HANDBOOK OF HYDROLOGY

Other McGraw-Hill Books of Interest

American Water Works Association • WATER QUALITY AND TREATMENT
American Water Works Association • WATER TREATMENT PLANT DESIGN
Brater & King • HANDBOOK OF HYDRAULICS
Brock • FIELD INSPECTION HANDBOOK
Corbitt • STANDARD HANDBOOK OF ENVIRONMENTAL ENGINEERING
Gaylord & Gaylord • STRUCTURAL ENGINEERING HANDBOOK
Herbich • HANDBOOK OF DREDGING ENGINEERING
Merritt • STANDARD HANDBOOK FOR CIVIL ENGINEERS
Parmley • FIELD ENGINEER'S MANUAL
Parmley • HYDRAULICS FIELD MANUAL
Zipparro & Hasen • DAVIS'S HANDBOOK OF APPLIED HYDRAULICS

HANDBOOK OF HYDROLOGY

David R. Maidment
Editor in Chief

Professor of Civil Engineering
University of Texas at Austin

McGRAW-HILL, INC.

New York San Francisco Washington, D.C. Auckland Bogotá
Caracas Lisbon London Madrid Mexico City
Milan Montreal New Delhi San Juan
Singapore Sydney Tokyo Toronto

Library of Congress Cataloging-in-Publication Data

Maidment, David R.
Handbook of hydrology / David R. Maidment, editor in chief.
p. cm.
Includes bibliographical references. Includes index.
ISBN 0-07-039732-5
1. Hydrology—Handbooks, manuals, etc. I. Title.
GB662.5.M35 1992
551.48—dc20 92-1 8193
CIP

2 3 4 5 6 7 8 9 0 DOC/DOC 9 8 7 6 5 4 3

ISBN 0-07-039732-5

The sponsoring editor for this book was Harold B. Crawford, the editing supervisor was Peggy Lamb, and the production supervisor was Pamela A. Pelton. It was set in Times Roman by Progressive Typographers Inc.

Printed and bound by R. R. Donnelley & Sons Company.

This book is printed on acid-free paper.

In Memory of Ven Te Chow
(1919–1981)

Scholar, Teacher, Author, Friend

CONTENTS

Part 1 Hydrologic Cycle

Part 2 Hydrologic Transport

Part 3 Hydrologic Statistics

Part 4 Hydrologic Technology

CONTRIBUTORS

Lajpat R. Ahuja *Soil Physicist and Research Leader, USDA Agricultural Research Service, Fort Collins, Colorado*

Mary P. Anderson *Professor of Geology and Geophysics, University of Wisconsin, Madison, Wisconsin*

Thomas O. Barnwell, Jr. *Civil Engineer, Environmental Research Laboratory, U.S. Environmental Protection Agency, Athens, Georgia*

Philip B. Bedient *Professor of Environmental Science and Engineering, Rice University, Houston, Texas*

Donald L. Brakensiek *Research Associate, Department of Agricultural Engineering, University of Maryland, College Park, Maryland*

Ian R. Calder *Head of Land-Use Section, Institute of Hydrology, Wallingford, United Kingdom*

Randall J. Charbeneau *Professor of Civil Engineering, University of Texas at Austin, Austin, Texas*

Malcolm K. Cleaveland *Assistant Professor of Geography, University of Arkansas, Fayetteville, Arkansas*

Timothy A. Cohn *Hydrologist, U.S. Geological Survey, Reston, Virginia*

Ian Cordery *Associate Professor of Civil Engineering, University of New South Wales, Kensington, New South Wales, Australia*

David E. Daniel *Professor of Civil Engineering, University of Texas at Austin, Austin, Texas*

Johannes J. DeVries *Regional Coordinator, Water Resources Center, University of California, Davis, California*

Robert E. Dickinson *Professor of Atmospheric Sciences, University of Arizona, Tucson, Arizona*

Roy D. Dodson *President, Dodson and Associates, Inc., Houston, Texas*

Edwin T. Engman *Branch Head, Hydrological Sciences Branch, Laboratory for Hydrospheric Processes, NASA Goddard Space Flight Center, Greenbelt, Maryland*

Efi Foufoula-Georgiou *Associate Professor of Civil and Mineral Engineering, University of Minnesota, Minneapolis, Minnesota*

Danny L. Fread *Director, Hydrologic Research Laboratory, National Weather Service, NOAA, Silver Spring, Maryland*

Edward J. Gilroy *Mathematical Statistician, U.S. Geological Survey, Reston, Virginia*

Donald M. Gray *Chairman, Division of Hydrology, University of Saskatchewan, Saskatoon, Saskatchewan, Canada*

Dennis R. Helsel *Chief, Branch of Systems Analysis, U.S. Geological Survey, Reston Virginia*

Robert M. Hirsch *Assistant Chief Hydrologist for Research and External Coordination, U.S. Geological Survey, Reston, Virginia*

T. V. Hromadka II *Principal Engineer, Boyle Engineering Corp., Newport Beach, California*

Wayne C. Huber *Professor and Chairman, Department of Civil Engineering, Oregon State University, Corvallis, Oregon*

Pierre Y. Julien *Associate Professor of Civil Engineering, Colorado State University, Fort Collins, Colorado*

Peter K. Kitanidis *Professor of Water Resources, Stanford University, Stanford, California*

John E. Kutzbach *Professor of Meteorology, University of Wisconsin, Madison, Wisconsin*

Eric G. Lappala *Senior Vice President, Harding, Lawson and Associates, Princeton Junction, New Jersey*

Vito J. Latkovich *Chief, Hydrologic Instrumentation Facility, U.S. Geological Survey, Stennis Space Center, Mississippi*

George H. Leavesley *Hydrologist, U.S. Geological Survey, Lakewood, Colorado*

Dennis P. Lettenmaier *Research Professor of Civil Engineering, University of Washington, Seattle, Washington*

David R. Maidment *Professor of Civil Engineering, University of Texas at Austin, Austin, Texas*

James L. Martin *Environmental Engineer, AScI Corporation, Athens, Georgia*

Steve C. McCutcheon *Environmental Research Laboratory, U.S. Environmental Protection Agency, Athens, Georgia*

Alistair I. McKerchar *Scientist, National Institute for Water and Atmospheric Research, Christchurch, New Zealand*

Thomas A. McMahon *Professor of Agricultural Engineering, University of Melbourne, Parkville, Victoria, Australia*

James W. Mercer *President, GeoTrans, Inc., Sterling, Virginia*

M. Paul Mosley *Hydrologist, National Institute for Water and Atmospheric Research, Wellington, New Zealand*

David H. Pilgrim *Professor and Head, Department of Water Engineering, University of New South Wales, Kensington, New South Wales, Australia*

Thomas A. Prickett *President, Thomas A. Prickett and Associates, Urbana, Illinois*

Terry D. Prowse *Research Scientist and Project Head, Cold Regions Hydrology and Ecology, National Hydrology Research Institute, Saskatoon, Saskatchewan, Canada*

Eugene M. Rasmusson *Senior Research Scientist, Department of Meteorology, University of Maryland, College Park, Maryland*

Walter J. Rawls *Hydrologist, U.S. Department of Agriculture, Beltsville, Maryland*

Hanadi S. Rifai *Research Associate, Department of Environmental Science and Engineering, Rice University, Houston, Texas*

Larry A. Roesner *National Technical Director Stormwater Practice, Camp Dresser & McKee, Inc., Orlando, Florida*

Jose D. Salas *Professor of Civil Engineering, Colorado State University, Fort Collins, Colorado*

Frank W. Schwartz *Ohio Eminent Scholar in Hydrogeology, Department of Geological Sciences, The Ohio State University, Columbus, Ohio*

Hsieh Wen Shen *Professor of Civil Engineering, University of California, Berkeley, California*

Adel Shirmohammadi *Associate Professor of Agricultural Engineering, University of Maryland, College Park, Maryland*

W. James Shuttleworth *Head, Hydrological Processes Division, Institute of Hydrology, Wallingford, United Kingdom*

James A. Smith *Assistant Professor of Civil Engineering and Operations Research, Princeton University, Princeton, New Jersey*

Leslie Smith *Professor of Geological Sciences, University of British Columbia, Vancouver, British Columbia, Canada*

Jery R. Stedinger *Professor of Civil and Environmental Engineering, Cornell University, Ithaca, New York*

Ben R. Urbonas *Chief, Master Planning Program, Urban Drainage and Flood Control District, Denver, Colorado*

Richard M. Vogel *Associate Professor of Civil Engineering, Tufts University, Medford, Massachusetts*

Richard K. Waddell *Vice President, GeoTrans, Inc., Boulder, Colorado*

David S. Ward *Vice President, Research and Development, GeoTrans, Inc., Sterling, Virginia*

Stephen W. Wheatcraft *Research Professor, Desert Research Institute, University of Nevada System, Reno, Nevada*

Eric F. Wood *Professor of Civil Engineering, Princeton University, Princeton, New Jersey*

INTERNATIONAL ADVISORY BOARD

PRACTITIONER ADVISORY BOARD

ABOUT THE EDITOR IN CHIEF

David R. Maidment is a Professor of Civil Engineering at The University of Texas at Austin where he has been on the faculty since 1981. He is a coauthor of *Applied Hydrology,* also published by McGraw-Hill, and an Editor of the *Journal of Hydrology.* He received his B.E. degree in Agricultural Engineering from the University of Canterbury, Christchurch, New Zealand, and his M.S. and Ph.D. degrees in Civil Engineering from the University of Illinois at Urbana-Champaign. He teaches graduate courses and conducts research in surface water hydrology, statistical methods in hydrology, and the use of geographic information systems in hydrology. In addition to his academic duties, Dr. Maidment serves as a consultant in hydrology and was for several years the drainage engineer for a small community.

PREFACE

The publication by McGraw-Hill in 1964 of the *Handbook of Applied Hydrology*, edited by Ven Te Chow, was a landmark in the development of hydrology. It coincided with the initiation of the International Hydrological Decade (1965–1974), the launching of many of the leading hydrologic journals, and the advent of the computer as a tool for data archiving and analysis. During the subsequent quarter century, hydrology has undergone a revolution in the methods available and in the degree of detail with which problems can be solved. Perhaps more than any other book, Chow's Handbook defined the subject of hydrology at the outset of this modern era and it stood as an influential reference for many years.

Following publication in 1988 of the text *Applied Hydrology*, which I coauthored with the late Ven Te Chow and Larry W. Mays, I was contacted by McGraw-Hill and asked to consider revising Chow's Handbook. I analyzed its contents and concluded that while an up-to-date *Handbook of Hydrology* is needed, it would be impractical to prepare that by revision of Chow's Handbook because so long has passed since its original material was prepared. Hence this volume: a new *Handbook of Hydrology* surveying the subject and synthesizing its established facts.

The need for a new Handbook is readily apparent. A large amount of knowledge of the various fields of hydrology has accumulated in journals, textbooks and reports but the basic facts have not been sifted down in one location. Many new methods and computer programs have been created for solving hydrologic problems, and practitioners need an authoritative source of guidance as to which methods are appropriate for a given task. The Handbook has been constructed to provide this information in a concise and readily accessible form.

Typical Handbook readers will be hydrologists or related professionals possessing a Bachelors degree in engineering or applied science and some years experience. They will turn to the Handbook to understand an unfamiliar term or a new subject area, to obtain a formula, to find the procedure for solving a problem, to assess what computer packages are available, or to obtain values from a map or table of data. The chapters will be valuable as reading material for students studying hydrology or other water-related subjects.

Although the Handbook is a compendium of hydrologic practice rather than a description of hydrologic science, sufficient scientific background is provided to enable to hydrologist to understand the hydrologic processes involved in a given problem, and to appreciate the limitations of the methods presented for solving it. The Handbook is not an encyclopedia attempting to catalog every known fact about the field: in many cases, a subject is simply introduced and references provided for the reader who seeks further detail. But a thorough effort has been made to ensure that the coverage of the field is as complete as space permits in a book of this size. Some 3000 references to the hydrology literature are provided.

The Handbook comprises four sections: Hydrologic Cycle, Chaps. 1–10, covering the flow of water through the phases of the hydrologic cycle; Hydrologic Transport, Chaps. 11–16, treating the motion of constituents and contaminants carried with the flow; Hydrologic Statistics, Chaps. 17–20, dealing with the analysis of hydrologic data and quantification of uncertainty; and Hydrologic Technology, Chaps. 21–29, describing computer programs, advanced data collection and forecasting methods, and procedures for hydrologic design. New fields of emphasis in recent years are given special attention, such as groundwater pollution control.

Space limitations did not permit the inclusion of individual chapters devoted to the sciences associated with hydrology such as geomorphology, or to the treatment of individual hydrologic environments, such as forest hydrology or wetlands; information on these subjects is contained in other chapters—for example, geomorphology is covered under streamflow, and forest hydrology and wetlands are treated in the chapter on the hydrologic effects of land use change. I have compiled a comprehensive index for these and related subjects to show where material on them can be found in the Handbook.

I am privileged to have had such a fine group of authors contributing the chapters to the Handbook. They are all experienced professionals in their respective fields, and I know that they have considered it an important responsibility to set down their hard-won knowledge for the benefit of our profession. Whenever you reference material from a particular chapter in this Handbook, I hope that you will attribute it to the authors of that chapter then simply cite me as the editor of the Handbook, rather than citing the Handbook as a whole by using my name alone as is sometimes done.

I am also grateful to the members of the two Advisory Boards listed at the front of the Handbook: the International Advisory Board, chaired by Stephen J. Burges, and the Practitioner Advisory Board, chaired by Leo R. Beard. I am especially indebted to Steve and Leo for the long hours they devoted to setting up and operating their Boards and for the wise counsel they provided on many occasions. Members of the Advisory Boards reviewed all the chapters to ensure that the material presented is sound, practical, and representative of the best information available in the United States and internationally.

In addition to the reviews provided by Board members, I solicited reviews of the chapters from specialists in each field, and I reviewed all the chapters in detail myself. A total of 230 chapter reviews were thus contributed, an average of 8 reviews per chapter, and the chapters were revised in light of these reviews. This extensive review procedure was designed to augment the careful attention to detail provided by the authors and to ensure that the information presented in each chapter is an authoritative survey that can be considered invaluable for reference and professional practice.

The assembly of this vast project, involving 58 authors in five countries, 65 Advisory Board members spread all over the world, and many additional technical reviewers, involved a considerable degree of organization. I wish to acknowledge the support I have received in this task from the Department of Civil Engineering and the Center for Research in Water Resources of the University of Texas at Austin. In particular, I am grateful to Brenda Nelson and her CRWR staff who directed the flow of material among all the authors, reviewers, myself, and McGraw-Hill. Without their assistance, this task would not have been completed. I have also appreciated the constant advice and encouragement of Harold Crawford, my editor at McGraw-Hill.

For many readers of this Handbook, hydrology is more than simply a subject to be learned, more than a science to be studied, more even than a profession we are proud to practice; it deals with natural phenomena of intrinsic beauty, whose technical intricacy and importance to humanity motivate our life's work. I have constantly felt this sense of dedication among the many members of the hydrology profession who contributed chapters and reviews to the Handbook. Preparing this Handbook has been a labor of love for me, and I trust that the result will serve our profession well in the years to come.

David R. Maidment

HANDBOOK OF HYDROLOGY

CHAPTER 1
HYDROLOGY

David R. Maidment
Department of Civil Engineering
University of Texas,
Austin, Texas

1.1 INTRODUCTION TO HYDROLOGY

Hydrology is concerned with the circulation of water and its constituents through the hydrologic cycle. It deals with precipitation, evaporation, infiltration, groundwater flow, runoff, streamflow, and the transport of substances dissolved or suspended in flowing water. Hydrology is primarily concerned with water on or near the land surface; ocean waters are the domain of oceanography and the marine sciences.

Hydrology as a Science. Hydrology is a science because it is concerned with a class of natural phenomena governed by general laws which the hydrologist seeks to understand and predict. A panel of the U.S. National Research Council[13] reviewed the evolution of hydrology as a science and presented the following definition of hydrology: *Hydrology is the science that treats the waters of the Earth, their occurrence, circulation and distribution, their chemical and physical properties, and their reaction with their environment, including their relation to living things. The domain of hydrology embraces the full life history of water on the Earth.* This definition was originally proposed in 1962 by an ad hoc committee of the U.S. Federal Council for Science and Technology, chaired by Walter Langbein.

As an earth science, hydrology is closely related to other natural sciences. Understanding precipitation and evaporation requires knowledge of climatology and meteorology; similarly, infiltration is connected to soil science, groundwater flow to geology, surface runoff to geomorphology, streamflow to fluid mechanics. Besides the flow of water, understanding the transport of constituents calls for a host of additional knowledge, drawn principally from chemistry and physics, to account for the decay, precipitation, dissolving, dispersion, and chemical reactions which change the concentration of its constituents as water flows along.

The touchstone of any science is its ability to predict the behavior of phenomena and then verify the predictions by observation. In hydrology, verification is complicated because the phenomena are driven by precipitation, which is an inherently random and uncertain process, and because the spatial extent and variability of the phenomena are so great as to defy exact prediction and measurement at every point. Thus statistical science must be employed to sift the critical numbers from large amounts of observed hydrologic data and to test the hypotheses being examined. Many hydrologic extremes are so unpredictable that the only way to assess their potential size and scope is through the assembly and analysis of historically recorded events. Hydrology is thus an observational science.

The extent, intricacy, and inherent uncertainty of hydrologic phenomena are so large as to cause some people to question whether they will ever be completely understood. The exciting reward of hydrologic research is the gradual unveiling of nature's processes by extension of observation and theory.

Hydrology as a Profession. Hydrology is a profession because the hydrologist seeks to apply hydrologic knowledge to solve problems and make life better for people. Webster's Third International Dictionary defines a profession as *A calling requiring specialized knowledge . . . which has as its prime purpose the rendering of a public service.*[19]

Traditionally, hydrologists have practiced their profession by specializing in hydrology but maintaining their primary affiliation with another discipline in engineering or applied science. In recent years, however, hydrology has become increasingly recognized as an independent profession. The American Institute of Hydrology, formed in 1981, has as one of its principal activities the certification of professional hydrologists and hydrogeologists if they meet specified standards of education and experience.[1]

The skills and methods used by the hydrologist rest in part on deductions from the basic laws governing hydrologic phenomena, and in part on professional experience and observed data. As knowledge evolves, a more rational scientific basis is created for hydrologic practice, but abstract principles alone do not guarantee accuracy of the result—ensuring that the input data for the analysis are realistic for local conditions is also a critical part of ensuring accuracy; there is a balance between a powerful model requiring detailed input data that are questionable, as compared with a simplified model whose inputs are reliable. Besides technical accuracy, the professional hydrologist must work within a limited budget and time frame which dictates choosing the most effective and practical approach that these resources will permit. Balancing these concerns and choosing the correct path through them is one of the skills a hydrologist acquires by experience.

The use of computer programs makes possible analyses of complex hydrologic phenomena of an extent and precision that are otherwise unattainable. Some computer programs are in such widespread use, and are required by government agencies to such a degree, that they have become standard methods of professional practice. Computer technology is evolving more rapidly than any other sphere of activity impinging on hydrologic practice. But computation is not an end in itself—the ultimate goal is to ensure that the result is correct. And computer technology, powerful though it is, carries within it the possibility of hidden flaws in the software or input data that may invalidate the computed result. Careful checking by independent methods is always necessary to ensure that computer output is correct and realistic.

What Hydrologists Do. The practicing hydrologist's task is to specify the inputs of water and constituents to a water resource system, such as a river, lake, or aquifer system, and to trace the motion of the water and constituents as they pass through the system. Hydrologists are concerned with three issues: water use, water control, and pollution control.

By *water use* is meant the withdrawal of water from lakes, rivers, and aquifers for water supply to cities, industries, and agriculture, the instream use of water for hydropower and recreation, and the protection of wildlife, both plant and animal life, which inhabits these water systems. The hydrologist is called upon to specify the inflows to the system for both normal and drought conditions and to predict how different withdrawal rates or instream flow policies would affect the flow through the system. Water quality properties, such as turbidity, temperature, and mineral and bacteriological content may be important in such situations. In these tasks, hydrology is an input into the broader field of *water resources planning and management,* in which the validity of the various uses for water is balanced through technical, legal, and political mechanisms.

By *water control* is meant control of hydrologic extremes, principally floods, and the erosion and sediment transport which occur during floods. The age-old human concern to be protected form the ravages of floods is manifested in flood protection works, such as levees and dams; in management schemes, such as floodplain delineation and policies to regulate development within floodplains; in erosion and sediment control works and in storm water detention and diversion projects. Again, the hydrologist's task is to specify the inputs to the system for given design conditions, and to trace the discharge of water and sediment through the system. Hydrology is especially dependent upon hydraulics in these tasks.

By *pollution control* is meant the prevention of the spread of pollutants or contaminants in natural water bodies, and the cleanup of existing pollution. Here the hydrologist must determine the sources and extent of pollution, how quickly and how far the pollution will spread, and where the pollutants will ultimately end up. *Point sources* of pollution, such as landfills and chemical waste dumps, must be located; *nonpoint sources* of pollution such as drainage or runoff of pesticides and fertilizers from agricultural lands must be identified; the various solutes and liquids (such as gasoline) which might flow from these sites must be determined; the streams and aquifers through which the contaminants will pass need to be studied; and designs are prepared either to prevent the pollutants from flowing into the natural water system at all or, if pollution occurs, to ensure that it is sufficiently limited so that the quality of the receiving water is not significantly damaged. In some cases, attempts may be made to extract polluted waters from the natural environment, such as by pumping out polluted groundwater, so that the water can be treated and discharged again. Hydrology relates closely to geology and environmental engineering in this type of work.

1.2 HYDROLOGIC CYCLE

The hydrologic cycle is the most fundamental principle of hydrology (Fig. 1.2.1). Water *evaporates* from the oceans and the land surface, is carried over the earth in atmospheric circulation as water vapor, *precipitates* again as rain or snow, is *intercepted* by trees and vegetation, provides *runoff* on the land surface, *infiltrates* into soils, *recharges* groundwater, *discharges* into streams, and ultimately, flows out into the oceans from which it will eventually evaporate once again. This immense water engine, fueled by solar energy, driven by gravity, proceeds endlessly in the presence or absence of human activity.

World Water Balance. The volumes of water flowing annually through the phases of the hydrologic cycle are shown in Fig. 1.2.1 in units relative to the annual precipitation on the land surface (119,000 km^3/year) which is set equal to 100 units in this diagram. The annual volume of evaporation from the ocean (424 units) is seven times larger than that from the land surface (61 units), making the oceans the primary

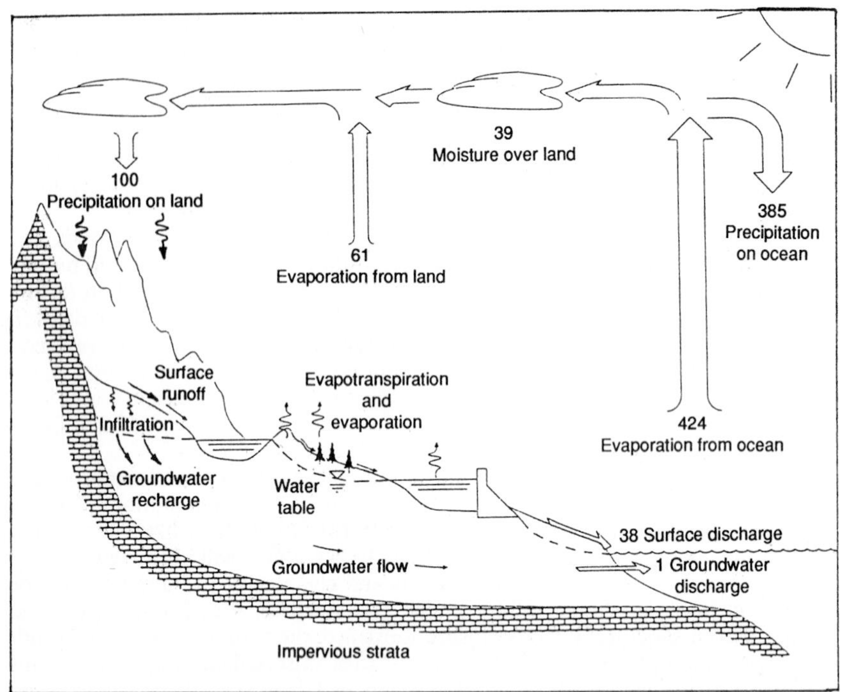

FIGURE 1.2.1 The hydrologic cycle with annual volumes of flow given in units relative to the annual precipitation on the land surface of the earth (119,000 km^3/year). *(Reproduced from Ref. 7 with permission.)*

source of precipitation over the earth's surface. The annual volume of discharge from the land surface to the oceans (39 units) is nearly all from surface water (38 units) and is counterbalanced by an equal net inflow of atmospheric water vapor from the oceans to the land areas.

When the annual quantities of the world water volumes are expressed in units of depth instead of volume (Fig. 1.2.2), annual precipitation on the land surface averages 800 mm (31 in), about two-thirds its value over the oceans; annual evaporation from the land surface is 480 mm (19 in), about one-third that over the oceans; and the remaining 320 mm (12 in) is runoff from the land to the oceans. Thus the dominance of the oceans in supplying atmospheric moisture arises in part from a more active hydrologic cycle over the oceans, and in part from their greater coverage of the earth's surface.

Earth's Water. The relative quantities of the earth's water contained in each of the phases of the hydrologic cycle are presented in Table 1.2.1. The oceans contain 96.5 percent of the earth's water, and of the 3.5 percent on land, approximately 1 percent is contained in deep, saline groundwaters or in saline lakes, leaving only 2.5 percent of the earth's water as fresh water. Of this fresh water, 68.6 percent is frozen into the polar ice caps and a further 30.1 percent is contained in shallow groundwater aquifers, leaving only 1.3 percent of the earth's fresh water mobile in the surface and atmospheric phases of the hydrologic cycle. The proportions of this water in the

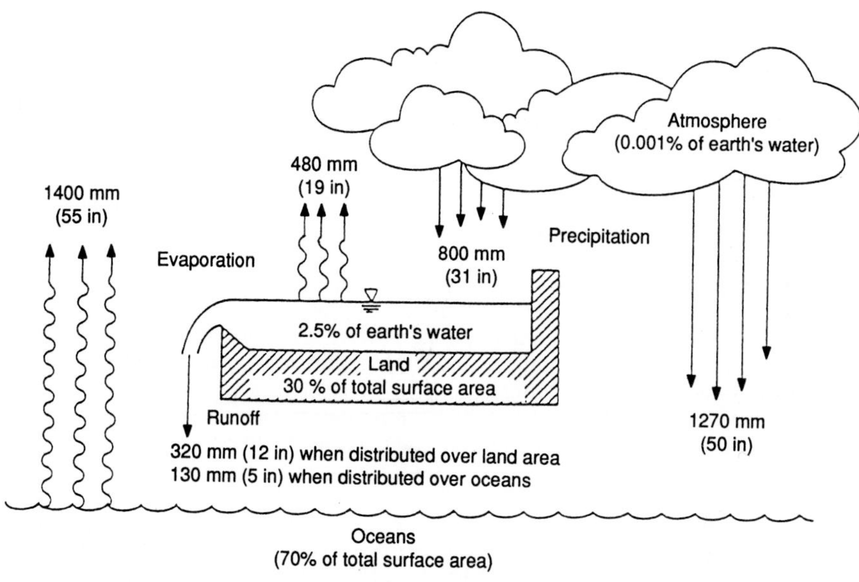

FIGURE 1.2.2 Annual quantities of the world water balance. *(Figure adapted from Bras.[5] Data from World Water Resources and Water Balance of the Earth, Copyright, UNESCO, 1978.)*

TABLE 1.2.1 Quantities of Water in the Phases of the Hydrologic Cycle

Item	Area, 10^6 km^2	Volume, km^3	Percent of total water	Percent of fresh water
Oceans	361.3	1,338,000,000	96.5	
Groundwater:				
Fresh	134.8	10,530,000	0.76	30.1
Saline	134.8	12,870,000	0.93	
Soil moisture	82.0	16,500	0.0012	0.05
Polar ice	16.0	24,023,500	1.7	68.6
Other ice and snow	0.3	340,600	0.025	1.0
Lakes:				
Fresh	1.2	91,000	0.007	0.26
Saline	0.8	85,400	0.006	
Marshes	2.7	11,470	0.0008	0.03
Rivers	148.8	2,120	0.0002	0.006
Biological water	510.0	1,120	0.0001	0.003
Atmospheric water	510.0	12,900	0.001	0.04
Total water	510.0	1,385,984,610	100	
Fresh water	148.8	35,029,210	2.5	100

Table adapted from World Water Balance and Water Resources of the Earth, Copyright, UNESCO, 1978.

atmosphere, soil moisture, and in lakes are similar, while that in rivers is less and that in snow and glacier ice is greater. A small amount of biological water remains fixed in the living tissues of plants and animals. All the data on the earth's waters cited here are taken from a comprehensive study of world water balance conducted in the Soviet Union during the International Hydrological Decade.[44] These values are estimates, and future studies made with more comprehensive data will lead to refinement of these values.

It is remarkable that the atmosphere, the driving force of the hydrologic cycle, contains only 12,900 km^3 of water, which is less than 1 part in 100,000 of all the waters of the earth. Atmospheric water would form a layer only 25 mm (1 in) deep if precipitated uniformly onto the earth's surface. The unpredictability of atmospheric motion, and the rapid recycling of water through the atmosphere (in which a water molecule resides on average for about 8 days before being precipitated again) create great uncertainty in weather patterns and random fluctuations in precipitation; these fluctuations feed forward into infiltration, runoff, groundwater flow, and streamflow and related hydrologic processes, making them random phenomena as well, though to a lesser degree than for precipitation. The unpredictability of its phenomena is a fascinating aspect of hydrology.

Transport Cycles. And not only is water moving, vast quantities of dissolved and suspended constituents are carried along in *transport cycles* associated with the water cycle. Ionic substances, principally chlorides and sulfates, are evaporated with ocean waters, dissolved in the atmosphere in rainwater along with nitrogen and sulfur compounds resulting from burning of fossil fuels, and precipitated on the land surface, producing acid rain. Water is a primary weathering agent for rocks and soils, breaking them down, dissolving them, and carrying the resulting sediments and dissolved solids down to the sea in an immense transport process.[4]

Once in the sea, many constituents settle or precipitate on the ocean floor, consolidate through time to form sedimentary rocks, and are redistributed over the earth's surface through the action of plate tectonics. Some sedimentary rocks sink deeper into the earth's mantle where increased temperature and pressure metamorphose them to new rock types. Small amounts of the rock in the mantle melt, flow under the mantle, and reemerge as igneous material either in volcanoes or in the deep ocean rift valleys whose upwelling from the earth's molten core drives plate tectonics. One can thus speak of a *rock cycle,* through which, over millennia, the earth's crust is continually being re-created.[17]

Origin of the Earth's Water. The origin of the earth's water is not known exactly, but the most likely source is *outgassing* of water vapor from the earth's interior as part of the extrusion of igneous material in volcanoes and ocean upwellings. Once released from the high temperature and pressure of the earth's core, this *juvenile water* condenses because of the peculiar fact that the prevailing temperature and pressure at the earth's surface are just right to permit water to exist as a liquid—on the other planets of our solar system the surface is either too hot, in which case the water dissociates into hydrogen and oxygen and escapes into space; or too cold, in which case the water freezes on the surface. Earth is unique among these planets in that water can exist in all three phases (solid, liquid, and gas) at the conditions of temperature and pressure prevailing at its surface.[10]

Some of the condensed water is combined into rocks and drawn down into the earth's core again, becoming *connate water,* much of which is expelled as metamorphosis proceeds, to reemerge as hot springs in volcanic regions or be absorbed into mineral veins.[23]

Water as a Physical Substance. The remarkable properties of water as a physical substance arise from its molecular structure. Filling the first electron shell around an atom requires two electrons, and filling the second shell requires eight more electrons; a hydrogen atom possesses one electron of its own, so it pairs up with an oxygen electron to complete its first shell; oxygen possesses eight electrons of its own, and by bonding with two hydrogen atoms it acquires the 10 electrons necessary to complete its second shell. Thus H_2O is created.

Two unshared electron pairs are left on the oxygen atom. These form, with the hydrogen atoms, a tetrahedral structure around the oxygen atom (Fig. 1.2.3). The two hydrogen vertices of this structure are slightly positively charged because their electron is preferentially located near the oxygen atom; the other two vertices containing oxygen's electron pairs are slightly negatively charged, thus creating a dipole between the vertices and permitting water molecules to form *polar* or *hydrogen bonds* with one another. These bonds are 10 to 50 times weaker than the covalent bonds between hydrogen and oxygen within the molecule,[4] but they are strong enough to cause water molecules to cluster in tetrahedral patterns, as shown for ice in Fig. 1.2.4*a*.

Changing State. Ice, though cold, possesses heat energy which is expressed by the vibration of the atoms and molecules in this fixed structure; as ice gets warmer, these vibrations increase to the point where the tetrahedral structure breaks down, ice melts, and its molecules mix slightly closer in water than they did in ice (Fig. 1.2.4*b*). As a result, water is slightly more dense than ice at its melting point, and ice floats on the surface of lakes, thus preventing deeper waters from freezing and protecting the aquatic life within them. Were it not for this property, ice formed at the surface of lakes would sink to the bottom until eventually many lakes would freeze solid. In the colder climates of the earth, repeated daily cycles of water freezing and thawing are an important factor in rock weathering: water percolates into rock cracks and expands as it freezes, thus generating a pressure of as much as 30,000 psi (207,000 kPa), sufficient to crack even the most durable rock.

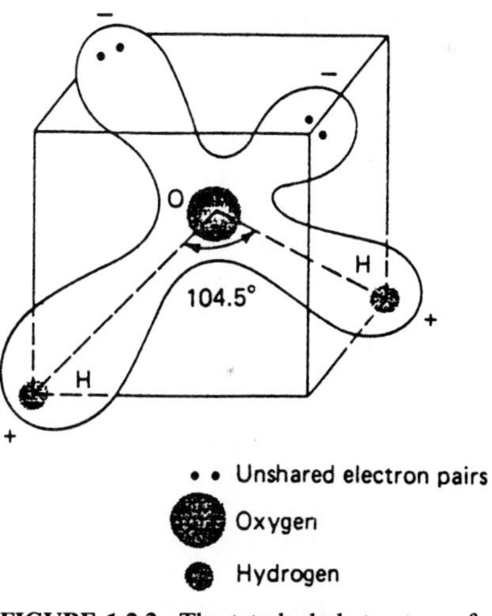

FIGURE 1.2.3 The tetrahedral structure of a water molecule. (Source: *Horne, R. A.*, "Marine Chemistry," © *1969 by John Wiley and Son, Inc. Reprinted by permission of the publisher.*)

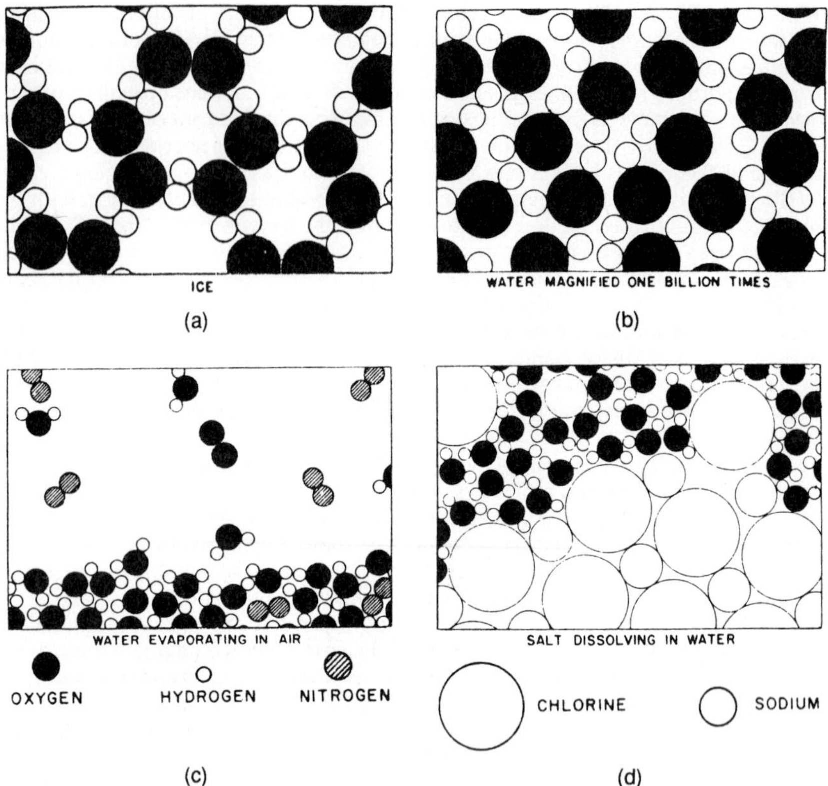

FIGURE 1.2.4 Water as a physical substance. (*Source: R. P. Feynman/R. B. Leighton,* "The Feynman Lectures on Physics, Vol. I," © *1963 by Addison Wesley Publishing Company, Inc. Reprinted by permission of the publisher.)*

The molecules in liquid water vibrate faster as the temperature rises,[14] some vibrations being so large as to cause a molecule to be thrown entirely off the water surface into the gaseous or vapor phase (Fig. 1.2.4*c*). Evaporating water consumes a large amount of energy, its *latent heat of vaporization,* an amount which is about 8 times larger than is necessary to melt ice, and about 600 times larger than its *heat capacity,* which is the energy required to raise water temperature by 1°C. This makes evaporation the dominant component of the energy balance associated with the hydrologic cycle. In fact, about 23 percent of the solar radiation reaching the earth is absorbed by evaporating water.

The latent heat of vaporization of water is larger than for any other liquid; its heat capacity is also larger than for any other liquid except liquid ammonia; these properties combine to permit the oceans to stabilize the temperature variation of coastal and island climates.[11] Even the human body, approximately 70 percent water by weight, is maintained at constant temperature in part through water vaporization, which is why we sweat when we exercise.

Water as a Solvent. Polar bonding also accounts for water's remarkable properties as a solvent—more substances dissolve in greater amounts in water than in any other liquid. A salt dissolves as water molecules surround its ions, the collective strength of the polar bonds between water and the ions being sufficient to break the bonds between the ions. The relatively small size of the water molecule helps: Fig. 1.2.4*d* shows how water molecules cluster around sodium and chlorine atoms as they dissociate.

Hydrogen bonds also cause water to have a high surface tension, which produces capillary rise in soils and causes rain to form into spherical droplets rather than into shapeless masses of particles.

It is indeed fortunate that water is so abundant on the earth and that it possesses these unique physical properties; without them, life on earth would not exist.

1.3 DEVELOPMENT OF HYDROLOGY

The origin and circulation of the waters of the earth has been a subject of speculation since ancient times, but the year 1850 might be regarded as marking the beginning of the development of methods in current use in hydrologic practice. In 1851, Mulvaney[34] first described the concept of the time of concentration that now forms the backbone of the rational method of runoff computation, and he also designed a primitive form of rain gauge that would record time-varying rainfall intensity during a storm. Five years later, Darcy[8] established the basic law of groundwater motion. During the following decades, knowledge gradually accumulated: In 1871 Saint-Venant[40] derived the equations of one-dimensional surface water flow, in 1891 Manning[32] developed his equation for open channel velocity, in 1908 the first watershed level measurement of the hydrologic effects of land-use change was done at Wagon Wheel Gap, in 1911 Green and Ampt[18] produced their infiltration model, and in 1925 Streeter and Phelps[41] developed the dissolved oxygen sag curve for rivers.

Figure 1.3.1 summarizes some significant contributions to the development of hydrologic knowledge from 1930 to 1980 under the classification of hydrology used in this handbook, namely, the hydrologic cycle, transport, statistics, and technology. Included in this table are discoveries of basic principles, such as the Theis equation in 1935; hydrologic computer models, such as HEC-1 in 1965; and books about hydrology, such as Meinzer's *Hydrology*. In some cases, developments in related fields which have subsequently become important in hydrology are noted, such as Box-Jenkins time series analysis methods. The citations in Fig. 1.3.1 are keyed to the reference lists in this chapter and the other chapters of the handbook. Study of these references is a fruitful effort for the serious hydrologist because the original description of a work often illuminates its fundamental assumptions and thinking more clearly than the concise statements of its results presented in later publications. The information presented in Fig. 1.3.1 is limited to the period 1930–1980 because it is difficult to assign a correct historical significance to more recent developments.

In scanning Fig. 1.3.1, it can be seen that there has been an acceleration in the growth of hydrologic knowledge since about 1950, and of computer-related technology since about 1960. Hydrologic principles and working equations can be derived from the fundamental laws of physics, or they can result from the synthesis of field observations of a phenomenon. Once an adequate process description has been constructed, it is often incorporated into computer models whose parameters are determined using local data for each application. While much has been accomplished, there is still a great deal that is unknown or uncertain about the functioning of hydrologic processes. And some of the existing knowledge of these processes has yet to be synthesized into a form in which it can be applied in a generalized fashion. Gradually clarifying these mists of uncertainty is the exciting challenge of hydrologic research, which leads in turn to improved hydrologic practice through the development of better methods and more precise parameter values.

FIGURE 1.3.1 Important contributions to the development of hydrology from 1930 to 1980. The reference given by each entry indicates the [chapter number, reference number] where the citation for this entry is located in the handbook.

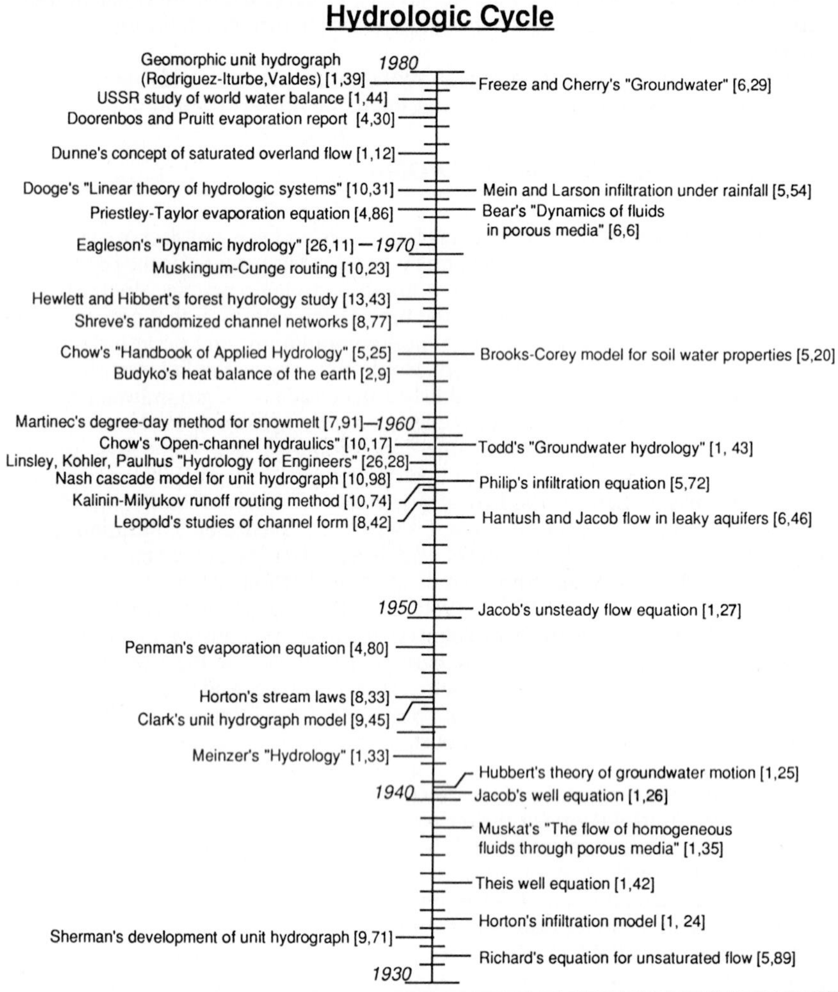

1.4 HYDROLOGIC UNITS

The evolution of hydrologic knowledge and the accumulation of hydrologic data over the past several decades have occurred during a period of transition in units of measure. Most countries have now standardized on SI (Système International d'Unites) units which evolved from the metric system of units developed in France about 200 years ago. The United States, however, still uses the traditional English

FIGURE 1.3.1 *(Continued)* The [Chap. No., Ref. No.] gives location of the citation in the Handbook.

1980
Fischer's "Mixing in Inland and Coastal Waters" [14,37]
Yotsukura and Sayre's transverse mixing in open channels [14,92]
Ackers and White formula for sediment transport in streams [12,3]
Stumm and Morgan's "Aquatic Chemistry" (Ist Ed.) [11,63]
1970
Mathematical analysis of aqueous equilibria (Barnes and Clark) [1,3]
Toffaleti's formula for sand discharge in rivers [12,125]
Fischer's dispersion studies in streams [14,35]
Bagnold's sediment transport equation [12,11]
Dispersion coefficients in porous media (Scheidegger) [16,72]
1960
Hem's "Study and Interpretation of the Chemical Characteristics of Natural Waters" (Ist Ed.) [11,28]
Groundwater retardation factor (Higgins) [1,22]
1-D dispersion in open channels (Elder) [1,15]
Random walk dispersion model in porous media (De Josselin de Jong) [1,9]
Blench's regime theory of channel form [8,6]
Lelliavsky's "Introduction to Fluvial Hydraulics" [1,31]
Taylor's studies of dispersion in pipes [14,83]
1950
Einstein's bed-load formula [12,36]
pE - pH diagram (Pourbaix) [1,38]
Meyer-Peter and Muller formula for sediment transport in streams [12,95]
Vanoni's suspended sediment measurements [1,45]
Piper diagram for groundwater quality [1,37]
1940
Calcium carbonate balance in water (Langelier) [1,30]
Shield's criterion for incipient sediment motion [12,119]
1930

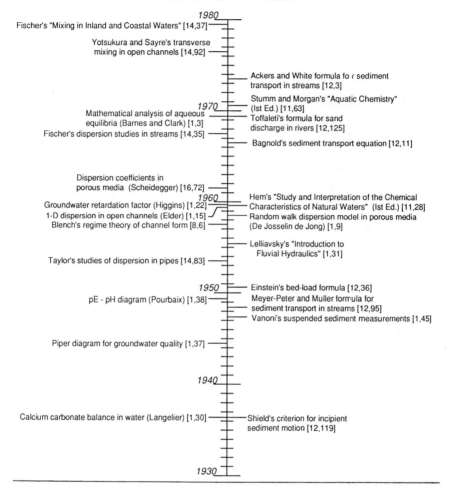

system of units for many purposes. Even when metric units are used, there are variations depending on whether one follows the strict SI system or retains convenient units from earlier metric systems. As a result, hydrologic equations, parameters, and data are expressed in a variety of units which may require conversion from one form to another.

SI System. The SI system contains seven *fundamental units:* the *meter* (m) for length, the *kilogram* (kg) for mass, the *second* (s) for time, the *ampere* (A) for electric current, the *kelvin* (K) for temperature, the *mole* (mol) for amount of substance, and the *candela* (cd) for luminous intensity. In addition to these, there are two supple-

FIGURE 1.3.1 *(Continued)* The [Chap. No., Ref. No.] gives location of the citation in the Handbook.

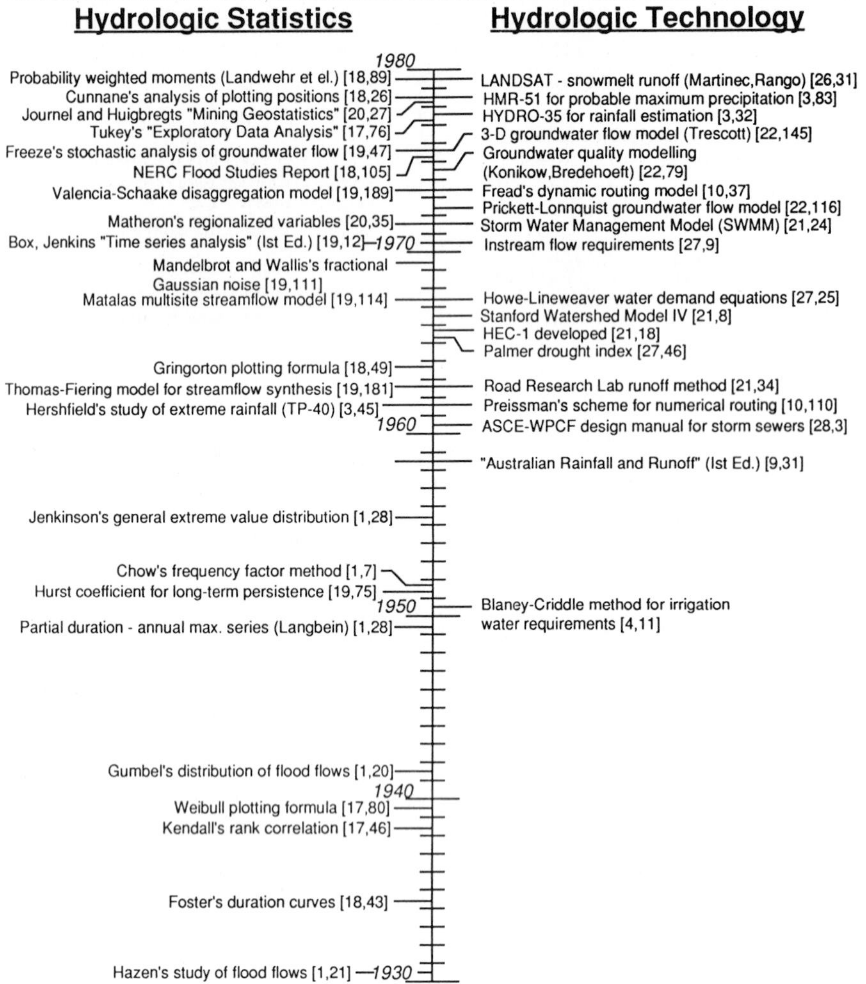

mentary units for angular measure: the *radian* (rad) for plane angles, and the *steradian* (sr) for solid angles.[2] All other units can be expressed as combinations of these quantities. The definitions of these fundamental units have been refined by periodic meetings of the General Conference on Weights and Measures and expressed in terms of invariant physical measures. For example, the second is defined to be "the duration of 9 192 631 770 periods of the radiation corresponding to the transition between the two hyperfine levels of the ground state of the cesium-133 atom."[36]

Units for other quantities are derived as products, ratios, or powers of the base units. These include the *newton* (N) for force, which is the force required to accelerate a 1-kg mass at 1 m/s^2; the *pascal* (Pa) for pressure or stress, where one pascal is equal to a force of 1 N distributed over an area of 1 m^2; the *joule* (J) for work or energy, which is the work done by a force of 1 N moving through a distance of 1 m; and the *watt* (W) for power, in which 1 W is an energy flow rate of 1 J/s. Raising a unit to a power of 10 is indicated by attaching a prefix to the unit. Thus, for example, 1 kW is a thousand watts, or $W \times 10^3$.

U.S. Customary Units. The traditional English system of units is referred to here as the U.S. Customary system, because the United States is the main country still using them. Mass, length, and time are the most important quantities, and of these, both the U.S. and SI systems share the second as the basic unit of time. Exact conversions have been established by the American Society for Testing and Materials and other standard setting organizations to permit conversion of mass and length units between the SI and the U.S. Customary systems (Ref. 46, p. 257). These exact conversions are:

$$1 \text{ inch (in)} = 2.54 \text{ cm} \tag{1.4.1}$$

$$1 \text{ pound mass } (\text{lb}_m) = 0.45359237 \text{ kg} \tag{1.4.2}$$

Consequently, 1 foot = 12 in = 30.48 cm = 0.3048 m exactly, and 1 mile = 5280 ft = 1609.344 m = 1.609344 km exactly. Units of force in the U.S. system are usually expressed in terms of *pounds force* (lb_f), which is the force required to accelerate 1 lb_m at the rate of acceleration due to gravity g. The value of the gravitational acceleration is not constant over the earth, varying slightly with latitude and altitude. A standard reference value of g has been adopted:[36]

$$g = 9.80665 \text{ m/s}^2 \tag{1.4.3}$$

By knowing conversions between SI and U.S. fundamental units and the relationship of derived units to the fundamental units, conversions between derived units in the U.S., SI, and other metric systems can be calculated to any desired degree of accuracy. In Appendix A at the end of the Handbook are presented conversion tables for commonly used units for length, area, volume, velocity or rate, discharge, mass, force, pressure or stress, energy or work, power, energy flux (power per unit area), temperature, and time.

Besides variations in units among countries, there are also variations in terminology. A glossary of hydrologic terms has been prepared by WMO and UNESCO.[47]

REFERENCES

1. American Institute of Hydrology, Brochure describing certification and registration of professional hydrologists and professional hydrogeologists, American Institute of Hydrology, 3416 University Ave, S.E., Minneapolis, Minn. (undated).
2. American Society of Mechanical Engineers, *ASME Orientation and Guide for Use of SI (Metric) Units,* 8th ed., United Engineering Center, New York, 1978.
3. Barnes, I., and F. E. Clark, "Chemical Properties of Groundwater and Their Corrosion and Encrustation Effects on Wells," *U.S. Geol. Surv. Prof. Pap.* 498D, 1969.
4. Berner, E. K., and R. A. Berner, *The Global Water Cycle,* Prentice-Hall, Englewood Cliffs, N.J., 1987.
5. Bras, R. L., *Hydrology,* Addison-Wesley, Reading, Mass., 1990.

6. Chow, V. T., "A General Formula for Hydrologic Frequency Analysis," *Trans. Am. Geophys. Union,* vol. 32, no. 2, pp. 231–237, 1951.
7. Chow, V. T., D. R. Maidment, and L. W. Mays, *Applied Hydrology,* McGraw-Hill, New York, 1988.
8. Darcy, H., *Les fontaines publiques de la ville de Dijon,* V. Dalmont, Paris, 1856.
9. De Josselin de Jong, G., "Longitudinal and Transverse Diffusion in Granular Deposits," *Trans. Am. Geophys. Union,* vol. 39, pp. 67–74, 1958.
10. Dooge, J. C. I., "On the Study of Water," *Hydrol. Sci. J.,* vol. 28, pp. 23–48, 1983.
11. Driscoll, F. G., ed., *Groundwater and Wells,* 2d ed., chap. 1, Johnson Division, St. Paul, Minn., 1986.
12. Dunne, T., T. R. Moore, and C. H. Taylor, "Recognition and Prediction of Runoff-Producing Zones in Humid Regions," *Hydrol. Sci. Bull.,* vol. 20, no. 3, pp. 305–327, 1975.
13. Eagleson, P. S., ed., *Opportunities in the Hydrologic Sciences,* National Academy Press, Washington, D.C., 1992.
14. Eisenberg, D., and W. Kauzmann, *The Structure and Properties of Water,* Oxford University Press, New York, 1969.
15. Elder, J. W., "The Dispersion of Marked Fluid in a Turbulent Shear Flow," *J. Fluid Mech.,* vol. 5, part 4, pp. 544–560, 1959.
16. Feynman, R. P., R. B. Leighton, and M. Sands, *The Feynman Lectures on Physics,* vol. 1, pp. 1-2 to 1-6, Addison-Wesley, Reading, Mass., 1963.
17. Flint, R. F., *The Earth and Its History,* W. W. Norton, New York, 1973.
18. Green, W. H., and G. A. Ampt, "Studies on Soil Physics," *J. Agric. Sci.,* vol. 4, part 1, pp. 1–14, 1911.
19. Gove, P. B., ed., *Webster's Third New International Dictionary,* Merriam, Springfield, Mass, p. 1811, 1968.
20. Gumbel, E. J., "The Return Period of Flood Flows," *Ann. Math. Stat.,* vol. 12, no. 2, pp. 163–190, 1941.
21. Hazen, A., *Flood Flows, a Study of Frequencies and Magnitudes,* Wiley, New York, 1930.
22. Higgins, G. H., "Evaluation of the Groundwater Contamination from Underground Nuclear Explosions," *J. Geophys. Res.,* vol. 64, pp. 1509–1519, 1959.
23. Holmes, A., *Principles of Physical Geology,* rev. ed., Ronald Press, New York, p. 412, 1965.
24. Horton, R. E., "The Role of Infiltration in the Hydrologic Cycle," *Trans. Am. Geophys. Union,* vol. 14, pp. 446–460, 1933.
25. Hubbert, M. K., "The Theory of Groundwater Motion." *J. Geol.* vol. 48, pp. 784–944, 1940.
26. Jacob, C. E., "On the Flow of Water in an Elastic Artesian Aquifer," *Trans. Am. Geophys. Union,* part 2, pp. 574–586, 1940.
27. Jacob, C. E., "Flow of Groundwater," in H. Rouse, ed., *Engineering Hydraulics,* Wiley, New York, pp. 321–386, 1950.
28. Jenkinson, A. F., "The Frequency Distribution of Annual Maximum (or Minimum) Values of Meteorological Elements," *Q. J. R. Met. Soc.,* vol. 81, pp. 158–171, 1955.
29. Langbein, W. B., "Annual Floods and the Partial Duration Flood Series," *Trans. Am. Geophys. Union,* vol. 30, pp. 879–881, 1949.
30. Langelier, W. F., "The Analytical Control of Anticorrosion Water Treatment," *J. Am. Water Works Assoc.,* vol. 28, pp. 1500–1521, 1936.
31. Lelliavsky, S., *An Introduction to Fluvial Hydraulics,* Constable, London, England, 1955.
32. Manning, R., "On the Flow of Water in Open Channels and Pipes," *Trans. Inst. Civil Eng. Ireland,* vol. 20, pp. 161–207, 1891; Supplement vol. 24, pp. 179–207, 1895.
33. Meinzer, O. E., ed., *Hydrology,* Dover, New York, 1942.

34. Mulvaney, T. J., "On the Use of Self-Registering Rain and Flood Gauges in Making Observations of the Relations of Rainfall and of Flood Discharges in a Given Catchment," *Proc. Inst. Civil Eng. Ireland,* vol. 4, pp. 18–31, 1851.

35. Muskat, M., *The Flow of Homogeneous Fluids Through Porous Media,* McGraw-Hill, New York, 1937.

36. National Institute of Standards and Technology, *The International System of Units (SI),* NIST Special Publication 330, 1991 ed., B. N. Taylor, ed., National Institute of Standards and Technology, Gaithersburg, Md., pp. 17, 28, 1991.

37. Piper, A. M., "Graphic Procedure in the Geochemical Interpretation of Water Analyses," *Trans. Am. Geophys. Union,* vol. 25, pp. 914–923, 1944.

38. Pourbaix, M. J. N., *Thermodynamics of Dilute Aqueous Solutions,* Edward Arnold, London, 1949.

39. Rodriguez-Iturbe, I., and J. B. Valdes, "The Geomorphologic Structure of Hydrologic Response," *Water Resour. Res.,* vol. 15, no. 6, pp. 1409–1420, 1979.

40. Saint-Venant, Barre de, "Theory of Unsteady Water Flow, with Application to River Floods and to Propagation of Tides in River Channels," *French Academy of Science,* vol. 73, pp. 148–154, 237–240, 1871.

41. Streeter, H. W., and E. B. Phelps, "A Study of the Pollution and Natural Purification of the Ohio River," *U.S. Public Health Bull.,* 146, February 1925.

42. Theis, C. V., "The Relation Between Lowering of the Piezometric Surface and the Rate and Duration of Discharge of a Well Using Groundwater Storage," *Trans. Am. Geophys. Union,* 16th annual meeting, part 2, pp. 519–524, 1935.

43. Todd, D. K., *Groundwater Hydrology,* Wiley, New York, 1959.

44. USSR Committee for the International Hydrological Decade, "World Water Balance and Water Resources of the Earth," English translation, *Studies and Reports in Hydrology,* vol. 25, UNESCO, Paris, 1978.

45. Vanoni, V. A., "Transportation of Suspended Sediment by Water," *Trans. Am. Soc. Civil Eng.,* vol. 111, pp. 67–133, 1946.

46. Wandmacher, C., *Metric Units in Engineering—Going SI,* Industrial Press, New York, 1978.

47. WMO-UNESCO, *International Glossary of Hydrology,* 2d ed., WMO Rept. No. 385 World Meteorological Organization, Geneva, Switzerland, 1992.

CHAPTER 2
CLIMATOLOGY

Eugene M. Rasmusson
Department of Meteorology
University of Maryland
College Park, Maryland

Robert E. Dickinson
Department of Atmospheric Sciences
University of Arizona
Tucson, Arizona

John E. Kutzbach
Department of Meteorology
University of Wisconsin
Madison, Wisconsin

Malcolm K. Cleaveland
Department of Geography
University of Arkansas
Fayetteville, Arkansas

2.1 THE GLOBAL CLIMATE SYSTEM: AN OVERVIEW

Definition of Climate. Although the same physical laws apply to both climate and weather, climatology is more than simply a branch of meteorology. *Weather* is the condition of the atmosphere at a particular time. *Climate* is the average state of the atmosphere during a period of time (weeks, decades, years, or millennia). Solar radiation is the major external energy source for the climate system, but there are other external factors, both natural, e.g., volcanic eruptions, and human-induced, e.g., land surface changes, which also affect, to some degree, the earth's climate.

The interactive response of the oceans and atmosphere, the cryosphere (ice and snow), and the land surface and its biomass to the external energy sources determines the earth's climate. The circulation of the oceans and the atmosphere is the powerful transport mechanism that moves heat over the long distances required to maintain the energy balance of the planet. The thermal capacity of the oceans provides seasonal heat storage that moderates what would otherwise be much larger summer to winter temperature changes.

Climate is not a static feature that can be described once and for all by averages of weather variables over a specified length of time. Rather, it is continuously evolving on a variety of temporal and spatial scales as a result of a complex system of internal dynamic interactions. In terms of hydrologic applications, the oft-posed question, "How long a record do I need in order to define 'the climate'?" is usually better stated as "What is the best estimate of the climate during some future period of interest?" The simple assumption that the statistics of the hydrologic cycle over the past few decades will hold for the next few decades could be significantly in error. A notable example of this is provided by the dramatic shift to a more arid climate over the African Sahel during the past quarter century.[46]

Climate Prediction. Accurate *climate predictions* are the obvious answer to this problem. At present some very modest predictive skill can be attained for periods of a month or season using empirical-statistical methods.[52] However, further progress in this area requires a better understanding of the climate system.

The modern climatologists' substitute for controlled laboratory experiments of the climate system is a *numerical climate model* (see Sec. 2.1.5). Currently, the ability of these models to simulate climate is limited by gaps in our knowledge of the hydrologic cycle, including cloudiness, the behavior of the ocean, small-scale processes in general, available computer power, and last but not least, the inherent unpredictability of many aspects of climate. The ultimate goal of these numerical models is the prediction of climate scenarios of sufficient realism to be useful for decision makers.

2.1.1 Global Heat and Water Balance

Radiation Balance of the Earth. The primary external energy source for the climate system is *solar radiation.* Figure 2.1.1 shows the annual average radiative fluxes

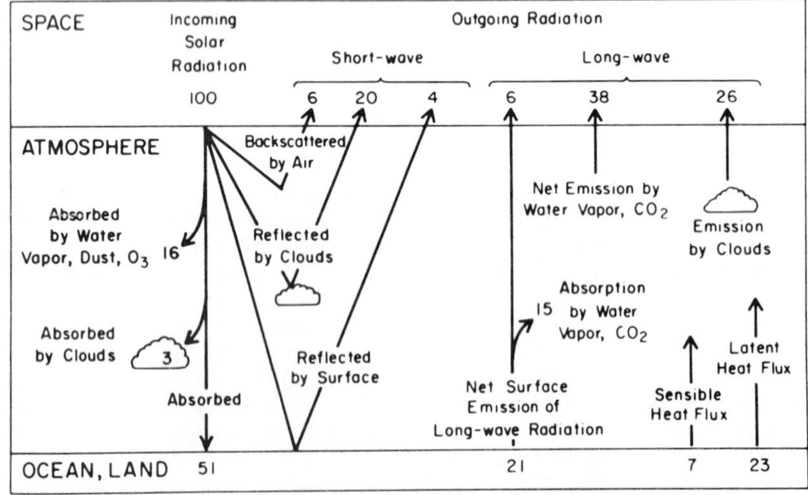

FIGURE 2.1.1 The global average components of the earth's energy balance. *(Adapted from Fig. 1.2, Ref. 56.)*

within the climate system. Average incoming radiation from the sun to the earth is represented by 100 units.

The wavelength of radiation is inversely proportional to the temperature of the emitter. Thus solar radiation from the hot sun is referred to as *short-wave radiation* while the radiation from the cooler earth is referred to as *long-wave radiation.* Of the 100 units of incoming solar radiation, 30 are reflected from the atmosphere and the earth's surface and returned to space as short-wave radiation. The remaining 70 units are absorbed, 19 units by the atmosphere and 51 at the earth's surface. Because the earth is in approximate *thermal equilibrium* (no long-term net heating), this 70 units is eventually reradiated back to space as long-wave radiation. However, before it returns to space, this energy passes through a complex recycling between the earth's surface and the atmosphere as shown on the right of Fig. 2.1.1. This recycling involves both radiant energy, *sensible heat* flux (heat transport by conduction and convection) and *latent heat* flux (the energy absorbed in evaporating water).

Meteorologists divide the lowest 30 km (18.6 mi) of the atmosphere into two layers based on the vertical gradient of temperature (the atmospheric *lapse rate*). In the lower layer, the *troposphere,* the temperature generally decreases with height. In the upper layer, the *stratosphere,* the temperature generally increases more gradually with height. The troposphere and stratosphere are separated by the *tropopause,* a temperature minimum which varies in height from around 16 km near the equator to 9 km near the poles.[35]

Since temperature decreases with altitude within the troposphere, the temperature of the radiating atmospheric molecules is lower than that of the earth's surface. The gases which absorb long-wave radiation, water vapor, carbon dioxide, and ozone are minor constituents of the atmosphere. These gases intercept radiation from lower layers of the atmosphere and from the surface and reradiate thermally according to their temperature.

Data from the earth radiation budget experiment (ERBE) satellite indicate a value of 235 to 240 W/m^2 for outgoing thermal radiation from the earth,[2] which corresponds to an emission temperature of 254 to 255 K (−18 to −19°C). However, observed global average surface temperature is 288 K (15°C), indicating that the earth's surface is warmer by about 33 to 34°C than it would be without an atmosphere. This is due to what is popularly known as the *greenhouse effect.*

Heat Transport. The difference between incoming and outgoing radiation varies with latitude. There is net radiative heating at low latitudes (near the equator) and net cooling at high latitudes (near the poles) (Fig. 2.1.2). This imbalance is offset by a poleward transport of energy; thus a fundamental coupling exists between the radiation budget of the earth and the *general circulation* of the atmosphere and oceans. A very significant part of the poleward transport of heat in both hemispheres is accomplished by ocean currents, particularly in the subtropics and lower middle latitudes.[10] In the atmosphere, heat is transported in the form of sensible heat, which is the heat associated with the temperature of the air parcel, and the latent heat of the water vapor contained in the air parcel. Approximately 600 cal (2.5 kJ) of heat must be supplied by the ocean or land surface to evaporate a gram of water. This latent or "hidden" energy is carried by the evaporated water vapor until it is released to the atmosphere upon vapor condensation in regions of upward atmospheric motion, cloud formation, and precipitation.

Latent heat (water vapor) transport by the atmosphere is a major contributor to the heat balance of the earth. It constitutes a vital link between the hydrologic cycle and the global energy balance.

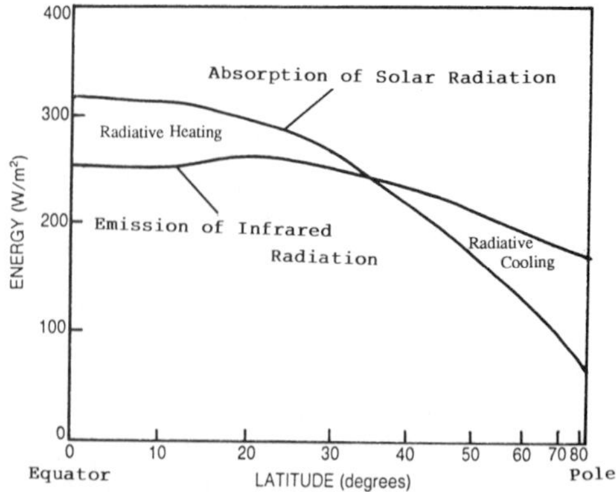

FIGURE 2.1.2 Mean annual latitude averaged absorbed solar radiation and outgoing long-wave radiation at the top of the atmosphere. The values are based on data obtained by satellites. *(Adapted from Fig. 3, Ref. 81. Used with permission.)*

2.1.2 Atmospheric Circulation

Low-Level Circulation. Figure 2.1.3 shows the mean sea-level pressure distribution and the low-level atmospheric circulation during January and July. Note particularly the following fundamental aspects of the circulation.

1. Even though there are large regional variations in the mean low-level circulation, a distinct pattern emerges when the data are zonally averaged, i.e., averaged around a latitude circle. This is schematically illustrated by Fig. 2.1.4. Beginning at low latitudes, there is a distinct zonally averaged convergence of the easterly *trade-wind flow* from the two hemispheres into the equatorial low-pressure zone (the *equatorial trough*). Reference to Fig. 2.1.3 shows that this circulation pattern is most distinct over the Atlantic and Pacific but is greatly modified by the monsoon circulation over the Indian Ocean–Indonesian sector.

2. Poleward of the equatorial trough is the belt of *subtropical high pressure.* Progressing to higher latitudes, one finds a low-pressure belt associated with the *extratropical storm tracks,* where the eastward-migrating cyclonic (low-pressure) and anticyclonic (high-pressure) weather systems are concentrated. Finally, there is a tendency for high pressure over the cold polar cap.

3. There is a pronounced seasonal variation in the atmospheric circulation, particularly in the northern hemisphere where the ocean-continent contrast is greater. During the northern summer, the oceanic subtropical highs are strongly developed and expand northward. Relatively low pressure appears over the warm continents, notably over South Asia, where the monsoon low is well developed.

During the northern winter (Fig. 2.1.3*a*) storminess increases over the North Atlantic and North Pacific, which is reflected in the appearance of pronounced monthly averaged low-pressure areas in the northern latitudes and a weak subtropical high-pressure belt. One notes a seasonal reversal from summer in the overall distribution of pressure between continents and oceans in the middle and high latitudes, with relatively high pressure over the cold continents and relatively low pressure over the oceans.

4. Outside the deep tropics, the rotation of the earth is a fundamental factor in determining the relationship between circulation and pressure. In a nonrotating

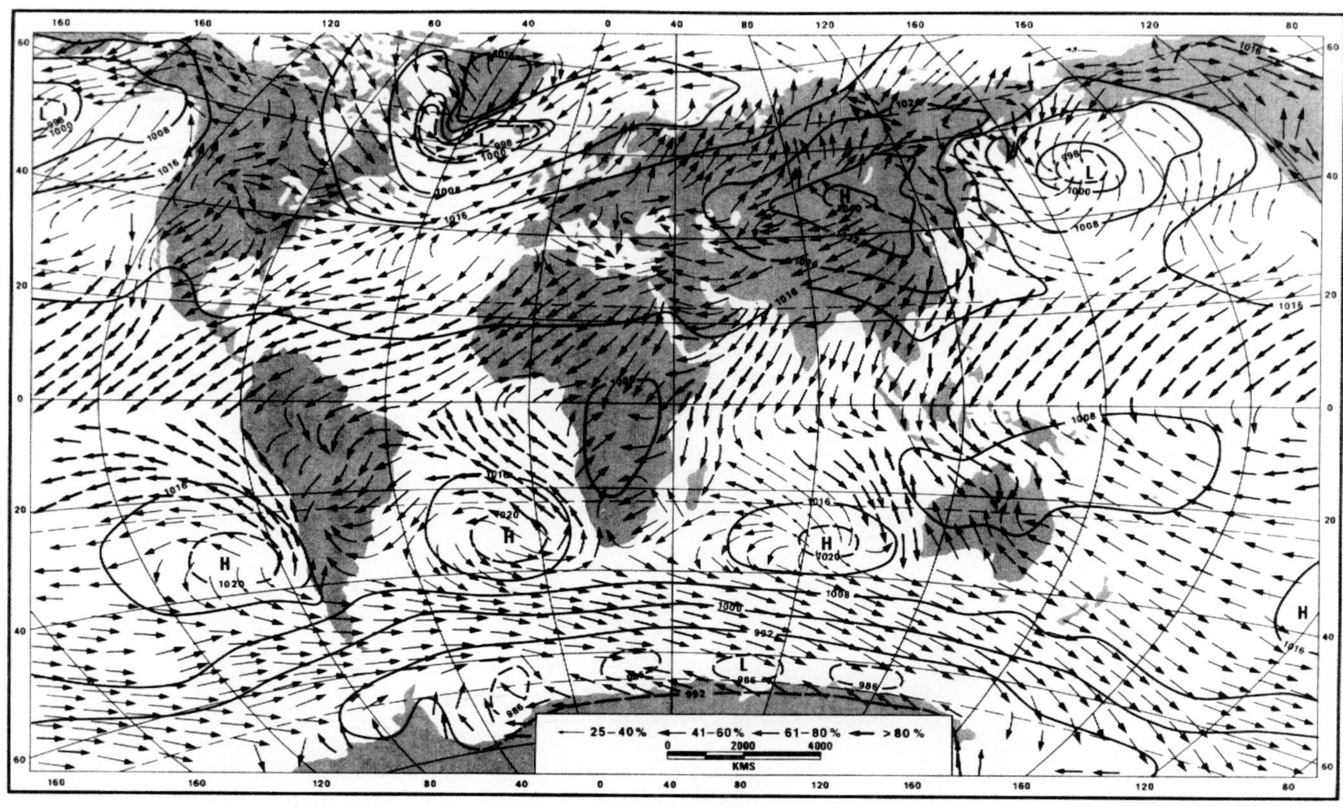

FIGURE 2.1.3 Average sea-level pressures (in millibars) and prevailing surface winds. Width of arrow indicates consistency of wind direction. Part *(a)* January. *(From Figs. 6.2 and 6.3, Ref. 55.)*

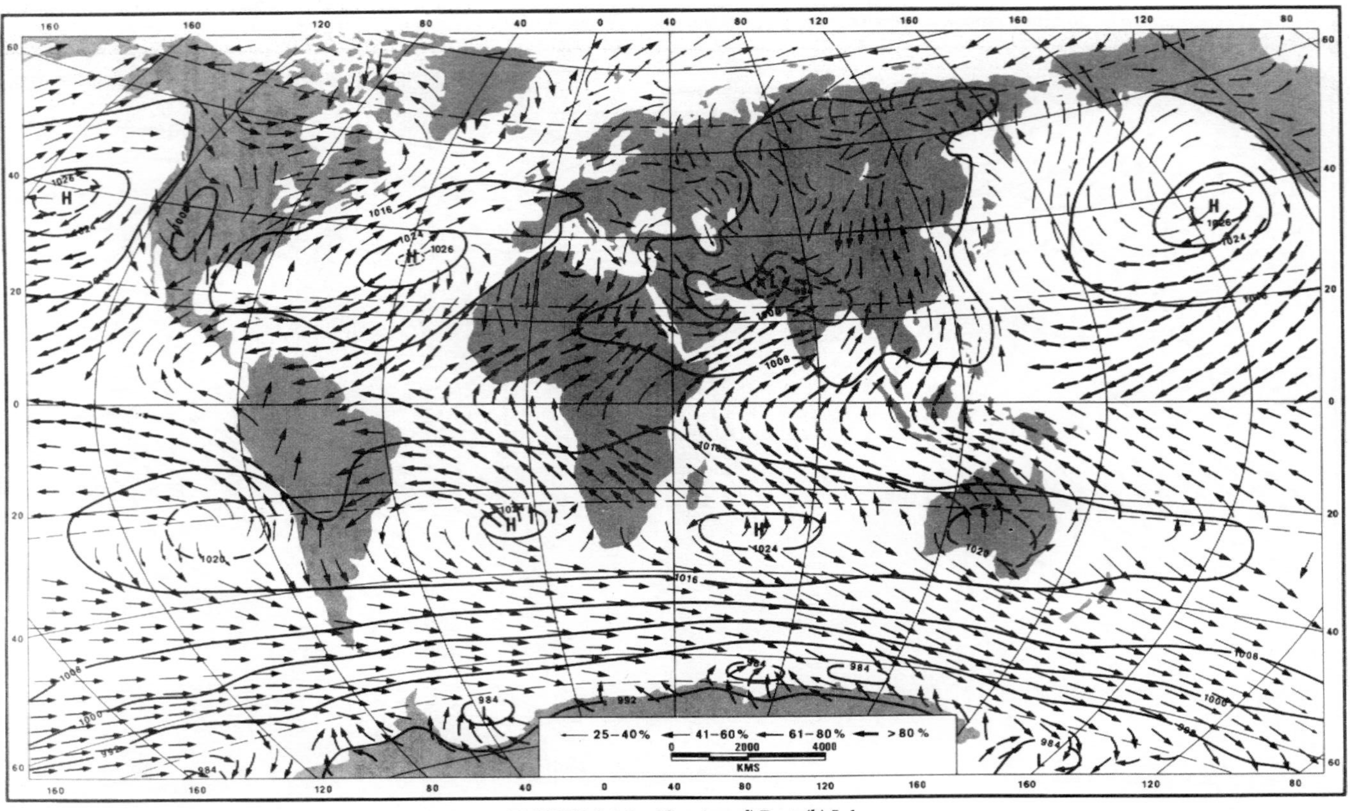

FIGURE 2.1.3 *(Continued)* Part *(b)* July.

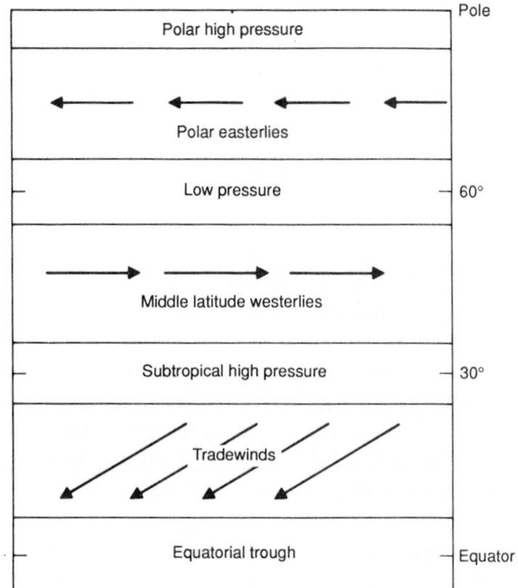

FIGURE 2.1.4 Simplified representation of the latitude averaged surface wind and pressure belts. *(Adapted from Fig. 4.9, Ref. 30. Used with permission.)*

system, the flow is largely a direct response to the pressure gradient and is essentially directed from areas of high pressure to areas of lower pressure. However, the earth's rotation gives rise to an additional apparent *Coriolis force* which results in a strong *geostrophic component* of flow. In the northern hemisphere, this appears as a clockwise circulation around high-pressure (anticyclonic) systems and a counterclockwise circulation around low-pressure (cyclonic) systems. The reverse relationship holds in the southern hemisphere. This feature of the circulation is clearly evident on the monthly mean charts of Fig. 2.1.3.

Upper-Level Circulation. The atmospheric circulation generally intensifies and simplifies upward through the troposphere. This is primarily due to the *thermal wind* effect. Simply put, the decrease in pressure with height is greater in the colder air of polar latitudes than it is in the warmer air of the subtropics. This results in an increasing north-south horizontal pressure gradient with height. Because of the geostrophic relationship between pressure and circulation previously noted, this results in stronger westerly winds with height in the subtropics and middle latitudes. The axis of strongest westerlies in the upper troposphere is called the *jet stream.*

Jet streams are continually developing, meandering, and decaying as part of the wavy, planetary-scale circulation systems of the upper troposphere. These upper tropospheric circulation features are associated with the ever-changing regional weather regimes of the extratropics.[35] The presence of these transient *synoptic features* of the middle-latitude circulation is implied by the relatively low consistency in wind direction in midlatitudes shown in Fig. 2.1.3, but their importance as a fundamental element of the general circulation of the atmosphere is not adequately represented by maps of the time-average circulation.

General Circulation. The first physically based theory of the general circulation was presented in 1735 by George Hadley.[53] Hadley reasoned that the equatorward direction of the low-level easterly winds of the vast oceanic trade-wind belts of the subtropics (Fig. 2.1.4) reflects the lower branch of a hemisphere-wide thermal convection cell, with rising motion in the warm equatorial belt and sinking motion in the subtropics and colder temperate zones. Such a circulation would be driven by the difference in solar heating between the tropics and high latitudes. Hadley considered the transient circulations associated with day-to-day weather changes, along with the mean differences in the circulation around a latitude band, and the annual cycle of temperature and other meteorological elements to be irrelevant "ornaments of the circulation," which are uncoupled from the fundamental processes of the general circulation. According to this theory, the general circulation could be described once and for all as a longitudinally uniform, steady circulation.

It is now known that the general circulation is far more complex than the simple one-cell *mean meridional circulation* pattern in each hemisphere envisioned by Hadley. Furthermore, the more recent picture of the general circulation illustrated in many monographs and textbooks that are widely used as references by hydrologists is also often oversimplified and several decades out of date. The major features of the observed zonally averaged circulation, as revealed by global observations during the past few decades, are shown in Fig. 2.1.5. The observations reveal a three-cell mean meridional circulation pattern (Fig. 2.1.5), but only in the tropics is the mean meridional circulation of central importance. The strong thermally driven circulation of the tropics is known as a *Hadley cell.* However, the tropical Hadley circulation cannot be adequately described by an annual average, for it changes seasonally. Except for the short periods during the equinox seasons (spring, autumn) the mean meridional circulation of the tropics is dominated by a single summer hemisphere Hadley cell. The zone of rising air shifts seasonally, being in the southern hemisphere from December to February (the southern summer) (Fig. 2.1.5*a*) and in the northern hemisphere from June to August. This zone of rising air is balanced by descending air on the opposite (winter) side of the equator. The center of the zone of rising air coincides with the warmest surface temperatures (thermal equator), and it shifts seasonally, north and south, following the sun.

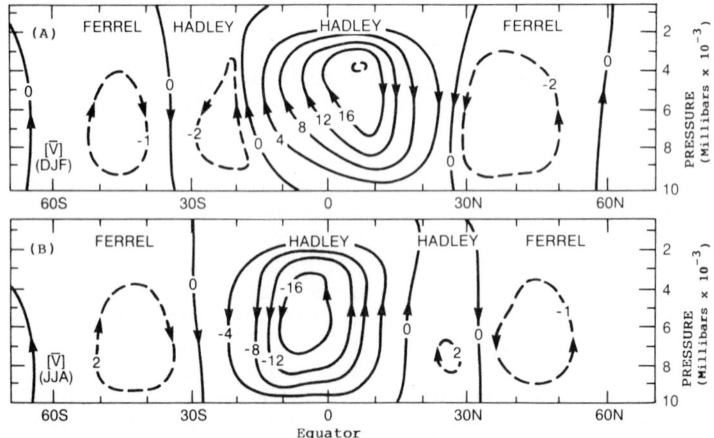

FIGURE 2.1.5 Mean meridional (latitude averaged) circulation of the atmosphere. (*a*) December–February, (*b*) June–August. Values on the streamlines indicate total mass circulation (10^{10} kg/s) between that streamline and the zero streamline. *(Data from Ref. 72.)*

Outside the tropics, the components of the mean meridional circulation are quite weak. They consist of weak indirect Ferrel cells in middle latitudes and even weaker direct polar cells at high latitudes. At these latitudes it is not the mean meridional circulation features but rather the "eddy transport" by the transient synoptic weather systems that is of central importance for the general circulation. It is these transient features rather than the latitudinally averaged features which transport warm, moist air poleward and cold, dry air equatorward, and account for most of the wintertime poleward transport of heat and water vapor.

2.1.3 Ocean Circulation

With 71 percent of its surface covered by water, the earth is largely an ocean planet. Because of its huge heat capacity, the ocean acts as the principal heat storage or "memory" component of the climate system. In contrast to the atmosphere, which is largely heated from below (see Sec. 2.3), the ocean is primarily heated and cooled at its upper surface. This results in an oceanic thermal structure that is fundamentally different from that found in the atmosphere. In terms of temperature, the tropical and temperate latitude oceans can be crudely divided into an upper zone and a deep zone (Fig. 2.1.6). The upper zone extends from the surface to 200 to 1000 m (656 to 3280 ft) depth. Within this zone is an upper *mixed layer* in which temperatures are similar to those at the surface. Below the mixed layer is a transition layer, called the *thermocline,* in which the temperature decreases rapidly with depth. Below the thermocline, in the deep ocean, temperatures decrease very slowly with depth, and temperature changes are relatively slow and small.

Except in polar waters, a "main" or "permanent" thermocline is present at all times. In middle latitudes, the temperature of the upper layer shows seasonal varia-

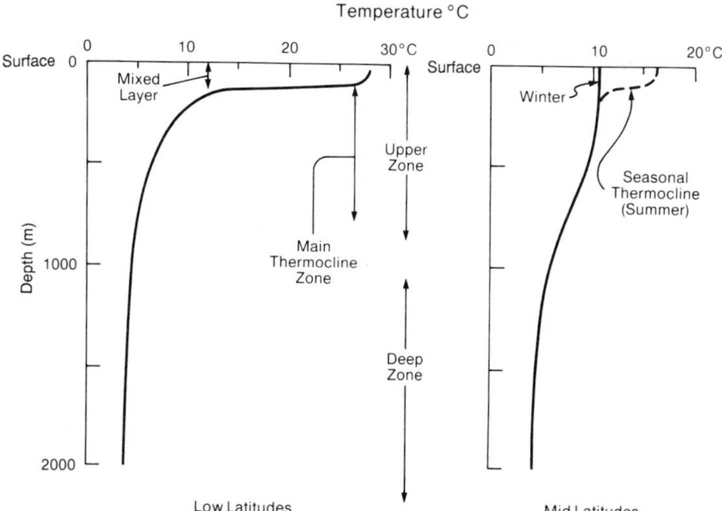

FIGURE 2.1.6 Typical ocean temperature vertical profiles for the tropical and temperature latitudes. Unit conversion: 1 meter = 3.28 feet. *(Adapted from Fig. 4.5, Ref. 77. Used with permission of Pergamon Press PLC.)*

tions associated with the summer storage and winter release of heat. In winter the surface temperature is low, and the mixed layer is deep and may extend to the main thermocline. In summer the surface temperature rises, the upper layer becomes more stably stratified and resistant to mixing, and a shallow "seasonal" thermocline often develops.

The mean annual sea surface temperature field is shown on Fig. 2.1.7. Mean annual open ocean surface temperatures range from more than 29°C (302 K) in parts of the tropics to near −1.8°C (271.4 K) (freezing point of salt water) at the ice edge. The SST isotherms are oriented roughly east-west over much of the ocean but deviate from this pattern near the coasts and in the vicinity of the equator, where ocean currents and upwelling processes play an important role in the temperature distribution.

The circulation systems of the ocean and atmosphere are coupled through energy and momentum exchanges across the air-sea interface. The atmosphere "feels" the ocean at its lower boundary by means of the sea surface temperature. The ocean, in turn, feels the atmosphere through the surface wind stress, which produces *wind-driven circulations* in the upper ocean. The ocean also responds to surface heat fluxes and to evaporation and precipitation, which change the ocean temperature and salinity. This results in density changes of the surface water that drive the deep *thermohaline circulations* of the ocean.

Thermohaline Circulation. Temperature and salinity determine the density of sea-water. The thermohaline circulation of the oceans refers to the slow circulation of water through the oceans arising from density changes. The thermohaline circulation originates in polar latitudes as a vertical flow sinking to middepth or even to the ocean bottom, followed by horizontal flow.[77] It is initiated by an increase of density at the upper surface, either directly by cooling and/or salinity increases or indirectly when ice freezes out, ejecting salt and thus increasing the salinity of the remaining water. While the downward branches of the thermohaline circulation occur over relatively restricted regions of the high-latitude ocean, the compensating upwelling branches are spread over large areas of the world oceans.

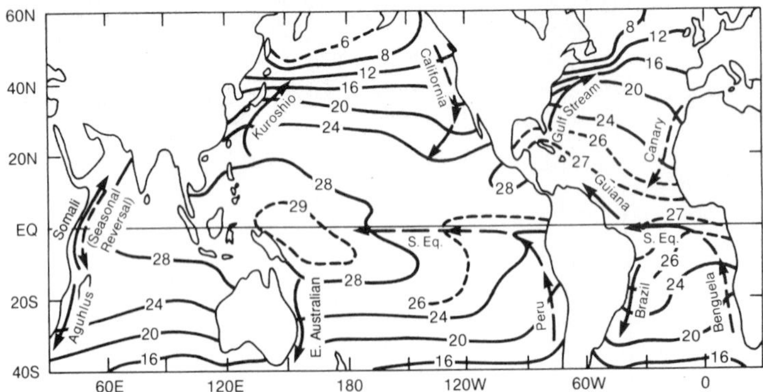

FIGURE 2.1.7 Annual mean sea surface temperature (SST) (°C). Data are from the period 1970–1988. Also shown schematically are several extratropical surface ocean currents. (See Refs. 14 and 20 for discussions of upwelling dynamics.) Unit conversion: K = °C + 273.2. *(SST data provided by R. Reynolds, NOAA Climate Analysis Center. Currents adapted from charts in Ref. 77.)*

The global scale redistribution of ocean water by thermohaline circulations operates on time scales of decades to centuries. On average, it takes around 1600 years for one complete cycle, i.e., for water to circulate from the surface to the deep ocean and back to the surface.[8] However, there is also evidence of rapid changes in the thermohaline circulation of the North Atlantic over periods as short as a decade or two,[51] suggesting that the thermohaline circulation could be a significant factor in climate variability on decadal time scales.

Wind-Driven Circulation. In contrast to the thermohaline circulation, the wind-driven circulation is primarily horizontal in nature and is limited to the upper few hundred meters of the ocean. Some of the regional components of the wind-driven circulation are indicated schematically in Fig. 2.1.7. Because of the complexity and seasonal variability of the circulation, larger-scale regional charts are required for a detailed description. These can be found in oceanographic texts and atlases.[50,77,93,96]

At the more southerly latitudes of the southern hemisphere (not shown in Fig. 2.1.7) where there are strong and persistent surface westerlies and no continental barrier, the *antarctic circumpolar current* is continuous around the latitude circle. Otherwise, outside the deep tropics, the largest-scale wind-driven circulation systems of the Atlantic and Pacific Oceans are the basin-scale *subtropical anticyclonic gyres.* They arise as a result of the prevailing winds, the continental boundaries, and the rotation of the earth.[77] They include relatively narrow and intense warm poleward current systems on the west side of the ocean basins, such as the Gulf Stream in the North Atlantic Ocean and the Kuroshio current east of Japan, which transport huge amounts of heat poleward, and cold, upwelling equatorward currents on the east sides of the basins (Fig. 2.1.7) such as the California current in the North Pacific Ocean and the Peru current in the South Pacific Ocean, which transport colder water equatorward.

The peculiar ocean dynamics in the vicinity of the equator give rise to a circulation regime quite different from that at higher latitudes.[28] This low-latitude circulation is characterized by a system of east-west, or zonal, currents, some only a few hundred kilometers in width. The complexity and small north-south extent of these currents preclude their inclusion in Fig. 2.1.7. Major changes in these equatorial current systems may be relatively rapid compared with those at higher latitudes, and thus they are more in tune with the time scale of year-to-year atmospheric variability, a crucial factor in the El Niño–southern oscillation (ENSO) cycle (see Sec. 2.1.4).

In contrast to the equatorial Atlantic and Pacific Oceans, the pronounced seasonal changes in the monsoon circulation (see Sec. 2.1.4) result in a seasonal reversal of the major wind-driven circulation systems of the low-latitude Indian Ocean.

2.1.4 Coupled Ocean and Atmosphere Processes

Annual Cycle. Through the storage of heat during summer and its poleward transport and release during winter, the oceans moderate the seasonal cycle of land surface temperatures. Consequently, the amplitude of the annual climate cycle in the southern hemisphere (81 percent ocean) is considerably smaller than that in the northern hemisphere (61 percent ocean).

Around 90 percent of the northern hemisphere seasonal heat storage takes place in the ocean.[73] Absorption of solar radiation and northward heat transport from the tropics increase the middle-latitude heat content of the oceans during the summer months. During winter, the high-latitude continents lose heat through radiation and are the source regions where cold arctic air masses form. These cold continental air masses regularly surge southward and eastward from Asia and North America over the warm waters of the western Pacific and Atlantic behind transient cyclones and associated cold fronts. The large contrast in temperature and moisture content between the cold continental air and the warm ocean surface results in huge wintertime fluxes of latent and sensible heat from the ocean to the atmosphere. The heat flux from the ocean quickly modifies the cold continental air mass as it moves eastward across the ocean, with the surface air taking on maritime characteristics long before it reaches the west coast of North America or Europe. In Europe, which has no substantial north-south mountain barriers, the maritime influence extends far inland, while over North America it is largely confined to the area west of the major coastal mountain barriers.

Monsoon Circulations. Since the heat capacity of the soil is small compared with that of the upper ocean, the amplitude of the annual cycle in land surface temperatures is far greater than that occurring in tropical sea surface temperature. The resulting annual cycle of land-ocean surface temperature difference along with the seasonal reversal in the sea surface temperature difference between the hemispheres drives the atmospheric *monsoon circulations* of the tropics.

The *Asian-Australian monsoon* system is the dominant monsoon circulation of the earth[32,80] (Fig. 2.1.8). During the winter phase of the monsoon, there is a low-level flow of dry, cool air from the cold continent to warmer ocean, and precipitation over the land is light. During summer, there is a flow of atmospheric moisture from the tropical ocean to the warmer land, where the upward motion of the heated air produces the heavy rains of the monsoon season.

Similar but less pronounced monsoon-type circulations also occur over west Africa and portions of Mexico and Central America, where they extend as far north as the southwestern United States. Monsoon regions exhibit a summer wet season,

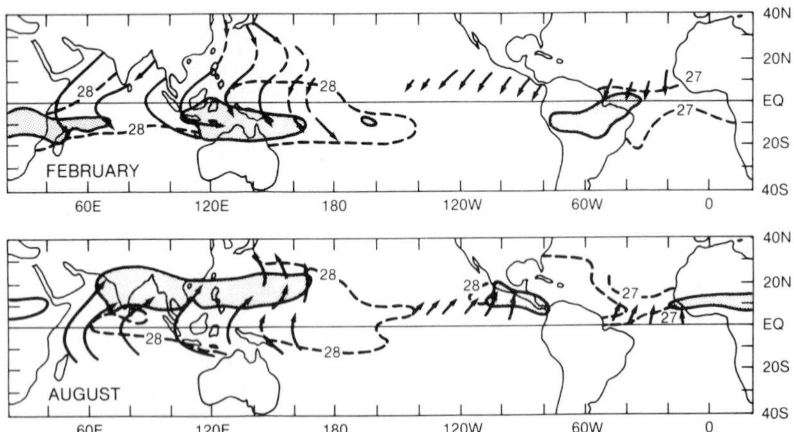

FIGURE 2.1.8 Monsoon component of atmospheric surface circulation. This diagram shows the departure of the monthly mean surface circulation during February and August from its annual mean value. Also shown are the 28°C (301.2 K) and 27°C (302.2 K) (Atlantic) sea surface temperature isotherms for the midsummer month of the respective hemisphere. The hatched areas delineate the regions where satellite data indicate much heavier precipitation than the average for all months. The charts are based on data provided by the NOAA Climate Analysis Center.

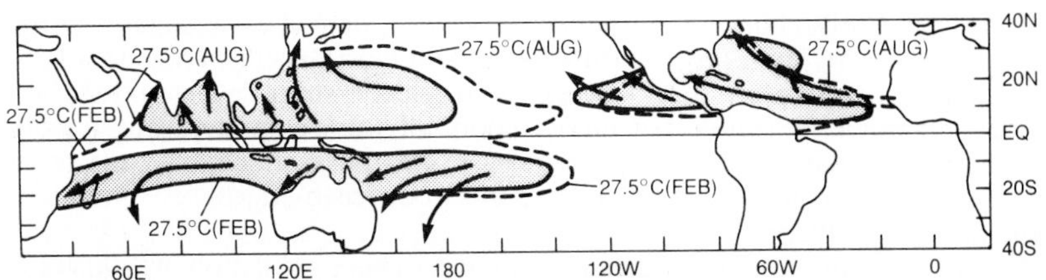

FIGURE 2.1.9 Primary paths and regions of tropical storm genesis (hatched). Also shown is the 27.5°C (300.9 K) sea surface temperature isotherm for August (northern hemisphere) and February (southern hemisphere). *(Adapted from Figs. 4 and 8 of Ref. 29. Used with permission.)*

and most have a winter dry season, but beyond this common characteristic, regional monsoon precipitation regimes vary widely.

Tropical Cyclones. The transient synoptic features of the tropics—depressions and tropical storms—each contain an area of intense convection, generally referred to as a *cloud cluster* because of its appearance on meteorological satellite images. These cloud clusters are the building blocks which in aggregate produce the large-scale, time-averaged precipitation patterns associated with the monsoons and oceanic convergence zones.

Tropical cyclones are the most energetic transient weather systems of the tropics. A mature tropical cyclone is a disturbance whose central core is considerably warmer than the surroundings. Tropical storm intensity is reached when a disturbance has sustained winds of 17.5 m/s (40 mi/h), and hurricane intensity is reached when winds reach 33 m/s (75 mi/h). *Hurricane* and *typhoon* are regional names for the same phenomenon.

Three necessary conditions have been identified for the formation of intense tropical cyclones.[29,73] First, a warm ocean surface is necessary to provide the required fluxes of water vapor and sensible heat from the ocean to the atmosphere. The lowest water temperature over which warm-core cyclones are likely to develop is about 26 to 27°C (299 to 300 K). Second, since strong rotation can be generated only in regions of significant Coriolis force (Sec. 2.1.2), these storms cannot form within about 5° to 8° of the equator, even though ocean temperatures may be high enough. Finally, a small change of wind with height is required if the storm is to survive.

Figure 2.1.9 shows the primary regions of tropical storm genesis. Most of the tropical storm activity occurs during the summer and autumn seasons over areas where sea surface temperature is 27.5°C (300.5 K) or higher. The absence of tropical storm activity over the South Atlantic and the eastern South Pacific is a consequence of the relatively low sea surface temperatures over those regions. Since the ENSO cycle is associated with changes in both sea surface temperature and vertical wind shear, it affects tropical storm activity.

El Niño-Southern Oscillation (ENSO) Cycle. This phenomenon is the most important source of year-to-year climate variability in the tropics and is a significant source of interannual climate variability in some regions of the higher latitudes as well.[82] It has a pronounced effect on the intensity of the Asian-Australian monsoon, particularly on the year-to-year variations in monsoon rainfall over India, Indonesia, and northern Australia.

El Niño refers to an anomalous warming of the upper ocean along the northern coast of Peru which typically begins after Christmas, hence the Spanish name El

Niño, or "the Christ child." The *southern oscillation,* named to distinguish it from the *northern oscillations* of the North Atlantic and North Pacific, refers to a large-scale seesaw in surface pressure between the tropical eastern Pacific and the west Pacific–Indian Ocean region, but more importantly, it is associated with a global pattern of year-to-year climate variations. The two phenomena are parts of the *ENSO mode* of climate variability which results from ocean-atmosphere interaction in the tropical Pacific.[82]

The major features of this coupled ocean-atmosphere oscillation are illustrated schematically by Fig. 2.1.10, which describes the situation in the central equatorial Pacific during the warm phase of the oscillation. We arbitrarily begin the description of this interactive feedback loop with branch A.

A. Above normal sea surface temperatures lead to enhanced, convective rainfall. The latent heat released to the atmosphere drives both local and distant circulation anomalies.

B. The remote circulation anomalies *(teleconnections)* result in changes in seasonal temperature and precipitation regimes over many regions of both the tropics and extratropics.

C. Locally, the enhanced convection results in westerly surface wind anomalies near the equator.

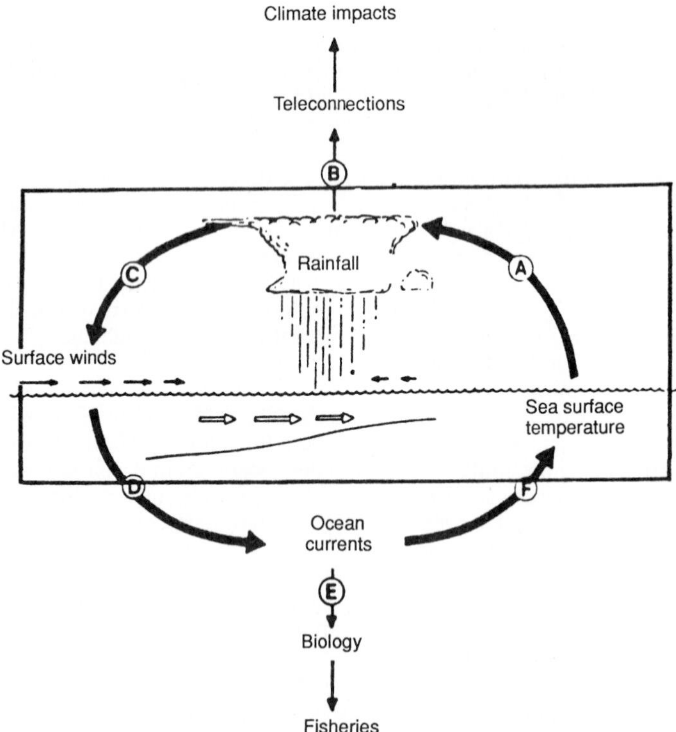

FIGURE 2.1.10 Schematic representation of the El Niño–southern oscillation (ENSO) cycle interactions. See text for discussion. *(Reprinted with permission from "U.S. Participation in the TOGA Program: A Research Strategy," 1986. Published by the National Academy Press, Washington, D.C.)*

D. The change in the surface wind results in changes in the ocean circulation which result in an eastward extension of the relatively warm water located over the western tropical Pacific (see Fig. 2.1.7).

E. The changes in surface wind over the central equatorial Pacific also lead to a remote ocean response in the eastern equatorial Pacific. Here, the upwelling of cold water decreases, leading to a warming of the upper layer of the ocean. This may seriously disrupt the oceanic food chain, leading to a dramatic decrease in the fish population. The warm water also brings heavy rains to the otherwise desert regions of coastal northern Peru and southern Ecuador.

F. Changes in the equatorial ocean currents lead to a further increase in sea surface temperature, resulting in further enhancement of convection, thus closing the feedback loop.

This feedback cycle typically exists for around a year, after which it reverses and the cold phase of the cycle begins. The time scale of the oscillation between warm and cold phases appears to be largely controlled by the complex ocean dynamics of the equatorial Pacific.[82]

2.1.5 Climate Models

Types of Models. Before the era of digital computers, a *climate model* was either a qualitative, conceptual model based on subjective interpretation of the governing equations or a highly simplified and often unrealistic simplification of these equations that could be solved analytically. While such analytic models are still a useful tool for theoreticians, a modern climate model is a computer program for integrating a set of equations which approximate the behavior of the climate system.

Models can be ranked in a hierarchy with respect to both type and degree of time-space resolution.[4] Simpler models may compute the spatial distribution at equilibrium of a single key climate parameter, usually temperature, as a function of latitude, elevation in the atmosphere, or both. Such models are primarily concerned with changes in gross climate conditions that occur when the climate system's near equilibrium between energy gain and loss is disturbed. Such models are quite simple in design but may give useful information about the sensitivity of the climate system to changes in variables such as solar radiation and *planetary albedo,* the percent of incoming solar radiation reflected back to space. Also, they help in interpreting the results of more complex models.

The most elaborate models depict climatic quantities for all three spatial dimensions and indicate their variations in time. These models are commonly referred to as *general circulation models* (GCMs) since they explicitly simulate the major features of the atmospheric circulation, and for interactive models, the ocean circulation as well, using dynamic principles. General circulation models predict future conditions from given initial and boundary conditions that describe a particular state of the climate system.

Whatever the resolution of the model, a representation of each climate process, such as watershed-scale hydrology, that is smaller in space or time scale than the resolution of the model grid, must be incorporated implicitly within the model unless it can be determined that such a process is unimportant for the particular model or application. These representations, which link such *subgrid scale processes* to those resolved by the model, are referred to as *parameterizations.* The realism of the climate model's results depends strongly on the success of these parameterizations.

Grid cells of around 5° latitude by 5° longitude were typical of early GCM climate simulations, but as computational power increases, the size of the cell is being decreased. Thus the mismatch between resolution of the model and the scales of interest to hydrologists continues to narrow.

There are components of the total climate variation that result from physical processes that cannot or need not be modeled deterministically. Some models take these variations into account by assuming that they are random about their mean values. These *stochastic models* yield probabilistic as well as deterministic predictions about the climate.[4]

2.2 DOCUMENTING THE RECENT CLIMATE

Various national networks have been established to collect climate data. *Climatological stations* are categorized by the World Meteorological Organization (WMO) as "principal and ordinary; precipitation stations; stations for specific purposes."[99] In the United States, official climatological records are collected from a variety of stations, and observations are taken by both paid and unpaid observers. Daily accumulated precipitation and maximum and minimum temperature are measured at almost all stations. In 1990, there were 910 manned stations which took observations hourly or four times daily, 11,330 climatological substations, and 170 automatic stations reporting to the U.S. National Weather Service (NWS).[75] Because of cost considerations, the number of automatic stations was expected to reach around 1000 by the mid-1990s. In many cases they will replace existing manual observation stations.

The major climate-observing networks now operated by the U.S. National Oceanic and Atmospheric Administration (NOAA) include the following:

1. The *principal climatological station network for local climatological data (LCD) stations* defines climate for about 275 major cities and towns.
2. The co-op network defines the baseline climate for the United States using about 5200 temperature stations and 7400 precipitation stations.
3. The *hourly precipitation data (HPD) network* of about 3000 stations establishes baseline rainfall records for hydrologic purposes.
4. The *historical climate network (HCN)* of 1215 temperature and precipitation stations is the definitive network for documenting climate variations and change. Figure 2.2.1 shows the location of these stations. Data from gauges in this network have been carefully checked for errors and missing values, and those data are the most reliable basis for hydrologic studies which require climatic data in the United States.
5. The newest NOAA *reference climate network* (RCN) serves two purposes:
   - ***a.*** *264 LCD stations (LCD RCN)* observe many climate elements. They are the baseline climate stations for elements other than temperature and precipitation. This is also the baseline network for monitoring climate for many major cities and towns.
   - ***b.*** *640 co-op stations (co-op RCN),* the best of stations in the HCN plus Hawaii, Alaska, and U.S. island stations, form a high-quality baseline network for monitoring temperature and precipitation climate trends.

Efforts have been made in some countries to identify high-quality *benchmark data sets* which are believed to be minimally affected by nonnatural climate factors. A good example is the hydro-climatic data network (HCDN) of discharge data for the United States identified by the U.S. Geological Survey.[48]

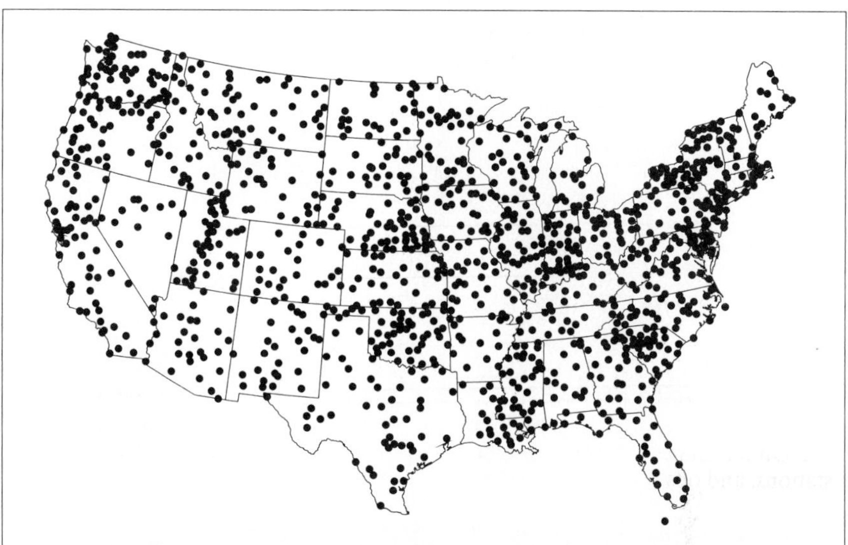

FIGURE 2.2.1 Distribution of climate observation stations in the U.S. historical climate network (HCN). Each dot covers an area of approximately 1620 km^2 (625 mi^2), which is considered a desirable density for climate observations. *(Chart provided by the National Climatic Data Center, Asheville, N.C.)*

2.2.1 Instrumentation

The World Meteorological Organization (WMO) has established requirements for *instrument shelters* in which temperature and humidity sensors are housed.[99] Figure 2.2.2 shows the standard wooden Cotton Region instrument shelter used in the United States. The National Weather Service has described this shelter,[70] and provided instructions for the exposure and installation of its instruments.[71]

A comprehensive review of sensors used for routine and special meteorological observations and an extensive bibliography on instruments and observational procedures can be found in Ref. 62. In addition to the standard mercury thermometer and maximum and minimum *thermometers,*[64] temperature sensors are available that use thermocouples, resistance elements, thermistors, and electric diodes and transistors. Before 1986, observers at over 90 percent of the U.S. climatological stations simply recorded the daily maximum and minimum temperature obtained from standard maximum and minimum thermometers housed in a standard wooden instrument shelter. The remainder used other types of thermometers, *hygrothermographs* (an instrument that records both temperature and humidity), and *thermographs.* Between 1986 and 1991, the National Weather Service replaced over half of its liquid-in-glass maximum and minimum thermometers in the standard shelters with thermistor-based maximum-minimum temperature systems (MMTSs) housed in the smaller plastic shelter,[79] which is also shown in Fig. 2.2.2.

Wind Measurement. Wind is simply air in motion relative to the surface of the earth. The standard level for *wind measurements* is 10 m (32.8 ft) above the ground over level open terrain, but many observations are taken at nonstandard levels. This creates serious problems of homogeneity, since the wind typically increases rapidly through the lowest few meters of the atmosphere. Equipment available for measuring

FIGURE 2.2.2 Standard medium-sized Cotton Region instrument shelter for housing temperature and humidity sensors. Height of the bottom of the shelter is 4 ft (1.12 m). Height of the thermometers is approximately 5.5 ft (1.68 m). Also shown on the right of the Cotton Region shelter and to the left of the rain gauge is the smaller plastic shelter which houses the newer thermistor-based maximum-minimum temperature system (MMTS) of the U.S. National Weather Service. *(Figure provided by the National Climatic Data Center, Asheville, N.C.)*

the horizontal wind speed and direction includes cup anemometers and standard wind vanes, propeller anemometers and vanes, and many others, each with its own dynamic characteristics.[62] The choice of the measurement technique depends on the application.

Radiation Measurement. Surface measurements of radiation are far less common and more difficult than measurements of the standard meteorological parameters. Consequently, the quality of much of the climatological radiation data is less reliable.[41] Instruments for measuring radiation are designed to measure the intensity of radiant energy over broad or narrow spectral bands. The radiation fluxes of interest may be divided into fluxes with wavelength less than 4.0 μm (4×10^{-6} m) *(short-wave solar radiation)* and fluxes whose wavelength is more than 4.0 μm *(long-wave terrestrial radiation).* The short-wave solar radiation can in turn be subdivided into *ultraviolet* (uv, between 0.2 and 0.38 μm), *visible* (0.38 to 0.75 μm), and *infrared* (greater than 0.75 μm radiation).

The generic term *radiometer* refers to any instrument for measuring radiation, regardless of its spectral characteristics and its mode of operation. The WMO classification of radiation instruments identifies five basic types.[99]

Pyranometer measures global solar radiation—that is, the short-wave radiation received on a horizontal surface from sun and sky (Fig. 2.2.3). Use of a masking device to screen out the sun's direct beam allows measurement of the diffuse component of solar radiation. (Previously called a *pyrheliometer.)*

Pyrheliometer measures only the direct beam component of the solar radiation, i.e., the solar radiation received on a horizontal surface.

Pyrradiometer measures the total short-wave and long-wave radiation incident on a horizontal plane.

Net pyrradiometer or *net radiometer* measures total (short-wave and long-wave) net radiative flux through a horizontal plane.

Pyrgeometer measures the long-wave atmospheric radiation on a horizontal upward-facing black surface at ambient air temperature.

In addition to these five general terms, trade names are often attached to specific instruments.

To ensure the reliability and compatibility of radiation measurements on an international and national scale, the WMO has established a system of world, regional, and national centers that provide facilities, maintain equipment, and conduct comparisons. The primary North American center is located at the NOAA Environmental Research Laboratories Solar Facility, 325 Broadway, Boulder, Colo., 80303.

FIGURE 2.2.3 A pyranometer for measuring solar radiation. This instrument is screened from the sun's direct beam by a shade ring in order to measure diffuse radiation. Without the ring, the instrument measures total solar radiation received on a horizontal surface. *(Photograph provided by U.S. National Weather Service.)*

2.2.2 Data Homogeneity

In using climatological data, one needs to be aware of the many types of *biases* which may appear in the data sets. These can arise because of:[41]

1. More (less) precise or accurate measurements
2. Changes of temporal and/or spatial sampling and/or processing
3. Microclimatic changes of the local sampling environment

Precipitation Measurement. Perhaps the best example of inhomogeneities due to changes in accuracy and precision is the measurement of precipitation. It is well known that current and past measurement practices underestimate the true precipitation.[85] This has led scientists, operational network managers, and system designers to strive for improved methods of measuring precipitation. Of particular importance is the change from *unshielded* to *shielded precipitation gauges.* Figure 2.2.4 shows the effect of changes in gauging methods on the average measured amount of liquid and solid precipitation in a number of countries. The biases are not necessarily uniform throughout the country. For example, prior to 1948 there were no shields on United States gauges. By 1980, 44 percent of all the contiguous United States precipitation gauges had shields installed around the orifice of the gauge.[41] It is apparent that the magnitudes of these changes are such that they make the detection of climate change from these records tenuous at many stations within these countries. Information on changes in the type of gauge used is found in the United States in the detailed manuscript *station histories* available from the National Climatic Data Center, Asheville, N.C.

Radiation Measurement Error. Another example of instrument accuracy involves the change in the measurement of sunshine in the United States. Three major changes in instrumentation have made the analysis of twentieth century changes in sunshine very difficult.[41]

An important climatological element which has been subject to biases associated with poor calibration and instrument drift is the measurement of solar radiation at the earth's surface. In the United States these measurements of solar radiation began during the 1950s, but calibration and maintenance problems have existed during much of the period of record.[40] In 1976, a Department of Energy (DOE) program was initiated to correct solar radiation instrument bias in the archived record. Although this has improved the data base for many climatological applications,[68] the radiation data are not recommended for trend analysis. Calibration procedures since the mid-1970s have been quite rigorous, and the data will be useful for documentation of climate change if the instruments are well maintained.[41] The WMO is actively coordinating a global *baseline radiation network* which includes a significant quality-control effort.[98]

Temperature Measurement Error. Changes in instrumentation often result in changes in the character and accuracy of the measurements which require adjustments in the data to homogenize the record. The recent change from liquid-in-glass maximum and minimum thermometers to the thermistor-based maximum-minimum temperature system (MMTS) in the United States has had such an effect. Comparisons between the two systems[79] show a mean daily minimum temperature change of +0.3°C (+0.6°F), a mean daily maximum temperature change of −0.4°C. While the change in mean daily temperature was only −0.1°C, the reduction of 0.7°C in the daily temperature range is quite substantial for climate change studies and requires adjustment of the record.

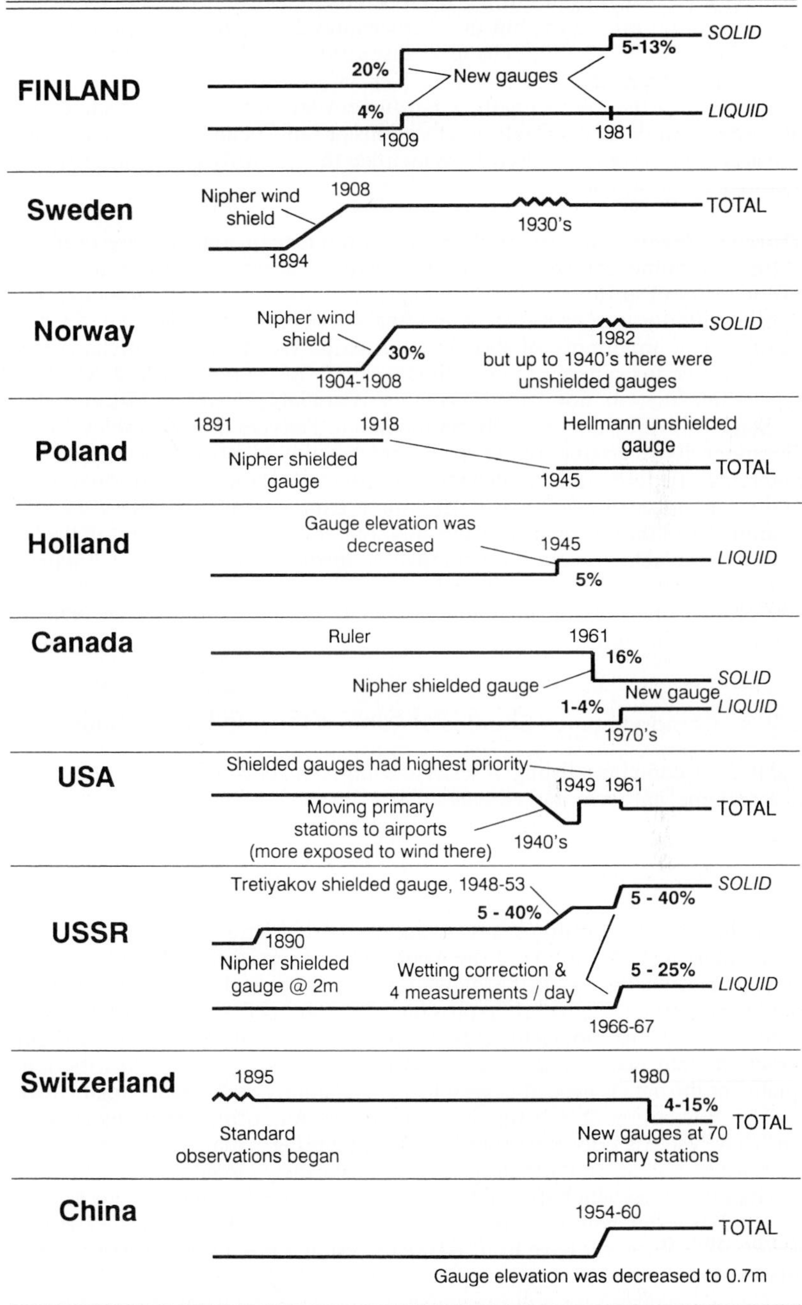

FIGURE 2.2.4 Discontinuities in the time series of measured precipitation arising from changes in gauging methods. (See Ref. 41 for more detailed discussion.) *(Adapted from Fig. 3, Ref. 41.)*

In the measurement of temperature, problems arise because of differences in the method for calculating mean daily temperature.[41] Virtually every country in the world has a set procedure, but these procedures differ from one another and they often change with time. Such changes cause serious problems of interpretation of the thermometric records.

Changes in the microclimatic environment around a sampling site can occur abruptly, as in the case of station relocations or rapid changes in land use, but they can also occur more gradually, as is sometimes the case with the transition from rural to urban environments.

Detecting Inhomogeneities in Records. A number of methods are available for detecting inhomogeneities in climate records in the absence of information about the history of the observations.[1,60,78] Although extremely useful, these methods should be considered only as a last resort, or as a final check of the homogeneity of a station's records. The importance of specific information on the timing of changes in the station observing system or microclimate cannot be overemphasized. Much of this type of information, now called *metadata,* is available in station histories.

Station histories contain information about the types of instruments used, their elevations above ground and exposure, information about the local surroundings, observing schedules, and maintenance procedures. Without such information, many of the inhomogeneities which exist in climatic data cannot be identified or corrected. A summary of the station history information for surface observations in the United States is available on computer-compatible media at the National Climatic Data Center, Asheville, N.C. Unfortunately, some station histories are far from complete. The lack of information on station history is considerably worse in many other countries of the world.

2.2.3 Comprehensive Sources of Climate Data and Information

Table 2.2.1 contains a listing of selected sources of general information on climate observations, data sets, and climate atlases.

2.2.4 Climate Charts

Since climate varies on almost any time scale one may wish to choose, a climate chart or summary is strictly valid only for a particular period of time. To be most useful, a climatic summary possesses the following characteristics: (1) all aspects of the climate presented must relate as nearly as possible to the same period of time, (2) the period of time should be long enough to properly average the shorter-term weather and climate variations that occur, and (3) a knowledge of the geographical distribution and quality of the data is needed to establish the reliability of regional climatic features.

The distribution of climatic data is highly nonuniform, and many areas of the earth have yet to be adequately sampled. In attempting to map the global climate, one often has little choice but to extrapolate between widely spaced stations and make use of data summaries which span different periods of time. Parameters such as evaporation and heat flux will never be directly measured except on a local, experimental basis; their values will have to be estimated using a mix of physical and empirical relationships.

The global distribution of precipitation over the continents is presented as Fig. 3.1.1. It is not possible to provide in this chapter the variety and detail of climate

TABLE 2.2.1 General Sources of Information on Climate Data Sets

Data catalogs and information	Comments
WMO, INFOCLIMA, Catalog of Climate System Data Sets, WMO/TD No. 293, World Meteorological Organization, C. Postale, No. 5, Geneva, 508 pp., 1989	Includes individual observations and summaries of meteorological and hydrologic data sets held at various data centers around the world
Wallis, A. L., Jr., Climate Research Data Catalog, available from U.S. National Climatic Data Center (NCDC), Asheville, N.C. 28801, 60 pp., 1988	The Air Force Data Center (USAF-ETAC), which is co-located with NCDC, has many sets of digitized observations for periods prior to 1965 that are not available elsewhere
Olsen, L. M., Data Set Availability through NASA's Climate Data System (NCDS). NASA/GSFC, Code 634, Greenbelt, Md. 20771, 9 pp., 1990. (For information contact NSSDC, Code 633, NASA Goddard Space Flight Center, Greenbelt, Md. 20771.)	Describes a U.S. National Aeronautics and Space Administration (NASA) information system which provides the user community with an online computer search to locate data sets for climate and other disciplines. This central catalog describes about 1000 data sets in broad terms suitable for judging their usefulness for a specific purpose
Jenne, R. L., Data Availability at NCAR, National Center for Atmospheric Research, P.O. Box 3000, Boulder, Colo. 80307, 1989	Included are data from climate model simulations of the "greenhouse" response, organized for easy input into climate assessment studies. This publication also includes references to catalogs available from other centers
U.S. DOE, U.S.-DOE Carbon Dioxide Research Program, publications and other documents (and numerical data), Department of Energy (DOE) Carbon Dioxide Information Center, Oak Ridge National Laboratories, P.O. Box 2008, Oak Ridge, Tenn. 37831-6335, 1989	A number of climate and paleoclimate data series are also available. Some volcanic information is included
CD ROMs and related media	CD ROMs are now available from many sources and are rapidly becoming an important medium for the distribution of climate data. Information on the current status of these efforts is available from Data Support Section, NCAR, Boulder, Colo. 80307 or the National Climatic Data Center, Asheville, N.C. 28801

Surface heat and water balance information	Comments
Baumgartner, A., and E. Reichel, The World Water Balance, 1975[3]	Maps, tabulations, and discussion of the world water balance
Budyko, M. I. (ed.), Atlas Teplovogo Balansa Zemnogo Shara (Atlas of the Heat Balance of the Earth), 1963[9]	Global maps of monthly averaged heat balance parameters and a map of mean annual precipitation
Henning, D., Atlas of the Surface Heat Balance of the Continents, 1989[34]	Contains continental-scale maps, and includes a discussion and comparison of different methods for estimating surface heat-balance components

TABLE 2.2.1 *(Continued)*

Surface heat and water balance information	Comments
Kessler, A., Heat Balance Climatology. World Survey of Climatology, vol. 1A, 1985[42]	Heat-balance maps and an extensive discussion of the heat-balance climatology
Korzoun, V. I. (editor in chief), World Water Balance and the Water Resources of the Earth, 1978[43]	Documents the results from the most comprehensive effort to date to document the world water balance
Jaeger, L., Monatskarten des Niederschlags für Ganze Erde, 1976[39]	Mean monthly global precipitation maps

Global climate atlases	Comments
Crutcher, H. L., and J. M. Merserve, Selected Level Heights, Temperatures and Dew Points for the Northern Hemisphere, NAVAIR 50-1C-52, U.S. Naval Weather Service Command, Washington, D.C., 1970[19]	Mean monthly upper air climatology for the northern hemisphere
Taljaard, J. J., H. van Loon, H. L. Crutcher, and R. L. Jenne, Climate of the Upper Air, Southern Hemisphere, vol. 1, Temperatures, Dew Points and Heights at Selected Pressure Levels, NAVAIR 50-1C-55. U.S. Naval Weather Service, Washington, D.C., 1969[95]	Mean monthly upper air climatology for the southern hemisphere
Rudloff, W., World Climates, CRC Press, 2000 Corporate Blvd., N.W., Boca Raton, Fla. 33431, 1988[84]	Monthly averages and extreme values for surface climate parameters for the years 1931–1960 for 1474 locations throughout the world
National Oceanic and Atmospheric Administration, 1983: Climatic Atlas of the United States (for sale from National Climatic Data Center, Federal Bldg., Asheville, N.C. 28801, Attn: Publications)[69]	A variety of climatic charts which describe the surface climatology of the United States

Surveys and handbooks	Comments
World Survey of Climatology[100]	A comprehensive source of information on the climates of the world. There are 15 volumes, the first published in 1969
Handbook of Applied Meteorology[31]	This volume is an excellent general reference on meteorology, including such topics as measurements and data. The chapter on data includes a listing of many climatological data sources and publications as well as sources of microfilm copies of weather maps. It also describes the services available from various data centers in the United States and Canada. There is a chapter on climatology as well as tabulations of primary monthly climate statistics for about 250 stations around the world

charts and tabulations required for the broad range of hydrologic applications. The reader is referred to the *climate atlases* listed in the previous section which provide regional and global charts and tabulations needed for many applications. Particular attention is called to Ref. 3 and 43, which include detailed global charts of mean precipitation, evaporation, and runoff.

2.3 ASPECTS OF LAND SURFACE CLIMATE

2.3.1 Scales of Variability

The processes which determine the local land surface climate operate on a variety of spatial scales. Table 2.3.1 shows a general classification of climate features by scale. Regional determinants which modify the planetary-scale north-south variation in climate associated with the variation of incoming solar radiation include (1) regional variations in absorbed solar radiation, (2) land-ocean contrast, (3) effects of large-scale topography, and (4) small-scale land surface variations.

2.3.2 Coupled Land-Atmosphere Interactions

The land surface and atmosphere are coupled through the exchange of energy and water and thus must be treated as interacting components of the climate system. The primary external source of energy for the climate system is solar radiation. Part of this radiation is reflected from clouds and the earth's surface. Averaged over the planet, this amounts to around 30 percent (Fig. 2.1.1). Two-thirds of the remaining radiation is absorbed at the earth's surface. This surface heat source is balanced by three surface energy sinks: net outgoing long-wave radiation, sensible heat transfer to the lower atmosphere, and evaporation (latent heat flux). These three surface energy sinks are energy sources for the atmosphere. To adequately describe and model the

TABLE 2.3.1 Scales of Climate Variability (Examples)

Planetary ($\geq 10^4$ km)	Macroscale (10^3 km)	Mesoscale (10^0-10^2 km)	Microscale ($\leq 10^{-1}$ km)
• Incoming solar radiation • Mean meridional circulations • Land-ocean contrast • Major mountain ranges • Subtropical highs • Trade-wind systems • Middle-latitude storm tracks • Jet streams	• Regional mountain ranges • Large lakes • Major vegetative features (forests, deserts, etc.) • Middle-latitude weather systems • Tropical storms	• Local terrain features • Squall lines • Land-sea breeze • Mountain valley winds • Urban effects	• Soil characteristics • Small-scale variations in vegetation • Individual structures

Note: Conversion: 1 km = 0.621 mile.

coupled interactions between land and atmosphere, one must model the processes which control the movement of heat and moisture between the earth's surface and the underlying soil, between the earth's surface and the atmospheric planetary boundary layer, and finally from the planetary boundary layer into the overlying atmosphere.

Surface-Atmosphere Exchange. The energy storage (heat and moisture) in the land surface layer is exceedingly small compared with that of the ocean. For averages over a few days or more, the land surface layer storage change is also small compared with the energy fluxes absorbed and emitted by the land surface; thus a near steady-state energy balance exists at the land surface.

In moist regions the change of water to its vapor phase through evaporation and the upward transport of this "latent heat" into the atmosphere by turbulence is by far the dominant energy sink that balances the net radiative heating of the land surface. Conversely, over a completely dry surface the net radiative heating is balanced by sensible heat transfer to the air and heat conduction into the soil.

It is natural to regard the *net radiation* at the land surface (the sum of the absorbed solar radiation and downward thermal infrared radiation from the atmosphere minus the upward thermal infrared radiation from the surface) as being provided by the atmosphere (or given by some measurement), and therefore the net radiation is equal to the sum of the sensible and latent heat fluxes from the land surface. Table 2.3.2 shows typical average values for the different energy-balance terms at the earth's surface. This partitioning is simple only for daily averaging and the case of the dry surface previously alluded to. Otherwise, it depends on the details of how the surface is modeled.

A model of the surface-atmosphere exchange must partition the energy balance into the individual fluxes. Modeling these fluxes presents a problem analogous to that of the observationist who wishes to estimate surface fluxes from near surface observations of the relevant atmospheric parameters. The atmosphere is typically modeled as a series of vertical layers and it is necessary to express the removal of heat and moisture from the surface in terms of the temperature, humidity, and wind in the lowest model layer. For a discussion of physical relationships used in modeling, see Ref. 20, 54, 57, and 92.

Planetary Boundary Layer. One soon learns when flying in an airplane that the air near the ground is much bumpier than that at higher levels. This low-level *turbulence*

TABLE 2.3.2 Examples of Surface Energy Fluxes, W/m²

Example	Absorbed solar radiation	Net long-wave emitted radiation	Sensible heat flux	Latent heat flux
Midsummer, midday dry desert*	600	150	300	0
Equatorial rain forest, (daily average)	180	60	20	100

* The difference between absorbed solar radiation and the three sink terms is the rate of change of energy storage in the desert's land surface layer (150 W/m²).

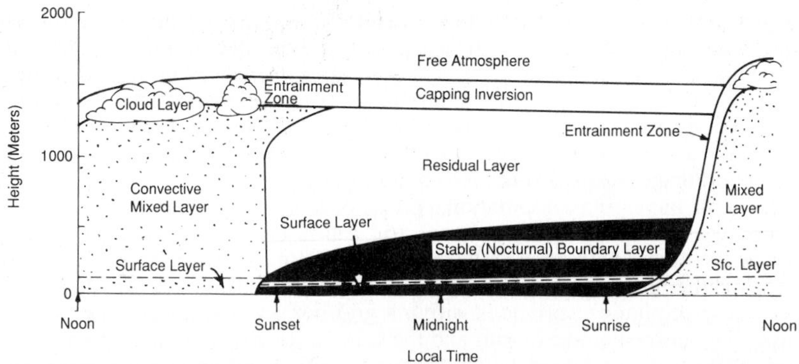

FIGURE 2.3.1 The planetary boundary layer in high-pressure regions over land. The planetary boundary layer consists of three major parts: a very turbulent mixed layer; a less turbulent residual layer containing former mixed-layer air; and a nocturnal stable boundary layer of sporadic turbulence. The mixed layer can be subdivided into a cloud layer and a subcloud layer. Unit conversion: 1 m = 3.28 ft. *(Adapted from Fig. 1.7, Ref. 92. Used with permission.)*

is a minimum at night and a maximum in the afternoon (Fig. 2.3.1). During the nighttime minimum, the turbulence is due to the mechanical stirring of air as it moves over the rough surface of the earth. The taller, more porous, and more unevenly distributed the land surface topography, the more vigorous is the mechanical mixing. Forests, for example, are much rougher than a flat surface with short vegetation. During the day, the relatively hot land surface which underlies a relatively cool atmosphere results in *thermal convection* and more vigorous stirring.

Atmospheric pressure decreases with increasing altitude. Consequently, stirring of an atmospheric layer causes rising parcels of air to cool by *adiabatic expansion* (because their pressure has decreased) and sinking parcels to correspondingly warm by compression. The net effect of such stirring is a vertical decrease in temperature of about 9.8°C per kilometer (5.4°F per 1000 ft), which is the *adiabatic lapse rate,* g/C_p [where g is the acceleration of gravity and C_p is the specific heat of the atmosphere at constant pressure (1004 J K^{-1} kg^{-1})]. The atmospheric layer near the surface is sufficiently mixed to maintain a lapse rate close to or slightly larger than the adiabatic. This is called the *planetary boundary layer.* Above the planetary boundary layer the atmosphere is generally kept quite stable (lapse rate less than adiabatic) by latent heat release associated with condensation and precipitation, together with upward heat transport by large-scale atmospheric motions.

As illustrated by Fig. 2.3.1, a model for the transport through the planetary boundary layer must include (1) a daytime atmospheric layer which extends from near the surface to a height of 1 km or more in which the stratification is nearly neutrally buoyant or slightly unstable, and (2) a nighttime surface layer shrinking to a thickness of a few hundred meters or so in which a stable lapse rate exists. The daytime layer can remove heat and moisture from the surface much more rapidly than the nighttime layer can give off heat and accumulate or give up water vapor by dew formation. Clouds with roots in the planetary boundary layer that extend into the free atmosphere are an additional mechanism for vertical transport. There are several modeling approaches for describing how heat and moisture are transported vertically through the planetary boundary layer.[54,92]

Surface Hydrology in Global Climate Models. Even with the most powerful digital computers, the resolution of the model, i.e., the minimum size of the grid boxes used for characterizing the climatic parameters, is limited by the vast number of computations required to perform a climate simulation over a period of decades or more. In the past, the typical grid box has been several hundred kilometers on a side. This has resulted in a severe mismatch between the model computational grid and the much smaller drainage basins of interest to hydrologists.

With the increasing computational power of each new generation of digital computers, the size of the computational grid cell is being decreased. For example, mesoscale models with grids of 30 km (18.6 mi) are now routinely used for short-range weather prediction. In addition, methods to statistically account for land surface and precipitation variations within a grid box are being developed. Thus the mismatch between model output and the scales of interest to hydrologists will continue to decrease, and the detail and accuracy with which surface hydrologic processes can be modeled will continue to increase.

Among the questions being addressed by coupled land surface–atmosphere climate models are: (1) What are the effects on hydrology of changing the net radiation at the surface? (2) What are the effects of modifying the partitioning between sensible and latent fluxes? (3) How do either of these classes of change feed back on the atmospheric hydrologic cycle? More specifically, these studies have examined the possible role of land surface processes in the dramatic decrease in rainfall over the African Sahel[12,86] and the effects on climate of Amazon deforestation.[21,49,87] Recent studies are also concerned with improving approaches to the underlying physics that determines the surface energy balance and atmospheric feedbacks, including their effect on these balances.

2.3.3 Small-Scale Climatic Variability

Because of a variety of local controls the climate of a small area can differ significantly from that of the larger surrounding region. Local differences in terrain (slope, elevation), land surface characteristics (lakes, forests, cultivated land, urban areas), and air pollution affect the airflow, cloudiness, temperature, and even precipitation through their effects on surface roughness, and the surface heat and water balance. These small-scale variations in climate introduce significant uncertainties in the characterization and mapping of climate from scattered point observations.

Topography and Elevation. Topography gives rise to a variety of mesoscale and microscale climate variations.[27] Regionally, the blocking effect of mountain ranges results in forced ascent, condensation, and heavy precipitation as the air moves up the windward slopes. To the lee of the crest, the subsiding air is relatively dry owing to the downward movement (subsidence) of the air and the upstream loss of moisture. This *rain shadow effect* of mountains may extend for hundreds or even thousands of kilometers downstream. Major mountain barriers such as the Rocky Mountains also channel the large-scale outflow of cold air from the polar regions.

Because of the atmospheric lapse rate, elevation has a profound effect on temperature and precipitation type. The dry adiabatic lapse rate of approximately 9.8°C per kilometer (5.4°F per 1000 ft) represents an upper limit on the rate of temperature decrease with height in a stable atmosphere. If the actual temperature gradient is greater than this value, the air becomes unstable for vertical displacements, and rapid mixing will bring the lapse rate back to near the dry adiabatic value.

The atmospheric lapse rate varies with climate zone, season, time of day, and weather conditions. At one extreme, the summertime afternoon lapse rate over arid regions such as the Sahara Desert may be near dry adiabatic through the lowest several thousand meters of the atmosphere. At the other extreme, a strong low-level temperature inversion, i.e., increase of temperature with height, often develops at night over middle- and high-latitude continental regions during winter when clear skies and light winds prevail. With little incoming solar radiation, such wintertime inversions may persist throughout the day in the arctic, leading to surface temperatures in the valleys which may be tens of degrees lower than the temperature a few hundred meters up the mountain slopes.

Local Circulations. *Local wind systems* are prevalent in many areas. They may be gravity-driven (mainly downslope winds in mountain regions and off ice fields and glaciers), thermally driven by differential surface heating,[23] or mechanically driven (due to isolated hills or mountains). On the larger scale, mountain lee wave phenomena can result in strong surface *foehn winds* that go by various regional names, such as *chinook winds* in the western United States.

Thermally induced local circulations are most pronounced in the tropics and in middle latitudes during the warmer months when the large-scale temperature gradient and circulation are weak. Thermally driven diurnal wind systems include mountain-valley winds that result from the larger daytime heating and nighttime cooling of the mountain slopes relative to the valley, land-lake and land-sea breezes, which result from the larger diurnal heating cycle over the land,[27,30] and urban-rural contrasts, which result from the urban heat island effect.[47]

Urban Climate. The climate of most urban areas differs from that of the surrounding countryside.[47] This may be partly due to the bias in the location of cities with respect to topography (usually in valleys or depressions), but it primarily results from the distinct land surface characteristics and air quality in an urban area. One particularly notable feature of an urban area is the *heat island effect,* which leads to higher urban temperatures, particularly at night.[47]

Climatic Effects of Large Lakes. For large bodies of water such as the *Great Lakes,* which have a large heat capacity, the seasonal change in surface water temperature lags that of the land; e.g., the water is colder than the land in spring and early summer and warmer in autumn and early winter. This results in seasonally varying land-lake wind effects. During early summer, the land-lake winds are from the lake to the land, inhibiting precipitation over the water. For the Great Lakes, the lake effect is particularly pronounced in early winter, when cold arctic air flows southeastward across the warm water surface. The large fluxes of heat and moisture from the warm lake surface into the cold air along with the different frictional effects over the water and the land lead to intense *lake-effect snowstorms* over the snowbelt regions immediately to the lee of lakes Erie and Ontario.

2.4 CLIMATE VARIABILITY AND CLIMATE CHANGE

Pronounced climate fluctuations such as the U.S. Great Plains droughts of the 1930s and 1950s, and the multidecadal downturn in Sahelian rainfall, which began in the 1960s and has extended through the 1980s,[46] are a pervasive feature of climate time series. Attempts to fully describe these modes of climate variability and relate them to causal mechanisms are seriously hindered by the short length of instrument records, which usually span less than a century. Tree-ring data can sometimes be used to extend the instrument time series. For example, the 300-year tree ring record from Iowa analyzed by Duvick and Blasing[22] clearly shows the recurrent nature of long-lasting severe drought epochs in that area.

Efforts to predict long-term variations in precipitation due to natural causes will continue to be hazardous and highly controversial until the controlling processes are better understood.[83] Many, perhaps most, aspects of multiyear variability derive from internal climate system dynamics, which are often associated with variations in sea surface temperature on a global rather than local scale.[24] While external factors such as volcanoes and solar variability may also be involved, their importance remains an unresolved and controversial issue. The suggested possibility of relationships between drought occurrence and the 11-year sunspot cycle or the 22-year solar magnetic variation[65] has been difficult to verify. A similar situation exists regarding the existence of a significant relationship between climate and volcanic eruptions.[61]

2.4.1 Drought

Like "bad weather," drought is a relative term. It is generally associated with a sustained period of significantly lower soil moisture levels and water supply relative to the normal levels around which the local environment and society have stabilized. Drought is a frequent and often catastrophic feature of semiarid climates, is less frequent and disruptive in humid regions, and is not a meaningful concept when applied to deserts. Drought in one region may represent normal conditions in a more arid region, or during a more arid climatic period.

Natural factors commonly assessed to determine drought presence include weather conditions, soil moisture, water-table conditions, water quality, and streamflow.[11] Their interactions and the areas of impact caused by drought are illustrated in Fig. 2.4.1. Some of the impacts affecting people involve water storage systems, the

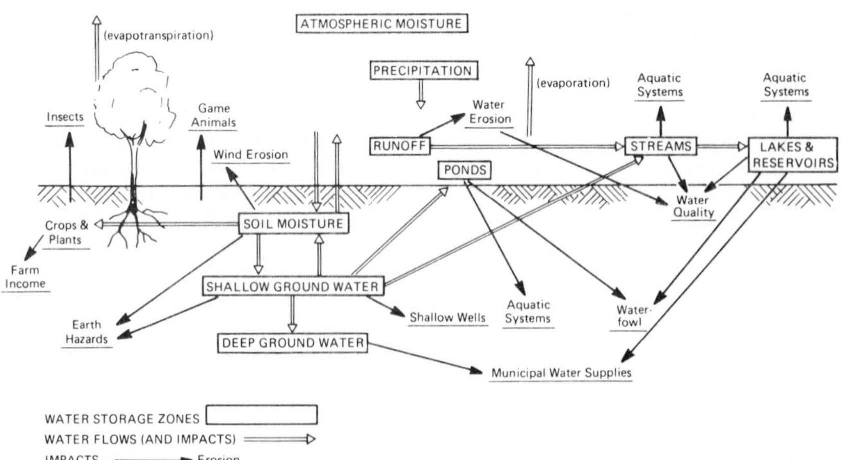

FIGURE 2.4.1 Hydrologic conditions affected by droughts, and related impacts. *(Adapted from Fig. 1, Ref. 11. Used with permission.)*

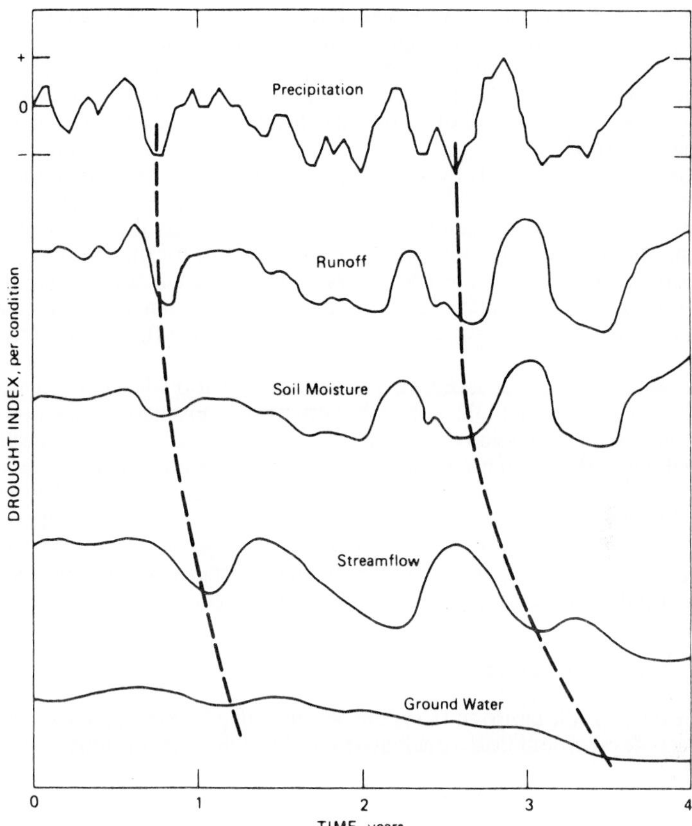

FIGURE 2.4.2 An idealized representation of how precipitation deficiencies during a hypothetical 4-year period are translated in delayed fashion, over time, through other components of the hydrologic cycle. *(Adapted from Fig. 2, Ref. 11. Used with permission.)*

availability of groundwater in shallow wells, decreased water use per capita, decreased water services, and a myriad of economic considerations.[11] Consequently, the particular criteria used in defining drought may include one or more of the following: precipitation, streamflow, runoff, evapotranspiration, groundwater levels, water supply, and water needs.

The temporal complexity of drought and its impacts on parts of the physical system are represented schematically in Fig. 2.4.2. This diagram shows how hypothetical fluctuations in precipitation over a 4-year period are translated, in delayed form, to runoff and then to soil moisture and groundwater. Deficiencies in each phase of the hydrologic cycle develop and end at different times.

Definitions of Drought. The most commonly used drought definitions are based on (1) meteorological and/or climatological conditions, (2) agricultural problems, (3) hydrologic conditions, and (4) economic considerations.[11]

Meteorological drought is defined as an interval of time, generally of the order of months or years, during which the actual moisture supply at a given place cumulatively falls short of climatically appropriate moisture supply.[74]

Agricultural drought is typically defined as a period when soil moisture is inadequate to meet evapotranspirative demands so as to initiate and sustain crop growth. Another facet of agricultural drought is a deficiency of water for livestock or other farming activities.[11]

Hydrologic drought typically refers to periods of below-normal streamflow and/or depleted reservoir storage.[5] Closely related to the concept of low flow in defining drought is the use of the number of consecutive months that a streamflow was deficient, that is, within the lowest 50 percent of record for the monthly record.

Economic droughts are a result of physical processes but concern the economic areas of human activity affected by drought.[11] The human effects, including the losses and benefits in the local and regional economy, are often a part of this definition.

The chain of drought causality can be examined at various levels. For a meteorologist, it begins with the identification of the persistent atmospheric circulation features associated with a suppression of precipitation. A more fundamental understanding is achieved when the forcing mechanisms which initiate and maintain these atmospheric anomalies are identified. The next step is an understanding of these processes in the context of coupled interactions between climate system components, for example, ocean and atmosphere.[83] Finally, a knowledge of the factors which determine the time scales, amplitudes, and frequency of occurrence of droughts provides insight into the physical processes behind drought regimes.

2.4.2 Climate Change

Climate involves the entire earth system, as illustrated by Fig. 2.4.3. Existing instrument records provide at least a qualitative picture of the present climate over most of

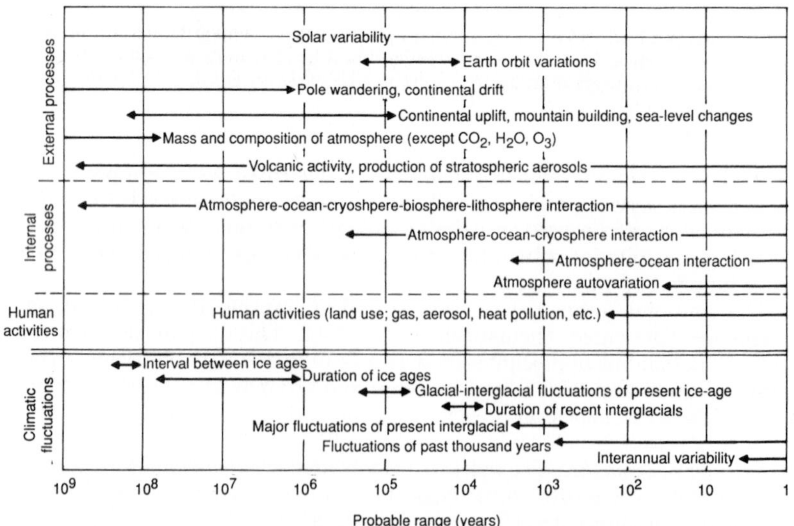

FIGURE 2.4.3 Time scales at which various forces act to change climate, and time scales of climatic fluctuations. Both natural and anthropogenic processes are given. *(Adapted from Fig. 2, Ref. 4.)*

the earth. However, few of these records span even a century, and they may not provide an adequate basis for long-range planning, since conditions during recent decades may not be typical of the longer-term climate.

Climate conditions several centuries in the past can sometimes be inferred from documentary material such as annals, chronicles, diaries, government papers, farm accounts, and ship logs, which serve as an important secondary source of climate information. The historical record is particularly complete for the last 1000 years in Europe but is less adequate in other regions and over longer periods.[45]

Paleoclimatic Data. For information on the longer time scales, we must interpret the proxy data recorded in the natural features of our planet.[33] *Proxy data* come from various paleoclimatic sensors that record the climatic responses of some natural system in a datable form. Each proxy source is a natural archive that has preserved the record of various paleoclimatic indicators.

Pollen records obtained from a network of coring sites have been used to reconstruct regional vegetation maps for the past 18,000 years.[16] Analyses of trace chemicals found in *lake sediments* are used to estimate the overall hydrologic balance of precipitation, evaporation, and runoff within the lake basin.[91] The physical dimensions of ancient lake basins, as estimated from raised or submerged paleoshorelines, provide yet another tool for calculating the hydrologic changes of the past.[44,91] The oxygen-isotope record in *marine sediments* indicates that there have been about ten major *glacial-interglacial cycles* during the past million years.

There are at least three examples of regional- to global-scale analyses that have synthesized a variety of *paleoclimatic indicators:* (1) the COHMAP group[16] produced estimates of the climate of the last 18,000 years at 3000-year intervals; (2) the CLIMAP group[14,15] produced estimates of the land surface, ocean temperature, and ice sheet topography of the glacial-age world; and (3) the University of Chicago Paleogeographic Atlas Project[101] is producing estimates of geography, topography, and vegetation for selected times in the past several hundred million years.

Dendroclimatology. Tree-ring chronologies, or *dendrochronology,* are the best source of precisely dated, high-resolution data related to hydrology over the past 1000 years. Long-lived trees suitable for compilation of climate proxy series are widely available in temperate and boreal climates worldwide.[26,89] Dendrochronologists have correlated tree growth with many environmental variables, and long tree-ring chronologies can be used to reconstruct the past history of snowpack,[97] precipitation,[6,89] temperature,[7,37] drought indices,[88,89] lake levels,[63] and streamflow.[13,18,89]

Proxy tree-ring series have been applied to several interesting hydrologic problems. For example, tree-ring data suggest that the long-term mean annual discharge of the Colorado River at Lee Ferry is as much as 2.0 million acre-ft (maf) (2.5 km^3) less than the 15.5 maf (19.1 km^3) that the 1922 Compact allocated (Fig. 2.4.4). The gauging data available from 1906 to 1922, which yielded the 15.5 maf figure, just happened to fall within the longest period of continuously high runoff in the last 450 years.[89] Reconstructed Potomac River annual summer low flow data since 1730[18] show that the period of observed streamflow data was not representative of the whole series and that flows exceeded the long-term median flow of the gauging records for long periods. A 281-year reconstruction of annual runoff in the White River of Arkansas indicated that longer periods of consecutive deficit and surplus runoff occurred before 1900 than in the 1900–1980 gauged record, and that both reconstructed and gauged runoff extremes showed significant interannual persistence.[13] In addition, double mass analysis with tree-ring chronologies detected inhomogeneity in the gauging record, a new practical use of tree-ring data.

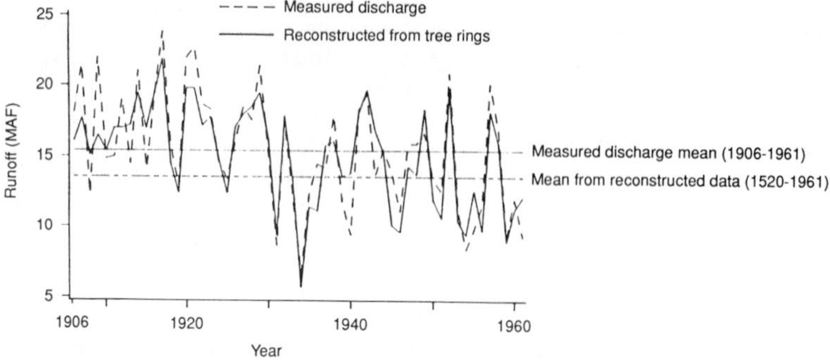

FIGURE 2.4.4 Reconstructed (solid) and gauged (dashed) series of Colorado River annual runoff at Compact Point (Lee Ferry), 1906–1961. The solid horizontal line is the mean of the gauged data and the dashed line is the long-term reconstructed mean, 1520–1961. Note that the gauged data before June 1921 are interpolated from other gauges in the basin and may contain substantial error. Gauging records for the three major sources of runoff first became available in 1914. The reconstruction ends in 1961 because that is the last year common to all 30 tree-ring chronologies used in the reconstruction, allowing for 3 years of lagged data. Ordinate is million acre-feet. Unit conversion: 1.0 acre-ft equals 1233 m^3. *(Data furnished by C. W. Stockton and D. M. Meko, Laboratory of Tree-Ring Research, Tucson, Ariz.)*

Measuring Tree Rings. Tree-ring reconstruction of environmental variables depends on compilation of long, well-replicated tree-ring chronologies sensitive to climate. The best chronologies are based on sampling many trees from a single species growing on ecologically homogeneous and undisturbed forest sites. Although many tree-ring chronologies come from dry sites,[25,90] some wetland species also exhibit sensitivity to drought.[13,88]

The cross-sectional surface of small samples removed from living trees (Fig. 2.4.5) is polished until the microscopic cellular structure becomes clear.[76,90,94] All rings are then carefully dated by matching patterns of wide and narrow rings.[25,26,76,90,94] This "crossdating" procedure ensures the exact chronological placement of each ring, an absolute necessity for the application of tree rings to high-resolution climatic or hydrologic reconstruction. The ring measurements are then transformed and the resulting series averaged into a site chronology to improve statistical characteristics and enhance climatic information.[17,25,26] The connection between tree-ring widths and streamflow or climatic data is established by statistical regression during the common data period. The tree-ring series and the regression equation are then used to reconstruct the hydrologic variable during the ungauged period.

Limitations of Tree-Ring Chronologies. While tree-ring chronologies provide the best source of annual climatic information over the last 1000 years, they do have important limitations as proxy climatic variables. When moisture-stressed trees are sampled, physiological limitations usually cause estimates of very wet conditions to have more error than drought reconstructions.[25,26] For this reason, droughts will usually be reconstructed more accurately than wet periods. Another limitation that must be appreciated is that ring widths integrate climate over relatively long periods and do not reflect short-term meteorological anomalies, such as those that produce flash floods.

One of the most critical limitations of tree-ring reconstruction of hydroclimatic variability is imposed by the period of dormancy in each annual cycle. Although

FIGURE 2.4.5 Preliminary examination of a tree-ring core sample just taken from a pine tree. *(Photograph by M. K. Cleaveland.)*

pre-growing-season climate may affect trees through physiological effects and soil moisture storage, tree growth usually responds most strongly to climate during the growing season.[25,26] As a consequence, reconstructions of growing-season climate are usually more accurate than reconstructions of dormant-period climate. Many of the above limitations can be partially overcome by using a mixture of species with somewhat different growing seasons and climate responses[25,26] or by deriving more information from each ring, e.g., by analyzing density variation within rings.[7]

Tree-ring data for the last three to five centuries (or longer in some cases) are readily available. Figure 2.4.6 is a map of some United States tree-ring sites. The International Tree-Ring Data Bank (ITRDB) is a repository for "raw" tree-ring data sets (i.e., untransformed ring width and density measurements) from many of these sites, but chronologies compiled from the raw data accompany most ITRDB submissions. Data are available from at least 650 sites on all continents (except Antarctica), with at least 400 from the United States. The National Oceanic and Atmospheric Administration maintains the ITRDB [ITRDB, National Geophysical Data Center, 325 Broadway (E/GC), Boulder, Colo. 80303.]

Many opportunities remain to develop long, climate-sensitive chronologies worldwide.[89] Research in the tropics shows promise of finding species suitable for dendrochronology,[37] which would open up exciting new possibilities for paleoclimatic investigation.

2.4.3 Anthropogenic Effects on Climate

To natural climate variations must now be added the *anthropogenic* or human factor. With increasing global population and advancing technology, human activity has become a fundamental component of climate change. Changes in land use alter

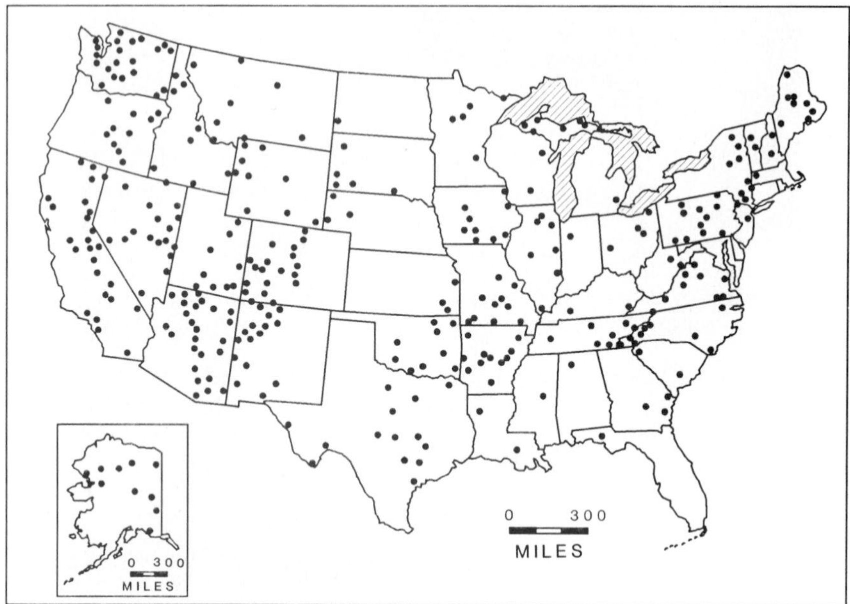

FIGURE 2.4.6 Map of some of the tree-ring site chronologies in the United States (dots). None of the chronologies begin more recently than 1750 and all end after 1960. Many are available through the International Tree-Ring Data Bank. The map is based on many sources, including Ref. 89 and unpublished information. Unit conversion: 1.0 mi equals 1.61 km.

surface reflectivity, surface temperatures, evaporation, water retention, and runoff. These changes impact the local energy and water balance. On the global scale, industrialized societies have been adding radiatively active trace gases and aerosols to the atmosphere at an ever-increasing rate.

The impact on climate of changes in atmospheric composition is at present only dimly perceived, but it is abundantly clear that today's climate statistics cannot be confidently projected into the twenty-first century. Thus increased uncertainty regarding the impact of climate on water supply, the structural integrity of facilities, and hydrograph statistics including extreme storm events will be a factor in long-term hydrologic planning and management for the foreseeable future.

Greenhouse Warming. Although humans are disturbing their environment in many ways, by far the biggest current issue is the effect on global temperature of increases in atmospheric trace gases arising from human activities. The primary *radiatively active gases* in the atmosphere are carbon dioxide (CO_2), methane (CH_4), and water vapor (H_2O), with lesser contributions from nitrous oxides, chlorfluorocarbons, and other gases. Although their concentrations are small, these gases consist of triatomic or multiatomic molecules, as compared with diatomic structure of the more abundant nitrogen (N_2) and oxygen (O_2) molecules. When radiation strikes a multiatomic molecule, the atoms are set in motion, vibrating in resonance to the frequency of the radiation, absorbing and reradiating energy. Rather than being reradiated directly to space, the long-wave radiation emitted from the earth's surface and lower atmosphere is partially intercepted by the radiatively active gases and reradiated back to earth again. In order to be in radiative balance, the temperature of the earth must increase over what it would be without the greenhouse gases, until the outgoing long-wave radiation of the planet balances the incoming solar short-wave radiation.

The disturbance of the global radiation balance by increases in atmospheric trace gases is both large and readily quantifiable. A sense of "large" is provided by climate models and the record of past climate variations. A good scale is the variability in energy output from the sun, the *solar constant.* Solar radiation has been observed to vary by 0.1 percent over a year. The consequences of this variation are small, and it requires a record many hundreds of years in length to detect these variations above the natural climate fluctuations. Any number of human activities, such as changing land-use patterns, could well have a comparable effect on the amount of solar radiation absorbed by the earth. In contrast, climate models suggest that a change of 1 percent in solar output would eventually lead to a global warming of at least 1°C (2°F).

The radiative effect of the observed changes in atmospheric carbon dioxide and methane over the last hundred years is already equivalent to a 1 percent change in the solar constant. Figure 2.4.7 shows the level of atmospheric carbon dioxide recorded at the Mauna Loa Observatory in Hawaii. From 1958 to 1990 this level increased from about 315 to 350 ppm at an increasing rate. The buildup in atmospheric carbon dioxide is due primarily to the burning of fossil fuels but is only about half of that produced by this source. The other half is presumed to have been absorbed by the upper layers of the ocean, or possibly in increased biomass. Because of the long lifetime of CO_2 in the atmosphere these changes are nearly irreversible within a human lifetime. The annual cycle of CO_2 shown in Fig. 2.4.7 results from the seasonal variation in plant photosynthesis, which is greatest during the northern hemisphere summer.

The rate of global warming is limited primarily by the several decades or more required to increase the water temperature of the surface and intermediate layers of the ocean. If the current level of greenhouse gases is to ultimately produce a 1°C warming, which is thought to be the case, the observed temperature increase to date should be about half of its equilibrium value, that is, about 0.5°C. The best attempts

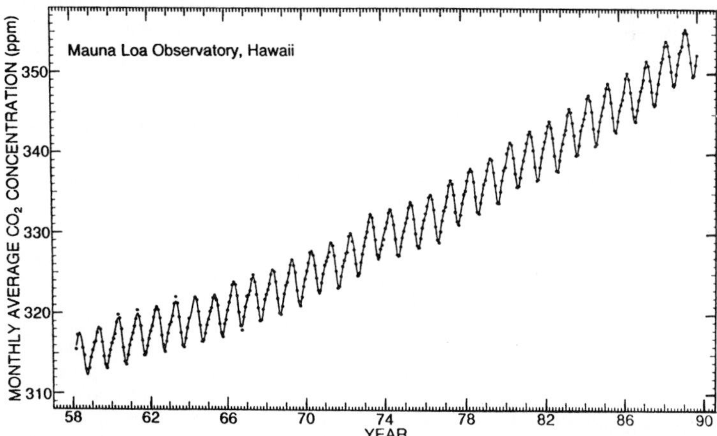

FIGURE 2.4.7 Observations of carbon dioxide concentration at Mauna Loa Observatory, Hawaii. *(Data courtesy of C. D. Keeling, Scripps Institution of Oceanography, La Jolla, Calif.)*

to construct global temperature records over the last century also show a half degree warming. However, natural variations in the climate system also have time scales of many decades or more and amplitudes of up to half a degree. Thus agreement between prediction and observation is interesting but far from scientifically conclusive.

The most disconcerting feature in the global temperature record is the evidence that global averaged temperatures around 1940 were nearly as high as they were in the late 1980s. During the intervening period, the temperature fell until the late 1970s, then rose rapidly. If the second temperature increase continues at a comparable rate during the 1990s, the observational evidence for greenhouse warming will become much more persuasive. For the present, the primary evidence of global warming is the unambiguous rise in greenhouse gas levels in the atmosphere and the consequences of this rise that are predicted by climate theory and modeling.

Energy Arguments. Table 2.4.1 shows the dominant greenhouse gas concentrations during preindustrial times, and during 1990, together with the implied trapping of thermal radiation caused by the increase in their concentrations. Future projections are based on societal assumptions and therefore are more problematical. The largest and most rapid growth is projected for the less developed countries, and attempts to curb future use of fossil fuels will be largely ineffective without their participation.

The question of global warming can be simply considered using the following model:

$$C\frac{\partial T}{\partial t} + \lambda T = \Delta Q \tag{2.4.1}$$

where T is global average temperature, C represents ocean heat capacity, ΔQ is the heat added, as indicated in Table 2.4.1, and λ is a feedback parameter that gives the dependence of thermal emission on temperature. The feedback parameter is generally referenced to what blackbody thermal emission would be at 255 K, about 4 W/m^2/K.

Water vapor is an important greenhouse gas. Assuming fixed atmospheric relative humidity, it has been shown[59] that the water vapor change with changing global temperature would reduce λ to about 2 W/m^2/K; i.e., there would be a positive feedback. Many other feedbacks have been discussed and some are generated implic-

TABLE 2.4.1 Change of Greenhouse Gas Forcing Since Preindustrial Times

Gas	Current (1990) concentration	Preindustrial concentration	ΔQ, W/m^2
Carbon dioxide	355 ppm	275 ppm	1.5
Methane	1.7 ppm	0.7 ppm	0.5
Nitrous oxide	310 ppb	285 ppb	0.2
CF11	0.28 ppb	0	0.06
CF12	0.48 ppb	0	0.14
Tropospheric ozone and other CFCS			0.1
Total			~2.4

Note: CF is chlorofluorocarbon; ΔQ is the increased atmospheric heating due to changes in greenhouse gas concentrations.

itly in current three-dimensional climate models. However, the present uncertainty as to how to treat clouds seems to dwarf the possible contributions of other terms. Several models have suggested a strong positive feedback from clouds that would reduce λ to 1 $W/m^2/K$, but other arguments suggest that changes in cloud properties might act to reduce the magnitude of global warming.

Implications of Climate Modeling Studies. The global warming scenario is the most quantitative information about the effect of increasing greenhouse gases. From a hydrologic viewpoint, however, an overall global temperature increase may be a relatively insignificant feature of the possible climate change. More important is the likelihood that the warming will be accompanied by significant regional changes in climate elements such as temperature, rainfall, and soil moisture.[36] While quantification of these changes is impossible without adequate treatments of clouds in climate models, studies up to now have suggested some specific major changes that must be considered in addition to the global temperature increases.

High-Latitude Surface Warming. All models suggest that high-latitude temperatures will increase, typically twice as much as the global average during the late fall and winter seasons. This enhancement has long been suggested on physical grounds as an implication of positive feedbacks by the snow and ice cover. Modeling studies show that the large heat capacities of the open ocean minimize this warming during summer, but it is effective during autumn, delaying sea ice formation, and in winter, when increased heat flow occurs through thinner ice and maximum positive feedback is expected from the temperature stratification.

Enhanced Hydrologic Cycle. The Clausius-Clapeyron equation gives the slope of the saturation vapor pressure curve as a function of temperature, i.e., the rate at which saturation vapor pressure increases with increasing temperature.[35] If surface and air temperatures are increased by the same amount, and relative humidities remain fixed, then global evaporation (precipitation) will simply increase because of the Clausius-Clapeyron relationship. In other words, the hydrologic cycle is expected to speed up with global warming. However, in one recent simulation the calculated increase in rainfall turned out to be less than half that inferred from simple thermodynamics.[66] This is because enhanced evaporation (heat loss) from the earth's surface cannot exceed the increased radiative flux. Hence surface relative humidities must increase enough to reduce evaporation to that which can be supported by the available radiative fluxes.

Enhanced Summer Drying. Changes in the hydrologic cycle are likely to have a significant impact on water resources. There could be large percentage changes in river flow, especially in arid regions, either increases or, with somewhat higher probability, decreases.[58] Some model simulations show increased midsummer dryness in midcontinental regions.[58] This is due to a poleward shift in the belt of maximum rainfall, a longer period where evapotranspiration exceeds precipitation, and the increased rates of evapotranspiration with warmer temperatures which occur in the model. This effect is most pronounced in areas where winter and early spring snowmelt and precipitation saturate the soil, after which further water inputs are relatively small. Increases in winter water storage or increases in summer precipitation could reverse this conclusion. For example, the snowpack might increase in mountainous terrain. The role of mountains in climate is not adequately resolved with current models.

Sea-Level Changes. Predictions of the effect of global warming on *sea level* are severely limited by a lack of understanding as to whether the Antarctic and Greenland ice sheets would store more or less water. Sea level could either rise or fall depending on whether the increase in temperature (melting) or related increase in precipitation (snowfall) would dominate. Current expectations are for a rise in sea level of several tenths of a meter by the middle of the next century.[36] Much larger increases in sea level are of low risk but would have an exceedingly large impact.

Biosphere Effects. Another area of major concern and equally large ignorance is the effect on the biosphere. Since human agricultural systems are opportunistic and CO_2 is an essential fertilizer, overall increases in productivity might be expected eventually, though there could be many severe dislocations at the regional and national level. Natural ecosystems, which are not so readily moved or helped in other ways, could undergo severe stresses. Thus much of the current efforts at species preservation through establishment of preserves could be wasted.

ACKNOWLEDGMENTS

We gratefully acknowledge the assistance of our colleagues as follows: A. Oort for data and information in Sec. 2.1; R. Ellingson, R. Jenne, R. Quayle, T. Karl, and R. Rignay for information in Sec. 2.2; H. C. Fritts, D. W. Stockton, D. A. Graybill, and D. M. Meko for information; D. W. Stahle and P. L. Chaney for editorial assistance and drafting of Fig. 2.4.4 and 2.4.6; and C. D. Keeling and W. Elliot for information in Sec. 2.4.4.

REFERENCES

1. Alexanderson, H., "A Homogeneity Test Applied to Precipitation Data," *J. Clim.*, vol. 6, pp. 661–675, 1986.
2. Barkstron, B. R., E. F. Harrison, and R. B. Lee, III, "Earth Radiation Budget Experiment, Preliminary Seasonal Results," *Eos,* vol. 297, pp. 304–305, 1990.
3. Baumgartner, A., and E. Reichel, *The World Water Balance,* Elsevier, Amsterdam, 1975.
4. Bergman, K. H., A. D. Hecht, and S. H. Schneider, "Climate Models," *Phys. Today,* vol. 34, (October), pp. 45–51, 1981.
5. Beron, M. A., and J. A. Radier, *Hydrological Aspects of Drought,* UNESCO, Paris, 1985.
6. Blasing, T. J., D. W. Stahle, and D. N. Duvick, "Tree Ring-Based Reconstruction of Annual Precipitation in the South-Central United States from 1750 to 1980," *Water Resour. Res.*, vol. 24, pp. 163–171, 1988.
7. Briffa, K. R., P. D. Jones, and F. H. Schweingruber, "Summer Temperature Patterns over Europe: A Reconstruction from 1750 A.D. Based on Maximum Latewood Density Indices of Conifers," *Quat. Res.*, vol. 30, pp. 36–52, 1988.
8. Broecker, W. S., *Chemical Oceanography,* Harcourt Brace Jovanovich, New York, 1974.
9. Budyko, M. I., ed., *Atlas Teplovogo Balansa Zemnogo Shara (Atlas of the Heat Balance of the Earth),* Mezhed. Geofiz. Komitet, Moskva, 1963.
10. Carissimo, B. C., A. H. Oort, and T. H. Vonder Haar, "Estimating the Meridional Energy Transports in the Atmosphere and Ocean," *J. Phys. Oceanogr.*, vol. 15, pp. 82–91, 1985.
11. Changnon, S. A. "Detecting Drought Conditions in Illinois," *Circ.* 169, *State of Illinois,*

Department of Energy and Natural Resources, State Water Survey Division, Champaign, Ill., 1987.

12. Charney, J., W. J. Quirk, S.-H. Chow, and J. Kornfield, "A Comparative Study of the Effects of Albedo Change on Drought in Some Arid Regions," *J. Atmos. Sci.,* vol. 34, pp. 1366–1385, 1977.
13. Cleaveland, M. K., and D. W. Stahle, "Tree Ring Analysis of Surplus and Deficit Runoff in the White River, Arkansas," *Water Resour. Res.,* vol. 25, pp. 1391–1401, 1989.
14. CLIMAP Project Members, "The Surface of the Ice Age Earth," *Science,* vol. 191, pp. 1131–1136. 1976.
15. CLIMAP Project Members, "Seasonal Reconstructions of the Earth's Surface at the Last Glacial Maximum," *Geological Society of America Map Chart Series,* MC-36, 1981.
16. COHMAP Members, "Climatic Changes of the Last 18,000 Years: Observations and Model Simulations," *Science,* vol. 241, pp. 1043–1052, 1988.
17. Cook, E. R., "The Decomposition of Tree-Ring Series for Environmental Studies, *Tree-Ring Bull.,* vol. 47, pp. 37–59, 1987.
18. Cook, E. R., and G. C. Jacoby, "Potomac River Streamflow Since 1730 as Reconstructed by Tree Rings," *J. Clim. Appl. Meteorol. (*now *J. Clim.),* vol. 22, pp. 1659–1672, 1983.
19. Crutcher, H. L., and J. M. Merserve, "Selected Level Heights, Temperatures and Dew Points for the Northern Hemisphere," *NAVAIR* 50-1C-52, U.S. Naval Weather Service Command, Washington, D.C., 1970.
20. Dickinson, R. E., "Modeling Evapotranspiration for Three-Dimensional Global Climate Models," *Climate Processes and Climate Sensitivity,* Geophysical Monograph 29, Maurice Ewing vol. 5, American Geophysical Union, Washington, D.C., pp. 58–72, 1984.
21. Dickinson, R. E., and A. Henderson-Sellers, "Modelling Tropical Deforestation: A Study of GCM Land-Surface Parameterizations," *Q. J. R. Meteorol. Soc.,* vol. 114, no. 13, pp. 439–462, 1988.
22. Duvick, D. N., and T. J. Blasing, "A Dendroclimatic Reconstruction of Annual Precipitation in Iowa Since 1680," *Water Resour. Res.,* vol. 17, pp. 1183–1189, 1981.
23. Flohn, H., "Local Wind Systems," in H. Flohn, ed., *World Survey in Climatology,* vol. 2, Elsevier, Amsterdam, pp. 139–171, 1969.
24. Folland, C. K., T. N. Palmer, and D. E. Parker, "Sahel Rainfall and Worldwide Sea Temperatures," *Nature,* vol. 320, pp. 602–607, 1986.
25. Fritts, H. C., and T. W. Swetnam, "Dendroecology: A Tool for Evaluating Variations in Past and Present Forest Environments," *Adv. Ecol. Res.,* vol. 19, pp. 111–188, 1989.
26. Fritts, H. C., *Tree Rings and Climate,* Academic Press, London, 1976.
27. Geiger, R., "Topoclimates," in H. Flohn, ed. *World Survey in Climatology,* vol. 2, pp. 105–138, Elsevier, Amsterdam, 1969.
28. Gill, A. E., *Atmospheric-Ocean Dynamics,* Academic Press, New York, pp. 429–492, 1982.
29. Gray, W. M., "Hurricanes, Their Formation, Structure, and Likely Role in the Tropical Circulation," in D. B. Shaw, ed. *Meteorology Over the Tropical Oceans,* Royal Meteorological Society, London, pp. 151–218, 1979.
30. Griffiths, J. F., and D. M. Driscoll, *Survey of Climatology,* Charles E. Merris Pub. Co., Columbus, Ohio, 1982.
31. *Handbook of Applied Meteorology,* D. Houghton, ed., Wiley, New York, 1985.
32. Hastenrath, S., *Climate and Circulation of the Tropics,* D. Reidel, Dordrecht, 1985.
33. Hecht, A. D., "Paleoclimatology: A Retrospective of the Past 20 Years, in A. D. Hecht, ed., *Paleoclimate Analysis and Modelling,* Wiley, New York, pp. 1–26, 1985.
34. Henning, D., *Atlas of the Surface Heat Balance of the Continents,* Gebrüder Borntraeger, Stuttgart, 1989.

35. Holton, J. R. *An Introduction to Dynamic Meteorology,* 2d ed., Academic Press, New York, 1979.

36. Intergovernmental Panel on Climate Change, *Climate Change* (J. T. Houghton, G. J. Jenkins, and J. J. Ephraums, eds.), Cambridge University Press, Cambridge, UK, 1990.

37. Jacoby, G. C., "Overview of Tree-Ring Analysis in Tropical Regions," *IAWA Bull.,* vol. 10, pp. 99–108, 1989.

38. Jacoby, G. C., and R. D'Arrigo, "Reconstructed Northern Hemisphere Annual Temperature Since 1671 Based on High Latitude Tree-Ring Data from North America, *Climatic Change,* vol. 14, pp. 39–59, 1989.

39. Jaeger, L., *Monatskarten des Niederschlags für Ganze Erde, Beitrage Deutsch. Wetterdeinst,* vol. 78, no. 139, 1976.

40. Jenne, R. L., and T. B. McKee, "Solar Data Sets," in D. D. Houghton, ed., *Handbook of Applied Meteorology,* Wiley, New York, pp. 1227–1230, 1985.

41. Karl, T. R., R. G. Quayle, and P. Ya Groisman, "Detecting Climate Variations and Change: New Challenges for Operational Observing and Data Management Systems," *J. Climate,* vol. 5 (in press), 1993.

42. Kessler, A., "Heat Balance Climatology," in O. M. Essenwanger, ed., *World Survey of Climatology,* vol. 1A, Elsevier, Amsterdam, 1985.

43. Korzoun, V. I. (editor in chief), *World Water Balance and the Water Resources of the Earth,* UNESCO, Paris, 1978.

44. Kutzbach, J. E., "Estimates of Past Climate at Paleolake Chad, North Africa, Based on a Hydrological and Energy Balance Model," *Quat. Res.,* vol. 14, no. 2, pp. 210–223, 1980.

45. Lamb, H. H., *The Changing Climate,* Methuen and Co., London, 1966.

46. Lamb, P. J., and R. A. Peppler, "West Africa," *Teleconnections Linking Worldwide Climate Anomalies* (M. Glantz, R. W. Katz, and N. Nicholls, eds.), Cambridge University Press, Cambridge, U. K. pp. 121–189, 1991.

47. Landsberg, H. E., "City Climate," *World Survey in Climatology,* vol. 3 (H. E. Landsberg, ed.), Elsevier, Amsterdam, pp. 299–333, 1981.

48. Landwehr, J. M., and J. R. Slack, "HDCN (Hydro-Climatic Data Network): A U.S. Geological Survey Discharge Data Set for Climatological Impact Analysis," *Proceedings, Special Sessions on Climate Variations and Hydrology,* American Meteorological Society, Boston, Mass., pp. 122–123, 1990.

49. Lean, J., and D. Warrilow, "Simulation of the Regional Climatic Impact of Amazon Deforestation," *Nature,* vol. 342, pp. 411–413, 1990.

50. Levitus, S., *Climatological Atlas of the World Ocean,* NOAA Professional Paper 13, U.S. Dept. of Commerce, Washington, D.C., 1982.

51. Levitus, S., "Interpentadal Variability of Temperature and Salinity in the Deep North Atlantic, 1970–1974 versus 1955–1959," *J. Geophys. Res.,* vol. 94, pp. 16125–16131, 1989.

52. Livezey, R. E., "Variability of Skill of Long-Range Forecasts and Implications for Their Use and Values," *Bull. Am. Meteorol. Soc.,* vol. 71, pp. 300–309, 1990.

53. Lorenz, E. N., *The Nature and Theory of the General Circulation of the Atmosphere,* World Meteorological Organization, Geneva, 1967.

54. Louis, J.-F., "A Parametric Model of Vertical Eddy Fluxes in the Atmosphere," *Boundary Layer Meteorol.,* vol. 17, pp. 187–202, 1979.

55. Lydolph, P. E., *The Climate of the Earth,* Rowman and Allanheld, Tatowa, N.J., 1985.

56. MacCracken, M. C., "Carbon Dioxide and Climate Change: Background and Overview," *Projecting the Climatic Effects of Increasing Carbon Dioxide* (M. C. MacCracken and F. M. Luther, eds.), Report DOE/ER-0237, U.S. Department of Energy, Washington, D.C., 1985.

57. Manabe, S., "Climate and Ocean Circulation: I. The Atmospheric Circulation and the Hydrology of the Earth's Surface," *Mon. Weather Rev.*, vol. 97, pp. 739–774, 1969.

58. Manabe, S., R. T. Weatherald, and R. J. Stouffer, "Summer Dryness Due to an Increase of Atmospheric CO_2 Concentration," *Climatic Change*, vol. 3, pp. 347–386, 1981.

59. Manabe, S., and R. T. Weatherald, "Thermal Equilibrium of the Atmosphere with a Given Distribution of Relative Humidity," *J. Atmos. Sci.*, vol. 24, pp. 241–250, 1967.

60. Maronna, R., and V. J. Yohai, "A Bivariate Test for the Detection of a Systematic Change in the Mean," *J. Am. Stat. Assoc.*, vol. 73, pp. 640–645, 1987.

61. Mass, C. F., and D. A. Portman, "Major Volcanic Eruptions and Climate: A Critical Evaluation," *J. Clim.*, vol. 2, pp. 566–593, 1989.

62. Mazzarella, D. A., "Measurements Today," in D. Houghton, ed., *Handbook of Applied Meteorology*, Wiley, New York, pp. 283–328, 1985.

63. Meko, D. M., and C. W. Stockton, "Tree-Ring Inferences on Historical Changes in the Level of Great Salt Lake," *Problems and Prospects for Predicting Great Salt Lake Levels*, (P. A. Kay and H. F. Diaz, eds.) University of Utah, Salt Lake City, 1988.

64. Middleton, W. E. K., and A. F. Spilhaus, *Meteorological Instruments*, University of Toronto Press, Toronto, 1953.

65. Mitchell, J. M., W. C. Stockton, and D. Meko, "Evidence of a 22-Year Rhythm of Drought in the Western United States Related to the Hale Solar Cycle Since the 17th Century," *Solar-Terrestrial Influences on Weather and Climate* (B. M. McCormac and T. A. Seliga, eds.), D. Reidel, Dordrecht, Holland, pp. 125–143, 1979.

66. Mitchell, J. F. B., C. A. Wilson, and W. M. Cunningham, "On CO_2 Climate Sensitivity and Model Dependence of Results," *Q. J. R. Meteorol. Soc.*, vol. 113, pp. 293–322, 1987.

67. National Academy of Sciences, "U.S. Participation in the TOGA Program," National Academy Press, Washington, D.C., 1986.

68. National Oceanic and Atmospheric Administration, "Hourly Solar Radiation-Surface Meteorological Observations," SOLMET, vol. 2, *Final Report* ID-9724, National Climatic Center, 1979.

69. National Oceanic and Atmospheric Administration, *Climatic Atlas of the United States* (for sale from National Climatic Data Center, Federal Bldg., Asheville, N.C. 28801, Attn: Publications.), 1983.

70. NOAA, *NOAA-NWS Specification* P300-SP004. National Weather Service Engineering Division, Washington, D.C., 1980.

71. NOAA, *Substation Observations*, NWS Observing Handbook, Government Printing Office, Washington, D.C., 1989.

72. Oort, A. H., and J. P. Peixoto, "Global Angular Momentum and Energy Balance Requirements from Observations," *Adv. Geophys.*, vol. 25, Academic Press, New York, pp. 355–490, 1983.

73. Palmén, E., and C. W. Newton, *Atmospheric Circulation Systems*, Academic Press, New York, pp. 27–64, 1969.

74. Palmer, W. C., "Meteorological Drought," *Research Paper 45*, U.S. Weather Bureau, Washington, D.C., 1965.

75. Personal communication, G. F. O'Brien, Surface Observations Section, National Weather Service, NOAA, 1990.

76. Phipps, R. L., "Collecting, Preparing, Crossdating, and Measuring Tree Increment Cores," *Water-Resources Investigations Report* 85-4148, U.S. Geological Survey, Reston, Va., 1985.

77. Picard, G. L., and W. J. Emery, *Descriptive Physical Oceanography*, 4th ed., Pergamon Press, New York, 1982.

78. Potter, K. W., "Illustration of a New Test for Detecting a Shift in Mean Precipitation Series," *Mon. Weather Rev.*, vol. 109, pp. 2040–2045, 1981.

79. Quayle, R. G., D. R. Easterling, T. R. Karl, and P. Y. Hughes, "Effects of Recent Thermometer Changes in the Cooperative Station Network," *Bull. Am. Meteorol. Soc.*, vol. 72, pp. 1718–1723, 1991.

80. Ramage, C. S., *Monsoon Meteorology,* Academic Press, New York, 1971.

81. Ramanathan, V., B. R. Barkstrom, and E. F. Harrison, "Climate and the Earth's Radiation Budget," *Phys. Today,* vol. 42 (May), pp. 22–32, 1989.

82. Rasmusson, E. M., "El Niño and Variations in Climate," *Am. Sci.,* vol. 73, pp. 168–177, 1985.

83. Rasmusson, E. M., "The Prediction of Drought: A Meteorological Perspective," *Endeavor,* vol. 11, pp. 175–182, 1987.

84. Rudloff, W., *World Climates,* CRC Press, Boca Raton, Fla., 1988.

85. Sevruk, B., "Methods of Correcting for Systematic Error in Point Precipitation Measurement for Operational Use," *Oper. Hydrol. Rep.* 21, *Pub.* 589, World Meteorological Organization, Geneva, Switzerland, 1982.

86. Shukla, J., and Y. Mintz, "Influence of Land-Surface Evapotranspiration on the Earth's Climate," *Science,* vol. 215, pp. 1498–1502, 1982.

87. Shukla, J., C. Nobre, and P. Sellers, "Amazon Deforestation and Climate Change," *Science,* vol. 247, pp. 1322–1325, 1990.

88. Stahle, D. W., M. K. Cleaveland, and J. G. Hehr, "North Carolina Climate Changes Reconstructed from Tree Rings: A.D. 372–1985," *Science,* vol. 240, pp. 1517–1519, 1988.

89. Stockton, C. W., W. R. Boggess, and D. M. Meko, "Climate and Tree Rings" in A. D. Hecht, ed., *Paleoclimate Analysis and Modeling,* chap. 3, Wiley, New York, 1985.

90. Stokes, M. A, and T. L. Smiley, *An Introduction to Tree-Ring Dating,* University of Chicago Press, Chicago, Ill., 1968.

91. Street-Perrott, F. A., and S. P. Harrison, "Lake Levels and Climatic Reconstruction," in A. D. Hecht, ed., *Paleoclimate Analysis and Modeling,* Wiley, New York, pp. 291–340, 1985.

92. Stull, R. B., *An Introduction to Boundary Layer Meteorology,* Kluwer Academic Publishers, Dordrecht, 1988.

93. Sverdrup, H. U., M. W. Johnson, and R. H. Fleming, *The Oceans, Their Physics, Chemistry and General Biology,* Prentice-Hall, New York, 1946.

94. Swetman, T. W., M. A. Thompson, and E. K. Sutherland, "Using Dendrochronology to Measure Radial Growth of Defoliated Trees," *Agriculture Handbook* 639, U.S. Department of Agriculture, Washington, D.C., 1985.

95. Taljaard, J. J., H. van Loon, H. L. Crutcher, and R. L. Jenne, *Climate of the Upper Air Southern Hemisphere, vol. 1, "Temperatures, Dew Points and Heights at Selected Pressure Levels," NAVAIR* 50-1C-55, U.S. Naval Weather Service, Washington, D.C., 1969.

96. Tchernia, P., *Descriptive Regional Oceanography,* Pergamon, Oxford, 1978.

97. Tunnicliff, B. M., "The Historical Potential of Snowfall as a Water Resource in Arizona," M. S. Thesis, University of Arizona, Tucson, 1975.

98. World Climate Research Program, "Radiation and Climate," Workshop on Implementing of the Baseline Surface Radiation Network, *WMO/TD* 406, World Meteorological Organization, Geneva, Switzerland, 1991.

99. World Meteorological Organization (WMO), *Guide to Meteorological Instruments and Observing Practices,* WMO 8TP.3, WMO, Geneva, 1969 (and subsequent updates).

100. *World Survey of Climatology,* (H. Landsberg, editor-in-chief), Elsevier, Amsterdam (15 vols), 1981.

101. Ziegler, A. M., "Phytogeographic Patterns and Continental Configurations during the Permian Period, in W. S. McKerrow and C. R. Scotese, eds., *Palaeozoic Palaeogeography and Biogeography,* Geological Society Memoir 12, pp. 363–379, 1990.

CHAPTER 3
PRECIPITATION

James A. Smith
Department of Civil Engineering and Operations Research
Princeton University
Princeton, New Jersey

3.1 INTRODUCTION

Precipitation exhibits marked variability in time and space. The spatial distribution of global continental precipitation at an annual time scale is shown in Fig. 3.1.1. Short-term fluctuations in precipitation over small areas are equally striking (Fig. 3.1.2). The dramatic consequences of precipitation variability, droughts and extreme floods have determined broad features of human settlement and commerce since ancient times. Describing and predicting the variability of precipitation is a fundamental requirement for a wide variety of human activities.

Extended treatment of precipitation can be found in a number of texts.[14,20,21,37,97,108] Wallace and Hobbs[102] present a detailed and accessible development of precipitation processes from a hydrometeorological perspective.

3.2 ATMOSPHERIC PROCESSES

In this section, the physical modeling framework for atmospheric processes is described. The set of equations that embody the physical laws of atmospheric motion is referred to as the *primitive equations.* These equations play a direct role in several areas of precipitation analysis, including the meteorological approach to probable maximum precipitation (PMP) (see Sec. 3.10) and in precipitation forecasting (see Sec. 3.11). They also provide a unified framework for all aspects of precipitation analysis.

The principal variables used to describe atmospheric dynamics are *density, pressure,* and *temperature.* The primitive equations relate these variables to atmospheric velocity[26,71] through a system of six equations (conservation of mass, conservation of energy, ideal gas law, and three conservation of momentum equations, one for each of the three components of velocity) in six unknowns (pressure, temperature, density, and three velocity components). In applications involving precipitation, equations governing the conservation of mass for water are typically added.

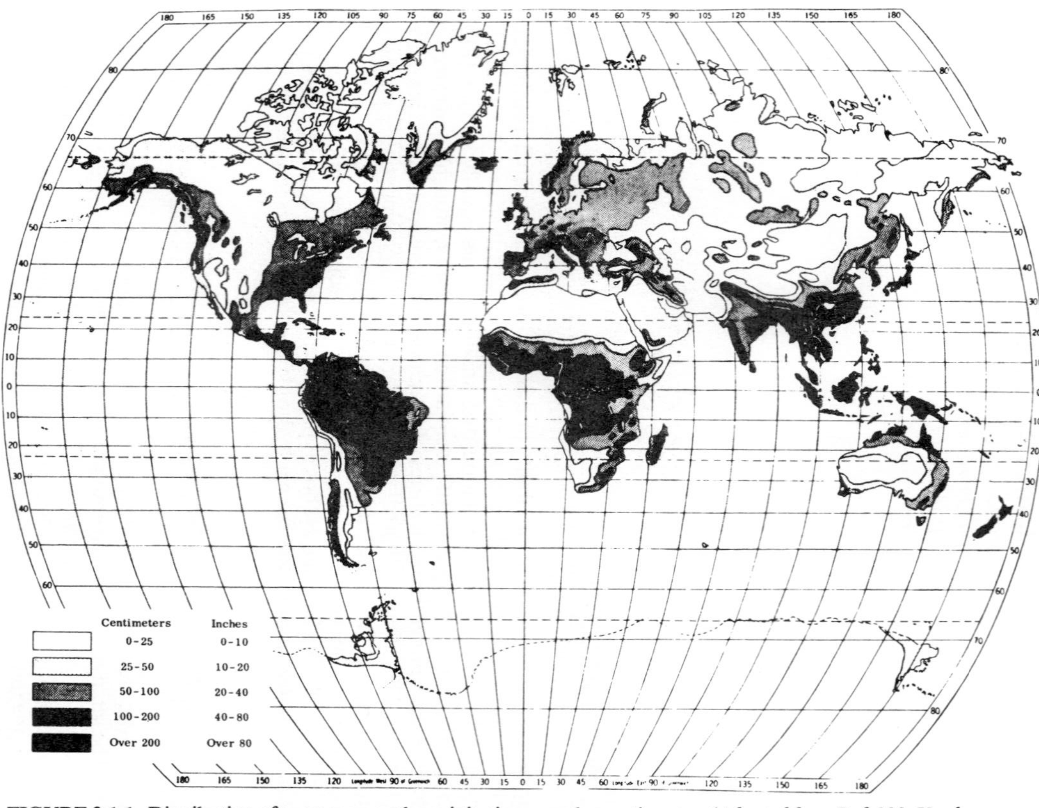

FIGURE 3.1.1 Distribution of average annual precipitation over the continents. *(Adapted from Ref. 102. Used with permission.)*

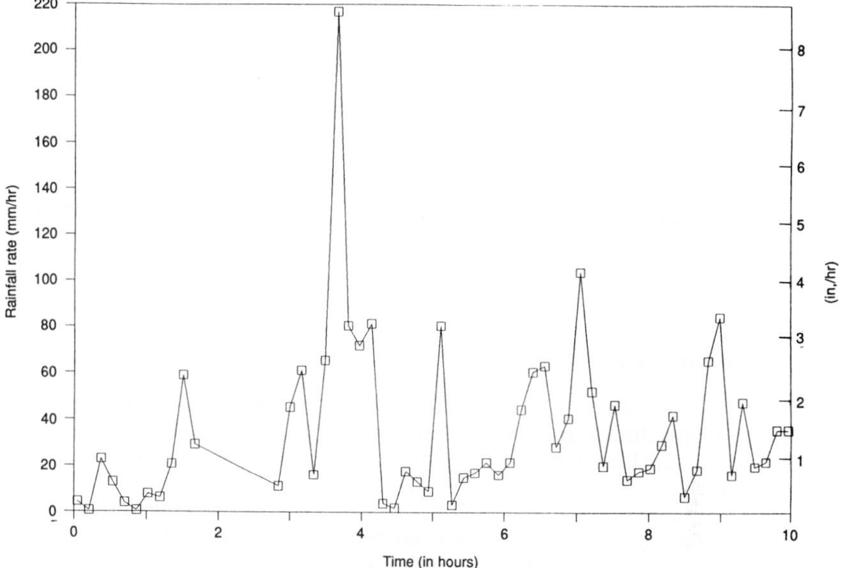

FIGURE 3.1.2 Precipitation rate time series for an extreme storm in Oklahoma on May 27, 1987, as measured by the National Service Storms Laboratory radar (values are for a 2- by 2-km area).

The vertical profiles of temperature, pressure, and density play an important role in precipitation processes. The rate of change of temperature in the vertical is termed the *lapse rate.* For unsaturated air parcels a typical value of lapse rate is approximately 10 K/km. For saturated conditions a typical value is 5.5 K/km. The rate of change of pressure in the vertical can be approximated for a stable atmosphere by the *hydrostatic equation*

$$\frac{dp}{dz} = -\rho g \tag{3.2.1}$$

where p is pressure, ρ is density, and g is the acceleration of gravity. The hydrostatic equation underlies computations of atmospheric moisture used for a variety of hydrologic applications.

3.3 ATMOSPHERIC MOISTURE

Atmospheric moisture content can be expressed in a number of ways. The ratio of the mass m_v of water vapor to the mass m_d of dry air is called the *mixing ratio* and is denoted r, that is,

$$r = \frac{m_v}{m_d} \tag{3.3.1}$$

The mixing ratio is generally expressed in grams of water vapor per kilogram of air. Typical values of mixing ratio at sea level range from 2 g/kg in middle latitudes to 20 g/kg in the tropics. *Specific humidity* is the ratio of the mass of water vapor to the total mass of the parcel:

$$q = \frac{m_v}{m_v + m_d} \tag{3.3.2}$$

Because of the relatively small contribution of atmospheric water vapor to the total mass of the parcel, the specific humidity is generally very close to the mixing ratio.

3.3.1 Air Saturation

A volume of air is "saturated" when it contains all the water vapor it can carry at the prevailing temperature; if more water vapor enters the air, a corresponding amount will condense out. Under these conditions, the partial pressure of water vapor is called the *saturation vapor pressure* and is denoted e_s. Experimental results show that the saturation vapor pressure depends only on temperature. The temperature dependence of saturation vapor pressure is described by the *Clausius-Clapeyron equation,*[102] for which a useful approximate solution is given by

$$e_s = 611 \exp\left(\frac{17.27T}{237.3 + T}\right) \tag{3.3.3}$$

(for T in degrees Celsius and e_s in pascals).

The *saturation mixing ratio* r_s is the ratio of the mass m_{vs} of water vapor in a volume of saturated air to the mass m_d of dry air:

$$r_s = \frac{m_{vs}}{m_d} \tag{3.3.4}$$

The relative humidity RH is the ratio (expressed as a percentage) of the actual mixing ratio to the saturation mixing ratio:

$$\mathrm{RH} = 100\,\frac{r}{r_s} \tag{3.3.5}$$

It can also be expressed as the ratio of the actual vapor pressure e to the saturation vapor pressure e_s at the prevailing temperature.

3.3.2 Dew-Point Temperature

The *dew point* is the temperature to which air must be cooled at a constant pressure in order for it to become saturated. In other words, the dew point is the temperature at which the actual mixing ratio equals the saturation mixing ratio. The *wet-bulb temperature* T_w is the temperature to which a parcel of air is cooled by evaporating water into it at a constant pressure until the air is saturated. The wet-bulb temperature is measured directly with a thermometer, the bulb of which is covered with a moist cloth over which air is drawn. The temperature of an evaporating cloud droplet or a raindrop is equal to the wet-bulb temperature.

3.3.3 Precipitable Water

The total mass of water vapor in a column of unit area extending to the top of the atmosphere is called the *precipitable water* (or *equivalent liquid water*). A useful approximation of the precipitable water W can be obtained by dividing the atmosphere vertically into N discrete intervals:

$$W = 0.0002 \sum_{n=1}^{N} (p_{n-1} - p_n) \frac{q_{n-1} + q_n}{2} \tag{3.3.6}$$

where p_n is air pressure (in millibars) for the nth layer from the surface, q_n is specific humidity (in g/kg), and W is in inches (1 in = 25.4 mm). Using the equation of state for an ideal gas, specific humidity can be expressed as follows:

$$q = 0.622 \frac{e_s \mathrm{RH}}{p} \tag{3.3.7}$$

The amount of precipitable water can be computed from Eqs. (3.3.6) and (3.3.7) using upper-air observations of temperature, pressure, and relative humidity. These observations are obtained by miniature weather stations, called *radiosondes,* which are carried aloft by meteorological balloons.[69] Mean precipitable water for the United States is illustrated in Fig. 3.3.1. In Fig. 3.3.2 the maximum observed precipitable water values are shown (together with the period of record for the stations used to derive the values). Note that maximum values of precipitable water seldom exceed 60 mm (2.1 in). To sustain high precipitation rates in an area, inflow of moisture is clearly required.

3.3.4 Atmospheric Moisture Flow

The flow of atmospheric water vapor is strongly dependent on atmospheric circulation features.[74,75] Figure 3.3.3 shows the vertically integrated water vapor flux field for the northern hemisphere from 1958. From Fig. 3.3.3 it can be seen from the thickness of the arrows that the primary source of atmospheric water vapor in the northern hemisphere is in the latitude belt from 15° north to 35° north. At low latitudes near the equator the atmospheric vapor transport is westward, reflecting the low-latitude easterly winds which blow from east to west. Maximum westward transport occurs at 10° north. At the midlatitudes, transport is dominantly eastward, reflecting the midlatitude westerlies blowing from west to east. Maximum eastward transport occurs at 40° north. The atmospheric moisture inflow for North America reflects two major points of inflow, one along the Pacific coast and one associated with the western Atlantic–Gulf of Mexico region. There is significantly higher moisture inflow along the southeastern United States than along the Pacific coast. Precipitation in the United States is largely fueled by the remote sources of moisture from the Gulf of Mexico and the Pacific Ocean.

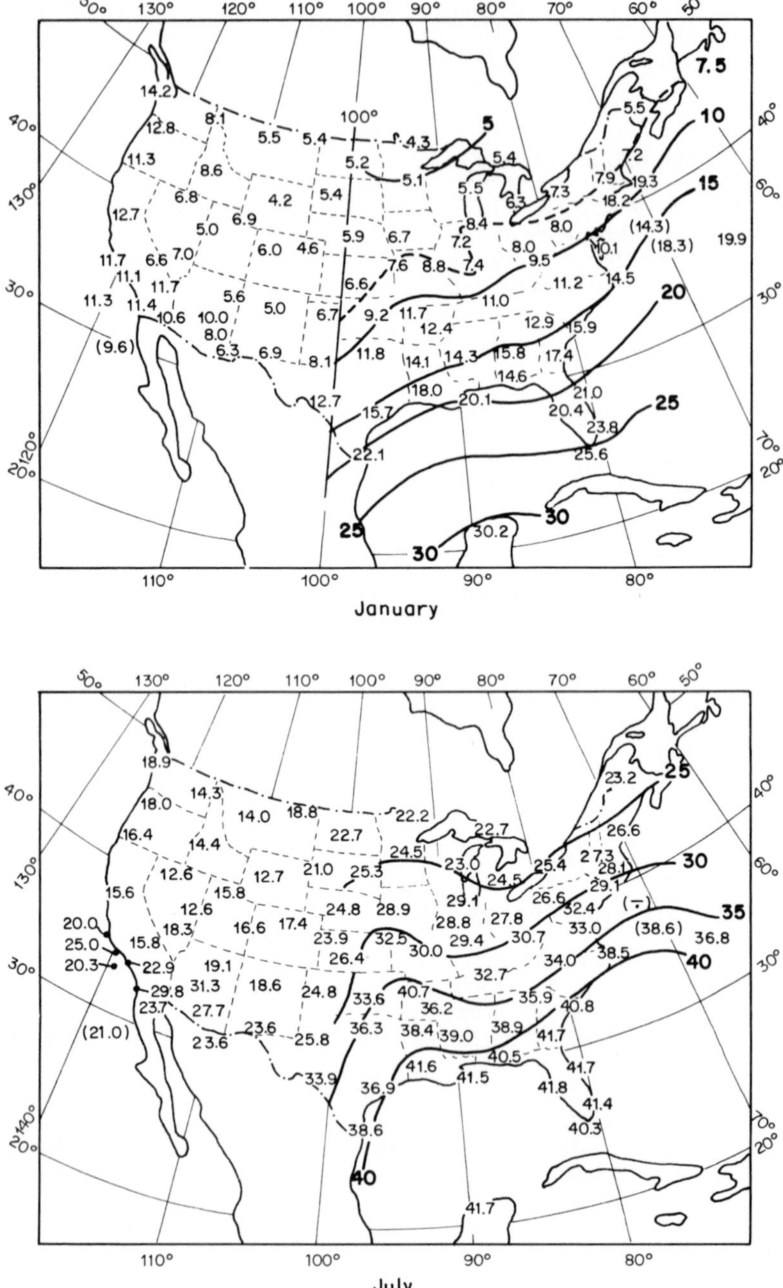

FIGURE 3.3.1 Mean precipitable water for the United States during January and July (in millimeters). *(From F. P. Ho and J. T. Riedel, "Precipitable Water over the United States," Nat. Weather Serv. Tech. Rept. 20, 1979.)*

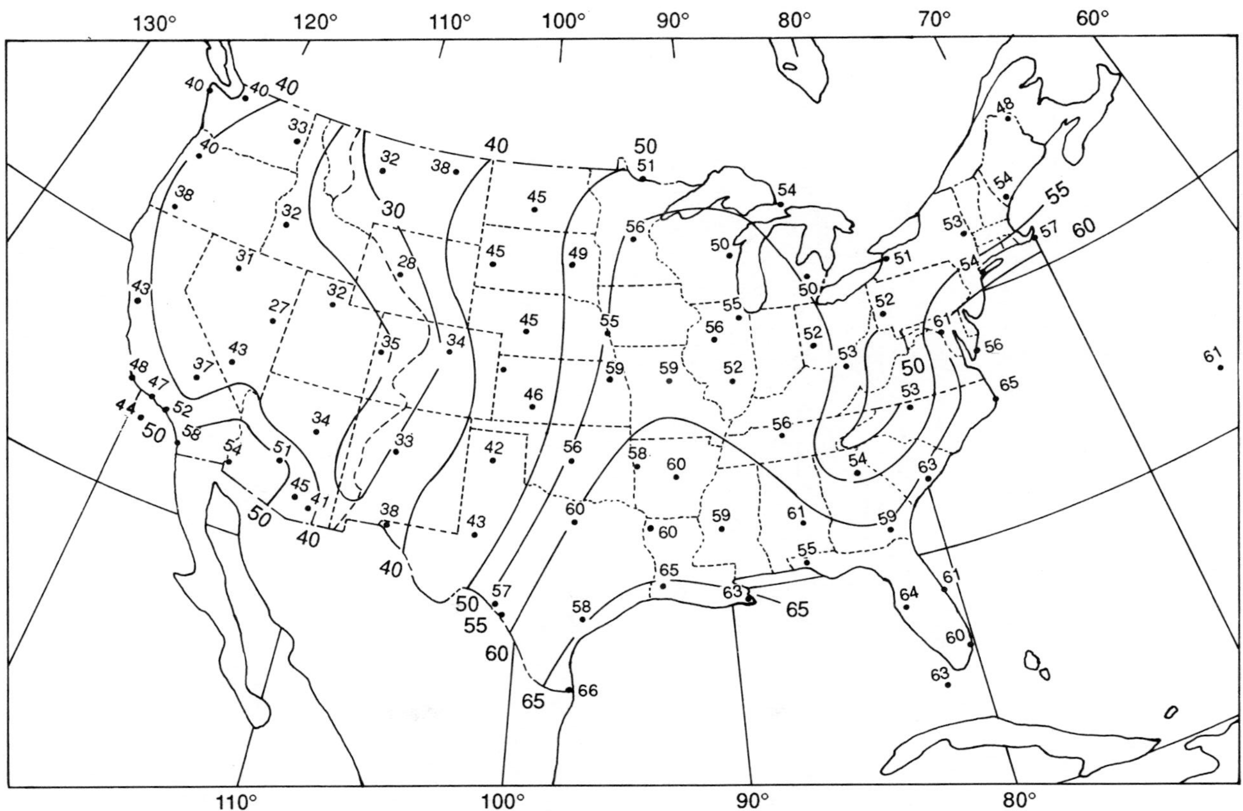

FIGURE 3.3.2 Maximum observed precipitable water for upper air stations in the United States (in millimeters). *(From F. P. Ho and J. T. Riedel, "Precipitable Water over the United States," Nat. Weather Serv. Tech. Rept. 20, 1979.)*

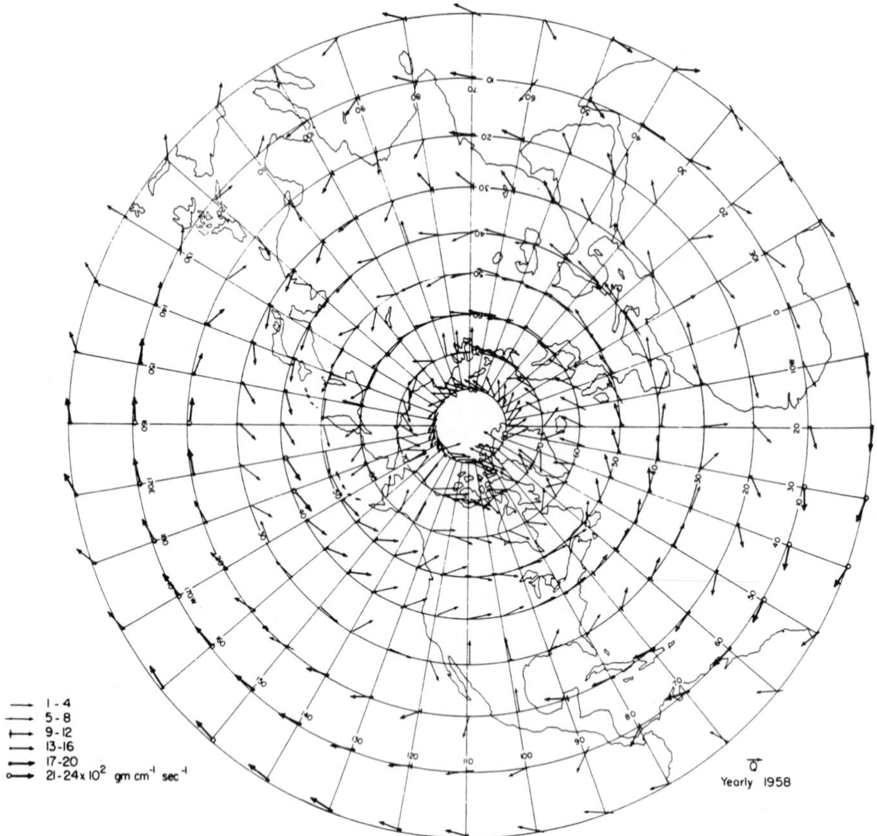

FIGURE 3.3.3 Mean vertically integrated water vapor flux field for the northern hemisphere during 1958. *(Adapted from Ref. 74.)*

3.4 PRECIPITATION PROCESSES

The formation of precipitation is studied in the field of *cloud physics* (see Mason[63] for an introduction). Numerous physical processes are involved in precipitation formation, and these have connections with a diverse collection of issues ranging from environmental quality to climate change.

3.4.1 Cloud Formation

Clouds form when air becomes *supersaturated,* that is, when the actual vapor pressure exceeds the saturation vapor pressure e_s. Supersaturation occurs through expansion and cooling of ascending air parcels, causing water vapor to condense on atmospheric particles, a process called *nucleation.* Atmospheric *aerosol,* which is a suspension of solid or liquid matter with small settling velocity, plays an important role in promoting condensation by providing nucleation sites for water vapor. Diameters of atmospheric aerosol range from 0.01 to 100 μm. Aerosol which serve as the nuclei upon which water vapor condenses in the atmosphere are called *cloud condensation nuclei* (CCN). The most effective CCN are large aerosol particles with high water solubility. Continental air masses are significantly richer in CCN than marine air masses, but marine air masses have higher concentrations of large CCN (Fig. 3.4.1), which means that marine air masses are more likely to produce precipitation than the continental air masses.[102]

Condensed water droplets form clouds. Two types of clouds can be distinguished: *cold clouds* and *warm clouds.* Clouds which extend above the 0°C level are called cold clouds. In cold clouds, water can be present in the form of ice particles, supercooled droplets, or water vapor. Clouds which contain both ice particles and supercooled droplets are termed *mixed clouds.* Ice particles can grow by three major processes: (1) growth from vapor condensation, (2) growth by collisions with supercooled water droplets, and (3) growth by aggregation with other ice particles. Growth of ice particles from the vapor phase results in *snow* crystals. Growth by collisions with supercooled water droplets produces *hailstones* and *graupel.* Growth by aggregation can occur for each type of ice particle.

3.4.2 Cloud Structure

In the early twentieth century Wegener discovered that in mixed clouds growth of ice crystals from vapor condensation is a common mechanism of precipitation formation, implying that precipitation often forms aloft as snow. If the 0°C isotherm lies between the level at which precipitation forms and the land surface, then the snow crystals melt and surface precipitation is in the form of rain. The development of

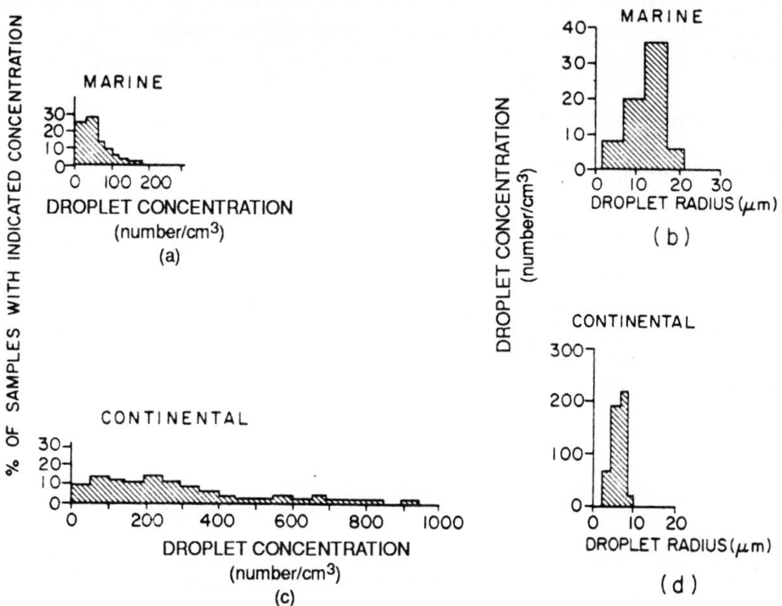

FIGURE 3.4.1 Concentrations of cloud condensation nuclei for continental and marine air masses.[102]

weather radar in the 1940s greatly expanded understanding of the vertical structure of precipitation in cold clouds. Radar observations from cold clouds are characterized by a region of high reflectivity, termed the *bright band,* at the elevation of the melting layer. The vertical profile of droplet fall speeds is also dominated by the melting layer, with low fall speeds associated with snow particles above the melting layer and relatively high fall speeds associated with water droplets below the melting layer. These features of clouds play an important role in developing procedures for estimating rainfall from radar and satellite sensors.

Clouds which lie completely below the 0° isotherm are called warm clouds. Only liquid water droplets can form in warm clouds. Water droplets grow in warm clouds by condensation, collision, and coalescence. Tropical cloud systems are often composed of warm clouds. Warm cloud processes are also important below the melting layer of cold clouds.

3.4.3 Cloud Seeding

Attempts to modify clouds and precipitation have occurred since ancient times. Battlefield and aviation considerations during World War II brought the topic under close scientific scrutiny. Since that time numerous "cloud seeding" programs have been undertaken to increase precipitation (or suppress hail; see Ref. 24 for historical notes). Attempts to increase precipitation can be distinguished by the dominant precipitation mechanisms discussed above. For warm clouds the most prominent procedure has been injection of large hygroscopic particles or water drops into the cloud base. This stimulates growth of water drops by the collision-coalescence mechanism. For cold clouds seeding by artificial ice nuclei (including dry ice, lead iodide, and cupric sulfide) is the most common means of attempting to induce precipitation. Dennis[24] reviews the results of many cloud seeding experiments.

3.4.4 Clouds and Climate

Clouds and precipitation play a central role in climate processes (see Chap. 2). An important role of precipitation is associated with transport of latent heat from the tropics poleward. The energy released in condensation of 1 kg of water vapor is approximately 2.5 million joules, so motion of water vapor effectively means motion of latent heat. Thus a general picture of latent heat transport in the northern hemisphere can be inferred from Fig. 3.3.3 corresponding to the mean vertically integrated water vapor flux field for the northern hemisphere. In addition to their role in latent heat transport, clouds also play a significant role in modifying the radiant energy fluxes of the atmosphere.

3.5 PRECIPITATING CLOUD SYSTEMS

The structural organization of precipitating cloud systems can be characterized by the relative dominance of *convective* or *stratiform* precipitation mechanisms. These two basic precipitation mechanisms differ in the times for growth of precipitation particles and the magnitudes of vertical air motions associated with the precipitating clouds. In stratiform precipitation vertical air motions are weak, precipitation particles are initiated near the top of the cloud system, and the time until precipitation develops can be quite long (hours). For convective precipitation, vertical air motions are locally strong, growth of precipitation particles initiates at cloud base at the time of cloud formation, and the time until precipitation develops is very short (approximately 45 min). A third precipitation mechanism, which can have both stratiform and convective components, is produced by *orographic lifting* of air masses over mountains or hills. Five major classes of precipitation-producing systems are discussed below.

3.5.1 Extratropical Cyclones

The atmospheric circulation is made up of air masses or streams which normally move in fairly stable and well-defined patterns over the earth. The temperature and moisture of air masses depends heavily on their source area; polar continental air masses are cold and dry; tropical oceanic air masses are warm and moist. Regions outside the tropics often experience contrasting air masses.

Two air masses or streams flowing in parallel with one another, but possessing different temperatures, may become coupled by an initially local instability near their interface that grows as the air masses intertwine and start to rotate together in an immense spiral called a *cyclone*. Warmer air, being lighter, rises over the cooler, denser air, forming *fronts* which are characterized by sharp gradients in temperature, pressure, and often moisture. A *cold front* occurs when cold air flows beneath slower-moving warm air (left-hand side of Fig. 3.5.1), while a *warm front* occurs when faster-moving warm air overtakes slow-moving cold air (right-hand side of Fig. 3.5.1).

Fronts associated with extratropical cyclones curve outward from the storm center for thousands of kilometers. Vertical lifting in extratropical cyclones is associated with fronts and is generally less than 0.1 km/h. Much of the precipitation in extratropical cyclones is thus dominated by the stratiform mechanism. The pattern of precipitation in cyclonic storms reflects the broad pattern of the front (Fig. 3.5.1). The detailed substructure within rainfall patterns has been described by Austin and Houze[10] and Houze.[46] An important feature is the organization of rainfall into more intense regions termed *rain bands* in which the convective precipitation mechanism is often prominent.

3.5.2 Midlatitude Thunderstorms

Just as extratropical cyclones represent the best example of stratiform precipitation, so midlatitude thunderstorms represent the best example of *convective* rainfall. Midlatitude thunderstorms are distinguished by the organization of thunderstorm cells.

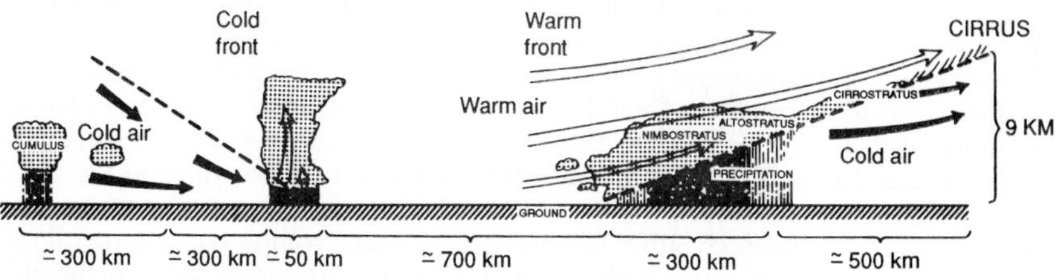

FIGURE 3.5.1 Conceptual model of an extratropical cyclone. *(Adapted from Ref. 102. Used with permission.)*

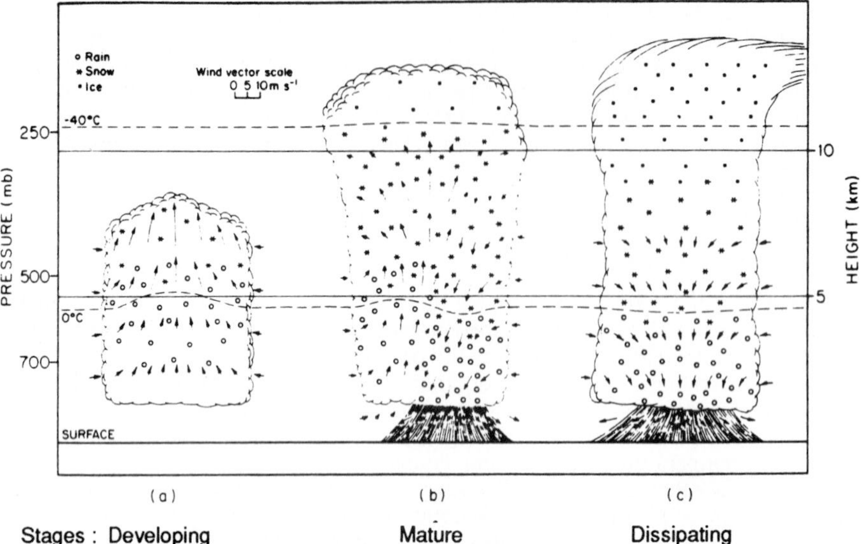

Stages : Developing Mature Dissipating

FIGURE 3.5.2 Stages of development of an air-mass thunderstorm. *(Adapted from Ref. 17.)*

Air-mass thunderstorms form in convectively unstable air masses with relatively large amounts of low-level moisture and little wind shear. The spatial structure of rainfall is characterized by organization into random patterns of thunderstorm cells. The Thunderstorm Project[17] of the late 1940s initiated the study of midlatitude thunderstorms and focused on air-mass thunderstorms. An important feature observed during the Thunderstorm Project was that each cell of a thunderstorm has a characteristic life cycle (Fig. 3.5.2), composed of (1) a developing "cumulus" stage, in which precipitation particles develop at the cloud base but do not reach the ground because of strong updrafts, (2) a "mature" stage in which the drag of precipitation particles creates a downdraft and precipitation particles begin reaching the ground, and (3) a "dissipating" stage in which light rain continues to fall. Air-mass thunderstorms generally do not produce large amounts of rain over extensive areas.

Organized thunderstorms, termed *mesoscale convective systems* (MCS), are a primary cause of flooding in many areas. Two broad categories of mesoscale convective systems can be distinguished: *mesoscale convective complexes* and *squall lines.* A mesoscale convective complex is characterized by a circular cloud shield that attains an area of about 100,000 km^2.[59] Squall lines are characterized by a linear organization of convective elements. Maddox et al.[60] have developed a classification system of flash-flood-producing storms for the United States and have shown that mesoscale convective complexes play a dominant role, particularly in the central United States. Fritsch et al.[33] note that mesoscale convective complexes provide more than half the growing-season precipitation for the Great Plains of the United States. A key feature of extreme flood-producing thunderstorms is that they repeatedly regenerate and cross the same area. The temporal pattern of rainfall rate is illustrated in Fig. 3.1.2. Chappell[19] terms midlatitude thunderstorms of this type "quasi-stationary convective events" (which can be either mesoscale convective complexes or squall lines). The general structure of quasi-stationary convective events was described in the early 1960s by Huff and Changnon.[49]

3.5.3 Tropical Cloud Clusters

Figure 3.5.3 illustrates that globally averaged precipitation is greatest in the tropics. The maximum in precipitation there is associated primarily with *cloud clusters* which occur in the zone of trade-wind convergence. Cloud clusters are, like all tropical cloud systems, convective in origin. Although tropical cloud systems cover a broad range of spatial scales, the bulk of precipitation is due to cloud clusters with rain areas of up to 50,000 km^2. Tropical rainfall plays an important role in global circulation and has important connections with atmospheric circulation anomalies such as El Niño.[88] Unfortunately, information on tropical rainfall is incomplete owing to the difficulties in accurate measurement of rainfall over the oceans. The Tropical Rainfall Measuring Mission (TRMM; see Ref. 88) and the Global Precipitation Climatology Project (GPCP; see Ref. 5) will greatly improve capabilities for assessing global precipitation in the tropics.

3.5.4 Monsoon Rainfall

The largest precipitation accumulations for periods greater than 24 h are associated with the Asian *monsoon.* India and southeast Asia are the prime locations of monsoonal rain during the Asian summer. Indonesia and Malaysia experience extreme monsoonal rain during the Asian winter.

The term monsoon is derived from the Arabic word for season. The characteristic feature of monsoon climates is a pronounced seasonal reversal of wind regime. Large-scale reversal of wind regimes occurs throughout India, much of southeastern Asia, and east Africa. A striking feature of monsoon climates is that most precipitation generally occurs during a relatively small fraction of the year. In Bombay, India,

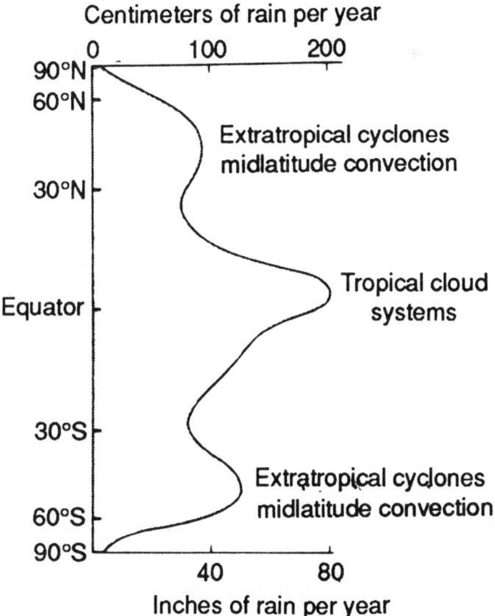

FIGURE 3.5.3 Annual precipitation by latitude bands. *(Reproduced with permission from Ref. 85.)*

for example, approximately 80 percent of the mean annual rainfall of 80 in (200 cm) occurs during the months of June, July, and August.

3.5.5 Hurricanes

"Hurricanes," or "typhoons" as they are known in the west Pacific, are responsible for extreme rainfall totals in coastal regions of the Atlantic and Pacific. Rainfall patterns associated with hurricanes often undergo significant change following landfall.[87] Extreme rainfall can be widespread and produce large-stream flooding, as Hurricane Agnes did in 1972, or be locally intense, as was the case with Hurricane Camille in 1969.[84] Hurricane rainfall is organized about an eye-wall rain band and outer spiral bands (Fig. 3.5.4). Hurricane rain bands exhibit both convective and stratiform characteristics. The mechanisms which initiate and sustain hurricanes are still the subject of scientific debate.[87]

3.5.6 Orographic Precipitation

Orographic influences can modify each of the basic storm types discussed above. A detailed discussion of orographic control of precipitation is given in Smith.[95] Smith distinguishes three principal orographic mechanisms: (1) convective initiation, (2) large-scale lifting, and (3) low-level growth.

Lifting of air as it flows over hills and mountains can trigger *convective instabilities*

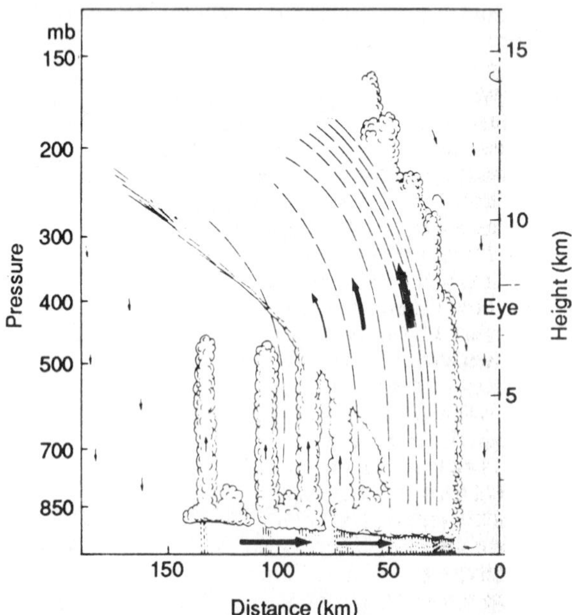

FIGURE 3.5.4 Idealized radial cross section through a hurricane. Arrows indicate radial and vertical mass fluxes. *(Adapted from Ref. 102. Used with permission.)*

that produce or enhance rainfall. Topographic features are clearly linked with a number of extreme storms in the United States including Hurricane Camille[84] and the Big Thompson storm.[19] Large-scale air-mass lifting over mountain ranges, which can produce precipitation by both stratiform and convective mechanisms, is an important element of water supply in much of the western United States. *Low-level growth* of precipitation can result from entrainment of liquid water droplets near the ground and from reduced evaporation losses for precipitation reaching the ground in topographically elevated regions. The relative importance of individual orographic mechanisms depends primarily on the size of the terrain feature and the storm type.

3.6 PRECIPITATION AT THE EARTH'S SURFACE

Two physical factors play a dominant role in determining the rate of precipitation, the *fall velocity* of raindrops (or frozen precipitation) and the *size distribution* of raindrops (or frozen precipitation). These two factors also influence what happens to the precipitation once it has reached the land surface.

3.6.1 Velocity of Raindrops

The terminal velocity of a rigid, spherical raindrop is proportional to the square root of the diameter of the raindrop.[20] As a water droplet falls through the air, however, it is affected by aerodynamic forces which cause the droplet to vibrate and deform. Drops of diameter less than 0.35 mm are essentially spherical and drops up to 1 mm in diameter have a shape well approximated by an oblate spheroid. Larger drops have a progressively flattened and then concave base (Fig. 3.6.1). For large droplets, vibrations and deformations frequently break up the droplet.

Precise measurements of terminal velocity of water droplets in stagnant air at sea level were made by Gunn and Kinzer.[40] Atlas and Ulbrich[6] show that the terminal velocity $v(D)$ (in m/s) of a raindrop of diameter D (in mm) can be accurately represented by

$$v(D) = 3.86D^{0.67} \tag{3.6.1}$$

for drop diameters between 0.8 and 4.0 mm.

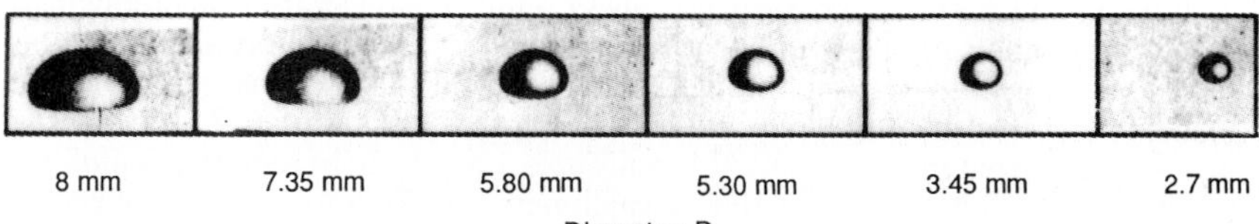

FIGURE 3.6.1 Typical shape of large raindrops at terminal velocity. *(Adapted from Ref. 25.)*

3.6.2 Drop-Size Distributions

Drop-size distributions in a volume of the atmosphere are characterized by relating drop density (in drops per cubic meter) and the distribution of drop sizes (in mm). (See Fig. 3.4.1.) Drop-size distributions are typically specified by a function $N(D)$ which represents the density of drops as a function of drop diameter. The most commonly used drop-size distribution is the Marshall-Palmer distribution,[62] which is given by

$$N(D) = N_0 \exp(-\Lambda D) \tag{3.6.2}$$

where $N(D)$ and N_0 have units of drops per cubic meter per mm of drop diameter and Λ has units mm^{-1}. A typical value of N_0 is 8000 $\text{m}^{-3}\text{mm}^{-1}$.[64] Marshall and Palmer relate the parameter Λ to rainfall rate through the formula

$$\Lambda = 4.1R^{-0.21} \tag{3.6.3}$$

where R is rainfall rate in mm/h. The Marshall-Palmer distribution has proved useful for a variety of applications. Numerous authors have noted, however, that observed drop-size distributions have characteristics that are often not well represented by the Marshall-Palmer distribution.[53] Alternative models are based on assumptions that drop diameters have either a lognormal or gamma distribution.

Several automated devices have been developed for measuring drop-size distributions, including the *distrometer*[52] and *raindrop camera.*[18] These devices provide the number of drops M within the sample volume and the diameters $D_1, \ldots, D_M$ of the raindrops. A number of important variables related to rainfall can be computed from these observations. Rainfall rate (in mm/h) is given by

$$R = (6\pi 10^{-4})V^{-1} \sum_{i=1}^{M} D_i^3 v(D_i) \tag{3.6.4}$$

where D is drop diameter (in mm), $v(D)$ is terminal velocity [in m/s; see Eq. (3.6.1)], and V is the volume of the sample element in cubic meters. Similarly, one can define the rainfall energy flux, i.e., kinetic energy per unit time (in joules/s/m²) by

$$\text{KE} = \left(\frac{\pi}{12} 10^{-9}\rho_w\right)V^{-1} \sum_{i=1}^{M} D_i^3 v(D_i)^3 \tag{3.6.5}$$

where ρ_W is the density of water (1000 kg/m³ at 0°C). The radar reflectivity factor (in mm⁶/m³) is defined by

$$Z = V^{-1} \sum_{i=1}^{M} D_i^6 \tag{3.6.6}$$

The radar reflectivity factor can be inferred from measurements taken by weather-surveillance radars (see Sec. 3.7.2).

Equations (3.6.4) and (3.6.5) are used by Wischmeier and Smith[109] in producing a regression relationship between the kinetic energy of rainfall and rainfall rate; this relationship forms the basis for development of the universal soil loss equation. The relationship between rainfall rate and rainfall energy flux (in watts/m², i.e., joules/s/m²) is summarized in Table 3.6.1.[92] From Table 3.6.1 it can be seen that a uniform rainfall rate of 25 mm/h over a 10 km² catchment translates to 1.29 MW of energy flux (i.e., 1.29×10^6 W). Rainfall energy flux increases rapidly with increasing rainfall rate. A rainfall rate of 10 mm/h results in a rainfall energy flux of 0.0439 W/m². A rainfall rate of 100 mm/h results in a rainfall energy flux of 0.664 W/m². Increasing

TABLE 3.6.1 Relationship between Rainfall Energy Flux and Rainfall Rate. (Data from Coweeta Hydrologic Research Laboratory, North Carolina.)

Rainfall rate, mm/h	Rainfall energy flux, W/m^2
0.25	0.000565
0.5	0.00128
1.0	0.00290
2.5	0.00855
5.0	0.0194
10.0	0.0439
25.0	0.129
50.0	0.293
100.0	0.664
250.0	1.96

Source: Adapted from Ref. 92.

rainfall rate by a factor of 10 increases rainfall energy flux by a factor of 15.

3.6.3 Interception

The disposition of precipitation at the ground surface depends on the characteristics of the land surface, especially the type and density of vegetation and cover of buildings, pavement, or roads. *Interception* is the term that covers a variety of processes that result from the temporary storage of precipitation by vegetation or man-made cover. Intercepted precipitation can be either evaporated to the atmosphere or ultimately transmitted to the ground surface. The main components of interception by vegetation are *throughfall, stemflow,* and *interception loss.*[103] Throughfall occurs either when precipitation falls through spaces in the vegetation canopy or when precipitation drips from leaves and twigs. Stemflow designates water that flows along twigs and branches with its ultimate delivery to the ground surface at the main stem or trunk. Interception loss accounts for precipitation that is retained by plant surfaces and later evaporated or absorbed by the plant.

Interception loss by vegetation depends both on characteristics of rainfall and on the vegetation cover. Interception loss is greatest at the beginning of a storm when interception storage, i.e., ability of the vegetation surfaces to collect and retain falling precipitation, is highest. It follows that rainfall frequency is more significant than duration or amount of rainfall in determining interception losses. Stemflow is a residual term that is approximately an order of magnitude smaller than throughfall and is very difficult to measure accurately. Stemflow is, however, quite important because it concentrates precipitation at the root base of vegetation.

3.6.4 Precipitation Scavenging

Precipitation scavenging is a principal mechanism of removing aerosol from the atmosphere and delivery of atmospheric pollutants to the land surface. Precipitation scavenging occurs as raindrops fall through the air and capture aerosol particles in their fall path. It is estimated that over 50 percent of atmospheric moisture acidity is

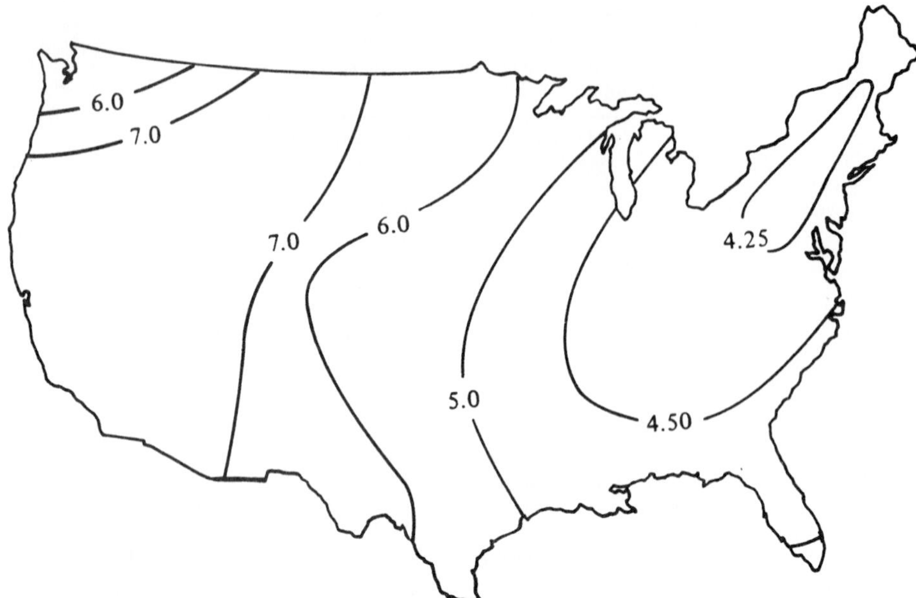

FIGURE 3.6.2 Typical pattern of rainfall pH in the continental United States (pH = 7.0 is neutral and decreasing numbers denote increasing acidification). *(Adapted from Ref. 61.)*

associated with precipitation scavenging. Figure 3.6.2 shows the geographic pattern of typical pH values of precipitation for the United States. The pattern is strongly correlated with industrial activity and mean wind patterns. The dominant source of acid rain is sulfuric acid, which arises from scavenging of sulfur particles produced by burning coal and oil.

3.7 PRECIPITATION MEASUREMENT

3.7.1 Rain Gauges

A variety of measurement devices have been developed for precipitation. *Rain gauges* are of two types: *recording* and *nonrecording.* A recording gauge automatically records rainfall accumulation at temporal resolutions down to 1 min or less. Recording rain gauges are often equipped with *telemetry* to allow for real-time transmission and utilization for water management. Three major types of recording rain gauges are the *weighing* type, the *float and siphon* type, and the *tipping-bucket* type. Figure 3.7.1 illustrates a weighing rain gauge. Nonrecording gauges consist simply of a cylindrical container and a calibrated measuring stick, which may be a part of the gauge. Primitive rain gauges were used in several cultures more than 2000 years ago. The invention of the "modern" nonrecording rain gauge is attributed to the Koreans and dates back to the sixteenth century.

Virtually all rain gauges suffer from errors due to modification of the wind field by the gauge. The magnitude of errors depends heavily on wind speed, siting characteristics, type of precipitation (rain or snow), and temperature.[28,57,70,78,79,86] Rain gauge measurement is difficult in a variety of settings, including mountain ridges, forests,

FIGURE 3.7.1 Illustration of a weighing rain gauge.[20]

and water bodies. Measurement errors for snow are typically much larger than for rain and are generally in the form of catch deficiencies; that is, less snow is measured than actually occurs (see Fig. 3.7.2).

3.7.2 Spatial Estimation of Rainfall from Rain Gauges

For hydrologic applications it is often necessary to compute estimates of *mean areal precipitation* for a catchment from rain gauge observations. Most of the procedures that are used for computing mean areal precipitation can be expressed as linear combinations of the observations. If n gauges, with values $P_1, \ldots, P_n$, are available for estimating precipitation in a catchment, then the estimate of mean areal precipitation is

$$\overline{P} = \sum_{i=1}^{n} a_i P_i \qquad (3.7.1)$$

where the *station weights* $a_1, \ldots, a_n$ are nonnegative constants that sum to 1.

The simplest method for computing mean areal precipitation is the *arithmetic mean* method in which the station weights are all $1/n$. This method is appropriate for

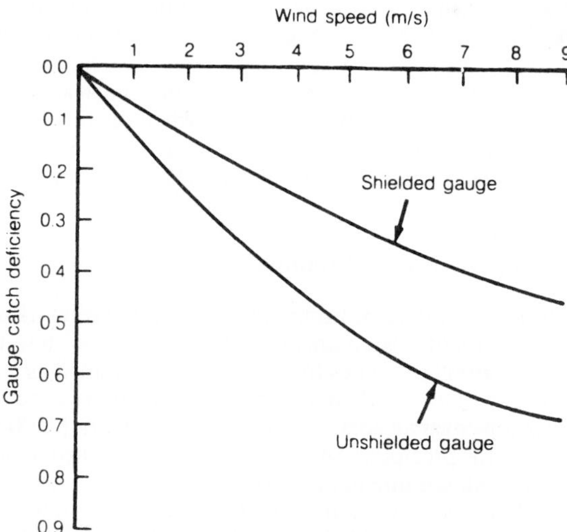

FIGURE 3.7.2 Mean rain gauge catch deficiency for snowfall of shielded and unshielded gauges in the United States as a function of wind speed.[57]

dense networks with uniform gauge locations. The *Thiessen polygon* method is based on the assumption that for any point in the watershed rainfall is equal to the observed rainfall at the closest gauge. The station weights are specified by the relative areas of the Thiessen polygon network, the boundaries of the polygons being formed by the perpendicular bisectors of the lines joining adjacent gauges. The Thiessen polygon method is well suited to graphical determination of weights.

For computer implementation, it is often convenient to compute mean areal precipitation by first interpolating rainfall at gauge sites onto a rectangular grid and then estimating areal precipitation by summing estimates from grid boxes within the catchment. The *inverse distance-squared* method is commonly used for this purpose. The estimate for the *j*th grid box is

$$\bar{P}_j = a \sum_{i=1}^{n} d_{ij}^{-2} P_i \tag{3.7.2}$$

where d_{ij} is the distance from gauge i to the center of grid box j and a is the inverse of the sum of the inverse distance-squared values for all gauges:

$$a = \left(\sum_{i=1}^{n} d_{ij}^{-2} \right)^{-1} \tag{3.7.3}$$

If m grid boxes cover the catchment, the mean areal precipitation is the arithmetic mean of the m estimates obtained from (3.7.2). The station weights in Eq. (3.7.1) for the inverse distance-squared method are given by

$$a_i = \frac{1}{m} \sum_{j=1}^{m} \frac{a}{d_{ij}^2} \tag{3.7.4}$$

If the gauge network does not vary with time, this formula can be used for computing mean areal precipitation of the catchment, eliminating the need for repeatedly computing precipitation at each grid point.

Another procedure which can be used for grid-based estimation of mean areal precipitation is "kriging."[23,56] For this procedure, the correlation properties of rainfall fields are used to specify the weights used in (3.7.2). Singh and Chowdhury[89] reviewed the performance of procedures for estimating mean areal precipitation and showed that all methods give comparable results when the time period is long, as in computation of mean annual rainfall on a watershed. As the time period diminishes, as for daily rainfall, the differences in results among the various methods increase.

3.7.3 Radar Measurement of Rainfall

The most important advantage of using radar for precipitation measurement is the coverage radar provides of a large area with high spatial and temporal resolution. Radar can provide rainfall estimates for time intervals as small as 5 min and spatial resolution as small as 1 km^2. With an effective range of approximately 200 km (130 mi), a single radar can cover an area of more than 10,000 km^2 (4500 mi^2). Systems of weather radars used for precipitation estimation are either active or planned in the United States, England, continental Europe, and Japan.

In the United States a network of more than 120 high-quality radars, the Next Generation Weather Radar system (NEXRAD), will be deployed by 1996 (locations are shown in Fig. 3.7.3). The first NEXRAD unit, which is located in Oklahoma City, Okla., began operation in 1991. The NEXRAD systems are termed WSR-88D,

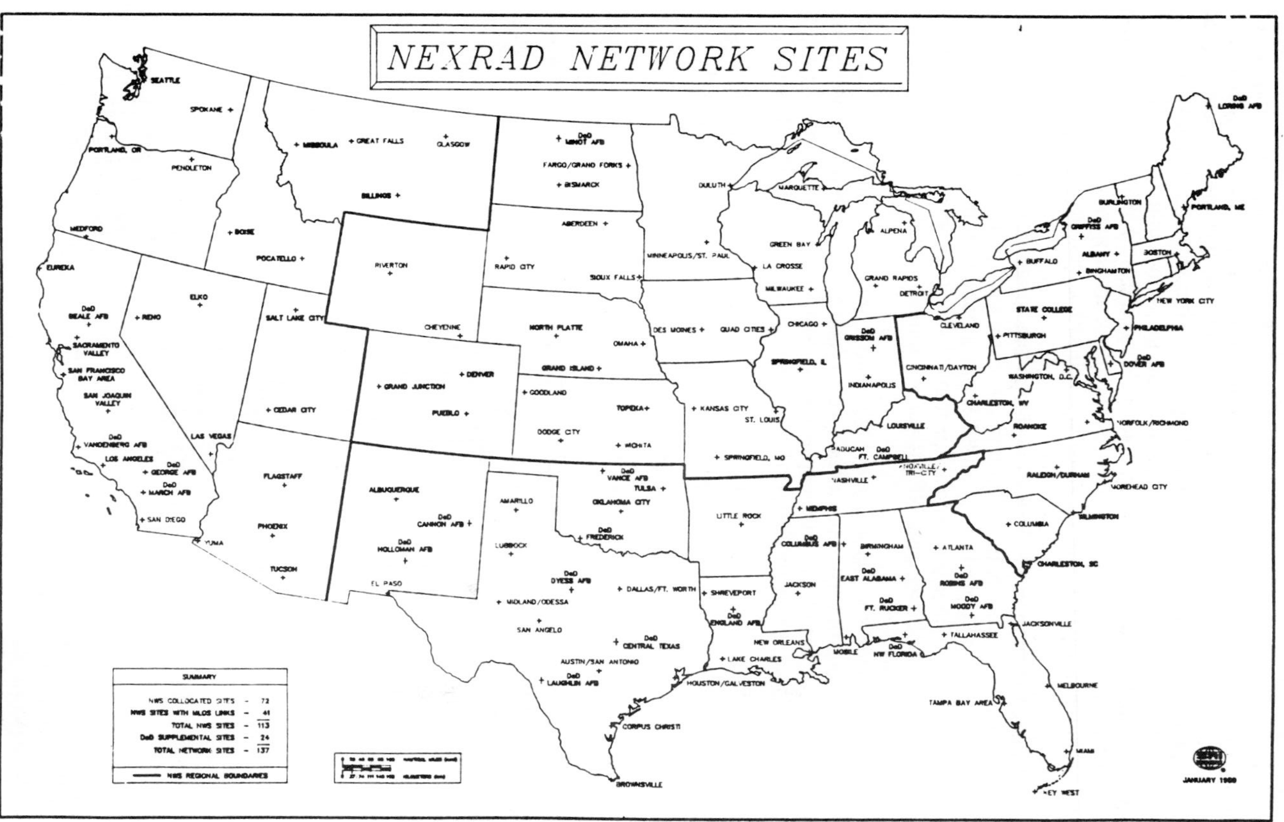

FIGURE 3.7.3 Proposed sites for the next generation weather radar (NEXRAD) network.

where WSR denotes "weather surveillance radar" and D denotes Doppler. Figure 3.7.4 illustrates the principal components of the WSR-88D radar. The centerpiece of the system is a 10-cm "radar data acquisition" unit. Each WSR-88D system also contains a computer processing system for converting radar data to meteorological products and a graphical display system. Figure 3.7.5 illustrates one of the WSR-88D products, storm total precipitation accumulation, for a storm that occurred in central Oklahoma on May 27, 1987. The data were obtained from a WSR-88D prototype radar in Norman, Okla. A time series of rainfall accumulation during the May 27 storm is shown in Fig. 3.1.2 for the radar grid cell with largest storm total accumulation.

The spatial resolution of WSR-88D digital precipitation products is approximately a 4 by 4 km square. In addition to the storm total precipitation product the WSR-88D system will produce 1- and 3-h precipitation accumulation products. Precipitation data at finer time and space resolutions can be archived by the WSR-88D. Algorithms for developing NEXRAD precipitation products are discussed in Refs. 2, 47, and 94.

"Electromagnetic" or "radio waves" from radar are electric and magnetic force fields that propagate through space at the speed of light and interact with matter along their paths.[25] The distance or time between successive wave peaks defines the wavelength λ (in cm) or the wave period f (in hertz). The two are related to the speed of light c by

$$c = \lambda f \tag{3.7.5}$$

Microwaves are electromagnetic forces having spatial wavelengths between 0.1 and 10 cm. Rainfall measurements are typically made with radar systems operating at wavelengths between 1 and 10 cm. The NEXRAD system, as noted above, employs a 10-cm radar.

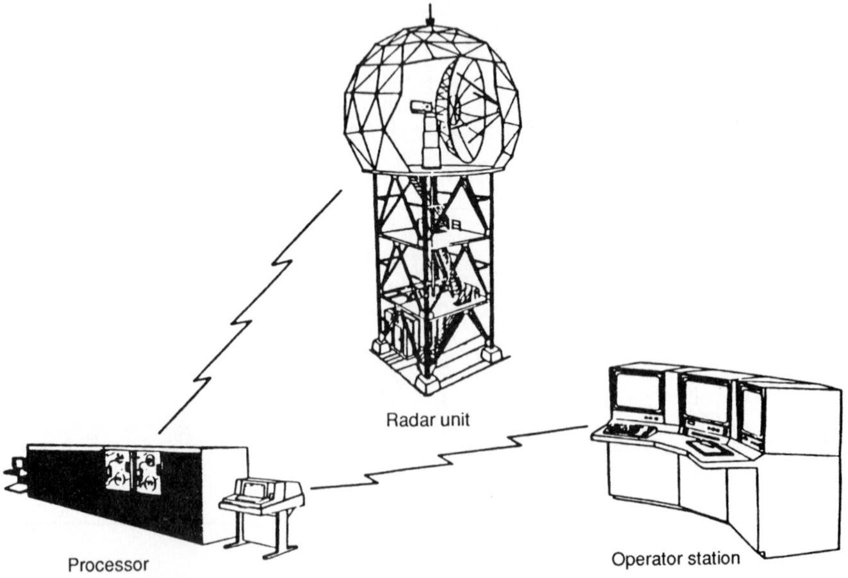

FIGURE 3.7.4 Schematic illustration of a WSR-88D (NEXRAD) system.

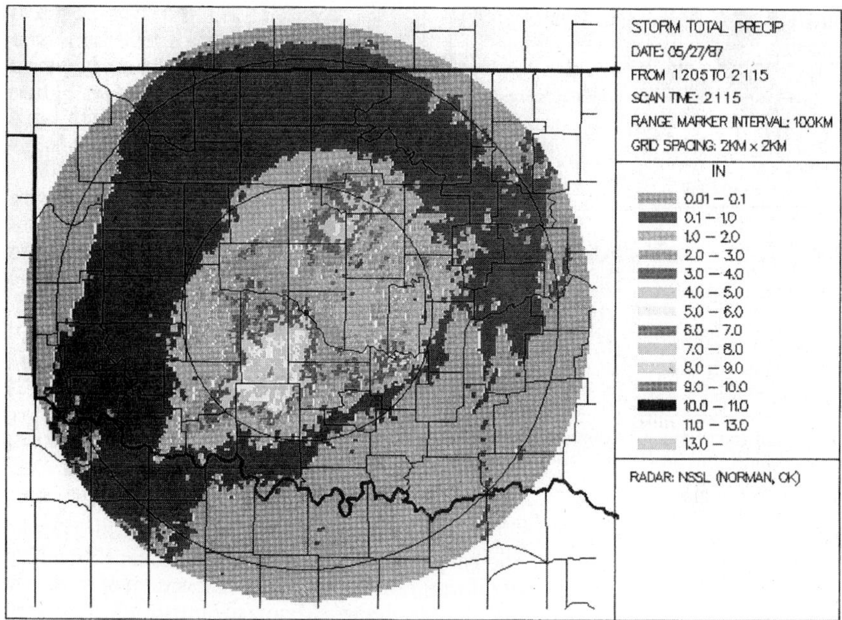

FIGURE 3.7.5 Storm total rainfall accumulation for a storm on May 27, 1987, in Oklahoma obtained from the National Severe Storms Laboratory radar.

The measurement taken by radar is of backscattered power. The utility of the measurement for rainfall estimation derives from the radar equation[12] which relates radar measured power to characteristics of the radar and to characteristics of the precipitation targets. From the radar equation the average power P received from a volume of rain-filled atmosphere at range r can be represented as follows:

$$P = \frac{CLZ}{r^2} \tag{3.7.6}$$

where C is a constant dependent upon radar design characteristics (including beamwidth, antenna gain, frequency, and pulse duration; see Table 3.7.1, which contains design parameters of the NEXRAD system), L represents fractional signal losses by attenuation, and Z is the radar reflectivity factor [Eq. (3.6.6)]. Estimates of the reflectivity factor Z obtained from the radar equation are termed equivalent reflectivities.

TABLE 3.7.1 NEXRAD (WSR-88D) System Design Parameters

Transmit frequency range	2.7–3.0 GHz
Peak power output	1.0 MW
Average power output	2.0 kW
Antenna beamwidth	0.95°
Antenna gain	45.5 dB
Receiver bandwidth	2.7–3.0 GHz
Dynamic range	95 dB

Power law models of the form

$$R = aZ^b \tag{3.7.7}$$

are used to estimate rainfall rate from reflectivity. These models are referred to as *Z-R relationships.* Parameters for

TABLE 3.7.2 Power Law Parameters for Relating Reflectivity Factor and Rainfall Rate in Eq. (3.7.7)

	b	a
Florida	0.74	0.013
North Carolina	0.72	0.019
New Jersey	0.71	0.018
Illinois	0.70	0.015
Arizona	0.67	0.013
Alaska	0.64	0.026
Oregon	0.62	0.028
Marshall Islands	0.76	0.016
Panama	0.75	0.012
Indonesia	0.70	0.017

Note: Parameters are developed from drop-size data obtained with the Illinois State Water Survey raindrop camera (see Cataneo and Stout[18]).

power law models can be estimated using paired samples of reflectivity and rainfall rate obtained from observed drop-size distributions using Eqs. (3.6.4) and (3.6.6). Table 3.7.2 lists estimated parameter values for a number of sites around the world.

Precipitation estimates derived from radar, like those obtained from rain gauges, are subject to errors arising from a number of possible sources. Detailed analyses of errors in radar rainfall estimates can be found in Refs. 9, 22, and 53.

3.7.4 Satellite Measurement of Rainfall

Satellite sensors provide the only systematic means of measuring rainfall over three quarters of the earth, the region covered by oceans. Satellite systems are the principal source of information for long-term assessment of trends in precipitation associated with global climate change. Major satellite monitoring programs for rainfall include the global precipitation climatology project (GPCP[5]) and the tropical rainfall measuring mission (TRMM[88]).

Infrared imagery has provided the primary information for routinely estimating rainfall from satellites. Satellite infrared images are composed of measurements of radiant energy originating from the atmosphere, land, or water. Measurements of infrared energy can be converted to temperature of the medium using the Stefan-Boltzmann law.[11] These measurements are termed *brightness temperatures.* Given a temperature lapse rate, brightness temperature observations can be used to infer cloud-top heights. Low brightness temperatures imply high cloud tops, which implies large thickness and high probability of rain. High brightness temperatures imply low cloud tops (or no clouds) and low probability of rain. Difficulties in estimating rainfall from infrared imagery follow from the fact that rainfall is inferred from cloud properties. Direct physical relationships between cloud properties, such as cloud-top height, and rainfall have not been established.

A standard procedure for estimating rainfall from satellite infrared imagery is the *temperature threshold method.*[4,76] The method has been used to produce rainfall estimates on a grid system of 2.5° longitude by 2.5° latitude and a time interval of 1 month. Rainfall in a given 2.5° grid cell is estimated by the following formula:

$$R = (3 \text{ mm/h})(\text{FRAC})(\text{HOURS}) \tag{3.7.8}$$

where R is rainfall in mm, FRAC is the fractional coverage of cloud-top temperature less than 235 K (−38°C) for the desired 2.5 grid cell, and HOURS is the number of hours in the observation period. This procedure has served as the basis of a climatological precipitation data base developed from operational satellite imagery under the GPCP.[50] Figure 3.7.6 shows mean annual oceanic rainfall for the period 1986–1989 derived for the GPCP. Other procedures for estimating rainfall from satellite-based observations are reviewed in Refs. 1, 7, and 11.

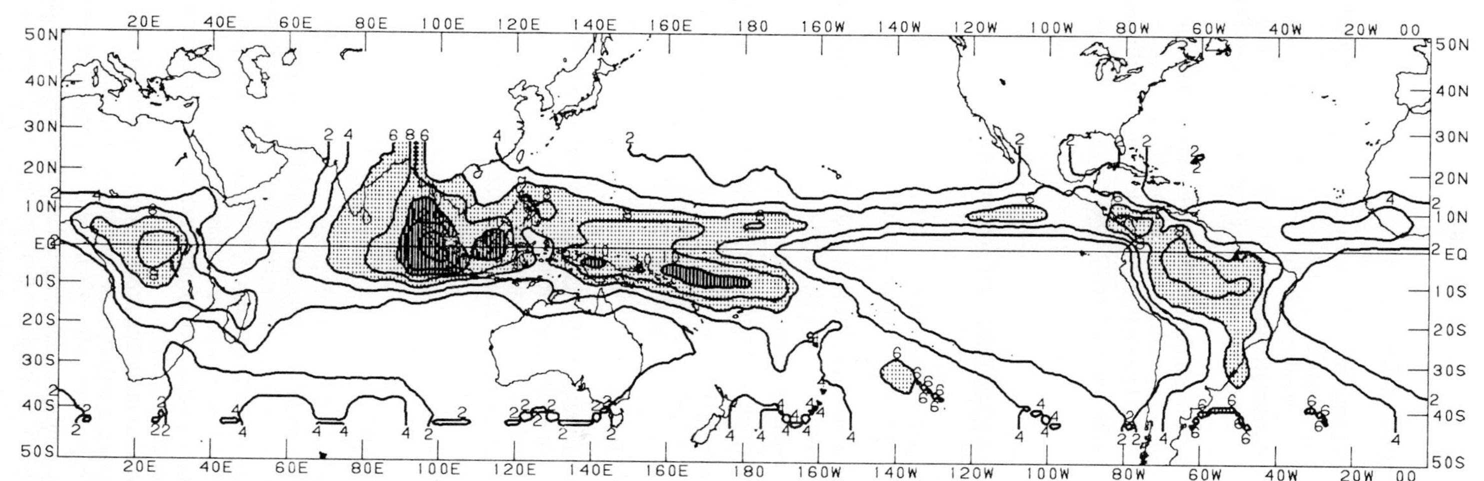

FIGURE 3.7.6 Mean daily rainfall (in millimeters per day) for the period 1986–1989 as determined by remote sensing of cloud temperatures by satellite.[50]

3.7.5 Multisensor Rainfall Estimation

Satellite, radar, and rain gauge procedures for estimating precipitation developed along largely independent paths. Increasingly, it is being recognized that observations from multiple sensor systems should be utilized jointly to form "multisensor" estimates of rainfall. This approach underlies the FRONTIERS system in the U.K.[15] and the precipitation processing systems associated with NEXRAD in the United States.[47]

Objective procedures are useful for combining data from different types of sensor systems. Krajewski[56] develops a framework for objective multisensor estimation of rainfall (see also Ref. 23). The key problem in implementing multisensor estimation procedures is characterizing the joint error structure of rainfall estimates for the different sensor systems.

3.8 STATISTICAL MODELING OF RAINFALL

Statistical models of rainfall can be distinguished by their representation of rainfall in space and time into the following three general classes of models:

1. *Spatial models,* which represent the spatial distribution of accumulated rainfall over a specified duration
2. *Temporal models,* which represent rainfall accumulation at a point over time
3. *Space-time models,* which represent both the spatial and temporal evolution of rainfall

3.8.1 Spatial Models

Spatial models of rainfall are generally used to represent the spatial distribution of storm total rainfall. Two basic types of models are generally used: (1) *gaussian random field models*[14] and (2) *cluster models.*[27,82,93] Model applications include precipitation frequency analysis[93] and design of precipitation sensor sampling strategies.[13]

3.8.2 Temporal Models

Two basic types of temporal models of rainfall occurrence are used, *discrete* and *continuous.* For discrete models the time scale is divided into intervals of fixed length (often daily or hourly). Markov chains and their generalizations have played a prominent role in modeling rainfall occurrence.[30,91,96] In continuous models of rainfall occurrence, the time interval between events is not constrained to fall into discrete intervals. Poisson processes and their generalizations are prominent in continuous modeling of rainfall occurrences.[54,99]

Two basic issues arise in modeling precipitation amounts associated with rainfall occurrences: (1) selecting a distribution function and (2) specifying the temporal correlation of precipitation amounts. A variety of distributional assumptions have been made for precipitation amounts, including exponential, gamma, lognormal, and mixed distributions.[104,111] Similarly, a variety of assumptions have been made on temporal correlation of precipitation amounts. As Rodriguez-Iturbe[80] notes, the appropriate assumptions are closely linked with the time scale of the model, because models with a fine time scale, such as minutes or hours, need to incorporate more information on interevent correlation than do those on a coarser time scale, such as daily rainfall.

3.8.3 Space-Time Models

Space-time models of rainfall have evolved from the cluster model framework introduced by LeCam.[3,58,105] In this framework, model rainfall is constructed from rain cells which are organized into larger rain bands and which have individual life cycles and trajectories. These models have been used for assessing sensor design[101] and assessment of the role of spatial variability of rainfall in determining spatial characteristics of infiltration.[90]

Statistical models of rainfall which include dynamic meteorological variables have also been developed. A principal motivation for these models has been precipitation forecasting.[35,73]

An important issue in modeling precipitation is accurate representation of precipitation over a range of space and time scales. Scale problems are particularly troublesome when models are needed at different time and space scales from those for which data are available. Extensive discussion of scale issues in precipitation modeling can be found in Refs. 106 and 113. The statistical structure of rainfall is still sufficiently unknown that it is not yet possible to develop a statistical model of rainfall for a small area, such as 1 km^2, and a small time interval, such as 5 min, and accumulate the results of this model over a larger area, such as 100 km^2, or a longer time interval, such as a day. Understanding the correlations of rainfall patterns in time and space, linking these patterns to the physical causes which produce them, and expressing these mechanisms mathematically are exciting challenges in rainfall research.

3.9 PRECIPITATION FREQUENCY ANALYSIS

Precipitation frequency analyses are used extensively for design of engineering works that control storm runoff. These include municipal storm sewer systems, highway and railway culverts, and agricultural drainage systems. Precipitation frequency analyses also play an important role in a diverse range of nonstructural problems involving natural hazards associated with extreme rainfall events.[110]

The precipitation frequency analysis problem is to compute the amount of precipitation y falling over a given area in a duration of x min with a given probability of occurrence in any given year. If an event of specified duration and intensity has probability of occurrence p for a year, the return interval T is the expected time, in years, between events and is given by $T = 1/p$. Precipitation frequency analysis studies provide rainfall accumulation depths for a specified duration and return interval.

Early precipitation frequency studies focused on design problems for small basins. For these studies, the spatial distribution of rainfall was typically ignored under the assumption that the areas being considered were small enough that point representations were sufficiently close to areally averaged values. The standard source of precipitation frequency procedures and results in the United States for many years was the work by Hershfield,[45] which is commonly referred to as TP-40.

Extensions to TP-40 include the works by Frederick et al.,[32] Miller et al.,[64] and U.S. Weather Bureau.[100] The report by Frederick et al. is typically referred to as HYDRO-35. These reports provide point precipitation frequency maps for durations ranging from 5 min to 10 days, and return intervals ranging from 2 to 100 years. Figure 3.9.1 illustrates HYDRO-35 precipitation frequency maps for durations of 5, 15, and 60 min for return intervals of 2 and 100 years. The standard steps in producing these maps are the following:

1. Estimate return intervals for N-min intensities using the Gumbel distribution and the fitting procedure of Gumbel[39] (see Chap. 18 for a discussion of statistical procedures for precipitation frequency analysis).
2. Apply corrections for conversion to partial duration series results, as needed.[32]
3. Apply corrections for conversion from clock-hour to 60-min values, as needed.[32]
4. Develop isopluvial maps from frequency results for gauge sites, using interpolation procedures and subjective judgment.

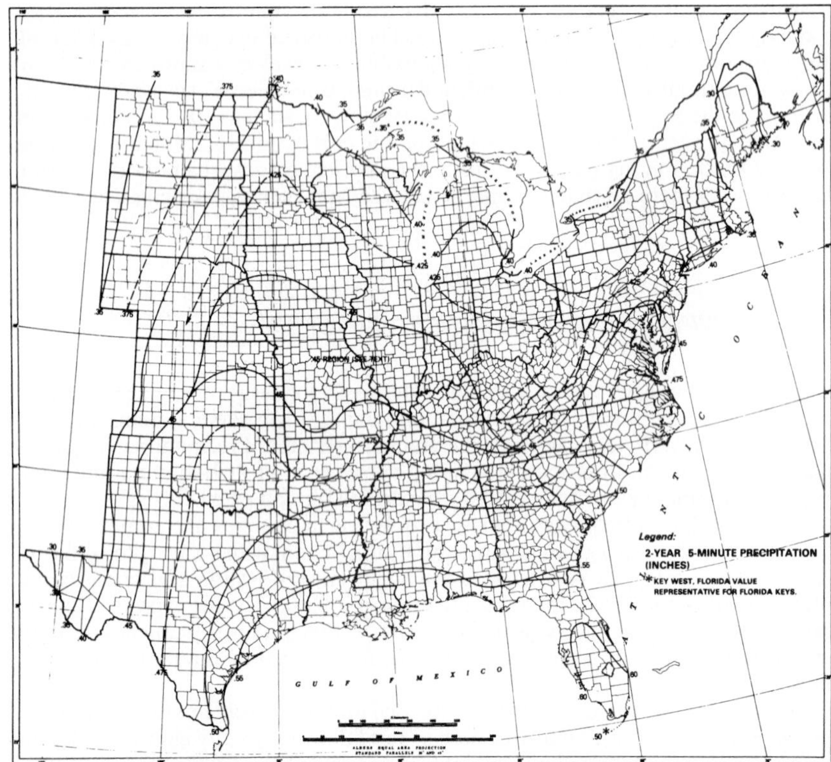

FIGURE 3.9.1 Precipitation frequency maps for the eastern United States from HYDRO-35.

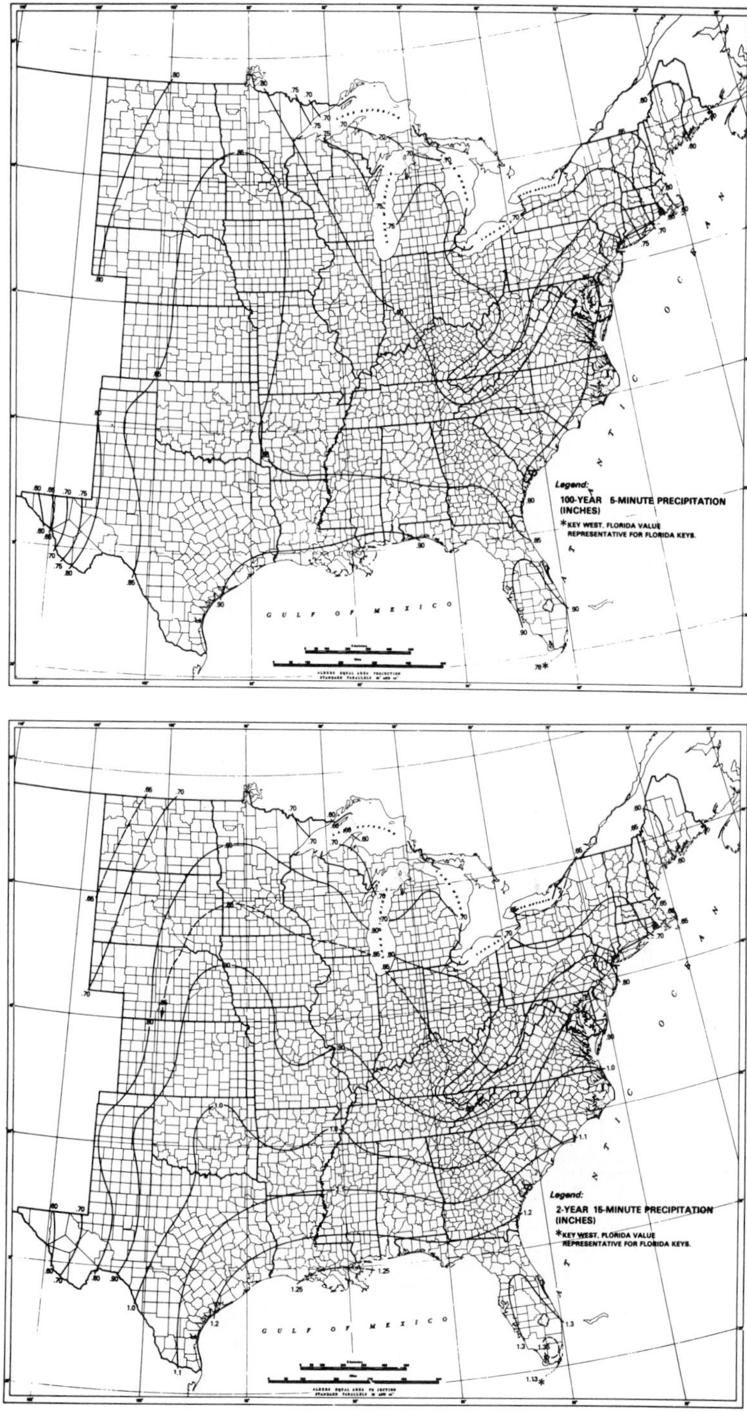

FIGURE 3.9.1 *(Continued)*

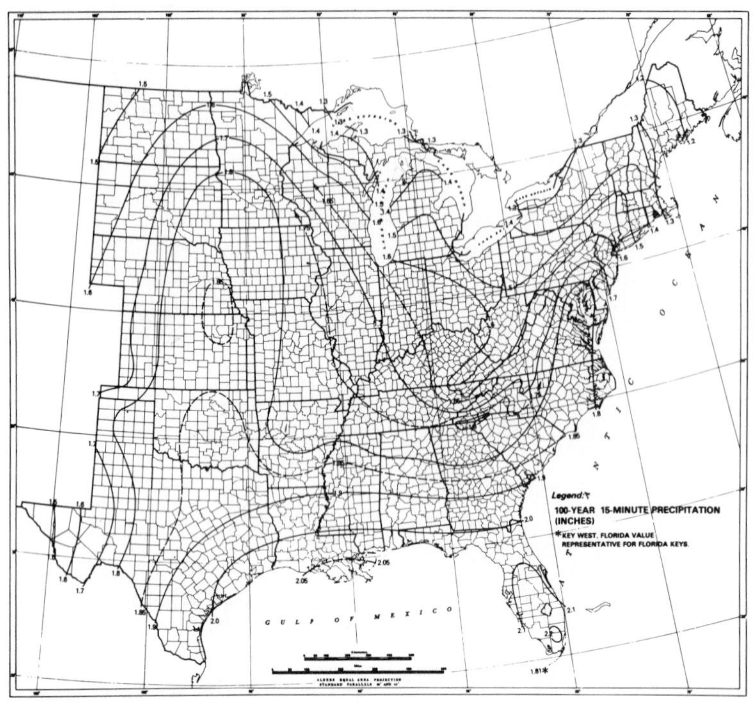

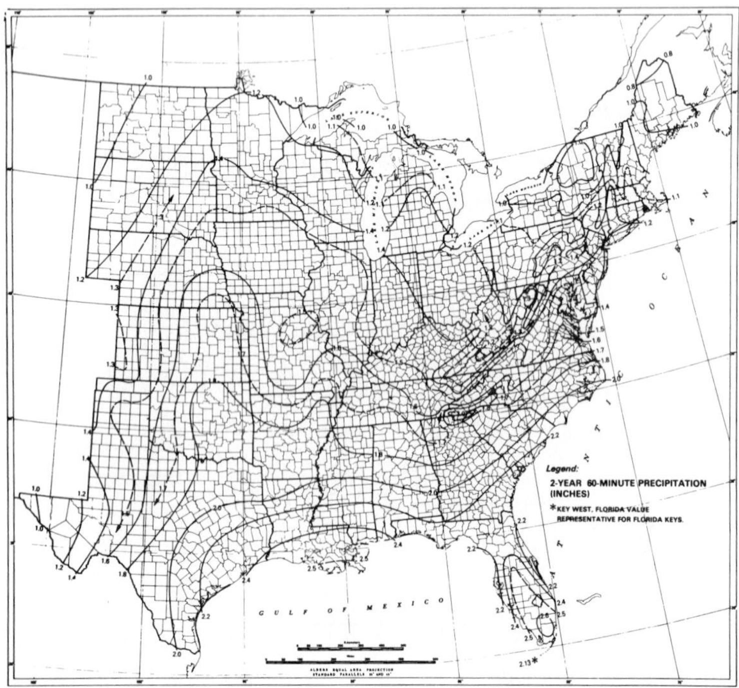

FIGURE 3.9.1 *(Continued)*

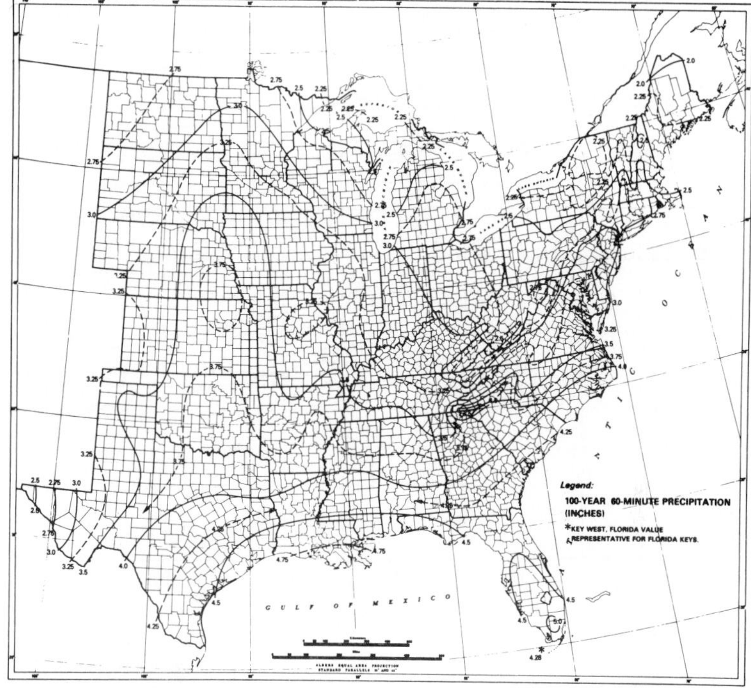

FIGURE 3.9.1 *(Continued)*

Empirical relationships have been developed for relating HYDRO-35 map information to precipitation frequency results at other durations and return intervals.[32] Precipitation frequency values for 10-min accumulations can be represented in terms of 5- and 15-min precipitation frequency values as follows:

$$P_{10\,\text{min}} = 0.41P_{5\,\text{min}} + 0.59P_{15\,\text{min}} \tag{3.9.1}$$

Similarly, precipitation frequency values for 30-min accumulations can be represented in terms of 15- and 60-min values as follows:

$$P_{30\,\text{min}} = 0.51P_{15\,\text{min}} + 0.49P_{60\,\text{min}} \tag{3.9.2}$$

Precipitation frequency values for return intervals other than 2 or 100 years can be obtained from the 2- and 100-year values by the following equation:

$$P_{T\,\text{yr}} = aP_{2\,\text{yr}} + bP_{100\,\text{yr}} \tag{3.9.3}$$

Table 3.9.1 contains parameters a and b for a range of return intervals.

Precipitation frequency values for basin-averaged rainfall depend on basin size. Typically all other aspects of basin geometry are ignored in precipitation frequency studies. Foufoula-Georgiou[31] shows how to include the effects of basin geometry. Precipitation frequency

TABLE 3.9.1 Frequency Parameters for Use with HYDRO-35 Maps and Eq. (3.9.3)

Return period T, years	a	b
5	0.674	0.278
10	0.496	0.449
25	0.293	0.669
50	0.146	0.835

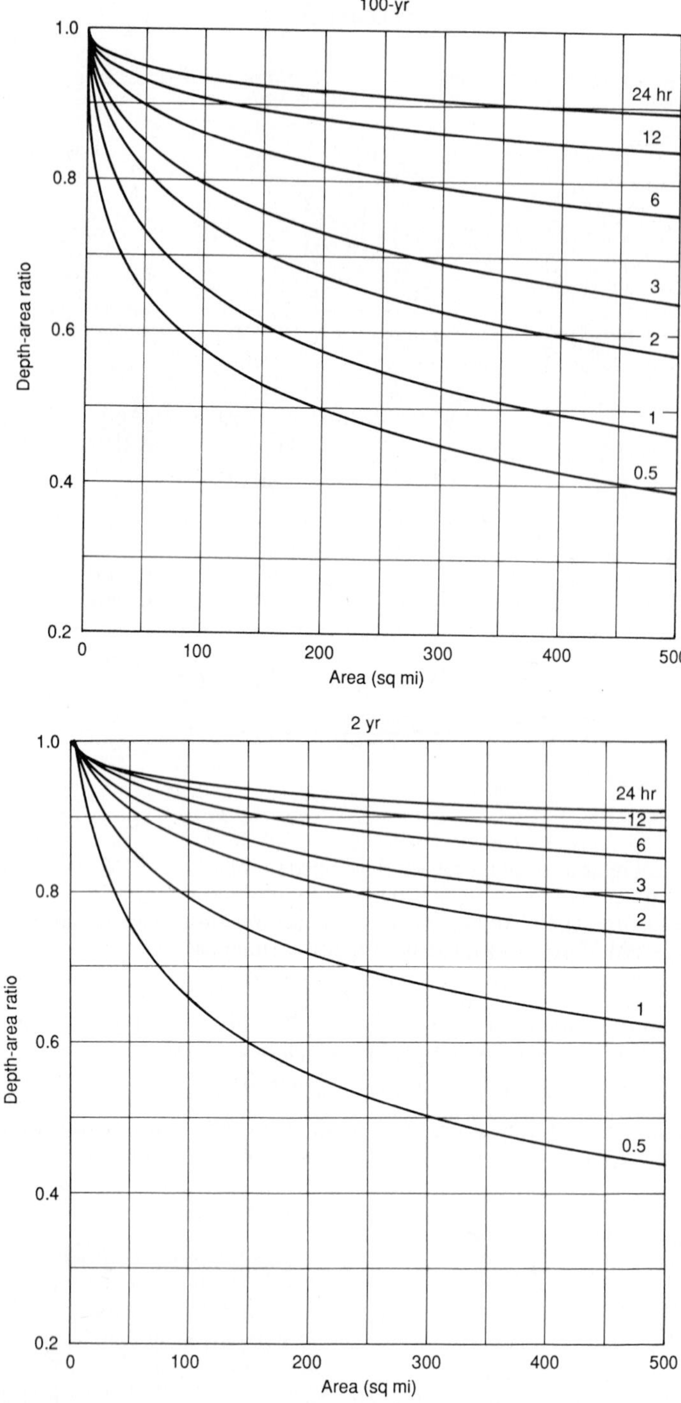

FIGURE 3.9.2 Depth area correction ratios for Chicago, Ill., by duration for 100- and 2-year return intervals.[67]

values for drainage basins are generally developed from point precipitation frequency values using a correction factor for basin area. Dense networks of rain gauges have been used to develop depth-area-duration correction factors.[67,68,81,114] Figure 3.9.2 illustrates area correction factors developed by Myers and Zehr[67] for the Chicago metropolitan area. Significant errors can arise from use of area correction factors in climates different from those in which they were developed. Radar rainfall data will ultimately provide a data source from which basin-averaged rainfall can be directly derived.[32]

For design applications it is often necessary to specify a temporal pattern associated with rainfall depths of a given duration and frequency. For example, the 5-year 60-min rainfall depth may be used in the form of 5-min increments in hydrologic models. Approaches to specify temporal patterns of design rainfall depths are given in Refs. 48, 49, 72, and 107. Pilgrim and Cordery[72] warn that approaches that overly smooth the temporal patterns of rainfall are unsuited for design applications, because the time variability of rainfall intensity often has a significant effect on the design hydrograph. Procedures for developing patterns incorporating the average variability of storm rainfall are described in Pilgrim and Cordery.[72] Two important points noted by Pilgrim and Cordery[72] and Huff and Changnon[49] are that the time variability of rainfall intensity increases with the storm return period and that the majority of extreme storms have multiple peaks of high rainfall intensity. Multiple peaks in rainfall intensity are typical of the flash-flood-producing storms described by Chappell[19] as quasi-stationary convective events (see Fig. 3.1.2).

3.10 PROBABLE MAXIMUM PRECIPITATION

For design of high-hazard structures such as spillways on large dams it is necessary to use precipitation values with very low risk of exceedance. Ideally a hydrologist would like to choose design storms for which there is no risk of exceedance. A theoretical problem that has plagued the search for such a storm is determining whether there is indeed an upper limit on rainfall amount. The conclusion of Gilman in 1964[37] that the existence of an upper limit on rainfall amount is both mathematically and physically realistic remains valid. The spatial and temporal context of the upper bound on rainfall amount is incorporated into the definition of *probable maximum precipitation* (PMP), which is defined by the World Meteorological Organization[112] as "theoretically the greatest depth of precipitation for a given duration that is physically possible over a given size storm area at a particular geographical location at a certain time of year." A more troublesome problem than ascertaining whether an upper bound exists is determining what it is.

Observed rainfall totals worldwide provide a broad indication of maximum possible rainfall totals at a point as a function of duration (see Fig. 3.10.1 and Table 3.10.1). Table 3.10.2 illustrates the dependence of maximum rainfall accumulation for the United States on duration and basin area.

The estimation of probable maximum precipitation was developed and has evolved in the United States as a hydrometeorological procedure.[66] Three meteorological components determine maximum possible precipitation: (1) amount of precipitable water, (2) rate of convergence, and (3) vertical motion. Meteorological models representing all three components have been developed,[108] but it has proved quite difficult to specify maximum rates of areal convergence and vertical motion for the models. In the standard approach to estimating PMP[112] observed storm totals for extreme storms are used as indicators of the maximum values of convergence and

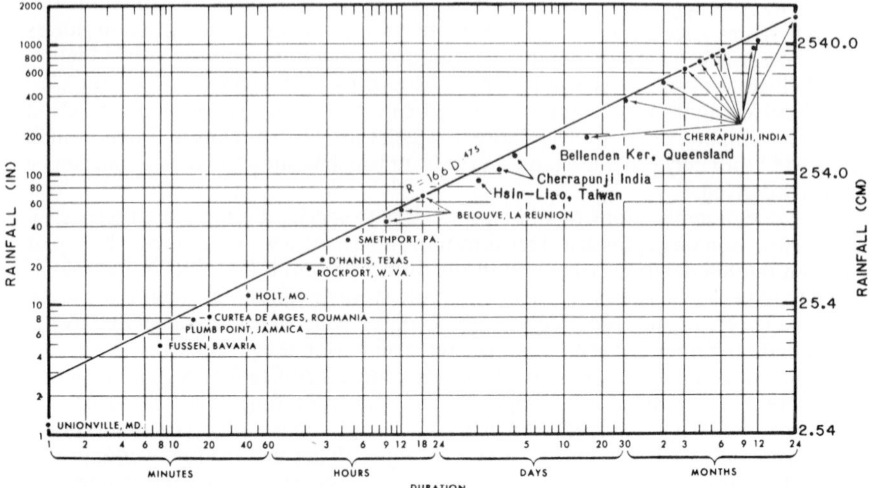

FIGURE 3.10.1 Maximum observed rainfall as a function of duration throughout the world. *(Adapted from Ref. 112.)*

vertical motion. The two major components of the standard approach to the PMP computation are *moisture maximization* for observed storms and *storm transposition.* In the first step of moisture maximization, the goal is to increase storm rainfall amounts to reflect the maximum possible moisture availability. In the storm transposition step, it is determined whether a given storm, which occurred in a broad region around the basin of interest, can be transposed to represent rainfall over the basin.

The principal data required for standard PMP computations are (1) a catalog of extreme storms and (2) surface dew-point temperature observations. The storm catalog for the United States[112] contains storm date, location, and rainfall accumulation about the storm center for increasing areas. Additional meteorological information, including surface and upper-air maps, is typically used subjectively in determining storm transposition regions and for other quality-control purposes.[112] Surface dew point is used as a moisture index for moisture maximization. Precipitable water can be computed from surface dew-point values using Eq. (3.3.7) under the assumption that saturation levels extend to the ground. In WMO[112] it is recommended that maximum moisture content is obtained from highest recorded dew point over a period of 50 or more years.

Storm transposition is based on the assumption that for a given storm meteorologically homogeneous regions exist over which the storm is equally likely to occur. The transposition procedure involves meteorological analysis of the storm to be transposed, determination of transposition limits, and application of adjustments for changes in storm location. Meteorological analysis provides a characterization of key aspects of storm type. Transposition limits are determined from a long series of daily weather charts by identifying the boundary of the region over which meteorologically similar storm types have occurred. Adjustments account for differences in moisture maxima for the storm location and transposition sites. Adjustments are sometimes also made for topographic effects, although objective procedures for determining orographic adjustments are not widely accepted. Having obtained a series of storms,

TABLE 3.10.1 World's Greatest Observed Point Rainfalls

Duration	Depth in	Depth mm	Location	Date
1 min	1.50	38	Barot Guadeloupe	Nov. 26, 1970
8 min	4.96	126	Fussen, Bavaria	May 25, 1920
15 min	7.80	198	Plumb Point, Jamaica	May 12, 1916
20 min	8.10	206	Curtea-de-Arges, Romania	July 7, 1947
42 min	12.00	305	Holt, Mo.	June 22, 1947
2 h 10 min	19.00	483	Rockport, W.V.	July 18, 1889
2 h 45 min	22.00	559	D'Hanis, Tex (17 mi NNW)	May 31, 1935
4 h 30 min	30.8+	782	Smethport, Pa.	July 18, 1942
9 h	42.79	1,087	Belouve, La Réunion	Feb. 28, 1964
12 h	52.76	1,340	Belouve, La Réunion	Feb. 28–29, 1964
18 h 30 min	66.49	1,689	Belouve, La Réunion	Feb. 28–29, 1964
24 h	71.85	1,825	Foc Foc, La Réunion	Mar. 15–16, 1952
2 days	88.94	2,259	Hsin-Liao-Taiwan	Oct. 17–18, 1967
3 days	108.62	2,759	Cherrapunji, India	Sept. 12–14, 1974
4 days	146.50	3,721	Cherrapunji, India	Sept. 12–15, 1974
8 days	151.46	3,847	Bellenden Ker, Queensland	Jan. 1–8, 1979
15 days	188.88	4,798	Cherrapunji, India	June 24–July 8, 1931
31 days	366.14	9,300	Cherrapunji, India	July 1861
2 months	502.63	12,767	Cherrapunji, India	June–July 1861
3 months	644.44	16,369	Cherrapunji, India	May–July 1861
4 months	737.70	18,738	Cherrapunji, India	April–July 1861
5 months	803.62	20,412	Cherrapunji, India	April–August 1861
6 months	884.03	22,454	Cherrapunji, India	April–September 1861
11 months	905.12	22,990	Cherrapunji, India	January–November 1861
1 year	1041.78	26,461	Cherrapunji, India	August 1860–July 1861
2 years	1605.05	40,768	Cherrapunji, India	1860–1861

Source: Ref. 112.

PMP is determined by *envelopment.* Envelopment entails selection of the storm which has the largest maximized storm rainfall for a given time interval. The envelopment process is used because a single historical storm is generally not the critical event over the entire range of time scales required.

3.10.1 PMP Estimation Using HMR-51 and HMR-52

PMP estimates can be derived either for specific basins or as generalized maps for larger regions. Figure 3.10.2 illustrates a site-specific PMP analysis for the Tennessee River basin.[112] Development of generalized maps entails use of objective analysis

TABLE 3.10.2 Maximum Observed Rainfall Depth, Area, and Duration Data for the United States.

Average rainfall in inches (millimeters)

Area	Duration, h 6	12	18	24	36	48	72
10 mi^2 26 km^2	24.7a (627)	29.8b (757)	36.3e (922)	38.7e (983)	41.8e (1062)	43.1e (1095)	45.2e (1148)
100 mi^2 259 km^2	19.6b (498)	26.3e (668)	32.5e (826)	35.2e (894)	37.9e (963)	38.9e (988)	40.6e (1031)
200 mi^2 518 km^2	17.9b (455)	25.6e (650)	31.4e (798)	34.2e (869)	36.7e (932)	37.7e (958)	39.2e (996)
500 mi^2 1,295 km^2	15.4b (391)	24.6e (625)	29.7e (754)	32.7e (831)	35.0e (889)	36.0e (914)	37.3e (947)
1,000 mi^2 2,590 km^2	13.4b (340)	22.6e (574)	27.4e (696)	30.2e (767)	32.9e (836)	33.7e (856)	34.9e (886)
2,000 mi^2 5,180 km^2	11.2b (284)	17.7e (450)	22.5e (572)	24.8e (630)	27.3e (693)	28.4e (721)	29.7e (754)
5,000 mi^2 12,950 km^2	8.1bh (206)	11.1b (282)	14.1b (358)	15.5e (394)	18.7i (475)	20.7i (526)	24.4i (620)
10,000 mi^2 25,900 km^2	5.7h (145)	7.9j (201)	10.1k (257)	12.1k (307)	15.1i (384)	17.4i (442)	21.3i (541)
20,000 mi^2 51,800 km^2	4.0h (102)	6.0j (152)	7.9k (201)	9.6k (244)	11.6i (295)	13.8i (351)	17.6i (447)
50,000 mi^2 129,000 km^2	2.5em (64)	4.2n (107)	5.3k (135)	6.3k (160)	7.9k (201)	9.9r (251)	13.2r (335)
100,000 mi^2 259,000 km^2	1.7m (43)	2.5om (64)	3.5k (89)	4.3k (109)	6.0p (152)	6.7p (170)	8.9q (226)

Storm	Date	Location of Center	
a	July 17–18, 1942	Smethport, Pa	
b	Sept. 8–10, 1921	Thrall, Tex.	
e	Sept. 3–7, 1950	Yankeetown, Fla.	Hurricane
i	June 27–July 1, 1899	Hearne, Tex.	
k	Mar. 13–15, 1929	Elba, Ala.	
q	July 5–10, 1916	Bonifay, Fla.	Hurricane
n	Apr. 15–18, 1900	Eutaw, Ala.	
m	May 22–26, 1908	Chattanooga, Okla.	
o	Nov. 19–22, 1934	Millry, Ala.	
h	June 27–July 4, 1936	Bebe, Tex.	
j	Apr. 12–16, 1927	Jefferson Parish, La.	
r	Sept. 19–24, 1967	Cibolo Ck., Tex	Hurricane
p	Sept. 29–Oct. 3, 1929	Vernon, Fla.	Hurricane

Source: Ref. 112.

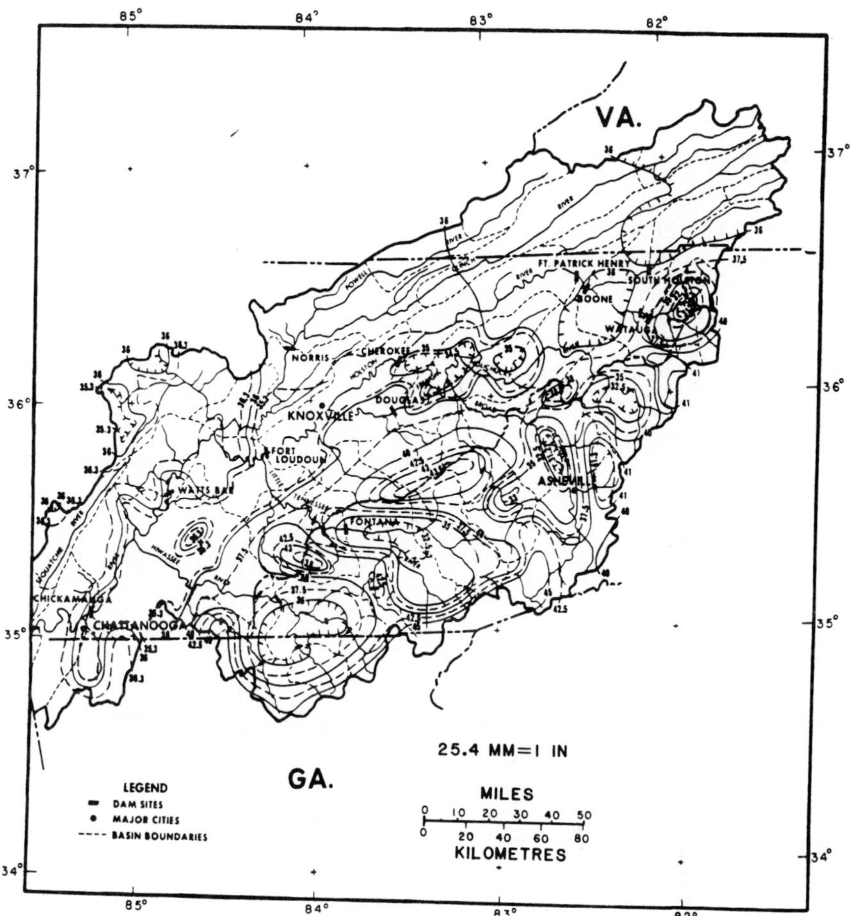

FIGURE 3.10.2 Site-specific Probable Maximum Precipitation for the Tennessee River basin. *(Adapted from Ref. 112.)*

procedures for mapping PMP estimates to a grid system.[112] Generalized maps have been developed for a variety of regions.[36,42,43,55,65,83] The report by Schreiner and Riedel[83] is referred to as HMR-51; the report by Hansen et al.[42] is referred to as HMR-52. Figure 3.10.3 shows a generalized PMP map for the eastern United States obtained from HMR-52.

To calculate PMP over a watershed from generalized maps, like those in HMR-52, involves five steps:

1. Development of depth-area-duration curves for PMP at the site
2. Describing the spatial distribution of storm rainfall by an elliptical isohyetal pattern
3. Application of an "orientation adjustment factor" to allow for regional patterns of atmospheric moisture flow
4. Determination of a "critical storm area" for the watershed

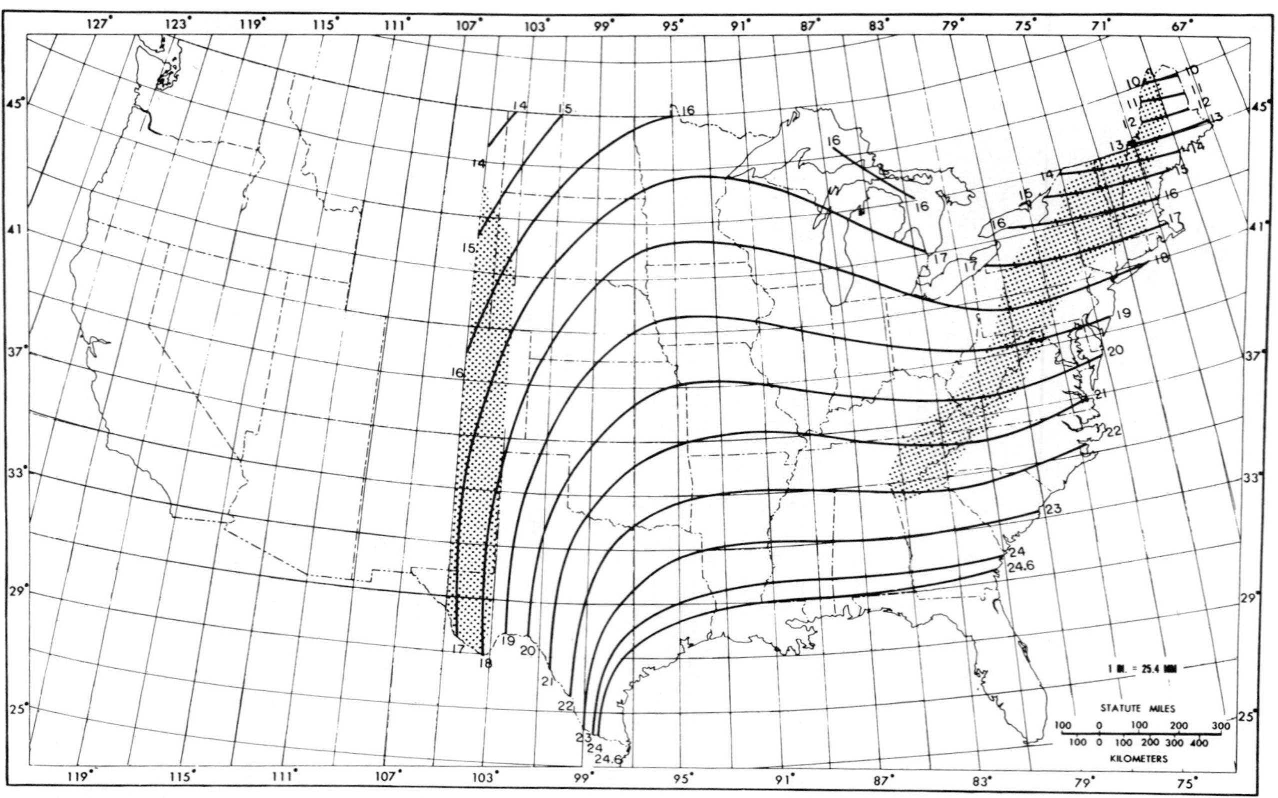

FIGURE 3.10.3 All-season PMP estimates for the eastern United States from HMR-52 for 6-h duration and 200-mi² area. *(From Ref. 42.)*

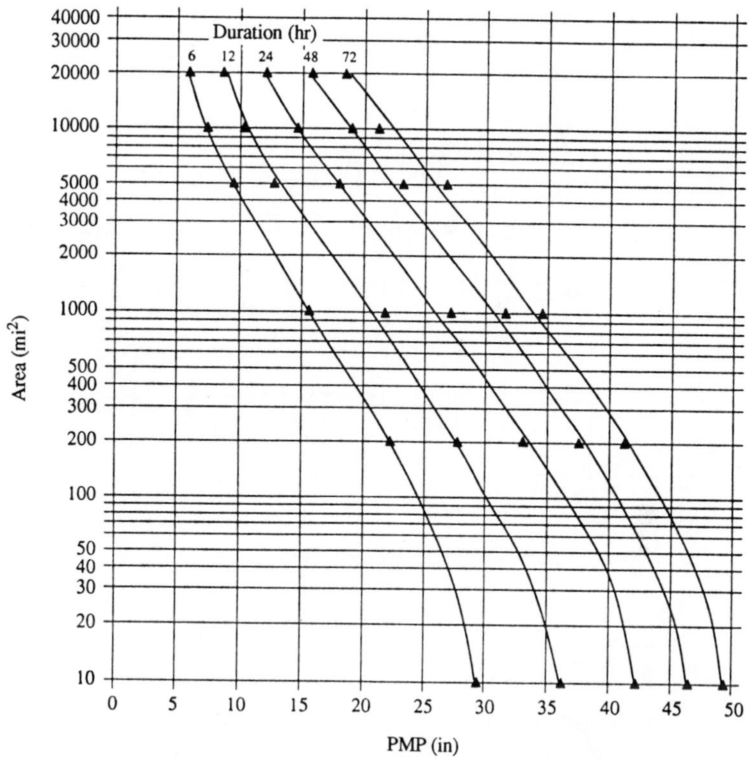

FIGURE 3.10.4 Depth-area-duration curves derived from HMR-52 for the Leon River in Texas. *(From Ref. 42.)*

5. Application of an "isohyetal adjustment factor" to determine the design rainfall as a function of area

These steps are described in detail in Chow et al.[20] HMR-52 contains figures and charts needed for carrying out the computations needed for steps 2 to 5, as well as the generalized PMP maps needed for step 1. An important component of a PMP analysis is a set of depth-area-duration curves. Figure 3.10.4 illustrates depth-area-duration curves derived from HMR-52 for the Leon River in Texas.

3.10.2 Validation of PMP Estimates

One method that has been used for assessing performance of PMP procedures is comparison of PMP values with maximum observed storm totals. Riedel and Schreiner[77] carried out a detailed analysis of PMP estimates for the United States. In Fig. 3.10.5, PMP estimates from HMR-51 and HMR-52 are compared with maximum station accumulations at 41 sites. On average, observed extreme rainfall was 60 percent of the PMP estimate at these sites.

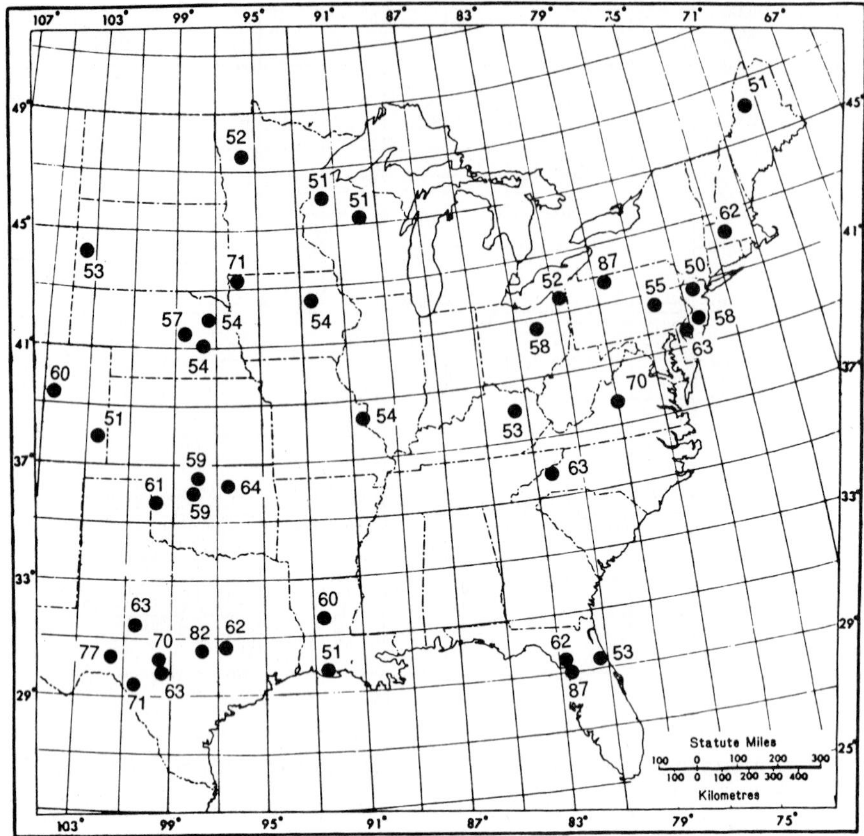

FIGURE 3.10.5 Maximum observed station rainfall expressed as a percent of PMP.[77]

3.10.3 PMP Estimates in Mountainous Regions

The standard approach to PMP computation cannot be readily applied in mountainous regions. Several approaches have been used for computing orographic PMP estimates. The *orographic separation method*[112] separately estimates the orographic and nonorographic components of PMP and combines them to produce a final PMP estimate. The nonorographic PMP component is computed using storms from valley and foothill locations and the standard PMP approach. Haiden et al.[41] describe a deterministic approach for estimating the orographic component of PMP. In the *orographic adjustment method*[112] corrections to standard PMP estimates are developed based on climatological precipitation characteristics. In the Tennessee River basin, for example, mean areal precipitation is used as a tool for adjusting PMP estimates for orographic influence.[112]

3.10.4 PMP Estimates and Precipitation Frequency Analysis

PMP procedures have a strong connection with the frequency analysis procedures discussed in Sec. 3.9. They can be interpreted as providing estimates of the upper bound of the distribution function for rainfall accumulation. The upper bound of a distribution function is the smallest value that has no probability of exceedance. Alternatively, procedures based on *stochastic storm transposition*[29,31] have been developed to estimate the frequency of extreme storms (i.e., storms with recurrence intervals greater than 100 years). In stochastic storm transposition, the frequency of occurrence of storms in the transposition region provides the link for obtaining frequency estimates of extreme storm magnitudes.

3.11 PREDICTION OF RAINFALL

Rainfall prediction plays an important role in management of water resources facilities such as flood-control dams and levees. For hydrologic applications rainfall forecasts with lead times between 0 and 12 h are generally of greatest use. For short-period spatial forecasts, radar plays a prominent role by virtue of its ability to measure both quantitative rainfall over large areas and short-term changes in rainfall. For applications to streamflow forecasting it is useful to link precipitation forecast models with surface runoff models into an integrated forecast system.[34] This allows precipitation and streamflow observations to be used in an efficient manner to partition prediction errors and update the appropriate system states.

For lead times of less than 2 h, rainfall-prediction procedures are often based on radar rainfall data and the assumptions that (1) the storm is moving with constant speed and direction and (2) the rainfall pattern remains relatively constant over the time period. Procedures of this type are often referred to as *precipitation projection* procedures or *nowcasting* procedures. Hydrometeorological forecast systems in several countries utilize procedures of this type, notably the United Kingdom,[15] Japan,[98] and the United States.[2] The key problem of rainfall projection procedures is determining the velocity of storm movement. Physically unreasonable forecasts can be obtained with these procedures unless suitable quality-control steps are taken.[2,8] Projection procedures can also be successfully applied using data from rain-gauge networks.[51]

Georgakakos and Bras[35] have developed a physically based prediction model for rainfall at a site. The model produces hourly forecasts based on surface observations of temperature, pressure, and dew point. The model is based on treatment of the storm-cloud system as a reservoir of condensed water. The state of the model is the equivalent liquid moisture content of the cloud column (per unit area). The principal equation governing the forecast model is conservation of liquid water.

Figure 3.11.1 illustrates the format of a standard quantitative precipitation forecast (QPF) product of the U.S. National Weather Service. The product is a 6-h forecast of probability of heavy precipitation. Similar forecast products are available for probability of precipitation. These forecasts are developed using synoptic-scale forecast models, which are based on the primitive equations described in Sec. 3.2. The output of these models does not, however, include precipitation. QPF products are developed using statistical procedures in which model output variables are input to regression models for the precipitation variables of interest. These procedures are referred to as model output statistics (MOS) procedures and are described in Glahn and Lowry.[38]

Improved meteorological observing systems will enable meteorological forecast models to run at smaller spatial scales in the future.[71] Significant improvements in precipitation forecasting for the 0- to 6-h time range should be realized. It is likely that

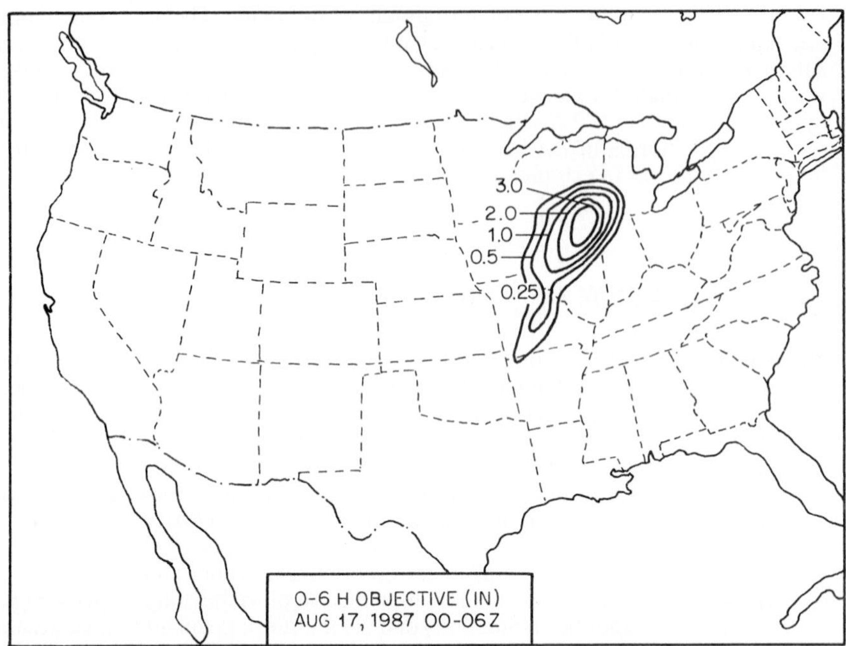

FIGURE 3.11.1 0- to 6-h quantitative precipitation forecast (QPF) product of the U.S. National Weather Service for Aug. 17, 1987.

future precipitation forecast systems will blend radar-based nowcasting procedures with simple meteorological models to produce hydrologically useful precipitation forecasts that cover the 0- to 6-h forecast period.

REFERENCES

1. Adler, R., and A. Negri, "A Satellite Infrared Technique to Estimate Tropical Convective and Stratiform Rainfall," *J. Appl. Meteorol.*, vol. 27, pp. 30–51, 1988.
2. Ahnert, P., M. Hudlow, E. Johnson, and D. Greene, "Proposed On-Site Precipitation Processing System for NEXRAD," Preprint Volume: 21st Conference on Radar Meteorology, American Meteorological Society, Boston, pp. 378–385, 1983.
3. Amorocho, J., and B. Wu, "Mathematical Models for the Simulation of Cyclonic Storm Sequences and Precipitation Fields," *J. Hydrol.*, vol. 32, pp. 329–345, 1977.
4. Arkin, P., "The Relationship between Fractional Coverage of High Cloud and Rainfall Accumulations during GATE over the B-Scale Array," *Mon. Weather Rev.*, vol. 92, pp. 14204–14216, 1979.
5. Arkin, P., and P. Ardanuy, "Estimating Climatic-Scale Precipitation from Space: A Review," *J. Clim.*, vol. 2, no. 11, pp. 1229–1238, 1989.
6. Atlas, D., and C. Ulbrich, "Path and Area-Integrated Rainfall Measurement by Microwave Attenuation in the 1-3 cm Band," *J. Appl. Meteorol.*, vol. 16, pp. 1322–1331, 1977.
7. Atlas, D., D. Rosenfeld, and D. Short, "The Estimation of Convective Rainfall by Area Integrals, 1, The Theoretical and Empirical Basis," *J. Geophys. Res.*, vol. 95(D3), pp. 2153–2160, 1990.

8. Austin, G., and A. Bellon, "The Use of Digital Weather Radar Records for Short-Term Precipitation Forecasting," *J. Appl. Meteorol.*, vol. 100, p. 658, 1974.

9. Austin, P. M., "Relation between Measured Radar Reflectivity and Surface Rainfall," *Mon. Weather Rev.*, vol. 115, pp. 1053–1070, 1987.

10. Austin, P. M., and R. A. Houze, "Analysis of the Structure of Precipitation Patterns in New England," *J. Appl. Meteorol.*, vol. 11, pp. 926–935, 1972.

11. Barrett, E. C., and D. W. Martin, *The Use of Satellite Data in Rainfall Monitoring,* Academic Press, San Diego, 1981.

12. Battan, L. J., *Radar Observation of the Atmosphere,* University of Chicago Press, Chicago, 1973.

13. Bell, T., "A Space-Time Stochastic Model of Rainfall for Satellite Remote-Sensing Studies," *J. Geophys. Res.*, vol. 92(D8), pp. 9361–9644, 1987.

14. Bras, R., and I. Rodriguez-Iturbe, *Random Functions and Hydrology,* Addison-Wesley, Reading, Mass., 1985.

15. Browning, K., "The FRONTIERS Plan: A Strategy for Using Radar and Satellite Imagery for Very-Short-Range Precipitation Forecasting," *Meteorol. Mag.*, vol. 108, pp. 161–184, 1979.

16. Browning, K., and C. Collier, "Nowcasting of Precipitation Systems," *Rev. Geophys.*, vol. 27, no. 3, pp. 345–370, 1989.

17. Byers, R., and R. Braham, *The Thunderstorm,* Government Printing Office, Washington, D.C., 1949.

18. Cataneo, R., and G. E. Stout, "Raindrop-Size Distributions in Humid Continental Climates and Associated Rainfall Rate-Reflectivity Relationships," *J. Appl. Meteorol.*, vol. 7, pp. 901–907, 1968.

19. Chappell, C., "Quasi-Stationary Convective Events," in P. Ray, ed., *Mesoscale Meteorology,* American Meteorological Society, Boston, pp. 289–310, 1989.

20. Chow, V. T., D. R. Maidment, and L. W. Mays, *Applied Hydrology,* McGraw-Hill, New York, 1988.

21. Collier, C., *Applications of Weather Radar Systems,* Ellis Horwood Limited, Chichester, 1989.

22. Collier, C., "Accuracy of Rainfall Estimates by Radar; Calibration by Telemetering Raingauges," *J. Hydrol.*, vol. 83, pp. 207–223, 1986.

23. Creutin, J., G. Delrieu, and T. Lebel, "Rain Measurement by Raingage-Radar Combination: A Geostatistical Approach," *J. Atmos. Ocean. Technol.*, vol. 5, pp. 102–115, 1988.

24. Dennis, A. S., *Weather Modification by Cloud Seeding,* Academic Press, New York, 1980.

25. Doviak, R., and D. Zrnic, *Doppler Radar and Weather Observations,* Academic Press, New York, 1984.

26. Dutton, J. A., *The Ceaseless Wind,* Dover, New York, 1986.

27. Eagleson, P. S., N. M. Fennessey, W. Qinliang, and I. Rodriguez-Iturbe, Application of Spatial Poisson Models to Air Mass Thunderstorm Rainfall, *J. Geophys. Res.*, vol. 92(D8), pp. 9661–9678, 1987.

28. Essery, C. I., and D. N. Wilcock, "Variation in Rainfall Catch from Standard U.K. Meteorological Office Raingages: A Twelve Year Case Study," *Hydrol. Sci. J.*, vol. 36, no. 1, pp. 23–34, 1990.

29. Fontaine, T. A., and K. W. Potter, "Estimating Probabilities of Extreme Rainfalls," *J. Hydraul. Eng.*, vol. 115, no. 11, pp. 1562–1575, 1989.

30. Foufoula-Georgiou, E., and D. Lettenmaier, "A Markov Renewal Model for Rainfall Occurrences," *Water Resour. Res.*, vol. 23, no. 5, pp. 875–884, 1987.

31. Foufoula-Georgiou, E., "A Probabilistic Storm Transposition Approach for Estimating Exceedance Probabilities of Extreme Precipitation Depths," *Water Resour. Res.*, vol. 25, no. 5, pp. 799–815, 1989.

32. Frederick, R. H., V. A. Meyers, and E. P. Auciello, "Five- to 60-Minute Precipitation Frequency for the Eastern and Central United States," *NOAA Tech. Mem. NWS HYDRO-35,* Washington D.C., 1977.

33. Fritsch, J. M., R. J. Kane, and C. R. Chelius, "The Contribution of Mesoscale Convective Weather Systems to the Warm Season Precipitation in the United States," *J. Clim. Appl. Meteorol.,* vol. 25, pp. 1333–1345, 1986.

34. Georgakakos, K. P., "A Generalized Stochastic Hydrometeorological Model for Flood and Flash-Flood Forecasting 1. Formulation," *Water Resour. Res.,* vol. 22, no. 13, pp. 2083–2095, 1986.

35. Georgakakos, K. P., and R. L. Bras, "A Hydrologically Useful Station Precipitation Model," *Water Resour. Res.,* vol. 20, no. 11, pp. 1585–1610, 1984.

36. German Water Resources Association, "Contributions to the Choice of the Design Flood and to Probable Maximum Precipitation," Publication 62, Hamburg (available in German only), 1983.

37. Gilman, C. S., "Rainfall," in V. T. Chow, ed., *Handbook of Applied Hydrology,* McGraw-Hill, New York, pp. 9.1–9.68, 1964.

38. Glahn, H. R., and D. A. Lowry, "The Use of Model Output Statistics in Objective Weather Forecasting," *J. Appl. Meteorol.,* vol. 11, p. 1203, 1972.

39. Gumbel, E. J., *Statistics of Extremes,* Columbia University Press, New York, 1958.

40. Gunn, R., and G. Kinzer, "The Terminal Velocity of Fall for Water Droplets in Stagnant Air," *J. Meteorol.,* vol. 6, pp. 243–248, 1949.

41. Haiden, T., P. Kahlig, M. Kerschbaum, and F. Nobilis, "On the Influence of Mountains on Large-Scale Precipitation: A Deterministic Approach toward Orographic PMP," *Hydrol. Sci. J.,* vol. 35, no. 5, pp. 501–510, 1990.

42. Hansen, E. M., L. C. Schreiner, and J. F. Miller, "Application of Probable Maximum Precipitation Estimates—United States East of the 105th Meridian," *Hydrometeorological Report* 52, National Weather Service, NOAA, U.S. Department of Commerce, Washington, D.C., 1982.

43. Hansen, E. M., F. K. Schwarz, and J. T. Riedel, "Probable Maximum Precipitation Estimates—Colorado River and Great Basin Drainages," *Hydrometeorological Report* 49, National Weather Service, NOAA, U.S. Department of Commerce, Silver Spring, Md., 1977.

44. Hansen, E. M., "Probable Maximum Precipitation for Design Floods in the United States," *J. Hydrol.,* vol. 96, pp. 267–278, 1987.

45. Hershfield, D. M., "Rainfall Frequency Atlas of the United States for Durations from 30 Minutes to 24 Hours and Return Periods from 1 to 100 Years," *U.S. Weather Bureau Technical Paper* 40, Washington, D.C., 1962.

46. Houze, R. A., "Structure of Atmospheric Precipitation Systems: A Global Survey," *Radio Sci.,* vol. 16, pp. 671–689, 1981.

47. Hudlow, M. D., R. G. Greene, P. Ahnert, W. Krajewski, T. Sivaramakrishnan, M. Dias, and E. Johnson, "Proposed Off-Site Precipitation Processing System for NEXRAD," Preprints, 21st Radar Meteorology Conference, American Meteorology Society, pp. 394–403, 1983.

48. Huff, F., "Time Distribution of Rainfall in Heavy Storms," *Water Resour. Res.,* vol. 3, no. 4, pp. 1007–1019, 1967.

49. Huff, F. A., and S. A. Changnon, Jr., "A Model 10-Inch Rainstorm," *J. Appl. Meteorol.,* vol. 3, pp. 587–599, 1964.

50. Janowiak, J., and P. Arkin, "Rainfall Variations in the Tropics During 1986–89, as Estimated from Observations of Cloud-Top Temperature," *J. Geophys. Res.,* vol. 96, pp. 3359–3373, 1991.

51. Johnson, E. R., and R. L. Bras, "Multivariate Short-Term Rainfall Prediction," *Water Resour. Res.,* vol. 16, no. 1, pp. 173–185, 1980.

52. Joss, J., and A. Waldvogel, "Ein Spektrograph für Niederschlagstropfen mit automatischer Auswertung," *Pure Appl. Geophys.*, vol. 68, pp. 240–246, 1967.

53. Joss, J., and A. Waldvogel, "Precipitation Measurement and Hydrology," in D. Atlas, ed., *Radar and Meteorology*, American Meteorology Society, Boston, pp. 577–606, 1990.

54. Kavvas, M. L., and J. Delleur, "A Stochastic Cluster Model of Daily Rainfall Sequences," *Water Resour. Res.*, vol. 17, no. 4, pp. 1151–1160, 1981.

55. Kennedy, M. R., "The Estimation of Probable Maximum Precipitation in Australia—Past and Current Practice," Proceedings of the Workshop on Spillway Design, 7–9 October 1981, *Conference Series* 6, Australian Water Resources Council, Australian Department of National Development and Energy, Australian Government Printing Office, Canberra, pp. 26–52, 1982.

56. Krajewski, W. F., "Cokriging of Radar and Rain Gage Data," *J. Geophys. Res.*, vol. 92(D8), pp. 9571–9580, 1987.

57. Larson, L., and E. Peck, "Accuracy of Precipitation Measurements for Hydrologic Modeling," *Water Resour. Res.*, vol. 10, no. 4, 1974.

58. LeCam, L., "A Stochastic Description of Precipitation," in *Fourth Berkeley Symposium on Mathematics, Statistics, and Probability*, vol. 3, University of California, Berkeley, Calif., pp. 165–186, 1961.

59. Maddox, R. A., "Mesoscale Convective Complexes," *Bull. Am. Meteorol. Soc.*, vol. 61, no. 11, pp. 1374–1387, 1980.

60. Maddox, R. A., C. F. Chappell, and L. R. Hoxit, "Synoptic and Meso-alpha Scale Aspects of Flash Flood Events," *Bull. Am. Meteorol. Soc.*, vol. 60, no. 2, pp. 115–123, 1979.

61. Manahan, S. E., *Environmental Chemistry*, 4th ed., Willard Grant Press, Boston, Mass., 1984.

62. Marshall, J. S., and W. Palmer, "The Distribution of Raindrops with Size," *J. Meteorol.*, vol. 5, pp. 165–166, 1948.

63. Mason, B. J., *The Physics of Clouds*, Oxford University Press, London, 1971.

64. Miller, J. F., R. H. Frederick, and R. J. Tracey, *Precipitation Frequency Analysis of the Western United States*, vol. I: Montana, vol. II: Wyoming, vol. III: Colorado, vol. IV: New Mexico, vol. V: Idaho, vol. VI: Utah, vol. VII: Nevada, vol. VIII: Arizona, vol. IX: Washington, vol. X: Oregon, vol. XI: California, NOAA Atlas 2, National Weather Service, NOAA, U.S. Department of Commerce, Silver Spring, Md., 1973.

65. Miller, J. F., E. M. Hansen, D. D. Fenn, L. C. Schreiner, and D. T. Jensen, "Probable Maximum Precipitation Estimates—United States between the Continental Divide and the 103rd Meridian," *Hydrometeorological Report* 55, National Weather Service, NOAA, U.S. Department of Commerce, Washington, D.C., 1984.

66. Myers, V., "Meteorological Estimation of Extreme Precipitation for Spillway Design Floods," *ESSA Tech. Mem. WBTM HYDRO-5*, U.S. Weather Bureau, Washington D.C., 1967.

67. Myers, V. A., and R. M. Zehr, "A Methodology for Point-to-Area Rainfall Frequency Ratios," *NOAA Technical Report NWS 24*, National Weather Service, NOAA, U.S. Department of Commerce, Washington, D.C., 1980.

68. Osborn, H. B., L. V. Lane, and V. A. Meyers, "Rainfall Watershed Relationships for Southwestern Thunderstorms," *Trans. Am. Soc. Agric. Eng.*, vol. 23, no. 1, pp. 82–87, 1979.

69. Parker, S., ed., *Meteorology Source Book*, McGraw-Hill, New York, 1988.

70. Peck, E. L., "Design of Precipitation Networks," *Bull. Am. Meteorol. Soc.*, vol. 61, no. 5, pp. 894–902, 1980.

71. Pielke, R. A., *Mesoscale Meteorological Modeling*, Academic Press, Orlando, 1984.

72. Pilgrim, D., and I. Cordery, "Rainfall Temporal Patterns for Design Floods," *J. Hydraul. Engr.*, vol. 101(HY1), pp. 81–95, 1975.

73. Ramirez, J., and R. Bras, "Clustered or Regular Cumulus Cloud Fields: The Statistical Character of Observed and Simulated Cloud Fields," *J. Geophys. Res.,* vol. 95(D3), pp. 2035–2045, 1990.

74. Rassmusson, E. M., "Atmospheric Water Vapor Transport and the Hydrology of North America," *MIT Department of Meteorology Report* A1, Cambridge, Mass., 1966.

75. Rassmusson, E. M., "Atmospheric Water Vapor and the Water Balance of North America," *Mon. Weather Rev.,* vol. 95, pp. 403–426, 1967.

76. Richards, F., and P. Arkin, "On the Relationship between Satellite-Observed Cloud Cover and Precipitation," *Mon. Weather Rev.,* vol. 109, pp. 1081–1093, 1981.

77. Riedel, J., and L. Schreiner, "Comparison of Generalized Estimates of Probable Maximum Precipitation with Greatest Observed Rainfalls," *NOAA Technical Report NWS* 25, Washington, D.C., 1980.

78. Rodda, J., "The Systematic Error in Rainfall Measurement," *J. Inst. Water Engs.,* vol. 21, pp. 173–177, 1967.

79. Rodda, J., and W. Smith, "The Influence of Systematic Error in Rainfall Measurement for Assessing Wet Deposition," *Atmos. Environ.,* vol. 20, no. 5, pp. 1059–1064, 1986.

80. Rodriguez-Iturbe, I., "Scale of Fluctuation of Rainfall Models," *Water Resour. Res.,* vol. 22, pp. 15S–37S, 1986.

81. Rodriguez-Iturbe, I., and J. Mejia, "On the Design of Rainfall Networks in Time and Space," *Water Resour. Res.,* vol. 10, no. 4, pp. 713–728, 1974.

82. Rodriguez-Iturbe, D. Cox, and P. Eagleson, "Spatial Modeling of Total Storm Rainfall," *Proc. R. Soc. London A,* vol. 403, pp. 27–50, 1986.

83. Schreiner, L. C., and J. T. Riedel, "Probable Maximum Precipitation Estimates, United States East of the 105th Meridian," *NOAA Hydrometeorological Report 51,* National Weather Service, Washington, D.C., June 1978.

84. Schwarz, F. K., "The Unprecedented Rains in Virginia Associated with the Remnants of Hurricane Camille," *Mon. Weather Rev.,* vol. 98, no. 11, pp. 851–859, 1970.

85. Sellers, W. D., *Physical Climatology,* University of Chicago Press, Chicago, Ill., 1965.

86. Sevruk, B., "International Workshop on Precipitation Measurement: Preface," *Hydrol. Processes,* vol. 5, pp. 229–232, 1991.

87. Simpson, R. H., and H. Riehl, *The Hurricane and Its Impact,* Louisiana State University Press, Baton Rouge, 1981.

88. Simpson, J., R. Adler, and G. North, "A Proposed Tropical Rainfall Measuring Mission (TRMM) Satellite," *Bull. Am. Meteorol. Soc.,* vol. 69, pp. 278–295, 1988.

89. Singh, V. P., and P. K. Chowdhury, "Comparing Some Methods of Estimating Mean Areal Rainfall," *Water Resour. Bull.,* vol. 22, no. 2, pp. 275–282, 1986.

90. Sivapalan, M., and E. Wood, "A Multidimensional Model of Nonstationary Space-Time Rainfall at the Catchment Scale," *Water Resour. Res.,* vol. 23, no. 7, pp. 1289–1299, 1987.

91. Smith, J. A., "Statistical Modeling of Daily Rainfall Occurrences," *Water Resour. Res.,* vol. 23, no. 5, pp. 885–893, 1987.

92. Smith, J. A., and R. D. De Veaux, "The Temporal and Spatial Variability of Rainfall Power," *Environmetrics,* vol. 3, no. 1, pp. 29–53, 1992.

93. Smith, J. A., and A. F. Karr, "A Statistical Model of Extreme Storm Rainfall," *J. Geophys. Res.,* vol. 95(D3), pp. 2083–2092, 1990.

94. Smith, J. A., and W. F. Krajewski, "Estimation of the Mean Field Bias of Radar Rainfall Estimates," *J. Appl. Meteorol.,* vol. 30, no. 4, pp. 397–412, 1991.

95. Smith, R. B., "The Influence of Mountains on the Atmosphere," *Adv. Geophys.,* vol. 21, pp. 87–230, 1979.

96. Stern, R., and R. Coe, "A Model Fitting Analysis of Daily Rainfall Data," *J. R. Stat. Soc. A,* vol. 147, no. 1, pp. 1–34, 1984.

97. Sumner, G., *Precipitation: Process and Analysis,* Wiley, Chichester, 1988.

98. Takasao, T., and M. Shiiba, "Development of Techniques for On-Line Forecasting of Rainfall and Flood Runoff," *Nat. Disaster Sci.,* vol. 6, p. 83, 1984.

99. Todorovic, P., and V. Yevjevich, "Stochastic Point Process Model of Precipitation," *Hydrology Paper 35,* Colo. State Univ., Fort Collins, 1969.

100. U.S. Weather Bureau, "Rainfall Intensities for Local Drainage Design in the Western United States for Durations of 20 Minutes to 24 Hours and 1 to 100 Year Return Periods," *Technical Paper* 28, U.S. Department of Commerce, Washington, D.C., 1956.

101. Valdes, J., I. Rodriguez-Iturbe, and V. Gupta, "Approximations of Temporal Rainfall from a Multidimensional Model," *Water Resour. Res.,* vol. 21, no. 8, pp. 1259–1270, 1985.

102. Wallace, J. M., and P. V. Hobbs, *Atmospheric Science: An Introductory Survey,* Academic Press, San Diego, 1977.

103. Ward, R. C., and M. Robinson, *Principles of Hydrology,* McGraw-Hill, London, 1990.

104. Waymire, E., and V. Gupta, "The Mathematical Structure of Rainfall Representations, 1. A Review of the Stochastic Rainfall Models," *Water Resour. Res.,* vol. 17, pp. 1261–1272, 1981.

105. Waymire, E., V. Gupta, and I. Rodriguez-Iturbe, "A Spectral Theory of Rainfall Intensity at the Meso-β Scale," *Water Resour. Res.,* vol. 20, pp. 1453–1465, 1984.

106. Waymire, E., and V. Gupta, "On Lognormality and Scaling in Spatial Rainfall Averages," in S. Lovejoy, ed., *Scaling, Fractals and Non-linear Variability in Geophysics,* D. Reidel Publishing Co., Dordrecht, The Netherlands, 1990.

107. Wenzel, H. G., "Rainfall for Urban Stormwater Design," in D. F. Kibler, ed., *Urban Storm Water Hydrology,* Water Resources Monograph 7, AGU, Washington, D.C., 1982.

108. Wiesner, C. J., *Hydrometeorology,* Chapman and Hall, London, 1970.

109. Wischmeier, W. H., and D. D. Smith, "Rainfall Energy and Its Relationship to Soil Loss," *Trans. Am. Geophys. Union,* vol. 39, no. 2, pp. 285–291, 1958.

110. Wolman, M. G., and R. Gerson, "Relative Scales of Time and Effectiveness of Climate and Watershed Geomorphology," *Earth Surface Processes,* vol. 3, pp. 189–208, 1978.

111. Woolhiser, D., and J. Roldan, "Stochastic Daily Precipitation Models, 2. A Comparison of Amounts," *Water Resour. Res.,* vol. 18, no. 5, pp. 1461–1468, 1982.

112. World Meteorological Organization, "Manual for Estimation of Probable Maximum Precipitation," *Operational Hydrology Report* 1, WMO No. 332, 2d ed., Secretariat of the World Meteorological Organization, Geneva, Switzerland, 1986.

113. Zawadzki, I., "Fractal Structure and Exponential Decorrelation in Rain," *J. Geophys. Res.,* vol. 92(D8), pp. 9586–9590, 1987.

114. Zehr, R. M., and V. A. Myers, "Depth-Area Ratios in the Semi-Arid Southwest United States," *NOAA Technical Memorandum NWS HYDRO*-35, National Weather Service, NOAA, U.S. Department of Commerce, Silver Spring, Md., 1984.

CHAPTER 4
EVAPORATION

W. James Shuttleworth*
Hydrological Processes Division
Institute of Hydrology
Wallingford, United Kingdom

4.1 INTRODUCTION

Evaporation occurs when water is converted into water vapor. The rate is controlled by the availability of energy at the evaporating surface, and the ease with which water vapor can diffuse into the atmosphere. Different physical processes are responsible for the diffusion, but the physics of water vapor loss from open-water surfaces and from soils and crops is essentially identical. In this chapter evaporation is defined as *the rate of liquid water transformation to vapor from open water, bare soil, or vegetation with soil beneath.* Unless otherwise stated, this rate is in millimeters of evaporated water per day. In the case of vegetation growing in soil, transpiration is defined as *that part of the total evaporation which enters the atmosphere from the soil through the plants.*

The rate of evaporation has traditionally been estimated using meteorological data from climate stations located at particular points within a region, and it has been assumed that the evaporating area is sufficiently small that the evaporation has no effect on regional climate or air movement. In reality, this simplified approach approximates a more complex situation in which local evaporation is a function of both local climate and regional air movement. With the advance of scientific studies of evaporation and the availability of remotely sensed measurements of regional variables such as temperature and radiation, it is to be expected that eventually evaporation estimation for local areas will be made a function of both local and regional variables; but for the moment, such estimates are possible only on intensively instrumented experimental sites.

4.1.1 Standard Evaporation Rates

Two standard evaporation rates are defined, *potential evaporation* and *reference crop evaporation,* and used as the basis for evaporation estimates. These rates are concep-

*Now at Department of Hydrology and Water Resources, University of Arizona, Tucson, Arizona.

tual in the sense that they represent idealized situations. In particular they ignore the fact that meteorological parameters near the surface are influenced by upwind surface energy exchange; that is, evaporation introduces water into the air and can remove energy from it, which changes the atmospheric humidity deficit and may alter the evaporation at downwind locations.

Potential Evaporation E_0 (in Millimeters per Day). Potential evaporation is here defined as *the quantity of water evaporated per unit area, per unit time from an idealized, extensive free water surface under existing atmospheric conditions.* This is a conceptual entity which measures the meteorological control on evaporation from an open water surface.

Reference Crop Evaporation E_{rc} (in Millimeters per Day). Reference crop evaporation is here defined as *the rate of evaporation from an idealized grass crop with a fixed crop height of 0.12 m, an albedo of 0.23, and a surface resistance of 69 s m^{-1}.* In terms of its evaporation rate, such a crop closely resembles previous definitions of a reference crop, namely, an extensive surface of short green grass cover of uniform height, actively growing, completely shading the ground, and not short of water.

4.2 PHYSICS OF EVAPORATION AND TRANSPIRATION

4.2.1 Surface Exchanges

Latent Heat. The molecules in liquid water are held close together by attractive intermolecular forces. In water vapor the molecules are at least ten times farther apart than in the liquid phase, so the intermolecular force is very much smaller. During evaporation, separation between molecules increases greatly, work is done against the attractive intermolecular force, and energy is absorbed. The energy required is called the *latent heat of vaporization of water* λ, and it decreases slightly with increasing water temperature since the initial separation of the molecules increases with temperature. If T_s is the surface temperature of the water in degrees Celsius, the latent heat of vaporization is given[51] by

$$\lambda = 2.501 - 0.002361\, T_s \qquad \text{MJ kg}^{-1} \tag{4.2.1}$$

which means about 2.5 million joules are required to evaporate a kilogram of water, where the joule is the standard SI unit of energy.

Water Molecule Movement between Water Surfaces and Air. Natural evaporation occurs by exchange of water molecules between air and a free water surface. This water surface could be a lake or a river, or inside plant leaves, adhering to soil particles, or on soil or vegetation surfaces during or just after rain.

Figure 4.2.1 illustrates the exchange of water molecules between liquid and vapor. Some of the vapor molecules hitting the surface rebound, but the capture rate of molecules by the surface is proportional to their rate of surface collision, and therefore to e, the vapor pressure adjacent to the water surface. A molecule must have a minimum energy if it is to leave the surface, and the number of such molecules is related to the surface temperature.

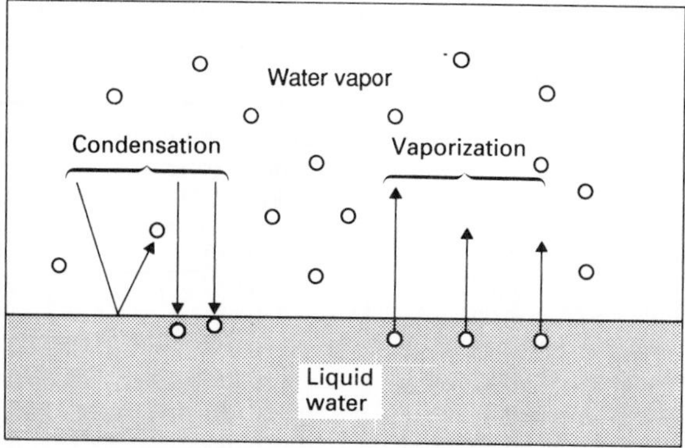

FIGURE 4.2.1 Molecular exchange between liquid water and water vapor. Not all the molecules hitting the surface are captured, but some condense at a rate which is proportional to the vapor pressure of the moist air: molecules with enough energy vaporize at a rate determined by the surface temperature.

Saturated Vapor Content of Air. Evaporation is the difference between two rates,[102] a *vaporization rate* determined by temperature, and a *condensation rate* determined by vapor pressure. If molecules can diffuse away from the surface, vapor pressure remains low, and the difference between these two rates is positive, so evaporation continues. If, on the other hand, the air above the water is thermally insulated and enclosed, the vapor pressure increases until the rates of vaporization and condensation are equal and there is no more evaporation. The air is then said to be *saturated.* At a given temperature this equilibrium occurs for a particular vapor pressure e_s. This is called the *saturated vapor pressure* and is related to temperature; if e_s is in kilopascals and T is in degrees Celsius, an approximate equation[118] is

$$e_s = 0.6108 \exp\left(\frac{17.27\ T}{237.3 + T}\right) \quad \text{kPa} \tag{4.2.2}$$

It is important in building physically based models of evaporation that not only is e_s a known function of temperature, but so is Δ, the gradient of this function, de_s/dT. This gradient is given by

$$\Delta = \frac{4098\ e_s}{(237.3 + T)^2} \quad \text{kPa °C}^{-1} \tag{4.2.3}$$

Table 4.2.1 lists values of e_s and Δ as a function of temperature.

Sensible Heat. A portion of the radiant energy input to the earth's surface is not used for evaporation; rather it warms the atmosphere in contact with the ground and then moves upward. We speak of the associated flow of energy as "sensible" heat flux because it changes air temperature, a property of air that can be measured or sensed. The temperature change is proportional to the product $(c_p\rho_a)$, where ρ_a is the density of air and c_p is the specific heat of air at constant pressure, taken as 1.01 kJ kg^{-1} K^{-1}. The density of (moist) air can be calculated from the ideal gas laws, but it is ade-

TABLE 4.2.1 Temperature Dependence of Saturated Vapor Pressure e_s, Its Temperature Gradient Δ, Together with the Psychrometric Constant at Standard Atmospheric Pressure

Temperature, °C	Saturated vapor pressure e_s, kPa	Gradient of saturated vapor pressure Δ, kPa °C^{-1}	Psychrometric constant γ, kPa °C^{-1}
0	0.611	0.044	0.0654
1	0.657	0.047	0.0655
2	0.706	0.051	0.0656
3	0.758	0.054	0.0656
4	0.814	0.057	0.0657
5	0.873	0.061	0.0658
6	0.935	0.065	0.0659
7	1.002	0.069	0.0659
8	1.073	0.073	0.0660
9	1.148	0.078	0.0660
10	1.228	0.082	0.0661
11	1.313	0.087	0.0661
12	1.403	0.093	0.0662
13	1.498	0.098	0.0663
14	1.599	0.104	0.0663
15	1.706	0.110	0.0664
16	1.819	0.116	0.0665
17	1.938	0.123	0.0665
18	2.065	0.130	0.0666
19	2.198	0.137	0.0666
20	2.339	0.145	0.0667
21	2.488	0.153	0.0668
22	2.645	0.161	0.0668
23	2.810	0.170	0.0669
24	2.985	0.179	0.0670
25	3.169	0.189	0.0670
26	3.363	0.199	0.0671
27	3.567	0.209	0.0672
28	3.781	0.220	0.0672
29	4.007	0.232	0.0673
30	4.244	0.243	0.0674
31	4.494	0.256	0.0674
32	4.756	0.269	0.0675
33	5.032	0.282	0.0676
34	5.321	0.296	0.0676
35	5.625	0.311	0.0677
36	5.943	0.326	0.0678
37	6.277	0.342	0.0678
38	6.627	0.358	0.0679
39	6.994	0.375	0.0670

quately estimated from

$$\rho_a = 3.486 \frac{P}{275 + T} \qquad \text{kg m}^{-3} \tag{4.2.4}$$

where P is the atmospheric pressure in kPa and T is air temperature in degrees Celsius.

The sensible heat flux, or heat flow per unit area H, is commonly upward from the ground during the day, but it is usually downward at night to support radiant energy loss from the land surface. Locally the flux of sensible heat can be downward during the day if the vegetation is wet and evaporation is rapid.

4.2.2 Radiation Balance at Land Surfaces

In the absence of restrictions due to water availability at the evaporative surface, the amount of radiant energy captured at the earth's surface is the dominant control on regional evaporation rates. As a monthly average, the radiant energy at the ground may be the most "portable" meteorological variable involved in evaporation estimation, in the sense that it is driven by astronomical rather than local climate conditions. Understanding surface radiation balance, and how to quantify it, is therefore crucial to understanding and quantifying evaporation.

Net Short-Wave Radiation. Figure 4.2.2 illustrates the radiation balance at the earth's surface. The sun is the main source of radiant energy. It is equivalent to a radiator of about 6000°C, but the input of extraterrestrial short-wave radiation S_0 is modified by absorption by atmospheric gases, particularly water vapor, through scattering by air molecules and dust particles in clear sky conditions, and additionally by clouds when these are present. Much of the radiation has short wavelengths, 0.3 to

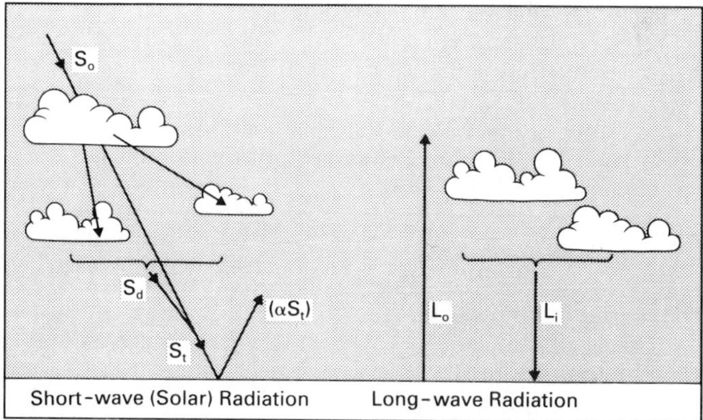

FIGURE 4.2.2 Radiation balance at the earth's surface. A proportion S_t of the solar radiation incident at the top of the atmosphere S_0 reaches the ground, some S_d indirectly after scattering by air and cloud. A proportion α, the albedo, is reflected. Outward long-wave radiation L_0 is partly compensated by incoming long-wave radiation L_i. S_t is typically 25 to 75 percent of S_0, while S_d can vary between 15 and 100 percent of S_t: both these proportions are influenced by cloud cover. α is typically 0.23 for land surfaces and 0.08 for water surfaces.

3.0 μm, the spectrum varying with the fraction of the total short-wave energy input S_t reaching the ground in the direct solar beam. Some of the short-wave radiation S_d reaches the surface in a diffuse, i.e., multidirectional form after scattering by atmospheric particles and clouds. This fraction is typically 15 to 25 percent in clear sky conditions but approaches 100 percent in overcast conditions.

Part of the short-wave radiation is reflected. The reflection coefficient or short-wave *albedo* α depends on transient features, such as the direction of the solar beam, and the proportion of diffuse radiation, but it also changes with land cover, since taller vegetation usually reflects less solar radiation than does shorter vegetation. The scientific literature reporting measured short-wave albedo is extensive, and reported values vary greatly. However, Table 4.2.2 gives the author's recommendation on plausible values of albedo for broad land cover classes. The value $\alpha = 0.23$ is a good overall average value for grassland and a range of agricultural crops, and $\alpha = 0.08$ is a reasonable value for open water surfaces.

The *net short-wave radiation* S_n is that portion of the incident short-wave radiation captured at the ground taking into account losses due to reflection, and is given by

$$S_n = S_t (1 - \alpha) \qquad \text{MJ m}^{-2}\text{ day}^{-1} \tag{4.2.5}$$

Solar radiation is measured in specialized agrometeorological stations with radiometers. These instruments require careful calibration and maintenance, however, and measured solar radiation data are usually not available at standard stations. The total incoming short-wave radiation can in most cases be estimated[16] from measured sunshine hours according to the following empirical relationship:

$$S_t = \left(a_s + b_s \frac{n}{N}\right) S_0 \qquad \text{MJ m}^{-2}\text{ day}^{-1} \tag{4.2.6}$$

where a_s = fraction of extraterrestrial radiation S_0 on overcast days ($n = 0$)
$a_s + b_s$ = fraction of extraterrestrial radiation S_0 on clear days
n/N = cloudiness fraction
n = bright sunshine hours per day, h
N = total day length, h
S_0 = extraterrestrial radiation, MJ m^{-2} day^{-1}

TABLE 4.2.2 Plausible Values for Daily Mean Short-Wave Solar Radiation Reflection Coefficient (Albedo) for Broad Land Cover Classes

Land cover class	Short-wave radiation reflection coefficient α
Open water	0.08
Tall forest	0.11–0.16
Tall farm crops (e.g., sugarcane)	0.15–0.20
Cereal crops (e.g., wheat)	0.20–0.26
Short farm crops (e.g., sugar beet)	0.20–0.26
Grass and pasture	0.20–0.26
Bare soil	0.10 (wet)–0.35 (dry)
Snow and ice	0.20 (old)–0.80 (new)

Note: Albedo can vary widely with time of day, season, latitude, and cloud cover.

In the absence of knowledge on crop cover the value $\alpha = 0.23$ is recommended.

Available local measurements of S_t can be used to carry out a regression analysis to determine the angstrom coefficients a_s and b_s by comparing with S_0 on overcast days to give a_s, and on days with bright sunshine to give $(a_s + b_s)$. Depending on atmospheric conditions (humidity, dust) and solar declination (latitude and month), the values of a_s and b_s will vary. When no actual solar radiation data are available, and no calibration has been carried out for improved a_s and b_s parameters, the following values are recommended for average climates:

$$a_s = 0.25 \qquad \text{and} \qquad b_s = 0.50$$

Net Long-Wave Radiation. There is a significant exchange of radiant energy between the earth's surface and the atmosphere in the form of radiation at longer wavelengths, i.e., in the range 3 to 100 μm (see Fig. 4.2.2). Both the ground and the atmosphere emit black-body radiation with a spectrum characteristic of their temperature. Since the surface is on average warmer than the atmosphere, there is usually a net loss of energy as thermal radiation from the ground.

The exchange of long-wave radiation L_n between vegetation and soil on the one hand, and atmosphere and clouds on the other, can be represented by the following radiation law:

$$L_n = L_i - L_o = -f\epsilon' \sigma (T + 273.2)^4 \qquad \text{MJ m}^{-2}\,\text{day}^{-1} \tag{4.2.7}$$

where L_o = outgoing long-wave radiation (ground to atmosphere), MJ m^{-2} day^{-1}
L_i = incoming long-wave radiation (atmosphere to ground), MJ m^{-2} day^{-1}
f = adjustment for cloud cover
ϵ' = net emissivity between the atmosphere and the ground
σ = Stefan-Boltzmann constant (4.903×10^{-9} MJ m^{-2} °K^{-4} day^{-1})
T = mean air temperature, °C

The *net emissivity* ϵ' can be estimated[3,14] from

$$\epsilon' = a_e + b_e \sqrt{e_d} \tag{4.2.8}$$

where e_d = vapor pressure, kPa
a_e = correlation coefficient
b_e = correlation coefficient

a_e lies in the range 0.34 to 0.44 and b_e in the range -0.14 and -0.25, but for average conditions the following indicative values can be taken:[30]

$$a_e = 0.34 \qquad \text{and} \qquad b_e = -0.14$$

When humidity measurements are not available, the dew point at minimum temperature can be taken to estimate average vapor pressure. Alternatively net emissivity can be estimated[53] from average temperature (in degrees Celsius) according to the equation

$$\epsilon' = -0.02 + 0.261 \exp(-7.77 \times 10^{-4}\, T^2) \tag{4.2.9}$$

The *cloudiness factor* f in Eq. (4.2.7) can be estimated[135] using solar radiation data from

$$f = a_c \frac{S_t}{S_{to}} + b_c \tag{4.2.10}$$

where S_t = measured solar radiation
S_{to} = solar radiation for clear skies [Eq. (4.2.6) with $n/N = 1$]
a_c, b_c = long-wave radiation coefficients for clear skies (sum = 1.0)

a_c and b_c parameters are calibration values to be determined through specialized local studies which involve measuring long-wave radiation values. The following indicative values are recommended:[30]

$$a_c = 1.35 \qquad b_c = -0.35 \qquad \text{(arid areas)}$$

$$a_c = 1.00 \qquad b_c = 0.00 \qquad \text{(humid areas)}$$

When data on sunshine hours n are available, the cloudiness factor for partly cloudy skies can be determined by substituting the relevant terms of Eq. (4.2.6) into Eq. (4.2.10) to give

$$f = \left(a_c \frac{b_s}{a_s + b_s}\right)\frac{n}{N} + \left(b_c + \frac{a_s}{a_s + b_s}\, a_c\right) \tag{4.2.11}$$

For $a_c = 1.35$, $b_c = -0.35$, $a_s = 0.25$, and $b_s = 0.50$, this becomes

$$f = 0.9\frac{n}{N} + 0.1 \tag{4.2.12}$$

Net Radiation. The *net radiation* R_n is the net input of radiation at the surface, i.e., the difference between the incoming and reflected solar radiation, plus the difference between the incoming long-wave radiation and outgoing long-wave radiation (see Fig. 4.2.2). It is given by

$$R_n = S_n + L_n \qquad \text{MJ m}^{-2}\text{ day}^{-1} \tag{4.2.13}$$

Net radiation is comparatively simple to measure using instrumentation, and indirect measurement is increasingly possible using satellite data,[91] albeit with considerable uncertainty. Often, however, practicing hydrologists are required to provide estimates of evaporation from data records which do not include net radiation. In these circumstances S_n must be estimated from Eqs. (4.2.5) and (4.2.6), and L_n from Eq. (4.2.7) and associated equations. Using the indicative values given in previous sections, for general purposes when only sunshine, temperature, and humidity data are available, net radiation (in MJ m^{-2} day^{-1}) can be estimated by the following equation:

$$R_n = \left(0.25 + 0.5\frac{n}{N}\right)S_0 - \left(0.9\frac{n}{N} + 0.1\right)(0.34 - 0.14\sqrt{e_d})\sigma T^4 \tag{4.2.14}$$

Because of the strong link between energy and evaporation (through the latent heat of vaporization; see Sec. 4.2.1), R_n can be expressed as an equivalent depth of evaporated water in mm by dividing R_n by $\rho_w\lambda$ where ρ_w (in kg m^{-3}) is the density of water, and λ is the latent heat of vaporization (in MJ kg^{-1}) from Eq. (4.2.1). It is convenient that with R_n in MJ m^{-2} day^{-1} the numerical value (R_n/λ) gives equivalent water depth in mm day^{-1}, since $\rho_w \approx 1000$ kg m^{-3}.

4.2.3 Energy Budget for a Unit Area

When describing the evaporation process it is usual to draw up an energy budget for a volume of defined vertical extent and unit area in the horizontal plane. In the case of

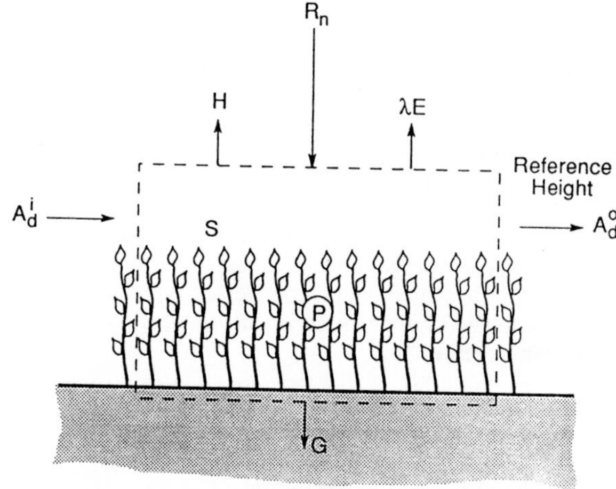

FIGURE 4.2.3 The components of the energy balance for a volume extending from just below the soil surface to the height at which the net radiation balance is determined.

a crop this volume extends from just below the soil surface, where energy lost by heat conduction to the soil is measured or estimated, to that level above the canopy at which the radiation balance described in the previous section is defined.

Figure 4.2.3 illustrates the components of the whole canopy energy balance for daytime conditions, which, ranked in approximate order of their magnitude, comprise:

R_n = net incoming radiant energy

λE = outgoing energy as evaporation

H = outgoing sensible heat flux

G = outgoing heat conduction into the soil

S = energy temporarily "stored" within the volume, and often neglected except for forests,[116] proportional to temperature changes in the vegetation, air, and shallow soil layer; and to changes in atmospheric humidity

P = energy absorbed by biochemical processes in the plants, typically taken as 2 percent of net radiation[116]

A_d = loss of energy associated with horizontal air movement;[119] significant in an "oasis" situation, but generally neglected otherwise ($A_d = A_d^o - A_d^i$ in Fig. 4.2.3)

It is usual to collect the energy terms and define an entity A, *the available energy,* which is the energy available for partitioning into latent or sensible heat. For the case of a complete canopy

$$A = \lambda E + H \qquad (4.2.15)$$

and

$$A = R_n - G - S - P - A_d \qquad \text{MJ m}^{-2}\ \text{day}^{-1} \qquad (4.2.16)$$

Conduction is the main mechanism for heat transfer in soils. In the idealized case of a uniform soil, and assuming the temperature of the soil-air interface oscillates sinusoidally over a day or a year, the heat flow is maximized when the rate of change of the soil surface temperature is greatest, one-eighth of a cycle before the peak temperature, i.e., 3 h earlier in the case of the diurnal wave and 1½ months earlier in the case of the annual wave. The amplitude of the daily cycle in soil surface temperature can be significantly greater than that in air temperature, with timing mainly linked to radiation input, and soil heat conduction G can be large, typically 30 percent of the net radiation exchange at the soil surface. With dense vegetation, little radiation reaches the ground and heat storage in the soil can often be neglected.[109]

The effective soil depth to which heat is transferred is greater for longer temperature cycles. To estimate the change in soil heat content for a given period, the following equation can be used:[126]

$$G = c_s d_s \frac{T_2 - T_1}{\Delta t} \quad \text{MJ m}^{-2}\ \text{day}^{-1} \tag{4.2.17}$$

where T_2 = temperature at the end of the period, °C
T_1 = temperature at the beginning of the period, °C
Δt = length of period, days
c_s = soil heat capacity (2.1 MJ m^{-3} °C^{-1}) for average moist soil
d_s = estimated effective soil depth, m

For *daily* temperature fluctuations (effective soil depth typically 0.18 m) the above formula[136] becomes

$$G = 0.38\,(T_{\text{day }2} - T_{\text{day }1}) \quad \text{MJ m}^{-2}\ \text{day}^{-1} \tag{4.2.18}$$

For *monthly* temperature fluctuations (effective soil depth typically 2.0 m)[56] it becomes

$$G = 0.14\,(T_{\text{month }2} - T_{\text{month }1}) \quad \text{MJ m}^{-2}\ \text{month}^{-1} \tag{4.2.19}$$

Since the magnitude of daily soil heat flux over 10- to 30-day periods is relatively small, it can often be neglected in hydrologic applications.

Heat transfer to depth in a water body is by conduction and thermal convection, and by the penetration of radiation below the surface. Its calculation is complex, and it is easier to measure it from successive temperature profile surveys.[27] The sensible heat transferred into a lake by water inflow and outflow A_h may be significant in the energy budget of a whole lake. The total advection rate per unit lake area can be approximated as follows:

$$\left.\begin{aligned} A_h &= \rho_w c_w (q_i T_i - q_o T_o + PT_p) \\ &= 4.19 \times 10^{-3} (q_i T_i - q_o T_o + PT_p) \end{aligned}\right\} \quad \text{MJ m}^{-2} \tag{4.2.20}$$

where ρ_w is the density of water, c_w is its specific heat; q_i and q_o are the rate of inflow and outflow per unit area of lake, P is the rate of precipitation (all in mm); and T_i, T_o, and T_p are the temperatures (in °C) of the inflow, outflow and precipitation water, respectively.

4.2.4 Diffusion through the Air

The extent to which the energy available at the ground is used to evaporate water is determined by the processes controlling vapor diffusion through the air. Movement occurs where there are variations in vapor concentration, and because the molecules making up the air are in permanent, random motion, either individually or in coherent groups as turbulent eddies. Eddy circulation raises moist air from near the land surface and replaces it with drier air from higher up, resulting in a net upward moisture movement in proportion to the vapor concentration difference between the upper and lower air masses.

The flows of water vapor λE and sensible heat H are proportional to differences in the vapor pressure e and temperature T, respectively. The constant of proportionality is related to the transport *resistance,* which measures the restriction placed on the movement by the diffusion process.

Molecular Diffusion. Individual air molecules are in rapid, haphazard motion at normal temperatures. The transfers associated with such motion control the movement of water vapor and warmed air at or near the leaves and stems within vegetation, and near the surface of the underlying soil. Air moving within vegetation, for instance, can be envisaged as interacting with leaves through the boundary layer of slowly moving air which surrounds each leaf. The rate of vapor flow is controlled by a *boundary-layer resistance.*[24,57,119] Similarly, a simple representation of evaporation within soil might assume, for instance,[71] that water vapor moves by molecular diffusion in the air between the soil particles in a progressively deepening layer of dry soil, thus producing a *soil surface resistance.*[112]

However, the most important resistance associated with molecular diffusion is that which controls the movement of water vapor from inside plant leaves to the air outside through small apertures in the surface of the leaves which are called *stomata.* Figure 4.2.4 illustrates this *transpiration* process. The air inside the stomatal cavity beneath the leaf surface is nearly saturated, while that outside is usually less. Water vapor movement is controlled by the plant, which opens or closes the stomatal aperture in response to atmospheric moisture demand and the amount of water in the soil. In this way, plants control their water loss to the atmosphere, and seek to ensure their survival when water is limited.

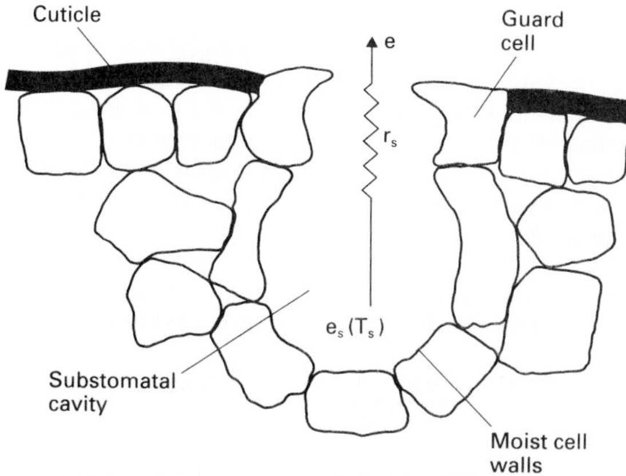

FIGURE 4.2.4 Transpiration by molecular diffusion of water vapor through the stomatal aperture of dry leaves. Air inside the substomatal cavity is saturated at the temperature of the leaf, and the water vapor diffuses through the stomatal opening to the less saturated atmosphere against a stomatal resistance which, for the whole canopy, is called the surface resistance r_s.

The stomatal resistance of the whole canopy, referred to as the *surface resistance* r_s, is less when more leaves are present since there are then more stomata through which transpired water vapor can diffuse. Empirical relationships exist between the surface resistance and leaf cover, soil water status, and environmental variables (e.g., Ref. 106) which can provide relationships suitable for practical application in estimating evaporation.

An analogy may be drawn with electrical resistance. The current i through a wire of resistance R is related to the potential difference V between the ends of the wire by Ohm's law, $V = iR$, which may be rewritten for the current i as $i = V/R$. Now in vapor transport, the measure of potential is the vapor pressure and the equivalent of current is the vapor flux rate E. Thus the vapor flux rate can be approximately estimated for leaf stomata as

$$E = \frac{k\,(e_s - e)}{r_s} \tag{4.2.21}$$

where k is a constant to account for units. The dimensions of vapor flow resistance r_s are $(T\,L^{-1})$ usually measured in s m^{-1}.

One approximation[2] for r_s is

$$r_s = \frac{200}{L} \qquad \text{s m}^{-1} \tag{4.2.22}$$

If h_c is the mean height of the crop, then the leaf area index L can be estimated[3] by

$$\begin{aligned} L &= 24\,h_c \qquad \text{(clipped grass with } 0.05 < h_c < 0.15 \text{ m)} \\ L &= 5.5 + 1.5 \ln (h_c) \qquad \text{(alfalfa with } 0.10 < h_c < 0.50 \text{ m)} \end{aligned} \tag{4.2.23}$$

The *surface resistance of the reference crop* of clipped grass 0.12 m high (see Sec. 4.1.1) is estimated as

$$r_s^{rc} = 69 \qquad \text{s m}^{-1} \tag{4.2.24}$$

Turbulent Diffusion. The wind blowing horizontally over natural surfaces is retarded by interaction with the ground and vegetation. This interaction creates random and haphazard air motion in which portions of air, of varying size, move in an ill-defined yet coherent way during their transient existence. This phenomenon, known as *turbulence,* is a much more efficient transport mechanism than molecular diffusion and is the primary process responsible for exchange between air close to the ground and that at higher levels in the atmosphere.

The rate of water vapor transfer away from the ground by turbulent diffusion is controlled by *aerodynamic resistance* r_a, which is inversely proportional to wind speed and changes with the height of the vegetation covering the ground, as

$$r_a = \frac{\ln\,[(z_u - d)/z_{om}]\,\ln\,[(z_e - d)/z_{ov}]}{(0.41)^2\,U_z} \qquad \text{s m}^{-1} \tag{4.2.25}$$

where z_u and z_e are the respective heights of the wind speed and humidity measurements, and U_z is the wind speed. Estimates can be made of r_a by assuming[15] that $z_{om} = 0.123\,h_c$ and $z_{ov} = 0.0123\,h_c$, and[71] that $d = 0.67\,h_c$, where h_c is the mean height of the crop.

If wind speed and humidity measurements are made at a height of (say) 2 m above the top of the vegetation, then for a measured wind speed of 5 m s^{-1}, Eq. (4.2.25) gives values of 45 s m^{-1}, 18 s m^{-1}, and 6.5 s m^{-1} for the aerodynamic resistance for grass ($h_c = 0.1$ m), agricultural crops ($h_c = 1.0$ m), and forest ($h_c = 10$ m), respectively. The rate of diffusion of water vapor is greater where the resistance to vapor diffusion is less, so the aerodynamic exchange of taller crops is more efficient than for shorter crops. In consequence, water caught on their leaves during rainstorms or overhead irrigation evaporates more quickly from taller crops as compared to shorter crops.

The *aerodynamic resistance of the reference crop* (see Sec. 4.1.1) with a crop height of 0.12 m, and for measurements of temperature and humidity at a standardized height of 2.0 m, is given by

$$r_a^{rc} = \frac{208}{U_2} \quad \text{s m}^{-1} \tag{4.2.26}$$

where U_2 is the wind speed in m s^{-1}, also measured at 2 m.

4.2.5 Simulation by Resistance Networks

In advanced models of evaporation, the diffusion of energy in the form of sensible heat or water vapor away from the plants or soil into the atmosphere is represented by a network of resistances of the types described in the previous section. Such models can be complex, with the plant canopy broken into separate layers and evaporation rates calculated from each of these individually.[113,126–128]

It is also possible to build physically realistic descriptions of sparse canopies using equivalent resistance networks which consider the evaporation from the plants separately to that from the soil.[45,61,111,112] Such models do not yet, however, have widespread use among practicing hydrologists.

Penman-Monteith Equation. Currently the most advanced resistance-based model of evaporation used in hydrologic practice assumes that all the energy available for evaporation is accessible by the plant canopy, and water vapor diffuses first out of the leaves against the surface (or stomatal) resistance r_s and then out into the atmosphere above against the aerodynamic resistance. Meanwhile the sensible heat, which originates outside rather than inside the leaves, only has to diffuse upward against the aerodynamic resistance r_a. This model is represented diagrammatically in Fig. 4.2.5.

Solving the equations describing the diffusion processes represented in Fig. 4.2.5 produces the Penman-Monteith equation. This equation allows the calculation of evaporation from meteorological variables and resistances which are related to the stomatal and aerodynamic characteristics of the crop, and has the form[72]

$$E = \frac{1}{\lambda}\left[\frac{\Delta A + \rho_a c_p D/r_a}{\Delta + \gamma(1 + r_s/r_a)}\right] \quad \text{mm day}^{-1} \tag{4.2.27}$$

in which Δ is given by Eq. (4.2.3), A by Eq. (4.2.16); D is the vapor pressure deficit $(e_s - e)$ (in kPa) measured at the height z_e for which r_a is calculated from Eq. (4.2.25); and r_s is the surface resistance of the land cover. The *psychrometric constant* γ is defined by the equation

$$\gamma = \frac{c_p P}{\epsilon\lambda} \times 10^{-3} = 0.0016286\,\frac{P}{\lambda} \quad \text{kPa °C}^{-1} \tag{4.2.28}$$

where c_p is the specific heat of moist air (= 1.013 kJ kg^{-1} °C^{-1}), P is the atmospheric

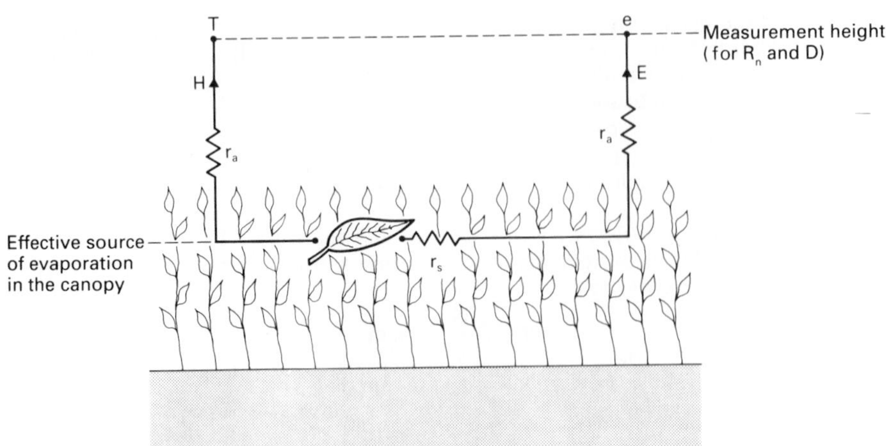

FIGURE 4.2.5 The equivalent network of diffusion resistances representing evaporation from an established crop canopy which intercepts almost all of the sun's radiant energy. It is assumed to share the available energy A as evaporation E or sensible heat H in the same way as would a single equivalent "big leaf" positioned at the effective source height $(d + z_o)$ within the crop.

pressure (in kPa), ϵ is the ratio of the molecular weight of water vapor to that for dry air (=0.622), and λ is the latent heat of vaporization of water (MJ kg^{-1}) given by Eq. (4.2.1). Values of γ as a function of air temperature are given in Table 4.2.1.

Wet Canopies. When vegetation is wet, which might be the case during or immediately after rain or following sprinkler irrigation, the source of water vapor is no longer inside the leaves; rather it is the water on the plants' surface. The stomata are no longer effective in restricting evaporation, and the surface resistance $r_s = 0$. For short crops, such as grass, r_a is large (see Sec. 4.2.4) and the consequence of setting $r_s = 0$ in Eq. (4.2.27) is less for grass than it is for tall vegetation such as forest.

The enhanced rate of evaporation of the water intercepted by forest during rain is an important aspect of the hydrologic effects of land-use change involving afforestation or deforestation (see Chap. 13). To make realistic estimates of forest evaporation it is necessary to describe the transpiration and the evaporation of rainwater separately. The latter component, the rainfall interception loss, can be estimated using models which calculate a running water balance for the storm water stored on the vegetation (e.g., Refs. 98, 99) from frequently sampled measurements of meteorological variables, including hourly rainfall.

Potential Evaporation Equation. Since potential evaporation occurs from an extensive free water surface, it follows that $r_s = 0$ is the appropriate value of surface resistance for estimating potential evaporation from Eq. (4.2.27). Further, the relevant value of energy supply A is given by replacing the smaller terms in Eq. (4.2.16) by A_h in Eq. (4.2.20), so that $A = (R_n + A_h)$. The appropriate form of r_a for open water evaporation was first determined empirically,[80,81] but its physical basis is now better understood,[121] and it can be estimated from

$$r_a^p = \frac{4.72\,[\ln (z_m/z_o)]^2}{1 + 0.536\,U_2} \quad \text{s m}^{-1} \tag{4.2.29}$$

where z_m (in m) is the height at which meteorological variables are measured, and z_o (in m) is the aerodynamic roughness of the surface.

For a standardized measurement height for wind speed, temperature, and humidity measurements of 2 m, and adopting the value $z_0 = 0.00137$ m which, according to Thom and Oliver[121] is that implicitly assumed by Penman,[80,81] *the equation here recommended for estimating potential evaporation* E_p (in mm day^{-1}) *is*

$$E_p = \frac{\Delta}{\Delta + \gamma}(R_n + A_h) + \frac{\gamma}{\Delta + \gamma}\frac{6.43\,(1 + 0.536\,U_2)D}{\lambda} \qquad (4.2.30)$$

where R_n = net radiation exchange for the free water surface, mm day^{-1}
A_h = energy advected to the water body, mm day^{-1}, if significant
U_2 = wind speed measured at 2 m, m s^{-1}
D = vapor pressure deficit $e_s - e$, kPa

and λ, Δ, and γ are given by Eqs. (4.2.1), (4.2.3), and (4.2.28), respectively.

Reference Crop Evaporation Equation. In this chapter the reference crop is precisely defined in terms of parameters appropriate to Eq. (4.2.27); see Sec. 4.1.1. Combining Eqs. (4.2.4), (4.2.16), (4.2.24), (4.2.27), and (4.2.28) [but neglecting the minor terms in Eq. (4.2.16)], and for a standardized measurement height for wind speed, temperature, and humidity of 2 m, *the equation here recommended for estimating reference crop evaporation,* in mm day^{-1}, *is*

$$E_{rc} = \frac{\Delta}{\Delta + \gamma^*}(R_n - G) + \frac{\gamma}{\Delta + \gamma^*}\frac{900}{T + 275}U_2 D \qquad (4.2.31)$$

where R_n = net radiation exchange for the crop cover, mm day^{-1}
G = measured or estimated soil heat flux, mm day^{-1}
T = temperature, °C
U_2 = wind speed at 2 m, m s^{-1}
D = vapor pressure deficit, kPa

and

$$\gamma^* = \gamma(1 + 0.33\,U_2) \qquad (4.2.32)$$

It should be noted that Eq. (4.2.31) is *not the Penman-Monteith equation as such,* since this equation has broader applicability. Rather Eqs. (4.2.30) and (4.2.31) are both implementations of the Penman-Monteith equation, in which the several resistances are here assigned particular values for specific, well-defined reference surfaces.

4.2.6 Empirical Estimation Equations

The physically-based equations for estimating evaporation rates for open water and reference crops recommended in the previous section, which are based on equivalent resistance networks, are currently the most physically realistic equations available for hydrologic application. Historically, conceptually simpler equations have been proposed, which are less fundamental and therefore necessarily have greater empirical content. Viewed in retrospect, these can be considered as simplifications of Eqs. (4.2.30) and (4.2.31) through the introduction of empirical relationships between meteorological variables derived from local data.

All the equations given below provide estimates of evaporation for a reference crop, and this is their primary purpose. In general they can also be considered to provide an approximate estimate of the potential evaporation rate for open water surfaces. However, in this case those equations which explicitly contain the available energy appropriate to a reference crop, i.e., $(R_n - G)$, should have this replaced by $(R_n + A_h)$, this being the equivalent form appropriate to a free water surface [see Eqs. (4.2.30) and (4.2.31)].

Combination Equations. Penman[80] was the first to derive an equation, given earlier as Eq. (4.2.30), which combines the energy required to sustain evaporation and an empirical description of the diffusion mechanism by which energy is removed from the surface as water vapor. Because it combines energy and diffusion features, this equation became known as a *combination equation,* and it spawned sibling equations which are given by "tuning" the empiricism in the description of atmospheric diffusion to better represent particular sets of data or local conditions, e.g., Refs. 30, 82, and 135.

When analyzing the link between the Penman and the Penman-Monteith equations, Thom and Oliver[121] demonstrated that Penman's implicit empirical function for the aerodynamic resistance [see Eq. (4.2.29)] not only provides some allowance for the effect of atmospheric buoyancy but also, when applied to a reference crop, makes approximate allowance for the absence of a surface resistance in the denominator, comparing Eqs. (4.2.27) and (4.2.30).

Studies of the comparative performance of several different forms of the combination equation[56] suggest that "tuning" the representation of the diffusion component has little universal advantage, though the introduction of some empirical seasonal dependence may be beneficial. This conclusion is consistent with Thom and Oliver's interpretation. The seasonal adjustment currently favored is based on data from Kimberly, Idaho, and the resulting Kimberly Penman equation[56,134] has the same nomenclature as Eq. (4.2.30) and takes the form

$$E_{rc} = \frac{\Delta}{\Delta + \gamma}(R_n - G) + \frac{\gamma}{\Delta + \gamma}\frac{6.43\ W_f D}{\lambda} \qquad \text{mm day}^{-1} \tag{4.2.33}$$

where

$$W_f = a_w + b_w U_2 \tag{4.2.34}$$

$$a_w = 0.4 + 1.4 \exp\{-[(J - 173)/58]^2\} \tag{4.2.35}$$

in which J is the Julian day number with, for northern latitudes,

$$b_w = 0.605 + 0.345 \exp\{-[(J - 243)/80]^2\} \tag{4.2.36}$$

and, for southern latitudes, J set to $J' = (J - 182)$ for $J \geq 182$, and set to $J' = (J + 183)$ for $J < 182$.

Radiation-Based Equations. The first term in Eq. (4.2.33) frequently exceeds the second term by a factor of about 4, and this suggests the possibility of a simpler empirical relation between reference crop evaporation rate and radiation called the Priestley-Taylor equation with the general form[86]

$$E_{rc} = \alpha \frac{\Delta}{\Delta + \gamma}(R_n - G) \qquad \text{mm day}^{-1} \tag{4.2.37}$$

In fact there is now more substantial evidence supporting such an empirical relationship, at least on a regional average (as opposed to a crop-specific) basis, for regions with uniform vegetation cover, or with land cover which is heterogeneous at the scale of a few kilometers.[104,107] Simple models of the evaluation of the near-surface atmospheric boundary layer[13,67,68] overviewed in Sec. 4.2.7 suggest that its partial containment by a thermal inversion layer in the atmosphere yields near surface vapor pressures in the second term of Eq. (4.2.33) which support a value of $\alpha \simeq 1.26$ in Eq. (4.2.37) for $r_s = 69$ s m^{-1}.

Recent evaluation[56] of Eq. (4.2.37) with $\alpha = 1.26$ confirms its applicability in humid climates. However, in arid climates the value of $\alpha = 1.74$ provides better estimates (Ref. 56, Table 7.16). On the basis of these results, and in view of its inherent simplicity, *the radiation-based equations here recommended for reference crop evaporation estimation are*

$$E_{rc} = 1.74 \frac{\Delta}{\Delta + \gamma} (R_n - G) \qquad \text{mm day}^{-1} \tag{4.2.38}$$

for arid locations, with relative humidity less than 60 percent in the month having peak evaporation, and

$$E_{rc} = 1.26 \frac{\Delta}{\Delta + \gamma} (R_n - G) \qquad \text{mm day}^{-1} \tag{4.2.39}$$

at all other (humid) locations.

Certain other empirical radiation-based equations remain in common use. *In humid climates,* the Turc equations[125] have been shown[52] to perform well. These equations have the form for RH < 50 percent

$$E = 0.31 \frac{T}{T + 15} (S_n + 2.09) \left(1 + \frac{50 - \text{RH}}{70}\right) \qquad \text{mm day}^{-1} \tag{4.2.40}$$

and for RH > 50 percent

$$E = 0.31 \frac{T}{T + 15} (S_n + 2.09) \qquad \text{mm day}^{-1} \tag{4.2.41}$$

where T is the average temperature in °C, S_n is the water equivalent of net solar radiation in mm day^{-1}, and RH is the relative humidity in percent. The general similarity in form between these earlier equations and Eqs. (4.2.38) and (4.2.39) is to be expected.

In arid climates, on the other hand, the radiation-based equation of Doorenbos and Pruitt,[30] which takes the form

$$E = -0.3 + b_{dp} \frac{\Delta}{\Delta + \gamma} S_n \qquad \text{mm day}^{-1} \tag{4.2.42}$$

has been shown[56] to provide estimates of reference crop evaporation, with[35]

$$\begin{aligned} b_{dp} = {} & 1.066 - 0.0013\,(\text{RH}_{\text{mean}}) + 0.045\,(U_d) - 0.0002\,(\text{RH}_{\text{mean}})\,(U_d) \\ & - 0.000315\,(\text{RH}_{\text{mean}})^2 - 0.0011\,(U_d)^2 \end{aligned} \tag{4.2.43}$$

where RH_{mean} is mean relative humidity in percentage, U_d is the mean daytime wind speed in m s^{-1}, and S_n is the net solar radiation in mm day^{-1}; see Eq. (4.2.5).

Temperature-Based Equations. The physical basis for estimating evaporation using temperature alone is that both terms in Eq. (4.2.31) are likely to have some relationship with temperature. Since the first (radiation-dependent) term is generally much the larger of the two, it is the correlation between radiation and temperature which is most important. The yearly temperature cycle is delayed with respect to the yearly radiation cycle and the empiricism in some past formulas has included allowance for this thermal lag.

In general, the only justification for using estimation equations of this type is that prediction of evaporation is required on the basis of existing data in which temperature is the only available variable measurement, and even in this case it is unwise to make evaporation estimates for less than a monthly averaging period. Certain relationships merit mention either because their relationship to radiation-based estimates is more explicit and plausible or because their empiricism is very broadly based and they are in very common use. Only the Hargreaves equation and Blaney-Criddle method are described below. Other temperature-based methods are not recommended.

The Hargreaves equation[47–50] is here taken as

$$E_{rc} = 0.0023\, S_o\, \bar{\delta}_T\, (T + 17.8) \qquad \text{mm day}^{-1} \tag{4.2.44}$$

where S_o is the water equivalent of extraterrestrial radiation in mm day^{-1} for the location of interest [see Eq. (4.4.4)], T is the temperature in °C, and $\bar{\delta}_T$ is the difference between mean monthly maximum and mean monthly minimum temperatures. This equation has been shown[56] to provide at least reasonable estimates of reference crop evaporation. Presumably this is because it contains an explicit link to solar radiation through S_o; some measure of the extent to which this radiation reaches the surface and warms the air near the ground, through the factor $\bar{\delta}_T$; while the temperature variation in $(T + 17.8)$ approximates the value of $\Delta/(\Delta + \gamma)$.

The Blaney-Criddle method[11,30] is well known and still in common use, and it is included here for this reason alone. In its most modern complex form[4,30,35] it contains much empiricism, and it is now hard to consider it merely a temperature-based method. The currently preferred form of the equation is

$$E = a_{BC} + b_{BC} f \tag{4.2.45}$$

with

$$f = p\,(0.46T + 8.13) \tag{4.2.45a}$$

$$a_{BC} = 0.0043\ \text{RH}_{\min} - (n/N) - 1.41 \tag{4.2.45b}$$

$$b_{BC} = 0.82 - 0.0041\,(\text{RH}_{\min}) + 1.07\,(n/N) + 0.066\,(U_d) - 0.006\,(\text{RH}_{\min})\,(n/N) - 0.0006\,(\text{RH}_{\min})\,(U_d) \tag{4.2.45c}$$

where p is the ratio of actual daily daytime hours to annual mean daily daytime hours expressed as a percent, T is mean air temperature in °C, (n/N) is the ratio of actual to possible sunshine hours, $\text{RH}_{\min}$ is the minimum daily relative humidity in percentage, and U_d is the daytime wind at 2 m height in m s^{-1}. The complexity of this equation is a tribute to the loyalty of its proponents, but precludes its ready interpretation in terms of a physically realistic equivalent.

4.2.7 Atmospheric Feedbacks

Changes in surface energy exchange alter the air as the atmosphere passes over variable land cover. Initially this change is immediately adjacent to the surface; then it moves progressively upward through the near-surface turbulent layer, finally permeating the whole planetary boundary layer and then the atmosphere above. In so doing, those properties of the air which control surface evaporation rate can be altered, and a feedback may occur to moderate the influence of changes in surface cover. This modification of the atmosphere happens at all horizontal scales, from the very small scale of the leaf or an evaporation pan, through the intermediate scale of the lake or irrigated field, and then at the regional and even continental scale.

Evaporation from Small Areas (Oasis Effect). Empirical and theoretical studies have been made of the evaporation from uniformly moist surfaces of limited extent, such as evaporation pans or lakes (e.g., Refs. 18, 19, 20, and 46) or moist grass[90] surrounded by dry ground. Although the effect of a change in surface cover propagates upward into the atmosphere fairly slowly, studies with numerical models of atmospheric turbulence suggest that the adjustment of the surface evaporation rate into a moving airstream as the air passes on to a different type of evaporative surface occurs quickly, within (say) 5 to 10 m of the boundary between the two surfaces (see Fig. 4.2.6).

In the case of water bodies the area of the water surface changes the effective size of the aerodynamic resistance (see Sec. 4.2.4) between the surface of the water, where the vapor pressure is e_{sf}, and the standardized height of 2 m at which the average vapor pressure and wind speed are e and U_2, respectively. The evaporation rates (in mm day^{-1}) for a water surface of area A (in m^2) are adequately described in Ref. 16 for pans, with 0.5 m $< A^{0.5} <$ 5 m, by

$$E = 3.623\, A^{-0.066}\, (e_{sf} - e)\, U_2 \qquad (4.2.46)$$

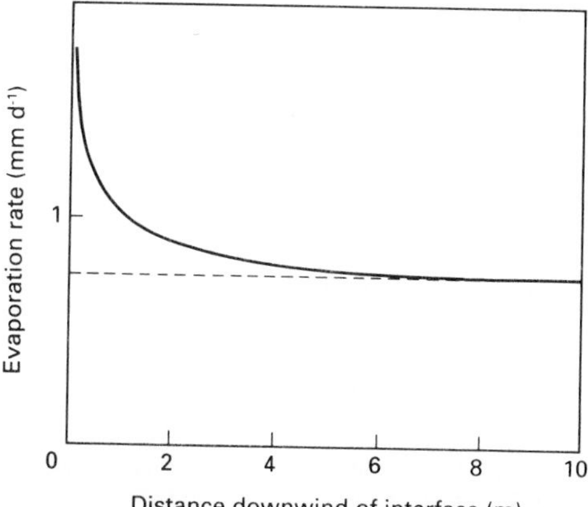

FIGURE 4.2.6 Predicted change in *surface* evaporation rate downwind of an interface between a dry, smooth surface and a well-watered, grassy area (reference crop) on a sunny day. *(Redrawn from Rao et al.[90] Used with permission.)*

and for lakes, with 50 m $< A^{0.5} <$ 100 km, by

$$E = 2.909\, A^{-0.05}\, (e_{sf} - e)\, U_2 \tag{4.2.47}$$

in which e and e_{sf} are in kPa, and e_{sf} can be assumed to be the saturated vapor pressure corresponding to the surface temperature of the water [see Eq. (4.2.2)].

Regional Scale Evaporation. There has long been an intuitive belief among hydrologists that atmospheric feedback mechanisms intervene at the regional scale to moderate the effect of surface controls on evaporation. It is the basis of the hypothetical potential evaporation rates defined in Sec. 4.1.1.

Faith in the concept of potential evaporation rates grew so strong that Bouchet[12] and Morton[73-76] postulated that the equilibrium around the hypothetical, potential evaporation λE_{po} which *would* have been applicable had water been freely available might provide a means of estimating actual evaporation. They postulate that the *excess* potential demand above this "preferred" potential rate as calculated from estimates of the potential evaporation λE_o using near-surface weather variables is directly related to the "short fall" in the actual evaporation λE due to restricted water availability, that is,

$$\lambda E_{po} - \lambda E = \lambda E_o - \lambda E_{po} \tag{4.2.48}$$

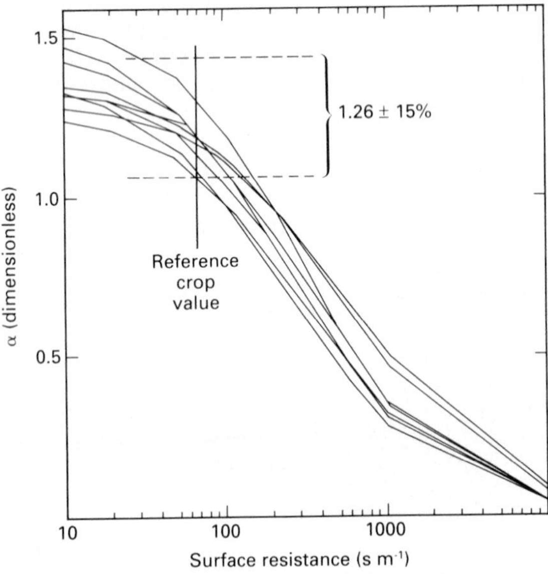

FIGURE 4.2.7 The effective daytime average value of the Priestley-Taylor parameter α in Eq. (4.2.37) as synthesized from a simple one-dimensional planetary boundary layer model. *(From McNaughton and Spriggs.[68] Used with permission.)* The several different lines correspond to model initializations with different radiosonde ascents drawn from data collected in the Netherlands.[31] Also shown is the value of surface resistance for a reference crop, and the preferred value of $\alpha = 1.26$ for humid climates [see Eq. (4.2.39)] with the range ± 15 percent around this value.

Brutsaert and Stricker[17] had some success with this concept by substituting E_p calculated from Eq. (4.2.30) for λE_{po}, and λE_{rc} from Eq. (4.2.39) for λE_o, but recent modeling studies[68] have raised doubt about the hypothesis.

Over extensive uniform surfaces, or where surface changes occur at a small scale and in a haphazard way so that mixing of the air makes them appear uniform to the atmosphere, regional scale atmospheric feedback processes can be adequately represented by one-dimensional models. Such models[13,67,68] have provided better though still incomplete understanding, with broad support for a regional evaporation rate, though they highlight that the concept of a potential rate is only approximate and that it has significant dependency on the control exerted by vegetation and soil.

Currently, such modeling is limited to simulating the development of a one-dimensional atmospheric boundary in daylight conditions, with field data from radiosonde ascents[31] used to initiate the models and define the meteorological variables aloft. There is therefore no simulation of clouds or precipitation in these models. Nonetheless such models suggest that both the Penman equation [Eq. (4.2.30)] and the Priestley-Taylor equation [Eq. (4.2.39)] can provide reasonable simulation of *daytime* surface evaporation, and that the value $\alpha = 1.26$ in this last equation is acceptable to an accuracy of 15 percent for a range of area-average surface resistance typical of pasture and agricultural crops, including the reference crop ($r_s = 69$ s m^{-1}; see Sec. 4.1.1). Figure 4.2.7 shows the simulated variation in the preferred daytime average Priestley-Taylor parameter α as a function of the area-average surface resistance. The Bouchet[12] "complementary evaporation" hypothesis is not, however, supported[68] by these studies.

4.3 MEASUREMENT OF EVAPORATION

Natural evaporation can be measured either as the rate of loss of liquid water from the surface or as the rate of gain of water vapor by the atmosphere. Measurements in the liquid phase either assume or create a closed system, such as an evaporation pan or lysimeter, and deduce evaporation as the net loss of water from that closed system over a given time, the measurement being one of *discrete changes* in total quantity of water in the system. Measurements in the vapor phase most commonly assume that the atmosphere is an open system and determine evaporation as an integration of the *rate of flow* of water vapor (or equivalently latent heat) into that open system through the turbulent boundary layer near the land surface. Measurements of the net change in the vapor content of the air over large areas using balloons can provide estimates of regional evaporation rate,[66,79,92,114] but this approach does not yet form part of hydrologic practice. Most of the techniques described below provide measurements of the local evaporation rate, with the exception of catchment water balances which estimate area-average evaporation for confined watersheds.

4.3.1 Measurement of Liquid Water Loss

Measurements of this type draw up a mass or volume balance for the water in a specified volume of soil or in a body of liquid water. The surface area of this sample is a necessary part of the measurement, while its depth can either be well defined, as in lysimetric measurements, or poorly defined but large enough for vertical drainage to be either neglected or computed, as in watershed experiments. This difference influences the time scale over which the results are applicable. In each case the measurement reduces to determining the terms in a basic water balance equation which is applied over a particular time interval.

$$E = P - (V_R + V_S + V_L)/A \quad \text{mm} \tag{4.3.1}$$

where E = net evapotranspiration loss from the specified volume per unit area, mm
P = net precipitation (or irrigation) input to the specified volume per unit area, mm
V_R = net volume of liquid water entering or leaving the specified volume as measured inflow or outflow both above and below the surface, liters
V_S = change in liquid water stored within the specified volume, liters
V_L = "leakage," i.e., that total volume of liquid water leaving the specified volume which is not, or cannot be, measured, and which therefore represents an error in the method, liters
A = effective area of the sample volume at the land surface, m^2

All water budget measurements share the problem that the error in the evaporation calculated from Eq. (4.3.1) is an accumulation of the errors in the other measured variables.

Evaporation Pans. Because of its apparent simplicity, the evaporation pan is probably the instrument used most widely to estimate *potential* evaporation. However, Gangopadhyaya et al.[37] list 27 different designs of evaporation pans and suggest that their list "is undoubtedly far from complete."

Sunken pans (e.g., Colorado, USSR-GGI, USDA-BPI) are sometimes preferred in crop water requirement studies, since these pans have a water level at soil height and give a better direct prediction of reference crop evaporation than other pan designs.[88] The Colorado pan is 0.92 m square and 0.46 m deep. It is made of galvanized iron, set in the ground with the rim 0.05 m above ground level. The water level inside the pan is maintained at or slightly below ground level.

The U.S. Weather Bureau Class A pan is shown in Fig. 4.3.1. It is circular, 1.21 m in diameter and 0.255 m deep, and is made of galvanized iron (22 gauge) or monel metal (0.8 mm). The pan is mounted on a wooden open-frame platform with its bottom 0.15 m above ground level. The soil is built up to within 0.05 m of the bottom of the pan. The pan must be level. It is filled with water to 0.05 m below the rim, and the water level should not drop to more than 0.075 m below the rim. In semiarid countries it is quite common to cover the exposed water surface with mesh to stop animals from drinking the water. This lowers the evaporation measurement by 10 to 20 percent,[28] depending on the dimensions of the mesh.

The evaporation from a pan can differ significantly from that from an adjacent lake or surrounding vegetation. It is necessary to accommodate these sometimes large differences using empirical *pan coefficients.* Evaporation from pans is generally greater than from adjacent large areas of water or well-watered vegetation, but pan coefficients vary significantly with siting and pan design as well as with climatic factors.

Since pan data are widely available and much used for estimating crop water use for irrigation purposes, empirical pan coefficients have been derived, particularly for the U.S. Class A pan. Table 4.3.1 gives suggested values[30] for k^A_{pan} in the equation

$$E_{rc} = k^A_{\text{pan}} E^A_{\text{pan}} \qquad \text{mm day}^{-1} \tag{4.3.2}$$

for a range of conditions, and for pans sited in cropped fields (case A) and noncropped, dry-surface fields (case B). In Eq. (4.3.2), E_{rc} is reference crop evaporation and E^A_{pan} is the measured Class A pan evaporation.

FIGURE 4.3.1 U.S. Weather Bureau Class A evaporation pans with screen in the foreground, and without screen in the background.

The wide range of coefficients given in Table 4.3.1 tends to overemphasize the shortcomings of pan data,[56] since most areas only occasionally experience the dry, strong wind conditions which give small k^A_{pan} values. Mean monthly E estimates based on pan evaporation should be within ± 10 percent in most climates.

Water Balances of Watersheds and Lakes. River runoff is arguably the most accurate hydrologic measurement and is a valuable, direct determination of the available surface water resource. Careful gauging can provide stream-flow measurements accurate to about 2 percent. But considerable difficulties are involved in using the measured runoff from closed catchments to provide a worthwhile, indirect measurement of evaporation.[26,59,62,83,84,96,130,131] This is exacerbated in the case of lakes by the additional need to gauge inflow as well as outflow, and by enhanced difficulties in estimating subsurface seepage. Notwithstanding the above, carefully selected and well-managed paired watersheds can provide valuable and convincing evidence of the consequences of land-use change on evaporation (e.g., Ref. 59); see Chap. 13.

Accurate estimates of area average precipitation [P in Eq. (4.3.1)] are problematic,[132] not just because of the real likelihood of systematic gauging errors (e.g., Ref. 97) but also because of spatial variability associated with topographic and surface features for watersheds, and the need to estimate P from lakeside gauges in the case of precipitation over water bodies. Errors in estimating P, typically 5 to 10 percent, can increase markedly if a proportion of the precipitation falls as snow.

A systematic uncertainty in the evaporation loss deduced from a catchment water balance arises from the possibility that the unmeasured leakage V_L forms a significant part of the total water balance. Considerable skill is required in selecting natural catchments without leakage, and it is important to recognize that the subterranean groundwater contours play an important, perhaps definitive, role in specifying catchment boundaries and that surface topography is not necessarily a reliable reflection of subsurface flow direction.

TABLE 4.3.1 Suggested Values for the Pan Coefficient k^A_{pan}, Which Relates Reference Crop Evaporation E_{rc} to Measured Class A Pan Evaporation E^A_{pan}

		Case A: Pan surrounded by short green crop				Case B: Pan surrounded by dry, bare area		
		Mean relative humidity, %				Mean relative humidity, %		
Wind	Upwind fetch of green crop, m	Low <40	Med 40–70	High >70	Upwind fetch of dry fallow, m	Low <40	Med 40–70	High >70
Light (<1 m/s)	0	0.55	0.65	0.75	0	0.7	0.8	0.85
	10	0.65	0.75	0.85	10	0.6	0.7	0.8
	100	0.7	0.8	0.85	100	0.55	0.65	0.75
	1000	0.75	0.85	0.85	1000	0.5	0.6	0.7
Moderate (2–5 m/s)	0	0.5	0.6	0.65	0	0.65	0.75	0.8
	10	0.6	0.7	0.75	10	0.55	0.65	0.7
	100	0.65	0.75	0.8	100	0.5	0.6	0.65
	1000	0.7	0.8	0.8	1000	0.45	0.55	0.6
Strong (5–8 m/s)	0	0.45	0.5	0.6	0	0.6	0.65	0.7
	10	0.55	0.6	0.65	10	0.5	0.55	0.65
	100	0.6	0.65	0.7	100	0.45	0.5	0.6
	1000	0.65	0.7	0.75	1000	0.4	0.45	0.55
Very strong (>8 m/s)	0	0.4	0.45	0.5	0	0.5	0.6	0.65
	10	0.45	0.55	0.6	10	0.45	0.5	0.55
	100	0.5	0.6	0.65	100	0.4	0.45	0.5
	1000	0.55	0.6	0.65	1000	0.35	0.4	0.45

Source: After Doorenbos and Pruitt (Ref. 30). Used with permission.

Notes: Mean relative humidity is the average maximum and minimum daily relative humidities.

Case A: For pans surrounded by cropped fields or wet soils, with very dry soil beyond the prescribed fetch.

Case B: For pans surrounded by very dry soil, with cropped fields or wet soil beyond the prescribed fetch.

The change in storage term V_S in Eq. (4.3.1) is difficult to measure reliably in extensive natural catchments and will usually provide the most important error in a weekly or monthly evaporation measurement. Its significance becomes less for an annual determination, when the error from this component can become comparable with those in precipitation and runoff.

In the light of the very real possibility of significant error in the bulk evaporation loss deduced from a watershed, it is advisable to supplement any such derivations of evaporation with parallel and independent meteorological or lysimetric measurements.

Soil Moisture Depletion. Given sufficient measurements of soil moisture content to account for spatial variability in water storage, and provided that drainage is insignificant or can be quantified, measurements of change in V_S in Eq. (4.3.1) allow evaporation to be calculated when precipitation is either absent or separately determined. The technique requires the repeated, in situ measurement of soil water content, which is now possible with neutron probes, capacitance sounders, or time-domain reflectometers.[7–9,29,43,123,124]

When applying this method and these instruments, considerable care is required to avoid disturbing the plant canopy or altering the density, aeration, and infiltration characteristics of the soil surface.[87] Drainage losses from the soil sample can be significant.[58,77,95,117] When subterranean water movement occurs via unsaturated flow, frequent simultaneous measurement of soil water tension profiles[8] can significantly enhance the method by helping to distinguish between the relative proportions of net water loss due to evaporation and drainage. Figure 4.3.2 illustrates the determination of a plane of zero potential gradient, the "zero flux" plane, and its use in better defining the proportion of stored water lost by evaporation during a simple soil "dry-down."

Lysimeters. A lysimeter is a device in which a volume of soil, typically 0.5 to 2.0 m in diameter, which may be planted with vegetation, is isolated hydrologically so that leakage $V_L = 0$ in Eq. (4.3.1). It either permits measurement of drainage V_R or makes it zero and, in the case of a weighing lysimeter, the change in water storage V_S is determined by weight difference. Though difficult and expensive to install, lysimeters have extensive and long-established use, primarily in research applications to test alternative measurement techniques or in the calibration of empirical equations to estimate evaporation. Excellent texts on the technical details of lysimeter design and on the value and shortcomings of lysimeter use exist in the literature (e.g., Refs. 1, 117, and particularly 89), to which the reader is referred for more comprehensive description.

Figure 4.3.3 shows an example of a well-designed weighing lysimeter in which vegetation is growing. If evaporation from the lysimeter is to be representative of the surrounding area, it should contain an undisturbed sample of the soil and vegetation.

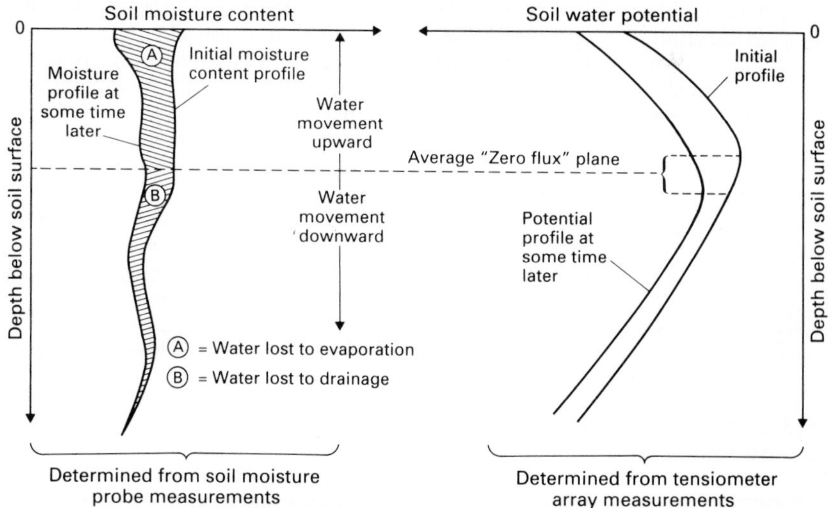

FIGURE 4.3.2 Illustration of the measurement of evaporation using soil moisture depletion supplemented with the determination of an average "zero flux" plane to discriminate between (upward) evaporation and (downward) drainage.

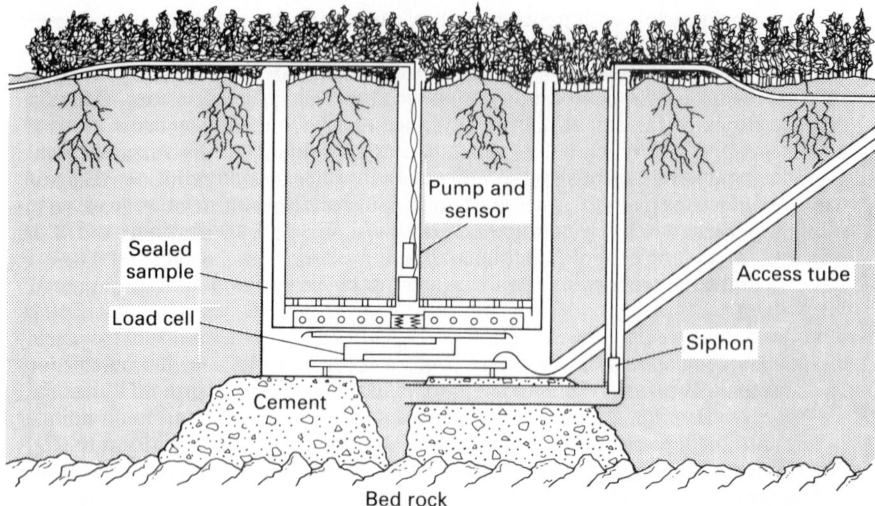

FIGURE 4.3.3 An example of a well-designed weighing lysimeter *(redrawn from Wright.*[133] *Used with permission.)* employing an undisturbed representative sample about 1 m in diameter, with the water status of the soil maintained similar to that of the surrounding area by pumping drainage water.

This means that steps must also be taken to ensure that the thermal, hydrologic, and mechanical properties of the soil are similar to those of surrounding soil, and to establish that the vegetation sample is representative of surrounding vegetation in terms of height, density, and physiological well-being.

Plant Physiological Techniques. Transpiration is the primary component (perhaps 95 percent) of evaporation from dense canopies where little direct solar radiation reaches the ground, or conversely, for sparse vegetation in arid or semiarid climates when the soil surface is very dry. In these particular cases, direct measurement of transpiration gives a worthwhile and cost-effective estimate of total evaporation.

Successful field measurements have been made of transpiration by cutting the stems of the vegetation under water, immersing the cut end in a tank of water, and noting the water uptake,[94] though care is necessary to check that the physiological status of the plant remains representative of the surrounding crop. Measurements of the rate of water flow which do not require cutting the stem are preferable, so methods of observing water flow through plant stems have been developed. These usually involve introducing "tracers" into the stem of the vegetation, sometimes as a pulse of heat to determine sap velocity (e.g., Ref. 25) or as deuterium in the "dilution gauging" method.[22] Success has also been reported[6,100] in deducing the rate of sap flow through its impact on thermal conduction in the stems of herbaceous plants.

4.3.2 Micrometeorological Measurements of Water Vapor Flow into the Atmosphere

Micrometeorological methods of measurement determine evaporation as the flux of water vapor through the air from the evaporating water surface, vegetation, or soil. The measurements are made in the atmosphere, within the turbulent air close to the ground, so that the measured vapor flow rate is a very good approximation to the surface evaporation rate.

There are two broad classes of micrometeorological evaporation measurement: those based on measurements of gradients and those based on measurements of fluctuations. Both rely on the fact that turbulent exchange is the dominant exchange mechanism within the near-surface atmosphere. Since micrometeorological measurements are necessarily made some distance above the ground, and the atmosphere is almost always moving horizontally, the measurements obtained at a particular location are representative of an area some distance upwind. This can be an advantage in that the upwind turbulent mixing helps to produce a value representing the average evaporation over a fairly large area. However, if the measured evaporation is meant to be representative of the particular uniform crop surrounding the instruments, it is necessary that there should be an extensive "fetch" of evaporating surface with essentially identical properties extending upwind from the measurement site for a considerable distance. To test this, the proportion F of the measured evaporation originating from the crop within a specified upwind fetch X_F can be estimated,[41,101] albeit only approximately. In (unstable) daytime conditions

$$F \approx \exp\left(\frac{-6\,(z-d)\,\{\ln[(z-d)/z_o]-1+z_o/(z-d)\}}{X_F[1-z_o/(z-d)]}\right) \tag{4.3.3}$$

here z is the measurement height and $d \simeq 0.67\,h_c$, $z_o \simeq 0.123\,h_c$, and h_c is the average height of the crop. All these heights are in meters.

Methods Based on Measurements of Mean Gradients. The assumption behind these techniques is that above an extensive homogeneous surface the transfer of vapor, momentum, and sensible heat can be described by similar vertical, one-dimensional diffusion equations. It is further assumed that the turbulent diffusion coefficients relating the fluxes of water vapor, sensible heat, and momentum to the respective vertical gradients of humidity, temperature, and wind speed are related to each other in a way which is not determined by characteristics of the surface but rather by characteristics of the turbulent boundary layer itself, and that this relationship is universal, i.e., not dependent on location or crop.

Describing the vertical transport in this way also assumes that evaporation is occurring at a steady rate. In fact this is rarely the case, but the rate of change of evaporation with respect to time is sufficiently slow that average gradients over 20 to 60 min can be successfully used in practice.

Aerodynamic Methods. The basic principle of the aerodynamic method is to assume that the *aerodynamic resistance* (see Sec. 4.2.4), which relates the flow of water vapor to the difference in vapor pressure at heights z_1 and z_2 above the ground, is related in a universally defined and known way to the equivalent aerodynamic resistance to momentum flow between the same two levels. Further, it is assumed that the equation describing the wind-speed profile, and hence momentum flow, above the ground is well known, so that the required difference in aerodynamic resistance between levels z_1 and z_2 can be calculated, as with Eq. (4.2.25).

In practice, the aerodynamic method is rarely applied in this precise way since, were a measurement of the difference in vapor pressure between two heights to be made, it would be preferable and simpler to make an equivalent differential measurement of air temperature over the same height interval, rather than a differential measurement of wind speed. It would then be possible to apply the more accurate energy-balance method described in the next section.

However, the aerodynamic method is used to provide measurement of sensible heat transport, and evaporation is then deduced indirectly from the energy balance using Eq. (4.2.15). This technique is less reliable than other micrometeorological methods, but it has the advantage that the required sensors, two for temperature, two for wind speed, and one for net radiation, are easily available and reasonably cheap and reliable.

Given the required measurements of wind speed u_1 and u_2 and of temperature T_1 and T_2, at the heights z_1 and z_2 *measured relative to the height of the zero plane displacement of the crop,* about 67 percent of crop height; and given the additional measurement of net radiation R_n and perhaps soil heat flux G, it has been shown[54,55] that it is possible to calculate sensible heat adequately using a few simple functions of the differences in temperature, wind speed, and height between z_1 and z_2.

Energy Balance Method. The turbulent diffusion processes responsible for the transport of water vapor and sensible heat through the atmosphere are very similar, but they both differ from those responsible for the transport of momentum. It is therefore a plausible assumption that the aerodynamic resistance which restricts the flow of water vapor and relates that flow rate to the difference in vapor pressure (Δe), between two particular heights, is numerically equal to the resistance which relates the flow of sensible heat to the temperature difference (ΔT) between the same two levels. The *Bowen ratio* β, which is the ratio of the sensible heat H to the latent heat λE, is therefore directly related to the ratio of the differences in temperature and humidity measured between any two heights; thus

$$\beta = \frac{H}{\lambda E} = \gamma \frac{\Delta T}{\Delta e} \tag{4.3.4}$$

the constant γ [see Eq. (4.2.28)] being necessary to account for units.

Using Eq. (4.3.4) it is therefore possible to calculate the ratio of H and λE from differential measurements of temperature and humidity. But, at the same time, by measuring net radiation R_n and soil heat flux G and expressing them as water equivalents, it is possible to also know the sum of these same two energy fluxes from Eq. (4.2.15) by neglecting or estimating the minor terms in Eq. (4.2.16). Solving these two simultaneous equations for the two unknown energy fluxes, the evaporation rate can be calculated from the equation

$$E = \frac{R_n - G}{1 + \gamma(\Delta T / \Delta e)} \qquad \text{mm day}^{-1} \tag{4.3.5}$$

which is sometimes called the *Bowen ratio* method.

Because of the inherent similarity between the diffusion of vapor and sensible heat, the Bowen ratio–energy balance technique is more robust with respect to changes in roughness and topography than is the aerodynamic technique since the profiles of temperature and vapor pressure are affected equally. Moreover, modern systems have been developed[10,36,69] which mechanically and regularly interchange the two temperature and the two humidity sensors between measurement levels, thereby allowing any systematic offset in their calibration to cancel out when the temperature and humidity differences are taken.

The primary practical problem with the Bowen ratio–energy budget technique is that the sign of H often changes in the evening and morning. (Sensible heat flux is usually upward during the day and downward at night.) H, and therefore ΔT, is then zero at the changeover times, so that Eq. (4.3.5) is not defined and applicable at these times. Further, the combination of low (usually negative) values of $(R_n - G)$ at night, and the fact that the two energy fluxes are commonly then in opposite directions, throws suspicion on the usefulness of nighttime measurements of evaporation by the Bowen ratio method.

Eddy Correlation Measurements. Near the surface the mean wind is parallel to the ground, so that the mean vertical wind is zero. However, turbulent eddies within the body of the moving air cause fluctuating vertical movements, both toward and away from the land surface. On average, of course, these fluctuations produce no net vertical movement of the air, but momentarily the air has a vertical velocity w'. If the average specific humidity of the air (in kg of water/kg of air) is $\bar{q}$, then there are similar turbulent fluctuations q' from this mean $\bar{q}$, q' positive meaning greater than average specific humidity and the converse for q' negative.

If w' and q' are simultaneously positive, moister than normal air will be carried away from the ground, and if w' and q' are simultaneously negative, then drier than normal air will move toward the ground. These conditions of positive correlations between w' and q' occur during evaporation; when w' and q' are negatively correlated condensation occurs. The technique for measuring w' and q' and for measuring evaporation from these is called *eddy correlation,* and the evaporation rate E is found as

$$E = 86{,}400\, \overline{\rho_a\, w'\, q'} \qquad \text{mm day}^{-1} \tag{4.3.6}$$

where the overbar denotes a time average. The analogous equation for sensible heat takes the form

$$H = \overline{\rho_a c_p\, w'\, T'} \qquad \text{W m}^2\,\text{s}^{-1} \tag{4.3.7}$$

where T' is the fluctuation in air temperature.

Many practical problems are involved in using the eddy correlation technique. The fluctuations in wind speed, humidity, and temperature can occur over a broad range of frequencies, for instance, some lasting several minutes but others only fractions of a second, and the sensors used to measure w' and q' must have a rapid response. At the same time these sensors have to be co-located, so that they measure the same moving air, yet they should ideally have limited size and be carefully positioned so as not to interfere with the air movements they measure.

It is important to correctly identify the fluctuating components q' and w' and to compute the integrated flux, preferably in real time. Advances in sensor technology and particularly in microprocessors in recent years have allowed associated advances in instrumentation to provide direct evaporation measurement, and integrated measurement systems have evolved (e.g., Ref. 110). Notwithstanding the technical difficulties involved in applying it properly, the eddy correlation technique is the preferred micrometeorological technique on the grounds that it is a direct measurement with minimum theoretical assumptions. Good present-day eddy correlation systems can provide routine evaporation measurements with accuracy in the order of 5 to 10 percent.

Indirect Measurements from Turbulence Statistics. There has been some investigation into measuring evaporation indirectly through its relationships to certain statistical properties of atmospheric turbulence.[16] One such technique is gaining acceptance as viable and simple to apply in semiarid environments, where evaporation is low because of dry soils and difficult to measure otherwise. It can be shown[64,122,129,137] that during the day the standard deviation of air temperature σ_T measured at 5 to

10 m above the ground can be used to calculate the sensible heat flux from

$$H = \rho_a c_p \left[\left(\frac{\sigma_T}{0.92} \right)^3 \frac{0.41\, g\, z}{T + 273.2} \right]^{1/2} \qquad \text{W m}^{-2}\,\text{s}^{-1} \tag{4.3.8}$$

where σ_T is the standard deviation of air temperature (in °C) at height z (in m), g is the acceleration due to gravity, and ρ_a is the density of air (in kg m^{-3}) and c_p the specific heat of air (1.01 kJ kg^{-1} K^{-1}). Evaporation can then be deduced from the energy balance using Eqs. (4.2.15) and (4.2.16), given measurements of R_n and G, and neglecting S, P, and A_d.

4.4 METHODS FOR ESTIMATING EVAPORATION

4.4.1 Introductory Comments

There is some distinction between the way evaporation is estimated in hydrologic practice and in hydrologic research. Hydrologists interested in providing realistic models of well-instrumented research catchments, or concerned with the true-to-life description of evaporation in hydrometeorological research, create complex descriptions of the movement of energy and water in the soil-vegetation-atmosphere interface,[44,63,113,127,128] and sometimes also of movement in the lower levels of the atmosphere (e.g., Refs. 13, 67, and 68). Using such complex models generally requires a high level of input data in the form of frequently sampled meteorological variables, and the specification of an extensive set of soil and vegetation parameters for which general values are not yet readily available in the literature. Mainly because of this heavy demand for input data, but partly for historical reasons, such research models are not yet "tools of the trade" for hydrologic practitioners.

Since research models of evaporation do not have the extensive calibration of parameters (or indeed the widespread acceptance) required to recommend them for use by practitioners in a *Handbook of Hydrology,* the understanding generated by modern hydrometeorological research is best applied in a more conservative way. The approach adopted here is to select preferred methods from among the broad range of empirical techniques which practicing hydrologists have proposed over the years and, where relevant and practical, to recommend refinement of such empirical practice.

Accordingly, the conventional approach of estimating evaporation as a two-stage process is retained here. These two stages comprise first estimating a "standard evaporation rate" (see Sec. 4.1.1) and then introducing allowance for different crops and/or soil moisture status as multiplicative factors. Two standard evaporation rates are supported, one for open water surfaces, namely, *potential evaporation* E_o, and one for a well-specified crop, namely, *reference crop evaporation* E_{rc}.

There is one exception to the use of the "two-stage" procedure for estimating evaporation, this being for the case of tall forest vegetation. Although still an active research issue, extensive investigation over the last two decades has demonstrated that, in this case, the difference between evaporation rates when the vegetation is wet and when it is dry is very marked, and so special attention is called for. Forests are therefore treated separately.

4.4.2 Estimating the Energy Available for Evaporation

The energy available at the land surface is normally the primary control on evaporation from open water and a well-watered reference crop. Providing a preliminary estimate (or measurement) of this available energy is necessary for use of the preferred techniques to estimate evaporation described in later sections. How such an estimate of available energy is best provided depends on the type of observational data available.

Table 4.4.1 describes the recommended selection and computation sequence to calculate available energy depending on the data available. This table draws on equations and tables already given but also requires formulas (and associated tables) to provide estimates of the maximum possible sunshine hours N, and of the extraterrestrial solar radiation S_o.

The maximum possible daylight hours[32] can be calculated from

$$N = \frac{24}{\pi} \omega_s \tag{4.4.1}$$

where ω_s is the sunset hour angle (in radians) given by

$$\omega_s = \arccos(-\tan\phi \tan\delta) \tag{4.4.2}$$

where ϕ is the latitude of the site (positive for the Northern hemisphere, negative for the Southern hemisphere) and δ is the solar declination (in radians) given by

$$\delta = 0.4093 \sin\left(\frac{2\pi}{365} J - 1.405\right) \tag{4.4.3}$$

where J is the Julian day number.

The extraterrestrial solar radiation S_o can be calculated[32] from

$$S_o = 15.392\, d_r(\omega_s \sin\phi \sin\delta + \cos\phi \cos\delta \sin\omega_s) \quad \text{mm day}^{-1} \tag{4.4.4}$$

where d_r is the relative distance between the earth and the sun given by

$$d_r = 1 + 0.033 \cos\left(\frac{2\pi}{365} J\right) \tag{4.4.5}$$

Equations (4.4.1) and (4.4.4) provide estimates of the maximum possible daylight hours and extraterrestrial solar radiation which are good to about 0.1 h and 0.1 mm day^{-1}, respectively, for latitudes between 55°S and 55°N. As a check on computer code to evaluate N and S_o, their April 15 values ($J = 105$) at latitudes 30°N, 0, and 30°S are 12.7, 12.0, and 11.3 h, and 15.0, 15.1, and 11.2 mm day^{-1}, respectively.

4.4.3 Computing Other Inputs for Evaporation Estimates

Vapor-Pressure Deficit. Preferred techniques for estimating evaporation require a value of the vapor-pressure deficit $\overline{D}$, the difference between saturated vapor pressure e_s and ambient vapor pressure e, averaged over the period for which the estimate is made. On rare occasions the true average vapor-pressure deficit is available—based on hourly values from an automatic weather station perhaps—but more usually it has to be estimated from climatological records.

TABLE 4.4.1 Selection and Computation Sequence for Estimating the Energy Available for Evaporation

---

1 Are local measurements of net radiation (R_n in MJ m^{-2} day^{-1}) available?
- YES: (i) Divide by λ [Eq. (4.2.1)] to give R_n in mm day^{-1}.
  (ii) Go to 5(a).
- NO: Continue with 2(a).

2(a) Are local records of fractional cloud cover (n/N) available?
- YES: Go to 3(a).
- NO: Continue with 2(b).

2(b) Local records of sunshine hours (n) available?
- YES: (i) Compute (n/N); N from Eq. (4.4.1).
  (ii) Go to 3(a).
- NO: Continue with 2(c).

2(c) Can n or n/N be estimated from regional records?
- YES: (i) Proceed as 2(a) or 2(b).
  (ii) Recognize increased uncertainty in (n/N).
  (iii) Go to 3(a).
- NO: Select pan- or temperature-based evaporation estimate.

3(a) Local measurements of solar radiation (S_t in MJ m^{-2} day^{-1}) available?
- YES: (i) Divide by λ [Eq. (4.2.1)] to give water equivalent.
  (ii) Go to 3(c).
- NO: Continue with 3(b).

3(b) Locally calibrated angstrom coefficients (a_s, b_s) available?
- YES: Select a_s and b_s from available values.
- NO: Set $a_s = 0.25$; $b_s = 0.50$.
- THEN: (i) Obtain value of extraterrestrial radiation (S_o) from Eq. (4.4.4).
  (ii) Compute S_t from Eq. (4.2.6).
  (iii) Continue with 3(c).

3(c) Local measurements of land cover albedo (α) available?
- YES: Select value of α from available measurements.
- NO: Estimate α using Table 4.2.2.
- THEN: (i) Compute net solar radiation from Eq. (4.2.5).
  (ii) Continue with 4(a).

4(a) Locally calibrated emissivity coefficients [a_e, b_e; Eq. (4.2.8)] available?
- YES: Select a_e and b_e from available values.
- NO: Set $a_e = 0.34$; $b_e = -0.14$.
- THEN: Continue with 4(b).

4(b) Measurements of dew point temperature available?
- YES: (i) Obtain vapor pressure e_d at dew point temperature from Eq. (4.2.2) or Table 4.2.1.
  (ii) Compute ϵ' from Eq. (4.2.8).
  (iii) Go to 4(d).
- NO: Continue with 4(c).

4(c) Measurements of minimum air temperature available?
- YES: (i) Set dew point to minimum temperature; obtain e_d from Eq. (4.2.2) or Table 4.2.1.
  (ii) Compute ϵ' from Eq. (4.2.8).
- NO: Compute ϵ' from Eq. (4.2.9).
- THEN: Continue with 4(d).

4(d) Locally calibrated cloudiness coefficients [a_c, b_c; Eq. (4.2.10)] available?
- YES: Select a_c and b_c from available values.
- NO: Set $a_c = 1.35$; $b_c = -0.35$ in arid areas
  or $a_c = 1.00$; $b_c = 0.00$ in humid areas.

TABLE 4.4.1 Selection and Computation Sequence for Estimating the Energy Available for Evaporation *(Continued)*

	THEN:	(i)	Compute clear sky solar radiation S_{to} as the value given by Eq. (4.2.6) with (n/N) set to zero.
		(ii)	Compute the cloudiness factor (f) from Eq. (4.2.10).
		(iii)	Compute net long-wave radiation from Eq. (4.2.7).
		(iv)	Compute net radiation from Eq. (4.2.13).
		(v)	Continue with 5(a).
5(a)	Estimate of available energy for open water surface required?		
	YES:		Go to 5(b).
	NO:		Go to 5(c).
5(b)	Data to estimate advected energy [A_h; Eq. (4.2.20)] available?		
	YES:	(i)	Compute A_h from Eq. (4.2.20).
		(ii)	Energy available for evaporation $A = R_n + A_h$.
	NO:		Energy available for evaporation $A = R_n$.
	THEN:		Energy estimation complete.
5(c)	Measurements or data to estimate soil heat flux G available?		
	YES:	(i)	Obtain G from measurements, or estimate from Eq. (4.2.18) or (4.2.19).
		(ii)	Energy available for evaporation $A = R_n - G$.
	NO:		Energy available for evaporation $A = R_n$.
	THEN:		Energy estimation complete.

Since saturated vapor pressure is not a linear function of temperature [see Eq. (4.2.2)] the particular procedure used to estimate vapor-pressure deficit from such climatological records can affect the estimated value (by as much as 30 percent), particularly in arid environments where the daily temperature cycle is often large. In general terms, and if possible, computing vapor-pressure deficit at the measured maximum and minimum temperature separately, and then taking the average of these deficit values, provides a better estimate of the true daily average deficit than averaging temperatures (or relative humidities) first and then computing the deficit.

Climatological humidity data are reported either (1) as relative humidity (RH_{max} and RH_{min}, in percentage); or (2) as daily average dry- and wet-bulb temperatures (T_d and T_w, in °C); or (3) as a daily average dew-point temperature (T_{dew}, in °C). The time of measurement is important but is often not given. Fortunately the actual vapor pressure of the air is fairly constant, and even one measurement per day may suffice, particularly if this measurement is made early in the day.

If a true daily average vapor-pressure deficit $\bar{D}$ is not available and an estimate from climatological data is required, the following averaging procedures are recommended for the three different humidity data availabilities given above.

1. *Relative Humidity.* Data given: T_{max}, T_{min}, RH.

$$\bar{D} = \left[\frac{e_s(T_{max}) + e_s(T_{min})}{2}\right]\frac{(1 - \mathrm{RH})}{100} \quad \mathrm{kPa} \tag{4.4.6}$$

with $e_s(T_{max})$ and $e_s(T_{min})$ from Eq. (4.2.2) or Table 4.2.1. If more than one value of relative humidity is available, RH is the average value.

2. *Wet- and Dry-Bulb Temperature.* Data given: T_{max}, T_{min}, T_{dry}, T_{wet}.

$$\bar{D} = \frac{e_s(T_{max}) + e_s(T_{min})}{2} - e \quad \mathrm{kPa} \tag{4.4.7}$$

with $e_s(T_{max})$ and $e_s(T_{min})$ from Eq. (4.2.2) or Table 4.2.1; and e taken from Table 4.4.2 for *(a)* aspirated and *(b)* unaspirated psychrometers, respectively. If more

TABLE 4.4.2a Vapor Pressure (in kPa) from Dry- and Wet-Bulb Temperature Data (in °C) for an Aspirated Psychrometer

Wet-bulb depression, °C (altitude < 1000 m)												Dry bulb, °C	Wet-bulb depression, °C (altitude > 1000 m)											
0	2	4	6	8	10	12	14	16	18	20	22		0	2	4	6	8	10	12	14	16	18	20	22
7.38	6.49	5.68	4.92	4.22	3.58	2.98	2.43	1.92	1.44	1.01	0.60	40	7.38	6.52	5.71	4.98	4.30	4.18	3.10	2.56	2.07	1.62	1.20	0.81
6.63	5.81	5.05	4.36	3.71	3.11	2.56	2.05	1.58	1.14	0.73		38	6.63	5.82	5.09	4.41	3.79	3.67	2.68	2.18	1.73	1.32	0.92	0.57
5.94	5.19	4.49	3.84	3.25	2.69	2.18	1.71	1.27	0.86	0.49		36	5.94	5.21	4.52	3.90	3.33	3.21	2.30	1.84	1.43	1.04	0.68	0.35
5.32	4.62	3.98	3.38	2.83	2.32	1.84	1.40	1.00	0.62			34	5.32	4.64	4.01	3.44	2.91	2.41	1.96	1.54	1.15	0.80	0.46	0.15
4.75	4.11	3.51	2.96	2.45	1.98	1.54	1.13	0.70	0.40			32	4.75	4.13	3.55	3.02	2.53	2.07	1.66	1.26	0.91	0.58	0.26	
4.24	3.65	3.09	2.58	2.11	1.67	1.26	0.88	0.53				30	4.24	3.67	3.13	2.64	2.19	1.77	1.38	1.02	0.69	0.38	0.09	
3.78	3.23	2.72	2.24	1.80	1.40	1.02	0.67	0.34				28	3.78	3.25	2.75	2.30	1.89	1.49	1.14	0.80	0.49	0.21		
3.36	2.85	2.38	1.94	1.53	1.15	0.80	0.47	0.16				26	3.36	2.87	2.41	2.00	1.61	1.25	0.92	0.60	0.32	0.05		
2.98	2.51	2.07	1.66	1.28	0.93	0.60	0.29					24	2.98	2.53	2.11	1.72	1.39	1.03	0.72	0.43	0.16			
2.64	2.20	1.80	1.42	1.06	0.74	0.43	0.14					22	2.64	2.23	1.83	1.43	1.15	0.83	0.55	0.27	0.02			
2.34	1.93	1.55	1.20	0.87	0.56	0.27						20	2.34	1.95	1.59	1.26	0.95	0.66	0.39	0.13				
2.06	1.68	1.33	1.00	0.69	0.41	0.14						18	2.06	1.71	1.37	1.06	0.78	0.50	0.25	0.01				
1.82	1.46	1.14	0.83	0.54	0.27							16	1.82	1.49	1.17	0.89	0.62	0.36	0.13					
1.60	1.27	0.96	0.67	0.40	0.15							14	1.60	1.29	1.00	0.73	0.48	0.24	0.03					
1.40	1.09	0.81	0.53	0.28								12	1.40	1.12	0.84	0.59	0.36	0.14						
1.23	0.94	0.67	0.41	0.17								10	1.23	0.96	0.70	0.47	0.26	0.04						
1.07	0.80	0.55	0.31	0.08								8	1.07	0.82	0.58	0.37	0.16							
0.93	0.68	0.44	0.21									6	0.93	0.70	0.48	0.27	0.07							
0.81	0.57	0.34	0.16									4	0.81	0.60	0.38	0.18								
0.71	0.48	0.28	0.08									2	0.71	0.50	0.29	0.10								
0.61	0.40	0.20										0	0.61	0.41	0.21									

Source: After Doorenbos and Pruitt.[30] Used with permission.

TABLE 4.4.2b Vapor Pressure (in kPa) from Dry- and Wet-Bulb Temperature Data (in °C) for a Nonventilated Psychrometer

Wet-bulb depression, °C (altitude < 1000 m)												Dry bulb, °C	Wet-bulb depression, °C (altitude > 1000 m)											
0	2	4	6	8	10	12	14	16	18	20	22		0	2	4	6	8	10	12	14	16	18	20	22
7.38	6.47	5.62	4.84	4.12	3.44	2.82	2.24	1.70	1.20	0.74	0.30	40	7.38	6.49	5.67	4.91	4.20	3.56	2.96	2.41	1.89	1.41	0.98	0.56
6.63	5.78	5.00	4.28	3.60	2.98	2.40	1.86	1.36	0.90	0.46	0.06	38	6.63	5.80	5.05	4.34	3.69	3.10	2.54	2.03	1.55	1.11	0.70	0.32
5.94	5.16	4.44	3.76	3.14	2.56	2.02	1.52	1.06	0.62	0.22		36	5.94	5.18	4.48	3.83	3.23	2.68	2.12	1.69	1.25	0.83	0.46	0.10
5.32	4.59	3.92	3.30	2.72	2.18	1.68	1.22	0.78	0.38			34	5.32	4.61	3.97	3.37	2.81	2.30	1.82	1.39	0.97	0.59	0.24	
4.75	4.08	3.46	2.88	2.34	1.84	1.38	0.94	0.54	0.16			32	4.75	4.10	3.51	2.95	2.43	1.96	1.52	1.11	0.73	0.37	0.04	
4.24	3.62	3.04	2.50	2.00	1.54	1.10	0.70	0.32				30	4.24	3.64	3.09	2.57	2.09	1.66	1.24	0.87	0.51	0.17		
3.78	3.20	2.66	2.16	1.70	1.26	0.86	0.48	0.12				28	3.78	3.22	2.71	2.23	1.79	1.38	1.00	0.65	0.31			
3.36	2.82	2.32	1.86	1.42	1.02	0.64	0.28					26	3.36	2.84	2.37	1.93	1.51	1.14	0.78	0.45	0.14			
2.98	2.48	2.02	1.58	1.19	0.80	0.44	0.11					24	2.98	2.50	2.07	1.65	1.27	0.92	0.58	0.28				
2.64	2.18	1.74	1.34	0.96	0.60	0.27						22	2.64	2.20	1.79	1.41	1.05	0.72	0.41	0.12				
2.34	1.90	1.50	1.12	0.76	0.43	0.11						20	2.34	1.92	1.55	1.19	0.85	0.55	0.25					
2.06	1.66	1.28	0.92	0.59	0.27							18	2.06	1.68	1.33	0.99	0.68	0.39	0.11					
1.82	1.44	1.08	0.75	0.43	0.14							16	1.82	1.46	1.13	0.82	0.52	0.25						
1.60	1.24	0.91	0.59	0.30	0.01							14	1.60	1.26	0.96	0.66	0.38	0.13						
1.40	1.07	0.75	0.46	0.17								12	1.40	1.09	0.80	0.52	0.26	0.03						
1.23	0.91	0.61	0.33	0.07								10	1.23	0.93	0.67	0.40	0.16							
1.07	0.77	0.49	0.23									8	1.07	0.79	0.54	0.30	0.06							
0.93	0.65	0.39	0.15									6	0.93	0.67	0.44	0.20								
0.81	0.55	0.29	0.09									4	0.81	0.57	0.34	0.11								
0.71	0.45	0.23										2	0.71	0.47	0.25	0.03								
0.61	0.37	0.15										0	0.61	0.38	0.17									

Source: After Doorenbos and Pruitt.[30] Used with permission.

than one pair of dry- and wet-bulb temperatures are available, e is the average value derived from these.

3. *Dew-Point Temperature.* Data given: $T_{\max}$, $T_{\min}$, T_{dew}.

$$\overline{D} = \frac{e_s(T_{\max}) + e_s(T_{\min})}{2} - e(T_{\text{dew}}) \qquad \text{kPa} \tag{4.4.8}$$

with $e_s(T_{\max})$, $e_s(T_{\min})$, and $e_s(T_{\text{dew}})$ from Eq. (4.2.2) or Table 4.2.1.

Adjustments for Measurement Heights. The preferred techniques for estimating evaporation given in later sections are appropriate for measured vapor-pressure deficit $\overline{D}$ and wind speed U_2 at a height of 2 m. For reference crop evaporation E_{rc} the equivalent value of the aerodynamic resistance r_a alters if the measurement height of either humidity or wind speed is other than 2 m. However, this can be allowed for by adjusting the effective value of wind speed used in Eq. (4.2.31) and the estimating equations below to U_2', this value being given by the equation

$$\frac{U_2'}{U_2} = \frac{34.9648}{\ln\left(\dfrac{z_e - 0.08}{0.001476}\right)\ln\left(\dfrac{z_u - 0.08}{0.01476}\right)} \tag{4.4.9}$$

where z_e and z_u are the actual heights of the humidity and wind speed measurements, respectively. For example, $U_2'/U_2 = 1.116$ for $z_e = 1$ m and $z_u = 2$ m, and $U_2'/U_2 = 0.749$ for $z_e = 2$ m and $z_u = 10$ m.

Equation (4.4.9) may also be adequate to correct wind speeds for the purposes of estimating potential evaporation rates for open water surfaces, bearing in mind the empirical and approximate nature of Eqs. (4.2.30) and (4.2.33).

4.4.4 Estimating the Evaporation Rate from Open Water

The preferred method for estimating the rate of evaporation from open water is from Eq. (4.2.30), which is here rewritten in the form

$$E_p = F_p^1 A + F_p^2 \overline{D} \qquad \text{mm day}^{-1} \tag{4.4.10}$$

where A is the energy available for evaporation (in mm day^{-1}), given by $A = (R_n + A_h)$, and estimated following the procedure given in Sec. 4.4.2; and $\overline{D}$ is the (average) vapor-pressure deficit (in kPa) calculated as in Sec. 4.4.3.

The coefficient F_p^1 is a function of temperature and the elevation of the site:

$$F_p^1 = \frac{\Delta}{\Delta + \gamma} \tag{4.4.11}$$

in which Δ is given by substituting e_s from Eq. (4.2.2) into Eq. (4.2.3); and γ by substituting λ from Eq. (4.2.1) into Eq. (4.2.28); with the atmospheric pressure P (in kPa), estimated from

$$P = 101.3\left(\frac{293 - 0.0065Z}{293}\right)^{5.256} \tag{4.4.12}$$

where Z is the elevation of the site (in m).

TABLE 4.4.3 Values of Parameters in Evaporation Estimation Eqs. (4.4.10), (4.4.14), and (4.4.17) for Two Elevations above Sea Level, and Sample Values of Mean Air Temperature and Wind Speed

Elevation Z, m	Temperature T, °C	Wind speed U_2, m s^{-1}	F_p^1	F_p^2	F_{rc}^1	F_{rc}^2	F_{pt}^{humid}	F_{pt}^{arid}
0	10	3	0.553	3.028	0.383	2.937	0.696	0.962
0	10	6	0.553	4.895	0.293	4.495	0.696	0.962
0	30	3	0.781	1.505	0.643	1.588	0.985	1.360
0	30	6	0.781	2.433	0.546	2.697	0.985	1.360
1000	10	3	0.582	2.832	0.411	2.803	0.733	1.012
1000	10	6	0.582	4.578	0.318	4.336	0.733	1.012
1000	30	3	0.801	1.371	0.670	1.470	1.010	1.394
1000	30	6	0.801	2.216	0.575	2.524	1.010	1.394

The coefficient F_p^2 is a function of temperature, wind speed, and the elevation of the site:

$$F_p^2 = \left(\frac{\gamma}{\Delta + \gamma}\right) \frac{6.43(1 + 0.536\,U_2)}{\lambda} \tag{4.4.13}$$

in which Δ and γ are given as for Eq. (4.4.11); U_2 is the wind speed measured at 2 m [or the effective value calculated from Eq. (4.4.9) if necessary]; and λ is given by Eq. (4.2.1). Computer code to calculate F_p^1 and F_p^2 can be checked using the sample values given in Table 4.4.3.

The allowance for atmospheric diffusion (see Sec. 4.2.5) implicit in F_p^2 is relevant to open water surfaces with reasonable small surface area, and Eq. (4.4.10) is therefore expected to provide weekly or monthly estimates of the evaporation rate from shallow, ground-level evaporation pans (or small ponds or lakes) which are good to 5 to 10 percent or 0.5 mm day^{-1}, whichever is greater. The effective value of the aerodynamic resistance for much larger expanses of water is larger (see Sec. 4.2.7), and the evaporation rate is therefore reduced. Equation (4.4.10) might therefore systematically overestimate the evaporation for very large lakes by approximately 10 to 15 percent.

Of the alternative techniques for estimating open water evaporation, the Kimberly-Penman equation,[56,134] given earlier as Eq. (4.2.33), has been shown[56] to have marginal advantage over Eq. (4.4.10) in semiarid environments but is marginally worse in humid environments. If not all of the climatological data are available to allow a calculation of E_p from either Eq. (4.4.10) or Eq. (4.2.33), the use of the same secondary estimation techniques as for reference crop evaporation is recommended based on radiation, pan, or temperature data (see Sec. 4.4.5). Estimates based on radiation, however, should use measurements or estimates of the energy available for evaporation relevant to open water (as opposed to grassed) surfaces; see Sec. 4.4.2.

4.4.5 Estimating the Evaporation Rate of the Reference Crop

Preferred Method. The preferred method for estimating the rate of evaporation from the reference crop or short actively growing grass (see Sec. 4.1.1 for definition) is from Eq. (4.2.31), which is here rewritten in the form

$$E_{rc} = F_{rc}^1 A + F_{rc}^2 \overline{D} \qquad \text{mm day}^{-1} \tag{4.4.14}$$

where A is the energy available for evaporation (in mm day^{-1}), given by $A = (R_n + A_h)$, and estimated following the procedure given in Sec. 4.4.2; and $\overline{D}$ is the (average) vapor-pressure deficit (in kPa) calculated as in Sec. 4.4.3.

The coefficient F_{rc}^1 is a function of temperature, wind speed, and the elevation of the site:

$$F_{rc}^1 = \frac{\Delta}{\Delta + \gamma^*} \tag{4.4.15}$$

in which Δ is given by substituting e_s from Eq. (4.2.2) into Eq. (4.2.3); and γ^* is given by Eq. (4.2.32), with γ obtained by substituting Eq. (4.2.1) into Eq. (4.2.28) and P taken from Eq. (4.4.12).

The coefficient F_{rc}^2 is similarly a function of temperature, wind speed, and the elevation of the site:

$$F_{rc}^2 = \left(\frac{\gamma}{\Delta + \gamma^*}\right) \frac{900\ U_2}{T + 275} \tag{4.4.16}$$

the required inputs being derived as for Eq. (4.4.15). Computer code to calculate F_{rc}^1 and F_{rc}^2 can be checked using the sample values given in Table 4.4.3.

Errors in estimating the evaporation from the precisely defined reference crop using Eq. (4.4.14) do not primarily arise because of the empirical values involved, since these are related to the physically-based values which control evaporation rate in the Penman-Monteith equation. Rather, they arise through differences between these values and those relevant to the particular crop for which estimates are made. Studies[56] suggest differences of the order 5 to 7 percent or 0.5 mm day^{-1} may arise.

Radiation-Based Estimates. The preferred radiation-based method for estimating reference crop evaporation is from Eq. (4.2.37), which is here rewritten in the form

$$E_{rc} = F_{pt} A \qquad \text{mm day}^{-1} \tag{4.4.17}$$

$F_{pt} = \alpha[\Delta/(\Delta + \gamma)]$, with $\alpha = 1.74$ for arid regions and $\alpha = 1.26$ for all other (humid) locations [see Eqs. (4.2.38) and (4.2.39)]. Arid regions, in this context, are defined as having relative humidity of less than 60 percent in the month with peak evaporation. The value of $(R_n - G)$ is estimated following the procedure given in Sec. 4.4.2.

Computer code to calculate F_{pt} for humid and arid locations can be checked using the sample values given in Table 4.4.3. Estimates using this method are prone to errors of the order 15 percent or 0.75 mm day^{-1}, whichever is greater, and should be made only for periods of 10 days or longer.

Other radiation-based estimation methods, especially those given by Turc[125] and Doorenbos and Pruitt,[30] are also in common use and are described in Sec. 4.2.6.

Pan-Based Estimates. The preferred pan-based method for estimating reference crop evaporation is from Eq. (4.3.2), with values of pan coefficients for Class A pans from Table 4.3.1. Estimated errors in using this technique are typically of the order 10 to 15 percent or 1 mm day^{-1}, whichever is greater; for regions where dry winds predominate, or where the upwind fetches are low, the errors may well be twice as large.

Temperature-Based Estimates. Estimating evaporation from temperature data is not recommended, except when lack of other data means this is the only option available. In these conditions the Hargreaves equation [Eq. (4.2.44)] is the preferred technique and Ref. 56 may provide estimates with errors in the order 10 to 15 percent or 1 mm day^{-1}, whichever is greater.

Other more complex temperature-based techniques are still in common use, in particular the Doorenbos and Pruitt version of the Blaney-Criddle method; see Eq. (4.2.45). Comparison[56] suggests the additional complexity in this provides a marginally inferior estimate to that given by the Hargreaves equation.

4.4.6 Estimating the Evaporation Rate of Other Crops

It is the actual evaporation which is most often required and, in principle, this could be estimated directly from Eq. (4.2.27) if there are values of r_s and r_a available which are appropriate to the particular crop for which estimates are required. In practice this is rarely the case except in research application, and it is therefore common practice to estimate first the evaporation rate for the reference crop (for which r_s and r_a are prescribed), and then multiply this rate by an additional factor; thus

$$E = K_c E_{rc} \qquad \text{mm day}^{-1} \tag{4.4.18}$$

The factor K_c in this equation is called the *crop coefficient* and, from a comparison with Eqs. (4.2.27) and (4.2.31), it is clearly a complex factor. It contains a significant dependence on the effective average surface resistance of the actual crop (relative to the reference crop), and this is the primary influence in dry conditions, but it also depends on vegetation height through r_a. Moreover, K_c has some dependence on meteorological variables, i.e., on temperature (through Δ), on wind speed (through r_a), and on rainfall (through r_s, indirectly, depending on the amount of time the canopy is wet). Clearly the value of crop coefficient will depend not only on the crop and its stage of development, but also in part on the average climate in which any empirical calibration is carried out.

Irrigated Field Crops. In practice many evaporation estimation applications concern the irrigation of agricultural crops. The objective is usually to supply water which is adequate, in that it does not limit growth, but not excessive, so that the soil surface is not waterlogged. In this case the *potential crop coefficient* K_{co} is relevant, this being defined from the equation

$$E = K_{co} E_{rc} \qquad \text{mm day}^{-1} \tag{4.4.19}$$

It is likely that K_{co} is less variable than K_c in moving from one location to the next, since it is a purer measure of stomatal resistance, which in turn is less variable since the soil moisture deficit remains small. (It will of course still have some local meteorological dependence through Δ and r_a.)

Considerable research has been directed toward defining K_{co} as a function of time for different crops. As might be expected for annual agricultural crops, with which it is often used, there is a pronounced seasonal variation of the type illustrated schematically in Fig. 4.4.1. This figure can be used to provide estimates of the seasonal variation in potential crop coefficient[30] for the range of (annual) field crops listed in Table 4.4.4 as follows:

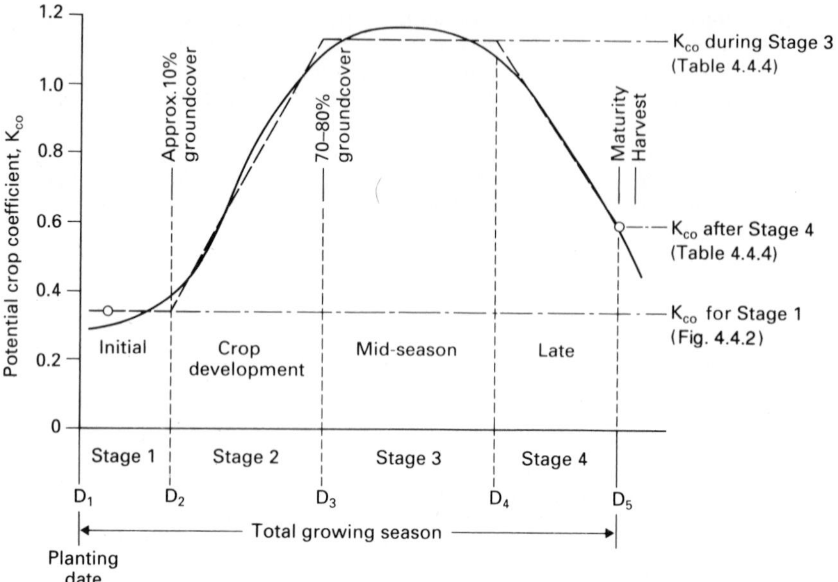

FIGURE 4.4.1 Schematic diagram of the seasonal changes in potential crop coefficient K_{co} for an irrigated field crop *(redrawn from Doorenbos and Pruitt.[30] Used with permission.)* illustrating four growth stages (initial, crop development, midseason, late) in the total growing season defined relative to the planting date D_1 and terminating at the dates D_2, D_3, D_4, and D_5, respectively. The value of K_{co} during the initial Stage 1 is taken from Fig. 4.4.2, and the values during Stage 3 and after D_5 are taken from Table 4.4.4. Intervening values during Stages 2 and 4 are determined by interpolation.

1. Establish the planting or sowing date and, if possible, determine the total growing season and length of the individual crop development stages from local information; otherwise estimates of the typical growing season and of the proportion of this at each growth stage are given in Table 4.4.4. Use these to determine the dates D_2, D_3, D_4, and D_5 for the time series of K_{co} (see Fig. 4.4.1) relative to the planting date D_1.
2. The (constant) value of K_{co} for Stage 1 is a function of the reference crop evaporation at the time of sowing and the average interval between rain or irrigation at this time. Determine this value from Fig. 4.4.2.
3. For a given crop growing in a given climate (humidity and wind), the (constant) value of K_{co} for Stage 3 (i.e., midseason, from date D_3 to D_4) is given in Table 4.4.4, as is the value of K_{co} at maturity or harvest (after Stage 4, at time D_5).
4. The full seasonal variation in K_{co} is then obtained by interpolating between the values at D_2 and D_3, and those at D_4 and D_5.

An estimate of the seasonal variation in E, and thus of the irrigation demand, can be made through the season by introducing these time-dependent estimates of K_{co} into Eq. (4.4.19), along with estimates of E_{rc}.

Irrigated Grasslike Crops. For irrigated alfalfa, grasses, clover, or grass-legumes successively cut for fodder throughout the season, or for well-irrigated and well-fertilized (grass, grass-legumes, or alfalfa) pastures, it is generally sufficient[30] to use

TABLE 4.4.4 For a Range of Irrigated Field Crops, the Typical Total Growing Season in Days; Representative Proportions of This Growing Season at Each of the Growth Stages Illustrated in Fig. 4.4.1; and Recommended Values of K_{co} During Stage 3 and After Stage 4 in Different (Wind and Humidity) Climates

						K_{co} during Stage 3				K_{co} after Stage 4			
	Typical growing season,	Fraction of stage time at growth stage				$RH_{min} > 70\%$		$RH_{min} < 20\%$		$RH_{min} > 70\%$		$RH_{min} < 20\%$	
						Wind, m s^{-1}		Wind, m s^{-1}		Wind, m s^{-1}		Wind, m s^{-1}	
Crop	days	1	2	3	4	0–5	5–8	0–5	5–8	0–5	5–8	0–5	5–8
Artichokes (perennial)	310–360	0.09	0.12	0.70	0.09	0.95	0.95	1.00	1.05	0.90	0.90	0.95	1.00
Barley	120–150	0.12	0.20	0.44	0.24	1.05	1.10	1.15	1.20	0.25	0.25	0.20	0.20
Beans (green)	75–90	0.22	0.33	0.33	0.12	0.95	0.95	1.00	1.05	0.85	0.85	0.90	0.90
Beans (dry)/Pulses	95–110	0.16	0.25	0.40	0.19	1.05	1.10	1.15	1.20	0.30	0.30	0.25	0.25
Beets (table)	70–90	0.24	0.35	0.29	0.12	1.00	1.00	1.05	1.10	0.90	0.90	0.95	1.00
Carrots	100–150	0.18	0.27	0.39	0.16	1.00	1.05	1.10	1.15	0.70	0.75	0.80	0.85
Castor beans	180	0.14	0.22	0.36	0.28	1.05	1.10	1.15	1.20	0.50	0.50	0.50	0.50
Celery	125–180	0.16	0.27	0.46	0.11	1.00	1.05	1.10	1.15	0.90	0.95	1.00	1.05
Corn (sweet)	80–110	0.23	0.29	0.37	0.11	1.05	1.10	1.15	1.20	0.95	1.00	1.05	1.10
Corn (grain)	125–180	0.17	0.28	0.33	0.22	1.05	1.10	1.15	1.20	0.55	0.55	0.60	0.60
Cotton	180–195	0.16	0.27	0.31	0.26	1.05	1.15	1.20	1.25	0.65	0.65	0.65	0.70
Crucifers	80–95	0.24	0.38	0.26	0.12	0.95	1.00	1.05	1.10	0.80	0.85	0.90	0.95
Cucumber	105–130	0.19	0.28	0.38	0.15	0.90	0.90	0.95	1.00	0.70	0.70	0.75	0.80
Eggplant	130–140	0.21	0.32	0.30	0.17	0.95	1.00	1.05	1.10	0.80	0.85	0.85	0.90
Flax	150–220	0.15	0.21	0.39	0.25	1.00	1.05	1.10	1.15	0.25	0.25	0.20	0.20
Grain (small)	150–165	0.14	0.20	0.40	0.26	1.05	1.10	1.15	1.20	0.30	0.30	0.25	0.25
Lentil	150–170	0.14	0.20	0.41	0.25	1.05	1.10	1.15	1.20	0.30	0.30	0.25	0.25
Lettuce	75–140	0.26	0.37	0.27	0.10	0.95	0.95	1.00	1.05	0.90	0.90	0.90	1.00
Melons	120–160	0.20	0.28	0.37	0.15	0.95	0.95	1.00	1.05	0.65	0.65	0.75	0.75
Millet	105–140	0.14	0.23	0.39	0.24	1.00	1.05	1.10	1.15	0.30	0.30	0.25	0.25
Oats	120–150	0.12	0.20	0.44	0.24	1.05	1.10	1.15	1.20	0.25	0.25	0.20	0.20
Onion (dry)	150–210	0.10	0.17	0.49	0.24	0.95	0.95	1.05	1.10	0.75	0.75	0.80	0.85

(Continued)

TABLE 4.4.4 For a Range of Irrigated Field Crops, the Typical Total Growing Season in Days; Representative Proportions of This Growing Season at Each of the Growth Stages Illustrated in Fig. 4.4.1; and Recommended Values of K_{co} During Stage 3 and After Stage 4 in Different (Wind and Humidity) Climates *(Continued)*

Crop	Typical growing season, days	Fraction of stage time at growth stage				K_{co} during Stage 3				K_{co} after Stage 4			
						$RH_{min} > 70\%$		$RH_{min} < 20\%$		$RH_{min} > 70\%$		$RH_{min} < 20\%$	
						Wind, m s^{-1}		Wind, m s^{-1}		Wind, m s^{-1}		Wind, m s^{-1}	
		1	2	3	4	0–5	5–8	0–5	5–8	0–5	5–8	0–5	5–8
Onion (green)	70–95	0.29	0.45	0.17	0.09	0.95	0.95	1.00	1.05	0.95	0.95	1.00	1.05
Groundnuts	130–140	0.22	0.30	0.30	0.18	0.95	1.00	1.05	1.10	0.55	0.55	0.60	0.60
Peas	90–100	0.21	0.26	0.37	0.16	1.05	1.10	1.15	1.20	0.95	1.00	1.05	1.10
Peppers (fresh)	120–125	0.22	0.29	0.33	0.16	0.95	1.00	1.05	1.10	0.80	0.85	0.85	0.90
Potato	105–145	0.21	0.25	0.33	0.21	1.05	1.10	1.15	1.20	0.70	0.70	0.75	0.75
Radishes	35–40	0.20	0.27	0.40	0.13	0.80	0.80	0.85	0.90	0.75	0.75	0.80	0.85
Safflower	125–190	0.17	0.27	0.35	0.21	1.05	1.10	1.15	1.20	0.25	0.25	0.20	0.20
Sorghum	120–130	0.16	0.27	0.33	0.24	1.00	1.05	1.10	1.15	0.50	0.50	0.55	0.55
Soybeans	135–150	0.14	0.21	0.46	0.19	1.00	1.05	1.10	1.15	0.45	0.45	0.45	0.45
Spinach	60–100	0.27	0.31	0.34	0.08	0.90	0.95	1.00	1.05	0.90	0.90	0.95	1.00
Squash	90–100	0.24	0.34	0.26	0.16	0.90	0.90	0.95	1.00	0.70	0.70	0.75	0.80
Sugar beet	160–230	0.18	0.27	0.33	0.22	1.05	1.10	1.15	1.20	0.90	0.95	1.00	1.00
Sunflower	125–130	0.17	0.28	0.36	0.19	1.05	1.10	1.15	1.20	0.40	0.40	0.35	0.35
Tomato	135–180	0.20	0.28	0.33	0.19	1.05	1.10	1.20	1.25	0.60	0.60	0.65	0.65
Wheat	120–150	0.12	0.20	0.44	0.24	1.05	1.10	1.15	1.20	0.25	0.25	0.20	0.20

Source: Derived from Doorenbos and Pruitt (Ref. 30, Tables 21 and 22). Used with permission.

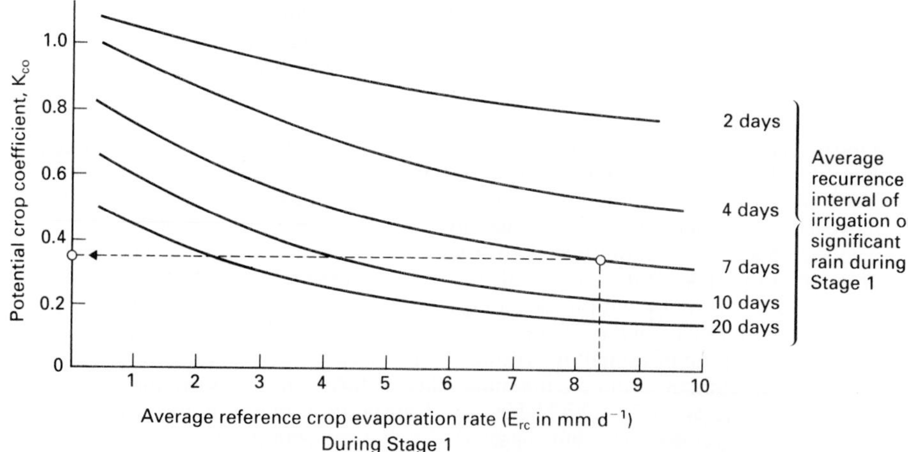

FIGURE 4.4.2 The potential crop coefficient K_{co} to be applied during Stage 1 (see text and Fig. 4.4.1) based on the average reference crop evaporation rate E_{rc} during this time and the estimated recurrence interval between significant rainfall or irrigation *(redrawn from Doorenbos and Pruitt.[30] Used with permission.)*. The example shown is for $E_{rc} = 8.4$ mm day^{-1} and an irrigation interval of 7 days, yielding the estimate $K_{co} = 0.35$.

the average values of potential crop coefficients given in Table 4.4.5 to estimate monthly average evaporation rate from Eq. (4.4.19). However, in the case of repeatedly cut fodder crops, there will be dramatic shorter-term variations in evaporation during successive cut and regrowth cycles.

Other Irrigated Crops. Values of K_{co} for particular field crops with growth season extending over several seasons (e.g., bananas, sugarcane); values for deciduous fruit and nut crops (e.g., apples, cherries, peaches, apricots, pears, plums, walnuts, citrus, grapes, coffee, tea); and values for specialized crops, such as rice, have been tabulated by Doorenbos and Pruitt.[30] Readers interested in such crops are referred to that source.

TABLE 4.4.5 Recommended Average Values of Potential Crop Coefficient K_{co} for Successively Cut Fodder Crops and Grazed Pastures Which Are Well-Watered and Fertilized

	RH > 70%		RH < 30%	
	Wind, m s^{-1}		Wind, m s^{-1}	
	0–5	5+	0–5	5+
Fodder crops:				
Alfalfa	0.85	1.05	0.95	1.05
Grasses	0.80	1.00	0.90	1.00
Clover/grass-legumes	1.00	1.10	1.05	1.10
Pasture (grass, grass-legumes, alfalfa)	0.95	1.05	1.00	1.05

Source: Derived from Doorenbos and Pruitt (Ref. 30, Table 23). Used with permission.

For tall perennial fruit crops the size of K_{co} can be significantly influenced by agricultural practice, in particular the absence of a weed understory can lower K_{co} by 20 to 30 percent. Equally the use of sprinkler irrigation can alter the evaporation loss from tree crops, from significantly less than E_{rc} to greater than E_{rc}; see next section.

Forests. The evaporation rate from forests is more difficult to describe and estimate than for other vegetation types, and the reader is referred to Chap. 13 for greater detail. Most of this difficulty arises because turbulent diffusion in the atmosphere above forests is much more efficient than for other crops. For this reason the rate of evaporation when the canopy is wet can be very much greater than when it is dry (e.g., Refs. 108, 115). It is not therefore realistic to represent the average effect of controlling processes operating within the canopy in terms of a single (effective) surface resistance as in the case of the reference crop. It is necessary to consider transpiration and the evaporation of rainfall intercepted by forest canopy separately.

Perhaps in compensation for the more efficient turbulent diffusion, the surface resistance of forests tends to be higher, and transpiration rates are lower. Present evidence is not yet definitive but suggests[106,108] that the *transpiration rate* of well-watered forests is perhaps 80 ± 10 percent of reference crop evaporation [Eq. (4.4.14)] provided this has been calculated with the appropriate (forest) value of albedo; see Table 4.2.2.

The rate of evaporation of intercepted water from a wet canopy, on the other hand, commonly exceeds the potential evaporation rate for open water surfaces [Eq. (4.4.10)], and indeed often the energy locally available to support it. The additional energy is withdrawn from the atmosphere from warmer, drier air upwind. The net interception loss is typically 10 to 30 percent of rainfall, and depends partly on the amount of water which the canopy can store, i.e., *the interception storage capacity S;* but partly on the nature of the rainfall, in particular the intensity and duration of the rainstorms since up to half the evaporation occurs during the storm itself. For forests with complete canopy cover, intense, short-lived, convective storms, more common in tropical regions, are associated with a lower fractional interception loss of say 10 to 18 percent of precipitation[65,105,106] while storms associated with frontal rainfall, which may be less intense but last longer, tend to give a higher fractional interception loss of say 20 to 30 percent of precipitation.[23,42] For deciduous forests, the loss of leaves in winter reduces the fractional interception loss, typically by a factor of 2 to 3.

An approximate estimate of forest evaporation averaged over a month might therefore be made from

$$E^{\text{forest}} = 0.8\, E_{rc}^{\text{forest}} + \alpha_i\, P \qquad \text{mm day}^{-1} \tag{4.4.20}$$

where the first term is an estimate of transpiration, with E_{rc}^{forest} the value of reference crop evaporation rate calculated from Eq. (4.4.14) but with net radiation appropriate to a forest; and the second term is interception loss, with α_i the fractional interception loss (see previous paragraph for typical values). Equation (4.4.20) is (at best) accurate to 10 to 15 percent, depending on the accuracy of α_i, and is prone to overestimate evaporation rates. The estimate it gives should be limited to 0.9 R_n for extensive midcontinental forests or to R_n for less extensive forests within (say) 200 km of the sea.[106]

If detailed short-term rainfall data are available, it may be possible[40,42] to provide an improved estimate of the interception loss term in Eq. (4.4.20) from

$$\alpha_i\, P = \frac{0.95\, N_s\, (S + 0.2\, \tau_s)}{N_d} \qquad \text{mm day}^{-1} \tag{4.4.21}$$

where N_s is the number of storms in the averaging period of N_d days and τ_s is their average duration (in h), while S is the interception storage capacity of the canopy (in mm), with $S = 1.2$ mm typical for coniferous canopies and $S = 0.8$ mm typical of broadleaf canopies in full leaf.[106] Equation (4.4.21) implicitly assumes an evaporation rate of 0.2 mm h^{-1} during storms.[65] The factor 0.95 in Eq. (4.4.21) allows partly for the suppression of transpiration during rainstorms and partly for the possibility that a small proportion of storms[65] involve a total rainfall less than S.

4.4.7 Soil Moisture Restrictions

If estimates of evaporation are required for nonirrigated crops, the water status of the soil can be important, acting through the surface resistance. The value of K_c is reduced from K_{co} by including a factor $K_s(\theta)$ related to the volumetric soil moisture content θ:

$$E = K_c\, E_{rc} = K_s(\theta)\, K_{co}\, E_{rc} \qquad \text{mm day}^{-1} \tag{4.4.22}$$

The amount of accessible soil water available to the plants depends on their rooting depth, which can of course change as the vegetation grows, a point of particular significance for annual crops. The most realistic models of the soil water restriction on evaporation rate simulate plant extraction from a series of moisture stores arranged vertically above each other in the soil through the rooting zone.[34,85] They perform a running water balance for each moisture store, with some water (from rain or irrigation) infiltrating into the top store and draining from the bottom store; some moving between stores; and meanwhile some being extracted at each level in proportion to the product of E_{rc}, the fraction of root in each store, and a moisture extraction function $f(\theta)$.

Studies have been made of the variation in $f(\theta)$ in response to decreasing soil water: Dyck[33] lists the formulas given in Table 4.4.6. Results differ in detail, as might be expected for an empirical soil-related function of this type, but many workers (e.g., Refs. 34, 38, 39, 52, 85) are in broad agreement that the overall behavior during a drying cycle follows the general pattern illustrated in Fig. 4.4.3.

Soil *saturated* by rain or irrigation, with moisture content θ_s, first drains until the remaining water held by surface tension on the soil particles is in equilibrium with the gravitational forces causing drainage. It is then said to be at *field capacity,* with a moisture content θ_f. The drying proceeds with little soil moisture restriction until the soil moisture falls to θ_d, when the moisture content is typically 50 to 80 percent of θ_f; then the hydraulic conductivity K_s and the transpiration rate start to fall. They continue to fall until a *wilting point* is reached, where the soil moisture content is θ_w, and when, it is assumed, $K_s = 0$. This behavior seems to be similar for both crops and soil. In conditions of prolonged drought the crop begins to die and the evaporation rate is no longer controlled by meteorological conditions, but by soil characteristics, especially by hydraulic conductivity.

When modeling the change in evaporation in response to soil water status it is often convenient to work in terms of the soil water content, but plants are more sensitive to soil water potential ψ_s. Some of the variability in the functional form of $f(\theta)$ reported in the literature may therefore be caused by the variability in the relationship between θ_s and ψ_s for different soils. In some hydrologic applications the precise form of $f(\theta)$ is not too important. In particular, in the case of modeled water budgets the cumulative error in the calculated water store is set to zero each time the soil dries completely or when it is completely wetted by heavy rain.

TABLE 4.4.6 Recommended Forms of the Soil Moisture Stress Function $f(\theta)$ as cited by Dyck.[33] Used with permission.

Soil moisture stress function	Reference
Daily values	
$f(\theta) = [1 - \exp(-\gamma\theta)]\,[1 - 2\exp(-\gamma\theta_f) + \exp(-\gamma\theta)]^{-1}$	70
$f(\theta) = [1 + (\theta_c/\theta)^{bk}]^{-1}$; with $k = 2.69 \exp(-0.09\,E_{rc})^{-0.62}$	78
$f(\theta) = \left[\frac{(\theta - \theta_w)}{(\theta_f - \theta_w)}\right] F_s$	5
$f(\theta) = (1 - 0.533\,V)^{-1} \left[\frac{(\theta - \theta_w)}{(\theta_f - \theta_w)}\right]$	60
5-day values	
$f(\theta) = 0.2 + 2\left[\frac{(\theta - \theta_w)}{(\theta_f - \theta_w)}\right] - 1.2\left[\frac{(\theta - \theta_w)}{(\theta_f - \theta_w)}\right]^2$	93
Monthly values	
$f(\theta) = \left[\frac{(\theta - \theta_w)}{(\theta_f - \theta_w)}\right]$	21

where θ = soil moisture content (variable)
θ_w = soil moisture content when soil at "wilting point" ($E/E_{rc} = 0$)
θ_c = soil moisture content when $E/E_{rc} = 0.5$
θ_f = soil moisture content when soil at "field capacity"
V = vegetation canopy density
F_s = adjustment factor for functional form of Fig. 4.4.3

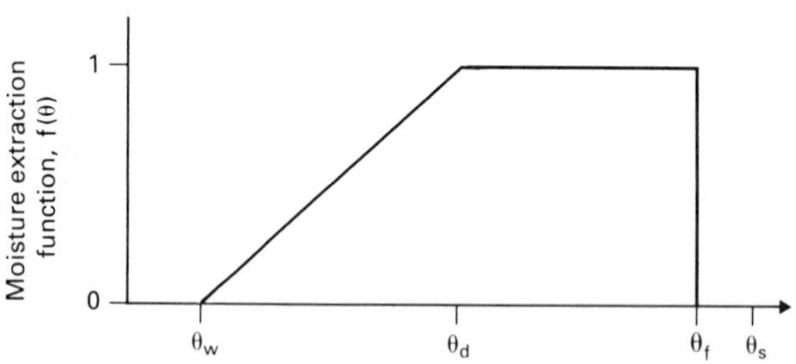

FIGURE 4.4.3 Typical variation of the moisture extraction function $f(\theta)$ which modifies the potential crop coefficient in response to changes in the volumetric soil moisture content θ in (portions of) the plants' rooting zone. θ_s, θ_f, and θ_w are the values of θ at saturation, field capacity, and wilting point, respectively, and (θ_d/θ_f) is typically 0.5 to 0.8. These values are determined by soil type.

4.4.8 Improving Evaporation Estimation

The adoption in this chapter of a reference crop evaporation rate which is precisely defined in terms of the physics of the evaporation process exploits research understanding gathered since the *Handbook of Applied Hydrology* was published in 1964. The significance of this development should not be underestimated in respect of the greater clarity and, on the basis of experimental evidence (e.g., Ref. 56), the improved accuracy it brings. At the same time the value of this step forward is compromised by the need to retain a "two-step" evaporation estimation procedure involving the poorly defined crop factor K_c, which has implicit meteorological dependence.

It is important that in punctuating the progress of hydrologic practice this *Handbook of Hydrology* should be a "semicolon," not a "full stop." With this in mind, I conclude with a recommendation. In order to provide a purer measure of crop control less contaminated by local climate, future field research into crop water requirements and the retrospective analysis of existing field data should seek to derive and report values of r_s^c, the effective average value of the surface resistance for the crop, rather than K_c.

If this becomes common practice, and in this way a relevant literature is created, then the author of this chapter in the next *Handbook of Hydrology* will be able to make a further significant step in advancing evaporation estimation practice by developing a "one-step" estimation procedure.

REFERENCES

1. Aboukhaled, A., A. Alfaro, and M. Smith, "Lysimeters," FAO Irrigation and Drainage Paper 39, Rome, 1982.
2. Allen, R. G., "A Penman for All Seasons," *J. Irrig. Drain. Eng.*, vol. 112, no. 4, pp. 348–368, 1986.
3. Allen, R. G., M. E. Jensen, J. L. Wright, and R. D. Burman, "Operational Estimates of Evapotranspiration," *Agron. J.*, vol. 81, pp. 650–662, 1989.
4. Allen, R. G., and W. O. Pruitt, "Rational Use of the FAO Blaney-Criddle Formula," *J. Irrig. Drain. Eng.*, vol. 112, no. IR2, pp. 139–155, 1986.
5. Baier, W., and G. W. Robertson, "A New Versatile Soil Moisture Budget," *Can. J. Plant Sci.*, vol. 46, pp. 299–315, 1966.
6. Baker, J. M., and C. H. M. Van Baval, "Measurement of Mass Flow of Water in the Stems of Herbaceous Plants," *Plant Cell Environ.*, vol. 10, pp. 777–782, 1987.
7. Bell, J. P., "A New Design Principle for Neutron Soil Moisture Gauges: The 'Wallingford' Neutron Probe," *Soil Sci.*, vol. 108, pp. 160–164, 1969.
8. Bell, J. P., "Neutron Probe Practice," *IH Report 19*, Institute of Hydrology, Wallingford, UK, 1987.
9. Bell, J. P., T. J. Dean, and M. G. Hodnett, "Soil Moisture Measurement by an Improved Capacitance Technique, part II: Field Techniques, Evaluation and Calibration," *J. Hydrol.*, vol. 93, pp. 79–90, 1987.
10. Black, T. A., and K. G. McNaughton, "Psychrometric Apparatus for Bowen Ratio Determination over Forests," *Boundary Layer Meteorol.*, vol. 2, pp. 246–254, 1971.
11. Blaney, H. F., and W. D. Criddle, "Determining Water Requirements in Irrigated Areas from Climatological and Irrigation Data," USDA (SCS) TP-96, p. 48, 1950.
12. Bouchet, R. J., "Evapotranspiration réelle et potentielle, signification climatique," in *Int. Assoc. Sci. Hydrol., Proceedings,* Berkeley, Calif., Symp. Publ. 62, pp. 134–142, 1963.

13. Bruin de, H. A. R., "A Model for the Priestley-Taylor Parameter," *J. Appl. Meteor.*, vol. 22, pp. 572–578, 1983.

14. Brunt, D., "Notes on Radiation in the Atmosphere," *Q. J. R. Meteorol. Soc.*, vol. 58, pp. 389–418, 1932.

15. Brutsaert, W., "Comments on Surface Roughness Parameters and the Height of Dense Vegetation," *J. Meteorol. Soc. Japan*, vol. 53, pp. 96–97, 1975.

16. Brutsaert, W., "Evaporation into the Atmosphere," D. Reidel Pub. Co., Dordrecht, Holland, 1982.

17. Brutsaert, W., and H. Stricker, "An Advection-Aridity Approach to Estimating Actual Regional Evaporation," *Water Resour. Res.*, vol. 15, pp. 443–450, 1979.

18. Brutsaert, W., and G. T. Yeh, "Evaporation from an Extremely Narrow Wet Strip at Ground Level," *J. Geophys. Res.*, vol. 74, pp. 3431–3433, 1969.

19. Brutsaert, W., and G. T. Yeh, "Implications of a Type of Empirical Evaporation Formula for Lakes and Pans," *Water Resour. Res.*, vol. 6, pp. 1202–1208, 1970.

20. Brutsaert, W., and S. L. Yu, "Mass Transfer Aspects of Pan Evaporation," *J. Appl. Meteorol.*, vol. 7, pp. 563–566, 1968.

21. Budyko, M., and L. I. Zubenok, "Opredelenie isparenija poverchnosti sushi," *AN SSSR Ser. Geogr.*, Booklet No. 6, 1961.

22. Calder, I. R., M. N. Narayanswamy, N. V. Srinivasalu, W. G. Darling, and A. J. Lardner, "Investigation into the Use of Deuterium as a Tracer for Measuring Transpiration from Eucalypts," *J. Hydrol.*, vol. 78, pp. 261–278, 1986.

23. Calder, I. R., and M. D. Newson, "Land Use and Upland Water Resources in Britain—A Strategic Look," *Water Resour. Bull.*, vol. 16, pp. 1628–1639, 1979.

24. Choudhury, B. L., and J. L. Monteith, "A Four-Layer Model for the Heat Budget of Homogeneous Land Surfaces," *Q. J. R. Meteorol. Soc.*, vol. 114, pp. 373–398, 1988.

25. Cohen, Y., G. C. Green, and M. Fuchs, "Improvement of the Heat Pulse Method for Determining Sap Flow in Trees," *Plant Cell Environ.*, vol. 4, pp. 425–431, 1987.

26. Coursey de, D. G., "The Goodwin Creek Research Catchment, Part I: Design Philosophy," *Proc. Symp. Hydrol. Res. Basins*, Bern, Sept. 21–23, 1982.

27. Crow, F. R., and S. D. Hottman, "Network Density of Temperature Profile Stations and Its Influence on the Accuracy of Lake Evaporation Calculations," *Water Resour. Res.*, vol. 9, pp. 895–899, 1973.

28. Dagg, M., "Evaporation Pans in East Africa," *Proc. Fourth Specialist Meeting on Applied Meteorology* in East Africa, Nairobi, 1968.

29. Dean, T. J., J. P. Bell, and A. J. B. Baty, "Soil Moisture Measurement by an Improved Capacitance Technique, Part I: Sensor Design and Performance," *J. Hydrol.*, vol. 93, pp. 67–78, 1987.

30. Doorenbos, J., and W. O. Pruitt, "Crop Water Requirements," *Irrigation and Drainage Paper,* 24, UN Food and Agriculture Organization, Rome, Italy, 1977.

31. Driedonks, A. G. M., "Dynamics of the Well-Mixed Atmospheric Boundary Layer," *Sci. Report W. R. 81-2,* K.N.M.I., de Bilt, The Netherlands, 1981.

32. Duffie, J. A., and W. A. Beckman, *Solar Engineering of Thermal Processes,* Wiley, New York, pp. 1–109, 1980.

33. Dyck, S., "Overview on the Present Status of the Concepts of Water Balance Models," *IAHS Publ.* 148, Wallingford, pp. 3–19, 1983.

34. Feddes, R. A., P. J. Kowalik, and H. Zaradny, "Simulation of Field Water Use and Crop Yield," *PUDOC,* Wageningen, 1978.

35. Frevert, D. K., R. W. Hill, and B. C. Braaten, "Estimation of FAO Evapotranspiration Co-efficients," *J. Irrig. Drain. Eng.*, vol. 109, no. IR2, pp. 265–270, 1983.

36. Fritschen, L. J., and J. R. Simpson, "Surface Energy and Radiation Balance Systems: General Description and Improvements," *J. Appl. Meteorol.*, vol. 28, pp. 680–689, 1989.

37. Gangopadhyaya, M., V. A. Uryvaev, M. H. Omar, T. J. Nordenson, and G. E. Harbeck, "Measurement and Estimation of Evaporation and Evapotranspiration," *WMO Technical Note,* no. 83, Geneva, Switzerland, 1966.

38. Gardner, W. R., and C. F. Ehlig, "The Influence of Soil Water on Transpiration of Plants," *J. Geophys. Res.*, vol. 68, pp. 5719–5724, 1963.

39. Gardner, W. R., and D. I. Hillel, "The Relation of External Evaporative Conditions to the Drying of Soils," *J. Geophys. Res.*, vol. 67, pp. 4319–4325, 1962.

40. Gash, J. H. C., "Comment on the Paper by A. S. Thom and H. R. Oliver, 'On Penman's Equation for Estimating Regional Evaporation'," *Q. J. R. Meteorol. Soc.*, vol. 104, pp. 532–533, 1978.

41. Gash, J. H. C., "A Note on Estimating the Effect of Limiting Fetch on Micrometeorological Evaporation Measurements," *Boundary Layer Meteorol.*, vol. 35, pp. 409–413, 1986.

42. Gash, J. H. C., I. R. Wright, and C. R. Lloyd, "Comparative Estimates of Interception Loss from Three Coniferous Forests in Great Britain," *J. Hydrol.*, vol. 48, pp. 89–105, 1980.

43. Greacon, E. L., ed., "Soil Water Assessment by the Neutron Method," CSIRO, Australia, 1981.

44. Halldin, S., and A. Lindroth, "Pine Forest Microclimate Simulation Using Different Diffusivities," *Boundary Layer Meteorol.*, vol. 35, pp. 103–123, 1986.

45. Ham, J. M., and J. L. Heilman, "Aerodynamic and Surface Resistances Affecting Energy Transport in a Sparse Crop," *Agric. For. Meteorol.*, vol. 53, pp. 267–284, 1991.

46. Harbeck, G. E., Jr, "A Practical Field Technique for Measuring Reservoir Evaporation Using Mass-Transfer Theory," *U.S. Geol. Surv. Prof. Paper,* 272-E, pp. 101–105, 1962.

47. Hargreaves, G. H., "Moisture Availability and Crop Production," *Trans. Am. Soc. Agric. Eng.*, vol. 18, no. 5, pp. 980–984, 1975.

48. Hargreaves, G. H., and Z. A. Samani, "Estimating Potential Evapotranspiration," Tech. Note, *J. Irrig. Drain Eng.*, vol. 108, no. 3, pp. 225–230, 1982.

49. Hargreaves, G. H., and Z. A. Samani, "Reference Crop Evapotranspiration from Temperature," *Appl. Eng. Agric.*, vol. 1, no. 2, pp. 96–99, 1985.

50. Hargreaves, G. L., G. H. Hargreaves, and J. P. Riley, "Agricultural Benefits for Senegal River Basin," *J. Irrig. Drain. Eng.*, vol. 111, no. 2, pp. 113–124, 1985.

51. Harrison, L. P., "Fundamental Concepts and Definitions Relating to Humidity," in A. Wexler, ed., *Humidity and Moisture,* vol. 3, Reinhold, New York, 1963.

52. Homes, R. M., "Discussion of 'A Comparison of Computed and Measured Soil Moisture under Snap Beans'," *J. Geophys. Res.*, vol. 66, pp. 3620–3622, 1961.

53. Idso, S. B., and R. D. Jackson, "Thermal Radiation from the Atmosphere," *J. Geophys. Res.*, vol. 74, pp. 5397–5403, 1969.

54. Itier, B., "Une méthode simplifée pour la mesure des flux de chaleur sensible," *Rech. Atmos.*, vol. 14, no. 1, pp. 17–34, 1980.

55. Itier, B., "Une méthode simple pour la mesure de l'evapotranspiration réelle a' l'echelle de la parcelle," *Agronomie,* vol. 1, no. 10, pp. 869–876, 1981.

56. Jensen, M. E., R. D. Burman, and R. G. Allen, "Evapotranspiration and Irrigation Water Requirements," *ASCE Manual* 70, p. 332, 1990.

57. Jones, H. G., *Plants and Microclimate,* Cambridge University Press, New York, 1984.

58. King, F. H., *Physics of Agriculture,* University of Wisconsin, Madison, Wis., 1910.

59. Kirby, C., and M. D. Newson, "The Plynlimon Catchments," *IH Report 109,* Institute of Hydrology, Wallingford, UK, 1990.

60. Koitzsch, R., and W. Golf, "Algorithmus zur Berechnung der realen Verdunstung, cited in 'Overview on the Present Status of the Concepts of Water Balance Models'," *IAHS Publ.* no. 148, Wallingford, pp. 3–19, 1983.

61. Lafleur, P. M., and W. R. Rouse, "Application of an Energy Combination Model for Evaporation from Sparse Canopies," *Agric. For. Meteorol.*, vol. 49, pp. 135–154, 1990.

62. Langbein, W. B., "Overview of AGU Chapman Conference on the Design of Hydrological Networks," *Water Resour. Res.*, vol. 15, no. 6, pp. 1867–1871, 1979.

63. Lhomme, J-P, "Extension of Penman's Formulae to Multi-Layer Models," *Boundary Layer Meteorol.*, vol. 42, pp. 281–291, 1988.

64. Lloyd, C. R., A. D. Culf, A. J. Dolman, and J. H. C. Gash, "Estimates of Sensible Heat Flux from Observations of Temperature Fluctuations," *Boundary Layer Meteorol.*, vol. 57, pp. 311–322, 1991.

65. Lloyd, C. R., J. H. C. Gash, W. J. Shuttleworth, and A. de O. Marques, "The Measurement and Modelling of Rainfall Interception by Amazonian Rainforest," *Agric. For. Meteorol.*, vol. 43, pp. 277–294, 1988.

66. Malhotra, G. P., and P. Bock, "Hydrologic Budget of North America and Sub-regions Formulated Using Atmospheric Vapor Flux Data," in *World Water Balance,* Studies and Reports in Hydrology, Report 11, UNESCO, Paris, 1972.

67. McNaughton, K. G., and T. W. Spriggs, "A Mixed-Layer Model for Regional Evaporation," *Boundary Layer Meteorol.*, vol. 34, pp. 243–263, 1986.

68. McNaughton, K. G., and T. W. Spriggs. "An Evaluation of the Priestley-Taylor Equation and the Complementary Relationship Using Results from a Mixed-Layer Model of the Convective Boundary Layer," *IAHS Publ.* no. 177, Wallingford, 1989.

69. McNeil, D. D., and W. J. Shuttleworth, "Comparative Measurements of Energy Fluxes over a Pine Forest," *Boundary Layer Meteorol.*, vol. 7, pp. 297–313, 1975.

70. Minhas, B. S., K. W. Parikh, and T. N. Srinavasan, "Toward the Structure of a Production Function for Wheat Yields with Dated Input of Irrigated Water," *Water Resour. Res.*, vol. 10, no. 3, pp. 383–393, 1974.

71. Monteith, J. L., "Evaporation and Surface Temperature," *Q. J. R. Meteorol. Soc.*, vol. 107, pp. 1–27, 1981.

72. Monteith, J. L., "Evaporation and the Environment," *Symp. Soc. Expl. Biol.*, vol. 19, pp. 205–234, 1965.

73. Morton, F. I., "Potential Evaporation and River Basin Evaporation," *J. Hydraul. Eng.*, vol. 91, no. HY6, pp. 67–97, 1965.

74. Morton, F. I., "Potential Evaporation as a Manifestation of Regional Evaporation," *Water Resour. Res.*, vol. 5, pp. 1244–1255, 1969.

75. Morton, F. I., "Estimating Evaporation and Transpiration from Climatological Observations," *J. Appl. Meteorol.*, vol. 14, pp. 488–497, 1975.

76. Morton, F. I., "Operational Estimates of Areal Evaporation and Their Significance to the Science and Practice of Hydrology," *J. Hydrol.*, vol. 66, pp. 1–76, 1983.

77. Nixon, P. R., and S. P. Lawless, "Translocation of Moisture with Time in Unsaturated Soil Profiles," *J. Geophys. Res.*, vol. 65, pp. 655–661, 1960.

78. Norero, A. L., "A Formula to Express Evapotranspiration as a Function of Soil Moisture and Evaporation Demands of the Atmosphere," Utah State University thesis, 1969.

79. Palmen, E., "Evaluation of Atmospheric Moisture Transport for Hydrological Purposes," *World Meteorological Organization IHD/Report* 1, International Hydrological Decade, 1967.

80. Penman, H. L., "Natural Evaporation from Open Water, Bare Soil and Grass," *Proc. R. Soc. London,* vol. A193, pp. 120–145, 1948.

81. Penman, H. L., *Vegetation and Hydrology,* Tech. Comm. 53, Commonwealth Bureau of Soils, Harpenden, England, 1963.

82. Penman, H. L., "Evaporation: An Introductory Survey," *Netherlands J. Agric. Sci.*, vol. 1, pp. 9–29, 87–97, 151–153, 1956.

83. Pereira, H. C., *Land Use and Water Resources in Temperate and Tropical Climates*, Cambridge University Press, 1973.

84. Pereira, H. C., *Policy and Practice in the Management of Tropical Watersheds*, Westview Press, Boulder, Colo., 1989.

85. Peschke, G., V. Dunger, and J. Gurtz, "Changes in Soil Moisture by Infiltration and Evapotranspiration," *IAHS Publ.* no. 156, 1986.

86. Priestley, C. H. B., and R. J. Taylor, "On the Assessment of Surface Heat Flux and Evaporation Using Large Scale Parameters," *Mon. Weather Rev.*, vol. 100, pp. 81–92, 1972.

87. Pruitt, W. O., "Prediction and Measurement of Crop Water Requirements: The Basis of Irrigation Scheduling," Faculty of Agriculture, University of Sydney, Sydney, NSW, 1986.

88. Pruitt, W. O., and J. Doorenbos, "Empirical Calibration, A Requisite for Evapotranspiration Formulae Based on Daily or Longer Mean Climatic Data?" *Proc. ICID International Round Table Conference on Evapotranspiration*, Budapest, Hungary, May 26–27, 1977.

89. Pruitt, W. O., and Lourence, F. J., "Experience in Lysimetry for ET and Surface Drag Measurements," in Advances in Evaporation, *ASAE Publ.* 14-85, American Society of Agricultural Engineers, St. Joseph, Mich., pp. 51–69, 1985.

90. Rao, K. S., J. C. Wyngaard, and O. R. Cote, "Local Advection of Momentum, Heat and Moisture in Micrometeorology," *Boundary Layer Meteorol.*, vol. 7, pp. 331–348, 1974.

91. Raphael, C., and J. E. Hay, "An Assessment of Models Which Use Satellite Data to Estimate Solar Irradiance at the Earth's Surface," *J. Clim. Appl. Meteorol.*, vol. 23, pp. 823–844, 1984.

92. Rasmussen, E. M., "Hydrological Application of Atmospheric Vapour-Flux Analyses," *WMO Operational Hydrology Report* 11, WMO-476, Geneva, Switzerland, 1977.

93. Renger, M., O. Strebel, and W. Giesel, "Beurteilung bodenkundlicher, kulturtechnischer und hydrolischer Fragen mit Hilfe von klimatischer Wasserbilanz und boden physikalischen Kennwerten," *Z. Kulturtech. Flurbereinig.*, vol. 15, pp. 148–160, 1974.

94. Roberts, J. M., "The Use of Tree Cutting Techniques in the Study of Water Relations of Mature *Pinus sylvestris L.*, Part I: The Technique and Survey of the Results," *J. Exp. Bot.*, vol. 28, pp. 751–767, 1977.

95. Robins, J. S., W. O. Pruitt, and W. H. Gardner, "Unsaturated Flow in Field Soils and Its Effect on Soil Moisture Investigations," *Soil Sci. Soc. Am. Proc.*, vol. 18, pp. 344–347, 1957.

96. Rodda, J. C., R. A. Downing, and F. M. Law, *Systematic Hydrology*, Butterworth, London, 1976.

97. Rodda, J. C., and S. W. Smith, "The Significance of the Systematic Error in Rainfall Measurement for Assessing Wet Deposition," *Atmos. Environ.*, vol. 20, pp. 1059–1064, 1986.

98. Rutter, A. J., K. A. Kershaw, P. C. Robins, and A. J. Morton, "A Predictive Model of Rainfall Interception in Forests. I: Derivation of the Model from Observations in a Plantation of Corsican Pine," *Agric. Meteorol.*, vol. 9, pp. 367–384, 1971.

99. Rutter, A. J., A. J. Morton, and P. C. Robins, "A Predictive Model of Rainfall Interception in Forests. II: Generalisation of the Model and Comparison with Observations in Some Coniferous and Hardwood Stands," *J. Appl. Ecol.*, vol. 12, pp. 367–380, 1975.

100. Sakuratani, T., "A Heat Balance Method for Measuring Water Flux in the Stem of Intact Plants," *J. Agric. Meteorol.*, vol. 37, no. 1, pp. 9–17, 1981.

101. Schuepp, P. H., M. Y. Leclerc, J. I. MacPherson, and R. L. Desjardin, "Footprint Prediction of Scalar Fluxes from Analytical Solutions of the Diffusion Equation," *Boundary Layer Meteorol.*, 1990.

102. Shuttleworth, W. J., "The Concept of Intrinsic Surface Resistance: Energy Budgets at a Partially Wet Surface," *Boundary Layer Meteorol.*, vol. 8, pp. 81–99, 1975.

103. Shuttleworth, W. J., "A Simplified One-Dimensional Theoretical Description of the Vegetation-Atmosphere Interaction," *Boundary Layer Meteorol.*, vol. 14, pp. 3–27, 1978.

104. Shuttleworth, W. J., "Macrohydrology—The New Challenge for Processes Hydrology," *J. Hydrol.*, vol. 100, pp. 31–56, 1988.

105. Shuttleworth, W. J., "Evaporation from Amazonian Rainforest," *Prof. R. Soc. London*, vol. B233, pp. 321–346, 1988.

106. Shuttleworth, W. J., "Micrometeorology of Temperate and Tropical Forest," *Phil. Trans. R. Soc. London*, vol. B324, pp. 299–334, 1989.

107. Shuttleworth, W. J., "Insight from Large-Scale Observational Studies of Land/Atmosphere Interactions," in Eric F. Wood, ed., "Land Surface Atmosphere Interactions: Parameterization and Analysis for Climate Modelling," *Surv. Geophys.*, vol. 12, pp. 3–39, 1990.

108. Shuttleworth, W. J., and I. R. Calder, "Has the Priestley-Taylor Equation Any Relevance to Forest Evaporation?" *J. Appl. Meteorol.*, vol. 18, pp. 634–638, 1979.

109. Shuttleworth, W. J., J. H. C. Gash, C. R. Lloyd, C. J. Moore, J. Roberts, A. de O. Marques, G. Fisch, V. de P. Silva Filho, M. N. G. Ribeiro, L. C. B. Molion, J. C. Nobre, L. D. A. de Sa, O. M. R. Cabral, S. R. Patel, and J. C. Mordas, "Observations of Radiation Exchange Above and Below Amazonian Rainforest," *Q. J. R. Meteorol. Soc.*, vol. 110, pp. 1163–1169, 1984.

110. Shuttleworth, W. J., J. H. C. Gash, C. R. Lloyd, D. D. McNeil, C. J. Moore, and J. S. Wallace, "An Integrated Micrometeorological System for Evaporation Measurement," *Agric. For. Meteorol.*, vol. 43, pp. 295–317, 1988.

111. Shuttleworth, W. J., and R. J. Gurney, "The Theoretical Relationship between Foliage," *Q. J. R. Meteorol. Soc.*, vol. 116, pp. 497–519, 1990.

112. Shuttleworth, W. J., and J. S. Wallace, "Evaporation from Sparse Crops—An Energy Combination Theory," *Q. J. R. Meteorol. Soc.*, vol. 111, pp. 839–855, 1984.

113. Sinclair, T. R., C. E. Murphy, and K. R. Knoerr, "Development and Evaluation of Simplified Models Simulating Canopy Photosynthesis and Transpiration," *J. Appl. Ecol.*, vol. 13, pp. 813–829, 1976.

114. Starr, V. P., and J. P. Peixoto, "On the Global Balance of Water Vapor and the Hydrology of Deserts," *Tellus*, vol. 10, pp. 188–194, 1958.

115. Stewart, J. B., "Evaporation from the Wet Canopy of a Pine Forest," *Water Resour. Res.*, vol. 13, no. 6, pp. 915–921, 1977.

116. Stewart, J. B., and A. S. Thom, "Energy Budgets in Pine Forest," *Q. J. R. Meteorol. Soc.*, vol. 99, pp. 154–170, 1973.

117. Tanner, C. B., "Measurement of Evapotranspiration," in *Irrigation of Agricultural Lands*, American Society of Agronomy Monograph 11, Madison, Wis., pp. 534–574, 1967.

118. Tetens, O., "Über einige meteorologische Begriffe," *Z. Geophys.*, vol. 6, pp. 203–204, 1930.

119. Thom, A. S., "Momentum, Mass and Heat Exchange of Vegetation," *Q. J. R. Meteorol. Soc.*, vol. 193, pp. 345–357, 1972.

120. Thom, A. S., "Momentum, Mass and Heat Exchange of Plant Communities," Sec. 3 in J. L. Monteith, ed., *Vegetation and the Atmosphere*, Academic Press, London, pp. 57–109, 1975.

121. Thom, A. S., and H. R. Oliver, "On Penman's Equation for Estimating Regional Evaporation," *Q. J. R. Meteorol. Soc.*, vol. 193, pp. 345–357, 1977.

122. Tillman, J. E., "The Indirect Determination of Stability, Heat and Momentum Fluxes in the Atmospheric Boundary Layer from Simple Scalar Variables During Dry, Unstable Conditions," *J. Appl. Meteorol.*, vol. 11, pp. 783–792, 1972.

123. Topp, G. C., J. L. Davis, and A. P. Annan, "Electromagnetic Determination of Soil-Water Content; Measurement in Coaxial Transmission Lines," *Water Resour. Res.*, vol. 16, pp. 574–582, 1980.

124. Topp, G. C., J. L. Davis, W. G. Bailey, and W. D. Zebchuk, "The Measurement of Soil Water Content Using a Portable TDR Hand Probe," *Can. J. Soils Sci.*, vol. 64, pp. 313–321, 1984.

125. Turc, L., "Evaluation des besoins en eau d'irrigation, evapotranspiration potentielle, formule climatique simplifice et mise a jour," *Ann. Agron.*, vol. 12, pp. 13–49, 1961.

126. Waggoner, P. E., *Vegetation and the Atmosphere,* vol. I, J. L. Monteith, ed., Academic Press, London, pp. 205–228, 1975.

127. Waggoner, P. E., and R. W. Reifsnyder, "Simulation of the Temperature Humidity and Evaporation Profiles in a Leaf Canopy," *J. Appl. Meteorol.,* vol. 7, pp. 400–409, 1968.

128. Waggoner, P. E., and N. C. Turner, "Comparison of Simulated and Actual Evaporation from Maize and Soil in a Lysimeter," *Agric. Meteorol.,* vol. 10, pp. 113–123, 1972.

129. Wesley, M. L., "Use of Variance Techniques to Measure Dry Air-Surface Exchange," *Boundary Layer Meteorol.,* vol. 44, pp. 13–31, 1988.

130. WMO, "Casebook on Hydrological Network Design Practice," *WMO Publ.* no. 324, Geneva, Switzerland, 1972.

131. WMO, "Hydrological Network Design and Information Transfer," *Operational Hydrology Report* 8, WMO Publ. 433, Geneva, Switzerland, 1976.

132. WMO, "Methods of Correction for Systematic Error in Point Precipitation Measurements for Operational Use," *Operational Hydrology Report* 21, WMO Publ. 589, Geneva, Switzerland, 1982.

133. Wright, I. R., "A Lysimeter for the Measurement of Evaporation from High Altitude Grass," in *Improved Methods of Hydrological Measurements in Mountain Areas,* IAHS Press, Wallingford, 1990.

134. Wright, J. L., "New Evapotranspiration Crop Co-efficients," *J. Irrig. Drain. Eng.,* vol. 108, no. IR2, pp. 57–74, 1982.

135. Wright, J. L., and M. E. Jensen, "Peak Water Requirements of (15) Crops in Southern Idaho," *J. Irrig. Drain. Eng.,* vol. 24, no. IR1, pp. 193–201, 1972.

136. Wyjk van, W. R., and D. A. de Vries, "Periodic Temperature Variations in Homogeneous Soil," in W. D. van Wijk, ed., *Physics of the Plant Environment,* North-Holland Publishing Co., Amsterdam, pp. 102–143, 1963.

137. Wyngaard, J. C., O. R. Cote, and Y. Izumi, "Local Free Convection, Similarity, and the Budgets of Shear Stress and Heat Flux," *J. Atmos. Sci.,* vol. 28, pp. 1171–1182, 1971.

CHAPTER 5
INFILTRATION AND SOIL WATER MOVEMENT

Walter J. Rawls
U.S.D.A. Agricultural Research Service
Beltsville, Maryland

Lajpat R. Ahuja
U.S.D.A. Agricultural Research Service
Fort Collins, Colorado

Donald L. Brakensiek
Agricultural Engineering Department
University of Maryland
College Park, Maryland

Adel Shirmohammadi
Agricultural Engineering Department
University of Maryland
College Park, Maryland

Infiltration is the process of water entry into a soil from rainfall, snowmelt, or irrigation. Soil water movement is the process of water flow from one point to another within the soil. The two processes cannot be separated as the rate of infiltration is controlled by the rate of soil water movement below the surface and the soil water movement continues after an infiltration event, as the infiltrated water is redistributed. The soil water movement also controls the supply of water for plant uptake and for evaporation at the soil surface. Infiltration and soil water movement play a key role in surface runoff, groundwater recharge, evapotranspiration, soil erosion, and transport of chemicals in surface and subsurface waters.

5.1 PROPERTIES AFFECTING SOIL WATER MOVEMENT

The soil properties affecting soil water movement are hydraulic conductivity (a measure of the soil's ability to transmit water) and water-retention characteristics (the ability of the soil to store and release water). These soil water properties are closely related to soil physical properties.

5.1.1 Soil Physical Properties

Particle-Size Properties. Particle-size properties are determined from the size distribution of individual particles in a soil sample. Soil particles smaller than 2 mm are divided into three soil texture groups: *sand, silt,* and *clay.* Figure 5.1.1[47] shows the particle size, sieve dimension, and the defined size class for the U.S. Department of Agriculture (USDA),[104] Canadian Soil Survey Committee,[53] the International Soil Science Society,[122] and the American Society for Testing and Materials[75] systems of classification. All mention of particle-size groups in this chapter refers to the USDA classification scheme. Figure 5.1.2 shows the limits for the basic USDA soil texture classes. Various methods for determining particle-size properties are given in Gee and Bauder.[32]

The particle-size properties which have the greatest effect on soil water retention[86] are the percentages of sand, silt, clay, fine sand, coarse sand, very coarse sand, and coarse fragments (>0.2 cm).

Morphological Properties. The morphological properties having the greatest effect on soil water properties are bulk density, organic matter, and clay type. These properties are closely related to soil structure and soil surface area. *Bulk density* is defined as the ratio of the weight of dry solids to the bulk volume of the soil. The bulk volume includes the volume of the solids and the pore space. Soil *porosity* (total volume occupied by pores per unit volume of soil) is computed from bulk density and *particle density* (normally assumed to be equal to 2.65 g/cm³) as follows:

$$\phi = 1 - \frac{BD}{PD} \tag{5.1.1}$$

where ϕ = total porosity (volume)
BD = soil bulk density, g/cm³
PD = particle density, g/cm³

As bulk density increases, water retention and hydraulic conductivity near saturation decrease. Also, water retention increases as the amount of soil organic matter (1.74 times the percent of organic carbon) increases.

In soil containing a large percentage of clay (> 10 percent) the clay mineralogy or clay type has a significant effect on soil water properties. For example, expandable clays such as montmorillonite have a significantly lower hydraulic conductivity and higher water retention than nonexpandable clays such as kaolinite.

Other morphological properties such as the thickness of the soil horizon and soil structure are derived from soil survey descriptions. A quantitative description of the effects of these properties on soil water movement has not been determined.

Chemical Properties. Chemical properties of the soil are also important because they affect the integrity of the soil aggregates (groups of soil particles bound together).

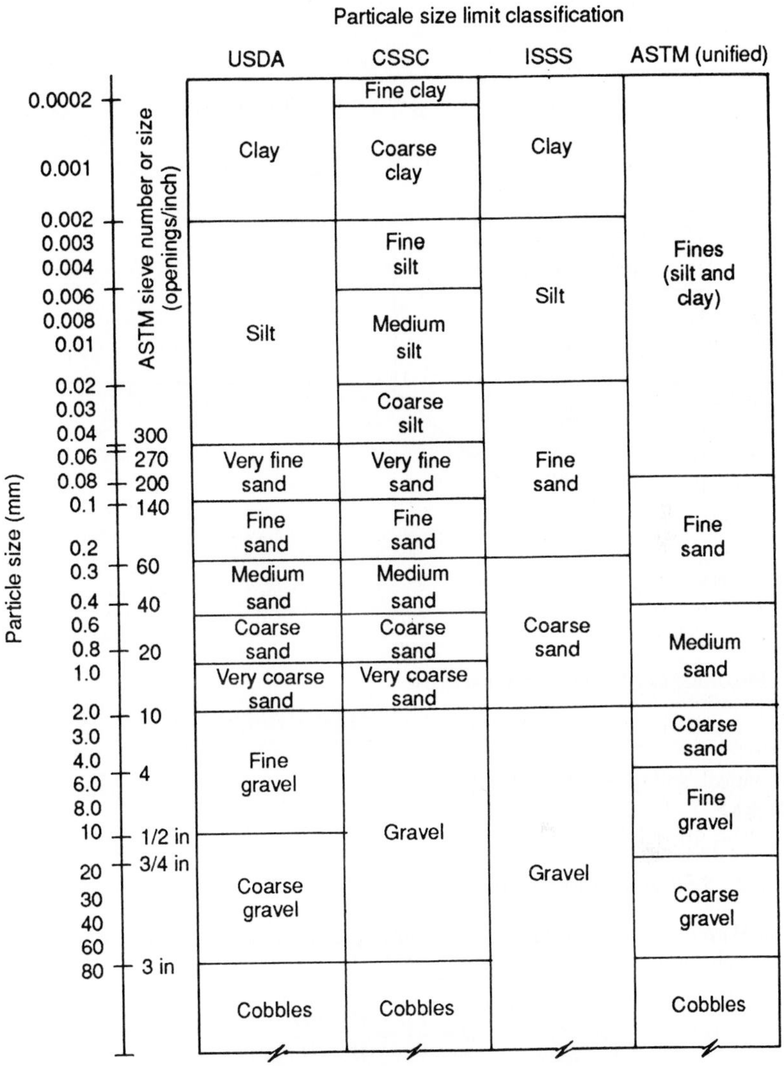

Particale size limit classification

Particle size (mm)	ASTM sieve number or size (openings/inch)	USDA	CSSC	ISSS	ASTM (unified)
0.0002		Clay	Fine clay	Clay	Fines (silt and clay)
0.001			Coarse clay		
0.002		Silt	Fine silt	Silt	
0.003					
0.004					
0.006			Medium silt		
0.008					
0.01					
0.02			Coarse silt	Fine sand	
0.03					
0.04	300				
0.06	270	Very fine sand	Very fine sand		
0.08	200				Fine sand
0.1	140	Fine sand	Fine sand		
0.2				Coarse sand	
0.3	60	Medium sand	Medium sand		
0.4	40				Medium sand
0.6		Coarse sand	Coarse sand		
0.8	20				
1.0		Very coarse sand	Very coarse sand		
2.0	10	Fine gravel	Gravel	Gravel	Coarse sand
3.0					
4.0					
6.0	4				Fine gravel
8.0					
10	1/2 in	Coarse gravel			
20	3/4 in				Coarse gravel
30					
40					
60					
80	3 in	Cobbles	Cobbles		Cobbles

USDA (104) U.S. Department of Agriculture
CSSC (53) Canadian Soil Survey Committee
ISSS (122) International Soil Science Society
ASTM (7) American Society for Testing and Materials

FIGURE 5.1.1 Particle-size limits according to several current classification schemes. *(Reproduced from Ref. 32 with permission.)*

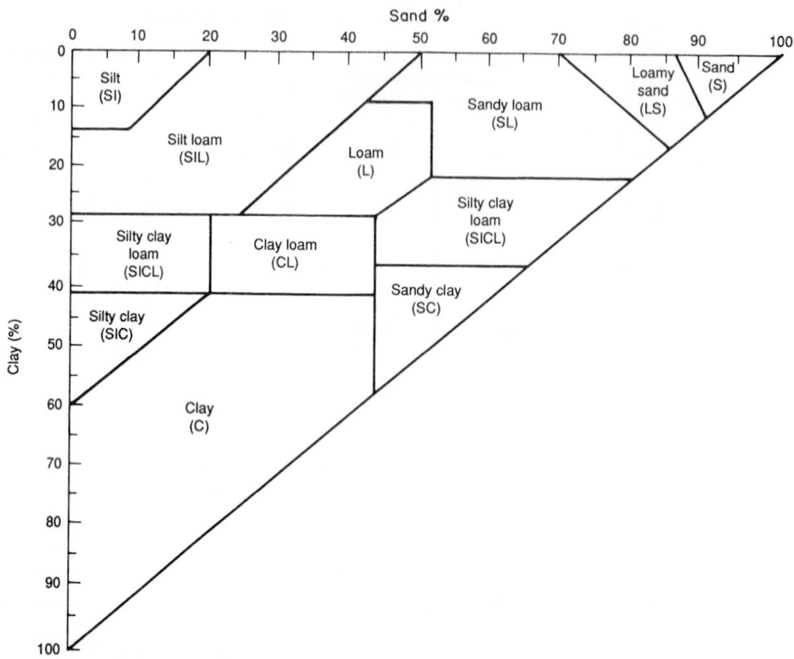

FIGURE 5.1.2 USDA soil textural triangle.

Chemical soil properties should be considered when they are outside normal ranges.[86]

5.1.2 Soil Water Properties

Soil Water Content. In most hydrologic applications soil water content is expressed on a volumetric basis as

$$\theta = \frac{V_w}{V_t} = \frac{W_w}{W_d}\frac{BD}{D} \tag{5.1.2}$$

where θ = volumetric water content, $cm^3\ cm^{-3}$
V_w = volume of water, cm^3
V_t = total volume of soil, cm^3
BD = bulk density of the soil, $g\ cm^{-3}$
W_w = weight of water, g
W_d = weight of dry soil, g
D = density of water (normally assumed equal to 1 $g\ cm^{-3}$)

Another commonly used water content term is percent saturation defined as

$$\text{Percent saturation} = 100\,\frac{\theta}{\phi} \tag{5.1.3}$$

Water-Retention Characteristic. The water-retention characteristic of the soil describes the soil's ability to store and release water and is defined as the relationship between the soil water content and the soil suction or matric potential. The water-retention characteristic is also known as the *moisture characteristic* or *capillary pressure-saturation curve.* Other terms that are synonymous with matric potential but may differ in sign or units are soil water suction, capillary potential, capillary pressure head, matric pressure head, tension, and pressure potential. *Matric potential* is the measure of the energy status of water in the soil and is a component of total soil water potential, which is described as

$$H_t = h_g + h_p + h_o + \cdots \tag{5.1.4}$$

where H_t is the total potential, h_g is the gravitational potential, h_p is the matric potential, and h_o is the osmotic potential; other types of potential are also possible.[37] The total soil water potential measures the energy of soil water at an elevation h_g subject to the suction pressure head h_p relative to water at atmospheric pressure located at elevation zero. Energy head is represented by units of length which is equivalent to the amount of energy possessed by a unit weight of water. Since unsaturated soil water pressures are less than atmospheric, the capillary pressure and matric potential are negative numbers.

Soil water content and matric potential have a power function relationship. The models most frequently used to describe this relationship are those proposed by Brooks and Corey,[20] Campbell,[22] and van Genuchten.[114] Table 5.1.1 describes these mathematical relationships. The model proposed by van Genuchten[114] permits a representation of the total water-retention curve, whereas Campbell[22] and Brooks and Corey[20] describe only the portion of the curve for matric potentials less than the bubbling pressure, or pressure at which air will enter the soil. Table 5.1.1 also presents the relationship between the model parameters.[80] Figure 5.1.3*a* illustrates the water-retention relationship for two contrasting soil textures. Note that the sandy loam soil retains less water than the clayey soil.

Hydraulic Conductivity. The hydraulic conductivity is a measure of the ability of the soil to transmit water and depends upon both the properties of the soil and the fluid.[47,99] Total porosity, pore-size distribution, and pore continuity are the important soil characteristics affecting hydraulic conductivity. Fluid properties affecting hydraulic conductivity are viscosity and density. The hydraulic conductivity at or above the saturation point ($h \geq 0$) is referred to as *saturated hydraulic conductivity,* and for water contents below saturation ($h < 0$), it is called the *unsaturated hydraulic conductivity.*

The hydraulic conductivity is a nonlinear function of volumetric soil water content, and varies with soil texture as is shown in Fig. 5.1.3*b*. For a sandy loam soil, the hydraulic conductivity at its saturated water content (point *b*1) is higher than that for a clay soil (point *b*2), even though the porosity is higher in the clay soil. As the soil water content decreases, the hydraulic conductivity of both soils decreases rapidly. Figure 5.1.3*c* presents K as a function of h. This shows that the hydraulic conductivity of the sandy loam soil decreases more rapidly with decrease in h than that of the clay soil, such that at lower h values (or higher suctions) the hydraulic conductivity of the clay soil is higher. The rate of change of h in the sandy soil also decreases much more rapidly than that of the clay soil.

Hysteresis. Figure 5.1.4 illustrates the matric potential water content $h(\theta)$ relationship as the soil was *desorbing,* or draining, and *absorbing,* or wetting. It has com-

TABLE 5.1.1 Soil Water Retention and Hydraulic Conductivity Relationships

Hydraulic soil characteristic	Parameters	Parameter correspondence
	Brooks and Corey[20]	
Soil water retention	λ = pore-size index	$\lambda = \lambda$
$\frac{\theta - \theta_r}{\phi - \theta_r} = \left(\frac{h_b}{h}\right)^\lambda$	h_b = bubbling capillary pressure	$h_b = h_b$
	θ_r = residual water content	$\theta_r = \theta_r$
	ϕ = porosity	$\phi = \phi$
Hydraulic conductivity	K_s = fully saturated conductivity ($\theta = \phi$)	$K_s = K_s$
$\frac{K(\theta)}{K_s} = \left(\frac{\theta - \theta_r}{\phi - \theta_r}\right)^n = (S_e)^n$	$n = 3 + \frac{2}{\lambda}$	
	Campbell[22]	
Soil water retention	ϕ = porosity	$\phi = \phi$
$\frac{\theta}{\phi} = \left(\frac{H_b}{h}\right)^{1/b}$	H_b = scaling parameter with dimension of length	$H_b = h_b$
	b = constant	$b = \frac{1}{\lambda}$
Hydraulic conductivity	$n = 3 + 2b$	
$\frac{K(\theta)}{K_s} = \left(\frac{\theta}{\phi}\right)^n$		
	Van Genuchten[114]	
Soil water retention	ϕ = porosity	$\phi = \phi$
$\frac{\theta - \theta_r}{\phi - \theta_r} = \left[\frac{1}{1 + (\alpha h)^n}\right]^m$	θ_r = residual water	$\theta_r = \theta_r$
	α = constant	$\alpha = (h_b)^{-1}$
	n = constant	$n = \lambda + 1$
	m = constant	$m = \frac{\lambda}{\lambda + 1}$
Hydraulic conductivity		
$\frac{K(\theta)}{K_s} = \left(\frac{\theta - \theta_r}{\phi - \theta_r}\right)^{1/2} \left\{1 - \left[1 - \left(\frac{\theta - \theta_r}{\phi - \theta_r}\right)^{1/m}\right]^m\right\}^2$		

θ = water content; h = capillary suction, cm; $K(\theta)$ = hydraulic conductivity for given water content, cm/h

monly been observed that the desorbing and absorbing relationships differ, and this phenomenon is called *hysteresis,* which is caused by entrapment of air in pockets connecting different size pores during wetting. The hysteresis in the $K(\theta)$ relationship is often small[67] compared with that in the $h(\theta)$ relationship. The hysteresis in $h(\theta)$ (Fig. 5.1.4) also shows that a soil undergoing cycles of wetting and drying exhibits secondary hysteretic loops. This nonuniqueness of the $h(\theta)$ causes considerable difficulty in modeling soil water movement in nature. Hysteresis can be avoided if one considers either just a wetting cycle or just a draining cycle. For practical applications, hysteresis has mostly been neglected.

5.1.3 Spatial Variability of Soil Water Properties

Natural field soils encompass considerable spatial variability in their properties even within a given soil type. During the last two decades, a large number of studies have

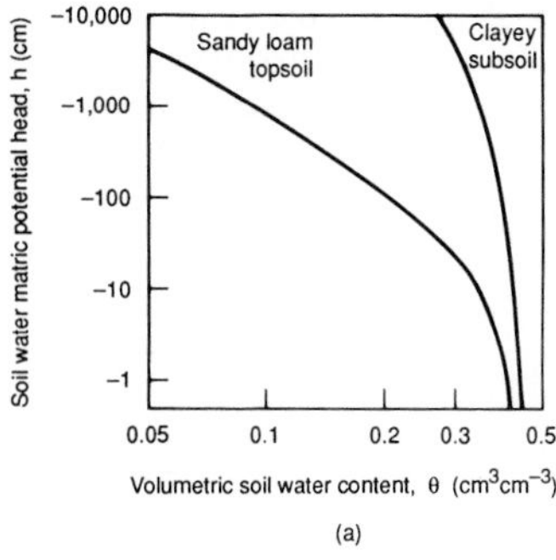

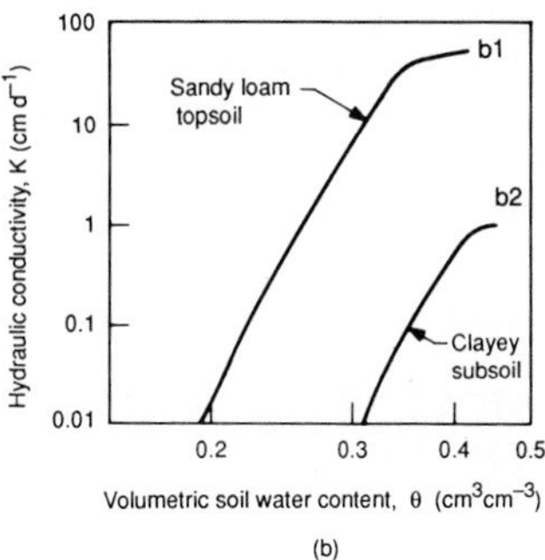

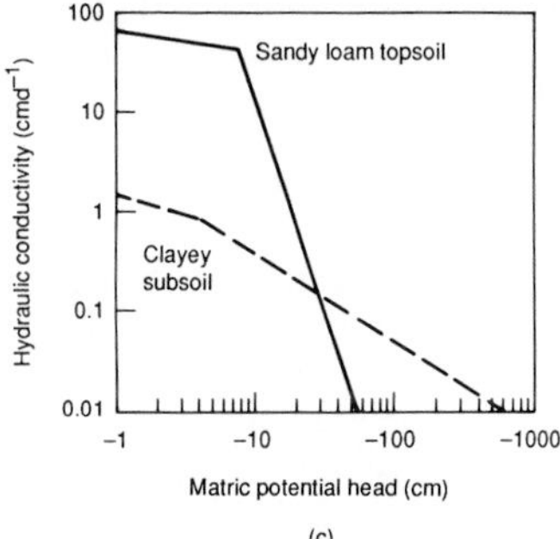

FIGURE 5.1.3 *(a)* The $h(\theta)$, *(b)* $K(\theta)$, and *(c)* $K(h)$ relationships of sandy loam and clayey horizons.

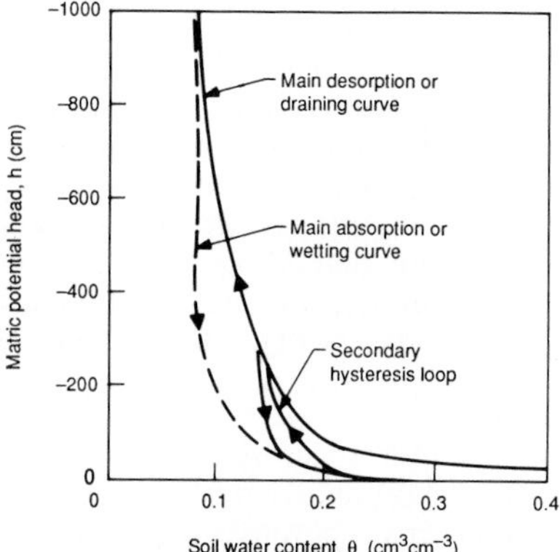

FIGURE 5.1.4 The effect of hysteresis on the $h(\theta)$ function of sandy loam topsoil during wetting and drying.

addressed the nature and statistical structure of spatial variability in soil water properties. The methods used to represent variability include frequency distributions, spatial trends or drift, spatial autocovariance and autocorrelation, and similar-media or related types of scaling.[6,113,116,117,119] At this time these techniques are not being widely used in practical applications.

5.2 MEASUREMENT OF SOIL WATER PROPERTIES

5.2.1 Soil Water Content

Soil water content can be determined by direct or indirect methods. The gravimetric method is the primary direct method. Indirect methods involve inferring the soil water content from the measurement of a soil property or a property of an object placed in the soil that is affected by soil water content. Indirect methods include radiological methods, electrical resistance, time-domain reflectometry, nuclear magnetic resonance, and remote-sensing techniques. The detailed description of the methods including installation and operation is given in Gardner[31] and Brakensiek et al.[17] The following is a brief description of the methods.

Gravimetric Method. The gravimetric method of soil water determination is the standard method to which all other methods are compared. This method consists of taking soil samples weighing between 100 and 200 g. The samples are put in a container of known weight, weighed (subtracting the weight of the container gives the weight of the wet soil), and sealed airtight. The samples are then taken to the lab and opened, put in an oven, and dried at 105°C until all the water has evaporated (usually about 24 h). After drying the samples, they are weighed again (subtracting the weight of the container gives the weight of the dry soil). The moisture content on a weight basis is the difference between the wet and dry weights divided by the dry weight. For more details see Brakensiek et al.[17] and Gardner.[31] The gravimetric method of soil water content determination requires simple equipment but is a time-consuming and destructive procedure.

Radiological Methods. Two radiological methods are available for measurement of soil water content. One is the *neutron thermalization method,* which is based on the interaction between high-energy neutrons (fast neutrons) and the nuclei of hydrogen atoms in the soil. The neutron method measures soil water content of a 7- to 15-cm-diameter sphere. The other is the *gamma attenuation,* based on the attenuation and backscattering of gamma rays as they pass through the soil. The gamma method measures a 1-cm-wide band. Both methods use portable equipment for taking measurements at permanent observation sites. These methods are therefore nondestructive and have the additional advantage of obtaining data from the same location at each observation.[17]

Electrical Resistance Method. Electrical resistance methods use electrodes encased in some porous medium (e.g., gypsum, nylon fabric, and fiberglass fabric) which is in contact with the soil. The water in the porous casing of the electrodes tends toward an equilibrium with the soil moisture. These methods are not sensitive at high soil water contents; however, they are nondestructive and yield data from the same location at each observation.[17]

Time-Domain Reflectometry (TDR) Method. This method determines the dielectric constant of soil by measuring the transmission time of an electromagnetic pulse between a pair of parallel, metallic rods of known length embedded in the soil. The soil water content is calculated from a unique relationship between the dielectric constant and water content.[28,110–112] This method has proved very useful in the field to make continuous measurements on the same area by installing permanent lines.

Nuclear Magnetic Resonance Method. This measurement approach depends on the interaction between hydrogen nuclear magnetic moments and a magnetic field. Discrimination of water signals from signals from other hydrogen-bearing material is based on nuclear relaxation times, which reflect the molecular environment and structural bonding characteristics of water molecules in the soil. This technique is capable of measuring soil water on static samples continuously along a transect.[68]

Remote Sensing. Measurement of soil moisture using remote-sensing techniques depends upon the type of reflected or emitted radiation.[97] See Chap. 24 for details. The primary advantage over the other soil moisture measurement techniques is that remote sensing produces a soil moisture measurement averaged over a large area rather than a point measurement.

5.2.2 Water-Retention Characteristic

Cassel and Klute[23] provide a detailed discussion of field and laboratory methods for measuring water potential measurement using *tensiometers* (porous ceramic cups connected to a vacuum or manometer). The theory and use of tensiometers has been discussed by Richards.[90]

Gypsum blocks are another relatively widely used apparatus for measuring the soil water matric potential. Gypsum blocks are cheap but are inaccurate pressure potential indicators, especially in the wet range.[99] However, gypsum blocks are good indicators in dry soil water range (potentials less than 100 kPa). Thus tensiometers and gypsum blocks are often used together, the tensiometers for wet conditions and gypsum blocks for dry conditions. Other procedures such as the psychrometric method have also been used in research for in situ measurement of matric potential.[77] In areas where salinity is a problem, salinity sensors have been used extensively.[99]

The $h(\theta)$ relationships for different soil horizons are determined in the field by simultaneously measuring h and θ at a number of times during redistribution of water. This field determination is usually restricted to the wet range, matric potentials above -33 kPa (-333 cm), using tensiometers and a neutron probe or gravimetric θ sampling. More commonly, the $h(\theta)$ relationship is measured on undisturbed soil cores brought from the field to the laboratory. The soil cores are slowly saturated with water and then drained to equilibrium with a sequence of negative-pressure steps during which the outflow of water and the change in weight of the soil cores is measured. Air-pressure chambers equipped with ceramic pressure plates or pressure membranes are used for this measurement, and they are restricted to matric potentials less than -10 kPa. Hanging water columns are commonly used to attain different pressures in the wet range (>-10 kPa).[23]

5.2.3 Hydraulic Conductivity

There are many field and laboratory methods for determining soil hydraulic conductivity. The choice of a method depends upon many factors, including the nature of the soil, availability of equipment and expertise, soil water content range for which measurements are needed (saturated or unsaturated), and the purpose for which the measured values are to be used.[47]

Saturated Hydraulic Conductivity—Field and Laboratory Measurements. Amoozegar and Warrick[8] provide a detailed discussion of the several field techniques used both below and above the water table. Bouwer[13] discusses the selection of an appropriate field method as it relates to the soil and the dominant flow direction.

The auger hole and piezometer methods are two widely used methods of measuring saturated hydraulic conductivity below a water table. The *auger hole method* as described by Bouwer and Jackson[14] can be used to determine the saturated hydraulic conductivity of each layer in a stratified profile by drilling auger holes of different depths.

The *piezometer method* is based on the measurement of the flow into an unlined cavity at the lower end of a lined hole. This method may be used to measure either horizontal or vertical hydraulic conductivity.

To determine the hydraulic conductivity for shallow water table conditions the *drain outflow measurement technique* is useful.[40,98]

The *Guelph permeameter*[88] is the most commonly used technique to measure saturated hydraulic conductivity in the field. This method is popular because it uses a smaller volume of water as compared with the other techniques.

Saturated hydraulic conductivity is usually measured in the laboratory using constant-head and falling-head methods on undisturbed soil cores.[37,47] Both of these methods are based on the direct application of Darcy's law through the saturated soil column of a uniform cross-sectional area.

Unsaturated Hydraulic Conductivity—Field and Laboratory Measurements. Green et al.[33] provide a detailed discussion of the field measurement of unsaturated hydraulic conductivity. They present both "unsteady drainage-flux methods" and "steady flux methods."

Steady flux methods involve establishment of a known flux of water at the soil surface rather than using the conservation of mass principle at each depth increment as in the unsteady drainage-flux technique.[33] Hillel and Gardner[39] and Bouma et al.[12] used a crust boundary to establish a steady water flux at the soil surface. Sprinkler applications of water through rainfall simulators have also been used to establish steady flux at the soil surface.[38,123]

Both the steady-state and non-steady-state principles have been used for measurement of unsaturated hydraulic conductivity in the laboratory. Steady-state methods include the establishment of boundary conditions to obtain steady, one-dimensional flow under different adjustable pressure heads.[37,47,99] Arya et al.[10] uses an evaporation-based method to determine the unsaturated hydraulic conductivity.

5.3 Estimation of Soil Water Properties

5.3.1 Water-Retention Characteristics

The simplest method for estimating $h(\theta)$ is to use soil texture reference curves (Fig. 5.3.1) or to relate soil properties to the soil water held at specific matric potentials

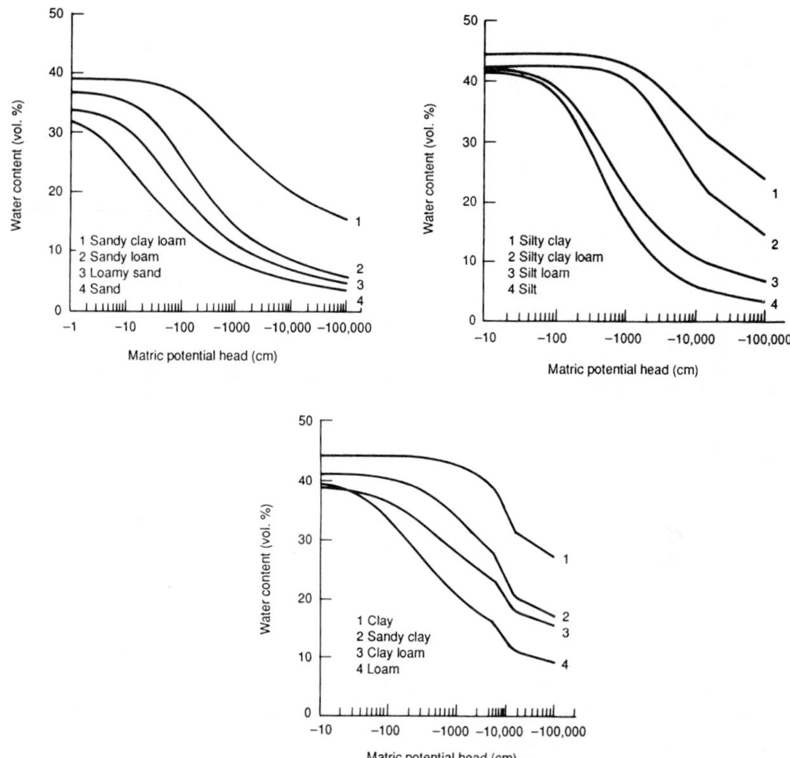

FIGURE 5.3.1 Water-retention curves for USDA soil textures.

using regression analysis. Rawls and Brakensiek[79] incorporated three levels of information for predicting the soil water at specific matric potentials (Table 5.3.1). The first set of equations is based on the particle size and organic matter data. The second level incorporates one measured point, −1500 kPa water content, with the particle size and organic matter data, and the third level incorporates two measured points, −33 and −1500 kPa water contents. A modification of this approach is to use a log-log linear interpolation between −33 and −1500 kPa water contents.[3] Also, the soil water-retention curve can be estimated from bulk density, −33 kPa water content, and a reference soil water-retention curve (Fig. 5.3.1).

A second method estimates parameters for models describing the $h(\theta)$ or $\theta(h)$ relationships such as the Brooks and Corey,[20] Campbell,[22] and Van Genuchten[114] models (Table 5.1.1). Several approaches relating the model parameters to soil properties have been used. One method is to use average parameter values for soil texture classes (Table 5.3.2).[84] Another approach relates the water-retention parameters to soil properties by regression analysis. Regression equations developed by Rawls and Brakensiek[80] for the Brooks and Corey model are given in Table 5.3.3. Using the correspondence between model parameters given in Table 5.1.1, the equations in Table 5.3.3 can be used for any of the water-retention models given in Table 5.1.1. Also, using the parameters, the complete water-retention curve can be developed (Fig. 5.3.1).

TABLE 5.3.1 Linear Regression Equations for Predicting Soil Water Content at Specific Suctions

The equations have the form $\theta_x = a + b \times$ sand (%) $+ c \times$ silt (%) $+ d \times$ clay (%) $+ e \times$ organic matter (%) $+ f \times$ bulk density (g/cm^3) $+ g \times 0.33$ bar moisture (cm^3/cm^3) $+ h \times 15$ bar moisture (cm^3/cm^3), where θ_x = predicted water retention (cm^3/cm^3) for a given suction, a–h = regression coefficients.

Matric potential, kPa	Intercept	Sand, %	Silt, %	Clay, %	Organic matter, %	Bulk density, g/cm^3	−33 kPa water retention, cm^3/cm^3	−1500 kPa water retention, cm^3/cm^3	Correlation coefficient R
	Regression coefficients								
	a	b	c	d	e	f	g	h	
	0.7899	−0.0037			0.0100	−0.1315			0.58
−4	0.6275	−0.0041			0.0239			−0.08	0.57
	0.1829				−0.0246	−0.0376	1.89	−1.38	0.77
	0.7135	−0.0030		0.0017		−0.1693			0.74
−7	0.4829	−0.0035			0.0263			0.25	0.74
	0.0888	−0.0003			−0.0107		1.53	−0.81	0.91
	0.4118	−0.0030		0.0023	0.0317				0.81
−10	0.4103	−0.0031			0.0260			0.41	0.81
	0.0619	−0.0002			−0.0067		1.34	−0.51	0.95
	0.3121	−0.0024		0.0032	0.0314				0.86
−20	0.3000	0.0024			0.0235			0.61	0.89
	0.0319	−0.0002					1.01	−0.06	0.99
	0.2576	−0.0020		0.0036	0.0299				0.87
−33	0.2391	−0.0019			0.0210			0.72	0.92
	0.2065	−0.0016		0.0040	0.0275				0.87
−60	0.1814	−0.0015			0.0178			0.80	0.94
	0.0136					−0.0091	0.66	0.39	0.99
	0.0349		0.0014	0.0055	0.0251				0.87

−100	0.1417	−0.0012			0.0151		0.85	0.96
	−0.0034				0.0022	0.52	0.54	0.99
	0.0281		0.0011	0.0054	0.0200			0.86
−200	0.0986	−0.0009			0.0116		0.90	0.97
	−0.0043				0.0026	0.36	0.69	0.99
	0.0238		0.0008	0.0052	0.0190			0.84
−400	0.0649	−0.0006			0.0085		0.93	0.98
	−0.0038				0.0026	0.24	0.79	0.99
	0.0216		0.0006	0.0050	0.0167			0.81
−700	0.0429	−0.0004			0.0062		0.94	0.98
	−0.0027				0.0024	0.16	0.86	0.99
	0.0205		0.0005	0.0049	0.0154			0.81
−1000	0.0309	−0.0003			0.0049		0.95	0.99
	−0.0019				0.0022	0.11	0.89	0.99
−1500	0.0260			0.0050	0.0158			0.80

Sand (%) + silt (%) + clay (%) = 100, Sand = 2.0–0.05 mm, Silt = 0.05–0.002 mm, Clay < 0.002 mm.
Source: Reproduced from Ref. 80 by permission of ASCE.

TABLE 5.3.2 Water-Retention Properties Classified by Soil Texture

Texture class	Sample size	Total porosity ϕ, cm^3/cm^3	Residual water content θ_r, cm^3/cm^3	Effective porosity ϕ_e, cm^3/cm^3	Bubbling pressure h_b: Geometric,[†] mean, cm	Pore-size distribution λ: Arithmetic mean	Water retained at −33 kPa, cm^3/cm^3	Water retained at −1500 kPa, cm^3/cm^3
Sand	762	0.437* (0.374–0.500)	0.020 (0.001–0.039)	0.417 (0.354–0.480)	7.26 (1.36–38.74)	0.694 (0.298–1.090)	0.091 (0.018–0.164)	0.033 (0.007–0.059)
Loamy sand	338	0.437 (0.368–0.506)	0.035 (0.003–0.067)	0.401 (0.329–0.473)	8.69 (1.80–41.85)	0.553 (0.234–0.872)	0.125 (0.060–0.190)	0.055 (0.019–0.091)
Sandy loam	666	0.453 (0.351–0.555)	0.041 (−0.024–0.106)	0.412 (0.283–0.541)	14.66 (3.45–62.24)	0.378 (0.140–0.616)	0.207 (0.126–0.288)	0.095 (0.031–0.159)
Loam	383	0.463 (0.375–0.551)	0.027 (−0.020–0.074)	0.434 (0.334–0.534)	11.15 (1.63–76.40)	0.252 (0.086–0.418)	0.270 (0.195–0.345)	0.117 (0.069–0.165)
Silt loam	1206	0.501 (0.420–0.582)	0.015 (−0.028–0.058)	0.486 (0.394–0.578)	20.76 (3.58–120.4)	0.234 (0.105–0.363)	0.330 (0.258–0.402)	0.133 (0.078–0.188)
Sandy clay loam	498	0.398 (0.332–0.464)	0.068 (−0.001–0.137)	0.330 (0.235–0.425)	28.08 (5.57–141.5)	0.319 (0.079–0.559)	0.255 (0.186–0.324)	0.148 (0.085–0.211)
Clay loam	366	0.464 (0.409–0.519)	0.075 (−0.024–0.174)	0.390 (0.279–0.501)	25.89 (5.80–115.7)	0.242 (0.070–0.414)	0.318 (0.250–0.386)	0.197 (0.115–0.279)
Silty clay loam	689	0.471 (0.418–0.524)	0.040 (−0.038–0.118)	0.432 (0.347–0.517)	32.56 (6.68–158.7)	0.177 (0.039–0.315)	0.366 (0.304–0.428)	0.208 (0.138–0.278)
Sandy clay	45	0.430 (0.370–0.490)	0.109 (0.013–0.205)	0.321 (0.207–0.435)	29.17 (4.96–171.6)	0.223 (0.048–0.398)	0.339 (0.245–0.433)	0.239 (0.162–0.316)
Silty clay	127	0.479 (0.425–0.533)	0.056 (−0.024–0.136)	0.423 (0.334–0.512)	34.19 (7.04–166.2)	0.150 (0.040–0.260)	0.387 (0.332–0.442)	0.250 (0.193–0.307)
Clay	291	0.475 (0.427–0.523)	0.090 (−0.015–0.195)	0.385 (0.269–0.501)	37.30 (7.43–187.2)	0.165 (0.037–0.293)	0.396 (0.326–0.466)	0.272 (0.208–0.336)

* First line is the mean value. Second line is ± one standard deviation about the mean.
[†] Antilog of the log mean.
Source: Reproduced from Ref. 80 by permission of ASCE.

TABLE 5.3.3 Estimation Equation for Brooks-Corey Parameters

h_b—Brooks-Corey bubbling pressure, cm

$$h_b = \exp[5.3396738 + 0.1845038(C) - 2.48394546(\phi) - 0.00213853(C)^2 - 0.04356349(S)(\phi) - 0.61745089(C)(\phi) + 0.00143598(S)^2(\phi^2) - 0.00855375(C^2)(\phi^2) - 0.00001282(S^2)(C) + 0.00895359(C^2)(\phi) - 0.00072472(S^2)(\phi) + 0.0000054(C^2)(S) + 0.50028060(\phi^2)(C)]$$

λ—Brooks-Corey pore-size distribution index

$$\lambda = \exp[-0.7842831 + 0.0177544(S) - 1.062498(\phi) - 0.00005304(S^2) - 0.00273493(C^2) + 1.11134946(\phi^2) - 0.03088295(S)(\phi) + 0.00026587(S^2)(\phi^2) - 0.00610522(C^2)(\phi^2) - 0.00000235(S^2)(C) + 0.00798746(C^2)(\phi) - 0.00674491(\phi^2)(C)]$$

θ_r—Brooks-Corey residual water content (volume fraction)

$$\theta_r = -0.0182482 + 0.00087269(S) + 0.00513488(C) + 0.02939286(\phi) - 0.00015395(C^2) - 0.0010827(S)(\phi) - 0.00018233(C^2)(\phi^2) + 0.00030703(C^2)(\phi) - 0.0023584(\phi^2)(C)$$

C = percent clay ($5 < \% < 60$)
S = percent sand ($5 < \% < 70$)
ϕ = porosity (volume fraction)
Source: Reproduced from Ref. 80 by permission of ASCE.

5.3.2 Hydraulic Conductivity

Selection of hydraulic conductivity prediction techniques depends upon the availability and the level of information on physical and hydraulic properties of soil.[64] The most common technique is to apply the Brooks and Corey, Campbell, or van Genuchten hydraulic conductivity relationships (Table 5.1.1). All parameters needed for the $K(\theta)$ equations can be obtained from those of the $\theta(h)$, except for the saturated hydraulic conductivity. If only soil texture classes are available the saturated hydraulic conductivities and corresponding unsaturated hydraulic conductivity curves can be obtained from Fig. 5.3.2.[96] If specific soil texture information is available, the saturated hydraulic conductivity for undisturbed conditions can be obtained from Fig. 5.3.3. Another technique which is more specific was developed by Ahuja et al.,[3–5] who related the saturated hydraulic conductivity to an effective porosity (ϕ_e, total porosity obtained from soil bulk density minus the soil water content at -33 kPa matric potential) by the following generalized Kozeny-Carman equation:

$$K_s = B\phi_e^n \tag{5.3.1}$$

where n can be set equal to 4 and B equals 1058 when K_s has units of cm h^{-1}. For soils with sand greater than 65 percent and/or clays greater than 40 percent the saturated hydraulic conductivities may vary by an order of magnitude or greater.

Coarse fragments (>2 mm in size) in the soil in addition to their effect in reducing porosity (Sec. 5.5.3) also affect the saturated hydraulic conductivity of the soil. The saturated hydraulic conductivity of the soil matrix should be multiplied by the following correction for coarse fragments:[15]

$$\text{Coarse fragment correction} = 1 - \frac{\%\text{ weight coarse fragments}}{100} \tag{5.3.2}$$

Freezing the soil will also affect the saturated hydraulic conductivity. Rawls and Brakensiek[80] generalized the work of Lee[49] and developed a saturated hydraulic

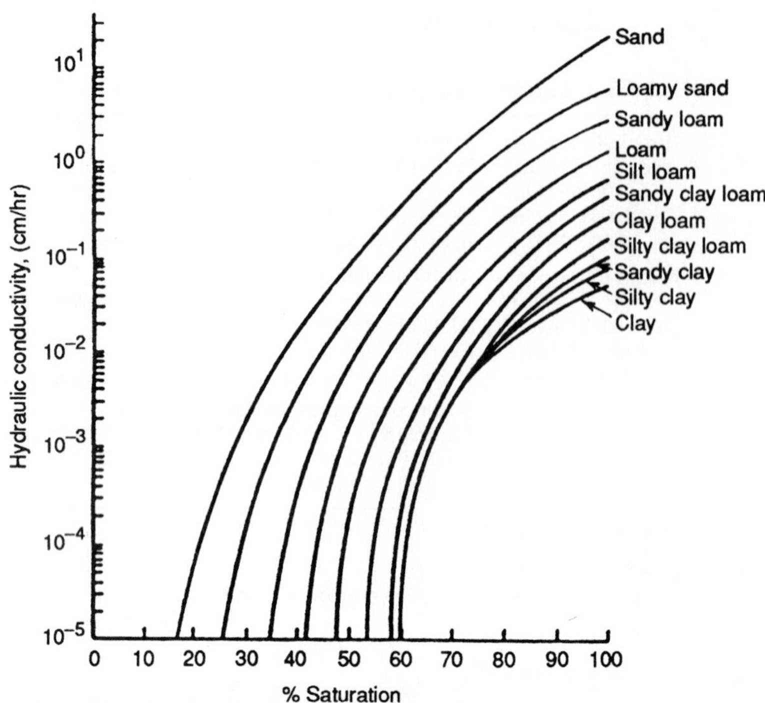

FIGURE 5.3.2 Hydraulic conductivity sorted by soil texture. *(Reproduced from Ref. 83 with permission.)*

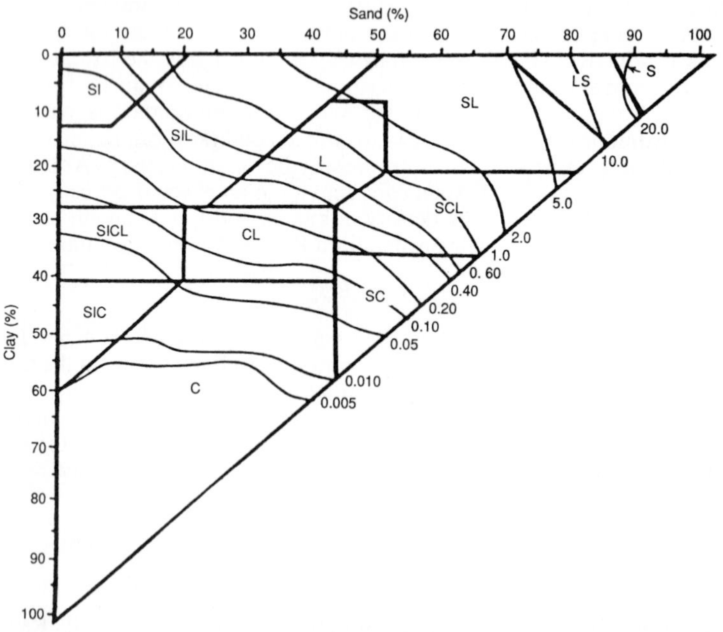

FIGURE 5.3.3 Saturated hydraulic conductivity for USDA soil texture triangle. *(Reproduced from Ref. 80 by permission of ASCE.)*

conductivity correction factor for frozen soil which is

$$\text{FSC} = 2.0 - 1.9\,\frac{\theta_f}{\theta_{33}} \tag{5.3.3}$$

where FSC = frozen soil hydraulic conductivity correction
θ_f = percent volume soil water at time of freezing
θ_{33} = percent volume soil water held at −33 kPa matric potential

When θ_f is greater than θ_{33}, FSC equals 0.001.

5.4 PRINCIPLES OF SOIL WATER MOVEMENT AND INFILTRATION

5.4.1 Forces Governing Water Flow in Saturated Soils—Darcy's Law

The work of Hagen in 1839 and of Poiseuille in 1841 on laminar flow in pipes established that water, while flowing from a point of higher energy head to a point of lower energy, loses energy due to friction along the flow path in proportion to the velocity of water.[27] Darcy[26] was familiar with their work and recognized a similarity between water flow through a water-saturated soil and laminar flow in pipes. His experiments on flow through saturated sand beds indicated that the rate of flow was indeed proportional directly to head loss and inversely to the length of flow path, with a constant proportionality factor. This finding is now well established as *Darcy's law.* It may be expressed as

$$Q = KA\,\frac{\Delta H}{\Delta l} \tag{5.4.1}$$

where Q is the volumetric discharge or flow rate ($m^3\ s^{-1}$) through a cross-sectional area A (m^2), ΔH (m) is the difference in hydraulic heads of water between two ends of the soil column of length Δl (m), and K ($m\ s^{-1}$) is the proportionality constant, called the *hydraulic conductivity.* The hydraulic conductivity is a characteristic of the soil as well as water properties. In differential form, Eq. (5.4.1) for steady flow may be written as

$$q = -K\,\frac{dH}{dz} \tag{5.4.2}$$

where q is the specific discharge or the flow rate per unit cross-sectional area of soil ($m\ s^{-1}$), often called the *Darcy velocity,* or *flux,* and z is the distance in the direction of flow. The *hydraulic head H,* the energy per unit weight of water, is expressed as

$$H = h + z \tag{5.4.3}$$

where h is the soil water pressure head and z is the elevation above an arbitrary datum. The soil water pressure head h is equal to p/γ, where p ($J\ m^{-3}$) is the soil water pressure and $\gamma = \rho g$ ($kg\ m^{-2}\ s^{-2}$) is the specific weight of water. Equation (5.4.2) is written for flow in one dimension, but this can be easily generalized to describe flow in two or three dimensions. The minus sign in Eq. (5.4.2) arises because the gradient dH/dz in the direction of flow is negative, whereas the q in that direction is considered positive. Darcy's law applies to a saturated soil which is *homogeneous* (soil water properties do not vary with location) and *isotropic* (soil water properties do not vary with direction). For a layered soil, it can be applied separately for each layer, if each soil layer is assumed to be homogeneous and isotropic within itself. The value of K and the gradient dH/dz vary from layer to layer, but there is a continuity of H and q across the interfaces between layers.

5.4.2 Flow in Unsaturated Soils—Buckingham-Darcy Equation

Soils are seldom completely saturated with water, although saturation may occur at certain depths for brief periods of time. Most of the time water flows in soil under unsaturated conditions; i.e., the soil pore space is occupied by both air and water. Strictly speaking, unsaturated flow is a process of simultaneous flow of two immiscible fluids, air and water. More commonly, however, it is assumed that the air phase is interconnected and continuous, so that air can easily escape as the water moves in, and thus offers negligible resistance to water flow. Most of the theoretical and experimental developments concerning infiltration and soil water movement in the literature, and presented in this chapter, are based on these assumptions. However, there are situations when the air cannot escape ahead of the infiltrating water, or even where it can, air movement may offer appreciable resistance to water flow at high soil water contents near saturation.[60]

It is also commonly assumed that unsaturated flow is *isomotic* and *isothermal*, so that the effects of *salt* and *temperature* variations in soil on liquid water movement are neglected. The vapor transport of water in soil is also generally neglected. However, the overall influence of temperature on soil hydraulic conductivity is taken into account.

Under these simplifying assumptions, Buckingham[21] modified Eq. (5.4.2) to describe unsaturated flow by generalizing the concepts of the K and h terms originally defined for saturated conditions. He reasoned that in an unsaturated soil the hydraulic conductivity K is a function of the volumetric soil water content θ, i.e., $K = K(\theta)$, and called it the *capillary conductivity*. The parameter K is now called the *unsaturated hydraulic conductivity*.[106] Buckingham also reasoned that the soil water potential h in an unsaturated soil would be negative, because of the presence of capillary suction forces, since it is a function of θ, $h = h(\theta)$. The $h(\theta)$ is often referred to as the *soil water matric potential head,* and its absolute value is called the *matric suction head* $\tau(\theta)$.[106] With introduction of these concepts, Eq. (5.4.2) for one-dimensional, unsteady, vertical flow is modified to become

$$q = -K(\theta)\left[\frac{\partial h(\theta)}{\partial z} - 1\right] \tag{5.4.4}$$

where z is the soil depth, taken positive downward. Buckingham[21] further introduced a term $D(\theta) = K(\theta)\, dh/d\theta$, now generally called the *soil water diffusivity,* which transforms Eq. (5.4.4) to

$$q = -\left[D(\theta)\frac{\partial \theta}{\partial z} - K(\theta)\right] \tag{5.4.5}$$

Equation (5.4.5) with θ as the dependent variable is not very convenient for applications in a layered soil, because θ becomes discontinuous at layer interfaces. Equation (5.4.4) is therefore more convenient. The term $K(\theta)$ can also be written as $K(h)$, when θ is assumed to be a unique function of h, the matric potential head.

5.4.3 Two-Phase Flow

As stated in Sec. 5.4.2, the effects of air movement on water flow are normally assumed negligible. When air can escape ahead of the infiltration wetting front, its effects on infiltration are small.[69] However, if air cannot escape, an air pressure will build up below the wetting front to decrease the infiltration rate. Also, when the air can easily escape, the air resistance can decrease infiltration. Morel-Seytoux and Noblanc[62] presented equations that consider the simultaneous movement of both air and water, and the theory has been further elaborated on in Refs. 60, 61, and 63. The theory shows that separate equations are written for flow rate, or Darcy velocity, of water v_w and air v_a. Assuming the density of air relative to that of water is very small, these equations are simplified to

$$v_w = -K(\theta)\frac{\partial h_w}{\partial z} + K(\theta) \tag{5.4.6}$$

$$v_a = -K_s \frac{\mu_w}{\mu_a} k_{ra} \frac{\partial h_a}{\partial z} \tag{5.4.7}$$

where h_w is the pressure of water and h_a that of air, both in equivalent water heights, μ_w is the viscosity of water, μ_a that of air, and k_{ra} is the relative permeability of air. The h_w is equal to $h + h_a$, where h is the soil water matric potential head as defined earlier. The k_{ra} becomes small when the soil water content is near saturation, and then the air flow is impeded, and so is water flow. The k_{ra} is a function of soil water content, and μ_w/μ_a parameters are the additional information needed for this two-phase flow approach. The difficulties in parameterizing the two-phase model hinder it from being readily used in practical applications.

5.4.4 The Mass-Conservation Equation of Soil Water Movement

Combining Darcy's law [Eq. (5.4.4)] with the law of conservation of mass results in a partial differential equation of one-dimensional vertical flow in an unsaturated soil:

$$\frac{\partial \theta\,(z, t)}{\partial t} = \frac{\partial}{\partial z}\left[K(\theta, z)\frac{\partial h(\theta, z)}{\partial z} - 1\right] \tag{5.4.8}$$

where t is time and the nomenclature $K(\theta, z)$, $h(\theta, z)$ allows for variation of $K(\theta)$ and $h(\theta)$ with depth z, as in a layered soil. This equation has two dependent variables θ and h. Under the assumption that θ is a unique function of h, we can write $\partial\theta/\partial t$ as $(d\theta/dh)\partial h/\partial t = C(h)\,\partial h/\partial t$, and obtain

$$C(h, z)\frac{\partial h(z, t)}{\partial t} = \frac{\partial}{\partial z}\left[K(h, z)\frac{\partial h(z, t)}{\partial z} - 1\right] \tag{5.4.9}$$

in which $C(h)$ is called the *specific moisture capacity*. Such an equation in a general three-dimensional form was first derived by Richards[89] and is therefore commonly called the *Richards equation*.

In portions of the soil horizons that become saturated for brief periods of time during water flow, the term $C(h)$ becomes zero and $K(h, z)$ becomes a constant saturated value $K_s(z)$. Equation (5.4.9) is then reduced to the Laplace equation:

$$\frac{\partial}{\partial z}\left[K_s(z)\frac{\partial h(z, t)}{\partial z} - 1\right] = 0 \qquad (5.4.10)$$

During redistribution or drainage of water between rainfall or irrigation events, root uptake of water is an important factor. This is added to the equation as a sink term S_w expressing the rate of uptake per unit volume of soil:

$$\partial\theta\frac{(z, t)}{\partial t} + S_w(z, t) = \frac{\partial}{\partial z}\left[K(\theta, z)\frac{\partial h(\theta, z)}{\partial z} - 1\right] \qquad (5.4.11)$$

5.4.5 Richards Equation in Relation to Infiltration

The time-dependent rate of infiltration into a soil is governed by Richards equation, subject to given antecedent soil moisture conditions in the soil profile, the rate of water application at the soil surface, and the conditions at the bottom of the soil profile. In general, the initial soil water potential will vary with soil depth. The initial conditions ($t = 0$) can be expressed as a profile of matric potential head varying with depth.

The boundary condition at the soil surface will depend upon the rate of water application. For a rainfall event with intensities less than or equal to the saturated hydraulic conductivity of the soil profile, all the rain will infiltrate into the soil without generating any runoff. For higher rainfall intensities, all the rain will infiltrate into the soil during early stages until the soil surface becomes saturated ($\theta = \theta_s$, $h \geq 0$, $z = 0$). After this point, the *ponding time*, the infiltration is less than the rain intensity and the runoff begins. These conditions may be expressed as

$$-K(h)\frac{\partial h}{\partial z} + 1 = R \qquad \theta(0, t) \leq \theta_s \qquad t \leq t_p \qquad (5.4.12)$$

$$h = h_o \qquad \theta(0, t) = \theta_s \qquad t > t_p \qquad (5.4.13)$$

where R is the rainfall intensity, h_o is a small positive ponding depth on the soil surface, and t_p is the ponding time. These conditions also accommodate time-varying rainfall intensities, as well as when rainfall intensity is smaller than K_s of the soil throughout the storm. For a surface ponded-water irrigation condition Eq. (5.4.13) will apply from time zero on.

The surface boundary conditions [Eqs. (5.4.12) and (5.4.13)] apply at any point in the field during rainfall. In a long sloping field, some infiltration may continue to occur in lower parts of the field even after the rainfall stops, owing to continued overland flow from upper parts. During this phase, conditions described by Eqs. (5.4.12) and (5.4.13) still apply after R is replaced by the overland flow per unit area at the point of interest. To obtain the overland flow rates, hydrodynamic equations of overland flow need to be solved interactively with infiltration.[94]

The lower boundary condition depends upon the depth of the unsaturated profile. For a deep profile, a unit-gradient flux condition is commonly applied at a depth L below the infiltration-wetted zone:

$$q(L, t) = K(\theta, L) \qquad t > 0 \qquad (5.4.14)$$

For a shallow profile, a constant pressure head is assumed at the water-table depth L:

$$h(L, t) = 0 \qquad t > 0 \tag{5.4.15}$$

5.4.6 Numerical Computer Solutions for General Conditions

The Richards equation (5.4.8) subject to the general conditions described in Eqs. (5.4.12) to (5.4.15) in a layered soil profile does not have any known analytical or closed-form solutions for infiltration. The solutions can, however, be obtained by using finite-difference or finite-element numerical methods.[54,87,92,95,115,121] Some solutions for infiltration rates and soil water content profiles with time for different rainfall intensities and soil conditions are illustrated in Figs. 5.4.1 and 5.4.2.

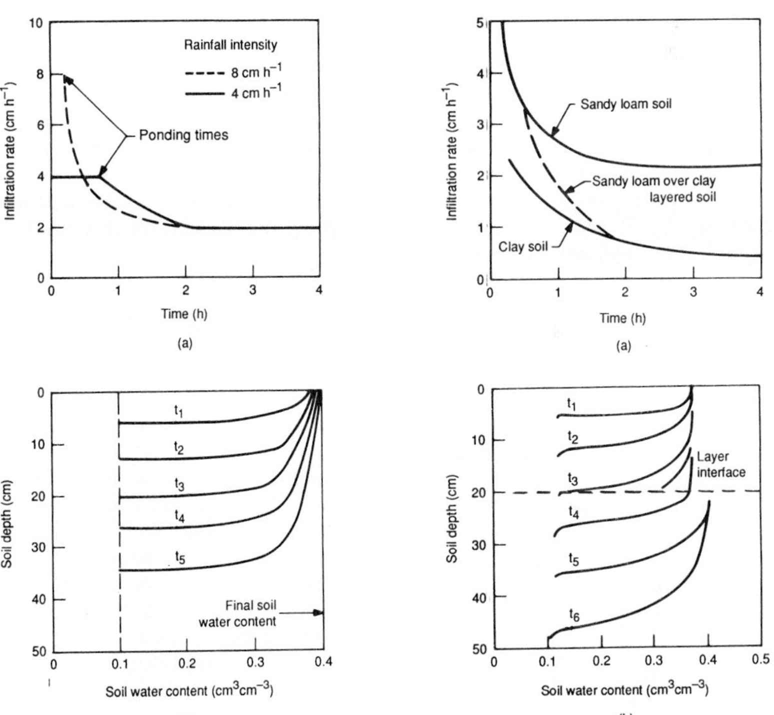

FIGURE 5.4.1 Infiltration curves and water-content profiles for a homogeneous soil. *(a)* Infiltration curves for two constant rainfall rates. *(b)* Soil water content profiles with initial moisture content $\theta = 0.1$.

FIGURE 5.4.2 Infiltration rate curves and water content profiles for a layered soil under ponded conditions. *(a)* Infiltration rate curves. *(b)* Soil water content profiles.

5.4.7 Analytical Solutions for Homogeneous Soils and Limited Conditions

For nonlayered soils, uniform initial soil moisture distribution, and limited surface boundary conditions, some closed-form solutions are available. The pioneering work of Philip[72] provided a series solution for vertical infiltration into a semi-infinite homogeneous soil, with a constant initial moisture content θ_i and a constant matric potential h_o maintained at the soil surface:

$$q(t) = 1/2\, St^{-1/2} + A_2 + A_3 t^{1/2} + A_4 t + \cdots + A_n t^{n/2-1} \qquad (5.4.16)$$

where $q(t)$ is the infiltration rate and S (commonly called the sorptivity), A_2, A_3, . . . , A_n are constants. These constants are, however, functions of soil hydraulic properties $K(\theta)$ and $h(\theta)$ as well as initial and boundary conditions. During early stages, when the effect of gravity is negligible, Eq. (5.4.16) reduces to an extensively used equation:

$$q(t) = \frac{1}{2} St^{-1/2} \qquad (5.4.17)$$

For early to intermediate times, the first two terms of the series should be used:

$$q(t) = \frac{1}{2} St^{-1/2} + A_2 \qquad (5.4.18)$$

where A_2 is commonly assumed to equal the saturated hydraulic conductivity. A limitation of Eq. (5.4.16) is that it does not hold valid for extended time periods, because the series diverges after a certain time. Philip[73,75] provides guidance as to the approximate time limit of its applicability, and also a separate large-time solution with which the series solution in Eq. (5.4.16) can be connected. At large times, the infiltration rate in a uniform soil becomes constant, equal to the saturated hydraulic conductivity K_s. Recently, Swartzendruber[107] presented a solution that holds for both small, intermediate, and large times.

5.4.8 Computing Redistribution, Drainage, Evaporation, and Transpiration of Water

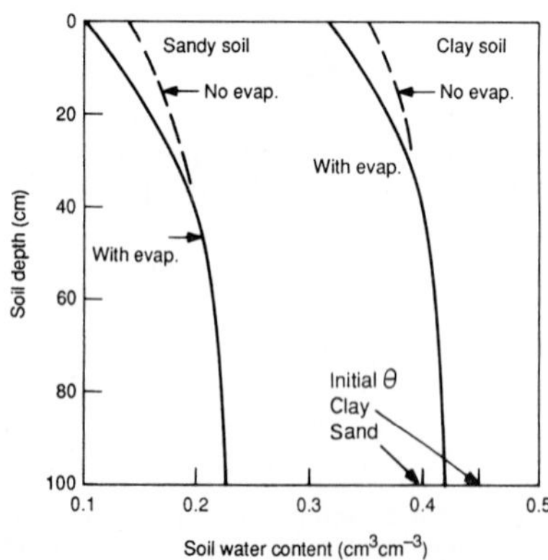

FIGURE 5.4.3 Redistribution of soil water in a clay and a sandy soil after 2 days of wetting at soil water contents of 0.45 and 0.40 cm³ cm⁻³, respectively.

After a rainfall or an irrigation, the infiltrated water is subject to downward redistribution and drainage, as well as plant uptake and evaporation at the soil surface. These changes and losses can be computed by using Eq. (5.4.11) subject to the known initial conditions at the cessation of infiltration and boundary conditions at the soil surface. The root sink term depends upon the evaporative demand of the atmosphere as well as on the extent of plant leaf cover over the ground, root density, and the ability of soil water movement to supply water to the root surface. A simple approach is to assume that the transpiration per unit leaf-covered area is equal to the demand rate as long as the "available water" in the root zone is less than 25 to 50 percent depleted, after which time the uptake decreases in proportion to the further depletion of the available water.[91] The available water is defined as the difference in soil water contents at h values of -33 kPa ($\frac{1}{3}$ bar) and -1500 kPa (15 bar). The transpiration rate is distributed equally throughout the assumed effective root zone. An example of redistributed water for wetted profiles with and without evaporation is illustrated in Fig. 5.4.3.

5.5 OPERATIONAL INFILTRATION MODELS

Models to characterize infiltration for field applications usually employ simplified concepts which predict the infiltration rate or cumulative infiltration volume assuming that surface ponding begins when the surface application rate exceeds the soil surface infiltration rate. The evolution of infiltration modeling has taken three directions, the empirical, the approximate (meaning approximation to the physically based models), and the physical approach. Most of the empirical and approximate models treat the soil as a semi-infinite medium with the soil saturating from the surface down. Physically based models specify appropriate boundary conditions and normally require detailed data input. The Richards equation is the physically based infiltration equation used for describing water flow in soils. Solving this equation mathematically is extremely difficult for many flow problems, and until recent advances in numerical methods and in personal computing power it was not feasible to use the Richards equation for practical applications. Ross[93] developed a software package using the Richards equation for simulating water infiltration and movement of water in soils where water is added as precipitation and removed by runoff, drainage, evaporation from the soil surface, and transpiration by vegetation. The practical use of the program is enhanced by the development of procedures for predicting soil hydraulic properties described in Sec. 5.3.

In addition to the above strictly infiltration-based models many consider the rainfall excess models which lump all losses (infiltration, depression storage, interception) together as infiltration models (Fig. 5.5.1).

5.5.1 Rainfall Excess Models

Rainfall excess is that part of the rainfall which is not lost to infiltration, depression storage, and interception. While a number of models have been proposed for estimating rainfall excess, the most commonly used models are the index models and the USDA Soil Conservation Service (SCS) runoff curve number model.

Index Models. Index models are relatively simple methods that can be useful when performing gauged analysis (i.e., when rainfall and runoff data are available) or when a simple method is commensurate with the data available for estimating loss values. The most commonly used index models are (1) constant fraction, (2) initial and constant loss rate, and (3) constant loss rate. The advantage of the index methods is that they require only one or two parameters. The disadvantages are that they require rainfall-runoff records and are dependent on watershed and storm conditions from which they are determined. For more information on these models see Chap. 9.

SCS Runoff Curve Number Model. The SCS runoff curve number (CN) method is described in detail in Chap. 4 of the Soil Conservation Service *National Engineering*

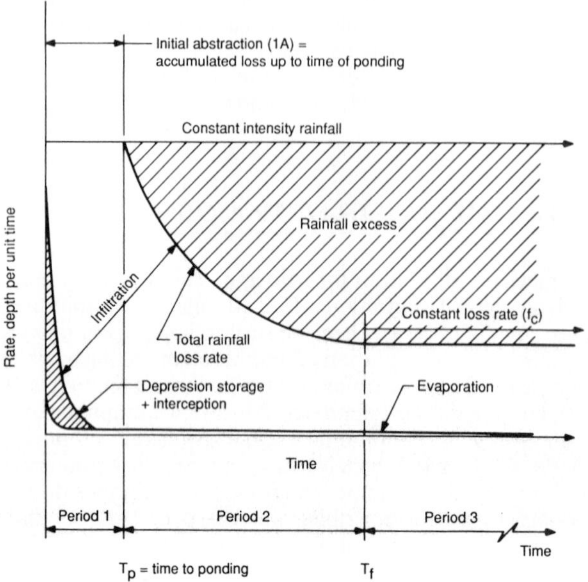

FIGURE 5.5.1 Schematic of rainfall excess components.

Handbook.[103] The SCS runoff equation is

$$Q = \frac{(P - I_a)^2}{(P - I_a) + S} \tag{5.5.1}$$

where Q = runoff, in
P = rainfall, in
S = potential maximum retention after runoff begins, in
I_a = initial abstraction, in

Initial abstraction is all losses before runoff begins. It includes water retained in surface depressions, water intercepted by vegetation, evaporation, and infiltration. I_a is highly variable but from data from many small agricultural watersheds, I_a was approximated by the following empirical equation:

$$I_a = 0.2S \tag{5.5.2}$$

By eliminating I_a as an independent parameter, this approximation allows use of a combination of S and P to produce a unique runoff amount. Substituting Eq. (5.5.2) into Eq. (5.5.1) gives

$$Q = \frac{(P - 0.2S)^2}{P + 0.8S} \tag{5.5.3}$$

where the parameter S is related to the soil and cover conditions of the watershed through the curve number CN. CN has a range of 30 to 100, and S is related to CN by

$$S = \frac{1000}{\text{CN}} - 10 \tag{5.5.4}$$

Equation (5.5.4) calculates S in units of inches. The major factors that determine CN are the hydrologic soil group, cover type, treatment, hydrologic condition, and antecedent runoff condition. The values of CN in Table 5.5.1 represent average antecedent runoff conditions for urban, cultivated agricultural, other agricultural, and arid and semiarid range land uses.[105] The following sections explain how to determine factors affecting the CN.

The SCS has classified all soils into four *hydrologic soil groups* (A, B, C, and D) according to their infiltration rate, which is obtained for bare soil after prolonged wetting. The four groups are defined as follows.

> Group A soils have low runoff potential and high infiltration rates even when thoroughly wetted. They consist chiefly of deep, well to excessively drained sands or gravels. The USDA soil textures normally included in this group are sand, loamy sand, and sandy loam. These soils have a transmission rate greater than 0.76 cm/h.
>
> Group B soils have moderate infiltration rates when thoroughly wetted and consist chiefly of moderately deep to deep, moderately well to well drained soils with moderately fine to moderately coarse textures. The USDA soil textures normally included in this group are silt loam and loam. These soils have a transmission rate between 0.38 and 0.76 cm/h.
>
> Group C soils have low infiltration rates when thoroughly wetted and consist chiefly of soils with a layer that impedes downward movement of water and soils with moderately fine to fine texture. The USDA soil texture normally included in this group is sandy clay loam. These soils have a transmission rate between 0.13 and 0.38 cm/h.
>
> Group D soils have high runoff potential. They have very low infiltration rates when thoroughly wetted and consist mainly of clay soils with a high swelling potential, soils with a permanent high water table, soils with a claypan or clay layer at or near the surface, and shallow soils over a nearly impervious material. The USDA soil textures normally included in this group are clay loam, silty clay loam, sandy clay, silty clay, and clay. These soils have a very low rate of water transmission (0.0 to 0.13 cm/h). Some soils are classified in group D because of a high water table that creates a drainage problem; however, once these soils are effectively drained, they are placed into another group.

A list of most of the soils in the United States and their respective hydrologic soil group classification is given by the Soil Conservation Service.[105] Maps and soil reports are available on a county basis for most of the United States and can be obtained from the library or SCS offices.

Treatment is a cover-type modifier used in Table 5.5.1*b* to describe the effect on CN of the management of cultivated agricultural lands. It includes mechanical practices, such as contouring and terracing, and management practices, such as crop rotations and reduced or no tillage.

Hydrologic condition indicates the effects of cover type and treatment on infiltration and runoff and is generally estimated from density of plant and residue cover on sample areas. A good hydrologic condition indicates that the soil usually has a low runoff potential for the given hydrologic soil group, cover type, and treatment. Some factors to consider in estimating the effect of cover on infiltration and runoff are (1) canopy or density of lawns, crops, or other vegetative areas; (2) amount of year-round cover; (3) amount of grass or close-seeded legumes in rotations; (4) percent of residue cover; and (5) degree of surface roughness. Several factors, such as the percentage of

TABLE 5.5.1 SCS Runoff Curve Numbers

a. Urban areas					
Cover description		Curve numbers for hydrologic soil group			
Cover type and hydrologic condition	Average % impervious area*	A	B	C	D
Fully developed urban areas (vegetation established):					
Open space (lawns, parks, golf courses, cemeteries, etc.)†					
Poor condition (grass cover < 50%)		68	79	86	89
Fair condition (grass cover 50 to 75%)		49	69	79	84
Good condition (grass cover > 75%)		39	61	74	80
Impervious areas:					
Paved parking lots, roofs, driveways, etc. (excluding right-of-way)		98	98	98	98
Streets and roads:					
Paved; curbs and storm sewers (excluding right-of-way)		98	98	98	98
Paved; open ditches (including right-of-way)		83	89	92	93
Gravel (including right-of-way)		76	85	89	91
Dirt (including right-of-way)		72	82	87	89
Western desert urban areas:					
Natural desert landscaping (pervious areas only)‡		63	77	85	88
Artificial desert landscaping (impervious weed barrier, desert shrub with 1- to 2-in sand or gravel mulch and basin borders)		96	96	96	96
Urban districts:					
Commercial and business	85	89	92	94	95
Industrial	72	81	88	91	93
Residential districts by average lot size:					
1/8 acre or less (town houses)	65	77	85	90	92
1/4 acre	38	61	75	83	87
1/3 acre	30	57	72	81	86
1/2 acre	25	54	70	80	85
1 acre	20	51	68	79	84
2 acres	12	46	65	77	82
Developing urban areas:					
Newly graded areas (previous areas only, no vegetation§		77	86	91	94
Idle lands (CNs are determined using cover types similar to those in Table 5.5.1*c*)					

* The average percent impervious area shown was used to develop the composite CNs. Other assumptions are as follows: impervious areas are directly connected to the drainage system, impervious areas have a CN of 98, and pervious areas are considered equivalent to open space in good hydrologic condition. CNs for other combinations of conditions may be computed.

† CNs shown are equivalent to those of pasture. Composite CNs may be computed for other combinations of open space cover type.

‡ Composite CNs for natural desert landscaping should be computed based on the impervious area percentage (CN = 98) and the pervious area CN. The pervious area CNs are assumed equivalent to desert shrub in poor hydrologic condition.

§ Composite CNs to use for the design of temporary measures during grading and construction should be based on the degree of development (impervious area percentage) and the CNs for the newly graded pervious areas.

Source: Ref. 105.

TABLE 5.5.1 SCS Runoff Curve Numbers *(Continued)*

b. Cultivated agricultural areas

Cover description			Curve numbers for hydrologic soil group			
Cover type	Treatment*	Hydrologic condition†	A	B	C	D
Fallow	Bare soil	—	77	86	91	94
	Crop residue cover (CR)	Poor	76	85	90	93
		Good	74	83	88	90
Row crops	Straight row (SR)	Poor	72	81	88	91
		Good	67	78	85	89
	SR + CR	Poor	71	80	87	90
		Good	64	75	82	85
	Contoured (C)	Poor	70	79	84	88
		Good	65	75	82	86
	C + CR	Poor	69	78	83	87
		Good	64	74	81	85
	Contoured and terraced (C&T)	Poor	66	74	80	82
		Good	62	71	78	81
	C&T + CR	Poor	65	73	79	81
		Good	61	70	77	80
Small grain	SR	Poor	65	76	84	88
		Good	63	75	83	87
	SR + CR	Poor	64	75	83	86
		Good	60	72	80	84
	C	Poor	63	74	82	85
		Good	61	73	81	84
	C + CR	Poor	62	73	81	84
		Good	60	72	80	83
	C&T	Poor	61	72	79	82
		Good	59	70	78	81
	C&T + CR	Poor	60	71	78	81
		Good	58	69	77	80
Close-seeded or broadcast legumes or rotation meadow	SR	Poor	66	77	85	89
		Good	58	72	81	85
	C	Poor	64	75	83	85
		Good	55	69	78	83
	C&T	Poor	63	73	80	83
		Good	51	67	76	80

* *Crop residue cover* applies only if residue is on at least 5 percent of the surface throughout the year.

† Hydrologic condition is based on combination of factors that affect infiltration and runoff, including (1) density and canopy of vegetative areas, (2) amount of year-round cover, (3) amount of grass or close-seeded legumes in rotations, (4) percent of residue cover on the land surface (good ≥ 20%), and (5) degree of surface roughness.

Poor: Factors impair infiltration and tend to increase runoff.

Good: Factors encourage average and better than average infiltration and tend to decrease runoff.

Source: Ref. 105.

TABLE 5.5.1 SCS Runoff Curve Numbers *(Continued)*

c. Other agricultural areas

Cover description		Curve numbers for hydrologic soil group			
Cover type	Hydrologic condition	A	B	C	D
Pasture, grassland, or range—continuous forage for grazing*	Poor	68	79	86	89
	Fair	49	69	79	84
	Good	39	61	74	80
Meadow—continuous grass, protected from grazing and generally mowed for hay	—	30	58	71	78
Brush—brush-weed-grass mixture with brush the major element†	Poor	48	67	77	83
	Fair	35	56	70	77
	Good	30	48	65	73
Woods-grass combination (orchard or tree farm)‡	Poor	57	73	82	86
	Fair	43	65	76	82
	Good	32	58	72	79
Woods§	Poor	45	66	77	83
	Fair	36	60	73	79
	Good	30	55	70	77
Farmsteads—buildings, lanes, driveways, and surrounding lots	—	59	74	82	86

* *Poor:* <50% ground cover or heavily grazed with no mulch.
Fair: 50 to 75% ground cover and not heavily grazed.
Good: >75% ground cover and lightly or only occasionally grazed.

† *Poor:* <50% ground cover.
Fair: 50 to 75% ground cover.
Good: >75% ground cover.

‡ CNs shown were computed for areas with 50% woods and 50% grass (pasture) cover. Other combinations of conditions may be computed from the CNs for woods and pasture.

§ *Poor:* Forest litter, small trees, and brush are destroyed by heavy grazing or regular burning.
Fair: Woods are grazed but not burned, and some forest litter covers the soil.
Good: Woods are protected from grazing, and litter and brush adequately cover the soil.

Source: Ref. 105.

d. Arid and semiarid range areas

Cover description		Curve numbers for hydrologic soil group			
Cover type	Hydrologic condition*	A†	B	C	D
Herbaceous—mixture of grass, weeds, and low-growing brush, with brush the minor element	Poor		80	87	93
	Fair		71	81	89
	Good		62	74	85
Oak-aspen—mountain brush mixture of oak brush, aspen, mountain mahogany, bitter brush, maple, and other brush	Poor		66	74	79
	Fair		48	57	63
	Good		30	41	48
Piñon-juniper—piñon, juniper, or both: grass understory	Poor		75	85	89
	Fair		58	73	80
	Good		41	61	71
Sagebrush with grass understory	Poor		67	80	85
	Fair		51	63	70
	Good		35	47	55

TABLE 5.5.1 SCS Runoff Curve Numbers *(Continued)*

d. Arid and semiarid range areas

Cover description		Curve numbers for hydrologic soil group			
Cover type	Hydrologic condition*	A†	B	C	D
Desert shrub—major plants include saltbush, greasewood, creosotebush, blackbrush, bursage, palo verde, mesquite, and cactus	Poor	63	77	85	88
	Fair	55	72	81	86
	Good	49	68	79	84

* *Poor:* <30% ground cover (litter, grass, and brush overstory).
Fair: 30 to 70% ground cover.
Good: >70% ground cover.
† Curve numbers for group A have been developed only for desert shrub.
Source: Ref. 105.

impervious area and the means of conveying runoff from impervious areas to the drainage system, should be considered in computing CN for urban areas.

Antecedent runoff condition (ARC) is an index of runoff potential for a storm event. The ARC is an attempt to account for the variation in CN at a site from storm to storm. CN for the average ARC at a site is the median value as taken from sample rainfall and runoff data. The CNs in Table 5.5.1 are for the average ARC, which is used primarily for design applications. The Soil Conservation Service[105] and Rallison[76] give a detailed discussion of storm-to-storm variation and a demonstration of upper and lower enveloping curves.

Curve numbers describe average conditions that are useful for design purposes. If the rainfall event used is a historical storm that departs from average conditions, the modeling accuracy decreases. The runoff curve number equation should be applied with caution when recreating specific features of an actual storm. The equation does not contain an expression for time and therefore does not account for rainfall duration or intensity, although Eq. (5.5.3) can be applied to the cumulative rainfall at a number of points within the cumulative rainfall hyetograph and thus generate an excess rainfall hyetograph for the storm.

The user should understand the assumptions reflected in the initial abstraction term I_a and should ascertain that the assumption applies to their situation. I_a, which consists of interception, initial infiltration, surface depression storage, evapotranspiration, and other factors, was generalized as $0.2S$ based on data from agricultural watersheds (S is the potential maximum retention after runoff begins). This approximation can be especially important in an urban application because the combination of impervious areas with pervious areas can imply a significant initial loss that may not take place. The opposite effect, a greater initial loss, can occur if the impervious areas have surface depressions that store some runoff. To use a relationship other than $I_a = 0.2S$, one must use rainfall-runoff data to establish new S or CN relationships for each cover and hydrologic soil group. Runoff from snowmelt or rain on frozen ground cannot be estimated using these procedures.

The CN procedure is less accurate when runoff is less than 0.5 in. As a check, one should use another procedure to determine runoff. The SCS runoff procedures apply only to direct surface runoff; do not overlook large sources of subsurface flow or high groundwater levels that contribute to runoff. These conditions are often related to hydrologic soil group A soils and forest areas that have been assigned relatively low

CN values in Table 5.5.1. Good judgment and experience based on stream gauge records are needed to adjust CNs as conditions warrant. When the weighted CN is less than 40, use another procedure to determine runoff.

5.5.2 Empirical Infiltration Models

The empirical models generally relate infiltration rate or volume to elapsed time modified by certain soil properties.[99] Parameters used in these models are commonly estimated from measured infiltration rate–time relationships for a given soil condition. The three most common empirical equations are Kostiakov's model,[48] Horton's model,[44] and Holtan's model.[41]

Kostiakov Model. Kostiakov[48] proposed a simple infiltration model relating the infiltration rate f_p to time t, which was presented by Skaggs and Khaleel[100] as

$$f_p = K_k t^{-\alpha} \tag{5.5.5}$$

where K_k and α are constants which depend on the soil and initial conditions and may be evaluated using the observed infiltration rate–time relationship.

The limitations of using Kostiakov's model are its need for a set of observed infiltration data for parameter evaluation; thus, it cannot be applied to other soils and conditions which differ from the conditions for which parameters K_k and α were determined. The Kostiakov model has primarily been used for irrigation applications.

Horton Model. A three-parameter empirical infiltration model was presented by Horton[44] and has been widely used in hydrologic modeling. He found that the infiltration capacity f_p to time t relationship may be expressed as

$$f_p = f_c + (f_o - f_c)\, e^{-\beta t} \tag{5.5.6}$$

where f_o is the maximum infiltration rate at the beginning of a storm event and reduces to a low and approximately constant rate of f_c as the infiltration process continues and the soil becomes saturated. The parameter β controls the rate of decrease in the infiltration capacity.[100] Horton's equation is applicable only when effective rainfall intensity i_e is greater than f_c.[50] Parameters f_o, f_c, and β must be evaluated using observed infiltration data. Generalized parameter estimates are given in Table 5.5.2.

TABLE 5.5.2 Parameter Estimates for Horton Infiltration Model

Soil and cover complex	f_o, mm h^{-1}	f_c, mm h^{-1}	β, min^{-1}
Standard agricultural (bare)	280	6–220	1.6
Standard agricultural (turfed)	900	20–290	0.8
Peat	325	2–29	1.8
Fine sandy clay (bare)	210	2–25	2.0
Fine sandy clay (turfed)	670	10–30	1.4

Source: Reproduced from Ref. 100 with permission.

Wide-scale application of this model is limited because of the dependence of the parameters on specific soil and moisture conditions. These parameters can be related to the physically based parameters of the Green-Ampt equation.[58,59]

Holtan Model. Holtan[41] developed an empirical equation on the premise that soil moisture storage, surface-connected porosity, and the effect of root paths are the dominant factors influencing the infiltration capacity. Holtan and Lopez[42] modified the equation to be

$$f = \mathrm{GI}\, A\, S_a^{1.4} + f_c \tag{5.5.7}$$

where f is the infiltration rate (in/h), GI is the growth index of crop in percent maturity varying from 0.1 to 1.0 during the season, A is the infiltration capacity (in h^{-1}) per $(in)^{1.4}$ of available storage and is an index representing surface-connected porosity and the density of plant roots which affect infiltration (Table 5.5.3), S_a is the available storage in the surface layer (A horizon) in inches, and f_c is the constant infiltration rate when the infiltration rate curve reaches asymptote (steady infiltration rate). Musgrave[65] related f_c to different hydrologic soil groups (Table 5.5.4).

TABLE 5.5.3 Estimates of Vegetative Parameter A in Holtan Infiltration Model

Land use or cover	Basal area rating* Poor condition	Basal area rating* Good condition
Fallow†	0.10	0.30
Row crops	0.10	0.20
Small grains	0.20	0.30
Hay (legumes)	0.20	0.40
Hay (sod)	0.40	0.60
Pasture (bunch grass)	0.20	0.40
Temporary pasture (sod)	0.20	0.60
Permanent pasture (sod)	0.80	1.00
Woods and forests	0.80	1.00

* Adjustments needed for "weeds" and "grazing."
† For fallow land only, poor condition means "after row crop" and good condition means "after sod."
Source: Ref. 30.

TABLE 5.5.4 Final Infiltration Rates by Hydrologic Soil Groups for Holtan Infiltration Model

Hydrologic soil group	f_c, cm/h
A	0.76
B	0.38–0.76
C	0.13–0.38
D	0.0–0.13

Source: Ref. 65.

Holtan's equation computes the infiltration rate based on the actual available storage S_a of the surface layer (A horizon). This equation is easy to use for prediction of rainfall infiltration, and the values for the input parameters can be obtained from tables for known soil type and land use. A major difficulty with the use of Holtan's equation is the evaluation of the depth of the top layer (control depth). Huggins and Monke[45] showed that the control depth is highly dependent on cultural practices and surface conditions. Holtan et al.[43] suggested using plow layer or depth to the impeding layer as the control depth.

5.5.3 Approximate Theory-Based Infiltration Models

Most approximate models give results similar to each other; however, the major problem is estimating their parameters. The Green-Ampt[34] and Philip[74] models are the most used.

Green-Ampt Model. The Green-Ampt[34] model is an approximate model utilizing Darcy's law. The original model was developed for ponded infiltration into a deep homogeneous soil with a uniform initial water content. Water is assumed to infiltrate into the soil as piston flow resulting in a sharply defined wetting front which separates the wetted and unwetted zones as shown in Fig. 5.5.2. Neglecting the depth of ponding at the surface, the Green-Ampt rate equation is

$$f = K\left[1 + \frac{(\phi - \theta_i)S_f}{F}\right] \tag{5.5.8}$$

and its integrated form is

$$Kt = F - S_f(\phi - \theta_i)\ln\left[1 + \frac{F}{(\phi - \theta_i)S_f}\right] \tag{5.5.9}$$

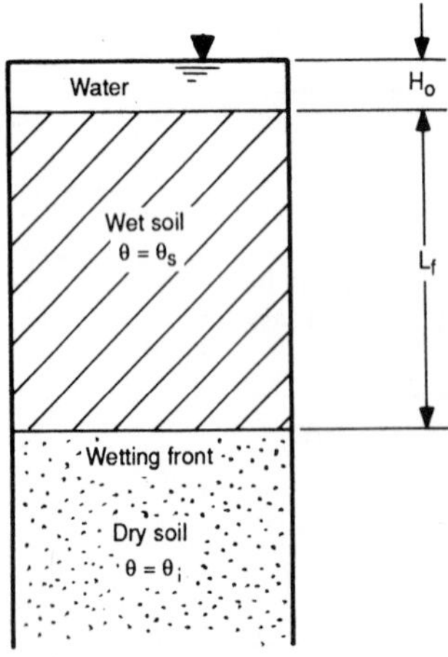

FIGURE 5.5.2 Green-Ampt model.

where K is the effective hydraulic conductivity, $[L/T]$; S_f is the effective suction at the wetting front, $[L]$; ϕ is the soil porosity, $[L^3/L^3]$; θ_i is the initial water content, $[L^3/L^3]$; F is accumulated infiltration, $[L]$; and f is infiltration rate, $[L/T]$. Equation (5.5.9) assumes a ponded surface so that infiltration rate equals infiltration capacity.

Mein and Larson[54] developed the following system for applying the Green-Ampt model to infiltration under rainfall. Just prior to surface ponding the rainfall rate R, $[L/T]$ equals the infiltration rate f, and the cumulative infiltration at time to ponding F_p equals the rainfall rate times the time to surface ponding t_p. Thus the infiltration rate for steady rainfall is

$$f = R \qquad \text{for } t \leq t_p \tag{5.5.10}$$

$$f = K + \frac{K S_f(\phi - \theta_i)}{F} \qquad t > t_p \tag{5.5.11}$$

where $t_p = F_p/R$ and $F_p = [S_f(\phi - \theta_i)]/(R/K - 1)$. The integrated form which is analogous to Eq. (5.5.14) is

$$K(t - t_p + t_p') = F - S_f(\phi - \theta_i) \ln \left[1 + \frac{F}{(\phi - \theta_i)S_f}\right] \tag{5.5.12}$$

where t_p' is the equivalent time to infiltrate volume F_p, under initially surface ponded conditions, and can be calculated from Eq. (5.5.9). Generally the Green and Ampt model is applied by incrementing F and solving for t in Eq. (5.5.12) and then for f using Eq. (5.5.11).

The Green-Ampt equations (5.5.8) and (5.5.9) for homogeneous soils can be extended to describe infiltration into layered soils, when the hydraulic conductivity of the successive layers decreases with depth.[24,35] As long as the wetting front is in the top layer, the equations remain the same. After the wetting front enters the second layer, the effective hydraulic conductivity K is set equal to the harmonic mean $K_h = \sqrt{K_1 K_2}$ for wetted depths of the first and second layers and the capillary head is set equal to S_f of the second layer. This principle is then carried through to the third and succeeding layers.

For a layered soil, in which the saturated hydraulic conductivity of a subsoil layer is greater than that of a layer above (typically a crusted soil), the above Green-Ampt equations cannot be used after the wetting front enters the higher-K layer. For such cases, it may be assumed that infiltration through the higher-K layer continues to be governed by harmonic mean K of the upper layers.[57]

One of the most common forms of soil layering is the formation of a crust on the soil surface caused by raindrop impact. The thickness of the crust is generally very small, e.g., 1.5 to 3.0 mm,[52] and usually develops during the first 10 cm of rainfall. Ahuja[2] developed a Green-Ampt approach based on physical principles to handle a developing crust; however, for practical applications it can be assumed that a stable crust exists on a bare soil.

Macroporosity comprises large noncapillary pores or voids in soil, such as worm holes, decayed root channels, and structural cracks, that are open at the soil surface and capture the free water available at the surface during rainfall or runoff, and conduct it downward very quickly, bypassing most of the soil matrix. A practical Green-Ampt approach to modeling infiltration with macropores is to adjust the effective hydraulic conductivity. A conceptual approach is to consider the soil porosity as two domains, the macropores and the soil matrix with an interaction between the two.

Under rainfall, the flow into macropores begins only after the start of surface ponding of the soil matrix. Infiltration into the soil matrix, before and after ponding, is modeled by the Green-Ampt approach as described earlier. After ponding, the free water available at the soil surface is allowed to flow into macropores. The flow in macropores is assumed to be fast and unrestricted by size of the holes. However, the macropore water can be absorbed by the drier soil matrix, below the transient wetting front generated by the continuing vertical infiltration into the soil matrix, by radial or lateral infiltration. The lateral infiltration starts from the uppermost point below the matrix wetting and proceeds downward, until the available water within a given time step is all absorbed or the lowest depth of interest is reached, below which the excess water freely drains away.

For infiltration under ponded conditions, the soil in the wetted zone is nearly saturated. The wetted zone then develops a viscous resistance to air flow, which reduces infiltration rate. To account for this effect, Morel-Seytoux and Khanji[61] introduced a correction factor to the Green-Ampt equation for a homogeneous soil. The correction factor varies with the soil type and ponding depth, ranging from 1.1 to 1.7 with an average of 1.4, and should be used to reduce the Green-Ampt infiltration rate (dividing the rate by the correction factor) when entrapped air in the soil is a problem.

Green-Ampt Parameter Estimation. To apply the Green-Ampt model the effective hydraulic conductivity K, the wetting front suction S_f, the porosity ϕ, and the initial moisture content θ_i need to be measured or estimated. These parameters can be determined by fitting to experimental infiltration data; however, for specific application purposes it is easier to determine the parameters from readily available data such as soils and land-use data.

Average values for the Green-Ampt wetting front suction S_f, saturated hydraulic conductivity K_s, and porosity ϕ are given in Table 5.5.5 for the 11 USDA soil

TABLE 5.5.5 USDA Soil Texture Green-Ampt Infiltration Parameters

Soil texture class	Porosity ϕ	Wetting front soil suction head S_f, cm	Saturated hydraulic conductivity K_s,* cm/h
Sand	0.437 (0.374–0.500)	4.95 (0.97–25.36)	23.56
Loamy sand	0.437 (0.363–0.506)	6.13 (1.35–27.94)	5.98
Sandy loam	0.453 (0.351–0.555)	11.01 (2.67–45.47)	2.18
Loam	0.463 (0.375–0.551)	8.89 (1.33–59.38)	1.32
Silt loam	0.501 (0.420–0.582)	16.68 (2.92–95.39)	0.68
Sandy clay loam	0.398 (0.332–0.464)	21.85 (4.42–108.0)	0.30
Clay loam	0.464 (0.409–0.519)	20.88 (4.79–91.10)	0.20
Silty clay loam	0.471 (0.418–0.524)	27.30 (5.67–131.50)	0.20
Sandy clay	0.430 (0.370–0.490)	23.90 (4.08–140.2)	0.12
Silty clay	0.479 (0.425–0.533)	29.22 (6.13–139.4)	0.10
Clay	0.475 (0.427–0.523)	31.63 (6.39–156.5)	0.06

* A method for determining the Green-Ampt K from K_s is given following Eq. (5.5.16). For bare ground conditions K can be taken as $K_s/2$.

Source: Reproduced from Ref. 81 with permission.

textures. These values can be used as a first estimate; however, if more detailed soil properties are available, more refined estimates can be made using the following prediction equations.

The porosity ϕ can be determined from measured bulk density or estimated bulk density determined from Fig. 5.5.3 using percent sand, percent clay, and percent organic matter available from most soil analyses. Also, if the cation-exchange capacity of the clay, which is an indicator of the shrink-swell capacity of the clay, is available, the bulk density at the water content for 33 kPa tension can be estimated by the following:

$$\text{BD} = 1.51 + 0.0025(S) - 0.0013(S)(\text{OM}) - 0.0006(C)(\text{OM}) - 0.0048(C)(\text{CEC}) \tag{5.5.13}$$

where BD = bulk density of <2-mm material, g/cm^3
S = percent sand
C = percent clay
OM = percent organic matter [1.7 × (percent organic carbon)]
$\text{CEC} = \dfrac{\text{cation-exchange capacity of clay}}{\text{percent clay}}$

CEC ranges from 0.1–0.9

Bulk density can be converted to porosity using Eq. (5.1.1).

Coarse fragments (>2 mm in size) in the soil affect the porosity, and adjustments

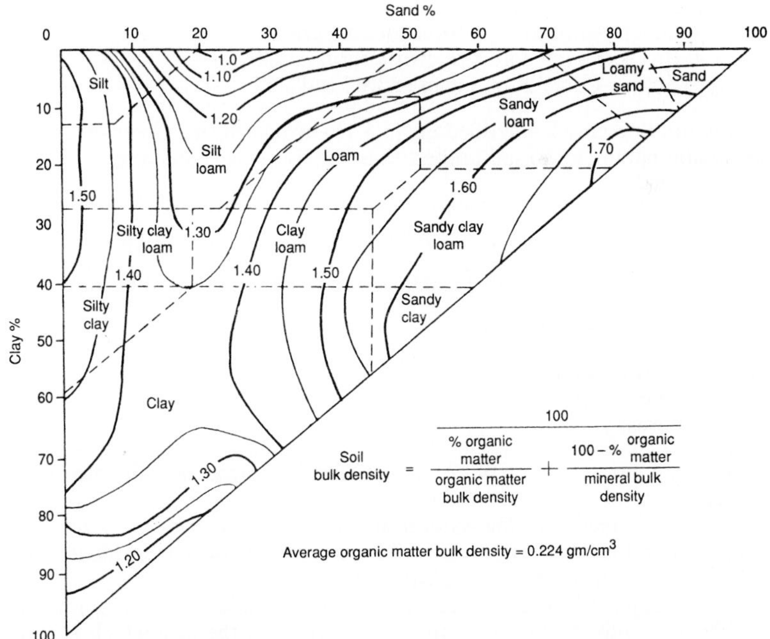

FIGURE 5.5.3 Mineral bulk density in gm/cm^3. The contents in the triangular chart give mineral bulk density. A value is selected from this chart and substituted into the equation to the right of the chart to determine the soil bulk density. *(Reproduced from Ref. 78 with permission. © by Williams and Wilkins.)*

should be made using the following equation:[15]

$$\phi_c = \phi \text{ CFC} \tag{5.5.14}$$

where ϕ_c = porosity of the soil with coarse fragments
ϕ = porosity of the soil without coarse fragments
$\text{CFC} = 1 - \dfrac{\text{VCF}}{100}$
VCF = volume coarse fragments (>2 mm) computed from
$= \dfrac{\text{WCF}}{2.65}\left[\dfrac{100}{(100 - \text{WCF})\text{BD}} + 1\right]$
WCF = percent weight of coarse fragments
BD = bulk density of the soil fraction less than 2 mm, g/cm^3

The initial water content θ_i should be measured, or it can be estimated from moisture-retention relationships.[81] A good estimate of wet, average, and dry initial water contents is the water content held at −10 kPa, −33 kPa, and −1500 kPa, respectively; however, this is dependent on location; e.g., in the western rangeland the average value is −1500 kPa water content while in the eastern part of the United States it is closer to the −33 kPa water content.

The *Green-Ampt wetting front suction parameter* S_f can be estimated from the Brooks-Corey parameters (Secs. 5.2.2 and 5.3.1)[81] as

$$S_f = \frac{2 + 3\lambda}{1 + 3\lambda}\frac{h_b}{2} \tag{5.5.15}$$

where S_f = Green-Ampt wetting front suction, cm
λ = Brooks-Corey pore-size distribution index
h_b = Brooks-Corey bubbling pressure head

Rawls and Brakensiek amplified Eq. (5.5.15) by relating the Green-Ampt wetting front suction parameter to soil properties in the following equation:

$$\begin{aligned} S_f = \exp\,[&6.53 - 7.326(\phi) + 0.00158(C^2) + 3.809(\phi^2) \\ &+ 0.000344(S)(C) - 0.04989(S)(\phi) + 0.0016(S^2)(\phi^2) \\ &+ 0.0016(C^2)(\phi^2) - 0.0000136(S^2)(C) - 0.00348(C^2)(\phi) \\ &+ 0.000799(S^2)(\phi)] \end{aligned} \tag{5.5.16}$$

where S = percent sand
C = percent clay
ϕ = porosity

A graphical representation of Eq. (5.5.16) for the normal porosity distribution is shown in Fig. 5.5.4.

It can be assumed that the natural and management factors except for bulk density do not influence the wetting front suction and all the management effects are incorporated into the *conductivity parameter K*.

To incorporate the effects of land cover on infiltration, it is recommended to divide the area into the following three categories: (1) the area which is bare and outside of canopy cover, (2) the area which has ground cover, and (3) the bare area under canopy, and to develop an effective hydraulic conductivity for each area. Compute the infiltration separately for each area and then sum the three infiltration

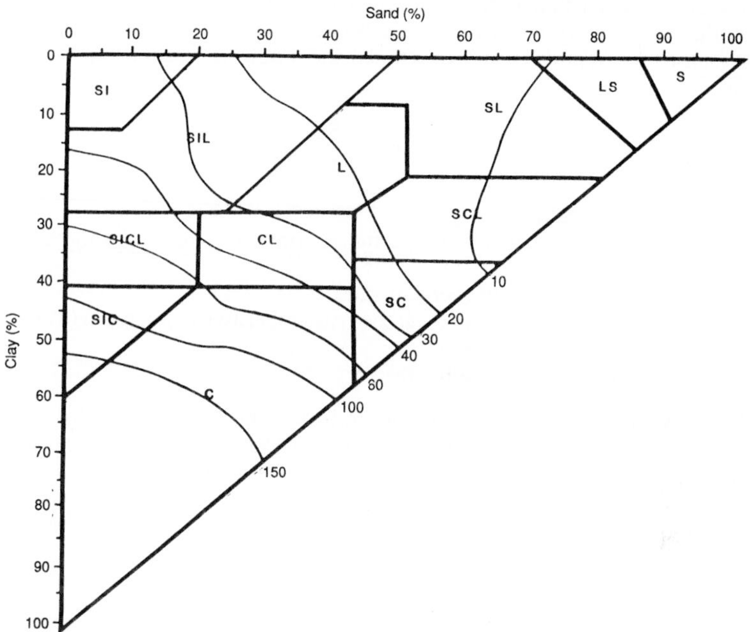

FIGURE 5.5.4 Green-Ampt wetting front suction for the USDA soil texture triangle. *(Reproduced from Ref. 81 with permission.)*

amounts weighted according to their areal cover to obtain the infiltration for the area. This method of determining the infiltration assumes that the three areas do not cascade. If the areas do cascade, this method overpredicts infiltration.

For the area which is bare under canopy the effective hydraulic conductivity can be assumed to be equal to the saturated hydraulic conductivity K_s of the soil.

The area which has ground cover is assumed to contain macroporosity, and the effective hydraulic conductivity is equal to the saturated hydraulic conductivity K_s times a macroporosity factor A. Rawls et al.[82] and Brakensiek and Rawls[16] developed two macroporosity factors for areas which do not undergo mechanical disturbance on a regular basis, for example, rangeland, and one for areas which do undergo mechanical disturbance on a regular basis, for example, agricultural areas. The prediction equation for macroporosity factor A for undisturbed rangeland areas is

$$A = \exp(2.82 - 0.099\,S + 1.94\,\mathrm{BD}) \tag{5.5.17}$$

and for undisturbed agricultural areas is

$$A = \exp(0.96 - 0.032(S) + 0.04(C) - 0.032\,(\mathrm{BD})] \tag{5.5.18}$$

where S = percent sand
C = percent clay
BD = bulk density of the soil (<2 mm), g/cm^3

The macroporosity factor in Eqs. (5.5.17) and (5.5.18) is constrained to be not less than 1.0.

The area which is bare outside canopy is assumed to be crusted and the effective hydraulic conductivity is equal to the saturated hydraulic conductivity K_s times a crust factor CRC. Rawls et al.[85] developed the following relationship for the crust factor:

$$\mathrm{CRC} = \frac{\mathrm{SC}}{1 + (\psi_i/L)} \tag{5.5.19}$$

where CRC = crust factor
SC = correction factor for partial saturation of the soil subcrust (Table 5.5.6)
= 0.736 + 0.0019 (percent sand)
ψ_i = matric potential drop at the crust-subcrust interface, cm (Table 5.5.6)
= 45.19 − 46.68 (SC)
L = wetting front depth, cm

The crust factor in Eq. (5.5.19) incorporates the effects of a crust into a one-layer Green-Ampt model. Also, it does not predict a transient crust conductivity as it assumes that a steady-state crust conductivity can be predicted by

$$K_c = \frac{K_s \, \mathrm{SC} \, Z_c}{\psi_i + Z_c} \tag{5.5.20}$$

where Z_c = crust thickness, cm (can be assumed equal to 0.5 cm)
K_s = saturated hydraulic conductivity of the soil, cm/h

Philip Model. Philip[74] proposed that the first two terms of his series solution could be used as an infiltration model. The equation is

$$f = \frac{1}{2} S t^{1/2} + A \tag{5.5.21}$$

TABLE 5.5.6 Mean Steady-State Matric Potential Drop ψ_i Across Surface Seals by Soil Texture

Soil texture	Matric potential drop ψ_i, cm	Reduction factor for subcrust conductivity (SC)
Sand	2	0.91
Loamy sand	3	0.89
Sandy loam	6	0.86
Loam	7	0.82
Silt loam	10	0.81
Sandy clay loam	5	0.85
Clay loam	8	0.82
Silty clay loam	10	0.76
Sandy clay	6	0.80
Silty clay	11	0.73
Clay	9	0.75

Source: Reproduced from Ref. 85 with permission.

where f is the infiltration rate, cm/h, t is time from ponding, S is the sorptivity, and A is a parameter with dimension of conductivity. Parameters in Eq. (5.5.21) can be evaluated from experimental infiltration data using regression analysis; however, the parameters can be estimated from soils data using the following approximations. S can be approximated using the following equation developed by Youngs:[123]

$$S = \sqrt{2(\phi - \theta_i)\, KS_f} \tag{5.5.22}$$

where S_f is the Green-Ampt effective wetting front suction, cm, which can be estimated from soils data using Eq. (5.5.15); ϕ is the total porosity, which can be estimated from the soil bulk density using Eqs. (5.5.13) and (5.5.14); the initial water content θ_i is measured or estimated from water-retention data according to the degree of wetness; K, cm/h, is the effective conductivity, which is estimated using procedures given for estimating the effective conductivity in the Green-Ampt model. The A parameter in Eq. (5.5.21) was found by Youngs[123] to range from 0.33 K_s to K_s, with K_s being the recommended value.

5.6 MEASUREMENT OF INFILTRATION

Infiltration is a very complex process which can vary temporally and spatially. Selection of measurement techniques and data-analysis techniques should consider these effects. Infiltration measurement techniques can be categorized by their spatial dimensions.

5.6.1 Areal Measurement

Areal infiltration estimation is accomplished by analysis of rainfall-runoff data from a watershed.[25]

5.6.2 Point Measurement

Point infiltration measurements are normally made by applying water at a specific site to a finite area and measuring the intake rate of the soil. There are four types of infiltrometers: the *ponded-water ring* or *cylinder* type, the *sprinkler* type, the *tension* type, and the *furrow* type. An infiltrometer should be chosen that replicates the system being investigated. For example, ring infiltrometers should be used to determine infiltration rates for inundated soils such as flood irrigation or pond seepage. Sprinkler infiltrometers should be used where the effect of rainfall on surface conditions influences the infiltration rate. Tension infiltrometers are used to determine the infiltration rates of the soil matrix in the presence of macropores. Furrow infiltrometers are used when the effect of flowing water is important, as in furrow irrigation.

Infiltrometers are primarily used for measuring the infiltration rate as it varies with time; however, depending on the specified goals, the following auxiliary data should also be collected: (1) physical plot characteristics (length, width, area, aspect, landscape position, slope, elevation, soil series, soil profile description, physical and chemical properties of the soil such as those reported in Sec. 5.1); (2) tillage information (date, type, implement, depth, speed, direction, soil moisture condition, random roughness, bulk density, amount and energy of rain between tillages, general tillage history for previous 3 years); (3) crop conditions (crop type, description, grain yield, residue yield for previous 3 years); (4) cover information (date, type and percent canopy cover, canopy height, leaf area index, percent surface cover, rock cover, and weight of surface cover); and (5) surface soil conditions (date, crust thickness, amount of surface cracks, amount of pores greater than 1 mm, depth of frozen ground).

Ring or Cylinder Infiltrometers. These infiltrometers are usually metal rings with a diameter of 30 to 100 cm and a height of about 20 cm. The ring is driven into the ground about 5 cm, water is applied inside the ring with a constant-head device, and intake measurements are recorded until a steady infiltration rate is observed. To help eliminate the effect of lateral spreading use a double-ring infiltrometer, which is a ring infiltrometer with a second larger ring around it. Both rings have water applied to them, and measurements are taken inside the inner ring. The outer ring serves as a buffer to help minimize the lateral spreading effect. ASTM standards for constructing ring infiltrometers are given in Lukens.[51] The advantages of the ring infiltrometer are that only a small area is needed for measurements, it is inexpensive to construct and simple to run, and it does not have a high water requirement. This method usually produces higher steady-state infiltration rates compared with sprinkler infiltrometers.[36]

Sprinkler Infiltrometers. Sprinkler infiltrometers are designed to emulate aspects of natural rainfall such as drop-size distribution, drop velocity at impact, range of intensities, angle of impact, nearly continuous and uniform raindrop application, and capability to reproduce rainstorm intensity distribution and duration.[55,66] In many instances simulated rain is the only way to obtain results in a reasonable time period and under controlled conditions. The size of the plot to represent the conditions to be evaluated, the water requirement, portability, operation unaffected by wind, operation in sloping terrain, and purpose of the investigation influence the selection of the appropriate rainfall simulator. Agassi et al.[1] report that the quality of the water should be the same as rain since the water applied can have a significant effect on the surface seal formation. Peterson and Bubenzer[71] inventoried the different rainfall simulators reported in the literature and categorized them according to whether the simulated rain was produced by a nozzle or by a drop former. Their inventory includes the appropriate reference, type of nozzle, nozzle pressure, nozzle movement and spray pattern, drop size, intensity range, plot size, and details for constructing a rainfall simulator.

Tension Infiltrometer. The tension infiltrometer, or disk infiltrometer as it is sometimes called, is made up of three main components (Fig. 5.6.1). The first component is a tension control tube which contains three air entry ports which independently determine the amount of tension in the system by varying the distance between the air entry point and the water level. The second component is a large tube with a scale used for measuring the change in water level. The third component is a contact plate at the base of the device with a wire mesh, outer nylon mesh, and a micromesh pulled tight around the base. The diameter of the base plate is normally the same size as the inner ring of the double-ring infiltrometer. The large tube is filled with water by submerging the base plate in a container of water and drawing water up to the tube by a means of a vacuum pump fastened to a port at the top of the tube.[9,70,118] The advantage of the tension infiltrometer is that by varying the tension pores of a certain size can be eliminated from the flow process; thus the effect of pore distribution on infiltration can be determined. The pore-size distribution is very important when macropores are present.

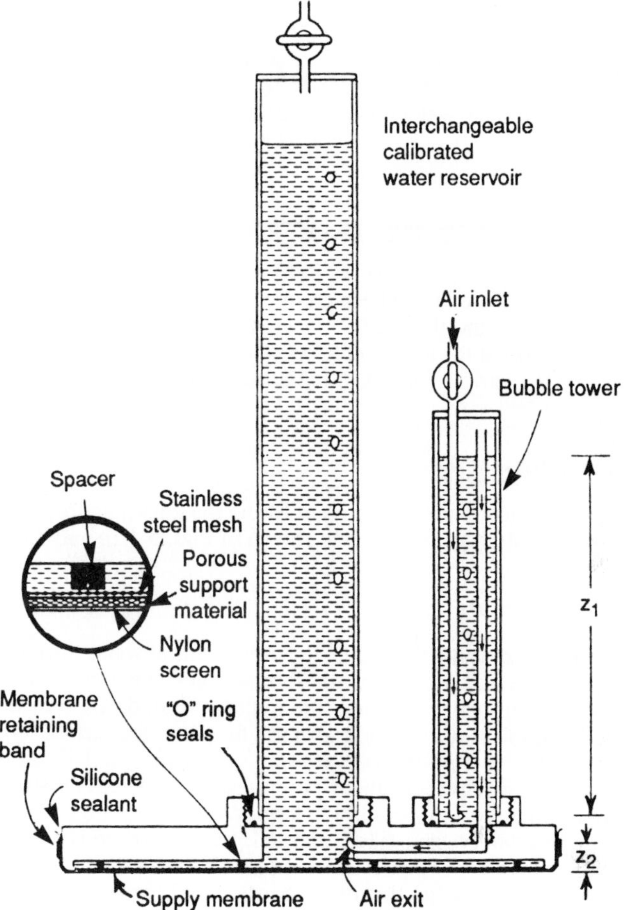

FIGURE 5.6.1 Tension infiltrometer. *(Reproduced from Ref. 70 with permission.)*

Furrow Infiltrometer. Infiltration measurements for furrow irrigation systems can be made with a blocked furrow infiltrometer, recirculating furrow infiltrometer, or inflow-outflow measurement.[46] The following is a brief description of the three furrow infiltration methods.

Blocked Furrow Infiltrometer. This infiltrometer consists of blocking off a section of furrow, usually up to 5 m in length, and then ponding water in the furrow section to a depth approximating the observed depth during an irrigation event. The water in the furrow is kept at a relatively constant level by continually adding water until a constant infiltration rate is achieved or the time exceeds the expected duration of the irrigation event.

Recirculating Furrow Infiltrometer. This infiltrometer is a modification of the blocked furrow infiltrometer with water being continuously recycled over the furrow segment in an attempt to better simulate the flow conditions of an actual irrigation.

Inflow-Outflow Method. The infiltration rates along a furrow segment are determined by taking the difference in measured flow rates at the beginning and end of a segment. Flumes or other suitable flow-measuring devices are used for these measurements. The length of segment depends on the infiltration characteristics of the furrow. Also the volume of water going into storage may have to be considered in the analysis.

5.7 FACTORS AFFECTING INFILTRATION

Factors affecting infiltration have been grouped into soil, soil surface, management, and natural categories, and if any of these factors are important it is important that their effect be accounted for in the infiltration model. Published material illustrates the effects which the various factors have on infiltration.

5.7.1 Soil Factors

Soil factors include the soil physical properties and the soil water properties discussed in Sec. 5.1 for soil water movement and shown in Figs. 5.4.1*a* and 5.4.2*a*.

5.7.2 Surface Factors

The surface factors are those that affect the movement of water through the air-soil interface. Cover materials protect the soil surface. Lack of cover, a bare soil, leads to the formation of a surface crust under the impact of raindrops[108] or other factors, which break down soil structure and move soil fines into the surface or near-surface pores. Once formed, a crust impedes infiltration. Figure 5.7.1 illustrates that the removal of the surface cover (straw or burlap) reduces the steady-state infiltration rate from approximately 3 to 4 cm h^{-1} to less than 1 cm h^{-1}. Figure 5.7.2 illustrates the differences between crusted, tilled, and sod (grass) cover soils on the infiltration curve. The bare tilled soil has higher infiltration than a crusted soil initially; however, its steady-state rate approaches that of the crusted soil because a crust is developing. Also, the grass-covered soil has a higher rate than a crusted soil partially because the grass protects the soil from crusting.

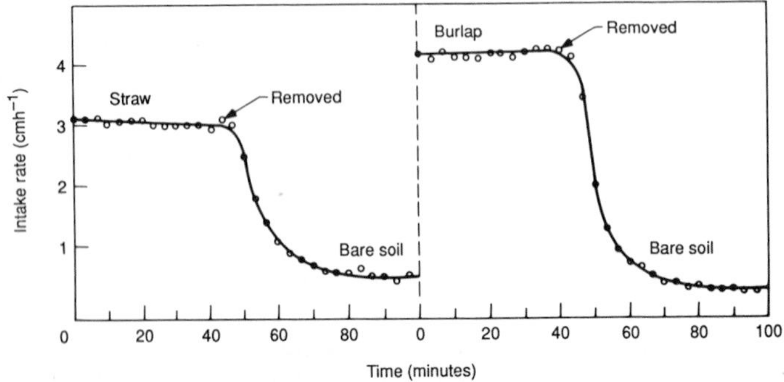

FIGURE 5.7.1 Effect of covered and bare soil on infiltration rates.[56]

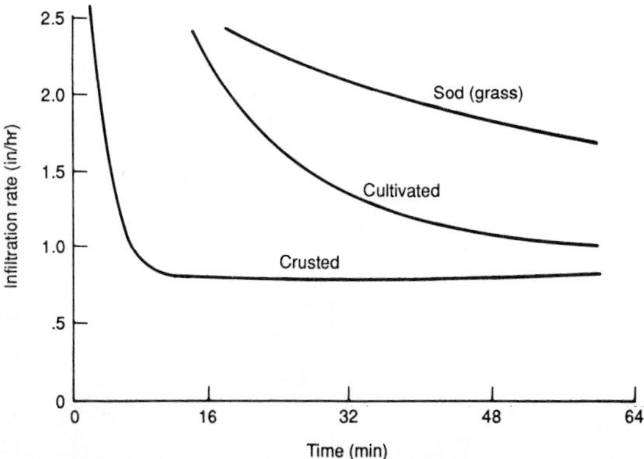

FIGURE 5.7.2 Effect of surface sealing and crusting on infiltration rates. *(Reproduced from Ref. 100 with permission.)*

Changes in soil surface configurations can be caused by natural processes such as erosion or man-made processes such as tillage. Thurow et al.[109] found that the area around plants maintained higher infiltration compared with the area between plants. Also, Freebairn[29] reported that there was a tendency for infiltration steady-state rates to increase with surface roughness generated by tillage.

5.7.3 Management Factors

Agricultural management systems involve different types of tillage, vegetation, and surface cover. Figure 5.7.3 illustrates the influence of tillage practices (moldboard plow, chisel plow, and no till) on infiltration. Brakensiek et al.[16] reported that moldboard plowing would increase soil porosity from 10 to 20 percent depending on soil texture and would increase infiltration rates over nontilled soils. Rawls[78] reported that increasing the organic matter of the soil lowers the bulk density, increases

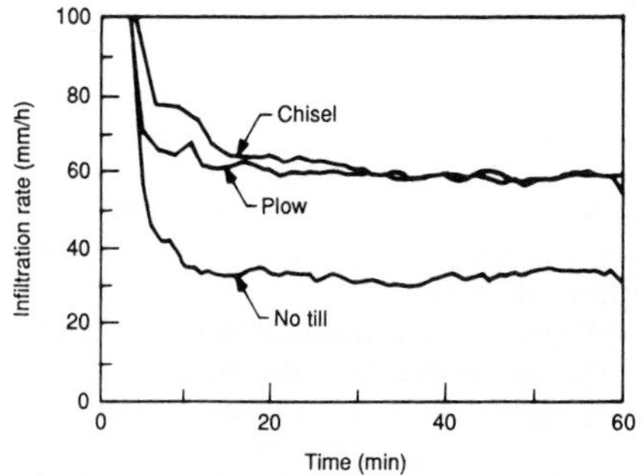

FIGURE 5.7.3 Infiltration rates for Port Byron silt loam, 2 months after chisel and moldboard plow tillage and 4 months after planting under no tillage. *(Reproduced from Ref. 29 with permission.)*

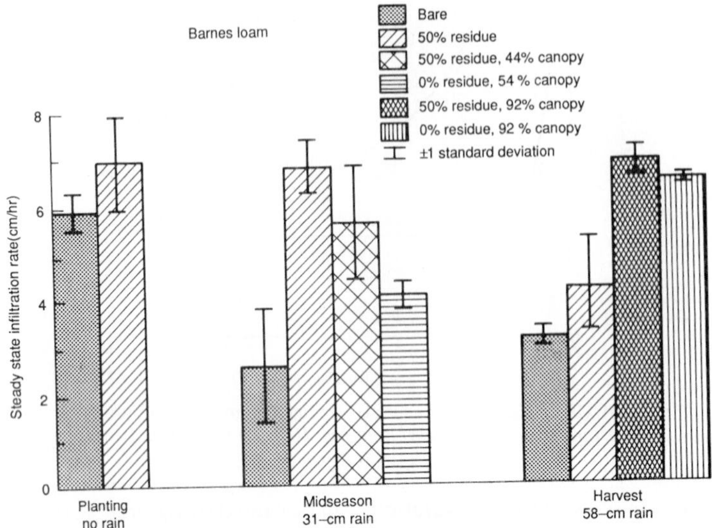

FIGURE 5.7.4 Temporal effects of soybean canopy and residue on the steady-state infiltration rate on Barnes loam.

porosity, and hence increases the infiltration. Figure 5.7.4 gives the steady-state infiltration rate of bare ground, soybean vegetation, and residue cover interactions for planting, midseason, and harvest. The bare-soil steady-state infiltration rate decreases between planting and midseason and then stays stable primarily as a result of crusting. Residue maintains a high steady-state rate until harvest while the canopy and canopy residue combination increase the steady-state rate.

Management practices on rangelands usually change the types of vegetation or the grazing practices. The type of vegetation on rangelands has a significant effect on infiltration. Grazing practices also influence infiltration, as is shown in Fig. 5.7.5. The effects of rangeland management systems on runoff are summarized by Branson et al.[19]

5.7.4 Natural Factors

Natural factors include natural processes such as precipitation, freezing, seasons, temperature, and moisture which vary with time and space and interact with other factors in their effect on infiltration.

Soil temperature influences infiltration through its effect on the viscosity of water. Lee[49] found freezing the soil with a high moisture content decreases infiltration to almost zero, while freezing the soil at a low moisture content increases infiltration to twice its normal rate.

The effect of cumulative antecedent rainfall on exposed and 50 percent residue covered agricultural soil is shown in Fig. 5.7.4, indicating a decrease in the steady-state infiltration rate with the continued exposure to the action of rainfall. For bare soil it seems that a stable steady-state infiltration rate is achieved between planting and midseason, indicating that a stable crust is achieved early in the growing season and maintained thereafter. The increasing steady-state infiltration rate with increases in canopy cover (Fig. 5.7.4) demonstrates the temporal variability in infiltration caused by a growing crop. Also, Fig. 5.7.4 indicates that canopy cover and residue cover do not cause additive increases in the steady-state infiltration rate.

Increases in rainfall intensity increase surface disturbance caused by the raindrops and the building up of a ponding head, which usually increases the bare soil infiltration, and for bare soil with canopy cover this intensity effect is dissipated by the canopy of the growing crop.

The effect of spatial variability on infiltration is a measure of the difference between "point" infiltration and apparent infiltration rates and amounts associated with composite areas or watersheds. The variability of soil, surface, and management factors over an area is a result of some of the practices or factors just discussed. Smith[102] states, "There is no set of parameters which, when used with point infiltration models will produce the same response as the net from the area."

Smith and Hebbert[101] studied the effect of a linear variation of hydraulic conductivity along a plane "watershed" on runoff for two rates of uniform rainfall. Their results indicate that different spatial hydraulic conductivity patterns resulting in the same average hydraulic conductivity will produce significantly different runoff hydrographs. They also found that higher rates of rainfall would produce a runoff

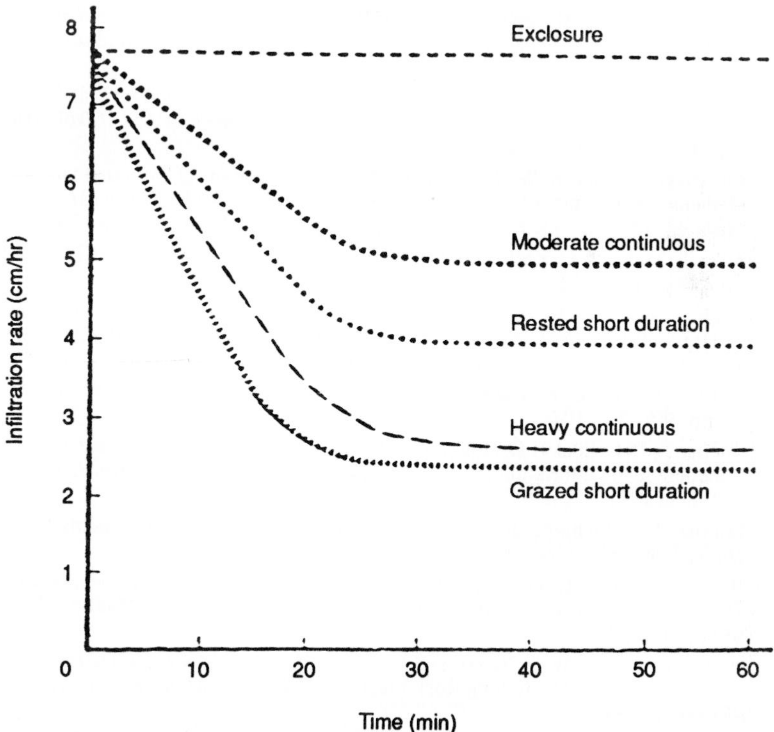

FIGURE 5.7.5 Mean infiltration rates for various grazing treatments at Fort Stanton, N. Mex. *(Reproduced from Ref. 120 with permission.)*

hydrograph similar to that produced by a uniform pattern of infiltration on the plane. However, the spatial variability in infiltration parameters has more effect at lower rainfall rates (near the saturated hydraulic conductivity of the soil).

REFERENCES

1. Agassi, M., I. Shainberg, and J. Morin, "Effects on Seal Properties of Changes on Drop Energy and Water Salinity During a Continuous Rainstorm," *Aust. J. Soil Res.*, vol. 26, pp. 1–10, 1988.
2. Ahuja, L. R., "Modeling Infiltration into Crusted Soils by the Green-Ampt Approach," *Soil Sci. Soc. Am. J.*, vol. 47, pp. 412–418, 1983.
3. Ahuja, L. R., J. W. Naney, and R. D. Williams, "Estimating Soil Water Characteristics from Simpler Properties or Limited Data," *Soil Sci. Soc. Am. J.*, vol. 49, pp. 1100–1105, 1985.
4. Ahuja, L. R., R. R. Bruce, D. K. Cassel, and W. J. Rawls, "Simpler Field Measurement and Estimation of Soil Hydraulic Properties and Their Spatial Variability for Modeling," *Proc. Modeling Agricultural Forest and Rangeland Hydrology,* ASAE Publ. 07-88, pp. 19–33, 1988.
5. Ahuja, L. R., D. K. Cassel, R. R. Bruce, and B. B. Barnes, "Evaluation of Spatial Distribution of Hydraulic Conductivity Using Effective Porosity Data," *Soil Sci.*, vol. 148, pp. 404–411, 1989.
6. Ahuja, L. R., and D. R. Nielsen, "Field Soil Water Relations," in *Irrigation of Agricultural Crops—Agronomy Monograph* 30 American Society of Agronomy, Madison, Wis., pp. 143–190, 1990.
7. American Society for Testing and Materials, "Standard Test Method for Classification of Soils for Engineering Purposes, d 2487-83, *1985 Annual Book of ASTM,* vol. 4.08, pp. 395–408, Philadelphia, Pa., 1985.
8. Amoozegar, A., and A. W. Warrick, "Hydraulic Conductivity of Saturated Soils—Field Methods," in A. Klute, ed., *Methods of Soil Analysis, Part I—Physical and Mineralogical Methods,* American Society of Agronomy Monograph 9, 2d ed., pp. 735–770, 1986.
9. Ankeny, M. D., T. C. Kaspar, and R. Horton, "Design for an Automated Tension Infiltrometer," *Soil Sci. Soc. Am. J.*, vol. 52, pp. 893–896, 1988.
10. Arya, L. M., D. A. Farrell, and G. B. Blake, "A Field Study of Soil Water Depletion Patterns in Presence of Growing Soybean Roots. I. Determination of Hydraulic Properties of the Soil," *Soil Sci. Soc. Am. J.*, vol. 39, pp. 424–430, 1975.
11. Boast, C. W., and D. Kirkham, "Auger Hole Seepage Theory," *Soil Sci. Soc. Am. J.*, vol. 35, pp. 365–373, 1971.
12. Bouma, J., D. I. Hillel, F. D. Hole, and C. R. Amerman, "Field Measurement of Hydraulic Conductivity by Infiltration through Artificial Crusts," *Soil Sci. Soc. Am. J.*, vol. 35, pp. 362–364, 1971.
13. Bouwer, H., "Planning and Interpreting Soil Permeability Measurements," *J. Irrig. Drain. Eng.*, vol. 95, no. IR3, pp. 391–402, 1969.
14. Bouwer, H., and R. D. Jackson, "Determining Soil Properties," in J. van Schilfgaarde, ed., *Drainage for Agriculture,* American Society of Agronomy, no. 7, Madison, Wis., pp. 611–672, 1974.
15. Brakensiek, D. L., W. J. Rawls, and G. R. Stephenson, "A Note on Determining Soil Properties for Soils Containing Rock Fragments," *J. Range Manage.*, vol. 39, no. 5, pp. 408–409, 1986.
16. Brakensiek, D. L., and W. J. Rawls, "Effects of Agricultural and Rangeland Management Systems on Infiltration," in *Modeling Agricultural, Forest, and Rangeland Hydrology,* American Society of Agricultural Engineers, St. Joseph, Mich. p. 247, 1988.

17. Brakensiek, D. L., H. B. Osborn, and W. J. Rawls, "Field Manual for Research in Agricultural Hydrology," *USDA Handbook* 224, 1979.

18. Brakensiek, D. L., W. J. Rawls, and G. R. Stephenson, "Determining the Saturated Hydraulic Conductivity of Soil Containing Rock Fragments," *Soil Sci. Soc. Am. J.*, vol. 50, no. 3, pp. 834–835, 1984.

19. Branson, F. A., G. F. Gifford, K. G. Renard, and R. F. Hadley, *Rangeland Hydrology,* Kendall/Hunt, Dubuque, Iowa, chap. 4, pp. 47–72, 1981.

20. Brooks, R. H., and A. T. Corey, "Hydraulic Properties of Porous Media," *Hydrology Paper* 3, Colorado State University, Fort Collins, Colo. 1964.

21. Buckingham, E., "Studies on the Movement of Soil Moisture," *USDA Bur. Soils Bull.* 38, 1907.

22. Campbell, G. S., "A Simple Method for Determining Unsaturated Conductivity from Moisture Retention Data," *Soil Sci.,* vol., 117, pp. 311–314, 1974.

23. Cassel, D. K., and A. Klute, "Water Potential: Tensiometry," in A. Klute, ed., *Methods of Soil Analysis, Part I—Physical and Mineralogical Methods,* American Society of Agronomy Monograph 9, 2d ed., pp. 563–596, 1986.

24. Childs, E. C., and M. Bybordi, "The Vertical Movement of Water in Stratified Porous Material, 1. Infiltration," *Water Resour. Res.,* vol. 5, no. 2, 1969.

25. Chow, V. T., *Handbook of Applied Hydrology,* McGraw-Hill, New York, pp. 12-1 – 12-30, 1964.

26. Darcy, H., Les fontaines publiques de la ville de Dijon, Dalmont, Paris, 1856.

27. De Wiest, R. J. M., "Fundamental Principles of Groundwater Flow," in R. J. M. De Wiest, ed., *Flow through Porous Media,* Academic Press, New York, pp. 1–52, 1969.

28. Drungil, C. E. C., K. Abt, and T. J. Gish, "Soil Moisture Determination in Gravelly Soils with Time Domain Reflectometry," *Trans. Am. Soc. Agric. Eng.,* vol. 32, no. 1, pp. 177–180, 1989.

29. Freebairn, D. M., S. C. Gupta, C. A. Onstad, and W. J. Rawls, "Antecedent Rainfall and Tillage Effects upon Infiltration," *Soil Sci. Soc. Am. Jour.,* vol. 53, pp. 1183–1189, 1989.

30. Frere, M. H., C. A. Onstad, and H. N. Holtan, "ACTMO, An Agricultural Chemical Transport Model," *USDA-ARS. ARS-H-3,* 1975.

31. Gardner, W. H., "Water Content," in A. Klute, ed., *Methods of Soil Analysis, Part I—Physical and Mineralogical Methods,* American Society of Agronomy Monograph 9, 2d ed., pp. 493–544, 1986.

32. Gee, G. W., and J. W. Bauder, "Particle Size Analysis," in A. Klute, ed., *Methods of Soil Analysis, Part I—Physical and Mineralogical Methods,* American Society of Agronomy Monograph 9, 2d ed., pp. 383–411, 1986.

33. Green, R. E., L. R. Ahuja, and S. K. Chong, "Hydraulic Conductivity, Diffusivity, and Sorptivity of Unsaturated Soils—Field Methods," in A. Klute, ed., *Methods of Soil Analysis, Part I—Physical and Mineralogical Methods,* American Society of Agronomy Monograph 9, 2d ed., pp. 771–798, 1986.

34. Green, W. H., and G. A. Ampt, "Studies on Soil Physics: 1. Flow of Air and Water through Soils," *J. Agric. Sci.,* vol. 4, pp. 1–24, 1911.

35. Hachum, A. Y., and J. F. Alfaro, "Rain Infiltration into Layered Soils: Prediction," *J. Irrig. Drain. Eng.,* vol. 106, pp. 311–321, 1980.

36. Hawkins, R. D., "Interpretations of Source Area Variability in Rainfall-Runoff Relations," in V. P. Singh, ed., *Rainfall-Runoff Relationships Water Resources Publications, Colorado,* pp. 303–325, 1981.

37. Hillel, D., *Soil and Water Physical Principles and Processes,* Academic Press, New York, 1971.

38. Hillel, D., and Y. Benyamini, "Experimental Comparison of Infiltration and Drainage Methods for Determining Unsaturated Hydraulic Conductivity of a Soil Profile in Situ:

Isotope and Radiation Techniques in Soil Physics and Irrigation Studies 1973," International Atomic Energy Agency, Vienna, pp. 271–275, 1974.

39. Hillel, D., and W. R. Gardner, "Measurement of Unsaturated Conductivity and Diffusivity by Infiltration through an Impeding Layer," *Soil Sci.*, vol. 109, pp. 149–153, 1970.

40. Hoffman, G. J., and G. O. Schwab, "Tile Spacing Prediction Based on Drain Outflow," *Trans. Am. Soc. Agric. Eng.*, vol. 7, pp. 444–447, 1964.

41. Holtan, H. N., "A Concept for Infiltration Estimates in Watershed Engineering," *USDA Bull.*, 41–51, 1961.

42. Holtan, H. N., and N. C. Lopez, "USDAHL-70 Model of Watershed Hydrology," *USDA Tech. Bull.* 1435, 1971.

43. Holtan, H. N., G. J. Stilner, W. H. Henson, and N. C. Lopez, USDAHL-74 Model of Watershed Hydrology," *USDA Tech. Bull.* 1518, 1975.

44. Horton, R. E., "An Approach toward a Physical Interpretation of Infiltration-Capacity," *Soil Sci. Soc. Am. J.*, vol. 5, pp. 399–417, 1940.

45. Huggins, L. F., and E. J. Monke, "The Mathematical Simulation of the Hydrology of Small Watersheds," *TR1, Purdue Water Resources Research Center,* Lafayette, Ind., 1966.

46. Kincaid, D. C., "Intake Rate: Border and Furrow," in A. Klute, ed., *Methods of Soil Analysis, Part I—Physical and Mineralogical Methods,* American Society of Agronomy Monograph 9, 2d ed., pp. 871–887, 1986.

47. Klute, A., and C. Dirksen, "Hydraulic Conductivity and Diffusivity—Laboratory Methods," in A. Klute, ed., *Methods of Soil Analysis, Part I—Physical and Mineralogical Methods,* American Society of Agronomy Monograph 9, 2d ed., pp. 687–734, 1986.

48. Kostiakov, A. N., "On the Dynamics of the Coefficient of Water-Percolation in Soils and on the Necessity for Studying It from a Dynamic Point of View for Purposes of Amelioration," *Trans. Sixth Comm. Intern. Soil Sci. Soc. Russian,* part A, pp. 17–21, 1932.

49. Lee, H. W., "Determination of Infiltration Characteristics of a Frozen Palouse Silt Loam Soil under Simulated Rainfall," Ph.D. Dissertation, University of Idaho Graduate School, 1983.

50. Linsley, R. K., Jr., M. A. Kohler, and J. L. H. Paulhus, *Hydrology for Engineers,* 2d ed., McGraw-Hill, New York, 1975.

51. Lukens, R. P., ed., *Annual Book of ASTM Standards,* part 19, *Soil and Rock, Building Stones,* pp. 509–514, 1981.

52. McIntyre, D. S., "Permeability Measurements of Soil Crusts Formed by Raindrop Impact," *Soil Sci.*, vol. 85, pp. 185–189, 1958.

53. McKeague, J. A., ed., *Manual on Soil Sampling and Methods of Analysis,* Canadian Society of Soil Science, Ottawa, Canada, 1978.

54. Mein, R. G., and C. L. Larson, "Modeling Infiltration During a Steady Rain," *Water Resour. Res.*, vol. 9, pp. 384–394, 1973.

55. Meyer, L. D., "Methods for Attaining Desired Rainfall Characteristics in Rainfall Simulations," *Proceedings of the Rainfall Simulator Workshop, USDA-ARS-ARM,* no. 10, pp. 35–48, 1979.

56. Miller, D. E., and R. O. Gifford, "Modification of Soil Crusts for Plant Growth," chap. 7 in *Soil Crusts, Agric. Exp. Sta. Univ. Ariz., Tech. Bull.* 214, p. 9, Tucson, Ariz., 1974.

57. Moore, I. D., and J. D. Eigel, "Infiltration into Two-Layered Soil Profiles," *Trans. Am. Soc. Agric. Eng.*, vol. 24, pp. 1496–1503, 1981.

58. Morel-Seytoux, H. J., *Unsaturated Flow in Hydrologic Modeling: Theory and Practice,* Kluwer Academic, Boston, 1989.

59. Morel-Seytoux, H. J., "Recipe for a Simple but Physically Based Approach to Infiltration under Variable Rainfall Conditions," *1988 Hydrology Days,* no. 1005, Ft. Collins, Colo. pp. 226–247, 1988.

60. Morel-Seytoux, H. J., and J. A. Bilica, "A Two-Phase Numerical Model for Prediction of Infiltration: Applications to a Semi-infinite Soil Column," *Water Resour. Res.*, vol. 21, pp. 607–615, 1985.

61. Morel-Seytoux, H. J., and J. Khanji, "Derivation of an Equation of Infiltration," *Water Resour. Res.*, vol. 10, no. 4, pp. 795–800, 1974.

62. Morel-Seytoux, H. J., and A. Noblanc, "Infiltration Prediction by a Moving Strained Coordinates Method," in A. Hadas et al., eds., *Physical Aspects of Soil, Water, and Salts in Ecosystems*, Springer-Verlag, New York, 1972.

63. Morel-Seytoux, H. J., and M. Vauclin, "Superiority of Two-Phase Formulation for Infiltration," *Advances in Infiltration*, American Society of Agricultural Engineering, vol. 11-83, pp. 34–47, 1983.

64. Mualem, Y., "Hydraulic Conductivity of Unsaturated Soils—Prediction and Formulas," in A. Klute, ed., *Methods of Soil Analysis, Part 1—Physical and Mineralogical Methods*, American Society of Agronomy Monograph 9, 2d ed., pp. 799–823, 1986.

65. Musgrave, G. W., "How Much of the Rain Enters the Soil?" in *USDA Water Yearbook of Agriculture*, Washington, D.C., pp. 151–159, 1955.

66. Neff, E. L., "Why Rainfall Simulation?" *Proceedings of the Rainfall Simulator Workshop, USDA-ARS-ARM*-10, pp. 3–8, 1979.

67. Nielsen, D. R., and J. W. Biggar, "Measuring Capillary Conductivity," *Soil Sci.*, vol. 92, pp. 192–193, 1961.

68. Paetzold, R. F., G. A. Matzkanin, and A. De Los Santos, "Surface Soil Water Content Measurement Using Pulsed Nuclear Magnetic Resonance Techniques," *Soil Sci. Soc. Am. J.*, vol. 49, pp. 537–540, 1985.

69. Parlange, J. Y., and D. E. Hill, "Air and Water Movement in Porous Media: Compressibility Effects," *Soil Sci.*, vol. 127, pp. 257–263, 1979.

70. Perroux, K. M., and I. White, "Designs of Disc Permeameters," *Soil Sci. Soc. Am. J.*, vol. 52, pp. 1205–1215, 1988.

71. Peterson, A., and G. Bubenzer, "Intake Rate Sprinkler Infiltrometer," in A. Klute, ed., *Methods of Soil Analysis, Part I—Physical and Mineralogical Methods*, American Society of Agronomy, 2d ed., pp. 845–870, 1986.

72. Philip, J. R., "The Theory of Infiltration: 1. The Infiltration Equation and Its Solution," *Soil Sci.*, vol. 83, pp. 345–357, 1957.

73. Philip, J. R., "The Theory of Infiltration: 2. The Profile at Infinity," *Soil Sci.*, vol. 83, pp. 435–448, 1957.

74. Philip, J. R., "The Theory of Infiltration. 4. Sorptivity and Algebraic Infiltration Equations," *Soil Sci.*, vol. 84, pp. 257–264, 1957.

75. Philip, J. R., "Theory of Infiltration," *Adv. Hydrosci.*, vol. 5, pp. 215–305, 1969.

76. Rallison, R. E., "Origin and Evolution of the SCS Runoff Equation," *Watershed Manage. Symp.*, ASCE, pp. 912–914, 1980.

77. Rawlins, S. L., "Measurement of Water Content and the State of Water in Soils," in T. T. Koslowski, ed., *Water Deficits and Plant Growth*, vol. IV, Academic Press, New York, pp. 1–55, 1976.

78. Rawls, W. J., "Estimating Soil Bulk Density from Particle Size Analysis and Organic Matter Content," *Soil Sci.*, vol. 135, no. 2, pp. 123–125, 1983.

79. Rawls, W. J., and D. L. Brakensiek, "Estimating Soil Water Retention from Soil Properties," *J. Irrig. Drain. Eng.*, vol. 108, no. IR2, pp. 166–171, 1982.

80. Rawls, W. J., and D. L. Brakensiek, "Prediction of Soil Water Properties for Hydrologic Modeling," *Watershed Management in the Eighties*, ASCE, pp. 293–299, 1985.

81. Rawls, W. J., and D. L. Brakensiek, "A Procedure to Predict Green Ampt Infiltration Parameters," *Adv. Infiltration, Am. Soc. Agric. Eng.*, pp. 102–112, 1983.

82. Rawls, W. J., D. L. Brakensiek, and R. Savabi, "Infiltration Parameters for Rangeland Soils," *J. Range Mange.*, vol. 42, no. 2, pp. 139–142, 1989.

83. Rawls, W. J., D. L. Brakensiek, and K. E. Saxton, "Estimation of Soil Water Properties," *Trans. Am. Soc. Agric. Engrs.*, vol. 25, no. 5, pp. 1316–1330, 1328, 1982.

84. Rawls, W. J., D. L. Brakensiek, and B. Soni, "Agricultural Management Effects on Soil Water Processes: Part I. Soil Water Retention and Green-Ampt Parameters," *Trans. Am. Soc. Agric. Engrs.*, vol. 26, no. 6, pp. 1747–1752, 1983.

85. Rawls, W. J., D. L. Brakensiek, J. R. Simanton, and K. D. Kohl, "Development of a Crust Factor for a Green Ampt Model," *Trans. Am. Soc. Agric. Engrs.*, vol. 33, no. 4, pp. 1224–1228, 1990.

86. Rawls, W. J., T. J. Gish, and D. L. Brakensiek, "Estimating Soil Water Retention from Soil Physical Properties and Characteristics," *Adv. Soil Sci.*, vol. 16, pp. 213–234, 1991.

87. Remson, I., G. M. Hornberger, and F. Molz, *Numerical Methods in Subsurface Hydrology,* Wiley, New York, 1971.

88. Reynolds, W. D., D. E. Elrick, and G. C. Topp, "A Reexamination of the Constant Head Well Permeameter Method for Measuring Saturated Hydraulic Conductivity above the Water Table," *Soil Sci.*, vol. 136, pp. 250–268, 1983.

89. Richards, L. A., "Capillary Conduction of Liquids in Porous Mediums," *Physics,* vol. 1, pp. 318–333, 1931.

90. Richards, L. A., "Physical Conditions of Water in Soil," in C. A. Black, ed., *Methods of Soil Analysis,* American Society of Agronomy, Monograph 9, pp. 128–151, 1965.

91. Ritchie, J. T., E. Burnett, and R. C. Henderson, "Dryland Evaporative Flux in a Subhumid Climate: III. Soil Water Influence," *Agron. J.*, vol. 64, pp. 168–173, 1972.

92. Ross, P. J., "Efficient Numerical Methods for Infiltration Using Richards' Equation," *Water Resour. Res.*, vol. 26, pp. 279–290, 1990.

93. Ross, P. J., "SWIM—A Simulation Model for Soil Water Infiltration and Movement (Reference Manual)," *CSIRO Division of Soils,* Davies Laboratory, Townsville, Queensland, Australia, 1990.

94. Rovey, E. W., D. A. Woolhiser, and R. E. Smith, "A Distributed Kinematic Model of Upland Watersheds," Hydrology Paper 93, *Colorado State University,* Ft. Collins, Colo., 1977.

95. Rubin, J., and R. Steinhardt, "Soil Water Relations During Rain Infiltration: 1. Theory," *Soil Sci. Soc. Am. J.*, vol. 27, pp. 246–251, 1963.

96. Saxton, K. E., W. J. Rawls, J. S. Romberger, and R. I. Papendick, "Estimating Generalized Soil Water Characteristics from Texture," *Trans. Am. Soc. Agric. Engrs.*, vol. 50, no. 4, pp. 1031–1035, 1986.

97. Schmugge, T. J., T. J. Jackson, and H. L. McKim, "Survey of Methods for Soil Moisture Determination," *Water Resour. Res.*, vol. 16, no. 6, pp. 961–979, 1980.

98. Skaggs, R. W., "Determinations of the Hydraulic Conductivity-Drainable Porosity Ratio from Water Table Measurements," *Trans. Am. Soc. Agric. Engrs.*, vol. 19, no. 1, pp. 73–80, 84, 1976.

99. Skaggs, R. W., D. E. Miller, and R. H. Brooks, "Soil Water, Part I—Properties," in *Design and Operation of Farm Irrigation Systems,* Monograph 3, American Society of Agricultural Engineers, St. Joseph, Mich., pp. 77–123, 1980.

100. Skaggs, R. W., and R. Khaleel, "Infiltration," in C. T. Haan, ed., *Hydrologic Modeling of Small Watersheds,* Monograph 5, American Society of Agricultural Engineers, St. Joseph, Mich., pp. 4–166, 1982.

101. Smith, R. E., and R. H. B. Hebbert, "A Monte Carlo Analysis of the Hydrologic Effects of Spatial Variability of Infiltration," *Water Resour. Res.*, vol. 15, no. 2, pp. 419–429, 1979.

102. Smith, R. E., "Flux Infiltration Theory for Use in Watershed Hydrology," in *Adv. Infiltration,* American Society of Agricultural Engineers, St. Joseph, Mich., pp. 313–323, 1983.

103. Soil Conservation Service, *SCS National Engineering Handbook,* Sec. 4, Hydrology, USDA, 1972.

104. Soil Conservation Service, "Procedures for Collecting Soil Samples and Methods of Analysis for Soil Survey," *Soil Survey Investigations Report 1,* Washington, D.C., 1982.

105. Soil Conservation Service, *Urban Hydrology for Small Watersheds,* Technical Release 55, pp. 2.5–2.8, 1986.

106. Swartzendruber, D., "The Flow of Water in Unsaturated Soils," in R. J. M. De Wiest, ed., *Flow through Porous Media,* Academic Press, New York, pp. 215–292, 1969.

107. Swartzendruber, D., "A Quasi Solution of Richards' Equation for Downward Infiltration of Water into Soil," *Water Resour. Res.,* vol. 5, pp. 809–817, 1987.

108. Tackett, J. L., and R. W. Pearson, "Some Characteristics of Soil Crusts Formed by Simulated Rainfall," *Soil Sci.,* vol. 99, pp. 407–413, 1965.

109. Thurow, T. L., W. H. Blackburn, and C. A. Taylor, Jr., "Hydrologic Characteristics of Vegetation Types as Affected by Livestock Grazing Systems, Edwards Plateau, Texas," *J. Range Manage.,* vol. 39, no. 6, p. 506, 1986.

110. Topp, G. C., J. L. Davis, and A. P. Annan, "Electromagnetic Determination of Soil Water Content: Measurement in Coaxial Transmission Lines," *Water Resour. Res.,* vol. 16, no. 3, pp. 574–582, 1980.

111. Topp, G. C., et al., "The Measurement of Soil Water Content Using a Portable TDR Hand Probe," *Can. J. Soil Sci.,* vol. 64, pp. 313–321, 1984.

112. Topp, G. C., and J. L. Davis, "Measurement of Soil Water Content Using Time Domain Reflectometry (TDR): A Field Evaluation," *Soil Sci. Soc. Am. J.,* vol. 49, pp. 19–24, 1985.

113. Trangmar, B. B., R. S. Yost, and G. Uehara, "Application of Geostatistics to Spatial Studies of Soil Properties," *Adv. Agron.,* vol. 38, pp. 45–94, 1985.

114. Van Genuchten, M. Th., "A Closed-Form Equation for Predicting the Hydraulic Conductivity of Unsaturated Soils," *Soil Sci. Soc. Am. J.,* vol. 44, pp. 892–898, 1980.

115. Wang, F. C., and V. Lakshminarayana, "Mathematical Simulation of Water Movement through Unsaturated Nonhomogeneous Soils," *Soil Sci. Soc. Am. J.,* vol. 32, pp. 329–334, 1968.

116. Warrick, W. A., and D. R. Nielsen, "Spatial Variability of Soil Physical Properties in the Field," in D. Hillel, ed., *Applications of Soil Physics,* Academic Press, New York, pp. 319–344, 1980.

117. Warrick, W. A., D. E. Myers, and D. R. Nielsen, "Geostatistical Methods Applied to Soil Science," in A. Klute, ed., *Methods of Soil Analysis, Part I—Physical and Mineralogical Methods,* American Society of Agronomy, 2d ed., Monograph 9, pp. 53–82, 1986.

118. Watson, K. W., and R. J. Luxmoore, "Estimating Macroporosity in a Forest Watershed by Use of a Tension Infiltrometer," *Soil Sci. Soc. Am. J.,* vol. 50, pp. 578–587, 1986.

119. Webster, R., "Quantitative Spatial Analysis of Soil in the Field," *Adv. Soil Sci.,* vol. 3, pp. 1–70, 1985.

120. Weltz, M., and M. K. Wood, "Short Duration Grazing in Central New Mexico: Effects on Infiltration Rates," *J. Range Manage.,* vol. 39, no. 4, p. 366, 1986.

121. Whisler, F. D., K. K. Watson, and S. J. Perrens, "The Numerical Analysis of Infiltration into Heterogeneous Porous Media," *Soil Sci. Soc. Am. J.,* vol. 36, pp. 868–874, 1972.

122. Yong, R. N., and B. P. Warkentin, *Introduction to Soil Behavior,* Macmillan, New York, 1966.

123. Youngs, E. G., "An Infiltration Method Measuring the Hydraulic Conductivity of Unsaturated Porous Materials," *Soil Sci.,* vol. 97, pp. 307–311, 1964.

CHAPTER 6
GROUNDWATER FLOW

Leslie Smith
Department of Geological Sciences
University of British Columbia
Vancouver, British Columbia, Canada

Stephen W. Wheatcraft
Desert Research Institute
University of Nevada System
Reno, Nevada

6.1 INTRODUCTION

This chapter focuses on groundwater flow in saturated porous media. The framework within which we summarize both the principles of groundwater flow and techniques for data interpretation is shown in Fig. 6.1.1. The figure outlines the steps in a *site investigation program* designed to characterize a groundwater flow system, for purposes of reaching a management decision. The management decision may involve, for example, the issue of whether or not groundwater withdrawals can meet a specified demand. Characterization of a hydrogeologic system begins with the formulation of a *conceptual model* of both the geologic environment and the subsurface flow system. The conceptual model can be viewed as a hypothesis describing the main features of the geology, the hydrologic setting, and site-specific relations between geologic structure and patterns of fluid flow. Mathematical modeling can be thought of as a process of hypothesis testing, leading to refinement of the conceptual model and its expression in the more quantitative framework of a *hydrogeologic simulation model.* Before any field measurements are taken, it can be very helpful to make preliminary calculations of subsurface fluid flux using a simplified hydrogeologic simulation model, which may be no more complicated than a one-dimensional analytic solution. Using the conceptual model and initial calculations as a guide, a field program is planned to refine the description of site geology, to measure water levels in boreholes, and to characterize hydraulic properties of each geologic unit. Data interpretation is facilitated by the development of an updated hydrogeologic simulation model, which subsequently is used in a predictive analysis. For example, the updated model may be a two-dimensional computer simulation of the drawdown in water levels in response to pumping from multiple wells. Model predictions are evaluated in light of the project needs, and a decision is made either to proceed to

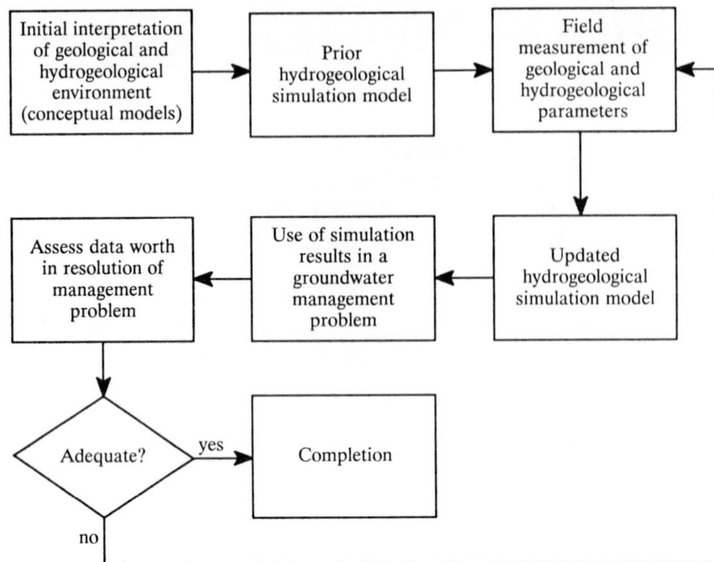

FIGURE 6.1.1 Decision flow diagram outlining the steps in a site investigation program designed to characterize a groundwater flow system for input into a management problem.

project completion or to cycle through additional sequences of data collection and hydrogeologic simulation in order to reduce uncertainties in the analysis.[30]

Introductory-level textbooks on groundwater hydrology include Refs. 17, 19, 27, and 29. Bibliographies include Ref. 71. An extensive overview of the major hydrogeologic regions within North America is available in Ref. 5. Standard texts in geology include Refs. 60 and 110.

6.2 A GEOLOGIC FRAMEWORK

6.2.1 The Geologic Setting

An analysis of subsurface fluid flow cannot be separated from consideration of the geologic setting. If there is a common thread to errors or misinterpretations in groundwater investigations, it can often be traced to an inadequate characterization of the geologic setting and/or an incomplete analysis of its influence on subsurface fluid flow. A description of the geologic setting includes documentation of the different sediment and rock types in the study area, spatial and temporal relationships among the different geologic units, some sense of the scale of spatial variability, and any structural features (e.g., folds, faults).

Geologists map rock units in terms of rock type and age. Table 6.2.1 presents the *geologic time scale* in terms of its broadest divisions, eras, and periods. The geologic time scale is the yardstick by which geologists organize relations between rock units. *Geologic eras* are defined based on the dominant life forms found in the corresponding rocks. The Paleozoic era marks the progression from marine invertebrates to fish

TABLE 6.2.1 Geologic Time Scale

Era	Period	Epoch	Duration	Before present
			Millions of years	
Cenozoic	Quaternary	Holocene	0.01	
		Pleistocene	3	3
	Tertiary	Pliocene	4	7
		Miocene	19	26
		Oligocene	12	38
		Eocene	16	54
		Paleocene	11	65
Mesozoic	Cretaceous		71	136
	Jurassic		54	190
	Triassic		35	225
Paleozoic	Permian		55	280
	Pennsylvanian		45	325
	Mississippian		20	345
	Devonian		50	395
	Silurian		35	430
	Ordovician		70	500
	Cambrian		70	570
Precambrian				>570

and reptiles; in the Mesozoic era, dinosaurs were the dominant vertebrates on land, while the Cenozoic era corresponds to the dominance of mammals. *Geologic periods* provide a finer division of geologic eras and are named either on a geographic basis or on the basis of the characteristics of the rocks in the region they were first defined. *Epochs* in the Tertiary period are defined by grouping rocks on the basis of the percentage of their fossils that are represented by still-living species. The Pleistocene epoch encompasses the repeated cycles of continental glaciation over the past three million years.

6.2.2 Geologic Properties Important to Hydrologic Analysis

A *mineral* is a naturally formed solid substance having a definite chemical composition and characteristic crystal structure. A *rock* is an aggregate mass of mineral matter.

Igneous Rocks. Igneous rocks are formed either by cooling and solidification of subsurface magma (intrusives) or by solidification of magma that erupts and flows across the surface of the earth (extrusives). Because rates of cooling are slower for *intrusive rocks,* mineral grains are large enough to be distinguished in hand specimens (phaneritic). With *extrusive rocks,* rates of cooling are more rapid, and individual mineral grains cannot be seen in a hand specimen (aphanitic). Extrusive rocks also have a variable character that depends upon whether they solidified from a lava flowing across the ground surface or from volcanic ash that was ejected into the atmosphere before settling on the ground surface.

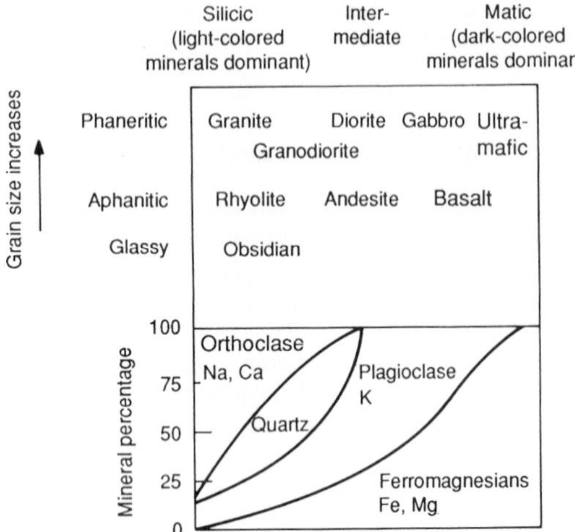

FIGURE 6.2.1 Classification of igneous rocks on the basis of grain size and mineral content. (*From Ref. 60. Used with permission.*)

Three main mineral groups occur in igneous rocks: quartz (SiO_2), feldspars {orthoclase ($KAlSi_3O_8$) or plagioclase [$(Ca,Na)(Al_2Si)\ AlSi_2O_8$]}, and ferromagnesium silicates (olivine, hornblende, pyroxene). Igneous rocks are classified by the relative proportion of each of these minerals (Fig. 6.2.1).

Sedimentary Rocks. *Clastic sedimentary rocks* are formed by the *lithification* of sediments (compaction and cementation) produced by weathering, transport, and deposition of rock fragments. *Chemical sedimentary rocks* are formed from either chemical precipitates (e.g., calcite) or the accumulation of biochemical material (e.g., shell fragments, coal). Table 6.2.2 presents a general classification of sedimentary rocks. The nature of the depositional environment is the key factor in determining rock type. The extent of compaction and cementation determines the degree to which the original porosity of the sediment is preserved as sediments are lithified.

Metamorphic Rocks. Metamorphic rocks form when minerals recrystallize owing to increased temperature and/or pressure. The magnitude of the temperature and pressure increase determines the suite of minerals which form from the parent rock. Geologists refer to the *metamorphic grade* to quantify the magnitude of pressure and/or temperature changes that drive metamorphic reactions. A major division of metamorphic rocks is based upon the presence or absence of *foliation,* a layered appearance due to parallel alignment of elongated mineral grains that have recrystallized normal to the direction of greatest pressure. Table 6.2.3 lists diagnostic features of common metamorphic rocks.

Unconsolidated Sediments. Unconsolidated sediments usually cover the bedrock surface. These sediments originate from in-place weathering of the parent rock mass, forming a *regolith;* or by transportation to the site by fluvial processes, downslope movement, or glacial processes. Sediments overlying bedrock vary with the nature of the sediment source, the mode and distance of sediment transport, and the mode of

TABLE 6.2.2 Classification of Sedimentary Rocks

Origin	Texture	Grain size or mineral type	Rock name
Detrital	Clastic*	Gravel and larger	Conglomerate
		Sand	Sandstone
		Silt	Siltstone
		Clay	Shale
Chemical inorganic	Clastic or nonclastic	Calcite ($CaCO_3$)	Limestone
		Dolomite [$CaMg(CO_3)_2$]	Dolomite
		Calcite admixed with clay	Chalk
		Halite (NaCl)	Salt
		Anhydrite ($CaSO_4$)	
Biochemical organic		Calcite (shells)	Limestone
		Silica (diatoms)	Chert
		Plant remains	Coal

* Composed of fragments.
Source: Adapted from Ref. 60. Used with permission.

sediment deposition. Spatial continuity of bedding structures and the architecture of clay, silt, and sand bodies are key properties of sedimentary deposits. Pleistocene deposits of glacial origin occur in most areas north of 40° latitude. Table 6.2.4 summarizes the characteristics of the main types of glacial deposits.

Weathering is the in situ physical disintegration and chemical decomposition of rocks under the environmental conditions occurring at the earth's surface. Depend-

TABLE 6.2.3 Classification of Metamorphic Rocks

Foliated metamorphic rocks
Slate: Produced by low-grade metamorphism of shale, under conditions of directed pressure but relatively low temperature, imparting a distinct cleavage within the rock mass
Phyllite: A higher grade of metamorphism than slate, producing a coarser-grained texture, with the formation of flaky mineral grains along cleavage planes that gives the rock a silky luster
Schist: A higher grade of metamorphism, coarser-grained than either slate or phyllite. Surfaces of cleavage planes are rough. Formed from many types of igneous and sedimentary rocks. Schists are classified according to the name of the most abundant minerals within the rock.
Gneiss: A coarse-grained, coarsely banded rock, consisting of alternating bands of light- and dark-colored minerals, reflecting a high grade of metamorphism. Gneisses are formed from silicic igneous rocks and various types of sedimentary rocks
Nonfoliated metamorphic rocks
Quartzite: Formed by recrystallization of quartz sandstone, consisting of a dense mass of nonplaty quartz grains
Marble: Formed by recrystallization of limestone or dolomite, producing large interlocking grains of calcite or dolomite

TABLE 6.2.4 Classification of Glacial Deposits

Till: Nonsorted, nonstratified sediment deposited beneath, from within, or from the top of glacial ice. Matrix is dominated by clay and silt, often containing boulders. Basal till, laid down at the base of an ice sheet, is dense and compact. Supraglacial (ablation) till, which accumulates as glacial ice melts in place, is less dense than basal till and is more variable in grain size and sorting

Glaciofluvial: Coarse-grained sediment deposited by meltwater in front of a glacier (outwash), or adjacent to ice bodies (ice-contact stratified deposits). Depending upon environmental conditions and distance from the glacial front, deposits can range from a well-sorted unit (uniform grain size) to a very heterogeneous mixture of grain sizes

Glaciolacustrine sediments: Well-sorted, clays, silts, and fine sands, deposited within glacial lakes, often showing cyclic banding due to annual fluctuations in sediment loading (varves).

Loess: Primarily silt-sized sediments composed of quartz and feldspar, deposited by winds (eolian), loosely packed.

Glacial-marine sediments: Clay-rich sediments deposited where glacial ice advances to the coastline.

Source: Adapted from Ref. 115. Used with permission.

ing upon the local balance between rates of weathering and erosion, the regolith can vary from a discontinuous skin to a zone 30 or 40 m in thickness. Changes in physical properties with depth in the regolith are normally gradual rather than abrupt.

6.3 A HYDRAULIC FRAMEWORK

6.3.1 Hydraulic Potential and Fluid Flux

Groundwater flows through interconnected void spaces, along microcracks between grain boundaries and in larger-scale fractures. Groundwater moves in response to differences in fluid pressure and elevation. The driving force is measured in terms of *hydraulic head* $h[L]$,* where

$$h = z + \frac{p}{\rho g} \tag{6.3.1}$$

Here z is the elevation of the measurement point above datum, p is the pressure of a fluid with constant density ρ, and g is the acceleration due to gravity. Hydraulic head (also referred to as the *piezometric* or *potentiometric head*) is equal to the mechanical energy per unit weight of the fluid. Contributions from kinetic energy to hydraulic head can be neglected in almost all cases because groundwater velocities are so low. Groundwater flows from regions where the hydraulic head is higher toward regions where it is lower. By convention, pressure is expressed in terms of values above atmospheric pressure (gauge pressure). Defining the *pressure head* as

$$h_p = \frac{p}{\rho g} \tag{6.3.2}$$

* Quantities in brackets are the dimensions of the variable indicated, $[L]$ = length.

leads to

$$h = z + h_p \tag{6.3.3}$$

where z is the *elevation head.*

Above the *water table,* in the capillary fringe and vadose zones, water is held in tension and the fluid pressure is less than atmospheric (i.e., $h_p < 0$). Below the water table, in the saturated zone, fluid pressure exceeds atmospheric pressure ($h_p > 0$). The water table is defined as the surface on which the pressure head is equal to zero. Given a datum of sea level, the hydraulic head at a point on the water table is equal to the elevation of the water table above mean sea level at that point.

Hydraulic head is measured in a *piezometer* (Fig. 6.3.1). Standpipe piezometers are constructed by installing plastic or metal casing inside a borehole, with an open screen connected to the end of the casing to permit inflow. The diameter of the casing is normally in the range of 4 to 6 cm (2 to 3 in), with a screen length from 1 to 3 m (3 to 9 ft). Sand is placed around the screen, and the remaining annular space between the borehole wall and the casing above the screen is backfilled to the surface with either cement or bentonite clay. In this way, an isolated measurement of fluid pressure is obtained. Equation (6.3.3) is used to calculate hydraulic head. By convention, the midpoint elevation of the piezometer screen equals the elevation head. Multiport piezometers are described in Ref. 19 (p. 617).

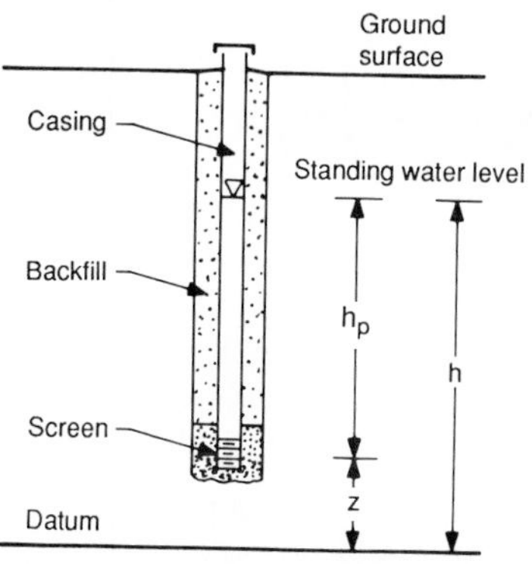

FIGURE 6.3.1 Schematic diagram of a piezometer, illustrating the relationships among hydraulic head, pressure head, and elevation head.

Contour maps of hydraulic head (*piezometric* or *potentiometric maps*) are used to infer directions of subsurface fluid flow since flow will be everywhere normal to the head contours in an isotropic medium (see Sec. 6.3.3). A contour map of water-table elevation is a particular kind of potentiometric map (Fig. 6.3.2). In constructing or interpreting a potentiometric map, it is important to recognize the underlying assumption of horizontal flow implicit in a two-dimensional map representation. In this light, hydraulic head measurements from piezometers completed at different depths, or in different geologic units, should be considered on an individual basis before they are added to the data base used to produce a potentiometric map.

The volumetric flow of groundwater is calculated using *Darcy's law:*

$$Q = -KA\frac{\partial h}{\partial x} = qA \tag{6.3.4}$$

where Q is the *volumetric flow rate* [L^3T^{-1}] in the x direction, A is the cross-sectional area for flow, K is the hydraulic conductivity [LT^{-1}], $\partial h/\partial x$ is the gradient in hydraulic head, and q is the *specific discharge* (flow rate per unit area [LT^{-1}]). The minus sign indicates that fluid moves in the direction of decreasing hydraulic head. The specific discharge is sometimes referred to as the *Darcy flux, Darcy velocity,* or *bulk velocity.*

The *pore water velocity* v [LT^{-1}] is related to the specific discharge by

$$v = \frac{q}{n} \tag{6.3.5}$$

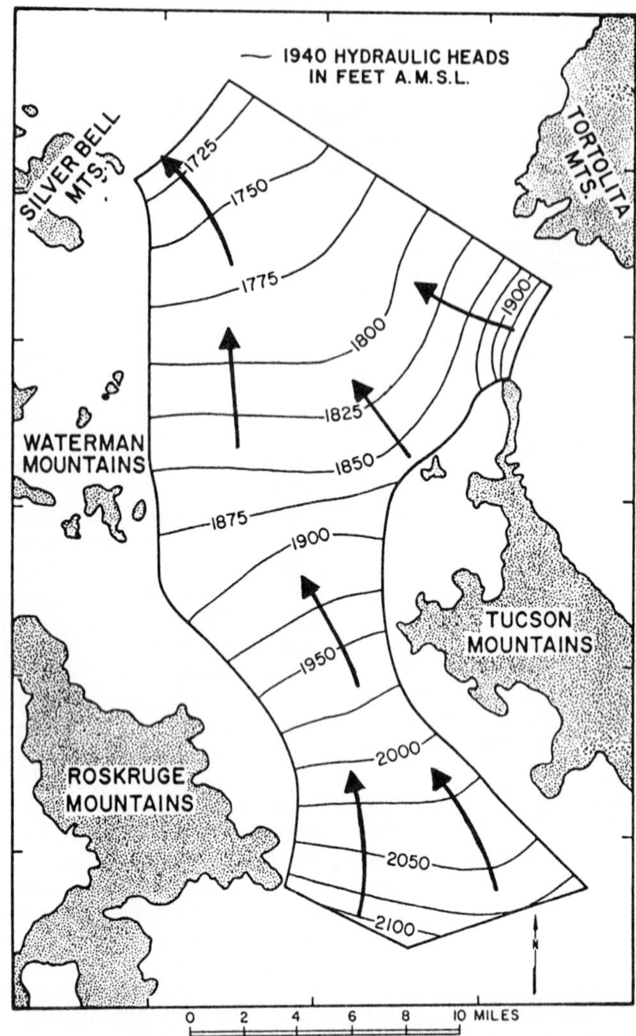

FIGURE 6.3.2 Example of a potentiometric map illustrating hydraulic head contours and the direction of groundwater flow in the uppermost aquifer of Avra Valley, west of Tucson, Ariz. Data from 1940, prior to large-scale aquifer development. (*From Ref. 12.*)

where n is the effective porosity (see Sec. 6.3.2). The pore water velocity (also called the *seepage velocity*) is the average velocity for transport of nonreactive solutes dissolved in groundwater. Under natural flow conditions, in near-surface permeable sediments, groundwater velocities ranging from ten to several hundred meters per year are typical. In media with lower hydraulic conductivity, groundwater velocities are correspondingly lower.

Darcy's law is valid for laminar flow, a condition met in the great majority of hydrogeologic settings. Possible exceptions include karstic limestones, basalts with flow tubes, and rock masses with large-aperture fractures subject to high hydraulic gradients. In these instances, the relation between specific discharge and the hydraulic gradient may be nonlinear.[24] In certain geologic settings, gradients in concentration or temperature contribute to the total driving force causing fluid flow (Ref. 19, p. 563). Additional terms are added to Darcy's law to account for these driving forces; however, coefficient values are poorly documented.

6.3.2 Hydraulic Properties of a Porous Medium

Porosity. Porosity is defined as the fraction of void space per unit volume of porous medium. Porosity is a dimensionless number less than 1, although it is frequently reported as a percentage. Table 6.3.1 summarizes a representative range of porosity values for various geologic media. *Effective porosity* includes only that void space that forms part of the interconnected flow paths through the medium and excludes void space in isolated or dead-end pores. Differences between total and effective porosity reflect lithologic controls on pore structure. In unconsolidated sediments coarser than silt size, effective porosity can be less than the total porosity by 2 to 5 percent (e.g., 0.28 rather than 0.30).

Primary porosity refers to the void space between grains; *secondary porosity* is due either to fracturing or to chemical dissolution of the mineral framework. Although fracture porosity may be on the order of 0.1 percent or less, a well-connected fracture network can have a large impact on hydraulic conductivity.

Hydraulic Conductivity and Permeability. *Hydraulic conductivity* is a measure of the ability of a fluid to move through the interconnected void spaces in the sediment or rock. Hydraulic conductivity is a function of both the medium and the fluid. To separate the effects of the medium from those of the fluid, the *permeability* (k) is defined in the following expression:

$$K = \frac{k \rho g}{\mu} \tag{6.3.6}$$

TABLE 6.3.1 Representative Values of Porosity

Sediment or rock type	Porosity
Clays	0.40–0.60
Silts	0.35–0.50
Fine sands	0.20–0.45
Coarse sands	0.15–0.35
Shales (near-surface, weathered)	0.30–0.50
Shales (at depth)	0.01–0.10
Sandstones	0.05–0.35
Limestones	See Table 6.5.3
Bedded salt	0.001–0.005
Unfractured igneous rocks	0.0001–0.01
Fractured igneous rocks	0.01–0.10
Basalts	0.01–0.25

TABLE 6.3.2 Representative Values of Hydraulic Conductivity and Permeability*

Sediment or rock type	Hydraulic conductivity,†,‡ m/day	Permeability,§ m^2
Clays	$10^{-7}-10^{-3}$	$10^{-19}-10^{-15}$
Silts	$10^{-4}-10^{0}$	$10^{-16}-10^{-12}$
Fine to coarse sands	$10^{-2}-10^{+3}$	$10^{-14}-10^{-9}$
Gravels	$10^{+2}-10^{+5}$	$10^{-10}-10^{-7}$
Glacial till	See Table 6.5.1	
Shales (matrix)	$10^{-8}-10^{-4}$	$10^{-20}-10^{-16}$
Shales (fractured and weathered)	$10^{-4}-10^{0}$	$10^{-16}-10^{-12}$
Sandstones (well-cemented)	$10^{-5}-10^{-2}$	$10^{-17}-10^{-14}$
Sandstones (friable)	$10^{-3}-10^{0}$	$10^{-15}-10^{-12}$
Carbonates	See Table 6.5.3	
Salt	$10^{-10}-10^{-8}$	$10^{-22}-10^{-20}$
Anhydrite	$10^{-7}-10^{-6}$	$10^{-19}-10^{-18}$
Unfractured igneous and metamorphic rocks	$10^{-9}-10^{-5}$	$10^{-21}-10^{-17}$
Fractured igneous and metamorphic rocks	$10^{-5}-10^{-1}$	$10^{-17}-10^{-13}$
Basalts	See Table 6.5.2	

* Values have been compiled from numerous sources and must be used with caution in site-specific applications. The table identifies a commonly observed range in hydraulic conductivity for each of the sediment and rock types. It does not identify bounding values, nor does it differentiate between laboratory and in situ measurements.

† Assuming water properties common to shallow groundwater.

‡ To convert values from m/day to ft/day, multiply by 3.28; to convert to ft/s, multiply by 3.8×10^{-5}.

§ To convert values from m^2 to ft^2, multiply by 10.75. One darcy is equal to a permeability of 9.87×10^{-13} m^2.

where μ is dynamic viscosity of the fluid and ρ is fluid density. Permeability $[L^2]$ is a property of the medium only. Table 6.3.2 lists representative ranges in hydraulic conductivity and permeability for a variety of sediment and rock types.

A hydrogeologic unit is *homogeneous* if its hydraulic properties are the same at every location. However, because of spatial and temporal variability in the geologic processes that create and modify rocks and sediments, no unit is truly homogeneous. *Heterogeneity* occurs on a range of spatial scales. Figure 6.3.3 shows an idealized section through a region where glacial deposits overlie bedrock. The variation in hydraulic conductivity among the depositional units could amount to five or six orders of magnitude; within each unit it can easily be several orders of magnitude. The analyst decides whether or not homogeneity approximations are adequate, and if not, at what scale or level of detail the heterogeneity must be specified.

Given the normal variation in a set of hydraulic conductivity measurements, the question arises as to how best to average a set of measurements to estimate a single value for the *effective hydraulic conductivity* (K_e) when using a homogeneous medium approximation. Unfortunately, averaging rules are not well defined.[38] The effective hydraulic conductivity depends upon properties of the porous medium and the pattern of fluid flow. For steady-state flow, with a spatially uniform hydraulic gradient, the following averaging rules apply:

1. Perfectly stratified medium, n layers, with layer thickness d_i and hydraulic conductivity K_i

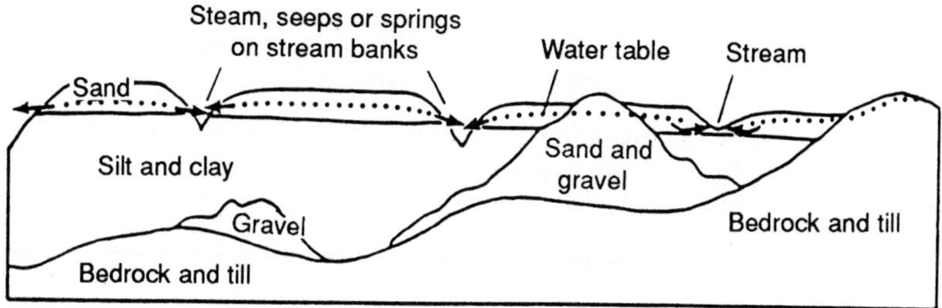

Lake or marine-dominated lowland

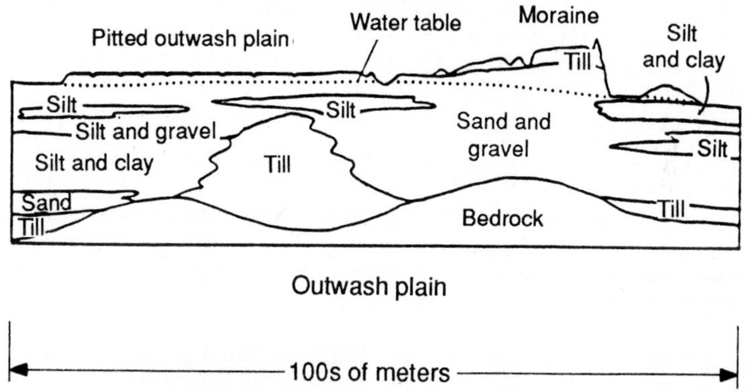

Outwash plain

100s of meters

FIGURE 6.3.3 Schematic diagram of geologic units in glaciated terrain, illustrating the potentially complex nature of the stratigraphy. (*From Ref. 101. Used with permission.*)

- Flow parallel to layering (arithmetic mean)

$$K_e = K_A = \sum_{i=1}^{n} \frac{d_i K_i}{\sum_{i=1}^{n} d_i}$$

- Flow perpendicular to layering (harmonic mean)

$$K_e = K_H = \sum_{i=1}^{n} \frac{d_i}{\sum_{i=1}^{n} \frac{d_i}{K_i}}$$

2. Heterogeneous medium, nonstratified, m measurements
   - Two-dimensional models (geometric mean)

$$K_e = K_G = (K_1 K_2 \cdots K_m)^{1/m}$$

   - Three-dimensional models

$$K_e = K_G (1 + \sigma_y^2/6)$$

where σ_y^2 is the variance of the natural logarithms of the hydraulic conductivity measurements. If the hydraulic gradient is not constant, as in radial flow toward a pumping well, no general averaging rules are available.

To account for the temperature dependence of fluid density and viscosity in converting from permeability to hydraulic conductivity, or vice versa, the following expressions are useful:

$$\rho_w = \rho_o\{1 - [3.17 \times 10^{-4}(T - T_o)] - [2.56 \times 10^{-6}(T - T_o)^2]\} \quad (6.3.7)$$

$$\mu = 2.39 \times 10^{-5}(10^{248.37/(T+133.15)}) \quad (6.3.8)$$

Density is in kg/m^3, viscosity in Pa · s. Equation (6.3.7) should be used to interpolate between tabulated values of ρ. Within $\pm$ 10°C from a tabulated value (ρ_o at T_o), the formula is accurate to 0.1 percent. Fluid density at 10, 20, 30 and 40°C is 999.7, 998.2, 995.7, and 992.2 kg/m^3, respectively. Equation (6.3.7) does not capture the density maximum at 4°C. Equation (6.3.8) is most accurate in the temperature range 15 to 20°C, with an absolute error of 0.8 percent at 5°C, and 0.5 percent at 40°C. The viscosity of water at 20°C is 1.005×10^{-3} Pa · s.

If fluid density is a function of total dissolved solids (TDS), then

$$\rho = \rho_o [1 + \beta_c (C - C_0)] \quad (6.3.9)$$

where C is the concentration of TDS (g/L), and C_0 is the total dissolved solids (g/L) in fresh water corresponding to ρ_o. β_c is equal to 7.14×10^{-4} L/g within the salinity range from fresh water to seawater. Seawater with a density of 1025 kg/m^3 is equivalent to a TDS of 35 g/L. For $\rho_o = 1000$ kg/m^3, $C_0 = 0$.

Specific Storage. When fluid pressures decline within a porous medium, two responses occur: (1) the fluid volume expands under the lower value of fluid pressure; and (2) the pore space decreases as an additional fraction of the overburden pressure must be carried by the solid matrix. The magnitude of the first response is controlled by the *compressibility of water* (β), the magnitude of the second by the *compressibility of the porous medium* (α). The compressibility of a porous medium is defined as the ratio of the change in volume ($-\Delta V/V$) to the change in effective stress ($\Delta\sigma_e$), where $\Delta\sigma_e = -\Delta p$. The compressibility is the inverse of the bulk modulus.

It is convenient to have a single parameter that characterizes the volume of water released from a porous medium with a decline in fluid pressure. *Specific storage* [L^{-1}] is defined as the volume of water that a unit volume of porous medium releases from storage per unit change in hydraulic head. Specific storage is calculated as

$$S_s = \rho g (\alpha + n\beta) \quad (6.3.10)$$

Representative values of compressibility and specific storage are given in Table 6.3.3.

Transmissivity and Storage Coefficient. In the case of horizontal flow through a layer of thickness b, two derived parameters are commonly used. *Transmissivity, T,* [L^{2}T^{-1}] is the product of hydraulic conductivity and the layer thickness and the *storage coefficient S* (or *storativity*) is the product of specific storage and thickness:

$$T = K\, b \quad (6.3.11)$$

$$S = S_s\, b \quad (6.3.12)$$

It should be emphasized that S and T are specifically defined for two-dimensional, horizontal analyses, having the aquifer thickness included within their values.

TABLE 6.3.3 Representative Values of Compressibility and Specific Storage

Sediment or rock type	Compressibility α, Pa^{-1}	Specific storage* S_S, m^{-1}
Unconsolidated clays	$10^{-6}-10^{-8}$	$10^{-2}-10^{-4}$
Unconsolidated sands	$10^{-7}-10^{-9}$	$10^{-3}-10^{-5}$
Unconsolidated gravels	$10^{-8}-10^{-10}$	$10^{-4}-10^{-6}$
Compacted sediments	$10^{-9}-10^{-11}$	$10^{-5}-10^{-7}$
Igneous and metamorphic rocks	$10^{-9}-10^{-11}$	$10^{-5}-10^{-7}$
Water*	4.4×10^{-10}	

* Assuming water properties common to shallow groundwater.

6.3.3 Anisotropic Hydraulic Conductivity

If the hydraulic conductivity of a sediment or rock is independent of the direction of measurement, the porous medium is *isotropic.* Fluid flow is in the same direction as the hydraulic gradient. If the value of hydraulic conductivity is dependent upon the direction of measurement, the medium is *anisotropic.* In an anisotropic medium, flow lines are not aligned with the direction of the hydraulic gradient but instead are rotated toward the direction of higher hydraulic conductivity. In sedimentary deposits, the hydraulic conductivity in a direction perpendicular to stratification, is usually less than that parallel to the stratification. Anisotropic ratios of conductivity measured on core samples are typically on the order of 2:1 to 10:1. At the field scale, anisotropic ratios of conductivity in layered geologic media may be as large as 100:1 or greater (Ref. 29, p. 33).

The specific discharge vector (q_x, q_y, q_z) is related to the components of the hydraulic gradient using the equation

$$\begin{Bmatrix} q_x \\ q_y \\ q_z \end{Bmatrix} = -\begin{bmatrix} K_{xx} & K_{xy} & K_{xz} \\ K_{yx} & K_{yy} & K_{yz} \\ K_{zx} & K_{zy} & K_{zz} \end{bmatrix} \begin{Bmatrix} \dfrac{\partial h}{\partial x} \\ \dfrac{\partial h}{\partial y} \\ \dfrac{\partial h}{\partial z} \end{Bmatrix} \tag{6.3.13}$$

The off-diagonal terms are symmetric (e.g., $K_{xy} = K_{yx}$). A simple interpretation of the off-diagonal terms is that K_{xy}, for example, is the hydraulic conductivity value determining the specific discharge in the x direction due to a hydraulic gradient directed along the y coordinate axis. It is useful to think of the direction dependence of hydraulic conductivity in terms of a *hydraulic conductivity ellipsoid* (or an ellipse in two dimensions). The major and minor axes of the ellipse, with lengths given by $\sqrt{K_{xx}'}$ and $\sqrt{K_{zz}'}$ correspond to the directions of the maximum and minimum values of hydraulic conductivity (Fig. 6.3.4*a*). The major and minor axes are generally parallel and perpendicular to the stratification in a sedimentary deposit.

Equation (6.3.13) is simplified when the major and minor axes of the hydraulic conductivity ellipsoid are aligned with the (x, y, z) coordinate axes of the flow domain. In this case, the off-diagonal terms equal zero. This particular orientation of the ellipsoid defines the *principal coordinate system.* The off-diagonal terms are always

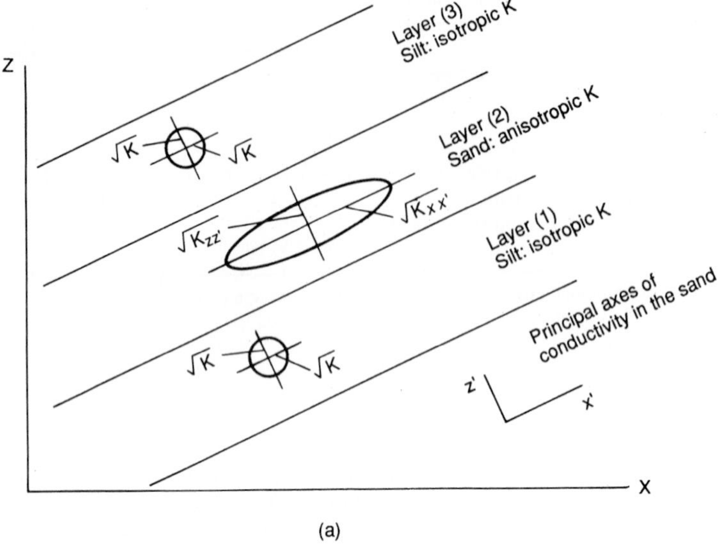

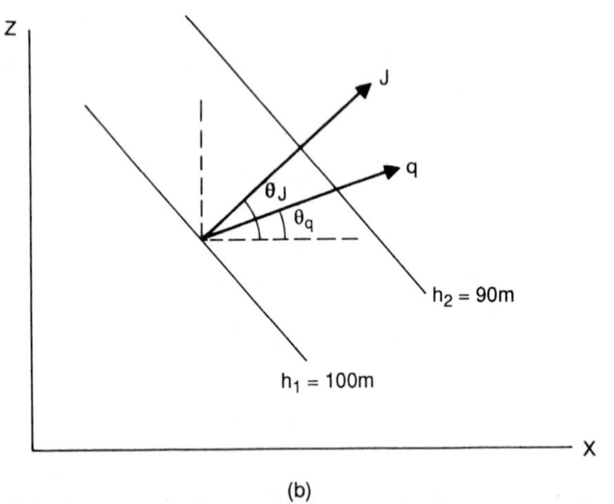

FIGURE 6.3.4 *(a)* A tilted sequence of layered sediments; illustrating the hydraulic conductivity ellipse for isotropic and anisotropic units, and the relation between the principal axes of the ellipse (x',z') and the coordinate axes of flow domain (x,z). *(b)* Relation between the direction of flow and the direction of the hydraulic gradient in an anisotropic porous medium.

zero, independent of the orientation of the coordinate system, in an isotropic medium.

If the values of hydraulic conductivity are known in the principal coordinate system (K_{xx}', K_{zz}' for the major and minor axes of the ellipse), then the values in a second coordinate system making an angle θ to the principal coordinate system are

$$K_{xx} = \frac{K_{xx}' + K_{zz}'}{2} + \frac{K_{xx}' - K_{zz}'}{2} \cos 2\theta$$

$$K_{xz} = K_{zx} = -\frac{K_{xx}' - K_{zz}'}{2} \sin 2\theta \qquad (6.3.14)$$

$$K_{zz} = \frac{K_{xx}' + K_{zz}'}{2} - \frac{K_{xx}' - K_{zz}'}{2} \cos 2\theta$$

Equation (6.3.14) is useful in cases where the principal directions of hydraulic conductivity differ among the geologic units within the flow domain (e.g., in structurally deformed sedimentary rocks, or with cross-cutting faults).

Assuming that x and z are the principal directions, the equation expressing the direction of the flow line as a function of the direction of the hydraulic gradient (θ_J) and the anisotropic ratio $K_r = K_{zz}'/K_{xx}'$ is (see Fig. 6.3.4*b*)

$$\theta_q = \tan^{-1}(K_r \tan \theta_J) \qquad (6.3.15)$$

The magnitude q of the specific discharge as a function of θ_J and the hydraulic gradient J is

$$q = J(K_{xx}'^2 \cos^2 \theta_J + K_{zz}'^2 \sin^2 \theta_J)^{1/2} \qquad (6.3.16)$$

These equations are useful in determining the direction of flow and fluid flux from a hydraulic head map when the porous medium is known to be anisotropic.

If the system is two-dimensional and anisotropic in the horizontal plane, transmissivity has the components (T_{xx}, T_{xy}, T_{yx}, T_{yy}), analogous to hydraulic conductivity.

6.3.4 Hydrogeologic Settings

Hydrogeologic models are developed on the basis of the hydraulic properties of sediments and rocks. Rocks or sediments with similar hydraulic properties are grouped to form *hydrostratigraphic units.* A terminology exists that allows for a simple classification of the major hydrogeologic features of a geologic setting. Thus, we define:

- *Vadose zone:* the partially saturated region between the ground surface and the water table.
- *Aquifer:* a permeable geologic unit that can transmit and store significant quantities of water.
- *Aquitard:* a less permeable geologic unit that stores but does not readily transmit water.
- *Unconfined aquifer:* a permeable geologic unit with the water table forming its upper boundary.
- *Confined aquifer:* a permeable geologic unit located beneath a saturated, less permeable unit (i.e., beneath an aquitard).
- *Perched zones:* a zone of limited areal extent, located above the main water table, that occurs when infiltrating water is impeded by a low permeability layer, creating saturated conditions above the impeding layer.

Unconfined aquifers respond to groundwater withdrawals differently than confined aquifers. Confined aquifers yield water to wells by the mechanisms of fluid volume expansion and compaction of pore volume [Eq. (6.3.10)]. By definition, a confined aquifer remains saturated. In addition to their elastic storage properties, unconfined aquifers yield water by desaturation of the pore space as the water table declines. Water released from storage in an unconfined aquifer greatly exceeds that of a confined aquifer for equal water level declines. To reflect this difference, the storativity of an unconfined aquifer is termed the *specific yield,* defined as the volume of water released from a unit area of the aquifer for a unit decline in the water table (Table 6.3.4). The specific yield is a significant fraction of the effective porosity.

TABLE 6.3.4 Representative Values of Specific Yield

Sediment or rock type	Specific yield, %
Clays	1–5
Silts	10–20
Fine sands	10–30
Sands and gravels	20–30
Sandstone	5–20
Shale	0.5–5
Limestone	0.5–20

One consequence of groundwater withdrawals in geologic settings that contain compressible sediments is *subsidence* of the land surface (e.g., the Central Valley of California and Mexico City). Most of the consolidation occurs not in the aquifers but in interstratified silts and clays that slowly release water from storage with the decline in hydraulic head.[48,98] While subsidence can be arrested by halting the decline in hydraulic head, because of the inelastic nature of many clays, subsidence is largely irreversible.

6.3.5 Concept of a Flow System

In developing a conceptual model of a flow system, it is important to consider the topographic setting. The key elements of a groundwater flow system are shown in Fig. 6.3.5. Topographically higher areas are typically *areas of recharge,* while topographically lower areas are *areas of discharge.* Flow directions are from recharge to discharge areas. The piezometers in Fig. 6.3.5 show that hydraulic head increases with depth in areas of discharge (*C* and *D*), while in areas of recharge, hydraulic head decreases with depth (*A* and *B*). Two piezometers located on the same equipotential, even though they may be separated by a substantial distance horizontally, will have the same water level (*E* and *B*). A topographic high may be coincident with a *groundwater divide* separating one flow system from another. An extended discussion of groundwater flow systems can be found in Ref. 19 (pp. 242–266).

Water levels in piezometers fluctuate on time scales ranging from a few minutes to hundreds of years, depending upon the nature of the processes that initiate the fluid pressure variations. Short-term fluctuations in confined aquifers can be caused by changes in barometric pressure of the atmosphere,[106] earth tides,[52] and seismic events.[105] Earth tides can lead to water-level changes of 1 or 2 cm; atmospheric pressure changes may cause fluctuations of several tens of centimeters, depending upon the elastic properties of the aquifer and the magnitude of change in atmospheric pressure. These types of water-level changes are damped in unconfined aquifers. However, fluctuations can occur in response to time-varying rates in consumptive use of water by plants whose roots penetrate to the water table. Yearly cycles reflect changing wet and dry seasons, and the resulting temporal distribution of groundwater recharge (e.g., Fig. 6.3.6). Multiyear cycles are tied to longer-term changes in precipitation patterns. It is possible to estimate values of the hydraulic properties of aquifers from the frequency response of the water-level fluctuations.[52,106]

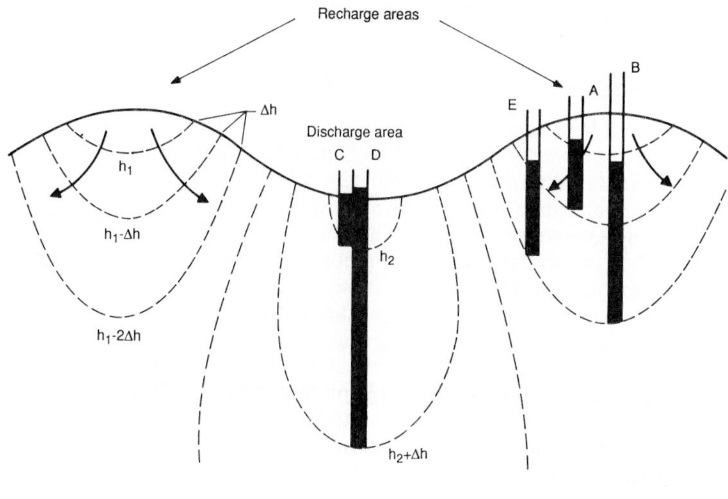

FIGURE 6.3.5 Conceptual model of a groundwater flow system. The dashed lines are lines of constant hydraulic head; head decreases with depth in recharge areas and increases with depth in discharge areas. The piezometers have small screens at the bottom of the borehole.

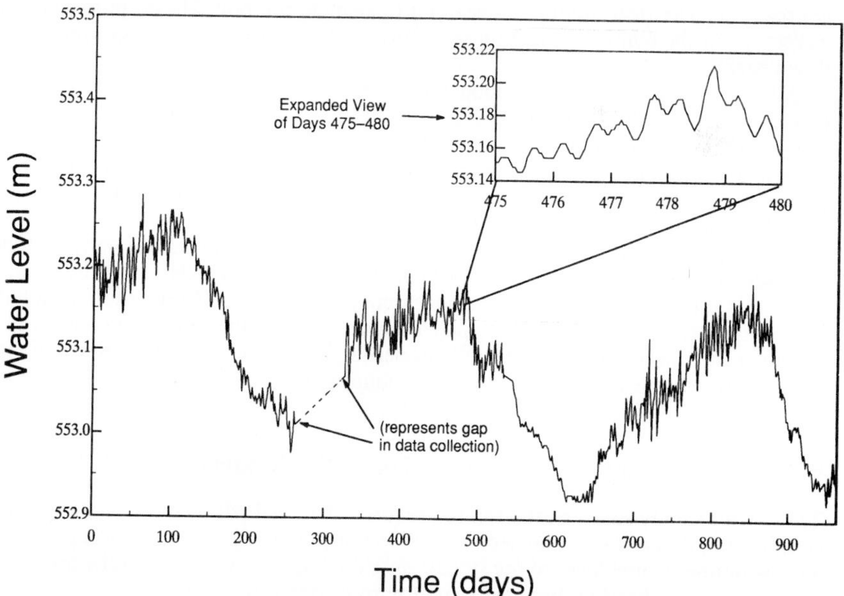

FIGURE 6.3.6 Example of water-level fluctuations in an unconfined aquifer. Day 0 is Jan. 1, 1989. The main graph has one data point per day; the expanded view has one data point per hour plotted.

6.4 HYDROGEOLOGIC SIMULATION MODELS

6.4.1 Concept of a Hydrogeologic Model

Figure 6.1.1 emphasizes a staged approach to site investigation, beginning with the development of a *conceptual model.* The conceptual model is based on a reconnaissance of the area of interest using geologic maps, air-satellite photos, field observations, other available data, and the skill and experience of the analyst. The conceptual model of the system consists of basic elements such as inflows, outflows, and system geometry. The conceptual model is a hypothesis of how the hydrogeologic system works. Mathematical modeling is a method of testing this hypothesis.

The *mathematical model* consists of differential equations for hydraulic head, together with specification of system geometry, boundary conditions, and for transient problems, initial conditions. Specification of appropriate boundary conditions on the flow domain is often the most difficult task in formulating the model, and one of the primary sources of uncertainty in model analysis. For the fluid flow problem, boundaries are normally either (1) impermeable, (2) specified flux, or (3) specified values of hydraulic head.

A *hydrogeologic simulation model* is a combination of a mathematical model, including values for its parameters, and a method for solving it. If the governing equations and boundary and initial conditions are simple enough, an analytic solution is possible.[47,122] Even for cases where the problem is too complex for an analytic solution, it is a good idea to simplify the conceptual model in the first instance to the point where an analytic solution is possible. Such a solution is usually easy to work with and can provide quick, valuable insight into how the system works. Analytic solutions vary in complexity from graphical solutions, such as flow nets, to solutions that are so cumbersome that it may be easier to solve the problem using a numerical approach such as the finite-difference or finite-element method. The emphasis in this chapter is on the common analytic solutions used in characterizing a groundwater flow system.

A hydrogeologic simulation model is developed to reach a management decision (Fig. 6.1.1). Simulation models serve at least three roles in this regard: (1) to provide insight to the hydrologic processes operative in the study area, (2) to develop predictions of system behavior under changed conditions, and (3) to test alternative hypotheses on system behavior to guide site investigation plans and to increase the confidence level in the management decision.

As the conceptual model is refined, a decision must be made about whether it is necessary to proceed with a more sophisticated mathematical model, or if the existing solution is adequate to resolve the management problem. If the decision is to proceed with a numerical solution, it is usually because of complex boundary conditions, because of spatially variable hydraulic properties, and/or because analytic solutions do not exist for the appropriate governing equations.

6.4.2 Steady-State Hydrogeologic Simulation Models

The use of a steady-state approximation requires that the hydrogeologic system be in equilibrium, that is, hydraulic head values do not change significantly through time. The two-dimensional form of the steady-state flow equation describing the distribution of hydraulic head within a saturated porous medium is

$$\frac{\partial}{\partial x}\left(K_{xx}b\frac{\partial h}{\partial x}+K_{xy}b\frac{\partial h}{\partial y}\right)+\frac{\partial}{\partial y}\left(K_{yx}b\frac{\partial h}{\partial x}+K_{yy}b\frac{\partial h}{\partial y}\right)=0 \qquad (6.4.1)$$

where b is the thickness of the medium perpendicular to the (x,y) plane. For a cross-sectional model, $b = 1$. For an areal flow model, b is equal to the aquifer thickness. If the coordinate axes of the flow system are aligned with the principal directions of the hydraulic conductivity ellipse, then $K_{xy} = K_{yx} = 0$.

Flow Nets. The simplest solution method is the graphical construction of a *flow net.* Flow nets are obtained through a trial-and-error process of drawing *streamlines* and *equipotentials.* Lines of equal hydraulic head, called equipotentials, are given the symbol Φ [L]. A streamline is a line that at each point is tangent to the groundwater flux vector. In a steady flow, a streamline traces the pathlines of fluid motion. Along a streamline, the *stream function* Ψ $[L^2T^{-1}]$ is constant. The stream function and equipotential are related[33]

$$K\frac{\partial \Phi}{\partial x} = \frac{\partial \Psi}{\partial y} \quad \text{and} \quad \frac{\partial \Psi}{\partial x} = -K\frac{\partial \Phi}{\partial y} \tag{6.4.2}$$

Lines of constant Φ and constant Ψ are orthogonal. Numerical values of the stream function map increments in the volumetric flow through the flow domain (per unit thickness) relative to a reference streamline where $\Psi = 0$.

The rules for constructing flow nets in an isotropic medium are (see Fig. 6.4.1):

- Streamlines are drawn in the direction of flow; they can be curvilinear, in response to system geometry. The streamlines form a set of *stream tubes.*
- Equipotentials are drawn perpendicular to the streamlines, to form *curvilinear squares.* The hydraulic head loss (Δh) between any two equipotentials is the same.
- Streamlines must be drawn perpendicular to constant-head boundaries and parallel to impermeable boundaries. Streamlines cannot intersect.
- Equipotential lines must intersect the water table at equal vertical intervals, such that Δz is constant.
- Equipotentials can be labelled knowing that where they intersect the water table, the hydraulic head is equal to the elevation of that point.

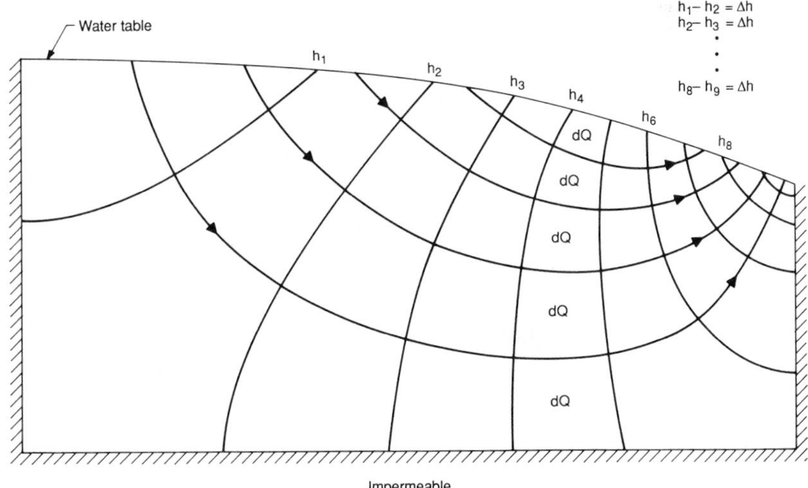

FIGURE 6.4.1 Example of construction of a flow net. Cross-sectional model of the saturated zone in a hill slope.

By drawing curvilinear squares, each stream tube carries the same quantity of flow ($dQ = K\Delta h$, per unit width normal to the diagram). Stream tubes will be smaller in the region of the flow domain where the specific discharge is higher. It always requires trial-and-error adjustment of the placement of the streamlines and equipotentials to meet these criteria. The solution by flow nets is approximate, at best. However, even with the ready access to computer models, flow nets are still a good starting place in the staged approach to understanding hydrogeologic systems. The process of drawing a flow net forces one to focus on a synthesis of the key features of the groundwater system.

When streamlines cross from a unit of hydraulic conductivity K_1 to a second unit of hydraulic conductivity K_2, streamlines refract according to the tangent law ($K_1/K_2 = \tan\theta_1/\tan\theta_2$), where θ_1 is the deviation of the flow line from the vertical in unit 1, and θ_2 is the deviation in unit 2. As a consequence of *flow-line refraction,* streamlines adjust to use higher-permeability layers or zones as preferential flow paths, and they tend to cross regions of lower permeability in the shortest distance possible.

Flow nets in anisotropic media can be drawn using the concept of a *transformed domain.*[29] For the case where $K_{xx} > K_{zz}$, the z axis of the flow domain is lengthened by the ratio $(K_{xx}/K_{zz})^{1/2}$, keeping the horizontal scale unchanged. The flow net is constructed in this equivalent isotropic medium using the rules outlined above. To calculate the volumetric flow within the transformed section, an equivalent hydraulic conductivity $(K_{xx}K_{zz})^{1/2}$ is used.

Confined, Steady-State Flow: Stream Functions. In the case where it is reasonable to approximate the porous medium as a homogeneous unit, there are a number of useful analytic relationships between the stream function and the equipotential. First, we define a new stream function $\psi = \Psi/K[\mathrm{L}]$, which is a normalization with respect to the homogeneous K value.

1. Uniform flow field with specific discharge q_0 (see Fig. 6.4.2*a*)

$$\Phi(x) = -\frac{q_0}{K}x \qquad \psi(y) = \frac{q_0}{K}y \tag{6.4.3}$$

2. Radial flow to a point source or sink (see Fig. 6.4.2*b*)

$$\Phi(r) = \frac{Q}{2\pi T}\ln r \qquad \psi(\theta) = \frac{Q}{2\pi T}\theta \tag{6.4.4}$$

   where Φ is the value of hydraulic head at a radial distance r from the source or sink, ψ is the cumulative flux between the streamlines defined by $\theta = 0°$ and the angle θ, and Q is the volumetric flow at the source or sink.

3. Point source or sink superimposed on a uniform flow field (see Fig. 6.4.2*c*). For a source or sink located at the origin

$$\begin{aligned} \Phi(r,\theta) &= \frac{-q_0}{K}r\cos\theta + \frac{Q}{2\pi T}\ln r \\ \psi(r,\theta) &= \frac{q_0}{K}r\sin\theta + \frac{Q}{2\pi T}\theta \end{aligned} \tag{6.4.5}$$

   If the uniform flow is moving from right to left, then the point source or sink is a sink (pumping well), and it captures a portion of the regional flow. The width of

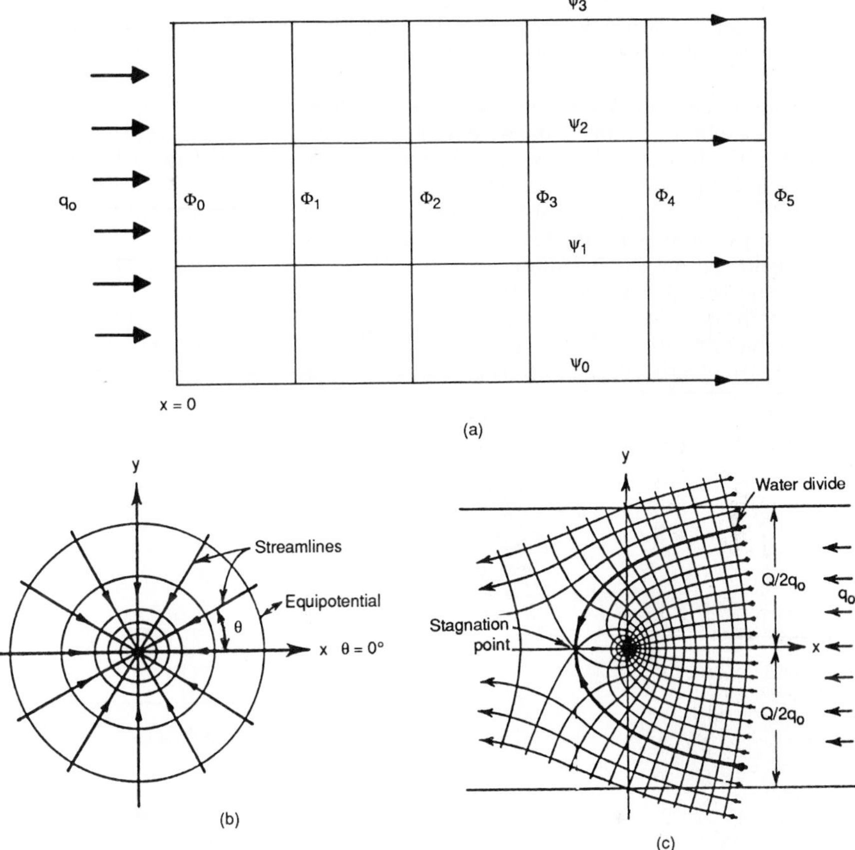

FIGURE 6.4.2 Streamlines and equipotentials for *(a)* a uniform flow field, *(b)* a point source or sink, *(c)* a combined uniform flow field and a point source or sink. *(Reproduced from Ref. 7 by permission of McGraw-Hill, Inc.)*

the capture zone y_c is given by

$$y_c = \frac{Q}{q_0\, b} \tag{6.4.6}$$

where b is the aquifer thickness. If the sign of q_0 is reversed, the uniform flow moves from left to right, and the point source or sink is a source (recharge well), and y_c represents the width of the zone of recharge water.

4. Point source or sink located at position (x_0, y_0) with respect to the origin

$$\Phi(x,y) = \frac{Q}{4\pi T} \ln\,[(x - x_0)^2 + (y - y_0)^2]$$

$$\psi(x,y) = \frac{Q}{2\pi T} \tan^{-1}\left(\frac{y - y_0}{x - x_0}\right) \tag{6.4.7}$$

The equations in this form are useful in calculating the flow field when more than one source and/or sink is present. The distribution of Φ and ψ is obtained by simply adding the solutions for each source and sink.

Unconfined, Steady-State Flow: Dupuit Assumptions. Because the upper boundary of an unconfined aquifer is the water table, with an elevation equal to the hydraulic head h, the boundary becomes part of the solution when predicting the hydraulic head distribution in the aquifer. The *Dupuit assumptions* simplify the problem to the point where some simple analytic results can be developed. The Dupuit assumptions are: (1) streamlines are assumed to be horizontal and the equipotentials vertical; (2) the hydraulic gradient is assumed to be equal to the slope of the water table and to be invariant with depth (Fig. 6.4.3).

For the problem shown in Fig. 6.4.3, the following solutions predict the height of the water table (h), the discharge per unit width (Q_x) at any distance x from the origin ([L^2T^{-1}]), and the location of the groundwater divide (d).

$$h^2 = h_1^2 - \frac{(h_1^2 - h_2^2)\,x}{L} + \frac{w}{K}(L - x)\,x \tag{6.4.8}$$

$$Q_x = \frac{K\,(h_1^2 - h_2^2)}{2\,L} - w\left(\frac{L}{2} - x\right) \tag{6.4.9}$$

$$d = \frac{L}{2} - \frac{K}{w}\,\frac{(h_1^2 - h_2^2)}{2\,L} \tag{6.4.10}$$

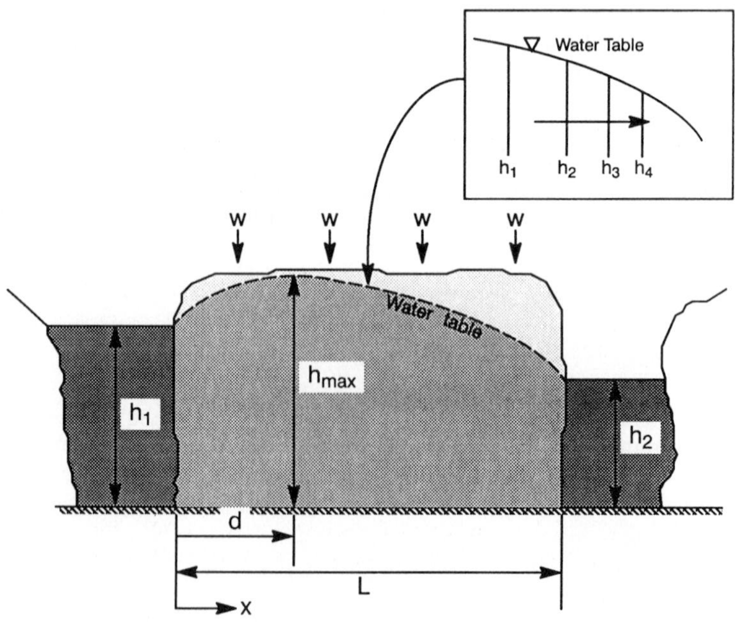

FIGURE 6.4.3 Conceptual model using the Dupuit approximation for unconfined flow to a stream in response to uniform recharge, Dupuit approximation. *(Modified from Ref. 47. Used with permission).*

The parameter w is the recharge rate ($[LT^{-1}]$) at the water table. Equations (6.4.8) and (6.4.9) can also be used to estimate the height of the water table, and volumetric flow, at an intermediate location between two monitoring wells where the water-table elevation is known. The Dupuit assumptions provide a good approximation for a water-table slope up to 10°. The magnitude of the error, for a water table with slope θ, is given by the percent difference between $\sin\theta$ and $\tan\theta$.

Radial Flow to a Well: Steady-State Models. The *Theim equation* describes steady-state *radial flow* to a well in a confined aquifer with transmissivity T. Given a well radius r_w, the well discharge (Q) is related to the transmissivity and drawdown by

$$Q = 2\pi T \frac{h - h_w}{\ln(r/r_w)} \tag{6.4.11}$$

where the hydraulic head at the well is h_w and h is the head at a distance r from the pumping well. This equation assumes the well fully penetrates the confined aquifer and that the drawdown cone has not intersected any hydraulic boundaries.

The analogous equation for an unconfined aquifer with hydraulic conductivity K is

$$Q = \pi K \frac{h_o^2 - h_w^2}{\ln(r_o/r_w)} \tag{6.4.12}$$

where h_o is hydraulic head at radial distance r from the fully penetrating well. If an observation well is not available, r_o is approximated by specifying the radius of influence of the well (i.e., the radial distance at which the drawdown approaches zero). In practice, calculations are not sensitive to this somewhat arbitrary estimate, because of the logarithmic dependence on r_o. Equation (6.4.12) does not provide accurate estimates of hydraulic head near the well because it is derived by neglecting vertical components of flow, but estimates of Q, given two hydraulic head measurements, are reasonable.

6.4.3 Transient Hydrogeologic Simulation Models

Three basic models are used in describing the time-dependent drawdown in hydraulic head in response to groundwater withdrawals (Fig. 6.4.4). The models represent a confined aquifer, a leaky aquifer system, and an unconfined aquifer, respectively.

The governing equation of transient flow, written for an areal (x, y) model, with the coordinate axes of the flow domain aligned with the principal directions of the hydraulic conductivity ellipse, is

$$\frac{\partial}{\partial x}\left(K_{xx} b \frac{\partial h}{\partial x}\right) + \frac{\partial}{\partial y}\left(K_{yy} b \frac{\partial h}{\partial y}\right) - Wb = \rho g b\,(\alpha + n\beta)\frac{\partial h}{\partial t} \tag{6.4.13}$$

where W is a volumetric flux per unit volume representing sources and sinks of water $[T^{-1}]$, n is porosity, α is the compressibility of the porous medium, β is the compressibility of water, and b is the vertical thickness of the model domain. Using the parameters transmissivity and storage coefficient, this equation is often seen in the more compact form

$$\frac{\partial}{\partial x}\left(T_{xx} \frac{\partial h}{\partial x}\right) + \frac{\partial}{\partial y}\left(T_{yy} \frac{\partial h}{\partial y}\right) - Wb = S\frac{\partial h}{\partial t} \tag{6.4.14}$$

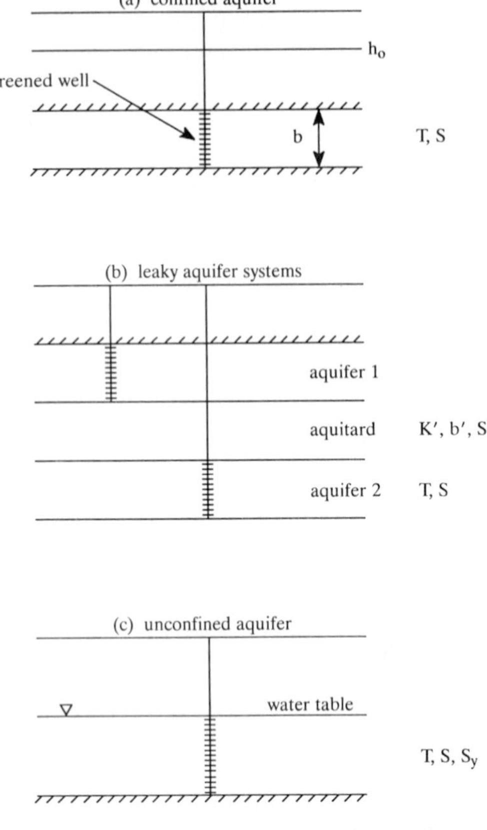

FIGURE 6.4.4 Conceptual models used in interpreting transient hydraulic tests, and the parameters obtained from these tests.

For transient flow in a hill slope, represented by a vertical cross-sectional slice, $b = 1$ and, commonly, $W = 0$.

Confined Aquifers. Figure 6.4.4*a* illustrates the model describing radial flow to a well fully penetrating a horizontal, confined aquifer of uniform thickness and infinite extent. The aquifer is represented as a homogeneous isotropic medium with transmissivity (T) and storage coefficient (S). No water enters the aquifer by leakage from adjacent strata. Storage of water in the well bore is ignored. The discharge at the well (Q) is constant through time. At $t = 0$, the hydraulic head in the aquifer is everywhere constant at a value h_0. The drawdown (s) is the difference $h - h_0$.

The analytic solution for the drawdown in hydraulic head [$s(r,t)$] is

$$s(r,t) = \frac{Q}{4\,\pi\,T}\int_u^\infty \frac{e^{-z}}{z}\,dz = \frac{Q}{4\pi T}\,W(u) \tag{6.4.15}$$

where $u = r^2S/4Tt$. Equation (6.4.15) is frequently referred to as the *Theis solution.*

It is valid for any consistent set of units. The integral expression in Eq. (6.4.15) is termed the well function $W(u)$. The following series expansion is useful for numerical evaluation of $W(u)$:

$$W(u) = -0.5772 - \ln u + u - \frac{u^2}{2\,2!} + \frac{u^3}{3\,3!} - \frac{u^4}{4\,4!} + \cdots \qquad (6.4.16)$$

If the series approximation is truncated to the form $W(u) = -0.5772 - \ln u$ (appropriate when $u < 0.05$), drawdown can be approximated by the Jacob equation:

$$s(r,t) = \frac{Q}{4\pi T} \ln \frac{2.25\,Tt}{r^2 S} \qquad (6.4.17)$$

Leaky Aquifer Systems. In a leaky aquifer system (Fig. 6.4.4*b*), water may enter the aquifer from adjacent lower-permeability units, providing an additional source of water and reducing drawdowns relative to those predicted by the Theis solution. Hantush and Jacob[46] developed a solution describing drawdown in a radially symmetric leaky confined aquifer separated from an overlying aquifer by a lower-permeability confining bed. Key simplifications are that the hydraulic head in the upper aquifer remains constant and that no water is released from storage in the aquitard. Leakage into the aquifer being pumped is a function of the vertical hydraulic conductivity (K') of the aquitard, its thickness (b'), and the difference in hydraulic head between the overlying aquifer and that in the aquifer being pumped.

Defining the dimensionless parameters:

$$u = \frac{r^2 S}{4Tt} \qquad \frac{r}{B} = r\left(\frac{K'}{Tb'}\right)^{1/2} \qquad (6.4.18)$$

the drawdown is given by:

$$s(r,t) = \frac{Q}{4\pi T} W\left(u, \frac{r}{B}\right) \qquad (6.4.19)$$

The well function $W(u,r/B)$ is tabulated in most hydrogeology texts.[19,27,29]

To account for the release of water from storage within the confining layer, the following approximate solution can be used:[43]

$$s(r,t) = \frac{Q}{4\pi T} W(u,\beta') \qquad (6.4.20)$$

where

$$u = \frac{r^2 S}{4Tt} \qquad \beta' = \frac{r}{4B}\left(\frac{S'}{S}\right)^{1/2} \qquad B = \left(\frac{Tb'}{K'}\right)^{1/2} \qquad (6.4.21)$$

and S' is the storage coefficient of the confining layer. The function $W(u,\beta')$ is tabulated in Ref. 27 (p. 553).

A more complete solution, taking into account storage in the confining bed and drawdown in the overlying aquifer, has been developed,[85] providing values of drawdown in the two aquifers and in the aquitard.

Unconfined Aquifers. In an unconfined aquifer (Fig. 6.4.4*c*), the water table declines and the zone above the water table desaturates. The drawdown, averaged over the depth of the unconfined aquifer, at a radial distance r from the pumping well, is given by[83]

$$s(r,t) = \frac{Q}{4\pi T} W(u_a, u_b, \Gamma) \tag{6.4.22}$$

where

$$u_a = \frac{r^2 S}{4Tt} \qquad u_b = \frac{r^2 S_y}{4Tt} \qquad \Gamma = \frac{r^2 K_z}{b^2 K_r} \tag{6.4.23}$$

and K_r and K_z are the horizontal and vertical hydraulic conductivities of the unconfined aquifer. The well function $W(u_a, u_b, \Gamma)$ is tabulated in Ref. 83. The parameter u_a describes the early-time response when the rate of drawdown is controlled by the elastic storage properties of the aquifer, while u_b describes the late-time response when the rate of drawdown is controlled by the specific yield. This simplified model of drawdown is valid if the specific yield (S_y) is much greater than the elastic storage coefficient (S) and the drawdown is substantially less than the initial saturated thickness.

Partial Penetration of Wells. The solutions above are based on the assumption that the pumping well is open to the full thickness of the aquifer. When this is not the case, there are significant vertical components of flow in the vicinity of the pumping well [for $r < 1.5b\,(K_r K_z)^{1/2}$]. Analytic solutions accounting for the partial penetration of wells have been developed for confined aquifers,[44] leaky aquifers,[45] and unconfined aquifers.[82] The geologic model of the site, which describes the thickness of aquifers and aquitards, is used in making a determination of whether partial-penetration effects need to be considered.

Multiple-Well Systems. In the case where more than one well is pumping water from a confined aquifer, the total drawdown in hydraulic head can be calculated as a summation of the drawdown due to each well operating independently. Thus the cumulative drawdown at an observation point for a system of n wells is

$$S_T = \frac{Q_1}{4\pi T} W(u_1) + \cdots + \frac{Q_n}{4\pi T} W(u_n) \qquad i = 1,2,3, \ldots, n \tag{6.4.24}$$

where $u_i = r_i^2 S/4Tt_i$, r_i is the radius vector from pumping location i to the observation point, and t_i is the time since pumping started at well i, at rate Q_i.

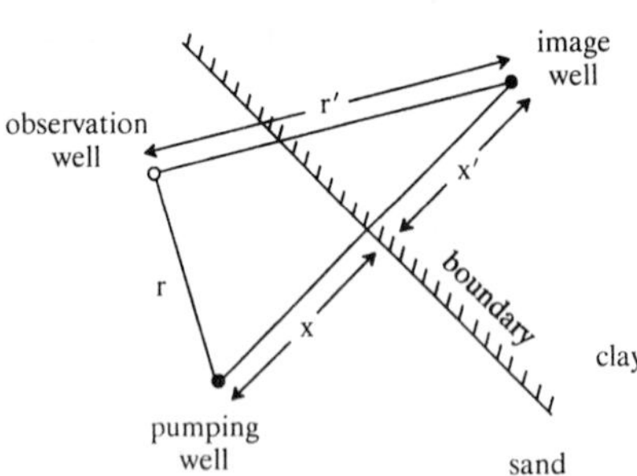

FIGURE 6.4.5 Placement of an image well for the case of a linear impermeable boundary.

Effects of Boundaries. Where wells are in close proximity to lateral boundaries of the aquifer, or the aquifer is of limited areal extent, it is necessary to account for the effect of those boundaries on cumulative drawdown. If the geometric configuration of the lateral boundaries is of a simple form, it is possible to apply the *principle of superposition* and *image well theory*. For a well located a perpendicular distance x from a linear impermeable boundary, an imaginary well pumping at the same rate Q is located at a perpendicular distance x' on the other side of the boundary, immediately opposite the real well (Fig. 6.4.5). The summation of the drawdown due to both wells accounts for the increased drawdown caused by the limited extent of the aquifer. If the boundary is a recharge boundary, such as a fully penetrating stream or river, and the streambed does not have a lower hydraulic conductivity than the aquifer, then the drawdown can be predicted in a similar manner, except that the image well injects water at rate Q. Often the geometry of aquifer boundaries is too complex to permit easy application of image theory. In this case, it is preferable to adopt numerical techniques to predict drawdown in hydraulic head (see Chap. 22).

6.5 FLOW SYSTEM CHARACTERISTICS FOR SELECTED GEOLOGIC SETTINGS

6.5.1 Intrusive Igneous and Metamorphic Terrain

The hydraulic properties of crystalline rock are strongly dependent upon the presence and extent of *fracturing*. *Faults* are fractures along which shear displacement has occurred. *Joints* are openings in which there has been no movement parallel to the fracture plane.[99] Joints typically have lengths on the order of centimeters to tens of meters and occur in sets defined on the basis of the orientation of joint planes. Because of these preferred orientations, the hydraulic properties of fractured rocks are commonly anisotropic.

Fractured media at a local scale can be extremely heterogeneous. Fracture properties controlling fluid flux include the number of fractures, spacing, fracture length, orientation of fracture sets, and fracture aperture. Techniques for mapping fracture geometries are given in Refs. 69 and 107. It should be recognized, however, that there may be a weak correlation between fracture density and permeability.[57] Fractures that are steeply dipping are poorly sampled with vertical boreholes; their presence and abundance may become apparent only during hydraulic testing. Alternatively, inclined boreholes could be considered.

Effective permeabilities of crystalline rock typically decrease by two or three orders of magnitude in the first several hundred meters below ground surface, as the number of fractures decreases and fractures close under increasing lithostatic load. However, isolated higher-permeability fracture zones can be encountered. Such higher-permeability zones potentially control the overall pattern of fluid flow at a site.[26]

Models to predict fluid flow in fractured rocks are grouped into three classes. In using an *equivalent porous-medium model,*[73] the connectivity of the fracture network must be sufficient, at the scale of observation, to represent fluid flow in terms of a hydraulic conductivity ellipse (Sec. 6.3.2). *Dual porosity models*[39] represent the fracture network and the intact rock matrix as two overlapping continua. *Discrete fracture models*[108] predict fluid flux by accounting for fluid flow on a fracture-by-fracture basis.

The permeability ellipse for a network of fractures can be estimated assuming fractures are of infinite length[114] or finite length.[89] Calculations based on an infinite-length approximation provide an upper bound. The greatest limitation in using such equations is to determine fracture aperture. Mechanical measurements of fracture opening are unreliable for prediction of fluid flux because they do not account for the effect of contact area and surface roughness within the plane of the fracture on hydraulic resistance. Complications arise because of the highly deformable nature of fractures.[36] Cross-hole interference tests provide an alternative to estimating the permeability tensor from geometric data.[53] At the scale of hundreds of meters (subbasin scale), hydrogeologic simulation models calibrated to estimates of recharge rates and water-table elevations may provide the most reliable approach to estimate a bulk or effective conductance of the rock mass.

Faults occur as an intermixed zone of clay-rich crushed rock *(gouge)* and highly fractured rock, grading outward into a less fractured rock mass. A single fault zone can vary spatially from a nearly planar feature to a complex set of connected fault segments. Permeability values for gouge range between 10^{-22} and 10^{-18} m^2,[81] while the permeability of open fracture zones may locally be up to five or six orders of magnitude higher.[16] If clay-rich gouge is predominant, the fault may be hydrologically similar to the intact rock mass, or it may act as a barrier to flow. Offset of geologic strata along faults may disrupt the lateral continuity of higher-permeability zones.

6.5.2 Volcanic Terrain

Hydraulic properties of volcanic rocks are largely determined by the mode of extrusion and geologic controls on porosity development and fracturing.[129] Repeated eruptions lead to a geologic unit constructed from numerous lava flows. Fracture development is lower in the middle of a flow. Volcanic rocks may be interstratified with fluvial sediments deposited between eruptive events. These *interflow deposits* exhibit varying degrees of lateral continuity; and where they are areally extensive and composed of coarse-grained material, they may be the most permeable units in the section.[72]

Basalt permeabilities are typically anisotropic with highest permeabilities in the direction that the basalt moved across the ground surface. Representative hydraulic conductivity values (Table 6.5.1) indicate that local values of permeability may differ by six or seven orders of magnitude within a sequence of basalt flows.

Ash flow tuffs are composed of glass fragments and shards. The extent to which

TABLE 6.5.1 Representative Hydraulic Conductivity and Permeability Values for Volcanic Rocks*

Rock type	Hydraulic conductivity,†,‡ m/day	Permeability,§ m^2
Cenozoic flood basalts:		
Dense, unfractured	$10^{-6}-10^{-3}$	$10^{-18}-10^{-15}$
Vesicular	$10^{-4}-10^{-3}$	$10^{-16}-10^{-15}$
Interbeds	$10^{-3}-10^{+3}$	$10^{-15}-10^{-9}$
Quaternary basalts:		
Vesicular	$10^{+1}-10^{+3}$	$10^{-11}-10^{-9}$
Tuffs:		
Densely welded (matrix)	$<10^{-6}$	$<10^{-18}$
Densely welded (fractured)	$10^{-6}-10^{+1}$	$10^{-18}-10^{-11}$
Nonwelded	$10^{-3}-10^{-2}$	$10^{-15}-10^{-14}$

* See footnote *, Table 6.3.2.

† Assuming water properties common to shallow groundwater.

‡ To convert values from m/day to ft/day, multiply by 3.28; to convert to ft/s, multiply by 3.8×10^{-5}.

§ To convert values from m^2 to ft^2, multiply by 10.75.

TABLE 6.5.2 Representative Values of Hydraulic Conductivity for Glacial Deposits*,†,‡

	Unweathered, m/day	Weathered, m/day	Fractured, m/day
Basal till	$10^{-6}-10^{-2}$	$10^{-4}-10^{-1}$	$10^{-4}-10^{0}$
Supraglacial till	$10^{-4}-10^{0}$	$10^{-4}-10^{0}$	$10^{-4}-10^{0}$
Glaciolacustrine	$10^{-8}-10^{-4}$		$10^{-6}-10^{-3}$
Loess	$10^{-6}-10^{0}$	$10^{-5}-10^{-2}$	
Glaciofluvial	$10^{-2}-10^{2}$		

* See footnote *, Table 6.3.2.
† Assuming water properties common to shallow groundwater.
‡ To convert values from m/day to ft/day, multiply by 3.28; to convert to ft/s, multiply by 3.8×10^{-5}.
Source: Adapted from Ref. 115. Used with permission.

the glass fragments are *welded* determines their hydraulic properties, for there is an inverse relationship between degree of welding and development of cooling fractures. While an average value for total porosity of welded tuff is 15 percent, and that for nonwelded tuff is 30 percent, the permeability of welded tuffs can be two or three orders of magnitude higher than that of nonwelded tuff (Table 6.5.1).

6.5.3 Glaciated Terrain

Many glacial deposits exhibit a complex stratigraphy (Fig. 6.3.3). During the Pleistocene, there were repeated glacial advances and retreats, and within each major glacial period, there were many minor advances and retreats, producing a sequence of interstratified glacial deposits of variable character and thickness (see Table 6.2.4). Table 6.5.2 lists representative values of hydraulic conductivity for various glacial deposits. Laboratory measurements of the hydraulic conductivity of till and glaciolacustrine sediments commonly yield values that are lower by two or three orders of magnitude than in situ field measurements. Tills may contain numerous vertical fractures, especially within 10 m of the ground surface.

6.5.4 Carbonate Terrain

The hydraulic properties of carbonate rocks are dependent on three key factors: (1) the nature of the original carbonate sediment; (2) the degree to which the initial porosity of the sediment has been lost due to postdepositional compaction, pressure solution, and cementation; and (3) the extent to which fresh water flowing through the carbonate matrix has created new porosity and permeability by chemical dissolution (secondary porosity). Representative values of hydraulic conductivity values for carbonate rocks are listed in Table 6.5.3. General features of a number of carbonate terrains in North America are summarized in Ref. 9.

Carbonate rocks have a diverse hydrologic character; they can be extremely productive aquifers, or tight confining beds. *Solution channels* and *karst features,* leading to a highly permeable, anisotropic rock mass, are favored if (1) surface topography and drainage, bedding features, and/or jointing promote flow localization which

TABLE 6.5.3 Representative Values of Porosity, Hydraulic Conductivity, and Permeability of Carbonate Rocks*

Lithology	Porosity	Hydraulic conductivity,†,‡ m/day	Permeability,§ m^2
Carbonate mud	0.40–0.70	10^{-3}–10^{-1}	10^{-15}–10^{-13}
Dolomite	0.001–0.15	10^{-4}–10^{0}	10^{-16}–10^{-12}
Tertiary limestone	0.20–0.35	10^{-4}–10^{0}	10^{-16}–10^{-12}
Paleozoic limestone	0.001–0.10	10^{-4}–10^{0}	10^{-16}–10^{-12}
Oolitic limestone	0.01–0.25	10^{-2}–10^{-1}	10^{-14}–10^{-13}
Holocene coral limestone	0.30–0.50	10^{2}–10^{4}	10^{-10}–10^{-8}
Karstified limestone	0.05–0.50	10^{-1}–10^{7}	10^{-13}–10^{-5}
Chalk	0.15–0.45	10^{-3}–10^{0}	10^{-15}–10^{-12}

* See footnote *, Table 6.3.2.
† Assuming water properties common to shallow groundwater.
‡ To convert values from m/day to ft/day, multiply by 3.28; to convert to ft/s, multiply by 3.8×10^{-5}.
§ To convert values from m^2 to ft^2, multiply by 10.75.
Source: Adapted from Ref. 9. Used with permission.

focuses the solvent action of circulating groundwater; and (2) well-connected pathways are maintained between the recharge area and the discharge area, favoring higher groundwater velocities. In the Floridan aquifer system, increases in permeability and conduit development have been greater in regions where the aquifer is unconfined, in comparison with regions where the aquifer is confined beneath thick clayey deposits.[56] Permeability and storage typically decrease with depth as karstification becomes less prominent. Ancient karst (paleokarst), now deeply buried, can be highly permeable.

Quantitative analysis of fluid flow in karst terrain is difficult, with considerable uncertainty to be expected in any hydrogeologic simulation model. Karst consists of two hydraulic domains, with a network of conduits transmitting water through a porous and fissured matrix. The extent of permeability development for each domain, and the coupling between them, determines the nature of flow. Four approaches are used to identify the hydrologic properties of a karstic flow system:[28] (1) water-balance estimation, (2) borehole investigations, (3) water tracing, and (4) spring hydrograph analysis. Ford and Williams[28] emphasize the importance of studying karst springs. Spring hydrograph analysis examines the form and rate of a recession curve following a recharge event to infer the storage and structural characteristics of a karst aquifer. Chemical fluctuations in the spring discharge are also used to aid in hydrograph separation.[20] Water tracing with fluorescent dyes can be used to delimit catchment boundaries, identify recharge areas, and estimate groundwater flow velocities (Ref. 28, p. 232).

Two questions must be asked if conduit development is suspected. First, can the aquifer, at the scale of analysis, be represented as a porous medium? Second, can flow be represented by Darcy's law, or is flow in the larger fissures and conduits turbulent? Conduit flow may be laminar in conduits up to about 0.5 m in diameter, provided fluid velocities do not exceed 3 m/h.[28] For purposes of groundwater resource management, porous medium approximations are often acceptable generalizations.[28] If a porous medium approximation is inappropriate, it is necessary to construct a model that incorporates the hydraulic equations of flow within individual conduits and to specify the geometry of the conduit pathways. Tracer tests, in conjunction with flow measurements, have been successful in defining the general character of conduit systems.[100,112,113] A statistical approach to quantify the residence time of water in a karst aquifer is given in Ref. 21.

6.5.5 Weathered Horizons

In regions underlain by lower-permeability rocks, a surficial weathered horizon is sometimes identified as a separate hydrostratigraphic unit. Breakdown of the rock matrix, and a greater amount of fracturing, can create a zone where the permeability is two to three orders of magnitude higher than in the parent rock. In nonglaciated regions, the weathered horizon may be 10 or more meters thick. Key factors in determining the hydraulic properties of a weathered horizon are the nature of the parent rock and the relative importance of mechanical and chemical weathering. If chemical weathering is dominant, and if clay minerals form within the weathered zone, then permeability of the weathered zone may be similar to that of the parent rock.

6.6 GROUNDWATER HYDROLOGY OF ISLAND AND COASTAL REGIONS

6.6.1 Introduction

Aquifers in island and coastal areas are prone to *seawater intrusion.* Because seawater is denser than fresh water, it will invade aquifers which are hydraulically connected to the ocean. Under natural conditions, fresh water recharge forms a lens that floats on top of a base of seawater. This equilibrium condition can be disturbed by changes in recharge and/or induced conditions of pumpage and artificial recharge. Seawater intrusion must be addressed in managing groundwater resources in island and coastal areas. The *fresh-water lens* is very susceptible to contamination, and once contaminated, it can be very difficult to restore to pristine conditions. Note that the term intrusion does not necessarily imply dynamic conditions; it has historically been used to simply indicate the presence of seawater.

The occurrence of seawater in coastal aquifers can be visualized with a simplified conceptual model (Fig. 6.6.1*a*). The essential elements are fresh-water recharge from inland sources, seawater intrusion from the ocean side, and an interface or transition zone between the two types of water. The *transition zone* is a zone of mixing between the fresh water and the salt water. Under equilibrium conditions, the *interface* will remain fixed, and fresh water will discharge along the *seepage face.* Examples of site-specific studies of seawater intrusion include the Nile Delta in Egypt,[59] Florida,[2] Israel,[77] and Micronesia.[4] A conceptual model of an island aquifer intruded with seawater is shown in Fig. 6.6.1*b*. In a coastal aquifer, recharge is predominantly due to lateral inflow, whereas in an island aquifer, it is due to vertical recharge. The conceptual models in Fig. 6.6.1 are analyzed in different ways. Coastal aquifers are usually modeled in the vertical plane. Island systems are also analyzed in the vertical plane when the thickness of the fresh-water lens is a significant fraction of the horizontal width of the lens, a situation common in atoll islands. When the lens thickness is small compared with its horizontal extent, the system can often be treated in the horizontal plane.

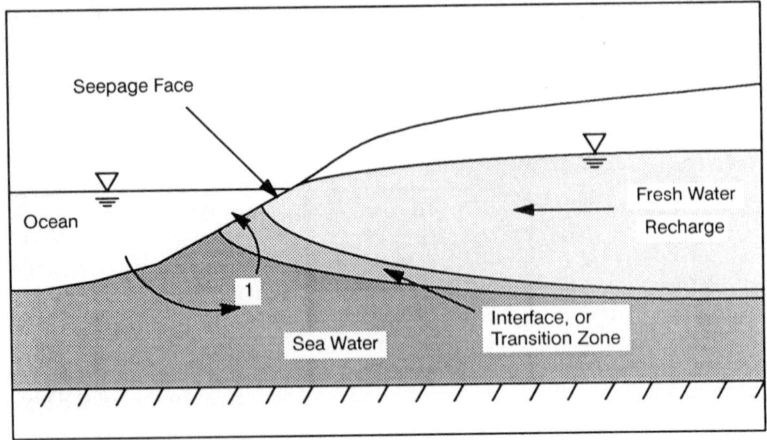

(a)

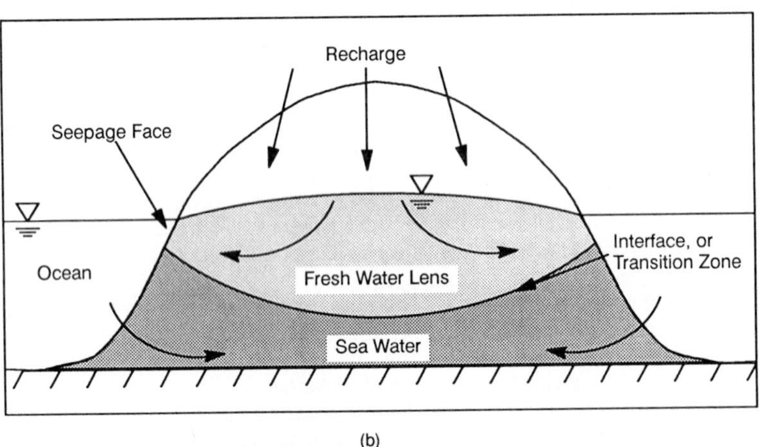

(b)

FIGURE 6.6.1 Occurrence of seawater intrusion in *(a)* coastal aquifers, *(b)* island aquifers.

6.6.2 Hydrogeologic Simulation Models of Seawater Intrusion

The first question in developing a simulation model is whether it is appropriate to treat the problem as *miscible* or *immiscible* flow. If the transition zone is thin relative to the thickness of the fresh-water lens and it is immobile, then it is appropriate to assume that the fresh water and salt water do not mix (immiscible), and the transition zone is considered to be a *sharp interface.* There are many methods for dealing with seawater intrusion as a two-phase, immiscible flow problem. These methods are generally known as *interface methods.* Under very dynamic conditions, or in cases where the fresh-water lens is relatively thin compared with the transition zone, it may be necessary to adopt to miscible flow models.

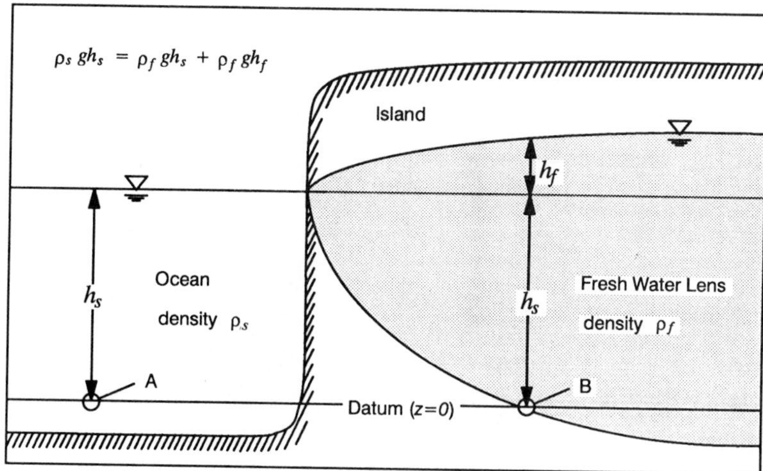

FIGURE 6.6.2 The Ghyben-Herzberg lens.

Interface Methods—Hydrostatic. The simplest way to model a fresh-water lens is to assume that the fresh water, interface, and underlying salt water are in hydrostatic equilibrium. This assumption is known as the *Ghyben-Herzberg relationship* (Fig. 6.6.2). At $z = 0$, under conditions of hydrostatic equilibrium, the pressure at A must equal the pressure at B. As a consequence:

$$h_s = \frac{\rho_f}{\rho_s - \rho_f} h_f \approx 40\, h_f \tag{6.6.1}$$

assuming a seawater density ρ_s of 1025 kg/m^3, and a fresh-water density ρ_f of 1000 kg/m^3. This relationship predicts that there will be approximately 40 times as much fresh water in the lens beneath sea level as there is fresh water above sea level. For many coastal and island regions, the fresh-water lens is potentially a major source of potable water.

Because of its simplicity, it is common practice to apply the Ghyben-Herzberg relationship to any study of aquifers subject to seawater intrusion. However, this same simplicity limits its accuracy, and it is only a general rule of thumb. Because it is a hydrostatic approximation, Eq. (6.6.1) neglects motion of the fresh water, salt water, or the interface, nor does it account for a transition zone.

Interface Methods—Steady State. An improvement over the Ghyben-Herzberg relationship is to allow steady recharge to the fresh-water lens, resulting in a *stationary interface* and a seepage face to permit fresh-water discharge to the ocean. The simplest stationary interface solution (Fig. 6.6.3) is the method proposed by Glover.[37] The equipotentials Φ and stream functions ψ are given by

$$\begin{aligned}\Phi &= \left(\frac{\delta Q}{K}\right)^{1/2} \left[x + (x^2 + y^2)^{1/2}\right]^{1/2} \\ \psi &= \left(\frac{\delta Q}{K}\right)^{1/2} \left[-x + (x^2 + y^2)^{1/2}\right]^{1/2}\end{aligned} \tag{6.6.2}$$

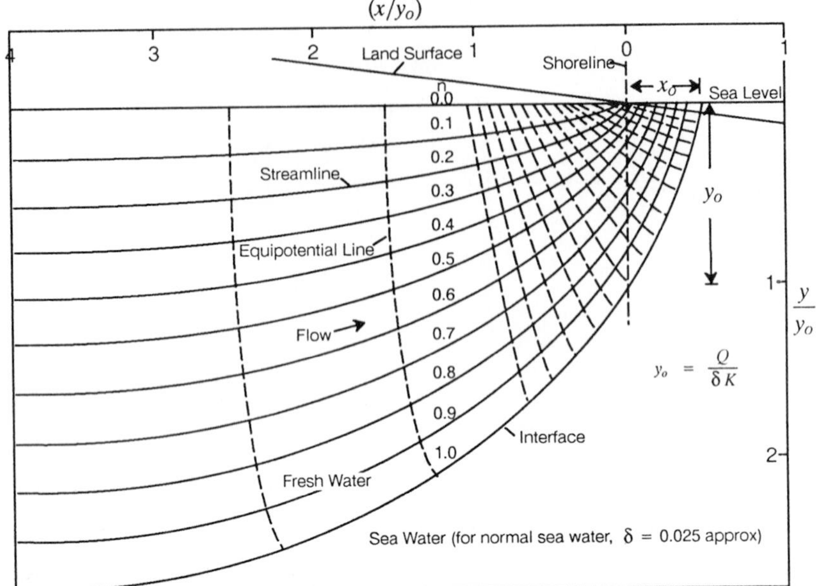

FIGURE 6.6.3 The Glover static interface solution for fresh-water discharge to the ocean. *(From Ref. 37.)*

where Q is the fresh-water recharge per unit length of shoreline [L^2T^{-1}] and $\delta = (\rho_s - \rho_f)/\rho_f$. The vertical thickness of the fresh-water zone as a function of the distance from the shoreline is

$$y(x) = \left(\frac{2\,Q}{\delta\,K}\,x\right)^{1/2} + \frac{Q}{\delta\,K} \tag{6.6.3}$$

The width of the region through which fresh water discharges to the ocean is

$$x_0 = -\frac{Q}{2\,\delta\,K} \tag{6.6.4}$$

If there is a way to estimate x_0, then Eq. (6.6.4) provides a simple method to calculate recharge (Q) to the coastal aquifer. One approach that has been used with some success to estimate x_0 is by infrared air or satellite photos, taking advantage of the fact that the fresh water escaping to the ocean may be of a different temperature than the near-shore ocean water.[1] The Glover solution does not allow for a seepage face above sea level at the shoreline (Fig. 6.6.3). This assumption has the effect of underestimating the thickness of the fresh-water lens, especially near the ocean. Also, the solution assumes that the seawater is not moving and that pumping operations are small compared with the recharge rate through the fresh-water lens.

Stationary interface solutions are available for numerous cases of interest. Strack[121] presents several solutions for problems of salt water upconing beneath a pumping well. Vacher[123] and Bear[7] present solutions for island and coastal lenses, respectively. Numerical simulations for the southern Oahu, Hawaii, fresh-water lens were conducted by Eyre.[25]

Interface Methods—Dynamic. If conditions of variable recharge and/or significant pumping exist, the interface will move in response to the nonequilibrium conditions. For this problem, the hydraulic head and fluid velocity in both the fresh and saline portions of the aquifer must be considered. Since these quantities are unknown, the problem becomes one of coupling differential equations for motion of the fresh and saline waters, and solving for the unknown position of the interface between them. There are no analytic solutions for the *moving interface problem,* and one must resort to numerical methods.[7,51,75,109]

Factors That Affect the Transition Zone. Under certain conditions, the transition zone may become so diffuse that it will no longer be appropriate to use the sharp interface approximations. Often the most disruptive influence is the effect of pumping activities. Withdrawal of water from the fresh-water lens via pumping wells creates *salt-water upconing,* which is the migration of the fresh-salt interface toward a well in response to pumping withdrawals. Solutions for salt-water upconing can be found in Refs. 7 and 121. Ocean tidal influences also contribute to the spread of the transition zone in coastal aquifers and small island aquifers.[126,128] Figure 6.6.4 provides several examples of transition zones in Hawaii. These transition zones range from the very narrow zone at Punaluu, which would be modeled as a sharp interface,

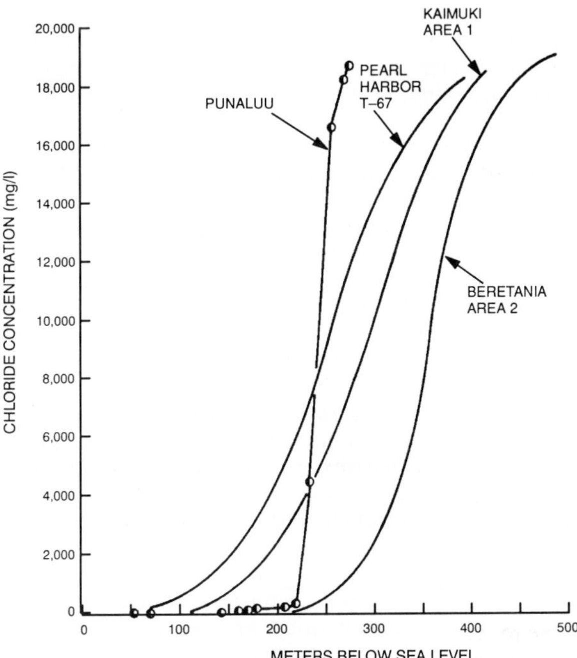

FIGURE 6.6.4 Examples of transition zones between fresh and salt water from Oahu, Hawaii. *(From Ref. 70.)*

to the very broad zone at Beretania. The shape of the transition zone can be estimated by electric logging[68] when wells are present or by geophysical methods.[117,118]

Miscible Seawater Intrusion. When the sharp interface approximation is inappropriate owing to the large thickness of the transition zone relative to the fresh-water lens, it is advisable to treat the fresh water and salt water as miscible. There are two elements to the problem: (1) a single fluid phase with variable density; and (2) a solute representing the total dissolved solids in seawater, which affects the fluid density. As a result, there are four unknown dependent variables: fluid specific discharge, fluid pressure, salt concentration, and fluid density, requiring the solution of four equations.[32,54,124] There are no analytic solutions to this problem because of the nonlinear nature of the governing equations. Henry[49,50] provides an approximate analytic solution which is often used for validating numerical codes.

6.7 CHEMISTRY OF NATURALLY OCCURRING SOLUTES

6.7.1 Geochemical Processes

As groundwater moves through the subsurface, it changes chemically as it comes in contact with different minerals and proceeds toward *chemical equilibrium.* Typically, the species Na^+, Ca^{2+}, Mg^{2+}, HCO_3^-, SO_4^{2-}, and Cl^- constitute more than 90 percent of the total dissolved solids in a sample of groundwater. At a given point along the flow system, the water chemistry will be a function of the chemical characteristics of the recharge water, the solubilities of minerals encountered along the flow path, the order in which minerals are encountered, and reaction rates relative to the groundwater velocity.

Many graphical methods of displaying chemical data have been developed to aid in compressing large volumes of numerical data into a form suited for interpretation. Most approaches are based on identification of *hydrochemical facies.* Figure 6.7.1 shows a common method, the *Piper trilinear plot.* Cations and anions are plotted separately in the lower triangular regions; a composite chemistry is determined by projection of those points into the central, diamond-shaped region. Ionic concentrations are plotted in terms of percent equivalents per liter, calculated separately for the cations and anions. Equivalents per liter are calculated by dividing the value in mg/L by the atomic weight, and multiplying this number by the valence of the ion.

The Piper plot can be used to describe qualitatively how the chemical characteristics of groundwater change along a flow path. By noting the position of each sample point in the diamond-shaped region of the Piper plot relative to that sample's position within the flow system, spatial trends in solute composition may become apparent. Chemical patterns are helpful in ensuring that interpretations of flow patterns based on hydraulic data are consistent with the "chemical signatures" of groundwater samples.

Chemical Equilibrium and Solubility. Chemical equilibrium is governed by the *law of mass action,* stated for the reaction ($aA + bB = cC + dD$) as

$$K = \frac{[C]^c\,[D]^d}{[A]^a\,[B]^b} \tag{6.7.1}$$

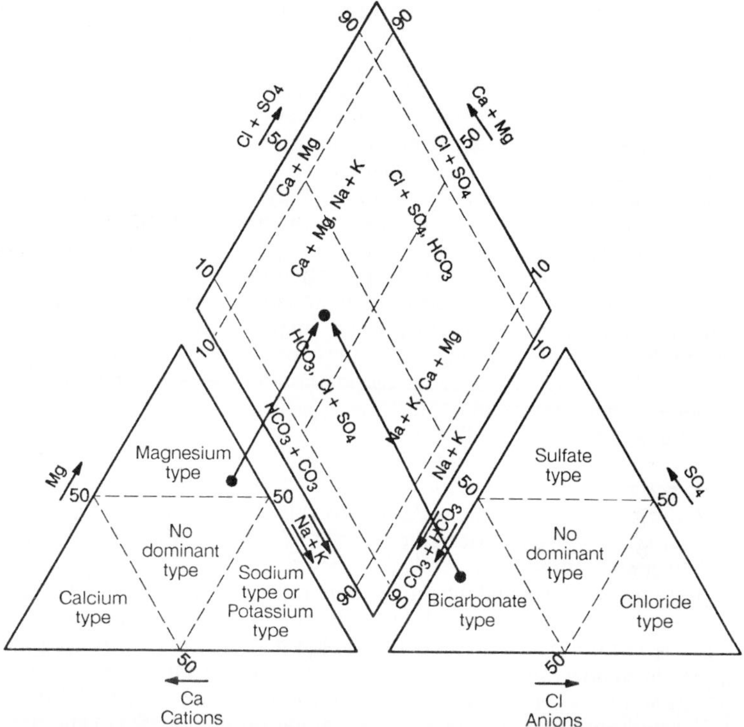

FIGURE 6.7.1 Piper trilinear plot representing chemical characteristics of groundwater. Cation and anion concentrations are plotted in the lower triangles and their locations are projected into the upper diamond to specify hydrochemical characteristics of the groundwater sample. *(From Ref. 29. R. Allan Freeze and John A. Cherry, "Groundwater," ©1979. Reprinted by permission of Prentice Hall, Englewood Cliffs, New Jersey.)*

where K is the equilibrium constant, $a, b, c,$ and d are the moles of constituents, and terms in square brackets refer to *chemical activities.* The activity of solid species is equal to 1. Chemical activity for an ionic species is equal to its molality times the *activity coefficient* of that species. Values of the activity coefficient for an ionic species, for a groundwater of a given ionic strength, can be calculated using the extended Debye-Huckel equation (Ref. 19, p. 395). Equilibrium constants for the common mineral dissolution and precipitation reactions of importance in groundwater studies are tabulated in Ref. 19 (p. 433).

The *saturation index* for a precipitation-dissolution reaction is defined ($S_i = IAP/K$) where IAP is the *ion activity product* calculated from the law of mass action [Eq. (6.7.1)] based on species concentrations determined in a water sample and K is the equilibrium constant for the reaction. For samples where the value of the total dissolved solids is less than 10,000 mg/L, concentrations in mg/L can be converted to molality by dividing the value in mg/L by ($10^3 \cdot$ formula weight). If S_i is less than 1, then at the point in the flow system from which the sample was taken, the groundwater is undersaturated with respect to the mineral species in question. A value of 1 indicates a state of equilibrium. Values greater than 1 signify supersaturation with respect to the mineral species.

Mineral dissolution and precipitation often proceed at a rate that is slow in comparison with rates of groundwater flow. Chemical equilibrium, rather than providing a methodology for predictive assessments of solute concentrations, is best viewed as providing a framework within which the overall character of species concentrations can be interpreted. Geochemical speciation models are used in a limited way to evaluate specific suites of mineral reactions.[40]

Oxidation-Reduction Reactions. Microbially catalyzed redox reactions are important in determining the mobility and ionic form of compounds involving iron, nitrogen, sulfur, and carbon, in addition to trace metals such as copper, mercury, cadmium, and chromium. Oxidation of organic matter is a key factor in determining dissolved oxygen levels in unconfined aquifers, and in controlling the evolution from oxidizing to reducing conditions. The primary interpretative tool is chemical equilibrium and the law of mass action, used in the construction of *pE-pH diagrams.* pE is a measure of the oxidizing or reducing potential of the pore water. These plots indicate stability fields for the major dissolved species and solid phases. An example of a pE-pH diagram is given in Ref. 19 (p. 456).

Surface Reactions. Surface reactions involve the transfer of mass between groundwater and the surfaces of solids, primarily clay minerals, metal oxides, and organic carbon. *Cation exchange* describes the transfer of exchangeable cations within an adsorbed layer adjacent to a negatively charged mineral surface. Transfer of cations between the adsorbed layer and the mobile pore water can be represented in terms of a law of mass action.[29] Clay minerals have differing exchange capacities, values are greater for montmorillonite and decrease for chlorite, illite, and kaolinite. For naturally occurring solutes, surface reactions are important in modifying ratios of calcium, magnesium, and sodium along a flow path, and in influencing mobility of trace metals. Formation of neutrally charged *ion complexes* (e.g., $CaSO_4^{o}$) and the presence of *colloids* can have a strong influence on the degree to which dissolved cations take part in surface reactions (Ref. 19, p. 434).

The clay mineral kaolinite, and metal oxides, have a surface charge due to adsorped hydroxyl groups. Their surface charge changes with the pH of the pore water. At low pH, these surfaces act as anion exchangers, having a net positive charge on the mineral surface. The *isoelectric point,* at which the net charge on the surface is zero, is specific to each mineral.[22] At higher pH values, surfaces act as cation exchangers. In these cases, the capacity of surface reactions to modify pore water chemistry is pH-dependent.

6.7.2 Environmental Isotopes

Isotopes of a particular element have the same number of protons but a different number of neutrons. Major applications of isotropic data include the differentiation of water masses with unique isotopic signatures and estimation of flow directions, travel times, and ground water recharge rates. The environmental isotopes in common use are the stable isotopes deuterium (^2H), oxygen-18 (^{18}O), and carbon-13 (^{13}C), and radioisotopes tritium (^3H), and carbon-14 (^{14}C). See Refs. 34 and 35 for a comprehensive summary.

The stable isotopes *oxygen*-18 and *deuterium* behave as nonreactive (conservative) tracers under near-surface conditions, moving at the average velocity of the groundwater. Measurements are reported in *delta δ units per mil,* where $\delta = (R/R_{std} - 1.0) \times 1000$, R is the measured isotopic ratio, and R_{std} is the isotopic ratio in a reference standard, in this case standard mean ocean water (Ref. 22, p. 368). When R differs from R_{std} by 1 part per thousand, this defines one δ unit. Accuracy for ^{18}O measurements is $\pm 0.2\ \delta$, and for ^2H it is $\pm 2\ \delta$. Data are often analyzed in graphical form as a plot of δ^2H versus $\delta^{18}O$. Samples are compared with the *meteoric water line*, an empirically derived relationship for continental precipitation, given as

$$\delta^2H = 8\ \delta^{18}O + 10 \tag{6.7.2}$$

Isotopic fractionation of $^2H/^1H$ and $^{18}O/^{16}O$ during phase changes of water enriches one isotope relative to the other. Fractionation is temperature-dependent. For example, winter precipitation is depleted in ^{18}O and ^2H compared with summer precipitation. Precipitation at the beginning of a storm is often higher in ^{18}O and ^2H compared with that at the end of a storm, as the heavier isotopes are selectively removed from the vapor phase. Water that has been subject to evaporation is enriched in ^2H relative to ^{18}O because of its lower atomic weight; sample points will lie off the meteoric water line. These processes provide water masses with unique signatures that can be used to aid in determining recharge areas, degrees of mixing between waters of differing origin, and in hydrograph separation.[111] Isotopic data are also useful in documenting anthropogenic inputs of sulfur and nitrogen to shallow groundwater systems.[104]

Natural production of radionuclides within the atmosphere by interaction of gases with cosmic radiation and major increases in radionuclides above naturally occurring values due to above-ground testing of thermonuclear weapons provide a means of estimating the subsurface residence time of groundwater since it was isolated from the atmosphere and soil gas. Calculation of the *residence time* ("groundwater age") is possible by noting the decrease in concentration that has taken place by radioactive decay, provided the initial concentration of the radionuclide in the recharge water can be estimated.

Dating of groundwater using *carbon-14* techniques (half-life of 5730 years) is a useful technique for water with residence times up to 30,000 years.[14,15] Hydrogeologic settings with residence times of this magnitude include large-scale regional flow systems and systems in thick, low-permeability sediments. Significant uncertainties are introduced into age determinations because of dilution by dead carbon dissolved from minerals encountered by the pore water along its flow path. A number of correction techniques exist. Phillips et al.[95] review six methods and apply age dating as a tool in modeling groundwater flow in the San Juan Basin.

From 1957 to 1963, large amounts of *tritium* (half-life of 12.26 years) were introduced into the upper atmosphere by above-ground testing of thermonuclear weapons. The tritium content of precipitation exceeded natural background levels by approximately one thousand in 1963 and 1964. Measurements are reported in terms of tritium units (TU), where 1 TU is equal to 1 tritium atom per 10^{18} atoms of hydrogen (about 3.2 pCi/L or 0.118 Bq/kg). Measurements can be made to a detection limit of 0.1 TU on enriched samples, with a counting error of ± 0.1 to 0.3; and on unenriched samples, the detection limit is 6 TU, with a counting error of ± 6 to 12 TU. Recharge rates can be estimated by locating the depth of the 1963 tritium peak.[103] To be useful as a quantitative tool, the time history of tritium input to the flow system must be known. A qualitative interpretation of tritium concentrations, for sites within the northern hemisphere, is given in Table 6.7.1. There has been recent interest in the use of ^{36}Cl to estimate recharge rates of modern water.[94]

Groundwater age estimates are best viewed as one more input in the formulation of a hydrogeologic simulation model, rather than in providing an exact determination of residence time. Processes such as hydrodynamic dispersion, diffusion of radionuclides from fractures into matrix blocks, and mixing of waters from different

TABLE 6.7.1 Interpretation of Tritium Concentrations in Groundwater in the Northern Hemisphere

Concentration, TU	Interpretation
Less than 0.2	Water is older than 50 years
Less than 2.0	Water is older than 30 years
Between 2 and 10	Interpretation is difficult; water is likely at least 20 years old
Between 10 and 100	Water is less than 35 years old; may be modern
More than 100	Probably related to water from peak fall-out period, 1960–1965

Source: From Ref. 15.

stratigraphic horizons in samples taken from boreholes with long completion intervals all act to modify the calculated age of the groundwater. Differences between residence times based on an estimated pore water velocity [Eq. (6.3.5)] and those based on atmospheric radionuclides can normally be expected to be in the range of 20 to 100 percent.[15]

6.8 ESTIMATION OF HYDROGEOLOGIC PARAMETERS

6.8.1 Porosity

Laboratory measurements of porosity are made with a gas pycnometer.[13] A simpler technique involves saturation of a sample of known volume under a vacuum, weighing of the sample, air drying at 105°C, and reweighing. Differences in weight can be converted to the volume of water filling the pore space. Field methods are based on borehole geophysical logging (see Sec. 6.9.4) or tracer tests.[55,90] Borehole methods measure total porosity, whereas tracer tests measure effective porosity. Tracer tests are expensive and time-consuming, and data interpretation can be difficult. Reliable estimates of effective porosity for fractured rock masses are difficult to obtain. Measurements on a core sample will not reveal the influence of connected pathways through fractures, nor will geophysical borehole logs.

6.8.2 Hydraulic Conductivity and Storage Properties

Laboratory-Scale Measurements. Approaches to estimating hydraulic conductivity include:

Constant-Head Test (see Fig. 6.8.1a)

$$K = \frac{Q}{A\,(H/L)} \tag{6.8.1}$$

where Q is the volumetric flow, given the constant hydraulic gradient (H/L). Because it is difficult to control fluid leakage and measure low flow rates for samples with hydraulic conductivities less than approximately 10^{-6} m/s, there are advantages in these cases in using transient tests.

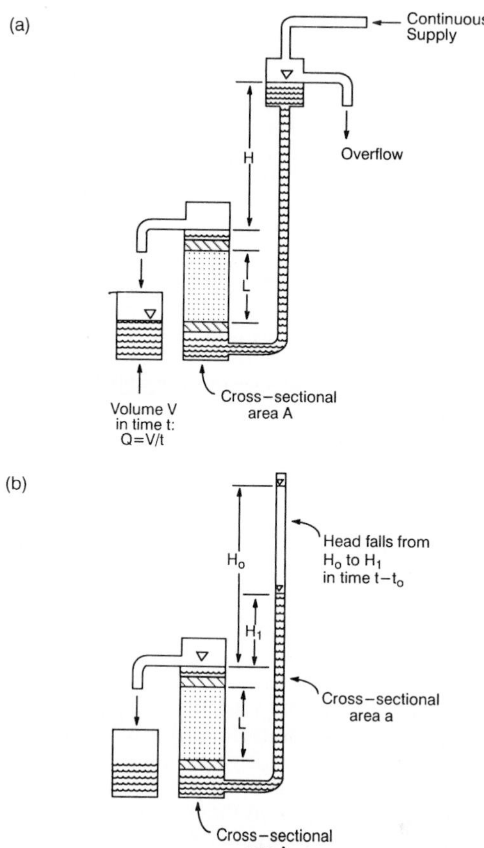

FIGURE 6.8.1 Schematic diagrams of permeameters: *(a)* Constant-head apparatus, *(b)* Falling-head apparatus. *(From Ref. 29. R. Allan Freeze and John A. Cherry, "Groundwater," ©1979. Reprinted by permission of Prentice Hall, Englewood Cliffs, New Jersey.)*

*Falling-Head Test (see Fig. 6.8.1*b*)*

$$K = \frac{aL}{A(t - t_0)} \ln \left(\frac{H_0}{H_1} \right) \tag{6.8.2}$$

where $(t - t_0)$ is the time required for the water level in the standpipe to decline from H_0 to H_1.

Transient Pressure Response. Specialized techniques, based on monitoring the transmission of a pressure pulse through the sample and matching the observed pressure-time data to numerically derived pressure histories, are described in Ref. 80.

Estimates Based on Empirical Correlations with Grain Size. Order-of-magnitude estimates of hydraulic conductivity for sands include

1. Hazen's equation:

$$K = Cd_{10}^2 \tag{6.8.3}$$

Here K is in units of cm/s and grain size is measured in millimeters. The variable d_{10} (determined by a sieve analysis) is the grain size for which 10 percent of the particles are finer by weight. Values of C range from 0.4 to 0.8 for fine sands, 0.8 to 1.2 for medium to coarse sand, to 1.2 to 1.5 for a well-sorted, coarse sand.[27]

2. Kozeny-Carmen equation:

$$K = \frac{\rho g}{\mu} \frac{n^3}{(1-n)^2} \frac{d_m^2}{180} \tag{6.8.4}$$

Here n is porosity and d_m the mean grain size.[6]

Values of hydraulic conductivity estimated in the laboratory are representative only of a core-sized volume of rock under the environmental conditions created during testing. For heterogeneous porous media, and fractured media, laboratory-scale measurements must be interpreted with caution, as they potentially lead to a poor estimate of field-scale fluid flux.

Porous medium compressibility (α) can be measured in the laboratory by a consolidation test.[8] Given the value of α, and porosity, Eq. (6.3.10) provides a laboratory-scale estimate of specific storage.

Single-Borehole Tests—Piezometer Tests. A common in situ method for estimating hydraulic conductivity is the *piezometer test* (also known as a *slug* or *bail test*). The test is based on recording the rate of recovery of water level within a piezometer, following disturbance of the equilibrium water level. In a slug test, water is added to the standing water column to raise the water level above its equilibrium level. In a bail test, water is instantaneously withdrawn from the piezometer (see Fig. 6.8.2*a*). The scale of the hydraulic conductivity measurement is of the order of 0.5 to 2 m beyond the tip of the piezometer, depending upon the hydraulic properties of the medium. Piezometer tests are cost-effective and can often be carried out in a minimum of time.

Two methods of interpreting piezometer tests are possible; one neglects changes in fluid storage in the medium surrounding the piezometer tip and the other takes it into account.

1. A plot is made of the normalized unrecovered head difference ($H\text{-}h/H\text{-}H_o$) versus time, with the unrecovered head difference plotted on a logarithmic scale; the data should plot as a straight line. The time (T_o) corresponding to an ($H\text{-}h/H\text{-}H_o$) value of 0.37 is calculated from the best-fitting straight line. When ($L/R > 8$), hydraulic conductivity is estimated from:

$$K = \frac{r_w^2 \ln (L/R)}{2LT_o} \tag{6.8.5}$$

Additional equations for estimating K for piezometer intakes of different geometry are given in Ref. 11.

2. The second approach is based on a solution describing transient, radial flow outward from the piezometer screen.[92] Type curves tabulated in Refs. 27 and 92 can be used to estimate values of hydraulic conductivity and specific storage. For media where the true value of specific storage S_s is less than approximately 0.005, computed values of S_s are uncertain to several orders of magnitude because of the

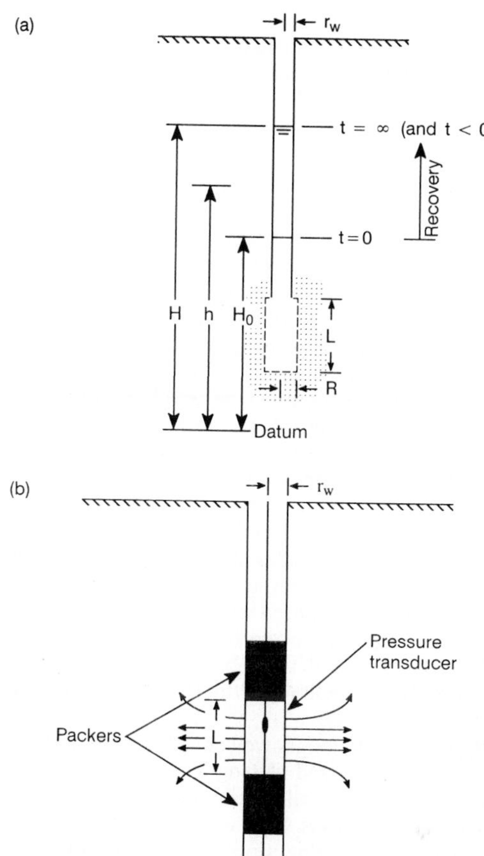

FIGURE 6.8.2 Single-borehole tests: *(a)* piezometer test, *(b)* packer test. *(From Ref. 29. R. Allan Freeze and John A. Cherry, "Groundwater," ©1979. Reprinted by permission of Prentice Hall, Englewood Cliffs, New Jersey.)*

similar shape of the type curves in this parameter range. For compressible media such as silts and clays, only the product KS_s can be reliably determined.[10]

In lower-permeability media, piezometer tests may be a more effective technique for estimating hydraulic conductivity than a pumping test. However, piezometer tests are limited in clayey deposits and tight rocks because recovery time may be on the order of months. In this case, *pressure slug tests* should be considered.[10,86,96] Here the piezometer is rapidly pressurized by injecting water into the test interval, which is then shut in, and the decline in pressure is monitored. If a lower-permeability "skin" occurs on the borehole wall (e.g., because of drilling mud penetrating the sediment or rock), then the effect of this zone on the hydraulic response should be considered.[79]

Single-Borehole Tests—Packer Tests. In a *packer test,* a length L of borehole is isolated between inflatable packers and fluid is injected into the test interval (Fig.

6.8.2*b*). Injection is continued until the flow rate stabilizes and data are interpreted using steady-state approximations. The hydraulic head within the injection zone and the flow rate (Q) are measured. The excess hydraulic head above the pretest measurement is calculated (Δh_w). Then K is determined by

$$K = \frac{Q}{2\pi L \Delta h_w} \ln\left(\frac{L}{r_w}\right) \qquad L \gg 2r_w \tag{6.8.6}$$

Cross-Borehole Tests—Pumping Test. A *pumping test* is the most common cross-hole procedure for estimating hydraulic properties. Water is pumped from one well, while the drawdown in hydraulic head is recorded in one or more observation wells. In comparison with single-borehole tests, cross-borehole tests have the advantages of sampling a larger volume of the porous medium, of yielding more reliable estimates of specific storage, in providing indications of aquifer boundaries, and in dealing with anisotropic media. The advantages must be weighed against the higher costs of carrying out these tests, and the value of the additional information gained.

In designing an aquifer test (Ref. 19, p. 185), distances between observation wells and the pumping well, and the pumping rate, can be chosen from calculations based on prior estimates of aquifer parameters, given the geologic model of the site, and any existing laboratory or single-borehole test data. For example, a pumping rate could be selected that yields a drawdown, after 24 h, of 1 to 2 m in an observation well located tens of meters from the pumping well. The observed decline in hydraulic head in the observation well is compared with type curves developed for the mathematical problem that corresponds to the field-test configuration. Selection of the mathematical model is based on both the behavior of the drawdown response and the conceptual model of the hydrogeologic setting that is being used to guide the site investigation (Fig. 6.1.1).

Drilling methods, water-well design, well development, and procedures for carrying out pumping tests are discussed at length in Ref. 23. Manuals and tests[66,125] provide discussions and example calculations for a wide range of hydrogeologic settings and pumping configurations. Interactive software for computer-aided interpretation of aquifer tests is becoming more widely available (e.g., AQTESOLV by Geraghty and Miller, Reston, Va; Graphical Well Analysis Package, Groundwater Graphics, San Diego, Calif.)

Confined Aquifers. The *Theis solution* describes radial flow to a well pumping water at a constant rate Q from a horizontal confined aquifer [Eq. (6.4.15)] Figure 6.8.3*a* illustrates the method for estimating the transmissivity and storage coefficient from the drawdown recorded in an observation well at a distance r from the pumping well. If the aquifer thickness is known, hydraulic conductivity and specific storage can be calculated. The measured time-drawdown values are plotted on a loglog scale at the same scale as the type curve $W(u)$ vs. $1/u$ [Eq. (6.4.15)]. The two curves are superimposed to get the best match between the field data and type curve, keeping the coordinate axis parallel. An arbitrary match point is selected and paired values $W(u)$, $1/u$, s, and t are determined. Then, using any consistent set of units, transmissivity and storage coefficient are calculated:

$$T = \frac{Q\,W(u)}{4\pi s} \qquad S = \frac{4uTt}{r^2} \tag{6.8.7}$$

The *Jacob method* is an approximate solution to the same boundary-value problem as the Theis method [see Eq. (6.4.17)]. The method is illustrated in Fig. 6.8.3*b*. Drawdown versus time data from the observation well are plotted using a semilog

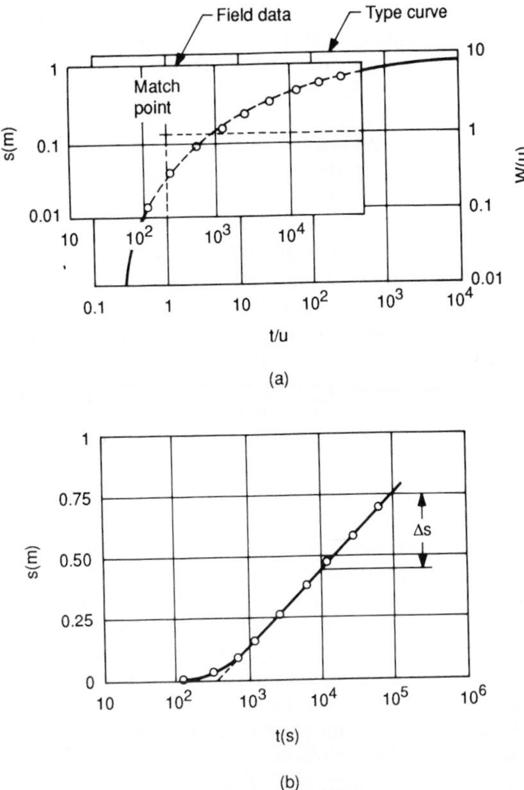

FIGURE 6.8.3 *(a)* Theis loglog method to estimate transmissivity and storage coefficient of a confined aquifer. *(b)* Jacob semilog approximation to estimate transmissivity and storage coefficient of a confined aquifer. *(From Ref. 29. R. Allan Freeze and John A. Cherry, "Groundwater," ©1979. Reprinted by permission of Prentice Hall, Englewood Cliffs, New Jersey.)*

scale, and the best-fitting straight line is obtained. This curve is projected to the zero-drawdown axis to determine the intercept t_0. The drawdown (Δs) for one log cycle of time is calculated. Using any consistent set of units, transmissivity and storage coefficient are calculated:

$$T = \frac{2.3\,Q}{4\,\pi\,\Delta s} \qquad S = \frac{2.25\,T\,t_0}{r^2} \tag{6.8.8}$$

For the Jacob approximation to be valid, the dimensionless parameter $u = r^2S/4Tt$ should be less than 0.05.

The *specific capacity* (C_s) of a pumping well is equal to ($Q/\Delta h_w$), where Q is the pumping rate and Δh_w is the drawdown in the pumping well. Drawdown in a pumping well is a function of both the hydraulic head in the aquifer immediately adjacent to the well screen and losses in head due to turbulent flow through the well screen and to the pump intake. In cases where there may not be observation wells near all pumping wells, it is possible to develop correlations between specific capacity and transmissivity, if both parameters can be estimated at some wells.

An estimate of transmissivity can be obtained using *recovery data* in a pumping well. During recovery, there are no well losses from pumping that limit the extrapolation of water levels in the pumping well to values of hydraulic head in the aquifer immediately adjacent to the well bore. A semilog plot is made of residual drawdown versus log $(1 + t_p/t)$, where t_p is the duration of pumping and t is the time since pumping stopped. Equation (6.8.8), based on the Jacob approximation, is then used to estimate transmissivity from the slope of the recovery curve.

An impermeable boundary not detected during initial geologic investigations may become apparent during a pumping test. An impermeable boundary appears as a break in slope on the drawdown plot. Boundaries are more easily detected on a semilog rather than a loglog plot of drawdown. The aquifer parameters T and S are estimated from the time-drawdown response before boundary effects are noted.[23]

Leaky Aquifers. Field data are plotted on loglog graph paper of the same scale as the leaky-well function $W(u,r/B)$ [Eq. (6.4.19)], and the best match is obtained between the field curve and the family of type curves, keeping the coordinate axis parallel. A match point is selected and paired values of $W(u,r/B)$ and S, and $1/u$ and t are determined. Substitution in Eqs. (6.4.19) and (6.4.18) provides estimates of the transmissivity and storage coefficient of the confined aquifer. The vertical hydraulic conductivity of the confining layer is estimated using Eq. (6.4.18).

Using the approximate solution that accounts for the storage properties of the confining layer [Eq. (6.4.20)], it is possible to estimate values for T and S of the aquifer, and the product $K'S'$ of the confining bed [substitution of match point values in Eqs. (6.4.20) and (6.4.21)]. If additional piezometers are installed within the confining bed and overlying aquifer, it is possible to obtain estimates of the hydraulic conductivity and storage properties of those units.[85]

Unconfined Aquifers. Field data are plotted on loglog graph paper of the same scale as the unconfined aquifer well function $W(u_a,u_b,\Gamma)$ [Eq. (6.4.22)]. The late-time data are used to estimate values of Γ, T, and S_y [Eq. (6.4.23)]. The beginning of the pumping test is used to estimate S [Eq. (6.4.23)]. The vertical hydraulic conductivity can be determined from the value of Γ [Eq. (6.4.23)]. Anisotropic effects, and the effects of delayed gravity drainage due to a falling war table, are more prominent if the aquifer test is designed using partially penetrating wells.[78]

Lower-Permeability Media. In fractured crystalline rock, storativity is typically quite low and it may be possible to carry out cross-hole, pulse interference tests.[58] The approach is identical to single-hole, pressure slug tests described earlier, except that the pressure response is also recorded in an isolated section of a second piezometer located a distance r from the test piezometer.

Extensions. A number of factors modify the water-level response that is observed in pumping tests from that predicted in the models discussed above. Where significant, it is necessary that these modifications be introduced in the parameter estimation procedure. Extensions include anisotropy in the horizontal plane,[84] the effects of water storage in larger-diameter well bores,[91] and the presence of a lower-permeability skin lining the well bore. Novakowski[87] summarizes analytic solutions accounting for these effects and describes a computer program to generate appropriate type curves.

Field-Scale Model Calibration. In the past decade, the approach to estimation of hydrogeologic parameters has shifted from one with a strong reliance on pumping test analysis to a more broadly based approach in which single borehole tests and pumping tests are viewed as one element of the model-building process that includes *parameter measurement, model calibration,* and *model validation.* Parameter estimates derived from calibration of a larger-scale simulation model have the following advantages over an approach based solely on pumping test analysis. First, estimates of hydraulic conductivity and storage properties can be obtained at the scale at which the analyst is attempting to make hydrologic predictions, incorporating in the estimation procedure the influence of assumptions on domain boundaries, geologic structure, and boundary conditions. Second, model calibration provides a framework for integrating measurements made at different scales (cores, single-well tests, pumping tests), and/or a suite of spatially distributed measurements obtained at the same scale. Third, greater insight can usually be obtained into the magnitude of uncertainties in the parameter estimates, and the effects of these uncertainties on the model results.

Estimation of hydraulic parameters, fluid fluxes across domain boundaries, or other model parameters is most commonly carried out by trial-and-error calibration. More formalized inverse simulation techniques are available, but they have been slow to be adopted as a practitioner's tool.

Reliability of Parameter Estimates. A number of factors influence the reliability of parameter estimates obtained from single borehole and cross-hole pumping tests. First, a correspondence is presumed between the mathematical model and the field conditions. It is often difficult to prove that such a correspondence exists. Second, considerable reliance is placed on hydrogeologic judgment when interpreting data from hydraulic tests. Third, a good match between observed and computed changes in hydraulic head does not guarantee that the correct mathematical model was formulated. Several models may fit the field data, yielding a range of parameter estimates. A subjective assessment of hydraulic testing procedures, in terms of scale, uncertainty in parameter estimates, ease of application, and cost is given in Ref. 93. It is difficult to justify more than two significant digits in reporting measurements of hydraulic conductivity.

6.9 GEOPHYSICS

6.9.1 Introduction

Geophysics as used in hydrologic applications can be defined as the collection and interpretation of physical measurements for the purpose of deducing or inferring geologic structure and the composition and properties of subsurface materials or providing insight to the moisture content distribution, water quality, and water-table elevation. The general objective in using geophysical methods is to aid in developing and refining a conceptual model during the process of building a hydrogeologic simulation model[88] (Fig. 6.1.1). There are three major types of geophysical data: acoustic (or seismic), potential field, and radioactive. Geophysical methods are either passive or active. *Passive methods* involve the collection and interpretation of naturally present, or naturally generated, signals or data. Examples include seismic waves, gravity methods, and electrical methods such as tellurics and self-potential. *Active methods* include seismic methods in which an acoustic energy source is activated and the resulting acoustic reflections and refractions are recorded, and radioactive methods such as neutron logging.

Because of their limited vertical resolution, *surface geophysical methods* are generally more useful for providing geological information for conceptual model development. *Borehole geophysical methods* have good vertical resolution compared with surface methods and are more useful for determining spatial variability of hydrogeologic parameters, and the distribution of moisture content with depth.

6.9.2 Acoustic Methods

Acoustic methods can be divided into two categories, reflection and refraction. The basic principle of *seismic reflection* is that seismic waves are reflected at interfaces between geologic units with different seismic velocities. Seismic velocities depend upon the elastic constants of a porous medium. The time required (arrival time) for the acoustic signal to travel from the source to a reflecting interface in the subsurface and back to the surface is measured by a geophone. Seismic reflection methods are capable of producing detailed information on subsurface structure.[41,64] However, interpretation of seismic reflection data requires analysis by trained geophysicists; moreover the collection of reflection data requires expensive field equipment and large field teams.

The *seismic refraction* method involves an acoustic energy source and at least one receiving geophone.[42] The basic principle is that different rock types have characteristic acoustic velocities. The source is activated and the geophone records the arrival time of a wave that has been refracted along the interface between two geologic units and returned to the surface. This method is simple to use and is particularly effective for identifying high-velocity layers. The acoustic velocity of a medium saturated with water is greatly increased in comparison with velocities in the vadose zone; seismic refraction can thus be used to determine the depth to the water table. The disadvantage of seismic refraction is that it is less accurate in determining vertical distances than seismic reflection, and lower-velocity layers beneath higher-velocity layers cannot be detected. For instance, in looking for the water table, this method is most commonly defeated when a higher-velocity layer (e.g., a hardpan) occurs in the vadose zone.

6.9.3 Potential Field Methods

Gravity methods are used primarily to provide information on the areal extent and thickness of a permeable alluvial deposit over a relatively dense, impermeable bedrock.[116] *Electromagnetic methods* include magnetic methods; electrical methods such as direct current (dc) and complex resistivity; telluric and magnetotelluric methods; induced polarization; and self-potential. Most of these methods have been developed in both surface and borehole configurations.[18,61]

Electrical methods exploit the fact that different geological materials have different capacities for transmitting an electric current. Saturated porous media have much lower resistance to electric current than similar dry materials. *Archie's law* is widely used to relate resistivity to porosity under saturated conditions

$$\rho_r = c\rho_w n^{-m} \tag{6.9.1}$$

where ρ_r is the bulk resistivity of the rock (ohm-m), ρ_w is the resistivity of the pore water, n is the porosity, and c and m are empirical constants,[61] whose values lie in the ranges $0.5 < c < 2.5$ and $1.5 < m < 2.5$, respectively. The ratio ρ_r/ρ_w, known as the formation factor,[67] is assumed constant for a given porosity. The presence of clay minerals and/or a pore fluid with very low electrical conductivity are key factors which may cause Archie's law to yield erroneous results.

Surface DC Electrical Resistivity. The elements of surface dc electrical resistivity methods are a current source, electrodes to allow current to flow into the subsurface, and potential electrodes to measure the voltage difference across two points on the surface. The measured resistivity will be an apparent resistivity ρ_a which is an average of the resistivities encountered by the current. The most common objective in surface electrical methods is to do a sounding in which resistivity measurements are taken for several different electrode spacings. As the spacings increase, information regarding vertical changes in resistivity is obtained. The two most common configurations are the Wenner and Schlumberger arrays.[18]

Field data of apparent resistivity and electrode spacing are plotted on a loglog axis. The analyst must decide whether the curve represents a one-, two-, or multilayer system. Master type curves are available[130] that, for a two-layer system, allow the analyst to determine the true resistivities of both layers and the thickness of the first layer. Master type curves for systems with more than two layers are also available, but the interpretation of these systems requires an experienced analyst. Computer programs which invert the field data to obtain a best fit for various layered models are routinely used.[65] It is always best to use resistivity measurements in conjunction with other geophysical and geological data, as the layered solutions are not unique. Examples of dc resistivity surveys applied to groundwater investigations can be found in Refs. 31 and 120.

Transient Electromagnetic Sounding. Transient electromagnetic methods (TEMs) generate a direct current which is abruptly terminated. The termination induces a transient eddy current in the earth which is sensed with a receiving coil. Measurement of strength versus time of the secondary magnetic field generated by the eddy currents yields information on the vertical variation of the electrical conductivity. TEM soundings are useful for determining salinity as a function of depth in coastal aquifers.[119] Although the equipment is more expensive than dc resistivity equipment, surveys are quicker and use a smaller field crew, and it is easier to achieve deeper soundings with this method.

6.9.4 Borehole Methods, Including Radioactivity Techniques

Borehole methods include all techniques in which measurements of physical parameters are collected in the borehole as a function of depth. Borehole data can provide information on physical and chemical characteristics of borehole fluids, formation porosity, moisture content, lithology, and groundwater flow rates and directions. Extended discussions can be found in Refs. 63 and 127.

Spontaneous Potential Logs. SP logs are a record of dc voltage differences between a moving electrode in the borehole and a fixed electrode on the surface. These potential differences arise from differences in salinities between formation fluids and borehole fluids. SP logs can be obtained only in uncased boreholes and are useful for delineating sand-shale sequences, which often correspond to permeable-impermeable zones (Ref. 3, pp. 28–40).

Induction Logs. Induction methods consist of a borehole tool containing a transmitter coil and a receiver coil. A very high frequency alternating current is passed through the transmitter coil. The magnetic field associated with this current induces electric current flow in the formation which is sensed by the receiving coil. These methods provide a measure of formation conductivity used in determining water content, porosity, and formation fluid chemistry. Induction logs can be run in PVC-cased or open boreholes.

Natural Gamma Logs. Natural gamma methods use a scintillation counter to measure the natural gamma radiation that is emitted in the vicinity of the borehole tool. They can be used in cased or uncased, liquid- or air-filled boreholes. Their use is primarily in determining lithology and stratigraphic correlation. Clay minerals are often enriched in potassium-40, a strong gamma emitter, allowing detection of low-permeability zones. Gamma logs can help determine if a low-resistivity zone reflects water chemistry or lithologic factors.

Active Gamma Logs. The active gamma log (density log) uses a concentrated source of gamma rays, usually cobalt-60 or cesium-137, and a gamma detector. The back-scattered gamma intensity is a function of formation density. The active gamma log is often combined with a neutron log. It can be run in open holes above or below the water table.

Neutron Logs. A neutron logging tool consists of a neutron source and a detector. Because the mass of a hydrogen atom has approximately the same mass as a neutron, a collision between a neutron (from the source) and a hydrogen atom will cause maximum energy loss of the neutron. Hence the energy of the detected neutron response is interpreted for moisture content in the vadose zone and for porosity in the saturated zone.[97]

TABLE 6.9.1 Borehole Methods and Their Use in Groundwater Investigations

Objective	Subsurface methods
1. Location of zones of saturation	Spontaneous potential log Temperature log Neutron log Gamma–gamma log
2. Physical and chemical characteristics of fluids	Spontaneous potential log Temperature log Fluid conductivity log Specific ion electrodes DO, Eh, pH probes
3. Stratigraphy and porosity	Formation resistivity log Induced polarization log Natural gamma log Active gamma log Resistance log Acoustic Neutron log Induction log Spontaneous potential log
4. Flow and direction	Flowmeter Tracer Water level

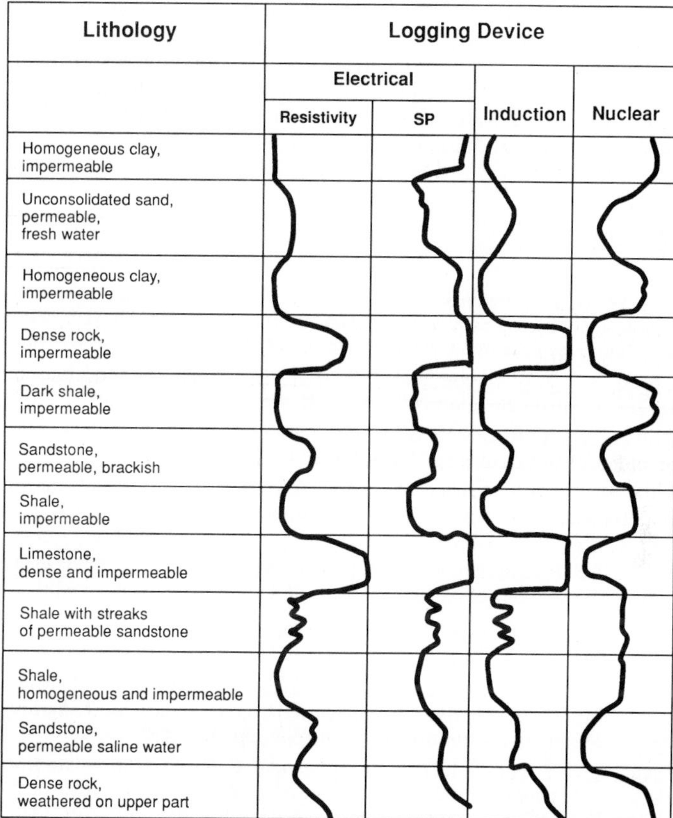

Lithology
Logging Device
Electrical
Resistivity
SP
Induction
Nuclear
Homogeneous clay, impermeable
Unconsolidated sand, permeable, fresh water
Homogeneous clay, impermeable
Dense rock, impermeable
Dark shale, impermeable
Sandstone, permeable, brackish
Shale, impermeable
Limestone, dense and impermeable
Shale with streaks of permeable sandstone
Shale, homogeneous and impermeable
Sandstone, permeable saline water
Dense rock, weathered on upper part

FIGURE 6.9.1 Conceptual example of interpretation of subsurface lithology using a suite of borehole geophysical methods. *(From Ref. 127.)*

Groundwater Flowmeters. Groundwater flowmeters provide measurements of groundwater velocity, hydraulic conductivity, and flow direction. There are two basic types: *thermal impulse* and *vertical impeller.* Using a circular array of thermistors, the thermal impulse flowmeter measures the speed and direction of a heat pulse as it is advected away from the center of a borehole.[76] It provides a relative measure of pore velocity and a direct measure of flow direction. The vertical impeller borehole flowmeter measures the horizontal radial discharge to a well as a function of the depth in the borehole.[102] From these measurements, the horizontal hydraulic conductivity as a function of depth is obtained.

Numerous other logs are available for use in groundwater investigations (Table 6.9.1). The use of several borehole methods together provides a more complete interpretation of subsurface conditions than the use of any individual method. One method often serves to confirm or elaborate on the findings of another. An example of this is shown in Fig. 6.9.1, which is a hypothetical interpretation of subsurface lithology from a suit of borehole methods. The nuclear (neutron) log reads low for both "sandstone, permeable saline water" and "limestone, dense and impermeable." Other logs, such as resistivity, provide a distinction between the two layers: the impermeable limestone has high resistivity, whereas the permeable sandstone has a medium resistivity value. Keys[62] has developed microcomputer programs for the analysis of borehole logs. MacCary[74] presents an example of the use of borehole logs in carbonate aquifers. Kwader[67] uses borehole logs to determine porosity, water quality, and hydraulic conductivity.

REFERENCES

1. Adams, W. M., F. L. Peterson, S. P. Mathur, L. K. Lepley, C. Warren, and R. D. Huber, "A Hydro-Geophysical Survey from Kawaihae to Kailua-Kona, Hawaii," *Tech. Rept.* 32, Water Resources Research Center, University of Hawaii, Honolulu, Hawaii, 1969.
2. Andersen, P. F., J. W. Mercer, and H. O. White, Jr., "Numerical Modeling of Salt-Water Intrusion at Hallandale, Florida," *Ground Water,* vol. 26, no. 5, pp. 619–630, 1988.
3. Asquith, G., and C. R. Gibson, *Basic Well Log Analysis for Geologists,* American Association of Petroleum Geologists, Tulsa, Okla., 1982.
4. Ayers, J. F., and H. L. Vacher, "Hydrogeology of an Atoll Island: A Conceptual Model from Detailed Study of a Micronesian Example," *Ground Water,* vol. 24, no. 2, pp. 185–198, 1986.
5. Back, W., J. S. Rosenshein, and P. R. Seaber, eds., *Hydrogeology,* Geological Society of America, The Geology of North America, vol. O-2, Boulder, Colo., 1988.
6. Bear, J., *Dynamics of Fluids in Porous Media,* Elsevier, New York, 1972.
7. Bear, J., *Hydraulics of Groundwater,* McGraw-Hill, New York, 1979.
8. Bradford, J. M., and S. C. Gupta, "Compressibility," in A. Klute, ed., *Methods of Soil Analysis. Part 1. Physical and Mineralogical Methods,* 2d ed., no. 7, Agronomy Series, American Society of Agronomy, Madison, Wis., pp. 479–492, 1986.
9. Brahana, J. V., J. Thrailkill, T. Freeman, and W. C. Ward, "Carbonate Rocks," in W. Back, J. S. Rosenshein, and P. R. Seaber, eds., *Hydrogeology,* Geological Society of America, The Geology of North America, vol. O-2, Boulder, Colo., pp. 333–352, 1988.
10. Bredehoeft, J. D., and S. S. Papadopoulos, "A Method for Determining the Hydraulic Properties of Tight Formations," *Water Resour. Res.,* vol. 16, no. 1, pp. 233–238, 1980.
11. Cedergren, H. R., *Seepage, Drainage, and Flow Nets,* 3d ed., Wiley, New York, 1989.
12. Clifton, P., and S. P. Neuman, "Effects of Kriging and Inverse Modeling on Conditional Simulation of the Avra Valley Aquifer in Southern Arizona," *Water Resour. Res.,* vol. 18, no. 4, pp. 1215–1234, 1982.
13. Danielson, R. E., and P. L. Sutherland, "Porosity," in A. Klute, ed., *Methods of Soil Analysis. Part 1. Physical and Mineralogical Methods,* 2d ed., no. 7, Agronomy Series, American Society of Agronomy, Madison, Wis., pp. 443–461, 1986.
14. Davis, S. N., and H. W. Bentley, "Dating Groundwater," a Short Review, in L. A. Curie, ed., *Nuclear and Chemical Dating Techniques: Interpreting the Environmental Record,* American Chemical Society Symposium Series 176, pp. 187–222, 1982.
15. Davis, S. N., and E. Murphy, "Dating Groundwater and the Evaluation of Repositories for Radioactive Waste," U.S. Nuclear Regulatory Commission, NUREG/CR-4912, 1987.
16. Davison, C. C., and E. T. Kozak, "Hydrogeologic Characteristics of Major Fracture Zones in a Large Granite Batholith of the Canadian Shield," *Proc. 4th Canadian-American Conference on Hydrogeology,* National Water Well Association, Dublin, Ohio, 1988.
17. de Marsily, G., *Quantitative Hydrogeology,* Academic Press, New York, 1986.
18. Dobrin, M. B., *Introduction to Geophysical Prospecting,* McGraw-Hill, New York, 1976.

19. Domenico, P.A., and F. W. Schwartz, *Physical and Chemical Hydrogeology,* Wiley, 1990.

20. Dreiss, S., "Regional Scale Transport in a Karst Aquifer. 1. Component Separation of Spring Flow Hydrographs," *Water Resour. Res.,* vol 25, no. 1, pp. 117–125, 1989.

21. Dreiss, S., "Regional Scale Transport in a Karst Aquifer 2. Linear Systems and Time Moment Analysis," *Water Resour. Res.* vol. 25, no. 1, pp. 126–134, 1989.

22. Drever, J. I., *The Geochemistry of Natural Waters,* 2d ed., Prentice-Hall, Englewood Cliffs, N.J., 1988.

23. Driscoll, F. G., *Groundwater and Wells,* 2d ed., Johnson Division, St. Paul, Minn., 1986.

24. Dybbs, A., and R. V. Edwards, "A New Look at Porous Media Fluid Mechanics—Darcy to Turbulent," in J. Bear and M. Y. Corapcioglu, eds., *Fundamentals of Transport Phenomena in Porous Media,* NATO ASI Series, Martinus Nijhoff Publishers, Boston, pp. 199–256, 1984.

25. Eyre, P. R., "Simulation of Ground-Water Flow in Southeastern Oahu, Hawaii," *Ground Water,* vol. 23, no. 3, pp. 325–330, 1985.

26. Farvolden, R. N., O. Pfannkuch, R. Pearson, and P. Fritz, "Region 12, Precambrian Shield," in W. Back, J. S. Rosenshein, and P. R. Seaber, eds., *Hydrogeology,* Geological Society of America, The Geology of North America, vol. O-2, Boulder, Colo., pp. 101–114, 1988.

27. Fetter, C. W., *Applied Hydrogeology,* 2d ed., Merrill Publ., Columbus, Ohio, 1988.

28. Ford, D. C., and P. W. Williams, *Karst Geomorphology and Hydrology,* Unwin Hyman, London, 1989.

29. Freeze, R. A., and J. A. Cherry, *Groundwater,* Prentice-Hall, Englewood Cliffs, N.J., 1979.

30. Freeze, R. A., J. Massmann, L. Smith, T. Sperling, and B. James, "Hydrogeological Decision Analysis: 1. A Framework," *Ground Water,* vol. 28, no. 5, pp. 738–766, 1990.

31. Fretwell, J. D., and M. T. Stewart, "Resistivity Study of a Coastal Karst Terrain, Florida," *Ground Water,* vol. 19, no. 2, pp. 156–162, 1981.

32. Frind, E. O., "Seawater Intrusion in Continuous Coastal Aquifer-Aquitard Systems," *Adv. Water Resour.,* vol. 5, 1982.

33. Frind, E. O., and G. B. Matanga, "The Dual Formulation of Flow for Contaminant Transport Modeling. 1. Review of Theory and Accuracy Aspects," *Water Resour. Res.,* vol. 21, no. 3, pp. 159–169, 1985.

34. Fritz, P., and J. Ch. Fontes, *Handbook of Environmental Isotope Geochemistry,* vol. 1, Terrestrial Environment A, Elsevier, Amsterdam, 1980.

35. Fritz, P., and J. Ch. Fontes, *Handbook of Environmental Isotope Geochemistry,* vol. 2, Terrestrial Environment B, Elsevier, Amsterdam, 1986.

36. Gale, J. E., "Assessing the Permeability Characteristics of Fractured Rock," in T. N. Narasimhan, ed., *Recent Trends in Hydrogeology,* Special Paper 189, Geological Society of America, Boulder, Colo., pp. 163–181, 1982.

37. Glover, R. E., "The Pattern of Fresh Water Flow in a Coastal Aquifer," *J. Geophys. Res.,* vol. 64, pp. 439–475, 1959.

38. Gomez-Hernandez, J. J., and S. Gorelick, "Effective Groundwater Model Parameter Values: Influence of Spatial Variability of Hydraulic Conductivity, Leakance, and Recharge," *Water Resour. Res.,* vol. 25, no. 3, pp. 405–420, 1989.

39. Gringarten, A. C., "Flow Test Evaluation of Fractured Reservoirs," in T. N. Narasimhan, ed., *Recent Trends in Hydrogeology,* Special Paper 189, Geological Society of America, Boulder, Colo., pp. 237–264, 1982.

40. Grove, D. B., and K. G. Stollenwerk, "Chemical Reactions Simulated by Groundwater Quality Models," *Water Resour. Bull.,* vol. 23, pp. 601–615, 1987.

41. Haeni, F. P., "Application of Continuous Seismic Reflection Methods to Hydrologic Studies," *Ground Water,* vol. 24, no. 1, pp. 23–31, 1986.

42. Haeni, F. P., "Application of Seismic Refraction Methods in Groundwater Modeling Studies in New England," *Geophysics,* vol. 51, no. 2, pp. 236–249, 1986.
43. Hantush, M. S., "Modification of the Theory of Leaky Aquifers," *J. Geophys. Res.,* vol. 65, pp. 3713–3725, 1960.
44. Hantush, M. S., "Aquifer Test on Partially Penetrating Wells," *Proc. ASCE,* vol. 87, pp. 171–195, 1961.
45. Hantush, M. S., "Hydraulics of Wells," *Adv. Hydrosci.,* vol. 1, pp. 281–432, 1964.
46. Hantush, M. S., and C. E. Jacob, "Nonsteady Radial Flow in an Infinite Leaky Aquifer," *Trans. Am. Geophys. Union,* vol. 36, pp. 95–100, 1955.
47. Harr, M. E., *Groundwater and Seepage,* McGraw-Hill, New York, 1962.
48. Helm, D. C., "One-Dimensional Simulation of Aquifer System Compaction Near Pixley, California, 1. Constant Parameters," *Water Resour. Res.,* vol. 11, no. 3, pp. 465–478, 1975.
49. Henry, H. R., "Salt Intrusion into Fresh-Water Aquifers," *J. Geophys. Res.,* vol. 64, no. 11, pp. 1911–1919, 1959.
50. Henry, H. R., "Effects of Dispersion on Salt Encroachment in Coastal Aquifers, Sea Water in Coastal Aquifers," *U.S. Geol. Surv. Water Supply Paper* 1613-C, pp. 70–84, 1964.
51. Henry, H. R., "Interfaces between Salt Water and Fresh Water in Coastal Aquifers, Sea Water in Coastal Aquifers, *U.S. Geol. Surv. Water Supply Paper* 1613-C, pp. 35–70, 1964.
52. Hsieh, P. A., J. D. Bredehoeft, and J. M. Farr, "Estimation of Aquifer Transmissivity from Phase Analysis of Earth Tide Fluctuations of Water Levels in Artesian Wells," *Water Resour. Res.,* vol. 23, no. 10, pp. 1824–1832, 1987.
53. Hsieh, P. A., and S. P. Neuman, "Field Determination of the Three-Dimensional Hydraulic Conductivity Tensor of Anisotropic Media, 1. Theory," *Water Resour. Res.,* vol. 21, no. 11, pp. 1655–1666, 1985.
54. Huyakorn, P. S., P. F. Anderson, J. W. Mercer, and H. O. White, Jr., "Saltwater Intrusion in Aquifers: Development and Testing of a Three-Dimensional Finite Element Model," *Water Resour. Res.,* vol. 23, no. 2, pp. 293–312, 1987.
55. Javandel, I., "On the Field Determination of Effective Porosity," *Proceedings of Conference on New Field Techniques for Quantifying the Physical and Chemical Properties of Heterogeneous Aquifers,* National Water Well Association, Dublin, Ohio, pp. 155–172, 1989.
56. Johnson, R. H., and J. A. Miller, "Region 24, Southern United States," in W. Back, J. S. Rosenshein, and P. R. Seaber, eds., *Hydrogeology,* Geological Society of America, The Geology of North America, vol. O-2, Boulder, Colo., pp. 229–236, 1988.
57. Jones, J. W., E. S. Simpson, S. P. Neuman, and W. S. Keys, "Field and Theoretical Investigation of Fractured Crystalline Rocks Near Oracle, Arizona," Top. Rep. NUREG/CR-3736, U.S. Nuclear Regulatory Commission, Washington, D.C., 1985.
58. Karasaki, K., J. C. S. Long, and P. A. Witherspoon, "Analytical Models of Slug Tests," *Water Resour. Res.,* vol. 24, no. 10, pp. 115–126, 1988.
59. Kashef, A. A. I., "Salt-Water Intrusion in the Nile Delta," *Ground Water,* vol. 21, no. 2, pp. 160–167, 1983.
60. Kehew, A. E., *General Geology for Engineers,* Prentice-Hall, Englewood Cliffs, N.J., 1988.
61. Keller, G. V., and F. C. Frischknecht, *Electrical Methods in Geophysical Prospecting,* Pergamon Press, New York, 1966.
62. Keys, W. S., "Analysis of Geophysical Logs of Water Wells with a Microcomputer," *Ground Water,* vol. 24, no. 6, pp. 750–760, 1986.
63. Keys, W. S., "Borehole Geophysics Applied to Groundwater Investigations," National Water Well Assoc., Dublin, Ohio, 1989.
64. Knapp, R. W., and D. W. Steeples, "High-Resolution Common-Depth-Point Seismic Reflection Profiling: Instrumentation," *Geophysics,* vol. 51, no. 2, pp. 276–282, 1986.

65. Kofoed, O., *Geosounding Principles,* vol. 1, Elsevier, New York, 1979.

66. Kruseman, G. P., and N. A. De Ridder, "Analysis and Evaluation of Pumping Test Data," 3d ed., International Institute for Land Reclamation and Improvement, Wageningen, The Netherlands, 1976.

67. Kwader, T., "Estimating Aquifer Permeability from Formation Resistivity Factors," *Ground Water,* vol. 23, no. 6, pp. 762–766, 1985.

68. Kwader, T., "The Use of Geophysical Logs for Determining Formation Water Quality," *Ground Water,* vol. 24, no. 1, pp. 11–15, 1986.

69. La Pointe, P. R., and J. A. Hudson, "Characterization and Interpretation of Rock Mass Joint Patterns," Geological Society of America, Special Paper 199, 1985.

70. Lau, L. S., "Seawater Encroachment in Hawaiian Ghyben-Herzberg Systems," in *Proceedings of the National Symposium on Ground-Water Hydrology,* American Water Resources Association, pp. 259–271, 1967.

71. Leeden, F. van der, *Geraghty and Miller's Groundwater Bibliography,* 4th ed., Water Information Center, Plainview, N.Y., 1987.

72. Lindholm, G. F., and J. J. Vaccaro, "Columbia Lava Plateau," in W. Bach, J. S. Rosenshein, and P. R. Seaber, eds., *Hydrogeology,* Geological Society of America, The Geology of North America, vol. O-2, Boulder, Colo., pp. 333–352, 1988.

73. Long, J. C. S., and P. A. Witherspoon, "The Relationship of the Degree of Interconnection to Permeability in Fracture Networks," *J. Geophys. Res.,* vol. 90(B4), pp. 3087–3098, 1985.

74. MacCary, L. M., "Geophysical Logging in Carbonate Aquifers," *Ground Water,* vol. 21, no. 3, pp. 334–342, 1983.

75. McElwee, C. D., "A Model Study of Salt-Water Intrusion to a River Using the Sharp Interface Approximation," *Ground Water,* vol. 23, no. 4, pp. 465–475, 1985.

76. Melville, J. G., F. J. Molz, and O. Guven, "Laboratory Investigation and Analysis of a Ground-Water Flowmeter," *Ground Water,* vol. 23, no. 4, pp. 486–495, 1985.

77. Mercado, A., "The Use of Hydrogeochemical Patterns in Carbonate Sand and Sandstone Aquifers to Identify Intrusion and Flushing of Saline Water," *Ground Water,* vol. 23, no. 5, pp. 635–645, 1985.

78. Mock, P., and J. Merz, "Observations of Delayed Gravity Response in Partially Penetrating Wells,"*Ground Water,* vol. 28, no. 1, pp. 11–16, 1990.

79. Moench, A. F., and P. A. Hsieh, "Analysis of Slug Test Data in a Well with Finite Thickness Skin," *Inter. Assoc. Hydrogeol., Mem. Proc.,* vol. 17, no. 1, pp. 17–29, 1985.

80. Morin, R. H., and H. W. Olsen, "Theoretical Analysis of the Transient Pressure Response from a Constant Flow Rate Hydraulic Conductivity Test," *Water Resour. Res.,* vol. 23, no. 8, pp. 1461–1470, 1987.

81. Morrow, C. A., L. Q. Shi, and J. Byerlee, "Permeability of Fault Gouge Under Confining Pressure and Shear Stress," *J. Geophys. Res.,* vol. 89(B5), pp. 3193–3200, 1984.

82. Neuman, S. P., "Effects of Partial Penetration on Flow in Unconfined Aquifers Considering Delayed Gravity Response," *Water Resour. Res.,* vol. 9, pp. 303–312, 1974.

83. Neuman, S. P., "Analysis of Pumping Test Data from Anisotropic Unconfined Aquifers Considering Delayed Gravity Response," *Water Resour. Res.,* vol. 11, pp. 329–342, 1975.

84. Neuman, S. P., G. R. Walter, H. W. Bentley, J. J. Ward, and D. D. Gonzalez, "Determination of Horizontal Aquifer Anisotropy with Three Wells," *Ground Water,* vol. 22, no. 1, pp. 66–72, 1984.

85. Neuman, S. P., and P. A. Witherspoon, "Field Determination of the Hydraulic Properties of Leaky Multiple-Aquifer Systems," *Water Resour. Res.,* vol. 8, pp. 1284–1298, 1972.

86. Neuzil, C. E., "On Conducting the Modified Slug Test in Tight Formations," *Water Resour. Res.,* vol. 18, no. 2, pp. 439–441, 1982.

87. Novakowski, K. S., "Analysis of Aquifer Tests Conducted in Fractured Rock: A Review of the Physical Background and the Design of a Computer Program for Generating Type Curve," *Ground Water,* vol. 28, no. 1, pp. 99–107, 1990.

88. O'Brien, K. M., and W. J. Stone, "Role of Geological and Geophysical Data in Modeling a Southwestern Alluvial Basin," *Ground Water,* vol. 22, no. 6, pp. 717–727, 1984.

89. Oda, M., Y. Hatsuyama, and Y. Ohnishi, "Numerical Experiments on Permeability Tensor and Its Application to Joined Rock at Stripa Mine, Sweden," *J. Geophys. Res.,* vol. 92 (B8), pp. 8037–8048, 1987.

90. Olhoeft, G. R., and G. R. Johnson, "Densities of Rocks and Minerals," in R. S. Carmichael, ed., *Practical Handbook of Physical Properties of Rocks and Minerals,* CRC Press, Boca Raton, Fla., pp. 141–176, 1989.

91. Papadopoulos, I. S., and H. H. Cooper, "Drawdown in a Well of Large Diameter," *Water Resour. Res.,* vol. 3, no. 1, pp. 241–244, 1967.

92. Papadopoulos, I.S., J. D. Bredehoeft, and H. H. Cooper, Jr., "Type Curves for Slug Test in a Well of Finite Diameter," *Water Resour. Res.,* vol. 9, pp. 1087–1089, 1973.

93. Peck, A., S. Gorelick, G. De Marsily, S. Foster, and V. Kovalevsky, "Consequences of Spatial Variability in Aquifer Properties and Data Limitations for Groundwater Modeling Practice," International Association Hydrological Science, Publ. 175, IAHS Press, Wallingford, England, 1988.

94. Phillips, F. M., J. L. Mattick, T. A. Duval, D. Elmore, and P. W. Kubik, "Chlorine 36 and Tritium from Nuclear Weapons Fallout as Tracers for Long-Term Liquid and Vapor Movement in Desert Soils," *Water Resour. Res.,* vol. 24, no. 11, pp. 1877–1891, 1988.

95. Phillips, F. M., M. K. Tansey, L. A. Peeters, S. Cheng, and A. Long, "An Isotopic Investigation of Groundwater in the Central San Juan Basin, New Mexico: Carbon 14 Dating as a Basis for Numerical Flow Modeling," *Water Resour. Res.,* vol. 25, no. 10, pp. 2259–2273, 1989.

96. Pickens, J. F., G. E. Grisak, J. D. Avis, D. W. Belanger, and M. Thury, "Analysis and Interpretation of Borehole Hydraulic Tests: Principles, Model Development, and Applications," *Water Resour. Res.,* vol. 23, no. 7, pp. 1341–1375, 1987.

97. Poeter, E. P., "Perched Water Identification with Nuclear Logs," *Ground Water,* vol. 26, no. 1, pp. 15–21, 1988.

98. Poland, J. F., ed., "Guidebook to Studies of Land Subsidence Due to Groundwater Withdrawals," UNESCO Studies and Reports in Hydrology, no. 40, 1984.

99. Pollard, D. D., and A. Aydin, "Progress in Understanding Jointing over the Past Century," *Geol. Soc. Am. Bull.,* vol. 100, pp. 1181–1204, 1988.

100. Quinlan, J. F., "Groundwater Basin Delineation with Dye-Tracing, Potentiometric Surface and Cave Mapping, Mammoth Cave Region, Kentucky, USA," *Beitr. Geol. Schweiz, Hydrologie,* vol. 28, pp. 177–189, 1982.

101. Randall, A. D., R. Francis, M. Frimpter, and J. Emery, "Region 19, Northeastern Appalachians," in W. Back, J. S. Rosenshein, and P. R. Seaber, eds., *Hydrogeology,* Geological Society of America, The Geology of North America, vol. O-2, Boulder, Colo., pp. 301–314, 1988.

102. Rehfeldt, K.R., "Application of the Borehole Flowmeter Method to Measure Spatially Variable Hydraulic Conductivity at the MADE Site," in F. J. Molz, J. G. Melville, and O. Guven, eds., *Proceedings of the Conference on New Field Techniques for Quantifying the Physical and Chemical Properties of Heterogeneous Aquifers,* National Water Well Association, Dublin, Ohio, pp. 419–443, 1989.

103. Robertson, W. D., and J. A. Cherry, "Tritium as an Indicator of Recharge and Dispersion in a Groundwater System in Central Ontario," *Water Resour. Res.,* vol. 25, no. 6, pp. 1097–1109, 1989.

104. Robertson, W. D., J. A. Cherry, and S. L. Schiff, "Atmospheric Sulfur Deposition 1950–1985 Inferred from Sulfate in Groundwater," *Water Resour. Res.,* vol. 25, no. 6, pp. 1111–1124, 1989.

105. Roeloffs, E., "Hydrologic Precursors to Earthquakes: A Review," *Pure Appl. Geophys.*, vol. 126, pp. 177–209, 1988.

106. Rojstaczer, S., "Determination of Fluid Flow Properties from the Response of Water Levels in Wells to Atmospheric Loading," *Water Resour. Res.*, vol. 24, no. 11, pp. 1927–1938, 1988.

107. Rouleau, A., and J. E. Gale, "Statistical Characterization of the Fracture System in the Stripa Granite, Sweden," *Int. J. Rock Mech. Min. Sci.*, vol. 22, no. 6, pp. 353–367, 1985.

108. Rouleau, A., and J. E. Gale, "Stochastic Discrete Fracture Simulation of Groundwater Flow into an Underground Excavation," *Int. J. Rock Mech. Min. Sci.*, vol. 24, pp. 99–112, 1987.

109. Shamir, U., and G. Dagan, "Motion of the Seawater Interface in Coastal Aquifers: A Numerical Solution," *Water Resour. Res.*, vol. 7, no. 3, pp. 644–657, 1971.

110. Skinner, B. J., and S. C. Porter, *Physical Geology*, Wiley, New York, 1987.

111. Sklash, M. G., "Environmental Isotope Studies of Storm and Snowmelt Runoff Generation," in M. G. Anderson and T. P. Burt, eds., *Surface and Subsurface Processes in Hydrology*, Wiley, Sussex, England, 1990.

112. Smart, C. C., "Artificial Tracer Techniques for the Determination of the Structure of Conduit Aquifers," *Ground Water*, vol. 26, no. 4, pp. 445–453, 1988.

113. Smart, C. C., "Quantitative Tracing of the Maligne Karst System, Alberta, Canada," *J. Hydrol.*, vol. 98, pp. 185–204, 1988.

114. Snow, D. T., "Anisotropic Permeability of Fractured Media," *Water Resour. Res.*, vol. 5, no. 6, pp. 1273–1289, 1969.

115. Stephenson, D. A., A. H. Fleming, and D. M. Mickelson, "Glacial Deposits," in W. Back, J. S. Rosenshein, and P. R. Seaber, eds., *Hydrogeology*, Geological Society of America, The Geology of North America, vol. O-2, Boulder, Colo., pp. 301–314, 1988.

116. Stewart, M. T., "Gravity Survey of a Deep Buried Valley," *Ground Water*, vol. 18, no. 1, pp. 24–30, 1980.

117. Stewart, M. T., "Evaluation of Electromagnetic Methods for Rapid Mapping of Salt-Water Interfaces in Coastal Aquifers," *Ground Water*, vol. 20, no. 5, pp. 538–545, 1982.

118. Stewart, M., "Electromagnetic Mapping of Fresh-Water Lenses on Small Oceanic Islands," *Ground Water*, vol. 26, no. 2, pp. 187–191, 1988.

119. Stewart, M., and M. C. Gay, "Evaluation of Transient Electromagnetic Soundings for Deep Detection of Conductive Fluids," *Ground Water*, vol. 24, no. 3, pp. 351–356, 1986.

120. Stewart, M., M. Layton, and T. Lizanec, "Application of Resistivity Surveys to Regional Hydrogeologic Reconnaissance," *Ground Water*, vol. 21, no. 1, pp. 42–48, 1983.

121. Strack, O. D. L., "A Single-Potential Solution for Regional Interface Problems in Coastal Aquifers," *Water Resour. Res.*, vol. 12, no. 6, pp. 1165–1174, 1976.

122. Strack, O. D. L., *Groundwater Mechanics*, Prentice-Hall, Englewood Cliffs, N.J., 1989.

123. Vacher, H. L., "Dupuit-Ghyben-Herzberg Analysis of Strip-Island Lenses," *Geol. Soc. Am. Bull.*, vol. 100, pp. 580–591, 1988.

124. Voss, C. I., and W. R. Souza, "Variable Density Flow and Solute Transport Simulation of Regional Aquifers Containing a Narrow Freshwater-Saltwater Transition Zone," *Water Resour. Res.*, vol. 23, no. 10, pp. 1851–1866, 1987.

125. Walton, W. C., *Groundwater Pumping Tests—Design and Analysis*, Lewis Publishers, Chelse, Mich., 1987.

126. Wheatcraft, S. W., and R. W. Buddemeier, "Atoll Island Hydrology," *Ground Water*, vol. 19, no. 3, pp. 311–320, 1981.

127. Wheatcraft, S. W., K. C. Taylor, J. W. Hess, and T. M. Morris, "Borehole Sensing Methods for Ground-Water Investigations at Hazardous Waste Sites," University of Nevada System Desert Research Institute, Water Resources Center Publ. 41099, 1986.

128. Williams, J. A., "Diffusivity, Storativity, and the Dupuit Assumptions for Periodic Flow in a Vertical Hele-Shaw Model," *Water Resour. Res.*, vol. 18, no. 4, pp. 925–930, 1982.

129. Wood, W. W., and L. A. Fernandez, "Volcanic Rocks," in W. Back, J. S. Rosenhein, and P. R. Seaber, eds., *Hydrogeology,* Geological Society of America, The Geology of North America, vol. O-2, Boulder, Colo., pp. 353–365, 1988.

130. Zohdy, A. A. R., G. P. Eaton, and D. R. Mabey, "Application of Surface Geophysics to Ground-Water Investigations," Techniques of Water Resources Investigations of the U.S. Geological Survey, 1974.

CHAPTER 7
SNOW AND FLOATING ICE

Don M. Gray
Division of Hydrology
University of Saskatchewan
Saskatoon, Saskatchewan, Canada

Terry D. Prowse
National Hydrology Research Institute
Saskatoon, Saskatchewan, Canada

7.1 INTRODUCTION

The objective of this chapter is to familiarize practicing hydrologists with the physical characteristics of seasonal snow cover and floating ice and the processes affecting them. Volumetrically, these components form only a small fraction of the world's fresh water, but their hydrologic importance is immense. Furthermore, they seasonally affect large portions of the landscape. In North America, for example, snow annually attains depths of more than 30 cm as far south as 38°N and floating ice makes waterways unnavigable for 100 days per year at 42°N.[11]

In high and midlatitudes, where precipitation is slight, melt of the seasonal snow cover is the most significant hydrologic event of the year. In these regions, runoff from the shallow snow covers often provides 80 percent or more of the annual surface runoff, augments soil water reserves, and recharges groundwater supplies. Even at lower latitudes, particularly in alpine regions, snowmelt is a primary source of water. For example, in California as much as 85 to 90 percent of annual streamflow comes from spring snowmelt.

The significance of floating ice to hydrology stems not from its melt contribution but from its influence on flow, water levels, and storage in lakes and in river channels. Extreme events in cold environments, such as low flows and floods, are often more a result of channel-ice effects than they are of landscape runoff processes. While peak spring flows in cold regions are primarily a function of snowmelt, the maximum water level in a stream is usually determined by the resistance to flow created by ice obstructions. Moreover, such ice-affected levels are frequently the annual water level peaks, exceeding those produced by higher discharge rates under open-water conditions.

Other forms of ice are more significant volumetric components of the hydrosphere. Glaciers and permafrost comprise approximately 75 percent of the world's fresh water supply, although they contribute only marginally to the annual hydrologic cycle. Within specific hydrologic regions, however, perennial ice forms dominate the water cycle. Treatment of these subjects is beyond the scope of this chapter, and readers are referred to general texts on these subjects.[110,134,162,168,169]

7.2 SNOW—FORMATION, INTERCEPTION, DISTRIBUTION, AND MEASUREMENT

7.2.1 Snowfall

Physics of Formation. An appreciation of the mechanics of snowfall formation is important because of the effects the formative processes have on the temporal and spatial character of snowfall and the structural and geometrical configuration of snow crystals. The latter affect the strength, water content, thermal, hydraulic, and other properties of snow.

The atmospheric requirements for snowfall formation are the presence of water vapor and ice nuclei and an ambient temperature below 0°C. *Ice nuclei* are particles that cause ice crystals to form through either direct freezing of cloud droplets or freezing of water deposited on the particle surface as a vapor. The major nuclei in the atmosphere are dust particles, products of combustion from industrial plants and forest fires, and organic matter from the earth's surface. At any time, there are billions of aerosol particles in the atmosphere; most have sizes ranging between 0.001 and 1 μm. A very small fraction of these particles are active as ice nuclei, only about 1 particle in 10^9 at −10°C, with the number increasing approximately by a factor of 10 with each decrease of 4°C down to −30°C.[136] Once ice crystals form, they may splinter and create large numbers of nuclei to aid the precipitation process.

Continued growth of an ice crystal leads to the formation of a *snow crystal.* This is a large particle, having a very complex shape, and of such size that it is visible to the naked eye. The size and shape of snow crystals may change as they fall to earth because of freezing and accretion of water droplets with which they collide. This results in *rimed crystals* and, in the extreme, *snow pellets.* A *snowflake* is an aggregation of snow crystals and may also grow in size during its fall owing to the adhesion of colliding snow crystals. Snow scientists classify atmospheric snow crystals by their shapes and growth processes.[82] A different system is used to classify seasonal snow on the ground.[32]

Whether a snowflake formed in the atmosphere arrives at the earth's surface as snow or rain depends primarily on the extent and temperature of layers of air through which it falls. Figure 7.2.1 shows the relationship between the frequency of occurrence of rain and snow and air temperature observed at Donner summit, California.[158]

7.2.2 Interception

Interception of snowfall by a vegetative canopy plays a major role in the snow hydrology of forests. Many questions remain unanswered as to the effects of interception by a forest canopy on the energy balance at the ground surface, the amount of snowfall lost to evaporation and sublimation, and the distribution of snow within a forest and forest openings.

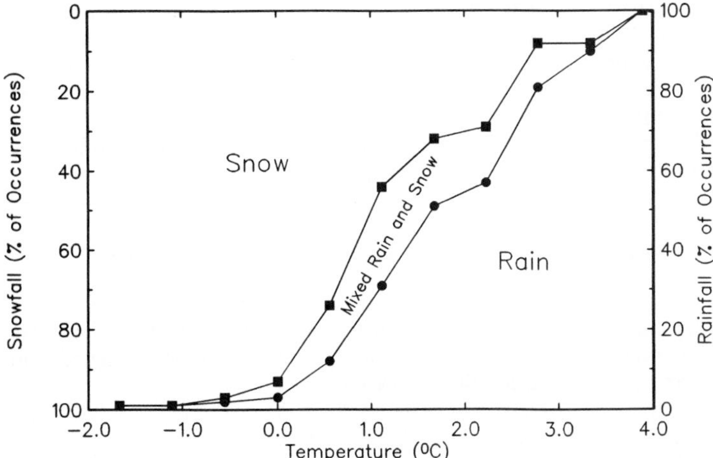

FIGURE 7.2.1 Relationship between frequency of occurrence of rain and snow events and air temperature monitored at Donner Summit, California, elevation 2195 m. *(Adapted from Ref. 158.)*

Early in a storm, snowflakes fall through branches and needles until small bridges form at narrow openings. These bridges increase the collection area, and snow is retained on the bridges by cohesion. At some point a tree reaches its maximum holding capacity, when snow retention from subsequent snowfall is roughly balanced by the loss of intercepted snow falling to the ground. Branch strength and flexure, needle configuration and orientation, mass and surface area of vegetation, and tree age and density are some of the properties which affect the amount of snow interception. Interception by coniferous forests is much greater than by deciduous forests because deciduous trees lose their leaves during winter. A factor of 2 difference in interception capacity between conifers and hardwoods has been suggested.[143] The intercepting surface of a mature spruce forest is about 20 times greater than the ground surface.[35]

Effect of Wind on Interception. The morphological characteristics of trees of similar size and species are generally of lesser importance to interception than meteorological factors, particularly air temperature and wind speed. The highest interception rates during storms have been reported to occur at air temperatures between −23 and 0°C.[71] Wind affects interception by disrupting the bridging process, by reducing the amount of snow intercepted because of vibration of the vegetation, by erosion of intercepted snow, and by retaining snow particles in suspension. Interception is small at wind speeds greater than 2 m/s.[102] Because of mechanical removal of snow due to vibration and increased erosion due to increased turbulence, high winds may decrease the amount of snow intercepted along exposed windward edges of a forest and at the top of the canopy. The flux of snow at the canopy top and the percentage of snow particles entering a canopy which fall to a lower level increase with increasing wind speed during snowfall.[155] Thus high winds may decrease the amount of interception near the top of a canopy while increasing accumulation within the canopy.

The amount of intercepted snow continuously changes because of wind erosion, evaporation and sublimation, melting, and avulsion. It is difficult to quantify the amount of intercepted snow and the losses due to problems in measuring the various components. Most of the information available is site-specific and the results from studies vary widely. For example, the annual evaporation losses of intercepted snow from pine forests, expressed as a percentage of seasonal snowfall, may range from 5.2 to 32 percent, depending on measurement location and year.[104,135,167] In central Colorado it has been reported that one-third of the snow intercepted by an artificial conifer sublimed during a 6-h warming period.[139]

7.2.3 Snow Cover Depth, Density, and Water Equivalent

The physical properties of a snowpack most important to hydrologists are depth, density, and water equivalent. *Snow water equivalent* (SWE), the equivalent depth of water of a snow cover, is not usually measured directly but is calculated from snow depth d_s and density ρ_s using the expression

$$\mathrm{SWE} = 0.01 d_s \rho_s \tag{7.2.1}$$

in which SWE is in mm when d_s is in cm and ρ_s is in kg/m^3.

An average density for new snowfall equal to 100 kg/m^3 is often assumed; this gives 1 unit of water for each 10 units of snow depth. It is not appropriate to use a fixed value of the ratio for all environments. The density of freshly fallen snow varies widely depending on the amount of air contained within the lattice of the snow crystals. Densities in the range of 50 to 120 kg/m^3 are common. Lower values are generally found in snowfalls formed under dry, cold conditions; higher values are found in the wet snows at warm temperatures. Figure 7.2.2 shows that the density of new-fallen snow decreases exponentially as air temperature decreases below freezing.

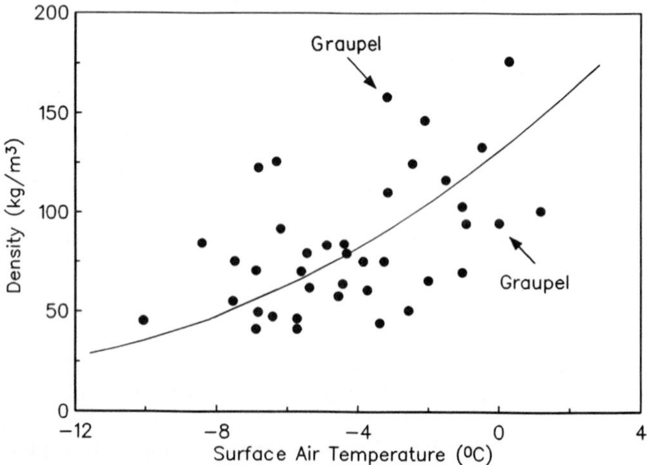

FIGURE 7.2.2 Relationship between density of new-fallen snow and air temperature.[158] Measurements were made at Central Sierra Snow Laboratory at an elevation of 2104 m in the Sierra Nevada Mountains of California. Air temperatures were taken at a height of about 1.22 m above the snow surface at about the same time as the density measurements.

TABLE 7.2.1 Densities of Dry Snow Covers

Snow type	Density, kg/m^3
Wild snow	10–30
Ordinary new snow, immediately after falling in still air	50–65
Settling snow	70–190
Settled snow	200–300
Very slightly wind toughened snow immediately after falling	63–80
Average wind-toughened snow	280
Hard wind slab	350

Source: Reproduced by courtesy of the International Glaciological Society from *Snow Structure and Ski Fields*, G. Seligman, 1980.

Time Variation of Snow Density. Following deposition, the density of new snow increases rapidly owing to *metamorphism*—changes in the size, shape, and bonding of snow crystals due to temperature and water-vapor gradients, settlement, and wind packing, with the rate of change being controlled primarily by meteorological conditions. Typical densities for dry snow covers are given in Table 7.2.1.

The average density of a snowpack also varies seasonally (see Fig. 7.2.3). The seasonal increase usually is largest in subarctic regions where densities may increase by more than 100 kg/m^3 as the snow ripens. During snowmelt, the density of a snowpack may vary extensively because of the storage and loss of meltwater. Densities commonly range between 350 and 500 kg/m^3. Lower densities occur in the morning, and density increases during the day as a snowpack becomes primed by infiltrating meltwater.

7.2.4 Snow Cover Distribution

The areal variability of snow cover is studied at three spatial scales:

1. *Macroscale or Regional:* Areas up to 10^6 km^2 with characteristic distances of 10 to 1000 km depending on latitude, elevation, orography, and the presence of large water bodies. At this scale, dynamic meteorological effects such as standing waves in the atmosphere, the directional flow around barriers, and lake effects are important.
2. *Mesoscale or Local (within Region):* Characteristic linear distances of 100 m to 1 km in which redistribution of snow along relief features may occur because of wind or avalanches, and deposition and accumulation of snow may be related to terrain variables and to vegetative cover.
3. *Microscale:* Characteristic distances of 10 to 100 m over which differences in accumulation patterns result from variations in airflow patterns and transport.

Effect of Wind on Distribution. Most hydrologic studies of wind transport have focused on the redistribution of snow and the consequential effect on the spatial variability in snow water equivalent. Wind hardens and compacts snow because of the drag forces exerted on the surface by the moving air and by impacting particles. Transported snow crystals undergo changes to their shape and size and form drifts and banks of higher density than the parent material.

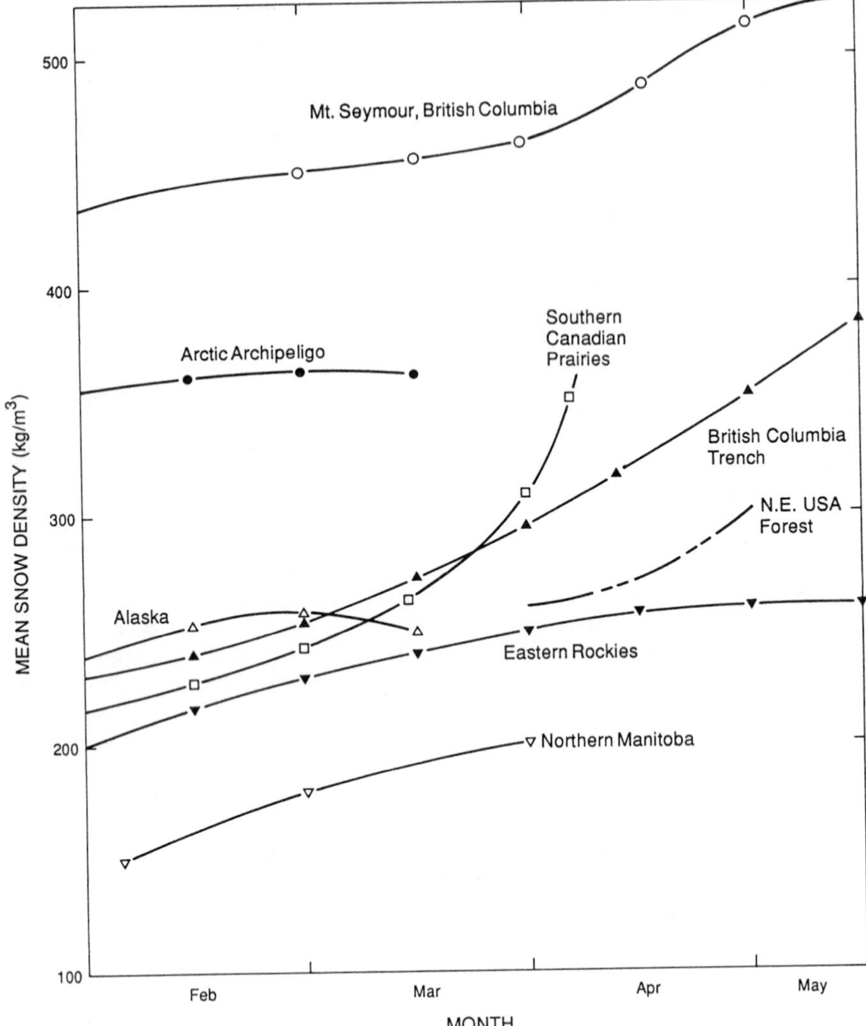

FIGURE 7.2.3 Seasonal variation in average snow cover density. *(Adapted from Ref. 94. Used with permission.)*

The physics of snow transport, erosion, and deposition by wind is summarized in Refs. 96, 117, 138, and 152. Three modes of movement are recognized:

1. *Creep:* Movement of heavy particles by rolling along the surface. Creeping particles usually comprise a very small part of the total amount of snow transported by wind, except at very low wind speeds.
2. *Saltation:* Movement of particles by skipping or jumping along the snow surface. The trajectories of saltating particles vary with wind speed, particle size, and

surface conditions. They follow arcs with abrupt, near-vertical ascents and flat, near-horizontal descents. A typical jump may be 1 cm high and 20 cm long. Most saltating particles attain a maximum trajectory height in the range from a few millimeters to 5 cm.

The transport rate of saltating snow depends on the vertical distribution of wind, the roughness and hardness of the snow surface, the size and density of roughness elements (e.g., vegetation) protruding above the snow surface, fetch distance, and other factors.

3. *Turbulent Diffusion (Suspension):* Movement of particles suspended in the turbulent airstream at a mean horizontal velocity close to that of the moving air. Saltating particles become *suspended* when the time-averaged drag force imposed by the upward-moving turbulent air overcomes the gravitational force. Because the drag force is proportional to the square of the particle diameter, whereas mass increases with the cube of the diameter, turbulent diffusion favors smaller particles. Suspended particles are therefore smaller than those moving in saltation near the snow surface.

The concentration of suspended snow is at a maximum just above the saltation layer and decreases with height at a rate that depends on the wind speed. The higher the wind speed the more uniform is the vertical distribution in the mean mass concentration of snow per unit volume of air. In strong winds and over terrain where the upstream fetch is several kilometers in length the layer of suspended snow may extend to heights of hundreds of meters. Although the concentrations may be small, the mass flux in suspension may be large because of the thickness of the suspended layer.

Empirical expressions for estimating the transport rate of saltating snow q_{salt}, the transport rate of suspended snow q_{susp}, and the total transport rate q_T (the sum of q_{salt} and q_{susp}) directly from a point measurement of wind speed are reported in the literature. Several of these equations are listed in Table 7.2.2. They assume an ade-

TABLE 7.2.2 Expressions for Calculating the Mass Fluxes of Blowing Snow in Saltation, q_{salt}, in Suspension, q_{susp}, and the Total Transport Rate q_T in Kilograms per Second per Meter of Width Perpendicular to the Wind. u is the Wind Speed in m/s at the Height Indicated by the Subscript in Meters

Equation	Height, m	Ref.	Comments
$q_{salt} = 0.00003(u_1 - 1.3)^3$		74	Prairies
$q_{salt} = (u_{10}^{1.30}/2100) - (1/17.37u_{10}^{1.30})$		118	fallow land
$q_{susp} = u_{10}^{4.13}/674{,}100$	5	117	$u_{10} > 6.5$, m/s
$\log q_T = -1.82 + 0.089u_{10}$	0.001–300	21	Antarctica
$q_T = 0.000077(u_{10} - 5)^3$	0–2	41	
$q_T = 0.0002u_1^{2.7}$	0–2	152	Old, firm snow
$q_T = 0.0000029u_1^{4.14}$	0–2	153	Settled old snow
$q_T = 0.0000022u_{10}^{4.04}$	0–5	117	Prairies fallow land $u_{10} > 6.5$, m/s

quate upwind supply of snow to support quasi-steady-state transport conditions. To apply them requires information on wind speed and the duration of blowing snow.

Sublimation of Blowing Snow. The loss of water by *sublimation,* the change in phase from ice to water vapor, during wind transport[40,117,137] has not received equal attention as the processes of saltation and suspension. The sublimation rate is a function of many factors. It increases with (1) increasing wind speed, (2) increasing ambient air temperature, (3) decreasing relative humidity, and (4) elevation (due to the decrease in vapor pressure and increase in solar radiation). The effects of wind speed, temperature, and humidity on the sublimation rate of blowing snow are illustrated in Fig. 7.2.4.

This loss of water during snow redistribution can be important in water-scarce regions where snow is a major source of manageable fresh water supply. Estimates of average annual sublimation were made using meterological observations monitored at 16 locations in the prairie provinces of Canada.[79,119] They show that on 1-km fetches of stubble and fallow: (1) the sublimation loss increases by ~7 percent with a change in land use from stubble to fallow, (2) the percentage of annual snowfall lost to sublimation is equal to or greater (up to 2.6 times) than the amount of snow transported from the fetch in saltation and suspension, and (3) up to 41 percent of annual snowfall on fallow land and 34 percent of annual snowfall on stubble land may be lost by sublimation during wind transport.

Snow Erosion and Deposition. Under invariant atmospheric and surface conditions and an adequate fetch of erodible snow, an *equilibrium* condition will develop where the mass of snow in saltation and suspension remains constant and the surface erosion rate is equal to the evaporation and sublimation rate. Atmospheric conditions, surface roughness, snow supply, and land use influence the length of fetch

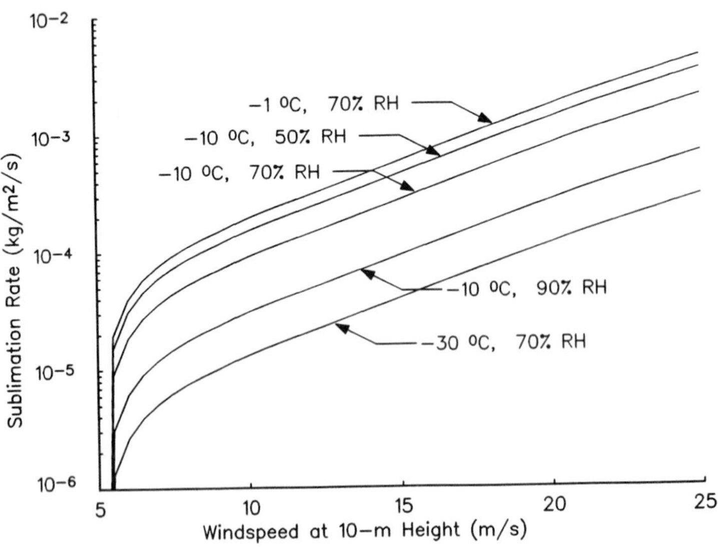

FIGURE 7.2.4 Effects of wind speed, temperature, and relative humidity on the sublimation rate of blowing snow.

TABLE 7.2.3 Relative Snow Deposition on Topographic Facets along a Traverse Across the Ural Mountains

Topographic facet	Length, km	Precipitation (relative)	Accumulation (relative)	Eddy
Flat	10	1.0	1.0	
Lower slope	2	1.75	1.2	
Upper slope	1	1.25	0.25	V*
Ridge	1.3	1.4	2.1	H†
Lee of main ridge	1	0.5	1.1	
Lee of second ridge	0.8	0.6	0.6	V
Downwind plateau	0.7	0.8	0.9	
Valley	0.7	0.7	1.3	
Windward slope	0.5	0.9	0.3	V
Ridge and lee slope	1.3	0.6	0.7	H
Lower lee slope	0.5	1.0	0.9	

Source: Reported in Ref. 103.
* V = vertical direction
† H = horizontal direction

required to attain *equilibrium* transport. This distance may vary between 150 and 500 m.[117,153] On fetches longer than the *equilibrium* fetch, whereas the saltation and suspension rates remain reasonably constant, the sublimation rate increases; e.g., the sublimation rate from a 1-km fetch increases by a factor of 1.4 for a fetch of 2 km and by a factor of 1.75 for a fetch of 4 km.

Deposition of snow occurs at locations where the wind decelerates because of changes in surface roughness.

Effect of Topography and Vegetation. In the absence of a complete understanding of the physical processes governing snow cover distribution, hydrologists use empirical relationships for estimating snow water equivalent and snow depth from topography, vegetation, and land use. This approach assumes consistent patterns in snow cover characteristics as related to landscape features and climate. Extrapolation of relationships outside the region where they are developed is not recommended.

Effect of Elevation, Slope, and Aspect. Although the amount of snowfall from an individual storm normally decreases with increasing altitude, the depth of seasonal snow cover usually increases with elevation because of the larger number of snowfall events and lower amounts of evaporation and melt. Thus, at a particular location in a mountainous region a strong linear association is often found between seasonal snow cover water equivalent and elevation within a selected elevation band.[158] However, even along specific transects the rate of increase in water equivalent with elevation may vary widely from year to year.[95] Elevation alone is not a causative factor in snow cover distribution, and a host of other variables such as slope, aspect, upwind topography, climatic factors, and characteristics of the parent weather systems must be considered to interpret distribution patterns accurately.

The wide variability in the amounts of snow accumulated on various topographic facets resulting from snow transport and other processes is demonstrated in the data reported for the Ural Mountains in Table 7.2.3. These data also illustrate the effects of slope and aspect on snow cover distribution. Snow depth decreases with distance along a slope oriented in the direction of prevailing winds, and major accumulations occur on lee slopes and in abrupt depressions. A primary influence of aspect on snow distribution patterns is its effect on the surface energy exchange processes and snowmelt.

Effect of Vegetation. Vegetation affects snow cover distribution through its influences on (1) surface roughness and wind speed, therefore on snow erosion, transport, and deposition; (2) the surface energy exchange; and (3) snowfall interception. The proportion of snowfall accumulated in a forest depends on *canopy density,* the proportion of the ground surface protected (shaded) by vegetation, and tree species. For example, Kuz'min[77] found the water equivalent of a snow cover in a fir forest decreased from 56 to 34 mm for cover densities of 0.1 and 1 in a winter with little snow. For a winter with heavy snow the corresponding decrease was from 230 to 150 mm. Because of the lack of leaves, deciduous forests have greater accumulations of snow on the ground than do coniferous forests.

Much of the research on snow cover distribution in forested environments has concentrated on the difference between the amounts of snow collected in forests and in openings, and has demonstrated that larger amounts of snow accumulate in clearings. An expression relating the snow water equivalent in a fir forest SWE_f and in a clearing SWE_c to forest cover density p (expressed as a decimal fraction) is[77]

$$SWE_f = SWE_c\,(1 - 0.37p) \tag{7.2.2}$$

The average depth of snow in a clearing varies with the size of the opening; increasing the size decreases the average depth.

The reason for the differences in snow accumulation between forests and forest openings is not clear, and researchers have postulated they may be due to (1) intercepted snow from adjacent trees being blown into the clearing, (2) the lack of interception over clearings, and (3) changes in the wind patterns produced by the clearing.

TABLE 7.2.4 Methods of Measuring Snowfall and Snow Cover Depth, Density, and Snow Water Equivalent

Parameter	Method	Reference
Point snowfall depth	Ruler, snow boards	56
Point snowfall water equivalent	Nonrecording precipitation gauge	56
	Weighing-type precipitation gauge	56
Point snow cover depth	Ruler, graduated rod, aerial markers	56
	Acoustic	57
Point snow cover density (water equivalent)	Gravimetric—snow tube: graduated hollow tube with a cutter fitted to end + weighing kit	56
	Radioisotope (nuclear) gauge	146
	Snow pillow	56
Areal snow cover depth and density (point sampling)	Snow survey	56
	Stratified (unitized) sampling	149, 150
Remote sensing (extent and areal coverage)	NOAA and Landsat satellites, thermal infrared band	67
Remote sensing (density)	Terrestrial survey: natural gamma radiation	25
Remote sensing (snow cover water equivalent)	Active and passive microwave	16

Recent studies suggest that because most of the increase in snow in an opening occurs during storms, that redistribution of intercepted snow is not the primary cause of the increased accumulations monitored in clearings. Wheeler[166] reported no significant difference between the average snow water equivalent monitored at sites in a forest located upwind and downwind of a clearing and concluded that interception was the major factor contributing to the difference between the amount of snow in the forest floor and in the clearing. The thesis that wind fields within clearings affect the snow distribution patterns in openings is supported by Refs. 36 and 47. They show larger snow accumulation on the windward side of a clearing than on the leeward side.

7.2.5 Snow Cover Measurement

Instruments and procedures for measuring snow depth, density, and water equivalent may vary widely with the user. Table 7.2.4 lists several methods used in North America and other parts of the world to measure these parameters. For additional material on snow measurement methods it is recommended that readers consult the *Proceedings* on the annual meetings of western and eastern snow conferences.

7.3 SNOWMELT

In many parts of the world, melt of the seasonal snow cover is the single most important event of the water year. Water produced by melting snow supplies reservoirs, lakes, and rivers and infiltrating meltwater recharges soil moisture and groundwater.

There are many models for forecasting snowmelt runoff. These use either an *energy balance* or a *temperature-index method* to compute melt. For extreme conditions and for short-term forecasts, physically based systems are recommended. Most index models are designed for simulation periods of a month or longer and give good results under average conditions. Irrespective of the choice of method, modeling of the spatial distribution of snow cover and melt usually is accomplished by dividing a watershed into a number of smaller land units based on topographic facets such as elevation bands and hill slopes or by geometrical subdivision into grid squares. Within individual land units the melt, internal energy changes, water transmission, and storage and other processes are parameterized or described by empirical formulas.

7.3.1 Energy Balance

Many studies have been directed toward an understanding of the energetics of snowmelt (e.g., Refs. 4, 64, 121, and 158). The energy-balance approach for calculating snowmelt applies the law of conservation of energy to a *control* volume. The control volume has as its lower boundary the snow-ground interface and as its upper boundary the snow-air interface. Use of a volume allows the energy fluxes into the snow to be expressed as internal energy changes.

The balance requires that the sum of the energy fluxes by radiation, convection, conduction, and advection plus the change in internal energy in the volume be zero. For a cube of snow of unit volume, assuming the energy transfers in the horizontal

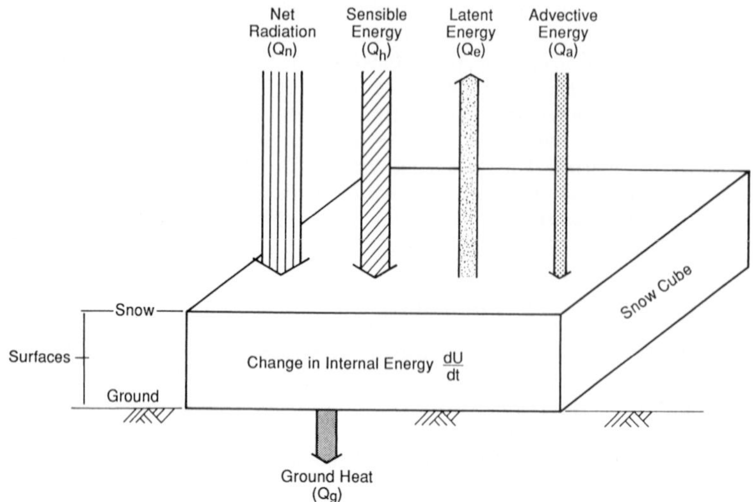

FIGURE 7.3.1 Schematic of the energy fluxes, in the vertical direction, during melting of a control volume of snow.

direction are negligible, the energy equation in the vertical direction gives the amount of energy available for melt Q_m as (see Fig. 7.3.1)

$$Q_m = Q_n + Q_h + Q_e + Q_g + Q_a - \frac{\Delta U}{\Delta t} \tag{7.3.1}$$

where Q_n = *net radiation,* the flux (energy per unit cross-sectional area per unit time) of energy at the surface due to the exchange of radiation

Q_h = *sensible energy,* the turbulent flux of energy exchanged at the surface due to a difference in temperature between the surface and overlying air

Q_e = *latent energy,* the turbulent flux of energy exchanged at the surface due to vapor movement as a result of a vapor pressure difference between the surface and overlying air; evaporation represents a loss, condensation a gain

Q_g = *ground heat,* the flux of energy exchanged between the volume and the underlying ground by conduction

Q_a = *advective energy,* energy derived from external sources, such as by rain, that is added to the volume

$\Delta U/\Delta t$ = rate of change of internal energy in the volume per unit surface area per unit time. U is often referred to as negative heat storage

In applying Eq. (7.3.1) fluxes into the volume are taken as positive; those out of the volume are negative. The time interval over which a balance is applied is governed by the data monitoring program, with 1-h, 6-h, 12-h, 24-h, and monthly periods being common. The amount of melt M is calculated using Q_m in the expression

$$M = 0.270 Q_m \tag{7.3.2}$$

in which M is in mm of meltwater per day when the average daily melt flux Q_m is in W/m^2.

Figure 7.3.2 shows the diurnal variation of various components of the energy balance during one day of melt of a shallow snow cover in an exposed environment. On this day net radiation dominated the process; the daily net radiation and melt fluxes were 33.1 and 32.6 W/m^2, respectively. The data also show: (1) the direction of the various energy terms may change over a day, that is sensible, latent, and ground components have the ability either to assist or to counteract net radiation and (2) the internal energy change may be a significant component in the energy balance of a shallow snow cover, which is usually not the case for deep snow.

One should not conclude from the above example that net radiation is *always* the dominant flux influencing melt. Table 7.3.1 demonstrates that the relative contributions of radiative and turbulent energy transfers over snow may vary widely and change from site to site and from year to year at the same site. These variations suggest that there is little basis for stating that one energy flux will control overwhelmingly or that another may be negligible in any particular region. Altitude, season, air mass, and vegetative cover affect the relative contributions of radiative and turbulent fluxes. Low-level, shallow snowpacks that melt quickly early in spring exhibit large scatter in these contributions because the rate of melt largely depends on the energy content of the air mass present and the presence of patches of bare ground increase the energy available for melt by advection of warmer air derived over bare ground. The effect of perturbations of air masses on deep snowpacks that melt late in the year and over longer periods tends to be masked by other factors. Also there is a trend for increased relative contributions of radiant energy later in the year.

Regarding the effect of altitude, below 2000 m elevation there is no obvious relationship between the relative contributions of the radiative and turbulent energy transfers, suggesting that other factors, such as topography or air-mass characteristics, are more important. Above the 2000-m level there is an apparent decrease in the relative importance of the turbulent fluxes.[78]

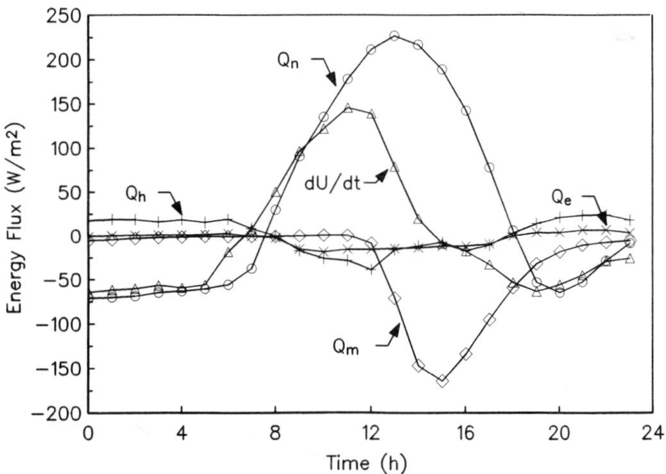

FIGURE 7.3.2 Diurnal variation in energy fluxes during melt of a shallow snow cover. *(Adapted from Ref. 58. Used with permission.)*

TABLE 7.3.1 Selected Results for the Relative Contributions of Radiative and Turbulent Energy Transfers over Snow

Observer	Site	Observation period	Percent contribution*		
			Q_n†	Q_h†	Q_e†
54	Open field (Ottawa)	Spring melt (14 days)	74.6	25.4	−74.1
165	Snow (Fairbanks)	Melt (18 days)	86	14	−24
58	Canadian prairies	Premelt (April)	100.0	−41.6	−35.3
		Early melt (March 1976)	25.9	74.1	−28.8
		Snowmelt (1974)	58.8	41.2	−9.6
		Snowmelt (1975)	95.2	4.8	−19.5
		Snowmelt (1976)	62.0	38.0	−4.2
2	Lower Meadow, Central Sierra Snow Laboratory	Snowmelt (1947)	19.4(70.7)‡	78.8(25.0)	1.8(4.3)
		Snowmelt (1948)	30.9(73.3)	66.2(20.9)	2.9(5.8)
		Snowmelt (1949)	28.7(76.9)	71.3(23.1)	−2.6(−1.4)
		Snowmelt (1950)	37.1(70.9)	59.1(22.5)	3.8(6.6)
		Snowmelt (1951)	36.6(71.4)	63.4(24.2)	−0.6(4.4)
154	Eastern U.S. and Canada	January melt, warm air-mass advection	17.2	46.9	35.9
33	Rangeland isolated drift	Melt	36	64	−48

Source: After Ref. 86.

* Values are percentage of the total energy input (the sum of the positive values) for the observation period in question. Negative numbers denote energy losses.

† Q_n = net radiation; Q_h = sensible energy; Q_e = latent energy.

‡ Numbers in parentheses refer to daytime values only.

7.3.2 Net Radiation

Net radiation Q_n, the sum of the net short-wave Q_{sn} and net long-wave Q_{ln} fluxes, is important to snowmelt in both open and forested environments. Since the net long-wave exchange is often a loss from the snow surface, Q_n is expressed as

$$Q_n = Q_{sn} - Q_{ln} \tag{7.3.3}$$

Net Short-Wave Radiation. Net short-wave radiation is equal to the incident short-wave flux received by the surface Q_s less the amount reflected Q_r, that is:

$$Q_{sn} = Q_s - Q_r \tag{7.3.4}$$

Incident Short-Wave Radiation. That portion of the electromagnetic radiation from the sun falling within the wavelength range from 0.2 to 4 μm is considered short-wave radiation. The intensity of the beam perpendicular to the sun's rays is equal to 1.35 kW/m^2, the *solar constant.*

While passing through the atmosphere, short-wave radiation is reflected by clouds; scattered diffusely by air molecules, dust, and other particles; and absorbed by ozone, water vapor, carbon dioxide, and nitrogen compounds. On average, only about 47 percent of the short-wave radiation received at the outer limit of the atmosphere reaches the earth's surface. The total short-wave flux incident to the earth's surface Q_s, often referred to as *global radiation,* is composed of a *direct-beam* and a *diffuse* component. The diffuse component is direct-beam radiation that has been reflected and scattered by clouds or other surfaces and diffused by atmospheric constituents.

In situations where land slope and aspect and field of view (exposure) affect snowmelt, it is advantageous to resolve the incident short-wave flux into its components. Direct-beam radiation received, reflected, and scattered is strongly affected by these factors, more so than diffuse radiation. Graphs of the effects of slope and aspect on the direct-beam component are provided by Ref. 84; they show that at a latitude of 50°N, clear-sky daily radiation in mid-April incident to a 20° north-facing slope is only about 40 percent of that received by a 20° south-facing slope. Procedures for estimating diffuse radiation to a slope and reduction to the flux due to surrounding topography are described by Refs. 39, 76, and 108.

Numerous models are available for estimating global radiation incident to the earth's surface.[46,59,80] One of the more practical approaches moderates the *extra-terrestrial short-wave flux* Q_A, the flux incident to a horizontal plane at the outer surface of the earth's atmosphere, by atmospheric transmittancy and cloud cover according to the expression[34]

$$Q_s = Q_A\left[a + b\left(\frac{n}{N}\right)\right] \tag{7.3.5}$$

in which a and b are empirically derived coefficients, n is the measured number of hours of bright sunshine, and N is the maximum possible number of hours of bright sunshine (see Ref. 80). A ratio of $n/N = 1$ represents a clear sky; $n/N = 0$ designates a completely overcast sky. Values of a in the range 0.18 to 0.4 (mean $\approx$ 0.27) and b in the range from 0.42 to 0.56 (mean $\approx$ 0.52) are reported in Ref. 34. (See also Sec. 4.2.2.) Because of the seasonal variation in atmospheric transmissivity, high in winter and low in summer, the intercept a tends to be higher and the slope b tends to be lower during the winter months. Table 7.3.2 gives Q_A on selected days at various latitudes.

Albedo. The greatest effect the presence of a snow cover has on the radiative exchange of the earth's surface is the change it produces in the albedo. *Albedo* is the ratio of the amount of short-wave radiation reflected by a surface to the global flux incident to the surface, i.e., $A = Q_r/Q_s$. The reflective properties of snow vary widely depending on wetness, impurities (dirt, chemicals), particle size, density and composition, surface roughness, and the spectral composition and direction of the illumi-

TABLE 7.3.2 Total Daily Solar Radiation to a Horizontal Surface at the Top of the Atmosphere in W/m^2

Latitude, °N	Approximate date: Oct. 16	Nov. 8	Nov. 30	Dec. 22	Jan. 13	Feb. 4	Feb. 26	Mar. 21	Apr. 13
0	429.18	421.43	412.22	408.35	414.16	425.30	434.02	433.54	422.88
10	404.47	382.29	362.33	355.07	364.27	385.10	409.32	427.24	434.51
20	368.14	331.81	303.72	293.06	305.17	334.72	372.50	407.38	433.05
30	321.16	273.20	238.32	225.73	239.29	275.14	324.55	375.41	419.01
40	264.00	207.81	168.57	153.55	169.54	210.23	267.87	332.30	390.91
50	200.54	138.54	98.82	85.25	99.30	139.99	202.96	278.53	354.58
60	132.24	70.72	34.88	23.74	35.36	70.72	133.69	216.53	307.59
70	62.49	11.63				11.63	63.46	148.71	254.31
80	3.39						3.39	75.08	204.90
90									204.90

Source: After Ref. 80.

nating beam.[76,97] Table 7.3.3 gives typical ranges in the surface albedo of various surfaces.

The albedo of a natural snow cover varies with time due to metamorphism, contamination by dust, forest litter, and other materials, and the occurrence of melt, rain, and snow events. Figure 7.3.3 shows a more rapid decay in the albedo of a deep, uncontaminated snow cover during melt than during the accumulation season. Presumably this is due to the presence of liquid water, which not only has a lower albedo than snow but also causes changes to the optical properties of a snow cover because of melting. The albedo of melting snow rapidly decreases from about 0.80 to 0.40 in about 15 days.

The albedo-depletion curves for shallow snow covers in open environments differ appreciably from those described above, the rate of decrease in albedo during ablation being much more rapid than that of a deep snowpack. In part, this is due to absorption of radiation by the snow cover and underlying ground. The albedo depletion of seasonal prairie snow covers between Feb. 1 and the end of ablation may be divided into three periods (Fig. 7.3.4). During *premelt,* from Feb. 1 up to the start of active melt, except for increase due to snowfall and decreases caused by periodic melt

TABLE 7.3.3 Albedo of Various Surfaces for Short-Wave Radiation

Surface	Typical range in albedo
New snow	0.80–0.90
Old snow	0.60–0.80
Melting snow—porous → fine-grained	0.40–0.60
Forests—conifers, snow	0.25–0.35
Forests—green	0.10–0.20
Water	0.05–0.15
Snow ice	0.30–0.55
Black ice: intact → candled → granulated	0.10 → 0.40 → 0.55

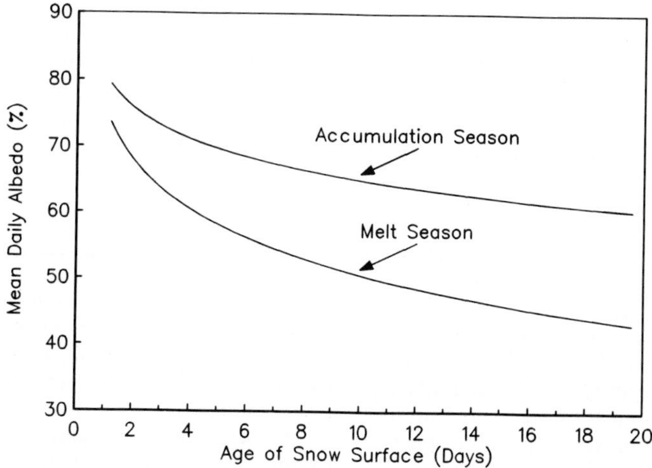

FIGURE 7.3.3 Variation in albedo of deep snow covers with time. *(Source: Ref. 158.)*

events, the rate of depletion is relatively constant in the range from 0.004 to 0.009 per day with an average of 0.0061 per day. During *melt,* the profile of the decay curve during continuous ablation is S-shaped, in which the period of rapid melt is preceded and followed by 1 to 2 days when the rate of change in albedo is lower. Over the period of rapid melt the decrease is about 0.071 per day and ablation spans only 4 to 7 days. During *postmelt,* following the disappearance of the seasonal snow cover, the

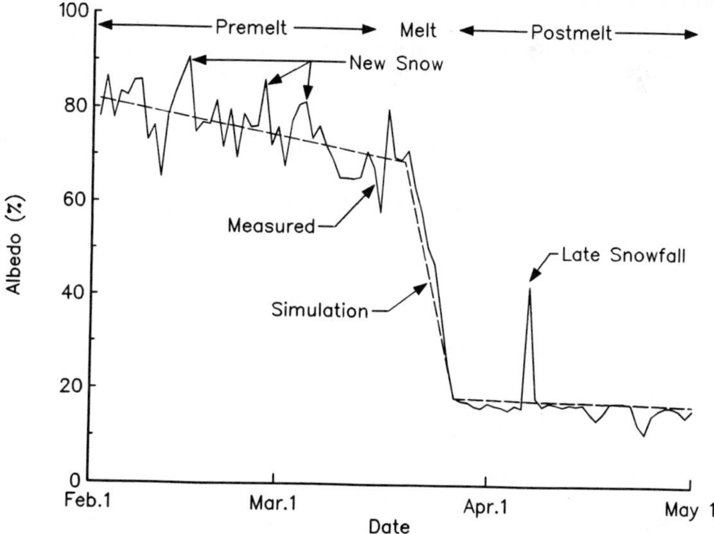

FIGURE 7.3.4 Decrease in albedo of a shallow prairie snow cover during premelt, melt, and postmelt periods. *(Source: Ref. 63.)*

albedo of the ground surface takes on a relatively constant value of 0.17. The decrease in albedo of late-occurring snows occurs at a rate of about 0.20 per day.

Correlation of Net All-Wave and Short-Wave Radiation. Global radiation frequently controls the radiative energy exchange at the snow surface under situations when net long-wave radiation is small. Therefore, researchers have attempted to use short-wave radiation as a predictor of net radiation. A few examples are given in Table 7.3.4. This regression approach is *not* recommended in forests with high canopy densities.

Net Long-Wave Radiation. Net long-wave radiation Q_{ln} is the sum of the long-wave radiation emitted downward by the atmosphere $Q_{l\downarrow}$, and the component emitted upward by the earth's surface $Q_{l\uparrow}$. Q_{ln} often is very important to snowmelt in a forest where a large portion of the incident short-wave radiation is absorbed by the canopy, then emitted downward to the snow surface as long-wave radiation, and when the presence of low-lying clouds inhibits the escape of long-wave radiation emitted by the ground surface. In open environments, cooling of a snow cover due to long-wave radiation emitted to the atmosphere during the nighttime hours affects the timing and release of snow cover runoff. Before significant runoff occurs, the snowpack must be isothermal at 0°C.

Long-wave radiation emitted by a body Q_l is a function of its surface temperature and is described by the Stefan-Boltzmann equation:

$$Q_l = \epsilon\sigma T_s^4 \tag{7.3.6}$$

where ϵ is the emissivity, σ is the Stefan-Boltzmann constant (5.67×10^{-8} W/m$^2 \cdot$ K^4), and T_s is the surface temperature, K. The emissivity of snow generally falls in the narrow range from 0.97 (dirty snow) to 0.99 (fresh snow). Thus long-wave radiation emitted by a melting snow surface is *locked in* to a maximum value of about 310 W/m^2 because of the upper limit in surface temperature of 0°C.

Atmospheric long-wave radiation is emitted by ozone, carbon dioxide, dirt, water vapor, and other contaminants from all levels; under clear skies the major portion occurs from the lowest 100-m layer. Because the largest flux is presumed to originate from the water vapor, investigators have expressed emissivity as a function of vapor

TABLE 7.3.4 Correlation of Net Radiation Q_n at the Snow Surface with Short-Wave Radiation Above the Canopy Q_s, Net Radiation Above the Canopy Q_{nc}, and Net Short-Wave Radiation Above the Surface Q_{sn}. Q_n, Q_s, Q_{nc}, and Q_{sn}, are in W/m^2

Equation	r*	Time interval	Canopy density†	Reference
$Q_n = -0.208 + 0.245Q_{sc}$	0.90	Hourly	0.162	116
$Q_n = -0.116 + 0.223Q_{sc}$	0.87	Hourly	0.163	116
$Q_n = 0.104 + 0.279Q_{sc}$	0.89	Hourly	0.504	116
$Q_n = 0.184 + 0.200Q_{sc}$	—	Daily	Leafless deciduous	69
$Q_n = 0.906 + 0.184Q_{nc}$	0.87	Hourly	Mixed forest	131
$Q_n = -25.92 + 0.651Q_{sn}$	0.87	Daily	Prairie	59

* r = correlation coefficient.
† Canopy density = fraction of snow surface shaded by vegetative canopy.

pressure. The most widely quoted of these relationships is the Brunt formula[20]

$$\epsilon = a + b\sqrt{e_a} \tag{7.3.7}$$

in which e_a is the vapor pressure of the air, Pa, and a and b are coefficients whose magnitudes change with location and climate. Values for a ranging from 0.30 to 0.66 and for b ranging from 0.0039 to 0.0127 are reported by Ref. 76. (See also Sec. 4.2.2.)

Net clear-sky long-wave radiation, calculated from Eqs. (7.3.6) and (7.3.7), should be corrected for field of view and cloud cover. Procedures for including a view factor in the calculations are discussed by Ref. 39. In general, cloud cover, by absorbing incoming solar radiation and emitted terrestrial thermal energy, increases the downward-directed atmospheric long-wave flux and decreases the amount of upward-directed long-wave radiation lost to universal space. The effects of cloud cover on the net long-wave flux Q_{ln} usually are accounted for by adjusting the clear-sky flux Q_{lno} for the degree of cloudiness n/N using the expression[114]

$$\frac{Q_{ln}}{Q_{lno}} = a + b\left(\frac{n}{N}\right) \tag{7.3.8}$$

Values for a usually fall in the range from 0.10 to 0.25 and for b in the range from 0.75 to 0.9. The absence of nighttime observations of cloud cover constitutes the largest practical drawback to this approach for calculating Q_{ln}.

Net Radiation—Forest Environment. The exchange of radiation with the snow surface in a forest differs appreciably from that experienced in open areas because the tree canopy, which shades the snow from global radiation, changes the albedo, emissivity, and roughness of the outer surface, influences the transmission (absorption) of short-wave radiation, and affects the net long-wave exchange. Crown coverage is often used to index the effects of trees on the various components of net radiation. The relationship between the amount of solar radiation transmitted and crown coverage is not linear because of repeated reflections of the radiation by the trees. Up to 90 percent of the incident short-wave flux may be absorbed by dense coniferous forests; leafless hardwoods, such as aspen and birch, may transmit half or more of the flux incident to the canopy. Diffuse sky radiation penetrates to a greater depth into the canopy than the direct beam, a result which is due in part to the variation in the shading effect of vegetation with the angle of incidence of the beam. Also, short-wave radiation reflected by the snowpack is susceptible to absorption by the overhanging branches, so that a large portion of this energy may remain within the canopy.

Radiation absorbed by the forest raises its temperature and is dissipated as long-wave radiation, sensible heat, and latent energy. Reradiation of long-wave radiation by the canopy significantly increases the flux incident to the snow surface. The presence of forest does not, however, appreciably affect the magnitude of the net long-wave exchange between the snowpack and the atmosphere.

Because of the various interactions, the net short-wave gain by the snow decreases as canopy coverage increases, whereas the net long-wave gain from the forest correspondingly increases. The changes are not compensating. Net radiation progressively decreases with increasing canopy coverage to a canopy closure of about 20 percent for a spruce-fir forest; from then on the flux increases up to total canopy cover, although at this coverage net radiation at the snow surface is only a little more than half of that without forest.[132] In general, melt in a forest is less than in the open, often in the range of 60 to 75 percent. These numbers vary widely, however, depending on surface albedo and the structure and density of the forest.

7.3.3 Sensible and Latent Energy Fluxes

Operational estimates of the turbulent transfer of sensible and latent heat over snow are usually obtained using measurements of air temperature, humidity, and wind speed in the aerodynamic formulas:

$$Q_h = D_h u_z (T_a - T_s) \tag{7.3.9}$$

and

$$Q_e = D_e u_z (e_a - e_s) \tag{7.3.10}$$

where D_h = bulk transfer coefficient for sensible heat transfer
u_z = wind speed, m/s, at a reference height z
T_a, T_s = temperatures of the air and snow surface, °C, respectively
D_e = bulk transfer coefficient for latent heat transfer
e_a, e_s = vapor pressures of the air and snow surface, Pa, respectively

Table 7.3.5 lists values for the transfer coefficients D_h and D_e reported in the literature. A review of the application of Eqs. (7.3.9) and (7.3.10)[4,85] indicates that both

TABLE 7.3.5 Values of the Bulk Transfer Coefficients for Turbulent Exchange Above Melting Snow

Author	$D_h \times 10^3$, kJ/m³ · °C	$D_e \times 10^3$ kJ/m³ · Pa	Measurement height, m			Comments
			u_z	T_a	e_a	
70	1.06	0.0515	3.2	2	2	Values based on eddy correlation measurements made in four 1-h periods under highly stable conditions with wind speeds less than 3.5 m/s
86	15	0.25	2	1.2	—	e_a measured 1.6 km from site. D_h and D_e are average values for a 2-week period. Assumed $D_h/D_e = 0.6$ (Bowen ratio)
170	3.56	0.0662	0.7	1.2	1.2	Values apply only for wind speeds greater than 2 m/s
158	1.68	0.080	1	1	1	Coefficients corrected to 1-m height using 1/6 power law
151	5.74	0.100	1	1	1	Coefficients corrected to 1-m height using 1/6 power law
158	—	0.114	1	1	1	Coefficients corrected to 1-m height using 1/6 power law
58	6.69	0.0217	1	1	1	Measurements taken above melting prairie snowpack. D_h = average for 6 days; D_e = average for 8 days, based on lysimeter results

Source: From D. H. Male and D. M. Gray, *Handbook of Snow,* Copyright 1981. Reprinted with permission of Macmillan Publishing Company, a division of Macmillan, Inc. (Ref. 84).

TABLE 7.3.6 Daily Evaporation Rates from Snow During Daytime Hours in mm/8 h

Site	Average	Maximum	Minimum
Open	0.30	0.76	0.02
Aspen stand	0.21	0.58	0.01
Conifer stand	0.14	0.17	0.07

Source: After Ref. 38.

expressions can result in large errors when applied in either exposed or forest environments.

Often the turbulent fluxes are ignored in a forest on the assumption that reduced wind speeds lead to negligible energy transfer. This assumption is questionable in maritime climates and at southern regions because snowmelt is often initiated by the movement of a warm air mass into a region, an indication that turbulent transfers are important. Sensible and latent energy transfers may increase or decrease the net energy available for melt. Table 7.3.1 suggests that over melting snow the net contribution of sensible energy usually is positive. The transfer of latent energy usually responds to the radiation flux following a cycle of evaporation during the day and condensation at night. Daily condensation is probable if nighttime radiation losses are large or a warm air mass is present, and can exceed evaporation over a 24-h period resulting in a net accretion to the snowpack.

A mean daily evaporation rate of 0.15 mm/day and maximum condensation rate of 0.54 mm/day during melt have been monitored in a prairie environment.[58] For northern Sweden, Nyberg[107] found a maximum evaporative loss of 1.6 mm/day in the presence of cold, dry arctic air and a maximum gain through condensation of 0.95 mm/day in the presence of a warm, moist maritime air. Table 7.3.6 lists evaporation rates monitored during the daytime hours in selected landscapes in Utah. They show evaporation was greatest in the open, intermediate in the aspen stand, and least under the conifer stand. The rate increased during the season at each site. Condensation gains during the night were found to be of the order of 20 to 30 percent of evaporation measured during the preceding day.

7.3.4 Ground Heat

The presence of snow, because of its low thermal conductivity, reduces the heat exchange between the ground and the atmosphere. In most environments, the ground heat flux Q_g is a very small component in the daily energy balance of a melting snowpack compared with radiation, sensible energy and latent energy, and its effects on total snowmelt can safely be ignored. Average daily values generally fall in the range of 0 to 4.6 W/m^2.[58,158] Over an extended melt season, however, the effects of ground heat on snowmelt runoff may be important. Q_g influences the temperature of the snow near the ground surface, contributes to conditioning of a snowpack before melt, and affects the state of the underlying soil.

The importance of ground heat to runoff in the arctic where completely frozen ground and permafrost impede the infiltration of meltwater and subzero soil temperatures persist throughout the duration of snow cover ablation is stressed by Marsh and Woo.[90] Meltwater produced during the day refreezes at night owing to the low ambient air temperatures and heat flow into the ground. This results in the formation of an ice layer at the ground surface. Melt of this ice, following the disappearance of the seasonal snow cover, is a major source of runoff.

Practical estimates of ground heat flow usually are obtained from soil temperatures taken near the surface in the expression for steady-state, one-dimensional heat flow by conduction:

$$Q_g = \lambda_g \left(\frac{\partial T_g}{\partial z} \right) \tag{7.3.11}$$

where λ_g is the thermal conductivity of the soil, T_g is the ground temperature, and z is the vertical directional coordinate. For unfrozen silt and clay soils values for λ_g typically range from 0.4 to 2.1 W/m · °C and for sand from 0.25 to 3.0 W/m · °C, depending on density and moisture content. Because the thermal conductivity of ice is approximately four times that of water, frozen soils have a higher thermal conductivity than unfrozen soils at the same moisture content.

7.3.5 Advective Energy

Rain. A common source of advective energy considered in snowmelt calculations is that supplied by rain. Heat transferred to a snow cover by rain is the difference between its energy content before falling on the surface and its energy content on reaching thermal equilibrium within the pack. Two cases are encountered:

1. Rainfall on a melting pack where the rain does not freeze
2. Rainfall on a pack with a temperature below 0°C where the water freezes and releases its heat of fusion

The first case can be approximated by the expression

$$Q_a = 4.2\,(T_r - T_s)\,P \tag{7.3.12}$$

where Q_a is the energy supplied to the pack by rain, MJ/m², T_r is the temperature of the rain (normally taken as the air temperature or the dew-point temperature, °C, T_s is the temperature of the control volume of snow, °C, and P is the depth of rain, mm.

The second case, where rain falls on snow that is at a temperature below 0°C, is complicated. Depending on the energy content of the pack, only part of the rain may freeze and release its heat of fusion. Usually, because of the lack of measurements, the amount of rain frozen is indeterminate and therefore not accounted for in the energy balance.

Patchy Snow Cover. Hydrologists studying shallow snow covers in exposed environments have long recognized that the maximum snowmelt and runoff rates occur when the land is only partially snow-covered, often less than 60 percent. Differences in rates under partial and complete snow cover in a prairie environment are demonstrated by Erickson et al.[42] and Granger.[58] They show maximum runoff rates from isolated snow patches ranging to 18.6 mm/°C · day, depending on land use, compared with maximum surface melt rates ranging from 8 to 12 mm/day for complete snow covers.

Large areas of bare ground within a snow field significantly alter the energy balance, and local advection of energy from the bare patches becomes increasingly important to melt as the snow cover dwindles. Averaged over a snow season the relative importance of radiative and turbulent energy sources to melt depends on the size of the snow field. Small snow patches are dominated by turbulent melt throughout the season or until they disappear, but the larger snow fields are dominated by radiation melt early in the season and turbulent melt late in the season as they decrease in area. It was estimated that on the prairies over a 6-day period, 44 percent of the energy supplied to an isolated melting snow patch was by sensible heat; during the same period the sensible heat supply to a complete snow cover was 7 percent.[65] A study of the energy balance of late-lying snowdrifts in Idaho reported that 71 percent of the energy available for snowmelt and evaporation came from the sensible heat transfer.[33]

Estimating Turbulent Transfers. A two-dimensional model for calculating the turbulent energy fluxes, which considers the development of the boundary layer beginning at the leading edge of a snowpack, is described by Weisman.[164] This model was applied to study melt from patchy snow cover in an alpine environment of the Front Range, Colorado, using average vertical profiles of temperature, horizontal wind, and humidity monitored over upwind tundra surfaces.[109] Their simulations showed that sensible heat comprised the major source of the total turbulent melt energy, although its relative importance vis-à-vis the latent heat depends upon ambient weather. Near the leading edge of a snow field, advection may contribute more than 347 W/m^2 of melt energy on very windy days and more than 139 W/m^2 on relatively windless days. The totals decrease following a power-law relationship to 57.8 and 23.1 W/m^2 at a distance of 1000 m from the leading edge of snow.

7.3.6 Internal Energy

In shallow snow covers and the upper layers of deep snow, melt produced in the late afternoon often refreezes at night owing to radiative cooling, causing large changes in the internal energy content of a snow cover. The nightly cooling delays the release of meltwater the following day until the snow cover becomes isothermal at 0°C.

The internal energy content U is calculated by the expression

$$U = d_s\,(\rho_i C_{pi} + \rho_\ell C_{p\ell} + \rho_v C_{pv})\,T_m \qquad (7.3.13)$$

where d_s is the depth of snow, ρ is the density ($\rho_i \approx 922$ kg/m^3; $\rho_\ell = 1000$ kg/m^3), C_p is the specific heat ($C_{pi} \approx 2.1$ kJ/kg · °C; $C_{p\ell} \approx 4.2$ kJ/kg · °C), and T_m is the mean snow temperature, and the subscripts i, ℓ, and v refer to ice, liquid, and vapor, respectively. In practical cases, the contribution of the vapor-phase term is assumed to be negligible and is omitted in computing changes in U.

Monitoring the changes in liquid water content is difficult and in the past has been accomplished by systematic sampling using calorimetric techniques which are unsuitable for operational practice. Advances in time-domain reflectometry for measuring the liquid water content in porous media and remote sensing of passive microwaves for estimating both the liquid and ice content of natural snow cover offer potential solutions to this measurement problem.

For deep snowpacks, such as those often encountered in alpine and forested regions, internal energy changes usually are small in relation to the other energy components, and therefore they are frequently neglected.

7.3.7 Temperature Index Methods

Although energy-balance models provide a physical basis for estimating snowmelt, the data required to solve the energy equation are so extensive that it will likely be some time before all physics will be employed in practice. Therefore, operational systems for snowmelt prediction substitute a *temperature-index approach.* The index procedure has the advantages: (1) it often gives melt estimates which are comparable with those determined by the energy balance, and (2) temperature is a reasonably easy variable to measure, extrapolate, and forecast. Air temperature often is a good index of the energy available for melt in a maritime climate and in forests; it is less reliable as an index in open, exposed areas because net short-wave radiation and the sensible and latent fluxes (none of which are related directly to air temperature) may exhibit wide variations in their relative importance to snowmelt. The temperature-index approach is least applicable during extremes, e.g., when net radiation and latent energy are large and air temperature is low. It is also unsuitable for monitoring

TABLE 7.3.7 Temperature-Index Expressions for Calculating Daily Snowmelt (M,mm) in Various Regions of North America

Location	Expression*		Reference
Western Canada mountains	$M = 3.0(T_m + \beta\{[(T_{max} - T_{min})/8] + T_{min}\})$ For $T_{min} \leq 0$ $\beta = 0$ For $T_{min} > 0$ $\beta = T_{min}/4.4$		130
Southern Manitoba–Red River	$M = (0.9 \rightarrow 2.7)T_m$		27
Southern Ontario	$M = (3.66 \rightarrow 5.7)T_m$		19, 81
Montana Rockies, variable forest cover (30–80% crown coverage)	$M^{\dagger} = 4.08T_m$ or $M = 1.10T_{max}$ $M^{\dagger} = 4.58T_m$ or $M = 1.42T_{max}$	(April) (May)	158
Western Cascades, Oregon, heavily forested	$M^{\dagger} = 1.70T_m$ or $M = 0.46T_{max}$ $M^{\dagger} = 3.30T_m$ or $M = 1.42T_{max}$	(April) (May)	158
Sierra Nevada, California, light open forest	$M^{\dagger} = 1.78T_m$ or $M = 0.96T_{max}$ $M^{\dagger} = 1.92T_m$ or $M = 1.14T_{max}$	(April) (May)	158
Eastern Canada forested basin	$M = 1.82(T_m + 2.4)$		129
Boreal forest	$M = 0.58T_m$ (midseason)		
	$M = 1.83(T_m - 3.5)$ (period of major melt)		
Taiga	$M = 0.91(T_m - 2.5)$ (midseason)		
	$M = 1.66T_m$ (period of major melt)		

* T_m = daily mean air temperature, °C
T_{min} = minimum daily air temperature, °C
T_{max} = maximum daily air temperature, °C
† M = basin snowmelt, mm.

diurnal variations in melt because air temperatures usually lag and attenuate short-term variations in net radiation.

The simplest and most common expression relating snowmelt to temperature is

$$M = M_f(T_i - T_b) \tag{7.3.14}$$

where M is the depth of meltwater produced in a selected interval of time, M_f is the melt factor, T_i is the index air temperature, and T_b is the base temperature. The most frequently used values for T_i and T_b are the mean daily temperature and 0°C, respectively. Therefore, the calculation is often referred to as the *degree-day method.* The choice of the index temperature and the base temperature is somewhat arbitrary and empirical. Where there is wide diurnal cycling of air temperature, setting T_i equal to the daily maximum or the mean temperature computed over that portion of the day when the temperature is above 0°C usually will provide the best results. The daily maximum has the advantage that it is an indicator of cloud cover. Examples of temperature-index models used to calculate daily snowmelt from measurements of air temperature in different regions of North America are given in Table 7.3.7.

An empirical expression for the degree-day melt factor is[91]

$$M_f = 0.011\rho_s \tag{7.3.15}$$

in which M_f is in mm per °C · day above 0°C and ρ_s is the snow density in kg/m³. Because the density of melting snow usually falls between 300 and 550 kg/m³, degree-day factors ranging from 3.5 to 6 mm/°C · day are common. Lower values are recommended for fresh snow and for snow cover under a forest canopy. The recommendation is consistent with the findings reported by Ref. 158 which shows mean values for the melt factor in the range from about 0.9 to 1.8 mm/°C · day for forested watersheds. Conversely, Ref. 61 shows melt factors for 6-h periods ranging from 1.5 to 7.0 mm/°C for prairie snow covers.

No single, universally applicable temperature index of snowmelt exists. Each index is unique to a specific basin and geographical location; indices vary with atmospheric conditions, time of year, vegetation and topography, physical properties of a snow cover, and other variables, This has led to attempts to relate M_f to the various factors affecting its magnitude. Most studies involving the temperature-index method show a trend for the melt factor to increase during the melt season. This may be due to several factors, e.g., increasing snow density, increasing solar radiation, decreasing albedo, decreasing internal energy, and other changes to the energy balance of a snow cover. Procedures used to account for this trend include increasing the melt factor and/or changing the base temperature. Figure 7.3.5 shows the seasonal variation in melt factors during nonrain periods for the contiguous United States and Alaska suggested by Anderson.[5] The effect of seasonal changes in internal energy and other factors on melt of shallow snow covers at northern latitudes is indexed by varying T_b;[93] in mid-January, $T_b > 4$°C; in mid-April, $T_b = 0$°C. When snow cover is incomplete it is necessary to include a measure of snow cover extent in the calculations of melt because less meltwater is produced on a basin than if 100 percent snow cover is assumed.[42,92] Martinec and Rango[92] warn against compensating a meltwater difference that arises from erroneous snow cover information by "optimizing" the melt factor.

The above discussion relates to snowmelt under rain-free conditions. A general equation for snowmelt during rain in a heavily forested area is given as[158]

$$M = (0.74 + 0.007P)(T_a - T_b) \tag{7.3.16}$$

in which M, the daily melt, and P, the 24-h precipitation, are in mm.

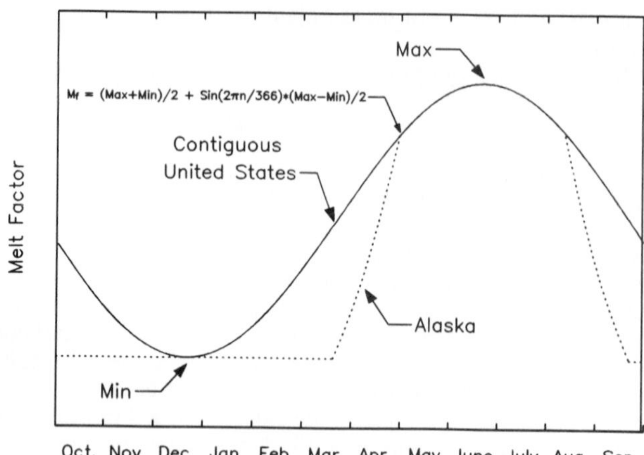

FIGURE 7.3.5 Seasonal variation in the melt factor during rainfree periods for the contiguous United States and Alaska. *(Source: Ref. 5.)*

7.4 MELTWATER MOVEMENT THROUGH SNOW

Movement of meltwater through snow governs the time between the occurrence of melt at the snow surface and the arrival of meltwater at the base of a snow cover. It is only within the past two decades that significant advances have been made in quantifying this process, although the similarity of water flow in "ripe" snow and through soil has been recognized for some time.[52] The major difference between snow and soil, as media for conducting water, is that following snowfall the size and shape of snow grains change continuously with time. Meltwater accelerates these changes, resulting in the formation of large, round grains. Refreezing of meltwater also may lead to the formation of relatively impermeable ice layers.

When meltwater moves into a homogeneous snow cover two zones are formed: (1) an upper layer of wet snow that is isothermal at 0°C and (2) a lower layer with a temperature below 0°C.[171] The rate at which a wetting front moves downward is controlled by refreezing of water onto grains to release sufficient heat of fusion to raise the temperature of the snow to 0°C and to increase the water content. By comparing the movement of wetting fronts in dry snow at −5°C and in wet snow at 0°C, Colbeck[29] showed that the meltwater needed to raise the temperature of the dry snow to 0°C and to fill the liquid storage capacity increases the time between the introduction of water at the surface and the initiation of runoff at the base and decreases the amount of runoff.

A wetting front does not advance as a uniform wave. Vertical leaders (flow fingers) usually develop which concentrate water that moves at a higher velocity than meltwater in the adjacent surrounding snow. These fingers are attributed to unstable flow. Instabilities in the flow regime usually are associated with inhomogeneities in the transmission properties of snow and tend to originate along horizontal boundaries. Meltwater channels may extend through numerous stratigraphic layers.

7.4.1 Velocity of Meltwater Movement

In free draining snow covers, where the effects of pressure (suction) forces on the vertical movement of meltwater are small in comparison with gravity, the average velocity of flow through snow can be described by[28]

$$V = \frac{\rho_w g}{\mu_w} k_s S^{*\epsilon} \tag{7.4.1}$$

where ρ_w is the density of water, g is the acceleration due to gravity, μ_w is the dynamic viscosity of water, k_s is the permeability of the snow matrix, S^* is the effective water saturation, and ϵ is the exponent. S^* is calculated by the expression

$$S^* = \frac{S_w - S_{wi}}{1 - S_{wi}} \tag{7.4.2}$$

in which S_w is the water saturation and S_{wi} is the irreducible water saturation, water that is permanently retained by surface (capillary) forces. Both terms are expressed as a fraction of the total pore volume.

An expression for k_s (m^2) is[144]

$$k_s = 0.077 d^2 \exp(-0.0078\, \rho_s) \tag{7.4.3}$$

in which d is the average grain diameter, m, and ρ_s is the snow density, kg/m^3. Equation (7.4.3) shows permeability increasing with increasing grain size and decreasing with increasing snow density.

Values for the exponent ϵ of Eq. (7.4.1) for snow range from 1.5 to 4.5. Its magnitude depends on grain size, density, stage of metamorphism, and other factors; however, since no clear relationship exists with any snow parameter, a value of $\epsilon = 3$ is usually adopted.

Equation (7.4.1) has been used in the continuity equation to develop an expression for the vertical rate of propagation of a meltwater flux through a relatively wet homogeneous snow cover.[30] During melting of a natural snow cover the melt rate may vary widely with time. This leads to situations where fast-moving fluxes overtake slower-moving fluxes and the development of a shock front. The rate of propagation of a shock front is described by Colbeck.[30]

A limitation of Eq. (7.4.1) for describing water movement through heterogeneous snow is the requirement for representative values of the permeability and the effective porosity of snow. As pointed out earlier, ice layers may cause ponding, enhance lateral flow, divert water downslope, and concentrate water on small areas or to major flow paths. Variations in flow may be modeled by extending the *gravity-flow theory* to describe simultaneous flows along different paths, i.e., through ice layers and through flow fingers.[31]

7.5 SNOWMELT INFILTRATION INTO FROZEN SOILS

In most northern regions, the single most important factor affecting the apportionment of snow water between direct runoff and soil water is the infiltrability of frozen ground. Snowmelt infiltration into a frozen soil is influenced by the thermal and hydrophysical properties of the soil, the soil temperature and moisture regimes, and the quantity and rate of release of water from the snowpack. If the depth of frost is small, the energy exchanged between the soil and meltwater may thaw the entire frozen profile, returning the infiltration characteristics of the soil to those of the unfrozen state. Studies in the former Soviet Union[75] suggest that a soil frozen to a depth less than 150 mm behaves the same as an unfrozen soil; once a soil is frozen to 600 mm, freezing to greater depth has no further effect on infiltration.

7.5.1 Modeling Frozen Ground Effects—Indirect Measures

Modification (or Calibration) of Soil Parameters. Most operational snowmelt runoff models use a soil moisture accounting routine to distribute infiltrating meltwater into some or all of the following components: upper zone storage, percolation, interflow, and groundwater recharge. Soil parameters and/or transfer coefficients are used to control the quantities of water distributed to each component. The parameters are varied to allow calibration of a model for unfrozen and frozen conditions.

Frost Index. A *frost index* is similar to a *degree-day factor* (see Sec. 7.3.7); it has units of temperature and is limited to temperatures below the freezing point. It is computed continuously from air temperature data and is affected by snow cover (insulating effect), the daily thaw rate from ground heat (thawing from below), and meltwater entering the soil. These models are used to identify the *state* of the ground. If a soil is frozen, the effects on infiltration are compensated by modifying parameters of the soil moisture accounting routine, runoff coefficients, or infiltration indices.

An example is described by Molnau and Bissell.[105] They use a continuous frozen ground index (CFGI) in which CFGI_i, the index on the *i*th day, is expressed as a function of the index on the preceding day, CFGI_{i-1}, the mean daily air temperature T_m and snow depth d_s as

$$\mathrm{CFGI}_i = A\,(\mathrm{CFGI}_{i-1}) - T_m e^{-0.4Kd_s} \tag{7.5.1}$$

in which A is a daily decay coefficient, K is a snow depth reduction coefficient, cm^{-1}, and d_s is the depth of snow on the ground, cm. With $A = 0.97$ and $K = 0.57$ cm^{-1}, it is found that CFGI values in the range of 56 to 83°C-days define the limits of the transition between unfrozen and frozen conditions in the Pacific northwestern region of the United States.

Depth of Frost. There are many temperature-driven models for predicting the depth of frost. An expression for frost depth z_f which includes the phase change and the specific heat of the soil is[43,157]

$$z_f = \frac{86{,}400\,\lambda_g F}{L_w + C_v\left(T_m + \dfrac{F}{2t}\right)} \tag{7.5.2}$$

where z_f = depth of frost, cm
λ_g = thermal conductivity of the soil, J/cm · s · K
F = frost index, number of degree-days the daily mean air temperature is above a base temperature, °C · days
L_w = latent heat of fusion of water in a unit volume of soil, J/cm
C_v = volumetric heat capacity of the soil matrix, J/cm^3 · K
T_m = mean annual temperature, °C
t = duration of the freezing period, days

Equation (7.5.2) does not account for the effects of heat flow from the unfrozen zone or migration of water on the advance of the freezing front. A practical equation for frost penetration which takes into account the exchange of heat by conduction is given by Ref. 55.

7.5.2 Completely Frozen Soils

In completely frozen soils meltwater always infiltrates against a freezing temperature. The effects of soil temperature at the time of melt, however, may be secondary to those of effective porosity (air-filled pores), which are inversely related to ice content.[62,148] Entry to and movement of water in frozen soils mainly occurs through large pores.

Following the investigations by Popov[120] concerned with long-range forecasting of spring runoff for lowland rivers in the former Soviet Union, Gray et al.[66] grouped completely frozen soils into the following infiltration classes:

Restricted: Infiltration is impeded by an ice lens on the surface or at shallow depth. The amount of meltwater infiltrating the soil is negligible, and most of the snow water goes to direct runoff and evaporation.

Limited: Infiltration is governed primarily by the snow cover water equivalent and the frozen water content of the 0- to 300-mm soil layer.

Unlimited: Soils containing many large, air-filled macropores which are capable of infiltrating most or all of the meltwater.

Figure 7.5.1 plots infiltration against snow water equivalent in which the three classes are shown. The 1 : 1 line represents the *unlimited* class where all the snow

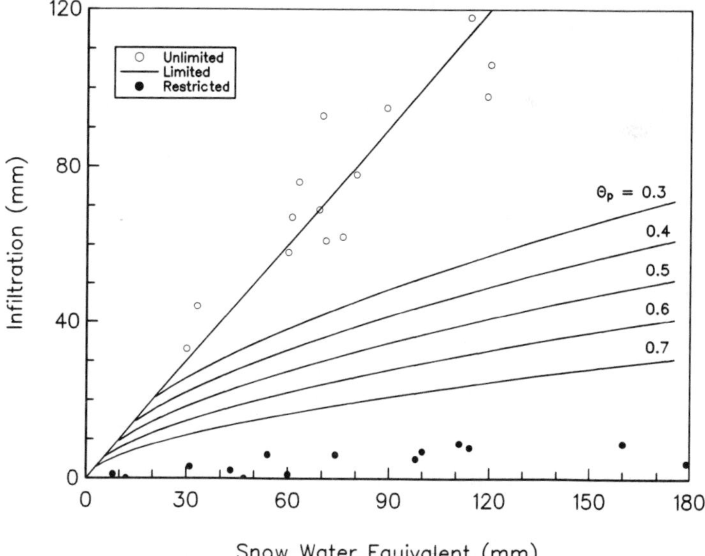

FIGURE 7.5.1 Relation between snowmelt infiltration INF, snow cover water equivalent SWE, and premelt moisture content of the 0- to 300-mm soil layer θ_p, for completely frozen soils of unlimited, limited, and restricted classes. *(Source: Ref. 60.)*

water infiltrates. Infiltration exceeds available snow water because of overland flow to and interflow within large soil cracks. The data points on and near the abscissa represent the *restricted* class. The family of curves, representing the *limited* class, are described by the following relationship between snowmelt infiltration INF, snow cover water equivalent SWE, and the premelt moisture content of the 0- to 300-mm soil layer θ_p:

$$\text{INF} = 5\,(1 - \theta_p)\,\text{SWE}^{0.584} \tag{7.5.3}$$

in which INF and SWE are in mm and θ_p is the degree of pore saturation, mm³/mm³.

An analogous expression for infiltration into severely cracked, heavy lacustrine clay soils which fall in the unlimited class (INFC in mm) is

$$\text{INFC} = 4.53\,\text{SWE}^{0.553} + 2.33\,(1 - \theta_p)\,\text{SWE}^{0.584} \tag{7.5.4}$$

Equation (7.5.4) was developed from measurements in soils in which the cracks had a mean length : area ratio at the surface of 1.75 m/m².

7.6 RUNOFF FROM SNOWMELT

Many of the snow processes described in the preceding systems have been incorporated into operational systems for forecasting catchment runoff. Among the better-known systems are the U.S. National Weather Service River Forecasting System—Snow Accumulation and Ablation Model,[3,161] the U.S. Army Corps of Engineers Streamflow Synthesis and Reservoir Regulation model,[159] and the HVB model of the Swedish Hydrological and Meteorological Institute.[16]

The manner in which the various models available for snowmelt simulate the processes varies widely in complexity, from single-variable indices of melt to complete energy balances. Each model is most reliable in the area where it was developed and may require extensive calibration when applied elsewhere. In 1986, the results of a comprehensive study undertaken by the World Meteorological Organization compared the performance of eleven models using data sets from six different regions.[168] This numerical assessment provides useful information on effects of basin size, relative elevation range, percentage forest cover, and other factors on model performance.

For the most part, forecasting systems have focused only on streamflow prediction for open-water conditions; operational flow models that consider the effects of ice are scarce (e.g., HEC-2 with ice option[160]). Moreover, such models are unable to simulate accurately the complex mix of processes that control the transitional periods of freeze-up and breakup. Research is being conducted in a host of cold-regions countries on these subjects, and the information is gradually being incorporated into new-generation hydraulic-flow models (e.g., RIVICE,[115] the Finnish Ice Research Project[72]). Comprehensive, predictive modeling of runoff processes in cold regions ultimately requires a linking of landscape snowmelt models and channel-ice hydraulic models. The following sections detail the unique hydrologic regime under ice-affected conditions.

7.7 ICE FORMATION AND FREEZE-UP PROCESSES

7.7.1 Cooling Water Surface

Where little flow exists, as in lakes or in the quiescent conditions of large head ponds with minimum throughput, the water column experiences a full circulation or overturn when the surface water cools to the temperature of maximum density at 4°C. This dense 4°C water "sinks" and is replaced at the surface by less-dense water from underneath. Without further mixing, the surface water will cool to the freezing point and develop an ice cover.

In most rivers, flow is turbulent and the associated vertical mixing usually precludes density stratification related to cooling. With effective vertical mixing, the entire water column cools at approximately the same rate. The heat fluxes controlling the water temperature are

$$Q^* = Q_{sn} + Q_{ln} + Q_h + Q_e + Q_p + Q_{gw} + Q_b + Q_f \tag{7.7.1}$$

where Q^* is the net heat flux to the water column, Q_{sn} is the net short-wave radiation heat flux, Q_{ln} is the net long-wave radiation heat flux, Q_h is the sensible heat flux from air, Q_e is the latent heat flux from water vapor, Q_p is the precipitation heat flux, Q_{gw} is the groundwater heat flux, Q_b is the combined geothermal and sediment heat flux, and Q_f is the heat from fluid friction (positive values represent a heat flux to the water). Water temperatures of the upper layer of a stably stratified lake can be calculated using the same equation but modified by deleting Q_f, adding a term to account for heat conduction from the underlying water, and by combining $Q_b + Q_{gw}$ for shallow lakes.

Q_{sn} frequently is the largest term in Eq. (7.7.1), but its role is often complicated by the presence of fog which develops over the water surface when the air is much cooler than the water. Cooling is primarily driven by negative values of the other three large heat fluxes Q_{ln}, Q_h, and Q_e which are respectively associated with a negative long-wave radiation balance, convection of heat away from the surface, and evaporation. The precipitation heat flux can be either negative or positive depending on the relative temperatures of the water and precipitation, and whether the latter is solid or liquid. Snow falling into the water produces the largest heat loss, not only because the snow is at or below freezing but because of the latent heat consumed in melting it.

The contributions of groundwater heat are highly variable and difficult to determine but can be significant, especially where the winter base flow is primarily derived from groundwater. Friction, sediment, and geothermal heat fluxes are small relative to other energy components and usually can be ignored in an open-water energy budget. Friction heat is most important in steep streams, especially where the open-water surface area is reduced by the growth of ice from shore. Similarly, once an ice cover has fully formed and reduced atmospheric exchanges, the heat flux from bottom sediments can become a relatively important heat source, especially in shallow lakes. Although these terms are usually less than approximately 4 W/m^2 and the resulting water heating is in the order of hundredths of a degree, the resulting heat flux in fast-flowing sections can keep small areas clear of ice.

7.7.2 0°C Conditions on Rivers

The larger the surface area to volume ratio, the more quickly a river exchanges heat with the atmosphere. Hence, given the same meteorological conditions, wide shallow streams cool most rapidly while deep rivers are usually the last to freeze. When the water reaches 0°C, additional cooling forms ice. For a river cooling in a downstream direction, the location at which the water is cooled to 0°C (0°isotherm) can be obtained from[8]

$$l_{T_w=0} - l_{T_{w,o}} = -\frac{\rho_w C_p V h_o}{C_o} \ln\left(\frac{-T_a}{T_{w,o} - T_a}\right) \tag{7.7.2}$$

where $l_{T_w=0}$ is the downstream location, m, at which the water temperature reaches 0°C; $l_{T_{w,o}}$ is the upstream location, m, of the reach with an initial water temperature of $T_{w,o}$, °C; ρ_w is the density of water, kg/m³; C_p is the specific heat of water, J/kg · °C; V is the mean flow velocity, m/s; h_o is the open-water flow depth, m; and T_a is the air temperature, °C. C_o is a heat-transfer coefficient (W/m² · °C) ideally reflecting all the major heat fluxes of Eq. (7.7.1). For example, average winter values of $C_o = 22$ to 28 W/m² · °C apply to reaches of the St. Lawrence River,[87,142] although this rate can more than double during periods of extreme cold.[87] Being a site-specific term, it should be calculated from energy-balance analyses for any new locations based on equations similar to those outlined for atmospheric exchanges over snow (Sec. 7.3) and specifically reviewed for cooling open water (e.g., Refs. 37 and 98).

7.7.3 Ice Formation

On a river or lake, the first ice usually forms along the margins or banks. Once bank material is cooled below its threshold temperature for nucleation, ice crystals form, and as long as flow and temperature conditions permit, ice continues to spread over the adjacent water surface. Conditions are most suitable for lateral ice growth where the flow is laminar, permitting early supercooling of the surface.

In turbulent sections of a river, water cooled at the surface is constantly mixed with the underlying flow. Before the supercooling necessary for nucleation can take place, the entire cross section has to reach 0°C. Then, with a slight degree of further cooling (usually not exceeding −0.1°C), small ice particles, referred to as *frazil,* begin to develop. Initial frazil particles are usually discoid but can assume a spicule shape in less turbulent water. Although their dimensions are small, typically < 1 mm in diameter and 0.05 to 0.5 mm thick, their concentration has been reported to be as high as 10^6/m³.[53] Such high concentrations have been suggested to affect the "equivalent viscosity" or sluggishness of river water and thereby the conveyance capability of a channel.[156]

Large quantities of flowing ice are a problem for facilities such as hydropower plants, especially if the frazil is in an active state (supercooled) and capable of adhering to objects. To avoid this, the common practice is to encourage the formation of an ice cover over the frazil-producing zone, normally by placing a boom across the river against which frazil will accumulate, or by slowing the water velocity to create quiescent conditions suitable for static ice growth.

While the water is supercooled, frazil crystals continue to grow and also adhere to and coat objects within the flow or on the bed, such as vegetation, boulders, and even areas of gravel and coarse sand. Given suitable hydrometeorological conditions, anchor ice can form blankets of ice on river bottoms. As long as conditions remain cold, anchor ice will remain firmly bonded to the riverbed, releasing only when its buoyancy is sufficient to lift the underlying bed material. Once the river begins to warm slightly, or when solar radiation penetrates to the bed, the bond between the riverbed and the overlying anchor ice disintegrates and the ice releases.

7.7.4 Ice Production

Given an abundant supply of freezing nuclei and cold weather conditions, the total volume of ice (both frazil and other forms) which can be produced in an open area A_o is estimated by integrating a simplified expression of the surface heat loss $C_o(T_w - T_a)$ over time:

$$V_f = \frac{A_o}{\rho_i L_i} \int_{t_1}^{t_2} C_o(T_w - T_a)dt \tag{7.7.3}$$

where V_f is the volumetric production of ice, m^3/s, A_o is the area of the open water reach, m^2, ρ_i is the density of ice, kg/m^3, L_i is the latent heat of fusion for ice, J/kg, T_w is mean water temperature, °C, and t is time, s.

Because the supercooling that controls frazil formation in rivers is not great, even slight changes to the thermal regime of a river, such as the frictional heat produced at the foot of rapids, can appreciably affect the production of frazil ice. In general, the hydrometeorological conditions most favorable for frazil production occur on cold clear nights and include a large heat loss by long-wave radiation to the atmosphere, and a strong wind accompanied by cold, dry air to produce large convective and evaporative heat losses. Under very cold winter conditions, frazil may be generated during the day, but under less severe circumstances, it will follow a diurnal cycle characterized by rapid growth at night and a cessation during the day, owing primarily to warming by solar radiation.

7.7.5 Floe Accumulation

Once formed, frazil crystals continue to grow and agglomerate first as small clusters and then as larger flocs (5 to 100 mm). Initially, the movement of frazil flocs is largely controlled by water turbulence. As the agglomerations continue to grow, they respond less to turbulence, remaining on the surface for longer periods to form *frazil slush, frazil pans, pancake ice,* and eventually large *frazil floes* (Fig. 7.7.1*a*). Flowing frazil can affect the growth of ice from the shore, by accumulating along or abrading the outer edge of previously formed *border ice.* A flow velocity of approximately 1.2 m/s is near the maximum for adherence of frazil.[101] In general, border-ice development by static ice growth and slush adhesion is a function of slush-ice concentration, heat flux from the water, and water velocity at the ice-water margin.

Under relatively constant meteorological conditions, the position of the 0°C isotherm may approximate that of the freeze-up ice edge. However, the formation of a stationary ice sheet requires first either that border ice completely covers the river or that the flowing ice is halted. In most situations, moving ice continues to grow and agglomerate in the flow until the surface concentration becomes sufficient for a bridge or arch to form across the channel (Fig. 7.7.1*b*). There is no reliable theory to predict bridging locations, although the susceptibility of a section to bridging is related to its transport capacity and to the characteristics of the incoming flow of ice. Bridging is common where border ice growth constricts the downstream passage of

FIGURE 7.7.1 *(a)* Pancake ice and ridges of border ice, North Saskatchewan River, Canada. *(Courtesy of R. Gerard.)* *(b)* Bridging of freeze-up accumulation between border ice, North Saskatchewan River, Canada. *(Courtesy of R. Gerard.)*

floating frazil pans and slush, or in low-velocity sections upstream of deeper water with a complete cover. In the latter case, incoming frazil is entrained underneath, or begins to accumulate against, the ice edge. Where the ice is not submerged, simple juxtaposition of floes can cause a cover (Fig. 7.7.1*b*) to rapidly progress upstream at rates as high as tens of kilometers per day.

7.7.6 Ice Cover Progression

Numerous laboratory studies have assessed the susceptibility of ice blocks to submerge or underturn as they arrive at an ice edge. However, since most of these studies used rectangular blocks, the resulting formulas can provide only rough approximations for the irregularly shaped porous slush which typifies the ice at freeze-up. For most practical applications, simple velocity criteria are used instead. In general, the entrainment of frazil flocs beneath a cover requires velocities in excess of 0.6 to 0.7 m/s,[7,147] while velocities as high as 2.0 m/s are required for the entrainment of well-developed ice pans.[98]

The velocity required to carry ice beneath the cover after its initial submergence is dependent on a number of parameters including the floe shape and porosity and the roughness of the bed and cover bottom. Although field data are scarce, critical transport velocities of 0.6 to 1.3 m/s are suggested.[100]

As the ice continues downstream, it will melt, reemerge in open-water sections, or deposit beneath the cover at a section where the transport capacity is reduced, thereby forming a *hanging dam*. Such an accumulation continues to thicken until the ice supply is depleted or the constriction it creates causes a resumption of ice transport.

7.7.7 Ice Jams—Hydraulic Control

Where ice submergence takes place, but the ice is deposited immediately downstream of the evolving ice edge, a cover will develop and progress upstream with a thickness given by[98,111]

$$\mathrm{Fr} = \frac{V}{\sqrt{gh_o}} = \sqrt{2\left(1 - \frac{\rho_i}{\rho_w}\right)(1 - p_j)\frac{t_j}{h_o}\left(1 - \frac{t_j}{h_o}\right)} \tag{7.7.4}$$

where Fr is the Froude number, V is the mean flow velocity, h_o is the open-water flow depth, ρ_i is the density of ice, ρ_w is the density of water, t_j is the accumulation thickness, g is the acceleration of gravity, and p_j is the porosity of the accumulation.

As the cover thickens, higher velocities are required to submerge floes. At a cover thickness to flow depth ratio of 0.33 the flow restriction becomes critical and Eq. (7.7.4) no longer applies. Inserting $t_j/h_o = 1/3$ into Eq. (7.7.4) and assuming that $\rho_i/\rho_w = 0.92$, a maximum critical Froude number (Fr_c) for this case is defined by

$$\mathrm{Fr}_c = 0.154\sqrt{1 - p_j} \tag{7.7.5}$$

from which the importance of ice porosity on river ice accumulation becomes apparent. Although Fr_c ranges from 0.06 to 0.12, a value of 0.08 is often used in the study of freeze-up accumulations.[99] The latter value translates into a porosity of approximately 0.73, indicating that freeze-up jams can be quite porous. Breakup jams comprised of solid ice floes are assumed to have a porosity of only 0.4 (e.g., Ref. 123).

7.7.8 Ice Jams—Mechanical Control

The preceding discussion assumes that the internal resistance of the accumulation withstands the increases in downstream forces created by the additional fluid shear stress and the gravitational-weight component as the cover lengthens upstream. To remain intact and stable, the internal strength of the accumulation must be sufficient to transfer these forces to the banks; otherwise it will compact to a larger thickness that provides sufficient internal strength. Such conditions also govern the behavior of breakup ice jams.

Figure 7.7.2 illustrates a typical ice jam where a cover has developed upstream of an obstruction. Maximum thickness is reached at the *toe* and water depths rise upstream in response to the increased flow resistance. Upstream of the *head,* water depths decrease until beyond the effect of any ice-related flow resistance. If the jam is long enough, an *equilibrium section* develops characterized by maximum water levels and uniform ice thickness and flow conditions. Example surface profiles are shown in Fig. 7.7.2*c* compared with that for comparable open-water flow.

Since the precise location of ice jams is difficult to predict, a conservative approach is to estimate potential flood hazard assuming the development of an equilibrium section within the reach of interest. A force-balance equation for the equilibrium reach (modified from the original by Pariset and Hausser[111]) can be written as

$$\mu\rho_i\left(1-\frac{\rho_i}{\rho_w}\right)gt_j^2+2ct_j=W_j\left(\rho_w gR_iS+\rho_i gt_jS\right) \qquad (7.7.6)$$

where μ is a dimensionless coefficient reflecting the internal friction and porosity of the accumulation and suggested to have an average value of 1.2,[12] t_j is the jam thickness, m, c is jam cohesion, W_j is the jam width, R_i is the hydraulic radius of the ice-controlled portion of the flow, S is slope, and the remaining terms are defined following Eq. (7.7.4). Downstream forces on the jam are represented by the two right-hand terms: the average shear stress on the base of the ice cover and the streamwise component of the weight of the ice jam per unit area, respectively. The two left-hand terms represent the frictional and cohesive resistance forces, respectively. Little is known about the cohesive strength of jams formed under cold conditions where interparticle freezing may occur, but it is commonly considered to be negligible under spring breakup conditions.

Solving Eq. (7.7.6) (after Ref. 12) for the equilibrium thickness (assuming $\rho_i/\rho_w = 0.92$ and no cohesion) yields

$$t_j = 6.25\,\frac{W_jS}{\mu}\left[1+\sqrt{0.174\mu\left(\frac{n_i}{n_o}\right)^{3/2}\frac{h_j}{W_jS}}\right] \qquad (7.7.7)$$

where h_j is the average flow depth, m, beneath the section given by

$$h_j = \left(\frac{1.59n_oQ}{S^{1/2}W_j}\right)^{3/5} \qquad (7.7.8)$$

In Eq. (7.7.8), Q is discharge and n_o is the composite Manning roughness, which can be obtained from the Beloken-Sabaneev formula

$$n_o = \left(\frac{n_i^{3/2}+n_b^{3/2}}{2}\right)^{2/3} \qquad (7.7.9)$$

where n_i and n_b are Manning's roughness coefficients for the ice and bed, respectively.

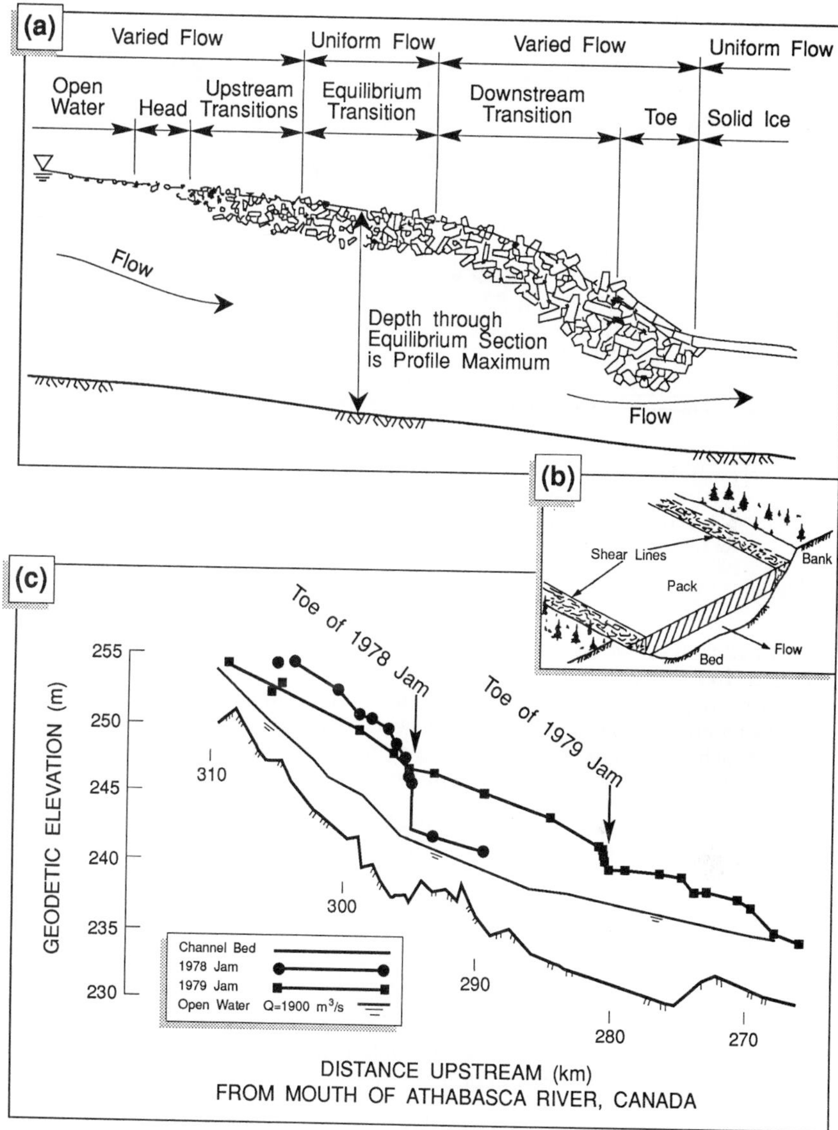

FIGURE 7.7.2 *(a, b)* Typical ice jam configuration. *(c)* Measured ice jam profiles. *(Adapted from Refs. 6 and 49 with permission.)*

The total water depth H_j, including the submerged portion of the pack thickness, then becomes

$$H_j = h_j + 0.92t_j \tag{7.7.10}$$

To derive estimates of peak water levels from the equilibrium jam equations requires some knowledge of the flow-resistance terms. Although n_i is known to range from

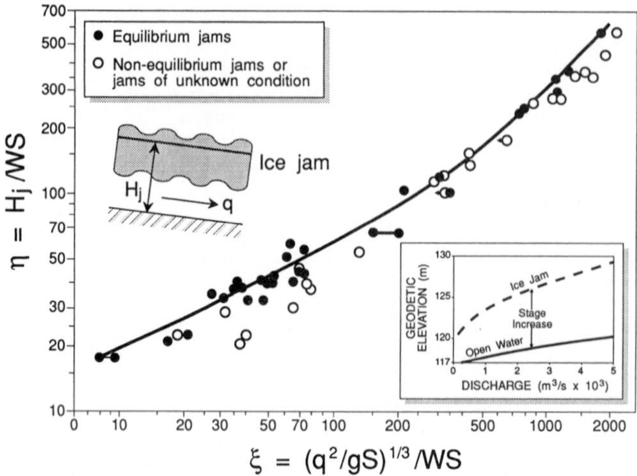

FIGURE 7.7.3 Dimensionless rating curve for the equilibrium reach in a fully developed ice jam *(Ref. 12, modified by Ref. 163)* and actual rating curve for Liard River, NWT, Canada.[122] *(Used with permission.)*

approximately 0.01 for a 0.1-m-thick slush layer to 0.1 for a 3-m-thick layer of ice floes (e.g., Ref. 106), current knowledge does not permit assignment of an accurate value of n_i to a particular jam. Based on some assumptions pertaining to resistance, a simplified method can be used to estimate water depths for the equilibrium reach.[12] The approach also assumes a reasonably prismatic channel and that width W does not vary with stage. A dimensionless discharge ξ is first computed from

$$\xi = \frac{(q^2/gS)^{1/3}}{WS} \tag{7.7.11}$$

where q is discharge per unit width, m²/s. Then a value for the dimensionless water depth η is obtained from Fig. 7.7.3 and an actual water depth from

$$H_j = \eta WS \tag{7.7.12}$$

Despite the reach geometry and resistance assumptions, the relationship has been found to describe satisfactorily most available field data. A sample ice-jam stage-discharge curve is shown in the inset of Fig. 7.7.3.

7.8 WINTER HEAT BUDGET OF ICE COVERS

7.8.1 Ice Growth

Once a surface ice cover becomes established, it thickens as the freezing front migrates vertically downward into the water column. The resultant ice is usually highly transparent with a columnar grain structure and is referred to as *blue* or *black* ice (Fig. 7.8.1*a*). At the base of the ice sheet, the rate of ice growth is determined by the

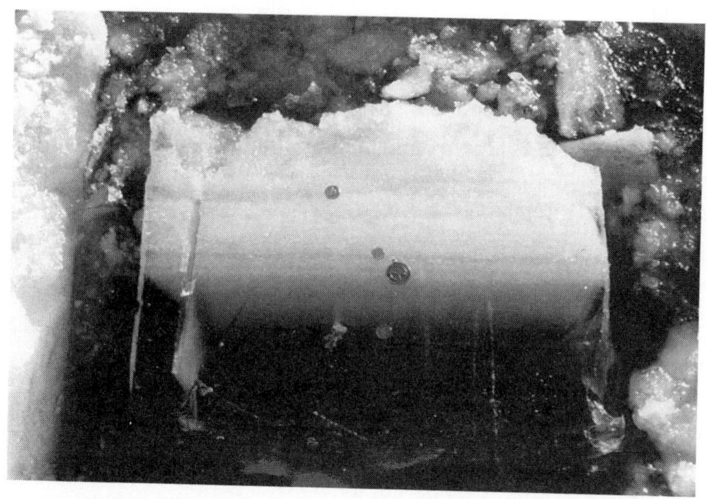

FIGURE 7.8.1 *(a)* Typical ice cover showing transparent black ice at the base overlain by white or snow ice. *(Photograph by T. D. Prowse.) (b)* Columnar ice cover showing effects of candling from radiation decay. *(Photograph by M. Demuth.)*

difference between heat exchanges at the ice surface and heat supplied to the base of the ice cover by the water:

$$\frac{dt_i}{dt}\rho_i L_i = \left(\frac{T_m - T_a}{\frac{t_i}{\lambda_i} + \frac{1}{C_a}}\right) - C_w(T_w - T_m) \quad (7.8.1)$$

where t_i and λ_i are the thickness, m, and thermal conductivity of ice, W/m · °C, and T_m is the basal ice temperature (commonly assumed to be 0°C), T_w is water temperature, T_a is air temperature, ρ_i is ice density, and L_i is the latent heat of fusion of ice (J/kg). C_a and C_w are the heat-transfer coefficients from ice to air and water to ice, respectively, W/m² · °C. Although typical values of C_a[9] approximate those given for C_o in Sec. 7.7.2, C_a should ideally be calculated from energy budget analyses of cold ice (e.g., Ref. 141) or snow as in Sec. 7.3. In the case of lakes and for periods of ice growth on rivers, the subsurface heat flux is ignored. Methods for calculation of C_w are reviewed by Ref. 89.

From Eq. (7.8.1) a simplified approach based on a degree-day function is

$$t_i = \alpha\sqrt{D_f} \quad (7.8.2)$$

where D_f is accumulated degree-days below freezing and α is a coefficient varied to account for conditions of exposure, surface insulation, and subsurface heat flux (Table 7.8.1). The index D_f is computed by summing the absolute values of daily mean air temperature when this temperature is below 0°C. When the temperature is above 0°C the value is not included in the cumulative degree-day index. Note, however, that when t_i is small (i.e., <10 cm), it increases in direct proportion to D_f rather than its square root.[9]

7.8.2 Snow-Cover and Frazil-Ice Effects

A surface snow cover, especially of newly fallen low-density snow, retards the growth of static ice by slowing the heat loss to the atmosphere. The insulating effect of snow is apparent in the lower values of α in Table 7.8.1 when snow is present and can be accounted for in Eq. (7.8.1) by adding an additional term t_s/λ_s to the denominator of the first term, where t_s and λ_s are the thickness, m, and thermal conductivity of snow, W/m · °C. Snow, however, also promotes the growth of superimposed ice, usually

TABLE 7.8.1 Coefficient α Used in Eq. (7.8.2) for Ice Growth under Varying Conditions of Exposure and Surface Insulation

	α, mm °C$^{1/2}$ · day$^{1/2}$
Theoretical maximum	34
Windy lakes with no snow	27
Average lake with snow	17–24
Average river with snow	14–17
Sheltered small river with rapid flow	7–14

Source: After Ref. 98.

referred to as *snow ice* or *white ice*. Typically, snow accumulating on the black-ice layer results in a gradual depression of the ice sheet and a positive (relative to the top of the ice sheet) hydrostatic water level. Water flows through fractures and cracks in the ice sheet to the new elevated level, flooding the lower layers of the snow cover. This water-ice mixture refreezes downward, forming a relatively opaque layer characterized by grains much smaller and less organized than those in black ice (Fig. 7.8.1*a*).

On lakes, snow ice formation is commonly related to the distribution of snow loads following a pattern of near-zero thickness in windswept central areas of a lake to maxima near the edges and in downwind accumulation zones. Insulation provided by the snow and snow ice retards the growth of black ice in these zones, whereas their absence leads to unhampered growth of black ice in the central portions of a lake. The net effect over a winter period is an equalization of ice thickness in different parts of a lake.[1] Similar processes can occur on large windswept rivers, but because of the added effect of water-level variations due to flow conditions, snow ice also can be promoted by water overflow, especially on regulated rivers with large variations in stage.

In rivers, growth rates at the base of the ice can also be affected by frazil deposits. As for snow ice, vertical migration of the freezing front into an underlying mixture of frazil and water is more rapid than into water alone, because a portion of the stratum already exists in the solid state. To account for frazil or snow ice in calculating ice growth, a correction should be made to the latent heat of fusion term L_i in Eq. (7.8.1) for such ice-water mixtures (e.g., Ref. 22).

7.8.3 Icings

Another form of surficial ice, termed *icings, aufeis,* or *naleds,*[24] develops from local spring or groundwater or when the underlying flow encounters some form of obstruction or resistance and is forced up through cracks and holes onto the ice surface. Freezing of the water may be rapid without the insulation offered by the underlying snow and ice cover. Common icing locations include shallow braided streams and culvert and bridge openings where icing can cause major restrictions to spring snowmelt.

Where this process continues through the winter, streams can develop ice several times the maximum thickness of thermal ice covers or even the normal depth of summer flow. Knowledge of icing volumes and period of formation can be used to calculate average groundwater discharge rates.[162]

7.9 WINTER FLOW CONDITIONS

7.9.1 Winter Storage

An ice cover decreases the flow conveyance of a channel by reducing cross-sectional area and increasing resistance because of an enlarged wetted perimeter. Hence upstream water levels or backwater increase as compared with open-water conditions. For uniform flow, the ratio of ice to open-water stage for the same discharge can be approximated by (assuming the bed roughness n_b is the same for ice and open-water conditions)

$$\frac{H_i}{h_o} = \left[1 + \left(\frac{n_i}{n_b}\right)^{3/2}\right]^{2/5} + 0.92\,\frac{t_i}{h_o} \tag{7.9.1}$$

where H_i and h_o are the stage, m, for ice and open-water conditions, respectively, n_i is the ice cover roughness, and t_i is the ice-cover thickness, m. With similar ice cover and bed roughness, for example, the stage is approximately 30 percent higher under ice than open-water conditions. The greatest increases, however, are produced by the rough accumulations characterizing ice jams. n_i for such ice jams can easily be $\sim 3n_b$ or larger. Assuming $t_i \sim h_o$, such high resistance will produce an approximate threefold increase in stage.

The rise in upstream water levels due to ice-cover resistance means that water is put into storage (Fig. 7.9.1) and will not be released to the flow downstream until a reduction in resistance occurs. With ice growth during the winter, further reductions in cross-sectional area and increased flow resistance will result. Fluctuations in storage can also result from changes in ice roughness. In general, ice resistance tends to be high during the freeze-up period, decreases over the winter, and then increases again with the beginning of melt. Because of ice growth, frazil deposition in concave portions of the ice bottom, and thermal smoothing, ice cover roughness n_i decreases during the main winter period with typical early to midwinter values in the order of 0.008 to 0.012 for smooth subsurfaces. Notably, higher values can prevail in the presence of significant quantities of frazil, particularly if deposited in the form of a hanging dam. Ice roughness n_i often rises again during the prebreakup period when slight warming of the flow produces ice ripples, the n_i of which has been reported to be as high as 0.028.[23]

Elevations in water levels also can occur from the growth of anchor ice on the riverbed. Although stage rises with the growth of anchor ice, the overall flow resistance will change depending on the surface roughness. Values of n_i as high as 0.045 have been reported for well-developed deposits.[88]

Significant reductions in lake storage and modifications to winter flow regimes can result from winter snow loading of ice surfaces. Accumulation of deep snow depresses the ice cover, displaces stored water, and increases discharge. Heavy snowfalls in drainage systems with a large lake component commonly produce downstream pulses in discharge. These can be an important modifier of the winter flow regime, particularly in areas of low winter discharge.

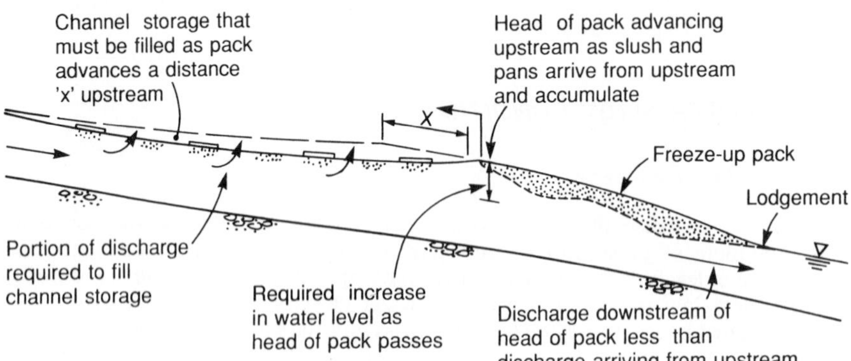

FIGURE 7.9.1 Channel storage resulting from advancement of freeze-up cover. *(Reproduced from Ref. 49 with permission.)*

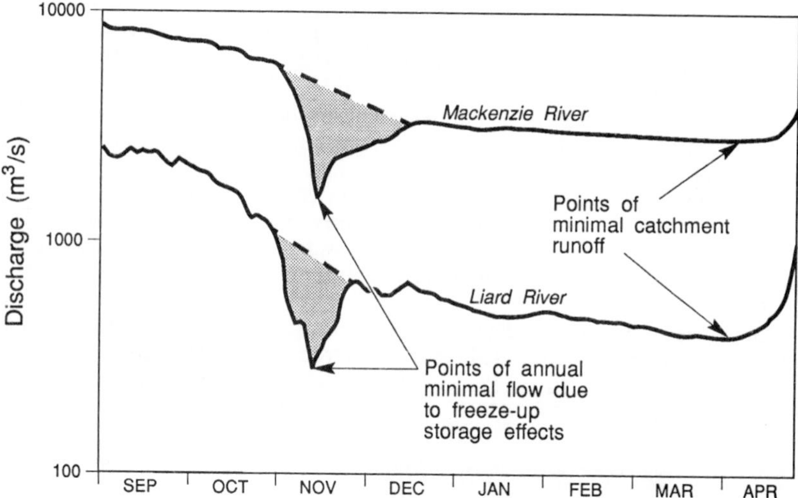

FIGURE 7.9.2 Winter hydrographs showing displacement of minimum flow due to freeze-up storage effects. *(Data source: 1990–1991 Water Survey of Canada.)*

7.9.2 Winter Low Flow

As flow is put into storage and abstracted for ice growth, a period of low flow can prevail. Flow withdrawal by these processes can shift the time of minimum flow on a river from the late-winter period, when the seasonal low flow derived from the landscape actually occurs, to the freeze-up period (Fig. 7.9.2). Frequency curves of low flows affected by such abstraction have a characteristic convex shape, the probability of the lowest flows falling below those for conditions or streams where abstraction for ice growth is not a significant factor.[48]

In high-northern climates, ice growth and low winter flow can be severe enough to cause complete freezing to the bed of small streams unless there is sufficient storage capacity, in the form of lakes and ponds, to supply flow throughout the winter. At shallow lake outlets, however, freezing to the bed also can isolate a lake from the winter flow system.

7.9.3 Flow Measurement

Because of seasonal variations in ice roughness, simple measurements of stage are inadequate for estimating winter flow. Standard practice is to conduct periodic discharge measurements and use a suitable technique for interpolation between measurements.[113] The reliability of a particular technique depends on river size and climatic conditions, small streams and temperate climates being the most problematic.[133]

During the period of stable ice cover, discharge is derived from velocity measurements obtained through holes at standard cross-section intervals. Standard practice in North America is to calculate an average velocity from measurements at 0.2 and 0.8 of the flow depth for flow depths in excess of 0.75 m. For shallower depths, measurements are made at the 0.5 or 0.6 level and corrected by multiplying by 0.88 or 0.92, respectively. Such methods provide good results on deep rivers, but special problems are created by slush accumulations and shallow streams.[112] Methods employed in other northern countries are reviewed by Pelletier.[113]

Calculating discharge is most complicated during periods of varied flow that often characterize freeze-up and breakup. Notably, breakup is also the time of peak annual discharge for many subarctic nival rivers.[26] The best method is to compute discharge based on flow calculations from upstream or downstream open-water stations. Site-specific measurements can be made with recently developed sonar-ranging equipment that permits above-ice measurements of water levels[49] and with a variety of velocity techniques that rely on shore-based and aerial surveys of ice floes or floating targets.[125]

Guidelines for conducting detailed river-ice measurement programs are outlined in Ref. 124 and, for extracting ice-breakup data from hydrometric station records in Ref. 14.

7.10 ICE DECAY AND BREAKUP

Lake and river ice breakup results from hydrothermal and hydromechanical processes, the relative importance of each being largely dependent on meteorological conditions. During the prebreakup period, even before air temperatures rise above freezing, snow on the ice and on land begins to melt under the influence of solar radiation. The onset of spring melt advances breakup in two ways. First, snowmelt runoff increases discharge, velocity, and water levels, and second, heat exchanges at both the snow-air and ice-water interfaces melt and decay the snow and ice cover, decreasing its thickness and mechanical strength.

7.10.1 Snow and Ice Ablation

Surface energy transfers of long-wave radiation and sensible and latent heat to the melting snow and ice covers of lakes and rivers occur much like those on land. The situation is more complicated, however, in the case of short-wave radiation because of the variability in optical properties of the ice sheet. Depending on the surface ice structure, varying amounts of short-wave radiation are reflected (see Table 7.3.3) or pass through and are attenuated within the ice. Polycrystalline ice, such as frazil and snow ice, has an albedo similar to that of aged snow but columnar ice is highly transparent with an albedo as low as 0.1, similar to that of open water. The reflective properties of the ice also influence the albedo of overlying thin snow, in the same way that underlying soil and vegetation influence land snow albedo. In general, a cover of highly reflective ice will melt and decay much more slowly, particularly where the surface energy balance is dominated by short-wave radiation.

Melt also may occur at the base of the ice sheet when the heat supplied by the underlying water exceeds the rate of heat conduction up through the ice. When an ice sheet becomes 0°C-isothermal (at 0°C throughout), conduction halts and any heat supplied to the water-ice boundary is consumed by melt. In the case of a river, sources of heat warming the water include that from local runoff, tributary flow, groundwater, radiation, bed sources (geothermal and heat released from sediments), and flow friction; the magnitude and relative importance of each source depend on local hydraulic and meteorological conditions. The greatest variability occurs in the radiation term because of seasonal and diurnal changes in incoming short-wave radiation and because of the decrease in albedo and radiation attenuation as the snow and snow-ice layers ablate. For lakes, where most bottom melt occurs from free (density-driven) convection, water temperatures of several degrees above freezing can develop from radiative warming beneath highly transparent ice covers.

In rivers, radiative warming of the flow also can be significant, but the general rise in temperatures is complicated by a more rapid heat transfer with water flow and by tributary inflow. Although the temperature of the river water may be only a fraction of a degree above freezing, the flow velocity promotes a rapid transfer of heat to the ice cover. The transfer is further assisted by the development of ice ripples which can increase the rate of heat transfer by over 50 percent compared with smooth ice.[10]

Because of the complexity and difficulty of evaluating the various heat fluxes, few attempts have been made to predict rates of ice thinning except those based on degree-day indices. Reductions in ice thickness during the prebreakup period, t_m (m), can be estimated from[18]

$$t_m = \beta D_b \tag{7.10.1}$$

where β is an empirical coefficient found to range between 0.004 and 0.01 m/C°-day for northern (>60°N) rivers, and D_b is the accumulated degree-days above a base daily mean air temperature of −5°C. There is evidence to suggest that the lower values of β also may apply to more southerly rivers.[126]

7.10.2 Reductions in Ice Strength and Resistance

General warming of an ice cover reduces its mechanical strength, although it remains structurally competent. Some of the warming is due to the absorption of short-wave radiation. If the rate of radiative warming exceeds the rate at which the heat can be conducted away, melt reduces intergranular contact, a process commonly termed *candling* for columnar-grained ice (Fig. 7.8.1*b*). This structural decay of the ice sheet reduces its mechanical strength and its ability to withstand loads such as those generated by wind on lakes or flood waves on rivers. Although not yet predictable, changes in strength occur most rapidly immediately after an ice cover becomes largely 0°C isothermal.[128]

7.10.3 Breakup Types

Breakup conditions can be divided into two general types as described by the quasi-synonymous terms: (1) *premature, mechanical,* or *dynamic* and (2) *overmature* or *thermal.* In the overmature case, which most closely approximates breakup conditions on lakes, spring discharge remains small while the ice sheet thins, detaches from the banks, and loses significant strength. For rivers, this is usually associated with low spring runoff. Eventually the ice sheet weakens until it poses almost no resistance to spring flow. For the premature case, conditions are reversed. Breakup of a strong, intact ice cover, which may still be attached strongly to the shore, is caused by a large spring flood wave created by rapid and extensive snowmelt, sometimes augmented by rainfall. With rising discharge, the downstream forces rupture and dislodge the intact ice sheet, driving it downstream.

The severity and pattern of breakup is influenced by the alignment of the river course relative to the local climate. On rivers in which snowmelt, runoff, and ice breakup proceed downstream with the seasonal advance of warm weather, the ice jam and flood risk is heightened because the spring flood wave is always pushing against an intact ice cover. The risk is lower on rivers flowing in a direction opposite to that of regional warming. Thermal ablation of the downstream cover prior to the start of significant upstream spring melt can greatly reduce the probability of ice jamming.

7.10.4 Breakup Prediction

Breakup at a site depends on a number of variables including cover thickness (ice and snow), ice strength, river geometry, flow velocity, and stage. Assuming that it is not generated by an ice-jam surge (see below), common methods of predicting breakup initiation (e.g., Ref. 15) stem from a technique originally proposed by Shulyakovskii[145] in which a relationship is established between two readily measurable variables: the rise in stage above the preceding freeze-up level ΔH and the total heat input to the cover ΣQ. Each is a surrogate index of physical changes in upstream forces associated with the spring flood wave and the decrease in downstream resistance offered by the ice cover. ΔH primarily reflects increases in (1) shear stress on the base of the ice cover produced by rising discharge, (2) channel flow width which permits greater downstream ice movement, and (3) shore-ice detachment. Decreases in ice thickness and strength are represented by ΣQ. Overmature breakups are represented by high ΣQ and low ΔH, the reverse being the case for premature breakups. Relations between these two terms remain empirical and site-specific. Figure 7.10.1 is an example of the approach. Points in the upper right corner represent thermal breakup conditions (large ΣQ and small ΔH) whereas those in the lower left are representative of dynamic breakups (lower heat input but potentially greater water level increases). Note that the vertical axis also shows the effect of ice thickness t_i—greater ice thickness requires more heat input to effect a breakup.

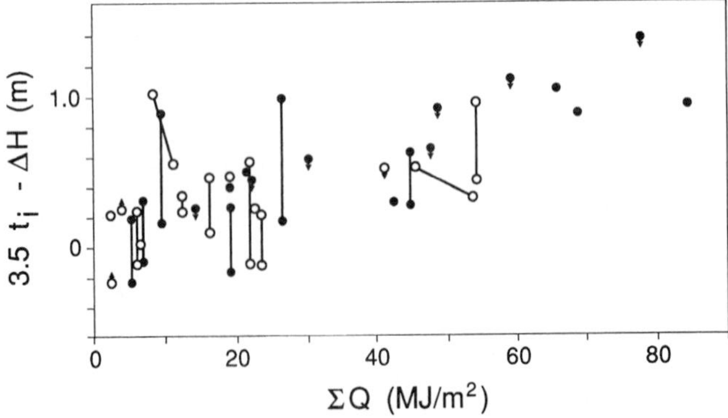

FIGURE 7.10.1 Relationship between ice thickness-stage increase and index of accumulated heat transfer for breakup initiation, Ganaraska River, Ontario, Canada. *(Adapted from Ref. 14. Used with permission.)*

FIGURE 7.10.2 Shear walls remaining after ice-jam release, Liard River, NWT, Canada. Walls are approximately 8 to 10 m high and extend 50 m from the shore. *(Photograph by T. D. Prowse.)*

7.10.5 Breakup Ice Jams

Once the intact ice cover has fractured, breakup ice jams can develop into thick and severe flow obstructions. Typical sites include those where ice passage is obstructed, such as by sharp bends, constrictions, islands, or bridge piers, and where the river slope rapidly decreases and the channel widens, particularly in the presence of a downstream intact cover. Mouths of rivers entering rivers or lakes or oceans that break up later are common examples of the latter type.

Although breakup jamming is a complex process, an estimate of potential water levels can be obtained from the approach detailed in Sec. 7.7.8. The thickness of breakup jams, as evidenced by remnant shear walls that line the banks after their release (Fig. 7.10.2), is often 5 to 10 m on large northern rivers. Representative values of ice roughness n_i for these thick accumulations are in the range 0.05 to 0.10.

Removal of an ice jam can result from a sudden collapse or from thermal ablation. During thermal ablation, the greatest source of melt is the downstream transfer of warm water that has been heated in upstream ice-clear reaches. Water temperatures several degrees above freezing immediately upstream of a jam head are not uncommon and are capable of producing rapid melting of the ice accumulation. Meltwater produced by such rapid melt has been reported to augment discharge by as much as 5 percent for several days.[123] Sudden collapse of an ice jam usually results from a combination of increases in discharge, the difference between upstream and downstream water levels, and instability at the jam toe.

7.10.6 Breakup Surges and Ice Runs

Collapse of an ice jam and the release of water in channel storage can produce a surge characterized by dramatic increases in downstream water levels and velocities. It is not uncommon in large rivers for *breaking fronts* (transition between intact and

FIGURE 7.10.3 Rubble breaking front moving at approximately 5 m/s, Liard River, NWT, Canada. *(Photograph by T. D. Prowse.)*

fragmented or fractured ice as shown in Fig. 7.10.3) to carren through downstream intact ice covers at speeds in excess of 5 m/s for tens of kilometers, and for rates of rise in water levels to be in the order of 1.0 m/min. The magnitude and rate of such surge characteristics represent a far more catastrophic flood potential than that possible under similar open-water flow conditions. Methods for estimating surge celerity and water velocity are described by Refs. 13 and 68.

Downstream progression of dynamic breakups is often characterized by breaking fronts produced by ice-jam surges. Types of cover failure associated with these fronts have been related to high- and low-energy conditions.[44] The former is characterized by large-scale fracturing of the ice sheet, the celerity of which approaches that of the surge. Matching of celerities for the surge and fracture front maximizes hydraulic forces on a cover.[45] In such a breakup, smaller-diameter brash ice is produced only as the cover is fragmented while being driven farther downstream, and hence the primary fracture front is separated from the ensuing rubble front. In lower-energy breakups the entire progression of breakup can be dominated by a rubble breaking front characterized by numerous ice-failure processes.[127] Resistance to the progression of breaking fronts is offered by channel resistance, as created by downstream hydraulic conditions (e.g., channel slope, roughness, flow depth, and velocity) and by ice conditions (e.g., bank-bed attachment, mechanical strength and thickness). Arresting of a breaking front can lead to the formation of another ice jam.

7.11 FLOOD FREQUENCY

Maximum annual stage on rivers subject to ice-cover formation can occur under ice or open-water conditions. The overall probability P_H of a given stage being equaled or exceeded in either circumstance is given by (assuming independent P_{Hi} and P_{Ho})

$$P_H = P_{Hi} + P_{Ho} - P_{Hi}P_{Ho} \tag{7.11.1}$$

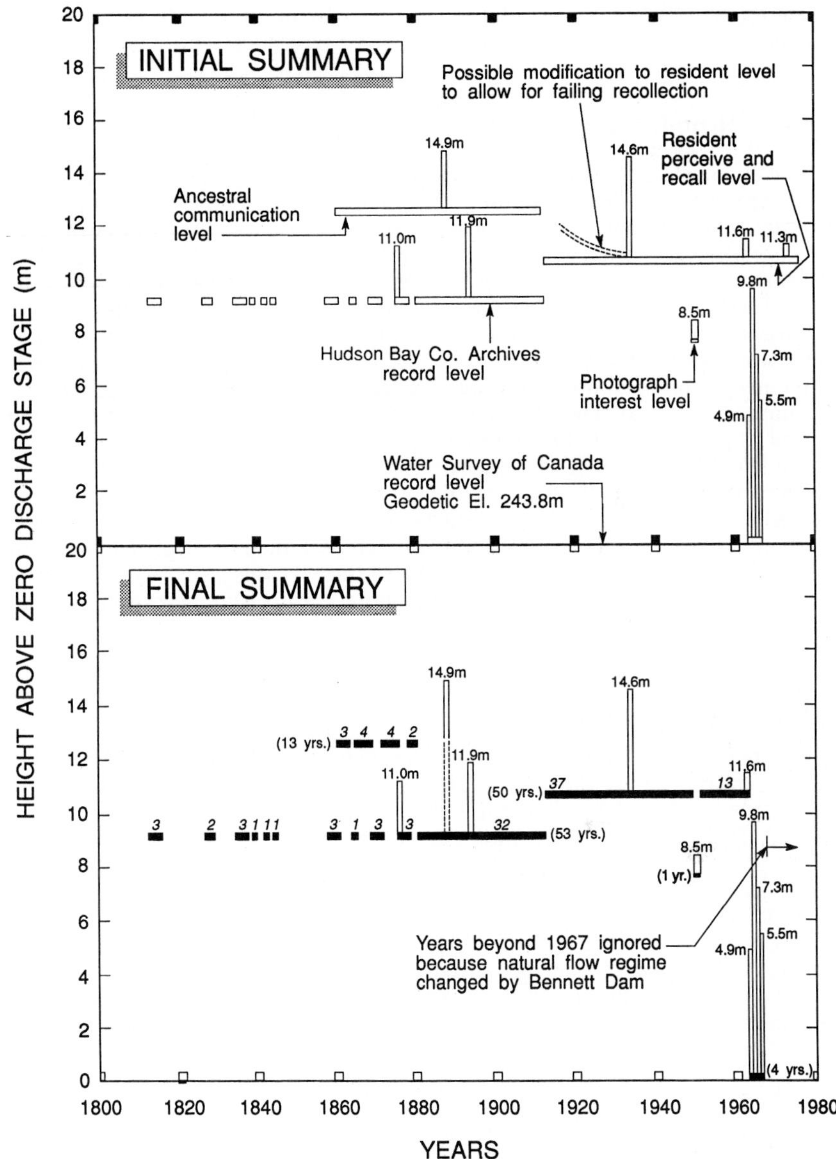

FIGURE 7.11.1 Observed ice-jam stages and assigned perception levels for the Peace River at Fort Vermilion, Alberta, Canada. *(Reproduced from Ref. 51 with permission.)*

where P_{Hi} and P_{Ho} are the probabilities that the maximum water levels for ice and open-water conditions, respectively, will equal or exceed a given water level.

7.11.1 Historical and Environmental Data

In rare circumstances, gauging records are available for the site of interest, but in most cases water-level records must be derived from other sources such as historical records, local accounts, and environmental evidence, especially through dendrochronological analysis of ice scars on vegetation. A sufficient length of record for probability estimates can usually be derived only by combining sources which are often biased toward major events. One method of accommodating this problem is to assign a perception stage to each source and time period.[51] Perception stage is defined as the water level below which the particular source probably would not have provided any documentation. The lowest perception stage for each source and the maximum ice-related stage then are determined for each documented year and plotted as in Fig. 7.11.1. Probability estimates then are obtained by treating the record length N associated with a particular event as the total number of years in which there are perception stages less than or equal to the event stage. (See also Sec. 18.6.3.) The rank of event m is established by considering higher events in this period of record N. The exceedance probability of each peak then is calculated. Standard probability distributions may not fit ice-related stage records primarily because ice-related stage-discharge relations are much more sensitive to overbank flow than open-water floods.

7.11.2 Jam-Discharge Frequency

In the absence of ice-jam stage data, maximum breakup discharges can be used to estimate the probability of ice-related stages (e.g., Refs. 50, 163). For locations where jam frequency data also are absent, only a range in potential water levels can be derived. The procedure first requires establishing ice-related stages for the range in discharge expected during the breakup period. A lower bound can be set by assuming no ice jam will form. Stage is calculated as a simple function of discharge using appropriate values for cover thickness and hydraulic roughness. To define an upper bound, it is assumed that a jam will develop and an equilibrium section will affect the site of interest. Stages then are obtained from an ice-jam rating curve defined by the equilibrium relationship in Sec. 7.7.8. The frequency of stages in either case is then determined by the expected or measured frequency of maximum discharges that prevail between the beginning of breakup and ice-clear conditions.

REFERENCES

1. Adams, W. P., and T. D. Prowse, "Evolution and Magnitude of Spatial Patterns in the Winter Cover of Temperate Lakes," *Fennia,* vol. 159, no. 2, pp. 343–359, 1981.
2. Anderson, E. A., "Development and Testing of Snow Pack Energy Balance Equations," *Water Resour. Res.,* vol. 4, no. 1, pp. 19–37, 1968.
3. Anderson, E. A., "National Weather Service River Forecast System—Snow Accumulation and Ablation Model," National Oceanic and Atmospheric Administration, Silver Springs, Md., *Tech. Memorandum* NES HYDRO-17, 1973.

4. Anderson, E. A., "A Point Energy and Mass Balance Model of a Snow Cover," *Tech. Memorandum* NWS 19 NOAA, 1976.

5. Anderson, E. A., "Snow Accumulation and Ablation Model," in operational forecast programs and improved data improvements for the National Weather Service River Forecast System Data Management Program Documentation. U.S. National Weather Service, Silver Springs, Md., 1978.

6. Andres, D. D., and P. F. Doyle, "Analysis of Break-up Ice Jams on the Athabasca River at Fort McMurray, Alberta," *Can. J. Civil Eng.,* vol. 11, no. 3, pp. 444–458, 1984.

7. Ashton, G. D., "River Ice," *Ann. Rev. Fluid Mech.,* vol. 10, pp. 369–392, 1978.

8. Ashton, G. D., ed., *River and Lake Ice Engineering,* Water Resources Publ., Littleton, Colo., 1986.

9. Ashton, G. D., "Thin Ice Growth," *Water Resour. Res.,* vol. 25, no. 3, pp. 564–566, 1989.

10. Ashton, G. D., and J. F. Kennedy, "Ripples on Underside of River Ice Covers," *J. Hydraul. Eng.,* vol. 98, pp. 1603–1624, 1972.

11. Bates, R. E., and M. A. Bilello, "Defining the Cold Regions of the Northern Hemisphere," U.S. Army Corps Eng. Cold Reg. Res. Eng. Lab., Hanover, N.H., *Tech. Rept.* 178, 1966.

12. Beltaos, S., "River Ice Jams: Theory, Case Studies and Applications," *J. Hydraul. Eng.,* vol. 109, no. 10, pp. 1338–1359, 1983.

13. Beltaos, S., *Monograph on River Ice Jams,* Chap. 3, Ice Jams Processes, Chap. 4, Theory, Submitted to Nat. Res. Council Can. Working Group on River Ice Jams, Nat. Water Res. Institute, Burlington, Ont., Canada, NHRI Contribution 88–50, 1988.

14. Beltaos, S., *Guidelines for Extraction of Ice-Break-up Data from Hydrometric Station Records,* Working Group on River Ice Jams, Field Studies and Research Needs, Environ. Can., Nat. Hydrol. Res. Institute, Saskatoon, Sask., NHRI Sci. Rep. 2, pp. 37–70, 1990.

15. Beltaos, S., "Fracture and Breakup of River Ice Cover," *Can. J. Civil Eng.,* vol. 17, pp. 173–183, 1990.

16. Bergström, S., "Spring Flood Forecasting by Conceptual Models in Sweden," in S. C. Colbeck and M. Ray, ed., *Proc. Model. Snow Cover Runoff,* U.S. Army Cold Reg. Res. Lab., Hanover, N.H., pp. 397–405, 1979.

17. Bernier, P. Y., "Microwave Remote Sensing of Snowpack Properties: Potential and Limitation," *Nordic Hydrol.,* vol. 18, pp. 1–20, 1987.

18. Bilello, M. A., *Maximum Thickness and Subsequent Decay of Lake, River and Fast Sea Ice in Canada and Alaska,* U.S. Army Cold Reg. Res. Eng. Lab. Rep. 80-6, Hanover, N.H., 1980.

19. Bruce, J. P., and R. H. Clark, *Introduction to Hydrometeorology,* Pergamon Press, Oxford, England, 1966.

20. Brunt, D., "Notes on the Radiation in the Atmosphere," *Q. J. R. Meteorol. Soc.,* vol. 58, pp. 389–420, 1932.

21. Budd, W. R., R. Dingle, and U. Radok, "The Byrd Snowdrift Project: Outline and Basic Results," *Studies in Antarctic Meteorology, Am. Geophys. Union, Antarctic Research Series 9,* pp. 71–134, 1966.

22. Calkins, D. J., *Accelerated Ice Growth in Rivers,* U.S. Army Cold Reg. Res. Eng. Lab., CRREL Report 79-14, Hanover, N.H., 1979.

23. Carey, K. L., "Observed Configuration and Computed Roughness of the Underside of River Ice, St. Croix River, Wisconsin," *U.S. Geol. Survey Professional Paper* 550-B, pp. B192–B198, 1966.

24. Carey, K. L., "Icings Developed from Surface Water and Groundwater," U.S. Army Cold Reg. Res. Eng. Lab, Cold Regions Sci. Eng. Monograph III-3, 1973.

25. Carroll, Thomas R., "Operational Airborne Measurements of Snow Water Equivalent and Soil Moisture Using Terrestrial Gamma Radiation," *Proc. Large Scale Effects of Seasonal Snow Cover. Int. Assoc. Sci. Hydrol.* 166, pp. 213–223, 1987.

26. Church, M., "Hydrology and Permafrost with Reference to North America," in Permafrost Hydrol., Proc. Workshop Seminar 1974, Can. Natl. Comm. Int. Hydrol. Decade, Ottawa, Ont., pp. 7–20, 1974.

27. Clark, R. H., "Predicting the Runoff from Snowmelt," *Eng. J.*, vol. 38, pp. 434–441, 1955.

28. Colbeck, S. C., "A Theory of Water Percolation in Snow," *J. Glaciol.*, vol. 11, no. 63, pp. 369–385, 1972.

29. Colbeck, S. C., "An Analysis of Water Flow in Dry Snow," *Water Resour. Res.*, vol. 12, no. 3, pp. 523–527, 1976.

30. Colbeck, S. C., "The Physical Aspects of Water Flow through Snow," *Adv. Hydrosci.*, vol. II, pp. 165–206, 1978.

31. Colbeck, S. C., "Thermodynamics of Snow Metamorphism Due to Variations in Curvature," *Cold Reg. Sci. Tech.*, vol. 1, pp. 37–45, 1979.

32. Colbeck, S., E. Akitaya, R. Armstrong, H. Gubler, J. Lafeuille, K. Lied, D. McClung, and E. Morris, *The International Classification for Snow on the Ground,* Int. Comm. Snow and Ice, Int. Assoc. Hydrol. Sci., World Data Center, University of Colorado, Boulder, Colo., pp. 1–27, 1990.

33. Cox, L. M., and J. F. Zuzel, "A Method for Determining Sensible Heat Transfer to Late-Lying Snowdrifts," *Proc. 44th Ann. Meet., West. Snow Conf.,* Calgary, Alta., pp. 23–28, 1976.

34. de Jong, B., *Net Radiation Received by a Horizontal Surface at the Earth,* Delft University Press, Rotterdam, The Netherlands, 1973.

35. Delfs, J., "Interception and Stemflow in Stands of Norway Spruce and Beech in West Germany," in W. E. Sopper and H. W. Lull, eds., *Forest Hydrology,* Pergamon Press, New York, pp. 179–183, 1967.

36. Dickinson, W. T., and F. G., Theakston, "Snow-Flume Modelling of Snow Accumulation in Forest Openings," in *Hydrological Processes of Forested Areas,* Proc. Can. Hydrol. Symp., Fredericton, N.B., Natl. Res. Council Canada 20548, pp. 113–129, 1982.

37. Dingman, S. L., W. F. Weeks, and Y. C. Yen, "The Effect of Thermal Pollution on River Ice Conditions," *Water Resour. Res.*, vol. 4, no. 2, pp. 349–362, 1968.

38. Doty, R. D., and R. S. Johnston, "Comparison of Gravimetric Measurements and Mass Transfer Computations of Snow Evaporation Beneath Selected Vegetation Canopies," *Proc. 37th Annual Meet. West. Snow Conf.,* pp. 57–62, 1969.

39. Dozier, J., "A Solar Radiation Model for a Snow Surface in Mountainous Terrain," in S. C. Colbeck and M. Ray, eds., *Proc. Modeling Snowcover Runoff,* U.S. Army Cold Reg. Res. Eng. Lab., Hanover, N.H., pp. 144–153, 1979.

40. Dyunin, A. K., "Fundamentals of the Theory of Snow Drifting," *Izvest. Sibirsk, Otdel. Akad. Nauk. U.S.S.R.* no. 12, pp. 11–24 (English translation by G. Belkov, Nat. Res. Counc. Can., Ottawa, Tech. Transl. 952, 1961), 1959.

41. Dyunin, A. K., and V. M. Kotlyakov, "Redistribution of Snow in Mountains under the Effect of Heavy Snow-Storms," *Cold Reg. Sci. Tech.,* vol. 3, pp. 287–294, 1980.

42. Erickson, D. E. L., W. Lin, and H. Steppuhn, "Indices for Estimating Prairie Runoff from Snowmelt," Paper presented to *Seventh Symp. Water Studies Inst. Appl. Prairie Hydrol.,* Saskatoon, Sask., 1978.

43. Farnsworth, R. K., "A Mathematical Model of Soil Surface Layers for Use in Predicting Significant Changes in Infiltration Capacity During Periods of Freezing Weather," Ph.D. Thesis, University of Michigan, Ann Arbor, Mich., 1976.

44. Ferrick, M. G., G. Lemieux, N. D. Mulherin, and W. Demont, "Controlled River Ice Cover Break-up: Part 2. Theory and Numerical Model Studies," *Proc. Int. Assoc. Hydraul. Res. Symp. Ice 1986,* Iowa City, Iowa, vol. III, pp. 293–305, 1986.

45. Ferrick, M. G., and N. D. Mulherin, *Framework for Control of Dynamic Ice Break-up by River Regulation,* U.S. Army Cold Reg. Res. Eng. Lab, Rep. 89-12, 1989.

46. Garnier, B. J., and A. Ohmura, "The Evaluation of Surface Variations in Solar Radiation Income," *Solar Energy,* vol. 13, pp. 21–34, 1970.

47. Gary, H. L., "Airflow Patterns and Snow Accumulation in a Forest Clearing," *Proc. 43rd Annual Meet. Western Snow Conf.,* pp. 106–113, 1975.

48. Gerard, R., "Regional Analysis of Low Flows: a Cold Region Example," *Proc. 5th Can. Hydrotech. Conf.,* Fredericton, N.B., pp. 95–112, 1981.

49. Gerard, R., "Hydrology of Floating Ice," in T. D. Prowse and C. S. L. Ommanney, eds., *Northern Hydrology, Canadian Perspectives,* Environ. Can., Nat. Hydrol. Res. Institute, Saskatoon, Sask., Sci. Rep. 1, pp. 103–134, 1990.

50. Gerard, R., and D. J. Calkins, "Ice-related Flood Frequency Analysis: Application of Analytical Estimates," *Proc. Cold Reg. Specialty Conf.,* Montreal, Que., pp. 85–101, 1984.

51. Gerard, R., and E. W. Karpuk, "Probability Analysis of Historical Flood Data," *J. Hydraul. Eng.,* pp. 1153–1165, 1979.

52. Gerdel, R. W., "The Dynamics of Liquid Water in Deep Snow-packs," *Trans. Am. Geophys. Union,* vol. 26, no. 1, pp. 83–90, 1945.

53. Gilfilian, R. E., W. L. Kline, T. E. Ostrkamp, and C. S. Benson, "Ice Formation in a Small Alaskan Stream," *Proc. Banff Symposia on the Role of Snow and Ice in Hydrology,* Banff, Alta., pp. 505–513, 1972.

54. Gold, L. W., and G. P. Williams, "Energy Balance During the Snow Melt Period at an Ottawa Site," *Int. Assoc. Sci. Hydrol. Publ.* 54, pp. 288–294, 1961.

55. Gooderich, L. E., "Efficient Numerical Technique for One-Dimensional Thermal Problems with the Phase Change," *Int. J. Heat Mass Transfer,* vol. 21, pp. 615–677, 1978.

56. Goodison, B. E., H. L. Ferguson, and G. A. McKay, "Measurement and Data Analysis," in D. M. Gray and D. H. Male, eds., *Handbook of Snow: Principles, Processes, Management and Use,* Pergamon Press, Toronto, pp. 191–274, 1981.

57. Goodison, B. E., B. Wilson, K. Wu, and J. Metcalfe, "An Inexpensive Remote Snow-Depth Gauge: An Assessment," *Proc. 52nd Ann. Meet. West. Snow Conf.,* pp. 188–191, 1984.

58. Granger, R. J., "Energy Exchange During Melt of a Prairie Snowcover," M.Sc. Thesis, University of Saskatchewan, Saskatoon, Sask., 1977.

59. Granger, R. J., and D. M. Gray, "A Net Radiation Model for Calculating Daily Snowmelt in Open Environments," *Nordic Hydrol.,* vol. 21, pp. 217–234, 1990.

60. Granger, R. J., and D. M. Gray, "The Impact of Frozen Soil on Prairie Hydrology," in K. R. Cooley, ed., *Proc. Int. Symp. Frozen Soil Impacts on Agricultural, Range and Forest Lands,* U.S. Army Cold Reg. Res. Eng. Lab., Hanover. N.H., Special Rep. 90-1 pp. 247–256, 1990.

61. Granger, R. J., and D. H. Male, "Melting of a Prairie, Snowpack," *J. Appl. Meteorol.* (now *J. Clim. Appl. Meteorol.*) vol. 17, no. 2, pp. 1833–1842, 1978.

62. Granger, R. J., D. M. Gray, and G. E. Dyck, "Snowmelt Infiltration to Frozen Prairie Soils," *Can. J. Earth Sci.,* vol. 21, no. 6, pp. 669–677, 1984.

63. Gray, D. M., and P. J. Landine, "Albedo Model for Shallow Prairie Snow Covers," *Can. J. Earth Sci.,* vol. 24, no. 9, pp. 1760–1768, 1987.

64. Gray, D. M., and P. J. Landine, "An Energy-Budget Snowmelt Model for the Canadian Prairies," *Can. J. Earth Sci.,* vol. 25, no. 9, pp. 1292–1303, 1987.

65. Gray, D. M., and A. D. J. O'Neill, "Applications of the Energy Budget for Predicting Snowmelt Runoff," *Adv. Concepts Tech. Study Snow Ice Resour.,* Natl. Acad. Sci., Washington, D.C., pp. 108–118, 1974.

66. Gray, D. M., P. G. Landine, and R. J. Granger, "Simulating Infiltration into Frozen Prairie Soils in Streamflow Models," *Can. J. Earth Sci.,* vol. 22, no. 3, pp. 464–474, 1985.

67. Hall, D. K., and J. Martinec, *Remote Sensing of Ice and Snow,* Chapman and Hall, New York, 1985.

68. Henderson, F. M., and R. Gerard, "Flood Waves Caused by Ice Jam Formation and Failure," *Proc. Int. Assoc. Hydraul. Res. Int. Symp. Ice,* Quebec, Canada, vol. 1, pp. 277–287, 1981.

69. Hendrie, L. K., and A. G. Price, "Energy Balance in a Deciduous Forest," in S. C. Colbeck and M. Ray, eds., *Proc. Modeling Snowcover Runoff,* U.S. Army Cold Reg. Res. Eng. Lab., Hanover N.H., pp. 211–221, 1979.

70. Hicks, B. B., and H. C. Martin, "Atmospheric Turbulent Fluxes Over Snow," *Boundary-Layer Meterol.,* vol. 2, pp. 496–502, 1972.

71. Hoover, M. D., and C. F. Leaf, "Process and Significance of Interception in Colorado Subalpine Forest," in W. E. Sopper and H. W. Lull, eds., *Forest Hydrology,* Pergamon Press, New York, pp. 213–223, 1967.

72. Huokuna, M., "The Finnish River Ice Research Project—The Numerical River Ice Model in Use," *Proc. Int. Assoc. Hydraul. Res. Symp. Ice,* Helsinki, Finland, 1990, pp. 215–230.

73. Kivisild, H. R., "Hydrodynamic Analysis of Ice Floods," vol. 3, *Proc. Int. Assoc. Hydraul. Res. 8th Congress,* Montreal, Que., 1959.

74. Kobayashi, D., "Studies of Snow Transport in Low-Level Drifting Snow," *Inst. Low Temperature Science, Series* A, no. 24, pp. 1–58, 1972.

75. Komarov, V. D., and T. T. Makarova, "Effect of the Ice Content, Cementation and Freezing Depth of the Soil on Meltwater Infiltration in a Basin," *Soviet Hydrology: Selected Pap.,* issue 3, pp. 243–249, 1973.

76. Kondratyev, K. Ya, *Radiation in the Atmosphere,* vol. 12. International Geophysical Series, Academic Press, New York, 1969.

77. Kuz'min P. P., *Formirovanie Snezhnogo Pokrova i Metody Opredeleniya Snegozapasov (Snow Cover and Snow Reserves)* (English translation by Israel Prog. Sci. Transl., Jerusalem), 1963.

78. Kuz'min, P. P., *Protsess Tayaniya Shezhnogo Pokrova (Melting of Snow Cover)* (English translation by Israel Prog. Sci. Transl., Jerusalem), 1972.

79. Landine, P. G., and D. M. Gray, *Snow Transport and Management,* Internal Report Division of Hydrology, University of Saskatchewan, Saskatoon, Sask., 1989.

80. List, R. J., *Smithsonian Meteorological Tables,* 6th ed., The Smithsonian Institution, Washington, D. C., 1968.

81. MacLaren Plansearch, "Snow Hydrology Studies Phases I and II: Study Methodology and Single Event Simulation," Ontario Ministry of Natural Resources, Toronto, Ont., 1984.

82. Magono, C., and C. Lee, "Meteorological Classification of Natural Snow Crystals," *J. Fac. Sci., Hokkaido Univ.,* Ser. VII, vol. 2, pp. 321–335, 1966.

83. Male, D. H., "The Seasonal Snowcover," in S.C. Colbeck, ed., *Dynamics of Snow and Ice Masses,* Academic Press, New York, pp. 305–395, 1980.

84. Male, D. H., and D. M. Gray, "Snowcover Ablation and Runoff," in D. M. Gray and D. H. Male, eds., *Handbook of Snow: Principles, Processes, Management & Use,* Pergamon Press, Toronto, pp. 360–436, 1981.

85. Male, D. H., and R. J. Granger, "Energy Mass Fluxes at the Snow Surface in a Prairie Environment," in S. C. Colbeck and M. Ray, eds., *Proc. Modeling Snowcover Runoff,* U.S. Army Cold Reg. Res. Eng. Lab., Hanover, N.H., pp. 101–124, 1979.

86. Male, D. H., and R. J. Granger, "Snow Surface Energy Exchange," *Water Resour. Res.,* vol. 17, no. 3, pp. 609–627, 1981.

87. Marcotte, N., *Heat Transfer from Open-Water Surfaces in Winter,* Nat. Res. Council Can. Tech. Memorandum 114, pp. 2–16, 1975.

88. Marcotte, N., "Anchor Ice in Lachine Rapids, Results of Observations and Analysis,"

Proc. Int. Assoc. Hydraul. Res. Int. Symp. Ice, Hamburg, Germany, vol. 1, pp. 151–159, 1984.

89. Marsh, P., and T. D. Prowse, "Water Temperature and Heat Flux at the Base of River Ice Covers," *Cold Reg. Sci. Technol.,* vol. 14, pp. 33–50, 1987.
90. Marsh, P., and M. K. Woo, "Wetting Front Advance and Freezing of Meltwater within a Snow Cover, 1. Observations in the Canadian Arctic," *Water Resour. Res.,* vol. 20, no. 12, pp. 1853–1864, 1984.
91. Martinec, J., "The Degree Day Factor for Snowmelt Runoff Forecasting," Int. Union Geodesy Geophys. Gen. Assembly of Helsinki, Int. *Assoc. Hydrol. Sci. Comm. Surface Waters,* IAHS Publ. 51, pp. 468–477, 1960.
92. Martinec, J., and A. Rango, "Parameter Values for Snowmelt Runoff Modelling," *J. Hydrol.,* 84, pp. 197–219, 1986.
93. McKay, G. A., "Relationships between Snow Survey and Climatological Measurements," Int. Union Geod. Geophys. Gen. Assem. Berkeley (Surface Water), *Int. Assoc. Hydrol. Sci.,* Publ. 63, pp. 214–227, 1963.
94. McKay, G. A., and H. A. Thompson, "Snowcover in the Prairie Provinces," Trans. Am. Soc. Agr. Eng., vol. 11, no. 6, pp. 812–815, 1968.
95. Meiman, J. R., "Snow Accumulation Related to Elevation, Aspect and Forest Canopy," *Proc. Workshop Seminar on Snow Hydrology,* Queen's Printer of Canada, Ottawa, pp. 35–47, 1970.
96. Mellor, M., "Blowing Snow," *CRREL Monograph,* Part III, Section A3c, U.S. Army Cold Reg. Res. Lab., Hanover, N.H., 1965.
97. Mellor, M., "Some Optical Qualities of Snow," Symp. Int. Aspects Scientifigues des Avalanches de Neige. 5–10 Avril, Davos, Suisse, *Int. Assoc. Sci. Hydrol. Publ.* 69, pp. 129–140, 1966.
98. Michel, B., *Winter Regime of Rivers and Lakes,* Cold Reg. Sci. Eng. Monograph III-Bla, U.S. Army Cold Reg. Res. Eng. Lab., Hanover, N.H., 1971.
99. Michel, B., *Ice Mechanics,* Les Presses de L'Université, Laval, Quebec, 1978.
100. Michel, B., "Comparison of Field Data with Theories on Ice Cover Progression in Large Rivers," *Can. J. Civil Eng.,* vol. 11, pp. 798–814, 1984.
101. Michel, B., N. Marcotte, F. Fonseca, and G. Rivard, "Formation of Border Ice in the Ste. Anne River," *Proc. 2nd Workshop on Hydraulics of Ice-Covered Rivers,* Edmonton, Alta., pp. 38–61, 1982.
102. Miller, D. H., "Interception Processes During Snow Storms," *USDA Forest Res. Serv., Paper* PSW-18, 1964.
103. Miller, D. H., "Spatial Interactions Produced by Meso-Scale Transports of Water in the Atmospheric Boundary Layer," Paper presented to the Annual Meeting, Association of American Geographers, New York, 1976.
104. Miner, N. H., and J. M. Trappe, "Snow Interception, Accumulation and Melt in Lodgepole Pine Forests in the Blue Mountains of Eastern Orgeon," *USDA Forest Res. Serv. For. Range Expt. Sta.,* Res. Note 143, 1957.
105. Molnau, M., and V. C. Bissell, "A Continuous Frozen Ground Index for Flood Forecasting," *Proc. 51st Ann. Meet. West. Snow Conf.,* pp. 109–119, 1983.
106. Nezhikhovskiy, R. A., "Coefficients of Roughness of Bottom Surface of Slush Ice Cover," *Soviet Hydrol. Selected Pap.,* 2, pp. 127–148, 1964.
107. Nyberg, A., "A Study of Evaporation and the Condensation at a Snow Surface," *Ark. Geofys.,* vol. 4, no. 30, pp. 577–590, 1965.
108. Obled, Ch., and H. Harder, "A Review of Snow Melt in the Mountain Environment," in S. C. Colbeck and M. Ray, eds., *Proc. Modeling Snowcover Runoff,* U.S. Army Cold Reg. Res. Eng. Lab., Hanover, N.H., pp. 179–204, 1979.
109. Olyphant, G. A., and S. A. Isard, "The Role of Advection in the Energy Balance of

Late-Lying Snowfields: Niwot Ridge, Front Range, Colorado," *Water Resour. Res.*, vol. 74, no. 11, pp. 1962–1968, 1988.

110. Ostrem, G., and M. Brugman, "Glacier Mass-Balance Measurements," *Environ. Can. Natl. Hydrol. Res. Inst., Saskatoon, Sask., Sci. Rep.* 4, 1991.

111. Pariset, E., and R. Hausser, "Formation and Evolution of Ice Covers on Rivers," *Trans. Eng. Inst. Can.*, vol. 5, no. 1, pp. 41–49, 1961.

112. Pelletier, P., "Uncertainties in the Single Determination of River Discharge: A Literature Review," *Can. J. Civil Eng.*, vol. 15, no. 5, pp. 834–850, 1988.

113. Pelletier, P., "A Review of Techniques Used by Canada and Other Northern Countries for Measurement and Computation of Streamflow under Ice Conditions," *Nordic Hydrol.*, vol. 21, pp. 317–340, 1990.

114. Penman, H. L., "Natural Evaporation from Open Water, Bare Soil and Grass," *Proc. R. Soc. London,* Series A, vol. 193, pp. 120–145, 1948.

115. Petryk, S., R. Saade, M. Sydor, and S. Beltaos, "Global Design of the Numerical River Ice Model RIVICE," *Proc. 6th Workshop on the Hydraulics of Ice-Covered Rivers,* Ottawa, Ont., 1991.

116. Petzold, D. R., and R. G. Wilson, "Solar and Net Radiation Over Melting Snow in the Subarctic," *Proc. 31st Ann. Meet. East. Snow Conf.*, Ottawa, Ont., pp. 51–59, 1974.

117. Pomeroy, J. W., "Wind Transport of Snow," Ph.D. Thesis, University of Saskatchewan, Saskatoon, Sask., 1988.

118. Pomeroy, J. W., and D. M. Gray, "Saltation of Snow," *Water Resour. Res.*, vol. 26, no. 7, pp. 1583–1594, 1990.

119. Pomeroy, J. W., D. M. Gray, and P. G. Landine, "Blowing Snow Transport and Sublimation on the Prairies," *Proc. East. Snow Conf.*, Guelph, Ont., 1991.

120. Popov, E. G., "Snowmelt Runoff Forecasts—Theoretical Problems," in The Role of Snow and Ice in Hydrology, *Proc. Banff Symp.*, UNESCO, WMO, IASH, vol. 2, pp. 829–839, 1972.

121. Price, A. G., and T. Dunne, "Energy Balance Computations of Snowmelt in a Subarctic Area," *Water Resour. Res.*, vol. 12, pp. 686–694, 1976.

122. Prowse, T. D., "Ice Jam Characteristics, Liard-Mackenzie Rivers Confluence," *Can. J. Civil Eng.*, vol. 13, no. 6, pp. 653–665, 1986.

123. Prowse, T. D., "Heat and Mass Balance of an Ablating Ice Jam," *Can. J. Civil Eng.*, vol. 17, no. 4, pp. 629–635, 1990.

124. Prowse, T. D., *Guidelines for River Ice Data Collection Programs,* Working Group on River Ice Jams, Field Studies and Research Needs, Environ. Can., Nat. Hydrol. Res. Inst., NHRI Sci. Rep. 2, pp. 1–36, 1990.

125. Prowse, T. D., J. C. Anderson, and R. L. Smith, "Discharge Measurement during River Ice Breakup," *Proc. 43rd East. Snow Conf.*, Hanover, N.H., pp. 55–69, 1986.

126. Prowse, T. D., S. Beltaos, B. Burrell, P. Tang, and J. Dublin, "Breakup of the Nashwaak River, New Brunswick," *Proc. 46th East. Snow Conf.*, Quebec City, Que., pp. 142–155, 1989.

127. Prowse, T. D., and M. N. Demuth, "Failure Modes Observed during River Ice Breakup," *Proc. 46th East. Snow Conf.*, Quebec City, Que., pp. 237–241, 1989.

128. Prowse, T. D., M. N. Demuth, and H. A. M. Chew, "The Deterioration of Freshwater Ice Due to Radiation Decay," *J. Hydraul. Res.*, vol. 28, no. 6, pp. 685–697, 1990.

129. Pysklywec, D. W., K. S. Davar, and D. I. Bray, "Snowmelt at an Index Plot," *Water Resour. Res.*, vol. 4, no. 5, pp. 937–946, 1968.

130. Quick, M. C., and A. Pipes, "U.B.C. Watershed Model,"*Proc. WMO/IASH Symp. Application Mathematical Models in Hydrol. Water Resour. Systems,* Bratislava, Czechoslovakia, 1975.

131. Rauner, Yu L., "On the Heat Budget of a Decidious Forest in Winter," *Izv. Acad. Nauk*

SSSR, vol. 4, pp. 83–90. (Transl. A. Nurklik, 1964, Can. Dept. Transport, Meteorol. Branch, Meteorol. Transl. 11, pp. 60–77), 1961.

132. Reifsnyder, W. E., and H. W. Lull, *Radiant Energy in Relation to Forests, USDA Tech. Bull.* 1344, 1965.

133. Rosenberg, H. B., and R. L. Pentland, "Accuracy of Winter Streamflow Records," *Proc. 23rd East. Snow Conf.,* Hartford, Conn., pp. 51–72, 1966.

134. Rothlisberger, H., and H. Lang, "Glacial Hydrology," in A. M. Gurnell and M. J. Clark, eds., *Glacio-Fluvial Sediment Transfer; an Alpine Perspective,* Wiley, New York, pp. 207–284, 1987.

135. Satterlund, D. R., and H. F. Haupt, "The Disposition of Snow Caught by Conifer Crowns," *Water Resour. Res.,* vol. 6, pp. 649–652, 1970.

136. Schemenauer, R. S., M. O. Berry, and J. B. Maxwell, "Snowfall Formation," in D. M. Gray and D. H. Male, eds., *Handbook of Snow, Principles, Processes Management and Use,* Pergamon Press, New York, pp. 129–152, 1981.

137. Schmidt, R. A., Jr., "Sublimation of Wind-Transported Snow—A Model," *Res. Paper* RM-90, USDA Forest Serv., Rocky Mtn. Forest and Range Expt. Sta., Fort Collins, Colo., 1972.

138. Schmidt, R. A., "Properties of Blowing Snow," *Rev. Geophys. Space Phys.,* vol. 20, pp. 39–44, 1982.

139. Schmidt, R. A., R. L. Jairell, and J. W. Pomeroy, "Measuring Snow Interception and Loss from an Artificial Conifer," *Proc. 56th Ann. Meet. West. Snow Conf.,* Kalispell, Mont., pp. 166–169, 1988.

140. Seligman, G., *Snow Structure and Ski Fields,* R. R. Clarke Ltd., Edinburgh, Scotland, 1936.

141. Shen, H. T., and L. A. Chiang, "Simulation of Growth and Decay of River Ice Cover," *J Hydraul. Eng.,* vol. 110, no. 7, pp. 958–971, 1984.

142. Shen, H. T., E. P. Foltyn, and S. F. Daly, *Forecasting Water Temperature Decline and Freeze-up in Rivers,* U.S. Army Eng. Cold Reg. Res. Eng. Lab., CRREL Rept 84-19, 1984.

143. Shevelev, N. N., "Interception of Vertical and Horizontal Precipitation in the Forests of the Central Urals," *Soviet Hydrol. Sel. Papers,* vol. 16, no. 4, pp. 313–318, 1977.

144. Shimizu, H., "Air Permeability of Deposited Snow," Institute of Low Temperature Science, Series A, no. 22, pp. 1–32, 1970.

145. Shulyakovskii, L. G., ed., *Manual of Forecasting Ice-Formation for Rivers and Inland Lakes,* Gidrometeorologicheskoe Izdatel'stvo, Leningrad, Israel Program for Scientific Translations, 1966.

146. Smith, J. L., H. G. Halverson, and R. A. Jones, "Central Sierra Profiling Snow Gauge: A Guide to Fabrication and Operation," *U.S. Atomic Energy Comm. Rep.* TID-25986, Natl. Tech. Inf. Serv., U.S. Department of Commer., Washington, D.C., 1972.

147. St. Lawrence Waterway Project, Report of Joint Board of Engineers on St. Lawrence Waterway Project, Appendix E, Ice Formation on the St. Lawrence and Other Rivers, Ottawa, pp. 406–422, 1927.

148. Steenhuis, T. S., G. D. Budenzer, and M. F. Walter, "Water Movement and Infiltration in a Frozen Soil: Theoretical and Experimental Considerations," *Am. Soc. Agr. Eng. Pap.* 77-2545, Winter Meet. Am. Soc. Agr. Eng., 1977.

149. Steppuhn, H. W., "Areal Water Equivalents for Prairie Snowcovers by Centralized Sampling," *Proc. 44th Ann. Meet. West. Snow Conf.,* pp. 63–68, 1976.

150. Steppuhn, H. W., and G. E. Dyck, "Estimating True Basin Snowcover," *Adv. Concepts Tech. Study Snow Ice Resour. Interdiscip. Symp.,* U.S. Natl. Acad. Sci., Washington, D.C., pp. 314–328, 1974.

151. Sverdrup, H. U., "The Eddy Conductivity in the Air over a Smooth Snow Field," *Geophysiske Publikasjoner,* vol. XI, no. 7, 1936.

152. Tabler, R. D., and R. A. Schmidt, "Snow Erosion, Transport, and Deposition in Relation to Agriculture," in H. Steppuhn and W. Nicholaichuk, eds., *Proc. Symp.: Snow Management for Agriculture.* Great Plains Agricultural Council, University of Nebraska, Lincoln, Neb., Publ. 120, pp. 11–58, 1986.

153. Takeuchi, M., "Vertical Profile and Horizontal Increase of Drift-Snow Transport," *J. Glaciology,* vol. 26, no. 94, pp. 481–492, 1980.

154. Treidl, R. A., "A Case Study of Warm Air Advection Over a Melting Snow Surface," *Boundary Layer Meteorol.,* vol. 1, pp. 155–168, 1970.

155. Troendle, C. A., R. A. Schmidt, and M. H. Martinez, "Snow Deposition Processes in a Forest Stand with a Clearing," *Proc. 56th Ann. Meet. West. Snow Conf.,* Kalispell, Mont., pp. 78–88, 1988.

156. Tsang, G., "Frazil Ice and Anchor Ice and Their Resistance Effect in Rivers," *Proc. Can. Hydrol. Symp.,* Vancouver, pp. 127–138, 1979.

157. U.S. Army Corps of Engineers, *Report on Frost Investigations, 1944–1945, Addendum* 1945–1947, New England Div. Frost Effects Lab., U.S. Army Corps of Engineers, 1949.

158. U.S. Army Corps of Engineers, *Snow Hydrology: Summary Report of the Snow Investigations,* North Pacific Division, Portland, Ore., 1956.

159. U.S. Army Corps of Engineers, *Program Description and Users Manual for SSARR Model Program 724-KJ-G0010,* North Pacific Division, Portland, Ore., 1972.

160. U.S. Army Corps of Engineers, *HEC-2 Water Surface Profiles: Users Manual,* Hydrol. Eng. Lab., U.S. Army Corps of Engineers, Davis, Calif., 1979.

161. Department of Commerce, *National Weather Service River Forecast System Forecast Procedures,* NOAA, Silver Springs, Md., Tech. Memo. NWS HYDRO-14, pp. 1.1–1.20, 1972.

162. van Everdingen, R. O., "Ground-Water Hydrology," in T. D. Prowse and C. S. L. Ommanney, eds., *Northern Hydrology, Canadian Perspectives,* Environ. Can. Nat. Hydrol. Res. Inst., Saskatoon, Sask., Sci. Rep. 1, pp. 63–76, 1990.

163. Watt, W. E., ed., *Hydrology of Floods in Canada: A Guide to Planning and Design,* Nat. Res. Council Can., Ottawa, Ont., 1989.

164. Weisman, R. W., "Snowmelt. A Two-Dimensional Turbulent Diffusion Model," *Water Resour. Res.,* vol. 13, no. 2, pp. 337–342, 1977.

165. Wendler, G., "The Heat Balance at the Snow Surface During the Melting Period (March–April, 1966) Near Fairbanks, Alaska," *Gerlands Beitr. Geophys.,* vol. 76, no. 6, pp. 453–460, 1967.

166. Wheeler, E., "Interception and Redistribution of Snow in a Subalpine Forest on a Storm-by-Storm Basis," *Proc. 55th Ann. Meet. West. Snow Conf.,* vol. 52, pp. 70–77, 1987.

167. Wilm, H. G., and E. F. Dunford, "Effect of Timber Cutting on Water Available for Streamflow from a Lodgepole Pine Forest," *USDA Tech. Bull.* 965, 1945.

168. World Meteorological Organization, *Intercomparison of Models of Snowmelt Runoff,* Operational Hydrol. Rep. 23, WMO 646, Geneva, 1986.

169. Woo, M. K., "Permafrost Hydrology," in T. D. Prowse and C. S. L. Ommanney, eds., *Northern Hydrology, Canadian Perspectives,* Environ. Can. Nat. Hydrol. Res. Instit., Saskatoon, Sask., Sci. Rep. 1, pp. 77–101, 1990.

170. Yoshida, S., "Hydrometeorological Study on Snowmelt," *J. Meteorol. Res.* vol. 14, pp. 879–899, 1962.

171. Yosida, Z., "Infiltration of Thaw Water into a Dry Snow Cover," *Low Temp. Sci.,* Series A, 1973.

CHAPTER 8
STREAMFLOW

M. Paul Mosley
National Institute for Water and Atmospheric Research
Wellington, New Zealand

Alistair I. McKerchar
National Institute for Water and Atmospheric Research
Christchurch, New Zealand

8.1 THE NATURE OF STREAMFLOW

8.1.1 Definition of Streamflow

Streamflow is the *flow rate,* or *discharge,* of water, in cubic feet per second (ft^3/s) or cubic meters per second (m^3/s), along a defined natural channel. It is the component of the hydrologic cycle which transfers water, originally falling as rain or snow onto a watershed, from the land surface to the oceans. Hence streamflow at a particular point on a channel system is contributed by runoff from the *watershed* (also known as the *catchment* or *drainage basin*) upstream of that point, and return flow from the groundwater aquifer.

Streamflow is generated (Fig. 8.1.1*a*) by a combination of (1) *baseflow* (return flow from groundwater), (2) *interflow* (rapid subsurface flow through pipes, macropores, and seepage zones in the soil), and (3) *saturated overland flow* from the surface of poorly permeable or temporarily saturated soil, or from permanently saturated zones near the channel system (Fig. 8.1.1*b* and *c*). Interflow and saturated overland flow together comprise *quickflow,* the rapid runoff during and after rainfall of "new" water. Quickflow and baseflow are conventionally separated, on a streamflow hydrograph, by a line extended from the foot of the rising limb of the hydrograph to the falling limb, or *recession.* The upward slope of the separation line (Fig. 8.1.1*a*) is assigned for a particular region by examination of hydrograph form, field observation of runoff processes, and trial and error. Isotope techniques are able to distinguish between "new" and "old" water, and indicate that there is not, in practice, a clear separation between quickflow and baseflow.

Streamflow may be (1) *perennial,* in a channel which never dries up, (2) *intermittent,* in a channel which at drier times of year may have some reaches with flowing water interspersed with other reaches in which the water flows below the surface, and

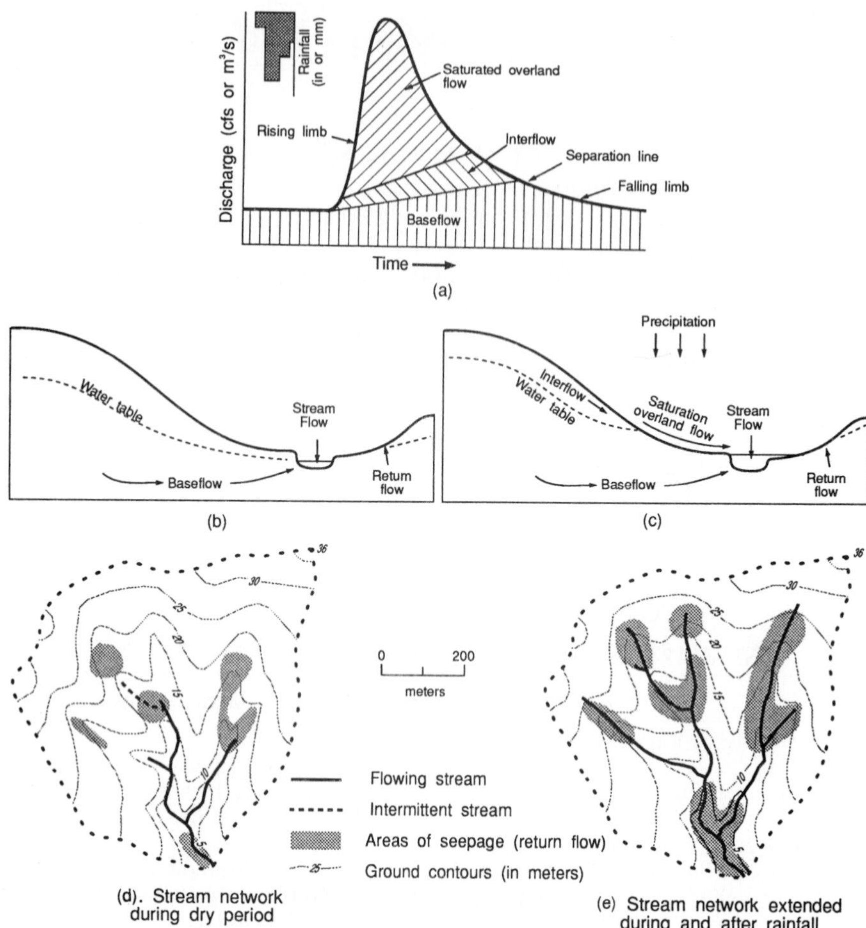

FIGURE 8.1.1 Separation of sources of streamflow on an idealized hydrograph *(a)*. Sources of streamflow on a hillslope profile during a dry period *(b)* and during a rainfall event *(c)*. The extent of a stream network during a dry period *(d)* and during a rainfall event *(e)*.

(3) *ephemeral,* in a channel which flows only after rainfall. For perennial streamflow to occur requires that the groundwater table intersect the streambed; otherwise seepage will cause the channel to dry up. Hence, as the water table rises and falls in response to rainfall (Fig. 8.1.1*b* and *c*), the flowing stream channel network may expand and contract (Fig. 8.1.1*d* and *e*).

8.1.2 Controls on Streamflow

Streamflow, as both a process and a storage component of the hydrologic cycle, reflects the volume of water supplied to a watershed as rain and other forms of precipitation, the rate of operation of other hydrologic processes (interception, infiltration, and evapotranspiration), and changes in the volume of other storages (lakes, aquifers, snowpack, soil moisture).

The streamflow rate at a particular point in time, at a particular point on a drainage system, integrates all the hydrologic processes and storages upstream of that point. The rate depends upon such factors as the sequence and size of rainfall events; the seasonal distribution and nature of precipitation; the extent, type, and transpiration of the vegetation cover; the infiltration capacity of the soil mantle; and the topography of the watershed. In many watersheds, streamflow is also strongly modified by human activity, such as by withdrawals for irrigation, interbasin water transfers, and discharges of wastewater.

In turn, streamflow creates and shapes the channel in which it is contained. In particular, there is a close relationship between the characteristics of an alluvial channel and the *bankfull discharge,* the flow which just fills the channel. Bankfull discharge has a recurrence interval on the order of once in 1.5 to 10 years[50,88] and is an index of those discharges which, over a period of years, are most effective in transporting sediment along the channel and in shaping its bed and banks.

The functional relationships between watershed characteristics, streamflow, and channel form may be expressed in terms of mathematical or statistical models which can be used to estimate streamflow. Hence the geomorphological context is an important starting point for considering streamflow.

8.2 GEOMORPHOLOGY AND STREAMFLOW

Geomorphology is the study of the landscape. The landscape system tends toward a dynamic equilibrium state in which the ability of flowing water to erode the land surface or streambed is balanced by the ability of those surfaces to resist erosion.

8.2.1 Morphometric Attributes of Waterways and Drainage Basins

The qualitative or descriptive attributes of the landscape are the subject of a vast literature, including textbooks such as those by Chorley et al.,[14] Dury,[17] Garner,[23] and Twidale.[84] The extensive research literature on drainage basins and rivers is summarized by Refs. 20, 25, 63, 65, 66, and 74, and others.

The quantitative approach to river and drainage basin morphology is well developed,[72,73] and many indices have been developed to describe landforms; selections are provided in Ref. 24 and in the above-mentioned texts. Some of these indices are duplicative, have been rarely used, or have little application for hydrologic purposes. Table 8.2.1 provides examples of the types of relationships which have been developed between hydrologic and geomorphologic variables, using statistical methods. These relationships have been used to enhance understanding of landscape formation but are also useful for estimating hydrologic parameters from measurements of landscape form. This is the basis of regional streamflow estimation (Sec. 8.5).

8.2.2 The Channel Cross Section

Morphometry of channel cross sections is well understood, particularly with respect to "regime theory" and *hydraulic geometry.*[6,43,65] The measurements commonly used are based on standard land surveying techniques or can be obtained from observations made for flow gauging purposes. The shape and size of alluvial channel

TABLE 8.2.1 Relationships between Geomorphology and Hydrology

Equation	Area	Reference
Discharge related to channel form		
1. $Q_{2.33} = 3.741\ X_S^{1.015} \cdot R_{asp}^{-0.515}$	South Island, New Zealand	49
2. $Q = 0.129\ X_s^{1.157} \cdot R_{asp}^{-0.781}$	South Island, New Zealand	49
3. $Q_{bf} = 8.913\ X_S^{1.27} \cdot \mathrm{P/R}^{-0.267}\ \mathrm{S}^{0.317}$	South Island, New Zealand	49
4. $Q_{2.33} = 0.00055\ \mathrm{L}^{1.96}$	South Island, New Zealand	50
5. $L = 395\ \mathrm{Q}_{2.33}^{0.48}\ \mathrm{M}^{-0.74}$	Australia and USA	70
Discharge related to drainage basin form		
6. $Q_{7,20} = 2.74 \times 10^{-8}\ A^{1.08} P^{3.93} F^{-0.61} I^{2.08}$	Potomac River, USA	83
7. $Q_{2.33} = 1.08\ A_{2.33}^{0.77} P^{2.92} D^{0.81}$	UK	68
8. $Q_{2.33} = 12\ A^{0.79}$	New Mexico	42
9. $Q_{2.33} = 0.0607\ A^{0.94} P_5^{1.03} I_{soil}^{1.23} FR^{0.27} I_{slope}^{0.16} (1 + L)^{-0.85}$	UK	59
10. $Q_{2.33} = 1.48 \times 10^{-4} A^{0.82} P_2^{2.18}$	North Island, New Zealand	5

$Q_{2.33}$ = mean annual flood, m³/s
Q = mean annual discharge, m³/s
Q_{bf} = bankfull discharge, m³/s

X_S = channel cross-section area, m²
R_{asp} = ratio of maximum depth/mean depth
P/R = ratio of wetted perimeter/hydraulic radius
S = channel slope, m/m
L = meander wavelength, m
M = Weighted silt-clay content and banks, %

Discharge:
$Q_{7.20}$ = minimum annual 7-day mean flow, 2-year return period, ft³/s
$Q_{2.33}$ = mean annual flood (ft³/s; m³/s in Eqs. 1–4, 9, 10)
Precipitation:
P = mean annual precipitation, in
P_5 = 5-year return, 24-h rainfall, with soil moisture deficit subtracted
$P_{2.33}$ = mean annual daily maximum rainfall, in
P_2 = 2-year return, 24-h rainfall, mm

Drainage basin:
A = drainage area (mi²; km² in Eqs. 1–4, 10)
D = drainage density, mi/mi²
F = forested area, %
FR = stream frequency, number/km²
L = fraction of area occupied by lake storage
I_{soil} = soil index
I = soil infiltration
I_{slope} = slope index

cross sections are closely related to the flows responsible for forming them. As flow increases in the downstream direction, channel width and mean depth tend to increase, while water surface slope decreases (Fig. 8.2.1). The rate of change may be broadly similar between rivers, although other factors—particularly the size and quantity of sediment load and the nature and vegetative cover of the banks—are also significant controls.

Streamflow indices may be estimated from channel form, which is measured more easily and cheaply than streamflow. Table 8.2.1 gives examples (which are applicable only to the locations from which the data were obtained) of regression relationships developed to estimate streamflow from channel measurements. These relationships were developed at sites for which streamflow records were already available, for estimating streamflow at nearby locations where records are not available.

Various indices of streamflow are used, including *mean annual flood* ($Q_{2.33}$), *most probable annual flood* ($Q_{1.58}$) (which have an average recurrence interval of 2.33, and 1.58 years, respectively), and *mean annual discharge* (Q). Because channel form is a function of the type and quantity of sediment load, and bed and bank material, as well as flows,[71] the relationship between channel morphometry and flow varies re-

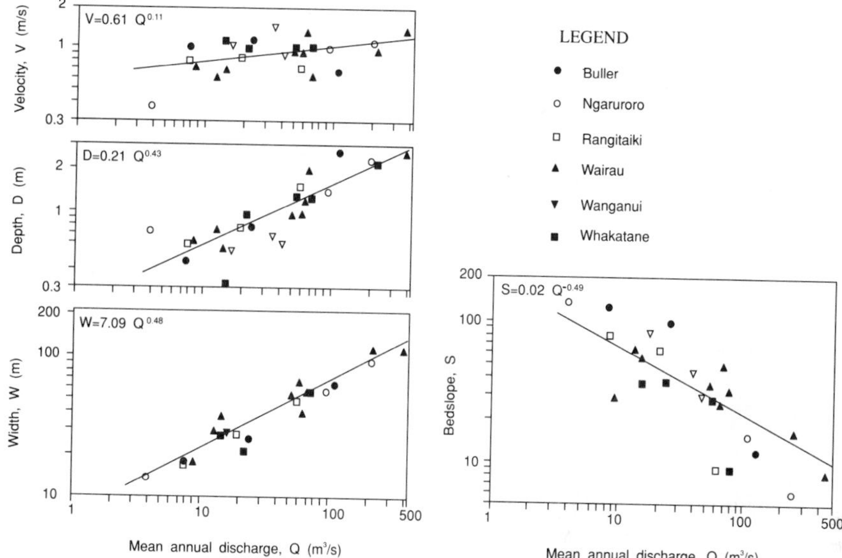

FIGURE 8.2.1 Hydraulic relationships downstream along six New Zealand gravel-bedded rivers: mean velocity, mean depth, water surface width, and channel bed slope as a function of mean annual discharge. *(After Griffiths.[26] Used with permission.)*

gionally, in response to sediment characteristics. Furthermore, large, infrequent floods may cause dramatic and long-lasting changes to a particular section of channel, so that a particular cross section may not be in equilibrium with those moderate flood events, occurring once a year or so, which are normally responsible for the long-term average channel form.

The concept of bankfull discharge as a channel-forming flow is not appropriate for all channels, such as those with bedrock banks or braided channels or those which are known to be actively degrading or aggrading. It may be applied in some channels without a floodplain or well-marked bank top by reference to vegetation lines, such as the boundary on a rock bank between the clean or slightly moss-covered rock and perennial grasses and shrubs, which can tolerate only infrequent and brief periods of inundation.

Channel cross-section shape and dimensions may vary markedly over short distances. This variation may be repeated in a quasi-regular fashion in a channel characterized by meanders, bends, and/or riffles and pools (Fig. 8.2.2), but in other channels the variation appears to be random. For purposes of quantitative analysis and comparison, it is therefore necessary to ensure that cross sections intended to be representative of different river reaches are selected in a consistent fashion, e.g., by always choosing a section at the crossover between two bends, or by making measurements at a number of sections in each reach, and calculating the mean and standard deviation of each morphometric variable in that reach. The number of sections depends on the variability displayed and the analyses to be used, but 25 to 30 or more are desirable. The latter approach is becoming more widely used, for instance, in studies of the stream channel environment for fish habitat assessment, because there is a need for information on frequency distributions of morphometric attributes rather than simple average values.[51]

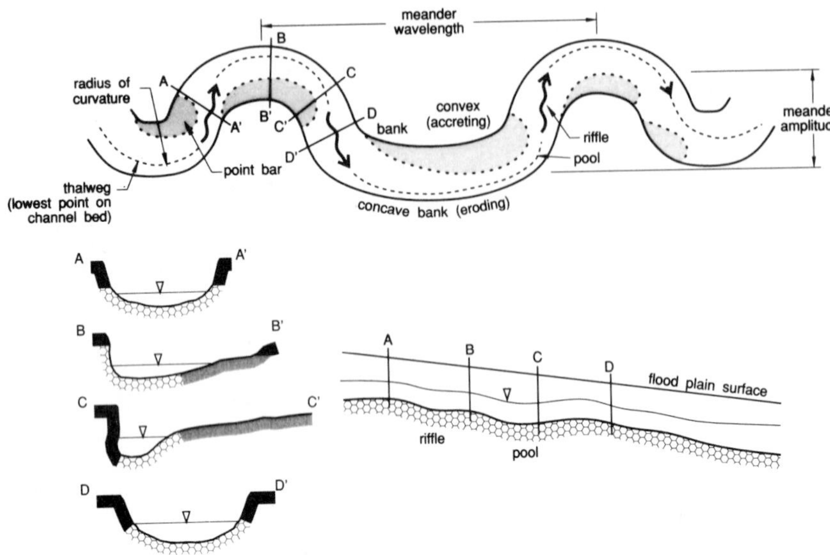

FIGURE 8.2.2 Definition sketch of the plan geometry of a meandering river.

8.2.3 The Channel Reach

Channel reaches are characterized by their gradient and plan geometry as shown in Fig. 8.2.3. Both are hydrologically significant; for example, channel gradient is a partial determinant of flow velocity, travel time, and sediment transport capacity, and planimetric indices such as meander wavelength are often closely correlated with discharge.

Channel gradient may be quite variable over short distances, and a field survey is desirable for detailed studies, particularly of streams and small rivers. However, for larger rivers, the floodplain gradient may be measured from 1:25,000 or 1:50,000 maps. Over distances greater than one or two wavelengths of riffles or bends, floodplain gradient becomes a close approximation of water surface or energy gradient, if the measurement is referenced to equivalent points on the repeating patterns of pools and riffles (bars or crossing, Fig. 8.2.2).

Several classifications of river channel planform or pattern have been proposed. Some of these, being based on the relationship between morphology, hydrologic characteristics, and sediment load characteristics, are useful in reconnaissance studies in assisting the hydrologist to make inferences about hydrology from the more readily observed morphometry (Fig. 8.2.3).[48,75] However, rivers form a continuum, rather than a series of exclusive classes.[52]

Quantitative measures of channel pattern, particularly those descriptive of meander form, have been much investigated. There is a strong relationship between meander wavelength and discharge, although few rivers have meanders which approach the regularity of shape needed to measure wavelength or related variables with great precision. *Sinuosity* is easily measured from good topographic maps or aerial photographs; it is more strongly related to sediment and topographic characteristics and to the energy of the waterway than to indices of streamflow.

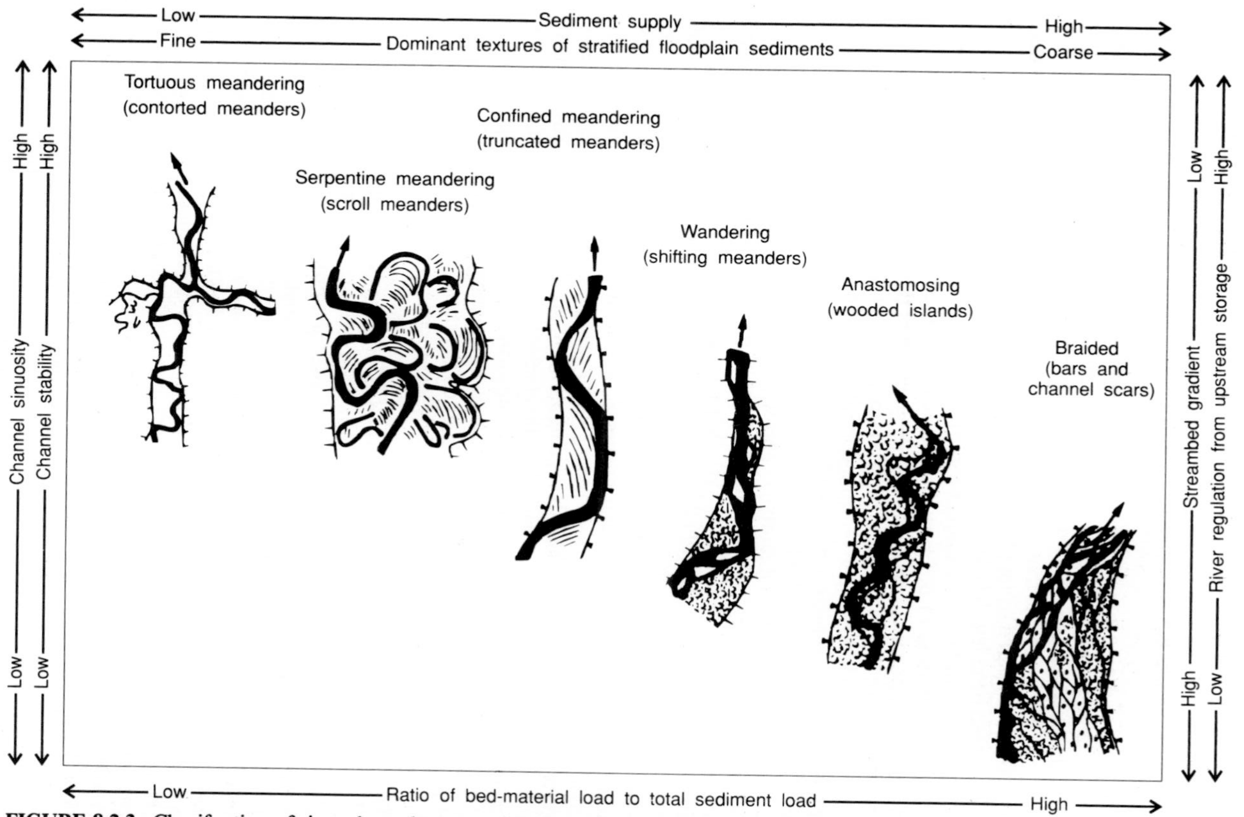

FIGURE 8.2.3 Classification of river channel types and controlling variables. *(After Mollard.[48] Used with permission.)*

8.2.4 The Channel Network

The shape of the channel network reflects hydrologic processes and also partially controls them, through the channel flow resistance. Network characteristics also strongly reflect lithology, topography, and the geologic structure and history of the area.

Analyses of drainage networks are very dependent on the quality of the data source. A large number of studies (many summarized by Goudie[24] and in Table 8.2.2) indicate the unreliability of standard topographic maps in many parts of the world. As a rule, it appears that only modern topographic maps at scales greater than 1 : 50,000 and prepared using photogrammetric techniques can be used confidently, and data taken from less detailed maps should be checked against aerial photographs. In comparative studies, data sources for different sites should be equally accurate.

The methods used in defining the extent of the network, and particularly the identification of fingertip tributaries and the location of their sources, are important. When using maps as a data source, marked streams ("blue lines") can be used which, depending on topographic convention, delimit the stream network at "normal summer flows" or some similar condition (Fig. 8.2.4*b*). Alternatively, the network can be extended into topographic hollows, marked by contour crenulations, on the assumption that surface flow during rainfall events will extend into locations where saturation occurs. When a network is being surveyed in the field, several definitions may similarly be adopted, for example: (1) channels with flow during the survey; (2) defined permanent channels, whether or not they are flowing at the time (ephemeral streams), and (3) topographic hollows with surface saturation or with vegetation characteristic of marshy ground, which could have ephemeral flow for short periods following rainfall (Fig. 8.2.4*a*). In some areas, the stream network may extend even further, in the form of subsurface pipes, which are particularly prevalent in peat-covered terrain.[40] Indeed, streamflow generation by rapid subsurface flow through macropores in the soil clouds even further the distinction between streamflow and its surface expression as stream channels and subsurface flow processes (Fig. 8.2.4*d*).

Once the drainage network has been defined and mapped, a variety of topological and geometric attributes can be measured, such as drainage density and area (Table 8.2.2; Fig. 8.2.4), and may be used in statistical models for estimating streamflow (Table 8.2.1, Fig. 8.2.5).

The topologic characteristics of a stream network may be described by the *stream ordering* systems developed by Strahler[79] and Shreve.[77] In the Strahler system, the "fingertip" tributaries are assigned order 1; two first-order channels combine to become, below their confluence, a second-order channel, and so on (Fig. 8.2.4*c*). A junction with a lower-order channel (e.g., a third order flowing into a fourth order) does not change the order of the higher-order stream. In the Shreve system, the fingertip tributaries are assigned magnitude 1; each downstream channel segment has a magnitude equal to the number of first-magnitude tributaries upstream.

Horton's laws of drainage network composition relate stream numbers, catchment areas, and lengths to stream order.[33] However, in general, efforts to establish relationships between streamflow and indices of network structure have not been very successful.

Drainage basin area is perhaps the easiest characteristic to relate to hydrology, although other variables such as basin slope are also correlated with drainage area, so that the significance of drainage area alone is not always easy to interpret. Indices of streamflow such as the mean annual flood or mean annual runoff generally increase with drainage area, but because of the restricted areal extent of intense precipitation, stormflow per unit area is inversely proportional to area.

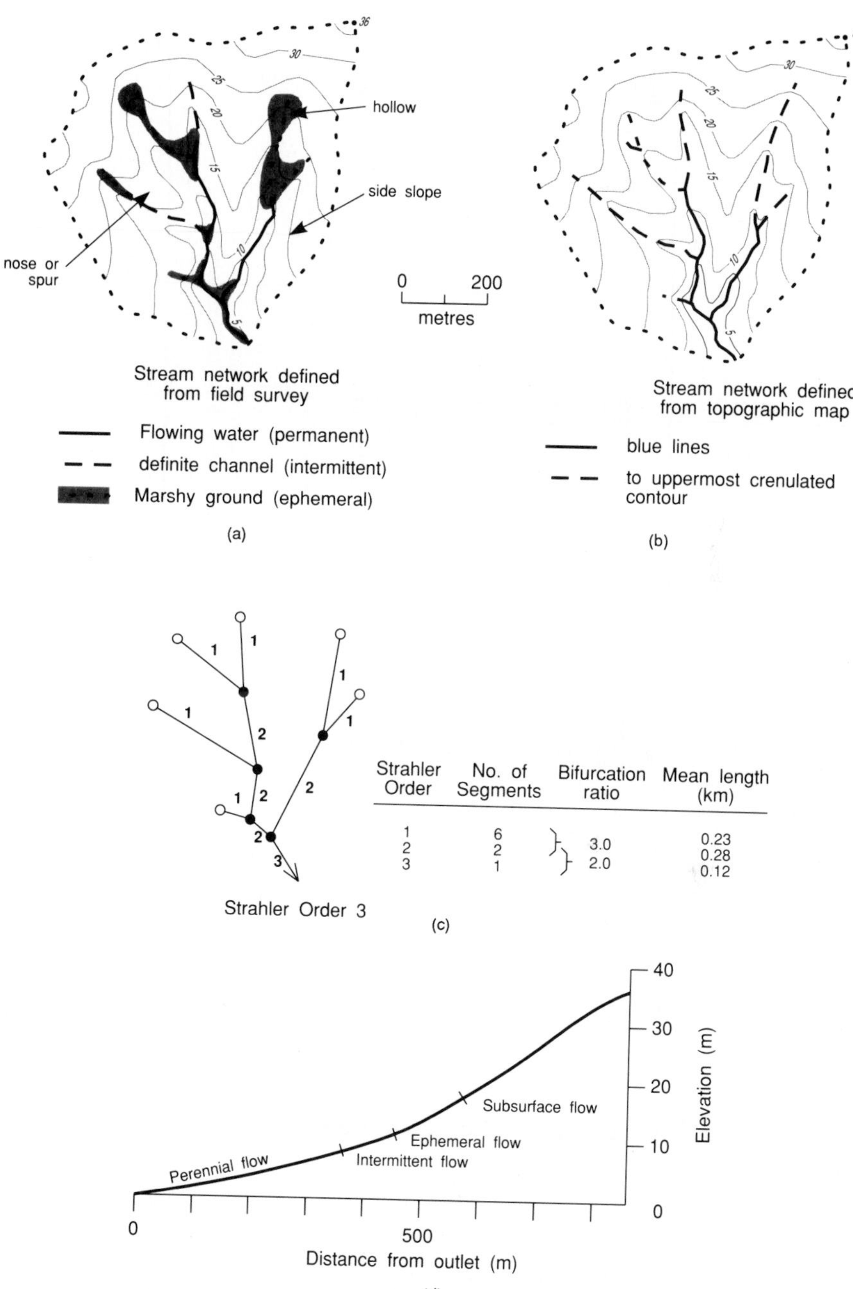

Strahler Order	No. of Segments	Bifurcation ratio	Mean length (km)
1	6	3.0	0.23
2	2	2.0	0.28
3	1		0.12

FIGURE 8.2.4 Definition sketches of network attributes, for an imaginary watershed. Stream network defined from field survey *(a)* and from a topographic map *(b)*; stream order and stream magnitudes defined using the Strahler system *(c)*. Longitudinal profile of the main stream in *(d)*.

TABLE 8.2.2 Drainage-Basin Morphometry

Variable	Symbol	Dimensions	Method of measurement	Comment	Value in Fig. 8.2.4b
Total channel length	C_t	Length	Opisometer, digitizer		2.08 km
Mainstream length	C_m	Length	Opisometer, digitizer	Needs subjective definition of the mainstream, to its source	0.86 km
Basin perimeter	P_b	Length	Opisometer, digitizer		2.44 km
Basin length	L_b	Length	Ruler, digitizer	Various definitions possible: to highest point; to point above source of mainstream; to most distant point	0.79 km
Basin area	A	Length^2	Planimeter, digitizer		0.38 km^2
Drainage density	D	Length^{-1}	C_t/A		5.47 km/km^2
Relief	H_b	Length	From contours or spot heights	To maximum elevation	36 m
Relief ratio	R	—	H_b/L_b		0.046
Elongation	E	—	$A^{0.5}/L_b$	E is smaller (<1.0) for basins which are more elongated	0.79
Circularity	E_c	—	$A/(P_b^2/4\,\pi)$	Ratio of basin area to that of a circle with same perimeter as the basin	0.80
Network diameter		—	From network map	Greatest number of links between outlet and source	5

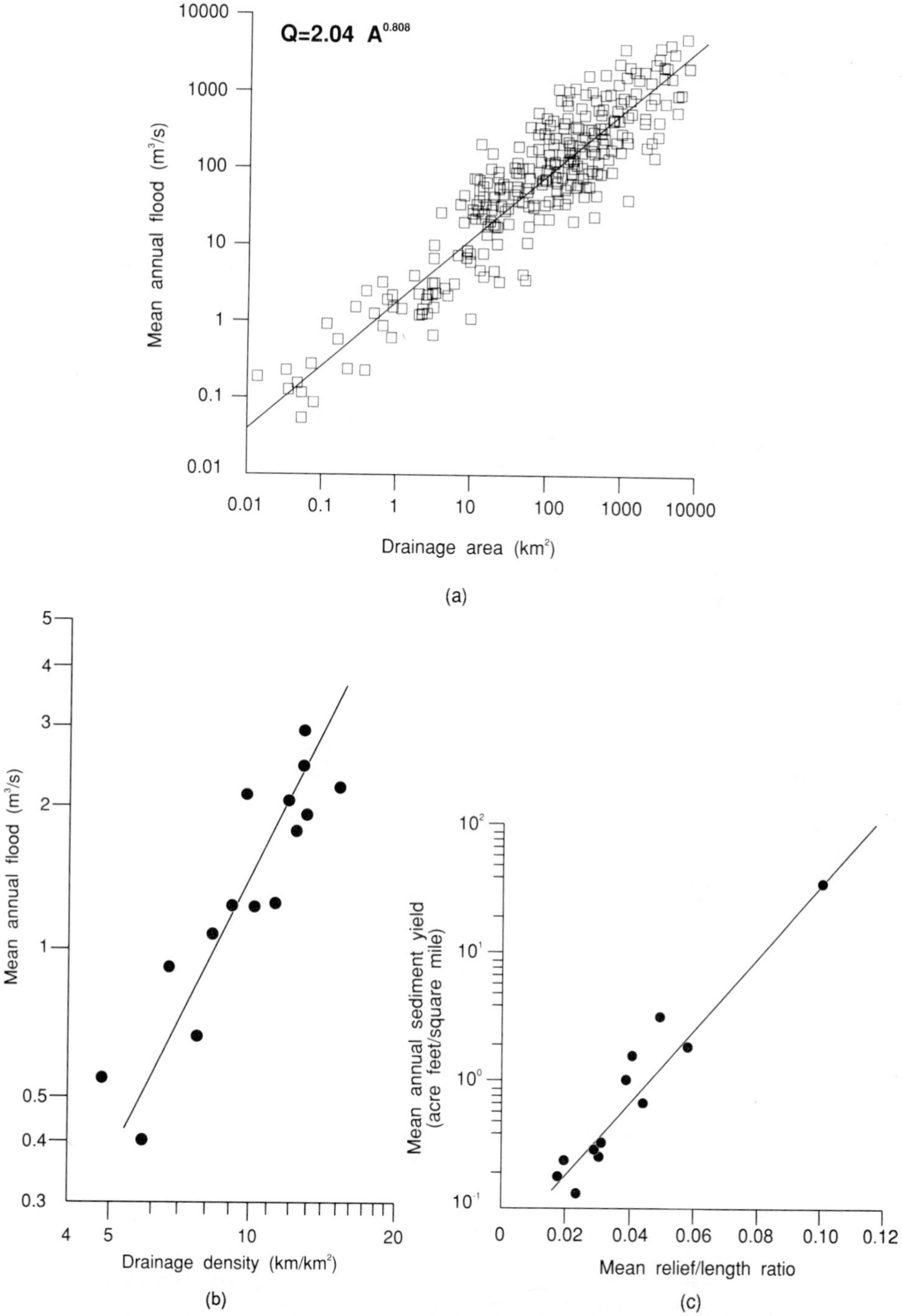

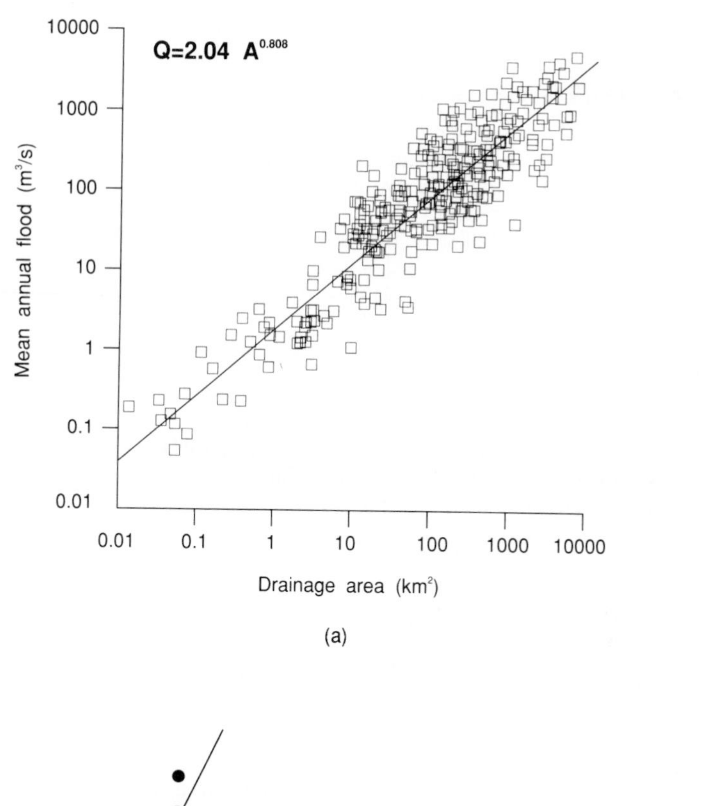

FIGURE 8.2.5 Representative examples of relationships between geomorphologic and hydrologic variables: *(a)* Mean annual flood and drainage area of drainage basins in New Zealand. *(b)* Mean annual flood and drainage basin density for northeast United States. *(After Carlston.[12])*. *(c)* Mean annual sediment yield and watershed relief-length ratio. *(After Hadley and Schumm.[27])*

8.3 STREAMFLOW MEASUREMENT

Streamflow or discharge measurement normally involves (1) obtaining a continuous record of water levels, or *stage* above a datum; (2) establishing the relationship between water level and discharge (the *stage-discharge relation*, or *rating curve*); and (3) transforming the record of stage into a record of discharge. A streamflow measurement station is commonly called a *gauging station*. Standard procedures for many aspects of streamflow measurement are summarized by the International Standards Organization.[36]

8.3.1 Operation of a Gauging Station

A recorder at a stream gauging station records water-surface elevation at regular intervals, usually every 15 min. The elevation datum is set with reference to at least three permanent reference marks or *benchmarks* located in stable ground separate from the recorder structure. In a natural channel, the rating curve is established manually over a period of years by measuring discharge and water level over the observed range of flow conditions. With weirs and flumes, the rating curve can be determined experimentally in a laboratory, but it is normally checked in the field.

Selection of a Site. The most important requirements for the location of a water-level recording station are that the downstream *hydraulic controls*, which control the water level at the station and determine the stage-discharge relation, are stable and sensitive to changes in discharge; that is, the stage-discharge relation should not change over time, and a measurable change in discharge should be matched by a measurable change in stage. Additional requirements for a good gauging station site are: the site should be accessible even during high flows, flows should be confined into a single channel at all times, and the hydraulic control should be unaffected by variable backwater conditions caused by tidal effects, downstream lakes, or inflowing tributaries.

Manual discharge measurement, or *gauging*, requires a location on a straight section of channel where the streamlines of flow are approximately parallel, and where water velocities are within the range that can be measured accurately with available current meters, typically 0.1 to 20 ft/s (0.03 to 6 m/s). The location need not be the same as that chosen for the water-level recorder but should be sufficiently close that the gauging is an accurate measurement of the discharge at the recorder.

8.3.2 Measurement of Stage

The Measurement Site and Control Section. A *section control* is a natural constriction in the channel caused by a rock outcrop, an artificial constriction such as at a bridge where the waterway width is narrower than the natural channel, or a downstream increase in channel slope, as encountered at the head of a waterfall. *Artificial controls*, such as groins or sills across the streambed, are often used to stabilize the channel. *Channel control* (or friction control) is provided when the natural roughness of the channel perimeter controls the velocity and hence the depth of flow. This is the case in a long straight channel with a uniform cross section.

Often a site in a natural channel has more than one control. A low flow or *primary control* may be a riffle of gravel or a rock bar; this may be drowned out by the action of other controls farther downstream at higher flows (Fig. 8.3.1).

Weirs and Flumes. Weirs (Fig. 8.3.2) or flumes may be deployed in smaller

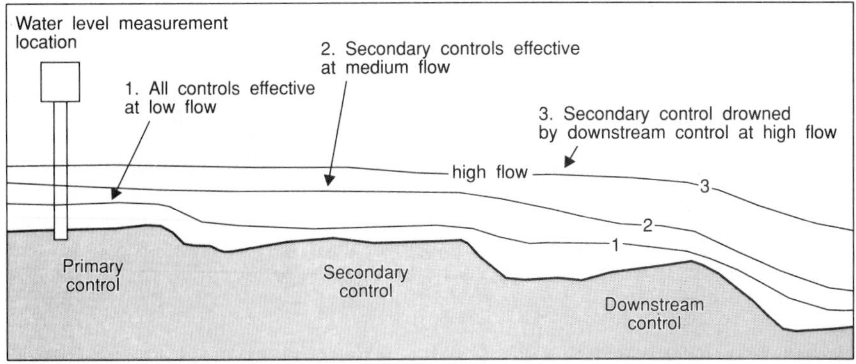

FIGURE 8.3.1 The influence of downstream controls on water level.

streams.[1] These artificial controls have predetermined rating curves, but it is advisable to calibrate the structure by carrying out discharge measurements. Some agencies, such as the U.S. Geological Survey, require in situ calibration. Continuing gaugings are also advisable to confirm the applicability of the rating curves, which could be altered by sediment movement and vegetation growth, or by structural settlement. Errors in extremely low flows may occur through imprecise leveling of the invert of the structure, and errors in flood discharges may occur if the rating curve is extended beyond the calibrated range.

Techniques for Stage Measurement. A traditional device for recording stage is a float in a stilling well, connected by a counterweighted tape to a chart or punched-tape recorder and ancillary equipment. Data are obtained either by digitizing charts or by

FIGURE 8.3.2 Streamflow measurement site equipped with V-notch weir and stilling well on Graham Creek near Nelson, New Zealand. *(Photo: M. J. Duncan.)*

machine-reading punched tapes, and after checking and processing, the data are stored in archives.

Float recorders are appropriate for streams with narrow, incised gravel-bed channels, for which the stilling well can be located close to the water. For wide sandy channels, the stilling well must be located on the stream bank some distance from the main body of flow and the pipes connecting the stilling well to the main stream are vulnerable to blockage by siltation. For these reasons, other installations common in the United States use *pressure transducers,* pressure bulbs, or *manometers.* Water-level data can be accumulated on-site with solid-state data loggers. *Telemetry,* using land-line, radio, or satellites, is a cost-effective means to retrieve data when the travel distance to the gauging station is large, and telemetry is particularly beneficial when real-time retrieval of data is required for water management and flood forecasting.[60]

Maintenance Requirements. Collection of data that conform to specified standards requires regular inspection and maintenance of gauging stations [ISO standard ISO 1100/1-1981(E), Sec. 7.1.7.1]. The records of inspections are important elements of a quality-assurance program for hydrologic data collection, because they provide evidence of the integrity of the archived data.

8.3.3 Measurement of Discharge

Velocity and Depth. Discharge at a given stage is computed from measurements of velocity and depth at a cross section near the recorder. Velocity is measured at locations or *verticals* spaced across the cross section, using a *current meter.* The spacing between velocity measurements should be such that not more than 10 percent of total flow is represented by any one vertical. The meter is suspended on a cable controlled by a winch (Fig. 8.3.3*b*), or in shallow water it is mounted on a measuring rod carried by the gauger (Fig. 8.3.3*a*). The depth at each vertical is measured using the cable or measuring rod. Other methods of measuring velocity use floats, a boat equipped with current meter and sonar, a velocity head rod, or ultrasonic equipment (Table 8.3.1). Velocity in a vertical approaches zero at the streambed and is greatest at or near the free surface (Fig. 8.3.4). The mean velocity typically is at 0.6 of the depth from the surface. Commonly, mean velocity in a vertical is estimated by measuring velocity at 0.6 depth in water less than 2.5 ft (0.76 m) deep, or by taking the mean of velocities measured at 0.2 and 0.8 depth in water more than 2.5 ft (0.76 m) deep. Where depth cannot be determined during floods, velocity may be measured just below the surface. The depth can be estimated after the flood has passed, and coefficients derived from the standard velocity curve (Fig. 8.3.4) can be used to estimate the mean velocity in the vertical.

Current Meters. A *current meter* (Fig. 8.3.3) contains a rotating element whose speed of rotation is proportional to the water velocity. Vertical-axis meters with rotating cups (*Price-type meter,* Fig. 8.3.3) are commonly used in the United States. Horizontal-axis rotating screw (or propeller) meters are less prone to fouling by weed. International standards specify the construction and maintenance of current meters (ISO standard ISO 2537-1974). Calibration tables for current meters, which relate the rotational speed to water velocity, are developed by towing the meter through still water in a tank at a series of known velocities (Ref. 31, ISO standard ISO 3455-1976).

Floats. When current meters are not available or cannot be used, *floats* provide a less accurate method of velocity measurement. A float is substantially submerged

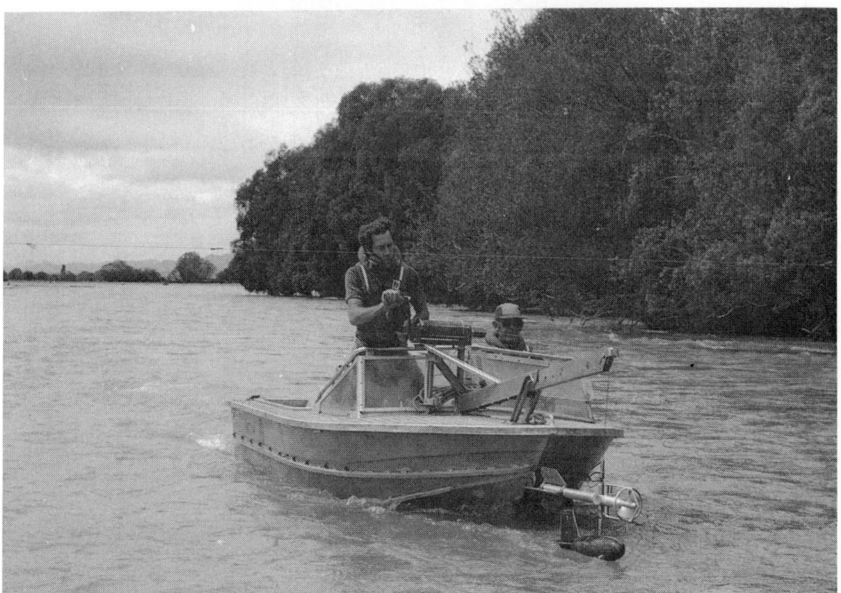

FIGURE 8.3.3 Current meter used for measuring velocity: *(a)* mounted on a measuring rod (Onyx River, Antarctica); and *(b)* suspended on a cable from the bow of a jet-boat (Hurunui River, New Zealand). *[Photo (a) M. P. Mosley; (b) J. K. Fenwick.]*

and therefore unaffected by wind. A distinctive color is helpful. The time of travel over a known distance is measured and velocity is computed as distance/time. The reach used should be straight with uniform flow, and the distance should be such that the time of travel is at least 20 s. Several floats should be used at intervals across the reach. If surface floats are used, measured velocity must be converted to an estimated

TABLE 8.3.1 Other Discharge Measurement Methods

Method	Description	Application	Features	References
Dilution gauging	Uses extent of dilution of a concentrated tracer to calculate discharge. Tracers include common salt, fluoroscene dyes	Steep mountain streams where mixing occurs readily	Some versions can be automated to measure discharge at a predetermined stage. Independent of velocity area methods, so provides useful check for systematic errors	31, 39, 89
Moving boat method	Echo sounder measures depths. Near-surface velocity deduced as upstream component of observed velocity vector when boat is driven across river between markers. Angles of apparent velocity measured with vane	Suited to large rivers where width exceeds 500 ft (150 m), depth exceeds 7 ft (2 m), velocity exceeds 1.5 ft/s (0.5 m/s). Suitable for measurement during tidal cycles	Requires motor boat, echo sounder, and vane to measure current angle	21, 78
Ultrasonic	Velocity is deduced after differencing the time of travel of ultrasonic pulses directed diagonally upstream and diagonally downstream between two transducers	Lake outflows, tidal channels, and other situations where a rating is undefined. Not suited to wide, shallow, or weedy channels	Continuous measurement. Requires no sediment load. Relatively expensive to install	Sec. 25.2.3
Electromagnetic	An electromagnetic field is generated in the water by a coil under the river. This creates a voltage difference between electrodes on opposite sides of the river which is proportional to flow velocity	Channels to about 100 ft (30 m) wide. Useful in weedy rivers, or rivers with silty or moving beds	Expensive installation—requires mains power and coil placed under impermeable membrane below the river	Sec. 25.2.2

Volumetric	Discharge measured by time taken to fill a container of known volume	Checking calibrations of weirs and flume. Needs at least 1.5 ft (0.5 m) free overall from notch of weir or base of flume. Used with low flows, e.g., less than 0.35 ft^3/s (10 L/s)		
Rising air float	Velocity profiles deduced from the surface profile of bubbles rising from a submerged perforated air hose laid across the cross section	Slow-flowing rivers and canals	Relatively undisturbed water surface necessary to view bubble profile	69
Velocity head rod	Height of standing wave is equated to velocity head	Small streams and irrigation canals that can be waded. Useful in weedy channels	Velocity range 1.5 to 8 ft/s (0.5 to 2.5 m/s). Simple, but less accurate method	76
Indirect measurement of peak discharge at width contractions	Solves energy equation for velocity using measured head loss through contractions	Used to estimate peak discharge where bridge abutments constrict flows	Gives peak discharge only. Peak levels can be measured with crest stage indicators	45
Indirect measurement of peak discharge at culverts	Uses measured water levels and culvert geometry to solve energy equation for velocity	Used at culverts	Gives peak discharge only. Peak levels can be measured with crest stage indicators	7
Indirect measurement of discharge over dams, crests, and embankments	Apply weir discharge equations	Used to estimate flow over dam crests and embankments		34

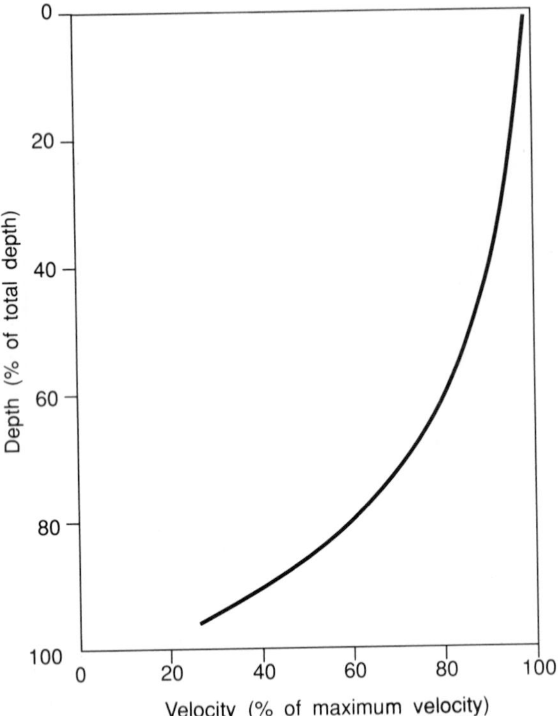

FIGURE 8.3.4 Typical vertical velocity curve. *(After Rantz and others.[64])*

mean velocity in the vertical (Fig. 8.3.4). Typically the mean velocity in a vertical is 0.86 of velocity measured at the surface. Separate soundings are necessary to determine the depth along the course of each float.

Velocity-Area Methods of Discharge Calculation. The *midsection method* of discharge calculation in velocity-area gauging assumes that the mean velocity for the vertical i represents the mean velocity in the cross section, from halfway to the preceding vertical $(i-1)$ to halfway to the next vertical $(i+1)$ (Fig. 8.3.5). Incremental discharge for each measurement is computed as

$$q_i = v_i\left(\frac{b_{i+1}-b_i}{2}+\frac{b_i-b_{i-1}}{2}\right)d_i$$
$$= v_i\left(\frac{b_{i+1}-b_{i-1}}{2}\right)d_i \tag{8.3.1}$$

and total discharge is obtained by summing the incremental discharges for all segments. This method is preferred because it is slightly more accurate than the alternative *mean section method.*

Standard gauging cards (Fig. 8.3.6) are used in the field to record gauging data, and computations of discharge may also be recorded on this form. The gauging card becomes an original record, which is part of the history of the station and is archived for future reference. Increasingly, gauging data are recorded on hand-held computers, which also do the discharge computations.

Standard procedures have been developed[31,64] to handle special situations: the stage is rapidly changing, and the mean stage must be calculated; the flow is deep and swift, and the sounding line does not hang vertically but is swept downstream; the measurement cross section is not normal to the direction of flow.

Sources of Uncertainty in Streamflow Gauging. Herschy[30] analyzed the sensitivity of velocity-area gaugings to *random uncertainties* in the constituent measurements of width b_i, depth d_i, and velocity v_i. The number of verticals and duration of the velocity measurements are the prime sources of uncertainty in velocity-area gauging. With 20 verticals and a Price-type current meter calibrated to ISO standards, gaugings in shallow water [less than 2.5 ft (760 mm) deep], where velocity is measured at 0.6 depth, have a coefficient of variation for random errors of approximately ± 4.3 percent. Gaugings in deeper water [more than 2.5 ft (760 mm)], with velocity measured at 0.2 and 0.8 of depth, have a coefficient of variation for random errors of ± 3.0 percent.[13] Gauging with at least 20 verticals is recommended in Sec. 8.1.2 of ISO standard ISO 748-1979(E).

Systematic uncertainty, or bias in velocity-area gauging, results from errors in calibrated tapes, cables, and winches for depth and width measurements, and from incorrect current meter calibration. Demonstration of systematic errors in gaugings requires alternative discharge measurements. For example, differences of 10 to 20 percent between American and Canadian winter discharge measurements in the Red River of the north were attributed to significant differences between a current meter rating on a rod and on a cable suspension.[61]

Slope-Area Method of Discharge Calculation. The slope-area method (Dalrymple and Benson,[16] ISO standard ISO 1070-1973) is applied in straight, uniform, or slightly convergent channels, to compute the peak discharge for a flood event. The maximum stage height for at least three cross sections along the channel is marked during a flood, measured with crest-stage indicators, or inferred from debris marks soon afterward. The cross-section areas and water surface slopes between these cross sections are surveyed, mean values are estimated, velocity is calculated from the Manning equation [Eq. (8.3.2)], and finally the associated maximum discharge is calculated.

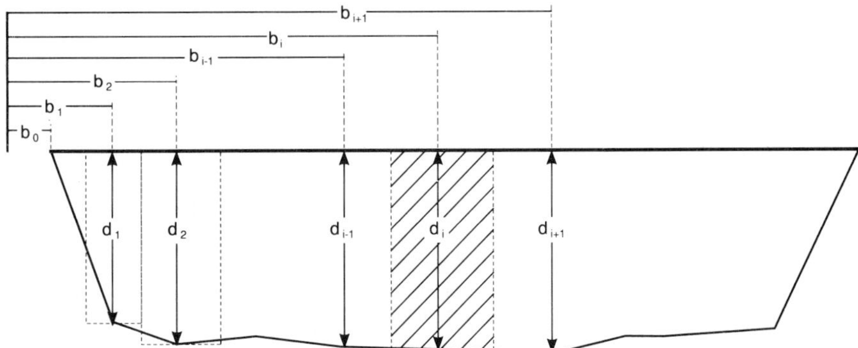

FIGURE 8.3.5 Delineation of a cross section for measurement of discharge by the velocity-area method.

Site No. 21801 WS Form 4A

DISCHARGE MEASUREMENT NO. 9484

MOHAKA River at RAUPUNGA

River Number 218000 Map Reference: N115: 542895

Party Smith / Fairlie Date: 9 Jan 1987

FIELD DATA:

Measured by: (Current Meter) ~~Floats/Slope Area/Dilution/Other~~

Meter Type: GURLEY No. TY0453 Rating No. 7 Date 8.5.86

Slope .... Constant ....

Spin Test: Before 270 secs. After 245 sec.

Used (Rod) Cable. Meter — mm above bottom of — kg weight

24 Verticals: 6 @ 0.6D, 18 @ 0.2D/0.8D

Measured from ~~slackline/cableway/boat/upstream/downstream side bridge~~ (wading.)

Measured 5 m (above) ~~below/at~~ Recorder tower

Wind 5-15 km/h (up) ~~down/across.~~ Angle of current (nil) variable/constant — degrees

Water Temp. 22.0 °C ~~Turbid~~ (Discoloured) ~~Clear~~

STAGE READINGS			
Time	Recorder	Well	Staff Gauge
1115	1.846	1.846	1.846
1252	Measurement began		
1300	1.845		
1315	1.845		
1330	1.841		
1345	1.842		
1351	Measurement ended		
1352	1.842	1.846	1.842
Mean G.H.	1.844		1.844

COMPUTED DATA:

Discharge: 24.082 m^3/s ~~l/s~~

Stage Ht. change ~~nil~~/ 0.004 m

Rate of ~~rise~~/(fall) mm/h

Area 37.125 m^2

Width 59.300 m

Max. Depth 0.860 m

Max. Surf. Vel. — m/s

Mean Vel. 0.649 m/s

Wet. Perim. 60.115 m

Hyd. Rad. 0.618 m

Slope ....

n ....

Specific Dis. ....

Water Level R.L.: ....

Sediment Conc. .... mg/*l*

Remarks: ....

Computed by: TF Checked by: TK

Supplied by Water Resources Support Centre DSIR Kainga, N.Z.

FIGURE 8.3.6 Example of gauging card, as used in New Zealand. Details of width, depth, and velocity measurements are listed in columns on the interior of the card. Calculations are normally done by computer and are no longer manually entered on the card.

Angle Coef.	Dist. from Init. Point	Obs. Depth	Method	Depth of Obs.	Revolu- tions	Time in Secs.	Velocity				Area	Mean Depth	Width (W)	Discharge
							At Point	Mean in Vert.	Adjusted for Horiz. Angle	Mean in Sect.				
	9.0	WELB		0										
	10.0	0.86	2	.172	2	52								
			8	.688	2	60								
	12.0	0.67	2	.134	25	41.6								
			8	.536	15	40.0								
	14.0	0.53	2	.106	40	40.2								
			8	.424	25	42.7								
	16.0	0.50	2	.1	45	44.3								
			8	.4	30	45.2								
	18.0	0.56	2	.112	40	40.5								
			8	.448	30	46.5								
	20.0	0.58	2	.116	45	45.2								
			8	.464	30	40.0								
	22.0	0.58	2	.116	40	40.0								
			8	.464	30	43.8								
	24.0	0.58	2	.116	40	43.0								
			8	.464	30	46.2								

(continued over page)...

FIGURE 8.3.6 (*Continued*).

The Manning equation is

$$v = \frac{C_1 R^{2/3} S^{1/2}}{n} \tag{8.3.2}$$

where v is mean velocity in the cross section, C_1 is 1.0 for SI units and 1.49 for U.S. units, S is slope, and R is hydraulic radius.

Information in Barnes,[4] Hicks and Mason,[32] and ISO standard ISO 1070-1973 assists in the estimation of Manning's n. In gravel-bed rivers where slope exceeds 0.002, an equation from Jarrett[37] may complement hydrologic judgment and experience. In U.S. units (R in ft) the equation is

$$n = 0.39\, S^{0.38} R^{-0.16} \tag{8.3.3a}$$

and in metric units (R in m) the equation is

$$n = 0.32\, S^{0.38} R^{-0.16} \tag{8.3.3b}$$

The standard error of estimate for n from this equation is about ±30 percent.

Other Methods of Discharge Measurement. Other discharge measurement methods are summarized in Table 8.3.1. New developments and recent applications of electronics to improve field data collection are reviewed in Chap. 25.

8.3.4 Computation of Streamflow Data

Stage-Discharge Rating Curves. The quality of the stage-discharge relation, or *rating curve* (Fig. 8.3.7), determines the quality of computed streamflow data. Hydraulic theory helps in determining the general form of the rating curve. In a long straight channel where channel friction control operates, a rating curve has the form

$$Q = C\,(h + a)^N \tag{8.3.4}$$

where Q = discharge
C and N = constants
h = stage
a = stage at which discharge is zero

Values of N for different cross-section shapes are:

Rectangular $N = 1.67$ (assuming width > 20 depth)
Parabolic $N = 2.17$ (assuming width > 20 depth)
Triangular $N = 2.67$

Because natural channels are often approximately parabolic in cross section, a value of about 2 for the exponent N is appropriate where there is channel friction control. Where there is a series of natural controls for different ranges of stage (Fig. 8.3.1), different values of C, a, and N may apply for each range of stage. Equation (8.3.4) also applies to weirs: for a V-notch thin-plate weir $N = 2.5$ and C is a function of notch angle; for a thin-plate rectangular weir $N = 1.5$.[1]

Figure 8.3.7*a* presents an example of a rating curve on arithmetic-scale paper for the Grey River at Dobson, New Zealand. Plots can also be constructed showing stage

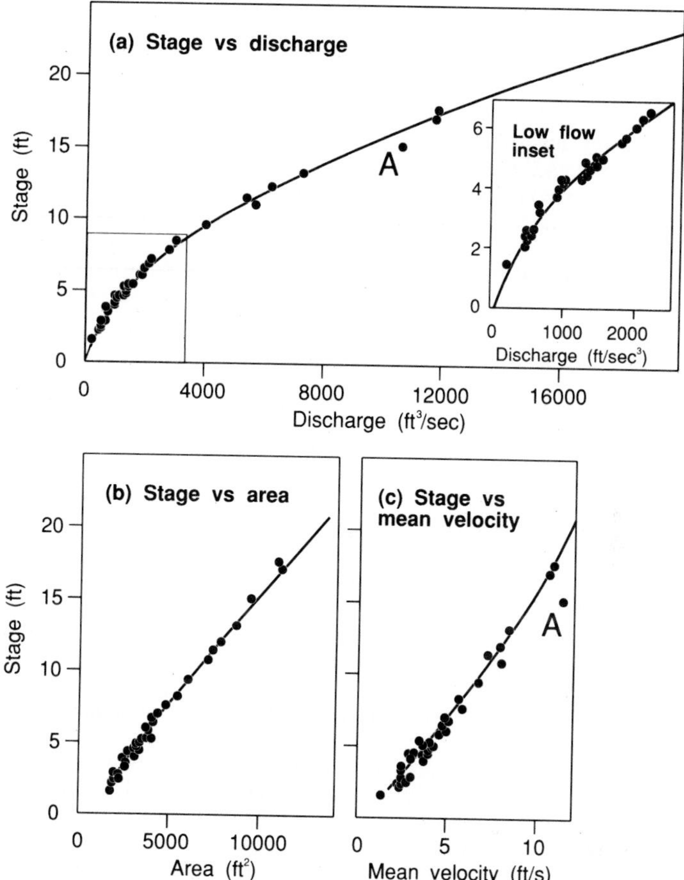

FIGURE 8.3.7 Rating curve for Grey River at Dobson, New Zealand. The stage-discharge rating *(a)* is extrapolated after extending curves through the stage-area *(b)* and stage-velocity *(c)* data to the maximum recorded stage of 20.4 ft. The extrapolated rating curve in *(a)* is drawn through points of area × mean velocity, read from curves *(b)* and *(c)*. Note that departure from the rating curve of the gauging marked *A* can be attributed to an incorrect mean velocity in *(c)*.

against cross-section area (Fig. 8.3.7*b*) and mean velocity (Fig. 8.3.7*c*). These plots are useful for:

1. Helping with extrapolation of ratings.
2. Helping to identify the causes of changes in the slope of the rating curve; e.g., the cross-section area vs. stage curve will flatten above bankfull stage.
3. Checking for mistakes. For example, when the cross-section area for a given gauging plots satisfactorily with other cross-section areas, a deviation in the plotted mean velocity may be caused by use of an incorrect current meter rating for the measurement (e.g., the measurement marked *A* in Fig. 8.3.7*a* and *c*).

4. Identifying when scour or aggradation has occurred, because the measured cross-section area for the gauging departs from other measured cross-section areas for the same stage.

Practice in the preparation of rating curves differs among agencies. In the U.S. Geological Survey, logarithmic plotting for high flow and normal (arithmetic) plotting at low flows is used initially to develop the general shape of the rating, which is subsequently displayed on logarithmic or arithmetic graph paper.[64] In arid Western Australia, which experiences especially large ranges between high and low discharge, plots of h against $Q^{2/5}$ on arithmetic paper are used. European and particularly British practice, where weirs and flumes are widely used, is to use logarithmic plots.[31]

A curve may be fitted to a series of stage-discharge measurements by eye or analytically. For analytical fitting, Eq. (8.3.4) can be written as

$$\log Q = \log C + N \log (h + a) \tag{8.3.5}$$

Where a cannot be measured accurately, it must be determined by various numerical methods (e.g., Ref. 31). With a determined, least-squares regression is used to estimate C and N (ISO[36]). By varying a and refitting the regression, the value of a can be determined which minimizes the error in the fitted equation. This method provides an objective result and an error of estimate, but in simple form it requires that one control operates over the whole of the discharge range. The operation of a second control, as may happen in a natural reach, is evident by a change in gradient: two equations must then be fitted and a transition point between them determined.

Alternatively, gaugings may be plotted and rating curves drawn by eye on arithmetic-scale paper. A separate larger-scale plot may be necessary for the lower part of the curve. Some hydrologic data archiving packages (e.g., Ref. 82) process ratings by using interpolation between pairs of h versus Q coordinates.

Extrapolation of Ratings. Extrapolation of ratings is necessary when a water level is recorded below the lowest or above the highest gauged level. Large errors can result if a function of the form of Eq. (8.3.4) is extrapolated beyond the range of gauged discharges without consideration of the cross-section geometry and controls. Where the cross section is stable, a simple method[36] is to extend the stage-area and stage-velocity curve and, for given stage values, take the product of velocity and cross-section area to give discharge values beyond the stage values that have been gauged.

The stage-area curve can be extended above the active channel by using standard land-surveying methods. Extrapolation of the stage-velocity curve requires understanding of the high stage control. Where there is a channel control, the Manning equation can be used to assist the extrapolation. In a wide approximately rectangular channel where width W exceeds 20 times mean depth d, $R = d$ approximately. The factor $C_1 S^{1/2}/n$ tends to a constant value, say C_S, at high stages, so that Eq. (8.3.2) becomes

$$v = d^{2/3} C_S \tag{8.3.6}$$

which specifies the shape of the velocity curve. However, this method is not reliable where Manning roughness varies with stage.

An upper bound on velocity is normally imposed by the Froude number

$$\mathrm{Fr} = \frac{v}{\sqrt{gd}} \tag{8.3.7}$$

where g is acceleration due to gravity. The Froude number Fr rarely exceeds unity in alluvial channels.[37]

The conveyance-slope method is a development of the use of the Manning equation, for the case where channel control applies and flow is uniform.[64] In this method, the Manning equation (8.3.2) is written as

$$Q = K\,S^{1/2} \tag{8.3.8}$$

where the conveyance $K = AR^{2/3}/n$ is determined from gauging measurements. The slope S, equal to $(Q/K)^2$, is plotted against stage and extrapolated, using the knowledge that slope tends to a constant value with increasing stage. The conveyance K is determined from the cross-section geometry and channel roughness n for a given stage.

Shifts in Rating Curves. The stage-discharge relationship can vary with time, in response to degradation, aggradation, or a change in channel shape at the control section; deposition of sediment causing increased approach velocities in a weir pond; vegetation growth; or ice accumulation. Shifts in rating curves are best detected from regular gaugings[35] and become evident when several gaugings deviate from the established curve. Sediment accumulation or vegetation growth at the control will cause deviations which increase with time, but a flood can flush away sediment and aquatic weed and cause a sudden reversal of the rating curve shift.

In gravel-bed rivers a flood may break up the armoring of the surface gravel material, leading to general degradation until a new armoring layer becomes established, and ratings tend to shift between states of quasi-equilibrium. It may then be possible to shift the rating curve up or down by the change in the mean bed level, as indicated by plots of stage and bed level versus time.[46] The technique is referred to by Liddell[44] as the Bolster method.

In rivers with gentle slopes, discharge for a given stage when the river is rising may exceed discharge for the same stage when the river is falling.[29] In such cases, adjustment factors must be applied in calculating discharge for rising and falling stages.[64]

Software for Calculating Streamflow. Traditionally, discharge data have been published in yearbooks, as annual series of daily mean discharges and other data derived therefrom, such as monthly mean discharges, plus summaries of annual extremes. In most countries, these data are now archived on computer tapes or disks and may therefore be processed by computer methods. Many national computer archives of hydrologic data continue to use daily mean discharge as the fundamental unit. Alternatively, the computer archive may contain the original measurements of stage and discharge, and the stage-discharge relations, and the software then calculates discharge at specified time steps at the time of enquiry. This alternative approach offers the advantage of being able to view the original stage data, examine flood hydrographs at a time resolution of less than one day, obtain flood peak data, and make retrospective adjustments to rating curves, using new gaugings. In such software (e.g., TIDEDA[82]) *data compression* is used to eliminate redundant values that can be estimated by interpolation, thereby achieving economies in data storage.

Several software programs are readily available for calculating streamflow and managing and archiving streamflow information, including TIDEDA (New Zealand), the VITUKI database management system (Hungary), the Water Survey of Canada system, and others. Such programs are available commercially or through the HOMS (Hydrological Operational Multipurpose System) of the World Meteorological Organization. HOMS[94] is a technology transfer system, which supplements normal commercial channels and provides easier access to technology such as software for processing, archiving, and analyzing streamflow data.

8.4 STREAMFLOW HYDROGRAPHS

8.4.1 Annual Hydrographs and Climatic Regime

Annual discharge hydrographs (Fig. 8.4.1) show the variation of discharge during a year. Figure 8.4.1*a* is for a *perennial stream* which flows continuously. Storm rainfall causes the high points or spikes of *quickflow*. *Baseflow* is evident in the dry January-February summer period. The second hydrograph (Fig. 8.4.1*b*) is for a seasonally *ephemeral stream* which yields runoff mostly during winter and after summer storms. As baseflow is absent during summer, there are long periods when stream-

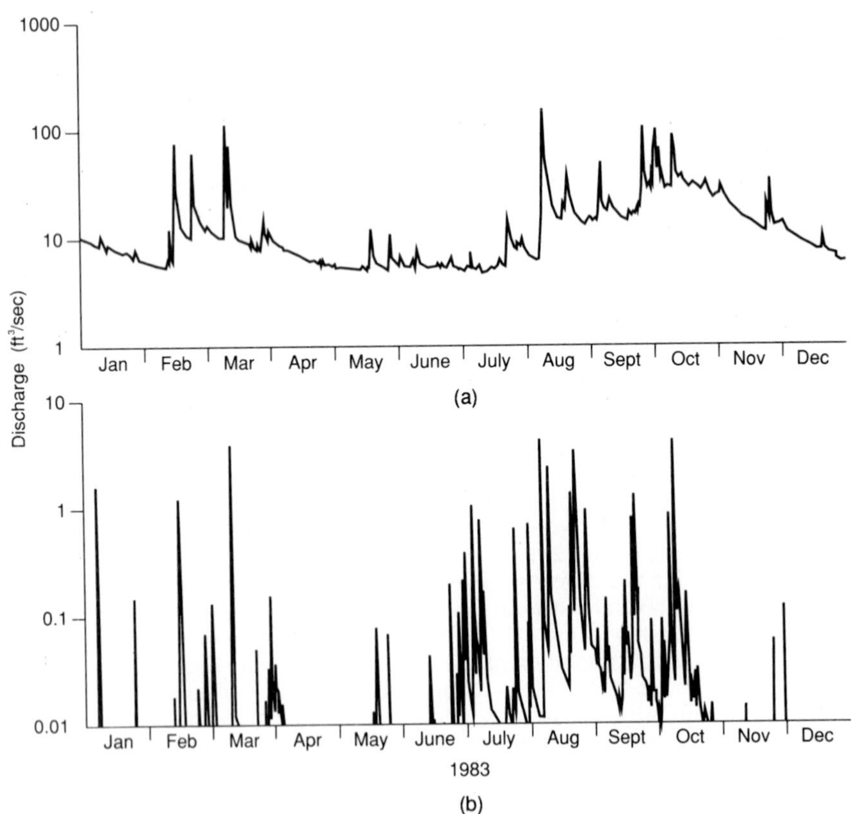

FIGURE 8.4.1 Examples of annual discharge hydrographs. *(a)* The perennial Dry Acheron Stream, New Zealand [mean discharge: 8.44 ft^3/s (0.239 m^3/s); mean annual runoff: 48.0 in (1220 mm)]. *(b)* The ephemeral Pakaraka Stream, New Zealand [mean discharge: 0.0322 ft^3/s (0.91 L/s); mean annual runoff 4.9 in (125 mm)].

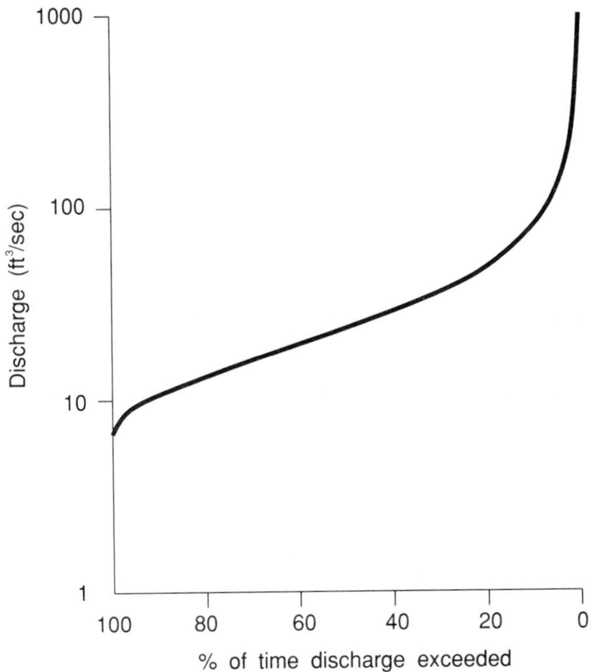

FIGURE 8.4.2 Flow duration curve for Punehu stream, 1970 through 1989.

flow is zero. Seasonal patterns in discharge are indicated by monthly mean discharge hydrographs. These have been classified into 15 regions for a global sample of 969 discharge records with an average length of 32 years.[28] The basis for classification is the seasons of the year in which discharge is greatest and is least. For example, much of western North America is classified as "moderate, early summer." Rivers in this regime show a strong broad peak in early summer months with a rapid decline to a low level in the late fall, winter, and early spring, a pattern also evident in portions of eastern and central Asia, southern Africa, and South America. The east and south of the United States is "early spring." Here discharge builds to a peak in early and midspring. The pattern can be produced by winter and early spring rainfall alone or by some snowmelt in temperate zones. Flow declines through the summer before beginning to increase again in the late fall. This pattern is shared with the Iberian peninsula, eastern Europe, Asia Minor, and portions of southern hemisphere countries.

8.4.2 Flow Duration Curves

The flow duration curve plots cumulative frequency of discharge, that is, discharge as a function of the percentage of time that the discharge is exceeded (Fig. 8.4.2). It is not a probability curve, because discharge is correlated between successive time intervals, and discharge characteristics are dependent on season of the year. Hence the probability that discharge on a particular day exceeds a specified value depends on the discharge on preceding days and on the time of year. Flow duration curves provide a compact graphical summary of streamflow variability. For example, they illustrate effects of river regulation upon low-flow percentiles, and indicate low-flow dependability for water resource investigations. The streamflow regime of rivers can be compared by overlaying flow duration curves for each river on one graph, after normalizing the curves by dividing by the mean discharge of each river. These curves are also used to estimate total suspended sediment load, by integration with sediment rating curves.

8.5 STREAMFLOW ESTIMATION

Where no records are available for a stream, streamflow must be estimated. Parameters required may include mean annual discharge and seasonal patterns, flood and low-flow quantiles, and percentiles of the flow duration curve. Alternatively, it may be necessary to estimate a hydrograph that would result from a particular sequence of rainfall or to estimate streamflow records over many years from rainfall records.

8.5.1 Estimation from Watershed Attributes

In the *regression method,* regions are first defined within which hydrologic regimes of watersheds are considered to be homogeneous. Regression equations of the type

$$Q = a\, B^b\, C^c\, D^d \tag{8.5.1}$$

are developed for gauged watersheds. In Eq. (8.5.1), Q is the streamflow parameter of interest; $a, b, c, d, \ldots$ are constants; $B, C, D, \ldots$ are the watershed characteristics which are expected to influence streamflow. The characteristics are assessed from maps, geographic information systems, meteorological statistics, etc. Examples of characteristics used are basin area, mean annual rainfall, and main channel slope (Table 8.2.2).

8.5.2 Low-Flow Estimation

Low flows are determined by the geology and lithology of the watershed, the losses to evaporation, and characteristics such as area and mean precipitation. Quantification of geological characteristics has enabled some success in prediction of regional low-flow values.[47]

The best method for estimating low-flow characteristics at a site with no continuous flow records requires gaugings of low flows at the site for at least 2 years. The gaugings are plotted against contemporaneous discharges for a stream in the vicinity that has a record long enough for low-flow quantiles to be defined (e.g., Fig. 8.5.1). This plot provides a means by which discharge values or the flow duration curve for the ungauged site can be estimated using discharge values or the flow duration curve for the site with the long record. To estimate less specific indices of low flow, such as the "normal summer low flow," gaugings at a large number of sites in a region can be carried out during a short period of stable low flows, and related to watershed characteristics such as area and percentage of forest cover.

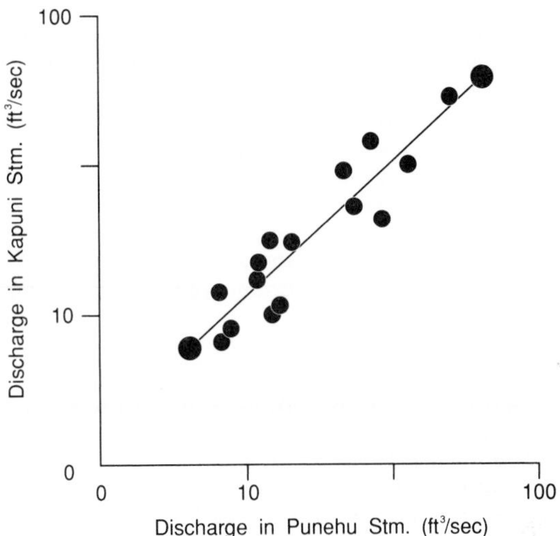

FIGURE 8.5.1 Low-flow correlations between two New Zealand Streams. Low flows gauged in the Kapuni stream are related to contemporaneous data for the Punehu stream, enabling low-flow characteristics for the Punehu (e.g., Fig. 8.4.2) to be transferred through the relation to the Kapuni.

8.5.3 Interpolating along River Channels

Where upstream and downstream gauging stations are operated, reliable estimates of discharge percentiles can be obtained by plotting discharge against location along the channel.[67]

In some cases where low flows are of interest, discharge may vary along a channel, through exchange with groundwater. Gaugings along the channel will identify the segments of channel where seepage losses (or gains) occur. In extreme cases, such as where a perennial river emerges from hills onto an alluvial fan, the losses can be sufficiently large that the river becomes ephemeral. Losses to streamflow along channels can be increased by riparian vegetation, especially those species called *phreatophytes,* with deep root systems which draw moisture from groundwater.[15,38]

8.6 STREAMFLOW INFORMATION

Information about streamflow is required to reduce uncertainty and to permit water-management decisions to be made with increased confidence. Streamflow data are used in combination with other types of information, particularly on meteorology, water quality, sediment loads, and land use, to facilitate integrated resource management.

8.6.1 Applications

Streamflow information is required in many areas of resource management and economic development. Applications can be classified according to (1) type of user (e.g., national government, consulting engineer, university researcher); (2) sector of the economy (e.g., energy, agriculture, transportation); (3) general field of application (e.g., flood mitigation, water resources management, design of structures); (4) type of investigation or analysis (e.g., flood estimation, storage reliability, waste loading capacity); and (5) type of information required (e.g., flow duration curves, basin attributes). Each application has different requirements for data quality, timeliness, form of presentation, and permissible cost, and these must be defined and reconciled. It is particularly important to consider future requirements, since data collected today cannot be remeasured.

UNESCO/WMO[85] provides a novel arrangement of elements of water resources projects, which indicates the relative importance of particular streamflow attributes (Table 8.6.1).

8.6.2 Objectives of a Streamflow Measurement Program

In designing a streamflow measurement program, it is necessary to set objectives which reflect user requirements.

User surveys are employed by hydrologic agencies (for example, Refs. 2, 3) to promote cost-effective provision of streamflow information. For example, an evaluation of the hydrometric network of New Brunswick, Canada, used a survey of users, in conjunction with technical studies, to develop and compare the cost-effectiveness of several data-collection strategies and make a series of recommendations for improving management of the network.[62] Similarly, a review of user needs in New Zealand led to proposal of a national hydrometric reference network, and specification of goals and objectives which provided the context within which data coverage and quality could be defined.[53]

8.6.3 Quality of Streamflow Information

Quality is defined as "the totality of features or characteristics of a product or service that bear on its ability to satisfy a given need."[9] In general, the usefulness of information may be considered in terms of 10 attributes which are aspects of quality (Ref. 11, p. 34): accessibility, comprehensiveness, accuracy, appropriateness, timeliness, clarity, flexibility, verifiability, freedom from bias, and quantifiability. These factors apply to the data or information themselves, the arrangements for storing them, and the means of delivering them to the user. The main determinants of the quality of streamflow information are (1) accuracy, continuity, and length of the record at each station, and (2) the spatial distribution of stations.

8.6.4 Quality Assurance of Streamflow Records

Comprehensive guidance on the design and implementation of a quality-assurance program is provided in the ISO 9000 series of International Standards Organization standards (ISO 9000 to ISO 9004).

The basic requirement is a quality manual which expresses the agency's commitment to quality, the various procedures it employs for monitoring quality and rectifying deficiencies, and staff responsibilities for all the different components of the program. A quality-assurance program can be set up for an entire streamflow observation program and for component parts, such as instrument servicing.

TABLE 8.6.1 UNESCO/WMO Recommendations for Types of Data Required for Water Resources Projects.

H = of high importance; M = of medium importance.

Element of water resource project	Example	River water levels: Time series	River water levels: Maximum	River water levels: Minimum	River flow: Time series	River flow: Maximum	River flow: Minimum	Quality	Channel character: Cross section	Channel character: Planform	Channel character: Velocity	Sediment
Modifiers of water balance	Increase or decrease precipitation, runoff, etc., through surface treatment, cloud seeding, etc.				H	M	M					
Redistribution of water in space	Canals, intakes, diversions, abstraction	H	H	H	H	H	H	H	H	H	H	H
Redistribution of water in time	Reservoirs	M	M	M	H	M	H	H				H
Extractors or supplies of energy	Hydroelectricity; waste heat disposed from thermal power station	H	H	H	H	M	H	M	M		M	H
Water confiners	Dams; floodbanks	H	H	H	M	H		M	M	M	M	M
Water relievers	Spillways	H	H	H	H	H		M	M	M	M	M
Quality improvers at source	Reduction of saline runoff; soil erosion control							H	M	M	M	H
Quality improvers at point of use	Water supply and sewage treatment plants; cooling towers				H	M	H	H	M	M	M	H
Water standards and legislation	Water-quality standards; minimum flows; flow allocations	M	M	M	M	M	M	M	M	M	M	M
Zoning	Floodplain zoning; wild and scenic rivers	M	M	M	M	M	M	M	M	M	M	M
Insurance	Flood damage; during construction		H	H	M	H	H	H				
Flow and level forecasting	Flood forecasting and working; reservoir operations	H	H	H	H	H	H	H	M		M	

Source: After *Water Resources Assessment Activities,* UNESCO/WMO, 1988. Used with permission.

Standards for Streamflow Data. Streamflow data of the highest practicable quality should be collected at general-purpose network stations. With current technology, an appropriate standard for stage data is provided by International Standards Organization standard ISO 4373-1979, clause 7:

> For the measurement of stage, in certain installations an uncertainty of ± 10 mm may be satisfactory; in others an uncertainty of ± 3 mm or better may be required; however, in no case should the uncertainty be worse than ± 10 mm or 0.1% of the range whichever is greater.

ISO standard 1100/2-1982, clause 7.1, provides standards for the stage-discharge rating curve required to derive discharge from stage data as specified in Table 8.6.2. However, there are no ISO standards for streamflow data themselves. Hence, to provide data of a quality which enables the objectives of its hydrometric network to be achieved (Sec. 8.6.2), the New Zealand Department of Scientific & Industrial Research (DSIR) defined the standards shown in Table 8.6.2.

Standard Procedures and Appropriate Technology. Procedures are to some extent constrained by the equipment used in a streamflow data-collection program, but many procedures are generally applicable with little modification. International sources for standard procedures include ISO *Handbook* 16,[36] the WMO *Technical Regulations,* vol. III,[93] and the WMO *Guide to Hydrological Practices.*[90] In many parts of the world the standard procedures promulgated by the U.S. Geological Survey, described in *Water Supply Paper* 2175,[64] and the relevant chapters of *Techniques of Water Resources Investigations of the United States Geological Survey* have been adopted as standard procedures. Other countries and agencies have prepared their own procedures to suit their particular types of rivers and equipment.

8.6.5 Network Design

Despite the development of a variety of objective and statistically based methods for streamflow and precipitation network design, judgment and experience are still indispensable. The WMO *Guide to Hydrological Practices* (Ref. 90, p. 3.18) makes recommendations for network density which provide a starting point (Table 8.6.3); there are several sources of additional guidance on the design of cost-effective networks.[10,55,56,86,87,91]

A general-purpose network must provide the ability to estimate hydrologic parameters over a wide area, such as by using a regional regression model. The sources of error in prediction using such a model are, apart from those due to operation of the gauging stations themselves, those due to sampling in time (a function of record length), sampling in space (a function of station density), and model error.

Record Length. Because of the temporal variability of meteorological and hydrologic phenomena, streamflow stations must be operated for a period of years. If a mean discharge is to be determined with standard error x, the length of record required is given by

$$x = \frac{s}{\sqrt{N}} \tag{8.6.1}$$

where s is standard deviation and N is length of record in years. For example, if mean annual discharge is to be determined with standard error $x = 10$ percent and the

TABLE 8.6.2 Standards Adopted by the Water Resources Survey, New Zealand Department of Scientific & Industrial Research

To provide hydrometeorological information of sufficient accuracy to meet the objectives of the National Hydrological Reference Network, the following standards have been specified:

NOTE:

Incidents outside the control of the Water Resources Survey (e.g., vandalism or major storms) may require that these standards are temporarily foregone for the period of affected record.

Water level (river, lake and groundwater) and river flow sites

Measurement of stage:

(a) Installed equipment and operating procedures shall be such as to ensure that 95% of instantaneous measurements are within ±3 mm or ±10 mm of the water level above the sensing device, depending on site instrumentation; specifications in clause 7 of ISO Standard ISO 4373-1979(E) should also be met.

(b) Instantaneous values shall be available on the Water Resources Archive within a maximum of six months of their being recorded.

(c) At any one measuring station, there shall be not more than 2% (approx. 7 days) missing record in a given calendar year, and not more than one calendar year in ten shall have any missing record.

(d) Field practice shall conform to standards specified in the Water Resources Survey Hydrologists' Field Manual.

Measurement of discharge, and rating curve construction

(a) Flow gaugings shall conform to the appropriate ISO standards as outlined in ISO Handbook 16 (ISO, 1983), and documented in the Water Resources Survey Field Hydrologists' Manual; in any case, 95% of discharges shall be measured to an accuracy of better than ±8% of the rated value, and a frequency specified by reference to flood event frequency, bed stability, and historical evidence.

(b) Stage discharge rating curves shall conform to specifications of ISO Standard ISO1100/2-1982, clause 7.1:
the curves shall invariably express the stage-discharge relation objectively and shall therefore be tested for absence from bias and goodness of fit in periods between shifts of control, and for shifts in control [Ref. 36].

(c) Flow gaugings and revised rating curves shall be available on the Water Resources Archive within a maximum of six months of the date of the gaugings.

(d) At any one measuring station, 95% of all flows estimated from a stage record with a rating applied shall be within ±8% of the actual values.

standard deviation of annual discharge $s = 30$ percent of the mean, then $N = 9$ years of recording is necessary. This assumes that the ratio of the standard deviation to mean is known before recording commences and that serial correlation between successive annual values is negligible. For at least some general-purpose stations, continuing operation is desirable, to provide information on hydrologic extremes, to define long-term trends and periodicities in streamflow and climate, and to detect changes resulting from changing land use.

Longer hydrologic records provide better data, in the sense that the decision maker has greater confidence in them. The resulting decisions may be more cost-effective. For example, the system of floodbanks (levees) constructed along the Waimakariri River, near Christchurch, New Zealand, were designed in 1960 on the basis of a design flood of 4300 m³/s, using data for 1930–1957. Recalculation using data for 1930–1987 gave a 100-year design flood of 3800 m³/s. Hence the protection

TABLE 8.6.3 Recommended Minimum Density of Hydrometric Stations

Type of region	Range of norms for minimum network, area, km² per station	Range of provisional norms tolerated in difficult conditions,* area, km² per station
I. Flat regions of temperate, mediterranean and tropical zones	1000–2500	3000–10,000
II. Mountainous regions of temperate, mediterranean, and tropical zones	300–1000	1000–5000§
Small mountainous islands with very irregular precipitation, very dense stream network	140–300	
III. Arid and polar zones†	5000–20,000‡	

* Last figure of the range should be tolerated only for exceptionally difficult conditions.
† Great deserts are not included.
‡ Depending on feasibility.
§ Under very difficult conditions this may be extended to 10,000 km².
Source: From *Guide to Hydrological Practices,* WMO, 1981. Used with permission.

afforded by the 1960 scheme is greater than originally planned (or the cost of the scheme was greater than necessary to provide the desired level of protection). The consequences would have been potentially more severe had the original design flood been an underestimate.

Spatial Coverage. The error in estimation of a streamflow statistic depends on its spatial variability and the spatial coverage of the monitoring network. It is possible to assess the change in accuracy that would be achieved by altering network density and to estimate the number of stations required to provide estimates of specified accuracy for a given region.[10,57,81,87] However, in practice, network density is in many areas constrained by other factors such as the availability of good sites for making measurements.

Network Analysis for Regional Information. Moss[54] has shown how spatial coverage and length of record can be traded off in designing a network to provide some specified level of accuracy; this approach is part of the design procedure known as Network Analysis for Regional Information (NARI).[57] NARI defines the combinations of number of stations and record length required to provide a specified standard error of estimate. The particular combination is then chosen on the basis of considering the costs of different operating strategies.

Network Analysis Using Generalized Least Squares. Another network design technique developed by the U.S. Geological Survey is known as Network Analysis Using Generalized Least Squares (NAUGLS).[80] To design a network that provides maximum information (minimum prediction error) for a specified budget and planning horizon, an approximate nonlinear procedure is used. A comparison of NAUGLS and NARI[58] indicates that NAUGLS estimates more precisely the mean square prediction error and provides a relatively unbiased estimate of mean streamflow.

8.6.6 Streamflow Data Archives

In the United States, the U.S. Geological Survey maintains two large-scale computerized systems for storing water data. The National Water Data Exchange (NAWDEX) program identifies sources of water data. The National Water Data Storage and Retrieval System (WATSTORE) provides processing, storage, and retrieval of surface water, groundwater, and water-quality data. NAWDEX is thus an index of the contents of WATSTORE. These are being integrated into a National Water Information System,[18,19] which will also integrate the National Water-Use Information Program and the Water Resources Scientific Information Center. In Canada, a *Surface Water Data Reference Index* (Environment Canada[21]) lists stations by province and drainage basin, and a separate publication[21] gives information on how to obtain data. North American data are also available on compact disk–read only memory (CD-ROM)[8] which can be purchased from EarthInfo, Inc., 5541 Central Ave., Boulder, Colo. 80301; Fax (303) 938-8183. A PC microcomputer equipped with a CD-ROM reader can then be used to study a continental database.

For other countries, the INFOHYDRO manual[92] provides information on where to obtain hydrologic data.

Streamflow Data Dissemination. A particularly valuable form of data presentation is a hydrologic or water atlas, such as that produced by the Federal Republic of Germany.[41] An atlas provides information in a highly interpreted form which is useful for many purposes, particularly where intercomparison with nonhydrologic and/or spatially distributed features is required. Technology for Geographic Information Systems (GIS) is developing at such a rate and becoming so accessible that hard-copy atlases, which are very expensive to produce, may be replaced by computer-compatible forms of data presentation and dissemination. GIS packages provide the opportunity for manipulating different forms of spatial information, including drainage networks, mean annual flows, sediment yields, and so forth, and have immense potential for application to streamflow analysis.

REFERENCES

1. Ackers P., W. R. White, J. A. Perkins, and A. J. M. Harrison, *Weirs and Flumes for Flow Measurement,* Wiley, New York, 1978.
2. Acres Consulting Services Ltd., "Economic Evaluation of Hydrometric Data," Acres Consulting Services Ltd., Report to Department of Fisheries and Environment, Canada, 1977.
3. Australian Water Resources Council, "The Importance of Surface Water Resources Data to Australia," *Water Management Series* 16, Australian Government Publishing Service, Canberra, 1988.
4. Barnes, H. H., Jr., "Roughness Characteristics of Natural Channels," *U.S. Geol. Surv. Water Supply Paper* 1849, 1967.
5. Beable, M. E., and A. I. McKerchar, "Regional Flood Estimation in New Zealand," *Water & Soil Tech. Publ.* 20, Ministery of Works & Development, Wellington, New Zealand, 1982.
6. Blench, T., *Regime Behaviour of Canals and Rivers,* Butterworths, London, 1957.

7. Bodhaine, G. L., "Measurement of Peak Discharge at Culverts by Indirect Methods," *Techniques of Water-Resources Investigations of the U.S. Geological Survey,* Book 3, Chap. A3, 1982.
8. Bradley, K., P. J. Call, J. B. Edwards, and E. Perry, "Disseminating Voluminous Resource Databases in an Era of Global Change," Proc. 1989 meeting, *Earth Observations and Global Change Decision Making: A National Partnership,* Krieger Publishing, 1989.
9. British Standards Association, *A Guide to Quality Assurance,* British Standard BS 4891, 1972.
10. Brown, J. A. H., "A Review of the Design of Water Resources Data Networks," *Australian Water Resources Council Tech. Paper* 52, 1980.
11. Burch, J. G., and F. R. Strater, *Information Systems: Theory and Practice,* Hamilton Publishing, Santa Barbara, 1974.
12. Carlston, C. W., "Drainage Density and Streamflow," *U.S. Geol. Surv. Prof. Paper* 422-C, 1963.
13. Carter, R. W., and I. E. Anderson, "Accuracy of Current Meter Measurements," *J. Hydraul. Eng.,* vol. 89, no. Hy4, pp. 105–115, 1963.
14. Chorley, R. J., S. A. Schumm, and D. E. Sugden, *Geomorphology,* Methuen, London, 1984.
15. Culler, R. C., R. L. Hanson, R. M. Myrick, R. M. Turner, and F. P. Kipple, "Evapotranspiration Before and After Clearing Phreatophytes, Gila River Floodplain, Graham County, Arizona," *U.S. Geol. Surv. Prof. Paper* 655-P, 1962.
16. Dalrymple, T., and M. A. Benson, "Measurement of Peak Discharge by Slope-Area Method," *Techniques of Water-Resources Investigations of the U.S. Geological Survey,* Book 3, Chap. A2, 1967.
17. Dury, G. H., *The Face of the Earth,* 5th ed., Allen & Unwin, London, 1986.
18. Edwards, M. D., "Plan for the Design, Development, Implementation, and Operation of the National Water Information System," *U.S. Geol. Surv. Open-File Report* 87-29, 1987.
19. Edwards, M. D., A. L. Putnam, and N. E. Hutchinson, "Conceptual Design for the National Water Information System," *U.S. Geol. Surv. Bull.* 1792, 1987.
20. Embleton, C., and J. Thornes, eds., *Process in Geomorphology,* Arnold, London, 1979.
21. Environment Canada, *Supplying Hydrometric and Sediment Data to Users,* 2d ed., Inland Waters Directorate, Water Resources Branch, Ottawa, 1980.
22. Environment Canada, *Surface Water Data Reference Index Canada 1988,* Inland Waters Directorate, Water Resources Branch, Ottawa, 1989.
23. Garner, H. F., *The Origin of Landscapes: A Synthesis of Geomorphology,* Oxford University Press, London, 1974.
24. Goudie, A., ed., *Geomorphological Techniques,* Allen & Unwin, London, 1981.
25. Gregory, K. J., and D. E. Walling, *Drainage Basin Form and Process,* Arnold, London, 1973.
26. Griffiths, G. A., "Hydraulic Geometry Relationships of Some New Zealand Gravel Bed Rivers," *J. Hydrol. (NZ),* vol. 19, pp. 106–118, 1980.
27. Hadley, R. F., and S. A. Schumm, "Sediment Sources and Drainage Basin Characteristics in Upper Cluyenne River Basin," *U.S. Geol. Surv. Water-Supply Paper* 1531-B, 1961.
28. Haines, A. T., B. L. Finlayson, and T. A. McMahon, "A Global Classification of River Regimes," *Appl. Geog.,* vol. 8, pp. 255–272, 1988.
29. Henderson, F. M., *Open Channel Flow,* Macmillan, New York, 1966.
30. Herschy, R. W., "Accuracy," chap. 10 in R. W. Herschy, ed, *Hydrometry: Principles and Practice,* Wiley, Chichester, pp. 353–397, 1978.
31. Herschy, R. W., *Streamflow Measurement,* Elsevier, 1985.

32. Hicks, D. M., and P. D. Mason, "Roughness Characteristics of New Zealand Rivers," Water Resource Survey, P.O. Box 14-901, Wellington, New Zealand, 1991.
33. Horton, R. E., "Erosional Development of Streams and Their Drainage Basins: Hydrophysical Approach to Quantitative Morphology," *Geol. Soc. Am. Bull.* 56, pp. 275–370, 1945.
34. Hulsing, H., "Measurement of Peak Discharge at Dams by Indirect Methods," Book 3, chap. A5 of *Techniques of Water-Resources Investigations of the United States Geological Survey,* 1967.
35. Ibbitt, R. P., and C. P. Pearson, "Gauging Frequency and Detection of Rating Changes," *Hydrol. Sci. J.,* vol. 32, no. 1, pp. 85–103, 1987.
36. ISO, "Liquid Flow Measurement in Open Channels," *Handbook* 16, International Standards Organization, 1983.
37. Jarrett, R. D., "Hydraulics of High-Gradient Streams," *J. Hydraul. Eng.,* vol. 110, pp. 1519–1539, 1984.
38. Johns, E. L., ed., "Water use by Naturally Occurring Vegetation: An Annotated Bibliography," A report prepared by the task committee on water requirements of natural vegetation, Committee on Irrigation Water Requirements, Irrigation and Drainage Division, American Society of Civil Engineers, 1989.
39. Johnstone, D. E., "Some Recent Developments of Constant Injection Salt Dilution Gaugings in Rivers," *J. Hydrol. (NZ),* vol. 27, no. 2, pp. 128–153, 1988.
40. Jones, J. A. A., "The Nature of Soil Piping: A Review of Research," *Br. Geomorph. Res. Group Monograph* 3, Geobooks, Norwich, 1981.
41. Keller, R., *Hydrological Atlas der Bundesrepublik Deutschland,* Deutsches Forschungsgemeinschaft, 1979.
42. Leopold, L. B., and J. P. Miller, "Ephemeral Streams—Hydraulic Factors and Their Relation to the Drainage Net," *U.S. Geol. Surv. Prof. Paper* 282A, 1956.
43. Leopold, L. B., M. G. Wolman, and J. P. Miller, *Fluvial Process in Geomorphology,* Freeman, San Francisco, 1964.
44. Liddell W. A., *Stream Gaging,* McGraw-Hill, New York, 1927.
45. Matthai, H. F., "Measurement of Peak Discharge at Width Contractions by Indirect Methods," *Techniques of Water-Resources Investigations of the United States Geological Survey,* Book 3, Chap. A4, 1967.
46. McKerchar, A. I., and R. D. Henderson, "Drawing and Checking State/Discharge Rating Curves," Publ. 11, Hydrology Centre, Christchurch, New Zealand, 1987.
47. McMahon, T. A., and R. G., Mein, *River and Reservoir Yield,* Water Resources Publications, 1986.
48. Mollard, J. D., "Airphoto Interpretation of Fluvial Features. Fluvial Processes and Sedimentation," in *Proceedings of Hydrology Symposium,* University of Alberta, Department of the Environment, Canada, pp. 341–380, 1973.
49. Mosley, M. P., "Prediction of Hydrologic Variables from Channel Morphology, South Island Rivers," *J. Hydrol. (N.Z.),* vol. 18, pp. 109–120, 1979.
50. Mosley, M. P., "Semi-determinate Hydraulic Geometry of River Channels, South Island, New Zealand," *Earth Surf. Proc. Landforms,* vol. 6, pp. 127–137, 1981.
51. Mosley, M. P., "Analysis of the Effects of Changing Discharge on Channel Morphology and Instream Uses of a Braided River, Ohau River, New Zealand," *Water Resour. Res.,* vol. 18, pp. 800–812, 1982.
52. Mosley, M. P., "The Classification and Characterization of Rivers," chap. 12 in K. S. Richards, ed., *River Channels: Environment and Process,* Blackwell, Oxford, pp. 295–320, 1987.
53. Mosley, M. P., and A. I. McKerchar, "Quality Assurance Programme for Hydrometric Data in New Zealand," *Hydrol. Sci. J.,* vol. 32, pp. 185–202, 1989.

54. Moss, M. E., "Space, Time, and the Third Dimension (Model Error)," *Water Resour. Res.*, vol. 15, pp. 1797–1800, 1979.
55. Moss, M. E., "Concepts and Techniques in Hydrological Network Design," *WMO Op. Hydrol. Report* 19, WMO, Geneva, 1982.
56. Moss, M. E., ed., "Integrated Design of Hydrological Networks," *IAHS Publ.* 158, 1986.
57. Moss, M. E., E. J. Gilroy, G. D. Tasker, and M. R. Karlinger, "Design of Surface Water Data Networks for Regional Information," *U.S. Geol. Surv. Water Supply Paper* 2178, 1982.
58. Moss, M. E., and G. D. Tasker, "An Intercomparison of Hydrological Network Design Technologies," *Hydrol. Sci. J.*, vol. 36, no. 3, pp. 209–221, 1991.
59. Natural Environment Research Council (NERC), "Flood Studies Report," vol. 1, *Hydrological Studies,* Natural Environment Research Council, London, 1975.
60. Nemec J., *Hydrological Forecasting,* D. Reidel, 1986.
61. Pelletier, P. M., "Uncertainties in Streamflow Measurement under Winter Ice Conditions, A Case Study: The Red River at Emerson, Manitoba, Canada," *Water Resour. Res.*, vol. 25, no. 8, pp. 1857–1868, 1989.
62. Perks, A., D. Ambler, Z. Davar, and B. Burrell, "New Brunswick Hydrometric Network Evaluation: Summary Report," Environment Canada, 1989.
63. Petts, G. E., *Impounded Rivers,* Wiley, Chichester, 1984.
64. Rantz, S. E., and others, *Measurement and Computation of Streamflow* (vol. 1, Measurement of Stage and Discharge; vol. 2, Computation of Discharge), *U.S. Geological Survey Water Supply Paper* 2175, 1983.
65. Richards, K. S., *Rivers, Form and Process in Alluvial Channels,* Methuen, London, 1982.
66. Richards, K. S., ed., *River Channels, Environment and Process,* Blackwell, Oxford, 1987.
67. Riggs, H. C., *Streamflow Characteristics,* Elsevier, 1985.
68. Rodda, J. C., "The Significance of Characteristics of Basin Rainfall and Morphology in a Study of Floods in the United Kingdom," UNESCO Symposium on Floods and Their Compilation, vol. 2, pp. 834–845, 1969.
69. Sargent, D. M., "The Development of a Viable Method of Streamflow Measurement Using the Integrating Float Technique," *Proc. Inst. Civil Engrs.*, pt. 2, 1981, p. 71.
70. Schumm, S. A., "River Adjustment to Altered Hydrologic Regimen Murrumbidgee River and Paleochannels, Australia," *U.S. Geol. Surv. Prof. Paper* 598, 1968.
71. Schumm, S. A., "River Metamorphosis," *J. Hydraul. Eng.*, vol. 95, pp. 255–273, 1969.
72. Schumm, S. A., *River Morphology,* Dowden, Hutchinson & Ross, Stroudsburg, Pa., 1972.
73. Schumm, S. A., *Drainage Basin Morphology,* Dowden, Hutchinson & Ross, Stroudsburg, Pa., 1977.
74. Schumm, S. A., *The Fluvial System,* Wiley, New York, 1977.
75. Schumm, S. A., "Evolution and Response of the Fluvial System: Sedimentologic Implications," *Soc. Econ. Paleontol. Min. Spec. Publ.* 31, pp. 19–29, 1981.
76. Service hydrologique national de Suisse, "Guide pour les jaugeages de cours d'eau," Service hydrologique national, Berne, Switzerland, 1983.
77. Shreve, R. L., "Statistical Law of Stream Numbers," *J. Geol.*, vol. 74, pp. 17–37, 1966.
78. Smoot, G. F., and C. C. Novack, "Measurement of Discharge by Moving Boat Method," *Techniques of Water Resources Investigations of the U.S. Geological Survey,* Book 3, Chap. A11, 1969.
79. Strahler, A. N., "Dynamic Basis of Geomorphology," *Geol. Soc. Am. Bull.*, vol. 63, pp. 923–938, 1952.
80. Tasker, G. D., "Generating Efficient Gaging Plans for Regional Information," *Int. Assoc. Hydrol. Sci. Publ.* 158, pp. 269–281, 1986.

81. Tasker, G. D., and M. E. Moss, "Analysis of Arizona Flood Data Network for Regional Information," *Water Resour. Res.,* vol. 15, pp. 1791–1796, 1979.

82. Taylor, M. E. U., S. M. Thompson, and M. W. Rodgers, "Micro-TIDEDA," pp. 127–139 in World Meteorological Organization, ed., *Microprocessors in Operational Hydrology,* D. Reidel, 1984.

83. Thomas, D. M., and M. A. Benson, "Generalization of Streamflow Characteristics from Drainage Basin Characteristics," *U.S. Geol. Surv. Water-Supply Paper* 1975, 1970.

84. Twidale, C. R., *Analysis of Landforms,* Wiley, New York, 1976.

85. UNESCO/WMO, *Water-Resources Assessment Activities; Handbook for National Evaluation,* UNESCO, Paris, 1988.

86. van der Made, J. W., "Design Aspects of Hydrological Networks," TNO Comm. Hydrol. Res., Proc. Inform. 35, TNO, The Hague, 1986.

87. van der Made, J. W., *Analysis of Some Criteria for Design and Operation of Surface Water Gauging Networks,* Rijkswaterstaat, The Hague, 1988.

88. Williams, G. P., "Bankfull Discharge of Rivers," *Water Resour. Res.,* vol. 14, pp. 1141–1154, 1978.

89. Wilson, J. F., E. C. Cobb, F. A. Kilpatrick, "Fluorometric Procedures for Dye Tracing," *Techniques of Water-Resources Investigations of the U.S. Geological Survey,* Book 3, Chap. A12, 1986.

90. WMO, *Guide to Hydrological Practices,* vol. I, *Data Acquisition and Processing,* WMO, Geneva, 1981.

91. WMO, "Hydrological Network Design—Needs, Problems and Approaches," *Operational Hydrology Rep.* 12, WMO, Geneva, 1982.

92. WMO, "INFOHYDRO Manual," *Operational Hydrology Rep.* 28, WMO, Geneva, 1987.

93. WMO, "Technical Regulations," vol. III, *Hydrology,* WMO, Geneva, 1988.

94. WMO, *HOMS Reference Manual,* 2d ed., WMO, Geneva, 1988.

CHAPTER 9
FLOOD RUNOFF

David H. Pilgrim and Ian Cordery
School of Civil Engineering
The University of New South Wales
Kensington, New South Wales, Australia

9.1 DESCRIPTION OF FLOODS

9.1.1 Introduction

Flood runoff results from short-duration highly intense rainfall, long-duration low-intensity rainfall, snowmelt, failure of dam or levee systems, or combinations of these conditions. Events such as earthquakes, landslides, ice blockages or releases, and high tides or storm surges can worsen flood conditions.

The best information on flood magnitudes that are likely to occur in the future is obtained from observed flow records—what has occurred in the past. The nature of the flood-producing system—the interaction of atmosphere, land geology and geomorphology, vegetation and soils, and the activities of people—is so complex that sole use of theoretical or modeling approaches can provide only generalized estimates of the flood regime of a stream or a region. Local information on observed floods is essential to calibrate models for valid application to particular drainage basins or regions. A valuable supplement to recorded streamflow data can be provided by historic flood information from old newspaper reports, long-term residents, local municipal bodies, and from road and railway authorities which usually keep records of damage to their installations. Information on more recent flood runoff events can be obtained from debris of flood marks on riverbanks and floodplains. Considerable work has also been carried out in recent years on the assessment of *paleofloods*. These are major floods that have occurred outside the historical record, but which are evidenced by geological, geomorphological, or botanical information. These are discussed in Chap. 18.

The need to utilize available local data for calibration cannot be overemphasized. A given storm may produce a large flood in one region, but in an apparently similar region, the same storm may produce little or no runoff.

9.1.2 Choice of Method of Flood Estimation

The choice of the method to be used is the first step in flood estimation. Unfortunately, the choice is often made on a largely subjective and intuitive basis. While some subjectivity is always involved, the following considerations provide a sound basis for choice of a flood estimation method:

- The form and structure of available methods, the factors they consider, their theoretical basis, and their relative accuracies
- Whether a *deterministic* or *probabilistic design* estimate is required, and whether a particular method and its parameter values are suited to this application
- Whether a method is capable of calibration with data recorded at the site or, if it is a regional method, whether it has been derived from data recorded in the region
- The type and importance of the work for which the estimate is required, the effects of inaccuracy and exceedance of the estimate, and whether a peak flow or complete hydrograph is required
- The time that can be spent in estimating the flood
- The available expertise, as more complex methods generally require greater expertise in their use and interpretation, without which results may be poorer than for simpler methods

For design at a site where observed flood data are available, a choice must be made between some form of flood frequency analysis or one of the methods based on design rainfall as described in this chapter. Flood frequency analysis gives a direct estimate of the flood of selected exceedance probability, but rainfall records are generally longer than flow records, are less variable over time, are available at more locations, and have greater spatial consistency in the surrounding region.

Very little quantitative guidance is available on choice of flood estimation methods, and rather arbitrary rules are recommended for most regions. *Bulletin* 17B of the Interagency Advisory Committee on Water Data[32] recommends that flood estimates from precipitation should be used only as an alternative method for estimating floods with exceedance probabilities of 1 percent or less if the length of available streamflow record is less than 25 years. The U.S. Bureau of Reclamation[79] gives a similar recommendation. The U.K. *Flood Studies Report*[53] recommends that a flood frequency curve should be extrapolated to a return period of $2N$ years only, where N years is the length of record. Beyond a return period of $4N$ years, a regional frequency curve is recommended up to a return period of 200 years, and even to 500 years with lower accuracy.

In Australia,[31,62] quantitative formulas involving the statistics and record lengths of rainfall and streamflow have been developed for the average recurrence interval where it is better to change from flood frequency analysis to an estimate computed from a design rainfall. These formulas are based on estimated confidence intervals of the frequency distributions of the two types of estimates at the site.

9.2 RAINFALL-RUNOFF RELATIONSHIPS

9.2.1 Runoff Processes

Flood runoff has often been considered to consist of surface runoff produced at the ground surface when the rainfall intensity exceeds the infiltration capacity. While this process, known as *Hortonian overland flow*, occurs in many situations, two other general storm runoff processes are now recognized, as a result of observations on natural basins during storm periods and many detailed studies of instrumented plots and small areas.

Saturated overland flow occurs when, on part of the drainage basin, the surface horizon of the soil becomes saturated as a result of either the buildup of a saturated zone above a soil horizon of lower hydraulic conductivity or the rise of a shallow water table to the surface. While this usually occurs in valley bottoms, in some regions saturated areas first occur on ridge tops where the surface soil horizon is thin.[61]

Throughflow is water that infiltrates into the soil and percolates rapidly, largely through macropores such as cracks and root and animal holes, and then moves laterally in a temporarily saturated zone, often above a layer of low hydraulic conductivity. It reaches the stream channel quickly and differs from other subsurface flow by the rapidity of its response and its relatively large magnitude.

Runoff processes operating at any location vary from time to time. Large variations in hydrologic characteristics, and therefore in runoff processes, also occur over small, apparently homogeneous areas to the extent that all three runoff processes discussed above may occur during a single storm event.[63] Associated with the recognition of several processes producing storm runoff has been the concept that storm runoff may be generated from only a small part of many drainage basins. In addition, this *source area* may vary in extent in different seasons and during the progress of a storm.

The type of runoff process and the location of source areas, whether close to the outlet and adjacent to stream channels or on ridges remote from the channels, has considerable influence on the resulting hydrographs. However, practical methods for estimating storm losses and runoff have not yet been developed to explicitly account for these differences. Uniform or average conditions, at least over subareas, are generally assumed. These new insights into physical processes do, however, give some guidance in assessing the validity and applicability of the practical methods.

9.2.2 Rainfall Excess and Direct Runoff

When precipitation begins, water is deposited on many different surfaces, and from these it travels by a large number of routes into the local stream, back to the atmosphere or to the local water table. The processes involved in the transfer of water from its point of impact to its point of disposal are shown schematically in Fig. 9.2.1. *Rainfall excess* or *direct runoff* is the water that eventually becomes flood runoff, being generated at the surface as Hortonian runoff and saturated overland flow, or having a subsurface route in throughflow. Three processes prevent some of the precipitation from becoming immediate runoff and are therefore considered as losses. *Interception* is water that is retained on vegetation and other surfaces (Sec. 3.6.3). *Depression storage* is water that is retained in innumerable depressions on the basin surface, ranging in size from lakes and swamps down to soil grain size cavities. Although the water held in these depressions may be continually exchanged during a storm, the volume stored is a loss to runoff and is finally removed by evaporation and infiltration. There may also be a small amount of evapotranspiration during a storm. *Infiltration,* the water passing into the surface of the soil, is the most important of the loss processes. With Hortonian runoff, infiltration is a direct loss, and it governs the volume of flood runoff and influences hydrograph shape. With saturated overland flow and throughflow, the effects of infiltration are more complex, as some infiltrating water becomes flood runoff or influences the runoff-producing areas.

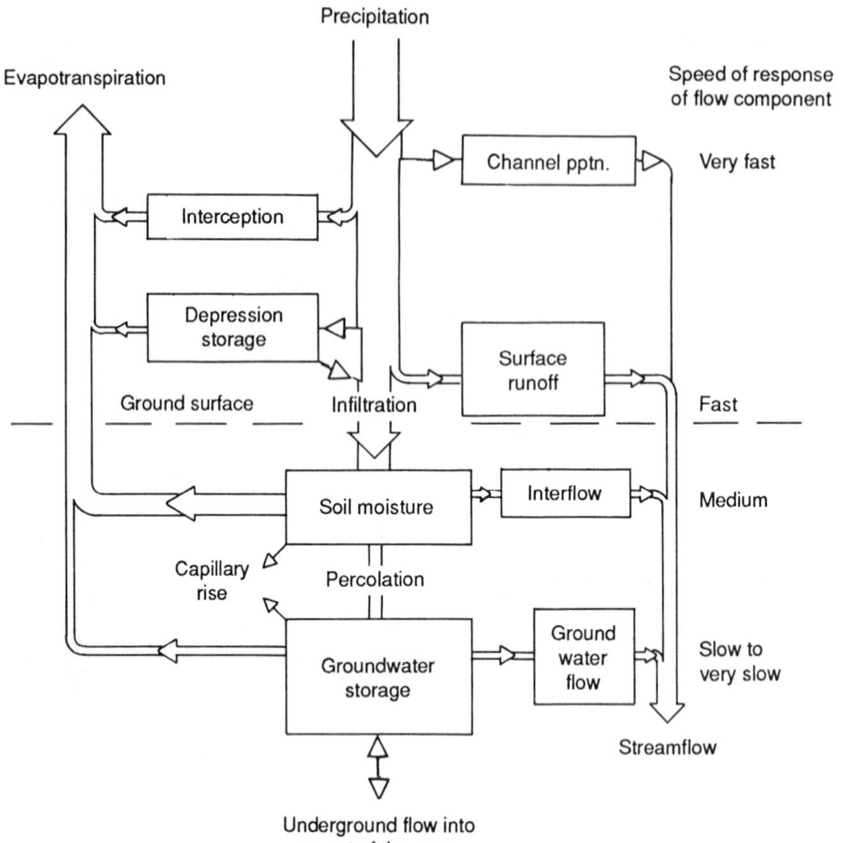

FIGURE 9.2.1 Land phase of the water cycle (width of the arrows indicates the average relative magnitudes of water transfer in humid zones). *(Reproduced from Ref. 75 with permission.)*

9.2.3 Separation of Baseflow and Direct Runoff Hydrographs

In flood analysis, it is often necessary to separate a flood hydrograph into different flow components. These component flows may be referred to as surface or direct runoff, rapid subsurface flow or interflow, and groundwater flow or baseflow, but it is only in intensive field studies that it is possible to distinguish paths by which water has reached the stream. In practice, separation of hydrograph components is made on the basis of travel or response times, which matches analytical procedures, rather than on the basis of physical processes. The emphasis here is on simple-to-use techniques, but other techniques may be useful, such as the linear filter approach.[9,44] The flow components are sometimes referred to as quickflow and baseflow, without any reference to the processes which produced them. Only two flow components generally need to be recognized in most practical procedures.

Separation of Component Flows. To separate components it is necessary to identify three features of the total hydrograph. These are the start and end of surface runoff and the shape of the baseflow hydrograph between these two points.

The *start of surface runoff* is usually easily identified as shown at A in Fig. 9.2.2, where, after rainfall begins, the hydrograph first diverges from the constant or steadily declining baseflow which prevailed previously. The time of ending of surface runoff and the shape of the baseflow hydrograph are more difficult to define.

For the *end of surface runoff*, two procedures have found practical application:

Semilogarithmic Plots of the Recession Curves. These are useful[24] for investigating the components of hydrograph recessions and for determining the end of surface runoff. Hydrograph recession curves plotted on semilogarithmic paper (discharge on the logarithmic scale) frequently approximate three straight-line segments. The end of surface runoff is generally assumed to be the point of intersection of the two lowest lines, as the different slopes represent different response times and hence runoff processes. The intersection of the two upper lines may be preferred in some cases, and the choice is a matter of judgment.

Each straight line is defined by

$$q_t = q_o K_r^t \tag{9.2.1}$$

where q_t = discharge at time t
q_o = initial discharge (both q_t and q_o are measured within the same straight segment)
K_r = recession constant whose value is dependent on the units of t

Typical values of K_r for the different flow components are shown below for basins ranging in size from 120 to 6500 mi^2 (300 to 16,000 km^2) in the United States, eastern Australia, and several other regions:[37]

Flow components	K_r hourly	K_r daily
Groundwater	0.998	0.95
Interflow	0.99	0.8–0.9
Surface runoff	0.95–0.99	0.3–0.8

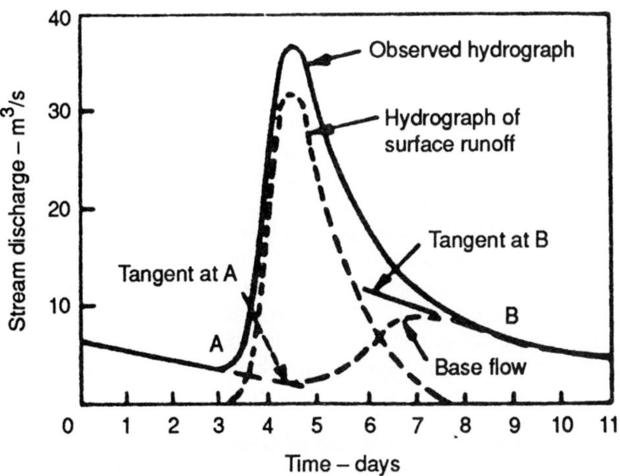

FIGURE 9.2.2 Typical baseflow separation. *(Reproduced from Ref. 31 with permission.)*

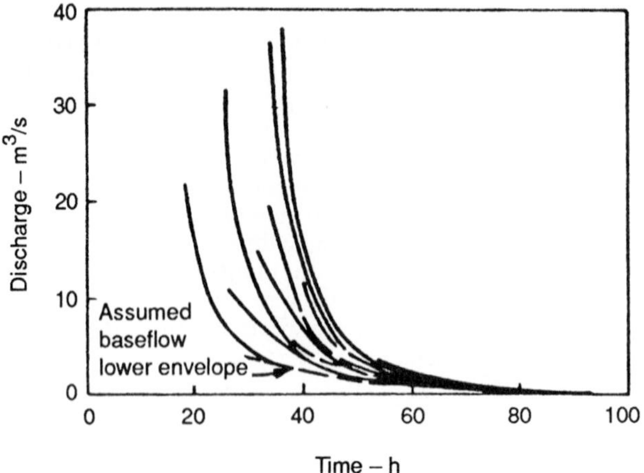

FIGURE 9.2.3 Hydrograph recession curves with assumed lower envelope baseflow recession. *(Reproduced from Ref. 31 with permission.)*

Larger basins may have K_r values at the upper end of the ranges shown and vice versa. These typical K_r values can be helpful in interpreting the analysis of streamflow recessions.

A similar approach involves plotting q_t versus q_{t+1} on linear paper, giving linear segments wherever K_r is constant. However, if consistent straight-line segments are not obtained in either procedure, it may be necessary to use the second approach below.

Typical Baseflow Recession Curve. This is obtained by superimposing the available hydrograph recessions (linear axes) for a basin in such a way that at their lower discharges they coincide as closely as possible, as shown in Fig. 9.2.3. The baseflow recession is then the lower envelope of all the recessions. By superimposing the baseflow recession curve on the hydrograph of interest using tracing paper, the time at which surface runoff ends can be determined as the time when the hydrograph recession takes up the shape of the baseflow recession. Variations of individual recessions (see Fig. 9.2.3) justify slight alterations to the hydrograph recession so that a definite time for the end of surface runoff is obtained.

Separation Hydrograph. The separation of baseflow from surface runoff is achieved by drawing tangents to the recession curves at the points of start and finish of direct runoff (A and B in Fig. 9.2.2) and assuming a shape for the baseflow hydrograph between these tangent points. A straight-line separation is often quite acceptable where the maximum baseflow discharge is well below 10 percent of the maximum discharge. A more realistic separation may be obtained by continuing the average baseflow recession forward in time from point A in Fig. 9.2.2 to a point approximately below the peak of the total hydrograph, and joining this extended baseflow recession to point B by a smooth curve. The latest possible occurrence of the peak of baseflow (to give a smooth curve) agrees with the slower travel of this component.

9.2.4 Factors Affecting the Relation of Storm Runoff and Rainfall

All runoff-producing processes shown in Fig. 9.2.1 are dependent on the length of time since the last precipitation, and the evapotranspiration during that time. Even for a given drainage basin, the rainfall-runoff relationship is thus very dependent on basin wetness. Interception storages empty in a few hours after precipitation, small depressions also empty in a few hours, but larger ones may empty only over months. Soil water is removed over a period of months, but the rate is dependent on the soil type and profile, season, and the type and state of the vegetation. The rainfall-runoff relationship is sensitive to surface soil water content and to the dominant runoff process operative on the basin.

9.2.5 Approaches to Estimating Storm Runoff

For practical estimation of rainfall excess or storm runoff volume, which are synonymous terms, numerical representations or models are required of rainfall and losses or of the relation of runoff to rainfall.

Rainfall is almost always represented by a *hyetograph,* which is a time pattern of rainfall intensity. Two distinct cases are used:

1. *Recorded rainfall* for simulating an actual hydrograph. The hyetograph may be assumed to be the same at all points over the basin (lumped), or if several gauges are available to provide definition, different hyetographs may be used for different subareas.
2. *Design rainfall* with average intensity obtained from published intensity-duration-frequency (IDF) information for the region. The design storm may be assumed to have a constant intensity or a more realistic variable intensity pattern over its duration. The storm model is usually lumped; i.e., it has the same hyetograph at all points on the basin. Point rainfalls estimated from IDF data may be reduced slightly as basin size increases (Sec. 3.9).

Models Used to Estimate Storm Runoff. A number of different models (Fig. 9.2.4) are used in practice,[31] some focusing on the runoff and others on the loss, the part of the rain which does not run off. The most frequently used models are:

1. Loss (and conversely runoff) is a constant fraction of rainfall in each time period, or if the storm has constant rainfall intensity, a simple proportion of the total rainfall. This is the runoff coefficient concept which is discussed further in Sec. 9.4.1.
2. Constant loss rate where the rainfall excess is the residual after a selected constant loss rate or infiltration capacity is satisfied.
3. Initial loss and continuing constant loss rate, which is similar to model 2 except that no runoff occurs until a given initial loss capacity has been satisfied, regardless of the rainfall rate. A variation of this model is to have an initial loss followed by a loss consisting of a constant fraction of the rainfall in the remaining time periods.
4. Infiltration curve or equation representing capacity rates of loss decreasing with time. This may be an empirical curve or a physically based model such as the Green and Ampt[23] equation.
5. Standard rainfall-runoff relation, such as the U.S. Soil Conservation Service relation.

Choice and validity of rainfall loss methods depend on the type of problem, the data available, and the runoff processes which are likely to be dominant. Some flood estimation methods specify the model to be used, such as the SCS method (model 5).

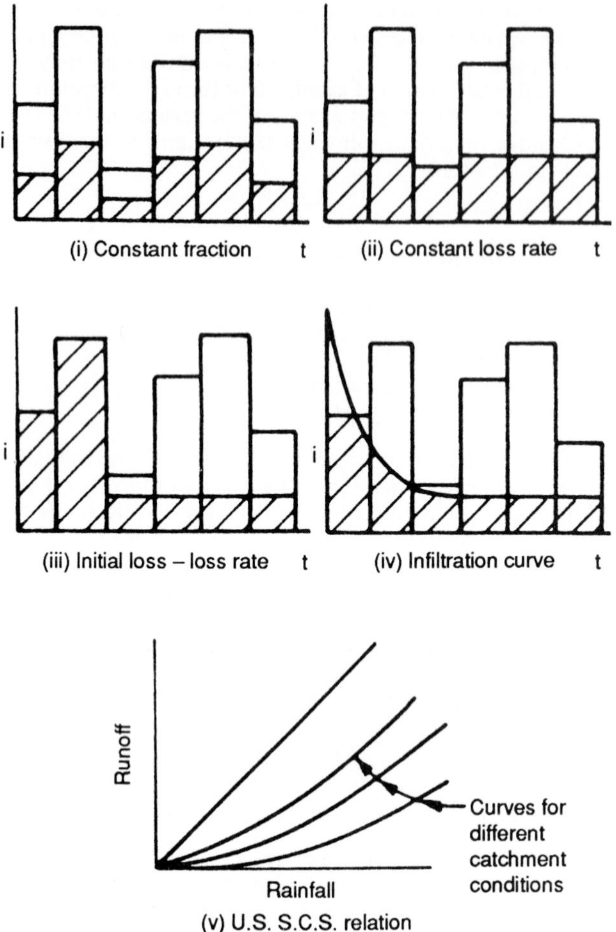

FIGURE 9.2.4 Loss models used to estimate rainfall excess. *(Reproduced from Ref. 31 with permission.)*

Where Hortonian runoff occurs from the whole drainage basin for the whole range of floods under consideration, models 2 and preferably 3 are likely to be most satisfactory. Model 4 is also good for these conditions but is more difficult to apply in practice. In some regions, the continuing loss rate in 2 and 3 is not constant but tends to increase with increasing rainfall intensity (e.g., Ref. 3), possibly reflecting the effects of runoff occurring from only part of the drainage basin. In other regions, probably where Hortonian runoff occurs over all or most of the drainage basin, the continuing loss rate has been found to decrease with increasing intensity. Where runoff occurs from only a saturated portion of a drainage basin, models 1 and 5 are likely to be satisfactory.

To estimate the rainfall excess from an actual storm, as distinct from the design situation, allowance must be made for the condition and wetness of the drainage basin immediately prior to the event. For this purpose rainfall-runoff relations such as 3, 4, or 5 above could be employed provided local data have been used in their calibration.

In fitting any of the models to an observed storm and resulting flood, the volume of storm runoff and the depth of rainfall excess (which is the volume divided by the basin area) are determined from the area under the direct runoff hydrograph after baseflow has been separated. In using the various loss models, the excess of rainfall over the losses (the *excess rainfall*) must equal the volume of storm runoff.

9.3 FUNDAMENTAL ISSUES IN ESTIMATING FLOOD DISCHARGES

The appropriateness and even validity of a flood estimate depends not only on the correct numerical application of a particular method but also on a number of basic issues that are often inadequately considered.

9.3.1 Importance and Approach to Different Types of Flood Estimates

The importance of estimates of flood magnitudes for management of their social and economic impacts is obvious in general terms, but it is helpful to consider more quantitatively the magnitude of economic investment. For the United States, Poertner[66] estimated that public investment in surface drainage improvements in developing urban areas involved an annual capital investment of about $4 billion as of 1978, and that annual flood damages in urban areas were about $3 billion. More comprehensive figures are available for Australia.[58] Approximate average annual expenditures on structures and works sized by flood estimates as of 1988 total $A700 million, about 0.3 percent of the gross domestic product (GDP). Similar relative figures should apply to most developed countries. In addition, average annual flood damages in riverine floodplains, and from urban drainage systems, are each estimated at approximately $A200 million, or 0.1 percent of GDP, in Australia as at 1988.

Several points are indicated by these figures:

1. The overall economic importance of flood estimation.
2. The average annual expenditure for works in small rural drainage basins, mainly for culverts, small bridges, and farm dams, amounts to 46 percent of the total, making flood estimation for these small drainage basins of greatest importance in terms of national expenditure.
3. Urban drainage design is next in importance with 26 percent of expenditure.
4. Both of these categories for small basins are of greater overall economic importance than flood estimation for large drainage basins where much of the remaining 28 percent of expenditure occurs, but for which most of the more sophisticated methods have been developed.
5. Many designers dealing with small drainage basins have limited hydrologic expertise.

9.3.2 Requirements of Methods to Be Used in Practice

Design Floods as Contrasted with Floods Resulting from Actual Storms. Much confusion has resulted from lack of recognition of the fundamental differences between these two types of problems where a flood is estimated from rainfall. The same computational procedures may be involved in each case, but valid application requires recognition of the differences.

Floods resulting from actual storms reflect the particular storm pattern and the moisture conditions prevailing at the time of the storm. For reproducing these floods, the flood estimation methods are used as deterministic models and the above factors must be represented as closely as possible. This type of estimate is required for flood forecasting, the estimation from recorded rainfall of missing flood records, and the calibration of flood models.

By contrast, the much more common practical requirement of the estimation of a design flood does not involve representation of the characteristics of a historical event but rather the probabilistic estimate of a flood whose magnitude will be equaled or exceeded with a given frequency (for example, 0.1, meaning that on average it will be exceeded 10 times in a 100-year interval). There is no unique combination of rainfall and drainage basin condition which produces the design flood. In this case, a flood estimation method (which may have the same mathematical form as for estimating a flood resulting from an actual storm) is used as a probabilistic model, and the designer's intention is to estimate a flood with a selected exceedance frequency from design rainfall.

Although the differences in these two types of problems are often not recognized, they have three important practical consequences, as illustrated further for the rational method in Sec. 9.4.1:

1. A particular procedure may be good for one case but quite unsuitable for the other.
2. If actual floods are to be estimated, values of parameters should be derived from calibration on individual observed events. For estimating design floods, values of parameters should be derived from probability analyses of data from many observed floods.
3. The parameters of methods are quite different conceptually in the two cases. For example, the common concept of the runoff coefficient as the ratio of the peak rate of runoff to the rainfall causing it in a particular event is correct in the very unusual case where the rational method is used to estimate an actual flood. However, this concept is incorrect and fundamentally misleading in the usual case where a design flood is to be estimated.

Regional Differences in Flood Characteristics. While the objective of many design procedures is that they should be applicable over large regions, there is growing recognition of regional differences in hydrologic characteristics that limit the transferability of design data and methods over large areas. These differences result from variations in storm runoff processes, partial and variable source area runoff production, and variations in the types and structure of soils and their infiltration characteristics. This highlights the danger of transposing procedures and design information from one country or region to another, and reinforces the need for the derivation of procedures and parameters from observed data in each region.

Linear or Nonlinear Response. A fundamental aspect of all methods for transforming rainfall excess into an outflow hydrograph is whether linear or nonlinear behavior is assumed and, if the latter, the form of nonlinearity. With linearity, an increase in the input results in a proportional increase in the output, and outputs from different inputs can be superimposed. In general, proportionality and superposition do not apply with nonlinear systems, although some systems fulfill the temporal superposition principle.[39] At least over part of the discharge range in drainage basins, nonlinearity is evidenced by the flood peak increasing at a faster than proportional rate as rainfall excess increases, and by earlier occurrence of the flood peak. The average flow velocity in the basin increases as the discharge increases, in contrast with constant average velocity for all discharges assumed with linear behavior. However, Wong and Laurenson[82] and Bates and Pilgrim[2] found decreasing average velocity with increasing discharge in river reaches, probably owing to floodplain storage.

Most practical flood estimation uses linear models, notably the unit hydrograph procedure and the rational method. Reasons include: (1) the simplicity of calculations in analysis and design and the availability of methodology; and (2) doubt regarding the worth of more complex procedures in view of the inaccuracies inherent in hydrologic data and lack of knowledge of actual runoff processes. Various methods have been developed to model nonlinear response.[17,39] The main methods that have been used in practice are the equations for unsteady flow in urban drainage, power function relations in cascades of conceptual storages and in kinematic wave models, and quasi-linear approaches.

While most of the models can be fitted to observed rainfall excess-runoff events, the choice of form of nonlinearity and parameter values can have a major effect on design estimates, especially where considerable extrapolation is involved. This is illustrated in results for a 35 mi² (90 km²) drainage basin near Sydney, Australia.[57] The RORB model (Sec. 9.6.2) consisting of a network of reservoirs was used with storage (S)-discharge (q) relations of the form

$$S = k\,q^m \tag{9.3.1}$$

Values of m of 0.8, 1.0 (linear), and 1.2 were used to calibrate the coefficient k in the model to a flood with average recurrence interval of 5 years. Each calibration enabled good reproduction of this flood. The calibrated models were then used to estimate floods for a design rainfall and for the probable maximum precipitation, whose rainfalls were, respectively, twice and approximately six times the magnitude of the rainfall for the calibration flood. The results were:

Flood	Average recurrence interval, years	Peak flow, m³/s		
		$m = 0.8$	$m = 1.0$	$m = 1.2$
1. Calibration	5	170	170	170
2. Design (twice case 1)	80	400	340	290
3. Probable maximum flood	—	1500	1020	760

The increasing differences in the estimated peaks with the degree of extrapolation illustrate the importance of the choice of the form and degree of nonlinearity, which are often selected arbitrarily in practice. Surprisingly little attention has been paid to the processes involved in nonlinearity of response of drainage basins. The usual approach has been to calibrate a nonlinear model using several observed floods, either with an assumed degree of nonlinearity, or including one or more parameters controlling the degree of nonlinearity among those to be evaluated. However, the design event to which the calibrated model is to be applied is usually much larger than the calibration events, and the validity of extrapolation may be open to serious question. Also, the relatively low accuracy of high discharge values in the calibration floods introduces uncertainty. The assumptions made in extrapolating rating curves may in themselves introduce a spurious degree of nonlinearity.

The physical characteristics of the stream channels of the drainage basin can provide helpful information through the likely effects of the variation of hydraulic radius and roughness on average flow velocities.[65] Considering hydraulic flow formulas such as Manning's equation, drainage basins where the design flow is retained within channels that are formed or have small floodplains are likely to respond in a highly nonlinear manner. In drainage basins with large floodplains and vegetation or other obstructions within the high banks and on overbank areas, average velocities are likely to remain fairly constant or even to decrease to some extent as flow rate increases. The classic and widely quoted evidence for nonlinearity in rainfall-runoff conversion is that of Minshall,[48] obtained mainly on a single 27-acre (10.9-ha) cultivated drainage basin. This basin would have had little if any floodplain. Subsidiary evidence on a larger 290-acre (117-ha) mixed-cover drainage basin showed less nonlinearity. Choice of form and degree of nonlinearity involves considerable uncertainty.

Need for Simple and Unambiguous Methods. Much design flood estimation for small drainage basins is carried out by designers with little hydrologic expertise. It is therefore important that adopted procedures should be relatively simple to apply, physically sound, based on locally observed data, and capable of giving reproducible answers by different designers. It is often assumed that more complex models with an apparently better physical and mathematical basis will give more accurate estimates. In a comparison of three types of methods, Loague and Freeze[43] found that accuracy decreased with increasing complexity, particularly in design where regression and unit hydrograph models were acceptable but a quasi physically based model performed poorly. Loague and Freeze claimed that the main difficulty was scale problems associated with the spatial variability of rainfall and soil properties, and that their results would also apply to similar types of models. It seems likely that for design, the use of an adequate base of observed data may lead to greater accuracy and is more important than using a complex model.

Design for Risk. All hydraulic works sized by a flood estimate are designed on a risk basis, and none is completely "safe." There is always a finite, and often a quite large, probability that the design flood will be exceeded. The design intention is that this will occur (albeit infrequently) but that any resulting damages will be socially and politically acceptable, there will be a low risk to life, and the average annual cost of works plus damages will be less than that of constructing works for larger design floods and lower risks.

As a consequence, failure or surcharging of minor and medium-sized structures should occur relatively frequently. Most minor bridges and culverts, farm dam spillways, and urban drainage systems for which the social and economic implications of surcharge are not great have been designed for floods with average recurrence intervals in the range of 2 to 100 years, with 20 years being an approximate mean. With thousands of these structures in a given region, several hundred (about 5 percent of the total number) should be surcharged every year, although high variability must be expected. If, on the average, these large numbers of failures do not occur, the design procedures are not accomplishing their objectives, and structures are being overdesigned relative to stated criteria. As surcharging is to be expected, one of the objectives in design should be to provide for passage of floods that exceed the design flood with a minimum of social, physical, and environmental damage.

9.3.3 Estimation of a Flood of Selected Probability from Rainfall

The intention in design generally is to estimate a flood of selected exceedance probability from rainfall intensity-duration-frequency data (Chap. 3) for the locality. The fundamental problem is that each part of the design model introduces some joint probability, obscuring the relation between the probabilities of the design rainfall and the flood estimated from it. There is no rigorous solution to the problem of how to maintain the selected probability, but four general approaches have been used, the last two being of greatest practical value.

1. *Frequency analysis of a synthetic streamflow record* generated by a continuous rainfall-runoff model from long rainfall records.
2. *Joint probability analysis* of the variables contributing to flood discharge, using transition probability matrices or a large number of simulations with random values drawn from assumed distributions.[4,53]
3. *Use of median values* of losses, baseflow, temporal pattern of rainfall, and hydrograph model parameters. Extreme values would convert a design rainfall of selected exceedance probability to a flood with a much smaller probability. If values of these variables are derived from several observed events, the probabilities of the occurrence of values higher and lower than the medians would be equal. Use of these median values in design should minimize the problem of joint probabilities and produce a flood estimate of similar probability to that of the design rainfall.
4. *Values derived from comparison of floods and rains of the same probability.* This is the most direct and simple procedure for regional methods. For each gauged drainage basin, floods of various probabilities are estimated by frequency analysis of observed floods. Values of design parameters or variables are then determined, which convert a design rainfall of a given probability to the flood of the same probability. The effects of the other variables are automatically accounted for. The probabilistic rational method (Sec. 9.4.1) developed for general design in Australia[31,60] is an example of this approach, and a more complex procedure is described by Beard.[3]

9.3.4 Pattern and Duration of Design Rainfall

Temporal Patterns. While these are discussed in more detail in Sec. 3.9, two points need to be recognized where design rainfalls are found from intensity-duration-frequency data. First, these rainfalls generally do not represent complete storms but are from intense bursts within these storms. The design temporal patterns therefore need to be appropriate for these intense bursts, and not for complete storms. Second, the pattern should be appropriate to achieve, as best as possible, the transformation of a rainfall of selected probability to a flood of the same probability. Approaches used to estimate design temporal patterns are described by Wenzel in Ref. 34 and in Ref. 59.

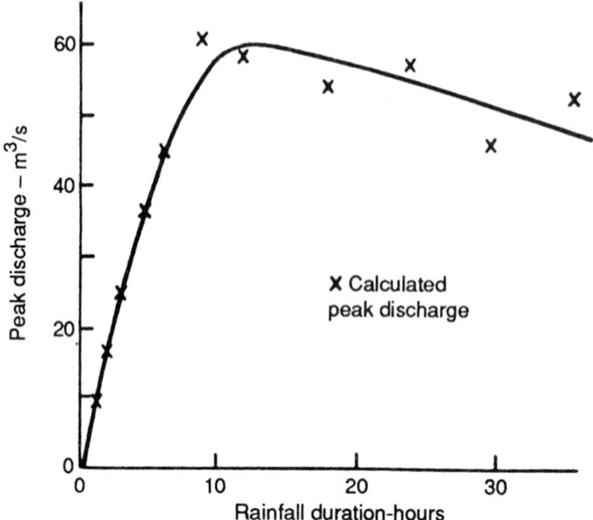

FIGURE 9.3.1 Critical duration of design rainfall from plot of peak discharge against duration. *(Reproduced from Ref. 31 with permission.)*

Critical Duration of Rainfall. For estimation of design floods of selected exceedance probability, there is a critical duration that gives the maximum peak discharge. Where the design procedure does not specify the duration, floods should be calculated from design rainfalls of several durations and the selected probability. The flood peaks are plotted against rainfall duration, and the design peak discharge and critical duration are obtained from the peak of a smooth curve drawn through the plotted points. Figure 9.3.1 shows a typical plot, the scatter of points reflecting minor variations resulting from differences in design temporal patterns for different durations. If outflow from a storage is required, the peaks of routed floods should be plotted.

9.4 PEAK FLOWS FOR SMALL TO MEDIUM-SIZED BASINS

The estimation of peak flows on small to medium-sized rural drainage basins is probably the most common application of flood estimation as well as being of greatest overall economic importance. These estimates are required for the design of culverts, small to medium-sized bridges, causeways and other drainage works, spillways of farm and other small dams, and soil conservation works. It is not possible to define precisely what is meant by "small" and "medium" sized, but upper limits of 10 mi^2 (25 km^2) and 200 mi^2 (500 km^2), respectively, can be considered as general guides. In almost all cases, no observed data are available at the design site, and little time can be spent on the estimate, precluding use of other data in the region.

Hundreds of different methods have been used for estimating floods on small drainage basins, most involving arbitrary formulas.[11] The three most widely used types of methods are the rational method, the U.S. Soil Conservation Service Method, and regional flood frequency methods. The last type is described in Chap. 18.

9.4.1 Rational Method

Traditional Approach. The rational method formula is

$$q = F\,C\,i\,A \tag{9.4.1}$$

where q is the peak discharge, C is the runoff coefficient which is dimensionless and loosely defined as the ratio of runoff to rainfall, i is the rainfall intensity, and A is the drainage basin area. F is a units conversion factor. When English units of ft^3/s, in/h, and acres are used, F equals 1.008 and is usually omitted from the formula. With SI units of m^3/s, mm/h, and km^2, F equals 0.278 or (1/3.6).

The average rainfall intensity i has a duration equal to the critical storm duration, normally taken as the time of concentration t_c. For design, i is estimated from the rainfall intensity-duration-frequency data for the location, with its frequency the same as that selected for the design flood. *Time of concentration* is an idealized concept and is defined as the time taken for a drop of water falling on the most remote point of a drainage basin to reach the outlet, where remoteness relates to time of travel rather than distance. Probably a better definition is that it is the time after commencement of rainfall excess when all portions of the drainage basin are contributing simultaneously to flow at the outlet.

The rational method is widely used around the world for flood estimation on small rural drainage basins and is the most widely used method for urban drainage design.[41,56] It is generally considered to be an approximate deterministic model representing the flood peak that results from a given rainfall, with the runoff coefficient being the ratio of the peak rate of runoff to the rainfall intensity. The formula was developed from a simplified analysis of runoff. On the drainage basin in Fig. 9.4.1, *isochrones* or lines of equal time of travel are drawn, with areas a, b, c, and d acres between them as shown. A constant rate of rainfall i in/h commences at time $t = 0$. The method theoretically assumes that there is no temporary storage of water on the surface of the drainage basin. The ratio of the peak rate of runoff to the rainfall intensity is then the same as the ratio of the volumes of runoff and rainfall. If the duration of the rainfall equals t_c, the hydrograph reaches an instantaneous peak of Ci expressed in units of runoff rate per unit area. However, if the rainfall duration is longer than t_c, the hydrograph remains constant after reaching a value of Ci for a time (rainfall duration $-\,t_c$). In both cases the times of rise and recession of the hydrograph

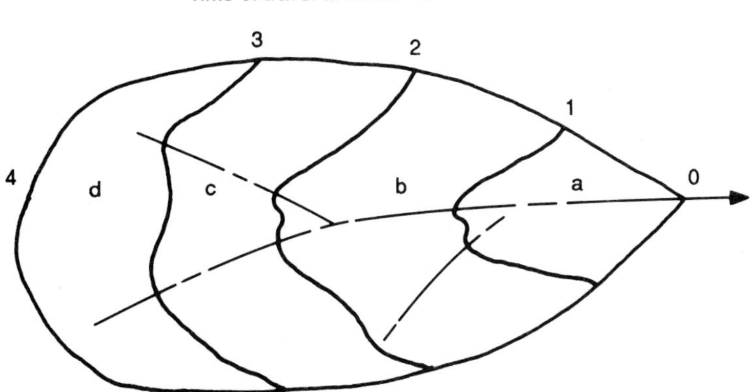

FIGURE 9.4.1 Isochrones of travel time and subareas (acres) of a drainage basin.

equal t_c. Considering the rainfall of i in/h commencing at $t = 0$, at time $t = 1$ h, an area a acres contributes to flow at the outlet, and scaling up the volumetric identity that 1 in/h on 1 acre gives 1.008 or 1 ft³/s, results in an outflow rate of $(Ci \cdot a)$. At $t = 2$ h, an area of $(a + b)$ acres contributes, and the outflow rate is $[Ci \cdot (a + b)]$ ft³/s. The area contributing increases with time until the time of concentration when the whole area A contributes, and the outflow rate is (CiA) ft³/s. This simplistic theoretical basis led to the name "rational method" compared with the prevailing arbitrary methods available when it was developed.

The basis enables assessment of the rational method as a description of the flood runoff process. The effects of rainfall and basin size are accounted for explicitly, and those of most other physical characteristics of the drainage basin are considered indirectly in the time of concentration and the runoff coefficient. The latter accounts for infiltration and other losses, though not in a physically realistic manner. Temporary storage and temporal and spatial variation of rainfall are neglected completely, and the method is reasonably valid only where their effects are small. This may be the case on urban and small rural drainage basins. In practice, the runoff coefficient must take account of these neglected factors as well as infiltration and other losses, and the effects of antecedent wetness.

The simple theoretical basis also indicates why the critical duration of rainfall is assumed to be the time of concentration. For durations less than t_c, the whole area does not contribute. For durations greater than t_c, there is no increase in contributing area and the rainfall intensity of a given frequency decreases, and therefore a duration of t_c gives the highest peak flow. It is assumed that for durations less than t_c, the effect of the reduction in contributing area is greater than that of increased rainfall intensity. This is not the case if the first or last increments of travel time (see Fig. 9.4.1) give only small additions of contributing area or flow, and use of a shorter-duration rainfall may then be desirable.

Design Values. In application of the rational method, design values of time of concentration and runoff coefficient must be estimated. For urban areas, values of t_c are normally calculated as length divided by velocity determined by hydraulic formulas or tabulated values. For rural drainage basins, t_c is generally estimated by means of an empirical formula. Some of these are listed by Chow et al.[13] and French et al.[21] One of the formulas often used in the United States was derived by Kirpich[36] based on data from six small agricultural drainage basins reported by Ramser.[68] The Kirpich formula is

$$t_c = 0.0078\, L^{0.77}\, S^{-0.385} \tag{9.4.2}$$

where t_c is in minutes, L is the length of channel from divide to outlet in feet, and S is the average channel slope in ft/ft. This formula has been found to give low values in Australia. Another formula that has been used in several countries was proposed by Bransby Williams.[7] A modified form with L in miles, A in mi², and S in ft/ft is

$$t_c = 21.3\, L\, A^{-0.1}\, S^{-0.2} \tag{9.4.3}$$

where the slope is that of a linear profile having the same area under it as the actual profile of the main stream. With units of L in km, A in km², and S in m/m, the coefficients in the above formulas are 3.97 and 14.6, respectively. The formula to be used in a particular region is a matter of judgment and is obviously a source of uncertainty. For routine design, it is important that a consistent procedure be used, even in details such as the slope measure. The formula adopted for a region should be assessed to ensure that it gives reasonable average velocities, and if possible estimated

t_c values should be compared with typical minimum times of rise of observed hydrographs.[64]

Estimating the value of the runoff coefficient is the greatest difficulty and the major source of uncertainty in application of the rational method. The coefficient must account for all the factors affecting the relation of peak flow to average rainfall intensity other than area and response time. Design values are normally obtained from tables of suggested values, graphs, or by means of the sum of "scores" given for each of several factors. A typical but rather simple example is given in Table 9.4.1[1] relating mainly to urban drainage. A common feature, and weakness, of most sets of design values is that they largely reflect judgment rather than hard data. All of the rather sparse analytical studies indicate that derived values of C on a given drainage basin vary widely from storm to storm, particularly with different antecedent wetness conditions. It is assumed that more stable values apply to the large events of interest

TABLE 9.4.1 Runoff Coefficients Recommended by American Society of Civil Engineers and Water Pollution Control Federation

	Runoff coefficients
Description of area:	
Business	
Downtown	0.70–0.95
Neighborhood	0.50–0.70
Residential	
Single-family	0.30–0.50
Multiunits, detached	0.40–0.60
Multiunits, attached	0.60–0.75
Residential (suburban)	0.25–0.40
Apartment	0.50–0.70
Industrial	
Light	0.50–0.80
Heavy	0.60–0.90
Parks, cemeteries	0.10–0.25
Playgrounds	0.20–0.35
Railroad yard	0.20–0.35
Unimproved	0.10–0.30
Character of surface:	
Pavement	
Asphaltic and concrete	0.70–0.95
Brick	0.70–0.85
Roofs	0.75–0.95
Lawns, sandy soil	
Flat, 2 percent	0.05–0.10
Average, 2–7 percent	0.10–0.15
Steep, 7 percent	0.15–0.20
Lawns, heavy soil	
Flat, 2 percent	0.13–0.17
Average, 2 to 7 percent	0.18–0.22
Steep, 7 percent	0.25–0.35

Source: Ref. 1. Used with permission.

in design. Often, the value of C is assumed to increase as average recurrence interval increases, thus allowing for nonlinearity of flood response. An example of a more detailed table of C values of this type is given by Chow et al.[13] Graber[22] describes a method for determining C at a given location from the Soil Conservation Service method runoff curve number CN. Unfortunately, there is considerable evidence that the one set of C values, even with detailed variations for different conditions, will not apply to regions with different hydrologic regimes or even uniformly within large regions. Considerable judgment and experience are required in selecting satisfactory values of C for design, and there is a need to check values against observed flood data in a given region, as discussed for the probabilistic approach.

Probabilistic Approach. As used in design, the rational formula [Eq. (9.4.1)] can be written in a more explicit form as

$$q(Y) = F \cdot C(Y) \cdot i(t_c, Y) \cdot A \tag{9.4.4}$$

where C as well as q and i is labeled with an average recurrence interval (ARI) of Y years. The intention in design is not to estimate the runoff from a particular storm. Rather, the intention is to estimate the discharge that would be given for this ARI by a frequency analysis of observed floods if a long and representative record of discharge was available at the site. Then the rational formula is a means to convert a rainfall with ARI of Y years into a peak discharge with the same ARI. Both the design rainfall and peak discharge are probabilistic values derived from frequency analyses of local data.

The nature of the rational method as used in design is best demonstrated by rearranging Eq. (9.4.4) as

$$C(Y) = \frac{q(Y)}{i(t_c, Y) \cdot F \cdot A} \tag{9.4.5}$$

If frequency curves of both rainfalls of duration t_c and peak flows are available at a site, Fig. 9.4.2 for a typical gauging station (Omadale Brook at Roma in eastern Australia, area 104 km^2) shows that with the appropriate scaling of $(F \cdot A)$, the value of $C(Y)$ is given by the ratio of $q(Y)$ to $i(t_c, Y)$. This value of $C(Y)$ will convert a design rainfall of duration t_c and ARI of Y years into the flood peak with ARI of Y years. The design situation is exactly fitted by this probabilistic interpretation and has little similarity to the traditional deterministic model of runoff resulting from a particular storm. Most of the criticism of the rational method has been based on the deterministic interpretation, and is not necessarily valid for the probabilistic design case.

The probabilistic rational method was developed by Horner and Flynt[30] but was then neglected until the work of Schaake et al.[70] with urban areas and French et al.[21] for rural drainage basins. In Australia, design procedures have now been derived for most regions using this approach.[31,58,64] Values of $C(Y)$ have been derived using Eq. (9.4.5) for all small to medium-sized gauged drainage basins, and a base value, usually $C(10)$, was either mapped or related by regression to physical variables. Figure 9.4.3 shows a map of 10-year runoff coefficient values for part of eastern New South Wales, Australia. The contours generally follow topographic features and rainfall contours. Values of C greater than 1 are possible with the probabilistic interpretation.[60] As shown by Fig. 9.4.2, the value of $C(Y)$ will vary in a regular manner with Y. Ratios $C(Y)/C(10)$ were derived for the rural regions, while $C(Y)$ was found to be independent of Y in urban areas. For Omadale Brook at Roma

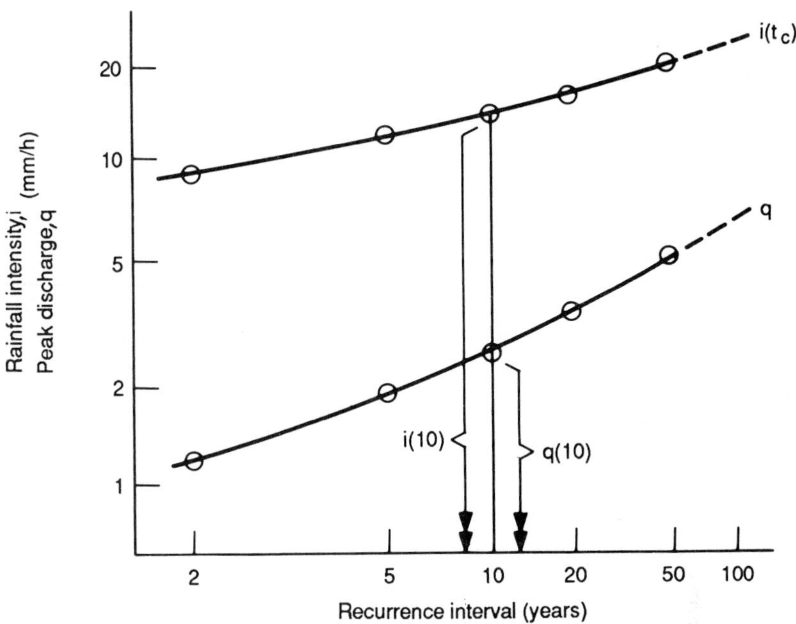

FIGURE 9.4.2 Frequency curves of rainfall and peak discharge for Omadale Brook, eastern Australia, for derivation of probabilistic runoff coefficients.

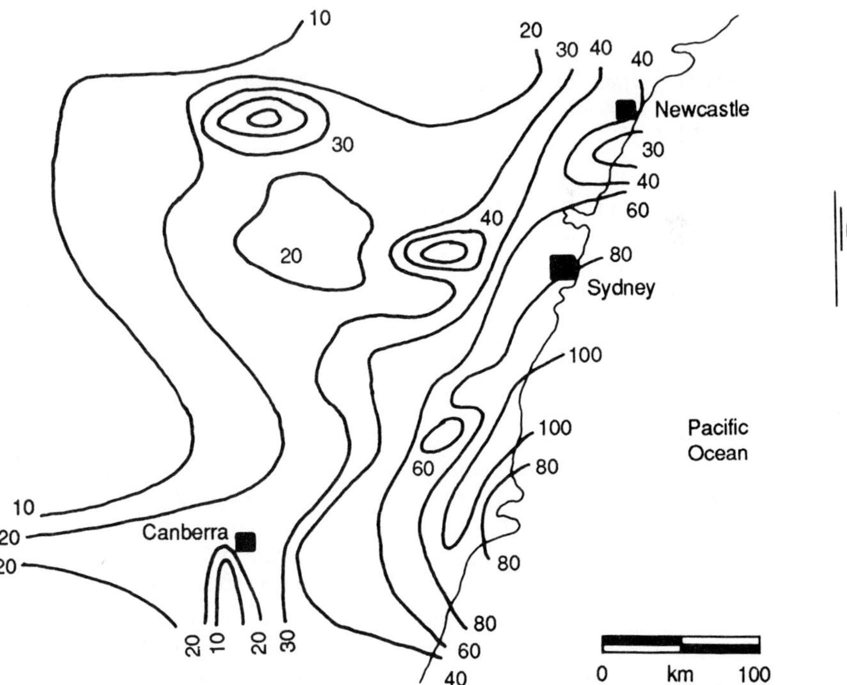

FIGURE 9.4.3 Design runoff coefficients for average recurrence interval of 10 years for part of eastern Australia (expressed as percentages).

illustrated in Fig. 9.4.2, values of $C(Y)$, derived ratios of $C(Y)/C(10)$, and average values of the ratio for the region are:

Average recurrence interval Y, years	2	5	10	20	50	100
$C(Y)$	0.13	0.16	0.19	0.22	0.26	0.30
$C(Y)/C(10)$	0.68	0.84	1.00	1.16	1.37	1.58
Average regional values of $C(Y)/C(10)$	0.70	0.86	1.00	1.14	1.33	1.50

Viewed in this way, the probabilistic rational method is an efficient regional flood frequency procedure, with rainfall intensity, a major determinant of flood response, being one of the independent variables. It is also simple and familiar to most designers.

Design procedures of this type should give much more accurate flood estimates than use of tabulated values of C in the traditional approach. In a study of 271 drainage basins in Australia,[60,64] the probabilistic approach was shown to use almost all the information contained in the observed data and gave flood estimates of good accuracy. More accurate estimates would be possible only with longer records. Estimates for the same drainage basins using the traditional approach, with C values determined from a graph depending on rainfall and soil and vegetation types, compared very poorly with values estimated from frequency analyses of observed floods, 63 percent of 271 basins having errors $> \pm 50$ percent and 42 percent with errors $> \pm 100$ percent. It is also noteworthy that the C values derived from observed data did not show large variations with different physical characteristics of drainage basins, contrary to the assumptions which underlie tabulations or graphs of suggested design values such as those in Refs. 1 and 13.

Derivation of Design Data for the Probabilistic Rational Method. The steps in deriving design data on a probabilistic basis for use of the rational method in a particular region are:

1. For each gauged basin, carry out a frequency analysis of observed floods to determine values of $q(Y)$ for a range of average recurrence intervals.
2. Select a design formula for time of concentration. This should be derived from or checked against observed data from basins in the region.[21,64] The adopted formula for t_c must be used throughout derivation of the design data and must be specified to be used in all applications of the derived procedure.
3. For each basin, values of the design rainfall intensity $i(t_c, Y)$ are determined for each recurrence interval for which flood values were derived in 1, from the design rainfall intensity-duration-frequency data.
4. Using the values from 1 and 3, values of $C(Y)$ are calculated for each basin by Eq. (9.4.5) for each recurrence interval Y.
5. A base value of $C(Y)$ is selected for relating to basin characteristics. The 2-year or 10-year values $C(2)$ or $C(10)$ are generally convenient as they are subject to relatively low sampling errors. The selected base values of $C(Y)$ are then related to basin characteristics by regression or are mapped over the region (see Fig. 9.4.3).
6. Regional average values of the ratio of $C(Y)$ to the coefficient value for the selected base ARI are then determined for each desired value of Y.
7. Application of the rational method for flood design for any basin in the region then involves use of the adopted formula for t_c, the rainfall intensity-duration-frequency data for the region, the base runoff coefficient from 5, and the relevant

frequency ratio from 6. The design information is valid only if the t_c formula and rainfall data used in derivation are also used in application of the method.

Range of Validity. Where design data have been derived from the available observed flood data in the region of interest using the probabilistic approach, the rational method is efficient and provides flood estimates of good accuracy over the range of drainage basin sizes used in derivation. Consistent flood estimates for drainage basins up to 100 mi^2 (250 km^2) have been derived in Australia.

9.4.2 Soil Conservation Service Method

The SCS method is widely used for estimating floods on small to medium-sized ungauged drainage basins. In the United States, it has replaced the rational method to a significant degree as a result of its wider apparent data base and the manner in which physical characteristics are considered in its application. It has been adopted as the required procedure by many municipal and regional authorities. The method has also been used to some extent in other countries. It was developed originally as a procedure to estimate runoff volume and peak discharge for design of soil conservation works and flood-control projects. Many variations of the procedure have been used. Current procedures are described in Refs. 80 and 81.

While there is an extensive literature on the method, little quantitative information is available on the data base from which it was developed, and the manner in which this data base was used in the development. Rallison[67] and Miller and Cronshey[47] briefly describe the origin and evolution of the method, which was derived largely from infiltrometer tests and measured rainfall and runoff on small plots and basins. The method is also reviewed in a series of papers in Ref. 72.

Volume of Runoff. The derivation of the basic equation for estimating the volume or depth of runoff is illustrated schematically in Fig. 9.4.4 for accumulated volumes during a storm. The general form of the relation is well established by both theory and

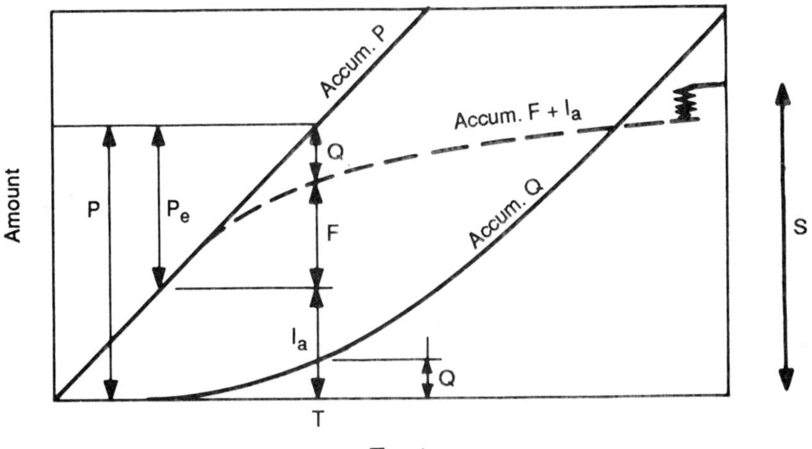

FIGURE 9.4.4 Accumulated rainfall, losses, and runoff during a uniform storm.

observation. No runoff occurs until rainfall equals an initial abstraction I_a. After allowing for I_a, the depth of runoff Q is the residual after subtracting F, the infiltration or water retained in the drainage basin (excluding I_a) from the rainfall P. The potential retention S is the value that $(F + I_a)$ would reach in a very long storm.

If P_e is the effective storm rainfall equal to $(P - I_a)$, the basic assumption in the method is

$$\frac{F}{S} = \frac{Q}{P_e} \tag{9.4.6}$$

This assumption seems to be quite arbitrary and has no theoretical or empirical justification, although it is obvious from Fig. 9.4.4 that the ratios will be of the same general order. Then as F equals $(P_e - Q)$,

$$Q = \frac{P_e^2}{P_e + S} = \frac{(P - I_a)^2}{P - I_a + S} \tag{9.4.7}$$

The empirical relation $I_a = 0.2\,S$ was adopted as the best approximation from observed data, and so

$$P_e = P - 0.2\,S$$

and

$$Q = \frac{(P - 0.2S)^2}{P + 0.8S} \tag{9.4.8}$$

For convenience and to standardize application of this equation, the potential retention is expressed in the form of a dimensionless runoff curve number CN, where (in English units, S in inches)

$$\text{CN} = \frac{1000}{S + 10} \tag{9.4.9}$$

Eliminating S from (9.4.8) and (9.4.9) gives the basic SCS relationship for estimating Q from P and CN shown in Fig. 9.4.5. This graphical relation and Eq. (9.4.8) have the advantage of having only one parameter. The value of CN depends on the soil, cover, and hydrologic condition of the land surface. These are described as:

- Soil, four classes
  - A: high infiltration, low runoff, as for deep sand or loess, aggregated silts
  - B: moderate infiltration, as for moderately fine to moderately coarse-textured soils such as sandy loam
  - C: slow infiltration, as for fine-textured soil such as clay loam, shallow sandy loam, soils low in organic content
  - D: very slow infiltration, such as swelling and plastic clays, and claypan
- Cover, relating to various types of vegetation and crops, land treatments and crop practices, paving and urbanization
- Hydrologic condition, relating to whether vegetation is dense and in good condition, and whether the soil is rich in organic matter and has a well-aggregated structure, resulting in high infiltration and low runoff

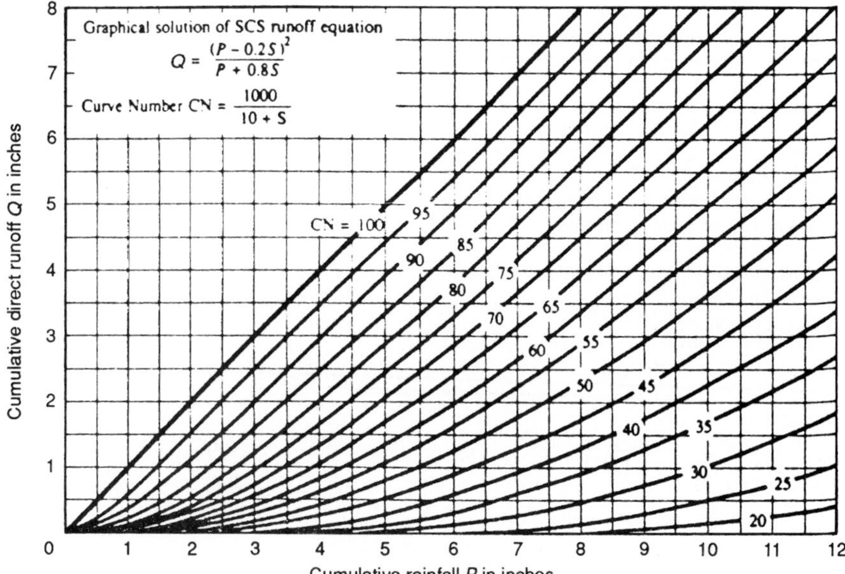

FIGURE 9.4.5 SCS relation between storm runoff and rainfall.[80]

CN also depends on the antecedent wetness of the drainage basin, and three classes of antecedent moisture condition (AMC) are defined, dry, average, and wet (AMC I, II, and III). Standard values of CN for AMC II are listed in Table 9.4.2 for various land uses and soil types and additional values are given in Table 5.5.1. More detailed values and values for AMC I and III are given in Ref. 80, and soils throughout the United States have been classified into the four hydrologic soil groups.

Where a drainage basin is made up of various soil types and land uses, or "hydrologic soil-cover complexes," a composite CN can be calculated. It should be noted that as a result of the nonlinearity of Eq. (9.4.7), the scale of subdivision and the spatial correlation of soil types will affect the overall value of CN and the resulting calculated runoff.[46]

Although not developed for this purpose, the SCS method is used to estimate the runoff from successive intervals in a storm, as well as the total runoff. For this, the accumulated Q to the end of each period is determined from the accumulated P to that time, and the runoff in each period is found as the difference between the accumulated Q at the end and start of the period.

Peak Discharge and Flood Hydrograph. The peak discharge in the SCS method is derived from the triangular approximation to the hydrograph shown in Fig. 9.4.6 resulting from a rainfall excess of duration D. Based on unpublished data, the lag L_a from the centroid of rainfall excess to the peak of the unit hydrograph is assumed to be 0.6 t_c. Then the time of rise T_p (h) to the peak of the hydrograph is

$$T_p = 0.5\ D + 0.6\ t_c \tag{9.4.10}$$

The volume of runoff under the hydrograph is $\frac{1}{2}\ q_p \cdot T_b$, where T_b is the base length of the hydrograph and is assumed equal to 2.67 T_p, based on the study of many

TABLE 9.4.2 Runoff Curve Numbers for Selected Agricultural, Suburban, and Urban Land Uses (Antecedent Moisture Condition II, $I_a = 0.2S$)

Land-use description		Hydrologic soil group			
		A	B	C	D
Cultivated land:*					
Without conservation treatment		72	81	88	91
With conservation treatment		62	71	78	81
Pasture or range land:					
Poor condition		68	79	86	89
Good condition		39	61	74	80
Meadow: Good condition		30	58	71	78
Wood or forest land:					
Thin stand, poor cover, no mulch		45	66	77	83
Good cover†		25	55	70	77
Open spaces, lawns, parks, golf courses, cemeteries, etc.:					
Good condition: grass cover on 75% or more of the area		39	61	74	80
Fair condition: grass cover on 50 to 75% of the area		49	69	79	84
Commercial and business areas (85% impervious)		89	92	94	95
Industrial districts (72% impervious)		81	88	91	93
Residential:‡					
Average lot size	Average % impervious§				
1/8 acre or less	65	77	85	90	92
1/4 acre	38	61	75	83	87
1/3 acre	30	57	72	81	86
1/2 acre	25	54	70	80	85
1 acre	20	51	68	79	84
Paved parking lots, roofs, driveways, etc.¶		98	98	98	98
Streets and roads:					
Paved with curbs and storm sewers¶		98	98	98	98
Gravel		76	85	89	91
Dirt		72	82	87	89

* For a more detailed description of agricultural land-use curve numbers, refer to Ref. 80, Chap 9.

† Good cover is protected from grazing and has litter and brush covered soil.

‡ Curve numbers are computed assuming the runoff from the house and driveway is directed toward the street with a minimum of roof water directed to lawns where additional infiltration could occur.

§ The remaining pervious areas (lawn) are considered to be in good pasture condition for these curve numbers.

¶ In some warmer climates of the country a curve number of 95 may be used.

Source: Reproduced by permission from Refs. 13 and 80.

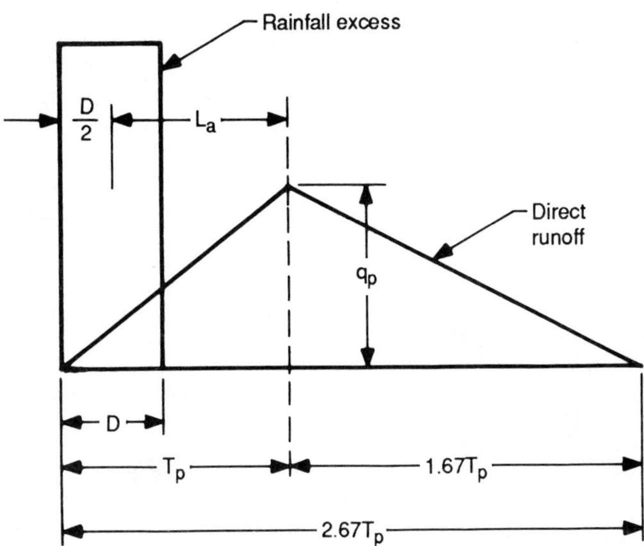

FIGURE 9.4.6 SCS triangular hydrograph.

unit hydrographs. Equating this volume to Q, the volume estimated from the rainfall, and then combining the above relations, rearranging, and adjusting for units, the peak discharge q_p in ft³/s is given by

$$q_p = \frac{484\,AQ}{0.5\,D + 0.6\,t_c} \tag{9.4.11}$$

where A is the area of the drainage basin in mi², Q is in inches, and D and t_c are in hours. In SI units, with q_p in m³/s, basin size in km², and Q in mm, the coefficient 484 is replaced by 0.208.

The estimate of q_p depends on t_c [see Eq. (9.4.11)]. For a given drainage basin, q_p therefore depends on the procedure used for estimating t_c. The Kirpich formula (9.4.2) has been the usual method, but the U.S. Soil Conservation Service[80,81] has proposed a method that depends on the value of CN, and a method involving separate estimates of times of sheet flow, shallow concentrated flow, and channel flow. In some approaches, the designer is free to choose the method of estimating t_c, and in others it is predetermined, resulting in different estimates of q_p for the one drainage basin.

Different methods are also available for estimating the complete hydrograph. Families of hydrographs have been precalculated from a curvilinear dimensionless unit hydrograph, together with several standard rainfall temporal patterns. These hydrographs represent 1 in of runoff from a 1 mi² basin. Scaling by Q and drainage area gives the required hydrograph.[81]

Applications and Validity. As noted earlier, the SCS method is probably the one most widely used in the United States for estimating floods on small rural drainage basins, and it is also used to a lesser extent in urban situations. The explicit considera-tion of the various factors that are thought to affect flood runoff makes the method attractive. The common basis of computed hydrographs makes the method useful for studies involving detention basins and other storages, and pollution management.

In practical use, application of the SCS method is simple and direct. Problems are caused by different analysts obtaining different answers as a result of choice of different procedures, especially for estimating t_c, and in choosing the curve number CN.

Confusion has resulted from application of the SCS method as a deterministic model for estimating the flood resulting from a particular rainfall, and as a probabilistic model to estimate a design flood. The method has been tested extensively in the deterministic sense with generally poor results. For example, Wood and Blackburn[83] using 1600 runoff plots in Nevada, Texas, and New Mexico found differences between observed and computed flood peaks of greater than ± 50 percent in 67 percent of the results. They found that the assumed antecedent moisture condition had a major effect and that results were better for bare soil or sparse vegetation than for dense vegetation. Similar findings on the effects of vegetation were reported by Steichen.[76] Hawkins[25] and Chen[10] give examples of the sensitivity of runoff estimates to variations in CN.

Some limited testing of the SCS method as a probabilistic model was carried out by Hiemstra[26] and Hjelmfelt.[27,28] Extensive testing was carried out by Hoesein et al.[29] using data from 139 drainage basins in eastern Australia in a procedure analogous to the probabilistic rational method. The CN values required to reproduce flood peaks estimated by frequency analyses of recorded flows depended on the method of estimating t_c and the average recurrence interval of the design flood. The required CN values depended to only a relatively minor extent on the characteristics of the drainage basins, in contrast with the assumption in the SCS method. There was a marked lack of agreement between the probabilistic CN values and those estimated by conventional means.

All of the above results cast some doubt on the accuracy and validity of the SCS method. Care is required in its application, and there is a need for checking of the method against observed flood data in each region where the method is applied. This is supported by the review of the current status of the method by Miller and Cronshey.[47] They report that the U.S. Soil Conservation Service recognizes that a single value of CN may not be applicable to similar conditions over the whole of the United States and that the median CN for given conditions varies from region to region.

9.5 UNIT HYDROGRAPH METHOD

9.5.1 Introduction

The unit hydrograph method of flood estimation was first proposed by Sherman[71] and has since found wide-ranging application for both design and estimation of actual floods where a hydrograph and reasonable accuracy are required.

A unit hydrograph is defined as the direct runoff hydrograph resulting from unit depth of excess rainfall produced by a storm of uniform intensity and specified duration. The unit depth of excess rainfall is taken to be 1 in or 1 mm over the entire drainage basin. Unit hydrographs are generally derived from streamflow data and estimates of temporal distribution of rainfall excess, or by synthesis using drainage basin physiographic data. To calculate a flood hydrograph, the unit hydrograph is applied to the hyetograph of rainfall excess to estimate the hydrograph of surface runoff, and baseflow is added to produce the flood hydrograph.

9.5.2 Unit Hydrograph Theory

A unit hydrograph is a specific type of hydrograph in that it represents the effects of the physical characteristics of the basin on the completely defined, standardized input rainfall excess. Two important assumptions are implicit in unit hydrograph theory. These are:

Basin Linearity. The linear assumptions of proportionality of outflow to inflow and the superposition of outputs from successive individual inputs are discussed in Sec. 9.3.2, together with the conditions under which the assumptions are likely to be valid. To minimize errors resulting from the assumption of linearity, unit hydrographs should be derived from floods of magnitudes as close as possible to those that will be calculated using the derived unit hydrograph. These floods are generally of sufficient magnitude to involve considerable overbank flow.

Basin as a Lumped System. The definition of the unit hydrograph indicates that the input rainfall excess is assumed to be uniform over the basin and is defined by a single hyetograph. In reality the depth of rainfall and the losses may act together to cause the real rate of runoff generation to vary widely from place to place at any instant. However, in some basins, particularly those with marked topographic features, the similarity of the spatial distribution of rainfall in all significant storms may mean that it is quite reasonable to characterize the basin rainfall by a single hyetograph. The unit hydrograph is really the basin response to a standard input.

An important feature of the unit hydrograph is its specified time period. This is the duration of rainfall excess that produced the unit hydrograph, and its duration must be included in the name of the unit hydrograph. For a given basin the 1-h unit hydrograph will be different from the 3-h unit hydrograph. In SI units, a 1-h unit hydrograph is produced by 1 mm of rainfall excess falling over the basin in 1 h at a rate of 1 mm/h, and a 3-h unit hydrograph by 1 mm of rainfall excess occurring uniformly during a 3-h period, at a rate of 1/3 mm/h.

In deriving a unit hydrograph from a burst of rainfall excess of approximately uniform intensity, the specified time period is fixed by the duration of the causative rainfall excess. When deriving a unit hydrograph from a multiperiod storm, or for application of a unit hydrograph for flood estimation, the specified time period may be chosen to suit the user's needs. Choosing a long specified time period reduces the labor of calculation. However, the duration needs to be short enough to provide good definition of both the rainfall excess hyetograph and the resulting unit hydrograph. Use of a period longer than a quarter of the unit hydrograph time of rise may result in large errors, especially at the hydrograph peak.

The instantaneous unit hydrograph (IUH) resulting from unit depth of rainfall excess in an infinitesimally short time interval is a useful concept for some synthetic unit hydrograph models. The T-h unit hydrograph (TUH) can be obtained from the IUH very simply, the ordinate at the end of any T-h interval being equal to the average ordinate of the IUH during that interval.

9.5.3 Derivation of Unit Hydrographs from Observed Data

Preparation of Data. It is desirable to derive unit hydrographs from several floods for which hydrograph and recording rain gauge data are available. If possible only floods with an average recurrence interval of 1 year or greater should be used to comply with the linearity requirements. Then for each flood:

1. Separate baseflow as described in Sec. 9.2.3.
2. Calculate the volume of direct runoff. This is equal to the volume of rainfall excess.
3. Estimate average depth and intensity of basin rainfall in each time period of the storm.
4. Calculate rainfall excess in each time period of the storm using an appropriate loss model to determine the loss in each time period.

Derivation of a Unit Hydrograph from a Single-Period Storm. A single-period storm is one in which all of the rainfall excess occurs at a reasonably uniform rate over a fairly short time period. A unit hydrograph is then derived by dividing the ordinates of the surface runoff hydrograph by the depth of surface runoff in inches or millimeters. The specified time period of the unit hydrograph is the duration of rainfall excess. If this specified time period is not a convenient interval, changing it by up to ± 25 percent is acceptable and will have little effect on any subsequent use of the unit hydrograph. The time axis is expressed in hours from the beginning of the rainfall excess. It is important that the position of the hydrograph in time relative to its causative rainfall should be preserved.

Derivation of Unit Hydrographs from Multiperiod Storms. Multiperiod storms are relatively long and have varying intensities of rainfall excess. A storm should be divided into a number of equal periods, each of which has a fairly constant rate of rainfall excess. The duration of the period will be the specified time period of the derived unit hydrograph.

The unit hydrograph can be defined in terms of the observed direct runoff hydrograph ordinates q_i and rainfall excess hyetograph depths P_i in a series of n equations in which there are k unknowns. Here

n = number of hydrograph ordinates

j = number of periods of rainfall excess in the hyetograph

k = number of unit hydrograph ordinates

These equations can be expressed conveniently in matrix form as set out, for example, in Refs. 13 and 31. The number of unknowns for which the equations are to be solved is $k = n - j + 1$. Whenever j is greater than unity, n will exceed k, and so there will be more equations than unknowns. Since all the known data contain errors (in observation, averaging rain over the basin, etc.), all the equations must also contain some error. In addition, unit hydrograph theory does not provide an absolutely correct model of basin behavior. There is therefore no unique solution to the equations.

Several analytical techniques are available for derivation of unit hydrographs from multiperiod storms. Some, such as the Collins method,[33] are suitable for manual use whereas most require computer solution, for example, the least-squares approach, which was first proposed by Snyder.[74]

Unit hydrographs derived using least-squares techniques and some other approaches often have a sawtooth appearance. These irregularities should be smoothed before making any use of the derived unit hydrograph. Severe oscillation of ordinates may result from lack of synchronization of the rainfall and runoff data, or from an incorrect choice of time distribution of rainfall excess. Small justifiable changes to the rainfall excess may largely eliminate wild fluctuations.

9.5.4 Changing the Specified Time Period of a Unit Hydrograph

The specified time period of the unit hydrograph can be changed easily, and two approaches are available, depending on whether a longer or shorter period is needed. These approaches are discussed in some detail in most engineering hydrology texts.[13,31,42]

9.5.5 Runoff Computations Using Unit Hydrographs

Selection of Design Unit Hydrograph. To develop a unit hydrograph for a drainage basin it is desirable to derive unit hydrographs from as many floods as possible. These should be examined to determine whether there is any trend in unit hydrograph characteristics with storm or other causative phenomena. For example, the unit hydrograph time to peak and peak discharge may vary with location of the storm center in the basin, the peak discharge of the flood, and the year of occurrence of the storm.

Suitable allowance for any trends should be made in selecting a design unit hydrograph. If the variation between unit hydrographs prepared from major flood events appears to be random, an average unit hydrograph should be prepared. This should not be done by taking a simple average of the unit hydrographs, since the resulting average could be quite different from any of the individual unit hydrographs. Rather, the average peak flow and the average time to peak flow should be computed. The average unit hydrograph is then sketched to have this peak flow and to conform with the general shape of the individual unit hydrographs. The total flow volume should be equivalent to 1 in (1 mm) depth over the basin.

Where snow cover or lakes affect the shape of the hydrograph, the unit hydrograph adopted should be one derived from storms with conditions similar to those of the design storm. If the design storm is spatially nonuniform, the design unit hydrograph should be derived from events which have similar spatial distributions of rainfall.

There are often quite large differences between unit hydrographs derived from several large floods. Differences in time to peak and peak discharge of 50 to 100 percent of the average values of these two properties can be expected.

Calculation of a Hydrograph. The excess rainfall and unit hydrograph are combined to form a hydrograph of surface runoff as shown in Table 9.5.1. This involves the summation of a number of hydrographs, each of which is produced by one period of rainfall excess. For example, in Table 9.5.1 the first period of rainfall excess, 15.8 mm, produces a hydrograph with ordinates (at intervals of the specified time period) equal to 15.8 times the unit hydrograph ordinates, as shown in column (4). To obtain the total flood hydrograph, a constant baseflow should be added to the adopted surface runoff hydrograph. An average baseflow for the basin should be used if possible, or if this is not known a regional value may be available. Values quoted for the United States, Australia, and New Zealand are of the order of 1.0 $ft^3/s/mi^2$ (0.011 $m^3/s/km^2$).[37,38] A convenient method for manually deriving a flood hydrograph is to use a tabulation for the calculation such as that given in Table 9.5.1.

In determining a design hydrograph, a range of rainfall durations must be used in this procedure to determine the critical duration and maximum peak discharge as discussed in Sec. 9.3.4.

TABLE 9.5.1 Flood Hydrograph Computations—Given the 1-h Unit Hydrograph

Time, h (1)	Excess rainfall, mm (2)	Unit hydrograph, m^3/s (3)	Flood hydrograph, m^3/s, due to rainfall excess in: First hour (4)	Second hour (5)	Third hour (6)	Base flow, m^3/s (7)	Total hydrograph, m^3/s (8)
1	15.8	1.89	29.9			4	34
2	3.6	5.87	92.7	6.8		4	103
3	13.0	10.43	164.8	21.1	24.6	4	215
4		14.45	228.3	37.6	76.3	4	346
5		11.28	178.2	52.0	135.6	4	370
6		7.04	111.2	40.6	187.9	4	344
7		4.39	69.4	25.3	146.6	4	245
8		2.74	43.3	15.8	91.5	4	155
9		1.72	27.2	9.9	57.1	4	98
10		1.07	16.9	6.2	35.6	4	63
11		0.67	10.6	3.9	22.4	4	41
12		0.42	6.6	2.4	13.9	4	27
13				1.5	8.7	4	14
14					5.5	4	10

9.5.6 Synthetic Unit Hydrographs

Synthetic unit hydrographs can be estimated for ungauged drainage basins by means of relationships between parameters of a unit hydrograph model and the physical characteristics of the drainage basin. These relationships are quite empirical and as such cannot be expected to be universally applicable. In general their application should be restricted to the region in which they were derived.

A number of synthetic unit hydrograph approaches are available, but the only ones to have found widespread use are those based on the models of Snyder,[73] Clark,[14] Nash,[49] and the U.S. Soil Conservation Service.[80] There are many similarities between the Snyder and Soil Conservation Service methods, but the Clark and Nash approaches are quite different.

Snyder's approach is based on a relationship of the form

$$t_p = C_t \, (L \, L_{CA})^n$$

where t_p = time from the centroid of rainfall excess to peak of hydrograph
L = length of main stream
L_{CA} = flow distance from center of area of basin to outlet
n = parameter

An improved formula suggested by the U.S. Corps of Engineers includes the square root of stream slope in the denominator of the term in parentheses.[42] Peak discharge is then given by

$$q_p = \frac{640 \, C_p}{t_p}$$

where C_p and C_t are coefficients. Equations relating the coefficients C_t and C_p to basin physical characteristics are given in Refs. 15 and 18. These show that the coefficients vary over a range of 10 times. In view of this, the method generally cannot be expected to provide accurate flood estimates. If the relations for the coefficients have been derived in the region of interest or can be rederived from observed data for several local drainage basins, the method may give reasonable flood estimates.

The Clark and Nash models involve routing of rainfall excess through concentrated storages which represent the storage effects of the basin, and the former is described in Sec. 9.6.2. To derive a unit hydrograph, the rainfall excess input must be 1 in (or 1 mm) occurring uniformly, and usually instantaneously, to give an IUH. The Clark model has been widely used, and its application is described in Refs. 31 and 33. For an instantaneous unit input of rainfall excess over the drainage basin, the inflow to the concentrated storage at the basin outlet is proportional to the time-area diagram, or graph of area between isochrones plotted against time of travel. Since velocity of flow throughout the channel system of a drainage basin in a given flood is approximately constant,[55] the time-area diagram may be constructed by measuring flow distances on a map. The two parameters of the model are C, the time base length of the time-area diagram, and K, the storage delay time of the linear concentrated storage at the outlet. For unit hydrograph synthesis, C and K must be related to drainage basin characteristics. Cordery and Webb[16] give an example and also provide standardized unit hydrograph shapes, which considerably reduce the labor involved in application of the method. This approach has the potential to provide better representations of unit hydrographs than the Snyder approach because it takes far more account of the hydrograph-forming processes that operate on a drainage basin.

The U.S. Soil Conservation Service[80] has presented a dimensionless unit hydrograph which can be applied to ungauged basins where the single parameter of the unit hydrograph L_a (basin lag) [or T_p (time to peak)] can be related to drainage basin characteristics. If accurate flood estimates are to be obtained, the relationships need to be rederived using local data. Further discussion of the Soil Conservation Service approach is given in Sec. 9.4.2.

A recently developed approach, the geomorphological unit hydrograph, has been reviewed by Bras and Rodriguez-Iturbe[8] but has not yet been applied in practice.

9.6 FLOOD HYDROGRAPH MODELS

9.6.1 Types of Models

Many types of models have been developed to estimate flood hydrographs from excess rainfall. The characteristic that distinguishes the models considered here from unit hydrographs and other transfer function procedures is that they attempt to represent the runoff processes in more detail. This can include consideration of the hydraulics of runoff, the routing of runoff through temporary storage within the drainage basin, and the arrangement or topology of the stream network. Computer programs are available for most models and are required for practical application. Some of these are described briefly in Chap. 21 and details are given in the user manuals. For valid application, it is important for the user to understand the principles embodied in the procedure, and the assumptions and simplifications involved.

9.6.2 Storage Routing Models

Nature and Modeling of Storage. In these models, the flood hydrograph is computed by routing rainfall excess through a model representing the storage in the drainage basin. As with all routing, the storage considered is only temporary storage, and not the water retained over a period that is long compared with the response time of the flood. Storage in this context is the total volume of water that is in transit to the basin outlet. It is primarily located in the channels throughout the drainage basin, which comprise the main stream, the tributary channels, and the myriad of rills and drainage lines that feed into the channels. There is also a component of storage in overland flow across the surface of the drainage basin, and/or in rapid subsurface flow.

The volume of storage S at any time in each element of the drainage basin system is related to a representative discharge q in that element at the same time. The most common form of assumed relation is

$$S = kq^m \tag{9.6.1}$$

where k and m are dimensional and dimensionless parameters, respectively. If $m = 1$, the system is linear. The usual range of m is 0.6 to 1.0. For concentrated, reservoir, or level-pool routing, the representative discharge is the outflow discharge.

From the description of storage above, it can be seen that it is highly distributed in a physical or spatial sense. At least four approaches have been used to model the distributed nature of storage.

Muskingum Method. Although it is the usual method for storage routing in river channels, the Muskingum method has received little use in hydrograph models, being employed in some cases for channel routing elements.

Lag and Route Method. The runoff is first delayed or translated without modification, and then routed through a concentrated storage. Artificial separation of these two effects of distributed storage is not really valid, but it provides a reasonable representation of the combined effect. This approach is widely used, especially in urban drainage models.

Series or Cascade of Concentrated Storages. In this, a long channel is divided into several short segments, and the volume of storage in each segment is considered to be concentrated at its midpoint. The flow is then successively routed through each of these storages in turn. Most of the distributed nature of the storage is accounted for, and only second-order effects are neglected because of concentrating subreach storages at their midpoints.

Hydraulic Equations of Unsteady Flow. The Saint Venant equations of unsteady flow, or simplified forms of these equations, automatically allow for the distributed nature of storage. The kinematic wave simplification is usually used in hydrograph models (Sec. 9.6.3).

A major advantage of most storage routing models is their ability to take account of the spatial variation of gross and excess rainfall, and the different amounts of storage through which rainfall excess on different parts of the drainage basin passes on its way to the outlet. The rainfall excess on the subarea most remote from the outlet is first routed to the next subarea, where rainfall excess for that area is added. This process is continued in sequence down the drainage basin to the outlet.

Simple Models. Figure 9.6.1 illustrates the simple model consisting of translation of flow over the drainage basin coupled to a concentrated storage at the outlet. The isochrones join points with equal time of travel to the outlet. Rainfall excess for each subarea is lagged or translated to the outlet with a delay equal to the average time of

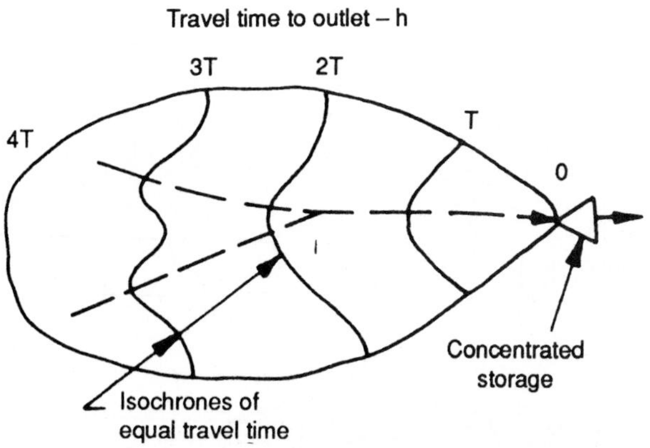

Travel time to outlet – h

3T 2T T 4T 0

Concentrated storage

Isochrones of equal travel time

FIGURE 9.6.1 Model comprising translation with concentrated storage at the outlet. *(Reproduced from Ref. 31 with permission.)*

travel of that subarea, but the discharge is not modified in magnitude. The total translated hydrograph is then routed through a concentrated storage conceptually located at the outlet. A variation of this approach is to use subbasins as the subareas, with travel times from the centroid of each subbasin to the outlet. This model, with a linear concentrated storage as proposed by Clark,[14] has found considerable use for synthesizing unit hydrographs for rural drainage basins. With a linear or nonlinear storage, the model is also widely used for urban drainage, as in the ILLUDAS model in Sec. 21.2. For relatively small urban areas, the concentrated storage at the outlet is often omitted.

Network Models. In these, the storages are arranged to represent the stream network of the drainage basin. The distributed nature of the storage is represented by separate series of concentrated storages or other forms of storage for the main stream and for major tributaries and their contributing areas, providing a degree of physical realism.

HEC-1 Model. The most widely used model in the United States, this is described by the U.S. Army Corps of Engineers.[78] The computer program is described in Sec. 21.2.

RORB Model. This is the most widely used of several storage routing models which have superseded unit hydrographs as the main practical method for estimating flood hydrographs in Australia. Interactive PC programs are available for each of the models. The principles of the various Australian models are described in Ref. 31. A detailed description of RORB is given by Laurenson and Mein[40] (available from Montech Pty Ltd, Monash University, Clayton, Vic., 3168, Australia.) As illustrated in Fig. 9.6.2*a*, the drainage basin is divided into between 5 and 20 (or possibly more) subareas based on watershed boundaries. Nodes, which are points of input and output within the model, are located at centroids of subareas for input of rainfall excess, at confluences of streams for addition of hydrographs, upstream and downstream of any reservoirs, at diversion sites, and at gauging stations and design sites. Nonlinear concentrated storages of the form of Eq. (9.6.1) are located between each pair of adjacent nodes and are also used to represent any lakes or reservoirs on the reach. The RORB model is efficient from a practical viewpoint, in that it embodies much of the physical nature of the drainage basin but has only one parameter to be calibrated, corresponding to k in Eq. (9.6.1).

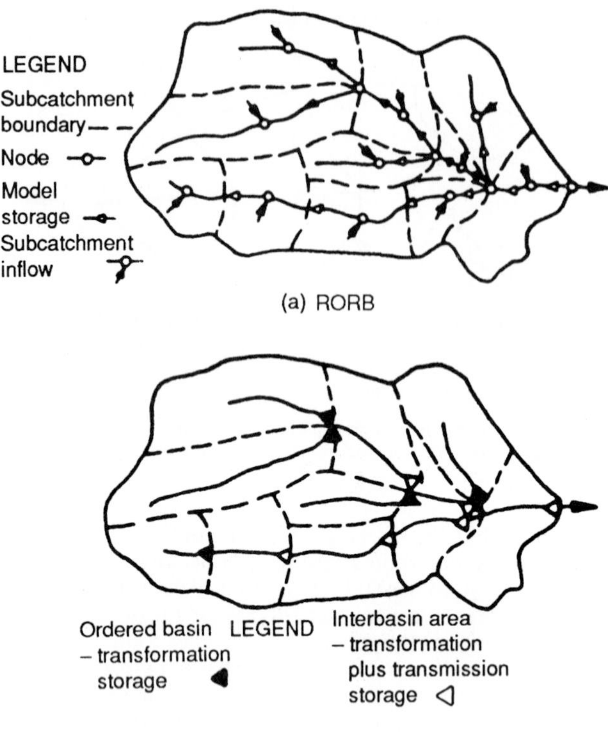

FIGURE 9.6.2 Australian network models. *(Reproduced from Ref. 31 with permission.)*

The Watershed Bounded Network Model (WBNM), described by Boyd et al.[5,6] (available from Unisearch, P.O. Box 1, Kensington, N.S.W., 2033, Australia), is similar in concept to RORB. The storages apply to subareas rather than to the channels between nodes as illustrated in Fig. 9.6.2*b*, and the model incorporates a more detailed consideration of basin geomorphology. Two different types of subareas are modeled, *ordered basins* where only rainfall excess is transformed to streamflow at the outlet, and *interbasin areas* where upstream runoff is transmitted through the subarea in a main channel in addition to the transformation of rainfall excess to streamflow.

Many other storage routing models have been developed, with varying amounts of practical use. Urban drainage has probably been the major field of application. Models of this type have also gained some practical acceptance in their country of origin, as in Japan and South Africa.

9.6.3 Kinematic Wave Models

Application of Kinematic Wave Routing. The full dynamic equations of unsteady flow are discussed in Chap. 10. Although it might be theoretically desirable to attempt to apply these equations to the analysis of flood runoff on drainage basins, they would generally be unnecessarily complex and computationally inefficient for this application. The kinematic wave simplification has usually been adopted as generally satisfactory. In this, the energy slope is identical with the bed slope, and the equation of motion can be replaced by

$$q = \alpha h^n \tag{9.6.2}$$

where q is the discharge per unit width, h is the local mean depth, α is a parameter that reflects the slope and hydraulic resistance, and the exponent n depends on whether the flow is laminar or turbulent. Kinematic wave routing involves solution of Eq. (9.6.2) with the equation of continuity and is generally considered to be of sufficient accuracy for analysis of overland flow and channel flow.

Many models have been developed using kinematic wave analysis of flow over plane and conic surfaces and channels of regular cross section to simulate overland and channel flow. They are often termed "physically based" models, but "mathematically based" might be a better term. The analysis of flow over the idealized sections is more rigorous than storage routing, but the grossly simplified physical elements lead to estimates on actual drainage basins that are no more realistic. Kinematic wave models have found considerable practical application in urban drainage models where the flow elements are of more regular shape, but they have found little practical application in natural drainage basins. Reviews of kinematic wave models are given by Overton and Meadows,[54] Woolhiser,[84] and Stephenson and Meadows.[77]

Simple Models. Several very simple models of drainage basins have been proposed. These generally involve simple geometric representations of the basin such as a converging overland flow model consisting of half an inverted cone, a folded rectangular plate, or combinations. All these models are grossly oversimplified in terms of representing a complete drainage basin, and as complete models are of more academic than practical interest.

Network Models. While the simple models described above may be of limited value by themselves, they are more useful in representing elements or subareas within a larger drainage basin. These elements can then be linked together by channel elements to represent the stream network of a drainage basin similar to the network-type storage routing models. Kinematic wave routing is usually used for the channels. Examples of this type of model are described in Refs. 19, 35, and 45.

9.6.4 Calibration of Models

In general, the ideal approach of estimating parameter values from physical considerations can be applied only partially. Since all models and their parameters are approximations to reality, there is a general need for calibration or checking with observed data. If data are not available at the site, this can only be accomplished using a regional relationship between parameter values and physical characteristics. Records at nearby basins can be used to recalibrate a relation of this type.

Where flood records are available at a site, one or more model parameters should be adjusted to give the best possible fit of the calculated and observed hydrographs for at least one flood. Where two or more parameters are adjusted, interaction of the values frequently occurs, so that different combinations of parameter values give very similar computed hydrographs for an observed flood but may give quite different results with design floods of larger magnitude. It is often desirable to select the values of all but the most important parameter from physical considerations or by judgment and to adjust the remaining parameter to give the best fit to the observed data.

Manual trial-and-error adjustment of parameters is often of sufficient accuracy for practical flood estimation and allows examination of the effects of changing parameter values. Alternatively, automatic optimization programs can be used, either of the search type or direct optimization.

Fitting Criteria. Various criteria may be used for judging the fit of a calculated to an observed hydrograph. Common measures are the differences between peak magnitudes, a measure of overall fit such as the sum of absolute values or squares of the differences of individual ordinates, or differences between lags or other time measures. The first two are most common and satisfactory. There is no one uniquely correct criterion, and the choice must be made by judgment, taking into consideration the objectives of the analysis, noting that use of different criteria will usually lead to different derived parameter values. Even when a single fitting criterion is used, parameter values derived from different floods will generally be different as a result of data errors and the fact that all models are only approximations to reality. The possibility of trends in derived values should always be examined.

Split-Sample Testing. If sufficient data are available, it is desirable to use split-sample testing. The model calibrated with about two-thirds of the data is tested by its ability to reproduce the remaining data, considering the effects of any trends. If the test results lie within acceptable limits, the data and model are accepted, and recalibration is carried out using all of the data. If the test results lie outside acceptable limits, the basic assumptions and data used in calibration should be thoroughly checked. If the test results are still outside acceptable limits, resort must be made to recalibrating with all of the data, but confidence in the results is then reduced.

9.7 OTHER FLOOD ESTIMATION METHODS

Only the main types of methods for calculating floods from rainfall have been described in this chapter. Several other methods have been referred to briefly, and many others have been developed but cannot be included. This applies particularly to methods derived for particular regions, such as those described by Chow[12] and Reich and Hiemstra.[69] In some cases simple regressions between rainfall and flood peaks may give useful results, but they often incorporate large scatter. The method developed in the U.K. *Flood Studies Report*[53] uses complex multiple regression to estimate the mean annual flood. The U.S. Geological Survey has developed similar regression models for most of the United States.

Another approach that can be very useful is based on consideration of the largest floods that have been observed in the region of interest. The usual procedure is to draw an *envelope curve* on a regional plot of maximum recorded flood at each gauging station against drainage basin area. Logarithmic values are normally plotted with discharge in ft^3/s per mi^2 (m^3/s per km^2). The graph provides a useful summary of flood experience in a region. Plotting and labeling of the maximum flood for each drainage basin makes the scatter of the data obvious. Trends in flood characteristics in a region can be examined, as with elevation, latitude, stream slope, distance from the ocean or other moisture source, or different record lengths. It may be possible to draw envelope curves for different subregions. However, as time proceeds, higher floods are recorded and the envelope curve moves to higher discharges. Probabilities of floods cannot be estimated objectively by this method. Except where data are sparse and other methods cannot be used, envelope curves are better employed for either checking that estimates by other methods are of the correct order of magnitude or providing preliminary estimates, rather than as a final design method.

Francou and Rodier[20] developed a systematic form of the envelope curve procedure, with a parameter to represent the flood potential of the region. It has been widely applied in French-speaking Africa and in South Africa.

9.8 EXTREME FLOODS

Large and extreme floods, ranging from those with annual exceedance probabilities of less than 1 in 100 up to the probable maximum flood (PMF), are required for the design of dam spillways, large detention basins, nuclear power stations, major bridges, and other major works. There is also a growing need for checking the adequacy of existing dams, and the effects of floods exceeding the design event in main drainage in urban areas. Four approaches are used in practice for the estimation of extreme floods:

- Extrapolation of a flood frequency curve, utilizing where possible historic and paleofloods and regionalization
- Extrapolation of rainfall frequency data to very low probabilities, and estimation of the flood from this
- Generation of a very long series of rainfalls from a stochastic rainfall model, estimating the floods from the largest rainfalls, and carrying out a frequency analysis of these large floods
- Estimation of the probable maximum flood, and use of either this or a smaller flood derived from it

The National Research Council[52] discusses each of these approaches and classifies the second and third methods listed above as subgroups of one class. Major assumptions are unavoidably involved in all the procedures. The PMF approach is the one most commonly used around the world in hydrologic design at present.

Computation of Probable Maximum Flood. In calculating extreme floods or the probable maximum flood from rainfall, the same procedures and principles are used as for the more usual design floods discussed in this chapter, but different values of design parameters are usually adopted. The temporal patterns of extreme design storms are generally more uniform than those for storms of higher probability. Antecedent wetness is generally assumed to be high, and adopted losses are assumed equal to or somewhat lower than the minimum in recorded large floods. Except for short storms where losses are negligible, the assumption of zero losses is too conservative if the probable maximum flood is regarded as a reasonable limiting value.

Where a linear hydrograph model is used, some arbitrary upward adjustment of flood peaks is generally made to allow for nonlinear effects. Linsley et al.[42] report adjustments of 5 to 20 percent for large floods in the United States, and the Institution of Engineers Australia[31] gives guidelines for 15 to 20 percent increases depending on characteristics of the drainage basin. The U.K. *Flood Studies Report*[53] recommends a 33 percent reduction in the time to peak of the unit hydrograph when it is applied in PMF computations. There is evidence[65] that nonlinear models may overestimate the PMF, and the Institution of Engineers Australia[31] provides guidance on adjustments.

Major structures that are classified as of medium or low hazard generally require a design flood which is based on but is smaller than the PMF. Two approaches are used to adjust from the PMF. The more usual and simpler is to adopt a percentage of the PMF discharge, based on recommendations such as those of the National Research Council.[50,51] A more consistent and intrinsically more logical approach is to design for specified very low probabilities. However, these floods are normally estimated by gross extrapolation of frequency curves. A method based on interpolation between the 100-year flood and the PMF is used in Australia.[31,62]

REFERENCES

1. American Society of Civil Engineers and Water Pollution Control Federation, *Design and Construction of Sanitary Storm Sewers,* ASCE Manuals and Reports on Engineering Practice no. 37 and WPCF Manual of Practice no. 9, 1969.
2. Bates, B. C., and D. H. Pilgrim, "Investigation of Storage-Discharge Relations for River Reaches and Runoff Routing Models," *Civ. Eng. Trans. Inst. Engrs. Aust.,* vol. CE25, pp. 153–161, 1983.
3. Beard, L. R., "Practical Determination of Hypothetical Floods," *J. Water Resour. Plan. Manage.,* vol. 116, no. WRPM3, pp. 389–401, 1990.
4. Beran, M. A., and J. V. Sutcliffe, "An Index of Flood-Producing Rainfall Based on Rainfall and Soil-Moisture Deficit," *J. Hydrol.,* vol. 17, pp. 229–236, 1972.
5. Boyd, M. J., D. H. Pilgrim, and I. Cordery, "A Storage Routing Model Based on Catchment Geomorphology," *J. Hydrol.,* vol. 42, pp. 209–230, 1979.
6. Boyd, M. J., B. C. Bates, D. H. Pilgrim, and I. Cordery, "WBNM: A General Runoff Routing Model—Programs and User Manual," *University of New South Wales, Water Res. Lab. Report* 170, Kensington, Australia, 1987.
7. Bransby Williams, G., "Flood Discharge and the Dimensions of Spillways in India," *The Engineer* (London), vol. 121, pp. 321–322, September 1922.
8. Bras, R. L., and I. Rodriguez-Iturbe, "A Review of the Search for a Quantitative Link between Hydrologic Response and Fluvial Geomorphology," in M. L. Kavvas, ed., *New Directions for Surface Water Modeling, IAHS Publ.* 181, pp. 149–163, 1989.
9. Chapman, T. G., Comment on "Evaluation of Automated Techniques for Baseflow and Recession Analyses," *Water Resour. Res.,* vol. 27, pp. 1783–1784, 1991.
10. Chen, C. L. "An Evaluation of the Mathematics and Physical Significance of the Soil Conservation Service Curve Number Procedure for Estimating Runoff Volume," in V. P. Singh, ed., *Rainfall-Runoff Relationship (Proc. Int. Symp. Rainfall-Runoff Modeling), Water Resources Publ.,* Littleton, Colo., pp. 387–415, 1982.
11. Chow, V. T., "Hydrologic Determination of Waterway Areas for the Design of Drainage Structures in Small Drainage Basins," *University of Illinois, Eng. Exp. Sta. Bull.,* 462, 1962.
12. Chow, V. T., "Hydrologic Design of Culverts," *J. Hydraul. Eng.,* vol. 88, no. HY2, pp. 39–55, 1962.
13. Chow, V. T., D. R. Maidment, and L. W. Mays, *Applied Hydrology,* McGraw-Hill, New York, 1988.
14. Clark, C. O., "Storage and the Unit Hydrograph," *Trans. Am. Soc. Civil Engrs.,* vol. 110, pp. 1419–1488, 1945.

15. Cordery, I., "Synthetic Unitgraphs for Small Catchments in Eastern New South Wales," *Civ. Eng. Trans. Inst. Engrs. Aust.*, vol. CE10, pp. 47–58, 1968.

16. Cordery, I., and S. N. Webb, "Flood Estimation in Eastern New South Wales—A Design Method," *Civ. Eng. Trans. Inst. Engrs. Aust.*, vol. CE16, pp. 87–93, 1974.

17. Diskin, M. H., "Nonlinear Hydrologic Models," in V. P. Singh, ed., *Rainfall-Runoff Relationship (Proc. Int. Symp. Rainfall-Runoff Modeling)*, Water Resources Publ., Littleton, Colo., pp. 127–146, 1982.

18. Espey, W. H., Jr., D. G. Altman, and C. B. Graves, "Nomographs for Ten-Minute Unit Hydrographs for Small Urban Watersheds," *Tech. Memo. 32, Urban Water Resources Research Prog., Am. Soc. Civil Engrs.*, New York, December 1977.

19. Field, W. G., and B. J. Williams, "A Generalized Kinematic Catchment Model," *Water Resour. Res.*, vol. 23, pp. 1693–1696, 1987.

20. Francou, J., and J. A. Rodier, "Essai de Classification des Crues Maximales," *Floods and Their Computation, Proc. Leningrad Symp.*, vol. 1, UNESCO, pp. 518–525, 1967.

21. French, R., D. H. Pilgrim, and E. M. Laurenson, "Experimental Examination of the Rational Method for Small Rural Catchments," *Civ. Eng. Trans. Inst. Engrs. Aust.*, vol. CE16, pp. 95–102, 1974.

22. Graber, S. D., "Relations between Rational and SCS Runoff Coefficients and Methods," in B. C. Yen, ed., *Channel Flow and Catchment Runoff*, Department of Civil Engineering, University of Virginia, Charlottesville, Va., pp. 111–120, 1989.

23. Green, W. H., and G. A. Ampt, "Studies on Soil Physics. 1. The Flow of Air and Water through Soils," *J. Agric. Sci.*, vol. 4, pp. 1–24, 1911.

24. Hall, F. R., "Baseflow Recessions—A Review," *Water Resour. Res.*, vol. 4, pp. 973–983, 1968.

25. Hawkins, R. H., "Infiltration and Curve Numbers: Some Pragmatic and Theoretic Relationships," *Proc. Symp. Watershed Management 1980, Boise, Idaho*, vol. II, ASCE, pp. 925–937, 1980.

26. Hiemstra, L. A. V., "Modifications on the 'Soil Conservation Service Method' for the Estimation of Design Floods on Very Small Catchments in the Arid Western Part of the United States of America," *Colo. State Univ., Civ. Eng. Dept., Tech. Report*, Contract 14-11-008-2812, 1968.

27. Hjelmfelt, A. T., "Empirical Investigation of Curve Number Technique," *J. Hydraul. Eng.*, vol. 106, pp. 1471–1476, 1980.

28. Hjelmfelt, A. T., "Curve Numbers: A Personal Interpretation," in J. Borrelli, V. R. Hasfurther, and R. D. Burman, eds., *Proc. Specialty Conf. Advances in Irrigation and Drainage: Surviving External Pressures*, ASCE, pp. 208–215, 1983.

29. Hoesein, A. A., D. H. Pilgrim, G. W. Titmarsh, and I. Cordery, "Assessment of the US Conservation Service Method for Estimating Design Floods," in M. L. Kavvas, ed., *New Directions in Surface Water Modeling, IAHS Publ.* 181, pp. 283–291, 1989.

30. Horner, W. W., and F. L. Flynt, "Relation between Rainfall and Runoff from Small Urban Areas," *Trans. Am. Soc. Civ. Engrs.*, vol. 101, pp. 140–183, 1936.

31. Institution of Engineers Australia, *Australian Rainfall and Runoff, A Guide to Flood Estimation*, D. H. Pilgrim, ed., Canberra, Australia, 1987.

32. Interagency Advisory Committee on Water Data, "Guidelines for Determining Flood Flow Frequency," *Bull.* 17B of the Hydrology SubCommittee, Office of Water Data Coordination, Geological Survey, U.S. Department of the Interior, Washington, D.C., 1982.

33. Johnstone, D., and W. P. Cross, *Elements of Applied Hydrology*, Ronald Press, New York, 1949.

34. Kibler, D. F., ed., *Urban Stormwater Hydrology*, American Geophysical Union, Water Resources Monograph 7, Washington, D.C., 1982.

35. Kibler, D. F., and D. A. Woolhiser, "The Kinematic Cascade as a Hydrological Model," *Hydrol. Papers*, 39, Colorado State University, Fort Collins, Colo., 1970.

36. Kirpich, Z. P., "Time of Concentration of Small Agricultural Watersheds," *Civ. Eng., Am. Soc. Civ. Engrs.*, vol. 10, p. 362, 1940.

37. Klaassen, B., and D. H. Pilgrim, "Hydrograph Recession Constants for New South Wales Streams," *Civ. Eng. Trans. Inst. Engrs. Aust.*, vol. CE17, pp. 43–49, 1975.

38. Knisel, W. G., "Baseflow Recession Analysis for Comparison of Drainage Basins and Geology," *J. Geophys. Res.*, vol. 68, pp. 3649–3653, 1963.

39. Kundzewicz, Z. W., and J. J. Napiórkowski, "Nonlinear Models of Dynamic Hydrology," *Hydrol. Sci. J.*, vol. 31, pp. 163–185, 1986.

40. Laurenson, E. M., and R. G. Mein, *RORB—Version 4, Runoff Routing Program, User Manual*, Department of Civil Engineering, Monash University, Clayton, Victoria, Australia, May 1988.

41. Linsley, R. K., "Flood Estimates: How Good Are They?" *Water Resour. Res.*, vol. 22(9, Supplement), pp. 159S–164S, 1986.

42. Linsley, R. K., M. A. Kohler, and J. L. H. Paulhus, *Hydrology for Engineers*, 3d ed., McGraw-Hill, New York, 1982.

43. Loague, K. M., and R. A. Freeze, "A Comparison of Rainfall-Runoff Modeling Techniques on Small Upland Catchments," *Water Resour. Res.*, vol. 21, pp. 229–248, 1985.

44. Lyne, V., and M. Hollick, "Stochastic Time-Variable Rainfall-Runoff Modelling," *Hydrol. Water Res. Symp.*, 1979, Inst. Engrs. Aust., Natl. Conf. Publ. 79/10, pp. 89–92, 1979.

45. Machmeier, R. E., and C. L. Larson, "Runoff Hydrographs for Mathematical Watershed Model," *J. Hydraul. Eng.*, vol. 94, pp. 1453–1474, 1968.

46. Mancini, M., and R. Rosso, "Using GIS to Assess Spatial Variability of SCS Curve Number at the Basin Scale," in M. L. Kavvas, ed., *New Directions in Surface Water Modeling, IAHS Pub.* 181, pp. 435–444, 1989.

47. Miller, N., and R. Cronshey, "Runoff Curve Numbers, the Next Step," in B. C. Yen, ed., *Channel Flow and Catchment Runoff*, Department of Civil Engineering, University of Virginia, Charlottesville, Va., pp. 910–916, 1989.

48. Minshall, N. E., "Predicting Storm Runoff on Small Experimental Watersheds," *J. Hydraul. Eng.*, vol. 86, pp. 17–38, 1960.

49. Nash, J. E., "A Unit Hydrograph Study with Particular Reference to British Catchments," *Proc. Inst. Civ. Engrs.*, vol. 17, pp. 249–282, 1960.

50. National Research Council, *Safety of Existing Dams: Evaluation and Improvement*, National Academy Press, Washington, D.C., 1983.

51. National Research Council, *Safety of Dams: Flood and Earthquake Criteria*, National Academy Press, Washington, D.C., 1985.

52. National Research Council, *Estimating Probabilities of Extreme Floods—Methods and Recommended Research*, National Academy Press, Washington, D.C., 1988.

53. Natural Environment Research Council, *Flood Studies Report*, vol. 1, Hydrological Studies, London, 1975.

54. Overton, D. E., and M. E. Meadows, *Stormwater Modeling*, Academic Press, New York, 1976.

55. Pilgrim, D. H., "Isochrones of Travel Time and Distribution of Flood Storage from a Tracer Study on a Small Watershed," *Water Resour. Res.*. vol. 13, pp. 587–595, 1977.

56. Pilgrim, D. H., "Bridging the Gap between Flood Research and Design Practice," *Water Resour. Res.*, vol. 22(9, supplement), pp. 165S–176S, 1986.

57. Pilgrim, D. H., "Estimation of Large and Extreme Floods," *Civ. Eng. Trans. Inst. Engrs. Aust.*, vol. CE28, pp. 62–73, 1986.

58. Pilgrim, D. H., "Regional Methods for Estimation of Design Floods for Small to Medium Sized Drainage Basins in Australia," in M. L. Kavvas, ed., *New Directions in Surface Water Modeling, IAHS Publ.* 181, pp. 247–260, 1989.

59. Pilgrim, D. H., and I. Cordery, "Rainfall Temporal Patterns for Design Flood Estimation," *J. Hydraul. Eng.*, vol. 100, no. HY1, pp. 81–95, 1975.

60. Pilgrim, D. H., and G. E. McDermott, "Design Floods for Small Rural Catchments in Eastern New South Wales," *Civ. Eng. Trans. Inst. Engrs. Aust.*, vol. CE24, pp. 226–234, 1982.

61. Pilgrim, D. H., I. Cordery, and B. C. Baron, "Effects of Catchment Size on Runoff Relationships," *J. Hydrol.*, vol. 58, pp. 205–221, 1982.

62. Pilgrim, D. H., I. A. Rowbottom, and D. G. Doran, "Development of Design Procedures for Extreme Floods in Australia," in V. P. Singh, ed., *Application of Frequency and Risk in Water Resources,* Reidel, pp. 63–77, 1987.

63. Pilgrim, D. H., D. D. Huff, and T. D. Steele, "A Field Evaluation of Subsurface and Surface Runoff, II, Runoff Processes," *J. Hydrol.*, vol. 38, pp. 319–341, 1978.

64. Pilgrim, D. H., G. E. McDermott, and G. E. Mittelstadt, "Development of the Rational Method for Flood Design for Small Rural Basins in Australia," in B. C. Yen, ed., *Channel Flow and Catchment Runoff,* Department of Civil Engineering, University of Virginia, Charlottesville, Va., pp. 51–60, 1989.

65. Pilgrim, D. H., B. C. Bates, M. J. Boyd, and I. Cordery, "Nonlinearity in Flood Estimation Models," in S. P. Simonovic, I. C. Goulter, D. H. Burn, and B. J. Lence, eds., *Water Resources Systems Application,* Department of Civil Engineering, University of Manitoba, Winnipeg, Canada, pp. 167–176, 1990.

66. Poertner, H. G., "Better Ways to Finance Stormwater Management," *Civ. Eng.*, vol. 51, no. 4, pp. 67–69, 1981.

67. Rallison, R. E., "Origin and Evolution of the SCS Runoff Equation," *Proc. Symp. on Watershed Management 1980, Boise, Idaho,* vol. II, ASCE, pp. 912–924, 1980.

68. Ramser, C. E., "Run-off from Small Agricultural Areas," *J. Agric. Res.*, vol. 34, pp. 797–823, 1927.

69. Reich, B. M., and L. A. V. Hiemstra, "Tacitly Maximized Small Watershed Flood Estimates," *J. Hydraul. Eng.*, vol. 91, pp. 217–245, 1965.

70. Schaake, J. G., J. C. Geyer, and J. W. Knapp, "Experimental Examination of the Rational Method," *J. Hydraul. Eng.*, vol. 93, pp. 353–370, 1967.

71. Sherman, L. K., "Streamflow from Rainfall by the Unit-graph Method," *Eng. News Record,* vol. 108, pp. 501–505, 1932.

72. Singh, V. P., ed., *Rainfall-Runoff Relationship (Proc. Int. Symp. Rainfall-Runoff Modeling),* Water Resources Publ., Littleton, Colo., 1982.

73. Snyder, F. F., "Synthetic Unit-Graphs," *Trans. Am. Geophys. Union,* vol. 19, pp. 447–454, 1938.

74. Snyder, W. M., "Hydrograph Analysis by the Method of Least Squares," *Proc. ASCE,* vol. 81, Separate 793, 1955.

75. Solomon, S. I., and I. Cordery, *Hydrometeorology,* vol. II, part 5, *Compendium of Meteorology, WMO Tech Publ.*, 364, Geneva, 1984.

76. Steichen, J. M., "Field Verification of Runoff Curve Numbers for Fallow Rotations," *J. Soil Water Conserv.*, vol. 38, no 6, pp. 496–499, 1983.

77. Stephenson, D., and M. E. Meadows, *Kinematic Hydrology and Modeling,* Developments in Water Science 26, Elsevier, Amsterdam, 1986.

78. U.S. Army Corps of Engineers, *HEC-1, Flood Hydrograph Package, Users Manual,* Hydrologic Engineering Center, 1981.

79. U.S. Bureau of Reclamation, *Criteria for Selecting and Accommodating Inflow Design Floods for Storage Dams and Guidelines for Applying Criteria to Existing Storage Dams,* U.S. Department of the Interior, Tech. Mem. 1, Asst. Commissioner—Engineering and Research, Denver, Colo., 1981.

80. U.S. Soil Conservation Service, *National Engineering Handbook,* Sec. 4, *Hydrology,* U.S. Department of Agriculture, Washington, D.C., 1985.

81. U.S. Soil Conservation Service, *Urban Hydrology for Small Watersheds,* Tech. Release 55, Washington, D.C., June 1986.

82. Wong, T. H. F., and E. M. Laurenson, "Wave Speed-Discharge Relations in Natural Channels," *Water Resour. Res.,* vol. 19, pp. 701–706, 1983.

83. Wood, M. K., and W. H. Blackburn, "An Evaluation of the Hydrologic Soil Groups as Used in the SCS Runoff Method on Rangelands," *Water Resour. Bull.,* vol. 20, pp. 379–389, 1984.

84. Woolhiser, D. A., "Physically Based Models of Watershed Runoff," in V. P. Singh, ed., *Rainfall-Runoff Relationship* (Proc. Int. Symp. Rainfall-Runoff Modeling), Water Resources Publ., Littleton, Colo., pp. 189–202, 1982.

CHAPTER 10
FLOW ROUTING

D. L. Fread
Hydrologic Research Laboratory
National Weather Service, NOAA
Silver Spring, Maryland

10.1 INTRODUCTION

10.1.1 General

Flow routing is a mathematical procedure for predicting the changing magnitude, speed, and shape of a flood wave as a function of time (i.e., the *flow hydrograph*) at one or more points along a *watercourse* (waterway or channel). The watercourse may be a river, stream, reservoir, estuary, canal, drainage ditch, or storm sewer. The flow hydrograph *(unsteady flow)* can result from precipitation runoff (rainfall and/or snowmelt), reservoir releases (spillway, gate, and turbine releases and/or dam failures), landslides into reservoirs, or tides (astronomical and/or wind-generated storm surges).

Many ways have been sought to predict the characteristic features of a flood wave in order to determine necessary actions for protecting life and property from the effects of flooding, and to improve the transport of water through natural or man-made watercourses for economic reasons. Commencing with investigations as early as the seventeenth century, mathematical techniques for flow routing have continually evolved over time. In 1871, Barré de Saint-Venant[116] formulated the basic theory for one-dimensional analysis of unsteady flow; however, owing to the mathematical complexity of the *Saint-Venant equations,* simplifications were necessary to obtain feasible solutions for the important characteristics of a flood wave and its movement. This resulted in the gradual development of many simplified flow routing methods. Only within the last four decades could the complete Saint-Venant equations be solved via computers with varying degrees of feasibility.

Flow routing may be classified as either lumped or distributed. In *lumped flow routing* or *hydrologic routing,* the flow is computed as a function of time at one location along the watercourse; however, in *distributed flow routing* or *hydraulic routing,* the flow is computed as a function of time simultaneously at several cross sections along the watercourse (see Fig. 10.1.1). Two methods for lumped flow routing and two for distributed flow routing are presented here along with their data requirements. The chapter concludes with several complexities which can be encountered in channel flow routing.

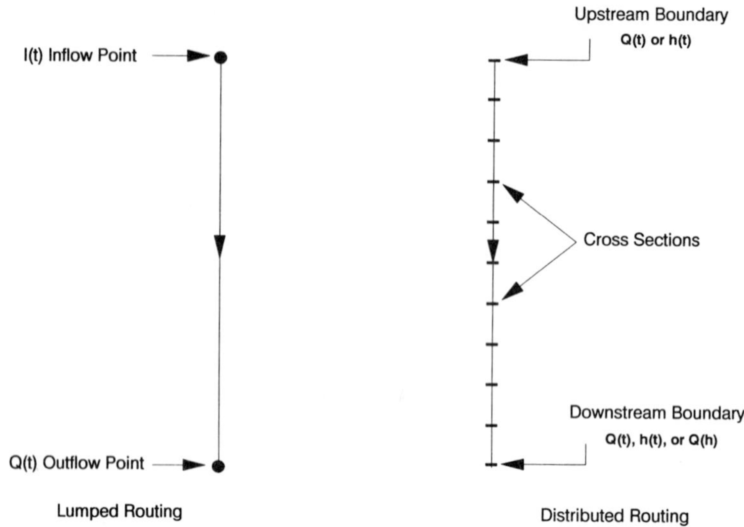

FIGURE 10.1.1 Schematic showing lumped and distributed flow routing system, where Q is discharge or flow rate and h is water surface elevation or stage.

10.1.2 Routing Model Selection

Flow routing has been an important type of hydrologic analysis, and its inherent complexity and computational requirements have resulted in the development of many routing models. The literature abounds with a wide spectrum of usable flow routing models,[44,88,94,95] which are sufficiently accurate when used within the bounds of their limitations. Selection of a flow routing model for a particular application is influenced by the relative importance placed on the following factors: (1) the model provides appropriate hydraulic information to answer the user's questions; (2) the model's accuracy; (3) the accuracy required in the flow routing application; (4) the type and availability of required data; (5) available computational facilities and costs; (6) familiarity with a given model; (7) extent of documentation, range of applicability, and availability of a "canned" or packaged routing model; (8) complexity of the mathematical formulation if the routing model is to be totally developed from "scratch" (coded for computer); and (9) capability and time available to develop a particular type of routing model. Taking the above factors into consideration and recognizing that the relative importance of each may change depending on the application, it is apparent there is no universally superior flow routing model; and a judicious selection of a model from among the several available models is necessary. The simplified routing models are appealing because of their computational simplicity; however, accuracy considerations can restrict their range of applicability.

Accuracy of Reservoir Routing Models. In reservoir applications, the accuracy of *level-pool routing* models (see Sec. 10.2.2) relative to the more accurate distributed *dynamic routing* models (see Sec. 10.3.3) is shown in Fig. 10.1.2, an extension of the author's previous work on this subject.[50] The error (in percent) associated with level-pool routing is expressed as a normalized error for the rising limb of the outflow hydrograph. The peak outflow is used as the normalizing parameter. The normalized

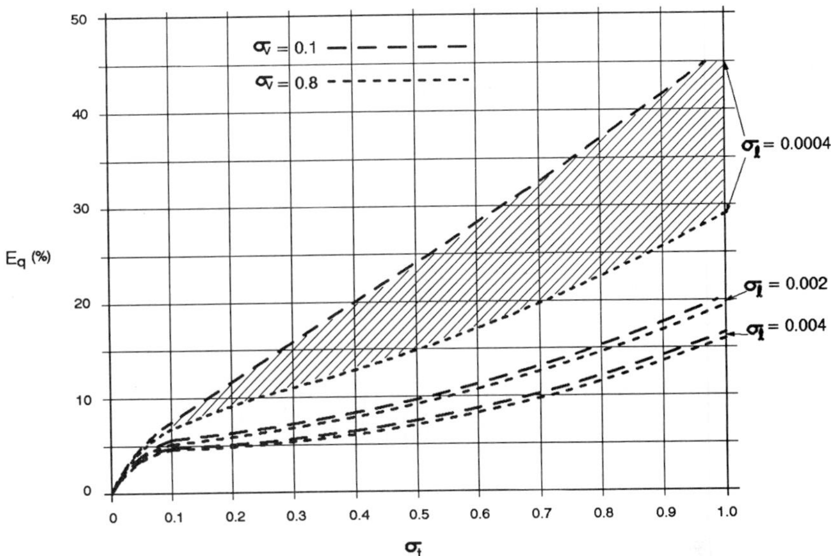

FIGURE 10.1.2 Level-pool routing compared with dynamic routing showing the normalized error E_q of the rising limb of the outflow hydrograph as a function of σ_t, σ_ℓ, and σ_v (dimensionless parameters that reflect the reservoir volume, length, depth, and the inflow hydrograph volume and time of rise).

error E_q is

$$E_q = \frac{100}{Q_{D_p}} \sqrt{\frac{\sum_{i=1}^{N'} (Q_{L_i} - Q_{D_i})^2}{N'}} \tag{10.1.1}$$

in which Q_{L_i} is the level-pool routed flow, Q_{D_i} is the dynamic routed flow, Q_{D_p} is the dynamic routed flow peak, and N' is the number of computed discharges comprising the rising limb of the routed hydrograph. Since level-pool routing is based on the assumption of a horizontal water surface along the length of the reservoir at all times, the error E_q associated with level-pool routing increases as (1) reservoir mean depth D_r decreases, (2) reservoir length L_r increases, (3) time of rise T_r of inflow hydrograph decreases, and (4) inflow hydrograph volume decreases. These effects can be represented by three dimensionless parameters, i.e., $\sigma_\ell = D_r/L_r$, $\sigma_t = L_r/(3600\ T_r \sqrt{g\ D_r})$ in which g is the gravity acceleration constant and T_r is the time, h, from beginning of rise until the peak of the hydrograph, and σ_v = hydrograph volume/reservoir volume. As shown in Fig. 10.1.2, E_q increases as σ_t increases and as σ_ℓ and σ_v decrease; also the influence of σ_v increases as σ_ℓ decreases.

Accuracy of River Routing Models. In river routing applications, the lumped as well as the kinematic-type and diffusion-type routing models offer the advantage of simplicity where there is an absence of significant backwater effects. Accuracy considerations restrict these models to applications where the depth-discharge relation is essentially single-valued, and the product of the time of rise of the hydrograph and the channel bottom slope is not small. An approximate criterion[42,107] which restricts kinematic-type routing relative to dynamic routing models errors to less than E percent is

$$E < \frac{\mu' \phi n^{1.2} q_p^{0.2}}{T_r S_o^{1.6}} \tag{10.1.2}$$

The error term E, expressed in percent, represents the energy slope ratio of kinematic models to dynamic models. A similar criterion (E') for the diffusion-type routing models is

$$E' < \frac{\mu'' \phi' q_p^{0.4}}{T_r S_o^{0.7} n^{0.6}} \tag{10.1.3}$$

where

$$\phi = \frac{(m+1)^2}{3m+5} \tag{10.1.4}$$

$$\phi' = \frac{m+3}{3m+5} \tag{10.1.5}$$

in which the units conversion factor μ' is 0.21 (U.S. units) or 0.43 (SI units) and μ'' is 0.0022 (U.S. units) or 0.0091 (SI units); T_r is the time of rise, h, of the inflow hydrograph; S_o is the channel bottom slope, ft/ft; q_p is a unit-width peak discharge, (ft^3/s or m^3/s); n is the Manning coefficient for flow resistance; and m is the cross-section shape factor, $0 \leq m \leq 2$, used to describe the channel top width B, as $B = ky^m$ in which y is depth of flow ($m = 0$ for rectangular, $m = 0.5$ for parabolic, and $m = 1.0$ for triangular channels).

An inspection of (10.1.2) and (10.1.3) shows the importance of the parameters T_r and S_o which can have a large range of possible values. The parameter q_p is not dominant owing to the power associated with it, and n has a rather restricted range of possible values, say 0.015 to 0.25. It is also apparent that diffusion models are applicable for a wider range of bottom slopes and hydrographs than the kinematic models. The simple diffusion-type Muskingum-Cunge method (see Sec. 10.3.2) can be used effectively in many applications where (10.1.3) is satisfied and backwater effects or reverse flows can be neglected.

In cases involving a gently sloping channel and rapidly rising flood wave, when the combination of S_o and T_r becomes small enough that (10.1.3) cannot be satisfied, dynamic routing models (see Sec. 10.3.3) based on the complete one-dimensional Saint-Venant equations are required. Dynamic routing models are required for (1) slowly rising flood waves in mild sloping channels, i.e., slopes less than about 0.10 percent; (2) situations where backwater effects are important owing to tides, significant tributary inflows, natural constrictions, dams, and/or bridges; and (3) situations where waves propagate upstream from large tides and storm surges or very large tributary inflows. As the trend for increased computer computational speed and storage capabilities at decreased costs continues, the economic feasibility of using dynamic routing models for a wider range of applications will increase, since dynamic models have the capability to correctly simulate the widest spectrum of wave types and waterway characteristics. Implicit dynamic routing models—the most efficient and versatile although the most complex of the dynamic routing models—will be increasingly utilized as improvements continue to be made in their computational robustness and reliability.

10.2 LUMPED FLOW ROUTING

10.2.1 General

A simplified description of unsteady flow along a watercourse (routing reach) depicts it as a lumped process, as shown in Fig. 10.1.1, in which the *inflow* I at the upstream end and the *outflow* Q at the end of the watercourse are functions of time, i.e., $I(t)$ and $Q(t)$. The principle of *mass conservation* requires the difference between the two flows (discharges) to be equal to the time rate of change of the storage S within the reach, i.e.,

$$I(t) - Q(t) = \frac{dS}{dt} \tag{10.2.1}$$

The *storage* S is related to I and/or Q by an arbitrary empirical storage function. The most simple is a single-valued function of outflow Q, i.e., $S = f(Q)$, or of water-surface elevation h, i.e., $S = f(h)$. This implies the water surface is level throughout the watercourse, usually a reservoir or lake. A more complex relationship exists for long narrow reservoirs or *open channels* (rivers and streams) where storage is a function of both inflow and outflow.

Solution of (10.2.1) for $Q(t)$ with various approximations for the storage constitutes lumped flow routing. Both graphical and mathematical techniques for solving (10.2.1) have been used. The attractiveness of lumped flow routing is its relative simplicity compared with distributed flow routing. However, lumped flow routing methods for rivers neglect backwater effects and are not accurate for rapidly rising hydrographs routed through mild to flat sloping rivers, and they are also inaccurate for rapidly rising hydrographs in long reservoirs. Lumped flow routing methods (several are listed in Table 10.2.1) can be categorized as (1) level-pool types which are used for reservoirs, assuming a level water surface at all times, and based on Eq. (10.2.1); (2) storage types used for rivers, considering the sloping water surface due to the passage of a flood wave, and are based on Eq. (10.2.1); and (3) *linear systems* types which assume the routing channel is composed of linear reservoirs connected by

TABLE 10.2.1 Lumped Flow Routing Methods

Model type	Name	References
Level-pool	Puls, Goodrich	17,18,44,55,61,91,94,113
Level-pool	Modified Puls	65
Level-pool	Runge-Kutta	19
Level-pool	Iterative trapezoidal integration	40,45
Storage routing	Kalinin-Miljukov	31,44,74,77,94
Storage routing	Lag and Route	44,91
Storage routing	Muskingum	17,18,19,32,44,91,92,94,96,104,122,139
Storage routing	SSARR	94,115
Storage routing	Tatum	44
Linear systems	Linear reservoir	19,31,44,98
Linear systems	SOSM	97
Linear systems	Linearized St. Venant	90
Linear systems	Multiple linearized	76,81
Linear systems	CLS	100

linear channels which may be uniquely characterized by a unit response function, and the inflow (input)–outflow (output) relationship is defined by a convolution integral.

10.2.2 Level-Pool Reservoir Routing

In this technique, the reservoir is assumed always to have a horizontal water surface throughout its length, hence level-pool. Unsteady flow routing in reservoirs which are not excessively long and in which the inflow hydrograph is not rapidly changing with time, as determined from Fig. 10.1.2, can be approximated by a simple technique known as *level-pool routing.* The water-surface elevation h changes with time t, and the outflow from the reservoir is assumed to be a function of $h(t)$. This is the case for reservoirs with *uncontrolled overflow spillways* such as the ogee-crested, broad-crested weir, and morning-glory types.[17–19,91] *Gate-controlled spillways* can be included in level-pool routing if the gate setting (height of the gate bottom above the gate sill) is a predetermined function of time, since the outflow is a function of h and the extent of gate opening. Several level-pool routing techniques have been proposed, many of them graphical or semigraphical; however, with computerization, nongraphical computational techniques are currently more prevalent. Since (10.2.1) is an ordinary differential equation, it can be solved by various numerical techniques such as a Runge-Kutta method[19] or an *iterative trapezoidal integration method*[40,45] which is presented here.

Iterative Trapezoidal Integration Method. In this solution method, the trapezoidal rule is used to integrate the conservation of mass equation (10.2.1). The time domain consists of time lines separated by Δt intervals, i.e., $t = 0, \Delta t, 2\Delta t, \ldots, j\Delta t, (j+1)\Delta t$. The time rate of change in storage is the product of reservoir surface area Sa and change of water-surface elevation h over the jth time step, i.e.,

$$\frac{dS}{dt} = \frac{0.5(Sa^j + Sa^{j+1})(h^{j+1} - h^j)}{\Delta t^j} \tag{10.2.2}$$

in which the surface area Sa is specified as a known tabular function of h. Using average values for $I(t)$ and $Q(t)$ over the Δt interval and substituting (10.2.2) into (10.2.1) yields the following:

$$0.5(I^j + I^{j+1}) - 0.5(Q^j + Q^{j+1}) - \frac{0.5(Sa^j + Sa^{j+1})(h^{j+1} - h^j)}{\Delta t^j} = 0 \tag{10.2.3}$$

The inflows I at times j and $j+1$ are known from the specified inflow hydrograph; the outflow Q at time j can be computed from the known water-surface elevation h^j and an appropriate spillway discharge equation. The surface area Sa^j can be determined from the known value of h^j. The unknowns in the equation consist of h^{j+1}, Q^{j+1}, Sa^{j+1}; the latter two are known nonlinear functions of h^{j+1}. Hence (10.2.3) can be solved for h^{j+1} by an *iterative method* such as *Newton-Raphson,* i.e.,

$$h_{k+1}^{j+1} = h_k^{j+1} - \frac{f(h_k^{j+1})}{f'(h_k^{j+1})} \tag{10.2.4}$$

in which k is the iteration counter and $f(h_k^{j+1})$ is the left-hand side of (10.2.3) evaluated with the first estimate for h_k^{j+1}, which for $k = 1$ is either h^j or a linearly extrapolated estimate of h^{j+1}; $f'(h_k^{j+1})$ is the derivative of (10.2.3) with respect to h^{j+1}.

It can be approximated by using a numerical derivative as follows: $f'(h_k^{j+1}) = [f(h_k^{j+1} + \epsilon) - f(h_k^{j+1} - \epsilon)]/[(h_k^{j+1} + \epsilon) - (h_k^{j+1} - \epsilon)]$ in which ϵ is a small value, say 0.1 ft (0.03 m). Using (10.2.4), only one or two iterations are usually required to solve (10.2.3) for h^{j+1}. Initially, the pool elevation h^j must be known to start the computational process. Once h^{j+1} is obtained, Q^{j+1} can be computed from the spillway discharge equation. $T_r/20 \le \Delta t \le T_r/10$ is usually the range for Δt.

Limitations. As shown in Fig. 10.1.2, level-pool routing is less accurate as reservoir length increases, reservoir mean depth decreases, and time of rise of the inflow hydrograph decreases. This inaccuracy can have significant economic effects on water-control management.[21,118]

10.2.3 Muskingum River Routing

The *Muskingum method,* developed by McCarthy,[92] is a popular lumped flow routing technique in the United States and elsewhere. It assumes a variable discharge-storage equation, i.e.,

$$S = K[X I + (1 - X) Q] \tag{10.2.5}$$

Assuming the stage is a single-valued function of discharge Q, the storage S in the routing reach is represented by (10.2.5) in which the prism storage in the reach is KQ, where K is a proportionality coefficient, and the volume of wedge storage is equal to $KX(I - Q)$, where X is a weighting factor having the range $0 \le X \le 0.5$ (most streams have X values between 0.1 and 0.3). The storage beneath a line parallel to the streambed is called *prism storage;* the water located between this line and the actual profile is *wedge storage.* In a channel, the storage relationship to discharge plots as a single or twisted loop if the storage is assumed to be related only to outflow; i.e., storage is greater for a given outflow during rising stages than during falling stages. This is caused by different backwater profiles existing at various times during passage of the flood wave.

The time rate of change of storage dS/dt in (10.2.1) is represented as follows:

$$dS/dt = \frac{S^{j+1} - S^j}{\Delta t^j} = \frac{K\{[X I^{j+1} + (1 - X) Q^{j+1}] - [X I^j + (1 - X) Q^j]\}}{\Delta t^j} \tag{10.2.6}$$

where the superscripts j and $j + 1$ denote the times separated by the interval Δt^j. Substituting (10.2.6) into (10.2.1) yields the following:

$$Q^{j+1} = C_1 I^{j+1} + C_2 I^j + C_3 Q^j \tag{10.2.7}$$

which is the Muskingum flow routing equation, where

$$C_1 = \frac{\Delta t - 2KX}{2K(1 - X) + \Delta t} \tag{10.2.8}$$

$$C_2 = \frac{\Delta t + 2KX}{2K(1 - X) + \Delta t} \tag{10.2.9}$$

$$C_3 = \frac{2K(1 - X) - \Delta t}{2K(1 - X) + \Delta t} \tag{10.2.10}$$

and where $C_1 + C_2 + C_3 = 1$, and $K/3 \le \Delta t \le K$ is usually the range for Δt.

Calibration of K and X. Values for K and X can be determined from observed inflow and outflow hydrographs.[18,19,91,92,96,122] Using (10.2.6) and the left side of (10.2.1), as expressed in (10.2.3), yields an equation for K, i.e.,

$$K = \frac{0.5\ \Delta t[I^{j+1} + I^j - (Q^{j+1} + Q^j)]}{X(I^{j+1} - I^j) + (1 - X)(Q^{j+1} - Q^j)} \tag{10.2.11}$$

If, at each time interval, values of the numerator are plotted against those of the denominator, a loop is formed. Iteratively varying X will tend to close the loop, and that value of X which causes the plot to be most nearly a single line is the correct value for the reach. Then K may be computed from the average value determined from (10.2.11) for the correct value of X. Lateral inflows can also be included in the calibration of the Muskingum method.[104] If observed inflow and outflow hydrographs are not available to compute K and X, these parameters may be estimated from the hydrograph and river cross-sectional and flow resistance characteristics as shown by Cunge[23] and others.[19,32,44,77,94] This technique, known as the *Muskingum-Cunge method,* is best utilized as a distributed routing method and is presented in Sec. 10.3.2.

Limitations. The Muskingum method sometimes produces unrealistic initial negative dips in the computed hydrograph;[23,99] however, it provides reasonably accurate results for moderate to slow rising floods propagating through mild to steep sloping watercourses. The Muskingum method is a kinematic-type routing model; its relative accuracy is approximately defined by (10.1.2). It is not suitable for rapidly rising hydrographs such as dam-break floods, and it neglects variable *backwater effects* due to downstream constrictions, bridges, dams, large tributary inflows, or tidal fluctuations. An index to indicate insignificant backwater effects is: The maximum volume that exceeds the peak normal depth and is stored in backwater pools throughout the routing reach should be insignificant compared with the volume of the rising limb of the inflow hydrograph.

10.3 DISTRIBUTED FLOW ROUTING

10.3.1 General

Unsteady flow in a watercourse is most accurately described as a distributed process because the flow rate, velocity, and depth (elevation) vary in space (at cross sections along the channel) as shown in Fig. 10.1.1. Estimates of these properties in a channel system can be obtained by using distributed flow routing based on the complete differential equations of one-dimensional unsteady flow (the Saint-Venant equations).[116] These equations allow the flow rate and water level to be computed as functions of space and time rather than time alone as in the lumped flow routing methods of Sec. 10.2. Distributed flow routing based on the complete Saint-Venant equations is known as *dynamic routing.* Also, simplified forms of the Saint-Venant equations, referred to as kinematic and diffusion (zero-inertia) equations, can be used for distributed flow routing.

Saint-Venant Equations. The original *Saint-Venant equations* are the *mass conservation equation,* i.e.,

$$\frac{\partial(AV)}{\partial x} + \frac{\partial A}{\partial t} - q = 0 \tag{10.3.1}$$

and the *momentum equation,* i.e.,

$$\frac{\partial V}{\partial t} + V\frac{\partial V}{\partial x} + g\left(\frac{\partial h}{\partial x} + S_f\right) = 0 \qquad (10.3.2)$$

in which t is time, x is distance along the longitudinal axis of the watercourse, A is cross-sectional area, V is velocity, q is *lateral inflow* or *outflow* distributed along the x axis of the watercourse (this term was not included in the original derivation), g is the gravity acceleration constant, h is the water-surface elevation above an arbitrary datum such that $\partial h/\partial x = \partial y/\partial x - S_o$ in which y is the flow depth and S_o is the bottom slope of the watercourse, and S_f is the *friction slope* which may be evaluated using a uniform, steady-flow empirical resistance equation such as Chezy's or Manning's.[17-19,61,91,96] Equations (10.3.1) and (10.3.2) are quasi-linear, hyperbolic partial differential equations with two dependent parameters (V and h) varying in one dimension only (the x direction) and two independent parameters (x and t). The area A and S_f are known functions of h and/or V. No analytical solutions of the complete equations for most practical applications are available. Derivations of the Saint-Venant equations[61,87,127,129,139] utilize the following basic assumptions: (1) the flow is essentially one-dimensional, (2) the stream length affected by the flood wave is many times greater than the flow depth, (3) the vertical accelerations are negligible and vertical pressure distribution in the wave is hydrostatic, (4) the water density is constant, (5) the channel bed and banks are fixed and not mobile, and (6) the channel bottom slope S_o is relatively small, less than about 15 percent.

Application of Distributed Flow Routing. Distributed flow routing models, which compute both the rate of flow Q and water-surface elevation h, are useful for determining floodplain depths, required heights of structures such as bridges or levees, and streamflow velocities affecting the transporting of pollutants. Distributed flow routing models can also be used for such applications as real-time forecasting of river floods, irrigation water deliveries through canals, inundation maps for dam-break contingency planning, transient waves created in reservoirs by gate or turbine changes, landslide-produced waves in reservoirs, and unsteady flow in storm sewer systems. The true flow process in each of these applications varies in all three space dimensions; e.g., velocity varies along a channel, across the channel, and from the water surface to the channel bottom. However, normally the spatial variation in the velocity across the channel and as a function of depth is negligible, so that the entire flow process can be approximated as varying in only one space dimension—the x direction along the flow channel. Thus the *one-dimensional* equations of unsteady flow are widely applicable.

Simplified Distributed Routing Models. Prior to the advent of computers, or more recently the feasible economical availability of such computational resources, the inability to obtain any solutions at all to the complete Saint-Venant equations resulted in the development of several simplified distributed routing models. They are based on the mass conservation equation (10.3.1) and various simplifications of the momentum equation (10.3.2).

Kinematic Wave Model. The simplest type of distributed routing model is the *kinematic wave model,* interest in which was stimulated by the work of Lighthill and Whitham.[90] It is based on the following simplified form of the momentum equation:

$$S_f - S_o = 0 \qquad (10.3.3)$$

in which S_o is the bottom slope of the channel (watercourse) and a component of the term $(\partial h/\partial x)$. This assumes that the momentum of the unsteady flow is the same as that of steady, uniform flow described by the Chezy, Manning equation or a similar expression in which discharge is a single-valued function of depth, i.e., $\partial A/\partial Q = dA/dQ = 1/c$. Also, since $\partial A/\partial t = \partial A/\partial Q \cdot \partial Q/\partial t$ and $Q = AV$, (10.3.1) can be expanded into the classical *kinematic wave equation,* i.e.,

$$\frac{\partial Q}{\partial t} + c\frac{\partial Q}{\partial x} - cq = 0 \tag{10.3.4}$$

in which the *kinematic wave velocity* or *celerity* (c) is defined as

$$c = k'V \tag{10.3.5}$$

where k' is the *kinematic ratio,* i.e., the kinematic wave celerity divided by the flow velocity V. If the Manning equation is used for the steady uniform flow, the kinematic ratio is given by the following expression:

$$k' = \frac{1}{V}\frac{dQ}{dA} = \frac{5}{3} - \frac{2}{3}\frac{A}{(BP)}\frac{dP}{dy} \tag{10.3.6}$$

in which B is the wetted top width, A the wetted area, P the perimeter of the wetted portion of the cross section, and dP/dy the derivative of P with respect to the water depth y. For flow in a wide, rectangular channel, $k' = 5/3$. Solution methods for the kinematic wave equation (10.3.4) consist of an analytical solution using the method of characteristics[19,90,127] or direct solution by finite-difference approximation techniques of either explicit or implicit types.[19,28,59,65,86,91,124] The kinematic wave equation does not theoretically account for hydrograph (wave) attenuation. It is only through the numerical error associated with the finite-difference solution that attenuation is achieved. Kinematic wave models are limited to applications where single-value, stage-discharge ratings exist—where there are no *loop ratings*—and where backwater effects are insignificant. Since, in kinematic wave models, flow disturbances can propagate only in the downstream direction, *reverse (negative) flows* cannot be predicted. Kinematic wave models are appropriately used as components of hydrologic watershed models[28,59] for overland flow routing of runoff; they are not recommended for channel routing unless the hydrograph is very slow rising, the channel slope is moderate to steep, and hydrograph attenuation is quite small. See Sec. 10.1.2 for the kinematic wave model's relative accuracy properties.

Diffusion Wave Model. Another simplified distributed routing model, known as the *diffusion wave (zero-inertia) model,* is based on (10.3.1) along with an approximation of the momentum equation that retains only the last two terms in (10.3.2), i.e.,

$$\frac{\partial h}{\partial x} + S_f = 0 \tag{10.3.7}$$

Finite-difference approximation techniques, both explicit[58] and implicit,[131] have been used to obtain simultaneous solutions to (10.3.1) and (10.3.7). This type of simplified routing model considers backwater effects but improperly distributes them instantaneously (in time) throughout the total routing reach; its accuracy is also deficient for very fast rising hydrographs, such as those resulting from dam failures, hurricane storm surges, or rapid intermittent reservoir releases, which propagate through mild to flat sloping watercourses. See Sec. 10.1.2 for the diffusion model's relative accuracy properties. Several types of distributed flow routing models (dynamic, diffusion, and kinematic) are listed in Table 10.3.1.

TABLE 10.3.1 One-Dimensional Distributed Flow Routing Models

Model type	Name	Reference(s)	Finite-difference scheme	Special features	Availability
Dynamic	BRANCH	119,120	4I	b,d,e,g,i,j	NP
Dynamic	BRASS	21	4I	a,d,e,f,g,i,j,n,r	NP
Dynamic	CARIMA (ONDYN)	25	4I	b,d,e,f,i,j	P
Dynamic	CHARIMA	63	4I	b,d,e,h,i,j	NP
Dynamic	DAMBRK	19,40,45,47	4I	c,d,e,f,i,j,k,m	NP
Dynamic	DWOPER	19,37,41	4I	b,d,e,f,g,i,j,l	NP
Dynamic	EXTRAN	64	E	a,d,l,n,p	NP
Dynamic	FEQ	35	4I	b,d,e,f,i,j,n,l	NP
Dynamic	FLDWAV	19,44,49	4I	b,c,d,e,f,g,h,i,j,k,l,m	NP
Dynamic	FLOSED	72	4IL	a,d,h,i	NP
Dynamic	FLUVIAL	15	4I	h,i	P
Dynamic	LORIS	123	4I	b,d,e,f,i,j	P
Dynamic	MOBED	80	4IL	h,i,j	NP
Dynamic	RICE	82,143	4I	j,q	P
Dynamic	RUBICON	138	4I	b,d,e,f,h,i,j,l	P
Dynamic	SOC/SOCJM	54,71	E	a,i	NP
Dynamic	S11	4,26	6I	b,c,d,e,f,h,i,j,k,l,n,p	P
Dynamic	UNET	11	4IFL	b,d,e,f,i,j	P
Diffusion	Muskingum-Cunge	15,19,23,44,94,101,108,112	4P	a,c,j	NP
Diffusion	PAB	133	CI/FB	b,c,f,h,j,n,p	NP
Diffusion	Zero-inertia	58,131	E		NP
Kinematic	HEC-1	65	E	k,n	NP
Kinematic	MITCAT	59	E	n	NP
Kinematic	Nonlinear	86,124	4I		NP

Legend

NP = nonproprietary
P = proprietary
4I = weighted 4-point nonlinear implicit
4IL = weighted 4-point linear implicit
4IFL = fully forward 4-point linear implicit
6I = 6-point linear implicit; optional iteration
E = explicit
4P = linear/nonlinear 4-point
CI/FB = convolution integral (flow)/finite difference backwater equation (depth)

Special features

a = dendritic system of interconnecting channels
b = dendritic/looped system of interconnecting channels
c = subcritical-supercritical mixed flow
d = assortment of external and internal boundaries
e = special treatment of floodplains
f = special treatment for overtopping of levees
g = automatic calibration of variable friction coefficient
h = sediment transport (mobile-bed) effects
i = dead storage effects
j = variable friction
k = dam-break flood generation
l = storm sewer capabilities including pressurized flow
m = non-Newtonian (mud) flow capabilities
n = hydrologic rainfall-runoff modeling capability
p = water quality modeling capability
q = ice effects
r = reservoir operations

10.3.2 Muskingum-Cunge Method

The Muskingum method can be modified by computing the routing coefficients in a particular way as shown by Cunge[23] and others,[32,77,94] which changes the kinematic-based Muskingum method to one based on the diffusion analogy which is capable of predicting hydrograph attenuation. This modified Muskingum method (known as the *Muskingum-Cunge method*) is most effectively used as a distributed flow routing technique.[19,44,77,94,101,108,112] The recursive equation applicable to each Δx_i subreach for each Δt^j time step is

$$Q_{i+1}^{j+1} = C_1\, Q_i^{j+1} + C_2\, Q_i^j + C_3\, Q_{i+1}^j + C_4 \tag{10.3.8}$$

which is similar to the Muskingum method (10.2.7), but expanded to include lateral inflow effects C_4. Q_i^{j+1} is the same as I_{j+1} in (10.2.7) while Q_i^j and Q_{i+1}^j are the same as I_j and Q_j, respectively. The coefficients C_1, C_2, and C_3 are positive values whose sum must equal unity; they are defined as in (10.2.8) to (10.2.10). The last term (C_4) in (10.3.8) accounts for the effect of time (Δt) and space (Δx) averaged lateral inflow $\bar{q}_i$, i.e.,

$$C_4 = \frac{\bar{q}_i\, \Delta x\, \Delta t}{2K(1 - X) + \Delta t} \tag{10.3.9}$$

in which K is a *storage constant* having dimensions of time, and X is a *weighting factor* expressing the relative importance inflow and outflow have on the storage. It can be shown[23,94] that (10.3.8) is a finite-difference representation of the classical kinematic wave equation (10.3.4); however, if X is expressed as a particular function of the flow properties, (10.3.8) can be considered an approximate solution of a combination of (10.2.1) and (10.3.7) known as the parabolic, diffusion analogy equation which accounts for wave attenuation but not for reverse (negative) flows or backwater effects. Its relative accuracy is approximately defined by (10.1.3). In the Muskingum-Cunge method, K and X are computed as follows:

$$K = \frac{\Delta x}{\bar{c}} \tag{10.3.10}$$

$$X = \frac{1}{2} - \frac{\bar{Q}}{2\bar{c}\,\bar{B}\,S_e\,\Delta x} = \frac{1}{2} - \frac{\bar{D}}{2\bar{k}'\,S_e\,\Delta x} \tag{10.3.11}$$

in which $\bar{c}$ is the kinematic wave celerity (10.3.5), $\bar{Q}$ is discharge, $\bar{B}$ is cross-sectional top width associated with $\bar{Q}$, S_e is the energy slope approximated by evaluating S_f in (10.3.2) for the initial flow condition, $\bar{D}$ is the hydraulic depth (A/B), and $\bar{k}'$ is the kinematic wave ratio (10.3.6). The bar indicates the variable is averaged over the Δx reach and over the Δt time step. For minimal numerical errors associated with the solution scheme, the time step Δt and distance step Δx should be selected as follows:[73]

$$\Delta t \le \frac{T_r}{M} \tag{10.3.12}$$

where $M \ge 5$, T_r is the time of rise of the hydrograph, and

$$\Delta x \simeq 0.5\, c\, \Delta t \left[1 + \left(1 + 1.5\, \frac{\bar{q}}{c^2\, S_o \Delta t}\right)^{1/2}\right] \tag{10.3.13}$$

in which $\bar{q}$ is the average unit-width discharge (Q/B) and S_o the bottom slope.

Solution Procedure. With coefficients defined by (10.2.8) to (10.2.10) and (10.3.9) to (10.3.11), the solution of (10.3.8) can be obtained by either linear or nonlinear (iterative) methods.[101,108,112] The coefficients are functions of Δx and Δt (the independent parameters), and D, c, and k' (the dependent variables) are also functions of water-surface elevations h. These may be obtained from a steady, uniform flow formula such as the *Manning equation*,[17–19,61,91,96] i.e.,

$$Q = \frac{\mu \, AR^{2/3} \, S_e^{1/2}}{n} \tag{10.3.14}$$

in which n is the *Manning roughness coefficient*, A is the cross-sectional area, R is the hydraulic radius given by A/P in which P is the wetted perimeter of the cross section, S_e is the energy slope approximated by S_o, and μ is a units conversion[17] factor (1.49 for U.S. and 1.0 for SI). Rather than using the Manning equation, a more accurate method[133] to determine the water-surface elevations h is to use a backwater solution technique (see Initial Conditions in Sec. 10.3.3). The h value at the most downstream section is computed from (10.3.14) in which S_e is evaluated as in (10.3.11). This provides a water-surface elevation from which the backwater solution proceeds section by section in the upstream direction for subcritical flows. For supercritical flows, h is computed from (10.3.14) at the most upstream section and the backwater solution proceeds in the downstream direction. The backwater method approximates the loop relation that exists between h and Q values when they are computed by dynamic routing methods.

In the *linear solution* procedure, the coefficients K and X are assumed constant for all time steps in each reach, or they are computed from the known flow properties, i.e., $Q(i, j)$, $Q(i, j+1)$, $Q(i+1, j)$, $h(i, j)$, $h(i, j+1)$, and $h(i+1, j)$. In the more accurate *nonlinear solution*, an estimated value of the unknown flow (Q_{i+1}^{j+1}) and its corresponding h value is also used to compute K and X. The estimated values are determined by extrapolation from previously computed values. The solution procedure is iterative and converges when computed and estimated values of h agree within a suitably small tolerance, say 0.01 ft (0.003 m).

Limitations. The nonlinear Muskingum-Cunge routing method, which uses a backwater technique for water-surface elevations, is limited to hydrographs that peak in more than about 2 h. Thus it is not suitable for fast-rising hydrographs such as those produced by dam failures (peak errors exceed 5 percent when $T_r < 0.002/S_o^{1.12}$); also, reverse flows and backwater effects are not accounted for in this method.

10.3.3 Dynamic Routing

General. If the complete Saint-Venant equations (10.3.1) and (10.3.2) are used, the routing model is known as a *dynamic routing model*. With the advent of high-speed computers, Stoker[126,127] first attempted in 1953 to use the complete Saint-Venant equations for routing Ohio River floods.[69] Since then, much effort has been expended on the development of dynamic routing models. Many models have been reported in the literature,[44,88,95] some of which are listed in Table 10.3.1.

Classification of Dynamic Routing Methods. Dynamic routing models can be categorized as characteristic and direct methods of solving the Saint-Venant equations. In the *characteristic methods*, these equations are first transformed into an equivalent set of four ordinary differential equations which are then approximated with finite differences to obtain solutions. Characteristic methods[1,6,10,61,83,88,128] have not proved advantageous over the direct methods for practical flow routing applications.

Direct methods can be classified further as either explicit or implicit. *Explicit schemes*[9,33,54,69,71,88,89,126,130,139] transform the differential equations into a set of algebraic equations which are solved sequentially for the unknown flow properties at each cross section at a given time. However, *implicit schemes*[4,7,10,11,19,25,26,33,35,37,38,40,41,44,45,49,63,72,75,80,88,91,109,114,119,120,130,137,138] transform the Saint-Venant equations into a set of algebraic equations which must be solved simultaneously for all Δx computational reaches at a given time; this set of simultaneous equations may be either linear or nonlinear, the latter requiring an iterative solution procedure.

Numerical Stability of Solution. Explicit methods, although simpler in application, are restricted by *numerical stability* considerations. Stability problems arise when inevitable errors in computational round-off and those introduced in approximating the partial differential equations via finite differences accumulate to the point that they destroy the usefulness and integrity of the solution, if not the total breakdown of the computations, by creating artificial oscillations of length about $2\Delta x$ in the solution. Because of stability requirements, explicit methods require very small computational time steps on the order of a few seconds or minutes depending on the ratio of the computational reach length Δx to the minimum dynamic wave celerity u, i.e., $\Delta t \leq \Delta x/u$. This is known as the *Courant condition,* and it restricts the time step to less than that required for an infinitesimal disturbance to travel the Δx distance. Such small time steps cause explicit methods to be very inefficient in the use of computer time.

Implicit finite-difference techniques, however, have no restrictions on the size of the time step due to mathematical stability; however, *numerical convergence* (accuracy) considerations require some limitation in time step size. Implicit techniques are generally preferred over explicit because of their computational efficiency. Rather than using finite-difference approximation techniques to solve the Saint-Venant equations, finite-element techniques[22,29,56] can be used; however, their greater complexity offsets any apparent advantages when compared with a weighted four-point implicit finite-difference scheme (described later) for solving the one-dimensional flow equations. Finite-element techniques are often applied to two- and three-dimensional flow computation.

Extended Saint-Venant Equations. More powerful and useful expressions of the Saint-Venant equations are their *conservation* or *divergent* form with additional terms to account for lateral flows,[87,127,129] off-channel storage areas,[33,87,129] and sinuosity effects.[29,30] The *extended Saint-Venant equations*[45] consist of the mass conservation equation, i.e.,

$$\frac{\partial Q}{\partial x} + \frac{\partial s_c(A + A_o)}{\partial t} - q = 0 \tag{10.3.15}$$

and the momentum equation, i.e.,

$$\frac{\partial(s_m Q)}{\partial t} + \frac{\partial(\beta Q^2/A)}{\partial x} + gA\left(\frac{\partial h}{\partial t} + S_f + S_{ec}\right) + L + W_f B = 0 \tag{10.3.16}$$

where h is the water-surface elevation, A is the active cross-sectional area of flow, A_o is the inactive *(off-channel storage)* cross-sectional area which may be preferred omitted[25] and its effect represented by a higher frictional resistance for that portion of the cross section, s_c and s_m are depth-weighted *sinuosity coefficients*[29,30,45] which correct for the departure of a sinuous in-bank channel from the x axis of the floodplain, x is the longitudinal mean flow-path distance measured along the center of the watercourse, t is time, q is the lateral inflow or outflow per lineal distance along the watercourse (inflow is positive and outflow is negative), β is the *momentum coefficient* for nonuniform velocity distribution within the cross section,[87,130] g is the gravity acceleration constant, S_f is the boundary friction slope, and S_{ec} is the expansion-contraction (large eddy loss) slope.[45,114]

Friction Slope. The boundary friction slope S_f is evaluated from Manning's equation for uniform, steady flow, i.e.,

$$S_f = \frac{n^2|Q|Q}{\mu^2 A^2 R^{4/3}} = \frac{|Q|Q}{K_c^2} \tag{10.3.17}$$

in which n is the Manning coefficient of frictional resistance, R is the hydraulic radius, μ is a units conversion factor (1.49 for U.S. units and 1.0 for SI), and K_c is the *channel conveyance* factor. The absolute value of Q is used to correctly account for reverse (negative) flows. The conveyance formulation is preferred (for numerical and accuracy considerations) for composite channels[25,45] having wide, flat overbanks or floodplains in which K_c represents the sum of the conveyance of the channel (which is corrected for sinuosity effects by dividing by s_m), and the conveyances of left and right overbank areas.

Expansion and Contraction Effects. The term S_{ec} is computed as follows:

$$S_{ec} = \frac{K_{ec}\Delta(Q/A)^2}{2g\,\Delta x} \tag{10.3.18}$$

in which K_{ec} is the *expansion and contraction coefficient* (negative for expansion, positive for contraction) which varies from -1.0 to 0.4 for an abrupt change in section geometry to -0.3 to 0.1 for a very gradual, warped transition between cross sections. The Δ represents the difference in the term $(Q/A)^2$ at two adjacent cross sections separated by a distance Δx. If the flow direction changes from downstream to upstream, K_{ec} can be automatically changed.[45]

Routing Parameters. The depth-weighted sinuosity coefficients are depth-dependent and computed from specified *sinuosity factors* which are ≥ 1; they represent the ratio of the flow-path distance along a meandering channel to the mean flow-path distance along the floodplain.

The momentum correction coefficient β for nonuniform velocity distribution is

$$\beta = \frac{K_{c_\ell}^2/A_\ell + K_{c_c}^2/A_c + K_{c_r}^2/A_r}{(K_{c_\ell} + K_{c_c} + K_{c_r})^2/(A_\ell + A_c + A_r)} \tag{10.3.19}$$

in which K_c is conveyance, A is wetted area, and the subscripts ℓ, c, and r denote left floodplain, channel, and right floodplain, respectively.[17,45] When floodplain properties are not separately specified and the total cross section is treated as a composite section, β can be approximated as $1.0 \leq \beta \leq 1.10$.

Lateral Flow Momentum. The term L in (10.3.16) is the momentum effect of lateral flows, and has the following form:[130] (1) lateral inflow, $L = -qv_x$, where v_x is the velocity of lateral inflow in the x direction of the main channel flow; (2) seepage lateral outflow, $L = -0.5qQ/A$; and (3) bulk lateral outflow, $L = -qQ/A$.

Wind Effects. The last term ($W_f B$) in (10.3.16) represents the resistance effect of wind on the water surface;[39,87] B is the wetted top width of the active flow portion of the cross section; and $W_f = V_r|V_r|c_w$, where the wind velocity relative to the water is $V_r = V_w \cos w - V$, V_w is the velocity of the wind, w is the acute angle the wind direction makes with the x axis, V is the velocity of the unsteady flow, and c_w is a wind friction coefficient.[33]

Mud or Debris Flows. Another term (S_i) can be included in the momentum equation (10.3.16) in addition to S_f to account for viscous dissipation of *non-Newtonian flows* such as mud or debris flows.[45] This effect becomes significant[103] only when the solids concentration of the flow is in the range of about 40 to 50 percent by volume. For concentrations of solids greater than about 50 percent, the flow behaves more as a landslide and is not governed by the Saint-Venant equations.

Implicit Four-Point, Finite-Difference Solution Technique. The extended Saint-Venant equations (10.3.15) and (10.3.16) constitute a system of partial differential equations with two independent variables x and t and two dependent variables h and Q; the remaining terms are either functions of x, t, h, and/or Q, or they are constants. The partial differential equations can be solved numerically by approximating them with a set of finite-difference algebraic equations; then the system of algebraic equations are solved in conformance with prescribed initial and boundary conditions.

Of various implicit, finite-difference solution schemes that have been developed, a *weighted four-point scheme* first used in 1961 by Preissmann[109] and more recently by many others[4,7,11,15,19,25,26,35,37–41,44,45,49,63,72,80,91,119,120,137,138] is most advantageous. It is readily used with unequal distance steps, its stability-convergence properties are conveniently modified, and boundary conditions are easily applied.

The Space-Time Plane. In the weighted four-point implicit scheme, the continuous x-t region in which solutions of h and Q are sought is represented by a rectangular grid of discrete points as shown in Fig. 10.3.1. The *x-t plane* (solution domain) is a

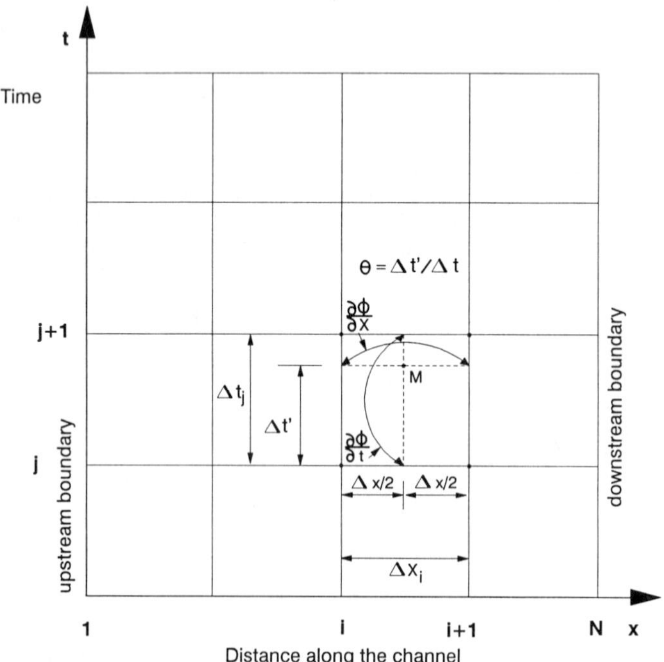

FIGURE 10.3.1 The x-t solution domain showing weighted four-point implicit scheme.

convenient device for visualizing relationships among the variables. The grid points are determined by the intersection of lines drawn parallel to the x and t axes. Those parallel to the t axis represent locations of cross sections; they have a spacing of Δx, which need not be the same between each pair of cross sections. Those parallel to the x axis represent time lines; they have a spacing of Δt, which also need not be the same between successive time points. Each point in the rectangular network can be identified by a subscript i which designates the x position or cross section and a superscript j which designates the particular time line.

Numerical Approximations of Derivatives. The time derivatives are approximated by a forward-difference quotient at point M (in Fig. 10.3.1) centered between the i and $i+1$ points along the x axis, i.e.,

$$\frac{\partial \phi}{\partial t} \simeq \frac{\phi_i^{j+1} + \phi_{i+1}^{j+1} - \phi_i^j - \phi_{i+1}^j}{2\,\Delta t_j} \tag{10.3.20}$$

where ϕ represents any dependent variable or functional quantity ($Q, s_c, s_m, A, A_o, q, h$). Spatial derivatives are approximated at point M by a forward-difference quotient located between two adjacent time lines according to weighting factors of θ (the ratio $\Delta t'/\Delta t$ shown in Fig. 10.3.1) and $1-\theta$, i.e.,

$$\frac{\partial \phi}{\partial x} \simeq \frac{\theta(\phi_{i+1}^{j+1} - \phi_i^{j+1})}{\Delta x_i} + \frac{(1-\theta)(\phi_{i+1}^j - \phi_i^j)}{\Delta x_i} \tag{10.3.21}$$

Nonderivative terms are approximated with weighting factors at the same time level (point M) where the spatial derivatives are evaluated, i.e.,

$$\phi \simeq \frac{\theta(\phi_i^{j+1} + \phi_{i+1}^{j+1})}{2} + \frac{(1-\theta)(\phi_i^j + \phi_{i+1}^j)}{2} \tag{10.3.22}$$

Stability of the Implicit Scheme. The weighted four-point implicit scheme is unconditionally linearly stable for $\theta \geq 0.5$; however, the sizes of the Δt and Δx steps are limited by the accuracy of the assumed linear variations of functions between the grid points in the x-t solution domain. Values of θ greater than 0.5 dampen parasitic oscillations which have wavelengths of about $2\Delta x$ that can grow enough to invalidate or destroy the solution. The θ weighting factor causes some loss of accuracy as it departs from 0.5, a *box scheme,*[7,69] and approaches 1.0, a *fully implicit scheme.*[10] This effect becomes more pronounced as the magnitude of the Δt step increases.[38] Usually, a θ weighting factor of 0.60 is used to minimize the loss of accuracy while avoiding the possibility of weak (pseudo) instability for θ values of 0.5 when frictional effects are minimal.[2,38,88]

Algebraic Routing Equations. Using the finite-difference operators of (10.3.20) to (10.3.22) to replace the derivatives and other variables in (10.3.15) and (10.3.16), the following weighted four-point, *implicit finite-difference equations* are obtained:

$$\theta\,\frac{Q_{i+1}^{j+1} - Q_i^{j+1}}{\Delta x_i} - \theta\, q_i^{j+1} + (1-\theta)\,\frac{Q_{i+1}^j - Q_i^j}{\Delta x_i} - (1-\theta)\, q_i^j$$
$$+ \frac{s_{c_i}^{j+1}\,(A+A_o)_i^{j+1} + s_{c_i}^{j+1}(A+A_o)_{i+1}^{j+1} - s_{c_i}^j(A+A_o)_i^j - s_{c_i}^j(A+A_o)_{i+1}^j}{2\,\Delta t_j} = 0 \tag{10.3.23}$$

$$\frac{(s_{m_i}Q_i)^{j+1} + (s_{m_i}Q_{i+1})^{j+1} - (s_{m_i}Q_i)^j - (s_{m_i}Q_{i+1})^j}{2\,\Delta t_j} + \theta\left[\frac{(\beta Q^2/A)^{j+1}_{i+1} - (\beta Q^2/A)^{j+1}_i}{\Delta x_i}\right.$$

$$+ g\,\bar{A}^{j+1}_i\left(\frac{h^{j+1}_{i+1} - h^{j+1}_i}{\Delta x_i} + \bar{S}^{j+1}_{f_i} + S^{j+1}_{ec_i}\right) + L^{j+1}_i + (W_f\bar{B})^{j+1}_i\Bigg]$$

$$+ (1-\theta)\left[\frac{(\beta Q^2/A)^j_{i+1} - (\beta Q^2/A)^j_i}{\Delta x_i} + g\,\bar{A}^j_i\left(\frac{h^j_{i+1} - h^j_i}{\Delta x_i} + \bar{S}^j_{f_i} + S^j_{ec_i}\right)\right.$$

$$\left. + L^j_i + (W_f\bar{B})^j_i\right] = 0 \tag{10.3.24}$$

where

$$\bar{A}_i = \frac{A_i + A_{i+1}}{2} \tag{10.3.25}$$

$$\bar{S}_{f_i} = \frac{n^2\,\bar{Q}_i|\bar{Q}_i|}{\mu^2\,\bar{A}_i^2\,\bar{R}_i^{4/3}} = \frac{\bar{Q}|\bar{Q}|}{\bar{K}^2_{c_i}} \tag{10.3.26}$$

$$\bar{Q}_i = \frac{Q_i + Q_{i+1}}{2} \tag{10.3.27}$$

$$\bar{R}_i = \frac{\bar{A}}{\bar{P}} \simeq \frac{\bar{A}}{\bar{B}} \tag{10.3.28}$$

$$\bar{B}_i = \frac{B_i + B_{i+1}}{2} \tag{10.3.29}$$

$$\bar{K}_{c_i} = \frac{K_{c_i} + K_{c_{i+1}}}{2} \tag{10.3.30}$$

the terms L and $W_f B$ are defined in (10.3.16); terms associated with the jth time line are known from initial conditions or previous time-step computations; and μ in (10.3.26) is defined in (10.3.17).

Solution Procedure. The flow equations are expressed in finite-difference form for all Δx reaches between the first and last (Nth) cross section ($i = 1, 2, \ldots, N$) along the watercourse and then solved simultaneously for the unknowns (Q and h) at each cross section. In essence, the solution technique determines the unknown quantities (Q and h at all specified cross sections along the watercourse) at various times into the future; the solution is advanced from one time to a future time over a finite time interval *(time step)* of magnitude Δt. Thus, applying (10.3.23) and (10.3.24) recursively to each of the $(N-1)$ rectangular grids in Fig. 10.3.1 between the upstream and downstream boundaries, a total of $(2N-2)$ equations with $2N$ unknowns are formulated. Then, prescribed *boundary conditions* for *subcritical flow (Froude number* less than unity, i.e., $\mathrm{Fr} = V/\sqrt{gD} < 1$), one at the upstream boundary and one at the downstream boundary, provide the two additional and necessary equations required for the system to be determinate. Since disturbances can propagate only in the downstream direction in *supercritical flow* ($\mathrm{Fr} > 1$), two upstream boundary conditions are required for the system to be determinate. The boundary conditions are described later. Because of the nonlinearity of (10.3.23) and (10.3.24) with respect to Q and h, an iterative, highly efficient quadratic solution technique, such as the

Newton-Raphson method[7,19,44,45,68] is frequently used. Other solution techniques linearize (10.3.23) and (10.3.24) via a Taylor-series expansion[11,25,26,109,120] or other means.[130] Convergence of the iterative technique is attained when the difference between successive solutions for each unknown is less than a relatively small prescribed tolerance. Convergence for each unknown at all cross sections is usually attained within about one to five iterations. A more complete description of the solution method may be found elsewhere.[7,19,25,39,44,91]

The solution of $2N \times 2N$ simultaneous equations requires an efficient technique for the implicit method to be feasible. One such procedure requiring $38N$ computational operations (+, −, *, /) is a *compact, pentadiagonal Gaussian elimination method*[36,39,44] which makes use of the banded structure of the coefficient matrix of the system of equations. This is essentially the same as the *double-sweep* elimination method.[2,25,33,72,88,110]

When flow is supercritical, the solution technique previously described can be somewhat simplified. Two boundary conditions are required at the upstream boundary[2,40] and none at the downstream boundary since flow disturbances cannot propagate upstream in supercritical flow. The unknowns h and Q at the most upstream cross section are determined from the two boundary equations. Then, cascading from upstream to downstream, (10.3.23) and (10.3.24) are solved for h_{i+1} and Q_{i+1} at each cross section by using Newton-Raphson iteration applied to a system of two nonlinear equations with two unknowns.

Initial Conditions. Values of water-surface elevation h and discharge Q for each cross section must be specified initially at time $t = 0$ to obtain solutions to the Saint-Venant equations. *Initial conditions* may be obtained from any of the following: (1) observations at gauging stations, or interpolated values between gauging stations for intermediate cross sections in large rivers; (2) computed values from a previous unsteady-flow solution (used in real-time flood forecasting); and (3) computed values from a steady-flow backwater solution. The backwater method is most commonly used, in which the steady discharge at each cross section is determined by

$$Q_{i+1} = Q_i + q_i \, \Delta x_i \qquad \cdots i = 1, 2, 3, \cdots, N-1 \tag{10.3.31}$$

in which Q_1 is the assumed steady flow at the upstream boundary at time $t = 0$ and q_i is the known average lateral inflow or outflow along each Δx reach at $t = 0$. The water-surface elevations h_i are computed according to the following steady-flow simplification of the momentum equation (10.3.24):

$$\left(\frac{Q^2}{A}\right)_{i+1} - \left(\frac{Q^2}{A}\right)_i + g\bar{A}_i (h_{i+1} - h_i + \Delta x_i \, \bar{S}_{fi}) = 0 \tag{10.3.32}$$

in which $\bar{A}_i$ and $\bar{S}_{fi}$ are defined by (10.3.25) and (10.3.26), respectively. The computations proceed in the upstream direction ($i = N - 1, \cdots, 3, 2, 1$) for subcritical flow (they must proceed in the downstream direction for supercritical flow). The starting water-surface elevation h_N can be specified or obtained from the appropriate downstream boundary condition for the discharge Q_N obtained via (10.3.30). The Newton-Raphson iterative solution method[48,68] for a single equation and/or a simple, less efficient, but more stable bisection iterative technique can be applied to (10.3.32) to obtain h_i. The steady water-surface profile can also be obtained from steady-flow backwater models such as HEC-2,[66] WSP2,[125] or WSPRO.[121] The initial conditions must be sufficiently accurate to result in convergence of the Newton-Raphson solution of the Saint-Venant finite-difference equations. Small initial errors will dampen out after several time steps owing to friction.

External Boundaries. Values for the unknowns at *external boundaries* (the upstream and downstream extremities of the routing reach) of the watercourse must be specified in order to obtain solutions to the Saint-Venant equations. In fact, in most unsteady-flow applications, the unsteady disturbance is introduced at one or both of the external boundaries.

Upstream Boundary. Either a specified discharge or water-surface elevation *time series (hydrograph)* can be used as the *upstream boundary condition.* The hydrograph should not be affected by downstream flow conditions.

Downstream Boundary. Specified discharge or water-surface elevation time series, or a tabular relation between discharge and water-surface elevation (single-valued rating curve) can be used as the *downstream boundary condition.*

Another downstream boundary condition can be a *loop-rating curve* based on the Manning equation.[39,40,44,45] The loop is produced by using the friction slope S_f rather than the channel bottom slope S_o. The friction slope exceeds the bottom slope during the rising limb of the hydrograph, while the reverse is true for the recession limb. The friction slope S_f is approximated by using (10.3.16) where L and W_f are assumed to be zero while s_m and β are assumed to be unity, i.e.,

$$S_f \approx -\frac{(Q^j_N - Q^{j-1}_N)}{gA^j_n\,\Delta t^j} - \frac{(Q^2/A)^j_N - (Q^2/A)^j_{N-1}}{gA^j_N\,\Delta x_{N-1}} - \frac{h^j_N - h^j_{N-1}}{\Delta x_{N-1}} \qquad (10.3.33)$$

The loop-rating boundary equation allows the unsteady wave to pass the downstream boundary with minimal disturbance by the boundary itself, which is desirable when the routing is terminated at an arbitrary location along the watercourse and not at a location of actual flow control such as a dam.

The downstream boundary condition can also be a *critical flow* section such as the entrance to a waterfall or a steep reach, i.e.,

$$Q = \sqrt{\frac{g}{B}}\,A^{3/2} \qquad (10.3.34)$$

Critical flow occurs when the bottom slope S_o exceeds the *critical slope* S_c, which can be easily computed as follows:

$$S_c = \frac{\hat{\mu}\,n^2}{D^{1/3}} \qquad (10.3.35)$$

where $\hat{\mu} = 14.6$ for U.S. units and $\hat{\mu} = 9.8$ for SI units.

When the downstream boundary is a stage-discharge relation (rating curve), the flow at the boundary should not be otherwise affected by flow conditions farther downstream. Although there are often some minor effects due to the presence of cross-sectional irregularities downstream of the chosen boundary location, these usually can be neglected unless the irregularity is so pronounced as to cause significant backwater or drawdown effects. Reservoirs or major tributaries located below the downstream boundary which cause backwater effects at the boundary should be avoided. When either of these situations is unavoidable, the routing reach should be extended downstream to the dam in the case of the reservoir or to a location downstream of where the major tributary enters. Sometimes the routing reach may be shortened by moving the downstream boundary to a location farther upstream where backwater effects are negligible.

Internal Boundaries. Along a watercourse, there are locations such as a dam, bridge, or waterfall (short rapids) where the flow is rapidly varied rather than gradually varied in space. At such locations *(internal boundaries),* the Saint-Venant equations are not applicable since gradually varied flow is a necessary condition for their derivation. Empirical water elevation–discharge relations such as weir flow are utilized for simulating rapidly varying flow. At internal boundaries, cross sections are specified for the upstream and downstream extremities of the section of watercourse where rapidly varying flow occurs. The Δx reach containing an internal boundary requires two internal boundary equations; since, as with any other Δx reach, two equations equivalent to the Saint-Venant equations are required. One of the required internal boundary equations represents conservation of mass with negligible time-dependent storage, i.e.,

$$Q_i - Q_{i+1} = 0 \tag{10.3.36}$$

The second equation is usually an empirical *rapidly varied flow* relation of the type $Q = f(h_i, h_{i+1})$. Several examples of rapidly varied flow internal boundary equations are listed in Table 10.3.2.

TABLE 10.3.2 Internal Boundary Conditions

Type of boundary	Equation	Reference
Critical flow	$Q_i = \sqrt{g/B}\, A^{3/2}$	17,44,61
Dam:		
Discharge hydrograph	$Q_i = Q(t)$	
Stage hydrograph	$h_i = h(t)$	
Rating curve	$Q_i = K_b Q_b + K_s Q_s + Q_g + K_{cs} Q_{cs} + Q_t$	44,45
Breach flow	$Q_v = 3.1b\,(h - h_{br})^{3/2} + 2.45z\,(h - h_{br})^{5/2}$	44,45,47
Spillway flow	$Q_s = C_s L_s\,(h - h_s)^{3/2}$	
Gate flow	$Q_g = \sqrt{2g}\, C_g A_g\,(h - \hat{h})^{1/2}$ $\quad \hat{h} = h_g \quad h_t \le h_g$; $\hat{h} = h_t \quad h_t > h_g$	44,45
Dam crest flow	$Q_{cs} = C_d L_d\,(h - h_d)^{3/2}$	
Turbine flow	$Q_t = Q(t)$	
Bridge:	$Q_i = Q_{bo} + K_{em} Q_{em} + K_b Q_b$	45
Bridge flow	$Q_{bo} = \sqrt{2g}\, C_b A_b\,[h - h_t + (Q/A)^2/2g]^{1/2}$	45
Embankment flow	$Q_{em} = C_{em} L_{em}\,(h - h_{em})^{3/2}$	45

A, A_g, A_b = wetted cross-sectional area, area of gate opening, wetted area of bridge opening

b = instantaneous bottom breach width = $b_o\,(t_{br}/\tau_{br})$ for $t_{br} \le \tau_{br}$; otherwise $b = b_o$ (Ref. 45,47)

b_o = final bottom breach width, ft = $14.3\,(V_r h_d)^{0.25} - 0.5\, z h_d$ (Ref. 45,47,53)

C_s, C_d, C_{em}, C_g = discharge coefficients for spillway, dam overtopping, embankment, bridge, or gate flow

h = elevation of water at the upstream side of structure

h_{br} = breach bottom elevation

h_d = top of dam elevation

h_s, h_g, h_{em} = elevation of spillway crest, centerline of gate opening, top of embankment

h_t = tailwater elevation

K_b, K_s, K_{cs}, K_{em} = submergence correction factor for breach, spillway, dam overtopping, bridge embankment overflow

L_s, L_d, L_{em} = length of spillway, dam, embankment

$Q_b, Q_{bo}, Q_{cs}, Q_{em}, Q_g, Q_i, Q_s, Q_t$ = discharge (breach, bridge opening, crest overflow, embankment overflow, gate, internal boundary, spillway, turbine)

t_{br} = time since beginning of breach formation, h

τ_{br} = formation of time of breach, h = $0.6\, V_r^{0.47}/h_d^{0.9}$ (Refs. 45,47,53)

V_r = volume of reservoir, acre-ft

z = side slope of breach, 1 : vertical to z : horizontal

***Selection of Computational* Δx *and* Δt.** The accuracy of the finite-difference solution is affected by the choice of the computational distance step Δx. For best accuracy, the maximum computational distance step Δx_m is selected as follows:[44,45,111,141]

$$\Delta x_m \leq \frac{c\,T_r}{20} \tag{10.3.37}$$

in which c is the *bulk wave celerity* (the celerity associated with an essential characteristic of the unsteady flow such as the peak of the hydrograph). In most applications, the wave speed is well approximated as a kinematic wave and c is estimated from (10.3.5) or from two or more observed flow hydrographs at different points along the channel. Since c can vary along the channel, Δx_m may not be constant along the channel.

Another criterion for selecting Δx_m is the restriction imposed by rapidly varying cross-sectional changes along the x axis of the watercourse. Such expansion and contraction is limited to the following inequality:[13,117]

$$0.635 \leq \frac{A_{i+1}}{A_i} < 1.576 \tag{10.3.38}$$

This condition results in the following approximation for the maximum computational distance step:[45]

$$\Delta x_m < \frac{L'}{1 + 2|A_i - A_{i+1}|/\hat{A}} \tag{10.3.39}$$

where $\hat{A} = A_{i+1}$ for a contracting reach and $\hat{A} = A_i$ for an expanding reach, and L' is the original distance step between adjacent specified cross sections at i and $i + 1$. The selection of Δx is determined by the minimum value of Δx_m from either (10.3.37) or (10.3.39).

Significant changes in the bottom slope of the watercourse also require small distance steps in the vicinity of the change. This is required particularly when the flow changes from subcritical to supercritical or conversely along the watercourse. Such changes can require computational distance steps in the range of 50 to 200 ft (15 to 61 m).

Selection of the Δt time step is governed by the following inequality:

$$\Delta t \leq \frac{T_r}{M} \tag{10.3.40}$$

where M is a numerical accuracy factor varying between 5 and 30; M can be estimated as $2.7\,[1 + \sqrt{g}\,n/(\mu D^{1/6} S_o^{1/2})]$ to provide desirable numerical error properties; μ is defined in (10.3.17); and T_r is the smallest time of rise of any wave that is routed.

10.4 DATA REQUIREMENTS FOR ROUTING MODELS

10.4.1 General

The data required for routing models are substantially different for the lumped models than for the distributed models. An inflow (discharge) hydrograph is always required for lumped models. Also, this is required for distributed models, although the diffusion (zero-inertia) and dynamic models can use a water-surface elevation time series as an alternate upstream boundary condition. The different data requirements for lumped and distributed models are shown in Table 10.4.1. The remaining portion of this subsection will provide additional information on cross-sectional, channel friction (Manning n), and lateral flow data.

10.4.2 Cross Sections

Much of the uniqueness of a specific distributed flow routing application is captured in the *cross sections* located at selected points along the watercourse as in Fig. 10.1.1.

Active Sections. That portion of the channel cross section in which flow occurs is called *active.* Cross sections may be of regular or irregular geometrical shape. As indicated in Fig. 10.4.1, each cross section is described by tabular values of channel top width and water-surface elevation which constitute a piecewise linear relationship. Generally about 4 to 12 sets of top widths and associated elevations provide a sufficiently accurate description of the cross section. Area-elevation tables can be generated initially from the specified top width–elevation data. Areas or widths associated with a particular water-surface elevation are linearly interpolated from the tabular values. Cross sections at gauging station locations are generally used as computational points. Such points are also specified at locations along the river where significant cross-sectional or flow-resistance changes occur or at locations where major tributaries enter. The spacing of cross sections can range from a few hundred feet to a few miles apart. Typically, cross sections are spaced farther apart for large rivers than for small streams, since the degree of variation in the cross-sectional characteristics is greater for the small streams. It is essential that the selected cross sections, with the assumption of linear variation between adjacent sections, represent

TABLE 10.4.1 Data Used in Lumped and Distributed Routing Models

	Routing models	
Data type	Lumped	Distributed
Observed inflow hydrograph, $I(t)$	a	c
Observed outflow hydrograph, $Q(t)$	a	c
Observed water-surface elevation time series, $h(t)$	—	c
Lateral inflow hydrograph, $q(t)$	—	d
Surface area—elevation table, $Sa(h)$	b	—
Cross-section top width—elevation table, $B(h)$	—	e
Friction coefficient—water-surface elevation or discharge table, $n(h \text{ or } Q)$	—	f
Expansion and contraction coefficients, K_{ec}	—	g
Sinuosity factors, s_c and s_m	—	h

a = required for calibration of storage and linear systems models
b = required for level-pool model
c = required for calibration
d = not always required
e = always required
f = can be obtained via calibration
g = can be assumed 0.0 for fairly uniform channels
h = can be assumed 1.0 for fairly straight channels

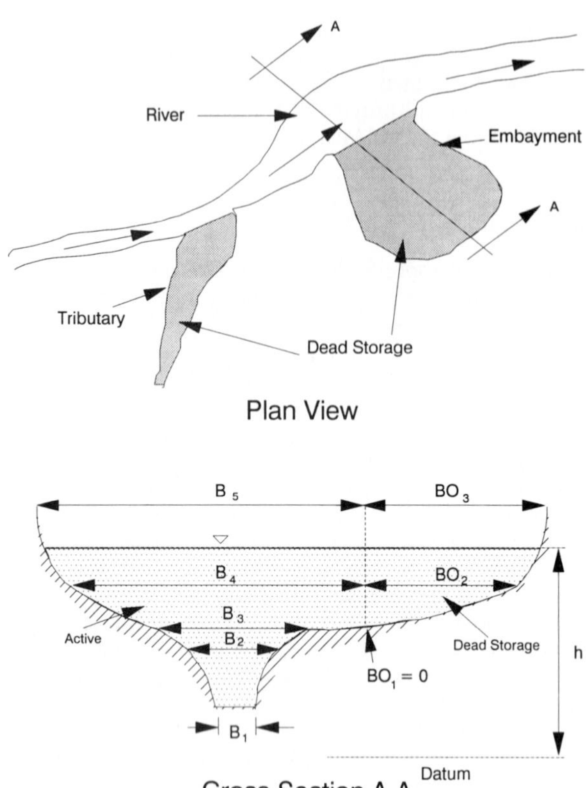

FIGURE 10.4.1 Plan view of river with active and dead storage areas, and cross section view.

the volume available to contain the flow along the watercourse. In some applications, particularly the routing of flood waves in large rivers with gradually varying cross sections, the number of required cross sections can be reduced by using a distance-weighted average section whose width is so computed that it replaces several (varying from a few to more than 50) intervening cross sections and yet conserves the volume within the reach.[39,44]

Inactive (Dead) Sections. There can be portions of a cross section where the flow velocity in the x direction is negligible relative to the velocity in the active portion. The inactive portion is called *off-channel (dead) storage;* it is represented by the term A_o in (10.3.15). Off-channel storage areas can be used to effectively account for adjacent embayments, ravines, or tributaries (see Fig. 10.4.1) which connect at some elevation with the flow channel but do not convey flow in the x direction; they serve only to store some of the passing flow. Sometimes, off-channel storage can be used to simulate a heavily wooded floodplain which primarily stores some of the floodwaters while conveying a very minimal portion of the flow. Dead storage cross-sectional properties are described by width (dead storage) vs. elevation tables.

10.4.3 Channel Friction

The resistance to flow[17–19,61,91,96,139] in a watercourse may be parameterized by the *Manning n, Chezy C, Darcy f,* or some other friction (roughness) coefficient which represents the effect of roughness elements of the channel bank and bed particles as well as form losses attributed to dynamic alluvial bed forms and vegetation of various types (grass, shrubs, field crops, brush, and trees) located along the banks and overbanks (floodplain). Also, small eddy losses due to mild expansion and contraction of cross-sectional reaches as well as river bend losses are often included as components of the Manning n.[17]

The Manning n varies with the magnitude of flow.[46] As the flow increases and more portions of the bank and overbank become inundated, the vegetation located at these elevations causes an increase in the resistance to flow.[79] Also, the n value may be larger for small floodplain depths than for larger depths because of flattening of the brush, thick weeds, or tall grass as the flow depths and velocities increase. This effect may be reversed in the case of wooded overbanks where, at the greater depths, the flow impinges against the leaved branches rather than only against the tree trunks, thus increasing Manning's n. The n values may also decrease with increasing discharge when the increase in the overbank flow area is relatively small compared with the increase of flow area within the banks, as in the case of wide rivers with levees situated closely along the natural riverbanks, or when floods remain confined within the channel banks.

Other conditions which can result in different n values for the same flow are: (1) change of season which affects the extent of vegetation, (2) change of water temperature which affects bed forms in some alluvial rivers,[20,135] (3) ice-cover effects,[17,102,136] and (4) man-made channel changes such as drift removal, channel straightening, bank stabilization, or paving.

The Manning n is defined for each channel reach and specified as a tabular piecewise linear function of stage or discharge, with linear interpolation used to obtain n values intermediate to the tabulated values.

Calibration. Results from dynamic routing or other distributed routing models are somewhat sensitive to the Manning n. Best results are obtained when n is adjusted to reproduce historical observations of stage and discharge. Such an adjustment process is known as *calibration,* which may be accomplished via either a trial-and-error or an automatic technique.[44,51,141] In the absence of observed flows and water-surface elevations, selection of the Manning n should reflect the influence of bank and bed materials, channel obstructions, irregularity of the riverbanks, and especially vegetation. Vegetation may cause the n values to vary considerably with flow as previously described. Some basic references for estimating n values are available.[12,17–19,61,91] Some references also describe the effects of urbanization of the floodplain,[8,60,140] wooded floodplains,[8] and steep rivers with cobble and boulder beds.[70] Manning n values for flows less than bankfull are approximately 0.015 to 0.035 for large rivers (Mississippi, Ohio, Missouri, Illinois), 0.03 to 0.04 for moderate-sized rivers and streams, 0.04 to 0.07 for mountain streams, and 0.04 to 0.25 for overbank flows. Some references provide estimates of the Darcy friction factor f, which is related to the Manning n as follows:

$$n = \tilde{\mu} f^{0.5} D^{0.17} \tag{10.4.1}$$

where D is hydraulic depth and $\tilde{\mu}$ is 0.093 (U.S. units) or 0.113 (SI units).

10.4.4 Lateral Flows

Specified unsteady flows associated with tributaries that are not dynamically routed can be added to the unsteady flow along the routing reach. This is accomplished via the term q in (10.3.15) and (10.3.16). The total tributary flow, which is a known function of time, i.e., $Q(t)$ which is a specified time series, is distributed along a single Δx reach, i.e., $q(t) = Q(t)/\Delta x_i$. Backwater effects of the routed flow on the tributary flow are ignored, and the *lateral flow* is usually assumed to enter perpendicular to the routed flow. Known outflows can be simulated by using a negative sign with the specified $Q(t)$. Numerical difficulties in solving the Saint-Venant equations sometimes arise when the ratio of lateral inflow to channel flow q_i/Q_i is too large; this can be overcome by increasing Δx_i for this reach.

10.5 FLOW ROUTING COMPLEXITIES

10.5.1 General

The routing of unsteady flows in natural waterways entails many complexities that require special treatment. These include network of channels, levee overtopping, mixed subcritical-supercritical flows, sediment transport effects, streamflow-aquifer interactions, ice effects, and two-dimensional flows.

10.5.2 Channel Networks

Flow routing is often required in natural waterways as well as man-made channels which are linked together forming a network of channels. The configuration may be *dendritic* (tree-type) and/or *looped* (islands, parallel channels connected by bypasses, etc). A network of channels presents complications in achieving computational efficiency when using implicit dynamic routing models. Necessary compatibility equations for flow conditions at the confluence of two channels produce a *coefficient matrix* in the solution procedure with elements which are not contained within the narrow band along the main diagonal of the matrix. Such *off-diagonal elements* produce a *sparse matrix* containing relatively few nonzero elements. Unless special matrix solution techniques are used for the sparse matrix, the computational time required to solve the matrix by conventional Gaussian elimination matrix solution techniques is too great and makes an implicit solution infeasible. Various algorithms have been developed to provide efficient computational treatment of channel networks. Most algorithms treat the channel junctions as internal boundaries consisting of three compatibility equations, usually a conservation of mass equation and two simplified momentum equations.[11,25,43,44,49,57,72] These are solved along with (10.3.23) and (10.3.24) applied to each Δx reach. The resulting sparse matrix is solved by various sparse matrix solution techniques of the *Gaussian elimination* or *double-sweep* variety.[25,36,43,44,57,72,75,88,120] Another type of network algorithm treats tributary flow at a junction as lateral flow q to the main-stem channel, and thereby eliminates the computational difficulty associated with off-diagonal elements; this is an iterative *relaxation* procedure[37,44] which is applicable only to first-order dendritic networks (a main-stem channel with one or more tributaries).

10.5.3 Supercritical and Subcritical Mixed Flow

Flow can change with either time or distance along the routing reach from supercritical to subcritical while passing through critical flow, or conversely. This *mixed flow* requires special treatment to prevent numerical instability when modeling it with the Saint-Venant equations. One method[43–45] of treating this is to avoid using the Saint-Venant equations at the point where mixed flow occurs; this is accomplished by dividing the routing reach at each time step into a series of subcritical and supercritical routing reaches (subreaches) and using appropriate external boundary conditions for each subreach. The Froude number is used to determine the supercritical reaches, for which $\mathrm{Fr} > 1$. At each time step, the solution commences with the most upstream subreach, and proceeds subreach by subreach in the downstream direction. Hydraulic jumps are allowed to move at the end of a time step according to the relative values of supercritical sequent depth and the adjacent downstream subcritical depth. Another more approximate method that has been used for mixed flows is to arbitrarily decrease the β coefficient in the second term of (10.3.16) so that it becomes zero as the flow approaches critical flow ($F_r = 1$). This rather crude approach helps to stabilize the numerical solution, but at the expense of accuracy since the second term $\partial(Q^2/A)/\partial x$ in (10.3.16) is thereby neglected.

10.5.4 Levee Overtopping

Flows which overtop *levees* located along either or both sides of a main-stem river and/or its principal tributaries can be treated as lateral outflow ($-q$) in (10.3.15) and (10.3.16) where the diverted lateral flow over the levee is computed as broad-crested weir flow.[25,43–45,49,138] This overtopping flow is corrected for submergence effects if the floodplain water-surface elevation sufficiently exceeds the levee crest elevation. The overtopping flow may reverse its direction when the floodplain elevation exceeds the river water-surface elevation, thus allowing flow to return to the channel after the flood peak passes. The overtopping broad-crested weir flow is computed according to the following:

$$q = -c_\ell K_s (h - h_c)^{3/2} \tag{10.5.1}$$

where

$$K_s = 1.0 \qquad \text{if } h_r \le 0.67 \tag{10.5.2}$$

$$K_s = 1.0 - 27.8\,(h_r - 0.67)^3 \qquad \text{if } h_r > 0.67 \tag{10.5.3}$$

$$h_r = \frac{h_{fp} - h_c}{h - h_c} \tag{10.5.4}$$

in which c_ℓ is the weir discharge coefficient ranging in value from 2.6 to 3.2 for U.S. units (1.4 to 1.8 for SI units). K_s is the submergence correction factor[44,45] similar to that used for internal boundaries (dams), h_c is the levee-crest elevation, h is the water-surface elevation of the river, and h_{fp} is the water-surface elevation of the floodplain. Flow in the floodplain can affect the overtopping flows via the submergence correction factor K_s. Flow may also pass from the waterway to the floodplain through a time-dependent *crevasse (breach)* in the levee[11,41,49] via a breach-flow equation similar to that shown in Table 10.3.2. The floodplain, which is separated from the principal routing channel (river) by the levee, may be treated as (1) a dead storage area[33,87,129] in the Saint-Venant equations; (2) a tributary which receives its inflow as lateral flow (the flow from the river which overtops the levee crest) which is dynamically routed along the floodplain;[41,44,49] and (3) flow and water-surface elevations which can be computed by using a level-pool routing method,[25,26,45,49,138] particularly if the floodplain is divided into compartments by levees or road embankments located perpendicular to the river levee.

10.5.5 Mobile-Bed Effects

A complex interaction of unsteady flow and sediment transport occurs in alluvial rivers with *mobile (movable) sand beds.*[15] The river bottom aggrades (raises) and degrades (lowers) during the passage of the flood wave. Also, the hydraulic resistance of the river bottom changes as the sand bed forms change their magnitude and shape. Routing models which account for the effects of *aggradation* and *degradation* have been developed.[15,16,63,67,72,80] In these models, the sediment conservation equation, i.e.,

$$(1-\lambda)\frac{\partial A_b}{\partial t}+\frac{\partial Q_s}{\partial x}-q_s=0 \tag{10.5.5}$$

is coupled to an implicit finite-difference solution of the Saint-Venant equations. In (10.5.5) λ is the porosity of the bed material, A_b is the cross-sectional area of the mobile portion of the channel bed, Q_s is the volumetric sediment transport rate computed by some appropriate steady-flow sediment transport technique,[5,34,93,134,142] and q_s is the lateral inflow rate of sediment per unit length. The coupling of (10.5.5) with (10.3.15) and (10.3.16) can be accomplished within the iteration of the nonlinear implicit solution, between these iterations, or simply between time steps of the flow computation; the particular type of coupling of water and sediment flow depends on the rate of change of A_b with respect to time. Also, the dynamic interaction between the changing sand bed forms and bed friction may be approximated[14,132] via a continuous modification of the roughness coefficient (Manning n). Additional information on sediment transport is presented in Chap. 12.

10.5.6 Streamflow-Aquifer Interaction

Interaction of streamflow and its adjacent *groundwater aquifer* for floods occurring in some channels, particularly those often located in arid regions, can be of sufficient magnitude to affect the river flow by attenuating the peak flow, reducing the wave peak celerity, and extending the recession limb of the river discharge hydrograph. The flow between the river and the aquifer can be simulated by coupling distributed flow routing models to either one- or two-dimensional *groundwater models.* The coupling occurs through the lateral flow term q in the Saint-Venant equations. Both explicit[52] and implicit[106] dynamic flow routing models have been coupled to the unsteady saturated or saturated-unsaturated porous media equation. Time steps required for the flow routing equations are usually smaller than those required for the groundwater (saturated equations); models can take advantage of this for greater efficiency by solving the groundwater equations periodically.

10.5.7 Ice Effects

The formation of an ice cover in a waterway affects the flow.[105] The ice cover floats except for those in narrow waterways or those with very thick covers located in extremely cold regions. The bottom of the ice cover causes an increase in the flow-resistance coefficient.[17] This varies over a wide range throughout the winter[102,136] as the ice cover begins to develop, reaches maturity, and then decays with increasing temperature. Localized hanging *ice dams* further increase flow resistance. Severe increases in water-surface elevation can occur when *ice jams* are formed during the breakup of the ice as air temperatures increase. The ice jam can considerably reduce the cross-sectional area of flow and act as a constricted flow-control section. Ice effects can be considered in dynamic flow routing models.[82,143] Ice effects are further discussed in Sec. 7.7.

10.5.8 Landslide-Generated Wave

Reservoirs are sometimes subject to landslides which move at high velocities into the reservoir; the slide displaces a portion of the reservoir contents and creates a very steep water wave which travels up and down the length of the reservoir,[27] and the wave can overtop the dam and cause its failure (breach). A *landslide-generated wave* can be simulated by using a dynamic routing model. Using known information on the volume of landslide mass, its porosity, and the time interval over which the landslide moves into the reservoir, the landslide volume is deposited in the reservoir during very small time steps in the dynamic routing computations; this rapid reduction in reservoir cross-sectional area[45,78] creates a landslide-generated wave in the solution of the Saint-Venant equations. *Wave runup* for near-vertical faces of concrete dams can be neglected; however, for earthen dams, the abrupt wave can advance up the sloping dam face to approximately 2.5 times the height of the wave.[96]

10.5.9 Two-Dimensional Flows

Two-dimensional unsteady-flow models which account for momentum conservation in the y direction (across the waterway perpendicular to the x axis) can be categorized as complete (all terms in the momentum equations are retained) or simplified (first two terms, inertial or acceleration terms in each momentum equation are neglected). Complete two-dimensional models[3,62,84,85] are used for unsteady flows in estuaries, bays, or lakes where environmental pollution concerns require knowledge of flow patterns (circulation) and velocities which are dominated by two-dimensional effects. Occasionally, simplified two-dimensional flow routing models have been used for unsteady flows in complex floodplains.[24] Two-dimensional models are generally much more expensive to calibrate and execute on computers than one-dimensional models. Generally, the additional accuracy gained does not justify their use to predict water-surface elevations and average flows in typical unsteady-flow applications having floodplains.[138]

REFERENCES

1. Abbott, M. B., *An Introduction to the Method of Characteristics,* American Elsevier, New York, 1966.

2. Abbott, M. B., *Computational Hydraulics,* Pitman, London, England, 1980.
3. Abbott, M. B., and J. A. Cunge, "Two-Dimensional Modeling of Tidal Deltas and Estuaries," in K. Mahmood and V. Yevjevich, eds. *Unsteady Flow in Open Channels,* vol. II, chap. 18, pp. 763–812, Water Resources Publications, Fort Collins, Colo., 1975.
4. Abbott, M. B., and J. A. Cunge, *Engineering Applications of Computational Hydraulics,* vol. 1, Pitman, London, England, 1982.
5. Ackers, P., and W. R. White, "Sediment Transport: A New Approach and Analysis," *J. Hydraul. Eng.,* vol. 99, no. HY11, pp. 2041–2060, 1973.
6. Amein, M., "Streamflow Routing on Computer by Characteristics," *Water Resour. Res.,* vol. 2, no. 1, pp. 123–130, 1966.
7. Amein, M., and C. S. Fang, "Implicit Flood Routing in Natural Channels," *J. Hydraul. Eng.,* vol. 96, no. HY12, pp. 2481–2500, 1970.
8. Arcement, G. J., Jr., and V. R. Schneider, *Guide for Selecting Manning's Roughness Coefficients for Natural Channels and Flood Plains,* Report RHWA-TS-84-204, U.S. Geological Survey for Federal Highway Administration, National Technical Information Service, PB84-242585, 1984.
9. Balloffet, A., "One-Dimensional Analysis of Floods and Tides in Open Channels," *J. Hydraul. Eng.,* vol. 95, no. HY4, pp. 1429–1451, 1969.
10. Baltzer, R. A., and C. Lai, "Computer Simulation of Unsteady Flow in Waterways," *J. Hydraul. Eng.,* vol. 94, no. HY4, pp. 1083–1117, 1968.
11. Barkow, R. L., *UNET One-Dimensional Unsteady Flow through a Full Network of Open Channels, Users Manual,* Hydrologic Engineering Center Rept CPD-66, U.S. Army Corps of Engineers, Davis, Calif., 1992.
12. Barnes, H. H., Jr., *Roughness Characteristics of Natural Channels,* Geological Survey Water-Supply Paper 1849, Government Printing Office, Washington, D.C., 1967.
13. Basco, D. R., "Improved Robustness of the NWS DAMBRK Algorithm," *Hydraulic Engineering (Proc. of the 1987 National Conference on Hydraulic Engineering),* ASCE, New York, pp. 776–781, August 1987.
14. Brownlie, W. R., "Flow Depth in Sand-Bed Channels," *J. Hydraul. Eng.,* vol. 109, HY7, pp. 959–990, 1983.
15. Chang, H. H., *Fluvial Processes in River Engineering,* Wiley, New York, 1988.
16. Chen, Y. H., and D. B. Simons, "Mathematical Modeling of Alluvial Channels," *Symp. Modeling Techniques,* vol. I, 2d Ann. Symp. Waterways, Harbors, and Coastal Eng. Div., ASCE, pp. 466–483, 1975.
17. Chow, V. T., *Open-Channel Hydraulics,* McGraw-Hill, New York, 1959.
18. Chow, V. T., *Handbook of Applied Hydrology,* secs. 7, 25-II, McGraw-Hill, New York, 1964.
19. Chow, V. T., D. R. Maidment, and L. W. Mays, *Applied Hydrology,* McGraw-Hill, New York, 1988.
20. Colby, B. R., and C. H. Scott, "Effects of Water Temperature on the Discharge of Bed Material," *USGS Professional Paper 4626,* Government Printing Office, Washington, D.C., 1965.
21. Colon, R., and G. F. McMahon, "BRASS Model: Application to Savannah River System Reservoirs," *J. Water Resour. Plann. Manage.,* vol. 113, no. 2, pp. 177–190, 1987.
22. Cooley, R. L., and S. A. Moin, "Finite Element Solution of Saint-Venant Equations," *J. Hydraul. Eng.,* vol. 102, no. HY6, pp. 759–775, 1976.
23. Cunge, J. A., "On the Subject of a Flood Propagation Computation Method (Muskingum Method)," *J. Hydraul. Res.,* vol. 7, no. 2, pp. 205–230, 1969.
24. Cunge, J. A., "Two-Dimensional Modeling of Flood Plains," in K. Mahmood and V. Yevjevich, eds., *Unsteady Flow in Open Channels,* vol. II, chap. 17, pp. 705–762, Water Resources Publications, Fort Collins, Colo., 1975.

25. Cunge, J. A., F. M. Holly, Jr., and A. Verway, *Practical Aspects of Computational River Hydraulics,* Pitman, Boston, Mass., 1980.

26. Danish Hydraulic Institute, *MIKE II Short Description,* Horsholm, Denmark, 1988.

27. Davidson, D. D., and B. L. McCartney, "Water Waves Generated by Landslides in Reservoirs," *J. Hydraul. Eng.,* vol. 101, no. HY12, pp. 1489–1501, 1975.

28. Dawdy, D. R., J. C. Schaake, Jr., and W. M. Alley, *Distributed Routing Rainfall-Runoff Model,* U.S. Geological Survey Water Resources Investigations 78–90, 1978.

29. DeLong, L. L., "Extension of the Unsteady One-Dimensional Open-Channel Flow Equations for Flow Simulation in Meandering Channels with Flood Plains," *Selected Papers in Hydrologic Science,* U.S. Geological Survey Water Supply Paper 2220, pp. 101–105, 1986.

30. DeLong, L. L., "Mass Conservation: 1-D Open Channel Flow Equations," *J. Hydraul. Eng.,* vol. 115, no. 2, pp. 263–268, 1989.

31. Dooge, J. C. I., *Linear Theory of Hydrologic Systems, Tech. Bull.* 1468, USDA Agricultural Research Service, Beltsville, Md., 1973.

32. Dooge, J. C. I., W. G. Strupczewski, and J. J. Napiorkowski, "Hydrodynamic Derivation of Storage Parameters of the Muskingum Model," *J. Hydrol.,* no. 54, pp. 371–387, 1982.

33. Dronkers, J. J., "Tidal Computations for Rivers, Coastal Areas, and Seas," *J. Hydraul. Eng.,* vol. 95, no. HY1, pp. 29–77, 1969.

34. Engelund, K., and E. Hansen, *A Monograph on Sediment Transport in Alluvial Streams,* Teknisk Vorlog, Copenhagen, Denmark, 1983.

35. Franz, D. D., *Unsteady Flow Solutions: FEQ/FEQUTL,* Linsley, Kraeger Associates, Ltd., Mountain View, Calif., 1991.

36. Fread, D. L., "Discussion of Implicit Flood Routing in Natural Channels," by M. Amein and C. S. Fang, *J. Hydraul. Eng.,* vol. 97, no. HY7, pp. 1156–1159, 1971.

37. Fread, D. L., "Technique for Implicit Dynamic Routing in Rivers with Tributaries," *Water Resour. Res.,* vol. 9, no. 4, pp. 918–926, 1973.

38. Fread, D. L., *Numerical Properties of Implicit Four-Point Finite Difference Equations of Unsteady Flow,* HRL-45, NOAA Tech. Memo NWS HYDRO-18, Hydrologic Research Laboratory, National Weather Service, Silver Spring, Md., 1974.

39. Fread, D. L., *Theoretical Development of Implicit Dynamic Routing Model,* HRL-113, Hydrologic Research Laboratory, National Weather Service, Silver Spring, Md., 1976.

40. Fread, D. L., "The Development and Testing of a Dam-Break Flood Forecasting Model," *Proc. Dam-Break Flood Modeling Workshop,* U.S. Water Resources Council, Washington, D.C., pp. 164–197, 1977.

41. Fread, D. L., "NWS Operational Dynamic Wave Model," *Verification of Mathematical and Physical Models,* Proceedings of 26th Annual Hydr. Div. Specialty Conf., ASCE, College Park, Md., pp. 455–464, 1978.

42. Fread, D. L., *Applicability Criteria for Kinematic and Diffusion Routing Models,* HRL-176, Hydrologic Research Laboratory, National Weather Service, Silver Spring, Md., 1983.

43. Fread, D. L., "Computational Extensions to Implicit Routing Models," *Proceedings of the Conference on Frontiers in Hydraulic Engineering,* ASCE, MIT, Cambridge, Mass., pp. 343–348, 1983.

44. Fread, D. L., "Channel Routing," in M. G. Anderson and T. P. Burt, eds., *Hydrological Forecasting,* Wiley, New York, chap. 14, pp. 437–503, 1985.

45. Fread, D. L., *The NWS DAMBRK Model: Theoretical Background/User Documentation,* HRL-256, Hydrologic Research Laboratory, National Weather Service, Silver Spring, Md., 1988.

46. Fread, D. L., "Flood Routing and the Manning *n*," in B. C. Yen, ed., *Proc. of the International Conference for Centennial of Manning's Formula and Kuichling's Rational Formula,* Charlottesville, Va., pp. 699–708, 1989.

47. Fread, D. L., "National Weather Service Models to Forecast Dam-Breach Floods," in O. Starosolszky and O. M. Melder, eds., *Hydrology of Disasters,* Proc. of the World Meteorological Organization Technical Conference, November 1988, Geneva, Switzerland, pp. 192–211, 1989.

48. Fread, D. L., and T. E. Harbaugh, "Open Channel Profiles by Newton's Iteration Technique," *J. Hydrol.,* vol. 13, pp. 70–80, 1971.

49. Fread, D. L., and J. M. Lewis, "FLDWAV: A Generalized Flood Routing Model," *Proc. National Conference on Hydraulic Engineering,* ASCE, Colorado Springs, Colo., pp. 668–673, 1988.

50. Fread, D. L., G. F. McMahon, and J. L. Lewis, "Limitations of Level-Pool Routing in Reservoirs," *Proceedings, Third Water Resources Operations and Management Workshop,* ASCE, Computational Decisions Support Systems for Water Managers, Fort Collins, Colo., 1988.

51. Fread, D. L., and G. F. Smith, "Calibration Technique for 1-D Unsteady Flow Models," *J. Hydraul. Eng.,* vol. 104, no. HY7, pp. 1027–1044, 1978.

52. Freeze, R. A., "Role of Subsurface Flow in Generating Surface Runoff, 1. Base Flow Contributions to Channel Flow," *Water Resour. Res.,* vol. 8, no. 3, pp. 609–623, 1972.

53. Froehlich, D. C., "Embankment-Dam Breach Parameters," *Proc. 1987 National Conf. on Hydraulic Engr.,* ASCE, New York, August, pp. 570–575, 1987.

54. Garrison, J. M., J. P. Granju, and J. T. Price, "Unsteady Flow Simulation in Rivers and Reservoirs," *J. Hydraul. Eng.,* vol. 95, no. HY5, pp. 1559–1576, 1969.

55. Goodrich, R. D., "Rapid Calculation of Reservoir Discharge," *Civ. Eng.,* vol. 1, pp. 417–418, 1931.

56. Gray, W. G., G. F. Pinder, and C. A. Brebbia, *Finite Elements in Water Resources,* Pentech Press, London, 1977.

57. Gunaratnam, D. J., and F. E. Perkins, "Numerical Solution of Unsteady Flows in Open Channels," *Report 127,* Department of Civil Engineering, MIT, Cambridge, Mass., 1970.

58. Harder, J. A., and L. V. Armacost, *Wave Propagation in Rivers,* Hydraulic Engng. Lab., Report 1, Series 8, University of California, Berkeley, 1966.

59. Harley, B. M., F. E. Perkins, and P. S. Eagleson, *A Modular Distribution Model of Catchment Dynamics,* Report 133, R. M. Parsons Lab for Water Resources and Hydrodynamics, MIT, Cambridge, Mass., 1970.

60. Heijl, H. R., "A Method for Adjusting Values of Manning's Roughness Coefficient for Floods in Urban Areas," *U.S. Geol. Surv. J. Res.,* vol. 5, no. 5, pp. 541–595, 1977.

61. Henderson, F. M., *Open Channel Flow,* Macmillan, New York, pp. 285–287, 1966.

62. Hinwood, J. B., and I. G. Wallis, "Review of Models of Tidal Waters," *J. Hydraul. Eng.,* vol. 101, no. HY11, pp. 1405–1421, 1975.

63. Holly, F. M., Jr., J. C. Yang, P. Schwerz, J. Schaefer, S. H. Hsu, and R. Einhellig, *CHARIMA-Numerical Simulation of Unsteady Water and Sediment in Multiple Connected Networks of Mobile-Bed Channels,* University of Iowa, IIHR 343 Iowa City, Iowa, 1990.

64. Huber, W. C., J. P. Heany, M. A. Meding, W. A. Peltz, H. Sheikh, and G. F. Smith, *Storm Water Management Model User's Manual,* Natl. Environ. Res. Ctr. Document EPA-670/2-75-017, U.S. Environmental Protection Agency, Cincinnati, Ohio, 1975.

65. Hydrologic Engineering Center, *HEC-1 Flood Hydrograph Package—Users Manual,* U.S. Army Corps of Engineers, Davis, Calif., 1981.

66. Hydrologic Engineering Center, *HEC-2 Water Surface Profiles Users Manual,* U.S. Army Corps of Engineers, Davis, Calif., 1982.

67. Interagency Ad Hoc Sedimentation Work Group, *Twelve Selected Computer Stream Sedimentation Models Developed in the United States,* Subcommittee on Sedimentation,

Interagency Advisory Committee on Water Data (S. Fan, ed.), Federal Energy Regulatory Commission, Government Printing Office: 1989-250-618/00854, 1988.

68. Isaacson, E., and H. B. Keller, *Analysis of Numerical Methods,* Wiley, New York, 1966.

69. Isaacson, E., J. J. Stoker, and A. Troesch, *Numerical Solution of Flood Prediction and River Regulation Problems* Report II/III, no. IMM-NYU-205/235, New York University Institute of Mathematical Science, New York, 1954.

70. Jarrett, R. D., "Hydraulics of High-Gradient Streams," *J. Hydraul. Eng.,* vol. 110, no. HY11, November, pp. 1519–1539, 1984.

71. Johnson, B. H., *Unsteady Flow Computations on the Ohio-Cumberland-Tennessee-Mississippi River System,* Tech. Report H-74-8, U.S. Army Engineer Waterways Experiment Station, Vicksburg, Miss., 1974.

72. Johnson, B. H., *Development of a Numerical Modeling Capability for the Computation of Unsteady Flow on the Ohio River and Its Major Contributaries,* Tech. Report HL-82-20, U.S. Army Engineer Waterways Experiment Station, Vicksburg, Miss., 1982.

73. Jones, S. B., "Choice of Space and Time Steps in the Muskingum-Cunge Flood Routing Method," *Proc. Inst. Civ. Eng.,* part 2, no. 71, pp. 759–772, 1981.

74. Kalinin, G. P., and P. I. Miljukov, "On the Computation of Unsteady Flow in Open Channels," *Meteorologiya i Gidrologiya Zhuzurnal,* vol. 10, Leningrad, U.S.S.R., 1957.

75. Kamphuis, J. W., "Mathematical Tidal Study of St. Lawrence River," *J. Hydraul. Eng.,* vol. 96, no. HY3, pp. 643–664, 1970.

76. Keefer, T. N., and R. S. McQuivey, "Multiple Linearization Flow Routing Model," *J. Hydraul. Eng.,* vol. 100, no. HY7, pp. 1031–1046, 1974.

77. Koussis, A. D., "Theoretical Estimations of Flood Routing Parameters," *J. Hydraul. Eng.,* vol. 104, no. HY1, pp. 109–115, 1978.

78. Koutitas, C. G., "Finite Element Approach to Waves Due to Landslides," *J. Hydraul. Eng.,* vol. 103, no. HY9, September, pp. 1021–1029, 1977.

79. Kouwen, N., "Field Estimation of the Biomechanical Properties of Grass," *J. Hydraul. Res.,* vol. 26, no. 5, pp. 559–568, 1988.

80. Krishnappan, B. G., "Unsteady Flow in Mobile Boundary Channels," Environ. Hydr. Sect., Hydr. Res. Div., National Water Res. Inst., *Canada Centre for Inland Waters,* Burlington, Ont. 1979.

81. Kundzewicz, Z. W., "Multilinear Flood Routing," *Acta Geophys. Pol.,* vol. 32, no. 4, pp. 419–445, 1984.

82. Lai, A. M. W., and H. T. Shen, "Mathematical Model for River Ice Process," *J. Hydraul. Eng.,* vol. 117, no. 7, pp. 851–867, 1991.

83. Lai, Chintu, "Numerical Modeling of Unsteady Open-Channel Flow," in V. T. Chow and B. C. Yen, eds., *Advances in Hydroscience,* vol. 14, pp. 161–333, Academic Press, Orlando, Fla., 1986.

84. Leenderste, J. J., "Aspects of a Computational Model for Long-Period Water Wave Propagation," *Rand Report RM-5294-PR,* Rand Corp., Santa Monica, Calif., 1967.

85. Leenderste, J. J., "Aspects of SIMSYS2D, a System for Two-Dimensional Flow Computation," *Rand Report R-3572-USGS,* Rand Corp., Santa Monica, Calif., 1987.

86. Li, R. M., D. B. Simons, and M. A. Stevens, "Nonlinear Kinematic Wave Approximation for Water Routing," *Water Resour. Res.,* vol. 11, no. 2, pp. 245–252, 1975.

87. Liggett, J. A., "Basic Equations of Unsteady Flow," in K. Mahmood and V. Yevjevich, eds., *Unsteady Flow in Open Channels,* vol. I, chap. 2, pp. 29–62, Water Resource Publ., Fort Collins, Colo., 1975.

88. Liggett, J. A., and J. A. Cunge, "Numerical Methods of Solution of the Unsteady Flow Equations," in K. Mahmood and V. Yevjevich, eds., *Unsteady Flow in Open Channels,* vol. I, chap. 4, pp. 89–182, Water Resources Publ., Fort Collins, Colo., 1975.

89. Liggett, J. A., and D. A. Woolhiser, "Difference Solutions of the Shallow-Water Equations," *J. Eng. Mech. Div., ASCE,* vol. 95, no. EM2, pp. 39–71, 1967.

90. Lighthill, M. J., and G. B. Whitham, "On Kinematic Floods—Flood Movements in Long Rivers," *Proc. R. Soc. London,* A220, pp. 281–316, 1955.

91. Linsley, R. K., M. A. Kohler, and J. L. H. Paulhus, *Hydrology for Engineers,* McGraw-Hill, New York, pp. 502–530, 1986.

92. McCarthy, G. T., "The Unit Hydrograph and Flood Routing," *Conf. North Atlantic Div.,* U.S. Corps of Engineers, New London, Conn., 1938.

93. Meyer-Peter, E., and P. Muller, "Formulas for Bed-Load Transport," *Proc. 2d Congress IAHR,* Paper 2, Stockholm, Sweden, pp. 39–64, 1948.

94. Miller, W. A., and J. A. Cunge, "Simplified Equations of Unsteady Flow," in K. Mahmood and V. Yevjevich, eds., *Unsteady Flow in Open Channels,* vol. I, chap. 5, pp. 183–257, Water Resources Publ., Fort Collins, Colo., 1975.

95. Miller, W. A., and V. Yevjevich, *Unsteady Flow in Open Channels,* Bibliography, vol. III, Water Resources Publ., Fort Collins, Colo., 1975.

96. Morris, H. M., and J. M. Wiggert, *Applied Hydraulics in Engineering,* Ronald Press, New York, 1972.

97. Napiórkowski, J. J., and P. O'Kane, "A New Non-Linear Conceptual Model of Flood Waves," *J. Hydrol.,* vol. 69, no. 4, pp. 43–58, 1984.

98. Nash, J. E., "The Form of the Instantaneous Unit Hydrograph," *Int. Assoc. Hydrol. Sci.,* no. 45, vol. 3–4, pp. 114–121, 1957.

99. Nash, J. E., "A Note on the Muskingum Method of Flood Routing," *J. Geophys. Res.,* vol. 64, pp. 1053–1056, 1959.

100. Natale, T., and E. Todini, "A Stable Estimator for Linear Models; 1. Theoretical Developments and Monte Carlo Experiments," *Water Resour. Res.,* vol. 12, no. 4, pp. 667–671, 1976.

101. Natural Environment Research Council, *Flood Studies Report,* vol. III, Flood Routing Studies, Institute of Hydrology, Wallingford, England, 1975.

102. Nezhikhovskiy, R. A., "Coefficients of Roughness of Bottom Surface of Slush Ice Cover," *Soviet Hydrology: Selected Papers,* no. 2, pp. 127–148, 1964.

103. O'Brien, J. S., and P. Julien, "Physical Properties and Mechanics of Hyper-Concentrated Sediment Flows," in D. S. Bowles, ed., *Delineation of Landslide, Flash Flood, and Debris Flow Hazards in Utah,* Utah State University, Utah Water Research Laboratory, Logan, Utah, General Series UWRL/G-85/03, pp. 260–279, 1984.

104. O'Donnell, T., C. Pearson, and R. A. Woods, "Improved Fitting for Three Parameter Muskingum Procedure," *J. Hydraul. Eng.,* vol. 114, no. 5, pp. 516–528, 1988.

105. Pariset, E., R. Hauser, and A. Gagnon, "Formation of Ice Covers and Ice Jams in Rivers," *J. Hydraul. Eng.,* vol. 92, no. HY6, pp. 1–24, 1976.

106. Pogge, E. C., and W. L. Chiang, *Further Development of a Stream-Aquifer System Model,* Kansas Water Resources Research Institute, University of Kansas, Lawrence, Kans., 1977.

107. Ponce, V. M., R. M. Li, and D. B. Simons, "Applicability of Kinematic and Diffusion Models," *J. Hydraul. Eng.,* vol. 104, no. HY3, pp. 353–360, 1978.

108. Ponce, V. M., and V. Yevjevich, "Muskingum-Cunge Method with Variable Parameters," *J. Hydraul. Eng.,* vol. 104, no. HY12, pp. 1663–1667, 1978.

109. Preissmann, A., "Propagation of Translatory Waves in Channels and Rivers," in *Proc. First Congress of French Assoc. for Computation,* Grenoble, France, pp. 433–442, 1961.

110. Preissmann, A., and J. A. Cunge, "Translatory Wave Calculations by Computer," *Proc. 9th Congress, IAHR,* Dubrovnik, pp. 656–664, 1961.

111. Price, R. K., "A Comparison of Four Numerical Methods for Flood Routing," *J. Hydraul. Eng.,* vol. 100, no. HY7, pp. 879–899, 1974.

112. Price, R. K., *FLOUT—A River Catchment Flood Model,* Report IT168, Hydraulics Research Station, Wallingford, England, 1977.

113. Puls, L. G., *Construction of Flood Routing Curves,* House Document 185, U.S. 70th Congress, 1st session, Washington, D.C., pp. 46–52, 1928.

114. Rajar, R., "Mathematical Simulation of Dam-Break Flow," *J. Hydraul. Eng.,* vol. 104, no. HY7, pp. 1011–1026, 1978.

115. Rockwood, D. M., "Columbia Basin Stream Flow Routing by Computer," *Trans. ASCE,* vol. 126, part 4, pp. 32–56, 1958.

116. Saint-Venant, Barré de, "Theory of Unsteady Water Flow, with Application to River Floods and to Propagation of Tides in River Channels," *Computes Rendus,* vol. 73, Acad. Sci., Paris, France, pp. 148–154, 237–240. (Translated into English by U.S. Corps of Engrs., no. 49-g, Waterways Experiment Station, Vicksburg, Miss., 1949.), 1871.

117. Samuels, P. G., *Models of Open Channel Flow Using Preissmann's Scheme,* Cambridge University, Cambridge, England, pp. 91–102, 1985.

118. Sayed, I., and D. C. Howard, "Application of Dynamic Backwater Modeling to Mactaquac Headpond—Saint John River, N.B.," *Proceedings of 6th Canadian Hydrotechnical Conference,* Canadian Society for Civil Engr., pp. 203–220, 1983.

119. Schaffranek, R. W., *Flow Model for Open Channel Reach or Network,* Professional Paper 1384, U.S. Geological Survey, 1987.

120. Schaffranek, R. W., R. A. Baltzer, and D. E. Goldberg, "A Model for Simulation of Flow in Singular and Interconnected Channels," *Book 7, Automated Data Processing and Computations, TWRI Series,* U.S. Geological Survey, chap. C3, 1981.

121. Shearman, J. O., *Users Manual for WSPRO: A Computer Model for Water-Surface Profile Computations,* Federal Highways Admin., Report FWHA/IP-89/027, 1990.

122. Singh, V. P., and R. C. McCann, "Some Notes on Muskingum Method of Flood Routing," *J. Hydrol.,* vol. 48, no. 3, pp. 343–361, 1980.

123. Slade, J. E., and P. G. Samuels, *Modeling Complex River Networks,* HR Pub. Paper 41, Hydraulics Research Limited, Wallingford, UK, 1980.

124. Smith, A. A., "A Generalized Approach to Kinematic Flood Routing," *J. Hydrol.,* vol. 45, pp. 71–89, 1980.

125. Soil Conservation Service, *WSP2 Computer Program,* Technical Release 61, Engineering Div., Soil Conservation Service, U.S. Department of Agriculture, 1976.

126. Stoker, J. J., *Numerical Solution of Flood Prediction and River Regulation Problems; Derivation of Basic Theory and Formulation of Numerical Methods of Attack,* Report I, no. IMM-NYU-200, New York University Institute of Mathematical Science, New York, 1953.

127. Stoker, J. J., *Water Waves,* Interscience, New York, pp. 452–455, 1957.

128. Streeter, V. L., and E. B. Wylie, *Hydraulic Transients,* McGraw-Hill, New York, pp. 239–259, 1967.

129. Strelkoff, T., "The One-Dimensional Equations of Open-Channel Flow," *J. Hydraul. Eng.,* vol. 95, no. HY3, pp. 861–874, 1969.

130. Strelkoff, T., "Numerical Solution of Saint-Venant Equations," *J. Hydraul. Eng.,* vol. 96, no. HY1, pp. 223–252, 1970.

131. Strelkoff, T., and N. D. Katopodes, "Border Irrigation Hydraulics with Zero Inertia," *J. Irrig. Drain. Eng.,* vol. 103, pp. 325–342, 1977.

132. Task Committee on Bed Configuration and Hydraulic Resistance of Alluvial Streams, *The Bed Configuration and Roughness of Alluvial Streams,* Committee on Hydraulics and Hydraulic Eng., The Japan Society of Civil Engineering, 1974.

133. Todini, E., and A. Bossi, *PAB (Parabolic and Backwater)—An Unconditionally Stable Flood Routing Scheme Particularly Suited for Real Time Forecasting and Control,* Publ. 1, Univ. di Bologna, Bologna, Italy, 1985.

134. Toffalleti, F. B., "Definitive Computations of Sand Discharge in Rivers," *J. Hydraul. Eng.,* vol. 95, no. HY1, pp. 225–246, 1969.

135. U.S. Army Corps of Engineers, "Missouri River Channel Regime Studies, Omaha District," Missouri River Div. Sediment Series 13B, Omaha, Neb., 1969.

136. Uzuner, M. S., "The Composite Roughness of Ice-Covered Streams," *J. Hydraul. Res.,* vol. 13, no. 1, pp. 79–102, 1975.

137. Vasiliev, O. F., M. J. Gladyshev, N. A. Pritvits, and V. G. Sudobicher, "Methods for the Calculation of Shock Waves in Open Channels," *Proc., IAHR Eleventh Int. Congress,* vol. 44, no. 3, Leningrad, U.S.S.R., 1965.

138. Verwey, A., and M. J. M. Haperen, "HD-System RUBICON—A User-Friendly Package for the Simulation of Unsteady Flow in Open Channel Networks," *Hydrosoft,* vol. 1, no. 1, pp. 3–12, 1988.

139. Viessman, W., Jr., J. W. Knapp, G. L. Lewis, and T. E. Harbaugh, *Introduction to Hydrology,* 2d ed., Intext Educational Publishers, New York, 1977.

140. Walton, R., and B. A. Christensen, "Friction Factors in Storm-Surges over Inland Areas," *J. Waterway, Port, Coastal and Ocean Div., ASCE,* vol. 106, WW2, pp. 261–271, 1980.

141. Wormleaton, P. R., and M. Karmegam, "Parameter Optimization in Flood Routing," *J. Hydraul. Eng.,* vol. 110, no. HY12, pp. 1789–1814, 1984.

142. Yang, C. T., "Unit Stream Power and Sediment Transport," *J. Hydraul, Eng.,* vol. 18, no. HY10, pp. 1805–1826, 1972.

143. Yapa, P. D., and H. T. Shen, "Unsteady Flow Simulation for an Ice-Covered River," *J. Hydraul. Eng.,* vol. 112, no. 11, pp. 1036–1049, 1986.

CHAPTER 11
WATER QUALITY

Steve C. McCutcheon
Environmental Research Laboratory
U.S. Environmental Protection Agency
Athens, Georgia

James L. Martin
AScI Corp.
Athens, Georgia

Thomas O. Barnwell, Jr.
Environmental Research Laboratory
U.S. Environmental Protection Agency
Athens, Georgia

In the early years of hydrology, water quality was relatively unimportant except in arid lands where salinization occurred. In arid regions, recognition of the problems of water quantity and quality dates back to the ancient civilizations in Egypt, Mesopotamia, and India. Since the beginning of the formal practice of hydrology in about 1930,[37] much has changed. Population growth has resulted in increased water demand and greater contamination of water from the disposal of wastes. In many areas, the use of water is limited by its quality rather than by the quantity available. Water quality limitations are defined by water use standards, and there are areas in the United States, western Europe, and Japan where these standards are violated. There are extensive areas in eastern Europe, the Soviet Commonwealth, and developing countries where reasonable water use criteria are exceeded. In addition, there are many other water bodies where water quality is impaired. This is especially true in the eastern United States, western Europe, Japan, South Korea, and a number of other regions. Furthermore, there are indications that rapidly developing, water-rich areas such as central eastern and southeastern China, and Brazil must soon begin to address water quality limitations on water use.

The availability of good, clean water is also a public health issue. Many human diseases, especially those causing cholera and other plagues that decimated the population of Europe during the Middle Ages, are serious problems. Regrettably, waterborne disease transmission remains a serious concern even in the United States where recent disease outbreaks have occurred. Poisoning by toxic chemicals and related chronic effects are the most recent public health concerns to arise over water quality. The chronic effects of man-made chemicals in water are not well known.

Water quality is important not only because of its linkage to the availability of water for various uses and its impact on public health, but also because water quality has an intrinsic value. The quality of life is often judged on the availability of pristine waters. Contamination of water deprives present and future generations of a birthright. There is also a need to preserve the aquatic habitats of fish, birds, and mammals. Preservation of habitats is rarely viewed in terms of simple economics but is often taken to be as important as ensuring the public health.

To assist the practicing hydrologist in planning for and adapting to limitations on the use of water and to aid in the protection of valuable water resources, this chapter covers the basic concepts of water chemistry, the physical properties of water, and its constituents or impurities. To aid in the interpretation of measurements, water quality standards and criteria for various uses are presented.

ACKNOWLEDGMENTS

Rather than cite the work of Hem,[28] other publications of the U.S. Geological Survey (e.g., Refs. 5, 23, 33, 39, 55, 61), publications of the EPA (e.g., Refs. 9, 10, 11, 12, 18, 27, 29, 31, 46, 51, 52, 57, 64, 70, 71, 72, 73, 74, 75, 76), *Standard Methods,*[15] Wunderlich,[81,82] and Stumm and Morgan[63] a number of times, we acknowledge these contributions to the understanding of water quality and refer the interested reader to these for more comprehensive reading. The reports by the U.S. Geological Survey are available through Books and Open File Reports Section, U.S. Geological Survey, Federal Center, Box 25425, Denver, Colo. Publications of the U.S. EPA and U.S. Geological Survey are available through the National Technical Information Service, 5285 Port Royal Road, Springfield, Va., 22161, USA, phone (703) 487-4650, and at repositories in the capitals of major countries across the globe.

11.1 PHYSICAL PROPERTIES AND BASIC CONCEPTS OF WATER CHEMISTRY

11.1.1 Physical Properties of Water

The temperature, density, specific weight, dissolved solids content, viscosity, surface tension, thermal capacity, enthalpy, vapor pressure, and heat of vaporization are important physical characteristics of natural waters. *Water temperature* normally ranges from 0 to 35°C but may be higher in warm springs, geysers, and ocean thermal vents. Many properties of water are related to water temperature, and these are listed in Table 11.1.1.

Density of Water. Water *density,* the mass per unit volume, is important in water quality investigations for several reasons. Water density ρ is the index used to determine fluid specific gravity (i.e., fluid specific gravity = density of the fluid/ρ). Density differences over the depth of water bodies lead to density stratification, which controls the flow of water and constituents through the water body. The density of water also controls buoyancy and the ease with which biota can remain in different aquatic habitats.

Density has SI units of kg m^{-3} and English customary units of slugs ft^{-3} or lb_m ft^{-3}. Pure water density has a maximum of approximately 1000 kg m^{-3} at 4°C and decreases to 994 kg m^{-3} at 35°C. Density decreases as temperature increases except between 0 and 4°C (see Fig. 11.1.1) where density increases with increasing temperature. At 0°C in pure water, the density is 999.87 kg m^{-3}. Density is best calculated as a function of temperature using the equation shown in Fig. 11.1.1. Since water is practically incompressible, pressure effects on density are normally ignored.

Specific Weight. *Specific weight* of water γ is the gravitational attractive force acting on a unit volume that derives from the density and acceleration of gravity ($\gamma = \rho g$). In SI units, specific weight is expressed as N m^{-3}. At normal water temperatures, the specific weight is 9789 N m^{-3} (at 20°C). In English customary units, specific weight is expressed as lb_f ft^{-3} and has the typical value of 62.4 lb_f ft^{-3} (at 15°C).

Dissolved Solids. Water density and other properties are also controlled by dissolved solids content. Dissolved solids are impurities that occur in all natural waters because of the weathering of rocks and soils, and also because of the solution of carbon dioxide from the atmosphere. As the dissolved solids content increases, water density increases. The density of water containing dissolved solids ρ_s is[82]

$$\rho_s = \frac{m_s + m_w}{V_s + V_w} = \rho_o \frac{1 + c}{1 + \dfrac{V_s}{V_w}} \tag{11.1.1}$$

where the mass ratio of constituents mixed together $c = \rho_{ds} V_s / \rho_o V_w$, ρ_{ds} = density of solids (approximately 2150 kg m^{-3} for natural salts), ρ_o = density of water normally determined by the equation given in Fig. 11.1.1, V_s = volume of solids, V_w = volume of water, $m_s = \rho_s V_s$ = mass of solids, $m_w = \rho_o V_w$ = mass of water.

Dissolved solids concentrations are expressed using at least three different units: (1) concentration as mg L^{-1}, (2) salinity in parts per thousand (‰ or g kg^{-1}), or (3) density in kg m^{-3}. Concentrations in mg per L of pure water are approximately the same as salinity × 1000 or concentration in units of parts per million (ppm) until concentrations exceed 7000 mg L^{-1}. Salinity is the mass of salt in grams per kg of seawater (mass of salt and water). Equation (11.1.1) can be used to relate pure water density, dissolved solids (salt) density, and concentration or salinity. In practice, however, empirical relationships between seawater density and salinity S are used. There are a number of approximate formulas, but the most useful is

$$\rho_s \,(\text{kg m}^{-3}) = \rho_o + AS + BS^{3/2} + CS^2 \tag{11.1.2}$$

where $A = 8.24493 \times 10^{-1} - 4.0899 \times 10^{-3}\, T + 7.6438 \times 10^{-5}\, T^2 - 8.2467 \times 10^{-7}\, T^3 + 5.3675 \times 10^{-9}\, T^4$
$B = -5.724 \times 10^{-3} + 1.0227 \times 10^{-4}\, T - 1.6546 \times 10^{-6}\, T^2$
$C = 4.8314 \times 10^{-4}$

T is in °C, S is in g kg^{-1}, and ρ_o is calculated from the equation in Fig. 11.1.1. Generally, density is computed to four or five significant figures.

Viscosity. *Viscosity* is an important water property controlling resistance to motion. Such motions include flow in open channels and in aquifers, flow in response to wind

TABLE 11.1.1 Physical Properties of Water

Temperature		Density		Specific weight		Absolute viscosity		Kinematic viscosity	
°F	°C	kg m^{-3}	slugs ft^{-3}	N m^{-3}	lb$_f$ ft^{-3}	kg m^{-1} s-1 *	slugs ft^{-1} s^{-1} †	m^2 s^{-1} ‡	ft^2 s^{-1}
32.0	0	999.87	1.940	9805.4	62.419	0.001787	3.732E-05	1.787E-06	1.924E-05
33.8	1	999.93	1.940	9805.9	62.423	0.001728	3.608E-05	1.728E-06	1.860E-05
35.6	2	999.97	1.940	9806.3	62.426	0.001671	3.491E-05	1.671E-06	1.799E-05
37.4	3	999.99	1.940	9806.6	62.427	0.001618	3.379E-05	1.618E-06	1.741E-05
39.2	4	1000.00	1.940	9806.6	62.428	0.001567	3.272E-05	1.567E-06	1.687E-05
41.0	5	999.99	1.940	9806.6	62.427	0.001518	3.171E-05	1.518E-06	1.634E-05
42.8	6	999.97	1.940	9806.3	62.426	0.001472	3.074E-05	1.472E-06	1.585E-05
44.6	7	999.93	1.940	9806.0	62.423	0.001428	2.982E-05	1.428E-06	1.537E-05
46.4	8	999.88	1.940	9805.4	62.420	0.001386	2.894E-05	1.386E-06	1.492E-05
48.2	9	999.81	1.940	9804.8	62.416	0.001346	2.810E-05	1.346E-06	1.449E-05
50.0	10	999.73	1.940	9804.0	62.411	0.001307	2.730E-05	1.308E-06	1.407E-05
51.8	11	999.63	1.940	9803.1	62.405	0.001270	2.653E-05	1.271E-06	1.368E-05
53.6	12	999.53	1.939	9802.0	62.398	0.001235	2.580E-05	1.236E-06	1.330E-05
55.4	13	999.41	1.939	9800.8	62.391	0.001202	2.510E-05	1.202E-06	1.294E-05
57.2	14	999.27	1.939	9799.5	62.382	0.001169	2.442E-05	1.170E-06	1.260E-05
59.0	15	999.13	1.939	9798.1	62.373	0.001139	2.378E-05	1.140E-06	1.227E-05
60.8	16	998.97	1.938	9796.6	62.363	0.001109	2.316E-05	1.110E-06	1.195E-05
62.6	17	998.80	1.938	9794.9	62.353	0.001081	2.257E-05	1.082E-06	1.164E-05
64.4	18	998.62	1.938	9793.2	62.342	0.001053	2.200E-05	1.055E-06	1.135E-05
66.2	19	998.43	1.937	9791.3	62.330	0.001027	2.145E-05	1.029E-06	1.107E-05
68.0	20	998.23	1.937	9789.3	62.317	0.001002	2.093E-05	1.004E-06	1.080E-05
69.8	21	998.02	1.936	9787.3	62.304	0.0009780	2.043E-05	9.799E-07	1.055E-05
71.6	22	997.80	1.936	9785.1	62.290	0.0009548	1.994E-05	9.570E-07	1.030E-05
73.4	23	997.57	1.936	9782.8	62.276	0.0009326	1.948E-05	9.349E-07	1.006E-05
75.2	24	997.33	1.935	9780.4	62.261	0.0009111	1.903E-05	9.136E-07	9.834E-06
77.0	25	997.08	1.935	9778.0	62.245	0.0008905	1.860E-05	8.931E-07	9.613E-06
78.8	26	996.81	1.934	9775.4	62.229	0.0008705	1.818E-05	8.733E-07	9.400E-06
80.6	27	996.54	1.934	9772.8	62.212	0.0008513	1.778E-05	8.543E-07	9.195E-06
82.4	28	996.26	1.933	9770.0	62.194	0.0008328	1.739E-05	8.359E-07	8.998E-06
84.2	29	995.98	1.933	9767.2	62.176	0.0008149	1.702E-05	8.182E-07	8.807E-06
86.0	30	995.68	1.932	9764.3	62.158	0.0007976	1.666E-05	8.011E-07	8.622E-06
87.8	31	995.37	1.931	9761.3	62.139	0.0007809	1.631E-05	7.845E-07	8.444E-06
89.6	32	995.06	1.931	9758.2	62.119	0.0007647	1.597E-05	7.686E-07	8.273E-06
91.4	33	994.73	1.930	9755.0	62.099	0.0007491	1.565E-05	7.531E-07	8.106E-06
93.2	34	994.40	1.929	9751.8	62.078	0.0007340	1.533E-05	7.382E-07	7.945E-06
95.0	35	994.06	1.929	9748.4	62.057	0.0007194	1.503E-05	7.237E-07	7.790E-06

* 1 kg m^{-1} s^{-1} = 10 poise = 1000 centipoise = 10 g cm^{-1} s^{-1} = 10 dynes s cm^{-2}. See Chap. 1 and Weast[78] for other conversions.

† 1 slug ft^{-1} s^{-1} = 1 lb$_f$ s ft^{-2}.

‡ 1 m^2 s^{-1} = 10,000 stokes = 10,000 cm^2 s^{-1} = 100,000 centistokes.

Notes: 1. Density computed from the Thiesen-Scheel-Diesselhorst equation.[66]

2. Specific weight = density × 9.806650 m s^{-2}.

3. Dynamic or absolute viscosity from U.S. National Bureau of Standards (Hardy and Cottington,[26] Swidells, unpublished data, and Weast[78]).

Surface tension		Thermal capacity,	Enthalpy	Heat of vaporization		Saturation vapor pressure		Saturation vapor pressure head	
N m^{-1}	lb$_f$ ft^{-1}	J g^{-1} °C^{-1}	J g^{-1}	J kg^{-1}	Btu lb^{-1}	Pa	psia	m	ft
0.076	0.00518	4.2177	0.1026	2.501E+06	1075.1	611	0.0886	0.062	0.204
		4.2141	4.3184	2.499E+06	1074.2	657	0.0952	0.067	0.220
		4.2107	8.5308	2.496E+06	1073.2	705	0.1023	0.072	0.236
		4.2077	12.7400	2.494E+06	1072.1	758	0.1099	0.077	0.253
		4.2048	16.9462	2.492E+06	1071.1	813	0.1179	0.083	0.272
0.075	0.00513	4.2022	21.1408	2.489E+06	1070.1	872	0.1265	0.089	0.292
		4.1999	25.5496	2.487E+06	1069.1	935	0.1356	0.095	0.313
		4.1977	29.5496	2.484E+06	1068.1	1001	0.1452	0.102	0.335
		4.1957	33.7463	2.482E+06	1067.1	1072	0.1555	0.109	0.359
		4.1939	37.9410	2.480E+06	1066.1	1147	0.1664	0.117	0.384
0.074	0.00509	4.1922	42.1341	2.477E+06	1065.0	1227	0.1780	0.125	0.411
		4.1907	46.3255	2.475E+06	1064.0	1312	0.1903	0.134	0.439
		4.1893	50.7041	2.473E+06	1063.0	1402	0.2033	0.143	0.469
		4.1880	54.7041	2.470E+06	1062.0	1497	0.2171	0.153	0.501
		4.1869	58.8916	2.468E+06	1061.0	1598	0.2317	0.163	0.535
0.073	0.00504	4.1858	63.0779	2.466E+06	1060.0	1704	0.2472	0.174	0.571
		4.1849	67.2632	2.463E+06	1058.9	1817	0.2636	0.186	0.609
		4.1840	71.4476	2.461E+06	1057.9	1937	0.2809	0.198	0.649
		4.1832	75.6312	2.459E+06	1056.9	2063	0.2992	0.211	0.691
		4.1825	79.8141	2.456E+06	1055.9	2196	0.3186	0.224	0.736
0.073	0.00498	4.1819	83.9963	2.454E+06	1054.9	2337	0.3390	0.239	0.783
		4.1813	88.1778	2.451E+06	1053.9	2486	0.3606	0.254	0.833
		4.1808	92.3589	2.449E+06	1052.9	2643	0.3833	0.270	0.886
		4.1804	96.5395	2.447E+06	1051.8	2809	0.4074	0.287	0.942
		4.1800	100.7196	2.444E+06	1050.8	2983	0.4327	0.305	1.001
0.072	0.00493	4.1796	104.8994	2.442E+06	1049.8	3167	0.4593	0.324	1.063
		4.1793	109.0788	2.440E+06	1048.8	3361	0.4874	0.344	1.128
		4.1790	113.2580	2.437E+06	1047.8	3565	0.5170	0.365	1.197
		4.1788	117.4369	2.435E+06	1046.8	3780	0.5482	0.387	1.269
		4.1786	121.6157	2.433E+06	1045.8	4006	0.5809	0.410	1.345
0.071	0.00488	4.1785	125.7943	2.430E+06	1044.7	4243	0.6154	0.435	1.426
		4.1784	129.9727	2.428E+06	1043.7	4493	0.6516	0.460	1.510
		4.1783	134.1510	2.425E+06	1042.7	4755	0.6897	0.487	1.599
		4.1783	138.3293	2.423E+06	1041.7	5031	0.7296	0.516	1.692
		4.1782	142.5078	2.421E+06	1040.7	5320	0.7716	0.546	1.790
		4.1782	146.6858	2.418E+06	1039.7	5624	0.8156	0.577	1.893

4. Kinematic viscosity = dynamic viscosity divided by density.
5. Surface tension values from Weast.[78]
6. Thermal capacity at 1 atm in absolute J from Weast.[78] Note that 1 J = 0.23885 cal = 9.4782×10^{-4} Btu.
7. Enthalpy values from Weast.[78]
8. Heat of vaporization values from $H_v = 2.501 \times 10^6 - 2361\ T$.
9. Vapor pressure from the Goff-Gratch formula used to compute values in the Smithsonian Meteorological Tables.[38] See Chap. 1 for conversion factors to other typical pressure units. The Goff-Gratch formula is different from the Smithsonian tables by 0.5 percent.

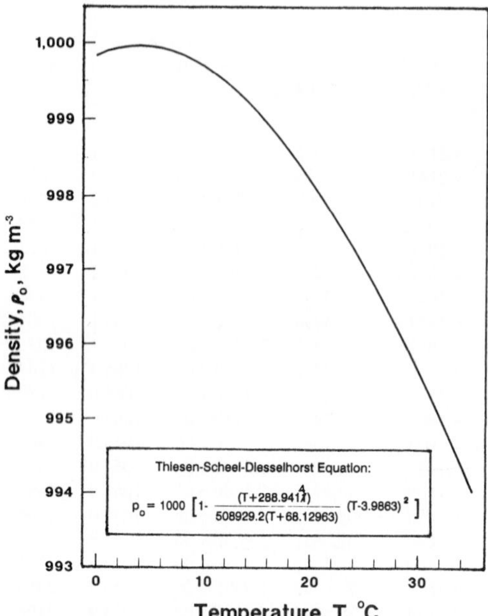

FIGURE 11.1.1 Pure water density as a function of temperature.

shear, flow driven by density differences, and displacement of water by the settling of particles. The *dynamic* or *absolute viscosity* μ is a measure of the water resistance to shear stress. For a Newtonian fluid like water, viscosity is the constant of proportionality relating shear stress τ and the strain rate or velocity gradient du/dy in Newton's law of viscosity

$$\tau = \mu \frac{du}{dy} \tag{11.1.3}$$

where u = horizontal velocity and y = vertical coordinate direction. Dynamic viscosity and density can be combined into one physical parameter referred to as the *kinematic viscosity* $\nu = \mu/\rho$.

In SI units, dynamic or absolute viscosity is expressed in the units $\mathrm{N \cdot s\ m^{-2}}$. The English common unit is $\mathrm{lb_f\ s\ ft^{-2}}$ or slugs $\mathrm{ft^{-1} \cdot s^{-1}}$. Note that $1\ \mathrm{lb_f \cdot s\ ft^{-2}} = 1$ slug $\mathrm{ft^{-1}\ s^{-1}}$. In SI units, kinematic viscosity is expressed in $\mathrm{m^2\ s^{-1}}$. The English unit is $\mathrm{ft^2\ s^{-1}}$.

Both absolute and kinematic viscosity decrease as molecular motion increases with increasing temperature as shown in Table 11.1.1. Dynamic viscosity values in Table 11.1.1 are derived from the empirical expressions of Hardy and Cottington[26] and Swidells (in Weast[78]) that are expressed in $\mathrm{N\ s\ m^{-2}}$ as

$$\log_{10}\left(\frac{\mu}{100}\right) = \frac{1301}{998.333 + 8.1855\,(T-20) + 0.00585\,(T-20)^2} - 1.30233$$

$$\text{for } T = 0 \text{ to } 20°\text{C} \tag{11.1.4}$$

$$\log_{10} \frac{\mu}{\mu_{20}} = \frac{1.3272\,(20 - T) - 0.001053\,(T - 20)^2}{T + 105} \qquad \text{for } T = 20 \text{ to } 100°\text{C} \quad (11.1.5)$$

where μ_{20} is the dynamic viscosity at 20°C = 0.001002 N s m^{-2}. Some compilations of viscosity values[4,56,62] are based on Bingham's formula,[6] which is out of date.

Surface Tension. *Surface tension* at the interface of water and air, or at the interface of any two immiscible fluids, results from the fluid forces holding individual molecules at the surface. At the air-water interface, surface tension exists because mild polarity allows water molecules to be orientated and weakly held in a surface layer. The obvious effects of the surface tension between water and air are seen in the convex surface of water in a glass extending above the rim without overflowing, in the ability of needles to float on a smooth water surface, and in the capability of insects to walk over a quiescent water surface. The surface tension force acts perpendicular to the free surface along a line such as at the edge of the meniscus on a glass tube. The force is proportional to the length of the line. The SI units for surface tension are N m^{-1}, while lb_f ft^{-1} are the customary English units. At 20°C, the surface tension of water is 0.073 N m^{-1} or 0.005 lb_f ft^{-1}. There is only a weak relationship between surface tension and temperature, as noted in Table 11.1.1.

Vapor Pressure. Both saturation vapor pressure and ambient vapor pressure must be known to accurately simulate evaporation and water temperature. For some dissolved gases, mass transfer between air and water can be related to water vapor exchange.

The *vapor pressure* of water vapor in air results from the kinetic energy of water molecules causing some molecules to break free of the surface and enter the air. Water molecules evaporate into the air until the air becomes saturated. At equilibrium, when the *saturation vapor pressure* is attained in the air above water, the kinetic exchange of molecules between water and air, and between air and water, is in balance. Perturbations from the original equilibrium caused by temperature changes in the air or water result in an increased flux from one medium to the other until equilibrium is again achieved. Vapor pressure increases as increased temperature forces more water molecules into the adjacent air. The change in saturated vapor pressure with water temperature is given in Table 11.1.1.

The SI units for vapor pressures are Pa or kN m^{-2}. The customary English units are lb_f ft^{-2}. Conversions involving inches or mm of mercury, ft of water, millibars, and other units are frequently a source of computational error that must be carefully checked (see Appendix A for conversion factors).

A change in water temperature, and some influence of air pressure, are the only effects known to cause vapor pressure and saturation vapor pressure to change. In some vapor pressure equations, the small effect of barometric pressure changes is ignored. A number of expressions are available to calculate saturation vapor pressure and vapor pressure but these have not been thoroughly investigated to determine which are the most practical.[44] Another drawback is that the various empirical formulations have not been converted to a consistent set of units. Typically a mixture of SI, cgs (cm g s), English, and other units is in use.

Table 11.1.1 gives saturated water vapor pressures for various temperatures that are equivalent to standard values compiled by List.[38] Values in Table 11.1.1 and the standard values are derived from the Goff-Gratch formulation. A simpler form is [37]

$$e_o = 3.38639\,[(0.00738T_s + 0.8072)^8 - 0.000019\,|1.8T_s + 48| + 0.001316] \quad (11.1.6)$$

where e_o = saturation vapor pressure in kPa, and T_s = water surface temperature in °C. These values are accurate to within 1 percent over the range of temperatures of 0 to 35°C (32 to 95°F).

Some investigators prefer the Magnus-Tetens equation that is expressed as

$$e_o\,(\text{Pa}) = 10^{7.5T_s/(T_s+237.3)+2.7858} \quad (11.1.7)$$

Further simplifications have been developed for water temperature simulations[9,10] and in reservoir studies,[64] but these equations do not seem to offer any advantage at present.

The Magnus-Tetens equation with appropriate coefficients also accurately describes saturated water vapor pressure over ice when expressed as

$$e_o\,(\text{Pa}) = 10^{9.5T_s/(T_s+265.5)+2.7858} \quad (11.1.8)$$

Ambient water vapor pressure is frequently computed as

$$e = e_o - 0.000661\,P_a(T_a - T_{wb})\,(1 + 0.00115T_{wb}) \quad (11.1.9)$$

where e and e_o have the units of kPa, T_a is the dry-bulb temperature of the air in °C, T_{wb} is the wet-bulb temperature in °C, and P_a is the barometric pressure in kPa. Saturation vapor pressure e_o is normally calculated from Eqs. (11.1.6) or (11.1.7).

Heat Energy. *Heat* quantities are normally measured in J (N m, SI unit), cal (cgs), or British thermal units (Btu, English system). *Heat capacity* is the quantity of heat energy required to increase the water temperature one degree. In SI units, 4186.8 J kg^{-1} $°C^{-1}$ is heat capacity of water while 1 cal g^{-1} $°C^{-1}$ is the heat capacity in the cgs system that is frequently used. The *calorie* is the heat required to raise the temperature of 1 g of water from 3.5 to 4.5°C (called a small calorie). The *normal calorie* is the heat required to raise the temperature of 1 g of water from 14.5 to 15.5°C. The *mean calorie* is one-hundredth of the amount of heat required to raise 1 g of water from 0 to 100°C. In English units, 4.860×10^{-6} Btu lb_m^{-1} $°F^{-1}$ is the heat capacity of water. The Btu is the heat required to raise the temperature of 1 lb of water from 60°F (maximum density) to 61°F.

Changes in the heat energy of water ΔH are related to water temperature changes ΔT, volume V, density ρ, and heat capacity c as $\Delta H = c\rho V\Delta T$. The heat flux is the amount of heat passing through a unit area. The most important heat fluxes in hydrology are the fluxes of solar and long-wave radiation through the water surface. McCutcheon[44] and Wunderlich[81] define the individual components of the water surface heat flux. Four important units are used to express the heat flux. The SI units are W m^{-2} (J s^{-1} m^{-2} or N s^{-1} m^{-1}). The English units are Btu h^{-1} ft^{-2} or Btu day^{-1} ft^{-2}. The cgs system units are kcal m^{-2} h^{-1} or langleys per day (1 langley = 1 cal cm^{-2}).

Heat of vaporization or evaporation (also latent heat of vaporization) is the amount of heat required to evaporate or condense a unit mass of water. The heat of vaporization can be accurately expressed over a range of 0 to 40°C (32 to 104°F) as $H_v = 2.501 \times 10^6 - 2361T$, where H_v is in units of J kg^{-1} and T, the water temperature, is in °C. Values are tabulated in Table 11.1.1. *Latent heat of fusion* is the amount of heat required to convert 1 g of ice to water. The value is 0.3337 MJ kg^{-1} or

79.7 cal (15°C) g^{-1} or approximately 1/7 of the latent heat of vaporization. The same amount of heat is released when 1 kg of water is converted to ice at a temperature of 0°C (32°F). *Latent heat of sublimation* is the amount of heat required to change 1 g of ice into water vapor or vice versa. At 0°C (32°F) the latent heat of sublimation is approximately 2.83 MJ kg^{-1}.

11.1.2 Ionization, Anions, Cations, and Dissolved Solids

Ionization and Speciation. *Ionization* has a significant impact on the behavior of a chemical in water and may be very important in mixing zones or where pH changes rapidly. The formation of free metal ions is especially important to toxicity to organisms. Ionization is also important for organic chemicals, especially when volatilization of the neutral species determines ultimate fate.[51] Volatilization is the exchange of contaminant between water and the atmosphere.[44,65,67] Ionization is important in the determination of ammonia toxicity where $NH_3 \rightleftarrows NH_4^- + H^+$.[50,75] Furthermore, ionization may be important when positively charged organic compounds are involved in cation exchange with solids.[47]

Solids that dissolve in water can *ionize* or react in water to form anions and cations by the separation of the original molecule into components. *Ionization* may also involve reaction with other ions, including the ionized components of the water molecule, $H_2O \rightleftarrows H^+ + OH^-$. *Anions* are negatively charged (e.g., OH^-) and *cations* are positively charged (e.g., H^+). The dissolved molecule (also known as the neutral species or the electrically neutral species) is normally in equilibrium with its cations and anions (i.e., the ionization reaction normally occurs very rapidly) as well as those of other neutral species so that a charge balance is maintained. The term *speciation* describes the relation between all ions and neutral species dissolved in water. One of the simplest ionization reactions in water (besides that of water itself) is that of common table salt or sodium chloride, where $NaCl \rightleftarrows Na^+ + Cl^-$. As in all such reactions, the positively charged sodium cation and negatively charged chloride anion species must be balanced so that the same number of negative charges and positive charges are present to maintain electrochemical neutrality in the solution. At least a trace amount of NaCl is present at all times, as ions combine and divide continuously. If another salt, such as potassium chloride, is added to the simple sodium chloride solution, the additional chloride ions achieve an equilibrium with both NaCl and KCl, which are in turn in equilibrium with Na^+ and K^+. The *species* in this more complex solution are NaCl, Na^+, Cl^-, KCl, and K^+.

Units of Concentration. Concentrations of dissolved solids are normally reported in mg L^{-1}, equivalent to g m^{-3} (instead of kg m^{-3}, which is the SI unit), or μg L^{-1} for trace amounts of metals and organic chemicals. Units of moles L^{-1} and millimoles L^{-1} are used by chemists, chemical engineers, and oceanographers. The conversion to mg L^{-1} is obtained by multiplying moles of dissolved material (mole L^{-1}) by the atomic or molecular weight in mg $mole^{-1}$. The atomic and molecular weights should be based on recent International Standard Tables referenced to carbon-12 (Fig. 11.1.2). Concentrations of milliequivalents (meq) L^{-1} are frequently reported and used in laboratory analyses as well. The *equivalent weight,* or combining weight, is the molecular weight of the ion dissolved in water divided by the charge of the ion (e.g., +1 for Na^+, +2 for Ca^{2+}, and −1 for Cl^{-1}). It is good practice to compute the meq of anions and meq of cations to be sure these are always equal. Every solution must be electrochemically neutral. Concentrations in meq L^{-1} are converted to mg

PERIODIC TABLE OF THE ELEMENTS

New notation / Previous IUPAC form / CAS version

KEY TO CHART

Atomic Number →	50 +2	← Oxidation States
Symbol →	Sn +4	
1983 Atomic Weight →	118.71	
	18 18 4	← Electron Configuration

1 Group IA	2 IIA	3 IIIA IIIB	4 IVA IVB	5 VA VB	6 VIA VIB	7 VIIA VIIB	8	9 VIIIA VIII	10	11 IB	12 IIB	13 IIIB IIIA	14 IVB IVA	15 VB VA	16 VIB VIA	17 VIIB VIIA	18 VIIIA	Orbit
1 H +1 −1 1.00794 1																	2 He 0 4.00260 2	K
3 Li +1 6.941 2-1	4 Be +2 9.01218 2-2											5 B +3 10.81 2-3	6 C +2 +4 −4 12.011 2-4	7 N +1 +2 +3 +4 +5 −1 −2 −3 14.0067 2-5	8 O −2 15.9994 2-6	9 F −1 18.9984 2-7	10 Ne 0 20.179 2-8	K-L
11 Na +1 22.9898 2-8-1	12 Mg +2 24.305 2-8-2											13 Al +3 26.9815 2-8-3	14 Si +2 +4 −4 28.0855 2-8-4	15 P +3 +5 −3 30.97381 2-8-5	16 S +4 +6 −2 32.06 2-8-6	17 Cl +1 +5 +7 −1 35.453 2-8-7	18 Ar 0 39.948 2-8-8	K-L-M
19 K +1 39.0983 -8-8-1	20 Ca +2 40.08 -8-8-2	21 Sc +3 44.9559 -8-9-2	22 Ti +2 +3 +4 47.88 -8-10-2	23 V +2 +3 +4 +5 50.9415 -8-11-2	24 Cr +2 +3 +6 51.996 -8-13-1	25 Mn +2 +3 +4 +7 54.9380 -8-13-2	26 Fe +2 +3 55.847 -8-14-2	27 Co +2 +3 58.9332 -8-15-2	28 Ni +2 +3 58.69 -8-16-2	29 Cu +1 +2 63.546 -8-18-1	30 Zn +2 65.39 -8-18-2	31 Ga +3 69.72 -8-18-3	32 Ge +2 +4 72.59 -8-18-4	33 As +3 +5 −3 74.9216 -8-18-5	34 Se +4 +6 −2 78.96 -8-18-6	35 Br +1 +5 −1 79.904 -8-18-7	36 Kr 0 83.80 -8-18-8	-L-M-N
37 Rb +1 85.4678 -18-8-1	38 Sr +2 87.62 -18-8-2	39 Y +3 88.9059 -18-9-2	40 Zr +4 91.224 -18-10-2	41 Nb +3 +5 92.9064 -18-12-1	42 Mo +6 95.94 -18-13-1	43 Tc +4 +6 +7 (98) -18-13-2	44 Ru +3 101.07 -18-15-1	45 Rh +3 102.906 -18-16-1	46 Pd +2 +4 106.42 -18-18-0	47 Ag +1 107.868 -18-18-1	48 Cd +2 112.41 -18-18-2	49 In +3 114.82 -18-18-3	50 Sn +2 +4 118.71 -18-18-4	51 Sb +3 +5 −3 121.75 -18-18-5	52 Te +4 +6 −2 127.60 -18-18-6	53 I +1 +5 +7 −1 126.905 -18-18-7	54 Xe 0 131.29 -18-18-8	-M-N-O
55 Cs +1 132.905 -18-8-1	56 Ba +2 137.33 -18-8-2	57* La +3 138.906 -18-9-2	72 Hf +4 178.49 -32-10-2	73 Ta +5 180.948 -32-11-2	74 W +6 183.85 -32-12-2	75 Re +4 +6 +7 186.207 -32-13-2	76 Os +3 +4 190.2 -32-14-2	77 Ir +3 +4 192.22 -32-15-2	78 Pt +2 +4 195.08 -32-16-2	79 Au +1 +3 196.967 -32-18-1	80 Hg +1 +2 200.59 -32-18-2	81 Tl +1 +3 204.383 -32-18-3	82 Pb +2 +4 207.2 -32-18-4	83 Bi +3 +5 208.980 -32-18-5	84 Po +2 +4 (209) -32-18-6	85 At (210) -32-18-7	86 Rn 0 (222) -32-18-8	-N-O-P
87 Fr +1 (223) -18-8-1	88 Ra +2 226.025 -18-8-2	89** Ac +3 227.028 -18-9-2	104 Unq +4 (261) -32-10-2	105 Unp (262) -32-11-2	106 Unh (263) -32-12-2	107 Uns (262) -32-13-2												O P Q

																Orbit
*Lanthanides	58 Ce +3 +4 140.12 -20-8-2	59 Pr +3 140.908 -21-8-2	60 Nd +3 144.24 -22-8-2	61 Pm +3 (145) -23-8-2	62 Sm +2 +3 150.36 -24-8-2	63 Eu +2 +3 151.96 -25-8-2	64 Gd +3 157.25 -25-9-2	65 Tb +3 158.925 -27-8-2	66 Dy +3 162.50 -28-8-2	67 Ho +3 164.930 -29-8-2	68 Er +3 167.26 -30-8-2	69 Tm +3 168.934 -31-8-2	70 Yb +2 +3 173.04 -32-8-2	71 Lu +3 174.967 -32-9-2		N O P
**Actinides	90 Th +4 232.038 -18-10-2	91 Pa +5 +4 231.036 -20-9-2	92 U +3 +4 +5 +6 238.029 -21-9-2	93 Np +3 +4 +5 +6 237.048 -22-9-2	94 Pu +3 +4 +5 +6 (244) -24-8-2	95 Am +3 +4 +5 +6 (243) -25-8-2	96 Cm +3 (247) -25-9-2	97 Bk +3 +4 (247) -27-8-2	98 Cf +3 (251) -28-8-2	99 Es +3 (252) -29-8-2	100 Fm +3 (257) -30-8-2	101 Md +2 +3 (258) -31-8-2	102 No +2 +3 (259) -32-8-2	103 Lr +3 (260) -32-9-2		O P Q

Numbers in parentheses are mass numbers of most stable isotope of that element

From *Chemical and Engineering News*, 63(5), 27, 1985. This format numbers the groups 1 to 18.

FIGURE 11.1.2 Periodic chart of elements. (*From* CRC Handbook of Physics and Chemistry, *67th ed., 1986. Used with permission.*)

L^{-1} by multiplying by the equivalent weight. Table 11.1.2 gives the reciprocals of the equivalent weights or atomic weights used to convert from mg L^{-1} to meq L^{-1} or millimoles L^{-1} for important ions found in natural water. For example, a concentration of 1 mg L^{-1} of Al^{3+} corresponds to 0.1119 meq L^{-1} or 0.03715 millimoles L^{-1} using the conversion factors listed in Table 11.1.2.

Occasionally, loads of dissolved solids carried by moving water are reported from water quality and discharge data. Loads are simply concentrations times the discharge (kg $m^{-3} \times m^3 s^{-1}$ = kg s^{-1} or kg day^{-1} or lb day^{-1}).

Anions, Cations, and Neutral Species in Water. Natural waters contain a complex mixture of cations and anions that include the primary cations calcium (Ca^{2+}), magnesium (Mg^{2+}), sodium (Na^+), and potassium (K^+); and the primary anions chloride (Cl^-), sulfate (SO_4^{2-}), carbonate (CO_3^{2-}), bicarbonate (HCO_3^-), fluoride (F^-), and nitrate (NO_3^-). These cations and anions, along with silicon (usually present in a nonionic form), are often present in water in concentrations in excess of 1 mg L^{-1}. Minor constituents that can occasionally achieve concentrations of 1 mg L^{-1} or higher include aluminum (Al^{3+}), boron (B), iron (Fe^{2+}, Fe^{3+}), manganese (Mn^{2+}), phosphate (PO_4^{3-}), organic carbon, ammonia (NH_3), nitrite (NO_2^-), organic nitrogen, and dissolved gases [carbon dioxide (CO_2), dissolved oxygen (O_2),

TABLE 11.1.2 Conversion Factors for Constituent Concentration

Element and reported species	F_1	F_2	Element and reported species	F_1	F_2
Aluminum (Al^{3+})	0.11119	0.03715	Magnesium (Mg^{2+})	0.08229	0.04114
Ammonium (NH_4^+)	0.05544	0.05544	Manganese (Mn^{2+})	0.03640	0.01820
Antimony (Sb)		0.00821	Mercury (Hg)		0.00499
Arsenic (As)		0.01334	Molybdenum (Mo)		0.01042
Barium (Ba^{2+})	0.01456	0.00728	Nickel (Ni)		0.01704
Beryllium (Be^{2+})	0.22192	0.11096	Nitrate (NO_3)	0.01613	0.01613
Bicarbonate (HCO_3)	0.01639	0.01639	Nitrite (NO_2)	0.02174	0.02174
Boron (B)		0.09250	Phosphate (PO_4^3)	0.03159	0.01053
Bromide (Br)	0.01252	0.01252	Phosphate (HPO_1^2)	0.02084	0.01042
Cadmium (Cd^{2+})	0.01779	0.00890	Phosphate (H_2PO_4)	0.01031	0.01031
Calcium (Ca^{2+})	0.04990	0.02495	Potassium (K^+)	0.02558	0.02558
Carbonate (CO_3^2)	0.03333	0.01666	Rubidium (Rb^+)	0.01170	0.01170
Cesium (Cs^+)	0.00752	0.00752	Selenium (Se)		0.01266
Chloride (Cl)	0.02821	0.02821	Silica (SiO_2)		0.01664
Chromium (Cr)		0.01923	Silver (Ag^+)	0.00927	0.00927
Cobalt (Co^{2+})	0.03394	0.01697	Sodium (Na^+)	0.04350	0.04350
Copper (Cu^{2+})	0.03147	0.01574	Strontium (Sr^{2+})	0.02283	0.01141
Fluoride (F^+)	0.05264	0.05264	Sulfate (SO_4^2)	0.02082	0.01041
Hydrogen (H^+)	0.99216	0.99216	Sulfide (S^2)	0.06238	0.03119
Hydroxide (OH^+)	0.05880	0.05880	Thorium (Th)		0.00431
Iodide (I^+)	0.00788	0.00788	Titanium (Ti)		0.02088
Iron (Fe^{2+})	0.03581	0.01791	Uranium (u)		0.00420
Iron (Fe^{3+})	0.05372	0.01791	Vanadium (V)		0.01963
Lead (Pb^{2+})	0.00965	0.00483	Zinc (Zn^{2+})	0.03059	0.01530
Lithium (Li^+)	0.14407	0.14407			

Based on 1975 atomic weights, referred to carbon-12. From Hem.[28]
Milligrams per liter × F_1 = milliequivalents per liter; milligrams per liter × F_2 = millimoles per liter.

and hydrogen sulfide (H_2S)]. Hydrogen ions (H^+) are normally present in smaller concentrations but are important in determining water quality and understanding the chemistry of a specific water.

The exact mixture of the anions and cations in water is controlled by the *dissolution* (dissolving of minerals into water), *precipitation* (formation of solid minerals from supersaturated dissolved materials in water), and *ionization* of the constituents present in a water. In addition, minerals and contaminants *sorb* to solids or are released into the water from solids. *Sorption* is the attachment of a portion of the dissolved phase of a substance to a solid by electrochemical or thermodynamic processes such as cation exchange or by absorption.

The form and concentration of dissolved substances present in a water depend on its history of contact with other geochemicals in the atmosphere, on the ground surface, within the soil profile, in groundwater aquifers, and in surface waters. Although concentrations can vary over a wide range, Tables 11.1.3 and 11.1.4 list typical concentrations found in surface and groundwaters. For comparison Table 11.1.5 lists concentrations in seawater. While the ion concentrations in oceans are approximately constant, concentrations can be higher in some isolated and semi-isolated seas and lakes and concentrations are variable in estuaries and coastal areas where freshwater mixing occurs. Groundwater concentrations are approximately the same over large areas because of the slow movement of water in aquifers but can be quite variable from one aquifer to another, or highly variable in aquifers affected by waste plumes. Surface water concentrations change as the water runs off and moves downstream. Contact time with soils and rocks from which material is dissolved is an important determinant of the dissolved concentration and the mixture of ions present.

Evaporation from surface waters is also important in controlling the concentration of dissolved solids. Rainfall has only limited concentrations of impurities, and runoff that quickly reaches streams is low in dissolved solids. However, when water remains in contact with soils longer or percolates into the ground, dissolved solids increase. Dissolved solids in streams increase downstream owing to evaporation and the inflow of groundwater. Sewage, industrial effluents, and urban runoff cause further increases in dissolved solids. As a result, it is possible to identify nonpoint sources or undocumented point sources by measuring dissolved solids (usually by measuring specific conductance; see Sec. 11.2.1) along a stream and noting where rapid increases in specific conductance occur. *Point sources* are discrete controlled flows of wastewater entering a stream through a pipe or well-defined channel. *Nonpoint sources* are diffuse flows through the stream or lake banks, or overland flows that enter the stream at many points. It is also possible to identify waste plumes in shallow groundwater because of the elevated dissolved solids content of the wastes. Where streams and aquifers are already polluted, it may not be possible to identify specific waste sources, especially when the concentrations of dissolved solids in the stream or aquifer and the waste inflow are similar. However, cleaner tributaries and inflows of more polluted waters can be identified by the differences in dissolved solids.

pH. The concentration of hydrogen ions is a critical water quality determinant in natural waters. Although hydrogen ions [H^+] are usually low in concentration compared with many other ions, their effect on water chemistry is far-reaching. Concentrations of [H^+] control speciation of many other geochemicals, influence dissolution and precipitation, and determine whether the water will support aquatic life. For simplicity, the concentrations of hydrogen ion are normally expressed as the negative log of the hydrogen-ion concentration in moles L^{-1} or $pH = -\log[H^+]$.

TABLE 11.1.3 Range and Typical Concentrations for Water Quality Parameters in Streams and Rivers

Water quality parameter	Typical value	Range of values observed[a]	Units	Alternative units[b]
Temperature	Variable	0–30	°C	°Fahrenheit Kelvin[c] °Rankine
pH	4.5–8.5	1–9	pH units $-\log[H^+]$	
Dissolved oxygen (O_2)	3–9	0–19	mg/L	% saturation[d]
Total nitrogen (N)	0.1–10	0.004–>100	mg/L	millimoles/L
Organic nitrogen	0.1–9	<0.2–20	mg/L	
Ammonia (NH_3-N)	0.01–10	<0.01–45	mg N/L	mg NH_3/L[e] moles/L
Nitrite (NO_2-N)	0.01–0.5	<0.002–10	mg N/L	mg NO_2/L[f]
Nitrate (NO_3-N)	0.23	0.01–250	mg N/L	mg NO_3/L[g] moles/L
Nitrogen gas (N_2)	0–18.4		mL/L	
Total phosphorus (P)	0.02–6	0.01–30	mg P/L	moles/L
Orthophosphate (PO_4)	0.01–0.5	<0.01–14	mg P/L	mg PO_4/L moles/L
Total organic carbon (C)	1–10	0.01–40	mg C/L	moles/L
Dissolved organic carbon (C)	1–6	0.3–32	mg C/L	moles/L
Volatile organic carbon	0.05		mg C/L	
Total organic matter	2–20	0.02–80	mg/L	
Inorganic carbon	50	5–250	mg $CaCO_3$/L	mg HCO_3/L moles/L meq/L
Carbon dioxide (CO_2)	0–5	0–50	mg CO_2/L	
Alkalinity (as $CaCO_3$)	150	5–250	mg $CaCO_3$/L	mg HCO_3/L moles/L meq/L
Acidity (as $CaCO_3$)		2.8–23.3	meq/L	
Biochemical oxygen demand (BOD_5)	2–15	<2–65	mg/L	
Chemical oxygen demand (COD)		<2–100	mg/L	
Hardness (as $CaCO_3$)	47–54	1–1000	mg $CaCO_3$/L	
Color	1–10	0–500	Color units	
Turbidity		0–3	NTU[h]	
Specific conductance	70	40–1500[i]	μS/cm at 25°C	mS/m at 25°C[j]
Dissolved solids (total)	73–89	5–317	mg/L	
Suspended solids	10–110	0.3–50,000	mg/L	
Total solids		20–1000	mg/L	
Cyanide	1–4		μg/L	
Phenol	<1	<1–6	μg/L	

TABLE 11.1.3 Range and Typical Concentrations for Water Quality Parameters in Streams and Rivers *(Continued)*

Water quality parameter	Typical value	Range of values observed[a]	Units	Alternative units[b]
Chloride (Cl^-)	8	≈0–158,000	mg/L	moles/L
Chlorine (Cl)		0.5–2	mg/L	moles/L
Iron (Fe)	0.04	0.01–2000	mg/L	
Manganese (Mn)	8.2	0.01–2200	μg/L	
Fluoride (F)	0.1–0.3	0–5	mg/L	
Bicarbonate (HCO_3^-) (as $CaCO_3$)	58.4	≈0–4467	mg/L	
Carbonate (CO_3^{2-})	≈0		mg/L	
Sulfate (SO_4)	8.3–11.2	0.13 to 3930	mg/L	
Hydrogen sulfide (H_2S)		<0.25	μg/L	
Calcium (Ca)	13–15	≈0–954	mg/L	
Magnesium (Mg)	4	0–379	mg/L	
Sodium (Na)	5.1–6.3	0.7–1220	mg/L	
Potassium (K)	1.3–2.3	0.02–189	mg/L	
Silicate (SiO_2)	10–14	0.15–101	mg/L	
Aluminum (Al)	50	7–4400	μg/L	
Boron (B)	18	0.7–840	μg/L	
Lithium (Li)	12	0.01–400	μg/L	
Rubidium (Rb)	1.5	0.3–7.4	μg/L	
Cesium (Cs)	0.035	0.004–0.2	μg/L	
Beryllium (Be)	0.013	0.01–1	μg/L	
Strontium (Sr)	60	6.3–<1500	μg/L	
Barium (Ba)	60	18–152	μg/L	
Titanium (Ti)	10	<0.01–107	μg/L	
Vanadium (V)	1	≈0–171	μg/L	
Chromium (Cr)	1	<0.01–84	μg/L	
Molybdenum (Mo)	0.5	≈0–145	μg/L	
Cobalt (Co)	0.2	<0.001–15	μg/L	
Nickel (Ni)	2.2	0.001–530	μg/L	
Copper (Cu)	10	0.05–>100	μg/L	
Silver (Ag)	0.30	0.03–2	μg/L	
Zinc (Zn)	30	≈0–<5000	μg/L	
Cadmium (Cd)	≈0–5	0.09–130	μg/L	
Mercury (Hg)	1[k]	<0.1–5[k]	μg/L	
Lead (Pb)	1	<0.01–55	μg/L	
Gold (Au)	0.002	<0.001–1	μg/L	
Tin (Sn)	≈0–2.1	<100	μg/L	
Bismuth (Bi)	<10		μg/L	
Thallium (Tl)	<10		μg/L	
Platinum (Pt)			μg/L	
Arsenic (As)	2	<0.10–1100	μg/L	
Antimony (Sb)	1	0.26–5.1	μg/L	
Selenium (Se)	0.20	0.11–2680	μg/L	
Bromine (Br)	20	0.5–4400	μg/L	
Iodine (I)	2 to 7	0.2–100	μg/L	
Uranium (U)	0.04	0.016–47	μg/L	
Thorium (Th)	0.1	0.044–10	μg/L	
Radon (Rn)	1		picocurie/L	
Radium (Ra)	0.3×10^{-7} to 4×10^{-7}	0.02×10^{-7} to 34×10^{-7}	μg/L	
Zirconium (Zr)		0.05–22.5	μg/L	

TABLE 11.1.3 Range and Typical Concentrations for Water Quality Parameters in Streams and Rivers *(Continued)*

Water quality parameter	Typical value	Range of values observed[a]	Units	Alternative units[b]
Scandium (Sc)	0.004	0.001–0.011	μg/L	
Gallium (Ga)	0.09	0.089–1	μg/L	
Tungsten (W)	0.03		μg/L	
Yttrium (Y)	0.7	<9.0	μg/L	
Lanthanum (La)	0.05	0.001–1.76	μg/L	
Cerium (Ce)	0.08		μg/L	
Praseodymium (Pr)	0.007		μg/L	
Neodymium (Nd)	0.04		μg/L	
Samarium (Sm)	0.008		μg/L	
Europium (Eu)	0.001		μg/L	
Gadolinium (Gd)	0.008		μg/L	
Terbium (Tb)	0.001		μg/L	
Dysprosium (Dy)	0.05		μg/L	
Holmium (Ho)	0.001		μg/L	
Thulium (Tm)	0.001		μg/L	
Ytterbium (Yb)	0.05	<0.9	μg/L	
Lutetium (Lu)	0.001		μg/L	
Erbium (Er)	0.004		μg/L	
Germanium (Ge)		n.d. to 3.7		

Note: From McCutcheon.[45] Many typical values taken from 1963 estimates by Livingstone[39] that may not fully incorporate more recent pollution effects, including effects of acid rain on increased leaching of geochemicals and from estimates by Martin and Meybeck[43] that excluded polluted rivers in the eastern United States and western Europe. Both investigators concentrated on the loads in large rivers draining directly into the oceans.

[a] High concentrations are observed at the mouth of rivers where mixing with seawater occurs. Concentrations approach maximum values found in seawater. Typical seawater concentrations representing the highest concentrations of many elements in streams are given in Table 11.1.5.

[b] See Table 11.1.2 for some conversions.

[c] SI unit.

[d] % saturation = (DO mg L^{-1}/DO_{sat}mg L^{-1}) × 100.

[e] Conversion is 1 mg N/L = 1.2159 mg NH_3/L.

[f] Conversion is 1 mg N/L = 3.2845 mg NO_2/L.

[g] Conversion is 1 mg N/L = 4.4268 mg NO_3/L.

[h] NTU = nephelometric turbidity units. A measure of light scatter in a hydrazine sulfate (concentration of 1.25 mg/L at 1 NTU) and hexamethylenetetramine (concentration of 12.5 mg/L at 1 NTU) solution that forms a formazine suspension. Before the development of modern light scattering measuring devices (nephelometers), turbidity was measured with the Jackson candle turbidimeter in Jackson turbidity units (JTU) which are not exactly convertible to NTU but are approximately the same, i.e., 40 NTU = 40 JTU.

[i] Extreme values have been observed in streams dominated by acid mine drainage (Spring Creek, Calif.; 350 μS/cm). Some industrial wastes may contain in excess of 10,000 μS/cm. Seawater has a conductivity of approximately 50,000 μS/cm. Some brines associated with halite may have conductivities of approximately 500,000 μS/cm. Precipitation may have as little as 2 μS/cm.

[j] SI units are millisiemens/m at 25°C (abbreviated mS/m). 1 millisiemen/m = 10 microsiemen/cm (abbreviated μS/cm). 1 μS/cm = 1 micromho/cm at 25°C (the typical English unit).

[k] It is suspected that many measurements of Hg are incorrect because of ubiquitous laboratory contamination.

TABLE 11.1.4 Examples of Groundwater with Relative High Concentrations of One or More Chemical Constituents, in mg/L

	Thermal		Igneous and crystalline rock					Sedimentary rock							Unconsolidated rock, alluvial fill, glacial outwash									
Sample no.	1	2	3	4	5	6	7	8	9	10	11	12	13	14	15	16	17	18	19	20	21	22	23	24
Aluminum	0.2	0.22				0.2					28			0.2	0.6					0.1	1.2			
Arsenic	1.5	4.0		1.3																				
Bicarbonate (HCO_3)		312	220	296	69	121	133	320	2080	285	0	146	241	241	30	85	202	153	161	101	100	402	412	440
Boron	4.4	48	0.08	4.6					0.4															
Bromide	1.5	1.5																						
Calcium	0.8	3.6	32	5	17	28	96	88	3	60	424	46	140	35	8.4	277	49	92	32	58	2.7	64	2.5	126
Carbonate (CO_3)			0	10	0				57		0		0	0			0	0	0	0		0	30	
Chloride	405	874	7.9	34	1.1	1.0	25	13	71	12	380	3.5	38	1	1.8	605	246	205	12	39	2.0	30	9.5	8.0
Color (units)			10		5		3	2				5		10	3		2		2		23		1	1
Copper						0.01														0.01				
Fluoride	25	2.6	0.2	0.8	1	0.1	0.4	0.3	2	0.5	1.8	0.0	0.8	0.9	0.1	0.2	0.1	0.6	0.7	0.0	0.1	0.1	1.7	0.7
Hardness (as $CaCO_3$)	2	9	129	17	49	78	318	250	38	337	1860	132	526	224	27	954	196	386	116	198	15	238	15	490
Iodide	0.3	0.6																						
Iron	0.06	0.52	0.01	0.1	0.33	2.7	1.0	0.02	0.15	0.37	0.88	0.04	0.01	0.39	11		0.00		0.0	0.04	2.9	0.28	0.2	2.3
Lithium	5.2	7																						
Magnesium	0.0	0.0	12	1	1.7	1.9	19	7.3	7.4	31	194	4.2	43	33	1.5	64	18	38	8.8	13	2.0	19	2.1	43
Manganese	0.0	0.00	0.00		0.00	0.22	0.01	0.00		0.05	9.6				0.32					1.3				0.00
Nitrate (NO_3)	1.8	2.7	2.9	0.7	0.0	0.2	0.4	4.6	0.2	0.8	3.1	7.3	4.1	1.2	0.4	35	2.2	83	0.6	0.6	0.6	60	0.6	0.2
pH (units)	9.6	8.9	7.8	8.5	7.1	6.9	7.8	7.5	8.3	7.6	4	7	7.4	8.2	6.3		7.7		7.9	7.0	7.4	7.4	8.7	7.6
Phosphate (PO_4)	1.3	0.24		2.6		0.0					0.0									0.1				
Potassium	24	65	5.2	1.2	*	4.2	1.5	2.8	2.4	4	11	0.8	*	1.3	3.6	*	*	*	*	2.8	1.7	9.5	*	2.1
Selenium																								
Silica (SiO_2)	363	314	49	10	29	31	15	24	16	8.7	98	8.4	13	18	7.9	74	22	23	71	10	12	27	22	20
Sodium	352	660	30	136	7.4*	6.8	18	19	857	12	416	1.5	21*	28	1.5	53*	168*	110*	42*	23	35	114	182*	13
Strontium		0.67								52														
Sulfate (SO_4)	23	108	11	10	6.9	1.4	208	6.7	1.6	111	2420	4	303	88	5.9	113	44	137	54	116	5.6	74	3.5	139

Total dissolved solids	1310	2360	257	363	98	137	468	322	2060	440	4190	139	701	329	44	1260	651	764	310	338	101	578	452	571
Zinc																				0.01				

* Sodium concentration includes potassium.

Sample no.	Source
1	Hot spring, Yellowstone National Park, Wyoming, temperature 94°C (high in silica and other elements)
2	Thermal well, 227 m deep. Washoe County, Nevada, bottom temperature 186°C (high in silica, arsenic, and other elements)
3	Well, 232 m deep, Umatilla County, Oregon; water from basalt of Columbia River Group (high in silica)
4	Well, 46 m deep, Lane County, Oregon; water from Fisher Formation (high in metals and other constituents)
5	Well, 122 m deep, Burke County, North Carolina; water from mica schist (high in silica)
6	Well, 51 m deep, Baltimore County, Maryland; water from granitic gneiss (high in aluminum and manganese)
7	Well, 37 m deep, Williamanset, Mass.; water from Portland Arkose (high in calcium)
8	Well, 30 m deep, Rice County, Arkansas; water from Dakota Sandstone (high in calcium)
9	Well, 152 m deep, Richland County, Montana; water from sandstone and shale (high in sodium)
10	Well, 581 m deep, Waukesha, Wis.; water from sandstone (high in metals)
11	Well, 6.7 m deep, Drew County, Arkansas; water from shale, sand, and marl (high in aluminum and manganese, also contained 1.7 pCi/L radium and 0.017 mg/L uranium)
12	Spring, Huntsville, Ala.; water from Tuscumbia Limestone (high in calcium)
13	Well, 257 m deep, Chaves County, New Mexico; water from San Andres Limestone (high in magnesium)
14	Well, 152 m deep, Milwaukee County, Wisconsin; water from Niagara Dolomite (high in magnesium)
15	Well, 64 m deep, Fulton, Miss.; water from Tuscaloosa Formation (high in iron)
16	Well, Maricopa County, Arizona; water from alluvial fill (high in calcium and nitrate)
17	Well, 152 m deep, Maricopa County, Arizona; water from alluvial fill (high in sodium)
18	Well, 84 m deep, Maricopa County, Arizona; water from alluvial fill (high in nitrate)
19	Average of seven wells, 76 218 m deep, Albuquerque, N.M.; water from alluvial fill (high in silica)
20	Two radial collector wells, 16 m deep. Parkersburg, W.V.; water from sand and gravel (high in manganese)
21	Well, 399 m deep, Memphis, Tenn.; water from Willcox Formation sand (high in iron)
22	Well, 10 m deep, Lincoln County, Kansas; water from alluvium (high in fluoride)
23	Well, 56 m deep, Raleigh-Durham, N.C.; water from Coastal Plain sediments (high in sodium)
24	Well, 36 m deep, Columbus, Ohio; water from glacial outwash (high in iron)

TABLE 11.1.5 Concentration and Range of Geochemicals in Seawater

Constituent	Concentration,* mg/L	Range,† mg/L
Oxygen (as H_2O)	883,000	
(dissolved O_2)	6.0	
Hydrogen (as H_2O)	110,000	
Hydrogen ions (H^+)‡	0.28	0.25–0.34
Chloride (Cl)	19,400	19,000–19,900
Sodium (Na)	10,800	10,500–11,070
Sulfate (SO_4)	2,700	2,650–2,790
Magnesium (Mg)	1,290	1,270–1,350
Calcium (Ca)	411	400–436
Potassium (K)	392	390–415
Bicarbonate (HCO_3^-)	142	140–157
Bromide (Br)	67.3	65–69.7
Nitrogen (dissolved N_2)	15.5	
(NH_4^+, NO_2^-, NO_3^-, and dissolved organic as N)	0.67	0–0.67
(Dissolved organic N)	0.2	
Strontium (Sr)	8.1	6.3–13
Silicate (SiO_2)	6.2	0–12
Boron (B)	4.5	4.1–4.7
Fluoride (F)	1.3	1.2–1.7
Organic carbon (C)	1.8	0.10–2.7
Argon (Ar)	0.45	
Lithium (Li)	0.17	0.10–0.195
Rubidium (Rb)	0.12	0.086–0.134
Phosphorus (P)	0.09	0–0.3
Iodine (I)	0.064	0.044–0.080
Barium (Ba)	0.021	<0.00004–0.093
Molybdenum (Mo)	0.010	0.0001–0.016
Nickel (Ni)	0.0066	0.0001–0.043
Zinc (Zn)	0.005	0.001–0.500
Iron (Fe)	0.0034	0–0.062
Uranium (U)	0.0033	0.0004–0.0047
Arsenic (As)	0.0026	0.00026–0.035
Vanadium (V)	0.0019	0.0003–0.007
Aluminum (Al)	0.001	0–0.01
Titanium (Ti)	0.001	0.0007–0.002
Copper (Cu)	0.0009	0.0002–0.027
Tin (Sn)	0.00081	0.0002–0.003
Cobalt (Co)	0.00039	<0.000005–0.045
Manganese (Mn)	0.0004	0.0002–0.0086
Antimony (Sb)	0.00033	0.00018–0.0011
Cesium (Cs)	0.00030	0.00027–0.00058
Silver (Ag)	0.00028	0.00004–0.0015
Krypton (Kr)	0.00021	0.00021–0.0003
Chromium (Cr)	0.0002	0.00004–0.0005
Mercury (Hg)	0.00015	0.00003–0.002
Neon (Ne)	0.00012	
Cadmium (Cd)	0.00011	0.00002–0.00025
Thallium (Tl)	0.0001	<0.00001–0.0001

TABLE 11.1.5 Concentration and Range of Geochemicals in Seawater *(Continued)*

Constituent	Concentration,* mg/L	Range,† mg/L
Tungsten (W)	<0.0001	0.00009–0.0001
Selenium (Se)	0.00009	0.000052–0.0005
Germanium (Ge)	0.00006	0.00005–0.00007
Xenon (Xe)	0.000047	0.000047–0.0001
Gallium (Ga)	0.00003	0.000023–0.000037
Lead (Pb)	0.00003	0.00002–0.0004
Zirconium (Zr)	0.000026	
Bismuth (Bi)	0.00002	0.000015–0.0002
Yttrium (Y)	0.000013	3×10^{-6}–16.3×10^{-6}
Niobium (Nb)	0.000015	0.00001–0.00002
Gold (Au)	0.000011	0.000004–0.000027
Rhenium (Re)	8.4×10^{-6}	
Hafnium (Hf)	$<8 \times 10^{-6}$	
Helium (He)	7.2×10^{-6}	5×10^{-6}–7.2×10^{-6}
Scandium (Sc)	$<4 \times 10^{-6}$	0.01×10^{-6}–1.8×10^{-6}
Lanthanum (La)	3.4×10^{-6}	0.7×10^{-6}–12×10^{-6}
Neodymium (Nd)	2.8×10^{-6}	0.2×10^{-6}–9.2×10^{-6}
Tantalum (Ta)	$<2.5 \times 10^{-6}$	
Cerium (Ce)	1.2×10^{-6}	0.6×10^{-6}–850×10^{-6}
Dysprosium (Dy)	9.1×10^{-7}	5.2×10^{-7}–14.0×10^{-7}
Erbium (Er)	8.7×10^{-7}	6.1×10^{-7}–12.4×10^{-7}
Ytterbium (Yb)	8.2×10^{-7}	4.8×10^{-7}–28×10^{-7}
Gadolinium (Gd)	7.0×10^{-7}	5.2×10^{-7}–11.5×10^{-7}
Ruthenium (Ru)	7×10^{-7}	
Praseodymium (Pr)	6.4×10^{-7}	4.1×10^{-7}–15.8×10^{-7}
Beryllium (Be)	6×10^{-7}	
Samarium (Sm)	4.5×10^{-7}	0.45×10^{-7}–17×10^{-7}
Thorium (Th)	4×10^{-7}	$<0.7 \times 10^{-7}$–500×10^{-7}
Holmium (Ho)	2.2×10^{-7}	1.2×10^{-7}–7.2×10^{-7}
Thulium (Tm)	1.7×10^{-7}	0.9×10^{-7}–3.7×10^{-7}
Lutetium (Lu)	1.5×10^{-7}	1.2×10^{-7}–7.5×10^{-7}
Terbium (Tb)	1.4×10^{-7}	0.6×10^{-7}–3.6×10^{-7}
Europium (Eu)	1.30×10^{-7}	0.9×10^{-7}–7.9×10^{-7}
Indium (In)	1×10^{-7}	1×10^{-7}–$<2 \times 10^{-2}$
Protactinium (Pa)	2×10^{-10}	5×10^{-11}–2×10^{-9}
Radium (Ra)	1×10^{-10}	0.3×10^{-10}–2.9×10^{-10}
Radon (Rn)	0.6×10^{-15}	

Note: Rhodium, palladium, tellurium, osmium, iridium, and platinum do not seem to have been measured in seawater.

* From McCutcheon.[45]

† Not based on an exhaustive review.

‡ Corresponds to a pH of 8.2 (8.0 to 8.3).

11.1.3 Ionization of Organic Acids and Bases

Using the law of mass action, the distribution of ionic forms ($[H^+]$, concentration of hydrogen ions in moles L^{-1}); ($[A^-]$, concentration of weak acid in moles L^{-1}, and neutral species [HA] in moles L^{-1}) is described by the *acid dissociation constant* K_a where

$$K_a = \frac{[H^+]\,[A^-]}{[HA]} \tag{11.1.10}$$

Similarly, the *base dissociation constant* relates the concentration of hydroxyl ions $[OH^-]$ in moles L^{-1}, and weak base concentration $[B^+]$ in moles L^{-1}, to the neutral species [BOH] in moles L^{-1}. That expression is

$$K_b = \frac{[OH^-]\,[B^+]}{[BOH]} \tag{11.1.11}$$

Values of K_a and K_b have been measured for a number of contaminants[12,18,51,78] and are reported as $pK_a = -\log_{10}(K_a)$ or $pK_b = -\log_{10}(K_b)$. K_a values for weak acids range from 10^{-3} to 10^{-9} while K_b values range from 10^{-3} to 10^{-10}. Strong bases have a K_b on the order of 1. The fraction of total contaminant that is in a neutral form, and thus subject to volatilization is, for acids

$$f_a = \frac{[HA]}{[HA] + [A^-]} = \frac{1}{1 + 10^{pH - pK_a}} \tag{11.1.12}$$

and for bases

$$f_b = \frac{[BOH]}{[BOH] + [B^+]} = \frac{1}{1 + 10^{pK_w - pK_b - pH}} = 1 - f_a \tag{11.1.13}$$

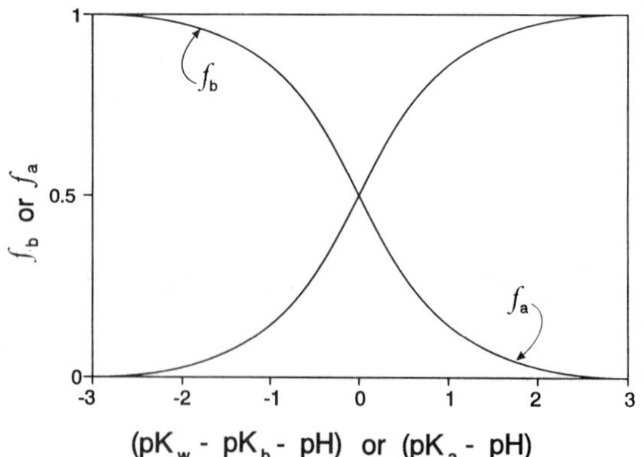

FIGURE 11.1.3 Fraction of weak acids and bases that remain un-ionized as a function of pH, the water dissociation constant (pK_w), and the dissociation constants for acids (pK_a) and bases (pK_b). Note that $f_a = 1 - f_b$ and $f_b = 1 - f_a$.

where pK_w is the log base 10 of $K_w = [H^+] \cdot [OH^-]$. Mills et al.[51] presents values of pK_w at different temperatures (e.g., $pK_w = 14.00$ at 25°C). Figure 11.1.3 shows the relationship between pK_a and f_a and pK_b and f_b. For example, if a total dissolved concentration of 2-nitrophenol of 20 μg L^{-1} is present in a water with a pH of 8.0 and the $pK_a = 7.21$, the fraction of the neutral species present is

$$f_a = \frac{1}{1 + 10^{(\mathrm{pH} - \mathrm{pK}_a)}} = \frac{1}{1 + 10^{(8.0-7.21)}} = 0.14 \tag{11.1.14}$$

or (0.14) 20 μg L^{-1} = 2.8 μg L^{-1} 2-nitrophenol un-ionized molecule will be present and subject to volatilization (loss to the atmosphere).

11.1.4 Geochemical Speciation

Equilibrium. Although the dissolution of minerals from rocks can be a slow process, once dissolution has occurred, the cations and anions in natural waters are generally at or near a state of equilibrium for reactions not involving a change in phase (solid, liquid, or gas). When solutions are at equilibrium, it is possible to calculate the distribution of elements among all the possible and feasible anions, cations, and complexes. The appendices in Allison et al.[1] list the complexes that are normally considered in the speciation analysis of natural waters.

Precipitation and dissolution reactions may or may not achieve an equilibrium state quickly. When these reactions control the form and amount of a dissolved solid present, an experienced geochemist or reference[28,34,79] should be consulted to determine if equilibrium calculations are appropriate.

Mass-Action Equations. Equilibrium calculations are based on the *law of mass action,* expressed as

$$\frac{[C]^c\,[D]^d}{[A]^a\,[B]^b} = K \tag{11.1.15}$$

where A and B are substances that react to form products C and D. The brackets indicate molal concentrations of moles per kg of water. For most natural waters, molal concentrations are essentially equivalent to molar concentrations of moles per L of solution (water plus dissolved solids). K is the *equilibrium* or *dissociation constant* for the reversible reaction

$$aA + bB \rightleftharpoons cC + dD \tag{11.1.16}$$

where a, b, c, and d are multiples required to balance the equation at electroneutrality. The equilibrium constant K changes with temperature and pressure, but changes in equilibrium constants due to pressure differences are usually negligible.

The law of mass action is strictly valid only in very dilute solutions. As ion concentrations increase, potential chemical interaction is lessened by electrostatic effects. A correction factor termed the *activity coefficient* must be calculated to convert measured ion concentrations to effective concentrations, or *activities.* Calculation of activities makes it possible to apply the law of mass action to natural waters. Typically, corrections for dissolved solids effects are based on the Debye-Huckel equation or the Davies equation.

The equilibrium constant K can be calculated using equations of the form

$$\log_{10} K_i = a + \frac{b}{T} + cT + d \log_{10} T \tag{11.1.17}$$

where T is absolute temperature in K, and a, b, c, and d are coefficients.[45]

Carbon Dioxide Dissolved in Water. As an example, speciation reactions and the mass-action equations for a simple system consisting of carbon dioxide gas in water can be written as

$$CO_2(gas) + H_2O(liq) \leftrightarrows H_2CO_3(aq) \leftrightarrows H^+ + HCO_3^- \leftrightarrows 2\,H^+ + CO_3^{2-} \tag{11.1.18}$$

where carbonic acid, H_2CO_3, ionizes into bicarbonate, HCO_3^- and hydrogen ions, and bicarbonate ions further ionize into a carbonate ion, CO_3^{2-} and an additional hydrogen ion under favorable pH conditions ($pH > 8.2$). Water ionizes as $H_2O \leftrightarrows H^+ + OH^-$. Where other dissolved materials are absent, as in rainwater, these solutions will be sufficiently dilute to permit assigning a value of 1.0 to the activity coefficients. The mass-action equations for equilibrium of the carbonate system at 25°C and 1 atmosphere pressure are

$$\frac{[H_2CO_3]}{P_{CO_2}} = K_H = 10^{-1.43} \tag{11.1.19}$$

$$\frac{[HCO_3^-]\,[H^+]}{[H_2CO_3]} = K_1 = 10^{-6.35} \tag{11.1.20}$$

$$\frac{[CO_3^{2-}]\,[H^+]}{[HCO_3^-]} = K_2 = 10^{-10.33} \tag{11.1.21}$$

and

$$[H^+]\,[OH^-] = K_W = 10^{-14.00} \tag{11.1.22}$$

where P_{CO_2} = the partial pressure of carbon dioxide in the atmosphere (decimal fraction of the total volume), K_1 = equilibrium constant for the dissociation of carbonic acid into bicarbonate ions, K_2 = equilibrium constant for the dissociation of bicarbonate ions into carbonate ions, K_W = equilibrium constant for the dissociation of water into hydrogen and hydroxyl ions, and K_H = the *Henry's law constant* relating the atmospheric partial pressure to the carbonic acid concentration in water. Henry's law is a simple concept relating the partial pressure of an atmospheric gas to the gas dissolved in water at equilibrium for gases that are only slightly soluble in water. Dissolved carbon dioxide is assumed to be almost completely hydrolyzed (reacts with water) to form carbonic acid, H_2CO_3. The brackets denote concentrations of the species in solution in moles per liter. The activity of water, $[H_2O]$, is normally assumed to be 1, especially for a dilute rainwater solution.

Because there must be an overall balance of negative and positive charges among the dissolved species, a charge balance equation can be derived from Eq. (11.1.18) and solved with Eqs. (11.1.19) through (11.1.22) to describe the entire carbonate system (see Sec. 11.2.2).

Figure 11.1.4 illustrates the change in species present at different pH values and three temperatures but is based only on Eqs. (11.1.22) and (11.1.23), assuming a fixed total alkalinity ($HCO^{3-} + CO_3^{2-}$) with no gas phase, and an activity coefficient = 1.0

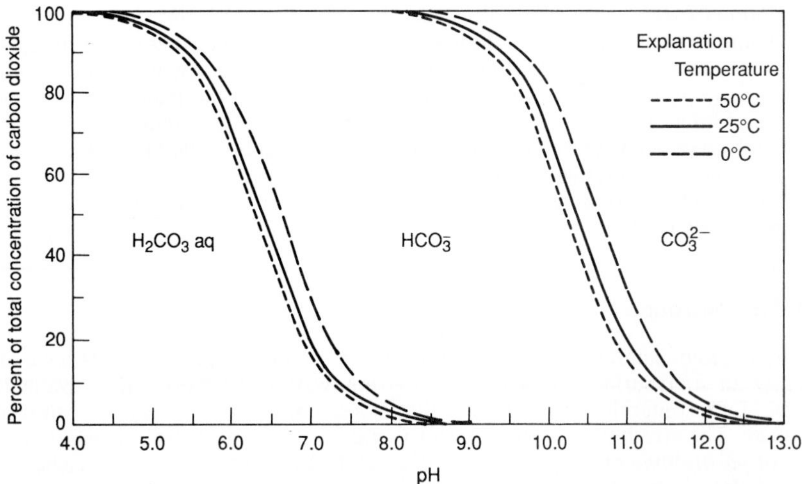

FIGURE 11.1.4 The distribution of dissolved carbon dioxide species concentrations (or activities) as a function of pH at 1 atmosphere pressure and over a range of temperatures of 0 to 50°C. The curves were calculated by Hem[28] from Eqs. (11.1.20) and (11.1.21). As pH increases, the water becomes increasingly alkaline. This deficit in hydrogen ions (H^+) causes the carbonic acid (H_2CO_3) to dissociate first into bicarbonate ions (HCO_3), and then into carbonate ions (CO_3^{2-}) as pH increases to greater than approximately 8.2. This *buffer capacity* tends to moderate pH changes in natural waters.

for all ions. The figure could be recomputed for any given partial pressure of atmospheric carbon dioxide and any specified temperature range (the reactions are assumed to be largely independent of atmospheric pressure changes) using the five equations (11.1.18) to (11.1.22) and six unknowns (P_{CO_2}, $[H_2CO_3]$, $[HCO_3^-]$, $[CO_3^{2-}]$, $[H^+]$, and $[OH^-]$) noted.

Saturation Index. Some speciation reactions in water involve solids precipitating or dissolving. The tendency to precipitate is normally expressed by an index value such as

$$\text{Saturation index} = \log\left(\frac{IAP}{K}\right) = \log\left(\frac{\dfrac{[C]^c[D]^d}{[A]^a[B]^b}}{K}\right) \tag{11.1.23}$$

where IAP = ion activity products defined by the left-hand side of Eq. (11.1.15) and K = equilibrium constant. Comparison of Eqs. (11.1.23) and (11.1.15) indicates that the saturation index is a simple restatement of the law of mass action when the solution is at saturation. When the saturation index is negative, the water is undersaturated with the solid component. When the saturation index is positive, the water is supersaturated. Saturation index values of zero indicate that the water is at saturation. This sign convention holds when the reaction involves a dissolving solid [either reactant A or B in Eq. (11.1.16)].

Some investigators define a saturation index as the simple ratio (IAP/K). When $IAP/K = 1$, the natural water is at saturation. At $IAP/K > 1$, supersaturation occurs, and at $IAP/K < 1$, the water is undersaturated with respect to the dissolved ion of interest.

Speciation Models. Geochemical speciation calculations are commonly performed with computer models based on the law of mass action and the other basic principles outlined in this section. These models simultaneously solve for chemical equilibrium from a large number of the mass-action equations for all the major geochemical species of interest. In this way, all the competing effects of different species can be taken into account. Computer models available are listed in Table 11.1.6. Except for highly saline or high-temperature waters where the model PHREEQE may be best, the best overall alternative may be the MINTEQA2 model,[1] which has the most extensive data base.

11.1.5 Sorption

Many organic chemicals, radionuclides, metals, and other geochemicals have an affinity for solid surfaces. In surface and groundwaters, the most readily available surface for a contaminant to sorb to is that of a sediment, soil, or very fine colloidal particles. The attachment of substances to solids is usually described as the *distribution* or *partitioning* of a chemical between the dissolved phase and the attached or sorbed phase associated with solids.

Sorption is a general term used to describe the attachment of dissolved molecules to solids or solid surfaces. The process is influenced by turbulent and molecular motion of the surrounding fluid that carries molecules to or near the surface, charges on the dissolved molecules and surfaces, and other chemical and thermodynamic properties of the water, contaminant, and solid that push, pull, or hold molecules on a surface or cause incorporation of the molecule into the structure of the solid. The process that describes attachment of molecules to the surface of a solid is *adsorption.* When dissolved molecules are incorporated within the structure of a solid, the process is known as *absorption.*

There are three categories of contaminants: (1) nonpolar or neutral organic chemicals such as PCB (polychlorinated biphenyls) and pesticides such as DDT (dichlorodiphenyltrichloroethane), (2) metals, and (3) ionizable organic chemicals. Nonpolar organic chemical sorption can be predicted when the fraction of organic carbon content of suspended or aquifer solids is known and the partitioning of the dissolved chemical between water and octanol has been measured. Partitioning of metals between the dissolved phase and solid phase can be predicted when the characteristics of the iron oxide and organic carbon coating on particles is known along with solution pH and other factors that influence the species of metal in solution. Ionizable organic chemicals are sorbed by organic carbon films and attach to charged sites depending on pH conditions.[29,30]

Sorption of Organic Chemicals. Many organic chemicals are *hydrophobic* (lacking affinity for water) and are pushed onto and held on solid surfaces by the random movement of water molecules and the contaminant molecule. Once at the surface, wetting effects and capillary action may lead to absorption. Toxic hydrophobic chemicals accumulate in organic films,[13,14,20,32] a ubiquitous feature of natural particles formed by sorption of natural organic materials. The organic carbon content of solids has been found to accurately measure the sorption capacity of organic films.

Natural particles are not perfect spheres with a solid outer surface. The outer surface is penetrated by numerous cracks, crevices, and channels into the interior of the particle, which is many times an aggregate of smaller particles. These interior surfaces become covered with organic material and represent extensive sorption sites for neutral organic chemicals and charged ions.

TABLE 11.1.6 Geochemical and Metals Speciation Models

Item†	$CaCO_3$ indices*: Basis for calculation of SI	$CaCO_3$ indices*: CCPP	Approximate temperature range, °C	Approximate limit of ionic strength	Ion pairs considered?	Alk_o considered?‡	Minimum equipment required*
1. Caldwell-Lawrence diagrams	pH_{sa}	P, D	2–25	0.030	No	No	Diagrams
2. ACAPP	RS	P, D	−10–110	6+	Yes	Yes	IBM-compatible PC, 512K bytes of RAM, MS DOS or PC DOS v.2.1 or higher
3. DRIVER	RS	P	7–65	2.5	Yes	Yes	Mainframe computer
4. INDEX C	pH_{sa}, pH_{sb}	P, D	0–50	0.5	No	No	Hewlett-Packard 41C calculator, with three memory modules
5. LEQUIL	RS	No	5–90	0.5	Yes	Yes	IBM-compatible PC, 256K RAM, Lotus 1-2-3 or work-alike, PC DOS or MS DOS v.2.0 or higher
6. MINTEQA2	RS	P, D	0–100	0.5	Yes	Yes	IBM-compatible PC, 512K bytes of RAM, PC DOS v.3.0 or higher, 10 megabyte hard disk drive, math coprocessor useful but not required. Also available for mainframe computers
7. PHREEQE standard	RS	P, D	0–100	0.5	Yes	Yes	IBM-compatible PC, known to work with 512K RAM, PC DOS or MS DOS v.2.11 or higher. Also available for mainframe computers

TABLE 11.1.6 Geochemical and Metals Speciation Models *(Contined)*

Item†	$CaCO_3$ indices*: Basis for calculation of SI	$CaCO_3$ indices*: CCPP	Approximate temperature range, °C	Approximate limit of ionic strength	Ion pairs considered?	Alk_o considered?‡	Minimum equipment required*
For high-salinity waters	RS	P, D	0–80	7–8	Yes	Yes	IBM-compatible PC, 640K RAM recommended, with math coprocessor, MS DOS v.3.2 or higher
8. SEQUIL	RS	P, D	7–65	2.5	Yes	Yes	IBM-compatible PC, 512K bytes of RAM, MS DOS or PC DOS v.2.1 or higher
9. SOLMINEQ.88	RS	P, D	0–350	6	Yes	Yes	IBM-compatible PC, 640K RAM, math coprocessor, PC DOS or MS DOS v.3.0 or higher. Also available for mainframe computers
10. WTRCHEM	pH_{sa}	P, D	0–100	0.5	No	No	Any PC equipped with a BASIC interpreter, 5K RAM
11. WATEQ4F	RS	No	0–100	0.5	Yes	Yes	IBM-compatible PC, known to work with 512K RAM, PC DOS or MS DOS, v.2.11 or higher

Source: From Clesceri et al.[15] Used with permission.

* SI = saturation index

CCPP = $CaCO_3$ precipitation potential

pH_{sa} = alkalinity-based pH_s

pH_{sb} = bicarbonate-based pH_s

P = calculates amount of $CaCO_3$ theoretically precipitated

D = calculates amount of $CaCO_3$ theoretically dissolved

RS = relative saturation

PC = personal computer

RAM = random access memory

†1. American Water Works Association, Denver, Colo., provides 30.5- by 38.1-cm diagrams (order number 20204), and documentation[4] (order number 20203); Loewenthal and Marais[40] provide 10.2- by 11.4-cm diagrams, with documentation; Merrill[48] provides 10.2- by 16.5-cm diagrams, with documentation.

2. Radian Corp., Austin, Tex. (software and documentation).

3. Power Computing Co., Dallas, Tex. (software and documentation EPRI[21])

4. Brown and Caldwell, Walnut Creek Calif., (software and documentation).

5. Illinois State Water Survey, Aquatic Chemistry Section, Champaign, Ill. (software and documentation).

6. Center for Exposure Assessment Modeling, Environmental Research Laboratory, Office of Research and Development, U.S. Environmental Protection Agency, Athens, Ga. (software and documentation) (Allison et al.[1])

7. U.S. Geological Survey, National Center, MS 437, Reston, Va. Chief of WATSTORE Program (provides software for mainframe version of standard code); U.S. Geological Survey, Water Resources Division, MS 420, Menlo Park, Calif. (provides software for personal computer version of standard code); National Water Research Institute, Canada Centre for Inland Waters, Burlington, Ont., Canada [provides software and documentation (Parkhurst et al.,[55] Crowe and Longstaff[16]) for personal computer versions of both standard and high-salinity codes]; U.S. Geological Survey, Oklahoma City, Okla. [provides documentation (Parkhurst et al.,[55] Fleming and Plummer[23]) for mainframe and personal computer versions of standard code].

8. Power Computing Company, Dallas, Tex. (software and documentation) (EPRI[22]).

9. U.S. Geological Survey, Water Resources Division, MS 427, Menlo Park, Calif. (software and documentation) (Kharaka et al.[33]).

10. Public Domain Software, Santa Clara, Calif. (code listing and documentation).

11. U.S. Geological Survey, Water Resources Division, MS 420, Menlo Park, Calif. (software and documentation) (Ball et al.[5]).

‡ Codes differ in the species included in Alk_o.

When sorption is controlled by surface or ion charges, there is a relationship between particle concentration and sorption. When organic chemicals are incorporated into organic films on particles, only the amount and type of organic carbon on the particle and the hydrophobicity of the dissolved substance are important.[32] Particle concentration does not affect sorption except that there is a relationship between organic carbon content and particle concentration.

Some investigators have distinguished between sorption onto a particle and *desorption* from a particle. However, since most of the sorption reactions are reversible, it is usually not necessary to distinguish between sorption and desorption. It is necessary to take the effects of colloids and diffusion into and out of complex aggregate particles into account.[24,25,32,80]

Solids and Particles in Natural Waters. Surface waters contain sand, silts, clays, organic detritus, and plankton (algae and bacteria). Sorption occurs preferentially onto the finer particles because of the high surface area to volume ratios and because they are normally present in greater quantities. These fine particles have a significant amount of organic carbon and oxide coatings that attract metal ions. Organic contaminants attach to organic material on mineral particles, detritus, plankton, fish, and other wildlife in proportion to the amount of organic carbon present. Many natural waters have significant quantities of colloidal material consisting of macromolecules of organic material, fragments of detritus, and fine inorganic particles, all less than 0.45 μm in diameter. These colloids are especially efficient in sorbing metals and organic chemicals and are frequently measured[15] as dissolved organic carbon.

Equilibrium. Sometimes sorption reaches equilibrium, but in many cases not before transport processes change the relationship between the dissolved substance and the solids in the flow. Equilibrium calculations are still useful, however, to describe the potential state of sorption and the tendency of dissolved and sorbed concentrations to increase or decrease. Various *sorption kinetics models* are available to describe the partitioning of chemicals between the dissolved phase and the sorbed phase as it changes over time.

Whether an equilibrium state has been reached for sorption depends on many site-specific factors. In general, nonpolar or neutral organic molecules like PCBs require 1 to 3 months to achieve full equilibrium. Some metals like zinc and copper require at least 24 h in streams.[46] If groundwater flow does not change significantly over several months, equilibrium partitioning may be a valid assumption. In stream flows that change conditions significantly from day to day, equilibrium partitioning may be a valid assumption for metals but not for organic chemicals consisting of large hydrophobic molecules, which may take longer to reach equilibrium.

Fugacity[41] is a thermodynamic term used to describe the tendency of chemicals to move between the dissolved phase and the sorbed phases associated with suspended particles, colloidal particles, plankton, and fish as shown in Fig. 11.1.5. Mass diffuses from regions of high fugacity to regions of low fugacity. If a chemical is not in equilibrium, measured concentrations of dissolved and sorbed chemical can be compared with equilibrium concentrations to determine if dissolved concentrations will be expected to increase or decrease. The same types of calculations can be made for body burdens of chemicals in fish and on other biota.

Because of the failure of many chemicals to reach equilibrium during experiments, coefficients that describe the partitioning between the dissolved and sorbed phases have not been precisely measured. Therefore, published data on partitioning must be carefully examined before use.

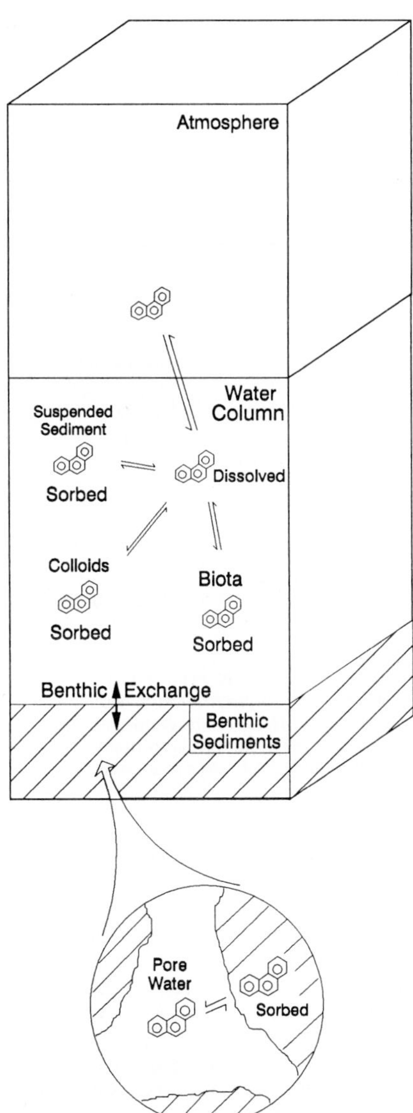

FIGURE 11.1.5 Illustration of the distribution of chemicals in the environment and fugacity. *(From McCutcheon.[45])*

Sorption Isotherms. A sorption isotherm describes the partitioning of chemicals between the dissolved phase and sorbed phase. Isotherms relate the concentration of chemical dissolved to the concentration sorbed. For organic chemicals, three types of isotherms are applicable: (1) the *Langmuir adsorption isotherm,* (2) the *Freundlich empirical isotherm,* and (3) the *linear isotherm* for dilute solutions. These relationships and the associated isotherm equations are given in Fig. 11.1.6. The sorption of some metals is described by an ion exchange adsorption model or by electrostatic models of various types.

For low concentrations of organic chemicals in natural waters, the linear isotherm is a very accurate representation of partitioning.[32] Only occasionally are concentrations high enough ($>10^{-3}$ molar) to warrant use of the Langmuir or Freundlich isotherms (e.g., landfill leachate and industrial processes that produce high-concentration wastes). For the linear isotherm, the relationship between the dissolved concentration C_w and the sorbed concentration C_s is $K_pC_w = C_s$ where K_p is the partitioning coefficient. C_w has units of mass of contaminant per volume of water ($mg_c\ L_w^{-1}$) and C_s has units of mass of contaminant per unit mass of sediment ($mg_c\ kg_s^{-1}$). Thus K_p has units of $L_w\ kg_s^{-1}$ or $mL\ g^{-1}$ (volume of water L_w per mass of sediment, kg_s). The subscripts *w*, *c*, and *s* are used to indicate that the units refer to L of water, mg of chemical, and kg of sediment.

There are no highly reliable compilations of partitioning coefficients. Many literature reports are based on duplicated citations of the few measurements available, and a number of measurements are highly suspect. At the U.S. EPA Environmental Research Laboratory in Athens, Ga., fate constants like the partitioning coefficient are derived from the literature by a knowledgeable chemist or are remeasured. Many values contained in other data bases are rejected. Access to this EPA data base is available through the EPA Center for Exposure Assessment Modeling, 960 College Station Road, Athens, GA 30613-0801.

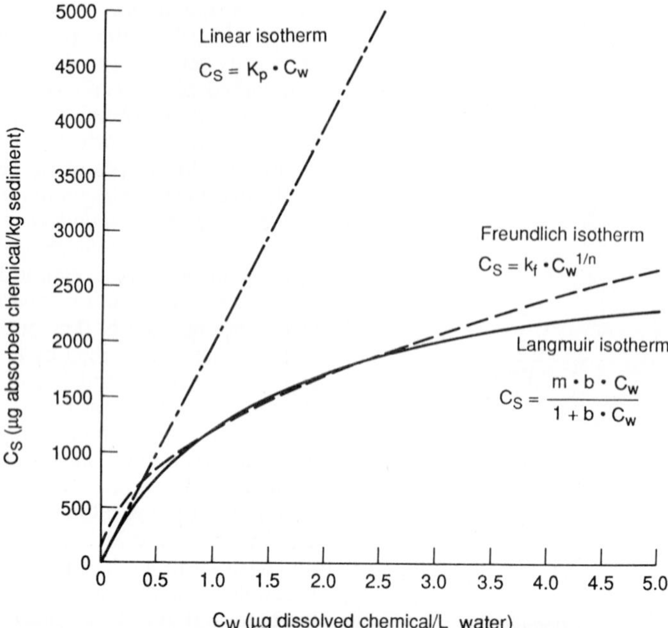

FIGURE 11.1.6 Sorption isotherms. For the Langmuir isotherm, m is the maximum sorption in moles L^{-1} or mg L^{-1}, b is the Langmuir sorption constant with units of moles L^{-1} or mg L^{-1} depending on the concentration units employed (moles L^{-1} vs. mg L^{-1}). k_f and n are Freundlich isotherm constants. n is normally assumed to be unitless and k_f has units of L kg^{-1} like the linear partitioning coefficient K_p. *(From Mills et al.[51])*

The best means of deriving K_p values involves normalizing values of the fraction of organic carbon f_{oc} present on sediments or aquifer solids and to an equivalent organic carbon partitioning coefficient K_{oc} as $K_p = K_{oc} f_{oc}$. f_{oc} is easily measured for sediments, soils, aquifer materials, and detritus. Furthermore K_{oc}, the partition coefficient normalized to the organic carbon content, has been related to the octanol-water partitioning coefficient K_{ow} for a number of classes of compounds. Table 16.5.4 gives a compilation of the empirical relationships between K_{oc} and K_{ow} for various classes of chemicals. The reader should refer to the original reference to determine if a compound of interest is included. Note that some relationships are based on a limited number of compounds in the class of compounds. Polynuclear aromatic compounds are a subclass of the class of aromatic hydrocarbons and both are crude oil or combustion-derived compounds. Karickhoff[32] reviews what is known about attachment of nonpolar chemicals when organic films are absent, and Jafvert[29,30] proposes methods for determining the partitioning of ionizable chemical compounds.

Partitioning in Natural Waters. One important result from partitioning calculations is the amount of chemical that remains in the dissolved form, available to impair the health of aquatic organisms, to volatilize into the air, to be degraded by bacteria and other organisms, and to react with other natural or man-made chemicals. The simplest relationships for the fraction that remains dissolved f, the fraction sorbed f_s, and relation between the two are

$$f = \frac{1}{1 + \frac{K_p S}{n}} \qquad f_s = \frac{\frac{K_p S}{n}}{1 + \frac{K_p S}{n}} \qquad \text{and} \qquad f + f_s = 1 \tag{11.1.24}$$

where S is the concentration of solids in $kg_s\ L^{-1}$ [multiply S in Eq. (11.1.24) by 10^{-6} when units of $mg\ L^{-1}$ are employed], and n is porosity (in $L_w\ L^{-1}$, volume of water per unit volume of water-sediment mixture). The porosity of natural waters is approximately 1 and may be about 0.5 for bedded sediments (ranging from 0.6 to 0.3, similar to porosity of soil). Natural systems are more complex in that partitioning calculations must also account for partitioning to colloids and plankton, and these calculations are covered in Refs. 2 and 25. See Ref. 2 for appropriate modeling techniques.

As an example, consider the equilibrium phase partitioning between the dissolved and solid phase (suspended solids) for 4,4′-dichlorobiphenyl (a PCB compound used as an insulating fluid in electrical equipment) having $K_p = 214\ L_w\ kg_s^{-1}$. Figure 11.1.7 shows the expected distribution of the chemical between the dissolved and sorbed phases for a range of sediment concentrations, including those characteristic of sediment beds. This makes it clear that, in a water column with typical amounts of suspended sediment (10 to 100 mg_s/L), the dissolved phase is the most important fraction. Only at higher sediment concentrations (including those typical of bed

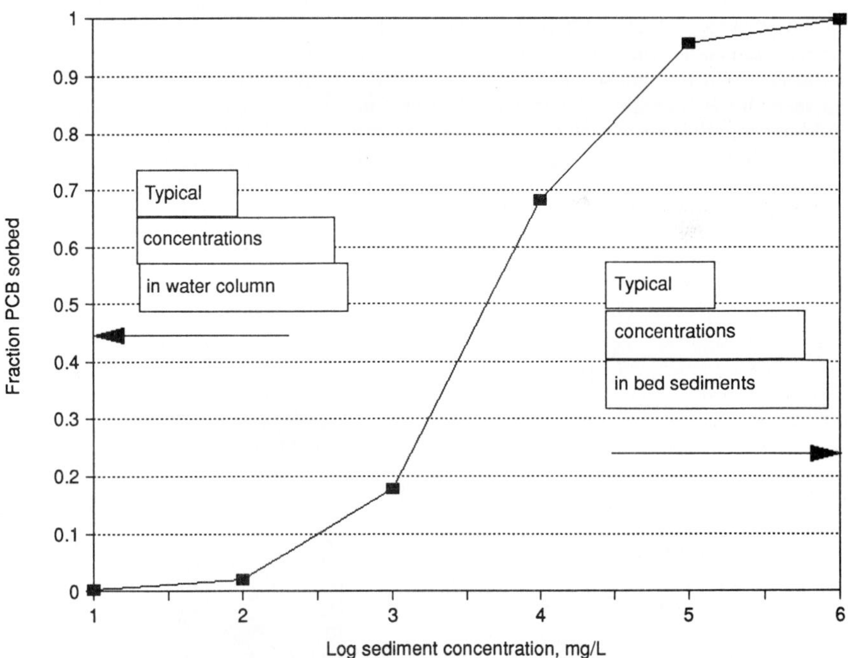

FIGURE 11.1.7 Relationship between dissolved and sorbed 4,4′-dichlorobiphenyl (PCB, electrical insulating fluid) for various solids concentrations. $K_p = 214\ L_w\ kg_s^{-1}$ and sediment specific gravity = 2.65.

sediments) is the sorbed phase important. The sorbed phase of pollutants on sediment is of importance only when high sediment concentrations are well mixed in the water column to facilitate equilibrium or when water is moving slowly through the bed to allow equilibrium. This changes dramatically for the most highly sorbed chemicals, however. For example, dioxin (a by-product of paper production) has $K_p = 10^5 L_w kg_s^{-1}$, and thus 50 percent (0.50) would be in the dissolved phase when as little as 10 mg_s/L of suspended sediments is present.

Sorption Kinetics. Sorption kinetics have been modeled using one and two compartments, and using interparticle diffusion as illustrated in Fig. 11.1.8. The mass transfer or chemical reaction viewpoint is typically adopted along with a conceptual definition of one or more boxes or compartments describing sorption sites or reservoirs (collection of sites). The mass transport approaches are essentially empirical and thus require calibration. The interparticle diffusion models may be sufficiently mechanistic to eventually not require calibration.

Sorption of Metals. Three important processes govern sorption of metals to solids: (1) the *physical adsorption* of metal ions in the external particle surface due to van der Waals forces (electrostatic forces between ions of different charges), (2) *chemical adsorption* due to chemical bonding (or reaction) between dissolved ions or molecules and particle surfaces,[31] and (3) *ion exchange* where there is an exchange of ions of like charges between the bulk solution and the surface. Since most natural particles have a negative surface charge, the exchange of ions usually involves dissolved cations and is thus called *cation exchange.* The capacity of a particular sediment or class of particles to exchange different cations can be measured in the laboratory and is referred to as the *cation exchange capacity,* which is normally measured in meq of cations per unit weight of solids (meq per 100 g or meq kg^{-1}). When ions attach to a surface, this is normally a reversible reaction. *Chemisorption,* where dissolved molecules react with surface molecules, also occurs. However, all of the processes affecting adsorption have not been well defined, and thus they are sometimes lumped together along with precipitation into a general process called *sorption* like that done for

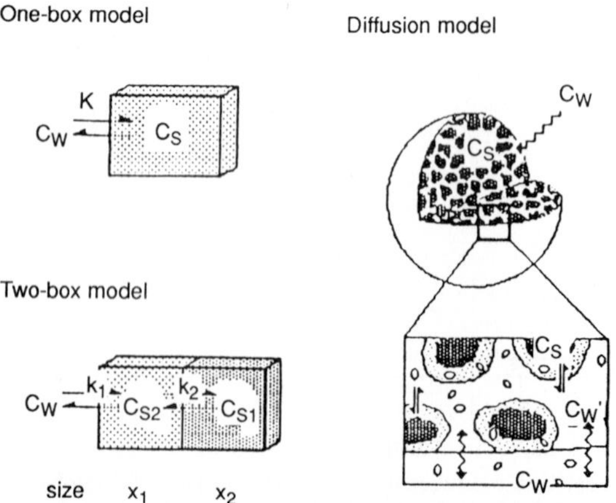

FIGURE 11.1.8 Different conceptual models of sorption kinetics. *(From Gschwend et al.[24].)*

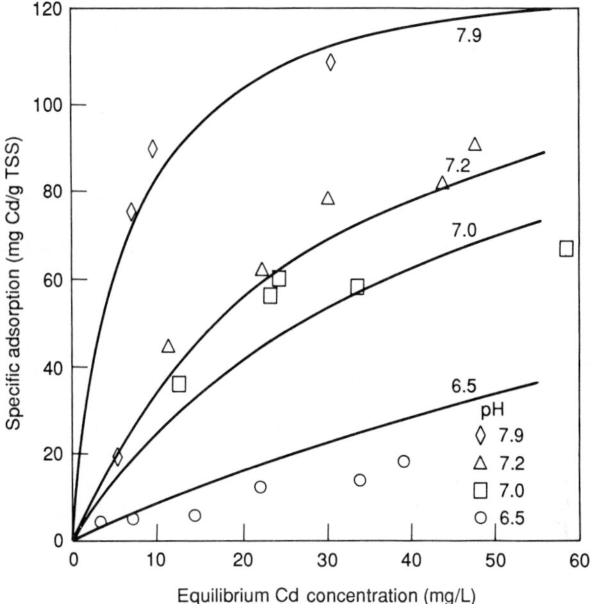

FIGURE 11.1.9 Measured distribution of cadmium between solid and dissolved phases for different pH values. The curves included represent predictions based on a surface complexion model of metal sorption described by Martin et al.[42] TSS is total suspended solids.

sorption of organic chemicals, which refers to the total distribution normally measured.

While changes in concentrations of a dissolved metal ion, competitive effects of other ions including other metal ions, effects of iron and manganese oxides, and effects of organic carbon have a significant influence on the extent of adsorption,[20,31] pH is the most important factor. Adsorption may vary from approximately 0 to 100 percent over a narrow range of pH that is often less than 2 pH units. The effect of pH variations on sorption isotherms for cadmium is illustrated in Fig. 11.1.9.

11.1.6 Mobility and Bioavailability of Metals

The ability of toxic metals to move through soils, sediments, aquifers, and surface waters depends on the geochemical conditions at a site as well as sorption to particles that may be present.[57] Although the mobility of metals is governed by the same basic principles (mass balance of metals and law of mass action) in groundwater and surface water, mobility is quite variable because of the sensitivity to variable geochemical conditions. The reactivity and toxicity of metals is also related to many of the same factors as mobility. Toxicity of algae to a metal is controlled by the free metal ion present rather than total metal present for most metals.

The mobility, reactivity, and *bioavailability* (ability of organisms to take up and concentrate a contaminant in water) of metals in natural waters depend on speciation and thus are more easily investigated using geochemical speciation computer models (see Table 11.1.6). These calculations are exactly like the general geochemical speciation calculations based on the law of mass action and the assumption that reactions of metals and natural geochemicals occur quickly. As with most natural major ions, most reactions of metals occur quickly enough to assume that equilibrium has been achieved except when phase reactions like precipitation and dissolution occur. When sorption reactions are also included, it is possible to make reliable predictions of mobility of metals in most natural waters.

Many sampling and analytical procedures, especially those appropriate for reconnaissance and screening-level studies, are limited to measuring the total amount of metals present but the mobility and bioavailability of metals cannot be determined from typical field measurements of total metals without conducting speciation calculations.

Metals may occur as (1) free ions; (2) metal-ligand complexes (e.g., $Al(OH)^{2+}$ or Cu-humate, where a *ligand* is simply an organic or inorganic molecule that will form a complex with a metal ion; humate is a natural organic compound that is found in almost all natural waters); (3) insoluble species (e.g., Ag_2S or $BaSO_4$ that precipitate); (4) adsorbed species (e.g., lead sorbed onto a ferric hydroxide surface of a mineral particle); (5) species held onto a surface by ion exchange (e.g., calcium ions on clay); and (6) species that differ by oxidation state (e.g., Mn^{2+} and Mn^{4+} and Fe^{2+} and Fe^{3+}). A number of factors influence the formation, mobility, and bioavailability of each of these species.

11.1.7 Other Processes Affecting Water Quality

Other processes affecting water quality include (1) hydrolysis reactions of chemicals with water molecules, (2) biodegradation of organic substances by bacteria, and (3) volatilization or gas exchange with the atmosphere (see Chap. 29).

Hydrolysis reactions are reactions with water that lead to transformations of pollutants by breaking a chemical bond and subsequent formation of a new bond. Hydrolysis is usually accelerated by acidic or basic conditions, although it can occur at neutral pH. Hydrolysis of metal ions is a complexation process. With organic pollutants, hydrolysis proceeds through the reaction of the compound with water followed by the formation of a new carbon-oxygen bond. The rate of the hydrolysis reaction for various organic compounds is dependent on pH and temperature with half-lives varying from a few seconds to 10^6 years.[7]

While many transformations of water constituents occur quickly, the *kinetics,* or time dependence, of some reactions are important determinants of water quality. These reactions primarily involve biological oxidation of carbonaceous compounds and nitrogen (oxygen demands), *biodegradation* of trace organic contaminants, and the uptake of nutrients for algae and bacteria growth (eutrophication). Biodegradation and eutrophication modeling are covered in Chap. 14.

In addition to microbiological mediation of redox reactions, it is possible to have significant abiotic redox reactions of importance. Synthetic organic contaminants also may be lost from water by reaction with an oxidant.

11.1.8 Acid Deposition

Since the 1970s, acid rain or, more appropriately, *acid deposition* (acid in dry dust depositing and rainwater) has become a public and scientific concern. A review of the acid deposition phenomenon is contained in the report *Acid Deposition: Long Term Trends.*[54]

The environmental effects of acid deposition include acidification of surface waters, poor health and productivity of forests and crops, deterioration of buildings and sculptures, and decline of human health. Figure 11.1.10 illustrates the relationship between the primary sources (emissions of sulfur and nitrogen oxides) and their effects in the atmosphere and biosphere. An equivalent amount of sulfur and nitrogen probably returns to the land surface as dry deposition as compared with the amount of acid in rainfall. Data on dry deposition are scarce, however. Acid deposition alone does not serve as an indicator of acidification because the role of forests, soils, and geologic materials must all be considered. Neutral rainwater infiltrating acidic soils will produce acidic stream flow.

The combustion of fossil fuels accounts for more than 90 percent of emissions of sulfur and nitrogen oxides into the atmosphere in eastern North America.[54] These oxides are transformed in the atmosphere into sulfates and nitrates in a complex series of reactions, and are subsequently removed from the atmosphere by wet and dry deposition.

The historical concern about the acidity or pH of precipitation has shifted to the deposition of sulfur, nitrogen, and other chemicals as understanding of the phenomena has improved. The acidity of wet deposition is the result of the concentration of these compounds as well as weak organic acids and alkaline compounds.

The geochemical characteristics of the watershed must also be considered in evaluating surface water impacts. Sulfate (SO_4^{2-}) is thought to be a major determinant of lake acidification. Sulfate moves through soils as a mobile, negatively charged anion. When this occurs, an equivalent amount of positively charged cations must be carried along to satisfy the charge balance. Thus, as the concentration of sulfate in solution increases, so does the concentration of cations. Watersheds composed of highly acidic soils contribute a higher proportion of acid cations (H^+ and Al^{3+}), thereby reducing surface water alkalinity while less acid soils will contribute a higher

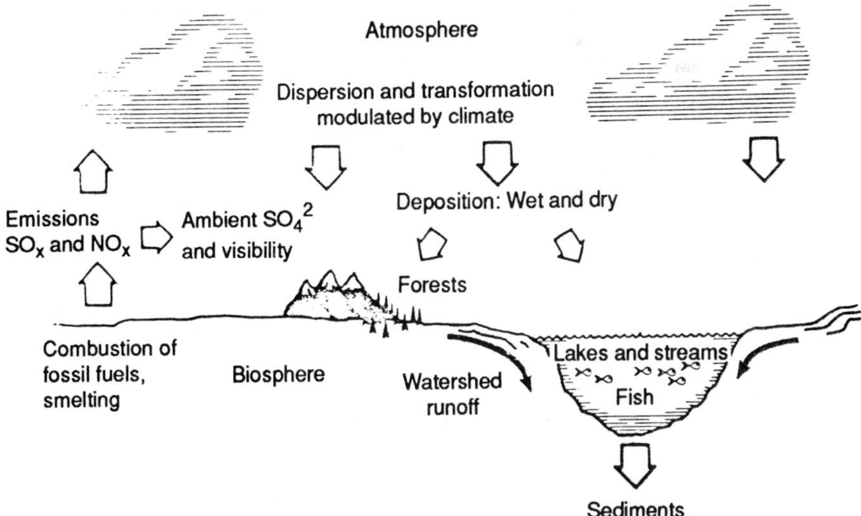

FIGURE 11.1.10 Acid deposition: sources and affected ecosystems. *(Reprinted with permission from* "Acid Deposition: Long-Term Trends," *1986. Published by National Academy Press, Washington, D.C.)*

proportion of base cations (Ca^{2+}, Mg^{2+}, or Na^{2+}). Sulfates are not mobile in all soils; watersheds may retain or release sulfate through biochemical reactions. Some watersheds contain sulfur-bearing minerals that can act as sulfur sources.

Nitrate fate and transport is complicated because nitrate is biologically active. Terrestrial nitrogen is incorporated in plant material during the growing season, and nitrogen entering surface waters may be rapidly consumed by aquatic biota. Thus excess nitrogen is incorporated into the nitrogen cycle, mitigating the impact of atmospheric emissions. However, for brief periods in the spring, nitric acid may be released from a watershed and cause a sharp, temporary decrease in aquatic pH.[54]

Differences in watershed geology, soil properties, and land-use practices in the watershed can cause surface waters to respond quite differently even when receiving equivalent atmospheric inputs of constituents. Figure 11.1.11 shows areas of sensitive soils and areas of present emissions of acidifying compounds.[59] See Fig. 3.6.2 for estimates of rainfall pH in the United States.

The biological effects of acid deposition can be considerable. Forests are affected by acid deposition in many industrialized countries, notably in eastern North America and Europe. Acid rain has caused a decline in fish populations in some freshwater lakes and streams in eastern North America, Canada, and the Scandinavian countries. Many species of fish cannot tolerate surface water pH below 5. Aluminum (Al^{3+}) can also reach toxic levels in acidified waters.

The relationship between emissions of sulfur and nitrogen compounds and terrestrial and aquatic effects is complicated and difficult to quantify. The scientific community has reached a consensus that acidic deposition is an important, but not the only, source of surface water acidification.[53,54]

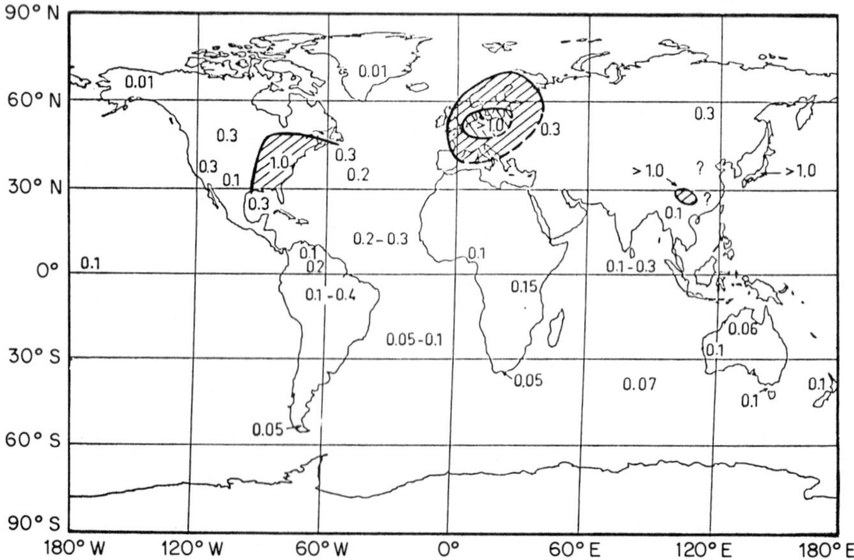

FIGURE 11.1.11 Areas of emissions and sensitive soils, showing present and potential problem areas. *(From Rodhe.[59] Used with permission.)*

11.2 CONSTITUENTS OF NATURAL AND POLLUTED WATERS

11.2.1 Dissolved Solids

Total Dissolved Solids (TDS). TDS is determined by evaporation of a filtered sample to obtain a residue whose weight divided by the sample volume is the total dissolved solids concentration. The TDS concentration of natural waters varies widely. Rainwater has a TDS concentration of < 10 mg L^{-1}; runoff from wet, well-drained regions can have a TDS concentration as little as 25 mg L^{-1}; seawater has approximately 35,000 mg L^{-1}; and a sodium chloride solution at saturation has in excess of 300,000 mg L^{-1} TDS. Saline waters are typically classified as follows:

Slightly saline	1000–3000 mg L^{-1}
Moderately saline	3000–10,000 mg L^{-1}
Very saline	10,000–35,000 mg L^{-1}
Briny	$>35,000$ mg L^{-1}

There are several reasons why TDS is an important indicator of water quality. Dissolved solids affect ionic strength, which has an impact on the mobility and transformation of metals and ionizable chemicals. In addition, TDS are a major determinant of aquatic habitat that affects both aquatic biota and irrigated plants. Many aquatic plants and animals are adapted to fresh waters or to salt waters and cannot survive when dissolved solids concentrations are too high or too low. Only a few of the higher-order animals can migrate between oceans and freshwater rivers. Some bacteria and other simple organisms survive in both fresh and brackish waters, but their metabolism is affected when moving from one to the other.

Monitoring of dissolved solids provides a gross measure of geochemical conditions and the load of constituents in rivers. The U.S. Geological Survey maintains an extensive U.S. network of sites where flows, dissolved solids, and other parameters are monitored. Their data may be used to estimate the rate of weathering in a basin and to track global balances of materials.

TDS is an important indicator of the suitability of water for drinking, irrigation, and industrial use. Dissolved solids increase as waters move over the land surface and through soils and aquifers. Use of the water for human consumption, industrial activities, and farming further increases TDS as the water is returned to surface waters. Eventually, TDS may increase until the waters can no longer be used or provide a proper wildlife habitat.

TDS affects saturation concentrations of dissolved oxygen and influences the ability of a water body to assimilate wastes. The rate of degradation of organic wastes is lower in salt waters than in fresh waters. Also eutrophication (the growth of algae in water) rates depend on total dissolved solids.

A mass balance of dissolved solids is useful to ensure that water quality models properly simulate the water balance. In estuary studies, salinity or TDS are good indicators of the mixing between fresh water and sea water.

TDS is determined by evaporating a sample of known volume to dryness and weighing the residue after a final oven-drying step. However, different oven-drying temperatures have been recommended so that it is appropriate to report the temperature used to evaporate the sample [e.g., mg L^{-1} (103°C)]. The ASTM[3] recommends drying for 1 h at 103°C or at 180°C. The higher-temperature evaporation is intended to drive off more of the water of crystallization for some residues, but Hem[28] notes that the water of crystallization is minor for dilute natural waters. *Standard Methods*[15] specifies drying at 103 to 105°C or at 180°C. The U.S. Geological Survey[61] recommends drying at 105°C overnight or at 180°C for 2 h.

TDS concentrations are used to check chemical analyses of individual major ions and constituents. The sum of the concentrations of individual constituents dissolved in the water sample, when compared with an independent measurement of total dissolved solids, should agree to within 1 or 2 percent for waters with moderate concentrations. Agreement to within 5 percent is generally adequate for speciation calculations and mobility estimates of metals. However, some constituents are affected by evaporative heating including bicarbonate ions, organic matter, nitrate, boron, and sulfate. When a water is high in calcium and sulfate, the water of crystallization retained in the residue may be equivalent to several hundred mg L^{-1}. When TDS exceeds 1000 mg L^{-1}, a calculated summation of constituent analytical values may be preferable to a residue-on-evaporation determination.

Specific Conductivity. The *specific conductivity* of water, or its ability to conduct an electric current, is related to the total dissolved ionic solids. Specific conductivity is defined as the reciprocal of the electrical resistance of a 1 cm^3 cube of a material at 25°C. To avoid any confusion with geophysical measurements of resistivity over a meter distance (ohms m^{-1}), it is appropriate to report standard specific conductivity measurements as millisiemens per m (mS m^{-1}, SI units) at the temperature in the field or corrected to 25°C. The standard unit of μmho cm^{-1} is 0.1 millisiemen m^{-1}. In the laboratory, conductance should be measured at 25°C to avoid the need for temperature corrections. Specific conductance approximately doubles in a 0.01 molar KCl solution from 0 to 35°C, at an approximate rate of 2 percent per °C.[28] However, the specific conductance of individual ions exhibits a different response to temperature changes so that temperature corrections for natural waters may be imprecise.

The specific conductivity of water varies from an absolute minimum of 0.05 microsiemen cm^{-1} (μS cm^{-1}) to 225,000 μS cm^{-1} for concentrated brines. Specific conductivities of precipitation range from approximately 2 to 100 μS cm^{-1}, surface and groundwaters range from 50 μS cm^{-1} up to the conductance of seawater of 50,000 μS cm^{-1}, potable waters normally range from 50 to 1500 μS cm^{-1}, and industrial wastes may approach 10,000 μS cm^{-1}. The specific conductivity of a water is often linearly related to dissolved solids as AC_s = TDS, where C_s = specific conductivity in μS cm^{-1} or μmho cm^{-1} at 25°C, TDS = total dissolved solids in mg L^{-1}, and A = a coefficient with units of mg cm L^{-1} μmho^{-1}. The value of A depends on the mixture of dissolved solids. The coefficient A has a typical value of 0.64 and a full range of 0.54 to 0.96. Natural waters can be represented by a value in the range of 0.55 to 0.75.[28] The higher values of A, up to approximately 1.0, are associated with high sulfate concentrations. Typically, long-term monitoring for several years of specific conductance and total dissolved solids is conducted at a site on a stream or in a well to derive a relationship between C_s and TDS. Figure 11.2.1 is a typical example for the Gila River in Arizona. Relationships between C_s and TDS have been developed for many U.S. Geological Survey monitoring sites and are reported in their annual data reports for each state. In addition, conductance is also linearly related to the concen-

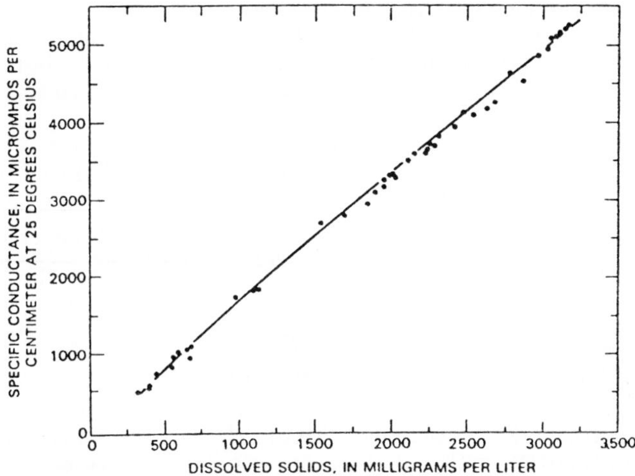

FIGURE 11.2.1 Example of a linear relationship between conductivity and total dissolved solids from daily composite samples collected at Bylas, Ariz., on the Gila River, from Oct. 1, 1943, to Sept. 30, 1944. *(From Hem.[28])*

trations of individual ions. The total concentrations of anions and cations in meq L^{-1} can be estimated as $0.01 \times C_s$ in $\mu S\ cm^{-1}$ at 25°C.[15]

Salinity. Salinity is another measure of dissolved solids, used primarily in oceanographic investigations and in studies of irrigation waters. Salinity is a concentration of the solids on a mass per mass basis (parts per thousand or g dissolved solids per kg of seawater, written as ‰) rather than the mass per volume basis that total dissolved solids in mg L^{-1} represents. Occasionally, brine concentrations are expressed as a weight percentage (i.e., 3.5 percent salt). The salinity of the oceans is fairly constant, ranging from 33 to 37‰ in over 95 percent of the world's oceans.

To be precise, salinity is defined as the inorganic matter in 1 kg of seawater after all halogens (Br and I^-) are converted to chlorine (Cl^-). Since salinity is difficult to measure directly, a practical salinity scale of 1978 was developed defining salinity relative to a standard KCl solution at 15°C. A mass of 32.4356 g of KCl in 1 kg of water was defined as a practical salinity of 35‰.[15]

Most standard field specific conductivity meters record salinity as well with a precision of at least ±0.01 ‰. Seawater salinities S in ‰ are related to specific conductivity C_s in $\mu S\ cm^{-1}$, as follows:[15]

$$S = a_o + a_1 R_T^{1/2} + a_2 R_T + a_3 R_T^{3/2} + a_4 R_T^2 + a_5 R_T^{5/2} + \Delta S \qquad (11.2.1)$$

where R_T is the ratio of C_s to conductivity of seawater at 35 g kg^{-1} (approx. 50,000 $\mu S\ cm^{-1}$) and the empirical coefficients a_i are $a_0 = 0.0080$, $a_1 = -0.1692$, $a_2 = 25.3851$, $a_3 = 14.0941$, $a_4 = -7.0261$, $a_5 = 2.7081$, and

$$\Delta S = \frac{T - 15}{1 + 0.0162\,(T - 15)}\,(b_0 + b_1 R_T^{1/2} + b_2 R_T + b_3 R_T^{3/2} + b_4 R_T^2 + b_5 R_T^{5/2}) \qquad (11.2.2)$$

T is in °C. The empirical coefficients b_i are $b_0 = 0.0005$, $b_1 = -0.0056$, $b_2 = -0.0066$, $b_3 = -0.0375$, $b_4 = 0.0636$, and $b_5 = -0.0144$. Equation (11.2.1) is valid for $S = 2$ to 42 ‰. For low salinities, the practical salinity scale has been extended to the salinity range of 0 to 40 ‰. The extended equation is written as

$$S = S_{PSS} - \frac{a_0}{1 + 1.5X + X^2} - \frac{b_0 f(T)}{1 + Y^{1/2} + Y^{3/2}} \tag{11.2.3}$$

where S_{PSS} is the value of salinity determined from the practical salinity scale given as Eq. (11.2.1), $a_0 = 0.008$, $b_0 = 0.0005$, $X = 400(C_s)$, $Y = 100(C_s)$, and the function $f(T)$ is

$$f(T) = \frac{T - 15}{1 + 0.0162\,(T - 15)} \tag{11.2.4}$$

It has been empirically determined that the relative proportion of the major ions in seawater is approximately constant. As a result, one of the major ions can be determined and used to estimate salinity. The major ion most often selected for this determination is chloride. *Chlorinity* is defined as the amount of chlorine equivalent to the total halide concentration (Br, CL) in seawater, as determined by the amount of silver (Ag), in grams, necessary to precipitate the Cl^- and Br^- (halogens) in 328.5233 g of seawater.[63] Originally, chlorinity was related to salinity as $S = 1.80655(Cl)$. With the definition of the practical salinity of 35 ‰, however, this is no longer precisely valid.

11.2.2 Inorganic Carbon and Buffering

Carbonate and bicarbonate are important to stream water quality because of the buffering capacity against changes in pH provided when dissolved carbon dioxide is consumed by reaction faster than it can be replaced by atmospheric gas exchange or bacterial decomposition. Phosphates and other chemical systems also provide buffering capacity in natural waters.

In addition to knowing the buffering capacity of a natural water, it is important also to understand the *buffering intensity.* Buffering intensity is a measure of how fast a system can react to maintain its pH value. Operationally, buffering intensity is defined as the slope of the titration curve at any point in the titration for a given solution (see Fig. 11.2.2). The speed of buffering reactions is important because the equilibrium of a natural water can involve many diverse interacting species, some of which involve precipitation and dissolution of solid phases. If the interactions with the solid phase are rapid, they assist in maintaining a certain pH level. If these reactions are slow or limited, less buffering capacity is available.

In most streams, inorganic carbon forms the bulk of the dissolved solids, and the bicarbonate concentration is generally held to within a moderate range by the carbonate equilibria. If a water is completely controlled by carbonate equilibria, then a pH of 6.4 is expected and approximately 160 mg L^{-1} of H_2CO_3 is expected. Typically, most streams contain less than 200 mg L^{-1} of H_2CO_3, but groundwaters may contain somewhat more. Rainwater typically contains less than 10 mg L^{-1} of H_2CO_3 and sometimes much less than 1.0 mg L^{-1}, depending on the pH. Typical values of bicarbonate concentrations are reported in Table 11.2.1.

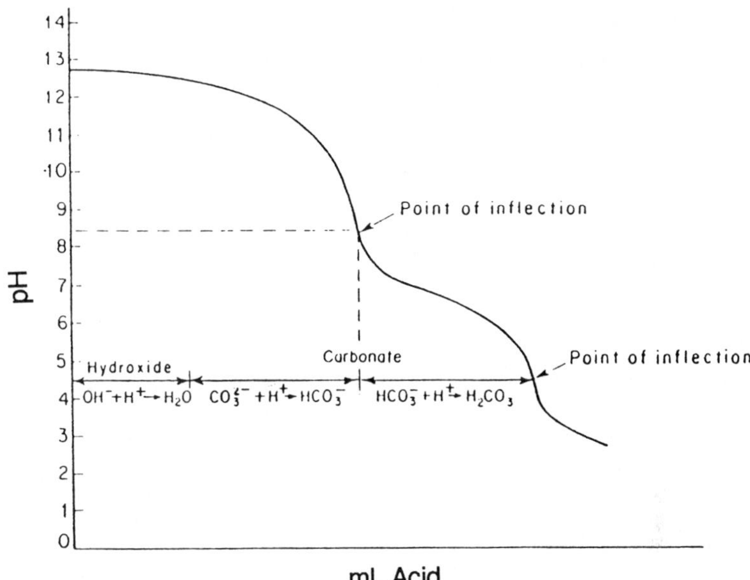

FIGURE 11.2.2 Typical alkalinity titration curve. *(From Sawyer and McCarty.[60] Used with permission.)*

11.2.3 Alkalinity and Acidity

Acidity and alkalinity are measures of the capacity of a water to neutralize added base or acid, respectively. As such, neither is a direct measurement of specific elements or compounds. Alkalinity is a measure of the net effect of cations and anions in a water. While it is not a quantitative measure of a specific chemical, alkalinity can be used with other chemical concentration measurements to estimate difficult parameters such as bicarbonate and carbonate. Acidity, when it is used, is somewhat qualitative.

Alkalinity. Alkalinity is a measure of buffering capacity defined using a proton (H^+) balance where the alkalinity plus $[H^+]$ is equal to the sum of the base cations C_B minus the sum of the anions of strong acids C_A in solution. The difference in the two sums is equal to the sum of the anions of weak acids in solution (derived from a mass and a charge balance[40]). Thus the total alkalinity may be defined as

$$\begin{aligned}\{\text{Alk}\} - [H^+] = C_B - C_A &= [Na^+] + [K^+] + 2[Ca^{2+}] + [Mg^{2+}] + [NH_4^+] \\ &+ [H^+] + \cdots - [Cl^-] - 2[SO_4^{2-}] - [NO_3^-] - 2[HPO_4^{2-}] \\ &- [HCO_3^-] - 2[CO_3^{2-}] - [OH^-] - \cdots \end{aligned} \quad (11.2.5)$$

or approximately,

$$\{\text{Alk}\} \approx [HCO_3^-] + 2[CO_3^{2-}] + [OH^-] - [H^+] \quad (11.2.6)$$

where the continuations (. . .) indicate other cations and anions of weak acids may be present and may affect alkalinity. Since this is a charge-balance equation, alkalin-

TABLE 11.2.1 Typical Concentrations of Inorganic Carbon and Alkalinity*

Location	Inorganic carbon concentrations†			Alkalinity	
	(moles L^{-1})	(mg $CaCO_3$ L^{-1})	(mg HCO_3 L^{-1})	(mg $CaCO_3$ L^{-1})	(mg HCO_3 L^{-1})
Rivers, average	10×10^{-4}	100	61	50	61
Range	1×10^{-4} to 50×10^{-4}	10–500	6–310‡	5–250	6–310
Loch Diuch, Scotland§					
Pore waters, aerobic sediments				0.098	0.12
anaerobic sediments				2.2	2.7
U.S. groundwaters, range	5×10^{-4} to 80×10^{-4}	50–800	30–500‡	5–250	6–310
Seawater	23×10^{-4}	230	140	125	150
Rainwater, typical	2×10^{-4}	20	10		
Range	0.1×10^{-4} to 0.5×10^{-4}	1–5	0.6–3	0–2	0–2.4

* From McCutcheon,[45] Stumm and Morgan,[63] and Hem.[28]

† Concentration of $[CO_2(aq)] + [H_2CO_3] + [HCO_3] + [CO_3^{2}]$.

‡ Extreme values of 1000 mg HCL_3^- L^{-1} have been observed in waters low in calcium and magnesium when sulfate reduction is releasing CO_2 (Hem[28]).

§From Thurman.[68] Originally from Krom and Sholkovitz.[36]

Note: Atmospheric $CO_2 = 0.033$ percent by volume in dry air or $P_{CO_2} = 3.3 \times 10^{-4}$ atmospheres; preindustrial $P_{CO_2} = 2.9 \times 10^{-4}$ atmospheres.[63]

ity is equal to the sum of the equivalent concentrations, which are the molar concentrations (bracketed values) multiplied by their valence (or charge), and alkalinity is expressed in equivalents L^{-1}. Listed in Eq. (11.2.6) are the major contributors to alkalinity normally found in natural waters. When carbonate species dominate all other ions, Eq. (11.2.5) reduces to the simpler form given in Eq. (11.2.6). Since carbon dioxide does not affect the net charge balance, changes in carbon dioxide alone do not affect alkalinity. In practice, the equivalent molar concentrations of the carbonate species ($[HCO_3^-]$ and $[CO_3^{2-}]$) are not known but can be calculated from the measurable quantities, pH and alkalinity. Therefore, the normal practice is to measure alkalinity by titration and to infer concentrations of $[HCO_3^-]$ and $[CO_3^{2-}]$.

The total alkalinity may alternatively be defined as a measure of the strong acid required to attain a pH by titration equal to that of a concentration of $[H_2CO_3] + [HCO_3^-] + [CO_3^{2-}]$ where carbonic acid is (H_2CO_3). Addition of this amount of acid results in all anions of weak acids (i.e., $[HCO_3^-] + [CO_3^{2-}] + [OH^-]$) being converted to dissolved carbon dioxide. Thus alkalinity can be determined by titrating a water sample with a strong acid (e.g., incremental addition of sulfuric acid, H_2SO_4) until a pH is achieved where virtually all of the solutes which contribute to alkalinity have reacted. For water samples where carbonates dominant alkalinity, this complete reaction occurs near a pH of 4.4 so that at this pH and below the concentrations of hydroxide, bicarbonate, and carbonate ions are approximately zero (see Fig. 11.1.4). The total carbonate is in the form of carbonic acid. The shift in carbonate species as pH changes, and the buffering capacity, is illustrated by a typical titration curve (Fig. 11.2.2) for a water sample whose alkalinity is dominated by carbonate species and with an initial pH above 12 and temperature of 20°C.

A variety of units are used to express alkalinity. Environmental engineers and water treatment specialists typically report concentrations in terms of mg $CaCO_3$ L^{-1}, whereas chemists and others prefer either meq L^{-1} or moles L^{-1}. In addition, units of mg HCO_3^- L^{-1} may be used. The various units and conversion factors are given in Table 11.1.2. Alkalinity, acidity, hardness, Ca^{2+}, and Mg^{2+}, are at times reported in units of mg $CaCO_3$ L^{-1}. The conversion factor from meq L^{-1} to mg L^{-1} is 1/50 when units of mg $CaCO_3$ L^{-1} are used. For example, 100 meq L^{-1} is 2 mg $CaCO_3$ L^{-1}.

As shown in Table 11.2.1, total alkalinity averages 123 mg $CaCO_3$ L^{-1} in rivers and ranges from 5 mg $CaCO_3$ L^{-1} to 250 mg $CaCO_3$ L^{-1}. For comparison, alkalinities of groundwaters, seawater, and rainwater are also given in Table 11.2.1. Groundwaters have more dissolved carbonates and other ions. Seawater is dominated by carbonate alkalinity but does have approximately 10 mg $CaCO_3$ L^{-1} of *noncarbonate alkalinity* (alkalinity caused by other anions besides carbonate and bicarbonate).

The alkalinity of a mixture of river waters, groundwaters, seawater, and rainwater cannot be determined from a simple algebraic mass balance. Even where equilibrium conditions can be assumed, the calculation requires the simultaneous solution of a number of mass balances and speciation reactions to determine the mixed water alkalinity. This is not a simple calculation but one best performed with a geochemical speciation model (see Table 11.1.6).

Acidity. *Acidity* in water is the "quantitative capacity of aqueous media to react with hydroxyl ions."[3,28] Acidity is produced by proton donors or acids, including water, carbonic acid, mineral acids, and others. *Mineral acidity* results from acids stronger than carbonic acid (H_2CO_3) that tend to displace the pH to values lower than 4.5.

The acidity of a water is determined in a manner very similar to alkalinity except that the sample is titrated with a strong base (e.g., NaOH) to a defined pH end point. For practical acidity determinations, the methyl orange dye and phenolphthalein dye are used to define how much base is required to achieve pH values of about 4.5 and 8.3. The amount of strong base required to achieve a pH of approximately 4.5 is a measure of *mineral acidity,* or the acidity due to the presence of acids stronger than carbonic acid. The amount of strong base required to increase the pH from 4.5 to 8.3 is a measure of the *carbon dioxide acidity.* Acidity, or *total acidity,* is determined by titration with a strong base to a pH of 8.3. Titration past a pH of 8.3 to about 10 may yield additional indications of acidity, but there are no rational ways of defining an end point for the titration.

The mineral acidity in a water sample is caused primarily by the reaction of metal salts (e.g., ferric chloride and other metallic compounds that ionize in water) with water molecules, oxidation of sulfates and iron pyrite, and ionization of organic acids from industrial discharges. These materials arise from metallurgical and synthetic chemical industries and drainage from mines, mine tailings, lean ore dumps, and "gob" piles (refuse piles of rock and other mining wastes[19,60]). Piles of tailings and mining wastes can cover 20 to 40 hectares (50 to 100 acres) or more. Some stream valleys in the mining regions of Colorado, Nevada, and Montana are completely covered with tailings. Other sources of acidity include fly ash ponds and coal piles at power plants. Salts of heavy metals, particularly trivalent metal ions such as Fe^{3+} and Al^{3+}, react with water to release hydrogen ions that are the source of mineral acidity.

Mineral acidity is also contributed by organic acids stronger than carbonic acids. In general, acidity may be contributed by hydrogen sulfide, organic acids, and polyvalent metal ions. Only when specific industrial effluents are present or mine drainage is occurring is mineral acidity important compared with carbon dioxide acidity. Normally carbon dioxide acidity is the dominant component in unpolluted waters. However, some waters (e.g., the Rio Negro in Brazil) are naturally acidic because of other causes.

Acidity measurements are important because of the corrosiveness of acidic water on steel, concrete, and other materials. Acidic waters are unpalatable so that they rarely present a human health problem.[60]

Acid Mine Drainage. Acid mine drainage and other mining wastes contain sulfur, sulfides, and iron pyrites that are converted to sulfuric acid and salts of sulfuric acid by sulfur oxidizing bacteria in aerobic waters. The oxidation of 1 mole of iron pyrite produces 4 equivalents of acidity, which cause the waters to reach a pH as low as 2.3, and occasionally 1.8.[19] Streams receiving acid mine drainage typically have distinct yellowish orange or yellow-brown iron deposits (ferric hydroxide) on the streambed.

11.2.4 Hardness

Originally *hardness* was defined as a measure of the ability of a water to precipitate soap, and water hardness is an ancient concept that predates modern water chemistry. Soap precipitation is controlled by the magnesium and calcium ions present and is computed from the sum of the two ions expressed in mg $CaCO_3$ L^{-1}. Hardness (in mg $CaCO_3$ L^{-1}) $= 2.497 \times$ Ca (in mg Ca L^{-1}) $+ 4.118 \times$ Mg (in mg Mg L^{-1}). Some European countries report hardness in "degrees." The French degree is 10 mg $CaCO_3$ L^{-1}; 1 German degree $= 17.8$ mg $CaCO_3$ L^{-1}; and the English or Clark degree $= 14.3$ mg $CaCO_3$ L^{-1}.

When hardness (sometimes referred to as *total hardness*) exceeds the sum of carbonate and bicarbonate alkalinity, that amount of hardness equal to the carbonate and bicarbonate alkalinity is referred to as *carbonate hardness.* The amount of hardness in excess of carbonate and bicarbonate alkalinity is referred to as *noncar-*

bonate hardness. Carbonate and noncarbonate hardness have also been referred to as *temporary and permanent hardness.* The terms *hard and soft waters* are somewhat arbitrary but may be classified based on hardness expressed in mg $CaCO_3$ L^{-1} as:

Soft	0–50 mg $CaCO_3$/L	Hard	121–180 mg $CaCO_3$/L
Moderately hard	51–120 mg $CaCO_3$/L	Very hard	>180 mg $CaCO_3$/L

The World Health Organization recommends that hardness not exceed 500 mg $CaCO_3$ L^{-1} but under ideal conditions, hardness should not exceed 80 mg $CaCO_3$ L^{-1}. However, the water is not particularly objectionable if hardness is less than 100 mg $CaCO_3$ L^{-1}.

In natural waters, hardness ranges from 0 to hundreds of mg $CaCO_3$ L^{-1}; 200 to 300 mg Ca CO_3 L^{-1} of hardness is common in many parts of the United States where the water has been in contact with limestone and gypsum. Groundwater from gypsum formations can exceed 1000 mg $CaCO_3$ L^{-1}.

Hardness is measured by (1) measuring magnesium and calcium ions and calculating the amount as noted above, and (2) titration with EDTA (ethylenediamine tetraacetic acid) which chelates, or forms a soluble complex, with ions such as calcium and magnesium. The first method is more accurate. Data reported before 1940 may be in terms of titration with soap. Since 1940, the chelation titration with EDTA or an equivalent agent is probably the basis for data reported. Most current data are based on the calculation from magnesium and calcium.

11.2.5 Color and Odor

Color and odor have no direct chemical significance, but they are qualitative indications of the chemical state of water. Both are usually related to organic matter in water, although odor can be a sign of pollution from industrial sources. Dark brown to yellow colored water normally results from the leaching of tannin and vegetable colors from plants and organic debris, but mineral sediments can lead to reddish or yellowish colors, as the names of many rivers and streams indicate. Odors arise from the decay of organic matter when dissolved oxygen is depleted, and also result from man-made pollutants such as phenols. *True color* is due to dissolved materials and *apparent color* is due to turbidity or suspended particles. Filtration is used to distinguish between the two types of color but some true color may be removed by filtration.

An arbitrary color standard of 500 units has been assigned to a solution of 1 g cobalt chloride, 1.245 g potassium chloroplatinate, and 100 mL concentrated hydrochloric acid in 1 L. Colored glass disks and tubes of a fixed length are used to judge color intensity in color units. Southeastern U.S. streams draining swampy areas may exceed 200 color units. Ten color units are barely noticeable to most people.

Pure water does not produce odor or taste. The exact physiological mechanism producing odor or taste is unknown. However, the sensation of odor is a powerful protective mechanism which warns humans of toxins. Odor, along with color, is one of the earliest indications that water pollution is occurring.

11.2.6 Sodium Adsorption Ratio

The hydraulic conductivity of soils containing clays is affected by the cations present in waters applied to the soil in irrigation. Of particular concern is the relationship between monovalent sodium and divalent calcium and magnesium. The relationship is usually represented by the sodium adsorption ratio (SAR) expressed as

$$\mathrm{SAR} = \frac{\{\mathrm{Na}^+\}}{\sqrt{\dfrac{\{\mathrm{Ca}^{2+}\} + \{\mathrm{Mg}^{2+}\}}{2}}} \tag{11.2.7}$$

where the brackets indicate that concentrations are expressed as meq L^{-1}. Hydraulic conductivity generally decreases with increasing sodium adsorption ratios. Values of SAR greater than 6 to 9 in a water used for irrigation indicate that a reduction of the hydraulic conductivity of the soil, and consequently the water infiltration rate, is to be expected if swelling, high-exchange-capacity clays are present. SAR values less than 6 are generally considered not to be a problem, while values greater than 9 may cause severe problems.[8] The reduction in infiltration capacity is also impacted by carbonate and bicarbonate concentrations. Carbonate deposition may remove Ca^{2+} and Mg^{2+} from the soil water, increasing the hazard due to sodium.

11.2.7 Dissolved Oxygen

Dissolved oxygen content is a measure of the ability of surface waters to support aquatic life. Oxygen is poorly soluble in water, and the *oxygen saturation concentration* is dependent on three primary parameters—temperature, atmospheric pressure, and dissolved solids. At 1 atm of pressure, the solubility of oxygen in fresh water ranges from 14.6 mg L^{-1} at 0°C to 6.4 mg L^{-1} at 40°C. See Table 14.4.5 and Eq. (14.4.12). Dissolved solids concentrations of 36.1 ‰ salinity reduce oxygen saturation concentration to 11.4 mg L^{-1} at 0°C and to 5.3 mg L^{-1} at 40°C.

The low solubility of oxygen in water is the primary factor that requires treatment of liquid wastes before discharge to a receiving water. The presence of dissolved oxygen determines whether waste material is degraded by aerobic (with oxygen) or anaerobic (without oxygen) processes. Aerobic processes use oxygen for the oxidation of organic matter in wastes and produce relatively innocuous end products. Anaerobic processes degrade wastes without oxygen in a slower process and the degradation products, such as hydrogen sulfide and methane, are often obnoxious. The critical conditions for dissolved oxygen deficiency typically occur during the late summer months when temperatures are high, saturation concentrations are low, biological processes are enhanced, and stream flow is low.

11.2.8 Biochemical Oxygen Demand (BOD) and Chemical Oxygen Demand (COD)

BOD is the amount of oxygen consumed by microorganisms while stabilizing or degrading carbonaceous and nitrogenous compounds under aerobic conditions. The BOD test is widely used as an indicator of the strength of municipal and industrial wastes and is an important test used in protecting aquatic life from oxygen deficiency. It is the primary regulatory tool used in limiting discharge of these wastes to water; and the BOD test is a key in defining the technology-based waste treatment requirements (e.g., secondary treatment or equivalent) required under the U.S. Clean Water Act. BOD modeling is reviewed in Chap. 14.

The COD test measures the total quantity of oxygen required for oxidation of a waste to carbon dioxide and water. It is sometimes used as an approximate substitute for BOD as the COD test can be performed in 3 h instead of the 5-day period required for the BOD test. During the COD test, all organic matter is converted to CO_2 and water in contrast to the BOD test, in which only biologically reactive carbon is oxidized. As a result, COD values are greater than BOD values. The primary limitation of the COD test is the inability to distinguish between biologically degradable and biologically refractory material (nonbiodegradable material).

11.2.9 Nitrogen

Occurrence in Natural Waters. The nitrogen cycle (Fig. 11.2.3) is important to water quality for several reasons. *Nitrification* (oxidation of ammonia and nitrite to

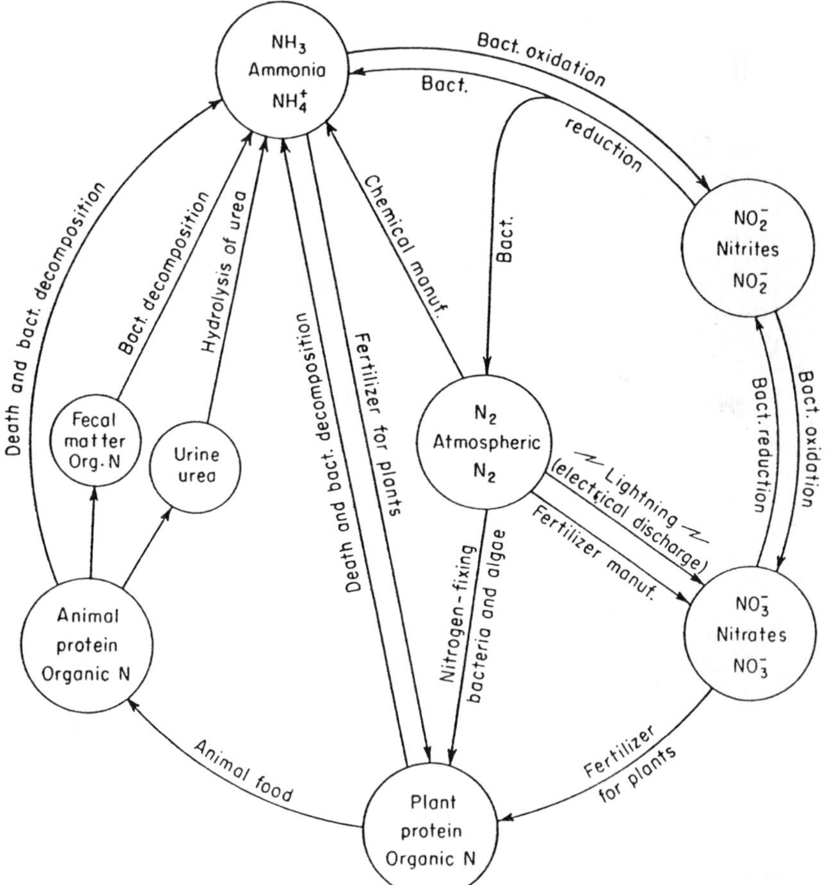

FIGURE 11.2.3 The nitrogen cycle. *(Reproduced from Ref. 60 with permission.)*

nitrate) consumes dissolved oxygen in the water column and streambed. Ammonia and nitrate are also important nutrients for the growth of algae and other plants. Excessive nitrogen can lead to *eutrophication.* Nitrogen is also an important component for cell building by bacteria and other aquatic animals.

Nitrogen gas in the atmosphere is a large reservoir that maintains equilibrium concentrations in surface waters. Like oxygen, atmospheric nitrogen varies little. At 20°C and 1 atmosphere of pressure, the saturation concentration of dissolved nitrogen gas is approximately 15 mg L^{-1}.[60] Like oxygen, nitrogen is only slightly soluble and is nonreactive with water.

Occasionally, supersaturation of nitrogen occurs owing to decreases of pressure. The most common occurrence is below high dams that release bottom waters in which the dissolved nitrogen gas is at equilibrium at lower temperatures and higher pressures. Fish get the "bends" when the nitrogen and other gases come out of solution to form bubbles in their blood vessels.

Humans influence the stream nitrogen balance in at least three related ways. Nitrogen gas is used to manufacture nitrate fertilizers, ammonia, and nitrogen-containing explosives.[71] Significant quantities of nitrogen oxides are released from automobiles, fossil fuel burning, and other industrial activities. Nitrogen is contained in wastewaters discharged into streams. The important forms of nitrogen include organic nitrogen, ammonia, nitrate, and cyanide (CN^-). Ammonia (NH_3 and NH_4^+) and cyanide are toxic to biota, especially when large concentrations are present. Ammonium and cyanide ions form complexes with metals, thus affecting the mobility of metals in water.

Forms of nitrogen are both oxidized and reduced in the stream environment, as shown in Fig. 11.2.3. In anaerobic stream sediments, denitrification produces nitrogen gas from the reduction of nitrate. Denitrification can also occur in anaerobic zones in the interior of particles suspended in the stream and along the stream bottom. The released nitrogen gas diffuses into the water column and influences the nitrogen gas exchange with the atmosphere or is occasionally lost to the atmosphere as bubbles. Under anaerobic conditions, organic and sometimes inorganic carbon is oxidized using nitrate as the oxygen source.

The principal water quality criteria for nitrogen focus on nitrate and ammonia.[74] The effects of nitrite have been investigated but stream concentrations rarely approach deleterious levels. Nitrate concentrations exceeding 10 mg N L^{-1} (44 mg L^{-1} as NO_3) present a potentially serious public health problem. Concentrations of 11 to 40 mg N L^{-1} of NO_3 may cause *methemoglobinemia* (blue babies) in infants under 6 months of age.[35]

Stream concentrations are on the order of 0.5 to 3 mg N L^{-1}. In the absence of wastewater discharges and significant nonpoint sources, concentrations of nitrates in streams are on the order of 0.1 to 0.5 mg N L^{-1}. Streams fed by shallow groundwater draining agricultural areas, however, may be more heavily polluted with nitrate (approaching 10 mg N L^{-1}). Farm animals produce large quantities of organic wastes that are oxidized to nitrate in the soil. Since nitrate is almost exclusively in a dissolved form, it is easily carried to streams with runoff and quickly leaches into shallow groundwater. Fertilizer use, especially on irrigated fields, contributes high concentrations of nitrate.

Ammonia concentrations may approach 5 mg N L^{-1} downstream of wastewater discharges in smaller streams, but generally concentrations are less than 3 to as low as 0.5 mg N L^{-1}. Concentrations higher than 0.5 mg N L^{-1} tend to cause significant ammonia toxicity to fish and other organisms. Concentrations may be as low as 0.01 mg L^{-1} in uncontaminated stream reaches. Ammonia concentrations in groundwaters are normally very low because of sorption of ammonia onto the soil.

Concentrations of ammonia in untreated sewage may exceed 30 mg L^{-1}. Ammonia concentrations of 6 mg N L^{-1} are typical of combined sewer outflows.

Organic nitrogen concentrations in streams are usually somewhat less. On average, organic nitrogen concentrations in untreated sewage are approximately 20 mg L^{-1} and higher. Concentrations may be as low as 0.2 mg L^{-1} in surface waters and below levels of concern in groundwater. As much as 20 mg L^{-1} occurs in treated wastewaters. Combined sewer overflows contain 1 to 2 mg L^{-1} of organic nitrogen.

Nitrite concentrations in many stream segments do not exceed 0.5 mg N L^{-1}. The highest concentrations are normally observed downstream of sewage discharges. Away from sewage outfalls and urban nonpoint sources, concentrations are lower and many times not detectable (i.e., less than 0.002 mg N L^{-1}).

Domestic sewage contains about 15 to 100 mg L^{-1} of *total nitrogen.* Approximately 50 to 60 percent of the wastewater total nitrogen is ammonia. The remainder is organic nitrogen. Nitrite and nitrate concentrations are usually negligible in wastewaters. Unless specifically designed to do so, wastewater treatment plants do not remove significant amounts of nitrogen.

Downstream of point sources of nitrogen, there is a distinctive sequence of change in the forms of nitrogen that occur as the nitrogen is oxidized. The initial concentrations of organic nitrogen and ammonia are high and these generally decrease in the downstream direction. Nitrite builds up to a maximum concentration and then decreases back to undetectable levels. Nitrate continually builds up until all of the nitrogen is converted to nitrate. Nitrate concentrations begin to decrease as phytoplankton uptake begins, but this normally occurs after the oxidation of organic nitrogen and ammonia has been largely completed. McCutcheon[45] discusses typical concentration profiles of organic nitrogen, ammonia, nitrite, and nitrate in streams.

Measurement of Nitrogen. There are a number of ways to measure the various forms of aquatic nitrogen. Figure 11.2.4 shows the relationship between the various types of measurements and the forms of nitrogen in water. Typical measurements of nitrogen for stream surveys include Kjeldahl nitrogen (organic nitrogen plus ammonia); ammonia; nitrite plus nitrate; nitrite; and nitrogen in plants as determined by chlorophyll *a* plus phytoplankton stoichiometry (relationship among carbon, oxygen, hydrogen, nitrogen, phosphorus, and chlorophyll *a*). Measurement of *dissolved nitrogen* involves filtration of water through a 0.45-μm filter whereas *total nitrogen* measurements are made with unfiltered water. Organic nitrogen is usually determined from the difference in Kjeldahl nitrogen and ammonia nitrogen measurements, but it can be measured directly. Organic nitrogen measurements are the same as the Kjeldahl measurement, but the ammonia originally in the sample is removed before organic nitrogen is converted to ammonia for the measurement. Nitrate is determined from the difference in nitrite plus nitrate minus nitrite.

Total nitrogen is composed of organic nitrogen, ammonia, nitrite, and nitrate, or measurements of Kjeldahl nitrogen and nitrite plus nitrate. Nitrogen gas is not considered in typical measurements. In addition, nitrogen oxides and other intermediate products of denitrification are not normally included.

Particulate organic nitrogen consists of detritus and other dead cell matter plus phytoplankton and other living protoplasm. *Dissolved organic nitrogen* consists of humic and fulvic acids, macromolecules, organic colloids, and other organic molecule fragments. There is a continuum of particle sizes that is operationally divided into dissolved and particulate matter according to what will pass through a 0.45-μm filter.

A separation of living and nonliving organic nitrogen is necessary but difficult to achieve with the measurements currently used. Both types represent very different

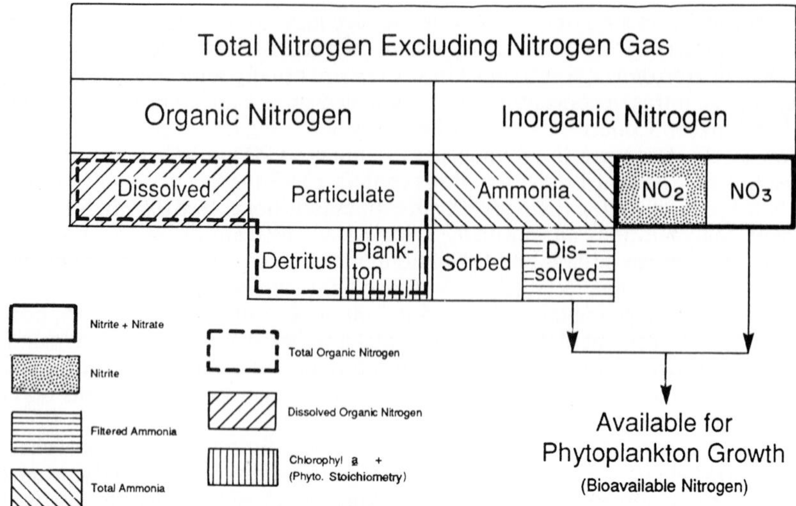

FIGURE 11.2.4 Forms of nitrogen present in water and the types of measurements available. *(From McCutcheon.[45])*

reservoirs of nitrogen that behave differently. Water quality models must track organic nitrogen contained in living phytoplankton and the nonliving organic nitrogen that serves as substrate for some bacteria. Current water quality models do not simulate the organic nitrogen in bacteria. In many cases, bacteria represent a negligible portion of the primary biological productivity in aquatic systems.[17,58]

Living phytoplankton are often a significant portion of the organic nitrogen present in surface waters, especially in eutrophic waters. Accounting for the amount of living biomass is important to determine the state of eutrophication and amount of nutrient uptake that is occurring in a water body.

The amount of organic nitrogen contained in the living biomass is often estimated by measuring the *chlorophyll a* present in phytoplankton cells. On average, phytoplankton cells contain 1.5 percent chlorophyll *a* on a dry weight basis.[15] Biomass is typically 6 to 7 percent nitrogen,[9,58] so the ratio of organic nitrogen to chlorophyll *a* is approximately 4 to 5 with a range of 2.7 to 29.

Measured concentrations of ammonia, nitrite, and nitrate are normally reported in terms of *equivalent nitrogen concentrations* and referred to as *ammonia-nitrogen, nitrite-nitrogen,* and *nitrate-nitrogen,* respectively. For environmental engineers and others dealing with polluted streams, the concentration units of mg N L^{-1} are preferred. Ecologists, geochemists, and others dealing with more pristine streams prefer the reporting concentrations in μg N L^{-1} or μg NH_3 L^{-1}, μg NO_2 L^{-1}, and μg NO_3 L^{-1}. Micromoles L^{-1} are also used in some cases. Reports in terms of equivalent concentrations are frequently confusing but quite useful in water quality studies. When all nitrogen species are expressed in terms of nitrogen, it is a simple matter to understand the degree of pollution and the distance the nitrogen may have traveled. In addition, a mass balance is easily performed when concentrations are expressed in terms of nitrogen. The conversions from ammonia, nitrite, and nitrate to equivalent nitrogen concentrations are 1 mg N L^{-1} = 4.427 mg NO_3^- L^{-1} = 3.285 mg NO_2^- L^{-1} = 1.215 mg NH_3^+ L^{-1}. See Table 11.1.2. For example, 5.12 mg NO_3^- L^{-1} = 1.16 mg N L^{-1}.

11.2.10 Phosphorus

Phosphorus is an important component of organic matter. As a constituent of nucleic acids, it is vital for all organisms. In many rivers, phosphorus is the limiting nutrient that prevents additional productivity (biological activity) from occurring. When this is the case, the control of phosphorus in sewage, industrial wastewaters, and non-point-source pollution is important. Critical levels of phosphorus causing excessive algae growth can be as low as 0.01 to 0.005 mg P L^{-1},[60,67] but concentrations are more frequently on the order of 0.05 mg P L^{-1}.[69] In contrast to streams and rivers, estuarine and coastal water productivity is most often limited by nitrogen.

The origin of phosphorus in streams is the mineralization of phosphates from the soil and rocks, or drainage containing fertilizer or other industrial products. The principal components of the phosphorus cycle, shown in Fig. 11.2.5, involve organic phosphorus and inorganic phosphorus, in the form of orthophosphate (PO_4^{3-}). Many bacteria and fungi break down organic phosphorus to release poly- or orthophos-

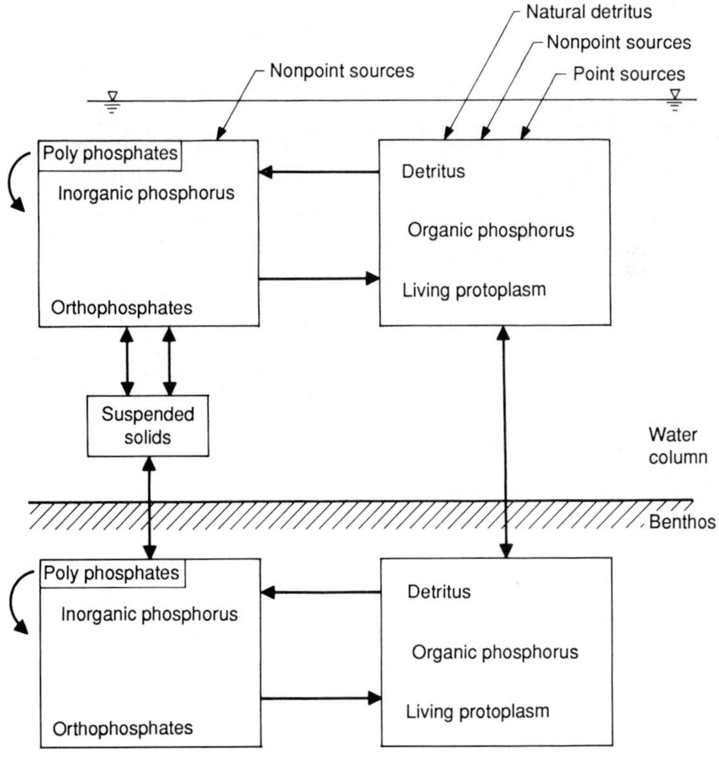

FIGURE 11.2.5 Phosphorus cycle in water. *(From McCutcheon.[45])*

phate. Orthophosphate is soluble and considered to be the only biologically available form of phosphorus. Hydrolysis slowly converts any polyphosphate into orthophosphate. Since phosphorus strongly associates with solid particles and is a significant part of organic material, sediments influence water column concentrations and are an important component of the phosphorus cycle in streams. In addition to sorption of phosphates, precipitation from solution at high pH is also important. At low pH, phosphates are mobilized into solution and often leach from bed sediments undergoing anaerobic decomposition.

The major sources of phosphorus in streams are natural organic material, organic and inorganic phosphorus in wastewaters and nonpoint sources, and phosphate detergents in wastewaters. In the 1950s and 1960s, sodium phosphate was used as a "builder" in household detergents to increase cleaning power. Starting in the 1960s, various states and cities banned phosphate detergents to avoid expensive treatment alternatives.

Phosphorus is generally present in streams in low concentrations—approximately 0.1 mg L^{-1} or less. Polluted segments of a stream may contain 1 mg L^{-1} or more. Treated sewage typically contains 5 mg P L^{-1} dissolved phosphorus.[69] However, phosphorus may be removed from wastewater streams by coagulation or advanced biological treatment. Meybeck[49] estimates that the natural background concentration of dissolved phosphates is approximately 0.01 mg P L^{-1}. Natural background concentration of total dissolved phosphorus is about 0.025 mg P L^{-1}.

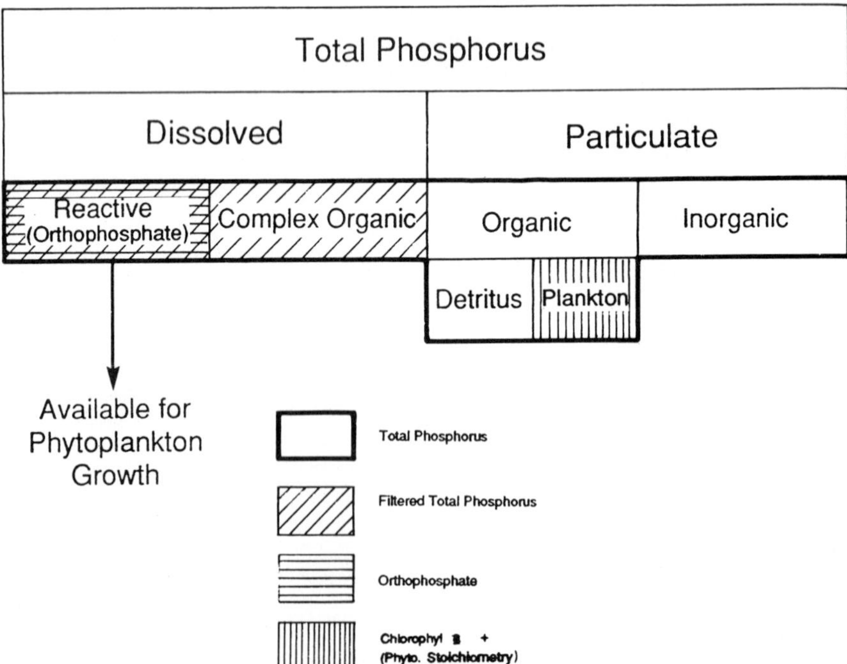

FIGURE 11.2.6 Forms of phosphorus present in water and the measurements available. *(From McCutcheon.[45])*

Particulate phosphorus is about 95 percent of the total phosphorus in most cases, however. Meybeck also estimates that waste disposal and other human activities cause a 10- to 100-fold increase in dissolved inorganic phosphorus concentrations in European and North American rivers. For example, a typical measurement of phosphorus in the Ohio River near its mouth was 0.58 mg L^{-1} during the 1970s. Upstream, the phosphate concentration was measured as 0.1 mg L^{-1}; 30 mg L^{-1} was observed in Peace Creek in Florida and this was related to phosphate mining in the area; 4.6 mg P L^{-1} of phosphates was observed in Powder River in Oregon (a tributary to the Snake River) and related to waste disposal in the stream. Orthophosphate concentrations tend to remain constant over long reaches of rivers, unless photosynthesis is occurring or until loads enter.

Important methods of measurement include detecting orthophosphate and total phosphorus. Figure 11.2.6 indicates that there are five forms of phosphorus in streams. These include orthophosphate (and the polyphosphates that are converted to orthophosphate), complex dissolved organic molecules, detritus and other filterable (0.45 μm filter) organic debris, plankton, and any suspended inorganic precipitates.

Phosphate concentrations are normally reported in mg L^{-1}, especially when polluted waters are sampled. Studies of more pristine streams may report concentrations in μg L^{-1} or micromoles L^{-1}. Environmental engineering studies normally report orthophosphate concentrations in terms of phosphorus using: 1 mg P L^{-1} = 3.066 mg PO_4^{3-} L^{-1}.

Measurement of total phosphorus, filtered total phosphorus, orthophosphate, and chlorophyll *a* (plus knowledge of the stoichiometric ratios of phosphorus to other cell constituents) allows definition of the important components of phosphorus. Only the amounts available in detritus, bacteria, and insoluble precipitates cannot be defined too well. However, for practical studies these measurements are adequate.

11.2.11 Microorganisms

Microorganisms are present in all surface waters but are typically in low concentrations in groundwater, where the filtering action of aquifer materials and long retention times reduce their concentrations. The significant microorganisms in water quality studies are viruses, bacteria, fungi, algae, and protozoa.

Standard Methods[15] includes a number of tests for microbial examination of water samples. These tests include the *coliform test,* which is generally accepted as the principal microbiological indicator of human contamination. Decades of experience have established the significance of the coliform group density as measure of the degree of pollution from sewage because *E. coli* is an abundant bacterial species in the human intestine. Coliform bacteria are present in the intestines of all warm-blooded animals and are excreted with feces. Their presence is an indication that other *pathogenic* (disease-causing) bacteria are present. Tests are available to differentiate the *Fecal coliform (F. coli)* bacteria from the larger coliform group (total coliforms) as a more specific indicator of warm-blooded animal contamination. This test is useful because the nonfecal members of the coliform group, which may have originated in soils, can be expected to live longer in the relatively unfavorable environment of surface waters. *Fecal streptococci* is also used as an indicator of fecal contamination.

Tests for coliform organisms and *F. strep.* include a multiple-tube fermentation test that results in a *most probable number* (MPN) index. This is an index of the number of coliform bacteria present, in contrast to the direct count of coliform colonies given by the more specific *membrane filter* procedure. In both procedures, coliform density is reported as either the MPN or membrane filter count per 100 mL.

Other indicator organisms used in the examination of recreational waters (swimming pools, whirlpools, and naturally occurring waters) include *Pseudomonas, Streptococcus, Staphylococcus,* and in rare cases, *Legionella.* Viruses may also be examined, as well as some spore- and cyst-producing bacteria. Routine testing for enteric viruses is not currently recommended in *Standard Methods,*[15] but special circumstances such as disease outbreaks or wastewater reclamation may dictate testing for these microorganisms. Iron and sulfur bacteria are of concern in surface water, and water treatment and distribution systems, especially those for industrial use. Iron and sulfur bacteria are also of concern in well waters where their presence may reduce pumping efficiency by plugging. *Actinomycetes* can cause difficult-to-remove earthy-musty odors in municipal water supplies. Though originally considered to be fungi, they have been identified as filamentous, branching bacteria. Traces of these microorganisms can be a problem in water supplies since certain of their metabolites can be detected by humans at concentrations below 20 ng L^{-1} and massive outbreaks can disrupt wastewater treatment processes.

Fungi, including yeasts and filamentous species or molds, may be found wherever nonliving organic matter occurs, but some species are pathogenic. Their water quality significance is that they can survive without sunlight and thus persist in water distribution systems where they can cause taste and odor problems at low densities and can clog systems at high densities.

Algae, or *phytoplankton,* are free-floating one-celled organisms that are ubiquitous in surface waters. Many are a primary food source for higher organisms. Most algae are photosynthetic, and the concentration of the photosynthetic pigment *chlorophyll a,* which constitutes 1 to 2 percent of their dry weight, is often used to estimate algal biomass. Algae have long been used as indicators of water quality. Some species flourish in highly eutrophic waters and develop noxious blooms, causing offensive tastes and odors. Excessive algal growth depletes dissolved oxygen and causes toxic conditions, resulting in fish kills, animal deaths, and human illness. Because of their short life cycle, algae respond rapidly to changing water quality. However, their usefulness as a indicator of water quality is limited by their transient nature and patchy distribution. In addition, algal response to pollution may occur at some distance from pollution sources.

Periphyton are a community of microscopic plants and animals that live on submerged surfaces. Unlike phytoplankton, which do not respond immediately to pollution sources, there are often dramatic changes in the periphyton community immediately downstream of waste sources. However, like phytoplankton, their patchy nature limits their use as a quantitative indicator.

Pathogenic *protozoa* of concern in drinking water supplies and wastewater are *Giardia lamblia, Entamoeba histolytica,* and *Cryptosporidium.* These organisms can cause diarrhea or gastroenteritis. *Giardia* is the most frequently identified agent in waterborne outbreaks. More than 28,000 cases of giardiasis were reported in the United States between 1965 and 1984. *E. histolytica* causes amoebic dysentery, resulting in an average of 28 deaths per year in the United States. *Cryptosporidium* can cause a cholera-like diarrhea and is of increasing concern as a human pathogen. *Schistosomiasis,* found in stream banks in Africa, Asia, and South America is a major cause of water-borne disease in the tropics.

11.3 WATER QUALITY STANDARDS

11.3.1 Standards, Criteria, and Water Quality Protection in the United States

Water is a limited resource subject to many demands. As the population and technology expands, the demands on available water resources increases as does the complexity of environmental issues. The intended use of water may be limited by both its quality and its quantity. Poor water quality can affect aquatic life as well as the use of water by humans for drinking, recreation, agriculture, and industry. Water quality conditions that affect the aesthetics (e.g., taste, odor) of these uses are also considered detrimental and often the subject of regulation. Water quality is regulated by the development of criteria and standards:

- A *water quality criterion* is that concentration, quality, or intensive measure (e.g., temperature) that, if achieved or maintained, will allow or make possible a specific water use. A criterion may be a concentration that, if not exceeded, will protect an organism, aquatic community, or designated use with an adequate degree of safety. A criterion may also be a narrative statement concerning some desirable condition. While water quality criteria are often the starting point in deriving standards, criteria do not have a direct regulatory impact because they relate to the effects of pollution rather than its causes.
- A *water quality standard* is the translation of a water quality criterion into a legally enforceable ambient concentration, mass discharge or effluent limitation expressed as a definite rule, measure, or limit for a particular water quality parameter. A standard may or may not be based on a criterion.

Standards may differ from criteria because of prevailing local quality conditions such as natural impairment of water quality. For example, some natural waters may exceed some water quality criteria even in the absence of anthropogenic pollution. Standards and criteria may also differ from one place to another owing to economic considerations, the perceived importance of a particular ecosystem, or the degree of safety required for a particular water.

Historically, the first effects of poor water quality that were the subject of criteria and standards were those affecting human health. For example, the control of infectious disease was the first goal of water quality regulations, followed by regulation of *toxic chemicals* (chemicals which, through their chemical or physical action, produce a harmful biological effect), then by regulation of those conditions that impact the convenience of use or aesthetics of water. For example, the quality of drinking water is impacted by its taste and odor, even though this taste or odor may not have serious health effects. The effect of water quality on the health of aquatic ecosystems is also an important consideration.

Presently, the primary U.S. federal agency responsible for the protection of water quality is the Environment Protection Agency (EPA), established in 1970. In addition to the EPA, over 40 other U.S. federal agencies and each state have authority over various aspects of water quality and quantity regulation.

In addition to permit authority, if authorized by the EPA, states are responsible for developing comprehensive state water management plans, including the establishment of water analysis (e.g., assessment, inventories, revisions of standards, and waste load allocations) and implementation programs (e.g., treatment works programs, use management programs, and regulatory programs).

11.3.2 U.S. Water Quality Criteria

In the United States, surface water quality criteria were established by the EPA and its predecessor agencies in order to guide the formulation of standards. These criteria were first published in 1968 in the report of the National Technical Advisory Committee to the Secretary of the Interior (the "Green Book," Federal Water Pollution Control Administration[77]). Section 304(a)(1) of the Clean Water Act of 1972 required the EPA to publish and periodically update ambient water quality criteria. Water quality criteria have been updated and published regularly,[52,70,74] and a series of individual water quality criteria documents were published for those pollutants listed as toxic under section 304(a)(1) of the Clean Water Act. In 1986, the EPA published "Quality Criteria for Water 1986" (the "Gold Book"[74]), which includes summaries of all contaminants for which EPA had established criteria recommendations up until that time. More recently, a number of updates have been published.

Human Health Criteria. Criteria developed for the protection of human health are based on the carcinogenic or toxic effects of pollutants. These criteria are based on human health assessments, the goal of which is to estimate ambient water concentrations that do not result in adverse health effects. In the case of suspected carcinogens, criteria are based on levels of cancer risk. Human health assessments contain four elements: *exposure, pharmacokinetics, toxic effects,* and *criteria formulation.*

Criteria formulation establishes a "safe" level of intake for the contaminant. Noncarcinogenic criteria calculations are usually based upon verified *reference dose* (Rfd) values (formerly *acceptable daily intake,* or ADI) in units of mg kg^{-1} day^{-1} (mg of contaminant per kg of body weight per day), which are derived using *no-observable-adverse-effect levels* (NOAEL) from human or animal toxicity studies. The reference dose is an estimate of the daily exposure to the human population that can occur without appreciable risk of deleterious effects over a lifetime. This dose is calculated using the NOAEL multiplied by a safety factor ranging from 10 to 1000, depending on the quality and quantity of data available. For carcinogenic risks there is no acceptable "safe" (risk-free) level for carcinogens other than zero concentration. Rather, guidelines are established from studies designed to establish relationships between dose (intake rates, mg kg^{-1} day^{-1}) and carcinogenic risk. "Acceptable" levels of risk are generally taken to be an incremental cancer risk of 10^{-5} to 10^{-7} (one additional case of cancer in a population ranging from 100,000 to 10,000,000, respectively). Once the "acceptable" level of risk is selected, the acceptable intake rate of contaminant can be computed.

Once the "safe" intake levels are established, the criterion may be computed from estimated exposure levels. *Exposure* is defined as the contact of an organism (humans, in the case of the health risk assessment) with the chemical agent. Determining human exposure includes a description of the pathway through which pollutants come in contact with humans, a mathematical expression that relates the pathway to an exposure level, and the determination of values to use for the exposure. Pathways include drinking, ingestion while swimming, eating contaminated fish, dermal contact with contaminated water or sediment, ingestion of contaminated sediments, and inhalation of volatilized contaminants. Present water quality criteria[74] are based on human ingestion of water and fish, or human ingestion of fish alone.

The mathematical expression for chemical exposure is given in Table 11.3.1 which, if the acceptable intake rate is known from the reference dose or carcinogenic risk factor, can be solved for the acceptable chemical concentration (the criterion). Calculations are usually made using the following basic exposure assumptions—a

TABLE 11.3.1 Generic Equation for Calculating Chemical Intake Rates for Chemical Exposure

$$I_R = \frac{CC_R E_{FD}}{B_W A_T}$$

I_R	Intake rate, the amount of chemical at the exchange boundary (mg contaminant/kg body weight/day)
	Chemical-related variables:
C	Chemical concentration, the average concentration contacted over the exposure period (e.g., mg L^{-1})
	Variables that describe the exposed population:
C_R	Contact rate; the amount of contaminated medium contacted per unit time or event (e.g., L/day)
E_{FD}	Exposure frequency and duration; describes how long and how often exposure occurs. Often calculated using $$E_{FD} = E_F E_D$$ where E_F = exposure frequency, days/year; E_D = exposure duration, years
B_W	Body weight; the average body weight over the exposure period, kg
	Assessment-determined variables:
A_T	Averaging time; period over which exposure is averaged (days) (Based on 70 years for lifetime or less where particular age groups are considered)

Source: From EPA.[76]

70-kg male consumes a daily average of 6.5 g day^{-1} of freshwater and estuarine fish and shellfish and an average of 2 L day^{-1} of water. For carcinogenic effects, a lifetime risk is computed from an exposure of 70 years.

Aquatic Health Criteria. Aquatic health criteria for the protection of aquatic organisms and their uses are derived from four major kinds of adverse effects: (1) *acute toxicity,* (2) *chronic toxicity,* (3) *toxicity to aquatic plants,* and (4) *bioaccumulation of contaminants. Acute effects* are short-term lethal effects and occur within 4 days of exposure. *Chronic effects* occur over a longer period and include changes in feeding, growth, metabolism, or reproduction in additional to eventual death. *Representative species* of aquatic organisms are used so that data available for these species indicate the sensitivity of untested species. The resulting toxicity criteria are intended to protect most species in a balanced, healthy aquatic community.

Limits on contaminants in plant and animal tissues (the *residue value*) are intended to protect wildlife which consume aquatic organisms and the marketability of aquatic organisms. The residue value is generally based on a *bioconcentration factor* (BCF) that relates concentrations of the contaminant in water to its residual concentration in tissue and a maximum permissible tissue concentration, which may be based on FDA (U.S. Federal Drug Administration) action levels or the results of chronic wildlife feeding studies. The permissible tissue concentration divided by the bioconcentration results in the criterion for the contaminant in water.

Toxicity is determined for individual contaminants (e.g., ammonia) or for mixtures of contaminants as found in waste discharges (often called *whole-effluent toxicity*). Whole-effluent toxicity tests are necessary because it is not cost-effective to identify every compound in a waste discharge and then determine a toxicity for that compound. In addition, whole-effluent toxicity tests identify the effects of chemical interaction that cannot be inferred from individual toxicity tests.

Aquatic health impacts of contaminants are a function of the magnitude, duration, and frequency of exposure to toxic materials. EPA's recommended criteria for both individual contaminant toxicity and whole-effluent toxicity are specified as two numbers:

1. *Criterion Continuous Concentration* (CCC):—The 4-day average concentration of a pollutant in ambient water that should not be exceeded more than once every 3 years on the average.
2. *Criterion Maximum Concentration* (CMC): The 1-h average concentration in ambient water which should not be exceeded more than once every 3 years on the average.

The frequency with which criteria may be exceeded can vary between specific sites.

U.S. Water Quality Criteria. A summary of the *1986 Water Quality Criteria* is found in Table 11.3.2, listing chemicals that are *priority pollutants* or suspected carcinogens. The term *priority pollutant* originally referred to a list of 129 pollutants agreed to in a court settlement between EPA and an environmental group but has come to refer to any pollutant listed in EPA court settlements and legislation. Aquatic life criteria are provided for acute and chronic effects, for both fresh and marine waters. Ambient water concentrations should not exceed these levels to ensure the health of aquatic populations. Human health effects are based on water and fish ingestion as well as fish consumption only. Not all U.S. states presently have aquatic life standards.

Bacteria. Bacterial standards are established primarily to protect human health. The criteria are based on an indicator bacteria that, if present, may suggest that other pathogenic species may be present. The indicator bacteria normally used is *Escherichia coli (E. coli)* or *enterococci.*

For waters supporting shellfish harvesting, the geometric mean fecal coliform bacterial concentration should not exceed 14 MPN per 100 mL, with not more than 10 percent of the samples exceeding 43 MPN per 100 mL. These standards are based on international agreements and data indicating no evidence of disease outbreak resulting from consumption of shellfish which were grown in waters meeting these bacteriological criteria.

Color. Criteria for the effect of color on the aquatic community are primarily due to its effect on light penetration. The Water Quality Criteria[74] state that waters shall be "virtually free from substances producing an objectionable color." Increased color should not reduce the depth of the compensation point for photosynthetic activity, or the depth where there is a balance between photosynthesis and respiration, by more than 10 percent from the seasonally established norm.

Total Dissolved Gases. To protect freshwater and marine organisms, the Water Quality Criteria[74] state that the total dissolved gas concentrations in water should not exceed 110 percent of the saturation value for gases at the existing atmospheric and hydrostatic pressures.

Oil and Grease. Domestic water supplies should be "virtually free from oil and grease," particularly from the tastes and odors that emanate from petroleum products.

Solids (Suspended, Settleable) and Turbidity. Criteria are established for solids and turbidity primarily due to their effect on productivity by reducing light penetration. The criterion for suspended solids is that the depth of the compensation point for photosynthetic activity should not be reduced by more than 10 percent from its seasonally established norm.

TABLE 11.3.2 Water Quality Criteria Summary

	Priority pollutant	Carcinogen	Concentrations, µg/L: Fresh acute criteria	Fresh chronic criteria	Marine acute criteria	Marine chronic criteria	Units per liter: Water and fish ingestion	Fish consumption
Acepthene	Y	N	1700*	520*	970*	710*		
Acrolein	Y	N	68*	21*	55*		320 µg	790 µg
Acrylonitrile	Y	Y	7550*	2600*			0.058 µg†	0.65 µg†
Aldrin	Y	Y	3.0		1.3		0.074 ng†	0.079 ng†
Alkalinity	N	N		20,000				
Ammonia	N	N	Criteria are pH and temperature dependent—see document					
Antimony	Y	N	9000*	1600*			146 µg	45,000 µg
Arsenic	Y	Y					2.2 ng†	17.5 ng†
Arsenic (pent)	Y	Y	850*	48*	2319*	13*		
Arsenic (tri)	Y	Y	380	190	69	36		
Asbestos	Y	Y					30 kf/L†	
Bacteria	N	N	For primary recreation and shellfish uses—see document					
Barium	N	N					1 mg	
Benzene	Y	Y	5300*		5100*	700*	0.65 µg†	40 µg†
Benzidine	Y	Y	2500*				0.12 ng†	0.53 ng†
Beryllium	Y	Y	130*	5.3*			6.8 ng†	117 ng†
BHC	Y	N	100*		0.34*			
Cadmium	Y	N	3.9‡	1.1‡	43	9.3	10 µg	
Carbon tetrachloride	Y	Y	35,200*		50,000*		0.4 µg†	6.94 µg†
Chlordane	Y	Y	2.4	0.0043	0.09	0.004	0.46 µg†	0.48 ng†
Chlorinated benzenes	Y	Y	250*	50	160*	129*		
Chlorinated naphthalenes	Y	N	1600*		7.5*			
Chlorine	N	N	19	11	13	7.5		
Chloroakyl ethers	Y	N	238,000*					
Chloroakyl ether (bis-2)	Y	Y					0.03 µg†	1.36 µg†
Chloroform	Y	Y	28,900*	1240*			0.19 µg†	15.7 µg†
Chlorosopropyl ether	Y	N					34.7 µg	4.36 mg
Chloromethyl ether	N	Y					0.00000376 mg†	0.00184 µg†

TABLE 11.3.2 Water Quality Criteria Summary *(Continued)*

	Priority pollutant	Carcinogen	Concentrations, µg/L: Fresh acute criteria	Fresh chronic criteria	Marine acute criteria	Marine chronic criteria	Units per liter: Water and fish ingestion	Fish consumption
Chlorophenol 2	Y	N	4380*	2000*				
Chlorophenol 4	N	N			29,700*			
Chlorophenoxy herbicides(2,4,5-TP)	N	N					10 µg	
Chlorophenoxy herbicides(2,4-D)	N	N					100 µg	
Chlorpyrifos	N	N	0.063	0.041	0.011	0.0056		
Chloro-4 methyl-3 phenol	N	N	30*					
Chromium (hex)	Y	N	16	11	1100	50	50 µg	
Chromium (tri)	N	N	1700‡	210‡	10,300*		170 mg	3433 mg
Color	N	N	Narrative statement—see document					
Copper	Y	N	18‡	12‡	2.9	2.9		
Cyanide	Y	N	22	5.2	1	1	200 µg	
DDT	Y	Y	1.1	0.001	0.13	0.001	0.024 ng†	0.024 ng†
DDT metabolites (DDE)	Y	Y	1050*		14*			
DDT metabolite (TDE)	Y	Y	0.06*		3.6*			
Demeton	Y	N		0.1		0.1		
Disutylphthalate	Y	N					35 mg	154 mg
Dichlorbenzenes	Y	N	1120*	763*	1970*		400 µg	2.6 mg
Dichlorobenzidines	Y	Y					0.01 µg†	0.020 µg†
Dichloroethane 1,2	Y	Y	118,000*	20,000*	113,000*		0.94 µg†	243 µg†
Dichloroethylenes	Y	Y	11,600*		224,000*		0.033 µg†	1.85 µg†
Dichlorphenol 2,4	N	N	2020*	365*			3.09 mg	
Dichloropropane	Y	N	23,000*	5700*	10,300*	3040*		
Dichloropropene	Y	N	6060*	244*	790*		87 µg	14.1 mg
Dieldrin	Y	Y	2.5	0.0019	0.71	0.0019	0.071 mg†	0.076 mg†
Diethylphthalate	Y	N					350 mg	1.8 g
Dimethylphenol 2,4	Y	N	2120*					

Dimethylphthalate	Y	N					313 mg	2.9 g
Dinitrotoluene 2,4	N	Y					0.11 μg†	9.1 μg†
Dinitrotoluene	Y	N					70 μg	14.3 mg
Dinitrotoluene	N	Y	330*	230*	590*	370*		
Dimitro-o-cresol 2,4	Y	N					13.4 μg	765 μg
Dioxin (2,3,7,8-TCDD)	Y	Y	0.01*	0.00001*			0.000013 mg†	0.000014 ng†
Diphenylhydrazine	Y	N					42 ng†	0.56 μg†
Diphenylhydrazine 1,2	Y	N	270*					
Di-2-ethylhexylphthalate	Y	N					15 mg	50 mg
Endosulfan	Y	N	0.22	0.056	0.034	0.0087	74 μg	159 μg
Endrin	Y	N	0.18	0.0023	0.037	0.0023	1 μg	
Ethylbenzene	Y	N	32,000*		430*		1.4 μg	3.26 mg
Fluoranthene	Y	N	3980*		40*	16	42 μg	54 μg
Gases, total dissolved	N	N	Narrative statement — see document					
Guthion	N	N		0.01		0.01		
Haloethers	Y	N	360*	122*				
Halomethanes	Y	Y	11,000*		12,000*	6400*	0.19 μg†	15.7 μg†
Heptachlor	Y	Y	0.52	0.0038	0.053	0.0036	0.28 ng†	0.29 ng†
Hexachloroethane	N	Y	980*	540*	940*		1.9 μg†	8.74 μg†
Hexachlorobenzene	Y	N					0.72 ng†	0.74 ng†
Hexachlorobutadiene	Y	Y	90*	9.3*	32*		0.45 μg†	50 μg†
Hexachlorocyclohexane	Y	Y	2.0	0.06	0.16			
Hexachlorocyclohexane-alpha	Y	Y					9.2 ng†	31 ng†
Hexachlorocyclohexane-beta	Y	Y					16.3 ng†	54.7 ng†
Hexachlorocyclohexane-gama	Y	Y					18.6 ng†	62.5 ng†
Hexachlorocyclohexane-technical	Y	Y					12.3 ng†	41.4 ng†
Hexachlorocyclopentadiene	Y	N	7*	5.2*	7*		206 μg	
Iron	N	N		1000			0.3 mg	
Isophorone	Y	N	117,000*		12,900*		5.2 mg	520 mg
Lead	Y	N	82‡	3.2‡	140	5.6	50 μg	
Malathion	N	N		0.1		0.1		
Manganese	N	N					50 μg	100 μg
Mercury	Y	N	2.4	0.012	2.1	0.025	144 ng	148 ng
Methoxychlor	N	N		0.03		0.03	100 μg	
Mirex	N	N		0.001		0.001		

TABLE 11.3.2 Water Quality Criteria Summary *(Continued)*

			Concentrations, $\mu g/L$				Units per liter	
	Priority pollutant	Carcinogen	Fresh acute criteria	Fresh chronic criteria	Marine acute criteria	Marine chronic criteria	Water and fish ingestion	Fish consumption
Monochlorobenzene	Y	N					488 μg	
Naphthalene	Y	N	2300*	620*	2350*			
Nickel	Y	N	1400‡	160‡	75	8.3	13.4 μg	100 μg
Nitrates	N	N					10 mg	
Nitrobenzene	Y	N	27,000*		6680*		19.8 mg	
Nitrophenols	Y	N	230*	150*	4850*			
Nitrosamines	Y	Y	5850*		3,300,000*			
Nitrosodisutylamine N	Y	Y					6.4 ng†	587 ng†
Nitrosodimethylamine N	Y	Y					0.8 ng†	1240 ng†
Nitrosodimethylamine N	Y	Y					1.4 ng†	16,000 ng†
Nitrosodiphenylamine N	Y	Y					4900 ng†	16,100 ng†
Nitrosopyrroladine N	Y	Y					16 ng†	91,900 ng†
Oil and grease	N	N	Narrative statement—see document					
Oxygen, dissolved	N	N	Warm water and cold water criteria matrix—see document					
Parathion	N	N	0.065	0.013				
PCBs	Y	Y	2.0	0.014	10	0.03	0.079 ng†	0.079 ng†
Penthachlorinated ethanes	N	N	7240*	1100*	390*	281*		
Penthachlorobenzene	N	N					74 μg	85 μg
Pentachlorophenol	Y	N	20§	13§	13	7.9*	1.01 mg	
Ph	N	N		6.5–9		6.5–8.5		
Phenol	Y	N	10,200*	2560*	5800*		3.5 mg	
Phosphorus elemental	N	N				0.1		
Phthalate esters	Y	N	940*	3*	2944*	3.4*		
Polynuclear aromatic hydrocarbons	Y	Y			300*		2.8 mg†	31.1 ng*
Selenium	Y	N	280	35	410	54	10 μg	
Silver	Y	N	4.1‡	0.12	2.3		50 μg	
Solids (dissolved) and salinity	N	N					250 mg	

Solids (suspended) and turbidity	N	N	Narrative statements—see document					
Sulfides—hydrogen sulfide	N	N		2		2		
Temperature	N	N	Species-dependent criteria—see document					
Tetrachlorinated ethanes	Y	N	9320*					
Tetrachlorobenzene 1,2,4,5	Y	N					38 μg	48 μg
Tetrachloroethane 1,1,2,2	Y	Y		2400*	9020*		0.17 μg	10.7 μg
Tetrachloroethanes	Y	N	9320*					
Tetrachloroethylene	Y	Y	5280*	840*	10,200*	450*	0.6 μg†	8.86 μg†
Tetrachlorephenol 2,3,4,8	Y	N				440*		
Thallium	Y	N	1400*	40*	2130*		13 μg	48 μg
Toluene	Y	N	17,500*		6300*	5000*	14.3 μg	424 mg
Toxaphene	Y	Y	0.73	0.0002	0.21	0.0002	0.71 ng†	0.73 ng†
Trichlorinated ethanes	Y	Y	18,000*					
Trichloroethane 1,1,1	Y	N		31,200*			18.4 mg	1.03 g
Trichloroethane 1,1,2	Y	Y		9400*			0.6 μg†	41.8 μg†
Trichloroethylene	Y	Y	45,000*	21,900*	2000*		2.7 μg†	80.7 μg†
Trichlorophenol 2,4,5	N	N					2600 μg	
Trichlorophenol 2,4,6	Y	Y		970*			1.2 μg	3.6 μg†
Vinyl chloride	Y	Y					2 μg†	525 μg†
Zinc	Y	N	120‡	110‡	95	86		

Source: From EPA,[74] excluding drinking water MCL.
Y = yes
N = no
* Insufficient data to develop criteria. Value presented is the LOEL (lowest observed effect level).
† Human health criteria for carcinogens reported for three risk levels, value presented in the 10^{-6} risk level.
‡ Hardness-dependent criteria (100 mg L^{-1} used).
§ pH-dependent criteria (7.8 pH used).

Temperature. Species-dependent temperature criteria are established to protect aquatic life. Two criteria are established—one for acute effects and another for chronic effects. For fresh waters, the acute criterion is a maximum temperature for short exposures of specific species. The chronic criterion for fresh waters is based on a weekly average temperature. The weekly average criterion may be based upon protection of aquatic species from rapid cooling during cooler months in the region of a plume with elevated temperatures, protection against excessively high temperatures during warmer months, or protection of aquatic life during reproductive seasons.

For marine waters, the maximum acceptable increase in the weekly average temperature is 1 °C during all months of the year (provided summer temperature maxima are not exceeded), and daily temperature cycle characteristics of the water body should not be altered in either amplitude or frequency. The summer thermal maximum defines the upper acceptable temperature limit and is established on a site-specific basis.

Aesthetic Criteria. Aesthetic criteria are associated with the perceived quality of water for a particular use. This perception of quality may be affected by taste and odor problems as well as the presence of nuisance aquatic plant growth and their associated animal pests. But taste, odor, and color perception should not be taken lightly as they represent the innate physiologic reactions. The Water Quality Criteria[74] list criteria for general aesthetic quality as well as criteria for color, phosphorus, and tainting substances. The criteria stipulate that all waters should be free from substances that settle to form objectionable deposits; float as debris, scum, oil, or other matter to form a nuisance; produce objectionable color, odor, taste, or turbidity; injure or are toxic, or produce adverse physiological responses in humans, animals, or plants; and produce undesirable or nuisance aquatic life.

Phosphorus. No U.S. criterion has been established for the control of phosphorus to prevent eutrophication. Limits on phosphate phosphorus are established primarily to control the development of plant nuisances and their associated animal pests in both flowing waters and downstream receiving water bodies. To prevent the development of biological nuisances, total phosphorus should not exceed 50 μg P L^{-1} in any stream or 25 μg P L^{-1} within a lake or reservoir.

Other Criteria. Other criteria have been established for agricultural and industrial use by EPA.[74]

11.3.3 Water Quality Standards

Water quality standards are the implementation of water quality criteria in an enforceable regulatory framework. All water discharges in the United States are controlled by state and federal standards. Standards may deviate from water criteria since they may consider factors other than health considerations. Water quality standards are typically established on the basis of one or more of the following factors:

1. Established and ongoing practice or experience
2. Existing criteria
3. Bioassays to establish new criteria
4. Ability to measure parameters reliably and other technical factors
5. Evidence derived from accidental human exposure
6. Epidemiological studies
7. Educated guess based on available information and judgment

8. Application of mathematical models (e.g., those that simulate health risks)
9. Economic attainability and impact on specific industries
10. Legal enforceability

Drinking water standards are intended primarily to protect human health, and secondarily to protect appearance or aesthetics of drinking waters. Drinking water standards have been established in the United States for some public water supplies since 1914, with the enactment of bacteriological standards (called Treasury standards since they were enforced by the Treasury Department). In 1974, the Safe Drinking Water Act was established. There are presently two types of standards. The maximum contaminant level goal (MCLG) is a best estimate of the concentration that protects against adverse human health effects and that allows an adequate margin of safety. However, this is a nonenforceable standard. The maximum contaminant levels (MCL) are the regulated standard and include other considerations in their formulation. The MCLs are the primary standard that cannot be exceeded for any waters delivered to any user of a U.S. public water system. U.S. drinking water standards went into effect in 1977 with the "Safe Drinking Water Act of 1977," Public Law 93-523. Existing and proposed MCL and MCLG levels in drinking water are provided in Table 11.3.3. Primary levels are enforceable, health-based standards while secondary standards are nonenforceable taste, odor, and appearance guidelines. Some MCLs correspond to practical quantitation limits (PQLs), which are a multiple of the detection limit (MDL) for those parameters. The water quality standards are also a part of the Water Quality Criteria and were omitted from Table 11.3.2 because of their inclusion here.

Prior to 1972, U.S. states were empowered to establish ambient water quality standards while the federal government could not. In 1972, the Federal Water Pollution Control Act (FWPCA) amendments were passed in order to provide a uniform national system of water quality standards, permits, and enforcement. These amendments gave broad powers to the EPA to define pollutants and develop effluent limitations. Two types of standards were placed into effect: ambient and effluent water standards. *Ambient standards* are based on the establishment of threshold values for a particular contaminant and consider the intended use of the water body as well as its ability to assimilate wastes. *Effluent standards* limit the amount of material that may be discharged regardless of the size of the receiving water body or the intended use of its waters. Effluent standards are often technology-based and may be imposed even if the level of contamination resulting from the effluent is less than that required to achieve ambient water quality standards. However, where the standard technologies are not sufficient to meet the ambient criteria, the use of more advanced technologies may be required.

The FWPCA was further amended in 1977 (known as the Clean Water Act). This act set a new class of effluent standards. The Clean Water Act also provided a variance provision for best available technology standards for nonconventional pollutants (e.g., chemical oxygen demand, fluoride, aluminum, sulfide, and ammonia) contained in Section 301(g). The Clean Water Act established two principal bases for effluent limitations: technology-based effluent controls and water quality standard–based controls. This law allows the EPA, with state approval, to modify effluent standards for nonconventional materials if the modifications do not interfere with water quality standards or public health. Under Sections 303 and 401 of the Clean Water Act, states are given primary responsibility for developing water quality standards and limits to those standards. The EPA's role is to review the state standards in order to ensure compliance with the act.

TABLE 11.3.3 Existing and Proposed U.S. EPA Maximum Contaminant Levels (MCL) in Drinking Water as of July 1991

Contaminant	Proposed MCLG, mg L^{-1}	Current MCL, mg L^{-1}	Proposed MCL, mg L^{-1}
	I. Primary levels		
A. Inorganic chemicals:			
Antimony	0.003		0.01/0.004
Arsenic		0.05	
Asbestos	7 million fibers/L		7 million fibers/L
Barium	5	1	5
Beryllium	0		0.001
Cadmium	0.005	0.01	0.005
Chromium	0.1	0.05	0.1
Copper	1.3		1.3
Cyanide	0.2		0.2
Fluoride		4	
Lead	0	0.05	0.005
Mercury	0.002	0.002	0.002
Nickel	0.1		0.1
Nitrate (as N)	10	10	10
Nitrite (as N)	1		1
Selenium	0.005	0.01	0.05
Silver		0.05	
Sulfate	400/500		400/500
Thallium	0.0005		0.002/0.001
B. Volatile organics			
Benzene		0.005	
Carbon tetrachloride		0.005	
o-Dichlorobenzene	0.6		0.6
p-Dichlorobenzene		0.075	
1,2-Dichloroethylene		0.005	
1,1-Dichloroethylene		0.007	
cis-1,2-Dichloroethylene	0.07		0.07
trans-1,2-Dichloroethylene	0.1		0.1
Dichloromethane (methylene chloride)	0		0.005
1,2-Dichloropropane	0		0.005
Ethylbenzene	0.7		0.7
Monochlorobenzene	0.1		0.1
Styrene	0/0.1		0.005/0.1
			0.005
Tetrachloroethylene	0		2
Toluene	2		
1,1,1-Trichloroethane		0.2	
1,1,2-Trichloroethane	0.003		0.005
Trichloroethylene		0.005	
Vinyl chloride		0.002	
Xylenes (total)	10		10
C. Pesticides/herbicides/PCBs/ base-neutral organics:			
Acrylamide	0		
Adipates [di(ethylhexyl)adipate]	0.5		0.5
Alachlor	0		0.002
Aldicarb	0.01		0.01
Aldicarb sulfoxide	0.01		0.01
Aldicarb sulfone	0.04		0.04

TABLE 11.3.3 Existing and Proposed U.S. EPA Maximum Contaminant Levels (MCL) in Drinking Water as of July 1991 *(Continued)*

Contaminant	Proposed MCLG, mg L^{-1}	Current MCL, mg L^{-1}	Proposed MCL, mg L^{-1}
	I. Primary levels		
Atrazine	0.003		0.003
Carbofuran	0.04		0.04
Chlordane	0		0.002
2,4-D	0.07	0.1	0.07
Dalapon	0.2		0.2
Dibromochloropropane	0		0.002
Dinoseb	0.007		0.007
Diquat	0.02		0.02
Endothall	0.1		0.1
Endrin	0.002	0.0002	0.002
Epichlorohydrin	0		
Ethylene dibromide	0		0.00005
Glyphosate	0.7		0.7
Heptachlor	0		0.0004
Heptachlor epoxide	0		0.0002
Hexachlorobenzene	0		0.001
Hexachloroyclopentadien (HEX)	0.05		0.05
Lindane	0.0002	0.004	0.0002
Methoxychlor	0.4	0.1	0.4
Oxamyl (Vydate)	0.2		0.2
PAHs [Benzo(a)pyrene]	0		0.0002
Polychlorinated biphenyls (PCBs)	0		0.0005
Pentachlorophenol	0.2		0.2
Phthalates [di(ethylhexyl) phthalate]	0		0.004
Picloram	0.5		0.5
Simazine	0.01		0.01
2,3,7,8-TCDD (Dioxin)	0		0.00000005
Toxaphene	0	0.005	0.005
2,4,5-TP (Silvex)	0.05	0.01	0.05
1,2,4-Trichlorobenzene	0.009		0.009
Trihalomethanes (total)		0.1	
D. Microbiological:			
Bacteria		4 per 100 mL	
Coliform		1 per 100 mL	
E. Physical:			
Turbidity in turbidity units (TU)		1 NTU (monthly avg.) 5 NTU avg. of 2 consecutive days	
F. Radionucleides:			
Gross alpha radiation		15 pCi/L	
Man-made beta radiation		4 millirems/year	
Radium 226 and 228		5 pCi/L	
	II. Secondary levels		
Aluminum			0.05
Chloride		250	
Color		15 color units	
Copper		1	

TABLE 11.3.3 Existing and Proposed U.S. EPA Maximum Contaminant Levels (MCL) in Drinking Water as of July 1991 *(Continued)*

Contaminant	Proposed MCLG, mg L^{-1}	Current MCL, mg L^{-1}	Proposed MCL, mg L^{-1}
	II. Secondary levels		
Corrosivity		Neither corrosive nor scale-forming	
o-Dichlorobenzene			
p-Dichlorobenzene			0.0100.005
Ethylbenzene			0.03
Fluoride		2	
Foaming agents		0.5	
Hexachlorocyclopentadiene			0.008
Iron		0.3	
Manganese		0.05	
Odor		3 T.O.N.	
Pentachlorophenol			0.03
pH		6.5–8.5 pH units	
Silver			0.09
Styrene			0.01
Sulfate		250	
Toluene			0.04
Total dissolved solids (TDS)		500	
Xylenes			0.02
Zinc		5	

Specific sections of the 1977 Clean Water Act include:

Section 301, which set standards for point sources that were not publicly owned treatment works (POTW). It required discharges to reduce emissions using the best practical control technology (BPCT) available in 1977 and the best available technology (BAT) economically available by 1983.

Section 302, which set ambient water quality standards. The standards used could be the federal or state standards, whichever were more stringent.

Section 306, which required all new sources to meet standards equivalent to the 1983 BAT standards.

Section 307, which covers toxic pollutants and requires that standards be developed based on public health and welfare and not technical feasibility.

Section 402, which empowers the federal government to create a National Pollution Discharge Elimination System (NPDES) that applies to any discharge.

Although applying primarily to drinking water, the maximum concentration limits (MCLs) are generally taken as the basis for ambient surface water standards as well as groundwater standards for aquifers used for drinking waters. The MCLs are also generally adopted as the appropriate standard for cleanup at superfund sites. States may implement their own standards, but they generally cannot be less than the MCL. However, differences may occur where the standards are implemented. For example, some states allow a mixing zone for materials discharged into a water body in which the standards may be violated.

The MCLs as listed do not include human health effects due to the ingestion of contaminated fish nor do they include the effects of aquatic toxicity. The criteria for human and aquatic health described earlier (Table 11.3.2) may be used as the basis for such standards. Aquatic health standards may also include the effects of the frequency and duration of exposure. For aquatic health standards for acute toxicity, the CMC should not exceed 0.3 acute toxic unit (TU_a) as measured by the most sensitive of three test species. For chronic protection, the CCC should not exceed 1.0 chronic toxic unit (TU_c) to the most sensitive of at least three test species. The allowable effluent concentration is determined by an exposure assessment that includes an analysis of how much of the water body is subject to the criteria being exceeded, and the frequency and duration of that exceedance. If mixing is not rapid and complete, then a mixing zone analysis may be performed (in those states which allow a mixing zone). Otherwise the concentrations may be estimated using mathematical far-field waste load allocation models.[73]

REFERENCES

1. Allison, J. D., D. S. Brown, and K. J. Novo-Gradac, MINTEQA2/PRODEFA2, "A Geochemical Assessment Model for Environmental Systems: Version 3.0 User's Manual," U.S. Environmental Protection Agency, Athens, Ga., 1990.
2. Ambrose, R. B., Jr., T. A. Wool, J. P. Connolly, and R. W. Schanz "WASP4, A Hydrodynamic and Water Quality Model—Model Theory, User's Manual, and Programmer's Guide," U.S. EPA Report EPA/600/3-87/039, Athens, Ga., 1988.
3. American Society for Testing and Materials, *Manual on Industrial Water and Industrial Waste Water,* Philadelphia, p. 364, 1964.
4. ASCE (American Society of Civil Engineers), Hydraulic Models, *Man. Eng. Pract.,* 25, New York, 1942.
5. Ball, J. W., D. K. Nordstrom, and D. W. Zachman, "WATEQ4F—A Personal Computer FORTRAN Translation of the Geochemical Model WATEQ2 with Revised Data Base, U.S. Geological Survey, Open-File Report 87-50, 1987.
6. Bingham, *Fluidity and Plasticity,* McGraw-Hill, New York, p. 340, 1922.
7. Bonzountas, M., "Soil and Groundwater Fate Modeling," in R. L. Swann, and A. Eschenroeder, eds. *Fate of Chemicals in the Environment,* American Chemical Society, Symposium Series 225, Washington, D.C., pp. 41–65, 1983.
8. Bouwer, H., *Groundwater Hydrology,* McGraw-Hill, New York, 1978.
9. Bowie, G. L., W. B. Mills, D. B. Porcella, C. L. Campbell, J. R. Pagenkopf, G. L. Rupp, K. M. Johnson, W. H. Chan, and S. A. Gherini, "Rates, Constants, and Kinetics Formulations in Surface Water Quality Modeling," 2d ed., Report EPA/600/3-85/040, U.S. Environmental Protection Agency, Athens, Ga., 1985.
10. Brown, L. C., and T. O. Barnwell, Jr., "The Enhanced Stream Water Quality Models QUAL2E and QUAL2E-UNCAS: Documentation and User Manual," EPA/600/3-87/007, U.S. EPA, Athens, Ga., 1987.
11. Burns, L. A., "Fate of Chemicals in Aquatic Systems: Process Models and Computer Codes," in R. L. Swann and A. Eschenroeder, eds., *Fate of Chemicals in the Environment,* pp. 25–40, Washington, D.C.: American Chemical Society Symp. Ser. 225, 1983.
12. Callahan, M. A., et al., "Water-Related Fate of 129 Priority Pollutants," U.S. EPA-440/4-79-029b, Washington, D.C., 1979.
13. Chiou, C. T., R. L. Malcolm, T. I. Brinton, and D. E. Kile, "Water Solubility Enhancement of Some Organic Pollutants and Pesticides by Dissolved Humic and Fulvic Acids," *Environ. Sci. Technol.,* vol. 20, no. 502, 1986.

14. Chiou, C. T., L. J. Peters, and V. H. Freed, "A Physical Concept for Soil-Water Equilibria for Nonionic Organic Compounds," *Science,* vol. 206, no. 16, p. 831, 1979.

15. Clesceri, L. S., A. E. Greenberg, and R. R. Trussell, eds., *Standard Methods for the Examination of Water and Wastewaters,* 17th ed., American Public Health Association, American Water Works Association, and Water Pollution Control Federation, Washington, D.C., 1989.

16. Crowe, A. S., and F. J. Longstaff, "Extension of Geochemical Modeling Techniques to Brines: Coupling of the Pitzer Equations to PHREEQE," *Proceedings of Solving Groundwater Problems with Models,* National Water Well Association, Denver, Colo., 1987.

17. Delwiche, C. C., "The Nitrogen Cycle," *Sci. Am.,* vol. 223, no. 3, p. 137, 1970.

18. Donigian, A. S., T. Y. R. Lo, and E. W. Shanahan, "Rapid Assessment of Potential Ground-Water Contamination under Emergency Response Conditions," U.S. EPA Report, 1983.

19. Dugan, P. R., *Biochemical Ecology of Water Pollution,* Plenum, New York, 1975.

20. Dzombak, D. A., and F. M. M. Morel, "Adsorption of Inorganic Pollutants in Aquatic Systems," *J. Hydraul. Eng.,* vol. 113, pp. 430–475, 1987.

21. Electrical Power Research Institute (EPRI), "Design and Operating Guidelines Manual for Cooling Water Treatment: Treatment of Recirculating Cooling Water," Sec. 4, Process Model Documentation and User's Manual, EPRI CS-2276, Palo Alto, Calif., 1982.

22. Electrical Power Research Institute (EPRI), "SEQUIL—An Inorganic Aqueous Chemical Equilibrium Code for Personal Computers," vol. 1, User's Manual/Workbook, Version 1.0, EPRI GS-6234, Palo Alto, Calif., undated.

23. Fleming, G. W., and L. N. Plummer, "PHRQINPT—An Interactive Computer Program for Constructing Input Data Sets to the Geochemical Simulation Program PHREEQE," U.S. Geological Survey Water Resources Investigation Report 83-4236, 1983.

24. Gschwend, P. M., S-C Wu, O. S. Madsen, J. Wilkins, R. A. Ambrose, Jr., and S. C. McCutcheon, "Modeling the Benthos-Water Column Exchange of Hydrophobic Chemicals," U.S. EPA Environmental Research Laboratory, EPA/600/3-86/044, Athens, Ga., 1986.

25. Gschwend, P. M., and S. C. Wu, "On the Constancy of Sediment-Water Partition Coefficients of Hydrophobic Organic Pollutants," *Environ. Sci. Technol.,* vol. 19, p. 90, 1985.

26. Hardy, R. C., and R. L. Cottington, *J. Res. Natl. Bureau Res.,* vol. 42, p. 573, 1949.

27. Hassett, J. J., J. C. Means, W. L. Banwart, and S. G. Wood, "Sorption Properties of Sediments and Energy-Related Pollutants," U.S. EPA Environmental Research Laboratory, EPA/600/3-80-041, Athens, Ga., 1980.

28. Hem, J. D., "Study and Interpretation of the Chemical Characteristics of Natural Water," Water Supply Paper 2254, 3d ed., U.S. Geological Survey, 1985; also see 2d ed., 1970.

29. Jafvert, C. T., "Assessing the Environmental Partitioning of Organic Acid Compounds," Environmental Research Brief EPA/600/M-89/016, U.S. EPA Environmental Research Laboratory, Athens, Ga., 1990.

30. Jafvert, C. T., "Sorption of Organic Acid Compounds to Sediments: Initial Model Development," *Environ. Toxicology Chem.,* vol. 9, p. 1259, 1990.

31. Johnson, R. L., C. D. Palmer, and W. Fish, "Subsurface Chemical Processes," chap. 5, in Transport and Fate of Contaminants in the Subsurface, EPA/625/4-89/019, U.S. Environmental Protection Agency, Office for Research Information, Cincinnati, Ohio, and Robert S. Kerr Environmental Research Laboratory, Ada, Okla., 1989.

32. Karickhoff, S. W., "Organic Pollutant Sorption in Aquatic Systems," *J. Hydraul. Eng.,* Vol. 110, p. 707, 1984.

33. Kharaka, Y. K., W. D. Gunter, P. K. Aggarwal, E. H. Perkins, and J. D. DeBraal, "SOLINEQ.88: A Computer Program for Geochemical Modeling of Water-Rock Interactions," U.S. Geological Survey, Water Resources Invest. Report 88-4227, 1988.

34. Krauskopf, K. B., *Introduction to Geochemistry,* McGraw-Hill, New York, 1979.

35. Krenkel, P. A., and V. Novotny, *Water Quality Management,* Academic Press, New York, 1980.

36. Krom, M. D., and E. R. Sholkovitz, "Nature and Reactions of Dissolved Organic Matter in the Interstitial Waters of Marine Sediments," *Geochimica et Cosmochimica Acta,* vol. 41, p. 1565, 1977.

37. Linsley, R. K., M. A. Kohler, and J. H. L. Paulhus, *Hydrology for Engineers,* 3d ed., McGraw-Hill, New York, 1982.

38. List, R. J., *Smithsonian Meteorological Tables,* Smithsonian Miscellaneous Collections, vol. 114, Publication 4041, 6th rev. ed., Washington, D.C., 1951.

39. Livingstone, D. A., "Chemical Composition of Rivers and Lakes, Data of Geochemistry," 6th ed., U.S. Geological Survey Professional Paper 440-G, Washington, D.C., G1-G64, 1963.

40. Lowenthal, R. E., and G. v. R. Marais, *Carbonate Chemistry of Aquatic Systems: Theory and Application,* Ann Arbor Science, Ann Arbor, Mich., 1976.

41. Mackay, D., "Finding Fugacity Feasible," *Environ. Sci. Tech.,* vol. 13, no. 10, p. 1218, 1979.

42. Martin, J. L., B. Batchelor, and S. C. Chapra, "Modification of a Metal Adsorption Model to Describe the Effect of pH," *J. Water Poll. Control Fed.,* vol. 57, no. 5, pp. 425–427, 1985.

43. Martin, J. M., and M. Meybeck, "Elemental Mass-Balance of Material Carried by Major World Rivers," *Mar. Chem.,* vol. 7, p. 173, 1979.

44. McCutcheon, S. C., *Water Quality Modeling,* vol. I, *Transport and Surface Exchange in Rivers,* CRC Press, Boca Raton, Fla., 1989.

45. McCutcheon, S. C., *Water Quality Modeling: Biogeochemical Cycles in Rivers,* CRC Press, Boca Raton, Fla., expected to be published 1993.

46. Medine, A. J., and B. R. Bicknell, "Case Studies and Model Testing of the Metals Exposure Analysis Modeling System (MEXAMS)," EPA-600/3-84-045, U.S. EPA, Athens, Ga., 1986.

47. Medine, A. J., and S. C. McCutcheon, "Fate and Transport of Sediment-Associated Contaminants," in J. Saxena, ed., *Hazard Assessment of Chemicals,* vol. 6, Hemisphere, New York, pp. 225–291, 1989.

48. Merrill, D. T., "Chemical Conditioning for Water Softening and Corrosion Control, in R. L. Sanks, ed., *Water Treatment Plant Design,* Ann Arbor Science Publishers, Ann Arbor, Mich., 1976.

49. Meybeck, M., "Carbon, Nitrogen, and Phosphorus Transport by World Rivers," *Am. J. Sci.,* vol. 282, p. 401, 1981.

50. Miller, D. C., S. Poucher, J. A. Cardin, and D. Hansen, "The Acute Toxicity of Ammonia to Marine Fish and a Mysid," *Arch. Environ. Contam. Toxicol.,* vol. 19, pp. 40–48, 1990.

51. Mills, W. B., D. B. Porcella, M. J. Ungs, S. A. Gherini, K. V. Summers, M. Lingfung, G. L. Rupp, and G. L. Bowie, "A Screening Procedure for Toxic and Conventional Pollutants," Parts 1 and 2, Reports EPA/600/6-85/002a and 002b, U.S. Environmental Protection Agency, Athens, Ga., 1985.

52. National Academies of Sciences and Engineering, Water Quality Criteria 1972, U.S. EPA/R3/73/033, Washington, D.C., 1973.

53. NAPAP (National Acid Precipitation Assessment Program), Annual Report, 1989, and Findings Update, National Acid Precipitation Assessment Program, 722 Jackson Place, NW, Washington, D.C., 1989.

54. NRC (National Research Council), *Acid Deposition: Long Term Trends,* National Academy Press, Washington, D.C., 1986.

55. Parkhurst, D. L., D. C. Thorstenson, and L. N. Plummer, "PHREEQE—A Computer

Program for Geochemical Calculations," U.S. Geological Survey, Water Resources Investigation 80-96, 1980.

56. Perry, R. H., C. H. Chilton, and S. D. Kirkpatrick, eds., *Perry's Chemical Engineers Handbook,* McGraw-Hill, New York, 1963.
57. Puls, R. W., and M. J. Barcelona, "Ground Water Sampling for Metals Analyses," EPA/540/4-89/001, U.S. Environmental Protection Agency, Office of Solid Waste and Emergency Response, Washington, D.C., 1989.
58. Rheinheimer, G., *Aquatic Microbiology,* 2d ed., Wiley, New York, 1980.
59. Rodhe, H., "Acidification in a Global Perspective," *Ambio,* vol. 18, no. 3, 1989.
60. Sawyer, C. N., and P. L. McCarty, *Chemistry for Sanitary Engineers,* McGraw-Hill, New York, 1967.
61. Skougstad, M. W., M. J. Fishman, L. C. Friedman, D. E. Erdmann, and S. S. Duncan, eds., "Methods for Determination of Inorganic Substances in Water and Fluvial Sediments," U.S. Geological Survey, Techniques of Water Resources Investigations, Book 5, chap. A1, Government Printing Office, Washington, D.C., 1979.
62. Streeter, V. L., and E. B. Wylie, *Fluid Mechanics,* 6th ed., McGraw-Hill, New York, 1975.
63. Stumm, W., and J. J. Morgan, *Aquatic Chemistry,* Wiley, New York, 1981.
64. Thackston, E. L., "Effect of Geographical Variation on Performance on Recirculating Cooling Ponds," U.S. EPA Rept. EPA-660/2-74-085, Corvallis, Ore., 1974.
65. Thibodeaux, L. J., *Chemodynamics,* Wiley, New York, 1979.
66. Tilton, L. W., and J. K. Taylor, "Accurate Representation of the Refractivity and Density of Distilled Water as a Function of Temperature," *J. Res. Natl. Bur. Std.,* vol. 18, pp. 205–214, 1937.
67. Thomann, R. V., and J. A. Mueller, *Principles of Surface Water Quality Modeling and Control,* Harper & Row, New York, 1987.
68. Thurman, E. M., *Organic Geochemistry of Waters,* Nijhoff and Junk Publishers, Dordrecht, The Netherlands, 1985.
69. Ulhmann, D., *Hydrobiology,* Wiley, New York, 1979.
70. U.S. EPA, "Water Quality Criteria," 1972, EPA R3-73-033, U.S. Environmental Protection Agency, Washington, D.C., 1973.
71. U.S. EPA, "Process Design Manual for Nitrogen Control," Technology Transfer Series, 1975.
72. U.S. EPA, "Methods for the Chemical Analysis of Water and Wastes," EPA-600/4-79-020. U.S. EPA Environmental Support Laboratory, 1979.
73. U.S. EPA, "Technical Support Document for Water Quality-Based Toxics Control," EPA-440/4-85-032, U.S. Environmental Protection Agency, Office of Water Regulations and Standards, Washington, D.C., 1985.
74. U.S. EPA, "Quality Criteria for Water," EPA-440/5-86-001, U.S. Environmental Protection Agency, Office of Water Regulations and Standards, Washington, D.C., 1987.
75. U.S. EPA, "Nitrogen-Ammonia/Nitrate/Nitrite; Water Quality Standards Criteria Summaries: A Compilation of State/Federal Criteria," EPA 440/5-88/029, Office of Water Regulations and Standards, Washington, D.C., 1988.
76. U.S. EPA, "Risk Assessment Guidance for Superfund: Human Health Evaluation Manual," Part A—Interim Final OSWER Directive 9285.7-Ola, U.S. EPA, Washington, D.C., 1989.
77. U.S. Federal Water Pollution Control Administration, Water Quality Criteria, National Technical Advisory Committee to the Secretary of the Interior, Washington, D.C., 1968.
78. Weast, R. C., ed., *CRC Handbook of Chemistry and Physics,* CRC Press, Boca Raton, Fla., 1986.

79. Wedepohl, K. H., ed., *Handbook of Geochemistry,* vols. I–II-4, Springer-Verlag, Berlin, 1969.

80. Wu, S. C., and P. M. Gschwend, "Sorption Kinetics of Hydrophobic Organic Compounds to Natural Sediments and Soils," *Environ. Sci. Technol.,* vol. 20, p. 717, 1986.

81. Wunderlich, W. O., "Heat and Mass Transfer between a Water Surface and the Atmosphere," Water Resources Research Lab. Rept. 14, TVA Rept. 0-6803, Norris, Tenn., 1972.

82. Wunderlich, W. O., "Heat Exchange between a Water Surface and the Atmosphere," to be published by W. O. Wunderlich, 3221 Essary Drive, Knoxville, Tenn., 37918, 1993.

CHAPTER 12
EROSION AND SEDIMENT TRANSPORT

Hsieh Wen Shen
Department of Civil Engineering
University of California
Berkeley, California

Pierre Y. Julien
Department of Civil Engineering
Colorado State University
Fort Collins, Colorado

Despite extensive research effort, knowledge of erosion and sediment transport still remains incomplete, and there is no generally accepted formula to be used for an accurate solution of the sediment transport rate and watershed sediment yield. However, significant progress has been made in recent decades and approximate solutions can be obtained.

This chapter describes current knowledge on erosion and sediment transport, compares well-known formulas, and presents guidelines for solving some problems. Knowledge of complex subjects such as cohesive sediment is limited and thus is not discussed here. For more complete analysis of sediment movement and related subjects, readers are referred to Graf,[59] Shen,[113,114,116] Yalin,[142] Vanoni,[113] Raudkivi,[105] Simons and Senturk,[119] Chien and Wan,[25] Garde and Ranga Raju,[50] Walling et al.,[135] Mehta,[90] Thorne et al.,[123] and Mehta et al.[88,89]

12.1 FUNDAMENTAL PROPERTIES OF SEDIMENT PARTICLES

Sediment transport characteristics are determined by the *size, shape, concentration, fall velocity,* and *bulk density* of sediment particles.

Size. Definitions: the *size* of a sediment particle is defined on three mutually perpendicular axes labeled *a, b, c.* The *a* axis is in the direction of the longest dimension of a sediment particle. Then the other two mutually perpendicular axes are chosen with the c axis close to the shortest dimension and b being the intermediate perpendicular axis. The selection of these three axes is somewhat arbitrary. Different people may choose different sets of axes.

The *mean diameter* d_m of a sediment particle is the arithmetic average of the three axes a, b, and c.

The *nominal diameter* d_n of a sediment particle is the diameter of a sphere having the same volume as the particle.

The *fall diameter* d_f of a sediment particle is the diameter of a sphere with a density of 2.65 and having the same terminal fall velocity in quiescent distilled water at 24°C as the particle.

The *sieve diameter* d_s of a sediment diameter is defined as the square size opening in a sieve which a given sediment particle will just pass through. Studies indicate that for most natural material $d_s = 0.9\ d_n$.

Sediment particles are divided into different groups such as boulders, cobbles, gravels, sand, silt, and clays according to their sizes as shown in Table 12.1.1.

Another size scale used by geologists is the phi (ϕ) scale, given by

$$\phi = -\log_2 d \qquad (12.1.1)$$

where d is the diameter in millimeters (mm) and the logarithm has a base of 2. The main advantage of using a ϕ scale is that the demarcating diameters in Table 12.1.1 become whole numbers on the ϕ scale. The minus sign is introduced so that sand sizes will have positive ϕ numbers.

Sediment Size-Distribution Measurements. Since both flow resistance and sediment movement are directly related to the sediment size distributions, it is important to obtain an accurate measurement of sediment size distributions in a stream (Vanoni[133]). The most reliable method to measure sediment size distribution is by *laboratory sieve analysis.* A sediment sample is placed in the top sieve of a series of sieves. The size of a sample is usually between 100 and 200 g. This series of sieves is placed in a Rotap mechanical shaker for 5 to 10 min to separate the sediment sample into various sizes. If the sieving time is too short, fine sediment particles may not reach the bottom pan of this series of sieves. If the sieving time is too long, sediment particles may be broken into smaller pieces. Sieve analysis can be applied only to a range of sediment sizes between 0.062 and 32 mm (from gravel to fine sand).

Another method to measure the sediment size distribution is to measure the different amounts of sediment settling through a *visual accumulation tube* (U.S. Inter-Agency Committee on Water Resources[129]) for sediment sizes between 0.062 and 2 mm. This instrument consists of an upper mixing chamber, a glass settling tube with tapered bottom, an eyepiece to measure the top of the sediment settling column at the bottom of the settling tube, and a recorder. The sediment sample is first well mixed in the mixing chamber. The sediment particles are then released from the mixing chamber downward to the settling tube through a valve. The eyepiece and the recorder give the top of sediment deposited in the settling tube as a function of time. With a calibrated chart, the sediment size distribution of the sample is determined.

Sampling Large Sediment. For sediment sizes larger than 16 mm, the number of sediment particles within a given size range may be counted in the field. These sediment particle samples must be selected randomly in order to obtain a representative sample. One method to select these sediment samples is to place a grid on top of the sediment bed surface and select sediment samples at the intercepts of vertical and horizontal grid lines. Another method is to select the sediment particles at the heel of

TABLE 12.1.1 Sediment Grade Scale and Sieve Size

	Size, mm		μm	Inches	Approximate sieve mesh openings per inch: Tyler	Approximate sieve mesh openings per inch: U.S. standard	Class
Boulders and cobbles:							
	4000–2000			160–80			Very large boulders
	2000–1000			80–40			Large boulders
	1000–500			40–20			Medium boulders
	500–250			20–10			Small boulders
	250–130			10–5			Large cobbles
	130–64			5–2.5			Small cobbles
Gravel:							
	64–32			2.5–1.3			Very coarse gravel
	32–16			1.3–0.6			Coarse gravel
	16–8			0.6–0.3	2 1/2		Medium gravel
	8–4			0.3–0.16	5	5	Fine gravel
	4–2			0.16–0.08	9	10	Very fine gravel
Sand:							
	2–1	2.00–1.00	2000–1000		16	18	Very coarse sand
	1–1 1/2	1.00–0.50	1000–500		32	35	Coarse sand
	1/2–1/4	0.50–0.25	500–250		60	60	Medium sand
	1/4–1/8	0.25–0.125	250–125		115	120	Fine sand
	1/8–1/16	0.125–0.062	125–62		250	230	Very fine sand
Silt:							
	1/16–1/32	0.062–0.031	62–31				Coarse silt
	1/32–1/64	0.031–0.016	31–16				Medium silt
	1/64–1/128	0.016–0.008	16–8				Fine silt
	1/128–1/256	0.008–0.004	8–4				Very fine silt
Clay:							
	1/256–1/512	0.004–0.0020	4–2				Coarse clay
	1/512–1/1024	0.0020–0.0010	2–1				Medium clay
	1/1024–1/2048	0.0010–0.0005	1–0.5				Fine clay
	1/2048–1/4096	0.0005–0.00024	0.5–0.24				Very fine clay

Source: From Lane.[78]

a person's shoe when this person is walking backward along a certain path. Thorne et al.[123] provide a detailed analysis of gravel bed sediment sampling.

The sediment size distributions in a streambed and banks can be analyzed by photographs, but the true sediment particle size distribution may be different from the apparent sediment sizes projected onto a plane surface on a photograph. Empirical correlations have been presented by Bray[18] and Adams[4] to correct this effect.

Sampling Fine Sediment. For sediment sizes finer than 0.062 mm, it is difficult to use sieves to separate small particles and fine sediment particles are usually separated by their fall velocities. The *pipette method* is a reliable indirect method to determine size distribution of sediment particles between 0.002 and 0.062 mm. Sediment size greater than 0.062 mm must be removed first by sieving. Clay particles may require special treatment to ensure that sediment particles fall singly during the analysis. The desirable sediment concentration is between 2000 and 5000 mg/liter for the pipette method. When the sediment concentration is greater than this upper limit, the sample may be diluted and split into several samples. The pipette method is based on the separation of different sediment sizes by their respective fall velocities. Sediment concentrations are determined in a vertical cylinder of water at a predetermined depth as a function of settling time. Particles having a settling velocity greater than that of the size at which separation is desired will settle below the point of withdrawal after a certain time lapse. One finds the size of sediment particle that should have settled to this location using Stokes' law as follows:

$$\omega = \frac{gd^2}{18\nu} \frac{\gamma_s - \gamma_f}{\gamma_f} \tag{12.1.2}$$

where ω, g, d, ν, γ_s, and γ_f are, respectively, the particle settling velocity, gravitational constant, particle diameter, kinematic viscosity of the liquid, specific weight of sediment, and specific weight of liquid. The *centrifugal method* can be used to separate extremely fine sediment particles. A centrifugal force several times stronger than the gravity can be generated by rotation of a sediment sample at a high speed to accelerate settling. After different rotation time periods, the various volumes of sediment accumulation can be measured to determine sediment size distribution.

The *bottom withdrawal* (BW) tube method can be used where the pipette equipment is not available. The size range to be measured by the BW tube is between 0.002 and 0.062 mm, which is the same as that for the pipette method. The desired sediment concentration range is between 1000 and 3500 mg/liter. After removal of sediment particles greater than 0.062 mm, the sediment sample (well-mixed with water) is placed in a 122-mm-long cylinder with an inside diameter between 25 and 26 mm. A clamp is installed at the bottom of this cylinder with a tapered lower section, and samples are withdrawn below the clamp at certain predetermined time periods.

The *hydrometer method* relies on the relationship between the progressive decrease in density that occurs at a given elevation in a well-dispersed sediment suspension and the sediment diameter of the particles. It is assumed that the difference in density between the suspension and pure fluid as measured with a hydrometer is proportional to the sediment concentration.

Cumulative Sediment Size Distribution. The *cumulative sediment size distribution* for a sediment sample is obtained by the following procedure. First, the weight of a total sediment sample is measured. Next, the sediment sample is separated into various size fractions by one or more procedures as described in the above sections and the weight of sediment in each size fraction is determined.

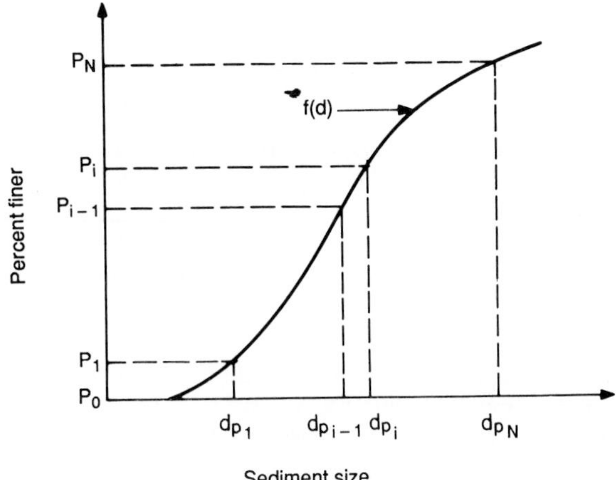

FIGURE 12.1.1 Cumulative sediment size distribution curve.

Let P_j be the percentage of the total sediment weight in each size fraction. The cumulative sediment size distribution $f(d)$ is plotted in Fig. 12.1.1. The d_{P_N} is the median sediment size for size fraction P_N and d_{p_i} is the sediment size for which i percent is finer. Thus the median size of a sediment sample is d_{50} and the mode size is the predominant size in a sediment sample.

The sediment *arithmetic mean size* d_a is defined by the following relationship:

$$d_a = \frac{\sum_{i=1}^{N} (P_i - P_{i-1}) \frac{d_{p_i} + d_{p_{i-1}}}{2}}{\sum_{i=1}^{N} (P_i - P_{i-1})} \tag{12.1.3}$$

If N covers the entire range between 0 and 100 percent,

$$d_a = \sum_{i=1}^{N} (P_i - P_{i-1}) \frac{d_{p_i} + d_{p_{i-1}}}{2} \tag{12.1.4}$$

The *geometric mean size* d_g is defined as

$$d_g = \sqrt[N]{\prod_{i=1}^{N} d_{p_i}} \qquad \text{or} \qquad \log d_g = \frac{1}{N} \sum_{i=1}^{N} \log d_{p_i} \tag{12.1.5}$$

The *nominal diameter* is the diameter of a sphere of the same volume as the particle and the *sorting coefficient* S_0 is defined as

$$S_0 = \sqrt{\frac{d_{75}}{d_{25}}} \tag{12.1.6}$$

The sorting coefficient for many natural sediment samples varies between 2.0 and 4.5. Often, the sediment sizes in natural sediment samples follow lognormal distribution. For this distribution, the geometric standard deviation σ_g is given by

$$\sigma_g = \frac{d_{84.1}}{d_{50}} = \frac{d_{50}}{d_{15.9}} \tag{12.1.7}$$

Shape of Sediment Particles. The shape of a sediment particle has a strong influence on its fall velocity. For gravel and cobbles in alluvial deposits, the flat faces are usually dipping upstream. The arrangement of sediment shapes strongly influences streambed stability. A shape factor K is defined as

$$K = \frac{c}{\sqrt{ab}} \tag{12.1.8}$$

where a, b, c are the particle's length, width, and height as defined previously. For well-rounded natural sediment particles, $K = 0.7$.

For more detailed discussion of shape factors, see Garde and Ranga Raju.[50]

Sediment Concentration. Sediment concentration measures the amount of sediment carried by the flow and is an important parameter in sediment movement studies.

The three commonly used terms to express sediment concentrations are:

S_v = sediment concentration by volume

= bulk volume of sediment solids/total solid-water volume $(V) = V_s/V$

$$= \frac{S_W}{S_G - S_W(S_G - 1)} \tag{12.1.9}$$

where S_G is the specific gravity of the sediment.

S_w = sediment concentration by weight (mass)

= weight (mass) by sediment $W_s(M_s)$/total sediment-water weight (mass) $W(M)$

$$= \frac{S_m}{\rho_W + (1 - 1/S_G)S_m} \tag{12.1.10}$$

where ρ_W is density of water.

S_m = sediment concentration by a mixture of weight (mass) and volume

= weight (mass) of sediment $W_s(M_s)$/total sediment-water volume V

$$= \frac{S_W \rho_W S_G}{S_G - (S_G - 1)S_W} \tag{12.1.11}$$

The sediment concentration is usually expressed as parts per million (ppm) by weight. In these cases, all the above three sediment concentrations are multiplied by a million, or 10^6. During annual floods, sediment concentrations in large rivers often reach between 1000 and 10,000 ppm by weight. The greatest sediment concentrations measured for a short duration in the United States (for the Rio Puerco River) exceeded 700,000 ppm, by weight.

Another sediment concentration measure is by kilograms (weight) per cubic meter (volume). For the Missouri and Colorado Rivers in the United States, the annual average sediment concentrations are, respectively, 3.54 and 27.5 kg/m³. The Yellow River in China is the greatest sediment-carrying stream in the world. Its annual average sediment concentration is 37.6 kg/m³ and the maximum measured sediment concentration has reached 911 kg/m³. The sediment concentration for 100 percent pure bulk sediment is about 1650 kg/m³.

For a Newtonian fluid μ, the dynamic viscosity of the water-sediment mixture varies with sediment concentration by volume as follows (Einstein and Chien[33]):

$$\mu = \mu_f(1 + 2.5S_v) \tag{12.1.12}$$

Krone[75] found that

$$\mu = \mu_f e^{2.5S_v} \tag{12.1.13}$$

where μ_f is the dynamic viscosity for the pure fluid and S_v is the sediment concentration in the fluid by volume.

Fall Velocity. The terminal fall velocity of a sediment particle depends on the effects of size, shape, and density of a sediment particle, the effects of fluid density, and turbulence. Figure 12.1.2 provides the fall velocity for various sediment sizes and shapes under different water temperatures. The set of curves for $K = 0.5$ has fall velocities in the range of 0.2 to 100 cm/s, respectively; the second set of curves for $K = 0.7$ (closed to well-rounded natural sediment particles) has fall velocities in the range of 0.1 to 100 cm/s. The third set of curves for $K = 0.9$ has fall velocities in the range of 0.1 to 50 cm/s, respectively.

The effect of particle shape on fall velocity is significant. Schulz, Wilde, and Albertson[109] found that the drag coefficient, which determines the fall velocity of a particle, varies greatly with the Reynolds number based on particle size.

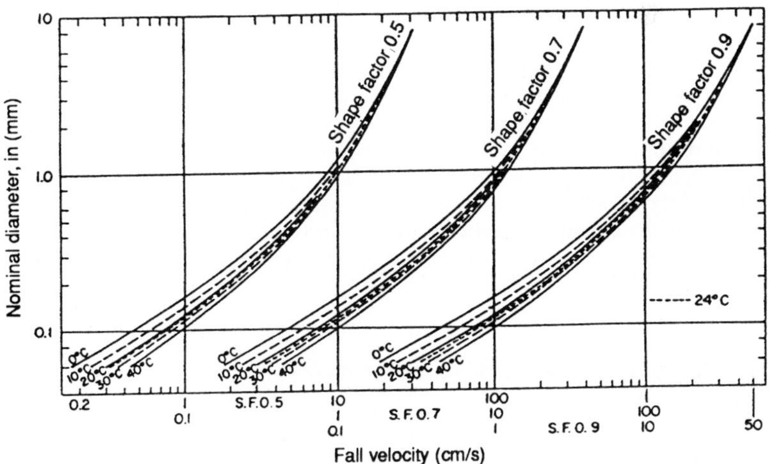

FIGURE 12.1.2 Sediment particle fall velocities.[129] Sediment fall velocities for three shape factors $K = 0.5, 0.7, 0.9$. Each set of curves uses a different horizontal scale for shape factor (S.F.).

Specific Weight and Bulk Specific Weight. The *specific weight* of a single sediment particle is the weight of this particle divided by its volume. The specific weight of a sediment particle can also be taken as the product of the specific gravity of the sediment particle and the weight of the water. Sediment particles coarser than clay consist mainly of quartz and feldspar, and the specific gravities of these particles are about 2.65. The specific weights of silt, sand, and gravel are about 2.65 g/cm^3 or 165 lbf/ft^3. Because of the voids occurring among the particles in a sediment deposit, the bulk specific weight of sediment deposits is always less than that for a single sediment particle. The *bulk specific weight* of a sediment deposit is defined as the dry weight of the sediment deposit divided by its bulk volume. The bulk specific weights of various sediment sizes are given in Table 12.1.2. The *porosity* is defined as the percentage of pore space in the total bulk volume of the sediment:

$$\text{Porosity} = 100\,\frac{\text{bulk volume} - \text{solid grain volume}}{\text{bulk volume}} \tag{12.1.14}$$

The bulk specific weights of clay and silt deposits can vary significantly over time owing to compaction. This is discussed in Sec. 12.8 (Reservoir Sedimentation) because it is an important factor in determining the sediment accumulation in a reservoir.

Heavy sediment concentration reduces the fall velocity for a sediment particle. The relation between the settling velocity ω of an individual particle in a fluid and ω_0, the settling velocity of the same particle immersed in a fluid with heavy sediment concentration, can be expressed as follows:

$$\frac{\omega_0}{\omega} = (1 - S_v)^m \tag{12.1.15}$$

where S_v is the sediment concentration by volume and the range of values of m varies from 2.25 to 7.0. Actually the value of m decreases with increase of the Reynolds number $\omega d/\nu$, where ν is the fluid kinematic viscosity. The most common m value for silt size is between 2.35 and 4.65 (see Chien and Wan[25]). Most of the recent research on heavy sediment concentration flows has been conducted in China (Chien and Wan[25]).

The effect of fluid turbulence on fall velocity is significant for a high turbulence level, low sediment density, and small sediment size. Jobson and Sayre[66] indicate that with normal turbulence in uniform flows in rivers and laboratory flumes with sediment of specific gravity of 2.65, the turbulent effect on fall velocity of sediment particles greater than fine sand particles can be neglected.

The settling velocity for clay particles is rather complex because these sediment particles may flocculate and settle as a group in hindered settling. Turbulence can also cause a breakup of flocculation. No general acceptable relationship for these effects is available.

12.2 FLUVIAL BED FORMS AND FLOW RESISTANCE

12.2.1 Bed Forms

Depending on the flow and sediment characteristics, sand bed streams can exhibit different bed forms. For sediment particles finer than about 0.6 mm, ripples form as soon as the sediment particles are moved by the flow. As flow increases, ripples are

TABLE 12.1.2 Porosity and Specific Weight for Sediments

Classification and range, mm	Fine sand, 1/8–1/4 mm	Fine sand, 1/4–1/2 mm	Medium sand, 1/2–1 mm	Coarse sand, 1–2 mm	Coarse sand, 2–4 mm	Gravelly sand, 4–8 mm	Fine gravel, 8–16 mm	Medium gravel, 16–32 mm	Coarse gravel, 32–64 mm	Coarse gravel, 64–128 mm	Coarse gravel and boulders, 128–256 mm
Porosity, %	44	43	41	39	37.5	34.5	33	27	23	18	17
Specific weight, kN/m^3	14	15	15	16	16	17	18	19	20	22	22
Specific weight, lbf/ft^3	93	94	98	101	103	108	111	121	127	130	137

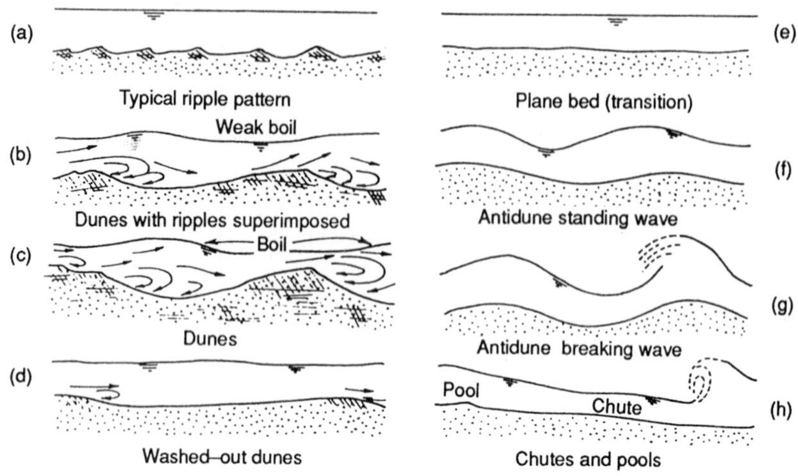

FIGURE 12.2.1 Sediment bed forms (flow increases consecutively from *a* to *h*).

changed into dunes, then go through a transition to plane beds, antidunes, and antidunes with hydraulic jumps (Fig. 12.2.1). Bed forms are generally classified into two groups: a lower regime for subcritical flows and an upper regime for supercritical flows. The dividing line between lower regime and upper regime flow is generally close to the critical flow where the flow Froude number is 1. The flow Froude number is defined as the flow velocity divided by the square root of the product of the flow depth and the gravitational constant. For gravel bed streams, the bed form is relatively flat with or without bars. A bar is usually formed by a large amount of sediment deposited over a large area of streambed. The vertical elevation of this bar does not vary greatly. Bars can occur in both sand and gravel streams.

For lower regime flows, the streambed can exhibit ripples and dunes with the water surface being approximately flat. For upper regime flows, the streambeds can exhibit antidune form, and the water surfaces are in phase with the bed forms. With antidunes, the bed form can move upstream, move downstream, or stand still. Antidunes can move upstream through sediment deposit on the upstream face of the dune and erosion from the downstream face of the dune. Thus the bed form may move upstream even if all sediment particles are essentially moving downstream.

For coarse sand with a size greater than 0.6 mm, Shen[117] found that ripples do not form at low sediment transport rates. Many attempts (see Garde and Albertson,[53] Simons and Richardson,[120] Garde and Ranga Raju,[52] Mather and Ranga Raju,[87] Vanoni,[134] Brownlie,[19] Engelund and Fredose[39]) have been made to predict the occurrence of different bed forms under various flow conditions for sediment particles between silt and coarse sand particles.

Figure 12.2.2*a* through *d* shows charts by Vanoni[134] which provide reasonable results on the type of bed form for flow depth D up to about 10 ft (or 3 m) and for d_{50} between 100 and 600 μm. One can use these charts to estimate the bed form of sandy bed channels under various flow regimes. For example, in a channel having a fine sandy bed ($d_{50} = 150\,\mu$m), a depth $D = 0.6$ m (2 ft) and a velocity of 0.7 m/s (2.4 ft/s), the depth-diameter ratio is $D/d_{50} = 0.6/15.0 \times 10^{-6} = 4000$, and the Froude number is $V/\sqrt{gD} = 0.7/\sqrt{9.81 \times 0.6} = 0.29$. Reading from Fig. 12.2.2*a*, the bed form for this flow regime is dunes.

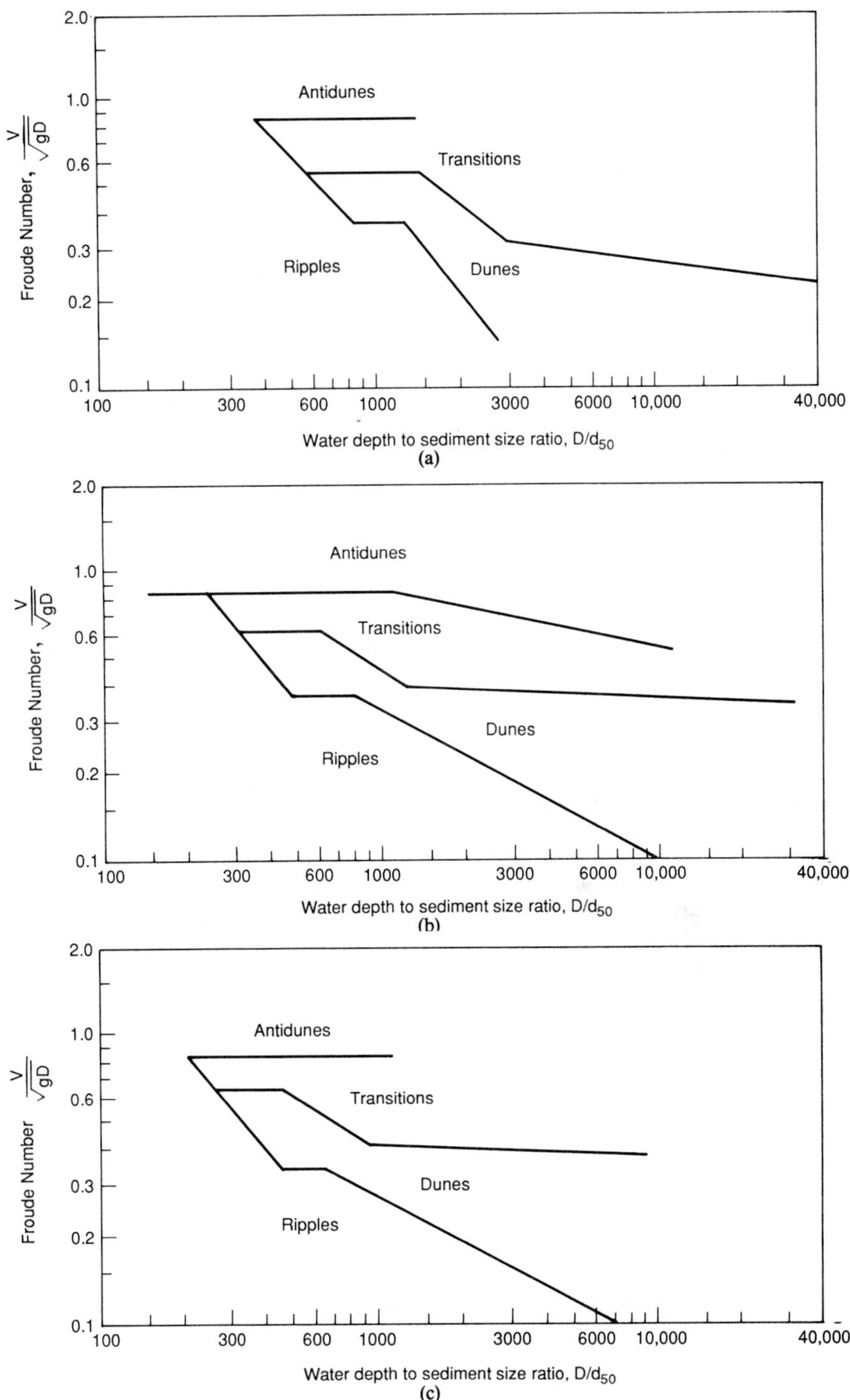

FIGURE 12.2.2 (*a*) Bed form chart for fine sand ($d_{50} = 100 \sim 200\ \mu m$). (*b*) Bed form chart for fine to medium sand ($d_{50} = 200 \sim 300\ \mu m$). (*c*) Bed form chart for medium sand ($d_{50} = 300 \sim 400\ \mu m$). (*d*) Bed form chart for medium to coarse sand ($d_{50} = 400 \sim 600\ \mu m$).

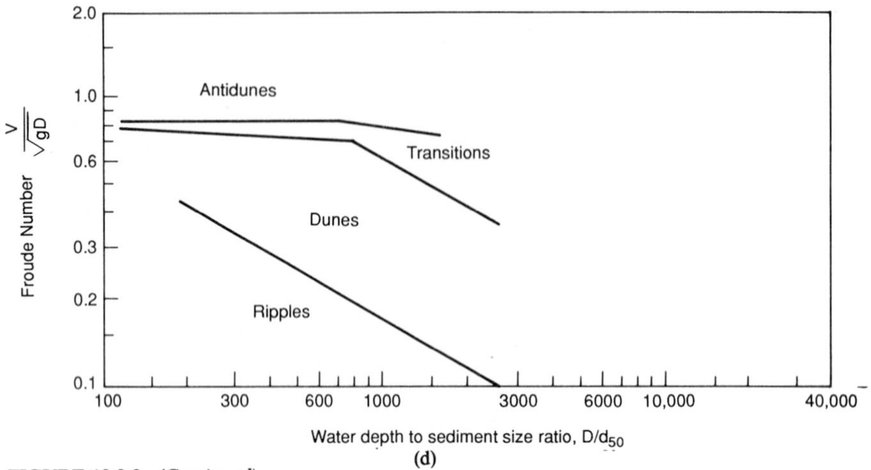

(d)

FIGURE 12.2.2 *(Continued)*

12.2.2 Flow Resistance

Different bed forms cause various amounts of flow resistance through their effect on bed roughness. This can be described by Manning's equation

$$U = \frac{1}{n} R^{2/3} S^{1/2} \tag{12.2.1}$$

where U, n, R, and S are, respectively, the average flow velocity in meters per second, the Manning's roughness coefficient, the hydraulic radius in meters, and the dimensionless energy slope. If the foot unit is used in Eq. (12.2.1) instead of the metric unit, a factor of 1 should be replaced by 1.486 on the right-hand side of the equation. The hydraulic radius is defined as the flow cross-sectional area divided by the wetted perimeter. The hydraulic radius R_a for the trapezoidal cross section with 45° bank slope and the circular cross section shown in Fig. 12.2.3 are given by:

$$R_a \text{ for trapezoidal shape} = \frac{\text{flow area } (BD + D^2)}{\text{wetted perimeter } (B + 2\sqrt{2}\,D)}$$

$$R_b \text{ for circular shape} = \frac{\text{flow area } \pi D^2/4}{\text{wetted perimeter } \pi D} = \frac{D}{4}$$

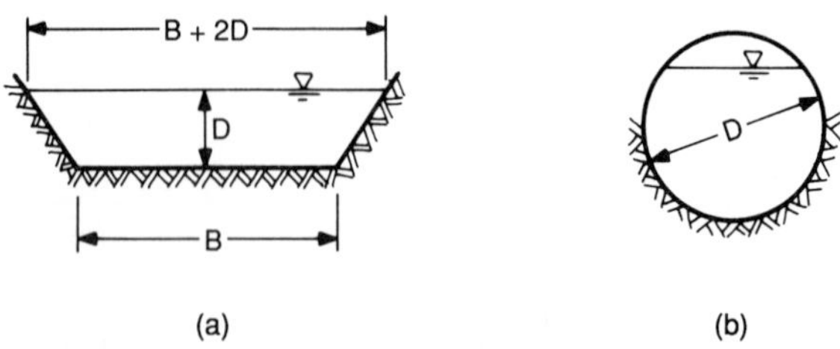

FIGURE 12.2.3 Hydraulic radius calculations.

For a wide channel (width $>20D$), $R(\approx BD/B) \approx D$. Thus the hydraulic radius for wide channel is approximately the same as flow depth.

Hydrologists are concerned about fluvial flow resistance because a change in flow resistance can alter a flood level significantly. Low regime flows with sediment size less than 0.6 mm correspond to a Manning's roughness coefficient between 0.015 and 0.045. For upper regime flows, Manning's roughness varies between 0.013 and 0.02. Figure 12.2.4 is an example of the effects of bed form on a stage-discharge curve.

Einstein and Barbarossa[35] were the first to separate total flow resistance into *skin friction resistance* and *form resistance.* They assumed that grain skin friction resistance can be related to the average flow velocity through the well-known logarithmic flow velocity distribution for turbulent flows. They then determined the form resistance by subtracting the grain skin friction resistance from the total resistance. From both field and laboratory data, they found that the Darcy-Weisbach form resistance factor is directly related to the bed-load transport rate. For detailed description, see Einstein,[36] Graf,[59] Raudkivi,[105] and Garde and Ranga Raju.[50]

Shen[117] and Da Cuhna[30] found that for sediment sizes greater than 0.6 mm, Einstein and Barbarossa's[35] form resistance curve must be corrected to include the phenomenon that ripple beds do not form with sediment coarser than 0.6 mm. Shen[117] also found that finer sediment has a stronger tendency to form high dunes than does coarser sediment. Engelund[41] found that the resistance to flow of a plane bed with sediment in motion was greater than that for a plane bed without sediment motion. Lovera and Kennedy[85] determined the variation of skin resistance from experimental studies conducted with movable plane bed; then Alam and Kennedy[5] subtracted this grain skin resistance from the total resistance to determine form resistance which was a function of the flow Froude number and relative roughness of the sediment particles. Brownlie[19] analyzed numerous field and laboratory data and found that the shear stress was a function of flow discharge, energy slope, sediment size, and bed form characteristics. His curves are shown in Fig. 12.2.5. Shen, Fehlman, and Mendoza[111] confirmed that the addition of grain resistance and form resistance to obtain the total resistance is feasible.

Current data indicate that none of the above approaches produces entirely satisfactory results for prediction of flow resistance because the true skin resistance of three-dimensional irregular and nonuniform dunes cannot be determined precisely so that the form resistance values obtained by subtracting an assumed skin resistance

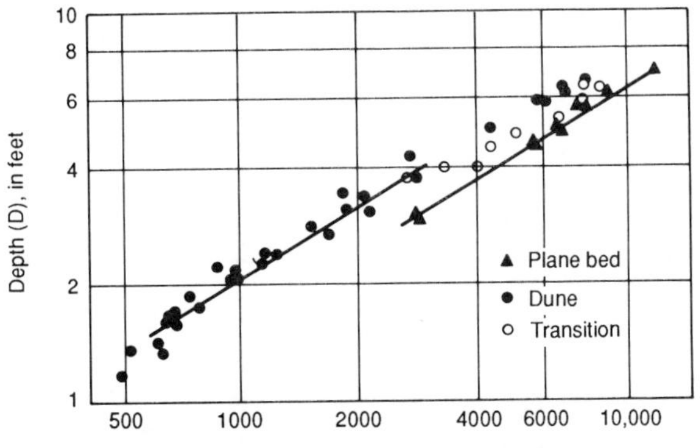

FIGURE 12.2.4 Stage-discharge curve affected by bed forms.

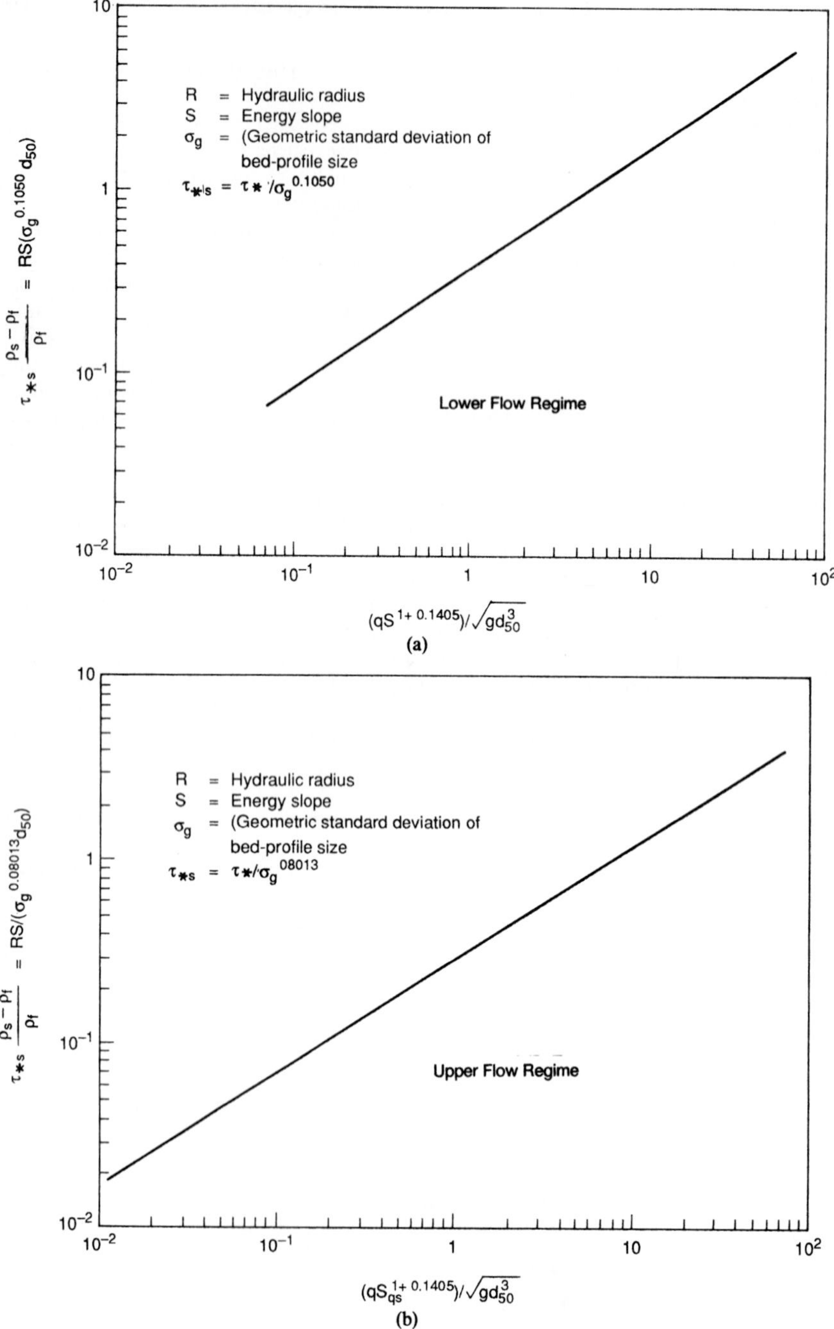

FIGURE 12.2.5 (*a*) Relationship between dimensionless shear stress τ_{*s} and flow discharge q per unit width and energy slope S for lower flow regime. (*b*) Relationship between dimensionless shear stress τ_{*s} and flow discharge q per unit width and energy slope S for upper flow regime. *(After Brownlie.[19])*

from the total resistance are also not precisely determined even in laboratory experiments.

12.2.3 Approach to Estimating Flow Resistance

To determine the Manning roughness or flow resistance for a desired discharge for a particular channel, one should obtain flow resistances from field measurements of flow velocity, flow depth, sediment size distribution, and energy slope for at least three different flow discharges. One of the flow discharges should be for as high a discharge as possible. Bed forms should be observed for all three discharges. First, compare the occurrences of bed forms with the charts in Fig. 12.2.2. For sediment sizes less than 0.6 mm, the bed form changes from ripples to dune to plane bed and to antidunes as the discharge increases. Thus the form resistance should continuously decrease as discharge increases. For sediment sizes greater than 0.6 mm, the bed form changes from plane bed to dunes back to plane bed and then to antidunes. The form resistance first increases and then decreases as flow discharge increases. For a wide channel, the skin resistance usually decreases as stage increases. The field measurements of Manning's roughness coefficient for the three or more flow discharges can be used to form an empirical relationship between roughness and river flow discharges, and this should be checked with the analyses presented by Einstein,[36] Engelund,[41] Alam and Kennedy,[5] and Brownlie.[19] One can also estimate the roughness value for a desired flood discharge by examining a combination of field data and results on the variation of bed forms found in previous research studies.

Table 12.2.1 gives reference values of Manning's roughness coefficient for known bed types based on Chow,[26] Barnes,[13] and other field data.

TABLE 12.2.1 Variation of Manning's Roughness Coefficients n with Bed Type

Bed characteristics	Reference Manning's roughness coefficient n
Sand:	
Plane bed	0.011–0.020
Ripple bed	0.018–0.035
Dune bed	0.020–0.035
Standing waves	0.014–0.025
Antidunes	0.015–0.035
Gravel and cobbles:	0.020–0.030
Boulder	Roughness varies greatly. Usually roughness increases with decreasing flow depth. n can reach 0.1
Vegetation	Roughness varies greatly with the changes of density, height, flexibility of vegetation, and the relative ratio between flow depth and vegetative elements
Bermuda, Kentucky, Buffalo grasses	Flow depth more than 5 times vegetation height—n between 0.03 and 0.06
	Flow depth the same or less than that of vegetation height. 0.01–0.2
Extremely dense vegetation	Vegetation height above flow depth. n can exceed 1
Natural sandy streams:	
Clean and straight	0.025–0.04
Winding and some weeds	0.03–0.05
Mountain streams with boulders	0.04–0.1
Floodplains:	
Short grass	0.02–0.04
High grass	0.03–0.05
Dense willow, brush, etc.	0.05–0.20

Strickler's equation can be used to estimate Manning's roughness for streambed and banks based on the prevailing sediment sizes on the bed and banks:

$$n = \frac{(d_{50})^{1/6}}{21.0} \tag{12.2.2}$$

where d_{50} is the median sediment size in meters, which is restated as:

$$n = \frac{(d_{50})^{1/6}}{25.6} \tag{12.2.3}$$

where the median sediment size is in feet.

12.3 THE BEGINNING OF SEDIMENT MOTION

The critical hydraulic conditions for the beginning of sediment motion have been extensively investigated because streambeds and banks are stable with hydraulic conditions below these critical values. Often these conditions are used for the selection of riprap size for bank protection, and they also affect sediment transport rate prediction equations. *Incipient motion* is the term commonly used to denote the beginning of sediment motion.

Sediment particles are moved by the flow whenever the magnitude of instantaneous fluid force acting on the sediment particle exceeds the resistance force for this particle to be moved. For laminar flow, the fluid force acting on the particle is constant in time, and thus incipient motion conditions can be readily determined in the laboratory. Laboratory experiments have demonstrated that a great majority of sediment particles on the streambed will be moved by the flow when the existing hydraulic condition exceeds the incipient motion condition for these sediment particles for laminar flow. However, laminar flows rarely occur in natural surface water flows except in the cases of thin layers of overland flow and flow through thick vegetation.

Most natural surface flows are turbulent. For turbulent flows, the magnitude of fluid forces acting on any single sediment particle fluctuates widely in time and thus there is random intermittent motion of sediment particles on the streambed near the point of incipient motion. Even in a low-flow condition, a few sediment particles will be moved by the flow if one waits long enough.

The determination of incipient motion conditions for turbulent flow is rather subjective. For practical purposes, the incipient motion criterion should be chosen to denote the conditions that no significant amount of sediment movement should occur, for all practical purposes. In certain cases, a few sediment particle movements at the bottom of the riprap may cause the collapse of the whole structure; one should then select the appropriate incipient motion condition or increase the safety factor to allow for this situation.

***Case 1* Nearly Uniform Cohesionless Sediment Particles.** As a first approximation, the two most important factors governing the movement of a sediment particle are (1) the fluid force acting on the particle and (2) the resistance of the particle to be moved. Let the fluid force F_1 acting on the particle be $\tau_0 k_1 d^2$ and F_2, the resistance force for motion, be $(\gamma_s - \gamma_f) k_2 d^3$, then

$$\frac{F_1}{F_2} = \frac{\tau_0 k_1 d^2}{(\gamma_s - \gamma_f) k_2 d^3} = \frac{\tau_0 K^*}{(\gamma_s - \gamma_t) d} \tag{12.3.1}$$

In the above two expressions, τ_0 is the fluid shear stress acting on the boundary which can be calculated as $\tau_0 = \gamma_f RS$ in a uniform steady flow, d is the average particle size, γ_s and γ_f are the specific weights of sediment particles and the fluid, respectively, k_1 and k_2 are related to the shape factors for the sediment particles, and $K^* = k_1/k_2$. F_1 includes a combination of lift and drag forces. F_2 is proportional to the submerged weight of the sediment particle, and k_2 is a combination of the volume shape factor of the sediment particle and the frictional coefficient. Equation (12.3.1) can be applied only to nearly uniform-sized cohesionless sediment particles. If one defines a term τ^* as the dimensionless ratio between τ_0 and $(\gamma_s - \gamma_t)d$, then

$$\tau^* = \frac{\tau_0}{(\gamma_s - \gamma_f)d} = \frac{1}{K^*}\frac{F_1}{F_2} \tag{12.3.2}$$

Actually, τ^* can also be expressed as (with U_* as the flow shear velocity and g the gravitational constant)

$$\tau^* = \frac{\gamma_f U_*^2}{(\gamma_s - \gamma_f)gd} \tag{12.3.3}$$

where U_*, the *shear velocity* is defined as

$$U_* = \sqrt{\frac{\tau_0}{\rho}} \tag{12.3.4}$$

and ρ is the fluid density.

Because τ^* is a ratio between the inertia force F_1 and the gravitational force F_2, it can be related to the square of a Froude number, U_*^2/gd, as Eq. (12.3.3) shows.

Shields[118] conducted a series of experiments with sediment of selected densities to establish the following criterion for incipient sediment motion as shown in Fig. 12.3.1 (v is the kinematic viscosity of the fluid in this figure). The dashed line in this figure is the critical condition for incipient motion to occur. For τ^* greater than this dashed line, sediment motion occurs.

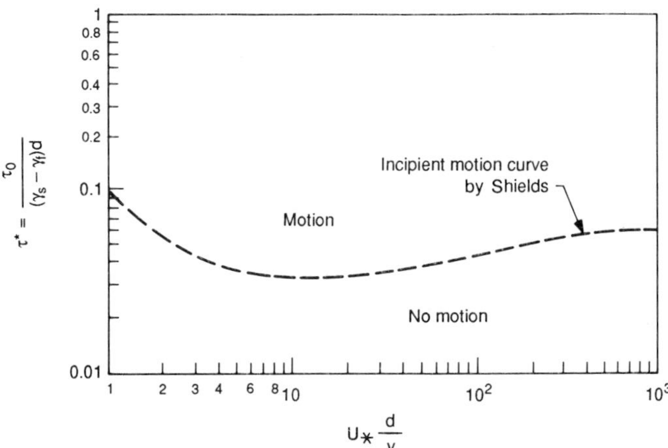

FIGURE 12.3.1 Incipient motion diagram. *(Shields.[118])*

In Fig. 12.3.1, v is the kinematic viscosity of the fluid, U_* is the shear velocity, τ_0 is the shear stress at the channel bed surface, γ_s and γ_f are the specific weights of sediment and fluid, respectively, and d is the representative sediment size.

EXAMPLE 12.1 Determining Incipient Motion in an Open Channel. Given:

$$\text{Rectangular channel with width} = 100 \text{ ft}$$

$$\text{Channel slope} = 0.001$$

$$\text{Sediment size on channel bed} = 3 \text{ mm } (=0.010 \text{ ft})$$

$$\text{Flow temperature} = 50°\text{F}$$

$$v, \text{ kinematic viscosity for water} = 1.4 \times 10^{-5} \text{ ft}^2/\text{s}$$

$$\text{Specific weight of water} = 62.4 \text{ lbf/ft}^3$$

$$\text{Specific gravity for sediment} = 2.65$$

$$\text{Density of water} = 1.94 \text{ slugs/ft}^3$$

Determine: At what flow velocity and flow discharge will the sediment particles on the channel bed be moved by the flow?

Solution: For a wide channel, hydraulic radius $R \approx D$ (flow depth). Assume:

$$\tau^* = \frac{\tau_0}{(\gamma_s - \gamma_f)d} \approx 0.04 \qquad \text{(to be checked later)}$$

so that

$$\tau_0 = 0.04(\gamma_s - \gamma_f)d$$

where shear bottom stress $\tau_0 = \gamma_f RS$. Hence

$$0.04(\gamma_s - \gamma_f)d = \gamma_f RS \qquad \text{and} \qquad R = 0.04\frac{(\gamma_s - \gamma_f)}{\gamma_f}\frac{d}{S}$$

$$= 0.04 \times 1.65 \times \frac{3}{305 \times 0.001} = 0.65 \text{ ft}$$

Now one should check the assumption of

$$\tau^* = \frac{\tau_0}{(\gamma_s - \gamma_f)d} \approx 0.04$$

$$\tau_0 = \gamma_f RS = 62.4 \times 0.65 \times 0.001 = 0.041 \text{ lbf/ft}^2$$

$$U_* = \sqrt{\frac{\tau_0}{\rho}} = \sqrt{\frac{0.041}{1.94}} = 0.145 \text{ ft/s}$$

$$\frac{U_* d}{v} = \frac{0.145 \times 3}{305 \times 1.4 \times 10^{-5}} = 102$$

As shown in Fig. 12.3.1, for $\dfrac{U_* d}{v} = 102$,

$$\frac{\tau_0}{(\gamma_s - \gamma_f)d} \text{ is } 0.04 \qquad \text{(on the incipient motion curve)}$$

Thus the previous assumption of $\tau_0/(\gamma_s - \gamma_f)d = 0.04$ is valid.

Use Strickler's equation, Eq. (12.2.3), to estimate Manning's n value:

$$n = \frac{d^{1/6}}{25.6} \text{ (where } d \text{ is in feet)} = \left(\frac{3}{305}\right)^{1/6} \frac{1}{25.6} = 0.018$$

$$U = \frac{1.49}{n} R^{2/3} S^{1/2} \qquad \text{with } R \text{ in feet and } U \text{ in ft/s}$$

$$= \frac{1.49}{0.018} (0.65)^{2/3} (0.001)^{1/2} = 1.96 \text{ ft/s}$$

$$Q = \text{flow velocity} \times \text{flow area} = 1.96 \times 100 \times 0.65 = 127 \text{ ft}^3\text{/s } (3.61 \text{ m}^3\text{/s})$$

Answer: When flow depth reaches 0.65 ft, flow velocity reaches 1.96 ft/s, and flow discharge reaches 127 ft^3/s, incipient sediment motion will occur in this channel.

Shields' criterion is often used to determine the riprap size d necessary for a stable bank under prescribed flow conditions (Wang and Shen[136]). In general, the Shields curve between motion and no motion can be treated as horizontal, and a conservative criterion for critical dimensionless shear stress is $\tau_c^* = 0.03$ for no motion. Since the definition of incipient motion for turbulent flow is subjective, one must choose an appropriate safety factor (normally between 2 and 4) to multiply this critical dimensionless shear stress according to the potential danger if objectionable sediment movement should occur. The Shields' diagram has been generally accepted as a good indication for incipient motion conditions of uniform-size cohesionless sediment particles, especially for plane beds. As discussed by Wang and Shen,[136] it is difficult to extend Shields' curve for a large Reynolds number (large sediment sizes) of $U_* d/\nu > 5000$.

Incipient Motion on Ripple and Dune Beds. For ripple and dune beds, Chabert and Chauvin[22] have modified Shields' diagram such that incipient motion begins for $\tau_c^* = 0.06$. Mantz[86] has modified Shields' diagram for disk-shaped particles.

Frequently, it is desirable to use a critical flow velocity which can be measured to indicate incipient sediment motion instead of flow shear stress, which cannot be measured easily. For a hydraulically rough turbulent boundary,

$$\frac{\overline{U}}{U_*} = 5.75 \log \frac{R}{k_s} + 6.25 \tag{12.3.5}$$

and for a hydraulically smooth turbulent boundary,

$$\frac{\overline{U}}{U_*} = 5.75 \log \left(\frac{U_* R}{\nu}\right) + 3.25 \tag{12.3.6}$$

where $\overline{U}$ is the depth averaged flow velocity, k_s is the representative grain roughness, which may be assumed to be d_{65}, and R is the hydraulic radius. By setting $\tau_c^* = 0.04$, the critical average flow velocity for incipient motion $\overline{U}_c$ can be obtained by combining Eqs. (12.3.2), (12.3.4), and (12.3.5) for a hydraulically rough boundary:

$$\frac{\overline{U}_c}{\sqrt{(\gamma_s - \gamma_f)d/\rho_f}} = \log \frac{R}{k_s} + 1.08 \tag{12.3.7}$$

Results from Eq. (12.3.7) agreed well with the equations obtained by Goncharov.

An equation similar to Eq. (12.3.7) for a smooth boundary can also be derived, but most natural channels are hydraulically rough.

Christensen[27] analyzed the incipient motion on cohesionless channel banks and showed that the ratio between the critical shear stress on the bank and the bed is a function of representative roughness to effective grain size (d_{35}) ratio.

Case 2 Mixture with Nonuniform Cohesionless Sediment Sizes. The movement of nonuniform sediment size distributions has received increasing attention in the past decade because nearly all streambeds consist of mixed-size sediment particles. The physical process of sediment movement with mixed sizes is complex. Sometimes larger particle sizes are moved by the flow at an earlier stage than smaller particles because larger particles intrude farther in the flow and face stronger turbulence. Also, smaller particles can hide behind larger particles and be protected from the flow. Egiazoroff,[32] Gessler,[54-57] Little and Meyer,[82] Ashida and Michiue,[9] Hayashi et al.,[61] and Misri et al.[96] have all investigated the critical traction stress of a mixture. Based on results from laboratory experiments conducted by Gessler,[57] Little and Meyer,[82] and field data collected by Lane and Carlson,[78] and Pemberton,[103] Shen and Lu[112] showed that for sand bed streams the bed is stable if

$$\tau^* = \frac{\tau_0}{(\gamma_s - \gamma_f)d_{30}} < 0.028 \tag{12.3.8}$$

Thus Shields' criterion of $\tau^* < 0.03$ as derived from uniform sediment size can also be used for nonuniform sand sizes, if d_{30} is substituted for d_{50}. The principle is that if the d_{30} size particle is stable, then all sizes greater than d_{30} should also be stable. All sizes smaller than d_{30} are sheltered behind larger particles and thus are also stable. Since Eq. (12.3.8) is derived based on limited data, an appropriate safety factor of 2 to 4 should be used with this equation for design purposes.

Case 3 Cohesive Sediment. Movement of cohesive sediment is much more complex than that of coarser cohesionless particles because of the action of physicochemical forces among cohesive sediment particles. Often the type of sediment, the concentration of different ions, and their cation-exchange capacities can be more important to erosion than the magnitude of fluid forces.

There are three major types of clay minerals: kaolinite, montmorillonite, and illite. Each type has its own distinct characteristics. Some cohesive soils contain a combination of different types of clay mineral. Two task committees[88,89,122] were appointed by the Hydraulic Division of the American Society of Civil Engineers to review current knowledge on cohesive sediments. No general relationship was found to define the incipient motion of cohesive material. Mehta[90] presents a detailed description of cohesive sediment behavior.

12.4 MECHANISMS OF SEDIMENT MOTION

Several commonly used terms to describe sediment motion are discussed in this section. Equations to predict sediment transport rates are given in the following sections.

The amount of sediment passing through a given stream cross section is a critical factor in (1) determining the amount of sediment deposited in a downstream reservoir and thus the useful life of a dam project; (2) investigating the stability of the streambed and stream banks; (3) influencing the water quality to be used for irrigation, city water supply, and recreation; (4) affecting the navigable depth of a stream; (5) determining the water level during floods; (6) studying the potential effects on stream ecology such as fisheries as well as other biological species; and (7) investigating the movement of pollutants attached to sediment particles.

Total Sediment Load. The term *sediment load* has been used to denote sediment transport rate, and the dimension of load can be expressed either as weight per unit time or as volume per unit time. Sediment concentration has been used to indicate the ratio of sediment transport rate and the fluid discharge rate. The sediment concentration can also be used to express either the weight of sediment per weight of fluid mixture, or volume of sediment per volume of fluid mixture. The fluid mixture includes both liquid and sediment particles.

In the United States, the sediment concentration is usually expressed as parts of sediment per million parts of mixture or milligrams per liter. In other parts of the world, the sediment concentration is frequently expressed as kilograms of sediment particles per cubic meter of fluid mixture.

If Q_s is the sediment load in mass per time, Q is the flow discharge in volume per time, and C is the sediment concentration in mass per volume, then

$$Q_s = QC \tag{12.4.1}$$

In U.S. customary units, with Q in ft³/s and C in mg/liter or ppm by mass, and Q_s in units of U.S. tons/unit time (1 U.S. ton = 2000 lb), then

$$\begin{aligned} Q_s &= 2.697 \times 10^{-3}\, QC && \text{(U.S. tons/day)} \\ &= 0.984\, QC && \text{(U.S. tons/year)} \end{aligned}$$

In SI units, with Q in m³/s and C in kg/m³, and Q_s in units of metric ton/unit time (1 metric ton = 1000 kg), then

$$\begin{aligned} Q_s &= 86.4\, QC && \text{(metric tons/day)} \\ &= 3.5 \times 10^4\, QC && \text{(metric tons/year)} \end{aligned}$$

The *total sediment load* is defined as the total amount of sediment passing through a given stream cross section. Depending on usage, total sediment load can be divided into the following categories:

1. According to the mechanism of movement

   Total sediment load = bed load + suspended load

2. According to the measurement capabilities:

   Total sediment load = sampled load + unsampled load

3. According to the method of calculation:

   Total sediment load = wash load + bed material load

   Bed material load = suspended bed material load + bed load

   Suspended load = wash load + suspended bed material load

The definitions of the above terms are given in the following sections, and Fig. 12.4.1 shows the various vertical zones for these classifications.

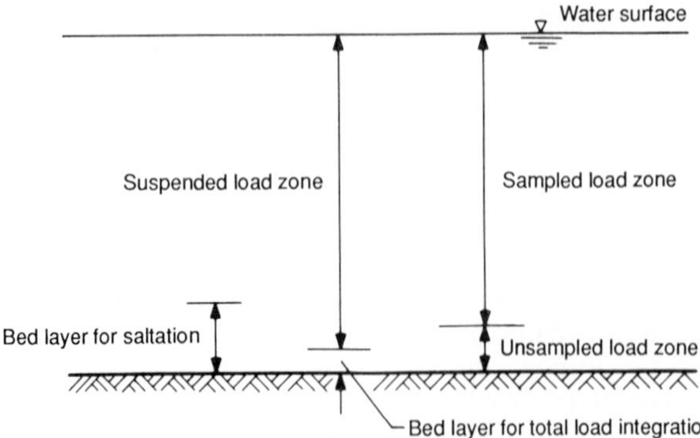

FIGURE 12.4.1 Vertical flow layers for sediment load classification.

Bed Load. *Bed load* is the rate of movement of sediment particles along the streambed in the processes of rolling, sliding, and/or hopping. In the hopping process, the flow turbulence picks up the sediment particle from the bed, then drops the particles back to the streambed. This hopping process is called *saltation.* There are two types of bed layers. A saltation bed layer is established by the height of saltation that exists in the flow (van Rijn[132]) and this is at least an order of magnitude greater than the sediment particle size. A second type of bed layer with a thickness of about two times the particle size is defined to be used for the integration of the vertical distribution of suspended sediment load. One must be careful in reading literature on the particular definition of bed layer being used. There is no general agreement on the thickness of bed layer for bed load.

Suspended Load. The *suspended load* is the rate of movement of sediment particles that are supported by the turbulent motion in the stream flow. For convenience, the suspended load is taken as the rate of sediment particle movements between the top of the bed layer (about twice the sediment particle size) and the water surface as shown in Fig. 12.4.1. The same sediment particles can be transported as suspended load and as bed load at different times.

Sampled Load. The *sampled load* is that part of the total sediment load that can be reasonably accurately measured by either a depth integrated sampler or a point integrated sampler. If samplers are placed too close to the streambed, the sampler changes the flow and sediment characteristics on top of the streambed and the measurements do not reflect true sediment rates at that location. The effective zone that can be measured reasonably accurately by a suspended load sampler is called the *sampled zone.* The minimum clearance for the sampler is about 10 to 20 cm above the streambed. Thus, in Fig. 12.4.1, the unsampled zone is within 10 to 20 cm vertically upward from the bed.

Unsampled Load. No instrument and procedure has been developed to obtain a reasonably accurate measurement of sediment movement rate in a flow zone within 10 to 20 cm from the top of the streambed. The rate of sediment movement in this unsampled zone is called the *unsampled load.*

Wash Load. Suspended load can be divided into two parts. One part of the suspended load is composed of the relatively coarse sediment particles that can be seen readily on the streambed surface. The amount of sediment movement of this size range is related to the flow condition and the sediment composition of the streambed and is called *suspended bed material load* because these particles appear on the streambed. A second part of the suspended load is composed of the relatively fine sediment particles called *wash load* that cannot easily be seen individually on the streambed surface. This load is not directly related either to the flow condition at the stream reach or to the sediment composition of the streambed. Einstein and Chien[34] found that wash load depends not only on the fluid and sediment characteristics but also on the upstream sediment supply rate. Einstein[36] suggested using the d_{10} size from the bed material size distribution to separate the wash load from the bed material load, where d_{10} is defined as the sediment size on the streambed surface of which 10 percent is finer. A more recent study (Chien and Wan[25]) indicated that d_5 may be a better indicator to divide wash load and bed material load. Wash load must be calculated from upstream supply rate (or watershed sediment yield) and/or by actual sediment rate measurements. Very little wash load is moved as bed load and thus wash load is a part of suspended load only. Watershed sediment yield is discussed in Sec. 12.10.

Bed Material Load. As discussed in the last section, *bed material load* is the total rate of sediment transport for those sediment particle sizes that are readily apparent on the surface of the streambed (or $d > d_5$ of the sediment bed material size). Movements of these particles are related to the flow and sediment characteristics of the bed. Thus bed material load can in principle be calculated from sediment transport equations. Bed material load can be divided into two parts. *Suspended bed material load* is the total rate of transport of sediment that is transported as suspended load with the sediment sizes appearing readily on the streambed surface. The second part of bed material load that moves along the streambed is *bed load.*

12.5 BED LOAD

As stated in the previous section, sediment particles can be moved by the flow as bed load along the streambed in the process of rolling, sliding, and saltation (hopping). With low flow rates just above the incipient motion condition, bed load motions are intermittent. Sediment particles may be moved by the flow for a short period of time and then rest on the streambed for a relatively long time period. In sand bed streams, as flow rate increases dunes appear in the streambed and sediment particles may be deposited on the downstream slopes of the dune and remain there for a long rest. For still greater flow rates, dunes may disappear and streambeds exhibit plane form; the entire layer or layers of the bed surface may move continuously as bed load. In this case, the analysis of duBoys[31] and that by Bagnold[10–12] may be important. A gravel streambed usually consists of two different sediment size populations; a coarse sediment sample, with gravels, pebbles, cobbles, and/or boulders, moves only during high flows; a second sediment sample, with sand and/or silts, moves on top of the gravel bed or fills in the void spaces among the coarse sediment material. Figure 12.5.1 shows four different types of stream bedding for fluvial gravels. In Fig. 12.5.1*a*, different layers of gravel and sand particles are deposited in response to the historical values of flow and incoming sediment characteristics.

In Fig. 12.5.1*b*, the fine sediment particles at the surface layer are either washed away by the flow during gravel movements or they are infiltrated down to the subsurface. In Fig. 12.5.1*c*, fine sediment particles begin to be deposited onto the upper layer of the sediment bed. In Fig. 12.5.1*d*, a mixture of fine and coarse sediment particles occurs underneath a layer of large sediment particles deposited on the streambed surface. These large particles in the stream may reduce or prevent the sediment particles below from being moved by the flow. The amounts of sediment movement for these four bed conditions are different even under the same flow condition. Thus no generally accepted gravel bed load equation is available. Several popular bed load equations for sand bed streams are described below.

Einstein's Bed Load Equation. This equation is introduced first because it is the most important comprehensive analysis available and many relationships derived by others can be converted into Einstein's final form for comparison (see Chien and Wan[25]). Einstein[36] conducted laboratory experiments to prove that under equilibrium conditions (streambed elevation does not change with time), the flow will erode certain amounts of sediment particles from the streambed and equal amounts of sediment particles will be deposited from the flow onto the streambed. Thus he developed the following relationships between the dimensionless *flow intensity parameter* Ψ and the dimensionless *sediment transport parameter* ϕ.

Einstein[36] developed a dimensionless transport rate as

$$\phi = \frac{q_B}{\gamma_s}\left(\frac{\rho_f}{\rho_s - \rho_f}\right)^{1/2}\left(\frac{1}{gd^3}\right)^{1/2} \tag{12.5.1}$$

where q_B is weight of sediment bed load per unit time and per unit width of channel.

The parameter ϕ has been used widely by many others as the dimensionless sediment transport rate for numerous studies.

Einstein developed a dimensionless flow intensity Ψ related to the hydraulic radius R_b':

$$\Psi = \frac{\gamma_s - \gamma_f}{\gamma_f}\frac{d_{35}}{R_b'S} \tag{12.5.2}$$

Einstein believed that only the grain resistance is related to the movement of sediment particles and thus R_b' is the hydraulic radius related to the grain resistance and d_{35} is the sediment size of which 35 percent is finer. A complex procedure was presented by Einstein[36] to obtain R_b'. Researchers should refer to this article for detailed description. For practicing hydrologists, R_b' can be estimated as follows.

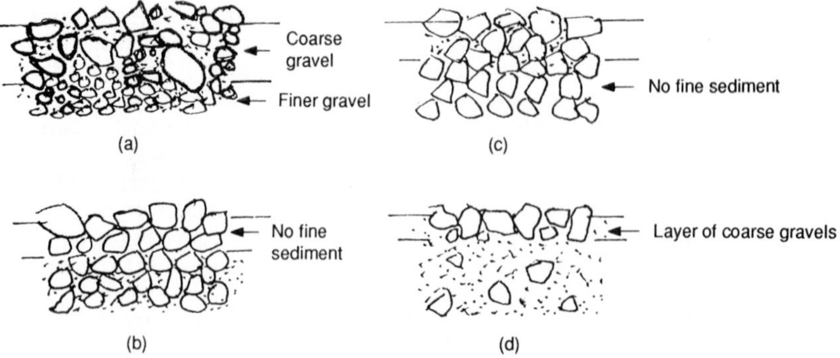

FIGURE 12.5.1 Some typical gravel bed compositions.

In a wide stream (width > 20 depth) with steady uniform flows: $R_b = R = D$ (flow depth), and R'_b is calculated as

$$R'_b \approx \frac{n'}{n} R_b \approx \frac{n'}{n} D \qquad (12.5.3)$$

where n is determined from Manning's equation, using measured values of U, h, and S [Eq. (12.2.1)] and n' is calculated from Strickler's equation as given in Eq. (12.2.2) or Eq. (12.2.3).

The formula for Ψ is just an inverse of the Shields' number τ^* as shown in Eq. (12.3.2) with $\tau_0 = \gamma_f R'_b S$.

From both field and laboratory data, Einstein[36] established the relationship between ϕ and Ψ as shown in Fig. 12.5.2.

For nonuniform sediment sizes, Einstein[36] introduced several correction factors. A correction factor Y was introduced to indicate the change in the lift coefficient due to the presence of various sediment sizes. Another correction factor was introduced to indicate the hiding of smaller particles behind large particles, for which Shen and Lu[112] suggested a modification by analyzing data collected by Gessler[57] and Little and Mayer.[82] For a detailed discussion on Einstein's procedure, the reader is referred to Einstein,[36] Shen,[114] Garde and Ranga Raju,[50] Chien and Wan,[25] Graf,[59] and many other reference texts.

Several other well-known bed load equations are discussed below.

Meyer-Peter and Muller Equation. This equation was introduced by Meyer-Peter and Muller.[94] Based on data collected with sediment sizes between 0.4 and 30 mm; flow depth 0.01 to 1.2 m; density of sediment bottom 1.25 to 4.22; and slope between

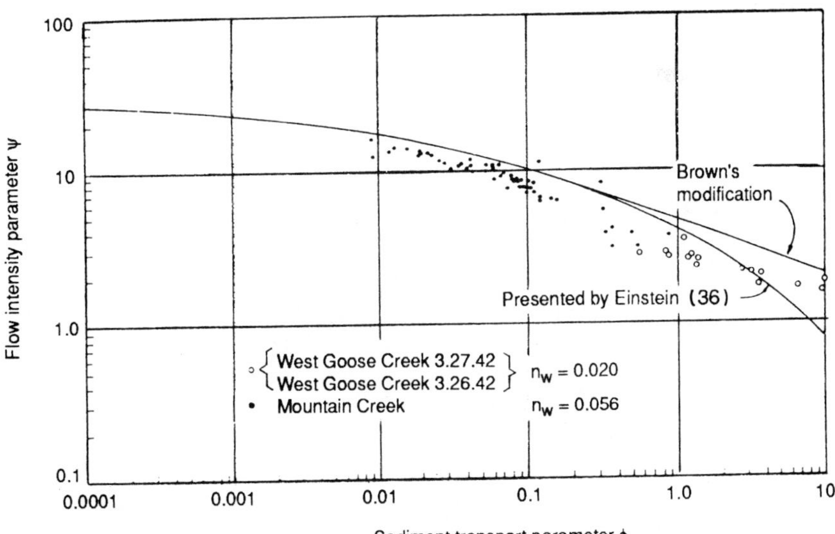

FIGURE 12.5.2 Einstein's[36] bed load relationship between the flow intensity parameter Ψ and the sediment transport parameter ϕ.

4×10^{-4} and 2×10^{-2}, the following empirical relationship was developed:

$$\left(\frac{n'}{n}\right)^{3/2} \frac{\gamma_f R_b S}{(\gamma_s - \gamma_f)d_a} = 0.047 + 0.25 \left(\frac{\gamma_f}{g}\right)^{1/3} \left(\frac{q_B}{\gamma_s}\right)^{2/3} \frac{1}{(\gamma_s - \gamma_f)^{1/3} d_a} \quad (12.5.4)$$

In this equation, n is the Manning's roughness, R_b is the hydraulic radius, S is the energy slope, γ_s and γ_f are the specific weights for solid and liquid, respectively, d_a is the arithmetic mean sediment size, g is the gravitational constant, and q_B is the bed load rate in weight per unit time.

In the above equation n' is the grain Manning's roughness and may be obtained as $n' = (d_{90}^{1/6})/26$. Since Einstein's[36] equation was developed based on a wide range of sediment sizes and Meyer-Peter's equation was developed based only on relatively coarse sediment, many engineers prefer to use Meyer-Peter's equation in dealing with relatively coarse sediment.

Chien and Wan[25] showed that Eq. (12.5.4) can be modified into the form

$$\left(\frac{n'}{n}\right)^{3/2} \tau^* = 0.047 + 0.25\, \phi^{2/3} \quad (12.5.5)$$

where ϕ is defined in Eq. (12.5.1). The factor $(n'/n)^{3/2}$ is the ratio of the skin shear stress to the total shear stress.

Chien and Wan[25] showed that Eq. (12.5.5) agrees with Einstein's curve (Fig. 12.5.2) reasonably well. Chien and Wan also plotted the Bagnold[11] equation with $d = 2$ mm, the Ackers and White[3] equation with $d > 2.5$ mm and $S = 0.01$ as well as $S = 0.001$, Yalin's[142] bed load equations, and the Engelund-Fredose[40] bed load equation, as shown in Fig. 12.5.3. These relationships are all rather similar.

Bagnold[10–12] Bed Load Formula. For high sediment transport rates, the intergranular solid friction force of bed material is important. Let tan α (ratio of tangential shear

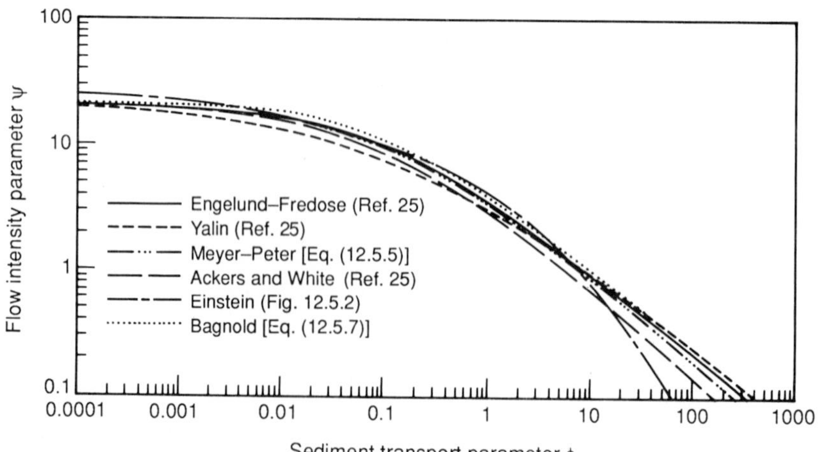

FIGURE 12.5.3 Comparison of various bed load equations. *(Chien and Wan.[25])*

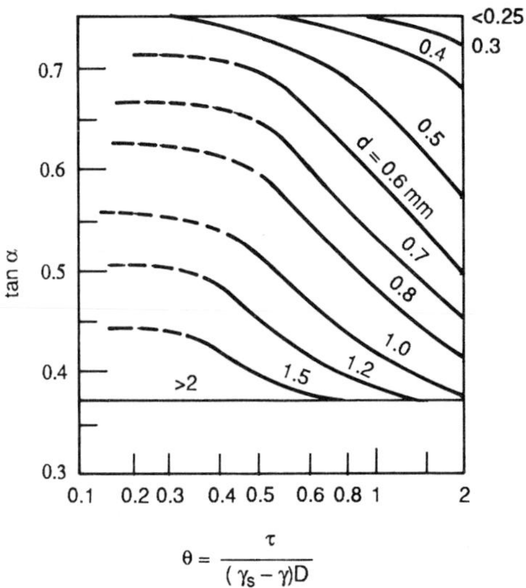

FIGURE 12.5.4 Bed load parameters. *(Bagnold.[10])*

to normal force) be the coefficient of this intergranular solid friction, Bagnold found that

$$q_B = \frac{\tau_0 U l_B}{\tan \alpha} \tag{12.5.6}$$

where τ_0 is the shear stress, U is the average flow velocity, and l_B is a bed load transport coefficient. Bagnold provided particle size data and Fig. 12.5.4 giving the variation of $\tan \alpha$.

Chien and Wan[25] determined that Eq. (12.5.6) can be approximated by

$$\phi = \frac{1}{\Psi}\left[\left(\frac{1}{\Psi}\right)^{1/2} - \left(\frac{1}{\Psi_c}\right)^{1/2}\right]\frac{1}{\tan \alpha}\left(5.75 \log 30.2 \frac{md}{k_s} - \frac{w}{U_*}\right) \tag{12.5.7}$$

where w is mean fall velocity and k_s the representative grain roughness.

$$m = 1.4\left(\frac{U_*}{U_{*C}}\right)^{0.6} \tag{12.5.8}$$

where U_{*c} is the critical shear stress for incipient motion.

Chien and Wan noted that the first part on the right-hand side of Eq. (12.5.7) varies more than the second part, thus

$$\phi \propto \frac{1}{\Psi}\left[\left(\frac{1}{\Psi}\right)^{1/2} - \left(\frac{1}{\Psi_c}\right)^{1/2}\right] \tag{12.5.9}$$

or

$$\phi \propto \tau^* \left(\tau^{*1/2} - \tau_c^{*1/2}\right) \tag{12.5.10}$$

Equation (12.5.10) was plotted by Chien and Wan[25] in Fig. 12.5.3 for $\tan \alpha = 0.63$, and sediment particle sizes of 0.2 and 2 mm. They found that Eq. (12.5.7) provided similar results to Einstein's curve.

Parker's[100] Relationship for Gravel Bed Streams. Based on data collected for gravel bed streams, Parker developed the following relationship for bed load transport of uniform sediment sizes:

$$q_B^* = 0.0218\,(\tau^{**})^{3/2}\,G(\Delta) \tag{12.5.11}$$

where

$$\Delta = \frac{\tau^{**}}{0.0386}$$

$$q_B = q_B^* \sqrt{Rgd}\,d \tag{12.5.12}$$

$$\tau^{**} = \frac{\tau}{Rgd} \tag{12.5.13}$$

and

$$G(\Delta) = \begin{cases} 5474\left(1 - \dfrac{0.853}{\Delta}\right)^{4.5} & \text{for } \Delta > 1.59 \\ \exp[14.2\,(\Delta - 1) - 9.28\,(\Delta - 1)^2] & 1 \le \Delta \le 1.59 \\ \Delta^{14.2} & \Delta < 1 \end{cases} \tag{12.5.14}$$

where d is sediment size; q_B is defined following Eq. (12.5.1).

The procedure for using the Parker method is as follows. The shear stress τ is computed using $\tau = \gamma_f RS$ in the same way as in Example 12.1, the value of τ^{**} is formed from Eq. (12.5.13), the value of Δ from $\Delta = \tau^{**}/0.0386$, the value of $G(\Delta)$ from Eq. (12.5.14), the value of q_B^* from Eq. (12.5.11), and finally the value of the sediment load q_B is found from Eq. (12.5.12).

For a high sediment transport rate

$$\frac{\tau^{**}}{0.0386} \gg 1.59 \qquad q_B \propto (\tau^{**})^{1.5} \tag{12.5.15}$$

Parker[100] also showed that with his relationship, data with $d = 28.6$ mm fall below Einstein's[36] curve and data with $d = 0.5$ mm fall above Einstein's[36] curve.

Assessment of Bed Load Transport Equations. It is difficult to determine a reliable bed load transport to be used in the field because of the lack of reliable data from streams. Thus many bed load relationships have been developed based on flume data. As shown by Chien and Wan,[25] many bed load relationships provide similar results for certain flow conditions. The procedure developed by Einstein[36] is still the most comprehensive one available. For transport of coarse sand, pebbles, and small gravel, Meyer-Peter and Muller's equation[94] is usually used because this relationship was developed based on relatively coarse sediment. For high sediment rates where intergranular forces are important, Bagnold's[12] relationship is used. Ackers and White's[3] equation, as described in the next section, has received increasing attention and should always be checked as a reference. It is difficult to use relationships presented by Wiberg and Smith[138] owing to the lack of knowledge on the variation of the representative dune dimensions. Parker[100] and his coworkers have presented extremely valuable knowledge on the movements of gravel particles, but the constants in their relationship should be calibrated with more data. Similarly, the relationships presented by Misri, Garde, and Ranga Raju[96] and van Rijn[132] should also be tested further by data.

As shown in Fig. 12.5.3, the bed load transport rate varies with different powers of the stream-flow shear and stress. For low transport rates, bed load rate varies with the 15th power of shear stress, and for high transport rates, bed load rate varies with the shear stress to the 1.5 power. Most investigators prefer to relate bed load rate to the grain shear stress, rather than the total shear stress because shear stress related to form roughness does not transport sediment bed load particles downstream.

The emphasis here is on the transport of uniform sediment size. For the transport of nonuniform sediment size, readers are referred to Einstein,[36] Ackers and White,[3] Garde and Ranga Raju,[50] and Parker.[100] More research is needed to obtain data for the transport of nonuniform sediment size in order to develop a generally acceptable equation. Perhaps the use of Einstein's equation with a modified hiding factor as introduced by Shen and Lu[112] will provide an answer as well as any other method.

12.6 VERTICAL SUSPENDED SEDIMENT CONCENTRATION DISTRIBUTION

Sediment suspended in a stream flow has a vertical distribution, less dense near the surface, more dense near the bed. The form of this distribution is determined by the balance between the rate at which particles are falling due to gravity and the rate at which they are being stirred up again by turbulent eddy motion. At equilibrium, these two rates are equal, so an equation can be written for concentration C as a function of depth y as

$$wC + \epsilon_m \frac{\partial C}{\partial y} = 0 \tag{12.6.1}$$

where w is the fall velocity of sediment particles, C is the sediment concentration, ϵ_m is the vertical mass transfer coefficient due to eddy motion, and y is the vertical direction.

By assuming a logarithmic vertical velocity distribution and the fluid shear stress relationship $\tau_0 = \epsilon_m \rho \, du/dy$, the solution to Eq. (12.6.1) is

$$\frac{C_y}{C_a} = \left(\frac{D-y}{y} \frac{a}{D-a} \right)^z \tag{12.6.2}$$

where C_a is the concentration at a reference level at elevation a above the bed, D is the depth of the stream, C_y is the concentration at elevation y above the bed, and

$$z = \frac{w}{U_* K} \tag{12.6.3}$$

and w is the fall velocity, U_* is the shear velocity, and K is the von Karman constant of 0.4 for clear water.

In order to solve the sediment concentration at any elevation y, one must know C at a given elevation. H. A. Einstein[36] suggested that the bed load transport rate q_B can be used to obtain C_a as follows:

$$C_a = C_{2d} = \frac{i_B q_B}{23.2 \, U'_* d} \tag{12.6.4}$$

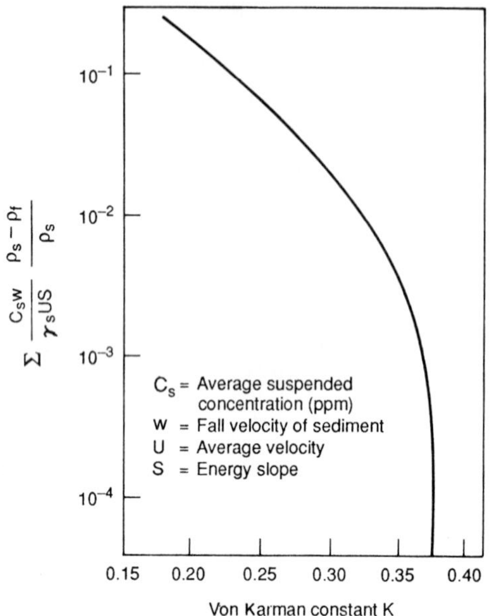

FIGURE 12.6.1 Variation of von Karman's *K*. *(Einstein and Chien.[33])*

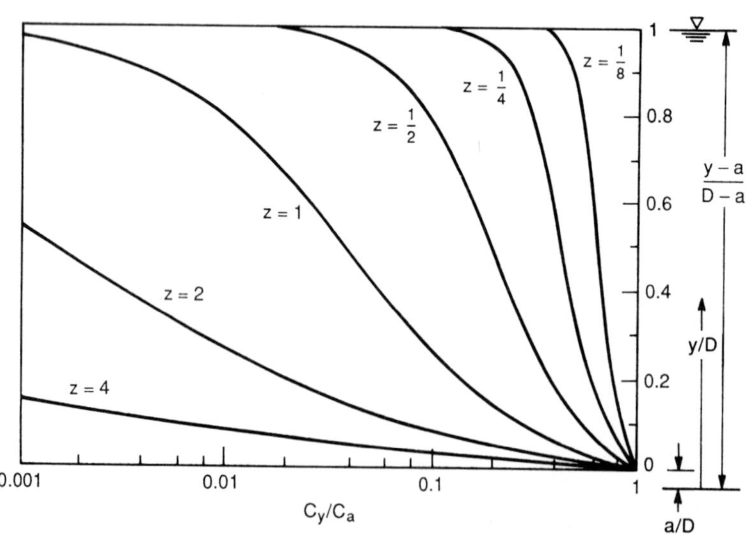

FIGURE 12.6.2 Vertical variations of sediment concentration for different values of parameter *z* in Eq. (12.6.2).

where $a = 2d$, C is in dry weight of sediment per unit volume, U'_* is the shear velocity resulting from shear stress due to skin friction alone, i_B is the fraction of the total sediment load in the bed load for a particular sediment size d, and q_B is the total bed load rate in weight per unit time and per unit width.

For high sediment concentration, the value of K may decrease from 0.4, which was derived for clear water. There is no entirely satisfactory method to obtain the K value, but perhaps the best-known curve to indicate the change of K with flow and sediment concentration is given in Fig. 12.6.1 by Einstein and Chien,[34] where C_s is the average sediment concentration.

Figure 12.6.2 shows the vertical variation of sediment concentration for different z values according to Eqs. (12.6.2) and (12.6.3).

As shown in Fig. 12.6.2 and Eq. (12.6.2), for $z = 0$, the sediment concentration is uniformly distributed vertically, and for $z \rightarrow 5$, most of the sediment transport is bed load. Thus if $z \rightarrow 5$, or $w/U_*K \rightarrow 5$ according to Eq. (12.6.3), or $w = 2U'_*$, bed load is the predominant mode of sediment transport rate. On the other hand, if $z \rightarrow 0$, the vertical suspended sediment concentration is uniformly distributed.

12.7 TOTAL SEDIMENT LOAD

Total sediment load is the total amount of sediment passing through a given stream cross section per unit time. As discussed in Sec. 12.4, total sediment load is the sum of wash load and bed material load. The wash load can be estimated by measurements as well as by watershed sediment yields. This will be discussed in the next section. The bed material load can be estimated as the sum of bed load and suspended bed material load. In this section, the bed material load is referred to as total sediment load due to common usage. Actually, the bed material load is that part of the total sediment load which can be calculated by a sediment transport rate relationship.

Einstein[36] presented the most comprehensive analysis on total sediment load. Unfortunately, the results from his final relationship were never verified by measurements and therefore will not be described here. Readers, especially researchers, should refer to Einstein's article for detailed description of sediment transport.

Sediment Load in Large Sand Bed Rivers (Flow Depth Greater than 10 ft or 3 m). Many Americans use the total sediment load relationship as described by Toffaleti[124] for large rivers because this relationship was developed by data collected from large rivers in the United States. Toffaleti[124] presented a procedure for the analytical determination of sediment transport based on the concepts of Einstein[36] and Einstein and Chien.[34] First, he established that Einstein's Ψ versus ϕ curve may be represented by the equivalent expression

$$\Psi = \frac{TA}{U^2} 10^4 d \qquad (12.7.1)$$

where Ψ is Einstein's flow intensity parameter, T (dimensions $[L/T^2]$) is a parameter that includes constants and those components of shear force that are functions of water temperature, A is a dimensionless correction factor to replace Einstein's correction factors for sediment of mixed sizes, $U[L/T]$ is the mean flow velocity, and $d[L]$ is the grain diameter. He divides the flow depth into three zones: the lower zone where the depth of flow is less than $R/11.24$ (R is the hydraulic radius), the middle zone depth between $R/2.5$ and $R/11.25$, and the upper zone where the depth is above

$R/2.5$. Toffaleti further states that the G_F or nucleus load in U.S. tons per day, computed for a 1-ft width in the lower zone (by assuming the bed is composed entirely of one size of sand), can be represented by

$$G_F = \frac{0.0600}{(TA/U^2)^{5/3}\,(d/0.00058)^{5/3}} \tag{12.7.2}$$

and for very fine sand $d < 1$ mm, this equation becomes

$$G_F = \frac{1.905}{(TA/U)^{5/3}} \tag{12.7.3}$$

These relationships were determined from field data collected from large rivers (such as the Atchafalaya River), taken over a 20-year period and covering a range of flow discharges of 20,000 to 500,000 ft³/s. (This information was obtained by private communication with Toffaleti.) The sediment concentration distribution is expressed as

$$C_y = C_a\left(\frac{y}{R}\right)^z \tag{12.7.4}$$

in which C_y is the sediment concentration at y, C_a is the sediment concentration at a level a, and y is the vertical elevation above the bed. For the middle zone

$$z = \frac{Uw}{C_z SR} \tag{12.7.5}$$

in which w is the fall velocity of the particle in feet per second, C_z is a temperature correction factor which equals $260.67 - 0.667\,T$ (T in °F), and S is the energy slope or water surface slope. The exponents z of the sediment concentration distribution in the lower and upper zones are, respectively, 0.756 and 1.5 times that of the middle zone.

After determining G_F and the sediment concentration distribution in the lower zone, one can proceed to determine the sediment concentration at the upper edge of the lower zone, then calculate the sediment concentration distribution in the middle zone and the total sediment load in the middle zone, and finally obtain the total sediment load in the upper zone. The summation of the total sediment load in the three zones is the total sediment load over the entire flow depth.

Readers are referred to the original article by Toffaleti[125] for a more detailed description of this method.

Sediment Load in Small Sand Bed Rivers (Flow Depth Less than 10 ft or 3 m). Colby[28] investigated the variations of sediment transport load as a function of the mean flow velocity, depth, viscosity, water temperature, and concentration of the fine sediment of the discharge of sand per foot of channel width. He recommended the diagrams shown in Figs. 12.7.1 and 12.7.2 for estimation of sediment load. In spite of many inaccuracies in the available data and uncertainties in the graphs, Colby found that ". . . about 75 percent of the sand discharges that were used to define the relationships were less than twice or more than half of the discharges that were computed from the graphs of average relationship. The agreement of computed and observed discharges of sands for sediment stations whose records were not used to define the graphs seemed to be about as good as that for stations whose records were used." Note that all curves of 100-ft depth, most curves of 10-ft depth, and part

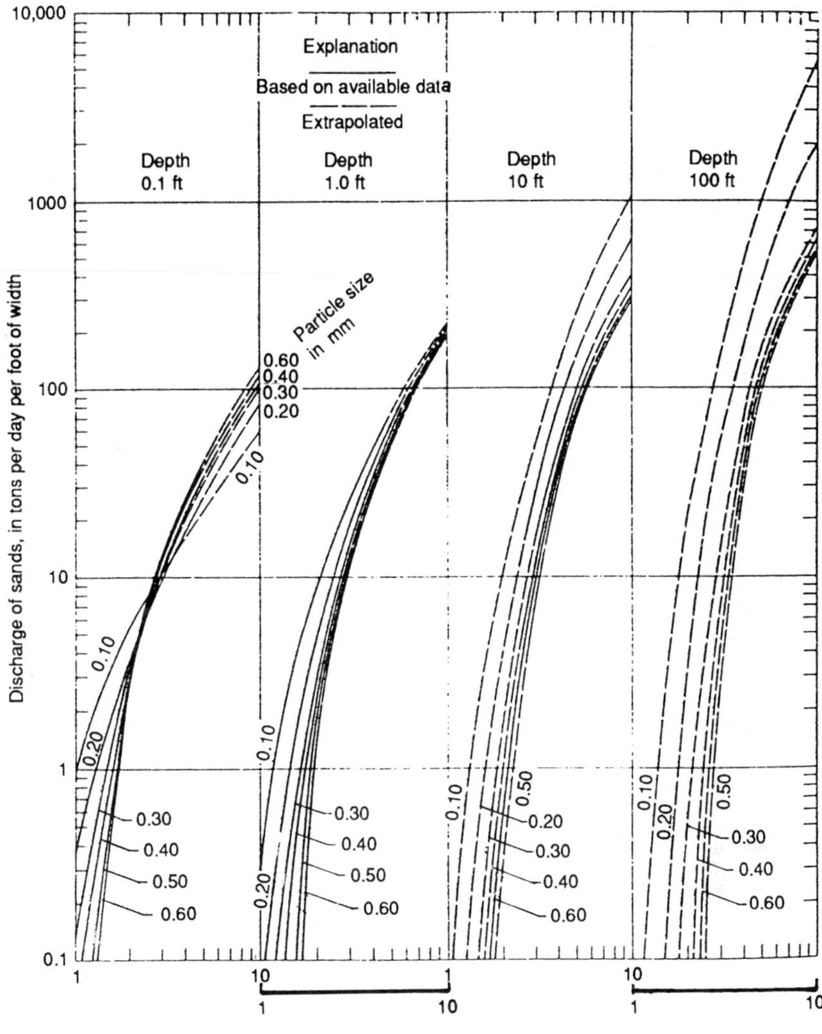

FIGURE 12.7.1 Relationships of discharge of sands to mean velocity for six median sizes of bed sands, four depths of flow, and a water temperature of 60°F. *(Colby.*[28]*)*

of the curves of 1.0 and 0.1 ft for Fig. 12.7.1 were not based on available data and were extrapolated. Thus Colby's diagrams should not be used for flow depth greater than 10 ft. The total sediment load L in U.S. tons/day/ft width is found by

$$L = L_0[1 + (k_1 k_2 - 1)k_3/100] \tag{12.7.6}$$

where L_0 is the expected load determined for a given flow depth, velocity, and particle size by interpolation from Fig. 12.7.1; k_1 is a proportionate factor to adjust the load

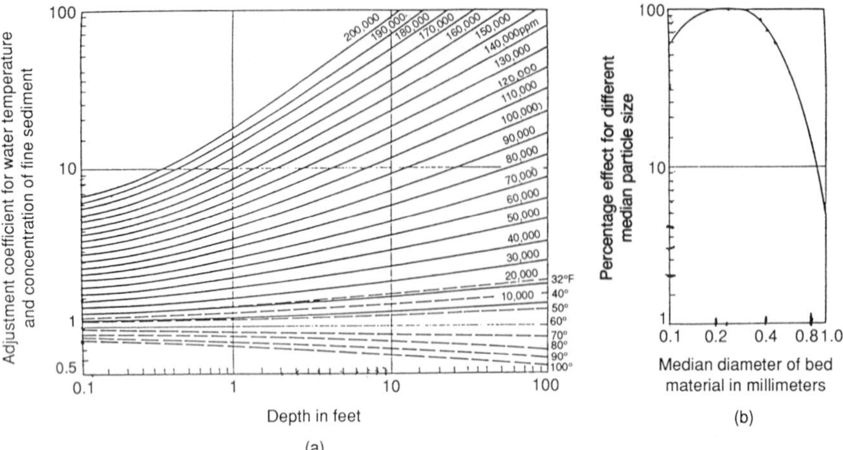

FIGURE 12.7.2 Approximate effect of water temperature and concentration of fine sediment on the relationship of discharge of sand to mean velocity. *(Colby.[28])* Graph (*a*) is based on sediment sizes 0.2 to 0.3 mm. For other sediment sizes, a correction factor is needed from graph (*b*).

for a water temperature different from 60°F, determined from the curves on the lower part of Fig. 12.7.2*a*; k_2 is a proportionate factor to adjust the load for the concentration of fine sediment, determined from the upper part of Fig. 12.7.2*a*; and k_3 is a proportionate factor to adjust the load for a median particle size different from the reference level for the method (0.2 to 0.3 mm), found from Fig. 12.7.2*b*. The conversion of sediment load to SI units is 1 U.S. tons/day/ft = 2.98 metric tons/day/m.

EXAMPLE 12.2 Colby[28] presented the following sample problem to illustrate his method. Selected data for example:

Item	Values
Mean velocity	6.5 ft/s
Depth	4.8 ft
Median size of bed sediment	0.43 mm
Water temperature	75°F
Concentration of fine sediment, mostly bentonite	33,000 ppm

Summary of the computations is as follows. From Fig. 12.7.1 for a velocity of 6.5 ft/s, the discharges of sand are about 92 and 150 U.S. tons/day/ft of width for depths of 1.0 and 10 ft, respectively. Hence about 130 U.S. tons/day/ft of width is interpolated for flow depth of 4.8 ft obtained by using linear interpolation between the logarithms of the loading and depth values. The adjustment coefficient from Fig. 12.7.2*a* for 75°F and a depth of 4.8 ft is 0.86. The adjustment coefficient for 33,000 ppm of fine sediment, mostly bentonite, is 1.92. From the same figure, the adjustment for the median diameter of 0.43 mm is 0.78. The 130 tons/day/ft multiplied by $[1 + (0.86 \times 1.92 - 1) \times 0.78]$, or 1.51, gives 196 tons/day/ft, which could well be rounded to 200 tons/day/ft because the discharge of sands ordinarily should not be stated to more than two significant figures.

Ackers and White Total Sediment Load Formula.[3] Ackers and White collected more than one thousand sets of data points and developed a total sediment load equation. This approach can be used to obtain a reference value for $\bar{C}$, the bed load concentration by weight.

$$\left(\frac{U_*}{U}\right)^{C_1} \frac{\gamma_f \bar{C} D}{\gamma_s d} = C_2 \left(\frac{F_1}{C_3} - 1\right)^{C_4} \tag{12.7.7}$$

where

$$F_1 = \left[\frac{U}{\sqrt{32} \log (10 D/d)}\right]^{1-c_1} \frac{U_*^{C_1}}{\sqrt{(\Delta\gamma_s/\rho_f)d}} \tag{12.7.8}$$

in which $\Delta\gamma_s = \gamma_s - \gamma_f$ and the values of C_1, C_2, C_3, and C_4 are given as a function of d_* by:

$$d_* = \left(\frac{\Delta\gamma_s}{\rho_f v^2}\right)^{1/3} d$$

For $1.0 < d_* \leq 60.0$

$$C_1 = 1.0 - 0.56 \log d_*$$

$$C_2 = 10\,[2.86 \log d_* - (\log d_*)^2 - 3.53]$$

$$C_3 = \frac{0.23}{d_*^{1/2}} + 0.14$$

$$C_4 = \frac{9.66}{d_*} + 1.34$$

For $d_* > 60.0$

$$C_1 = 0,\ C_2 = 0.025,\ C_3 = 0.17,\ C_4 = 1.50$$

Sediment Load in Laboratory Flume Studies (Flow Depth 1.5 ft or Less). For total sediment load studies in the laboratory flumes, either the Shen and Hung[115] or the Yang[144] curves can be used. Shen and Hung's results are presented in Fig. 12.7.3. In Fig. 12.7.3, the sediment concentration is in parts per million by weight, U is the average flow velocity in feet per second, w is the terminal fall velocity of the sediment particles in feet per second, and S is the energy slope.

To Obtain Unsampled Sediment Load (from Measurements Taken from the Sampled Zone). Because of the difficulty in collecting reliable sediment data in the unsampled zone (within a few inches from the top of the streambed), the sediment transport rate inside the unsampled zone can be estimated from measurements taken in the sampled zone based on the modified Einstein procedure. (See Colby and Humbree[29] and Schroeder and Humbree.[108])

The major changes used in the modified Einstein procedure from the original Einstein procedure are given below.

1. The calculation of the shear velocity U_* is based on a measured mean velocity [see Eqs. (12.3.5) and (12.3.6)] rather than a calculated flow velocity by Einstein's procedure.

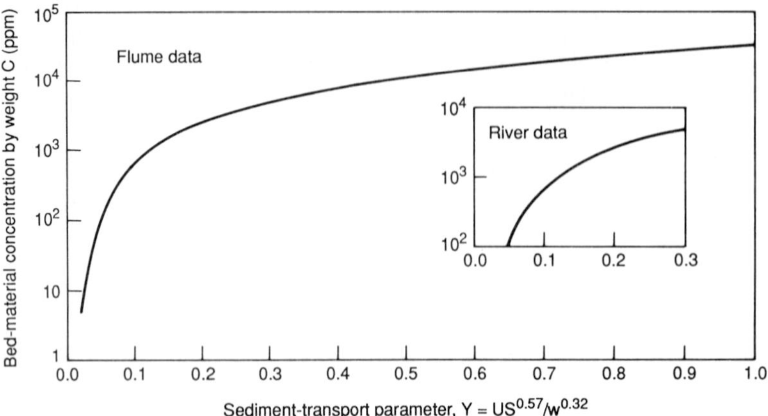

FIGURE 12.7.3 Bed-material load vs. flow. *(Shen and Hung.[115])*

2. The suspended load exponent z in Eq. (12.6.2) is determined from the observed z value for a dominant grain size. Values of z for other grain sizes are derived from that of the dominant size and are assumed to change with the 0.7 power of the settling velocity.
3. A slight change in the hiding factor is introduced.
4. The depth D is used to replace the hydraulic radius.
5. The value of Einstein's intensity of bed load transport is arbitrarily divided by a factor of 2 to fit the data more closely.

Since the modified Einstein procedure estimates the total sediment load from the measured load, it no doubt can give better agreement with field data than the Einstein procedure.

Recommendation. It is impossible to recommend a single approach to obtain total sediment load. The modified Einstein's method has been used widely to estimate the sediment transport load in the unsampled load from measurements taken in the sampled zone. However, this approach cannot provide a predicted total sediment load from a given set of flow conditions with data collected from these flow conditions. In order to obtain predicted total sediment load from flow conditions, Toffaleti's method[124] may be used for large rivers (flow depth greater than 10 ft or 3 m), Colby's method[28] for small rivers (flow depth smaller than 10 ft or 3 m), and Shen and Hung's method[115] for laboratory flume studies (flow depth smaller than 2 ft or 0.6 m), respectively. Many European hydrologists and some American hydrologists have used Ackers and White's method[3] because this method was derived based on a large amount of data. It is important to collect field data from the river basin of concern to check the suitability of using any of these methods. Often, a new set of equations may be derived for a particular river basin based on data collected from that basin. For gravel bed rivers without fine sediment, Meyer-Peter and Muller's equation[94] has been used widely to estimate the movement of gravel bed load. For a mixture of sand and gravel, no complete equation exists at this time.

12.8 RESERVOIR SEDIMENTATION

Estimating the effects of potential sediment accumulation in reservoirs is an important element in the planning of a dam project. Sediment accumulation in a reservoir may (1) reduce the useful storage volume for water in this reservoir, (2) change the water quality near the dam, (3) increase flooding level upstream of the dam due to sediment aggradation, (4) influence the stability of the stream downstream of the dam, (5) affect stream ecology in the dam region, and (6) cause other environmental impacts by changing the water quality. Sediment particles may also cause abrasion of turbine blades, tunnels, and gate recess. Sediment accumulations at various parts of the hydraulic structure such as sluices, and navigation channels may affect the operation of these structures. In the following sections, a commonly used method to estimate potential sediment accumulation in the reservoir is described, and methods to reduce sediment accumulation in reservoirs are discussed.

Estimating Potential Sediment Accumulation in a Reservoir. The first step is to construct a flow duration curve, which is the cumulative distribution curve of the stream runoff passing the dam (see also Sec. 8.4.2).

The next step is to construct a sediment rating curve, relating sediment concentration to stream discharge. The sediment rating curve may be obtained entirely by sediment concentration and flow discharge measurements, but this procedure is sometimes not adequate because sediment concentrations for high stream flows are difficult to obtain. Usually high flows carry disproportionate amounts of sediment concentration and it is difficult to extend the empirical curve to greater flow discharges. Thus the estimation of potential wash load from watershed sediment yields as described in Sec. 12.10 and bed material load by procedures described in Secs. 12.5 and 12.7 are needed. Bed material is defined as the total load minus the wash load. Of course, the measured sediment concentrations for various stream flow discharges at the potential dam site are very important values to be used for the calibration of both the watershed sediment yield estimations and the calculated bed material loads.

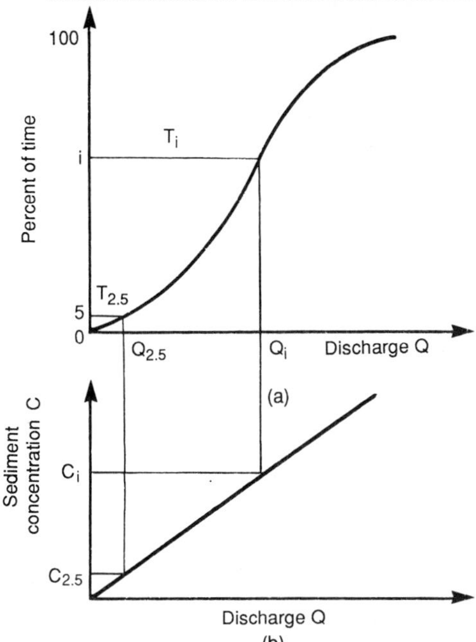

FIGURE 12.8.1 Flow and sediment discharge.

The flow duration curve and the sediment rating curve are plotted in Fig. 12.8.1. The flow duration curve is divided into several sections. For instance, one may divide the flow time duration into 20 equally spaced sections of 5 percent each (or $\Delta P = 0.05$). In each time section, the average Q_i is read from the flow duration curve. The sediment concentration C_i for each Q_i is read from the sediment rating curve. The average total sediment load in weight per unit time is given by the following equation:

$$q_t = \Sigma\, C_i Q_i\, \Delta P = \Delta P \Sigma C_i Q_i \qquad (12.8.1)$$

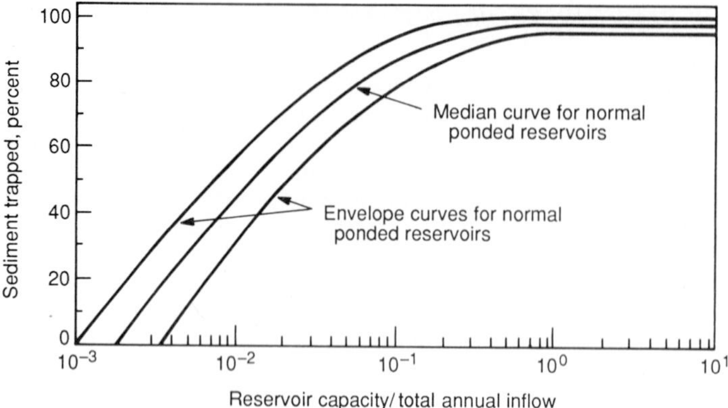

FIGURE 12.8.2 Sediment trap efficiency in a reservoir. *(After Brune.[21])*

One can convert the above average total sediment load into weight per year.

Not all sediment passing through the dam section will be deposited in the reservoir because part of this amount of sediment will pass through sluices, spillways, and other diversion flow releases from the reservoir. The percentage of sediment trapped in the reservoir is a function of the fall velocity of sediment and the allowable time provided for the sediment to settle. The relative size of the reservoir, the reservoir shape, the reservoir operation, and the sediment particle size are some of the more important factors to determine the amount of sediment trapped in the reservoir. Figure 12.8.2 developed by Brune[21] is the most commonly used relation for determining sediment trapping. The trap efficiency is defined as the ratio between sediment trapped in the reservoir and the total sediment entering the reservoir. The reservoir capacity is the reservoir volume at normal operation pool elevation for the period considered. The trap efficiency will change with time if reservoir capacity should change owing to sediment accumulation. For cohesive sediment deposits, the density of deposited sediment will also change with time owing to compaction.

There are three methods of reservoir operation as shown in Table 12.8.1, each of which has different sediment accumulation characteristics.

The total weight of sediment accumulation can be calculated by the following equation:

$$W_0 = p_c W_c + p_m W_m + p_s W_s \tag{12.8.2}$$

TABLE 12.8.1 Modes of Reservoir Operation

Operation no.	Mode of reservoir operation
1	Sediment always submerged or nearly submerged
2	Moderate to considerable reservoir drawdown occurs often
3	Reservoir normally empty

Source: IRTCES.[63]

where W_0, W_c, W_m, and W_s are the densities for total, clay, silt, and sand, respectively, in kilograms per cubic meter, and p_c, p_m, p_s are the percentage of the total sediment composition for clay, silt, and sand, respectively.

The initial specific weights of sediment deposited in different modes of operation are given in Table 12.8.2.

The average density of sediment accumulation after T_1 years of operation W_{T_1} is given by

$$W_{T_1} = W_0 + 0.4343\ K_0 \left[\frac{T_1}{T_1 - 1}(\ln T_1) - 1\right] \tag{12.8.3}$$

where W_0 is the initial specific weight and K_0 is a factor given by Table 12.8.3.

For more detailed analysis and example of this calculation, readers are referred to IRTCES[63] or Bruk.[20]

Methods to Reduce Sediment Accumulation in a Reservoir. As discussed by Bruk[20] and Fan,[42,43] there are three methods to reduce sediment inflow into the reservoir. These methods are reduction of sediment inflow by soil conservation, bypassing heavy sediment-laden flows, and trapping of sediment by a vegetation screen. Soil conservation is commonly used to reduce sediment yield from a watershed. A great deal of effort and time is needed to produce significant results. Bypassing heavy sediment-laden flows is an excellent procedure if an appropriate bypass system can be constructed. The retention of sediment by a vegetation screen can cause severe environmental problems at the sediment deposition site.

The trap efficiency of a reservoir may also be reduced by drawndown flushing during heavy sediment inflow season and by density current flushing. The first method is to open the bottom sluices during heavy sediment inflow seasons to increase flow velocity in the reservoir to flush sediment particles out of the reservoir. This can be very effective.

Density currents can occur along the top of the sediment deposit in the reservoir, and these currents can assist the movement of sediment particles toward the dam. These currents can carry significant amounts of sediment particles through the reservoir with appropriate vents constructed in the dam. The necessary conditions have generally been established for the occurrence of density currents (Fan[42,43]). But these conditions do not always produce density currents. It is generally believed that not more than 10 percent of annual sediment accumulation may be carried out of the reservoir by venting density currents.

An alternative way of reducing sediment accumulation in the reservoir is by

TABLE 12.8.2 The Initial Specific Weight of Sediment Deposits in Different Modes of Reservoir Operation

	Initial density, kg/m³		
Operation no.	w_c	w_m	w_s
1	416	1120	1550
2	561	1140	1550
3	641	1150	1550

Source: IRTCES[63] or Bruk.[20]

TABLE 12.8.3 Constant K_0 Based on the Types of Reservoir Operation and Sediment Size

	K_0 (for metric units)		
Operation no.	Sand	Silt	Clay
1	0	91	256
2	0	29	135
3	0	0	0

Source: IRTCES.[63]

removal of sediment through mechanical means by dredging, or by reservoir operation and siphoning. These are very costly procedures but have received increasing attention because of the gradual depleting of desirable dam sites.

For detailed discussions of all these methods, see Bruk[20] and Fan.[42,43]

12.9 SEDIMENT MEASUREMENTS IN STREAMS

Sediment measurements in streams are important for determining sediment characteristics and for calibrating sediment models. The U.S. Geological Survey (USGS) is the official U.S. federal agency for collecting sediment data, and readers should consult the U.S. Inter-Agency Committee stationed at the St. Anthony Falls Hydraulics Laboratory (Mississippi River and 3rd Ave., S.E., Minneapolis, Minn. 55414), University of Minnesota, for detailed information on the sediment measurement instruments and procedures. The USGS[130,131] presents detailed information on measurement instruments and procedures.

Site Selection. The sedimentation collection site should have the following requirements:

1. Similar to flow gauging stations:
   a. Accessibility.
   b. Stable streambed and banks and relatively uniform section.
   c. Relatively free of backwater effects.
   d. Existence of natural control or a suitable site for installing artificial control.
2. In addition to that for flow gauging stations:
   a. Close to a long-term gauging station for possible data extension.
   b. Long uniform and stable reach with stable streambed and banks to determine the friction slope when needed.
   c. Sediment characteristics that are useful to the particular investigation concerned. If the purpose is to study the change of streambed and bank, then the previous requirement may be waived.

Sediment Bed Material Sampler. Sediment size is an important element in sediment load calculation and may indicate the change of stream behavior. The USGS BM-54 and BMH-60, as shown in Fig. 12.9.1, are two popular grab bed material samples. BM is an indicator for bed material sampler and 54 indicates the year the sampler was developed. The BM-54 can be used to collect bed-material samples from a truck, a boat, or a bridge. It is 0.56 m long, made of cast steel, and streamlined with tail vanes for aligning itself with the flow velocity. The BM-54 is supported by a cable and its bottom has a scoop bucket. When the sampler touches the streambed, the tension in the cable is released and a bed material sample is collected from the top 6 cm of the streambed. The BMH-60 is a hand-held sampler that is lighter and can be operated from a hand line. The BMH-53 (not shown) is a piston-type hand sampler and is designed for a shallow stream that can be waded. A core of 6 cm diameter and 24 cm deep can be collected with the sampler. These two samplers are particularly suitable for sand particles. For detailed description of these two samplers, see U.S. Inter-Agency Report 14.[129] As shown in Fig. 12.5.1, sediment bed material size distributions vary greatly with depth. It is important to decide at what depth sediment size distribution is needed.

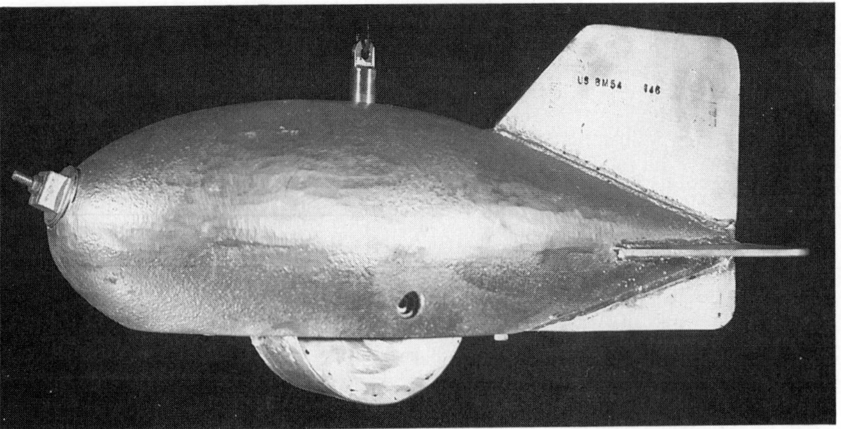

FIGURE 12.9.1 Sediment bed material sampler BM-54. *(Photo by U.S. Inter-Agency Committee on water resource, subcommittee on sedimentation.)*

Suspended Load Sampler. Two different types of suspended load sampler are available. The *point integrating sampler* is used to collect suspended load for a short duration at a given depth. The P-61 (weight 105 lb) is used to take a point sample with a maximum depth of 40 m. A heavier P-63 (weight 200 lb) is used for relatively high flow velocities. A special heavy 300-lb sampler P-62 can be specially ordered for high flow velocities. Figure 12.9.2*a* shows a P-61 sampler. A bottle is inserted in the sampler and the sediment with water enters the bottle through a nozzle. The size of nozzle should be greater than at least 1.5 times the largest sediment size to be collected. These samplers are streamlined, cast in bronze, with tail fins to orient the sampler into the flow. A solenoid-activated valve is used to equalize the pressure inside and outside the sampler to avoid an initial surge during the opening of this nozzle. One should never let the sampled bottle be more than two-thirds full to prevent the entrance of excess sediment particles.

The second type of suspended sampler is a *depth integrated sampler.* The sampler is lowered slowly down to about 10 cm from the top of the streambed to prevent the sampler from touching the streambed. Then the sampler is raised at the same speed to the water surface. During the round trip, suspended sediment continuously enters the sampler. The average sediment concentration collected in the sampler provides an indication of the average suspended sediment concentration in the stream. A D-74 sampler is shown in Fig. 12.9.2*b*. This has a cast bronze streamlined body with tail vanes. It is suitable for a maximum depth of 5 m with fluid velocity less than 2.1 m/s. The DH-59, as shown in Fig. 12.9.2*c*, is a hand-held depth integrated sampler. The nozzle size must be at least 1.5 times the largest sediment size expected to enter the sampler. The nozzle size must also be small enough so that the bottle will not be two-thirds full after a transverse across the stream flow depth. If these two criteria cannot be fulfilled, a point integrating sampler must be used.

Bed Load Sampler. As discussed in Sec. 12.4, bed load cannot be measured with a sampler satisfactorily. A Helly-Smith bed load sampler, as shown in Fig. 12.9.3, has been developed recently. This sampler provides reasonable data in certain limiting cases. The sample bag with a mesh screen is attached to the top of the frame. The

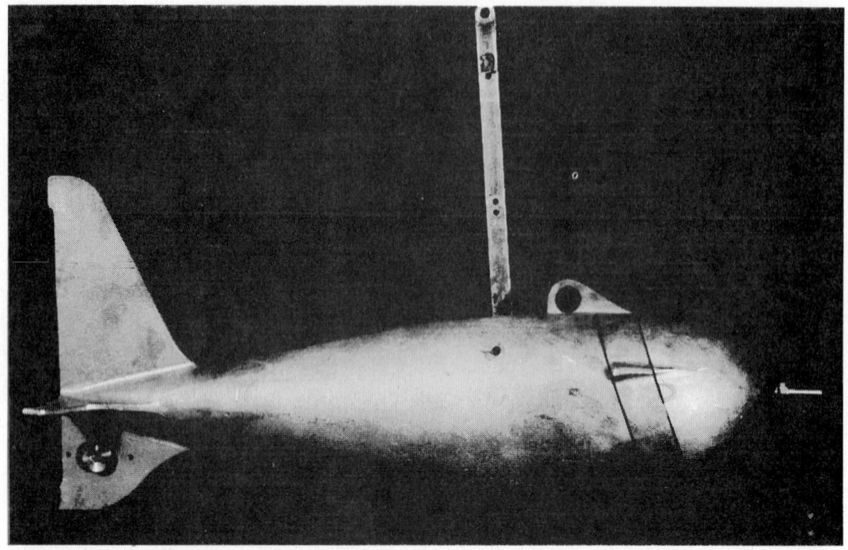

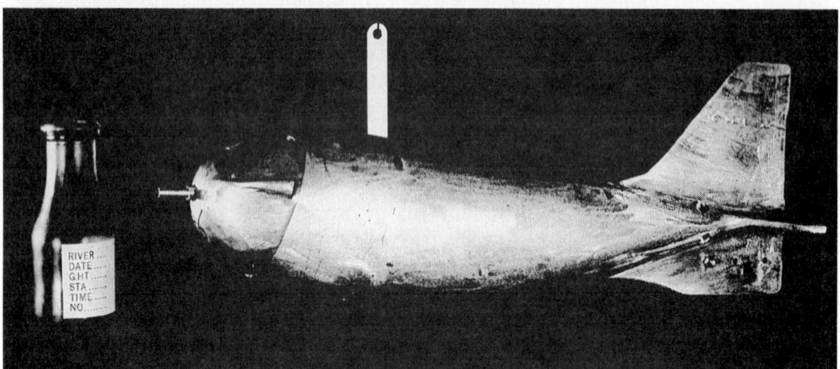

FIGURE 12.9.2 Suspended sediment samplers: *top,* P-61 sampler; *bottom,* D-74 sampler. *(Photos by U.S. Inter-Agency Committee on water resource, subcommittee on sedimentation.)*

screen in the sample bag must have openings less than the expected sediment size and the nozzle at the main entrance must be large enough to permit entrance of large sediment particles. The mesh opening, the bag size, and the orifice opening must be mutually adjusted to maintain normal flow profile in the entrance. The sampler should be placed on a relatively flat bed. Readers should contact the U.S. Inter-Agency Committee at St. Anthony Falls Hydraulics Laboratory, University of Minnesota, for current information on this sampler.

Computation of Fluvial-Sediment Discharge. The U.S. Geological Survey[130] provides a detailed discussion of the computation procedure for fluvial-sediment discharge, and the following is a brief description of this subject.

Two basic types of sediment records—daily and periodic—are published by the

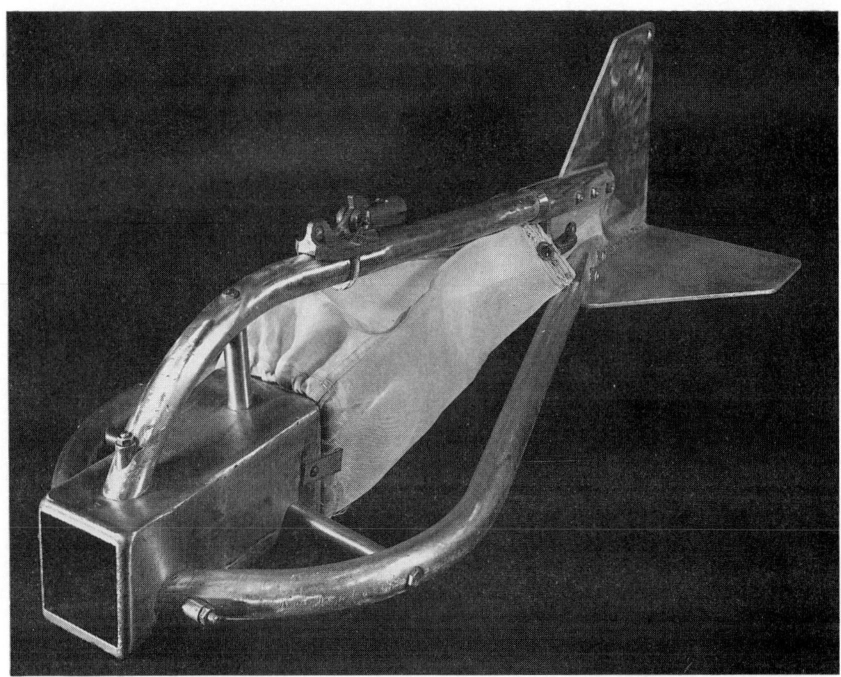

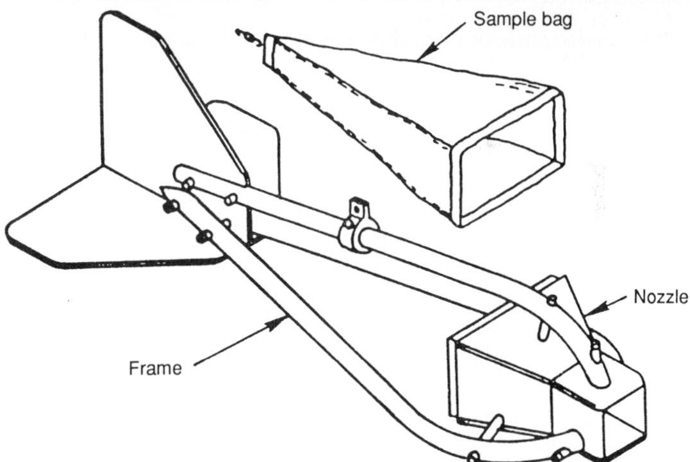

FIGURE 12.9.3 Helley-Smith bed load sampler. *(Photo by U.S. Inter-Agency Committee on water resource, subcommittee on sedimentation.)*

U.S. Geological Survey. In stations with daily records, the following items are normally available: daily mean sediment concentration, suspended sediment discharge, periodic determinations of sediment concentration, suspended sediment discharge, and periodic determinations of the particle size distributions of suspended sediment and bed material. These are combined with other water quality data and published annually in "Quality of Surface Water" reports by water year (October through the following September) from the U.S. Geological Survey. Periodic data are collected at special locations where daily data are not needed. All sediment concentrations are determined in parts per million and reported in milligrams per liter. The computation of sediment discharge is made according to the following equation:

$$Q_s = Q \times C_s \times k \qquad (12.9.1)$$

where Q_s = sediment discharge
Q = water discharge
C_s = concentration of suspended sediment
$k = 2.697 \times 10^{-3}$ when Q_s in U.S. tons/day, Q is in ft^3/s, and C is in mg/liter, $k = 86.4$ when Q_s is in metric tons/day, Q is in m^3/s, and C is in kg/m^3. Both these values of k incorporate a sediment specific gravity of 2.65.

$$C_s = \frac{\text{dry weight of sediment} \times 1{,}000{,}000}{\text{weight of water-sediment mixture}} \text{ (parts per million)} \qquad (12.9.2)$$

Sediment concentration in parts per million can be converted to milligrams per liter by a factor of C according to the following relation:

$$C_s \text{ (milligrams per liter)} = C \times C_s \text{ (parts per million)}$$

Table 12.9.1 gives values of C for different sediment concentration ranges.

TABLE 12.9.1 Conversion Factors C for Sediment Concentration: Parts per Million to Milligrams per Liter

Concentration range, ppm	C	Concentration range, ppm	C
0–15,900	1.00	322,000–341,000	1.26
16,000–46,800	1.02	342,000–361,000	1.28
46,900–76,500	1.04	362,000–380,000	1.30
76,600–105,000	1.06	381,000–399,000	1.32
106,000–133,000	1.08	400,000–416,000	1.34
134,000–159,000	1.10	417,000–434,000	1.36
160,000–185,000	1.12	435,000–451,000	1.38
186,000–210,000	1.14	452,000–467,000	1.40
211,000–233,000	1.16	468,000–483,000	1.42
234,000–256,000	1.18	484,000–498,000	1.44
257,000–279,000	1.20	499,000–514,000	1.46
280,000–300,000	1.22	515,000–528,000	1.48
301,000–321,000	1.24	529,000–542,000	1.50

The factors are based on the assumption that the density of water is 1.000 (plus or minus 0.005), the range of temperature is 0 to 29°C, the specific gravity of sediment is 2.65, and the dissolved solids concentration is less than 10,000 ppm.

For metric systems, the flow discharge Q is in cubic meters per second and the sediment discharge Q_2 is in metric tons per day. Thus the value of k is 0.0864, and C_s is in milligrams per liter.

12.10 WATERSHED SEDIMENT YIELDS

This section covers the general subject of erosion and sediment yields from watersheds. The natural processes of erosion, transport, and deposition of sediments have occurred throughout geologic times and have shaped the landscape of the world in which we live.[62,73,145] Erosion often causes serious damage to agricultural land by reducing the fertility and productivity of soils. Eroded soil is the largest pollutant of surface waters in the United States.[92] Sediment transport affects water quality and its suitability for consumption and industrial use. Soil eroded from upland areas is the source of most sediments transported by rivers to reservoirs.

Problems associated with the deposition of sediments are varied. Sediment deposition in stream channels reduces flood carrying capacity, resulting in greater flood damage to adjacent properties. Reservoirs not only trap the incoming sediment load but reservoir sedimentation also increases the flooding risks because of aggradation upstream of the reservoir. Upstream aggradation depends on the stream slope, the sediment size distribution, and the water-level fluctuations in the reservoir. Streams with low slope carrying large quantities of sediment may result in aggradation many miles upstream of the reservoir. Reservoir sedimentation results in loss of storage capacity for flood control and/or irrigation.

Downstream of reservoirs, erosion of channel bed and bank material usually occurs as a result of the action of the clear water released from the reservoir on the erodible channel bank and bed material. Lowering of the main channel elevation causes degradation of the tributary channels and initiates upstream degradation and head cutting through the tributary watersheds. Damage to agricultural lands through lowering the groundwater table and undermining of bridge piers can be severe. Bank material can be preferentially eroded when the bed material is more resistant and degrading streams then have a strong tendency to meander or braid.

Human activities have increased the rate of erosion over the *normal erosion rate,* also called the *geologic erosion rate.* The erodibility of natural soils can be altered by disturbing the soil structure through plowing and tillage. When compared with geologic erosion rates on the order of 0.1 ton/acre-year, *accelerated erosion rates* due to human activities can be more than 100 times in excess of the geologic erosion rate. Erosion rates of grazed areas can exceed 5 tons/acre-year, and average values of 40 to 50 tons/acre-year can reasonably be expected during urban development when the soil is not vegetated and is constantly reworked.[104] Human activities also influence the natural characteristics of channel flows through channel stabilization and hydraulic structures.

Physical Processes. The extent of erosion, specific degradation, and sediment yield from watersheds relates to a complex interaction between topography, geology, climate, soil, vegetation, land use, and man-made developments. Erosion is characterized by the detachment and entrainment of solid particles from the land surface or from the bed and banks of streams. Erosion has been observed to occur in various forms under the influence of water, gravity, wind, and ice. Particles detached by weathering or rock fracturation are usually angular in shape. These particles are called sediments after a complete cycle of erosion, transport, and sedimentation.

Abrasion of sediments during multiple cycles causes the sediments to become more round as they travel long distances over geological times.

Sheet and Rill Erosion. Water is the most widespread agent of erosion. Water erosion can be classified into two types: sheet erosion and channel erosion. *Sheet erosion* is the detachment of land surface material by raindrop impact and thawing of frozen grounds and its subsequent removal by overland flow. The impact of raindrops on the soil surface causes the detachment of soil particles. The finer particles detached further reduce infiltration by sealing the soil pores. The transport capacity of thin overland flow, usually called sheet flow, increases with field slope and flow discharge per unit width.[6,69,71] As sheet flow concentrates and the unit discharge increases, the increased sediment transport capacity scours microchannels called rills. Rill erosion is the removal of soil by concentrated sheet flow. Rills are small enough to be removed by normal tillage.

The surface erosion process begins when raindrops impact the ground and detach soil particles by splash.[37,92,98] The kinetic energy released by raindrop impact on the ground is sufficiently large to break the bonds between soil particles. The characteristics of raindrop splash depend on raindrop size and sheet flow depth; a crown-shaped crater forms a few milliseconds after impact. The impact shear stress can be as large as 100 times the base shear stress from sheet flow. For example, a 3-mm raindrop striking a 2-mm water layer exerts shear stresses exceeding 80 N/m^2.[60] This far exceeds the critical shear stress of 2.5 N/m^2 for eroding cohesive soils.[121] Ellison[38] experimentally measured the splash erosion rates from splash samplers and proposed a relationship between the measured amount of soil intercepted in splash samplers and the raindrop velocity and diameter, and the rainfall intensity.

Without surface runoff, the soil erosion losses from nearly level fields are very small. Surface erosion rates largely depend on unit discharge and surface slope. Musgrave[97] reported on soil-loss measurements for some 40,000 storms on experimental plots in the United States. His analysis indicates that soil erosion losses depend on soil erodibility, runoff length and slope, the maximum 30-min amount of rainfall, and a cover factor. Further developments led to the universal soil loss equation (USLE), which is designed to predict the long-term average soil losses in runoff from field areas under specified cropping and management systems.[127,141] The USLE computes the soil loss at a given site as a product of six major factors as follows:

$$A = RKLSCP \tag{12.10.1}$$

where A is the soil loss per unit area normally in tons per acre, R is the rainfall erosivity factor, K is the soil erodibility factor, usually in tons per acre, L is the field length factor normalized to a plot length of 72.6 ft, S is the field slope factor normalized to a field slope of 9 percent, C is the cropping-management factor normalized to a tilled area with continuous fallow, and P is the conservation practice factor normalized to straight-row farming up and down the slope.

The rainfall erodibility factor R can be evaluated for each storm (summed over hours) from

$$R = 0.01 \sum EI \tag{12.10.2}$$

where the summation is performed over the time increments of the storm and E is given by

$$E = (916 + 331 \log I) \tag{12.10.3}$$

in which E is the kinetic energy in foot-tons per acre-inch and I is the rainfall intensity in inches per hour. Soil erosion losses from single storms strongly correlate with the maximum 30-min rainfall intensity. On an annual basis, the annual rainfall erosion index in the United States is shown in Fig. 12.10.1. When using the values of R from this figure, average annual erosion losses are calculated from Eq. (12.10.1).

The soil-erodibility factor K describes the inherent erodibility of the soil expressed in the same units as the annual erosion losses in tons per acre. Numerous factors control the erodibility of cohesive soils such as grain size distribution, texture, permeability, and organic content. The evaluation of K can be obtained from the nomograph proposed by Wischmeier and Smith[140] given the percentage of sand, the percentage of silt and fine sand, the percentage of organic matter, the soil structure, and the soil permeability. Typical values of the factor K relate to the general triangular soil classification shown in Fig. 12.10.2. For each type of soil, approximate values of K can be found in Table 12.10.1 given the soil type and the percentage of organic matter.

The slope length–steepness factor LS is a topographic factor relating erosion losses from a field of given slope and length when compared with soil losses of a standard plot 72.6 ft long inclined at 9 percent slope. The values of LS are plotted in Fig. 12.10.3 given the length and slope of the area under consideration.

The cropping-management factor C for bare soils is taken as a standard value equal to unity. The factor C accounts for soils under different cropping and management combinations such as different vegetation, canopy during growth stage, before and after harvesting, crop residues, mulching, fertilizing, and crop sequence. Typical values of C are given in Table 12.10.2 for undisturbed forest land; pasture, range, and idle land; and construction slopes. Area-averaged values of C can be used when several vegetation types cover a given area.

The conservation practice factor P equals 1 for downslope rows, and typical values for contouring, strip cropping, and terracing are given in Table 12.10.3. Contour practices are most effective on slopes less than 12 percent, in which case P can be as low as 0.5. Contouring does not reduce erosion losses at slopes exceeding 24 percent. Strip cropping and terracing reduce erosion significantly on slopes less than 12 percent.

Wind Erosion. Wind erosion can be important in arid and semiarid areas. The rate of wind erosion primarily depends on the particle size distribution, wind velocity, soil moisture, surface roughness, and vegetation cover. The empirical equation proposed by Chepil and Woodruff[24] for estimating the rate of wind erosion provides rough estimates of wind erosion rates.

Snowmelt Erosion. Extremely high erosion rates have been recorded when large amounts of precipitation and snowmelt runoff occur during spring months in combination with minimal soil cover and freeze-thaw cycles that cause weathering. In cold regions, the major portion of the annual sediment yield from watershed is observed during snowmelt runoff.[68]

Gully Erosion. Gully erosion is the removal of soil in larger channels which cannot be obliterated by normal tillage. Gully erosion involves waterfall erosion at the gully head, channel erosion in the gully, and mass movement of bank material into the gully. As the gully widens and the gully head progresses upslope, the channel reaches a stable slope and vegetation begins to grow and stabilizes the gully.

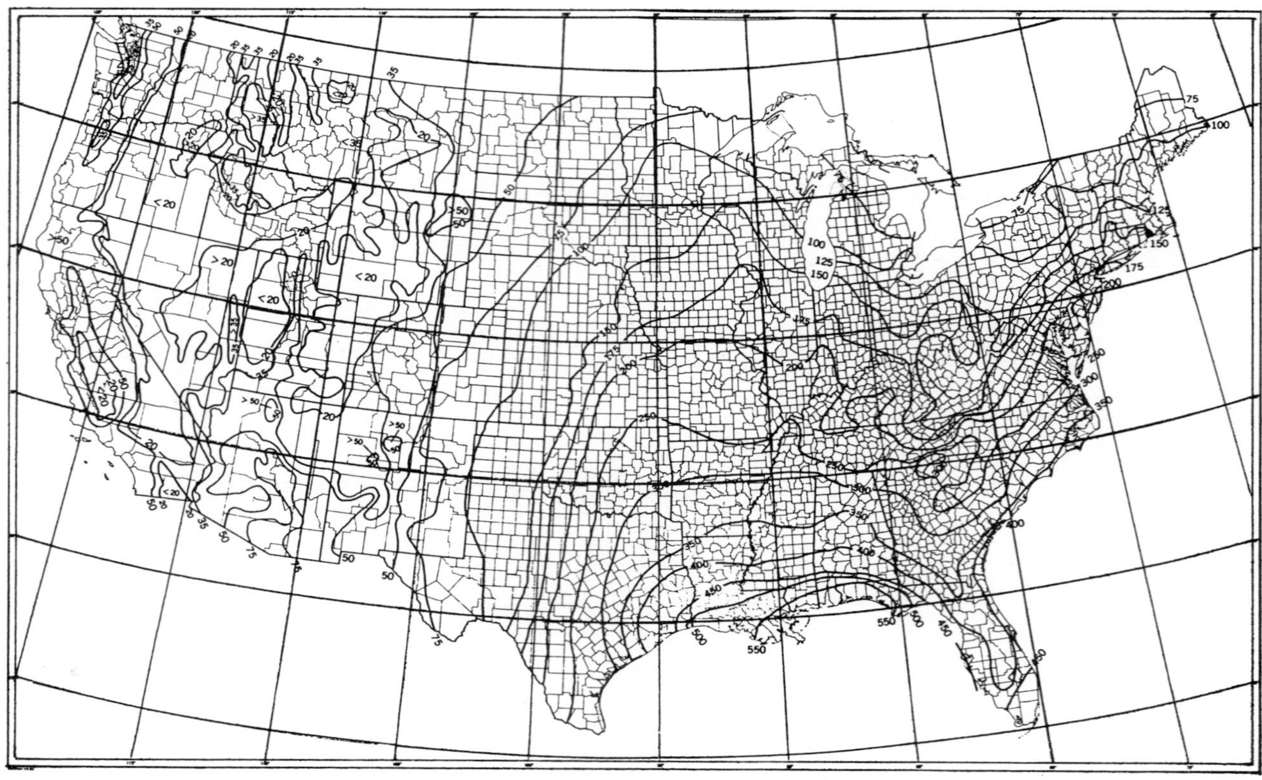

FIGURE 12.10.1 Average annual values of the rainfall erosion index *R*. (*Wischmeier and Smith.*[140])

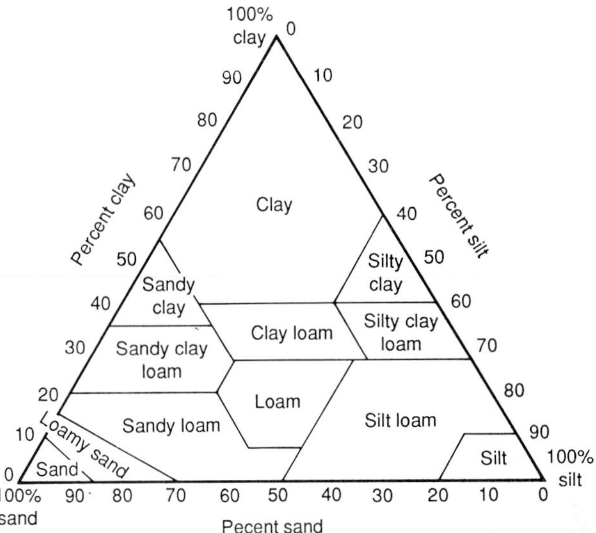

FIGURE 12.10.2 Triangular soil classification. *(From Brady.[17])*

Channel Erosion. Channel erosion, which includes bed and bank erosion, can be very significant in alluvial channels. The sediment transport capacity of a stream is generally proportional to water discharge and channel slope, and this capacity varies inversely with bed sediment size. A qualitative equilibrium relationship between hydraulic and sediment parameters has been proposed by Lane:[77]

$$QS_0 \sim Q_s d \tag{12.10.4}$$

TABLE 12.10.1 Soil Erodibility Factor K in tons/acre

	Organic matter content, %	
Textural class	0.5	2
Fine sand	0.16	0.14
Very fine sand	0.42	0.36
Loamy sand	0.12	0.10
Loamy very fine sand	0.44	0.38
Sandy loam	0.27	0.24
Very fine sandy loam	0.47	0.41
Silt loam	0.48	0.42
Clay loam	0.28	0.25
Silty clay loam	0.37	0.32
Silty clay	0.25	0.23

Source: From Schwab et al.[110]

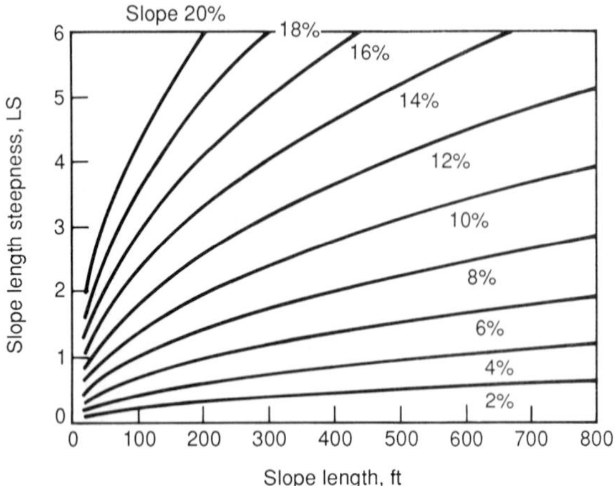

FIGURE 12.10.3 Slope length steepness factor LS. *(From Wischmeier.[141])*

in which Q is the stream discharge, S_0 is the downstream slope of the channel, Q_s is the bed sediment discharge, and d is the bed sediment size.

Quantitative estimates of channel erosion and deposition rates are obtained after careful consideration of the applied hydraulic forces and the resistive gravitational forces of sediment particles. A complete treatment involves the understanding of incipient motion of sediment particles, sediment discharge formulas, and bed form configurations under the lower and upper flow regime.[119,133] The beginning of motion of noncohesive sediment particles can be calculated from the Shields' diagram while the erosion of cohesive soils remains difficult to predict.[84,121] Channel bed erosion rates can be determined by comparing the actual cross-sectional geometry with earlier profiles. Channel bank erosion rates can be measured by comparisons of channel positions as detected from a pair of recent and old areal photographs.

Sediment Transport. The time variability of sediment concentration measurements in natural channels depends on many factors such as the location of the measurement, the magnitude of the flood, the source of water and sediments, and the seasonal watershed conditions prior to the flood. In general, the sediment concentration increases with discharge, although the sediment concentration at a given discharge may vary depending on the season, the source of sediment, and whether the discharge is increasing or decreasing.[49] The sediment flux at a given point is given by the product of the sediment concentration by the point velocity. Since the flow velocity is maximum at the surface while the sediment concentration is maximum near the bed, the sediment flux must be integrated over the entire cross-sectional area to obtain the total sediment discharge Q_s passing through a given cross section.

The *flux-averaged concentration* C is the ratio between the total sediment discharge Q_s and the total water discharge Q. On a daily basis, the *sediment load* is the amount of sediment passing a stream cross section given by

$$Q_s = 0.0864\,QC_s \qquad (12.10.5)$$

where Q_s is the daily sediment load in metric tons per day, C_s is the flux-averaged concentration in mg/liter, and Q is the daily discharge in m³/s (the units of the

TABLE 12.10.2 Cropping Management Factor *C*

Undisturbed forest land

Percent of area covered by canopy of trees and undergrowth	Percent of area covered by duff at least 2 in deep	Factor *C*
100–75	100–90	0.0001–0.001
70–45	85–75	0.002–0.004
40–20	70–40	0.003–0.009

Permanent pasture, range, and idle land*

Vegetative canopy		Cover that contacts the soil surface					
		Percent ground cover					
Type and height†	Type‡	0	20	40	60	80	95+
No appreciable canopy	G	0.45	0.20	0.10	0.042	0.013	0.003
	W	0.45	0.24	0.15	0.091	0.043	0.011
Tall weeds or short brush with average drop fall height of 20 in	G	0.17–0.36	0.10–0.17	0.06–0.09	0.032–0.038	0.011–0.013	0.003
	W	0.17–0.36	0.12–0.20	0.09–0.13	0.068–0.083	0.038–0.041	0.011
Appreciable brush or bushes, with average drop fall height of 6 1/2 ft	G	0.28–0.40	0.14–0.18	0.08–0.09	0.036–0.040	0.012–0.013	0.003
	W	0.28–0.40	0.17–0.22	0.12–0.14	0.078–0.087	0.040–0.042	0.011
Trees, but no appreciable low brush. Average drop fall height of 13 ft	G	0.36–0.42	0.17–0.19	0.09–0.10	0.039–0.041	0.012–0.013	0.003
	W	0.36–0.42	0.20–0.23	0.13–0.14	0.084–0.089	0.041–0.042	0.011

Construction slopes

Type of mulch	Mulch rate, tons/acre	Factor *C*
Straw	1.0–2.0	0.06–0.20
Crushed stone, 1/4–1.5 in	135	0.05
	240	0.02
Wood chips	7	0.08
	12	0.05
	25	0.02

Source: Adapted from Wischmeier and Smith.[140]

* The listed *C* values assume that the vegetation and mulch are randomly distributed over the entire area.

† Canopy height is measured as the average fall height of water drops falling from the canopy to the ground. Canopy effect is inversely proportional to drop fall height and is negligible if fall height exceeds 33 ft.

‡ G: cover at surface is grass, grasslike plants, decaying compacted duff, or litter at least 2 in. deep. W: cover at surface is mostly broadleaf herbaceous plants (as weeds with little lateral-root network near the surface) or undecayed residues or both.

TABLE 12.10.3 Conservation Practice Factor P for Contouring, Strip Cropping, and Terracing

	P value			
			Terracing	
Land slope, %	Farming on contour	Contour strip crop	*	†
2–7	0.50	0.25	0.50	0.10
8–12	0.60	0.30	0.60	0.12
13–18	0.80	0.40	0.80	0.16
19–24	0.90	0.45	0.90	0.18

Source: From Wischmeier.[141]
* For erosion-control planning on farmland.
† For prediction of contribution to off-field sediment load.

constant are 0.0864 metric ton second liter/day m³ mg). Equation (12.10.5) is related to Eq. (12.9.1), mentioned previously.

A *sediment rating curve* is obtained by plotting the flux-averaged sediment concentration C_s as a function of Q. The analysis of data points usually gives a power relationship of the form

$$C_s = a\,Q^b \tag{12.10.6}$$

where a and b are coefficients usually obtained by regression analysis.

The interannual sediment yield from a watershed can be obtained from the *flow duration–sediment rating curve method.* The method illustrated in Table 12.10.4 consists of dividing the flow duration curve in intervals becoming increasingly small at higher discharges as shown in column 1. The duration percentage of each interval and the duration midpoint are inserted in columns 2 and 3, respectively. The midpoint flow discharge in column 4 is then obtained from the flow duration curve given the duration midpoint from column 3. From this discharge in column 4, the corresponding total sediment concentration in column 6 is given from the sediment rating curve, or Eq. (12.10.6). For each interval, the water yield in column 5 is calculated from multiplying columns 2 and 6. Likewise, the annual sediment yield in column 7 is calculated from Eq. (12.10.5) given Δp, Q, and C_s from columns 2, 4, and 6. The interannual total sediment yield is finally obtained from the sum of column 7.

Sediment Yield. The total amount of sheet and channel erosion in a watershed is known as the gross erosion. All eroded particles in a watershed, however, do not reach the outlet of the watershed.[16,65,126,128] Particles detached in bare upland areas are trapped in vegetated areas farther downstream. Some material carried in natural streams is deposited in the channels to cause global channel aggradation, or is locally and temporarily trapped behind channel bed forms and point bars. Some material deposits on the floodplain during major floods and large amounts are permanently trapped in lakes and reservoirs. The total amount of sediment which is delivered to the outlet of the watershed is known as the *sediment yield.*

Sediment-delivery Ratio. The *sediment-delivery ratio* expresses the percentage of on-site eroded material that reaches a designated downstream location. The ratio between the sediment yield and the gross erosion in a watershed channel network to

TABLE 12.10.4 Long-Term Sediment Yield Calculated by Flow Duration Sediment Rating Curve Method

Interval $p_1 - p_2$, % (1)	Duration $\Delta p = p_2 - p_1$, % (2)	Midpoint $(p_1 + p_2)/2$, % (3)	Discharge Q, ft³/s (4)	$Q \times \Delta p$, ft³/s (5)	Concentration mg/liter C_s (6)	Sediment yield $Q_s \times \Delta p$, tons/year (7)
0.00–0.02	0.02	0.01	58,000	12	321	3,324*
0.02–0.1	0.08	0.06	52,000	42	279	10,363
0.1–0.5	0.4	0.3	43,000	172	219	33,632
0.5–1.5	1.0	1.0	33,000	330	155	45,670
1.5–5.0	3.5	3.25	21,000	737	87	57,094
5–15	10	10	10,640	1064	36	34,200
15–25	10	20	5,475	548	15	7,333
25–35	10	30	3,484	348	9	2,800
35–45	10	40	2,435	244	5	1,087
45–55	10	50	1,839	184	4	657
55–65	10	60	1,375	138	3	369
65–75	10	70	1,030	103	1.8	166
75–85	10	80	765	76	1.2	82
85–95	10	90	547	55	0.8	39
95–98.5	3.5	96.75	397	14	0.5	6
Total annual sediment yield						197,000

$$* \; Q_s \times \Delta p = 365.25 \frac{\text{days}}{\text{year}} \times 0.0864 \frac{\text{tons s liter}}{\text{day m}^3\text{mg}} \times \frac{58{,}000 \text{ ft}^3\text{m}^3 \text{ 321 mg}}{\text{s } 35.32 \text{ ft}^3 \times \text{liter}} \times \frac{0.02}{100} = 3324 \frac{\text{tons}}{\text{year}}.$$

convey eroded material to the outlet depends on drainage area, watershed slope, drainage density, and runoff as discussed by Gottschalk.[58] It appears that the probability of entrapment of particles increases with the size of the drainage area. Therefore, the sediment-delivery ratio SDR decreases primarily with drainage area A_t as shown in Fig. 12.10.4.[16,49]

Computer Models. Hillslope models are based on the concept that soil losses and sediment yield at the bottom of the slope are determined by the processes of rainfall

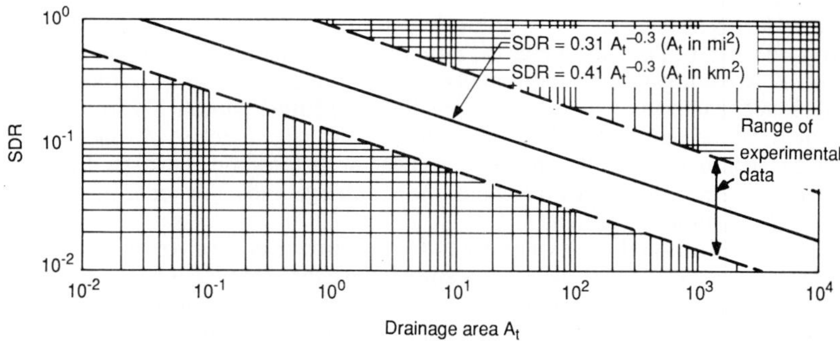

FIGURE 12.10.4 Sediment-delivery ratio (SDR) vs. drainage area.

and runoff erosivity combined with sediment detachment, transport, and deposition in overland flow. Recent advances in hydrology, soil science, erosion mechanics, and computer technology have provided the technological basis for a computerized process-based erosion prediction technology. The U.S. Department of Agriculture has initiated a national project called the USDA Water Erosion Prediction Project (WEPP) to develop a new generation of water erosion prediction technology applicable to "field-sized" areas up to a few hundreds of hectares. This project is based on substantial mathematical treatments of surface runoff and sediment detachment and transport.[15,46,47,73,76,81,91]

On small watersheds, physically based soil erosion and sediment yield models usually subdivide the watershed into square grid elements or simple open book planes. On each plane, infiltration losses are subtracted from the rainfall and the excess rainfall intensity is routed by numerically solving the Saint-Venant equations. Sediment transport capacity is usually calculated from the bed load and total load equations such as the equations of Meyer-Peter and Muller, Yalin, and others. Such models need to be calibrated with field data from single or multiple storms.[1,2,80,106]

On large watersheds, the calculation of erosion losses from very fine square grids requires large data bases on precipitation, soil types, topography, vegetation, land use, and conservation practices. A correction factor describing the influence of grid size on the calculation of soil erosion losses can be used to estimate sheet erosion losses from large watersheds given the average characteristics of the watershed.[48] Fine-meshed grids are best used to define the spatial distribution of erosion losses, while the average characteristics of the watershed can appropriately define the total upland erosion losses. In cold regions, most of the sediment yield occurs during snowmelt, and monthly calculations of soil erosion losses and sediment yield from both rainfall and snowmelt are possible.[67]

12.11 SEDIMENT EROSION CONTROL

The specific purpose of sediment control methods is to reduce the sediment yield from a watershed. Sediment control methods are very diversified,[62,72,145] and their cost and efficiency depend on the soil erosion type and countermeasure. Two types of methods are generally considered: (1) reduction of erosion losses at the source; and (2) reduction in sediment yield with sediment traps.

Reducing Erosion Losses. Some solutions attempt to solve the problem at the source by reducing erosion losses from upland areas. A review of the relative influence of the various terms of the universal soil-loss equation [Eq. (12.10.1)] indicates that relatively little can be done to change the rainfall erosivity factor R or the soil erodibility factor K. In some cases, the addition of organic residues increases the infiltration rate and reduces soil erosion losses from sheet flow and surface runoff. The runoff length factor L can be reduced only by increasing the drainage density, in which case the surface erosion losses may be partially reduced at the expense of a faster drainage of surface water resulting in an increased peak discharge and a greater sediment transport capacity in the channels.

One of the most effective methods to reduce surface erosion losses is to increase the vegetation cover which will significantly reduce the cropping management factor C. The presence of natural vegetation is most effective in reducing the detachment of soil particles induced by raindrop impact and also in stabilizing the soil in the root zone. The surface slope factor S can be reduced by forming terraces, although the cost of terracing may be prohibitive. Terraces[14,110] significantly reduce soil erosion losses, through the practice factor *P*, in reducing runoff length and slope steepness besides increasing infiltration. In agricultural areas, reduction of soil losses can be obtained by contouring, strip cropping, mulching, residues, and minimum tillage.[110]

Sediment Control Structures. Sediment control structures are designed with the specific purpose to trap sediments below eroded areas. Ponds and reservoirs are generally efficient in trapping coarse sediments but are not effective in trapping fine sediments. The sediment yield is reduced by settling of sediment particles in the reservoir and also by reducing the bed slope of the channel entering the reservoir.

Reducing channel erosion is also possible but usually expensive. Solution to local erosion problems can be economically feasible; the cost of bank stabilization methods to entire watersheds is, however, prohibitive. Local bank erosion can be reduced by appropriate streambank protection methods[14,107,125] such as vegetation, gabions, mattresses, trenches, soil cement, riprap, spurs, hardpoints, retards, dikes, jetties, fences, and drop structures. Fences and vegetation screens to increase channel roughness are also effective in trapping coarse material and reducing the slope of natural channels. The grassed waterway is one of the basic conservation practices on farmland.[14] Gully stabilization can usually be achieved by a combination of runoff reduction entering the gully and vegetative means to stabilize the gully surface. Drop spillways control the downstream bed elevation of eroding streams. These structures not only reduce the gradient but also induce deposition of transported material on the ponded reach upstream.

REFERENCES

1. Abbott, M. B., J. C. Bathurst, J. A. Cunge, P. E. O'Connell, and J. Rasmussen, "An Introduction to the European Hydrological System—Systeme Hydrologique Europeen, 'SHE'. 1. History and Philosophy of a Physically-Based, Distributed Modelling System," *J. Hydrol.,* vol. 87, pp. 45–59, 1986.
2. Abbott, M. B., J. C. Bathurst, J. A. Cunge, P. E. O'Connell, and J. Rasmussen, "An Introduction to the European Hydrological System—Systeme Hydrologique Europeen, 'SHE'. 2. History and Philosophy of a Physically-Based, Distributed Modelling System," *J. Hydrol.,* vol. 87, pp. 61–77, 1986.
3. Ackers, P., and W. R White, "Sediment Transport: New Approach and Analysis," *J. Hydraul. Eng.,* vol. 99, no. HY-11, pp. 2041–2060, 1973.
4. Adams, J., "Grain Size Analysis from Photographs," *J. Hydraul. Eng.,* vol. 105, pp. 1247–1285, 1979.
5. Alam, Z. U., and J. F. Kennedy, "Friction Factors for Flow in Sand-Bed Channels," *J. Hydraul. Eng.,* vol. 95, no. HY-6, pp. 1973–1992, 1969.
6. Alonso, C. V., W. H. Neibling, and G. R. Foster, "Estimating Sediment Transport Capacity in Watershed Modeling," *Trans. ASAE,* vol. 24, no. 5, pp. 1211–1220, 1981.
7. Andrews, E. D., and G. Parker, "Formulation of a Coarse Surface Layer as the Response to Gravel Mobility Sediment Transport in Gravel-Bed Rivers," C. Thorne et al., eds., *Gravel Bed Rivers,* Wiley, pp. 289–300, 1987.
8. Andrews, E. D., and D. C. Erman, "Persistence in the Size Distribution of Surficial Bed Material During an Extreme Snowmelt Flood," *Water Resour. Res.,* vol. 22, pp. 191–197, 1986.
9. Ashida, K., and M. Michiue, "An Investigation of River Degradation Downstream of a Dam," *Proc. IAHR 14th Congress,* vol. 3, pp. 247–255, Paris, 1971.

10. Bagnold, R. A., "The Nature of Saltation and of Bed Load Transport in Water," *Proc. R. Soc. London,* ser A, vol. 332, pp. 473–504, 1973.
11. Bagnold, R. A., "An Approach to the Sediment Transport Problem from General Physics," *U.S. Geol. Survey, Prof. Paper* 422-I, 1966.
12. Bagnold, R. A., "Flow of Cohesionless Grains in Fluids," *Phil. Trans. R. Soc. London,* no. 964, vol. 249, pp. 235–297, 1956.
13. Barnes, H. H., "Roughness Characteristics of Natural Channels," *U.S. Geological Survey Water Supply Paper* 1899, 1967.
14. Barnes, R. C., Jr., "Erosion Control Structures," chap. 28 in H. W. Shen, ed., *River Mechanics.* Water Resources Publication, Littleton, Colo., 1972.
15. Beasley, D. B., "ANSWERS: A Mathematical Model for Simulating the Effects of Land Use and Management on Water Quality," Ph.D. dissertation, Purdue University, West Lafayette, Ind., 1977.
16. Boyce, R., "Sediment Routing and Sediment-Delivery Ratios," in *Present and Prospective Technology for Predicting Sediment Yields and Sources,* USAD0ARS-S-40, pp. 61–65, 1975.
17. Brady, N. S., *The Nature and Properties of Soils,* 8th ed., Macmillan, New York, 1974.
18. Bray, D. I., "Generalized Regime Type Analysis of Alberta Rivers," Ph.D. Thesis, Department of Civil Engineering, University of Alberta, Canada, 1972.
19. Brownlie, W. R., "Prediction of Flow Depth and Sediment Discharge in Open Channels," Report KH-R43A, W. M. Keck Laboratory, California Institute of Technology, Pasadena, Calif., Nov., 1981.
20. Bruk, S., "Methods of Computing Sedimentation in Lakes and Reservoirs," UNESCO International Hydrological Programme, Paris, 1985.
21. Brune, G. M., "Trap Efficiency of Reservoirs," *Trans. Am. Geophys. Union,* vol. 34, no. 3, pp. 407–418, 1953.
22. Chabert, J., and J. L. Chauvin, "Formation des dunes et des rides dans las modeles fluviaux," *Bull. du Centre de Recherches et d'Essais de Chaton,* no. 4, 1963.
23. Chepil, W. S., "Dynamics of Wind Erosion II: Initiation of Soil Movement," *Soil Sci.,* vol. 60, pp. 397–411, 1945.
24. Chepil, W. S., and N. P. Woodruff, "Estimations of Wind Erodibility of Field Surfaces," *J. Soil Water Cons.,* vol. 9, no. 6, pp. 257–265, 1954.
25. Chien, N., and S. W. Wan, *Mechanics of Sediment Transport,* Science Press, Beijing, China (in Chinese), 1983.
26. Chow, V. T., *Open Channel Hydraulics,* McGraw-Hill, New York, 1959.
27. Christensen, B. A., "Incipient Motion on Cohesionless Channel Banks," in H. W. Shen, ed., *Sedimentation,* Water Resources Publications, Littleton, Colo., 1972.
28. Colby, B., "Practical Computations of Bed Material Discharge," *J. Hydraul. Eng.,* vol. 90, HY2, pp. 217–246, 1964.
29. Colby, B. R., and C. H. Humbree, "Computation of Total Sediment Discharge. Niobrara River Near Cody, Nebraska," *U.S.G.S. Water Supply Paper* 1357, 1955.
30. Da Cuhna, L. V., "About the Roughness in Alluvial Channels with Comparatively Coarse Bed Material," *Proc. 12th IAHR Congress,* Fort Collins, Colo., 1967.
31. duBoys, P., "Etudes du Regime du Rhone et l'Action Exercee par les Laux sur un Lit a Fond de Graviers Undefiniment Affouillable," *Ann. Ponts et Chausses,* ser. 5, vol. 18, pp. 141–195, 1879.
32. Egiazoroff, L. V., "Calculation of Non-Uniform Sediment Concentration," *J. Hydraul Eng.,* vol. 91, no. HY-4, pp. 225–248, 1965.
33. Einstein, H. A., and N. Chien, "Effect of Heavy Sediment Concentration Near Bed Motion Velocity and Sediment Distribution," *U.S. Army Corps of Engineers, Division Series* 8, 1955.

34. Einstein, H. A., and N. Chien, "Transport of Sediment Mixtures with Large Ranges of Grain Sizes," IER, MRD series, no. 2, 1953.

35. Einstein, H. A., and N. L. Barbarossa, "River Channel Roughness," *Trans. ASCE,* vol. 117, pp. 1121–1146, 1952.

36. Einstein, H. A., "The Bed Load Function for Sediment Transportation in Open Channel Flows," *USDA Tech. Bull.* 1026, 1950.

37. Ekern, P. C., "Raindrop Impact as the Force Initiating Soil Erosion," *Soil Sci. Soc. Am.,* vol. 15, pp. 7–10, 1950.

38. Ellison, W. D., "Some Effects of Raindrops and Surface Flow on Soil Erosion and Infiltration," *Trans. AGU,* vol. 26, no. 2, pp. 415–429, 1945.

39. Engelund, F., and J. Fredose, "Sediment Ripples and Dunes," *Ann. Rev. Fluid Mech.,* vol. 14, pp. 13–37, 1982.

40. Engelund, F., and J. Fredose, "A Sediment Transport Model for Straight Alluvial Channels," *Nordic Hydrol.,* vol. 7, no. 5, pp. 293–306, 1976.

41. Engelund, F., "Hydraulic Resistance of Alluvial Streams," *J. Hydraul. Eng.,* vol. 92, no. HY2, pp. 315–326, 1966.

42. Fan, J., and G. L. Morris, "Reservoir Sedimentation. I: Delta and Density Current Deposits," *J. Hydraul. Eng.,* vol. 118, no. 3, pp. 354–369, 1992.

43. Fan, J., and G. L. Morris, "Reservoir Sedimentation. II: Reservoir Desiltation and Long-Term Storage Capacity," *J. Hydraul. Eng.,* vol. 118, no. 3, pp. 370–384, 1992.

44. Fenton, J., and J. E. Abbot, "Initial Movement of Grains on a Stream Bed: The Effect of Relative Protrusions," *Proc. R. Soc. London,* series A-352, pp. 523–537, 1977.

45. Fisher, J. S., B. L. Sill, and D. F. Clark, "Organic Detritus Particles: Initiation of Motion Criteria in Sand and Gravel Beds," *Water Res.,* vol. 19, no. 6, pp. 1627–1637, 1983.

46. Foster, G. R., "Modeling of Soil Erosion Process," in *Hydrologic Modeling of Small Watersheds,* ASAE Monograph 5, ASAE, St. Joseph, Mo., pp. 298–380, 1982.

47. Foster, G. R., and L. D. Meyer, "Mathematical Simulation of Upland Erosion Using Fundamental Erosion Mechanics," in *Present and Prospective Technology for Predicting Sediment Yields and Sources,* USDA-ARS-40, pp. 190–207, 1975.

48. Frenette, M., and P. Y. Julien, "LAVSED-I-Un Modele pour predire l'erosion des bassins et le transfert de sediments fins dan les cours d'eau nordiques," *Can. J. Civ. Eng.,* vol. 13, no. 2, pp. 150–161 (in French), 1986.

49. Frenette, M., and P. Y. Julien, "Computer Modeling of Soil Erosion and Sediment Yield from Large Watersheds," *Int. J. Sediment Res.,* vol. 2, November, pp. 39–68, 1987.

50. Garde, R. J., and K. G. Ranga Raju, *Mechanics of Sediment Trasnportation and Alluvial Stream Problems,* 2d ed., Wiley, New York, 1985.

51. Garde, R. J., K. Al-Shaikh-Ali, and S. Diette, "Armouring Process in Degrading Streams," *J. Hydraul. Eng.,* vol. 103, no. 9, pp. 1091–1095, 1977.

52. Garde, R. J., and K. G. Ranga Raju, "Regime Criteria for Alluvial Streams," *J. Hydraul. Eng.,* vol. 89, no. HY-6, pp. 153–164, 1963.

53. Garde, R. J., and M. L. Albertson, "Sand Waves and Regimes of Flow in Alluvial Channels," *Proc. IAHR,* 8th Congress, Montreal, vol. 4, 1959.

54. Gessler, J., "Stochastic Aspects of Incipient Motion on River Beds," chap. 25 in H. W. Shen, ed., *Stochastic Approach to Water Resources,* vol. 2, Water Resources Publications, Littleton, Colo., 1976.

55. Gessler, J., "Beginning and Ceasing of Sediment Motion," chap. 7 in H. W. Shen, ed., *River Mechanics,* Water Resources Publications, Littleton, Colo., 1971.

56. Gessler, J., "Self-Stabilizing Tendencies of Sediment Mixtures with Large Range of Grain Sizes," *J. Waterway Harbor Div. ASCE,* vol. 96, no. WW2, pp. 235–249, 1970.

57. Gessler, J., "The Beginning of Bed Load Movement of Mixtures Investigated as Natural

Armouring in Channels," *Trans. T-5 Keck Laboratory,* California Institute of Technology, Pasadena, Calif., 1967.

58. Gottschalk, L. C., "Sedimentation, part I, Reservoir Sedimentation," Sec. 17-1 in V. T. Chow, ed., *Handbook of Applied Hydrology,* McGraw-Hill, New York, 1964.
59. Graf, W., *Hydraulics of Sediment Transport,* McGraw-Hill, New York, 1971.
60. Hartley, D. M., "Boundary Shear Stress Induced by Raindrop Impact," Ph.D. dissertation, Department of Civil Engineering, Colorado State University, Fort Collins, 1990.
61. Hayashi, T., S. Ozaki, and T. Ichibashi, "Study on Bed Load Transport of Sediment Mixture," *Proc. 24th Japanese Conference on Hydraulics,* 1980.
62. Hudson, N., *Soil Conservation,* 2d ed., Cornell University Press, 1981.
63. IRTCES, "Lecture Notes of the Training Course on Reservoir Sedimentation," from *International Research and Training Center on Erosion and Sedimentation,* P.O. Box 366, Beijing, China, 1985.
64. Iwagaki, Y., "Hydrodynamical Study on Critical Tractive Force," *Tran. J.S.C.E.,* no. 41 (in Japanese), 1956.
65. Jansson, M. B., "Land Erosion by Water in Different Climates," *UNGI Report* 57, Uppsala University, 1982.
66. Jobson, H. E., and W. W. Sayre, "An Experimental Investigation of the Vertical Mass Transfer of Suspended Sediments," *Proc. IAHR 13th Congress,* vol. 2, Kyoto, Japan, 1969.
67. Julien, P. Y., and M. Frenette, "Macroscale Analysis of Upland Erosion," *Hydrol. Sci. J.,* vol. 32, no. 3, pp. 347–358, 1987.
68. Julien, P. Y., and M. Frenette, "LAVSED-II-A Model for Predicting Suspended Load in Northern Streams," *Can. J. Civ. Eng.,* vol. 13, no. 2, pp. 162–170, 1986.
69. Julien, P. Y., and D. B. Simons, "Sediment Transport Capacity of Overland Flow," *Trans. ASAE,* vol. 28, no. 3, pp. 755–762, 1985.
70. Karim, M. F., and F. M. Holly, Jr., "Armouring and Sorting Simulation in Alluvial Rivers," *J. Hydraul. Eng.,* vol. 112, no. 8, pp. 705–715, 1986.
71. Kilinc, M. Y., and E. V. Richardson, "Mechanics of Soil Erosion from Overland Flow Generated by Simulated Rainfall," *Hydrology Paper* 63, Colorado State University, Fort Collins, 1973.
72. Kirkby, M. J., and R. P. C. Morgan, eds. *Soil Erosion,* Wiley, New York, 1980.
73. Knisel, W. G., ed., "CREAMS: A Field-Scale Model for Chemicals, Runoff, and Erosion from Agricultural Management Systems," *USDA Cons. Res. Rept.* 26, 1980.
74. Krone, R. B., "A Study of Rheological Properties of Estuarial Sediments," *Tech Bull.* 7, Committee on Tidal Hydraulics, U.S. Army Engineer Waterway Experiment Station, Vicksburg, Miss., 1963.
75. Kuhnle, R. A., "Bed-Surface Size Changes in Gravel Bed Channel," *J. Hydraul. Eng.,* vol. 115, no. 6, pp. 731–743, 1989.
76. Lane, L. J., and M. A. Nearing, "USDA-Water Erosion Prediction Project: Hillslope Profile Model Documentation," *USDA-ARS National Soil Erosion Research Laboratory Report* 2, 1989.
77. Lane, E. W., "The Importance of Fluvial Morphology in Hydraulic Engineering," *Proc. ASCE,* vol. 81, no. 745, 1955.
78. Lane, E. W., and E. J. Carlson, "Some Factors Affecting the Stability of Canals Constructed in Coarse Granular Materials," *Proc. 5th IAHR Congress,* University of Minnesota, 1953.
79. Lee, Hong-Yuan, and A. J. Odgaard, "Simulation of Bed Armoring in Alluvial Channels," *J. Hydraul. Eng.,* vol. 112, no. 9, pp. 794–801, September, 1986.
80. Li, R. M., D. B. Simons, W. T. Fullerton, K. G. Eggert, and B. O. Spronk, "Simulation of

Water Runoff and Sediment Yield from a System of Multiple Watersheds," *Proc. 18th Congress JAHR,* Cagliary, Italy, vol. 5, pp. 219–226, 1979.

81. Li, R. M., "Water and Sediment Routing from Watersheds," chap. 9 in H. W. Shen, ed., *Modeling of Rivers,* Wiley, New York, 1979.
82. Little, W. C., and R. G. Mayer, "The Role of Sediment Gradation in Channel Armouring," *Georgia Institute of Technology, Publ.* ERC-0672, 1972.
83. Liu, H. K., "Mechanics of Sediment Ripple Formation," *J. Hydraul. Eng.,* vol. 83, no. HY2, pp. 1197-1–1197-23, 1957.
84. Lovell, C. W., and R. L. Wiltshire, "Engineering Aspects of Soil Erosion, Dispersive Clays and Loess," *Geotech. Spec. Publ.* 10, ASCE, 1987.
85. Lovera, F., and J. F. Kennedy, "Friction Factors for Flat-Bed Flows in Sand Channels," *J. Hydraul. Eng.,* vol. 95, no. HY4, pp. 1227–1234, 1969.
86. Mantz, P. A., "Incipient Transport of Fine Grains and Flakes by Fluids-Extended Shields Diagram," *J. Hydraul. Eng.,* vol. 103, no. 6, pp. 601–616, 1977.
87. Mather, A. K., and K. G. Ranga Raju, "Prediction of Regimes of Flow," *Symposium on Modern Trends in Civil Engineering,* University of Roorkee, India, 1973.
88. Mehta, A. J., E. J. Hayter, W. R Parker, R. B. Krone, and A. M. Teeter, "Cohesive Sediment Transport, I: Process Description," *J. Hydraul. Eng.,* vol. 115, no. 8, pp. 1076–1093, 1989.
89. Mehta, A. J., E. J. Hayter, W. R. Parker, R. B. Krone, and A. M. Teeter, "Cohesive Sediment Transport, II: Application," *J. Hydraul. Eng.,* vol. 115, no. 8, pp. 1094–1112, 1989.
90. Mehta, A. J., ed., *Estuarine Cohesive Sediment Dynamics* (Lecture notes on coastal and estuarine studies), vol. 14, Springer-Verlag, Berlin, Germany, 1986.
91. Meyer, L. D., and W. H. Wischmeier, "Mathematical Simulation of the Process of Soil Erosion by Water," *Trans. ASAE,* vol. 12, no. 6, pp. 754–758, 1969.
92. Meyer, L. D., "Soil Erosion by Water on Upland Areas," Chap. 27 in H. W. Shen, ed., *River Mechanics,* Water Resources Publication, Littleton, Colo., 1972.
93. Meyer-Peter, E., H. Favre, and H. A. Einstein, "Neuere Veruschresultate Uber den Geschiebetrieb," *Schweiz, Bauzeitung,* vol. 103, no. 12, pp. 147–150, 1934.
94. Meyer-Peter, E., and R. Muller, "Formula for Bed Load Transport," *Proc. 2d Meeting, IAHR,* vol. 6, 1948.
95. Miller, R. L., and R. J. Byrne, "The Angle of Repose for a Single Grain on a Fixed Rough Bed," *Sedimentology,* vol. 6, pp. 303–314, 1966.
96. Misri, R. L., R. J. Garde, and K. G. Ranga Raju, "Bed Load Transport—An Experimental Study," *JIP,* India.
97. Musgrave, G. W., "The Quantitative Evaluation of Factors in Water Erosion: A First Approximation," *J. Soil Water Cons.,* vol 2, no. 3, pp. 133–138, 1947.
98. Mutchler, C. K., "Parameters for Describing Raindrop Splash," *J. Soil Water Cons.,* vol. 22, pp. 91–94, 1967.
99. Noyce, R., "Sediment Routing with Sediment-Delivery Ratios," in *Present and Prospective Technology for Predicting Sediment Yields and Sources,* USDA-ARS-S-40, pp. 61–65, 1975.
100. Parker, G., "Surface Based Bed Load Transport Relative to Grand Rivers," *J. Hydraul. Res. IAHR,* vol. 4, pp. 417–436, 1990.
101. Parker, G., S. Dhamotharan, and H. Stefan, "Model Experiments on Mobile, Paved Grave Bed Streams," *Water Resour. Res.,* vol. 18, no. 5, pp. 1395–1408, 1982.
102. Parker, G., P. C. Klingeman, and D. G. McLean, "Bedload and Size Distribution in Paved Gravel-Bed Streams," *J. Hydraul. Eng.,* vol. 108, no. 4, pp. 544–571, 1982.
103. Pemberton, E. L., "Einstein's Bed Load Function Applied to Channel Design and Degra-

dation," chap. 16 in H. W. Shen, ed., *Sedimentation* (Einstein), Water Resources Publications, Littleton, Colo., 1972.

104. Piest, R. F., and C. R. Miller, "Sediment Sources and Sediment Yields," chap. IV in V. Vanoni, ed., *Sedimentation Engineering,* ASCE-Manual and Reports on Engineering Practice 54, 1977.

105. Raudkivi, A. J., *Loose Boundary Hydraulics,* 3d ed., Pergamon Press, New York, 1990.

106. Renard, K. G., and J. R. Simanton, "Application of RUSLE to Rangelands," *Proc. Symp. Watershed Planning and Analysis in Action, ASCE Durango,* Colo., pp. 164–173, 1990.

107. Richardson, E. V., D. B. Simons, and P. Y. Julien, "Highways in the River Environment," Federal Highway Administration Publication FHWA-HI-90-016, February, 1990.

108. Schroeder, K. B., and C. H. Humbree, "Application of Modified Einstein Procedure for Computation of Total Sediment Load," *Trans. AGU.,* vol. 37 (see also its discussion in *Trans. AGU.,* vol. 38, no. 5, October 1957), 1956.

109. Schulz, E. F., R. H. Wilde, and M. L. Albertson, "Influence of Shape on the Fall Velocity of Sedimentary Particles," Mississippi River Sediment Series 5, 1954.

110. Schwab, G. O., R. K. Frevert, T. W. Edminster, and K. K. Barnes, *Soil and Water Conservation Engineering,* 3d ed., Wiley, New York, 1981.

111. Shen, H. W., H. M. Fehlman, and C. Mendoza, "Bed Form Resistance in Open Channel Flows," *J. Hydraul. Eng.,* vol. 116, no. 6, pp. 799–815, June 1990.

112. Shen, H. W., and J. Y. Lu, "Development and Prediction of Bed Armouring," *J. Hydraul. Eng.,* vol. 109, no. 4, pp. 611–629, 1983.

113. Shen, H. W., ed., *Modelling of Rivers,* Wiley, New York, 1976.

114. Shen, H. W., ed., *Environmental Impacts on Rivers,* Water Resources Publications, Littleton, Colo., 1973.

115. Shen, H. W., and C. S. Hung, "An Engineering Approach to Total Bed-Material Load by Regression," chap. 14 in H. W. Shen, ed., *Sedimentation* (Einstein), Water Resources Publications, Littleton, Colo., 1972.

116. Shen, H. W., ed., *River Mechanics,* vols. I, II, Water Resources Publications, Littleton, Colo., 1971.

117. Shen, H. W., "Development of Bed Roughness in Alluvial Channels, *J. Hydraul. Eng.,* Proc. Paper 3113, vol. 88, no. HY3, pp. 45–58, 1962.

118. Shields, A., "Anwendung der Aechichkeits-mechanik und der turbuleng forschung auf dir Geschiebewegung' Mitt Preussische," *Versuchsanstalt für Wasserbau and Schiffbau,* Berlin, 1936.

119. Simons, D. B., and F. Senturk, *Sediment Transport Technology,* Water Resources Publications, Littleton, Colo., 1977.

120. Simons, D. B., and E. V. Richardson, "A Study of the Variables Affecting Flow Characteristics and Sediment Transport in Alluvial Channels," *Proc. Federal Inter-Agency Sedimentation Conference,* USDA (Washington), Misc. Publ. 970, 1962.

121. Smerdon, E. T., and R. P. Beasley, "Critical Cohesive Forces in Cohesive Soils," *Agric. Eng.,* vol. 42, pp. 26–29, 1961.

122. Task Committee, "Erosion of Cohesive Sediment," *J. Hydraul. Eng.,* vol. 94, no. 4, 1969.

123. Thorne, C. R., J. C. Bathurst, and R. D. Hey, ed., *Sediment Transport in Gravel-Bed Rivers,* Wiley, New York, 1987.

124. Toffaleti, F. B., "Definite Computations of Sand Discharges in Rivers," *J. Hydraul. Eng.,* vol. 95, HY1, pp. 225–248, 1969.

125. U.S. Army Corps of Engineers, "The Streambank Erosion Control Evaluation and Demonstration Act of 1974," Section 32, Public Law 93-251: Final Report to Congress, Main Report and Appendices A to H, 1981.

126. U.S. Department of Agriculture, Soil Conservation Service, "Cropland Erosion," in the Second Natl. Water Assessment, Water Resources Council, Washington, D.C., 1977.

127. U.S. Department of Agricultural Research Service, "A Universal Equation for Predicting Rainfall-Erosion Losses," USDA-ARS Spec. Report 22-26, 1961.

128. U.S. Department of Agriculture Soil Conservation Service, Generalized Map of Soil Erosion in the United States, 1948.

129. U.S. Inter-Agency Committee on Water Resources, Subcommittee on Sedimentation, "Measurement and Analysis of Sediment Loads in Streams," Report no. 14, 1963.

130. U.S. Geological Survey, "Computation of Fluvial-Sediment Discharge," by G. Porterfield, chap. 3, book 3, 1972.

131. U.S. Geological Survey, "Field Methods for Measurement of Fluvial Sediment," by H. Guy and V. Norman, chap. C2, book 3, 1970.

132. van Rijn, L. C., "Sediment Transport: Part I: Bed Load Transport," *J. Hydraul. Eng.*, vol. 110, no. 12, pp. 1431–1456, 1984.

133. Vanoni, V. A., ed., *Sedimentation Engineering, ASCE Manual and Reports on Engineering Practice,* 54, New York, 1975.

134. Vanoni, V. A., "Factors Determining Bed Form of Alluvial Streams," *J. Hydraul. Eng.*, vol. 100, no. HY3, pp. 363–378, 1974.

135. Walling, D. E., R. F. Hadley, R. Lal, C. A. Onstad, and A. Yair, "Recent Developments in Erosion and Sediment Yield Studies," UNESCO, 1985.

136. Wang, S. Y., and H. W. Shen, "Incipient Sediment Motion and Riprap Design," *J. Hydraul. Eng.,* vol. 109, 1983.

137. White, C. M., "The Equilibrium of Grains on the Bed of Stream," *Proc. R. Soc. London,* Series A, Mathematical and Physical Sciences, no. 958, vol. 174, 1946.

138. Wiberg, P. L., and J. Dungan Smith, "Calculations of the Critical Shear Stress for Motion of Uniform and Heterogeneous Sediments," *Water Resour. Res.,* vol. 23, no. 8, pp. 1471–1480, 1987.

139. Wischmeier, W. H., C. B. Johnson, and B. V. Cross, "A Soil-Erodibility Monograph for Farmland and Construction Sites," *J. Soil and Water Cons.,* vol. 26, no. 5, pp. 189–193, 1971.

140. Wischmeier, W. H., and D. D. Smith, "Predicting Rainfall Erosion Losses—A Guide to Conservation Planning," *USDA Agriculture Handbook* 537, 1978.

141. Wischmeier, W. H., "Upslope Erosion Analysis," chap. 15 in H. W. Shen, ed., *Environmental Impact on Rivers,* Water Resources Publications, Littleton, Colo., 1972.

142. Yalin, M. S., *Mechanisms of Sediment Transport,* Pergamon Press, New York, 1972.

143. Yalin, M. S., and E. Karahan, "Inception of Sediment Transport," *J. Hydraul. Eng.,* vol. 105, HY11, pp. 1433–1443, 1979.

144. Yang, C. T., "Incipient Motion and Sediment Transport," *J. Hydraul. Eng.,* vol. 99, HY10, pp. 1679–1704, 1973.

145. Zachar, D., *Soil Erosion,* Developments in Soil Science No. 10, Elsevier, Amsterdam, 1982.

CHAPTER 13
HYDROLOGIC EFFECTS OF LAND-USE CHANGE

Ian R. Calder
Institute of Hydrology
Wallingford, United Kingdom

13.1 INTRODUCTION

13.1.1 Major Land-Use Changes and the Problems They Pose

We live in a changing world, and the effects of the changes are of interest to everybody. Land-use change can have local, regional, and global hydrologic consequences. It is the objective of this chapter to outline these effects and to indicate how they can be measured and predicted.

On a global scale, the largest change in terms of land area, and arguably also in terms of hydrologic effects, is from *afforestation* and *deforestation.* Forests are mostly planted or allowed to remain on areas unsuitable for agricultural use, often upland mountainous areas which are also the principal catchments for water supplies for cities and towns. Planting trees creates concern that they will intercept more rainfall and deplete downstream water supplies; stream acidification may also result. Cutting down trees raises concerns of erosion, siltation of streams, and increased leaching of soil nutrients.

Intensification of agriculture involving land drainage, the use of fertilizers and pesticides, and the stall and battery farming of animals and poultry represents another major land-use change which may have impacts on soil erosion and water quality and quantity.

The draining of wetlands is a third area where the hydrologic consequences of land-use change are important. Formerly, wetlands were regarded as being of little value; investment in wetlands meant drainage and conversion to forestry, agriculture, or industry. More recently the value of wetlands in providing biological diversity, aesthetics and recreation, fishery and wildlife products and through their hydrologic role of flood control, shoreline anchorage, and water purification is becoming more appreciated. This has provided an impetus for the conservation and protection of wetlands.

Urbanization is another land-use change with important hydrologic effects. These are discussed in Chap. 28.

A summary of the hydrologic impacts associated with land-use change is given in Table 13.1.1.

TABLE 13.1.1 Summary of the Major Hydrologic Effects of Land-Use Change

Land-use change	Component affected	Principal hydrologic processes involved	Geographic scale and likely magnitude of effect
Afforestation (deforestation has converse effect except where disturbance caused by forest clearance may be of overriding importance)	Annual flow	Increased interception in wet periods Increased transpiration in dry periods through increased water availability to deep root systems	Basin scale; magnitude proportional to forest cover, world average is 34 mm $year^{-1}$ reduction for 10% increase in forest cover
	Seasonal flow	Increased interception and increased dry period transpiration will increase soil moisture deficits and reduce dry season flow	Basin scale; can be of sufficient magnitude to stop dry season flows
		Drainage activities associated with planting may increase dry season flows through initial dewatering and also through long-term effects of the drainage system	Basin scale; drainage activities will increase dry season flows
		Cloud water (mist or fog) deposition will augment dry season flows	High-altitude basins only; increased cloud water deposition may have a significant effect on dry season flows
	Floods	Interception reduces floods by removing a proportion of the storm rainfall and by allowing buildup of soil moisture storage	Basin scale; effect is generally small but greatest for small storm events
		Management activities: cultivation, drainage, road construction, all increase floods	Basin scale; increased floods for all sizes of storm events
	Water quality	Leaching of nutrients is less from forests through reduced surface runoff and reduced fertilizer applications	Basin scale; variable but leaching can be an order of magnitude less than from agricultural land
		Deposition of most atmospheric pollutants is higher to forests because of reduced aerodynamic resistance	Basin scale; leads to acidification of catchments and runoff
	Erosion	High infiltration rates in natural, mixed forests reduce surface runoff and erosion	Basin scale; reduces erosion
		Slope stability is enhanced by reduced soil pore water pressure and binding of forest roots	Basin scale; reduces erosion

TABLE 13.1.1 Summary of the Major Hydrologic Effects of Land-Use Change *(Continued)*

Land-use change	Component affected	Principal hydrologic processes involved	Geographic scale and likely magnitude of effect
		Windthrow of trees and weight of tree crop reduce slope stability	Basin scale; increases erosion
		Soil erosion, through splash detachment, is increased from forests without an understory of shrubs or grass	Basin scale; increases erosion
		Management activities: cultivation, drainage, road construction, felling, all increase erosion	Basin scale; management activities are often more important than the direct effect of the forest
	Climate	Increased evaporation and reduced sensible heat fluxes from forests affect climate	Micro, meso, and global scale; forests generally cool and humidify the atmosphere; a 2°C increase in regional temperature is predicted for Amazonia if deforestation continues
Agricultural intensification			
	Water quantity	Alteration of transpiration rates affects runoff	Basin scale; effect is marginal
		Timing of storm runoff altered through land drainage	Basin scale; significant effect
	Water quality: fertilizers	Application of inorganic fertilizers	Basin scale; increased nutrient concentrations in surface and groundwaters
	Pesticides	Application of nonselective and persistent pesticides poses health risks to humans and animal life	Basin, regional, and global scale; effects can be long-lasting
	Farm wastes	Inadequate disposal of farm organic and inorganic water pollutes surface and groundwater bodies	Basin scale; effect on groundwater and surface waters
	Erosion	Cultivation without proper soil conservation measures and uncontrolled grazing on steep slopes increases erosion	Basin scale; effects are very site-dependent
Draining wetlands			
	Seasonal flow	Upland peat bogs, groundwater fens, and African dambos have little effect in maintaining dry season flows	Basin scale: drainage or removal of wetland will not reduce, and may increase dry season flows

TABLE 13.1.1 Summary of the Major Hydrologic Effects of Land-Use Change *(Continued)*

Land-use change	Component affected	Principal hydrologic processes involved	Geographic scale and likely magnitude of effect
		Lowering of the water table may induce soil moisture stress, reduce transpiration, and increase dry season flows	Basin scale; a reduction of water-table depth to minimum of 30 cm below surface is required
		Initial dewatering following drainage will increase dry season flows	Basin scale; effect may last from 1–2 years to decades
		The deeper flow outlet of the drainage system will lead to increased dry season flows	Basin scale; effects will be long-term
	Annual flow	Initial dewatering following drainage will increase annual flow	Basin scale; effect may last from 1–2 years to decades
		Afforestation following drainage will reduce annual flow	Basin scale; effects as for afforestation
	Floods	Drainage method, soil type, channel improvement, all affect flood response	Basin scale; open drains increase, closed drains reduce, flood peak
	Water quality	Redox potentials are altered leading to decomposition of peat, acidification, and increased organic loads in runoff	Basin scale; increases acidity
		Drainage systems intercepting mineral horizons will reduce acidity	Basin scale; reduces acidity
	Carbon balance	Accumulating peat bogs are a sink for atmospheric CO_2	Global scale effect; disturbance and drainage of bogs is such that the carbon balance of the world's peat bogs is neutral at present

13.1.2 Scale of Effects and Interactions

The hydrologic effects of land-use change are manifest in many ways and at different spatial and time scales. Most obvious are the immediate and direct effects on the quantity and quality of catchment runoff. For changes of sufficient areal scale, effects on climate may also occur. Climatic change, whether on the meso (local) or global scale, may produce secondary effects on local and global hydrology. Traditionally hydrologists have been concerned most with the direct and local effects of land-use change on hydrology. Increasingly hydrologists and water resource planners are becoming concerned with the secondary impacts such as the effects on water supply reliability resulting from climate change. On a global scale, a prominent issue is the effect of large-scale removal and burning of tropical rain forest which may enhance the greenhouse effect by releasing more carbon dioxide into the atmosphere and may also affect local and global climate by altering the energy balance and evaporation flux of the region in which the forest was located.

13.1.3 Controls on Water Use from Different Land-Use Systems

The evaporation from a particular form of land use is the result of the balance achieved between externally applied atmospheric and radiative demand, the availability of water at the evaporating surface, and the aerodynamic resistance to the transport of vapor between the surface and the atmosphere (Fig. 13.1.1). A change in land use is likely to alter the availability of water at the evaporating surface through changes in:

- The surface area of free water surfaces in streams and lakes
- The availability of soil water to plants (for example, when short-rooted agricultural crops replace deep-rooted trees, the availability of water will be reduced in dry periods or when drainage reduces soil moisture content in the rooting zone)
- Replacement of crops with different total leaf area per unit ground area (leaf area index, LAI), different stomatal resistance, and different stomatal responses to soil water and atmospheric humidity deficits.

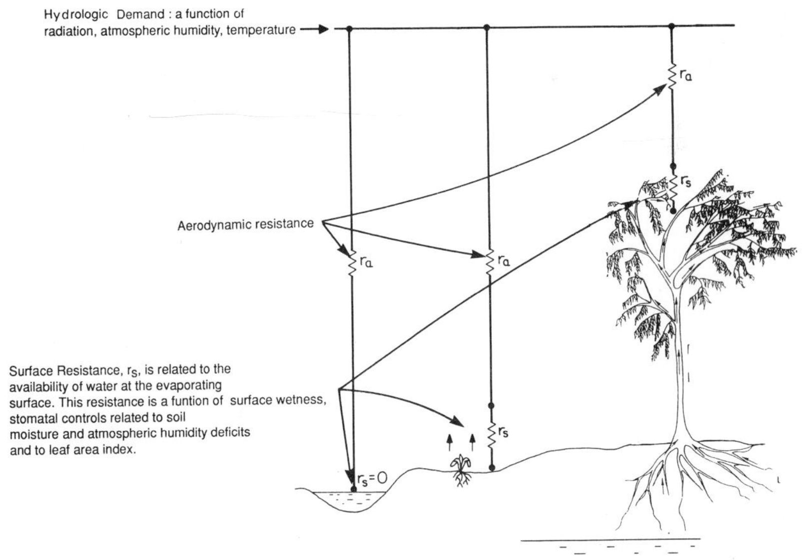

FIGURE 13.1.1 The evaporation rate is determined by the balance between meteorological demand and the availability of water at the evaporating surface.

Atmospheric Feedback Effects. Changes in water availability cause first-order changes in the evaporation rate which alter near surface (microclimate) atmospheric conditions. If the land-use change is of sufficient areal extent, changes to the regional (mesoscale) and global climate may follow. The altered atmospheric conditions, whether on the *micro, meso,* or *global* scale, exert a second-order change in the evaporation rate. This is a negative feedback process; a reduction in evaporation through restricted water availability will increase atmospheric demand for evaporation which will feed back as a (second-order) increase in evaporation rate.

Atmospheric feedback processes operate over a range of time scales. In many plants, an increase in atmospheric vapor pressure deficit will result in a reduction in stomatal aperture and a reduction in transpiration within minutes. The process may be considered to be first-order: the effect may be sufficient to place an upper limit on evaporation rate irrespective of increases in atmospheric demand. Limitations on evaporation as a result of soil water deficits are manifest over longer time periods, usually weeks or months.

The aerodynamic transport resistance to water movement between the evaporating surface and the atmosphere is determined by the aerodynamic roughness of the surface, and by the windspeed. Land-use change affects aerodynamic roughness; taller vegetation such as trees has a higher aerodynamic roughness and lower aerodynamic transport resistance compared with shorter vegetation such as grass.

The feedback, or "complementary," relationship between atmospheric demand and surface evaporation has been used by Brutsaert and Stricker[8] and Morton[61] to estimate regional evaporation from measurements of the regional climate and energy balance. The approach may be of value for obtaining approximate estimates of regional evaporation in certain circumstances but is not generally valid;[16] in regions with large-scale advection, as, for example, in some forested areas which have high interception losses, the approach cannot be applied.

The evaporation-atmosphere feedback processes operate on all scales and should be borne in mind in any detailed studies into the effects on evaporation of a change in land use, particularly if the change takes place over a large area.

13.1.4 Choice of Prediction Model

The hydrologic effects of land-use change can be predicted by synthesis of data from field experiments (empiricism) or by theoretical considerations. Both methods have their advantages. If the effect of a particular land-use change occurring under a given set of environmental conditions has been measured experimentally, such observations can be used to predict the effects of the same type of land-use change at another location where the conditions are similar—this is the philosophy of *catchment experiments.* Where predictions are required of the effects of land-use change under different conditions from those where measurements have been obtained, an understanding of the operation of the system is required (a system model). *Process studies* are then necessary to allow the development and calibration of a *process model* which will account for the different environmental conditions at the prediction site.

Effects on climate, whether global or mesoscale, and the secondary effects of climate change on hydrology can be studied only using process-based climate models. Process studies are required to define the interactions of the atmosphere with the surface hydrology and to measure parameter values for these models.

A model for predicting the effects of a land-use change should have:

1. Input data requirements which can be satisfied

2. A range of application which covers the problem being considered
3. Sufficient complexity to give the required prediction accuracy—use the simplest model which will give a sufficiently accurate result

13.2 METHODS AND MEASUREMENTS

13.2.1 Catchment Experiments

Catchment experiments are designed to investigate the effects of land-use change on a catchment scale. This requires the use of a *control,* which can be either the existing land use before the catchment land use is changed or a separate *paired catchment* or a combination of both. In a *single-catchment experiment,* the effect of a land-use change is measured by comparing measurements made before and after the change occurred. In a *paired-catchment experiment,* land use is held constant on the *control catchment* and changed on the *treatment catchment*—differences in hydrologic response with respect to annual runoff, flood, and low-flow response and water quality can then be compared for catchments experiencing the same weather patterns. In single-catchment experiments, it is difficult to separate the effects of land-use change from the effects of different weather patterns before and after the land-use change; paired-catchment experiments are more costly but provide more accurate results because the effects of change are being measured under the same weather pattern on the two catchments.

The first catchment experiments designed to investigate the hydrologic effects of forests began in the Alps in 1902, with the establishment of the Sperbelgraben and Rappengraben catchment experiments, and in the United States in 1909 with the studies initiated at Wagon Wheel gap, Colorado. Since then many experiments have been carried out in all continents, e.g., at Coweeta[30] in North Carolina, in East Africa,[71] at Mokobulaan in South Africa,[97] at Plynlimon in the United Kingdom,[47] in Malaysia, in New Zealand, at the Niligris in India, and at the Perth water supply catchments in Australia.

Error Analysis in Catchment Experiments. It is critical in the design of catchment experiments, particularly those concerned with the water balance, that the experimental error attached to the effect being studied is not larger than the effect itself. If the effects of a change in land use on evaporation are being studied using a paired-catchment experiment, the conventional *water-balance* equation can be written for the treatment catchment as

$$E = P - Q + \Delta S \tag{13.1.1}$$

and for the control catchment

$$E' = P' - Q' + \Delta S' \tag{13.1.2}$$

where E, P, Q, and ΔS refer to evaporation, precipitation, runoff, and storage change, respectively. Over a sufficiently long time period the storage change term becomes insignificant compared with the other terms and the equations reduce to

$$E = P - Q \tag{13.1.3}$$

$$E' = P' - Q' \tag{13.1.4}$$

If the *standard errors* of estimate of the rainfall and runoff are denoted by α and β, respectively, then the standard error of the measurement of evaporation γ is given by

$$\gamma = \sqrt{\alpha^2 + \beta^2} \tag{13.1.5}$$

under the assumption that the errors of estimate for rainfall and runoff are independent.

The standard error so calculated can be large and may approach or even exceed the estimated mean evaporation in situations where both P and Q are large in comparison with E and both have significant measurement errors.

The difference in evaporation between the two catchments is given by

$$E - E' = P - P' - Q + Q' \tag{13.1.6}$$

The standard error of $(E - E')$, assuming all errors are independent, is given by

$$\gamma_{E-E'} = \sqrt{\alpha^2 + \beta^2 + \alpha'^2 + \beta'^2} \tag{13.1.7}$$

and the corresponding *relative error* is given by

$$\frac{\gamma_{E-E'}}{E - E'} = \frac{\sqrt{\alpha^2 + \beta^2 + \alpha'^2 + \beta'^2}}{P - P' - Q + Q'} \tag{13.1.8}$$

The experimental errors attached to the measurement of precipitation and runoff have both *systematic* and *random* components.

Systematic or consistent errors in the precipitation measurement may occur through systematic undercatching because of wind effects and because of rain gauge siting. With a finite number of rain gauges in a network the network will, unless the gauges are randomly relocated at intervals, systematically record more or less than the true areal mean. Systematic error in stream-flow estimation results from uncertainties in the stage discharge relationship for flow measuring stations. The magnitude of systematic errors is usually difficult to determine, although the use of independent calibration techniques is of value in streamflow measurement. The determination of the systematic errors attached to areal precipitation measurement in mountainous or forested areas, particularly when snow forms a significant component of the annual precipitation, remains a major difficulty for the assessment and interpretation of results from catchment experiments.

Random errors of measurement in precipitation can arise from many sources including, for example, the precision of the measurement and spatial variability. For flow measurement the precision of the measurement and short-term fluctuations in the stage discharge relationship are sources of random error.

Where the experimental errors attached to the precipitation or stream-flow measurement are systematic in nature, the consequent relative error in the measurement of evaporation difference will not diminish with time. Where the errors are random in nature, extending the measurement period will reduce relative errors.

For the Plynlimon catchments the estimate of mean areal precipitation at ($\sim$20 gauges per catchment) has been calculated to be within 5 percent of the true areal mean as measured with an infinitely dense network of gauges.[48] Where rain gauge networks are fixed in time, that is, where they are not randomly relocated at intervals, this error represents a systematic error. If other systematic errors in the rain gauge method, for example, undercatching or overcatching due to wind effects, are discounted, the relative standard (systematic) error of precipitation measurement can be taken as 5 percent.

The relative standard (systematic) error in the streamflow measurement is 2.5

percent.[48] For upland mid-Wales, where the Plynlimon catchments experiments are located (Fig. 13.2.1) the annual rainfall is of the order of 2400 mm and the annual runoff is of the order of 2000 mm (see Table 13.2.1). The corresponding α and β values are therefore 120 and 50 mm and if assumed to be independent (there is no reason to expect that the measurement errors, either systematic or random, attached to the rainfall and runoff measurements are correlated or that there will be any correlation of these measurement errors between catchments), the standard error of the difference in evaporation γ (which can be calculated prior to the setting up of a catchment experiment) is given from Eq. (13.1.8) as 180 mm.

Where the experimental errors attached to the measurement of P and Q are random in nature, the relative error in the measurement of evaporation difference will diminish as the experiment is continued. If an experiment is carried on for n years, the mean $E - E'$ and standard deviation $\sigma_{E-E'}$ of n measured values of $E - E'$ can be calculated (Table 13.2.1). The variation of the term $E - E'$ arises from both true time variation in $E - E'$ and random errors in the measurement of the component precipitation and runoff. The value of $\sigma_{E-E'} = 90$ mm therefore represents an upper limit on the standard deviation due to random measurement error (assuming no true time variation of $E - E'$). For the 11-year measurement period the upper limit on the standard error due to random measurement error is then given by $\sigma_{E-E'}/\sqrt{n} = \sigma_{\overline{E-E'}} = 27$ mm, which is small compared with the estimate of the systematic error of 180 mm. Combining these by adding their variances gives the standard error of $E - E'$ due to both systematic and random component errors as $\sqrt{180^2 + 27^2} = 182$ mm.

The difference in evaporation between catchments can be judged to be significant with respect to random errors if $(\overline{E - E'})/\sigma_{\overline{E-E'}}$ is greater than the t-test critical value for a given probability level (see Chap. 17 for details of the t test). For the Plynlimon

FIGURE 13.2.1 The headwaters of the Severn catchment at Plynlimon. This is the site of a heavily instrumented land-use catchment experiment in Wales.

TABLE 13.2.1 Annual Values of Precipitation P, Runoff Q, Losses $P-Q$, and Difference in Losses for the Grassland Wye Catchment and the Forested Area of the Severn Catchment Only, for the Years 1970–1980

	Precipitation P		Runoff Q		$E=P-Q$ loss		Difference in losses (Severn − Wye)	Running values up to and including year shown				t-Test critical value at 99% level
					Severn E	Wye E'	$E-E'$					
	Wye	Severn	Wye	Severn	(3) − (5)	(2) − (4)	(6) − (7)	$\overline{E-E'}$	$\sigma_{E-E'}$	$\sigma_{\overline{E-E'}}$ (10)/$\sqrt{n}$	$(\overline{E-E'})/\sigma_{\overline{E-E'}}$ (9)/(11)	
(1)	(2)	(3)	(4)	(5)	(6)	(7)	(8)	(9)	(10)	(11)	(12)	(13)
1970	2869	2485	2415	1636	849	454	395	395	—	—	—	—
1971	1993	1762	1562	797	965	431	534	465	98	70	6.7	31.8
1972	2131	2124	1804	1342	782	327	455	461	70	40	11.5	7.0
1973	2606	2380	2164	1581	799	442	357	435	77	39	11.3	4.5
1974	2794	2703	2320	1785	918	474	444	437	68	30	14.6	3.7
1975	2099	2035	1643	1213	822	456	366	425	66	27	15.7	3.4
1976	1736	1645	1404	921	724	332	392	420	62	23	17.9	3.1
1977	2561	2573	2236	1638	935	325	610	444	88	31	14.2	3.0
1978*	2356	2367	2128	1668	699	228	471	447	83	28	16.2	2.9
1979*	2742	2683	2463	2016	667	279	388	441	80	25	17.3	2.8
1980	2695	2517	2377	1914	603	318	285	427	90	27	15.8	2.8
Mean	2417	2298	2047	1501	797	370	427					

Source: Reproduced with permission from Calder, Newson, and Walsh.[21]
Units: millimeters.
* Years in which snow made necessary the estimation of monthly precipitation total in winter.

catchment results the effect would be judged significant at the 99 percent level after a period of 3 years (Table 13.2.1). The accuracy of the experiment, with respect to random errors of measurement, increases in proportion to the square root of its duration; an experiment run for 4 years is twice as accurate as one run for 1 year, and a 9-year experiment will be 3 times more accurate. However, it will be noted from the above that when random errors are small compared with the likely systematic errors, extending the experiment over many years will have little effect on accuracy of the measurement.

Assessment of Catchment Experiments. Notwithstanding the error limitations outlined above, catchment experiments give the integrated effect of the treatment over all the processes operating within a catchment. Before such experiments are initiated, it is strongly recommended that a careful study be made to determine both the likely systematic and random errors of measurement and the number of years of data that will be required to measure the anticipated effect within the accuracy of the experimental data. Many catchment studies have given inconclusive results because the experimental error was the same size as the effect being measured, when that fact could have been anticipated before the studies were initiated if proper consideration of the measurement errors had taken place.

The principal disadvantage of catchment experiments is that without a knowledge of the processes taking place in the catchment the results are difficult to extrapolate elsewhere and the results are strictly applicable only to situations where the same set of processes is operating.

13.2.2 Process Studies, Methodology, and Theory

While catchment studies of the effects of a land-use change measure directly the integrated effect of the treatment, process studies must first identify the hydrologic processes which are affected by the treatment and then quantify them. The principal processes affecting water quantity and quality are:

1. *Aerodynamic transfer processes.* A change in the height or aerodynamic roughness properties of the land use will affect both *transpiration* and *interception* rates, affecting water quantity, and will also alter the *deposition rates* for atmospheric pollutants, affecting water quality.

2. *Plant physiological responses.* The primary plant physiological controls on transpiration are those related to leaf area and soil water availability (rooting depth). Secondary controls are related to the manner in which plants respond to changing environmental controls; these include the important "feedback" stomatal responses that many plant species exhibit, in which increases in atmospheric humidity deficit are associated with the closing of stomata. Plants which exhibit this response are thus able to limit their water use in times of increasing atmospheric evaporative demand. Plants which do not exhibit this response, if well supplied with water, are able to transpire at much higher rates than those that do. Deposition rates of gaseous pollutants which diffuse through plant stomata and are absorbed by plants (SO_2 and O_3) are also affected by stomatal response mechanisms.

3. *Erosion processes.* Erosion processes are affected by many factors, including soil type and slope, soil cultivation, land drainage, road construction, and other engineering works. Erosion rates are also related to vegetation type through raindrop size modification and splash detachment, through changes in pore water pressure, and through altered soil stability.

4. *Unsaturated and saturated flow processes.* Differences in the flood and low-flow response of catchments to a change in land use arise through alteration of the saturated and unsaturated flow processes. A knowledge of these processes is also required to assess the effects in time and in magnitude of agricultural applications of fertilizers and pesticides on groundwater and surface waters.

5. *Oxidation and reduction processes.* Cultivation and drainage of agricultural and forestry land alter oxidation and reduction reactions in soil profiles and redox potentials, which leads to the leaching of nutrients and acidification of streams.

6. *Ion exchange processes.* Catchments with high calcium content in soils and bedrock show less response to acidic pollutants, through neutralizing ion exchange processes; water acidity is reduced where drainage ditches or covered drains cut through basic mineral soil horizons.

Thus process studies must aim to identify and quantify the important processes operating in the catchment which influence the output variable under consideration. For studies of water quantity and quality the required measurements and methods may include the following:

Precipitation: Rain, snow, hail, mist, and fog and net precipitation beneath vegetation canopies.

Meteorological variables: Solar and net radiation, humidity, air temperature, windspeed, and wind direction.

Micrometeorological studies: Bowen ratio measurements of sensible and latent heat fluxes, and eddy correlation measurements of evaporation.

Plant physiological measurements: Leaf water potential, stomatal conductance, tissue osmotic potentials, and CO_2 assimilation rates.

Chemistry: Chemical concentrations in the atmosphere in the dry and wet form and in precipitation, net precipitation, soil water, plant water, and runoff.

Soil water status: Neutron probes, tensiometers, gypsum blocks, gamma transmission probes, capacitance probes, and time domain reflectrometry.

Tracing: Radioactive and stable isotopes, chemical tracers, heat as a tracer in flow measurements in plants and watercourses.

Processes occurring in the natural environment encompass many disciplines including meteorology, physics, chemistry, and plant physiology, and require multidisciplinary teams. Process studies are particularly valuable when associated with properly designed catchment experiments so that the reasons for the observed land-use change can be identified.

Effects of Land-Use Change on Aerodynamic Transport Processes. A change in land use can cause major changes in the aerodynamic transport processes operating above the surface, affecting both water quantity and quality. It is now recognized that for forests, higher evaporation rates in wet conditions (leading to high interception losses) and higher deposition rates of particulate, mist droplet, and some gaseous pollutants, as compared with those over grassland vegetation, are a direct result of the enhanced aerodynamic transport operating above the aerodynamically rough forest surface. This change in aerodynamic properties has major implications on both water quantity and quality. The interpretation and the application of the results from process studies on the effects of land-use change on rainfall interception, transpiration, and catchment acidification require a knowledge of aerodynamic transport theory.

Aerodynamic Transport Theory. The most widely used model for the transport of entities, whether momentum, gases, water vapor, particles, or heat, between the atmosphere and a surface involves a resistance analogy in which the flux of the entities is considered as analogous to the flow of electric current through a network of resistances (Fig. 13.2.2). With this Ohm's law analogy the flux is equivalent to current, concentration difference is equivalent to potential difference, and the resistance terms are equivalent to the resistance in a conductor. The transport process is usually considered as three stages: turbulent atmospheric transport between the well-mixed planetary boundary layer to the immediate vicinity of the surface against an *aerodynamic resistance* r_a, molecular diffusion through the surface boundary layer against a molecular *boundary layer resistance* r_b, and reaction or transport at or through the surface itself against a *surface resistance* r_s.

Aerodynamic Resistance. Aerodynamic transport within the planetary boundary layer occurs mainly through turbulent eddy diffusion. The aerodynamic resistance to transport by eddy diffusion can be calculated from a knowledge of the wind velocity and the aerodynamic roughness properties of the surface. In conditions without thermal buoyancy the aerodynamic resistance r_a to the transport of entities (assuming similar diffusion coefficients to momentum) between the surface and a

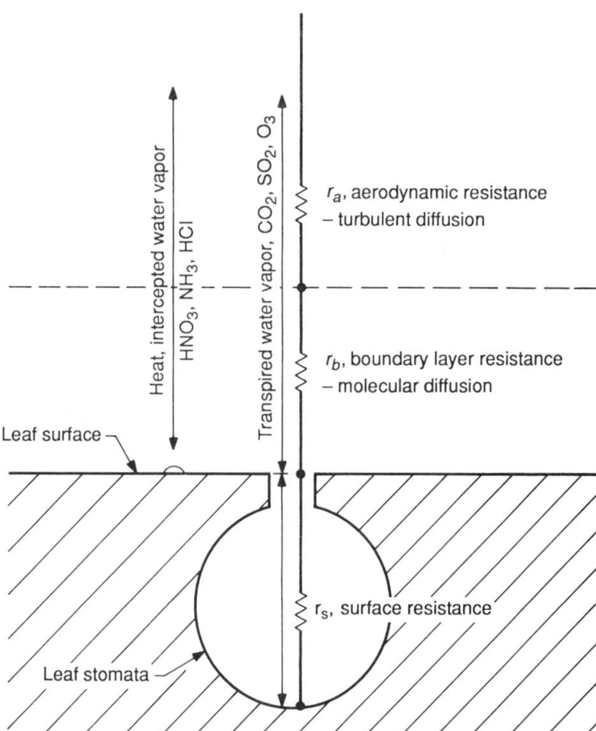

FIGURE 13.2.2 Schematic diagram of the resistances encountered in the transport of entities between the atmosphere and the outer leaf surface and from the atmosphere to within the stomatal cavity.

height z is given by[92,93]

$$r_a = \frac{[\ln(z-d)/z_0]^2}{k^2 u(z)} \tag{13.2.1}$$

where $u(z)$ is the wind velocity at height z; k is the dimensionless constant, von Karman's constant, the value of which has been determined experimentally as 0.41; and the terms d and z_0 relate to the height and roughness of the surface and are referred to, respectively, as the *roughness length* and the *zero plane displacement.* Approximate values for d and z_0 can be determined from the height of the vegetation h using the relation

$$d = 0.75h \tag{13.2.2}$$

and

$$z_0 = 0.1h \tag{13.2.3}$$

These equations imply an inverse relationship between the aerodynamic resistance and the mean windspeed and also a strong dependence with vegetation height such that the aerodynamic resistance for trees will be approximately an order of magnitude less than that for grass.

Boundary Layer Resistance. The different entities, including gases, particles, droplets, momentum, heat, and water vapor, are exchanged between the turbulent atmosphere and the surface through a viscous surface boundary layer within which there is laminar flow (Fig. 13.2.2). The boundary layer resistance has seldom been measured directly and is often estimated from empirical relations making allowance for the different diffusivities of gases and particles. Thom[92] has developed a theoretical basis for r_b and shown that the presence of "bluff body" forces allows momentum to be transferred through the boundary layer more effectively than heat or gases; that is, r_b is less for momentum than for other entities. For gases, transport through the viscous boundary layer of air close to the surface is by molecular diffusion and is therefore dependent on molecular weight.

For most entities, other than aerosol particles and some vapors, the boundary layer resistance is usually much smaller than the aerodynamic or the surface resistance and can then be ignored in comparison with these other, much larger, resistances.

Surface Resistance. Water vapor (transpiration) and CO_2 used in photosynthesis and other gases (SO_2 and O_3) are transferred from air to vegetation by diffusion through leaf stomata. This is a purely physiological resistance, termed the surface resistance r_s, and varies as leaf stomata open and close in response to changing environmental conditions.

Deposition Velocity. The flux F of pollutant particles or gases or other entities can be considered to occur down a concentration gradient between the atmosphere and the surface against an aerodynamic resistance, a boundary layer resistance, and a surface resistance acting in series:

$$F = \frac{\chi(z) - \chi(0)}{r_a + r_b + r_s} \tag{13.2.4}$$

where $\chi(z)$ = concentration of entity at height z.

The reciprocal of the total resistance to transfer is known as the *deposition velocity* $v_g(z)$ and is widely used to describe the deposition rates of pollutant reactive gases.

$$v_g(z) = \frac{1}{r_a + r_b + r_s} \tag{13.2.5}$$

When the surface concentration of the entity can be assumed to be zero, the deposition velocity may also be regarded as a flux normalized for concentration, i.e.,

$$v_g(z) = \frac{F}{\chi(z)} \tag{13.2.6}$$

A study of the transport processes operating both within and above the land surface can clarify the effects of a land-use change for both water quantity and water quality issues.

13.2.3 A Water Quantity Process Study

Plynlimon Process Studies. An example of the use of process studies in conjunction with catchment experiments to examine the effect on water use of land-use changes associated with forestry is provided by the studies carried out by the Institute of Hydrology (UK) at Plynlimon, Central Wales. The catchments are located at the source of the Wye and Severn rivers in steep upland topography. The precipitation is of the order of 2400 mm annually (Table 13.2.1) and is distributed throughout the year; most of the precipitation occurs as rain; the snow contribution per year is very variable but averages 5 percent. The rainfall is mostly of low intensity, generated from frontal systems enhanced by the orographic effect of the hills. Streamflow is perennial, although storm runoff forms a major proportion of the flow. The soils are predominantly peaty, overlying mudstone and shale drifts on the slopes, often with deep peat deposits of up to 3 m depth on the hilltops and in the valley bottoms where the peat overlies glacially deposited boulder clay. The Wye catchment is under grass cover; the Severn catchment has 70 percent coniferous afforestation, mostly Norway and Sitka spruce, and the remaining 30 percent is under moorland grass.

Measurements at Plynlimon. At Plynlimon the processes controlling transpiration were measured with a "natural" lysimeter (Fig. 13.2.3), together with neutron probe and tensiometric measurements of soil moisture,[11–13] plant physiological measurements of stomatal conductance and leaf water potential, and "tree cutting" measurements of water uptake from excised trees.[77] These experiments were located within the instrumented and forested Severn catchment at Plynlimon (Tables 13.2.1 and 13.2.2). Measurements of the processes controlling the interception of precipitation by the forest canopy were made using conventional throughfall troughs and stemflow gauges,[34] plastic-sheet net-rainfall gauges,[22] and gamma transmission measurements.[24]

Results at Plynlimon. These process studies conclusively demonstrated that at Plynlimon:

1. The reduced runoff per unit area from the forested Severn catchment compared with the grass-covered Wye catchment is principally the result of the increased interception losses from the forest. The higher interception losses are generated because of the increased turbulence and lower aerodynamic resistance to the transport of water vapor and heat between the forest surface and the atmosphere. This leads to higher evaporation rates from the forest in wet conditions compared with grassland. The enhanced evaporation rates occur both during rainfall and immediately afterward

FIGURE 13.2.3 A neutron probe soil moisture meter being used on the forest lysimeter, Plynlimon.

from the wetted vegetation surface; about half the interception loss by evaporation occurred during rainfall.

2. The transpiration from the forest was typically less by about 10 percent than that from grassland as a result of the lower stomatal conductance of the forest (at Plynlimon periods with soil moisture deficits sufficient to limit transpiration are not common).

Implications of the Plynlimon Process Studies. These two results: that interception losses from tall vegetation are likely to be higher than those from short vegetation

TABLE 13.2.2 Measurements from the Plynlimon Forest Lysimeter, February 1974–September 1976

Period	Precipitation	Interception	Transpiration
Feb. 6, 1974–Dec. 31, 1974	2328	685	289
Jan. 1, 1975–Dec. 31, 1975	2013	529	335
Jan. 1, 1976–Oct. 1, 1976	1103	366	277
Total	5444	1580	901

Units: millimeters.

The precipitation was recorded at the nearby Tanllwyth gauge within the Severn catchment.

Source: Reproduced with permission from Calder, Newson, and Walsh.[21]

and that, when soil moisture is nonlimiting, forest transpiration is likely to be similar to but less than "grass" have a general significance and can, with a few qualifications, explain the results from the majority of the world's "forest and grass" catchment experiments (see Hewlett and Hibbert[43]). A third generalization, applicable in more arid regions, where large soil moisture deficits occur, is that the greater rooting depth of forests, and greater soil water availability compared with grass and agricultural crops, leads to higher transpiration rates from forests. So, during drought periods evaporation losses from forests may also be higher than from grasslands, but for different reasons than during rainy periods.

The qualifications to these generalizations mainly concern the observations from four southeast Australian catchments[51] which, after forest fires, showed a decrease in runoff in subsequent years of 24 percent compared with runoff from a catchment that escaped the fire. This apparently anomalous result, one of the few examples of forest removal decreasing runoff, was explained by Greenwood[37] in terms of the unique forest structure. He pointed out that the forest, which was composed predominantly of exceptionally tall (98 m) *Eucalyptus regnans* (some of the world's tallest trees), had a minimal canopy and that this implied both low interception and transpiration losses. Following the fire the germination of seeds and the subsequent regrowth rapidly led to the leaf area per unit ground area exceeding that of the former forest canopy.

13.3 EFFECTS OF AFFORESTATION AND DEFORESTATION

13.3.1 Annual and Seasonal Flow, Measurement, and Prediction

Annual flow results from catchment experiments have been reviewed by Hewlett and Hibbert[43] and by Bosch and Hewlett.[5] From an analysis of results from 94 catchments worldwide Bosch and Hewlett concluded that:

1. Pine and eucalypt types cause an average change of 40 mm in annual flow for a 10 percent change in cover with respect to grasslands; that is, a 10 percent increase in forest cover will decrease annual flow by 40 mm, and a 10 percent decrease in cover will increase annual flow by the same amount.
2. The equivalent response on annual flow of a 10 percent change in cover of deciduous hardwood or scrub is 25 to 10 mm; that is, if 10 percent of a grassland catchment is converted to hardwood trees or scrub vegetation, the annual runoff will decrease by 10 to 25 mm.

Three generic methods, regression, semiempirical, and the more physically realistic methods based on the combination equation, are available for the prediction of annual flow differences following changes in forest cover. The selection of method depends upon the availability of input data and the accuracy and detail required in estimating the seasonal variation of effects.

Regression Approach. Making use of the analysis from 55 of the catchments reviewed by Bosch and Hewlett,[5] and from a further 10 catchments in the southern Piedmont, Trimble et al.,[94] using the regression method, showed that the annual reduction in stream flow in millimeters Y could be fitted to the percentage change of forest cover X via the constrained regression relationship:

$$Y = 3.26X \tag{13.3.1}$$

The fit gave an r^2 of 0.5, and the standard error of the estimate was 89 mm (Fig. 13.3.1).

Semiempirical Methods—Non-Soil-Water Limiting Conditions *Annual, Forest and Grass Model.* A semiempirical model was developed by Calder and Newson[17,19,20] for estimating both the annual and seasonal differences in runoff from afforested, upland catchments in the United Kingdom, which were previously under grass cover. These upland catchments receive rainfall throughout the year, and periods with large soil moisture deficits are uncommon. The model requires information on annual or daily rainfall, annual or daily Penman[69] E_T estimates of evaporation, and the proportion of the catchment with complete canopy coverage.

For the calculation of annual evaporation the assumptions inherent in the method are that:

1. Evaporation losses from grassland are equal to the annual Penman potential transpiration estimate for grass E_{Ta}.
2. Transpiration losses from forest are equal to the annual E_{Ta} value multiplied by the fraction of the year that the canopy is dry.
3. The annual interception loss from forest, with complete canopy coverage, is a simple function of the annual rainfall P_a (Fig. 13.3.2).
4. Soil moisture deficits are insufficient to limit transpiration from grass or trees in this (wet) area of the United Kingdom.

Annual evaporation is then given by

$$E_a = E_{Ta} + f(P_a\alpha - w_a E_{Ta}) \tag{13.3.2}$$

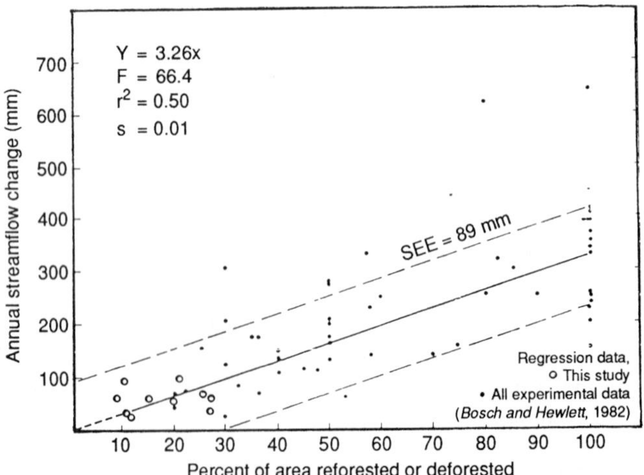

FIGURE 13.3.1 Reduction in stream flow against percentage increase of catchment forest area (or increase in flow against percentage decrease in catchment forest area) together with fitted regression. *(From Trimble et al.[94])*

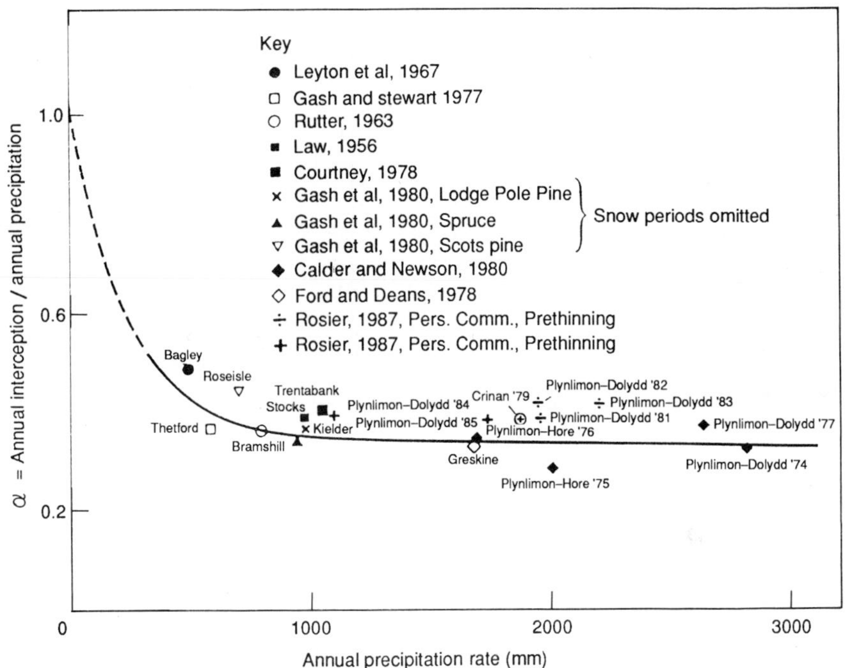

FIGURE 13.3.2 Observations of the annual fractional interception loss α from forests in the United Kingdom. *(Reproduced with permission from Calder.[17])*

where α = interception fraction (35 to 40 percent for regions of the United Kingdom where annual rainfall exceeds 1000 mm)

w_a = fraction of the year when the canopy is wet ($\sim 0.000122P_a$)

f = fraction of the catchment area under forest canopy cover

Use of aerial photographs has shown that, typically, for areas marked on maps as extensive forests, the f value is about 0.66, the remaining area comprising roads, gaps between forest blocks, riverbanks, clearings, and immature plantation with unclosed canopies.

The Calder-Newson model indicates that in the wet upland regions of the United Kingdom annual evaporation rates from forested catchments (with 75 percent of their area afforested, equivalent to 50 percent canopy coverage) may exceed those from grassland by 100 percent, and runoff will be reduced, typically by about 15 to 20 percent (Figs. 13.3.3 and 13.3.4).

This simple model was used[19,20] to investigate the effects on water supplies of afforesting the catchments of the major United Kingdom reservoirs (Fig. 13.3.4) and to provide information for the Centre for Agricultural Strategy's investigations into the feasibility of proposals to increase greatly the proportion of forestry in Britain. It has been used subsequently in many studies into the effects of afforestation on water resources in the United Kingdom.

Annual Forest and Grass Model—Worked Example. The Plynlimon catchments are used to provide an example of the use of the model.

The Severn catchment rainfall $P_a = 2298$ mm (from Table 13.2.1); the intercep-

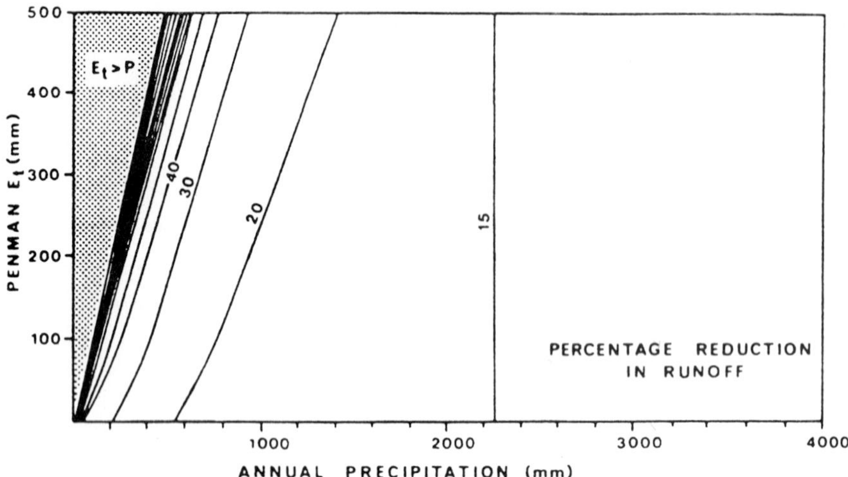

FIGURE 13.3.3 Predictions of percentage decrease in runoff as a result of afforestation to 50 percent canopy coverage plotted against the Penman E_T estimate and mean annual rainfall. *(Reproduced from Calder and Newson.[19] Used with permission.)*

tion fraction $\alpha = 0.35$ (from Fig. 13.3.2); then $w_a = 0.000122P_a = 0.281$; for the forested part of the Severn catchment, to which the data in Table 13.2.1 refer, the fraction of the catchment area under canopy coverage is taken as $f = 0.66$ (if the calculation was for the whole of the Severn catchment, with 70 percent forestry, 30 percent grass, then $f = 0.7 \times 0.66$).

The published value[58] of E_{Ta} for an altitude of 442 m in this area of Wales is 412 mm.

Calculation of Eq. (13.3.2) then gives for the forested part of the Severn catchment:

$$E_a = 867 \text{ mm, as compared with 797 mm measured (Table 13.2.1)}$$

and for the Wye catchment:

$$E_a = 412 \text{ mm, as compared with 370 mm measured (Table 13.2.1)}$$

The calculated difference in runoff between the forested and grassland catchments is $867 - 412 = 455$ mm, equivalent to 22 percent of the grassland catchment runoff, as compared with the measured value of 427 mm (21 percent of the grassland catchment runoff) given in Table 13.2.1. Thus good agreement is obtained between the predictions of the semiempirical model and the observations from a land-use catchment experiment.

Annual Forest, Heather, and Grass Model. Research on the evaporative characteristics of heather *Calluna vulgaris* in the United Kingdom uplands has established that transpiration losses are smaller but interception losses greater than those from grassland. These observations suggest[17,38] (Table 13.3.1) that the annual interception losses from heather can be estimated with an equation of the form

$$E_a = \beta E_{Ta}(1 - w_a) + \alpha P_a \tag{13.3.3}$$

where $\beta = 0.5$ and $\alpha = 0.2$.

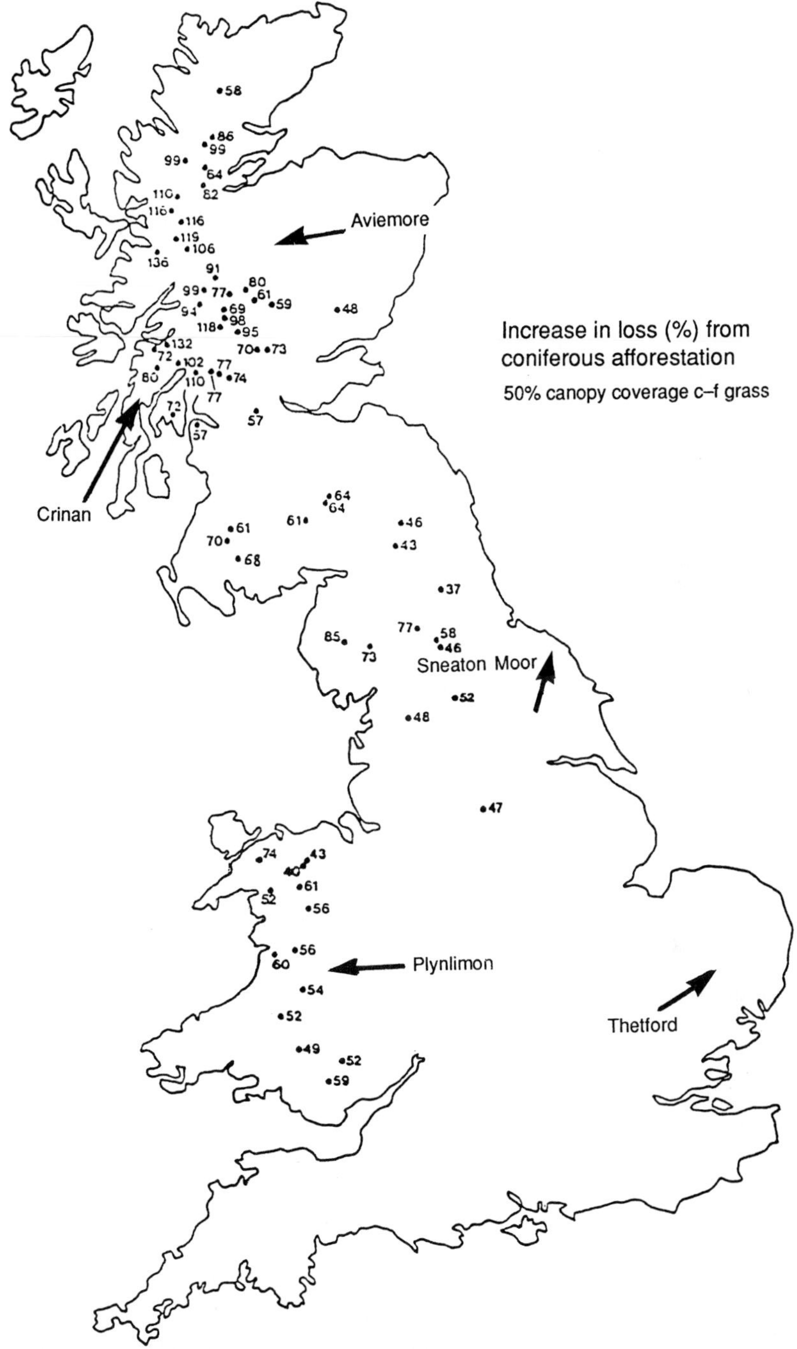

FIGURE 13.3.4 Predicted percentage increase in evaporation as a result of afforesting to 75 percent forest cover (equivalent to 50 percent canopy cover) the catchments supplying the major British upland reservoirs. Also shown are sites where forest, heather, and grass water use studies were carried out. *(Reproduced with permission from Calder.[17])*

TABLE 13.3.1 Interception and Transpiration Observations Summarized in Terms of the Average Interception Ratio α, the Daily Interception Model Parameters γ, δ, and the Ratio of Actual to Penman E_T Evaporation β

Source	Period	Interception Parameters			Transpiration fraction β
		α	γ, mm	δ (mm^{-1})	
Forest:					
All sites interception: Plynlimon, Dolydd, Crinan, and Aviemore		0.35	6.9	0.099	
Plynlimon forest lysimeter	1974–1976	0.30	6.1	0.099	0.9
Dolydd	1981–1983	0.39	7.6	0.099	—
Crinan	1978–1980	0.36	6.6	0.099	—
Aviemore	1982–1984*	0.45	7.1	0.099	—
Heather:					
Model estimate derived using automatic weather station data and measured interception parameters	1981	—	2.65	0.36	—
Crinan, neutron probe	1981–1983	—	—	—	0.58–0.67
Law's heather lysimeters	1964–1968	0.16	—	—	0.47
Sneaton moor lysimeter[100]	1980	0.19	—	—	0.25–0.5
Grass:					
Wye catchment Plynlimon indicates total annual evaporation consistent with					1.0

Source: Reproduced with permission from Calder.[17]
* Not including snow periods.

This equation indicates that similar increases in evaporation and reductions in runoff will be expected when forests replace either heather or grass moorland in regions which experience an annual rainfall of about 1250 mm. For annual rainfall greater than this, the higher interception losses from heather than from grass will outweigh the reduced transpiration, and the total annual evaporation from heather will be greater than that from grass; the converse is true for annual rainfall less than 1250 mm.

Seasonal Variation in Evaporation Differences. To investigate the seasonal variation of the effects of a land-use change among forest, heather, and grassland, the same approach may be adopted with the incorporation of an interception model operating on a daily time scale.

Interception is represented by a two-parameter exponential relationship of the form

$$I = \gamma[1 - \exp(-\delta P)] \tag{13.3.4}$$

where I is the daily interception loss, mm, P is the daily precipitation, mm, and with parameter values $\gamma = 6.91$ and $\delta = 0.099$ was found to give a good fit to coniferous forest interception losses recorded in the United Kingdom.[17]

Seasonal evaporative losses can then be obtained by summing daily evaporation estimates E_d as calculated by

$$E_d = \beta E_T(1 - w) + \gamma\,[1 - \exp(-\delta P)] \tag{13.3.5}$$

where the transpiration is estimated as the product of β, defined in Eq. (13.3.3), a climatologically derived daily Penman E_T estimate, and a term $(1 - w)$ which represents the fraction of the day that the canopy is dry and is able to transpire (where w is the fraction of the day the canopy is wet $= 0.045P$, for $P < 22$ mm; $w = 1$ for $P \geq 22$ mm, after Calder and Newson[20]). This equation has been validated by comparison with soil moisture and interception measurements at different sites in the uplands of the United Kingdom. Estimates of the α, β, γ, and δ parameters for the different vegetation types and the sources from which they were derived are shown in Table 13.3.1.

Applicability of the Seasonal Model. The seasonal model, as for the annual model, is appropriate for conditions similar to those of the British uplands where soil moisture stress is an infrequent occurrence. Furthermore the model as outlined is strictly applicable only to mature stands of vegetation in climates where snow is not a significant component of the annual precipitation. The treatment within the model of immature forests and the effects of snow is far from complete, but two working hypotheses have been proposed. For catchments with a high proportion of immature forest the fractional canopy coverage parameter f is assumed to be related to age on the basis of an S-shaped function with $f = 0.1$ for trees aged 0 to 5 years, 0.33 (6 to 10 years), 0.75 (11 to 15 years), 0.95 (16 to 20 years), and 1.0 for trees older than 20 years. The interception losses from forests in both snow and rain conditions are assumed to be the same (the data from which the interception parameters were derived included some snow periods). The equivalent "working assumption" for snow interception from heather or grass is that the average daily evaporation rate is zero.

Within their range of applicability the semiempirical methods may actually provide more accurate estimates of long-term evaporation losses than the more theoretical methods. This is because of deficiencies in the instrumentation used for measuring the meteorological variables which are used in the more theoretical models. This is particularly marked when interception losses are being estimated in wet conditions of low humidity deficits, as evaporation estimates calculated from the Penman-Monteith equation then become extremely sensitive to absolute errors in the measurement of wet-bulb depressions (which are used for the calculation of the atmospheric humidity deficit term[12]).

Semiempirical Methods—Soil Water Limiting Conditions. In many areas of the world, soil water limitations will have a profound effect on transpiration rates of crops. These effects may be taken into account using a moderator function or "root constant" where the actual evaporation is equal to the product of the moderator and the potential value. Penman[70] originally proposed that actual transpiration remained at the potential rate until the soil moisture deficit exceeded the root constant by 25 mm (the root constant for grass in southern England was taken as 75 mm); thereafter the transpiration was assumed to occur at one-twelfth of the potential rate. Various soil moisture moderator functions have since been proposed, the simplest of which assumes that the moderator is equal to the ratio of the water content to the available water capacity of the soil profile; the most complex functions assume the profile is composed of different layers with different available water capacity values and different moderator functions. The layer models have the advantage that following rainfall into a previously dry soil the model will allow evaporation at the potential rate if the surface layer is wet even though the total soil profile remains at a high deficit.

Soil moisture modelling studies under grassland in the dry east of England[18] showed that the choice of soil moisture regulating function and, in particular, the estimate of the maximum soil water available to the roots (available water capacity) were more important than the choice of method for estimating the potential transpiration.

The importance of soil moisture limitations in regulating evaporation from tree and agricultural crops has been demonstrated[23] in the drier areas (800 mm annual rainfall) of Karnataka state, southern India. Here the fundamental control on dry season evaporation is soil water availability. Indigenous forest and eucalypt plantations have deep root systems and are able to utilize virtually all of the seasonal rainfall, whereas annual agricultural crops with shallower root systems utilize typically only 60 percent of the rainfall.

Semiempirical Methods—Water Use of Immature Plantation. The estimation of interception losses from immature plantations has been discussed above. It is expected that transpiration from forest plantations will also be related to tree size and age. Measurements made on *Eucalyptus* plantations in Karnataka[23] have established a very close correlation (Fig. 13.3.5) between the transpiration rate of an individual tree and its stem cross-sectional area (a better correlation than was found with leaf area).

This relationship, when expressed in terms of the total stem cross-sectional area of the stand per hectare, and with the use of a suitable soil moisture regulating function, enables the stand evaporation to be calculated and has been used in models to predict the evaporation, the soil moisture deficit, and the volume growth. The only meteorological data required is daily rainfall. Meteorological demand, although providing the driving force for evaporation, is not thought to be a limiting factor during most of the year; the principal limitations on transpiration are thought to be soil moisture availability and tree size. These results from semiarid Karnataka, which indicate that

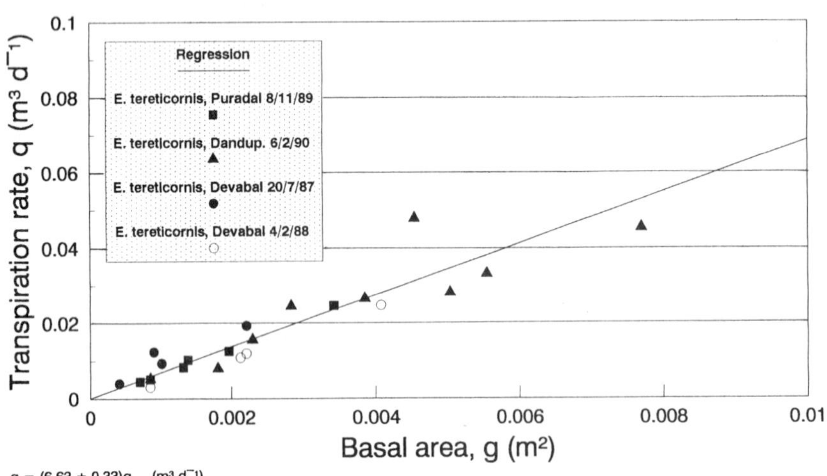

FIGURE 13.3.5 Transpiration rate of *Eucalyptus tereticornis* trees, measured using the deuterium tracing method at sites in southern India, plotted against the basal (stem) cross-sectional area of the tree measured at 1.2 m above ground level. The gradient of the regression line, 6.63 ± 0.33 m day^{-1}, has units of velocity and represents the Darcy velocity of sap flow in the stems. *(From Calder et al.*[23] *Used with permission.)*

evaporation is limited principally by soil water availability and plant physiological controls, are therefore in direct contrast to the observations from the wet uplands of the United Kingdom where evaporation is principally limited by atmospheric demand and physical, aerodynamic controls.

Penman-Monteith Approach. The third and most rigorous method of estimating the effects of a land-use change on evaporation involves the use of the combination equation (e.g., the Penman-Monteith[59] equation, see Chap. 4) in conjunction with models which take account of both the feedback effects of surface evaporation on atmospheric moisture content and the stomatal feedback effects between atmospheric moisture content and surface evaporation.

This approach gives the greatest insight into the mechanisms involved and is quite general in its applicability, but the detailed meteorological data that are required generally restrict its use to research applications. The method requires estimates of the aerodynamic resistance and surface resistance parameters in both wet and dry conditions together with (preferably hourly) measurements of net radiation, air temperature, vapor pressure deficit, and windspeed. Unless direct measurements are available of the surface resistance term from, for example, porometry measurements, it is necessary to obtain estimates using submodels which take into account the controlling environmental variables, particularly soil moisture deficit and atmospheric humidity deficit. When the method is used operationally, empirical relationships (similar to those described above for the semiempirical models) need to be incorporated. An example of the use of this approach, in which the feedback between atmospheric humidity and surface resistance is taken into account through a surface resistance submodel, is provided by the Plynlimon forest lysimeter studies[12,13] for which the hourly surface resistance r_s is described in relation to the day number of the year D and the hourly value of the vapor pressure deficit δe (kPa) as

$$r_s = \frac{74.5\left\{1 - 0.30\cos\left[2\pi\left(\dfrac{D-222}{365}\right)\right]\right\}}{1 - 0.457\delta e} \qquad \text{for } \delta e < 2.2 \text{ kPa}$$

$$r_s = \infty \qquad \text{for } \delta e \geq 2.2 \text{ kPa} \tag{13.3.6}$$

In wet conditions the surface resistance r_s is zero and an interception submodel is required to predict the duration of the wet conditions.[15,17,83]

13.3.2 Water Yield Management

Water yield management aims to control the seasonal and annual discharge from a catchment by managing the vegetation and its location within the catchment.

Annual Flow Increase. Increases in annual flow resulting from cutting trees or removing scrub vegetation will be most marked in both very high and very low rainfall areas. In very high rainfall areas forest evaporation will be higher than that from other land uses because of rainfall interception losses; in very low rainfall areas forest evaporation is likely to be higher than that from other crops because forests, with deep root systems, are better able to exploit soil water reserves.

Annual Flow Increase in High-Rainfall Areas. The upland areas of the United Kingdom are a prime example of high-rainfall areas. Their maritime climate is typified by high rainfall, a high number of rain days per year, and high windspeeds.

TABLE 13.3.2 Observations of the Annual Water and Energy Balance of Moist Tropical and Temperate Forests

Site	Rainfall, mm	Transpiration, mm	Interception, mm	Total evaporation, mm	Radiation, mm equivalent
Indonesia, West Java, Aug. 80–July 81[25]	2835	886	595 (21%)	1481 ± 12%	1543 ± 10%
Brazil, Amazonia, 1984, Reserve Ducke Forest[87]	2593	1030	363 (13%)	1393	1514
Wales, Plynlimon, 1975[13]	2013	335	529 (26%)	864	617

The Amazonia site experiences dry periods which may limit transpiration.

Total evaporative losses from forest can consume an amount of latent heat that easily exceeds the radiant energy input to the forest (Table 13.3.2); the extra source of energy is provided by advection, the extraction of heat from the air mass as it passes over the forest. In the highest rainfall areas (3000 mm year^{-1}), it has been calculated (using the Calder-Newson model) that from mixed-age wet evergreen coniferous forest (with 50 percent canopy coverage) the increased stream flow resulting from a conversion to grassland would typically be about 450 mm year^{-1}.

The effects of timber cutting on annual flow are most marked in areas where regrowth is slow. In the Rocky Mountains of the United States Anderson and colleagues[1] state that there is compatibility between the management of water and other resources; timber regrowth following cutting is slow, and as a result, increases in water yield from timber cutting last 3 to 5 times longer than in other parts of the United States.[45]

The wet evergreen forests of the tropics represent another land use where significant increases in yield, through reduction of interception losses, might be expected following conversion to grassland or agriculture. Here, although the forests are not able to make use of large-scale advection to support high rates of interception, as in the uplands of the United Kingdom (because the climate circulation patterns are different), there is evidence that humid rain forest is able to convert, on an annual basis, virtually the equivalent of all the net radiation (Table 13.3.2) into evaporation. It is therefore unlikely that any other land use will be able to evaporate at a higher rate, and conversion to other land uses will therefore result in increased annual flow.

Annual Flow Increase in Low-Rainfall Areas. In very low rainfall areas the principal limit on annual evaporation is likely to be soil water availability. Studies in Karnataka, southern India,[40] show that the available soil water capacity of both indigenous, dry deciduous forest and *Eucalyptus* plantation is of the order of 480 mm, whereas in the same region, the available water capacity for finger millet, an annual agricultural crop, is 150 mm. The annual evaporation from the indigenous and plantation forests is, within the errors of measurement (10 percent), equal to the rainfall of 800 mm year^{-1}; the evaporation from the finger millet, with a reduced soil water reservoir to exploit, is 500 mm year^{-1}. Conversion from forest to agricultural crops in this area will therefore increase annual flow (or catchment recharge) by this difference in annual evaporation.

Aspect and Location. The aspect and location of forestry within a catchment will have an effect on annual (and seasonal) flow patterns.

The evaporation from short crops, e.g., grass, is determined largely by the net radiation that they receive. Mountainous south-facing catchments in the northern hemisphere will therefore receive more radiation and have higher evaporation, and less annual flow, than those facing north. The evaporation from forests, because of their aerodynamically rough surfaces (and low aerodynamic resistance) is controlled more by the prevailing atmospheric humidity deficits than by net radiation (i.e., the aerodynamic term in the Penman-Monteith[59] equation is generally greater for forests than the radiation term) and will not, therefore, be affected greatly by aspect. Thus water yields from south-facing grassland catchments (in the northern hemisphere) are likely to be less than those from those facing north (roughly in proportion to the difference in the annual net radiation input) whereas for forested catchments the aspect effect on water yield will be much less.[57]

The location of forestry within the catchment may have important effects on base flow. On the mountainous Coweeta catchments in America, where base flow is derived from soil water drainage adjacent to streams,[42] conversion from forest to grass at the base of slopes is expected to increase base flows significantly,[53] whereas tree cutting in other parts of the catchment is expected to produce a much lesser effect.

Seasonal Flow. Afforestation may affect dry season flow through two principal mechanisms. First, the higher interception losses from forests in wet periods and increased transpiration losses in dry periods (because of deeper root systems) both tend to increase soil moisture deficits in dry periods compared with those under shorter crops. These increased deficits lead to reduced dry season flows where part, at least, of the dry season flow is derived from the soil moisture reservoir.[53,81] Second, land drainage operations, which are often part of the management associated with afforestation in wet, temperate climates, tend to increase flows as a result of both the initial dewatering (which may take a number of years) and the long-term effects of the alteration of the drainage regime (see Sec. 13.5.2). The two mechanisms are opposing and the net effect on low flows may result in either higher or reduced low flows.

For high-altitude forest, which is above the cloud base for a significant proportion of the year, or *cloud forest,* the deposition of cloud water onto the forest is likely to be a significant hydrologic process.[26] Because of the reduced aerodynamic resistance r_a and increased leaf area of forest, compared with shorter crops, the cloud deposition rates onto forest will be many times greater than those onto short vegetation. For cloud forest in locations such as the Andes, Hawaii, and Sri Lanka cloud water deposition may provide a significant component of the dry-season flow in rivers. The gains arising from the deposition of cloud water onto the forest may compensate or even exceed the losses from rainfall interception. In these situations removal of the forest may lead to a reduction in annual water yield.

In regions where high-intensity rainfall may induce surface runoff the higher infiltration rates under indigenous forest compared with other land uses may result in higher recharge to the aquifer. Afforestation may also have the additional benefit in steeply sloped areas of reducing landslips and preserving the soil aquifer which may be the source of dry-season flows. Both these effects of afforestation may therefore have beneficial effects on low flows, but these benefits are likely to be very site-specific and will accrue only in particular regions of the world.

No altogether satisfactory model for predicting the effects of forestry on low flows currently exists. Ideally it would be necessary to take into account the effects of the forest on the development of soil moisture deficits through, for example, the Calder-

Newson model, and the effects of these deficits on the contributing areas for base flow, in conjunction with the effects of any associated land drainage system. Uncertainty in the identification of contributing areas for base flow in any particular catchment is likely to limit the practical usefulness of such an approach in predicting the absolute magnitude of the effect, although it may be adequate, through sensitivity analysis, to indicate the range of possible effects.

13.3.3 Forestation and Floods

Processes Affecting Floods. Afforestation or deforestation affects flood generation both directly, through the presence or absence of the forest itself, and indirectly, as a result of associated forest management practices.

Tree interception mitigates flooding by removing a proportion of the storm-producing rainfall and by allowing, over periods with small rain events, the buildup of soil moisture deficits. Usually these effects are of minor importance in the largest storm and flood generation events. Paired-catchment studies in Bavaria[81] (Fig. 13.3.6) show clearly the magnitude of the effect: following tree planting there is a reduction in peak flows with forest growth; most of the reduction occurs over the first 10 years, and the reduction is proportionately much greater for the smaller storms.

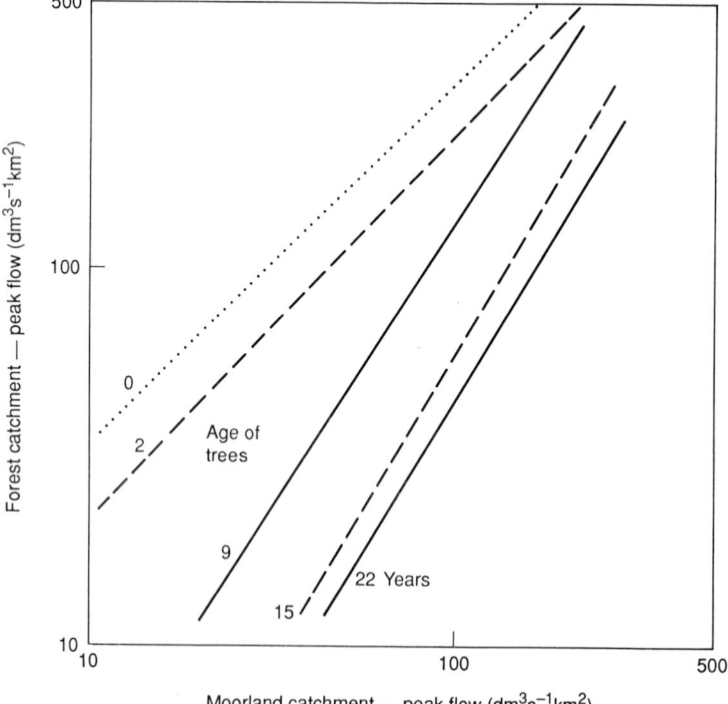

FIGURE 13.3.6 Comparison of peak instantaneous discharges for different ages of forest with an untouched moorland control catchment in the southern Chiemseemoors, Bavaria, Germany. *(Reproduced with permission from Robinson et al.[81])*

The management practices associated with forestry activities—the creation of roads, culverts, drainage ditches and logging and clearing activities and grazing which result in soil compaction and reduced infiltration rates of soils—are potentially more important in influencing flood generation than the presence or absence of the forest itself.

Douglas[30] reports that for the Coweeta catchments in the United States the effects of cutting and harvesting timber on the storm hydrograph are cumulative; stormflow volume, peak flow, and storm flow duration are all generally increased by timber cutting. The effect is greatest during the harvest and declines logarithmically with time as vegetation regrows.

Evidence of the effects on flood peaks of large-scale deforestation in the tropics is conflicting. Gentry and Parody[35] attributed an increase in the height of the flood crest for the Amazon at Iquitos during the period 1970–1978 to deforestation in the upper reaches of the basin. Nordin and Meade[66] contested this conclusion, attributing it to periodic (climatic) fluctuations unrelated to deforestation.

In the United Kingdom proved changes to the flood response in relation to forestry are associated almost entirely with ground preparation that involves cultivation and drainage of poorly drained soils.[63] The United Kingdom studies reveal a shorter time interval between rainfall and flood peaks after ground preparation than before, and because volumes are undiminished, peak rates of runoff are therefore higher. The results obtained by Robinson[79,80] showed that time to peak reduced from 2.1 to 1.6 h after (open) drainage operations and peak flows increased by 40 percent, but both effects diminished slightly 10 years after the implementation of drainage.

Flood Mitigation. The USDA Forest Service[1] recommends three methods to mitigate the effects on floods resulting from forest harvesting by:

1. Reducing the disturbed area by choice of a *yarding* method (i.e., logging in some areas, leaving other areas undisturbed)[75,76]
2. Rehabilitating logged areas that produce overland flow
3. Providing filter strips between bared areas and streams to keep soil out of stream channels and thus maintain channel capacity

There are also a number of management options for mitigating flooding from existing forests: reduced grazing diminishes soil compaction and overland flow; reducing forest fires decreases erosion and the clogging of river channels. Snowmelt flood events can also be reduced through forest management.[36] Forests on south-facing slopes (in the northern hemisphere), if managed to produce wide-open strips running north and south, will produce seasonally early runoff; conversely on a north-facing slope with continuous forest cover the melt will occur later. The snowmelt from different parts of the catchment can therefore be "desynchronized" so that runoff is spread out over a longer time period and flood peaks are diminished.

In principle, "desynchronization" of flood peaks could also be achieved through land drainage; a land drainage scheme near the catchment outflow would accelerate the arrival of flood peaks from this area and desynchronize them with the peak flow, arriving later, from the rest of the catchment.

13.3.4 Forests and Water Quality

Forests affect stream acidification through their interaction with atmospheric pollution and also through nutrient leaching.

Acidification. The sensitivity of a catchment to acid rain is related to the calcium content of its soils and bedrock; the higher the calcium content, the greater is the ability of the soil to neutralize the effects of acidic pollutants (acid rain). There is now evidence that from catchments with the same soil sensitivity, stream flow from those under afforestation is more acid and has higher concentrations of aluminum than does flow from catchments draining grassland or moorland.[64,74,90,95] The differences in water chemistry between land uses are most marked at high flows which typically show a difference of one unit of pH and a fourfold difference in aluminum concentration between forest and grassland. Afforested catchments also show more rapid variation in pH[62,102] with time. Concentrations of chloride, sulfate, and nitrate are also higher in forested catchments and are proportional to the area of the catchment under afforestation.[101]

The increased acidification of runoff from forested catchments is generally attributed to the increased deposition rates of pollutants onto the forest canopy. The greater leaf areas, per unit ground area, and smaller aerodynamic resistances of forests lead to higher capture rates of pollutants through impaction of cloud droplets and dry particles.

Acid Pollutants. The burning of fossil fuels, forests, field stubble, and industrial waste produces pollutants in the gaseous form as hydrocarbons, nitrous oxides, and sulfur dioxide and in the particulate form as smoke. Sulfate and nitrate are formed in the atmosphere by the oxidation of SO_2 and NO_2. The oxidation may occur through photochemical processes (Fig. 13.3.7) or it may take place in cloud or mist droplets where the acid nuclei are neutralized by metallic cations or ammonia. Sulfate and nitrate particles are hygroscopic and at high relative humidities form mist droplets. (Ammonium sulfate forms 60 percent of the total inorganic aerosol in country districts in Europe.) In conditions of high atmospheric humidity (high vapor pressure) hygroscopic aerosol particles grow into droplets.

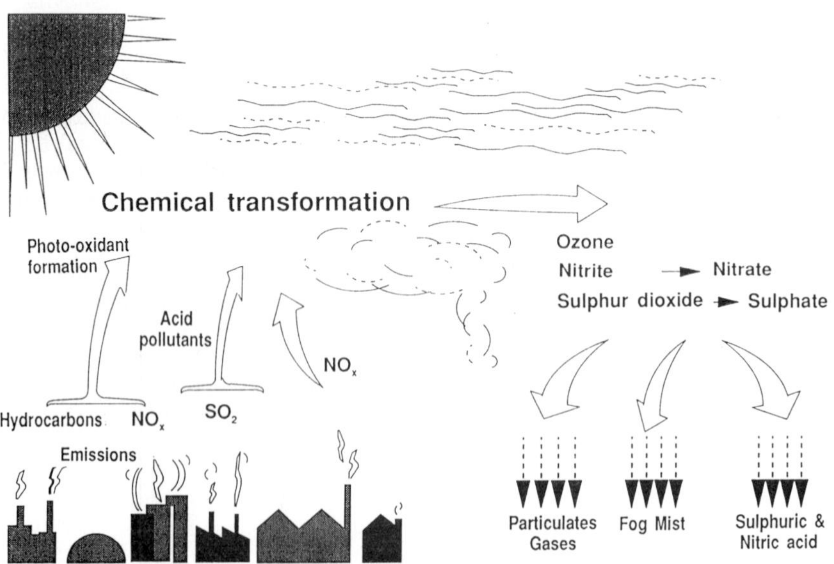

FIGURE 13.3.7 Sources and deposition pathways for acidic atmospheric pollutants.

Deposition Processes. Atmospheric deposition can occur in the wet form in rain, snowflakes, or cloud droplets, and in the dry form as aerosol particles or gases. Deposition rates of pollutants in rain or snow are largely independent of the surface over which they occur, while deposition rates of pollutants contained in particles, gases, and cloud droplets depend on many pollutant and surface properties including the particle or droplet size, the reactivity of the gas, the aerodynamic transport resistance in the atmosphere (which is itself dependent on the aerodynamic roughness of the surface and the windspeed), the molecular boundary layer resistance at the surface, and for some gases which are absorbed through leaf stomata, the leaf stomatal resistance (Fig. 13.2.2). Particle deposition rates on surfaces are also influenced by the processes of impaction, bounce-off, and blow-off which are all dependent on surface properties. Measurement difficulties still preclude the direct measurement of some of these forms of atmospheric deposition, and estimates of the effects of land-use change on deposition rates in the gaseous and particulate form and wet deposition in the mist droplet form are usually obtained through modeling studies and a knowledge of atmospheric concentrations. The recognized deposition processes leading to acidification are:

1. Wet deposition in precipitation of H^+, SO_4^{2-}, NO_3^-, NH_4^+, Na^+, Cl^-, Mg^{2+}, and Ca^{2+} ions
2. Cloud and fog precipitation involving the impaction of cloud droplets from ground-level clouds
3. Dry deposition of SO_2, NO_2, HNO_3, NH_3, HCl, and O_3 gases
4. Dry deposition of sulfate- and nitrate-containing particles

Wet Deposition in Rain and Snow. Wet deposition rates of pollutants in rain and snow precipitation are the rates most easily measured and most frequently reported, and these rates are independent of the surface over which the precipitation occurs.

Wet Deposition in Clouds and Fog. Cloud and fog water contain significantly larger ionic concentrations than rain (Table 13.3.3) with peak concentration up to 50 times greater.[96] Recent studies[32] indicate that for high-altitude forests in the United Kingdom (~500 m), altitudes sufficient for forests to frequently intercept cloud and mist droplets, the deposition of sulfur particles contained within cloud droplets (5 to 10 μm radius) may make a large contribution to the total annual deposition. Because cloud droplets, as opposed to submicron-sized dry particles, are efficiently captured by vegetation surfaces (the effective r_b is low), and as forests have lower aerodynamic resistances compared with shorter vegetation, deposition rates of cloud-borne pollutants onto forests will be greater than deposition onto shorter crops.

TABLE 13.3.3 Deposition of Reactive Gases HNO_3, NH_3, and HCl onto Different Vegetation Types

	Short grass	Moorland	Cereal	Forest
Height, m	0.1	0.4	1.0	10.0
Deposition velocity v_g, mm s^{-1}, at windspeed of 1 m s^{-1}	5.2	7.6	14.3	40.0
Deposition velocity v_g, mm s^{-1}, at windspeed of 4 m s^{-1}	23.5	33.3	50.0	100.0

Source: From the United Kingdom Review Group on Acid Rain.[96] Used with permission.

Dry Deposition of Reactive Gases. As the reactive gases HNO_3, NH_3, and HCl are readily absorbed by the outer leaf surfaces of vegetation the surface resistance r_s to their absorption is essentially zero and consequently the deposition velocity onto canopies is controlled by the aerodynamic resistance r_a [Fig. 13.2.2, Eq. (13.2.4)]. Deposition rates and deposition velocities will therefore be higher onto forests, with aerodynamically rough surfaces and low aerodynamic resistances, than onto shorter, smoother crops (Table 13.3.3).

Dry Deposition of Absorbed Gases. The gases SO_2 and O_3 are absorbed through stomata and are subject, during their transfer, to the extra stomatal (surface) resistance term r_s [Fig. 13.2.2, Eq. (13.2.4)]. Deposition rates and deposition velocities of absorbed gases will therefore vary less between vegetation types[31,33,44] than for the reactive gases as stomatal resistance varies less than aerodynamic resistance between vegetation types. The exchange of NO_2 and NO gases between the atmosphere and vegetation is poorly understood. Physiological processes within plants may produce

TABLE 13.3.4 Pollutant Concentrations in Air, Rain, and Cloud Droplets and Calculated Deposition Loads to Forest and Moorland at Kielder in Northern Britain

Annual average pollutant concentrations in air, parts in 10^9 by volume

SO_2	NO_2	HNO_3	NH_3	HCl^-
2.0	5.0	0.3	3.0	0.1

Rain and cloud droplet pollutant concentrations, μeq dm^{-3}

	SO_4^{2-}	NO_3^-(N)	NH_4^+(N)	Cl^-	Na^+	Mg^{2+}	Ca^{2+}
Rain	46	15	21	64	58	11	10
Cloud water	93	42	50	166	139	26	24

	Wet deposition	Dry deposition	Cloud water deposition	Total deposition
Deposition rates to forest, g m^{-2} year^{-1}				
SO_4^{2-}	1.31	0.31 (SO_2)	0.65	2.27
NO_3^-(N)	0.35	0.42 (NO_2, HNO_3)	0.09	0.86
NH_4^+(N)	0.45	0.93 (NH_3)	0.10	1.48
Cl^-	4.98	0.20	0.86	6.04
Na^+	2.81	—	0.46	3.27
Mg^{2+}	0.33	—	0.04	0.37
Ca^{2+}	0.62	—	0.07	0.69
Deposition rates to moorland, g m^{-2} year^{-1}				
SO_4^{2-}	1.31	0.31 (SO_2)	0.13	1.75
NO_3^-(N)	0.35	0.22 (NO_2, HNO_3)	0.02	0.59
NH_4^+(N)	0.45	0.18 (NH_3)	0.02	0.65
Cl^-	4.98	0.04	0.17	5.19
Na^+	2.81	—	0.09	2.90
Mg^{2+}	0.33	—	0.01	0.34
Ca^{2+}	0.62	—	0.01	0.63

Source: Reproduced from Ref. 32 with permission.

NO_2 and NO, so where atmospheric concentrations of these gases are low, deposition rates may be negative; that is, the net flux of the gases may be directed away from the surface.

Dry Deposition of Particles. Aerosol particles of SO_4^{2-}, NO_3^-, and NH_4^+ occur mainly in the size range 0.1 to 1.0 μm. For particles of this size deposition rates are limited by the high resistance encountered in transport through the viscous surface molecular boundary layer r_b (Fig. 13.2.2). The particles are too large to diffuse efficiently by Brownian diffusion and are too small to impact efficiently on most natural surfaces.[27] Dry deposition of these particles is now thought to be a relatively inefficient removal process.[96] The deposition of these aerosol particles is thought to occur mostly as wet deposition in rain or cloud drops subsequent to the nucleation of the cloud droplet.

Calculation of Deposition Rates of Atmospheric Pollutants. The measured pollutant concentrations and calculated deposition rates for forest and moorland at Kielder in northern Britain are shown in Table 13.3.4.[32] Inputs of nitrogen and sulfur to the moorland are calculated at 1.24 and 1.75 g m^2 $year^{-1}$, respectively.

Deposition rates to forests will be greater by 90 and 30 percent, respectively. For forest at Kielder, which is assumed to be in cloud for 1000 h per year (possibly an overestimate), the wet deposition of cloud drop sulfur is an important deposition process and is largely responsible for the predicted increased deposition onto forests compared with moorland. Recent studies have shown that although deposition rates onto forests are generally higher than onto shorter vegetation, actual rates vary quite widely between sites as a result of variation in pollutant concentrations and other site factors including altitude and annual rainfall (Fig. 13.3.8).

Nutrient Leaching. Studies of the effects of land use on water quality as characterized by the concentrations of the nutrients *N*, *P*, and *K* generally indicate much lower values in water emanating from forests than from agricultural lands (Table 13.3.5).[64] The differences between land uses are largely a reflection of the much higher fertilization rates and intensity of management on agricultural lands.

The most disruptive effects of forestry on water quality arise through intensive management practices associated with harvesting, site preparation, and site management. In particular, clear cutting can result in large increases in nutrient concentrations in watercourses. The highest concentrations reported in the United States are from forests in New Hampshire. Hornbeck et al.[46] and Pierce et al.[72] report increased nutrient values of 26 mg dm^{-3}. More commonly values of about 1 mg dm^{-3} have been reported for other forests in America.[3,54] The increased nutrient concentrations affect lake and stream eutrophication and increase the outbreaks of phytoplankton blooms (Table 13.3.5).

13.3.5 Afforestation and Erosion

Forestry operations are often associated with increased erosion. Land drainage operations prior to afforestation, the construction of access roads, and felling operations involving soil compaction and disturbance all increase erosion as they do flooding. Similar management practices (see Sec. 13.3.3) can be employed to mitigate these increases.

The presence of the forest also affects erosion, principally through the effects on slope stability and on splash detachment.

Slope Stability. O'Loughlin and Ziemer[67] state that the positive influences of forests on erosion depend upon the reduced soil pore water pressure caused by the forest

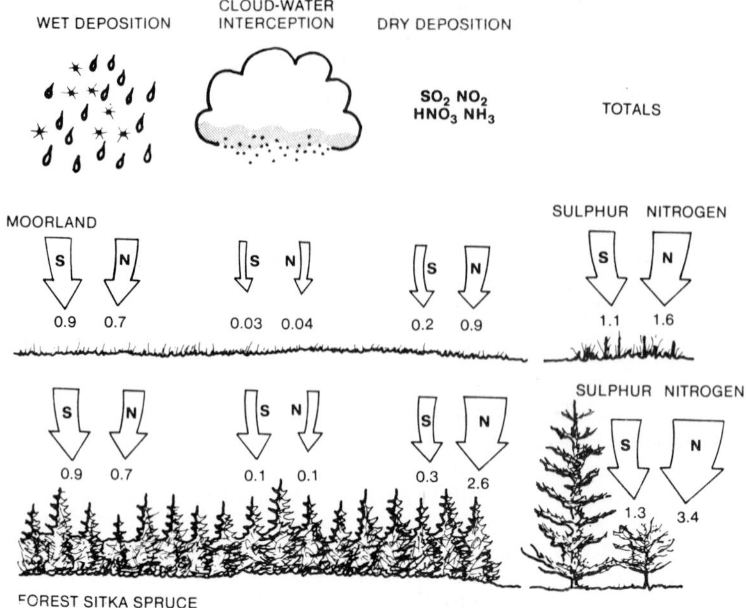

(a) ESKDALEMUIR 259 m (asl), 1400 mm annual precipitation

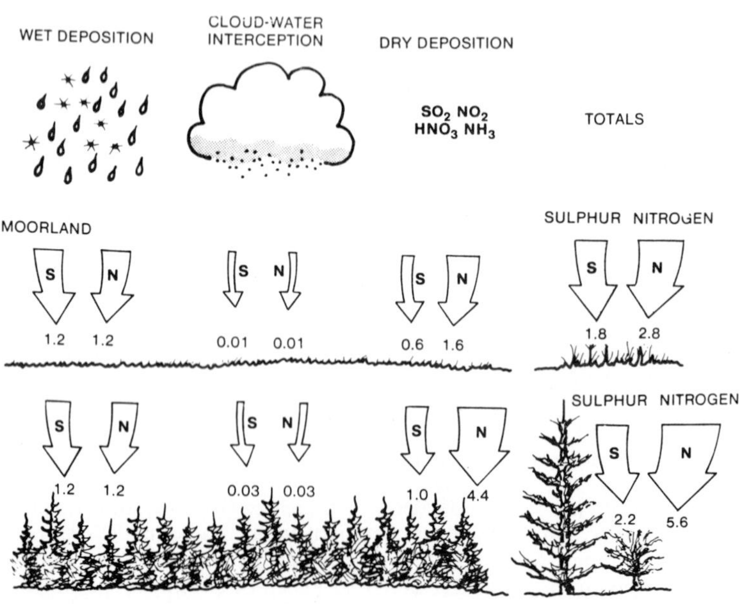

(b) HIGH MUFFLES 267 m(asl), 865 mm annual precipitation

FIGURE 13.3.8 Calculated deposition rates of atmospheric *S* and *N* (g m^{-2} $year^{-1}$) onto forest and moorland at Eskdalemuir and High Muffles sites in the United Kingdom. *(From Garland.[33] Used with permission.)*

TABLE 13.3.5 Effects of Different Land Use on Water Quality: Losses and Concentrations of *N*, *P*, and *K*

	N	*P*	*K*
Losses, kg ha^{-1} year^{-1}			
Forest:			
Slash and loblolly pine, U.S.	0.32	0.04	—
Deciduous hardwood, U.S.	2.0	—	—
Mixed coniferous, Canada	1.7*	0.0	—
Native evergreen, New Zealand	1.45	0.12	—
Eucalypt, Australia	0.12	0.004	—
Spruce, U.K.	16.1	—	5.7
Aspen, U.S.	0.3	0.1	4.4
Evergreen oak, Spain	0.03	0.01	0.3
Agriculture:			
Arable land, U.K.	4–13*	0.06*	—
Pasture, U.K.	8*	0.05*	—
Various intensive, U.S.	2–38*	0.2–1.2*	—
Pasture, U.S.	2–12*	0.1–4.6*	0.1–0.6
Maize, U.S.	2–62*	0.1–2.5*	0.4–26
Concentration in runoff, mg dm^{-3}			
Forest:			
Slash and loblolly pine, U.S.	0.08	0.01	—
Deciduous hardwood, U.S.	1.47	0.008	0.23
Mixed coniferous, Canada	0.23	0.01	—
Eucalypt, Australia	0.01	0.003	—
Rain forest, Amazon	0.004	0.01	0.15
Pine plantation, Florida	0.01	0.00	—
Agriculture:			
Grassland, U.K.	4.1*	0.09*	—
Arable, U.K.	9.0*	0.02–1.7*	—
Various intensive, U.S.	5–25*	0.13–0.33*	4.0–11
Threshold for phytoplankton blooms	0.3–0.65*	0.01–0.03*	—

Source: From Newton et al.[64] Used with permission.
* Total *N* or *P* (otherwise quoted as nitrate or phosphate ions).

evaporation, accumulation of an organic forest floor layer, and mechanical reinforcement of the soil by tree roots. Negative influences result from windthrow of trees and the weight of the tree crop itself.

Splash Detachment. Vegetation canopies influence splash detachment of soil particles through the modification of the natural raindrop size spectrum.

For storms with small raindrop sizes, usually low-intensity storms, canopies tend to amalgamate drops until vegetation elements are fully wetted and larger drops are released as net rainfall. Depending upon the height of the vegetation above the ground (drops of up to 6 mm diameter will reach terminal velocity within 12 m) drops may approach terminal velocity and acquire a *higher* kinetic energy than those in the natural rainfall.[7,60] The potential for greater splash detachment from bare mineral soils is therefore *greater* under tall forest canopies than under shorter vegetation.

Conversely, for storms with the largest drop sizes, usually the higher-intensity storms, vegetation canopies may break up the large drops and reduce both the mean drop size and the mean kinetic energy of the incident rain. Recent studies[39] have shown that vegetation canopies have characteristic net rainfall spectra. For *Pinus carribea,* irrespective of the drop size spectra of the incident rain the throughfall spectrum remains essentially unchanged (Fig. 13.3.9) and retains a "signature" characteristic of this particular vegetation type. For three tree species studied, *Pinus carribea, Eucalyptus camaldulensis,* and *Tectona grandis,* median volume drop diameters of the throughfall ranged from 2.3 to 4.2 mm (Fig. 13.3.10) while corresponding kinetic energies, assuming the drops reached terminal velocity, ranged by a factor of 7 with *Pinus carribea* having the least and *Tectona grandis* the greatest kinetic energies.

Splash detachment mobilizes soil particles which can be transported if there is surface runoff. These small soil particles can clog surface micropores and macropores, leading to an impermeable crust which itself reduces infiltration and enhances the production of runoff.

In natural, mixed forests, where a surface vegetation cover or a deep litter layer is usually present which helps protect the soil surface from raindrop impact, and where infiltration capacities are high, surface runoff and surface erosion are usually minimal.

For plantation forest the understory cover of vegetation is often reduced by shading or through competition for soil water or nutrients. For some plantations outbreaks of fire are a common occurrence which destroy both understorey vegetation and litter layers.

Plantations which have both tree species with large net raindrop spectra, such as

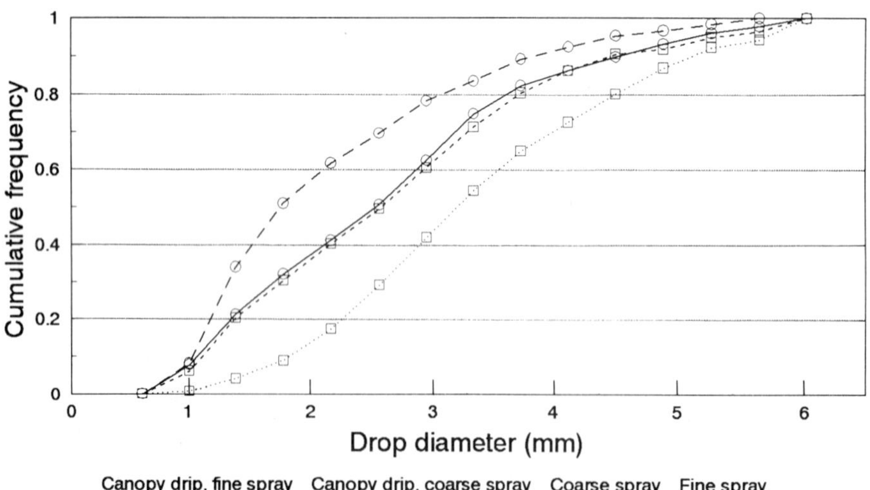

FIGURE 13.3.9 Cumulative frequency distribution of throughfall drop spectra beneath *Pinus carribea* subject to spray with median volume drop diameter (the drop diameter for which 50 percent of the volume was in drops less than this value) of 3.2 and 1.9 mm.

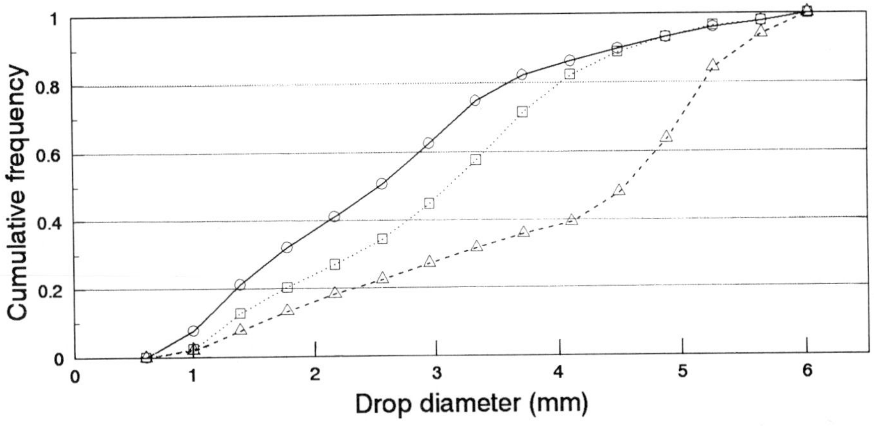

FIGURE 13.3.10 Cumulative frequency distribution of throughfall drop spectra for three tree species subject to spray with median volume drop diameter (the drop diameter for which 50 percent of the volume was in drops less than this value) of 3.2 mm. Median volume drop diameters of the throughfall spectra for *Pinus carribea* was 2.3 mm, for *Eucalyptus camaldulensis* was 2.8 mm and for *Tectona grandis* it was 4.2 mm.

Tectona grandis, and a lack of understorey or a litter layer, have the potential for particularly high rates of soil erosion.

The universal soil loss equation, USLE (see Chap. 12), is widely used for estimating sheet and rill erosion from agricultural lands. Dissmeyer and Foster[29] describe how this equation can be modified for forested lands.

13.3.6 Climate

Land use affects climate. Depending upon the scale of the land-use change the effect can occur on a micro, meso, or global scale. The effect occurs principally through the different inputs, into the atmosphere, of heat, water vapor, and radiation from the different land surfaces.

Microclimate. The variation with height of temperature, humidity, and windspeed close to a surface is the result of a balance between externally applied climate variables, the surface fluxes of heat and water vapor, and the aerodynamic resistance of the surface. *Profiles* of temperature, humidity, and windspeed show a logarithmic variation with height. For different land uses of small areal extent, the conditions at the planetary boundary layer which is typically at a height of a few kilometers (the height of the cloud base), can be considered to be a fixed boundary condition (for the different land uses) and conditions within the planetary boundary layer define the values for the externally applied climate variables.

McNaughton and Jarvis[56] have shown that, using an "up-down" procedure, microclimate changes resulting from a change in land use can be calculated by making use of climate measurements over the existing land use, calculating from these the conditions in the planetary boundary layer and then inferring from these conditions the surface fluxes and microclimate over the new land use.

Differences in the water availability at the evaporating surface will produce marked differences in microclimate as a result of altering the surface fluxes of heat and water vapor. An extreme example is the cool, moist microclimate found over a forest which has a deep root system and readily available soil water as compared with the hotter, drier microclimate found above a short-rooted crop or a bare soil where evaporative fluxes are much less.

Mesoclimate. For land uses or land-use changes occurring over areal extents with dimensions of the order of tens of kilometers it is no longer safe to assume that the conditions in the planetary boundary layer (at a height of a few kilometers) remain unchanged, and the "leakage" of the different fluxes through the planetary boundary layer then needs to be taken into account. Analytical models for doing this are described by McNaughton and Spriggs[57] and de Bruin.[28] Numerical (finite difference) mesoscale climate models, which are currently used for synoptic weather forecasting, may also be used to provide this information when the surface properties are suitably described. The scales on which these mesoscale processes operate are poorly understood at present, particularly for evaporation from wet forests, and are the subject of current research.

Rainfall. The question of whether the effects of a land-use change can alter rainfall is still controversial. Kittridge[49] concluded in 1948 that the influence of forests on rainfall generation is small, less than a 3 percent increase in temperate climates in rainfall over forests as compared with grassland, which is caused by the increased orographic effect resulting from the height of the trees raising the effective height of the topography. Some 40 years later it is possible to say little more on the effects of land use on rainfall generation on the mesoscale, although recent developments in mesoscale climate modeling indicate that the increased evaporation of intercepted water from forests can humidify the planetary boundary layer and can lead to a 5 to 10 percent increase in the regional rainfall. Further experimental and modeling studies are required to provide information on this important and contentious topic.

Silvicultural Practices for the Ameliorization of the Effects of Climate Change. The Australian Department of Conservation and Land Management and the Water Authority of Western Australia[89] are considering how silvicultural practices within the natural jarrah forest (*Eucalyptus* marginata Don ex Sm) may be used to mitigate the hydrologic influences of the greenhouse effect. One scenario is that stream flows and reservoir inflows will decrease as a result of a decrease in regional rainfall,[84] which Stoneman and Schofield suggest could be offset by selective thinning of forest blocks to increase stream flow by 47 percent.

13.4 EFFECTS OF AGRICULTURAL INTENSIFICATION

In the developed world, to increase yields and reduce costs, farmers have become increasingly reliant on intensive agricultural practices: the applications of fertilizers, pesticides, and herbicides, land drainage, and the feedlot or stall and battery farming of animals and poultry.

13.4.1 Hydrologic Effects

Intensive agricultural practices affect the quantity of runoff through the alteration of evaporation, the timing of runoff through changes in land drainage, and water quality, through soil erosion, nutrient leaching from disturbed soils, fertilizer and pesticide applications, and animal waste disposal. Water quality impacts are generally of most concern; where water quantity effects are important it is often through their secondary impact on quality such as in salinization induced by lack of drainage in irrigation schemes or where a rising water table, in response to changes in vegetation type, leaches out salts from the profile and produces salinity problems in streams and rivers. The removal of the jarrah forest in western Australia for bauxite mining is a good example of the latter situation; the replacement of the forest by grassland has led to rising water tables and salt being leached from the soil profile which has affected the streams providing the water supply for the city of Perth.

13.4.2 Water Quality: Nitrate Concentrations

Since the 1950s nitrate concentrations in the surface waters of agricultural catchments of most developed countries have been rising steadily. Although these increases can be attributed to many factors, it is now generally agreed that agricultural intensification is largely responsible, in particular the heavy application of inorganic fertilizers. Comparison of trends of nitrate concentration with time of four rivers in England (Fig. 13.4.1) illustrates the difference associated with different land use.[78] The river Tees in northern England drains mainly poor upland pasture and shows little increase in concentration; the Stour, Thames, and Ouse, which drain lowland catchments, subject predominantly to arable farmland, all showing a consistent rising trend in nitrate concentration which is approaching the World Health Organization's limit of 11.3 mg dm^{-3}.

High levels of nitrogen and phosphorus in watercourses lead to eutrophication and biological degradation of water bodies which can result in phytoplankton blooms (Table 13.3.5) causing problems to water authorities by blocking intake filters and by imparting unpleasant tastes and odors to the water.

13.4.3 Water Quality: Pesticides

Pesticides are, by design, biologically active chemicals; they act by interfering with biochemical reactions and so disrupting the normal chemical balance within the target organism. They pose possible risks to humans and wildlife. A wide range of pesticides is now in use with varying levels of toxicity, selectivity, and persistence. The ideal of a selective pesticide which is harmless to nontarget organisms, and which breaks down after achieving its purpose and before it reaches a watercourse, is far from being generally achieved.[82] Deposition of pesticides in rain and snow (wet deposition) has been widely reported. The usage of many of the pesticide compounds (e.g., DDT) was restricted in North America in the early 1970s, but their occurrence is still being reported there. The persistence of these chemicals with respect to chemical breakdown can only partially explain their continued occurrence. Aerial transport is regarded as the most likely mechanism for explaining both the widespread and the continued occurrence of these chemicals in many parts of the world. The *volume weighted rainfall concentration* of organic pollutants at Lake Superior, Canada, and

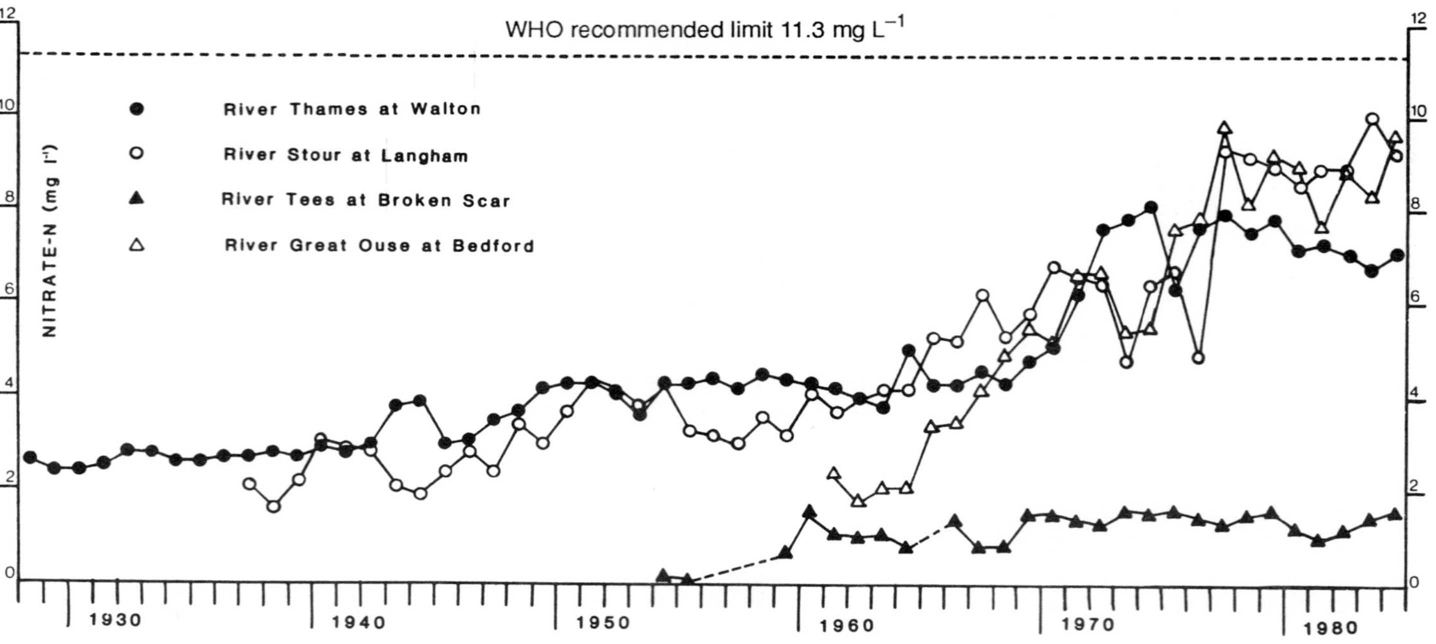

FIGURE 13.4.1 Trends in nitrate concentrations in four British rivers. *(Reproduced with permission from Roberts and Marsh.[78])*

TABLE 13.4.1 Estimates of Rainfall Loadings of Organics to Lake Superior

Compound	Volume weighted rain concentration, ng dm^{-3}	Loadings from rain and snow, kg $year^{-1}$
α-BHC	17.4	860
Lindane	5.9	290
Dieldrin	0.56	28.0
Endrin	0.085	4.2
p,p'-DDE	0.12	5.9
p,p'-DDT	0.11	5.4
p,p'-DDD	0.11	5.4

Source: Reproduced from Ref. 91 with permission of Pergamon Press, Ltd.

the annual loading of these pollutants to the lake are shown in Table 13.4.1. Of increasing concern is not only the toxicity of the individual compounds but also the compounding effects on toxicity of a combination of different pesticides appearing in watercourses at the same time.

13.4.4 Water Quality: Farm Wastes

Farm wastes, including manures and slurries from farm livestock, silage effluent, sheep dip liquors, and pesticide containers, are all potential sources of surface or groundwater pollution.

The concentration of animal production in feedlots leads to a concentration of waste products which are a potential source of water pollution for both surface and groundwater if they are not disposed of adequately. Animal wastes in the United States are estimated to be as much as 1.6 billion tonnes per year, of which 50 percent comes from feedlots.[86] Incidents of agricultural pollution are increasing. Beck[4] reports that the proportion of pollution incidents arising from agriculture in Yorkshire, United Kingdom, rose from zero in the mid-sixties to 12 percent by the late eighties; the Solway River Purification Board[88] in Scotland reported that 59 percent of pollution incidents were agricultural in origin.

13.4.5 Soil Erosion

The erosion of soils as a result of agricultural activities is a worldwide problem. Pimentel[73] has estimated that in the United States soil erosion on agricultural land occurs at a rate of about 30 tons ha^{-1} $year^{-1}$, approximately eight times the rate at which soil is being formed. Agricultural developments on steep slopes in the tropics, without proper soil conservation measures, and uncontrolled grazing on steep slopes lead to high soil losses. The hydrologic consequences include siltation of watercourses and reservoirs and an increase in flood peaks and a reduction in river low flows as the vegetation and soil are removed from hill lands.

13.5 EFFECTS OF DRAINING WETLANDS

13.5.1 Hydrologic Processes in Wetlands

Wetlands occur under conditions where (1) soils are saturated and water is ponded through inadequate drainage; (2) small catchments have no outlet, in hollows and in hill and mountainous areas; (3) river floods cause frequent inundation of the floodplains; (4) there is inadequate drainage to the sea in coastal areas. Wetlands cover an estimated 6 percent of the earth's land surface[55] in many forms including those found in the peat areas of the cold and temperate regions, the dambos at the headwaters of rivers in Africa, the floodplains of the Nile, the mangroves of Southeast Asia, and the Everglades in South Florida.

Water enters wetlands as precipitation, surface and groundwater inflows, and in coastal areas from tides. Water is lost from wetlands by evaporation, which because the surface is close to saturation, is near the potential rate from an open water surface; water is also lost through recharge to groundwater systems and through stream flow.

Wetlands are attributed with the generally beneficial hydrologic functions of:

1. Stream-flow moderation
2. Flood storage and peak mitigation, shoreline anchorage, and dissipation of the erosive forces of fast-flowing water
3. Sediment trapping and nutrient retention and removal
4. Carbon storage

13.5.2 Stream flow

Seasonal Flow. Wetlands have often been said to provide seasonal flow regulation, but recent studies indicate that this effect is marginal or nonexistent. Verry and Boelter[99] show that the flow duration curves from *upland peatbogs,* which occur in confined hollows, are very different from those from *groundwater fens* (Fig. 13.5.1). They argue, however, that neither the bogs nor the fens significantly regulate the

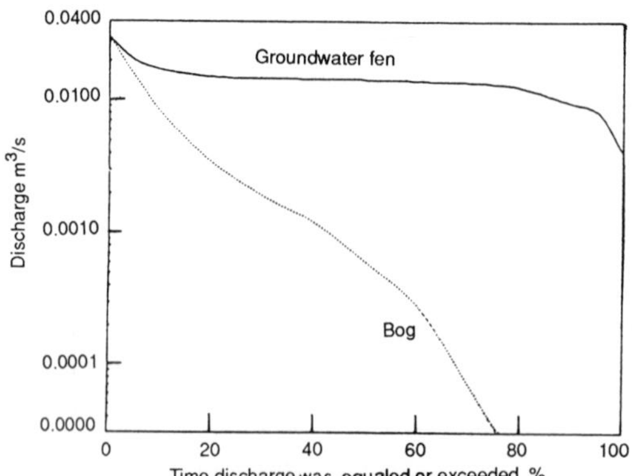

FIGURE 13.5.1 Stream-flow-duration curves for a groundwater fen and perched bog basin. *(From Verry and Boelter.[99])*

stream flow: the confined bogs only release stream flow sporadically during and shortly after storms, while the flat flow duration curve observed from groundwater fens is a consequence of the constant supply of water from the groundwater system to the fen. Studies of the flow duration curves from African catchments show little difference in relation to the proportion of the catchment with dambo wetlands,[9] implying that the dambos have no beneficial effect in sustaining dry season flows. It is argued that,[10] where dambos occur in permeable catchments which provide significant dry season flows, the increased evaporation from dambo wetlands has the detrimental effect, for water resources, of reducing the dry season flow.

The drainage of wetlands produces an initial dewatering while the water table is brought down toward the levels of the drains and drain outlets. Depending upon the type of wetland, this may take from 1 to 2 years to decades until a new quasi-equilibrium is achieved between net input and outflow for the wetland. During this time the dry season flows will be augmented by the "dewatering" flow.

When the new quasi-equilibrium has been reached and the "dewatering" flow has ceased, the increased hydraulic gradient for flow between the surface and the lowered drainage outlets is likely to result in long-term increases in the dry season flows from a drained wetland.

Annual Flow. The presence of a free water surface and the lack of soil water stress to vegetation in wetlands ensure generally high evaporation rates (a low surface resistance r_s) compared with other land uses. Annual runoff is therefore likely to be less than from other land uses.

The drainage of wetlands may increase annual runoff, particularly where:

1. The lowering of the water table through the use of closed or open channel drains depletes water stored in the previously saturated soil profile; for peat wetlands the process may take 1 to 2 years depending on the time of drain installation and the weather.
2. Interceptor ditches or main drains extend into the aquifer; in groundwater-fed wetlands the annual increase in stream flow can be as much as 30 percent and the increase can last for decades or even centuries as the water table in the main aquifer is lowered.[98]
3. The lowering of the water table in the wetland is sufficient to induce soil moisture stress to vegetation; transpiration rates will be reduced and annual stream flow will increase; a lowering to 30 to 40 cm beneath the land surface is regarded as the minimum for this effect to occur.

If a drained wetland is forested, stream flow may again decrease if interception and transpiration exceed that of the prior vegetation.[41]

13.5.3 Floods

The effects of draining wetlands on floods are determined principally by three factors: the type of drainage method employed, the soil type, and the fraction of the total catchment area which is drained.

The processes of water movement into open drainage ditches are different from those into closed (mole and tile) drains. Closed drains are fed by unsaturated or saturated flow in porous media while open drains also receive water by direct precipitation on the channels and from surface runoff. On a plot basis, storm-flow rates are typically an order of magnitude greater in open channel drainage systems than in closed drains;[68] the rates are higher from open channel drains even when the water table is below the surface (Fig. 13.5.2). For drainage systems in relatively impermeable soils, open drain peak flows are likely to be higher, while for closed drains peak flows will be less than for the undrained situation. Channel improvement through enlargement and straightening will also increase peak flows. Although the type of drainage system used affects the peak flow and the time to peak, it does not significantly affect the total quantity of storm period flow.[80]

The effects of drainage on flood response are also related to soil type; there is evidence that from clay soils, closed drainage systems reduce peak flows compared with undrained land while drainage of more permeable soils results in higher peak flows.[80]

13.5.4 Water Quality

Drainage of peat bogs fundamentally alters the chemical processes operating in the bog; redox potentials in the surface peat increase and processes of peat formation revert to peat decomposition. Where the drainage is by open ditches, erosion of the channels may occur.

Both open and closed drainage systems will result in increased flow through the soil profile, and where flow to mineral soils beneath the bog occurs there is likely to be increased leaching of iron, manganese, phosphorus, and base cations and a lowering of acidity (higher pH).

Potentially harmful effects of draining peat bogs, including increased organic loads and acidification as a result of the release of organic acids or of sulfate, have been recorded,[52] but the conditions under which acidification occurs are not well understood. Sallantaus[85] proposed that as the processes of cation and anion retention are reversible, the net effect of bog decomposition on acidity is the reverse of the behavior in natural aggrading peat bogs.

Afforestation on drained peats starts a new phase of aggrading biomass that takes up nutrients released by the decomposing peat; the net uptake of cations decreases the potential for neutralization due to peat decomposition.[62] Peat mining in Finland in which all the surface vegetation is destroyed usually results in the neutralization of runoff.[85]

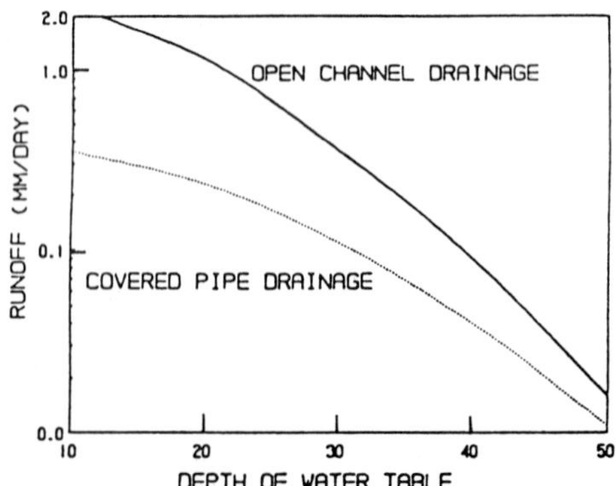

FIGURE 13.5.2 Dependence of runoff rates on water table depth for closed and open drains. *(From Paivenan.[68])*

13.5.5 Carbon Balance

Whereas most indigenous vegetation types are neutral with respect to the annual carbon balance, aggrading peat bogs represent a net sink of atmospheric carbon.

The world's peat deposits of 180 Gt are a significant component of the earth's carbon budget;[6,50] the atmospheric component is 600 to 700 Gt; plant biomass is 800 Gt, and the annual primary productivity value on land is 50 Gt.[103]

The full significance of the effects of draining peat bogs on the world's carbon balance is not yet known, but Armetano and Menges[2] have estimated that before the large-scale disturbance by humans over the last few centuries the net accumulation rate was 57 to 83 Mt of carbon per year. Following disturbance, they calculate a total annual shift in the carbon balance, as a result of the release of previously accumulated carbon and because of the loss of the sink for carbon, of 63 to 85 Mt per year, implying approximate neutrality in the carbon balance of the world's peat bogs at present.

REFERENCES

1. Anderson H. W., M. D. Hoover, and K. G. Reinhart, "Forests and Water: Effects of Forest Management on Floods, Sedimentation, and Water Supply," *USDA Forest Service General Technical Report* PSW-18, 1976.
2. Armetano, T. V., and E. S. Menges, "Patterns of Change in the Carbon Balance of Organic Soil-Wetlands of the Temperate Zone," *J. Ecol.,* vol. 74, pp. 755–774, 1986.
3. Aubertin G. M., and J. H. Patric, "Water Quality, after Clearcutting a Small Watershed in West Virginia," *J. Environ. Qual.,* vol. 3, pp. 243–249, 1974.
4. Beck, L., "Farm Waste Pollution," *Proc. Inst. Water Environ. Manage.,* annual symposium 1989, paper 5, 1989.
5. Bosch J. M., and J. D. Hewlett, "A Review of Catchment Experiments to Determine the Effects of Vegetation Changes on Water Yield and Evapotranspiration," *J. Hydrol.,* vol. 55, pp. 3–23, 1982.
6. Bramryd, T., "The Effects of Man on the Biogeochemical Cycle of Carbon in Terrestrial Ecosystems," in B. Bolin, E. T. Degens, S. Kemp, and P. Ketner, eds. *The Global Carbon Cycle,* Wiley, Chichester, pp. 183–218, 1979.
7. Brandt, J., "The Effect of Different Types of Forest Management on the Transformation of Rainfall Energy by the Canopy in Relation to Soil Erosion," in *Forest Hydrology and Watershed Management.* Proc. of the Vancouver symposium, August 1987, IAHS-AISH Publ. 167, 1987.
8. Brutsaert, W., and H. Stricker, "An Advection-Aridity Approach to Estimating Actual Regional Evaporation," *Water Resour. Res.,* vol. 15, pp. 443–450, 1979.
9. Bullock, A., "The Role of Dambos in Determining River Flow Regimes in Zimbabwe," *J. Hydrol.,* vol. 134, pp. 349–372, 1992.
10. Bullock, A., "Dambo Hydrology in Southern Africa—Review and Reassessment," *J. Hydrol.,* vol. 134, pp. 373–396, 1992.
11. Calder, I. R., "The Measurement of Water Losses from a Forested Area Using a 'Natural' Lysimeter," *J. Hydrol.,* vol. 30, pp. 311–325, 1976.
12. Calder, I. R., "A Model of Transpiration and Interception Loss from a Spruce Forest in Plynlimon, Central Wales," *J. Hydrol.,* vol. 33, pp. 247–265, 1977.
13. Calder, I. R., "Transpiration Observations from a Spruce Forest and Comparisons with Predictions from an Evaporation Model," *J. Hydrol.,* vol. 38, pp. 33–47, 1978.

14. Calder, I. R., "Do Trees Use More Water than Grass?" *Water Services,* vol. 83, pp. 11–14, 1979.
15. Calder, I. R., "A Stochastic Model of Rainfall Interception," *J. Hydrol.,* vol. 89, pp. 65–72, 1987.
16. Calder, I. R., "What Are the Limits on Forest Evaporation?—A Further Comment," *J. Hydrol.,* vol. 89, pp. 13–33, 1987.
17. Calder, I. R., *Evaporation in the Uplands,* Wiley, Chichester, England, 1990.
18. Calder, I. R., R. J. Harding, and P. T. W. Rosier, "An Objective Assessment of Soil-Moisture Deficit Models," *J. Hydrol.,* vol. 60, pp. 329–355, 1983.
19. Calder, I. R., and M. D. Newson, "Land Use and Upland Water Resources in Britain—A Strategic Look," *Water Resour. Bull.,* vol. 16, pp. 1628–1639, 1979.
20. Calder, I. R., and M. D. Newson, "The Effects of Afforestation on Water Resources in Scotland," in M. F. Thomas and J. T. Coppock, eds., *Land Assessment in Scotland,* Proceedings of the Royal Scottish Geographical Society, Edinburgh, May 1979. Aberdeen University Press, Aberdeen, Scotland, pp. 51–62, 1980.
21. Calder, I. R., M. D. Newson, and P. D. Walsh, "The Application of Catchment, Lysimeter and Hydrometeorological Studies of Coniferous Afforestation in Britain to Land-Use Planning and Water Management," *Proc. Symp. Hydrolog. Research Basins, Sonderh. Landeshydrologie,* Bern, pp. 853–863, 1982.
22. Calder, I. R., and P. T. W. Rosier, "The Design of Large Plastic Sheet Net Rainfall Gauges," *J. Hydrol.,* vol. 30, pp. 403–405, 1976.
23. Calder, I. R., M. H. Swaminath, G. S. Kariyappa, N. V. Srinivasalu, K. V. Srinivasa Murty, and J. Mumtaz, "Deuterium Tracing for the Estimation of Transpiration from Trees, 3. Measurements of Transpiration from *Eucalyptus* plantation, India," *J. Hydrol.,* vol. 130, pp. 37–48, 1992.
24. Calder, I. R., and I. R. Wright, "Gamma-Ray Attenuation Studies of Rainfall Interception from Sitka Spruce—Some Evidence for an Additional Transport Mechanism," *Water Resour. Res.,* vol. 22, pp. 409–417, 1986.
25. Calder, I. R., I. R. Wright, and D. Murdiyarso, "A Study of Evaporation from Tropical Rainforest—West Java," *J. Hydrol.,* vol. 89, pp. 13–33, 1987.
26. Cavelier, J., and G. Goldstein, "Mist and Fog Interception in Elfin Cloud Forests in Colombia and Venezuela," *J. Tropical Ecol.,* vol. 5, pp. 309–322, 1989.
27. Chamberlain, A. C., "The Movement of Particles in Plant Communities," in J. L. Monteith, ed., *Vegetation and the Atmosphere,* vol. 1, pp. 155–201, Academic Press, London, 1975.
28. de Bruin, H. A. R., "A Model for the Priestley-Taylor Equation," *J. Clim. Appl. Meteorol.,* vol. 22, pp. 572–578, 1983.
29. Dissmeyer, G. E., and G. R. Foster, "Modifying the Universal Soil Loss Equation for Forest Land," in S. A. El-Swaify, W. C. Moldenhauer, and A. Lo, eds., *Soil Erosion and Conservation,* Soil Conservation Society of America, pp. 480–495, 1985.
30. Douglas, J. E., "A Summary of Some Results from the Coweeta Hydrologic Laboratory," Appendix B, in L. S. Hamilton and P. N. King, authors, *Tropical Forested Watersheds,* Westview Press, Boulder, Colo., 1981.
31. Fowler, D., and M. H. Unsworth, "Turbulent Transfer of Sulphur Dioxide to a Wheat Crop," *Q. J. R. Meterol. Soc.,* vol. 105, pp. 767–783, 1979.
32. Fowler, D., J. N. Cape, and M. H. Unsworth, "Deposition of Atmospheric Pollutants on Forests," *Phil. Trans. R. Soc. Lond. B,* vol. 324, pp. 247–265, 1989.
33. Garland J. A., "The Dry Deposition of Sulphur Dioxide to Land and Water Surfaces," *Proc. R. Soc. Lond. A.,* vol. 354, pp. 245–268, 1977.
34. Gash, J. H. C., I. R. Wright, and C. R. Lloyd, "Comparative Estimates of Interception Loss from Three Coniferous Forests in Great Britain," *J. Hydrol.,* vol. 48, pp. 89–105, 1980.

35. Gentry, A. H., and J. L. Parody, "Deforestation and Increased Flooding of the Upper Amazon," *Science,* vol. 210, pp. 1354–1356, 1980.

36. Goodell, B. C., "Management of Forest Stands in Western United States to Influence the Flow of Snow-Fed Streams," *Int. Assoc. Sci. Hydrol. Publ.* 48, pp. 49–58, 1959.

37. Greenwood, E. A. N., "Deforestation, Revegetation, Water Balance and Climate," *Adv. Bioclimatology,* Springer-Verlag, in press.

38. Hall, R. L., "Processes of Evaporation from Vegetation of the Uplands of Scotland," *Trans. R. Soc. Edin. (Earth Sciences),* vol. 78, pp. 327–334, 1987.

39. Hall, R. L., and I. R. Calder, "Drop Size Modification by Forest Canopies—Measurements Using a Disdrometer," submitted to *J. Geophys. Res.,* 1992.

40. Harding, R. J., R. L. Hall, M. H. Swaminath, and K. V. Srinivasa Murthy, "The Soil Moisture Regimes beneath Forest and an Agricultural Crop in Southern India—Measurements and Modelling," in I. R. Calder, R. L. Hall, and P. G. Adlard, eds., *Growth and Water Use of Forest Plantations,* Proceedings of the International Symposium on the Growth and Water Use of Forest Plantations, Bangalore, Feb. 7–11, 1991, Wiley, Chichester, England, 1992.

41. Heikurainen, L., and J. Paivanen, "The Effect of Thinning, Clear Cutting, and Fertilization on the Hydrology of Peatland Drained for Forestry," *Acta For. Fenn.,* vol 104, pp. 1–23, 1970.

42. Hewlett, J. D., "Soil Moisture as a Source of Base Flow from Steep Mountain Watersheds," *USDA Forest Serv. Southeast,* Forest Expt. Sta. Pap. 132, 1961.

43. Hewlett, J. D., and A. R. Hibbert, "Factors Affecting the Response of Small Watersheds to Precipitation in Humid Areas," in W. E. Sopper, and H. W. Lull, eds., *International Symposium on Forest Hydrology,* pp. 275–290, Pergamon Press, Oxford, 1967.

44. Holland, P. K., J. Sugden, and K. Thornton, "The Direct Deposition of SO_2 from Atmosphere, part 2: Measurements at Rininglow Bog Compared with Other Results," *Central Electricity Generating Board North-West Region,* Report NW/SSD/RN/PL/1/74, 1974.

45. Hoover, M. D., "Watershed Management in Lodgepole Pine Ecosystems," in D. M. Baumgartner, eds., *Management of Lodgepole Pine Ecosystems Symp.,* Washington State University, Pullman, October 1973, pp. 569–580, 1973.

46. Hornbeck, J. W., R. S. Pierce, G. E. Likens, and C. W. Martin, "Moderating the Impact of Contemporary Forest Cutting on Hydrologic and Nutrient Cycles," in *Int. Symp. Hydrol. Characteristics of River Basins* (Tokyo, Japan, Dec. 8–11, 1975), Int. Assoc. Hydrol. Sci. Publ. 117, pp. 423–433, 1975.

47. Institute of Hydrology, "Water Balance of the Headwater Catchments of the Wye and Severn, 1970–75," *Inst. Hydrol. Rep.* 33, Wallingford, UK, 1976.

48. Institute of Hydrology, "Plynlimon Research: The First Two Decades," *Inst. Hydrol. Report* 109, Wallingford, UK, 1991.

49. Kittredge, J., *Forest Influences,* McGraw-Hill, New York, 1948.

50. Kivinen, E., and P. Pakarinen, "Geographical Distribution of Peat Resources and Major Peatland Complex Types in the World," *Ann. Acad. Sci. Fenn. Ser.* A, 132, 1981.

51. Langford, K. J., "Changes in Yield of Water Following a Bushfire in a Forest of *Eucalyptus regnans,*" *J. Hydrol.,* vol. 29, pp. 87–114, 1976.

52. Lundin, L., "Effects of Forest Drainage on the Acidity of Groundwater and Surface Water," in *International Symposium on Acidification and Water Pathways,* May 4–8, 1987, Bolkesjo, Norway. The Norwegian Committee for Hydrology in Cooperation with UNESCO and WMO, the IHP National Committee of Denmark, Finland and Sweden, vol. 11, pp. 229–238, 1987.

53. Lynch J. A., W. E. Sopper, and E. S. Corbert, "Watershed Behaviour under Controlled Simulated Rainfall," *Natl. Tech. Inform. Serv.,* Springfield, Va., Report PB-240921, 1974.

54. Lynch, J. A., E. S. Corbett, and D. W. Aurand, "Effects of Management Practices on Water Quality and Quantity: The Penn State Experimental Watersheds," in *Proc. Munic-*

ipal Watershed Manage. Symp. (Penn. State University, Sept. 11–12, 1973), USDA Forest Serv. Gen. Tech. Rept. 32–46, 1975.

55. Maltby, E., and R. E. Turner, "Wetlands of the World," *Geog. Mag.*, vol. 55, pp. 12–17, 1983.
56. McNaughton, K. G., and P. G. Jarvis, "Using the Penman-Monteith Equation Predictively," *Agric. Water Manage.*, vol. 8, pp. 263–278, 1984.
57. McNaughton, K. G., and T. W. Spriggs, "A Mixed-Layer Model for Regional Evaporation," *Boundary-Layer Meteorol.*, vol. 34, pp. 243–262, 1986.
58. Ministry of Agriculture, Fisheries and Food, *Technical Bulletin* 16, Potential Transpiration, Her Majesty's Stationery Office, U.K.
59. Monteith, J. L., *Principles of Environmental Physics,* Edward Arnold, London, U.K., 1973.
60. Morgan, R. P. C., "Establishment of Plant Cover Parameters for Modelling Splash Detachment," in S. A. El-Swaify, W. C. Moldenhauer, and A. Lo, eds., *Soil Erosion and Conservation,* Soil Conservation Society of America, pp. 377–383, 1985.
61. Morton, F. I., "Potential Evaporation as a Manifestation of Regional Evaporation," *Water Resour. Res.*, vol. 5, pp. 1244–1255, 1969.
62. Neal, C., P. G. Whithead, R. Neale, and B. J. Cosby, "Modelling the Effects of Acid Deposition and Conifer Afforestation on Stream Acidity in the British Uplands," *J. Hydrol.*, vol. 86, pp. 15–26, 1986.
63. Newson, M. D., and I. R. Calder, "Forests and Water Resources: Problems of Prediction on a Regional Scale," *Phil. Trans. R. Soc. Lond. B,* vol. 324, pp. 283–298, 1989.
64. Newton, A., M. G. R. Cannell, and I. R. Calder, "Trees and the Environment," *NERC Report,* Institute of Terrestrial Ecology, Institute of Hydrology, U.K., 1990.
65. Nillsson, I. S., H. G. Miller, and J. D. Miller, "Forest Growth as a Possible Cause of Soil and Water Acidification: An Examination of the Concepts," *Oikos,* vol. 39, pp. 40–49, 1982.
66. Nordin, C. F., and R. H. Meade, "Deforestation and Increased Flooding in the Upper Amazon," *Science,* vol. 215, pp. 426–427, 1982.
67. O'Loughlin, C. L., and R. R. Ziemer, "The Importance of Tree Root Strength and Deterioration Rates upon Edaphic Stability in Steepland Forest," in R. H. Waring, ed., *Carbon Uptake and Allocation: A Key to Management of Subalpine Ecosystems,* pp. 84–91, Corvallis, Ore., 1982.
68. Paivenan, J., "Effects of Different Types of Contour Ditches on the Hydrology of an Open Bog," *Proc. Fifth Int. Peat Congress,* Sept. 21–25, 1976, Poznan, Poland, Int. Peat Soc. Helsinki, vol. 1, pp. 93–106, 1976.
69. Penman, H. L., "Natural Evaporation from Open Water, Bare Soil and Grass," *Proc. R. Soc. Ser. A,* vol. 193, pp. 120–145, 1948.
70. Penman, H. L., "The Dependence of Transpiration on Weather and Soil Conditions," *J. Soil Sci.*, vol. 1, pp. 74–89, 1949.
71. Pereira, H. C., *Land Use and Water Resources in Temperate and Tropical Climates,* Cambridge University Press, London, 1973.
72. Pierce, R. S., W. C. Martin, C. C. Reeves, and others, "Nutrient Loss from Clearcuttings in New Hampshire," in *Natl. Symp. Watersheds Transition Proc.*, Am. Water Resour. Assoc., Urbana, Ill., pp. 285–295, 1972.
73. Pimentel, D., "Land Degradation: Effects on Food and Energy Resources," *Science,* vol. 194, pp. 149–155, 1976.
74. Reynolds, B., C. Neal, M. Hornung, and P. A. Stevens, "Baseflow Buffering of Streamwater Acidity in Five Mid Wales Catchments," *J. Hydrol.*, vol. 87, pp. 167–185, 1986.
75. Rice, R. M., "Forest Management to Minimize Landslide Risk," in Guidelines for Watershed Management, *FAO Conservation Guide* 1, Rome, pp. 271–287, 1977.

76. Rice, R. M., J. S. Rothacher, and W. F. Megahan, "Erosional Consequences of Timber Harvesting—An Appraisal," in *Proc. Symp. Watersheds in Transition,* Fort Collins, Colo., June 19–21, 1972, *Water Resour. Assoc. Ser.* 14, pp. 321–329, 1972.

77. Roberts, J. M., "The Use of the "Tree Cutting" Technique in the Study of the Water Relations of Norway Spruce [Picea abies (L.) Karst.]," *J. Exp. Bot.,* vol. 29, pp. 465–471, 1978.

78. Roberts, G., and T. Marsh, "The Effects of Agricultural Practices on the Nitrate Concentrations in the Surface Water Domestic Supply Sources of Western Europe," in *Water for the Future Hydrology in Perspective* (Proc. Rome Symposium, April 1987), IAHS Publ. 164, 1987.

79. Robinson, M., "Changes in Catchment Runoff Following Drainage and Afforestation," *J. Hydrol.,* vol. 86, pp. 71–84, 1986.

80. Robinson, M., "Impact of Improved Land Drainage on River Flows," *Institute of Hydrology Report* 113, Wallingford, U.K., 1990.

81. Robinson, M., B. Gannon, and M. Schuch, "A Comparison of the Hydrology of Moorland under Natural Conditions, Agricultural Use and Forestry," *Hydrol. Sci. J.,* vol. 36, pp. 565–577, 1991.

82. Royal Commission on Environmental Pollution, Her Majesty's Stationery Office, U.K., 1979.

83. Rutter, A. J., K. A. Kershaw, P. C. Robins, and A. J. Morton, "A Predictive Model of Rainfall Interception in Forests. I: Derivation of the Model from Observations in a Plantation of Corsican Pine," *Agric. Meteorol.,* vol. 9, pp. 367–384, 1971.

84. Sadler, B. S., G. W. Mauger, and R. A. Stokes, "The Water Resource Implications of a Drying Climate in South Western Australia," in G. I. Pearman, ed., Greenhouse Planning for Climate Change, *CSIRO Publ.,* Melbourne, pp. 296–311, 1988.

85. Sallantaus, T., "Water Quality of Peatlands and Man's Influence on It," *Proc. International Symposium on the Hydrology of Wetlands in Temperate and Cold Regions,* The Academy of Finland, Helsinki, pp. 80–98, 1988.

86. Sanders, W. M., "Nutrients," in R. T. Oglesby, C. A. Carlson, and J. A. McCann, eds., *River Ecology and Man,* Academic Press, New York, pp. 389–415, 1972.

87. Shuttleworth, W. J. "Evaporation from Amazonian Rainforest," *Proc. R. Soc. Lond.* B. 233, pp. 321–346, 1988.

88. Solway River Purification Board, *Report for the Year Ending 31st December 1987,* The Director, Clyde River Purification Board, East Kilbride, Glasgow, U.K., 1988.

89. Stoneman, G. L., and N. J. Schofield, "Silviculture for Water Production in Jarrah Forest of Western Australia: An Evaluation," *For. Ecol. Manage.,* vol. 27, pp. 273–293, 1989.

90. Stoner, J. H., and A. S. Gee, "Effects of Forestry on Water Quality and Fish in Welsh Rivers and Lakes," *J. Inst. Water Eng. Sci.,* vol. 39, pp. 27–45, 1985.

91. Strachan, W. M. J., "Organic Substances in the Rainfall of Lake Superior: 1983," *Environ. Toxicol. Chem.,* vol. 4, pp. 677–683, 1985.

92. Thom, A. S., "Momentum, Mass and Heat Exchange of Vegetation," *Q. J. R. Meteorol. Soc.,* vol. 98, pp. 124–134, 1972.

93. Thom, A. S., "Momentum, Mass and Heat Exchange of Plant Communities," in J. L. Monteith, ed., *Vegetation and the Atmosphere,* Academic Press, London, vol. 1, pp. 57–109, 1975.

94. Trimble, S. W., F. H. Weirich, and B. L. Hoag, "Reforestation and the Reduction of Water Yield on the Southern Piedmont Since Circa 1940," *Water Resour. Res.,* vol. 23, pp. 425–437, 1987.

95. United Kingdom Acid Waters Review Group, *Acidity in United Kingdom Fresh Waters,* Second report, Her Majesty's Stationery Office, U.K., 1989.

96. United Kingdom Review Group on Acid Rain, *Acid Deposition in the United Kingdom,*

1986–1988, Warren Springs Laboratory, Department of Trade and Industry, Stevenage, U.K., 1990.

97. Van Lill, W. S., F. J. Kruger, and D. B. Van Wyk, "The Effect of Afforestation with *Eucalyptus grandis* hill ex. maiden and *Pinus patula* schlect. et. cham. on Streamflow from Experimental Catchments at Mokobulaan, Transvaal," *J. Hydrol.,* vol. 48, pp. 107–118, 1980.

98. Verry, E. S., "The Hydrology of Wetlands and Man's Influence on It," *Proc. International Symposium on the Hydrology of Wetlands in Temperate and Cold Regions,* Joensuu, Finland, June 6–8, 1988.

99. Verry, E. S., and D. H. Boelter, "The Influence of Bogs on the Distribution of Streamflow from Small Bog-Upland Catchments," in *Hydrology of Marsh-ridden Areas,* Proc. Minsk Symp., June 1975, UNESCO Press, Paris, IAHS, pp. 469–478, 1975.

100. Wallace, J. S., J. M. Roberts, and A. M. Roberts, "Evaporation from Heather Moorland in North Yorkshire, England," *Proc. Symp. Hydrolog. Research Basins,* Bern, pp. 397–405, 1982.

101. Wells, D. E., and R. Harriman, "Acidification Studies in Scottish Catchments: The Effects of Deposition, Catchment Type and Runoff," in R. Perry et al., eds., *Acid Rain: Scientific and Technical Advances,* pp. 293–300, Selper, Ltd., U.K., 1987.

102. Welsh, W. T., and J. C. Burns, "The Loch Dee Project: Runoff and Surface Water Quality in an Area Subject to Acid Precipitation and Afforestation," *Trans R. Soc. Edin.: Earth Sciences,* vol. 78, pp. 249–260, 1987.

103. Woodwell, G. M., R. H. Whittaker, W. A. Reiners, G. E. Likens, C. C. Delwiche, and D. B. Botkin, "The Biota and World Carbon Budget," *Science,* vol. 199, pp. 141–146, 1978.

CHAPTER 14
CONTAMINANT TRANSPORT IN SURFACE WATER

Wayne C. Huber
Department of Civil Engineering
Oregon State University
Corvallis, Oregon

14.1 INTRODUCTION

14.1.1 Scope and Objectives

This chapter describes methods for evaluation of the transport and fate of water-quality constituents in surface waters. A quality *constituent* may be a pollutant or a relatively benign variable such as dissolved oxygen; hence, *constituent* is used generically and does not necessarily mean a polluting substance. The *fate* of a constituent depends both on its transport (movement) through the water and on sources, sinks, reactions, and decay mechanisms (sources and sinks). This chapter defines governing transport mechanisms of *advection* and *diffusion* (or diffusion-like mechanisms), and augments information on sources and sinks presented in Chap. 11.

Perhaps the most common hydrologic application of principles of fate and transport is in the prediction of dissolved oxygen (DO) in streams subject to input of biochemical oxygen demand (BOD) and subject to oxygen replenishment via surface reaeration. The well-known Streeter-Phelps equation[81] provides the steady-state, one-dimensional distribution of DO along a river for these conditions, and variants of this equation are still commonly applied in practice. Fate and transport calculations can also be applied to study phenomena such as lateral mixing of polluted water across a river, the distribution of constituents in an outfall plume, the downstream distribution of constituents from an accidental spill, the two- and three-dimensional distribution of constituents in a large lake or estuary, lake eutrophication, etc. The objective of this chapter is to present transport and fate principles, and sufficient information for an initial evaluation of problems such as these, with references for more detailed analysis. Stated another way, the objective is to predict $C(x, y, z, t)$, that is, to predict concentration as a function of location x, y, z and time t throughout the water body.

14.1.2 Definitions and Relationships among Flow, Concentration, and Load

The primary measure of the quantity of a constituent is its concentration C, defined as

$$C = \frac{\text{quantity (mass) of constituent}}{\text{volume of fluid}} \tag{14.1.1}$$

Most constituents are measured in terms of their mass, and C usually has units such as mg/L = g/m^3. Because the density of water is nearly 1.0 g/cm^3, units of μg/cm^3, mg/L, and g/m^3 are numerically equivalent to parts per million (ppm) by mass in water, a convenient coincidence, whereas μg/L (micrograms per liter) is numerically equivalent to parts per billion (ppb) by mass. Concentration may still be defined for variables not measured in mass units. For example, bacteria are often measured as a number (e.g., most probable number or MPN) per volume, radiation is measured in units of curies per volume, and heat is measured in units of calories or joules per volume.

The impact of constituents on a water body may be influenced by both the concentration and by the load. *Load* may mean either the total mass M in a volume V of water,

$$M = CV \tag{14.1.2}$$

or the *mass flow rate* L (mass/time) in water flowing with a rate Q (volume/time),

$$L = CQ \tag{14.1.3}$$

14.1.3 Problem Definitions

Steady-state analysis is commonly applied to discharges into water that are relatively constant in their concentration and flow rate (e.g., constant L), such as point-source discharges from wastewater treatment plants and industries. The analysis is generally easier mathematically, and constant (average) flow conditions in the receiving water are used. (The Streeter-Phelps equation is an example.) However, *transient analyses* must be applied to spills or instantaneous discharges and to problems for which receiving water conditions change rapidly, e.g., during a storm event.

Far-field problems are those in which the constituent concentration is influenced only by the transport conditions of the ambient fluid, that is, by the velocities, turbulence, and density of the receiving water body itself. The *near field* is the region near the outfall or discharge point within which the mixing and hydrodynamics are influenced by the nature of the outfall release, such as the region of a buoyant jet at an outfall pipe or diffuser. The extent of near-field phenomena establishes the *mixing zone,* which may be relatively small for the discharge of a pipe into a river, or larger for the discharge of sewage through an ocean outfall diffuser. The near-field region is important for establishment of initial dilution in the receiving water, but it is not considered in detail in this chapter. Extensive information on near-field analysis is provided by Fischer et al.[37] and Holley and Jirka.[44]

14.2 CONTAMINANT SOURCES

14.2.1 Point Sources

A *point source* is a regulatory term meaning a source that discharges through a pipe at a known location, such as from an industry or a wastewater treatment plant. Typical influent concentrations to municipal waste treatment plants are shown in Table 14.2.1. Effluent characteristics depend on the degree of treatment as shown in Table 14.2.2 for five important constituents (see Sec. 14.4 for further definitions of BOD_5 and COD). A more extensive summary is provided by Mills et al.[61] The values in Table 14.2.2 assume proper operation and maintenance of the plant. In fact, considerable variability occurs in the actual plant performance,[65] and monitoring is usually required to provide reliable effluent quality and quantity characteristics for specific dischargers. This statement is true for virtually every type of pollutant source!

As expected, the characteristics of industrial wastes depend heavily on the industry and degree of treatment. Representative concentrations for various industries are shown in Table 14.2.3. Additional general information is given in the U.S. Environmental Protection Agency (EPA) "Treatability Manual."[30]

All point-source dischargers in the United States are required to have permits according to the EPA National Pollutant Discharge Elimination System (NPDES)

TABLE 14.2.1 Typical Influent Municipal Waste Concentrations

	Concentration, mg/L		
Constituent	Strong	Medium*	Weak
Solids, total	1200	720	350
Dissolved, total	850	500	250
Fixed	525	300	145
Volatile	325	200	105
Suspended, total	350	220	100
Fixed	75	55	20
Volatile	275	165	80
Settlable solids, mL/L	20	10	5
Biochemical oxygen demand, 5-day, 20° (BOD_5-20°)	400	220	110
Total organic carbon (TOC)	290	160	80
Chemical oxygen demand (COD)	1000	500	250
Nitrogen (total as N)	85	40	20
Organic	35	15	8
Free ammonia	50	25	12
Nitrites	0	0	0
Nitrates	0	0	0
Phosphorus (total as P)	15	8	4
Organic	5	3	1
Inorganic	10	5	3
Chlorides†	100	50	30
Alkalinity (as $CaCO_3$)*	200	100	50
Grease	150	100	50

* In the absence of other data, use medium strength data for planning purposes.
† Values should be increased by amount in carriage water.
Source: Ref. 60.

TABLE 14.2.2 Characteristics of Treated Wastewater

	Effluent concentrations, mg/L, Total removal efficiencies*				
Scheme no.†	5-day BOD	Chemical oxygen demand	Total suspended solids	Total phosphorus	Total nitrogen
0 (Raw influent)	200(0%)	500(0%)	200(0%)	10(0%)	40(0%)
1	130(35%)	375(25%)	100(25%)	9(10%)	32(20%)
2	40(80%)	125(75%)	30(85%)	7.5(25%)	26(35%)
3	25(88%)	100(80%)	12(94%)	7(30%)	24(40%)
4	18(91%)	70(86%)	7(96%)	1(90%)	22(45%)
5	18(91%)	70(86%)	7(96%)	1(90%)	4(90%)
6	13(94%)	60(88%)	1(99.5%)	1(90%)	3(92%)
7	2(99%)	15(97%)	1(99.5%)	1(90%)	2(95%)

* Efficiencies for wastewater treatment are for the approximate BOD_5 concentration range of $100 \leq BOD_5 \leq 400$ mg/L.

† Scheme no. Process

Scheme no.	Process
0	Wastewater influent (no treatment)
1	Primary treatment
2	1 + activated sludge (secondary treatment)
3	2 + polishing filter (high efficiency or super secondary)
4	3 + phosphorus removal and recarbonation
5	4 + nitrogen stripping
6	5 + pressure filtration
7	6 + activated carbon adsorption

Source: Ref. 61.

TABLE 14.2.3 EPA 1977 Effluent Guidelines for Industrial Wastes

Industry	Daily max. BOD, mg/L	Daily max. TSS, mg/L
Dairy processing	0.05–0.5	0.07–1.0
Grain mills	0.04–150	0.02–150
Sugar beet processing	3.3	
Cane sugar processing	1.1–2.4	0.5–4.2
Plastics/synthetics	0.1–18	0.3–29
Soaps/detergents	0.03–3.6	0.01–2.2
Fertilizer manufacturing	50	
Petroleum refining	0.4–9.9	0.3–6.9
Petrochemicals	12.1	8.3
Iron/steel manufacturing		0.01–0.3
Tannery	3.2–9.6	4.0–11.6
Rubber	0.5–15	0.1–7
Timber processing	0.09–1.6	0.4–33
Textiles	1.4–22	1.4–35
Cement	0.005–50	
Organic chemicals	0.01–3.1	0.07–2.8
Pulp/paper	6.0–17.4	10–24
Meat products	0.02–0.74	0.04–0.9

TSS = total suspended solids

Ranges given reflect standards for different processes within an industry. Additional chemical constituents are specified for some industries.

Source: Ref. 64.

regulations. The permit requirements include information on the flow rate and concentrations at industries and waste-treatment plants, and this information is available from EPA regional offices and from state environmental regulatory agencies.

14.2.2 Nonpoint Sources

Nonpoint sources of constituents originate from flow distributed over the land surface. However, there are ambiguities in this definition, because storm-water flow, for example, may be collected in a sewer system and discharged to the receiving water through a pipe so that point-source discharges may have nonpoint-source origins. Some agencies classify storm water and *combined sewer overflows* (CSOs) as point sources, while others classify them as nonpoint sources. The following land-use types produce predominantly nonpoint sources: urban, agriculture, silviculture (forestry), mining, and undeveloped or natural. Atmospheric contributions may also be considered as a nonpoint source. Only urban runoff quality will be discussed briefly at this point.

Most newer cities in the United States have separate sewer systems; that is, the collection system for storm water is separate from the collection system for sanitary sewage. However, many older cities have combined sewer systems, in which the same pipe is used for sanitary sewage as for storm water. During dry weather, the sanitary base flow is diverted at a downstream regulator (e.g., an orifice or weir) to an interceptor sewer and then to a treatment plant, but during wet weather, the flow in excess of the regulator and/or interceptor capacity overflows directly to the receiving water, constituting a combined sewer overflow event.[51]

14.2.3 Nature of Surface Transport

Although point sources frequently exhibit daily and weekly variations, nonpoint-source flows originate from rainfall events and follow the temporal and spatial characteristics of rainfall to a large degree. A plot of concentration versus time is often called a *pollutograph,* and a plot of load (concentration × flow rate or mass/time) versus time is often called a *loadograph;* see Fig. 14.2.1. If the concentration during a storm were constant, the shape of the loadograph would exactly match that of the hydrograph (Fig. 14.2.1*a, b, c*). However, the pollutograph frequently exhibits considerably higher concentrations near the beginning of the storm (Fig. 14.2.1*b*). This is known as the *first flush* phenomenon and is due to higher rainfall intensities near the beginning of a storm that result in higher runoff and greater boundary shear stress and erosive potential, and to greater availability of solids that have *built up* on urban surfaces during dry weather. The *wash-off* of these solids is thus greater nearer the beginning of a storm (Fig. 14.2.1*c*). Urban runoff quality is often simulated by conceptual models of build-up and wash-off.[46] The first flush is most evident in combined sewers in which solids are deposited during dry weather and scoured during the beginning of a wet weather event. The first flush is least evident in highly urbanized urban cores, in which traffic and other factors tend to rapidly establish an equilibrium between factors that generate contaminants on the street surface, and those that remove them, including wind and planned activities such as street cleaning.

On pervious surfaces, build-up plays a lesser role, and entrainment of water quality constituents in runoff is due more to erosion and solution mechanisms.[46] Constituents may be adsorbed onto particulate matter and thus be subject to transport as solids.

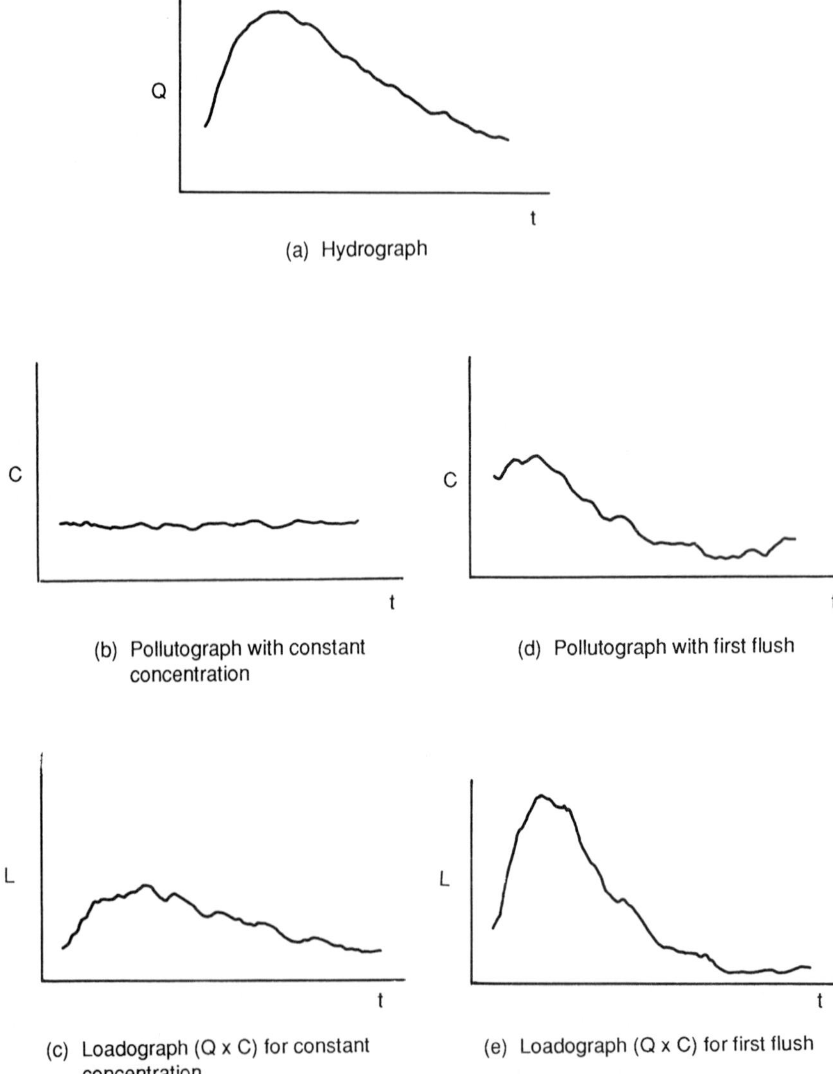

FIGURE 14.2.1 Effect of first flush on shapes of pollutograph and loadograph: (*a*) hydrograph (Q vs. T), (*b*) pollutograph (C vs. t), and (*c*) loadograph (load vs. t). Load $= Q \times C$.

14.2.4 Characterization of Urban Non-Point-Source Data

The *event mean concentration* (EMC) is the total storm load (mass) divided by the total runoff volume, although EMC estimates are usually obtained from a flow-weighted composite of concentration samples taken during a storm.[62] Mathematically,

$$\mathrm{EMC} \equiv \overline{C} = \frac{M}{V} = \frac{\int C(t)Q(t)\,dt}{\int Q(t)\,dt} \tag{14.2.1}$$

where $C(t)$ and $Q(t)$ are the time-variable concentration and flow measured during the runoff event, and M and V are the pollutant mass and runoff volume as defined in Eq. (14.1.2). It is clear that the EMC results from a *flow-weighted average,* not simply a time average of the concentration. Thus, when the EMC is multiplied by the runoff volume, an estimate of the loading to the receiving water is provided. As is evident from Fig. 14.2.1*b*, the instantaneous concentration during a storm can be higher or lower than the EMC, but the use of the EMC as an event characterization replaces the actual time variation of C versus t in a storm with a pulse of constant concentration having equal mass and duration as the actual event. This ensures that mass loadings from storms will be correctly represented.

Just as instantaneous concentrations vary within a storm, event mean concentrations vary from storm to storm, as illustrated in Fig. 14.2.2, and from site to site as well. The median or 50th percentile EMC at a site, estimated from a time series of the type illustrated in Fig. 14.2.2, is called the *site median EMC.* When site median EMCs from different locations are aggregated, their variability can be quantified by their median and *coefficient of variation* (CV = standard deviation divided by the mean) to achieve an overall description of the runoff characteristics of a constituent

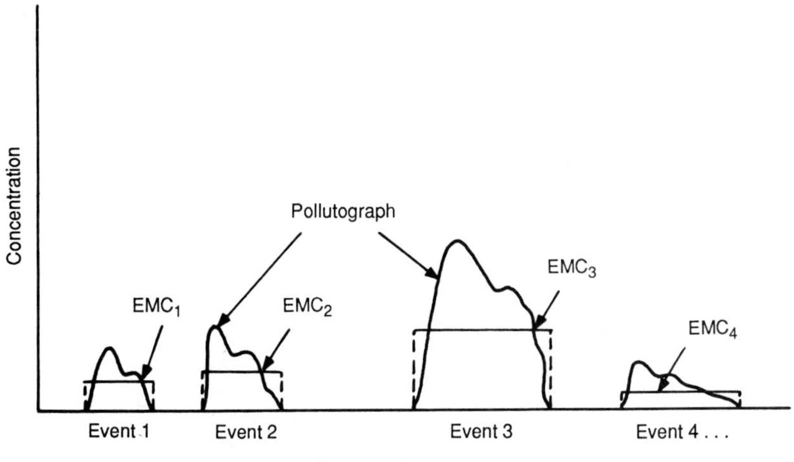

FIGURE 14.2.2 Interstorm variation of pollutographs and event mean concentrations.

TABLE 14.2.4 Variation in Event Mean Concentrations (EMCs) in Combined Sewer Overflows

Constituent	Site median concentration, mg/L: 10th Percentile	50th Percentile (median)	90th Percentile	Coefficient of variation* of EMCs
Total suspended solids	121	184	279	0.7
5-day biochemical oxygen demand	32	53	86	0.7
Chemical oxygen demand	62	132	283	0.8
Orthophosphorus	0.5	0.8	1.1	0.4
Total phosphorus	0.8	2.4	7.9	0.7
TKN†	4.1	6.5	10.3	0.6
NH_3-N†	0.9	1.9	3.9	0.8
NO_2-N†	0.1	0.1	0.1	0.6
NO_3-N†	0.2	1.0	4.5	0.5
Chrome	0.008	0.099	0.957	0.6
Copper	0.039	0.102	0.271	0.5
Lead	0.158	0.346	0.755	0.6
Zinc	0.223	0.348	0.544	0.6
Fecal coliform (10^6 MPN/100mL)	0.5	1		1.5

* Coefficient of variation = standard deviation/mean.

† Definitions: TKN = total Kjeldahl nitrogen = organic nitrogen + NH_3-N; NH_3-N = ammonia nitrogen; NO_2-N = nitrite nitrogen; NO_3-N = nitrate nitrogen.

Source: Reproduced from Ref. 26 with permission.

TABLE 14.2.5 Median Event Mean Concentrations (EMCs) for all EPA Nationwide Urban Runoff Program (NURP) Sites by Land Use Categories

		Residential		Mixed*		Commercial		Open/nonurban	
Constituent†	Units	Median	CV	Median	CV	Median	CV	Median	CV
5-day biochemical oxygen demand	mg/L	10.0	0.41	7.8	0.52	9.3	0.31	—	—
Chemical oxygen demand	mg/L	73	0.55	65	0.58	57	0.39	40	0.78
Total suspended solids	mg/L	101	0.96	67	1.14	69	0.85	70	2.92
Total lead	μg/L	144	0.75	114	1.35	104	0.68	30	1.52
Total copper	μg/L	33	0.99	27	1.32	29	0.81	—	—
Total zinc	μg/L	135	0.84	154	0.78	226	1.07	195	0.66
TKN	μg/L	1900	0.73	1288	0.50	1179	0.43	965	1.00
NO_2-N + NO_3-N	μg/L	736	0.83	558	0.67	572	0.48	543	0.91
Total phosphorus	μg/L	383	0.69	263	0.75	201	0.67	121	1.66
Soluble phosphorus	μg/L	143	0.46	56	0.75	80	0.71	26	2.11

* Mixed land use consists primarily of low- and medium-density residential and commercial land uses.

† See Table 14.2.4 for constituent definitions.

Source: Ref. 31.

across various sites. An indication of this variability is shown for combined sewer overflows in Table 14.2.4[26] and for storm water in Tables 14.2.5[26] and 14.2.6.[31] The CSO data (Table 14.2.4) are based on a database from 18 combined sewer catchments in seven U.S. and Canadian cities.[47] The storm-water data are from 81 catchments in 22 U.S. cities sampled as part of the U.S. Environmental Protection Agency's Nationwide Urban Runoff Program (NURP).[31,62]

TABLE 14.2.6 Water-Quality Characteristics of Typical Urban Runoff

Constituent*		Event-to-event variability in EMCs (Coefficient of variation)	Site median EMC: For median urban site	Site median EMC: For 90th percentile urban site
BOD_5	mg/L	0.5–1.0	9	15
COD	mg/L	0.5–1.0	65	140
TSS	mg/L	1–2	100	300
Total lead	μg/L	0.5–1.0	144	350
Total copper	μg/L	0.5–1.0	34	93
Total zinc	μg/L	0.5–1.0	160	500
TKN	μg/L	0.5–1.0	1500	3300
NO_2-N + NO_3-N	μg/L	0.5–1.0	680	1750
Total phosphorus	μg/L	0.5–1.0	330	700
Soluble phosphorus	μg/L	0.5–1.0	120	210

* See Table 14.2.4 for constituent definitions.
Source: Ref. 31.

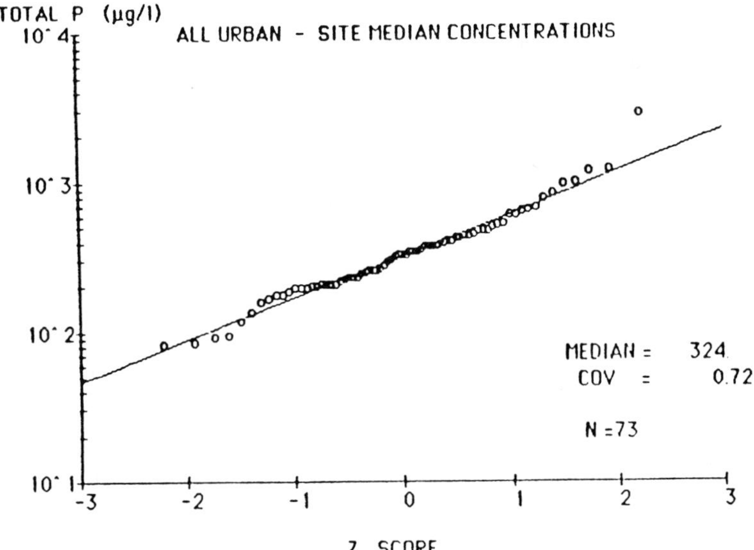

FIGURE 14.2.3 Lognormality of event mean concentrations. (*Reproduced from Ref. 26 with permission.*)

In general, storm-water contaminant concentrations are less than those for combined sewer overflows, but considerable variations occur. Although variations among land uses are shown in Table 14.2.5, the differences are generally not statistically significant, and the data may be combined to characterize a typical urban site, as indicated in Table 14.2.6. Driscoll and James[26] summarize the results of monitoring efforts as follows:

> Site median differences for individual sites are usually as great for sites (*a*) within a study area as for sites within different regions, and (*b*) within a land use as for sites with different land uses. In other words, the type of urban land use has, at best, only a minor influence on pollutant concentrations in runoff. However, land use does affect mass loads, since there are considerable differences in the percentage imperviousness between areas and thus in the volume of runoff.

The lognormal frequency distribution has been found to describe the variation among EMCs very well (e.g., an example of a good fit is shown in Fig. 14.2.3), both for storm water and CSOs.[25] This attribute has been used to develop predictive methods for the frequency distribution of EMCs and loads.[27] In addition, the properties of the lognormal distribution may be used to derive mean concentrations from the median and coefficient of variation as follows:

$$\text{Mean} = \text{median}(1 + \text{CV}^2)^{1/2} \tag{14.2.2}$$

For example, to compute the event mean concentration of BOD in runoff from a residential area from Table 14.2.5, the median concentration is 10 mg/L and CV = 0.41 for 5-day BOD, so the mean concentration for residential storm water is $10 \times (1 + 0.41^2)^{1/2} = 10.8$ mg/L.

14.2.5 Use of Regression for Concentration Estimates

As an alternative to site-specific concentration data, regression equations have been developed from the same Nationwide Urban Runoff Program database[62] for both event mean concentrations (EMCs) and total storm event loads.[28,82] The two most significant explanatory (independent) variables for most constituents were total storm rainfall and total contributing drainage area, followed by imperviousness, land use, and mean annual climatic characteristics. The resulting regression equations were of the form

$$Y = \beta_0 \times X_1^{\beta_1} \times X_2^{\beta_2} \cdots X_n^{\beta_n} \times \text{BCF} \tag{14.2.3}$$

where Y = event mean concentration, $X_1, \ldots, X_n$ = explanatory variables, $\beta_0, \ldots, \beta_n$ = regression coefficients, and BCF = bias correction factor. Regression coefficients for 11 water-quality parameters in three hydrologic regions are presented in Table 14.2.7 with descriptions and units for all variables. The hydrologic regions correspond to regions of less than 20 in mean annual rainfall (region I), between 20 and 40 in mean annual rainfall (region II), and greater than 40 in (region III). Regression coefficients identified as being not significant can be omitted from the calculation.

For example, suppose it is desired to calculate the expected event mean concentration for COD in region II. The required explanatory variables and assumed values are TRN (total storm rain) = 1.3 in; DA (drainage area) = 0.9 mi^2; MAR (mean annual rainfall) = 37 in. Thus,

$$\text{COD} = 0.254 \times 1.3^{-0.259} \times 0.9^{-0.054} \times 37^{1.556} \times 1.299 = 85 \text{ mg/L} \quad (14.2.4)$$

Driver and Tasker[28] and Tasker and Driver[82] provide similar equations for total storm and mean annual loads, plus information on the derivation and limitations of these equations.

14.3 TRANSPORT

14.3.1 Transport Mechanisms and Definitions

Flux. The *transport* of a constituent is measured by the *flux,* with units of quantity per area per time. For constituents measured in units of mass, typical units are $g/m^2 \cdot s$ or $lb/ft^2 \cdot day$, etc. Two fundamental transport mechanisms are of importance in fate and transport studies: *advection* and *diffusion.* (Radiation is a third fundamental transport mechanism that is not considered here.)

Advection. Advection (sometimes known as *convection*) is bodily transport due to the motion of the fluid. The constituent moves with the fluid velocity. Mathematically, the flux F due to advection equals velocity times concentration,

$$F = UC \quad (14.3.1)$$

where U is velocity and concentration C should have units of quantity per volume. For example, the mass flux of a constituent of concentration 2 g/m^3 in a fluid with velocity 3 m/s is 6 $g/m^2 \cdot s$. Similarly, the heat flux in water with temperature 25°C and velocity 3 m/s is

$$\begin{aligned} F &= U \rho c_p T \\ &= 3 \text{ m/s} \times 1000 \text{ kg/m}^3 \times 4.2 \times 10^3 \text{ J/kg} \cdot {}^\circ\text{C} \times 25^\circ\text{C} \\ &= 3.15 \times 10^8 \text{ J/m}^2 \cdot \text{s} \end{aligned}$$

where ρ is the density and c_p is the heat capacity of water.

Advection does not alter the shape (distribution of concentration versus distance) of the constituent distribution as long as the velocity distribution remains uniform. The center of mass of the constituent distribution in water moves with the mean velocity.

Diffusion. Diffusion is the process of transport of a constituent in the direction of decreasing concentration of the constituent. The flux is proportional to the gradient of concentration; e.g., in the x direction,

$$F_x = -E_x \frac{\partial C}{\partial x} \quad (14.3.2)$$

where concentration again should be expressed in units of quantity per volume and E_x is the diffusivity, with units of length²/time. Equation (14.3.2) is also known as Fick's first law of diffusion, and the negative sign indicates a positive flux in the direction of negative gradient (i.e., in the direction of decreasing concentration). *Molecular diffusion* occurs in still water or laminar flow, for which molecular diffusiv-

TABLE 14.2.7 USGS Regression Models for Storm-Runoff Event Mean Concentration

Response variable and region	Regression coefficients														
	$\hat{\beta}_0$	TRN, in	DA, sq mi	IA + 1, percent	LUI + 1, percent	LUC + 1, percent	LUR + 1, percent	LUN + 2, percent	PD, people per square mile	DRN, minutes	INT, inches	MAR, inches	MNL, pounds of nitrogen per acre	MJT, degrees Fahrenheit	BCF
COD I	5.035	−0.473	−0.087	—	0.388	0.012*	—	0.048*	—	—	—	0.855	—	—	1.163
COD II	0.254	−0.259	−0.054	—	0.0003*	0.025*	—	−0.033*	—	—	—	1.556	—	—	1.299
COD III	46.9	−0.179	−0.047	—	0.320	0.031	—	−0.169	—	—	—	—	—	—	1.270
SS I	2,041	0.143*	0.108	—	—	—	—	—	—	−0.370	—	—	—	—	1.543
SS II	734	0.132	−0.342	−0.329	—	—	—	—	0.041	—	—	—	—	−0.519	1.650
SS III	176	0.054*	0.286	—	0.168	0.072	—	−0.295	—	—	—	—	—	—	1.928
DS I	0.333	−0.402	0.469	0.445	—	—	—	—	—	—	—	1.497	—	—	1.352
DS II	2,398	−0.112	0.519	0.468	—	—	—	—	—	—	—	—		−1.373	1.179
TN I	3.52	−0.285	0.033*	—	0.512	0.017*	—	0.012*	—	—	—	−0.129	—	—	1.096
TN II	1.65	−0.204	0.065	0.176	—	—	—	—	—	—	—	—	−0.296	—	1.256
TN III	26,915	−0.253	−0.169	0.057*	—	—	—	—	—	—	—	−2.737	—	—	1.308
TKN I	1.282	−0.449	0.022*	—	0.426	−0.016*	—	−0.012*	—	—	—	—	0.347	—	1.167
TKN II	0.830	−0.224	−0.066	0.039*	—	—	—	—	—	—	—	—	0.106	—	1.321
TKN III	9,549	−0.157	−0.159	—	—	—	—	−0.086	—	—	—	−2.447	—	—	1.326
TP I	0.085	−0.232	−0.012*	—	0.552	−0.080	—	0.038*	—	—	—	0.530	—	—	1.261
TP II	0.022	−0.177	−0.133	0.006	—	—	—	—	—	—	2.019	—	—	—	1.521
TP III	2.630	−0.016	−0.107	—	—	0.053	0.184	−0.168	—	—	—	—	—	−0.710	1.365

DP I	0.352	−0.294	−0.013*	—	0.629	−0.136	—	−0.046*	—	—	—	−0.297*	—	—	1.266
DP II	0.003	−0.209	−0.174	0.245	—	—	—	—	—	—	1.514	—	—	—	1.567
DP III	0.060	0.189	−0.076	—	—	—	—	0.358	—	—	—	—	—	—	1.341
CD I	0.338	−0.256	0.025*	—	0.090*	0.033*	—	−0.110	—	—	—	0.481*	—	—	1.166
CD II	0.851	0.223*	0.189*	—	—	—	—	—	—	—	—	—	—	0.394*	1.284
CU I	11.3	−0.327	0.066*	—	0.237	0.048*	—	0.155	—	—	0.406*	—	—	—	1.297
CU II	9.683	−0.298	−0.151	0.157*	—	—	—	—	—	—	—	—	—	—	1.473
CU III	1,774	−0.104	−0.077	—	0.446	0.078	—	−0.204	—	—	−3.247	—	—	—	1.348
PB I	141	−0.347	0.145	—	−0.109	0.034*	—	−0.086	—	—	—	0.046*	—	—	1.304
PB II	0.487	−0.268	−0.359	—	—	0.099	0.152	−0.008*	—	—	—	1.088	—	—	1.433
PB III	39.8	−0.196	0.123	0.404	—	—	—	—	—	—	—	—	—	—	1.510
ZN I	199	−0.338	0.070*	—	—	0.029	0.114*	0.068*	—	—	—	−0.004*	—	—	1.242
ZN II	0.149	−0.238	−0.201	0.278	—	—	—	—	—	—	—	—	—	1.961	1.650
ZN III	1,879	−0.149	−0.061*	—	0.285	0.146	−0.078	—	—	—	—	—	—	−0.916	1.322

$\hat{\beta}_0$ is the regression coefficient that is the intercept in the regression model; TRN is total storm rainfall; DA is total contributing drainage area; IA is impervious area; LUI is industrial and use; LUC is commercial land use; LUR is residential land use; LUN is nonurban land use; PD is population density; DRN is duration of each storm; INT is maximum 24-h precipitation intensity that has a 2-year recurrence interval; MAR is mean annual rainfall; MNL is mean annual nitrogen load in precipitation; MJT is mean minimum January temperature; BCF is bias correction factor; COD is chemical oxygen demand in storm-runoff mean concentration, in milligrams per liter; I is region I, representing areas that have mean annual rainfall less than 20 in; II is region II, representing areas that have mean annual rainfall of 20 to less than 40 in; III is region III representing areas that have mean annual rainfall equal to or greater than 40 in; SS is suspended solids in storm-runoff mean concentration, in milligrams per liter; DS is dissolved solids in storm-runoff mean concentration, in milligrams per liter; TN is total nitrogen in storm-runoff mean concentration, in milligrams per liter; TKN is total ammonia plus organic nitrogen as nitrogen in storm-runoff mean concentration, in milligrams per liter; TP is total phosphorus in storm-runoff mean concentration, in milligrams per liter; DP is dissolved phosphorus in storm-runoff mean concentration, in milligrams per liter; CD is total recoverable cadmium in storm-runoff mean concentration, in micrograms per liter; CU is total recoverable copper in storm-runoff mean concentration, in micrograms per liter; PB is total recoverable lead in storm-runoff mean concentration, in micrograms per liter; ZN is total recoverable zinc in storm-runoff mean concentration, in micrograms per liter; RUN is storm-runoff volume, in cubic feet; dashes (—) indicate that the variable is not included in the model; asterisk (*) indicates the explanatory variable is not significant at the 5-percent level; equation form is given by Eq. (14.2.3).

Source: Ref. 28.

ities depend on the transporting fluid (water), the constituent, and on temperature; molecular diffusivities are reported in publications such as the *Handbook of Chemistry and Physics.*[89]

In surface waters, molecular diffusion is seldom an important process because it is overwhelmed by the process of *turbulent diffusion.* If velocity fluctuations due to turbulence could be described exactly as a function of time and space, then turbulent transport could be analyzed as an advective process. Unfortunately, turbulent fluctuations ordinarily can only be described statistically, but their net effect on transport is to generate a turbulent flux similar to the form of Eq. (14.3.2), in which E_x is the *turbulent diffusivity* in the x direction and C is interpreted as a time average over an interval long enough to average out turbulent fluctuations (from fractions of a second in turbulent pipe flow, to minutes or hours for turbulent eddies in large water bodies). The turbulent diffusivity for mass transport is analogous (and approximately equal to) the eddy diffusivity for momentum transport and depends only on flow conditions (e.g., Reynold's number, roughness and stratification). Thus, mass, heat, and momentum diffuse at approximately equal rates in turbulent flow. For *isotropic* conditions, the diffusivities are equal in the three coordinate directions ($E_x = E_y = E_z$), and for *homogeneous* turbulence, the diffusivity is not a function of location.

Although the use of Eq. (14.3.2) constitutes a convenient and very widely used method for describing turbulent transport, the diffusivities are usually unknown and must be estimated from tracer experiments or other empirical data. Turbulent diffusivities in surface water are several orders of magnitude greater than molecular diffusivities and are often on the order of 1 to 10,000 cm²/s (0.001 to 10 ft²/s).

In practice, "diffusion" often is used to account for all unknown factors in a problem, including undefined velocity fields, trapping of the constituent along boundaries, secondary currents, density effects, etc. The diffusivity thus becomes a calibration parameter, which is difficult to transfer from one situation to another. *Shear-flow dispersion* is another diffusive-type transport process that is very important in one-dimensional analysis in rivers and streams and in some two-dimensional problems.

14.3.2 Shear-Flow Dispersion

Consider a stream or river in which a one-dimensional analysis could be performed to find concentration as a function of longitudinal distance along the stream, x, and time t, that is, $C(x, t)$. Variations across the stream's cross section in the y (lateral or transverse) and z (vertical) directions are eliminated through the use of cross-sectional averages for velocity U and concentration $\overline{C}$,

$$U = \frac{Q}{A} = \frac{1}{A}\int_A u(y, z)\, dA \tag{14.3.3}$$

and

$$\overline{C} = \frac{1}{A}\int_A C(y, z)\, dA \tag{14.3.4}$$

where Q = stream flow rate, A = cross-sectional area, $u(y, z)$ = velocity distribution over the cross section, and $C(y, z)$ = concentration distribution over the cross section.

Consider the instantaneous input of a tracer (depicting a contaminant) uniformly across the stream section, as indicated in Fig. 14.3.1. If the velocity distribution is

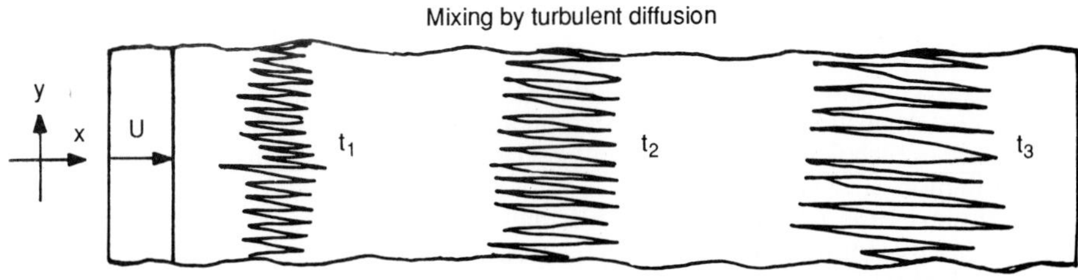

a. Plan view of stream with hypothetical uniform velocity distribution.

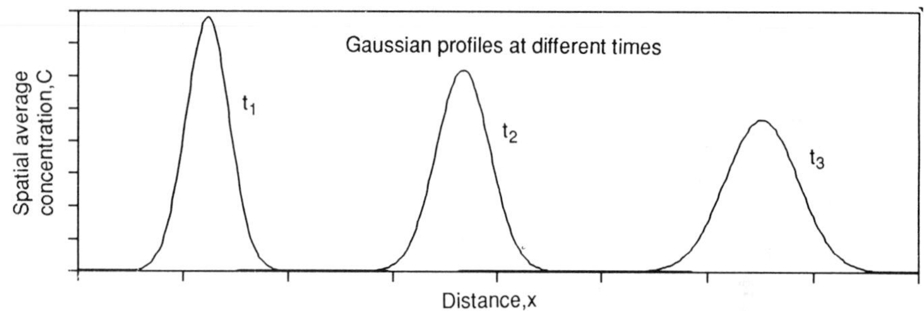

b. Concentration distribution due to instantaneous injection across cross section for Figure 14.3.1a.

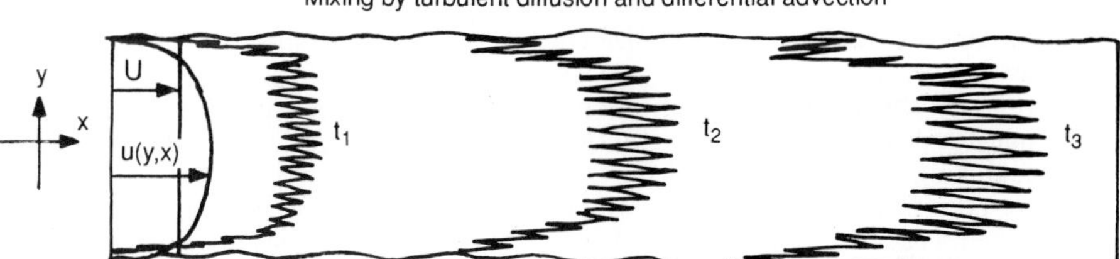

c. Plan view of stream with usual, non-uniform velocity distribution.

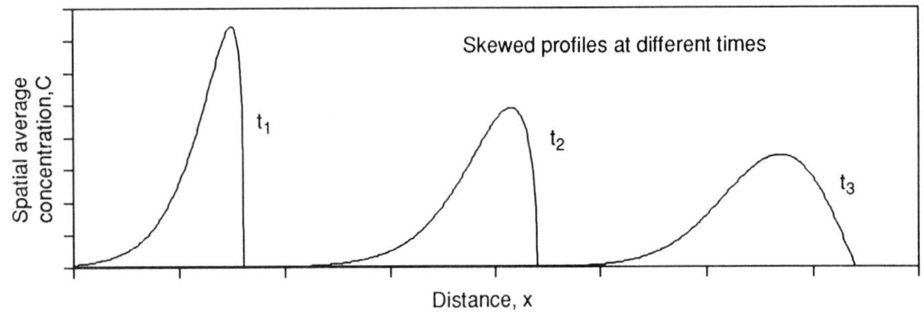

d. Concentration distribution due to instantaneous injection across cross section for Figure 14.3.1c.

FIGURE 14.3.1 Definition sketch for longitudinal dispersion: (*a*) plan view of stream with hypothetical uniform velocity distribution, (*b*) concentration distribution due to instantaneous injection across cross section of Fig. 4.3.1 (*a*), (*c*) plan view of stream with usual nonuniform velocity, (*d*) concentration distribution due to instantaneous injection across cross section for Fig. 14.3.1(*c*).

uniform (constant) over the cross section (Fig. 14.3.1*a*), the tracer center of mass is advected downstream, and mixing and spreading occur uniformly across the section because of turbulent diffusion. No difference occurs between $\overline{C}$ and $C(y, z)$ and the distribution of $\overline{C}(x)$ at a given time is theoretically gaussian (Fig. 14.3.1*b*).

However, the velocity distribution actually varies from zero at the stream boundaries to a maximum at some point in the cross section near the water surface. Hence, in the real stream (Fig. 14.3.1*c, d*), the tracer distribution would be similar to the velocity distribution near the point of release, and the spatial average concentration will be considerably stretched and decidedly skewed (non-Gaussian). There is an additional longitudinal mixing process that occurs, known as *shear-flow dispersion* or *longitudinal dispersion,*[41] because of differential advection and the consequent retardation of tracer near the stream boundaries.

Far enough downstream from the release point of the tracer, an equilibrium is achieved between transverse mixing and shear, and thereafter a one-dimensional analysis can be conducted. Near the release point, the analysis must be either two- or three-dimensional. When a one-dimensional analysis is valid, the longitudinal dispersion process obeys Eq. (14.3.2) with a dispersion coefficient E_L instead of a turbulent diffusion coefficient. Remarkably, the value of E_L depends only on the velocity distribution $u(y, z)$ and on the transverse and vertical turbulent diffusivities and can be predicted from hydraulic considerations alone.[37]

The term *dispersion* is commonly used to refer to diffusive-type mixing; however, for applications to natural waters, it is convenient to reserve the word strictly for shear-flow-type dispersion.[41] This phenomenon was first analyzed by Taylor[83,84] for laminar and turbulent flow in a pipe, followed by Elder[29] for two-dimensional flow down a plane surface. Fischer[35,36] and other investigators extended the analysis to natural streams; results are summarized by Fischer et al.,[37] Holly,[45] and Holley and Jirka.[44]

Although shear-flow dispersion can be predicted for natural streams, in practice the magnitude of E_L incorporates all the additional processes in a stream that enhance mixing. The magnitude of E_L is typically on the order of 100 to 1000 times greater than turbulent diffusivities. The effects of shear-flow dispersion will be encountered whenever an average is taken over one or two dimensions in which the velocity varies, for example, during a two-dimensional, vertically averaged analysis of a shallow lake or estuary.

14.3.3 Governing Advective-Diffusion Equation

The governing conservation of mass equation for a constituent can be derived by equating the change of mass in a control volume to the sum of the net (advective plus diffusive) flux through the control volume plus sources and sinks.[37,40,44,45,91] The most general three-dimensional form in cartesian coordinates is

$$\frac{\partial C}{\partial t} + u\frac{\partial C}{\partial x} + v\frac{\partial C}{\partial y} + w\frac{\partial C}{\partial z} = \frac{\partial}{\partial x}\left(E_x\frac{\partial C}{\partial x}\right) + \frac{\partial}{\partial y}\left(E_y\frac{\partial C}{\partial y}\right) + \frac{\partial}{\partial z}\left(E_z\frac{\partial C}{\partial z}\right) - KC \quad (14.3.5)$$

in which u, v, w = velocity components in the three coordinate directions, x, y, z; C = concentration in the turbulent flow; E_x, E_y, E_z = nonisotropic (a function of direction), nonhomogeneous (a function of location) turbulent diffusivities in the x, y, z directions; and first-order decay (with coefficient K) is assumed.

Equation (14.3.5) can be applied to two-dimensional problems by averaging in one coordinate direction. For example, if vertical averaging (in the z direction) is accomplished, the two-dimensional, nonisotropic, nonhomogeneous form in cartesian coordinates is[40,91]

$$\frac{\partial (hC)}{\partial t} + \frac{\partial (uhC)}{\partial x} + \frac{\partial (vhC)}{\partial y} = \frac{\partial}{\partial x}\left(hE'_x \frac{\partial C}{\partial x}\right) + \frac{\partial}{\partial y}\left(hE'_y \frac{\partial C}{\partial y}\right) - hKC \quad (14.3.6)$$

For this equation, the concentration and u and v velocities (in the x and y directions) are vertically averaged over the variable depth $h(x, y)$. Thus, E'_x and E'_y include a diffusive mixing component due to shear-flow dispersion. Equation (14.3.6) can be simplified if the depth is constant in space and time, since h can then be eliminated from each term.

The one-dimensional form of the advective-dispersion equation commonly applied in streams is[91]

$$\frac{\partial C}{\partial t} + \frac{\partial (UC)}{\partial x} = \frac{\partial}{A\,\partial x}\left(AE_L \frac{\partial C}{\partial x}\right) - KC \quad (14.3.7)$$

In Eq. (14.3.7), $U = Q/A$ is the average longitudinal velocity, $A(x)$ is the cross-sectional area at any location, and $E_L(x)$ is the longitudinal dispersion coefficient.

14.3.4 Useful Analytical Solutions

A multitude of analytical solutions is available for Eqs. (14.3.5) to (14.3.7).[14,17,37,40] One taxonomic breakdown is to consider instantaneous and continuous sources (in time), and point (three-dimensional), line (two-dimensional) and plane (one-dimensional) injections in space. For homogeneous, nonisotropic turbulence, the solutions may be reduced to the Gaussian forms shown in Table 14.3.1.

For the solutions of Table 14.3.1, a point source means that the contaminant release of mass M is concentrated at a single point and can diffuse in all three coordinate directions. A line source means that the contaminant release is spread uniformly over the length h of a line and diffuses only in the x-y plane perpendicular to the line. A plane source means that the contaminant release is spread uniformly over a plane surface of area A and diffuses only in the x direction perpendicular to the plane. These concepts are illustrated in Fig. 14.3.2. Mathematically, the concentration at the point, line, or plane source is infinite for all but the continuous plane source [Eq. (14.3.13)]. The instantaneous releases result in fully transient solutions. The continuous-source solutions assume steady-state conditions at which the concentration is not a function of time.

Equation (14.3.8) gives Gaussian profiles in the x, y, and z directions as an observer moves through the diffusing cloud produced by a point release of mass M at (x_1, y_1, z_1) (Fig. 14.3.2*a*). Equation (14.3.9) (Fig. 14.3.2*b*) gives Gaussian profiles in the x and y directions from an instantaneous line source in the z direction, and Eq. (14.3.11) (Fig. 14.3.2*d*) gives Gaussian profiles in the y and z directions as the observer moves through the diffusing cloud or plume from a continuous point source at (x_1, y_1, z_1). The distribution is also Gaussian in the y direction from the continuous line source [Eq. (14.3.12) and Fig. 14.3.2*e*]. Equation (14.3.10) (Fig. 14.3.2*c*) indicates that C versus x is Gaussian for the solution to Eq. (14.3.7) with an instantaneous source, e.g., "slug" load (Fig. 14.3.2*c*), although the plot of C versus t will always exhibit skew. Finally, the result for a continuous plane source (release in the y-z plane

TABLE 14.3.1 Solutions to Advective Diffusion Equation

Eq. (14.3.8), instantaneous point source [solution to Eq. (14.3.5)]:

$$C(x, y, z, t) = \frac{M}{(4\pi t)^{3/2}(E_x E_y E_z)^{1/2}} \exp - \left\{ \frac{[(x - x_1) - ut]^2}{4E_x t} + \frac{(y - y_1)^2}{4E_y t} + \frac{(z - z_1)^2}{4E_z t} + Kt \right\} \quad (14.3.8)$$

Eq. (14.3.9), instantaneous line source [solution to Eq. (14.3.6)]:

$$C(x, y, t) = \frac{M}{h 4\pi t (E'_x E'_y)^{1/2}} \exp - \left\{ \frac{[(x - x_1) - ut]^2}{4E'_x t} + \frac{(y - y_1)^2}{4E'_y t} + Kt \right\} \quad (14.3.9)$$

Eq. (14.3.10), instantaneous plane source [solution to Eq. (14.3.7)]:

$$C(x, t) = \frac{M}{A(4\pi E_L t)^{1/2}} \exp - \left\{ \frac{[(x - x_1) - Ut]^2}{4E_L t} + Kt \right\} \quad (14.3.10)$$

Eq. (14.3.11), continuous point source [solution to Eq. (14.3.5) with $\partial C/\partial t = 0$]:

$$C(x, y, z) = \frac{q}{4\pi x (E_y E_z)^{1/2}} \exp - \left\{ \frac{u(y - y_1)^2}{4E_y x} + \frac{u(z - z_1)^2}{4E_z x} + K\frac{x}{u} \right\} \quad (14.3.11)$$

Eq. (14.3.12), continuous line source [solution to Eq. (14.3.6) with $\partial C/\partial t = 0$]:

$$C(x, y) = \frac{q}{h(4\pi u x E'_y)^{1/2}} \exp - \left\{ \frac{u(y - y_1)^2}{4E_y x} + K\frac{x}{u} \right\} \quad (14.3.12)$$

Eq. (14.3.13), continuous plane source [solution to Eq. (14.3.7) with $\partial C/\partial t = 0$]:

$$C(x) = C_o \exp \left\{ \frac{x}{2E_L}(U \mp \Omega) \right\} \quad (14.3.13)$$

Inject at $x = 0$,
$-$for $+x$ (downstream),
$+$for $-x$ (upstream)

Note: For all solutions the velocity vector is parallel to the x axis ($u \neq 0$, $v = w = 0$) and water is initially clean ($C = 0$ at $t = 0$). Parameters:

M = mass injected instantaneously
q = mass rate of injection (mass/time)
h = depth of injection (length of line source)
A = cross-sectional area
x_1, y_1, z_1 = coordinates of injection location
u = x velocity
U = average velocity in stream
E_x, E_y, E_z = diffusivities in x, y, z directions
E'_y = depth-averaged diffusivity
E_L = longitudinal dispersion coefficient
$C_o = q/(A\rho\Omega) = q/(hw\Omega) \approx q/(A\rho U) \approx$ well-mixed initial concentation, Eq. (14.4.18)
ρ = water density
$\Omega = U(1 + 4KE_L/U^2)^{1/2}$
K = first-order decay coefficient (1/time)

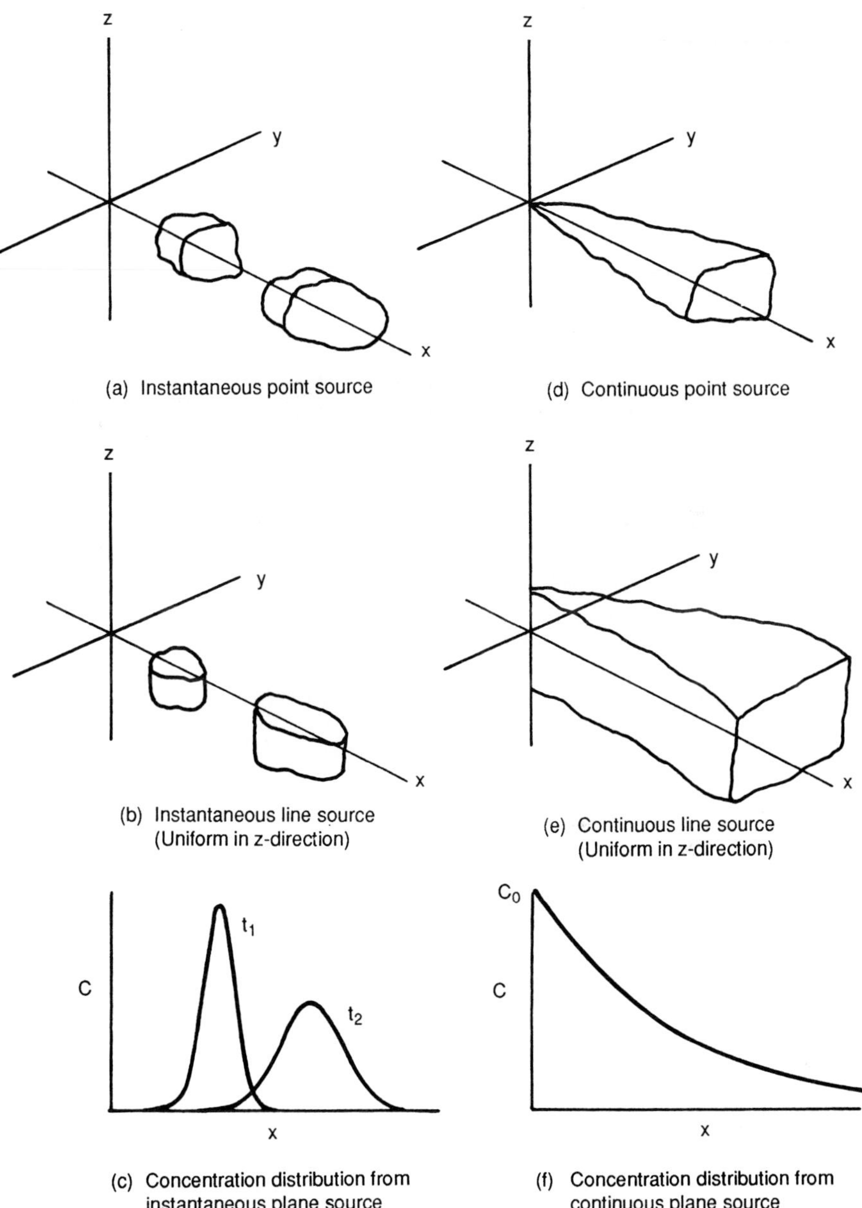

FIGURE 14.3.2 Concentration distributions downstream from release at origin.

whose motion is traced in the x-direction), Eq. (14.3.13) (Fig. 14.3.2f), shows a decrease in concentration only because of decay; the concentration otherwise would remain constant at the completely mixed value as the constituent moved downstream. Although Eq. (14.3.13) can be applied upstream of the injection point, this would ordinarily not be meaningful in rivers. In estuaries, however, the equations sometimes are applied for average conditions over a tidal cycle (Sec. 14.5.2), in which case diffusive transport would occur upstream of the injection point.

In general, the plumes described by these equations have a spatial variance σ^2 in any direction that is related to E by[34]

$$E = 0.5\frac{d\sigma^2}{dt} = 0.5\,u\frac{d\sigma^2}{dx} \tag{14.3.14}$$

which for constant E reduces to

$$\sigma^2 = 2Et + \sigma_o^2 = \frac{2Ex}{u} + \sigma_o^2 \tag{14.3.15}$$

where σ_o^2 is the initial spatial variance of the contaminant when released into the flow field. In this discussion, the x axis is assumed to be aligned with the velocity vector. The variance may be substituted into Eqs. (14.3.8) to (14.3.12) to derive equivalent common Gaussian forms for the plume concentration.[44,76]

Equations (14.3.8) to (14.3.13) assume an infinite flow field. If the source lies on a boundary, such as an instantaneous point source at the surface of a deep lake or a vertical line source on one bank of a wide stream, then the values of C given in the Table 14.3.1 solutions should be multiplied by 2 to account for the semi-infinite flow field. If the source lies near a boundary, e.g., a pipe discharging in the middle of a stream or a line source on one bank of a narrow river, the *method of images* (Fig. 14.3.3) can often be used to account for the boundaries.[37,44,70] Several image-source

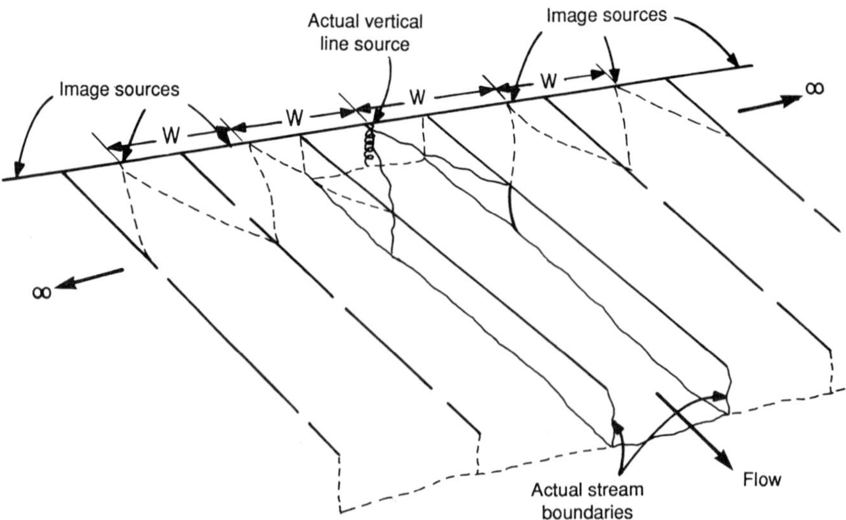

FIGURE 14.3.3 Method of images.

solutions are given by Holley and Jirka[44] for various source configurations in streams having approximately rectangular cross sections.

If boundaries are too irregular for application of the method of images, a numerical solution to the differential equation should be employed. The effect of boundaries is to retain the constituent within the defined flow field (e.g., within the stream, estuary, or lake). For example, for the situation of an instantaneous release in a stream, Eq. (14.3.8) would first apply to define the three-dimensional conditions near the source. After vertical mixing occurs, Eq. (14.3.9) could be used to describe two-dimensional lateral and longitudinal mixing. After the stream is mixed both vertically and laterally, Eq. (14.3.10) can be used to describe further mixing in the longitudinal direction.

In the six solutions shown in Table 14.3.1, first-order decay is simulated very simply by multiplication of the solution for a conservative pollutant by a factor $e^{-Kt} = e^{-Kx/u}$, where $t = x/u$ is the travel time. For the situation of a continuous, plane source [Eq. (14.3.13)], the only reason for the exponential decrease in concentration with x is this decay. If no decay occurred, a conservative pollutant would simply maintain a constant concentration downstream from the initial condition of complete mixing at the injection point.

In the one-dimensional solutions, Eqs. (14.3.10) and (14.3.13), the average velocity U and the longitudinal dispersion coefficient E_L are used to emphasize that E_L accounts for shear-flow dispersion encountered when using the average velocity for the stream. Equations (14.3.11) and (14.3.12) assume that longitudinal turbulent diffusion can be neglected compared to longitudinal advection. This is almost always true in natural streams. For one-dimensional analysis downstream from a continuous plane source [Eq. (14.3.13)], dispersion can be neglected relative to advection if

$$\frac{4KE_L}{U^2} \ll 1 \tag{14.3.16}$$

a condition satisfactorily fulfilled if $4KE_L/U^2 \leq 0.1$. This is usually true for rivers but may not be true in estuaries, where U is usually smaller and E_L is usually higher than for rivers.

When dispersion can be neglected compared to advection, Eq. (14.3.13) reduces to

$$C(x) = C_o e^{-Kx/U} \tag{14.3.17}$$

where C_o is the (well-mixed) concentration at $x = 0$, and it is seen that simple exponential decay occurs in proportion to travel time along the stream.

If mixing is complete in the system, then no concentration gradients occur and Eq. (14.3.5) reduces to

$$\frac{dC}{dt} = -KC \tag{14.3.18}$$

with the familiar solution for a *continuously stirred tank reactor* (CSTR),

$$C(t) = C_o e^{-Kt} \tag{14.3.19}$$

Clearly, Eqs. (14.3.17) and (14.3.19) represent similar phenomena if t is interpreted as travel time, $t = x/U$ (Fig. 14.3.2f).

14.3.5 Parameter Value Estimates

Methods for Evaluation. The solution of Eqs. (14.3.5) to (14.3.7) and the use of Eqs. (14.3.8) to (14.3.13) depend on knowledge of the velocity field, turbulent diffusivities for problems approximated by two- and three-dimensional solutions, and the longitudinal dispersion coefficient for one-dimensional problems. Velocities may be determined by a variety of conventional techniques, but diffusivities are much more difficult to measure. In some cases, the fundamental definition of the diffusion coefficient can be used [Eq. (14.3.2)], and the diffusivity found by

$$E = -\frac{\text{flux}}{\text{gradient}} \tag{14.3.20}$$

if both the flux and gradient can be measured. Most often, however, tracer experiments are conducted. For two- and three-dimensional analyses, the Gaussian solutions can be manipulated in various ways to evaluate E,[20] but the most common method is probably to use the relationship between diffusivity and variance, Eq. (14.3.14). The spatial variance is the second central moment, which can be evaluated from tracer data. For example, if a boat is used to take transverse samples across a plume downstream from a continuous point source, then the estimate S_y^2 of the variance of concentration in the transverse or y direction is

$$S_y^2 = \frac{\sum_i C_i y_i^2 \, \Delta y_i}{\sum_i C_i \, \Delta y_i} - \bar{y}^2 \tag{14.3.21}$$

where the first moment about the arbitrary origin is

$$\bar{y} = \frac{\sum_i C_i y_i \, \Delta y_i}{\sum_i C_i \, \Delta y_i} \tag{14.3.22}$$

As shown in Fig. 14.3.4, the two equations assume samples at variable distances across the plume, y_i, with corresponding variable intervals between the samples, Δy_i. When variances are computed at several distances or travel times downstream from the release point, Eq. (14.3.14) can be evaluated numerically.

Oceans and Large Lakes. Generalized results are available for very large water bodies such as the ocean and the Great Lakes, for which the diffusivity is found to depend on the length scale of mixing.[9,10,18,68] For example, Brooks[10] advocated the use of the 4/3 law for oceanic diffusion at the surface, downstream of locations of submerged diffusers of sewage effluent,

$$E = 0.01 \, L^{4/3} \tag{14.3.23}$$

where E = diffusivity in the horizontal plane (cm^2/s), and L = length scale (cm). The length scale L is usually a multiple of the square root of the spatial variance of the plume or cloud [Eq. (14.3.21)], e.g., $L = 3S_y$.[18,37] Okubo[68] notes that the coefficient may vary with the length scale and suggests

$$E = 0.01 L^{1.15} \tag{14.3.24}$$

as the best fit for all oceanic and Great Lakes data (again, units are cm^2/s for E and cm for L, and the coefficient 0.01 is dimensional). The 4/3 law is limited to locations far

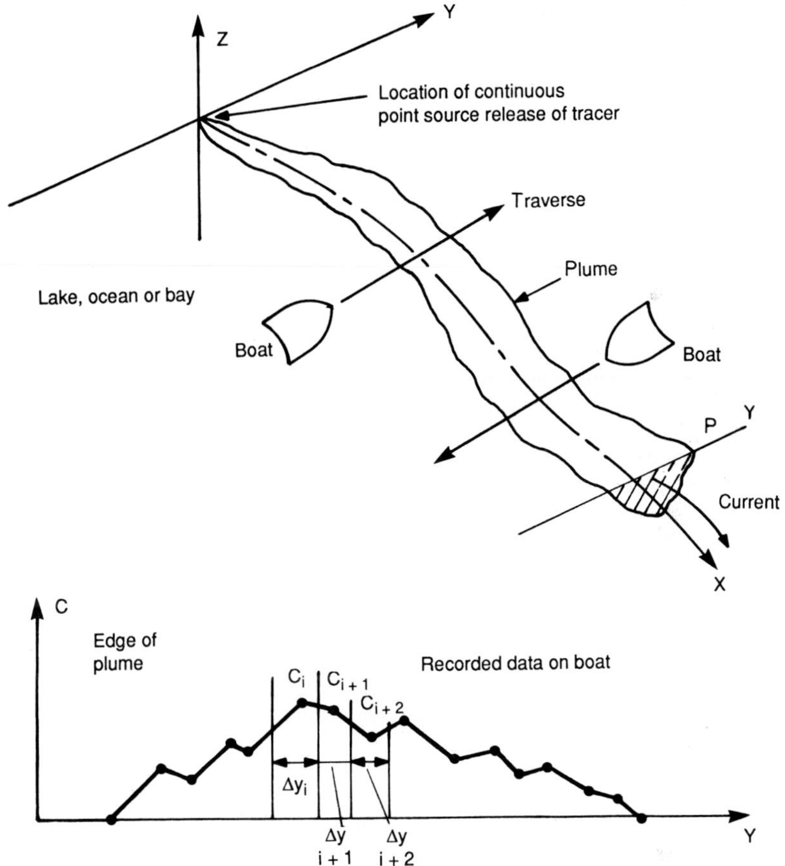

FIGURE 14.3.4 Data from tracer experiments.

from boundaries[37] and thus seldom applicable in streams, estuaries, and smaller lakes. For application to these locations, site-specific analysis must usually be performed to determine diffusivities, especially since they are influenced by wind, geometry, and all the mixing processes discussed earlier.

Rivers and Streams. Typical values for E_y (lateral or transverse diffusivity) for rivers have been compiled by Fischer et al.[37] and Bowie et al.[9] On the basis of these data, Fischer et al.[37] recommend for "relatively straight" natural channels

$$E_y \approx 0.6\bar{d}u_* \tag{14.3.25}$$

where $\bar{d}$ is the average depth and u_* is the shear or friction velocity,

$$u_* = \sqrt{\frac{\tau_o}{\rho}} = \sqrt{\frac{f}{8}}\,U = \sqrt{gRS} \tag{14.3.26}$$

where τ_o = boundary shear stress
ρ = fluid density
f = Darcy-Weisbach friction factor
U = average velocity
g = gravitational acceleration
R = hydraulic radius
S = slope of the energy grade line

As a very crude approximation, $u_* \approx 0.1U$, but such an approximation should be made only in the total absence of other hydraulic information.

The factor of 0.6 in Eq. (14.3.25) is stated as being within ±50 percent by Fischer et al.[37] and can be 2 to 8 times greater when the effect of bends, meanders, and other stream factors are considered.[19] Nonetheless, Eq. (14.3.25) is probably the most useful means for predicting transverse diffusivities in streams, short of the superior technique of site-specific measurements. Fischer et al.[37] and Holley and Jirka[44] provide techniques for approximating the additional mixing caused by bends in rivers. Yotsukura and Sayre[92] illustrate an alternative coordinate system for problems of transverse mixing in natural channels.

The depth-averaged vertical diffusivity E_z can be adequately approximated by

$$E_z = 0.067du_* \tag{14.3.27}$$

where d is the local depth.[37] Equation (14.3.27) is based on a logarithmic velocity profile and the assumption that the mass diffusivity equals the momentum diffusivity *(eddy diffusivity)*. In most streams and rivers, vertical diffusion is sufficient to produce vertically well-mixed conditions soon after entry of contaminants into the stream, from which point the problem becomes two-dimensional and lateral diffusion governs.

The longitudinal dispersion coefficient E_L varies widely in natural streams and rivers as demonstrated in tabulations by Fischer et al.[37] and Schnoor et al.,[74] with values of E_L ranging from approximately 30 to 15,000 ft²/s (3 to 1500 m²/s). Discounting the high value of 15,000 ft²/s (1,500 m²/s) for the Missouri River, the upper bound is still on the order of 7000 ft²/s (700 m²/s), although most values are in the range 30 to 700 ft²/s (3 to 70 m²/s). The range of values for the dimensionless dispersion coefficient,

$$E'_L = \frac{E_L}{u_* \bar{d}} \tag{14.3.28}$$

is still high, with approximate bounds of $10 < E'_L < 3000$ and most values within the range 100 to 500. Thus, it is difficult to select a value of E_L strictly on the basis of experiments in other streams and rivers.

Hence, the best method for finding E_L remains site-specific measurements, often utilizing Eq. (14.3.14) in finite-difference form. For studies in which a pollutant cloud from an upstream instantaneous injection is sampled as it moves past fixed downstream locations, Fischer[34] has shown that

$$\sigma_x^2 = U^2 \sigma_t^2 \tag{14.3.29}$$

where σ_x^2 is the variance of the C versus x distribution (units of length squared), and σ_t^2 is the variance of the C versus t distribution measured at a fixed location (units of time squared). Hence, data collected at fixed locations can be used to evaluate the temporal variances, and Eq. (14.3.14) can be used in the form

$$E_L = \frac{U^2}{2}\frac{d\sigma_t^2}{dt} = \frac{U^3}{2}\frac{d\sigma_t^2}{dx} \approx \frac{U^2}{2}\frac{(\sigma_{t_2}^2 - \sigma_{t_1}^2)}{(\bar{t}_2 - \bar{t}_1)} \tag{14.3.30}$$

where the variances (second central moments) are estimated from the temporal concentration distribution at the upstream (subscript 1) and downstream (subscript 2) stations, and $\bar{t}$ is the center of mass (first moment) of a temporal distribution. The numerical evaluation of σ_t^2 and $\bar{t}$ is analogous to Eqs. (14.3.21) and (14.3.22), respectively. The best estimate of average velocity from such studies is found from the time taken for the center of mass to traverse the reach:[49]

$$U = \frac{x_2 - x_1}{\bar{t}_2 - \bar{t}_1} \tag{14.3.31}$$

In lieu of tracer measurements, Fischer et al.[37] provide an estimate for E_L based strictly on the hydraulic properties of the stream cross section,

$$E_L \approx \frac{0.011 U^2 w^2}{\bar{d} u_*} \tag{14.3.32}$$

where $\bar{d}$ is the average depth and w is the stream width. Fischer et al.[37] report that Eq. (14.3.32) is generally within a factor of 4 of values of E_L measured from tracer experiments. Improvements are possible given more detailed hydraulic information, such as the cross-sectional velocity distribution.

Estuaries. In estuaries, generalizations for diffusivities and dispersion coefficients are even more difficult. Fischer et al.[37] report dimensionless transverse diffusivities [equivalent to the coefficient of 0.6 in Eq. (14.3.25)] of 0.4 to 1.6 for five estuaries and recommend site-specific measurements. However, Eq. (14.3.27) is probably adequate for the average vertical diffusivity. These authors also provide a method for approximating the longitudinal dispersion coefficient as a function of hydraulic parameters and the tidal period. Further discussion is contained in Sec. 14.5.2.

Lakes. In lakes, vertical diffusion is usually the primary flux of interest, and Bowie et al.[9] and Schnoor et al.[74] provide literature reviews. The diffusivity E_z is a function of wind shear, bottom roughness (for shallow lakes), lake morphometry, and degree of stratification. Average E_z values for whole lakes range from 10^{-5} to 1 ft^2/s (10^{-6} to 0.1 m^2/s) with smaller values during the summer period of stratification. Values within the *thermocline* (region of highest density difference and greatest stratification) range from 10^{-6} to 2.5×10^{-4} ft^2/s (10^{-7} to 2.5×10^{-5} m^2/s). Many of these values were determined by fitting one-dimensional models of thermal stratification to observed temperature profiles. For shallow lakes, Bowie et al.[9] provide summaries of theoretical relationships of E_z to wind stress and degree of stratification.

14.4 CONSTITUENT SOURCES AND SINKS

14.4.1 Kinetics and Rate Constants

The production or loss of a constituent, with or without interaction with another constituent, is a *kinetic process,* to be distinguished from the physical transport processes of advection and diffusion/dispersion. Examples of kinetic processes include decay of bacteria (loss), oxidation of carbonaceous material (loss), reaeration of dissolved oxygen across the air-water interface (gain), etc. More specifically, constituents may be affected by the production-degradation processes listed in Table 14.4.1.[61]

Details of the processes of Table 14.4.1 are beyond the scope of this chapter, except for discussion of selected first-order processes of special and historic concern for water pollution, e.g., oxygen demand and dissolved oxygen balance. Useful screening procedures to account for the remaining processes of Table 14.4.1 are provided by Mills et al.[61] and Schnoor et al.[74]

In general, quantification of such processes involves insertion of source/sink terms into the governing advective-diffusion equation. The general form of such terms is

$$\frac{dC_i}{dt} = f(C_i, C_j, T) \qquad j = 1, 2, \ldots \tag{14.4.1}$$

where the rate of change of concentration of constituent i may depend on C_i as well as on temperature T and concentrations of other constituents, C_j. A common assumption that performs adequately for many processes is that the decay (or production) of a constituent is *first order:*

$$\frac{dC}{dt} = -KC \tag{14.4.2}$$

where K is the first-order rate coefficient (time^{-1}), and the negative sign indicates decay or loss. A *second-order process* is one in which dC/dt is proportional to C^2, etc. A more complex formulation is shown in Fig. 14.4.1 for *Michaelis-Menten* or *Monod* kinetics in which, for decay,

$$\frac{dC}{dt} = -\frac{k_s C}{k_{1/2} + C} \tag{14.4.3}$$

TABLE 14.4.1 Physical, Chemical, and Biological Processes Affecting the Production and Degradation of Water-Quality Constituents

Process	Nature	First order	Comments
Sorption	Chemical-physical	No	Adsorption on sediment. Sediment transport governs constituent transport.
Volatilization	Chemical-physical	Yes	Governed by fluid turbulence (reaeration) and/or atmospheric mass transfer. Henry's law constant is key parameter.
Sedimentation	Physical	Sometimes	Sediment transport governs deposition and scour of particulates and thus adsorbed constituents.
Biodegradation	Biological	Usually	Decay through microbial decomposition. Mineralization is breakdown of organic matter to CO_2, N_2 and H_2O.
Photolysis	Chemical-physical	Usually	Photolysis rate is a complex function of light absorbance properties of compound.
Hydrolysis	Chemical	Yes	Influenced by pH.

Source: Based on Ref. 61.

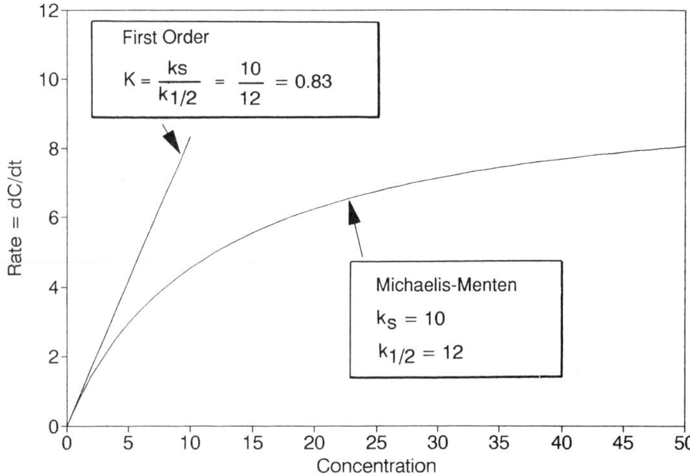

FIGURE 14.4.1 Michaelis-Menten kinetics.

where k_s (concentration/time) is the limiting reaction rate (when $C \gg k_{1/2}$), and $k_{1/2}$ is called the *half-saturation constant* because dC/dt is half the limiting value when $C = k_{1/2}$. As shown in Fig. 14.4.1, the behavior is first order when $C \ll k_{1/2}$, with $K = k_s/k_{1/2}$. Equation 14.4.3 is often applied to the kinetics of nutrients (nitrogen and phosphorus), especially in eutrophication studies.[73]

Generally, when one constituent decays, another may be formed or degraded. That is, the kinetics of several constituents may be coupled, leading to a set of simultaneous equations of the form of Eq. (14.4.1). Tables of first- and second-order decay coefficients for many chemicals can be found in Mills et al.,[61] Schnoor et al.,[74] and EPA.[32] When several first-order processes (Table 14.4.1) affect the fate of a constituent, the overall first-order rate coefficient is the sum of the first-order rate coefficients for the individual component processes.

The temperature dependence of rate coefficients is usually described with reference to their rate at 20°C; that is,

$$K(T) = K(20)\theta^{T-20} \tag{14.4.4}$$

where T is in degrees Celsius and the constant θ is typically in the range 1.01 to 1.10 (Table 14.4.2), depending on the constituent.[9,61] The key point is that reaction rates tend to increase with increasing temperature.

TABLE 14.4.2 Typical Temperature Correction Factors θ [Eq. (14.4.4)]

Constituent and process (defined in text)	Rate coefficient	θ	Reference
Oxidation of BOD	K_1	1.047	Camp[13]
Reaeration of DO	K_2	1.024	Camp[13]
Oxidation of NOD*	K_N	1.085	Bowie et al.[9]
Decay of total coliforms	K_s	1.07	Mancini[53]

* Nitrogenous oxygen demand.

If Eq. (14.4.2) [same as Eq. (14.3.18)] is solved for the case of a continuously stirred tank reactor, the solution is given by Eq. (14.3.19). Literature values for rate constants sometimes refer to base *e* logarithms implied by Eq. (14.3.19), and sometimes to base 10 logarithms (common logarithms), for which 10^{-Kt} would replace e^{-Kt} in Eq. (14.3.19). It is important to know the proper base when reviewing rate coefficients, because

$$K_{\text{base } e} = 2.303 K_{\text{base } 10} \tag{14.4.5}$$

Thus, Eq. (14.3.19) can be written in two forms:

$$C(t) = C_o e^{-K_{\text{base } e} t} = C_o 10^{-K_{\text{base } 10} t} \tag{14.4.6}$$

14.4.2 The Oxygen Balance

Dissolved oxygen is one of the most important water-quality factors because of its effect on aquatic flora and fauna, and especially on the ability of a water body to degrade organic wastes. All regulatory agencies within the United States have a DO standard for natural waters, such as the dissolved oxygen shall never be less than 4 mg/L, and the 24-h average DO $\geq$ 5 mg/L. The concentration 5 mg/L is often specified as the acceptable minimum in waste-load allocation studies (i.e., studies in which the degree of wastewater treatment is to be determined for an individual or group of wastewater dischargers). Source characteristics of oxygen demand are shown in Tables 14.2.2 and 14.2.3.

Various sources and sinks of DO are summarized in Table 14.4.3. If oxygen demand exceeds the rate of production of DO enough so that the DO in the water column is completely depleted, anaerobic conditions can produce fish kills and also lead to production of methane and other noxious gases. If there was no oxygen demand (clean water), natural waters would become saturated with dissolved oxygen because of reaeration across the water surface. When the contribution of sources of oxygen demand is equaled by the supply of oxygen, equilibrium conditions result, with a DO concentration lower than the saturation value.

The oxidation of BOD may be represented by a first-order process,

$$\frac{dL}{dt} = -K_1 L \tag{14.4.7}$$

where L = BOD concentration (mg/L) and K_1 is the deoxygenation coefficient (also denoted as K_d in the literature). Typical values of K_1 are in the range 0.1 to 4.0 day^{-1}, with larger values for untreated wastewater and smaller values for treated wastewater and natural waters.[90] Rate constants for natural waters are seldom the same as for wastewater effluent.[85] Inverse relationships with stream flow have been reported, e.g. Ref. 90,

$$K_1 = 10.3 Q^{-0.49} \tag{14.4.8}$$

where K_1 is in day^{-1} at 20°C and Q is in ft^3/s. With Q in m^3/s, the coefficient of 10.3 becomes 1.80.

Laboratory tests often report 5-day BOD, or BOD_5, that is, the amount of oxygen utilized by carbonaceous matter in 5 days. The relationship between BOD_5 and ultimate first-stage (carbonaceous) BOD, L_o, is based on Eq. (14.4.7) and Fig. 14.4.2,

$$L_o = \frac{\text{BOD}_5}{1 - e^{-5K_1}} \tag{14.4.9}$$

TABLE 14.4.3 Sources and Sinks of Oxygen and Oxygen Demand in Natural Waters

Produce oxygen
1. Reaeration across the water surface 2. Photosynthesis by plants (diurnal effect)
Remove oxygen
3. Biochemical oxygen demand = BOD, a measure of O_2 required by bacteria in the stabilization (oxidation) of organic (carbonaceous) matter 4. Nitrogenous oxygen demand = NOD, a measure of O_2 used by bacteria to oxidize organic and ammonia nitrogen to nitrite and nitrate nitrogen 5. Chemical oxygen demand = COD, laboratory measure of all oxidizable material, unlikely to be realized in most streams 6. Respiration of plants (diurnal effect) and animals 7. Purging of gases rising from the benthal layer (layer of decaying and often anaerobic matter on stream bottom) 8. Diffusion into benthal layer to satisfy O_2 demand in the aerobic zone of this layer. Numbers 7 and 8 are often combined into sediment oxygen demand = SOD.
Remove oxidation demand
9. Oxidation by DO 10. Sedimentation and adsorption
Add oxidation demand
11. Point sources, e.g., wastewater treatment plants, industries 12. Nonpoint sources and local runoff along reach, e.g., storm water and agricultural runoff 13. Scour of bottom sediments and diffusion of partly decomposed organic matter from the benthal layer into overlying water.

Either ultimate BOD or BOD_5 can be included in a conservation-of-mass equation, but it is more customary to use ultimate BOD as L in Eq. (14.4.7).

The *nitrogen cycle* as it affects dissolved oxygen is sketched in Fig. 14.4.3. Organic nitrogen (contained in organic waste and algae) is converted first to ammonia nitrogen that is then oxidized to nitrite and nitrate. Algae utilize nitrate nitrogen, and

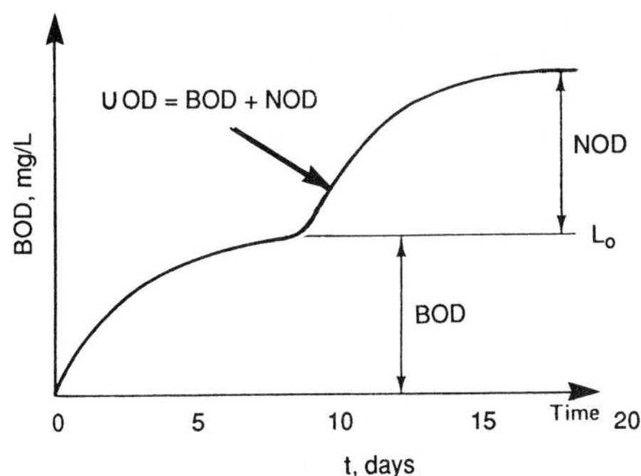

FIGURE 14.4.2 Carbonaceous (BOD), nitrogenous (NOD), and ultimate (UOD) oxygen demand curves.

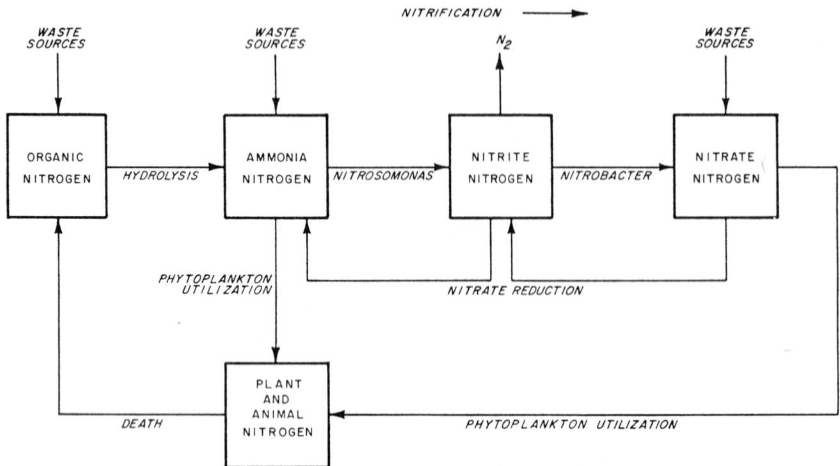

FIGURE 14.4.3 The nitrogen cycle. (*Reproduced from Ref. 8 with permission.*)

recycling occurs to the organic form as they die. In the literature, the notation NO_3-N is often encountered, meaning the nitrogen present in the nitrate form, etc. For example, 1 mg/L of NO_3 contains only 14/62 = 0.23 mg/L of nitrogen, based on the ratio of molecular weights of nitrogen and nitrate. Hence, it is important to know the chemical form of nitrogen, phosphorus, etc. that is reported in data summaries. The sum of organic nitrogen and ammonia nitrogen is called *total Kjeldahl nitrogen* (TKN) in a laboratory analysis procedure.

The various nitrogen components shown in Fig. 14.4.3 may be considered individually, with separate first-order or Michaelis-Menten kinetics for each component[85] or treated collectively as *nitrogenous oxygen demand,*

$$\text{NOD} = 4.57\text{TKN} + 1.14\text{NO}_2\text{-N} \tag{14.4.10}$$

where the stoichiometric constants 4.57 and 1.14 arise because 4.57 g oxygen is required to convert 1 g TKN to NO_3-N and 1.14 g oxygen is required to convert 1 g NO_2-N to NO_3-N.[80] (Conversion of 1 g TKN to NO_2-N requires the difference, or 3.43 g oxygen.) Most of the NOD is due to TKN (organic N + NH_3-N) because nitrite concentrations in most wastewater streams and ambient waters are very low, less than 0.1 mg/L. NOD is often treated with first-order kinetics with a rate constant K_N. Values of K_N for streams are in the range of 0.2 to 0.6 day^{-1} (Ref. 56), although much wider ranges (0.02 to 6 day^{-1}) have been reported for natural waters.[9]

An even simpler way in which to deal with both BOD and NOD is to combine them into a single *ultimate oxygen demand* (UOD),

$$\text{UOD} = \text{ultimate BOD} + \text{NOD} \tag{14.4.11}$$

with a single first-order rate coefficient, approximately similar in magnitude to a value for K_1. This approximation is most suited to shallow streams in which benthic (stream bottom) nitrifying organisms exert a nitrogenous oxygen demand simultaneously with carbonaceous oxygen demand.[16]

Oxygen is added to the water column as plants (e.g., algae) photosynthesize during daylight hours, and oxygen is removed during nighttime hours as plants respire.

TABLE 14.4.4 Sediment Oxygen Demand Ranges

Location	Range, g $O_2/m^2 \cdot$day	Source
Municipal sewage sludge, outfall vicinity	2–10	Thomann and Mueller[85]
Municipal sewage sludge, aged, downstream of outfall	1–2	Thomann and Mueller[85]
Estuarine mud	1–2	Thomann and Mueller[85]
Sandy bottom	0.2–1	Thomann and Mueller[85]
Mineral soils	0.05–0.1	Thomann and Mueller[85]
Measured in rivers and streams	0.02–44	Bowie et al.[9]
Measured in estuaries and ocean	0.1–11	Bowie et al.[9]
Measured in lakes and reservoirs	0.004–9	Bowie et al.[9]

Depending on algal concentrations, large (1 to 5 mg/L) diurnal swings in DO can occur.[85] The likely magnitude can be determined from light- and dark-bottle BOD tests[85] and simulated by using simple cosine source/sink functions[24] or more complex coupled models of algae and DO such as QUAL2E.[11]

Sediment oxygen demand is typically reported in units of grams O_2 per square meter per day. Thus, when multiplied by the surface area, required source/sink units [Eqs. (14.3.5) to (14.3.7)] of concentration per unit time result. Values in natural waters cover a very wide range (Table 14.4.4).

Oxygen concentrations may range from zero to a *saturation concentration*, C_s, that is a function of temperature and salinity as shown in Table 14.4.5. Saturation concentrations in fresh and saline water are closely approximated by[15]

$$C_s = \exp\left[c_0 + \frac{c_1}{T} + \frac{c_2}{T^2} + \frac{c_3}{T^3} + \frac{c_4}{T^4} + \text{Sal}\left(c_5 + \frac{c_6}{T} + \frac{c_7}{T^2}\right)\right] \quad (14.4.12)$$

TABLE 14.4.5 Sea-Level Values for Dissolved Oxygen Saturation Concentration, mg/L.

Temperature, °C	Salinity, ppt 0	10	20
0	14.621	13.636	12.714
5	12.770	11.947	11.175
10	11.288	10.590	9.934
15	10.084	9.485	8.921
20	9.092	8.572	8.081
25	8.263	7.807	7.376
30	7.559	7.155	6.773
35	6.950	6.590	6.249

Source: B. B. Bensen and D. Krause, Jr., "The Concentration and Isotopic Fractionation of Oxygen Dissolved in Freshwater and Seawater in Equilibrium with the Atmosphere," *Limnology and Oceanography,* vol. 29, no. 3, pp. 620–632, 1984. Reproduced with permission.

where C_s is in mg/L, T is in kelvins (K = °C + 273.15), Sal is salinity (related to chlorinity or chloride concentration) in parts per thousand (ppt), and the coefficients are given in Table 14.4.6.

TABLE 14.4.6 Coefficients for Dissolved Oxygen Saturation in Eq. (14.4.12)

Coefficient	Value
c_0	-139.34411
c_1	1.575701×10^5
c_2	-6.642308×10^7
c_3	1.243800×10^{10}
c_4	-8.621949×10^{11}
c_5	-0.017674
c_6	10.754
c_7	2140.7

Source: B. B. Benson and D. Krause, Jr., "The Concentration and Isotopic Fractionation of Oxygen Dissolved in Freshwater and Seawater in Equilibrium with the Atmosphere," *Limnology and Oceanography*, vol. 29, no. 3, pp. 620–632, 1984. Reproduced with permission.

Salinity is a close representation of total dissolved solids in water and is related to chlorinity by[15]

$$\text{Sal} = 1.80655\text{Chl} \tag{14.4.13}$$

with traditional units of parts per thousand for both salinity and chlorinity. Ocean salinity is approximately 32 ppt = chlorinity of 17.7 ppt. An approximate correction for altitude is[85]

$$C_s(Z) = C_s(0)(1 - 0.000035Z) \tag{14.4.14}$$

where Z is elevation (feet) above sea level ($Z = 0$), or a somewhat more accurate correction can be applied involving saturation vapor pressure.[9]

Oxygen is replenished by reaeration across the air-water interface with a rate

$$\frac{dC}{dt} = K_2(C_s - C) = K_2 D \tag{14.4.15}$$

where C is DO concentration, C_s is the saturation DO concentration, $D = C_s - C$ is called the *DO deficit,* K_2 (also called K_a in the literature) is the *reaeration coefficient* (units of time^{-1}).

The reaeration coefficient increases with increasing turbulence (i.e., with increasing velocity and slope) and decreases with increasing depth, for which many empirical and semiempirical formulations are available.[42] Two equations that performed well in a comparison between predicted and measured reaeration rates[72] are those by Padden and Gloyna,[69]

$$K_2 = 6.9U^{0.703}\bar{d}^{-1.054} \tag{14.4.16}$$

and Tsivoglou and Wallace,[86]

$$K_2 = 4161US \tag{14.4.17}$$

where K_2 has units of day^{-1} at 20°C, U is average stream velocity (ft/s), $\bar{d}$ is average stream depth (ft), and S is stream slope (dimensionless). For velocity in units of m/s and average depth in units of m, the coefficient 6.9 in Eq. (14.4.16) becomes 4.55, and the coefficient 4161 in Eq. (14.4.17) becomes 13,648. Reaeration coefficients are often found in the range of 1 to 10 day^{-1}.

For bodies of standing water such as lakes, lagoons, and bays, reaeration may be related to wind velocity through the liquid-film coefficient,

$$K_2 = \frac{K_L}{d} \tag{14.4.18}$$

where K_L has units of velocity and d is the local depth. A relationship between K_L and wind is given by Banks and Herrera,[6]

$$K_L = 0.728U_w^{1/2} - 0.317U_w + 0.037U_w^2 \tag{14.4.19}$$

where K_L has units of m/day and U_w is the wind speed in m/s at an elevation of 10 m above the ground. O'Connor[67] provides a more detailed investigation into wind effects on gas transfer at the air-water interface.

14.4.3 Streeter-Phelps Analysis of Dissolved Oxygen

The most common type of DO analysis originated with Streeter and Phelps[81] in a study of BOD and DO downstream of the discharge of a sewage treatment plant. Their steady-state, one-dimensional analysis is still useful today, except that numerical solutions are readily obtained, so that other sources, sinks, and mixing parameters can be included, as well as consideration of the nitrogen cycle, photosynthesis and respiration, and other oxygen-demanding reactions. In order to obtain simple analytical solutions, the oxygen-demanding parameters are considered first, with the assumption of a continuous point source discharging into a stream, at which point there is complete mixing with the contents of the stream. Thus, for BOD,

$$L_o = \frac{Q_u L_u + Q_s L_s}{Q_u + Q_s} \tag{14.4.20}$$

where L_o is the resulting well-mixed concentration in the river at the point of waste entry ($x = 0$), Q_u and L_u are the upstream river flow and ultimate BOD concentration, respectively, and Q_s and L_s are the flow and strength of the source (e.g., a treatment plant or industry). Other constituents (e.g., NOD, DO) can be similarly mixed to provide initial conditions.

BOD is then subject to advection and diffusion/dispersion as it progresses down the river. A sink term ($-K_1L$) is included in the advective-diffusion equation (14.3.7) with C replaced by L, and other source/sink terms can be added optionally. For the steady-state condition, $\partial L/\partial t = 0$ in Eq. (14.3.7), and the ordinary differential equation is readily solved to give

$$L(x) = L_o e^{mx} \tag{14.4.21}$$

where $L(x)$ is the BOD distribution downstream of the initial condition L_o, and the exponent m is

$$m = \left(\frac{U}{2E}\right)\left(1 - \sqrt{1 + \frac{4K_1E}{U^2}}\right) \tag{14.4.22}$$

where E is the dispersion coefficient and U is the average velocity. For most streams and rivers, dispersion can be neglected relative to advection if

$$\frac{4K_1E}{U^2} \ll 1 \tag{14.4.23}$$

for which in the limit as $E \Rightarrow 0$ the exponent m reduces to

$$m = -\frac{K_1}{U} \tag{14.4.24}$$

Thus, for most rivers and streams, Eq. (14.4.21) reduces to the simple

$$L(\mathrm{x}) = L_o e^{-K_1 x/U} = L_o e^{-K_1 t} \tag{14.4.25}$$

which describes a simple exponential decay as a function of travel time, $t = x/U$. If NOD is analyzed as a single demand, analogous equations to Eqs. (14.4.21) and (14.4.25) can be readily developed.

The term $-K_1L(x)$ becomes a sink term for DO. Reaeration [Eq. (14.4.15)] is a source term. The option remains of adding other terms to represent other sources and sinks due to NOD, photosynthesis, respiration, sediment oxygen demand, local runoff, etc. (e.g., Refs. 21 and 66). With only the BOD sink and reaeration source in Eq. (14.3.7), the solution for the steady-state, one-dimensional DO deficit [$D(x) = C_s - C(x)$] distribution is[59]

$$D(x) = \left(\frac{K_1 L_o}{K_2 - K_1}\right)\left[e^{mx} - \left(\frac{s_1}{s_2}\right) e^{rx}\right] + D_o e^{rx} \tag{14.4.26}$$

where D_o is the initial, well-mixed deficit at $x = 0$, and parameters m, r, s_1, and s_2 are defined below:

Parameter	Including dispersion	Without dispersion
s_1	$\sqrt{1 + 4K_1E/U^2}$	1
s_1	$\sqrt{1 + 4K_2E/U^2}$	1
m	$(U/2E)(1 - s_1)$	$-K_1/U$
r	$(U/2E)(1 - s_2)$	$-K_2/U$

As discussed above, dispersion can be neglected when $4KE/U^2 \gg 1 (4KE/U^2 \leq 0.1$ is adequate). This is usually the case for rivers, but not for estuaries, for which the dispersion coefficient is relatively high and the freshwater velocity (used for U) is low. Except for inclusion of dispersion, Eq. (14.4.26) is the same as derived by Streeter and Phelps in 1925 and is still in common use today, with the recognition that diurnal photosynthesis-respiration and other temporally and spatially varying sources and sinks are omitted. If NOD were included in the analysis, an analogous term to the first of Eq. (14.4.26) would be added to the equation to account for the additional deficit due to nitrification.[85]

Equation (14.4.26) describes the classical *dissolved oxygen sag equation,* sketched in Fig. 14.4.4*a* and *b*. The DO typically decreases (deficit increases) downstream of the source (as long as BOD oxidation K_1L exceeds reaeration K_2D) until a minimum (critical) concentration or maximum (critical) deficit is reached; after the minimum, the DO increases until saturation is reached (unless there are additional BOD sources or DO sinks included in the analysis). The location of the minimum DO is readily

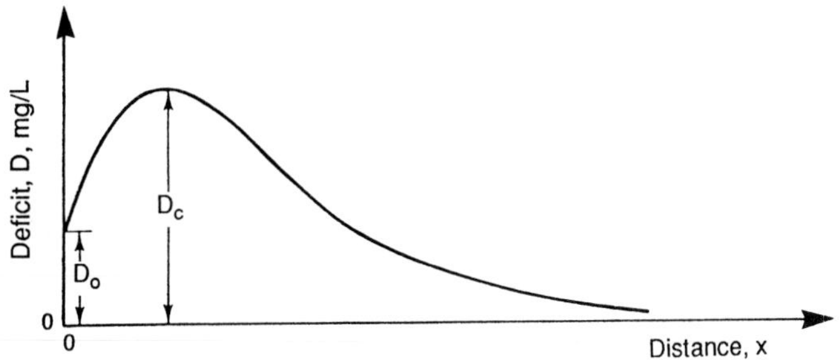

(a) Variation of DO deficit, $D = C_s - C$, with distance, x.

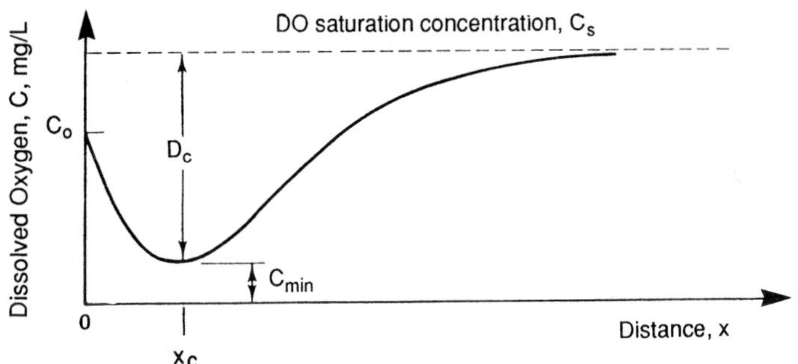

(b) Variation of DO concentration, C, with distance, x.

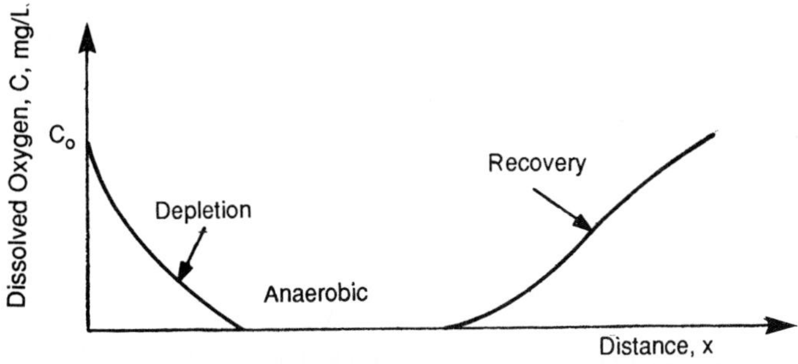

(c) Do variation when reaching anaerobic conditions.

FIGURE 14.4.4 Typical dissolved oxygen sag in rivers. Note: The variation is the same with travel time, $t = x/U$: (*a*) variation of DO deficit, $D = C_s - C$, with distance, x, (*b*) variation of DO concentration C with distance x, (*c*) DO variation when reaching anaerobic conditions.

found[59] by differentiation of Eq. (14.4.26) ($dD/dx = 0$),

$$x_c = \left(\frac{1}{m-r}\right) \ln \left\{\frac{r}{m}\left[\frac{s_1}{s_2} - \frac{(K_2 - K_1)D_o}{K_1 L_o}\right]\right\} \tag{14.4.27}$$

including dispersion, and

$$x_c = \frac{U}{K_2 - K_1} \ln \left\{\frac{K_2}{K_1}\left[1 - \frac{(K_2 - K_1)D_o}{K_1 L_o}\right]\right\} \tag{14.4.28}$$

without dispersion. The critical deficit then is obtained by substituting Eq. (14.4.27) or (14.4.28) into (14.4.26) or more simply by substituting into

$$D_c = \frac{K_1 L_o}{K_2} e^{m x_c} \tag{14.4.29}$$

obtained by setting $dD/dx = 0$ in the simplification of Eq. (14.3.7) (with $C = D$). For the possible situation in which $K_1 = K_2$, limits should be considered,[58] with the result that the critical deficit is simply

$$D_c = L_o e^{-(1 - D_o/L_o)} \tag{14.4.30}$$

The above analysis assumes that anaerobic conditions will not be reached, that is, that the DO always will be greater than or equal to 0. If the DO sag reaches zero (Fig. 14.4.4*c*), reaeration continues at its maximum rate ($dC/dt = -dD/dt = K_2 C_s$) and oxidation of BOD occurs at this limiting rate until the BOD is reduced enough that the reaeration rate exceeds the deoxygenation rate ($-K_1 L$), at which point the DO begins to rise above zero (Fig. 14.4.4*c*).

The DO analysis outlined here can be applied to individual stream reaches in order to account for variable flows, inputs, and rate coefficients. At each junction or source, complete mixing is performed, such that the downstream conditions for one reach become the initial conditions for the following (downstream) reach. In this manner, complex stream networks can be analyzed with spatially varying parameters and multiple sources. This is the method used by the QUAL2E model,[11] for example. Numerical solutions must be used in lieu of the analytical solutions shown herein for cases of time-varying sources or sinks, such as to account for photosynthesis and respiration (e.g., Ref. 24).

In addition to the hydrodynamic parameters of dispersion coefficient E_L and average velocity U discussed earlier, parameter estimation for Eqs. (14.4.21) and (14.4.26) involves the BOD and DO parameters K_1, K_2, L_o, and D_o. The last two parameters are usually given through knowledge of the source conditions, while the reaeration coefficient is obtained from an appropriate formula (Sec. 14.4.2). The deoxygenation coefficient K_1 includes effects of sedimentation and resuspension of BOD as well as other possible first-order processes and thus will not necessarily equal the bottle test result from tests of source effluent. Rather, K_1 should be obtained from tests of stream samples and calibration of the governing equations (14.4.21) and (14.4.26).

Examination of Eq. (14.4.26) indicates that the deficit is a nonlinear function of the initial loading L_o. For waste load allocation problems, the objective is often to determine the value of L_o that will produce an acceptable DO concentration (e.g., $C \geq 5$ mg/L). For this purpose, nomographs are available,[54] and linear approximations to the solution have been used in order to optimize the degree of wastewater treatment required.[5] Commonly for complex river networks and effluent conditions, trial-and-error analysis is performed using models such as QUAL2E[11] for steady-state hydrodynamics, or WASP4[3] or HSPF[23,48] for fully transient hydrodynamics and water-quality simulation.

The dissolved oxygen analysis discussed above has been the subject of countless articles and texts. For example, good reviews and assistance in parameter estimation can be found in Smith and Eilers,[77] Krenkel and Novotny,[50] Bowie et al.,[9] Mills et al.,[61] Steele,[79] Thomann and Mueller,[85] and McBride et al.[55] Current EPA models are reviewed by Ambrose et al.,[4] Ambrose and Barnwell,[2] and McCutcheon.[57]

14.4.4 Bacteria and Other Constituents

First-order decay coefficients for many constituents are summarized by Bowie et al.,[9] Mills et al.,[61] and Schnoor et al.[74] Coliform bacteria are often modeled as part of water-quality studies; first-order decay has been a very good assumption in many modeling studies, with coefficients ranging from 0.0004 to 1.1 h^{-1} (Ref. 9) but with most values in the range 0.02 to 0.1 h^{-1}, with a median rate of 0.04 h^{-1} for total coliforms. Mancini[53] shows that coliform mortality increases with percent seawater,

$$\frac{K_s}{K_f} = 0.8 + 0.006P \tag{14.4.31}$$

where K_s and K_f are first-order decay coefficients for saline and fresh water, respectively, and P is percent seawater (e.g., $P = 100$ percent in the ocean). The presence of light is also a factor.

14.4.5 Parameter Uncertainty

Estimation of parameters and coefficients for water-quality analysis is not a deterministic process in which the analyst determines exact numerical values from judicious evaluation of reference material. Rather, so many physical, chemical, and biological factors influence the concentration of a constituent in nature that its value can never be known with certainty. For example, see the discussion of event mean concentrations in Sec. 14.2.4, for which the uncertainty is quantified using the lognormal frequency distribution. As another example, first-order sensitivity analysis can be applied to deduce the variance of dissolved oxygen concentrations predicted by the Streeter-Phelps equation on the basis of the variances of the several input parameters.[12] A comprehensive review of issues and techniques is presented by Beck.[7]

14.5 APPLICATIONS TO NATURAL WATERS

14.5.1 Rivers and Streams

Contaminant transport in rivers and streams commonly can be considered as three-dimensional near the source, two-dimensional after vertical mixing has occurred a short distance downstream from the source, and one-dimensional after complete cross-sectional mixing has occurred. A *time scale* for mixing, τ, is

$$\tau = \frac{\beta l^2}{E} \tag{14.5.1}$$

where l is a characteristic length, E is the appropriate diffusivity (i.e., vertical, E_z, or transverse, E_y), and β is a coefficient that differs for different mixing criteria. For conditions where one can assume complete vertical mixing and for complete cross-sectional mixing, $\beta \approx 0.4$ to 0.5.[37,40] Larger β values apply for less uniform injections and in the presence of *dead zones,* that is, when tracer is trapped along the stream bank or bottom. Similarly, enhanced mixing of influents to the stream through diffusers, for example, can result in complete cross-sectional mixing much sooner. The concentration profile (C versus x) due to a plane source (one-dimensional), instantaneous injection approaches a gaussian distribution [Eq. (14.3.10)] when $\beta > 1.0$.[37] For vertical mixing, the characteristic length l is the depth. For lateral or transverse mixing, l is the distance from the location of maximum velocity to the farthest bank, or more conservatively, simply the width of the stream. The time scale τ can be converted to a distance x by multiplication by the velocity.

Thus to summarize, near the point of injection, three-dimensional solutions (or near-field jet discharge solutions) apply, e.g., Eqs. (14.3.8) and (14.3.11). After vertical mixing has occurred (i.e., $x \geq 0.4\ Ud^2/E_z$), two-dimensional solutions apply, e.g., Eqs. (14.3.9) and (14.3.12). The range of validity of three- and two-dimensional solutions can be enhanced with image sources.[44] One-dimensional analysis may be used for $x \geq 0.4\ Uw^2/E_y$ (often several miles downstream of the injection point), but skewness of the concentration profile from an instantaneous injection will remain until at least $x \geq Uw^2/E_y$, invalidating Eq. (14.3.10) until that point. Values of $x < 0.4\ Uw^2/E_y$ were said by Fischer[35] to be in the *convective period,* whereas larger values are in the *diffusive period.*

As an example, image sources are incorporated into two useful solutions for streams. Consider a continuous vertical line source in the center of an approximately rectangular cross-section river (Fig. 14.5.1*a*). Equation (14.3.12) can be used near the source, but this solution is unable to account for "reflection" off each bank of the

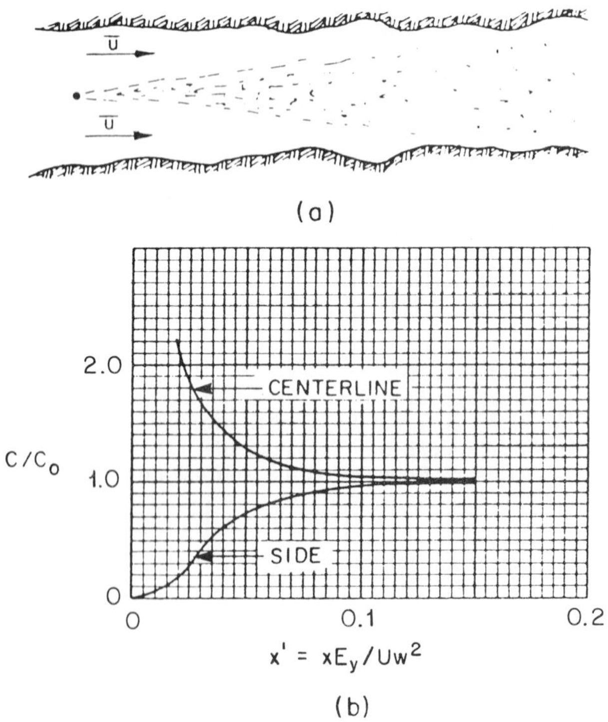

FIGURE 14.5.1 Continuous discharge at stream centerline. *(Reproduced from Ref. 37 with permission.)*

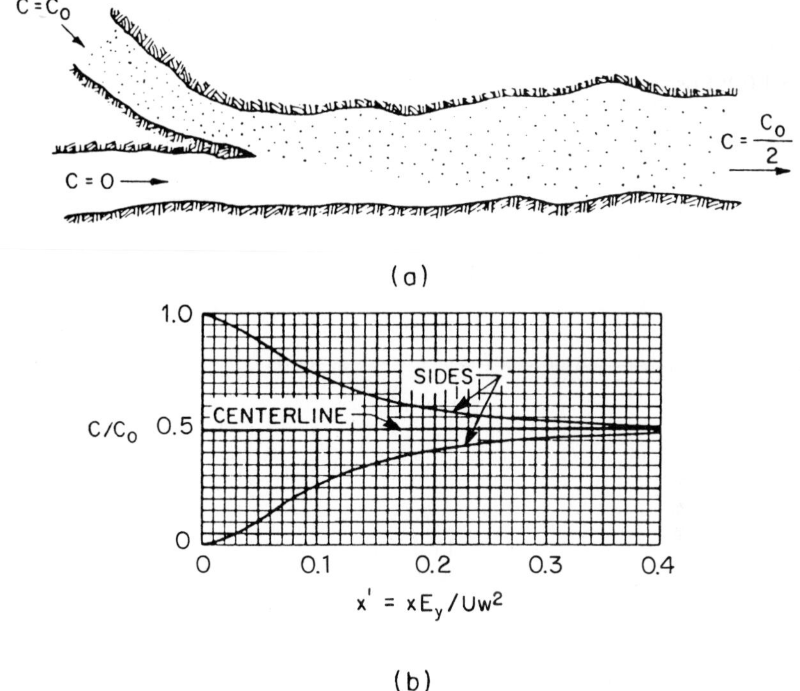

FIGURE 14.5.2 Blending of two streams of equal discharge. *(Reproduced from Ref. 37 with permission.)*

stream. By summing the results of an infinite series of line sources spaced w units apart from each other in the y direction (Fig. 14.3.3), the solution shown in Fig. 14.5.1*b* results. Dimensional values of x can be computed using the transverse diffusivity E_y from Eq. (14.3.25), the average velocity U, and the width w. In this solution, C_o is the well-mixed concentration far downstream of the injection point (Table 14.3.1). By the symmetry of the problem, Fig. 14.5.1*b* applies equally well to a vertical line source at the stream boundary, if the width w is replaced by $2w$. Then the bottom curve (marked *side*) is the concentration on the opposite bank, and the top curve (marked *centerline*) is the concentration along the bank with the discharge. For example, for a centerline discharge, concentration differences across the stream are within 5 percent at a dimensionless distance of 0.1, thus providing one definition of *complete mixing.*[37] For a side discharge, the analogous dimensionless distance would be $x' = 0.4$ if w is replaced by $2w$. This forms one basis for the beginning of the diffusive zone discussed earlier.

The solution for transverse mixing of two merging rivers is shown in Fig. 14.5.2. What will be the concentration on each bank? Because the constituent mass is reflected off each bank, the exact solution is a summation of the error function solution for mixing of two infinitely wide rivers[37] and is graphed in Fig. 14.5.2*b*. For example, the concentration at each bank would deviate from the average at the centerline by only 5 percent at a dimensionless distance x' of approximately 0.3.

A third additional solution is for the case of an instantaneous *volume source,* that

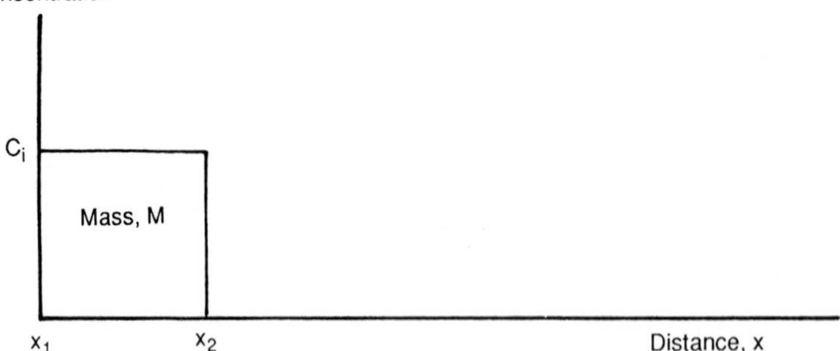

(a) Initial condition for volume source

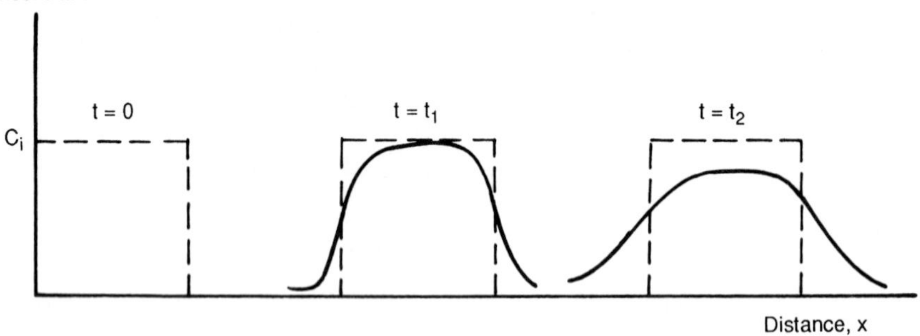

(b) Concentration distribution [Eq. (14.5.3)] showing advection and dispersion of volume source. Center of mass moves with mean velocity, U.

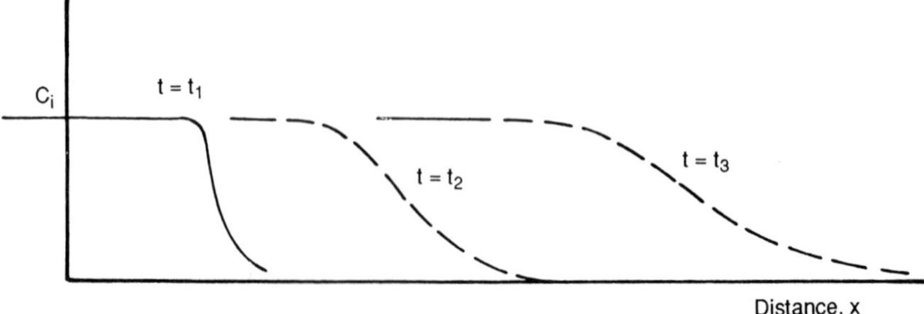

(c) Concentration distribution [Eq. (14.5.4)] for "break-through" of liquid of different concentration, C_i. Concentration $C_i/2$ moves with mean velocity.

FIGURE 14.5.3 Volume source solutions for a stream.

is, a mass M of constituent placed uniformly in a stream channel of cross-sectional area A between locations x_1 and x_2 (Fig. 14.5.3). The initial concentration in the reach is then

$$C_i = \frac{M}{A(x_2 - x_1)} \tag{14.5.2}$$

with the solution

$$C(x, t) = \frac{C_i}{2}\left[\operatorname{erf}\left(\frac{(x - x_1) - Ut}{\sqrt{4E_L t}}\right) - \operatorname{erf}\left(\frac{(x - x_2) - Ut}{\sqrt{4E_L t}}\right)\right] \tag{14.5.3}$$

The shape of the concentration distribution is sketched in Fig. 14.5.3 and approaches Gaussian [Eq. (14.3.10)] far downstream. For the special case in which $x_1 = -\infty$ and $x_2 = 0$, that is, the *breakthrough* of an interface between fluids of two different concentrations, $C = C_i$ upstream of $x = 0$ and $C = 0$ downstream of $x = 0$, Eq. (14.5.3) simplifies to

$$C = \frac{C_i}{2} \operatorname{erfc} \frac{x}{\sqrt{4E_L t}} \tag{14.5.4}$$

the solution to which is also sketched in Fig. 14.5.3. The error function and complementary error function [$\operatorname{erfc}(x) = 1 - \operatorname{erf}(x)$] are tabulated in many mathematics and statistics handbooks (e.g., Ref. 1). Equation (14.5.4) also arises in the breakthrough of a different fluid in column tests in porous media.[38] Both solutions, Eqs. (14.5.3) and (14.5.4), can be modified for first-order decay by multiplication by e^{-Kt}.

Design conditions for water-quality modeling in streams and rivers have traditionally been taken at low-flow conditions to represent minimum dilution, e.g., the 7-day, 10-year low flow, or 7Q10. Steady-state flow models such as QUAL2E have then been used for determination of in-stream concentrations and determination of required treatment at discharge points. With recognition of the importance of nonpoint sources (Sec. 14.2), the critical combination of loadings and in-stream flows is not obvious (e.g., wet-weather conditions could control), and continuous simulation with transient-water-quality models can be used to evaluate control measures, e.g., Medina,[58] with the option to control at any design frequency.[33]

14.5.2 Estuaries

Estuary-type flow is characterized by a velocity that is the sum of a freshwater contribution U_f and an oscillatory contribution due to tides. An idealized representation (Fig. 14.5.4) is

$$U(t) = U_f + U_T \sin \sigma t \tag{14.5.5}$$

where U_T is the amplitude of the tidal velocity component and $\sigma = 2\pi/T$, where T is the tidal period (typically 12.5 h). (A phase parameter can be included in the argument of the sine if zero tidal velocity does not correspond to $t = 0$.) The freshwater velocity is

$$U_f = \frac{Q_f}{A} \tag{14.5.6}$$

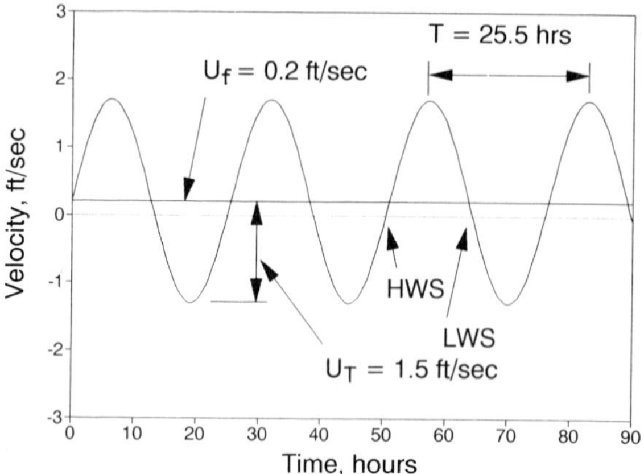

FIGURE 14.5.4 Estuary-type flow.

where U_f and Q_f are the freshwater velocity and flow, respectively, and A is the cross-sectional area.

If $U(t)$ of Eq. (14.5.5) is substituted into the one-dimensional advective dispersion equation, Eq. (14.3.7), it can be solved in closed form for the case of an instantaneous plane source and in the form of an integral for continuous injection.[39,43] The analytical solution requires that the dispersion coefficient, which actually varies with velocity, be treated as a constant average value over the tidal period E_A. The actual varying $E_L(x, t)$ can be used in numerical models. This is the so-called *real-time method* of analysis[63] because temporal variations in the advective flux are accounted for. In the freshwater portion of estuaries, values of E_A are the same order of magnitude as values of E_L in rivers, that is, 30 to 700 ft²/s (3 to 70 m²/s).

In lieu of the real-time analysis, many one-dimensional solutions for rivers and streams have been used in estuaries under one of two simplifying assumptions. The *tidal average method* assumes that the concentration in Eq. (14.3.7) is an average over the tidal cycle and that the advective velocity is that due only to freshwater flow. The dispersion coefficient then has a specific definition E_T for this set of assumptions, and $E_T \neq E_A$. A quasi-steady-state assumption then is made to the effect that concentration distributions are identical during the same portion of each tidal cycle, as evidenced by the salinity distribution that migrates up and down the estuary during the flood and ebb tides. The quasi-steady-state contaminant distribution then is determined, with $\partial C/\partial t = 0$ in Eq. (14.3.7) and U_f for the velocity, yielding solutions of the form of Eq. (14.3.13). Values of E_T range from 200 to 10,000 ft²/s (20 to 1000 m²/s), with most values in the range of 1000 to 5000 ft²/s (100 to 500 m²/s).[9]

The *slack tide method* examines concentrations only at high water slack (HWS) or low water slack (LWS) conditions, corresponding to the end of the flood and ebb tides, respectively, and corresponding to times of zero velocity in Fig. 14.5.4. Again, only the freshwater velocity component is used. For an instantaneous injection, the slack tide result is Eq. (14.3.10) with dispersion coefficient E_A and is the same as the real-time result evaluated at HWS or LWS. However, for continuous injection, the result of Eq. (14.3.13) requires a much larger dispersion coefficient E_S to produce results that correspond to the real-time solution upstream of the injection location.[39]

Any of the three methods can be applied to conditions in an estuary. For example, Fischer et al.[37] illustrate the tidal average method and show how the dispersion coefficient E_T can be evaluated from the salinity distribution. These authors also show how the salt balance in an estuary or bay can be used to infer the residence time and equilibrium concentration of constituents in an estuary or bay, assuming that the salinity will mix in the same way as will the constituents over several tidal cycles. Values of tidally averaged dispersion coefficients are summarized by Bowie et al.[9] Thomann and Mueller[85] demonstrate the slack tide method, and Najarian and Harleman[63] illustrate the real-time method in the context of numerical modeling. In general, instantaneous values of the dispersion coefficient, $E_L(x, t)$, are different from those for real-time analysis, E_A, tidal average analysis, E_T, and slack tide analysis, E_S. It is important to recognize which value is being reported in one-dimensional studies of estuaries.[39]

Considerable caution should be used in application of one-dimensional analysis to estuaries, because it is appropriate only under well-mixed conditions in the absence of vertical density gradients, such as in the freshwater portion of the estuary, or in a shallow estuary with relatively small freshwater inflows compared to the volume of tidal flow. The magnitude of the dispersion coefficient (discussed in Sec. 14.3.5) depends on the relative time scales of cross-sectional mixing, tidal period, and decay.[37] If a one-dimensional analysis is not suitable, then two- or three-dimensional methods must be used (e.g., Ref. 75).

14.5.3 Lakes and Reservoirs

For a large lake (not well-mixed, subject to concentration gradients), prediction of the concentration distribution due to effluent discharges is similar to that for discharges to the ocean or a bay; that is, the three-dimensional concentration distribution is influenced by the complex pattern of wind-driven currents and turbulence. The simple, closed-form solutions, Eqs. (14.3.8), (14.3.9), (14.3.11), and (14.3.12) can be applied to situations in which boundaries (shoreline and bottom) do not affect the concentration distribution. Otherwise, image sources and/or numerical methods must be used. Values of diffusivities for lakes for use in the closed-form solutions were discussed in Sec. 14.3.5. Complex modeling of lake dynamics is discussed by Fischer et al.,[37] Reckhow and Chapra,[73] and Mills et al.,[61] as is thermal stratification that can determine the vertical temperature distribution and affect diffusivities.

More commonly, perhaps, water-quality analysis of lakes involves questions of *eutrophication:* deleterious effects due to increased *nutrient* (nitrogen and phosphorus) supply. The deleterious effects as viewed by humans are increased life and growth of a lake's biota, especially algae and *macrophytes* (large aquatic plants), with a consequent increase in turbidity and color, possible reduction in dissolved oxygen, and change in nature of the fish population away from sportfishing species. Thus, in-lake nitrogen and phosphorus concentrations are a reflection of the *trophic status* of a lake, as are Secchi disk depth and chlorophyll *a* concentrations (Table 14.5.1).

TABLE 14.5.1 Trophic Status of Lakes

Water-quality variable	Oligotrophic	Mesotrophic	Eutrophic
Total phosphorus, μg/L	<10	10–20	>20
Chlorophyll *a*, μg/L	<4	4–10	>10
Secchi depth, m	>4	2–4	<2
Hypolimnetic oxygen, % saturation	>80	10–80	<10

Source: Ref. 85.

Oligotrophic refers to clear lakes with low biological productivity. *Mesotrophic* refers to lakes with intermediate biological productivity, and *eutrophic* refers to lakes with high biological productivity relative to natural levels.[85] The *hypolimnion* of a lake is the bottom region below the *thermocline* (region of steep temperature and density gradients).

An indication of trophic status is based on the relative amounts of nitrogen and phosphorus required by algae for growth. As an approximation, lakes are *nitrogen-limited* if the ratio of total nitrogen N to total phosphorus P is less than 13, *nutrient-balanced* if $13 \leq N/P \leq 21$, and *phosphorus-limited* if $21 < N/P$ (Refs. 78 and 85). The exact limits of these ranges (i.e., 13 and 21) depend on algal species and are not well-defined, but these numbers are typical. Some authors report the ratio of *available nitrogen* (NH_3-N + NO_2-N + NO_3-N) to *available phosphorus* (orthophosphorus or soluble reactive phosphorus), but the critical ratios change relatively little. *Phosphorus-limited* means that additional phosphorus is required to produce additional algal growth. Most but not all lakes are phosphorus-limited, especially those for which nonpoint-source runoff is the dominant source of phosphorus.

Simple eutrophication modeling is often based on the assumption that a lake behaves in a well-mixed fashion as a *continuously stirred tank reactor* on an annual average basis,[73] which at steady state ($dP/dt = 0$) yields

$$P = \frac{L'}{H(1/t_d + K)} = \frac{L'}{q + w_s} \tag{14.5.7}$$

where P = total phosphorus concentration
L' = areal loading rate (mass/area-time) = L/A_s
L = P load to lake (mass/time)
A_s = lake surface area
H = mean depth
t_d = detention time = V/Q
Q = lake outflow (excluding evaporation)
V = lake volume
K = first-order loss rate for total P

The hydraulic overflow rate $q = Q/A_s = H/t_d$ may alternatively be used in this equation. The decay coefficient K is related to the settling velocity w_s of particulate phosphorus,

$$K = \frac{w_s}{H} \tag{14.5.8}$$

but generally neither K nor w_s is readily known. Empirical approaches have led to Vollenweider's relationship[87] of K to mean depth,

$$K = \frac{10}{H} \tag{14.5.9}$$

with units of year^{-1} for K and meters for H. If H is measured in units of feet, the coefficient 10 becomes 32.8. Vollenweider[88] and Larsen and Mercier[52] relate K to detention time as

$$K = \sqrt{\frac{1}{t_d}} \tag{14.5.10}$$

with units of years for t_d and year^{-1} for K, and the coefficient 1 is dimensional. Most lakes have some nonevaporative outflow at least by seepage to groundwater, but if a lake is completely landlocked and sealed on the bottom, then Eq. (14.5.7) reduces to

$$P = \frac{L'}{HK} = \frac{L'}{w_s} \tag{14.5.11}$$

On the basis of comparisons of eutrophic and oligotrophic lakes, Vollenweider[87] defined eutrophic lakes as those with P concentrations greater than 20 $\mu g/L$, oligotrophic lakes as those with P concentrations less than 10 $\mu g/L$, and mesotrophic for $10 \leq P \leq 20$ $\mu g/L$. He then used Eqs. (14.5.7) and (14.5.9) to define acceptable combinations of loading rates, mean depth, and residence times (Fig. 14.5.5). Mills et al.[61] report that Eq. (14.5.10) is also acceptable for eutrophication screening and provide examples of simple eutrophication screening methods.

Another alternative for eutrophication screening is the use of *trophic state indices,* usually a regression equation that relates various combinations of Secchi disk readings and concentrations of phosphorus, nitrogen, and chlorophyll *a*. Many of these are site-specific and are reviewed by Reckhow and Chapra,[73] Mills et al.,[61] and Thomann and Mueller.[85]

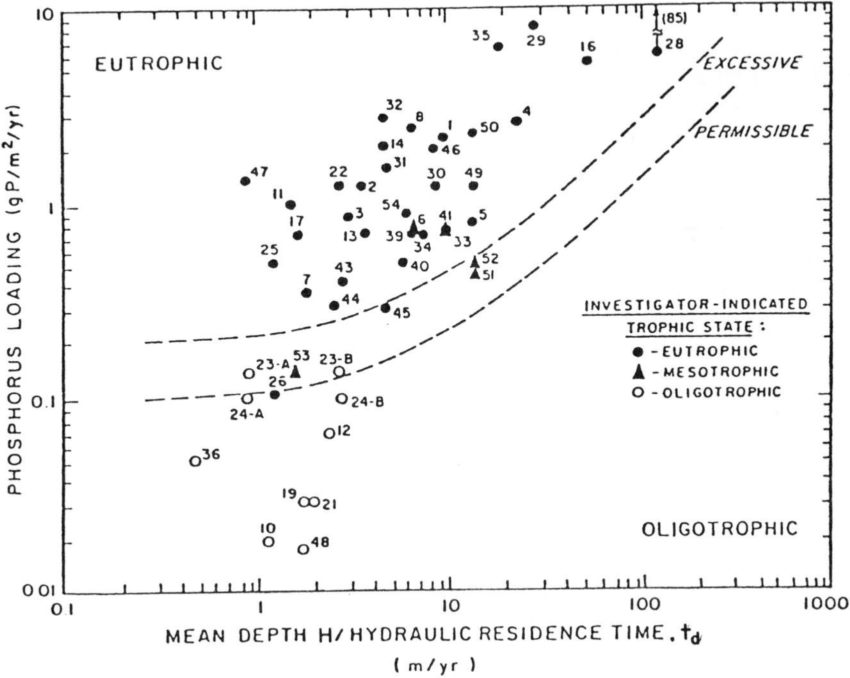

FIGURE 14.5.5 Vollenweider's relationship among phosphorus loading, mean depth, and residence time. (*Source: Ref. 71.*)

14.6 SUMMARY

The abbreviated presentation in this handbook cannot possibly cover all aspects of transport of contaminants in surface waters. Only the highlights have been presented, with emphasis on analytical methods and screening tools. A few reference sources already cited should be reiterated as useful sources of additional details on the topics of this chapter, in particular Fischer et al.[37] for the physics of mixing in natural waters; Yotsukura[91] and Harleman[40] for analytical solutions to the advective-dispersion equation; Krenkel and Novotny[50] and Thomann and Mueller[85] for a review of constituent characterization, sources, sinks, and interactions; McBride et al.[55] for dissolved oxygen analysis; Bowie et al.[9] for parameter estimation; and Mills et al.[61] and Schnoor et al.[74] for screening methods for combinations of constituents and water bodies most commonly encountered in practice.

Probably the best U.S. source of numerical models for many of the processes discussed in this chapter is the Center for Exposure Assessment Modeling, U.S. Environmental Protection Agency, College Station Road, Athens, Ga. 30613.[2,4] Additional sources include the Corps of Engineers Waterways Experiment Station in Vicksburg, Miss. and the Corps of Engineers Hydrologic Engineering Center in Davis, Calif. Other U.S. federal agencies that perform modeling of contaminant transport include the U.S. Geological Survey (USGS) and the National Oceanic and Atmospheric Administration (NOAA). A review of several non-point-source water-quality models distributed by various federal agencies is provided by Donigian and Huber.[22]

REFERENCES

1. Abramowitz, M., and I. A. Stegun, *Handbook of Mathematical Functions,* National Bureau of Standards, Applied Mathematics Series 55, U.S. Government Printing Office, Washington, D.C., 1965.
2. Ambrose, R. B., Jr., and T. O. Barnwell, Jr., "Environmental Software at the U.S. Environmental Protection Agency's Center for Exposure Assessment Modeling," *Environ. Software,* vol. 4, no. 2, pp. 76–93, 1989.
3. Ambrose, R. B., Jr., T. A. Wool, J. P. Connolly, and R. W. Schanz, *WASP4: A Hydrodynamic and Water Quality Model—Theory, User's Manual and Programmer's Guide,* EPA/600/3-87/039 (NTIS PB88-185095), Environmental Protection Agency, Athens, Ga., 1988*a*.
4. Ambrose, R. B., Jr., et al., "Waste Allocation Simulation Models," *J. Water Pollut. Control Fed.,* vol. 60, no. 9, pp. 1646–1655, September 1988*b*.
5. Arbabi, M., J. Elzinga, and C. ReVelle, "The Oxygen Sag Equation: New Properties and a Linear Equation for the Critical Deficit," *Water Resour. Res.,* vol. 10, no. 5, pp. 921–929, October 1974.
6. Banks, R. B., and F. F. Herrera, "Effect of Wind and Rain on Surface Reaeration," *J. Env. Eng. Div., Am. Soc. Civ. Eng.* (now *J. Environ. Eng.*), vol. 103, no. EE3, pp. 489–504, June 1977.
7. Beck, M. B., "Water Quality Modeling: A Review of the Analysis of Uncertainty," *Water Resour. Res.,* vol. 23, no. 8, pp. 1393–1442, August 1987.
8. Biswas, A. K., *Systems Approach to Water Management,* McGraw-Hill, New York, 1976.
9. Bowie, G. L., et al., *Rates, Constants and Kinetics Formulations in Surface Water Quality Modeling (Second Edition),* EPA/600/3-85/040, Environmental Protection Agency, Athens, Ga., June 1985.

10. Brooks, N. H., "Diffusion of Sewage Effluent in an Ocean Current," in E. A. Pearson (ed.), *Waste Disposal in the Marine Environment,* Pergamon Press, New York, pp. 246–267, 1960.

11. Brown, L. C., and T. O. Barnwell, Jr., *The Enhanced Stream Water Quality Models QUAL2E and QUAL2E-UNCAS: Documentation and User's Manual,* EPA/600/3-87/007 (NTIS PB87-202156), Environmental Protection Agency, Athens, Ga., 1987.

12. Burges, S. J., and D. P. Lettenmaier, "Probabilistic Methods in Stream Quality Management," *Water Resour. Bull.,* vol. 11, no. 1, pp. 115–130, February 1975.

13. Camp, T. R., *Water and its Impurities,* Reinhold, New York, 1963.

14. Carslaw, H. S., and J. C. Jeager, *Conduction of Heat in Solids,* 2d ed., Oxford University Press, New York, 1959.

15. Clesceri, L. S., A. E. Greenberg, and R. R. Trussell (eds.), *Standard Methods for the Examination of Water and Wastewater,* 17th ed. American Public Health Association, Washington, D.C., 1989.

16. Cooper, A. B., "Developing Management Guidelines for River Nitrogenous Oxygen Demand," *J. Water Pollut. Control Fed.* vol. 58, no. 8, pp. 845–852, August 1986.

17. Crank, J., *The Mathematics of Diffusion,* Oxford University Press, New York, 1956.

18. Csanady, G. T., *Turbulent Diffusion in the Environment,* D. Reidel, Boston, 1973.

19. Demetracopoulos, A. C., and H. G. Stefan, "Transverse Mixing in Wide and Shallow River: Case Study," *J. Env. Eng.,* vol. 109, no. 3, pp. 685–699, June 1983.

20. Diachishin, A. N., "Dye Dispersion Studies," *J. San. Eng. Div., Am. Soc. Civ. Eng.* (now *J. Environ. Eng.*) vol. 89, no. SA1, pp. 29–49, January 1963.

21. Dobbins, W. E., "BOD and Oxygen Relationships in Streams," *J. San. Eng. Div., Am. Soc. Civ. Eng.* (now *J. Environ. Eng.*), vol. 90, no. SA3, pp. 53–78, June 1964.

22. Donigian, A. S., Jr., and W. C. Huber, *Modeling of Nonpoint Source Water Quality in Urban and Non-Urban Areas,* EPA/600/3-91/039, Environmental Protection Agency, Athens, Ga., 1991.

23. Donigian, A. S., Jr., J. C. Imhoff, B. R. Bicknell, and J. L. Kittle, Jr., *Application Guide for the Hydrological Simulation Program—FORTRAN,* EPA/600/3-84/065 (NTIS PB84-215763), Environmental Protection Agency, Athens, Ga., 1984.

24. Dresnack, R., and W. E. Dobbins, "Numerical Analysis of BOD and DO Profiles," *J. San. Eng. Div., Am. Soc. Div. Eng.* (now *J. Environ. Eng.*), vol. 94, no. SA3, pp. 789–807, October 1968.

25. Driscoll, E. D., "Lognormality of Point and Nonpoint Source Pollutant Concentrations," *Proc. Stormwater and Water Quality Model Users Group Meeting,* Orlando, Fla. EPA/600/9-86/023 (NTIS PB87-117438/AS), Environmental Protection Agency, Athens, Ga., pp. 157–176, March 1986.

26. Driscoll, E.D., and W. James, "Evaluation of Alternatives," in W. James (ed.), *Pollution Control Planning, Proc. Ontario Ministry of the Environment Technology Transfer Workshop,* Toronto, published by CHI, Guelph, Ontario, Chap. 7, pp. 139–175, February 1987.

27. Driscoll, E. D., P. E. Shelley, and E. W. Strecker, *Pollutant Loadings and Impacts from Highway Stormwater Runoff,* vol. I, Design Procedure, FHWA-RD-88-006, Office of Engineering and Highway Operations R&D, Federal Highway Administration, McLean, Va., 1989.

28. Driver, N. E., and G. D. Tasker, *Techniques for Estimation of Storm-Runoff Loads, Volumes and Selected Constituent Concentrations in Urban Watersheds in the United States,* USGS Open File Report 88-191, Denver, 1988.

29. Elder, J. W., "The Dispersion of Marked Fluid in Turbulent Shear Flow," *J. Fluid Mechanics,* vol. 5, no. 4, pp. 544–560, May 1959.

30. Environmental Protection Agency, *Treatability Data, Vol. 1,* EPA-600/2-82-001a, Cincinnati, OH, 1982.

31. Environmental Protection Agency, *Results of the Nationwide Urban Runoff Program,* vol. I, Final Report, NTIS PB84-185552, Washington, D.C., 1983.

32. Environmental Protection Agency, *Superfund Public Health Evaluation Manual,* EPA/540/1-86/060, Washington, D.C., 1986.

33. Environmental Protection Agency, *Guidance for Water Quality-Based Decisions: The TMDL Process,* EPA/440/4-91/001, Washington, D.C., April 1991.

34. Fischer, H. B., "A Note on the One-Dimensional Dispersion Model," *Int. J. of Air and Water Pollut.,* vol. 10, pp. 443–452, June–July 1966.

35. Fischer, H. B., "The Mechanisms of Dispersion in Natural Streams," *J. Hydraul. Div., Am. Soc. Civ. Eng.* (now *J. Hydraul. Eng.*), vol. 93, no. HY6, pp. 187–216, November 1967.

36. Fischer, H. B., "Dispersion Prediction in Natural Streams," *J. San. Eng. Div., Am. Soc. Civ. Eng.* (now *J. Environ. Eng.*), vol. 94, no. SA5, pp. 927–943, October 1968.

37. Fischer, H. B., E. J. List, R. C. Y. Koh, J. Imberger, and N. H. Brooks, *Mixing in Inland and Coastal Waters,* Academic Press, New York, 1979.

38. Freeze, R. A., and J. A. Cherry, *Groundwater,* Prentice-Hall, Englewood Cliffs, N.J., 1979.

39. Harleman, D. R. F., "One-Dimensional Models," in Tracor, Inc., ed., *Estuarine Models, An Assessment,* 16070DZV02/71 (NTIS PB-206807), Environmental Protection Agency, Washington, D.C., February 1971.

40. Harleman, D. R. F., *Transport Processes in Environmental Engineering,* lecture notes, Parsons Hydrodynamics Laboratory, Massachusetts Institute of Technology, Cambridge, 1988.

41. Holley, E. R., "Unified View of Diffusion and Dispersion," *J. Hydraul. Div., Am. Soc. Civ. Eng.* (now *J. Hydraul. Eng.*), vol. 95, no. HY2, pp. 621–632, March 1969.

42. Holley, E. R., "Oxygen Transfer at the Air-Water Interface," in R. J. Gibbs, ed., *Transport Processes in Lakes and Oceans,* Plenum Press, New York, pp. 179–202, 1977.

43. Holley, E. R., and D. R. F. Harleman, *Dispersion of Pollutants in Estuary-Type Flows,* MIT Hydrodynamics Laboratory Tech. Rep. No. 74, Cambridge, Mass., January 1965.

44. Holley, E. R., and G. H. Jirka, *Mixing in Rivers,* Tech. Rept. E-86-11, Corps of Engineers, Waterways Experiment Station, Vicksburg, Miss., September 1986.

45. Holly, F. M., Jr., "Dispersion in Rivers and Coastal Waters—1. Physical Principles and Dispersion Equations," Chap. 1 in *Developments in Hydraulic Engineering—3,* P. Novak, editor, Elsevier, New York, 1985.

46. Huber, W. C., "Deterministic Modeling of Urban Runoff Quality," in H.C. Torno, J. Marsalek, and M. Desbordes eds., *Urban Runoff Pollution,* Series G: Ecological Sciences, vol. 10, Springer-Verlag, New York, pp. 167–242, 1985.

47. Huber, W. C., J. P. Heaney, D. A. Aggidis, R. E. Dickinson, K. J. Smolenyak, and R. W. Wallace, *Urban Rainfall-Runoff-Quality Data Base,* EPA-600/2-81-238 (NTIS PB82-221094), Environmental Protection Agency, Cincinnati, July 1982.

48. Johanson, R. D., J. C. Imhoff, J. L. Kittle, Jr., and A. S. Donigian, *Hydrological Simulation Program—FORTRAN (HSPF): User's Manual for Release 8.0,* EPA/600/3-84/066 (NTIS PB84-224385), Environmental Protection Agency, Athens, Ga., 1984.

49. Kirkpatrick, F. A., and J. F. Wilson, Jr., "Measurement of Time of Travel in Streams by Dye Tracing," Chap. A9, Book 3, *Techniques of Water-Resources Investigations of the United States Geological Survey,* USGS, Denver, 1989.

50. Krenkel, P. A., and V. Novotny, *Water Quality Management,* Academic Press, New York, 1980.

51. Lager, J. A., W. G. Smith, W. G. Lynard, R. M. Finn, and E. J. Finnemore, *Urban Stormwater Management and Technology: Update and User's Guide,* EPA-600/8-77-014 (NTIS PB-275654), Environmental Protection Agency, Cincinnati, 1977.

52. Larsen, D. P., and H. T. Mercier, "Phosphorus Retention Capacity of Lakes," *J. Fish. Res. Board Canada,* vol. 33, pp. 1731–1750, 1976.

53. Mancini, J. L., "Numerical Estimates of Coliform Mortality Rates under Various Conditions," *J. Water Pollut. Control Fed.*, vol. 50, no. 11, pp. 2477–2484, October 1978.

54. McBride, G. B., "Nomographs for Rapid Solutions for the Streeter-Phelps Equations," *J. Water Pollut. Control Fed.* vol. 54, no. 4, pp. 378–384, April 1982.

55. McBride, G. B., J. C. Rutherford, and R. D. Pridmore, "Modeling Organic Pollution of Streams," Chap. 22 in *Civil Engineering Practice*, vol. 5, *Water Resources/Environmental*, P. N. Cheremisinoff, N. P. Cheremisinoff, and S. L. Cheng (eds.), Technomic, Lancaster, Pa., 1988.

56. McCutcheon, S. C., "Laboratory and Instream Nitrification Rates for Selected Sites," *J. Environ. Eng.*, vol. 113, no. 3, pp. 628–646, June 1987.

57. McCutcheon, S. C., *Transport and Surface Exchange in Rivers*, vol. 1 of *Water Quality Modeling*, R. H. French and S. C. McCutcheon, eds., CRC, Boca Raton, Fla., 1989.

58. Medina, M. A., Jr., *Level III: Receiving Water Quality Modeling for Urban Stormwater Management*, EPA-600/2-79-100 (NTIS PB80-134406), Environmental Protection Agency, Cincinnati, August 1979.

59. Medina, M. A., Jr., W. C. Huber, J. P. Heaney, and R. Field, "River Quality Model for Urban Stormwater Impacts," *J. Water Resources Planning and Management Div., Am. Soc. Civ. Eng.* (now *J. Water Resources Planning and Management*), vol. 107, no. WR1, pp. 263–280, March 1981.

60. Metcalf and Eddy, Inc., *Wastewater Engineering*, McGraw-Hill, New York, 1979.

61. Mills, W. B., et al., *Water Quality Assessment: A Screening Procedure for Toxic and Conventional Pollutants in Surface and Ground Water (Revised 1985)*, Parts 1 and 2, EPA/600/6-85/002a,b (NTIS PB86-122496 and PB86-122504), Environmental Protection Agency, Athens, Ga., September 1985.

62. Mustard, M. H., N. E. Driver, J. Chyr, and B. G. Hansen, *U.S. Geological Survey Urban Stormwater Data Base of Constituent Storm Loads; Characteristics of Rainfall, Runoff, and Antecedent Conditions; and Basin Characteristics*, Water Resources Investigations Report 87-4036, U.S. Geological Survey, Denver, 1987.

63. Najarian, T. O., and D. R. F. Harleman, "Real Time Simulation of Nitrogen Cycle in an Estuary," *J. Environ. Eng. Div., Am. Soc. Civ. Eng.* (now *J. Environ. Eng.*), vol. 103, no. EE4, pp. 523–538, August 1977.

64. Nemerow, N. L., *Industrial Water Pollution*, Addison-Wesley, Reading, Mass., 1978.

65. Niku, S., E. D. Schroeder, G. Tchobanoglous, and F. J. Samaniego, *Performance of Activated Sludge Processes: Reliability, Stability and Variability*, EPA-600/2-82-227 (NTIS PB82-109604), Environmental Protection Agency, Cincinnati, December 1981.

66. O'Connor, D. J., "The Temporal and Spatial Distribution of Dissolved Oxygen in Streams," *Water Resour. Res.*, vol. 3, no. 1, pp. 65–79, 1967.

67. O'Connor, D. J., "Wind Effects on Gas-Liquid Transfer Coefficients," *J. Environ. Eng.*, vol. 109, no. 3, pp. 731–752, June 1983.

68. Okubo, A., "Oceanic Diffusion Diagrams," *Deep Sea Res.*, vol. 18, pp. 789–802, August 1971.

69. Padden, T. J., and E. F. Gloyna, *Simulation of Stream Processes in a Model River*, report no. EHE-70-23, CRWR-72, University of Texas, Austin, May 1971.

70. Prakash, A., "Convective-Dispersion in Perennial Streams," *J. Env. Eng. Div., Am. Soc. Civ. Eng.* (now *J. Environ. Eng.*), vol. 103, no. EE2, pp. 321–340, February 1977.

71. Rast, W., and G. F. Lee, *Summary Analysis of the North American (U.S. Portion) OECD Eutrophication Project*, EPA-600/3-78-008, Environmental Protection Agency, Corvallis, Ore. 1978.

72. Rathbun, R. E., "Reaeration Coefficients of Streams—State-of-the-Art," *J. Hydraul. Div., Am. Soc. Civ. Eng.* (now *J. Hydraul. Eng.*), vol. 103, no. HY4, pp. 409–424, April 1977.

73. Reckhow, K. H., and S. C. Chapra, *Engineering Approaches for Lake Management*, Butterworth Publishers, Woburn, Mass. two vols., 1983.

74. Schnoor, J. L., C. Sato, D. McKechnie, and D. Sahoo, *Process, Coefficients, and Models for Simulating Toxic Organics and Heavy Metals in Surface Waters,* EPA/600/3-87/015, Environmental Protection Agency, Athens, Ga., June 1987.

75. Sheng, Y. P., "Evolution of a Three-Dimensional Curvilinear Grid Hydrodynamic Model for Estuaries, Lakes and Coastal Waters: CH3D," in *Estuarine and Coastal Modeling,* Am. Soc. of Civ. Eng., New York, 1989.

76. Slade, D. H. (ed.), *Meteorology and Atomic Energy, 1968,* NTIS TID-24190, U.S. Atomic Energy Commission, Oak Ridge, Tenn. July 1968.

77. Smith, R., and R. G. Eilers, *Stream Models for Calculating Pollutional Effects of Stormwater Runoff,* EPA-600/2-78-148, Environmental Protection Agency, Cincinnati, August 1978.

78. Smith, V. H., "Nutrient Dependence of Primary Productivity in Lakes," *Limnol. Oceanogr.,* vol. 24, no. 6, pp. 1051–1064, 1979.

79. Steele, T. D., "Water Quality," Chap. 10 in *Hydrological Forecasting,* M.G. Anderson and T.P. Burt (eds.), Wiley, New York, 1985.

80. Stratton, F. D., and P. L. McCarty, "Prediction of Nitrification Effects on the Dissolved Oxygen Balance of Streams," *Environ. Science and Eng.,* vol. 1, no. 5, pp. 405–410, May 1967.

81. Streeter, H. W., and E. B. Phelps, *A Study of the Pollution and Natural Purification of the Ohio River,* U.S. Public Health Bulletin no. 146, February 1925 (reprinted by U.S. Dept. Health, Education and Welfare, 1958).

82. Tasker, G. D., and N. E. Driver, "Nationwide Regression Models for Predicting Urban Runoff Water Quality at Unmonitored Sites," *Water Resources Bull.* vol. 24, no. 5, pp. 1091–1101, October 1988.

83. Taylor, G. I., "Dispersion of Soluble Matter in Solvent Flowing Slowly Through a Tube," *Proc. Royal Soc. London (A),* vol. 219, pp. 186–203, 1953.

84. Taylor, G. I., "The Dispersion of Matter in Turbulent Flow Through a Pipe," *Proc. Royal Soc. London (A),* vol. 223, pp. 446–468, 1954.

85. Thomann, R. V., and J. A. Mueller, *Principles of Surface Water Quality Modeling and Control,* Harper and Row, New York, 1987.

86. Tsivoglou, E. C., and J. R. Wallace, *Characterization of Stream Reaeration Capacity,* EPA-R3-72-012 (NTIS PB-214649), Environmental Protection Agency, Washington, D.C., October 1972.

87. Vollenweider, R. A., "Input-Output Models with Special Reference to the Phosphorus Loading Concept in Limnology," *Schweiz. Z. Hydrol.* vol. 37, pp. 53–84, 1975.

88. Vollenweider, R. A., "Advances in Defining Critical Loading Levels for Phosphorus in Lake Eutrophication," *Mem. Ist. Ital. Idrobiol.,* vol. 33, pp. 53–83, 1976.

89. Weast, R. C., and M. J. Astel (eds.), *Handbook of Chemistry and Physics,* CRC Press, Boca Raton, Fla., 1981.

90. Wright, R. M., and A. J. McDonnell, "In-stream Deoxygenation Rate Prediction," *J. Environ. Eng. Div., Am. Soc. Civ. Eng.* (now *J. Environ. Eng.*), vol. 105, no. EE2, pp. 323–335, April 1979.

91. Yotsukura, N., "Derivation of Solute-transport Equations for a Turbulent Natural-channel Flow," *J. Research, U.S. Geological Survey,* vol. 5, no. 3, pp. 277–284, May–June 1977.

92. Yotsukura, N., and W. W. Sayre, "Transverse Mixing in Natural Channels," *Water Resour. Res.,* vol. 12, no. 4, pp. 695–704, August 1976.

CHAPTER 15
CONTAMINANT TRANSPORT IN UNSATURATED FLOW

Randall J. Charbeneau
David E. Daniel
Department of Civil Engineering
University of Texas
Austin, Texas

15.1 INTRODUCTION

The movement of contaminants in unsaturated soil is an important hydrologic problem. For example, there are about 2 million underground storage tanks located at over 700,000 facilities in the United States which contain petroleum products, and it is estimated that approximately 25 percent of these tanks are nontight and potentially leaking and polluting the subsurface environment.[152] Figure 15.1.1 shows a region of contaminated soil and the factors and processes that affect the fate and transport of chemicals. As rainfall percolates into the soil, it carries with it dissolved chemicals from wastes released near the land surface and hydrocarbons from surface washoff or underground storage tank leakage. Rainfall drives contamination into the soil through the *vadose zone* which extends from the ground surface to the water table, and then past the water table to the groundwater zone in which the chemicals may be transported laterally for distances of thousands of feet or meters. The presence of air in the soil complicates not only water flow but also flow of immiscible fluids such as hydrocarbons which may vaporize. In some cases *losses* through adsorption of the contamination on the soil, volatilization to the atmosphere, degradation by microorganisms, or through other physical, chemical, or biological processes may prevent the contamination from reaching the water table.

Table 15.1.1 shows the factors that affect transport and potential contamination in a number of hydrogeologic, environmental, geotechnical, and agricultural problems. These factors determine the ability of the soil to adsorb and degrade wastes (the soils' *assimilative capacity*) and whether chemicals are likely to accumulate within the soil profile or leach through the profile and contaminate groundwater. Understanding these factors helps in identifying proper waste disposal sites and determining suitable remediation methods for contaminated sites. In addition, these factors determine what happens to chemicals after old or abandoned waste sites are closed, and the type and quantity of chemical emissions to the atmosphere that may occur at

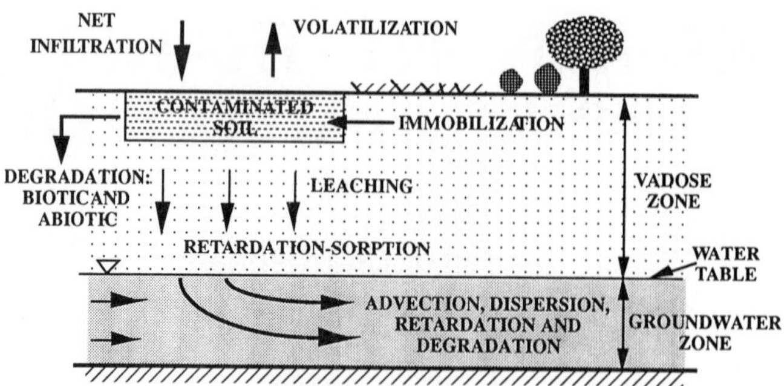

FIGURE 15.1.1 Processes which affect the movement and fate of contaminants in the subsurface environment.

sites which have recently been contaminated by volatile chemicals. Finally, these factors determine the appropriate mathematical models to predict transport and fate of chemicals in the unsaturated zone.[89] For protection of public health and the environment, particularly groundwater, it is desirable to enhance losses and retardation of contamination in the soil. This may be done through proper design and operation of a waste management site, or through development of structural and engineering controls at a site under closure or remediation.

15.2 PROCESSES OF SOLUTE TRANSPORT

15.2.1 Observations of Solute Transport through Soils

When water moves through the soil profile, it carries solutes dissolved in the water, and these chemicals are *partitioned* between soil particles and the water and air in voids, which means that some of the chemical migrates from the water to the soil particles or the soil air until an equilibrium is established among the various locations where the chemical may reside. One of the first problem areas where constituent

TABLE 15.1.1 Important Fate and Transport Processes in the Unsaturated Zone

Processes that affect losses:
- Degradation—biological, chemical, photochemical
- Volatilization

Processes that affect retardation:
- Immobilization
- Sorption
- Ion exchange

Processes that affect transport
- Advection
- Diffusion and dispersion
- Residual saturations
- Preferential flow

movement in soil water was investigated concerned salinity control in soils under irrigation. Irrigation water brings with it dissolved salt which builds up in the soil as evapotranspiration consumes soil water. If not controlled, salt can accumulate in the soil to the point where plants can no longer grow and the land becomes barren. To control soil salinization, additional irrigation water called a *leaching requirement* is added to flush or leach the salt through the crop root zone to the intermediate vadose zone and groundwater where the salt will remain in solution. Early estimates[153] of leaching requirements were based on the quantity of water leaving the root zone and the mass balance of salt in a given depth of soil. Formulas based on such simple assumptions neglect examination of how solute movement is related to water movement. Solutes are displaced differently in different soils and for different rates of water movement.[17,18,98]

Figure 15.2.1 presents an example showing differing solute displacement behavior under different rates of water movement as presented by Miller et al.,[98] who used three water application methods to displace chemically pure KCl applied on the soil surface (Panoche clay loam) at a loading of about 1 kg/m^2. Based on a solubility of 264 kg/m^3,[135] the KCl would dissolve with a 0.38-cm (0.15-in) application of water. The three water application methods used by Miller et al.[98] included continuous surface ponding, intermittent ponding with 6 in (15 cm) of water, and intermittent ponding with 2 in (5 cm) of water.

Once the chemical dissolved and entered the soil column, water application through *continuous surface ponding* required much more water than the other application methods to displace the peak chloride concentration to a given depth. Under continuous ponding conditions, the bulk of the water movement occurs in the larger pores, and the chloride contained within the smaller pores is bypassed. *Intermittent application* of water allows more time for the salt concentration to equilibrate across the range in pore sizes at a given depth, so the constituent moves more uniformly

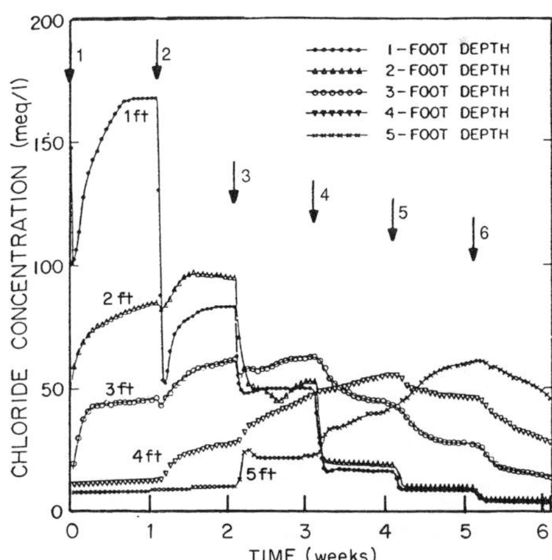

FIGURE 15.2.1 Chloride concentration vs. time for Panoche clay loam with 6-in (15-cm) water applications. Arrows represent the times for each subsequent application of 6 in of ponded water at the soil surface (1 ft = 30 cm). *(After Miller et al.[98])*

through the soil and the resulting displacement is much more efficient (less bypassing of salt contained in the smaller pores).

Figure 15.2.1 shows the concentration of chloride in the soil solution versus time at five 1-ft (30-cm) depth intervals for six periods of intermittent ponding with 6 in (15 cm) of water. Zero time represents the time when the first 6-in irrigation was initiated. The vertical arrows represent other times when additional 6-in applications were initiated. Of particular interest are the relative minimum or maximum concentrations reached at the initiation of each 6-in application. This is shown most clearly at about 1 week for the 1-ft depth, where the concentration decreases abruptly to 50 meq/L and then during the subsequent week rises to approximately 80 meq/L. This shows that when drainage occurs, the flow first comes primarily from the larger pore sequences which contain the recently ponded water, and the extracted solution may not be exactly the same as the average soil solution.

Figure 15.2.2 shows the concentrations of chloride in the soil solution versus time for 13 periods of intermittent ponding with 2 in (5 cm) of water. With the 2-in water application, the soil water content remains lower than for either the continuous ponding or 6-in water application methods, suggesting that the largest pores do not become filled with water. Figure 15.2.2 shows that, especially at the 1-ft (30-cm) depth, rapid displacement is associated with each 2-in water application. However, it appears that the concentrations remain nearly in equilibrium across the range of pore sizes that are filled with water at a given depth. It is also apparent that at the 3-ft (90-cm) depth and below, the effect of the individual water applications is lost, and the resulting concentration curves are representative of the average flow conditions even though the constituent concentration is still time-varying.

The field experiment of Miller et al.[98] shows that solute displacement behavior depends on the water application method. Attempts to flush constituents rapidly through the soil profile using large application rates result in rapid water movement through the larger pores, bypassing the constituent contained in the smaller pores. The resulting displacement is not very efficient because larger quantities of water are required to displace the constituent from the profile. In addition, the constituent is

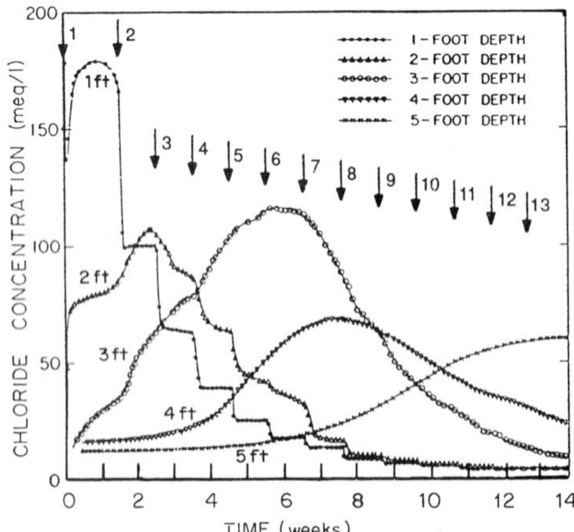

FIGURE 15.2.2 Chloride concentration vs. time for Panoche clay loam with 2-in ponded water applications (2 in = 5 cm; 1 ft = 30 cm). *(After Miller et al.[98])*

spread over a greater depth. Small application rates, which are more typical of infiltration from natural rainfall events, result in a more uniform displacement of the constituent through the soil profile. The smaller infiltration rate results in a lower water content and requires smaller quantities of water to displace the constituent from the profile.

The results of many other investigations have shown that the long-term displacement of solutes through many field soils may be modeled through use of average water flow rates along with average water contents, instead of modeling the transient transport of solutes through a heterogeneous soil profile in response to individual rainfall or infiltration events. The computational savings associated with the simplified approach are obvious. The important point is that simplified models can provide appropriate accuracy for many practical applications.

15.2.2 Solute Transport Models for a Conservative Species

It is useful to be able to estimate the travel time for solutes from the ground surface to the water table, and the rate at which leachate leaves the unsaturated zone to become groundwater contamination. The estimate is based on *advection* or simple displacement of constituent with water without considering dispersion within the soil column. Three pieces of information are required: the net groundwater recharge rate, the average soil water content, and the depth to the water table.

Net Groundwater Recharge Rate. Estimation of the net groundwater recharge rate requires that one perform a water balance at the ground surface. The water balance will be a function of the characteristics of the soil profile and the climate. The general equation describing the water balance may be written as

$$P + I = R + ET + G + \Delta S \tag{15.2.1}$$

The two terms on the left represent the precipitation P and the applied irrigation water I. The sum of these two terms represents the net addition of water to the soil surface. The terms on the right of Eq. (15.2.1) are the surface runoff R, the evapotranspiration ET, the deep percolation or groundwater recharge G, and the water storage change ΔS of the soil profile. Each of the terms in Eq. (15.2.1) represents water flows or storage changes measured in units of depth over some arbitrary time interval. To estimate the net groundwater recharge, Eq. (15.2.1) is applied over a long time interval (e.g., 1 year or more) and the water storage change is neglected. A number of computer models are available for computation of the water balance, including CREAMS[83] and HELP.[132] Such models provide estimates of the *groundwater recharge rate* W (which equals the net groundwater recharge divided by the time interval: $W = G/\Delta t$) as well as its variability through the year. See Secs. 21.3 and 21.5 for additional details on soil water balance models.

Average Water Content. The second piece of information required for estimation of the solute travel time from the ground surface to the water table is the average water content in the soil. If the average volumetric water content $\bar{\theta}$ is known, then the average solute displacement velocity, *pore water velocity,* or *seepage velocity* v is given by

$$v = \frac{q}{\bar{\theta}} = \frac{W}{\bar{\theta}} \tag{15.2.2}$$

where the *Darcy velocity* or *specific discharge* q is equal to the net groundwater recharge rate W. The travel time to the water table at a depth L, t_L is given by

$$t_L = \frac{L}{v} = \frac{L\bar{\theta}}{W} \tag{15.2.3}$$

Volumetric water contents are usually measured in the field using neutron attenuation or in the laboratory using gravimetric methods on soil samples.[52] When field data are not available, models may be used to estimate average soil water content. Near the ground surface, the water content is quite variable over time and both gravitational and capillary pressure forces are important. However, below the upper meter (3 ft) or so, the water content does not vary significantly with either depth or time for a homogeneous soil. A nearly uniform water content dictates a nearly constant capillary pressure, so that capillary pressure is nearly constant below the upper soil zone. The Darcy velocity is

$$q = K\left(\frac{\partial \psi}{\partial z} + 1\right) \tag{15.2.4}$$

where q is the vertical specific discharge, K is the hydraulic conductivity (which is a function of either water content or capillary pressure in the unsaturated zone), ψ is the capillary suction head, and z is the distance measured positive downward from the ground surface. Equation (15.2.4) is a statement of Darcy's law. Since the capillary pressure gradients are assumed to be small, one may replace Eq. (15.2.4) with

$$q = K(\theta) \tag{15.2.5}$$

where the hydraulic conductivity is taken as a function of the volumetric water content θ. The soil water content adjusts itself so that the hydraulic conductivity is sufficient to pass the volumetric flux q in the downward direction under the force of gravity, and the water content θ can thus be inferred from knowledge of q and $K(\theta)$. For example, Fig. 15.2.3 shows the soil water retention curve measured by Brooks and Corey[27] for fine sand, along with the fitted soil water retention models of Brooks and Corey[27] and van Genuchten[158] which are discussed in Chap. 5. The corresponding *relative permeability* curves, which show the ratio of the hydraulic conductivity at a given water content to its saturated value, are given in Fig. 15.2.4. With the Brooks and Corey[27] *power law model,* Eq. (15.2.2) gives

$$v = \frac{W}{\theta_r + (n - \theta_r)(W/K_s)^{\lambda/(3\lambda+2)}} \tag{15.2.6}$$

where $\bar{\theta}$ in the denominator is found by inverting Eq. (15.2.5) for the power law model. In Eq. (15.2.6), θ_r is the irreducible water content, n is the porosity, K_s is the saturated hydraulic conductivity, and λ is the Brooks and Corey pore size distribution index. For a mean recharge rate of 25 cm/year (10 in/year) and the fine sand shown in Figs. 15.2.3 and 15.2.4, Eq. (15.2.6) gives $v = 279$ cm/year (9.2 ft/year) and $\bar{\theta} = 0.090$. Use of the van Genuchten model from Chap. 5 for the same soil and recharge rate gives an average solute velocity of 316 cm/year (10.4 ft/year) and $\bar{\theta} = 0.079$.

Table 15.2.1 provides estimates of the average seepage velocity v from Eq. (15.2.6) for different soil textures and recharge rates. For each soil texture, the average parameters determined by Carsel and Parrish[33] were used to determine the tabulated values. These parameters are based on data obtained from measurements for all soils reported in SCS Soil Survey Information Reports, and were analyzed using a multiple

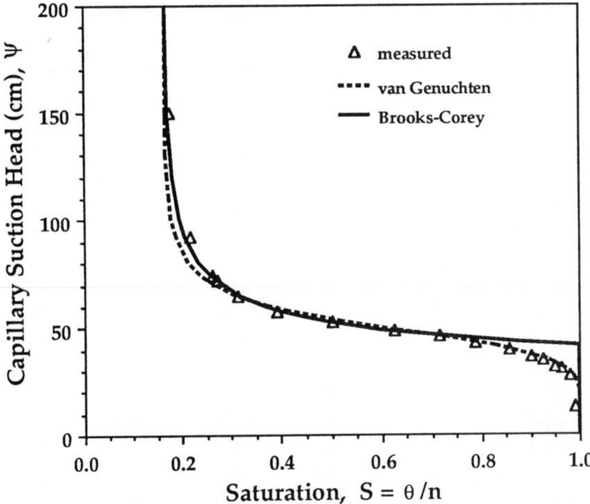

FIGURE 15.2.3 Capillary suction curve for fine sand and fitted models of Brooks and Corey[27] and van Genuchten.[158] The fitted parameters are porosity $n = 0.36$, irreducible water content, $\theta_r = 0.06$, pore size distribution index $\lambda = 3.74$, bubbling capillary pressure head $\psi_b = 41$ cm, saturated hydraulic conductivity $K_s = 2.8 \times 10^{-3}$ cm/s, van Genuchten parameter $m = 0.85$, and van Genuchten parameter $\alpha = 0.021$ cm^{-1}.

regression equation developed by Rawls and Brakensiek.[122] It is important to note that the velocities reported in Table 15.2.1 are quite small. This suggests that the average residence time for solutes in the unsaturated zone can be very long, especially at locations where the water table is found at great depths. It also should be noted that Eq. (15.2.6) and Table 15.2.1 provide a rough guide only. They are based on the assumption of steady recharge and uniform flow in a homogeneous soil profile. The

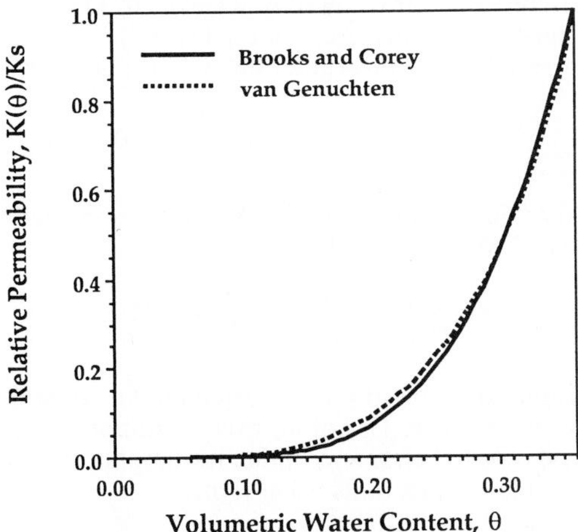

FIGURE 15.2.4 Relative permeability curve for fine sand corresponding to fitted capillary suction curve of Fig. 15.2.3. Porosity $n = 0.36$.

TABLE 15.2.1 Average Solute Displacement Velocities, cm/year

	Average annual recharge, cm			
Soil type	5	10	25	50
Clay	16	31	75	148
Clay loam	19	37	86	164
Loam	26	49	113	211
Loamy sand	53	99	225	416
Silt	21	39	88	164
Silt loam	22	41	93	174
Silty clay	16	30	74	145
Silty clay loam	16	30	72	137
Sand	68	127	286	527
Sandy clay	18	35	82	158
Sandy clay loam	25	48	112	212
Sandy loam	39	73	167	308

1 in = 2.54 cm

effects of solute mixing or dispersion, macropores, and spatial variability are not included, and these effects can be very dramatic.

Governing Advection-Dispersion Equation. Most analyses of transport in the unsaturated zone assume one-dimensional vertical flow. The basic flow equation is

$$\frac{\partial \theta}{\partial t} + \frac{\partial q}{\partial z} = 0 \tag{15.2.7}$$

where θ is the volumetric water content, t is time, z is depth measured positive downward, and q is the volumetric flow of water given by Darcy's law [Eq. (15.2.4)]. Water uptake by plant roots is neglected in Eq. (15.2.7).

For a conservative solute, the one-dimensional transport equation takes the form

$$\frac{\partial(\theta c)}{\partial t} + \frac{\partial}{\partial z}\left(qc - \theta D \frac{\partial c}{\partial z}\right) = 0 \tag{15.2.8}$$

where c is the solute concentration and D is the dispersion coefficient. Combining Eqs. (15.2.7) and (15.2.8), one obtains

$$\theta \frac{\partial c}{\partial t} + q \frac{\partial c}{\partial z} = \frac{\partial}{\partial z}\left(\theta D \frac{\partial c}{\partial z}\right) \tag{15.2.9}$$

which is the governing advection-dispersion equation for contaminant transport in both steady and transient flow. Use of an average uniform water content, which observations have shown to be an appropriate assumption for many applications, allows Eq. (15.2.9) to be written in its usual form

$$\frac{\partial c}{\partial t} + v \frac{\partial c}{\partial z} = D \frac{\partial^2 c}{\partial z^2} \tag{15.2.10}$$

where $v = q/\theta$ as in Eq. (15.2.2).

Dispersion Coefficient. The value of D reflects not only molecular diffusion arising from natural molecular motion of dissolved constituents but also mechanical dispersion resulting from the fact that local fluid velocities inside individual pores deviate from the average fluid flux in both magnitude and direction. Nielsen et al.[105] note that, in practice, D is used as an empirical parameter that includes all the solute spreading mechanisms that are not directly included in Eq. (15.2.9). These mechanisms include, for aggregated soils, the nonequilibrium concentrations between the inter- and intraaggregate pore space, the presence of macropores, and for nonconservative solutes, the effects of nonlinear sorption or decay. As such, D plays the role of a fudge factor incorporating those processes not directly considered in the simple advection-dispersion model formulation.

Nevertheless, the conceptual model is that D accounts for the two additive phenomena of molecular diffusion and mechanical dispersion, and in a one-dimensional system the dispersion coefficient D is written as

$$D = \tau D_m + \alpha |v|^m \tag{15.2.11}$$

where v is the pore water velocity given by $v = q/\theta$, τ is a *tortuosity factor* that depends on the water content[79] but not on the pore water velocity v, D_m is the *molecular diffusion coefficient,* and α and m are empirical constants. The tortuosity factor is discussed in detail in Sec. 15.2.5. For homogeneous, saturated soil, the exponent m is approximately unity,[130] and the parameter α is then known as the *dispersivity.* For unsaturated soils the value of α ranges from about 0.5 cm or less for laboratory-scale experiments involving disturbed soils, to about 10 cm or more for field-scale experiments.[19,75,157]

Solving for Concentration Profiles. Solving the differential equation (15.2.10) for the concentration as a function of depth and time $c(z, t)$ requires an *initial condition,* which is usually assumed to be $c(z, 0) = 0$ (an initially clean soil), and two boundary conditions, one at the top of the soil column and one at the bottom. At the bottom, it is usual to assume that there is no contaminant, $c(\infty, t) = 0$. The solution will then depend on the boundary condition at the soil surface. Many results are available, including those for a constant surface concentration value $c(0, t) = c_o$ (which is called a *type 1 boundary condition,* whose solution is presented by Ogata and Banks[107]) and for a constant surface flux

$$qc_o = qc - \theta D \left. \frac{\partial c}{\partial z} \right|_{z=0}$$

(which is a *type 3 boundary condition,* where the solution is given by Lindstrom et al.[88]). A simple approximate solution for both of these boundary conditions is given by

$$\frac{c}{c_o} = \frac{1}{2} \operatorname{erfc} \left(\frac{z - vt}{\sqrt{4Dt}} \right) \tag{15.2.12}$$

To use Eq. (15.2.12), one must be able to calculate the complementary error function erfc( ). This function is discussed in Abramowitz and Stegun,[1] who give an extensive tabulation. A brief listing is given in Table 15.2.2. For large arguments ($X > 3$) the complementary error function may be approximated by

$$\operatorname{erfc}(X) \cong \frac{\exp(-X^2)}{X\sqrt{\pi}} \tag{15.2.13}$$

TABLE 15.2.2 The Complementary Error Function

X	erfc(X)	X	erfc(X)
0.00	1.000000	1.60	0.023652
0.10	0.887537	1.70	0.016210
0.20	0.777297	1.80	0.010909
0.30	0.671373	1.90	0.007210
0.40	0.571608	2.00	0.004678
0.50	0.479500	2.10	0.002979
0.60	0.396144	2.20	0.001863
0.70	0.322199	2.30	0.001143
0.80	0.257899	2.40	0.000689
0.90	0.203092	2.50	0.000407
1.00	0.157299	2.60	0.000236
1.10	0.119795	2.70	0.000134
1.20	0.089686	2.80	0.000075
1.30	0.065992	2.90	0.000041
1.40	0.047715	3.00	0.000022
1.50	0.033895		

Equation (15.2.13) is accurate to within 5 percent for $X > 3$ and to within 1 percent for $X > 7$. For small arguments an approximation is

$$\text{erfc}(X) \cong 1 - \frac{2X}{\sqrt{\pi}} \tag{15.2.14}$$

which is accurate to within 5 percent for $X < 0.425$ and to within 1 percent for $X < 0.268$. Other approximations are presented in Abramowitz and Stegun.[1] For negative arguments ($X < 0$), one must use

$$\text{erfc}(X) = 2 - \text{erfc}(-X) \tag{15.2.15}$$

For example, erfc(-3) $= 2 -$ erfc(3).

Figure 15.2.5 shows the concentrations for a depth of 0.5 m predicted by the type 1 solution of Ogata and Banks,[107] the type 3 solution of Lindstrom et al.,[88] and Eq. (15.2.12), which shows that the approximate solution is very close to the exact solution under either Type 1 or Type 3 boundary conditions. The curves showing concentration versus time at a given depth are called *breakthrough curves.* The parameters used to derive the breakthrough curves in Fig. 15.2.5 are typical of those from laboratory experiments where a laboratory column is filled with soil and a chemical tracer is introduced at one end and its concentration is measured in the effluent at the other end. Figure 15.2.6 shows the model results predicted for a parameter set which may be more typical of a natural field condition. The simple advection model would predict a breakthrough time of 500 days for seepage at 0.01 m/day through the 5-m (16-ft) depth, while the models which include longitudinal dispersion show initial breakthrough as early as 200 days and complete breakthrough requiring more than 800 days.

15.2.3 Transport in Spatially Variable Soils

It is well known that soils vary significantly from point to point in their textural composition and hydraulic properties. This means that most of the parameters char-

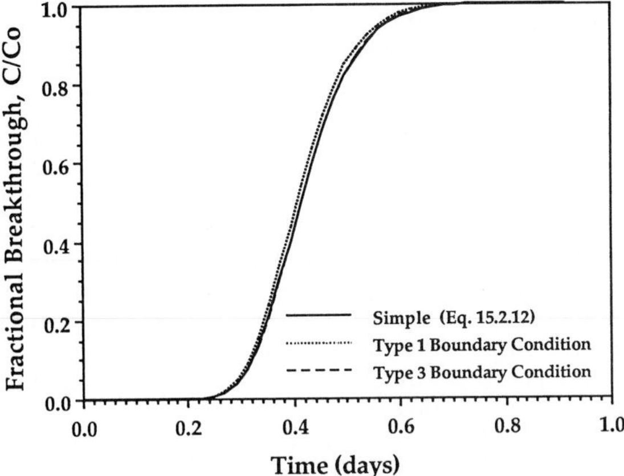

FIGURE 15.2.5 Fractional breakthrough curves for $v = 1.2$ m/day, $D = 0.012$ m^2/day, and $z = 0.5$ m, typical values for a laboratory soil column.

acterizing solute transport in the unsaturated zone vary both laterally and vertically. The spatial variability in soil water properties within various texture classes has been investigated on a national basis (for the United States) by Rawls and Brakensiek[121,123] and by Carsel and Parrish.[33] These authors present equations for estimating soil water retention and hydraulic properties from data on soil texture, bulk density, organic matter, and clay contents, and they provide information on the variability of these parameters within soil texture classes (see Chap. 5 for details). The results of these investigations are useful in application of generic models for solute transport in the unsaturated zone, but for many problems one is interested in the variability of transport over a field at a particular site.

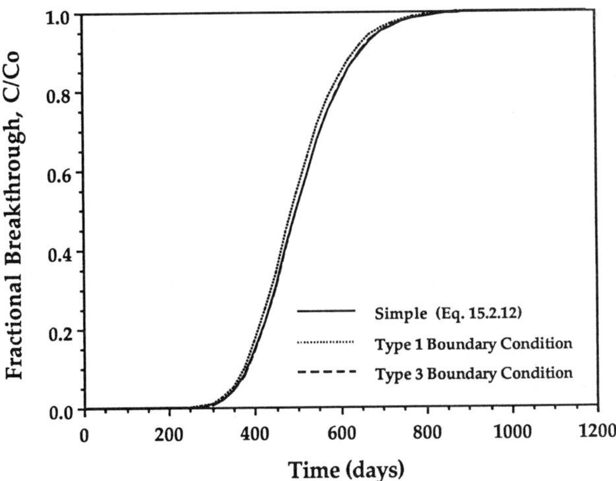

FIGURE 15.2.6 Fractional breakthrough curve for $v = 0.01$ m/day, $D = 1.1 \times 10^{-3}$ m^2/day, and $z = 5$ m, typical values for field conditions.

Statistical Variation of Soil Water Properties. The type of variability in transport behavior which is observed in the field is shown in the study of Van de Pol et al.,[157] who examined the rates of movement of chloride and tritium at several locations within a field plot and compared these solute movement rates with the rate at which water was applied to the plot. They used an 8 by 8 m plot and a uniform steady infiltration rate of 2 cm/day. The soil profile consisted of layers of clay, silty clay, silty loam, and medium to fine sand. The site was instrumented with 24 sampling locations to a depth of 1.5 m. For each sampling location, the concentration curves were fit to the type 3 equation of Lindstrom et al.,[88] which gives results similar to Eq. (15.2.12) (see Figs. 15.2.5 and 15.2.6), and the best-fit values of v and D were determined. Frequency histograms for v and D were analyzed, and it was found that for both chloride and tritium they were lognormally distributed.

If a variable X is lognormally distributed, then $Y = \ln(X)$ is normally distributed with a probability density function given by

$$f(Y) = \frac{\exp\left(-\frac{(Y - m_y)^2}{2s_y^2}\right)}{X\sqrt{2\pi s_y^2}} \tag{15.2.16}$$

where m_y and s_y are the mean and standard deviation of the natural logarithm of the variate (s_y^2 is the variance). For the X variate, its mean and variance are given by

$$m_x = \exp\left(m_y + \frac{s_y^2}{2}\right) \tag{15.2.17}$$

and

$$s_x^2 = [\exp(2m_y + s_y^2)][\exp(s_y^2) - 1] \tag{15.2.18}$$

The data from Van de Pol et al.[157] for chloride showed that the velocity v had log-mean and log-standard deviation values of $m_{\ln(v)} = 1.203$ and $s_{\ln(v)} = 0.564$, while the log-mean and log-standard deviation values of D were found to be $m_{\ln(D)} = 2.963$ and $s_{\ln(D)} = 1.132$, respectively. The lognormal velocity distribution for these parameters is shown in Fig. 15.2.7. This distribution has a mean velocity of $m_v = 3.90$ cm/day. A glance at Fig. 15.2.7 shows that while in most of the field the seepage

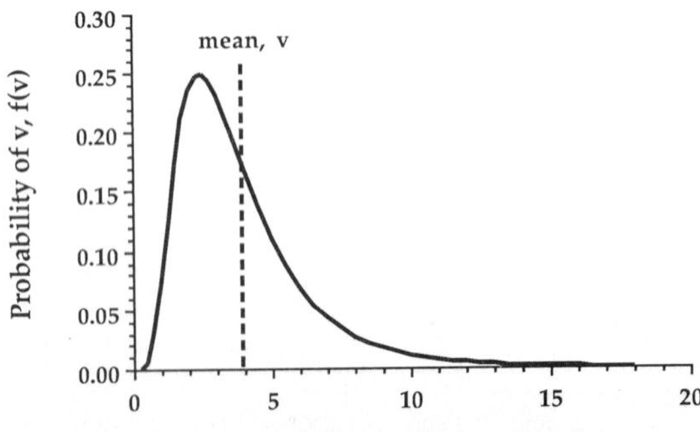

FIGURE 15.2.7 Lognormal velocity distribution from Van de Pol et al.[157] with a mean $m_v = 3.90$ cm/day and standard deviation $s_v = 2.39$ cm/day.

TABLE 15.2.3 Field Studies of the Spatial Variability of Saturated Hydraulic Conductivity

Mean, cm/day	$\sigma_{\ln K_s}$	Texture	Area, ha	Depth, cm	Number	Method of measurement
21	1.4	Clay loam	150	0–150	120	Ponded 6.2 × 6.2 m plots
168	1.2	Sandy loam	15	30–150	64	Undisturbed ponded lab cores
316	0.6	Sand	0.8	0–120	120	In situ permeameter
84	0.6	Loamy sand	0.4	0–20	12	Undisturbed ponded lab cores
4	0.5	Silty clay loam	*	30–90	33	Undisturbed ponded lab cores
19	0.9	Coarse	*	30–90	330	Undisturbed ponded lab cores
11	0.9	Fine	*	30–90	287	Undisturbed ponded lab cores
7	0.8	Silty clay	*	30–90	339	Undisturbed ponded lab cores
28	1.6	Very coarse	*	30–90	36	Undisturbed ponded lab cores
56	0.9	Coarse	*	30–90	352	Undisturbed ponded lab cores
71	0.9	Loamy sand	*	30–90	121	Undisturbed ponded lab cores
99	0.7	Loamy sand	91.6	0–15	5	Undisturbed ponded lab cores
24	1.2	Silty loam	91.6	120–135	5	Undisturbed ponded lab cores

* Composite single soil series data from Soil Conservation Service, Imperial County.
Source: After Jury et al.[73] Used with permission.

velocity is less than this mean value, there is a substantial part of the field where the pore water velocity is more than three times the mean value. The mean chloride dispersion coefficient is, from Eq. (15.2.17), $m_D = 36.74$ cm²/day. For tritium, the values of v and D were also found to be lognormally distributed with mean values of 3.78 cm/day and 36.65 cm²/day, respectively. From these data with pore water velocities varying over an order of magnitude at a single field site, it may be concluded that calculation of solute flux from $\bar{q}c$, where $\bar{q}$ is the mean infiltration rate and c is solute concentration, can lead to substantial errors in estimation of the amount of solute leaching past a given soil depth, and that spatial variability should be explicitly included in field estimates of solute transport.

Summary of Field Studies. Table 15.2.3, from Jury et al.,[73] summarizes the results from a number of field studies of the spatial variability of saturated hydraulic conductivity. While the mean value varies over nearly two orders of magnitude, the standard deviation of $\ln(K_s)$ lies within the range of 0.5 to 1.6. Table 15.2.4, also from

TABLE 15.2.4 Field Studies of the Spatial Variability of Infiltration Rate

Mean, cm/day	$\sigma_{\ln W}$	Texture	Area, ha	Number	Method of measurement
203	0.47	Silty loam	9.6	26	Double ring 0–30 min
15	1.18	Clay loam	150	20	Steady ponded 6.2 × 6.2 m plots
17	0.38	Loam	0.9	1280	Double ring steady state
7	0.64	Silty clay loam	0.004	625	5 transects of 5 cm diam. rings
9	0.52	Silty clay loam	0.004	125	5 transects of 25 cm diam. rings
9	0.23	Silty clay loam	0.004	25	5 transects of 125 cm diam. rings
47	0.78	7 series	100	20	Double ring
263	0.84	7 series	100	15	Inverse auger hole

Source: After Jury et al.[73] Used with permission.

Jury et al.,[73] shows the spatial variability of recharge rate W, which has less variability than does the saturated hydraulic conductivity. Both saturated hydraulic conductivities and infiltration rates appear to be lognormally distributed. On the other hand, values of the porosity appear to be normally distributed. For a field site, Nielsen et al.[102] report mean and standard deviation values of the porosity of 0.454 and 0.048, respectively. The variability in porosity is small compared with its mean value, so porosity may be considered constant, at least for this site.

Preferential Flow. *Preferential flow* refers to the rapid movement of solutes through fractures, root holes, and other heterogeneities, at rates much greater than expected from consideration of the porous medium as a whole. Preferential flow is much more important in vadose zone transport than in transport within saturated media. This is due, primarily, to the greater prevalence of preferential flow channels in the shallow subsurface. In addition, the hydraulic gradient driving flow is much greater in the vadose zone (the gravitational gradient is unity because water is going vertically downward) than in the zone of saturation (the gravitational gradient in groundwater is often 0.01 or less, and the water is traveling nearly horizontally).

Dyes and other chemical tracers have been used to study contaminant movement through the vadose zone.[54,92,143] Kung[85] notes that funneling due to heterogeneities may result in flow occurring through less than 1 percent of the soil. Unstable flow and fingering occurs in layered soils and may cause lateral spreading of contaminants. Factors such as these make flow and transport difficult to predict and measure: point samples can easily miss narrow fingers of solute. Similar problems occur with transport through unsaturated fractured rock.[48]

Unfortunately, there are no proven predictive models for vadose zone transport with preferential flow. However, most of the solutes which are transported below the root zone will eventually be transported to the water table because the loss rates by adsorption and volatilization decrease rapidly with depth in the vadose zone.

Monte Carlo Methods. Calculation of solute transport in the presence of spatial variability in processes and parameters requires the use of stochastic modeling methods. The simplest approach to stochastic modeling is through application of Monte Carlo methods described more fully in Chap. 19. The approach is to pick values of v and D from the estimated distributions, and then use analytical models such as Eq. (15.2.12) to predict the breakthrough curve at the depth of interest for the given parameters (random choice of v and D). The procedure is repeated many times, and the resulting breakthrough curves are averaged. As an alternative to considering v and D as independent variables, one may use models correlating their values. For example, from tracer experiments on 20 subplots of a 150-ha field to a depth of 6 ft, Biggar and Nielsen[19] have found the approximate relation

$$D = 0.6 + 2.93\, v^{1.11} \tag{15.2.19}$$

where D is in cm²/day and v in cm/day. With an equation such as (15.2.19), one could use a single estimated distribution of v to carry through the analysis. Carsel and Parrish[33] discuss the correlations between soil water retention and hydraulic parameters which are important for application of Monte Carlo methods.

The Monte Carlo approach was followed by Amoozegar-Fard et al.,[14] who used the lognormal distributions of v and D measured by Biggar and Nielsen[19] for Panoche soil. Their simulations showed that the average rate at which salt is leached from the field hinges on the variability of the pore water velocity and not on that of the apparent diffusion coefficient, especially for greater depths. The relative concentration remaining near the soil surface for a pulse input was also found to be dramatically different when a random pore water velocity was used rather than an average deterministic value, and the resulting concentration distribution compared favorably with the field averaged measurements of Biggar and Nielsen.[19]

There are a number of alternatives to Monte Carlo analysis for estimation of transport through spatially variable soils. Probabilistic models such as that presented by Dagan and Bresler[42] and transfer function models[69] have found some application. However, their use currently falls within the realm of scientific research rather than hydrologic practice.

15.2.4 Solute Partitioning in a Multiphase System

A solute is a chemical substance dissolved in a solution. In a soil several solutions may coexist, such as a chemical dissolved in water and in an immiscible oil, each one being a separate *phase* (a phase is a separate, homogeneous part of a heterogeneous system). A physical *interface* exists between each of the phases in contact, which is a dividing surface between the phases, which compounds can migrate across. Investigations of fate and transport of chemicals in the unsaturated zone must inherently deal with a multiphase system consisting of water, air, and soil. In addition, for certain applications, such as spills, leaking tanks, or land treatment of petroleum hydrocarbons, there also is a separate liquid hydrocarbon phase which is immiscible with water. The pore space is filled by the sum of the fluids present so

$$n = \theta_w + \theta_a + \theta_o \tag{15.2.20}$$

where n is the porosity, θ_w is the volumetric water content, θ_a is the volumetric air content, and θ_o is the volumetric content of the nonaqueous phase liquid (NAPL) or, alternatively, organic immiscible liquid (OIL). Individual chemical constituents partition themselves among the various phases according to thermodynamic equilibrium principles and mass transfer kinetic factors. The concentrations of a constituent in the water, air, and oil phases are designated c_w, c_a, and c_o, respectively, all on a mass per unit volume basis. The chemical located on soil particles, or *soil phase concentration,* is specified as mass of chemical sorbed per unit mass of soil, and is designated c_s. The *bulk concentration m,* which is the mass of constituent per bulk volume of soil, is then specified by

$$m = \theta_w c_w + \theta_a c_a + \theta_o c_o + \rho_b c_s \tag{15.2.21}$$

where ρ_b is the soil bulk density (mass of soil per unit bulk volume).

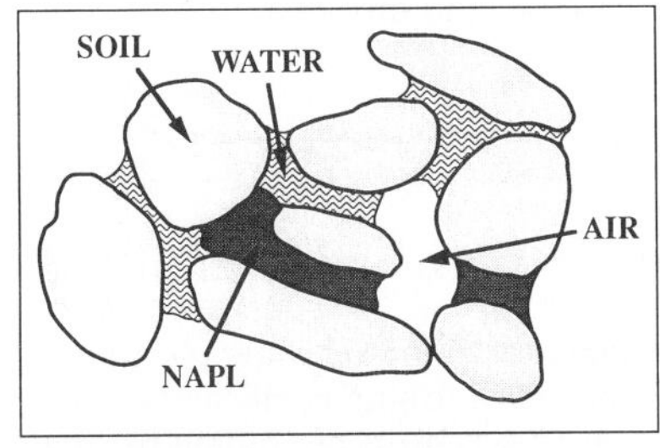

FIGURE 15.2.8 A hydrocarbon contaminated unsaturated zone has four physical phases (air, soil, water, NAPL). A chemical constituent may be present in any one, or all four phases.

Multiphase Equilibrium. Figure 15.2.8 shows a schematic representation of a hydrocarbon contaminated soil. Four phases are present: air, soil, water, and the hydrocarbon or NAPL phase. A petroleum hydrocarbon such as gasoline consists of more than one hundred chemical constituents. These constituents may dissolve or attach to any or all of the phases present. When considering the transport of the constituent within the multiphase system, a fundamental question concerns how the concentrations of constituents within the various

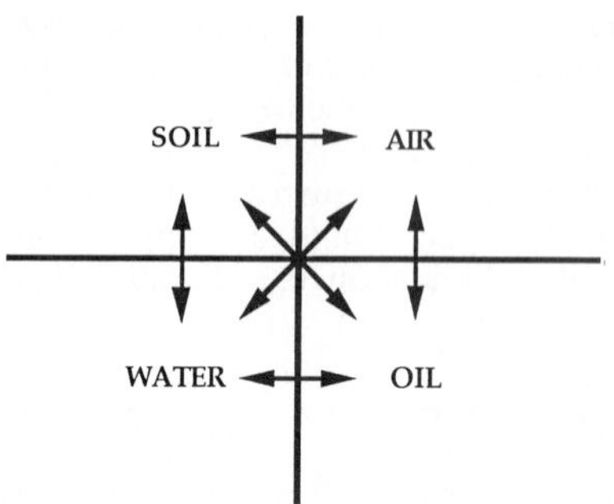

FIGURE 15.2.9 Partitioning in a multiphase system.

phases relate to each other. The simplest and most common approach assumes that the rate of mass transport through the soil within a phase is slow compared with the rate of mass transfer between phases in contact locally. Then the concentrations in adjacent phases remain in thermodynamic equilibrium, which is called the *local equilibrium assumption.*

The local equilibrium assumption assumes that the problem is *separable;* that is, even though a solute can exist in any one of four phases—air, soil, water, and OIL—at any point where two of these phases touch each other, the equilibrium set up at that interface is assumed to hold independent of the presence of the other phases. Thus it is being assumed that the presence of OIL does not affect the water-soil partitioning properties of a medium; rather, the total amount of material just gets shared among the phases. The equilibrium relationship for a multiphase system is shown schematically in Fig. 15.2.9. If the constituent of interest is lost from one phase, for example, by leaching of the water phase, then the other phases serve as a reservoir of contaminant which resupplies the phase which is losing mass while maintaining equilibrium partitioning.

Miller et al.[97] have experimentally investigated the dissolution characteristics of trapped NAPL and found that equilibrium between the water and NAPL phases is achieved rapidly, over a wide range of NAPL saturations and aqueous phase seepage velocities. Jennings and Kirkner,[66] Valocchi,[155] Parker and Valocchi,[111] and others have discussed the limitations of the local equilibrium assumption. The use of kinetic models for sorption is discussed by Karickhoff.[76]

Linear Partitioning Relations. For analysis and computation of solute transport in a multiphase system, it is convenient to be able to reference concentrations in any one phase to concentrations in another phase, or to the bulk concentration. For example, in the analysis of solute leaching, it is convenient to reference concentrations in other phases to the concentration in water, while for analysis of volatilization (vaporization of a solute with subsequent transport to the atmosphere), reference to the air phase concentration is more appropriate. The local equilibrium assumption allows one to express the bulk concentration in terms of the concentration of a single phase. The water phase serves as the reference phase, and one has

$$c_a = K_H c_w \tag{15.2.22}$$

$$c_o = K_o c_w \tag{15.2.23}$$

$$c_s = K_d c_w \tag{15.2.24}$$

In these equations, K_H is the Henry's law constant describing air-water partitioning, while K_o and K_d are the oil-water and soil-water partitioning coefficients, respectively. Both K_H and K_o are dimensionless, while K_d has units of volume per mass. Substituting Eqs. (15.2.22) through (15.2.24) in Eq. (15.2.21) gives

$$m = (\theta_w + \theta_a K_H + \theta_o K_o + \rho_b K_d) c_w = B_w c_w \tag{15.2.25}$$

where B_w is the *bulk water partition coefficient.* There are similar bulk partition coefficient definitions for the other phases. For example, one has

$$m = B_a c_a = \frac{B_w}{K_H} c_a \qquad (15.2.26)$$

$$m = B_o c_o = \frac{B_w}{K_o} c_o \qquad (15.2.27)$$

$$m = B_s c_s = \frac{B_w}{K_d} c_s \qquad (15.2.28)$$

for the air, hydrocarbon (OIL), and soil phases, respectively.

Henry's Law. Henry's law states that water-vapor partitioning is described by a linear relationship under equilibrium conditions. This is the relationship shown in Eq. (15.2.22). The Henry's law constant in Eq. (15.2.22) is dimensionless. This constant is often expressed in the literature as the ratio of the vapor pressure P_{vp} to the water solubility S for the constituent with units of atm-m^3/mol.

$$K'_H = \frac{P_{vp}}{S} \qquad (15.2.29)$$

The Henry's law constants in Eqs. (15.2.29) and (15.2.22) are related through

$$K_H = \frac{K'_H}{RT} \qquad (15.2.30)$$

where R is the gas constant (8.2×10^{-5} atm-m^3/mol-K) and T is the temperature in kelvins. The vapor pressure, solubility, and Henry's law constant for selected chemicals are presented in Table 15.2.5. The values in this table were selected from an extensive tabulation presented by Mercer et al.,[95] and Table 15.2.5 includes many of the chemical species which are identified at hazardous waste sites.

Sorption on Soil Organic Carbon: Hydrophobic Theory. Nonpolar organic compounds sorb onto the solid organic matter component of the soil matrix primarily because they do not like being in the water phase.[77] Such compounds are *hydrophobic,* and the resulting chemical interactions are weak and nonspecific (strong chemical bonds are not formed). When the organic compounds are present in trace concentrations, linear partitioning relationships (linear sorption isotherms) are often observed. A *sorption isotherm* is the relationship between the sorbed and solution concentrations. A linear isotherm plots as a straight line with a slope equal to the distribution coefficient K_d in Eq. (15.2.24). The distribution coefficient is a function of the hydrophobic character of the organic compound and the amount of organic matter present in the soil, and may be written[28,78,133]

$$K_d = K_{oc} f_{oc} \qquad (15.2.31)$$

where K_{oc} is the organic carbon partitioning coefficient and f_{oc} is the fraction of organic carbon within the soil matrix. Values of K_{oc} for various organic compounds are shown in Table 15.2.5. The fraction of organic carbon f_{oc} must be measured for a particular soil. Sorption partitioning coefficients, indexed to organic carbon (K_{oc}) are relatively invariant for natural sorbents, and the K_{oc}'s can be estimated from other

TABLE 15.2.5 Partitioning Characteristics for Selected Chemicals

	Water solubility, mg/L	Vapor pressure, atm	Henry's law K_H atm·m³/mol	Organic carbon K_{oc}, L/kg
Acetone	Infinite	3.55E−01*	2.06E−05	2.20E+00
Aldrin	1.80E−01	7.89E−09	1.60E−05	9.60E+04
Atrazine	3.30E+01	1.84E−09	2.59E−13	1.63E+02
Benzene	1.75E+03	1.25E−01	5.59E−03	8.30E+01
Bis-(2-ethylhexyl)phthalate	2.85E−01	2.63E−10	3.61E−07	5.90E+03
Chlordane	5.60E−01	1.32E−08	9.63E−06	1.40E+05
Chlorobenzene	4.66E+02	1.54E−02	3.72E−03	3.30E+02
Chloroethane	5.74E+03	1.32E+00	6.15E−04	1.70E+01
DDT	5.00E−03	7.24E−09	5.13E−04	2.43E+05
Diazinon	4.00E+01	1.84E−07	1.40E−06	8.50E+01
Dibutyl phthalate	1.30E+01	1.32E−08	2.82E−07	1.70E+05
1,1-Dichloroethane	5.50E+03	2.39E−01	4.31E−03	3.00E+01
1,2-Dichloroethane	8.52E+03	8.42E−02	9.78E−04	1.40E+01
1,1-Dichloroethene	2.25E+03	7.89E−01	3.40E−02	6.50E+01
1,2-Dichloroethene *(trans)*	6.30E+03	4.26E−01	6.56E−03	5.90E+01
Dieldrin	1.95E−01	2.34E−10	4.58E−07	1.70E+03
Ethyl benzene	1.52E+02	9.00E−03	6.43E−03	1.10E+03
Methylene chloride	2.00E+04	4.76E−01	2.03E−03	8.80E+00
Methyl parathion	6.00E+01	1.28E−08	5.59E−08	5.10E+03
Naphthalene	3.17E+01	3.03E−04	1.15E−03	1.30E+03
Parathion	2.40E+01	4.97E−08	6.04E−07	1.07E+04
Phenol	9.30E+04	4.49E−04	4.54E−07	1.42E+01
Tetrachloroethene (PERC)	1.50E+02	2.30E−02	2.59E−02	3.64E+02
Toluene	5.35E+02	3.70E−02	6.37E−03	3.00E+02
Toxaphene	5.00E−01	5.26E−04	4.36E−01	9.64E+02
1,1,1-Trichloroethane	1.50E+03	1.62E−01	1.44E−02	1.52E+02
Trichloroethene (TCE)	1.10E+03	7.60E−02	9.10E−03	1.26E+02
Trichloromethane (chloroform)	8.20E+03	1.99E−01	2.87E−03	4.70E+01
Vinyl chloride	2.67E+03	3.50E+00	8.19E−02	5.70E+01
o-Xylene	1.75E+02	9.00E−03	5.10E−03	8.30E+02

Source: Mercer et al.[95]
* 3.55E−01 = 3.55×10^{-1} = 0.355

physical properties of pollutants such as aqueous solubility or octanol-water partition coefficients (K_{ow}). For further details see Sec. 16.5.2.

Oil-Water Partitioning. Very little is known in detail about the magnitude of the oil-water partitioning coefficient K_o, except that it is not a constant but depends on the composition of the OIL phase. Since this composition changes with time as the pollutant ages, one may anticipate that K_o will change with time also. *Compositional models,* which simulate the behavior of the many components of the hydrocarbon phase, are used regularly in petroleum engineering, though their use in environmental studies has been limited.[15,16,39] For the partitioning between the aqueous and oil phases, Corapcioglu and Baehr[39] apply Raoult's law, which states that the aqueous phase concentration is equal to the aqueous phase solubility of the constituent in equilibrium with the pure constituent phase, multiplied by the mole fraction of the constituent in the OIL phase. The resulting compositional model leads to

$$K_o = \frac{\omega_k \sum_{j=1}^{N} (c_{oj}/\omega_j)}{S_k \gamma_k} \tag{15.2.32}$$

Equation (15.2.32) is written for a species k which is one out of N species which make up the OIL phase; ω_j is the molecular weight of the jth constituent (g/mol), c_{oj} is the concentration of the jth constituent in the OIL phase (g/L), S_k is the solubility of species k in water (g/L), and γ_k is the activity coefficient of the kth species (which equals 1 for ideal solutions). Equation (15.2.32) makes it apparent that K_o changes as the composition of the OIL phase changes, because of dissolution, volatilization, and degradation of constituents.

Example: Partitioning of BTX from Gasoline. A gasoline spill whose constituents are given in Table 15.2.6 occurs on a soil whose bulk density $\rho_b = 1.6$ kg/L and organic carbon fraction $f_{oc} = 0.01$. The porosity $n = 0.4$ is divided among the water, OIL, and air phases as $\theta_w = 0.15$, $\theta_o = 0.05$, $\theta_a = 0.20$. Determine the partitioning among soil, water, OIL, and air phases of the BTX compounds (benzene, toluene, and xylene).

Solution

1. The molar concentration for the gasoline mixture is calculated by dividing the constituent concentration (in g/L) in Table 15.2.6 by the corresponding molecular weight (in g/mol) and summing over all constituents:

$$\sum_{j=1}^{9} \frac{c_{oj}}{\omega_j} = 7.0 \, \frac{\text{mol}}{\text{liter}}$$

TABLE 15.2.6 Characteristics for Gasoline Spill Example—Gasoline Composition

Constituent	Constituent concentration, g/L	Molecular weight ω_j, g/mol
Benzene	8.2	78
Toluene	43.6	92
Xylene	71.8	106
1-Hexene	15.9	84
Cyclohexane	2.1	84
n-Hexane	20.4	86
Other aromatics	74.0	106
Other paraffins (C4–C8)	336.7	97.2
Heavy ends (>C8)	145.1	128
	717.8	

Baehr and Corapcioglu.[16]

2. The bulk concentration for each constituent is calculated from $m_j = \theta_o c_{oj}$. For benzene this gives $m_{\text{benzene}} = (0.05)(8.2) = 0.41$ g/L. The resulting bulk concentrations are

$$m_{\text{benzene}} = 0.41 \text{ g/L}$$

$$m_{\text{toluene}} = 2.18 \text{ g/L}$$

$$m_{\text{xylene}} = 3.59 \text{ g/L}$$

3. Using Eqs. (15.2.30), (15.2.31), and (15.2.32), the partitioning characteristics from Table 15.2.5, and an assumed temperature of 25°C (298 K), the resulting partitioning coefficients are shown below. The bulk water partitioning coefficients are calculated from Eq. (15.2.25).

	Partitioning coefficients			
Constituent	K_H	K_d, L/kg	K_o	B_w
Benzene	0.23	0.83	310	17
Toluene	0.26	3.0	1200	65
Xylene	0.21	8.3	4200	220

4. The BTX concentrations in the water, air, soil, and OIL phases are calculated from $c_w = m/B_w$, $c_a = m/B_a = K_H m/B_w$, $c_s = m/B_s = K_d m/B_w$ and $c_o = m/B_o = K_o m/B_w$. The corresponding percentages of the constituent in the water, air, soil, and OIL phases are calculated from percent $w = 100\ (\theta_w/B_w)$, percent $a = 100\ (\theta_a K_H/B_w)$, percent $s = 100\ (\rho_b K_d/B_w)$, and percent $o = 100\ (\theta_o K_o/B_w)$. The resulting concentrations and percentages are shown below.

	Phase concentration (percentage)			
Constituent	Water, mg/L	Air, mg/L	Soil, mg/kg	Oil, g/L
Benzene	24 (0.88)	5.5 (0.27)	20 (7.8)	7.5 (91)
Toluene	34 (0.23)	8.7 (0.08)	100 (7.4)	40 (92)
Xylene	16 (0.07)	3.4 (0.02)	130 (5.9)	67 (94)

15.2.5 Diffusion in Multiphase Systems

The diffusive transport of gases such as O_2 and volatile organic constituents in the soil occurs partly in the gaseous phase and partly in the liquid phase. Diffusion through the gaseous phase is primarily responsible for the transport of O_2 and CO_2 for maintenance of soil respiration and for supply of oxygen for aerobic biodegradation of organic soil contaminants. The vertical air-phase diffusive process is described by Fick's first law

$$J_{da} = -nD_{sa} \frac{\partial c_a}{\partial z} \tag{15.2.33}$$

where J_{da} is the air-phase diffusive mass flux (mass diffusing in the air phase across a unit area per unit time), D_{sa} is the *effective air-phase diffusion coefficient* in the soil, c_a is the air-phase concentration (mass per unit volume of air) of the diffusing substance, and $\partial c_a/\partial z$ is the vertical concentration gradient. The effective diffusion coefficient in the soil must be smaller than the bulk air-phase molecular diffusion coefficient D_a, owing to the limited fraction of the total volume occupied by continuous air-filled pores and also to the tortuous nature of these pores. One may write

$$nD_{sa} = \theta_a \tau_a D_a \tag{15.2.34}$$

where θ_a is the volumetric air content and τ_a is the air-phase *tortuosity factor* which accounts for the decreased cross-sectional area and increased path length induced by a porous medium, which is a function of the volumetric air space available for transport.[104] Different workers have over the years found different relations between D_{sa} and D_a, and various examples are presented in Table 15.2.7. Figure 15.2.10 shows the air-phase diffusion coefficient in soil predicted by these equations for a soil with a porosity of 0.45 and a bulk-air molecular diffusion coefficient D_a of 0.19 cm^2/s. The *Millington formula* has been shown to be very successful at representing the tortuosity effect on diffusion coefficients of pesticides in soil, even at water contents very close to saturation,[51,72,87,131,138] and it is probably the most widely used relation in the recent literature for prediction of diffusion coefficients in soil.

For a solute in a multiphase system under equilibrium conditions, the concentration gradients in each of the fluid phases are proportional, and diffusion will occur in each of the phases. Use of the Millington formula for the air, water, and OIL phases gives

$$nD_{sa} = \frac{\theta_a^{10/3}}{n^2} D_a \tag{15.2.35}$$

$$nD_{sw} = \frac{\theta_w^{10/3}}{n^2} D_w \tag{15.2.36}$$

$$nD_{so} = \frac{\theta_o^{10/3}}{n^2} D_o \tag{15.2.37}$$

TABLE 15.2.7 Air-Phase Diffusion Coefficients in Soil

Value of $\frac{n D_{sa}}{D_a} = \theta_a \tau_a$	Reference
$\kappa \theta_a^2$	Buckingham[30]
$0.66\ \theta_a$	Penman[112]
$0.62–0.8\ \theta_a$	Blake and Page[21]
$0.61\ \theta_a$	Van Bavel[156]
$\theta_a^{3/2}$	Marshall[91]
$\frac{\theta_a^{10/3}}{n^2}$	Millington[99]
$0.9\ \theta_a - 0.1$	Wesseling[160]

Source: From Hillel.[63]

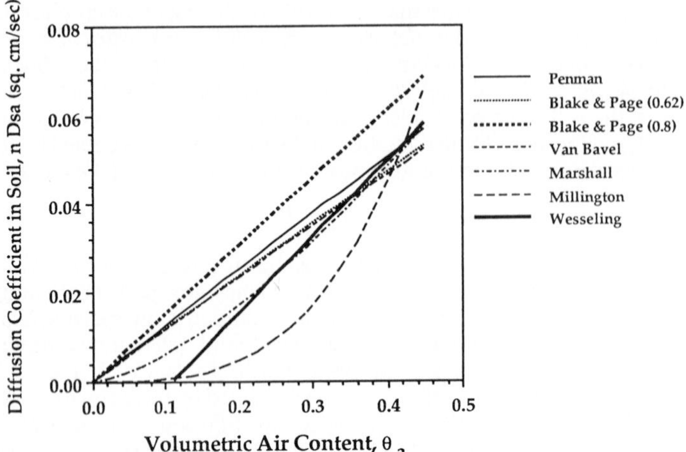

FIGURE 15.2.10 Comparison of air diffusion coefficients in soils, $n\,D_{sa}$. Values based on a porosity of $n = 0.45$ and an air molecular diffusion coefficient of $D_a = 0.19$ cm²/s.

where D_a is the air-phase molecular diffusion coefficient, D_w is the water-phase molecular diffusion coefficient, and D_o is the oil-phase molecular diffusion coefficient. The total vertical diffusive flux including transport in the three fluid phases is calculated from

$$J_d = -D_s \frac{\partial m}{\partial z} \qquad (15.2.38)$$

where D_s is the *effective soil diffusion coefficient* and m is the bulk concentration. The effective soil diffusion coefficient is given by

$$D_s = \frac{nD_{sa}}{B_a} + \frac{nD_{sw}}{B_w} + \frac{nD_{so}}{B_o} \qquad (15.2.39)$$

where B_a, B_w, B_o are the bulk partitioning coefficients given by Eqs. (15.2.26) to (15.2.28). The effective soil diffusion coefficient is a function of the molecular diffusion coefficients for each of the fluid phases, the phase volumetric contents, and the solute partitioning coefficients.

Molecular Diffusion Coefficients. The air-phase molecular diffusion coefficient D_a depends on the chemical substance and on the temperature and air pressure.[20] For a temperature of 25°C, D_a varies between 0.15 and 0.25 cm²/s (14 and 23 ft²/day) for low-molecular-weight gases (e.g., O_2, CO_2). The diffusion coefficient decreases with increasing size of the diffusing molecule. The water-phase molecular diffusion coefficient for gases which dissolve is about 10^{-4} smaller than the air-phase molecular diffusion coefficient.[90] Johnson and Babb[67] show that diffusion coefficients are on the same order of magnitude for a hydrocarbon phase as for the aqueous phase. For pesticides of intermediate molecular weight (100 to 300 g/mol), Jury et al.[74] recommend using an average value of about 0.05 cm²/s for the air-phase molecular diffusion coefficient, and a value of about 5×10^{-6} cm²/s for the water-phase molecular diffusion coefficient.

Figure 15.2.11 shows the effective soil diffusion coefficient for xylene, calculated from Eq. (15.2.39) as a function of the volumetric air content for a fixed oil content of 0.05 and porosity of 0.4. The effective soil diffusion coefficient is over two orders of magnitude larger for a dry soil than for a wet soil.

15.2.6 Volatilization Losses of Soil Contaminants

Contaminants may leave the soil by vaporizing into the soil air and then exiting to the atmosphere by diffusion in soil air to the ground surface, or by being carried in soil air as it is forced from the soil by vacuum extraction in wells. Since the mass must enter the atmosphere by traveling in soil air, it is often assumed that gaseous transport must dominate the mass transport process. However, if the mass transfer between phases is sufficiently fast so that local equilibrium conditions are achieved, the concentration in the air phase is proportional to that in the other phases, so the mass transfer in all phases may be important.

The rate of volatilization is affected by many factors, such as soil properties, chemical properties, and environmental conditions. Its rate is ultimately limited by the chemical vapor concentration which is maintained at the soil surface and by the rate at which this vapor is carried away from the soil surface to the atmosphere. The mechanisms of volatilization are similar to those of evaporation of soil water. The factors which control the rate of volatilization have been studied mostly for pesticides, but there is little that distinguishes pesticides from other organic chemicals, and one may assume that the observations based on pesticides are applicable to organic chemicals.[55,70,140,141,148]

The loss of chemicals from the soil profile due to volatilization may be calculated

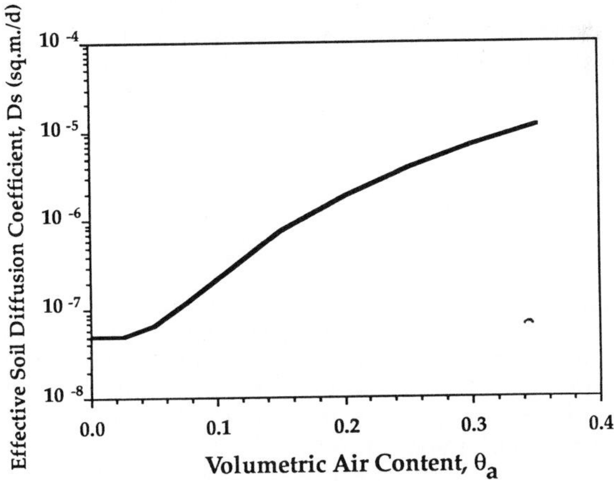

FIGURE 15.2.11 Effective soil diffusion coefficient D_s for xylene as a function of air content for a porosity $n = 0.40$ and a fixed volumetric oil content of $\theta_o = 0.05$. Values are based on molecular diffusion coefficients $D_a = 0.43$ m²/day and $D_w = D_o = 4.3 \times 10^{-5}$ m²/day.

using the theory of diffusion. One approach is to combine Fick's first law with the continuity equation giving Fick's second law, which is a partial differential equation called the *diffusion equation.*[40] If the soil profile initially contains a chemical at a uniform bulk concentration m_i, then the volatile flux (mass per area per unit time) from the soil profile is given by

$$J_{\text{vol}} = m_i \sqrt{\frac{D_s}{\pi t}} \tag{15.2.40}$$

where D_s is the effective soil diffusion coefficient and t is time. The cumulative volatilization loss (per unit surface area) is given by

$$Q_{\text{loss}} = m_i \sqrt{\frac{4D_s t}{\pi}} \tag{15.2.41}$$

Equations (15.2.40) and (15.2.41) come from the model presented by Hamaker,[60] except that they account for partitioning and diffusive transport in all of the phases.

Moving Boundary Method. An alternative formulation for computing the volatilization loss of a contaminant from the soil, which is easier to generalize for some applications, considers a sharp moving boundary separating a contaminated region below from an uncontaminated region of diffusive transport above. Thibodeaux and Hwang[147] refer to the region below the boundary as the "wet zone" and that above the boundary as the "dry zone" (see Sec. 15.2.8). The constituent concentration at the ground surface is assumed to be zero, and the volatile flux is proportional to the concentration difference ($m_i - 0$) across the dry zone. As vaporization occurs at the plane separating the wet and dry zones, the dry zone increases in depth. According to the moving boundary model, the volatilization flux is calculated by

$$J_{\text{vol}} = D_s \frac{m_i}{y} \tag{15.2.42}$$

where y is the depth to the interface separating the wet and dry zones. Application of the continuity principle leads to the result that $y = \sqrt{2D_s t}$ and the volatile flux is

$$J_{\text{vol}} = m_i \sqrt{\frac{D_s}{2t}} \tag{15.2.43}$$

while the cumulative volatile loss is

$$Q_{\text{loss}} \; m_i \sqrt{2D_s t} \tag{15.2.44}$$

A direct comparison of the predictions given by the classical diffusion model of Eqs. (15.2.40) and (15.2.41) with the moving boundary model given by Eqs. (15.2.43) and (15.2.44) shows that the moving boundary model differs from the classical Fickian diffusion model by a factor of $\sqrt{2/\pi} \cong 0.8$, which is much less than the accuracy with which most of the parameters can be measured in the field. The moving boundary model is also simpler.

15.2.7 Biodegradation Losses of Soil Contaminants

Biodegradation is an important environmental process that causes the breakdown of organic chemicals in soil. The chemical transformations are associated with, or *mediated*, through the activities of microorganisms which are naturally present. The transformation of organic carbon to inorganic carbon (CO_2) is accomplished through enzymatic oxidation. Under aerobic conditions, molecular oxygen is involved as the terminal electron acceptor, while under anaerobic conditions, the final electron acceptor is something other than molecular oxygen, such as sulfate or nitrate. *Mineralization* refers to the complete degradation of an organic chemical to inorganic products such as carbon dioxide, water, sulfate, nitrate, or ammonia. *Partial degradation* describes a level of degradation less than complete mineralization. The degradation products may be more or less toxic than their parent compounds. Chemical compounds that are not easily degraded are said to be *recalcitrant*, and they persist in the environment.[4,5,84,134,154]

The rate of biodegradation is determined by the number and type of microorganisms present, the toxicity of the parent compound or its daughter products to the microorganism population, the water content and temperature of the soil, the presence of electron acceptors and the oxidation-reduction potential, the soil pH, the availability of other nutrients for microbial metabolism, the water solubility of the chemical, and possibly other factors. Various models for the rate of biodegradation are reviewed by Alexander and Scow.[6] For most investigations, it is assumed that degradation can be described by first-order rate reactions in which the rate of loss of a chemical is proportional to the chemical concentration present in the soil, that is,

$$\frac{dm}{dt} = -\lambda m \tag{15.2.45}$$

where λ (time^{-1}) is the first-order rate constant, which is related to the half-life $T_{1/2}$ by

$$T_{1/2} = \frac{\ln(2)}{\lambda} \cong \frac{0.693}{\lambda} \tag{15.2.46}$$

Measured values of $T_{1/2}$ are quite variable between the laboratory and the field, and from site to site in the field. Representative values for selected chemicals are shown in Table 15.2.8.

15.2.8 Screening Models for Chemicals in Unsaturated Soil

The purpose of a screening model is to evaluate the behavior of a chemical in soil, and to determine the processes which control its fate and transport. The goal is to screen a wide variety of chemicals under a range of soil types and environmental conditions in a computationally efficient manner.

As an example, the screening model presented by Short[139] was developed for assessment of the movement of constituents from *oily wastes* during land treatment operations. This regulatory and investigative treatment zone (RITZ) model utilizes a *moving boundary formulation*. (RITZ is available through the Robert S. Kerr Environmental Research Laboratory, Ada, OK, 74820, USA.)[106] A feature of the RITZ model is that it considers an oil phase which is present at low saturations (*residual saturations*, so that it is immobile as a phase), and constituents from the oily phase may be lost owing to volatilization and biodegradation. The immiscible oily phase may also degrade over time.

The oily waste is initially incorporated over a depth L_o, as shown in Fig. 15.2.12*a*. The computations in the RITZ model are concerned with the fate and transport of constituents contained within the oily waste, for example, benzene, toluene, or xy-

TABLE 15.2.8 Half-life in Soils for Selected Organics*

Chemical	Half-life, days	Chemical	Half-life, days
Aldicarb	70	Methyl ethyl ketone	100
Aldrin	365	Methyl parathion	15
Atrazine	71	Naphthalene	40
Benzene	40	4-Nitrophenol	16
Benzo A anthracene	100	Parathion	18
Chlordane	100	PCE	1000
Chloroethane	30	Pentachlorophenol	10
Chloroethene	30	Phenanthrene	50
Chloroform	100	Phenol	20
Chloromethane	120	Pyrene	500
2-Chloronaphthalene	1440	Tetrachloroethene	10
DDT	3840	Toluene	5
Diazinon	32	Triallate	100
1,1-Dichloroethane	45	1,1,1-Trichloroethane	365
1,2-Dichloroethane	90	Trichloroethene	4
Dichloromethane	100	1,1-Trichloroethylene	730
2,4-Dichlorophenol	20	Trichloromethane	50
Dieldrin	868	24-D	15
Diuron	328	Xylene	110
Lindane	266		

Source: After Jury et al.,[73] Wilkerson et al.,[161] American Petroleum Institute,[8] and Rao et al.[119]

* Illustrative values which are conservative for a natural site. Half-lives can be reduced considerably for a properly engineered soil remediation site.

lene in gasoline. The model can simulate the behavior of only one constituent, or *pollutant,* of the oily waste at a time. The following assumptions are made in the RITZ model:[139] (1) the oily phase is immobile, (2) dispersion or diffusion is negligible, except as it contributes to volatilization, (3) equilibrium conditions are assumed for partitioning of the pollutant between the phases, (4) an effective degradation constant is used which accounts for degradation within the separate phases, and (5) the net infiltration of rainfall occurs at a constant rate. As time proceeds, a "dry zone" develops near the ground surface, marking the region influenced by volatilization. The base of the dry zone, shown at a depth y in Fig. 15.2.12*b*, proceeds downward at a rate controlled by the volatile flux from the profile and by the leaching rate of the pollutant. The pollutant is also leached from the wet zone and migrates downward into the region of the soil profile which does not contain the oily phase. The depth of the base of the zone of leachate is $z = L$. The concentration of the leachate from the wet zone changes with time, because both the pollutant concentration and the oil content vary because of first-order degradation losses.

The decay or degradation relations are assumed to be first-order for both the pollutant and the oily phase, with first-order rate constants of λ_p and λ_o, respectively. The bulk concentration and OIL content are given by

$$m = m_i e^{-\lambda_p t} \tag{15.2.47}$$

and

$$\theta_o(t) = \theta_o(0) e^{-\lambda_o t} \tag{15.2.48}$$

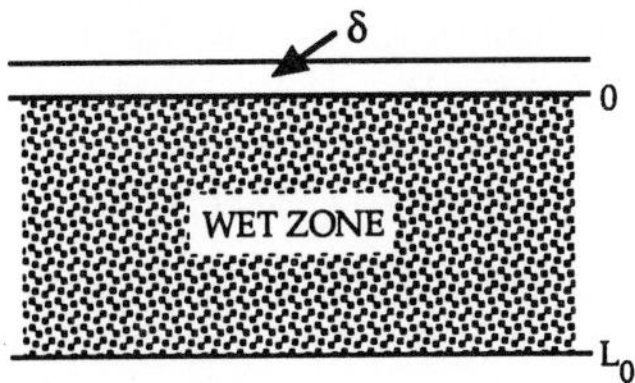

(a) Initial region of contamination to a depth L_O with atmospheric viscous sublayer of thickness δ.

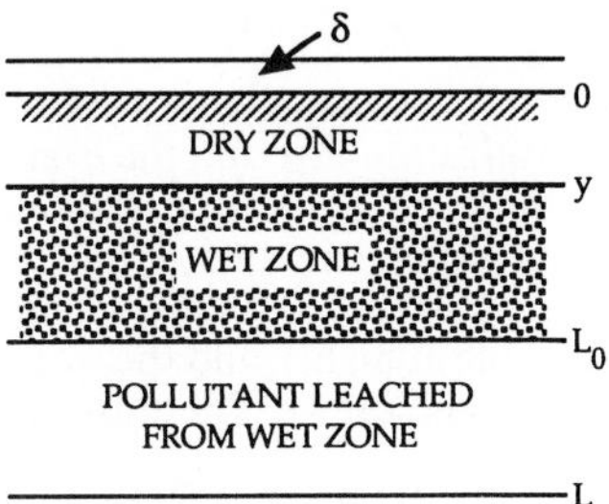

(b) Dry zone, wet zone, and leachate zone in a contaminated soil profile.

FIGURE 15.2.12 Soil profile modeled in the regulatory and investigative treatment zone (RITZ) model.

where the initial bulk pollutant mass concentration and OIL contents are m_i and $\theta_o(0)$, respectively. As the OIL degrades, this pushes the pollutant out into the other phases as determined by the multiphase partitioning relationships. These partitioning relations are the same as in Sec. 15.2.4, except that the bulk water partition coefficient defined in Eq. (15.2.25) is now time-dependent:

$$B_w(t) = \theta_w + \theta_a K_H + \theta_o(0)e^{-\lambda_o t}K_o + \rho_b K_d \tag{15.2.49}$$

The bulk water partition coefficient is a function of time within the wet and dry zones, but it is a constant in the leachate zone below L_o because the OIL phase is absent there. The aqueous pollutant concentration within the wet zone, which is the same as the concentration entering the leachate zone, is given by

$$c_w(t) = \frac{m_i e^{-\lambda_p t}}{B_w(t)} \tag{15.2.50}$$

Depending on the relative magnitudes of the decay constants for the pollutant and the OIL phases λ_p and λ_o, respectively, the aqueous concentration can either increase or decrease soon after the release occurs, though it must ultimately decrease exponentially.

The computational effort for the wet zone goes into tracking the depth of the moving boundary y as a function of time. Mass balance for this interface gives

$$\frac{dy}{dt} = v_s + \frac{D_s}{r+y} \tag{15.2.51}$$

where the first term on the right-hand side is the effective or *retarded velocity* of the constituent $v_s = q/B_w(t)$ (for a saturated medium without an OIL phase, the factor $R = B_w/\theta_w$ is called the *retardation factor*), and the second term accounts for volatilization with additional mass transfer resistance due to the presence of an atmospheric viscous sublayer of thickness δ. The viscous sublayer resistance factor r is calculated from

$$r = \frac{D_s B_a \delta}{D_a} \tag{15.2.52}$$

Jury et al.[74] note that the volatilization flux and the evaporation rate are limited by diffusion through a stagnant air layer of thickness δ above the soil surface, and they suggest that the magnitude of δ be estimated from the evaporation rate e. The steady-state daily evaporation rate is assumed to equal one-half of a typical evaporation rate (evaporation is negligible at night), and the water vapor density is assumed to equal its saturated value at the soil surface. These assumptions lead to the equation

$$\delta = \frac{D_{wv}\rho_{wv}(1 - \mathrm{RH})}{2e\rho_w} \tag{15.2.53}$$

where D_{wv} is the binary diffusion coefficient of water vapor in air (~2 m²/day), ρ_{wv} is the saturated water vapor density (~0.023 kg/m³ at 25°C), RH is the relative humidity, and ρ_w is the liquid water density (~1000 kg/m³). With an evaporation rate of 0.5 cm/day and a relative humidity of 0.5, Eq. (15.2.53) predicts a viscous sublayer thickness of 0.23 cm. While the resistance of this sublayer is small, it does prevent the model equation (15.2.51) from predicting an infinite volatilization flux immediately after the release.

For *nonvolatile* chemical species, K_H is small, and the volatilization term may be omitted from Eq. (15.2.51) to give

$$y = \frac{q}{B_w(\infty)} \left\{ t + \frac{1}{\lambda_o} \ln \left[\frac{B_w(t)}{B_w(0)} \right] \right\} \tag{15.2.54}$$

where $B_w(\infty) = \theta_w + \theta_a K_H + \rho_b K_d$. On the other hand, for a *highly volatile* chemical species with a large K_H, the volatile loss is very large at first and decreases rapidly as the dry zone depth increases because of the large diffusion resistance offered by transport through the soil zone. As long as the half-life of the OIL phase is not too small, θ_o is approximately constant during this period of significant volatilization. This is the same as the assumption that $\lambda_o \to 0$. With $\lambda_o = 0$, Eq. (15.2.51) may be integrated to find

$$y + \frac{D_s}{v_s} \ln \left[\frac{r v_s + D_s}{(r+y)v_s + D_s} \right] = v_s t \tag{15.2.55}$$

For the case with *negligible air resistance,* Eq. (15.2.51) with $r = 0$ gives

$$y - \frac{D^*}{q} \ln \left(1 + \frac{qy}{D^*} \right) = \frac{q}{B_w(\infty)} \left\{ t + \frac{1}{\lambda_o} \ln \left[\frac{B_w(t)}{B_w(0)} \right] \right\} \tag{15.2.56}$$

where $D^* = n(K_H D_{sa} + D_{sw} + K_o D_{so})$. Equations (15.2.54), (15.2.55), (15.2.56), and the more general equation (15.2.51), give y as a function of time. The maximum time for leachate generation from the initial region of oily waste contamination t_m may be found by substituting $y = L_o$ in these equations and solving for $t = t_m$. With $y(t)$ known from integration of Eq. (15.2.51) or from the approximate equations (15.2.54) to (15.2.56), the cumulative volatilization loss may be obtained through integration of

$$Q_{\text{vol}} = m_i D_s \int_0^t \frac{e^{-\lambda_p \tau}}{r + y(\tau)}\, d\tau \tag{15.2.57}$$

Likewise, the cumulative leachate from the wet zone is calculated from

$$Q_{\text{leach}} = m_i q \int_0^t \frac{e^{-\lambda_p \tau}}{B_w(\tau)}\, d\tau \tag{15.2.58}$$

Finally, the amount of degradation from the wet zone may be calculated from

$$Q_{\text{deg}} = m_i \lambda_p \int_0^t [L_o - y(\tau)] e^{-\lambda_p \tau} d\tau \tag{15.2.59}$$

Equations (15.2.57) to (15.2.59) provide the mass balance for the initial region of contamination by oil waste released into the soil, and they are valid so long as $y \le L_o$.

For land treatment systems and for other releases of contaminants, major questions concern when and at what concentration any leachate from the initial region of contamination will leave the *treatment zone* [in land treatment applications, the treatment zone consists of the upper 5 ft (1.5 m) of the soil profile] or enter an underlying aquifer at a depth L_{aq}. Within the leachate zone below the wet zone, the bulk water partition coefficient corresponds to that without an OIL phase present (B_{wlz}). The leading edge of the contaminated pulse will reach depth L_{aq} after at time

$$t_{\text{aq}} = \frac{B_{wlz}(L_{\text{aq}} - L_o)}{q} \tag{15.2.60}$$

For a given constituent, if the leading edge of the dry zone front catches the leachate front at a depth y before the time specified by Eq. (15.2.60), then this constituent will not be transported to the aquifer. All of the constituent will be lost to the volatilization and degradation processes.

The travel time from the region of initial concentration to the water table for any "parcel" of pollutant is also equal to t_{aq} (this is due to the assumed linearity of the sorption isotherm). Thus the concentration reaching the aquifer is equal to the concentration leaving the wet zone at a time t_{aq} earlier, multiplied by the factor $e^{-\lambda_p t_{\text{aq}}}$ to account for the degradation losses during transport from the wet zone to the water table. For the aqueous phase concentration reaching the aquifer at time t this gives

$$c_{w\text{aq}}(t) = \frac{m_i e^{-\lambda_p t}}{B_w(t - t_{\text{aq}})} \tag{15.2.61}$$

The time for the pollutant to be completely removed from the soil profile is the time at which the dry zone interface reaches the water table. The continuity equation for the region beneath the zone of incorporation is still given by Eq. (15.2.51), but now the "initial condition" is $y(t_m) = L_o$ and $B_w = B_{wlz}$. This equation may be integrated to find

$$y - L_o + \frac{D_s}{v_s} \ln \left[\frac{(r + L_o)v_s + D_s}{(r + y)v_s + D_s} \right] = v_s(t - t_m) \tag{15.2.62}$$

The maximum time for leachate from the vadose zone is found by setting $y = L_{aq}$ and solving for t in Eq. (15.2.62).

RITZ Model Example. As an example of the application of the RITZ model, consider the leaching of benzene from a gasoline spill where the gasoline has contaminated the upper meter of the soil profile and has a volumetric content of $\theta_o = 0.05$. The partitioning of benzene from the gasoline is presented in the example in Sec. 15.2.4. It is assumed that the average infiltration rate is 25 cm/year, that the depth to the water table is 5 m, and that degradation of both benzene and gasoline is negligible. As shown in Fig. 15.2.13, benzene first reaches the water table at 25 years after the release. Benzene is leached from the wet zone at 58 years, and it is removed from the vadose zone at 82 years. With a benzene concentration of 24 mg/L in water, the flux to the aquifer is 6 g/m^2/year over the time period from 25 to 82 years. The volatilization loss is calculated to be 15 percent of the total benzene (60 g/m^2). Figure 15.2.13 also shows the depth of the dry zone for a case with negligible volatilization. For the nonvolatile case, the constituent is not leached from the vadose zone until 93 years.

15.2.9 Evaluation of Contaminant Fate and Transport Models

There has been a considerable effort in development of comprehensive fate and transport models, especially for pesticides. Many of these models are designed to be quite general in their range of application, allowing for vertically heterogeneous soil profiles and for arbitrary meteorological conditions. An effort has been made to validate some of these more comprehensive fate and transport models, where valida-

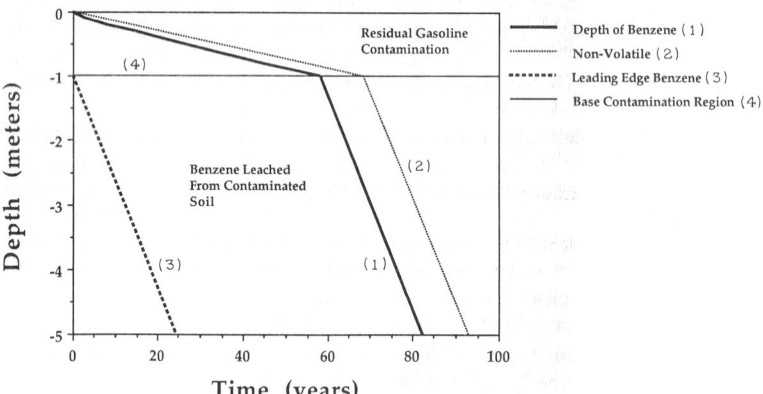

FIGURE 15.2.13 Simulation of benzene migration from gasoline-contaminated soil using the RITZ model. Upper 1 m of the soil profile is contaminated by a spill with gasoline immobilized by capillary forces at a volumetric content of $\theta_o = 0.05$. The recharge rate is 25 cm/year.

tion is generally taken to mean a favorable comparison of model results with numerical environmental data collected in the field or in laboratory observations. Hern et al.[61] note that complete model validation requires testing over the full range of conditions for which predictions are intended. At a minimum, this requires a series of validations in various climates and soil types with chemicals that typify the major fate and transport processes. Such a comprehensive testing program has not been undertaken to date.

Models may be evaluated through both field and laboratory studies. Melancon et al.[94] evaluated the *SESOIL, PRZM,* and *PESTAN* models through laboratory column leaching experiments. *PRZM* (pesticide root zone model) has also been tested in the field by Jones et al.,[68] who found satisfactory performance through favorable comparison of the model results with field data. In a comprehensive investigation Pennell et al.[113] have evaluated five pesticide simulation models using a data set developed for application of aldicarb in a citrus grove located over a well-drained, deep sandy soil in central Florida. The models employed and a description of the methods used in them for simulating environmental processes are shown in Table 15.2.9. Model performance was evaluated based on the ability to predict the depth of solute center of mass, solute dissipation, and solute concentration distributions within the soil profile. Three of the models (CMLS, PRZM, and LEACHMP) were found to provide satisfactory predictions of both solute center of mass and pesticide degradation, while none of the models accurately described measured solute concentration distributions. Based on their study, Pennell et al.[113] recommend CMLS as both a management and teaching tool, especially when consideration of metabolites is not important and concentration distributions are not desired. PRZM requires essentially the same input as CMLS but also can incorporate runoff and additional crop parameters. PRZM can predict concentration distributions (which CMLS cannot), and it is recommended as a management model for users who desire concentration distributions but are not interested in the behavior of metabolites. LEACHMP is the most complex of the models considered by Pennell et al.,[113] requiring considerably more input data, and it was also found to be the most difficult to execute. This model is recommended for use by scientists and experienced modelers who desire simulations of all pesticide components.

Pennell et al.[113] conclude that it is reasonable to expect that pesticide simulation models are able to predict the location of the solute center of mass and the amount of mass remaining in the soil profile within approximately 50 percent of the actual value of these variables. However, it may be unrealistic to expect deterministic pesticide simulation models to accurately predict solute concentration distributions.

15.3 MEASUREMENT OF SOIL PROPERTIES

Analysis of contaminant transport in unsaturated flow begins with an understanding of liquid fluxes. If one cannot accurately predict the direction of water movement or the flux, there is little hope of accurately predicting rates or patterns of contaminant movement. Also, the accuracy of predictions of contaminant transport in unsaturated flow is limited by one's ability to describe accurately the subsurface conditions and to quantify soil properties. Imperfect characterization of subsurface stratigraphy and inaccurate description of the properties of the subsoils (including the spatial distribution of those properties) usually are the most important factors that limit the overall accuracy of modeling.

TABLE 15.2.9 Description of Methods Used in Models to Simulate Environmental Processes

	Simulation model				
Process	CMLS (Version 4.0, 1987)	GLEAMS (Version 1.8.54, 1989)	LEACHMP (Version 1.0, 1987)	MOUSE (Version 7x, 1986)	PRZM (Release 1, 1985)
Water flow	Piston displacement of water. Instantaneous redistribution between field capacity and wilting point	Predicts water flow between soil layers based on a storage similar to the "tipping bucket" method	Solves Richards' equation. Requires moisture release curve data which must be fit to a modified Campbell's function	Water drainage occurs under unit hydraulic gradient. Unsaturated hydraulic conductivity is calculated using an exponential function	Water flow based on "tipping bucket" method. Operates between field capacity and wilting point. Instantaneous or time-dependent water redistribution
Runoff	Runoff not considered	Runoff based on SCS curve-number method. Erosion calculated using overland, channel, and impoundment elements, and soil particle characteristics	Runoff not considered	Runoff based on SCS curve-number method	Runoff based on SCS curve-number method. Erosion based on the Universal Soil Loss Equation
Solute transport	Piston displacement of solute	Convection transport of solute using water flow between soil layers. Solute can move upward by capillary flow	Solves the convective-dispersive transport equation	Piston displacement of solute	Convective transport of solute based on water flow between soil increments

Solute dispersion	Tracks a nondispersive solute point	Numerical dispersion, from convective transport equation, used to simulate actual solute dispersion	Calculates hydrodynamic dispersion	Calculates dispersion around midpoint of solute band using an error function	Numerical dispersion, from convective transport equation, used to simulate actual dispersion
Sorption	Inpute solute K_{oc}. Input organic carbon by soil horizon or enter K_d by soil horizon	Input K_{oc}'s for up to 10 solutes and metabolites. Input organic matter by soil horizon	Input K_{oc} for solute and two metabolites. Input organic carbon by soil increment	Input K_d for solute in the root zone and region below the root zone	Input solute K_d by soil horizon
Degradation	Input solute half-life by soil horizon	Input half-life for each solute or metabolite by horizon. Input one coefficient of transformation for each component	Input five degradation rate coefficients for three components by soil increment	Input solute half-life by soil increment	Input solute degradation rate coefficients by soil horizon
Evapotranspiration	Input daily PET. Water removed from wettest soil horizons in root zone first	Potential evaporation calculated from solar radiation and air temperature. Actual ET is then calculated using leaf area index and soil-water content	Input weekly PET total. Water removal based on root distribution, root resistance, and soil-water potential	Input daily PET. Water removal based on soil moisture content and ET input parameters	Input daily ET and crop ET coefficient. Water removal based on root distribution, and soil-water content
Roots	Input maximum rooting depth. Root biomass constant	Input maximum rooting depth. Water use is a function of depth based on an exponentially decreasing function	Root biomass can be constant or increasing. For constant root biomass, input relative root fraction by soil increment	Uniform root distribution to a rooting depth that can vary over time	Root biomass can be constant or increasing. Root distribution decreases linearly to maximum rooting depth

Source: After Pennell et al.[113]

Two critical relationships must be defined for all unsaturated, subsurface units that are to be modeled:

- The moisture characteristic curve (also called *soil water retention curve*), which is the relationship between capillary pressure head ψ and either volumetric water content θ or degree of water saturation (S, which is equal to θ/n, where n is the porosity of the soil)
- The relationship between hydraulic conductivity K and either pressure head or water content; for convenience, the relative permeability of the water phase (k_{rw}) is defined as

$$k_{rw} = \frac{K}{K_s} \tag{15.3.1}$$

such that

$$K = k_{rw} K_s \tag{15.3.2}$$

where K_s is the hydraulic conductivity at full saturation and k_{rw} is a function of either capillary pressure head $k_{rw}(\psi)$, volumetric water content $k_{rw}(\theta)$, or degree of water saturation $k_{rw}(S)$

The hydraulic conductivity at full saturation K_s is often evaluated independently from the other parameters. Typically, the moisture characteristic curve and relative permeability function are determined in the laboratory whereas the hydraulic conductivity at saturation is assessed with a combination of laboratory and in situ measurements. To define initial conditions, information is required on the initial capillary pressure heads that exist in the field soil; this information can be obtained by direct measurement of ψ (laboratory or field) or by measuring θ of the soil and estimating ψ from the moisture characteristic curve.

Additional soil properties may need to be measured, depending on site-specific conditions:

- Fluid-phase molecular diffusion coefficient D_m [used in Eq. (15.2.11)]
- Soil-water partition coefficient K_d [used in Eq. (15.2.31)]

15.3.1 Moisture Characteristic Curve

The moisture characteristic curve relates capillary pressure head ψ to either volumetric water content θ or degree of water saturation ($S = \theta/n$). Other terms used to describe this relationship in the literature include *water retention function, capillary pressure-saturation curve,* or just *moisture characteristic.*

The capillary pressure head is usually negative in unsaturated soil. To employ positive numbers rather than negative numbers, some people prefer to work with suction head, which is the negative of ψ, and prefer to report the moisture characteristic curve as suction head versus θ or S. The capillary suction head is used in Sec. 15.2.2. Any of the various forms of the moisture characteristic curve can easily be converted to a different form.

An important problem encountered when attempting to measure the moisture characteristic curve is the fact that different curves are measured for wetting and drying conditions. Figure 15.3.1 illustrates the difference in wetting and drying curves and depicts the hysteresis in moisture characteristic curves when the soil is cyclically wetted and dried. The two primary reasons why the wetting and drying curves are different may be explained as follows:

- The pores of the soil are not of uniform size, and variably sized pores drain and fill differently. The capillary pressure needed to drain a pore of variable radius is controlled by the minimum radius of the pore, whereas the capillary pressure needed to fill an empty pore with water is controlled by its maximum radius.
- When an unsaturated soil is wetted, air bubbles are inevitably trapped in the pores of the soil, which causes θ to be less than n at $\psi = 0$. Even wetting a column of soil with the base of the column immersed in a pool of water and the top subjected to a vacuum does not remove all the air bubbles from a soil.

If a soil is fully saturated with water and then dried, the ψ-θ curve (Fig. 15.3.1) is called the *initial drainage curve.* It is very difficult to fully saturate the soil and to purge all air bubbles from a soil. Approximately 10 percent of the total pore volume of the soil is usually occupied by air bubbles even in carefully soaked soil samples. If the soil is dried from an initial water content corresponding to a soaked but not fully saturated soil, the ψ-θ curve is called the *main drying curve.* The *main wetting curve* is obtained by wetting the soil from a very low water content. Intermediate curves corresponding to wetting-drying cycles are often called *scanning curves.* More complete discussion of hysteresis may be found in articles by Topp and Miller[151] and Mualem,[101] and a good summary of nomenclature is provided by Klute.[81] Common laboratory methods for measuring the moisture characteristic curve are listed in Table 15.3.1.

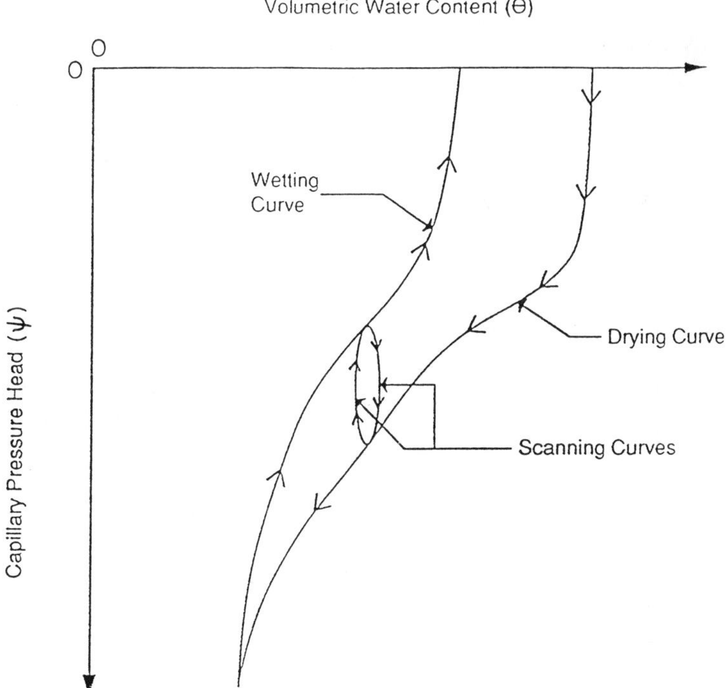

FIGURE 15.3.1 Moisture characteristic curve.

TABLE 15.3.1 Common Laboratory Methods for Measuring Moisture Characteristic (ψ-θ) Curve

Type of measurement	Name of method	Suction range, bars	Suggested references	Advantages	Disadvantages
Incremental wetting or drying	Pressure plate	0–15	Klute[81] ASTM[9]	Long history of use Equipment available commercially Little ambiguity in results ASTM method is available	Problems with air diffusion through porous plate Drying curve is easier to measure than wetting curve
	Pressure membrane	0–70	Richards[127] Coleman and Marsh[37] ASTM[10]	Can test at very negative water potential ASTM method is available	Not nearly as widely used as pressure plate
	Tensiometer	0–0.9	Cassel and Klute[34] Stannard[142]	Tensiometers are widely used Wetting or drying curve can be measured	Range is limited
	Thermocouple psychrometer	1–75	Rawlins and Campbell[120]	Very convenient and easy to use technique Wide range of water potential can be increased	Must control temperature carefully
	Filter paper method	Unlimited	McQueen and Miller[93] Al-Khafaf and Hanks[7]		
Dynamic wetting or drying	Neutron thermalization plus tensiometry	0–0.9	Topp et al.[150]	Very first testing	Higher degree of equipment sophistication is required

1 bar = 100 kPa.

To develop a moisture characteristic curve for a soil, the first step is to determine whether or not hysteresis will be taken into account. Hysteresis is usually ignored to simplify the analysis. However, even if hysteresis is ignored, it is important to recognize which condition (wetting or drying) is applicable to a particular problem to be modeled and to develop the moisture characteristic curve for that condition.

Three approaches can be taken to determine the moisture characteristic curve for a particular soil. The first technique is to estimate the curve from published data for similar soils. Three moisture characteristic curves for soils with different grain sizes are shown in Fig. 15.3.2. Information on the soil type can be used in conjunction with catalogs of published soil moisture characteristic curves, e.g., Mualem,[100] to estimate the moisture characteristic curve for a particular soil. Gupta and Larson[59] describe the use of grain-size distribution data, organic content, and bulk density to estimate the moisture characteristic curve. Rawls and Brakensiek[121] collected and analyzed data from 500 soils. Ahuja et al.[3] compared predicted soil moisture characteristic curves with measured ψ-θ relationships for 189 soil cores; the mean error in predicted water content (based on measured ψ) was 10 to 30 percent.

The second, and probably most common, technique is to assume an analytic function for $\psi(\theta)$, e.g., Brooks and Corey,[27] Campbell,[31] and van Genuchten[158] (see Chap. 5). The empirical coefficients in these functions are usually estimated based on

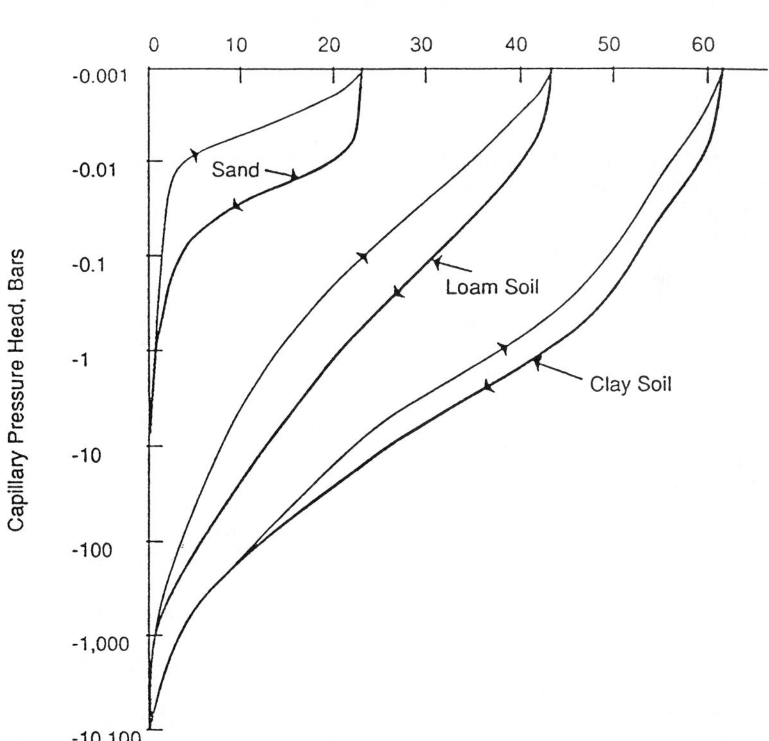

FIGURE 15.3.2 Moisture characteristic curves for three soils. *(After Yong and Warkentin.[162])*

correlations of various soil characteristics, e.g., soil classification or grain-size distribution. Because developers of closed-form solutions and computer codes for flow problems in unsaturated soils often assume that the $\psi(\theta)$ function follows one of the commonly used relationships, the hydrologist is usually compelled to force the assumed $\psi(\theta)$ relationship to fit the available data as well as possible.

The third approach for developing a soil moisture characteristic curve is to measure it directly. Several methods of measurement may be used. The methods may be divided into two categories:

- Incremental equilibrium methods
- Dynamic methods

With incremental equilibrium methods, the soil is allowed to come to equilibrium at some θ (or ψ), and then ψ (or θ) is measured. Next θ (or ψ) is changed, enough time is allowed for equilibrium to be established, and finally ψ (or θ) is measured. The process is repeated until a sufficient number of ψ-θ points have been measured to define the moisture characteristic curve. With dynamic methods, the soil is continuously wetted or dried, and both ψ and θ are measured directly. Incremental equilibrium methods are much more widely used than dynamic methods because of greater simplicity. However, dynamic methods yield a moisture characteristic curve much more quickly than incremental equilibrium methods.

Among the incremental equilibrium methods, by far the most commonly used method of measurement is the *pressure plate*. Procedures are described by the American Society for Testing and Materials[9] and by Klute.[81] Soil samples are placed in rings, soaked with water, and then placed in contact with a high-air-entry-value, saturated porous disk. A chamber is placed around the assembly, and a positive air pressure is applied to the chamber. The air pressure forces water out of the soil samples until the strength of the soil water tension holding water in the pores is equal to the applied air pressure; the outflow may be monitored to confirm equilibrium, which usually takes a few days to establish. The applied air pressure and any negative pressure in the porous disk are measured and are used to calculate ψ. After equilibrium is established, the chamber is disassembled and the soil samples are oven-dried to determine θ. The technique is used primarily for development of drying curves but can also be used to measure a wetting curve. The *pressure membrane* is a similar apparatus, but the very high air entry value of the membrane allows testing at even higher pressure.

Another incremental equilibrium technique involves sealing the soil in a cell and measuring ψ with tensiometers or thermocouple psychrometers; after equilibrium is established, the soil is wetted or dried by adding a small amount of water or allowing some evaporation, the cell is sealed, equilibrium is established after a few days, ψ is measured, and the cell is weighed to track changes in θ. After completion of several wetting or drying increments, the soil is oven-dried to determine the final θ. Daniel[43] gives examples of this technique, which is particularly convenient for determining the wetting curve. Other techniques, including vapor equilibrium methods and the null method, are described by Campbell and Gee[32] and Klute.[81]

With dynamic methods, θ must be measured nondestructively and ψ must be measured with instruments with a very fast response time, since ψ is continuously changing. Water content is typically measured with nuclear methods or flux integration, and ψ is measured with tensiometry.[114,150]

Moisture characteristic curves can also be measured in situ using techniques similar in principle to those described for dynamic measurements in the laboratory. Bruce and Luxmoore[29] summarize the technique. The reader will find additional discussion and references on this subject in Chap. 5, including a discussion of field methods of measuring soil-water properties.

15.3.2 Hydraulic Conductivity

Hydraulic conductivity in unsaturated soils is determined from the hydraulic conductivity at saturation K_s and the relative permeability k_{rw} [Eqs. (15.3.1) and (15.3.2)]. The relative permeability can be related to ψ, θ, or the saturation $S = \theta/n$. However, the $k_{rw}(\theta)$ or $k_{rw}(S)$ relation is usually preferred over $k_{rw}(\psi)$ because the $k_{rw}(\psi)$ relationship is hysteretic; in contrast, $k_{rw}(\theta)$ and $k_{rw}(S)$ exhibit virtually no hysteresis.

Relative Permeability. There are many examples in the literature of procedures to estimate $k_{rw}(\theta)$ or $k_{rw}(S)$ from other properties of the soil, including the moisture characteristic curve (e.g., Refs. 3, 47, 57, 100, 101). Estimating k_{rw} is far easier than measuring it.

Roughly a dozen techniques have been developed in the laboratory to measure k_{rw}. The techniques that have been most commonly used are summarized in Table 15.3.2. One of the main problems in measuring k_{rw} is that to induce flow of water through a soil column, a hydraulic gradient must exist; if a gradient exists, either (1) flow must take place vertically under gravity drainage at unit gradient, or (2) ψ must vary along the soil column (which means that θ and k_{rw}, which is the parameter to be measured, must also vary).

Gravity drainage through a vertical soil column in unsaturated soils is accomplished by placing a high-resistance porous plate on top of the soil column. Water is applied to the top of the plate. Gravity drainage through the plate drives a downward flux of water through the plate and soil column. The flux is limited by the hydraulic resistance of the porous plate. Once steady conditions are established, the flux is measured and either θ or ψ (or both) is measured in the soil. The high-resistance porous plate is replaced with a plate having a different resistance, and the experiment is repeated. Several repetitions lead to development of a k_{rw} curve. The procedure is described more fully by Ahuja.[2]

Steady-state methods induce steady flow through a soil column with a controlled ψ at both ends of the soil column and a small hydraulic gradient to drive flow. Procedures are described by Nielsen and Biggar[102] and Klute and Dirksen.[82] This technique is not practical for soils with low K because the flow rate resulting from the small hydraulic gradient is too small to be measured accurately.

Transient tests may be performed with a pressure plate *(pressure plate outflow test)*. Soil is placed in a pressure plate device (discussed earlier), a step increase in air pressure in the chamber is imposed, the rate of water flow out of the soil is measured as a function of time, and hydraulic conductivity is computed from the resulting data. Procedures and methods of calculation are reported by Gardner[53] and Olson and Daniel.[108]

Another method of transient testing is called the *instantaneous profile method*. Instantaneous profiling provides a powerful means for determining K in the laboratory or field. The soil is continuously wetted; by monitoring the rate of change of ψ and/or θ with position x and time t along a soil column, K can be calculated. The ψ-x curve is differentiated to obtain the hydraulic gradient, and the θ-x curve is integrated from a given point in the soil column to the downstream end of the column to determine the volume of water that exists downstream of a point, which is used to compute the rate of flow. Richards and Weeks,[128] Weeks and Richards,[159] and Daniel[43] give details of the method.

TABLE 15.3.2 Common Laboratory Methods for Measuring the Hydraulic Conductivity of Unsaturated Soil

Type of measurement	Name of method	Suction range, bars	Suggested references	Advantages	Disadvantages
Steady state	Constant head	0–0.9	Klute and Dirksen[82]	Analogous to tests on saturated soils Interpretation of results is straightforward	May be difficult to establish a significant hydraulic gradient without causing variation in water content in the soil column
	High-resistance	0–1	Ahuja[2]	Simple, straightforward test	Not as flexible as some test methods
	Steady flux	0–0.9	Klute and Dirksen[82]	Analogous to steady-state infiltration	Long time to establish steady flow in long column Need water content and water potential probes
Transient	Sorptivity through pressure plate	0–15	Dirksen[46] Olson and Daniel[108] Klute and Dirksen[82]	Relatively rapid testing	Several potential experimental problems, e.g., effect of resistance of porous plate
	Instantaneous profile	0–75	Weeks and Richards[159] Daniel[43]	Excellent range and flexibility	Somewhat complex measurements and data analysis

Klute and Dirksen[82] summarize several other methods of measurement of unsaturated hydraulic conductivity in the laboratory. Examples of field applications are provided by Clothier and White,[36] Green et al.,[56] Perroux and White,[115] and Clothier and Smettem.[35]

There are no simple guidelines as to which methods are best to use for determining relative permeability. Many testing techniques are available, and these types of tests (particularly for contaminant transport studies) tend to be performed by experienced, specialized laboratories. The main points are (1) simple analytic functions can be assumed for relative permeability, and in many cases no further effort is needed; (2) for important projects, the relative permeability function can be measured using one of several available techniques, and the measurements can be used to confirm an assumed relationship or to determine appropriate curve-fitting parameters for an analytic function; and (3) the relative permeability is multiplied by the hydraulic conductivity at saturation—there is no point in going to extremes to define the relative permeability function accurately unless the hydraulic conductivity at saturation K_s is well characterized. In nearly all practical situations, uncertainties in K_s will far outweigh uncertainties in k_{rw}, particularly for values of soil water saturation S greater than about 70 percent.

Hydraulic Conductivity at Saturation. The hydraulic conductivity at saturation K_s is a parameter that can vary over many orders of magnitude. Although much has been published about K_s's of many soils, it is inappropriate to guess a value of K_s in serious modeling work. Although other properties of the soil may be used to aid in estimating K_s (e.g., grain-size distribution data or the moisture characteristic curve)[96] it is best to measure directly the hydraulic conductivity at saturation for each significant soil unit at the site. One reason why it is best to measure K_s directly is that unpredictable secondary porosity features (e.g., cracks, fractures, or worm holes) can play a dominant role in controlling K_s and yet are not reflected in parameters such as grain-size distribution.

Laboratory permeameters for measuring K_s fall into two categories: rigid-wall permeameters and flexible-wall permeameters. With rigid-wall devices, the soil is contained in a rigid cylinder. One problem with rigid-wall cells is that spurious sidewall flow may occur with soils having low K_s. The problem of sidewall flow is overcome by confining the soil column with a flexible latex membrane in what is known as a *flexible-wall permeameter.* Hydraulic conductivity tests may be performed with a constant head, falling head, or constant flux. Recommended sources of information for testing details include Olson and Daniel,[108] Daniel et al.,[45] and Klute and Dirksen.[82]

Laboratory and field hydraulic conductivities at saturation often correlate poorly because the structural features that control flow in the field exist at a scale that is too large to be accurately reflected in small laboratory test specimens. Olson and Daniel[108] compared laboratory- and field-measured K_s's for 72 data sets involving clayey soils and found that the ratio of $(K_s)_{\text{field}}/(K_s)_{\text{laboratory}}$ ranged from 0.3 to 46,000. In 52 of the 72 data sets, the field-measured K_s was greater than the lab-measured value. The $(K_s)_{\text{laboratory}}$ was less than $(K_s)_{\text{field}}$ in only 13 of 72 cases. These trends are typical: one usually finds that the hydraulic conductivity increases with increasing scale of measurement owing to structural features of a soil that usually are measurable in a reliable manner only on a large scale (see, for example, Bradbury and Muldoon[26]).

Numerous methods of measurements of K_s in the field have been described in the literature. However, most of the literature describes testing of soils below the water table. The interest of this chapter is in methods of measurement of K_s in unsaturated soils. For unsaturated soils, there are essentially two methods of measurement: infiltration testing and borehole testing.

Infiltration tests are performed with single- or double-ring infiltrometers. The rate of infiltration and the depth of wetting front is measured as a function of time. The Green-Ampt method typically is used to calculate K_s; the wetting-front pressure head is either calculated, assumed, or measured with an air-entry permeameter. Bouwer,[23] Amoozegar and Warrick,[13] Amoozegar,[12] and Reynolds and Elrick[125] provide more information on the infiltration test.

The second type of test is a *borehole test.* A hole is drilled into the soil, the hole is filled with water, and the rate of inflow is measured, typically while a constant head is maintained in the borehole. Methods of calculating K_s are reviewed by Stephens and Neuman,[144] Philip,[118] Reynolds, Elrick, and Clothier,[126] Reynolds and Elrick,[124] and Herzog and Morse.[62]

15.3.3 Capillary Pressure Head ψ

For modeling purposes, the initial distribution of capillary pressure head ψ must be known in the flow domain. One way to estimate the initial distribution of ψ is to measure θ and determine the corresponding ψ from the moisture characteristic curve. Often, however, it is desirable to measure ψ directly to obtain more accurate information on initial conditions.

The capillary pressure head can be measured in two ways: (1) an undisturbed sample of soil can be obtained and ψ measured in the laboratory on that sample, or (2) the measurement can be obtained directly in the field. The most commonly used device for measuring ψ in both the laboratory and field is the tensiometer (Fig. 15.3.3). The tensiometer consists of a porous element that is inserted into the soil and a pressure-sensing device. The tensiometer is initially saturated with a liquid (usually water). When the saturated porous element is brought into contact with the soil, the soil will try to suck water out of the tensiometer. However, a negative pressure develops in the tensiometer. In a short period of time, equilibrium is established between the soil water and the water in the tensiometer. The negative pressure reading indicated on the tensiometer is equal to ψ, although a correction for the elevation difference between the pressure gauge and the porous element must be made. Tensiometry is described in detail by Cassell and Klute[34] and Stannard.[142]

One limitation of the tensiometer is that it cannot read ψ's more negative than approximately 0.9 bar owing to problems with cavitation of water. The thermocouple psychrometer[120] is a convenient device for laboratory measurements of total water potentials between -1 and -75 bar (-100 to -7500 kPa). The thermocouple psychrometer works as follows. A tiny drop of water is formed on the measuring junction of a miniature thermocouple. As the drop evaporates, the junction cools to the dew point. The thermocouple generates an electromotive force that is proportional to the difference between wet- and dry-bulb temperatures, which can be related to the relative humidity of the soil air. A microvoltmeter is used to measure the output from the thermocouple psychrometer. Provided that equilibrium exists, the relative humidity of the soil air is related to the total water potential. One problem with the thermocouple psychrometer is that it measures the total water potential, which includes capillary, osmotic, and adsorptive effects. Also, the thermocouple is sensitive to temperature fluctuations and is less accurate near the upper end of its range [in the range of about -3 to -1 bar (-300 to -100 kPa)].

15.3.4 Diffusion Coefficient D

The diffusion coefficient in either the liquid or gas phase may be measured directly. Diffusion coefficients in soil are not routinely measured for everyday projects, but

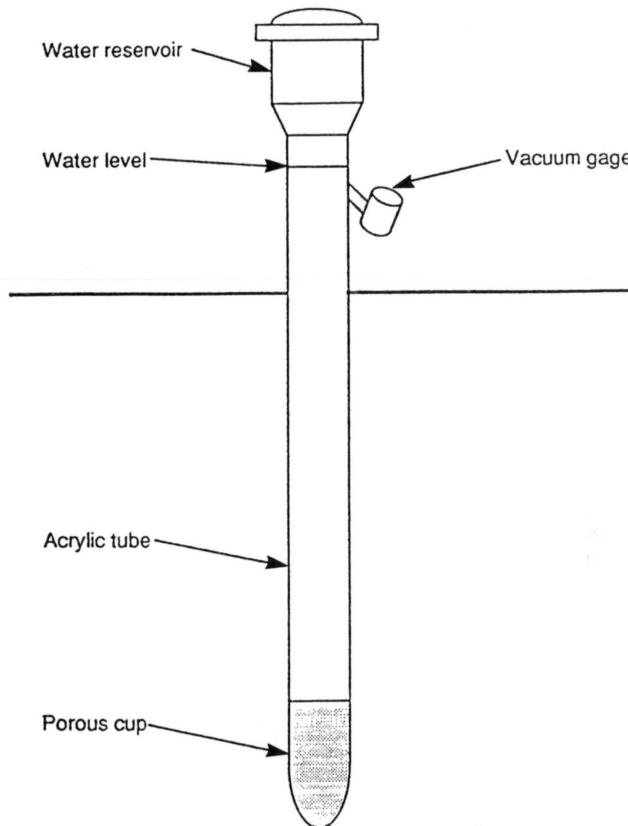

FIGURE 15.3.3 A soil water tensiometer.

researchers have developed measurement techniques. Shackelford[136] and Shackelford and Daniel[137] review techniques for measuring diffusion coefficients in saturated soil. Although little has been written about liquid-phase diffusion in unsaturated soil, some of the same measurement techniques may be applicable to both saturated and unsaturated soils.

Gas diffusion in unsaturated soil has been studied far more extensively than liquid-phase diffusion. Sources of information include Taylor,[146] Currie,[41] Lai et al.,[86] Jellick and Schnabel,[65] and Rolston et al.[129] Methods of measurement essentially are divided into steady-state procedures based on Fick's first law, and transient methods in which parameter fitting to an appropriate solution of Fick's second law is used to determine the diffusion coefficient.

15.3.5 Distribution Coefficient K_d

The distribution coefficient K_d is a measure of the tendency for a solute to be sorbed onto the soil. The larger K_d, the greater is the soil sorption.

The K_d is used to characterize solute partitioning between liquid and solid or between gas and solid. The usual test involves mixing the soil and relevant fluid (liquid or gas) containing one or more solutes of interest and measuring the solute concentration in the fluid before and after mixing. The mass of solute sorbed by the soil is computed from the difference in concentration before and after mixing. A graph is made of the mass of solute sorbed per unit mass of soil versus the solute concentration in the fluid phase. If this graph (termed the "isotherm") is linear, the slope is K_d. Nonlinear models, e.g., Freundlich isotherm, are used if appropriate. Batch absorption testing procedures are described by Griffin et al.[58] and ASTM.[11]

15.4 FIELD MONITORING

Monitoring of unsaturated soils in the field may be designed to sense various parameters:

- Amount of fluid stored in the soil
- Capillary pressure of soil water
- Chemistry of soil water

In addition, one may want to obtain samples of soil water for direct testing of water quality.

15.4.1 Water Content Monitoring

Virtually all techniques for measuring water content in field soils are indirect methods. The ones that are most commonly used are summarized in Table 15.4.1. Neutron thermalization has been the most commonly used technique for precise measurements, but advances in technology associated with time domain reflectometry (TDR) and nuclear magnetic resonance (NMR), and the recent development of commercial devices for these techniques, are likely to cause a shift toward these nonnuclear-source techniques.

15.4.2 Capillary Pressure

In the field, two measurement techniques are commonly used to measure soil water potential: tensiometers[34,142] and thermocouple psychrometers.[120] Tensiometers span the range of 0 to -0.9 bar of soil water potential, and thermocouple psychrometers work in the range of -1 to -75 bar.

15.4.3 Chemistry of Soil Water

Some of the techniques listed in Table 15.4.1 are sensitive not only to water content but also to soil salinity. The four-electrode probe is particularly useful as a simple, low-cost indication of the resistivity of the soil (resistivity decreases with increasing soil salinity).

TABLE 15.4.1 Field Monitoring Techniques for Water Content of Soil

Measurement technique	Principle measurement	Suggested references	Advantages	Disadvantages
Porous blocks (gypsum or nylon)	Electrical resistance of block changes with water content of block	Bouyoucos and Mick[25] Coleman and Hendrix[38] Tanner and Hanks[145] Bouyoucos[24] Bourget et al.[22] Gardner[52] Daniel et al.[45]	Extremely low cost probes Long history of use Convenient and simple to use	Must be calibrated with each soil Probes deteriorate over time Soil water salinity affects reading
Heat dissipation sensor (porous block)	Rate of dissipation of heat pulse varies with water content	Phene, Hoffman, and Rawlins[116] Phene, Rawlins, and Hoffman[117] Daniel et al.[45]	Low-cost probes Convenient and simple to use	Must be calibrated with each soil Probes deteriorate over time
Four-electrode probe	Resistivity of soil depends on water content of soil	Kirkham and Taylor[80] Daniel et al.[45]	Extremely low cost probe Array spacing can vary to adjust volume of soil monitored	Device provides relative (not absolute) indication of water content
Neutron thermalization	Thermalization of neutrons depends on density of hydrogen atoms	Gardner[52]	Very accurate and reproducible technique Can monitor multiple depths	Requires radioactive source and access tubes Extraneous sources hydrogen can affect measurements
Time domain reflectometry (TDR)	Dielectric constant of soil depends on water	Topp et al.[151]	Very accurate and reproducible method Probe can be buried in soil No radioactive sources	Measurements are affected by soil water chemistry
Pulsed nuclear magnetic resonance (PNMR)	Electromagnetic response depends on water content	Paetzold et al.[109]	Accurate and reproducible method No radioactive source	Sensitivity depends on measurement depth

15.4.4 Soil Water Sampling

The most common device used for sampling water from unsaturated soil is the pressure-vacuum lysimeter.[50,110] The device consists of a porous element (preferably with a high air entry pressure) attached to the end of a tube. The device is placed at the base of a borehole, and the space around the porous element is backfilled with a fine silica flour. A vacuum is drawn inside the lysimeter. If the pressure inside the lysimeter is less than the capillary pressure in the soil water, water will be drawn into the lysimeter. The lysimeter may be evacuated several times to capture a large sample of water. The device is pressurized to lift the water sample to the surface via tubing.

There are many potential sources of error in sampling with lysimeters. The most significant errors are loss of volatiles and sorption of solutes onto the porous element. Porous Teflon minimizes problems with sorption, but porous Teflon has a very low air entry value and is most applicable for sampling water from wet soils.

Pan and trench lysimeters[49] have also been used to sample pore fluids from unsaturated soil. Holder et al.[64] describe a capillary-wick pan lysimeter. Pan lysimeters are particularly useful for sampling soils located near the surface, e.g., for soil treatment systems.

REFERENCES

1. Abramowitz, M., and I. A. Stegun, *Handbook of Mathematical Functions,* Dover, New York, 1965.
2. Ahuja, L. R., "Unsaturated Hydraulic Conductivity from Cumulative Inflow Data," *Soil Sci. Soc. Am. Proc.,* vol. 38, no. 5, pp. 695–698, 1974.
3. Ahuja, L. R., J. W. Naney, and R. D. Williams, "Estimating Soil Water Characteristics from Simpler Properties or Limited Data," *Soil Sci. Soc. Am. J.,* vol. 45, no. 5, pp. 1100–1105, 1985.
4. Alexander, M., "Biodegradation: Problems of Molecular Recalcitrance and Microbial Infallibility," *Adv. Appl. Microbiol.,* vol. 7, pp. 35–80, 1965.
5. Alexander, M., *Introduction to Soil Microbiology,* 2d ed., Wiley, New York, 1977.
6. Alexander, M., and K. M. Scow, "Kinetics of Biodegradation in Soil," in B. L. Sawhney, and K. Brown, eds., *Reactions and Movement of Organic Chemicals in Soils,* SSSA Special Publication 22, Soil Science Society of America, Madison, Wis., 1989.
7. Al-Khafaf, S., and R. J. Hanks, "Evaluation of the Filter Paper Method for Estimating Soil Water Potential," *Soil Sci.,* vol. 117, pp. 194–199, 1974.
8. American Petroleum Institute, "The Land Treatability of Appendix VIII Constituents Present in Petroleum Industry Wastes," *API Publication* 4379, 1984.
9. American Society for Testing and Materials, "Test Method for Capillary-Moisture Relationships for Coarse- and Medium-Textured Soils by Porous Plate Apparatus," Method D-2325, *Annual Book of ASTM Standards,* vol. 04.08, pp. 302–308, 1992.
10. American Society for Testing and Materials, "Test Method for Capillary-Moisture Relationships for Fine-Textured Soils by Pressure-Membrane Apparatus," Method D-3152, *Annual Book of ASTM Standards,* vol. 04.08, pp. 431–436, 1992.
11. American Society for Testing and Materials, "Standard Test Method for Distribution Ratios by the Short-term Batch Method," Method D-4319, *Annual Book of ASTM Standards,* vol. 04.08, pp. 66–691, 1992.
12. Amoozegar, A., "A Compact Constant-Head Permeameter for Measuring Saturated Hydraulic Conductivity of the Vadose Zone," *Soil Sci. Soc. Am. J.,* vol. 53, no. 5, pp. 1356–1361, 1989.

13. Amoozegar, A., and A. W. Warrick, "Hydraulic Conductivity of Saturated Soils: Field Methods," in A. Klute, ed., *Methods of Soil Analysis,* Part 1, American Society of Agronomy, Madison, Wis., pp. 735–798, 1986.
14. Amoozegar-Fard, A., D. R. Nielsen, and A. W. Warrick, "Soil Solute Concentration Distributions for Spatially Varying Pore Water Velocities and Apparent Diffusion Coefficients," *Soil Sci. Soc. Am. J.,* vol. 46, pp. 3–9, 1982.
15. Baehr, A. L., "Selective Transport of Hydrocarbons in the Unsaturated Zone Due to Aqueous and Vapor Phase Partitioning," *Water Resour. Res.,* vol. 23, no. 10, pp. 1926–1938, 1987.
16. Baehr, A. L., and M. Y. Corapcioglu, "A Compositional Multiphase Model for Groundwater Contamination by Petroleum Products, 2. Numerical Solution," *Water Resour. Res.,* vol. 23, no. 1, pp. 201–214, 1987.
17. Biggar, J. W., and D. R. Nielsen, "Diffusion Effects in Miscible Displacement Occurring in Saturated and Unsaturated Porous Materials," *J. Geophys. Res.,* vol. 65, pp. 2885–2895, 1960.
18. Biggar, J. W., and D. R. Nielsen, "Miscible Displacement, 2, Behavior of Tracers," *Soil Sci. Soc. Am. Proc.,* vol. 26, pp. 125–128, 1962.
19. Biggar, J. W., and D. R. Nielsen, "Spatial Variability of the Leaching Characteristics of a Field Soil," *Water Resour. Res.,* vol. 12, no. 1, pp. 78–84, 1976.
20. Bird, R. B., W. E. Stewart, and E. N. Lightfoot, *Transport Phenomena,* Wiley, Madison, Wis., 1960.
21. Blake, G. R., and J. B. Page, "Direct Measurement of Gaseous Diffusion in Soils," *Soil Sci. Soc. Am. Proc.,* vol. 13, pp. 37–42, 1948.
22. Bourget, S. J., D. E. Elrick, and C. B. Tanner, "Electrical Resistance Measurements: Their Moisture Hysteresis, Uniformity, and Sensitivity," *Soil Sci.,* vol. 86, pp. 298–304, 1958.
23. Bouwer, H., "Rapid Field Measurement of Air Entry Value and Hydraulic Conductivity of Soil as Significant Parameters in Flow System Analysis," *Water Resour. Res.,* vol. 2, pp. 729–732, 1966.
24. Bouyoucos, G. J., "More Durable Plaster-of-Paris Moisture Blocks," *Soil Sci.,* vol. 76, pp. 447–451, 1953.
25. Bouyoucos, G. L., and A. H. Mick, "Improvements in the Plaster-of-Paris Absorption Block Electrical Resistance Method for Measuring Soil Moisture under Field Conditions," *Soil Sci.,* vol. 63, pp. 455–465, 1947.
26. Bradbury, K. R., and M. A. Muldoon, "Hydraulic Conductivity Determinations in Unlithified Glacial and Fluvial Materials," STP 1053, in D. M. Nielsen and A. I. Johnson, eds., *Water and Vadose Zone Monitoring,* American Society for Testing and Materials, Philadelphia, pp. 138–151, 1990.
27. Brooks, R. H., and A. T. Corey, "Hydraulic Properties of Porous Media," *Hydrol. Pap. 3,* Colorado State University, Fort Collins, 1964.
28. Brown, D. S., and E. W. Flagg, "Empirical Prediction of Organic Pollutant Sorption in Natural Sediments," *J. Environ. Qual.,* vol. 10, no. 3, pp. 382–386, 1981.
29. Bruce, R. R., and R. J. Luxmoore, "Water Retention: Field Methods," in A. Klute, ed., *Methods of Soil Analysis,* Part 1, American Society of Agronomy, Madison, Wis., pp. 663–686, 1986.
30. Buckingham, E., "Contributions to Our Knowledge of the Aeration of Soils," *Bulletin* 25, U.S. Department of Agriculture Bureau of Soils, Washington, D.C., 1904.
31. Campbell, G. S., "A Simple Method for Determining Unsaturated Conductivity from Moisture Retention Data," *Soil Sci.,* vol. 117, no. 6, pp. 311–314, 1974.
32. Campbell, G. S., and G. W. Gee, "Water Potential: Miscellanous Methods," in A. Klute, ed., *Methods of Soil Analysis,* Part 1, American Society of Agronomy, Madison, Wis., pp. 619–633, 1986.

33. Carsel, R. F., and R. S. Parrish, "Developing Joint Probability Distributions of Soil Water Retention Characteristics," *Water Resour. Res.*, vol. 24, no. 5, pp. 755–769, 1988.

34. Cassell, D. K., and A. Klute, "Water Potential: Tensiometry," in A. Klute, ed., *Methods of Soil Analysis,* Part 1, 2d ed. American Society of Agronomy, Madison, Wis., pp. 563–596, 1986.

35. Clothier, B. E., and K. R. J. Smettem, "Combining Laboratory and Field Measurements to Define the Hydraulic Properties of Soil," *Soil Sci. Soc. Am. J.*, vol. 54, no. 2, pp. 299–304, 1990.

36. Clothier, B. E., and I. White, "Measurement of Sorptivity and Soil Water Diffusivity in the Field," *Soil Sci. Soc. Am. J.*, vol. 45, no. 2, pp. 241–245, 1981.

37. Coleman, J. D., and A. D. Marsh, "An Investigation of the Pressure Membrane Method for Measuring the Suction Properties of Soil," *J. Soil Sci.*, vol. 12, pp. 343–360, 1961.

38. Colman, E. A., and T. M. Hendrix, "The Fiberglass Electrical Soil-Moisture Instrument," *Soil Sci.*, vol. 67, pp. 425–438, 1949.

39. Corapcioglu, M. Y., and A. L. Baehr, "A Compositional Multiphase Model for Groundwater Contamination by Petroleum Products, 1. Theoretical Considerations," *Water Resour. Res.*, vol. 23, no. 1, pp. 191–200, 1987.

40. Crank, J., *The Mathematics of Diffusion,* 2d ed., Oxford University Press, Oxford, 1975.

41. Currie, J. A., "Gaseous Diffusion in the Aeration of Aggregated Soils," *Soil Sci.*, vol. 92, no. 1, pp. 40–45, 1961.

42. Dagan, G., and E. Bresler, "Solute Dispersion in Unsaturated Heterogeneous Soil at Field Scale: I. Theory," *Soil Sci. Soc. Am. J.*, vol. 43, pp. 461–467, 1979.

43. Daniel, D. E., "Permeability Test for Unsaturated Soil," *Geotech. Testing J.*, vol. 6, no. 2, pp. 81–86, 1983.

44. Daniel, D. E., P. M. Burton, and S. D. Hwang, "Laboratory Tests on Four Probes Used for Vadose Zone Monitoring," in D. M. Nielsen and M. N. Sara, eds., *Current Practice for Ground Water and Vadose Zone Investigations,* STP 1118, American Society for Testing and Materials, Philadelphia, in press, 1992.

45. Daniel, D. E., S. J. Trautwein, S. S. Boynton, and D. E. Foreman, "Permeability Testing with Flexible-Wall Permeameters," *Geotech. Testing J.*, vol. 7, no. 3, pp. 113–122, 1984.

46. Dirksen, C., "Determination of Soil Water Diffusivity by Sorptivity Measurements," *Soil. Sci. Soc. Am. Proc.*, vol. 39, pp. 22–27, 1975.

47. Elzeftawy, A., and R. S. Mansell, "Hydraulic Conductivity Calculations for Unsaturated Steady State and Transient-State Flow in Sand," *Soil Sci. Soc. Am. Proc.*, vol. 39, no. 4, pp. 599–603, 1975.

48. Evans, D. D., and T. J. Nicolson, "Flow and Transport through Unsaturated Fractured Rock," *Geophysical Monograph,* AGU Geophysical Monograph Board, Washington, D.C., 1987.

49. Everett, L. G., "Soil Pore-Liquid Monitoring," J. S. Devinny, L. G. Everett, J. C. S. Lu, and R. L. Stollan, eds., *Subsurface Migration of Hazardous Wastes,* van Nostrand Reinhold, New York, pp. 306–336, 1990.

50. Everett, L. G., L. G. McMillion, and L. A. Eccles, "Suction Lysimeter Operation at Hazardous Waste Sites," in A. G. Collins and A. I. Johnson, eds., *Contamination: Field Methods,* ASTM STP 963, American Society for Testing and Materials, Philadelphia, pp. 304–327, 1988.

51. Farmer, W. J., M. S. Yang, J. Letey, and W. F. Spencer, "Hexachlorobenzene: Its Vapor Pressure and Vapor Phase Diffusion in Soil," *Soil Sci. Soc. Am. J.*, vol. 44, pp. 676–680, 1980.

52. Gardner, W. H., "Water Content," in Klute, A., ed., *Methods of Soil Analysis,* Part 1, Monograph 9, American Society of Agronomy, Madison, Wis., 1986.

53. Gardner, W. R., "Calculation of Capillary Conductivity from Pressure Plate Outflow Data," *Soil Sci. Soc. Am. Proc.*, vol. 20, pp. 317–320, 1956.

54. Ghodrati, M., and W. A. Jury, "A Field Study Using Dyes to Characterize Preferential Flow of Water," *Soil Sci. Soc. Am. J.*, vol. 54, p. 1558, 1990.

55. Glotfelty, D. E., and C. J. Schomburg, "Volatilization of Pesticides from Soil," in B. L. Sawhney and K. Brown, eds., *Reactions and Movement of Organic Chemicals in Soils,* SSSA Special Publication 22, Soil Science Society of America, Madison, Wis., 1989.

56. Green, R. E., L. R. Ahuja, and S. K. Chong, "Hydraulic Conductivity, Diffusivity, and Sorbtivity of Unsaturated Soils: Field Methods," in A. Klute, ed., *Methods of Soil Analysis,* Part 1, *Physical and Mineralogical Methods,* 2d ed., ASA Monograph 9, American Society of Agronomy, Madison, Wis., pp. 799-823, 1986.

57. Green, R. E., and J. C. Corey, "Calculation of Hydraulic Conductivity: A Further Evaluation of Some Predictive Methods," *Soil Sci. Soc. Am. Proc.*, vol. 35, no. 1, 3-8, 1971.

58. Griffin, R. A., W. A. Sack, W. R. Roy, C. C. Ainsworth, and I. G. Krapuc, "Batch Type 24-h Distribution Ratio for Contaminant Adsorption by Soil Materials," in D. Lorenzen, R. A. Conway, L. P. Jackson, A. Hamza, C. L. Perket, and W. J. Lacy, eds., *Hazardous and Industrial Solid Waste Testing and Disposal: Sixth Volume,* ASTM STP 933, American Society for Testing and Materials, Philadelphia, pp. 390-408, 1986.

59. Gupta, S. C., and W. E. Larson, "Estimating Soil Water Retention Characteristics from Particle Size Distribution, Organic Matter Percent, and Bulk Density," *Water Resour. Res.*, vol. 15, pp. 1633-1635, 1979.

60. Hamaker, J. W., "Diffusion and Volatilization," in C. A. I. Goring, and J. W. Hamaker, eds., *Organic Chemicals in the Soil Environment,* vol. 1, Marcel Dekker, New York, 1972.

61. Hern, S. C., S. M. Melancon, and J. E. Pollard, "Generic Steps in the Field Validation of Vadose Zone Fate and Transport Models," in S. C. Hern and S. M. Melancon, eds., *Vadose Zone Modeling of Organic Pollutants,* Lewis Publishers, Chelsea, Mich., 1986.

62. Herzog, B. L., and W. J. Morse, "Comparison of Slug Test Methodologies for Determination of Hydraulic Conductivity in Fine-Grained Sediments," in D. M. Nielsen and A. I. Johnson, eds., *Ground Water and Vadose Zone Monitoring,* ASTM STP 1053, American Society for Testing and Materials, Philadelphia, pp. 152-164, 1990.

63. Hillel, D., *Introduction to Soil Physics,* Academic Press, Orlando, 1982.

64. Holder, M., K. W. Brown, J. C. Thomas, D. Zabcik, and H. E. Murray, "Capillary-Wick Unsaturated Zone Pore Water Sampler," *Soil Sci. Soc. Am. J.*, vol. 55, no. 5, pp. 1195-1202, 1991.

65. Jellick, G. J., and R. R. Schnabel, "Evaluation of a Field Method for Determining the Gas Diffusion Coefficient in Soils," *Soil Sci. Soc. Am. J.*, vol. 50, no. 1, pp. 18-23, 1986.

66. Jennings, A. A., and D. J. Kirkner, "Criteria for Selecting Equilibrium or Kinetic Sorption Descriptions in Groundwater Quality Models," in H. T. Shen, ed., *Frontiers in Hydraulic Engineering,* American Society of Civil Engineers, New York, 1983.

67. Johnson, P. A., and A. L. Babb, "Liquid Diffusion in Non-electrolytes," *Chem. Revs.*, vol. 56, pp. 387-453, 1956.

68. Jones, R. L., G. W. Black, and T. L. Estes, "Comparison of Computer Model Predictions with Unsaturated Zone Field Data for Aldicarb and Aldoxycarb, *Environ. Toxicol. Chem.*, vol. 5, pp. 1027-1037, 1986.

69. Jury, W. A., "Simulation of Solute Transport Using a Transfer Function Model," *Water Resour. Res.*, vol. 18, no. 2, pp. 363-368, 1982.

70. Jury, W. A., "Volatilization from Soil," in S. C. Hern and S. M. Melancon, eds., *Vadose Zone Modeling of Organic Pollutants,* Lewis Publishers, Chelsea, Mich., 1986.

71. Jury, W. A., W. J. Farmer, and W. F. Spencer, "Behavior Assessment Model for Trace Organics in Soil. II. Chemical Classification Parameter Sensitivity," *J. Environ. Qual.* vol. 13, pp. 567-572, 1984.

72. Jury, W. A., R. Grover, W. F. Spencer, and W. F. Farmer, "Modeling Vapor Losses of Soil-Incorporated Triallate," *Soil Sci. Soc. Am. J.*, vol. 44, pp. 445-450, 1980.

73. Jury, W. A., D. Russo, G. Sposito, and H. Elabd, "The Spatial Variability of Water and

Solute Transport Properties in Unsaturated Soil: I. Analysis of Property Variation and Spatial Structure with Statistical Models, and II. Scaling Models of Water Transport," *Hilgardia,* vol. 55, no. 4, pp. 1–32, 33–56, 1987.

74. Jury, W. A., W. F. Spencer, and W. J. Farmer, "Behavior Assessment Model for Trace Organics in Soil: I. Model Description," *J. Environ. Qual.,* vol. 12, no. 4, pp. 558–564, 1983.

75. Jury, W. A., and G. Sposito, "Field Calibration and Validation of Solute Transport Models for the Unsaturated Zone," *Soil Sci. Soc. Am. J.,* vol. 49, no. 6, pp. 1331–1341, 1985.

76. Karickhoff, S. W., "Sorption Kinetics of Hydrophobic Pollutants in Natural Sediments," in R. A. Baker, ed., *Contaminants and Sediments,* vol. 2, Ann Arbor Science Publishers, Ann Arbor, Mich., pp. 193–205, 1980.

77. Karickhoff, S. W., "Organic Pollutant Sorption in Aquatic Systems," *J. Hydraul. Eng.,* vol. 110, no. 6, pp. 707–735, 1984.

78. Karickhoff, S. W., D. S. Brown, and T. A. Scott, "Sorption of Hydrophobic Pollutants on Natural Sediments," *Water Res.,* vol. 13, pp. 241–248, 1979.

79. Kemper, W. D., and J. C. van Schaik, "Diffusion of Salt in Clay-Water Systems," *Soil Sci. Soc. Am. Proc.,* vol. 30, no. 5, pp. 535–540, 1966.

80. Kirkham, D., and G. S. Taylor, "Some Tests of Four-Electrode Probe for Soil Moisture Measurement," *Soil Sci. Soc. Am. Proc.,* vol. 14, pp. 42–46, 1949.

81. Klute, A., "Water Retention: Laboratory Methods," in A. Klute, ed., *Methods of Soil Analysis,* Part 1, American Society of Agronomy, Madison, Wis., pp. 635–686, 1986.

82. Klute, A., and C. Dirksen, "Hydraulic Conductivity and Diffusivity: Laboratory Methods," in A. Klute, ed., *Methods of Soil Analysis,* Part 1, American Society of Agronomy, Madison, Wis., pp. 687–734, 1986.

83. Knisel, W. G., "CREAMS: A Field-Scale Model for Chemicals, Runoff, and Erosion from Agricultural Management Systems," U.S. Department of Agriculture, Conservation Research Report 26, 1980.

84. Kuhn, E. P., and J. M. Suflita, "Dehalogenation of Pesticides by Anaerobic Microorganisms in Soils and Groundwater—A Review," in B. L. Sawhney, and K. Brown, eds., *Reactions and Movement of Organic Chemicals in Soils,* SSSA Special Publication 22, Soil Science Society of America, Madison, Wis., 1989.

85. Kung, K-J. S., "Preferential Flow in a Sandy Vadose Zone: 1. Field Observation," *Geoderma,* vol. 46, pp. 51–58, 1990.

86. Lai, S. H., J. M. Tiedje, and A. E. Erickson, "In Situ Measurement of Gas Diffusion Coefficient in Soils," *Soil Sci. Soc. Am. J.,* vol. 40, no. 1, pp. 3–6, 1976.

87. Letey, J., and W. J. Farmer, "Movement of Pesticides in Soil," in W. D. Guenzi, ed., *Pesticides in Soil and Water,* Soil Science Society of America, Madison, Wis., 1974.

88. Lindstrom, F. T., R. Haque, V. H. Freed, and L. Boersma, "Theory on the Movement of some Herbicides in Soils: Linear Diffusion and Convection of Chemicals in Soils," *J. Environ. Sci. Tech.,* vol. 1, no. 7, pp. 561–565, 1967.

89. Loehr, R. C., and R. J. Charbeneau, "Understanding the Fate and Transport of Contaminants in the Unsaturated Zone of Soil," Groundwater Quality Protection Pre-Conference Workshop Proceedings, Water Pollution Control Federation 61st Annual Conference, Dallas, Tex., October 1988.

90. Lyman, W. J., W. F. Reehl, and D. H. Rosenblatt, *Handbook of Chemical Property Estimation Methods, Environmental Behavior of Organic Compounds,* McGraw-Hill, New York, 1982.

91. Marshall, T. J., "The Diffusion of Gases through Porous Media," *J. Soil Sci.,* vol. 10, pp. 79–82, 1959.

92. McCord, J. T., D. B. Stephens, and J. L. Wilson, "Toward Validating State-Dependent

Macroscopic Anisotropy in Unsaturated Media: Field Experiments and Modeling Considerations," *J. Contaminant Hydrol.*, vol. 7, p. 145, 1991.

93. McQueen, I. S., and R. R. Miller, "Calibration and Evaluation of a Wide Range Gravimetric Method of Measuring Moisture Stress," *Soil Sci.*, vol. 106, pp. 225–231, 1968.

94. Melancon, S. M., J. E. Polard, and S. C. Hern, "Evaluation of SESOIL, PRZM and PESTAN in a Laboratory Column Leaching Experiment," *Environ Toxic. Chem.*, vol. 5, pp. 865–878, 1986.

95. Mercer, J. W., D. C. Skipp, and D. Griffin, "Basic Pump-and-Treat Ground-Water Remediation Technology," R. S. Kerr Environmental Research Laboratory, Ada, Okla., U.S. Environmental Protection Agency, EPA/600/8-90/003, March 1990.

96. Messing, I., "Estimation of the Saturated Hydraulic Conductivity in Clay Soils from Soil Moisture Retention Data," *Soil Sci. Soc. Am. J.*, vol. 53, no. 3, pp. 665–668, 1989.

97. Miller, C. T., M. M. Poirier-McNeill, and A. S. Mayer, "Dissolution of Trapped Nonaqueous Phase Liquids: Mass Transfer Characteristics," *Water Resour. Res.*, vol. 26, no. 11, pp. 2783–2796, 1990.

98. Miller, R. J., J. W. Biggar, and D. R. Nielsen, "Chloride Displacement in Panoche Clay Loam in Relation to Water Movement and Distribution," *Water Resour. Res.*, vol. 1, no. 1, pp. 63–73, 1965.

99. Millington, R. J., "Gas Diffusion in Porous Media," *Science*, vol. 130, pp. 100–102, 1959.

100. Mualem, Y., *A Catalog of Hydraulic Properties of Unsaturated Soils*, Israel Institute of Technology, Haifa, Israel, 1976.

101. Mualem, Y., "Hydraulic Conductivity of Unsaturated Soils: Prediction and Formulas," in A. Klute, ed., *Methods of Soil Analysis*, Part 1, American Society of Agronomy, Madison, Wis., pp. 799–823, 1986.

102. Nielsen, D. R., and J. W. Biggar, "Measuring Capillary Conductivity," *Soil Sci.*, vol. 92, pp. 192–193, 1961.

103. Nielsen, D. R., J. W. Biggar, and K. T. Erh, "Spatial Variability of Field-Measured Soil-Water Properties," *Hilgardia*, vol. 42, no. 7, pp. 215–259, 1973.

104. Nielsen, D. R., R. D. Jackson, J. W. Cary, and D. D. Evans, *Soil Water*, American Society of Agronomy, Madison, Wis., 1972.

105. Nielsen, D. R., M. Th. van Genuchten, and J. W. Biggar, "Water Flow and Solute Transport Processes in the Unsaturated Zone," *Water Resour. Res.*, vol. 22, no. 9, pp. 89S–108S, 1986.

106. Nofziger, D. L., and J. R. Williams, "Interactive Simulation of the Fate of Hazardous Chemicals During Land Treatment of Oily Wastes: RITZ User's Guide," R. S. Kerr Environmental Research Laboratory, Ada, Okla., U.S. Environmental Protection Agency, EPA/600/8-88-001, January 1988.

107. Ogata, A., and R. B. Banks, "A Solution of the Differential Equation of Longitudinal Dispersion in Porous Media," *U.S. Geol. Surv. Prof. Paper* 411-A, 1961.

108. Olson, R. S., and D. E. Daniel, "Measurement of the Hydraulic Conductivity of Fine-Grained Soils," in T. F. Zimmie and C. O. Riggs, eds., *Permeability and Groundwater Contaminant Transport*, STP 746, American Society for Testing and Materials, Philadelphia, pp. 18–64, 1981.

109. Paetzold, R. F., A. de los Santos, and B. A. Matzkanin, "Pulsed Nuclear Magnetic Resonance Instrument for Soil-Water Content Measurement: Sensor Configurations," *Soil Sci. Soc. Am. J.*, vol. 51, no. 2, pp. 287–290, 1987.

110. Parizek, R. R., and B. E. Lane, "Soil Water Sampling Using Pan and Deep Pressure-Vacuum Lysimeters," *J. Hydrol.*, vol. 11, pp. 1–21, 1970.

111. Parker, J. C., and A. J. Valocchi, "Constraints on the Validity of Equilibrium and First-Order Kinetic Transport Models in Structured Soils," *Water Resour. Res.*, vol. 22, no. 3, pp. 399–408, 1986.

112. Penman, H. L., "Gas and Vapor Movements in the Soil. I. The Diffusion of Vapors through Porous Solids," *J. Agr. Sci.*, vol. 30, pp. 437–462, 1940.

113. Pennell, K. D., A. G. Hornsby, R. E. Jessup, and P. S. C. Rao, "Evaluation of Five Simulation Models for Predicting Aldicarb and Bromide Behavior under Field Conditions," *Water Resour. Res.*, vol. 26, no. 11, pp. 2679–2693, 1990.

114. Perroux, K. M., P. A. C. Raats, and D. E. Smiles, "Wetting Moisture Characteristic Curves Derived from Constant-Rate Infiltration into Thin Samples," *Soil Sci. Soc. Am. J.*, vol. 46, no. 2, pp. 231–234, 1982.

115. Perroux, K. M., and I. White, "Designs for Disc Permeameters," *Soil Sci. Am. J.*, vol. 52, no. 5, pp. 1205–1215, 1988.

116. Phene, C. J., G. L. Hoffman, and S. L. Rawlins, "Measuring Soil Matric Potential in Situ by Sensing Heat Dissipation within a Porous Body: I. Theory and Sensor Calibration," *Soil Sci. Soc. Am. Proc.*, vol. 35, pp. 27–33, 1971.

117. Phene, C. J., S. L. Rawlins, and G. L. Hoffman, "Measuring Soil Matric Potential in Situ by Sensing Heat Dissipation within a Porous Body: II. Experimental Results," *Soil Sci. Soc. Am. Proc.*, vol. 35, pp. 225–229, 1971.

118. Philip, J. R., "Approximate Analysis of the Borehole Permeameter in Unsaturated Soil," *Water Resour. Res.*, vol. 21, no. 7, pp. 1025–1033, 1985.

119. Rao, P. S. C., A. G. Hornsby, and R. E. Jessup, "Indices for Ranking the Potential for Pesticide Contamination of Groundwater," *Proc. Soil Crop Sci. Soc. Fla.*, vol. 44, pp. 1–8, 1985.

120. Rawlins, S. L., and G. S. Campbell, "Water Potential: Thermocouple Psychrometry," in A. Klute, ed., *Methods of Soil Analysis*, Part 1, American Society of Agronomy, Madison, Wis., pp. 597–618, 1986.

121. Rawls, W. J., and D. L. Brakensiek, "Estimating Soil Water Retention from Soil Properties," *J. Irrig. Drain. Eng.*, vol. 108, no. IR2, pp. 166–177, 1982.

122. Rawls, W. J., and D. L. Brakensiek, "Prediction of Soil Water Properties for Hydrologic Modeling," in *Proceedings of Symposium on Watershed Management*, pp. 293–299, American Society of Civil Engineers, New York, 1985.

123. Rawls, W. J., and D. L. Brakensiek, "Estimation of Soil Water Retention and Hydraulic Properties," in H. J. Morel-Seytoux, ed., *Unsaturated Flow in Hydrologic Modeling: Theory and Practice*, NATO ASI Series C: Mathematical and Physical Sciences, vol. 275, Kluwer Academic Publishers, Dordrecht, 1989.

124. Reynolds, W. D., and D. E. Elrick, "A Laboratory and Numerical Assessment of the Guelph Permeameter Method," *Soil Sci.*, vol. 144, pp. 282–299, 1987.

125. Reynolds, W. D., and D. E. Elrick, "Determination of Hydraulic Conductivity Using a Tension Infiltrometer," *Soil Sci. Soc. Am. J.*, vol. 55, no. 3, pp. 633–639, 1991.

126. Reynolds, W. D., D. E. Elrick, and B. E. Clothier, "The Constant Head Well Permeameter: Effort of Unsaturated Flow," *Soil Sci.*, vol. 139, no. 2, pp. 172–180, 1985.

127. Richard, L. A., "A Pressure-Membrane Extraction Apparatus for Soil Suction," *Soil Sci.*, vol. 51, pp. 377–386, 1941.

128. Richards, S. J., and L. V. Weeks, "Capillary Conductivity Values from Moisture Yield and Tension Measurements in Soil Columns," *Soil Sci. Soc. Am. Proc.*, vol. 17, pp. 206–209, 1953.

129. Rolston, D. E., R. D. Glauz, G. L. Grundmann, and D. T. Louie, "Evaluation of an In Situ Method for Measurement of Gas Diffusivity in Surface Soils," *Soil Sci. Soc. Am. J.*, vol. 55, no. 6, pp. 1536–1542, 1991.

130. Saffman, P. G., "A Theory of Dispersion in Porous Media," *J. Fluid Mech.*, vol. 6, no. 6, pp. 321–349, 1959.

131. Sallam, A., W. A. Jury, and J. Letey, "Measurement of Gas Diffusion Coefficient under Relatively Low Air-Filled Porosity," *Soil Sci. Soc. Am. J.*, vol. 48, pp. 3–6, 1984.

132. Schroeder, P. R., B. M. McEnroe, R. L. Peyton, and J. W. Sjostrom, "The Hydrologic Evaluation of Landfill Performance (HELP) Model," vol. 4, Documentation for Version 2, IA #DW 21931425-01-3, U.S. Army Engineer Waterways Experiment Station, Vicksburg, Miss., October 1989.

133. Schwarzenbach, R. P., and J. Westall, "Transport of Nonpolar Organic Compounds from Surface Water to Groundwater. Laboratory Sorption Studies," *Environ. Sci. Technol.*, vol. 15, no. 11, pp. 1360-1367, 1981.

134. Scow, K. M., "Rate of Biodegradation," in W. J. Lyman, W. F. Reehl, and D. H. Rosenblatt, eds., *Chemical Property Estimation Methods,* McGraw-Hill, New York, 1982.

135. Seidell, A., *Solubilities,* 1, 4th ed., American Chemical Society, D. van Nostrand, Princeton, N.J., 1958.

136. Shackelford, C. D., "Laboratory Diffusion Testing for Waste Disposal—A Review," *J. Contaminant Hydrol.*, vol. 3, no. 7, pp. 177-217, 1991.

137. Shackelford, C. D., and D. E. Daniel, "Diffusion in Saturated Soil: I. Background," *J. Geotech. Eng.*, vol. 117, no. 3, pp. 467-484, 1991.

138. Shearer, R. C., J. Letey, W. J. Farmer, and A. Klute, "Lindane Diffusion in Soil," *Soil Sci. Soc. Am. Proc.*, vol. 37, pp. 189-194, 1973.

139. Short, T. E., "Movement of Contaminants from Oily Wastes during Land Treatment," in E. J. Calabrese and P. J. Kostecki, eds., *Soils Contaminated by Petroleum: Environmental and Public Health Effects,* Wiley, New York, pp. 317-330, 1988.

140. Spencer, W. F., W. J. Farmer, and M. M. Cliath, "Pesticide Volatilization," *Residue Rev.*, vol. 49, pp. 1-47, 1973.

141. Spencer, W. F., W. J. Farmer, and W. A. Jury, "Review: Behavior of Organic Chemicals at Soil, Air, Water Interfaces as Related to Predicting the Transport and Volatilization of Organic Pollutants," *Environ. Toxicol. Chem.*, vol. 1, pp. 17-26, 1982.

142. Stannard, D. I., "Tensiometers—Theory, Construction, and Use," in D. M. Neilsen and A. I. Johnson, eds., *Ground Water and Vadose Zone Monitoring,* ASTM STP 1053, American Society for Testing and Materials, Philadelphia, pp. 34-51, 1990.

143. Steenhuis, T. S., W. Staubitz, M. S. Andreini, J. Surface, T. Richard, R. Paulsen, N. B. Pickering, J. R. Hagerman, and L. D. Geohring, "Preferential Movement of Pesticides and Tracers in Agricultural Soils," *J. Irrig Drain.*, vol. 116, p. 50, 1990.

144. Stephens, D. B., and S. P. Neuman, "Vadose Zone Permeability Tests: Summary," *J. Hydraul. Eng.*, vol. 108, no. 5, pp. 623-639, 1982.

145. Tanner, C. B., and R. J. Hanks, "Moisture Hysteresis in Gypsum Moisture Blocks," *Soil Sci. Soc. Am. Proc.*, vol. 16, pp. 48-51, 1952.

146. Taylor, S. A., "Oxygen Diffusion in Porous Media as a Measure of Soil Aeration," *Soil Sci. Soc. Am. Proc.*, vol. 14, no. 2, pp. 55-61, 1949.

147. Thibodeaux, L. J., and S. T. Hwang, "Landfarming of Petroleum Wastes—Modeling the Air Emission Problem," *Environ. Progr.*, vol. 1, no. 1, pp. 42-46, 1982.

148. Thomas, R. G., "Volatilization from Soil," in W. J. Lyman, W. F. Reehl, and D. H. Rosenblatt, eds., *Chemical Property Estimation Methods,* McGraw-Hill, New York, 1982.

149. Topp, G. C., J. L. Davis, and A. P. Annan, "Electromagnetic Determination of Soil Water Content: Measurement in Coaxial Transmission Lines," *Water Resour. Res.*, vol. 16, pp. 574-582, 1980.

150. Topp, G. C., A. Klute, and D. B. Peters, "Comparison of Water Content-Pressure Head Data Obtained by Equilibrium, Steady State, and Unsteady State Methods," *Soil Sci. Soc. Am. Proc.*, vol. 31, no. 3, pp. 312-314, 1967.

151. Topp, G. C., and E. E. Miller, "Hysteretic Moisture Characteristics and Hydraulic Conductivities for Glass-Bead Media," *Soil Sci. Soc. Am. Proc.*, vol. 30, no. 2, pp. 156-162, 1966.

152. U.S. Environmental Protection Agency, "Part II Underground Storage Tanks; Technical Requirements and State Program Approval; Final Rules," *Fed. Regist.*, vol. 53, no. 185, 1988.

153. U.S. Salinity Laboratory, "Diagnosis and Improvement of Saline and Alkali Soils," *U.S.D.A. Agr. Handbook* 60, 1954.

154. Valentine, R. L., and J. L. Schnoor, "Biotransformation," in S. C. Hern and S. M. Melancon, eds., *Vadose Zone Modeling of Organic Pollutants,* Lewis Publishers, Chelsea, Mich., 1986.

155. Valocchi, A. J., "Validity of the Local Equilibrium Assumption for Modeling Sorbing Solute Transport through Homogeneous Soils," *Water Resour. Res.*, vol. 21, no. 6, pp. 808–820, 1985.

156. Van Bavel, C. H. M., "Gaseous Diffusion and Porosity in Porous Media," *Soil Sci.*, vol. 73, pp. 91–104, 1952.

157. Van de Pol, R. M., P. J. Wierenga, and D. R. Nielsen, "Solute Movement in a Field Soil," *Soil Sci. Soc. Am. J.*, vol. 41, no. 1, pp. 10–13, 1977.

158. Van Genuchten, M. T., "A Closed-Form Equation for Predicting the Hydraulic Conductivity of Unsaturated Soil," *Soil Sci. Soc. Am. J.*, vol. 44, pp. 892–898, 1980.

159. Weeks, I. V., and S. J. Richards, "Soil-Water Properties Computed from Transient Flow Data," *Soil Sci. Soc. Am. Proc.*, vol. 31, pp. 721–735, 1967.

160. Wesseling, J., "Some Solutions of the Steady-State Diffusion of Carbon Dioxide through Soils," *Neth. J. Agr. Sci.*, vol. 10, pp. 109–117, 1962.

161. Wilkerson, M., D. Kim, and M. Nodell, "The Pesticide Groundwater Prevention Act: Setting Specific Numerical Values," California Department of Food and Agriculture Report, September 1984, State of California, Department of Food and Agriculture, Sacramento, Calif., 1984.

162. Yong, R. N., and B. P. Warkentin, *Soil Properties and Behavior,* Elsevier, Amsterdam, 1975.

CHAPTER 16
CONTAMINANT TRANSPORT IN GROUNDWATER

James W. Mercer
GeoTrans, Inc.
Sterling, Virginia

Richard K. Waddell
GeoTrans, Inc.
Boulder, Colorado

16.1 INTRODUCTION

The movement of dissolved constituents in groundwater is affected by three factors: (1) *advection* of the constituent with the water flowing through the aquifer, (2) *dispersion* of the constituent, and (3) *sources* and *sinks* of the constituent within the volume such as chemical reactions or adsorption onto the solid matrix of the aquifer. Computer models of solute transport are based on mass-balance equations that describe these factors. In these models, advection by pumping (q_{out}) or injection (q_{in}), as well as that due to the natural flow of water, is generally considered. Thus, the contaminant mass balance may be written:

Dispersion + advection by natural flow + advection by pumping or injection + other sources and sinks = rate of change of mass of contaminant stored in the aquifer (16.1.1)

Or, in more mathematical terms, the same equation is written as the *advection-dispersion equation:*

$$\bar{\nabla} \cdot \phi \bar{\bar{D}} \cdot \bar{\nabla} C - \bar{\nabla} \cdot \bar{q} C + q_{in} C^* - q_{out} C + R = \frac{\partial(\phi C)}{\partial t} \qquad (16.1.2)$$

where C = material concentration, M/L^3
C^* = concentration of the source term, M/L^3
$\bar{q}$ = Darcy velocity, L/t
q_{in} = volumetric flow rate of the water source, L^3/t
q_{out} = volumetric flow rate of the water sink, L^3/t
R = chemical source or sink, M/L^3t
ϕ = porosity (dimensionless)
$\overline{\overline{D}}$ = dispersion tensor, L^2/t
$\overline{\nabla}$ = gradient differential operator, $1/L$

On the left-hand side, the first term is the dispersive term, the second is the advective term, and the remaining terms are source/sink terms. R includes sorption, precipitation/dissolution, and decay. The term on the right-hand side is the accumulation, or storage, term.

Table 16.1.1 summarizes the various processes associated with solute transport. Many of these processes, including advection, dispersion, nonaqueous-phase liquid flow, retardation, and biodegradation, are discussed in more detail in the following sections. Following the process descriptions is a discussion on data collection which is necessary to understand and describe the processes occurring at a site.

TABLE 16.1.1 A Summary of the Processes Associated with Dissolved Solute Transport and their Impact[57]

Process	Definition	Impact on transport
	Solute transport	
Advection	Movement of solute as a consequence of groundwater flow.	Most important way of transporting solute away from source.
Diffusion	Solute spreading due to molecular diffusion in response to concentration gradients.	A second-order mechanism in most flow systems where advection and dispersion dominate.
Dispersion	Fluid mixing due to effects of unresolved heterogeneities in the permeability distribution.	A mechanism that reduces solute concentration in the plume. However, a dispersed plume is more widespread than a plume moving by advection alone.
	Biologically mediated mass transfer	
Biological transformations	Reactions involving the degradation of organic compounds and whose rate is controlled by the abundance of the microorganisms and redox conditions.	Important mechanism for solute reduction, but can lead to undesirable daughter products.

TABLE 16.1.1 A Summary of the Processes Associated with Dissolved Solute Transport and their Impact[57] *(Continued)*

Process	Definition	Impact on transport
	Solute transfer	
Radioactive decay	Irreversible decline in the activity of a radionuclide through a nuclear reaction.	An important mechanism for attenuation when the half-life for decay is comparable to or less than the residence time of the flow system. Also adds complexity by production of daughter products.
Sorption	Reaction between solute and the surfaces of solids causing the solute to bond (to varying degrees) to the surface.	An important mechanism that reduces the rate at which the solute is apparently moving. Makes it more difficult to remove solute at a site in a given time.
Dissolution/precipitation	The process of adding solutes to or removing them from solution by reactions dissolving or creating various solids.	Precipitation is an important attenuation mechanism that can limit the concentration in solution. Dissolution of these solids can slow aquifer restoration by acting as a continuing source.
Acid-base reactions	Reactions involving a transfer of protons (H^+).	Mainly exercises an indirect, but important, control on solute transport by controlling the pH of groundwater.
Complexation	Combination of cations and anions to form a more complex ion.	An important mechanism that commonly increases the mobility of metals by forming uncharged or negatively charged ions, or by increasing a metal's solubility.
Hydrolysis/substitution	Reaction of an organic compound with water or a component ion of water (hydrolysis) or with another anion (substitution).	Often hydrolysis/substitution reactions make an organic compound more susceptible to biodegradation and more soluble.
Redox reactions	Reactions that involve a transfer of electrons and include elements with more than one oxidation state.	An extremely important family of reactions affecting mobility of metals through changes in ionic charges and solubility and of organic compounds through degradation reactions.

16.2 ADVECTION

Advection (or *convection*) is the movement of solute (for instance, a contaminant) caused by the actual motion or flow of the fluid, or solvent (water). The bulk movement of water through an aquifer causes solute transfer via advection. Advection is the primary process by which solute moves in the subsurface. Processes of dispersion and chemical reactions normally tend to reduce the concentration of a contaminant as it moves with the groundwater flow.

Because of the difficulties involved in measuring groundwater velocity directly, groundwater velocities are routinely determined indirectly using measurements of hydraulic head, hydraulic conductivity, and Darcy's equation. *Darcy's equation* for a uniform density fluid in one dimension may be written as follows:

$$q = -K\frac{\partial h}{\partial s} \tag{16.2.1}$$

where q = the Darcy velocity, L/t
K = the hydraulic conductivity, L/t
$\partial h/\partial s$ = the hydraulic gradient, dimensionless

The *Darcy velocity* (also called *specific discharge*) is an average value representative of the entire cross-sectional area for which the hydraulic conductivity is determined. That is, Darcy's equation describes volumetric flow per unit cross-sectional

TABLE 16.2.1 Representative Values for Effective Porosity*

Soil textural classes	Effective porosity†
Unified soil classification system	
GC, GP, GM, GS SW, SP, SM, SC	0.20
ML, MH	0.15
CL, OL, CH, OH, PT	0.01‡
U.S. Department of Agriculture soil textural classes	
Clays, silty clays, sandy clays	0.01‡
Silts, silt loams, silty clay loams	0.10
All others	0.20
Rock units (all)	
Porous media (nonfractured rocks such as sandstone and some carbonates)	0.15
Fractured rocks (most carbonates, granites, etc.)	0.0001

* Effective porosity is the proportion of the total aquifer volume including the aquifer matrix that is actually available for water movement.

† These values are estimates. There may be differences between similar units.

‡ Assumes negligible secondary porosity. If fractures or soil structure are present, effective porosity should be 0.001 (0.1%), but hydraulic conductivity should be higher than that for intact material.

Source: Ref. 90.

area, which includes the solid matrix. It does not represent the average velocity of the water through the pores. To obtain the interstitial velocity (also called *average pore velocity, average linear velocity,* or *seepage velocity*), the Darcy velocity must be divided by the effective porosity:

$$v = \frac{q}{\phi_e} = -\frac{K}{\phi_e}\frac{\partial h}{\partial s} \tag{16.2.2}$$

where v = the interstitial velocity (L/t) and ϕ_e = effective porosity (dimensionless).

Because there are dead-end pores within the groundwater system, the entire pore space is not effective in transmitting fluid. Similarly, the pore space within the lower-permeability parts of an aquifer is less active in transmitting fluids than the pores in the higher-permeability parts. Therefore, the effective porosity ϕ_e is less than the total porosity of the aquifer material. Table 16.2.1 provides estimates for effective porosity. For an unfractured medium, the variability (or uncertainty) of the effective porosity is much less significant than the variability in the hydraulic conductivity in calculating the interstitial velocity. However, if flow occurs in fractures, the effective porosity of the medium is much lower because of dead-end fractures, and the importance of developing accurate measures of the effective porosity increases greatly. If matrix diffusion (see Sec. 16.3.1) is important, solute movement is controlled both by matrix porosity and by fracture porosity.[64]

16.3 DISPERSION

Mechanical dispersion refers to the spreading and mixing caused by variations in the velocity with which water moves. Dispersion occurs at different scales. *Microscopic dispersion* is dispersion that results from variation in velocity within pores (the velocity is higher in the center of the pore than next to the solid grains), and from the tortuous movement of water around the grains. This dispersion is observed in laboratory experiments. The term *macroscopic dispersion* refers to dispersion resulting from the interbedding or interfingering of materials of differing permeability. In most field situations, macroscopic dispersion is more important than microscopic dispersion. However, in some field experiments, it has been observed that it is possible to adequately represent the movement of solute by developing a model that incorporates the properties of the interbedded materials and using microscopic dispersion within each bed.[55]

Except where water velocities are very low (e.g., some clays with very low hydraulic conductivity), molecular diffusion (described by Fick's law) is small compared to mechanical dispersion and can be ignored as a dispersive mechanism. However, molecular diffusion can have an important role in retardation of solute movement if diffusion can occur from materials or fractures of high permeability into materials of low permeability.[6] This process is referred to as *matrix diffusion*[33] and is discussed in the next section.

The form of the *hydrodynamic dispersion tensor* is complex.[72] For isotropic media, the dispersion coefficient, written to incorporate molecular diffusion, is calculated as follows:

$$D_{ij} = a_{ij}\, v_i + D \tag{16.3.1}$$

where $i = 1, 2, 3$ and $j = 1, 2, 3$ represent summation over the coordinate system (x, y, z). D_{ij} (the dispersion coefficient, L^2/t) and a_{ij} (the dispersivity, L) are 3×3

matrices, if three-dimensional flow is considered. This is because both the direction i in which dispersion is occurring and the surface j on which it is occurring must be considered. The term v_i is the water velocity (L/t), and D is the molecular diffusion coefficient (L^2/t). The first term to the right of the equal sign is the mixing term. The dispersivity matrix is considered to be symmetric about the diagonal (which implies that the aquifer is isotropic with respect to dispersivity), so the number of unique components is reduced from 9 to 6. Further, if flow is in the principal directions (parallel to one of the coordinate axes), the off-diagonal terms become 0, and the number of terms is further reduced to three (one parallel to flow, a_L, and two perpendicular to flow, a_T). The dispersivity in the direction of flow (known as the *longitudinal dispersivity*) is typically larger than the dispersivities *(transverse dispersivities)* in the directions perpendicular to flow.

In general, the amount of dispersion is determined by model calibration, adjusting the dispersivities until a match with measurements of concentration is obtained. Anderson[5] discusses some of the difficulties in determining these parameters. Dispersion spreads initially localized contaminants, and the larger the dispersivity, the larger the spreading.

To illustrate the effects of dispersion, consider the one-dimensional distribution in Figure 16.3.1. The concentration front without dispersion appears as a sharp front

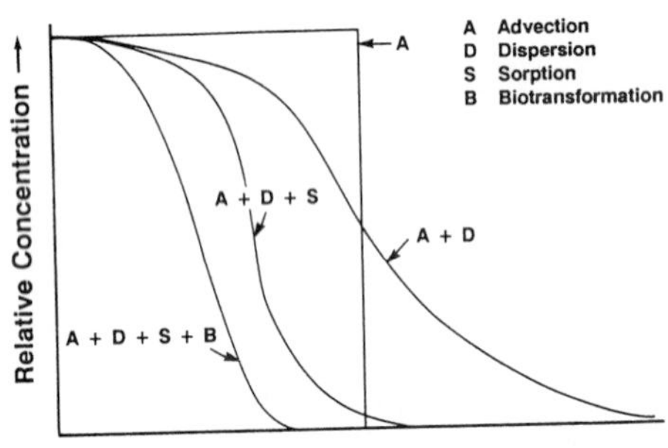

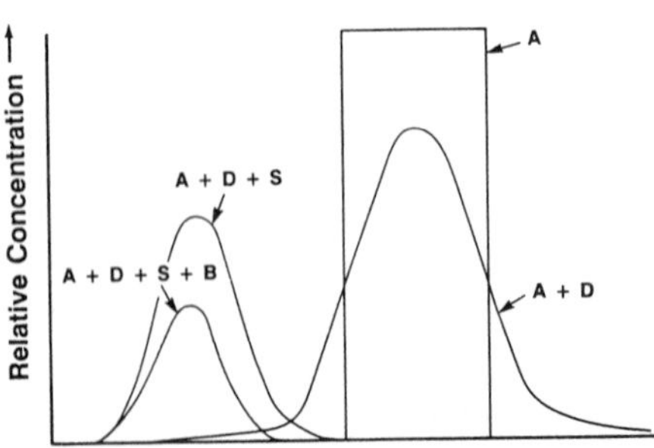

FIGURE 16.3.1 The influence of natural processes on levels of solute downgradient from continuous and slug-release sources. *(Reproduced from Ref. 40 with permission.)*

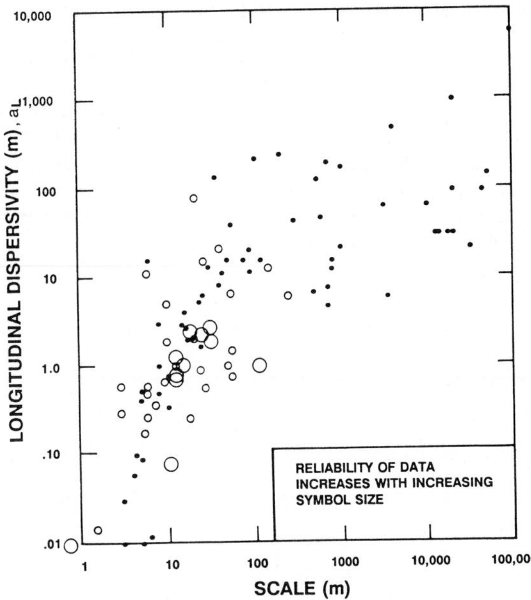

FIGURE 16.3.2 Compilation of longitudinal dispersivity values versus the scale of measurement and the reliability of the data. *(Copyright © 1985. Electric Power Research Institute: EPRI EA-4190. A Review of Field-Scale Physical Solute Transport Processes in Saturated and Unsaturated Porous Media. Reprinted with permission from Ref. 93.)*

(A) that moves at the average fluid velocity. With dispersion, however, the front is smeared out (A + D), so that part of the solute travels at a velocity faster than the average velocity.

The form of the dispersion term presented is the one most commonly used in applications. However, the nature of the dispersion phenomenon and its precise mathematical form are still sources of controversy. Values determined by observations from laboratory and field data are *scale-dependent.* Values of longitudinal dispersivity vary from a few centimeters for laboratory experiments, through a few meters for tracer studies, to tens or hundreds of meters for regional pollution problems. The commonly accepted explanation for this is that the longitudinal dispersivity coefficient is a measure of the scale of heterogeneity that is not included in the analysis (the larger the area, the larger the heterogeneity).

Dispersivities are obtained from regional studies, as well as tracer tests. Anderson[5] and Isherwood[38] reviewed the literature and summarized values of dispersivity. There is a considerable range of values, with longitudinal (parallel to the direction of groundwater flow) dispersivity generally being larger than transverse (perpendicular to the direction of flow) dispersivity. Figure 16.3.2 shows a compilation of dispersivity measurements. As the figure indicates, measured dispersivities seem to depend on the scale of the measurement. For more information on dispersion, see Refs. 5, 6, 29, 31, 51, 73, 74, 80, and 93.

16.3.1 Diffusion into Low-Permeability Material

Solute can be removed from the actively flowing part of a groundwater system by diffusion into parts of the flow system with lower permeability (referred to as *matrix diffusion*). An example is diffusion from a fracture (high permeability) into the rock matrix (lower permeability). This mechanism slows the movement of the solute front. This effect is modeled by coupling a one-dimensional diffusion equation into the advection-dispersion equation through the term R. The particular form of the equation and the additional boundary conditions will vary according to the geometry of the problem. An effective diffusion coefficient is given by

$$D_{\text{eff}} = \alpha\phi\left(\frac{L}{L_e}\right)^2 D \tag{16.3.2}$$

where α = constrictivity (an empirical correction factor)
ϕ = porosity of the matrix, dimensionless
L/L_e = macroscopic diffusion length divided by the effective diffusion length
D = molecular diffusion coefficient in the fluid, L^2/t

Matrix diffusion may also be a significant retardation mechanism in interbedded sands and silts/shales.

16.4 MOVEMENT OF NON-AQUEOUS-PHASE LIQUIDS

Non-aqueous-phase liquids (NAPLs) are associated with many contamination problems, often serving as sources for dissolved contamination. In addition, contaminants can dissolve into the NAPL and be transported with the moving NAPL. This may mobilize contaminants through a facilitated transport mechanism known as *cosolvent effects.* For that reason, a discussion of NAPLs is included in this chapter. NAPLs are grouped according to whether they are denser than water (DNAPL) or lighter than water (LNAPL). An example of a LNAPL is gasoline in sufficient quantity to form a separate fluid; similarly, examples of a DNAPL are chlorinated solvents, such as trichloroethene (TCE). It is the density of the NAPL that is important, rather than the density of the separate components of the NAPL. For example, a mixture of oil and TCE may be an LNAPL or a DNAPL, depending on the amount of TCE present and the resulting density of the mixture.

As a separate fluid, a NAPL moves in response to pressure and gravitational gradients. Because more than one fluid occupies the pore space (water and NAPL in the saturated zone; water, air, and NAPL in the vadose zone), multifluid flow concepts are used to characterize NAPL movement. Some of the important properties of the fluid and media are summarized in Table 16.4.1.

NAPL migration in the subsurface is affected by:[25] (1) volume of NAPL released, (2) area of infiltration, (3) time duration of release, (4) properties of the NAPL, (5) properties of the media, and (6) subsurface flow conditions. The cross-sectional schematic in Fig. 16.4.1*a* depicts the distribution of organic chemicals in multiple phases resulting from a release of lighter-than-water non-aqueous-phase liquid. When introduced into the subsurface, gravity causes the NAPL to migrate downward through the vadose zone as a distinct liquid. This vertical migration also is accompanied to some extent by a lateral spreading due to the effect of capillary forces[75] and

TABLE 16.4.1 Properties of NAPL and Media that Influence Contaminant Migration

Property	Definition	Range of values
Saturation	The volume fraction of the total void volume occupied by that fluid at a point.	0–1.
Residual saturation	Saturation at which NAPL becomes discontinuous and is immobilized by capillary forces.	Approximately 0.1–0.5.
Interfacial tension	The free surface energy at the interface formed between two immiscible or nearly immiscible liquids. Surface tension is the interfacial tension between a liquid and its own vapor.	Values of interfacial and surface tensions for NAPL-forming chemicals generally range between 0.015–0.5 N/m.
Wettability	Describes the preferential spreading of one fluid over solid surfaces in a two-fluid system; it depends on interfacial tension.	Determined through contact-angle studies. Commonly, in the vadose zone, NAPL is the wetting fluid; in the saturated zone, water is the wetting fluid.
Capillary pressure	The difference between the nonwetting fluid pressure and the wetting fluid pressure. It is a measure of the tendency of a porous medium to attract the wetting fluid and repel the nonwetting fluid.	Depends on the interfacial tension, contact angle, and pore size.
Relative permeability	The ratio of the effective permeability of a fluid at a fixed saturation to the intrinsic permeability.	0.0–1.0, depending on the fluid saturation.
Solubility	The maximum concentration of the chemical that will dissolve under a set of chemical and physical conditions.	Varies widely depending on chemical and aquifer conditions.
Volatilization	The transfer of matter from liquid and soil to the gaseous phase.	Depends on organic partitioning between water and air, and NAPL and air.
Density	Mass per unit volume of a substance. Specific gravity is the ratio of a substance's density to that of some standard substance, usually water.	Specific gravities of petroleum products may be as low as 0.7, whereas chlorinated aliphatic compounds can be as high as 1.2 to 1.5.
Viscosity	The internal friction within a fluid that causes it to resist flow.	Varies depending on fluid and temperature.

due to media spatial variability (layering), which is not shown in Fig. 16.4.1*a*. As the NAPL progresses downward through the vadose zone, it leaves residual liquid (at the residual saturation) trapped in the pore spaces. This entrapment is due to surface tension effects. In addition to the migration of the NAPL, some of the organics may volatilize and form a vapor extending beyond the NAPL.

If the NAPL release is sufficiently large, some of the NAPL will eventually reach saturated groundwater. Here, density influences flow. An LNAPL will spread laterally along the capillary fringe forming a complex distribution near the top of the water table. This distribution may be a lens that depresses natural groundwater levels. Soluble components may dissolve in the water and migrate because of the movement of the water. Denser-than-water non-aqueous-phase liquid (DNAPL) will displace water and, given sufficient volume, will continue its vertical migration until it encounters a layer of low permeability (Fig. 16.4.1*b*). Here it will move under pressure and gravity forces along the confining layer. As in the vadose zone, some of the DNAPL will be held in the pore space within the saturated zone. This residual DNAPL will serve as a source of contaminants to a dissolved plume, depending on the aqueous solubility of the organic compounds. In the vadose zone, infiltrating rainwater may dissolve organic vapors or the residual NAPL and transport these organic components to the saturated region.

For more information on NAPLs, see Ref. 53. Field techniques applied to NAPL sites are discussed in Ref. 92. Modeling approaches for NAPL problems are reviewed in Ref. 1.

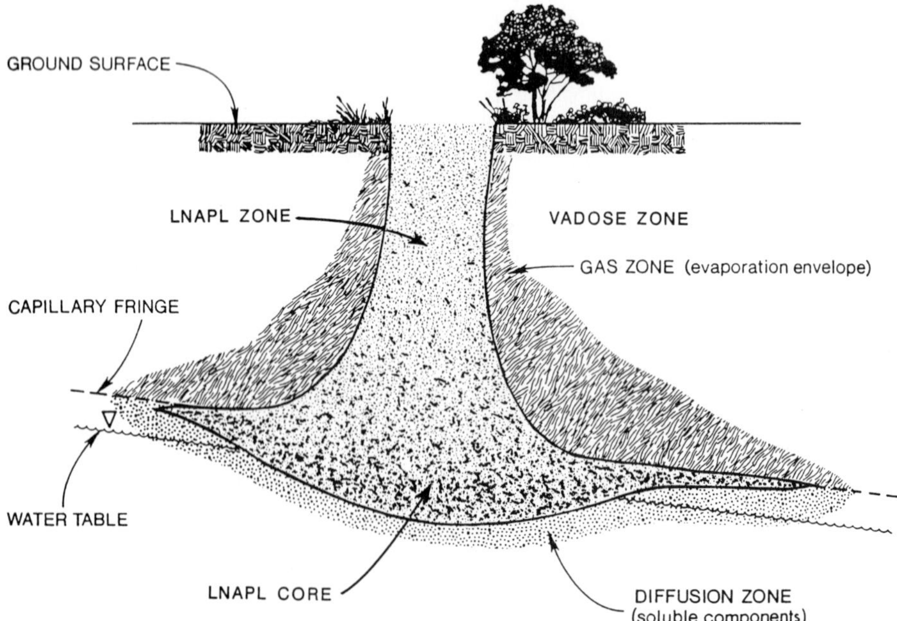

FIGURE 16.4.1*(a)* LNAPL infiltration schematic. *(Copyright © 1988. Electric Power Research Institute. EPRI EA-5976. Multiple Flow and Transport Models for Organic Chemicals. A Review and Assessment. Reprinted with permission from Ref. 1.)*

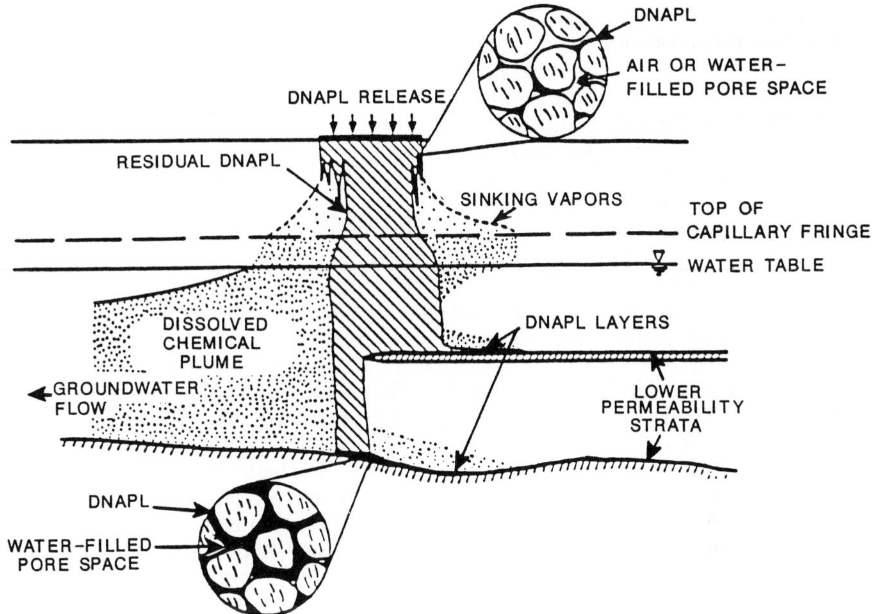

FIGURE 16.4.1*(b)* DNAPL infiltration schematic.[25]

16.5 RETARDATION

Removal of molecules of solute from the active part of the groundwater system results in retardation of the movement of the front (see A + D + S in Fig. 16.3.1). There are several mechanisms by which retardation occurs. These include precipitation and/or coprecipitation, sorption, ion exchange, diffusion into low-permeability material (discussed previously), and transformation. Although retardation mechanisms generally cause the contaminant to be immobile, under certain conditions small solid particles (colloids) may be transported with the moving water. This type of facilitated transport can mobilize contaminants that are thought to be immobile. Retardation processes are discussed below for inorganic and organic chemicals, and include precipitation/dissolution, sorption, and adsorption kinetics.

16.5.1 Inorganic Chemicals

Precipitation/Dissolution. Precipitation and dissolution of minerals can change the chemistry of water and provide a mechanism for retardation of the movement of solute. The solubility of minerals is a function of many factors; therefore, numerical guidelines pertinent to predicting the removal of solute are beyond the scope of this handbook. These factors include pH, the oxidation-reduction environment, concentrations of major and minor ions and gases, and temperature. Several computer codes are available for investigating the solubility of minerals; these include

WATEQ[85] and its progeny, PHREEQE,[58] MINTEQ,[26] and EQ3/EQ6.[102] These codes use thermodynamic information and information on the above factors to calculate the concentrations and activities of ions in the water, and the state of saturation of the water for a large suite of minerals. Many versions of these programs are available, and reflect changes in the computer programming and/or the thermodynamic database. Several of these codes also allow various chemical reactions, such as precipitation or dissolution of minerals, ion exchange, and sorption, to be simulated. These codes assume that equilibrium has been established, and the code output should be interpreted with this assumption in mind.

Table 16.5.1 lists complexes that may constitute a significant proportion of the total amount of various metals in solution. The distribution of the metal among these complexes, and perhaps others which may be significant, will affect the solubility of the metal. Many of the complexes are hydration products; thus it is necessary to know the pH of the solution. Other complexes are also pH-sensitive. For example, CO_3^{-2} is a common complexing agent. Its activity depends, among other things, on the amount of inorganic carbon present and the solution pH. It is therefore important to have data on concentrations of major ions in the water, as well as on significant complexing ions, when evaluating the state of the water relative to solubility controls. Table 16.5.1 also provides the formulas of mineral phases known to limit or suspected of limiting the concentrations in water of various metals. Other solid phases may have lower solubilities than those solids listed, but the kinetics of precipitation appear to be too slow for these phases to provide a control on solute movement.

Sorption. In general terms, sorption refers to the removal of a solute from solution through association with a solid surface by electrostatic or chemical forces. The attraction between *sorbate* (the surface) and *sorbent* (material attaching to the surface) can result from a variety of forces, including coulombic attraction, London–van der Waals forces, orientation energy, induction forces, hydrogen bonding, and chemical forces.[65]

Specific sorption (*or chemisorption*) represents a chemical interaction between sorbate and sorbent and occurs at specific sites on the surface of the sorbate. In chemisorption the chemical bond between sorbate and sorbent usually is covalent. A molecule undergoing chemisorption may lose its identity as atoms are rearranged to form new compounds fulfilling the unsatisfied valences of the surface atoms.

Nonspecific sorption is the process in which solutes are attracted to the surface of soil particles, generally from London–van der Waals (polar-bond) or electrostatic forces. Adsorbed molecules are not fixed to a specific site but are free to move within the interface. This form of sorption is weaker and less specific than when sorption occurs through a true chemical reaction, as in specific sorption.

Exchange adsorption or *ion exchange* is a type of nonspecific sorption process in which ions of one substance are sorbed and displace other ions. An example is Ca^{++} replacing Na^+. The charge and size of the ion are the determining factors for exchange adsorption. Generally, a more highly charged and smaller ion is more strongly adsorbed.[35] The radius of the hydrated ion, however, actually determines the relative order of adsorption between competing ions. A measure of the capacity for ion exchange is the *cation-exchange capacity,* for which values for various clay minerals are provided in Table 16.5.2.

Adsorption isotherms are commonly used to describe sorption. These assume that adsorption is attained instantaneously and remains constant. These equilibrium methods are considered valid when (1) the rate of adsorption is much greater than the rate of change of solute concentration due to any other cause and/or (2) the rate of

TABLE 16.5.1 Common Complexes, Solubility-Controlling Phases, and Sorption Behavior of Several Metals

Metal	Valence state	Important complexes	Solubility control	Adsorption
Al	+3	SO_4^{-2}, F^-, OH^-	Gibbsite—moderate pH, alunite, basalunite—low pH and high SO_4^{-2}	Organic matter, clays; K > Al > Ca.
As	+3	$H_3AsO_3^0$, $H_2AsO_3^-$		Fe and Mn hydrous oxides, but less sorption than As^{+5}.
	+5	$H_2AsO_4^-$, $HAsO_4^{-2}$	$FeAsO_4$, $Pb_3(AsO_4)_2$, $Mn_3(AsO_4)_2$	Extractable hydrous oxides of Fe and Mn at pH < 7–9; competition by PO_4.
Ba	+2	$BaSO_4^0$, $BaCO_3^0$	$BaSO_4$, pH < ~9 $BaCO_3$, pH > ~9	Nonspecific, ion exchange (Ba > Sr > Ca > Mg). Less important specific sorption in oxides and hydrous oxides.
B	+3	$H_3BO_3^0$, pH < ~9.2 $B(OH)_4^-$, pH > ~9.2		Greatest on fresh Al oxides or hydroxides, maximum at pH 7.5 to 10. Competition from OH^-. Enhanced by Ca and Mg. Sorption is sometimes not reversible.
Cd	+2	$CdSO_4^0$, $CdCO_3$, $CdCl^+$	$CdCO_3$ at high pH $Cd(PO_4)_2$	Specific at < 10^{-5} M on oxides of Al, Fe, and Mn, and on calcite. Nonspecific at higher concentrations. Competition from Co, Pb, Zn, Ca, and Mg.
Cr	+3	Complexes of OH^-	$FeCr_2O_4$, Cr_2O_3, (Fe,Cr) $(OH)_3$	At pH < 4.5 and [Cr] < 10^{-6} M, specifically sorbed by Fe and Mn oxides. Sorption increases with increasing pH on silicate surfaces.
	+6	CrO_4^{-2} $HCrO_4^{-2}$	$PbCrO_4$ Cr_2O_3	Specifically sorbed by Al and Fe oxides, pH < 7. Competition by SO_4^{-2} and PO_4^{-2}.
Cu	+1 +2	$CuCl_2^-$ $CuSO_4^0$ $CuCl^+$ $Cu(OH)_2^0$	$Cu(OH)_2$ $CuFe_2O_4$	Organic matter important sorbent. Specific sorption to Fe, Al, Mn oxides at low concentrations. At high $Cu(CO_3)_2^{-2}$ concentrations, strong pH dependence. $CuOH^+$ more strongly sorbed than Cu^{+2}.

TABLE 16.5.1 Common Complexes, Solubility-Controlling Phases, and Sorption Behavior of Several Metals (*Continued*)

Metal	Valence state	Important complexes	Solubility control	Adsorption
Fe	+2	$FeSO_4^0$ $FeOH^+$	$Fe_3(OH)_8$ $FeCO_3$	
	+3	$Fe(OH)_2^+$, $Fe(OH)_4^-$ FeF_2^+	$Fe(OH)_3$	Humic acid, pH < 3; Clays
Pb	+2	$PbCO_3^0$, $Pb(CO_3)_2^{-2}$ $PbSO_4^0$ $PbCl^+$	Lead phosphates $PbCO_3$ $Pb(OH)_2$	At low concentrations, specific sorption on Mn and Fe oxides. Ion exchange at higher concentrations (> 10^{-4} M), with competition by Al, Fe, Ca, Mg.
Mn	+2, +3, +4 (+2 dominates in solution)	$MnSO_4^0$ $MnHCO^+$ $MnOH^+$	Nsutite (oxidizing) MnOOH, Mn_2O_3, Mn_3O_4 (reducing) $MnCO_3$	Poorly understood. Ion exchange may be important at > 10^{-4} M. Weakly specifically sorbed to iron oxide, stronger to Mn oxide. May sorb to calcite.
Hg	0 +1	Hg^0 HgI^0 $Hg(HS)_2^0$ HgS_2^2 $HgHS_2^-$	Hg(l) αHgS, βHgS	
	+2	HgI_2^0 $Hg(OH)_2^0$		Strongly sorbed by organics. Also sorbed manganese oxides. Cl^-, Br^-, and I^- inhibit sorption.
Mo	+6	MoO_4^{-2}, pH > 4 $H_2MoO_4^0$, pH < 4	$Fe_2(MoO_4)_3$ $PbMO_4$, $CaMoO_4$	Fe and Al oxides important. Marked decline in sorption with increased pH due to amphoteric nature of solids. Competition by PO_4. Sorption on iron oxides becomes irreversible with time, perhaps due to formation of iron molybdate.
Ni	+2	$NiSO_4^0$ $Ni(OH)_2^0$ $Ni(OH)_3^-$	$NiFe_2O_4$ NiS	Specific sorption by Fe and Mn oxides, clays. Adsorption edge, pH of 5.5 to 8.5. Sorption decreased by increasing ionic strength. Strong competition by Ca^{++}. Sorption slow, partially reversible at moderate to high pH, rapid and reversible at low pH.

Se	−2	H_2Se^0, pH < 3.8 HSe^-, pH > 3.8	Se $FeSe_2$		Not well-characterized by valence states. In general, sorption on crytocrystalline and amorphous SiO_2, Al_2O_3, Fe_2O_3. Decreasing sorption at higher pH, probably controlled by the surface charge of the sorbent. Competition by PO_4, SO_4.
	+4	H_2SeO_3, pH < 2.9 $HSeO_3^-$, 2.9 < pH < 8.4 SeO_3^{-2}, pH > 8.4	$Fe(OH)_4(SeO_3)$		
	+6	$HSeO_4^-$, pH < 2.0 SeO_4^{-2}, pH > 2.0			
Zn	+2	$ZnSO_4^0$ $Zn(CO_3)_2^{-2}$ $ZnCO_3^0$	$ZnCO_3 \cdot H_2O$ or $ZnCO_3$ $Zn(OH)_2$		Specific sorption on Fe and Mn oxides at low concentrations with marked pH dependency. At high concentrations, coulombic interactions become important, and sorption on clays occurs. Competition by high concentrations of Ca^{2+} and Mg^{2+}. Sorption enhanced by anions specifically adsorbed on Fe oxides.

Source: Copyright © 1984. Electric Power Research Institute. EPRI EA-3356. *Chemical Attenuation Rates, Coefficients, and Constants in Leachate Migration, Vol. 1: A Critical Review.* Reprinted with permission from Ref. 62.

TABLE 16.5.2 Cation-Exchange Capacity of Clay Minerals

Mineral	Exchange capacity, meq/100 g at pH 7
Kaolinite	3–15
Halloysite ($2H_2O$)	5–10
Halloysite ($4H_2O$)	40–50
Montmorillonite group	70–100
"Illites" (hydrous micas)	10–40
Vermiculite	100–150
Chlorite	10–40
Glauconite	11–20+
Palygorskite group	20–30
Allophane	70

Source: Ref. 30.

adsorption is much greater than the rate of groundwater flow.[84] In some instances, the assumption of equilibrium is not valid, and kinetic approaches are used. Kinetic methods may be needed when (1) the contaminant is first introduced into the groundwater system, (2) the contaminant source is in the form of a spike or a slug, and/or (3) the adsorption rate is too slow or the groundwater flow rate is too fast for equilibrium to be attained.[54] Kinetic models will be discussed briefly in a later section; only equilibrium methods are discussed here.

Because sorption is a surface reaction, it is a function of the surface area available. Minerals and solids with a high specific surface area include clays and oxides and hydrous oxides of iron and manganese. These materials often occur as coatings on larger particles in the soil so that sorption can be important even in coarse-grained materials.

When a solution contacts a solid, solute will tend to transfer from liquid to solid until the concentration of solute in solution is in equilibrium with the concentration of the sorbate attached to the solid. This process is not always completely reversible. Equilibrium data are determined from laboratory batch or column experiments and are usually represented at a given temperature by an adsorption isotherm. The adsorption isotherm is the relationship between the quantity sorbed per unit mass of sorbent and the concentration of solute, and is valid only for the chemical conditions (solid and solution) under which it was determined. Changes in pH or other parameters can greatly change sorption behavior. The most commonly used isotherms are discussed below.

Linear equilibrium adsorption assumes the sorbate concentration is linearly related to the solute concentration. It is described by

$$S = \frac{X}{M} = K_d C \tag{16.5.1}$$

where S = sorbate concentration, MM^{-1}
C = solute concentration, ML^{-3}
K_d = distribution coefficient, L^3M^{-1}
X = mass of solute sorbed, M
M = mass of sorbent, M

Note that the symbol M is used also to represent mass when defining units. The slope of the above equation is obtained from

$$\frac{dS}{dC} = K_d = \frac{X/M}{C} \tag{16.5.2}$$

The distribution coefficient K_d is defined as the ratio of the quantity of the sorbate per gram of solid, X/M, to the concentration of the solute, C, at equilibrium. It may be obtained by plotting the results from laboratory batch experiments as shown in Fig. 16.5.1. The importance of the distribution coefficient approach for organic chemicals is discussed in Sec. 16.5.2.

Freundlich equilibrium adsorption describes the sorbate concentration as a nonlinear function of the solute concentration. It may be described by

$$S = \frac{X}{M} = KC^n \tag{16.5.3}$$

where K and n are experimentally determined constants. (In some formulations, the exponent is written as $1/n$.) If $n = 1$, the above formula reduces to the linear equilibrium adsorption equation; n is typically less than or equal to 1. By taking the logarithm of both sides, this equation is converted to the following linear form:

$$\log S = \log K + n \log C \tag{16.5.4}$$

Plots of both forms of the equation are shown in Fig. 16.5.2. Note that a plot of log S versus log C gives a straight line. The slope of the equation is obtained by taking the derivative to yield the following:

$$\frac{dS}{dC} = nKC^{n-1} \tag{16.5.5}$$

Langmuir equilibrium adsorption also assumes a nonlinear function between the sorbed concentration and the solute concentration. The Langmuir isotherm is de-

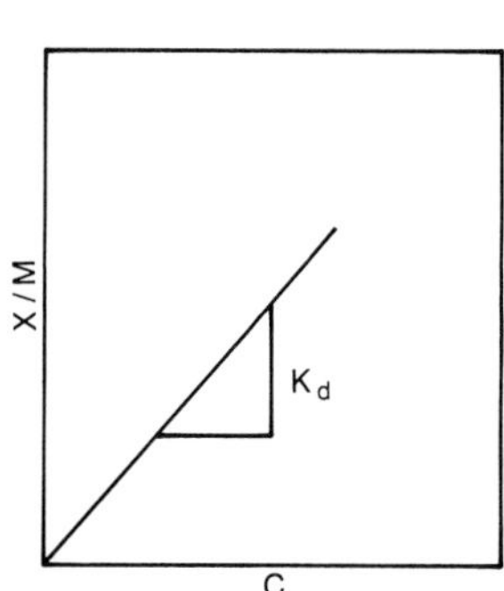

FIGURE 16.5.1 Linear adsorption isotherm showing the slope equal to the distribution coefficient.

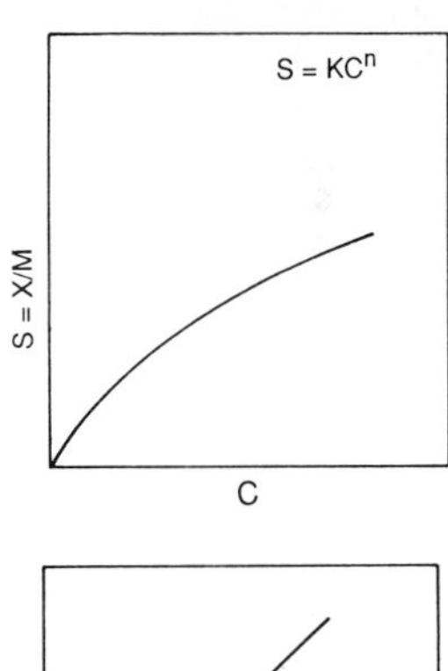

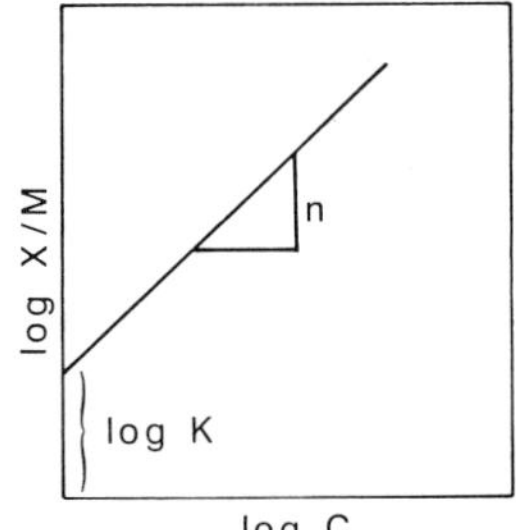

FIGURE 16.5.2 Freundlich equilibrium isotherm showing how constants are determined.

scribed by the following:

$$S = \frac{X}{M} = \frac{aC}{1 + bC} \tag{16.5.6}$$

where a and b are experimentally determined constants. Equation (16.5.6) is sometimes written with $a = bK$, where b is a constant related to the enthalpy of adsorption and K is the maximum adsorption capacity of the solid (mass solute adsorbed/mass adsorbent for a complete monolayer). The equation is based on the assumptions that: (1) sorption is limited to a single layer of solute molecules, (2) sorbed molecules are not free to move on the surface, and (3) the enthalpy of sorption is the same for all molecules.

There have been several more mechanistic sorption models developed, such as the electrostatic adsorption models (constant-capacitance, diffuse-layer, double-layer, and triple-layer models). These are discussed in Refs. 24, 79, and 96. The use of these models is not yet widespread outside the research community. Their usage will increase as the required thermochemical data are incorporated into computer models for geochemical speciation.

One-Dimensional Transport with Linear Adsorption. The preceding discussion describes sorption, but it does not discuss how sorption affects transport. The effects are examined in this section using a one-dimensional transport equation for illustrative purposes. The one-dimensional transport of a single sorbing solute in a homogeneous porous medium under saturated steady-state water-flow conditions is described by

$$D \frac{\partial^2 C}{\partial x^2} - v \frac{\partial C}{\partial x} = \frac{\partial C}{\partial t} + \frac{\rho_b}{\phi_e} \frac{\partial S}{\partial t} \tag{16.5.7}$$

where D = dispersion coefficient ($L^2 t^{-1}$)
v = interstitial velocity (Lt^{-1})
ρ_b = bulk mass density (ML^{-3})
ϕ_e = effective porosity (dimensionless)

The last term in the above equation represents the change in concentration in the fluid caused by adsorption. Using the chain rule, it may be expanded as

$$\frac{\partial S}{\partial t} = \frac{dS}{dC} \frac{\partial C}{\partial t} \tag{16.5.8}$$

For purposes of illustration, it is assumed that linear equilibrium sorption is valid and, therefore,

$$\frac{dS}{dC} = K_d \tag{16.5.9}$$

Substituting and rearranging terms yields the following:

$$D \frac{\partial^2 C}{\partial x^2} - v \frac{\partial C}{\partial x} = \left(1 + \frac{\rho_b}{\phi_e} K_d\right) \frac{\partial C}{\partial t} \tag{16.5.10}$$

The quantity $[1 + (\rho_b/\phi_e)K_d]$ is known as the *retardation factor.* The retardation of the concentration relative to the bulk mass of water is described by the following retardation equation:

$$v_c = \frac{v}{1 + (\rho_b/\phi_e)K_d} \tag{16.5.11}$$

where v_c is the velocity of the $C/C_o = 0.5$ point on the concentration profile of the retarded constituent.[28] The retardation concept is illustrated schematically in Fig.

LABORATORY REPRESENTATION

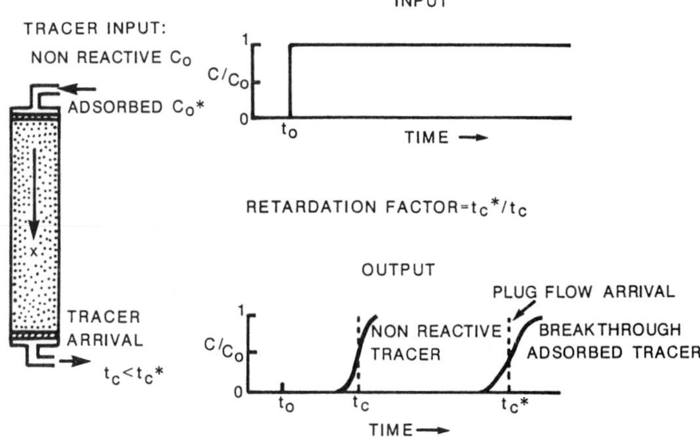

FIELD REPRESENTATION

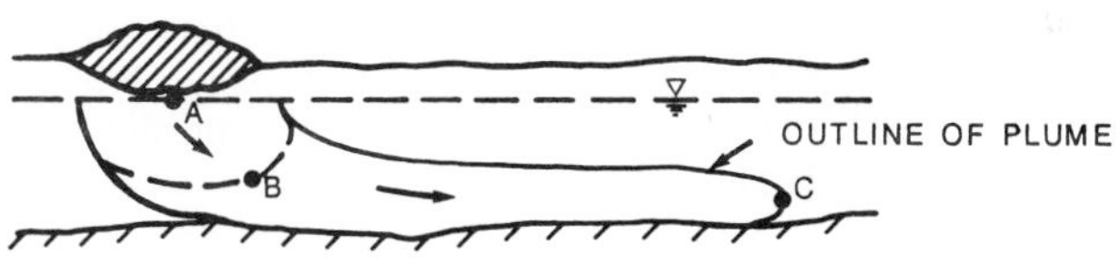

RETARDATION FACTOR=DISTANCE AC /DISTANCE AB

FIGURE 16.5.3 Schematic illustration of the retardation concept: the ideal laboratory case and a hypothetical field case. *(Reprinted with permission from Ref. 18: Groundwater Contamination, 1984. Published by the National Academy Press, Washington, D.C.)*

16.5.3 for laboratory conditions and for a hypothetical field situation. For the field representation shown in this figure, the groundwater velocity is assumed to have no spatial variability.

Limitations of the adsorption-isotherm approach are discussed in Ref. 18. One problem is that breakthrough curves that are simulated using the advection-dispersion model with linear isotherms are nearly symmetrical, whereas curves obtained from laboratory-column experiments are asymmetrical with extended tails, especially on the trailing side of the contaminant distribution. Reynolds[66] compiled data from several published studies in which tracers with linear batch isotherms were used to obtain breakthrough curves from column experiments. The results were scaled and plotted on the same graph of dimensionless concentration versus effective pore volumes (Fig. 16.5.4). The data for Fig. 16.5.4 represent the work of several investigators and a variety of sorbates and sorbents. All the breakthrough curves are asymmetrical and sit within a single narrow band on the dimensionless graph. This suggests the presence of a common factor that is currently not accounted for in the advection-dispersion formulation, perhaps the effect of kinetics.

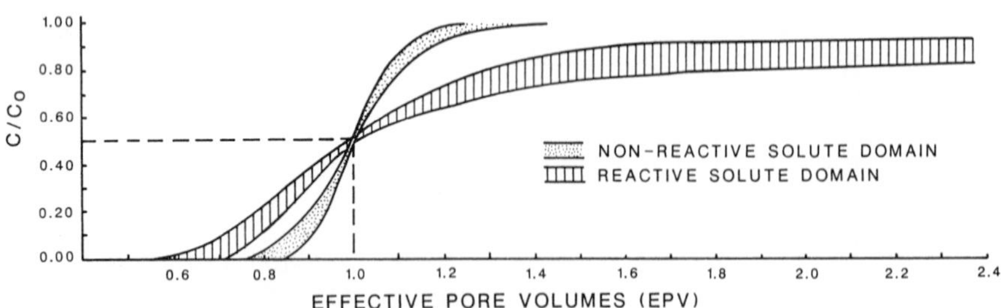

FIGURE 16.5.4 Breakthrough graphs for nonreactive and adsorbed tracers in column experiments with homogeneous porous media. Each graph represents numerous sets of experimental data. EPV is the volume of water contained in a unit of porous material. The horizontal axis represents the volume of water flushed through the system in units of EPV. *(Source: Ref. 32.)*

Adsorption Kinetics. Sorption is often treated as a process that achieves rapid equilibrium so that kinetic expressions are not needed. However, a brief discussion of sorption kinetics is included here for completeness. Studies of sorption kinetics are apparently limited; as a result, parameters required in rate expressions are ill-defined and applicable only under a specific set of conditions. Under these constraints, kinetics expressions are less practical for application unless the analyst can determine values of the rate constants which apply to the specific system being investigated.

Most kinetic expressions for sorption and desorption are assumed to be first-order. The equation for the change in solute concentration is

$$\frac{\partial C}{\partial t} = -k_r\left(C - \frac{S}{K}\right) \tag{16.5.12}$$

and that for the sorbate is

$$\frac{\partial S}{\partial t} = -k_r(S - KC) \tag{16.5.13}$$

where k_r = rate parameter (generally assumed to be the same for adsorption or desorption) and K = an experimentally determined parameter.

Commonly, sorption rates are much faster than desorption rates. Sorption of some metals does not appear to be reversible, or becomes less reversible with time. Several explanations of this behavior have been proposed. Crystallization of metal oxides can trap metals forming these oxides. In some instances, it is thought that sorption may cause crystallization of a phase less soluble than the sorbent, if a solid-solution series exists, or that the ion migrates into the crystal of the sorbent if it is of appropriate size and charge. In these cases, the above equations do not adequately describe the kinetic behavior.

16.5.2 Sorption of Organics

Sorption results from the affinity of a solute for a solid phase but may also occur from the lack of affinity of the solute for a solvent phase.[94] In aqueous systems, this latter phenomenon is called *hydrophobic adsorption.* The hydrophobic character of these compounds is also responsible for the low solubility of these compounds in water.

The distribution coefficient K_d is a valid representation of the partitioning between the solution phase and the solid phase only if the partitioning is fast (compared to the flow velocity) and reversible, and the isotherm is linear. Many organic compounds are reported to follow a linear adsorption isotherm. They include: several halogenated aliphatic hydrocarbons,[19] polynuclear aromatic hydrocarbons,[52] dibenzothiophene,[34] benzene,[70] and halogenated hydrocarbons and some substituted benzene compounds.[100] Given these data and the reported frequency of these compounds as groundwater contaminants, organic adsorption from groundwaters is most often assumed to follow a linear isotherm and to have a constant K_d. Sorption of organics is not always rapid enough to be in equilibrium. Karickhoff et al.[39] investigated the sorption and desorption of organic pollutants and found that a very rapid component of adsorption preceded a much slower component of adsorption. First-order kinetics were obeyed during each of the two periods. Approximately half of the sorptive equilibrium was realized within minutes, while the slower component required days or weeks to complete. The slower second period was visualized as diffusive transfer to sorption sites that were inaccessible to the bulk water.

Different methods have been used to determine the sorption-desorption behavior of contaminants. The most common approach is laboratory column and batch studies where effluent concentrations are measured in order to describe the overall interaction between the liquid phase and the solid matrix. The applicability of such laboratory tests to field conditions is often questionable. Another method is based on field measurement of contaminant concentrations in soil samples collected at various depths during drilling and in adjacent groundwater during subsequent monitor-well sampling. A third method calculates the distribution coefficient, based on the total organic carbon content of the soil and either the aqueous solubility of the compound or its octanol/water partition coefficient. This method is the topic discussed in this section.

The work of Lambert et al.[45–47] demonstrated that the sorption of neutral organic pesticides correlated with the organic carbon content of a given soil. Weed and Weber[95] also found that the retention of pesticides by soil is most closely related to the amount of organic matter in the soil.

Lambert et al. suggest that soil organic matter functions like organic solvent in a solvent extraction process. Thus, partitioning of a neutral organic compound between soil organic matter and water should correlate well with its partitioning between water and an immiscible organic solvent. Chiou et al.[19] also showed that the transfer of nonionic organic compounds from water to soil could be due to partitioning in the soil organic matter, and they found that linear adsorption isotherms applied over a wide range of concentrations.

The distribution coefficient K_d is related to organic carbon content fraction according to the equation

$$K_d = K_{oc} f_{oc} \tag{16.5.14}$$

where K_{oc} is the adsorption constant based on organic carbon content and f_{oc} is the fraction of total organic carbon content in terms of grams of organic carbon per gram of soil. This equation is usable when $f_{oc} > 0.1$ percent. A typical value of organic matter in mineral soils is 3.25 percent.[13] The amount of organic matter is approximately 1.9 times the amount of organic carbon; therefore, a typical value for organic carbon content is 1.7 percent. However, f_{oc} will vary from site to site. Several f_{oc} values are listed in Table 16.5.3.

The determination of K_{oc} usually is accomplished by using an equation with the

TABLE 16.5.3 Fraction of Organic Carbon (f_{oc})

Aquifer material	Value, %	Reference
Borden aquifer Canada (sand)	0.028	59
Silt clay	16.2	56
Sandy loam	10.8	56
Silt clay	1.7	56
Silt loam	1.0	56
Silt loam	1.9	20
Silt clay loam	2.6	70
Silt clay loam	1.8	70
Traverse City, Mich. (sand and gravel)	0.008	82
Granger, Ind. (sand)	0.1	82
Fine sand	0.087	100
Silt loam	1.6	83

following general form:

$$\log K_{oc} = a \log (S_w \text{ or } K_{ow}) + b \tag{16.5.15}$$

where S_w = aqueous solubility of a compound
K_{ow} = octanol-water partition coefficient (ratio of concentration of solute in octanol phase to concentration of solute in aqueous phase)
a = slope on log-log plot
b = value of $\log K_{oc}$ when $\log (S_w$ or $K_{ow})$ equals zero

Many studies of various chemical groups have been conducted. Karickhoff et al.[39] related K_{oc} to the octanol-water partition coefficient and to the water solubility by the following relationships:

$$K_{oc} = 0.63 K_{ow} \tag{16.5.16}$$

and

$$\log K_{oc} = -0.54 \log S_w + 0.44 \tag{16.5.17}$$

where S_w = aqueous solubility of a compound (expressed as a mole fraction).

The water solubilities of the compounds examined ranged from 1 ppb to 1000 ppm. Hassett et al.[34] found a similar relationship between K_{oc} and K_{ow} for organic energy-related pollutants.

Chiou et al.[19] also investigated the relationship between octanol-water partitioning and aqueous solubilities for a wide variety of chemicals, including aliphatic and aromatic hydrocarbons, aromatic acids, organochlorine and organophosphate pesticides, and polychlorinated biphenyls. Their results, shown in Fig. 16.5.5, cover more than eight orders of magnitude in solubility and six orders of magnitude in the octanol-water partition coefficient. The regression equation based on these data is

$$\log K_{ow} = -0.670 \log S_w + 5.00 \tag{16.5.18}$$

where S_w = aqueous solubility of a compound, mol/L.

Brown and Flagg[15] have extended the work of Ref. 39 by developing an empirical relationship between K_{ow} and K_{oc} for nine chloro-s-triazine and dinitroaniline com-

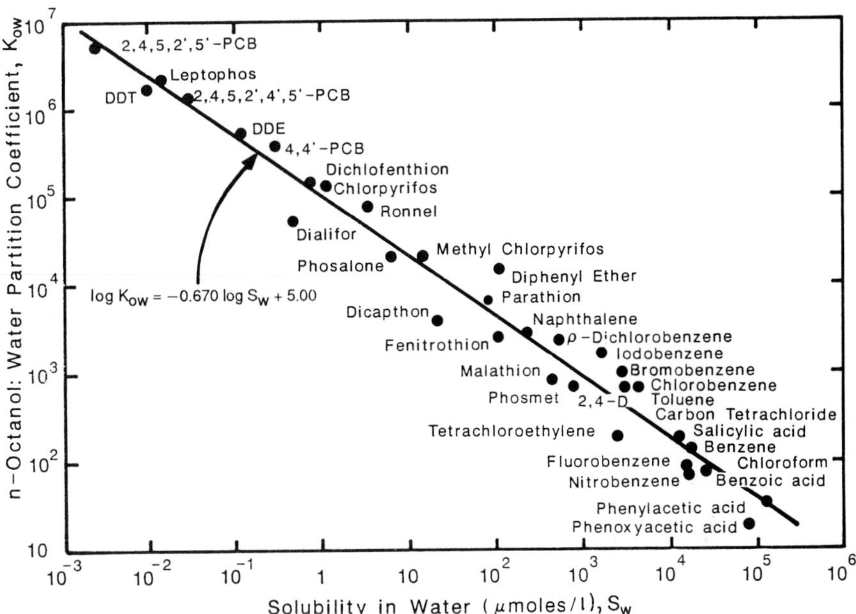

FIGURE 16.5.5 Correlation of aqueous solubility with octanol-water partition coefficient. *(Reproduced from Ref. 19 with permission.)*

pounds. They plotted their results, along with those of Karickhoff et al.,[39] as shown in Fig. 16.5.6. The combined data set produces the following correlation:

$$\log K_{oc} = 0.937 \log K_{ow} - 0.006 \qquad (16.5.19)$$

The correlation between K_{oc} and K_{ow} for the compounds studied by Brown and Flagg [15] has a larger factor of uncertainty than those studied by Karickhoff et al.[39]

The octanol-water partition coefficient and the solubility of an organic compound are generally available or may be calculated. Leo et al.[49] have compiled a list of octanol-water partition coefficients. For organic compounds, Ref. 91 provides many common properties. Lyman et al.[50] provide methods of calculating solubilities and octanol-water partition coefficients.

The theoretical method for calculating an adsorption coefficient K_d can be summarized as follows:

1. Find the aqueous solubility of the compound, S_w, or its octanol-water partitioning coefficient K_{ow} from chemical references. If S_w is available but not K_{ow}, convert S_w to K_{ow} by Eq. (16.5.18).
2. Convert K_{ow} to the organic carbon partitioning coefficient K_{oc} by Eq. (16.5.19).
3. Find the organic carbon fraction f_{oc} in the soil from Table 16.5.3.
4. Compute K_d by Eq. (16.5.14), $K_d = K_{oc} f_{oc}$.

The reliability of the theoretical method for calculating the adsorption coefficient depends on several factors. Regression equations for K_{oc} as a function of K_{ow} or S_w are

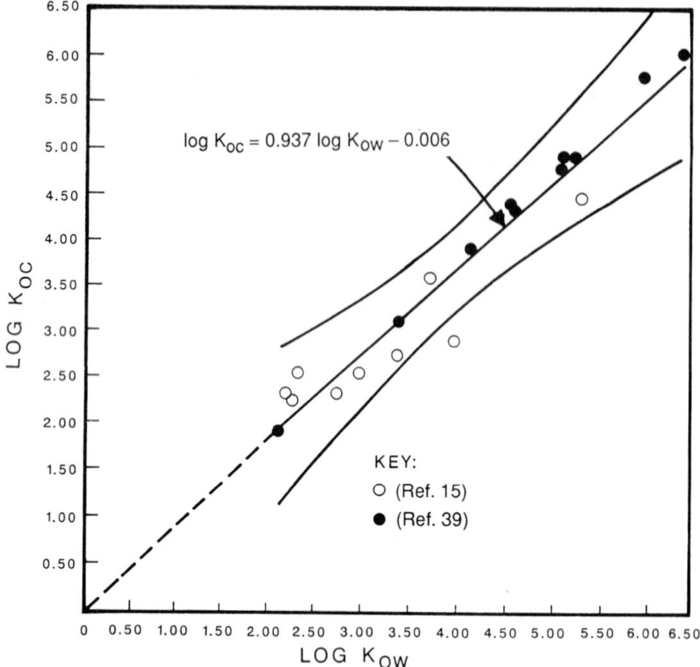

FIGURE 16.5.6 Relationship between K_{oc} and K_{ow} for coarse silt.

experimentally derived from a specific data set that represents particular classes of chemicals and ranges of parameters. Therefore, care must be taken when selecting a regression equation. Also, the regression equation method is only valid in porous media with greater than 0.1 percent organic carbon content.[50] Regression equation methods consider organic carbon as the primary adsorbent; however, below an organic carbon content of 0.1 percent, inorganic surfaces are the dominant adsorbent.[60]

Table 16.5.4, adapted from Ref. 50, includes examples of various regression equations, the source of each equation, and the classes of organic chemicals for which each equation was derived. Table 16.5.5, also adapted from Ref. 50, lists the parameters required for each equation, the range of values for each parameter, and the range of K_{oc} values calculated with each equation.

Lyman et al.[50] recommend selection of a regression equation for K_{oc} on the basis of data available about the chemical, the chemical classes covered by each regression equation, and the range of K_{oc} and input parameter values covered by each regression equation. Highest priority should be given to the most accurate data from actual measurements. If data are available for all input parameters, a regression equation using K_{ow} is preferred to an equation using S_w.

For selection of a regression equation on the basis of chemical classes, priority should be given to the equation which was derived from the same chemical class as

TABLE 16.5.4 Various Regression Equations for the Organic Carbon Partitioning Coefficient

Equation no.	Equation	Chemical classes represented
1	$\log K_{oc} = -0.55 \log S_w + 3.64$	Wide variety, mostly pesticides[41]
2	$\log K_{oc} = -0.54 \log S_w + 0.44$	Mostly aromatic or polynuclear aromatics; two chlorinated[39]
3	$\log K_{oc} = -0.557 \log S_w + 4.277$	Chlorinated hydrocarbons[19]
4	$\log K_{oc} = 0.544 \log K_{ow} + 1.377$	Wide variety, mostly pesticides[41]
5	$\log K_{oc} = 0.937 \log K_{ow} - 0.006$	Aromatics, polynuclear aromatics, triazines and dinitroaniline herbicides[15]
6	$\log K_{oc} = 1.00 \log K_{ow} - 0.21$	Mostly aromatic or polynuclear aromatics; two chlorinated[39]
7	$\log K_{oc} = 1.029 \log K_{ow} - 0.18$	Variety of insecticides, herbicides, and fungicides[63]
8	$\log K_{oc} = 0.524 \log K_{ow} + 0.855$	Substituted phenylureas and alkyl-N-phenylcarbamates[14]

Source: Reprinted with permission from Ref. 50. Copyright 1982, American Chemical Society.

TABLE 16.5.5 Parameters for Regression Equations in Table 16.5.4

Equation no.	Parameter required	Range of parameter	Range of K_{oc} values
1	S_w (mg/L)	0.0005 – 1,000,000	1 – 1,000,000
2	S_w (mole fraction)	$(0.03–410,000) \times 10^{-9}$	80 – 1,000,000
3	S_w (moles/L)	0.002 – 100,000	30 – 380,000
4	K_{ow}	0.001 – 4,000,000	10 – 1,000,000
5	K_{ow}	100 – 4,000,000	100 – 1,000,000
6	K_{ow}	100 – 4,000,000	100 – 1,000,000
7	K_{ow}	0.3 – 400,000	2 – 250,000
8	K_{ow}	3 – 2200	10 – 400

Source: Reprinted with permission from Ref. 50. Copyright 1982, American Chemical Society.

the chemical being modeled. However, if there is no clear match of chemical classes, Ref. 50 suggests using Eqs. (1) and (4) from Table 16.5.4 because they are derived from the widest variety of chemicals.

For selection of a regression equation on the basis of the range of K_{oc} and input parameter values, the values should be within the range originally covered by the regression equation. The use of the parameter values or estimation of K_{oc} values outside the range of the original data set will subject the estimated K_{oc} and, therefore, the estimated retardation coefficient to greater uncertainty. Data for a variety of organic compounds are provided in Refs. 50 and 86.

16.6 BIODEGRADATION REACTIONS

Biodegradation refers to biologically mediated processes that chemically alter a solute or substrate. *Primary biodegradation* refers to any biologically induced structural transformation in the parent compound that changes its molecular form. Microorganisms are responsible for biodegradation and are abundant in aquatic environments. Microbes influence chemical processes because of their ability to supply energy for reactions through metabolic processes and to catalyze reactions through enzymatic activity. Chemical reactions that proceed slowly or not at all in the absence of biota may occur and proceed faster in the presence of biological *enzymes* (organic substances produced by cells and causing changes in other substances by catalytic action).

Although microbial communities catalyze countless reactions, many of them fall into a few classes. Oxidative reactions make up one important class of biochemical reactions. The hydroxylation of aromatic compounds, such as benzene changed to catechol, is an example of an oxidative reaction that generates polar compounds from nonpolar ones. Microbes also catalyze reductive reactions. An example is the dehydrochlorination of DDT to produce DDE. Enzymes can catalyze otherwise slow hydrolytic reactions as well, e.g., malathion to malathion β-monacid.

Although each reaction causes the primary degradation of a compound, different reactions affect the toxicity of a compound in different ways.[3] *Mineralization* refers to the complete degradation of an organic compound to inorganic products. In many reactions, however, organic products remain. *Detoxication* reactions produce innocuous metabolites from a toxic substance. In *activation* reactions, microbes convert an innocuous compound into a toxic compound. The defusing of potentially hazardous compounds occurs when biota produce an innocuous compound before the parent compound's harmful form is generated. Finally, a toxic compound may be transformed chemically but retain its toxicity.

Reactions can take place either in the presence or in the absence of oxygen. Reactions in the presence of oxygen are *aerobic,* whereas those in the absence of oxygen are *anaerobic.* Some compounds, such as DDT, are transformed under both aerobic (forming DDE) and anaerobic (forming DDD) conditions. Examples of anaerobic transformations that occur in groundwater are[103]

Carbon tetrachloride → chloroform → methylene chloride
Tetrachloroethylene → trichloroethylene ⏋

$$\rightarrow \left\{ \begin{array}{l} \text{cis 1,2-dichloroethene} \\ \text{trans 1,2-dichloroethene} \\ \text{1,1-dichloroethene} \end{array} \right\} \rightarrow \text{vinyl chloride} \qquad (16.6.1)$$

Other compounds that are degraded in groundwater are shown in Table 16.6.1.

TABLE 16.6.1 Prospect of Biotransformation of Selected Organic Compounds in Groundwater

Class of compounds	Aerobic water, conc. of pollutant, μg/L		Anaerobic water
	>100	<10	
Halogenated aliphatic hydrocarbons			
Trichloroethylene	None	None	Possible*
Tetrachloroethylene	None	None	Possible*
1,1,1-trichloroethylene	None	None	Possible*
Carbon tetrachloride	None	None	Possible*
Chloroform	None	None	Possible*
Methylene chloride	Possible	Improbable	Possible
1,2-dichloroethane	Possible	Improbable	Possible
Brominated methanes	Improbable	Improbable	Probable
Chlorobenzenes			
Chlorobenzene	Probable	Possible	None
1,2-dichlorobenzene	Probable	Possible	None
1,4-dichlorobenzene	Probable	Possible	None
1,3-dichlorobenzene	Improbable	Improbable	None
Alkylbenzenes			
Benzene	Probable	Possible	None
Toluene	Probable	Possible	None
Dimethylbenzenes	Probable	Possible	None
Styrene	Probable	Possible	None
Phenol and alkyl phenols	Probable	Probable	Probable†
Chlorophenols	Probable	Possible	Possible
Aliphatic hydrocarbons	Probable	Possible	None
Polynuclear aromatic hydrocarbons			
Two and three rings	Possible	Possible	None
Four or more rings	Improbable	Improbable	None

* Possible, probably incomplete
† Probable, at high concentration
Source: Ref. 101. Reproduced with permission.

In all biochemical reactions, the number of electrons must be conserved. When one reaction product has carbon atoms in a higher oxidized state through the loss of electrons, then another product must be present in a reduced state by having more electrons per carbon atom. Very often the reduced product (electron acceptor) is not part of the original substrate. In aerobic reactions, oxygen is the electron acceptor and is reduced to water. In anaerobic systems, NO_3^- may be the electron acceptor and be reduced to NO_2^-, N_2O, or N_2. If SO_4^{-2} is present, it will also accept electrons and be reduced to H_2S. Even CO_2 is utilized as an electron acceptor by methane bacteria to form CH_4. Another anaerobic process is *fermentation,* which occurs in the absence of external electron acceptors. The general reduction reaction may be written as[81]

$$\text{Substrate} + \text{electron acceptor} \xrightarrow{\text{biomass}} \text{products} + CO_2 + H_2O + \text{biomass} + \text{reduced electron acceptor} \quad (16.6.2)$$

For any single microorganism, organic compounds can be divided into three groups according to their biodegradability: (1) usable immediately as an energy or nutrient source, (2) degraded slowly, (3) not degraded (recalcitrance). These groups or types of degradation are subsets of metabolism and are shown in Fig. 16.6.1. *Metabolism* refers to degradation of organic compounds to provide energy and carbon for growth. A fourth group also exists, consisting of compounds subject to cometabolic degradation. *Cometabolism* refers to degradation of organic compounds which are not themselves used as a nutrient or growth substrate, but are used concomitantly with other compounds which are. Figure 16.6.2 compares metabolic and cometabolic degradation rate curves. Because cometabolism does not provide a growth substrate, the population increase characteristic of metabolic degradation does not take place, and the rate of degradation often is slower.

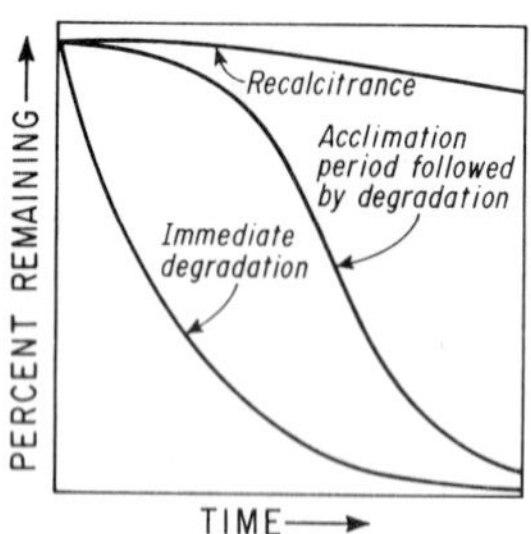

FIGURE 16.6.1 Types of degradation reactions based on organic compounds' biodegradability. *(Reprinted with permission from Ref. 50. Copyright 1982, American Chemical Society.)*

16.6.1 Factors Influencing Biodegradation Reactions

The variables that influence the rate of biodegradation fall into two categories: (1) those that determine the availability and concentration of the compound to be degraded (e.g., sorption) or that affect the microbial population size and activity and (2) those that directly control the reaction rate itself (e.g., temperature). Both direct and indirect variables can be classified as substrate-related, organism-related, or environment-related. Most of the information on these factors is qualitative, primarily derived from observations of degradation by specific species. Because of considerable variations in species, habitat, and chemical environment, not all variables will influence all situations in the same way.

Biodegradation has been studied in several different habitats including (1) aquatic systems, (2) wastewater treatment, (3) activated sludge, and (4) soil. The following discussion focuses on the subsurface environment. The majority of the microbial population in the subsurface is located in the top layer of soil because nutrient levels and oxygen availability are high there. Wilson and McNabb[101] provide similar data for depths of up to 6 m (20 ft), both above and below the water table. The numbers of organisms at these deeper locations are surprisingly high and do not decline drastically with depth. They report a uniformity of numbers for different seasons and in material from replicate bore holes at the same site.

16.6.2 Mathematical Representation of Biodegradation

Before the rate of biodegradation can be quantified and a rate constant can be calculated, a kinetic expression must be derived to describe the pattern of loss over time. Two general rate laws have been proposed to describe biodegradation: the power rate law and the hyperbolic rate law. These are discussed below in the context of aquatic systems.

The *power rate law* states that the rate of biodegradation is proportional to some power of the substrate concentration as follows:

$$-\frac{dC}{dt} = kC^n \tag{16.6.3}$$

where C = concentration of substrate in solution
k = biodegradation rate constant
n = order of reaction

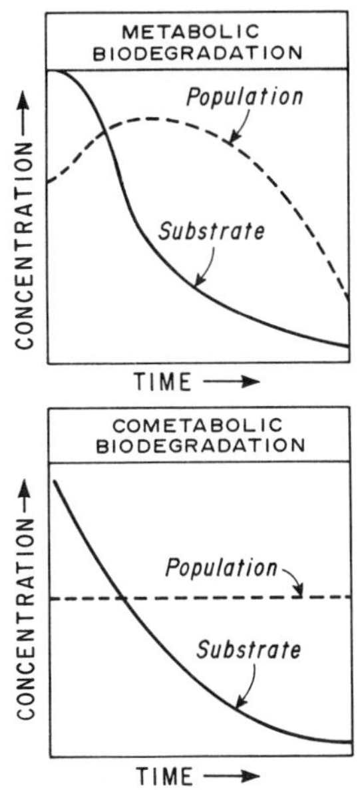

FIGURE 16.6.2 Population and substrate concentration behavior during metabolic and cometabolic biodegradation. *(Reprinted with permission from Ref. 50. Copyright 1982, American Chemical Society.)*

If first-order kinetics are assumed (i.e., $n = 1$), the rate is simply the product of the rate constant and the substrate concentration. The first-order decay equation is similar to that used for radionuclides, where the rate of decomposition is proportional to the amount of substance present. The rate law is

$$N = N_o e^{-\lambda t} \tag{16.6.4}$$

where N is the number of atoms remaining after time t and N_o is the original number present at time $t = 0$. The decay constant λ is analogous to k in Eq. (16.6.3) and has units of T^{-1}.

Information usually is given on half-lives, instead of on λ. The relationship between half-life and decay constant is straightforward. Because the half-life $t_{1/2}$ is the time at which the number of atoms remaining is one-half the original number, the above equation may be rewritten

$$\frac{N}{N_o} = \frac{1}{2} = e^{-\lambda t_{1/2}} \tag{16.6.5}$$

or

$$t_{1/2} = \frac{\ln 2}{\lambda}$$

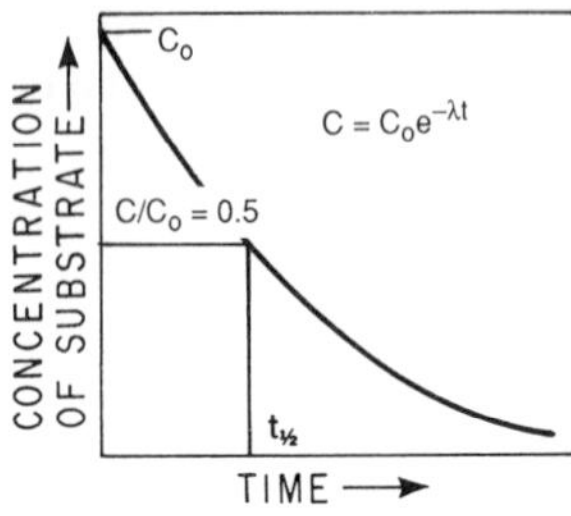

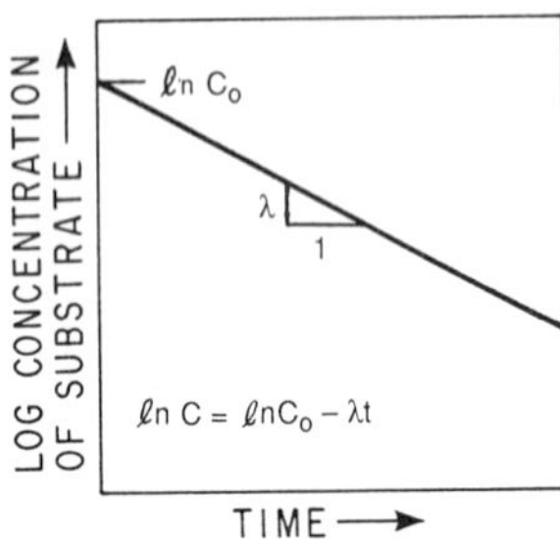

FIGURE 16.6.3 First-order disappearance curve of a chemical. *(Reprinted with permission from Ref. 50. Copyright 1982, American Chemical Society.)*

and finally

$$\lambda = \frac{0.693}{t_{1/2}} \tag{16.6.6}$$

Figure 16.6.3 shows a typical first-order decay curve due to radioactive decay or biodegradation; when the log of the concentration is plotted against time, the curve becomes a straight line. At low concentrations, the assumption of first-order kinetics for biodegradation is reasonable.[50] Huyakorn et al.[36] showed that the degradation of aldicarb in transport through soils on Long Island could be represented as a first-order decay process.

The *hyperbolic rate law* is commonly used to describe the growth of microbial populations. When associated with enzymatic processes, it is known as *Michaelis-Menten kinetics;* for microbial growth, it is known as *Monod kinetics.* Based on Monod kinetics, this law expresses the biodegradation rate as a hyperbolic saturation function of the substrate concentration. Although the measured rate generally refers to population growth, it can be converted to a term describing the disappearance of the substrate supporting growth. For further discussion on this, see Refs. 10, 11, 43, 67, 68, 69, 76, and 81.

16.7 PROCESS CHARACTERIZATION

The character of a solute transport problem is evident from the distribution of solutes and the processes controlling their transport. Therefore, characterizing a site involves gathering and analyzing data sets to help predict future conditions based on past behavior. In this section, data collection activities are discussed, including goals, sources of data, collection techniques, and strategies.

16.7.1 Data Collection Goals

Site characterization, a process following the scientific method, is performed in phases (Fig. 16.7.1). In the initial phase, a hypothesis about the site or system behavior is made. Based on the hypothesis, a data collection program is designed in the second phase. Data are collected and an analysis or assessment follows. An iterative

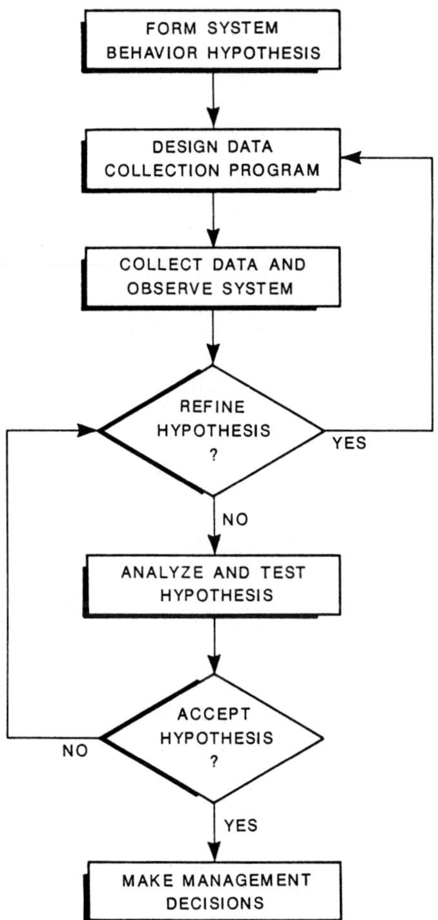

FIGURE 16.7.1 Site characterization phases. *(Reproduced from Ref. 12 with permission.)*

step of refining the hypothesis is performed by using the results of the analysis, and additional data may be collected. As the knowledge of the site increases and becomes more complex, the working hypothesis may take the form of either a numerical or analytical model. Data collection continues until the hypothesis is proven sufficiently. The proof forms the basis for decision making.

The ultimate goal of site characterization is to make informed decisions based on studies with clearly defined objectives:[17] to assess the background water quality; to establish the impact of certain facilities, practices, or natural phenomena on water quality; and to predict future groundwater quality trends under various conditions.

Whatever the objective, however, assessment of the groundwater flow system and of groundwater chemistry must be made to characterize the site. All too often, emphasis is placed on groundwater chemistry through intensive groundwater quality monitoring and, at many sites, groundwater quality data are abundant. Unfortunately, water level and stratigraphic data, which are used to determine direction and rate of groundwater movement, are limited. Ironically, these equally important data are easier and less expensive to collect than water-quality data.

16.7.2 Sources of Data

Flow Data. One of the key elements affecting solute transport is the characterization of the groundwater flow system. This includes the physical parameters of the overall hydrologic region influencing the contaminated site (e.g., hydraulic conductivity, storage coefficient, and aquifer thickness); system stresses (e.g., recharge and pumping rates); and other system characteristics (e.g., physical and hydraulic boundaries and groundwater flow directions and rates). By understanding where and the rates at which groundwater recharges and discharges (mass balance), the laws governing flow (e.g., Darcy's equation), and the geological framework through which this flow occurs, it is possible to determine these characteristics. Table 16.7.1 lists the information typically used to identify and quantify the important characteristics of a groundwater system. The methods for collecting these data are discussed later.

Because migrating miscible contaminants travel with moving groundwater, it is important to characterize groundwater flow. Groundwater flows from areas of re-

TABLE 16.7.1 Aspects of Site Hydrogeology

Geologic aspects
1. Type of water-bearing unit or aquifer (overburden, bedrock).
2. Thickness, areal extent of water-bearing units and aquifers.
3. Type of porosity (primary, such as intergranular pore space, or secondary, such as bedrock discontinuities, e.g., fracture or solution cavities).
4. Presence or absence of impermeable units or confining layers.
5. Depths to water table; thickness of vadose zone.
Hydraulic aspects
1. Hydraulic properties of water-bearing unit or aquifer (hydraulic conductivity, transmissivity, storativity, porosity, dispersivity).
2. Pressure conditions (confined, unconfined, leaky confined).
3. Groundwater flow directions (hydraulic gradients, both horizontal and vertical), volumes (specific discharge), rate (average linear velocity).
4. Recharge and discharge areas.
5. Groundwater or surface water interactions; areas of groundwater discharge to surface water.
6. Seasonal variations of groundwater conditions.
Groundwater use aspects
1. Existing or potential aquifers.
2. Existing or near-site use of groundwater.

Source: Ref. 89.

charge (commonly via rainfall, surface water bodies, or irrigation) to areas of discharge (surface water or wells). Along the way, subsurface heterogeneities (such as faults) influence its direction. The rate of groundwater flow is controlled by the porosity and hydraulic conductivity of the media through which it travels and by hydraulic gradients, which are influenced by recharge and discharge.[27,28] Often it is important to conduct a site characterization quickly; however, groundwater flow systems vary with time. Seasonal variations in water levels, which are often several feet in magnitude, can adversely impact remediation and should be characterized.

Contaminant information includes: (1) source characterization, (2) concentration distribution of contamination and naturally occurring chemicals, and (3) data associated with the processes that affect plume development. Source characterization consists of the following: (1) the chemical volume released, (2) the area infiltrated, and (3) the time duration of release. Often, the release occurred so long ago that information is difficult to obtain.

Chemical Data. Quantitative characterization of subsurface chemistry can include sampling the vadose and saturated zones to determine the concentration distributions in groundwater, soil, and vadose water. Vadose zone monitoring is discussed in Refs. 97, 98, and 99. A network of monitoring wells (also necessary for the hydrogeologic data) needs to be installed to collect depth-discrete groundwater samples.[87] Wells should be located in areas that will supply information on ambient (background) groundwater chemistry and on plume chemistry. At a minimum, soil and groundwater samples should be analyzed for the parameters of concern from the waste stream. A full-priority pollutant scan on the first round provides information on plume chemistry and may be useful in differentiating plumes that have originated from several different sources. On subsequent rounds, the parameter list may be refined on the basis of site-specific considerations.

If inorganic contaminants are of concern or interest, a full suite of major and minor ions should be analyzed. Field measurements of pH will be especially important. After analyzing the samples, the resulting concentration data should be mapped in three dimensions to determine the spatial distribution of contamination. These plume delineation maps and the results from aquifer tests will yield estimates of plume movement and identify locations for extraction wells.

Solute Transport Data. The data requirements for contamination characterization are presented in Table 16.7.2. Sorption-desorption and transformation processes are important in controlling the migration rate and concentration distributions. Some of these processes tend to retard the rate of contaminant migration and act as mechanisms for concentration attenuation. Because of their effects, the plume of a reactive contaminant generally expands more slowly and the concentration (relative to source concentration) is less than that of an equivalent nonreactive contaminant.

Chemical properties of the plume are necessary (1) to characterize the transport of the chemicals and (2) to evaluate the feasibility of remediation. The following properties influence the mobility of dissolved chemicals in groundwater and should be considered for plume migration and cleanup:

1. *Aqueous solubility:* Determines the degree to which the chemical will dissolve in water. Solubility indicates maximum possible concentrations. High solubility indicates low sorption tendencies.
2. *Chemical speciation:* Determines the charge and distribution of ions and, hence, the mobility of inorganic chemicals.

TABLE 16.7.2 Data Pertinent to Groundwater Contamination Characterization

General category	Specific data
Site physical framework	Estimates of hydrodynamic dispersion parameters Effective porosity distribution Natural (background) aquifer constituent concentration distributions Fluid density and relationship to concentrations
System stresses	Pollution source locations Pollutant releases
Chemical/biological framework	Mineralogy Organic content Groundwater temperature Solute properties Major ion chemistry Minor ion chemistry Eh-pH environment
Observable responses	Areal and temporal distributions of water, solid, and vapor phase contaminants Stream flow/quality information for associated surface water

Source: Ref. 12. Reproduced with permission.

3. *Henry's law constant (see Eq. (15.2.29)):* High values may signify volatilization from the aqueous phase as an important transport process. Used in conjunction with vapor pressure.
4. *Density:* For high concentrations, the density of the contaminated fluid may be greater than the density of pure water. This causes the downward vertical movement of contaminants.
5. *Octanol-water partition coefficient:* Indicates a chemical's tendency to partition between the groundwater and the soil. A large octanol-water partition coefficient signifies a highly hydrophobic compound, which indicates strong sorption. This provides similar information to that provided by solubility.
6. *Organic carbon partition coefficient:* Another indicator of a chemical's tendency to partition between groundwater and the soil. For certain chemicals, it is directly related to the distribution coefficient via the fraction of total organic carbon content.
7. *Half-life and degradation process:* This provides information regarding the persistence of the chemical and which, if any, transformation products might be expected.

These parameters for many chemicals may be obtained from references such as Refs. 21 and 50. Conducting a background data search reduces the amount of information that will have to be collected in the field. As indicated above, chemical-specific information is available in handbooks. Various sources of general information on specific sites are available as shown in Table 16.7.3. Other sources of information are listed in Ref. 89. Once the available data have been reviewed, it is possible to design an approach to collect the initial field data.

TABLE 16.7.3 Potential Sources of Information for Site Characterization

Problem specific	Federal or state geological surveys, university libraries, geology and engineering departments, state health departments, property owners, county records, well drillers, medical libraries, state or federal environmental protection agencies, state attorney general's office.
Site specific	Weather bureaus, state water resource boards, census bureaus, soil and water conservation districts, employment commissions, corporation commissions, Department of Agriculture, Forest Service, air photo services.

16.7.3 Data Collection Techniques

Subsurface conditions can be studied by indirect techniques or by using point data. Table 16.7.4 lists common data collection methods. References on monitoring wells include Refs. 16, 23, and 71; references on geophysical techniques include Refs. 22, 42, 44, and 77. Choosing the appropriate methods depends on the overall scope of the project. A conceptualization of the site and contamination problem should be made and updated as data become available. Throughout the study, it is essential to document all well construction details, sampling episodes, etc., to arrive at an accurate evaluation of the entire site. Understanding the hydrogeology and extent of contamination is important to a successful field study. Formulating adequate design plans ensures that wells are sited to a proper depth and stratigraphic layer so the extent of contamination is not increased by cross contamination of clean layers by leakage from contaminated layers through wells that penetrate both types of layers.

To determine flow directions and vertical and horizontal gradients, water levels are measured and converted to elevations relative to a datum, usually mean sea level. Water-level measurements may be taken by several different means including (1) chalk and tape, (2) electrical water-level probe, and (3) pressure transducer. These techniques are discussed in Refs. 2 and 78. Horizontal gradients are determined by using water-level data from wells that are open to the same hydrologic unit and/or at the same elevation but separated areally. Vertical gradients are determined by using water-level data from wells in the same location but screened at different elevations. The gradient is the difference in water levels (hydraulic head) divided by the distance between the measurement points. Because water levels often yield a complex three-dimensional surface, care must be taken in computing the hydraulic gradient. Groundwater velocity is determined by multiplying the gradient by hydraulic conductivity and dividing by effective porosity. Methods for determining hydraulic conductivity are discussed in Chap. 6.

For fractured media and karst formations, site characterization and remediation designs are more difficult. Techniques such as fracture trace analysis[48] and geophysical prospecting may be useful for locating the more permeable zones where contaminants are most likely to be located and, thus, where extraction wells might be placed. Other characterization techniques include continuous coring, aquifer tests, and tracer tests.[37] For more detailed discussion on flow in the special heterogeneous conditions of fractured media, see Ref. 78 and, for karst formations, see Refs. 9, 37 and 61.

To ensure proper quality assurance (QA) and quality control (QC) of groundwater samples, strict protocols are followed in the field. The pH, temperature, and specific conductance of a sample are measured. Ideally, before a sample is gathered,

TABLE 16.7.4 Data Collection Methods for Site Characterization

Category	Commonly used methods	Advantages/disadvantages
Geophysics (indirect data method)	Electromagnetics	Good for delineation of high conductivity plumes
	Resistivity	Useful in locating fractures
	Seismic	Limited use in shallow studies
	Ground-penetrating radar	Useful in very shallow soil studies
Drilling	Augering	Poor stratigraphic data
	Augering with split-spoon sampling	Good soil samples
	Air/water rotary	Rock sample information
	Mud rotary	Fills fractures—needs intensive development
	Coring	Complete details on bedrock
	Jetting/driving	No subsurface data
Groundwater sampling	Bailer	Allows escape of volatiles (operator-dependent)
	Centrifugal pump	Can produce turbid samples, increasing chance of misrepresented contamination
	Peristaltic/bladder pump	Gives more representative samples
	Core penetrometer	Can determine stratigraphy, and collect soil and groundwater samples
Soil boring		Restricted to shallow depths
Aquifer tests	Pumping test	Samples a large aquifer section
	Slug test	Does not require liquid disposal

water is extracted from the well until these parameters have stabilized. This ensures that the sample is from the formation rather than being a residual accumulation in the well. Proper sample storage and shipment to a qualified laboratory is also important. A sampling plan addresses issues such as sampling frequency, locations, and statistical relevance of samples.[88] For more details on sampling guidance, see Refs. 7, 8, and 17. For methods to determine partition coefficients from cores, see Ref. 81; for NAPL characterization, see Ref. 4.

16.8 SUMMARY

Emphasis in this chapter has been placed on explaining the processes associated with solute transport. In particular, retardation and biodegradation are discussed in detail because of their importance in contaminant hydrology. The approach in writing this chapter has been to provide tables of data that should be helpful to the practicing hydrogeologist. Finally, data collection activities have been discussed because they are critical to understanding the processes active at a particular site.

REFERENCES

1. Abriola, L. M., "Multiphase Flow and Transport Models for Organic Chemicals: A Review and Assessment," EPRI EA-5976, Electric Power Research Institute, Palo Alto, Calif. 1988.
2. Acker III, W. L., *Basic Procedures for Soil Sampling and Core Drilling,* Acker Drill Co., Scranton, Pa., 1974.
3. Alexander, M., "Biodegradation of Toxic Chemicals in Water and Soil," *Dynamics, Exposure, and Hazard Assessment of Toxic Chemicals,* R. Hague, editor, Ann Arbor Science, Ann Arbor, Mich., 1980.
4. American Petroleum Institute, A Guide to the Assessment and Remediation of Underground Petroleum Releases, *API Publication 1628* (2d ed.), 1989.
5. Anderson, M. P., "Using Models to Simulate the Movement of Contaminants through Ground-Water Flow Systems," *Critical Reviews in Environmental Control,* vol. 9, pp. 97–156, 1979.
6. Anderson, M. P., "Movement of Contaminants in Groundwater: Groundwater Transport-Advection and Dispersion," *Groundwater Contamination, Studies in Geophysics,* National Academy Press, Washington, D.C., pp. 37–45, 1984.
7. Barcelona, M. J., J. P. Gibb, and R. A. Miller, "A Guide to the Selection of Materials for Monitoring Well Construction and Ground-Water Sampling," Illinois State Water Survey Contract Report No. 327, USEPA-RSKERL, EPA-600/52-84-024, U.S. Environmental Protection Agency, Ada, Okla., 1983.
8. Barcelona, M. J., J. P. Gibb, J. A. Helfrich, and E. E. Garske, "Practical Guide for Ground-Water Sampling," Illinois State Water Survey Contract Report No. 374, USEPA-RSKERL under cooperative agreement CR-809966-01, U.S. Environmental Protection Agency, Ada, Okla., 1985.
9. Bogli, A., *Karst Hydrology and Physical Speleology,* Springer-Verlag, New York, 1980.
10. Borden, R. C., M.D. Lee, J. T. Wilson, C. H. Ward, and P. B. Bedient, "Modeling the Migration and Biodegradation of Hydrocarbons Derived from a Wood-Creosoting Process Waste," *Proceedings of Petroleum Hydrocarbons and Organic Chemicals in Ground Water,* National Water Well Association, Houston, pp. 130–143, 1984.
11. Bouwer, E. J., and P. L. McCarty, "Modeling of Trace Organics Biotransformation in the Subsurface," *Ground Water,* vol. 22, no. 4, pp. 433–440, 1984.
12. Bouwer, E. J., J. Mercer, M. Kavanaugh, and F. DiGiano, "Coping with Groundwater Contamination," *J. Water Pollut. Control Fed.,* vol. 60, no. 8, pp. 1414–1428, 1988.
13. Brady, N. C., *The Nature and Properties of Soils* (8th ed.), MacMillan, New York, 1974.
14. Briggs, G. G., "A Simple Relationship between Soil Adsorption of Organic Chemicals and Their Octanol/Water Partition Coefficients," *Proceedings of the 7th British Insecticide and Fungicide Conference,* vol. 1, Boots, Nottingham, Great Britain, 1973.
15. Brown, D. S., and E. W. Flagg, "Empirical Prediction of Organic Pollutant Sorption in Natural Sediments." *J. Environmental Quality,* vol. 10, no. 3, pp. 382–386, 1981.
16. Campbell, M. D., and J. H. Lehr, *Water Well Technology,* McGraw-Hill, New York, 1973.
17. Cartwright, K., and Shafer, J. M., "Selected Technical Consideration for Data Collection and Interpretation-Groundwater," *National Water Quality Monitoring and Assessment,* National Academy Press, Washington, D.C., pp. 33–56, 1987.
18. Cherry, J. A., R. W. Gillham, and J. F. Barker, "Contaminants in Groundwater: Chemical Processes, Studies in Geophysics," *Groundwater Contamination,* National Academy Press, Washington, D.C., pp. 46–64, 1984.
19. Chiou, C. T., V. H. Freed, D. W. Schmedding, and R. L. Kohnert, "Partition Coefficient and Bioaccumulation of Selected Organic Compounds," *Environ. Sci. Tech.,* vol. 11, pp. 475–478, 1977.
20. Chiou, C. T., P. E. Porter, and D. W. Schmedding, "Partition Equilibria of Non-Ionic

Organic Compounds between Soil Organic Matter and Water," *Environmental Science and Technology,* vol. 17, pp. 227–231, 1983.

21. CRC Press, *Handbook of Chemistry and Physics* (46th ed.), Boca Raton, Fla., 1990.
22. Dobrin, M. B., *Introduction to Geophysical Prospecting* (3d ed.), McGraw-Hill, New York, 1976.
23. Driscoll, F. G., *Ground Water and Wells* (2d ed.), Johnson Division, UOP, St. Paul, Minn., 1986.
24. Dzombak, D. A., *Toward a Uniform Model for the Sorption of Inorganic Ions on Hydrous Oxides,* Ph.D. thesis, Massachusetts Institute of Technology, Cambridge, Mass., 1986.
25. Feenstra, S., and J. A. Cherry, "Subsurface Contamination by Dense Non-Aqueous Phase Liquids (DNAPL) Chemicals," International Groundwater Symposium, International Association of Hydrogeologists, Halifax, Nova Scotia, May 1–4, 1988.
26. Felmy, A. R., D. C. Girvin, and E. A. Jenne, "MINTEQ, A Computer Program for Calculating Aqueous Geochemical Equilibrium," Report EPA-600/3-84-032, U.S. Environmental Protection Agency, Athens, Ga., 1983.
27. Fetter, C. W., Jr., *Applied Hydrogeology,* Charles E. Merrill, Columbus, Ohio, 1980.
28. Freeze, R. A., and J. A. Cherry, *Groundwater,* Prentice Hall, Englewood Cliffs, N.J., pp. 402–413, 1979.
29. Freyberg, D. L., "A Natural Gradient Experiment on Solute Transport in a Sand Aquifer: II. Spatial Moments and the Advection and Dispersion of Non-Reactive Tracers," *Water Resour. Res.,* vol. 22, no. 13, pp. 2031–2046, 1986.
30. Garrels, R. M., and C. L. Christ, *Solutions, Minerals, and Equilibria,* Harper & Row, New York, 1965.
31. Gelhar, L. W., "Stochastic Subsurface Hydrology from Theory to Applications," *Water Resour. Res.,* vol. 22, no. 9, pp. 135s–145s, 1986.
32. Gillham, R. W., and J. A. Cherry, "Contaminant Transport by Groundwater in Nonindurated Deposits," *Recent Trends in Hydrogeology* (T. N. Narasimhan, editor), Geological Society of America Special Publication 189, Boulder, Colo. pp. 31–62, 1982.
33. Grisak, G. E., and J. F. Pickens, "Solute Transport through Fractured Media, 1. The Effect of Matrix Diffusion," *Water Resour. Res.,* vol. 16, no. 4, pp. 719–730, 1980.
34. Hassett, J. J., J. C. Means, W. L. Bonwart, and S. G. Wood, "Sorption Properties of Sediments and Energy-Related Pollutants," Report EPA-600/3-80-041, U.S. Environmental Protection Agency, Athens, Ga., 1980.
35. Hounslow, A. W., "Adsorption and Movement of Organic Pollutants," *Proceedings of the Third National Symposium of Aquifer Restoration and Ground Water Monitoring,* National Water Well Association, Worthington, Ohio, pp. 334–346, 1983.
36. Huyakorn, P. S., J. W. Mercer, and D. S. Ward, "Finite Element Matrix and Mass Balance Computational Schemes for Transport in Variably Saturated Porous Media," *Water Resour. Res.,* vol. 21, no. 3, pp. 346–358, 1985.
37. International Association of Hydrological Sciences, "Karst Hydrogeology and Karst Environment Protection," *IAHS Publication 176,* 1988.
38. Isherwood, D., *Geoscience Data Base Handbook for Modeling a Nuclear Waste Repository,* vol. 1, Lawrence Livermore Laboratory Report UCRL-52719, Livermore, Calif., 1981.
39. Karickhoff, S. W., D. S. Brown, and T. A. Scott, "Sorption of Hydrophobic Pollutants on Natural Sediments," *Water Res.,* vol. 13, pp. 241–248, 1979.
40. Keely, J. F., M. D. Piwoni, and J. T. Wilson, "Evolving Concepts of Subsurface Contaminant Transport," *J. Water Pollut. Control Fed.,* vol. 58, p. 349, 1986.
41. Kenaga, E. E., and C. A. I. Goring, "Relationship between Water Solubility, Soil Sorption, Octanol-Water Partitioning, and Concentration of Chemicals in Biota," *Aquatic Toxicology,* ASTM STP 707, J. G. Eaton, P. R. Parrish, and A. C. Hendricks, eds., American Society for Testing and Materials, pp. 78–115, 1980.

42. Keys, W. S., and L. M. MacCary, "Application of Borehole Geophysics to Water-Resources Investigations," in *Techniques of Water-Resources Investigations,* U.S. Geological Survey, Book 2, Chap. E1, 1971.

43. Kobayashi, H., and B. E. Rittmann, "Microbial Removal of Hazardous Organic Compounds," *Environmental Science and Technology,* vol. 16, no. 3, pp. 170A–182A, 1982.

44. Kwader, T., "The Use of Geophysical Logs for Determining Formation Water Quality," *Ground Water,* vol. 24, no. 1, pp. 11–15, 1986.

45. Lambert, S. M., P. E. Porter, and H. Schieferstein, "Movement and Sorption of Chemicals Applied to Soil," *Weeds,* vol. 13, pp. 185–190, 1965.

46. Lambert, S. M., P. E. Porter, and H. Schieferstein, "Functional Relationship Between Sorption in Soil and Chemical Structure," *J. Agric. Food Chemistry,* vol. 15, no. 4, pp. 572–576, 1967.

47. Lambert, S. M., P. E. Porter, and H. Schieferstein, "Omega, a Useful Index of Soil Sorption Equilibrium," *J. Agric. Food Chemistry,* vol. 16, no. 2, pp. 340–343, 1968.

48. Lattman, L. H., and R. R. Parizek, "Relationship between Fracture Traces and the Occurrence of Ground Water in Carbonate Rocks," *J. Hydrol.,* vol. 2, pp. 73–91, 1964.

49. Leo, A., C. Hansch, and D. Elkins, "Partition Coefficients and Their Uses," *Chemical Review,* vol. 71, pp. 525–621, 1971.

50. Lyman, W. J., W. F. Reehl, and D. H. Rosenblatt (eds.), *Handbook of Chemical Property Estimation Methods: Environmental Behavior of Organic Compounds,* McGraw-Hill, New York, 1982.

51. Mackay, D. M., D. L. Freyberg, P. V. Roberts, and J. A. Cherry, "A Natural Gradient Experiment on Solute Transport in a Sand Aquifer, 1: Approach and Overview of Plume Movement," *Water Resour. Res.,* vol. 22, no. 13, pp. 2017–2029, 1986.

52. Means, J. C., S. G. Wood, J. J. Hassett, and W. L. Banwart, "Sorption of Polynuclear Aromatic Hydrocarbons by Sediments and Soils," *Environmental Science and Technology,* vol. 14, no. 12, pp. 1524–1528, 1980.

53. Mercer, J. W., and R. M. Cohen, "A Review of Immiscible Fluids in the Subsurface: Properties, Models, Characterization, and Remediation," *J. Contaminant Hydrol.,* vol. 6, no. 2, pp. 107–163, 1990.

54. Miller, C. T., and W. J. Weber, Jr., "Modeling Organic Contaminants Partitioning in Ground-Water Systems," *Ground Water,* vol. 22, no. 5, pp. 584–592, 1984.

55. Moltyaner, G. L., "Stochastic versus Deterministic: A Case Study," *Hydrogeologie,* vol. 2, pp. 183–196, 1986.

56. Nathwani, J. S., and C. R. Phillips, "Absorption-Desorption of Selected Hydrocarbons in Crude Oil on Soils," *Chemosphere,* vol. 6, pp. 157–162, 1977.

57. National Research Council, *Ground Water Models: Scientific and Regulatory Applications,* National Academy Press, Washington, D.C., 1989.

58. Parkhurst, D. L., D. C. Thorstenson, and L. N. Plummer, "PHREEQE—A Computer Program for Geochemical Calculations," U.S. Geological Survey Water Resources Investigation 80–96, 1980.

59. Patrick, G. C., C. J. Ptacek, R. W. Gillham, J. F. Barker, J. A. Cherry, D. Major, C. I. Mayfield, and R. D. Dickhout, "The Behavior of Soluble Petroleum Product Derived Hydrocarbons in Groundwater," Pace Phase I Report No. 85-3, Petroleum Association for Conservation of the Canadian Environment, Ottawa, 1985.

60. Pennington, D., "Retardation Factors in Aquifer Decontamination of Organics," *Proceedings of the Second Annual Symposium and Exposition on Aquifer Restoration and Ground Water Monitoring,* National Water Well Association, Dublin, Ohio, pp. 1–5, 1982.

61. Quinlan, J. F., and R. O. Ewers, "Ground Water Flow in Limestone Terrains: Strategy Rationale and Procedure for Reliable, Efficient Monitoring of Ground Water Quality in Karst Areas," *Proceedings from the Fifth National Symposium and Exposition on Aquifer*

Restoration and Ground Water Monitoring, Columbus, Ohio, National Water Well Association, Dublin, Ohio, pp. 197–234, 1985.

62. Rai, D., J. M. Zachara, A. P. Schwab, R. L. Schmidt, D. C. Girvin, and J. E. Rogers, *Chemical Attenuation Rates, Coefficients, and Constants in Leachate Migration,* vol. 1: *A Critical Review,* EPRI EA-3356, Electric Power Research Institute, Palo Alto, Calif., 1984.
63. Rao, P. S. C., and J. M. Davidson, "Estimation of Pesticide Retention and Transformation Parameters Required in Nonpoint Source Pollution Models," *Environmental Impact of Nonpoint Source Pollutants,* M. R. Overcash and J. M. Davidson, eds., Ann Arbor Science, Ann Arbor, Mich., pp. 23–67, 1980.
64. Reeves, M., V. A. Kelley, and J. F. Pickens, "Regional Double-Porosity Solute Transport in the Culebra Dolomite: An Analysis of Parameter Sensitivity and Importance at the Waste Isolation Pilot Plant (WIPP) Site," Sandia National Laboratories Report SAND87-7105, Albuquerque, N.M., 1987.
65. Reinbold, K. A., J. J. Hassett, J. C. Means, and W. L. Banwart, "Adsorption of Energy-Related Organic Pollutants: A Literature Review," Environmental Research Laboratory, EPA-600/3-79-086, U.S. Environmental Protection Agency, Athens, Ga., 1979.
66. Reynolds, W. D., *Column Studies of Strontium and Cesium Transport through a Granular Geologic Porous Medium,* M.Sc. thesis, University of Waterloo, Ontario, Canada, 1978.
67. Rittmann, B. E., P. L. McCarty, and P. V. Roberts, "Trace Organics Biodegradation in Aquifer Recharge," *Ground Water,* vol. 18, no. 3, pp. 236–243, 1980.
68. Rittmann, B. E., P. L. McCarty, "Model of Steady-State Biofilm Kinetics," *Biotechnology and Bioengineering,* vol. 22, pp. 2343–2357, 1980a.
69. Rittmann, B. E., P. L. McCarty, "Evaluation of Steady-State Biofilm Kinetics," *Biotechnology and Bioengineering,* vol. 22, pp. 2359–2373, 1980*b*.
70. Rogers, R. D., J. C. McFarlane, and A. J. Cross, "Adsorption and Desorption of Benzene in Two Soils and Montmorillonite Clay," *Environmental Science and Technology,* vol. 14, pp. 457–460, 1980.
71. Scalf, M. R., S. F. McNabb, W. I. Dunlap, R. L. Cosby, and I. Frybenber, *Manual of Ground Water Quality Sampling Procedures,* Robert S. Kerr Environmental Research Laboratory, U. S. EPA, Ada, Ok., 1981.
72. Scheidegger, A. E., "General Theory of Dispersion in Porous Media," *J. Geophys. Res.,* vol. 66, pp. 3273–3278, 1961.
73. Schwartz, F. W., "On Radioactive Waste Management: An Analysis of the Parameters Controlling Subsurface Contaminant Transport," *J. Hydrol.,* vol. 27, pp. 51–71, 1975.
74. Schwartz, F. W., "Macroscopic Dispersion in Porous Media: The Controlling Factors," *Water Resour. Res.,* vol. 13, no. 4, pp. 743–752, 1977.
75. Schwille, F., *Dense Chlorinated Solvents in Porous and Fractured Media,* Lewis, Chelsea, Mich., 1988.
76. Srinivasan, P., and J. W. Mercer, "Simulation of Biodegradation and Sorption Processes in Ground Water," *Ground Water,* vol. 26, no. 4, pp. 475–487, 1988.
77. Stewart, M., M. Layton, and T. Lizanec, "Application of Resistivity Surveys to Regional Hydrogeologic Reconnaissance," *Ground Water,* vol. 21, no. 1, pp. 42–48, 1983.
78. Streltsova, T. D., *Well Testing in Heterogenous Formations,* Wiley, New York, 1988.
79. Stumm, W., and J. J. Morgan, *Aquatic Chemistry,* Wiley, New York, 1981.
80. Sudicky, E. A., "A Natural Gradient Experiment on Solute Transport in a Sand Aquifer: Spatial Variability of Hydraulic Conductivity and its Role in the Dispersion Process," *Water Resour. Res.,* vol. 22, no. 13, pp. 2069–2082, 1986.
81. Sundstrom, D. W., and H. E. Klei, *Wastewater Treatment,* Prentice-Hall, Englewood Cliffs, N.J., 1979.
82. Thomas, J. M., G. L. Clark, M. B. Thomson, P. O. Bedient, H. S. Rifai, and C. H. Ward,

"Environmental fate and attenuation of gasoline components in the subsurface," American Petroleum Institute, Washington, D.C., 1988.

83. Thurman, E. M., *Organic Geochemistry of Natural Waters,* Martinus Nijhoff/Dr. W. Junk, Boston, 1985.

84. Travis, C. C., and E. L. Etnier, "A Survey of Sorption Relationships for Reactive Solutes in Soil," *J. Environmental Quality,* vol. 10, no. 1, pp. 8–17, 1981.

85. Truesdell, A. H., and B. F. Jones, "WATEQ: A Computer Program for Calculating Chemical Equilibria of Natural Waters," *U.S. Geological Survey Journal of Research,* vol. 2, pp. 233–248, 1974.

86. U.S. Environmental Protection Agency, Superfund Public Health Evaluation Manual, EPA/540/1-86-060, Washington, D.C., 1986a.

87. U.S. Environmental Protection Agency, "RCRA Ground-water Monitoring Technical Enforcement Guidance Document," OSWER-9950.1, Washington, D.C., 1986b.

88. U.S. Environmental Protection Agency, *Handbook of Ground Water,* EPA/625/6-87/016, Cincinnati, Ohio, 1987.

89. U.S. Environmental Protection Agency, "Guidance for Conducting Remedial Investigations and Feasibility Studies Under CERCLA," Interim Final, EPA/540/6-89/004, Office of Emergency and Remedial Response, Washington, D.C., 1988.

90. U.S. Environmental Protection Agency, Interim Final RCRA Facility Investigation (RFI) Guidance—Volumes I and II, OSWER Directive 9502.00-6D, EPA/530/SW-89-031, Washington, D.C., 1989.

91. Verschueren, K., *Handbook of Environmental Data on Organic Chemicals,* Van Nostrand Reinhold, New York, 1983.

92. Villaume, J. F., "Investigations at Sites Contaminated with Dense, Non-Aqueous Phase Liquids (NAPLs)," *Ground Water Monitoring Review,* pp. 60–75, 1985.

93. Waldrop, W. R., L. W. Gelhar, A. Mantoglou, C. Weltry, and K. R. Rehfeldt, "A Review of Field-Scale Physical Solute Transport Processes in Saturated and Unsaturated Porous Media," EPRI EA-4190, Electric Power Research Institute, Palo Alto, Calif., 1985.

94. Weber, W. J., Jr., *Physicochemical Processes for Water Quality,* Wiley-Interscience, New York, 1972.

95. Weed, S. B., and J. W. Weber, "Pesticide-Organic Matter Interactions," In Guenzi, W. D., ed., *Pesticides in Soil and Water,* Soil Science Society of America, Madison, Wisc., 1974.

96. Westall, J. C., and H. Hohl, "A Comparison of Electrostatic Models for the Oxide/Solution Interface," *Adv. Coll. Inter. Science,* vol. 12, pp. 265–294, 1980.

97. Wilson, L. G., "Monitoring in the Vadose Zone, Part I: Storage Changes," *Ground Water Monitoring Review,* vol. 1, no. 3, p. 32, 1981.

98. Wilson, L. G., "Monitoring in the Vadose Zone, Part II," *Ground Water Monitoring Review,* vol. 2, no. 4, p. 31, 1982.

99. Wilson, L. G., "Monitoring in the Vadose Zone, Part III," *Ground Water Monitoring Review,* vol. 3, no. 4, p. 155, 1983.

100. Wilson, J. T., et al., "Transport and Fate of Selected Organic Pollutants in a Sandy Soil," *J. Environmental Quality,* vol. 10, p. 501–506, 1981.

101. Wilson, J. T., and J. F. McNabb, "Biological Transformation of Organic Pollutants in Groundwater," *EOS, 64*(33), American Geophysical Union, 1983.

102. Wolery, T. J., "Calculations of Chemical Equilibrium between Aqueous Solutions and Minerals: The EQ3/6 Software Package," UCRL-52658, Lawrence Livermore Laboratory, Livermore, Calif., 1979.

103. Wood, P.R., R. F. Lang, and I. L. Payan, "Anaerobic Transformation, Transport, and Removal of Volatile Chlorinated Organics in Ground Water," *Ground Water Quality,* C. H. Ward, W. Giger, and P. L. McCarty, eds., Wiley-Interscience, New York, pp. 493–511, 1985.

CHAPTER 17
STATISTICAL ANALYSIS OF HYDROLOGIC DATA

Robert M. Hirsch
Dennis R. Helsel
Timothy A. Cohn
Edward J. Gilroy
U.S. Geological Survey
Reston, Virgina

17.1 INTRODUCTION

17.1.1 Hydrology and Chance

Many hydrologic processes are subject to chance in the sense that they exhibit substantial variability that cannot be adequately accounted for by physical laws. The difficulty in explaining or predicting hydrologic variables arises for three reasons. The first is the inherent randomness of the driving variables (predominantly precipitation) and the randomness of the hydrologic system (including topographic, aquifer, and soil characteristics). The second is sampling error—the measurements that hydrologists have to work with are only a small sample from a large or potentially infinite population. For example, precipitation data or soil property data are generally collected at only a few points over an entire watershed, or water-quality samples from a river are collected only infrequently, although the water quality is changing constantly. These small samples have limited accuracy in describing the quantities of interest. As the number of samples grows, the accuracy typically improves. The third source of uncertainty is a result of incorrect understanding of the processes involved. This means that, even if sampling errors were eliminated, there would still be errors in estimating or predicting system outputs from system inputs.

Most of the chapters of this handbook are concerned with the proper understanding of the hydrologic processes. It is primarily in this chapter and in Chaps. 18, 19, and 20 that the handbook addresses statistical error. A guideline for statistical analysis is that the hydrologist should first fully explore the potential use of process understanding before proceeding to the use of statistical methods. The extent to which process understanding can be used is generally dictated by the availability of data on the inputs to the system and the characteristics of the system. The extent to which one uses process-based explanation covers a wide range:

1. Mathematical models incorporating directly the fundamental laws of physics, chemistry, and biology
2. Conceptual models, for example, cascades of linear reservoirs
3. Regression-based models that incorporate rather obvious hydrologic truisms, e.g., more precipitation results in more stream flow
4. No model at all, where the hydrologic process is viewed as being entirely random

The methods described in this chapter focus on the development of models of the third and fourth types mentioned above. However, the methods developed here can be used for characterizing the data and its variability, and assessing questions of differences among groups or changes over time for all four kinds of models.

The material presented here emphasizes the statistical characteristics common to many types of hydrologic data that are not generally addressed in introductory statistical texts. These characteristics are

1. Nonnormal, skewed distribution functions, often having a lower bound of zero
2. Outliers—a few values that are much larger or much smaller than the bulk of the data
3. Serial correlation, or lack of independence among observations
4. Distributions dependent on other variables
5. Censoring—data concentration values reported as less than a detection limit, or discharge values less than the discharge at which flood damage occurs
6. Seasonal patterns; statistical characteristics tend to vary with the seasons

Nearly all of the techniques that are addressed in this chapter can be found in the statistics literature. However, many results found in the standard statistical literature are not appropriate for data with the characteristics mentioned above. It is the purpose of this chapter to identify statistical methods that are appropriate for use with typical hydrologic data. Nonetheless, it is the duty of the hydrologist to identify the characteristics of the data being analyzed, and to consider the validity of any proposed statistical technique for data with these characteristics.

17.1.2 Basic Concepts of Probability and Statistics

Where repeated independent experiments are performed, as in flipping coins or casting a die, the relative frequency of particular events often appears to approach a limit even though the events themselves continue to be unpredictable. For example, if one repeatedly tosses a die, and records the outcomes on a bar chart, the shape of the bar chart will assume a specific form. This effect is generally called *statistical regularity.* In laboratory experiments one can develop confidence that statistical regularity is present by repeating experiments under nearly identical circumstances. However, much of the data that arise in hydrology are *observational* rather than experimental: for these data statistical regularity cannot be demonstrated through repetition of an experiment. The hydrologist cannot repeat the experiment of a great flood or drought. Thus, the justification for the use of statistics and probability in hydrology rests, in most cases, on the insight that statistical methods provide into the expected magnitude and variability of future observations.

Definition of Statistics. Statistics deals with methods for drawing inferences about the properties of a *population* based on the properties of a sample from that population. However, statistics goes beyond simply describing the population. It also offers some measure of the uncertainty in knowledge about the population. Statistics quantifies the value of the information, measured in terms of a decrease in uncertainty, that can be obtained from collecting additional data. Knowledge of the magnitude of the uncertainties is essential to identifying those areas where it will be worthwhile to collect additional data.

The fundamental concept in statistics is the *population,* which refers to a collection of objects whose measurable properties are of interest. Populations of interest can be finite and enumerated explicitly. For example, one might be interested in the population of rivers in the world that drain more than 1000 km^2. Populations can also be infinite and abstractly defined, as in the population of all future flows that might occur on a river.

Probability provides a theoretical underpinning for statistical methods. Probability deals with methods for calculating the likelihood of an event (e.g., observing a given sample value) given known population characteristics. Conversely, statistics deals with methods for drawing inferences about the properties of a population, based on a given sample. For example, one might postulate that the concentration of chloride in an aquifer is approximately 10 mg/L. This might be interpreted to mean that the probability is one-half that an individual sample will exceed 10 mg/L. Given this assumption, one can calculate the likelihood that each of three measurements selected randomly (C_1, C_2, C_3) would exceed 10 mg/L. The relevant characteristics of the system—i.e., that the median concentration is 10 mg/L—is assumed to be known exactly, and is used to derive the probability of the compound event: three samples, all of whose concentrations exceed 10 mg/L.

One could recast this example in terms of statistics: suppose one had collected three samples of water from the aquifer and found that the chloride concentrations were (2, 4, 17). What could one say about the probability that a future measurement from the aquifer will be less than 10 mg/L? With what degree of confidence would one be willing to make a statement about likely future concentrations? In this case the object of interest is the characteristic—in this case the median—of the population, and the question relates to *estimation* of the properties of the population. One could also ask whether, on the basis of the information in the sample, the median of the population differed from 10 mg/L, which would be a *hypothesis test.*

17.1.3 Populations versus Samples

Where a population consists of a finite number of elements, one can sometimes measure the characteristics of every element and thus define the exact properties of the population. For example, if stream flow is measured in a continuous fashion (based on gauge heights recorded on a strip chart), then the total stream flow for a year can be computed as the integral of the continuous stream-flow record (computed from the gauge-height record and the station rating curve). The mean flow is this total volume divided by the length of the year. When computed this way, the mean flow is said to be a *statistic* based on a complete census.

Often, however, one is limited to a *sample* consisting of a fraction of the elements of the *population.* To learn about the characteristics of the population, one must understand how the properties of the sample relate to the properties of the population. Various types of samples have been defined:

1. *Random sampling:* The likelihood of selection of each member of the population is equal. This can be achieved by using a random number generator to determine which element is selected.

2. *Stratified random sampling:* The population is divided into groups. Inside each group, random sampling is used (see Ref. 12, pp. 89–149; Ref. 24, pp. 45–57; Ref. 74.
3. *Uniform sampling:* Data are selected according to a strict rule so that the sampled points are uniformly far apart in time or space.
4. *Convenience sampling:* The data are collected at the convenience of the experimenter. For example, some hydrologists do not like to work in the rain or the cold, and thus collect data on rainless summer days.

Ideally one hopes to have a random sample, a stratified random sample, or a uniform sample. Uniform sampling often offers some logistical advantages, and in many cases is more efficient than random sampling because it minimizes the effect of serial dependence on sampling variability.

Biased or *nonrepresentative* sampling methods can lead to biased estimates or conclusions. In particular, convenience sampling schemes often result in the *sampling population*—the population of elements with greater-than-zero likelihood of inclusion in the sample—differing substantially from the *target population.* If one knows the relationship between the sampling population and the target population, one can draw inferences about the latter from samples drawn from the former. Otherwise, inferences drawn from biased samples apply only to the sampling population.

Stratified random sampling is used in cases where the target population can be divided into groups. This is done because the different groups may have different amounts of variability, and it may be more efficient to sample the highly variable groups more intensively than the less variable groups. It is then necessary to take this sampling preference into account when computing an overall estimate.

A common form of *sampling bias* can arise where one uses a stratified approach for selecting samples, but then assumes the sampling was based on an unstratified random method. For example, suppose one measured the chloride concentrations in a river basin. If most of the samples were collected in the urban areas, such a data set might not be *representative* of the river as a whole if there were extensive use of road salt in the urban area. However, if the river basin is divided into two strata, an urban-area stratum and a nonurban stratum, and the size of each stratum is known, and each sample is known to belong to a given stratum, then unbiased estimates are possible. This issue arises when studies addressing a particular question employ data collected to address a different question. For example, historical water-quality data are sometimes not representative of typical conditions, because water-quality studies are often conducted after a water-quality problem has been identified.

Estimators and Estimates. A procedure for describing the properties of a population on the basis of a sample is called an *estimator;* a particular number computed by an estimator is called an *estimate* or *statistic.* The process of ranking a data set and selecting the median is an *estimator;* for example, the number 5 is an *estimate* of the median of a population based on the sample (3, 5, 15). A biased estimator is one that, over repeated uses, results in estimates that, on average, depart from the true population value of the parameter one is attempting to estimate. A sample statistic is an estimate, made on the basis of the available data, of some characteristic parameter of the population.

17.1.4 Categories of Analysis: Single Variable versus Covariation

Many statistical analyses consider the distribution of a single variable in isolation. For example, one may be interested in the distribution of sediment concentration in a stream. In other cases it is worthwhile to consider how a variable changes with respect to a collection of other variables. This change is called *covariation.* For example, one might choose to relate sediment concentrations in a river to the stream flow or to the land-use practices in the drainage basin.

17.1.5 Exploratory versus Confirmatory Statistics

Classical statistics was developed with a specific concept of how science is conducted: a conjecture is hypothesized; then it is either rejected on the basis of statistical tests or one concludes that there is not sufficient evidence to justify rejection. Classical statistics limits itself to providing methods for obtaining unbiased estimates, confidence intervals, and hypothesis tests, assuming that the investigator has a specific model or hypothesis in mind before collecting the data. This subject is called *confirmatory* statistics.

The advent of low-cost computing has substantially changed the way in which data are analyzed. Investigators now commonly use statistical and graphical methods to help explore and understand their data before any hypotheses are formed. Graphical display of data serves two purposes: it reveals characteristics of a distribution or relationships among the variables that may not otherwise be discovered, and it illustrates important concepts when the results are presented to others. The first of these tasks has been called *exploratory data analysis* (EDA), an approach to statistics developed by John Tukey.[39,50,78] EDA procedures provide a "first look" at data. Patterns and theories of how a system behaves are developed by graphing the data. These are inductive procedures—the data are summarized rather than tested. The results provide guidance for the selection of appropriate deductive hypothesis-testing procedures.

However, one must be cautious about statistical results based on such analyses. The validity of the traditional *confirmatory* tests is based on an assumption that the investigator developed the hypothesis *prior* to examining the data.[43,59]

17.2 CHARACTERIZATION OF A SINGLE VARIABLE

17.2.1 Summary Statistics

Examples of *summary statistics* are the sample mean, median, interquartile range, and standard deviation. In many cases, the statistics that are used to describe samples correspond to *parameters* that are used to describe populations, and this can cause confusion. For example, the term *median* is used to refer to both the 50th percentile of a population (which is fixed, although it may be unknown), and the middle observation of a sample, which is a random variable. Table 17.2.1 defines various parameters and statistics that are used to describe populations and samples.

17.2.2 Graphical Display of Data

Histograms. Histograms are plots of bars whose height is the number n_i, or fraction n_i/n, of data falling into one of several intervals of equal width. Iman and Conover[44] suggest that for a sample size of n, the number of intervals k should be the smallest integer such that $2^k \geq n$. Though histograms are popular graphics, their deficiencies are rarely pointed out. Figure 17.2.1 shows two histograms of annual stream flow.

TABLE 17.2.1

Concept	Population value, discrete case	Population value, continuous case	Sample value
Cumulative distribution function (cdf)	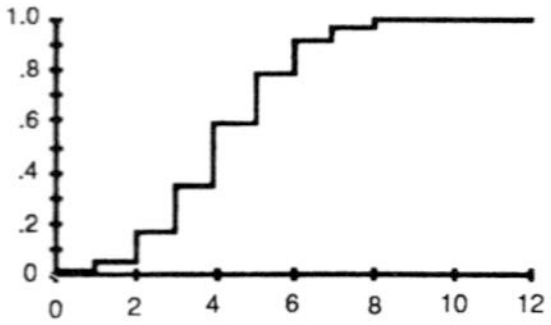 Describes the probability that a random variable is less than or equal to a specified value x	 Describes the probability that a random variable is less than or equal to a specified value x	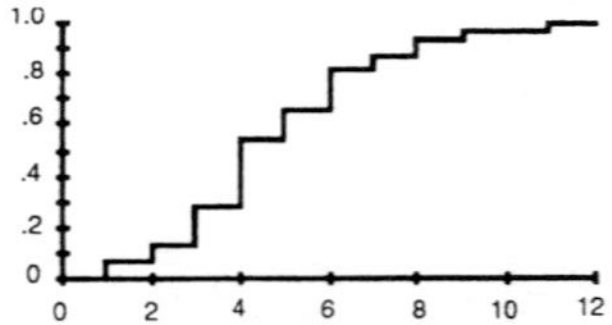 Empirical distribution function (edf): describes the observed frequency of a random variable being less than or equal to a specified value x
Probability mass function (pmf) and probability density function (pdf)	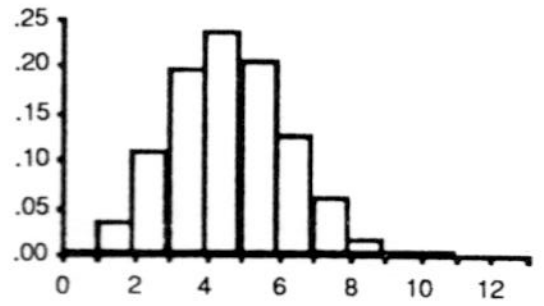 pmf: the probability that X is equal to k	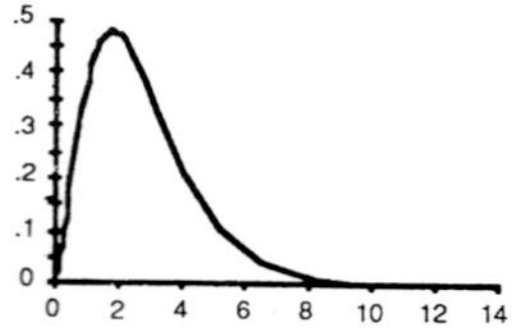 pdf: first derivative of the cumulative distribution function $f(x) \equiv \dfrac{dF(x)}{dx}$	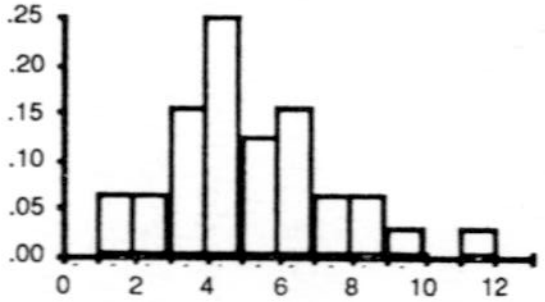 Histogram: observed frequency with which random variable X falls into the assigned ranges
Mean, average, or expected value	$\mu \equiv \sum_{i=1}^{\infty} P(X = x_i)x_i$	$\mu \equiv \int_{-\infty}^{\infty} xf(x)\,dx$	$\overline{X} \equiv \sum_{i=1}^{n} \dfrac{X_i}{n}$

Variance	$\sigma^2 \equiv \sum_{i=1}^{\infty} P(X = x_i)(x_i - \mu)^2$	$\sigma^2 \equiv \int_{-\infty}^{\infty} (x - \mu)^2 f(x)\, dx$	$S^2 \equiv \sum_{i=1}^{n} \frac{(X_i - \overline{X})^2}{n - 1}$
kth central moment	$M_k \equiv \sum_{i=1}^{\infty} P(X = x_i)(x_i - \mu)^k$	$M_k \equiv \int_{-\infty}^{\infty} (x - \mu)^k f(x)\, dx$	$\tilde{M}_k \equiv \sum_{i=1}^{n} \frac{(X_i - \overline{X})^k}{n}$
Standard deviation	$\sigma \equiv \sqrt{\sigma^2}$		$S \equiv \sqrt{S^2}$
Coefficient of variation or relative standard deviation (if $\mu \neq 0$)	$\text{CV} \equiv \frac{\sigma}{\mu}$		$\text{CV} \equiv \frac{S}{\overline{X}}$
Coefficient of skew (a measure of asymmetry)	$\gamma \equiv \frac{M_3}{\sigma^3}$		$G \equiv \frac{\tilde{M}_3}{S^3}$
Quantiles	x_p is any value of X that has the properties that $P(X < x_p) \leq p$ $P[X > x_p] \leq 1 - p$		$\hat{X}_p$ is the pth quantile of EDF
Median (useful for describing central tendency regardless of skewness)	$x_{0.5}$ Any value of X that has the property that $P[X < x_p] \leq 0.5$ $P[X > x_p] \leq 0.5$		$\hat{X}_{0.5}$ The middle observation in a sorted sample, or the average of the two middle observations if the sample size is even.
Upper quartile, lower quartile, and hinges	Upper quartile $\equiv x_{0.75}$ Lower quartile $\equiv x_{0.25}$		Upper hinge $\equiv \hat{X}_{0.75}$ This is an approximation to the sample upper quartile; it is defined as the median of all sample values of $X \geq x_{0.50}$. The lower hinge, $\hat{X}_{0.25}$, is defined analogously.
Interquartile range (useful for describing spread of data regardless of symmetry)	$x_{0.75} - x_{0.25}$ Width of central region of population containing probability of 0.5		$\hat{X}_{0.75} - \hat{X}_{0.25}$ Width of central region of data set encompassing approximately half the data

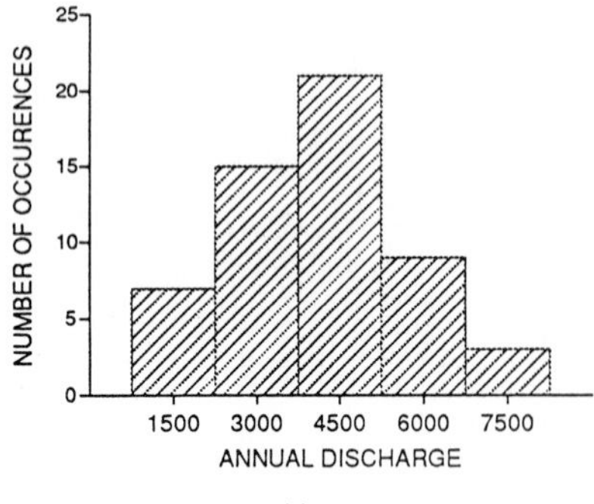

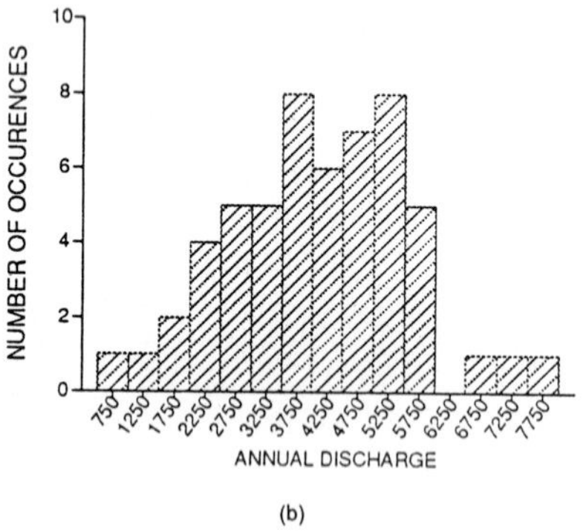

FIGURE 17.2.1 Histogram of annual stream flow for the Licking River at Catawba, Ky., 1929–1983. Parts *a* and *b* are from the same data, but the histograms use different interval widths.

The data are the same but the graphical impression is quite different because the representation in part *b* uses too many intervals and obscures the overall pattern of the data distribution.

If the class intervals of the histogram are large, important information about the distribution of the variable can be lost, especially in the first and last intervals. If the intervals are small, there is an excessive amount of class-to-class variation which is not meaningful information. The arbitrary selection of class intervals is a drawback to their use as a statistical tool.

Quantile Plots. Quantile plots or cumulative distribution functions (cdf's) portray the quantiles (percentiles/100) of the distribution of sample data. They are also called *empirical distribution functions* (edf's). Quantiles of importance such as the median are easily discerned from the plot. The spread and skewness of the data, as well as any bimodal character it possesses, can also be examined. Quantile plots have the advantage that arbitrary intervals are not required, as with histograms.

Figure 17.2.2 is a quantile plot of the stream-flow data from Fig. 17.2.1. Attributes of the data such as the gap between 6000 and 6800 ft^3/s (indicated by the nearly horizontal line segment) are evident. The percent of data in the sample less than a given value can be read from the graph with much greater accuracy than from a histogram.

Variations of quantile plots are used in hydrology for three purposes: (1) to compare two or more data distributions, (2) to compare data to a theoretical distribution such as the normal distribution (a probability plot), and (3) to calculate frequencies of exceedance (a flow-duration curve).

To construct a quantile plot, the data $X_{(i)}$ are ranked from smallest to largest, the smallest being denoted $X_{(1)}$ and the largest, $X_{(n)}$. For each value a sample estimate of the probability of exceedance p_i is computed. This p_i is known as the *plotting position.* There is no single all-purpose formula for the plotting position. The plotting position formulas have the following form:

$$p_i = \frac{i - a}{n + 1 - 2a}$$

where p_i is the cumulative frequency, the probability of a value being less than the ith smallest observation in the data set of n observations. The parameter a is a constant that can take on values in the range 0 to 0.5. The most commonly used formulas are presented in Table 18.3.1.

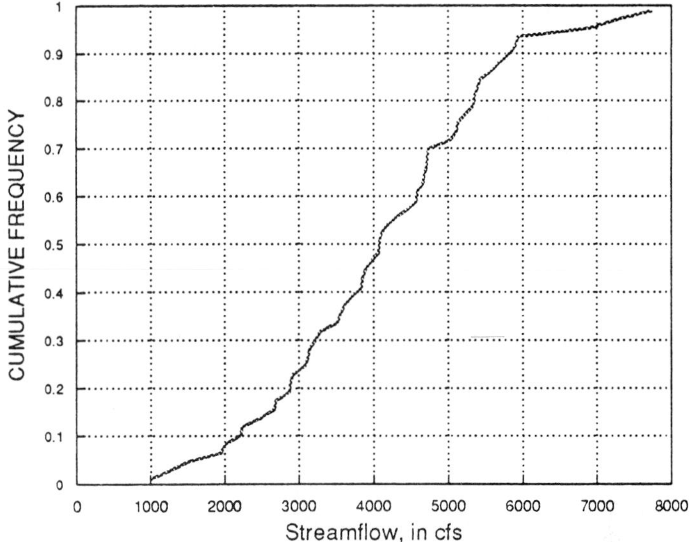

FIGURE 17.2.2 Quantile plot or cumulative distribution function of the annual stream flow, Licking River at Catawba, Ky., 1929–1983.

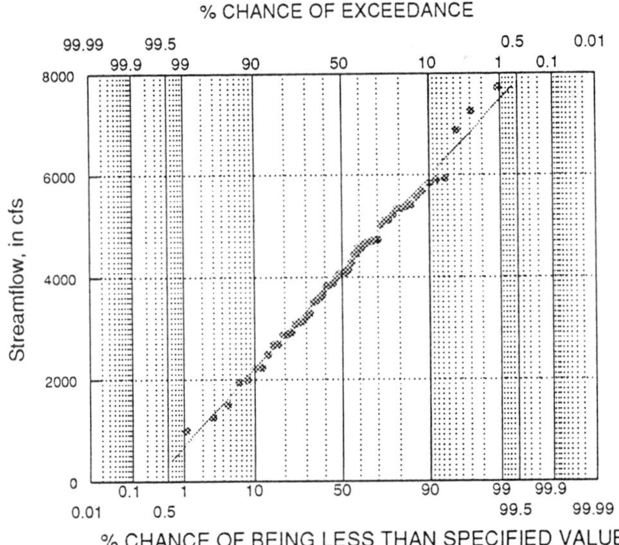

FIGURE 17.2.3 Probability plot of the annual stream flow, Licking River at Catawba, Ky., 1929–1983, on normal probability paper.

A specialized form of the quantile plot is the probability plot using specially designed graph paper such that a given theoretical probability distribution function would plot as a straight line. Figure 17.2.3 is an example of such a curve for the Licking River discharge data. The figure uses paper designed for the normal distribution. The percent chance values are simply the p_i values of Fig. 17.2.2 multiplied by 100. The fact that the sample data plot close to a straight line is an indication that these data are approximately normal. The straightness of such plots can be exploited to create a hypothesis test for the validity of a distribution. See Sec. 18.3.3.

Box Plots. A box plot is a concise graphical display for summarizing the distribution of a data set (Fig. 17.2.4).[9,56,76] Box plots provide visual summaries that provide at a glance an idea of the central tendency of the data, the variability, the symmetry, and the presence of outliers. Box plots are useful for examining the characteristics of a single data set, and, when plotted side by side on the same scale, they are particularly useful for comparing several related data sets (such as concentrations by season or annual floods on different rivers). See Fig. 17.2.5.

The interpretation of a box plot is as follows (see Table 17.2.1 for definitions):

1. The middle line of the box is the sample median
2. The bottom of the box is the lower hinge, which is approximately the lower quartile (25 percent value)
3. The top of the box is the upper hinge, which is approximately the upper quartile (75 percent value)
4. The height of the box is called the H spread; it is approximately equal to the interquartile range
5. The step size is defined as 1.5 times the H spread
6. The line that extends above (below) the box is called the upper (lower) whisker

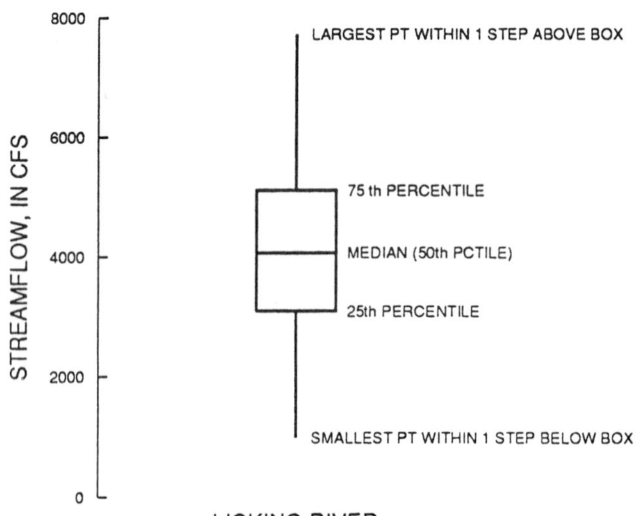

FIGURE 17.2.4 Box plot of the annual stream flow, Licking River at Catawba, Ky., 1929–1983.

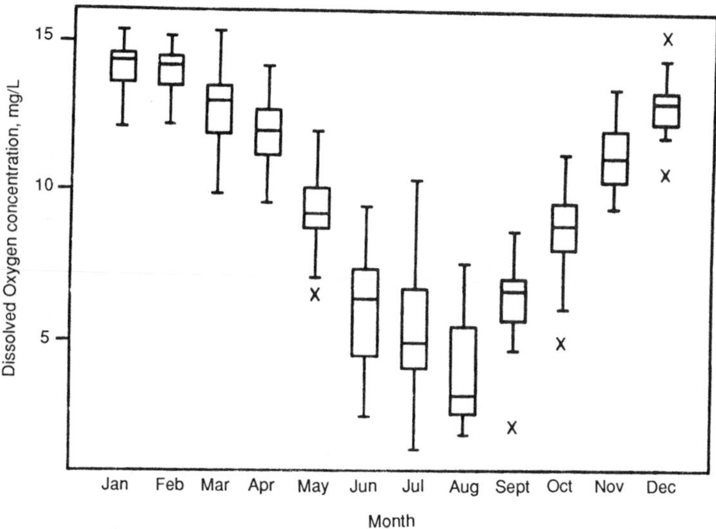

FIGURE 17.2.5 Side-by-side box plots of dissolved oxygen concentrations (in mg/L) measured at Conowingo Dam on the Susquehanna River, 1979–1989, by month of the year.

7. The upper (lower) whisker extends to the largest (smallest) value that is less than or equal to one step away from the box
8. *Outside* values are those observations that are between 1 and 2 steps away from the box; they are each marked with a ×
9. *Far-outside* values are those that are more than two steps away from the box; they are each marked with a ○

Box plots provide a ready visual check on the normality of a data set. For data from a normal distribution, the frequency of outside values is less than 1 in 100 and the frequency of far-outside values is less than 1 in 300,000. If outside or far-outside values occur substantially more frequently than this, it is indicative that the data do not originate from a normal distribution. If the high frequency of outside or far-outside values persists even after the data have been transformed to make the box nearly symmetrical, one should give serious consideration to using the nonparametric procedures described later in this chapter when conducting further analysis of the data.

17.2.3 Hypothesis Testing

The descriptive and exploratory methods described in Secs. 17.2.1 and 17.2.2 illustrate the characteristics of a sample of hydrologic data. The hydrologist's investigations may go further, asking questions such as: Does water in this aquifer typically violate drinking-water standards? Have annual flood peaks increased over time as a result of basin development? Has the concentration of a pollutant declined as a result of the installation of a new treatment plant? Are hydraulic conductivities different in the upper and lower units of this aquifer?

These practical questions, often involving an assumed causative agent such as the building of a treatment plant or the progress of basin development, must be translated into rigorous statistical hypotheses. Table 17.2.2 specifies the procedure for

TABLE 17.2.2 Definitions for Use in Hypothesis Testing

Term	Definition	Comments
Null hypothesis, H_0	The hypothesis being tested. Usually a hypothesis of no change or no difference.	Example: the distributions of hydraulic conductivities are identical in two aquifer units, or concentration is not related to flow.
Alternative hypothesis H_1	Some departure from H_0 that might be expected. The departure that one expects to see.	Example: the distributions of hydraulic conductivities are different, or concentration is related to flow.
Test statistic T	A random variable, defined as some function of the sample being tested.	Test statistics are always dimensionless. The probability distribution of T, given H_0, is a function of sample size only.
One-sided test	A hypothesis test where H_1 is a departure from H_0 in only one direction.	Example: the hydraulic conductivities in aquifer unit A tend to be higher than in aquifer unit B, or concentration increases with flow.
Two-sided test*	A hypothesis test where H_1 is a departure from H_0 in either direction.	Example: the mean hydraulic conductivity in aquifer unit A differs from the mean in aquifer unit B, or concentration changes monotonically with flow.
Decision rule for the test	Reject H_0 if $T < T_L$ or if $T > T_U$	T_L and T_U are the critical values of the test statistic T. Their values are established on the basis of H_0 and the sample size.

Type I error	Rejecting H_0 (using the decision rule) in the case where H_0 is true.	
Significance level α	The probability of a Type I error, assuming H_0.	Prob (reject $H_0 \| H_0$ true) using the decision rule.
Attained significance level p	The smallest α level for which H_0 would be rejected, given the observed value of T.	If $p < \alpha$, reject H_0. The smaller the value of p, the stronger the evidence against H_0.
Type II error	Failure to reject H_0 when H_0 is false	
Power	The probability that H_0 will be rejected, given a particular kind and magnitude of departure from H_0. It is equal to 1-Prob (Type II error).	To the extent that the kind and magnitude of departures from H_0 which are expected can be postulated, one can estimate the power of a given test.
Parametric test	Any hypothesis test that includes the specification of the type of data distribution.	In most cases, the assumption is that the data are normal. Examples: t test, equal-variance tests, tests related to linear regression
Nonparametric or distribution free test	Any hypothesis test that does not include the specification of the type of data distribution.	Tests computed by comparing only relative magnitudes (ranks) of observations. No requirement that the data follow a specified distribution. Examples: Kendall's tau, rank-sum test, Kruskal-Wallis test.

* In this chapter all tests will be assumed to be two-sided.

hypothesis testing. There is a null hypothesis, denoted H_0, which is a nominal or simplest case, and an alternative hypothesis H_1, which is based on the kind of departure from H_0 that the hydrologist expects to see. The process of hypothesis testing is a rigorous way of determining if the available data provide strong enough evidence to lead one to reject H_0 in favor of H_1. It can not be used to prove that H_0 is correct. Hypothesis testing functions by examining whether some characteristics of the data (quantified by the *test statistic*) depart from what one would expect if H_0 were true, to an extent that is very unlikely to arise by chance alone.

A sample will, in almost all cases, depart by some amount from conforming perfectly with the expected situation under H_0. Table 17.2.3 describes the various possible outcomes of a hypothesis test when compared to the true state of nature. Hypothesis tests answer the question: is this lack of conformity large enough to have occurred by chance alone if H_0 were actually true? H_0 is an abstraction, and conducting the test does not imply that one believes that H_0 may be an accurate conception of the true state of nature. Rather, hypothesis tests provide a convenient, theoretically founded, convention for measuring the strength of statistical evidence. They do not, however, provide proof of causality or proof that some population has a particular characteristic (e.g., normality, zero mean, or linear trend with time).

Figure 17.2.6 illustrates the concepts of the probability distribution of T and the significance level α and the attained significance level p. Table 17.2.4 provides some often-used critical values of four types of distributions.

Descriptions of the tests for differences between two or more groups of data are given in Sec. 17.3. The tests for relationships between continuous variables are given in Sec. 17.4. Tests for changes over time (trends) are given in Sec. 19.2. However, in many cases, tests for trends can be carried out with the procedures in Sec. 17.4 by using time as the explanatory variable. Another kind of trend testing arises where there are two distinct periods of time to be compared—typically before and after some modification of the hydrologic system. In these cases trend tests are identical to the tests used to compare any two populations (as discussed in Sec. 17.3.2).

Parametric and Nonparametric Tests. In many situations, both parametric and nonparametric tests are available. If the distributional assumptions of the parametric test were correct, the power (for a given α level) of the parametric test would be slightly higher than the power of the nonparametric alternative. In many cases the differences in power between the parametric and nonparametric tests become smaller as the sample size increases. If the data distributions depart substantially from the assumed distribution, then the nonparametric test can be much more

TABLE 17.2.3 The Possible Outcomes of a Hypothesis Test and Their Probabilities

	States of nature	
Decisions	H_0 is true	H_0 is false
Do not reject H_0	Correct decision. Probability of this outcome is $1 - \alpha$.	Type II error. Probability of this outcome is β.
Reject H_0	Type I error. Probability of this outcome is α (the significance level).	Correct decision. Probability of this outcome is $1 - \beta$ (the power of the test).

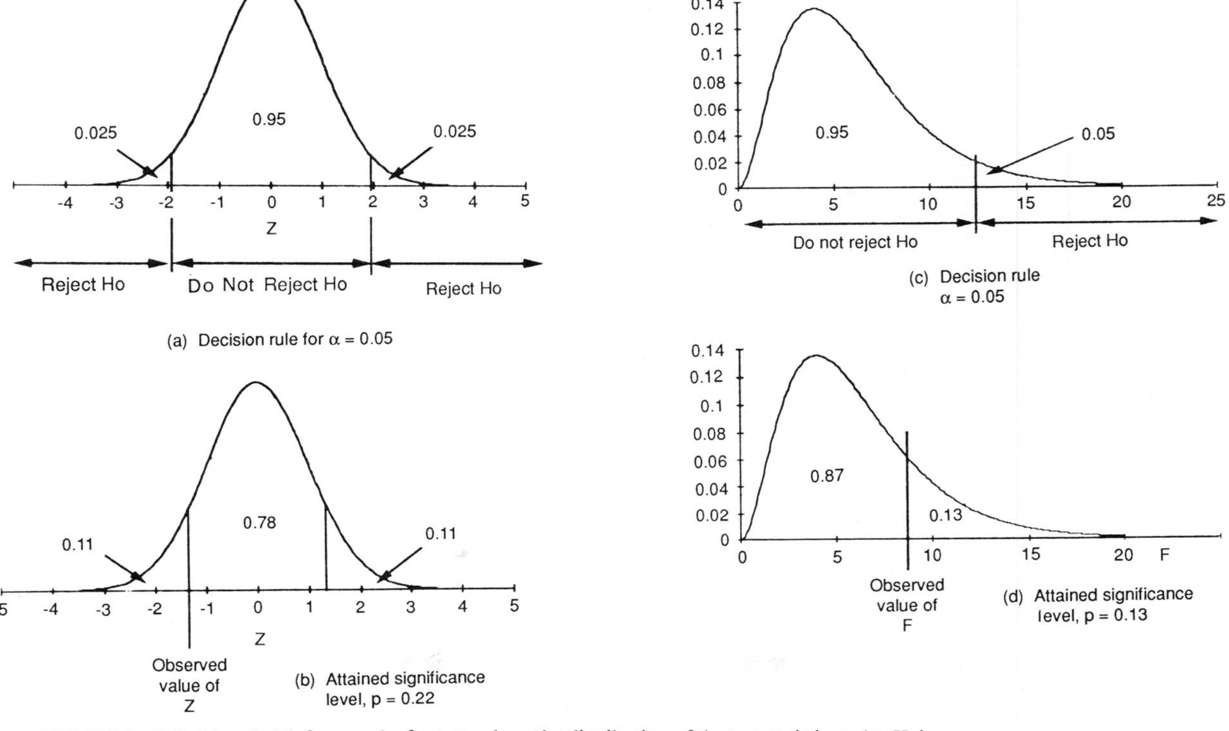

FIGURE 17.2.6 Definition sketch for α and p for tests where the distribution of the test statistic under H_0 is a normal or t distribution (a, b) or a chi-square or F distribution (c, d). Critical values and α are illustrated in a and c, and attained significance levels (p) are illustrated in b and d.

powerful than the parametric test. Unfortunately, procedures for testing the fit of data to probability distributions are often insensitive to departures that are large enough to cause a serious loss of power for the parametric tests. Thus, unless there is a great deal of prior knowledge of the distributional characteristics of the data, it is risky to use a test that depends on a specific distribution. For parametric tests that depend on the normal distribution, the significance levels are reasonably accurate regardless of the actual distribution of the data. However, the power of the tests may be severely diminished if the data are truly nonnormal.

Many hydrologic data sets are positively skewed and have variability proportional to the magnitude of the data. Logarithmic or power transformations can be effective in making these distributions approximately symmetrical and more nearly normal so that parametric tests can be used. Transformations will not necessarily eliminate the "heavy tails" (a high frequency of extreme outliers) of a distribution. Nonparametric tests are particularly well-suited to populations that have heavy tails because the test statistics, and associated estimators, are highly resistant to the effect of these outliers as compared to parametric tests.

Parametric tests are more attractive when dealing with multivariate analyses. Historically, statistical practice has been dominated by parametric procedures because of their general versatility and their computational elegance. With widespread access to digital computers and incorporation of nonparametric methods into statistical software, it is both possible and wise to try both the parametric and nonparametric approaches to any given problem. If the two analyses give similar results, this should add confidence to the conclusions. If they do not, this should lead to more data analysis to understand the causes of the discrepancy and to attempts to refine judgments about the appropriateness of the two procedures for the data.

1. Test of the Probability of Exceeding Some Value. The null hypothesis is H_0: $\text{Prob}(X > X_0) = P_0$ versus H_1:$\text{Prob}(X > X_0) \neq P_0$, where X is a random variable, X_0 is a threshold value of interest, and P_0 is some probability of exceedance $(0 < P_0 < 1)$. This is a nonparametric test.

An example is when X is the annual flood, X_0 is the threshold above which flood damages occur, and P_0 is the design probability (for example, 0.01 for the 100-year design or 0.1 for the 10-year design). Another example is when X is the concentration of some contaminant in a random sample from the river or aquifer, X_0 is a drinking water standard, and P_0 is the frequency of exceedance of the standard that constitutes a health hazard.

The test statistic T is the number of exceedances of X_0 which occur in a sample of size n.

The decision rule is: reject H_0 if $T < T_L$ or $T > T_U$. For small sample sizes (where $n < 20$ or $nP_0 < 5$ or $n(1 - P_0) < 5$), T_L and T_U are computed in exact form from the binomial distribution (see Ref. 34, pp. 77–84). For large sample sizes the normal approximation is used. Note that this does not imply that X is normal; rather, it implies that the binomial distribution converges to the normal distribution as n becomes large.

$$T_L = nP_0 + Z_{\alpha/2}\sqrt{nP_0(1 - P_0)}$$

where $Z_{\alpha/2}$ is the $\alpha/2$ quantile from the standard normal distribution.

$$T_U = nP_0 + Z_{1-\alpha/2}\sqrt{nP_0(1 - P_0)}$$

where $Z_{1-\alpha/2}$ is the $1 - \alpha/2$ quantile of the standard normal distribution. Table 17.2.4 gives the quantiles (z) of the standard normal distribution for various values of α. A value of α that is commonly used in statistical tests is 0.05 (probability of a Type

TABLE 17.2.4 Often-Used Critical Values of Four Distributions

The table entry x_p *is the value for which Prob* $(X < x_p) = p$.

p	0.9	0.95	0.975	0.99	0.995
		t distribution			
df = 1*	3.08	6.31	12.71	31.82	63.66
df = 2	1.89	2.92	4.30	6.96	9.92
df = 3	1.64	2.35	3.18	4.54	5.84
df = 4	1.53	2.13	2.78	3.75	4.60
df = 5	1.48	2.01	2.57	3.36	4.03
df = 7	1.41	1.89	2.36	3.00	3.50
df = 10	1.37	1.81	2.23	2.76	3.17
df = 15	1.34	1.75	2.13	2.60	2.95
df = 20	1.33	1.72	2.09	2.53	2.85
df = 25	1.32	1.71	2.06	2.49	2.79
df = 30	1.31	1.70	2.04	2.46	2.75
		Normal distribution			
	1.28	1.64	1.96	2.33	2.58
		Chi-square distribution†			
df = 1	2.71	3.84	5.02	6.64	7.88
df = 2	4.60	5.99	7.38	9.21	10.60
df = 3	6.35	7.82	9.35	11.34	12.84
df = 4	7.78	9.49	11.14	13.28	14.86
df = 5	9.24	11.07	12.83	15.09	16.75
df = 7	12.02	14.07	16.01	18.48	20.28
df = 10	15.99	18.31	20.48	23.21	25.19
df = 15	22.31	25.00	27.49	30.58	32.80
df = 20	28.41	31.41	34.17	37.57	40.00
df = 30	40.26	43.77	46.98	50.89	53.67
		F distribution‡			
df = 1, *n*					
df = 2, 10	2.92	4.10	5.46	7.56	9.43
df = 2, 20	2.59	3.49	4.46	5.85	6.99
df = 2, 30	2.49	3.32	4.18	5.39	6.35
df = 2. 60	2.39	3.18	3.92	4.98	5.79
df = 2, 120	2.36	3.07	3.80	4.79	5.54
df = 2, ∞	2.30	3.00	3.69	4.60	5.30
df = 3, 10	2.73	3.71	4.83	6.55	8.08
df = 3, 20	2.38	3.10	3.86	4.94	5.82
df = 3, 30	2.28	2.92	3.59	4.51	5.24
df = 3, 60	2.18	2.76	3.34	4.13	4.73
df = 3, 120	2.13	2.68	3.23	3.95	4.50
df = 3, ∞	2.08	2.60	3.12	3.78	4.28
df = 4, 10	2.60	3.48	4.47	5.99	7.34
df = 4, 20	2.25	2.87	3.52	4.43	5.17
df = 4, 30	2.14	2.69	3.25	4.02	4.62
df = 4, 60	2.04	2.52	3.01	3.65	4.14
df = 4, 120	1.99	2.45	2.89	3.48	3.92
df = 4, ∞	1.94	2.37	2.68	3.32	3.72

* df = degrees of freedom.

† For the chi-square distribution a good approximation of the cdf is $x_p = 0.5(z_p + \sqrt{2\text{df} - 1})^2$, where z_p is the p percentage point on the standard normal distribution (read from the table above).

‡ For the *F* distribution with 1 and *n* degrees of freedom, the F value for a given *p* level is equal to the *t* value squared for *n* degrees of freedom and the same *p* level.

I error is 1 in 20). If this is the case, then the values needed from Table 17.2.4 are -1.96 and 1.96 for $Z_{0.025}$ and $Z_{0.975}$, respectively.

2. Test of Hypothesis about Sample Mean. The null hypothesis for this test is H_0: population mean $\mu = \mu_0$. The alternative hypothesis is H_1: $\mu \neq \mu_0$. The sample is of size n with mean value $\bar{X}$. The assumptions are that the observations are normal, independent, and identically distributed. If $n > 30$ the assumption of normality can be relaxed and the significance level of the test will remain accurate, but the power may be reduced as compared to a nonparametric approach such as the median test.

One situation in which the test would be used is where some independent analysis indicates that the population mean is μ_0. The test then becomes a test of that independent analysis. Another example is the case where a model is postulated to predict individual values of some random variable from some other variable (for example runoff as a function of precipitation). X can be defined as observed minus predicted value. Setting $\mu_0 = 0$ makes this a test for bias of the estimator.

The test statistic is

$$T = \frac{\bar{X} - \mu}{s/\sqrt{n}}$$

The decision rule is: reject H_0 if $|T| > t(1 - \alpha/2, n - 1)$, where $t(1 - \alpha/2, n - 1)$ is the $1 - \alpha/2$ quantile of a Student's t distribution with $n - 1$ degrees of freedom. For example if $\alpha = 0.05$, then $1 - \alpha/2 = 0.975$, and if n is large (say $n > 30$), then the critical t value is approximately 2 (see Table 17.2.4).

3. Test for the Population Median. The statement of the null hypothesis H_0 is that the population median $= M$, versus H_1: population median $\neq M$. The assumptions are that the data consist of n independent and identically distributed samples drawn from a single distribution. The test is more powerful than the test for the population mean in cases where the distribution departs substantially from the normal. The test statistic T is the number of observations that are less than or equal to M. The decision rule is: reject H_0 if $|T - \frac{n}{2}| \geq z_{1-\alpha/2}\frac{\sqrt{n}}{2}$. This applies only for $n > 20$ (see Ref. 51). For example, if $\alpha = 0.05$, then $z_{1-\alpha/2} = 1.96$. If $n = 100$, then the null hypothesis would be rejected for $T \leq 40$ or $T \geq 60$.

17.3 COMPARISON OF GROUPS

17.3.1 Paired Comparisons

Paired comparisons occur where there is a logical pairing of observations between two groups. The question is whether there is a tendency for one group to have higher values than the other. Differences between the groups are seen clearly when tests are performed on the differences between data pairs.

For paired observations (x_i, y_i), $i = 1, 2, \ldots, n$, the null hypothesis that x_i and y_i are from the same population implies that the mean or median of their differences $D_i = x_i - y_i$ is zero. Examples of situations in hydrology where paired tests are appropriate are

1. Nitrate concentrations are measured in a set of shallow wells (x_i) at the end of a growing season when a standard fertilizer application rate was employed. The same wells are resampled (y_i) after another growing season when reduced applica-

tions of fertilizer were made. The test is used to see if the different levels of fertilizer application affect groundwater nitrate concentrations.

2. Two methods (x_i and y_i) of measuring stream flow are compared. Both methods are applied simultaneously at various times at a given station. D_i is the difference in the two measurements at a given time. It may be that the errors are multiplicative rather than additive (i.e., if discharge is low, the differences are small, and if discharge is high, the differences are large). In such cases, a more powerful test may be conducted by using the logarithms of the discharge data rather than the data themselves.

The two tests for paired comparisons listed in Table 17.3.1 are both computed from the variable D_i. The choice of test depends on the assumptions that can safely be made about the data.

The Signed-Rank Test. The null hypothesis H_0 for the signed-rank test is that the x's and y's come from the same population. The alternative H_1 is that they differ in level (their means or medians are different). There is no requirement that the x and y distributions be normal or even that they be symmetric. The computational steps for the signed-rank test are as follows:

1. Eliminate all $D_i = 0$ from the data set (the cases where $x_i = y_i$).
2. Define n as the number of nonzero D_i values.
3. Rank the D_i according to their absolute values. Largest $|D_i|$ has rank n, smallest $|D_i|$ has rank 1.
4. Let $R^+ =$ the sum of the ranks corresponding to $D_i > 0$.
5. Let $R^- =$ the sum of the ranks corresponding to $D_i < 0$.
6. Compute the test statistic W^+, the minimum of $\{R^+, R^-\}$.
7. For the sample size n, compute the theoretical mean and standard deviation of W^+ under H_0. They are, respectively,

$$\mu_{W^+} = \frac{n(n+1)}{4} \quad \text{and} \quad \sigma_{W^+} = \sqrt{\frac{n(n+1)(2n+1)}{24}}$$

TABLE 17.3.1 Tests for Differences between Sets of Paired Samples

Test	H_0	Comments
Paired t test	The differences $(x_i - y_i)$ are normally distributed with zero mean.	Most powerful test if differences are normal. If the data are skewed or have outliers, it should not be used.
Signed-rank test	The differences $(x_i - y_i)$ have a symmetric distribution with zero median.	More powerful than the t test if the data are nonnormal.

8. Compute the standardized form of the test statistic Z_{sr^+}

$$Z_{sr^+} = \begin{cases} \dfrac{W^+ - \frac{1}{2} - \mu_{W^+}}{\sigma_{W^+}} & \text{if } W^+ > \mu_{W^+} \\ 0 & \text{if } W^+ = \mu_{W^+} \\ \dfrac{W^+ + \frac{1}{2} - \mu_{W^+}}{\sigma_{W^+}} & \text{if } W^+ < \mu_{W^+} \end{cases}$$

9. Reject H_0 if $|Z_{sr^+}| > Z_{1-\alpha/2}$, where $Z_{1-\alpha/2}$ is the $1 - \alpha/2$ point on the standard normal probability distribution (see Table 17.2.4). For example, if $\alpha = 0.05$, then H_0 would be rejected if $|Z_{sr^+}| > 1.96$. For $n < 15$, this decision rule is not accurate and an exact test should be used. Tables for the exact test can be found in Ref. 63, pp. 199–202, and Ref. 34, pp. 142–145.

Paired t test. The paired t test is the most commonly used test for evaluating matched pairs of data. However, it assumes that the differences D_i follow a normal distribution. The computation of the paired t test is as follows:

1. Compute the $D_i = x_i - y_i$ from all n pairs of data.
2. Compute the mean difference $\overline{D}$, the average of the D_i.
3. Compute the sample variance

$$S^2 = \sum_{i=1}^{n} \frac{(D_i - \overline{D})^2}{n - 1}$$

4. Compute the standardized test statistic

$$t_p = \frac{\overline{D}}{\sqrt{S^2/n}}$$

5. Reject H_0 if $|t_p| > t_{1-\alpha/2,n-1}$

Graphical approaches can be used to illustrate matched-pair test results in a manner similar to those already given in Sec. 17.2.2 for illustrating a single data set, as the differences between matched pairs constitute a single data set. A probability plot of the paired differences, for example, shows whether or not the differences follow a normal distribution. The best method for directly illustrating the results of the signed-rank or paired t tests is a box plot of the differences. The number of data above and below zero and the nearness of the mean or median difference to zero are clearly displayed, as is the degree of symmetry of the D_i.

Though a box plot is an effective and concise way to illustrate the characteristics of the differences, it will not show the characteristics of the original data. This can be better done with a *scatter plot* [a graph of (x, y) data pairs]. Similarity between the two groups of data is illustrated by their closeness to an $x = y$ line. If x is generally greater than y, most of the data will fall below the line. When y exceeds x, the data will lie largely above the $x = y$ line. This relationship can be made clearer for large data sets by plotting a smoothed curve representing the x, y relationship using the LOWESS technique introduced in Sec. 17.4.9. Data points (or their smooth fitted curve) falling generally parallel to the $x = y$ line on the scatter plot indicate an additive difference between the (x, y) data pairs. The line $x = y + d$ can be plotted on the figure to illustrate the magnitude of the difference between x and y, where d is the mean difference between x and y.

Estimators of Differences between Paired Groups. After testing for differences between matched pairs, a measure of the magnitude of that difference is usually desirable. The mean difference $\bar{D}$ is appropriate whenever the paired t test is valid. When outliers or nonnormality are suspected, a more robust estimator is necessary, such as a Hodges-Lehmann estimator.[40] This estimator is computed by taking the median of the averages of all possible pairs of differences (i.e., if there are $n(x, y)$ pairs, then the estimate is the median of $n(n-1)/2$ averages, $(x_i + y_j)/2$, $i \neq j$).

17.3.2 Comparisons between Two Independent Data Groups

Tests to compare two independent groups of data are performed to determine whether one group tends to contain larger values than the other. The data are independent in the sense that there are no pairings of data between observation 1 of group A and observation 1 of group B, etc. An example of such a situation is a study in which chloride concentrations at base flow during winter are collected in a random sample of stream reaches in rural basins and a random sample of reaches in basins with substantial urban land use. Even if the number of samples in each group were equal, there is no natural pairing of the data from the two groups and the paired sample methods are not appropriate. Two methods are presented in Table 17.3.2, one parametric and one nonparametric.

The Rank-Sum Test. The rank-sum test, the Wilcoxon rank-sum test, the Mann-Whitney test, and the Wilcoxon-Mann-Whitney rank-sum test are all the same test. The null hypothesis H_0 is that the two groups are identically distributed. The alternative H_1 is that one group tends to produce larger observations than the second group. No assumptions are made about how the data are distributed. The test's results are not affected by monotonic transformations of the data (such as the log transformation). This is an advantage over the t test. It may be necessary to transform a skewed data set in order to apply the t test properly. The choice of transformation is an additional step of data manipulation. The use of such manipulations can lead critics

TABLE 17.3.2 Assumptions and Estimators Associated with Tests between Two Independent Groups

Test	H_0	Comments
Two-sample t test	The means for the two groups are equal	Assumes that the two populations are both normal with equal variances. Substantial departures from normality seriously decrease the power of the test.
Rank-sum test	The medians of the two groups are equal	Assumes that the two populations are identically distributed.

of the result to argue that the arbitrary manipulation was used to obtain the desired results. The rank-sum test avoids the need for such transformations. The computational procedure for the rank-sum test is:

1. Assign ranks from 1 (smallest) to N (largest) to all of the data. $N = n + m$, where n is the sample size of the smaller sample, and m is the sample size for the larger of the two samples. In the case of ties (equal data values) use the average of the ranks for the tied values.
2. Compute the test statistic W as the sum of the ranks of the n observations in the *smaller* group.
3. For the sample size, compute the theoretical mean and standard deviation of W under H_0. They are, respectively,

$$\mu = \frac{n(N+1)}{2} \qquad \sigma = \sqrt{\frac{nm(N+1)}{12}}$$

The standardized form of the test statistic Z_{rs} is computed as

$$Z_{rs} = \begin{cases} \dfrac{W - ½ - \mu}{\sigma} & \text{if } W > \mu \\ 0 & \text{if } W = \mu \\ \dfrac{W + ½ - \mu}{\sigma} & \text{if } W < \mu \end{cases}$$

4. Reject H_0 if $|Z_{rs}| > Z_{1-\alpha/2}$, where $Z_{1-\alpha/2}$ is the $1 - \alpha/2$ point on the standard normal probability distribution (see Table 17.2.4). For example, if $\alpha = 0.05$, then H_0 would be rejected if $|Z_{rs}| > 1.96$. If the sample size of one or both groups is less than 10, this approximate decision rule is inaccurate and an exact test should be used. Tables for the exact test can be found in Refs. 34 and 51. When more than a few ties occur, a correction to σ must be used. The formula to use when there are tied ranks is

$$\sigma = \sqrt{\frac{nm}{N(N-1)} \sum_{k=1}^{N} R_k^2 - \frac{nm(N+1)^2}{4(N-1)}}$$

where R_k is the rank of the kth value.

Two-Sample t Test. The t test is the most widely used method for comparing two independent groups of data. H_0 is that the means of both groups are equal. H_1 is that the means are unequal. Two issues make the t test less applicable for general use than the rank-sum test. These are (1) it lacks power when applied to nonnormal data and (2) it is powerful for the case where the two groups differ by an additive constant but is not powerful for the case where the two cases differ by a multiplicative factor. Generally, the rank-sum test should be used instead of the t test when the data are substantially skewed or nonnormal. The computational procedure for the t test is as follows:

1. Compute the sample means and sample standard deviations of the data in both groups $\bar{x}$, s_x and $\bar{y}$, s_y, where x denotes the first group (sample size n) and y denotes the second group (sample size m).
2. Compute the degrees of freedom for the test. If the null hypothesis that $s_x = s_y$ cannot be rejected, then df $= n + m - 2$. If the standard deviations are signifi-

cantly different, then compute the degrees of freedom from the following formula (rounding to the nearest integer)

$$\mathrm{df} = \frac{\left(\frac{s^2_x}{n} + \frac{s^2_y}{m}\right)^2}{\frac{(s^2_x/n)^2}{n-1} + \frac{(s^2_y/m)^2}{m-1}}$$

3. Compute the test statistic t. If the variances are equal, then

$$t = \frac{\bar{x} - \bar{y}}{\sqrt{\frac{s^2_x}{n} + \frac{s^2_y}{m}}}$$

4. Reject H_0 if $|t| > t_{1-\alpha/2,\,\mathrm{df}}$.

Graphical Presentations of Test Results. Side-by-side box plots of the variables x and y are well-suited to both describing the results of these hypothesis tests and visually allowing a judgment of whether the data fit the assumptions of the test being employed.

Estimates of Differences between Two Groups. If a hypothesis test indicates that two groups of data are different, the next step is to determine by how much the two groups differ. A common approach, related to the two-sample t test, is to compute the difference between the two group means $(\bar{x} - \bar{y})$. A more robust alternative, related to the rank-sum test, is one of a class of nonparametric estimators known as Hodges-Lehmann estimators $\hat{\Delta}$ (Refs. 40; 42, p. 75–77). It is the median of the nm possible pairwise differences between the x and y values.

17.3.3 Comparisons among Several Data Groups

Analysis of variance, an extension of the t test, is a parametric test which makes comparisons among more than two groups of data. A nonparametric method more appropriate for the frequent situations where data do not follow a normal distribution is the Kruskal-Wallis test.

These tests are applicable with continuous data, such as concentrations or water levels, when the distribution of the data in one group may differ from one or more of the other groups. Examples of groups include: season of the year, aquifer type, land-use type, storm type, or hydrologic condition (e.g., rising limb, falling limb, or base flow).

Tests for Differences Due to One Factor. The data consists of a set of k groups, with each data point belonging in one of the k groups. Within each group there are n_j observations. Observation y_{ij} is the ith of n_j observations in group j, so that $i = 1, \ldots, n_j$ for the jth of k groups $j = 1, \ldots, k$. The total number of observations N is thus

$$N = \sum_{j=1}^{k} n_j$$

which simplifies to $N = kn$ when the sample size is equal to n for all k groups.

Analysis of Variance. Analysis of variance (ANOVA) determines whether all groups have identical mean values by comparing two estimates of the overall variance. If the null hypothesis H_0 is true, group means will differ only slightly from the overall mean. The total variance in the data will therefore be very similar to the variance within a group around that group mean. If group means are dissimilar, some of them will be sufficiently different from the overall mean that the variance within groups no longer equals the total variance.

The within-group variance is estimated by the mean square error (MSE). The treatment mean square (MST) estimates the sum of the within-group and among-group variances. Their computation is shown in Table 17.3.3. The treatment and error mean squares are computed as their sum of squares divided by their degrees of freedom (df). The computations and results of an ANOVA are usually organized into an ANOVA table.

The formulas for computing MST and MSE are

$$\text{MST} = \frac{\sum_{j=1}^{k} n_j(\bar{y}_j - \bar{y})^2}{k-1} = \frac{\text{SST}}{k-1}$$

$$\text{MSE} = \frac{\sum_{j=1}^{k}\sum_{i=1}^{n_j} (y_{ij} - \bar{y}_j)^2}{N-k} = \frac{\text{SSE}}{N-k}$$

where $\bar{y}_j$ is the jth group mean, n_j is the number of observations in group j, and $\bar{y}$ is the overall mean.

The test statistic is $f = \text{MST/MSE}$. The decision rule for the test is reject H_0 if $f \geq F(1-\alpha, k-1, N-k)$, the $(1-\alpha)$ quantile of the F distribution with $k-1$ and $N-K$ degrees of freedom (see Table 17.2.4). For example, if $\alpha = 0.05$, with 5 groups and a total of 100 observations the critical value would be $F(0.95, 4, 95)$, which is 2.47. (In this example, 4 is the number of numerator degrees of freedom and 95 is the denominator degrees of freedom.) Thus H_0 would be rejected if $f \geq 2.47$.

Violation of either the normality or constant variance assumptions (all groups have the same variance even though their means may be different) results in a loss of ability to identify differences between means, which is a loss of power (for a given α level). When ANOVA is conducted on nonnormal data, a conclusion of no difference between groups may result from the lack of power of the ANOVA, and not from a true equivalence of means. The Kruskal-Wallis test (see below) is close in power to ANOVA when data follow a normal distribution, and may be far more powerful for nonnormal data. In addition, the transformation of data to approximate normality prior to performing ANOVA is more problematic than for the t test, as there are more than two groups to transform. It may be difficult to find a single transformation which, when applied to all groups of data, will result in each becoming normal with equal variance.

TABLE 17.3.3 Table for a One-Way ANOVA

Source	df	Sum of squares (SS)	Mean square (MS)	F	p value
Treatment	$k-1$	SST	MST	MST/MSE	p
Error	$N-k$	SSE	MSE		
Total	$N-1$	Total SS			

The Kruskal-Wallis Test. The null hypothesis of the Kruskal-Wallis test is that all k groups have identical distributions. The alternative hypothesis is that at least one group has a different distribution. As with the other nonparametric tests, the results are not affected by a monotonic transformation of the data. The test procedure is as follows:

1. Assign ranks to all of the data (1 is lowest value, N is highest value). Use average ranks in the event of ties.
2. For each of the k groups of data, compute the average of the rank ($\bar{R}_j$) of the n_j data values in that group.
3. Compute the test statistic KW:

$$\mathrm{KW} = \frac{12}{N(N+1)} \sum_{j=1}^{k} n_j \left(\bar{R}_j - \frac{N+1}{2} \right)^2$$

4. Reject H_0 if $\mathrm{KW} \geq \chi^2(1-\alpha), k-1$, the $1-\alpha$ quantile of the chi-square distribution, with $k-1$ degrees of freedom. For example, if $\alpha = 0.05$ and there are 6 groups ($k = 6$), then reject H_0 if $\mathrm{KW} \geq 11.07$. See Table 17.2.4 for the chi-square distribution. For small sample sizes (3 groups with sample sizes of 5 or less per group, or with 4 or more groups of size 4 or less per group), see Ref. 51 for the exact critical values for KW.

Tests for the Effects of More Than One Factor. More than one factor may simultaneously be influencing the magnitudes of observations. As the effect of one factor may depend on another, sequential one-factor tests will not adequately measure the respective influences of each factor. Even when only one factor is actually influencing the data and a one-way ANOVA for that factor soundly rejects H_0, a second one-way test for a related factor may erroneously reject H_0 simply because of the association between the two factors. The test for the second factor should remove the effect of the first before establishing that the second has any influence. By evaluating all factors simultaneously, the influence of one factor can be measured while compensating for the others. This is the objective of a multifactor analysis of variance and of its nonparametric analog.

The effects of two or more factors may be simultaneously evaluated using a factorial ANOVA design. A factorial ANOVA occurs when none of the factors is a subset of the others. If subsetted factors do occur, the design includes "nested" factors and the equations for computing the F test statistics will differ from those presented here. More than two factors can also be simultaneously tested, but the equations are beyond the scope of this handbook. See Ref. 62 for more detail on higher-way and nested analysis of variance. The nonparametric equivalents to these tests are generally conducted by substituting the ranks for the data values and then performing the ANOVA procedure on the ranks (see Ref. 16). The parametric and nonparametric versions of multifactor ANOVA are beyond the scope of this handbook. See Ref. 34 for details.

Multiple Comparison Tests. In most cases an analyst is interested not only in whether the groups differ, but which ones differ from others. This information is not supplied by the previous tests, but by methods called *multiple comparison tests.* Multiple comparison tests compare all possible pairs of treatment group means (or medians), and are performed only after the null hypothesis of "all means (or medians) identical" has been rejected. Of interest is the pattern of group means (or medians), for example:

$$\text{Group A} \approx \text{group B} \ll \text{group C}$$

Multiple comparison tests are not efficient methods for contrasting specific sets of groups known to be of interest before an ANOVA or Kruskal-Wallis test is done, such as a treatment versus a control. Other tests are available for making specific contrasts. Instead, multiple comparison tests compare all possible combinations of treatment group centers, ranking the centers in order and indicating which are similar or different from others. Stoline[70] reviews the many types of parametric multiple comparison tests. Campbell and Skillings[8] discuss nonparametric multiple comparisons.

Parametric Multiple Comparisons. ANOVA should always be performed first as the appropriate test for determining whether any differences occur between group means. If no differences occur, stop there. If differences occur, then multiple comparison tests can follow to determine which group means differ from the others. These tests require the same assumptions as does ANOVA—data within each treatment group must be normally distributed, and with equal variance. Violations of these assumptions will generally result in a loss of power to detect differences between means which are actually present. Multiple comparison tests calculate a *least significant range,* the distance between any two means which must be exceeded in order for the two groups to be considered significantly different at a significance level α. The least significant range is a function of the estimated group variance and sample size. Some tests use the stated α level for each pairwise comparison (α_p = pairwise error rate). When there are multiple comparisons each having a pairwise error rate of α_p, the overall probability of declaring at least one false difference (the overall error rate α_o) is much greater than α_p. This overall error rate is the error rate for the "pattern" of group means, and is more often of interest than a pairwise error rate. For example, for six group means, there are $(6 \cdot 5)/2 = 15$ pairwise comparisons. If $\alpha_p = 0.05$ is used for each test, the overall Type I error rate is approximately $\alpha_o = 1 - (1 - \alpha_p)^{15} = 0.54$. That is, if all six groups actually had the same population mean, the probability of rejecting at least one H_0 (that two means are equal) is 0.54, rather than the desired level of 0.05.

Pairwise rates are sometimes incorrectly presented as if they were overall rates. This gives a false sense of security in the results. When the primary interest is in the overall pattern and its accuracy, methods which set the error rate equal to the overall α, such as Tukey's test (see Ref. 60, pp. 73–78), should be performed.

17.4 COVARIATION: CONTINUOUS VARIABLES

This section addresses the analysis of relationships of two (or more) continuous variables. Throughout this discussion the variable of interest (the response variable) is referred to as the *dependent variable* and is denoted by y. Sections 17.4.1 to 17.4.4 deal with only one *explanatory variable* (denoted x). Many of the techniques described in this section can be extended for use with multiple explanatory variables. For extensive detail in dealing with multiple explanatory variables, refer to standard texts on regression.[21,61]

The explanatory variable can be a random variable, typically a measure of something that is a driving force to the system being studied. For example: x may be rainfall and y stream flow or x may be stream flow and y a solute concentration. In other cases, x is nonrandom; for example, it may be a measure of time or some geographic coordinate (e.g., distance downstream, or, in a multiple regression, distances north and east of some datum). In these cases, the analysis of the relationship between x and y becomes an analysis of temporal or spatial trend.

The section is organized according to approach. The approaches are nonparametric descriptions of a linear relationship (Sec. 17.4.1), regression-based description and estimation assuming a linear relationship (Secs. 17.4.2 to 17.4.8), curve fitting and estimation with no assumption of linearity (Sec. 17.4.9), and finally techniques for estimating multiple missing values in a hydrologic record (Sec. 17.4.10). Table 17.4.1 outlines the topics within the section according to the type of question they attempt to answer.

17.4.1 Nonparametric Approaches to Correlation

Correlation is a mathematical description of the strength of the relationship between two variables. All of the measures of correlation ρ have the characteristic of being dimensionless and scaled to lie in the range $-1 \le \rho \le 1$. In the case where $\rho = 0$, the data are said to be uncorrelated. Typically, when correlation is evaluated by a hy-

TABLE 17.4.1 List of Topics in Sec. 17.4

Question	Method	Section
Are the two variables related in a monotonic fashion, and how strong is the relationship?	Kendall's tau	17.4.1
Are the two variables related in a linear fashion, and how strong is the relationship?	Pearson's correlation	17.4.2
Describe the variation in the dependent variable using a linear model.	Linear regression	17.4.2, 5, 6, 7, 8
Describe the variation in the dependent variable with no restriction that the relationship be linear.	Locally weighted scatter plot smoothing (LOWESS)	17.4.9
Do two or more linear models differ from each other?	Analysis of covariance	17.4.5
Estimate a single value of the dependent variable using a linear model (prediction).	Regression analysis	17.4.2, 3, 5
Estimate a single value of the dependent variable using a linear model in which the dependent variable is transformed.	Transformation bias correction	17.4.4
Estimate a single value of the dependent variable with no requirement that the model be linear.	LOWESS	17.4.9
Graphically display the relationship of the dependent and independent variables.	LOWESS	17.4.9
Compute residuals from the relationship to remove some unwanted source of variation before proceeding to some further analysis.	Regression or LOWESS	17.4.2, 5, 8, 9
Estimate multiple values of the dependent variable to fill in a set of missing values.	Maintenance of variance extension (MOVE)	17.4.10

pothesis test, the null hypothesis H_0 is $\rho = 0$ and the alternative H_1 is $\rho \neq 0$. Graphical methods (a scatter plot) are the best approach to identifying the presence of the type of dependence that exists: nonmonotonic versus monotonic and linear versus nonlinear. *Monotonic* means that there are no reversals in the slope of the relationship. *Linear* is a special case of monotonic for which the slope of the relationship is constant over the range of x.

Correlation coefficients provide a means for quantifying and testing the strength of monotonic relationships between two variables. However, correlation does not provide evidence for causal relationship between two variables. They may be correlated because one causes the other; for example, precipitation produces runoff. They may also be correlated because they both share the same cause; for example, two solutes measured at a variety of times or a variety of locations may both be influenced by variations in the source of the water.

Kendall's Correlation Coefficient. An effective and general measure of correlation between two variables is Kendall's correlation coefficient, generally known as Kendall's τ (tau).[46,47] It is a rank-based procedure and is therefore resistant to the effect of extreme values and to deviations from a linear relationship. Thus, it is well-suited to use with dependent variables for which the variation around the general relationship exhibits a high degree of skewness or kurtosis. Examples include dependent variables such as river discharge, concentration or transport rates of sediment, and concentrations or transport rates of substances that are transported in association with suspended sediment.

The following is the method of computing Kendall's τ and conducting Kendall's test for correlation (the null hypothesis H_0 is that the distribution of y does not change as a function of x):

1. The n data pairs $(x_1, y_1), (x_2, y_2), \ldots, (x_n, y_n)$ are indexed according to the magnitude of the x value, such that $x_1 \leq x_2 \leq \ldots \leq x_n$ and y_i is the dependent variable value that corresponds to x_i.
2. Examine all $n(n-1)/2$ ordered pairs of y_i values. Let P be the number of cases where $y_i > y_j$ $(i > j)$, and let M be the number of cases where $y_i < y_j (i > j)$.
3. Define the test statistics $S = P - M$.
4. For $n > 10$, the test is conducted using a normal approximation. The standardized test statistic Z is computed,

$$Z = \begin{cases} \dfrac{S-1}{\sqrt{\text{Var}(S)}} & S > 0 \\ 0 & S = 0 \\ \dfrac{S+1}{\sqrt{\text{Var}(S)}} & S < 0 \end{cases}$$

and $\text{Var}(S) = n(n-1)(2n+5)/18$.

5. The null hypothesis is rejected at significance level α if $|Z| > Z_{(1-\alpha/2)}$, where $Z_{(1-\alpha/2)}$ is the value of the standard normal distribution with a probability of exceedance of $\alpha/2$. For example, if $\alpha = 0.05$, then the null hypothesis would be rejected for $|Z| > 1.96$. In cases where some of the x and/or y values are tied, this formula for Var (S) is modified (see below). If the sample size is less than 10, then it is necessary to use tables for the S statistic.[47]

6. The Kendall correlation coefficient τ is defined as

$$\tau = \frac{S}{n(n-1)/2}$$

As with other types of correlation coefficients, τ can only take on values between -1 and 1, its sign indicates the sign of the slope of the relationship, and the absolute value indicates the strength of the relationship.

Because the test is based only on the ranks of the data, it can be implemented even in cases where some of the data are censored. This is an important feature of the test for application in hydrology (see Sec. 17.5 for a discussion of censored sampling situations in hydrology). The correction in the formula for the variance of S to account for ties follows; this correction is important for situations with censoring, in that all of the censored values must be considered to be tied:

$$\text{Var}(S) = \frac{n(n-1)(2n+5) - \sum_{i=1}^{n} t_i i(i-1)(2i+5)}{18}$$

where t_i is the number of ties of extent i. For example, in the data set 5, 5, 6, 7, 8, 8, 8, 10, 10, 11, 12, 12 the t_i values are as follows: $t_1 = 3$ [three untied values (6, 7, 11)], $t_2 = 3$ [three ties of extent two (5, 10, 12)], $t_3 = 1$ [one tie of extent three (8)], and for all higher values of i, $t_i = 0$.

The test can not be employed where there are multiple censoring thresholds in the data set (e.g., one value is reported as "<10 μg/L" and another as "<1 μg/L") because the values cannot be unambiguously ranked.

The Kendall S statistic has had wide application to the analysis of time trends in hydrology.[38] The test of correlation becomes a test for trend if the x values are time. One variation on this test that is particularly useful is the seasonal Kendall test.[36] In this test all of the data are grouped by season (the year can be divided into any number of seasons such as 2, 4, 6, or 12, depending on data availability). A Kendall S statistic can be computed for each season; the variance of each of these is known from the sample size in each season. The individual S statistics provide a test of trend in each season. The S statistics can be summed $S' = \sum_i s_i$, and their variances can be summed $\text{Var}(S') = \sum_i \text{Var}(s_i)$, so that a standardized variate Z can be computed by using S' and $\text{Var}(S')$ in place of S and $\text{Var}(S)$ to solve for Z, the standard form of the statistic. This method provides an overall test of trend for the entire data set in a manner that is not confounded by the typical differences in distribution among the seasons. A drawback to the procedure is that it is insensitive to cases where some seasons exhibit positive trend slopes while other seasons exhibit negative trend slopes. There are related nonparametric trend analysis procedures that address this issue.[52,77]

17.4.2 Linear Regression

Another approach to describing the linear relationship between two variables is ordinary least squares linear regression. The approach has a long and rich history because it was possible to make the necessary computations, even for large data sets, without the aid of a digital computer. The advent of the computer has made it possible to go beyond what is possible with regression methods, although they remain quite useful and form the basis for many of the newer, more flexible or robust techniques. The linear regression model is

$$y_i = \beta_0 + \beta_1 x_i + \epsilon_i \qquad i = 1, 2, \ldots, n$$

where y_i = ith observation of the response (or dependent) variable
x_i = ith observation of the explanatory variable
β_0 = intercept
β_1 = slope
ϵ_i = random error or residual for the ith observation
n = sample size

It is further assumed that ϵ_i is a random variable which is independent of x_i and has a mean of zero and a constant variance σ^2, which does not depend on x. When hypothesis tests or confidence intervals are computed, the ϵ_i are also assumed to follow a normal distribution. Table 17.4.2 defines the variables used in regression modelling.

Another way of describing the linear regression model is in terms of a conditional mean and variance of y, given x_0. The conditional mean of y given x_0 is $E[y|x_0] = \beta_0 + \beta_1 x_0$, and the conditional variance of y, given x_0, is $\mathrm{Var}(y|x_0) = \sigma^2$. Note that σ^2 is not a function of x_0.

The Correlation Coefficient. The Pearson (or product moment) correlation coefficient r is specifically a measure of linear association. If a data set were such that all of the x, y pairs plotted exactly on a straight line, then r would equal $+1$ or -1 exactly ($+1$ if upward sloping or -1 if downward sloping). Data sets that follow some nonlinear monotonic function exactly will have $|r| < 1$. In contrast to this, the rank-based alternative, the Kendall τ, would have $|\tau| = 1$ if the data followed a linear or nonlinear monotonic function exactly.

As with Kendall's τ, the correlation coefficient r can form the basis of a statistical test of independence. The null hypothesis is that the y_i are independent and identically distributed normal random variables, not dependent on the x_i. The test statistic t is defined: $t = \dfrac{r\sqrt{n-2}}{\sqrt{1-r^2}}$. The null hypothesis is rejected if $|t| > t_{\mathrm{crit}}$, where t_{crit} is the point on the Student's t distribution with $n - 2$ degrees of freedom that has a probability of exceedance of $\alpha/2$.

Developing a Regression Model. The ease with which linear regression model parameters can be estimated with digital computers leads to a temptation to apply the procedure without carefully exploring the modeling choices. The following is a set of general procedures that should be used in developing and testing a regression model. The procedures are summarized in Table 17.4.3.

1. There should be a physically plausible argument for selecting the explanatory variable x to use to estimate the dependent variable y. The scatter plot of y versus x should be made and two questions addressed: Does the relationship appear to be linear or curved? Does the variability of y vary for different levels of x? A relationship where the variance is constant is called *homoscedastic.*

There are a wide variety of transformations that can be considered. The *ladder of powers* is a widely used family of transformations that is suitable for most purposes. The ladder of powers takes the form $z = x^p$, where p typically takes on values such as

TABLE 17.4.2 Formulas Used in Linear Regression

Formula	Name
$\overline{X} = \frac{1}{n} \sum_{i=1}^{n} x_i$	Mean of x
$\overline{Y} = \frac{1}{n} \sum_{i=1}^{n} y_i$	Mean of y
$S_{yy} = \sum_{i=1}^{n} (y_i - \overline{Y})^2 = \sum_{i=1}^{n} y_i^2 - n(\overline{Y})^2$	Sums of squares y
$S_{xx} = \sum_{i=1}^{n} (x_i - \overline{X})^2 = \sum_{i=1}^{n} x_i^2 - n(\overline{X})^2$	Sums of squares x
$Sxy = \sum_{i=1}^{n} (x_i - \overline{X})(y_i - \overline{Y}) = \sum_{i=1}^{n} (x_i y_i) - n\overline{X}\,\overline{Y}$	Sums of cross products
$b_1 = \frac{S_{xy}}{S_{xx}}$	Estimate of β_1 (slope)
$b_0 = \overline{Y} - b_1 \overline{X}$	Estimate of β_0 (intercept)
$\hat{y}_i = b_0 + b_1 x_i$	Estimate of y given x_i
$e_i = y_i - \hat{y}_i$	Estimated residual
$s^2 = \frac{S_{yy} - b_1 S_{xy}}{n-2} = \frac{\sum_{i=1}^{n} e_i^2}{n-2}$	Estimate of σ^2, also called mean square error
$S_{ee} = \sum_{i=1}^{n} e_i^2 = \text{SSE}$	Error sum of squares
$s = \sqrt{s^2}$	Standard error of regression
$r = \frac{S_{xy}}{\sqrt{S_{xx} S_{yy}}} = b_1 \sqrt{\frac{S_{xx}}{S_{yy}}}$	Correlation coefficient
$R^2 = \frac{S_{yy} - s^2(n-2)}{S_{yy}} = 1 - \frac{S_{ee}}{S_{yy}}$	Coefficient of determination fraction of the variance explained by regression

TABLE 17.4.3 Steps That Should Be Taken in Regression Model Building, Depending on the Appearance of the Scatter Plot

Appearance	Linear	Curved
Homoscedastic (variance constant over range of x)	Proceed with linear regression	Transform the x variable to achieve linearity
Heteroscedastic (variance changes over range of x)	Transform the y and possibly the x variables to achieve a linear and homoscedastic relationship	Transform the y and possibly the x variables to achieve a linear and homoscedastic relationship

$-2, -1, -0.5, -0.33, 0, 0.33, 0.5, 1, 2$. It is convenient to define the case where $p = 0$ as the log transformation [$z = \ln(x)$]. The changes in curvature of the z versus y plot are relatively gradual as p is varied. A good choice of p can be selected by eye, although objective procedures exist.[6,21,71] The natural logarithm transformation has some properties that make it particularly appropriate in many hydrologic applications and some special characteristics of this transformation are discussed in Sec. 17.4.4. Base 10 logarithms can be used, but the mathematics of some of the formulas presented below become more complex. Hence, natural logarithms are used in this chapter. The decision to transform y should be based on the appearance of plots and residuals (see step 4 below). It should not be based on considerations of R^2. Comparisons of R^2 between a model using y and another using a transform of y are meaningless.

2. Estimate the parameters of the model, b_0 and b_1. If they result in estimates of y that are unreasonable for a reasonable value of x, then the model is probably misspecified and one or both variables should be transformed. This is particularly so where y is a physical quantity that cannot attain negative values (flows, concentrations, or fluxes) and yet reasonable values of x result in estimates of y that are negative.

3. Use the t statistics to determine the significance of the estimated slope b_1. For a single x variable, the test of significance of b_1 is the same as the test of significance of r, the correlation coefficient given above. If it is not possible to reject the null hypothesis that $\beta_1 \neq 0$ then the regression model should not be used, and the sample mean of y should be considered the best estimate of y. As an approximation, a regression coefficient is significant if the absolute value of its t statistic is greater than 2, that is, $t < -2$ or $t > 2$.

4. Compute the residuals from the regression model $e_i = y_i - \hat{y}_i = y_i - b_0 - b_1 x_i$. Make a scatter plot of e_i versus x_i. This plot should appear as a cloud of data centered around the line $e = 0$ (Fig. 17.4.1). If it takes on the shape of a U or inverted U this means that there was some nonlinearity that was not identified in step 1; see Fig. 17.4.2. If it takes on a funnel shape this means that there was some heteroscedasticity that was not identified in step 1; see Fig. 17.4.3. If either of these is the case, return to step 1 and continue to modify transformations until the desired condition is met. In some cases there is no simple transformation that can remove the nonlinearity. In

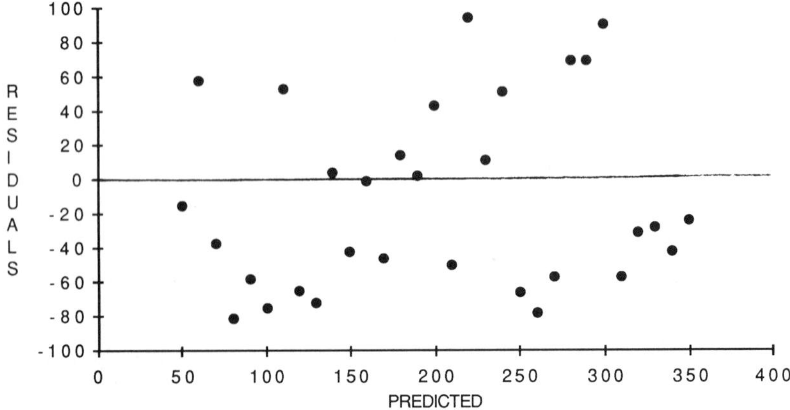

FIGURE 17.4.1 Example of regression residuals. No problems are indicated—neither curvature nor heteroscedasticity.

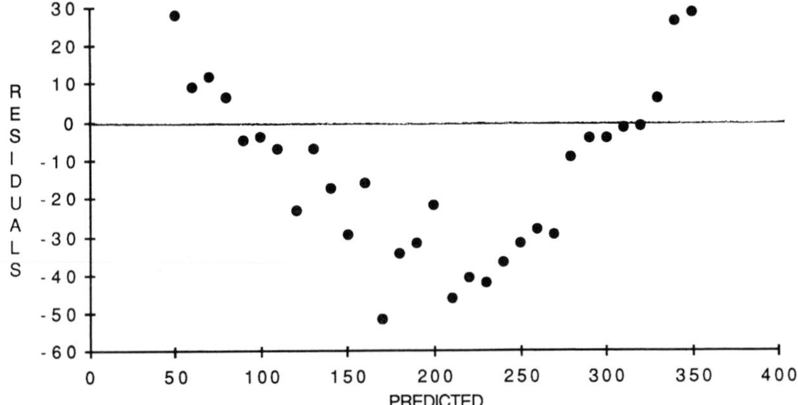

FIGURE 17.4.2 Example of regression residuals. Poor fit is indicated by curvature.

these cases polynomial-based multiple regression models are appropriate. The approach to constructing these models is described below in Sec. 17.4.5. If homoscedasticity cannot be achieved, an appropriate technique to use may be weighted least squares, described below in Sec. 17.4.8.

5. Examine the marginal distribution of the residuals using techniques such as a box plot, probability plot, or histogram. If they depart from a normal distribution, then one must be cautious about relying on the inferences described below (such as confidence intervals and prediction intervals) which assume normality of residuals.

6. Examine the residuals as a function of space or time or logical category of the observations. By examining plots of residuals versus time, one can evaluate time trends in the relationship between x and y. Maps of residuals can be contoured or evaluated for spatial trend. The residuals can also be categorized into groups and

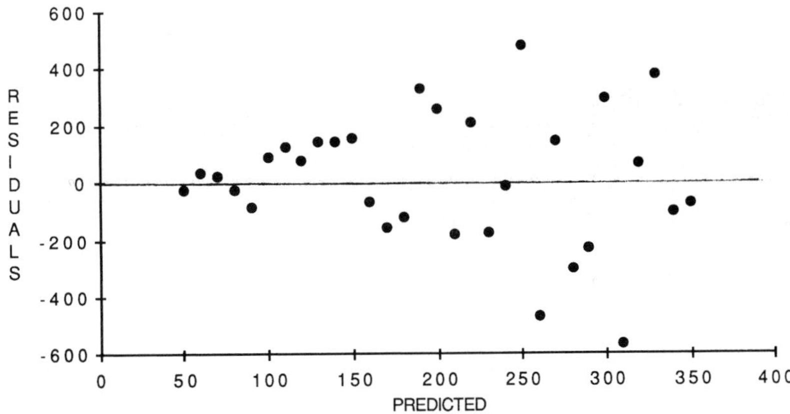

FIGURE 17.4.3 Example of regression residuals. Heteroscedasticity is indicated by funnel-shaped pattern.

presented as side-by-side box plots (see Fig. 17.2.5). The categories can be seasons or some description of hydrologic conditions such as rising limb versus falling limb of the hydrograph, or days with rain versus days without rain. Residual analysis which shows some predictable variation in the x-y relationship can lead to the use of multiple regression techniques to model simultaneously all of the known sources of variation.

17.4.3 Confidence Intervals and Prediction Intervals

If the true relationship is linear and the residuals are independent, normal random variables with constant variance, then the $(1-\alpha)\cdot 100\%$ confidence intervals (Cls) for the individual parameters β_0 and β_1 can be defined. That is, $\alpha = 0.05$ corresponds to a 95 percent Cl. The meaning of the Cl is that, in repeated applications of the technique, the frequency with which the Cl would not contain the true parameter value is α.

For the slope β_1 the Cl is

$$\left(b_1 - \frac{ts}{\sqrt{S_{xx}}},\ b_1 + \frac{ts}{\sqrt{S_{xx}}}\right)$$

where s is the standard error of the regression and t is the point on Student's t distribution with $n-2$ degrees of freedom with a probability of exceedance of $\alpha/2$.

For the intercept β_0 the Cl is

$$\left(b_0 - ts\sqrt{\frac{1}{n} + \frac{\bar{x}^2}{S_{xx}}},\ b_0 + ts\sqrt{\frac{1}{n} + \frac{\bar{x}^2}{S_{xx}}}\right)$$

where t is defined as above.

For a given x_0, the estimate of the *conditional mean* of y is $\hat{y} = b_0 + b_1 x_0$. The $(1-\alpha)\cdot 100\%$ confidence interval for the conditional mean of y is

$$\left(\hat{y} - ts\sqrt{\frac{1}{n} + \frac{(x_0-\bar{x})^2}{S_{xx}}},\ \hat{y} + ts\sqrt{\frac{1}{n} + \frac{(x_0-\bar{x})^2}{S_{xx}}}\right)$$

The *prediction interval* is a representation of the range of values that an individual y might take on for a given x_0. It incorporates the parameter uncertainty as well as the unexplained variability of y. The $(1-\alpha)\cdot 100\%$ prediction interval for a single response, given x_0, is

$$\left(\hat{y} - ts\sqrt{1 + \frac{1}{n} + \frac{(x_0-\bar{x})^2}{S_{xx}}},\ \hat{y} + ts\sqrt{1 + \frac{1}{n} + \frac{(x_0-\bar{x})^2}{S_{xx}}}\right)$$

If the sample size is large (say $n > 30$) and if x_0 is close to $\bar{x}$, a very good approximation to the prediction interval is $(\hat{y} - ts,\ \hat{y} + ts)$. Figure 17.4.4 illustrates the confidence interval and prediction interval for a regression model.

17.4.4 Transformation of y

In those cases where y is transformed in order to achieve homoscedasticity of the residuals, special consideration must be given to the interpretation of the results of the regression analysis. In particular, the regression estimate $\hat{y}$ is defined as the

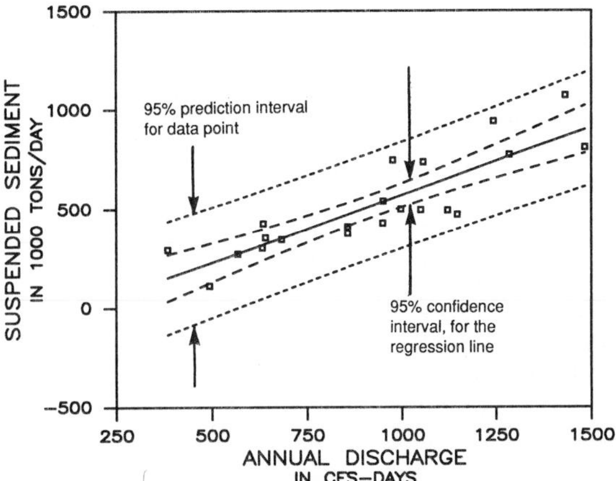

FIGURE 17.4.4 Example of a regression estimate and the 95 percent confidence intervals and 95 percent prediction intervals. Data are annual suspended sediment load and annual discharge for the Green River at Munfordville, Ky.

conditional mean of y given a particular value of x. However, if y is a transformed variable, then it is incorrect to simply do the inverse transformation to compute the conditional mean of the original (untransformed) variable. This problem is of special concern in estimating the mass or flux of a hydrologic variable and adding these over time.

To illustrate the issues in interpreting the results of a transformed regression model, consider the following case. The model is

$$\ln L = \beta_0 + \beta_1 \ln Q + \epsilon$$

where L is constituent load (tons/day), and Q is discharge (cubic feet per second). Assume that the ϵ values are normal with mean zero and variance σ^2.

Figure 17.4.5 illustrates a data set typical of such L versus Q data, shown here in a log-log plot. The curves are results from a linear regression computed in log units. The middle line is the regression line and the 50 and 95 percent prediction intervals are shown. Note that, because of the normality assumption, the prediction intervals are symmetric about the regression line. For any given Q value the five lines on the graph represent five different percentage points on the conditional distribution of $\ln L$. They are the 2.5, 25, 50 (median), 75, and 97.5 percentage points. The 50th percentage point is both the mean and the median for $\ln L$ because normality is assumed.

Figure 17.4.6 replots these data points and curves in real units (L versus Q). The five curves remain the 2.5, 25, 50, 75, and 97.5 percentage points on the conditional distribution, but now it is a distribution of L conditional on Q. Note, however, that the symmetry of these curves is lost. The distribution of L conditional on Q is not normal, it is lognormal. For a lognormal distribution, the mean of the untransformed variable is not equal to its median. In fact, the conditional mean of L will always lie above the central line, which remains the conditional median following transformation.

Assume that the objective is estimating the amount of sediment, nutrient, or contaminant entering a lake, reservoir, or estuary over some period of time such as a

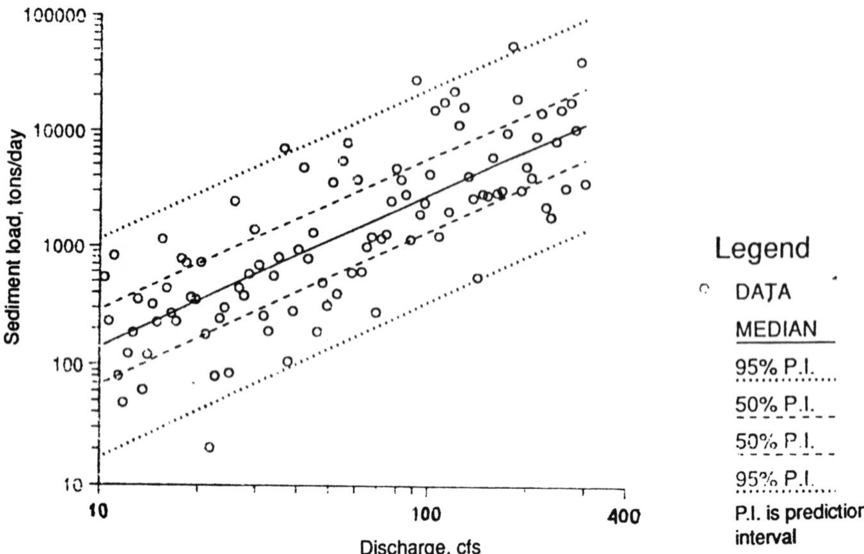

FIGURE 17.4.5 Example of the fit of a regression model fitted to the log of load. Solid line is the regression indicating both the mean and the median of the conditional distribution (assuming log data are normal). The dashed line is the 50 percent prediction interval. The dotted line is the 95 percent prediction interval.

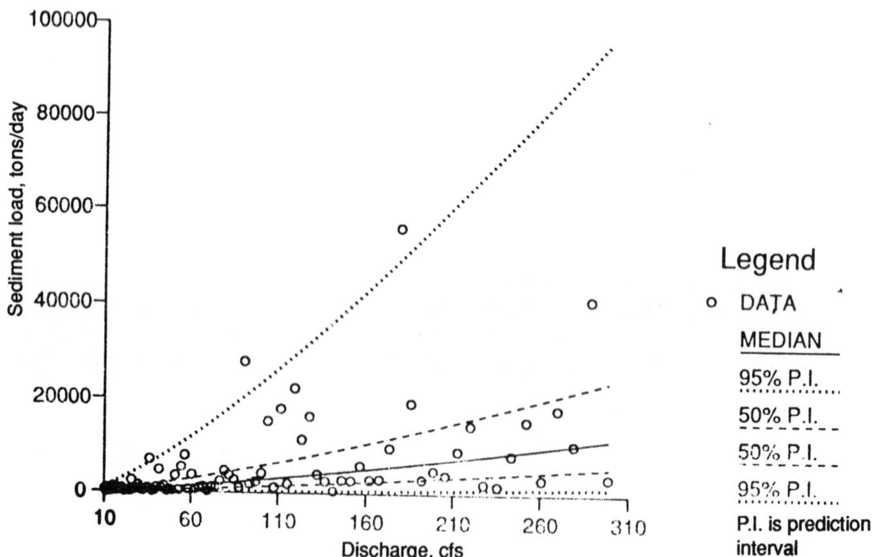

FIGURE 17.4.6 Example of the fit of a regression model fitted to the log of load (same data as shown in Fig. 17.4.5). Solid line is the regression indicating the median of the conditional distribution. The dashed line is the 50 percent prediction interval. The dotted line is the 95 percent prediction interval. Note the asymmetry of the conditional distribution for load. The mean lies substantially above the median line.

season, year, or decade. The purpose of such an analysis is to estimate the accumulation of mass in the water body, as opposed to finding instantaneous concentrations in the inflow. Because these inputs are typically strongly related to discharge, the objective is best accomplished by estimating the expected loading for each of many short time periods (such as an hour or day) as a function of discharge and perhaps time of year and other explanatory variables, and then summing the individual estimates to arrive at an expected value for the long time period. This is appropriate because the sum of the means is the mean of the sum. However, the sum of the medians is not the median of the sum. Simply transforming the regression equation back to real space provides a median estimate of L and not a mean. Therefore the sum of these estimates will not provide an unbiased estimate of the mean (or median) of L over the long time period.

Using the equation $\hat{L} = \exp(b_0 + b_1 \ln Q)$ will result in underestimates of the mean of as much as 50 percent in some hydrologically reasonable cases.[23] If the residuals in log space were normal and the parameters of the model $(\beta_0, \beta_1, \sigma^2)$ were known without error, then it would be possible to use the results from the theory of the lognormal distribution[1] to remove the bias.

One approach is to treat the estimates as if they were the true parameters, resulting in the following estimator:[23]

$$\hat{L}_F = \exp(b_0 + b_1 \ln Q_0 + \tfrac{1}{2}s^2)$$

When the residuals follow a normal distribution, n is large (>30), and σ is small (<0.5), this is a very good approximation. However, in other cases, this estimator can greatly overestimate the true mean (it overcompensates for the bias).

There is a minimum-variance unbiased solution to this problem.[7] This still holds only where the residuals from the regression on $\ln L$ are truly normal. See Refs. 15 and 26 for its application to hydrologic problems and for the necessary computational formulas.

An alternative, the so-called *smearing estimate*[22] requires only the assumption that the residuals are independent and identically distributed (homoscedastic), but they can follow any distribution. In the case of the log transform the smearing estimate of the mean is

$$\hat{L}_D = \exp(b_0 + b_1 \ln Q_0) \frac{\sum_{i=1}^{n} \exp(e_i)}{n}$$

If the residuals are normal, the smearing estimator performs nearly as well as unbiased estimator.[7] It avoids the overcompensation of the approach used in Ref. 23. It is the preferred approach unless the distribution of the residuals is known to be normal, because it is insensitive to the form of the distribution of residuals. Crawford[19] provides a comparison of the techniques.

The smearing estimator can be generalized to any transformation. If $y = f(Y)$, where Y is the raw dependent variable and f is the transformation function (e.g., square root, reciprocal, or log), then

$$\hat{Y}_D = \frac{\sum_{i=1}^{n} f^{-1}(b_0 + b_1 X_0 + e_i)}{n}$$

where b_0 and b_1 are the coefficients of the fitted regression, e_i are the residuals from the fitted regression, f^{-1} is the inverse of the transformation that was used (e.g., square, reciprocal, or exponential, respectively), and X_0 is the specific value of X (the explanatory variable) for which the estimate of Y is being made.

17.4.5 Multiple Regression

Commonly the hydrologist knows, from first principles, from experience, or from analysis of regression residuals, that improved explanation of the variation in y may be achieved by simultaneously considering the effects of more than one explanatory variable. In such cases, multiple linear regression should be used rather than simple linear regression, as described above.

The multiple linear regression model takes the form:

$$y = \beta_0 + \beta_1 x_1 + \beta_2 x_2 + \cdots + \beta_k x_k + \epsilon$$

where there are k explanatory variables $x_1, x_2, x_3, \ldots, x_k$. There are assumed to be n observations of the k explanatory variables and of the dependent variable. The subscript i ($i = 1, 2, \ldots, n$) has been omitted.

A common type of multiple linear regression model in hydrology is the stream-flow basin characteristics model.[66,68,73] Some stream-flow statistic (such as the mean flow or the 10-year flood) is estimated as a function of drainage basin area, average basin altitude, and percentage of basin forested. A similar approach is the basin yield–basin characteristics model[64] in which, for example, basin dissolved solids yield is estimated as a function of average rainfall, percentage of basin area underlain by carbonate rocks, and basin population.

Multiple regression analysis is also an effective tool in trend analysis where one or more explanatory variables are used to account for the temporal or spatial trend in some relationship. In the case of temporal trend, time is used as an explanatory variable. In the case of spatial trend, some geographical coordinates are used as explanatory variables. These may be expressed as distances east and north of an arbitrary datum and may include cross products or higher-order terms.

An example of a model for time trend is

$$C = \beta_0 + \beta_1 \ln Q + \beta_2 (\ln Q)^2 + \beta_3 T + \epsilon$$

where C is concentration of some solute in a river, $\ln Q$ is the natural log of discharge, and T is time in years. If such a model can be justified by the data (in particular if it can be shown that $\beta_3 \neq 0$), then one can state that the relationship between C and Q has been changing over time. This is a more powerful approach for identifying and describing trends than simply examining the trend in C as a function of T, because the simultaneous use of discharge and time as explanatory variables can remove a great deal of the variability of C.

Functions of the same variable can be used as multiple explanatory variables. A polynomial of some explanatory variable can be used to account for the curvature in the relationship between that variable and y. When this is done, it is advisable to *center* the original explanatory variable (subtract its mean) before computing the powers that are to be used as the additional explanatory variables. An example of this is the model

$$y = \beta_0 + \beta_1 (x - \bar{x}) + \beta_2 (x - \bar{x})^2 + \beta_3 (x - \bar{x})^3 + \epsilon$$

This centering is done to avoid multicollinearity, which can lead to numerical problems in the regression and to difficulties in interpreting regression statistics. One can use orthogonal polynomials (Ref. 48, vol. 2, pp. 380–384) to achieve the desired result of completely eliminating multicollinearity, but generally the centering of these variables will be sufficient.

Periodic functions of a time variable can be used to explain some cyclic behavior. Typically this is done to account for diurnal cycles, tidal cycles, or annual cycles. If the cycle is well-described by a single sine function, then the model can be expressed as

$$y = \beta_0 + \beta_1 \sin(2\pi t) + \beta_2 \cos(2\pi t) + \epsilon$$

where t is in units equivalent to the cycle length (e.g., days for a diurnal cycle, years for an annual cycle). In cases where the cycle is not a simple sine function, additional periodic functions with frequencies that are some integer multiple of the primary frequency can be used. For example,

$$y = \beta_0 + \beta_1 \sin(2\pi t) + \beta_2 \cos(2\pi t) + \beta_3 \sin(4\pi t) + \beta_4 \cos(4\pi t) + \epsilon$$

These periodic functions can be used in conjunction with time variables (for long-term trends) and with other explanatory variables in a multiple regression model to estimate the transport of constituents in a stream. For example, the loads of several nutrient species entering Chesapeake Bay through its tributaries have been described by models of the form:[14]

$$\ln C = \beta_0 + \beta_1 \ln Q + \beta_2 \ln^2 Q + \beta_3 T + \beta_4 T^2 + \beta_5 \sin(2\pi T) + \beta_6 \cos(2\pi T) + \epsilon$$

In general, this simple linear model has been found to fit the observed data reasonably well. The same six-variable model is used for predicting the concentrations of each constituent, regardless of whether or not each coefficient is significantly different from zero. Because all the explanatory variables are measured continuously, the model can be used to provide continuous estimates of nutrient loads entering the bay. Using such a model facilitates the computation of standard errors of the estimated annual loads.[15,26]

Analysis of Covariance. Yet another type of explanatory variable is the *dummy variable:* one that takes on only the value 0 or 1. The use of such variables in regression is called *analysis of covariance.* It is just a special case of multiple linear regression and, as such, is described as a part of this section on multiple linear regression. It is used in cases where the relationship between x and y changes according to some factor that is best thought of as a categorical variable rather than a continuous variable. Examples include: day versus night, the rising limb of the hydrograph versus the falling limb, the absence of a facility (such as a sewage treatment plant or dam) versus the presence of the facility. The dummy variable is defined as 0 if the observation is in one group and 1 if in the other. New variables can be formed as the products of these dummy variables and some other explanatory variables.

Examples of two types of models are

$$y = \beta_0 + \beta_1 x + \beta_2 Z + \epsilon \qquad (1)$$

$$y = \beta_0 + \beta_1 x + \beta_2 Z + \beta_3 xZ + \epsilon \qquad (2)$$

where Z is the dummy variable, defined to be either 0 or 1 depending on which group the observation belongs in, and x is some explanatory variable which is continuous. Model (1) describes a shift in intercept—the relationship of x and y shifts by an amount β_2 for cases in the category where $Z = 1$, but the slope of the relationship is constant. Model 2 describes changes in both slope and intercept. For $Z = 0$ the slope is β_1 and the intercept is β_0; for $Z = 1$ the slope is $\beta_1 + \beta_3$ and the intercept is $\beta_0 + \beta_2$. The reason for using model 2 (rather than splitting the data into two separate groups and doing analyses of each) is that with it, statistical tests on the coefficients of the terms containing Z can determine if the groups are, in fact, distinct in terms of their x, y relationships. Analysis of covariance is a highly useful tool in hydrology because it can be used to test if the relationship between variables has changed (e.g., runoff as a function of rainfall, concentration as a function of stream flow). This is generally much more informative (and more powerful in terms of hypothesis testing) than simply exploring the question of changes in the variable of interest, without regard to some driving variables.

17.4.6 Evaluation of a Multiple Regression Equation

In multiple regression problems, plots of the residuals versus predicted values should always be made and examined for curvature and heteroscedasticity. Also, plots should be made of residuals versus time sequence or location or other candidate explanatory variables not in the model. For a thorough discussion of such plots see Ref. 21.

The standard form in which to report a multiple regression equation is

$$\begin{array}{llllll} \hat{y} = b_0 + & b_1 x_1 & + b_2 x_2 & + \cdots & + b_k x_k \\ (t_0) & (t_1) & (t_2) & \cdots & (t_k) \end{array}$$

$$R^2 = \qquad , s =$$

where the t_i terms are the t statistics on the individual coefficients b_i and s is the standard error of the regression.

Multicollinearity. One of the most serious problems encountered in multiple regression is *multicollinearity*. This is the condition where at least one explanatory variable is highly correlated with another explanatory variable or with some linear combination of other explanatory variables. With multicollinearity, regression coefficients can be highly unstable and unreliable (a small change in the data would result in large changes in some of the β's). One diagnostic for this problem is the variance inflation factor (VIF).[54] For variable j the VIF is $\text{VIF}_j = 1/(1 - r_j^2)$, where r_j^2 is the r^2 from a regression of the jth explanatory variable on all of the other explanatory variables. Ideally one would like to use orthogonal predictor variables, for which $\text{VIF}_j = 1$. Serious problems can result if regressions contain one or more variables with $\text{VIF}_j > 10$, and if the regression equation is being used to make predictions that are somewhat beyond the range of the observed data or if some conclusions are being drawn about the regression coefficients themselves. There are several possible solutions to problems of multicollinearity.

1. If the model contains functions (e.g., powers and cross products) of other explanatory variables, the problem may be solved by centering those explanatory variables (subtracting their mean) before computing the various functions.

2. Eliminate variables in instances where two of the variables describe essentially the same thing. It is often the case that in selecting explanatory variables for use in a regression, the hydrologist has a general idea of what the important influences are, but the available data provides several choices of explanatory variables that roughly describe that influence. If all of the choices are used and the VIFs are high, then some of these alternative variables should be removed from the model. A good rule for removal is to attain the highest model R^2, subject to there being no VIF > 10. The hydrologist's judgment about the most important influences and most appropriate representation of those influences should play a significant role in this process.

3. Another approach is to collect more data. For example, the explanatory variables might be percentage of basin in forest and basin altitude. These are likely to be closely related in a mountainous region (forests are at higher altitudes). If it is possible to add a few basins that are atypical of the general pattern (e.g., low altitude–heavily forested) then it may be possible to eliminate the colinearity problem.

4. In some situations one can use ridge regression[41] or principal components regression.[57,65] These techniques are beyond the scope of this handbook. They are described in many textbooks on regression.

Regression Diagnostics. There are several diagnostics that relate to individual observations in the data set. The formulas presented here for the diagnostics are for the case of only one explanatory variable. For multiple regression see Ref. 4 or other regression texts published since about 1980.

The first of these is the *leverage statistic* which describes how far the data point is from the center of mass of all of the data points. It is based only on the x values and not on the y. It is defined as

$$h_i = \frac{1}{n} + \frac{(x_i - \bar{x})^2}{S_{xx}}$$

An observation can be considered to have high leverage when $h_i > 2p/n$, where p is the number of coefficients in the model (for simple linear regression $p = 2$). High leverage alone is not a problem, but it does create the potential for a given value exerting excessive influence on the regression, depending on the value of y_i. The leverage statistic can be used to identify *hidden extrapolation,* the situation where predictions are being made beyond the range of the original data used to estimate the equation. See Ref. 61, p. 142, for a discussion of hidden extrapolation.

Another diagnostic is the *standardized residual* e_{si}. It is the residual e_i $(= y_i - \hat{y}_i)$, standardized by its standard error:

$$e_{si} = \frac{e_i}{s\sqrt{1 - h_i}}$$

An extreme outlier is one for which $|e_{si}| > 3$. There should be only an average of 3 of these in 1000 observations if the residuals are normally distributed. $|e_{si}| > 2$ should occur about 5 times in 100 observations if normally distributed.

An even more useful form of residual computation is the *prediction residual* $e_{(i)}$, defined as $e_{(i)} = y_i - \hat{y}_{(i)}$, where $\hat{y}_{(i)}$ is the regression estimate of y_i based on a regression equation computed leaving the ith observation out. The prediction residual can, however, be computed without having to solve n different regression problems. The computational formula is

$$e_{(i)} = \frac{e_i}{1 - h_i}$$

An excellent measure of the quality of a regression equation is the *PRESS statistic,* the prediction error sum of squares:

$$\text{PRESS} = \sum_{i=1}^{n} e_{(i)}^2$$

It is a validation-type estimator of error, but it does not require withholding observations from the estimation data set. In multiple regression, it is a very useful estimate of the quality of various possible regression models. When selecting among several alternative regression models, selection based on minimizing PRESS is a good approach.

A highly influential observation is one that, if left out of the data set, would lead to a substantially different set of coefficients. To have high influence, the observation must have high leverage and a large residual.

The influence of the *i*th observation is defined as[18]

$$D_i = \frac{e_i^2 \, h_i}{ps^2(1 - h_i)^2}$$

where p = number of coefficients in the model. It is suggested that the *i*th observation has high influence if $D_i > F_{p,n-p}$ at $\alpha = 0.5$. Note that, for simple linear regression with more than about 30 observations, the critical value for D_i would be about 0.7, and for multiple linear regression with several explanatory variables, the critical value would be in the range of 0.8 to 0.9.

Finding an observation with high Cook's *D* should lead to very careful examination of that datum value for possible errors or special conditions which might have prevailed at the time it occurred. If it can be shown that an error occurred, the value should be corrected if possible or deleted if it cannot be corrected. If error cannot be shown, then a more complex model which explains the point better is one option to consider if the other data also support this. The other option is to use a more robust procedure such as weighted least squares (see Sec. 17.4.8).

17.4.7 Model Selection

First and foremost, models need to be conceptually reasonable for the situation at hand. For example, a model for stream flow should have drainage area and rainfall depth as multiplicative terms rather than additive terms. Thus, one could use log flow as the dependent variable and log area and log precipitation depth as explanatory variables. Also, the model should not produce unrealistic estimates of the dependent variables, given reasonable values of the explanatory variables. For example, there should be no negative values predicted for a variable that must be positive.

Selection of a "best" regression model should begin with initial selection of candidate explanatory variables, each of which has some theoretical justification for being used in the model. The number of explanatory variables considered should generally be no more than about $n/10$. Inclusion of additional explanatory variables (even a set of random numbers) will always result in some decrease in the standard error of estimate S_{ee} and some increase in R^2. What is needed is a determination of whether the β's for additional variables are significantly different from zero and whether a reduction in prediction error is likely to be realized.

To start a search for a best regression model, most statistical software packages have a simple procedure to fit all possible regressions. If there are *k* possible explanatory variables, then there will be 2^k possible models (including the no-variable model). However, when models using polynomial terms are being considered, they should not include models that contain higher-order terms without the lower-order terms. Similarly, if periodic functions are used, the model should not contain the sine without the cosine (or the cosine without the sine) for any given frequency. Also, if analysis of covariance models is being used, the cross product variable should not be considered without the dummy variable being used.

Having fit all the candidate models (leaving out these special cases and leaving out models with high multicollinearity), the hydrologist can then focus on a set of k potential models: the best k-variable model, the best $(k-1)$-variable model, . . . , down to the best 1-variable model and the no-variable model. *Best* (for a fixed number of explanatory variables) is defined here as minimum S_{ee} or maximum R^2 (the result is identical). To optimize for prediction accuracy, select from among these k best models the model with the lowest PRESS statistic. Another statistic that is also useful in model selection is *Mallows' Cp.*[53] Model selections based on PRESS and those based on Cp are likely to be the same except where there are highly influential observations.[27]

Multiple regression can be used to determine whether some effect is significant. Examples include: Is there a significant time trend in the relationship of solute concentration to streamflow? Is there a seasonal cycle in the stage-versus-discharge relationship? Did the rainfall runoff relationship change after the dam was built? These kinds of questions can be addressed by the use of the F test on nested models. A simple model is nested in a complex model if all of the variables in the simple model are contained in the complex model and the complex model has at least one more variable than the simple model.

Let model s be the simpler model:

$$y = \beta_0 + \beta_1 x_1 + \beta_2 x_2 + \ . \ . \ . \ + \beta_k x_k + \epsilon$$

It has $k + 1$ parameters and $n - (k + 1)$ residual degrees of freedom (df_s). Its sum of squared errors is SSE_s.

Let model c be the more complex model:

$$y = \beta_0 + \beta_1 x_1 + \beta_2 x_2 + \cdots + \beta_k x_k + \beta_{k+1} x_{k+1} + \cdots + \beta_m x_m + \epsilon$$

It has $m + 1$ parameters and $n - (m + 1)$ residual degrees of freedom (df_c). Its sum of squared errors is SSE_c. The null hypothesis is $H_0: b_{k+1} = b_{k+2} = \cdots = b_m = 0$, versus the alternative H_1: at least one of these $m - k$ coefficients is not equal to zero.

The test statistic is

$$F = \frac{(SSE_s - SSE_c)/(df_s - df_c)}{(SSE_c/df_c)}$$

where $df_s - df_c = m - k$. If F exceeds the tabulated value of the F distribution with $(df_s - df_c)$ and df_c degrees of freedom for the selected α (such as $\alpha = 0.05$), then one can reject H_0. That is: choose the more complex model in preference to the simpler model.

Note that this result does *not* mean that every variable in the model has a coefficient that is significantly different from zero. There may be a simpler model that one would choose over this model if one were to conduct another nested F test with it.

In the special case of nested tests where there is only one additional variable in the complex model (a "partial" F test) then the test results are equivalent to a t test on the one variable in question. Thus, an easy way to establish whether any given model coefficient is significantly different from zero (given that all of the other variables are in the model) is to examine the t statistics on the coefficients that are provided in most statistical packages. If $|t| >$ the $1 - \alpha/2$ point on the Student's t distribution with $n - k$ degrees of freedom, reject the null hypothesis that the coefficient is equal to zero.

Overview of Procedures for Building Regression Models

1. If the question is "Should the dependent variable (y) be transformed?," the procedure to follow is to select the best untransformed model and best transformed model and compare them graphically in terms of three things: (1) homoscedasticity, (2) normality of residuals, and (3) curvature in a plot of residual versus predicted. Statistics such as R^2, SSE, and PRESS are *not* appropriate for making this comparison because the errors are measured in different metrics.

2. If the question is "What, if any, transformation should be taken on one or more of the explanatory variables (x)?," then one can get considerable help from a statistic such as R^2 (maximize), or PRESS (minimize). One can screen many possible transformations rapidly with such statistics. Once some good candidate models are identified, then one should look at a residual-versus-predicted plot before making a final model selection.

3. If the question is "Which of several models, each with the same dependent variable and the same number of explanatory variables (all of which are justifiable on first principles), should be selected?," then use of R^2 or PRESS is appropriate, but the choice should be checked with residuals plots.

4. If the question is "Which of several nested models should be selected?," then the F test of any pair of models may be used to find the best model. One may also select on the basis of minimum PRESS. The nested F test is most appropriate if the interest is in the model and its coefficients; the PRESS statistic is most appropriate if the interest is in the accuracy of predictions.

5. If the question is "Which of several different models, each with the same dependent variables but not necessarily nested, should be selected?," then minimum PRESS would be the appropriate selection criterion.

17.4.8 Weighted and Generalized Least Squares

Weighted least squares and *generalized least squares* are two techniques that extend the regression approach and which can be quite useful in hydrology. The concepts will only be introduced here. A key assumption in ordinary regression (with a single explanatory variable or multiple explanatory variables) is that the true residuals from the model have the same variance and are independent of each other. There are many practical situations in which these conditions do not hold. For example: in some cases the data set may contain observations of y with substantially different measurement errors. This can arise because a different measurement technique was used or because the observations were averages of replicate measurements and in some cases more replicates were used. The variance of these residuals will be the sum of a true natural variance and the measurement variance. There are other cases where there are inherent differences in variability due to some characteristics in the location or situation.

If the error variances are known (even approximately), it can be useful to use them in the estimation process. Intuitively, it is clear that one would like the fitting technique to give more weight to those observations that have a lower variance, and this is precisely what weighted least squares accomplishes. Thus, if a reasonable model (either continuous or discrete) of the error variance as a function of some explanatory variable can be developed, then that variance can be used in weighted least squares estimation. The weights on the ith individual observations should be equal to $1/\sigma_i^2$, where σ_i^2 is the error variance of the ith observation. See Ref. 21 or 45 for detailed discussions of weighted least squares (WLS) techniques.

The other assumption mentioned above is that the residuals are independent of each other. An example of where this is likely to be violated is when the observations are closely spaced, either temporally or spatially, such that they are dependent on each other. For instance, if the data set comprises pairs of concentration and discharge values that are spaced several hours or several days apart in time, then it is likely that the errors would be dependent on each other. The correlation among the errors is likely to decrease as the time spacing gets longer. Another example is where the data are estimates of the 10-year flood and a set of basin characteristics at a set of stream gauges located around a region. These at-site estimates of the 10-year flood are likely to be correlated as a function of the distance between the gauges and also as a function of the degree of temporal overlap of the flow records that are used to make the estimates.

Generalized least squares is the technique that should be applied in such situations. It provides optimal estimates of the model coefficients and variances in cases where the correlations among observations are nonzero. Like weighted least squares, it does not require that the variances all be equal. Thus, to use generalized least squares it is necessary to have some technique to estimate all of the $n(n-1)/2$ covariances and the n variances for the regression errors. Examples of generalized least squares applications to regional flood frequency estimation can be found in Refs. 68, 69, and 72.

17.4.9 LOWESS

All of the techniques described above require an assumption that the relationship of the dependent variable to some explanatory variable is either linear or else some relatively simple nonlinear function (such as a transformation of x or a polynomial of x). It is rarely the case that there are theoretical grounds to justify use of a single simple function over the entire range of the variable. Furthermore, when the data set is large it is commonly the case that one can visually identify departures from the simple function that could be accommodated in a regression model only by using a number of high-order terms. This is generally an unwise procedure because the high-order terms can create highly unstable behavior near or beyond the minimum and maximum values of x used to fit the regression model.

An alternative approach to describing the relationship between y and some x variable is the iterative procedure called *locally weighted scatter-plot smoothing* (LOWESS).[10,11] The idea of LOWESS is to allow the data to dictate the location of a smooth curve that goes through the middle of the data set. It selects the $\hat{y}$ value for any given x_o based only on the data that are relatively close to x_o. It uses weighted least squares to apply progressively stronger weights the closer an observation is to x_o.

The technique is computationally intensive. It involves the fitting of at least $2n$ separate weighted least squares models. In general the procedure for finding the $\hat{y}$ for any arbitrary x_o is to estimate a linear relationship for y as a function of x for all the data in the range $(x_o - D_x, x_o + D_x)$, where D_x is the *half-window width*. The smoothness of the curve is a function of D_x. Selection of an appropriate value for D_x

is subjective, and generally depends on the amount of data available and the data set's inherent smoothness. The computational procedure can be found in the references listed above and in Ref. 34.

Figure 17.4.7 is an example of LOWESS applied to a large data set of the natural log of sand transport versus natural log of discharge for the Colorado River at Lees Ferry, Ariz., during 1949–1952. It shows that LOWESS provides an excellent graphical means of depicting the relationship without having to resort to linear or higher-order models that really do not fit the pattern of the data. The technique eliminates the undue influence from the two very extreme low outliers and reflects that the x versus y relationship is generally linear but with a change in slope at about 6000 ft^3/s.

Examples of situations in which it would be particularly appropriate to use LOWESS are

1. cases where one wants to compute a set of residuals from some relationship (say concentration versus discharge) and then plot these residuals versus time or test them for trend over time. The LOWESS can be used as the estimates from which the residuals are calculated.

2. LOWESS can be used to aid in the display of several large x, y data sets in one figure. Plotting all of the data points using different symbols for each set is very cumbersome and often not very informative. If the LOWESS curves themselves (and not the data) are plotted, then the similarities and differences of the curves can be discerned. Also, if there is an interest in a possible abrupt shift in the relationship between x and y, then two LOWESS curves can be developed and the curves can be plotted to see how the relationship has shifted from one time period to the other. Other examples include chemical data from different aquifer types, the LOWESS of concentration of a metal versus pH,[81] and LOWESS of time series of atmospheric deposition at various sampling locations to see if the data follow similar temporal patterns or if they vary independently.[67]

3. These two applications can be combined: using LOWESS to describe the relationship between two variables and then using LOWESS again to smooth the time trend of the residuals or departures of the data from the relationship. For example, in the case of the Colorado River data shown in Fig. 17.4.7, the residuals from the LOWESS could be plotted as a function of time and a LOWESS smooth curve could

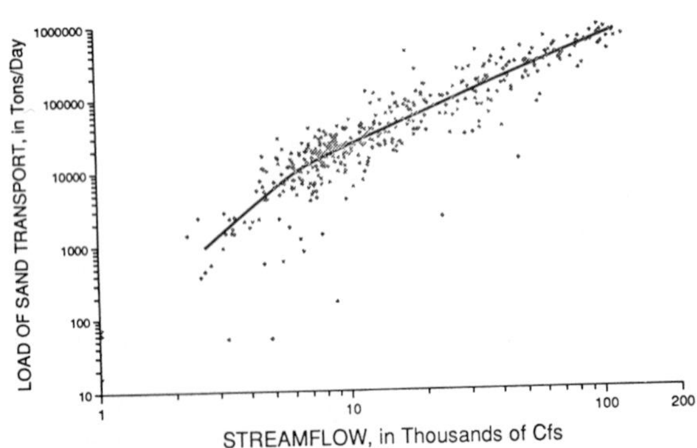

FIGURE 17.4.7 Sand size fraction for the Colorado River at Lees Ferry, Ariz. (1949–1952) as a function of stream flow. LOWESS curve is shown with the data.

be plotted to examine changes over time in the relationship between sand transport and stream flow at this site.

There are some additional variants of LOWESS that are also quite useful. One is the *upper and lower smooths*. These are two additional curves that can be plotted to represent the upper and lower quartiles of the conditional distribution of y as a function of x. They are constructed by computing the LOWESS curve of all the positive LOWESS residuals and that of all the negative LOWESS residuals and then adding these curves to the original LOWESS plot. Such a combined LOWESS plot is useful for showing how the spread and/or symmetry of the conditional distribution of y changes as a function of x. Another variant of LOWESS is the polar smooth,[11] which can be used to plot the general shape and location of groups of data so that the groups can be contrasted with each other.

17.4.10 Record Extension

In many cases, the random variable that is of greatest practical interest has been observed only for a short period of time, but other random variables that are closely related have been observed for much longer periods. For example, the random variable of interest might be the stream flow entering a reservoir. There may have been a stream gauge record (Y_i, $i = 1, 2, \ldots, n$) near the inflow point for only a few years. However, some tributary or nearby stream may have a very long record (X_i, $i = 1, 2, \ldots, N$). It will be assumed that $n < N$ and that records for both stations exist for the period $i = 1, 2, \ldots, n$. The presentation of the problem here assumes that the period of length n is at the beginning of the longer record but the methods are generalizable to overlaps that cover any portion of the period $i = 1, 2, \ldots, N$. The methods are general in that the time step could be years, months, days, or any other appropriate period.

In conducting studies of the probability of water-supply shortage in the reservoir, one would like to be able to simulate operations over some long hydrologic record. Suppose n were 5 years and N were 100 years and the events of interest (extreme, long-term droughts) may be 0.05 or 0.1 probability events. It is clear that having a 100-year record would provide for much more accurate results than could be achieved with only 5 years of record.

Thus one wishes to exploit the correlation between X and Y to extend the length of the Y record by estimating a set of $N - n$ flows for the period during which there was an X record but no Y record. The obvious approach to this problem is to use regression to estimate Y as a function of X. If the relationship of X and Y meets the assumptions for regression (linear and homoscedastic) then the regression estimate of Y_i is the lowest-variance estimator of Y_i. However, it is not a particular value of Y_i that is important, but rather the full collection of estimates of Y. What is needed is a set of Y values that possess the correct statistical properties of Y. These properties include the mean and variance of Y and may also include its serial correlation structure.

The variance of the regression estimate of Y, $\hat{Y}_i$ ($i = 1, \ldots, N - n$) is biased downward[55] because regression estimates lie on the regression line and the actual data are scattered about the regression line and are thus more variable. Matalas and Jacobs[55] proposed a means of adding the lost variance back into the data by adding a random error term to the regression estimates. This approach achieves the desired variance but suffers from two problems. One is the arbitrary nature of the results—repeated analyses with different random errors added result in different Y estimates.

The other is that the addition of the random term modifies the serial correlation structure of the record. A better approach to this problem is a class of methods known as maintenance of variance extension (MOVE).[2,35,79]

The principle of the MOVE techniques is to derive a set of coefficients for the model that transforms the X data to estimates of Y in such a way that the expected values of the statistics of the estimated Y values equal the true, but unknown, values of the statistical properties of Y. The simplest version, MOVE.1, preserves the mean and variance of Y using the equation

$$\tilde{Y}_i = a + bX_i$$

where $b = \sqrt{S_{yy}/S_{xx}}\ \mathrm{sgn}\ (r)$, $a = \bar{y} - b\bar{x}$, and sgn (r) is the sign of r (the correlation coefficient). In statistics this equation is known as the line of *organic correlation.*[37,50] Although the MOVE.1 technique is designed only to preserve the mean and variance, it preserves the distribution shape and serial correlation structure rather well if the two stations are in similar hydrologic settings. The technique has been extended to the simultaneous extension of multiple records, explicitly preserving many aspects of the correlation structure of the records.[30]

17.5 TECHNIQUES FOR USE WITH CENSORED VARIABLES

17.5.1 Examples of Censored Variables

Some hydrologic data are not reported as a specific value, but are reported as less than or greater than some threshold value. Some examples of this are (1) chemical analyses reported as being less than the detection limit, (2) flood discharges known from historical reports to have been greater than some threshold level which caused inundation of an area but the depth of inundation is not known, and (3) water table levels may be known to be below some altitude at some location because a well drilled there did not reach the water table. These kinds of data series are referred to as *censored.* It is always preferable to obtain uncensored data if possible, but because of laboratory practices or because of the imprecise nature of historical information, a censored record may be all that is available. The purpose of the procedures described in this section is to make the most effective use of such data. For a general overview of censored data techniques in water quality, see Ref. 33.

17.5.2 Estimation of Summary Statistics of Censored Data

An effective method for estimating summary statistics for censored data, including cases where there are multiple detection limits (e.g., different chemical analysis methods were used, each with a different detection limit), is to use regression to extend the empirical distribution function on a logarithmic probability plot to make estimates of the data values that have been censored.[25,32] The sample mean, standard deviation, median, and quartiles can then be estimated from the estimated empirical distribution which combines the known data values and the estimated data values for the censored observations.

17.5.3 Hypothesis Tests

Parametric and nonparametric methods can be used for hypothesis testing of censored data. These are described below. Censored data values should not be deleted before hypothesis testing, nor should arbitrary values be substituted for the censored data. The parametric tests all involve the assumption of the form of the probability distribution of the data. Various tests that are conceptually similar to the common parametric tests can be conducted by using the Tobit regression procedure (described in Sec. 17.5.4).

Nonparametric Tests on Censored Data. In nonparametric tests, all censored data are considered to be tied. The following is an example of the application of the rank-sum test described in Sec. 17.3.2 to a data set with extensive censoring. A sampling device with a detection limit of 10 units is used to obtain the two sets of samples, group A and group B, each with 10 data values. If the actual level is less than 10, then the data value is recorded as <10. The 20 data values from groups A and B are pooled and ranked, with the rank assigned to ties being the average rank of the tied set. For example, the lowest nine values in the pooled data set are each reported as <10, so they are all assigned the rank 5, the average of ranks 1 through 9. The next three values are each equal to 20 units, so they are assigned rank 11, the average of ranks 10 through 12, and so on. After ranking of the pooled data set, the values from groups A and B are separated to produce the following ranked lists:

Group A		Group B	
Data	Rank	Data	Rank
<10	5	<10	5
<10	5	<10	5
<10	5	20	11
<10	5	20	11
<10	5	30	13
<10	5	40	14.5
<10	5	50	16.5
20	11	60	18
40	14.5	90	19
50	16.5	100	20

The sum of the ranks, W, is 77.0 for group A. For the formulas presented in Sec. 17.3.2, with $n = 10$, $m = 10$, and $N = 20$: $\mu = 105$ and $\sigma = 12.59$. Given the test statistic $W = 77$, the standardized form of the test statistic $z = -2.18$. Thus if $\alpha = 0.05$, we can reject the null hypothesis that the two groups are identically distributed because the test statistic value of -2.18 is farther away from 0 than $z = -1.96$, which is the lower limit for the normal distribution with $\alpha = 0.05$.

When severe censoring occurs, say 50 percent or more of the data values are censored, the conventional nonparametric tests will have little power to detect differences between groups. An alternative is the use of *contingency tables* to test for differences in the proportion of data above the reporting limit in each group.[17] This nonparametric test can be used when the data are reported only as above or below some threshold. It may also be used when response data can be categorized into three or more groups such as below detection, detected but below some health standard, and exceeding standards. Contingency tables determine whether the proportion of data falling into each response category differs between groups (e.g., different sites or land-use categories). An example of its application might be that there are two sets of wells: one upgradient from a landfill and one group downgradient from the landfill. The question that this test would attempt to answer is: is the frequency of detections of the contaminant higher in one group than in the other? For the test procedure see Ref. 44, p. 304.

Hypothesis Testing with Multiple Reporting Limits. More than one reporting limit is often present in hydrologic data. When this occurs, hypothesis tests such as comparisons between data groups are more complicated. One approach which can always be used is to censor all data at the highest reporting limit, and then perform the appropriate test. Thus the data set

<1 <1 <1 5 7 8 <10 <10 <10 12 16 25

would become

<10 <10 <10 <10 <10 <10 <10 <10 <10 12 16 25

and a rank-sum test can be performed to compare this with another data set. This results in a loss of information which may be severe enough to obscure actual differences between groups. However, for some situations this is the best that can be done.

17.5.4 Correlation and Regression for Censored Data

With censored data one cannot use product-moment correlation coefficients or regression models (as described in Sec. 17.4.2). Correlation coefficients, and regression slopes and intercepts, cannot be computed without the actual values for the censored observations. Substituting fabricated values may produce coefficients strongly dependent on the arbitrarily selected values that are substituted. There are alternative methods that can be used with censored observations. The choice of method for regression depends on the amount of censoring present, as well as on the purpose of the analysis (Table 17.5.1).

Tobit regression is a widely accepted method for estimating a regressionlike model when there are censored data.[45,75] The Tobit procedure employs a maximum likelihood estimator. At present, the use of Tobit procedures in hydrology is in the experimental stage. Tobit procedures should be limited to those cases where one has at least 10 above-threshold observations for each parameter to be estimated. A bias correction for the method has been derived.[13] Tobit models are appropriate for the estimation of transport of a trace level constituent in streams, as a function of stream flow, where some of the concentrations are below the analytical detection limit. They can also be used to carry out the equivalent of a two-sample t test, by fitting a Tobit model

TABLE 17.5.1 Regression Methods for Censored Data

× denotes method appropriate for use

Method	Below 20% censoring	20–50% censoring	>50% censoring
Tobit regression	×	×	
Logistic regression		×	×
Contingency tables			×

in which the only explanatory variable is a dummy variable (set to 0 or 1, depending on which group the data come from). The fitted coefficient becomes a measure of difference between groups and the significance level of this coefficient is a measure of the significance of the difference between groups.

In *logistic regression*[3] the response variable is categorical. Instead of predicting, for example, the concentration, the probability of being in discrete binary categories such as above or below the reporting limit is estimated. Other examples of such categories are annual floods which cause damage versus ones that do not, or contaminant concentrations that violate a health standard versus ones that do not. One response (above, for example) is assigned a value of 1, and the second response, a 0. The probability of being in one category versus the second is tested to see if it differs as a function of continuous explanatory variable(s). An example would be to predict the probability of a detectable concentration of some pesticide being found in a groundwater sample based on the nitrate concentration in the same sample. Predictions of detection probabilities fall between 0 and 1, and are interpreted as the probability p of observing a response of 1 (pesticide detection). Therefore $1 - p$ is the probability of a 0 response (no pesticide detection).

The form of the logistic regression model is

$$\ln\left(\frac{p}{1-p}\right) = b_0 + b_1X_1 + b_2X_2 + \cdots$$

which is equivalent to

$$p = \frac{\exp(b_0 + b_1X_1 + b_2X_2 + \cdots)}{1 + \exp(b_0 + b_1X_1 + b_2X_2 + \cdots)}$$

where $X_1, X_2, \ldots$ are the explanatory variables, which have a continuous range.

Correlation. Correlation coefficients measure the strength of relationship between two variables without determining an equation describing that relationship. The product-moment correlation coefficient (defined in Sec. 17.4.2) cannot be computed for censored data without substituting fabricated values for the censored data, and so is not applicable. Two rank correlation methods are available, Kendall's tau (Sec. 17.4.1) and Spearman's rho.[17] When data are multiply censored, contingency tables can provide the phi coefficient as a measure of correlation (Ref. 44, p. 308).

Rank Correlation Coefficients. The rank-based correlation coefficient Kendall's tau (Sec. 17.4.1) can be computed when one or both variables are censored. All values below the reporting limit for a single variable are assigned tied ranks, as in the example in Sec. 17.5.3. For single detection limits these comparisons are unequivocal (a < 1 is less than a 5, etc.). Thus it is possible to do a test for trend over time using the Kendall S statistic (Sec. 17.4.1). For example, suppose the data set in time order is

$$<10, <10, 20, <10, <10, 30, 20, <10, 50, 40, 70, <10, 60$$

The corresponding Kendall's tau statistics are $P = 48$, $M = 14$, $S = 34$. The variance of S is 239.33 (based on $n = 13$ and $t_1 = 5$, $t_2 = 1$, and $t_6 = 1$, and all other $t_i = 0$). The standardized test statistic $z = 2.133$. Therefore, one could reject H_0 (that the data are identically distributed) at an $\alpha = 0.05$ (for which the critical value of the test statistic is $z = 1.96$) and conclude that there is a monotonic (upward) trend in the data.

REFERENCES

1. Aitchison, J., and J. A. C. Brown, *The Lognormal Distribution,* Cambridge University Press, Cambridge, England, 1981.
2. Alley, W. M., and A. W. Burns, "Mixed-Station Extension of Monthly Streamflow Records," *J. Hydraul. Div. Am. Soc. Civ. Eng.* (now *J. Hydraul. Eng.*) vol. 109, no. 10, pp. 1271–1284, 1983.
3. Amemiya, T., "Qualitative Response Models: A Survey, *J. Economic Literature,* vol. 19, pp. 1483–1536, 1981.
4. Belsley, D. A., E. Kuh, and R. E. Welsch, *Regression Diagnostics,* Wiley, New York, 1980.
5. Blom, G., *Statistical Estimates and Transformed Beta Variables,* Wiley, New York, pp. 68–75, 143–146, 1958.
6. Box, G. E. P., and D. R Cox, "An Analysis of Transformations," *J. Royal Statistical Soc.,* vol. *B-26,* no. 2, pp. 211–52, 1964.
7. Bradu, D., and Y. Mundlak, "Estimation in Lognormal Linear Models," *J. Am. Stat. Assoc.,* vol. 65, pp. 198–211, 1970.
8. Campbell, G., and J. H. Skillings, "Nonparametric Stepwise Multiple Comparison Procedures," *J. Am. Stat. Assoc.,* vol. 80, pp. 998–1003, 1985.
9. Chambers, J. M., W. S. Cleveland, B. Kleiner, and P. A. Tukey, *Graphical Methods for Data Analysis,* Duxbury, Boston, 1983.
10. Cleveland, W. S., "Robust Locally Weighted Regression and Smoothing Scatterplots," *J. Am. Stat. Assoc.,* vol. 74, pp. 829–836, 1979.
11. Cleveland, W. S., and R. McGill, "The Many Faces of a Scatterplot," *J. Am. Stat. Assoc.,* vol. 79, pp. 807–822, 1984.
12. Cochran, W. G., *Sampling Techniques* (3d ed.), Wiley, New York, 1977.
13. Cohn, T. A., "Adjusted Maximum Likelihood Estimation of the Moments of Lognormal Populations from Type I Censored Samples," *U.S. Geological Survey Open-File Report* 88-350, 1988.
14. Cohn, T. A., D. L. Caulder, E. J. Gilroy, L. D. Zynjuk, and R. M. Summers, "The Validity of a Simple Statistical Model for Estimating Fluvial Constituent Loads: An Empirical Study Involving Nutrient Loads Entering Chesapeake Bay," *Water Resour. Res.,* vol. 28, no. 9, pp. 2353–2364, 1992.
15. Cohn, T. A., L. L. DeLong, E. J. Gilroy, R. M. Hirsch, and D. Wells, "Estimating Constituent Loads," *Water Resour. Res.,* vol. 25, no. 5, pp. 937–942, 1989.
16. Conover, W. J., and R. L. Iman, "Rank Transformation as a Bridge Between Parametric and Nonparametric Statistics," *Am. Stat.,* vol. 35, no. 3, pp. 124–129, 1981.
17. Conover, W. L., *Practical Nonparametric Statistics* (2d ed.), Wiley, New York, 1980.
18. Cook, R. D., "Detection of Influential Observations in Linear Regression," *Technometrics,* vol. 19, pp. 15–18, 1977.
19. Crawford, C. G., "Estimation of Suspended-Sediment Rating Curves and Mean Suspended-Sediment Loads," *J. Hydrol.,* vol. 129, pp. 331–348, 1991.
20. Cunnane, C., "Unbiased Plotting Positions—A Review," *J. Hydrol.,* vol. 37, pp. 205–222, 1978.
21. Draper, N. R., and H. Smith, *Applied Regression Analysis* (2d ed.), Wiley, New York, 1981.
22. Duan, N., "Smearing Estimate: A Nonparametric Retransformation Method," *J. Am. Stat. Assoc.,* vol. 87, pp. 605–610, 1983.
23. Ferguson, R. I., "River Loads Underestimated by Rating Curves," *Water Resour. Res.,* vol. 22, pp. 74–76, 1986.
24. Gilbert, R. O., *Statistical Methods for Environmental Monitoring,* Van Nostrand Reinhold, New York, 1987.

25. Gilliom, R. J., and D. R. Helsel, "Estimation of Distributional Parameters for Censored Trace Level Water Quality Data—1. Estimation Techniques," *Water Resour. Res.*, vol. 22, pp. 1201–1206, 1986.
26. Gilroy, E. J., R. M. Hirsch, and T. A. Cohn, "Mean Square Error of Regression-Based Constituent Transport Estimates," *Water Resour. Res.*, vol. 26, no. 9, pp. 2069–2077, 1990.
27. Gilroy, E. J., and G. D. Tasker, "Multicollinearity and influential observations in hydrologic model selection," in K. Berk and L. Malone, eds., *Computing Science and Statistics, Proceedings of the 21st Symposium on the Interface,* pp. 350–354, 1990.
28. Gringorten, I. I., "A Plotting Rule for Extreme Probability Paper," *J. of Geophys. Res.*, vol. 68, pp. 813–814, 1963.
29. Groggel, D. J., and J. H. Skillings, "Distribution-Free Tests for Main Effects in Multifactor Designs," *Am. Stat.*, vol. 40, pp. 99–102, May 1986.
30. Grygier, J. C., J. R. Stedinger, and H. Yin, 1989, "A Generalized Maintenance of Variance Extension Procedure for Extending Correlated Series," *Water Resour. Res.*, vol. 25, no. 3, pp. 345–349, 1989.
31. Hazen, A., "Storage to Be Provided in the Impounding Reservoirs for Municipal Water Supply," *Trans. Am. Soc. Civ. Eng.*, vol. 77, pp. 1547–1550, 1914.
32. Helsel, D. R., and T. Cohn, "Estimation of Descriptive Statistics for Multiply-Censored Water-Quality Data," *Water Resour. Res.*, vol. 24, no. 12, pp. 1997–2004, 1988.
33. Helsel, D. R., "Less than Obvious: Statistical Treatment of Data below the Detection Limit," *Environ. Sci. Technol.*, vol. 24, pp. 1766–1774, 1990.
34. Helsel, D. R., and R. M. Hirsch, *Statistical Methods in Water Resources,* Elsevier, Amsterdam, 1992.
35. Hirsch, R. M., "A Comparison of Four Streamflow Record Extension Techniques," *Water Resour. Res.*, vol. 18, pp. 1081–1088, 1982.
36. Hirsch, R. M., J. R. Slack, and R. A. Smith, "Techniques of Trend Analysis for Monthly Water-Quality Data," *Water Resour. Res.*, vol. 18, pp. 107–121, 1982.
37. Hirsch, R. M., and E. J. Gilroy, "Methods of Fitting a Straight Line to Data: Examples in Water Resources," *Water Resour. Bull.*, vol. 20, no. 5, pp. 706–711, 1984.
38. Hirsch, R. M., R. B. Alexander, and R. A. Smith, "Selection of Methods for the Detection and Estimation of Trends in Water Quality," *Water Resour. Res.*, vol. 27, pp. 803–813, 1991.
39. Hoaglin, D. C., F. Mosteller, and J. W. Tukey, *Understanding Robust and Exploratory Data Analysis,* Wiley, New York, 1983.
40. Hodges, J. L., Jr., and E. L. Lehmann, "Estimates of Location Based on Rank Tests," *Annals Math. Stat.*, vol. 34, pp. 598–611, 1963.
41. Hoerl, A. E., and R. W. Kennard, "Ridge Regression: Biased Estimation for Nonorthogonal Problems," *Technometrics,* vol. 12, pp. 55–67, 1970.
42. Hollander, M., and D. A. Wolfe, *Nonparametric Statistical Methods,* Wiley, New York, 1973.
43. Hurvich, C. M., and C. L. Tsai, "The Impact of Model Selection on Inference in Linear Regression," *Am. Stat.*, vol. 44, no. 3, pp. 214–216, 1990.
44. Iman, R. L., and W. J. Conover, *A Modern Approach to Statistics,"* Wiley, New York, 1983.
45. Judge, G. G., W. E. Griffiths, R. C. Hill, H. Lutkepohl, and T. C. Lee, "Qualitative and Limited Dependent Variable Models," Chap. 18 in *The Theory and Practice of Econometrics,* Wiley, New York, 1985.
46. Kendall, M. G., "A New Measure of Rank Correlation," *Biometrika,* vol. 30, pp. 81–93, 1938.
47. Kendall, M. G., *Rank Correlation Methods,* Charles Griffin, London, 1975.

48. Kendall, M. G., and Alan Stuart, *The Advanced Theory of Statistics,* Oxford University Press, Oxford, England, 1979.

49. Kleiner, B., and T. E. Graedel, "Exploratory Data Analysis in the Geophysical Sciences," *Reviews of Geophysics and Space Physics,* vol. 18, pp. 699–717, 1980.

50. Kruskal, W. H., "On the Uniqueness of the Line of Organic Correlation," *Biometrics,* vol. 9, pp. 47–58, 1953.

51. Lehmann, E. L., *Nonparametrics: Statistical Methods Based on Ranks,* Holden-Day, Oakland, Calif., 1975.

52. Lettenmaier, D. P., E. R. Hooper, C. Wagoner, and K. B. Faris, "Trends in Stream Quality in the Continental United States 1978–1987," *Water Resour. Res.,* vol. 27, pp. 327–340, 1991.

53. Mallows, C. L., "Some Comments on Cp," *Technometrics,* vol. 15, pp. 661–675, 1973.

54. Marquardt, D. W., "Generalized Inverses, Ridge Regression, Biased Linear Estimation, and Nonlinear Estimation," *Technometrics,* vol. 12, pp. 591–612, 1970.

55. Matalas, N. C., and B. Jacobs, "A Correlation Procedure for Augmenting Hydrologic Data," U.S. Geological Survey Professional Paper 434-E, E1–E7, 1964.

56. McGill, R., J. W. Tukey, and W. A. Larsen, "Variations of Box Plots," *Am. Stat.,* vol. 32, pp. 12–16.

57. Meyers, R. H., *Classical and Modern Regression with Applications,* Duxbury, Boston, 1986.

58. Millard, S. P., and S. J. Deverel, Nonparametric Statistical Methods for Comparing Two Sites Based on Data With Multiple Nondetect Limits," *Water Resour. Res.,* vol. 24, no. 12, pp. 2087–2098, 1988.

59. Miller, A. J., "Selection of Subsets of Regression Variables" (with discussion), *J. Royal Statistical Soc.,* Series A, vol. 147, pp. 389–425, 1984.

60. Montgomery, D. C., *Design and Analysis of Experiments* (3d ed.), Wiley, New York, 1991.

61. Montgomery, D. C., and E. A. Peck, *Introduction to Linear Regression Analysis,* Wiley, New York, 1982.

62. Neter, J., W. Wasserman, and M. H. Kutner, *Applied Linear Statistical Models* (2d ed.), Irwin, Homewood, Ill., 1985.

63. Ott, L., *An Introduction to Statistical Methods and Data Analysis,* Kent, Boston, 1988.

64. Peters, N. E., "Evaluation of Environmental Factors Affecting Yields of Major Dissolved Ions in Streams of the United States," U.S. Geological Survey Water Supply Paper 2228, 1984.

65. Rawlings, J. O., *Applied Regression Analysis: A Research Tool,* Wadsworth and Brooks, Pacific Grove, Calif., 1988.

66. Sauer, V. B., W. O. Thomas, V. A. Strickler, and K. V. Wilson, "Flood Characteristics of Urban Watersheds in the United States," U.S. Geological Survey Water Supply Paper 2207, 1983.

67. Schertz, T. L., and R. M. Hirsch, "Trend Analysis of Weekly Acid Rain Data—1978–83," U.S. Geological Survey Water Resources Investigations Report 85-4211, 1985.

68. Stedinger, J. R., and G. D. Tasker, "Regional Hydrologic Analysis 1. Ordinary, Weighted, and Generalized Least Squares Compared," *Water Resour. Res.,* vol. 21, no. 9, pp. 1421–1432, 1985.

69. Stedinger, J. R., and G. D. Tasker, "Regional Hydrologic Analysis 2: Model-Error Estimators, Estimation of Sigma, and Log-Pearson Type 3 Distributions," *Water Resour. Res.,* vol. 22, no. 10, pp. 1487–1499, 1986.

70. Stoline, M. R., "The Status of Multiple Comparisons: Simultaneous Estimation of All Pairwise Comparisons in One-Way ANOVA Designs," *Am. Stat.,* vol. 35, no. 3, pp. 134–141, 1981.

71. Stoline, M. R., "An Examination of the Lognormal and Box and Cox Family of Transformations in Fitting Environmental Data," *Environmetrics,* vol. 2, no. 1, pp. 85–106, 1991.

72. Tasker, G. D., and J. R. Stedinger, "An Operational GLS Model for Hydrologic Regression," *J. Hydrol.,* vol. 111, pp. 361–375, 1989.

73. Thomas, D. M., and M. A. Benson, "Generalization of Streamflow Characteristics from Drainage-Basin Characteristics," U.S. Geological Survey Water Supply Paper 1975, 1970.

74. Thomas, R. B., "Estimating Total Suspended Sediment Yield with Probability Sampling," *Water Resour. Res.,* vol. 21, no. 9, pp. 1381–1388, 1985.

75. Tobin, J., "Estimation of Relationships for Limited Dependent Variables," *Econometrica,* vol. 26, pp. 24–36, 1958.

76. Tukey, J. W., *Exploratory Data Analysis,* Addison-Wesley, Reading, Mass., 1977.

77. van Belle, G., and J. P. Hughes, "Nonparametric Tests for Trend in Water Quality," *Water Resour. Res.,* vol. 20, pp. 127–136, 1984.

78. Velleman, P. F., and D. C. Hoaglin, *Applications, Basics, and Computing of Exploratory Data Analysis,* Duxbury, Boston, 1981.

79. Vogel, R. M., and J. R. Stedinger, "Minimum Variance Streamflow Record Augmentation Procedures," *Water Resour. Res.,* vol. 21, no. 5, pp. 715–723, 1985.

80. Weibull, W., *"The Phenomenon of Rupture in Solids," Ingeniors Vetenskaps Akademien Handlinga,* vol. 153, p. 17, 1939.

81. Welch, A. H., M. S. Lico, and J. L. Hughes, "Arsenic in Ground Water of the Western United States," *Ground Water,* vol. 26, no. 3, pp. 333–338, 1988.

CHAPTER 18
FREQUENCY ANALYSIS OF EXTREME EVENTS

Jery R. Stedinger
School of Civil and Environmental Engineering
Cornell University
Ithaca, New York

Richard M. Vogel
Department of Civil Engineering
Tufts University
Medford, Massachusetts

Efi Foufoula-Georgiou
Department of Civil and Mineral Engineering
University of Minnesota
Minneapolis, Minnesota

18.1 INTRODUCTION TO FREQUENCY ANALYSIS

Extreme rainfall events and the resulting floods can take thousands of lives and cause billions of dollars in damage. Flood plain management and designs for flood control works, reservoirs, bridges, and other investigations need to reflect the likelihood or probability of such events. Hydrologic studies also need to address the impact of unusually low stream flows and pollutant loadings because of their effects on water quality and water supplies.

The Basic Problem. Frequency analysis is an information problem: if one had a sufficiently long record of flood flows, rainfall, low flows, or pollutant loadings, then a frequency distribution for a site could be precisely determined, so long as change over time due to urbanization or natural processes did not alter the relationships of concern. In most situations, available data are insufficient to precisely define the risk of large floods, rainfall, pollutant loadings, or low flows. This forces hydrologists to use practical knowledge of the processes involved, and efficient and robust statistical techniques, to develop the best estimates of risk that they can.[115] These techniques are generally restricted, with 10 to 100 sample observations to estimate events exceeded with a chance of at least 1 in 100, corresponding to exceedance probabilities of 1 percent or more. In some cases, they are used to estimate the rainfall exceeded with a chance of 1 in 1000, and even the flood flows for spillway design exceeded with a chance of 1 in 10,000.

The hydrologist should be aware that in practice the true probability distributions of the phenomena in question are not known. Even if they were, their functional representation would likely have too many parameters to be of much practical use. The practical issue is how to select a reasonable and simple distribution to describe the phenomenon of interest, to estimate that distribution's parameters, and thus to obtain risk estimates of satisfactory accuracy for the problem at hand.

Common Problems. The hydrologic problems addressed by this chapter primarily deal with the magnitudes of a single variable. Examples include annual minimum 7-day-average low flows, annual maximum flood peaks, or 24-h maximum precipitation depths. These annual maxima and minima for successive years can generally be considered to be independent and identically distributed, making the required frequency analyses straightforward.

In other instances the risk may be attributable to more than one factor. Flood risk at a site may be due to different kinds of events which occur in different seasons, or due to risk from several sources of flooding or *coincident events,* such as both local tributary floods and large regional floods which result in backwater flooding from a reservoir or major river. When the magnitudes of different factors are independent, a *mixture model* can be used to estimate the combined risk (see Sec. 18.6.2). In other instances, it may be necessary or advantageous to consider all events that exceed a specified threshold because it makes a larger data set available, or because of the economic consequences of every event; such *partial duration series* are discussed in Sec. 18.6.1.

18.1.1 Probability Concepts

Let the upper case letter X denote a *random variable,* and the lower case letter x a possible value of X. For a random variable X, its *cumulative distribution function* (cdf), denoted $F_X(x)$, is the probability the random variable X is less than or equal to x:

$$F_X(x) = P(X \leq x) \tag{18.1.1}$$

$F_X(x)$ is the nonexceedance probability for the value x.

Continuous random variables take on values in a continuum. For example, the magnitude of floods and low flows is described by positive real values, so that $X \geq 0$. The *probability density function* (pdf) describes the relative likelihood that a continuous random variable X takes on different values, and is the derivative of the cumulative distribution function:

$$f_X(x) = \frac{dF_X(x)}{dx} \tag{18.1.2}$$

Section 18.2 and Table 18.2.1 provide examples of cdf's and pdf's.

18.1.2 Quantiles, Exceedance Probabilities, Odds Ratios, and Return Periods

In hydrology the *percentiles* or *quantiles* of a distribution are often used as design events. The $100p$ percentile or the pth quantile x_p is the value with cumulative probability p:

$$F_X(x_p) = p \tag{18.1.3}$$

The $100p$ percentile x_p is often called the $100(1-p)$ percent exceedance event because it will be exceeded with probability $1-p$.

The *return period* (sometimes called the *recurrence interval*) is often specified rather than the exceedance probability. For example, the annual maximum flood-flow exceeded with a 1 percent probability in any year, or chance of 1 in 100, is called the 100-year flood. In general, x_p is the T-year flood for

$$T = \frac{1}{1-p} \tag{18.1.4}$$

Here are two ways that return period can be understood. First, in a fixed T-year period the expected number of exceedances of the T-year event is exactly 1, if the distribution of floods does not change over that period; thus on average one flood greater than the T-year flood level occurs in a T-year period.

Alternatively, if floods are independent from year to year, the probability that the first exceedance of level x_p occurs in year k is the probability of $(k-1)$ years without an exceedance followed by a year in which the value of X exceeds x_p:

$$P\,(\text{exactly } k \text{ years until } X \geq x_p) = p^{k-1}\,(1-p). \tag{18.1.5}$$

This is a geometric distribution with mean $1/(1-p)$. Thus the *average* time until the level x_p is exceeded equals T years. However, the probability that x_p is not exceeded in a T-year period is $p^T = (1 - 1/T)^T$, which for $1/(1-p) = T \geq 25$ is approximately 36.7 percent, or about a chance of 1 in 3.

Return period is a means of expressing the exceedance probability. Hydrologists often speak of the 20-year flood or the 1000-year rainfall, rather than events exceeded with probabilities of 5 or 0.1 percent in any year, corresponding to chances of 1 in 20, or of 1 in 1000. Return period has been incorrectly understood to mean that one and only one T-year event should occur every T years. Actually, the probability of the T-year flood being exceeded is $1/T$ in every year. The awkwardness of small probabilities and the incorrect implication of return periods can both be avoided by reporting *odds ratios:* thus the 1 percent exceedance event can be described as a value with a 1 in 100 chance of being exceeded each year.

18.1.3 Product Moments and their Sample Estimators

Several summary statistics can describe the character of the probability distribution of a random variable. Moments and quantiles are used to describe the location or central tendency of a random variable, and its spread, as described in Sec. 17.2 and Table 17.2.1. The *mean* of a random variable X is defined as

$$\mu_X = E[X] \tag{18.1.6}$$

The second moment about the mean is the *variance,* denoted Var (X) or σ_X^2 where

$$\sigma_X^2 = \text{Var}\,(X) = E[(X - \mu_X)^2] \tag{18.1.7}$$

The standard deviation σ_X is the square root of the variance and describes the width or scale of a distribution. These are examples of *product moments* because they depend upon powers of X.

A dimensionless measure of the variability in X, appropriate for use with positive random variables $X \geq 0$, is the *coefficient of variation,* defined in Table 18.1.1. Table 18.1.1 also defines the *coefficient of skewness* γ_X, which describes the relative asymmetry of a distribution, and the *coefficient of kurtosis,* which describes the thickness of a distribution's tails.

Sample Estimators. From a set of observations $(X_1, \ldots, X_n)$, the moments of a distribution can be estimated. Estimators of the mean, variance, and coefficient of skewness are

$$\hat{\mu}_X = \bar{X} = \sum_{i=1}^{n} \frac{X_i}{n}$$

$$\hat{\sigma}_X^2 = S^2 = \left[\frac{\sum_{i=1}^{n} (X_i - \bar{X})^2}{n-1}\right] \tag{18.1.8}$$

$$\hat{\gamma}_X = G = \frac{n \sum_{i=1}^{n} (X_i - \bar{X})^3}{(n-1)(n-2)S^3}$$

TABLE 18.1.1 Definitions of Dimensionless Product-Moment and L-Moment Ratios

Name	Denoted	Definition
	Product-moment ratios	
Coefficient of variation	CV_X	σ_x/μ_x
Coefficient of skewness*	γ_X	$\dfrac{E(X-\mu_x)^3}{\sigma_X^3}$
Coefficient of Kurtosis†		$\dfrac{E(X-\mu_x)^4}{\sigma_X^4}$
	L-moment ratios	
L-coefficient of variation‡	L-CV, τ_2	λ_2/λ_1
L-coefficient of skewness	L-skewness, τ_3	λ_3/λ_2
L-coefficient of kurtosis	L-kurtosis, τ_4	λ_4/λ_2

* Some texts define $\beta_1 = [\gamma_X]^2$ as a measure of skewness.
† Some texts define the kurtosis as $\{E[(X-\mu_x)^4]/\sigma_X^4 - 3\}$; others use the term *excess kurtosis* for this difference because the normal distribution has a kurtosis of 3.
‡ Hosking[72] uses τ instead of τ_2 to represent the L-CV ratio.

Some studies use different versions of $\hat{\sigma}_X^2$ and $\hat{\gamma}_X$ that result from replacing $(n-1)$ and $(n-2)$ in Eq. (18.1.8) by n. This makes relatively little difference for large n. The factor $(n-1)$ in the expression for $\hat{\sigma}_X^2$ yields an unbiased estimator of the variance σ_X^2. The factor $n/[(n-1)(n-2)]$ in expression for $\hat{\gamma}_X$ yields an unbiased estimator of $E[(X-\mu_X)^3]$, and generally reduces but does not eliminate the bias of $\hat{\gamma}_X$ (Ref. 159). Kirby[84] derives bounds on the sample estimators of the coefficients of variation and skewness; in fact, the absolute value of both S and G cannot exceed $\sqrt{n}$ for the sample product-moment estimators in Eq. (18.1.8).

Use of Logarithmic Transformations. When data vary widely in magnitude, as often happens in water-quality monitoring, the sample product moments of the logarithms of the data are often employed to summarize the characteristics of a data set or to estimate parameters of distributions. A logarithmic transformation is an effective vehicle for normalizing values which vary by orders of magnitude, and also for keeping occasionally large values from dominating the calculation of product-moment estimators. However, the danger with use of logarithmic transformations is that unusually small observations (or *low outliers*) are given greatly increased weight. This is a concern if it is the large events that are of interest, small values are poorly measured, small values reflect rounding, or small values are reported as zero if they fall below some threshold.

18.1.4 L Moments and Probability-Weighted Moments

L moments are another way to summarize the statistical properties of hydrologic data.[72] The first L-moment estimator is again the mean:

$$\lambda_1 = E[X] \tag{18.1.9}$$

Let $X_{(i|n)}$ be the ith-largest observation in a sample of size n ($i = 1$ corresponds to the largest). Then, for any distribution, the second L moment is a description of scale based on the expected difference between two randomly selected observations:

$$\lambda_2 = \tfrac{1}{2}\, E[X_{(1|2)} - X_{(2|2)}] \tag{18.1.10}$$

Similarly, L-moment measures of skewness and kurtosis use

$$\begin{aligned} \lambda_3 &= \tfrac{1}{3}\, E[X_{(1|3)} - 2X_{(2|3)} + X_{(3|3)}] \\ \lambda_4 &= \tfrac{1}{4}\, E[X_{(1|4)} - 3X_{(2|4)} + 3X_{(3|4)} - X_{(4|4)}] \end{aligned} \tag{18.1.11}$$

as shown in Table 18.1.1.

Advantages of L Moments. Sample estimators of L moments are linear combinations (hence the name *L moments*) of the ranked observations, and thus do not involve squaring or cubing the observations as do the product-moment estimators in Eq. (18.1.8). As a result, L-moment estimators of the dimensionless coefficients of variation and skewness are almost unbiased and have very nearly a normal distribution; the product-moment estimators of the coefficients of variation and of skewness in Table 18.1.1 are both highly biased and highly variable in small samples. Both Hosking[72] and Wallis[163] discuss these issues. In many hydrologic applications an occasional event may be several times larger than other values; when product moments are used, such values can mask the information provided by the other observations, while product moments of the logarithms of sample values can overemphasize small values. In a wide range of hydrologic applications, L moments provide simple and reasonably efficient estimators of the characteristics of hydrologic data and of a distribution's parameters.

L-Moment Estimators. Just as the variance, or coefficient of skewness, of a random variable are functions of the moments $E[X]$, $E[X^2]$, and $E[X^3]$, L moments can be written as functions of *probability-weighted moments* (PWMs),[48,72] which can be defined as

$$\beta_r = E\{X\,[F(X)]^r\} \tag{18.1.12}$$

where $F(X)$ is the cdf for X. Probability-weighted moments are the expectation of X times powers of $F(X)$. (Some authors define PWMs in terms of powers of $[1-F(X)]$.) For $r=0$, β_0 is the population mean μ_X.

Estimators of L moments are mostly simply written as linear functions of estimators of PWMs. The first PWM estimator b_0 of β_0 is the sample mean $\overline{X}$ in Eq. (18.1.8).

To estimate other PWMs, one employs the ordered observations, or the *order statistics* $X_{(n)} \leq \cdots \leq X_{(1)}$, corresponding to the sorted or ranked observations in a sample $(X_i | i = 1, \cdots, n)$. A simple estimator of β_r for $r \geq 1$ is

$$b_r^* = \frac{1}{n} \sum_{j=1}^{n} X_{(j)} \left[1 - \frac{(j-0.35)}{n}\right]^r \tag{18.1.13}$$

where $1-(j-0.35)/n$ are estimators of $F(X_{(j)})$. b_r^* is suggested for use when estimating quantiles and fitting a distribution at a single site; though it is biased, it generally yields smaller mean square error quantile estimators than the unbiased estimators in Eq. (18.1.14) below.[68,89]

When unbiasedness is important, one can employ unbiased PWM estimators

$$\begin{aligned}
b_0 &= \overline{X} \\
b_1 &= \sum_{j=1}^{n-1} \frac{(n-j)X_{(j)}}{n(n-1)} \\
b_2 &= \sum_{j=1}^{n-2} \frac{(n-j)(n-j-1)X_{(j)}}{n(n-1)(n-2)} \\
b_3 &= \sum_{j=1}^{n-3} \frac{(n-j)(n-j-1)(n-j-2)X_{(j)}}{n(n-1)(n-2)(n-3)}
\end{aligned} \tag{18.1.14}$$

These are examples of the general formula

$$\hat{\beta}_r = b_r = \frac{1}{n} \sum_{j=1}^{n-r} \frac{\binom{n-j}{r} X_{(j)}}{\binom{n-1}{r}} = \frac{1}{(r+1)} \sum_{j=1}^{n-r} \frac{\binom{n-j}{r} X_{(j)}}{\binom{n}{r+1}} \tag{18.1.15}$$

for $r = 1, \ldots, n-1$ [see Ref. 89, which defines PWMs in terms of powers of $(1-F)$]; this formula can be derived using the fact that $(r+1)\beta_r$ is the expected value of the largest observation in a sample of size $(r+1)$. The unbiased estimators are recommended for calculating L moment diagrams and for use with regionalization procedures where unbiasedness is important.

For any distribution, L moments are easily calculated in terms of PWMs from

$$\begin{aligned}\lambda_1 &= \beta_0\\ \lambda_2 &= 2\beta_1 - \beta_0\\ \lambda_3 &= 6\beta_2 - 6\beta_1 + \beta_0\\ \lambda_4 &= 20\beta_3 - 30\beta_2 + 12\beta_1 - \beta_0\end{aligned} \tag{18.1.16}$$

Estimates of the λ_i are obtained by replacing the unknown β_r by sample estimators b_r from Eq. (18.1.14). Table 18.1.1 contains definitions of dimensionless L-moment coefficients of variation τ_2, of skewness τ_3, and of kurtosis τ_4. L-moment ratios are bounded. In particular, for nondegenerate distributions with finite means, $|\tau_r| < 1$ for $r = 3$ and 4, and for positive random variables, $X > 0$, $0 < \tau_2 < 1$. Table 18.1.2 gives expressions for λ_1, λ_2, τ_3, and τ_4 for several distributions. Figure 18.1.1 shows relationships between τ_3 and τ_4. (A library of FORTRAN subroutines for L-moment analyses is available;[74] see Sec. 18.11.)

Table 18.1.3 provides an example of the calculation of L moments and PWMs. The short rainfall record exhibits relatively little variability and almost zero skewness. The sample product-moment CV for the data of 0.25 is about twice the L-CV $\hat{\tau}_2$ equal to 0.14, which is typical because λ_2 is often about half of σ.

L-Moment and PWM Parameter Estimators. Because the first r L moments are linear combinations of the first r PWMs, fitting a distribution so as to reproduce the first r sample L moments is equivalent to using the corresponding sample PWMs. In fact, PWMs were developed first in terms of powers of $(1 - F)$ and used as effective statistics for fitting distributions.[89,90] Later the PWMs were expressed as L moments which are more easily interpreted.[72,128] Section 18.2 provides formulas for the parameters of several distributions in terms of sample L moments, many of which are obtained by inverting expressions in Table 18.1.2. (See also Ref. 72.)

18.1.5 Parameter Estimation

Fitting a distribution to data sets provides a compact and smoothed representation of the frequency distribution revealed by the available data, and leads to a systematic procedure for extrapolation to frequencies beyond the range of the data set. When flood flows, low flows, rainfall, or water-quality variables are well-described by some family of distributions, a task for the hydrologist is to estimate the parameters Θ of that distribution so that required quantiles and expectations can be calculated with the "fitted" model. For example, the normal distribution has two parameters, μ and σ^2. Appropriate choices for distribution functions can be based on examination of the data using probability plots and moment ratios (discussed in Sec. 18.3), the physical origins of the data, previous experience, and administrative guidelines.

Several general approaches are available for estimating the parameters of a distribution. A simple approach is the *method of moments,* which uses the available sample to compute an estimate $\hat{\Theta}$ of Θ so that the theoretical moments of the distribution of X exactly equal the corresponding sample moments described in Sec. 18.1.3. Alternatively, parameters can be estimated using the sample L moments discussed in Sec. 18.1.4, corresponding to the *method of L moments.*

Still another method that has strong statistical motivation is the *method of maximum likelihood.* Maximum likelihood estimators (MLEs) have very good statistical

TABLE 18.1.2 Values of L Moments and Relationships for the Inverse of the cdf for Several Distributions

Distribution and inverse cdf	L moments
Uniform:	$\lambda_1 = \dfrac{\beta+\alpha}{2} \quad \lambda_2 = \dfrac{\beta-\alpha}{6}$
$x = \alpha + (\beta - \alpha)F$	$\tau_3 = \tau_4 = 0$
Exponential:*	$\lambda_1 = \xi + \dfrac{1}{\beta} \quad \lambda_2 = \dfrac{1}{2\beta}$
$x = \xi - \dfrac{\ln[1-F]}{\beta}$	$\tau_3 = \dfrac{1}{3} \quad \tau_4 = \dfrac{1}{6}$
Normal†	$\lambda_1 = \mu \quad \lambda_2 = \dfrac{\sigma}{\sqrt{\pi}}$
$x = \mu + \sigma\Phi^{-1}[F]$	$\tau_3 = 0 \quad \tau_4 = 0.1226$
Gumbel: $x = \xi - \alpha \ln[-\ln F]$	$\lambda_1 = \xi + 0.5772\,\alpha \quad \lambda_2 = \alpha \ln 2$ $\tau_3 = 0.1699 \quad \tau_4 = 0.1504$
GEV:	$\lambda_1 = \xi + \dfrac{\alpha}{\kappa}\{1 - \Gamma[1+\kappa]\} \quad \lambda_2 = \dfrac{\alpha}{\kappa}(1 - 2^{-\kappa})\,\Gamma(1+\kappa)$
$x = \xi + \dfrac{\alpha}{\kappa}\{1 - [-\ln F]^{\kappa}\}$	$\tau_3 = \left\{\dfrac{2(1-3^{-\kappa})}{(1-2^{-\kappa})} - 3\right\}$
	$\tau_4 = \dfrac{1 - 5(4^{-\kappa}) + 10(3^{-\kappa}) - 6(2^{-\kappa})}{1 - 2^{-\kappa}}$
Generalized Pareto:	$\lambda_1 = \xi + \dfrac{\alpha}{1+\kappa} \quad \lambda_2 = \dfrac{\alpha}{(1+\kappa)(2+\kappa)}$
$x = \xi + \dfrac{\alpha}{\kappa}\{1 - [1-F]^{\kappa}\}$	$\tau_3 = \dfrac{1-\kappa}{3+\kappa} \quad \tau_4 = \dfrac{(1-\kappa)(2-\kappa)}{(3+\kappa)(4+\kappa)}$
Lognormal	See Eqs. (18.2.12), (18.2.13)
Gamma	See Eqs. (18.2.30), (18.2.31)

* Alternative parameterization consistent with that for Pareto and GEV distributions is: $x = \xi - \alpha \ln[1-F]$ yielding $\lambda_1 = \xi + \alpha$; $\lambda_2 = \alpha/2$.
† Φ^{-1} denotes the inverse of the standard normal distribution (see Sec. 18.2.1).
Note: F denotes cdf $F_X(x)$.
Source: Adapted from Ref. 72, with corrections.

properties in large samples, and experience has shown that they generally do well with records available in hydrology. However, often MLEs cannot be reduced to simple formulas, so estimates must be calculated using numerical methods.[85] MLEs sometimes perform poorly when the distribution of the observations deviates in significant ways from the distribution being fit.

A different philosophy is embodied in Bayesian inference, which combines prior information and regional hydrologic information with the likelihood function for available data. Advantages of the Bayesian approach are that it allows the explicit modeling of uncertainty in parameters, and provides a theoretically consistent framework for integrating systematic flow records with regional and other hydrologic information.[3,88,127,150]

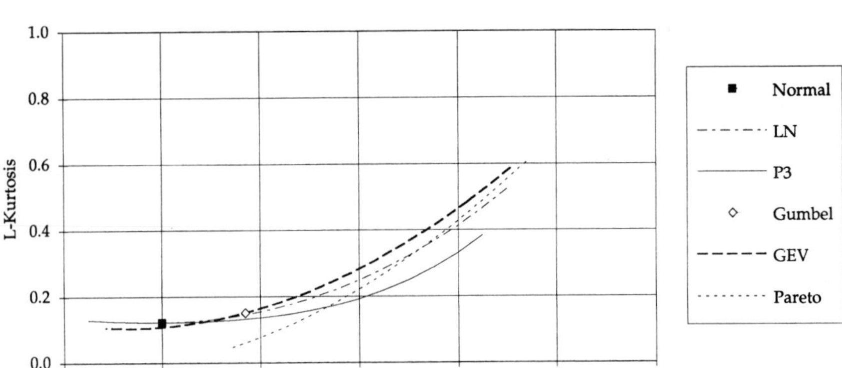

FIGURE 18.1.1 This L-moment diagram illustrates the relationship between the L-kurtosis τ_4 and the L-skewness τ_3 for the normal, lognormal (LN), Pearson type 3 (P3), Gumbel, generalized extreme value (GEV), and Pareto distributions.

TABLE 18.1.3 Example of Calculation of L Moments Using Ranked Annual Maximum 10-min Rainfall Depths for Chicago, 1940–1947

Year	1943	1941	1944	1945	1946	1947	1942	1940
Depth, in*	0.92	0.70	0.66	0.65	0.63	0.60	0.57	0.34
Rank	1	2	3	4	5	6	7	8

Biased PWMs [Eq. (18.1.13)]

$\bar{x} = b_0^* = 0.6338$; $b_1^* = 0.3434$; $b_2^* = 0.2355$

Unbiased PWMs [Eq. (18.1.14)]

$$\bar{x} = b_0 = 0.6338;\ b_1 = \frac{0.92 + (6/7)*(0.70) + (5/7)*(0.66) + \cdots + 0*(0.34)}{8} = 0.3607$$

$$b_2 = \frac{0.92 + (30/42)*(0.70) + (20/42)*(0.66) + \cdots + 0*(0.57) + 0*(0.34)}{8} = 0.2548$$

L moments using unbiased PWMs [Eq. (18.1.16)]

$\hat{\lambda}_1 = b_0 = 0.6338$; $\hat{\lambda}_2 = 2b_1 - b_0 = 0.0877$; $\hat{\lambda}_3 = 6b_2 - 6b_1 + b_0 = -0.0016$

L-CV and L skewness (Table 18.1.1)

$\hat{\tau}_2 = 0.138$; $\hat{\tau}_3 = -0.018$

Product moments for comparison [Eq. (18.1.8)]

$n = 8$, $\bar{x} = 0.6338$; $s = 0.160$; CV $= 0.252$, $G = -0.088$

* 1 in = 25.4 mm.

Occasionally *nonparametric methods* are employed to estimate frequency relationships. These have the advantage that they do not assume that floods are drawn from a particular family of distributions.[2,7] Modern nonparametric methods have not yet seen much use in practice and have rarely been used officially. However, curve-fitting procedures which employ plotting positions discussed in Sec. 18.3.2 are nonparametric procedures often used in hydrology.

Of concern are the *bias, variability,* and *accuracy* of parameter estimators $\hat{\Theta}[X_1, \ldots, X_n]$, where this notation emphasizes that an estimator $\hat{\Theta}$ is a random variable whose value depends on observed sample values $\{X_1, \ldots X_n\}$. Studies of estimators evaluate an estimator's bias, defined as

$$\text{Bias}\,[\hat{\Theta}] = E[\hat{\Theta}] - \Theta \tag{18.1.17}$$

and sample-to-sample variability, described by Var $[\hat{\Theta}]$. One wants estimators to be nearly unbiased so that on average they have nearly the correct value, and also to have relatively little variability. One measure of accuracy which combines bias and variability is the *mean square error,* defined as

$$\text{MSE}\,[\hat{\Theta}] = E[(\hat{\Theta} - \Theta)^2] = \{\text{Bias}\,[\hat{\Theta}]\}^2 + \text{Var}\,[\hat{\Theta}] \tag{18.1.18}$$

An *unbiased* estimator (Bias $[\hat{\Theta}] = 0$) will have a mean square error equal to its variance. For a given sample size n, estimators with the smallest possible mean square errors are said to be *efficient.*

Bias and mean square error are statistically convenient criteria for evaluating estimators of a distribution's parameters or of quantiles. In particular situations, hydrologists can also evaluate the expected probability and under- or overdesign, or use economic loss functions related to operation and design decisions.[112,124]

18.2 PROBABILITY DISTRIBUTIONS FOR EXTREME EVENTS

This section provides descriptions of several families of distributions commonly used in hydrology. These include the normal/lognormal family, the Gumbel/Weibull/generalized extreme value family, and the exponential/Pearson/log-Pearson type 3 family. Table 18.2.1 provides a summary of the pdf or cdf of these probability distributions, and their means and variances. (See also Refs. 54 and 85.) The L moments for several distributions are reported in Table 18.1.2. Many other distributions have also been successfully employed in hydrologic applications, including the five-parameter Wakeby distribution,[69,75] the Boughton distribution,[14] and the TCEV distribution (corresponding to a mixture of two Gumbel distributions[119]).

18.2.1 Normal Family: N, LN, LN3

The normal (N), or Gaussian distribution is certainly the most popular distribution in statistics. It is also the basis of the lognormal (LN) and three-parameter lognormal (LN3) distributions which have seen many applications in hydrology. This section describes the basic properties of the normal distribution first, followed by a discussion of the LN and LN3 distributions. Goodness-of-fit tests are discussed in Sec. 18.3 and standard errors of quantile estimators in Sec. 18.4.2.

The Normal Distribution. The normal distribution is useful in hydrology for describing well-behaved phenomena such as average annual stream flow, or average annual pollutant loadings. The *central limit theorem* demonstrates that if a random variable X is the sum of n independent and identically distributed random variables with finite variance, then with increasing n the distribution of X becomes normal regardless of the distribution of the original random variables.

The pdf for a normal random variable X is

$$f_X(x) = \frac{1}{\sqrt{2\pi\sigma_X^2}} \exp\left[-\frac{1}{2}\left(\frac{x-\mu_X}{\sigma_X}\right)^2\right] \tag{18.2.1}$$

X is unbounded both above and below, with mean μ_X and variance σ_X^2. The normal distribution's skew coefficient is zero because the distribution is symmetric. The product-moment coefficient of kurtosis, $E[(X-\mu_X)^4]/\sigma^4$, equals 3. L moments are given in Table 18.1.2.

The two moments of the normal distribution, μ_X and σ_X^2, are its natural parameters. They are generally estimated by the sample mean and variance in Eq. (18.1.8); these are the maximum likelihood estimates if $(n-1)$ is replaced by n in the denominator of the sample variance. The cdf of the normal distribution is not available in closed form. Selected points z_p for the *standard normal distribution* with zero mean and unit variance are given in Table 18.2.2; because the normal distribution is symmetric, $z_p = -z_{1-p}$.

An approximation, generally adequate for simple tasks and plotting, for the standard normal cdf, denoted $\Phi(z)$, is

$$\Phi(z) = 1 - 0.5 \exp\left[-\frac{(83z+351)z+562}{703/z+165}\right] \tag{18.2.2}$$

for $0 < z \leq 5$. An approximation for the inverse of the standard normal cdf, denoted $\Phi^{-1}(p)$ is

$$\Phi^{-1}(p) = z_p = \frac{p^{0.135}-(1-p)^{0.135}}{0.1975} \tag{18.2.3a}$$

or the more accurate expression valid for $10^{-7} < p < 0.5$

$$\Phi^{-1}(p) = z_p = -\sqrt{\frac{y^2[(4y+100)y+205]}{[(2y+56)y+192]y+131}} \tag{18.2.3b}$$

where $y = -\ln(2p)$. [Eqs. (18.2.2) and (18.2.3*b*) are from Ref. 35.]

Lognormal Distribution. Many hydrologic processes are positively skewed and are not normally distributed. However, in many cases for *strictly positive random variables* $X > 0$, their logarithm

$$Y = \ln(X) \tag{18.2.4}$$

is well-described by a normal distribution. This is particularly true if the hydrologic variable results from some multiplicative process, such as dilution. Inverting Eq. (18.2.4) yields

$$X = \exp(Y)$$

TABLE 18.2.1 Commonly Used Frequency Distributions in Hydrology (see also Table 18.1.2)

Distribution	pdf and/or cdf	Range	Moments
Normal	$f_X(x) = \frac{1}{\sqrt{2\pi\sigma_X^2}} \exp\left[-\frac{1}{2}\left(\frac{x-\mu_X}{\sigma_X}\right)^2\right]$	$-\infty < x < \infty$	μ_X and σ_X^2; $\gamma_X = 0$
Lognormal*	$f_X(x) = \frac{1}{x\sqrt{2\pi\sigma_Y^2}} \exp\left[-\frac{1}{2}\left(\frac{\ln(x)-\mu_Y}{\sigma_Y}\right)^2\right]$	$x > 0$	$\mu_X = \exp\left(\mu_Y + \frac{\sigma_Y^2}{2}\right)$
			$\sigma_X^2 = \mu_X^2[\exp(\sigma_Y^2) - 1]$
			$\gamma_x = 3\text{CV}_x + \text{CV}_x^3$
Pearson type 3	$f_X(x) = \lvert\beta\rvert[\beta(x-\xi)]^{\alpha-1} \frac{\exp[-\beta(x-\xi)]}{\Gamma(\alpha)}$	$\alpha > 0$	$\mu_x = \xi + \frac{\alpha}{\beta}$; $\sigma_X^2 = \frac{\alpha}{\beta^2}$
	$\Gamma(\alpha)$ is the gamma function	for $\beta > 0$: $x > \xi$	and $\gamma_x = \frac{2}{\sqrt{\alpha}}$
	(for $\beta > 0$ and $\xi = 0$: $\gamma_x = 2\,\text{CV}_x$)	for $\beta < 0$: $x < \xi$	and $\gamma_x = \frac{-2}{\sqrt{\alpha}}$
log-Pearson type 3	$f_X(x) = \lvert\beta\rvert\{\beta[\ln(x) - \xi]\}^{\alpha-1} \frac{\exp\{-\beta[\ln(x)-\xi]\}}{x\Gamma(\alpha)}$	for $\beta < 0$, $0 < x < \exp(\xi)$; for $\beta > 0$, $\exp(\xi) < x < \infty$	See Eq. (18.2.34)
Exponential	$f_X(x) = \beta \exp[-\beta(x-\xi)]$	$x > \xi$ for $\beta > 0$	$\mu_X = \xi + \frac{1}{\beta}$; $\sigma_X^2 = \frac{1}{\beta^2}$
	$F_X(x) = 1 - \exp\{-\beta(x-\xi)\}$		$\gamma_X = 2$

Gumbel	$f_X(x) = \frac{1}{\alpha}\exp\left[-\frac{x-\xi}{\alpha} - \exp\left(-\frac{x-\xi}{\alpha}\right)\right]$	$-\infty < x < \infty$	$\mu_x = \xi + 0.5772\alpha$
	$F_X(x) = \exp\left[-\exp\left(-\frac{x-\xi}{\alpha}\right)\right]$		$\sigma_X^2 = \frac{\pi^2\alpha^2}{6} \approx 1.645\alpha^2;\ \gamma_x = 1.1396$
GEV	$F_X(x) = \exp\left\{-\left[1 - \frac{\kappa(x-\xi)}{\alpha}\right]^{1/\kappa}\right\}$	(σ_x^2 exists for $\kappa > -0.5$)	$\mu_x = \xi + \left(\frac{\alpha}{\kappa}\right)[1 - \Gamma(1+\kappa)]$
	when $\kappa > 0,\ x < \left(\xi + \frac{\alpha}{\kappa}\right);\ \kappa < 0,\ x > \left(\xi + \frac{\alpha}{\kappa}\right)$		$\sigma_X^2 = \left(\frac{\alpha}{\kappa}\right)^2\{\Gamma(1+2\kappa) - [\Gamma(1+\kappa)]^2\}$
Weibull	$f_X(x) = \left(\frac{k}{\alpha}\right)\left(\frac{x}{\alpha}\right)^{k-1}\exp\left[-\left(\frac{x}{\alpha}\right)^k\right]$	$x > 0;\ \alpha,\ k > 0$	$\mu_x = \alpha\,\Gamma\left(1 + \frac{1}{k}\right)$
	$F_X(x) = 1 - \exp[-(x/\alpha)^k]$		$\sigma_X^2 = \alpha^2\left\{\Gamma\left(1+\frac{2}{k}\right) - \left[\Gamma\left(1+\frac{1}{k}\right)\right]^2\right\}$
Generalized Pareto	$f_X(x) = \left(\frac{1}{\alpha}\right)\left[1 - \kappa\frac{(x-\xi)}{\alpha}\right]^{1/\kappa - 1}$	for $\kappa < 0,\ \xi \le x < \infty$	$\mu_X = \xi + \frac{\alpha}{(1+\kappa)}$
	$F_X(x) = 1 - \left[1 - \kappa\frac{(x-\xi)}{\alpha}\right]^{1/\kappa}$	for $\kappa > 0,\ \xi \le x \le \xi + \frac{\alpha}{\kappa}$	$\sigma_X^2 = \alpha^2/[(1+\kappa)^2(1+2\kappa)]$
		(γ_X exists for $\kappa > -0.33$)	$\gamma_X = \frac{2(1-\kappa)(1+2\kappa)^{1/2}}{(1+3\kappa)}$

* Here $Y = \ln(X)$. Text gives formulas for three-parameter lognormal distribution, and for two- and three-parameter lognormal with common base 10 logarithms.

TABLE 18.2.2 Quantiles of the Standard Normal Distribution

p	0.5	0.6	0.75	0.8	0.9	0.95	0.975	0.99	0.998	0.999
z_p	0.000	0.253	0.675	0.842	1.282	1.645	1.960	2.326	2.878	3.090

If X has a lognormal distribution, the cdf for X is

$$F_X(x) = P(X \le x) = P[Y \le \ln(x)] = P\left[\frac{Y-\mu_Y}{\sigma_Y} \le \frac{\ln(x)-\mu_Y}{\sigma_Y}\right]$$

$$= \Phi\left[\frac{\ln(x)-\mu_Y}{\sigma_Y}\right] \qquad (18.2.5)$$

where Φ is the cdf of the standard normal distribution. The lognormal pdf for X in Table 18.2.1 is illustrated in Fig. 18.2.1. $f_X(x)$ is tangent to the horizontal axis at $x = 0$. As a function of the coefficient of variation CV_X, the skew coefficient is

$$\gamma_X = 3\text{CV}_X + \text{CV}_X^3$$

As the coefficients of variation and skewness go to zero, the lognormal distribution approaches a normal distribution.

Table 18.2.1 provides formulas for the first three moments of a lognormally distributed variable X in terms of the first two moments of the normally distributed variable Y. The relationships for μ_X and σ_X^2 can be inverted to obtain

$$\sigma_Y = \left[\ln\left(1 + \frac{\sigma_X^2}{\mu_X^2}\right)\right]^{1/2} \qquad \text{and} \qquad \mu_Y = \ln(\mu_X) - \tfrac{1}{2}\sigma_Y^2 \qquad (18.2.6)$$

These two equations allow calculation of the method of moments estimators of μ_Y and σ_Y, which are the natural parameters of the lognormal distribution.

Alternatively, the logarithms of the sample (x_i) are a sample of Y's: $[y_i = \ln(x_i)]$.

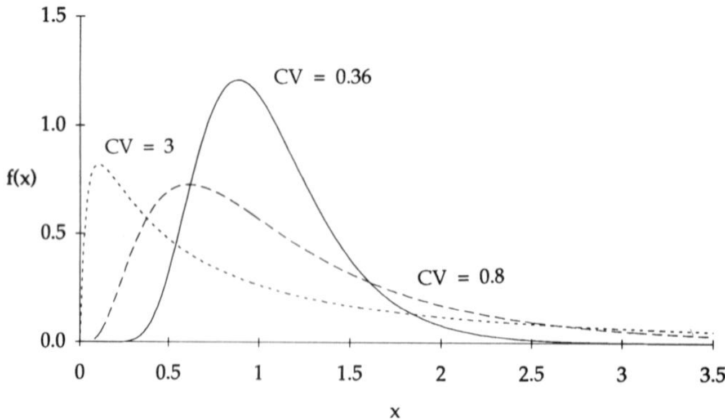

FIGURE 18.2.1 The probability density function of the lognormal distribution with coefficients of variation CV = 0.36, 0.8, and 3, which have coefficients of skewness $\gamma_X = 1.13$, 2.9, and 33 (corresponding to $\mu_Y = 0$ and $\sigma_Y =$ 0.35, 0.7, and 1.5 for base e logarithms, or $\mu_W = 0$ and $\sigma_W = 0.15, 0.30$, and 0.65 for common base 10 logarithms.)

The sample mean and variance of the observed (y_i), obtained by using Eq. (18.1.8), are the maximum-likelihood estimators of the lognormal distribution's parameters if $(n-1)$ is replaced by n in the denominator of s_Y^2. The moments of the y_i's are both easier to compute and generally more efficient than the moment estimators in Eq. (18.2.6), provided the sample does not include unusually small values;[126] see discussion of logarithmic transformations in Sec. 18.1.3.

Hydrologists often use common base 10 logarithms instead of natural logarithms. Let W be the common logarithm of X, $\log(X)$. Then Eq. (18.2.5) becomes

$$F_X(x) = P\left[\frac{W-\mu_W}{\sigma_W} \le \frac{\log(x)-\mu_W}{\sigma_W}\right] = \Phi\left[\frac{\log(x)-\mu_W}{\sigma_W}\right]$$

The moments of X in terms of those of W are

$$\mu_X = 10^{\mu_W + \ln(10)\sigma_W^2/2} \quad \text{and} \quad \sigma_X^2 = \mu_X^2\,(10^{\ln(10)\sigma_W^2} - 1) \tag{18.2.7}$$

where $\ln(10) = 2.303$. These expressions may be inverted to obtain:

$$\sigma_W = \left[\frac{\log(1+\sigma_X^2/\mu_X^2)}{\ln(10)}\right]^{1/2} \quad \text{and} \quad \mu_W = \log(\mu_X) - \tfrac{1}{2}\ln(10)\,\sigma_W^2 \tag{18.2.8}$$

Three-Parameter Lognormal Distribution. In many cases the logarithms of a random variable X are not quite normally distributed, but subtracting a lower bound parameter ξ before taking logarithms may resolve the problem. Thus

$$Y = \ln(X-\xi) \tag{18.2.9a}$$

is modeled as having a normal distribution, so that

$$X = \xi + \exp(Y) \tag{18.2.9b}$$

For any probability level p, the quantile x_p is given by

$$x_p = \xi + \exp(\mu_Y + \sigma_Y z_p) \tag{18.2.9c}$$

In this case the first two moments of X are

$$\mu_X = \xi + \exp(\mu_Y + \tfrac{1}{2}\sigma_Y^2) \quad \text{and} \quad \sigma_X^2 = [\exp(2\mu_Y + \sigma_Y^2)]\,[\exp(\sigma_Y^2) - 1] \tag{18.2.10a}$$

with skewness coefficient

$$\gamma_X = 3\phi + \phi^3$$

where $\phi = [\exp(\sigma_Y^2) - 1]^{0.5}$. If common base 10 logarithms are employed so that $W = \log(X-\xi)$, the value of ξ and the formula for γ_X are unaffected, but Eq. (18.2.10a) becomes

$$\mu_X = \xi + 10^{\mu_W + \ln(10)\,\sigma_W^2/2} \quad \text{and} \quad \sigma_X^2 = (\mu_X - \xi)^2\phi^2 \tag{18.2.10b}$$

with $\phi = (10^{\ln(10)\sigma_W^2/2} - 1)^{0.5}$.

Method-of-moment estimators for the three-parameters lognormal distribution are relatively inefficient. A simple and efficient estimator of ξ is the *quantile-lower-bound estimator:*

$$\hat{\xi} = \frac{x_{(1)}x_{(n)} - x_{\text{median}}^2}{x_{(1)} + x_{(n)} - 2x_{\text{median}}} \tag{18.2.11}$$

when $x_{(1)} + x_{(n)} - 2x_{\text{median}} > 0$, where $x_{(1)}$ and $x_{(n)}$ are, respectively, the largest and smallest observed values; x_{median} is the sample medium equal to $x_{(k+1)}$ for odd sample sizes $n = 2k + 1$, and $\frac{1}{2}(x_{(k)} + x_{(k+1)})$ for even $n = 2k$. [When $x_{(1)} + x_{(n)} - 2x_{\text{median}} <$ 0, the formula provides an upper bound so that $\ln(\xi - x)$ would be normally distributed.] Given ξ, one can estimate μ_Y and σ_Y^2 by using the sample mean and variance of $y_i = \ln(x_i - \hat{\xi})$, or $w_i = \log(x_i - \hat{\xi})$. The quantile-lower-bound estimator's performance with the resultant sample estimators of μ_Y and σ_Y^2 is better than method-of-moments estimators and competitive with maximum likelihood estimators.[67,126]

For the two-parameter and three-parameter lognormal distribution, the second L moment is

$$\lambda_2 = \exp\left(\mu_Y + \frac{\sigma_Y^2}{2}\right) \operatorname{erf}\left(\frac{\sigma_Y}{2}\right) = 2\exp\left(\mu_Y + \frac{\sigma_Y^2}{2}\right)\left[\Phi\left(\frac{\sigma_Y}{\sqrt{2}}\right) - \frac{1}{2}\right] \quad (18.2.12)$$

The following polynomial approximates, within 0.0005 for $|\tau_3| < 0.9$, the relationship between the third and fourth L-moment ratios, and is thus useful for comparing sample values of those ratios with the theoretical values for two- or three-parameter lognormal distributions:[73]

$$\tau_4 = 0.12282 + 0.77518\,\tau_3^2 + 0.12279\,\tau_3^4 - 0.13638\,\tau_3^6 + 0.11368\,\tau_3^8 \quad (18.2.13)$$

18.2.2 GEV Family: Gumbel, GEV, Weibull

Many random variables in hydrology correspond to the maximum of several similar processes, such as the maximum rainfall or flood discharge in a year, or the lowest stream flow. The physical origin of such random variables suggests that their distribution is likely to be one of several *extreme value (EV) distributions* described by Gumbel.[51] The cdf of the largest of n independent variates with common cdf $F(x)$ is simply $F(x)^n$. (See Sec. 18.6.2.) For large n and many choices for $F(x)$, $F(x)^n$ converges to one of three extreme value distributions, called types I, II, and III. Unfortunately, for many hydrologic variables this convergence is too slow for this argument alone to justify adoption of an extreme value distribution as a model of annual maxima and minima.

This section first considers the EV type I distribution, called the *Gumbel distribution.* The *generalized extreme value distribution* (GEV) is then introduced. It spans the three types of extreme value distributions for maxima popularized by Gumbel.[68,80] Finally, the *Weibull distribution* is developed, which is the extreme value type III distribution for minima bounded below by zero. Goodness-of-fit tests are discussed in Sec. 18.3 and standard errors of quantile estimators in Sec. 18.4.4.

The Gumbel Distribution. Let $M_1, \ldots, M_n$ be a set of daily rainfall, stream flow, or pollutant concentrations, and let the random variable $X = \max(M_i)$ be the maximum for the year. If the M_i are independent and identically distributed random variables unbounded above, with an "exponential-like" upper tail (examples include the normal, Pearson type 3, and lognormal distributions), then for large n the variate X has an extreme value type I distribution, or Gumbel distribution.[3,51] For example, the annual-maximum 24-h rainfall depths are often described by a Gumbel distribution, as are annual maximum stream flows.

The Gumbel distribution has the cdf, mean, and variance given in Table 18.2.1, and corresponding L moments are given in Table 18.1.3. The cdf is easily inverted to obtain

$$x_p = \xi - \alpha \ln[-\ln(p)] \quad (18.2.14)$$

The estimator of α obtained by using the second sample L moment is

$$\hat{\alpha} = \frac{\hat{\lambda}_2}{\ln(2)} = 1.443\,\hat{\lambda}_2 \tag{18.2.15}$$

If the sample variance s^2 from Eq. (18.1.8) were employed instead, one obtains

$$\hat{\alpha} = \frac{s\sqrt{6}}{\pi} = 0.7797\,s \tag{18.2.16}$$

The corresponding estimator of ξ in either case is

$$\hat{\xi} = \bar{x} - 0.5772\,\hat{\alpha} \tag{18.2.17}$$

L-moment estimators for the Gumbel distribution are generally as good or better than method-of-moment estimators when the observations are actually drawn from a Gumbel distribution, though maximum likelihood estimators are the best in that case.[89] However, L-moment estimators have been shown to be robust, providing more accurate quantile estimators than product-moment and maximum-likelihood estimators when observations are drawn from a range of reasonable distributions for flood flows.[90]

The Gumbel distribution's density function is very similar to that of the lognormal distribution with $\gamma = 1.13$ in Fig. 18.2.1. Changing ξ and α moves the center of the Gumbel pdf, and changes its width, but does not change the shape of the distribution. The Gumbel distribution has a fixed coefficient of skewness $\gamma = 1.1396$. For large x, the Gumbel distribution is asymptotically equivalent to the exponential distribution with cdf $\{1 - \exp[-(x - \xi)/\alpha]\}$.

The Generalized Extreme Value Distribution. This is a general mathematical form which incorporates Gumbel's type I, II, and III extreme value distributions for maxima.[68,80] The GEV distribution's cdf can be written

$$F(x) = \exp\left\{-\left[1 - \frac{\kappa(x-\xi)}{\alpha}\right]^{1/\kappa}\right\} \qquad \text{for } \kappa \neq 0 \tag{18.2.18}$$

The Gumbel distribution is obtained when $\kappa = 0$. For $|\kappa| < 0.3$, the general shape of the GEV distribution is similar to the Gumbel distribution, though the right-hand tail is thicker for $\kappa < 0$ and thinner for $\kappa > 0$.

Here ξ is a location parameter, α is a scale parameter, and κ is the important shape parameter. For $\kappa > 0$ the distribution has a finite upper bound at $\xi + \alpha/\kappa$ and corresponds to the EV type III distribution for maxima that are bounded above; for $\kappa < 0$, the distribution has a thicker right-hand tail and corresponds to the EV type II distribution for maxima from thick-tailed distributions like the generalized Pareto distribution in Table 18.2.1 with $\kappa < 0$.

The moments of the GEV distribution can be expressed in terms of the gamma function, $\Gamma(\bullet)$. For $\kappa > -1/3$, the mean and variance are given in Table 18.2.1, whereas

$$\gamma_X = \text{Sign}(\kappa)\,\frac{-\Gamma(1+3\kappa) + 3\Gamma(1+\kappa)\,\Gamma(1+2\kappa) - 2\Gamma^3(1+\kappa)}{[\Gamma(1+2\kappa) - \Gamma^2(1+\kappa)]^{3/2}} \tag{18.2.19}$$

where Sign (κ) is plus or minus 1 depending on the sign of κ, and $\Gamma(\,)$ is the gamma function for which an approximation is supplied in Eq. (18.2.21). For $\kappa > -1$, the order r PWM β_r of a GEV distribution is

$$\beta_r = (r+1)^{-1}\left\{\xi + \frac{\alpha}{\kappa}\left[1 - \frac{\Gamma(1+\kappa)}{(r+1)^{\kappa}}\right]\right\} \qquad (18.2.20)$$

L moments for the GEV distribution are given in Table 18.1.2.

For $0 \le \delta \le 1$, a good approximation of the *gamma function,* useful with Eqs. (18.2.19) and (18.2.20) is

$$\Gamma(1+\delta) = 1 + \sum_{i=1}^{5} a_i \delta^i + \epsilon \qquad (18.2.21)$$

where $a_1 = -0.574\,8646$
$a_2 = 0.951\,2363$
$a_3 = -0.699\,8588$
$a_4 = 0.424\,5549$
$a_5 = -0.101\,0678$

with $|\epsilon| \le 5 \times 10^{-5}$ [Eq. (6.1.35) in Ref. 1]. For larger arguments one can use the relationship $\Gamma(1+w) = w\Gamma(w)$ repeatedly until $0 < w < 1$; for integer w, $\Gamma(1+w) = w!$ is the factorial function.

The parameters of the GEV distribution in terms of L moments are[68]

$$\kappa = 7.8590c + 2.9554c^2 \qquad (18.2.22a)$$

$$\alpha = \frac{\kappa\lambda_2}{\Gamma(1+\kappa)\,(1-2^{-\kappa})} \qquad (18.2.22b)$$

$$\xi = \lambda_1 + \frac{\alpha}{\kappa[\Gamma(1+\kappa)-1]} \qquad (18.2.22c)$$

where

$$c = \frac{2\lambda_2}{\lambda_3 + 3\lambda_2} - \frac{\ln(2)}{\ln(3)} = \frac{2\beta_1 - \beta_0}{3\beta_2 - \beta_0} - \frac{\ln(2)}{\ln(3)}$$

The quantiles of the GEV distribution can be calculated from

$$x_p = \xi + \frac{\alpha}{\kappa}\{1 - [-\ln(p)]^{\kappa}\} \qquad (18.2.23)$$

where p is the cumulative probability of interest. Typically $|\kappa| \le 0.20$.

When data are drawn from a Gumbel distribution ($\kappa = 0$), using the biased estimator b_r^* in Eq. (18.1.13) to calculate the L-moment estimators in Eq. (18.2.22), the resultant estimator of κ has a mean of 0 and variance[68]

$$\mathrm{Var}(\hat{\kappa}) = \frac{0.5633}{n} \qquad (18.2.24)$$

Comparison of the statistic $Z = \hat{\kappa}\sqrt{n/0.5633}$ with standard normal quantiles allows construction of a powerful test of whether $\kappa = 0$ or not when fitting a GEV distribution.[68,72] Chowdhury et al.[22] provide formulas for the sampling variance of the sample L-moment skewness and kurtosis $\hat{\tau}_3$ and $\hat{\tau}_4$ as a function of κ for the GEV distribution so that one can test if a particular data set is consistent with a GEV distribution with a regional value of κ.

Weibull Distribution. If W_i are the minimum stream flows in different days of the year, then the annual minimum is the smallest of the W_i, each of which is bounded below by zero. In this case the random variable $X = \min(W_i)$ may be well-described by the EV type III distribution for minima, or the Weibull distribution. Table 18.2.1 includes the Weibull cdf, mean, and variance. The skewness coefficient is the negative of that in Eq. (18.2.19) with $\kappa = 1/k$. The second L moment is

$$\lambda_2 = \alpha(1 - 2^{-1/k})\,\Gamma\left(1 + \frac{1}{k}\right) \tag{18.2.25}$$

Equation (18.2.21) provides an approximation for $\Gamma(1 + \delta)$.

For $k < 1$ the Weibull pdf goes to infinity as x approaches zero, and decays slowly for large x. For $k = 1$ the Weibull distribution reduces to the exponential distribution in Fig. 18.2.2 corresponding to $\gamma = 2$ and $\alpha_{P3} = 1$ in that figure. For $k > 1$, the Weibull density function is like a Pearson type 3 distribution's density function in Fig. 18.2.2 for small x and $\alpha_{P3} = k$, but decays to zero faster for large x. Parameter estimation methods are discussed in Refs. 57 and 85.

There are important relationships between the Weibull, Gumbel, and GEV distributions. If X has a Weibull distribution, then $Y = -\ln[X]$ has a Gumbel distribution. This allows parameter estimation procedures [Eqs. (18.2.15) to (18.2.17)] and goodness-of-fit tests available for the Gumbel distribution to be used for the Weibull; thus if $+\ln(X)$ has mean $\lambda_{1,(\ln X)}$ and L-moment $\lambda_{2,(\ln X)}$, X has Weibull parameters

$$k = \frac{\ln(2)}{\lambda_{2,(\ln X)}} \qquad \text{and} \qquad \alpha = \exp\left(\lambda_{1,(\ln X)} + \frac{0.5772}{k}\right) \tag{18.2.26}$$

Section 18.1.3 discusses use of logarithmic transformations.

If Y has a EV type III distribution (GEV distribution with $\kappa > 0$) for maxima bounded above, then $(\xi + \alpha/\kappa) - Y$ has a Weibull distribution with $k = 1/\kappa$; thus for $k > 0$, the third and fourth L-moment ratios for the Weibull distribution equal $-\tau_3$ and τ_4 for the GEV distribution in Table 18.1.2. A three-parameter Weibull distribution can be fit by the method of L moments by using Eq. (18.2.22) applied to $-X$.

18.2.3 Pearson Type 3 Family: Pearson Type 3 and Log-Pearson Type 3

Another family of distributions used in hydrology is that based on the Pearson type 3 (P3) distribution.[13] It is one of several families of distributions the statistician Pearson proposed as convenient models of random variables. Goodness-of-fit tests are discussed in Sec. 18.3, and standard errors of quantile estimators, in Sec. 18.4.3.

The pdf of the P3 distribution is given in Table 18.2.1. For $\beta > 0$ and lower bound $\xi = 0$, the P3 distribution reduces to the *gamma distribution* for which $\gamma_X = 2\mathrm{CV}_X$. In some instances, the P3 distribution is used with $\beta < 0$, yielding a negatively skewed distribution with an upper bound of ξ.

Figure 18.2.2 illustrates the shape of the P3 pdf for various values of the skew coefficient γ. For a fixed mean and variance, in the limit as the shape parameter α goes to infinity and the skew coefficient γ goes to zero, the Pearson type 3 distribution converges to the normal distribution. For $\alpha < 1$ and skew coefficient $\gamma > 2$, the P3 pdf goes to infinity at the lower bound. For $\alpha = 1$ and $\gamma = 2$, the two-parameter *exponential distribution* is obtained; see Table 18.2.1.

The moments of the P3 distribution are given in Table 18.2.1. The moment equations can be inverted to obtain

$$\alpha = 4/\gamma_X^2$$
$$\beta = \frac{2}{\sigma_X \gamma_X} \tag{18.2.27}$$
$$\xi = \mu_X - \frac{\alpha}{\beta} = \mu_X - 2\frac{\sigma_X}{\gamma_X}$$

which allows computation of method-of-moment estimators. The method of maximum likelihood is seldom used with this distribution; it does not generate estimates of α less than 1, corresponding to skew coefficients in excess of 2.

A closed-form expression for the cdf of the P3 distribution is not available. Tables or approximations must be used. Many tables provide *frequency factors* $K_p(\gamma)$ which are the pth quantile of a standard P3 variate with skew coefficient γ, mean zero, and variance 1.[20,79] For any mean and standard deviation, the pth P3 quantile can be written

$$x_p = \mu + \sigma K_p(\gamma) \tag{18.2.28}$$

With this parameterization, it is not necessary to estimate the underlying values of α and β when the method of moments is used because the quantiles of the fitted distribution are written as a function of the mean, standard deviation, and the frequency factor. Tables of frequency factors are provided in Ref. 79. The frequency factors for $0.01 \le p \le 0.99$ and $|\gamma| < 2$ are well-approximated by the Wilson-Hilferty

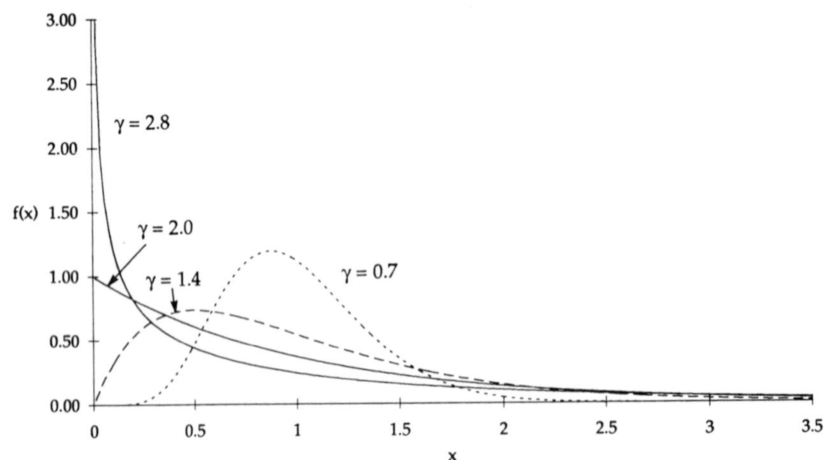

FIGURE 18.2.2 The probability density function for the Pearson type 3 (P3) distribution with lower bound $\xi = 0$, mean $\mu = 1$, and coefficients of skewness $\gamma = 0.7$, 1.4, 2.0, and 2.8 (corresponding to a gamma distribution and shape parameters $\alpha = 8$, 2, 1, and 0.5, respectively).

transformation

$$K_p(\gamma) = \frac{2}{\gamma}\left(1 + \frac{\gamma z_p}{6} - \frac{\gamma^2}{36}\right)^3 - \frac{2}{\gamma} \tag{18.2.29}$$

where z_p is the pth quantile of the zero-mean unit-variance standard normal distribution in Eq. (18.2.3). (Reference 83 provides a better approximation; Ref. 21 evaluates several approximations.)

For the P3 distribution, the first two L moments are

$$\lambda_1 = \xi + \frac{\alpha}{\beta} \quad \text{and} \quad \lambda_2 = \frac{\Gamma(\alpha + 0.5)}{\sqrt{\pi}\,\beta\,\Gamma(\alpha)} \tag{18.2.30}$$

An approximation which describes the relationship between the third and fourth L-moment ratios, accurate to within 0.0005 for $|\tau_3| < 0.9$, is[73]

$$\tau_4 = 0.1224 + 0.30115\,\tau_3^2 + 0.95812\,\tau_3^4 - 0.57488\,\tau_3^6 + 0.19383\,\tau_3^8 \tag{18.2.31}$$

Log-Pearson Type 3 Distribution. The log-Pearson type 3 distribution (LP3) describes a random variable whose logarithms are P3-distributed. Thus

$$Q = \exp\,[X] \tag{18.2.32}$$

where X has a P3 distribution with shape, scale, and location parameters α, β, and ξ. Thus the distribution of the logarithms X of the data is described by Fig. 18.2.2, Eqs. (18.2.27) to (18.2.29), and the corresponding relationships in Table 18.2.1.

The product moments of Q are computed for $\beta > r$ or $\beta < 0$ by using

$$E[Q^r] = e^{r\xi}\left(\frac{\beta}{\beta - r}\right)^{\alpha} \tag{18.2.33}$$

yielding

$$\mu_Q = e^{\xi}\left(\frac{\beta}{\beta - 1}\right)^{\alpha} \qquad \sigma_Q^2 = e^{2\xi}\left[\left(\frac{\beta}{\beta - 2}\right)^{\alpha} - \left(\frac{\beta}{\beta - 1}\right)^{2\alpha}\right] \tag{18.2.34}$$

and

$$\gamma_Q = \frac{E[Q^3] - 3\,\mu_Q\,E[Q^2] + 2\,\mu_Q^3}{\sigma_Q^3}$$

The parameter ξ is a lower bound on the logarithms of the random variable if β is positive, and is an upper bound if β is negative. The shape of the real-space flood distribution is a complex function of α and β.[11–13] If one considers W equal to the common logarithm of Q, log (Q), then all the parameters play the same roles, but the new β' and ξ' are smaller by a factor of 1/ln (10) = 0.4343.

This distribution was recommended for the description of floods in the United States by the U.S. Water Resources Council in Bulletin 17[79] and in Australia by their Institute of Engineers;[110] Sec. 18.7.2 describes the Bulletin 17 method. It fits a P3 distribution by a modified method of moments to the logarithms of observed flood series using Eq. (18.2.28). Section 18.1.3 discusses pros and cons of logarithmic transformations. Estimation procedures for the LP3 distribution are reviewed in Ref. 5.

18.2.4 Generalized Pareto Distribution

The generalized Pareto distribution (GPD) is a simple distribution useful for describing events which exceed a specified lower bound, such as all floods above a threshold or daily flows above zero. Moments of the GPD are described in Tables 18.1.2 and 18.2.1. A special case is the 2-parameter exponential distribution (for $\kappa = 0$).

For a given lower bound ξ, the shape κ and scale α parameters can be estimated easily with L-moments from

$$\kappa = \frac{\lambda_1 - \xi}{\lambda_2} - 2 \qquad \text{and} \qquad \alpha = (\lambda_1 - \xi)(1 + \kappa) \tag{18.2.35}$$

or the mean and variance formula in Table 18.2.1. In general for $\kappa < 0$, L-moment estimators are preferable. Hosking and Wallis[70] review alternative estimation procedures and their precision. Section 18.6.3 develops a relationship between the Pareto and GEV distributions. If ξ must be estimated, the smaller observation is a good estimator.

18.3 PROBABILITY PLOTS AND GOODNESS-OF-FIT TESTS

18.3.1 Principles and Issues in Selecting a Distribution

Probability plots are extremely useful for visually revealing the character of a data set. Plots are an effective way to see what the data look like and to determine if fitted distributions appear consistent with the data. Analytical goodness-to-fit criteria are useful for gaining an appreciation for whether the lack of fit is likely to be due to sample-to-sample variability, or whether a particular departure of the data from a model is statistically significant. In most cases several distributions will provide statistically acceptable fits to the available data so that goodness-of-fit tests are unable to identify the "true" or "best" distribution to use. Such tests are valuable when they can demonstrate that some distributions appear inconsistent with the data.

Several fundamental issues arise when selecting a distribution.[82] One should distinguish between the following questions:

1. What is the true distribution from which the observations are drawn?
2. What distribution should be used to obtain reasonably accurate and robust estimates of design quantiles and hydrologic risk?
3. Is a proposed distribution consistent with the available data for a site?

Question 1 is often asked. Unfortunately, the true distribution is probably too complex to be of practical use. Still, L-moment skewness-kurtosis and CV-skewness diagrams discussed in Secs. 18.1.4 and 18.3.3 are good for investigating what simple families of distributions are consistent with available data sets for a region. Standard goodness-of-fit statistics, such as probability plot correlation tests in Sec. 18.3.2, have also been used to see how well a member of each family of distributions can fit a sample. Unfortunately, such goodness-of-fit statistics are unlikely to identify the actual family from which the samples are drawn—rather, the most flexible families generally fit the data best. Regional L-moment diagrams focus on the character of sample statistics which describe the "parent" distribution for available samples, rather than goodness-of-fit. Goodness-of-fit tests address Question 3.

Question 2 is important in hydrologic applications and has been the subject of many studies (Ref. 29; examples include Refs. 32, 87, 90, 112, 124). At one time the distribution that best fitted each data set was used for frequency analysis at that site, but this approach has now been largely abandoned. Such a procedure is too sensitive to sampling variations in the data. Operational procedures adopted by different national flood-frequency studies for use in their respective countries should be based on a combination of regionalization of some parameters and split-sample/Monte Carlo evaluations of different estimation procedures to find distribution-estimation procedure combinations which yield reliable flood quantile and risk estimates. Such estimators are called *robust* because they perform reasonably well for a wide range of cases. In the United States, the log-Pearson type 3 distribution with weighted skew coefficient was adopted; an index-flood GEV procedure was selected for the British Isles (see Secs. 18.7.2 and 18.7.3). This principle also applies to frequency analyses of other phenomena.

18.3.2 Plotting Positions and Probability Plots

The graphical evaluation of the adequacy of a fitted distribution is generally performed by plotting the observations so that they would fall approximately on a straight line if a postulated distribution were the true distribution from which the observations were drawn. This can be done with the use of special commercially available probability papers for some distributions, or with the more general technique presented here, on which such special papers are based.[30] Section 17.2.2 also discusses the graphical display of data.

Let $\{X_i\}$ denote the observed values and $X_{(i)}$, the ith largest value in a sample, so that $X_{(n)} \le X_{(n-1)} \le \ldots \le X_{(1)}$. The random variable U_i defined as

$$U_i = 1 - F_X[X_{(i)}] \tag{18.3.1}$$

corresponds to the *exceedance probability* associated with the ith largest observation. If the original observations were independent, in repeated sampling the U_i have a *beta distribution* with mean

$$E[U_i] = \frac{i}{n+1} \tag{18.3.2}$$

and variance

$$\text{Var}(U_i) = \frac{i(n-i+1)}{(n+1)^2(n+2)} \tag{18.3.3}$$

Knowing the distribution of the exceedance probabilities U_i, one can develop estimators q_i of their values which can be used to plot each $X_{(i)}$ against a probability scale.

Let $G(x)$ be a proposed cdf for the events. A visual comparison of the data and a fitted distribution is provided by a plot of the ith largest observed event $X_{(i)}$ versus an estimate of what its true value should be. If $G(x)$ is the distribution of X, the value of $X_{(i)} = G^{-1}(1 - U_i)$ should be nearly $G^{-1}(1 - q_i)$, where the *probability-plotting position* q_i is our estimate of U_i. Thus the points $[G^{-1}(1 - q_i), X_{(i)}]$ when plotted would, apart from sampling fluctuation, lie on a straight line through the origin. Such a plot would look like Fig. 18.3.1, which actually displays $[\Phi^{-1}(1 - q_i), \log X_{(i)}]$. The exceed-

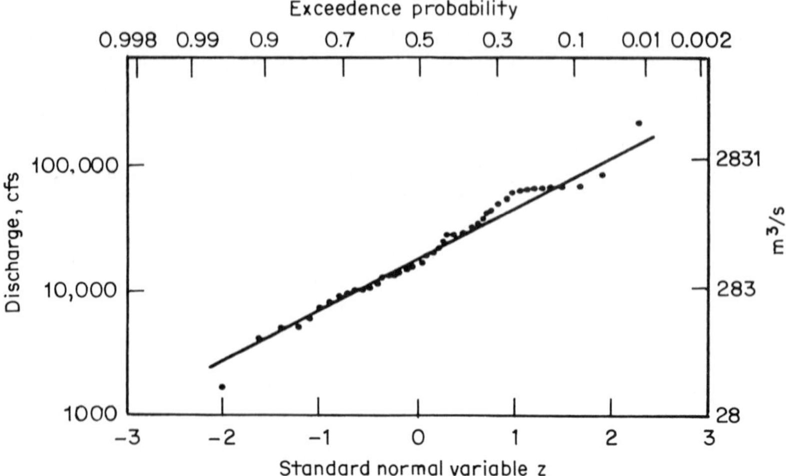

FIGURE 18.3.1 A probability plot using a normal scale of 44 annual maxima for the Guadalupe River near Victoria, Texas. *(Reproduced with permission from Ref. 20, p. 398.)*

ance probability of the ith-largest event is often estimated using the *Weibull plotting position:*

$$q_i = \frac{i}{n+1} \tag{18.3.4}$$

corresponding to the mean of U_i.

Choice of plotting position. Hazen[59] originally developed probability paper and imagined the probability scale divided into n equal intervals with midpoints $q_i = (i - 0.5)/n$, $i = 1, \ldots, n$; these served as his plotting positions. Gumbel[51] rejected this formula in part because it assigned a return period of $2n$ years to the largest observation (see also Harter[58]); Gumbel promoted Eq. (18.3.4).

Cunnane[26] argued that plotting positions q_i should be assigned so that on average $X_{(i)}$ would equal $G^{-1}(1 - q_i)$; that is, q_i would capture the mean of $X_{(i)}$ so that

$$E[X_{(i)}] \approx G^{-1}(1 - q_i) \tag{18.3.5}$$

Such plotting positions would be almost quantile-unbiased. The Weibull plotting positions $i/(n+1)$ equal the average exceedance probability of the ranked observations $X_{(i)}$, and hence are probability-unbiased plotting positions. The two criteria are different because of the nonlinear relationship between $X_{(i)}$ and $U_{(i)}$.

Different plotting positions attempt to achieve almost quantile-unbiasedness for different distributions; many can be written

$$q_i = \frac{i - a}{n + 1 - 2a} \tag{18.3.6}$$

which is symmetric so that $q_i = 1 - q_{n+1-i}$. Cunanne recommended $a = 0.40$ for obtaining nearly quantile-unbiased plotting positions for a range of distributions.

Other alternatives are Blom's plotting position ($a = \frac{3}{8}$), which gives nearly unbiased quantiles for the normal distribution, and the Gringorten position ($a = 0.44$) which yields optimized plotting positions for the largest observations from a Gumbel distribution.[49] These are summarized in Table 18.3.1, which also reports the return period, $T_1 = 1/q_1$, assigned to the largest observation. Section 18.6.3 develops plotting positions for records that contain censored values.

The differences between the Hazen formula, Cunanne's recommendation, and the Weibull formula is modest for i of 3 or more. However, differences can be appreciable for $i = 1$, corresponding to the largest observation (and $i = n$ for the smallest observation). Remember that the actual exceedance probability associated with the largest observation is a random variable with mean $1/(n + 1)$ and a standard deviation of nearly $1/(n + 1)$; see Eqs. (18.3.2) and (18.3.3). Thus all plotting positions give crude estimates of the unknown exceedance probabilities associated with the largest (and smallest) events.

A good method for illustrating this uncertainty is to consider quantiles of the beta distribution of the actual exceedance probability associated with the largest observation $X_{(1)}$. The actual exceedance probability for the largest observation $X_{(1)}$ in a sample is between $0.29/n$ and $1.38/(n + 2)$ nearly 50 percent of the time; and between $0.052/n$ and $3/(n + 2)$ nearly 90 percent of the time. Such bounds allow one to assess the consistency of the largest (or, by symmetry, the smallest) observation with a fitted distribution better than does a single plotting position.

Probability Paper. It is now possible to see how probability papers can be constructed for many distributions. A probability plot is a graph of the ranked observations $x_{(i)}$ versus an approximation of their expected value $G^{-1}(1 - q_i)$. For the nor-

TABLE 18.3.1 Alternative Plotting Positions and their Motivation*

Name	Formula	a	T_1	Motivation
Weibull	$\dfrac{i}{n+1}$	0	$n+1$	Unbiased exceedance probabilities for all distributions
Median†	$\dfrac{i-0.3175}{n+0.365}$	0.3175	$1.47n+0.5$	Median exceedance probabilities for all distributions
APL	$\dfrac{i-0.35}{n}$	~0.35	$1.54n$	Used with PWMs [Eq. (18.1.13)]
Blom	$\dfrac{i-3/8}{n+1/4}$	0.375	$1.60n+0.4$	Unbiased normal quantiles
Cunnane	$\dfrac{i-0.40}{n+0.2}$	0.40	$1.67n+0.3$	Approximately quantile-unbiased
Gringorten	$\dfrac{i-0.44}{n+0.12}$	0.44	$1.79n+0.2$	Optimized for Gumbel distribution
Hazen	$\dfrac{i-0.5}{n}$	0.50	$2n$	A traditional choice

* Here a is the plotting-position parameter in Eq. (18.3.6) and T_1 is the return period each plotting position assigns to the largest observation in a sample of size n.

† For $i = 1$ and n, the exact value is $q_1 = 1 - q_n = 1 - 0.5^{1/n}$.

mal distribution

$$G^{-1}(1 - q_i) = \mu + \sigma\, \Phi^{-1}(1 - q_i) \tag{18.3.7}$$

Thus, except for intercept and slope, a plot of the observations $x_{(i)}$ versus $G^{-1}[1 - q_i]$ is visually identical to a plot of $x_{(i)}$ versus $\Phi^{-1}(1 - q_i)$. The values of q_i are often printed along the abscissa or horizontal axis. Lognormal paper is obtained by using a log scale to plot the ordered logarithms log $(x_{(i)})$ versus a normal-probability scale, which is equivalent to plotting log $(x_{(i)})$ verus $\Phi^{-1}(1 - q_i)$. Figure 18.3.1 illustrates use of lognormal paper with Blom's plotting positions.

For the Gumbel distribution,

$$G^{-1}(1 - q_i) = \xi - \alpha \ln [-\ln (1 - q_i)] \tag{18.3.8}$$

Thus a plot of $x_{(i)}$ versus $G^{-1}(1 - q_i)$ is identical to a plot of $x_{(i)}$ versus the *reduced Gumbel variate*

$$y_i = -\ln [-\ln (1 - q_i)] \tag{18.3.9}$$

It is easy to construct probability paper for the Gumbel distribution by plotting $x_{(i)}$ as a function of y_i; the horizontal axis can show the actual values of y or, equivalently, the associated q_i, as in Fig. 18.3.1 for the lognormal distribution.

Special probability papers are not available for the Pearson type 3 or log Pearson type 3 distributions because the frequency factors depend on the skew coefficient. However, for a given value for the coefficient of skewness γ one can plot the observation $x_{(i)}$ for a P3 distribution, or log $(x_{(i)})$ for the LP3 distribution, versus the frequency factors $K_{p_i}(\gamma)$ defined in Eq. (18.2.29) with $p_i = 1 - q_i$. This should yield a straight line except for sampling error if the correct skew coefficient is employed. Alternatively for the P3 or LP3 distributions, normal or lognormal probability paper is often used to compare the $x_{(i)}$ and a fitted P3 distribution, which plots as a curved

TABLE 18.3.2 Generation of Probability Plots for Different Distributions

Normal probability paper. Plot $x_{(i)}$ versus z_{p_i} given in Eq. (18.2.3), where $p_i = 1 - q_i$. Blom's formula ($a = 3/8$) provides quantile-unbiased plotting positions.

Lognormal probability paper. Plot ordered logarithms log $[x_{(i)}]$ versus z_p. Blom's formula ($a = 3/8$) provides quantile-unbiased plotting positions.

Exponential probability paper. Plot ordered observations $x_{(i)}$ versus $\xi - \ln (q_i)/\beta$ or just $-\ln (q_i)$. Gringorten's plotting positions ($a = 0.44$) work well.

Gumbel and Weibull probability paper. For Gumbel distribution plot ordered observations $x_{(i)}$ versus $\xi - \alpha \ln [-\ln (1 - q_i)]$ or just $y_i = -\ln [-\ln (1 - q_i)]$. Gringorten's plotting positions ($a = 0.44$) were developed for this distribution. For Weibull distribution plot ln $[x_{(i)}]$ versus $\ln (\alpha) + \ln [-\ln (q_i)]/k$ or just $\ln [-\ln (q_i)]$. (See Ref. 154.)

GEV distribution. Plot ordered observations $x_{(i)}$ versus $\xi + (\alpha/\kappa)\{1 - [-\ln (1 - q_i)]^\kappa$, or just $(1/\kappa)\{1 - [-\ln (1 - q_i)]^\kappa\}$. Alternatively employ Gumbel probability paper on which GEV will be curved. Cunnane's plotting positions ($a = 0.4$) are reasonable.[52]

Pearson type 3 probability paper. Plot ordered observations $x_{(i)}$ versus $K_{p_i}(\gamma)$, where $p_i = 1 - q_i$. Blom's formula ($a = 3/8$) is quantile-unbiased for normal distribution and makes sense for small γ. Or employ normal probability paper. (See Ref. 158.)

Log Pearson type 3 probability paper. Plot ordered logarithms log $[x_{(i)}]$ versus $K_{p_i}(\gamma)$ where $p_i = 1 - q_i$. Blom's formula ($a = 3/8$) makes sense for small γ. Or employ lognormal probability paper. (See Ref. 158.)

Uniform probability paper. Plot $x_{(i)}$ versus $1 - q_i$, where q_i are the Weibull plotting positions ($a = 0$). (See Ref. 154.)

line. Table 18.3.2 summarizes how probability plots may be constructed for these and other distributions.

18.3.3 Goodness-of-Fit Tests and L-Moment Diagrams

Rigorous statistical tests are available and are useful for assessing whether or not a given set of observations might have been drawn from a particular family of distributions, as discussed in Sec. 18.3.1. For example, the *Kolmogorov-Smirnov test* provides bounds within which *every observation* on a probability plot should lie if the sample is actually drawn from the assumed distribution; it is useful for evaluating visually the adequacy of a fitted distribution. Stephens[136] gives critical Kolmogorov-Smirnov values for the normal and exponential distributions (reproduced in Ref. 95, p. 112); Chowdhury et al.[22] provide tables for the GEV distribution.

The *probability plot correlation test* discussed below is a more powerful test of whether a sample has been drawn from a postulated distribution; a test with greater *power* has a greater probability of correctly determining that a sample is not from the postulated distribution. *L-moment tests* are also relatively powerful and can be used to determine if a proposed Gumbel, GEV, or normal distribution is consistent with the data. L-moment diagrams are useful as a guide in selecting an appropriate family of distributions for describing a set of variables, such as flood distributions in a region.

Probability Plot Correlation Coefficient Test. A simple but powerful goodness-of-fit test is the probability plot correlation test developed by Filliben.[41] The test uses the correlation r between the ordered observations $x_{(i)}$ and the corresponding fitted quantiles $w_i = G^{-1}(1 - q_i)$, determined by plotting positions q_i for each $x_{(i)}$. Values of r near 1.0 suggest that the observations could have been drawn from the fitted distribution. Essentially, r measures the linearity of the probability plot, providing a quantitative assessment of fit. If $\bar{x}$ denotes the average value of the observations and $\bar{w}$ denotes the average value of the fitted quantiles, then

$$r = \frac{\Sigma\,(x_{(i)} - \bar{x})\,(w_i - \bar{w})}{[\Sigma\,(x_{(i)} - \bar{x})^2\,\Sigma\,(w_i - \bar{w})^2]^{0.5}} \tag{18.3.10}$$

Table 18.3.3 gives critical values of r for the normal distribution, or the logarithms of lognormal variates, based on a plotting position with $a = 3/8$. Values for the Gumbel distribution are reproduced in Table 18.3.4 for use with $a = 0.44$; the table also applies to logarithms of Weibull variates (see Table 18.3.2 and Sec. 18.2.2). Other tables are available for the uniform,[154] the GEV,[22] the Pearson type 3,[158] and exponential and other distributions.[30]

L-Moment Diagrams and Ratio Tests. Figure 18.1.1 provides an example of an L-moment diagram.[72,163] Sample L moments are less biased than traditional product-moment estimators, and thus are better suited for use in constructing moment diagrams. (See Sec. 18.1.4.) Plotting sample statistics on such diagrams allows a choice between alternative families of distributions (Ref. 29). L-moment diagrams include plots of τ_3 versus τ_2 for choosing among two-parameter distributions, or of τ_4 versus τ_3 for choosing among three-parameter distributions. Chowdhury et al.[22] derive the sampling variance of $\hat{\tau}_2$, $\hat{\tau}_3$, and $\hat{\tau}_4$ as a function of κ for the GEV distribution to provide a powerful test of whether a particular data set is consistent with a GEV distribution with a regionally estimated value of κ, or a regional κ and CV. Equation (18.2.24) provides a very powerful test for the Gumbel versus a general

TABLE 18.3.3 Lower Critical Values of the Probability Plot Correlation Test Statistic for the Normal Distribution Using $p_i = (i - 3/8)/(n + 1/4)$

	Significance level		
n	0.10	0.05	0.01
10	0.9347	0.9180	0.8804
15	0.9506	0.9383	0.9110
20	0.9600	0.9503	0.9290
30	0.9707	0.9639	0.9490
40	0.9767	0.9715	0.9597
50	0.9807	0.9764	0.9664
60	0.9835	0.9799	0.9710
75	0.9865	0.9835	0.9757
100	0.9893	0.9870	0.9812
300	0.99602	0.99525	0.99354
1000	0.99854	0.99824	0.99755

Source: Refs. 101, 152, 153. Used with permission.

TABLE 18.3.4 Lower Critical Values of the Probability Plot Correlation Test Statistic for the Gumbel and Two-Parameter Weibull Distributions Using $p_i = (i - 0.44)/(n + 0.12)$

	Significance level		
n	0.10	0.05	0.01
10	0.9260	0.9084	0.8630
20	0.9517	0.9390	0.9060
30	0.9622	0.9526	0.9191
40	0.9689	0.9594	0.9286
50	0.9729	0.9646	0.9389
60	0.9760	0.9685	0.9467
70	0.9787	0.9720	0.9506
80	0.9804	0.9747	0.9525
100	0.9831	0.9779	0.9596
300	0.9925	0.9902	0.9819
1000	0.99708	0.99622	0.99334

Source: Refs. 152, 153. See also Table 18.3.2.

GEV distribution using the sample L-moment estimator of κ. Similarly, if observations have a normal distribution, then $\hat{\tau}_3$ has mean zero and Var $[\hat{\tau}_3] = (0.1866 + 0.8/n)/n$, allowing construction of a powerful test of normality against skewed alternatives[72] using $Z = \hat{\tau}_3/\sqrt{(0.1866/n + 0.8/n^2)}$.

18.4 STANDARD ERRORS AND CONFIDENCE INTERVALS FOR QUANTILES

A simple measure of the precision of a quantile estimator is its variance Var (x_p), which equals the square of the *standard error*, SE, so that $\text{SE}^2 = \text{Var}(\hat{x}_p)$. Confidence intervals are another description of precision. Confidence intervals for a quantile are often calculated using the quantile's standard error. When properly constructed, 90 or 99 percent confidence intervals will, in repeated sampling, contain the parameter or quantile of interest 90 or 99 percent of the time. Thus they are an interval which will contain a parameter of interest most of the time.

18.4.1 Confidence Intervals for Quantiles

The classic confidence interval formula is for the mean μ_X of a normally distributed random variable X. If sample observations X_i are independent and normally distributed with the same mean and variance, then a $100(1-\alpha)\%$ confidence interval for μ_X is

$$\bar{x} - \frac{s_X}{\sqrt{n}} t_{1-\alpha/2,n-1} \leq \mu_X \leq \bar{x} + \frac{s_X}{\sqrt{n}} t_{1-\alpha/2,n-1} \tag{18.4.1}$$

where $t_{1-\alpha/2,n-1}$ is the upper $100(\alpha/2)\%$ percentile of Student's t distribution with $n-1$ degrees of freedom. Here $s_X/\sqrt{n}$ is the estimated *standard error* of the sample mean; that is, it is the square root of the variance of the estimator $\bar{X}$ of μ_X. In large samples ($n > 40$), the t distribution is well-approximated by a normal distribution, so that $z_{1-\alpha/2}$ from Table 18.2.2 can replace $t_{1-\alpha/2,n-1}$ in Eq. (18.4.1).

In hydrology, attention often focuses on quantiles of various distributions, such as the 10-year 7-day low flow, or the rainfall depth exceeded with a 1 percent probability. Confidence intervals can be constructed for quantile estimators. Asymptotically (with increasingly large n), most quantile estimators $\hat{x}_p$ are normally distributed. If $\hat{x}_p$ has variance Var $(\hat{x}_p)$ and is essentially normally distributed, then an approximate $100(1-\alpha)\%$ confidence interval based on Eq. (18.4.1) is

$$\hat{x}_p - z_{1-\alpha/2}\sqrt{\text{Var}(\hat{x}_p)} \quad \text{to} \quad \hat{x}_p + z_{1-\alpha/2}\sqrt{\text{Var}(\hat{x}_p)} \tag{18.4.2}$$

Equation (18.4.2) allows calculation of approximate confidence intervals for quantiles (or parameters) of distributions for which good estimates of their standard errors, $\sqrt{\text{Var}(\hat{x}_p)}$, are available.[85]

18.4.2 Results for Normal/Lognormal Quantiles

For a normally distributed random variable, the traditional estimator of x_p is

$$\hat{x}_p = \bar{x} + z_p s_X \tag{18.4.3}$$

Asymptotically, the variance of this estimator is

$$\mathrm{Var}\,(\hat{x}_p) = \frac{s_X^2}{n}\left(1 + \frac{1}{2}\,z_p^2\right) \tag{18.4.4}$$

Thus, an approximate $100(1-\alpha)\%$ confidence interval for x_p is

$$(\bar{x} + z_p s_X) \pm z_{1-\alpha/2}\sqrt{\frac{s_X^2}{n}\left(1 + \frac{1}{2}\,z_p^2\right)} \tag{18.4.5}$$

These results can also be used to obtain confidence intervals for quantiles of the two-parameter lognormal distribution. If X is lognormally distributed, then $Y = \ln(X)$ is normally distributed and

$$x_p = \exp\,(\mu_y + z_p\,\sigma_y) \tag{18.4.6}$$

The maximum likelihood estimator of x_p is essentially $\exp\,(\bar{y} + z_p s_y)$; for this estimator,[85,129]

$$\mathrm{Var}\,(\hat{x}_p) \approx x_p^2\left[\frac{\sigma_y^2}{n}\left(1 + \frac{1}{2}z_p^2\right)\right] \tag{18.4.7}$$

A simple but approximate $100(1-\alpha)\%$ confidence interval for the lognormal quantile x_p is

$$\exp\left[(\bar{y} + z_p s_y) \pm z_{1-\alpha/2}\sqrt{\frac{s_y^2}{n}\left(1 + \frac{1}{2}\,z_p^2\right)}\right] \tag{18.4.8}$$

Confidence intervals obtained by substituting Eq. (18.4.7) into Eq. (18.4.2) are not as good as Eq. (18.4.8).[129]

For the normal (and lognormal) distribution, it is also possible to calculate exact confidence intervals using the noncentral t distribution. Let $\zeta_{\alpha,p}$ and $\zeta_{1-\alpha,p}$ denote the 100α and $100(1-\alpha)$ percentiles of the noncentral t distribution. Then an exact $100(1-2\alpha)\%$ confidence interval for $x_p = \mu + z_p\sigma$ when X has a normal distribution is

$$\bar{x} + \zeta_{\alpha,p} s_X < x_p < \bar{x} + \zeta_{1-\alpha,p}\, s_X \tag{18.4.9}$$

Stedinger[129] and App. 9 in Ref. 79 provide tables of percentage points of the ζ distribution. An approximation for $\zeta_{\alpha,p}$ is

$$\zeta_{\alpha,p} \approx \frac{z_p + z_\alpha\sqrt{\dfrac{1}{n} + \dfrac{z_p^2}{2(n-1)} - \dfrac{z_\alpha^2}{2n(n-1)}}}{1 - \dfrac{z_\alpha^2}{2(n-1)}} \tag{18.4.10}$$

which is reasonably accurate for $n \geq 15$ and $\alpha \geq 0.05$ (Ref. 21). $\zeta_{1-\alpha,p}$ is obtained using Eq. (18.4.10) by replacing z_α by $z_{1-\alpha}$ which equals $-z_\alpha$.

Less work has been done on formulas for the variances of lognormal quantiles when three parameters are estimated. Formulas for maximum likelihood, moment, and moment/quantile-lower-bound estimators are evaluated in Ref. 67.

18.4.3 Results for Pearson/Log-Pearson Type 3 Quantiles

Confidence intervals for normal quantiles can be extended to obtain approximate confidence intervals for Pearson Type 3 (P3) quantiles y_p for known skew coefficient γ by using a scaling factor η, obtained from a first-order asymptotic approximation of the P3/normal quantile variance ratio:

$$\eta = \left[\frac{\text{Var}(\hat{y}_p)}{\text{Var}(\hat{x}_p)}\right]^{1/2} \cong \sqrt{\frac{1 + \gamma K_p + \tfrac{1}{2}(1 + \tfrac{3}{4}\gamma^2) K_p^2}{1 + \tfrac{1}{2} z_p^2}} \tag{18.4.11}$$

where K_p is the standard P3 quantile (frequency factor) in Eqs. (18.2.28) and (18.2.29) with cumulative probability p for skew coefficient γ;[129] z_p is the frequency factor for the standard normal distribution in Eq. (18.2.3) employed to compute $\hat{x}_p$ in Eq. (18.4.3). An approximate $100(1 - 2\alpha)\%$ confidence interval for the pth P3 quantile is

$$\hat{y}_p + \eta(\zeta_{1-\alpha,p} - z_p)\, s_y < y_p < \hat{y}_p + \eta(\zeta_{\alpha,p} - z_p) s_y \tag{18.4.12}$$

where $\hat{y}_p = \bar{y} + K_p s_y$.

Chowdhury and Stedinger[21] show that a generalization of Eq. (18.4.12) should be used when the skew coefficient γ is estimated by the at-site sample skew coefficient G_s, a generalized regional estimate G_g, or a weighted estimate of G_s and G_g. For example, if a generalized regional estimate G_g of the coefficient of skewness is employed, and G_g has variance $\text{Var}(G_g)$ about the true skew coefficient, then the scaling factor in Eq. (18.4.12) should be calculated as[21]

$$\eta = \sqrt{\frac{1 + \gamma K_p + \tfrac{1}{2}(1 + \tfrac{3}{4}\gamma^2) K_p^2 + n\, \text{Var}(G_g)\, (\partial K_p/\partial \gamma)^2}{1 + \tfrac{1}{2} z_p^2}} \tag{18.4.13}$$

where, from Eq. (18.2.29),

$$\frac{\partial K_p}{\partial \gamma} \approx \frac{1}{6}(z_p^2 - 1)\left[1 - 3\left(\frac{\gamma}{6}\right)^2\right] + (z_p^3 - 6z_p)\frac{\gamma}{54} + \frac{2}{3} z_p \left(\frac{\gamma}{6}\right)^3 \tag{18.4.14}$$

for $|\gamma| \le 2$ and $0.01 \le p \le 0.99$.

18.4.4 Results for Gumbel and GEV Quantiles

For the Gumbel distribution with two parameters estimated by the method of moments, the variance of the pth quantile is asymptotically[105]

$$\text{Var}(\hat{x}_p) = \frac{\alpha^2(1.11 + 0.52y + 0.61y^2)}{n} \tag{18.4.15}$$

for a sample of size n where $y = -\ln[-\ln(p)]$ is the Gumbel reduced variate.

For unbiased L-moment estimators[38,109]

$$\text{Var}(\hat{x}_p) = \frac{\alpha^2[(1.1128 - 0.9066/n) - (0.4574 - 1.1722/n)y + (0.8046 - 0.1855/n)y^2]}{n - 1} \tag{18.4.16}$$

Equation (18.4.16) also provides a reasonable estimate of Var (x_p) for use with biased PWMs. These values can be used in Eq. (18.4.2) to obtain approximate confidence intervals. Reference 109 provides formulas for Var $(\hat{x}_p)$ when maximum likelihood estimators are employed.

GEV Index Flood Procedures. The Gumbel and GEV distributions are often used as normalized regional distributions or regional growth curves, as discussed in Sec. 18.5.1. In that case the variance of $\hat{x}_p$ is given by Eq. (18.5.3).

GEV with Fixed κ. The GEV distribution can be used when the location and scale parameters are estimated by using L moments via Eqs. (18.2.22*b*) and (18.2.22*c*) with a fixed regional value of the shape parameter κ, corresponding to a two-parameter index flood procedure (Sec. 18.5.1). For fixed κ the asymptotic variance of the pth quantile estimator with unbiased L-moment estimators is

$$\text{Var}\,(\hat{x}_p) = \frac{\alpha^2(c_1 + c_2 y + c_3 y^2)}{n} \tag{18.4.17}$$

where $y = 1 - [\ln (p)]^\kappa$ when $\kappa \neq 0$ and c_1, c_2, c_3 are coefficients which depend on κ. The asymptotic values of c_1, c_2, c_3 for $-0.33 < \kappa < 0.3$ are well-approximated by[96]

$$c_1 = 1.1128 - 0.2384\kappa + 0.0908\kappa^2 + 0.1084\kappa^3$$

where, for $\kappa > 0$,

$$c_2 = 0.4580 - 3.0561\kappa + 1.1104\kappa^2 - 0.4071\kappa^3$$
$$c_3 = 0.8046 - 2.8890\kappa + 8.7874\kappa^2 - 10.375\kappa^3$$

and, for $\kappa < 0$,

$$c_2 = 0.4580 - 7.5124\kappa + 5.0832\kappa^2 - 11.623\kappa^3 + 2.250 \ln (1 + 2\kappa)$$
$$c_3 = 0.8046 - 2.6215\kappa + 6.8989\kappa^2 + 0.003\kappa^3 - 0.1 \ln (1 + 3\kappa)$$

For $\kappa = 0$, use Eq. (18.4.16).

Estimation of Three GEV Parameters. All three parameters of the GEV distribution can be estimated with L moments by using Eq. (18.2.22).[68] Asymptotic formulas for the variance of three-parameter GEV quantile estimators are relatively inaccurate in small samples;[96] an estimate of the variance of the pth quantile estimator with

TABLE 18.4.1 Coefficients for an Eq. (18.4.18) That Approximates Variance of Three-Parameter GEV Quantile Estimators

	Cumulative probability level p						
Coefficient	0.80	0.90	0.95	0.98	0.99	0.998	0.999
a_0	−1.813	−2.667	−3.222	−3.756	−4.147	−5.336	−5.943
a_1	3.017	4.491	5.732	7.185	8.216	10.711	11.815
a_2	−1.401	−2.207	−2.367	−2.314	−0.2033	−1.193	−0.630
a_3	0.854	1.802	2.512	4.075	4.780	5.300	6.262

Source: Ref. 96.

unbiased L-moment estimators for $-0.33 < \kappa < 0.3$ is

$$\text{Var}(\hat{x}_p) = \frac{\exp[a_0(p) + a_1(p)\exp(-\kappa) + a_2(p)\kappa^2 + a_3(p)\kappa^3]}{n} \qquad (18.4.18)$$

with coefficients $a_i(p)$ for selected probabilities p in Table 18.4.1 based on the actual sampling variance of unbiased L-moment quantile estimators in samples of size $n = 40$; the variances provided by Eq. (18.4.18) are relatively accurate for sample sizes $20 \leq n \leq 70$ and $\kappa > -0.20$.

18.5 REGIONALIZATION

Frequency analysis is a problem in hydrology because sufficient information is seldom available at a site to adequately determine the frequency of rare events. At some sites no information is available. When one has 30 years of data to estimate the event exceeded with a chance of 1 in 100 (the 1 percent exceedance event), extrapolation is required. Given that sufficient data will seldom be available at the site of interest, it makes sense to use climatic and hydrologic data from nearby and similar locations.

The National Research Council (Ref. 104, p. 6) proposed three principles for hydrometeorological modeling: "(1) 'substitute space for time'; (2) introduction of more 'structure' into models; and (3) focus on extremes or tails as opposed to, or even to the exclusion of, central characteristics." One substitutes space for time by using hydrologic information at different locations to compensate for short records at a single site. This is easier to do for rainfall which in regions without appreciable relief should have fairly uniform characteristics over large areas. It is more difficult for floods and particularly low flows because of the effects of catchment topography and geology. A successful example of regionalization is the index flood method discussed below. Many other regionalization procedures are available.[28] See also Secs. 18.7.2 and 18.7.3.

Section 18.5.2 discusses regression procedures for deriving regional relationships relating hydrologic statistics to physiographic basin characteristics. These are particularly useful at *ungauged sites*. When floods at a short-record site are highly correlated with floods at a site with a longer record, the *record augmentation* procedures described in Sec. 18.5.3 can be employed. These are both ways of making use of regional hydrologic information.

18.5.1 Index Flood

The index flood procedure is a simple regionalization technique with a long history in hydrology and flood frequency analysis.[31] It uses data sets from several sites in an effort to construct more reliable flood-quantile estimators. A similar regionalization approach in precipitation frequency analysis is the station-year method, which combines rainfall data from several sites without adjustment to obtain a large composite record to support frequency analyses.[15] One can also smooth the precipitation quantiles derived from analysis of the records from different stations.[63]

The concept underlying the index flood method is that the distributions of floods at different sites in a region are the same except for a scale or index-flood parameter which reflects the size, rainfall, and runoff characteristics of each watershed. Generally the mean is employed as the index flood. The problem of estimating the pth quantile x_p is then reduced to estimation of the mean for a site μ_X, and the ratio x_p/μ_X of the pth quantile to the mean. The mean can often be estimated adequately with the record available at a site, even if that record is short. The indicated ratio is estimated by using regional information. The British *Flood Studies Report*[105] calls these normalized regional flood distributions *growth curves.* The index flood method was also found to be an accurate and reproducible method for use at ungauged sites.[107]

At one time the British attempted to normalize the floods available at each site so that a large composite sample could be constructed to estimate their growth curves;[105] this approach was shown to be relatively inefficient.[69] Regional PWM index flood frequency estimation procedures that employ PWM and L moments, and often the GEV or Wakeby distributions, have been studied.[71,81,91,112,162] These results demonstrate that L-moment/GEV index flood procedures should in practice with appropriately defined regions be reasonably robust and more accurate than procedures that attempt to estimate two or more parameters with the short records often available at many sites. Outlined below is the L-moment/GEV version of the algorithm initially proposed by Landwehr, Matalas, and Wallis (personal communication, 1978), and popularized by Wallis and others.[69,160,162]

Let there be K sites in a region with records $[x_t(k)]$, $t = 1, \ldots, n_k$, and $k = 1, \ldots, K$. The L-moment/GEV index-flood procedure is

1. At each site k compute the three L-moment estimators $\hat{\lambda}_1(k)$, $\hat{\lambda}_2(k)$, $\hat{\lambda}_3(k)$ using the unbiased PWM estimators b_r.
2. To obtain a normalized frequency distribution for the region, compute the regional average of the normalized L moments of order $r = 2$ and 3:

$$\hat{\lambda}_r^R = \frac{\sum_{k=1}^{K} w_k\,[\hat{\lambda}_r(k)/\hat{\lambda}_1(k)]}{\sum_{k=1}^{K} w_k} \qquad \text{for } r = 2, 3 \tag{18.5.1}$$

For $r = 1$, $\hat{\lambda}_1^R = 1$. Here w_k are weights; a simple choice is $w_k = n_k$, where n_k is the sample size for site k. However, weighting by the sample sizes when some sites have much longer records may give them undue influence. A better choice which limits the weight assigned to sites with longer records is

$$w_k = \frac{n_k n_R}{n_k + n_R}$$

where n_k are the sample sizes and $n_R \approx 25$; the optimal value of the weighting parameter n_R depends on the heterogeneity of a region.[138,141]

3. Using the average normalized L moments $\hat{\lambda}_1^R$, $\hat{\lambda}_2^R$, and $\hat{\lambda}_3^R$ in Eqs. (18.2.22) and (18.2.23), determine the parameters and quantiles $\hat{x}_p^R$ of the normalized regional GEV distribution.
4. The estimator of the $100p$ percentile of the flood distribution at any site k is

$$\hat{x}_p(k) = \hat{\lambda}_1^k\, \hat{x}_p^R \tag{18.5.2}$$

where λ_1^k is the at-site sample mean for site k:

$$\hat{\lambda}_1^k = \frac{1}{n_k} \sum_{t=1}^{n_k} x_t(k)$$

Of value is an estimate of the precision of flood quantiles obtained with Eq. (18.5.2). Across the region of interest, let the variance of the differences $\tilde{x}_p^S - \hat{x}_p^R$ between the actual normalized quantile $\tilde{x}_p^S$ for a random site and the average regional estimator $\hat{x}_p^R$ be denoted ϑ^2; ϑ^2 describes the heterogeneity of a region. Then the variance of the error associated with the flood quantile estimator $\hat{x}_p$, equal to $\hat{\lambda}_1 \hat{x}_p^R$ for at-site sample mean $\hat{\lambda}_1$, can be written

$$\mathrm{Var}\,(\hat{x}_p) = E[\hat{\lambda}_1\, \hat{x}_p^R - \lambda_1\, \tilde{x}_p^S]^2 = \mathrm{Var}\,(\hat{\lambda}_1)\; E[(\hat{x}_p^R)^2] + (\lambda_1)^2\, \vartheta^2 \qquad (18.5.3)$$

The expected error in $\hat{x}_p$ is a combination of sampling error in site k's sample mean

$$\mathrm{Var}\,(\hat{\lambda}_1) = \mathrm{Var}\,[\bar{x}(k)] = \frac{\sigma_x^2}{n_k}$$

and the precision ϑ^2 of the regional flood quantile $\hat{x}_p^R$ as an estimator of the normalized quantile $\tilde{x}_p^S$ for a site in the region. In practice ϑ^2 is generally difficult to estimate. The generalized least squares regional regression methodology in Sec. 18.5.2 addresses the relevant issues and can provide a useful estimator.

A key to the success of the index flood approach is identification of reasonably similar sets of basins to keep the error in the regional quantiles ϑ^2 small.[94] Basins can be grouped geographically, as well as by physiographic characteristics including drainage area and elevation. Regions need not be geographically contiguous. Each site can potentially be assigned its own unique region consisting of sites with which it is particularly similar,[17] or regional regression equations can be derived to compute normalized regional quantiles as a function of a site's physiographic characteristics, or other statistics.[120]

For regions which exhibit a large ϑ^2, or when the record length for a site is on the order of 40 or more, then a two-parameter index flood procedure that uses the regional value of κ with at-site estimates of the GEV distribution's ξ and α parameters becomes attractive.[94] Chowdhury et al.[22] provide goodness-of-fit tests to assess whether a particular dimensionless regional GEV distribution, or a specified regional κ, is consistent with the data available at a questionable site.

18.5.2 Regional Regression

Regression can be used to derive equations to predict the values of various hydrologic statistics (including means, standard deviations, quantiles, and normalized regional flood quantiles) as a function of physiographic characteristics and other parameters. Such relationships are needed when little or no flow data are available at or near a site. Figure 18.5.1 illustrates the estimated prediction errors for regression models of low-flow, mean annual flows, and flood flows in the Potomac River Basin in the United States. Regional regression models have long been used to predict flood quantiles at ungauged sites, and in a nationwide test this method did as well or better than more complex rainfall-runoff modeling procedures.[107]

Consider the traditional log-linear model for a statistic y_i which is to be estimated by using watershed characteristics such as drainage area and slope:

$$y_i = \alpha + \beta_1 \log\,(\text{area}) + \beta_2 \log\,(\text{slope}) + \ldots + \epsilon \qquad (18.5.4)$$

A challenge in analyzing this model and estimating its parameters with available records is that one only obtains sample estimates, denoted $\hat{y}_i$, of the hydrologic statistics y_i. Thus the observed error ϵ is a combination of: (1) the time-sampling error

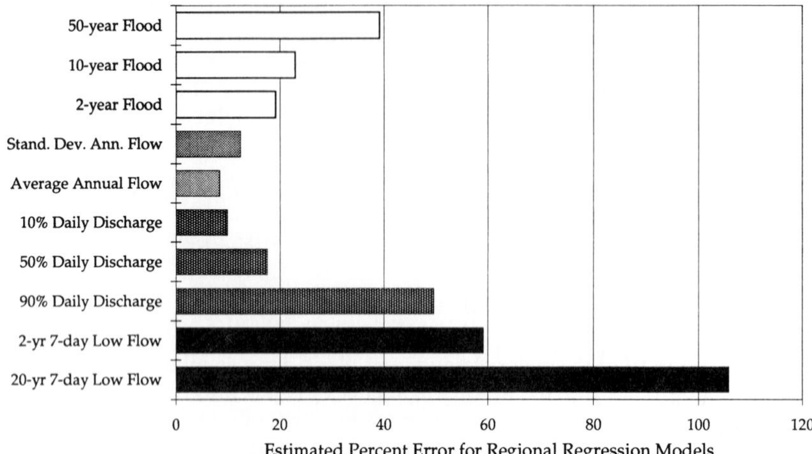

FIGURE 18.5.1 Percentage error for regional regression estimators of different statistics in the Potomac River Basin in the United States. *(From Ref. 142.)*

in sample estimators of y_i (these errors at different sites can be cross-correlated if the records are concurrent) and (2) underlying model error (lack of fit) due to failure of the model to exactly predict the true value of the y_i's at every site. Often these problems have been ignored and standard ordinary least squares (OLS) regression has been employed.[142] Stedinger and Tasker[130–132] develop a specialized generalized least squares (GLS) regression methodology to address these issues. Advantages of the GLS procedure include more efficient parameter estimates when some sites have short records, an unbiased model-error estimator, and a better description of the relationship between hydrologic data and information for hydrologic network analysis and design;[130,141] see also Sec. 17.4.8. Examples are provided by Tasker and Driver,[140] Vogel and Kroll,[156] and Potter and Faulkner.[111]

18.5.3 Record Augmentation and Extension

One can fill in missing observations in a short record by using a longer nearby record with which observations in the short record are highly correlated. Such cross-correlation can be used to fill in a few scattered missing observations, to extend the shorter record, or to improve estimates of the mean and variance of the events at the short-record site. For this third purpose it is not necessary to actually construct the extended series; one needs only the improved estimates of the moments. This idea is called *record augmentation* (Ref. 97, Ref. 105, App. 7 in Ref. 79).

Let x and y denote the flow record at the long- and short-record sites, respectively; let subscript 1 denote sample means and variances calculated for the period of concurrent record and subscript 2 denote the sample mean and variance for the long-record x site calculated using only the observations for which there is no corresponding y. The Matalas-Jacobs augmented-record estimator of the mean is

$$\hat{\mu}_Y = \bar{y}_1 + \frac{n_2}{n_1 + n_2}\, b\,(\bar{x}_2 - \bar{x}_1) \qquad n_1 \geq 4 \tag{18.5.5}$$

where n_1 is the number of concurrent observations and n_2 is the number of additional observations available at the x site. Their estimator of the variance is essentially

$$\hat{\sigma}_Y^2 = s_{y1}^2 + \frac{n_2}{n_1 + n_2} b^2 (s_{x2}^2 - s_{x1}^2) \qquad n_1 \geq 6 \tag{18.5.6}$$

except for several negligible adjustments; here

$$b = \hat{\rho}_{xy} \frac{s_{y1}}{s_{x1}} \tag{18.5.7}$$

is the standard linear regression estimator of change in y from a change in x. Equation (18.5.5) is relatively effective at improving estimates of the mean when the cross-correlation is 0.70 or greater; Eq. (18.5.6) transfers less information about the variance, which generally requires a cross-correlation of at least 0.85 to be worthwhile.[151] If the observations are serially correlated, considerably less information is transferred.[137,157]

In some cases one actually wants to create a longer series that will be used in simulation or archived as described in Sec. 17.4.10. In such cases it would be preferable if the extended series y_t had the same variance as the original series and was not smoothed by the process of regressing one record on another. This idea of *record extension* is developed in Refs. 64, 65, and 151 and, for the multivariate case with cross-correlation, in Ref. 50.

18.6 PARTIAL DURATION SERIES, MIXTURES, AND CENSORED DATA

This section discusses situations where data describing hydrologic events are not a simple series of annual values. Partial duration series and mixture models discussed in Secs. 18.6.1 and 18.6.2 describe hydrologic events by more than an average or a single annual maximum or minimum. These are examples of the idea of stochastic structure discussed in the introduction to Sec. 18.5. Section 18.6.3 discusses methods for dealing with censored data sets that occur when some observations fall below a recording threshold.

18.6.1 Partial Duration Series

Two general approaches are available for modeling flood, rainfall, and many other hydrologic series. Using an *annual maximum series,* one considers the largest event in each year; using a *partial duration series* (PDS) or peaks-over-threshold (POT) approach, the analysis includes all peaks above a truncation or threshold level. An objection to using annual maximum series is that it employs only the largest event in each year, regardless of whether the second largest event in a year exceeds the largest events of other years. Moreover, the largest annual flood flow in a dry year in some arid or semiarid regions may be zero, or so small that calling them floods is misleading.

Partial duration series analyses avoid such problems by considering all independent peaks which exceed a specified threshold. Fortunately one can estimate annual exceedance probabilities from the analysis of PDS with Eq. (18.6.4), below, or empirical relationships.[63] Arguments in favor of PDS are that relatively long and reliable PDS records are often available, and if the arrival rate for peaks over the threshold is large enough (1.65 events/year for the Poisson arrival with exponential exceedance model), PDS analyses should yield more accurate estimates of extreme quantiles than the corresponding annual-maximum frequency analyses.[105,118,145] Still, a drawback of PDS analyses is that one must have criteria to identify only independent peaks (and not multiple peaks corresponding to the same event); thus PDS analysis can be more complicated than analyses using annual maxima.

Partial duration models are applicable to modeling flood or rainfall events that exceed some threshold depth, or the occurrence of runoff carrying non-point-pollution loads. Partial duration models, perhaps with parameters that vary by season, are often used to estimate expected damages from hydrologic events when more than one damage-causing event can occur in a season or within a year.[108]

Two issues arise in modeling PDS. First, one must model the arrival rate of events larger than the threshold level; second, one must model the magnitudes of those events. For example, a Poisson distribution is often used to model the arrival of events, and an exponential distribution to describe the magnitudes of peaks which exceed the threshold.[16] For large-return-period events, the actual probabilistic model for arrivals is not important, provided different models yield the same average number of arrivals per year.[27,105]

There are several general relationships between the probability distribution for annual maximum and the frequency of events in a partial duration series. For a PDS, let λ be the arrival rate, equal to the average number of events per year larger than a threshold x_0; let $G(x)$ be the probability that events when they occur are less than x, and thus falls in the range (x_0, x). Then the arrival rate for any level x, with $x \geq x_0$, is

$$\lambda^* = \lambda[1 - G(x)] \tag{18.6.1}$$

The cdf $F_a(x)$ for the corresponding annual maximum series is the probability that the annual maximum for a year will not exceed x. For independent events, the probability of *no exceedances of* x over a 1-year period is given by the Poisson distribution, so that

$$F_a(x) = \exp(-\lambda^*) = \exp\{-\lambda[1 - G(x)]\} \tag{18.6.2}$$

[This relationship can be derived by dividing a year into m intervals, each with arrival rate λ^*/m. Then for small λ^*/m, the probability of no arrivals in a year is essentially $(1 - \lambda^*/m)^m$. Equation (18.6.2) is obtained in the limit as $m \to \infty$.]

Equation 18.6.2 reveals the relationship between the cdf for the annual maximums, and the arrival rate of and distribution for partial duration peaks. If the annual exceedance probability $1 - F_a(x)$ is denoted $1/T_a$, for an annual return period T_a (denoted as T elsewhere in the chapter) and the corresponding exceedance probability $[1 - G(x)]$ for level x in the partial duration series is denoted q_e, then Eq. (18.6.2) can be written

$$\frac{1}{T_a} = 1 - \exp(-\lambda\, q_e) = 1 - \exp\left(-\frac{1}{T_p}\right) \tag{18.6.3a}$$

where $T_p = 1/\lambda q_e$ is the average return period for level x in the PDS. Equation (18.6.3*a*) can be solved for T_p to obtain

$$T_p = -\frac{1}{\ln(1 - 1/T_a)} \tag{18.6.3b}$$

T_p is less than T_a because more than one event can occur per year in a PDS.

Equation (18.6.3*a*) transforms the *average arrival rate* λq_e for events larger than x into the *annual exceedance probability* $1/T_a$ in the annual maximum series. For levels x with $T_a > 10$, corresponding to infrequent events, the annual exceedance probability $1/T_a$ essentially equals the average arrival rate $\lambda q_e = \lambda[1 - G(x)]$ for the PDS, so that $T_a = T_p$ (Ref. 93). [See also Eq. (18.10.1).]

Consider a useful application of Eq. (18.6.2). Suppose a generalized Pareto distribution (Sec. 18.2.4) describes the distribution $G(x)$ of the magnitude of events larger than a threshold x_0:

$$G(x) = 1 - \left[1 - \kappa\left(\frac{x - x_0}{\alpha}\right)\right]^{1/\kappa} \qquad \text{for } \kappa \neq 0 \tag{18.6.4}$$

For positive κ this cdf has upper bound $x_{\max} = x_0 + \alpha/\kappa$; for $\kappa < 0$, an unbounded and thick-tailed distribution results; $\kappa = 0$ yields a two-parameter exponential distribution. Substitution of Eq. (18.6.4) for $G(\cdot)$ into Eq. (18.6.2) yields a GEV distribution for the annual maximum series greater than x_0 if $\kappa \neq 0$:[34,70,125]

$$F_a(x) = \exp\left[-\left(1 - \kappa\frac{x - \xi}{\alpha^*}\right)^{1/\kappa}\right] \qquad \kappa \neq 0 \tag{18.6.5a}$$

and a Gumbel distribution for $\kappa = 0$:[16]

$$F_a(x) = \exp\left[-\left(\frac{x - \xi}{\alpha}\right)\right] \tag{18.6.5b}$$

when $x \geq x_0$; the transformed parameters ξ and α^* are defined by

$$\begin{aligned} &\xi = x_0 + \frac{\alpha(1 - \lambda^{-\kappa})}{\kappa} \qquad \alpha^* = \alpha\lambda^{-\kappa} \qquad \kappa \neq 0 \\ &\xi = x_0 + \alpha \ln(\lambda) \qquad\qquad\qquad\qquad\;\; \kappa = 0 \end{aligned} \tag{18.6.6}$$

This general Poisson-Pareto model is a flexible and physically reasonable model of many phenomena. It has the advantage that regional estimates of the GEV distribution's shape parameter κ from annual maximum and PDS analyses can be used interchangeably.

In practice the arrival rate λ can simply be estimated by the average number of exceedances of x_0 per year. For either the exponential or generalized Pareto distributions in Table 18.1.2 for $G(x)$, the lower bound (denoted ξ in Table 18.1.2) equals x_0. The other parameters in Eqs. (18.6.5) and (18.6.6) can be estimated by substituting sample estimators into the inverse of the relationships in Table 18.1.2:

$$\begin{aligned} &\text{For } \kappa \neq 0: \qquad \kappa = \frac{\mu - x_0}{\lambda_2} - 2; \ \alpha = (\mu - x_0)(1 + \kappa) \\ &\text{For fixed } \kappa = 0: \qquad \frac{1}{\beta} = \alpha = \mu - x_0 \end{aligned} \tag{18.6.7}$$

where $\mu = \lambda_1$ is the mean of X, λ_2 is the second L moment, and β is the exponential distribution's scale parameter in Tables 18.1.2 and 18.2.1.

18.6.2 Mixtures

A common problem in hydrology is that annual maximum series are composed of events that may arise from distinctly different processes. Precipitation series may correspond to different storm types in different seasons (such as summer thunderstorms, winter frontal storms, and remnants of tropical hurricanes). Floods arising from different types of precipitation events, or from snow melt, may have distinctly different probability distributions.[168]

The annual maximum series M can be viewed as the maximum of the maximum summer event S and the maximum winter event W:

$$M = \max(S, W) \tag{18.6.8}$$

Here S and W may be defined by a rigidly specified calendar period, a loosely defined climatic period, or the physical characteristics of the phenomena.

Let the cdf of the S and W variables be $F_S(s)$ and $F_W(w)$. Then, if the magnitudes of the summer and winter events are *statistically independent,* meaning that knowing one has no effect on the probability distribution of the other, the cdf for M is

$$F_M(m) = P[M = \max(S, W) \leq m] = F_S(m)\, F_W(m) \tag{18.6.9}$$

because M will be less than m only if *both* S and W are less than m. If two or more independent series of events contribute to an annual maximum, the distribution of the maximum is the product of their cdfs.

An important question is when it is advisable to model several different component precipitation or flood series separately, and when it is better to model the composite annual maximum series directly. If several series are modeled, then more parameters must be estimated, but more data are available than if the annual maximum series (or the partial duration series) for each type of event is employed. Fortunately, the distributions of large events caused by different mechanisms can be relatively similar.[62] Modeling the component series separately is most attractive when the annual maximum series is composed of components with distinctly different distributions which are individually easy to model because classical two-parameter Gumbel or lognormal distributions describe them well, and such a simple model provides a poor description of the composite annual maximum series.

18.6.3 Analysis of Censored Data

In some water-quality investigations, a substantial portion of reported values of many contaminants is below limits of detection. Likewise, low-flow and sometimes flood-flow observations are rounded to or reported as zero. Such data sets are called *censored samples* because it is as if the values of observations in a *complete sample* that fell above or below some level were removed, or censored. Several approaches are available for analysis of censored data sets, including probability plots and probability-plot regression, weighted-moment estimators, maximum likelihood estimators, and conditional probability models.[55,57,61] See also Section 17.5.

Probability-plot methods for use with censored data are discussed below. They are relatively simple and efficient when the majority of values are observed, and unobserved values are known to be below (above) some detection limit or perception threshold which serves as an upper (lower) bound. In such cases, probability-plot regression estimators of moments and quantiles are as accurate as maximum likelihood estimators, and almost as good as estimators computed with complete samples.[33,60] Partial PWMs are the expectation of $xF(x)^r$ for x values above a threshold; they are conceptually similar to probability-plot regression estimators and provide a useful alternative for fitting some distributions.[164]

Weighted moment estimators are used in flood frequency analyses with data sets that include both a complete gauged record and a historical flood record consisting of all events above a perception threshold.[79,133,165] (See Sec. 18.7.4.) Weighted moment estimators weight values above and below the threshold levels so as to obtain moment estimators consistent with a complete sample. These methods are reasonable when a substantial fraction of the observations remain after censoring (at least 10 percent), and a value is either observed accurately or falls below a threshold and thus is censored.

Maximum likelihood estimators are quite flexible, and are more efficient than plotting and weighted moment estimators when the frequency with which a threshold was exceeded represents most of the sample information.[23,133] They allow the recorded values to be represented by exact values, ranges, and various thresholds that either were or were not exceeded at various times; this can be particularly important with historical flood data sets because the magnitudes of many historical floods are not recorded precisely, and it may be known that a threshold was never crossed or was crossed at most once or twice in a long period.[23] (See Sec. 18.7.4.) In these cases maximum likelihood estimators are perhaps the only approach that can make effective use of the available information.[33]

Conditional probability models are appropriate for simple cases wherein the censoring occurs because small observations are recorded as zero, as often happens with low-flow and some flood records. An extra parameter describes the probability p_0 that an observation is zero. A continuous distribution $G(x)$ is derived for the strictly positive nonzero values of X; the parameters of the cdf G can be estimated by any procedure appropriate for complete samples. The unconditional cdf $F(x)$ for any value $x > 0$ is then

$$F(x) = p_0 + (1 - p_0)\, G(x) \tag{18.6.10}$$

Equations (18.7.6) to (18.7.8) provide an example of such a model.

Plotting Positions for Censored Data. Section 18.3.2 discusses plotting positions useful for graphical fitting methods, as well as visual displays of data. Suppose that among n samples a detection limit or perception threshold is exceeded by water-quality observations or flood flows r times. The natural estimator of the exceedance probability q_e of the perception threshold is r/n. If the r values which exceeded the threshold are indexed by $i = 1, \ldots, r$, reasonable plotting positions approximating the exceedance probabilities within the interval $(0, q_e)$ are

$$q_i = q_e\left(\frac{i-a}{r+1-2a}\right) = \frac{r}{n}\left(\frac{i-a}{r+1-2a}\right) \tag{18.6.11}$$

where a is a value from Table 18.3.1. For $r \gg (1 - 2a)$, q_i is indistinguishable from $(i - a)/(n + 1 - 2a)$ for a single threshold. Reasonable choices for a generally make little difference to the resulting plotting positions.[60]

The idea of an exceedance probability for the threshold is important when detection limits change over time, generating multiple thresholds. In such cases, an exceedance probability should be estimated for each threshold so that a consistent set of plotting positions can be computed for observations above, below, or between thresholds.[60,66] For example, consider a historical flood record with an h-year historical period in addition to a complete s-year gauged flood record. Assume that during the total $n = (s + h)$ years of record, a total of r floods exceeded a perception threshold (censoring level) for historical floods. These r floods can be plotted by using Eq. (18.6.11).

Let e be the number of gauged-record floods that exceeded the threshold and hence are counted among the r exceedances of that threshold. Plotting positions within $(q_e, 1)$ for the remaining $(s - e)$ below-threshold gauged-record floods are

$$q_j = q_e + (1 - q_e)\left(\frac{j - a}{s - e + 1 - 2a}\right) \qquad (18.6.12)$$

for $j = 1$ through $s - e$, where again a is a value from Table 18.3.1. This approach directly generalizes to several thresholds.[60,66] For records with an r of only 1 or 2, Ref. 166 proposes fitting a parametric model to the gauged record to estimate q_e; these are cases when nonparametric estimators of q_e and q_i in Eq. (18.6.11) are inaccurate,[66] and MLEs are particularly attractive for parameter estimation.[23,133]

Probability-Plot Regression. Probability-plot regression has been shown to be a robust procedure for fitting a distribution and estimating various statistics with censored water-quality data.[60] When water-quality data is well-described by a lognormal distribution, available values $\log [X_{(1)}] \geq \ldots \geq \log [X_{(r)}]$ can be regressed upon $\Phi^{-1}[1 - q_i]$ for $i = 1, \ldots, r$, where the r largest observation in a sample of size n are available, and q_i are their plotting positions. If regression yields constant m and slope s, a good estimator of the pth quantile is

$$\hat{x}_p = 10^{m + s z_p} \qquad (18.6.13)$$

for cumulative probability $p > (1 - r/n)$. To estimate sample means and other statistics, one can fill in the missing observations as

$$X_{(i)} = 10^{y(i)} \qquad \text{for } i = r + 1, \ldots, n \qquad (18.6.14)$$

where $y(i) = m + s\ \Phi^{-1}(1 - q_i)$ and an approximation for Φ^{-1} is given in Eq. (18.2.3). Once a complete sample is constructed, standard estimators of the sample mean and variance can be calculated, as can medians and ranges. By filling in the missing small observations, and then using complete-sample estimators of statistics of interest, the procedure is made relatively insensitive to the assumption that the observations actually have a lognormal distribution.[60]

18.7 FREQUENCY ANALYSIS OF FLOODS

Lognormal, Pearson type 3, and generalized extreme value distributions are reasonable choices for describing flood flows using the fitting methods described in Sec. 18.2. However, as suggested in Sec. 18.3.3, it is advisable to use regional experience to select a distribution for a region and to reduce the number of parameters estimated for an individual site. This section describes sources of flood flow data and particular procedures adopted for flood flow frequency analysis in the United States and the United Kingdom, and discusses the use of historical flood flow information.

18.7.1 Selection of Data and Sources

A convenient way to find information on United States water data is through the U.S. National Water Data Exchange (NAWDEX) assistance centers. [For information contact NAWDEX, U.S. Geological Survey (USGS), 421 National Center, Reston, Va. 22092; tel. 703-648-6848.] Records are also published in annual U.S. Geological Survey water data reports. Computerized records are stored in the National Water Data Storage and Retrieval System (WATSTORE). Many of these records (climate data, daily and annual maximum stream flow, water-quality parameters) have been put on compact disc read-only memories (CD-ROMs) sold by EarthInfo Inc. (5541 Central Ave., Boulder, Colo. 80301; tel. 303-938-1788; fax 303-938-8183) so that the data can be accessed directly with personal computers. The WATSTORE peak-flow records contain annual maximum instantaneous flood-peak discharge and stages, and dates of occurrence as well as associated partial duration series for many sites. USGS offices also publish sets of regression relationships (often termed *state equations*) for predicting flood and low-flow quantiles at ungauged sites in the United States.

18.7.2 Bulletin 17B Frequency Analysis

Recommended procedures for flood-frequency analyses by U.S. federal agencies are described in Bulletin 17B.[79] Bulletin no. 15, "A Uniform Technique for Determining Flood Flow Frequencies," released in December 1967, recommended the log-Pearson type 3 distribution for use by U.S. federal agencies. The original Bulletin 17, released in March 1976, extended Bulletin 15 and recommended the log-Pearson type 3 distribution with a regional estimator of the log-space skew coefficient. Bulletin 17A followed in 1977. Bulletin 17B was issued in September 1981 with corrections in March 1982. Thomas[143] describes the development of these procedures.

The Bulletin 17 procedures were essentially finalized in the mid-1970s, so they did not benefit from subsequent advances in multisite regionalization techniques. Studies in the 1980s demonstrated that use of reasonable index flood procedures should provide substantially better flood quantile estimates, with perhaps half the standard error.[91,112,162] Bulletin 17 procedures are much less dependent on regional multisite analyses than are index flood estimators, and Bulletin 17 is firmly established in the United States, Australia, and other countries. However, App. 8 of the bulletin does describe a procedure for weighting the bulletin's at-site estimator and a regional regression estimator of the logarithms of a flood quantile by the available record length and the effective record length, respectively. The resulting weighted estimator reflects a different approach to combining regional and at-site information than that employed by index flood procedures.

Bulletin 17B recommends special procedures for zero flows, low outliers, historic peaks, regional information, confidence intervals, and expected probabilities for estimated quantiles. This section describes only major features of Bulletin 17B. The full Bulletin 17B procedure is described in that publication and is implemented in the HECWRC computer program discussed in Sec. 18.11.

The bulletin describes procedures for computing flood flow frequency curves using annual flood series with at least 10 years of data. The recommended technique fits a Pearson type 3 distribution to the common base 10 logarithms of the peak discharges. The flood flow Q associated with cumulative probability p is then

$$\log (Q_p) = \overline{X} + K_p S \tag{18.7.1}$$

where $\overline{X}$ and S are the sample mean and standard deviation of the base 10 logarithms, and K_p is a frequency factor which depends on the skew coefficient and selected exceedance probability; see Eq. (18.2.28) and discussion of the log-Pearson type 3 distribution in Sec. 18.2.3. The mean, standard deviation, and skew coefficient of station data should be computed by Eq. (18.1.8), where X_i are the base 10 logarithms of the annual peak flows. Section 18.1.3 discusses advantages and disadvantages of logarithmic transformations.

The following sections discuss three major features of the bulletin: generalized skew coefficients, outliers, and the conditional probability adjustment. Expected probability adjustments are also discussed. Confidence intervals for Pearson distributions with known and generalized skew coefficient estimators are discussed in Sec. 18.4.3. Use of historical information is discussed in Sec. 18.7.4, mixed populations in Sec. 18.6.2, and record augmentation in Sec. 18.5.3.

Generalized Skew Coefficient. Because of the variability of at-site sample skew coefficients in small samples, the bulletin recommends weighting the station skew coefficient with a generalized coefficient of skewness, which is a regional estimate of the log-space skewness. In the absence of detailed studies, the generalized skew coefficient G_g for sites in the United States can be read from Plate I in the bulletin. Assuming that the generalized skew coefficient is unbiased and independent of the station skew coefficient, the mean square error (MSE) of the weighted estimate is minimized by weighting the station and generalized skew coefficients inversely proportional to their individual mean square errors:

$$G_w = \frac{G_s/\text{MSE}(G_s) + G_g/\text{MSE}(G_g)}{1/\text{MSE}(G_s) + 1/\text{MSE}(G_g)} \tag{18.7.2}$$

Here G_w is the weighted skew coefficient, G_s is the station skew coefficient, and G_g is the generalized regional estimate of the skew coefficient; MSE( ) is the mean square error of the indicated variable. When generalized regional skew coefficients are read from its Plate I, Bulletin 17 recommends using $\text{MSE}(G_g) = 0.302$.

From Monte Carlo experiments,[159] the bulletin recommends that $\text{MSE}(G_g)$ be estimated using the bulletin's Table 1, or an expression equivalent to

$$\text{MSE}(G_s) = \frac{10^{a+b}}{n^b} \tag{18.7.3}$$

where

$$\begin{aligned} a &= -0.33 + 0.08|G_s| && \text{if} \quad |G_s| \le 0.90 \\ &= -0.52 + 0.30|G_s| && \text{if} \quad |G_s| > 0.90 \\ b &= 0.94 - 0.26|G_s| && \text{if} \quad |G_s| \le 1.50 \\ &= 0.55 && \text{if} \quad |G_s| > 1.50 \end{aligned}$$

$\text{MSE}(G_s)$ is essentially $5/n$ for small G_s and $10 \le n \le 50$. G_g should be used in place of G_s in Eq. (18.7.3) when estimating $\text{MSE}(G_s)$ to avoid correlation between G_s and the estimate of $\text{MSE}(G_s)$ (Ref. 138). McCuen[98] and Tasker and Stedinger[138] discuss the development of skew-coefficient maps, and regression estimators of G_g and $\text{MSE}(G_g)$.

Outliers. Bulletin 17B defines outliers to be "Data points which depart significantly from the trend of the remaining data." In experimental statistics an outlier is often a rogue observation which may result from unusual conditions or observational or recording error; such observations are often discarded. In this application low outliers are generally valid observations, but because Bulletin 17 uses the logarithms of the observed flood peaks to fit a two-parameter distribution with a generalized skew coefficient, one or more unusual low-flow values can distort the entire fitted frequency distribution. Thus detection of such values is important and fitted distributions should be compared graphically with the data to check for problems.

The thresholds used to define high and low outliers in log space are

$$\overline{X} \pm K_n S \tag{18.7.4}$$

where $\overline{X}$ and S are the mean and standard deviations of the logarithms of the flood peaks, excluding outliers previously detected, and K_n is a critical value for sample size n. For normal data the largest observation will exceed $\overline{X} + K_n S$ with a probability of only 10 percent; thus Eq. (18.7.4) is a one-sided outlier test with a 10 percent significance level. Values of K_n are tabulated in Bulletin 17B; for $5 \leq n \leq 150$, K_n can be computed by using the common base 10 logarithm of the sample size

$$K_n = -0.9043 + 3.345\sqrt{\log(n)} - 0.4046\log(n) \tag{18.7.5}$$

Flood peaks identified as low outliers are deleted from the record and a conditional probability adjustment is recommended. High outliers are retained unless historical information is identified showing that such floods are the largest in an extended period.

Conditional Probability Adjustment. A conditional probability procedure is recommended for frequency analysis at sites whose record of annual peaks is truncated by the omission of peaks below a minimum recording threshold, years with zero flow, or low outliers. The bulletin does not recommend this procedure when more than 25 percent of the record is below the truncation level. Section 18.6.3 discusses other methods.

Let $G(x)$ be the Pearson type 3 (P3) distribution fit to the r logarithms of the annual maximum floods *that exceeded the truncation level* and are included in the record, after deletions of zero, low outliers, and other events. If the original record spanned n years ($n > r$), then an estimator of the probability the truncation level is exceeded is

$$q_e = \frac{r}{n} \tag{18.7.6}$$

Flood flows exceeded with a probability $q \leq q_e$ in any year are obtained by solving

$$q = q_e[1 - G(x)] \tag{18.7.7}$$

to obtain

$$G(x) = 1 - \frac{q}{q_e} = 1 - q\left(\frac{n}{r}\right) \tag{18.7.8}$$

Bulletin 17 uses Eq. (18.7.8) to calculate the logarithms of flood flows which will be exceeded with probabilities of $q = 0.50$, 0.10, and 0.01. These three values are used to define a new Pearson type 3 distribution for the logarithms of the flood flows which reflects the unconditional frequency of above threshold values. The new Pearson type 3 distribution is defined by its mean M_a, variance S_a^2, and skew coefficient

G_a, which are calculated as

$$G_a = -2.50 + 3.12 \frac{\log (Q_{0.99}/Q_{0.90})}{\log (Q_{0.90}/Q_{0.50})}$$

$$S_a = \frac{\log (Q_{0.99}/Q_{0.50})}{K_{0.99} - K_{0.50}} \tag{18.7.9}$$

$$M_a = \log (Q_{0.50}) - K_{0.50}\, S_a$$

for log-space skew coefficients between -2.0 and $+2.5$. The Pearson type 3 distribution obtained with the moments in Eq. (18.7.9) should not be used to describe the frequency of flood flows below the median $Q_{0.50}$. Fitted quantiles near the threshold are likely to be particularly poor if the P3 distribution $G(x)$ fit to the above threshold values has a lower bound less than the truncation level for zeros and low outliers, which is thus a lower bound for x.

Expected Probability. A fundamental issue is what a hydrologist should provide when requested to estimate the flood flow exceeded with probability $q = 1/T$ using short flood flow records. It is agreed that one wants the flood quantile x_{1-q} which will be exceeded with probability q. An unresolved question is what should be the statistical characteristics of estimators $\hat{x}_{1-q}$. Most estimators in Sec. 18.2 yield $\hat{x}_{1-q}$ that are almost unbiased estimators of x_{1-q}:

$$E[\hat{x}_{1-q}] \approx x_{1-q} \tag{18.7.10}$$

and which have a relatively small variance or mean square error. However, an equally valid argument suggests that one wants $\hat{x}_{1-q}$ to be a value which in the future will be exceeded with probability q, so that

$$\mathrm{P}(X > \hat{x}_{1-q}) \approx q \tag{18.7.11}$$

when both X and $\hat{x}_{1-q}$ are viewed as random variables. If one had a very long record, these two criteria would lead to almost the same design value x_{1-q}. With short records they lead to different estimates because of the effect of the uncertainty in the estimated parameters.[8,113,127]

For normal samples, App. 11 in Bulletin 17B[79] (see also Ref. 20) provides formulas for the probabilities that the almost-unbiased estimator $\hat{x}_p = \bar{x} + z_p s$ of the 100p percentile will be exceeded. For $p = 0.99$ the formula is

$$\text{Average exceedance probability for } \hat{x}_{0.99} = 0.01\left(1 + \frac{26}{n^{1.16}}\right) \tag{18.7.12}$$

For samples of size 16, estimates of the 99 percentile will be exceeded with a probability of 2 percent on average. Bulletin 17B notes that for lognormal or log-Pearson distributions, the equations in its App. 11 can be used to make an expected probability adjustment.

Unfortunately, while the expected probability correction can eliminate the bias in the expected exceedance probability of a computed T-year event, the corrections would generally increase the bias in estimated damages calculated for dwellings and economic activities located at fixed locations in a basin.[4,127] This paradox arises because the estimated T-year flood is a (random) level computed by the hydrologist based on the fitted frequency distribution, whereas the expected damages are calculated for human and economic activities at fixed flood levels. Expected probability issues are related to Bayesian inference.[127]

18.7.3 British Frequency Analysis Procedures

The *Flood Studies Report*[105] contains hydrological, meteorological, and flood routing studies for the British Isles. The report concluded that the GEV distribution provided the best description of British and Irish annual maximum flood distributions and was recommended for general use in those countries. The three parameter P3 and LP3 distribution also described the data well (Ref. 105, pp. 241, 242).

A key recommendation was use of an index flood procedure. The graphically derived normalized regional flood distributions were summarized by dimensionless GEV distributions called *growth curves*. Reevaluation of the original method showed that the L-moment index flood procedure is to be preferred.[69] (See Sec. 18.5.1.) The report distinguishes between sites with less than 10 years of record, those with 10 to 25 years of record, and those with more than 25 years of record (Ref. 105, pp. 14 and 243):

Sites with $n < 10$ Years. The report recommends a regional growth curve with $\overline{Q}$ obtained from catchment characteristics (see Sec. 18.5.2), or at-site data extended if possible by correlation with data at other sites (see Sec. 18.5.3).

Sites with $10 \leq n \leq 25$ Years. Use either the index flood procedure with at-site data to estimate $\overline{Q}$ or, if the return period $T < 2n$, the Gumbel distribution.

Sites with $n > 25$ Years. Use either an index flood estimator with at-site data to estimate $\overline{Q}$ or, if the return period $T < 2n$, GEV distribution (see Sec. 18.2.2).

For $T > 500$. Use $\overline{Q}$ with a special country-wide growth curve.

18.7.4 Historical Flood Information

Available at-site systematic gauged records are the traditional and most obvious source of information on the frequency of floods, but they are of limited length. Another source of potentially valuable at-site information is historical and paleoflood records. Historical information includes written and other records of large floods left by human observers: newspaper accounts, letters, and flood markers. The term paleoflood information describes the many botanical and geophysical sources of information on large floods which are not limited to the locations of past human observations or recording devices.[6,25,134] Botanical data can consist of the systematic interpretation of tipped trees, scars, and abnormal tree rings along a water course providing a history of the frequency with which one or more thresholds were exceeded.[77,78] Recent advances in physical paleoflood reconstruction have focused on the use of slack-water deposits and scour lines, as indicators of paleoflood stages, and the absence of large flows that would have left such evidence; such physical evidence of flood stage along a water course has been used with radiocarbon and other dating techniques to achieve a relatively accurate and complete catalog of paleofloods in favorable settings with stable channels.[86]

Character of Information. Different processes can generate historical and physical paleoflood records. A flood leaving a high-water mark, or known to be the largest flood of record from written accounts, is the largest flood to have occurred in some period of time which generally extends back beyond the date at which that flood occurred.[66] In other cases, several floods may be recorded (or none at all), because they exceed some perception level defined by the location of dwellings and economic activities, and thus sufficiently disrupted people's lives for their occurrence to be noted, or for the resultant botanical or physical damage to document the event. In statistical terms, historical information represents a *censored sample* because only the largest floods are recorded, either because they exceeded a threshold of perception for the occupants of the basin, or because they were sufficiently large to leave physical evidence which was preserved. To correctly interpret such data, hydrologists should understand the mechanisms or reasons that historical, botanical, or geophysical records document that floods of different magnitudes either did, or did not, occur. The historical record should represent a complete catalog of all events that exceeded various thresholds so that it can serve as the basis for frequency analyses.

Estimation Procedures. A general discussion of estimation techniques with censored data is provided in Sec. 18.6.3, including plotting positions and curve fitting based on a graphical representation of systematic and historical flood peaks. Bulletin 17B[79] recommends a historically weighted moments procedure. A similar partial PWM method has been developed.[165] Curve fitting and weighted moments require that historical flood peaks above the perception level be assigned specific values. Even when the magnitudes of the few observed historical floods are available, historically weighted moments are not as efficient as maximum likelihood estimators.[92,133] The value of historical information using maximum likelihood estimation techniques is well-documented.[23,81,133] Maximum likelihood estimation is quite flexible and allows the historical record to be represented by thresholds that were not exceeded and by flood events whose magnitude is known only to have exceeded a threshold, to lie within some range, or which can be described by a precise value.[135]

18.8 FREQUENCY ANALYSIS OF STORM RAINFALL

The frequency of rainfall of various intensities and durations is used in the hydrologic design of structures that control storm runoff and floods, such as storm sewers, highway culverts, and dams. Precipitation frequency analysis typically provides rainfall accumulation values at a point for a specified exceedance probability and various durations. Basin-average rainfall values are usually developed from point rainfall by using a correction factor for basin areas greater than 10 mi^2 (25.9 km^2), as shown in Fig. 3.9.2.[100,103]

18.8.1 Selection of Data and Sources

United States precipitation data are published in *Climatological Data* and *Hourly Precipitation Data* by the National Oceanic and Atmospheric Administration (NOAA) from their National Climatic Data Center (NCDC); precipitation records and publications can be obtained directly from the center (NCDC, Federal Building, Asheville, NC 28801; tel. 704-259-0682; fax 704-259-0876). Climatic data have been put on CD-ROMs sold by EarthInfo Inc. (5541 Central Ave., Boulder, Colo. 80301; tel. 303-938-1788; fax 303-938-8183). NRC[104] discusses the availability and interpretation of United States rainfall data. Other national and regional agencies publish their own precipitation records.

The user of precipitation data should be aware of possible errors in data collection caused by wind effects, changes in the station environment, and observers. Users should check data for outliers and consistency. Interpretation of data is needed to account for liquid precipitation versus snow equivalent and observation time differences. Stations submitting data to NCDC are expected to operate standard equipment and follow standard procedures with observations taken at standard times.[169]

Rainfall frequency analysis is usually based on annual maximum series or partial duration series at one site (at-site analysis) or several sites (regional analysis). Since rainfall data are usually published for fixed time intervals, e.g., clock hours, they rarely yield the true maximum amounts for the indicated durations. For example, true annual maximum 24-h rainfalls for the United States are on the average 13 percent greater than annual maximum daily values corresponding to a fixed 24-h period.[63] Adjustment factors are usually employed with the results of a frequency analysis of annual maximum series. Such factors depend on the number of observational reporting times within the duration of interest. (See Ref. 172, p. 5-36).

Another source of data which has been used to derive estimates of the probable maximum precipitation, and to a lesser extent for rainfall frequency analysis, is the U.S. Army Corps of Engineers catalog of extreme storms. The data collection and processing were a joint effort of the U.S. Army Corps of Engineers and the U.S. Weather Bureau. Currently, a total of 563 storms, most of which occurred between 1900 and 1940, have been completed and published in Ref. 146; see also Refs. 104 and 123. There are problems associated with the use of this catalog for frequency analysis. It may be incomplete because the criteria used for including a storm in the catalog are not well-defined and have changed. Also, the accuracy in the estimation of the storm depths varies.

18.8.2 Frequency Analysis Studies

The *Rainfall Frequency Atlas,*[63] known as TP-40, provides an extended rainfall frequency study for the United States from approximately 4000 stations. The Gumbel distribution (Sec. 18.2.2; see also Ref. 172) was used to produce the point precipitation frequency maps of durations ranging from 30 min to 24 h and exceedance probabilities from 10 to 1 percent. The report also contains diagrams for making precipitation estimates for other durations and exceedance probabilities. The U.S. Weather Bureau, in a publication called TP-49,[149] published rainfall maps for durations of 2 to 10 days. Isohyetal maps (which partially supersede TP-40) for durations of 5 to 60 min are found in Ref. 46, known as HYDRO-35, and for 6 to 24 h for the western United States in *NOAA Atlas 2.*[100] Examples of frequency maps can be found in Chap. 3.

For a site for which rainfall data are available, a frequency analysis can be performed. Common distributions for rainfall frequency analysis are the Gumbel, log-Pearson type 3, and GEV distributions with $\kappa < 0$, which is the standard distribution used in the British Isles.[105]

Maps presented in TP-40 and subsequent publications have been produced by interpolation and smoothing of at-site frequency analysis results. Regional frequency analysis, which uses data from many sites, can reduce uncertainties in estimators of extreme quantiles (Refs. 15 and 161; see Sec. 18.5.1). Regional analysis requires selection of reasonably homogeneous regions. Schaefer[120] found that rainfall data in Washington State have CV and γ which systematically vary with mean areal precipitation. He used mean areal precipitation as an explanatory variable to develop a regional analysis methodology for a heterogeneous region, thereby eliminating boundary problems that would be introduced if subregions were defined.

Models of daily precipitation series (as opposed to annual maxima) are constructed for purposes of simulating some hydrologic systems. As Chap. 3 discusses, models of daily series need to describe the persistence of wet-day and dry-day sequences. The mixed exponent distribution, and the Weibull distribution with $k =$ 0.7 to 0.8, have been found to be good models of daily precipitation depths on rainy days, though an exponential distribution has often been used.[122,173]

18.8.3 Intensity-Duration-Frequency Curves

Rainfall intensity-duration-frequency (IDF) curves allow calculation of the average design rainfall intensity for a given exceedance probability over a range of durations. IDF curves are available for several U.S. cities; two are shown in Fig. 18.8.1.[148] When an IDF curve is not available, or a longer data base is available than in TP-25 or TP-40, a hydrologist may need to perform the frequency analyses necessary to construct an IDF curve (see p. 456 in Ref. 20).

IDF curves can be described mathematically to facilitate calculations. For example, one can use

$$i = \frac{c}{t^e + f} \tag{18.8.1}$$

where i is the design rainfall intensity (inches per hour), t is the duration (minutes), c is a coefficient which depends on the exceedance probability, and e and f are coefficients which vary with location.[170] For a given return period, the three constants can be estimated to reproduce i for three different t's spanning a range of interest. For example, for a 1 in 10 year event, values for Los Angeles are $c = 20.3$, $e = 0.63$, and $f = 2.06$, while for St. Louis $c = 104.7$, $e = 0.89$, and $f = 9.44$.

More recently, generalized intensity-duration-frequency relationships for the United States have been constructed by Chen[19] using three depths: the 10-year 1-h rainfall (R_1^{10}), the 10-year 24-h rainfall (R_{24}^{10}), and the 100-year 1-h rainfall (R_1^{100}) from TP-40. These depths describe the geographic pattern of rainfall in terms of the depth-duration ratio (R_1^T/R_{24}^T) for any return period T, and the depth-frequency ratio (R_t^{100}/R_t^{10}) for any duration t. Chen's general rainfall IDF relation for the rainfall depth R_t^T in inches for any duration t (in minutes) and any return period T (in years) is

$$R_t^T = \frac{a_1 R_1^{10} \, [x - 1)\log(T_p/10) + 1] \left(\dfrac{t}{60}\right)}{(t + b_1)^{c_1}} \tag{18.8.2}$$

where $x = (R_1^{100}/R_1^{10})$, T_p is the return period for the partial duration series (equal to the reciprocal of the average number of exceedances per year), and a_1, b_1, and c_1 are coefficients obtained from Fig. 18.8.2 as functions of (R_1^{10}/R_{24}^{10}) with the assumption that this ratio does not vary significantly with T. Chen uses $T_p = -1/\ln(1 - 1/T)$ to relate T_p to the return period T for the annual maximum series (see Sec. 18.6.1); for $T > 10$ there is little difference between the two return periods. The coefficients obtained from Fig. 18.8.2 are intended for use with TP-40 rainfall quantiles.

For many design problems, the time distribution of precipitation (hyetograph) is needed. In the design of a drainage system the time of occurrence of the maximum rainfall intensity in relation to the beginning of the storm may be important. Design hyetographs can be developed from IDF curves following available procedures (see Chap. 3).

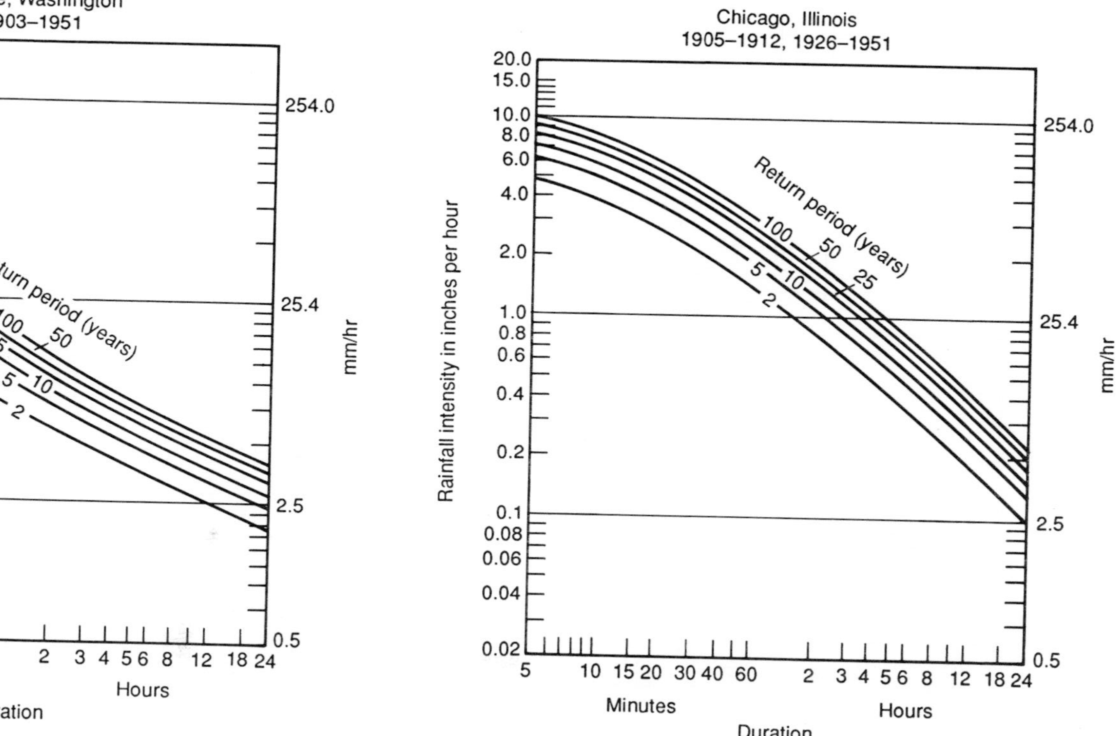

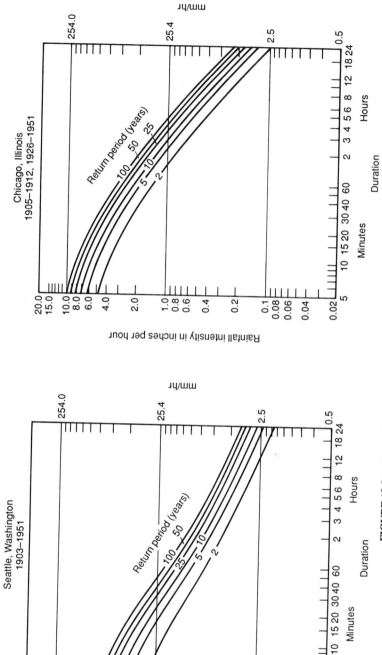

FIGURE 18.8.1 Typical intensity-duration-frequency curves. *(From Ref. 148.)*

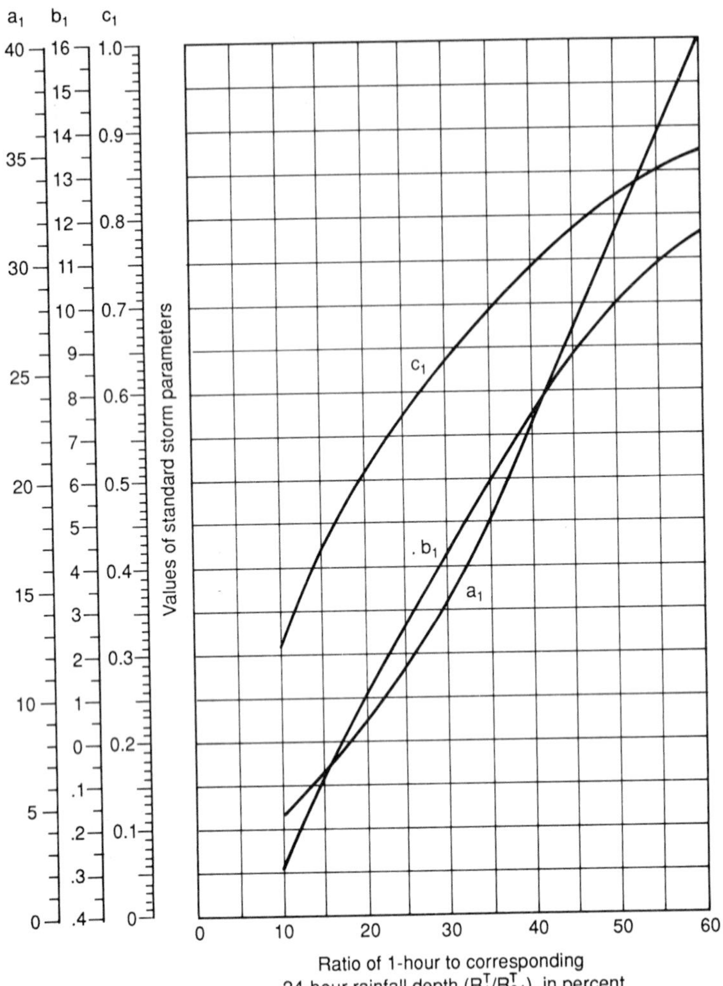

FIGURE 18.8.2 Relationship between standard storm parameters and ratio of 1-h to corresponding 24-h rainfall depth. *(Reproduced from Ref. 19 with permission.)*

18.8.4 Frequency Analysis of Basin Average Extreme Storm Depths

The stochastic storm transposition (SST) methodology has been developed for very low frequency rainfall (exceedance probabilities less than 1 in a 1000). SST provides estimates of the annual exceedance probability of the average catchment depth, which is the storm depth deposited over the catchment of interest. The estimate is based on regionalized storm characteristics and estimation of the joint probability distribution of storm characteristics and storm occurrences within a region.[42,44,106,171]

The SST method first selects a large climatologically homogeneous area (called the *storm transposition area*). Extreme storms of record within that area are employed to estimate the joint probability distribution of selected storm characteristics (such as maximum storm-center depth, storm shape parameters, storm orientation, storm depth spatial variability, etc.). Then, the probability distribution of the position of the storm centers within the transposition area is determined. This distribution is generally not uniform because the likelihood of storms of a given magnitude will vary across a region.[45] Integration over the distribution of storm characteristics and storm locations allows calculation of annual exceedance probabilities for various catchment depths. An advantage of the SST method is that it explicitly considers the morphology of the storms including the spatial distribution of storm depth and its relation to the size and shape of the catchment of interest.

18.9 FREQUENCY ANALYSIS OF LOW FLOWS

Low-flow quantiles are used in water-quality management applications including waste-load allocations and discharge permits, and in siting treatment plants and sanitary landfills. Low-flow statistics are also used in water-supply planning to determine allowable water transfers and withdrawals. Other applications of low-flow frequency analysis include determination of minimum downstream release requirements from hydropower, water-supply, cooling plants, and other facilities.

18.9.1 Selection of Data and Sources

Annual-Event-Based Low-Flow Statistics. Sources of streamflow and low-flow data are discussed in Sec. 18.7.1. The most widely used low-flow index in the United States is the one in 10-year 7-day-average low flow, denoted $Q_{7,0.10}$ (Ref. 117). In general, $Q_{d,p}$ is the annual minimum d-day consecutive average discharge not exceeded with probability p.

Prior to performing low-flow frequency analyses, an effort should be made to "deregulate" low flow series to obtain "natural" streamflows. This includes accounting for the impact of large withdrawals and diversions including water- and wastewater-treatment facilities, as well as urbanization, lake regulation, and other factors. Since low flows result primarily from groundwater inflow to the stream channel, substantial year-to-year carryover in groundwater storage can cause sequences of annual minimum low flows to be correlated from one year to the next (see Fig. 9 in Ref. 116 or Fig. 2 in Ref. 157). Low-flow series should be subjected to trend analysis so that identified trends can be reflected in frequency analyses.

The Flow-Duration Curve. Flow-duration curves are an alternative to analysis of annual minimum d-day averages (see Sec. 8.6.1). A flow-duration curve is the empirical cumulative distribution function of all the daily (or weekly) streamflow recorded at a site. A flow-duration curve describes the fraction of the time over the entire record that different daily flow levels were exceeded. Flow-duration curves are often used in hydrologic studies for run-of-river hydropower, water supply, irrigation planning and design, and water-quality management.[36,40,43,53,121] The flow-duration curves should not be interpreted on an annual event basis, as is $Q_{d,p}$, because the flow-duration curve provides only the fraction of the time that a stream flow level was exceeded; it does not distinguish between regular seasonal variation in flow levels, and random variations from seasonal averages.

18.9.2 Frequency Analysis Methods for Low Flows and Treatment of Zeros

Estimation of $Q_{d,p}$ from stream flow records is generally done by fitting a probability distribution to the annual minimum d-day-average low-flow series. The literature on low-flow frequency analysis is relatively sparse. The extreme value type III or Weibull distribution (see Sec. 18.2.2) is a theoretically plausible distribution for low flows. Studies in Canada and the eastern United States have recommended the three-parameter Weibull, the log-Pearson type 3, and the two-parameter and three-parameter lognormal distributions based on apparent goodness-of-fit.[24,139,154] Fitting methods for complete samples are described in Sec. 18.2.

Low-flow series often contain years with zero values. In some arid areas, zero flows are recorded more often than nonzero flows. Stream flows recorded as zero imply either that the stream was completely dry or that the actual stream flow was below a recording limit. At most U.S. Geological Survey gauges there is a lower stream flow level (0.05 ft^3/s) below which any measurement is reported as a zero. This implies that low-flow series are censored data sets, discussed in Sec. 18.6.3. Zero values should not simply be ignored, nor do they necessarily reflect accurate measurements of the minimum flow in a channel. Based on the hydraulic configuration of a gauge, and knowledge of the rating curve and recording policies, one can generally determine the lowest discharge which can be reliably estimated and would not be recorded as a zero.

The plotting-position method or the conditional probability model in Sec. 18.6.3 are reasonable procedures for fitting a probability distribution with data sets containing recorded zeros. The graphical plotting position approach without a formal statistical model is often sufficient for low-flow frequency analyses. One can define visually the low-flow frequency curve or estimate the parameters of a parametric distribution using probability-plot regression.

18.9.3 Regional Estimates of Low-Flow Statistics

Regional regression procedures are often employed at ungauged sites to estimate low-flow statistics by using basin characteristics. If no reliable regional regression equations are available, one can also consider the drainage area ratio, regional statistics, or base-flow correlation methods described below.

Regional Regression Procedures. Many investigators have developed regional models for the estimation of low-flow statistics at ungauged sites using physiographic basin parameters. This methodology is discussed in Sec. 18.5.2. Unfortunately, most low-flow regression models have large prediction errors, as shown in Fig. 18.5.1, because they are unable to capture important land-surface and subsurface geological characteristics of a basin. In a few regions, efforts to regionalize low-flow statistics have been improved by including basin parameters which in some manner describe the geohydrologic response of each watershed.[9,18,156] Conceptual watershed models can be used to formulate regional regression models of low-flow statistics.[155]

Drainage Area Ratio Method. Perhaps the simplest regional approach for estimation of low-flow statistics is the drainage area ratio method, which would estimate a low-flow quantile y_p for an ungauged site as

$$y_p = \left(\frac{A_y}{A_x}\right) x_p \tag{18.9.1}$$

where x_p is the corresponding low-flow quantile for a nearby gauging station and A_x and A_y are the drainage areas for the gauging station and ungauged site, respectively. Seepage runs, consisting of a series of discharge measurements along a river reach during periods of base flow, are useful for determining the applicability of this simple linear drainage-area discharge relation.[116] Some studies employ a scaling factor $(A_y/A_x)^b$ to allow for losses by using an exponent $b < 1$ derived by regional regression. (See Sec. 18.5.2.)

Regional Statistics Methods. One can sometimes use a gauging station record to construct a monthly streamflow record at an ungauged site using

$$y(i, j) = M(y_i) + \frac{S(y_i)\,[x(i, j) - M(x_i)]}{S(x_i)} \tag{18.9.2}$$

where $y(i, j)$ and $x(i, j)$ are monthly stream flows at the ungauged and nearby gauged sites, respectively, in month i and year j; $M(x_i)$ and $S(x_i)$ are the mean and standard deviation of the observed flows at the gauged site in month i; and $M(y_i)$ and $S(y_i)$ are the corresponding mean and standard deviation of the monthly flows at the ungauged site obtained from regional regression equations, discussed in Sec. 18.5.2. Hirsch[64] found that this method transferred the characteristics of low flows from the gauged site to the ungauged site.

Base Flow Correlation Procedures. When *base flow measurements* (instantaneous or average daily values) can be obtained at an otherwise ungauged site, they can be correlated with concurrent stream flows at a nearby gauged site for which a long flow record is available.[114,116] Estimators of low-flow moments at the ungauged site can be developed by using bivariate and multivariate regression, as well as estimates of their standard errors.[144] This is an extension to the record augmentation idea in Sec. 18.5.3. Ideally the nearby gauged site is hydrologically similar in terms of topography, geology, and base flow recession characteristics. For a single gauged record, if regression of concurrent daily flows at the two stations yields a model

$$y = a + bx + \epsilon \qquad \text{with Var}\,(\epsilon) = s_\epsilon^2 \tag{18.9.3}$$

estimators of the mean and variance of annual minimum d-day-average flows y are

$$\begin{aligned} M(y) &= a + b\,M(x) \\ S^2(y) &= b^2\,S^2(x) + s_\epsilon^2 \end{aligned} \tag{18.9.4}$$

where $M(x)$ and $S^2(x)$ are the estimators of the mean and variance of the annual minimum d-day averages at the gauged x site. Base flow correlation procedures are subject to considerable error when only a few discharge measurements are used to estimate the parameters of the model in Eq. (18.9.3), as well as error introduced from use of a model constructed between base flows for use in relating annual minimum d-day averages. (Thus the model R^2 should be at least 70 percent; see also Refs. 56 and 144.)

18.10 FREQUENCY ANALYSIS OF WATER-QUALITY VARIABLES

In the early 1980s, most U.S. water-quality improvement programs aimed at obvious and visible pollution sources resulting from direct *point discharges* of sewage and wastewaters to surface waters. This did not always lead to major improvements in the quality of receiving waters subject to *non-point-source pollution* loadings, corresponding to storm-water runoff and other discharges that carry sediment and pollutants from various distributed sources. The analyses of point- and nonpoint-source water-quality problems differ. Point sources are often sampled regularly, are less variable, and frequently have severe impacts during periods of low stream flow. Nonpoint sources often occur only during runoff-producing storm events, which is when nonpoint discharges generally have their most severe effect on water quality.

18.10.1 Selection of Data and Water-Quality Monitoring

Water-quality problems can be quantified by a number of variables, including biodegradable oxygen demand (BOD) and concentrations of nitrogenous compounds, chlorophyll, metals, organic pesticides, and suspended or dissolved solids. Water-quality monitoring activities include both design and actual data acquisition, and the data's utilization.[102,167] Monitoring programs require careful attention to the definition of the population to be sampled and the sampling plan.[47] A common issue is the detection of trends or changes that have occurred over time because of development or pollution control efforts, as discussed in Chap. 17.

The statistical analysis of water-quality data is complicated by the facts that quality and quantity measurements are not always made simultaneously, the time interval between water-quality samples can be irregular, the precision with which different constituents can be measured varies, and base flow samples (from which background levels may be derived) are often unavailable in studies of non-point-source pollution. The U.S. National Water Data Exchange provides access to data on ambient water quality; see Sec. 18.7.1. A list of other non-point-source water-quality data bases appears in Ref. 76.

18.10.2 Frequency Analysis Methods and Water-Quality Data

It is often inappropriate to use the conventional approach of selecting a single design flow for managing the quality of receiving waters. The most critical impact of pollutant loadings on receiving water quality does not necessarily occur under low flow conditions; often the shock loads associated with intermittent urban storm-water runoff are more critical.[99] Nevertheless, water-quality standards are usually stated in terms of a maximum allowable d-day average concentration.[147] The most common type of design event for the protection of aquatic life is based on the one in T-year d-day average annual low stream flow.[10]

For problems with regular data collection programs yielding continuous or regularly spaced observations, traditional methods of frequency analysis can be employed to estimate exceedance probabilities for annual maxima, or the probability that monthly observations will exceed various values. Event mean concentrations (EMC) corresponding to highway storm-water runoff, combined sewer overflows, urban runoff, sewage treatment plants, and agricultural runoff are often well approximated by lognormal distributions,[39] which have been a common choice for constituent concentrations.[47] Section 18.1.3 discusses advantages and disadvantages of logarithmic transformations. Procedures in Sec. 18.3 for selecting an appropriate probability distribution may be employed.

Investigations of the concentrations associated with trace substances in receiving waters are faced with a recurring problem: a substantial portion of water sample concentrations are below the limits of detection for analytical laboratories. Measurements below the detection limit are often reported as "less than the detection limit" rather than by numerical values, or as zero.[47] Such data sets are called *censored data* in the field of statistics. Probability-plot regression and maximum likelihood techniques for parameter estimation with censored data sets are discussed in Sec. 18.6.3.[60,61]

For intermittent loading problems the situation is more difficult and corresponds roughly to partial duration series, discussed in Sec. 18.6.1. In the context of urban storm-water problems, the average recurrence interval (in years) of a design event has been estimated as

$$T = \frac{1}{NP(C \geq C_0)} \tag{18.10.1}$$

where N is the average number of rainfall runoff events in a year, C is a constituent concentration, and the probability $P(C \geq C_0)$ that an observation C exceeds a standard C_0 in a runoff event is obtained by fitting a frequency distribution to the concentrations measured in observed runoff events.[37] This corresponds to Eq. (18.6.3) of Sec. 18.6.1 with event arrival rate $\lambda = N$. Models such as SWMM (see Chap. 21) can also be used to estimate T directly.

18.11 COMPUTER PROGRAMS FOR FREQUENCY ANALYSIS

Many frequency computations are relatively simple and are easily performed with standard functions on hand calculators, spreadsheets, or general-purpose statistical packages. However, maximum likelihood estimators and several other procedures can be quite involved. Water management agencies in most countries have computer packages to perform the standard procedures they employ. Four sets of routines for flood frequency analyses are discussed below.

U.S. Army Corps of Engineers Flood Flow Frequency Analysis (HECWRC). The U.S. Army Corps of Engineers has developed a library of 60 FORTRAN routines to support statistical analysis on MS-DOS personal computers. The library includes routines for performing the standard Bulletin 17B analyses, as well as general-purpose functions including general statistics, time series, duration curves, plotting positions, and graphical display. Information can be obtained by contacting Flood Frequency Investigations, Department of the Army, COE 51 Support Center, Hydrologic Engineering Center, 609 Second St., Davis, Calif. 95616-46897. HECWRC and other HEC software, with some improvements in the user interface and user support, are actively marketed by several private vendors. The U.S. Geological Survey also provides a program for Bulletin 17B analyses (Chief Hydrologist, U.S. Geological Survey, National Center, Mail Stop 437, Reston, Va. 22092).

British Flood Studies Software. Micro-FSR is a microcomputer-based implementation of the flood-frequency analysis methods developed by the Institute of Hydrology.[105] It also contains menu-driven probable maximum precipitation, unit hydrograph, and reservoir routing calculations for personal computers running MS-DOS. The package and training information can be obtained from Software Sales, Institute of Hydrology, Maclean Building, Crowmarsh Gifford, Wallingford, Oxfordshire

OX10 38800, United Kingdom; phone, 0491 38800; telex, 849365 HYDROL G; fax, 0491 32256

Consolidated Frequency Analysis (CFA) Package. The package incorporates FORTRAN routines developed by Environment Canada for flood-frequency analyses in that country with MS-DOS computers. Routines allow fitting of three-parameter lognormal, Pearson, log-Pearson, GEV, and Wakeby distributions using moments, maximum likelihood, and sometimes probability-weighted moment analyses. Capabilities are provided for nonparametric trend and tests of independence, as well as employing maximum likelihood procedures for historical information with a single threshold. Contact Dr. Paul Pilon, Inland Waters Directorate, Water Resources Branch, Ottawa, Ontario K1A 0E7, Canada.

FORTRAN Routines for Use with the Method of L Moments. Hosking[74] describes a set of FORTRAN-77 subroutines useful for analyses employing L moments, including subroutines to fit 10 different distributions. Index-flood procedures and regional diagnostic analyses are included. Contact Dr. J. R. M. Hosking, Mathematical Sciences Dept., IBM Research Division, T. J. Watson Research Center, Yorktown Heights, N.Y. 10598. The routines are available through STATLIB, a system for distribution of statistical software by electronic mail. To obtain the software send the message "send lmoments from general" to the e-mail address: statlib@lib.stat.cmu.edu.

REFERENCES

1. Abramowitz, M., and I. A. Stegun, *Handbook of Mathematical Functions,* Appl. Math. Ser. 55, National Bureau of Standards, U.S. Government Printing Office, Washington, D.C., 1964.
2. Adamowski, K., "A Monte Carlo Comparison of Parametric and Nonparametric Estimation of Flood Frequencies," *J. Hydrol.,* vol. 108, pp. 295–308, 1989.
3. Ang, A. H-S., and W. H. Tang, *Probability Concepts in Engineering, Planning and Design,* vol. I, *Basic Principles,* Wiley, New York, 1984.
4. Arnell, N., "Expected Annual Damages and Uncertainties in Flood Frequency Estimation," *J. Water Resour. Plann. Manage.,* vol. 115, no. 1, pp. 94–107, 1989.
5. Arora, K., and V. P. Singh, "A Comparative Evaluation of the Estimators of the Log Pearson Type (LP) 3 Distribution," *J. Hydrol.,* vol. 105, no. 1/2, pp. 19–37, 1989.
6. Baker, V. R., R. C. Kochel, and P. C. Patton, eds., *Flood Geomorphology,* Wiley, New York, 1988.
7. Bardsley, W. E., "Using Historical Data in Nonparametric Flood Estimation," *J. Hydrol.,* vol. 108, no. 1–4, pp. 249–255, 1989.
8. Beard, L. R., "Impact of Hydrologic Uncertainties on Flood Insurance," *J. Hydraul. Div., Am. Soc. Civ. Eng.,* (now *J. Hydraul. Eng.*), vol. 104, no. HY11, pp. 1473–1483, 1978.
9. Bingham, R. H., "Regionalization of Low-Flow Characteristics of Tennessee Streams," U.S. Geological Survey, Water Resources Investigations report 85-4191, 1986.
10. Biswas, H., and B. A. Bell, "A Method for Establishing of Site-Specific Stream Design Flows for Wasteload Allocations," *J. Water Pollut. Control Fed.,* vol. 56, no. 10, pp. 1123–1130, 1984.
11. Bobée, B., "The Log Pearson Type 3 Distribution and Its Applications in Hydrology," *Water Resour. Res.,* vol. 14, no. 2, pp. 365–369, 1975.

12. Bobée, B., and R. Robitaille, "The Use of the Pearson Type 3 Distribution and Log Pearson Type 3 Distribution Revisited," *Water Resour. Res.*, vol. 13, no. 2, pp. 427–443, 1977.

13. Bobée, B., and F. Ashkar, *The Gamma Distribution and Derived Distributions Applied in Hydrology,* Water Resources Press, Littleton, Colo., 1991.

14. Boughton, W. C., "A Frequency Distribution for Annual Floods," *Water Resour. Res.*, vol. 16, no. 2, pp. 347–354, 1980.

15. Buishand, T. A., "Bivariate Extreme-Value Data and the Station-Year Method," *J. Hydrol.*, vol. 69, pp. 77–95, 1984.

16. Buishand, T. A., "Bias and Variance of Quantile Estimates from a Partial Duration Series," *J. Hydrol.*, vol. 120, no. 1/4, pp. 35–49, 1990.

17. Burn, D. H., "Evaluation of Regional Flood Frequency Analysis with a Region of Influence Approach," *Water Resour. Res.*, vol. 26, no. 10, pp. 2257–2265, 1990.

18. Cervione, M. A., R. L. Melvin, and K. A. Cyr, "A Method for Estimating the 7-Day 10-Year Low Flow of Streams in Connecticut," Connecticut Water Resources Bulletin No. 34, Department of Environmental Protection, Hartford, Conn., 1982.

19. Chen, C., "Rainfall Intensity-Duration-Frequency Formulas," *J. Hydraul. Eng.*, vol. 109, no. 12, pp. 1603–1621, 1983.

20. Chow, V. T., D. R. Maidment, and L. W. Mays, *Applied Hydrology,* McGraw-Hill, New York, 1988.

21. Chowdhury, J. U., and J. R. Stedinger, "Confidence Intervals for Design Floods with Estimated Skew Coefficient," *J. Hydraul. Eng.*, vol. 117, no. 7, pp. 811–831, 1991.

22. Chowdhury, J. U., J. R. Stedinger, and L-H. Lu, "Goodness-of-Fit Tests for Regional GEV Flood Distributions," *Water Resour. Res.*, vol. 27, no. 7, pp. 1765–1776, 1991.

23. Cohn, T. A., and J. R. Stedinger, "Use of Historical Information in a Maximum-Likelihood Framework," *J. Hydrol.*, vol. 96, no. 1-4, pp. 215–223, 1987.

24. Condie, R., and G. A. Nix, "Modeling of Low Flow Frequency Distributions and Parameter Estimation," Int. Water Resour. Symp. on Water for Arid Lands, Teheran, Iran, Dec. 8–9, 1975.

25. Costa, J. E., "A History of Paleoflood Hydrology in the United States, 1800–1970," *EOS,* vol. 67, pp. 425–430, 1986.

26. Cunnane, C., "Unbiased Plotting Positions—A Review," *J. Hydrol.*, vol. 37, no. 3/4, pp. 205–222, 1978.

27. Cunnane, C., "A Note on the Poisson Assumption in Partial Duration Series Models," *Water Resour. Res.*, vol. 15, no. 2, pp. 489–494, 1979.

28. Cunnane, C., "Methods and Merits of Regional Flood Frequency Analysis," *J. Hydrol.*, vol. 100, pp. 269–290, 1988.

29. Cunnane, C., "Statistical Distributions for Flood Frequency Analysis, Operational Hydrology Report No. 33, World Meteorological Organization WMO—No. 718, Geneva, Switzerland, 1989.

30. D'Agostino, R. B., and M. A. Stephens, *Goodness-of-Fit Procedures,* Marcel Dekker, New York, 1986.

31. Dalrymple, T., "Flood Frequency Analysis," U.S. Geological Survey, *Water Supply Paper* 1543-A, 1960.

32. Damázio, J. M., Comment on "Quantile Estimation with More or Less Floodlike Distributions," by J. M. Landwehr, N. C. Matalas, and J. R. Wallis, *Water Resour. Res.*, vol. 20, no. 6, pp. 746–750, 1984.

33. David, H. A., *Order Statistics,* 2d ed., Wiley, New York, 1981.

34. Davison, A. C., "Modeling Excesses over High Thresholds, with an Application," in *Statistical Extremes and Applications,* J. Tiago de Oliveira, ed., D. Reidel, Dordrecht, Holland, pp. 461–482, 1984.

35. Derenzo, S. E., "Approximations for Hand Calculators Using Small Integer Coefficients," *Math. Computation,* vol. 31, no. 137, pp. 214–225, 1977.

36. Dingman, S. L., "Synthesis of Flow-Duration Curves for Unregulated Streams in New Hampshire," *Water Resour. Bull.,* vol. 14, no. 6, pp. 1481–1502, 1978.

37. DiToro, D. M., and E. D. Driscoll, "A Probabilistic Methodology for Analyzing Water Quality Effects of Urban Runoff on Rivers and Streams," U.S. Environmental Protection Agency, Washington, D.C., 1984.

38. Downton, F., "Linear Estimates of Parameters of the Extreme Value Distribution," *Technometrics,* vol. 8, pp. 3–17, 1966.

39. Driscoll, E. D., "Lognormality of Point and Non-Point Source Pollutant Concentrations," *Urban Runoff Quality,* Proceedings of an Engineering Foundation Conference, New England College, Henniker, N.H., pp. 438–458, 1986.

40. Fennessey, N., and R. M. Vogel, "Regional Flow-Duration Curves for Ungauged Sites in Massachusetts," *J. Water Resour. Plann. Manage.,* vol. 116, no. 4, pp. 530–549, 1990.

41. Filliben, J. J., "The Probability Plot Correlation Test for Normality," *Technometrics,* vol. 17, no. 1, pp. 111–117, 1975.

42. Fontaine, T. A., and K. W. Potter, "Estimating Probabilities of Extreme Rainfalls," *J. Hydraulic Eng. ASCE,* vol. 115, no. 11, pp. 1562–1575, 1989.

43. Foster, H. A., "Duration Curves," *Trans. Am. Soc. Civ. Eng.,* vol. 99, pp. 1213–1267, 1934.

44. Foufoula-Georgiou, E., "A Probabilistic Storm Transposition Approach for Estimating Exceedance Probabilities of Extreme Precipitation Depths," *Water Resour. Res.,* vol. 25, no. 5, pp. 799–815, 1989.

45. Foufoula-Georgiou, E., and L. L. Wilson," In Search of Regularities in Extreme Rainstorms," *J. Geophys. Res.,* vol. 95, no. D3, pp. 2061–2072, 1990.

46. Frederick, R. H., V. A. Meyers, and E. P. Auciello, "Five and Sixty Minute Precipitation Frequency for Eastern and Central United States," NOAA tech. memorandum Hydro-35, Silver Springs, Md., 1977.

47. Gilbert, R. O., *Statistical Methods for Environmental Pollution Monitoring,* Van Nostrand Reinhold, New York, 1987.

48. Greenwood, J. A., J. M. Landwehr, N. C. Matalas, and J. R. Wallis, "Probability Weighted Moments: Definitions and Relation to Parameters of Several Distributions Expressible in Inverse Form," *Water Resour. Res.,* vol. 15, no. 6, pp. 1049–1054, 1979.

49. Gringorten, I. I., "A Plotting Rule for Extreme Probability Paper," *J. Geophys. Res.,* vol. 68, no. 3, pp. 813–814, 1963.

50. Grygier, J. C., J. R. Stedinger, and H. B. Yin, "A Generalized Move Procedure for Extending Correlated Series," *Water Resour. Res.,* vol. 25, no. 3, pp. 345–350, 1989.

51. Gumbel, E. J., *Statistics of Extremes,* Columbia University Press, New York, 1958.

52. Guo, S. L., "A Discussion on Unbiased Plotting Positions for the General Extreme Value Distribution," *J. Hydrol.,* vol. 121, no. 1-4, pp. 33–44, 1990.

53. Gupta, R. S., *Hydrology and Hydraulic Systems,* Prentice-Hall, Englewood Cliffs, N.J., pp. 363–367, 1989.

54. Haan, C. T., *Statistical Methods in Hydrology,* Iowa State University Press, Ames, Iowa, 1977.

55. Haas, C. N., and P. A. Scheff, "Estimation of Averages in Truncated Samples," *Environ. Sci. Technol.,* vol. 24, no. 6, pp. 912–919, 1990.

56. Hardison, C. H., and M. E. Moss, "Accuracy of Low-Flow Characteristics Estimated by Correlation of Base-flow Measurements," U.S. Geological Survey, *Water Supply Paper* 1542-B, pp. 350–355, 1972.

57. Harlow, D. G., "The Effect of Proof-Testing on the Weibull Distribution," *J. Material Science,* vol. 24, pp. 1467–1473, 1989.

58. Harter, H. L., "Another Look at Plotting Positions," *Commun. Statistical-Theory and Methods,* vol. 13, no. 13, pp. 1613–1633, 1984.
59. Hazen, A., "Discussion on 'Flood flows' by W. E. Fuller," *Trans. Amer. Soc. Civ. Eng.,* vol. 77, pp. 526–632, 1914.
60. Helsel, D. R., and T. A. Cohn, "Estimation of Descriptive Statistics for Multiply Censored Water Quality Data," *Water Resour. Res.,* vol. 24, no. 12, pp. 1997–2004, 1988.
61. Helsel, D. R., "Less Than Obvious: Statistical Treatment of Data Below the Detection Limit," *Environ. Sci. Technol.,* vol. 24, no. 12, pp. 1767–1774, 1990.
62. Hershfield, D. M., and W. T. Wilson, "A Comparison of Extreme Rainfall Depths from Tropical and Nontropical Storms," *J. Geophys. Res.,* vol. 65, pp. 959–982, 1960.
63. Hershfield, D. M., "Rainfall Frequency Atlas of the United States for Durations from 30 Minutes to 24 Hours and Return Periods from 1 to 100 Years," tech. paper 40, U.S. Weather Bureau, Washington, D.C., 1961.
64. Hirsch, R. M., "An Evaluation of Some Record Reconstruction Techniques," *Water Resour. Res.,* vol. 15, no. 6, pp. 1781–1790, 1979.
65. Hirsch, R. M., "A Comparison of Four Record Extension Techniques," *Water Resour. Res.,* vol. 18, no. 4, pp. 1081–1088, 1982.
66. Hirsch, R. M., and J. R. Stedinger, "Plotting Positions for Historical Floods and Their Precision," *Water Resour. Res.,* vol. 23, no. 4, pp. 715–727, 1987.
67. Hoshi, K., J. R. Stedinger, and S. Burges, "Estimation of Log Normal Quantiles: Monte Carlo Results and First-Order Approximations," *J. Hydrol.,* vol. 71, no. 1/2, pp. 1–30, 1984.
68. Hosking, J. R. M., J. R. Wallis, and E. F. Wood, "Estimation of the Generalized Extreme-Value Distribution by the Method of Probability Weighted Moments," *Technometrics,* vol. 27, no. 3, pp. 251–261, 1985.
69. Hosking, J. R. M., J. R. Wallis, and E. F. Wood, "An Appraisal of the Regional Flood Frequency Procedure in the UK Flood Studies Report," *Hydrol. Sci. Jour.,* vol. 30, no. 1, pp. 85–109, 1985.
70. Hosking, J. R. M., and J. R. Wallis, "Parameter and Quantile Estimation for the Generalized Pareto Distribution," *Technometrics,* vol. 29, no. 3, pp. 339–349, 1987.
71. Hosking, J. R. M., and J. R. Wallis, "The Effect of Intersite Dependence on Regional Flood Frequency Analysis," *Water Resour. Res.,* vol. 24, no. 4, pp. 588–600, 1988.
72. Hosking, J. R. M., "L-Moments: Analysis and Estimation of Distributions Using Linear Combinations of Order Statistics," *J. Royal Statistical Soc., B,* vol. 52, no. 2, pp. 105–124, 1990.
73. Hosking, J. R. M., "Approximations for Use in Constructing L-moment Ratio Diagrams," research report RC-16635, IBM Research Division, T. J. Watson Research Center, Yorktown Heights, N.Y., March 12, 1991.
74. Hosking, J. R. M., "Fortran Routines for Use with the Method of L-Moments," research report RC-17097, IBM Research Division, T. J. Watson Research Center, Yorktown Heights, N.Y., August 21, 1991.
75. Houghton, J. C., "Birth of a Parent: The Wakeby Distribution for Modeling Flood Flows," *Water Resour. Res.,* vol. 14, no. 6, pp. 1105–1109, 1978.
76. Huber, W. C., and J. P. Heaney, "Urban Rainfall-Runoff-Quality Data Base," EPA-600/8-77-009 (NTIS PB-270 065), Environmental Protection Agency, Cincinnati, 1977.
77. Hupp, C. R., "Botanical Evidence of Floods and Paleoflood History," *Regional Flood Frequency Analysis,* V. P. Singh, ed., D. Reidel, Dordrecht, pp. 355–369, 1987.
78. Hupp, C. R., "Plant Ecological Aspects of Flood Geomorphology and Paleoflood History," Chap. 20 in *Flood Geomorphology,* V. R. Baker, R. C. Kochel and P. C. Patton, Wiley, New York, 1988.
79. Interagency Advisory Committee on Water Data, *Guidelines for Determining Flood Flow*

Frequency, Bulletin 17B, U.S. Department of the Interior, U.S. Geological Survey, Office of Water Data Coordination, Reston, Va., 1982.

80. Jenkinson, A. F., "Statistics of Extremes," in "Estimation of Maximum Floods," WMO No. 233, TP 126 (tech. note no. 98), pp. 183–228, 1969.
81. Jin, M., and J. R. Stedinger, "Flood Frequency Analysis with Regional and Historical Information," *Water Resour. Res.,* vol. 25, no. 5, pp. 925–936, 1989.
82. Kelman, J., "Cheias E Aproveitamentos Hidrelétricos," *Revista Brasileria de Engenharia,* Rio de Janeiro, 1987.
83. Kirby, W., "Computer Oriented Wilson-Hilferty Transformation That Preserves the First Three Moments and the Lower Bound of the Pearson Type 3 Distribution," *Water Resour. Res.,* vol. 8, no. 5, pp. 1251–1254, 1972.
84. Kirby, W., "Algebraic Boundness of Sample Statistics," *Water Resour. Res.,* vol. 10, no. 2, pp. 220–222, 1974.
85. Kite, G. W., *Frequency and Risk Analysis in Hydrology,* Water Resources Publication, Littleton, Colo., 1988.
86. Kochel, R. C., and V. R. Baker, "Paleoflood Analysis Using Slackwater Deposits," Chap. 21 in *Flood Geomorphology,* V. R. Baker, R. C. Kochel and P. C. Patton, eds., Wiley, New York, 1988.
87. Kuczera, G., "Robust Flood Frequency Models," *Water Resour. Res.,* vol. 18, no. 2, pp. 315–324, 1982.
88. Kuczera, G., "Effects of Sampling Uncertainty and Spatial Correlation on an Empirical Bayes Procedure for Combining Site and Regional Information," *J. Hydrol.,* vol. 65, no. 4, pp. 373–398, 1983.
89. Landwehr, J. M., N. C. Matalas, and J. R. Wallis, "Probability Weighted Moments Compared with Some Traditional Techniques in Estimating Gumbel Parameters and Quantiles," *Water Resour. Res.,* vol. 15, no. 5, pp. 1055–1064, 1979.
90. Landwehr, J. M., N. C. Matalas, and J. R. Wallis, "Quantile Estimation with More or Less Floodlike Distributions," *Water Resour. Res.,* vol. 16, no. 3, pp. 547–555, 1980.
91. Landwehr, J. M., G. D. Tasker, and R. D. Jarrett, "Discussion of Relative Accuracy of Log Pearson III Procedures, by J. R. Wallis and E. F. Wood," *J. Hydraul. Eng.,* vol. 111, no. 7, pp. 1206–1210, 1987.
92. Lane, W. L., "Paleohydrologic Data and Flood Frequency Estimation," in *Regional Flood Frequency Analysis,* V. P. Singh, ed., D. Reidel, Dordrecht, pp. 287–298, 1987.
93. Langbein, W. B., "Annual Floods and the Partial Duration Flood Series," EOS, *Trans. AGU,* vol. 30, no. 6, pp. 879–881, 1949.
94. Lettenmaier, D. P., J. R. Wallis, and E. F. Wood, "Effect of Regional Heterogeneity on Flood Frequency Estimation," *Water Resour. Res.,* vol. 23, no. 2, pp. 313–324, 1987.
95. Loucks, D. P., J. R. Stedinger, and D. A. Haith, *Water Resource Systems Planning and Analysis,* Prentice-Hall, Englewood Cliffs, N.J., 1981.
96. Lu, L., and J. R. Stedinger, "Variance of 2- and 3-Parameter GEV/PWM Quantile Estimators: Formulas, Confidence Intervals and a Comparison," *J. Hydrol.,* vol. 138, no. 1/2, pp. 247–268, 1992.
97. Matalas, N. C., and B. Jacobs, "A Correlation Procedure for Augmenting Hydrologic Data," *U.S. Geological Survey,* prof. paper 434-E, E1–E7, 1964.
98. McCuen, R. H., "Map Skew???," *J. Water Resour. Plann. Manage. Div., Am. Soc. Civ. Eng.* (now) *J. Water Res. Plann. Manage.,* vol. 105, no. WR2, pp. 265–277 [with Closure, vol. 107, no. WR2, p. 582, 1981], 1979.
99. Medina, M. A., "Level III: Receiving Water Quality Modeling for Urban Stormwater Management," EPA-600/2-79-100, Environmental Protection Agency, Cincinnati, 1979.
100. Miller, J. F., R. H. Frederick, and R. J. Tracey, "Precipitation-Frequency Atlas of the Conterminous Western United States (by states)," *NOAA Atlas 2,* National Weather Service, Silver Spring, Md., 1973.

101. Minitab, *Minitab Reference Manual,* Release 5, Minitab, Inc., State College, Pa., December 1986.

102. Montgomery, R. H., and T. G. Sanders, "Uncertainty in Water Quality Data," in *Statistical Aspects of Water Quality Monitoring,* A. H. El-Sharwi and R. E. Kwiatkowski, eds., Elsevier, New York, pp. 17–29, 1985.

103. Myers, V. A., and R. M. Zehr, "A Methodology for Point-to-Area Rainfall Frequency Ratios," *NOAA,* National Weather Service tech. rep. 24, 1980.

104. National Research Council, Committee on Techniques for Estimating Probabilities of Extreme Floods, *Estimating Probabilities of Extreme Floods, Methods and Recommended Research,* National Academy Press, Washington, D.C., 1988.

105. Natural Environmental Research Council, *Flood Studies Report,* vol. I, Hydrological Studies, London, 1975.

106. Newton, D. W., "Realistic Assessment of Maximum Flood Potentials," *J. Hydraul. Eng.,* ASCE, vol. 109, no. 6, pp. 905–918, 1983.

107. Newton, D. W., and J. C. Herrin, "Assessment of commonly used flood frequency methods," Transportation Research Record Series TRR896, Washington, D.C., 1982.

108. North, M., "Time-dependent Stochastic Model of Floods," *J. Hydraul. Div., Am. Soc. Civ. Eng.* (now *J. Hydraul. Eng.*), vol. 106, no. HY5, pp. 649–665, 1980.

109. Phien, H. N., "A Method of Parameter Estimator for the Extreme Value type I Distribution," *J. Hydrol.,* vol. 90, pp. 251–268, 1987.

110. Pilgrim, D. H., ed., *Australian Rainfall and Runoff, A Guide to Flood Estimation,* vol. I, The Institute of Engineers, Barton ACT, Australia, (rev. ed.), 1987.

111. Potter, K. W., and E. B. Faulkner, "Catchment Response Times as a Predictor of Flood Quantiles," *Water Resour. Bull.,* vol. 23, no. 5, pp. 857–861, 1987.

112. Potter, K. W., and D. P. Lettenmaier, "A Comparison of Regional Flood Frequency Estimation Methods Using a Resampling Method," *Water Resour. Res.,* vol. 26, no. 3, pp. 415–424, 1990.

113. Rasmussen, P. F., and D. Rosbjerg, "Risk Estimation in Partial Duration Series," *Water Resour. Res.,* vol. 25, no. 11, pp. 2319–2330, 1989.

114. Riggs, H. C., "Estimating Probability Distributions of Drought Flows," *Water and Sewage Works,* vol. 112, no. 5, pp. 153–157, 1965.

115. Riggs, H. C., Chap. 3, "Frequency Curves," U.S. Geological Survey Surface Water Techniques Series, Book 2, U.S. Geological Survey, Washington, D.C., 1966.

116. Riggs, H. C., "Low Flow Investigations," *U.S. Geological Survey Techniques of Water Resources Investigations,* Book 4, U.S. Geological Survey, Washington, D.C., 1972.

117. Riggs, H. C., et al., "Characteristics of Low Flows, Report of an ASCE Task Committee," *J. Hydraul. Eng.,* vol. 106, no. 5, pp. 717–731, 1980.

118. Rosbjerg, D., "Estimation in Partial Duration Series with Independent and Dependent Peak Values," *J. Hydrol.,* vol. 76, no. 1, pp. 183–196, 1985.

119. Rossi, F., M. Fiorentino, and P. Versace, "Two Component Extreme Value Distribution for Flood Frequency Analysis," *Water Resour. Res.,* vol. 20, no. 7, pp. 847–856, 1984.

120. Schaefer, M. G., "Regional Analyses of Precipitation Annual Maxima in Washington State," *Water Resour. Res.,* vol. 26, no. 1, pp. 119–131, 1990.

121. Searcy, J. K., "Flow-Duration Curves," U.S. Geological Survey, *Water Supply Paper* 1542-A, 1959.

122. Selker, J. S., and D. A. Haith, "Development and Testing of Single-Parameter Precipitation Distributions," *Water Resour. Res.,* vol. 26, no. 11, pp. 2733–2740, 1990.

123. Shipe, A. P., and J. T. Riedel, "Greatest Known Areal Storm Rainfall Depths for the Contiguous United States," NOAA tech. memo. Hydro-33, National Weather Service, 1976.

124. Slack, J. R., J. R. Wallis, and N. C. Matalas, "On the Value of Information in Flood Frequency Analysis," *Water Resour. Res.,* vol. 11, no. 5, pp. 629–648, 1975.

125. Smith, R. L., "Threshold Methods for Sample Extremes," in *Statistical Extremes and Applications,* J. Tiago de Oliveira, ed., D. Reidel, Dordrecht, Holland, pp. 621–638, 1984.

126. Stedinger, J. R., "Fitting Log Normal Distributions to Hydrologic Data," *Water Resour. Res.,* vol. 16, no. 3, pp. 481–490, 1980.

127. Stedinger, J. R., "Design Events with Specified Flood Risk," *Water Resour. Res.,* vol. 19, no. 2, pp. 511–522, 1983.

128. Stedinger, J. R., "Estimating a Regional Flood Frequency Distribution," *Water Resour. Res.,* vol. 19, no. 2, pp. 503–510, 1983.

129. Stedinger, J. R., "Confidence Intervals for Design Events," *J. Hydraul. Div., Am. Soc. Civ. Eng.* (now *J. Hydraul. Eng.*) vol. 109, no. HY1, pp. 13–27, 1983.

130. Stedinger, J. R., and G. D. Tasker, "Regional Hydrologic Regression, 1. Ordinary, Weighted and Generalized Least Squares Compared," *Water Resour. Res.,* vol. 21, no. 9, pp. 1421–1432, 1985.

131. Stedinger, J. R., and G. D. Tasker, "Correction to 'Regional Hydrologic Analysis, 1. Ordinary, Weighted and Generalized Least Squares Compared,'" *Water Resour. Res.,* vol. 22, no. 5, p. 844, 1986.

132. Stedinger, J. R., and G. D. Tasker, "Regional Hydrologic Analysis, 2. Model Error Estimates, Estimation of Sigma, and Log-Pearson Type 3 Distributions," *Water Resour. Res.,* vol. 22, no. 10, pp. 1487–1499, 1986.

133. Stedinger, J. R., and T. A. Cohn, "Flood Frequency Analysis with Historical and Paleoflood Information," *Water Resour. Res.,* vol. 22, no. 5, pp. 785–793, 1986.

134. Stedinger, J. R., and V. R. Baker, "Surface Water Hydrology: Historical and Paleoflood Information," *Reviews of Geophysics,* vol. 25, no. 2, pp. 119–124, 1987.

135. Stedinger, J., R. Surani, and R. Therivel, "Max Users Guide: A Program for Flood Frequency Analysis Using Systematic-Record, Historical, Botanical, Physical Paleohydrologic and Regional Hydrologic Information Using Maximum Likelihood Techniques," Department of Environmental Engineering, Cornell University, Ithaca, N.Y., 1988.

136. Stephens, M., "E.D.F. Statistics for Goodness of Fit," *J. Amer. Statistical Assoc.,* vol. 69, pp. 730–737, 1974.

137. Tasker, G. D., "Effective Record Length for the T-Year Event," *J. Hydrol.,* vol. 64, pp. 39–47, 1983.

138. Tasker, G. D., and J. R. Stedinger, "Estimating Generalized Skew with Weighted Least Squares Regression," *J. Water Resour. Plann. Manage.,* vol. 112, no. 2, pp. 225–237, 1986.

139. Tasker, G. D., "A Comparison of Methods for Estimating Low Flow Characteristics of Streams," *Water Resour. Bull.,* vol. 23, no. 6, pp. 1077–1083, 1987.

140. Tasker, G. D., and N. E. Driver, "Nationwide Regression Models for Predicting Urban Runoff Water Quality at Unmonitored Sites," *Water Resour. Bull,* vol. 24, no. 5, pp. 1091–1101, 1988.

141. Tasker, G. D., and J. R. Stedinger, "An Operational GLS Model for Hydrologic Regression," *J. Hydrol.,* vol. 111, pp. 361–375, 1989.

142. Thomas, D. M., and M. A. Benson, "Generalization of Streamflow Characteristics from Drainage-Basin Characteristics," U.S. Geological Survey, *Water Supply Paper* 1975, 1970.

143. Thomas, W. O., "A Uniform Technique for Flood Frequency Analysis," *J. Water Resour. Plann. Management,* vol. 111, no. 3, pp. 321–337, 1985.

144. Thomas, W. O., and J. R. Stedinger, "Estimating Low-Flow Characteristics at Gaging Stations and Through the Use of Base-Flow Measurements," pp. 197–205, in *Proceedings of the US-PRC Bilateral Symposium on Droughts and Arid-Region Hydrology,* W. H. Kirby and W. Y. Tan, eds., open-file report 81-244, U.S. Geological Survey, Reston, Va., 1991.

145. Todorovic, P., "Stochastic Models of Floods," *Water Resour. Res.*, vol. 14, no. 2, pp. 345–356, 1978.

146. U.S. Army Corps of Engineers, "Storm Rainfall on the United States," Office of the Chief Engineer, Washington, D.C., ongoing publication since 1945.

147. U.S. Environmental Protection Agency, "Quality Criteria for Water—1986," EPA 440/5-86-001, Office of Water Regulations and Standards, Washington, D.C., 1986.

148. U.S. Weather Bureau, "Rainfall Intensity-Duration Frequency Curves (for Selected Stations in the United States, Alaska, Hawaiian Islands, and Puerto Rico)," tech. paper no. 25, Washington, D.C., 1955.

149. U.S. Weather Bureau, "Two-to-Ten-Day Precipitation for Return Period of 2 to 100 Years in the Contiguous United States," tech. paper no. 49, Washington, D.C., 1964.

150. Vicens, G. J., I. Rodríguez-Iturbe, and J. C. Schaake, Jr., "A Bayesian Framework for the Use of Regional Information in Hydrology," *Water Resour. Res.*, vol. 11, no. 3, pp. 405–414, 1975.

151. Vogel, R. M., and J. R. Stedinger, "Minimum Variance Streamflow Record Augmentation Procedures," *Water Resour. Res.*, vol. 21, no. 5, pp. 715–723, 1985.

152. Vogel, R. M., "The Probability Plot Correlation Coefficient Test for Normal, Lognormal, and Gumbel Distributional Hypotheses," *Water Resour. Res.*, vol. 22, no. 4, pp. 587–590, 1986.

153. Vogel, R. M., "Correction to 'The Probability Plot Correlation Coefficient Test for Normal, Lognormal, and Gumbel Distributional Hypotheses," *Water Resour. Res.*, vol. 23, no. 10, p. 2013, 1987.

154. Vogel, R. M., and C. N. Kroll, "Low-Flow Frequency Analysis Using Probability-Plot Correlation Coefficients," *J. Water Resour. Plann. Manage., ASCE,* vol. 115, no. 3, pp. 338–357, 1989.

155. Vogel, R. M., C. N. Kroll, and K. M. Driscoll, "Regional Geohydrologic-Geomorphic Relationships for the Estimation of Low-Flows," *Proceedings of the International Conference on Channel Flow and Catchment Runoff,* B. C. Yen, ed., University of Virginia, Charlottesville, pp. 267–277, 1989.

156. Vogel, R. M., and C. N. Kroll, "Generalized Low-Flow Frequency Relationships for Ungauged Sites in Massachusetts," *Water Resour. Bull,* vol. 26, no. 2, pp. 241–253, 1990.

157. Vogel, R. M., and C. N. Kroll, "The Value of Streamflow Augmentation Procedures in Low Flow and Flood-Flow Frequency Analysis," *J. Hydrol.*, vol. 125, pp. 259–276, 1991.

158. Vogel, R. M., and D. E. McMartin, "Probability Plot Goodness-of-Fit and Skewness Estimation Procedures for the Pearson Type III Distribution," *Water Resour. Res.*, vol. 27, no. 12, pp. 3149–3158, 1991.

159. Wallis, J. R., N. C. Matalas, and J. R. Slack, "Just a Moment!," *Water Resour. Res.*, vol. 10, no. 2, pp. 211–221, 1974.

160. Wallis, J. R., "Risk and Uncertainties in the Evaluation of Flood Events for the Design of Hydraulic Structures," in *Piene e Siccita,* E. Guggino, G. Rossi, and E. Todini, eds., Fondazione Politecnica del Mediterraneo, Catania, Italy, pp. 3–36, 1980.

161. Wallis, J. R., "Probable and Improbable Rainfall in California," res. rep. RD9350, IBM, Armonk, N.Y., 1982.

162. Wallis, J. R., and E. F. Wood, "Relative Accuracy of Log Pearson III Procedures," *J. Hydraul. Eng.*, vol. 111, no. 7, pp. 1043–1056 [with discussion and closure, *J. Hydraul. Eng.*, vol. 113, no. 7, pp. 1205–1214, 1987], 1985.

163. Wallis, J. R., "Catastrophes, Computing and Containment: Living in Our Restless Habitat," *Speculation in Science and Technology,* vol. 11, no. 4, pp. 295–315, 1988.

164. Wang, Q. J., "Estimation of the GEV Distribution from Censored Samples by Method of Partial Probability Weighted Moments," *J. Hydrol.*, vol. 120, no. 1-4, pp. 103–114, 1990*a*.

165. Wang, Q. J., "Unbiased Estimation of Probability Weighted Moments and Partial Proba-

bility Weighted Moments for Systematic and Historical Flood Information and Their Application to Estimating the GEV Distribution," *J. Hydrol.*, vol. 120, no. 1-4, pp. 115–124, 1990.

166. Wang, Q. J., "Unbiased Plotting Positions for Historical Flood Information," *J. Hydrol.*, vol. 124, no. 3-4, pp. 197–206, 1991.

167. Ward, R. C., J. C. Loftis, and G. B. McBride, *Design of Water Quality Monitoring Systems*, Van Nostrand Reinhold, New York, 1991.

168. Waylen, P., and M-K. Woo, "Prediction of Annual Floods Generated by Mixed Processes," *Water Resour. Res.*, vol. 18, no. 4, pp. 1283–1286, 1982.

169. WB-ESSA, Weather Bureau Observing Handbook No. 2, Substation Observations, Office of Meteorological Operations, Weather Bureau, Environmental Sciences Administration, Silver Spring, Md., 1970.

170. Wenzel, H. G., "Rainfall for Urban Stormwater Design," in *Urban Storm Water Hydrology*, D. F. Kibler, ed., Water Resources Monograph 7, AGU, Washington, D.C., 1982.

171. Wilson, L. L., and E. Foufoula-Georgiou, "Regional Rainfall Frequency Analysis via Stochastic Storm Transposition," *J. Hydraul. Eng.*, vol. 116, no. 7, pp. 859–880, 1990.

172. World Meteorological Organization, *Guide to Hydrological Practices*, vol. II, *Analysis, Forecasting, and Other Applications*, WMO no. 168 (4th ed.), Geneva, 1983.

173. Woolhiser, D. A., and J. Roldan, "Stochastic Daily Precipitation Models, 2. A Comparison of Distributions of Amounts," *Water Resour. Res.*, vol. 18, no. 5, pp. 1461–1468, 1982.

CHAPTER 19
ANALYSIS AND MODELING OF HYDROLOGIC TIME SERIES

Jose D. Salas
Engineering Research Center
Colorado State University
Fort Collins, Colorado

Time-series analysis has become a major tool in hydrology. It is used for building mathematical models to generate synthetic hydrologic records, to forecast hydrologic events, to detect trends and shifts in hydrologic records, and to fill in missing data and extend records. This chapter includes definitions related to the stochastic structure of hydrologic series; time-series analysis principles; stochastic models for single and multiple series, disaggregation models, Markov chains, and point process models; methods for filling in missing data and extension of records; data generation and simulation principles; and a summary of software available for analysis, modeling, and generation of hydrologic series. Applications of stochastic models to forecasting are included in Chap. 26.

19.1 STOCHASTIC STRUCTURE OF HYDROLOGIC TIME SERIES

19.1.1 Hydrologic Time Series

In general, hydrologic processes such as precipitation and runoff evolve on a continuous time scale. For instance, a recording gauging station in a stream provides a continuous record of water stage and discharge $y(t)$ through time. A plot of the flow hydrograph $y(t)$ versus time t constitutes a *stream-flow time series* in continuous time or a *continuous time series* (see Fig. 19.1.1*a* for illustration). However, most hydrologic processes of practical interest are defined in a discrete time scale. A *discrete time series* may be derived by sampling the continuous process $y(t)$ at discrete points in time, or by integrating the continuous time series over successive time intervals as shown in Fig. 19.1.1*a*. For example, a daily stream-flow series may be derived by sampling the flows of a stream once daily or by integrating the continuous-flow hydrograph on a daily basis. Most hydrologic series are defined on hourly, daily,

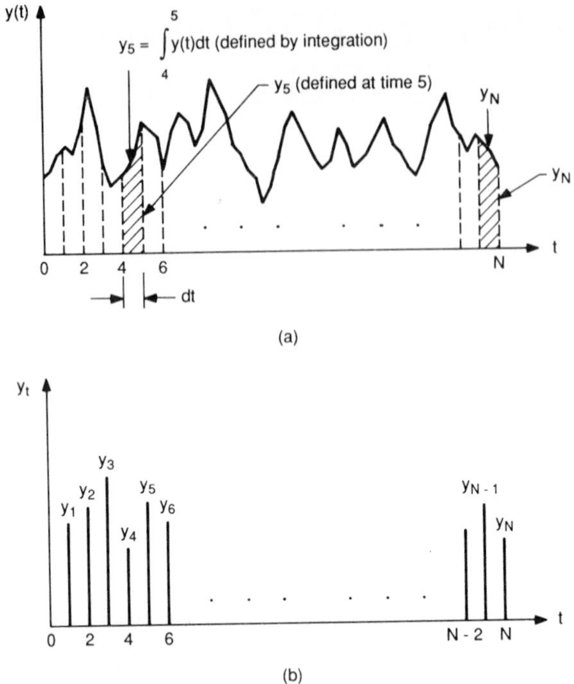

FIGURE 19.1.1 (*a*) A continuous time series $y(t)$. (*b*) A discrete time series y_t derived from the continuous series.

weekly, monthly, bimonthly, quarterly, and annual time intervals. The term *seasonal time series* is used for series defined at time intervals which are fractions of a year (usually multiples of a month). Figure 19.1.1*b* plots a discrete time series in the form of a bar or stick diagram; however, because of convenience or preference, the series is often plotted in the form of a continuous line by successively connecting the tops of the sticks. A continuous plot of a discrete time series should not be confused with a continuous time series.

Hydrologic time series may be classified into several categories depending on a number of factors. Each of these categories is defined below.

Single Time Series. A *single time series* (or *univariate series*) is simply a time series of one hydrologic variable at a given site. Consider a basin with five precipitation gauges and a stream network system with three stream-flow gauging stations. The precipitation time series measured at each site is a single time series. Likewise, the series resulting from the areal average of the five precipitation series is also a single time series. Similarly, the flow time series at any given site of the stream network system is a single time series.

Multiple Time Series. Consider the basin and flow network referred to above. The set of five precipitation time series of the basin represents a multivariate time series. Likewise, the set of three flow time series represents a multivariate series. In general, a set of two or more time series constitutes a *multiple time series* or a *multivariate time series*. Furthermore, multiple time series may be a set of time series of different processes. For instance, the flow time series at sites 1, 2, and 3 and the corresponding precipitation time series at gauges 1 through 5 constitute a multiple time series. Additionally, a multiple time series may arise at a stream-flow gauging station when the station measures different variables such as discharge, flow depth, water temperature, and sediment transport or at a given weather station when it measures variables such as precipitation, air temperature, evaporation, and humidity.

Uncorrelated and Correlated Time Series. Figure 19.1.2 shows a single time series x_t. If the x's at time t depend (linearly) on the x's at time $t-k$, for $k = 1, 2, \ldots$, then the time series is called *autocorrelated, serially correlated*, or *correlated in time*. Otherwise, it is *uncorrelated*. An uncorrelated series is also called an *independent series*. Autocorrelation or *dependence* in some hydrologic time series such as stream flow usually arises from the effect of storage, such as surface, soil, and groundwater storages, which causes the water to remain in the system through subsequent time periods. For instance, basins with significant surface storage in the form of lakes, swamps, or glaciers produce stream-flow series showing significant autocorrelation. Likewise, subsurface storage, especially groundwater storage, produces significant autocorrelation in the stream-flow series derived from groundwater outflow. Conversely, time series of monthly or annual precipitation and time series of annual maximum flows (flood peaks) are usually uncorrelated, although in cases that a time series is *nonhomogeneous*, significant serial correlation may occur.[136]

Refer to the two series of Fig. 19.1.2. If the y's at time t depend (linearly) on the x's at time $t-k$, for $k = 0, 1, \ldots$ —then the two time series are *cross-correlated*. Several combinations of autocorrelation and cross-correlation exist. For instance, it is possible that both series y_t and x_t are uncorrelated in time, yet are cross-correlated with one another. Likewise, it is possible that each series can be autocorrelated, yet

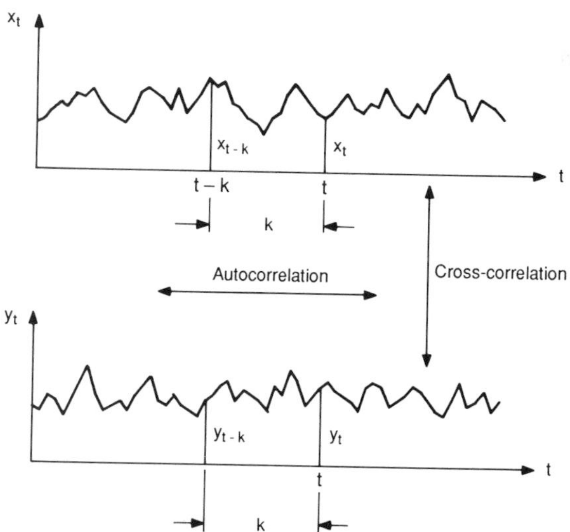

FIGURE 19.1.2 A pair of single time series x_t and y_t each having autocorrelation. Cross-correlation is between the two series.

there is no cross-correlation between them. Just as there are physical reasons why some hydrologic time series are autocorrelated, there are also physical reasons why two or more series are cross-correlated. Examples are precipitation series at two nearby sites and stream flow at two nearby gauging stations. In both cases, one would expect that the time series will be cross-correlated because the sites are relatively close to each other and therefore subject to similar climatic and hydrologic events. As the sites considered become father apart, their cross-correlation decreases. Likewise, one would expect a significant cross-correlation between stream-flow time series and the corresponding areal average precipitation time series over the same basin. One of the problems in hydrology is searching for significant correlation among time series.

Intermittent Time Series. Hydrologic time series are *intermittent* when the variable under consideration takes on nonzero and zero values throughout the length of record. For instance, the precipitation observed in a recording rain gauge is an intermittent continuous time series. Likewise, a discrete time series derived by integrating an intermittent continuous precipitation time series can be intermittent when the time interval of integration is relatively small. Thus, hourly, daily, and weekly rainfall are typically intermittent time series, while monthly and annual rainfall are usually nonintermittent. However, in semiarid and arid regions, even monthly and annual precipitation may be intermittent as well. Figure 19.1.3 shows a six-hourly rainfall series for a given gauging station and the corresponding stream flow series at two gauging stations. The rainfall series is intermittent, displaying a sequence of nonzero and zero rainfalls, while the stream-flow series is nonintermittent, with nonzero flows throughout the record. Stream-flow time series are often intermittent in semiarid and arid regions.

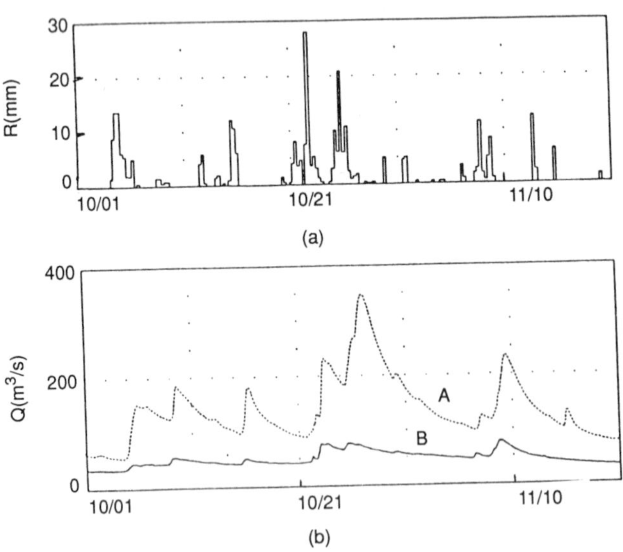

FIGURE 19.1.3 (*a*) Six-hour rainfall *R*, mm, during the period Oct. 1–Nov. 15, 1983, at site Te Haroto, Mohaka, New Zealand. (*b*) Corresponding stream flow *Q*, m^3/s, on the Mohaka River at Raupunga (A) and Glenfalls (B), New Zealand. *(Provided by S. M. Thompson.)*

Counting Time Series. The variable of interest may be the result of counting the occurrence of certain hydrologic events. An example is the count of rainy days for each month throughout the period of record. The resulting sequence of integer numbers $d_1, d_2, \ldots, d_n$ is a *counting time series.*

Regularly and Irregularly Spaced Time Series. Most time series are defined on a *regularly spaced* time interval; i.e., the value of the variable has been determined every hour, every day, or every week, etc., throughout the record. This is the case for most variables which are of interest in hydrology. However, in some cases, data may be collected at irregular time intervals. This is commonly true of water-quality measurements. Nearly all methods of time-series analysis require regularly spaced data, but some methods, such as the use of regression for trend analysis, can also be applied to irregularly spaced data. In this chapter it is assumed that the time series under consideration has been defined on a regular time scale.

Stationary and Nonstationary Time Series. A hydrologic time series is *stationary* if it is free of trends, shifts, or periodicity (cyclicity). This implies that the statistical parameters of the series, such as the mean and variance, remain constant through time. Otherwise, the time series is *nonstationary.* Generally, hydrologic time series defined on an annual time scale are stationary, although this assumption may be incorrect as a result of large-scale climatic variability, natural disruptions like a volcanic eruption, and human-induced changes such as the effect of reservoir construction on downstream flow. Hydrologic time series defined at time scales smaller than a year, such as monthly series, are typically nonstationary, mainly because of the annual cycle.

19.1.2 Partitioning of the Time-Series Structure

Hydrologic time series exhibit, in various degrees, trends, shifts or jumps, seasonality, autocorrelation, and nonnormality. These attributes of hydrologic time series are referred to as *components.* A time series can be *partitioned* or *decomposed* into its component series.

Trends and Shifts. In general, natural and human-induced factors may produce gradual and instantaneous trends and shifts (jumps) in hydrologic series. For example, a large forest fire in a river basin can immediately affect the runoff, producing a shift in the runoff series, whereas a gradual killing of a forest (for instance by an insect infestation that takes years for its population to build up) can result in gradual changes or trends in the runoff series (see Chap. 13 for further details). A large volcanic explosion such as the 1980 Mount St. Helens explosion, or a large landslide, can produce sudden changes in the sediment transport series of a stream. Trends in non-point-source water-quality series may be the result of long-term changes in agricultural practices and agricultural land development. Likewise, shifts in certain water-quality constituents may be caused by agricultural activities such as sudden changes in the use of certain types of pesticides. An important source of trends and shifts in stream-flow series arises from changes in land use and the development of reservoirs and diversion structures. The current concern about global warming and climatic changes is making hydrologists more aware of the occurrence of trends and shifts in hydrologic time series. Figure 19.1.4 shows the monthly series of water levels of Lake Victoria at Entebbe (Uganda), which has a significant upward shift.

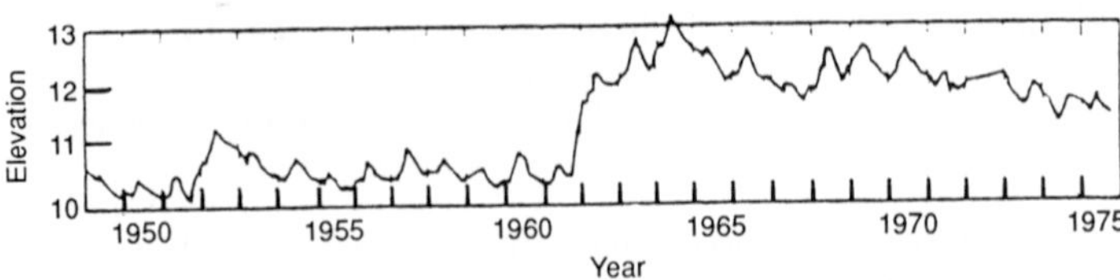

FIGURE 19.1.4 Time series of monthly levels of Lake Victoria at Entebbe, Uganda, for the period 1949–1975 showing an upward shift.[154]

Removing Trends. A hydrologic time series may exhibit shifts in one or more of its statistical characteristics. The most common ones are trends in the mean and in the variance. The partitioning of a time series with a simple trend is schematically shown in Fig. 19.1.5. A linear trend in the mean is shown in Fig. 19.1.5*a*. The trend $\bar{y}_t$ can be removed by the difference $y_t - \bar{y}_t$ as shown in Fig. 19.1.5*b*. The variance of such difference series, expressed by s_t^2, may be either a function of time (in which case there is a trend in the variance) or may be a constant, as shown graphically in Fig. 19.1.5*c*. The trend in the variance can be removed by $(y_t - \bar{y}_t)/s_t$ (the process of

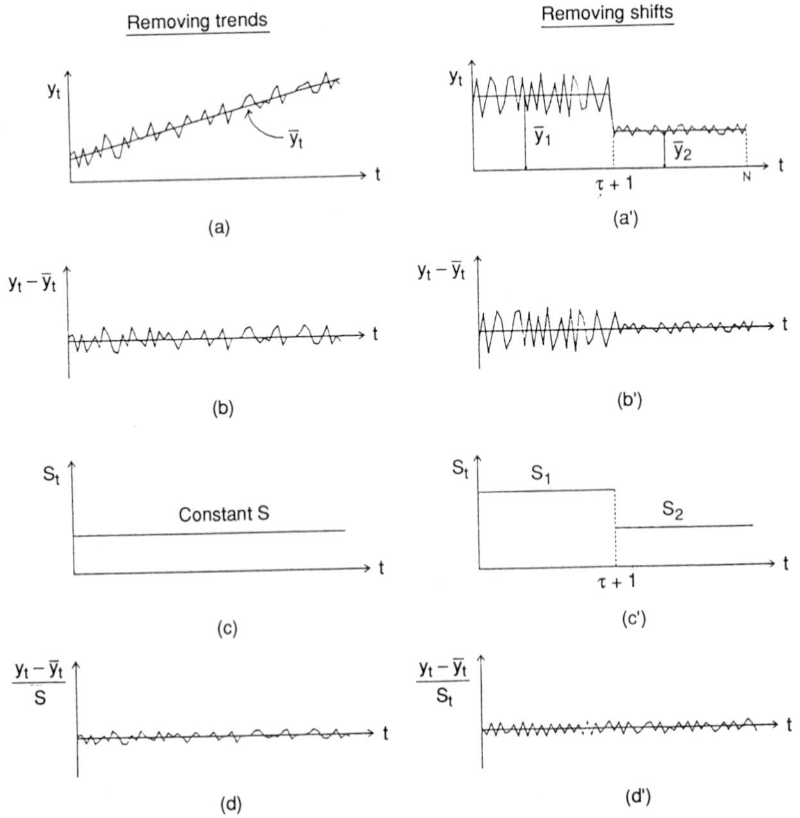

FIGURE 19.1.5 Partitioning an annual time series with trends and partitioning an annual series with shifts.

constructing a new series by subtracting the mean and dividing by the standard deviation is called *standardization*). The residual series in Fig. 19.1.5*d* may still have other properties such as correlation structure which can be decomposed or removed.

Removing Shifts. A hydrologic series may also exhibit shifts in one or more of its statistical characteristics. Positive (upward) or negative (downward) shifts in the mean and variance are most commonly analyzed. Figure 19.1.5*a*′ to *d*′ gives examples of removing sudden downward shifts from an annual series. In Fig. 19.1.5*a*′, shifts in the mean and variance occur at time $\tau + 1$. The shift in the mean is removed by $y_t - \bar{y}_t$ as shown in Fig. 19.1.5*b*′, and the shift in the variance is removed by $(y_t - \bar{y}_t)/s_t$. The residual series plotted in Fig. 19.1.5*d*′ now shows a series with mean zero and variance one and may further exhibit other properties such as autocorrelation.

Seasonality. Hydrologic series defined at time intervals smaller than a year (such as monthly series) generally exhibit distinct *seasonal* (or *periodic*) patterns. These result from the annual revolution of the earth around the sun which produces the annual cycle in most hydrologic processes. Some series of interest to hydrology, such as daily series of urban water use or daily series of hydropower generation, may also exhibit a *weekly pattern* due to variations of demands within a week. Likewise, hourly time series may have a distinct *diurnal pattern* due to the variations of demands within a day. Summer hourly rainfall series or certain water-quality constituents related to temperature may also exhibit distinct diurnal patterns due to the daily rotation of the earth which causes variations within the day of net radiation.

Seasonal or periodic patterns of hydrologic series translate into statistical characteristics which vary within the year (or within a week or a day as the case may be). Generally seasonal or periodic variations in the mean, variance, covariance, and skewness are important. Figure 19.1.6 shows how seasonal series are partitioned into basic components (the annual series is also shown for comparison). A part of the original time series y_t is plotted in Fig. 19.1.6*a*′, in which the seasonal (periodic) pattern is evident. It is a *periodic-stochastic* series since, in addition to the periodic pattern, a random pattern is also observed. This periodic-stochastic pattern repeats through time in a similar fashion. In contrast, the annual series in Fig. 19.1.6*a* does not show a periodic pattern; it simply varies about a constant mean $\bar{y}$ (see Fig. 19.1.6*b*). The fact that the series in Fig. 19.1.6*a*′ behaves in a cyclic fashion implies that the mean of the series is also cyclic or periodic, as shown in Fig. 19.1.6*b*′. For instance, for monthly stream-flow series, each month will have its own mean $\bar{y}_t$ (refer to Sec. 19.2 for the definition of seasonal statistics).

Removing Seasonality in the Mean and Variance. Removing the seasonality in the mean is accomplished by taking the difference $y_t - \bar{y}_t$, where $\bar{y}_t$ is the monthly mean for January, February, . . . , if t is a monthly index. When this difference is plotted in Fig. 19.1.6*c*′, the series fluctuates about zero with a particular pattern. The variability of the series is initially small, then increases, and then decreases. This pattern repeats in the second year and subsequent years throughout the record. If such variability is measured by the variance s_t^2 in each time interval in the year (for instance, one variance for each month in the year for a monthly series) and s_t is plotted as in Fig. 19.1.6*d*′, it will exhibit a seasonal (periodic) pattern similar to that of the mean in Fig. 19.1.6*b*′. In contrast, for the annual series $y_t - \bar{y}_t$ in Fig. 19.1.6*c*, the variance s^2 is a constant. The seasonality in the variance can be removed by $(y_t - \bar{y}_t)/s_t$. This operation is also called *seasonal standardization* and often is referred to in literature as *deseasonalizing* the original series. Actually, this latter term

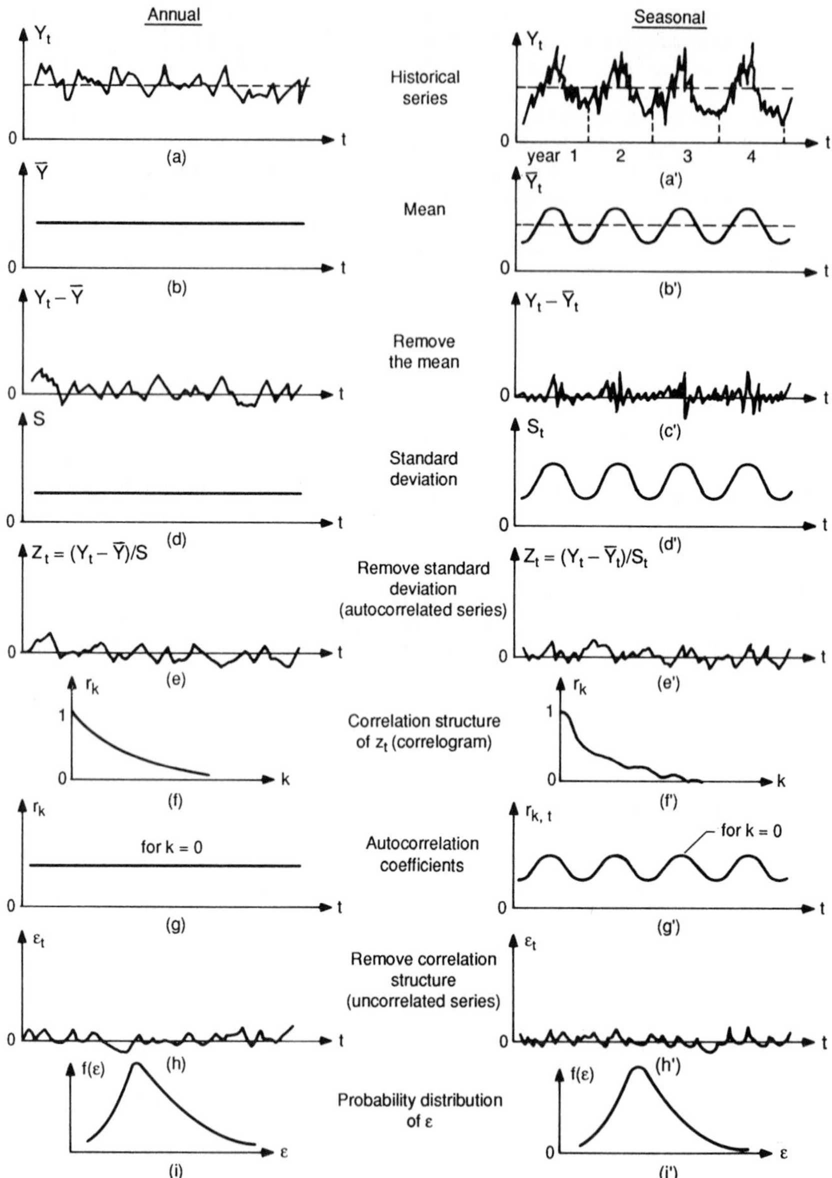

FIGURE 19.1.6 Partitioning an annual and a seasonal time series into their various components.

is a misnomer since it may imply that $z_t = (y_t - \bar{y}_t)/s_t$ is free of seasonality. However, other seasonalities may still be present in z_t.

Removing Seasonality in the Correlation. Unlike the seasonality in the mean and the variance, other remaining seasonalities are difficult to observe from the plot of the time series. Further analysis must be done to detect them. The series z_t in Fig. 19.1.6*e* and *e′* may have a dependence structure or autocorrelation. This is described by the *correlogram* r_k, which is a plot of the autocorrelation coefficient as a function of lag k, as shown in Fig. 19.1.6*f* and *f′*. However, a better way of analyzing the dependence structure of series z_t in Fig. 19.1.6*e′* is by determining the correlation on a season-by-season basis. For instance, for monthly series, one would correlate the February observations for all years of record with the corresponding January observations to obtain $r_{1,2}$, similarly for other months to obtain $r_{1,1}, \ldots, r_{1,12}$ and, in general, $r_{k,1}, \ldots, r_{k,12}$. The plot of $r_{k,t}$ for $k > 0$ may, depending on the hydrologic series under consideration, exhibit a seasonal or periodic pattern as in Fig. 19.1.6*g′*. In contrast, for annual series, the correlation coefficient r_k is assumed to remain constant, as depicted in Fig. 19.1.6*g*.

Regardless of whether the autocorrelation is constant or periodic, removing such correlation structure from the series requires a mathematical model to represent the correlation. A simple model that may be considered for series z_t in Fig. 19.1.6*e* is $z_t = r_1 z_{t-1} + \varepsilon_t$, the *lag-one autoregressive process.* The residual series $\varepsilon_t = z_t - r_1 z_{t-1}$ is uncorrelated or free of autocorrelation. A similar operation, although it uses a different model, can be used to remove the periodic correlation structure of z_t in Fig. 19.1.6*e′*. Thus, the original seasonal series y_t has been partitioned into components, periodic mean, periodic variance, and periodic correlation, and these components have been removed from y_t, yielding a residual series ε_t free of periodicities. The residual series ε_t of Fig. 19.1.6*h* or *h′* is represented by a probability distribution function $f(\varepsilon)$ which may be normal or nonnormal.

19.2 TIME-SERIES ANALYSIS PRINCIPLES

This section addresses the estimation of a number of statistical properties of annual and seasonal hydrologic time series. In addition, the detection and estimation of trends, shifts, seasonality, and nonnormality are briefly discussed, as are some procedures for transforming a skewed series into a normally distributed series.

19.2.1 Statistical Properties of Time Series

Overall Sample Statistics. The *mean* and the *variance* of a time series y_t are estimated by

$$\bar{y} = \left(\frac{1}{N}\right) \sum_{t=1}^{N} y_t \tag{19.2.1}$$

$$s^2 = \frac{1}{N-1} \sum_{t=1}^{N} (y_t - \bar{y})^2 \tag{19.2.2}$$

respectively, where N = sample size. The square root of the variance is the *standard deviation* s, and $c_v = s/\bar{y}$ is the *coefficient of variation.* Likewise, the *skewness coeffi-*

cient is estimated by

$$g = \frac{N \sum_{t=1}^{N} (y_t - \bar{y})^3}{(N-1)(N-2)s^3} \tag{19.2.3}$$

in which $\bar{y}$ and s are as defined above. Since the sample statistics $\bar{y}$, s^2, and g are *estimators* of the population statistics μ, σ^2, and γ, respectively, sometimes the notations $\hat{\mu}$, $\hat{\sigma}^2$, and $\hat{\gamma}$ are used to represent the sample statistics.

The *sample autocorrelation function* r_k (or autocorrelation coefficients) of a time series may be estimated by[80]

$$r_k = \frac{c_k}{c_0} \tag{19.2.4}$$

$$c_k = \left(\frac{1}{N}\right) \sum_{t=1}^{N-k} (y_{t+k} - \bar{y})(y_t - \bar{y}) \qquad k \geq 0 \tag{19.2.5}$$

The estimator r_k of Eq. (19.2.4) is an estimator of the population autocorrelation coefficient ρ_k. The plot of r_k versus k is the *correlogram.* Sometimes, the correlogram is used for choosing the type of stochastic model to represent a given time series. The lag-one serial correlation coefficient r_1 is a simple measure of the degree of time dependence of a series. When the correlogram decays rapidly to zero after a few lags, it may be an indication of small persistence or *short memory* in the series, while a slow decay of the correlogram may be an indication of large persistence or *long memory.* This short or long memory is shown schematically in Fig. 19.2.1.

Corrections for Bias. An estimator of a population statistic is *biased* when its mean value is different from the population statistic. The estimator r_k of Eq. (19.2.4) is a biased (downward) estimator of ρ_k; that is, the average value of r_k estimated by the formula is less than the true value ρ_k. Several procedures have been suggested to correct the values of r_k from limited samples.[195,206] Wallis and O'Connell[195] suggested the following correction to obtain an unbiased estimator of ρ_1

$$\hat{\rho}_1 = \frac{r_1 N + 1}{N - 4} \tag{19.2.6}$$

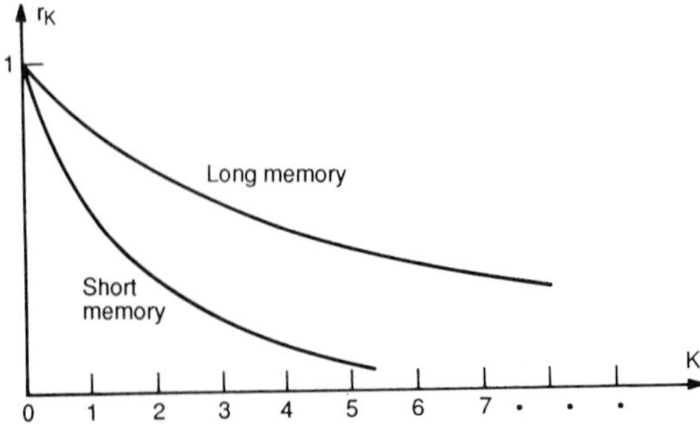

FIGURE 19.2.1 Schematic representation of a correlogram with short and long memory.

where N is the sample size. Likewise, s^2 of Eq. (19.2.2) is an unbiased estimator of σ^2 if the series is uncorrelated. When the series is autocorrelated, the effective sample size N is reduced. In this case, an unbiased estimator of the variance can be determined by[114,130]

$$\hat{\sigma}^2 = \frac{(N-1)\, s^2}{N-K} \tag{19.2.7}$$

where

$$K = \frac{[N(1-\hat{\rho}_1^2) - 2\,\hat{\rho}_1\,(1-\hat{\rho}_1^N)]}{[N(1-\hat{\rho}_1)^2]} \tag{19.2.8}$$

and s^2 and $\hat{\rho}_1$ are given by Eqs. (19.2.2) and (19.2.6), respectively.

The estimator g of Eq. (19.2.3) is an unbiased estimator of γ if the time series is uncorrelated and normally distributed. For nonnormal series, Bobee and Robitaille[9] proposed the unbiased estimator

$$\hat{\gamma}_0 = \frac{Lg[A + Bg^2(L^2/N)]}{\sqrt{N}} \tag{19.2.9a}$$

$$A = 1 + 6.5N^{-1} - 20.2N^{-2} \tag{19.2.9b}$$

$$B = 1.48N^{-1} + 6.77N^{-2} \tag{19.2.9c}$$

$$L = \frac{N-2}{\sqrt{(N-1)}} \tag{19.2.9d}$$

where $\pm L$ is the theoretical limit on skewness from a sample of size N.[88] Furthermore, for gamma correlated series, Fernandez and Salas[39] gave the unbiased estimator

$$\hat{\gamma} = \frac{\hat{\gamma}_0}{1 - 3.12\hat{\rho}_1^{3.7}N^{-0.49}} \tag{19.2.10}$$

in which $\hat{\gamma}_0$ and $\hat{\rho}_1$ are determined by Eqs. (19.2.9*a*) and (19.2.6), respectively.

Observed Annual Statistics. The overall sample statistics $\bar{y}$, s^2, c_v, g, and r_k are usually determined for annual hydrologic time series. Coefficients of variation c_v of annual flows are typically smaller than one, although they may be close to one or greater than one in streams in arid and semiarid regions. From an analysis of the annual flows of 126 rivers, McMahon and Mein[117] report a median value of c_v of 0.25. Coefficients of skewness g of annual flows are typically greater than zero. In some streams, small values of g are found, suggesting that annual flows are approximately normally distributed. On the other hand, in some streams of arid and semiarid regions, g can be greater than one. A range of g between -0.4 and about 2.0 and a median value of 0.40 has been reported.[117] Similarly, r_1 of annual flows are generally small but positive, although in some cases, because of sample variability, the r_1's are negative. It is quite typical to find values of r_1 in the range of $+0.0$ to 0.4 for annual stream-flow series. Yevjevich[202] found that, for a large number of rivers worldwide, the average value of r_1 was about 0.15, while McMahon and Mein[117] found a range of r_1 between -0.2 and 0.8 with a mean value of 0.23. Large values of r_1 for annual flows can be found for a number of reasons, including the effect of natural or manmade surface storage such as lakes, reservoirs, or glaciers, the effect of slow groundwater storage response, and the effect of nonstationarities. Figure 19.2.2 shows a slow-decaying correlogram r_k for the annual flows of the White Nile River at Mongalla and a fast decaying r_k for the Blue Nile River at Khartoum, while the r_k for the Nile River at Aswan lies between the other two.

Seasonal Sample Statistics. Seasonal hydrologic time series, such as monthly flows, may be better described by considering statistics on a seasonal basis. Let the seasonal time series $y_{\nu,\tau}$, in which ν = year; τ = season; $\nu = 1, \ldots, N$; and $\tau = 1, \ldots, \omega$, with N and ω denoting the number of years of record and the number of seasons per year, respectively. The seasonal mean $\bar{y}_\tau$ is obtained by applying Eq. (19.2.1) for each season τ as

$$\bar{y}_\tau = \frac{1}{N}\sum_{\nu=1}^{N} y_{\nu,\tau} \qquad \tau = 1, \ldots, \omega \tag{19.2.11}$$

Likewise, Eqs. (19.2.2) and (19.2.3) can be applied on a seasonal basis to determine the seasonal variance s_τ^2 and the seasonal skewness coefficient g_τ, respectively, for $\tau = 1, \ldots, \omega$. Furthermore, the season-to-season correlation coefficient $r_{k,\tau}$ is determined by

$$r_{k,\tau} = \frac{c_{k,\tau}}{(c_{0,\tau}\, c_{0,\tau-k})^{1/2}} \tag{19.2.12}$$

$$c_{k,\tau} = \frac{1}{N}\sum_{\nu=1}^{N} (y_{\nu,\tau} - \bar{y}_\tau)\,(y_{\nu,\tau-k} - \bar{y}_{\tau-k}) \tag{19.2.13}$$

For instance, for monthly stream-flow time series, $r_{1,4}$ represents the correlation between the flows of the fourth month with those of the third month.

Description of Seasonal Statistics. Each of the statistics $\bar{y}_\tau$, s_τ, g_τ, and $r_{k,\tau}$ may be plotted versus time $\tau = 1, \ldots, \omega$ to observe whether they exhibit a seasonal pat-

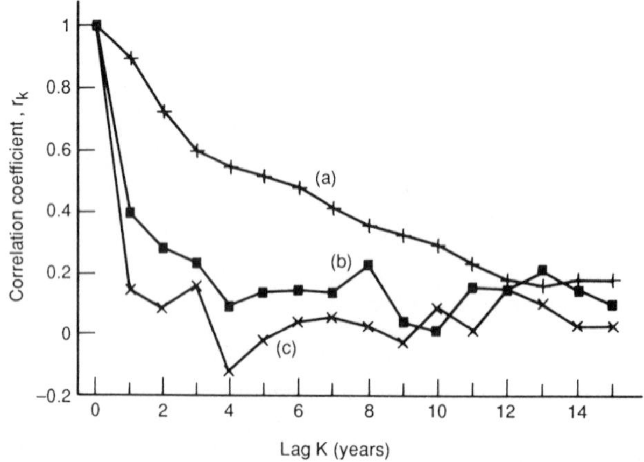

FIGURE 19.2.2 Correlogram of annual flows of (*a*) the White Nile River at Mongalla (1914–1983), (*b*) the Nile River at Aswan (1871–1989), and (*c*) the Blue Nile River at Khartoum (1912–1989).

tern (see Sec. 19.2.4 for testing for seasonality). These statistics may be fitted by Fourier series. This technique is especially effective with weekly and daily data.[155,206] Generally, for seasonal stream-flow series, $\bar{y}_\tau > s_\tau$, although for some streams $\bar{y}_\tau$ may be smaller than s_τ for the low-flow seasons. Furthermore, for intermittent stream-flow series, generally $\bar{y}_\tau < s_\tau$ throughout the year. Likewise, values of the skewness coefficient g_τ for the dry season are generally larger than those for the wet season, indicating that data in the dry season depart more from normality than those in the wet season. Values of the skewness for intermittent hydrologic series are usually larger than skewness for similar nonintermittent series. Seasonal correlations $r_{k,\tau}$ for stream flow during the dry season are generally larger than those for the wet season, and seasonal correlations $r_{k,\tau}$ for monthly precipitation are generally not significantly different from zero.[148] The data interval must be less than a month (typically, less than a week) to find significant autocorrelation in precipitation series.

Figure 19.2.3 shows $\bar{y}_\tau$, s_τ, g_τ, and $r_{k,\tau}$, $\tau = 1, \ldots, 12$ for the monthly flows of the Nile River at Aswan. A well-defined seasonality is shown in each statistic, meaning that the monthly statistics for the low-flow season are significantly different from those of the high-flow season. Furthermore, the seasonal patterns of $\bar{y}_\tau$ and s_τ are in phase while g_τ is out of phase relative to $\bar{y}_\tau$ and s_τ. Also, a well-defined seasonality in $r_{1,\tau}$ is evident with the larger values in dry months and smaller values in wet months. Most of the correlations $r_{12,\tau}$ are significant, with seasonality in phase with $r_{1,\tau}$. The correlations of Fig. 19.2.3 indicate the complex, long-term dependence (long-memory) structure of monthly flows of the Nile River. The significant correlations $r_{12,\tau}$ shown for this river are not typical for monthly stream-flow series. In fact, these correlations usually will be small or not significant. Conversely, rivers that exhibit long-term autocorrelation in seasonal flows will exhibit long-term autocorrelation in annual flows.

Sample Statistics for Multiple Series. For multiple time series, in addition to the above-defined sample statistics such as the mean, variance, skewness, and autocorrelation, the cross-correlation between each pair of time series can be determined. Consider the time series $y_t^{(i)}$ and $y_t^{(j)}$ at sites i and j, respectively. The sample lag-k cross-correlation coefficient is

$$r_k^{ij} = \frac{c_k^{ij}}{(c_0^{ii}\, c_0^{jj})^{1/2}} \qquad (19.2.14)$$

$$c_k^{ij} = \left(\frac{1}{N}\right) \sum_{t=1}^{N-k} (y_{t+k}^{(i)} - \bar{y}^{(i)})\,(y_t^{(j)} - \bar{y}^{(j)}) \qquad k \geq 0 \qquad (19.2.15)$$

and c_k^{ij} is the *cross-covariance* between the two series. The coefficient r_k^{ij} is an estimator of the population cross-correlation coefficient ρ_k^{ij}. The plot of r_k^{ij} versus k is the *cross-correlogram* or sample *cross-correlation function*. For instance, r_k^{ij} for annual flows of the Nile River at Aswan and the Blue Nile River at Khartoum is shown in Fig. 19.2.4. The lag-zero cross-correlation has a high value $r_0^{ij} = 0.85$. For n time series, the values r_k^{ij}, $i = 1, \ldots, n$ and $j = 1, \ldots, n$ are the elements of the cross-correlation matrix $\hat{\mathbf{M}}_k$ (n by n matrix). The lag-zero cross-correlation matrix $\hat{\mathbf{M}}_0$ is a symmetric matrix with 1's in the diagonal. The diagonal elements of the lag-k cross-correlation matrix $\hat{\mathbf{M}}_k$ for $k > 0$ are the lag-k serial correlation coefficients for each site.

Likewise, for seasonal time series, the season-to-season cross-correlation between sites can be determined. Consider the seasonal time series $y_{\nu,\tau}^{(i)}$ and $y_{\nu,\tau}^{(j)}$ for sites i and j, respectively. The lag-k seasonal cross-correlation coefficient $r_{k,\tau}^{ij}$ between both series

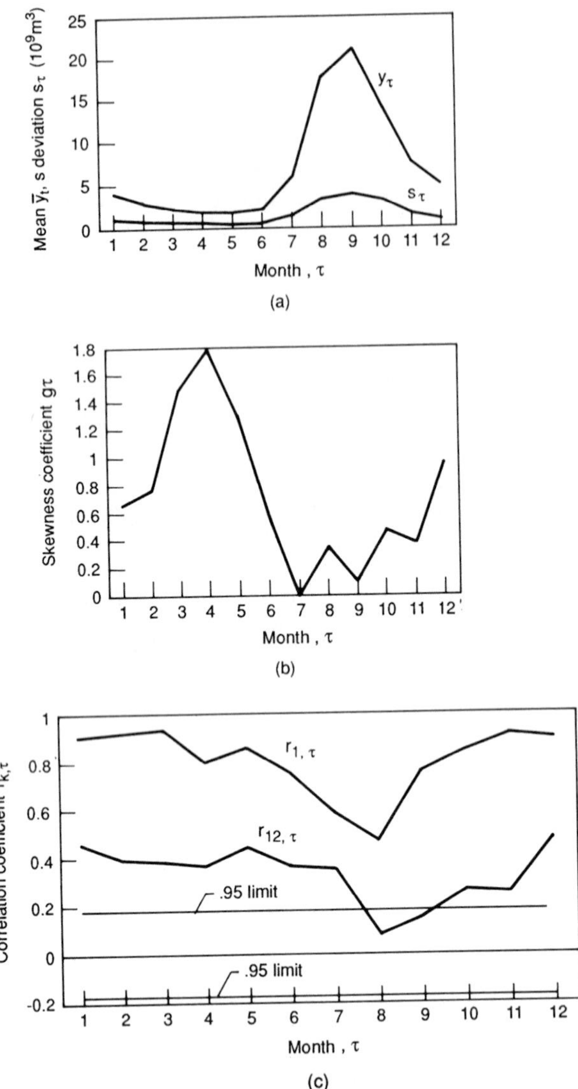

FIGURE 19.2.3 (*a*) Monthly mean $\bar{y}_\tau$ and standard deviation s_τ, (*b*) monthly skewness coefficient g_τ, and (*c*) lag-1 and lag-12 month-to-month correlations for monthly stream-flow series of the Nile River at Aswan (1871–1989).

can be determined by applying Eqs. (19.2.14) and (19.2.15) on a seasonal basis. For instance, for monthly stream flow, $r^{ij}_{1,4}$ is determined by correlating the stream-flow series of the fourth month of site i with those of the third month of site j.

Storage-Related Statistics. In modeling hydrologic time series for simulation studies of reservoir systems, storage-related statistics may be particularly important. Consider the hydrologic time series y_t, $t = 1, \ldots, N$ and a subsample $y_1, \ldots, y_n$

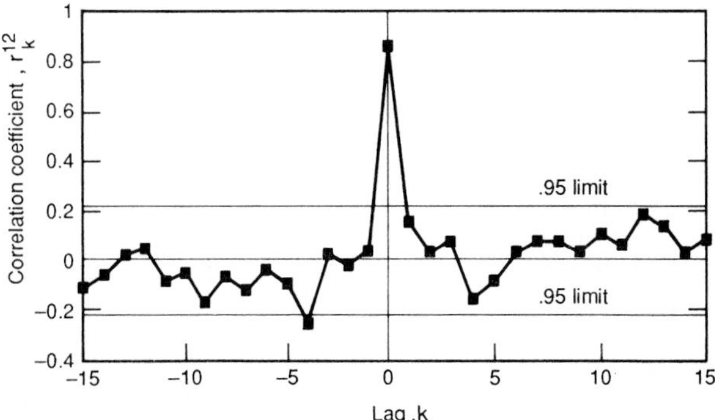

FIGURE 19.2.4 Cross-correlogram r_k^{12} of annual flows of the Nile River at Aswan (1) and the Blue Nile River at Khartoum (2), for the concurrent period 1912–1989.

with $n \leq N$. Form the sequence of partial sums S_i as

$$S_i = S_{i-1} + (y_i - \bar{y}_n) \qquad i = 1, \ldots, n \tag{19.2.16}$$

where $S_0 = 0$ and $\bar{y}_n$ is the sample mean. For instance, in the case of stream-flow time series, S_i represents the cumulative departure from the mean flow $\bar{y}_n$. The plot of S_i versus i, $i = 1, \ldots, n$ is the typical mass curve from which the minimum *storage capacity* D_n^* of a reservoir to deliver $\bar{y}_n$ through the time period n can be obtained.[41] For instance, Fig. 19.2.5 shows a mass curve for which $D_n^* = \max(D_1, \ldots, D_5)$. Related to D_n^* are the *range* R_n^* and the *rescaled range* R_n^{**} defined by

$$R_n^* = \max(S_0, S_1, \ldots, S_n) - \min(S_0, S_1, \ldots, S_n) \tag{19.2.17}$$

$$R_n^{**} = \frac{R_n^*}{s_n} \tag{19.2.18}$$

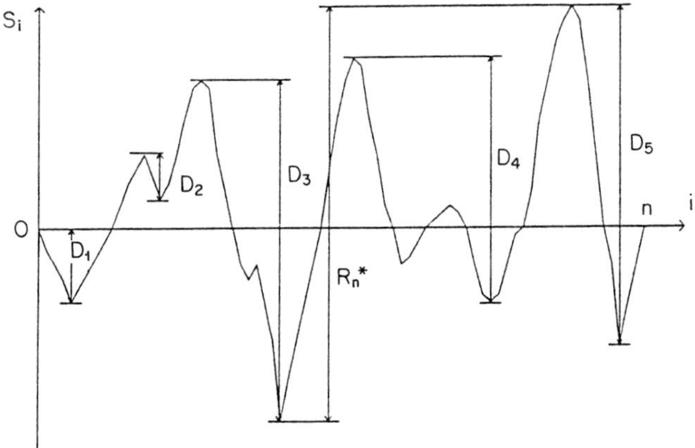

FIGURE 19.2.5 Mass curve S_i of Eq. (19.2.16), sequence of deficits $D_1, \ldots, D_5$, and range R_n^*.

in which s_n is the sample standard deviation. Thus, for a given sample of size N the ranges $R_2^*, \ldots, R_N^*$ and rescaled ranges $R_2^{**}, \ldots, R_N^{**}$ will result. Both R_n^* and R_n^{**} have been widely used in literature as measures of long-term dependence and for comparing alternative models of hydrologic series.[68,196,204]

In particular, Hurst[75] showed that for a large number of geophysical time series such as stream-flow, precipitation, temperature, and tree-ring series, the mean rescaled range $\overline{R}_n^{**}$ is proportional to n^h with $h > ½$. The values of h obtained for different series gave a mean of about 0.73 and a standard deviation of 0.09. Theoretical results for normal, independent processes[37] and for autoregressive processes[110] indicated that asymptotically $h = ½$ for these processes. The discrepancy between theoretical results stating that $h = ½$ and Hurst's empirical finding that $h > ½$ has become known as the *Hurst phenomenon.* Several estimators of h have been proposed and used in stochastic hydrology such as the original Hurst estimator K:[75]

$$K = \frac{\log (\overline{R}_n^{**})}{\log (n/2)} \tag{19.2.19}$$

It has been shown that the estimators of h are transient, meaning they depend on n, and, as $n \to \infty$, they generally converge to a limiting value, equal to ½ for many time-series models.[153] One interpretation of the Hurst phenomenon has been to associate $h = ½$ with short-memory models possessing short-term dependence structure and $h > ½$ with long-memory models possessing long-term dependence. This is a valid interpretation in the asymptotic sense. However, such interpretation as a criterion for selecting stochastic models for hydrologic time-series simulation is not practical because a number of models, including ARMA processes (see Sec. 19.3 for definition of ARMA processes) can have long-term dependence structure, yet asymptotically their value of $h = ½$. Nevertheless, estimates of h can be useful for comparing the performance of alternative modeling strategies and alternative estimation procedures.

Drought-Related Statistics. Drought-related statistics are also important in modeling hydrologic time series. Consider a hydrologic time series y_t, $t = 1, \ldots, N$ and a *demand level d* (also called a *crossing level*). Assume that y_t is an annual series and d is a constant (equal to the sample mean $\bar{y}$ or a fraction of $\bar{y}$) as shown in Fig. 19.2.6. A deficit occurs when $y_t < d$ consecutively during one or more years until $y_t > d$ again. Such a deficit can be defined by its duration L, by its magnitude M, and by its intensity $I = M/L$.[205] Since a number of deficits can occur in a given hydrologic

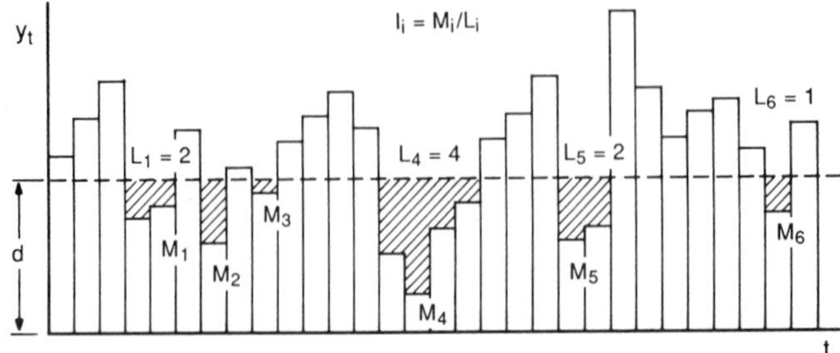

FIGURE 19.2.6 Definition of deficit length L_i, deficit magnitude M_i, and deficit intensity I_i.

sample (Fig. 19.2.6), the maximum deficit duration *(critical drought)* in a given sample $L^* = \max(L_1, \ldots, L_m)$ is often of most interest, where m is the number of deficits. Similar critical drought properties in relation to the magnitude and intensity are $M^* = \max(M_1, \ldots, M_m)$ and $I^* = \max(I_1, \ldots, I_m)$, respectively.

19.2.2 Determination and Testing of Trends

A number of parametric and nonparametric tests for trends have been suggested in the literature. This section includes one parametric and one nonparametric test. See also Sec. 17.2.3 of Chap. 17 for a discussion on the criteria for selecting parametric and nonparametric tests and Sec. 17.4 for trend analysis.

Detection and Estimation of Linear Trends. Assume that y_t, $t = 1, \ldots, N$ is an annual time series and $N =$ sample size. A simple linear trend can be written as

$$y_t = a + bt \tag{19.2.20}$$

where a and b are the parameters of the regression model (see Sec. 17.4.2). Rejection of the hypothesis $b = 0$ can be considered as a detection of a linear trend. The hypothesis that $b = 0$ is rejected if

$$T_c = \left| \frac{\sqrt{N-2}}{r\sqrt{1-r^2}} \right| > T_{1-\alpha/2,\nu} \tag{19.2.21}$$

in which r is the cross-correlation coefficient between the sequences $y_1, \ldots, y_N$ and $1, \ldots, N$, and $T_{1-\alpha/2,\nu}$ is the $1-\alpha/2$ quantile of the Student t distribution with $\nu = N-2$ degrees of freedom. See Sec. 17.4.5 for the use of polynomial regression for nonlinear trends.

Mann-Kendall Test for Trends. This is a nonparametric test which tests for a trend in a time series without specifying whether the trend is linear or nonlinear. Consider the annual time series y_t, $t = 1, \ldots, N$. Each value $y_{t'}$, $t = 1, \ldots, N-1$ is compared with all subsequent values y_t, $t = t'+1, t'+2, \ldots, N$, and a new series z_k is generated by

$$\begin{aligned} z_k &= 1 && \text{if } y_t > y_{t'} \\ z_k &= 0 && \text{if } y_t = y_{t'} \\ z_k &= -1 && \text{if } y_t < y_{t'} \end{aligned} \tag{19.2.22}$$

in which $k = (t'-1)(2N-t')/2 + (t-t')$. The Mann-Kendall statistic is given by the sum of the z_k series[71]

$$S = \sum_{t'=1}^{N-1} \sum_{t=t'+1}^{N} z_k \tag{19.2.23}$$

This statistic represents the number of positive differences minus the number of negative differences for all the differences considered.

The test statistic for $N > 40$ may be written as[71]

$$u_c = \frac{S+m}{\sqrt{V(S)}} \tag{19.2.24}$$

$$V(S) = \frac{1}{18}\left[N(N-1)(2N+5) - \sum_{i=1}^{n} e_i(e_i-1)(2e_i+5)\right] \qquad (19.2.25)$$

where $m = 1$ if $S < 0$ and $m = -1$ if $S > 0$, n is the number of tied groups, and e_i is the number of data in the ith (tied) group. The statistic u_c is assumed to be zero if $S = 0$. Then the hypothesis of an upward or downward trend cannot be rejected at the α significance level if $|u_c| > u_{1-\alpha/2}$, where $u_{1-\alpha/2}$ is the $1 - \alpha/2$ quantile of the standard normal distribution. Kendall[86] indicated that this test can be used even for N as low as 10 if there are not too many tied values. Hirsch et al.[71] extended this test to seasonal time series.

Final Remarks. The Mann-Kendall test can be extended to test whether different seasons exhibit trends in the same direction and of the same magnitude.[53,71] This test has been applied to testing trends in water-quality time series.[190] Other nonparametric tests for trends useful for hydrologic time series have been suggested, such as the Hotelling-Pabst test[27] and the Sen test,[53] and there are also tests for seasonally correlated data.[72] An excellent reference for nonparametric tests is Hollander and Wolfe.[73]

19.2.3 Determination and Testing of Shifts (Jumps)

A number of parametric and nonparametric tests are available for testing and determining shifts in statistical properties of time series such as the mean and variance.[53,164,173] One parametric and one nonparametric test are included here. The reader is referred to Sec. 17.3.2 for further discussion concerning the assumptions and applicability of tests for shifts.

t Test for Shift in the Mean. Suppose that y_t, $t = 1, \ldots, N$ is an annual hydrologic series which is uncorrelated and normally distributed with mean μ and standard deviation σ and N = sample size. The series is divided into two subseries of sizes N_1 and N_2 such that $N_1 + N_2 = N$. The first subseries y_t, $t = 1, 2, \ldots, N_1$, has mean μ_1 and standard deviation σ, and the second subseries y_t, $t = N_1 + 1, N_1 + 2, \ldots, N$, is assumed to have mean μ_2 and standard deviation σ. The simple t test can be used to test the hypothesis $\mu_1 = \mu_2$ when the two subseries have the same standard deviation σ. Rejection of the hypothesis can be considered as a detection of a shift. The test statistic in this case is given by[106,173]

$$T_c = \frac{|\bar{y}_2 - \bar{y}_1|}{S\sqrt{\dfrac{1}{N_1} + \dfrac{1}{N_2}}} \qquad (19.2.26)$$

$$S = \sqrt{\frac{(N_1 - 1)s_1^2 + (N_2 - 1)s_2^2}{N - 2}} \qquad (19.2.27)$$

where $\bar{y}_1$ and $\bar{y}_2$ and s_1^2 and s_2^2 are the estimated means and variances of the first and the second subseries, respectively. The hypothesis $\mu_1 = \mu_2$ is rejected if $T_c > T_{1-\alpha/2,\nu}$ where $T_{1-\alpha/2,\nu}$ is the $1 - \alpha/2$ quantile of the Student's t distribution with $\nu = N - 2$ degrees of freedom and α is the significance level of the test. Modifications of the test are available when the variances in each group are different[173] and when the data exhibit some significant serial correlation.[106]

Mann-Whitney Test for Shift in the Mean. Suppose that y_t, $t = 1, \ldots, N$ is an annual hydrologic series that can be divided into two subseries $y_1, \ldots, y_{N_1}$ and $y_{N_1+1}, \ldots, y_N$ of sizes N_1 and N_2, respectively, such that $N_1 + N_2 = N$. A new series, z_t, $t = 1, \ldots, N$, is defined by rearranging the original data y_t in increasing order of magnitude. One can test the hypothesis that the mean of the first subseries is equal to the mean of the second subseries by using the statistic[173]

$$u_c = \frac{\sum_{t=1}^{N_1} R(y_t) - N_1(N_1 + N_2 + 1)/2}{[N_1 N_2(N_1 + N_2 + 1)/12]^{1/2}} \tag{19.2.28}$$

where $R(y_t)$ is the rank of the observation y_t in ordered series z_t. The hypothesis of equal means of the two subseries is rejected if $|u_c| > u_{1-\alpha/2}$, where $u_{1-\alpha/2}$ is the $1 - \alpha/2$ quantile of the standard normal distribution and α is the significance level of the test. Equation (19.2.28) can be modified for the case of groups of values that are tied.[53]

Final Remarks. The foregoing tests are for a shift in the mean of a sample series when the point of change is known. However, when the point of change is not known, bayesian analysis can be used to detect the point of change and its amount.[105] Likewise, the foregoing tests assumed that the underlying series is uncorrelated. Although adjustments to the t test can be made for serial correlation,[106] intervention analysis is a more effective approach.[66] The F test has been used for changes in the variance. Furthermore, the tests included in this section assume a single series. For multiple series, double mass analysis can be used and the significance of a change in the slope can be tested. In the case of testing the mean of a series versus the mean of another series (or the mean of a group of series), the Mann-Whitney test can also be applied.[173] Furthermore, in the case of testing for equality in means for a group of series, one-way and two-way analysis of variance can be used.[58]

19.2.4 Testing for Seasonality

In most analysis and modeling of hydrologic time series, testing for seasonality in statistics is done by using simple procedures, mostly by observing the plot of the statistic under consideration versus the season τ. Figure 19.2.3 plots the mean $\bar{y}_\tau$ versus $\tau = 1, \ldots, 12$. The plot suggests that the $\bar{y}_\tau$'s during the low-flow season are quite different from those of the high-flow season. Thus, even though some of the $\bar{y}$'s in the low-flow season are similar to each other, one would conclude overall that $\bar{y}_\tau$ is a seasonal statistic. A similar argument can be made in relation to seasonality in other statistics such as s_τ, g_τ, and $r_{k,\tau}$.

More rigorous procedures for testing for seasonality can be made, although they are rarely applied. For instance, since $\bar{y}_\tau$ is an estimator of the population mean μ_τ, one could test the hypothesis that $\mu_\tau = \mu_{\tau'}$ (for any $\tau \neq \tau'$) versus $\mu_\tau \neq \mu_{\tau'}$. Under the assumption that the underlying variables are independent and normally distributed, one could apply the t test of Sec. 19.2.3. The independence assumption may not be quite true for seasonal stream-flow series, except if τ and τ' are far apart, while it is usually satisfied by seasonal precipitation series for any $\tau \neq \tau'$. This test can be extended to consider the hypothesis $\mu_1 = \cdots = \mu_\omega$ versus $\mu_1 \neq \cdots \neq \mu_\omega$ by applying *one-way analysis of variance.*[124] Likewise, one can test the hypothesis $\sigma_\tau^2 = \sigma_{\tau'}^2$ versus $\sigma_\tau^2 \neq \sigma_{\tau'}^2$ by using the two-tailed F test. Furthermore, extensions to test the hypothesis $\sigma_1^2 = \cdots = \sigma_\omega^2$ versus $\sigma_1^2 \neq \cdots \neq \sigma_\omega^2$ can be made by using an approximate *generalized likelihood-ratio test.*[124]

19.2.5 Testing for Normality

Several of the models and approaches included in this chapter assume that the variable under consideration is normally distributed. Therefore, it is usual practice to test the data for normality before further analysis. A widely used method for judging whether a certain data set is normally distributed is to plot the empirical frequency distribution of the data on normal probability paper (see also Sec. 17.2.2 of Chap. 17). A straight-line probability plot indicates that the data are normally distributed. Otherwise, the data are not normal and a transformation may be needed to make them normal. Given the availability of plotting packages for personal and workstation computers, the graphical test for normality is quite attractive in practice. More importantly, a powerful correlation plot test procedure is available.[94,95,193] However, other tests are also used, such as the chi-square test, the Kolmogorov-Smirnov test,[7,124] and procedures based on testing the hypothesis that the skewness coefficient is equal to zero or the kurtosis coefficient is equal to three.[173] See also Sec. 18.3.2 of Chap. 18 for details of normality tests.

19.2.6 Transformations to Normal

A widely used method for transforming data to normal is based on the *logarithmic transformation.* For instance, if x_t is the original series, $y_t = \log(x_t - c)$ is normally distributed provided that x_t is lognormal with lower-bound parameter c. Often the simple log transformation with $c = 0$ works approximately, even if the original variable x_t is not lognormal. Likewise, *power transformations* such as $y_t = (x_t - c)^b$ with $b < 1$ (usually ½ or ⅓) is an alternative. An equivalent method is the *Box-Cox transformation* given by[10]

$$y_t = \begin{cases} \dfrac{(x_t^\lambda - 1)}{\lambda} & \lambda \neq 0 \qquad (19.2.29a) \\ \ln(x_t) & \lambda = 0 \qquad (19.2.29b) \end{cases}$$

in which λ is a parameter which must be estimated so that the skewness of the y_t becomes zero.

19.3 TIME-SERIES MODELING

The concepts and principles discussed in the previous sections are used in this section for representing hydrologic time series by mathematical models. A number of *stochastic models* are presented here along with parameter estimation methods and model testing procedures. The models can be applicable for many hydrologic processes, particularly stream-flow and precipitation processes. The models included in this section belong to the class of autoregressive (AR), autoregressive with moving average terms (ARMA), and disaggregation modeling schemes. These models should be able to reproduce the most important statistical features of the hydrologic time series under consideration. Alternative models such as fractional gaussian noises,[111] broken line models,[31,118] and shifting level models[154] are not included in this section because AR and ARMA models can accommodate most typical cases. This is not to say that alternative models may not be useful. For instance, shifting-level models may be useful to capture the effect of climatic shifts in hydrologic series. Likewise, modeling of short-term rainfall processes has been developing rapidly in recent years.

The theory and modeling schemes included here simply introduce some basic concepts around which some of the recent rainfall models have been developing.

19.3.1 Modeling of Single Stationary Series

AR Models. A time series defined as[15,155]

$$y_t = \mu + \sum_{j=1}^{p} \phi_j (y_{t-j} - \mu) + \varepsilon_t \tag{19.3.1}$$

is called an *autoregressive model* of order p in which ε_t is an uncorrelated normal random variable (also referred to as *noise, innovation, error term,* or *series of shocks*) with mean zero and variance σ_ε^2; it is uncorrelated with $y_{t-1}, \ldots, y_{t-p}$. Since ε_t is normally distributed, then also y_t is normal. The parameters of the model are μ, $\phi_1, \ldots, \phi_p$, and σ_ε^2. The model, Eq. (19.3.1), is often denoted as the AR(p) model or simply the AR model. The AR(1) model takes the simple form

$$y_t = \mu + \phi_1 (y_{t-1} - \mu) + \varepsilon_t \tag{19.3.2}$$

Low-order AR models such as Eq. (19.3.2) have been widely used for modeling annual hydrologic time series (Refs. 42, 61, 114, 155, and 203, among others), and seasonal and daily series after seasonal standardization.[138,207]

The mean, variance, and autocorrelation function of the AR(p) process are[15,155]

$$E(y) = \mu \tag{19.3.3}$$

$$\text{Var}(y) = \sigma^2 = \frac{\sigma_\varepsilon^2}{\left(1 - \sum_{j=1}^{p} \phi_j \rho_j\right)} \tag{19.3.4}$$

$$\rho_k = \phi_1 \rho_{k-1} + \cdots + \phi_p \rho_{k-p} \tag{19.3.5}$$

respectively. The last expression is known as the *Yule-Walker equation.* The three foregoing equations are useful for determining the properties of a model, given the model parameters, and for estimating the parameters of the model given a set of observations $y_1, \ldots, y_N$. For the AR(1) model, Eqs. (19.3.4) and (19.3.5) give respectively

$$\sigma^2 = \frac{\sigma_\varepsilon^2}{1 - \phi_1^2} \tag{19.3.6}$$

$$\rho_k = \phi_1 \rho_{k-1} = \phi_1^k \tag{19.3.7}$$

ARMA Models. A more versatile model than the AR is the *autoregressive moving average model*[12]

$$y_t = \mu + \sum_{j=1}^{p} \phi_j (y_{t-j} - \mu) + \varepsilon_t - \sum_{j=1}^{q} \theta_j \varepsilon_{t-j} \tag{19.3.8}$$

with p autoregressive parameters $\phi_1, \ldots, \phi_p$, and q moving average parameters $\theta_1, \ldots, \theta_q$. The model is also denoted as an ARMA(p, q) model or simply as ARMA. Note that an ARMA(p, 0) model is the same as an AR(p) model and the ARMA(0, q) model is the same as the *moving average model* MA(q). The noise ε_t in

Eq. (19.3.8) is an uncorrelated normal process with mean zero and variance σ_ε^2 and is also uncorrelated with $y_{t-1}, \ldots, y_{t-p}$. A simple version of the ARMA(p, q) model is the ARMA(1, 1) as

$$y_t = \mu + \phi_1(y_{t-1} - \mu) + \varepsilon_t - \theta_1\varepsilon_{t-1} \qquad (19.3.9)$$

Low-order ARMA models such as Eq. (19.3.9) are useful for operational hydrology in general, especially for modeling annual series[23,67,68,107,115,129,155,156] and for seasonal series after seasonal standardization.[33,116]

ARMA models must fulfill the *stationarity* and *invertibility* conditions, which imply certain constraints of the parameters. These are specified by the solution of the *characteristic equations*[12]

$$u^p - \phi_1\, u^{p-1} - \cdots - \phi_p = 0 \qquad (19.3.10)$$

$$u^q - \theta_1\, u^{q-1} - \cdots - \theta_q = 0 \qquad (19.3.11)$$

whose roots in each case must lie within the unit circle. For instance, for the ARMA(1, 1) model, the constraints $-1 < \phi_1 < 1$ and $-1 < \theta_1 < 1$ arise from the foregoing equations.

The variance and the lag-1 autocorrelation coefficient of the ARMA(1, 1) model are

$$\sigma^2 = \frac{1 - 2\,\phi_1\theta_1 + \theta_1^2}{1 - \phi_1^2}\,\sigma_\varepsilon^2 \qquad (19.3.12)$$

$$\rho_1 = \frac{(1 - \phi_1\theta_1)\,(\phi_1 - \theta_1)}{1 - 2\,\phi_1\theta_1 + \theta_1^2} \qquad (19.3.13)$$

respectively. Furthermore, the autocorrelation function is

$$\rho_k = \phi_1\rho_{k-1} = \rho_1\phi_1^{k-1} \qquad k > 1 \qquad (19.3.14)$$

Comparing Eqs. (19.3.7) and (19.3.14), one may observe that ρ_k of the AR(1) process is less flexible than that of the ARMA(1, 1) process, since the former depends on the sole parameter ϕ_1 while the latter depends on ϕ_1 and θ_1. Figure 19.3.1 gives some examples of typical correlograms for both the AR(1) and the ARMA(1, 1) processes. In relation to modeling hydrologic processes, one may say that AR processes are short-memory processes and ARMA processes are long-memory processes.[130,153]

GAR Models. Skewed hydrologic processes must be transformed into normal processes before AR and ARMA models are applied. However, a direct modeling approach which does not require a transformation may be a viable alternative. The *gamma autoregressive process* (GAR process) offers such an alternative. It is defined as[103]

$$y_t = \phi y_{t-1} + \varepsilon_t \qquad (19.3.15)$$

where ϕ is the autoregressive coefficient, ε_t is a random component, and y_t has a three-parameter gamma marginal distribution. The noise ε_t can be obtained as a function of ϕ and the parameters of the gamma distribution λ, α, and β (the location, scale, and shape parameters, respectively) as

$$\varepsilon = \lambda\,(1 - \phi) + \eta \qquad (19.3.16)$$

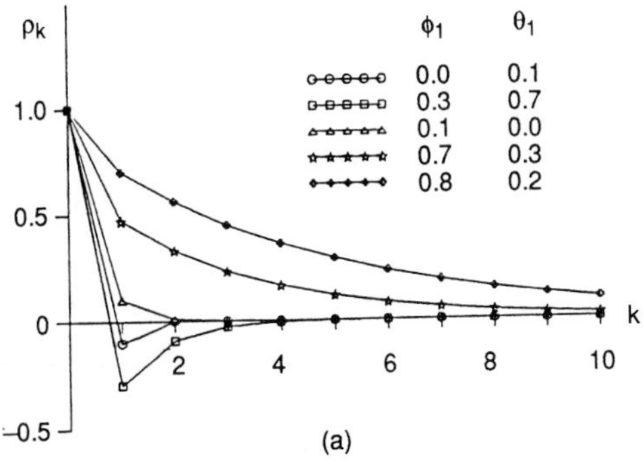

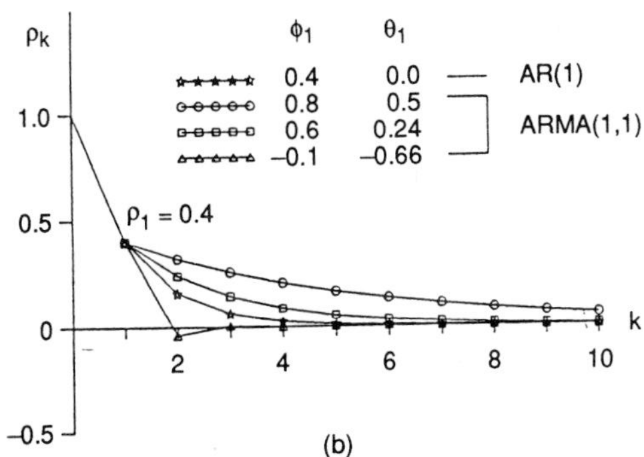

FIGURE 19.3.1 Correlograms ρ_k for (*a*) the ARMA(1, 1) process for various sets of parameters ϕ_1 and θ_1 and (*b*) the AR(1) and ARMA(1, 1) processes for which $\rho_1 = 0.4$.

$$\eta = 1 \qquad \text{if } M = 0$$

$$\eta = \sum_{j=1}^{M} E_j \phi^{U_j} \qquad \text{if } M > 0 \tag{19.3.17}$$

in which M is an integer random variable, Poisson-distributed, with mean $-\beta \ln(\phi)$, the set $U_1, \ldots, U_M$ are independent identically distributed (iid) random variables with uniform distribution (0, 1) and the set $E_1, \ldots, E_M$ are iid random variables exponentially distributed with mean $1/\alpha$. The GAR model has been applied to modeling annual stream-flow series.[39]

Product Models. While AR, ARMA, and GAR models are useful for modeling many hydrologic processes such as stream-flow processes in perennial rivers, they are impractical for *intermittent processes* such as stream flow in some ephemeral streams. Intermittent processes can be modeled as the product[79]

$$y_t = B_t z_t \tag{19.3.18}$$

where y_t is a nonnegative intermittent hydrologic variable, B_t is a discrete autocorrelated variable, z_t is a positive-valued continuous autocorrelated variable, and B_t and z_t are assumed to be mutually uncorrelated. B_t may be represented by a dependent (1, 0) *Bernoulli process* and z_t by an AR(1) process. Thus, the resulting product process y_t is intermittent and autoregressive. Product models have been applied for modeling short-term rainfall processes[17,24] and intermittent stream-flow processes.[159]

Parameter Estimation. Parameter estimation methods generally fall into three categories: *method of moments, method of maximum likelihood,* and *method of least squares.* The method of moments is based on taking as many moment equations as the number of parameters, substituting the population moments by the sample moments, and solving the equations simultaneously for the parameters. In the method of maximum likelihood, the likelihood function is first determined (this function is a function of the parameters given the observations), then the function (or its logarithm) is maximized and the parameters corresponding to such a maximum are the maximum likelihood estimators. In the method of least squares, the parameters which minimize the sum of square residuals $\Sigma\, \varepsilon_t^2$ are the least squares estimators; this is particularly useful for ARMA models. *Moment estimators* are generally available for all models. The method of maximum likelihood is the most efficient estimation method in a mean square error sense, although biases may be a problem, especially for small samples. The method of moments has been attractive in practice, since it is easier to apply and corrections for bias are available for some models. The method of least squares is an approximation to the method of maximum likelihood, and generally both methods require a numerical solution.

Moment estimators of the parameters of the AR(p) model can be obtained from Eqs. (19.3.3), (19.3.4), and (19.3.5). For instance, for the AR(1) model, the estimators $\hat{\mu}$, $\hat{\sigma}_\varepsilon^2$, and $\hat{\phi}_1$ are respectively

$$\hat{\mu} = \bar{y} \tag{19.3.19}$$

$$\hat{\sigma}_\varepsilon^2 = s^2\,(1 - r_1^2) \tag{19.3.20}$$

$$\hat{\phi}_1 = r_1 \tag{19.3.21}$$

in which $\bar{y}$, s^2, and r_1 are the sample mean, variance, and lag-1 autocorrelation coefficient determined by Eqs. (19.2.1), (19.2.2), and (19.2.4), respectively. Corrections for bias may be made by using $\hat{\rho}_1$ of Eq. (19.2.6) instead of r_1 and $\hat{\sigma}^2$ of Eq. (19.2.7) instead of s^2.

Moment estimators for the ARMA(p, q) process can also be derived. For the ARMA(1, 1) model, the moment estimator of μ is also $\bar{y}$ and the estimators of σ_ε^2, ϕ_1, and θ_1 are respectively

$$\hat{\sigma}_\varepsilon^2 = \frac{s^2\,(1 - \hat{\phi}_1^2)}{(1 - 2\,\hat{\phi}_1\hat{\theta}_1 + \hat{\theta}_1^2)} \tag{19.3.22}$$

$$\hat{\phi}_1 = \frac{r_2}{r_1} \tag{19.3.23}$$

$$\hat{\theta}_1 = \frac{-b \pm \sqrt{b^2 - 4\,(r_1 - \hat{\phi}_1)^2}}{2\,(r_1 - \hat{\phi}_1)} \tag{19.3.24}$$

in which $b = 1 - 2\,\hat{\phi}_1 r_1 + \hat{\phi}_1^2$. O'Connell[130] provides a bias correction for s^2, and procedures for estimating ϕ_1 and θ_1 so that the Hurst slope is preserved.

Maximum likelihood estimators of ARMA models are generally found approximately by minimizing the sum $S(\underline{\phi}, \underline{\theta}) = \Sigma\ \varepsilon_t^2$, in which $\underline{\phi}$ and $\underline{\theta}$ represent the parameter sets $\phi_1, \ldots, \phi_p$ and $\theta_1, \ldots, \theta_q$, respectively. The parameters $\hat{\underline{\phi}}$ and $\hat{\underline{\theta}}$ corresponding to the minimum sum will be the maximum likelihood estimators. Then, the estimator of σ_ε^2 is found by $\hat{\sigma}_\varepsilon^2 = (1/N)S(\hat{\underline{\phi}}, \hat{\underline{\theta}})$. A number of algorithms are available for minimizing the sum $S(\underline{\phi}, \underline{\theta})$.[12,15]

Fernandez and Salas[39] gave the moment estimators of the parameters of the GAR model as

$$\hat{\beta} = \frac{4}{\hat{\gamma}^2} \tag{19.3.25}$$

$$\hat{\alpha} = \frac{\sqrt{\hat{\beta}}}{\hat{\sigma}} \tag{19.3.26}$$

$$\hat{\lambda} = \hat{\mu} - \frac{\hat{\beta}}{\hat{\alpha}} \tag{19.3.27}$$

$$\hat{\phi} = \hat{\rho}_1 \tag{19.3.28}$$

in which $\hat{\rho}_1$, $\hat{\sigma}^2$, and $\hat{\gamma}$ are the unbiased estimators of Eqs. (19.2.6), (19.2.7), and (19.2.10), respectively and $\hat{\mu}$ is the unbiased sample mean of Eq. (19.2.1). Given that $\hat{\beta}$, $\hat{\alpha}$, $\hat{\lambda}$, and $\hat{\phi}$ are determined, then the noise term ε of Eq. (19.3.15) is completely specified. Parameter estimation for product models such as Eq. (19.3.18) has been suggested.[17,24,159]

Model Testing and Selection. One of the basic tests for AR and ARMA models is in regard to the assumptions of normality and independence of the noise ε_t. Once the model parameters are determined, the residuals (noise) are found and tested for normality as in Sec. 19.2.5. A common test to determine if the ε's derived from an ARMA(p, q) model are independent is by the *Portemanteau lack of fit test.*[11] It uses the statistic

$$Q = N \sum_{k=1}^{L} r_k^2(\varepsilon) \tag{19.3.29}$$

where $r_k(\varepsilon)$ is the sample autocorrelation function of ε_t and L is the number of lags considered (for instance, $L = 0.25N$), Q is approximately chi-square distributed with $v = L - p - q$ degrees of freedom, and N is the sample size. If $Q < \chi^2_{1-\alpha,v}$, then ε_t is uncorrelated and the model from which the ε's were derived is judged to be an adequate model (α is the significance level of the test). Often, the bounds $\pm\ 1.96/\sqrt{N}$ are determined and one would like to see the $r_k(\varepsilon)$'s within the limits, especially for small k.

The adequacy of a time-series model is often examined by comparing the historical statistics with those derived from the model. The statistics considered are the mean, variance, skewness, and autocovariance, although more thorough comparisons include storage and drought-related statistics as well. If the method of moments is used for parameter estimation, certain statistics such as the mean and variance should be preserved or reproduced by the model. However, this may not necessarily be the case if transformations to normal are used in the modeling process or are due to biases of estimators. Matalas[114] showed that biases resulting from transformations may be important and suggested an estimation scheme to avoid such biases. Likewise, basic historical moments may not be reproduced by the model if maximum likelihood estimates are used. Sometimes, the comparison of the historical and model correlograms is used as the basis of judging the adequacy of a model. For instance, if the ARMA(1, 1) model is the model to be tested, the model correlogram ρ_k of Eq. (19.3.14) is compared with the historical or sample correlogram r_k. One would like to see that the model correlogram resembles the historical correlogram for the model to be adequate. Computer simulation is often used for comparing historical and model statistics. Stedinger and Taylor[176] offer a procedure to follow in using data generation studies for comparing historical and model-generated statistics.

Finally, comparison among models and model selection can be made by using the *Akaike information criteria* (AIC) suggested by Akaike.[1] A modified AIC called AICC has been suggested:[15]

$$\text{AICC}(p, q) = N \ln (\hat{\sigma}_\varepsilon^2) + \frac{2\,(p+q+1)N}{(N-p-q-2)} \tag{19.3.30}$$

in which $\hat{\sigma}_\varepsilon^2$ is the maximum likelihood estimator of the noise variance. The model which minimizes the AICC is selected.

19.3.2 Modeling of Single Periodic Series

PAR Models. Assume that a periodic hydrologic process is represented by $y_{\nu,\tau}$, in which ν defines the year and τ defines the season, such that $\tau = 1, \ldots, \omega$ and ω is the number of seasons in the year. Without loss of generality, τ could represent a day, week, month, or season. A time series defined as[155]

$$y_{\nu,\tau} = \mu_\tau + \sum_{j=1}^{p} \phi_{j,\tau}(y_{\nu,\tau-j} - \mu_{\tau-j}) + \varepsilon_{\nu,\tau} \tag{19.3.31}$$

is called a *periodic autoregressive model* of order p, in which $\varepsilon_{\nu,\tau}$ is an uncorrelated normal variable with mean zero and variance $\sigma_\tau^2(\varepsilon)$, and it is also uncorrelated with $y_{\nu,\tau-1}, \ldots, y_{\nu,\tau-p}$. The model parameters are μ_τ, $\phi_{1,\tau}, \ldots, \phi_{p,\tau}$ and $\sigma_\tau^2(\varepsilon)$ for $\tau = 1, \ldots, \omega$, and the model is often denoted as the PAR(p) or simply the PAR model. Note that in Eq. (19.3.31), if $\tau - j \le 0$, then $y_{\nu,\tau-j}$ becomes $y_{\nu-1,\omega+\tau-j}$ and $\mu_{\tau-j}$ becomes $\mu_{\omega+\tau-j}$. The PAR(1) model arises by making $p = 1$ in Eq. (19.3.31) as

$$y_{\nu,\tau} = \mu_\tau + \phi_{1,\tau}(y_{\nu,\tau-1} - \mu_{\tau-1}) + \varepsilon_{\nu,\tau} \tag{19.3.32}$$

Low-order PAR models such as Eq. (19.3.32) have been widely used in hydrology. For instance, the PAR(1) model was used by Hannan[64] for modeling monthly rainfall series and by Thomas and Fiering[181] for monthly stream-flow simulation. Likewise, PAR(1), PAR(2), and PAR(3) models have been used for simulation of seasonal hydrologic processes.[28,29,32,151]

PARMA Models. One may extend the PAR model [Eq. (19.3.31)] to include periodic moving average parameters. Such a model is the *periodic autoregressive moving average model* or PARMA(p, q) model. Low-order PARMA models are useful for modeling periodic hydrologic time series. For instance, the PARMA(1, 1) model is simply written as

$$y_{\nu,\tau} = \mu_\tau + \phi_{1,\tau}(y_{\nu,\tau-1} - \mu_{\tau-1}) + \varepsilon_{\nu,\tau} - \theta_{1,\tau}\varepsilon_{\nu,\tau-1} \tag{19.3.33}$$

This model has been applied to monthly stream-flow series.[69,155,180] Likewise, PARMA(2, 1) and PARMA(2, 2) models[5] and more complex *multiplicative PARMA models*[121,160] may be needed for stream-flow modeling and simulation when preservation of both seasonal and annual statistics are desired.

Periodic GAR Model. Consider that $y_{\nu,\tau}$ is an autocorrelated variable with three-parameter gamma marginal distribution with location λ_τ, scale α_τ, and shape β_τ parameters varying with τ, $\tau = 1, \ldots, \omega$. The new variable $z_{\nu,\tau} = y_{\nu,\tau} - \lambda_\tau$ is a two-parameter gamma and can be represented by[38]

$$z_{\nu,\tau} = \phi_\tau z_{\nu,\tau-1} + (z_{\nu,\tau-1})^{\delta_\tau} w_{\nu,\tau} \tag{19.3.34}$$

where $z_{\nu,0} = z_{\nu-1,\omega}$, ϕ_τ is a periodic autoregressive coefficient, δ_τ is a periodic autoregressive exponent, and $w_{\nu,\tau}$ is the noise process. The *periodic GAR model* [Eq. (19.3.34)] has a periodic autocorrelation structure equivalent to that of the PAR(1) process. Refer to Fernandez and Salas[38] for properties and applications of the periodic GAR model.

Periodic Product Models. Intermittent hydrologic time series which are periodic and correlated can be modeled by[159]

$$y_{\nu,\tau} = B_{\nu,\tau} z_{\nu,\tau} \tag{19.3.35}$$

where $y_{\nu,\tau}$ is an intermittent periodic autocorrelated process, $B_{\nu,\tau}$ is a periodic autocorrelated Bernoulli (1, 0) process, and $z_{\nu,\tau}$ may be either an uncorrelated or correlated periodic process with a specified marginal distribution. Furthermore, the processes B and z are assumed to be mutually uncorrelated. Refer to Salas and Chebaane[159] and Chebaane et al.[25] for properties and applications of such *periodic product models* for stream-flow modeling and simulation.

Parameter Estimation. Estimation by method of moments and approximate maximum likelihood is available for the various periodic models included in this section. For instance, for the PAR(1) model of Eq. (19.3.32), the parameters μ_τ, ϕ_τ, and $\sigma_\tau^2(\varepsilon)$ may be estimated by

$$\hat{\mu}_\tau = \bar{y}_\tau \tag{19.3.36}$$

$$\hat{\phi}_{1,\tau} = \left(\frac{s_\tau}{s_{\tau-1}}\right) r_{1,\tau} \tag{19.3.37}$$

$$\hat{\sigma}_\tau^2(\varepsilon) = s_\tau^2 - s_{\tau-1}^2\, r_{1,\tau}^2 \tag{19.3.38}$$

where $\bar{y}_\tau$, s_τ, and $r_{1,\tau}$ are the sample seasonal mean, seasonal standard deviation, and lag-1 season-to-season correlation coefficient, respectively.

Likewise, for the PARMA(1, 1) model of Eq. (19.3.33), the moment estimators are[157]

$$\hat{\phi}_{1,\tau} = \frac{c_{2,\tau}}{c_{1,\tau-1}} \tag{19.3.39}$$

$$\hat{\theta}_{1,\tau} = \hat{\phi}_{1,\tau} + \frac{(s_\tau^2 - \hat{\phi}_{1,\tau} c_{1,\tau})}{(\hat{\phi}_{1,\tau} s_{\tau-1}^2 - c_{1,\tau})} - \frac{(\hat{\phi}_{1,\tau+1} s_\tau^2 - c_{1,\tau+1})}{(\hat{\phi}_{1,\tau} s_{\tau-1}^2 - c_{1,\tau})\hat{\theta}_{1,\tau+1}} \tag{19.3.40}$$

$$\sigma_\tau^2(\varepsilon) = \frac{\hat{\phi}_{1,\tau+1} s_{\tau-1}^2 - c_{1,\tau+1}}{\hat{\theta}_{1,\tau+1}} \tag{19.3.41}$$

for $\tau = 1, \ldots, \omega$, where $c_{k,\tau}$ is determined by Eq. (19.2.13). Equation (19.3.40) is a system of equations that must be solved simultaneously to obtain $\theta_{1,\tau}$, $\tau = 1, \ldots, \omega$. Note that in Eqs. (19.3.37) through (19.3.41), $\tau - 1$ is to be interpreted as ω when $\tau = 1$ and as $\tau + 1$ when $\tau = \omega$. Moment estimation procedures for higher-order PAR and PARMA models are also available.[155,157]

In the method of maximum likelihood, the sum of the square residuals $S = \sum_{\nu=1}^{N} \sum_{\tau=1}^{\omega} \varepsilon_{\nu,\tau}^2$ in which $\varepsilon_{\nu,\tau}$, is the noise term of the PAR or PARMA model under consideration, is minimized to obtain the approximate maximum likelihood estimators $\hat{\phi}_{1,\tau}, \ldots, \hat{\phi}_{p,\tau}$ and $\hat{\theta}_{1,\tau}, \ldots, \hat{\theta}_{q,\tau}$. Then, the noise variance $\sigma_\tau^2(\varepsilon)$ can be obtained by $\hat{\sigma}_\tau^2(\varepsilon) = (1/N) \sum_{\nu=1}^{N} \varepsilon_{\nu,\tau}^2$, $\tau = 1, \ldots, \omega$ in which the noises are evaluated from the PAR or PARMA model equations, as the case may be.

Moment estimators of the parameters of the periodic GAR model of Eq. (19.3.34) are available.[38] Likewise, parameter estimation procedures for periodic product models are also available.[25,159]

Model Testing. Testing of PAR and PARMA models can be done by testing the basic assumptions of the models, i.e., that the noise $\varepsilon_{\nu,\tau}$ is uncorrelated and normal. After the residuals $\varepsilon_{\nu,\tau}$ are determined, the season-to--season correlations $r_{1,\tau}(\varepsilon)$, $\tau = 1, \ldots, \omega$ can be obtained from Eq. (19.2.12) and it can be verified that the r's fall within the bounds $\pm 1.96/\sqrt{N}$. Likewise, the residuals can be tested for normality on a seasonal basis. The model may be tested for adequacy to preserve certain historical statistics such as $\bar{y}_\tau$, s_τ^2, g_τ, *and* $r_{k,\tau}$ $(k = 1, \ldots)$. If the method of moments is used to estimate the parameters and the original data are approximately normally distributed, then the model must be able to reproduce such basic historical statistics. However, if transformations are used to make the original data normally distributed, then the model based on the method of moments generally will not reproduce the original statistics $\bar{y}_\tau$, s_τ^2, g_τ, and $r_{k,\tau}$, although it will reproduce similar statistics in the transformed domain. However, for the log-transformation and the PAR(1) model, it is possible to reproduce the basic statistics in the original domain by a procedure suggested by Burges.[18] Likewise, one may be interested to see the capability of a given model to reproduce statistics at aggregated time scales, typically at the annual time scale. Such statistics may be the annual correlation structure and storage- and drought-related statistics. In this case, data generation experiments are generally made.[155,176]

19.3.3 Modeling of Multiple Stationary Series

Analysis and modeling of multiple time series are widely needed in hydrology. For instance, one may like to model precipitation series at several sites in a river basin, stream-flow data recorded at several gauging stations in the stream network, or a mix of precipitation with stream-flow data recorded at various sites. Analysis and modeling procedures for multiple series are more involved than for a single series. In analyzing multiple series, vector and matrix notations are needed. However, the basic principles are similar, and we will still be referring to means, variances, and covariances, but in vector and matrix forms. We will start in this section with modeling of stationary series and the following section will deal with periodic series.

Multivariate AR and Multivariate ARMA Models. Consider a multiple time series $\mathbf{Y}_t$, a column vector with elements $y_t^{(1)}, \ldots, y_y^{(n)}$ in which n is the number of series (number of sites or number of variables) under consideration. The *multivariate AR(1) model* suggested by Matalas[114] is defined as

$$\mathbf{Z}_t = \mathbf{A}_1\mathbf{Z}_{t-1} + \mathbf{B}\boldsymbol{\varepsilon}_t \tag{19.3.42}$$

in which $\mathbf{Z}_t = \mathbf{Y}_t - \boldsymbol{\mu}$, $\mathbf{A}_1$ and $\mathbf{B}$ are n- by n-parameter matrices and $\boldsymbol{\mu}$ is a column parameter vector with elements $\mu^{(1)}, \ldots, \mu^{(n)}$. The noise term $\boldsymbol{\varepsilon}_t$ is also a column vector of noises $\varepsilon_t^{(1)}, \ldots, \varepsilon_t^{(n)}$, each with zero mean such that $\mathbf{E}(\boldsymbol{\varepsilon}_t\boldsymbol{\varepsilon}_t^T) = \mathbf{I}$, where $\mathbf{T}$ denotes the transpose of the matrix and $\mathbf{I}$ is the identity matrix, and $\mathbf{E}(\boldsymbol{\varepsilon}_t\boldsymbol{\varepsilon}_{t-k}^T) = 0$ for $k \neq 0$. In addition, it is assumed that $\boldsymbol{\varepsilon}_t$ is uncorrelated with $\mathbf{Z}_{t-1}$ and $\boldsymbol{\varepsilon}_t$ is normally distributed. Model (19.3.42) has been widely used in operational hydrology. Higher-order multivariate AR models are also available.[134,155] Likewise, the *multivariate ARMA(1, 1) model* is written as[130,155]

$$\mathbf{Z}_t = \mathbf{A}_1\mathbf{Z}_{t-1} + \mathbf{B}\boldsymbol{\varepsilon}_t - \mathbf{C}_1\boldsymbol{\varepsilon}_{t-1} \tag{19.3.43}$$

in which $\mathbf{C}_1$ is an additional n- by n-parameter matrix.

Contemporaneous AR and ARMA Models. Using the full multivariate AR and multivariate ARMA models as described above often leads to complex parameter estimation, especially for the last model. Thus, model simplifications have been suggested. For instance, a simpler model will result from Eq. (19.3.42) if $\mathbf{A}_1$ is assumed to be a diagonal matrix.[114] In general, a *contemporaneous ARMA* (CARMA) model results if the matrices $\mathbf{A}_1$ and $\mathbf{C}_1$ of Eq. (19.3.43) are considered to be diagonal matrices.[22,155,158] In this case, model (19.3.43) implies a contemporaneous relationship in that only the dependence of concurrent values of the y's are considered important. Furthermore, the diagonalization of the parameter matrices allows model decoupling into component univariate models so that the model parameters do not have to be estimated jointly, and univariate modeling procedures can be employed.

To illustrate the foregoing concept, let a multivariate ARMA(p, q) process be

$$\mathbf{Z}_t = \sum_{j=1}^{p} \mathbf{A}_j\mathbf{Z}_{t-j} + \boldsymbol{\varepsilon}_t - \sum_{j=1}^{q} \mathbf{C}_j\boldsymbol{\varepsilon}_{t-j} \tag{19.3.44}$$

Assuming that matrices $\mathbf{A}_j$ and $\mathbf{C}_j$ are diagonals, model (19.3.44) can be decoupled into the model components

$$z_t^{(i)} = \sum_{j=1}^{p} a_j^{(i)} z_{t-j}^{(i)} + \varepsilon_t^{(i)} - \sum_{j=1}^{q} c_j^{(i)}\, \varepsilon_{t-j}^{(i)} \tag{19.3.45}$$

for $i = 1, \ldots, n$. Thus, the model components at each site are simply univariate ARMA(p, q) models where each $\varepsilon_t^{(i)}$, $i = 1, \ldots, n$, is uncorrelated, but are contemporaneously correlated with a variance-covariance matrix $\mathbf{G}$. Thus, the parameters a and c in each model, can be estimated by using univariate estimation procedures and the ε's can be modeled by

$$\boldsymbol{\varepsilon}_t = \mathbf{B}\boldsymbol{\xi}_t \tag{19.3.46}$$

in which $\boldsymbol{\xi}$ is normal, such that $\mathbf{E}(\boldsymbol{\xi}_t\boldsymbol{\xi}_t^T) = \mathbf{I}$ and $\mathbf{E}(\boldsymbol{\xi}_t\boldsymbol{\xi}_{t-k}) = \mathbf{0}$ and $k \neq 0$. Note that one does not have to consider the same univariate ARMA(p, q) model for each site.

Parameter Estimation. Moment estimators for the multivariate AR(1) model (19.3.42) are[114]

$$\hat{\mathbf{A}}_1 = \hat{\mathbf{M}}_1\hat{\mathbf{M}}_0^{-1} \tag{19.3.47}$$

$$\hat{\mathbf{B}}\,\hat{\mathbf{B}}^T = \hat{\mathbf{M}}_0 - \hat{\mathbf{A}}_1\hat{\mathbf{M}}_1^T \tag{19.3.48}$$

in which $\hat{\mathbf{M}}_0$ and $\hat{\mathbf{M}}_1$ are the lag-zero and lag-one cross-covariance matrices of the multivariate series Z_t whose elements are determined by Eq. (19.2.15). Equation (19.3.48) gives estimates of the product $\hat{\mathbf{B}}\,\hat{\mathbf{B}}^T = \hat{\mathbf{D}}$ where $\hat{\mathbf{D}}$ is the right-hand side of (19.3.48). This matrix equation can be solved for $\hat{\mathbf{B}}$ by *principal component analysis*[114] or by *square root procedure.*[208] Matrices $\hat{\mathbf{A}}_1$ and $\hat{\mathbf{B}}$ can be solved if $\hat{\mathbf{M}}_0$ and $\hat{\mathbf{M}}_1$ satisfy certain conditions. First of all, $\hat{\mathbf{M}}_0$ must be a positive definite matrix. This is generally satisfied when the sample sizes for all sites are the same. A technique that ensures that $\hat{\mathbf{M}}_0$ will be positive definite when the sample sizes are different is available.[30] In addition, matrix $\hat{\mathbf{B}}\,\hat{\mathbf{B}}^T = \hat{\mathbf{M}}_0 - \hat{\mathbf{A}}_1\hat{\mathbf{M}}_1^T$ must be positive definite. Estimators of $\hat{\mathbf{M}}_0$ and $\hat{\mathbf{M}}_1$ which ensure that $\hat{\mathbf{B}}\,\hat{\mathbf{B}}^T$ will be positive definite are also available.[30] With large matrices and with transformed data, numerical errors may still give an inconsistent $\mathbf{B}\,\mathbf{B}^T$ matrix. In such cases $\mathbf{B}\,\mathbf{B}^T$ can be adjusted.[13,119]

One can solve the matrix equation $\mathbf{B}\mathbf{B}^T = \mathbf{D}$ by using the square root method.[58] This method assumes that $\mathbf{B}$ is a *lower triangular matrix* (above diagonal elements are zero) and requires that $\mathbf{D}$ is a positive definite matrix. In such case the elements b^{ij} of $\mathbf{B}$ are

$$b^{ij} = \frac{d^{ji}}{b^{ij}} \qquad j = 1, i = 1, \ldots, n \tag{19.3.49a}$$

$$b^{ii} = \left[d^{ii} - \sum_{k=1}^{i-1} (b^{ik})^{1/2}\right]^{1/2} \qquad i = 2, \ldots, n \tag{19.3.49b}$$

$$b^{ij} = \frac{d^{ij} - \sum_{k=1}^{j-1} b^{ik}\, b^{jk}}{b^{jj}} \qquad j = 2, \ldots, n-1, i > j \tag{19.3.49c}$$

and $b^{ij} = 0$ for $i < j$, in which d^{ij}, $i, j = 1, \ldots, n$, are the elements of $\mathbf{D}$. In cases where $\mathbf{D}$ is a positive semidefinite matrix, a method based on principal components is needed.[13] An alternative method for solving $\mathbf{B}\mathbf{B}^T = \mathbf{D}$ which works for $\mathbf{D}$ positive definite or positive semidefinite is also available.[155]

The estimation difficulties normally encountered with the full multivariate ARMA models can be overcome by using CARMA models. Since the CARMA model (19.3.44) can be decoupled as in Eq. (19.3.45), univariate estimation procedures by the method of moments or maximum likelihood can be used to estimate the a's and c's. The elements of matrix $\hat{\mathbf{G}}$ for the CARMA(1, 1) model can be estimated as[178]

$$\hat{g}^{(ij)} = \frac{m_0^{(ij)}\,(1 - \hat{a}^{(i)}\,\hat{a}^{(j)})}{1 - \hat{a}^{(i)}\,\hat{c}^{(j)} - \hat{a}^{(j)}\,\hat{c}^{(i)} + \hat{c}^{(i)}\,\hat{c}^{(j)}} \tag{19.3.50}$$

for $i, j = 1, \ldots, n$, in which $m_0^{(ij)}$ is the ijth element of $\hat{\mathbf{M}}_0$, the lag-zero cross-covariance matrix of $\mathbf{Z}_t$. Finally, $\mathbf{B}$ of Eq. (19.3.46) will be estimated by solving $\hat{\mathbf{B}}\hat{\mathbf{B}}^{\mathrm{T}} = \hat{\mathbf{G}}$.

Model Testing. Model testing depends on the type of multivariate model considered. In the case of the full multivariate AR(p) and ARMA(p, q) models, one may test the assumption of normality of residuals $\boldsymbol{\varepsilon}_t$ and the assumptions $\mathbf{E}(\boldsymbol{\varepsilon}_t\boldsymbol{\varepsilon}_t^T) = \mathbf{I}$ and $\mathbf{E}(\boldsymbol{\varepsilon}_t\boldsymbol{\varepsilon}_{t-k}^T) = \mathbf{0}$ for $k \neq 0$. Thus, once the residuals $\varepsilon_t^{(i)}$, $i = 1, \ldots, n$, are found, one should test that $\hat{\mathbf{M}}_0(\varepsilon) = \mathbf{I}$ and that, at least, $\hat{\mathbf{M}}_1(\varepsilon) = \mathbf{0}$. The elements of these matrices [except the diagonal of $\mathbf{M}_0(\varepsilon)$] must be within the limits of $\pm 1.96/\sqrt{N}$. Likewise, in the case of CARMA models, one can make similar tests as above for the residual variable ξ of Eq. (19.3.46). Testing whether a given model reproduces historical statistics is often performed by data generation procedures.[155]

19.3.4 Modeling of Multiple Periodic Series

Multivariate PAR and Multivariate PARMA Models. The *multivariate PAR(1) model* is given by[152]

$$\mathbf{Z}_{\nu,\tau} = \mathbf{A}_\tau \mathbf{Z}_{\nu,\tau-1} + \mathbf{B}_\tau \boldsymbol{\varepsilon}_{\nu,\tau} \tag{19.3.51}$$

in which $\mathbf{Z}_{\nu,\tau} = \mathbf{Y}_{\nu,\tau} - \boldsymbol{\mu}_\tau$, $\mathbf{A}_\tau$ and $\mathbf{B}_\tau$ are n- by n-parameter matrices, and $\boldsymbol{\mu}_\tau$ is a column parameter vector with elements $\mu_\tau^{(1)}, \ldots, \mu_\tau^{(n)}$. All parameters $\boldsymbol{\mu}_\tau$, $\mathbf{A}_\tau$, and $\mathbf{B}_\tau$ are periodic. The noise term $\boldsymbol{\varepsilon}_{\nu,\tau}$ is a column vector normally distributed with mean zero, and $\mathbf{E}(\boldsymbol{\varepsilon}_{\nu,\tau}\boldsymbol{\varepsilon}_{\nu,\tau}^T) = \mathbf{I}$ and $\mathbf{E}(\boldsymbol{\varepsilon}_{\nu,\tau}\boldsymbol{\varepsilon}_{\nu,\tau-k}^T) = \mathbf{0}$ for $k \neq 0$. In addition, it is assumed that $\boldsymbol{\varepsilon}_{\nu,\tau}$ is uncorrelated with $\mathbf{Z}_{\nu,\tau-1}$. This model has been widely used for generating seasonal hydrologic processes. Likewise, the *multivariate PARMA(1, 1) model* is written as[63,158]

$$\mathbf{Z}_{\nu,\tau} = \mathbf{A}_\tau \mathbf{Z}_{\nu,\tau-1} + \mathbf{B}_\tau \boldsymbol{\varepsilon}_{\nu,\tau} - \mathbf{C}_\tau \boldsymbol{\varepsilon}_{\nu,\tau-1} \tag{19.3.52}$$

in which $\mathbf{C}_\tau$ is an additional n by n periodic matrix parameter.

Contemporaneous PAR and PARMA Models. Simplifications of the foregoing models can be made. For instance, $\mathbf{A}_\tau$ of Eq. (19.3.51) can be made diagonal. In the case of the multivariate PARMA(1, 1) model of Eq. (19.3.52), it is more convenient to write the model as[4]

$$\mathbf{Z}_{\nu,\tau} = \mathbf{A}_\tau \mathbf{Z}_{\nu,\tau-1} + \boldsymbol{\varepsilon}_{\nu,\tau} - \mathbf{C}_\tau \boldsymbol{\varepsilon}_{\nu,\tau-1} \tag{19.3.53}$$

and consider that $\mathbf{A}_\tau$ and $\mathbf{C}_\tau$ are diagonal matrices. Then, the model can be decoupled into univariate models for each site. However, to maintain cross-correlation among sites, the vector $\boldsymbol{\varepsilon}_{\nu,\tau}$ will be assumed to have a variance-covariance matrix $\mathbf{G}_\tau$ or $\mathbf{E}(\boldsymbol{\varepsilon}_{\nu,\tau}\boldsymbol{\varepsilon}_{\nu,\tau}^T) = \mathbf{G}_\tau$ and $\mathbf{E}(\boldsymbol{\varepsilon}_{\nu,\tau}\boldsymbol{\varepsilon}_{\nu,\tau-k}^T) = 0$ for $k \neq 0$. Then, $\boldsymbol{\varepsilon}_{\nu,\tau}$ can be modeled as

$$\boldsymbol{\varepsilon}_{\nu,\tau} = \mathbf{B}_\tau \boldsymbol{\xi}_{\nu,\tau} \tag{19.3.54}$$

such that $\mathbf{E}(\boldsymbol{\xi}_{\nu,\tau}\boldsymbol{\xi}_{\nu,\tau}^T) = \mathbf{I}$ and $\mathbf{E}(\boldsymbol{\xi}_{\nu,\tau}\boldsymbol{\xi}_{\nu,\tau-k}^T) = 0$ for $k \neq 0$. The foregoing modeling scheme is a *contemporaneous PARMA(1, 1) model.* Similar simplifications can be made for higher-order models.

Parameter Estimation. Moment estimators of the parameters of the multivariate PAR(1) model (19.3.51) are[155]

$$\hat{\mathbf{A}}_\tau = \hat{\mathbf{M}}_{1,\tau} \hat{\mathbf{M}}_{0,\tau-1}^{-1} \tag{19.3.55}$$

$$\hat{\mathbf{B}}_\tau \hat{\mathbf{B}}_\tau^T = \hat{\mathbf{M}}_{0,\tau} - \hat{\mathbf{A}}_\tau \hat{\mathbf{M}}_{1,\tau}^T \tag{19.3.56}$$

in which $\hat{\mathbf{M}}_{k,\tau}$ $(k = 0, 1)$ are sample season-to-season covariance matrices of $\mathbf{Z}_{\nu,\tau}$. Matrix $\hat{\mathbf{A}}_\tau$ is determined directly while $\hat{\mathbf{B}}_\tau$ must be found by the square root method. Estimation for the multivariate PARMA(1, 1) model is more complex.[63,158]

An alternative is to use the contemporaneous PARMA(1, 1) model of Eq. (19.3.53). Since it allows decoupling, univariate procedures can be used to estimate the parameters $a_\tau^{(i)}$ and $c_\tau^{(i)}$, $i = 1, \ldots, n$ (elements of $\mathbf{A}_\tau$ and $\mathbf{C}_\tau$). Then, the elements g_τ^{ij} of matrix $\mathbf{G}_\tau$ may be obtained as[63]

$$g_\tau^{(ij)} = e_\tau^{(ij)} + f_\tau^{(ij)} g_{\tau-1}^{(ij)} \tag{19.3.57a}$$

$$e_\tau^{(ij)} = m_{0,\tau}^{(ij)} - a_\tau^{(i)} m_{0,\tau-1}^{(ij)} a_\tau^{(j)} \tag{19.3.57b}$$

$$f_\tau^{(ij)} = a_\tau^{(i)} c_\tau^{(j)} + a_\tau^{(i)} c_\tau^{(i)} - c_\tau^{(i)} c_\tau^{(j)} \tag{19.3.57c}$$

where $m_{0,\tau}^{(ij)}$, $i, j = 1, \ldots, n$ are the elements of $\hat{\mathbf{M}}_{0,\tau}$. Once matrix $\mathbf{G}_\tau$ is determined, then $\mathbf{B}_\tau$ of Eq. (19.3.54) can be determined from $\hat{\mathbf{B}}_\tau \mathbf{B}_\tau^T = \mathbf{G}_\tau$.

Model Testing. Model testing for multivariate periodic models is similar to testing of multivariate stationary models except that periodicity must be considered. For instance, for full multivariate PAR or PARMA models, one can verify that the residuals $\boldsymbol{\varepsilon}_{v,\tau}$ are normally distributed and check the assumptions $\mathbf{E}(\boldsymbol{\varepsilon}_{v,\tau}\, \boldsymbol{\varepsilon}_{v,\tau}^T) = \mathbf{I}$ and $\mathbf{E}(\boldsymbol{\varepsilon}_{v,\tau}\, \boldsymbol{\varepsilon}_{v,\tau-k}^T) = \mathbf{0}$ for $k \neq 0$. Likewise, for contemporaneous PARMA models, one can make similar tests for the residuals $\boldsymbol{\xi}_{v,\tau}$ of Eq. (19.3.54). Additionally, one can do further testing by using data generation procedures.

19.3.5 Disaggregation of Annual to Seasonal Series

Generally, modeling of seasonal hydrologic time series is geared to preserving seasonal statistics only, while statistics at other levels of aggregation, such as annual statistics, may not be preserved. For instance, if the PAR(1) model is used to generate monthly flows, the historical monthly statistics are usually preserved, yet if such generated monthly flows are aggregated to obtain the corresponding annual flows, there is no assurance that the historical annual statistics will be preserved. *Disaggregation models* have been developed for reproducing statistics at more than one level of aggregation. Disaggregation models can be used for both temporal and spatial disaggregation; however, the models in this section are mostly described in terms of temporal disaggregation.

Traditional Valencia-Schaake Model. Assume that $\mathbf{X}$ and $\mathbf{Y}$ are normalized variables with mean zero. The basic form of the disaggregation model suggested by Valencia and Schaake is[189]

$$\mathbf{Y} = \mathbf{AX} + \mathbf{B}\boldsymbol{\varepsilon} \tag{19.3.58}$$

where $\mathbf{X}$ is an n vector of annual values at n sites, $\mathbf{Y}$ is an $n\omega$ vector of seasonal values in which ω is the number of seasons in the year, $\mathbf{A}$ and $\mathbf{B}$ are $n\omega$- by $n\omega$-parameter matrices, and $\boldsymbol{\varepsilon}$ is an $n\omega$ vector of independent standard normal variables. A desirable property of disaggregation models is additivity, i.e., the sum of the seasonal values must add up to the annual values. The parameters $\mathbf{A}$ and $\mathbf{B}$ may be estimated by[189]

$$\hat{\mathbf{A}} = \mathbf{S}_{YX}\mathbf{S}_{XX}^{-1} \tag{19.3.59}$$

$$\hat{\mathbf{B}}\hat{\mathbf{B}}^T = \mathbf{S}_{YY} - \mathbf{A}\mathbf{S}_{XY} \tag{19.3.60}$$

in which $\mathbf{S}_{UV}$ represents the sample covariance of the vectors $\mathbf{U}$ and $\mathbf{V}$. In the foregoing formulation, it is assumed that $\mathbf{X}$ (say the annual series) has been previously generated by a specified model such as the AR(1) or ARMA(1, 1) process.

Valencia-Schaake's model does not preserve the covariances of the first season of a year and any preceding season. To circumvent this, Eq. (19.3.58) is modified as[120]

$$\mathbf{Y} = \mathbf{AX} + \mathbf{B}\boldsymbol{\varepsilon} + \mathbf{CZ} \tag{19.3.61}$$

where **C** is a new parameter matrix and **Z** is a vector of seasonal values from the previous year for each site. Usually, **Z** is a vector containing only the last season of the previous year in which case **C** is an $n\omega$ by n matrix. **A**, **B**, and **C** may be estimated by[99]

$$\hat{\mathbf{A}} = (\mathbf{S}_{YX} - \mathbf{S}^*_{YZ}\mathbf{S}^{-1}_{ZZ}\mathbf{S}^*_{ZX})\,(\mathbf{S}_{XX} - \mathbf{S}^*_{XZ}\mathbf{S}^{-1}_{ZZ}\mathbf{S}^*_{ZX})^{-1} \tag{19.3.62a}$$

$$\hat{\mathbf{C}} = (\mathbf{S}^*_{YZ} - \hat{\mathbf{A}}\mathbf{S}^*_{XZ})\mathbf{S}^{-1}_{ZZ} \tag{19.3.62b}$$

$$\hat{\mathbf{B}}\hat{\mathbf{B}}^T = \mathbf{S}_{YY} - \hat{\mathbf{A}}\mathbf{S}_{XY} - \hat{\mathbf{C}}\mathbf{S}^*_{ZY} \tag{19.3.62c}$$

where $\mathbf{S}^*_{XZ} = \mathbf{S}_{XX'}\mathbf{S}^{-1}_{XX}\mathbf{S}_{X'Z}$, $\mathbf{S}^*_{YZ} = \mathbf{S}_{YZ} + \mathbf{S}_{YX}\mathbf{S}^{-1}_{XX}(\mathbf{S}^*_{XZ} - \mathbf{S}_{XZ})$ and $\mathbf{X}'$ is the vector of the previous year. With **A**, **B**, and **C** thus estimated, model (19.3.61) preserves the covariances $\mathbf{S}_{YY}$ and $\mathbf{S}_{YX}$ as well as the additivity property. However, Eqs. (19.3.62) assume an annual model which reproduces $\mathbf{S}_{XX}$ and $\mathbf{S}_{XX'}$.

On the other hand, a scheme which does not depend on the annual model's structure can be formulated as[177]

$$\mathbf{Y}_\tau = \mathbf{AX}_t + \boldsymbol{\varepsilon}_t \tag{19.3.63a}$$

$$\boldsymbol{\varepsilon}_t = \mathbf{C}\boldsymbol{\varepsilon}_{t-1} + \boldsymbol{\xi}_t \tag{19.3.63b}$$

in which the subscript t is now introduced, ε_t is independent of $\mathbf{X}_t$, and $\boldsymbol{\xi}_t$ is a random component with covariance matrix parameter $\mathbf{S}_{\xi\xi}$. The parameters may be estimated by

$$\hat{\mathbf{A}} = \mathbf{S}_{YX}\mathbf{S}^{-1}_{XX} \tag{19.3.64a}$$

$$\hat{\mathbf{C}} = \mathbf{S}_{\varepsilon\varepsilon'}\mathbf{S}^{-1}_{\varepsilon\varepsilon} \tag{19.3.64b}$$

$$\mathbf{S}_{\xi\xi} = \mathbf{S}_{\varepsilon\varepsilon} - \hat{\mathbf{C}}\mathbf{S}^T_{\varepsilon\varepsilon}\hat{\mathbf{C}}^T \tag{19.3.64c}$$

in which $\mathbf{S}_{\varepsilon\varepsilon} = \mathbf{S}_{YY} - \hat{\mathbf{A}}\mathbf{S}_{XY}$ and $\boldsymbol{\varepsilon}' = \boldsymbol{\varepsilon}_{t-1}$. The model scheme reproduces the moments $\mathbf{S}_{YY}$, $\mathbf{S}_{YX}$, and $\mathbf{S}_{XX}$.

Lane's Condensed Model. The foregoing disaggregation models have too many parameters, a problem which may be significant, especially when the number of sites is large and the sample size is small. Lane[98] sets to zero some parameters of model (19.3.61). Thus

$$\mathbf{Y}_\tau = \mathbf{A}_\tau\mathbf{X} + \mathbf{B}_\tau\boldsymbol{\varepsilon} + \mathbf{C}_\tau\mathbf{Y}_{\tau-1} \qquad \tau = 1, \ldots, \omega \tag{19.3.65}$$

is a model in which the number of parameters is reduced considerably (ω sets of parameters $\mathbf{A}_\tau$, $\mathbf{B}_\tau$, and $\mathbf{C}_\tau$). However, a shortcoming is that the additivity property is lost because the model is applied separately to each season. This shortcoming can be avoided by adjusting the seasonal values so that they add exactly to the annual values at each site. The estimation of model parameters and approximate adjustments can be found in Lane.[98]

Santos-Salas Step Disaggregation Model. In some of the disaggregation approaches, it is necessary to solve $\hat{\mathbf{B}}\hat{\mathbf{B}}^T = \hat{\mathbf{D}}$ for $\hat{\mathbf{B}}$. Since $\hat{\mathbf{B}}\hat{\mathbf{B}}^T$ should be positive semidefinite (because of the additivity property), the principal components technique is usually followed. However, when the matrices involved are large (which is typical for multisite seasonal disaggregation), the solution for $\hat{\mathbf{B}}$ usually deteriorates, and large computer storage capacity is required.[13] The disaggregation problem can be made computationally more amenable if it can be done in steps (stages or cascades); thus the size of the matrices involved will decrease and, consequently, so will the number of parameters. For instance, Fig. 19.3.2*a* schematically shows that annual flows are disaggregated into monthly flows directly in one step (this is the usual approach), while Fig. 19.3.2*b* shows such disaggregation is performed in two steps, into quarterly flows in the first step, then each quarterly flow is further disaggregated into monthly flows in the second step. However, even in the latter approach, large matrices will result when the number of seasons and the number of sites are large.

Santos and Salas[165,166] proposed a step disaggregation in which, at each step, the disaggregation is always into two parts or two seasons. For instance, Fig. 19.3.3 shows that the yearly value $\mathbf{X}_\nu$ is disaggregated into 12 monthly values by first disaggregating the year into the first month $\mathbf{Y}_{\nu,1}$ and the sum of the remaining 11 months $\sum_{\tau=2}^{12} \mathbf{Y}_{\nu,\tau}$. Then, this latter sum is disaggregated into the second month $\mathbf{Y}_{\nu,2}$ and the sum of the remaining 10 months $\sum_{\tau=3}^{12} \mathbf{Y}_{\nu,\tau}$ and so on until the months $\mathbf{Y}_{\nu,11}$ and $\mathbf{Y}_{\nu,12}$ are obtained in the eleventh disaggregation step. This stepwise disaggregation scheme leads to a maximum parameter matrix size of 2×2 for single site disaggregation and $2n \times 2n$ for multisite disaggregation. Note that the seasonal covariance between flows is preserved if model (19.3.61) is applied in each step. The step disaggregation model as

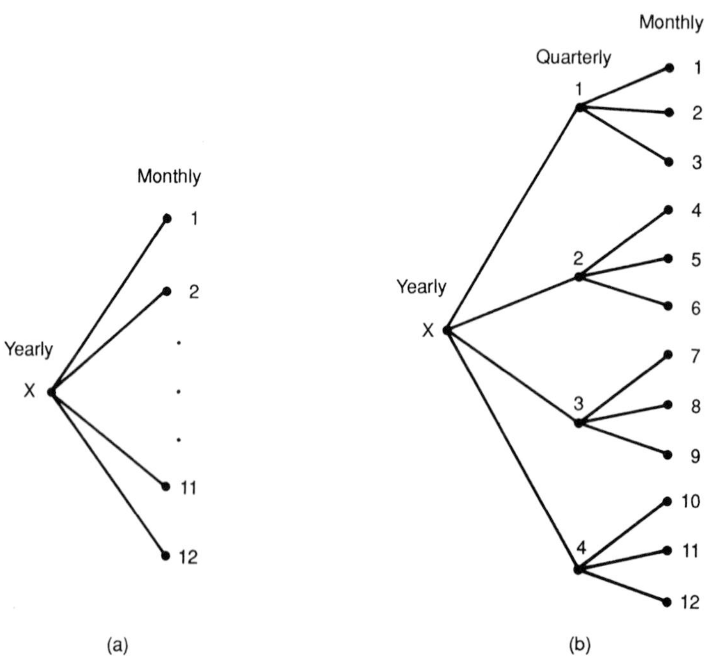

FIGURE 19.3.2 Disaggregation of annual flows X into monthly flows (*a*) in one step, and (*b*) in two steps, first into quarterly flows, then into monthly flows.

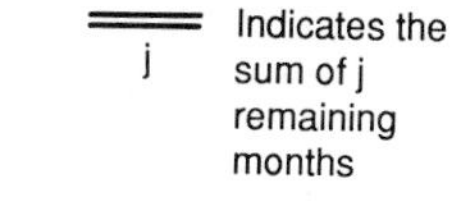

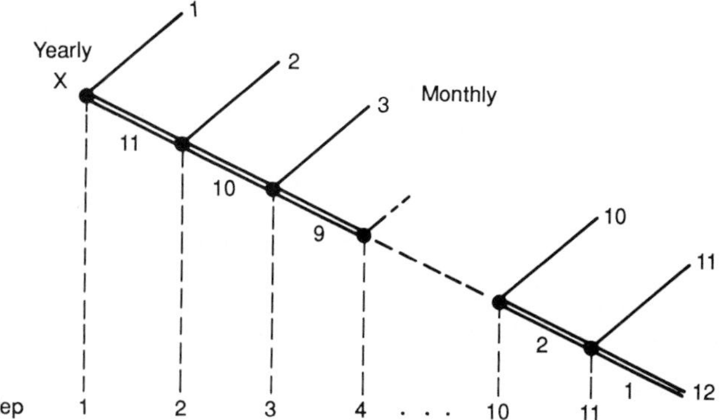

FIGURE 19.3.3 Step disaggregation of annual flows X into monthly flows in 11 steps. First step disaggregates into month 1 and the sum of months 2–12, the second step pertains to month 2 and sum of months 3–12, and so on.[166]

suggested above has the advantage over the previous models in that it has the minimum size of matrices involved, thus the smallest number of parameters, while keeping the additivity property. However, it has the same drawback of all previous models when the underlying variables are not normally distributed. However, the approach proposed by Todini[182] can be implemented in conjunction with the step disaggregation model for skewed variables.

Stedinger, Pei, and Cohn's Stagewise Models. Stedinger et al.[179] suggested a single site disaggregation model which reproduces seasonal statistics and the covariance of seasonal flows with annual flows, assuming lognormally distributed seasonal flows and lognormally distributed annual flows. Consider that $y_{v,\tau} = \log (Q_{v,\tau} - q_\tau)$ is normally distributed seasonal flows, $Q_{v,\tau}$ is the original seasonal flows, and q_τ is the lower bound for season τ. In addition, consider that $x_v = \log (Q_v - q)$ is also normally distributed, Q_v is the original annual flows, and q is the lower bound. The model to generate $y_{v,\tau}$ may be written as[179]

$$y_{v,1} = a_1 + b_1 x_v + \varepsilon_{v,1} \tag{19.3.66a}$$

$$y_{v,2} = a_2 + b_2 x_v + d_2 y_{v,1} + \varepsilon_{v,2} \tag{19.3.66b}$$

$$y_{v,\tau} = a_\tau + b_\tau x_v + c_\tau y_{v,\tau-1} + d_\tau \sum_{j=1}^{\tau-1} w_\tau y_{v,j} + \varepsilon_{v,\tau} \quad \tau = 3, \ldots, \omega \tag{19.3.66c}$$

where $\varepsilon_{v,\tau}$ are uncorrelated zero-mean normal random variables. The term $\Sigma\, w_\tau y_{v,j}$ allows for reproduction of the variance of the first-order approximation of Q_v, the actual lognormal annual flow, and $w_\tau = \exp(\mu_\tau + 0.5\,\sigma_\tau^2)$, in which μ_τ and σ_τ^2 are the mean and variance of $y_{v,\tau}$. The seasonal variables $y_{v,\tau}$ from Eq. (19.3.66) will produce a generated annual flow equal to $Q_v^* = \sum_{\tau=1}^{\omega} [q_\tau + \exp(y_{v,\tau})]$ which will not be equal to the original generated annual flow Q_v. Then, the seasonal flows can be adjusted as

$$Q_{v,\tau} = \left(\frac{Q_v}{Q_v^*}\right)[q_\tau + \exp(y_{v,\tau})] \tag{19.3.67}$$

in which case the sum of $Q_{v,1}, \ldots, Q_{v,\omega}$ will be exactly equal to Q_v.

Model (19.3.66) essentially amounts to adding an extra term to Lane's model (19.3.65). Likewise, model (19.3.66), if applied to real space flows, is similar to Santos-Salas' step model. Model (19.3.66) relates each monthly flow with the yearly flow explicitly, and the additivity property is preserved by including an extra term which represents the sum of the previously generated monthly flows. On the other hand, in Santos and Salas' step model, the relationship between each monthly flow and the annual flow is not explicit. However, if such a step model is written in regression form, then a term representing the sum of the remaining flows in the year will appear. Thus, the models are different, but accomplish essentially the same thing in the real space flows. The advantage of model (19.3.66) is that it can generate lognormally distributed seasonal flows which, with some adjustment, will add up to a lognormally distributed annual flow. A shortcoming of the model is that there is no provision for preserving the covariance of flows of the first season of this year with flows of the last season of the previous year. The multisite version of model (19.3.66) is also available.[59,61]

19.3.6 Markov Chains

The models included in the previous sections are applicable for continuous variables. However, various processes in hydrology can be formulated as discrete-valued processes or continuous processes can be discretized for computational convenience. In these cases, the theory of *Markov chains* may be applicable. Markov chains have been used in hydrology for modeling processes such as precipitation, stream flow, soil moisture, and water storage in reservoirs. Gabriel and Neumann[50] developed a Markov chain model for the occurrence of dry and wet days in daily rainfall. Many others have used Markov chains for modeling precipitation processes[3,26,40,62,122,149,172,184] and water storage processes.[51,56,109,126,210]

Definition and Properties. Consider $X(t)$ to be a discrete-valued process which started at time 0 and developed through time t. The values that the $X(t)$ process takes on are denoted by x_t, $t = 0, 1, \ldots$. Then

$$P[X(t) = x_t | X(0) = x_0, X(1) = x_1, \ldots, X(t-1) = x_{t-1}] \tag{19.3.68}$$

is the probability of the process being equal to x_t at time t, given its entire history. If the foregoing probability simplifies to

$$P[X(t) = x_t | X(t-1) = x_{t-1}] \tag{19.3.69}$$

it means that the outcome of the process at time t can be defined by using only the outcome at time $t-1$. A process which has this property is a *first-order Markov chain* or a *simple Markov chain*. Higher-order Markov chains can be formulated; however, only simple Markov chains will be considered here. Furthermore, the notation $X(t) = j$, $j = 1, \ldots, r$ will be used instead of x_t, which means that $X(t)$ is at *state j*, and r is the number of states. For instance, in modeling daily rainfall, one may consider two states, $j = 1$ for a dry day (no rain) and $j = 2$ for a wet day. Figure 19.3.4 shows schematically the definition of states.

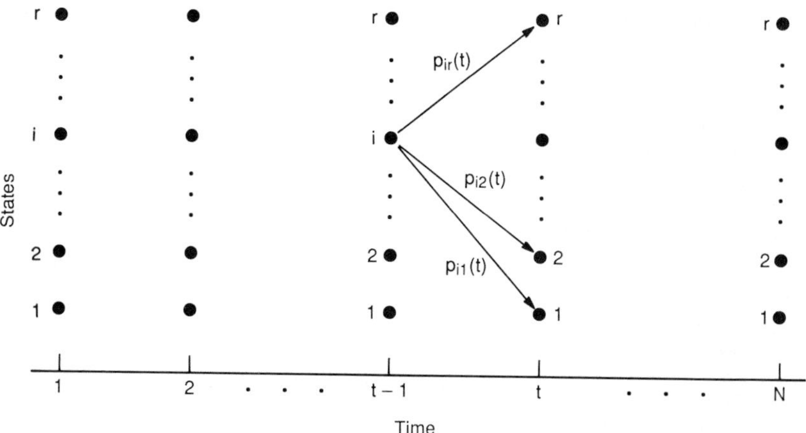

FIGURE 19.3.4 Definition of states for a Markov chain $X(t)$ and the corresponding transition probabilities $p_{ij}(t)$, $j = 1, \ldots, r$.

Transition Probability Matrix. A simple Markov chain is defined by its *transition probability matrix* $\mathbf{P}(t)$, which is a square matrix with elements $p_{ij}(t)$ given by

$$p_{ij}(t) = P[X(t) = j | X(t-1) = i] \tag{19.3.70}$$

for all i, j pairs. Figure 19.3.4 shows that the chain may go from state i at time $t-1$ to states $1, \ldots, r$ at time t, with corresponding transition probabilities $p_{i1}(t), \ldots, p_{ir}(t)$. Then,

$$\sum_{j=1}^{r} p_{ij}(t) = 1 \qquad i = 1, \ldots, r$$

Furthermore, if the transition probability matrix $\mathbf{P}(t)$ does not depend on time, the Markov chain is a *homogeneous chain* or a *stationary chain.* In this case, the notations P and p_{ij} are used. For the rest of this section, a homogeneous Markov chain is assumed.

n-Step Probability. Assume that the chain is now in state i and after n time steps it is in state j. The transition probability from i to j in n steps, denoted by $p_{ij}^{(n)}$, is given by[133]

$$p_{ij}^{(n)} = \sum_{k=1}^{r} p_{ik}^{(n-1)} p_{kj} \qquad n > 1 \tag{19.3.71}$$

and $p_{ij}^{(1)} = p_{ij}$. Thus, $p_{ij}^{(n)}$, $i, j = 1, \ldots, r$ are elements of the n-step *transition probability matrix* $\mathbf{P}^{(n)}$. It may be shown that $\mathbf{P}^{(n)}$ can be found by multiplying the one-step transition probability matrix $\mathbf{P}$ by itself n times.

Marginal Distribution. The probability distribution of the chain being at any state j at time t, denoted by $q_j(t) = P[X(t) = j]$, $j = 1, \ldots, r$, is called the *marginal distribution* of the process. Thus, $q_j(0)$ is the distribution of the initial states. The marginal state probability $q_j(t)$, given that $q_j(0)$ is known, may be determined as[133]

$$q_j(t) = \sum_{i=1}^{r} q_i(0) p_{ij}^{(t)} \tag{19.3.72}$$

Also, $\mathbf{q}(t) = \mathbf{q}(0)\,\mathbf{P}^t$, in which $\mathbf{q}(t)$ denotes the (row) vector of marginal state probabilities.

Steady-State Probabilities. The steady-state probability vector $\mathbf{q}^*$ with elements $q_1^*, \ldots, q_r^*$ represents the average fraction of time the chain is in states $1, \ldots, r$, respectively. It can be found by estimating $\mathbf{P}^{(t)}$ for large t until it converges. Also, the elements q_i^*, $i = 1, \ldots, r$ can be found by solving the system of equations

$$q_i^* = \sum_{k=1}^{r} q_k^* p_{ki} \qquad i = 1, \ldots, r \tag{19.3.73a}$$

$$\sum_{i=1}^{r} q_i^* = 1 \tag{19.3.73b}$$

Example. Assume that daily rainfall for a given site is represented by a simple Markov chain with two states, $j = 1$ for dry and $j = 2$ for wet, and a transition probability matrix $\mathbf{P}$ with elements $p_{11} = 0.6$, $p_{12} = 0.4$, $p_{21} = 0.3$, and $p_{22} = 0.7$. Assume also that initially the day is dry or $j = 1$ at $t = 0$. This also means that the initial marginal state probability vector is $\mathbf{q}(0) = [1, 0]$. Find: (1) the probability that the next day will be a dry day, (2) the probability that after 2 days, the day will be wet, (3) the probability of states dry and wet after 3 days, and (4) the probability of states dry and wet at any given day (regardless of the initial state). Since initially the day is dry, then $p_{11}^{(1)} = p_{11} = 0.60$ and Eq. (19.3.71) gives $p_{12}^{(2)} = p_{11}p_{12} + p_{12}p_{22} = 0.6 \times 0.4 + 0.4 \times 0.7 = 0.52$. The probabilities of states dry and wet after 3 days are determined by

$$\mathbf{q}(3) = \mathbf{q}(0)\begin{bmatrix} 0.6 & 0.4 \\ 0.3 & 0.7 \end{bmatrix}^3 = [1 \quad 0]\begin{bmatrix} 0.444 & 0.556 \\ 0.417 & 0.583 \end{bmatrix} = [0.444 \quad 0.556]$$

Finally, the probabilities of states dry and wet regardless of the initial state (long-run probabilities) are obtained by solving the system of Eqs. (19.3.73). Alternatively, it may be obtained from $\mathbf{P}^{(t)}$ where t is large. For example, for $t = 8$, it may be shown

$$\mathbf{P}^{(8)} = \begin{bmatrix} 0.429 & 0.571 \\ 0.429 & 0.571 \end{bmatrix}$$

Therefore, $\mathbf{q}^* = [0.429 \quad 0.571]$ with approximation to the third decimal figure.

Estimation and Testing. Estimation for a simple Markov chain amounts to estimating the elements p_{ij} of the transition probability matrix. For instance, consider modeling weekly rainfall for a period of ω weeks during the summer and denote by state 1 when it does not rain (dry week) and by state 2 when it rains (wet week). Thus, a sequence of rainfall states for a given summer may appear as 2112211222212 in which $\omega = 13$ (weeks). Thus, for a homogeneous simple Markov chain, the probabilities p_{11}, p_{12}, p_{22}, and p_{22} will be estimated. Assume that the total sample, considering N years of data, is ωN. Then n_{11} = number of times a dry week is followed by another dry week, $\mathbf{n}_{12}$ = number of times a dry week is followed by a wet week, n_{21} = number of times a wet week is followed by a dry week, and n_{22} = number of times a wet week is followed by a wet week. Furthermore, denote $n_1 = n_{11} + n_{12}$ and $n_2 = n_{21} + n_{22}$. Then, $\hat{p}_{ij} = n_{ij}/n_i$, $i = 1, 2$ and $j = 1, 2$. This algorithm can be extended for r states.

To test whether a simple Markov chain is an adequate model to describe the process under consideration, one can check some of the assumptions of the model and check whether it is able to reproduce some relevant properties of the process. For instance, one should check whether Eq. (19.3.68) simplifies to Eq. (19.3.69). Statistical methods for such tests are available.[40] In addition, one can compare the n-step transition probability of Eq. (19.3.71) with that obtained from the observed data. For $t = 2$ and $r = 2$, one can compare $p_{21}^{(2)} = p_{21}p_{11} + p_{22}p_{21}$ with $\hat{p}_{21}^{(2)}$ obtained from the data. Likewise, for a two-state Markov chain, the probability that the chain remains in state 1 during k steps is $P(L_1 = k) = p_{11}^{k-1}p_{12}$ and the probability that the chain stays in state 2 during k steps is $P(L_2 = k) = p_{22}^{k-1}p_{21}$. These probabilities can be compared with the corresponding probabilities obtained from the historical data.

Remarks. The emphasis of this section has been on homogeneous, first-order (simple) Markov chains. Although in some cases this model may be adequate to model the hydrologic process under consideration, often more complex models may be necessary. For instance, in modeling daily rainfall processes, the parameters of the Markov chain are often assumed to vary with time across the year.[40,149,201] Higher-order Markov chains may be necessary in other cases.[26] In addition, maximum likelihood estimation of parameters has been suggested.[149,201] Furthermore, selection of the order of Markov chain models can be based on the Akaike information criteria.[26,52,149,185]

19.3.7 Point Process Modeling

Simple Point Process. The theory of *point processes* has also been suggested for modeling time series of short-term rainfall since Le Cam[104] and Todorovic and Yevjevich[183] suggested that the occurrence of rainfall showers can be modeled by a *Poisson process.* Assume that storm arrivals are governed by a Poisson process. This means that the number of storms $N(t)$ in a time interval $(0, t)$ is Poisson-distributed with parameter λt, or

$$P[N(t) = n] = \frac{(\lambda t)^n}{n!} \exp(-\lambda t) \qquad n = 0, 1, \ldots \tag{19.3.74}$$

in which λ represents the storm arrival rate (per unit time). Referring to Fig. 19.3.5*a*, n storms arrived in the interval $(0, t)$ at times $t_1, \ldots, t_n$. The number of storms in any time interval T is also Poisson-distributed with parameter λT. The second assumption is that a *white noise* (random) rainfall amount R is associated with a storm arrival. For instance, R can be gamma-distributed with scale parameter μ and shape parameter δ. In addition, $N(t)$ and R are assumed to be independent. Thus, in Fig. 19.3.5*a*, rainfall amounts $r_1, \ldots, r_n$ are associated with storms occurring at times $t_1, \ldots, t_n$. Such a rainfall-generating process, called *Poisson white noise* (PWN), is a simple example of a point process.[35]

Likewise, the cumulative rainfall (mass curve process) in the interval $(0, t)$ is given by

$$Z(t) = \sum_{j=1}^{N(t)} R_j \tag{19.3.75}$$

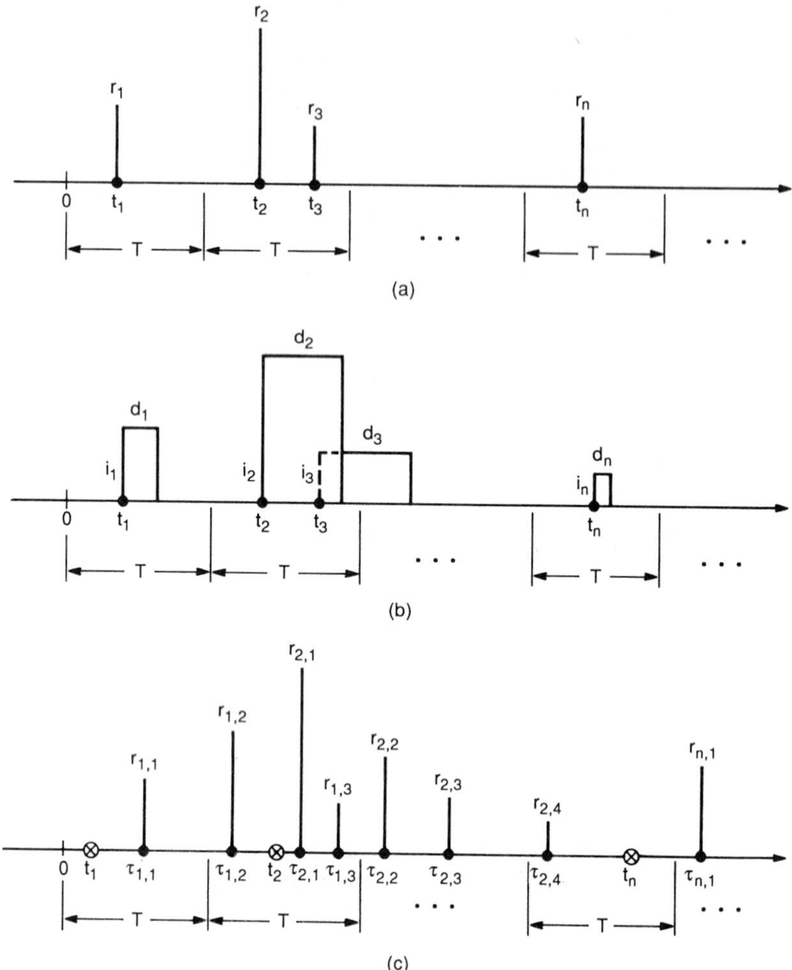

FIGURE 19.3.5 Schematic representation of (*a*) Poisson white noise, (*b*) Poisson rectangular pulse, and (*c*) Neyman-Scott white noise processes.

Such cumulative rainfall $Z(t)$ is called a *compound Poisson process.*[133] While the PWN model and $Z(t)$ describe the rainfall process in continuous time, a related process is the cumulative rainfall Y_i over nonoverlapping time intervals T as depicted in Fig. 19.3.5 (for instance, $T = 1$ h). Thus, the rainfall process Y_i in discrete time is defined by

$$Y_i = Z(iT) - Z(iT - T) \qquad i = 1, \ldots \tag{19.3.76}$$

The basic statistical properties of this process have been studied.[21,35,133] However, the rainfall process Y_i does not reproduce some of the important features of observed rainfall patterns. For instance, the lag-one serial correlation coefficients for hourly and daily precipitation at Denver Airport station for the months of June and December based on the 1948–1983 record are 0.446 and 0.172, respectively, while the process Y_i derived from the PWN model has zero autocorrelation.[133] Despite this shortcoming, even such a simple model can produce useful results for predicting the distribution of annual precipitation[35] and the distribution of extreme precipitation events.[20]

In the PWN model, the rainfall is assumed to occur instantaneously, so the storms have zero duration, which is unrealistic. Instead, one may consider that rainfalls occur with finite durations (rectangular pulses) as schematically shown in Fig. 19.3.5*b*. Each rainfall occurrence has a random intensity I and a random duration D. This is called the *Poisson rectangular pulse* (PRP) model. I and D can be assumed independent, and each exponentially distributed. Figure 19.3.5*b* shows a PRP process with n storms in the interval $(0, t)$ occurring at times $t_1, \ldots, t_n$ and associated with them are storms of intensities and durations $(i_1, d_1), \ldots, (i_n, d_n)$. In this case, storms can overlap, and, as a result, the aggregated process Y_i will be correlated. Properties and estimation procedures for such PRP process are available.[2,128,144,147] The PRP model is better conceptualized than the PWN model, but it is still limited when applied to rainfall data. Thus, more complex models have been suggested such as those based on the concept of clusters.

Cluster Processes. The concept of *clusters* was originally suggested by Neyman and Scott[127] in modeling the spatial distribution of galaxies. Le Cam[104] and Kavvas and Delleur[84,85] applied this concept of space clustering to model daily rainfall. Further developments have been made.[43,77,141,142,144–147] Here the concept of clusters as applied to modeling rainfall processes at a point is briefly described. The *Neyman-Scott cluster process* can be described as a two-level mechanism for modeling rainfall. First, storm-generating mechanisms (systems), or simply storms, arrive governed by a Poisson process with parameter λt. With reference to Fig. 19.3.5*c*, assume that in the period $(0, t)$, n storms arrive at points $t_1, \ldots t_n$. Then, associated with each storm, there are M precipitation bursts which are Poisson or geometrically distributed with parameter ν. In Fig. 19.3.5*c* there are three precipitation bursts associated with the storm that arrived at time t_1, four precipitation bursts associated with the storm that arrived at time t_2, and, in general, m_j precipitation bursts associated with a storm that arrived at time t_j. In addition, the time of occurrence, τ, of bursts relative to the storm origin t_j may be assumed to be exponentially distributed. For instance, the three bursts arising from the first storm are located at times $\tau_{1,1}$, $\tau_{1,2}$, and $\tau_{1,3}$ relative to time t_1. Finally, if the precipitation burst is described by an instantaneous random precipitation of depth R, then the resulting precipitation process is known as *Neyman-Scott white noise* (NSWN), while if the precipitation burst is described by a rectangular pulse of random intensity I and random duration D, then the precipitation process is known as the *Neyman-Scott rectangular pulse* (NSRP).

Properties and estimation of parameters for Neyman-Scott models are available.[36,43,45,77,84,85,128,141,142,147] The usual estimation approach has been the method of moments, although maximum likelihood has been suggested.[43] An apparent major problem is that parameters estimated based on one level of aggregation, say hourly data, are inconsistent with those estimated from another level of aggregation, say daily.[43,128,144] The problem seems to be that, as data are aggregated, information is lost and corresponding second-order statistics do not have enough information to give reliable estimates of the parameters of the generating process (model). As a consequence, the parameters of the generating process become significantly biased and have large variance. Estimation based on weighted moments of various time scales in a least squares fashion is an alternative.[19,36,77] Physical considerations may be useful in setting up constraints in some of the parameters and for initializing the estimates to be based on statistical considerations.[77,78] Koepsell and Valdes[92] applied this concept, based on the space-time cluster model suggested by Waymire et al.[200] for modeling rainfall in Texas, and pointed out the difficulty in estimating the parameters even when using physical reasoning as well as statistical analysis.

Remarks. The models included in this section essentially involved concepts of point processes in one dimension as applied to modeling precipitation processes at a point or a site within the class of Poisson processes and Neyman-Scott cluster processes. However, other alternative (somewhat related) temporal precipitation models have been suggested such as those based on Cox processes,[168,169] renewal processes,[44,57,143] and Barlett-Lewis processes.[146] Likewise, alternative space-time multidimensional precipitation models have been suggested.[77,96,170,171] For both temporal and space-time categories, which class of models is best is still open to question. Even small differences in modeling a certain component of a Neyman-Scott process can lead to significant differences in inferring the rest of the model parameters from actual data.[43] Furthermore, all precipitation models proposed to date are limited in some respects; for example, they do not include the daily periodicity observed in actual convective rainfall processes.[128]

19.4 FILLING IN MISSING OBSERVATIONS AND EXTENSION OF RECORDS

Records of hydrologic processes such as precipitation and stream flow are usually short and often have missing observations. Therefore, one of the first steps in any hydrologic data analysis is to *fill in missing values* and to *extend short records.* Observations may be missing for a number of reasons such as interruption of measurements because of equipment (mechanical, electrical, or electronic) failure, effects of extreme natural phenomena such as hurricanes or landslides or of human-induced factors such as wars and civil disturbances, mishandling of observed records by field personnel, or accidental loss of data files in the computer system. Likewise, hydrologic records are generally short; however, no matter the length of the record at a given site, if there are nearby sites with longer records, it may be possible to extend shorter records.

This section includes a number of information transfer techniques which can be applicable to both filling in missing observations and extending records. Some classical methods for filling in missing hydrologic data, such as the normal ratio method and the weighted distance interpolation method, are not included here. The methods included here are based on linear regression and time-series analysis.

19.4.1 Methods Based on Simple Linear Regression

Simple linear regression is most commonly applied for transferring hydrologic information between two gauging stations. Consider a short and a long sequence of a pair of hydrologic random variables represented by y_t and x_t, respectively. For instance, one variable can represent flows, and the other can also represent flows or rainfall. Assume that N_1 is the length of the short sequence and $(N_1 + N_2)$ is the length of the long sequence. The length N_1 also denotes the concurrent period of record. Without loss of generality, the sequences y and x may be represented as

$$y_1, y_2, \ldots, y_{N_1}$$
$$x_1, x_2, \ldots, x_{N_1}, x_{N_1+1}, \ldots, x_{N_1+N_2}$$

Then, a simple linear regression model may be established to extend the short sequence y_t.

Mathematical Model. A *simple linear regression model* between variables y_t and x_t may be generally represented as[113]

$$y_t = a + bx_t + \alpha\theta(1-\rho^2)^{1/2}\sigma_y\varepsilon_t \tag{19.4.1}$$

where y_t = dependent variable (short record)
x_t = independent variable (long record)
a, b = population parameters of the regression
α = coefficient
θ = 1 when the noise ε_t is added; $\theta = 0$ when ε_t is not added
ρ = population cross-correlation coefficient between y_t and x_t
σ_y = population standard deviation of y_t
ε_t = normal uncorrelated variable with mean zero and variance one which is uncorrelated with x_t

Estimation of Parameters. The estimators of a and b are given by[113]

$$\hat{a} = \bar{y}_1 - \hat{b}\bar{x}_1 \tag{19.4.2}$$

$$\hat{b} = \hat{r}\frac{s_1(y)}{s_1(x)} \tag{19.4.3}$$

where $\bar{y}_1$ and $\bar{x}_1$ are estimated means of the variables y_t and x_t, respectively, based on the concurrent record of size N_1, $s_1(y)$ and $s_1(x)$ are the corresponding estimated standard deviations of y_t and x_t, and $\hat{r}$ is the sample cross-correlation coefficient (refer to Sec. 19.2.1 for estimation of these statistics). In addition, σ_y of Eq. (19.4.1) is estimated by $s_1(y)$ and α is given by

$$\alpha = \left[\frac{N_2(N_1-4)(N_1-1)}{(N_2-1)(N_1-3)(N_1-2)}\right]^{1/2} \tag{19.4.4}$$

Model (19.4.1) can be used to extend the short sequence y_t; i.e., the values $\hat{y}_{N_1+1}, \ldots, \hat{y}_{N_1+N_2}$ are estimated from the concurrent values $x_{N_1+1}, \ldots, x_{N_1+N_2}$. Therefore, the new mean $\bar{y}$ and the new variance $s^2(y)$ of the extended sequence $y_1, \ldots, y_{N_1}, \hat{y}_{N_1+1}, \ldots, \hat{y}_{N_1+N_2}$ are[113]

$$\bar{y} = \bar{y}_1 + \frac{N_2}{(N_1+N_2)}\hat{b}(\bar{x}_2 - \bar{x}_1) \tag{19.4.5}$$

$$s^2(y) = \frac{1}{(N_1+N_2-1)}\Bigg[(N_1-1)s_1^2(y) + (N_2-1)\hat{b}^2 s_2^2(x) + \frac{N_1N_2}{(N_1+N_2)}\hat{b}^2(\bar{x}_2-\bar{x}_1)^2 + (N_2-1)\theta^2\alpha^2(1-\hat{r}^2)s_1^2(y)\Bigg] \tag{19.4.6}$$

Model (19.4.1) without noise ε_t ($\theta = 0$) may be used for filling in missing data only when just a few records are missing. For a significant number of missing records or for extension of short records in general, model (19.4.1) with $\theta = 0$ causes the variance of the extended record to be underestimated.[113] While this problem is eliminated by considering model (19.4.1) with noise ($\theta = 1$), this requires generating random numbers, which does not lead to a unique extended sequence. Hirsch[70] proposed a method known as *maintenance of variance extension* in which model (19.4.1) with $\theta = 0$ is considered in such a way that the mean $\bar{y}$ and the variance $s^2(y)$ of Eqs. (19.4.5) and (19.4.6), respectively, are maintained. Section 17.4.10 in Chap. 17 discusses some of the principles behind this method. Following Hirsch, model (19.4.1) with $\theta = 0$ may be used for record extension in which the parameters a and b are determined by[192]

$$\hat{a}_1 = \frac{(N_1 + N_2)\,\bar{y} - N_1\,\bar{y}_1}{N_2} - \hat{b}_1\bar{x}_2 \tag{19.4.7}$$

$$\hat{b}_1 = \left[\frac{(N_1 + N_2 - 1)s^2(y) - (N_1 - 1)s_1^2(y) - N_1(\bar{y}_1 - \bar{y})^2 - N_2\,(\hat{a} - \bar{y})^2}{(N_2 - 1)s_2^2(x)} \right]^{1/2} \tag{19.4.8}$$

where $\bar{y}$ and $s^2(y)$ are estimated by Eqs. (19.4.5) and (19.4.6), respectively; $\bar{y}_1$ and $s_1^2(y)$ are the sample mean and the variance of the original short record; and $\bar{x}_2$ and $s_2^2(x)$ are the sample mean and variance of the longer record $x_{N_1+1}, \ldots, x_{N_1+N_2}$. Thus, Eqs. (19.4.7) and (19.4.8), when used for estimating $\hat{y}_{N_1+1}, \ldots, \hat{y}_{N_1+N_2}$, will produce a record $y_1, \ldots, y_{N_1}, \hat{y}_{N_1+1}, \ldots, \hat{y}_{N_1+N_2}$ which will have mean $\bar{y}$ and variance $s^2(y)$.

Criteria for Improving Estimators of Parameters. In using correlation analysis to extend the short record y_t on the basis of a longer record x_t, a question arises whether the combined record of y_t consisting of N_1 recorded values and N_2 estimated values improves the estimates of the parameters such as the mean and variance. The criteria to be briefly described here assumes model (19.4.1) with noise $\theta = 1$. The variance of the mean $\bar{y}$ based on the longer record is[113]

$$\text{Var}\,(\bar{y}) = \frac{\sigma_y^2}{N_1}\left[1 - \frac{N_2}{N_1 + N_2}\left(r^2 - \frac{1 - r^2}{N_1 - 3}\right)\right] \tag{19.4.9}$$

For $\bar{y}$ (extended record) to be a better estimator of the population mean μ_y than $\bar{y}_1$ (short record), Var ($\bar{y}$) must be smaller than Var ($\bar{y}_1$). This occurs if

$$|r| > \left(\frac{1}{N_1 - 2}\right)^{1/2} \tag{19.4.10}$$

The right side of Eq. (19.4.10) is called the critical minimum correlation coefficient for improving the estimate of the mean. Such critical correlation is shown in Table 19.4.1 under $m = 1$, column (1) for various values of N_1.

Following the same concept described above, critical minimum correlation coefficients for improving the estimator of the variance can be found. Table 19.4.1, under $m = 1$, column (2), gives such critical correlations for $\theta = 1$, N_1 varying from 8 to 60 and $N_2 = 60$. For all practical purposes, critical values in the table can be applied for any value of N_2.

TABLE 19.4.1 Critical Correlations for Improving the Estimates of the Mean (1) and the Variance (2) for $m = 1, 2, 3, 4, 5$, Values of N_1 From 8 Through 60, and $N_2 = 60$

The assumed models are Eqs. (19.4.1) and (19.4.11) with $\theta = 1$ (m = number of concurrent records used)

	$m=1$		$m=2$		$m=3$		$m=4$		$m=5$	
N_1	(1)	(2)	(1)	(2)	(1)	(2)	(1)	(2)	(1)	(2)
8	0.408	0.720	0.577	0.835	0.707	0.914	0.816		0.913	
10	0.354	0.650	0.500	0.763	0.612	0.841	0.707	0.799	0.791	0.945
12	0.316	0.597	0.447	0.707	0.548	0.785	0.632	0.845	0.707	0.892
14	0.289	0.556	0.408	0.661	0.500	0.739	0.577	0.199	0.645	0.847
16	0.267	0.522	0.380	0.624	0.463	0.700	0.535	0.759	0.598	0.808
18	0.250	0.494	0.354	0.592	0.433	0.666	0.500	0.725	0.559	0.774
20	0.236	0.469	0.333	0.565	0.408	0.637	0.471	0.695	0.527	0.744
25	0.209	0.422	0.295	0.510	0.361	0.578	0.417	0.634	0.466	0.681
30	0.189	0.386	0.267	0.469	0.327	0.533	0.378	0.587	0.423	0.632
35	0.174	0.359	0.246	0.436	0.302	0.498	0.348	0.548	0.389	0.592
40	0.162	0.336	0.229	0.410	0.281	0.468	0.324	0.517	0.363	0.559
45	0.152	0.317	0.216	0.387	0.264	0.443	0.305	0.490	0.341	0.531
50	0.144	0.301	0.204	0.368	0.250	0.422	0.289	0.467	0.323	0.506
55	0.137	0.288	0.194	0.352	0.238	0.404	0.275	0.447	0.307	0.485
60	0.131	0.276	0.186	0.338	0.227	0.387	0.263	0.429	0.294	0.466

Note: Values of the critical correlation for improving the estimate of the variance are for $N_2 = 60$, but they can be used for any N_2, since the critical correlation does not vary significantly as N_2 varies.

19.4.2 Methods Based on Multiple Linear Regression

Multiple linear regression for transferring information to a site with a short record may be applied when two or more nearby sites with longer records are available. Assume that the short record of size N_1 is represented by y_t and the m longer records of size $N_1 + N_2$ are represented by the vector $\mathbf{x}_t$ as[54]

$$
\begin{array}{l}
y_1,\ y_2,\ \ldots,\ y_{N_1} \\
x_1^{(1)},\ x_2^{(1)},\ \ldots,\ x_{N_1}^{(1)},\ x_{N_1+1}^{(1)},\ \ldots,\ x_{N_1+N_2}^{(1)} \\
\vdots \\
x_1^{(m)},\ x_2^{(m)},\ \ldots,\ x_{N_1}^{(m)},\ x_{N_1+1}^{(m)},\ \ldots,\ x_{N_1+N_2}^{(m)}
\end{array}
$$

It is also assumed that the concurrent observations are drawn from a multivariate normal population with parameters $\mu_x^{(i)}$, μ_y, $\sigma_x^{2(i)}$, σ_y^2, and R, where $\mu_x^{(i)}$ and $\sigma_x^{2(i)}$ denote the population mean and variance of $x_t^{(i)}$, respectively, for sites $i = 1, \ldots,$ m; μ_y and σ_y^2 are the population mean and variance of y_t, respectively; and R is the population *multiple correlation coefficient.* In addition, it is assumed that at each site the observations are serially uncorrelated. The problem is to transfer information from the m sites with records of length $N_1 + N_2$ to the site with a short record.

Mathematical Model. The short record y_t of length N_1 may be related to the m records x_t by the multiple linear regression model[54]

$$y_t = a + \sum_{i=1}^{m} b_i x_t^{(i)} + \alpha\theta(1 - R^2)^{1/2} \sigma_y \varepsilon_t \tag{19.4.11}$$

where $\theta = 1$ if noise is added, otherwise $\theta = 0$; a and b are estimated by

$$\hat{a} = \bar{y}_1 - \sum_{i=1}^{m} \hat{b}_i \bar{x}_1^{(i)} \tag{19.4.12}$$

$$\hat{b}_i = \sum_{j=1}^{m} \hat{d}_1^{(ij)} \hat{c}_1^{(j)} \qquad i = 1, \ldots, m \tag{19.4.13}$$

with $\bar{y}_1$ and $\bar{x}_1^{(i)}$, $i = 1, \ldots, m$, the sample means of y_t and $x_t^{(i)}$, respectively, based on the sample of size N_1; $\hat{d}_1^{(ij)}$ are the elements of the inverse of the matrix whose elements are the lag-zero cross-covariances between $x_t^{(i)}$ and $x_t^{(j)}$, $i, j = 1, \ldots, m$; and $\hat{c}_1^{(j)}$ are the lag-zero cross-covariances between $x_t^{(j)}$ and y_t, $j = 1, \ldots, m$ [these cross-covariances may be determined from Eq. (19.2.15)]. The multiple-correlation coefficient R is estimated from the N_1 concurrent observations as

$$\hat{R} = \left[\frac{N_1 \sum_{i=1}^{m} \hat{b}_i \hat{c}_1^{(i)}}{\sum_{t=1}^{N_1} (y_t - \bar{y}_1)^2} \right]^{1/2} \tag{19.4.14}$$

and the coefficient α is given by

$$\alpha = \left[\frac{N_2(N_1 - 2m - 2)(N_1 - 1)}{(N_2 - 1)(N_1 - m - 2)(N_1 - m - 1)} \right]^{1/2} \tag{19.4.15}$$

Then, the new estimators of the mean μ_y and of the variance σ_y^2 are[54]

$$\bar{y} = \bar{y}_1 + \frac{N_2}{N_1 + N_2} \sum_{i=1}^{m} \hat{b}_i [\bar{x}_2^{(i)} - \bar{x}_1^{(i)}] \tag{19.4.16}$$

$$\begin{aligned} s^2(y) = \frac{1}{N_1 + N_2 - 1} \Bigg\{ &(N_1 - 1)\, s_1^2(y) + N_2 \sum_{i=1}^{m} \sum_{j=1}^{m} \hat{b}_i \hat{b}_j \hat{g}_2^{(ij)} \\ &+ \frac{N_1 N_2}{N_1 + N_2} \left[\sum_{i=1}^{m} \hat{b}_i \, (\bar{x}_2^{(i)} - \bar{x}_1^{(i)}) \right]^2 + (N_2 - 1)\, \alpha^2 \theta_2 \, (1 - \hat{R}^2)\, s_1^2(y) \Bigg\} \end{aligned} \tag{19.4.17}$$

where $\bar{y}_1$ and $s_1^2(y)$ are the sample mean and variance of the short series y_t; $\bar{x}_1^{(i)}$ and

$\bar{x}_2^{(i)}$ are the sample means of the short sample $x_1^{(i)}, \ldots, x_N^{(i)}$ and the additional sample $x_{N_1+1}^{(i)}, \ldots, x_{N_1+N_2}^{(i)}$, respectively; and $\hat{g}_2^{(ij)}$, $i, j = 1, \ldots, m$, are the lag-zero cross-covariances between $x_t^{(i)}$ and $x_t^{(j)}$ for the additional samples of size N_2. The maintenance-of-variance extension method described in Sec. 19.4.1 under simple linear regression has been extended to the multivariate case.[60]

Criteria for Improving Estimators of Parameters. The criterion for improving estimates of parameters of the short record for the multiple linear regression model is based on comparing the variances of the original and new estimators. The mean of the short sample y_t is improved if[54]

$$|R| > \left(\frac{m}{N_1 - 2}\right)^{1/2} \tag{19.4.18}$$

Table 19.4.1, under columns (1), gives the critical minimum multiple correlation coefficient for various values of N_1 and m. Likewise, the variance of the short sample is improved if the estimated multiple correlation coefficient R is larger than a critical value.[54,125] Such critical values are given in Table 19.4.1 under columns (2) for $m = 1$ through 5 and $N_1 = 8$ through 60.

19.4.3 Methods Based on Time-Series Models

Missing observations can be filled in and records can be extended by using many of the time-series models described in Sec. 19.3.

Use of AR(1) and PAR(1) Models. These models can be used to fill in missing observations for a given site when no other nearby sites with concurrent information are available. Assume the AR(1) model (19.3.2):

$$y_t = \mu + \phi(y_{t-1} - \mu) + \varepsilon_t \tag{19.4.19}$$

where the model parameters can be estimated from available data. If an observation at time t is missing, but y_{t-1} is known, then y_t can be determined from Eq. (19.4.19) using $\varepsilon_t = 0$ (the mean of ε_t). Since model (19.4.19) assumes stationarity, nonstationary data must be made stationary before the model is applied. A more convenient model for seasonal data is the PAR(1) model (19.3.32):

$$y_{v,\tau} = \mu_\tau + \phi_{1,\tau}(y_{v,\tau-1} - \mu_\tau) + \varepsilon_{v,\tau} \tag{19.4.20}$$

which can be used to fill in missing seasonal data such as monthly observations. The foregoing models should not be used to fill in successive missing observations.

Use of Multivariate Models. Multivariate models can be used for filling in missing data and extension of records. Suppose that y_t is the site with missing records and that data at sites $x_t^{(1)}$ and $x_t^{(2)}$ are available. Then, a multivariate model can be formulated as

$$y_t = a + \sum_{j=1}^{p} b_j y_{t-j} + \sum_{j=0}^{p_1} b_j^{(1)} x_{t-j}^{(1)} + \sum_{j=0}^{p_2} b_j^{(2)} x_{t-j}^{(2)} + \varepsilon_t \tag{19.4.21}$$

in which a, b_j, $j = 1, \ldots, p$; $b_j^{(1)}$, $j = 0, \ldots, p_1$ and $b_j^{(2)}$, $j = 0, \ldots, p_2$ are the parameters to be estimated from the data. Estimation of parameters and testing can be made by least squares.[34] Usually, p, p_1, and p_2 are small—of the order of 1 or 2. Note that model (19.4.21) falls in the category of *ARMAX* (ARMA with exogenous variables) and *transfer function models.* Applications of multivariate models can be found in Beauchamp et al.[6] and Kottegoda and Elgy.[93]

19.5 MONTE CARLO SIMULATION*

19.5.1 Introduction

Basic Concepts of Monte Carlo Simulation. Consider a *hydrologic system,* in which I represents the *input* and O represents the *output.* The system can be simple or complex, and the input and output of each can be either a single variable or a vector of several variables or any combination of these. For instance, the system may be a watershed system in which the input is simply the average precipitation series over the basin and the output is the stream-flow series at the outlet of the basin. In general, *Monte Carlo simulation* is a method for obtaining the probability distribution of output O given the probability distribution of the input I. Thus, in Monte Carlo simulation studies, three steps are usually required, namely, determining the input, transforming the input into the output, and then analyzing the output.

The input to be used in Monte Carlo simulation studies may be the historical hydrologic records, or artificially or synthetically generated records. In fact, this is one of the purposes of the stochastic models that are included in Sec. 19.3. Further discussion on how to generate synthetic records based on such models is given in Secs. 19.5.2 and 19.5.3 below. The transformation of the input into the output is made by means of a mathematical model which represents the behavior of the physical system under study. In the case of a reservoir system in which the input is the set of inflows to the reservoir and the output is the set of reservoir outflows, the inflows are transformed into the outflows by operating the reservoir according to the reservoir operating rule, a set of constraints, and the mass balance equation of the reservoir. Thus, the transformation process involves routing the input through the system to obtain the output. Put another way, the system input drives the system, which transforms it into the system output. Finally, the system output is analyzed statistically so that it can be used for decision making. The analysis of the output may consist of determining basic output statistics, such as the mean and variance, box plots to observe the variance of the output graphically, and the overall frequency distribution of the output variables under consideration. Details of such analysis can be found in Secs. 17.2 and 17.3 of Chap. 17.

Applications of Monte Carlo Simulation Studies. Some examples of typical applications of Monte Carlo simulation studies in hydrology are included here for illustration. The applications selected are purposely presented in a simplified schematic manner in order to illustrate the underlying concepts only. These concepts can then be extended and applied to more complex cases.

Design the Capacity of a Reservoir. Assume that the capacity of a reservoir for water supply will be determined so that a given water demand d will be delivered from the reservoir throughout a specified planning horizon. N years of historical stream-flow record (inflows to the reservoir) are available; in this case, for simplicity, it is assumed that N coincides with the planning horizon. One possible solution may be to deter-

*Part of the material in this section was contributed by Fidel Saenz de Ormijana.

mine the needed reservoir capacity assuming that the historical record will be identically repeated in the future; this capacity is the output of the system shown as $O(h)$ in Fig. 19.5.1. However, such identical realization of inflows is unlikely to occur in the future. An alternative approach for determining the reservoir capacity is Monte Carlo simulation. Thus, a mathematical model of the inflows x_t is determined and then used to generate a large number of possible sequences that may occur in the future. Then, for each sequence $x_t(i)$, also denoted as $I(i)$ in Fig. 19.5.1, the reservoir capacity $O(i)$ is determined, yielding the set $O(1), \ldots, O(m)$ where m is the number of realizations considered (usually large). This also means that the uncertain occurrence of future inflows is translated into an uncertain reservoir capacity. The set $O(1), \ldots, O(m)$ can be analyzed statistically to provide the hydrologist with additional information to make a decision on what reservoir capacity to use. This use of Monte Carlo analysis for reservoir design has been widely suggested in the literature.[41,107,108]

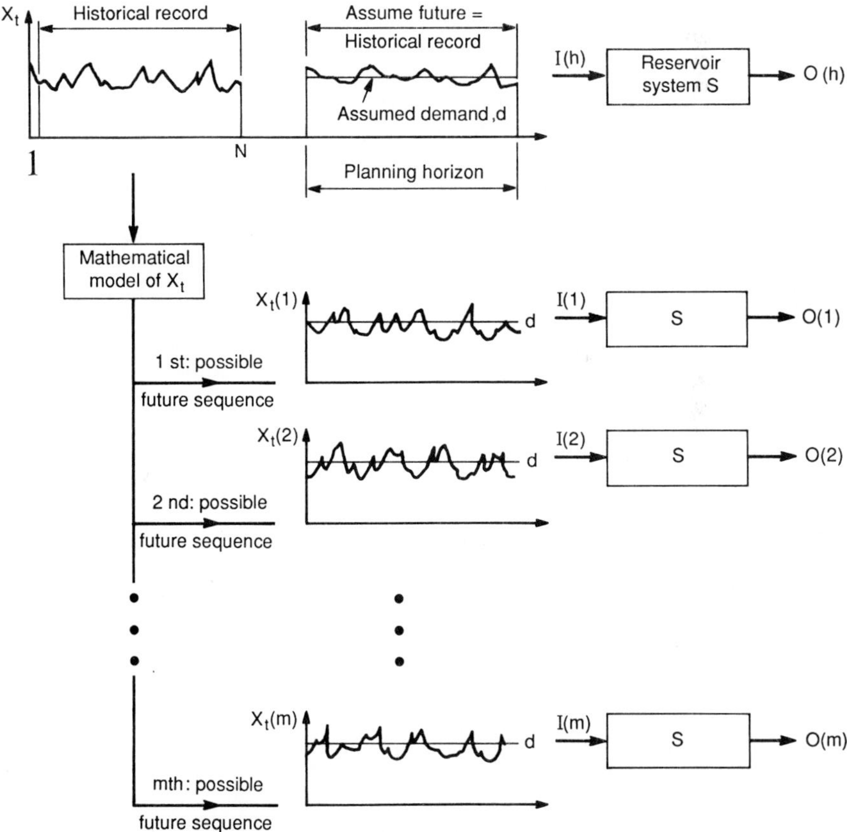

FIGURE 19.5.1 Schematic representation of a hydrologic system in which historical records $I(h)$ and synthetic records $I(i)$, $i = 1, \ldots, m$ are used as inputs for determining historical output $O(h)$ and simulated outputs $O(i)$, $i = 1, \ldots, m$, respectively.

Evaluating the Performance of a Reservoir of Given Capacity. The performance of a reservoir of specified capacity, operating rule, and projected water demands can be evaluated to determine, for instance, the reliability of the system to meet target demands, the total shortages or how often the reservoir may run dry, how often it may spill, and the duration and magnitude of those episodes. This problem can be approached by Monte Carlo simulation studies in which the set of inflows to the reservoir is the system input I and the system output O can be any desired set of performance measures of the reservoir system under consideration. As shown in Fig. 19.5.1, the simulation study can be made by using the historical record alone or by using synthetically generated inflow records. For reservoirs with seasonal regulation, and short operation planning horizon as compared to the historical record, the use of historical records alone may be sufficient. However, since these requirements are rarely met, synthetic records are typically used.

Evaluating the Performance of a Reservoir of Specified Operating Rules. In this case, the capacity of a reservoir and projected water demands are known, but the effect of an operating rule in the performance of the reservoir is being evaluated. A reservoir may have been operated historically under a certain operating rule and an alternative operating rule may need to be considered. In this case, Monte Carlo simulation studies are conducted in which the same inputs (synthetic records) are used for each operating rule. Then the two sets of outputs (performance measures) are compared statistically to determine whether the alternative operating rule gave significantly different results than the original rule.

Evaluating the Performance of Irrigation Water Delivery Systems. The assessment of alternatives to improve irrigation water delivery systems can be made by considering the system's performance under uncertainty. The objective of water delivery systems may be defined in terms of the desire to best meet the water requirements at the farm level. Performance measures related to such objectives can be based on a number of system state variables, for instance, the amounts of water required at a number of diversion points in the system and the amounts of water actually delivered to the diversion points, all variables defined throughout the irrigation season.[65] Monte Carlo simulation studies of system inputs, such as reservoir inflows, effective precipitation, and crop evapotranspiration, can be generated by multivariate models (such as those referred to in Sec. 19.3) for a length equal to the irrigation season. These inputs can then be routed through the irrigation system simulation model to obtain the system's state variables. This process is repeated a large number of times to provide an array of performance measures which can be analyzed statistically.

Determining the Dependable Capacity of Hydropower Systems. A common approach in the analysis of hydropower systems is to use the critical period of record for determining project dependable capacity. However, this approach usually underestimates the power capacity actually available for marketing purposes. An alternative approach is to use Monte Carlo simulation by which inflows to a reservoir system can be synthetically generated for a length equal to the operational horizon (usually several years long), then routed through the system to obtain the hydropower output (power capacity and energy). The process can be repeated several times to obtain an array of hydropower outputs $O(i)$, $i = 1, \ldots, m$ as indicated in Fig. 19.5.1. The output is then subject to statistical analysis to assist in defining a marketable hydropower output.[97]

Determining the Drought Properties of Water-Supply Systems. Drought properties of various return periods are needed to assess the degree to which a water-supply system will be able to cope with future droughts and, accordingly, to plan alternative water-supply strategies. The estimation of long-term droughts, for instance, the drought length, magnitude, and intensity as defined in Sec. 19.2, based on the historical record alone, are not reliable when the return periods are of the order of the historical record or they cannot be determined when the historical record is smaller than the return period. In these cases, drought properties can be determined by synthetically generating inflows at key points in the water-supply system under consideration, of length equal to the return period of interest, routing such inflows through the system to obtain the flows at the diversion point of the water-supply system. These inflows in turn can be analyzed to obtain the drought property of interest, denoted by $O(i)$ in Fig. 19.5.1. The process is repeated several times to find the array of drought properties $O(1), \ldots, O(m)$ whose average is an estimate of the T-year drought.[48,49]

Other Applications of Monte Carlo Simulation. A number of other applications of Monte Carlo simulation can be found in literature. For instance, for water systems in which a short record of stream flow and a longer record of precipitation are available, one may use the concurrent precipitation–stream flow records to calibrate a watershed model and generate a number of sequences of synthetic precipitation records which then are routed through the watershed model to generate synthetic streamflow records. In turn, these records can be used for design, evaluation of operational rules, or for any other purpose at hand. In some cases, Monte Carlo simulation has been used for estimating floods of long return periods.[46] Hourly rainfall and daily pan evaporation can be synthetically generated and routed through a previously calibrated watershed model to produce a long record of hourly stream flow. Annual flood peaks and volumes can be obtained from such synthetic flow traces, and frequency analysis will provide the flood estimates for the desired return periods.

Monte Carlo simulation has been used for studying the impacts of global climate change on the operation of water resources projects[102] and for determining the variability of the system's output as a result of the uncertainty in the parameters of the system's model. For instance, consider a watershed model used for transforming precipitation into runoff which involves a set of parameters, one of which (for the sake of simplicity) is assumed to be uncertain. Based on previous applications of the model in similar watersheds, one may estimate or assume the distribution of such a parameter. One can randomly sample values of the parameter and for each value run the watershed model to find the runoff sequence corresponding to a given precipitation input sequence. Thus, m values of the parameter will produce m sets of runoff sequences which can be analyzed statistically to determine the effect of parameter uncertainty on the simulated runoff.

Monte Carlo simulation studies are also used for many problems in groundwater hydrology, typically for deriving the distribution of the underlying output variable of a groundwater flow equation, given the distribution of the parameters and boundary conditions. For instance, one can use a steady-state groundwater flow equation for a two-dimensional isotropic nonuniform medium and study the effect of spatial variability of hydraulic conductivity K on the hydraulic head (system output). The variability of K can be modeled by a given probability distribution function and covariance function. Thus, a large number m of realizations of random spatially correlated parameters K can be generated and the groundwater flow equation solved for each realization to find the set of m hydraulic heads at various points in space, from which one can find the distribution of hydraulic heads.[47,55,83] One can extend the foregoing concept to study the effect of variability of other parameters such as porosity, the effect of variability on boundary conditions, and the effect of variability of model inputs such as groundwater recharge. Finally, an important application of Monte

Carlo simulation is to establish the uncertainty in travel time and spread of pollutants in porous media as a function of the uncertainty in the parameters of the groundwater contamination transport model. For this purpose, one can follow the random sampling and simulation approach described previously.

19.5.2 Generation of Random Inputs

In the early days of Monte Carlo simulation, random inputs were generated by randomly drawing cards from a stack, by sequentially reading values from tables of random numbers, or by using devices such as coins or dice. However, with the advent of digital computers and the development of mathematical and statistical techniques, the use of these procedures has become obsolete. Instead, mathematically based approaches for generating random inputs for Monte Carlo simulation studies have become the state of the art. This section includes some basic models for generating uniform and normal random numbers.

Generation of Uniform Random Numbers. Random numbers from a uniform distribution between the bounds 0 and 1 are considered here. These random numbers, referred to as $u(0, 1)$ are widely used for generating random numbers from other distributions. The most popular generators are called *linear congruential* generators, which are integer algorithms of the type $x_i = (ax_{i-1} + b) \bmod c$ where a, b, and c are integers and the mod c notation indicates that x_i is the remainder after dividing $(ax_{i-1} + b)$ by c. This procedure may also be expressed as

$$x_i = ax_{i-1} + b - c\,\mathrm{Int}\left(\frac{ax_{i-1} + b}{c}\right) \tag{19.5.1}$$

For example, if $a = 13$, $b = 3$, $c = 11$, and $x_0 = 3$, then the first five random numbers generated are 9, 10, 1, 5, and 2. This algorithm produces integer numbers between 0 and $c - 1$. The values of the generated numbers depend on the constants a, b, and c and on the initial value x_0 called the *seed number.* Linear congruential generators always have a finite cycle length; i.e., the sequence repeats itself exactly after generating a certain number of values. There are conditions for a, b, and c, under which the generator yields all the values between 0 and $c - 1$ before repeating.[91]

To obtain numbers uniformly distributed between zero and one, simply let $u_i = x_i/c$. Since u must not be zero or one exactly, c is always a very large value. For certain choices of the parameters a, b, and c, the variable u is approximately uniformly distributed between zero and one. For instance, a common choice of parameters, supported by the IMSL[76] subroutine library, is $a = 16807$, $b = 0$, and $c = 2^{31} - 1$, which may be used with 32-bit-word computers. For computers with smaller word size, a and c should be chosen so that their product does not exceed the machine word size, otherwise an integer overflow will occur. Likewise, the seed number x_0 must be provided in order to start the generation. It may be obtained through some random mechanism, for instance, by reading the computer clock. However, debugging a computer code and checking the output is easier if repeated simulations give the same results. In these cases, it is convenient to use the same starting seed. Table 19.5.1 gives 30 starting seeds for generating uniform random numbers by the linear congruential generator supported by IMSL, as noted above. Each seed value x_0 can be used to generate a set of 131,072 uniform random numbers and all sets will be independent of each other. Additional choices of a, b, and c and alternative generators are available.[14,137]

TABLE 19.5.1 Starting Seeds x_0 for Generating Uniform Random Numbers Based on Eq. (19.5.1) with $a = 16807$, $b = 0$, and $c = 2^{31} - 1$

748932582	250756106	431442774
1985072130	1025663860	1659181395
1631331038	186056398	400219676
67377721	522237216	1904711401
366304404	213453332	263704907
1094585182	1651217741	350425820
1767585417	909094944	873344587
1980520317	2095891343	1416387147
392682216	203905359	1881263549
64298628	2001697019	1456845529

Source: From Bratley et al.[14]

Generally, random number generators such as the linear congruential generator described above produce *pseudo-random numbers* because they follow a deterministic sequence even though the successive values of that sequence are uncorrelated. There are ways to improve the randomness of generated random numbers. One can do *shuffling,* i.e., temporarily store an array of random numbers, using each new generated random number to obtain one from the stored array, then replacing the numbers in the array by newly generated numbers as they are used. Shuffling can do no harm (it has no effect on a string of perfectly generated random numbers) and it improves significantly the sequences produced by a good generator. Press et al.[137] give FORTRAN, Pascal, and C codes for shuffling. *Combining* two or more generators to generate a set of random numbers is another approach.[14] Finally, one can use shuffling and combining simultaneously.[112]

Generation of Normal Random Numbers. A fast and simple method for generating standard normal random numbers ε (i.e., with zero mean and unit standard deviation) is the approximation based on the lambda distribution[82,132,139,140,167,186] as $\epsilon = 4.91\,[u^{0.14} - (1-u)^{0.14}]$, in which u is a uniform (0, 1) random number. The accuracy (the difference between the true standard normal variable and that obtained by the approximation) of this approximation is 0.0032 for $|\epsilon| < 2$, and 0.0038 for $2 < |\epsilon| < 3$, and increases rapidly for $|\epsilon| > 3$. The variable range is limited to $|\epsilon| < 4.91$ (the probability of ϵ being outside this interval is less than 10^{-6}). The variance and kurtosis of the lambda approximation are 0.997 and 2.972, close to the theoretical values 1 and 3, respectively, for a normal distribution. If greater accuracy is desired or the simulation focuses on tail behavior (such as for extreme value analysis), a better approximation may be used.

Several approximations for the normal distribution based on polynomial equations can be found.[209] A rational approximation with relative accuracy 10^{-6} is[131]

$$\epsilon(u) = t + \frac{p_0 + p_1 t + p_2 t^2 + p_3 t^3 + p_4 t^4}{q_0 + q_1 t + q_2 t^2 + q_3 t^3 + q_4 t^4} \qquad 0.5 \leq u < 1 \qquad (19.5.2a)$$

$$\epsilon(u) = -\epsilon(1-u) \qquad 0 < u < 0.5 \qquad (19.5.2b)$$

where $t = [-2 \ln (1-u)]^{1/2}$, u is a uniform (0, 1) random number, and the coefficients p_i and q_i, $i = 0, 1, \ldots, 4$ are given by $p_0 = -0.322232431088$, $p_1 = -1$,

$p_2 = -0.342242088547$, $p_3 = -0.0204231210245$, $p_4 = -0.0000453642210148$, $q_0 = 0.0993484626 06$, $q_1 = 0.588581570495$, $q_2 = 0.531103462366$, $q_3 = 0.103537752285$, and $q_4 = 0.0038560700634$.

Two other methods are often used to generate standard normal random numbers. The sum of uniform random numbers $\epsilon = \sum_{i=1}^{n} u_i - n/2$, where n is commonly 12, yields an approximate normal number with the same mean, variance, and skewness as the standard normal, but kurtosis is 2.9, against a kurtosis of 3.0 for the normal distribution. The range of ϵ is limited to $|\epsilon| < 6$, and the difference between the actual standard normal value and the value obtained by the above approximation is less than 0.009 for $|\epsilon| < 2$, but it can be as high as 0.9 for $2 < |\epsilon| < 3$. Likewise, the Box-Muller method gives two standard normal numbers based on $\epsilon_1 = (-2 \ln u_1)^{1/2} \cos(2\pi u_2)$ and $\epsilon_2 = (-2 \ln u_1)^{1/2} \sin(2\pi u_2)$ in which u_1 and u_2 are two uniform (0, 1) random numbers. This method is slightly faster than the above rational approximation, but the random numbers produced are not independent when the required uniform random numbers are generated by a linear congruential generator.

Remarks. In general, one can generate random numbers from any continuous distribution with cumulative distribution function $F(x)$ by the *inversion method.* The cumulative distribution function $F(x)$ varies between 0 and 1, so if a uniform (0, 1) random number u is generated, this value is made equal to $F(x)$ and the distribution function inverted to find the value of x, then $x = F^{-1}(u)$ is a random number from the distribution $F(x)$. This method can be extended to generate discrete random numbers as well.

19.5.3 Generation of Correlated Inputs

Autocorrelated and cross-correlated inputs are often needed for Monte Carlo simulation studies. For this purpose one can use a number of alternatives including procedures based on the available historical time series and on mathematical models such as those presented in Sec. 19.3. Some of these methods are discussed here in some detail for illustration.

Use of Historical Data. Using historical data is a widespread method of assessing alternative inputs when the available historic series of the input variables are long enough to define a sufficient approximation to the input distribution. For example, in simulating reservoir performance, if the historic stream-flow series is long compared to the simulation horizon, it may be possible to break down the historic stream-flow series into a number of subseries of length equal to the length of the simulation horizon, simulate the system operation with each subseries, and obtain the corresponding output statistics (such as supply reliability) over the simulation horizon. Yet another alternative in using the historic series is to work with the N *wrapped-around series* $\{x_2, x_3, \ldots, x_{N-1}, x_N, x_1\}$, etc., obtained by circular permutation of the historic values. This procedure is also known as *index-sequential.* A major drawback with this procedure is that the resulting set of N input series yields N outputs which are not independent and, as a consequence, the outputs have less precision. However, this approach has been used in some cases.[74,87,97]

Use of Univariate Time-Series Models. Suppose that we need to generate N consecutive observations of an ARMA(p, q) model with known or estimated parameters.

The ARMA(p, q) model defined by Eq. (19.3.8) is rewritten here as

$$z_t = \phi_1 z_{t-1} + \cdots + \phi_p z_{t-p} + \epsilon_t - \theta_1 \epsilon_{t-1} - \cdots - \theta_q \epsilon_{t-q} \qquad (19.5.3)$$

where $z_t = y_t - \mu$ represents the deviations of the process from its mean, $\phi_1, \ldots, \phi_p$ and $\theta_1, \ldots, \theta_q$ are autoregressive and moving average coefficients, respectively, and ϵ_t is normally distributed with mean zero and standard deviation σ_ϵ. To generate the z_t's, one can simply generate ϵ_t as needed by the procedures described in Sec. 19.5.2. The problem is that to generate the first value z_1, one needs to known in advance the values $z_{-p+1}, \ldots, z_0$. A number of procedures can be used to get around this problem. A convenient procedure is to use a *warm-up period* in which $r + N$ values $z_{-r+1}, \ldots, z_0, z_1, \ldots, z_N$ are generated setting $z_t = 0$ (its mean) for $t \le 0$, the first r values are deleted, and only the remaining N are used. If r is large enough, this will remove the bias introduced by taking $z_t = 0$ for $t \le 0$ and the effect on $z_1, \ldots, z_N$ will be negligible. A value of r around 50 has been recommended[42] for generations based on low-order ARMA models. In addition, for some low-order ARMA models, the initialization of the generation can be made directly without using the warm-up period.[107,161]

Finally, the ARMA(p, q) normal series are generated by $y_t = \mu + z_t$. If the original data x_t has been previously transformed by using $y_t = g(x_t)$, in which $g(\cdot)$ is a transformation function [for instance, $y_t = \log(x_t)$], then the generated series in the real or original domain will be $x_t = g^{-1}(y_t)$, the inverse transformation function [for instance, $x_t = \exp(y_t)$ if the logarithmic transformation was used]. Likewise, the generation of periodic AR and periodic ARMA time series can be made by following similar procedures except that the periodic parameters must be considered in the generation algorithm.

Use of Multivariate Time-Series Models. Generally the principles involved in univariate generation can be extended to multivariate generation. For instance, one can generate a set of n cross-correlated normal variables $\boldsymbol{\epsilon}_t$ with variance-covariance matrix $\mathbf{G} = \mathbf{B}\mathbf{B}^T$ by using the simple stationary multivariate model (19.3.46) $\boldsymbol{\epsilon}_t = \mathbf{B}\boldsymbol{\xi}_t$ in which $\mathbf{B}$ is an n by n lower triangular matrix and $\boldsymbol{\xi}$ is an n by 1 vector of uncorrelated standard normal variables. First generate the set $\xi_1^{(1)}, \ldots, \xi_1^{(n)}$ of uncorrelated standard normal random numbers. Then, generate successively the set of cross-correlated numbers by expanding the multivariate model as

$$\begin{aligned}
\epsilon_1^{(1)} &= b^{11}\xi_1^{(1)} \\
\epsilon_1^{(2)} &= b^{21}\xi_1^{(1)} \qquad + b^{22}\xi_1^{(2)} \\
&\vdots \\
\epsilon_1^{(n)} &= b^{n1}\xi_1^{(1)} + \cdots + b^{nn}\xi_1^{(n)}
\end{aligned}$$

Finally, other sets of cross-correlated numbers $\boldsymbol{\epsilon}_t$ for $t = 2, \ldots$ can be generated by repeating the same procedure.

Likewise, one can generate series which are simultaneously autocorrelated and cross-correlated by using, for instance, multivariate AR(1) model (19.3.42), multivariate ARMA(1, 1) model (19.3.43), or the CARMA model (19.3.44). In any case, initial values are required to start the generation. The initialization can be made by using the warm-up period approach as in the univariate generation; however, a direct method is available for generation based on the multivariate ARMA(1, 1) model.[178] Finally, the foregoing generation procedures can be extended to the case of multivariate autocorrelated and cross-correlated periodic series by using the PAR(1) model (19.3.51), and PARMA(1, 1) model (19.3.52), or the contemporaneous PAR or

PARMA model (19.3.53). The generation procedure is similar to the cases of stationary models except that periodic parameters are used.[155,161]

Use of Disaggregation Models. Consider here as an illustration a generation based on the traditional Valencia-Schaake model. For ease of reference, rewrite model (19.3.58) as $\mathbf{Y}' = \mathbf{AX}' + \mathbf{B}\boldsymbol{\epsilon}$, where $\mathbf{Y}'$ is the vector of the series being generated (for instance, seasonal series), $\mathbf{X}'$ is the series to be disaggregated (for instance, annual series), $\boldsymbol{\epsilon}$ represents a vector of uncorrelated standard normal variables, and **A** and **B** are the parameters of the model. It is assumed that all random variables are normally distributed with mean zero, i.e., $\mathbf{Y}' = \mathbf{Y} - \bar{Y}$ and $\mathbf{X}' = \mathbf{X} - \bar{X}$. The parameter matrices obtained for a case of disaggregating annual data at one site into three seasons are

$$\mathbf{A} = \begin{bmatrix} 0.4821 \\ 0.4837 \\ 0.0342 \end{bmatrix} \qquad \mathbf{B} = \begin{bmatrix} 17.6242 & 0 & 0 \\ -14.9866 & 4.2467 & 0 \\ -2.6376 & -4.2467 & 0 \end{bmatrix}$$

Note that the column in matrix **A** adds to unity and all the columns in matrix **B** add to zero in order to preserve additivity of the seasonal flows to form the correct annual flow. In addition, the analysis of annual and seasonal data gives $\bar{x} = 461.04$, $\bar{y}_1 = 168.68$, $\bar{y}_2 = 269.00$, and $\bar{y}_3 = 23.36$. Likewise, the model for annual data is an AR(1) as $X'_t = 0.3X_{t-1} + \xi_t$ in which ξ_t is normal with mean zero and standard deviation 209.5, and $X_t = X'_t + 461.04$, where X_t is the annual flow value. Thus, seasonal series are generated by

$$\begin{bmatrix} y'_{v,1} \\ y'_{v,2} \\ y'_{v,3} \end{bmatrix} = \begin{bmatrix} 0.4821 \\ 0.4837 \\ 0.0342 \end{bmatrix} X'_v + \begin{bmatrix} 17.6242 & 0 & 0 \\ -14.9866 & 4.2467 & 0 \\ -2.6376 & -4.2467 & 0 \end{bmatrix} \begin{bmatrix} \epsilon_{v,1} \\ \epsilon_{v,2} \\ \epsilon_{v,3} \end{bmatrix} \tag{19.5.5}$$

where v denotes the year.

The step-by-step generating procedure is

1. Generate a sequence of annual values. For instance, three generated values are $X'_1 = 262.9$, $X'_2 = -287.5$, and $X'_3 = 90.9$.
2. Disaggregate the generated annual values into seasonal values. For the first year, generate three normal (0, 1) values such as $\epsilon_{1,1} = -0.319$, $\epsilon_{1,2} = 0.994$, and $\epsilon_{1,3} = 0.662$. Then Eq. (19.5.5) gives:

   $$y'_{1,1} = 0.4821 \times 262.9 + 17.6242 \times (-0.319) = 121.12$$
   $$y'_{1,2} = 0.4837 \times 262.9 - 14.9866 \times (-0.319) + 4.2467 \times 0.994 = 136.16$$
   $$y'_{1,3} = 0.0342 \times 262.9 - 2.6376 \times (-0.319) - 4.2467 \times 0.994 = 5.61$$

   In a similar manner, the rest of the seasonal values can be computed from the other annual values.
3. Add the seasonal means to the generated seasonal series as $y_{1,1} = 168.68 + 121.12 = 289.80$, $y_{1,2} = 269.00 + 136.16 = 405.16$, and $y_{1,3} = 23.36 + 5.61 = 28.97$.

19.5.4 Length and Number of Simulations

In any analysis involving Monte Carlo simulation, the questions arise of how long the sample size of the generated sequences should be and how many samples should be generated. Answers to such questions vary from one analysis to another depending on the problem under consideration. This section attempts to give some practical guides for some typical problems in hydrology.

Length of Generated Samples. The length of the generated sample can be the same, shorter, or longer than the length of the historical sample, depending on the particular situation. For instance, consider the case of evaluating a stochastic model in regard to its ability to reproduce historical statistics. The model can be verified theoretically to some extent, especially in relation to statistics such as the mean, standard deviation, and correlations. But the interest may be in regard to more complex statistics such as storage- and drought-related statistics which must be verified by simulation. In any case, the generated sample length should be equal to the length of the historical sample. Now, consider that data generation is required for design of a reservoir. Clearly, in this case, the length of the generated sample must be equal to the planning horizon or economic life of the reservoir being designed.

Likewise, reservoir operational studies require similar considerations. If the purpose of the generation is to test alternative operating rules, the length of the simulation should be equal to the length of the operational planning horizon. In the case that a reservoir is operated under a well-established rule and the purpose is to determine, for instance, the hydropower output, the length of the generation should also coincide with the operational planning horizon (the Western Area Power Administration of the U.S. Department of Energy typically uses 5 to 10 years for this purpose[17]). Finally, if the purpose of the simulation is to determine extreme droughts of a specified return period T, the length of the generation must coincide with T.

Number of Samples to Generate. Enough samples should be generated so that the required output statistics are estimated accurately. To a large extent, the number of samples depends on what statistic of the Monte Carlo simulation output is of interest for the problem at hand. For example, to determine the mean of the output O with a given accuracy, one can use the normal approximation to establish the $1-\alpha$ confidence limits on the population mean $\mu(O)$, from which one can write $P[-u_{1-\alpha/2}\ \sigma(O)/\sqrt{m} < \bar{O} - \mu(O) < u_{1-\alpha/2}\ \sigma(O)/\sqrt{m}] = 1-\alpha$ where $\bar{O}$ = sample mean of the output, $\sigma(O)$ = population variance, $u_{1-\alpha/2} = 1-\alpha/2$ quantile of the normal distribution with mean zero and variance one, $1-\alpha$ = confidence level, and m = sample size. Thus, if $\bar{O}$ must be within 0.1 $\sigma(O)$ of $\mu(O)$ with a probability $1-\alpha = 0.95$, then $u_{0.975} = 1.96$ and the sample size required is given by 1.96 $\sigma(O)/\sqrt{m} = 0.1\sigma(O)$, which gives $m = 384$. Likewise, for an accuracy of $0.2\sigma(O)$, $m = 96$. A better approximation may be generally obtained by using instead the confidence limits based on the t distribution. In this case, the number of samples is obtained, for instance, by solving $t_{1-\alpha/2,m-1} = 0.1\sqrt{m}$ for m [for $0.1\sigma(O)$ accuracy], where $t_{1-\alpha/2,m-1}$ is the $1-\alpha/2$ quantile of the t distribution with $m-1$ degrees of freedom.

Likewise, to determine the standard deviation of the output O with a given accuracy, one can establish the confidence limits on the population variance $\sigma^2(O)$. Thus, one can write

$$P\left(\frac{\sqrt{m-1}}{\sqrt{\chi^2_{1-\alpha/2,m-1}}} < \frac{\sigma(O)}{S(O)} < \frac{\sqrt{m-1}}{\sqrt{\chi^2_{\alpha/2,m-1}}}\right) = 1-\alpha$$

in which $S(O)$ = sample standard deviation and $\chi^2_{\beta,m-1}$ = β-quantile of the chi-square distribution with $m-1$ degrees of freedom. For example, for $m = 384$ and $1-\alpha = 0.95$, $P[0.934 < \sigma(O)/S(O) < 1.077] = 0.95$, which means that with a sample size of 384 one can determine the sample standard deviation of the output such that its ratio with the population standard deviation is within about 15 percent.

Furthermore, one may be interested in determining the distribution of the output O with a prescribed accuracy. In this case, one can use the confidence limits on the true distribution F based on the Kolomogorov theorem.[68] Thus, $P(\max|\hat{F} - F| < 1.36/\sqrt{m}) = 0.95$ states that the maximum absolute difference between the sample distribution $\hat{F}$ and the population distribution F is less than $1.36/\sqrt{m}$ with probability 0.95. For instance, for $m = 5000$, the error in estimating the distribution is less than 0.019 with probability 0.95.

In addition, some practical guides have been offered. For example, when data generation is required for designing a reservoir, if annual data are used, as many as 1000 samples may be needed to accurately define the probability distribution of the maximum storage required.[108] On the other hand, if monthly flows are used, fewer samples may be adequate. In general, the number of samples varies from about 300 for streams that exhibit low variability to the order of 1000 for streams with high variability.[108] Obviously, these practical guides are less precise than the statistical criteria given above.

19.5.5 Model and Parameter Uncertainty

Much of the material of the previous sections was presented without making explicit reference to model uncertainty and parameter uncertainty. However, the importance of such uncertainties for many applications of Monte Carlo simulation must be recognized.[81,89,90,194] Model uncertainty arises because hydrologic processes are inherently complex, so alternative mathematical formulations have been proposed to reproduce the historical record in a statistical sense. Thus, for a particular problem at hand, the hydrologist is faced with the problem of selecting a model among the several alternatives. For instance, in the case of stream-flow modeling, a number of alternative short-memory and long-memory models have been proposed. On the other hand, parameter uncertainty arises because the model parameters are estimated from historical samples which usually are small. Naturally, both model uncertainty and parameter uncertainty are closely related.[90]

Model uncertainty can be alleviated in a number of ways. One can use physical arguments in deriving the model structure. For instance, in the case of stream-flow modeling, conceptual arguments have been proposed to justify the use of ARMA and PARMA processes.[5,41,155] Likewise, in the case of modeling precipitation processes, conceptual arguments have been proposed to justify the family of cluster processes.[197–199] Given that a family of models has been selected, the problem is still to define the type of model within the family. Statistical considerations may be used to assist in identifying the type of model. For instance, in the case of ARMA models, the type of model is selected by using diagnostic checks and other criteria such as the Akaike information criteria.[1] Likewise, model selection can be made by Monte Carlo simulation experiments to compare the performance of competing models in reproducing historical statistics, especially those statistics which have not been used in parameter estimation and those which are pertinent to the study at hand. Furthermore, in complex systems involving several sites, the issue of model selection is really more in terms of selecting among model strategies involving an array of univariate, multivariate, and disaggregation models. In these cases experience will likely provide the best solution.

TABLE 19.6.1 Summary of General-Purpose Programs for Time-Series Analysis, Modeling, and Forecasting

Name	Brief description	Reference
1. BMDP	Time-series analysis, modeling, and forecasting, estimation of missing data, analysis of variance, and nonparametric tests. Time-series modules include univariate and bivariate spectral analysis, intervention analysis, and ARMA and transfer-function models. Estimation of missing data includes modules based on regression analysis. Nonparametric analysis modules include several tests for detection of shifts.	BMDP[8] BMDP Statistical Software 1440 Sepulveda Blvd., Suite 316 Los Angeles, CA 90025 U.S.A.
2. IMSL	Time-series analysis including modeling simulation and forecasting, analysis of variance, and nonparametric tests. Modeling, simulation, and forecasting are based on ARMA models. Forecasting includes Kalman filtering. Programs for nonparametric tests include tests for detection of trends and shifts, tests of randomness, and tests of goodness of fit.	IMSL[76] IMSL 14141 Southwest Freeway, Suite 3000 Sugarland, TX 77478-3498 U.S.A.
3. ITSM	Time-series analysis and modeling and forecasting based on ARMA models.	Brockwell and Davis[16] ITSM Statistical Dept. Colorado State University Fort Collins, CO 80523 U.S.A.
4. MINITAB	Time-series analysis and modeling based on ARMA models. It includes programs for analysis of variance and for nonparametric tests to detect trends and shifts and test for randomness.	Minitab[123] Minitab Inc. 3081 Enterprise Dr. State College, PA 16801-3008 U.S.A.
5. SAS/ETS	Time-series analysis and modeling and forecasting based on ARMA models.	SAS/ETS[150] SAS Institute Inc. SAS Campus Dr. Cary, NC 27513 U.S.A.
6. SPSS	Time-series analysis, modeling and forecasting, analysis of variance and nonparametric tests. Modeling and forecasting are based on ARMA models. Programs for nonparametric tests include tests for detection of shifts, tests of randomness and tests of goodness of fit.	SPSS[174] SPSS Inc. 444 N. Michigan Ave. Chicago, IL 60611-3962 U.S.A.

TABLE 19.6.1 Summary of General-Purpose Programs for Time-Series Analysis, Modeling, and Forecasting *(Continued)*

Name	Brief description	Reference
7. Statgraphics	Time-series analysis and modeling, analysis of variance, and nonparametric tests. Time-series modeling is based on ARMA models. Modules on nonparametric methods include tests for detection of shifts, randomness, and goodness of fit.	Statgraphics[175] STSC Inc. 2115 E. Jefferson St. Rockville, MD 20852 U.S.A.

Parameter uncertainty is a major issue in stochastic modeling and simulation studies.[90,91] Procedures have been developed to deal with parameter uncertainty,[116,155,179,188,191] although in practice they are rarely applied. For instance, consider the AR(1) model of Eq. (19.3.2). To include parameter uncertainty in data generation studies based on this model, one needs to know the distribution of the parameters μ, ϕ_1, and σ_ϵ^2 given a historical sample of size N. In this case the sample distributions are[116,155] $\mu \sim N\{\hat{\mu}, \hat{\sigma}_\epsilon^2/[(1-\hat{\phi}_1)^2 N]\}$, $\phi_1 \sim N[\hat{\phi}_1, (1-\hat{\phi}_1^2)/(N-1)]$ and $\sigma_\epsilon^2 \sim N[\hat{\sigma}_\epsilon^2, 2\hat{\sigma}_\epsilon^2/N]$, where $\hat{\mu}$, $\hat{\phi}_1$, and $\hat{\sigma}_\epsilon^2$ denote the estimates of μ, ϕ_1, and σ_ϵ^2 obtained from the historical sample and $\sim$ denotes *distributed as.* The procedure to simulate m sequences of length n which includes parameter uncertainty is:[116,155] (1) follow steps 2 and 3 for $i = 1, \ldots, m$; (2) for sequence i, generate the parameter set $\mu(i)$, $\phi_1(i)$, and $\sigma_\epsilon^2(i)$ from the above normal distribution, respectively; and (3) using model (19.3.2) and the parameters $\mu(i)$, $\phi_1(i)$, and $\sigma_\epsilon^2(i)$, generate the sequence $y_1(i)$, $y_2(i)$, . . . , $y_n(i)$ following the approach suggested in Sec. 19.5.3. Similar procedures for univariate ARMA and PAR models are also available.[116,155,179] For multivariate models the problem is more complex; however, bayesian procedures are available.[188] In complex modeling and simulation studies involving seasonal data and multiple sites, the consideration of parameter uncertainty is generally complex. However, since such complex systems usually involve an array of models, it is always possible to include parameter uncertainty at least in some parts of the modeling process.[155,179]

19.6 COMPUTER PROGRAMS FOR TIME-SERIES ANALYSIS

Analysis, modeling, and simulation of hydrologic time series can be done effectively with the aid of computers. Several alternative software packages are available. Some of them are general-purpose programs for analysis of any kind of time series and others are specifically oriented for hydrologic time series. General-purpose program packages, such as those included in Table 19.6.1, are attractive because usually they are well-documented, have more statistical features, and are accompanied by good

TABLE 19.6.2 Summary of Special Programs for Hydrologic Time-Series Analysis, Modeling, and Simulation

Name	Brief description	Reference
1. HEC-4	Analysis, modeling, and simulation of multiple time series of monthly flows based on multiple linear regression models. It can also generate synthetic monthly flows at ungauged sites.	U.S. Army Corps of Engineers[187] Dept. of the Army Hydrologic Engineering Center 609 2d St. Davis, CA 95616-4687 U.S.A.
2. LAST	Analysis, modeling, and simulation of multiple annual and seasonal streamflow data. The main features of the approach are (1) preservation of annual serial correlation and annual cross-correlations, (2) generation of "key" stations annual flows and disaggregation of these values into component substations on an annual basis, and (3) disaggregation of annual flows into seasonal flows preserving both season-to-season correlations and cross-correlation between sites. Modeling and simulation are based on univariate and multivariate AR(1) and AR(2) models and disaggregation models.	Lane,[98] Lane and Frevert[100, 101] LAST Bureau of Reclamation, D-5077 Earth Sciences Div. P.O. Box 25007 Denver, CO 80225-007 U.S.A.
3. SPIGOT	Three modeling and generation schemes are used: (1) aggregated annual flow for the entire basin is generated by a univariate model, then is disaggregated into basin (aggregated) monthly flows, which in turn are disaggregated into key site monthly values; (2) aggregated annual flow for the entire basin is generated by a univariate model, then is disaggregated into monthly flows at key sites in a single step; and (3) annual flows at key sites are generated by a multivariate model and are disaggregated into monthly flows by a multivariate disaggregation model. In all schemes, modeling and simulation of annual flows are based on either univariate or multivariate AR(0) or AR(1) models, as the case may be.	Grygier and Stedinger[61] Dr. J. R. Stedinger School of Civil Engineering Hollister Hall Cornell University Ithaca, NY 14853-3501 U.S.A.

TABLE 19.6.2 Summary of Special Programs for Hydrologic Time-Series Analysis, Modeling, and Simulation *(Continued)*

Name	Brief description	Reference
4. CSUPAC1	Consists of programs CSU001 and CSU002 for modeling and generation of single-site hydrologic series and programs CSU003 and CSU004 for modeling and generation of multisite series. Univariate modeling is based on PAR(0), PAR(1), PAR(2), and PARMA(1, 1) models, and multivariate modeling is based on low-order contemporaneous PAR or contemporaneous PARMA models. Options for alternative transformations and Fourier series analysis are included.	Dr. J. D. Salas Engineering Research Center Colorado State University Fort Collins, CO 80523 U.S.A.
5. WASIM	Consists of programs WASIM1 and WASIM2 for modeling and generation of hydrologic time series based on stationary ARMA models, and program WASIM3, which includes parameter uncertainty in the generation.	McLeod and Hipel[116] Dr. Angus I. McLeod Statistics and Actuarial Science Group University of Western Ontario London, Ontario Canada N6A5B9

graphical display capabilities. However, generally, they are expensive. Furthermore, most general-purpose packages, for the most part, do not consider the model structures that one normally finds in hydrologic time series (an exception is the family of stationary ARMA models); estimation based on short samples, which is a typical case in hydrology; or periodicity in the covariance structure, which is also typical in seasonal hydrologic time series. Nor do they consider aggregation and disaggregation schemes on direct approaches to deal with nonnormal series.

On the other hand, programs developed specifically for hydrologic time series, such as those included in Table 19.6.2, have the advantage that the underlying models and estimation procedures involve features that are unique to hydrologic time series. However, no single package can handle all cases that may arise in practice, yet most packages can be applicable to the typical cases. Likewise, most packages have been developed for modeling seasonal and annual hydrologic series, such as stream-flow series, while packages for modeling and generation of short-term processes such as hourly rainfall are not readily available.

Generally, any computer package must be used with care. This is especially true in programs that do estimation and generation all at once. In this section, general-purpose programs and special programs for hydrologic time-series analysis, modeling and generation are presented in summarized form in Tables 19.6.1 and 19.6.2. The intent here is to make hydrologists aware of what computer packages are available without attempting detailed comparisons among them.

REFERENCES

1. Akaike, H., "A New Look at the Statistical Model Identification," *IEEE Trans. Automat. Contr.*, vol. AC-19, pp. 716–723, 1924.
2. Bacchi, B., P. Burlando, and R. Rosso, "Extreme Value Analysis of Stochastic Models of Point Rainfall," 3d Scientific Assembly of the IAHS, Poster Session, Baltimore, May 10–19, 1989.
3. Bardossy, A., and E. J. Plate, "Modeling Daily Rainfall Using a Semi-Markov Representation of Circulation Pattern Occurrence," *J. Hydrol.*, vol. 122, pp. 33–47, 1991.
4. Bartolini, P., and J. D. Salas, "Properties of Multivariate Periodic ARMA(1, 1) Models," Int. Symp. Multivariate Analysis in Hydrologic Processes, Colorado State University, Fort Collins, Colo., July 15–17, 1985.
5. Bartolini, P., and J. D. Salas, "Modeling of Streamflow Processes at Different Time Scales," submitted to *Water Resour. Res.*, 1992.
6. Beauchamp, J. J., D. J. Downing, and S. F. Railsback, "Comparison of Regression and Time Series Methods for Synthesizing Missing Streamflow Records," *Water Resour. Bull.*, vol. 25, no. 5, pp. 961–975, 1989.
7. Benjamin, J. R., and C. A. Cornell, "Probability, Statistics and Decisions for Civil Engineers," McGraw-Hill, New York, 1970.
8. BMDP, *BMDP Statistical Software,* University of California Press, 1981.
9. Bobee, B., and R. Robitaille, "Correction of Bias in the Estimation of the Coefficient of Skewness," *Water Resour. Res.*, vol. 11, no. 6, pp. 851–854, 1975.
10. Box, G. E. P., and D. R. Cox, "An Analysis of Transformations," *J. Royal Stat. Soc.*, Series B, vol. 26, pp. 211–252, 1964.
11. Box, G. E. P., and Pierce, D. A., "Distribution of Autocorrelations in Autoregressive Integrated Moving Average Time Series Models," *J. Am. Stat. Assoc.*, vol. 65, pp. 1509–1526, 1970.
12. Box, G. E. P., and G. M. Jenkins, "Time Series Analysis Forecasting and Control," Holden-Day, San Francisco, 1976.
13. Bras, R. L., and I. Rodriguez-Iturbe, *Random Functions and Hydrology,* Addison-Wesley, Reading, Mass., 1985.
14. Bratley, P., B. L. Fox, and L. E. Schrage, *A Guide to Simulation* (2d ed.), Springer-Verlag, New York, 1987.
15. Brockwell, P. J., and R. A. Davis, *Time Series: Theory and Methods* (2d ed.), Springer-Verlag, New York, 1991*a*.
16. Brockwell, P. J., and R. A. Davis, *ITSM: An Interactive Time Series Modeling Package for the PC,* Springer-Verlag, New York, 1991*b*.
17. Buishand, T. A., "Stochastic Modeling of Daily Rainfall Sequences," Dept. of Mathematics and Dept. of Land Water Use, Agricultural University, Wageningen, The Netherlands, 1977.
18. Burges, S. J., "Some Problems with Lognormal Markov Runoff Models," *J. Hydraul. Div., Am. Soc. Cir. Eng.* (now *J. Hydraul. Eng.*), vol. 98, no. HY9, pp. 1487–1496, 1972.
19. Burlando, P., "Modelli Stochastici per la Simulazione e la Previsione della Precipitazione nel Tempo," Ph.D. thesis, Institute of Hydraulics, Politecnico di Milano, Italy, 1989.
20. Burlando, P., and R. Rosso, "Stochastic Models of Temporal Rainfall, Reproducibility, Estimation and Prediction of Extreme Events," NATO Advanced Study Institute on Stochastic Hydrology in Water Resources Systems: Simulation and Optimization, Peñiscola, Spain, Sept. 18–29, 1989.
21. Cadavid, L. G., J. D. Salas, and D. C. Boes, "Disaggregation of Short-Term Precipitation

Records," water resources paper 106, Colorado State University, Fort Collins, Colo., 1992.

22. Camacho, F., A. I. McLeod, and K. W. Hipel, "Developments in Multivariate ARMA Modeling in Hydrology," Int. Symp. on Multivariate Analysis of Hydrologic Processes, Colorado State University, Fort Collins, Colo., July 15–17, 1985.
23. Carlson, R. F., A. J. MacCormick, and D. G. Watts, "Application of Linear Models to Four Annual Streamflows Series," *Water Resour. Res.*, vol. 6, pp. 1070–1078, 1970.
24. Chang, T. J., M. L. Kavvas, and J. W. Delleur, "Daily Precipitation Modeling by Discrete Autoregressive Moving Average Processes," *Water Resour. Res.*, vol. 20, no. 5, pp. 565–580, 1984.
25. Chebaane, M., J. D. Salas, and D. C. Boes, "Product Autoregressive Process for Modeling Intermittent Monthly Streamflows," paper submitted to *Water Resour. Res.*, 1992.
26. Chin, E. H., "Modeling Daily Precipitation Occurrence Process with Markov Chain," *Water Resour. Res.*, vol. 13, no. 6, pp. 949–956, 1977.
27. Conover, W. J., "Practical Non-Parametric Statistics," Wiley, New York, 1971.
28. Croley, T. E., II, "Sequential Stochastic Optimization in Water Resources," Ph.D. dissertation, Department of Civil Engineering, Colorado State University, Fort Collins, Colo., 1972.
29. Croley, T. E., and K. N. Rao, "A Manual for Hydrologic Time Series Deseasonalization and Serial Dependence Reduction," rep. 199, Iowa Institute of Hydraulic Research, University of Iowa, Iowa City, 1976.
30. Crosby, D. S., and T. Maddock, III, "Estimating Coefficients of a Flow Generator for Monotone Samples of Data," *Water Resour. Res.*, vol. 6, no. 4, pp. 1079–1086, 1970.
31. Curry, K., and R. L. Bras, "Theory and Applications of the Multivariate Broken Line, Disaggregation and Monthly Autoregressive Streamflow Generators to the Nile River," TAP report 78-5, MIT, Cambridge, Mass., 1978.
32. Delleur, J. W., P. C. Tao, and M. L. Kavvas, "An Evaluation of the Practicality and Complexity of Some Rainfall and Runoff Time Series Models," *Water Resour. Res.*, vol. 12, no. 5, pp. 953–970, 1976.
33. Delleur, J. W., and M. L. Kavvas, "Stochastic Models for Monthly Rainfall Forecasting and Synthetic Generation," *J. App. Meteor.*, (now *J. Clim. Appl. Meteorol.*), vol. 17, no. 10, pp. 1528–1536, 1978.
34. Draper, N. R., and H. Smith, *Applied Regression Analysis,* (2d ed.), Wiley, New York, 1981.
35. Eagleson, P., "Climate, Soil and Vegetation: 2. The Distribution of Annual Precipitation Derived from Observed Storm Sequences," *Water Resour. Res.*, vol. 14, no. 5, pp. 713–721, 1978.
36. Entekhabi, D., I. Rodriguez-Iturbe, and P. S. Eagleson, "Probabilistic Representation of the Temporal Rainfall Process by a Modified Neyman-Scott Rectangular Pulse Model: Parameter Estimation Validation," *Water Resour. Res.*, vol. 25, no. 2, pp. 295–302, 1989.
37. Feller, W., "The Asymptotic Distribution of the Range of Sums of Independent Random Variables," *Annals Math. Stat.*, vol. 22, pp. 427–432, 1951.
38. Fernandez, B., and J. D. Salas, "Periodic Gamma Autoregressive Processes for Operational Hydrology," *Water Resour. Res.*, vol. 22, no. 10, pp. 1385–1396, 1986.
39. Fernandez, B., and J. D. Salas, "Gamma-Autoregressive Models for Streamflow Simulation," *J. Hydraul. Eng.*, vol. 116, no. 11, pp. 1403–1414, 1990.
40. Feyerherm, A. M., and L. D. Bark, "Statistical Methods for Persistent Precipitation Patterns," *J. App. Meteorol.*, (now *J. Clim. Appl. Meteorol.*), vol. 4, pp. 320–328, 1964.
41. Fiering, M. B., *Streamflow Synthesis,* Harvard University Press, Cambridge, Mass., 1967.

42. Fiering, M. B., and B. B. Jackson, "Synthetic Streamflows," water resources monograph 1, American Geophysical Union, Washington D.C., 1971.

43. Foufoula-Georgiou, E., and P. Guttorp, "Compatibility of Continuous Rainfall Occurrence Models with Discrete Rainfall Observations," *Water Resour. Res.*, vol. 22, no. 8, pp. 1316–1322, 1986.

44. Foufoula-Georgiou, E., and D. P. Lettenmaier, "A Markov Renewal Model of Rainfall Occurrences," *Water Resour. Res.*, vol. 23, no. 5, pp. 875–884, 1987.

45. Foufoula-Georgiou, E., and P. Guttorp, "Assessment of a Class of Neyman-Scott Models for Temporal Rainfall," *J. Geophys. Res.*, vol. 92, no. D8, pp. 9679–9682, 1987.

46. Franz, D. D., B. A. Kraeger, and R. K. Linsley, "A System for Generating Long Streamflow Records for Study of Floods of Long Return Period," NUREG/CR-4496, vol. 2, Linsley, Kraeger Associates, Ltd., 1988.

47. Freeze, R. A., "A Stochastic-Conceptual Analysis of One-Dimensional Groundwater Flow in Nonuniform Homogeneous Media," *Water Resour. Res.*, vol. 11, no. 5, pp. 725–741, 1975.

48. Frevert, D. K., M. S. Cowan, and W. L. Lane, "Use of Stochastic Hydrology in Reservoir Operation," *J. Irrig. Drain. Eng.*, vol. 115, no. 3, pp. 334–343, 1989.

49. Frick, D. M., D. Bode, and J. D. Salas, "Effect of Drought on Urban Water Supplies: I. Drought Analysis," *J. Hydraul. Eng.*, vol. 116, no. 6, pp. 733–753, 1990.

50. Gabriel, K. R., and J. Neumann, "A Markov Chain Model for Daily Rainfall Occurrences at Tel Aviv," *Royal Meteorol. Soc. Quart. J.*, vol. 88, pp. 90–95, 1962.

51. Gani, J., "Recent Advances in Storage and Flooding Theory," *Adv. Appl. Prob.*, vol. 1, pp. 90–110, 1969.

52. Gates, P., and H. Tong, "On Markov Chain Modeling to Some Weather Data," *J. Appl. Meteorol.*, (now *J. Clim. Appl. Meteorol.*) vol. 15, pp. 1145–1151, 1976.

53. Gilbert, R. O., *Statistical Methods for Environmental Pollution Monitoring,* Van Nostrand Reinhold, New York, 1987.

54. Gilroy, E. J., "Reliability of a Variance Estimate Obtained from a Sample Augmented by Multivariate Regression," *Water Resour. Res.*, vol. 6, no. 6, pp. 1595–1600, 1970.

55. Gomez-Hernandez, J. J., and S. M. Gorelick, "Effective Groundwater Model Parameter Values: Influence of Spatial Variability of Hydraulic Conductivity, Leakage and Recharge," *Water Resour. Res.*, vol. 25, no. 3, pp. 405–419, 1989.

56. Gomide, F. S., "Range and Deficit Analysis Using Markov Chains," hydrology paper 79, Colorado State University, Fort Collins, Colo., 1975.

57. Grace, R., and P. Eagleson, "The Synthesis of Short-Time Increment of Rainfall Sequences," MIT Hydrodynamics Laboratory report 91, MIT Press, Cambridge, Mass., 1966.

58. Graybill, F. A., "Theory and Application of the Linear Model," Wadsworth & Brooks/Cole Advanced Books and Software, Pacific Grove, Calif., 1976.

59. Grygier, J. C., and J. R. Stedinger, "Condensed Disaggregation Procedures and Conservation Corrections for Stochastic Hydrology," *Water Resour. Res.*, vol. 24, no. 10, pp. 1584, 1988.

60. Grygier, J. C., J. R. Stedinger, and H. B. Yin, "A Generalized Maintenance of Variance Extension Procedure for Extending Correlated Series," *Water Resour. Res.*, vol. 25, no. 3, pp. 345–349, 1989.

61. Grygier, J. C., and J. R. Stedinger, "SPIGOT, A Synthetic Streamflow Generation Software Package," technical description, version 2.5, School of Civil and Environmental Engineering, Cornell University, Ithaca, N.Y., 1990.

62. Haan, C. T., D. M. Allen, and J. O. Street, "A Markov Chain Model for Daily Rainfall," *Water Resour. Res.*, vol. 12, no. 3, pp. 443–449, 1976.

63. Haltiner, J. P., and J. D. Salas, "Development and Testing of a Multivariate Seasonal ARMA(1, 1) Model," *J. Hydrol.,* vol. 104, pp. 247–272, 1988.

64. Hannan, E. J., "A Test for Singularities in Sydney Rainfall," *Aust. J. Phys.* vol. 8, no. 2, pp. 289–297, 1955.

65. Heyder, W. E., T. K. Gates, D. G. Fontane, and J. D. Salas, "Multicriterion Strategic Planning for Improved Irrigation Delivery: II. Application," *J. Irrig. Drain. Eng.,* vol. 117, no. 6, pp. 914–934, 1991.

66. Hipel, K. W., W. C. Lennox, T. E. Unny, and A. I. McLeod, "Intervention Analysis in Water Resources," *Water Resour. Res.,* vol. 11, no. 6, pp. 855–861, 1975.

67. Hipel, K. W., A. I. McLeod, and W. C. Lennox, "Advances in Box-Jenkins Modeling—1. Modeling Construction," *Water Resour. Res.,* vol. 13, no. 3, pp. 567–575, 1977.

68. Hipel, K. W., and A. I. McLeod, "Preservation of the Rescaled Adjusted Range: 2. Simulation Studies Using Box-Jenkins Models," *Water Resour. Res.,* vol. 14, no. 3, pp. 509–516, 1978.

69. Hirsch, R. M., "Synthetic Hydrology and Water Supply Reliability," *Water Resour. Res.,* vol. 15, no. 6, pp. 1603–1615, 1979.

70. Hirsch, R. M., "A Comparison of Four Record Extension Techniques," *Water Resour. Res.,* vol. 18, no. 4, pp. 1081–1088, 1982.

71. Hirsch, R. M., J. R. Slack, and R. A. Smith, "Techniques of Trend Analysis for Monthly Water Quality Data," *Water Resour. Res.,* vol. 18, no. 1, pp. 107–121, 1982.

72. Hirsch, R. M., and S. R Slack, "A Nonparametric Trend Test for Seasonal Data with Serial Dependence," *Water Resour. Res.,* vol. 20, no. 6, pp. 727–732, 1984.

73. Hollander, M., and D. A. Wolfe, "Nonparametric Statistical Methods," Wiley, New York, 1973.

74. House, P. M., and Ungvari, J. L., "Indexed Sequential Modeling of Hydropower System Operations," pp. 1046–1055, *Proc. Water Power '83 Inter. Conf. on Hydropower,* vol. 2, Knoxville, Tenn., 1983.

75. Hurst, H. E., "Long Term Storage Capacities of Reservoirs," *Trans. Am. Soc. Civ. Eng.,* vol. 116, pp. 776–808, 1951.

76. IMSL, *IMSL User's Manual,* IMSL Inc., Houston, Texas, 1984.

77. Islam, S., R. L. Bras, and I. Rodriguez-Iturbe, "Multidimensional Modeling of Cumulative Rainfall: Parameter Estimation and Model Adequacy Through a Continuum of Scales," *Water Resour. Res.,* vol. 24, no. 7, pp. 985–992, 1988.

78. Islam, S., R. L. Bras, and I. Rodriguez-Iturbe, "Comment on 'On Parameter Estimation of Temporal Rainfall Models,' by J. T. B. Obeysekera et al.," *Water Resour. Res.,* vol. 25, no. 4, p. 764, 1989.

79. Jacobs, P. A., and P. A. W. Lewis, "Discrete Time Series Generated by Mixtures—3. Autoregressive Process (DAR(P))," tech. rep. NPS 55-78-022, Naval Postgraduate School, Monterey, Calif., 1978.

80. Jenkins, G. M., and D. G. Watts, "Spectral Analysis and its Applications," Series in Time-Series Analysis, Holden-Day, San Francisco, 1969.

81. Jettmar, R. U., and G. K. Young, "Hydrologic Estimation and Economic Regret," *Water Resour. Res.,* vol. 11, no. 5, pp. 648–655, 1975.

82. Joiner, B. L., and J. R. Rosenblath, "Some Properties of the Range in Samples from Tukey's Symmetric Lambda Distributions," *J. Amer. Stat. Assoc.,* vol. 66, pp. 394–399, 1971.

83. Jones, L., "Explicit Monte Carlo Simulation Head Moment Estimates for Stochastic Confined Groundwater Flow," *Water Resour. Res.,* vol. 26, no. 6, pp. 1145–1153, 1990.

84. Kavvas, M. L., and J. W. Delleur, "The Stochastic and Chronological Structure of Rainfall Sequences—Application to Indiana," technical report 57, Purdue University, Water Resources Center, 1975.

85. Kavvas, M. L., and J. W. Delleur, "A Stochastic Cluster Model of Daily Rainfall Sequences," *Water Resour. Res.*, vol. 17, no. 4, pp. 1151–1160, 1981.

86. Kendall, M. G., "Rank Correlation Methods," Charles Griffin, London, 1975.

87. Kendall, D. R., and Dracup, J. A., "A Comparison of Index-Sequential and AR(1) Generated Hydrologic Sequences," *J. Hydrol.*, vol. 122, pp. 335–352, 1991.

88. Kirby, W., "Algebraic Boundedness of Sample Statistics," *Water Resour. Res.*, vol. 10, no. 2, pp. 220–222, 1974.

89. Klemes, V., "The Unreliability of Reliability Estimates of Storage Reservoir Performance Based on Short Streamflow Records," pp. 127–136. *Proc. Int. Symp. on Risk and Reliability in Water Resources Systems,* University of Waterloo, Waterloo, Ontario, Canada, vol. 1, 1978.

90. Klemes, V., R. Srikanthan, and T. A. McMahon, "Long-Memory Flow Models in Reservoir Analysis: What Is Their Practical Value?" *Water Resour. Res.*, vol. 17, no. 3, pp. 737–751, 1981.

91. Knuth, D. E., *Seminumerical Algorithms,* (2d ed.), vol. 2, *The Art of Computer Programming,* Addison-Wesley, Reading, Mass., 1981.

92. Koepsell, R. W., and J. B. Valdes, "Multidimensional Rainfall Parameter Estimation from a Sparse Network," *J. Hydraul. Eng.*, vol. 117, no. 7, pp. 832–850, 1991.

93. Kottegoda, N. T., and J. Elgy, "Infilling Missing Flow Data," pp. 60–73, Proc. Fort Collins 3d Int. Symp. on Applied and Theoretical Hydrology, Water Resources Publications, Littleton, Colo., 1979.

94. Kottegoda, N. T., "Investigation of Outliers in Annual Maximum Flow Series," *J. Hydrol.*, vol. 72, pp. 105–137, 1984.

95. Kottegoda, N. T., "Assessment of Non-Stationarity in Annual Series Through Evolutionary Spectra," *J. Hydrol.*, vol. 76, pp. 381–402, 1985.

96. Krajewski, W. F., and J. A. Smith, "Sampling Properties of Parameter Estimators for a Storm Field Rainfall Model," *Water Resour. Res.*, vol. 25, no. 9, pp. 2067–2075, 1989.

97. Labadie, J. W., D. Fontane, G. Q. Tabios, and N. F. Chou, "Stochastic Analysis of Dependable Hydropower Capacity," *J. Water Resour. Plann. Manage.*, vol. 113, no. 3, pp. 422–437, 1987.

98. Lane, W. L., "Applied Stochastic Techniques (Last Computer Package); User Manual," Division of Planning Technical Services, U.S. Bureau of Reclamation, Denver, Colo., 1979.

99. Lane, W. L., "Corrected Parameters Estimates for Disaggregation Schemes," in *Statistical Analysis of Rainfall and Runoff,"* V. Singh, ed., Water Resources Publications, Littleton, Colo., 1982.

100. Lane, W. L., and D. K. Frevert, "Applied Stochastic Techniques, User Manual," 6th Revision, Earth Sciences Division, U.S. Bureau of Reclamation, Denver, Colo., 1989.

101. Lane, W. L., and D. K. Frevert, "Applied Stochastic Techniques, Personal Computer Version 5.2, User's Manual," Earth Sciences Division, U.S. Bureau of Reclamation, Denver, Colo., 1990.

102. Lane, W. L., "Synthetic Streamflows for Global Climate Change," USGS open file report 91-244, *Proc. US-PRC Bilateral Symp. on Droughts and Arid Region Hydrology,* Tucson, Ariz., Sept. 16–20, 1991.

103. Lawrance, A. J., and P. A. W. Lewis, "A New Autoregressive Time Series Model in Exponential Variables [NEAR(1)]," *Adv. Appl. Prob.*, vol. 13, no. 4, pp. 826–845, 1981.

104. Le Cam, L. A., "A Stochastic Description of Precipitation," *Proc. 4th Berkeley Symp. Mathematics, Statistics and Probability,* J. Newman, ed., pp. 165–186, University of California Press, Berkeley, 1961.

105. Lee, A. F. S., and S. M. Heghinian, "A Shift of the Mean Level in a Sequence of Indepen-

dent Normal Random Variables—a Bayesian Approach," *Technometrics,* vol. 19, pp. 503–506, 1977.

106. Lettenmaier, D. P., "Detection of Trends in Water Quality Data from Records with Dependent Observations," *Water Resour. Res.,* vol. 12, pp. 1037–1046, 1976.

107. Lettenmaier, D. P., and S. J. Burges, "Operational Assessment of Hydrologic Models of Long-Term Persistence," *Water Resour. Res.,* vol. 13, no. 1, pp. 113–124, 1977.

108. Linsley, R. K., M. A. Kohler, and J. L. H. Paulhus, *Hydrology for Engineers* (3d ed.), McGraw-Hill, New York, 1982.

109. Lloyd, E. H., "A Probability Theory of Reservoirs with Serially Correlated Inputs," *J. Hydrol.,* vol. 1, pp. 99–128, 1963.

110. Mandelbrot, B. B., and J. W. Van Ness, "Fractional Brownian Motions, Fractional Noises and Applications," *SIAM Rev.,* vol. 10, no. 4, pp. 422–437, 1968.

111. Mandelbrot, B. B., and J. R. Wallis, "Computer Experiments with Fractional Gaussian Noises: Part 1, Averages and Variances," *Water Resour. Res.,* vol. 5, no. 1, pp. 228–241, 1969.

112. Marsaglia, G., and T. A. Bray, "On Line Random Number Generators and Their Use in Combinations," *Comm. Assoc. Comp. Mach.,* vol. 11, pp. 757–759, 1968.

113. Matalas, N. C., and B. Jacobs, "A Correlation Procedure for Augmenting Hydrologic Data," *U.S. Geological Survey professional paper* 434-E, pp. E1–E7, 1964.

114. Matalas, N. C., "Mathematical Assessment of Synthetic Hydrology," *Water Resour. Res.,* vol. 3, no. 4, pp. 937–945, 1967.

115. McLeod, A. I., K. W. Hipel, and W. C. Lennox, "Advances in Box-Jenkins Modeling—2. Applications," *Water Resour. Res.,* vol. 13, no. 3, pp. 577–586, 1977.

116. McLeod, A. I., and K. W. Hipel, "Simulation Procedures for Box-Jenkins Models," *Water Resour. Res.,* vol. 14, no. 5, pp. 969–975, 1978.

117. McMahon, T. A., and R. G. Mein, "River and Reservoir Yield," Water Resources Publications, Littleton, Colo., 1986.

118. Mejia, J. M., I. Rodriguez-Iturbe, and D. R. Dawdy, "Streamflow Simulation: 2. The Broken Line Process as a Potential Model for Hydrologic Simulation," *Water Resour. Res.,* vol. 8, no. 4, pp. 931–941, 1972.

119. Mejia, J. M., and J. Millan, "Una Metodologia para Tratar el Problema de Matrices Inconsistentes en la Generation Multivariada de Series Hidrologicas," VI Congreso Latino Americano de Hidraulica, Bogota, Colombia, 1974.

120. Mejia, J. M., and J. Rousselle, "Disaggregation Models in Hydrology Revisited," *Water Resour. Res.,* vol. 12, no. 2, pp. 185–186, 1976.

121. Mendonca, A., J. D. Salas, and M. W. Abdelmohsen, "Multiplicative PARMA Model for Seasonal Streamflow Simulation," submitted to *Water Resour. Res.,* 1992.

122. Mimikou, M., "Daily Precipitation Occurrences Modelling with Markov Chain of Seasonal Order," *Hydrol. Sci. J.,* vol. 28, nos. 2, 6, pp. 221–232, 1983.

123. Minitab, "Minitab Reference Manual," release 7, Minitab Inc., State College, Pa., 1989.

124. Mood, A. M., F. A. Graybill, and D. C. Boes, *Introduction to the Theory of Statistics* (3d ed.), McGraw-Hill, New York, 1974.

125. Moran, M. A., "On Estimators Obtained from a Sample Augmented by Multiple Regression," *Water Resour. Res.,* vol. 10, no. 1, pp. 81–85, 1974.

126. Moran, P. A. P., "A Probability Theory of Dams and Storage Systems," *Austral. J. Appl. Sci.,* vol. 5, pp. 116–124, 1954.

127. Neyman, J., and E. L. Scott, "Statistical Approach to Problems of Cosmology," *J. Roy. Statistical Soc.,* Ser. B, vol. 20, no. 1, pp. 1–43, 1958.

128. Obeysekera, J. T. B., G. Tabios, and J. D. Salas, "On Parameter Estimation of Temporal Rainfall Models," *Water Resour. Res.,* vol. 23, no. 10, pp. 1837–1850, 1987.

129. O'Connell, P. E., "A Simple Stochastic Modeling of Hurst's Law," in *Mathematical Models in Hydrology,* Symposium, Warsaw (IAHS Pub. No. 100, 1974), vol. 1, pp. 169–187, 1971.

130. O'Connell, P. E., "Stochastic Modeling of Long-Term Persistence in Streamflow Sequences," unpublished Ph.D. dissertation, Imperial College of Science and Technology, University of London, England, 1974.

131. Odeh, R. E., and J. O. Evans, "Algorithm AS 70: Percentage Points of the Normal Distribution," *Appl. Stat.,* vol. 23, pp. 96–97, 1974.

132. Ozturk, A., and R. F. Dale, "Least Squares Estimation of the Parameters of the Generalized Lambda Distribution," *Technometrics,* vol. 27, pp. 81–84, 1985.

133. Parzen, E., *Stochastic Processes,* Series in Probability and Statistics, Holden-Day, San Francisco, 1962.

134. Pegram, G. G. S., and W. James, "Multilag Multivariate Autoregressive Model for the Generation of Operational Hydrology," *Water Resour. Res.,* vol. 8, no. 4, pp. 1074–1076, 1972.

135. Pegram, G. G. S., "Factors Affecting Draft from a Lloyd Reservoir," *Water Resour. Res.,* vol. 10, no. 1, pp. 63–66, 1974.

136. Potter, K. W., "Annual Precipitation in Northwest United States: Long Memory, Short Memory, or No Memory," *Water Resour. Res., 15*(2):340–346, 1979.

137. Press, W. H., B. P. Flannery, S. A. Teukolsky, and W. V. Vetterling, *Numerical Recipes: The Art of Scientific Computing,* Cambridge University Press, New York, 1986.

138. Quimpo, R. G., "Stochastic Model of Daily River Flow Sequences," hydrology paper 18, Colorado State University, Fort Collins, Colo., 1967.

139. Ramberg, J. S., and B. W. Schmeiser, "An Approximate Method for Generating Symmetric Random Variables, *Comm. Assoc. Comput. Mach.,* vol. 15, pp. 987–990, 1972.

140. Ramberg, J. S., and B. W. Schmeiser, "An Approximate Method for Generating Asymmetric Random Variables," *Comm. Assoc. Comput. Mach.,* vol. 17, pp. 78–82, 1974.

141. Ramirez, J. A., and R. L. Bras, "Optimal Irrigation Control Using Stochastic Cluster Point Process for Rainfall Modeling and Forecasting," Ralph M. Parsons Laboratory, report 275, MIT, Cambridge, Mass., 1982.

142. Ramirez, J. A., and R. L. Bras, "Conditional Distributions of Neyman-Scott models for Storm Arrivals and Their Use in Irrigation Control," *Water Resour. Res.,* vol. 21, no. 3, pp. 317–330, 1985.

143. Rodhe, H. and J. Grandell, "Estimates of Characteristic Times for Precipitation Scavenging," *J. Atmospher. Sci.,* vol. 38, no. 2, pp. 370–386, 1981.

144. Rodriguez-Iturbe, I., V. K. Gupta, and E. Waymire, "Scale Considerations in the Modeling of Temporal Rainfall," *Water Resour. Res.,* vol. 20, no. 11, pp. 1611–1619, 1984.

145. Rodriguez-Iturbe, I., "Scale of Fluctuation of Rainfall Models," *Water Resour. Res.,* vol. 1, no. 4, pp. 489–498, 1986.

146. Rodriguez-Iturbe, I., D. R. Cox, and V. Isham, "Some Models for Rainfall Based on Stochastic Point Processes," *Proc. Royal Soc. London* (A), vol. 410, pp. 269–288, 1987*a*.

147. Rodriguez-Iturbe, I., B. Febres de Power, and J. B. Valdes, "Rectangular Pulses Point Process Models for Rainfall: Analysis of Empirical Data," *J. Geophys. Res.,* vol. 92, no. D8, pp. 9645–9656, 1987*b*.

148. Roesner, L. A., and V. Yevjevich, "Mathematical Models for Time Series of Monthly Precipitation and Monthly Runoff," hydrology paper 15, Colorado State University, Fort Collins, Colo., 1966.

149. Roldan, J., and D. A. Woolhiser, "Stochastic Daily Precipitation Models: 1. A Comparison of Occurrence Processes," *Water Resour. Res.,* vol. 18, no. 5, pp. 1451–1459, 1982.

150. SAS/ETS, "SAS/ETS User's Guide," version 6, 1st ed., SAS Inc., Cary, N.C., 1988.

151. Salas, J. D., and V. Yevjevich, "Stochastic Modeling of Water Use Time Series," hydrology paper no. 52, Colorado State University, Fort Collins, Colo., 1972.

152. Salas, J. D., and G. G. S. Pegram, "A Seasonal Multivariate Multilag Autoregressive Model in Hydrology," *Proc. Third Int. Symp. on Theoretical and Applied Hydrology,* Colorado State Univ., Fort Collins, Colo., 1977.

153. Salas, J. D., D. C. Boes, V. Yevjevich, and G. G. S. Pegram, "Hurst Phenomenon as a Pre-Asymptotic Behavior," *J. Hydrol.,* vol. 44, no. 1, pp. 1–15, 1979.

154. Salas, J. D., and D. C. Boes, "Shifting Level Modelling of Hydrologic Series," *Adv. Water Resour.,* vol. 3, pp. 59–63, 1980.

155. Salas, J. D., J. R. Delleur, V. Yevjevich, and W. L. Lane, "Applied Modeling of Hydrologic Time Series," Water Resources Publications, Littleton, Colo., 1980.

156. Salas, J. D., and J. T. B. Obeysekera, "ARMA Model Identification of Hydrologic Time Series," *Water Resour. Res.,* vol. 18, no. 4, pp. 1011–1021, 1982.

157. Salas, J. D., D. C. Boes, and R. A. Smith, "Estimation of ARMA Models with Seasonal Parameters," *Water Resour. Res.,* vol. 18, no. 4, pp. 1006–1010, 1982.

158. Salas, J. D., G. Tabios, and P. Bartolini, "Approaches to Multivariate Modeling of Water Resources Time Series," in special AWRA monograph, *Time Series in Water Resources,* K. W. Hipel (ed.), 1985.

159. Salas, J. D., and M. Chebaane, "Stochastic Modeling of Monthly Flows in Streams of Arid Regions," *Proc. Int. Symp. HY&IR Div.,* Am. Soc. Civ. Eng., San Diego, Calif., pp. 749–755, 1990.

160. Salas, J. D., and M. W. Abdelmohsen, "Determining Streamflow Drought Statistics by Stochastic Simulation," paper presented at the U.S.-PRC Bilateral Symposium on Droughts and Arid Region Hydrology, Tucson, Ariz., September 16–20, 1991.

161. Salas, J. D., and M. W. Abdelmohsen, "Initialization for Generating Single Site and Multisite Low Order PARMA Processes," submitted to *Water Resour. Res.,* 1992.

162. Salas, J. D., R. A. Smith, and M. Markus, "Modeling and Generation of Univariate Seasonal Hydrologic Data (Programs CSU001 and CSU002), "technical report 2, Computing Hydrology Laboratory, Engineering Research Center, Colorado State University, Fort Collins, Colo., 1992.

163. Salas, J. D., and M. Markus, "Modeling and Generation of Multivariate Seasonal Hydrologic Data (Program CSU003 and CSU004)," technical report 3, Computing Hydrology Laboratory, Engineering Reseach Center, Colorado State University, Fort Collins, Colo., 1992.

164. Salas, J. D., R. A. Smith, G. Tabios, and J. H. Heo, *Statistical Computer Techniques in Hydrology and Water Resources,* draft of forthcoming book, 1993.

165. Santos, E., and J. D. Salas, "A Parsimonious Step Disaggregation Model for Operational Hydrology," AGU Fall Meeting, San Francisco, *AGU Abstract,* EOS, vol. 64, no. 45, p. 706, 1983.

166. Santos, E., and J. D. Salas, "Stepwise Disaggregation Scheme for Synthetic Hydrology," *J. Hydraul. Eng.,* vol. 118, no. 5, pp. 765–784, 1992.

167. Shapiro, S. S., and M. B. Wilk, "An Analysis of Variance Test for Normality (complete samples)," *Biometrika,* vol. 52, pp. 591–611, 1965.

168. Smith, J. A., and A. F. Karr, "A Point Process Model of Summer Season Rainfall Occurrences," *Water Resour. Res.,* vol. 19, no. 1, pp. 95–103, 1983.

169. Smith, J. A., and A. F. Karr, "Statistical Inference for Point Process Models of Rainfall," *Water Resour. Res.,* vol. 21, no. 1, pp. 73–79, 1985*a.*

170. Smith, J. A., and A. F. Karr, "Parameter Estimation for a Model of Space-Time Rainfall," *Water Resour. Res.,* vol. 21, no. 8, pp. 1251–1258, 1985*b.*

171. Smith, J. A., and W. F. Krajewski, "Statistical Modeling of Space-Time Rainfall Using Radar and Rain Gage Observations," *Water Resour. Res.,* vol. 23, no. 10, pp. 1893–1900, 1987.

172. Smith, R. E., and H. A. Schreiber, "Point Process of Seasonal Thunderstorm Rainfall: 2. Rainfall Depth Probabilities," *Water Resour. Res.,* vol. 10, no. 3, pp. 418–426, 1974.

173. Snedecor, G. W., and W. G. Cochran, *Statistical Methods,* The Iowa State University Press, Ames, Iowa, 1980.

174. SPSS, *SPSS Update 7-9,* C. H. Hull and N. H. Nie, eds., McGraw-Hill, New York, 1981.

175. Statgraphics, "Statgraphics User Manual," Statistical Graphics Corp., Rockville, Md., 1984.

176. Stedinger, J. R., and M. R. Taylor, "Synthetic Streamflow Generation: 1. Model Verification and Validation," *Water Resour. Res.,* vol. 18, no. 4, pp. 909–918, 1982.

177. Stedinger, J. R., and R. M. Vogel, "Disaggregation Procedures for Generating Serially Correlated Flow Vectors," *Water Resour. Res.,* vol. 20, no. 1, pp. 47–56, 1984.

178. Stedinger, J. R., D. P. Lettenmaier, and R. M. Vogel, "Multisite ARMA(1, 1) and Disaggregation Models for Annual Streamflow Generation," *Water Resour. Res.,* vol. 21, pp. 497–509, 1985*a.*

179. Stedinger, J. R., D. Pei, and T. A. Cohn, "A Condensed Disaggregation Model for Incorporating Parameter Uncertainty into Monthly Reservoir Simulations," *Water Resour. Res.,* vol. 21, no. 5, pp. 665–675, 1985*b.*

180. Tao, P. C., and J. W. Delleur, "Seasonal and Nonseasonal ARMA Models in Hydrology," *J. Hydraul. Div., Am. Soc. Civ. Eng.* (now *J. Hydraul. Eng.*), vol. 2, no. HY10, pp. 1591–1559, 1976.

181. Thomas, H. A., Jr., and M. B. Fiering, "Mathematical Synthesis of Streamflow Sequences for Analysis of River Basins by Simulation," in *The Design of Water Resources Systems,* A. Maas et al., Harvard University Press, Cambridge, Mass., pp. 459–493, 1962.

182. Todini, E., "The Preservation of Skewness in Linear Disaggregation Schemes," *J. Hydrol.,* vol. 47, pp. 199–214, 1980.

183. Todorovic, P., and V. Yevjevich, "A Particular Stochastic Process as Applied to Hydrology," *Proc. Int. Hydrol. Symp.,* Fort Collins, Colo., September 6–8, vol. 1, pp. 298–303, 1967.

184. Todorovic, P., and D. A. Woolhiser, "Stochastic Structure of the Local Pattern of Precipitation," in *Stochastic Approaches to Water Resources,* H. W. Shen, ed., vol. II, H. W. Shen, Fort Collins, Colo., pp. 15.1–15.37, 1976.

185. Tong, H., "Determination of the Order of a Markov Chain by Akaike's Information Criterion," *J. Appl. Prob.,* vol. 12, pp. 488–497, 1975.

186. Tukey, J. W., "The Practical Relationship Between the Common Transformations of Counts and of Amounts," technical report 36, Statistical Techniques Research Group, Princeton University, Princeton, N.J., 1960.

187. U.S. Army Corps of Engineers, "HEC-4 Monthly Streamflow Simulation," Hydrologic Engineering Center, Davis, Calif., 1971.

188. Valdes, J. B., I. Rodriguez-Iturbe, and G. J. Vicens, "Bayesian Generation of Synthetic Streamflows: 2. Multivariate Case," *Water Resour. Res.,* vol. 13, no. 2, pp. 291–295, 1977.

189. Valencia, R. D., and J. C. Schaake, Jr., "Disaggregation Processes in Stochastic Hydrology," *Water Resour. Res.,* vol. 9, no. 3, pp. 580–585, 1973.

190. van Belle, G., and J. P. Hughes, "Nonparametric Tests for Trend in Water Quality," *Water Resour. Res.,* vol. 20, no. 1, pp. 127–136, 1984.

191. Vicens, G. J., I. Rodriguez-Iturbe, and J. C. Schaake, "Bayesian Generation of Synthetic Streamflows," *Water Resour. Res.,* vol. 11, no. 6, pp. 827–838, 1977.

192. Vogel, R. M., and J. R. Stedinger, "Minimum Variance Streamflow Record Augmentation Procedures," *Water Resour. Res.,* vol. 21, no. 5, pp. 715–723, 1985.

193. Vogel, R. M., "The Probability Plot Correlation Coefficient Test for the Normal, Lognor-

mal and Gumbel Distribution Hypotheses," *Water Resour. Res.*, vol. 22, no. 4, pp. 587–590, 1986.

194. Vogel, R. M., and J. R. Stedinger, "The Value of Stochastic Streamflow Models in Overyear Reservoir Design Applications," *Water Resour. Res.*, vol. 24, no. 9, pp. 1483–1490, 1988.

195. Wallis, J. R. and P. E. O'Connell, "Small Sample Estimation of ρ_1," *Water Resour. Res.*, vol. 8, no. 3, pp. 707–712, 1972.

196. Wallis, J. R., and E. O'Connell, "Firm Reservoir Yield—How Reliable Are Hydrological Records?" *Hydrol. Sci. Bull.* (now *Hydrol. Sci. Jour.*), vol. 18, pp. 347–365, 1973.

197. Waymire, E., and V. K. Gupta, "The Mathematical Structure of Rainfall Representations: 1. A Review of Stochastic Rainfall Models," *Water Resour. Res.*, vol. 17, no. 5, pp. 1261–1272, 1981*a*.

198. Waymire, E., and V. K. Gupta, "The Mathematical Structure of Rainfall Representations: 2. A Review of the Theory of Point Processes," *Water Resour. Res.*, vol. 17, no. 5, pp. 1273–1285, 1981*b*.

199. Waymire, E., and V. K. Gupta, "The Mathematical Structure of Rainfall Representations: 3. Some Applications of the Point Process Theory to Rainfall Processes," *Water Resour. Res.*, vol. 17, no. 5, pp. 1287–1294, 1981*c*.

200. Waymire, E., V. K. Gupta, and I. Rodriguez-Iturbe, "A Spectral Theory of Rainfall Intensity at the Meso-β Scale," *Water Resour. Res.*, vol. 20, no. 10, pp. 1453–1465, 1984.

201. Woolhiser, D. A., and G. G. S. Pegram, "Maximum Likelihood Estimation of Fourier Coefficients to Describe Seasonal Variations of Parameters in Stochastic Daily Precipitation Models," *J. Appl. Meteorol.*, vol. 18, pp. 34–42, 1979.

202. Yevjevich, V., "Fluctuation of Wet and Dry Years, Part I, Research Data Assembly and Mathematical Models," hydrology paper 1, Colorado State University, Fort Collins, Colo., 1963.

203. Yevjevich, V., "Fluctuations of Wet and Dry Years—Part II, Analysis by Serial Correlation," hydrology paper 4, Colorado State University, Fort Collins, Colo., 1964.

204. Yevjevich, V., "The Application of Surplus, Deficit and Range in Hydrology," hydrology paper 10, Colorado State University, Fort Collins, Colo., 1965.

205. Yevjevich, V., "An Objective Approach to Definitions and Investigations of Continental Hydrologic Droughts," hydrology paper 23, Colorado State University, Fort Collins, Colo., 1967.

206. Yevjevich, V., "Structural Analysis of Hydrologic Time Series," hydrology paper 56, Colorado State University, Fort Collins, Colo., 1972.

207. Yevjevich, V., "Generation of Hydrologic Samples, Case Study of the Great Lakes," hydrology paper 72, Colorado State University, Fort Collins, Colo., 1975.

208. Young, G. K., and Pisano, W. C., "Operational Hydrology Using Residuals," *J. Hydraul. Div., Am. Soc. Civ. Eng.* (now *J. Hydraul. Eng.*), vol. 94, no. HY4, pp. 909–923, 1968.

209. Zelen, M., and N. C. Severo, "Probability Functions," in *Handbook of Mathematical Functions,* M. Abramowitz and I. A. Stegun (eds.), pp. 925–995, Dover, New York, 1964.

210. Zsuffa, I., and A. Gálai, *Reservoir Sizing by Transition Probability: Theory, Methodology and Applications,* Water Resources Publications, Littleton, Colo., 1987.

CHAPTER 20
GEOSTATISTICS

Peter K. Kitanidis
Department of Civil Engineering
Stanford University
Stanford, California

20.1 INTRODUCTION

20.1.1 Typical Applications

Geostatistics[35] is a set of statistical estimation techniques involving quantities which vary in space; popular in mining, it has found many applications in hydrology as well as other areas of geophysics. Examples:

- From measurements of solute concentration in a number of irregularly spaced locations in a geologic formation, plot the contour map of the concentration. Before employing a contouring computer program, the concentration on the nodes of a fine grid needs to be estimated (*interpolation* problem).
- From measurements of precipitation at a number of rain gauges, infer the mean areal precipitation and evaluate the accuracy of the estimate (*averaging* or *integration* problem).
- From measurements of the hydraulic head at a number of wells and other information, infer the values of transmissivity using a groundwater model (*inverse* problem in groundwater modeling).
- Select the location of new monitoring wells at a site of subsurface contamination so that the concentration of mass of pollutants can be evaluated with sufficient accuracy or it can be ascertained that a water-quality standard is met (*monitoring network design* problem).

Common characteristics of these applications are (1) they involve spatial variables such as precipitation, snow-pack thickness, or moisture content, water-table elevation, or aquifer transmissivity; (2) these variables must be estimated or inferred from measurements; and (3) the spatial variability and the type of measurements are such that deterministic (error-free) estimation is not possible. For instance, even if the head has been measured at two wells in an aquifer, the head cannot be predicted with certainty at half the distance between the wells.

Statistical methods are well-suited for the solution of such problems of estimation with incomplete information; they can be used to: (1) produce estimates of the unknown quantities which are best in a logical and well-defined way, (2) evaluate the reliability of the estimates, and (3) predict the effects on estimation accuracy of an additional set of measurements, as in the evaluation of proposed sampling networks. Statistical estimation is a broad term which refers to all types of problems in which unknowns must be calculated approximately from measurements. The emphasis in this chapter is on *linear estimation methods,* which are by far the most practical and widely used; these methods are related to ARIMA time-series methods (Chap. 19) and to Kalman filtering (Chap. 26).

20.1.2 Why Use Geostatistical Estimation Techniques?

Often manual and ad hoc methods are applied in hydrology for the solution of interpolation and averaging problems. For example, contour maps are often drawn manually, inverse distance weighting is used in interpolation, and the method of Thiessen polygons is applied for the determination of mean areal precipitation. Advantages of geostatistical techniques include:

1. They are amenable to automation and the results are reproducible.
2. They rely on an explicit model of spatial variability which is confirmed with data and other pertinent information. Thus, in computing mean areal precipitation, the area of influence of each gauge is determined from an analysis of the structure of precipitation over the gauged area instead of following an inflexible and arbitrary procedure, such as Thiessen polygons.
3. They provide "best" (in a well-defined sense) estimates and measures of the reliability of these estimates, which means that geostatistics is particularly useful in evaluating the worth of additional measurements and in designing monitoring networks.
4. Estimation techniques are useful in combination with mathematical models of groundwater flow and transport for the solution of parameter estimation problems and for evaluation of the accuracy of model predictions. Geostatistical methods are less labor-intensive and more objective as well as potentially more accurate than manual calibration techniques.

20.1.3 The Role of Probability Theory

The estimation of the contour map of precipitation over an area from a few point measurements relies on the analyst's understanding of how this variable is distributed in space, i.e., on the spatial-variability model or spatial law. However, hydrologic and hydrogeologic variables are distributed in space in complex and inadequately understood ways. In most applications, an empirical model must be developed from the analysis of the site-specific data and other information, such as past experience and mathematical models. The model usually involves the concept of probability in the sense that spatial variability is described only partially by using averages. For example, the model may specify that the precipitation fluctuates about some mean value and may provide a formula to calculate the correlation coefficient of the fluctuations at two locations from their separation distance.

Probability theory is thus introduced because of inadequate information. For example, if there is not enough information to specify the actual value of the hydraulic head in a well, the next best thing is to consider all possible head values and to assign a probability to each of them being the right one. For the sake of illustration assume that the only possible solutions are $\phi = 4, 6$, or 11 m with equal probability, ⅓. From this description one can calculate the *mean value* $(4 + 6 + 11)/3 = 7$; this estimate is *best* in the sense that its mean square difference from the actual value (4, 6, or 11) is the smallest possible. For distributions that are nearly symmetric, the mean value is usually the most reasonable *point* (single-number) *estimate*. In addition to calculating a best estimate, one may come up with a measure of its accuracy in predicting the actual value. A reasonable way is through the *estimation variance* or *mean square error:* $[(4 - 7)^2 + (6 - 7)^2 + (11 - 7)^2]/3 = 8.66$. Note that when the mean is used as a best estimate, the mean square error is given by the *variance* of the distribution. As a measure of error, the square root of the mean square error, called the *standard error*, is used: $\sqrt{8.66} = 2.94$.

20.1.4 Steps in a Geostatistical Analysis

A geostatistical analysis of spatial data can be conveniently subdivided into two phases: *structural analysis* and *best linear unbiased estimation (BLUE)*. *Structural analysis* is the selection of a model of spatial variability. In basic geostatistics, it boils down to the selection of a semivariogram, and for this reason it is sometimes called *variography*. More generally in linear estimation, it is equivalent to selection of the first two moments of the spatial functions of interest and is called *second-moment characterization*. Selection of a model is based on the analysis of data and other information, including experience with data at similar sites and geologic and hydrologic information. Typically, model selection is an iterative procedure consisting of (1) *exploratory data analysis* on the basis of which a model is tentatively selected, (2) *parameter estimation*, such as selection of numerical values for the parameters of the semivariogram, and (3) *model validation or diagnostic checking*, which is a careful examination of the performance of the model in test cases. *Best linear unbiased estimation* deals with finding, as linear functions of the data, the estimates of the unknowns which are unbiased and have minimum variance, using the model developed during structural analysis.

The basic idea is that we proceed to figure out an unknown function (e.g., the distribution of rainfall depth over a basin) in two stages. During the first stage, structural analysis, the choice is narrowed down to the functions which share certain characteristics, collectively known as *structure*. This structure is usually quantified through relatively simple probabilistic models which are described in Sec. 20.2. During the second stage, BLUE, the choice is further narrowed down by eliminating the functions which are not consistent with available observations.

This chapter is a collection of methods and formulas used in applied linear geostatistics and should be used in a way that is consistent with the reader's background and experience.

Most practitioners deal with a single variable and simple variogram models; they should focus on Secs. 20.2.1, 20.2.2, 20.2.3, 20.3.1, 20.3.2, 20.3.5, 20.3.7, and all of Sec. 20.4. An example of structural analysis of a single variable is included in Sec. 20.4. Users who want to utilize models with variable mean and kriging with generalized covariance functions should additionally consult Secs. 20.2.4 and 20.3.3.

Users who want to analyze jointly two or more variables should consult Refs. 1, 2, 4, 5, 9–11, 16, 20–23, 31, 33, 38, 43, and 46. For simulation (how to generate a set of sample functions which are plausible answers given the available information) see Refs. 6, 15, 27, 34, 39, and 45.

20.2 BASIC METHODS AND MODELS

20.2.1 The First Two Moments

Consider a spatially variable quantity, or *regionalized variable, z*, such as precipitation or transmissivity; this quantity is a function of the spatial coordinates and may be represented by $z(x_1)$, $z(x_1, x_2)$, or $z(x_1, x_2, x_3)$ depending on whether it varies in one, two, or three dimensions. For brevity, the notation $z(\mathbf{x})$ will be used to include all three cases, where $\mathbf{x}$ is the location (a vector with one, two, or three components.) The function $z(\mathbf{x})$ is not known everywhere but needs to be estimated from available observations and, perhaps, additional information.

It is postulated that the actual unknown $z(\mathbf{x})$ function is one out of a collection (or *ensemble*) of functions $z(\mathbf{x}; 1)$, $z(\mathbf{x}; 2)$, . . . , and that the probability that $z(\mathbf{x}) = z(\mathbf{x}; i)$ is P_i. The ensemble of functions defines a *random field* (or *random function* or *spatial stochastic process.*) *Expectation* is the process of computing a probability-weighted average over the ensemble and is denoted by the symbol E. Thus, the expected value of z at location $\mathbf{x}$ is

$$E[z(\mathbf{x})] = P_1\, z(\mathbf{x}; 1) + P_2\, z(\mathbf{x}; 2) + \cdots = \sum_i P_i z(\mathbf{x}; i) \tag{20.2.1}$$

Instead of specifying all possible solutions separately, it is more convenient to specify and work with ensemble averages or statistical moments. The first two *statistical moments* of the random field are defined as:

1. The mean function (first moment) which gives the expected value at any point $\mathbf{x}$:

$$m(\mathbf{x}) = E[z(\mathbf{x})] \tag{20.2.2}$$

2. The covariance function (second moment), which is the covariance for any pair $\mathbf{x}$ and $\mathbf{x}'$:

$$R(\mathbf{x}, \mathbf{x}') = E\{[z(\mathbf{x}) - m(\mathbf{x})][z(\mathbf{x}') - m(\mathbf{x}')]\} \tag{20.2.3}$$

When $\mathbf{x}$ and $\mathbf{x}'$ are the same location, the covariance reduces to the variance

$$\sigma^2(\mathbf{x}) = R(\mathbf{x}, \mathbf{x}) \tag{20.2.4}$$

and $\sigma(\mathbf{x})$, the square root of the variance, is the standard deviation. The correlation coefficient between $z(\mathbf{x})$ and $z(\mathbf{x}')$ is

$$\rho(\mathbf{x}, \mathbf{x}') = \frac{R(\mathbf{x}, \mathbf{x}')}{\sigma(\mathbf{x})\, \sigma(\mathbf{x}')} \tag{20.2.5}$$

The correlation coefficient satisfies $-1 \leq \rho(\mathbf{x}, \mathbf{x}') \leq 1$. The significance of covariance or correlation in linear estimation is illustrated through the following example: without any measurements, the best estimate of $z(\mathbf{x}')$ is $m(\mathbf{x}')$ and the mean square error is $\sigma^2(\mathbf{x}')$; however, when $z(\mathbf{x})$ is observed, the estimate of $\mathbf{z}(\mathbf{x}')$ may be corrected if the value of $z(\mathbf{x})$ conveys information about $z(\mathbf{x}')$. Using a correction which is linear in the observation, the updated estimate of $z(\mathbf{x}')$ is

$$m(\mathbf{x}') + \rho(\mathbf{x}, \mathbf{x}')\, \frac{\sigma(\mathbf{x}')}{\sigma(\mathbf{x})}\, [z(\mathbf{x}) - m(\mathbf{x})]$$

with mean square estimation error reduced to $[1 - \rho^2(\mathbf{x}, \mathbf{x}')]\sigma^2(\mathbf{x}')$. If ρ is ± 1, $z(\mathbf{x}')$ can be computed with certainty (i.e., zero mean square error) from $z(\mathbf{x})$ through a linear relation. On the other extreme, if ρ is zero or $z(\mathbf{x})$ and $z(\mathbf{x}')$ are uncorrelated, observation of $z(\mathbf{x})$ cannot improve the accuracy of estimation of $z(\mathbf{x}')$ through a correction which is linear in the observation. Thus, the estimation accuracy depends critically on the degree of correlation between the unknown and the observations.

20.2.2 Stationarity and Isotropy

In practice, the first two moments are not known but must be deduced from the data. *Stationarity* and *isotropy* are simplifications useful in the development of empirical models of spatial variability. In the context of linear estimation methods, *stationarity* signifies that the mean and the variance are the same everywhere and the correlation between any two observations $z(\mathbf{x})$ and $z(\mathbf{x}')$ depends only on their *relative* location in space, i.e., $\mathbf{x} - \mathbf{x}'$.

$$E[z(\mathbf{x})] = m \tag{20.2.6}$$

$$E[(z(\mathbf{x}) - m)(z(\mathbf{x}') - m)] = R(\mathbf{x} - \mathbf{x}') \tag{20.2.7}$$

The notation $R(\mathbf{x} - \mathbf{x}')$ implies that R depends on all components of the separation vector $\mathbf{x} - \mathbf{x}'$. For example, for three-dimensional variability, $R(\mathbf{x} - \mathbf{x}') = R(x_1 - x_1', x_2 - x_2', x_3 - x_3')$ where $\mathbf{x} = (x_1, x_2, x_3)$ and $\mathbf{x}' = (x_1', x_2', x_3')$. The variance $R(\mathbf{0})$ is known as the *sill.*

Another simplification useful in empirical-model building is *isotropy:* the covariance function depends only on the separation distance, not the direction of the separation vector

$$E[(z(\mathbf{x}) - m)(z(\mathbf{x}') - m)] = R(|\mathbf{x} - \mathbf{x}'|) \tag{20.2.8}$$

where $|\mathbf{x} - \mathbf{x}'|$ stands for the length of vector $\mathbf{x} - \mathbf{x}'$,

$$|\mathbf{x} - \mathbf{x}'| = \sqrt{(x_1 - x_1')^2 + (x_2 - x_2')^2 + (x_3 - x_3')^2} \tag{20.2.9}$$

The correlation for stationary functions vanishes at a distance which is known as the *range.* The spatial extent of fluctuations of function z about its mean value is proportional to the range.

A generalization of the stationary model is the *intrinsic* model. The random function $z(\mathbf{x})$ is intrinsic if the following two conditions are met:

$$E[z(\mathbf{x}) - z(\mathbf{x}')] = 0 \tag{20.2.10}$$

$$2\gamma(\mathbf{h}) = E\{[z(\mathbf{x}) - z(\mathbf{x}')]^2\} \tag{20.2.11}$$

where $\mathbf{h} = \mathbf{x} - \mathbf{x}'$ is the separation vector. The function $2\gamma(\mathbf{h})$ is known as the *variogram* of the random field and $\gamma(\mathbf{h})$ as the *semivariogram.* [However, it is not uncommon to find $\gamma(\mathbf{h})$ referred to as the variogram.] Note that the definition of the variogram does not make use of the mean of $z(\mathbf{x})$ and that it does not assume that the variance of the fluctuations of $z(\mathbf{x})$ is necessarily finite. In the isotropic case, γ is a function only of the length of the separation vector, $h = |\mathbf{h}|$.

Stationary functions are also intrinsic. For a stationary process, the variogram is related to the covariance function through

$$2\gamma(\mathbf{h}) = 2R(\mathbf{0}) - 2R(\mathbf{h}) \tag{20.2.12}$$

and the semivariogram at large distances takes a value equal to the sill (variance), $\gamma(\infty) = R(0) = \sigma^2$. However, not all intrinsic functions are stationary. As a practical rule, an intrinsic function is nonstationary if its semivariogram tends to infinity as h tends to infinity. For example, the $\gamma(\mathbf{h}) = h$ semivariogram characterizes an intrinsic nonstationary function.

20.2.3 Commonly Used Models

Not all functions can be used as a covariance function of semivariogram (as explained in Sec. 20.2.4.) In modeling practice, covariance functions or semivariograms describing the spatial structure of a function are formed by combining a small number of simple, mathematically acceptable expressions or models. This section contains a list of such covariance functions R and semivariograms γ, some of which are plotted in Fig. 20.2.1.

Stationary Models

1. *Gaussian model*

$$R(h) = \sigma^2 \exp\left(-\frac{h^2}{L^2}\right) \quad \text{and} \quad \gamma(h) = \sigma^2\left[1 - \exp\left(-\frac{h^2}{L^2}\right)\right] \qquad (20.2.13)$$

where $\sigma^2 > 0$ and $L > 0$ are the two parameters of this model. Because the covariance function decays asymptotically, the range α is defined in practice as the distance at which the correlation is 0.05; i.e., $\alpha \approx 7L/4$. The gaussian model is the only covariance in this list with parabolic behavior at the origin [$\gamma(h) \sim h^2$ for small h, where $\sim$ stands for *proportional to*] indicating that it represents a regionalized variable which is smooth enough to be differentiable (i.e., the slope between two points tends to a well-defined limit as the distance between these points vanishes). See Figs. 20.2.1 and 20.2.2.

2. *Exponential model*

$$R(h) = \sigma^2 \exp\left(-\frac{h}{l}\right) \quad \text{and} \quad \gamma(h) = \sigma^2\left[1 - \exp\left(-\frac{h}{l}\right)\right] \qquad (20.2.14)$$

where $\sigma^2 > 0$ and $l > 0$. The range is $\alpha \approx 3l$. It is a model popular in hydrologic applications, because it is versatile and has a simple analytical form. See Figs. 20.2.1 and 20.2.3.

3. *Spherical model*

$$R(h) = \begin{cases} \left(1 - \dfrac{3}{2}\dfrac{h}{\alpha} + \dfrac{1}{2}\dfrac{h^3}{\alpha^3}\right)\sigma^2 & \text{for } 0 \le h \le \alpha \\ 0 & \text{for } h > \alpha \end{cases}$$
$$\gamma(h) = \begin{cases} \left(\dfrac{3}{2}\dfrac{h}{\alpha} - \dfrac{1}{2}\dfrac{h^3}{\alpha^3}\right)\sigma^2 & \text{for } 0 \le h \le \alpha \\ \sigma^2 & \text{for } h > \alpha \end{cases} \qquad (20.2.15)$$

where σ^2 = variance and α = range. Models 2 and 3 exhibit linear behavior at the origin; i.e., $\gamma(h) \sim h$ for small h. The realizations of a random field with such a variogram are continuous but not differentiable; i.e., they are less smooth than the realizations of a random field with a gaussian covariance function.

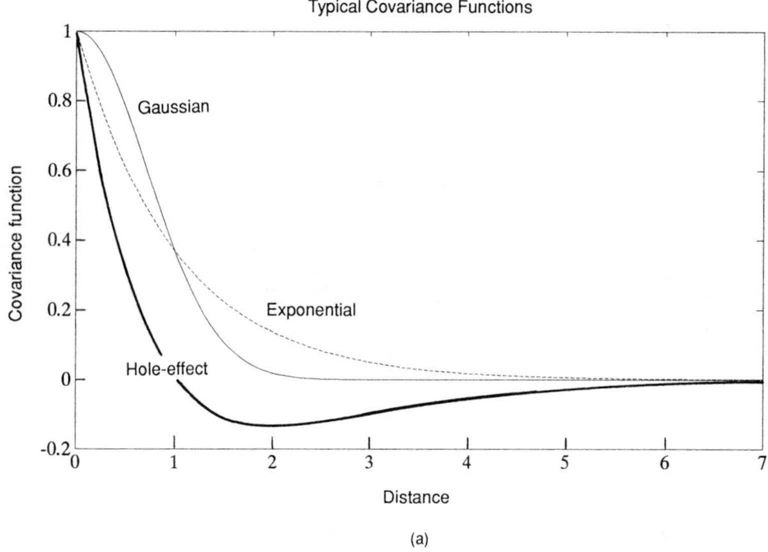

(a)

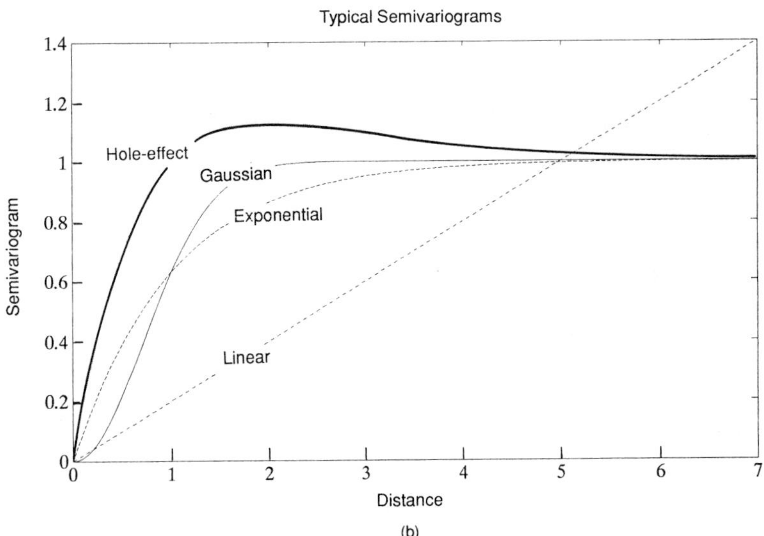

(b)

FIGURE 20.2.1 *(a)* The shape of covariance functions for the gaussian, exponential, and hole-effect models. *(b)* Semivariograms for the gaussian, exponential, hole-effect, and linear models.

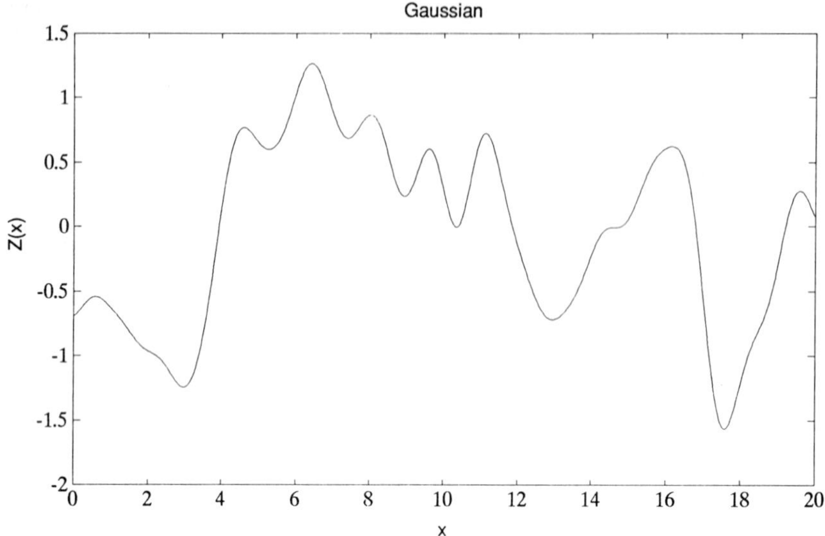

FIGURE 20.2.2 Sample function $z(x)$ with gaussian covariance function.

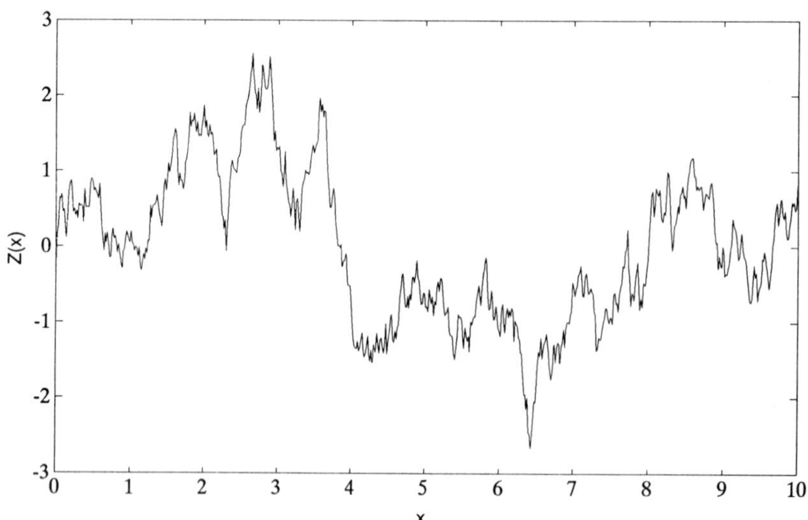

FIGURE 20.2.3 Sample function $z(x)$ with exponential covariance function.

4. *Hole-effect model.* In this case, the covariance function does not decay monotonically with distance. It is used to represent some type of pseudo-periodicity. (At the extreme, when the covariance is periodic, the regionalized variable is exactly periodic.) A model that has been used in hydrology[5] is

$$R(h) = \sigma^2 \left(1 - \frac{h}{L}\right) \exp\left(-\frac{h}{L}\right) \quad \text{and}$$

$$\gamma(h) = \sigma^2 \left[1 - \left(1 - \frac{h}{L}\right) \exp\left(-\frac{h}{L}\right)\right] \tag{20.2.16}$$

This model describes processes for which excursions above the mean tend to be compensated by excursions below the mean (see Fig. 20.2.4).

5. *Nugget-effect model*

$$R(h) = C_0\, \delta(h) = \begin{cases} 0 & h > 0 \\ C_0 & h = 0 \end{cases} \tag{20.2.17}$$

$$\gamma(h) = C_0\, [1 - \delta(h)] = \begin{cases} C_0 & h > 0 \\ 0 & h = 0 \end{cases}$$

where $C_0 > 0$ is the nugget variance and the symbol $\delta(h)$ stands for 1 if $h = 0$ and for 0 in all other cases. Note the discontinuity at the origin. The realizations of this random field are not continuous; i.e., $z(\mathbf{x})$ can be different from $z(\mathbf{x}')$ no matter how small the distance $|\mathbf{x} - \mathbf{x}'|$ that separates them.

The nugget effect represents microvariability as well as random measurement error. Microvariability is variability at a scale smaller than the separation distance

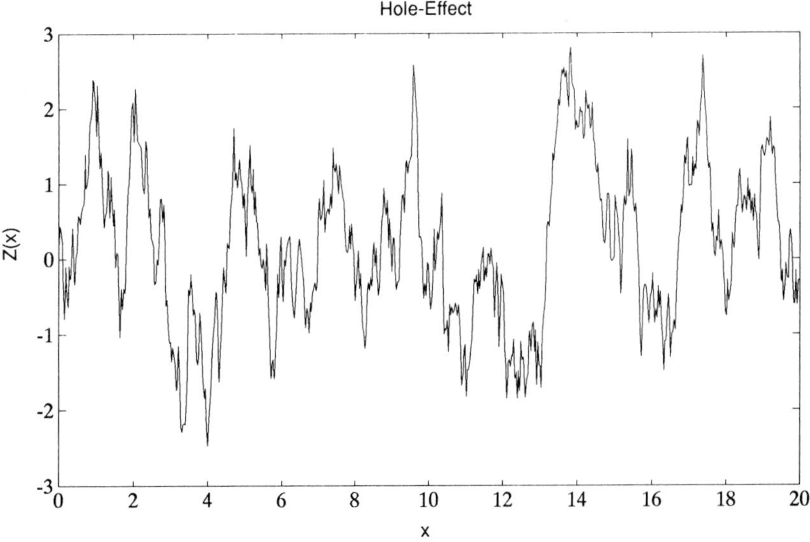

FIGURE 20.2.4 Sample function $z(x)$ with hole-effect covariance function.

between the closest measurement points. For example, if rain gauges are located with a typical spacing of 1 km, rainfall variability at the scale of 10 m or 100 m reduces the correlation between rainfall depths measured at the various gauges. The nugget semivariogram can be visualized as the limit of the exponential semivariogram $\gamma(h) = C_0\,[1 - \exp\,(-h/l)]$ when the typical distance h is much larger than the inherent scale of the fluctuations of the phenomenon, l. Discontinuous behavior can also be attributed to random measurement error which produces observations which vary from gauge to gauge in an unstructured or random way.

Intrinsic Nonstationary Models

6. *Power model*

$$\gamma(h) = \theta h^s \tag{20.2.18}$$

where $\theta > 0$ and $0 < s < 2$.

7. *Linear model*

$$\gamma(h) = \theta h \tag{20.2.19}$$

where the positive parameter θ = slope of the variogram. Although a special case of the power model (for $s = 1$) it is mentioned separately because of its usefulness in applications. See Figs. 20.2.1 and 20.2.5.

Note that in models 1 to 7, $\gamma(0) = 0$.

8. *Logarithmic model*

$$\gamma(h) = A\log\,(h) \tag{20.2.20}$$

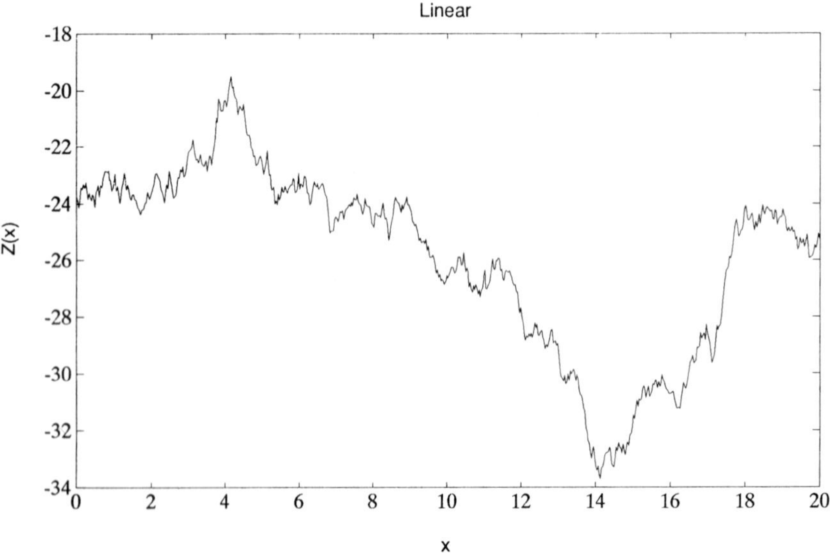

FIGURE 20.2.5 Sample function $z(x)$ with linear semivariogram.

where $A > 0$. This model, which originated in mining geostatistics, can be used only for integrals over finite volumes and *cannot* be used with point values of the regionalized variable. For example, it can be used to estimate solute mass over a finite volume given measurements of mass in samples also of finite volume. (See Sec. 20.3.8.)

Isotropic versus Anisotropic Models. In Eqs. (20.2.13) to (20.2.20), h represents a measure of the distance between two points $\mathbf{x}$ and $\mathbf{x}'$. A distinction should be made between two cases:

1. *Isotropic,* when

$$h = |\mathbf{x} - \mathbf{x}'| = \sqrt{(x_1 - x_1')^2 + (x_2 - x_2')^2 + (x_3 - x'_3)^2} \qquad (20.2.21)$$

2. *Anisotropic,* when h depends not only on the length but also on the orientation of the separation vector. It is common to adopt a model of geometric anisotropy in which anisotropic variability can be turned into isotropic variability through a linear transformation of the coordinates (rotation and shrinkage of the cartesian system, similar to the method used by groundwater hydrologists to deal with anisotropic conductivity). For example, one can select the x_1, x_2, x_3 directions along the main axes of correlation anisotropy and use

$$h = |\mathbf{x} - \mathbf{x}'| = \sqrt{\left(\frac{x_1 - x_1'}{c_1}\right)^2 + \left(\frac{x_2 - x_2'}{c_2}\right)^2 + \left(\frac{x_3 - x_3'}{c_3}\right)^2} \qquad (20.2.22)$$

where c_1, c_2, and c_3 are dimensionless numbers. This h is then used with any of the models 1 to 8. For example, combining this expression with the exponential model[19] gives

$$R(x_1 - x_1', x_2 - x_2', x_3 - x_3') = \sigma^2 \exp\left\{-\left[\left(\frac{x_1 - x_1'}{l_1}\right)^2 + \left(\frac{x_2 - x_2'}{l_2}\right)^2 + \left(\frac{x_3 - x_3'}{l_3}\right)^2\right]^{1/2}\right\}$$

a model with four parameters.

Superposition of Basic Models. By adding variograms 1 to 8 from the list in this section, one can obtain other mathematically acceptable variograms. For example, combining the linear and nugget effect semivariograms, we obtain another useful model:

$$\gamma(h) = \begin{cases} C_0 + \theta h & h > 0 \\ 0 & h = 0 \end{cases} \qquad (20.2.23)$$

with two parameters, $C_0 \geq 0$ and $\theta \geq 0$. In this way, one can find a model which adequately represents the structure of the spatial variable which is to be estimated.

20.2.4 Generalized Covariance Functions and Variable Mean

In linear estimation, the mean square error of estimation of an unknown quantity depends on the covariance function or the variogram. Because the mean square error cannot be negative, the covariance function or variogram cannot be chosen arbitrarily. Furthermore, the weights used in linear estimation are chosen by minimizing the mean square error of estimation. A well-defined minimum is not possible with an arbitrary covariance function. For example, in simple kriging (see Sec. 20.3) the mean m and the covariance function $R(\mathbf{x}-\mathbf{x}')$ of $z(\mathbf{x})$ are known. The value of z at location $\mathbf{x}_0$ is estimated from a weighted average of the mean and the observations. For a unique solution and a nonnegative minimum mean square error, the covariance function must meet some requirements which are met by the gaussian, exponential, spherical, hole-effect, and nugget-effect models of Sec. 20.2.3. An example of a function which does not meet the requirements is $R(\mathbf{x}-\mathbf{x}') = -\alpha\,|\mathbf{x}-\mathbf{x}'| + C$, where α and C are constants.

When the mean m of the random function $z(\mathbf{x})$ is unknown, as is common in applications, it is advantageous to select the weights so that the mean error of estimation vanishes independently of the actual value of m (thus achieving unbiasedness.) This is what ordinary kriging does: it imposes a restriction on the possible values of weights, thus allowing more freedom in the choice of covariance functions. For example, in addition to the exponential and the other covariance functions which one can use in simple kriging, one can use the $R(\mathbf{x}-\mathbf{x}') = -\alpha\,|\mathbf{x}-\mathbf{x}'| + C$ covariance function (where α and C are positive coefficients) and still obtain a unique solution and nonnegative mean square error. Furthermore, the result does not change with the addition of an arbitrary constant; that is, the covariance function $-\alpha|\mathbf{x}-\mathbf{x}'| + C$ gives the same best estimate and mean square error as $-\alpha|\mathbf{x}-\mathbf{x}'|$.

Thus, because of the unbiasedness condition, one needs to know the covariance function within a constant. This covariance within an arbitrary constant is called a *generalized covariance function.* The negative variogram $-\gamma(\mathbf{x}-\mathbf{x}')$ is a generalized covariance function. It must be chosen to satisfy requirements which are less stringent than the requirements that need to be satisfied in the case of covariance functions used in simple kriging. The concept of the generalized covariance functions is even more useful in the case of functions $z(\mathbf{x})$ with variable mean, as will be seen next.

To model the spatial structure of a hydrologic variable which has a well-defined trend, as is often the case with piezometric head at a regional scale or with precipitation in the presence of an orographic effect, the following model is used:

$$z(\mathbf{x}) = m(\mathbf{x}) + \xi(\mathbf{x}) \tag{20.2.24}$$

where $\mathbf{x}$ stands for the spatial coordinates, $m(\mathbf{x})$ is a deterministic function of spatial coordinates called the *drift* (or trend) and represents the expected value of $z(\mathbf{x})$, and $\xi(\mathbf{x})$ is a random function with zero mean.

For convenience in statistical model building and estimation, the drift is commonly represented as the weighted sum of known functions of the coordinates

$$m(\mathbf{x}) = \sum_{k=1}^{p} f_k(\mathbf{x})\beta_k \tag{20.2.25}$$

where $f_1(\mathbf{x}), \ldots, f_p(\mathbf{x})$ are known functions of the spatial coordinates $\mathbf{x}$, called *base functions;* $\beta_1, \ldots, \beta_p$ are constants, the *drift coefficients.* Because the mean function is linear in the drift coefficients, Eqs. (20.2.24) and (20.2.25) constitute what is known in statistics as the *linear model* (not to be confused with the linear variogram).

In applications, monomials are commonly chosen for base functions. In this context, Ref. 36 extends the idea of intrinsic functions to intrinsic functions of higher order: the drift of an *intrinsic function of* kth order is a k-order polynomial. For example, in the two-dimensional case:

$$\mathrm{k} = 0\colon p = 1, f_1(x_1, x_2) = 1 \qquad \text{(a constant plane)} \tag{20.2.26a}$$

$$k=1: \quad p=3, f_1(x_1,x_2)=1, f_2(x_1,x_2)=x_1, f_3(x_1,x_2)=x_2 \quad \text{(a sloping plane)} \tag{20.2.26b}$$

$$k=2: \quad p=6, f_1(x_1,x_2)=1, f_2(x_1,x_2)=x_1, f_3(x_1,x_2)=x_2, f_4(x_1,x_2)=x_1x_1, f_5(x_1,x_2)=x_2x_2, f_6(x_1,x_2)=x_1x_2 \quad \text{(a quadratic surface)} \tag{20.2.26c}$$

The intrinsic function is a special case of this model (order $k=0$).

Usually, the weights applied to the observations for the computation of the estimate are selected so that the estimation error is independent of the drift coefficients. This is achieved by enforcing that the weights satisfy as many linear equations (called *unbiasedness constraints*) as the unknown drift coefficients. By analogy with the intrinsic case, the covariance function needs to be known within a function with arbitrary coefficients. The covariance within a function is called *generalized* and must satisfy requirements which depend on the unbiasedness constraints. The essence of the generalized covariance is that it is the part of the actual covariance function that matters when performing this type of kriging.

When the model is constructed from data, $\xi(x)$ is commonly postulated stationary, in which case the (generalized) covariance function is $K(\mathbf{h})$, where $\mathbf{h}=\mathbf{x}-\mathbf{x}'$, and is often postulated isotropic, in which case $K(h)$, where $h=|\mathbf{h}|$. The models listed in Sec. 20.2.3 can be used for the covariance. However, generalized covariance functions characterize a very broad class of functions, called *stationary-increment* processes, which include stationary and intrinsic functions as special cases. A useful isotropic model uses the polynomial generalized covariance function combined with the nugget effect for intrinsic functions of kth order.[14,36] Extension to nonisotropic cases is straightforward if Eq. (20.2.22) is used.

20.3 BEST LINEAR UNBIASED ESTIMATION (BLUE)

20.3.1 What Is BLUE?

An estimate of the value of z at some unmeasured location $\mathbf{x}_0$ from observations $z(\mathbf{x}_1), \ldots, z(\mathbf{x}_n)$ is obtained by using a linear estimator:

$$\hat{z}(\mathbf{x}_0)=\sum_{i=1}^{n}\lambda_i z(\mathbf{x}_i) \tag{20.3.1}$$

where ^ stands for *estimate* and the λ's are weights. This type of estimator is quite common in hydrology. Consider, for example, that $z(\mathbf{x}_i)$, $i=1, \ldots, n$ are rainfall measurements and $z(\mathbf{x}_0)$ is rainfall at an ungauged location. An estimate is obtained through a linear interpolator, such as Eq. (20.3.1); the λ weights are usually chosen so that they add up to 1 and the λ_i weight is inversely proportional to some power of the distance between $\mathbf{x}_0$ and $\mathbf{x}_i$. In many cases, the method for selecting the weights is arbitrary. A more objective approach is to select the weights on the basis of the spatial structure of rainfall fluctuations. For example, in case *a* of Fig. 20.3.1, the weights should be approximately equal, all measurements being about equally informative, because rainfall varies so much over short distances. However, in case *b* of the same figure, the scale of rainfall fluctuations is large compared to the distance between the closest gauges; it is reasonable to select larger weights for the measurements which are

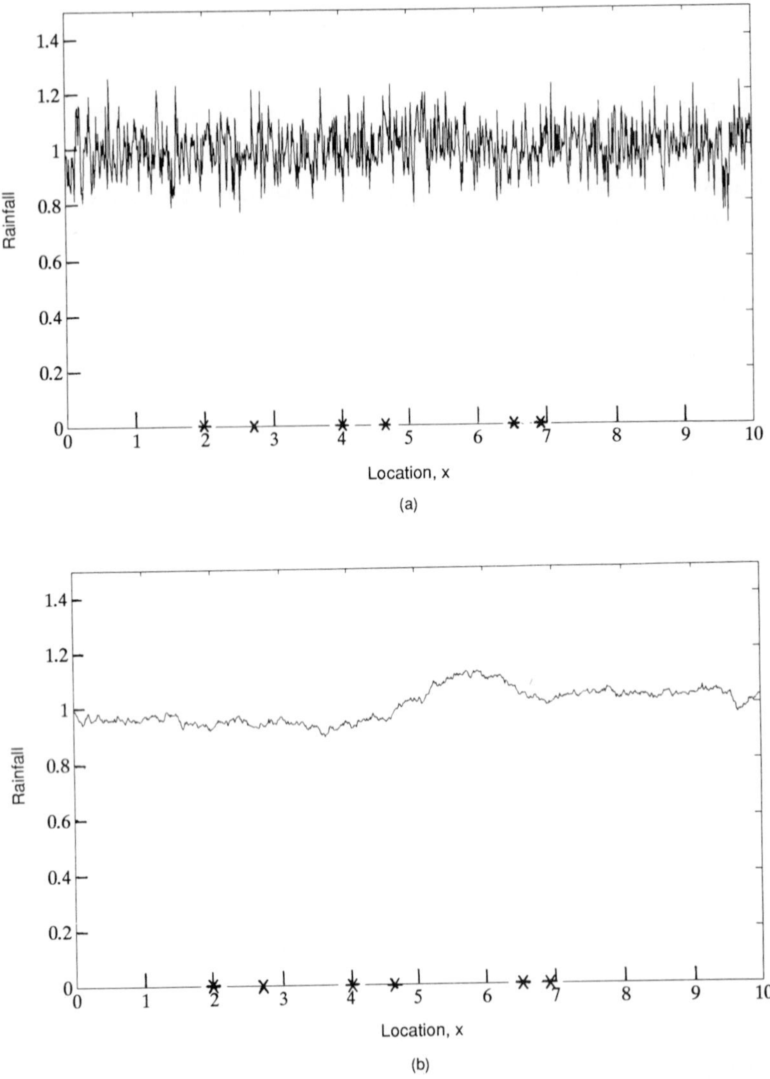

FIGURE 20.3.1 *(a)* The scale of rainfall fluctuations are of the order of 0.01. *(b)* The scale of rainfall fluctuations are of the order of 3. The location of gauges is indicated by * and the location of the unknown by x.

near the location of the unknown. Geostatistics provides a systematic procedure to select weights based on the variogram (or generalized covariance function) or rainfall.

In geostatistics, the weights are chosen from the specifications that the estimation error [estimate $\hat{z}(\mathbf{x}_0)$ minus true unknown $z(\mathbf{x}_0)$] must be on the average zero *(unbiasedness)* and must have the smallest possible mean square error *(minimum variance)*. From the minimization of the variance subject to the constraint of unbiasedness, a system of linear equations (the *kriging system*) is obtained. Its solution gives the weights. The same general approach applies if the unknown or measurements are not point values of z.

Regarding unbiasedness, a distinction can be made between: (1) unbiasedness for any value and (2) unbiasedness for a given value of a drift coefficient (such as the mean). The first case, known as *ordinary kriging,* is the most common. The second case, known as *simple kriging,* is appropriate when the drift coefficients are somehow known; the equations for obtaining the coefficients are a simple modification of the method appropriate for ordinary kriging. A generalization which includes partially known drift coefficients is given in Ref. 28.

20.3.2 Common Applications of Estimation from Point Measurements

A list of useful formulas for linear estimation from point observations $\mathbf{x}_1, \mathbf{x}_2, \ldots, \mathbf{x}_n$ will be presented for stationary and intrinsic cases.

Stationary with Unknown Mean. For ordinary kriging with the covariance function $R(\mathbf{x}_i - \mathbf{x}_j)$. *Interpolation,* i.e., estimation of $z(\mathbf{x}_0)$.

Estimator:

$$\hat{z}(\mathbf{x}_0) = \sum_{i=1}^{n} \lambda_i z(\mathbf{x}_i) \tag{20.3.2}$$

A kriging system of $n + 1$ linear equations with $n + 1$ unknowns $(\lambda_1, \ldots, \lambda_n, \nu)$ is

$$\sum_{j=1}^{n} \lambda_j R(\mathbf{x}_i - \mathbf{x}_j) + \nu = R(\mathbf{x}_i - \mathbf{x}_0) \qquad i = 1, 2, \ldots, n \tag{20.3.3}$$

$$\sum_{j=1}^{n} \lambda_j = 1 \tag{20.3.4}$$

Mean square error:

$$E\{[\hat{z}(\mathbf{x}_0) - z(\mathbf{x}_0)]^2\} = -\nu - \sum_{i=1}^{n} \lambda_i R(\mathbf{x}_i - \mathbf{x}_0) + R(0) \tag{20.3.5}$$

The ν coefficient is the Lagrange multiplier associated with the unbiasedness constraint [Eq. (20.3.4)].

Spatial Averaging. Estimate the value of

$$z_A = \frac{1}{|A|} \int_A z(\mathbf{x})\, d\mathbf{x}$$

where A is region, $|A|$ indicates the size (length, area, or volume) of A, and $\int_A$ is shorthand notation for a single, double, or triple integral over the domain A.

Estimator:

$$\hat{z}_A = \sum_{i=1}^{n} \lambda_i z(\mathbf{x}_i) \tag{20.3.6}$$

Kriging system:

$$\sum_{j=1}^{n} \lambda_j R(\mathbf{x}_i - \mathbf{x}_j) + \nu = R_{Ai} \qquad \text{for } i = 1, 2, \ldots, n \tag{20.3.7}$$

$$\sum_{j=1}^{n} \lambda_j = 1 \tag{20.3.8}$$

Mean square error:

$$E[(\hat{z}_A - z_A)^2] = -\nu - \sum_{i=1}^{n} \lambda_i R_{Ai} + R_{AA} \tag{20.3.9}$$

where R_{Ai} is the average covariance of $z(\mathbf{x}_i)$ with all $z(\mathbf{u})$ in the domain A and R_{AA} is the average covariance of all pairs of points in the domain:

$$R_{Ai} = \frac{1}{|A|} \int_A R(\mathbf{x}_i - \mathbf{u})\, d\mathbf{u} \qquad \text{and} \qquad R_{AA} = \frac{1}{|A|^2} \int_A \int_A R(\mathbf{u} - \mathbf{v})\, d\mathbf{u}\, d\mathbf{v}$$

See Sec. 20.3.6 regarding these integrals. For estimation of a slope or gradient, see Ref. 41.

Stationary Case, Known Mean (Simple Kriging). The formulas for ordinary kriging apply with the modifications:

1. In the estimator, replace every appearance of z with $z - m$.
2. The unbiased constraint is eliminated.
3. The Lagrange multiplier ν is set to zero.

For example, consider the interpolation problem:

Estimator:

$$\hat{z}(\mathbf{x}_0) - m = \sum_{i=1}^{n} \lambda_i [z(\mathbf{x}_i) - m] \qquad \text{or} \qquad \hat{z}(\mathbf{x}_0) = m + \sum_{i=1}^{n} \lambda_i [z(\mathbf{x}_i) - m] \tag{20.3.10}$$

Kriging system of n equations with n unknowns:

$$\sum_{j=1}^{n} \lambda_j R(\mathbf{x}_i - \mathbf{x}_j) = R(\mathbf{x}_i - \mathbf{x}_0) \qquad i = 1, 2, \ldots, n \tag{20.3.11}$$

Mean square error:

$$E\{[\hat{z}(\mathbf{x}_0) - z(\mathbf{x}_0)]^2\} = -\sum_{i=1}^{n} \lambda_i R(\mathbf{x}_i - \mathbf{x}_0) + R(0) \tag{20.3.12}$$

Intrinsic Case (Using Variograms). In ordinary kriging, because of the unbiasedness constraint, Eqs. (20.3.1) to (20.3.9) can be expressed through the variogram γ instead of the covariance function R through the substitution $R(\mathbf{h}) = -\gamma(\mathbf{h})$. [For numerical reasons, sometimes the substitution is $R(\mathbf{h}) = c - \gamma(\mathbf{h})$, where c is a large positive number.] For example, for interpolation:

Kriging system:

$$-\sum_{j=1}^{n} \lambda_j \gamma(\mathbf{x}_i - \mathbf{x}_j) + \nu = -\gamma(\mathbf{x}_i - \mathbf{x}_0) \qquad i = 1, 2, \ldots, n \tag{20.3.13}$$

$$\sum_{j=1}^{n} \lambda_j = 1 \tag{20.3.14}$$

Mean square error:

$$E\{[\hat{z}(\mathbf{x}_0) - z(\mathbf{x}_0)]^2\} = -\nu + \sum_{i=1}^{n} \lambda_i \, \gamma(\mathbf{x}_i - \mathbf{x}_0) \tag{20.3.15}$$

Note that the diagonal elements of the value of the matrix of coefficients of the kriging system $\gamma(\mathbf{x}_i - \mathbf{x}_i)$ are zero (because γ at lag exactly zero is zero even when a nugget is present).

Example. A one-dimensional function was the variogram

$$\gamma(h) = \begin{cases} 1 + h & h > 0 \\ 0 & h = 0 \end{cases}$$

and three measurements, at locations $x_1 = 0, x_2 = 1, x_3 = 3$. Estimate the value of $\mathbf{z}$ location $x_0 = 2$. The kriging system of equations, which gives the weights λ_1, λ_2, and λ_3, is

$$\begin{bmatrix} 0 & -2 & -4 & 1 \\ -2 & 0 & -3 & 1 \\ -4 & -3 & 0 & 1 \\ 1 & 1 & 1 & 0 \end{bmatrix} \begin{bmatrix} \lambda_1 \\ \lambda_2 \\ \lambda_3 \\ \nu \end{bmatrix} = \begin{bmatrix} -3 \\ -2 \\ -2 \\ 1 \end{bmatrix}$$

And the mean square estimation error is $-\nu + 3\,\lambda_1 + 2\,\lambda_2 + 2\,\lambda_3$.

Solving the system, obtain $\lambda_1 = 0.1304$, $\lambda_2 = 0.3913$, $\lambda_3 = 0.4783$, $\nu = -0.304$. The mean square estimation error is MSE = 2.43.

For $x_0 = 0$, i.e., coinciding with an observation point, the kriging system is

$$\begin{bmatrix} 0 & -2 & -4 & 1 \\ -2 & 0 & -3 & 1 \\ -4 & -3 & 0 & 1 \\ 1 & 1 & 1 & 0 \end{bmatrix} \begin{bmatrix} \lambda_1 \\ \lambda_2 \\ \lambda_3 \\ \nu \end{bmatrix} = \begin{bmatrix} 0 \\ -2 \\ -4 \\ 1 \end{bmatrix}$$

By inspection, one can verify that the only possible solution is $\lambda_1 = 1$, $\lambda_2 = \lambda_3 = 0$, $\nu = 0$. The mean square estimation error is MSE = 0, indicating error-free prediction. The kriging estimate at a measurement location is the measurement itself.

20.3.3 Extension to Variable Mean

Linear minimum-variance unbiased estimation is readily extended to the variable-mean case which follows the linear model (refer to Sec. 20.2.4); it is known in geostatistics by the name *universal kriging* or *kriging with a drift.*

Stationary Fluctuations with Covariance R. The problem is to estimate an unknown quantity y (point value, average, slope, etc.) from point measurements, as in Sec. 20.3.1. The estimator is again

$$y = \sum_{i=1}^{n} \lambda_i z(\mathbf{x}_i) \tag{20.3.16}$$

When the drift coefficients are unknown, there are p constraints. The kth constraint ($k = 1, \ldots, p$), needed to ensure that the result does not depend on the drift coefficient, β_k, is of the form

$$\sum_{i=1}^{n} f_k(\mathbf{x}_i)\lambda_i = f_k[y] \tag{20.3.17}$$

where for the common cases:

Interpolation:

$$f_k[y] = f_k(\mathbf{x}_0) \tag{20.3.18}$$

Averaging:

$$f_k[y] = \frac{1}{|A|} \int_A f_k(\mathbf{x})\, d\mathbf{x} \tag{20.3.19}$$

Furthermore, to the kth unbiasedness constraint there corresponds a Lagrange multiplier ν_k, $k = 1, \ldots, p$. The system of $n + p$ equations with the $n + p$ unknowns $\lambda_1, \ldots, \lambda_n, \nu_1, \ldots, \nu_p$ is

$$\sum_{j=1}^{n} R(\mathbf{x}_i - \mathbf{x}_j)\lambda_j + \sum_{k=1}^{p} f_k(\mathbf{x}_i)\nu_k = R_{yi} \qquad i = 1, \ldots, n \tag{20.3.20}$$

$$\sum_{j=1}^{n} f_k(\mathbf{x}_j)\lambda_j = f_k[y] \qquad k = 1, \ldots, p \tag{20.3.21}$$

where R_{yi} is the covariance of y and the measurement i. For the two common cases:

Interpolation:

$$R_{yi} = R(\mathbf{x}_0 - \mathbf{x}_i) \tag{20.3.22}$$

Averaging:

$$R_{yi} = \frac{1}{|A|} \int_A R(\mathbf{x}_0 - \mathbf{x}) d\mathbf{x} \tag{20.3.23}$$

Mean square estimation error:

$$E[(\hat{y} - y)^2] = -\sum_{k=1}^{p} f_k[y]\, \nu_k - \sum_{i=1}^{n} \lambda_i R_{yi} + R_{yy} \tag{20.3.24}$$

where R_{yy} is the variance of y. For the common cases:

Interpolation:

$$R_{yy} = R(0) \tag{20.2.25}$$

Averaging:

$$R_{yy} = \frac{1}{|A|^2} \int_A \int_A R(\mathbf{x} - \mathbf{y})\, d\mathbf{x}\, d\mathbf{y} \tag{20.3.26}$$

Example. Interpolation in two dimensions with first-order (linear) drift:

$$f_1 = 1$$

$$f_2 = x_{i1}$$

$$f_3 = x_{i2}$$

where x_{i1} and x_{i2} are the two cartesian coordinates of location $\mathbf{x}_i$. The kriging system is

$$\sum_{j=1}^{n} \lambda_j R(\mathbf{x}_i - \mathbf{x}_j) + v_1 + v_2\, \mathbf{x}_{i1} + v_3\, x_{i2} = R(\mathbf{x}_i - \mathbf{x}_0) \qquad i = 1, \ldots, n$$

$$\sum_{j=1}^{n} \lambda_j = 1$$

$$\sum_{j=1}^{n} x_{j1}\lambda_j = x_{01}$$

$$\sum_{j=1}^{n} x_{j2}\lambda_j = x_{02}$$

Mean square estimation error:

$$E\{[\hat{z}(\mathbf{x}_0) - z(\mathbf{x}_0)]^2\} = -v_1 - x_{01}v_2 - x_{02}v_2 - \sum_{i=1}^{n} \lambda_i R(\mathbf{x}_i - \mathbf{x}_0) + R(\mathbf{0})$$

Note that, if the drift coefficients were known, there would be no need for the unbiasedness constraints or the Lagrange multipliers. However, $z(\mathbf{x}) - m(\mathbf{x})$ should be used instead of $z(\mathbf{x})$ in the expression for the estimator.

Using Generalized Covariance Functions. Instead of the covariance function, the generalized covariance function may be used; this allows extension of these formulas to cases where the fluctuations are not stationary but belong to the much broader class of processes with stationary increments. The polynomial generalized covariance function, which describes k-order intrinsic functions, is often used in applications (see Sec. 20.2.4.)

20.3.4 Estimation of the Continuous Part

The interpolators presented in Sec. 20.3.2 are "exact" in the sense that they reproduce the observations exactly. That is, when the location of the unknown, $\mathbf{x}_0$, coincides with the location of an observation, $\mathbf{x}_i$, the estimate is $\hat{z}(\mathbf{x}_0) = z(\mathbf{x}_i)$ with zero mean square estimation error. If the covariance function is discontinuous at the origin $R(\mathbf{h}) = R_c(\mathbf{h}) + C_0\delta(\mathbf{h})$, where $R_c(\mathbf{h})$ is continuous at the origin and $C_0\delta(\mathbf{h})$ corresponds to the pure nugget effect, then the estimate at a location at a small distance from a measurement can be quite different; consequently the estimator is discontinuous.

One may consider z as the sum of two functions, $z_c(\mathbf{x})$ and $\eta(\mathbf{x})$. $z_c(\mathbf{x})$ is continuous with covariance R_c and represents variability at the scale of typical distances between measurements; $\eta(\mathbf{x})$ is a discontinuous function with zero mean and no spatial correlation (pure nugget effect) and stands for random measurement error and microvariability. In some cases, it is appropriate to view the continuous part as the "signal" of interest while the measurements contain the signal plus "noise." Measurements are taken from z, but the objective is to estimate z_c.

The formulas of Secs. 20.3.1 to 20.3.3 apply for the estimation of z_c with the following modifications:

- In the right-hand side of the kriging equations and in the expression for the mean square error, R_c is used instead of R.
- The left-hand side of the kriging system remains unchanged. This estimator is continuous. The mean square estimation error of z_c is smaller than that of z by C_0, unless the estimation point coincides with a measurement point.

20.3.5 Neighborhood Kriging

In interpolation, it is common in applications where many measurements are available to use only a few measurement points (usually by fixing the number to about 10 or by using all points within a small radius) in the neighborhood of the estimation point. Advantages of this approach are that computational cost and memory requirements are reduced, results do not depend on the value of the variogram at large distances (which is hard to infer from data), and the effect of a spatially variable mean is reduced. Disadvantages of this approach include that the estimator is discontinuous, especially if a sill is present, and the neighborhood radius or the number of points is like any other model parameter and needs to be calibrated. The method is recommended only as a means of reducing computational cost, especially for variograms without a sill for which the remaining measurements contain little additional information about the unknown.

20.3.6 Integrations for Volume Averaging

Finding the kriging coefficients involves the calculation of $\int_A R(\mathbf{x}_i - \mathbf{u})\, d\mathbf{u}$ and the estimation error involves $\int_A \int_A R(\mathbf{u} - \mathbf{v})\, d\mathbf{v}\, d\mathbf{u}$. Note that in the calculation of these integrals only the continuous part of the covariance function is used; i.e., $\int_A R(\mathbf{x}_i - \mathbf{u}) d\mathbf{u} = \int_A R_c(\mathbf{x}_i - \mathbf{u})\, d\mathbf{u} + \int_A C_0\delta(\mathbf{x}_i - \mathbf{u})\, d\mathbf{u} = \int_A R_c(\mathbf{x}_i - \mathbf{u})\, d\mathbf{u}$.

A straightforward and general numerical method is to divide the total region into m small subregions $A_1, \ldots, A_m$. The kth subregion A_k is represented through point $\mathbf{u}_k$, located near its center. Then

$$|A| = \sum_{k=1}^{m} |A_k| \tag{20.3.27}$$

$$\int_A R_c(\mathbf{x}_i - \mathbf{u})\, d\mathbf{u} = \sum_{k=1}^{m} R_c(\mathbf{x}_i - \mathbf{u}_k)\, |A_k| \tag{20.3.28}$$

$$\int_A \int_A R_c(\mathbf{u} - \mathbf{v})\, d\mathbf{u}\, d\mathbf{v} = \sum_{k=1}^{m} \sum_{l=1}^{m} R_c(\mathbf{u}_k - \mathbf{v}_l)\, |A_k|\, |A_l| \tag{20.3.29}$$

Although not necessarily the most efficient, this method is usually adequate. There is no general rule on the size of m; it depends on desired accuracy, available computer, the size and shape of the region, and the covariance function. The author has found that $m = 50$ is adequate for most practical purposes. Thus, start with $m = 50$ and then check whether the results are significantly affected when $m = 100$. (More efficient integration techniques are available for special cases, see Ref. 27, pp. 95–147.)

20.3.7 Confidence Interval

The 95 percent confidence interval of estimation of $z(\mathbf{x}_0)$ is the interval which contains the actual value of $z(\mathbf{x}_0)$ with probability 95 percent. It depends on the probability distribution of $z(\mathbf{x})$. The common approach is to assume normality, i.e., gaussian distribution of the estimation errors, in which case the confidence interval is approximately $(z - 2\sigma,\ \hat{z} + 2\sigma)$, where z is the estimate and σ is the standard error of estimation. Even when normality is not explicitly mentioned, this method of determining confidence intervals is the conventional way of obtaining confidence intervals in linear geostatistics.

20.3.8 Finite Support Volume

In some cases, the ith measured value is a volume average

$$\frac{1}{V_i} \int z(\mathbf{u})\, d\mathbf{u}$$

over volume (or area or length) V_i centered at $\mathbf{x}_i$. In this case, instead of $R(\mathbf{x}_i - \mathbf{x}_j)$, use

$$\bar{R}(\mathbf{x}_i - \mathbf{x}_j) = \frac{1}{V_i V_j} \int_{V_i} \int_{V_j} R(\mathbf{u} - \mathbf{v})\, d\mathbf{u}\, d\mathbf{v}$$

Similarly, for estimation at point $\mathbf{x}_0$, instead of $R(\mathbf{x}_i - \mathbf{x}_0)$, use

$$\bar{R}(\mathbf{x}_i - \mathbf{x}_0) = \frac{1}{V_i} \int_{V_i} R(\mathbf{u} - \mathbf{x}_0)\, d\mathbf{u}$$

20.4 STRUCTURAL ANALYSIS OF A SPATIAL VARIABLE

In applications, best linear unbiased estimation is preceded by structural analysis, i.e., the process of selecting a model and its parameters from the data. (However, the kriging formulas were presented first because they are used in model calibration and testing.) Geostatistics is similar to any other field of endeavor in that the inductive reasoning by which an empirical model is developed from data cannot be reduced to a cookbook recipe. The iterative approach which follows is a practical and well-established way to develop a model adequate for most applications. A flowchart of this approach is shown in Fig. 20.4.1. *Exploratory analysis* leads to the tentative selection of a model. The model involves some parameters which are fitted to the data in the *parameter estimation* or *calibration* step. Then the model is subjected to a number of tests. If it passes, the model is accepted. Otherwise, it is adjusted (e.g., replace the assumption of constant mean with that of a linear trend) and the new model is recalibrated and retested. The focus of the discussion is on a single variable (e.g., precipitation) and simple models which assume that correlation between two measurements depends on their relative location.

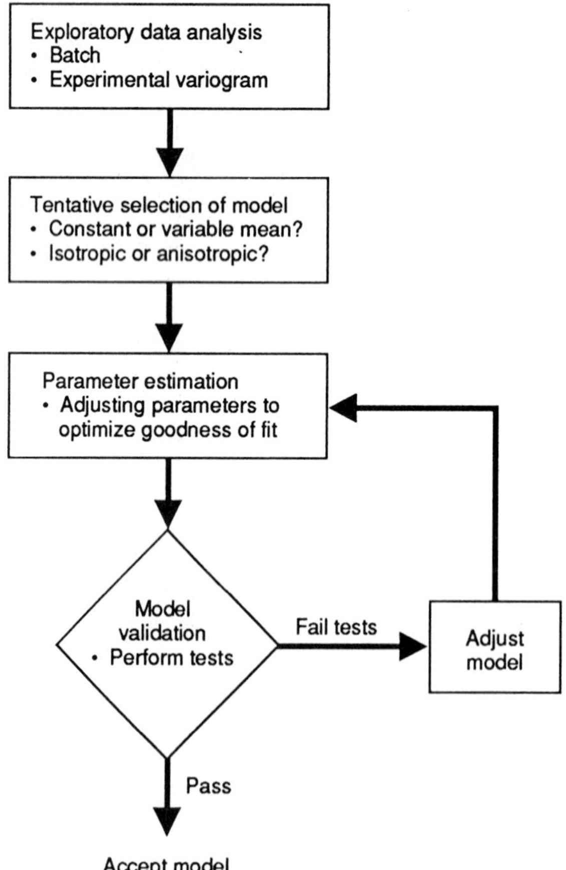

FIGURE 20.4.1 Flowchart of geostatistical model development (structural analysis).

20.4.1 Exploratory Analysis

The objective of exploratory analysis is to familiarize the analyst with the data and, in the process, to assist in the selection of an appropriate model. At this stage, the analyst relies on graphical methods and simple summary statistics to detect patterns of variability and peculiar data. Typically, the data consist of a set of observations $z(\mathbf{x}_1)$, $z(\mathbf{x}_2)$, . . . , $z(\mathbf{x}_n)$ at irregularly distributed points. For organizational purposes, one may distinguish between *batch* and *covariation* data analyses.

Batch analysis is the analysis of observations without consideration of their location in space. The applicable summary statistics (median, mean, mode, quartiles, interquartile range, variance, standard deviation, coefficient of variation, skewness coefficient) and graphical tools (mainly the box plot and histogram) are described in Chap. 17.

Can batch analysis determine the univariate distribution of the data? No final conclusion can be reached without examination of the spatial covariation of the observations. In particular, certain characteristics of the histogram, such as bimodality, may be explained away when spatial correlation is taken into account. Batch analysis is most informative about the univariate distribution of the data when the data can reasonably be assumed as independent and identically distributed. (This is the case for data with a nugget-effect variogram or some appropriately chosen residuals, as will be seen in Sec. 20.4.3.)

Covariation analysis displays the data in ways that reveal the correlation structure of the regionalized variable. The most important tool is the *experimental variogram,* described in the next subsection. However, one should not neglect other means of exploring spatial structure, such as plots of z versus each of its spatial coordinates. For two dimensions, a particularly useful way is plotting on a map the location of each measurement with symbols indicating whether it is above or below the median, whether it is a stray value, and other pertinent information.

20.4.2 The Experimental Variogram

Plot the square difference $[z(\mathbf{x}_i) - z(\mathbf{x}'_i)]^2$ against the separation distance $|\mathbf{x}_i - \mathbf{x}'_i|$ for all measurement pairs. For n measurements, there are $n(n-1)/2$ such pairs which form a scatter plot known as the *raw variogram.* The experimental variogram is, in a sense, a smooth line through the scatter plot.

In the common method of plotting the experimental variogram, the axis of separation distance is divided into consecutive intervals. The kth interval is (h_k^l, h_k^u) and contains N_k pairs of measurements $z(x_i)$ and $z(x'_i)$ for which $h_k^l < |x_i - x'_i| \leq h_k^u$. This interval is represented by a single point h_k. Then, compute

$$\hat{\gamma}(h_k) = \frac{1}{2N_k} \sum_{i=1}^{N} [z(\mathbf{x}_i) - z(\mathbf{x}'_i)]^2 \tag{20.4.1}$$

where index i refers to each pair of measurements $z(\mathbf{x}_i)$ and $z(\mathbf{x}'_i)$ for which

$$h_k^l < |x_i - x'_i| < h_k^u \tag{20.4.2}$$

Take h_k equal to the midpoint of the interval; i.e.,

$$h_k = \frac{|h_k^u - h_k^l|}{2} \tag{20.4.3}$$

or, even better, the average value found by

$$h_k = \frac{1}{N_k} \sum_{i=1}^{N_k} |x_i - x_i'| \tag{20.4.4}$$

Next, these points are connected to form a polygon, which is the experimental variogram. (For figures of the raw and experimental variograms, see the example of Sec. 20.4.7.) Numerous modifications to this basic approach have been proposed to improve its robustness or to account for other effects.[3,8,13,40]

The plotted experimental variogram depends partially on the selected intervals. It is unprofitable to spend too much time at this stage fiddling with the intervals in a pointless search for the "best" experimental variogram. (This time could be used more productively to test the model or adjust its parameters.) In selecting the length of an interval, keep in mind that, by increasing the length of the interval, you average over more points, thus decreasing the variance (associated with the raw variogram), but you may smooth out the curvature of the variogram. Some useful guidelines to obtain a reasonable experimental variogram are (1) use three to six intervals, (2) make sure you have at least 10 pairs in each interval, and (3) include more points (use longer intervals) at distances where the raw variogram is spread out. These rules are usually adequate for exploratory analysis purposes and preliminary variogram estimation.

The experimental variogram presented is a measure of spatial correlation independent of orientation. In some cases, however, better predictions can be made by taking into account the anisotropy in the structure of the unknown function; for example, conductivities in a layered medium are more correlated in a horizontal direction than in the vertical. The variogram should then depend on the orientation as well as the separation distance (anisotropic model.) The *directional experimental variogram* is a tool useful in exploring the anisotropy of the data. To plot it, consider the raw variogram. For two dimensions, in addition to the separation distance, compute the direction angle ϕ_{ij}, ranging from $-90°$ to $90°$ (because $110°$ is the same as $-70°$) for any pair of measurements. The data pairs are grouped with respect to the orientation. Then a separate directional experimental variogram is plotted for each group of direction angles. That is, plot one experimental variogram for those in the interval $-60°$ to $-30°$ and another experimental variogram for those in the interval $30°$ to $60°$. Commonly, in two dimensions, two orthogonal (orientations differ by $90°$) directions are selected. Significant differences between directional variograms indicate anisotropy and the orthogonal directions where the contrast is the sharpest may be taken as the principal directions of anisotropy. However, the same differences may be indicative of a drift in the mean value of the variable.

When using a model with a drift, one method is to fit, perhaps through least squares regression, this drift to the data and then to determine the variogram of the residuals (data minus the fitted drift.) This approach is used in exploratory analysis but is not appropriate for final determination of the model or its parameters. For final estimates, the methods of Sec. 20.4.4 should be used instead.

In most applications, the intrinsic model is appropriate, in which case structural analysis becomes equivalent to finding the variogram. However, the experimental variogram should be viewed as only a preliminary estimate of the variogram and should not be confused with the variogram itself. The first adjustment to be made on the experimental variogram is to replace it with a mathematical model obtained by combining models from the list of Sec. 20.2 (for reasons explained in Sec. 20.2.4). The mathematical model should resemble the experimental variogram, particularly at short separation distances. Examination of the experimental variogram near the origin provides clues about the continuity and differentiability of $z(\mathbf{x})$. In some cases, the variogram does not seem to tend to 0 as the separation tends to zero, which indicates discontinuity in the spatial variable z. In this case, the model should include a nugget effect. When the variogram tends to zero, the variable z appears to be continuous (no nugget effect), but its degree of smoothness depends on whether the variogram converges linearly, parabolically, etc. If it varies linearly, one should use a model with linear behavior at the origin (such as exponential, spherical, or linear.) If it varies parabolically, one should use a model with parabolic behavior (e.g., the gaussian model). The behavior of the variogram at large distances reveals whether stationarity is appropriate. If the experimental variogram indicates the presence of a sill, use models corresponding to stationary functions, such as the nugget effect, the exponential, and the gaussian models. If the experimental variogram indicates continual increase with distance, then use a variogram corresponding to a nonstationary intrinsic case, such as the linear variogram.

Waste no time trying to achieve a close visual fit between the model and the experimental variogram. As a general rule, the experimental variogram at large distances is corrupted by sampling variability so that it is pointless to take great pains to reproduce it in the mathematical model. For a linearly increasing variogram, make no effort to reproduce the experimental variogram at distances larger than about half the maximum separation distance.

With few exceptions, start with the simplest model possible (intrinsic and isotropic) and a variogram represented with one, two, or, for large data sets, three parameters. Subsequently, the model can be improved in two ways: (1) by calibrating the parameter and (2) by testing the model and, if inadequacies are found, by modifying it.

20.4.3 Model Validation

Assume that a model (including the numerical value of its parameters) has been chosen. For example, that $z(\mathbf{x})$ is intrinsic with variogram

$$\gamma(h) = \begin{cases} 0 & \text{if } h = 0 \\ 1 + h & \text{if } h > 0 \end{cases}$$

Is this an appropriate model? Find out by evaluating its performance in test cases. In practice, testing involves the evaluation of *residuals* or estimation errors which generally do a good job of revealing whether the model is adequate. We describe the use of the commonly used *standardized residuals* and their improved version, the *orthonormal residuals.*

Using Standardized Residuals. Assume that the sample consists of n point measurements, $z(\mathbf{x}_1), \ldots, z(\mathbf{x}_n)$. Drop one measurement, $z(\mathbf{x}_k)$. Then, using the other measurements and the assumed variogram, estimate the value of z at location $\mathbf{x}_k$ and its mean square estimation error, $\hat{z}_k$ and σ_k^2, respectively. Estimation is through kriging with the same procedure to be used in prediction. Calculate the difference between the actual observation $z(\mathbf{x}_k)$ and the predicted value using the other measurements, $\hat{z}_k$, divided by the standard error of estimation, σ_k:

$$e_k = \frac{z(x_k) - \hat{z}_k}{\sigma_k} \qquad k = 1, \ldots, n \tag{20.4.5}$$

Repeat the same procedure for all measurements, dropping one measurement at a time. This way, obtain the *standardized* residuals $e_1, \ldots, e_n$. These residuals should not contradict the behavior expected on the basis of the model. Calculate the following statistics:

$$S_1 = \frac{1}{n} \sum_{k=1}^{n} e_k \tag{20.4.6}$$

$$S_2 = \frac{1}{n} \sum_{k=1}^{n} e_k^2 \tag{20.4.7}$$

If the model is consistent with the data, the first number must be near zero, while the second must be near one. This is the "correct" answer; if the computed answers are not in agreement, there are good reasons to question the validity of the assumed model. There are no simple and reliable rules to judge whether the computed S_1 is close enough to 0 and S_2 is close enough to 1 for the model to be accepted. (It seems to be common practice to reject a model if the computed S_1 and S_2 differ from their "correct" values by 0.15 to 0.20.) Another limitation of the standardized residuals is that they are not necessarily uncorrelated.

Orthonormal Residuals. Another method to test the validity of the assumed model involves the formulation of uncorrelated *orthonormal* residuals.[30]

The n measurements are treated as if they were ordered. Then, calculate the kriging estimate of z at the kth point, $\mathbf{x}_k$, using *only* the first $k-1$ measurements, $\mathbf{x}_1, \ldots, \mathbf{x}_{k-1}$. Here, $k = p+1, \ldots, n$; p = number of drift coefficients; e.g., $p = 1$ for the intrinsic case. Calculate the actual error $z(\mathbf{x}_k) - \hat{z}_k$ and normalize it by using the standard error of estimation σ_k:

$$\epsilon_k = \frac{z(x_k) - \hat{z}_k}{\sigma_k} \qquad k = p+1, \ldots, n \tag{20.4.8}$$

This way, obtain $n-p$ residuals which, if the right model has been used, have zero mean and unit variance, and are uncorrelated (pure nugget effect):

$$E[\epsilon_k \epsilon_l] = \begin{cases} 1 & \text{if } k = l \\ 0 & \text{if } k \neq l \end{cases} \qquad k, l = p+1, \ldots, n \tag{20.4.9}$$

Compute

$$Q_1 = \frac{1}{n-p} \sum_{k=p+1}^{n} \epsilon_k \tag{20.4.10}$$

$$Q_2 = \frac{1}{n-p} \sum_{k=p+1}^{n} \epsilon_k^2 \tag{20.4.11}$$

Q_1 must be near zero and Q_2 must be near one. Approximating the sampling distribution of Q_1 by a normal with zero mean and variance $1/(n-p)$, there is a 95 percent probability that

$$|Q_1| < \frac{2}{\sqrt{n-p}} \tag{20.4.12}$$

The distribution of $(n-p)Q_2$ is approximated by a chi square with $n-p$ degrees of freedom; then there is a 95 percent probability that

$$U > Q_2 > L \tag{20.4.13}$$

where the values of U and L can be found from Table 20.4.1.

Thus, if Q_1 and Q_2 are outside these intervals, Eqs. (20.4.12) and (20.4.13), there is reason to question the validity of the model. The choice of 95 percent, although customary in statistics, is not the only confidence value possible. In special cases, one may adjust the test on the basis of prior confidence in the model.

The ϵ residuals are also used to test the hypothesis that they follow a gaussian distribution and to locate outliers. Also, one may compare the experimental variogram of $\epsilon_{p+1}(\mathbf{x}_{p+1}), \ldots, \epsilon_n(\mathbf{x}_n)$ with the pure nugget effect variogram.

In addition to these diagnostic tests, there are measures of overall model accuracy, or goodness of fit of the model to the data, based on the orthonormal residuals. The basic idea is that a good model should give accurate predictions when used in kriging. The first criterion is the actual mean square error:

$$M = \frac{1}{n-p} \sum_{k=p+1}^{n} [z(x_k) - \hat{z}_k]^2 = \frac{1}{n-p} \sum_{k=p+1}^{n} \epsilon_k^2 \sigma_k^2 \tag{20.4.14}$$

The smaller the value of M, the better the fit provided by the model. Another measure of fit, which is less sensitive to the largest values of the kriging error squares and is thus more stable than M, is the geometric mean of the kriging variances of the orthonormal residuals adjusted for Q_2:

$$cR = Q_2 \exp\left[\frac{1}{n-p} \sum_{k=p+1}^{n} \ln(\sigma_k^2)\right] \tag{20.4.15}$$

TABLE 20.4.1 The 0.025 and 0.975 Percentiles of the Distribution of Q_2 Statistic

$n-p$	L	U
1	0.001	5.02
2	0.025	3.69
3	0.072	3.12
4	0.121	2.78
5	0.166	2.56
10	0.325	2.05
15	0.417	1.83
20	0.479	1.71
25	0.524	1.62
30	0.560	1.57
35	0.589	1.52
40	0.610	1.48
45	0.631	1.45
50	0.648	1.43
75	0.705	1.34
100	0.742	1.30
>100	$1-a$*	$1+a$

* $a = \dfrac{2.8}{\sqrt{n-p}}$

20.4.4 Parameter Estimation

In the intrinsic isotropic case, the variogram may be obtained by combining graphical fitting to the experimental variogram, judgment, and the results of validation tests. This approach is less useful as the complexity of the problem increases, particularly for variable mean cases. Computational methods, reviewed in Ref. 29, are very useful in estimating the variogram parameters and should be used when possible. These methods find the numerical values of the parameters which optimize a measure of goodness of fit of the model to the data.

Some of the available methodologies are weighted least squares,[14] minimum-variance unbiased quadratic estimation, and restricted maximum likelihood (RML).[32] Of those, RML is the most accurate method but also has the highest computational cost.

The effect on predictions (kriging) of uncertainty in the estimates of the parameters can be taken into account,[18,28] but is usually neglected in routine applications.

20.4.5 Some Comments on Model Selection

Development from data of a model, i.e., the form of the drift and the covariance function or variogram, is a trial-and-error procedure in which a series of assumptions are tentatively proposed and then tested. The following comments are a practical guide in model selection:

1. Choose the simplest model consistent with the data. It is usually reasonable to start with the assumption that the regionalized variable is intrinsic and isotropic and to determine its experimental variogram. This model should, however, be put to the test (see Sec. 20.4.3); if model inadequacies are detected, relax the assumption of constant mean or of isotropy. Do so gradually, keeping the model as simple as possible.

Complicated drift models should be used only when there is sufficient evidence for them derived from observations and other information. At best, estimating an unknown drift coefficient corresponds to losing one measurement; thus, unless the new coefficient significantly improves the fit, it must be avoided. Similarly, there should be good reasons for adopting anisotropy; for example, the anisotropic model should give significantly better values of the goodness-of-fit measures M or cR of Sec. 20.4.3 than the simpler isotropic model. A useful rule of thumb is to avoid covariance models which involve more than two unknown parameters. Empirical models with many parameters are usually less accurate in making predictions (using kriging) than simpler models which have been calibrated and tested with care.

2. For practical estimation purposes, the right model is the one which represents available information about the structure of a spatial variable. The model is developed to solve a specific prediction problem. It serves no practical purpose to keep searching for the model which represents the "actual" structure when a reasonable simple model has been calibrated and tested. Occasionally, however, one may come up with two or more models which appear equally plausible, given the available information. One should calculate the predictions using each of these models and then compare the results.

3. Make use of available information. The model describes the structure of a concrete hydrologic variable (e.g., precipitation or hydraulic head) and should be consistent with what is known about this variable. For example, the presence of anisotropy of a hydrogeologic parameter should be given a geological explanation; or the presence of a linear trend in precipitation depths should be related to orographic effect or some other hydrometeorological mechanism. Such process understanding can be invaluable in distinguishing actual model improvements from overmodeling (incorporating spurious features) which does nothing to improve the accuracy of predictions. It may also reduce the number of parameters which need to be fitted to the data; for example, knowledge of stratification can suggest the principle directions of anisotropy.

20.4.6 Variable Transformation and Outside Values

There are practical advantages in working with unimodal and nearly symmetric probability distributions: a single number can be used to represent the central value in the batch, because the mode, median, and arithmetic mean are very near each other. If the distribution resembles the normal distribution, then the standard deviation is a good measure of the spread about the mean.

When a data set does not have these properties, it is reasonable to search for a simple transformation which will make its distribution easier to manage. The so-called power transformation (which includes the logarithmic transformation) is the most common for positive data:

$$y = \begin{cases} z^\theta & \text{if } \theta > 0 \\ \log(z) & \text{if } \theta = 0 \\ -z^\theta & \text{if } \theta < 0 \end{cases} \tag{20.4.16}$$

where θ is a parameter. Transmissivity and conductivity data are often distributed approximately lognormally; thus, θ is chosen as zero. Another example is pollutant concentrations which may vary over many orders of magnitude at a hazardous-waste site; $\theta = 1/2$ or $1/4$ often is sufficient to make linear estimation methods applicable.

Transformations are a useful way of extending the applicability of linear estimation methods; however, *the estimates are in terms of the transformed variables* and care must be taken in making the back transformation. Here is an example of how to make the transformation back to the original variable: in analyzing measurements of the concentration $c(\mathbf{x})$ of a pollutant in an aquifer the variable transformation $c_t(\mathbf{x}) = [c(\mathbf{x})]^{1/4}$ was made. Then, a geostatistical analysis was performed for the transformed variable $c_t(\mathbf{x})$ and a model was selected, calibrated, and validated. Next, the model was used in kriging to obtain the best estimate $\hat{c}_t(\mathbf{x}_0)$ and mean square error $\sigma_t(\mathbf{x}_0)^2$ at a given location $\mathbf{x}_0$. What conclusions can be drawn about the original variable, the concentration $c(\mathbf{x}_0)$, at the same location? The most likely value of $c(\mathbf{x}_0)$ is $[\hat{c}_t(\mathbf{x}_0)]^4$ and the 95 percent confidence interval ranges from $[\hat{c}_t(\mathbf{x}_0) - 2\sigma_t(\mathbf{x}_0)]^4$ to $[\hat{c}_t(\mathbf{x}_0) + 2\sigma_t(\mathbf{x}_0)]^4$. [The idea is that we treat the distribution of $c_t(\mathbf{x}_0)$ conditionally on the measurements as normal with mean $\hat{c}_t(\mathbf{x}_0)$ and variance $\sigma_t(\mathbf{x}_0)^2$ and then find the distribution of the $c(\mathbf{x}_0)$ through the transformation. No general and practical method exists for finding the minimum-variance estimate or conditional mean of the original variable.]

Sometimes, even after a transformation, the box plot of the orthonormal residuals may call attention to values which stand out from the bulk of data. Typically, we consider orthonormal residuals larger in absolute value than 3.5 or 4 as stray values. It would be unwise at a preliminary stage to lump those stray or unusual values with the other data. However, it would be equally unwise to mark these measurements as outliers and to discard them without a thorough study because these are often legitimate observations which may be more important than the rest of the data. Before deciding what to do with an outside value, the analyst must go back to the source of the data and to the understanding of the physical processes which are involved. One should make a reasonable effort to verify that the measurement was taken, interpreted, and transcribed correctly. If it is verified that it is a legitimate measurement, it should be reported separately instead of being lumped together with the other data.

20.4.7 Illustrative Example of Structural Analysis

This example presents the overall procedure and illustrates the application of the most important principles and methods we have discussed. Space limitations do not allow discussion of all tools which could (and occasionally should) enrich a geostatistical study.

Consider 100 measurements of, say, log conductivity (logarithm of conductivity) in a borehole.

Data Exploration

1. *Use whatever graphics software is available to plot the raw data.* In this case, plotting the observations against the location (Fig. 20.4.2) indicates that the scale of fluctuations is comparable to the largest separation distance between observations, suggesting that a stationary model might not be appropriate.

2. *Perform batch analysis.* This includes computing the univariate statistics (Table 20.4.2) and plotting the box plot (see Fig. 20.4.3.) The experimental distribution of the data is somewhat skewed but no outliers are present.

3. *Plot the raw variogram and the experimental variogram.* We start with a visual inspection of the raw variogram (Fig. 20.4.4.) We note that the scatter (in the

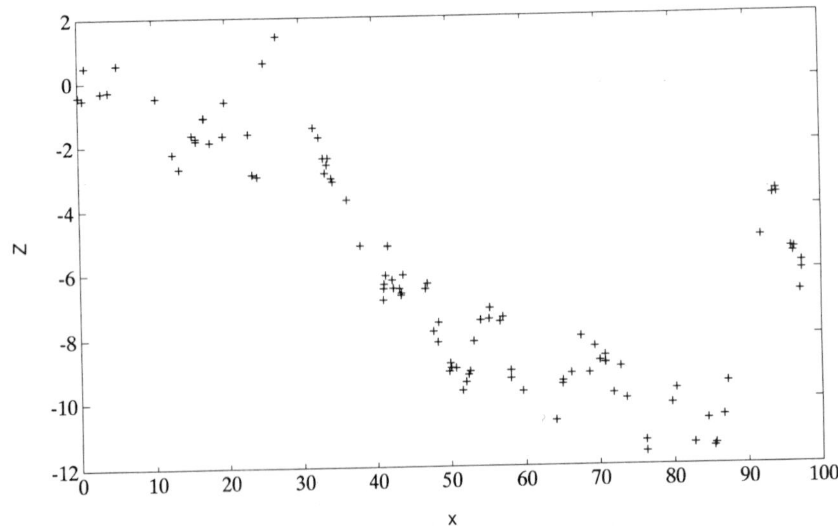

FIGURE 20.4.2 A plot of the data versus their location used in the example.

TABLE 20.4.2 Experimental Statistics of the Data Batch

Number of measurements	100.
Minimum value	−11.6
First quartile	−9.08
Median	−6.51
Third quartile	−2.89
Maximum value	1.4
Mean	−5.97
Variance	12.3
Skew coefficient	0.31

vertical direction) is small at low separation distances and the points tend to zero near the origin. Most points are at about half the maximum separation distance. At large distances, there are fewer points and they are widely scattered. Based on the raw variogram, we choose five intervals of unequal length (see Table 20.4.3) and we connect the representative points to form the experimental variogram (see Fig. 20.4.5.) The experimental variogram indicates linear behavior near the origin and

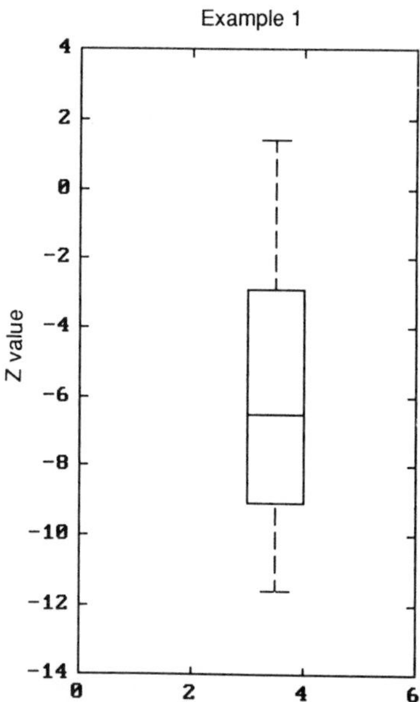

FIGURE 20.4.3 Box plot of data. A graphic presentation of the experimental distribution: the continuous lines show the quartiles and the median, thus the box contains 50 percent of the data; the discontinuous line contains 95 percent of the data.

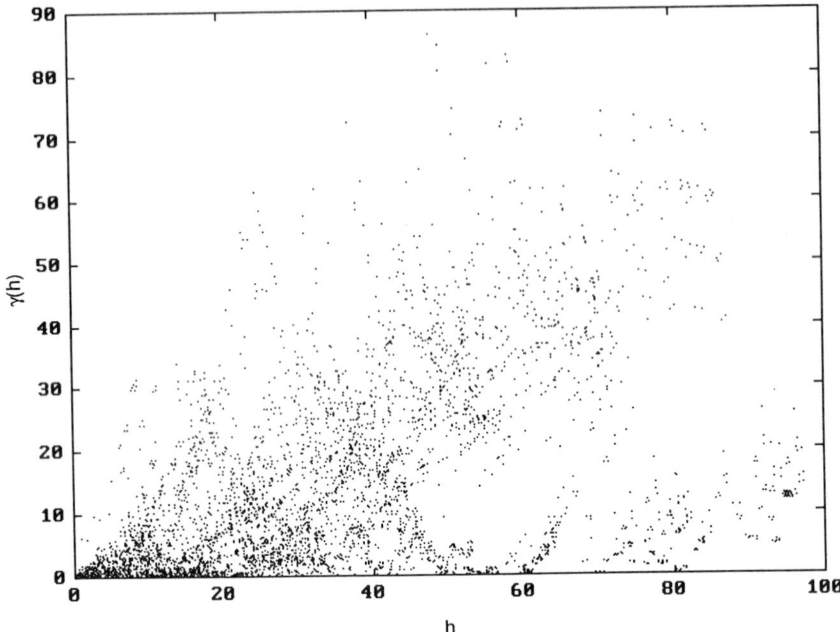

FIGURE 20.4.4 The raw semivariogram.

the absence of a well-defined sill. All these considerations lead us to adopt tentatively a linear variogram model.

Parameter Estimation. *Select numerical values for model parameters.* The parameter θ of the model semivariogram $\gamma(h) = \theta h$, where h is the separation distance, can be found by graphically fitting the model variogram to the experimental variogram. In general, the fit should be closer near the origin (at short separation distances) particularly if the nugget-effect variance is small. In our example, where the nugget-effect variance appears to be negligible, the fit must be very close near the origin. We make no effort to achieve a good fit at large separation distances.

Another approach is to determine the variogram parameter from a parameter

TABLE 20.4.3 Characteristics of the Experimental Variogram

Interval	Number of pairs	h	$\gamma(h)$
0–2	224	1.0	0.4
2–5	269	3.4	1.2
5–10	504	7.6	3.7
10–30	1732	19.8	7.6
30–100	2221	50.6	20.5

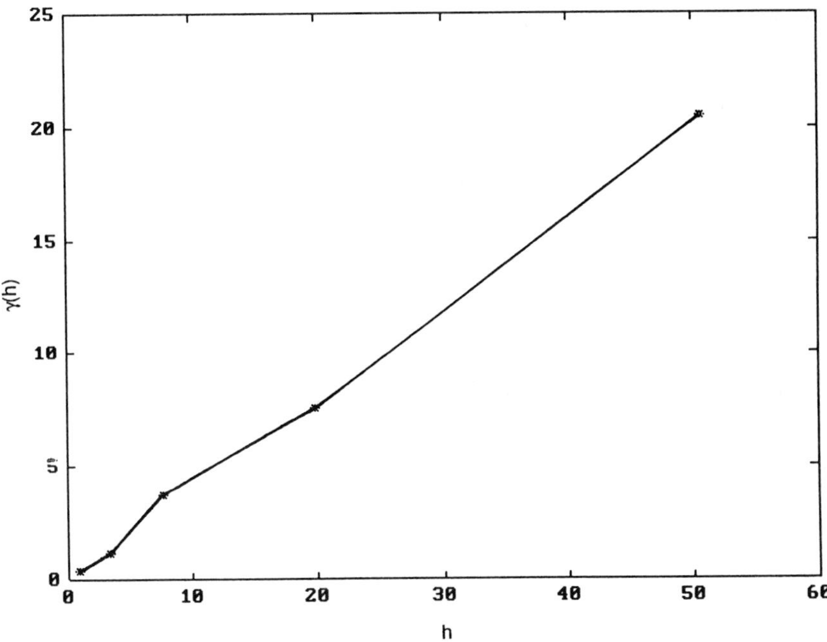

FIGURE 20.4.5 The experimental semivariogram.

estimation procedure of Sec. 20.4.4. We use the RML method (which in this case is equivalent to setting the value of $Q_2 = 1$), obtaining the estimate $\hat{\theta} = 0.501$. Figure 20.4.6 shows the RML-fitted linear variogram and the nodes of the experimental semivariogram.

Validation through Orthonormal Residuals. *Perform on the orthonormal residuals a complete exploratory analysis and diagnostic tests looking for indications that the orthonormal residuals may not: be uncorrelated, have zero mean, have variance 1, and be normally distributed.* The statistics of the 99 orthonormal residuals are shown on Table 20.4.4. (In our example, Q_2 is not useful for validation purposes, since the parameter was adjusted so that Q_2 is exactly 1.) The box plot of the orthonormal residuals is shown on Fig. 20.4.7. A statistical test (with 5 percent probability of rejecting normality when normality is appropriate) did not reject the normality of the orthonormal residuals. Regarding spatial structure, a plot of the orthonormal residuals against their location (Fig. 20.4.8) shows no spatial correlation structure. This is confirmed by their experimental variogram (see Fig. 20.4.9) which is practically indistinguishable from the nugget effect variogram. Thus, the examination of the orthonormal residuals suggests no reason to reject this model.

Examination of Alternative Models. *To assure that a good model has been chosen, compare with other models.* Should we include a nugget effect? Is a variogram with a sill more appropriate? Is the mean variable; for example, does it have a linear drift? To answer these questions, we calibrate and test three other models and then compare them with the base model: (1) intrinsic with variogram linear plus nugget effect, [Eq. (20.2.23)]; (2) exponential variogram; and (3) linear drift and linear variogram

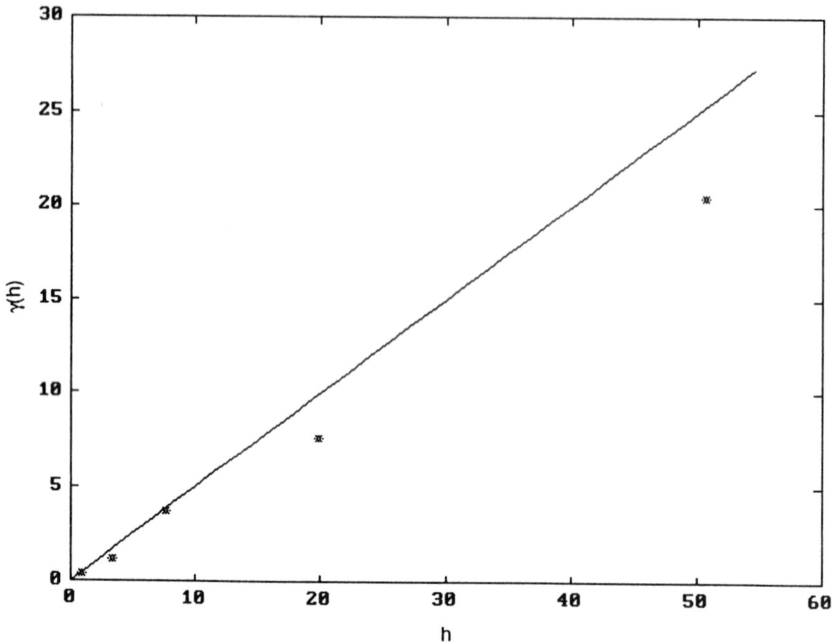

FIGURE 20.4.6 The fitted linear semivariogram (continuous line) and the nodes of the experimental semivariogram (*).

[an order $k = 1$ intrinsic function (see Sec. 20.2.4)]. In all these models, parameters are fitted with the RML method and additional sensitivity analysis is done by varying the parameters and examining the residuals. For purposes of comparison, we use the cR criterion of Eq. (20.4.15). In model (1), using a large nugget effect increased the value of cR (indicating poor fit); the RML method estimated a practically zero nugget effect. In the case of model (2), the relative reduction in the value of cR compared to the base model is about 0.004. However, the better fit of the exponential model is not sufficient to justify the use of a more complicated model—the exponential has two

TABLE 20.4.4 Statistics of Orthonormal Residuals

Number of measurements	99
Minimum value	−3.53
First quartile	−0.69
Median	0.02
Third quartile	0.75
Maximum value	1.96
Mean (Q_1)	−0.01
Variance (Q_2)*	1.00
Skew coefficient	−0.47

* Parameter is fitted so that Q_2 is 1.

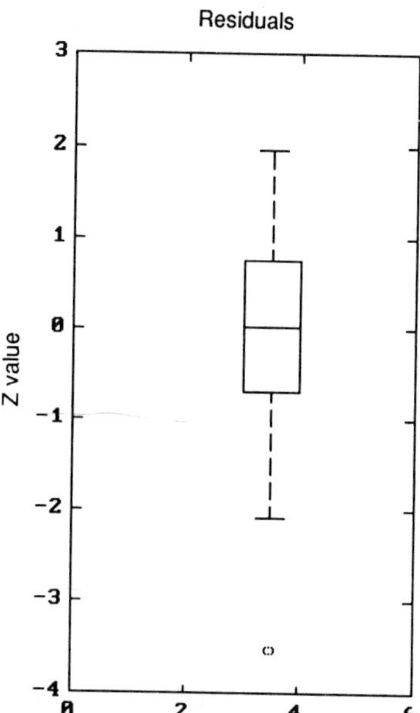

FIGURE 20.4.7 Box plot of the orthonormal residuals.

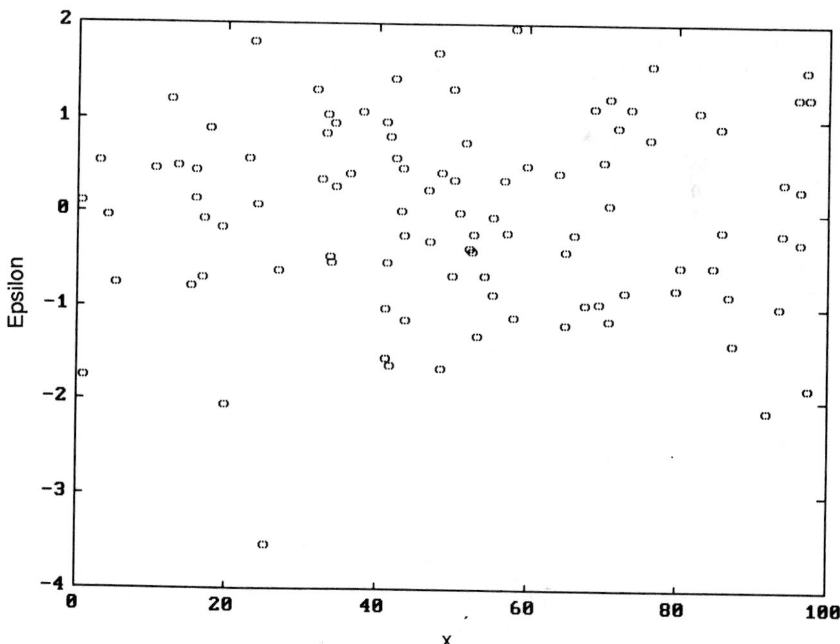

FIGURE 20.4.8 Plot of the orthonormal residuals versus their location.

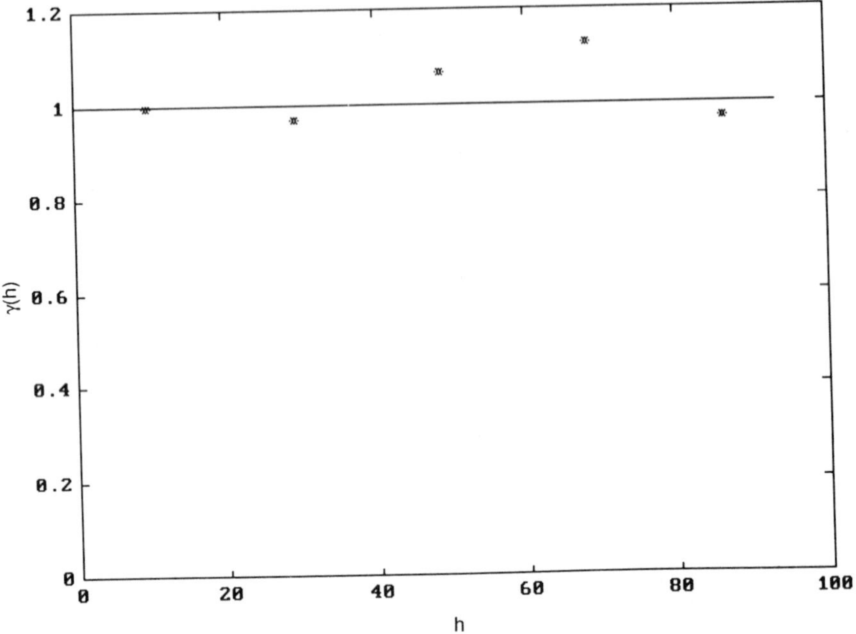

FIGURE 20.4.9 Experimental semivariogram for the orthonormal residuals.

parameters while the linear has only one. A rule of thumb is not to introduce an extra parameter unless the relative reduction in cR is at least $1/(n - p)$, where $n - p$ is the number of orthonormal residuals (in our example, we have 99 orthonormal residuals). As for model 3, its cR is actually greater by 5 percent than the cR of the base model, indicating that in this case adding a drift to the model does not promise to improve the predictions.

20.5 LIMITATIONS AND ALTERNATIVES

In linear geostatistics the estimates are obtained as linear functions of the observations with weights determined from the first two moments. If the probability of each weighted average of sampled values is normal or could be approximated as such (multivariate gaussian case), then kriging is the estimator with the smallest mean square error. Otherwise, kriging is the minimum-variance estimator among those that are linear functions of the data; it is possible that a nonlinear function of the data may further reduce the mean square error. However, such an estimator must rely on information additional to the first two moments, and its application is usually considerably more cumbersome.

As already mentioned, the most common practical approach is to transform the variable of interest and then to analyze the transformed variable by using linear geostatistics. A case that has been examined in some depth is that of the logarithmic transformation, known as *log-kriging*.[4]

Disjunctive kriging[37,42,47] is particularly useful in estimating probability of exceedance of a threshold level. Another useful alternative to linear geostatistics is indicator and probability kriging.[26,44,]

20.6 SELECTED READING AND SOFTWARE

Textbooks with emphasis on mining, petroleum, and geological applications are Refs. 7, 12, 24, 25, 27, 35. A two-part paper, Ref. 4, is an introduction to geostatistics and its applications in geohydrology with an extensive list of references. Reference 16 has a chapter on geostatistics and stochastic analysis in geohydrology. Reference 11 is a good reference on stochastic groundwater methods. Public domain geostatistical software packages include GEO-EAS[17] (which can be obtained from Evan J. Englund, U.S. EPA EMSL-LV, EAD, P.O. Box 93478, Las Vegas, NV 89193-3478) and GEOPACK[48] (which can be obtained from Robert S. Kerr Environmental Research Laboratory, Office of Research and Development, U.S. EPA, Ada, OK 74820).

REFERENCES

1. Abourifassi, M., and M. A. Marino, "Cokriging of Aquifer Transmissivities from Field Measurements of Transmissivity and Specific Capacity," *J. Math. Geology,* vol. 16, no. 1, pp. 19–35, 1984.
2. Ahmed, S., and G. deMarsily, "Comparison of Geostatistical Methods for Estimating Transmissivity Using Data on Transmissivity and Specific Capacity," *Water Resour. Res.,* vol. 23, no. 9, pp. 1717–1737, 1987.
3. Armstrong, M., "Improving the Estimation and Modeling of the Variogram," in *Geostatistics for Natural Resources Characterization,* G. Verly et al. (eds.), vol. 1, pp. 1–20, Reidel, Dordrecht, The Netherlands, 1984.
4. ASCE Task Committee on Geostatistical Techniques in Geohydrology, "Review of Geostatistics in Geohydrology," Parts I and II, *J. Hydraul. Eng.,* vol. 116, no. 5, pp. 612–658, 1990.
5. Bakr, A., L. W. Gelhar, A. L. Gutjahr, and J. R. MacMillan, "Stochastic Analysis of Spatial Variability of Subsurface Flows—1. Comparison of One- and Three-Dimensional Flows," *Water Resour. Res.,* vol. 14, no. 2, pp. 263–271, 1978.
6. Bras, R. L., and I. Rodriguez-Iturbe, *Random Functions in Hydrology,* Addison-Wesley, Reading, Mass., 1985.
7. Clark, I., *Practical Geostatistics,* Applied Science, London, 1979.
8. Cressie, N., and D. Hawkins, "Robust Estimation of the Variogram," *J. Math. Geology,* vol. 12, no. 2, pp. 115–126, 1980.
9. Dagan, G., "Stochastic Modeling of Groundwater Flow by Unconditional and Conditional Probabilities—1. Conditional simulation and the direct problem," *Water Resour. Res.,* vol. 18, no. 4, pp. 813–833, 1982.
10. Dagan, G., "Stochastic Modeling of Groundwater Flow by Unconditional and Conditional Probabilities: The Inverse Problem," *Water Resour. Res.,* vol. 21, no. 1, pp. 65–72, 1985.
11. Dagan, G., *Flow and Transport in Porous Media,* Springer-Verlag, Berlin, 1989.
12. David, M., *Geostatistical Ore Reserve Estimation,* Elsevier, Amsterdam, 1977.
13. David, M., *Handbook of Applied Advanced Geostatistical Ore Reserve Estimation,* Elsevier, Amsterdam, 1988.
14. Delfiner, P., "Linear Estimation of Nonstationary Spatial Phenomena," in *Advanced*

Geostatistics in the Mining Industry, M. Guarascio, M. David, and C. Huijbregts (eds.), pp. 49–68, D. Reidel, Bingham, Mass., 1976.

15. Delhomme, J. P., "Kriging in the Hydrosciences," *Adv. Water Resour.,* vol. 1, no. 5, pp. 251–266, 1978.
16. deMarsily, G., *Quantitative Hydrogeology,* Academic Press, New York, 1986.
17. Englund, E., and A. Sparks, "GEO-EAS (Geostatistical Environmental Assessment Software) User's Guide," U.S. EPA 600/4-88/033a, Las Vegas, Nev., 1988.
18. Feinerman, E., G. Dagan, and E. Bresler, "Statistical Inference of Spatial Random Function," *Water Resour. Res.,* vol. 22, no. 6, pp. 935–942, 1986.
19. Gelhar, L. W., and C. L. Axness, "Three-Dimensional Stochastic Analysis of Macrodispersion," *Water Resour. Res.,* vol. 19, no. 1, pp. 161–180, 1983.
20. Graham, W., and D. McLaughlin, "Stochastic Analysis of Nonstationary Subsurface Solute Transport—2. Conditional Moments," *Water Resour. Res.,* vol. 25, no. 11, pp. 2331–2355, 1989.
21. Gutjahr, A. L., and L. W. Gelhar, "Stochastic Models of Subsurface Flows: Infinite Versus Finite Domains and Stationarity," *Water Resour. Res.,* vol. 17, no. 2, pp. 337–351, 1981.
22. Hoeksema, R. J., and P. K. Kitanidis, "An Application of the Geostatistical Approach to the Inverse Problem in Two-Dimensional Groundwater Modeling," *Water Resour. Res.,* vol. 20, no. 7, pp. 1003–1020, 1984.
23. Hoeksema, R. J., et al., "Cokriging Model for Estimation of Water Table Elevation," *Water Resour. Res.,* vol. 25, no. 3, pp. 429–438, 1989.
24. Hohn, M. E., *Geostatistics and Petroleum Geology,* Van Nostrand Reinhold, New York, 1988.
25. Isaaks, E. H., and R. M. Srivastava, *Applied Geostatistics,* Oxford University Press, New York, 1989.
26. Journel, A., "The Place of Non-Parametric Geostatistics," in *Geostatistics for Natural Resources Characterization,* G. Verly et al. (eds.), Reidel, Dordrecht, The Netherlands, 1984.
27. Journel, A. G., and Ch. J. Huigbregts, *Mining Geostatistics,* Academic Press, New York, 1978.
28. Kitanidis, P. K., "Parameter Uncertainty in Estimation of Spatial Functions: Bayesian Analysis," *Water Resour. Res.,* vol. 22, no. 4, pp. 499–507, 1986.
29. Kitanidis, P. K., "Parameter Estimation of Covariances of Regionalized Variables," *Water Resour. Bull.,* vol. 23, no. 4, pp. 557–567, 1987.
30. Kitanidis, P. K., "Orthonormal Residuals in Geostatistics: Model Criticism and Parameter Estimation," *Mathematical Geology,* vol. 23, no. 5, pp. 741–758, 1991.
31. Kitanidis, P. K., and E. G. Vomvoris, "A Geostatistical Approach to the Inverse Problem in Groundwater Modeling (Steady State) and One-Dimensional Simulations," *Water Resour. Res.,* vol. 19, no. 3, pp. 677–690, 1983.
32. Kitanidis, P. K., and R. W. Lane, "Maximum Likelihood Parameter Estimation of Hydrologic Spatial Processes by the Gauss-Newton Method," *J. Hydrol.,* vol. 79, pp. 53–79, 1985.
33. Krajewski, W. F., "Co-Kriging of Radar-Rainfall and Rain Gage Data," *J. Geophys. Res.,* vol. 92, no. D8, pp. 9571–9580, 1987.
34. Mantoglou, A., and J. L. Wilson, "The Turning Bands Method for Simulation of Random Fields Using Line Generation by a Spectral Method," *Water Resour. Res.,* vol. 18, no. 5, pp. 1379–1394, 1982.
35. Matheron, G., *The Theory of Regionalized Variables and Its Applications,* École de Mines, Fontainbleau, France, 1971.
36. Matheron, G., "The Intrinsic Random Functions and Their Applications," *Adv. Appl. Prob.,* vol. 5, pp. 439–468, 1973.

37. Matheron, G., and M. Armstrong, "Disjunctive Kriging Revisited," parts I and II, *Mathematical Geology,* vol. 18, no. 8, pp. 711–742, 1986.

38. Myers, D. E., "Matrix Formulation of Cokriging," *Mathematical Geology,* vol. 14, no. 3, pp. 249–257, 1982.

39. Neuman, S. P., "Role of Geostatistics in Subsurface Hydrology," in *Geostatistics for Natural Resources Characterization,* G. Verly et al. (eds.), part I, pp. 787–816, Reidel, Dordrecht, The Netherlands, 1984.

40. Omre, E., "The Variogram and Its Estimation," in *Geostatistics for Natural Resources Characterization,* G. Verly et al. (eds.), vol. 1, pp. 107–125, Reidel, Dordrecht, The Netherlands, 1984.

41. Philip, R. D., and P. K. Kitanidis, "Geostatistical Estimation of Hydraulic Head Gradients," *Ground Water,* vol. 27, no. 6, pp. 855–865, 1989.

42. Rendu, J. M., "Disjunctive Kriging: A Simplified Theory," *Mathematical Geology,* vol. 12, pp. 306–321, 1980.

43. Rubin, Y., and G. Dagan, "Stochastic Identification of Transmissivity and Effective Recharge in Stready Groundwater Flow," parts I and II, *Water Resour. Res.,* vol. 23, no. 7, pp. 1185–1200, 1987.

44. Sullivan, J., "Conditional Recovery Estimation Through Probability Kriging—Theory and Practice," in *Geostatistics for Natural Resources Characterization,* G. Verly et al. (eds.), Reidel, Dordrecht, The Netherlands, 1984.

45. Tompson, A. F. B., R. Ababou, and L. W. Gelhar, "Implementation of the Three-Dimensional Turning Bands Random Field Generator," *Water Resour. Res.,* vol. 25, no. 10, pp. 2227–2243, 1989.

46. Vauclin, M., S. R. Vieira, G. Vachaud, and D. R. Nielsen, "The Use of Cokriging with Limited Field Soil Information," *Soil Sci. Proc. Am. J.,* vol. 47, no. 2, pp. 175–184, 1983.

47. Yates, S. R., A. W. Warrick, and D. E. Meyers, "Disjunctive Kriging, Parts I and II—Overview of Estimation and Conditional Probability," *Water Resour. Res.,* vol. 22, pp. 615–630, 1986.

48. Yates, S. R., and M. V. Yates, "Geostatistics for Waste Management: User's Manual for GEOPACK Version 1.0," Kerr Environmental Research, Ada, Okla., 1989.

CHAPTER 21
COMPUTER MODELS FOR SURFACE WATER

Johannes J. DeVries
Water Resources Center
University of California
Davis, California

T. V. Hromadka
Boyle Engineering Corp.
Newport Beach, California

21.1 PURPOSE OF CHAPTER

The purpose of this chapter is to describe computer program packages in hydrology that are generally available to hydrologists with access to standard computer systems such as personal computers (PCs) using the DOS or UNIX operating systems or to mainframes and minicomputers using standard operating systems and compilers. New programs are being written each year; some are variations of previous programs, while others use completely new techniques and provide the user with the great power of modern desktop computers, both in making a large number of computations very rapidly and in being able to deal with large volumes of data. It is not possible to include every existing model in a single chapter, and the discussion here is limited to several models of various types that have been found useful by practicing hydrologists and hydraulic engineers. Sufficient information is provided to permit the reader to make a judgment about the suitability of the program for specific hydrologic applications.

The programs covered in this chapter deal with surface water quantity and quality. The models are grouped into the following classes: (1) single-event rainfall-runoff and routing models, (2) continuous-stream-flow simulation models, (3) flood-hydraulics models, and (4) water-quality models.

The basic format of this chapter is as follows:

1. An overview of each model is provided.
2. The types of hydrologic problems and applications that the model is intended to deal with are discussed.
3. The required input for the model is briefly described.

4. Computer requirements are given.
5. Information on how the program can be obtained is provided.

21.1.1 General Remarks on Models in Hydrology

The models discussed in this section span the range from very simple, black-box-type models, the basic concepts of which were developed decades ago, through models that have been released for general use in just the last few years. The last group includes physically based distributed-parameter models, such as the SHE (European Hydrological System—Système Hydrologique Européen) model.

The data requirements of the models also vary over a wide range. In many design studies, standard design rainfall distributions and volumes are applied to catchments for which runoff response is characterized by synthetic unit hydrograph and generalized loss-rate functions. For physically based distributed-parameter models, large amounts of data are required for calibration of the model. However, the data are usually not available, because data collection is expensive and difficult. Also, as discussed by Abbott et al.[1,2] and Loague and Freeze,[22] a primary barrier to successful application of physically based models is the scaling problem. Field measurements are made at the point or local scale, while the application in the model is at the larger scale of the grid used to represent the hydrologic processes. This is especially true in the characterization of the spatial variability of precipitation and soil properties. In general, the larger the area considered, the larger is the variability of properties within that area.

The accuracy of model results is a function of the accuracy of the input data and the degree to which the model structure correctly represents the hydrologic processes appropriate to the problem. Complex models require complex data, and if the required data can be only roughly estimated, it may be better to use a model whose input data requirements are in tune with available data sources. Likewise, the modeling of the various hydrologic processes should be in balance. For example, continuous-stream-flow simulation requires accounting for both surface runoff and subsurface flow to streams in a balanced manner. A highly accurate model of surface runoff may not yield good stream-flow estimates if it is combined with a very approximate model of subsurface flows.

21.1.2 Characterization of Models

Simple, Single-Event, Rainfall-Runoff Models. Models that fall into this category include the Corps of Engineers HEC-1 model, the Soil Conservation Service TR-20 model, and similar models. Calculations proceed from upstream to downstream in the watershed, and the general modeling sequence is the following:

1. Subbasin average precipitation
2. Determination of precipitation excess from time-varying losses
3. Generation of the direct surface runoff hydrograph from precipitation excess
4. Addition of a simplified base flow to the surface runoff hydrograph
5. Routing of stream flow
6. Reservoir routing
7. Combination of hydrographs

In these models, the primary interest is the flood hydrograph, so it is not necessary to calculate evapotranspiration, soil moisture changes during and between storms, or detailed base flow processes.

Continuous-Stream-Flow Simulation Models. A second major class of hydrologic models is one that includes models that continuously account in time for all precipitation that falls on a watershed and the movement of water through the catchment to its outlet. During periods in which there is no precipitation, the main concern is the depletion of water stored in the watershed, with consequent emphasis on soil moisture accounting, evapotranspiration, and subsurface flows in the unsaturated and saturated groundwater zones.

Models which provide simulations of this type are called *continuous-stream-flow simulation models.* These models range in complexity from the very simple—antecedent precipitation index (API) and tank models—to the very complex—distributed-parameter models such as the HSPF and SHE models discussed below.

Physically based distributed-parameter models describe the major hydrologic processes which govern water movement through a catchment. These processes include, but are not limited to:

1. Canopy interception
2. Evapotranspiration
3. Snowmelt
4. Interflow
5. Overland flow
6. Channel flow
7. Unsaturated subsurface flow
8. Saturated subsurface flow

The spatial variation of these processes is represented by the spatial variation of precipitation, catchment parameters, and hydrologic response. Variability is modeled by representing the catchment by individual subbasins or by developing a grid of individual elements and prescribing the hydrologic characteristics of each element. Vertical variability is represented by subsurface zones or vertical layers of soil for each grid element.

Flood-Hydraulics Models. These models compute water surface profiles in open channels and represent flows in natural channels such as rivers and streams, where the geometry of the cross section changes from section to section. Programs in this group usually analyze flow through bridges and culverts, as well as on floodplains adjacent to the main river system.

Steady-flow models (such as HEC-2 and WSPRO) employ conventional step-backwater analyses. These programs assume that the flow is one-dimensional, gradually varied steady flow in the direction of the stream centerline. In situations where the flow is actually two-dimensional in nature or is rapidly varied (as at a bridge), hydraulic equations involving empirical head-loss coefficients are used to represent approximately these complicated flows. Both subcritical and supercritical flow profiles can be analyzed.

One class of unsteady-flow models is represented by the National Weather Service program FLDWAV and its predecessors DWOPER and DAMBRK. These programs are based on the one-dimensional St. Venant equations. The models provide many capabilities in addition to water surface profile calculation, such as dam breach simulation, embankment overtopping, and representation of structures and their operation schedules.

Some of the models which have been developed to analyze water-quality problems have detailed procedures for analyzing time-varying flows. For example, the EXTRAN block of the SWMM model described below has been frequently used for unsteady-flow analysis in stream systems. The MIKE11 model has similar capabilities.

Water-Quality Models. This group of models link the determination of water quantity and analysis of water quality. These models vary a great deal in their complexity. In general, the models require that relations between water-quality loading and the hydraulic features of the system be established. In most cases, empirical relations for defining chemical and biological reactions are used. Both lumped-parameter and distributed-parameter models are available, and the models may be single-event models or continuous models. Distributed-parameter models provide detailed information about local conditions, while lumped-parameter models are useful when large-scale systems are being studied. Single-event models are useful for evaluating water-quality conditions occurring with extreme flow events, while continuous models provide sequences of water-quality events that can be subjected to frequency analyses.

21.2 RAINFALL-RUNOFF AND ROUTING MODELS (SINGLE-EVENT MODELS)

21.2.1 HEC-1 Flood Hydrograph Package

Overview of Program. HEC-1 is a computer model for rainfall-runoff analysis developed by the Hydrologic Engineering Center of the U.S. Army Corps of Engineers.[11,18] This program develops discharge hydrographs for either historical or hypothetical events for one or more locations in a basin. The basin can be subdivided into many subbasins. Uncontrolled reservoirs and diversions can also be accommodated. Figure 21.2.1 shows an example river basin and the schematic for the division of the basin into subbasins and routing elements for modeling by HEC-1.

The available program options include the following: calibration of unit hydrograph and loss-rate parameters, calibration of routing parameters, generation of hypothetical storm data, simulation of snowpack processes and snowmelt runoff, dam safety applications, multiplan/multiflood analysis, flood damage analysis, and optimization of flood-control system components.

The 1990 version of HEC-1 also has the capability to communicate with a data storage system (DSS) file. DSS is a database system that is designed for efficient storage and utilization of time series data between the various Hydrologic Engineering Center programs.[19] For example, observed precipitation and discharge data can be stored in a DSS file and retrieved automatically during an execution of HEC-1. Additionally, simulation results can be written to a DSS file. Utility programs develop graphs or tabulations of data. DSS also has programs for loading data from WATSTORE and other data sources, for data editing, and for producing reports. DSS provides a convenient means for transfer of data generated with one program (say HEC-1) to input to another HEC computer program (such as the reservoir system simulation program HEC-5).

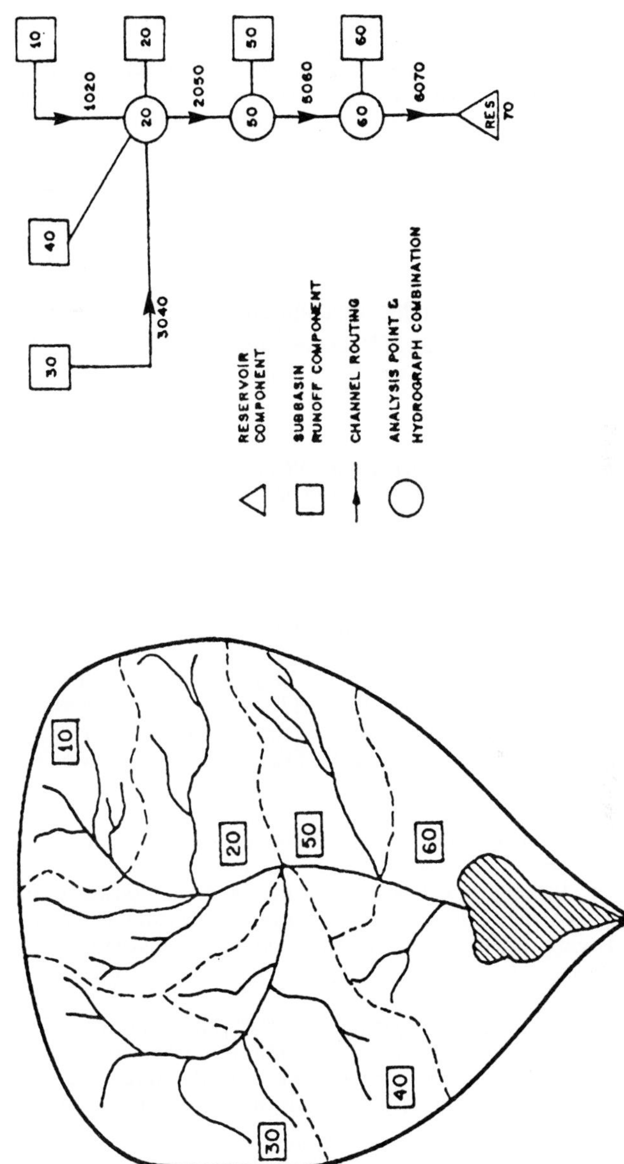

FIGURE 21.2.1 Example river basin and basin schematic for HEC-1 *(Source: Hydrologic Engineering Center.)*

HEC-1 allows a wide variety of options for specifying precipitation, losses, base flow, runoff transformation, and routing. A description of these options follows.

Precipitation. Spatially averaged precipitation can be determined externally and supplied as program input. As an alternative, precipitation for individual recording and nonrecording gauges can be specified, along with weighting factors to calculate the average precipitation for each subbasin. The basin-average precipitation can be further adjusted if the gauges from which it is determined have a normal annual rainfall systematically different from the basin as a whole, for example, if the gauges are in the valleys and the precipitation is greater in the hills.

Losses. Losses can be computed from:

1. An initial loss and constant loss rate
2. A four-parameter exponential loss function unique to HEC-1
3. The Soil Conservation Service (SCS) curve number (with an optional initial loss)
4. The Holtan formula
5. The Green and Ampt method

Rainfall Excess to Runoff Transformation. Precipitation excess can be transformed to direct runoff using either unit hydrograph or kinematic wave techniques. Several unit hydrograph options are available. A unit hydrograph may be supplied directly or the unit hydrograph may be expressed in terms of Clark, Snyder, or Soil Conservation Service unit hydrograph parameters. The kinematic wave option permits depiction of subbasin runoff with elements representing one or two overland-flow planes, one or two collector channels, and a main channel.

Base Flow. Base flow is specified by means of three input variables: (1) a starting discharge at the beginning of the simulation, (2) an exponential recession rate term, and (3) a recession threshold discharge for the recession limb of the hydrograph. Once the discharge drops below this threshold, the discharge is based solely on the recession rate.

Routing through Stream Channels. The Muskingum-Cunge, kinematic wave, Muskingum, modified Puls, and normal depth methods can be used for stream-flow routing. For the Muskingum-Cunge and normal depth methods, a routing reach is specified in terms of length, slope, and Manning's n values (for the main channel and for the left and right overbanks), and the cross section is defined with eight pairs of x-y coordinate values.

Reservoir Routing. Storage routing techniques are used for routing of flows through uncontrolled reservoirs. The reservoir outflow may be computed by specifying spillway and low-level outlet hydraulic characteristics, or a discharge rating curve for the reservoir can be given. Program output includes a time history of water storage and water surface elevation.

Parameter Calibration Capabilities. A very useful option of HEC-1 is the ability to employ an automatic parameter calibration procedure for single basins (basins that are not subdivided) when both discharge data and precipitation data are available for historical flood events. Unit hydrograph and loss-rate parameters are optimized by using a univariate gradient procedure.

There is also an option to optimize routing parameters, given historical inflow and outflow hydrographs for a routing reach. This requires specification of a time distribution for lateral inflow (flow that enters the routing reach between the inflow and outflow locations). The optimized values for routing parameters are usually very sensitive to the time distribution of lateral inflow, especially when this flow is a substantial proportion of the total outflow.

Hypothetical Storm Generation. HEC-1 has capabilities to generate frequency-based hypothetical storms as well as U.S. Army Corps of Engineers standard project storms for regions in the continental United States east of the Rocky Mountains. Frequency-based storms can be developed from generalized rainfall criteria such as that found in various technical publications of the U.S. Weather Service of the National Oceanographic and Atmospheric Administration (NOAA). Point-to-area adjustments to average precipitation are made automatically. Adjustments from partial duration series to annual series data can also be made.

Snowpack/Snowmelt Simulation. Snow accumulation is computed within elevation zones specified for the subbasin. Air temperatures (specified for the bottom of the lowest zone) and a lapse rate are used to determine whether precipitation occurs as rain or snow. Two snowmelt routines are available: (1) an energy budget method which requires meteorological data such as solar radiation, wind speed, and dew point and (2) a simple temperature-index method requiring only air temperatures. Air temperatures are required at constant time increments and are supplied as time series input. The data do not have to be provided at the same time increment used for program computations; HEC-1 will interpolate time series data as required.

Dam Safety Applications. The reservoir routing procedures can also include the calculation of flows which overtop the dam, including the formation of a breach described by its final geometry after the breach is complete. Flow through the breach is calculated in addition to the flows over the top of the dam, through the spillway, and through the low-level outlet. The various components of dam outflow are combined and routed downstream. Flooding elevations downstream from the dam are computed by the hydrologic routing techniques in the program.

Multiplan/Multiflood Analysis. This option is useful for analyzing alternative future land use and/or project conditions. Various alternative sets of values for modeling parameters can be specified and runoff calculated in a single application of the program. Also, ratios can be applied to hydrographs or to historical or hypothetical storm hyetographs for evaluation of runoff response under a range of conditions. For example, in urban hydrology applications, three conditions can be simulated in a single run of the model: runoff from the existing watershed, runoff after development has occurred, and outflow from the developed watershed with flood detention structures.

Flood Damage Analysis. A flood-peak versus return-period curve can be specified for existing conditions at each basin outlet. Using the multiplan option, alternative land use or project proposals can be simulated and their effect on the flood frequency curve determined. By combining the flood frequency curve with a curve of damage cost versus flood discharge, the expected annual flood damage can be calculated for each alternative plan.

Optimization of the Size of Flood Control System Components. The required size or capacity of various components such as uncontrolled reservoirs, diversions, and pumping plants can be obtained by optimization. Data requirements include cost and capacity data for each component as well as discharge frequency and damage data for damage centers.

Computer Requirements. HEC-1 is written in FORTRAN 77. The MS-DOS version of the program is distributed for the PC in two forms: (1) the standard program, which permits up to 300 ordinates for each computed hydrograph, and (2) a large-array version which permits up to 2000 ordinates to be used. The standard version occupies 475 Kbytes of random-access memory (RAM) on the PC. It can run with or without a math coprocessor chip (8087, 80287, or 80387). The large-array version requires extended memory for the PC (2.5 Mbytes of RAM). This version requires a 80386 system with math coprocessor or an 80486 central processing unit (CPU). The program source code is also available for installation on PC, mainframe, or mini-computer systems. HEC-1 is also available for Macintosh systems.

The personal computer version of HEC-1 is generally supplied as part of a package that includes a menu program with which execution of HEC-1 and associated tasks can be performed, an editor for creating and editing input files, and a utility program for viewing program output.

How Program Can Be Obtained. HEC-1 is available from program vendors who supply the compiled program or source code and also provide various degrees of program support. Some vendors provide modified versions of the program as well as their own data editors and plotting utilities. HEC-1 is available directly from the Hydrologic Engineering Center only to U.S. government agencies. A list of program vendors is available from the Hydrologic Engineering Center, 609 Second St., Davis, CA 95616.

21.2.2 TR-20—U.S. Soil Conservation Service Model

Overview of Program. The U.S. Soil Conservation Service TR-20 computer model is a single-event rainfall-runoff model that is normally used with a design storm as rainfall input.[30] The program computes runoff hydrographs, routes flows through channel reaches and reservoirs, and combines hydrographs at confluences of the watershed stream system. Runoff hydrographs are computed by using the SCS runoff equation and the SCS dimensionless unit hydrograph. Computed flows are routed through channel reaches and reservoirs. Multiple passes through the data file can be made to permit several alternatives to be evaluated in a single computer run.

The TR-20 model is a computer program which utilizes the Soil Conservation Service methods given in the Hydrology section of the *National Engineering Handbook*.[29] The name TR-20 is an acronym for the report (technical release number 20) which describes the program's use.

The watershed is usually divided into subbasins with similar hydrologic characteristics and which are based on the location of control points through the watershed. Control points are locations of tributary confluences, a structure, a reservoir, a diversion point, a damage center, or a flood-gauge location.

Historical or synthetic storm data are used to compute surface runoff from each subbasin. Precipitation excess is based on losses computed by the SCS curve number technique. The unit hydrograph used is the standard SCS dimensionless unit hydrograph. If synthetic rainfall is used, the rainfall excess hyetograph is determined from the effective rainfall and a specified rainfall distribution. This rainfall is then applied to the unit hydrograph to generate the subbasin runoff hydrograph. Base flow can be treated as a constant flow or as a triangular hydrograph. A linear routing procedure is used to route flow through stream channels. The modified Puls method (storage-indication routing) is used for reservoir routing. As many as 200 channel reaches and 99 reservoirs or water-retarding structures can be used.

TR-20 uses land-use information and soil maps indicating soil type to define the SCS curve number for specific land areas. The SCS dimensionless unit hydrograph is defined by a single parameter, the watershed lag, and the subbasin area. Standard procedures are available for determining the lag.[30] The method is not only simple to apply, but similar results are obtained by different hydrologists applying the SCS procedures, since the methodology is standardized.

TR-20 has been widely used by SCS engineers in the United States for urban and rural watershed planning, for flood insurance and flood hazard studies, and for design of reservoirs and channel projects. The SCS methodology is accepted by many local agencies also. McCuen[23] describes a variety of applications of the model and provides a detailed guide to its use. TR-55 is a simplified version of TR-20 that does rainfall-runoff modeling for a single watershed.

Required Input. The TR-20 model requires:

1. Watershed characteristics for each subbasin: area, SCS curve number, antecedent moisture condition, time of concentration, channel reach length
2. Hydrograph data: SCS dimensionless unit hydrograph, a given unit hydrograph, and base flow
3. Rainfall data: cumulative rainfall depths at constant intervals or dimensionless storm distribution with total depth and duration
4. Structure data: reservoir or structure data
5. Stream cross section data: if these are specified, flow routing coefficients are determined from them (otherwise routing coefficients must be specified)

Computer Requirements. The TR-20 program is available in different formats. There are versions of TR-20 for the PC. Other versions exist for mainframe and microcomputers.

How Program Can Be Obtained. The microcomputer (PC) version of the program can be obtained from local SCS offices. The various versions of the program for other types of computers can be obtained from the Soil Conservation Service, U.S. Department of Agriculture, Washington, D.C. The TR-20 report[30] can be obtained as PB-8818-4122 from National Technical Information Service, 5285 Port Royal Rd., Springfield, VA 22161.

21.2.3 ILLUDAS—Illinois Urban Drainage Simulator

Overview of Program. The Illinois Urban Drainage Simulator (ILLUDAS) was developed by the Illinois State Water Survey[32] for hydrologic design of storm drainage systems in urban areas and the hydrologic analysis of watersheds. ILLUDAS is based on the British Road Research Laboratory (RRL) model[34] and was modified to simulate U.S. conditions.[31] It computes runoff from directly connected paved areas in the same manner as the RRL model and also includes runoff from grassed and other pervious surfaces.

ILLUDAS uses observed rainfall or a rainfall distribution pattern as its primary input. The basin is divided into subbasins, one for each design point in the basin. For each subbasin, the program computes a paved area hydrograph and a grassed-area supply rate, leading to a separate hydrograph for the grassed area. The impervious and grassed-area hydrographs are combined, and the result is then combined with hydrographs from other tributaries. If the program is in the design mode, the required pipe size is computed. Flow is then routed to the next combining point. Results are printed, and the program moves to the next subbasin.

Since the program requires only a little more design information than the rational method, it may serve as a next step beyond the rational method for the determination of urban runoff. Terstriep and Stall[32] give the results of ILLUDAS applications for 21 urban basins and two rural basins. The basins were all less than 8.5 mi^2 in size.

Required Input. ILLUDAS requires:

1. Identification information and run type
2. Basin parameters—paved- and grassed-area abstractions, minimum pipe size and Manning's n
3. Rainfall parameters—measured rainfall data or rainfall distribution, duration and return period, and antecedent moisture condition
4. Reach data—type of branch channel, section dimensions, and available storage
5. Subbasin data—subbasin area for paved and grassed areas, contribution area, slopes and path lengths, and hydrologic soil group.

Computer Requirements. ILLUDAS is written in FORTRAN IV. The program consists of over 700 lines of code, and on a mainframe computer the program requires 220 Kbytes of core. The program has also been compiled for operation on the IBM PC and compatible computers. The current version is ver. 2.16.

How Program Can Be Obtained. The ILLUDAS program is available to users in the State of Illinois from the Illinois State Water Survey, 2204 Griffith Dr., Champaign, IL 61820. Others can obtain the program from CE Software, Box 2472, Station A, Champaign, IL 61820.

21.2.4 DRM3—USGS Rainfall-Runoff Model

Overview of Program. The Distributed Routing Rainfall-Runoff Model (DR3M) was developed by the U.S. Geological Survey.[3] DR3M provides a detailed simulation of runoff from rainfall for user-selected storm periods. In DRM3 a drainage basin (or basin subarea) is represented by an overland-flow element, a channel element, and (optionally) reservoirs. Kinematic wave procedures are used to route overland flow over areas that are defined as contributing areas. Flow through the channel segments is also routed by using kinematic wave procedures.

Subareas can be arranged into a basin network to permit simulation of complex hydrologic basins. The model is intended primarily for simulation of urban watersheds (for which the kinematic wave procedures are most applicable), but the model has also been successfully applied to rural watersheds.

The currently available version of DR3M is version II. Program documentation is provided by Alley and Smith.[3]

Required Input. Input consists of short-interval precipitation and discharge data, daily precipitation and evaporation totals, subcatchment areas, roughness, and hydraulic data.

Computer Requirements. DRM3 is written in FORTRAN and can be run on PC systems as well as minicomputers and mainframe computer systems.

How Program Can Be Obtained. The DRM3 model can be obtained from the U.S. Geological Survey. For information contact the U.S. Geological Survey, WRD, 415 National Center, Reston, VA 22092.

21.3 CONTINUOUS-STREAM-FLOW SIMULATION MODELS

21.3.1 SWRRB—Simulator for Water Resources in Rural Basins

Overview. Simulator for Water Resources in Rural Basins (SWRRB) simulates hydrologic and related processes in rural (agricultural) basins. This computer model was developed by the U.S. Department of Agriculture[4] to predict the effect of various types of watershed management procedures on water and sediment yields in ungaged rural basins. The major processes which are included in the model are surface runoff, evapotranspiration, transmission losses, pond and reservoir evaporation, sedimentation, and crop growth. There is also a special component of the program that simulates the runoff of pesticides.

The model deals with large basins which are subdivided into as many as 10 subbasins, each of which can have a different rainfall input. There is no limitation on basin area. The soil profile can be divided into as many as 10 layers. The upper layer has a fixed thickness of 10 mm (0.4 in); the other layers are of variable thickness.

The model is physically based and is intended to be used for situations in which calibration data are not available. Periods of many years of daily flows can be simulated by SWRRB.

SWRRB has three major components: hydrology, weather, and sediment yield. These are discussed in the following paragraphs. A flowchart showing the major elements of the program is given in Fig. 21.3.1.

Hydrology Component. The SWRRB hydrology model is based on the water balance equation, and the change in soil water content is computed from rainfall, runoff, evapotranspiration, percolation, and return flow. Basins are subdivided to reflect differences in hydrologic characteristics, such as the different evapotranspiration rate for different crops, soils, and other factors. The runoff from each subbasin is computed separately.

Surface runoff is computed from daily rainfall values. Runoff volume is determined by using the Soil Conservation Service curve number approach. Peak discharge is estimated by using a modification of the rational formula. Rainfall intensity is related to the subbasin time of concentration, which is based on an average overland flow time of concentration computed from average overland and channel flow lengths, slopes, and Manning n values. If snow is present, it is assumed to melt on days for which the average temperature exceeds 0°C.

The percolation component of SWRRB uses a storage routing technique combined with a crack-flow model to predict flow through soil layers. Water which percolates below the root zone is moved to groundwater or shows up as return flow in downstream basins. The crack-flow model permits percolation of infiltrated rainfall for conditions where the soil water content is below field capacity.

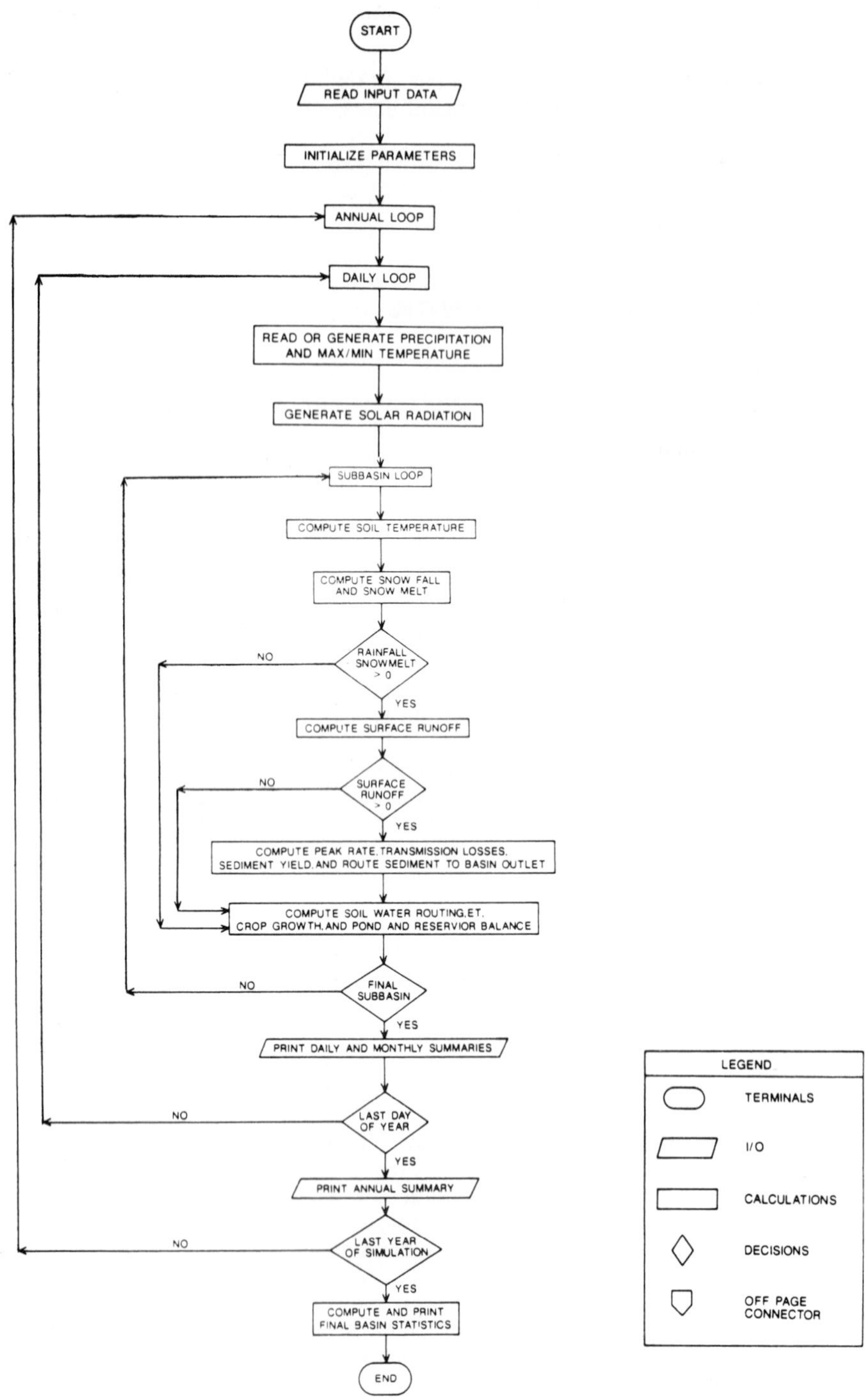

FIGURE 21.3.1 Schematic of the SWRRB program. *(Source: Arnold et al.[4] Used with permission.)*

Lateral subsurface flow is calculated to occur simultaneously with percolation, and lateral flow travel time is adjusted to account for variations in the soil water content and subsurface flow characteristics.

Potential evapotranspiration is related to daily solar radiation, mean air temperatures, crop cover and snow cover (if present). Soil and plant evaporation are computed separately. Soil evaporation is determined by the water content of the top 30 cm (12 in) of the soil. If soil water is limited, plant evaporation will be reduced accordingly.

Transmission losses in alluvial channels can be computed from stream channel data (width, depth, length, and the hydraulic conductivity of the channel alluvium).

Water yield in SWRRB is related to storage, seepage, evaporation, and outflow from ponds and reservoirs. The water balance equation is used for ponds and is based on the percentage of area occupied by ponds in the catchment employing an empirical pond storage-volume–surface-area relationship to compute daily inflows, outflows, seepage, and evaporation. The reservoir component is similar to the pond component. The reservoir calculations can also include flow from spillways.

Weather Component. SWRRB uses precipitation, air temperatures, and solar radiation for driving the weather component of the model. Two methods of providing precipitation data are available: daily precipitation can be used as a direct input, or precipitation can be simulated as a first-order Markov chain process which uses the probabilities of receiving precipitation if the preceding day was dry or wet. When no wet-dry probabilities are available, probabilities are estimated from the average number of days of precipitation in a month. Air temperatures and solar radiation for each day are generated from daily statistics of these variables.

Sediment Yield Component. Sediment yield for each subbasin is computed from the modified universal soil loss equation.[35] Sediment yield is computed from the surface runoff volume, the peak discharge, a soil erodibility factor, a crop management factor, an erosion control management factor, and a slope-length–steepness factor. The sediment yield from subbasins is used as an input to ponds and reservoirs.

Sediment is routed through the system, first through ponds and reservoirs, and then through stream channels by using a stream power concept described by Williams.[36]

SWRRB is used to predict the effects of various types of land uses on water yield (on a monthly or annual basis), on sediment production in the watershed, and on pollution. For example, sediment yields from a watershed can be compared for alternative management practices of fall plowing, conservation tilling, or no tilling. The model has been used to illustrate how various mixes of crops on the watershed and alternative management practices can affect water yield from the basin. SWRRB has been used to demonstrate the effects of reservoir storage on water and sediment yield in Oklahoma, and in another study it was used to determine the effect of urbanization on water and sediment entering a Texas lake.[4]

Required Input. The data used by the SWRRB program includes the following (specific input data formats are given in the program users manual):[4]

1. Program control codes governing the total length of simulation (1 to 100 years), number of subbasins, output codes, etc.
2. General watershed data such as rain-gauge correction factors, base flow parameters, and initial soil water storage

3. Subbasin centroid coordinates
4. General weather data
5. Monthly temperatures
6. Monthly solar radiation
7. Monthly rainfall
8. Parameters for generation of daily rainfall
9. Basin data such as SCS curve numbers, soil albedo, snow cover at start of simulation, overland and channel data, and surface erosion parameters
10. Channel routing data (including channel erosion parameters)
11. Pond data
12. Reservoir data
13. Soils data
14. Crop data
15. Irrigation data
16. Daily temperature and rainfall parameters

SWRRB provides access to meteorological statistics compiled for about a hundred first-order weather stations in the United States, and SWRRB contains a very extensive database of soil properties developed by the U.S. Department of Agriculture.

Computer Requirements. The SWRRB is written in FORTRAN 77. It runs on PC systems with 256 Kbyte RAM. Fortran source code is available. An interactive data editor is available for the program also.

How Program May Be Obtained. The program may be obtained from the Grassland, Soil, and Water Research Laboratory, Agricultural Research Service, U.S. Department of Agriculture, 808 East Blackland Rd., Temple, TX 76502.

21.3.2 PRMS—Precipitation-Runoff Modeling System

Overview. The Precipitation-Runoff Modeling System (PRMS) is a program developed by the U.S. Geological Survey (USGS) to simulate watershed response over time periods that are longer than those used with the USGS DR3M distributed-parameter watershed model discussed above. It was developed to evaluate the effects of various combinations of precipitation, climate, and land use on watershed response. The model provides simulations on both a daily and a storm time scale by using variable time steps. In the storm simulation mode, the program simulates selected hydrologic components using time increments that can be as small as 1 min. Before and after the storm period a daily time step is used and a daily average or daily total is simulated for the hydrologic components. Stream flow is computed as mean daily flow.

Watersheds are partitioned into units based on specific characteristics. The USGS has developed maps for the United States that show watershed delineation into *hydrologic response units* (HRU). Parameters of these units include surface slope, aspect, elevation, soil type, vegetation type, and distribution of precipitation. Each hydrologic response unit is assumed to have homogeneous hydrologic characteristics. A water balance and energy balance are computed daily for each HRU. The sum of the responses of all HRUs gives the daily watershed response.

During storm periods, a second level of partitioning is used to permit the short-term response of the watershed to be determined. For this the watershed is assumed to be composed of a series of interconnected flow planes and channel segments. Each HRU can be represented by a single flow plane or by a series of flow planes. The watershed drainage network is represented by a system of channel segments, reservoirs, and junctions in a manner similar to the DR3M model. Kinematic flow routing is used for routing flows over flow planes and through the channel segments.

PRMS is used in conjunction with two other programs developed by the U.S. Geological Survey. A program named ANNIE provides computer data management and analysis capabilities for PRMS. Hydrologic forecasting capabilities are provided by a modified version of the National Weather Service program Extended Streamflow Prediction (ESP). PRMS, ANNIE, and ESP, when used together, form a complete watershed modeling system that gives the user the capability to reduce, analyze, and prepare data for model applications; to simulate and forecast watershed response; and to analyze model results statistically and graphically.

Required Input. Short interval precipitation and discharge data, daily precipitation and evaporation totals, subcatchment areas, roughness and hydraulic data.

Computer Requirements. The DRM3 program is written in FORTRAN and can be run on PC systems as well as on minicomputers and mainframe computer systems.

How Program Can Be Obtained. The DRM3 model can be obtained from the U.S. Geological Survey. For information contact the Office of Surface Water, U.S. Geological Survey, WRD, 415 National Center, Reston, VA 22092.

21.3.3 SHE—European Hydrologic System Model (Système Hydrologique Européen)

Overview. SHE is a physically based, distributed-parameter catchment modeling system that was produced jointly by the Danish Hydraulic Institute, the U.K. Institute of Hydrology, and SOGREAH (France). The development work was financially supported by the Commission of the European Communities, and the development of SHE was aimed at providing a strong European capability in hydrologic modeling. The background for the development of SHE is given by Abbott et al.[1] The process equations and model structure description can be found in Abbott et al.[2] and Bathurst.[5,6]

The model considers the major hydrologic processes of which govern water movement through a catchment, namely: snowmelt, canopy interception, evapotranspiration, overland flow, channel flow, and unsaturated and saturated subsurface flow. Spatial variability of hydrologic processes is described by using a rectangular grid of (x, y) points in the horizontal plane with vertical variation in properties represented by a series of horizontal planes of various depths.

SHE is applicable to a wide range of hydrologic processes and can be applied to a variety of hydrologic problems, including irrigation schemes, determination of land-use changes, water development studies, groundwater contamination, erosion and sediment transport, and flood prediction. See Table 21.3.1 for details.

Computer Requirements. SHE applications have been performed on mainframe computers. Applications of the program on PCs are limited because of the large number of computations that must be made.

TABLE 21.3.1 Suggested Applications for SHE at Different Operational Scales

Topic	Primary hydrologic process*	Possible scale of operation
Irrigation schemes:		
Irrigation water requirement	ET/UZ	Field
Crop production	ET/UZ	Project
Waterlogging	ET/UZ	Field
Salinity/irrigation management	UZ	Field
Land-use changes:		
Forest clearance	ET/UZ/SZ	Catchment
Agricultural practices	ET/UZ/SZ	Field/catchment
Urbanization	ET/UZ/SZ	Catchment
Water developments:		
Groundwater supply	SZ	Catchment
Surface-water supply	ET/UZ/SZ	Catchment
Irrigation	UZ/SZ	Project/catchment
Stream-flow depletion	SZ/OC	Catchment
Surface water/groundwater interaction	ET/UZ/SZ	Project/catchment
Groundwater contamination:		
Industrial and municipal waste disposal	UZ/SZ	Field/catchment
Agricultural chemicals	UZ/SZ	Field/project/catchment
Erosion/sediment transport	OC/UZ	Project/catchment
Flood prediction	OC/UZ	Catchment

* ET = evapotranspiration; UZ = unsaturated zone; SZ = saturated zone; OC = overland and channel flow.

Source: Abbott.[2] Used with permission.

How Program Can Be Obtained. The program can be obtained through one of the sponsoring agencies: (1) The Danish Hydraulic Institute, Agern Alle 5, DK-2970 Horsholm, Denmark; (2) Institute of Hydrology, Maclean Building, Crowmarsh Gifford, Wallingford, Oxon OX10 8BB, United Kingdom; (3) Ministère de l'Environnement et du Cadre de Vie, 14, Bd. du General Leclerc, 95521 Neuilly-sur-Seine Cedex, France. Training in the use of the program is also available through these institutions.

21.4 FLOOD-HYDRAULICS MODELS

21.4.1 HEC-2 Water Surface Profiles

Overview. HEC-2 was developed by the Hydrologic Engineering Center of the U.S. Army Corps of Engineers to compute steady-state water surface elevation profiles in natural and constructed channels.[20] Its primary use is for natural channels with complex geometry such as rivers and streams. The program analyzes flow through bridges, culverts, weirs, and other types of structures. Water surface elevation profiles can be computed for cases where there is a loss of flow from the stream because of levee or embankment overflows, overtopping of watershed divides, or flow through diversion structures (the split-flow option). Cross-section modifications can be easily analyzed by specifying a template for cut and fill to the cross section using the channel improvement option.

The encroachment computation option has been widely used in the analysis of flood-plain encroachments for the U.S. Federal Emergency Management (FEMA) flood insurance program. There are several types of encroachment calculation procedures, including the specification of encroachments with fixed dimensions and the designation of target values for water surface increases associated with floodplain encroachments.

HEC-2 uses the standard step method for water surface profile calculations, assuming that flow is one-dimensional, gradually varied steady flow. Either subcritical or supercritical flow profiles can be analyzed. The input data records for subcritical profile calculations are specified from downstream to upstream. The cross sections must be placed in reverse order in the data file for supercritical profile runs.

HEC-2 will compute up to 14 individual water surface elevation profiles in a given run. Usually a different discharge is used for each profile, although when the encroachment or channel improvement options are used, the section dimensions are changed rather than the discharge. The discharge can be changed at each cross section to reflect tributaries, lateral inflows, or diversions.

The water surface elevation at the starting cross section can be specified in one of four ways: as a given elevation, as critical depth, by a rating curve, or as a computation by the program, by the slope-area method, when the energy grade line slope is specified. The water surface elevation associated with critical flow is computed for conditions of minimum energy at the cross section.

The energy-loss term in the energy equation is computed from two factors: boundary resistance (friction loss) and eddy loss between sections (expansion or contraction loss). Friction losses are computed by using Manning's equation. Eddy losses are calculated by using a head-loss coefficient multiplied by the change in the velocity head between cross sections.

HEC-2 computes water surfaces as either a subcritical flow profile or a supercritical profile. Mixed subcritical and supercritical profiles are not computed simultaneously. If the computations indicate that the profile should cross critical depth, the water surface elevation used for continuing the computations to the next cross section is the critical water surface elevation.

The kinetic energy coefficient α is included in the energy equation and is calculated from a weighted sum of kinetic energies of the main channel and the overbank areas. The overbanks can be subdivided into individual flow sections by using the divisions established by the assigned distribution of Manning's n across the full section. The program requires that three flow path distances be used between cross sections: a channel length and left and right overbank lengths. The reach length used in friction-loss calculations is a discharge-weighted length based on the relative amounts of flow in each portion of the cross section. Friction loss is computed as the product of the reach length and the mean friction slope for the reach. The mean friction slope can be computed from the average conveyance, the average slope, the geometric mean slope, or the harmonic mean slope. An alternative method for defining frictional resistance is to use a *roughness height* based on the Colebrook-White approach. The roughness height can be varied horizontally across the section in the same manner as Manning's n.

Bridge Hydraulics. Head loss at bridges and other river structures is also computed by HEC-2. Bridge head loss is assumed to be due to two factors: contraction and expansion losses before and after the structure, and loss at the structure itself. Structure losses can be computed by either the *special bridge method*, which uses hydraulic formulas for pier losses, pressure flow when the bridge opening is submerged, and weir flow when the bridge and roadway embankment are overtopped, or by the *normal bridge method,* which computes bridge head loss as the sum of increased friction plus expansion and contraction losses at the structure. The special bridge method is usually easier to apply, since fewer cross sections are required. In bridge calculations, the weir may be described by using a fixed width and elevation or the weir crest may be defined using *x-y* coordinates to more closely represent the weir flow section produced by bridge approach roadway embankments and the bridge deck.

Culvert Flows. Single- or multiple-barrel box or circular culverts may be modeled. Culvert inlet control flow computations are based on Federal Highway Administration (FHWA) nomographs. The program uses hydraulic formulas for outlet control calculations (inlet and outlet losses plus friction in the culvert). Flows which pass over the roadway embankment are treated as weir flow in the same manner as in the special bridge method. There are limitations in the HEC-2 culvert procedures. Only circular or box culverts can be analyzed, and, when multiple culverts are used, they must all be identical.

Encroachment Analyses. The six methods for encroachment analysis of flood plains available in HEC-2 are listed in Table 21.4.1.

Required Input. The program is designed to operate in a batch mode, and a data file is prepared by using an editing program. The Corps of Engineers editor COED is supplied with the HEC-2 program package for the PC. COED has on-line help information for all input variables used by HEC-2. Individual data records have a two-character identifier at the beginning of each record. Multiple title records can be used, and comments can be inserted at any point in the data file. In addition to running identification data, the program requires job control information, discharge and loss data, and cross-section geometry data. The cross-section data make up the major portion of the input and include cross-section numbering, reach lengths, geometry data, and modifications to the basic cross-section data (points added to cross section, filling of all low areas to a specified elevation, blocking out of ineffective

TABLE 21.4.1 HEC-2 Floodplain Encroachment Calculation Methods

Method No.	Description
1	Encroachment elevations and stations specified
2	Top width of water surface specified and centered on the channel
3	Encroachment stations automatically computed for a specified percent reduction in conveyance
4	Encroachment stations automatically computed corresponding to a target difference in the water surface elevations of natural and encroached flows
5	Encroachment stations are optimized to achieve a target difference in the water surface elevations of natural and encroached flows
6	Encroachment stations automatically computed corresponding to a target difference in the energy grade line elevations of natural and encroached flows

Source: Feldman.[11] Used with permission.

flow areas). Manning's n values, and expansion and contraction loss coefficients, may also be changed at each cross section.

Data which do not change from section to section can be repeated from the previous cross section. Cross-section dimensions can be expanded or reduced in size and adjusted in elevation at the repeated cross section. The data can be entered in free format and transformed to fixed format by a program utility.

Program Output. The amount of output produced by an HEC-2 run can be controlled by input codes. HEC-2 can generate large volumes of output because detailed information about the computations is normally printed by the program. Various summary tables can be requested; for example, summary tables for bridge and culvert computations, for the channel improvement option, and for encroachment data are available. Special flood insurance study tables for displaying encroachment output that match the table format required by the U.S. Federal Emergency Management Agency's guidelines and specifications for flood insurance studies can be printed. User-defined tables can be produced which display any variable computed by the program.

The interactive screen plotting program PLOT2 can produce profile plots, cross-section plots, rating curves, and plots of other computed data. The plots can be sent to a graphics printer or to plotters which use the Hewlett-Packard (HP) graphics language format.

Computer Requirements. The HEC-2 program is written in ANSI-standard FORTRAN 77. Versions of the program are available in compiled form for personal computers and as source code for PCs, minicomputers, and mainframe computers. The PC source code package contains information for compiling the program. Some program utilities (such as PLOT2) are not written in FORTRAN and are supplied only in executable form. A math coprocessor is not required for the PC version, but it speeds up program operation significantly. A graphics monitor and hard disk are not mandatory, but make the program much easier to use effectively.

How Program Can be Obtained. HEC-2 is available directly from HEC only to U.S. government agencies. HEC provides lists of program vendors for the United States and other countries (Hydrologic Engineering Center, 609 Second St., Davis, CA, 95616).

21.4.2 WSPRO—A Model for Water Surface Profile Computations

Overview. WSPRO is a computer program which computes steady-state water surface profiles in open channels. It is intended for use with natural channels such as rivers and streams, where the geometry of the channel changes from section to section. The program analyzes flow through bridges and culverts, through multiple-opening stream crossings, and embankment overflows. Two Federal Highway Administration reports[27,28] provide the documentation for this model.

Conventional step-backwater analyses are used in this program. The program assumes that the flow is one-dimensional, gradually varied steady flow. Both subcritical and supercritical flow profiles can be analyzed. The input data records (which are usually ordered from downstream to upstream for a subcritical profile analysis) do not have to be rearranged for a supercritical profile run.

The program can compute 1 to 20 individual water surface elevation profiles in a given run. Usually a different discharge is used for each profile. The discharge can be changed at each cross section. The water surface elevation at the starting cross section can either be specified by the user or be computed by the program. If a slope is specified, the program will compute the water surface elevation corresponding to normal depth using the slope-conveyance method. If neither elevation nor slope is specified, the program will assume the starting depth to be critical depth. The water surface elevation associated with critical flow conditions is based on occurrence of minimum energy at the cross section.

WSPRO allows simultaneous variation of bed roughness both across the cross section and with water depth. This type of roughness variation is fairly common in rivers;[28] however, this computational technique is not available in other models of this type. Friction-loss computations are based on specified flow lengths between cross sections. The user can select the technique used by the program for computing the average friction slope. Coefficients for energy losses associated with expansion and contraction of the flow may be specified as input.

Flow through Bridges. Flow through a bridge is treated as either free surface flow or pressure flow. When the water surface does not have significant contact with the underside of the bridge and the flow is subcritical or critical, a free surface flow profile is computed by techniques that are based on the USGS contracted-opening method,[26] in which a coefficient of discharge, reflecting the characteristics of bridge geometry and the flow, is determined for the bridge opening. The energy equation is then solved by using a minimum of three cross sections: an exit section located at least one bridge length downstream from the opening, a section at the bridge opening, and an upstream approach section. If spur dikes upstream from the bridge are present, a fourth section can be used.

Pressurized flow will exist when the water surface is in contact with the lower members of the bridge structure. WSPRO computes pressure flow conditions by one of two orifice-type discharge equations.[7] One equation represents unsubmerged orifice flow in which the water is in contact with only the upstream lower portion of the bridge. The second equation is used when the flow is in contact with both the upstream and downstream low-chord elements of the bridge.

Culvert Flows. Culvert flow computations in WSPRO are based on Federal Highway Administration design procedures.[33] Single- or multiple-barrel configurations of box, circular, and pipe-arch culverts of concrete, corrugated metal, and aluminum may be simulated. Culverts may also be included in the procedure for analyzing multiple openings in a roadway embankment described below. However, analyses are limited to using culvert data to compute headwater elevation on the basis of input data for discharge and downstream water surface elevations. The program does not determine a continuous water surface profile through the culvert.

Embankment Overflow. Roadway embankment overflow at a bridge is computed by treating the overflow sections as broad-crested weirs. Embankment overflow may occur in combination with either free surface flow through the bridge or with pressurized flow. In either case, a trial-and-error procedure is used.

For free surface flow, the following steps are followed:

1. An upstream water surface elevation is assumed.
2. The embankment overflow based on this elevation is computed.
3. Embankment overflow is subtracted from the total flow to give the flow passing through the bridge opening.

4. An upstream water surface elevation for the required flow through the bridge is computed.
5. The computed elevation from step 4 is compared with the assumed elevation of step 1.

The five steps are repeated until the elevation difference of step 5 is within an acceptable tolerance. The sign and magnitude of the difference is used to select the new assumed elevation in step 1 of the next iteration of the procedure.

A similar procedure is followed for pressurized flow. As for the free surface flow case:

1. An upstream water surface elevation is assumed.
2. The embankment overflow based on this elevation is computed.
3. In this case, however, pressure flow through the bridge opening is computed from the upstream elevation.
4. The computed flows from steps 2 and 3 are added.
5. The computed total discharge from step 4 is compared with the given discharge. Again, the five steps are repeated until the difference in step 5 is within an acceptable tolerance.

Multiple Openings. The WSPRO program is unique among programs of its type in that it provides capabilities for analyzing road crossings of streams in which there are two or more bridges or culverts at the crossing. In this analysis, flow is apportioned among the individual openings, and a water surface profile is computed for each individual opening using a representative strip of the valley. Flow apportionment is based on both the flow area of the openings and the distribution of flow conveyance across the total cross section. The valley strips are determined from flow division points which are based on the relative flow areas of adjacent openings. Iterations are made until the flow computed for each opening and a conveyance-weighted water surface elevation at a common upstream section do not change significantly on successive iterations.

Encroachment Analyses. Two options for channel encroachment analysis are available in the program. One method involves the analysis of fixed limit encroachments; the second determines encroachments based on conveyance removal to obtain a target rise in water surface elevation. In this latter method equal conveyance can be removed from each side of a cross section, with or without specified constraints. The procedures permit the analysis of the majority of encroachment problems encountered.

The WSPRO program was initially developed to provide bridge designers with a tool for analyzing alternative bridge openings and embankment configurations. Because of its usefulness for general stream profile computations, it is widely used in highway design, floodplain mapping, flood insurance studies, and developing stage-discharge relationships.

Required Input. The program is designed to operate in a batch process, in which a data file is generated by using an editing program. The individual data records use one- or two-character identifiers at the beginning of each record (this aspect of the program is similar to the input formats for HEC-2). The order in which the input records can be placed in the file is quite flexible. Also, the data can be entered in free format, if desired.

Types of input data used by the program are (1) title information, (2) job parameters, (3) profile control data, (4) cross-section definition, and (5) data display commands. Title information is used only for output identification. Up to three title information records can be used per run. Job parameters are provided on two records in each run to define error tolerances and test values, and to provide parameters for tables. Profile control data are discharges, starting water surface elevations or energy grade line slopes, information about subcritical or supercritical profile, and end of input.

Cross-section data make up the major portion of the input and include information on the type of cross section (unconstricted section, bridge, culvert, spur dike, etc.), cross-sectional geometry data, roughness data, and flow length data. Special data records for bridges, approach sections, road grades, and culverts are also used.

Data display commands are used to generate tables of cross-section properties and velocity/conveyance distributions and to produce plots. Data which do not change from section to section can be coded only for the first section to which they apply. Values which are not supplied are taken from the next downstream cross section. Template cross sections which can be expanded or reduced in size and adjusted in elevation can also be used.

Default values are provided for the parameters which govern the computational procedures, such as test values for computational tolerances.

Program Output. WSPRO generates a relatively large amount of output since it provides a detailed record of the processing of the input data and the results of all profile computations. This output is written to a file which can either be printed directly or be viewed on a terminal or personal computer screen by using a utility program. Three general types of information can be generated: (1) printer plots of cross sections, (2) cross-section property information, and (3) user-defined tables of computed quantities. The user can define which plots are to be included in the output. The model offers no option to suppress any output. However, an editing utility program could be used to produce smaller printed files. A wide selection of key input parameters and computed results are stored in a machine-readable direct-access file, and users can develop utility routines to access this file and generate additional tabular or plotted output.

Computer Requirements. The WSPRO program is written in American National Standards Institute (ANSI) standard FORTRAN 77. Program length is about 8000 lines of FORTRAN source code. Versions of the program are available for personal computers, minicomputers, and mainframe computers. Program execution requires about 200 to 250 Kbytes of memory on mainframe computers and about 400 Kbytes on PC systems. Three printer-compatible output files are automatically generated. Two direct-access files in machine-readable format are used. Additional output files can also be generated by the user.

How Program Can Be Obtained. The program can be obtained from the U.S. Geological Survey, WRD, 415 National Center, Reston, VA 22092, or from the Federal Highway Administration, U.S. Department of Transportation, Washington, D.C. A number of vendors of hydrologic models will also supply the program, either in compiled form for use on personal computers or in ASCII format for other types of computer systems.

21.4.3 FLDWAV—NWS National Weather Flood Wave Model

Overview. The FLDWAV program is a generalized unsteady-flow simulation model for open channels. It was developed by D. Fread of the U.S. National Weather Service (NWS) and replaces the DAMBRK, DWOPER, and NETWORK models, combining their capabilities and providing new hydraulic simulation procedures within a more user-friendly model structure.[14]

The FLDWAV program is intended to be used on PC-type computers, minicomputers, or mainframe computers. It simulates a wide range of unsteady-flow applications for a dendritic (tree-shape) system of waterways subject to backwater effects, including real-time flood forecasting; dam breach analysis and inundation mapping; design of waterway structures such as levees, off-channel detention, etc.; floodplain mapping; analysis of irrigation systems with flows regulated by gates; analysis of storm sewer systems which can operated with both free-surface and pressure flows; and unsteady flows due to hydropower operations.

FLDWAV can simulate the failure of dams caused by either overtopping or piping failure of the dam. The program can also represent the failure of two or more dams located sequentially on a river.

The program is based on the complete equations for unsteady open-channel flow (St. Venant equations). Various types of external and internal boundary conditions are programmed into the model. At the upstream and downstream boundaries of the model (external boundaries), either discharges or water surface elevations which vary with time can be specified.

FLDWAV uses an expanded form of the original St. Venant equations to include the following hydraulic effects:

1. Lateral inflows and outflows
2. Nonuniform velocity distribution across the flow section
3. Expansion and contraction losses
4. Off-channel storage (which is referred to as *dead storage*)
5. Procedures for representing flow path differences between the sinuous main channel and the flood plain
6. Surface wind shear effects
7. Representation of internal viscous dissipation effects that occur in mud and debris flows

Special Features. FLDWAV's special features include:

1. A subcritical/supercritical mixed-flow solution algorithm
2. Levee overtopping calculations
3. Interaction between channel flow and the floodplain flow
4. Calibration of Manning's roughness parameters
5. Analysis of combined free-surface and pressure flow conditions
6. Automatic selection of time and distance steps.

Internal Boundaries. The St. Venant equations apply to gradually varied flow with a continuous profile. If features which control or interrupt the water surface profile exist along the main stem of the river or its tributaries, internal boundary conditions are required in the program. These features can include dams, bridges, roadway embankments, falls, short steep rapids, weirs, etc. If a bridge is being simulated, the program uses a coefficient of discharge for flow through the bridge and another coefficient of discharge for flow over the roadway embankment. This latter flow is treated as flow over a broad-crested weir with a correction for submergence of the weir, if required.

If the structure is a dam, the total discharge past the structure is the sum of spillway flow, flow over the top of the dam, gated-spillway flow, flow through turbines, and flow through a breach in the dam, if a breach occurs. The spillway flow and dam overtopping are treated as weir flow, with corrections for submergence. The gated outlet can represent a fixed gate or one in which the gate opening can vary with time. These flows can also be specified by rating curves which define discharge passing through the dam as a function of upstream water surface elevation.

The turbine discharge can be either a fixed discharge or time-dependent. The dam breach flow is computed from a relationship expressing the time-varying characteristics of the dam breach. If the internal boundary is a waterfall or rapids, the water surface elevation at this location can be determined from a computation of critical depth.

External Boundaries. External boundary condition relationships must be specified at the upper and lower end of all unconnected stream reaches. The external boundary condition is the main factor driving the unsteady-flow conditions in many cases. The upstream boundary condition may be either a discharge hydrograph specified as a time series of flows or a stage hydrograph (a time series of water surface elevations). Downstream, the boundary condition could be a discharge or stage hydrograph, a rating curve which defines discharge as a function of water surface elevation, or a channel control loop-rating relationship based on Manning's equation in which the dynamic energy slope is related to the flow conditions at the last two cross sections at the downstream end of the river. Initial conditions (discharge and water surface elevation throughout the stream system at time zero) can be automatically obtained by using a steady-flow solution based on an initial discharge, or unsteady-flow conditions at time zero may be input directly.

If the river consists of a main stem and tributaries, the set of equations describing this case is solved by an iterative method[12] in which the flow at the confluence from or to the tributary is taken as the lateral inflow/outflow term in the St. Venant equations. If the river has bifurcations (such as islands) or is a dendritic system with tributaries connected to tributaries, a network solution scheme is used.[13]

Subcritical/Supercritical Mixed-Flow Solution Algorithm. When this algorithm is used, it automatically divides the routing reach into subreaches in which only subcritical or supercritical flow occurs. The Froude number is used to determine if the flow at a particular section is subcritical or supercritical. The transition locations where the flow changes from subcritical to supercritical are treated as external boundary conditions. At each time step, the solution begins with the farthest-upstream subreach and proceeds, subreach by subreach, in the downstream direction. The internal subreach boundary conditions are based on hydraulic conditions in the adjacent subreaches. Hydraulic jumps are allowed to move upstream or downstream prior to advancing to the next time step; this is done by comparing computed sequent water surface elevations with computed elevations in each section in the vicinity of the hydraulic jump. Using the subcritical/supercritical algorithm increases the computational time by about 20 percent.

Levee Overtopping Calculations. The program will compute the amount of flow which overtops levees along either or both sides of the mainstream of the river or its principal tributaries. The overtopping is represented by the lateral flow term in the St. Venant equations (a discharge per unit channel length). The levee is treated as a broad-crested weir in the computation of the amount of flow leaving the river. Because the amount of flow passing over the levee is affected by the water depth in the adjacent floodplain, the FLDWAV program has three options for analyzing the flow. In the first option the presence of the floodplain is ignored. For the second option, the receiving floodplain is treated as a storage area having a user-specified storage-elevation relationship, and level-pool routing procedures are used to compute the flow into and out of the floodplain. In the third option, the floodplain is treated as a tributary and the St. Venant equations are used to determine water surface elevations and discharges. In each option, a breech in the levee may be specified at some location.

Interaction between Channel Flow and the Floodplain Flow. The program also allows the floodplain to be separated into compartments where there are levees or roadway embankments perpendicular to the flow paths. The flow from compartment to compartment is computed by using the broad-crested weir equation. If the upstream water levels drop below the downstream levels, flow reversals are computed.

Calibration of Manning's Roughness Parameters. An optional feature of the FLDWAV program allows an automatic determination of Manning's n values when measured stage hydrographs are available. The program computes the Manning n as a function of either discharge or water surface elevation within river subreaches bounded by water-level recording stations.

Computer Requirements. The FLDWAV program is written in standard FORTRAN IV. Versions of the program are available for personal computers, minicomputers, and mainframe computers. The PC version requires a minimum of 600 Kbytes of RAM and requires a math-coprocessor chip for the program to run.

The program has more than 80 subroutines. Arrays are coded with a variable dimensioning technique within a single large array which is the only array of fixed size. When the program is executed, the large array is automatically partitioned into individual variable arrays required for a particular application. The size of each array is determined by the input which describes the application. In this way the maximum utilization of computer memory is assured, since arrays which are not used in a specific application are not assigned to memory.

Related Programs. DAMBRK, DWOPER, and other NWS models have been very widely used and are still in general distribution. DWOPER is the original form of the program. It was modified into the NETWORK program to deal with stream networks, such as those encountered in branching river systems, rivers with islands, and canal networks.

DAMBRK, the Dambreak Flood Forecasting Model (latest revision in 1988), is an unsteady-flow dynamic-routing model which uses the one-dimensional St. Venant equations to route reservoir outflow and dambreak floods through the downstream river valley. The program develops an outflow discharge hydrograph due to spillway and/or dam-failure flows.

SMPDBK, the Simplified Dam-Break Model, is an interactive simplified dam-break model which computes the peak discharge, water surface elevation, and time of occurrence of flooding at selected cross sections downstream from a breached dam.

BREACH, the Breach Erosion Model, is a deterministic model of the erosion-formed breach in an earthen dam caused by overtopping of the dam or initiated by a piping failure of the dam. The dam may be a manmade structure or formed by a landslide. BREACH computes the outflow and the breach parameters used in NETWORK and SMPDBK.

CROSS, a Cross-Section Reduction Plot Program, is for development of data to be used with the NWS models FLDWAV, NETWORK, and DAMBRK. It computes top-width–elevation tables from cross-section data in x-y coordinate format and also computes distance-weighted average cross-section data. CROSS can be used to convert data that are in HEC-2 format to the NWS format used by FLDWAV, NETWORK, and DAMBRK.

How Program Can Be Obtained. The FLDWAV program can be obtained from the National Weather Service, Hydrologic Research Laboratory, 1325 East-West Highway, Silver Spring, MD 20910. FLDWAV is available in executable form for PCs and as FORTRAN source code for other computer systems. A number of vendors of hydrologic models also supply the programs described above, either in compiled form for use on personal computers or as source code for other types of computer systems.

21.4.4 DHM—Diffusion Hydrodynamic Model

Overview. The DHM program is a coupled topographic and channel flow simulation model for two-dimensional floodplain analysis. It was developed for a two-dimensional dam-break floodplain study of a hypothetical failure of a large dam, and it was subsequently extended to include channel flow analysis and the exchange of water between stream channels and the topography. Hromadka and Yen[15] describe the application of the DHM to one-dimensional open-channel floodwave analysis; two-dimensional floodwave analysis; rainfall-runoff modeling of dam breaks, reservoirs, and estuaries; and coupled topographic and open-channel flow modeling.

The DHM is intended for use on PCs and larger systems. Recent applications of the DHM include river overflow floodplain studies, alluvial fan analysis, water reservoir dam-break analysis, and regional flood control deficiency analysis. Most of these studies were accomplished on PCs.

The DHM is based on the noninertial form of the St. Venant equations for two-dimensional flow. Various types of external and internal boundary conditions are available in the model such as spatially distributed rainfalls, hydraulic parameters, stage-discharge curves, and critical flow control. Kinematic wave techniques for flood routing calculations can be used in place of the diffusion routing calculations if desired.

Special Features. DHM's special features include:

1. Two-dimensional unsteady-flow topographic-flow modeling
2. Open-channel unsteady-flow modeling
3. Coupling of flow exchange between the open-channel and topographic-flow models
4. Modeling of backwater and storage effects by the use of the noninertial form of the St. Venant equations
5. The model can be switched to the kinematic wave technique
6. A small amount of program code

7. Availability of various output forms such as flow depth versus time or discharge versus time

The model reduces to a set of two-dimensional point estimates in the limit as the grid area approaches zero. Each grid element is connected to other grids by a north, south, east, west local coordinate system. Parameters required are area-averaged local elevations, Manning's *n* factor, effective area ratio (i.e., ratio of grid area with respect to area available to store water), and initial flow depth.

The open-channel flow model permits one-dimensional unsteady-flow routing of an interconnected channel network through the topographic model. Overflow from the open channels or drainage into the channel is modeled as a simple source-sink term within the two-dimensional topographic model.

The program advances the topographic and open-channel flow models forward in time independently, using an interface between models defined at prescribed intervals by the user. The interface algorithm simply provides coincident water surfaces at grid points where channel flows occur, while conserving mass. Excess channel flows spill into the surrounding grid as a source term contribution, whereas water stored in a grid element spills into a channel as a sink term with respect to the grid.

The model is initiated by flow versus time inputs at grid or open-channel locations, or by runoff distributed over the topography. Generally, small time steps are used because of the nonlinearity of the flow equations and model stability considerations. Using a 80486-type PC processor, a 60-mi^2 DHM simulation, with 1000-ft by 1000-ft grids and channel link network, requires about 1 h to simulate a 24-h storm event.

Computer Requirements. The DHM program is written in standard FORTRAN IV. The program runs on most modern PC-type systems. Because the model can utilize a large database, it can be very effectively used in geographic information system (GIS) applications.

How Program Can Be Obtained. The DHM program can be obtained from the Computational Hydrology Institute, 1510 Red Hill Ave., Tustin, CA 92680.

21.5 WATER-QUALITY MODELS

21.5.1 SWMM—Storm Water Management Model

Overview. The Storm Water Management Model (SWMM) was originally developed for the Environmental Protection Agency in 1971 by Metcalf and Eddy Inc., Water Resources Engineers Inc., and the University of Florida.[24] It was initially designed as a single-event model for the simulation of both runoff quantity and water-quality processes associated with urban runoff and in combined sewer systems for prediction of flows, stages, and pollution concentrations. The model has more recently been adapted to permit continuous simulation of urban storm water flows in addition to single-event simulations.[17,24]

The SWMM program consists of a number of segments or *blocks*. These include:

1. *RUNOFF Block.* Generates runoff from rainfall and routes flows to combining points. Water which infiltrates through the ground surface may also be routed as subsurface flow.

2. *TRANSPORT block.* Routes flow through watershed channels using the kinematic wave method.
3. *EXTRAN block.* Routes channel flow using an explicit finite-difference solution of the St. Venant equations. This is the only block in which water quality cannot be simulated along with runoff.
4. *STORAGE/TREATMENT block.* Routes flow through reservoir-type storages using a storage-routing procedure.
5. *STATISTICS block.* Separates the continuous hydrograph record and pollutographs (concentration as a function of time) into independent storm events. It also calculates statistics and performs frequency analyses.

SWMM permits simulation of a wide range of features of urban hydrology and water-quality processes including rainfall, snowmelt, surface runoff, subsurface contributions to runoff, flow routing, storage, and treatment of flows. SWMM deals with the movement of pollutants from the land surface of the modeled area to combined sewers or storm drainage outfalls. Hydrographs developed in the hydrologic portions are input to the water-quality part of the model. Output is in the form of pollutographs for each pollutant modeled. The hydrographs and the pollutographs are read into the TRANSPORT block where they are combined with the dry weather and infiltrated flow components to produce outflow graphs of water quality and quantity. SWMM predicts concentrations of suspended solids, nitrates, phosphates, and other pollutants in storm water runoff. For each time step, the runoff rate is computed in the hydrologic part of the model. The amount of pollutant removed by runoff is also computed for the time interval, and this can be related to the quantity of runoff to produce a pollutograph. Calibration data are considered essential to permit credible simulations of pollutographs. Without calibration, the computed pollutographs should be considered to provide only relative comparisons between control approaches.

Types of Problems. SWMM has been applied to nearly all aspects of urban hydrology. Huber, Heaney, and Cunningham[16] have published a bibliography of SWMM usage. A SWMM users group holds annual conferences on applications of the SWMM model.

In planning studies, the model can be used to simulate a period of a number of years using long-term precipitation data. The STATISTICS block can be used for a frequency analysis of the long-term record of hydrographs and pollutographs.

The EXTRAN block has been frequently used as a stand-alone modeling element for hydraulic analysis of drainage flows.[25] Hydrographs generated by another model can be supplied as input, and the unsteady-flow behavior of flood control system elements like pumping plants, storage reservoirs, and conveyance channels and conduits can be modeled in detail.

Computer Requirements. SWMM version 4 is distributed for use primarily on PC systems. A math coprocessor is recommended, but not mandatory. For simulations of large problems, execution times can be lengthy, and an 80386- or 80486-based computer is recommended for these applications. The program is written in FORTRAN 77, which permits it to be compiled on mainframe computers and microcomputers.

How Program Can Be Obtained. The program can be obtained from the Center for Exposure Assessment Modeling, Environmental Research Laboratory, Environmental Protection Agency, College Station Rd., Athens, GA 30613. Various software vendors also supply the program.

21.5.2 HSPF—Hydrologic Simulation Program—Fortran

Overview. The Hydrologic Simulation Program—Fortran (HSPF) model simulates both watershed hydrology and water quality.[21] It allows an integrated simulation of land and soil contaminant runoff processes with in-stream hydraulic and sediment-chemical interactions. The program provides a time history of runoff rate, sediment load, and nutrient and pesticide concentration, along with a time history of water quality and quantity at specific points in a watershed. HSPF simulates sand, clay, and silt sediments and a single organic chemical and transformation products of that chemical. Transfer and reaction products modeled are hydrolysis, oxidation, biodegradation, volatilization, and sorption. Resuspension and settling of silts and clays are based on the computed shear stress at the sediment-water interface. Resuspension and settling of sand is determined from the difference between the sand in suspension and the stream's total transport capacity for sand. Calibration of the model requires data from each of the sediment types. Exchanges of chemicals between benthic deposits (bottom sediments) and the overlying water column are also allowed.

The water-quantity routines in HSPF are a FORTRAN version of the Hydrocomp Simulation Program which was developed from the Stanford Watershed Model (SWM) originally developed in 1959. The model has undergone a great deal of modification since its initial development.[8] A commercial version of the Stanford Watershed Model was developed by Hydrocomp, Inc.[9] The most recent versions of the HSPF model contain all of the basic computational routines of the SWM-IV as well as various routing routines and water-quality simulation routines.

HSPF computes a continuous hydrograph of stream flow at the basin outlet. Input is a continuous record of precipitation and evaporation data. Rainfall is distributed into interception loss, rainfall on impervious areas which contributes directly to runoff, and an infiltrated portion. The infiltration is divided into (1) surface runoff and interflow which moves through the upper soil zone to channel flow and (2) flow into the lower soil zone or groundwater storage which contributes to active and inactive groundwater storage. The model utilizes three soil moisture zones: an upper soil zone, a lower soil zone, and a groundwater storage zone. Rapid runoff is accounted for in the upper zone. Both the upper and lower zones influence factors such as overland flow, infiltration, and groundwater storage. Water that is computed as moving into the lower zone can move into deep groundwater storage, some of which can become base flow to the stream. Total stream flow is a combination of overland flow, interflow, and groundwater flow.

The program user must supply parameters for each of the various processes. More than 20 parameters are needed to describe merely the hydrologic parameters, some of which cannot be directly measured (such as the various soil moisture parameters). Without calibration data, it can be difficult to verify the flows computed by this model.

Computer Requirements. The HSPF model is available for use on PC systems with 640 Kbytes of RAM and a hard disk. A math coprocessor is required to run the program. Because the program is written in FORTRAN 77, it can be compiled on mainframe computers and microcomputers.

How Program Can Be Obtained. The current public domain version of the program (HSPF) was developed for the U.S. Environmental Protection Agency. The program can be obtained from the EPA's Center for Exposure Assessment Modeling, Environmental Research Laboratory, Athens, GA 30613. Various software vendors also supply the program.

21.5.3 QUAL2E—Stream Water-Quality Model

Overview. QUAL2E is the Enhanced Stream Water Quality Model and is the latest in a series of water-quality management models initially developed by the Texas Water Development Board in the 1960s. QUAL-I was required by the Environmental Protection Agency during the 1970s for the development of basin-specific water-quality models. Several improved versions of QUAL were developed as part of this effort, and the QUAL-II series of models have been widely used after extensive review and testing. Present support is by the Environmental Protection Agency's Center for Exposure Assessment Modeling (CEAM).

QUAL2E simulates several water-quality constituents in branching stream systems. The model uses a finite-difference solution of the advective-dispersive mass transport and reaction equation. A stream reach is divided into a number of subreaches, and for each subreach a hydrologic balance in terms of discharge, a heat balance in terms of temperature, and a materials balance in terms of concentration is written. Both advective and dispersive transport processes are considered in the materials balance. Mass is gained or lost from the subreach by transport processes and by waste discharges and withdrawals. Mass can also be gained or lost by internal processes such as benthic sources or biological transformations.

The program simulates changes in conditions in time by computing a series of steady-flow water surface profiles, that is, the stream is conceived as a series of reaches, with water passing from one reach to the next, and water-quality processes take place in each reach separately from its neighbors upstream and downstream. The basis for mass transport calculations is the stream-flow rate; velocity, cross-sectional area, and water depth are computed from the flow, and the mass-flow rate of constituent moving from one tank to the next is found as the product of the stream-flow rate and the mass concentration of the constituent in the water. Time-varying computations are made by the program from climatological variables that primarily affect temperature and algal growth. QUAL2E simulates the major interactions of the nutrient cycles, algal production, benthic and carbonaceous demand, atmospheric reaeration, and their effect on the dissolved oxygen balance. The program determines mass balances for conservative minerals, coliform bacteria, and nonconservative constituents such as radioactive substances. QUAL2E uses chlorophyll *a* as the indicator of planktonic algae biomass. The nitrogen cycle is divided into four compartments: organic nitrogen, ammonia nitrogen, nitrite nitrogen, and nitrate nitrogen. In a similar manner, the phosphorus cycle is modeled by using two compartments. The primary internal sink of dissolved oxygen in the model is biochemical oxygen demand (BOD). The major sources of dissolved oxygen are surface reaeration, algal photosynthesis, and atmospheric reaeration.

Description of a stream network in QUAL2E requires dividing the streams into headwaters, reaches, and junctions. For each reach, as many as 26 physical, chemical, and biological parameters must be defined. The modeler must define more than 100 individual inputs when developing model input data. Model calibration and evaluation is not an easy task with such a complex model, and the developers of the program have used the principles of uncertainty analysis to assist users in the model calibration process. A version of the program named QUAL2E-UNCAS incorporates three uncertainty analysis techniques: sensitivity analysis, first-order error analysis, and Monte Carlo simulation. Selected input variables are changed by the program, and the user selects the specific variables and locations on the stream where the uncertainty analysis is to be applied.

Computer Requirements. The QUAL2E model is designed to be used on PC systems with 640 Kbytes of RAM and a hard disk. The program is written in FORTRAN 77, which permits it to be compiled on mainframe computers and microcomputers.

How Model May Be Obtained. The QUAL2E model is available from the U.S. Environmental Protection Agency, Center for Exposure Assessment Modeling (CEAM), Environmental Research Laboratory, Athens, GA 30613.

21.5.4 WASP4—Water-Quality Simulation Program

Overview. WASP4 is a simulation program for modeling contaminant fate and transport in surface waters. There has been a series of models designated as WASP—WASP4 is the latest in this series. WASP4 is designed for use by modelers who have a background in water-quality modeling. It is a sophisticated model that permits a good deal of flexibility in its application. WASP4 can be used to simulate one-, two-, or three-dimensional flows. The program is designed to allow users to substitute their own subroutines in the program. WASP4 input can also be linked to other models, such as HSPF, whose output files can be reformatted and read by WASP.

Two water-quality models are provided with WASP: (1) TOXI4, a simulation of transport and transformation of toxic substances and (2) EUTRO4, a model of dissolved oxygen and phytoplankton dynamics affected by nutrients and organic material.

WASP represents a body of water as a series of computational segments. Environmental properties and chemical concentrations are assumed to be spatially constant within each segment. Segment volumes and type (surface water, subsurface water, surface benthic, and subsurface benthic) and hydraulic coefficients must be specified. WASP4 uses several mechanisms for describing transport: advection and dispersion in the water column; advection and dispersion in pore water; settling, resuspension, and sedimentation of one to three classes of solids; evaporation; and precipitation.

The simulation of advection requires specification of each inflow or circulation pattern for the flow routed through each water-column segment. The flow can vary with time. Dispersion requires cross-section areas for model segments, characteristic mixing lengths, and dispersion coefficients to be specified.

The user must also specify loads, boundary concentrations, and initial concentrations. Only particulate concentrations are transported as solids, and only dissolved concentrations can be transported as pore water.

Advection and dispersion between each segment are computed for each variable by the model, and exchange with surfical benthic segments is determined. Sorbed or particulate fractions can settle through the water column and deposit to or erode from the surfical benthic segments. Dissolved materials may migrate downward or upward through the bed by percolation and pore water diffusion. Sorbed materials may migrate downward or upward as a result of sedimentation or erosion.

The TOXI4 Component. TOXI4 simulates transport and transformation of one to three chemicals and one to three types of particulate material. The model is composed of as many as six systems—three chemical and three solid—for which the WASP4 mass-balance equation is solved. The chemicals may be independent, or they may be linked through chemical reactions, such as a parent-compound–daughter product sequence.

Transfer processes defined in the model include sorption, ionization, and volatilization. Transformation processes include biodegradation, hydrolysis, photosynthesis, and chemical oxidation. Sorption and ionization are treated as equilibrium reactions. All processes are described using rate equations which may be described either by first-order constants or by second-order chemical-specific constants and time-varying environment-specific parameters that also vary in space.

Sediment is treated as a conservative constituent that is advected and dispersed among water segments. It may settle to or erode from benthic segments and may move between benthic segments through erosion and deposition.

The EUTRO4 Component. EUTRO4 simulates the transport and transformation of as many as eight state variables in the water column and sediment bed, including dissolved oxygen, carbonaceous biochemical oxygen demand, phytoplankton carbon and chlorophyll *a*, ammonia, nitrate, organic nitrogen, organic phosphorus, and orthophosphate. The model can be used to simulate any or all these variables and the interactions between them. Each variable may exist in both dissolved and particulate phases in each stream segment, as specified by the user. Settling and resuspension of organic solids, phytoplankton solids, and inorganic solids may be specified.

The program can deal with simulations at various levels of complexity. The simplest level allows the computation of final BOD, dissolved oxygen (DO), and sediment oxygen demand (SOD). At the next level, BOD is divided into carbonaceous (CBOD) and nitrogenous (NBOD) fractions. The third level is a linear dissolved oxygen balance influenced by photosynthesis and respiration of phytoplankton and nitrification of ammonia to nitrate besides SOD and CBOD. The fourth level adds the phosphorus cycle and simulates phytoplankton dynamics. Level 5 adds benthic interactions.

WASP4 and its predecessors have been used in a wide range of studies for the U.S. Environmental Protection Agency. Some of the applications have been validated by using field data or verified by model experiments.

Computer Requirements. The WASP4 model is written in FORTRAN 77, and versions are maintained both for PC systems (available on diskette) and for DEC and VAX systems using the VMS operating system (on nine-track magnetic tape). To use the model on PC systems, 640 Kbytes of RAM and a hard disk are required. A math coprocessor is desirable.

How Model May Be Obtained. The WASP4 models are available from the U.S. Environmental Protection Agency, Center for Exposure Assessment Modeling (CEAM), Environmental Research Laboratory, Athens, GA 30613.

21.5.5 AGNPS—Agricultural Non-Point-Source Pollution Model

Overview. The Agricultural Non-Point Source (AGNPS) model[37] simulates runoff water quality from agricultural watersheds. It was developed by the U.S. Agricultural Research Service (ARS) in cooperation with the Minnesota Pollution Control Agency and the U.S. Soil Conservation Service (SCS).[38] Other models of this type (such as SWRRB discussed above) are limited in the size of watershed which can be used because of the large amount of input data required. AGNPS is designed to be a simple, easy-to-use model which can run on a PC and deal with watersheds ranging in size from a few acres to 50,000 acres (20,000 hectares) or larger. Estimates of runoff water quality can be made with AGNPS to evaluate potential pollution problems for a watershed, and remedial measures can be recommended on the basis of an assessment of the effects of applying alternative management practices. Data representing these management alternatives can be used as program input, and the resulting watershed responses can be evaluated and compared.

AGNPS is an event-based model which uses geographic data cells of 1 to 40 acres (0.4 to 16 hectares) to represent land surface conditions. Within the framework of the cells, runoff characteristics and transport processes for sediment, nutrients, and chemical oxygen demand (COD) are simulated for each cell. Flows and pollutants are routed through the channel system to the basin outlet. Point source inputs (such as nutrient COD from animal feedlots) can also be simulated and combined with the non-point-source contributions.

Basic model components include hydrology, erosion, sediment transport, and chemical transport. In the hydrology component, runoff volume is calculated by the SCS curve number procedure. Peak flow rate is estimated by using an empirical equation which takes into account drainage area, channel slope, runoff volume, and watershed length-width ratio. Erosion is computed from a modified form of the universal soil loss equation. Soil loss is calculated for each cell of the watershed. Eroded soil and sediment yield are subdivided into particle size classes; sediment routing is based on the effective transport capacity of the stream channels. In the chemical transport part of the model the transport of nitrogen, phosphorus, and COD is calculated throughout the watershed. Chemical transport calculations are divided into soluble and sediment-adsorbed phases. COD is assumed to be soluble and to accumulate without losses.

In AGNPS applications, a uniform grid of cells is placed over the watershed. Storm rainfall and runoff produce erosion through sheet and rill erosion processes, soil loss and sediment yield is calculated for each cell, and the upland transport of sediment and nutrients is determined for each cell outlet. Calculations for AGNPS occur in three loops: initial calculations for all cells are made in the first loop, runoff volume is calculated for cells containing impoundments and sediment yields for cells that no other cells drain into are computed in the second loop, and sediments and nutrients are routed through the watershed in the third loop.

Required Input. Model input consists of watershed data (area, number of cells, precipitation) and cell parameter data. Twenty-two cell parameters are used. These include: SCS curve number, average land slope, slope shape factor, field slope length, channel data (slope, length, side slope, and roughness), universal soil-loss equation data (erodibility, cropping, and practice factors), soil texture, fertilization level, point-source indicator, gully source parameters, chemical oxygen demand factor, and a channel indicator for the presence of a defined channel in a cell.

Computer Requirements. AGNPS is written in FORTRAN 77 and consists of about 2000 lines of source code. On a minicomputer system, the model requires 400 Kbytes of storage memory for a watershed containing 3200 cells. The PC version of the program operates from a shell which provides access to a full-screen data editor, help screens, viewing of output on the computer monitor, and control of printed output. Graphical output display is also available, and the program supports several types of graphics printers and plotters.

How Program Can Be Obtained. The program is available from the North Central Soil Conservation Research Laboratory, Soil Conservation Service, U.S. Department of Agriculture, Morris, MN 56267. The AGNPS developers publish a quarterly AGNPS newsletter and conduct several training sessions throughout the United States each year.

21.5.6 MIKE11—Microcomputer-Based Modeling System—Rivers and Channels

Overview. MIKE11 was developed by the Danish Hydraulic Institute for the simulation of flows, water levels, and transport of sediment and dissolved or suspended materials.[10] MIKE11 is a general-purpose microcomputer-based model that simulates not only rainfall-runoff processes, but also river hydraulics, sediment transport, and water quality. MIKE11 can be used in design, management, and operation of river systems and channel networks.

MIKE11 is based on the Danish Hydraulic Institute program System 11, which provides a similar set of modeling system capabilities for the mainframe environment.

The MIKE11 model consists of several individual modules, allowing the user to add specific modules for various types of hydrologic simulation as the need for these features arises in the application. The model is menu-driven. It is configured with a core component termed the *basis module* plus a series of other add-on modules. The basis module includes the menu portion that deals with data handling and program execution; a catchment database that includes river cross-section data; a database for rainfall time series and water level and discharge data; computational modules for rainfall-runoff simulation and for river flow; and a module that permits plotting of input and output data. Figure 21.5.1 shows a simplified flow chart of the system.

The catchment and stream channel network system is modeled by the rainfall-runoff module. Complex river systems can be simulated. Runoff computations are based on a lumped-conceptual-type model that continuously accounts for the moisture content in four storage zones: (1) surface storage, (2) lower zone storage, (3) upper groundwater storage, and (4) lower groundwater storage. Runoff to stream channels is assumed to consist of overland flow, interflow, and base flow. The river

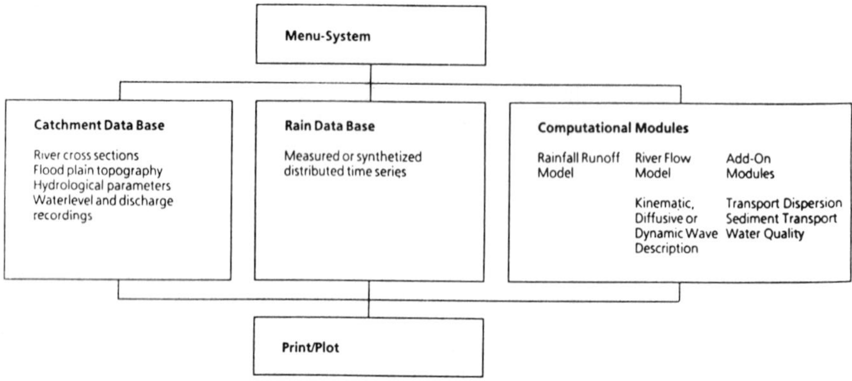

FIGURE 21.5.1 Simplified flow chart of the MIKE11 program. *(Source: Danish Hydraulic Institute.[10] Used with permission.)*

flow module permits the use of a variety of computational procedures. The full nonlinear one-dimensional unsteady-flow equations are normally used, while simplified channel routing equations (kinematic wave or diffusion wave equations) can be employed as deemed suitable in specific parts of the full model. Complex channel configurations can be accommodated, including looping channels. Channel computations can also include lateral discharges, free and submerged flow at weirs, flooding and drying of overflow areas, flow over embankments, and two-dimensional floodplain flows. Culverts and other stream structures can also be simulated. Irregular cross-section geometry can be used, flow-related roughness and local head losses can be employed, and the model can deal with both subcritical and supercritical flow conditions.

Sediment Transport Module. A sediment budget accounts for erosion and deposition, and the resulting changes in model geometry are used in the hydrodynamic calculations by the model. The model can account for sediment transfer between the river and floodplains. MIKE11 also has an option for computing bedform dimensions and the resistance coefficients associated with bedform roughness.

Transport-Dispersion Module. This module solves the one-dimensional conservation of mass equation. Transport and dispersion are simulated for any dissolved or suspended material. The behavior of conservative materials with linear decay characteristics can be simulated. Cohesive sediment transport can also be simulated.

Water-Quality Module. This module is an extension of the transport-dispersion module. It simulates the reaction processes of multicompound systems, and models a wide variety of biochemical interaction processes, ranging from simple BOD and DO computations to multicompound simulations including nutrients, macrophytes, and plankton.

MIKE11 has been used in the analysis of a number of river systems throughout the world. Applications include flood control planning, reservoir operation, design of river structures, irrigation system hydraulics, studies of tides and storm surges in rivers and estuaries, dam-break simulation, and detailed design of channel systems.

Required Input. Required data for the catchment and river data module include hydrologic parameters, river cross sections, floodplain topography, and discharge and water level records. For the rain database, either measured rainfall time series data or synthesized rainfall time series are required.

Computer Requirements. MIKE11 is specifically designed for the PC environment and is supplied in compiled form to run as a single-user program on 16-bit and 32-bit systems under MS-DOS and the CPM/86 operating systems or in a multiuser system under UNIX. MIKE11 supports plotters using HP graphics language conventions.

How Program Can Be Obtained. The program can be purchased from the Danish Hydraulic Institute, Agern Alle 5, DK-2970 Horsholm, Denmark.

21.6 COMPUTER MODELS AVAILABLE THROUGH THE HOMS PROGRAM

The Hydrological Operational Multipurpose System (HOMS) of the World Meteorological Organization (WMO) provides a program for the transfer of technology used by hydrologists throughout the world. A number of computer programs for hydrologic analysis are included in the series of HOMS *components.* Also available as HOMS components are technical manuals and descriptions of hydrologic instruments. The HOMS Reference Manual provides guidance in the use of the components and user requirements. The manual provides what are termed *sequences,* or logical aggregations of components that are compatible with each other and may be used together.

There are about 430 components available; a number of these are hydrologic models of the type discussed in this chapter. Summaries are available for each component, either from the HOMS Office, WMO, Geneva, Switzerland (address below), or from HOMS National Reference Centers (HNRC) located in over 100 countries. The address of the HNRC for a particular country can be obtained from the WMO Secretariat.

There are a number of models which are described by HOMS. HOMS lists hydrologic models in two categories: (1) hydrologic models for forecasting and design and (2) analysis of data for planning, design, and operation of water resources systems. A list of some of the models available through HOMS follows, using the HOMS classification number for the component.

Section J: Models for Forecasting Stream Flow

J04.1.01 Tank model (Japan)

J04.1.04 Snowmelt-runoff model (United States)

J04.1.06 Microcomputer-based flood forecasting system (Philippines)

J04.2.01 Conceptual watershed model for flood forecasting (China)

J04.2.02 Conceptual watershed model (the HBV model) (Sweden)

J04.2.03 Model to forecast rainfall floods (Russia)

J04.3.01 Sacramento soil moisture accounting model (SAC-SMA) (U.S. National Weather Service)

J04.3.03 Snow accumulation and ablation model (SNOW-17) (U.S. National Weather Service)

J15.2.01 Streamflow synthesis and reservoir regulation (SSARR) (U.S. Army Corps of Engineers)

J15.2.02 Nonlinear cascade hydrologic model (Czechoslovakia)

J15.2.03 CLSX (constrained linear system extended) model (Italy)

J15.3.02 Multipurpose unsteady-flow simulation system (MUFSYS 3) (Czechoslovakia)

J15.3.05 General-purpose flood forecasting modeling system—NAMS11/FF (Denmark)

Section K: Hydrologic Analysis Models

K22.1.04 Computer program for structure site analysis—DAMS2 (United States)

K22.1.05 Model CA HYDRO (Colombia)

K22.2.02 Semiconceptual watershed model (Italy)

K22.2.05 Rainfall-runoff model for medium-sized urban basins (France)

K22.3.03 General-purpose rainfall-runoff model—NAM (Denmark)

K35.3.07 WSP2 computer program for water surface profiles (United States)

K35.3.09 Microcomputer modeling package for rivers and estuaries—MIKE11 (Denmark)

K55.2.06 Water-quality simulation model—WATQUAL (United States)

K55.3.01 Storage, treatment, overflow, runoff model—STORM (United States)

K55.3.02 Water quality for river-reservoir systems—WQRRS (United States)

How Models May Be Obtained. For further information on HOMS and the HOMS components, including the address of the HOMS National Reference Centers in your country, write to: HOMS Office, Hydrology and Water Resources Department, WMO Secretariat, Case postale No. 2300, *CH-1211 GENEVA 2,* Switzerland. In the United States the HNRC is located at the National Weather Service, NOAA, 1325 East-West Highway, Silver Spring, MD 20910.

REFERENCES

1. Abbott, M. B., et al., "An Introduction to the European Hydrological System—Système Hydrologique Européen, 'SHE'—1: History and Philosophy of a Physically-Based, Distributed Modelling System," *J. Hydrol.,* vol. 87, pp. 45–59, 1986*a.*
2. Abbott, M. B., "An Introduction to the European Hydrological System—Système Hydrologique Européen, 'SHE'—2: Structure of a Physically-Based, Distributed Modelling System," *J. Hydrol.,* vol. 87, pp. 61–77, 1986*b.*
3. Alley, W. M., and P. E. Smith, "Distributed Routing Rainfall-Runoff Model—Version II," U.S. Geological Survey open file report 82-344, 1982.
4. Arnold, J. G., J. R. Williams, A. D. Nicks, and N. B. Sammons, *SWRRB—A Basin Scale Model for Soil and Water Resources Management,* Texas A&M University Press, College Station, Texas, 1990.
5. Bathurst, J. C., "Physically-Based Distributed Modelling of an Upland Catchment using the Système Hydrologique Européen." *J. Hydrol.,* vol. 87, pp. 79–102, 1986.
6. Bathurst, J. C., "Sensitivity Analysis of the Système Hydrologique Européen for an Upland Catchment," *J. Hydrol.,* vol. 87, pp. 79–102, 1986.
7. Bradley, J. N., "Hydraulics of Bridge Waterways," U.S. Federal Highway Administration, Hydraulic Design Series No. 1, U.S. Department of Transportation, Washington, D.C., 1970.
8. Crawford, N. H., and R. K. Linsley, Jr., "Digital Simulation in Hydrology: Stanford Watershed Model IV," Department of Civil Engineering, Stanford University, tech. report no. 39, July 1966.
9. Crawford, N. H., "Studies in the Application of Digital Simulation to Urban Hydrology," Hydrocomp International, Palo Alto, Calif., September 1971.
10. Danish Hydraulic Institute, "MIKE11—a Microcomputer Based Modelling System for Rivers and Channels, Users Manual," Horsholm, Denmark, 1990.
11. Feldman, A. D., "HEC Models for Water Resources System Simulation: Theory and Experience," in *Advances in Hydroscience,* vol. 12, V. T. Chow (ed.), Academic Press, New York, pp. 297–423, 1981.
12. Fread, D. L., "A Technique for Implicit Flood Routing in Rivers with Major Tributaries," *Water Resour. Res.,* vol. 9, no. 4, pp. 918–926, 1973.
13. Fread, D. L., "Channel Routing," chap. 14 in *Hydrological Forecasting,* M. G. Anderson and T. P. Burt (eds.), Wiley, New York, pp. 437–503, 1985.

14. Fread, D. L., and J. M. Lewis. "FLDWAV: A Generalized Flood Routing Model," *Proceedings of National Conference on Hydraulic Engineering,* ASCE, Colorado Springs, Colo., 1988.
15. Hromadka, T. V., and C. C. Yen, "A Diffusion Hydrodynamic Model," U.S. Geological Survey Water Resources Investigations report 87-4137, U.S. Government Printing Office, Washington, D.C., 1987.
16. Huber, W. C., J. P. Heaney, and B. A. Cunningham, "Storm Water Management Model (SWMM) Bibliography," EPA/600/3-85/077, Environmental Protection Agency, Athens, Ga., September 1988.
17. Huber, W. C., et al., "Storm Water Management Model User's Manual Version III," EPA-600/2-84-109a, Environmental Protection Agency, Athens, Ga., November 1981.
18. Hydrologic Engineering Center, "HEC-1 Flood Hydrograph Package," *Program Users Manual,* U.S. Army Corps of Engineers, Davis, Calif., 1990*a.*
19. Hydrologic Engineering Center, "HECDSS User's Guide and Utility Program Manuals," U.S. Army Corps of Engineers, Davis, Calif., 1990*b.*
20. Hydrologic Engineering Center, "HEC-2 Water Surface Profiles," *Program Users Manual,* rev. February 1991, U.S. Army Corps of Engineers, Davis, Calif., 1991.
21. Johanson, R. C., J. C. Imhoff, and H. Dana, Jr., "Users Manual for Hydrological Simulation Program—Fortran (HSPF)," EPA/9-80-015, Environmental Protection Agency, Athens, Ga., April 1980.
22. Loague, K. M., and R. A. Freeze, "A Comparison of Rainfall-Runoff Modeling Techniques on Small Upland Catchments," *Water Resour. Res.,* vol. 21, pp. 229–248, 1985.
23. McCuen, R. H., *Hydrologic Analysis and Design,* Prentice-Hall, Englewood Cliffs, N.J., 1989.
24. Metcalf and Eddy, "Storm Water Management Model," vol. 1. Environmental Protection Agency, Water Resources Engineers, University of Florida, Gainsville, Fla., 1971.
25. Roesner, L. A., R. P. Shubinski, and J. A. Aldrich, "Storm Water Management Model User's Manual Version III: Addendum I, EXTRAN," EPA-600/2-84-109b, Environmental Protection Agency, Cincinnati, Ohio, November 1981.
26. Schneider, V. T., et al., "Computation of Backwater and Discharge at Width Constrictions of Heavily Vegetated Flood Plains," U.S. Geological Survey Water-Resources Investigations 76-129, U.S. Department of the Interior, Washington, D.C., 1977.
27. Shearman, J. O., "Users Manual for WSPRO: A Model for Water Surface Profile Computations," U.S. Geological Survey, Federal Highway Administration, U.S. Department of Transportation, Washington, D.C., 1988.
28. Shearman, J. O., W. H. Kirby, V. R. Schneider, and H. N. Filippo, "Bridge Waterways Analysis Model," U.S. Federal Highway Administration report no. FHWA/RD86-108, U.S. Department of Transportation, Washington, D.C., 1986.
29. Soil Conservation Service, "Hydrology," Supplement A to Sec. 4, *National Engineering Handbook,* U.S. Department of Agriculture, Washington, D.C., August 1972.
30. Soil Conservation Service, "Computer Program for Project Formulation Hydrology," technical release no. 20, U.S. Department of Agriculture, June 1973.
31. Stall, J. B., and M. L. Terstriep, "Storm Sewer Design—An Evaluation of the RRL Method," Environmental Protection Agency technology series EPA-R2-72-068, October 1972.
32. Terstriep, M. L., and J. B. Stall, "The Illinois Urban Drainage Simulator, ILLUDAS," bulletin 58, Illinois State Water Survey, Urbana, Ill., 1974.
33. U.S. Federal Highway Administration, "Hydraulic Charts for the Selection of Highway Culverts," hydraulic engineering circular no. 5, U.S. Department of Transportation, Washington, D.C., 1980.
34. Watkins, L. H., "The Design of Urban Sewer Systems," road research technical paper no.

55, Department of Scientific and Industrial Research. Her Majesty's Stationery Office, London, 1962.

35. Williams, J. R., and H. D. Berndt, "Sediment Yield Prediction Based on Watershed Hydrology," *Trans. Am. Soc. Agric. Eng.,* ASAE vol. 20, no. 6, pp. 1100–1104, 1977.
36. Williams, J. R., "SPNN, A Model for Sediment, Phosphorous, and Nitrogen Yield From Agricultural Basins," *Water Resour. Bull.,* vol. 16, no. 5, pp. 843–848, 1980.
37. Young, R. A., C. A. Onstad, D. D. Bosch, and W. P. Anderson, "AGNPS, Agricultural Non-Point Source Pollution Model, A Watershed Analysis Tool," conservation research report 35, Agricultural Research Service, U.S. Department of Agriculture, Washington, D.C., 1987.
38. Young, R. A., C. A. Onstad, D. D. Bosch, and W. P. Anderson, "AGNPS, A Non-Point Source Pollution Model for Evaluating Agricultural Watersheds," *J. Soil Water Conserv.,* vol. 44, no. 2, pp. 168–173, 1989.

CHAPTER 22
COMPUTER MODELS FOR SUBSURFACE WATER

Mary P. Anderson
University of Wisconsin
Madison, Wisconsin

David S. Ward
Geotrans, Inc.
Sterling, Virginia

Eric G. Lappala
Harding Lawson Associates
Princeton Junction, New Jersey

Thomas A. Prickett
Thomas A. Prickett and Associates
Urbana, Illinois

22.1 INTRODUCTION

22.1.1 Types of Models

Computer models are essential for analyzing complex problems of subsurface flow and transport because they provide a quantitative framework for synthesizing a large set of parameters that describe spatial variability within the subsurface environment, as well as spatial and temporal trends in hydrologic parameters and stresses, and historical trends in water levels and concentrations. While some decisions can be made using best engineering or best geologic judgment, human reasoning alone is inadequate to synthesize the many factors involved in most complex problems.

Computer modeling refers to the application of a computer program, or *code,* to solve a set of equations that forms a *mathematical model* of the physical and chemical processes occurring in porous media. The code is generic whereas a model includes a set of *boundary* and *initial conditions* as well as a site-specific nodal grid and site-specific parameter values and stresses. A code is written once but a new model is designed for each application.

The codes used most often are those that simulate flow, with or without solute transport, in *continuous porous media* (in contrast to fractured media). These codes are the focus of this chapter. Other types of models sometimes used in subsurface analysis include those that simulate *multiphase flow, fracture flow,* or complex chemistry *(geochemical models)*. Other specialized models include *optimization models,* and models used to solve the *inverse problem*. These specialized models are discussed briefly in Sec. 22.7.

To help provide information on the large number of codes developed since the 1960s, the U.S. Environmental Protection Agency (U.S. EPA) established a clearinghouse known as the International Ground Water Modeling Center (IGWMC), which conducts short courses on modeling and distributes selected software. Other distributors of software include the Scientific Software Group, the U.S. Geological Survey (USGS), and the Energy Science and Technology Software Center (ESTSC). Addresses of these and other distributors of groundwater modeling software are given in Sec. 22.8.

Selection of Codes. The flow and solute transport codes discussed in this chapter are divided into five groups. *Groundwater flow codes* (Sec. 22.2) simulate the flow of groundwater. *Codes for contaminant transport in groundwater* (Sec. 22.3) simulate the advection, dispersion, and chemical reactions of constituents dissolved in groundwater. *Particle-tracking codes* (Sec. 22.4) trace flow paths. *Variably saturated flow codes* (Sec. 22.5) simulate the flow of water in both the unsaturated and saturated zones. *Codes for contaminant transport in variably saturated porous media* (Sec. 22.6) simulate advection, dispersion, and chemical reactions of constituents in water in the continuum representing the unsaturated and saturated zones.

Representative numerical models were selected in each of these five classes for presentation in this chapter. Selection of just a few codes was a difficult task. The featured codes are not necessarily the best codes in their class, but they are among the codes that are used most often in professional practice. In those cases where codes that were developed recently are featured, the selection was made because the code interfaces with another code that has proved popular. Furthermore, the selection of codes was focused on those that are well-accepted by the research and consulting communities, are either user-friendly or relatively well-documented, and are readily available for $1000 or less. The numerical codes featured in this chapter are listed in Table 22.1.1. All except BIO1D are available with source code. The USGS provides only the source code for the models it distributes. Other distributors listed in Sec. 22.8 typically furnish both source code and executable code for IBM-compatible and/or Macintosh microcomputers.

Some attention was paid to selecting families of codes so that users familiar with the structure of one code could move more easily to a related code that builds on this structure. Our purpose was not to compile an exhaustive list of codes, inasmuch as this type of information is available elsewhere.[95,148,149] The selection of featured codes is necessarily biased toward those with which the authors are most familiar. A few representative codes that compute analytical solutions and other numerical codes are also mentioned within each section. These codes do not necessarily meet the restrictions set for featured codes.

The degree of support provided for the featured codes varies widely and depends partly on the distributor. For example, the USGS distributes and maintains its codes but does not provide technical support in the form of consultation on problems encountered when using the code. Codes distributed by others may come with some degree of technical support. The code MODFLOW, for example, is available from several distributors including the USGS, each of whom provides a different level of

TABLE 22.1.1 Connections Among the Numerical Codes Featured in This Chapter

	Families of codes				
Groundwater flow		PLASM		MODFLOW	
Transport in groundwater		RNDWALK	MT3D		
Particle tracking	FLOWPATH			PATH3D	MODPATH
Variably saturated flow		FEMWATER			VS2D
Variable saturated transport		FEMWASTE			VS2DT
	Single codes				
Groundwater flow	AQUIFEM-1				
Transport in groundwater	BIO1D, USGS MOC, SWIFT/386				
Variably saturated flow	UNSAT2				
Transport in variably saturated media	SUTRA				

support. The user who anticipates needing technical support should check the current status of support provided before purchasing the code.

The summaries of codes in this chapter are intended to provide the reader with enough information to determine whether the code might be useful. Summaries of featured codes are arranged roughly in order of increasing complexity and/or sophistication. It must be emphasized that this software is very different from a word processor or a spreadsheet in that intelligent use of this software requires more than reading the user's manual. Correct application of these codes requires a general background in modeling[7,14,64,65,71,157] as well as a background in subsurface hydrology. Use of codes that are not interactive will also require careful study of the user's manual. Training courses in groundwater modeling are offered through the IGWMC, the National Ground Water Association [6375 Riverside Dr., Dublin, OH; tel. (614) 761-1711; fax (614) 761-3446.], and the E^3 Institute [Columbus, OH; tel. (614) 792-0005; fax (614) 792-0006.]

22.1.2 Mathematical Formulation

Governing equations are discussed and derived in other chapters of the handbook, but for convenience are presented here in the form they are typically used in the codes discussed in this chapter. Boundary and initial conditions, as well as parameter values, are also discussed elsewhere in this handbook.

Groundwater Flow. In three dimensions, the governing equation is

$$\frac{\partial}{\partial x}\left(K_x \frac{\partial h}{\partial x}\right) + \frac{\partial}{\partial y}\left(K_y \frac{\partial h}{\partial y}\right) + \frac{\partial}{\partial z}\left(K_z \frac{\partial h}{\partial z}\right) = S_s \frac{\partial h}{\partial t} + W \qquad (22.1.1)$$

where h is head (L); K_x, K_y, K_z, denote hydraulic conductivity (L/T) in the x, y, and z directions (L); S_s is specific storage ($1/L$); and W is a general term for sources and sinks of water ($1/T$). In two dimensions, the governing equation is usually written with two components of the transmissivity tensor, T_x and T_y, allowing the equation to be applied to both confined and unconfined conditions by using the Dupuit assumptions, which require that $T_x = K_x h$. These concepts are discussed further in Chap. 6.

Variably Saturated Flow. In three dimensions, the governing equation is

$$\frac{\partial}{\partial x}\left[K(\theta)\frac{\partial h}{\partial x}\right]+\frac{\partial}{\partial y}\left[K(\theta)\frac{\partial h}{\partial y}\right]+\frac{\partial}{\partial z}\left[K(\theta)\frac{\partial h}{\partial z}\right]=C(\theta)\frac{\partial h}{\partial t}+W \quad (22.1.2)$$

where h is head (L); $K(\theta)$ is the unsaturated hydraulic conductivity (L/T), which is a function of moisture content θ; $C(\theta)$ is the specific moisture capacity ($1/L$) equal to $d\theta/d(\psi)$, where ψ is pressure head; and W is a general sink/source term ($1/T$) or forcing function, as in Eq. (22.1.1). These concepts are discussed further in Chap. 5.

Solute Transport. A solute transport model requires that the velocity distribution in the problem domain be known. For this reason, solute transport codes generally include a flow code which calculates the head distribution, from which seepage velocities are calculated using Darcy's law:

$$v_x = -\frac{K_x}{n_e}\frac{\partial h}{\partial x} \quad (22.1.3a)$$

$$v_y = -\frac{K_y}{n_e}\frac{\partial h}{\partial y} \quad (22.1.3b)$$

$$v_z = -\frac{K_z}{n_e}\frac{\partial h}{\partial z} \quad (22.1.3c)$$

where v_x, v_y, and v_z are the components of the average linear velocity and n_e is the effective porosity. In unsaturated porous media, hydraulic conductivity is $K(\theta)$ and θ is used instead of n_e.

The advection-dispersion equation is used to simulate the transport of solutes influenced by advection, dispersion, and chemical reactions, through both saturated and unsaturated porous media. The three-dimensional form of this equation is

$$\frac{\partial}{\partial x}\left(D_{xx}\frac{\partial c}{\partial x}+D_{xy}\frac{\partial c}{\partial y}+D_{xz}\frac{\partial c}{\partial z}\right)+\frac{\partial}{\partial y}\left(D_{yx}\frac{\partial c}{\partial x}+D_{yy}\frac{\partial c}{\partial y}+D_{yz}\frac{\partial c}{\partial z}\right)+\frac{\partial}{\partial z}\left(D_{zx}\frac{\partial c}{\partial x}+D_{zy}\frac{\partial c}{\partial y}+D_{zz}\frac{\partial c}{\partial z}\right)-\frac{\partial}{\partial x}(cv_x)-\frac{\partial}{\partial y}(cv_y)-\frac{\partial}{\partial z}(cv_z)+\frac{W(c-C')}{n_e}-R_d\lambda c=R_d\frac{\partial c}{dt} \quad (22.1.4)$$

where c is concentration (M/L^3), D with various subscripts represents components of the dispersion coefficient (L^2/T), and C' is a known source or sink concentration. In applications to unsaturated media, n_e in Eq. (22.1.4) is equal to θ. The chemical reaction terms include a retardation factor R_d, and a first-order rate constant λ. The technical background to this material is presented in Chap. 15 for transport in the unsaturated zone and in Chap. 16 for transport in the saturated zone.

Particle Tracking. Codes that track the movement of solute solely by advection are known as *particle-tracking codes.* These codes typically do not compute solute concentrations; they simply trace path lines. Numerical particle-tracking codes require input of heads from a flow model. Velocities at nodal points are computed by using Eq. (22.1.3). The simulation begins when imaginary particles are introduced into the flow field. Velocities are estimated at particle positions by interpolation of nodal velocities.[7] Particles are tracked through the flow field to delineate path lines by solving

$$\frac{dx}{dt} = v_x \tag{22.1.5a}$$

$$\frac{dy}{dt} = v_y \tag{22.1.5b}$$

$$\frac{dz}{dt} = v_z \tag{22.1.5c}$$

Numerical codes are designed to remove particles as they arrive at exit boundaries and sinks. There are four integration methods[7] commonly used to solve Eq. (22.1.5): semianalytical, Euler, Runge-Kutta, and Taylor series expansion. The semianalytical scheme avoids numerical errors associated with discretization of time. The other three methods are numerical schemes, of which the Runge-Kutta method is usually preferred.

Particle-tracking codes are popular because they are much easier to use than solute transport codes, but still allow calculation of contaminant pathways and travel times. Chemical reactions that cause retardation can be simulated by using a retardation factor R_d to adjust velocities before input to the particle tracking code. These codes, however, do not account for other types of chemical reactions nor do they consider dispersion or dilution of contaminant. Particle-tracking codes provide an alternative to transport modeling, when the uncertainty associated with the selection of dispersivity values and chemical reaction terms is large.

22.1.3 Numerical Solution Techniques

The two most widely used numerical techniques for solving both flow and transport models are *finite differences* and *finite elements*. Choosing between a finite difference and a finite element model depends on the problem to be solved and on the preference of the user. Details on the methods of finite differences and finite elements as applied to modeling flow and transport in the subsurface may be found in a number of textbooks.[14,64,65,71,157]

Both methods solve for the dependent variable at each node in a grid superimposed over the problem domain. The governing equation is approximated by a set of algebraic equations which are solved by iteration, matrix solution, or some combination of these methods. A simple iterative solution of the set of algebraic equations resulting from application of the method of finite differences can be obtained with the aid of a spreadsheet.[106,108] Spreadsheet modeling is a useful learning tool but it is doubtful that spreadsheet solutions offer any advantages over standard computer codes. The time required to set up and test a complex spreadsheet model is likely to be equal to or greater than the time needed to set up and run a standard flow code.

Other numerical methods include *integrated finite differences*, the *boundary integral equation method*,[91] and *analytic elements*,[136,137] which is an emerging technique that may prove very powerful in practical applications. It is based on the superposition of analytic solutions, but, unlike traditional application of analytical solutions, this numerical technique can handle heterogeneous conditions and multiple sources and sinks. Unlike conventional finite-difference and finite-element methods, analytic elements can generate detailed solutions of flow locally as well as regionally within the same model. This is possible because the problem domain is not discretized. Thus the solution can be viewed in two or more "windows" of different scales.[43] SLAEM[138] is a two-dimensional steady-state analytic element code that is interactive with graphics, but it is also costly, priced at over $6000 with a postprocessor. QuickFlow® from Geraghty and Miller Inc. integrates analytic elements from Strack[137] with pre- and postprocessing.

22.1.4 Modeling Protocol

There are numerous pitfalls into which the modeler can fall when using even some of the simpler models discussed in this chapter. For this reason, it is essential that the modeler be conversant with basic principles of geology and groundwater hydrology and have a thorough understanding of the relevant physics that govern the movement of fluids and solutes in the subsurface. In addition, familiarity with basic principles of modeling[14,64,65,71,157] and model application[7] is essential. Even with all of these skills, it will be necessary to acquire modeling experience by regular use of models.[5] Following a good modeling protocol will help direct the modeling effort. *Modeling protocol* refers to an accepted procedure for applying models. While an official modeling protocol does not yet exist, standards are currently being formulated by the American Society for Testing and Materials.[46] The modeling protocol presented by Anderson and Woessner[7] is discussed here.

1. The first step in the modeling protocol is to define the purpose of the model. In most modeling projects, the purpose is to *predict* the consequences of a proposed action. However, models can also be used in an *interpretive sense* to gain insight into the controlling parameters or as a framework for assembling and organizing field data and formulating ideas about system dynamics. Models can also be used to study processes in *generic geologic settings* with the objective of formulating regional regulatory guidelines or of using the model as a screening tool to identify regions suitable for some proposed action.[56] Once the purpose has been clearly defined, it may be obvious that the problem can be solved by using a simple analytical solution[94] or is more appropriately answered by additional field investigations.

2. When the purpose dictates that a numerical model is necessary, a *conceptual model* is designed. During this step, the geologic setting is described and important hydrologic and hydrogeochemical processes are identified. Field data are assembled, and, if possible, the site is visited.

3. A mathematical model is designed, based on the conceptual model, and an appropriate computer code is selected drawing on information presented in this chapter. Documentation of the code should provide evidence that the code has been *verified* by comparisons between numerical solutions generated by the code and one or more analytical solutions. Design of the model will require good geologic and engineering judgment as well as careful study of the user's manual. Model design includes construction of the grid, setting boundary and initial conditions, and preliminary selection of parameter values from field data. During this stage, errors in the field data are assessed and quantified and *calibration targets* are defined.

4. During model *calibration,* parameters are adjusted within a predetermined range of uncertainty until the model produces results that approximate the set of field measurements selected as calibration targets. In practice, this is typically done by manual *trial-and-error* adjustment of parameters, but *automated parameter estimation codes* may be used instead (Sec. 22.7). As part of the calibration process, the effects of uncertainty in aquifer parameters, sink/source terms, and possibly boundary and initial conditions are tested. This analysis of uncertainty is known as a sensitivity analysis.

Calibration results should be presented in a tabular listing of calibration targets versus simulated values. The *residual error,* i.e., the difference between field-measured values and simulated values should also be tabulated. It is helpful to display these data in a *scatter plot,* which is a linear plot of field-measured values on the horizontal axis versus simulated values on the vertical axis. Some average measure of the residual such as the *mean absolute error* (the average of the absolute values of the residuals), or the *root mean squared error* (the average of the squares of the residuals) should be calculated and presented. A contour map showing the distribution of residual errors should also be presented. Residual errors should be randomly distributed over the model grid.

5. In *model verification,* the calibrated parameters are used in another simulation, often involving a change in sink or source terms or boundary conditions, and the results are compared to a second independent set of field data. The model is said to be verified if it produces an acceptable match to the verification targets without changing parameter values that were set during calibration. When an independent set of field data does not exist, as is often the case, this step is omitted. Model verification is an attempt to demonstrate the accuracy of the calibrated model and should not be confused with *verification of the governing equations,* which consist of a demonstration that the equations used in the model accurately describe the processes of flow and/or transport in porous media, or with *code verification,* which is a demonstration that the computer program accurately solves the equations that constitute the mathematical model.

6. A *predictive simulation* requires estimating future stresses such as groundwater recharge rates as influenced by fluctuation in precipitation, future pumpage based on projected use rates, and future waste volumes and leaching rates. Review of simulations performed around 20 years ago indicates that the failure of a model to predict the future is often a result of an inadequate conceptual model and/or a failure to anticipate future stresses accurately.[2,78,81,90] In general a good guideline for reliable predictions is that the simulation should not be extended into the future more than twice the period for which calibration data are available.[38] However, some environmental problems require predicting the system response many years, perhaps as many as 10,000 years, into the future.[99] In view of the uncertainties about future stresses, it is advisable to perform a *predictive sensitivity analysis* to test the effects of ranges in estimated future stresses on the model's predictions.

7. *Model validation* refers to the process of demonstrating that a given site-specific model is capable of making accurate predictions. Ideally, the model should be validated by means of a *postaudit* performed several years after the modeling study is completed and the model's prediction can be evaluated. A postaudit should occur long enough after the prediction was made to allow adequate time for the system to move far from the calibrated solution. When new field information is available, the model should be redesigned, if necessary, so that it can be continually improved and used on a regular basis.

The issue of validation is mainly a regulatory one, not a scientific one. From a scientific standpoint a model can never be proved valid, or stated in other words, "a valid model is an unattainable goal of model validation."[99] Model validation generally is not a fruitful exercise because of the uncertainties inherent in the conceptual model of the system and the uncertainties involved in assessing future stresses to the system. Nevertheless, it is essential to assess the degree of confidence that can be placed in model predictions. The confidence level depends largely on the results of the calibration, sensitivity analyses, and verification.

8. Finally, modeling results are presented in a written report. Information used to construct the model and all modeling results should be archived so that someone else can reconstruct the model and duplicate the results. A modeling journal that records the thought process used during model design and execution should be part of this documentation.

22.2 GROUNDWATER FLOW CODES

22.2.1 Analytical Codes

Most of the codes that compute analytical solutions are designed to aid in aquifer test analysis. These include AQTESOLV from Geraghty and Miller Inc., an interactive code that analyzes data from pumping tests and slug tests through both visual and statistical curve matching. Other programs for the analysis of pumping test data include TECTYPE from the Scientific Software Group, and PUMPTEST, THCVFIT, and VARQ from the IGWMC. TGUESS, from the IGWMC, is a program for analyzing specific capacity data.

Other groundwater flow problems are usually solved with a numerical code. However, in some applications, an analytical solution may be appropriate. For example, an analytical solution for two-dimensional steady flow in cross section is included in the U.S. EPA's CANSAZ code.[140] GWFLOW from the IGWMC is a package of seven frequently used analytical solutions for groundwater flow. Analytical solutions for flow are also used in particle tracking (Sec. 22.1.4).

22.2.2 Featured Numerical Codes

Introduction. Three numerical codes are featured in this section—PLASM, MODFLOW, and AQUIFEM-1 (Table 22.2.1). PLASM, a two-dimensional finite-

TABLE 22.2.1 Featured Numerical Groundwater Flow Codes

Name	Type	Selected applications (see "References")
PLASM	2D, FD	34, 67, 113, 116, 135, 154
MODFLOW	3D, FD	Many; see 7 for selected citations
AQUIFEM-1	2D, FE(T)	143

FD = finite differences, FE(T) = finite elements (triangular).

Required input: Spatial distribution of hydraulic conductivity or transmissivity, and storage coefficient; values of all hydrologic stresses (withdrawal and recharge rates) and boundary and initial conditions.

Output: Distribution of head in space and time; water balance information.

difference code introduced in 1971, was the first well-documented flow code to be readily available. MODFLOW, a three-dimensional finite-difference code released in 1988, is the latest in a series of two- and three-dimensional flow codes developed by the U.S. Geological Survey.[145,146] AQUIFEM-1 is a two-dimensional finite-element code originally developed at the Massachusetts Institute of Technology; a three-dimensional version was developed recently by one of its original authors.

All three codes have similar capabilities and can be expected to produce similar results. AQUIFEM-1, being a finite-element model, allows more flexibility in grid design. MODFLOW, being a three-dimensional code, allows simulation of more complex flow fields. AQUIFEM-1 and MODFLOW allow more flexibility in specifying time-dependent boundary conditions than does PLASM. MODFLOW and AQUIFEM-1 provide more detailed water balance information than does PLASM, although PLASM could be modified to provide similar detail.

PLASM is recommended as a first code for inexperienced modelers because it is interactive and easiest to operate. MODFLOW is an extremely popular and versatile code, but this versatility causes the assembly of input files to be more involved than for PLASM. The practitioner should allow time to master the relatively complex file structure of MODFLOW and to assimilate the lengthy user's manual. AQUIFEM-1 has relatively complex data files and an excellent user's manual.

Types of Problems. All three codes treat transient and steady-state conditions in both confined and unconfined aquifers, including regional flow problems and problems of local flow such as flow to a well. Unconfined conditions are treated using the Dupuit assumptions. The codes can also simulate leaky confined aquifers, flow to partially penetrating rivers and lakes, spring flow, flow to drains, and two-dimensional flow in a cross section.

All three codes allow irregularly spaced nodes and variable time steps as well as specified-head, specified-flow, and head-dependent boundary conditions. In a head-dependent boundary condition, the head specified at the boundary and the head in the aquifer just inside the boundary are used to compute a head gradient which is used to compute a boundary flux. Anderson and Woessner[7] discuss the design and application of groundwater flow models in detail, with emphasis on PLASM, MODFLOW, and AQUIFEM-1.

Required Input. The spatial distribution of aquifer properties (hydraulic conductivity or transmissivity, and storage parameters) and sink and source terms representing recharge and withdrawal are required.

PLASM. PLASM stands for the Prickett-Lonnquist Aquifer Simulation Model. This two-dimensional finite-difference code was originally written in FORTRAN[116] for the Illinois State Water Survey. An interactive version, written in BASICA, was developed by T. A. Prickett and Associates.[114]

Types of Problems. PLASM has been applied to numerous problems in consulting and regulatory work, but only a few of these applications are reported in the published literature (Table 22.2.1).

Special Features. Transmissivity values are specified for areas between nodes, rather than for a finite-difference cell, as is typical of other block-centered models. Input of transmissivities in this manner allows the user to adjust the transmissivities in order to "trick" PLASM into placing flow boundaries directly on the node rather than on the edge of the block as is done in other block-centered finite-difference codes.[7] The BASICA version of PLASM includes preprocessors, debugging routines, restart capabilities, a water balance computation, and a simplified water-level contouring routine. Output data may be saved in a format suitable for direct input to contouring packages such as SURFER (produced by Golden Software).

Computer Requirements. The BASICA version of PLASM runs on an IBM-compatible microcomputer. Output is automatically directed to a printer and the code, as written, will not run without connection to a printer. The code is easily modified, however, to direct output to a file instead of the printer.

How to Obtain the Code. The BASICA version can be obtained from T. A. Prickett and Associates. The IGWMC distributes a version based on the original documentation by the Illinois State Water Survey.[116] It includes a preprocessor (PREPLASM) to create or edit input files. PLASM has also been modified and renamed by users other than the original developer.[113,156]

AQUIFEM-1. AQUIFEM-1[143,166] is a two-dimensional finite-element model that simulates flow in one layer. The multilayer version is called AQUIFEM-N.[142] AQUIFEM-1 and -N are unrelated to the codes AQUIFEM[110] and AQUIFEM-SALT.[150]

Special Features. The grid is designed with linear triangular elements. Aquifer properties may be assigned to either the node or the element. The code has special features to simulate leakage to or from narrow rivers and provides a complete water balance including a list of fluxes at all nodes.

Computer Requirements. AQUIFEM-1 and -N run on IBM-compatible microcomputers.

How to Obtain the Code. AQUIFEM-1, with a graphics postprocessor to contour output, can be obtained from the GEOCOMP Corp. AQUIFEM-N, including grid generation and graphics packages, is available from Dr. Lloyd Townley. Versions for IBM-compatible computers, including 386-based machines, and for Macintosh computers may be requested.

MODFLOW. MODFLOW[93] is modular three-dimensional finite-difference flow code developed by the USGS.

Types of Problems. MODFLOW can simulate fully three-dimensional systems and quasi-three-dimensional systems in which flow in aquifers is horizontal and flow through confining beds is vertical. The code can also be used to simulate flow in two dimensions either in one horizontal layer or in a cross section. MODFLOW and its predecessors have been used in numerous applications, many of which are documented in USGS Water-Supply Papers, Professional Papers, and Water-Resources Investigations Reports. Some of these applications are cited in Ref. 7.

Special Features. The code permits the user to select a series of packages (or modules) to be used during a given simulation. The packages include three equation-solver packages,[58,59,93] stream packages,[93,118] a recharge package, and packages to simulate pumping or injection wells, drains, and evapotranspiration from the water table. A package to consider release of water from clay lenses is also available.[86,87]

Options set in an output control module allow the user to specify the print format of the head and/or drawdown arrays and to request printout of the water balance. Water balance output includes a summary statement of total flows to and from the system. If desired, flows to specified head nodes and discharges via pumping and evapotranspiration can also be printed. Preprocessors available for MODFLOW include PM and MODINP from the Scientific Software Group, PREPRO3FLO from GeoTrans Inc. and Model-CAD from Geraghty and Miller Inc. Postprocessors include PM and MODGRAF from the Scientific Software Group, a statistical package[127] from the USGS, contouring packages provided with some versions of the code, and the particle-tracking codes MODPATH and PATH3D (Table 22.1.1 and Sec. 22.4).

Computer Requirements. MODFLOW is written in strict FORTRAN 77. It runs on all computers that have a FORTRAN 77 compiler. It has been run without modification on many different types of mainframes, minicomputers, and workstations. It also has been run without modification on IBM and Apple personal computers and their clones. The allowable size of a problem, as measured by the number of finite-difference cells, depends on the size of MODFLOW's *X* array, which can be set at compilation time. On 8088 and 80286 computers, the practical limit is about 8000 cells. Extended-memory versions have no theoretical limits. Problems using grids of 60,000 cells are common.

How to Obtain the Code. Source code can be obtained from the USGS, McDonald Morrissey Associates Inc., and the IGWMC. Versions compiled for IBM-compatible computers are available from McDonald Morrissey Associates Inc., Geraghty and Miller Inc., and the Scientific Software Group. A version for the Macintosh computer is available from Geraghty and Miller Inc.

22.2.3 Other Numerical Codes

FE3DGW[48,50,51] from the ESTSC is a three-dimensional finite-element code. It was originally written for a VAX minicomputer but with some effort can be adapted for other systems such as Sun workstations. It includes supporting graphics software for visualization of well logs and mesh as well as contour plots of heads. The code has also been extended to include transport (CFEST, Sec. 22.3.3).

FREESURF,[101] from Dr. Shlomo Neuman, is a two-dimensional code utilizing deformable finite elements to handle problems involving a moving water table and seepage faces. FLONET from the Scientific Software Group, is a two-dimensional, finite-element, steady-state model designed to simulate flow in cross sections. It computes potentials, streamlines, and groundwater velocities, and includes deformable elements at the water table. Other flow codes include SLAM from Lewis Publishers, which is a five-layer finite-element code for two-dimensional steady-state[8] and transient[9] conditions. MICRO-FEM,[55] from the IGWMC, is an interactive four-layer steady-state finite-element code with on-screen graphics that uses triangular elements and up to 3000 nodes. INTERSAT, from the Scientific Software Group, is an interactive three-dimensional finite-difference code. The United Nations software package[68] includes numerical models for confined and unconfined aquifers and a model that uses the Ghyben-Herzberg assumption to compute water levels, location of the freshwater-saltwater interface, and thickness of the freshwater lens in a small oceanic island.

22.3 CODES FOR CONTAMINANT TRANSPORT IN GROUNDWATER

All of the codes discussed in this section solve the advection-dispersion equation [Eq. (22.1.4)] for transport of only one chemical constituent.

22.3.1 Analytical Codes

Most analytical solutions assume that the flow velocity is constant in space and time, which is an unrealistic assumption in most field applications. Nevertheless, analytical solutions are generally useful as preliminary screening tools. AT123D,[168] available from the IGWMC, is a collection of analytical solutions for one-, two-, and three-dimensional contaminant transport including advection, dispersion, and decay. Many other analytical codes are available, including a package developed by the USGS[165] and MYGRT^{TM}[141] from the Electric Power Research Institute (EPRI). SOLUTE, PLUME, AGU-10,[66] and EPA-VHS[32] are distributed by the IGWMC. Computer codes for analytical solutions are also presented in textbooks.[33,155]

22.3.2 Featured Numerical Codes

Four numerical codes are featured in this section (Table 22.3.1). BIO1D is conceptually the simplest physical model but it offers the most complex chemistry of the featured codes. RNDWALK and USGS MOC simulate two-dimensional transient flow and dispersion, with options for adsorption and decay. Both codes solve similar types of problems. MT3D allows for three-dimensional transient flow and dispersion and includes linear and nonlinear adsorption and first-order decay. SWIFT/386 is a complex model that allows variable-density transport in fractured and nonfractured porous media.

RNDWALK and BIO1D are interactive and are good choices for beginning modelers. USGS MOC offers somewhat more versatility than RNDWALK, since later versions of the code include nonlinear adsorption,[144] density-dependent flow,[125] and biodegradation.[123] MT3D is a relatively advanced model requiring a complex assembly of input files, which is facilitated by an excellent user's manual. Use of MT3D also requires familiarity with a three-dimensional block-centered flow model like MODFLOW (Sec. 22.2.2). All of these codes are finite-difference-based codes that solve Eq. (22.1.4) by decoupling the advective and dispersive portions of the equation, thereby minimizing the possibility of numerical dispersion. SWIFT/386 is an advanced three-dimensional finite-difference code that requires a sophisticated user.

BIO1D *Types of Problems.* BIO1D[132] assumes constant velocity, one-dimensional (longitudinal) dispersion, linear or nonlinear adsorption, and aerobic or anaerobic

TABLE 22.3.1 Featured Numerical Codes for Contaminant Transport in Groundwater

Name	Type	Selected applications (see "References")
BIO1D	1D flow and dispersion; FD	133
RNDWALK	2D; FD with MOC and DPRW	4, 56, 75, 154
USGS MOC	2D; FD with MOC and PIC	1, 16, 24, 36, 47, 61, 62, 69, 77, 79, 81, 88, 124, 139, 167
MT3D	3D; FD with MOC, MMOC, and HMOC, all with PIC	
SWIFT/386	3D; FD	19, 53, 97, 159, 160, 162

FD = finite differences, MOC = method of characteristics, DPRW = discrete parcel random walk, MMOC = modified method of characteristics, PIC = particle in cells, HMOC = hydrid method of characteristics.

Required input: See text for details.

Output: Distribution of heads and concentrations in space and time; water and mass balance information.

degradation. The code can be used to simulate reactive species such as organic solvents and petroleum products as well as inorganic solutes including radionuclides. BIO1D is ideally suited for deciding which transport processes are important in a given problem.

Required Input. BIO1D requires input of velocity, longitudinal dispersion coefficient, adsorption rates, decay rates, source concentration, and initial and boundary conditions.

Special Features. Two finite-difference equations are solved iteratively for oxygen and solute concentration. Adsorption is described by linear, Freundlich, or Langmuir isotherms. The process of degradation is based on Monod kinetics, but the nonlinear isotherms do not include kinetic reactions. A user-friendly front-end module helps guide the user through input screens. The time step, number of iterations, and mass balance are monitored as execution proceeds. The resulting concentration profiles or breakthrough curves can be viewed and plotted.

Computer Requirements. The code is written for an IBM-compatible microcomputer with a graphics card.

How to Obtain the Code. The code is available from GeoTrans Inc. Source code is not available.

RNDWALK. RNDWALK is the Random-Walk model, originally developed by the Illinois State Water Survey.[117] An interactive version[115] was developed by T. A. Prickett and Associates. A three-dimensional version, RD3D,[76] for use with the flow code MODFLOW (Sec. 22.2.3) is also available.

Types of Problems. RNDWALK is designed to simulate transient contaminant movement in a two-dimensional horizontal plane for both regional and local flow problems, under confined and unconfined conditions. The code includes a version of the flow code PLASM (Sec. 22.2.2) for calculating heads, from which a velocity distribution is computed. Options for simulating sinks and sources of contaminants include: injection of contaminated water into wells, vertically averaged saltwater fronts, leachate from landfills or leakage from underground storage tanks, leakage from overlying and underlying source beds, surface water sources such as contaminated lakes and streams, and diffuse leakage of contaminants from nonpoint sources. Retardation and decay and spatially dependent dispersivity can also be included in the simulation.

Required Input. RNDWALK requires the same input as PLASM (Sec. 22.2.1) as well as boundary and initial conditions and sink and source concentrations for the transport simulation. Transport parameters include longitudinal and transverse dispersivities, effective porosity, retardation factor, and half-life of constituents.

Special Features. An advantage of this code is that particles are introduced only where contaminants are present. A mass is assigned to each particle, and solute concentrations are calculated from the distribution of particles present in each cell of the grid used to solve the flow problem. Because calculation of concentrations is done independently of the tracking of particles, concentrations need not be calculated at every time step.

A disadvantage is that a high density of particles may be needed for accurate solutions. Also the solution is visually "bumpy" owing to the random movement of particles; successive runs of the model for an identical problem will yield slightly different, yet theoretically equivalent, results. Difficulties may arise for strongly heterogeneous porous media where the velocity varies markedly within the system. In this case, particles may become stuck in zones of low velocity unless small time increments are used or the code is modified to include a correction formula.[72]

Computer Requirements. RNDWALK runs on IBM-compatible microcomputers.

How to Obtain the Code. The interactive version of RNDWALK can be obtained from T. A. Prickett and Associates. IGWMC distributes another version with a preprocessor. A guidance document prepared by others is also available.[98]

USGS MOC *Types of Problems.* USGS MOC[44,80] was designed for analysis of two-dimensional transient solute transport in confined aquifers or in unconfined aquifers when the transmissivity can be considered to be constant in time. This model is not suited for problems involving radial flow.[36] Numerous applications have been documented (Table 22.3.1).

Required Input. The code has the same input requirements as RNDWALK. It is not interactive, but the assembly of data files is relatively simple. A preprocessor for data input is available.

Special Features. The code uses particles that track concentration (not mass), so particles are placed in every cell of the finite-difference grid, even where contaminants are not present. Relative to other solute transport models based on particle-tracking solutions, USGS MOC uses a lower density of tracer particles; particles are initially placed where concentrations are zero and the same low density of particles is also used where concentrations are high. Nodes must be uniformly spaced, i.e., Δx and Δy must be constants. Errors associated with tracking the particles sometimes cause poor mass balance, with errors exceeding 10 percent. Advantages include numerous case studies (Table 22.3.1) and the availability of more advanced versions, including MOCDENSE,[125] which treats variable-density fluids, and BIOPLUME II,[123] which allows parallel computation of an organic plume with biodegradation and oxygen concentrations. A decision support system is available for use with BIOPLUME II on a Macintosh microcomputer.[103,104]

Computer Requirements. USGS MOC runs on IBM-compatible microcomputers, including versions for 286 and 386 machines. A version for Macintosh computers is also available.

How to Obtain the Code. The code is available from the USGS and the IGWMC, who include a preprocessor. Geraghty and Miller Inc. sell a version for the Macintosh computer and a version for 386 and 486 IBM-compatible machines. MOCDENSE is available from the IGWMC and the Scientific Software Group.

MT3D. MT3D, Modular Transport Model in Three Dimensions, was recently developed by Papadopulos and Associates Inc.[176] for the U.S. Environmental Protection Agency (R. S. Kerr Environmental Research Laboratory).

Types of Problems. MT3D was developed to simulate three-dimensional solute transport without the complications of heat and density effects included in some other three-dimensional transport codes (Sec. 22.3.3). The model allows three-dimensional dispersion and chemical reactions involving linear or nonlinear adsorption and first-order decay. There are provisions for sinks and sources of contaminants compatible with the sink and source options in MODFLOW (Sec. 22.2.2). Its grid is also compatible with MODFLOW. Irregularly spaced nodes are allowed, i.e., Δx and Δy need not be constants.

Required Input. This code requires the same type of input as RNDWALK, described above, but in three dimensions, including three dispersivity components.

Special Features. MT3D was designed to accept input from MODFLOW, but can be modified to accommodate output from any block-centered flow model. It offers four solution methods, including a finite-difference option and three decoupled solution procedures. The MOC option is similar to the solution procedure used in USGS MOC and requires several particles in each finite-difference cell. Another procedure (MMOC) requires only one particle per cell, thereby requiring much less computer memory and computational time. This method, however, may introduce some numerical dispersion, especially for problems with sharp concentration fronts. An option for combining both solution procedures (HMOC) is also provided.

Computer Requirements. MT3D runs on IBM-compatible microcomputers.

How to Obtain the Code. The code can be obtained from S. S. Papadopulos and Associates Inc.

SWIFT/386 *Types of Problems.* SWIFT/386[158] solves finite-difference equations for density-dependent fluid flow and transport of solutes, heat, and radionuclides. While the code can be applied to continuous porous media, it was especially designed for fractured media (Sec. 22.7).

Required Input. SWIFT/386 requires the same type of input as MT3D, but also requires compressibility of water and of porous media. Nuclide decay series and fracture permeability and porosity are required, if those options are used.

Special Features. The code considers several radionuclide species in an arbitrary decay-chain sequence. The flow portion of the code considers both confined and unconfined aquifers using pressure as the dependent variable. Submodels include the representation of injection and production wells, incorporation of aquifer influence functions,[73] and leaching of radionuclides from a geologic repository.

Computer Requirements. SWIFT/386 was designed for the 386- and 486-based IBM-compatible microcomputers.

How to Obtain the Code. SWIFT/386 is available from GeoTrans Inc. Another version (SWIFT II) is available from the Scientific Software Group. The code is well-documented, with numerous problems for self-education and code verification.[121,122,161]

22.3.3 Other Numerical Codes

The featured codes are all finite-difference-based codes. Two-dimensional finite-element transport codes include SEFTRAN from GeoTrans Inc. and AQUA, an interactive menu-driven code with on-screen graphics distributed by the Scientific Software Group. The Golder Groundwater Computer Package from Golder Associates Inc. is a two- and quasi-three-dimensional finite-element code.

The three-dimensional, finite-difference model FTWORK from GeoTrans Inc., was designed as a management tool for the U.S. Department of Energy's Savannah River Site.[3] It solves the same types of problems as MT3D (Sec. 22.3.2). INTERSTAT, available from the Scientific Software Group, is an interactive three-dimensional, transient random-walk model. RD3D[76] is a three-dimensional version of RNDWALK (Sec. 22.3.2) linked to MODFLOW (Sec. 22.2.2). VARDEN,[82,83,84] from the USGS, is a three-dimensional finite-difference model with density effects.

HST3D,[74], a Heat- and Solute-Transport code for three-dimensional simulations, is a finite-difference code developed by the USGS to solve the same types of problems as SWIFT. HST3D is available from the USGS, the IGWMC, and the Scientific Software Group, which sells versions for IBM-compatible 386-based computers and the Macintosh computer.

CFEST (Coupled Fluid, Energy, and Solute Transport), from the ESTSC, is a three-dimensional finite-element code[49] to simulate coupled fluid flow, heat, and contaminant transport. It is similar in function to SWIFT and HST3D. CFEST is an extension of the flow code FE3DGW (Sec. 22.2.3) and, like FE3DGW, was developed for a VAX computer, requiring reprogramming to run on other systems.

Results from benchmarking studies[70,130] indicate that transport simulations using different codes are comparable when applied to the same or a similar conceptual model. Hence, selection of a code from among those in a similar class is mainly determined by user preference and the need for special features specific to one code.

22.4 PARTICLE-TRACKING CODES

22.4.1 Analytical Codes

Analytical solutions used in particle tracking usually rely upon a two-dimensional, steady-state solution of the flow equation for a homogeneous, isotropic confined aquifer. Under these conditions, path lines are streamlines and the flow equation may be written to solve directly for the stream function.[13] Linear adsorption represented by a retardation factor is easily included.

Nelson[100] was an early advocate of the use of particle-tracking codes, and his work led to the development of the analytical code S-PATHS.[105] DREAM,[17] from Lewis Publishers, consists of a set of analytical solutions for groundwater flow and includes options for calculating pathlines. PAT, available from the Scientific Software Group, is another analytical code that computes path lines and travel times.

WHPA[15] is a particle-tracking package, developed for the U.S. EPA and distributed by the IGWMC, that includes options for computing path lines analytically. There is also a numerical option that accepts input from a finite-difference or finite-element numerical model. WHPA was designed for problems involving the analysis of capture zones for delineation of wellhead protection areas, from which the code gets its name.

RESSQ,[66] which is included in the AGU-10 package of codes (Sec. 22.3.1), calculates the location of contaminant fronts around sources at various times and the concentration of solute at a production well as a function of time based on the arrival of streamlines. RESSQ is distributed by the IGWMC with an interactive pre- and postprocessor code, RESCUE.

22.4.2 Featured Numerical Codes

Three numerical codes are feature (Table 22.4.1). FLOWPATH includes a flow code adapted from PLASM (Sec. 22.2.2) and is a good choice for beginners because it is interactive. MODPATH and PATHD3D are three-dimensional particle-tracking codes developed for use with MODFLOW (Sec. 22.2.2).

TABLE 22.4.1 Featured Numerical Particle-Tracking Codes

Name	Type	Selected applications (see "References")
FLOWPATH	2D, steady flow	
MODPATH	3D, steady flow; transient possible	20
PATH3D	3D, steady and transient flow	6, 163, 177

Required input: Spatial distribution of heads, effective porosity.
Output: Path lines and travel times of imaginary tracer particles.

FLOWPATH *Types of Problems.* FLOWPATH[41] solves for path lines in two-dimensional steady-state flow fields. It can be used as a postprocessor to flow simulations to provide a picture of the flow field. This is done by placing particles around the perimeter of the grid and tracking each particle through the flow field.[10] It also can be used to track contaminants by advection, as influenced by linear adsorption, and to delineate capture zones for wellhead protection and remedial investigations.

Required Input. FLOWPATH has the same input requirements as PLASM (Sec. 22.2.2) and also requires input data on effective porosity.

Special Features. FLOWPATH uses Euler integration (Sec. 22.1.3) to move particles along path lines. Numerical errors introduced by Euler integration tend to be large unless small tracking steps are used. The size of a tracking step is controlled by an error criterion set by the user. FLOWPATH is interactive, with excellent on-screen graphics.

Computer Requirements. FLOWPATH runs on an IBM-compatible microcomputer with a graphics card.

How to Obtain the Code. FLOWPATH is available from Waterloo Hydrogeologic Software and the Scientific Software Group.

MODPATH *Types of Problems.* MODPATH[112] was designed for delineating path lines in a steady-state three-dimensional flow field. It can solve the same types of problems as FLOWPATH but in three dimensions. The code permits backward tracking,[128] whereby particles can be introduced into a sink such as a pumping well, and the particles are moved backward along the reverse path lines to their source. Backward tracking is particularly useful in defining capture zones and helping to identify the location of a contaminant source.

Required Input. MODPATH requires input of heads from MODFLOW (Sec. 22.2.2). It also requires effective porosity. Preassembled input files can be entered interactively or in batch mode.

Special Features. MODPATH is based on a semianalytical particle-tracking scheme,[111,112] whereby an analytical expression is used to calculate the flow path within the cell. This method of tracking particles is efficient because only one calculation is required to move a particle between faces of a finite-difference block. The semianalytical technique ensures that the solution is free from numerical errors associated with discretizing tracking steps. However, the algorithm is much less efficient in transient simulations because heads in a finite-difference block may change before a particle leaves the block, making it impossible to allow the particle to traverse an entire block in one time step unless extremely small time steps are used. The code keeps track of the coordinate locations of each marker particle as it moves through the flow field. Time of travel marks can be used to produce a visual record of travel times on path line plots.

How to Obtain the Code. The mainframe version of MODPATH, available from the USGS, includes a plotting routine (MODPATH-PLOT) that uses the DISSPLA graphics subroutine library (produced by Computer Associates). A version of MODPATH for IBM-compatible microcomputers is available from the IGWMC and the Scientific Software Group. Geraghty and Miller Inc. and the Scientific Software Group sell versions with graphics for 386- and 486-based computers.

PATH3D. PATH3D was developed by Zheng for the Wisconsin Geological and Natural History Survey and later revised and improved for Papadopulos and Associates Inc.[175]

Types of Problems. PATH3D solves the same type of problems as MODPATH, except that it was designed for both steady-state and transient problems.

Required Input. PATH3D has the same input requirements as MODPATH. Preassembled input files can be entered interactively or in batch mode.

Special Features. PATH3D uses a fourth-order Runge-Kutta tracking solution (Sec. 22.1.3) that allows a particle to take several small tracking steps while moving across a finite-difference block. A particle may also traverse several blocks in a single tracking step. To minimize numerical errors associated with the use of tracking steps, PATH3D automatically adjusts the step sizes according to an error criterion set by the user.

PATH3D produces an output file containing a particle identification number along with the *x*, *y*, *z* coordinates and travel times. The output file can be used directly as input to most commercial graphics packages such as SURFER (Golden Software) to obtain a plot of travel paths. Particle velocities and nodal coordinates can also be printed.

How to Obtain the Code. The code can be obtained from S. S. Papadopulos and Associates Inc.

22.4.3 Other Numerical Codes

Other particle-tracking codes include GWPATH,[11,129] an interactive particle tracking code for two-dimensional flow with graphics, WHPA (Sec. 22.4.1), and STLINE, a three-dimensional code from GeoTrans Inc.

22.5 VARIABLY SATURATED FLOW CODES

22.5.1 Analytical Codes

Solution of problems involving unsaturated media generally requires a numerical code, but a few analytical codes are available. For example, INFIL, from the IGWMC, solves for transient one-dimensional flow in a deep homogeneous soil profile with infiltration from ponded water at the surface.

22.5.2 Featured Numerical Codes

This section includes a description of three numerical codes, FEMWATER, UNSAT2, and VS2D (Table 22.5.1), for simulation of variably saturated flow using a two-dimensional form of Eq. (22.1.2).

TABLE 22.5.1 Featured Numerical Flow Codes for Variably Saturated Media

Name	Type	Selected applications (see "References")
FEMWATER	2D, FE(Q)	40
UNSAT 2	2D, FE(T)	39, 52, 134, 164
VS2D	2D, IFD	

FE(Q) = finite elements (quadrilateral), FE(T) = finite elements (triangular), IFD = integrated finite differences.

Required input: See text for details.

Output: Distribution of pressure head (or total head) and moisture content in space and time; water balance information.

Types of Problems. The models described here solve for total head or pressure head under both unsaturated and saturated conditions in two-dimensional flow under steady-state and transient conditions. They are used for problems in which it is important to describe water movement between the land surface and the water table. These models assume that the pressure of the air phase under unsaturated conditions is equal to atmospheric pressure, and that the air phase is immobile. If this is not the case, it may be necessary to use a multiphase flow code (Sec. 22.7).

Required Input. Solving Eq. (22.1.2) requires that the dependence of the hydraulic conductivity $K(\theta)$ and specific moisture capacity $C(\theta)$ on the pressure head be specified. The relation between θ and ψ is also specified so that moisture contents can be calculated from pressure heads.

FEMWATER. FEMWATER[120,169,173] is a two-dimensional finite-element model of water flow. A three-dimensional version is also available.[170,171]

Types of Problems. FEMWATER solves the types of problems described above in the general discussion of variably saturated flow codes with pressure head as the dependent variable. The code is not recommended for highly nonlinear problems such as those involving sharp infiltration fronts into dry soils; thus the code is limited to problems characterized by wet to moderately wet soils. The code does not have options for extraction of water by plant roots.

Special Features. FEMWATER discretizes the spatial domain using quadrilateral finite elements and/or triangular elements to allow flexibility in grid design. It computes velocities at the end of each time step. It also allows local anisotropy in hydraulic conductivity within elements so that the principal hydraulic conductivities may be oriented at any desired angle with respect to the coordinate axes of the problem. Later versions of the code[169] allow for radial coordinates.

Computer Requirements. FEMWATER was designed to run on an IBM 360 mainframe computer, but a version for IBM-compatible microcomputers is available.

How to Obtain the Code. The mainframe version is available from the ESTSC. The version for IBM-compatible microcomputers, 2-D FEMWATER, is available for PCs, XTs, or ATs, and for 386-based machines, from Environmental Consulting Engineers (ECE), who include a graphical pre- and postprocessor, NAG (Numerical Approximation Graphics), which plots the grid, labeling nodes and elements, and also generates plots of velocity vectors and a variety of contour and time series plots. The three-dimensional version of FEMWATER, 3DFEMWATER, is available from Dr. George Yeh.

UNSAT2. UNSAT2 is a two-dimensional finite-element code[31,102] that uses pressure head as the dependent variable and triangular elements.

Required Input. In addition to the data described in the general discussion of input for variably saturated flow codes, UNSAT2 requires a root effectiveness function to simulate uptake of water by plant roots.

Special Features. Like FEMWATER, UNSAT2 allows for axisymmetric flow problems using radial coordinates and includes local anisotropy in hydraulic conductivity within elements so that the principal hydraulic conductivities may be oriented at any desired angle with respect to the coordinate axes of the problem. However, UNSAT2 allows uptake of water by plant roots. Output includes a water balance, which prints fluxes to and from the system. There are two versions of UNSAT2. Version 1 includes an option for describing soil characteristics and version 2 has automatic gridding and time-step options.

Computer Requirements. UNSAT2 runs on 386-based IBM-compatible microcomputers.

How to Obtain the Code. UNSAT2 is available from Dr. Shlomo Neuman and from ECE, which provides a graphical pre- and postprocessor (NAG).

VS2D. VS2D is a two-dimensional model developed by the USGS.[85] It uses total hydraulic head h as the dependent variable and integrated finite differences to discretize the problem domain.

Types of Problems. VS2D is especially designed to handle relatively difficult nonlinear problems including sharply defined wetting fronts caused by infiltration into very dry, highly conductive soils, and fluid movement in highly heterogeneous and/or highly anisotropic porous media.

Required Input. Input is generally the same as for UNSAT2.

Special Features. VS2D can simulate domains described by either rectangular or cylindrical coordinate systems. It tabulates pressure head, total head, fluid saturation, moisture content, and a water balance for each time step. The user may select a variety of output files for input to postprocessing programs that generate contour maps or history plots of pressure head, total head, saturation and/or moisture content, water balance components, and flux across specified surfaces.

The principal limitation of VS2D arises from the method used to discretize the spatial domain. The requirement to use fine discretization in regions that undergo rapid and/or frequent changes in fluid contents, combined with the use of a rectangular mesh, can result in a large number of finite-difference cells.

Computer Requirements. VS2D was written for a PRIME mainframe computer, but versions for IBM-compatible microcomputers are available.

How to Obtain the Code. The mainframe version of the code is distributed by the USGS. A version for IBM-compatible microcomputers with an interactive preprocessor is available from the IGWMC. ECE also sells a version with a graphical pre- and postprocessor (NAG) for IBM-compatible computers.

22.5.3 Other Numerical Codes

Other variably saturated flow codes include TRUST and FLUMP. The IGWMC provides information on these codes.[149]

22.6 VARIABLY SATURATED FLOW AND TRANSPORT CODES

22.6.1 Introduction

Many models have been developed during the past several years to simulate the transport and fate of chemicals in subsurface environments,[57,107] but there is much debate over the validity of these models when applied to field situations. The codes described in this section should be used with the realization that they have had less field testing than the codes described in preceeding sections of this chapter.

Models for variably saturated flow and mass transport include compartment and continuous-domain models. Compartment models simulate one-dimensional movement through a soil column routing water and chemicals between discrete compartments in the column. Continuous domain models solve Eqs. (22.1.2) and (22.1.4). Continuous domain models are discussed below; compartment models are discussed in Chap. 15.

22.6.2 Featured Numerical Codes

All of the codes featured in this section (Table 22.6.1) are continuous-domain models. FEMWASTE and VS2DT have companion codes that solve for flow (Table 22.1.1 and Sec. 22.5.2) and are intended for simulations that are not complicated by the effects of density dependent flow or heat transport. SUTRA has its own flow code that includes density effects. This code also considers heat transport.

FEMWASTE. FEMWASTE and a related code LEWASTE are two-dimensional finite-element codes developed at Oak Ridge National Laboratory.[35,174] The former uses conventional finite-element methods and the latter uses hybrid Lagrangian-Eulerian finite-element methods to minimize numerical oscillation, which can cause negative concentrations, and numerical dispersion. Three-dimensional versions, 3DFEMWASTE and 3DLEWASTE,[172] are also available.

Types of Problems. FEMWASTE and LEWASTE consider transport of a dissolved chemical in soil water and/or groundwater, as influenced by advection, dispersion, linear or nonlinear adsorption using Langmuir or Freundlich isotherms, and first-order decay.

Required Input. FEMWASTE and LEWASTE require input of flow velocities from FEMWATER (Sec. 22.5.2) as well as longitudinal and transverse dispersivities, effective porosity, source and sink concentrations, distribution coefficient or Langmuir or Freundlich adsorption parameters, decay constant, and boundary and initial conditions for the transport problem.

Special Features. FEMWASTE and LEWASTE use quadrilateral and/or triangular finite elements and output from the flow code FEMWATER (Sec. 22.5.2).

Computer Requirements. The code is designed to run on an IBM 360 mainframe computer but can be executed on a 386-based IBM compatible microcomputer or on a workstation.

How to Obtain the Code. FEMWASTE for the mainframe computer is available from the ESTSC. ECE sells a version with a graphical pre- and postprocessor for IBM-compatible computers. LEWASTE and the three-dimensional versions, 3DFEMWASTE and 3DLEWASTE, are available from Dr. George Yeh.

TABLE 22.6.1 Featured Numerical Codes for Transport in Variably Saturated Media

Name	Type	Selected applications (see "References")
FEMWASTE	2D, FE(Q)	40
VS2DT	2D, IFD	
SUTRA	2D, IFD & FE(Q)	18, 25, 152

FE(Q) = finite elements (quadrilateral), FE(T) = finite elements (triangular), IFD = integrated finite differences.

Required input: See text for details.

Output: Distribution of heads (or pressure) and concentrations in space and time.

VS2DT. VS2DT is VS2D (Sec. 22.5.2) modified by the USGS[54] for solute transport. Several minor changes were made to the VS2D code and a transport code was added to create VS2DT.

Types of Problems. VS2DT solves for concentration of solute in the types of problems described for VS2D.

Required Input. VS2DT requires the same input as VS2D plus input similar to that required by FEMWASTE.

Special Features. VS2DT solves the two-dimensional version of the transport equation [Eq. (22.1.4)] using a finite-difference approximation. It includes first-order decay and adsorption as described by Freundich or Langmuir isotherms and ion exchange.

Computer Requirements. VS2DT was written in FORTRAN 77 for a PRIME mainframe computer and will require some reprogramming for other types of systems. A version for IBM-compatible microcomputers is also available.

How to Obtain the Code. The source code is distributed by the USGS. ECE sells a version with a graphical pre- and postprocessor.

SUTRA. SUTRA (Saturated-Unsaturated Transport), developed by the USGS,[151] is a complex code that requires a sophisticated user.

Types of Problems. SUTRA simulates two-dimensional, density-dependent flow and transport of either a dissolved solute or heat under variably saturated conditions. It can simulate variable density leachate, salt-water intrusion, thermal energy storage, thermal pollution of aquifers, and problems involving geothermal reservoirs.

SUTRA is intended mainly for the simulation of flow and solute or heat transport in fully saturated systems. The capability to simulate variably saturated flow is included to assist in the analysis of mildly nonlinear problems that also may involve thermal energy or solute transport. SUTRA is not specialized for the nonlinearities of unsaturated flow and consequently requires fine spatial and temporal discretization for simulations involving unsaturated porous media.

Required Input. SUTRA requires all the input needed for a variably saturated flow model (for example, see discussion under FEMWATER) and the input for a companion transport code (see discussion under FEMWASTE) as well as the density and viscosity of the fluid. It also requires the thermal conductivity of the porous medium and information on boundary and initial conditions to solve the heat-transport equation, if heat transport is simulated.

Special Features. SUTRA uses a numerical approximation method that combines integrated finite differences and quadrilateral finite elements utilizing either a cartesian or cylindrical coordinate system. SUTRA solves two partial differential equations—one for fluid flow and another for either concentration or temperature. Postprocessing packages include NAG (see information under FEMWATER) and packages developed by the USGS[131] and Geraghty and Miller Inc.

Computer Requirements. SUTRA runs on mainframes and on 386-based IBM-compatible and Macintosh microcomputers.

How to Obtain the Code. The mainframe version of SUTRA is available from the USGS and the IGWMC. Versions for IBM-compatible microcomputers are available with interactive postprocessors from Geraghty and Miller Inc., the Scientific Software Group, and ECE. Geraghty and Miller Inc. also sells a version for the Macintosh Plus computer.

22.6.3 Other Numerical Codes

Other numerical codes for transport in variably saturated media include VAM2D, VAM3D, FLAMINCO, and SATURN. The IGWMC provides information about these codes.[149]

22.7 OTHER TYPES OF CODES

Other types of codes used in subsurface analysis include multiphase flow models, fracture flow models, geochemical models, optimization models, and inverse models. Currently these models are more often used in research than in professional practice, but, with additional testing, may eventually be widely used by practitioners.

Multiphase Flow. Flow of fluids of different density may involve miscible fluids, which mix and combine readily, or immiscible fluids such as nonaqueous-phase liquids (NAPLs), which do not mix with water. Miscible fluids are simulated using the solute transport models discussed in Sec. 22.3 and 22.6. Multiphase flow of immiscible fluids is discussed in Chap. 15 and 16.

SWANFLOW is a three-dimensional finite-difference code[37] from GeoTrans Inc. and the IGWMC. ARMOS and MOTRANS, available from the Scientific Software Group, are two-dimensional finite-element codes which simulate the flow of water and a NAPL within and below the unsaturated zone. TOUGH, a multiphase flow code that can also be used for fractured media, is discussed below.

Fracture-Flow Codes. Flow and contaminant transport in fractured rock have always been and will continue to be of interest to hydrologists. Although much recent literature has been devoted to the problem of simulating the flow of water and solutes through fractured rock,[99,126] there is no consensus on the most appropriate modeling technique. If the fracture network is dense and continuous, an equivalent porous media model such as those discussed in Sec. 22.2 and 22.5 may be used. If the fracture network consists of discrete fractures, models can be developed to simulate flow through each fracture individually. If the rock matrix surrounding the fractures is permeable, a dual-porosity model is used. In this type of model, flow through the fractures is accompanied by diffusion into the surrounding porous rock matrix.

SWIFT/386 (Sec. 22.3.2) was especially designed for fractured media. It includes both dual-porosity and discrete-fracture conceptualizations. TOUGH,[119,147] available from the ESTSC, is a multiphase fracture-flow code that simulates the coupled three-dimensional transport of liquid water, water vapor, air, and heat in variably saturated fractured media or continuous porous media. Other fracture flow codes include TRAFRAP-WT available from the IGWMC, NEFTRAN[92] from the ESTSC, and MAFIC2D/3D from Golder Associates Inc.

Geochemical Codes. Geochemical codes calculate the concentrations of ions in solution in water at chemical equilibrium by solving a mass balance problem.[12] Some of these codes, e.g., PHREEQE,[109] available from the USGS and the IGWMC, can also simulate the change in speciation that will occur with the addition or removal of a chemical to the system, a procedure known as *reaction path modeling.*[12] There have been a few attempts to incorporate geochemical codes into contaminant transport modeling,[63,95] but their application to practical problems is still limited.

Optimization Models. Optimization or management models link a code for flow and/or solute transport with an optimization code that considers decision variables related to economics.[45] Groundwater management models may consider hydraulics or aquifer remediation problems, such as the optimal placement of wells, or problems of policy evaluation and water allocation. AQMAN[89] is an optimization code developed by the USGS. MODMAN from GeoTrans Inc. is an optimization code for use with MODFLOW.

Inverse Models. Most modeling applications are directed toward solving the forward problem. That is, given information on characteristics of the subsurface geology and hydrology, the code solves for head or concentration. The uncertainty associated with specifying parameter values forces the modeler into a trial-and-error calibration of the forward model. During calibration, parameter values are adjusted until the simulated results approximately match observed heads or concentrations (Sec. 22.1.4). An automated parameter estimation or inverse model accepts the observed heads or concentrations as input and calculates the set of parameter values that yields the best match between simulated and observed heads or concentrations.[21–23,96]

FTWORK (Sec. 22.3.2) contains an inverse routine for calibration of steady-state flow problems. A two-dimensional inverse code for steady-state groundwater flow was developed by the USGS[29] and several applications have been reported.[26–28,30,42,153] The USGS recently released a three-dimensional inverse code based on MODFLOW (Sec. 22.2.2) and known as MODFLOWP.[60] The Scientific Software Group also distributes a three-dimensional inverse code for groundwater flow known as MODINV.

22.8 SOFTWARE DISTRIBUTORS FOR SUBSURFACE FLOW AND TRANSPORT CODES

ECE: Environmental Consulting Engineers, Inc., P.O. Box 22668, Knoxville, TN 37933. Tel: (615) 966-6622; fax: (615) 966-0123.

EPRI: Electric Power Research Institute, P.O. Box 10412, Palo Alto, CA 94303. Tel: (415) 855-2150; fax: (415) 855-1069.

ESTSC: Energy Science and Technology Software Center, P.O. Box 1020, Oakridge, TN 37831-1020. Tel: (615) 576-2606; fax: (615) 576-2685.

GEOCOMP Corporation, 66 Commonwealth Ave., Concord, MA 01742. Tel: (508) 369-8304; fax (508) 369-4392.

GeoTrans Inc., 46050 Manekin Plaza, Suite 100, Sterling, VA 20166. Tel (703) 444-7000; fax: (703) 444-1685.

Geraghty and Miller Inc., Groundwater Modeling Group, 10700 Parkridge Blvd., Suite 600, Reston, VA 22091. Tel: (703) 758-1200; fax: (703) 758-1201.

Golder Associates Inc., 4104 148th Ave., NE, Redmond, WA 98052. Tel: (206) 883-0777; fax: (206)882-5498.

IGWMC/Europe: International Groundwater Modeling Center Europe, P.O. Box 6012, 2600 JA Delft, The Netherlands. Tel: 31.15.697214; fax: 31.15.564800.

IGWMC/USA: International Groundwater Modeling Center USA, Colorado School of Mines, Golden, CO 80401. Tel: (303) 273-3103; fax (303) 273-3278.

Lewis Publishers Inc., 2000 Corporate Blvd. N.W., Boca Raton, FL 33431. Tel: (313) 475-8619; fax (303) 475-8650.

McDonald Morrissey Associates Inc., 11305 Taffrail Court, Reston, VA 22091. Tel: (703) 758-9031; fax: (703) 758-7998, or Box 180 South Road, Hopkinton, N.H. 03229. Tel: (603) 746-6195; fax: (603) 746-5224.

Neuman: Dr. Shlomo P. Neuman, Department of Hydrology and Water Resources, University of Arizona, Tucson, AZ 85721. Tel: (602) 621-7114; fax: (602) 621-1422.

S. S. Papadopulos and Associates Inc., 7944 Wisconsin Ave., Bethesda, MD 20814. Tel: (301) 468-5760; fax: (301) 881-0832.

T. A. Prickett and Associates, 6 G. H. Baker Dr., Urbana, IL 61801. Tel: (217) 384-0615; fax: (217) 384-0518.

Scientific Software Group, P.O. Box 23041, Washington, D.C. 20026-3041. Tel: (703) 620-9214; fax: (703) 620-6793.

Townley: Dr. Lloyd Townley, CSIRO Division of Water Resources, Western Australian Laboratories, Private Bag, PO Wembley, WA 6014, Australia. Tel: (09) 387 0200; fax 61-9-387-8211.

USGS: U.S. Geological Survey, Watstore Program Office, 437 National Center, Reston, VA 22092. Tel: (703) 648-5695; fax: (703) 648-5295.

Waterloo Hydrogeologic Software, Attn.: Dr. Nilson Guiguer, 200 Candlewood Cr., Waterloo, Ontario, Canada N2L5Y9. Tel/fax: (519) 746-1798.

Yeh: Dr. George T. Yeh, Department of Civil Engineering, Pennsylvania State University, 213 Sackett Bldg., State College, PA 16802. Tel: (814) 863-3014; fax: (814) 865-3287.

22.9 REFERENCES

1. Al-Layla, R., H. Uazicigil, and R. de Jong, "Numerical Modeling of Solute Transport Patterns in the Dammam Aquifer," *Water Res. Bull.,* vol. 24, no. 1, pp. 77–85, 1988.
2. Alley, W. M., and P. A. Emery, "Groundwater Model of the Blue River Basin, Nebraska —Twenty Years Later," *J. Hydrol.,* vol. 85, pp. 225–250, 1986
3. Andersen, P. F., P. N. Sims, D. H. Davis, S. C. Hughes, *A Numerical Model of the Hydrogeological System Underlying the Savannah River Plant,* GeoTrans Inc., Sterling, Va, 1989.
4. Anderson, G., R. Stein, D. Kock, and P. Mattejet, "Three-Dimensional Ground Water Quality Modeling in Support of Risk Assessment at the Louisiana Army Ammunition Plant," *Superfund '90,* Washington, D.C., pp. 896–900, 1990.
5. Anderson, M. P. and W. W. Woessner, "Comments on: Who are these Manuals For? The Model Documentation Needs of Practitioners," *Ground Water,* vol. 28, no. 1, pp. 113, 1990.
6. Anderson, M. P., "Aquifer Heterogeneity—a Geological Perspective," in *Parameter Identification and Estimation for Aquifer and Reservoir Characterization,* National Water Well Association, Dublin, Ohio, pp. 3–22, 1991.
7. Anderson, M. P., and W. W. Woessner, *Applied Groundwater Modeling: Simulation of Flow and Advective Transport,* Academic Press, Orlando, Fla., 1992.
8. Aral, M. M., *Ground Water Modeling in Multilayer Aquifers: Steady Flow,* Lewis Publishers, Boca Raton, Fla., 1989.
9. Aral, M. M., *Ground Water Modeling in Multilayer Aquifers: Unsteady Flow,* Lewis Publishers, Boca Raton, Fla., 1990.
10. Bair, E. S., R. A. Sheets, S. M. Eberts, "Particle-Tracking Analysis of Flow Paths and Travel Times from Hypothetical Spill Sites within the Capture Area of a Wellfield," *Ground Water,* vol. 28, no. 6, pp. 884–892, 1990.
11. Bair, E. S., A. E. Springer, G. S. Roadcap, "Delineation of Travel Time-Related Capture Areas of Wells using Analytical Flow Models and Particle-Tracking Analysis," *Ground Water,* vol. 29, no. 3, pp. 397, 1991.
12. Bassett, R. L., and D. Melchior, "Chemical Modeling of Aqueous Systems: An Over-

view," in *Chemical Modeling of Aqueous Systems II,* D. C. Melchior and R. L. Bassett (eds), American Chemical Society, pp. 1–15, 1989.

13. Bear, J., *Dynamics of Fluids in Porous Media,* American Elsevier, New York, 1972.
14. Bear, J., and A. Verruijt, *Modeling Groundwater Flow and Pollution,* D. Reidel, Dordrecht, The Netherlands, 1987.
15. Blandford, T. N., and P. S. Huyakorn, *WHPA: A Modular Semi-Analytical Model for the Delineation of Wellhead Protection Areas,* U.S. Environmental Protection Agency, Office of Ground-Water Protection, 1990.
16. Bond, L. D., and J. D. Bredehoeft, "Origins of Seawater Intrusion in a Coastal Aquifer—A Case Study of the Pajaro Valley, California," *J. Hydrol.,* vol. 92, pp. 363–388, 1987.
17. Bonn, B. A., and S. A. Rounds, *DREAM—Analytical Ground Water Flow Programs,* Lewis Publishers, Boca Raton, Fla., 1990.
18. Bush, P. W., "Simulation of Saltwater Movement in the Floridan Aquifer System, Hilton Head Island, South Carolina," Water-Supply Paper 2331, U.S. Geological Survey, 1988.
19. Buss,D. R., B. H. Lester, and J. W. Mercer, "A Numerical Simulation Study of Deep-Well Injection," *Current Practices in Environmental Sciences and Engineering,* vol. 2, pp. 93–117, 1986.
20. Buxton, H. T., T. E. Reilly, D. W. Pollock, and D. A. Smolensky, "Particle Tracking Analysis of Recharge Areas on Long Island, New York," *Ground Water,* vol. 29, no. 1, pp. 63–71, 1991.
21. Carrera, J., "State of the Art of the Inverse Problem Applied to the Flow and Solute Transport Equations," in *Groundwater Flow and Quality Modelling,* E. Custodio et al. (eds.), D. Reidel, Dordrecht, The Netherlands, pp. 549–583, 1988.
22. Carrera, J., and S. P. Neuman, "Estimation of Aquifer Parameters Under Transient and Steady State Conditions—3. Application to Synthetic and Field Data," *Water Resour. Res.,* vol. 22, no. 2, pp. 228–242, 1986.
23. Carrera, J., et al., "Three-Dimensional Modeling of Saline Pond Leakage Calibrated by INVERT-3, a Quasi Three-Dimensional, Transient Parameter Estimation Program," in *Practical Applications of Ground Water Models,* National Water Well Association, Columbus, Ohio, pp. 547–569, 1984.
24. Chapelle, F. H., "A Solute-Transport Simulation of Brackish-Water Intrusion near Baltimore, Maryland," *Ground Water,* vol. 24, no. 3, pp. 304–311, 1986.
25. Cherkauer, D. S., and P. F. McKereghan, "Ground-Water Discharge to Lakes: Focusing in Embayments," *Ground Water,* vol. 29, no. 1, pp. 72–81, 1991.
26. Connell, J. F. and Z. C. Bailey, "Statistical and Simulation Analysis of Hydraulic-Conductivity Data for Bear Creek and Melton Valleys," Oak Ridge Reservation, Tennessee, Water-Resources Investigations Report 89-4062, U.S. Geological Survey, 1989.
27. Cooley, R. L., "A Method of Estimating Parameters and Assessing Reliability for Models of Steady State Groundwater Flow," *Water Resour. Res.,* vol. 15, no. 3, pp. 603–617, 1979.
28. Cooley, R. L., L. F. Konikow, and R. L. Naff, "Non-Linear-Regression Groundwater Flow Modeling of a Deep Regional Aquifer System," *Water Resour. Res.,* vol. 22, no. 13, pp. 1759–1778, 1986.
29. Cooley, R. L., and R. L. Naff, "Regression Modeling of Ground-Water Flow," Techniques of Water-Resources Investigations O3-B4, U.S. Geological Survey, 1990.
30. Czarnecki, J. B., and R. K. Waddell, "Finite-Element Simulation of Ground-Water Flow in the Vicinity of Yucca Mountain, Nevada-California, Water-Resources Investigations Report 84-4349, U.S. Geological Survey, 1984.
31. Davis, L. A., and S. P. Neuman, "Documentation and User's Guide to UNSAT2-Variably Saturated Flow Model," NUREG/CR-3390, WWL/TM-1791-1, U.S. Nuclear Regulatory Commission, 1983.

32. Domenico, P. A., and V. V. Palciauskas, "Alternative Boundaries in Solid Waste Management," *Ground Water,* vol. 20, no. 3, pp. 303–311, 1982.

33. Domenico, P. A. and F. W.Schwartz, *Physical and Chemical Hydrogeology,* Wiley, New York, 1990.

34. Doran, F. J. and J. E. Thresher, Jr., "Prediction of the Fate of Chromium from a Proposed Municipal Sanitary Landfill," in *Solving Ground Water Problems with Models,* National Water Well Association, Dublin, Ohio, pp. 253–284, 1987.

35. Duguid, J. O., and M. Reeves, "Material Transport through Porous Media: a Finite-Element Galerkin Model", Oak Ridge National Laboratory, ORNL-4928, Environmental Sciences Division, Publication 733, 1976.

36. El-Kadi, A. I., "Applying the USGS Mass-Transport Model (MOC) to Remedial Actions by Recovery Wells," *Ground Water,* vol. 26, no. 3, pp. 281–288, 1988.

37. Faust, C. R., J. H. Guswa, and J. W. Mercer, "Simulation of Three-Dimensional Flow of Immiscible Fluids within and below the Unsaturated Zone," *Water Resour. Res.,* vol. 25, no. 12, pp. 2449–2464, 1989.

38. Faust, C. R., L. R. Silka, and J. W. Mercer, "Computer Modeling and Ground-Water Protection," *Ground Water,* vol. 19, no. 4, pp. 362–365, 1981.

39. Feddes, R. A., E. Bresler, and S. P. Neuman, "Finite Element Analysis of Two-Dimensional Flow in Soils Considering Water Uptake by Roots: II. Field Applications," *Soil Sci. Soc. Amer. Proc.,* vol. 39, pp. 231–237, 1975.

40. Flavelle, P., S. Nguyen, and W. Napier, "Lessons Learned from Model Validation—A Regulatory Perspective," GEOVAL-1990: *Symposium on Validation of Geosphere Flow and Transport Models,* Organization for Economic Co-operation and Development (OEDC), Nuclear Energy Agency, Paris, pp. 580–588, 1991.

41. Franz, T., and N. Guiguer, *FLOWPATH, Two-Dimensional Horizontal Aquifer Simulation Model,* Waterloo Hydrogeologic Software, Waterloo, Ontario, 1990.

42. Garabedian, S. P., "Application of a Parameter-Estimation Technique to Modeling the Regional Aquifer Underlying the Eastern Snake River Plain, Idaho," Water-Supply Paper 2278, U.S. Geological Survey, 1986.

43. GeoTrans, Inc., "Analytic Element Modeling of Groundwater Flow," *GeoTrans Newsletter,* Spring 1991.

44. Goode, D. J., and L. F. Konikow, "Modification of a Method-of-Characteristics Solute Transport Model to Incorporate Decay and Equilibrium-Controlled Sorption or Ion Exchange," Water-Resources Investigations Report 89-4030, U.S. Geological Survey, 1989.

45. Gorelick, S. M., "A Review of Distributed Parameter Groundwater Management Modeling Methods," *Water Resour. Res.,* vol. 19, no. 2, pp. 305–319, 1983.

46. "An Update on ASTM Subcommittee D-18.21 on Ground Water and Vadose Zone Investigations," *Ground Water Monitoring Review,* vol. X, no. 2, pp. 74–77, 1990.

47. Gupta, A. D., and S. Sabanathan, "Saltwater Transport in a Heterogeneous Formation: A Case Study," *Transport in Porous Media,* vol. 3, no. 3, pp. 217–256, 1988.

48. Gupta, S. K., C. R. Cole, F. W. Bond, and A. M. Monti, "Finite-Element Three-Dimensional Ground-Water (FE3GW) Flow Model: Formulation, Computer Source Listings, and User's Manual," BMI/ONWI-548, Office of Nuclear Waste Isolation, Battelle Memorial Institute, Columbus, Ohio, 1984.

49. Gupta, S. K., C. R. Cole, C. T. Kincaid, and A. M. Monti, "Coupled Fluid, Energy and Solute Transport (CFEST) Model: Formulation and User's Manual," BMI/ONWI-660, Office of Nuclear Waste Isolation, Battelle Memorial Institute, Columbus, Ohio, 1987.

50. Gupta, S. K., C. R. Cole, and G. F. Pinder, "A Finite-Element Three Dimensional Ground-Water (FE3DGW) Model for a Multiaquifer System," *Water Resour. Res.,* vol. 20, no. 5, pp. 553–563, 1984.

51. Gupta, S. K., K. K. Tanji, and J. N. Luthin, "A Three-Dimensional Finite Element Groundwater Model," California Water Research Center, contrib. 152, 1975.

52. Guzman, A. G., L. G. Wilson, S. P. Neuman, and M. D. Osborn, "Simulating Effect of Channel Changes on Stream Infiltration," *J. Hydraul. Eng.,* vol. 115, no. 12, pp. 1631–1645, 1988.

53. Haug, A., "A Modeling Study of the Culebra Dolomite," in *Solving Ground Water Problems with Models,* National Water Well Association, Dublin, Ohio, pp. 827–852, 1987.

54. Healy, R. W., "Simulation of Solute Transport in Variably Saturated Porous Media with Supplemental Information of Modifications to the U.S. Geological Survey's Computer Program VS2D," Water-Resources Investigations Report 90-4025, U.S. Geological Survey, 1990.

55. Hemker, C. J., and H. van Elburg, *Micro-Fem, Version 2.0, User's Manual,* Hemker and van Elburg, Elandsgracht 83, 1016TR, Amsterdam, The Netherlands, 1987.

56. Hensel, B. R., D. A. Keefer, R. A. Griffin, R. C. Berg, "Numerical Assessment of a Landfill Compliance Limit," *Ground Water,* vol. 29, no. 2, pp. 218–224, 1991.

57. Hern, S. C., and S. M. Melancon, *Vadose Zone Modeling of Organic Pollutants,* Lewis Publishers Inc., Boca Raton, Fla., 1986.

58. Hill, M. C., "Preconditioned Conjugate-Gradient 2 (PCG2), a Computer Program for Solving Ground-Water Flow Equations," Water-Resources Investigations Report 90-4048, U.S. Geological Survey, 1990, 43 p.

59. Hill, M. C., "Solving Groundwater Flow Problems by Conjugate-Gradient Methods and the Strongly Implicit Procedure," *Water Resour. Res.,* vol. 26, no. 9, pp, 1961–1970, 1970.

60. Hill, M. C., "MODFLOWP: A Computer Program for Estimating Parameters of a Transient, Three-Dimensional Ground-Water Flow Model Using Nonlinear Regression, Open-File Report 91-484, U.S. Geological Survey, 1990.

61. Holmes, K. J., W. Chu, and D. R. Erickson, "Automated Calibration of a Contaminant Transport Model for a Shallow Sand Aquifer," *Ground Water,* vol. 27, no. 4, pp. 501–508, 1989.

62. Hossain, M. A., and M. Y. Corapcioglu, "Modifying the USGS Solute Transport Computer Model to Predict High-Density Hydrocarbon Migration," *Ground Water,* vol. 26, no. 6, pp. 717–723, 1988.

63. Hostetler, C. J., R. L. Erikson, J. S. Fruchter, and C. T. Kincaid, "Overview of FASTCHEM™ Code Package: Application to Chemical Transport Problems," report EQ-5870-CCM, vol. 1, Electric Power Research Institute, Palo Alto, Calif., 1988.

64. Huyakorn, P. S., and G. F. Pinder, *Computational Methods in Subsurface Flow,* Academic Press, New York, 1983.

65. Istok, J., *Groundwater Modeling by the Finite Element Method,* Monograph 13, American Geophysical Union, Washington, D.C., 1989.

66. Javendel, I., C. Doughty, and C. F. Tsang, *Groundwater Transport: Handbook of Mathematical Models,* Water-Resources Monograph 10, American Geophysical Union, Washington, D.C., 1984.

67. Karanjac, J., M. Altankaynak, and G. Oval, "Mathematical Model of Uluova Plain, Turkey—a Training and Management Tool," *Ground Water,* vol. 15, no. 5, pp. 348–357, 1977.

68. Karanjac, J., and D. Braticevic, "Ground Water Software Series (Overivew)," United Nations, Department of Technical Co-operation for Development, Natural Resources and Energy Division Water Resources Branch, New York, 1989.

69. Khaleel, R., and D. L. Reddell, "MOC Solutions of Convective-Dispersion Problems," *Ground Water,* vol. 24, no. 6, pp. 798–810, 1986.

70. Kincaid, C. T., et al., *Geohydrochemical Models for Solute Migration,* vol. 2: *Preliminary Evaluation of Selected Computer Codes,* EA-3417, Electric Power Research Institute, 1984.

71. Kinzelbach, W., *Groundwater Modelling: An Introduction with Sample Programs in BASIC,* Developments in Water Science 25, Elsevier, New York, 1986.

72. Kinzelbach, W., "Methods for Simulation of Pollutant Transport in Ground Water—A Model Comparison," in *Solving Problems with Ground Water Models,* National Water Well Association, Dublin, Ohio, pp. 656–675, 1987.

73. Kipp, K. L., Jr., "Adaptation of the Carter-Tracy Water Influx Calculation to Groundwater Flow Simulation," *Water Resour. Res.,* vol. 22, no. 3, pp. 423–428, 1986.

74. Kipp, K. L., Jr., "HST3D: A Computer Code for Simulation of Heat and Solute Transport in Three-Dimensional Ground-Water Systems," Water-Resources Investigations Report 86-4095, U.S. Geological Survey, 1987.

75. Koch, D., et al., "Using a Three-Dimensional Solute Transport Model to Evaluate Remedial Actions for Groundwater Contamination at the Picatinny Arsenal, New Jersey," *Superfund '89,* Washington, D.C., pp. 152–156, 1989.

76. Koch, D. L., and T. A. Prickett, *User's Manual for RD3D, a Three-Dimensional Mass Transport Random Walk Model Attachment to the USGS MODFLOW Three-Dimensional Flow Model,* Joint Engineering Technologies Associates and T. A. Prickett and Associates, Scientific Publication 3, Ellicott City, Md., 1989.

77. Konikow, L. F., "Modeling Chloride Movement in the Alluvial Aquifer at the Rocky Mountain Arsenal, Colorado," Water-Supply Paper 2044, U.S. Geological Survey, 1977.

78. Konikow, L. F., "Predictive Accuracy of a Ground-Water Model—Lessons from a Postaudit," *Ground Water,* vol. 24, no. 2, pp. 173–184, 1986.

79. Konikow, L. F., and J. D. Bredehoeft, "Modeling Flow and Chemical Quality Changes in an Irrigated Stream-Aquifer System," *Water Resour. Res.,* vol. 10, no. 3, pp. 546–562, 1974.

80. Konikow L. F., and J. D. Bredehoeft, "Computer Model of Two-Dimensional Solute Transport and Dispersion in Ground Water," Techniques of Water-Resources Investigations, Book 7, Chap. C2, U.S. Geological Survey, 1978.

81. Konikow, L. F., and M. Person, "Assessment of Long-Term Salinity Changes in an Irrigated Stream-Aquifer System," *Water Resour. Res.,* vol. 21, no. 11, pp. 1611–1624, 1985.

82. Kontis, A. L., and R. J. Mandle, "Modifications of a Three-Dimensional Ground-Water Flow Model to Account for Variable Water Density and Effects of Multiaquifer Wells," Water-Resources Investigations Report 87-4265, U.S. Geological Survey, 1988.

83. Kuiper, L. K., "A Numerical Procedure for the Solution of the Steady-State Variable Density Groundwater Flow Equation," *Water Resour. Res.,* vol. 19, no. 1, pp. 243–240, 1983.

84. Kuiper, L. K., "Documentation of a Numerical Code for the Simulation of Variable Density Ground-Water Flow in Three Dimensions," Water-Resources Investigations Report 84-4302, U.S. Geological Survey, 1985.

85. Lappala, E. G., R. W. Healy, and E. P. Weeks, "Documentation of Computer Program VS2D to Solve the Equations of Fluid Flow in Variably Saturated Porous Media," Water-Resources Investigations Report 83-4099, U.S. Geological Survey, 1987.

86. Leake, S. A., "Interbed Storage Changes and Compaction in Models of Regional Groundwater Flow," *Water Resour. Res.,* vol. 26, no. 9, pp. 1939–1950, 1990.

87. Leake, S. A., and D. E. Prudic, "Documentation of a Computer Program to Simulate Aquifer-System Compaction Using the Modular Finite-Difference Model," Open File Report 88-482, U.S. Geological Survey, 1988.

88. LeBlanc, D. R., "Sewage Plume in a Sand and Gravel Aquifer, Cape Cod, Massachusetts," Water-Supply Paper 2281, U.S. Geological Survey, 1984.

89. Lefkoff, L. J., and S. M. Gorelick, "AQMAN: Linear and Quadratic Programming Matrix Generator Using Two-Dimensional Ground-Water Flow Simulation for Aquifer Management Modeling," Water-Resources Investigations Report 87-4061, 1987.

90. Lewis, B. D., and F. S. Goldstein, "Evaluation of a Predictive Ground-Water Solute-Transport Model at the Idaho National Engineering Laboratory, Idaho," Water-Resources Investigations Report 82-85, U.S. Geological Survey, 1982.

91. Liggett, J. A., and P. L.-F. Liu, *The Boundary Integral Equation Method for Porous Media Flow,* Allen and Unwin, London, 1983.

92. Longsine, D. ., E. J. Bonano, and C. P. Harlan, "User's Manual for the NEFTRAN Computer Code," NUREG/CR-4766, Nuclear Regulatory Commission, Washington, D.C., 1987.

93. McDonald, M. G., and A. W. Harbaugh, "A Modular Three-Dimensional Finite-Difference Ground-Water Flow Model," *Techniques of Water--Resources Investigations,* 06-A1, U.S. Geological Survey, 1988.

94. McLaughlin, D., and W. K. Johnson, "Comparison of Three Groundwater Modeling Studies," *J. Water Resour. Plann. Manage.,* vol. 113, no. 3, pp. 405–421, 1987.

95. Mangold, D. C., and C-F Tsang, "A Summary of Subsurface Hydrological and Hydrochemical Models," *Reviews of Geophysics,* vol. 29, no. 1, pp. 51–79, 1991.

96. Menke, W., *Geophysical Data Analysis: Discrete Inverse Theory,* Academic Press, Orlando, Fla., 1989.

97. Mercer, J. W., B. H. Lester, S. D. Thomas, and R. L. Bartel, "Simulation of Saltwater Intrusion in Volusia County, Florida," *Water Res. Bull.,* vol. 22, no. 6, pp. 951–965, 1986.

98. National Council of the Paper Industry for Air and Stream Improvement Inc., "A Review and Further Documentation of the Groundwater Solute Transport Model Random Walk and Guidance for its Use," Technical Bulletin 505, 1986.

99. National Research Council, *Ground Water Models: Scientific and Regulatory Applications,* National Academy Press, Washington, D.C., 1990.

100. Nelson, R. W., "Evaluating the Environmental Consequences of Groundwater Contamination," *Water Resour. Res.,* vol. 14, no. 3, pp. 409–428, 1978.

101. Neuman, S. P., *User's Guide for FREESURF I,* Department of Hydrology and Water Resources, University of Arizona, Tuscon, 1976.

102. Neuman, S. P., R. A. Feddes, and E. Bresler, *Finite Element Simulation of Flow in Saturated–Unsaturated Soils Considering Water Uptake by Plants,* Technion, Hydrodynamics and Hydraulic Engineering Laboratory Report, Haifa, Israel, 1974.

103. Newell, C. J., et al., "OASIS: Parameter Estimation System for Aquifer Restoration Models, User's Manual Version 2.0," EPA/600/S8-90/039, U.S. Environmental Protection Agency, Washington, D.C., 1990.

104. Newell, C. J., J. F. Haasbeek, and P. B. Bedient, "OASIS: A Graphical Decision Support System for Ground-Water Contaminant Modeling," *Ground Water,* vol. 28, no. 2, pp. 224–234, 1990.

105. Oberlander, P. L. and R. W. Nelson, "An Idealized Ground-Water Flow and Chemical Transport Model (S-PATHS)," *Ground Water,* vol. 22, no. 4, pp. 441–449, 1984.

106. Olsthoorn, T. N., "The Power of the Electronic Worksheet: Modeling without Special Programs," *Ground Water,* vol. 27, no. 2, pp. 193–201, 1985.

107. Oster, C. A., "Review of Ground-Water Flow and Transport Models in the Unsaturated Zone," U.S. Nuclear Regulatory Commission, NUREG/CR-2917, PNL-4427, 1982.

108. Ousey, J. R., Jr., "Modeling Steady-State Groundwater Flow Using Microcomputer Spreadsheets," *J. Geol. Educ.,* vol. 34, pp. 305–311, 1986.

109. Parkhurst, D. L., D. C. Thorstenson, and L. N. Plummer, "PHREEQE—A Computer Program for Geochemical Calculations," Water-Resources Investigations Report 80-96, U.S. Geological Survey, 1980.

110. Pinder, G. F., and C. I. Voss, *AQUIFEM, A Finite Element Model for Aquifer Evaluation (Documentation),* Department of Water Resource Engineering, Royal Institute of Technology, Stockholm, Sweden, Report 7911, TRITA-VAT-3806, 1979.

111. Pollock, D. W., "Semianalytical Computation of Path Lines for Finite Difference Models," *Ground Water,* vol. 26, no. 6, pp. 743–750, 1988.

112. Pollock, D. W., "Documentation of Computer Programs to Compute and Display Pathlines using Results from the U.S. Geological Survey Modular Three-Dimensional Finite-Difference Ground-Water Model," Open-File Report 89-381, U.S. Geological Survey, 1989.

113. Potter, S. T., and W. J. Gburek, "Seepage Face Simulation Using PLASM," *Ground Water,* vol. 25, no. 6, pp. 722–732, 1987.

114. Prickett, T. A., *User's Manual for the Standard PLASM Two-Dimensional Groundwater Flow Model for Desk-Top Computers,* version 6B, T. A. Prickett and Associates, Urbana, Ill., 1990.

115. Prickett, T. A., *User's Manual for the Standard RNDWALK Two-Dimensional Mass Transport Model for Desk-Top Computers with Geographic Overlays, Version 6C,* T. A. Prickett and Associates, Urbana, Ill., 1990.

116. Prickett, T. A., and C. G. Lonnquist, "Selected Digital Computer Techniques for Groundwater Resource Evaluation," *Ill. State Water Surv. Bull.,* vol. 55, 1972.

117. Prickett, T. A., T. G. Naymik, and C. G. Lonnquist, "A Random-Walk Solute Transport Model for Selected Groundwater Quality Evaluations," *Ill. State Water Surv. Bull.,* vol. 65, 1981.

118. Prudic, D. E., "Documentation of a Computer Program to Stimulate Stream-Aquifer Relations Using a Modular, Finite-Difference, Ground-Water Flow Model, Open-File Report 88-729, U.S. Geological Survey, 1989.

119. Pruess, K., "TOUGH User's Guide," NUREG/CR-4645, SAND86-7104, LBL-20700, U.S. Nuclear Regulatory Commission, 1987.

120. Reeves, M., and J. O. Duguid, "Water Movement through Saturated-Unsaturated Porous Media: A Finite-Element Galerkin Model," Oak Ridge National Laboratory, ORNL-4927, 1976.

121. Reeves, M., D. S. Ward, P. A. Davis, and E. J. Bonano, "SWIFT II Self-Teaching Curriculum: Illustrative Problems for the Sandia Waste-Isolation Flow and Transport Model for Fractured Media," NUREG/CR-3925, SAND84-1586, Sandia National Laboratories, Albuquerque, N.M., 1987.

122. Reeves, M., D. S. Ward, N. D. Johns, and R. M. Cranwell, "Theory and Implementation for SWIFT II, The Sandia Waste-Isolation Flow and Transport Model for Fractured Media, Release 4.84," NUREG/CR-3328, SAND83-1159, Sandia National Laboratory, Albuquerque, N.M., 1986.

123. Rifai, J. S., P. B. Bedient, R. C. Borden, and F. F. Haasbeek, "BIOPLUME II—Computer Model of Two-Dimensional Transport under the Influence of Oxygen Limited Biodegradation in Ground Water," EPA/600/8-88/093a, Environmental Protection Agency, Washington, D.C., 1988.

124. Robson, S. G., "Feasibility of Digital Water-Quality Modeling Illustrated by Application at Barstow, California," Water-Resources Investigations Report 46-73, U.S. Geological Survey, 1974.

125. Sanford, W. E., and L. F. Konikow, "A Two-Constituent Solute-Transport Model for Ground Water having Variable Density," Water-Resources Investigations Report 85-4279, U.S. Geological Survey, 1985.

126. Schmelling, S. G., and R. R. Ross, "Contaminant Transport in Fractured Media: Models for Decision Makers," EPA/540/4-89/004, U.S. Environmental Protection Agency, 1989.

127. Scott, J. C., "A Statistical Processor for Analyzing Simulations Made Using the Modular Finite-Difference Ground-Water Flow Model," Water-Resources Investigations Report 89-4159, U.S. Geological Survey, 1989.

128. Shafer, J. M.,"Reverse Pathline Calculation of Time-Related Capture Zones in Nonuniform Flow," *Ground Water,* vol. 25, no. 3, pp. 283–289, 1987.

129. Shafer, J. M., GWPATH, *version 4.0,* J. M. Shafer, 1013 Devonshire Dr., Champaign, Il 61821, 1990.

130. Sims, P. N., P. F. Andersen, D. E. Stephenson, and C. R. Faust, "Testing and Benchmarking of a Three-Dimensional Groundwater Flow and Solute Transport Model," *Proc. Solving Ground Water Problems with Models Conf.,* International Ground Water Modeling Center and Association of Ground Water Scientists and Engineers, Indianapolis, Ind., pp. 821–841, 1989.

131. Souza, W. R., "Documentation of a Graphical Display Program for the Saturated-Unsaturated Transport (SUTRA) Finite-Element Simulation Model," Water-Resources Investigations Report 87-4245, 1987.

132. Srinivasan, P., and J. W. Mercer, "BIO1D One-Dimensional Model for Comparison of Biodegradation and Adsorption Processes in Contaminant Transport," GeoTrans, Sterling Va., 1987.

133. Srinivasan, P., and J. W. Mercer, "Simulation of Biodegradation and Sorption Processes in Ground Water," *Ground Water,* vol. 26, no. 4, pp. 475–487, 1988.

134. Stephens, D. B., "Groundwater Flow and Implications for Groundwater Contamination North of Prewitt, New Mexico, U.S.A.," *J. Hydrol.,* vol. 61, 391–408, 1983.

135. Stevens, W. S., A. M. Klock, and R. T. Leigh, "Remedial Investigation and Design for a Mine-Tailings Wastewater Facility in Southwestern Wyoming," in *Solving Ground Water Problems with Models,* National Water Well Association, Dublin, Ohio, pp. 285–312, 1987.

136. Strack, O. D. L., "The Analytic Element Method for Regional Groundwater Modeling," in *Solving Ground Water Problems with Models,* National Water Well Association, Dublin, Ohio, pp. 929–941, 1987.

137. Strack, O. D. L., *Groundwater Mechanics,* Prentice Hall, Englewood Cliffs, N.J., 1988.

138. Strack, O. D. L., *Technical Summary: SLAEM Single Layer Analytic Element Model,* 23 Black Oak Rd., North Oaks, MN 55127, 1990.

139. Strecker, E. W., and W-S. Chu, "Parameter Identification of a Ground-Water Contaminant Transport Model," *Ground Water,* vol. 24, no. 1, pp. 56–62, 1986.

140. Sudicky, E. A., and HydroGeoLogic, Inc., *CANSAZ: Combined analytical numerical code for stimulating flow and contaminant transport in the saturated zone module of EPA Composite Model for Surface Impoundments (EPACMS),* prepared for U.S. EPA, Office of Solid Waste, Washington, D.C., 1988.

141. Summers, K. V., et al., "MYGRT® Code Version 2.0: An IBM Code for Simulating Migration of Organic and Inorganic Chemicals in Groundwater," EN-6531 Project 2879-2, Electric Power Research Institute, Palo Alto, Calif., 1989.

142. Townley, L. R., *AQUIFEM-N, A Multi-Layered Finite Element Aquifer Flow Model,* CSIRO, Western Australian Laboratories, Wembley, 1990.

143. Townley, L. R., and J. L. Wilson, "Description of and User's Manual for a Finite Element Aquifer Flow Model AQUIFEM-1," MIT Ralph M. Parsons Laboratory for Water Resources and Hydrodynamics, Technology Adaptation Program, report no. 79-3, Cambridge, Mass., 1980.

144. Tracy, J. V., "Users Guide and Documentation for Adsorption and Decay Modifications to the USGS Solute Transport Model," NUREG/CR-2502, U.S. Nuclear Regulatory Commission, 1982.

145. Trescott, P. C., "Documentation of Finite-Difference Model for Simulation of Three-Dimensional Ground-Water Flow," Open File Report 75-438, U.S. Geological Survey, 1975.

146. Trescott, P. C., G. F. Pinder, and S. P. Larson, "Finite-Difference Model for Aquifer Simulation in Two Dimensions with Results of Numerical Experiments," *Techniques of Water-Resources Investigations,* Book 7, U.S. Geological Survey, 1976.

147. U.S. Nuclear Regulatory Commission, "Comparison of Strongly Heat-Driven Flow Codes for Unsaturated Media," NUREG/CR-5367, 1989.

148. van der Heijde, P. K. M., et al., *Groundwater Management: the Use of Numerical Models* (2d ed.), American Geophysical Union, Washington, D.C., 1985.

149. van der Heijde, P. K. M., A. I. El-Kadi, and S. A. Williams, "Groundwater Modeling: An Overview and Status Report," EPA/600/2-89/028, U.S. Environmental Protection Agency, 1988.

150. Voss, C. I., "AQUIFEM-SALT: A Finite-Element Model for Aquifers Containing a Seawater Interface," Water-Resources Investigations Report 84-4263, U.S. Geological Survey, 1984.

151. Voss, C. T., "A Finite-Element Simulation Model for Saturated-Unsaturated, Fluid-Density-Dependent Groundwater Flow with Energy Transport or Chemically-Reactive Single-Species Solute Transport," Water-Resources Investigations Report 84-4369, U.S. Geological Survey, 1984.

152. Voss, C. I., and W. R. Souza, "Variable Density Flow and Solute Transport Simulation of Regional Aquifers Containing a Narrow Freshwater-Saltwater Transition Zone," *Water Resour. Res.*, vol. 23, no. 10, pp. 1851–1866, 1987.

153. Waddell, R. K., "Two-Dimensional, Steady-State Model of Ground-Water Flow, Nevada Test Site and Vicinity, Nevada-California," Water-Resources Investigations Report 82-4085, U.S. Geological Survey, 1982.

154. Walton, W. C., *Practical Aspects of Ground Water Modeling,* National Water Well Association, Dublin, Ohio, 1985.

155. Walton, W. C., *Analytical Groundwater Modeling: Flow and Contaminant Migration,* Lewis Publishers, Boca Raton, Fla., 1989.

156. Walton, W., *Numerical Groundwater Modeling: Flow and Contaminant Migration,* Lewis Publishers, Boca Raton, Fla., 1989.

157. Wang, H. F., and M. P. Anderson, *Introduction to Groundwater Modeling: Finite Difference and Finite Element Methods,* W. H. Freeman, San Francisco, 1982.

158. Ward, D. S., "Data Input Guide for SWIFT/386, Version 2.50," *GeoTrans Technical Report,* Sterling, Va., 1991.

159. Ward, D. S., D. R. Buss, J. W. Mercer, and S. S. Hughes, "Evaluation of a Groundwater Corrective Action at the Chem-Dyne Hazardous Waste Site using a Telescopic Mesh Refinement Modeling Approach," *Water Resour. Res.,* vol. 23, no. 4, pp. 603–617, 1987.

160. Ward, D. S., D. R. Buss, D. M. Morganwalp, and T. D. Wadsworth, "Waste Confinement Performance of Deep Injection Wells," *Solving Ground Water Problems with Models,* National Water Well Association, Dublin, Ohio, pp. 764–781, 1987.

161. Ward, D. S., M. Reeves, and L. E. Duda, "Verification and Field Comparison of the Sandia Waste-Isolation Flow and Transport Model (SWIFT). NUREG/CR-3316, SAND83-1154, Sandia National Laboratories, Albuquerque, N.M., 1984.

162. Ward, D. S., T. D. Wadsworth, D. R. Buss, and J. W. Mercer, "Analysis of Potential Failure Mechanisms Pertaining to Hazardous Waste Injection in the Texas Gulf Coast Region," *J. Underground Injection Practices Council,* vol. 1, pp. 120–152, 1986.

163. Weaver, T. R., and J. M. Bahr, "Geochemical Evolution in the Cambrian-Ordovician Sandstone Aquifer, Eastern Wisconsin, 2. Correlation between Flow Paths and Ground-Water Chemistry," *Ground Water,* vol. 29, no. 4, pp. 510–515, 1991.

164. Wei, C. Y. and W. Y. J. Shieh, "Transient Seepage Analysis of Guri Dam," *J. Tech. Council ASCE,* vol. 105(TCI), pp. 135–147, 1979.

165. Wexler, E. J., "Analytical Solutions for One-, Two-, and Three-Dimensional Solute Transport in Ground Water Systems with Uniform Flow, Open-File Report 89-56, U.S. Geological Survey, 1989.

166. Wilson, J. L., L. R. Townley, and A. Sa da Costa, "Mathematical Development and Verification of a Finite Element Aquifer Flow Model AQUIFEM-1," MIT Technology Adaptation Program, report 79-2, Cambridge, Mass., 1979.

167. Yager, R. M., and M. P. Bergeron, "Nitrogen Transport in a Shallow Outwash Aquifer at

Olean, Cattaraugus County, New York," Water-Resources Investigations Report 87-4043, 1988.

168. Yeh, G. T., "AT123D, Analytical Transient One-, Two-, and Three-Dimensional Simulation of Waste Transport in the Aquifer System," ORNL-5602, Oak Ridge National Laboratory, Oak Ridge, Tenn., 1981.

169. Yeh, G. T., "FEMWATER: A Finite Element Model of WATER Flow through Saturated-Unsaturated Porous Media—First Revision," Oak Ridge National Laboratory, Environmental Science Division, publication no. 2943, Oak Ridge, Tenn., 1987.

170. Yeh, G. T., "3DFEMWATER: A Three-Dimensinal Finite Element Model of WATER flow through Saturated-Unsaturated Media," Oak Ridge National Laboratory, Environmental Sciences Division, publication no. 2904, 1987.

171. Yeh, G. T., *Users' Manual: A Three-Dimensional Finite Element Model of WATER Flow through Saturated-Unsaturated Media,* Pennsylvania State University, State College, 1990.

172. Yeh, G. T., *Users' Manual of a Hybrid Lagrangian-Eulerian Finite Element Model of WASTE Transport through Saturated-Unsaturated Media,* Pennsylvania State University, State College, 1990.

173. Yeh, G. T., and D. S. Ward, "FEMWATER: A Finite-Element Model of Water Flow through Saturated-Unsaturated Porous Media," ORNL-5567, Oak Ridge National Laboratory, Oak Ridge, Tenn., 1979.

174. Yeh, G. T., and D. S. Ward, "FEMWASTE: A Finite-Element Model of Waste Transport through Saturated-Unsaturated Porous Media," ORNL-5601, Oak Ridge National Laboratory, Oak Ridge, Tenn., 1981.

175. Zheng, C., *PATH3D a Ground-Water Path and Travel-Time Simulator, Version 2.0 User's Manual,* S. S. Papadopulos and Associates Inc., Bethesda, Md., 1989.

176. Zheng, C., *MT3D, a Modular Three-Dimensional Transport Model,* S. S. Papadopulos and Associates, Bethesda, Md., 1990.

177. Zheng, C., G. D. Bennett, and C. B. Andrews, "Analysis of Ground Water Remedial Alternatives at a Superfund Site," *Ground Water,* vol. 29, no. 6, pp. 838–848, 1991.

CHAPTER 23
ADVANCES IN HYDROLOGIC COMPUTATION

Roy D. Dodson
Dodson & Associates, Inc.
Houston, Texas

This chapter provides an overview of several advances in computational technology that are beginning to affect the practice of hydrology. Many technologies described in this chapter are not yet widely applied among practicing hydrologists. This contrasts with the other chapters of this book, which primarily deal with accepted practices and methods.

Yogi Berra once said "Prediction is hard—especially about the future." This chapter makes predictions about rapidly changing technologies and trends. Because of this, it is likely to become obsolete more quickly than the remainder of this book. Therefore, the emphasis of this chapter is on introducing important ideas and terminology that should have lasting importance.

The practice of hydrology has become increasingly computational during the past several years. This trend is largely attributable to the widespread availability of digital computers with sufficient speed and storage capacity for many hydrologic computations. It has become commonplace for hydrologic computer models originally developed for use on large computer systems to be successfully operated on desktop computers.

The most important advance in hydrologic computation that will occur over the next several years is the continuing performance improvements of common computer systems. This chapter explores that trend and its expected consequences for practicing hydrologists. However, this chapter also discusses several other advances in computation that will affect the practice of hydrology:

- User interface improvements
- Database management systems
- Digital mapping and other applications of computer graphics
- Expert systems and artificial intelligence
- Software engineering

23.1 ADVANCES IN COMPUTER HARDWARE

Many inventions may rightfully claim to have revolutionized certain industries or even society as a whole. However, no other invention has undergone so many revolutions in its own capabilities as the digital computer. The production of the ENIAC computer in 1946 initiated a series of breakthroughs that continue even today. Most computer engineers feel that the present rate of improvement in computer performance, cost, and reliability will continue, or improve through at least the year 2000.[3]

Within the next 10 years or so, most practicing hydrologists and engineers will have access to a computer system with the following characteristics:

- Very high computational performance—equivalent to today's fastest multimillion dollar computers
- Advanced user input devices such as voice input, scanners, and video
- Photographic-quality color displays and printers
- Large volumes of data storage on each computer—enough to store huge quantities of recorded hydrologic values, maps, and other information
- High-speed links to data storage on other computers, even at other locations

Hydrologists will receive some benefits from computer improvements by operating existing hydrologic computer models more frequently and with more data. However, the practice of hydrology will probably benefit much more from new computer software that will become possible because of the hardware improvements.

23.1.1 Computer Terminology

In approaching the subject of computer performance, it is important to understand some basic computer terms:

Hardware: the physically tangible portion of a computer system. The electrical and electronic components that make up the computer, screen, keyboard, printer, etc.

Software: the computer instructions and other information that control the operation of the computer hardware.

Central processing unit (CPU): the primary portion of the computer unit that processes the instructions.

Microprocessor: an integrated electronic circuit that can process instructions. Most computer systems are based on microprocessors.

Random-access memory (RAM): a portion of the computer that provides short-term storage of instructions or information for use by the CPU during processing. We measure RAM by the number of characters *(bytes)* that may be stored. One *megabyte (MB)* of RAM memory will store 1,048,576 characters of information. (This seemingly arbitrary number is exactly 2^{20}. It is called a megabyte because it is close to 1 million bytes.)

File storage device: a portion of the computer that provides long-term storage of instructions or information. File storage devices are most commonly disk drives. File storage capacity is also measured in bytes.

Disk drive: an electromechanical device for the storage and retrieval of computer data. The information is generally stored as millions of tiny magnetic fields in a coating on a disk-shaped piece of aluminum, glass, or plastic. These disks rotate to allow access to the data. Most computers are equipped with two types of disk drives: diskette drives and hard disks. Disk drives continue to store data even with the power off.

Diskette drive: a file storage device that uses removable plastic disks enclosed in a plastic housing. The removable diskettes (floppy disks) are portable, inexpensive, and fairly durable, so they are appropriate for transferring files from one computer system to another. They may also serve as a backup device for hard disks.

Hard disk: a file storage device that uses nonremovable aluminum or glass disks enclosed in an airtight case. Hard disks operate much more quickly than diskette drives. Hard disk drives can have a very large data storage capacity. One *gigabyte (GB)* of file storage will hold 1,073,741,824 (2^{30}) bytes, or 1024 MB. This is sufficient capacity for over 250,000 printed pages.

Megahertz (MHz): the speed of a CPU, as measured by the number of instruction cycles performed each second. Microprocessors commonly operate at 5 to 50 MHz or more. The speed of two different CPUs may not be proportional to the megahertz ratings of the CPUs.

Million instructions per second (MIPS): the performance of a CPU as measured by the time required to complete various test programs.

Million floating-point operations per second (MFLOPS): a measure of the performance of a CPU by the number of real number computations performed per second. For computationally intensive applications such as hydrology, the MFLOPS value is a reasonable indication of computer performance.

Parallel processing: a type of computer design in which a set of computer instructions is divided into different portions for simultaneous execution by different CPUs. Programs must generally be written specifically for parallel processing to get significant gains in performance.

Personal computer: a low-cost general-purpose computer designed for use by one individual at a time.

Workstation: a high-performance computer designed for use by one individual at a time, but with extensive capabilities for network communication with other computer systems. Workstations generally have special enhancements that improve their performance in engineering and graphics applications.

Supercomputer: the fastest type of computer system that is technologically feasible at a given time. Supercomputers use very expensive technologies, such as supercooled circuits, to operate as quickly as possible.

Operating system: a special type of computer program that controls the basic operations of the computer, including the file storage and communication devices. The operating system provides a link between the computer hardware and other, more specialized computer programs called *application programs.* Every computer is controlled by an operating system at all times. Most personal computers use the MS-DOS operating system developed by Microsoft Inc. Most workstations and some larger computers use the Unix operating system developed by AT&T. MS-DOS was designed for a single-user computer system whereas Unix and many other operating systems were designed for use by several people at the same time. Application programs written for use under one operating system must generally be modified for use under other operating systems.

CD-ROM: an acronym for compact disk read-only memory. A data storage device based on laser technology originally developed for audio compact disks.

CD-ROMs provide very large data capacity and are durable and inexpensive. The CD-ROM manufacturer stores information on the disk only during the original manufacturing process. Individual users cannot change this information or add more information later.

Modem: an abbreviation for modulator-demodulator. A device that transmits computer data by telecommunications links such as phone lines. Modem speed is measured in baud, which is approximately 8 times the number of bytes which are transferred per second. For example, a 2400-baud modem can transfer about 300 bytes of data per second. This is much slower than retrieving data from a file storage device such as a hard disk.

Scanner: a device that generates an image of a document or drawing in a computer-readable format. Scanners can be used to read printed material, paper drawings, or photographs into a computer data file. The computer can then process the data to identify characters on the printed page, or the individual elements of a drawing.

23.1.2 Projected Computer System Performance

In a comprehensive study of computer technology, Cutaia[3] projected the improvements in computer system technology likely to occur by the year 2000. He also modeled the effects of these technological improvements on total system performance. The study emphasized midrange multiuser business computer systems, rather than engineering workstations or personal computers. However, all of these computer systems share much of the same technology. Cutaia's projections considered many computer system components. However, three particularly important areas are CPU, RAM memory, and file storage. Figure 23.1.1 illustrates the projected performance improvements of a $6000 computer system compared with 1987 base conditions. As shown, the CPU performance (in terms of MIPS) is expected to improve by a factor of about 80 by the year 2000. The $6000 desktop computer of the year 2000 will be about as fast as the Cray XMP supercomputer of 1990. Such a machine would complete in less than a minute the hydrologic computations that required an hour to complete on a typical 1987 system.

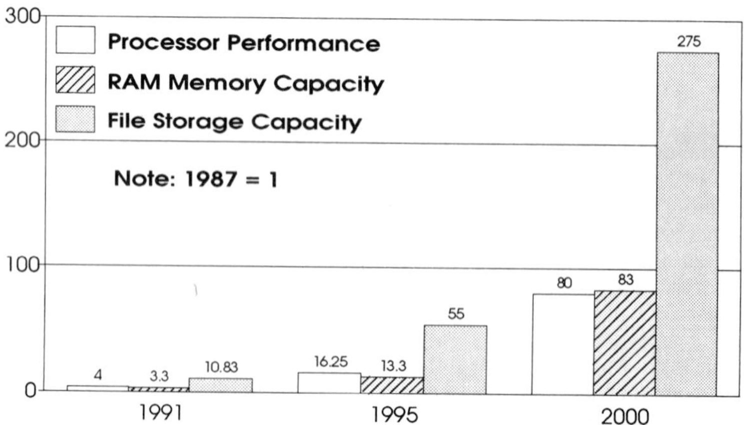

FIGURE 23.1.1 Improvements in desktop computer systems. *(Source of data: Ref. 3.)*

The CPU performance projections in Figure 23.1.1 do not consider parallel processing. Effective use of parallel processing could increase performance by an additional factor of 6, at the same price. This performance increase would reduce the 1-min computation time of the model mentioned above to less than 10 seconds. At this performance level, truly interactive hydrologic modeling becomes feasible.

As shown in Figure 23.1.1, available RAM memory capacity is expected to increase by a factor of more than 80 from 1987 to the year 2000. The RAM memory capacity of the typical $6000 desktop computer in the year 2000 will match or exceed the RAM memory available on the largest computers available in 1987.

This increase in memory capacity will reduce the existing memory constraints on hydrologic models. Hydrologists will analyze watersheds in more detail and with shorter computation intervals. Programs will offer a wider variety of optional computations.

Certain data-intensive types of software perform much better when sufficient RAM memory is available. Examples of such programs include finite-difference algorithms, detailed graphics, digital terrain models, and expert systems. Large increases in the RAM memory capacity of common computer systems will lead to increased use of such programs.

File storage capacity is expected to increase by a factor of about 275 between 1987 and the year 2000. This increase is even more dramatic than the increase in processor performance or RAM capacity. Future storage devices will also become much more reliable. Improvements in data backup systems will increase the total system reliability.

Other storage devices will also affect hydrologic computations. Laser read-only memory devices (such as CD-ROMs) will provide immediate access to huge quantities of data that do not require frequent updates. Stream-flow and rainfall records are already available on CD-ROM from EarthInfo Inc., Boulder, Colo. In the future, various types of maps should also be available on CD-ROM. In addition, improvements in high-speed communication links will allow individual workstations to access data stored on larger computer systems. Therefore, the total quality of data readily available will increase even more dramatically than Figure 23.1.1 suggests.

Changes in software will partially offset improvements in hardware performance. The average computer operation in the year 2000 may involve about 30 times more information than today's average operation. Therefore, the improvement in processing speed and storage capacity may not be as dramatic as the performance numbers alone seem to suggest. However, the new software capabilities should be more valuable than the raw processing power required to implement them.

23.2 USER INTERFACE AND GRAPHICS IMPROVEMENTS

The *user interface* is the total interaction between a computer system and the person using the system. The first computer systems had extremely crude user interfaces. The user had to enter all commands and data using switches on the computer's front panels. Later, punched cards and paper tapes stored computer instructions and data. Engineers developed special devices to read the data from the punched cards or paper tape directly into the computer.

Within a few years, computer engineers adapted the display technology of the television industry to provide computer monitors or screen displays. Designers also adapted Teletype machines to provide the first computer keyboards. With various enhancements, video monitors and keyboards are still the predominant user interface hardware components for computer systems.

Traditionally, the user interface includes text input provided by the user through the keyboard, and text output provided by the computer on the display screen or printer. Over the years, computer programmers have developed several methods of interaction to increase the productivity of computer users. These include command-line interfaces, full-screen interfaces, graphical user interfaces, and others.

In a *command-line interface,* the user prepares and submits a single line of data for input to the computer system. The user then waits for the computer system to respond with a result or an error message. Using this type of interface is like viewing a room through the mail slot in the door. However, the command-line interface can become very productive for skilled typists who memorize the commands required to operate the computer system. The popular MS-DOS operating system uses a command-line interface.

A *full-screen interface* allows the user to view and modify at least one screen full of data at a time. The computer system provides some intermediate responses while the user modifies the data. This type of interface allows much more flexibility than a strict command-line interface. For example, the computer may display menus to prompt for the user selection of commands or options. Well-designed interfaces allow users to rely on menus when learning the system, then progress to more efficient command-line interaction later.

The *spreadsheet interface* is a variation of the full-screen interface designed for the entry and display of numeric data in a grid pattern on the screen. A series of horizontal menus generally control the data entry operations. The spreadsheet interface is especially productive for analysts accustomed to working with numbers.

The *graphical user interface (GUI)* generally relies on a device such as a mouse to point to different portions of the screen. The computer interprets the pointing actions according to the current context, and executes commands accordingly. The screen display consists of graphic images, which may include representations of text. Several studies have shown that many users can learn a new computer program that employs a graphical user interface more quickly than other programs. A graphical user interface is generally superior for applications involving the production or processing of graphic images. As computers have become powerful enough to begin to perform serious graphic processing, the importance of the GUI has increased accordingly.

The emerging *pen-based interface* involves moving a penlike pointing device on the surface of the computer screen. In addition to using the pen for pointing, the user can also perform intuitive gestures such as striking through unwanted text or drawing a caret where text is to be inserted. Obviously, the pen would also be a convenient device for handwritten input, and manufacturers are making significant advances toward reliable computer recognition of handwritten or handprinted text.

The widespread use of the pen-based interface may provide a much more natural interaction with the computer system, because most people already know how to use a pen. Hydrologists have traditionally used maps, diagrams, sketches, flowcharts, and other types of graphics to represent the condition or operation of large natural or manmade systems. A user interface that deals directly with these types of graphical representations could make hydrologists more productive.

On the horizon, many anticipate that the *vocal interface* will form a major turning point in the use of computer systems. The technology for reliable voice recognition is still being developed.

An important element in the user interface is the output generated by the computer system. More flexible methods of output are becoming important, including:

- *High-resolution (photographic quality) color graphics display screens and printers.* The broader availability of such devices will improve the quality and usefulness of graphical output. The high-quality output will also facilitate new applications such as computer storage and display of photographs, construction drawings, and detailed mapping data.
- *Graphic animation and full-motion video.* Computer animation will allow hydrologists to better understand and communicate computer-simulated hydrologic events. Full-motion video capabilities will allow computer simulations to be compared with sequences of recorded video. Recorded video may also provide a realistic backdrop for animated images. Sound and voice storage and generation will complement the animated sequences for dramatic presentations.

23.3 SOFTWARE ENGINEERING

Software engineering is the application of traditional engineering principles of planning and design to the production of computer software. Engineers and computer scientists have invested many years of work in designing, programming, documenting, and using existing computer programs for hydrologic analysis. Several programs have been in use for 20 years or more, and should continue to be used, with periodic maintenance and upgrades, for many more years.

Engineering designers and analysts rely on information provided by hydrologic computer programs in making billions of dollars in public and private expenditures each year. For this reason, these computer programs are really a part of the public infrastructure. Like any other constructed entity, they have a manufacturing or construction cost, maintenance and upgrade costs, risks, and costs of failure. Eventually, they also have a replacement or overhaul cost. Major engineering computer programs used in this way must be considered as "engineered facilities," which should be designed according to proven engineering principles.

23.3.1 Computer Languages and Programming Styles

Computer programs are sequences of instructions that are carried out by the computer. Computer programmers use programming languages that combine words and mathematical symbols to produce a complete specification for a sequence of operations. A very early computer language was FORTRAN (FORmula TRANslator). FORTRAN is efficient at performing mathematical computations. Therefore, many hydrology computer programs use FORTRAN.

The first important application of engineering principles to software came in a careful review of programming languages themselves. Niklaus Wirth, a Swiss computer scientist, created a computer programming language called Pascal in 1969. Pascal differed from FORTRAN and most other contemporary languages in three important ways:

1. *Modularization.* Pascal requires the program to be divided into small single-purpose modules. Wirth called these modules *procedures* and *functions.* They are similar to FORTRAN subroutines.
2. *Strict sequencing.* Pascal controls the sequence of program operations according to strict rules. The program execution cannot jump from one statement to another in a remote portion of the program.

3. *Explicit declaration.* Pascal forces the programmer to list and describe all variables prior to any use.

Experience has proven that Wirth's structured approach allows computer programs to be developed and tested more quickly and to operate more reliably. Structured programs are also easier to understand and maintain. Most computer languages developed since 1969 allow a structured approach to program development. Currently, the most common language for professional software development is the C language developed at the AT&T Bell Laboratories research facility. C provides powerful and efficient programming while still allowing a structured approach to program development. A later version of FORTRAN, called FORTRAN-77, also allows a more structured programming style.

Many of the first hydrologic computer programs were not written by structured programming methods. These programs generally had large multipurpose modules and complex control-flow sequences. Fortunately, many programs have been partially rewritten in FORTRAN-77 to incorporate structured programming concepts. For example, the U.S. Army Corps of Engineers Hydrologic Engineering Center has rewritten a portion of the HEC-1 watershed runoff model. In the new version of HEC-1, the programmers have divided large multipurpose modules into smaller single-purpose modules.

23.3.2 Object-Oriented Programming (OOP)

Objects are computer data structures that are defined by their interaction with one another. Object-oriented programming began with Simula-67, a language designed for use in simulating physical systems. Several object-oriented languages are now available, as are object-oriented versions or derivations of popular languages such as C and Pascal.

An object-oriented system provides *data encapsulation,* in which a well-defined interface is used to access data. *Methods* are the interface functions that provide access to the encapsulated data. The data and methods together make up the object. Object-orientation also requires *inheritance,* which is the mechanism for constructing complex objects from simpler ones.

The primary advantage of using an object-oriented system is that it provides a natural way to divide data into individual modules. Once a particular type of object is developed and successfully tested, it can be safely used again in subsequent programming tasks. The judicious use of inheritance allows complex new objects to be built from simple existing objects. A properly designed object-oriented system can cut development time and improve software reliability.

Since object-oriented programming was originally developed to model physical systems, it can be applied to hydrology. For example, a watershed could be thought of as an "object" that may be described and represented by a combination of data and methods. Watershed data might include drainage area, soil types, development characteristics, and much more. Watershed methods might include loss rate functions and runoff methods.

23.3.3 Software Reliability and Testing

A major concern of hydrologists who use computer programs for analysis and design is the reliability of the computer software. The issue of software reliability has been a topic of extensive research among computer scientists as well. Computer scientists have shown mathematically that it is not possible to prove the correctness of any nontrivial computer program. In other words, we can never be absolutely sure that a computer program is providing correct solutions under all circumstances. For hydrologists, however, what is important is whether a computer program provides adequate solutions under a sufficiently wide range of conditions. This can be established by a comprehensive program of testing.[8]

23.4 DATABASE MANAGEMENT SYSTEMS

A major problem which hydrologists and drainage engineers will face during the next several years is *information overload*—the inability to deal with available information using present methods. As discussed previously, many hydrologists enjoy using the spreadsheet interface, in which they view and edit one screen full of data at a time. In effect, technical professionals often use the spreadsheet as a data management system. This approach is effective enough with small data sets, but not with the massive increases in data storage that are beginning to become available.

By about the year 2000, the data storage capacity of the desktop computer will grow to about 33 GB or more.[3] Today's average computer screen displays about 2000 text characters at a time. At a rate of one screen of data every 5 s, it would require about 11 years of typical 8-h workdays just to review 33 GB of data! Obviously, effectively using large data sets will require a more sophisticated approach. A database management system offers such an approach. The primary advantages of a database management system include:

- *Data storage and access.* Database management systems provide storage and access capabilities for large amounts of data.
- *Data revision.* The typical hydrology program application involves several analyses using only slight variations of the input data set. Proper use of a data management system eliminates the need for maintaining several highly redundant input data files.
- *Data reuse.* Users want to employ the same hydrologic data for many types of analyses and applications programs. Storing the data in a database management system with adequate access capabilities allows the data to be reused without extensive reformatting.
- *Data reporting.* The typical hydrology program user requires comparisons of the results of several similar analyses. Storing results using a data management system simplifies comparative reporting.

Figure 23.4.1 illustrates the interrelationships between the database management system (DBMS), the database, the hydrology application program, and the user. A hydrologist may access the database either through the hydrology application program or through DBMS utilities. Input and output may be via standard input (keyboard) and output (video display) devices, a printer, or secondary storage media (disk files, etc.).

Types of hydrologic data stored in a database management system commonly include:

- *Time-series data,* such as precipitation records, stream flow, and gauge records. Time-series data represent sequences of real-world observations or calculations.

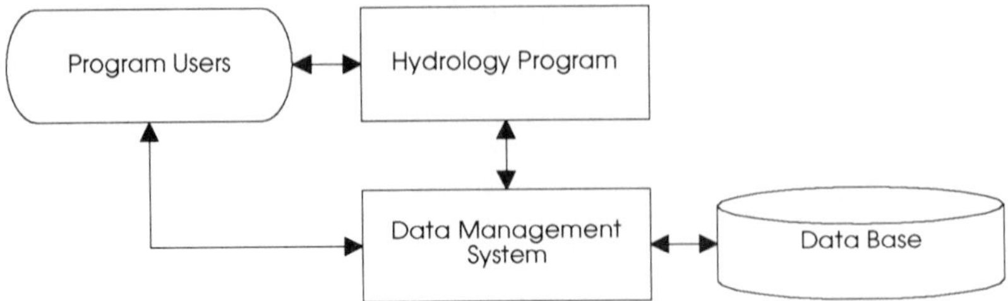

FIGURE 23.4.1 Hydrology program data management.

For regular interval time-series data, the time intervals between data values are all the same. Time-series data may also be available at random time intervals.

- *Geographic, topographic, and other data* from hydrologic and graphic applications programs.[20]
- *Graphic image data,* such as photographs and scanned documents.

Each of these types of data has unique characteristics. It may not be possible to select a single DBMS to handle all types of data efficiently.

23.4.1 Development of Database Technology

Databases have their roots in the 1960s, when corporate programmers began to see a need to centralize data storage. Instead of spreading inventory and personnel information through many files, it made more sense to collect it into common files available to all programs. A successful centralized database system performs two basic processes logically and consistently:

- *Data storage* as determined by the logical structure of the database—its *data model.* There are presently four major data models, each applicable to a different class of problem. These are the *hierarchical, network, inverted list,* and *relational models.* A fifth major model, the *object-oriented model,* is also under development. A major issue in hydrologic data storage is *data compression,* which reduces the required data storage space.
- *Data retrieval,* in which the user *queries* the database using a data query language. The Structured Query Language (SQL) is the best-known data query language. Some data query languages can be embedded within programs written in other programming languages such as FORTRAN and C.

Some database management systems are *data-dependent.* In a data-dependent system, each computer program which uses the data must directly control the details of data storage and data retrieval.[4] If you change the data storage structure or retrieval method, you will generally have to change all the computer programs that use the data. Some modern database management systems provide *data independence.* In a data-independent system, you can change the structure of stored data without having to revise all the programs that use the data.

23.4.2 Relational and Inverted-List Database Systems

A relational database system is

- ***Based on Tables.*** The user sees the data as a set of rectangular tables. Each table consists of *rows,* or *records,* that are separate instances of the same kind of information, and *columns,* or *fields,* which are different kinds of information about the feature. The data records are not listed in any particular order in each table.
- ***Capable of Generating New Tables.*** The DBMS data manipulation operators generate new tables from old ones.[4] Database access in relational systems should be data-independent. The user does not need to know anything about the physical order of the data to use the system.

The relational model has become the standard for most applications. Relational database management systems are appropriate for storing common business data and records of business transactions.

Despite the dominance of the relational DBMS for general applications, there are some disadvantages in using this technology for hydrology application programs. With a relational DBMS, time-series data would be stored as items in a table. The date and time of each data value would be stored individually with the data value. The date and time are necessary to allow the relational DBMS to distinguish between records, because the physical order of the records is not important in a relational DBMS. The storage of date and time values with every record would add greatly to the storage area needed for regular interval time-series data.

The inverted-list model is similar to the relational model because it also stores data as a collection of tables. However, in an inverted-list system, the stored tables and the physical access paths to the stored tables are visible to the user.[4] Further, the data in the tables have a definitive order that is also "visible" to the user. The order of the data is important to the data manipulation operators in an inverted-list system. Therefore, inverted-list DBMS applications are data-dependent. Inverted-list models are more efficient than relational models for storage of regular interval time-series data, because it is not necessary to store the date and time values separately for each record.

23.4.3 Hierarchic and Network Database Systems

The hierarchic data model uses a strict hierarchy of data structures, like a data tree.[4] Data trees begin with a single node (the *root*). Each node may have one or more *child* nodes, though each child node has only one parent. To manipulate data stored in a hierarchic DBMS, you must traverse paths up and down the trees.[4] Hierarchic DBMS applications are data-dependent.

Any treelike data set is an excellent candidate for a hierarchic data model. Regular interval time-series data sets are naturally hierarchic because smaller time intervals can be combined into larger time intervals. For example, hourly observations can be grouped into months. It is not necessary to store the time associated with each observation value. It suffices to store the starting time of the series and the interval between observations. The individual time values can then be derived.

A network database is an extended form of the hierarchic model.[4] In a network DBMS, a child can have multiple parents. This allows more complex relationships. Data manipulation operators for a network DBMS process data records and links between records. Network DBMS applications are data-dependent. Network databases can be very complex. Sophisticated databases require large numbers of intricately related nodes.

23.4.4 Data Compression

Data compression relies on the predictability of data values to reduce storage by truncating redundant data values. There are three types of data compression schemes important in dealing with time-series hydrologic data:

1. A *repeat counter* that flags duplicate values. Some hydrologic time-series data consist of many repeated values in sequence, such as zero rainfall readings or constant water elevation levels. These repeated sequential values can be replaced by a single value and a repeat counter.
2. A *difference method* that records the differences between each value and the minimum values in the record. The difference method works well where the difference between the maximum value and minimum value is not too large, and the precision of the values is known. Reservoir elevation and precipitation data can often be effectively compressed by using a difference method.
3. A *significant digits method* that records a significant number of digits for each value. Flow data can often be effectively compressed by this method.

Other, more sophisticated methods of data compression are available for other types of data, especially graphical data. Because of the very large quantities of data involved in detailed hydrologic analyses, automatic compression and decompression of data values are important attributes of an efficient database management system. However, most commercially available systems do not have this capability. By their very nature, data compression schemes must be somewhat data-dependent to be effective. The HEC Data Storage System (HECDSS), a database management system developed by the U.S. Army Corps of Engineers Hydrologic Engineering Center, can use data compression in storing time-series hydrologic data.[2] Tests indicate that the data compression routines in HECDSS achieve about an 80 percent reduction in storage requirements, without reducing the quality of the data or increasing the time required to access the data.

23.4.5 Using Databases with Hydrology Programs

Hydrology computer programs typically perform several types of computations while processing time-series data:[5]

Type I: Simple sums (or differences). For example, computing the total inflow volume from a time-series of inflow values would be a Type I computation.

Type II: Calculations using a time-varying state variable. For example, a flood routing computation that computes pond surface water elevation at time i from the pond elevation at time $i - 1$ is a Type II computation.

Type III: Calculations requiring a subiteration. For example, a hydrograph convolution, which includes a complete iteration of up to 150 or more terms for each time interval, would represent a Type III computation.

Database query languages such as SQL provide powerful tools for data manipulation, including insert, update, delete, and query. Database query languages are also called *fourth-generation languages* (4GL), and are generally associated with relational databases. For example, a Type I calculation would be represented in SQL as:

```
SELECT SUM (INFLOW1) FROM TABLE_X
WHERE DATE > MINDATE AND DATE < MAXDATE
```

The result of this query would be the sum of the inflow data (INFLOW1) between a predefined MINDATE and MAXDATE. This is a *nonprocedural* database access, since the method used to determine the result of the required calculation is not specified.

To perform the same calculation as the SQL example shown above by using a data-dependent system (HECDSS in this example), with FORTRAN as the host language, the following commands are required:

```
NVALS = MAXDATE − MINDATE − 1
CALL ZRRTS (PATH1 . . ., MINDATE, . . ., NVALS, INFLOW1, . . .)
SUM = 0
DO 200 I = 1, NVALS
SUM = SUM + INFLOW1 (I)
200 CONTINUE
```

This is a *procedural* database access since the host language actually performs the calculation. For this type of hydrologic calculation (Type I), the nonprocedural database access method is simpler and potentially more reliable than the procedural method. However, nonprocedural database access queries can be inefficient. For data-independent systems, the user cannot take advantage of the existing organization of the data to speed database access.

While nonprocedural database query capability would be superior to procedural queries for Type I calculations, Type II and Type III calculations are difficult or impossible to formulate in a nonprocedural language. Type II and Type III calculations are naturally procedural. Therefore, procedural database access would generally be simpler and more reliable for Type II and Type III calculations.

Because the database query languages available in most relational DBMS are inefficient or cumbersome for many calculations involving time-series data, and because relational database management systems do not store time-series data efficiently, a relational database may not be the most efficient choice for most hydrologic time-series data. The data independence provided by a relational DBMS may reduce the impact of changes in data storage on the hydrologic application program. However, data independence may prevent the hydrologic application program from taking advantage of the physical storage method to improve access efficiency. It also may prevent the use of the most effective methods of data compression to improve storage efficiency.

In spite of these technical disadvantages in handling time-series data, however, relational DBMS enjoy very widespread use in all types of applications, including many hydrologic applications. High-quality relational DBMS software is available on almost all computer systems, and most database users have some familiarity with the terminology and common operations used on most relational DBMS.

23.5 GRAPHICS APPLICATIONS FOR HYDROLOGY

Graphics applications make up probably the fastest-growing set of computational tools for hydrologists. Several different types of graphics applications are familiar to many hydrologists:

- *Computer-assisted drafting and design (CADD)* programs, which provide powerful tools for the input of project design data and field survey data. CADD programs produce high-quality two- or three-dimensional images of existing or designed facilities. CADD programs generally deal with *vector data.* Vector data are low-level geometric entities such as points, lines, and arcs, that are located according to a coordinate system. These low-level geometric entities are used to assemble a complex drawing.
- *Digital terrain models (DTM),* which can conveniently receive, store, and display large numbers of ground elevation and location coordinates. DTMs can show existing and proposed ground contours or grids for a watershed or project area.
- *Digital elevation models (DEM),* which store and process ground elevation values measured at the intersections of horizontal grid lines. Gridded elevation data is an example of *raster data,* which is an array (generally two-dimensional) of values measured at evenly spaced locations throughout an area. Raster data values are important because they can generally be obtained at low cost by using automated data collection equipment such as high-altitude cameras and other remote-sensing devices.
- *Geographic information systems (GIS),* which allow the input, storage, analysis, and representation of map-related data. GISs may deal with vector data such as points, lines, and polygons (closed two-dimensional figures, such as parcels of land). However, some GISs are based on raster data.
- *Image processing systems,* which store, enhance, analyze, and display digitized images such as those received from high-altitude photography and other forms of remote sensing. Image processing systems deal primarily with raster data.
- *Paint and rendering systems,* which include a broad category of tools primarily intended to produce high-quality illustrations and images. These systems generally deal with raster data from digital scanners or direct user input. They also may be used to provide more realistic images of vector-based drawings generated by CADD systems or three-dimensional terrain and landscape features from a DTM.[10]

Figure 23.5.1 illustrates these various categories of graphical software. As the figure also illustrates, there are increasing connections and areas of overlap among all categories.

23.5.1 Computer-Assisted Drafting and Design Programs

The most widely used graphics applications programs among hydrologists and engineers are CADD programs such as AutoCAD or Intergraph. These programs are often used for a wide variety of graphics applications, such as mapping and general technical illustration purposes, as well as traditional drafting and design functions. Because of the widespread use of these programs, and because they provide very

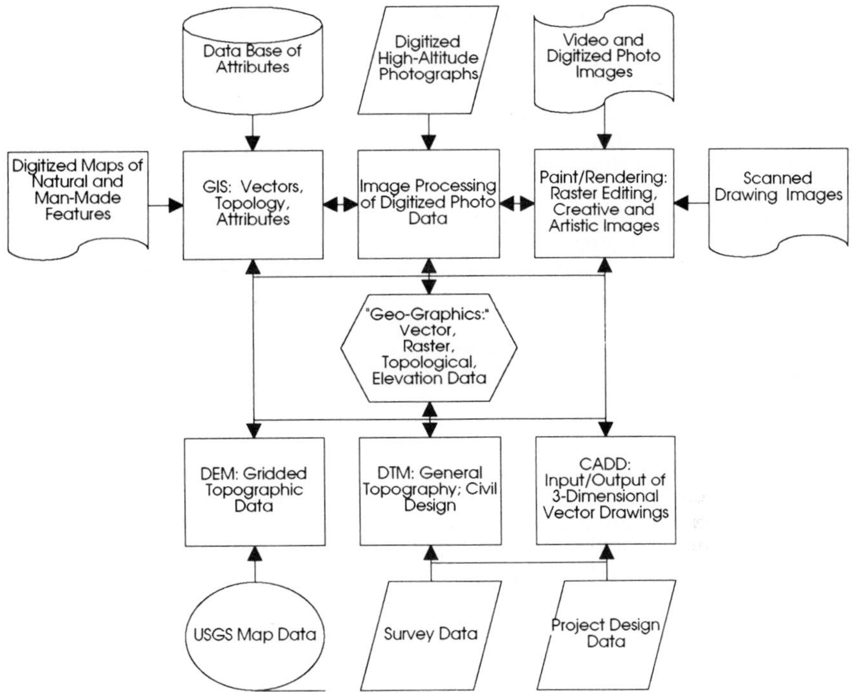

FIGURE 23.5.1 Relationships among graphics applications.

powerful tools for drawing, some of the other more specialized graphics applications are often based on CADD programs. For example, a geographic information system or digital terrain model may be built using a CADD program to provide the necessary graphics.

Other graphics applications programs are often compatible with CADD graphics file formats. The most widely accepted file format is the AutoCAD Data Interchange Format (DXF), which is recognized by almost all vector-based graphics application programs.

CADD programs have generally provided vector-based drawing tools, but have not included capabilities for displaying or editing raster data, such as data files from scanners. However, CADD programs are increasingly able to deal with combinations of raster and vector data. The CADD program may provide one or more of the following operations:

1. *Raster backdrops.* The CADD program may simply display the raster image as a backdrop for a vector drawing. Some programs may store the vector drawing data with the raster image to form a hybrid raster/vector drawing.
2. *Raster editing.* The CADD program may allow editing of the raster image. Using these editing functions, the user may clean up or modify the raster image directly.
3. *Raster vectorizing.* The CADD program may vectorize the raster image and recognize the lines, arcs, text, and other objects much as a skilled human drafter might do.

23.5.2 Digital Terrain Models and Digital Elevation Models

Digital terrain and digital elevation models provide graphical representations of land surface elevations. A DEM generally deals with a regular array of elevations, normally created in a square grid or hexagonal pattern over the ground surface.[11] The shortcoming of this regular grid-based approach is that the distribution of data points is not related to the characteristics of the terrain itself.[17] By contrast, a DTM can usually accept elevations at random locations. As such, a DTM can preserve important topographic features such as ridge lines, stream banks, and stream-flow lines. Generally, the DTM approach is superior for local areas in which data for specific features are available. The DEM approach is superior for analysis of very large areas where fully automated data collection is much more cost-effective.

In the United States, the National Mapping Division of the U.S. Geological Survey (USGS) and the Defense Mapping Agency (DMA) are both actively converting existing maps to digital form.[18] Two different categories of DEM data are now available:

1. 7.5- × 7.5-min blocks corresponding to the standard 1:24,000 scale USGS quadrangle maps. The data have a resolution (grid interval) of 100 ft (30 m) or 0.05 in (1.25 mm) at map scale. They are available for only a portion of the United States. Because the elevations are based on stereo measurements from high-altitude (40,000 ft), 1:80,000 scale aerial photographs, the root mean square error is generally ±20 ft (7 m).[18] A typical 7.5-ft quadrangle map includes about 190,000 elevation values.
2. 1° × 1° blocks representing one-half of the standard 1:250,000 scale 1° × 2° quadrangle maps. These data values are produced by the DMA and distributed by USGS, and are available for the entire United States except Alaska. The data have a resolution of 3 arc-seconds, which corresponds to about 90 m in the north-south direction and about 60 m in the east-west direction (at 50° north latitude). The accuracy of these elevations varies according to type of terrain: ±50 ft (15 m) for flat terrain, ±100 ft (30 m) for moderate terrain, and ±200 ft (60 m) for steep terrain.

Canada and most western European nations are undertaking similar programs for the development of DEM data from existing topographic data. In particular, Sweden is producing DEM data for the entire country with a resolution of 50 m and an accuracy of ±2.5 m (Ref. 18).

Most digital terrain models use a *triangulated irregular network (TIN)* model of the land surface. A TIN is built up by linking available ground surface points into triangles that form a network covering the site. The ground surface points from the corners of the triangles. Therefore, the TIN uses every measured data point directly. Important features such as ridge lines and stream banks are easily incorporated into the TIN as edges of adjacent triangles.

Two widely used algorithms for constructing the TIN are the *Delaunay* method and the *radial sweep* algorithm. Both of these generate a unique set of triangles—the Delaunay triangulation—that are as equilateral as possible and have minimum side lengths.[17] The computation time for both methods is linearly related to the number of data points, and computer memory requirements are not large.[14]

Figure 23.5.2 illustrates a set of spot elevations and (in dotted lines) the TIN constructed by using Delaunay triangulation on these points. Figure 23.5.2 also illustrates ground surface contours generated from the TIN by linear interpolation. Note the angular appearance of the contour lines. These contours may be smoothed by using a cubic spline method. The smoothed contours are more realistic looking, but are actually less accurate.

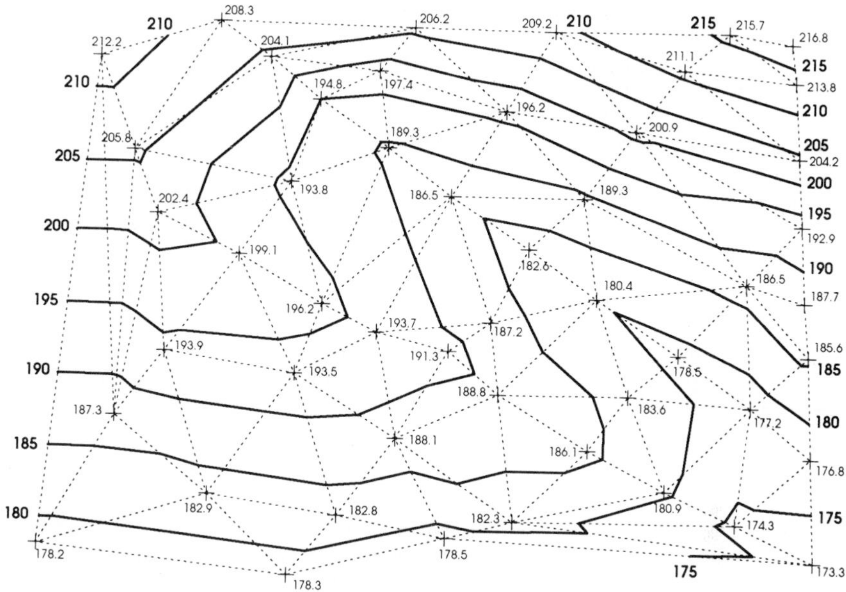

FIGURE 23.5.2 A triangulated irregular network and interpolated contours. *(From Ref. 15. Used with permission.)*

Some DTM systems have the capability of extracting ground geometry and analyzing the terrain for physical attributes other than simple elevations. Such systems may be called intelligent ground models (IGM) or intelligent terrain models.[6] Of particular interest to hydrologists are systems such as McDonnell-Douglas GDS-SITES that can display watershed boundaries and the direction of runoff.[9] Some systems can also generate *slope maps,* that depict isolines of ground slope in place of ground elevation contours.[20] Such computations are based on the slope and aspect of each individual triangle. The *aspect* of the slope is the direction (with respect to north) that runoff would move under the influence of gravity down the angle of maximum slope.[22]

Detailed digital terrain models can provide channel and floodplain cross sections for hydraulic studies. Floodplain boundaries can also be determined and displayed by using a DTM. However, this is not widely practiced because detailed DTMs are generally not available along significant reaches of stream channels. In addition, the most widely used river hydraulic models are one-dimensional. The cross sections in these models do not have the full three-dimensional data necessary to allow automated floodplain mapping.

Digital terrain models form the basis for many useful civil engineering computer programs that superimpose engineered facilities such as roadways, fill, and other structures onto the existing ground surface. These programs can compute important information such as cut and fill quantities for earthwork. They generally also include sophisticated tools for engineering design. For example, coordinate geometry (COGO) computations are generally provided to assist in precisely locating survey points and in laying out engineered facilities.

DTM programs and data will become increasingly available and important for hydrology. The detailed analysis of flow paths possible with a digital model of the ground surface will encourage the development and use of more detailed algorithms and approaches to hydrologic analysis. At the same time, existing methods that rely on lumped watershed parameters such as time of concentration will probably be modified to allow automatic computation of these parameters from the digital model. For example, many hydrologic methods relate the runoff characteristics of a subbasin to the average basin slope. Presently, this slope is typically computed by the formula[23]

$$\text{Basin slope} = \frac{(\text{elevation at } 0.85L) - (\text{elevation at } 0.10L)}{0.75L} \tag{23.5.1}$$

where L = the length of the longest channel extended from the basin mouth to the drainage divide.

In other words, the average slope of the entire basin is estimated from the elevation of only two points. If a contour map of the watershed were available from a digital terrain model, however, the following formula could be used:[23]

$$\text{Basin slope} = \frac{(\text{contour interval}) \times (\text{total length of contours})}{\text{drainage area}} \tag{23.5.2}$$

This formula considers the entire subbasin area in computing the average basin slope, and should therefore provide a more representative value. The difficulty in using this formula is that it is time-consuming to measure the total length of all contours within the subbasin. With a DTM, however, such computations are much more feasible. In fact, the DTM could avoid dealing with contours altogether by computing the area and slope of each individual triangle in a TIN model and using these values to compute the average watershed slope.

23.5.3 Geographic Information Systems

A *geographic information system* is an electronic system of maps connected to tables of data that describe the features on the maps. A vector-based GIS describes map features as *points, lines,* or *polygons.* Points can represent storm water inlets, groundwater wells, or any item located at some fixed point in space. Lines can represent streams, pipes, road segments, property boundaries, or similar linear items. A polygon is a closed set of connected lines. Polygons can represent fields, counties, watersheds, tracts of land, or similar areal entities.

A GIS can store and analyze the *topology* of the data, which includes the spatial relationships between the various points, lines, and polygon entities. GISs also store attributes describing each entity, and can analyze or group entities according to these attributes. Each spatial feature in a GIS has a unique geographic location specified by its coordinates and a unique identifying number by which it is connected to descriptive data in a relational database.

Each record in the database also has a unique identifying number. In a GIS, there is a one-to-one relationship between records in the database and features on the map. The GIS generally uses Structured Query Language or a similar nonprocedural database query language to find information in the database.

A *coverage* is a GIS map and its associated data tables. A GIS can have many coverages. For example, soil maps can form a coverage. A road map might be used to develop another coverage for the same area.

GIS originated in automated mapping systems created about 1970, but recent advances in computer hardware and operating systems have greatly expanded its use.

The leading vector-based GIS software program is Arc/Info, a product of the Environmental Systems Research Institute (ESRI) in Redlands, Calif. Another popular GIS program is MicroStation GIS from Intergraph Inc., Huntsville, AL.

Within the military and some other government agencies, a raster-based GIS called GRASS (Geographical Resources Analysis Support System) is widely used. GRASS was developed by the Construction Engineering Research Laboratory of the U.S. Army Corps of Engineers. In a purely raster-based GIS, the only topological entity available is the grid cell. The size and shape of each cell are uniform, and the geographic location can be easily computed. Numerous attributes can be attached to each grid cell location. Raster-based GIS systems are well-suited to the analysis of remotely sensed data. Raster structures are simple because entities are represented implicitly, whereas vector entities must be explicitly stored in a database. In vector-based systems, much effort is expended in defining polygons for overlaid data layers.[16] However in raster-based systems, the level of detail is limited by the grid cell size, and the requirement to store all attributes for each grid cell may greatly increase the data storage requirements.

A significant expansion of the Arc/Info system has been the linkage to the ERDAS image processing software.[7] The ERDAS software package processes satellite images such as LANDSAT and SPOT. The ERDAS image processing system and Arc/Info may exchange data files.[21] This provides users of the Arc/Info vector-based GIS access to common forms of remotely sensed raster data.

Many local government agencies use vector-based GIS to create digital maps of all streets, tax parcels, legal lots, utilities, etc. These maps then form a computerized information base for government operations.

Government agencies and industries use GIS to map natural resource features such as streams, oil and gas reservoirs, soils, and forestry. They then construct more complex maps using overlays. These overlays might define, for example, the likely habitat area of a particular kind of animal that requires a certain kind of vegetation, access to water, and hilly terrain.

GIS databases are tedious and expensive to build. The process of building a GIS database often exposes the incompleteness and inaccuracy of existing data. For example, many municipal GIS users first implement a database of all property on the tax rolls. After this is complete, they may digitize utility maps, only to find numerous inconsistencies between the existing utility maps and the basic GIS coverage. These inconsistencies often were not apparent (or were not important) while the tax maps and utility maps were maintained separately. However, they must be resolved when these maps are combined.

In spite of the problems, a completed GIS is powerful because it constitutes a digital information base of a region, which can be analyzed repeatedly in many different ways. A GIS performs sophisticated manipulations and analyses that may include:

- *Map overlays.* The GIS can create overlays of two or more coverages to identify areas having certain characteristics. For example, an overlay of soil coverage and roads could be used to identify all of the roadways built on a particular type of soil.
- *Buffer generation.* The GIS can generate buffers around points, lines, or polygons to identify all features within the buffer area. The GIS can then produce a tabular report listing all of the identified features.
- *Boundary dissolve.* The GIS can regroup and reclassify existing entities to form new entities or coverages.
- *Tabular data analyses.* Because the GIS is based on a full-featured relational

database, it can generate reports of tabular data from the database. For example, a GIS used to map all taxable property in a certain area could generate a report of the total tax base, grouped into categories.

- *Network analyses.* The GIS uses network analysis methods to identify routes through networks. An example application for the network analysis tool might be finding the quickest way from one point in a city to another through the street network.
- *Digital terrain model.* Some GISs provide a TIN model of the surface terrain as an additional coverage of the study area.
- *COGO.* This is a coordinate geometry package which helps to locate features accurately on maps.

A complete GIS system contains computer mapping and display capabilities for generating high-quality cartographic products. The user can specify size and scale to produce the desired map outputs.

As GIS databases are built up for most metropolitan areas, GIS technology will extensively influence the practice of hydrology. Initially, GIS will serve primarily as a front end or a back end to existing hydrologic models:

- *Front-end applications* include the computation of watershed parameters for existing lumped-parameter hydrologic models. "It is probably true that the factor most limiting hydrologic modelling is not the ability to mathematically characterize hydrologic processes, or to solve the resulting equations, but rather the ability to accurately specify the values of the model parameters representing the flow environment. GIS will help overcome that limitation."[13]
- *Back-end applications* include the cartographic display of computed hydrologic data. For example, water surface elevations computed at stream cross sections could be displayed as a map of floodplain boundaries with a GIS.

The digital terrain model coverage of a GIS should allow computation of some watershed parameters from basin topography, as described for stand-alone DTMs earlier in this chapter. Depending on the other coverages available within the GIS, additional parameters may be computed. For example, an important factor in urban hydrology is the percentage of impervious cover within a subbasin. Possible methods for computing the percentage impervious using a GIS may include:

1. *Direct computation.* If the GIS provides coverages for various types of urban development, then the percentage impervious cover may be computed by overlaying the watershed boundaries with each of the urban development coverages, to compute the watershed area within each category of urban development. Then, each acreage is multiplied by a percentage impervious factor and totaled to produce the total impervious cover within the subbasin. This is simply an automated version of the method most commonly used for percentage impervious computations.

2. *Road density.* The GIS may not provide coverages for each type of urban development, but the GIS will generally provide a road network coverage. In such a case, the percentage impervious cover can be estimated from the road density. The GIS can provide the total length of roadways within each subbasin. This value, divided by the subbasin area, gives the road density. This should be proportional to the level of urban development, and thus to the percentage impervious cover. The relationship between road density and percentage impervious cover will vary, but a relationship could probably be established through analysis of a few representative watersheds.

A major limitation to the application of GIS to hydrologic parameter computation is the lack of sufficient data on important characteristics such as soils and land uses. The USGS has a set of digital files, called the LULC (Land Use Land Cover) files, covering the United States with a 1 : 100,000 or 1 : 250,000 scale classification of land uses according to Level II of the Anderson classification system.[13] However, these classifications have not yet been related to runoff potential in a systematic manner. In addition, the available data may not be sufficiently detailed for urban hydrology studies.

There is no consistent national set of soil maps available in digital form. Many U.S. Department of Agriculture Soil Conservation Service state offices are in the process of digitizing the boundaries between different soils as indicated on the individual County Soil Survey maps.[16]

GIS coverages that include manmade drainage facilities such as storm sewers will also provide valuable analytical possibilities for hydrologists. It is possible to perform the functional equivalent of the rational method in hydrology, using the network modeling methods available in current vector-based GIS.[13]

Extending the hydrologic applications of GIS beyond those identified above will require modifications to the existing GIS and/or hydrologic modeling applications:

- *Modifications to hydrologic models.* Existing hydrologic models generally depend on parameters that are average values for an entire watershed or catchment. By building hydrologic models based on polygon coverages in GIS, much better results might be obtained.[13] For example, a polygon-based hydrologic model should recognize that the hydrologic response of a watershed will vary greatly not only according to the total amount of impervious area within the watershed, but also according to the *location* of the impervious area.
- *Extensions to GIS.* Many hydrologic processes are time-dependent, and hydrologic models generally require time-series input data and generate time-series output. Present GIS are not equipped to store or manipulate time-series data. Therefore, the ability to store, retrieve, and perform operations on time-series data will be crucial for implementing serious hydrologic modeling within a GIS.[13]

23.6 ARTIFICIAL INTELLIGENCE

Artificial intelligence (AI) describes the application of computer resources to the imitation of human reasoning ability. AI really includes a group of related technologies including natural language processing and expert systems.

Natural language processing is the ability of a computer to process language that humans use in ordinary conversation.[19] Natural language processing, which is still under development, is the technology behind the vocal user interface described earlier in this chapter.

An *expert system* is a program that attempts to mimic how the human mind reasons about a particular topic. Expert systems are also called *knowledge-based systems* or *rule-based systems.* Expert systems are based on sets of *production rules.* A rule is simply a statement with the following format:

IF (conditions) THEN . . .

Some advanced expert systems (including the Nexpert Object program from Neuron Data, Palo Alto, Calif.) use an augmented rule format that allows more complex simulations:

IF (conditions) THEN DO (actions)

Sets of these rules can be strung together so that the hypothesis of one rule is included in the conditions of another. This forms a *rule network* that permits the computer to simulate complex reasoning processes. The rule network contains the *procedural knowledge* of the expert system.[12]

A rule network requires facts to reason on. These facts are sometimes called *objects*. Objects may be incorporated into the conditions of a rule or its actions. They can change in value as the reasoning process proceeds. The object network contains the *declarative knowledge* of the expert system.[12]

The connected rule and object networks form the *knowledge base* of the expert system. The main characteristic that differentiates expert systems from conventional programs is that the knowledge base is separated from the way in which the knowledge is utilized by application programs. Therefore, the knowledge base should be independent in the same way that a relational database should be independent. Just as a database can be revised without changing the application programs that use the database, the rules or facts in a knowledge base can be revised or extended without redesigning the programs using the knowledge base. A conventional program can contain facts and rules which match the operation of the knowledge base, but those facts or rules cannot be modified or extended without changing the program itself.

Expert systems generally use either of two types of reasoning. Reasoning by *forward chaining* starts with a set of data in the object network and reasons forward to search out the consequences of those particular data values. Forward chaining is best suited for design or evaluation problems. Reasoning by *backward chaining* starts from a final hypothesis and deduces what conditions would have been necessary to make this hypothesis true. In backward chaining, the expert system works backward through all the conditions of rules leading to this hypothesis. A backward strategy is most appropriate in determining if the desired results are met, and if not, for what reasons. Backward chaining is mostly used for diagnosis and decision problems.[1]

Some expert systems are designed as individual programs. Stand-alone expert systems have been designed for diagnosis of certain medical conditions, for configuration of large computer systems, for quick approval of credit by financial institu-

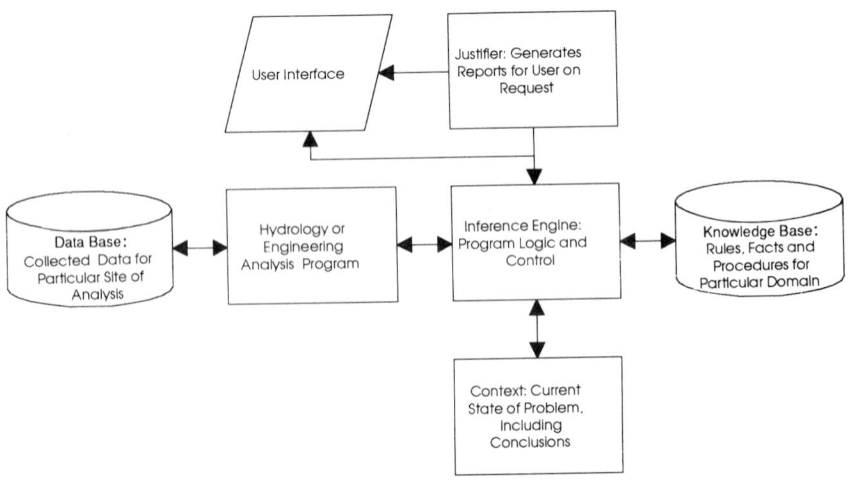

FIGURE 23.6.1 General architecture of an expert system. *(From Ref. 12. Reproduced by permission of ASCE.)*

tions, and for many other purposes. Stand-alone expert systems are useful where there is no need for extensive arithmetic analysis, and where much of the system input is provided interactively. Special programming languages, such as LISP and PROLOG, have been developed to make it easy to process knowledge bases.

Increasingly, however, expert systems are embedded within other programs to make them operate in a more intelligent fashion. Figure 23.6.1 illustrates the structure of a typical expert system which interacts with a hydrology applications program.

Figure 23.6.1 illustrates the relationship of the hydrology applications program to the database as well as the knowledge base. The expert system uses an inference engine to process the knowledge base according to forward-chaining or backward-chaining reasoning. The *justifier* is a part of the expert system that generates reports for the user. These reports can provide details on how the expert system reaches its conclusions. An *expert system shell* is a computer program that provides a general-purpose inference engine and usually a justifier, along with the tools necessary for the expert system shell to interact with external applications programs and databases. Some applications programs now contain embedded inference engines.

Expert systems should have extensive applications in hydrology. These applications will include the structure of the model, calibration of hydrologic models, analysis of hydrologic model output, and possibly preparation of hydrologic (engineering) designs.

REFERENCES

1. Baffaut, C., and J. W. Delleur, "Calibrating SWMM Runoff Quality Model with an Expert System." *Computing in Civil Engineering: Computers in Engineering Practice,* American Society of Civil Engineers, New York, 1989.
2. Charley, W. J., *DSS Data Compression Schemes for Regular Interval Time Series Data,* U.S. Army Corps of Engineers Hydrologic Engineering Center, November 1988.
3. Cutaia, A., *Technology Projection Modeling of Future Computer Systems,* Prentice-Hall, Englewood Cliffs, N.J., 1990.
4. Date, C. J., *An Introduction to Database Systems* (4th ed.), vol. 1. Addison-Wesley, Reading, Mass., 1986.
5. Dodson & Associates Inc., *Recommended Data Management Approach for the Interior Flood Hydrology (IFH) Software Package,* prepared for the U.S. Army Corps of Engineers Hydrologic Engineering Center, Davis, Calif., January 6, 1989.
6. Durrant, A. M., "The Use of a Digital Terrain Model Within a Geographical Information System for Stimulating Overland Hydraulic Mine Waste Disposal," Chap. 18 in *Terrain Modelling in Surveying and Civil Engineering,* G. Petrie and T. J. M. Kennie, eds., McGraw-Hill, New York, 1991.
7. ERDAS, *Image Processing System User's Guide,* ver. 7.3, Atlanta, Ga., 1988.
8. Howden, W. E., *Functional Program Testing and Analysis,* McGraw-Hill, New York, 1987.
9. Kennie, T. J. M., "Software Packages for Terrain Modelling Applications," Chap. 11 in *Terrain Modelling in Surveying and Civil Engineering,* G. Petrie and T. J. M. Kennie, eds., McGraw-Hill, New York, 1991.
10. Kennie, T. J. M. and R. A. McLaren, "Visualization for Planning and Design," Chap. 19 in *Terrain Modelling in Surveying and Civil Engineering,* G. Petrie and T. J. M. Kennie, eds., McGraw-Hill, New York, 1991.
11. Kennie, T. J. M. and G. Petrie, "Introduction to Terrain Modelling—Application Fields and Terminology," Chap. 1 in *Terrain Modelling in Surveying and Civil Engineering,* G. Petrie and T. J. M. Kennie, eds., McGraw-Hill, New York, 1991.

12. Knapp, G. M., H. P. Wang, K. C. Change, and G. C. Lee, "GABLE: An Expert System for Seismic Performance Evaluation," *Computing in Civil Engineering: Computers in Engineering Practice,* American Society for Civil Engineers, New York, 1989.

13. Maidment, D. R., "GIS and Hydrologic Modeling," Ch. 13 in *Geographic Information Systems and Environmental Modelling,* M. F. Goodchild, B. O. Parks, and L. F. Steyaert, eds., Oxford Univ. Press, New York (in press), 1993.

14. McCullagh, M. J., "Digital Terrain Modelling and Visualization," Chap. 9 in *Terrain Modelling in Surveying and Civil Engineering,* G. Petrie and T. J. M. Kennie, eds., McGraw-Hill, New York, 1991.

15. Milne, P. H., "Survey Software for Digital Terrain Modelling," Chap. 12 in *Terrain Modelling in Surveying and Civil Engineering,* G. Petrie and T. J. M. Kennie, eds., McGraw-Hill, New York, 1991.

16. Moore, I. D., et al., "GIS and Land Surface-Subsurface Process Modelling," in *Geographic Information Systems and Environmental Modelling,* M. F. Goodchild, B. O. Parks, and L. F. Steyaert, eds., Oxford Univ. Press, New York (in press), 1993.

17. Petrie, G., "Modelling, Interpolation and Contouring Procedures," Chap. 8 in *Terrain Modelling in Surveying and Civil Engineering,* G. Petrie and T. J. M. Kennie, eds., McGraw-Hill, New York, 1991.

18. Petrie, G., "Terrain Data Acquisition and Modelling from Existing Maps," Chap. 7 in *Terrain Modelling in Surveying and Civil Engineering,* G. Petrie and T. J. M. Kennie, eds., McGraw-Hill, New York, 1991.

19. Spence, L., and R. N. Palmer, "Inlet: Access to Water Resources Management Data through a Natural Language Interface." *Computing in Civil Engineering: Computers in Engineering Practice,* American Society of Civil Engineers, New York, 1989.

20. Steidler, F., et al., "Digital Terrain Models and Their Incorporation in a Database Management System," Chap. 15 in *Terrain Modelling in Surveying and Civil Engineering,* G. Petrie and T. J. M. Kennie, eds., McGraw-Hill, New York, 1991.

21. Terstriep, M. L., and M. T. Lee, "Regional Stormwater Modeling, Q-Illudas and Arc/Info," *Computing in Civil Engineering: Computers in Engineering Practice,* American Society of Civil Engineers, New York, 1989.

22. Thorpe, L. W., "Graphical Display and Manipulation in Two- and Three-Dimensions of Digital Cartographic Data," Chap. 22 in *Terrain Modelling in Surveying and Civil Engineering,* G. Petrie and T. J. M. Kennie, eds., McGraw-Hill, New York, 1991.

23. Van Haveren, B. P., *Water Resources Measurements: A Handbook for Hydrologists and Engineers,* American Water Works Association, Denver, Colo., 1986.

CHAPTER 24
REMOTE SENSING

Edwin T. Engman
NASA/Goddard Space Flight Center
Greenbelt, Maryland

24.1 INTRODUCTION

Remote sensing applications to hydrology are relatively new but are rapidly becoming an important information source for practicing hydrologists. To date, most of the reported applications have been scattered throughout the scientific literature, which has made it difficult to assimilate them into practice. This chapter attempts to pull together examples of current uses of remote sensing information in hydrology and to give the reader information on how to pursue this subject further. For a more in-depth coverage of remote sensing applications to hydrology the reader is referred to a recent UNESCO publication, "Advances in remote sensing for hydrology and water resources management"[55] and a recent book on the subject, *Remote Sensing in Hydrology.*[27]

Remote sensing uses measurements of the electromagnetic spectrum to characterize the landscape, or infer properties of it, or, in some cases, to actually measure hydrologic state variables. Aerial photography in the visible wavelengths is a remote sensing technique frequently used by hydrologists; however, modern remote sensing is centered around satellite systems and much of the discussion in this chapter emphasizes the use of satellite data. Over the years, remote sensing techniques have expanded to the point where they now include most of the electromagnetic spectrum. Different sensors can provide unique information about properties of the surface or shallow layers of the earth. For example, measurements of the reflected solar radiation give information on albedo, thermal sensors measure surface temperature, and microwave sensors measure the dielectric properties and, hence, the moisture content, of surface soil or of snow. Remote sensing and its continued development has added new techniques that hydrologists can use in a large number of applications. However important remote sensing data can be, it must be emphasized that this is just another tool in the hydrologist's tool kit and for the most part should be considered an addition to, rather than a replacement for, familiar hydrologic procedures.

24.2 BASIC PRINCIPLES

The electromagnetic spectrum, which is shown in Fig. 24.2.1, is the basis for all remote sensing. Remote sensing measures reflected or emitted energy from the earth's surface and takes advantage of the unique signal in specific regions of the spectrum.

An important feature of the electromagnetic spectrum is its interaction with the atmosphere and how this limits which parts of the spectrum can be used. Specific gases in the atmosphere block the amount of energy that is transmitted, and the regions of the electromagnetic spectrum that efficiently transmit the energy are known as *atmospheric windows*. Figure 24.2.1 illustrates these atmospheric windows as the solid black portions of the relative transmission graph. These are the parts of the spectrum which can be used for remote sensing by satellite imagery.

24.2.1 Sensors

The type of sensor dictates which part of the spectrum can be used for remote sensing applications. Carefully matching the sensor to the problem ensures that the study results will be useful and easily quantifiable. The commonly used sensors are briefly described below.

Gamma Radiation. Gamma-radiation remote sensing uses the difference between natural terrestrial gamma radiation and that attenuated by soil water or a layer of snow.[18] Most soils emit gamma radiation from the radioisotopes naturally occurring in soils (see γ-Rays in Fig. 24.2.1).

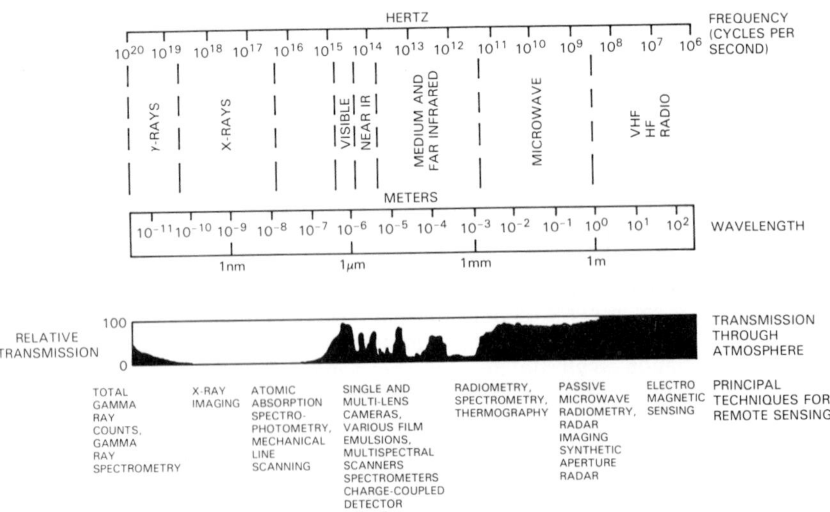

FIGURE 24.2.1 The electromagnetic spectrum, atmospheric windows, and principal techniques used in remote sensing. *(Reproduced from Ref. 27 with permission.)*

Aerial Photography. This has a long history of use in hydrology.[58] At first, aerial photographs used the visible portion of the electromagnetic spectrum, but advances in photography and films have led to the ability to make images of other parts of the spectrum, especially in the near infrared and thermal regions. The metric cameras used in aerial photography have been designed for a high degree of distortion correction. An additional remote sensing instrument analogous to the camera is the *vidicon,* or video recorder that can be used to record different regions of the spectrum with the use of filters.

Multispectral Scanners. Multispectral scanners are instruments that measure the spectral reflectance of the landscape in two or more narrow wavelength bands and record this information electronically. Multispectral scanners have two basic designs; one uses a motorized scanning mirror to sample successive strips of the earth beneath the sensor (Fig. 24.2.2*a*) so that the incoming radiation is separated into several spectral channels by a prism or diffraction grating. The other uses a "push broom" concept that projects the radiances directly onto an array of detectors (Fig. 24.2.2*b*). Thus, the reflected or emitted radiation is measured in discrete wavebands with detectors sensitive to specific regions of the spectrum.

Thermal Sensors. Thermal remote sensors directly measure the emitted thermal energy of the earth's surface. Surface temperature changes are the result of the balance of radiant, latent, sensible, and ground heat fluxes as well as artificial heat sources such as urban areas and industrial discharges. Thermal sensors generally operate on the line-scan principal of the multispectral scanner.

Microwave Sensors. Microwave remote sensing directly measures the dielectric properties of the earth's surface. The physical relationships between moisture, dielectric properties, and microwave response, along with the ability of microwave sensors to penetrate cloud cover, make microwave sensors a useful all-weather sensor to measure the moisture content of the earth's surface. There are two basic types of microwave sensors: *passive (radiometers)* and *active (radar)* instruments. Passive instruments measure the emitted energy in that particular region of the spectrum with a radiometer. Active systems, or radar, send out a pulse of energy and measure the portion that is reflected by the land surface.

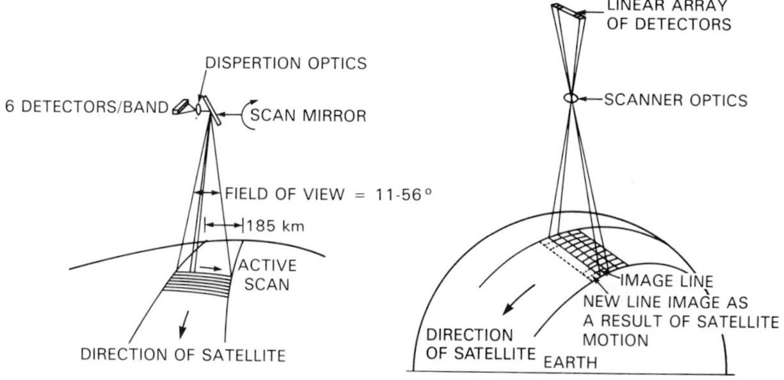

FIGURE 24.2.2 Schematic of *(a)* multispectral scanner and *(b)* pushbroom type of scanner. *(Part (b) is reproduced by permission of SPOT Image Corp., Reston, Va.)*

A special microwave instrument is the side-looking airborne radar (SLAR) that is used for rapid survey of terrain features. The advantage of SLAR is that it will penetrate cloud cover as well as much of the vegetation; as a result, areas have been mapped by SLAR that can not be mapped with more conventional aerial photography.

Lasers. Laser remote sensing is a rapidly expanding research area with potential application to hydrology. The principle involves projecting a narrow beam of coherent visible or near-infrared light and measuring the reflected energy with a photomultiplier tube to measure the distance between the laser system and the object of interest, or to measure the concentration of aerosols, or the elevation or roughness of the land surface.

24.2.2 Platforms and Satellite Systems

Platforms for mounting remote sensors include ground-based supports, aircraft, and satellites. Generally, truck-mounted ground-based systems and aircraft systems are used in sensor development to verify design characteristics and to learn how the sensor responds to the target characteristics. Aircraft are effective for coverage of relatively small areas and generally nonrepetitive missions. Aircraft are typically used for aerial photography and increasingly for multispectral-scanner, thermal, and side-looking airborne radar surveys.

Satellites provide an ideal platform for remote sensing because they can provide coverage nearly everywhere (if they are *polar-orbiting*) or continuous coverage (if they are *geostationary*). Polar-orbiting satellites are generally at relatively low earth altitude and provide complete, high-resolution coverage over a period of days. Geostationary satellites orbit with the earth's rotation so that they observe the same area of the globe continuously. The geostationary satellites have a relatively high orbit and provide coarse spatial resolution data. Although the choice of satellite orbit and design is beyond the scope of this chapter, a practicing hydrologist should understand the basics of orbit paths and repeat coverage intervals for the major satellite systems because they can influence the choice of data. Also, the practicing hydrologist will, for the most part, need to know only about the accessibility and characteristics of the major commercial satellites, the U.S. NASA Landsat and TM series, the U.S. NOAA satellites, and the French Systeme Probatoire d'Observation de la Terre (SPOT). Addresses for each of these data sources are listed in Table 24.2.1 for those wishing

TABLE 24.2.1 Sources of Satellite Data

EOSAT 4300 Forbes Blvd. Lanham, MD 20706 Telephone (301) 552-0547 Fax (301) 344-9933	SPOT Image Corp. 1897 Preston White Dr. Reston, VA 22091-4368 Telephone (703) 620-2200 Fax (703) 648-1813
Satellite Data Services Branch D543 Environmental Data Service NOAA World Weather Building, Room 606 Washington, DC 20233 Telephone (301) 763-8111	EROS Data Center Attn: User Services Sioux Falls, SD 57198 Telephone (605) 594-6151

Additional information on sources of aerial photography as well as other sources of remote sensing data can be found in Ref. 3.

TABLE 24.2.2 Spectral Characteristics of Satellite Systems Available for Hydrologic Applications

	Spectral range, μm			
Band number	NOAA-AVHRR*	Landsat	TM	SPOT
1	0.55–0.68		0.45–0.52	0.50–0.59
2	0.73–1.10		0.52–0.60	0.61–0.68
3	3.55–3.93		0.63–0.69	0.79–0.89
4	10.50–11.50	0.5–0.6	0.76–0.90	0.51–0.73†
5	11.50–12.50	0.6–0.7	1.55–1.75	
6		0.7–0.8	10.40–12.50	
7		0.8–1.1	2.08–2.35	
8‡		10.4–12.6		

* AVHRR = advanced very high resolution radiometer
† Panchromatic mode
‡ Only on Landsat 3

further information. These systems are emphasized because they are operational and data are readily available, and because there is a developed infrastructure of computer systems and software for analysis of data from these systems.

The existing satellite systems provide very good coverage of the earth and give the hydrologist a number of options for meeting data needs. Since 1972, the NASA Landsat (originally ERTS) satellites have been providing four spectral band measurements of the Earth's surface with the multispectral scanner (MSS), and more recently with the thematic mapper (TM) (seven spectral bands). Since 1986, the French SPOT satellites (four spectral bands) have also been in orbit. Table 24.2.2 lists the spectral characteristics of these four systems.

The choice of which satellite system to use depends greatly on the requirements for the data, which translate into the need for specific spectral bands, spatial requirements, temporal coverage, and the possible need for stereo coverage. Table 24.2.3 lists the principal applications for different spectral bands that are commonly available.

TABLE 24.2.3 Remote Sensing Applications for Different Spectral Bands (Approximate)

Band, μm	Applications
Blue (0.45–0.50)	Water penetration, land use, soil and vegetation characteristics, sediment
Green (0.50–0.60)	Green reflectance of healthy vegetation
Red (0.60–0.70)	Vegetation discrimination because of red chlorophyll absorption
Panchromatic (0.50–0.75)	Mapping, land use
Reflective (0.75–0.90)	Biomass, crop identification, soil-crop, land-water boundaries
Mid-infrared (1.5–1.75)	Plant turgidity, droughts, clouds—snow-ice discrimination
Mid-infrared (2.0–2.35)	Geology, rock formation
Thermal infrared (10–12.5)	Relative temperature, thermal discharges, vegetation classification, moisture studies

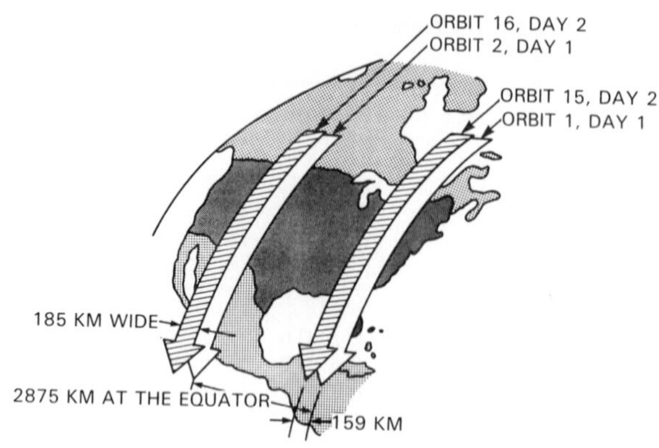

FIGURE 24.2.3 Illustration of orbit coverage of the earth with Landsat satellites.

The satellite orbits may limit the choice if the time interval between multiple passes is an important consideration. All remote sensing is sun-synchronous, and in polar or near-polar orbits. Thus the instruments "paint" a swath on the earth as they pass over the earth. For example, the Landsat satellites make 14 orbits each day, and because of the earth's rotation, their orbit paths are shifted westward so that, after 18 days, the entire earth is covered. Figure 24.2.3 illustrates the orbit paths and how they shift. Table 24.2.4 lists the swath widths, spatial resolutions, and repeat intervals for the major satellites. An added feature of the French satellite is that SPOT has the ability to point its sensor off-nadir by up to 27° from its central position. This allows tracking of transient phenomena (e.g., floods) as well as stereo coverage.

24.3 DATA ANALYSIS

Once data have been collected from a remote sensing system, the user must then interpret the data for the specific application. Generally, the data are available in three forms: as an *image* analogous to an aerial photograph, in an *analog format,* and in *digital format,* typically as a computer-compatible tape (CCT).

The conversion from imagery or analog data to digital data, and vice versa, is

TABLE 24.2.4 Spatial and Orbital Characteristics of Satellite Systems Available for Hydrologic Applications

Characteristic	NOAA-AVHRR	Landsat	TM	SPOT
Spatial resolution	1.1 km	79 m	30 m	20 m (10 m)*
Swath width	2700 km	185 km	185 km	117 km (425 km)†
Days for global coverage (repeat interval)	1	18	18	26 (2–3)†

* Panchromatic mode
† Off-nadir pointing capability

based on separating the measured intensity of reflectance into increments, usually based on byte word lengths, with 0 representing the lowest (darkest) level of reflectance and 255 the highest (brightest) reflectance.

There are several methods for converting images to digital data so that they can be analyzed with digital analysis equipment. One of the most common is the *densiometer,* a device that measures the average density or brightness of a small area of an image. The densiometer scans the image by rows and columns to create gray-level pixels. Imagery can also be digitized by a video camera, which makes the analog-to-digital conversion. A third technique has been developed relatively recently that uses a precisely scanning array of photodiodes.

Once in digital form, there are many instruments and computer systems that can perform a myriad of useful analyses to help the user interpret the data. Computer systems can also produce various statistics of the scene, and the data may be included in numerical hydrologic models. Scene preprocessing may be necessary to account for normalization for sun angle, correction for atmospheric conditions, and georeferencing the scene to a chosen map scale and coordinate system.

There are a number of image-analysis systems and very good commercial software packages for just about any hardware configuration. Currently, most of the newer systems are developed around the PC and workstation environments, although minicomputers and mainframes are still used. In addition, there are a number of "freeware" packages that are quite powerful and versatile, although support and user friendliness may be lacking. Figure 24.3.1 illustrates the components of a typical image-analysis system, which consists of a tape reader or some other input device, the computer, a monitor (preferably a high-density color system), and a joystick or tracking ball for moving around in the image. Disk storage, a separate monitor for instruction input, and various printing or high-quality color film writers are additional options. The costs vary with the sophistication of the hardware and software but can be expected to vary from a couple of thousand dollars to close to a hundred thousand dollars for a sophisticated system.

For more details on these aspects of remote sensing, see Refs. 6, 21, 36, 41, and 62.

Information on data analysis systems is frequently reviewed by journals; see, for example, a review of hardware and software used by forestry schools.[54] Vendors of software and hardware are also often listed.[2]

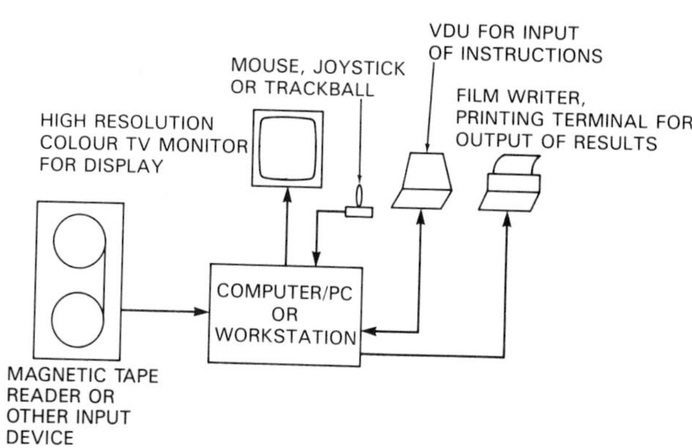

FIGURE 24.3.1 Components of a typical digital image processing system.

Once the data have been obtained and reduced to a usable form, their interpretation for hydrology is based on interpreting measurable variations in *spectral, temporal,* and *spatial* characteristics of the earth:

1. *Spectral characteristics.* Spectral characteristics (or *signature*) of the target are the unique spectral reflectances for specific earth features. Figure 24.3.2 shows typical spectral reflectance curves for three basic types of earth features: healthy green vegetation, dry bare soil (gray-brown loam), and clear lake water. By comparing the vegetation curve with the hatched boxes at the top of Fig. 24.3.2, it can be seen that some of the Landsat TM wave bands have been selected to fall within those wavelengths in which vegetation is especially reflective of incoming radiation.

Note how distinctive the curves are for each feature. Although the reflectance of individual features varies considerably above and below the average, these curves demonstrate the fundamental concept concerning spectral reflectance: that different surfaces produce different reflectances, and these vary over different parts of the spectrum.

2. *Temporal characteristics.* Temporal effects result from factors that change the spectral characteristics of a feature over time. For example, the spectral characteristics of many species of vegetation are in a nearly continuous state of change throughout a growing season. These changes often influence when sensor data are collected for a particular application. Moreover, temporal effects also may be the keys to gleaning the information sought in an analysis because, by comparing summer scenes with winter ones, for example, the spatial extreme of deciduous compared to evergreen vegetation can be determined.

3. *Spatial characteristics.* Spatial characteristics involve shapes and relative sizes as well as absolute sizes of objects. Typical examples of spatial characteristics are roads, rivers, and buildings. Geologic and geomorphic analysis relies on the identification of spatial features, including linear features such as faults or lineaments on the land surface that indicate geologic fracturing below the earth's surface.

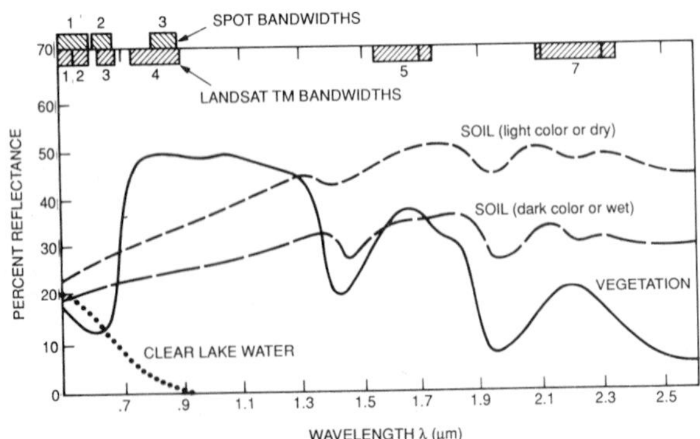

FIGURE 24.3.2 A schematic representation of spectral signatures for common earth targets. Landsat TM and SPOT bands and bandwidths are shown across the top of the figure. *(Reproduced from Ref. 27 with permission.)*

24.4 PRECIPITATION

Recognizing the practical limitations of rain gauges for measuring spatially averaged rainfall over large areas, hydrologists have increasingly turned to remote sensing as a possible means for quantifying the precipitation input, especially in areas where there are few surface gauges. Because the fundamental approach to measuring rainfall and snow are different with respect to remote sensing, snow measurement is discussed separately in this chapter.

Direct measurement of rainfall from satellites has not been generally feasible because the presence of clouds prevents observation of the precipitation directly with visible, near-infrared, and thermal-infrared sensors. However, improved analysis of rainfall can be achieved by combining satellite and conventional rain gauge data. Satellite data are most useful in providing information on the spatial distribution of potential rain-producing clouds, and gauge data are most useful for accurate point measurements. Ground-based radar has also proved useful for locating regions of heavy rain and for estimating rainfall rates (see Sec. 3.7).

Useful data can be derived from satellites used primarily for meteorological purposes, including polar orbiters such as the NOAA-N series and the Defense Meteorological Satellite Program, and from geostationary satellites such as GOES, GMS, and Meteosat, but their visible and infrared images can provide information only about the cloud tops rather than cloud bases or interiors. However, since these satellites provide frequent observations (even at night, with thermal sensors), the characteristics of potentially precipitating clouds and the rates of changes in cloud area and shape can be observed. From these observations, estimates of rainfall can be made which relate cloud characteristics to instantaneous rainfall rates and cumulative rainfall over time. For the practicing hydrologist, satellite rainfall methods are most valuable where there are no or very few surface gauges for measuring rainfall.

24.4.1 Visible and Infrared Techniques

The availability of meteorological and Landsat satellite data has produced a number of techniques for inferring precipitation from the visible and/or infrared (VIS/IR) imagery of clouds.[9,26] These techniques have led to the development of three approaches: a *cloud-indexing* approach, a *thresholding* approach, and a *life-history* approach. Cloud indexing,[44] which is time-independent, identifies different types of rain clouds and estimates the rainfall from the number and duration of clouds or their area. Thresholding[7] techniques consider that all clouds with low upper surface temperatures are likely to be rain clouds. Life history[56] methods are time-dependent and consider the rates of change in individual convective clouds or in clusters of convective clouds. Creutin et al.[22] used cokriging to combine ground rainfall measurements and satellite data to estimate spatial rainfall in the Middle East. All such spatial rainfall estimation methods are empirical in that they use statistical coefficients based on historical cloud and ground-measured rainfall to characterize the cloud-rainfall relationship.

24.4.2 Microwave Radiometry

Microwave techniques have great potential for measuring precipitation because at some microwave frequencies clouds are essentially transparent, and the measured microwave radiation is directly related to the raindrops themselves.

Microwave radiometry or passive microwave techniques react to rain in two ways: by *emission/absorption* and by *scattering.* With the emission/absorption approach, rainfall is observed by the emission of thermal energy by the raindrops. With the scattering approach, the rain attenuates upwelling radiation from the earth's surface and scatters or reflects cold, cosmic background radiation to the radiometer antenna. Considerable effort is now being spent on the development of passive microwave rainfall algorithms using dual or multifrequency principles, or polarizations at a single frequency. The results are promising,[60] and, although this method is still in the research phase, it appears that the most successful operational methods for rainfall sensing will combine passive microwave data with visible and infrared data.[8]

24.5 SNOW HYDROLOGY

Snow is a form of precipitation; however, in hydrology it is treated somewhat differently because of the lag between when it falls and when it produces runoff and groundwater recharge. Remote sensing is a valuable tool for obtaining snow data for predicting snowmelt runoff. For more in-depth coverage of this subject, the reader is referred to the book *Remote Sensing of Ice and Snow.*[31]

Nearly all regions of the electromagnetic spectrum provide useful information about the snowpack. Ideally, one would like to know the areal extent of the snow, its water equivalent, and the snow *condition* (grain size, density, and presence of liquid water). Although no single region of the spectrum provides all these properties, certain regions of the spectrum can be used to measure individual properties.

24.5.1 Gamma Radiation

The water content of some snowpacks is measured from low-elevation aircraft carrying sensitive gamma-radiation detectors. This method takes advantage of the natural emission of low-level gamma radiation from the soil. Aircraft flights over the same flight line before and during snow cover measure the attenuation of soil gamma radiation resulting from the snow layer, which is empirically related to an average snow water equivalent for that site.[20] This approach is limited to low aircraft altitudes (approximately 150 m) because the atmosphere attenuates a significant portion of the gamma radiation. In the U.S. National Oceanic and Atmospheric Administration (NOAA) operational airborne gamma-radiation snow water mapping program, procedures for correcting for the soil moisture are included in the system. Currently, this operational program covers over 1400 flight lines annually in the United States and Canada.[17] This method is effective for measuring snow cover in open plains, such as those in Kansas or Iowa, but is less effective in more hilly terrain or where there is extensive forest cover, such as in northern Wisconsin.

24.5.2 Visible/Near Infrared

The *albedo* or reflectance of the snow surface is the property most easily measured by remote sensing. Typically, new snow has an albedo of 90 percent or greater, whereas older snow that has been weathered and has accumulated dust and litter can have an albedo as low as 40 percent.[29] The reflectivity of new snow decreases as it ages, particularly in the infrared region of the spectrum.

Snow can readily be identified and mapped with the visible bands of satellite imagery because of its high reflectance in comparison to nonsnow areas. Generally this is done using the NOAA advanced very high resolution radiometer (AVHRR) visible channel, Landsat MSS channels 4 or 5, SPOT channels 1 to 4, or Landsat TM channels 2 and 4. Although snow can be detected at longer wavelengths, i.e., in the near-infrared region, the contrast between snow and nonsnow areas is considerably reduced in the infrared region as compared to the visible region of the spectrum. However, the contrast between clouds and snow is greater in Landsat TM band 5 (in the infrared region), and this band's data serve as a useful discriminator between clouds and snow.[25]

Use of satellite data for snow mapping has become operational in several regions of the world. Currently, NOAA develops daily snow cover maps for 56 river basins in the U.S. for use in stream-flow forecasting.[19] NOAA also produces maps of mean monthly snow cover for the northern hemisphere[43] using visible data in the lower resolution range (4 to 8 km). A handbook to assist potential users of satellite data in the mapping of snow-covered areas is available.[13]

24.5.3 Thermal Infrared

Thermal data are perhaps the least useful of the common remote sensing products for measuring snow and its properties; however, thermal data can be useful for helping identify snow/no-snow boundaries and discriminating between clouds and snow with AVHRR data.

24.5.4 Microwave

Microwave remote sensing offers great promise for future applications to snow hydrology. This is because the microwave data can provide information on the snowpack properties of most interest to hydrologists, i.e., snow cover area, snow water equivalent (or depth), and the presence of liquid water in the snowpack, which signals the onset of melt.[40] With the availability of microwave data from Special Sensor/Microwave Imager (SS/MI), this utility is being demonstrated.[28]

24.6 EVAPOTRANSPIRATION

Remote sensing techniques cannot measure evaporation or evapotranspiration directly. However, remote sensing does have two potentially very important roles in estimating evapotranspiration. First, remotely sensed measurements offer methods for extending empirical evapotranspiration relationships to much larger areas, including those areas where gauged temperature data may be sparse.[37,47,64] Secondly, remotely sensed measurements may be used to measure variables in the energy and moisture balance models of evapotranspiration. However, there has been little progress made in the direct remote sensing of the atmospheric parameters which affect evapotranspiration such as near-surface air temperature, near-surface water vapor gradients, and near-surface winds.

The question of how to use the spatial nature of remote sensing data to extrapolate point evapotranspiration measurements to a more regional scale has been addressed by Jackson,[34] and Gash[30] has formalized an analytical framework relating the horizontal changes in evaporation to horizontal changes in surface temperature.

Several variables related to evapotranspiration can be measured by remote sensing. Incoming solar radiation can be estimated from satellite observations of cloud cover, primarily from geosynchronous satellites.[14,63] For clear sky conditions, the surface albedo may be estimated by measurements covering the entire visible and near-infrared wave band, while measurements in narrow spectral bands can be used to determine vegetative cover.[15,34] Surface temperature is estimated from measurements in thermal infrared wavelengths, that is, the 10.5 to 12.5 μm wave band, assuming a surface emissivity close to unity for natural surfaces. Using the temperature sounders on the meteorological satellites in a linear regression model, Davis and Tarpley[24] estimated shelter temperatures with an error of about 2 K for clear or partly cloudy conditions.

Price[48] used thermal data from the Heat Capacity Mapping Mission (HCMM) to estimate regional scale evapotranspiration rates which were found to be comparable to pan evaporation data. Figure 24.6.1 illustrates the sensitivity of the thermal data to surface conditions, especially the irrigated agricultural areas that show up as cool in Fig. 24.6.1*b* because of the cooling effects of water evaporation in them.

One formulation of potential evapotranspiration that lends itself to remote sensing inputs is that developed by Priestley and Taylor.[49] Their evaporation formula is

$$LE = \alpha \frac{\Delta}{\Delta + \gamma} (R_n - G) \tag{24.6.1}$$

where L = latent heat of vaporization
E = evaporation rate
R_n = net radiation
G = ground heat flux
Δ = slope of vapor pressure versus temperature curve
α = empirical evaporation constant, determined to be 1.26

Barton[10] and Davies and Allen[23] have modified this formula for evaporation from an unsaturated land surface by treating α as a function of the surface layer soil moisture. Barton used airborne microwave radiometers to sense soil moisture remotely in his study of evaporation from bare soils and grasslands. Equation (24.6.1) is the basis for the model which Kanemasu et al.[38] used for estimating evapotranspiration with satellite data.

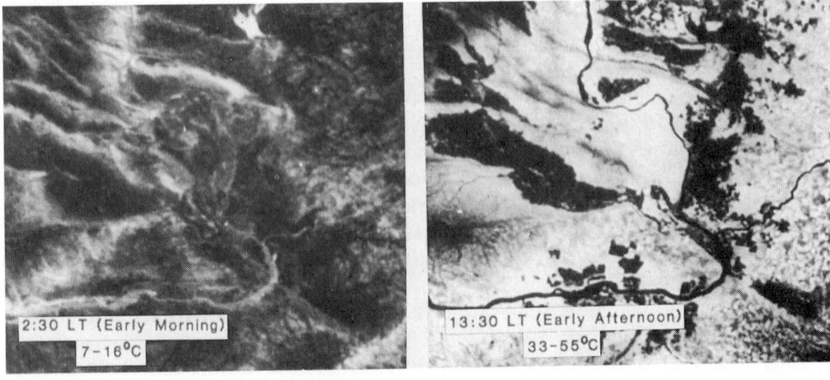

FIGURE 24.6.1 *(a)* Early morning and *(b)* early afternoon thermal imagery of eastern Washington State, U.S.A. Dark areas are cold and light areas are warm. The irrigated areas are identifiable in the afternoon scene *(b)* by their darker tone. *(Reproduced by permission of John C. Price.)*

24.7 RUNOFF

Runoff cannot be directly measured by remote sensing techniques. The role of remote sensing in runoff calculations is to provide a source of input data estimating equation coefficients and model parameters. There are three general areas where remote sensing is used: (1) to determine watershed geometry, drainage network, and other map-type of information; (2) to provide input data for empirical flood peak, annual runoff, or low flow equations; and (3) to delineate land-use classes.

24.7.1 Watershed Geometry

Remote sensing data can be used to obtain almost any information that is typically obtained from maps or aerial photography. In many regions of the world, remotely sensed data, and particularly Landsat, TM, or SPOT data, are the only source of good cartographic information. Drainage basin areas and the stream network are easily obtained from good imagery, even in remote regions. There have also been a number of studies to extract quantitative geomorphic information from Landsat imagery.[32]

Selection of imagery is important if one is to obtain the maximum possible information. Vegetation state is an important consideration. In the case of Landsat, TM, or SPOT, the choice of imagery with a low sun angle will enhance topographic and drainage features. Landsat MSS bands 5 and 7; TM bands 3, 4, and 5; and SPOT bands 2, 3, and 4 have proven to be the best choices for discerning physiographic features. The visible red band (MSS band 5, TM band 3, and SPOT band 2) is best for showing stream channel networks when their size is too small to be detected directly. This band is also good for separating vegetation types and for delineating nonvegetated areas. MSS band 7, TM bands 4 and 5, and SPOT band 3 show the most contrast between water and land areas. In cloud-covered and tropical regions, side-looking airborne radar penetrates dense vegetation and produces an image that reveals topographic features and drainage patterns.

24.7.2 Empirical Relationships

Empirical flood formulas are useful for making quick estimates of peak flow when there is very little other information available. Generally these equations are restricted in application to the size range of the basin and the climatic/hydrologic region of the world in which they were developed.

Most of the empirical flood formulas relate peak discharge to the drainage area of the basin. See United Nations Flood Control Series No. 7.[65] Remote sensing is used to estimate drainage basin area.

Landsat data are used to improve empirical regression equations of various runoff characteristics. For example, Allord and Scarpace[1] have shown how the addition of Landsat-derived land cover data can improve regression equations based on topographic maps alone.

24.7.3 Runoff Models

Landsat and SPOT data are used to determine both urban and rural land use for estimating runoff coefficients. Land use is an important characteristic of the runoff process that affects infiltration, erosion, and evapotranspiration. Thus, almost any physically based hydrologic model uses some form of land-use data or parameters based on these data. Distributed models, in particular, need specific data on land use and its location within the basin. Most of the work on adapting remote sensing to hydrologic modeling has involved the Soil Conservation Service (SCS) runoff curve number model.[67]

The empirical SCS models have widespread appeal because the runoff curve number depends only on land use and soil type and does not require hydrologic data for calibration of the rainfall-runoff relationship. Relating the curve numbers to land

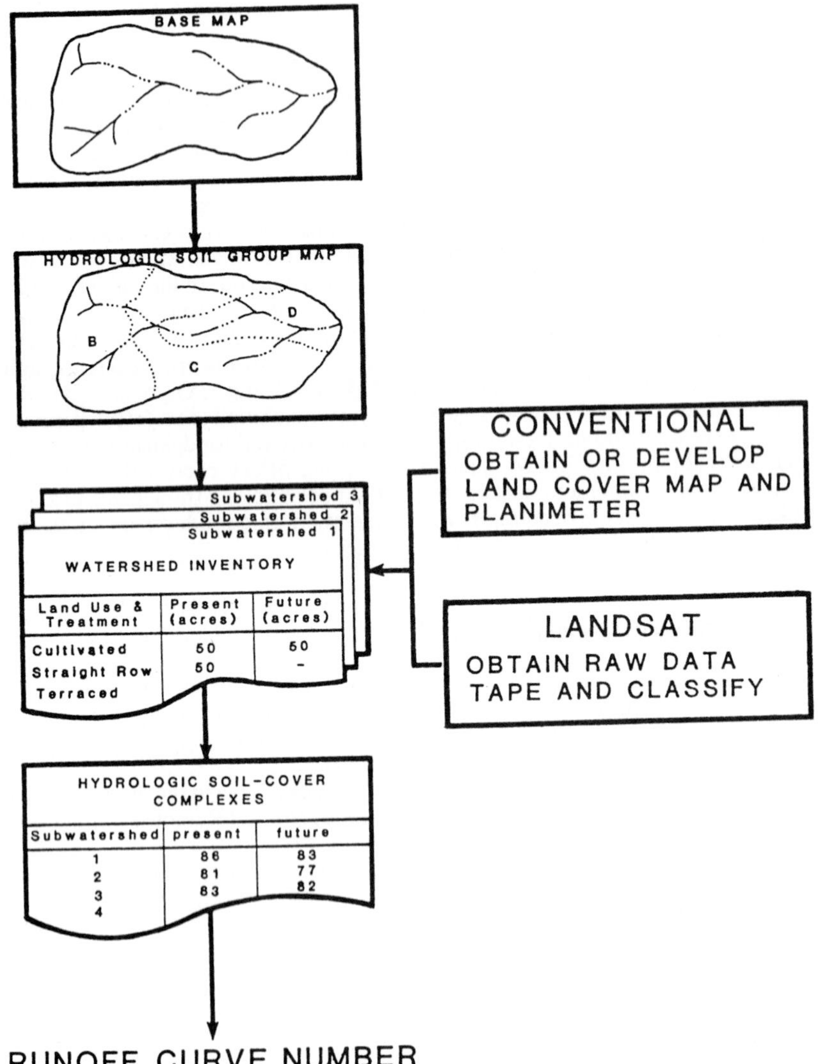

FIGURE 24.7.1 Flow diagram illustrating how Landsat data can replace conventional land-use data in the curve number procedure. *(After Ref. 68.)*

TABLE 24.7.1 Curve Numbers Table for Use with Landsat Data in Pennsylvania[12]

	Hydrologic soil group			
Land cover	A	B	C	D
Woods	25	55	70	77
Agriculture	64	75	83	87
Residential	60	74	83	87
Highly impervious	90	93	94	95
Water	98	98	98	95

use allows the examination of alternative forms of land development and management and assessment of their hydrologic impact.

Remote sensing data are used as a substitute for land cover maps obtained by conventional means, as shown in Fig. 24.7.1.

The linear resolution of the Landsat TM data is about 30 m which is equivalent in area to ¼ acre (0.1 ha). SPOT data are somewhat better (linear resolution of 10 or 20 m), but this spatial resolution is insufficient to provide the detail necessary to use the published SCS land-use tables (see Tables 5.5.1 and 9.4.2). Consequently, one must develop a land cover table analogous to the published SCS tables, but compatible to satellite data scales. An example of such a table, developed for some watersheds in Pennsylvania,[12] is shown in Table 24.7.1.

Such tables have much less detail than the published SCS land-use tables. One must develop a table suitable for the specific watershed and design a classifier to delineate the chosen land-use classes on that watershed.

In a study of the Upper Anacostia River Basin in Maryland, Ragan and Jackson[50] compared Landsat-derived flood-frequency parameters with those produced by conventional procedures employing low-level aerial photography to determine land use (Table 24.7.2).

In remote sensing applications, one seldom duplicates detailed land-use statistics exactly, but flood discharges are not sensitive to small differences in these statistics.

TABLE 24.7.2 Comparison of Aerial Photograph and Landsat-Derived Discharges Computed with the SCS Runoff Curve Number Model

		Discharge, ft^3/s	
Return period, years	Precipitation, in	Aerial photographs	Landsat
2	3.0	3,850	3,490
5	3.3	6,064	5,140
10	5.4	7,580	6,900
25	5.8	9,300	8,759
50	6.7	10,400	9,900
100	7.3	11,806	11,100

Source: From Ref. 50.

Another study of the three complex basins in Pennsylvania demonstrates this point.[12] Table 24.7.3 compares the land-use statistics and the computed runoff curve numbers for conventional and Landsat classifications of land cover.

One can easily see that the individual watershed's land-use statistics may differ by a significant amount but the computed curve numbers are nearly the same. A study by the Corps of Engineers[53] estimated that an individual pixel may be incorrectly classified about one-third of the time. However, by aggregating land use over a significant area, the misclassification of land use can be reduced to about 2 percent. These studies were all based on Landsat multispectral scanner data (80-m resolution). One can expect better results with 30-m Landsat TM data or 20-m data from the SPOT satellites.

Studies have shown[35] that for planning studies the Landsat approach to land cover classification is cost-effective. The authors estimated that the cost benefits were on the order of 2.5 to 1 or 6 to 1 in favor of the Landsat approach, depending on the experience of the analysts and the availability of data and background information. These benefits increase for larger basins or for multiple basins in the same general hydrologic area.

Other types of runoff models that are not based only on land use are beginning to be developed. For example, Strubing and Schultz[61] have developed a runoff regression model that is based on Barrett's[5] indexing technique. The cloud area and temperature are the satellite variables used to develop a temperature-weighted cloud cover index. This index is then transformed linearly to mean monthly runoff. Their results from the River Baise are very promising.

24.7.4 Integration with GIS

The pixel format of digital remote sensing data makes it ideal for merging with Geographical Information Systems (GIS). The US Army Corps of Engineers[66] developed a microcomputer-based system that combines remote sensing image processing and spatial data analysis through a GIS. The system provides options to use the SCS

TABLE 24.7.3 Summary of Conventional versus Landsat Land Cover Classification and Comparison of Curve Numbers for the Little Mahoney and Chickies Creek Watersheds[12]

	Land cover, %	
Classification	Conventional	Landsat
	Little Mahoney Creek	
Forest	76.0	82.3
Agriculture	6.9	0.1
Urban	16.0	14.3
Curve number	*72*	*70*
	Chickies Creek	
Forest	29.0	30.5
Agriculture	57.0	62.4
Urban	14.0	7.1
Curve number	*73*	*73*

curve number or the Snyder unit hydrograph to estimate runoff from single storm events. Remote sensing can be incorporated into the system in a variety of ways: as a measure of land use and impervious surfaces, for providing initial conditions for flood forecasting, and for monitoring flooded areas.[45]

24.7.5 Flood Monitoring

The area inundated by floods and floodplains can be mapped effectively with remotely sensed data. Satellite data such as those from Landsat can be used to define coverage of an entire river basin but may have some limitations on small basins because of the limited spatial resolution. Black-and-white photography, infrared photography, thermal infrared data, multispectral scanner data, and radar have all been used to map the areal extent of flooding. For most approaches using remotely sensed data, determining areas of inundation depends on measuring reductions in reflectivity caused by standing or flowing water, high soil moisture, moisture-stressed vegetation, and surface temperature changes induced by flooding. These effects may be detected for up to 2 weeks or longer after the passage of a flood; thus it may not be necessary to obtain data exactly during the flood peak.[4]

Cloud cover is frequently a problem in mapping floods with Landsat data because the 18-day cycle time between overpasses may not provide a clear image during or shortly after the flood. SPOT has a pointing capability to help overcome the cloudiness problem. The NOAA satellites have an advantage of more frequent coverage over the target area (twice daily). In spite of the coarser spatial resolution (approximately 1100 m versus 80 m for Landsat MSS and 30 m for the thematic mapper), the NOAA satellite thermal infrared sensor has proved effective in measuring areas of flood inundation.[11]

All-weather flood mapping is possible with microwave sensors. Radar systems are capable of higher spatial resolution than passive systems under similar situations and should be well-suited for this task. Lowry et al.[42] demonstrated that airborne synthetic-aperture radar (SAR) could provide all-weather flood area delineation. The shuttle imaging radar (SIR-B) was used to map flood boundaries and assess damage over areas of Bangladesh.[33]

24.7.6 Floodplain Mapping

Floodplains have been delineated by using remotely sensed data to infer the extent of the floodplain from vegetation changes, soils, or some other cultural features commonly associated with floodplains.[52]

Sollers et al.[59] produced flood and flood-prone maps at 1:24,000 and 1:62,500 scales using digital Landsat data. One can reasonably expect better results and more accurate delineation of flood-prone areas with the TM and SPOT data, although for many legal requirements it is necessary to map flood-prone areas from high-resolution aerial photography.

24.8 WATER QUALITY

Remote sensing has an important role in water-quality evaluation and management strategy. Sources of pollution are often easy to identify, especially where there are pipes or open channels discharging into a lake or river. Non-point-source pollution can perhaps be evaluated best by remote sensing, especially when the spatially distributed nature of non-point-source pollution is considered. The large-scale view provided by remote sensing gives an environmental scientist very different data from that which can be obtained with surface data collection and sampling. Monitoring large areas on a frequent basis can be achieved economically only with remote sensing. Unfortunately, remote sensing is largely limited to surface measurements of turbidity, suspended sediment, chlorophyll, eutrophication, and temperature. However, these characteristics of water quality can be used as indicators of more specific pollution problems. Effective use of remote sensing measurements is made when they are used as ancillary data with other data or information.

The region of the electromagnetic spectrum that includes visible and infrared light is useful for water-quality variables. Thermal infrared is also used for measuring water quality but it uses a direct measure of the emitted energy. The microwave region is not particularly useful for determining indicators of water quality because microwaves hardly penetrate water. Oil slicks or other surface contamination are well-detected by microwave radiation, however.

Because the intensity and color of light is modified by the volume of water and its contaminants, empirical relationships can be established between the reflectance measurement and water-quality variables.

There are a number of water-quality parameters that can be determined through remote sensing. Table 24.8.1 is a brief summary of some of these.[27]

Khorram[39] used Landsat data and samples taken at 50 sites in San Francisco Bay to study salinity, turbidity, suspended solids, and chlorophyllia. Khorram developed regression equations between these water-quality parameters and the mean radiance value of different Landsat bands. In San Francisco Bay, the saline water is generally more turbid than freshwater inflows, and there is thus a strong correlation between salinity and turbidity.

The temporal features of the satellite data may provide a time series of sediment distribution and concentration. It is possible to go back in time to generate a historical description of sedimentation activities in a given body of water. To do this, the Landsat MSS data may be obtained for the period from 1972 to the present and related to sediment concentration with current samples, assuming that the relationship between sediment characteristics and reflectance has not changed with time.

Large-scale imagery of computer maps of reflectance or surface temperature are

TABLE 24.8.1 Applications of Remote Sensing for Various Indicators of Water Quality

Water-quality indicator	Aquatic variable	Use
Color	Plankton, algae, dissolved chemicals	Food supply, food chain studies
Turbidity	Movement of water masses	Water body dynamics, bathymetry
Sediment	Concentration of suspended sediment	Lake sediment studies, erosion sources
Trophic state	Color, turbidity, temperature	Nutrient status, productivity
Thermal pollution	Temperature	Sources, thermal maps, evaluation of engineering structures

useful tools for planning *in situ* sampling programs and for locating sampling buoys and other permanently placed monitoring instrumentation.

24.9 FUTURE DEVELOPMENTS

New sensors, particularly in the microwave region, promise great potential for hydrologic applications. For example, remote sensing of soil moisture can be accomplished to some degree in all regions of the electromagnetic spectrum, but only the microwave region offers the potential for truly quantitative measurement from a spaceborne instrument.

Microwave techniques for measuring soil moisture include both the passive and active microwave approaches, with each having distinct advantages. The theoretical basis for measuring soil moisture by microwave techniques is based on the large contrast between the dielectric properties of liquid water and dry soil.

Existing hydrologic models have represented soil moisture in a way needed to make the model work but have not considered the possibility of independent determination of soil moisture levels or of soil parameters.[46]

Several new satellites are planned for launch over the next decade which will carry payloads and make measurements relevant to the land part of the hydrologic cycle. There are several satellites, such as ERS-1 recently launched by the European Space Agency, the J-ERS-1 to be launched by Japan, and Radarsat to be launched by Canada, which have primarily oceanographic applications but which will also be used for acquiring data over land. All will carry single-polarization, single-wavelength, synthetic-aperture radars, plus radiometers at various wavelengths. Continuing high-spatial-resolution data from the Landsat and SPOT satellites, passive microwave data from the Special Sensor/Microwave Imager, and continuing meteorological satellite coverage from the NOAA, GOES, GMS, and Meteosat series all mean that the remotely sensed techniques described in this chapter can continue to be employed and expanded. However, there are many other sensors and satellites being planned that will have considerable hydrologic interest, such as the Tropical Rainfall Measurement Mission (TRMM) and Earth Observational System (EOS). TRMM,[57] to be launched into an orbit inclined at about 30°, will have on board a rainfall radar that will be capable of estimating rainfall rates over land. The orbit is such that any point within the tropics will be sampled at every hour twice each month.

EOS[16] and its counterpart European and Japanese platforms will lead to considerable advances in the understanding of all the earth sciences, including hydrology. Although the draft manifest of instruments is impressive, so is the fact that the data on hydrology and other earth sciences is organized in an information system in which time series of all the data will be easily available. This data system will allow many types of data to be used simultaneously to calibrate or be assimilated into numerical models.

There are many new and exciting observations of the hydrologic cycle that are going to be available from new satellite systems, and concurrently, new models to allow the data to be analyzed and address previously intractable problems. Remote sensing can provide many of the necessary data to supplement conventional data and thereby expand hydrology in new and exciting directions. Remote sensing will also provide entirely new data types and forms that will help hydrologists tackle previously unsolvable questions.

REFERENCES

1. Allord, G. J., and F. L. Scarpace, "Improving Streamflow Estimates Through Use of Landsat," in *Satellite Hydrology,* 5th Annual William T. Pecora Memorial Symposium on Remote Sensing, Sioux Falls, S.D., pp. 284–291, 1979.
2. American Society of Photogrammetry and Remote Sensing, "Products and Services Index—and Corporate Descriptions," *Photogrammetric Engineering and Remote Sensing,* vol. 56, no. 5, May 1990.
3. ASTM, "Geotechnical Applications of Remote Sensing and Remote Data Transmission," ASTM special technical publication 967, A. I. Johnson and C. B. Pettersson, eds., American Society for Testing and Materials Symposium Proceedings, ASTM, Philadelphia, 1987.
4. American Water Resources Association, "Satellite Analysis," *Water Resour. Bull.,* vol. 10, no. 5, pp. 1023–1086, 1974.
5. Barrett, E. C., "The Estimation of Monthly Rainfall from Satellite Data," *Mon. Weather. Rev.,* vol. 98, pp. 322–327, 1970.
6. Barrett, E. C., and L. F. Curtis, *Introduction to Environmental Remote Sensing,* Chapman and Hall, London, 1982.
7. Barrett, E. C., G. D'Souza, C. H. Power, and C. Kidd, "Towards Trispectral Satellite Rainfall Monitoring Algorithms," in *Tropical Precipitation Measurements,* proc. int. symp., Tokyo, Japan, NASA/NASDA, J. S. Theon and J. Fugono, eds., A. Depak, Hampton, Va., pp. 285–292, 1988.
8. Barrett, E. C., and C. Kidd, "The Use of SMMR Data in Support of a VIR/IR Satellite Rainfall Monitoring Technique in Highly-Contrasting Climatic Environments," in *Passive Microwave Observing from Environmental Satellites, a Status Report,* J. C. Fischer, ed., NOAA technical report NESDIS 35, Washington D.C., pp. 109–123, 1987.
9. Barrett, E. C., and Martin D. W., *The Use of Satellite Data in Rainfall Monitoring,* Academic Press, London, 1981.
10. Barton, I. J., "A Case Study Comparison of Microwave Radiometer Measurements over Bare and Vegetated Surfaces," *J. Geophys. Res.,* vol. 83, pp. 3513–3517, 1978.
11. Berg, C. P., M. Matson, and D. R. Wiesnet, "Assessing the Red River of the North 1978 Flooding from NOAA Satellite Data," *Satellite Hydrology,* American Water Resources Association, Minneapolis, Minn., pp. 309–315, 1981.
12. Bondelid, T. R., T. J. Jackson, and R. H. McCuen, "Estimating Runoff Curve Numbers Using Remote Sensing Data," *Proc. Int. Symp. on Rainfall-Runoff Modeling, Applied Modeling in Catchment Hydrology,* Water Resources Publications, Littleton, Colo., pp. 519–528, 1982.
13. Bowley, C. J., J. C. Barnes, and A. Rango, *Applications Systems Verification and Transfer Project,* vol. VIII: "Satellite Snow Mapping and Runoff Prediction Handbook," NASA technical paper 1829, Goddard Space Flight Center, Greenbelt, Md., 1981.
14. Brakke, T. W., and E. T. Kanemasu, "Insolation Estimation from Satellite Measurements of Reflected Radiation," *Rem. Sens. Environ.,* vol. 11, pp. 157–167, 1981.
15. Brest, C. L., and Goward, S. N., "Deriving Surface Albedo Measurements from Narrow Band Satellite Data," *Int. J. Rem. Sens.,* vol. 8, pp. 351–367, 1987.
16. Butler, D., et al., "From Pattern to Process: The Strategy of the Earth Observing System," NASA, Washington, D.C., 1988.
17. Carroll, S. S., and T. R. Carroll, "Effect of Forest Biomass on Airborne Snow Water Equivalent Estimates Obtained by Measuring Terrestrial Gamma Radiation," *Rem. Sens. Environ.,* vol. 7, pp. 313–320, 1989.
18. Carroll, T. R., "Airborne Soil Moisture Measurements Using Natural Terrestrial Gamma Radiation," *Soil Science,* vol. 132, pp. 358–366, 1981.
19. Carroll, T. R., and M. Allen, "Airborne Gamma Radiation Snowwater Equivalent and Soil

Moisture Measurements and Satellite Aerial Extent of Snow Cover Measurements: A User's Guide," ver. 3.0, National Weather Service, NOAA, Minneapolis, Minn., 1988.

20. Carroll, T. R., and K. G. Vadnais, "Operational Airborne Measurement of Snow Water Equivalent Using Natural Terrestrial Gamma Radiation," *Proc. 48th Annual Western Snow Conf.,* Laramie, Wyom., pp. 97–106, 1980.

21. Colwell, R. N., ed., *Manual of Remote Sensing,* vols. I and II, American Society of Photogrammetry, Falls Church, Va., 1983.

22. Creutin, J. D., G. Delrieu, and T. Lebel, "Rain Measurement by Raingage-Radar Combination: A Geostatistical Approach." *J. Atmospher. Ocean Tech.,* vol. 5, no. 1., pp. 102–115, 1988.

23. Davies, J. A., and C. D. Allen, "Equilibrium, Potential and Actual Evaporation from Cropped Surfaces in Southern Ontario," *J. Appl. Meteorol.* (now *J. Clim. Appl. Meteorol.*) vol. 12, pp. 649–657, 1973.

24. Davis, P. A., and J. D. Tarpley, "Estimation of Shelter Temperatures from Operational Satellite Sounder Data," *J. Clim. Appl. Meteorol.,* vol. 22, pp. 369–376, 1983.

25. Dozier, J., "Snow Reflectance from Landsat-4 Thematic Mapper," *IEEE Trans. Geosci. Rem. Sens.,* vol. GE-22, no. 3, pp. 323–328, 1984.

26. D'Souza, G., and E. C. Barrett, "A Comparative Study of Candidate Techniques for U.S. Heavy Rainfall Monitoring Operations Using Meteorological Satellite Data," final report to U.S. Department of Commerce: cooperative agreement no. NA86AA-H-RA001, amendment no. 3, University of Bristol, Bristol, U.K., 1988.

27. Engman, E. T., and R. J. Gurney, *Remote Sensing in Hydrology,* Chapman and Hall, London, 1991.

28. Foster, J. L., A. T. C. Chang, D. K. Hall, and A. Rango, "Derivation of Snow Water Equivalent in Boreal Forests Using Microwave Radiometry," *Arctic,* vol. 44, supp. 1, pp. 147–152, 1991.

29. Foster, J. L., D. K. Hall, and A. T. C. Chang, "Remote Sensing of Snow," *EOS,* vol. 68, no. 32, pp. 681–684, 1987.

30. Gash, J. H. C., "An Analytical Framework for Extrapolating Evaporation Measurements by Remote Sensing Surface Temperature," *Int. J. Rem. Sens.,* vol. 8, no. 8, pp. 1245–1249, 1987.

31. Hall, D. K., and J. Martinec, *Remote Sensing of Ice and Snow,* Chapman and Hall, London, 1985.

32. Haralick, R. M., S. Wang, L. G. Shapiro, and J. B. Campbell, "Extraction of Drainage Networks by Using a Consistent Labeling Technique," *Rem. Sens. Environ.,* vol. 18, pp. 163–175, 1985.

33. Imhoff, M. L., et al., "Monsoon Flood Boundary Delineation and Damage Assessment Using Space Borne Imaging Radar and Landsat Data," *Photogramm. Eng. Rem. Sens.,* vol. 53, pp. 405–413, 1987.

34. Jackson, R. D., "Estimating Evapotranspiration at Local and Regional Scales," *IEEE Trans. Geosci. Rem. Sens.,* vol. GE-73, pp. 1086–1095, 1985.

35. Jackson, T. J., R. M. Ragan, and W. N. Fitch, "Test of Landsat-Based Urban Hydrologic Modeling," *J. Water Resour. Plann. Manage. Div., Am. Soc. Civ. Eng.* (now *J. Water Resour. Plann. Manage.*), vol. 103, no. WR1, proc. papers 12950, pp. 141–158, 1977.

36. Jensen, J. R., *Introductory Digital Image Processing: A Remote Sensing Perspective.* Prentice-Hall, Englewood Cliffs, N.J., 1986.

37. Jensen, M. E., and H. R. Haise, "Estimating Evapotranspiration from Solar Radiation," *Proc. Am. Soc. Civ. Eng., J. Irrigation and Drainage Division,* vol. 89, pp. 15–41, 1963.

38. Kanemasu, E. T., L. R. Stone, and W. L. Powers, "Evapotranspiration Model Tested for Soybeans and Sorghum," *Agron. J.,* vol. 68, pp. 569–572, 1977.

39. Khorram, S., "Development of Water-Quality Models Applicable Throughout the Entire San Francisco Bay and Delta," *Photogramm. Eng. Rem. Sens.,* vol. 51, pp. 53–62, 1985.

40. Kunzi, K. F., S. Patil, and H. Rott, "Snow-Covered Parameters Retrieved from NIMBUS-7 SMMR Data," *IEEE Trans. Geosci. Rem. Sens.*, vol. GE-20, pp. 452–467, 1982.

41. Lillesand, T. M., and R. W. Kiefer, *Remote Sensing and Image Interpretation*, Wiley, New York, 1979.

42. Lowry, R. T., E. J. Langham, and N. Mudry, "A Preliminary Analysis of SAR Mapping of the Manitoba Flood, May 1979," *Satellite Hydrology*, American Water Resources Association, Minneapolis, Minn., pp. 316–323, 1981.

43. Matson, M., and D. R. Wiesnet, "New Data Base for Climate Studies," *Nature*, vol. 289, pp. 451–456, 1981.

44. Moses, J. F., and E. C. Barrett, "Interactive Procedures for Estimating Precipitation from Satellite Imagery," *Hydrologic Applications of Space Technology* (proc. Cocoa Beach, Fl., Workshop, August 1985) IAHS publication no. 160, pp. 25–29, 1986.

45. Neumann, P., W. Fett, and G. A. Schultz, "A Geographic Information System as Data Base for Distributed Hydrological Models," *Proc. International Symposium, Remote Sensing and Water Resources*, August 1990, Enschede, The Netherlands, pp. 781–791, 1990.

46. Peck, E. L., T. N. Keefer, and E. R. Johnson, "Strategies for Using Remotely Sensed Data in Hydrologic Models," NASA report no. CR-66729, Goddard Space Flight Center, Greenbelt, Md., 1981.

47. Penman, H. L., "Natural Evaporation from Open Water, Bare Soil and Grass," *Proc. Royal Soc. London*, vol. A193, pp. 129–145, 1948.

48. Price, J. C., "Estimation of Regional Scale Evapotranspiration Through Analysis of Satellite Thermal-Infrared Data," *IEEE Trans. Geosci. Rem. Sens.*, vol. GE-20, pp. 286–292, 1982.

49. Priestly, C. H. B., and R. J. Taylor, "On the Assessment of Surface Heat Flux and Evaporation Using Large Scale Parameters," *Mon. Weather Rev.*, vol. 100, pp. 82–92, 1972.

50. Ragan, R. M., and T. J. Jackson, "Runoff Synthesis Using Landsat and SCS Model," *J. Hydraul. Div., Am. Soc. Civ. Eng.* (now *J. Hydraul. Eng.*) vol. 106, no. HY5, pp. 667–678, 1980.

51. Rango, A., "Operational Applications of Satellite Snow Cover Observations," *Water Resour. Bull.*, vol. 16, no. 6, pp. 1066–1073, 1980.

52. Rango, A., and A. T. Anderson, "Flood Hazard Studies in the Mississippi River Basin Using Remote Sensing," *Water Resour. Bull.*, vol. 10, pp. 1060–1081, 1974.

53. Rango, A., A. Feldman, T. S. George, III, and R. M. Ragan, "Effective Use of Landsat Data in Hydrologic Models," *Water Resour. Bull.*, vol. 19, pp. 165–174, 1983.

54. Sader, S. A., and J. C. Winne, "Digital Image Analysis Hardware/Software Use at U.S. Forestry Schools," *Photogramm. Eng. Rem. Sens.*, vol. 57, no. 2, pp. 209–211, 1991.

55. Schultz, G. A., and E. C. Barrett, "Advances in Remote Sensing for Hydrology and Water Resources Management," *Technical Documents in Hydrology*, UNESCO, Paris, 1989.

56. Scofield, R. A., and V. J. Oliver, "A Scheme for Estimating Rain from Satellite Imagery," NOAA/NESS technical memo 86, U.S. Department of Commerce, NOAA, Washington D.C., 1977.

57. Simpson, J., R. F. Adler, and G. R. North, "A Proposed Tropical Rainfall Measuring Mission (TRMM) Satellite," *Bull. Am. Meteorol. Soc.*, vol. 69, pp. 278–295, 1988.

58. Slama, C. C., ed., *Manual of Photogrammetry*, American Society of Photogrammetry, Falls Church, Va., 1980.

59. Sollers, S. C., A. Rango, and D. L. Henninger, "Selecting Reconnaissance for Flood Plain Surveys," *Water Resour. Bull.*, vol. 14, pp. 359–373, 1978.

60. Spencer, R. W., H. M. Goodman, and R. E. Wood, "Precipitation Retrieval Over Land and Ocean with SSM/I, Part 1: Identification and Characteristics of the Scattering Signal," *J. Atmospher. Oceanic Tech.*, vol. 2, pp. 254–263, 1988.

61. Strubing, G., and A. Schultz, "Estimation of Monthly River Runoff Data on the Basis of

Satellite Imagery," *Proc. Hamburg Symposium,* IAHS publication no. 145, pp. 491–498, 1983.

62. Swain, P. H., and S. M. Davis, eds., *Remote Sensing: The Quantitative Approach,* McGraw-Hill, New York, 1978.

63. Tarpley, J. D., "Estimating Incident Solar Radiation at the Surface from Geostationary Satellite Data," *J. Appl. Meteorol.* (now *J. Clim. Appl. Meteorol.*), vol. 18, pp. 1172–1181, 1979.

64. Thornthwaite, C. W., "An Approach Toward a Rational Classification of Climates," *Geophys. Rev.,* vol. 38, pp. 55–94, 1948.

65. United Nations Economic Commission for Asia and the Far East, *Multipurpose River Basin Development,* Part 1: *Manual of River Basin Planning,* Flood Control Series No. 7. United Nations publication ST/ECAFE/SERF/7, New York, 1955.

66. U.S. Army Corps of Engineers, "Remote Sensing Technologies and Spatial Data Applications," research document no. 29, Hydrologic Engineering Center, Davis, Calif., pp. 1–5, 1987.

67. U.S. Department of Agriculture, Soil Conservation Service, *National Engineering Handbook,* Sec. 4, *Hydrology,* U.S. Government Printing Office, Washington, D.C., 1972.

68. Welle, P. I., and T. J. Jackson, "Application of Landsat Data and Computer Data Bases for Runoff Curve Number Estimation," ASAE paper 82-2097, American Society of Agricultural Engineers, Summer Meeting, Madison, Wis., 1982.

CHAPTER 25
AUTOMATED DATA ACQUISITION AND TRANSMISSION

Vito J. Latkovich
U.S. Geological Survey
Stennis Space Center, Mississippi

George H. Leavesley
U.S. Geological Survey
Lakewood, Colorado

Hydrologic data collection and dissemination involve a number of systematic, stepwise procedures[3] that range from sensing, recording, and transmission of selected data measurements to the editing, storage, and application of these data (Fig. 25.1.1). Advances in the fields of electronics and computer technology have enabled the development of cheaper, faster, more accurate, and more reliable sensors, data loggers, storage media, data transmission methods, and data analysis capabilities. These improvements have enhanced the ability to collect and disseminate reliable and timely water-resources information. The focus of this chapter is on the current data acquisition and transmission technologies used in hydrologic data collection and dissemination systems.

25.1 SENSING WATER LEVEL

Water level is the height of a water surface above an established datum or elevation. In stream-gauging applications, this height is termed *gauge height* or *stage.*[2] The datum of the gauge may be an established datum, such as mean sea level, or an arbitrary elevation chosen for convenience at the gauge site. Water-level data are monitored and collected by many types of mechanical, electromechanical, and electronic sensors. The three most commonly used types are *float-driven sensors, pressure sensors,* and *ultrasonic sensors.* These sensors are discussed in the following sections as they apply to monitoring water levels in surface water bodies. These sensors are also commonly used to monitor water levels in wells.

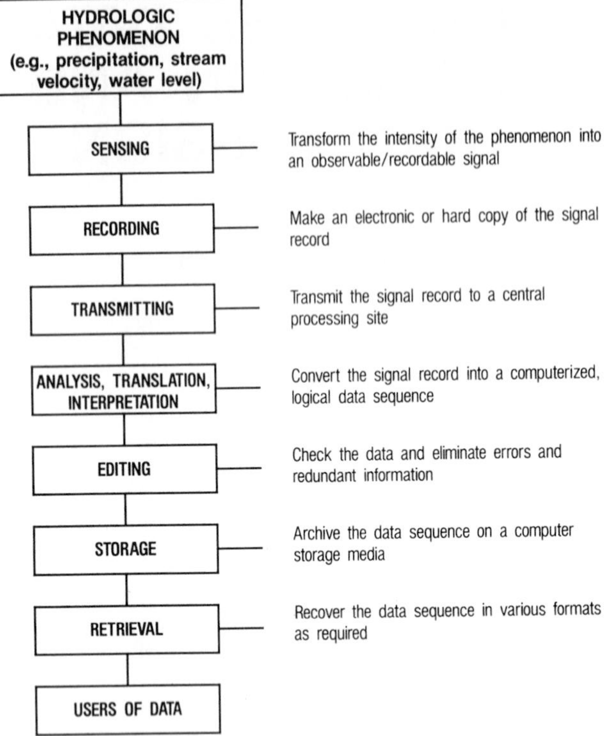

FIGURE 25.1.1 The hydrologic data-collection and dissemination sequence. *(Adapted from Chow, Maidment, and Mays.[3] Used with permission.)*

25.1.1 Float-Driven Sensors

The use of float-driven water-level sensors requires the installation of a *stilling well* in the stream or in the stream bank adjacent to the point of measurement (Fig. 25.1.2).[15] As its name implies, the stilling well protects the float and dampens the fluctuations in water-surface elevation due to wind and turbulence.[23] Detailed descriptions of stilling wells and floats, and their installation, are given by Herschy[13] and Rantz et al.[23]

In a typical installation of a float-driven water-level sensor, a float is connected to a sheave on a horizontal shaft by wire or a steel tape. The vertical movement of the float resulting from fluctuations in water level is translated to a mechanical movement or an electronic signal by the rotation of the shaft. Shaft rotation can be recorded mechanically by a pen trace on a strip chart or by a series of holes punched in a digital paper tape. Shaft rotation also can be recorded electronically through the use of an *encoder* or a *potentiometer*. Encoders are electromechanical devices that sense rotational shaft movement and convert or encode that movement into an electrical signal that can be recorded by an electronic data logger.

There are *absolute* and *incremental* encoders. In an absolute encoder, the position or number of rotations of the shaft has an absolute value associated with it. Absolute encoders usually have a limited life cycle, depending on their mechanical system.

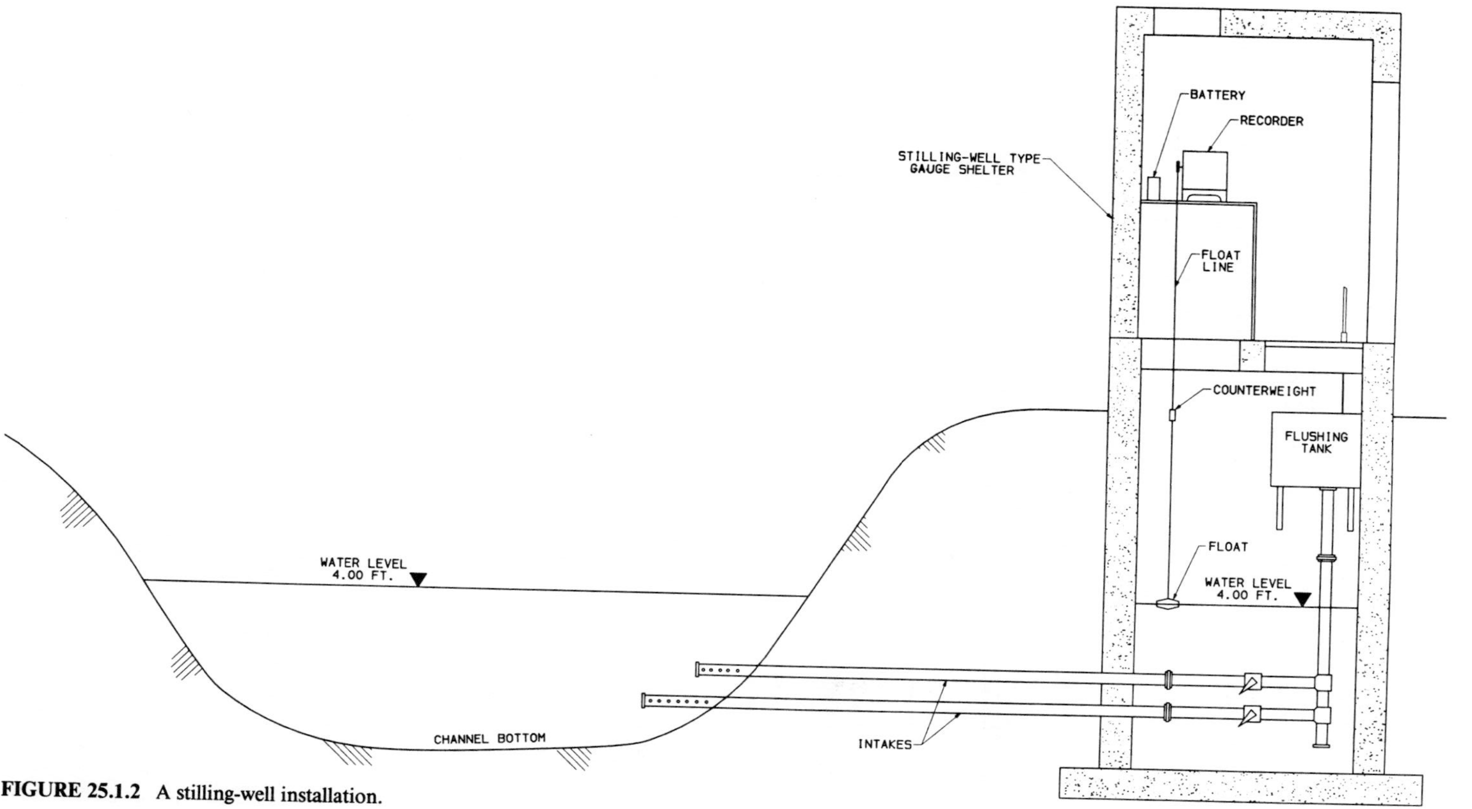

FIGURE 25.1.2 A stilling-well installation.

An incremental encoder uses an electronic counter to monitor and count increments of shaft rotation. Each increment normally represents 0.01 ft of stage change. The incremental changes are added to or subtracted from a base-stage value preprogrammed into the data logger. An incremental encoder is less expensive and has a longer life cycle than an absolute encoder.

A potentiometer is a resistive device with a rotating shaft input that can be placed in an electrical circuit to convert mechanical shaft rotation to a variable voltage output. The relation between resistance, as measured by voltage output, and stage is determined by calibration. This relation is then used to compute stage.

25.1.2 Pressure Sensors

The pressure exerted by water at any point in the water column is a function of water depth. This pressure can be transmitted through an air- or liquid-filled tube to a *manometer* or a *pressure transducer* for conversion to a mechanical or electrical signal that can be recorded. Pressure can also be sensed by a submersed transducer and converted to an electrical signal.

Manometers. A manometer converts a pressure-change signal to a mechanical movement of a rotating shaft. The shaft rotation is transferred to the shaft of a mechanical or electromechanical data recorder, or to an encoder, by a chain and sprocket assembly. Two types of manometers are commonly used—the *mercury manometer*[4] and the *balance beam manometer.*[13] Both devices require a gas-purge system that feeds gas through a tube whose orifice is permanently mounted below the water surface. The gas is bubbled freely through the orifice, and the gas pressure in the tube is equal to the hydraulic head or pressure on the orifice.

The mercury manometer employs two mercury reservoirs. The mercury in one reservoir is displaced by gas pressure changes resulting from water-level fluctuations over the orifice. This displacement activates a motor that moves the other reservoir to balance the pressure change. This movement is converted to a shaft rotation for recording. Figure 25.1.3 depicts a typical mercury manometer installation.[15]

A balance beam manometer employs a bellows system coupled with a balance beam and traveling weight. Pressure changes are transmitted to the bellows which moves the balance beam. This movement causes the traveling weight on the balance beam to adjust to a new position to put the beam back in balance. The change in position of the traveling weight corresponds to the change in water level over the manometer orifice and is converted to a shaft rotation for recording.

The use of manometers does not require the installation of stilling wells, as do conventional float-type gauges, and thus offers considerable savings in installation cost and setup time. The manometer shelter can be located hundreds of feet from the water's edge and the orifice may be easily moved to follow a stream channel that shifts location.[23] Manometers also are used on sand-channel streams because the gas bubbles tend to keep the orifice from being covered with sand.

Pressure Transducers. Pressure transducers convert pressure forces to electrical signals that can be recorded by a data logger. There are two general types of transducer systems, *submersible* and *nonsubmersible.* As the names imply, one is employed in the stream and the other is not.

The submersible pressure transducer consists of a pressure-sensitive diaphragm system which translates the pressure exerted on the diaphragm to an electrical signal. The signal is transmitted to an electronic data logger. The diaphragm system is

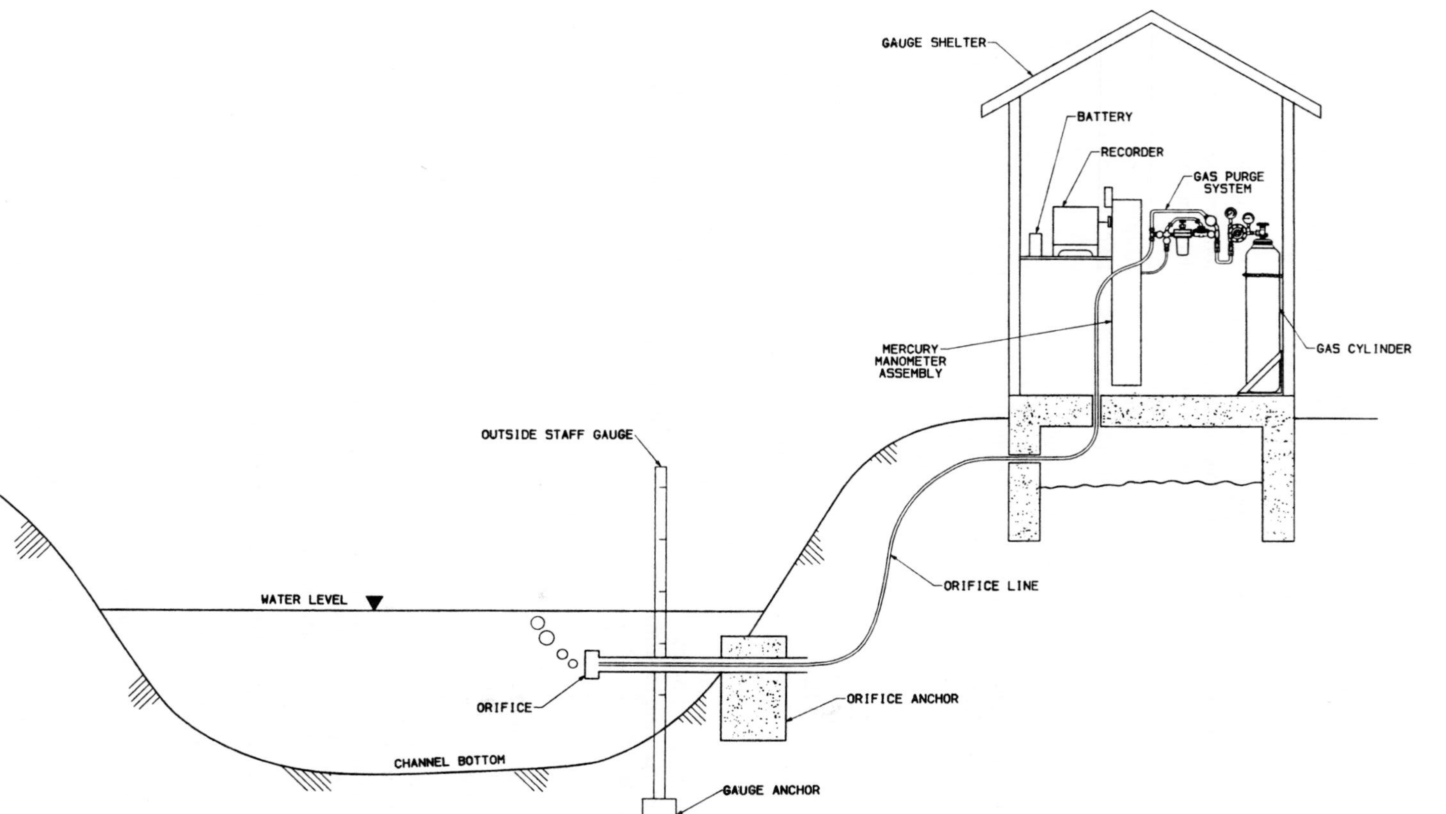

FIGURE 25.1.3 A typical mercury manometer installation.

vented to the atmosphere to ensure that pressure changes at the transducer are measured relative to atmospheric pressure. If the transducer is not vented, a barometric pressure measurement is required to correct the recorded data for changes in atmospheric pressure.

The nonsubmersible pressure transducer also consists of a pressure-sensitive diaphragm system but uses a gas-purge system just as the manometer does. This type of transducer is being used to replace the mercury manometer at many gauges. Concerns regarding the safety of the use of mercury have resulted in a shift away from the mercury manometer toward pressure transducers or other nonmercury manometers.

A variety of pressure-sensing technologies are used in transducers. The range of pressure that can be sensed varies among the technologies from a few pounds per square inch to many thousands of pounds per square inch. This translates to a range of water depths from a few feet to several hundreds of feet. Accuracies also vary by technology, ranging from ± 0.01 percent of full scale for certain vibrating quartz-crystal transducers to ± 1.0 percent for other devices. Transducers also are temperature sensitive, and thus temperature changes are normally monitored at the transducers to permit correction of the collected data.

Electrical signals provided for recording also vary by transducer technology. These signals can be an *analog signal* which is a voltage or current that varies continuously in relation to the pressure applied, a *frequency signal* where the frequency of a voltage waveform varies in relation to the pressure applied, or a *digital signal* where a voltage is switched between two distinct values in accordance with a coding scheme to represent the magnitude of the measured pressure.[1]

Pressure transducers require less service and maintenance than manometers, are small and compact, and vary greatly in cost, depending on the accuracy and type of sensing technology used. However, care must be taken to protect the transducer diaphragm from pressure beyond its certified limits and other mechanical damage.

25.1.3 Ultrasonic Sensors

Ultrasonic sensors use acoustic pulses to sense water levels either by contact or noncontact methods. The *contact ultrasonic sensor* (Fig. 25.1.4*a*) is mounted below the water surface and transmits acoustic pulses that are reflected back to the sensor by the water surface. The time interval between signal transmission and receipt, and the speed of sound in water, are used to compute the distance from the sensor to the water surface. This sensor is susceptible to errors caused by high water velocities, air entrainment in the water, and solids suspended in the water.

The *noncontact ultrasonic sensor* (Fig. 25.1.4*b*) is mounted above the water surface and sends acoustic pulses to the water surface that are reflected back to the sensor. The time between the transmission and receipt of the signal is used with the speed of sound in air to compute the distance from the sensor head to the water surface. The range of water levels that can be sensed by most commercial ultrasonic units of this type is approximately 1 to 33 ft (0.3 to 10 m).

The noncontact sensor measurements are very susceptible to changes in temperature in the air column through which the acoustic pulses travel; hence, overall accuracy is approximately ±0.1 ft (±0.03 m) over the full range of 33 ft. The device is an excellent backup sensor during floods and is used for measuring mud flows and for general reconnaissance investigations where greater accuracy is not required.

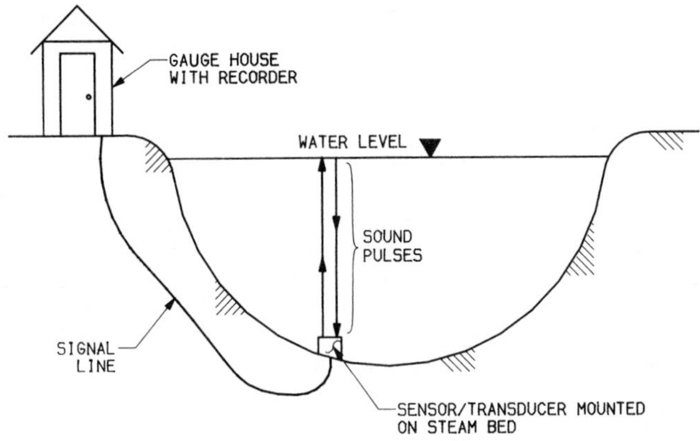

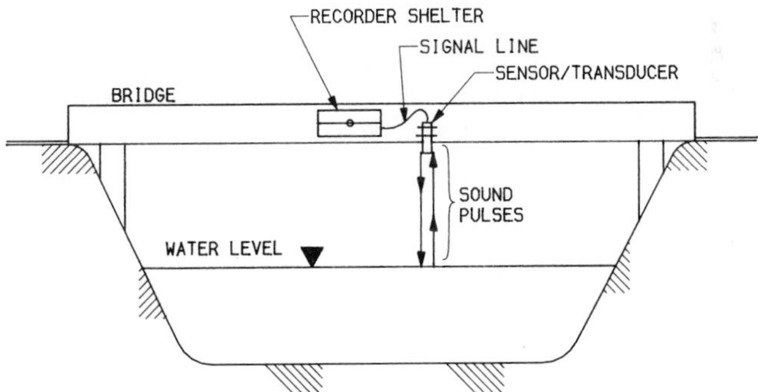

FIGURE 25.1.4 Upward/downward-looking ultrasonic sensor installations.

25.2 SENSING WATER VELOCITY

Velocity is one of the variables needed to compute water discharge from surface- and ground-water systems as well as the rates of transport of a variety of dissolved and suspended constituents in water. Velocity can be measured by a variety of techniques.

25.2.1 Mechanical Sensors

Most mechanical sensors for measuring water velocity are horizontal-axis impeller-type meters or vertical-axis bucket-wheel-type current meters. The meters commonly used in the United States are the *Price AA* and the *Price pygmy meters.*[23] These meters have a vertical axis rotor with six cone-shaped cups or buckets. The pygmy meter is used for flow depths less than 1.5 ft and the AA meter is used for larger flow depths. Both may employ either a mechanical or fiber-optic counting system to monitor the rotational speed of the buckets. The rotational speed is transmitted by wire to a standard headset or *current-meter digitizer.*

The current-meter digitizer,[25] which is a microprocessor-controlled system that monitors the number of bucket-wheel (or rotor) revolutions, measures the time during which a preset number of revolutions occurs, and uses that value to calculate velocity from an internal rating-table program. The digitizer then displays the resulting water velocity. The rating tables for various meters are preprogrammed into the digitizer. During floods, when more rapid measurements may be needed, the length of the counting period for each velocity measurement can be shortened to one-quarter or one-half the usual counting period. Recently developed digitizers have the ability to compute the cross-sectional area, velocity, and total discharge for each measurement section and for the total stream cross section.

25.2.2 Electromagnetic Sensors

The *electromagnetic current meter* measures water velocity by generating a magnetic field in the water around the sensor and using the water as the electrical conductor. The voltage measured by the meter at the sensor's electrodes is proportional to the magnitude of the water velocity in a direction perpendicular to the plane of the magnetic field. The meter is compact, fairly rugged, and has no moving parts. The meter can be used in a portable mode during periodic measurement visits or in a continuous mode when permanently mounted at the gauging site. Problems encountered with these meters include fouling of electrodes and interference from electrical and magnetic fields in the area.

25.2.3 Ultrasonic Sensors

The *ultrasonic velocity meter* measures an integrated water velocity along a path across a flowing water body, be it a river or a closed conduit, by timing the transmission of sound pulses between a transducer and receiver. The time of travel of the pulses is correlated with the flow velocity. Operation and reliability of the meter are hindered by excessive sediment concentrations, air entrainment in the water, unstable channel bottoms, and obstructed horizontal line-of-sight acoustic visibility. In open channels where the flow may not be parallel to the streambank, a crossed-path transducer system (Fig. 25.2.1) is used to compensate for this variation in flow direction. Multiple sets of transducers and receivers can be installed in a river cross section if velocity measurements at several depths are needed, such as in tidal reaches where upstream flow may occur intermittently.

The *acoustic Doppler velocity meter* measures water velocity in flowing water bodies and in closed conduits by reflecting sound pulses off particles in the water. The meter also can be operated in a profiler mode and can simultaneously measure velocity at various depths in the profile. When used to measure velocity profiles in the water column, it can be pointed either upward from a stable bottom position or downward from a surface position such as a boat.

25.2.4 Thermal-Pulse Sensor

The *thermal-pulse flowmeter* measures water velocities up or down the water column in a borehole or well.[14] A thermal pulse is induced and the time of travel of this pulse

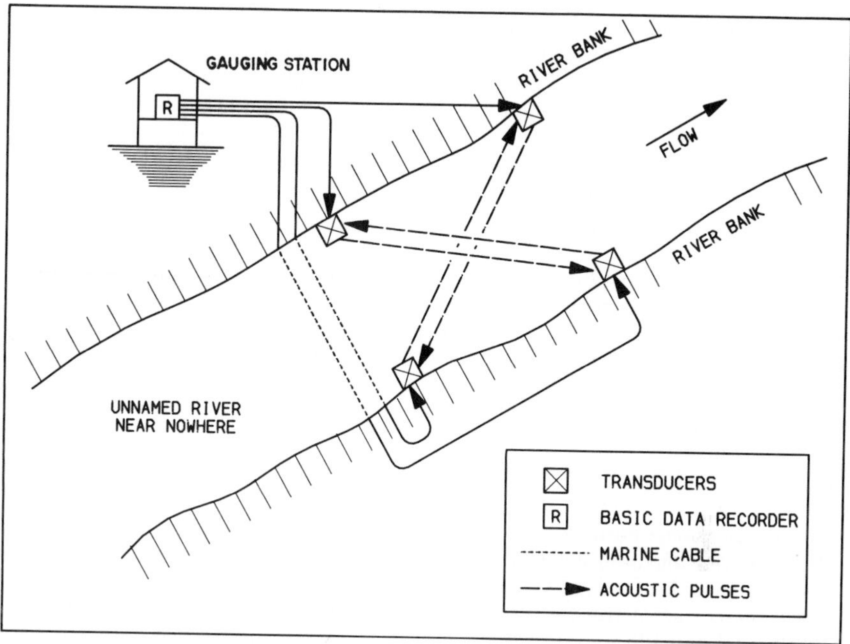

FIGURE 25.2.1 A crossed-path ultrasonic velocity meter installation.

up or down the column is monitored. The meter is capable of sensing and monitoring extremely small velocities in water moving between aquifer stratigraphic sections.

25.3 SENSING WATER QUALITY

Measurements of the chemical and physical characteristics of surface and ground waters provide data to assess the suitability of water for a variety of uses and to monitor changes in water quality over time. Many of the methods for measuring inorganic, organic, and radioactive substances in water use analytical procedures that can be performed only in a laboratory. However, a number of chemical and physical measurements can be made directly in the field at stream and well sites.

25.3.1 Sample Collection for Laboratory Analysis

Collection of water samples for transport to a laboratory is often done manually for single-purpose or infrequent sampling. However, for repetitive sampling at one or more sites over a range of hydrologic conditions, automatic *pumping samplers* provide a much more flexible and cost-effective alternative.

Pumping samplers can operate unattended to collect samples in preprogrammed time sequences or they can be triggered to sample on selected hydrologic events. Such events include the magnitude and rate of change in water levels beyond selected thresholds, the change in concentration of a selected chemical constituent, or the start of rainfall. Sampler features may include a wide range of numbers and capacities of sample bottles, capabilities to collect sequential or composite samples, refrigerated sample compartments, and operation on AC or DC power. Pump suction can lift water samples to heights of about 26 ft (8 m), which may limit the use of pumping samplers in some applications.

25.3.2 Field Measurements

The environmental conditions at most field sites limit the type and complexity of automated field measurements to a few basic physical properties and a small number of chemical constituents. Physical properties include temperature, pH, specific conductance, and dissolved-oxygen concentration. Some chemical ions can be measured using *selective-ion electrodes.* Electrodes are currently available to measure ammonia, bromide, calcium, chloride, cyanide, fluoride, iodide, lead, nitrate, sodium, and potassium ion activity. Field sensors for water-quality measurements can be classified as *portable, in-stream,* or *flow-through.*

Portable Sensors. Portable sensors are generally small, lightweight sensors that can be transported and used at field locations where continuous recording is not required. They are capable of sensing one or several water-quality constituents or properties. These sensors are useful for reconnaissance studies, routine operational measurements, and periodic measurements to evaluate trends in water quality. They are particularly effective for calibrating or verifying other field sensors in more permanent installations. However, portable sensors commonly are not well-protected from the environment and must be used with care.

In-Stream Sensors. An in-stream sensor is placed below the water surface of a stream at a selected point of measurement and is intended for continuous data collection. Most of the commercial in-stream sensors are environmentally rugged and may be powered by either alternating or direct current. In-stream sensors are submersible and either contain their own data storage devices or are connected to a data logger with data storage capabilities.

Fouling of sensors due to biological, chemical, and physical constituents in the water is a common problem and thus sensor accessibility is an important consideration. Good service and maintenance practices are required to minimize this problem. Other drawbacks include susceptibility to vandalism and damage by floods and waterborne debris or ice. Some limits also exist regarding the distance between the sensors and the data logger.

Flow-Through Sensors. Flow-through sensors are located away from the water body and require water to be pumped to sensor chambers. An AC-powered submersible pump is normally used to pump water from the lake or stream to the sensor chambers. Multiple sensor chambers may be used, depending on the number of constituents being measured.

Fouling is less of a problem for flow-through sensors than for in-stream installations because of the ease of service and maintenance and the ability to use automatic cleaning systems. Cleaning solutions such as bleach can be periodically injected into the sensor chambers prior to sensor operation to act as a cleaning agent. Sensor calibration is also easier and can be better controlled in the flow-through environment. Consequently, these sensor systems normally provide higher data reliability and accuracy than in-stream sensors.

Disadvantages of flow-through sensors include long pumping lines which increase the opportunity for sediment and biological buildup in the lines that could affect the constituent concentrations or physical properties being measured. High sediment concentrations in the water being pumped to the sensor chambers can damage the pump. The system is vulnerable to power outages, which may occur during large storms when data measurements are important. Systems must also be protected from freezing conditions.

25.4 DATA ACQUISITION AND STORAGE SYSTEMS

Recording of hydrologic data has evolved from manual reading of graduated staff plates and dipsticks to the use of various devices including pen-type graphical recorders, punched-paper-tape recorders, and electronic data loggers. Punched-paper-tape recorders offer an advantage over pen-type graphical recorders because the records they generate can be processed by automatic data processing techniques. These recorders have been used primarily in instances where the sensor output is mechanical motion. Punched-paper-tape recorders can be interfaced with some electronic sensors but may require modification. By contrast, most electronic data loggers or data-acquisition systems are easily interfaced with a large number of sensors for a wide variety of hydrometeorological measurements and data-transmission systems. Industry has set electrical interface standards, such as RS-232, which enable the user to program the system and retrieve data more easily with a minimum of peripheral equipment.

Data are collected, stored, and transported from field sites using a variety of technologies. Some data-acquisition systems store only a small amount of data but use one of several data transmission techniques to periodically send data to central data processing facilities using a variety of technologies that are discussed in more detail in Sec. 25.5. Other systems record and store data locally for manual retrieval and transport.

Most systems employ data memory internally or in *removable storage modules.* The internal data storage technology commonly uses *random-access memory* or *electronically programmable read-only memory* chips. Removable storage modules also use random-access memory and can be exchanged at the site or can be used to copy and store data from other storage modules at one or more sites for transport to the office for processing. Magnetic memory is not really practical in the field because of its susceptibility to mechanical damage, data loss from the effect of local variations in magnetic fields, and low-temperature problems.

Another option is the use of a *portable field computer* to retrieve data from a data logger or storage module. The field computer also provides the capability to review and analyze data at the site and thus assess sensor performance. However, environmental conditions at field sites necessitate the use of rugged systems that are protected from dust and moisture. These conditions may limit the use of some storage media such as floppy disks. Figure 25.4.1 depicts a data collection and retrieval system that includes a field computer component.

Data acquisition and storage systems vary in size and complexity. They can generally be classified according to applications as *dedicated systems* or *general-purpose systems.*

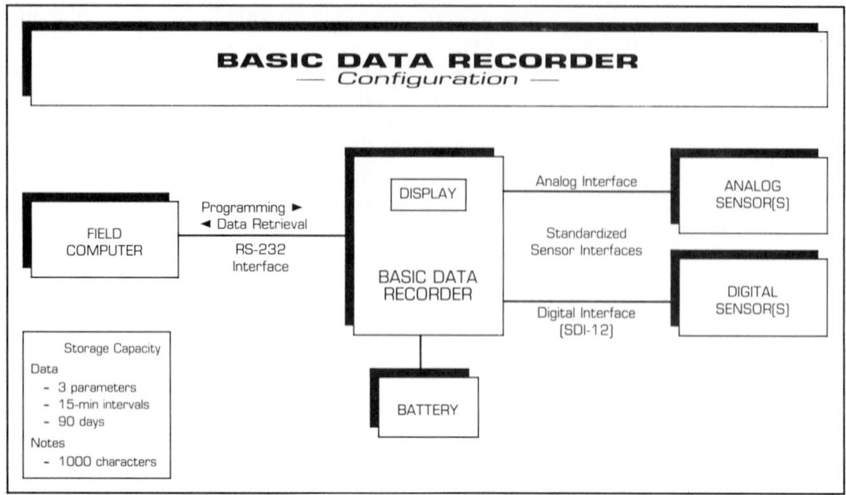

FIGURE 25.4.1 The U.S. Geological Survey's data-collection and retrieval configuration.

25.4.1 Dedicated Systems

Dedicated systems are designed and developed for use on single-purpose or limited tasks. They use a specific sensor or set of sensors and have relatively little flexibility for user interaction. All interfaces, protocols, sensor commands, and operating instructions are system-specific. User-programmable changes to system operations usually are limited to those affecting the frequency and number of samples collected.

An example of a dedicated system is the U.S. Geological Survey's R200 downhole recorder system.[17] The R200 was designed to measure water level in boreholes or wells that have a diameter of 2 in (5 cm) or larger. The system sensor is a pressure transducer. Several models with maximum pressure ranges from 5 to 30 lb/in^2 and an accuracy of approximately ±0.1 ft (±0.03 m) are available. This system measures the water level hourly, computes average daily water level, and stores this value in memory. Maximum and minimum water levels for the period of record and their time of occurrence are also computed and stored. The R200 operates unattended on six D-cell batteries for a minimum of 12 months.

25.4.2 General-Purpose Systems

General-purpose data-acquisition systems are flexible and expandable systems that permit the coupling of a wide variety of data loggers, sensors, and data-transmission methodologies to address hydrologic data-collection needs. The degree of flexibility is a function of the capabilities of both the data logger and the sensors.

Most data loggers can be classified as *programmable, menu-driven,* or *internally programmed.* Programmable systems provide selectable modes of operation and have options that allow the user to enter programs using the system's software programming language. This capability is useful for complex sensors and for reducing sensor signals to meaningful engineering units that can be read at the data logger. Menu-driven systems are geared toward routine data-collection activities where few changes are required. Menu options usually relate to sampling frequency and the number of samples to be collected at each sample interval. Internally programmed systems are similar to menu-driven systems but have few or no user-programming capabilities. These systems are preprogrammed by the manufacturer and may require return to the manufacturer for program changes.

Sensor operations may be controlled externally by a data logger or internally by a microprocessor within the sensor. Externally controlled sensors require multiwire lines to transfer control signals and data. Internally controlled sensors, however, need only a single signal line to transport the data signal to the data logger. A single signal-carrying line can carry signals to the data logger from a number of sensors. Internally controlled sensors can operate according to programmed instructions without tying up the data logger unnecessarily.

25.5 DATA TRANSMISSION

A variety of technologies are available to transmit data from a remote site to a central station for storage and processing. The choice of one technology over another depends on several factors, including the minimum allowable time between data collection and reporting, the spatial scale of the application, and the cost. A detailed overview of these technologies is provided by Flanders.[7]

25.5.1 Land Lines

Land lines involve the use of commercial telephone lines between the sensor or data logger and the point of storage and processing. Telephone lines provide a flexible transmission medium that is effectively unlimited in distance of application. The availability of *cellular communications* for telephone service now eliminates the need for cabling to the data-collection site in regions having cellular telephone service. Cellular telephone service uses a series of radio and microwave links to connect portable telephones to the full telephone network. Both cabling and cellular service, however, are most readily available in urban areas, and the cost of installing cabling to remote sites can be prohibitive.

The use of telephone lines requires a *modem* at the remote site and at the central data processing site. A modem modulates the data signal for transmission at the remote site and demodulates the signal, converting it back to a data signal, at the receiving site where the data are processed and stored. The central data collection instrument can be programmed to dial the remote site at a fixed time interval to retrieve, process, and store data. At remote sites having data loggers with microprocessor capabilities, hydrologic events, such as the start of precipitation or an increase in stream stage above a selected threshold, can be used to initiate data communication from the remote site and provide a real-time warning.

25.5.2 VHF/UHF and Microwave Radio Frequencies

VHF/UHF (30 to 3000 MHz) and *microwave* (3000 to 300,000 MHz) bands are the frequencies used for radio transmission of data. Transmission distances decrease with a decrease in available power and with an increase in radio frequency. The higher frequencies require that the path between transmitter and receiver be line of sight. Where transmission distances are limited by power or topography, *repeaters,* which receive and retransmit radio signals, can be used to extend transmission distances. However, repeaters increase system costs and add vulnerable links whose failure can disable the entire communications system. As with telephone systems, microprocessor-controlled radio communication can be initiated from a central collection station for fixed-time-interval sampling or from a remote site for event-based reporting for real-time warning systems.

25.5.3 Satellites

Satellites can be used as repeaters for transmission of data from remote data-collection sites. Data collected from remote sites can be transmitted to a communications satellite which then retransmits the data to ground stations for processing and dissemination to users. The use of communications satellites expands the scale of data transmission from local to regional and global. Two basic types of satellites are used for data collection and retransmission: *polar-orbiting* and *geostationary.* A description of these satellite systems and their application is provided by a number of authors.[8,10–12,20,29]

Polar-Orbiting Satellites. A major polar-orbiting satellite system used for data transmission is the Argos Data Collection and Location System (DCLS). Argos is administered jointly by Le Centre National d'Études Spatiales (CNES) of France and two agencies of the United States, the National Oceanic and Atmospheric Administration (NOAA) and the National Aeronautics and Space Administration (NASA). The DCLS is capable of acquiring environmental information from either fixed or moving data-collection platforms.

Polar-orbiting satellites have a nominal altitude of 870 km and each orbit takes 101 min to complete. The orbits are *sun-synchronous,* which means they cross the equatorial plane of the earth at the same time each day. The satellite radio view encompasses a circle with a diameter of 5000 km on the earth's surface. The time and frequency of satellite visibility varies with latitude. Two satellites are normally in operation, thus providing locations at the equator with six to eight satellite passes per day and sites near the poles with 28 passes per day.

Data are collected on the ground from up to 32 analog or digital sensors and stored by a unit termed a *platform transmitter terminal* (PTT) for transmission to the satellite as it passes over a site. Transmissions are limited to 256 bits of data and take less than a second to complete. Transmissions are repeated at 40- to 60-s intervals for moving PTTs requiring location computation and at 100- to 200-s intervals for fixed-site PTTs used only for data collection.

Data are stored on magnetic tape in the satellite. After each orbit, these data are relayed from the tape to a NOAA telemetry station where they are then transmitted by communication satellite to the NOAA and Argos processing centers. Processed data are distributed to users by telephone, telex, and mail. The average delay from time of data collection to availability of data for distribution to the user is 3 to 4 h. An example of the application of the Argos system for the Niger River in Africa is shown in Fig. 25.5.1.

Geostationary Satellites. Several geostationary satellites are located over the equator at an altitude of about 35,000 km. At this altitude, each satellite's orbit is synchronous with the rotation of the earth and thus the satellite appears stationary over a fixed point on the earth. The United States uses two Geostationary Operational

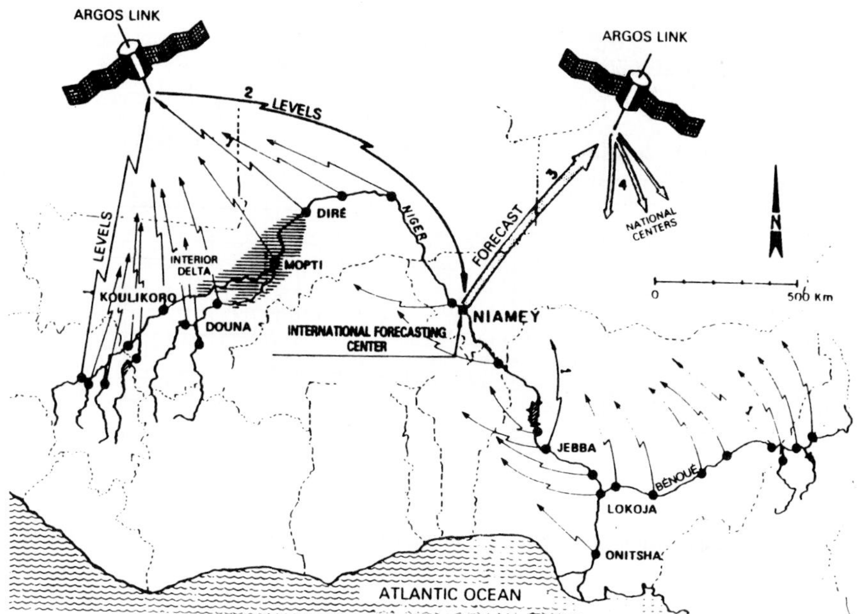

FIGURE 25.5.1 A flood forecasting and warning system for the Niger River, Africa, using the polar-orbiting Argos satellite. *(From Rodda.[24] Used with permission.)*

Environmental Satellite (GOES) platforms, which under normal operating circumstances are located at 75° W and 135° W longitude. Their effective range is from 75° N to 75° S latitude and from 0° longitude westward to about 150 to 180° E longitude.

The European Space Agency operates a geostationary satellite Meteosat stationed at 0° longitude for hydrometric data transmission over Europe and the African continent.[11] A number of other countries also have geostationary satellites, including Japan,[21] India,[16] and the former Soviet Union. The locations of these satellites and their telecommunications coverage are depicted in Fig. 25.5.2.

Data to be transmitted by satellite are collected from analog and digital sensors and stored for transmission by a unit termed a *data-collection platform* (DCP) (Fig. 25.5.3). DCPs can collect data at a short time interval, for example every 15 or 30 min, accumulate these data, and then transmit them at time intervals of 3 to 4 h. To prevent interference between transmissions from different DCPs, each DCP transmits on an assigned channel for a duration of about 1 min. The number of DCPs that can transmit simultaneously is limited to the number of available channels. Each GOES satellite has 233 channels for data relay and each Meteosat satellite has 66 channels.

There are three types of DCPs. The first is self-timed and sends data at a fixed time interval using an internal preset clock. The second is an interrogable DCP that transmits data only when interrogated from a ground control center via the satellite. The third type of DCP is one that reports in a random manner in response to preset sensor thresholds. Messages from all three types are repeated several times over a period of minutes to ensure receipt at the ground station.

Data from DCPs are relayed to a central command and data acquisition station

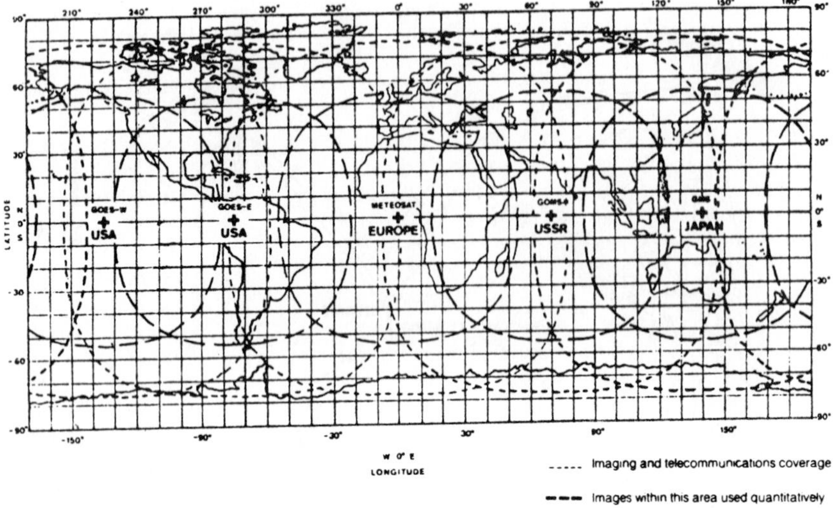

FIGURE 25.5.2 Locations and coverages of the geostationary satellites. *(From Herschy.[12] Used with permission.)*

operated by the National Earth Satellite Data and Information Service (NESDIS), or to user-operated direct-readout ground stations. Data received by NESDIS are disseminated to users via a variety of communication methods. Direct-readout ground stations commonly are collocated with and operated by the agency collecting the data. They do not communicate with the satellite but simply receive DCP messages, flag and disseminate alert messages, and store data for further processing.[26,27]

25.5.4 Meteor-Burst Telemetry

Meteor-burst telemetry uses ionized meteor trails as reflectors for VHF radio signals to overcome the line-of-sight limitations of standard VHF radio and microwave communications.[5,18] Billions of meteors enter the earth's atmosphere daily, burning

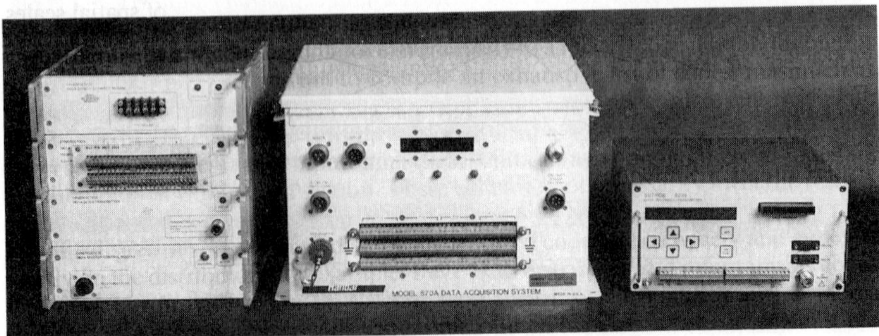

FIGURE 25.5.3 Data-collection platforms (DCPs) for GOES satellite telecommunications. *(Used with permission.)*

up and leaving an ionized trail of gases that remain for periods of a few microseconds to a few seconds. The altitude of useful trails is 80 to 120 km above the earth's surface, which limits the range of communications to about 2000 km. There are also diurnal and seasonal variations in meteor trail density which can affect transmission reliability.

A meteor-burst system is composed of one or more master stations and a number of remote sites. Remote sites are microprocessor-controlled data collection and transmission stations that collect data at preselected time intervals, for example every 15 min, and process and store these data for transmission to the master station one or more times per day. During retrieval of data from a remote site, the master station continuously transmits until a meteor trail occurs at the correct location to reflect the signal to the remote site. Upon receiving the master station signal, the remote site immediately transmits its data using the same meteor trail as a signal reflector. Remote sites can also be programmed to initiate communications to the master station when selected sensor output exceeds a specified threshold.

Advantages of meteor-burst technology include access to data in near real time, relatively low system costs, the ability to transmit at a common radio frequency, and communication security because of the random nature of the meteor trail location.[18] Disadvantages are related to the short duration of meteor trails which limit the message length, and the diurnal and seasonal variation in the density of meteor trails. The largest operational meteor-burst system is the Snow Telemetry (SNOTEL) system of the U.S. Department of Agriculture.[5] SNOTEL has more than 550 remote sites located in the mountainous regions of the western United States.

25.6 DATA ACQUISITION AND TRANSMISSION SYSTEMS

Data acquisition systems and transmission methodologies can be combined in a variety of ways to meet program objectives, time and space considerations, and budgetary constraints. If costs are disregarded, then the major considerations become the geographical size and topographical variability of the area of interest and the acceptable time limit between data sensing and data application. Requirements for real-time data acquisition are normally associated with smaller areas at a local scale where the monitoring of rapidly changing environmental conditions is important. Data networks at the regional scale usually are associated with monitoring and characterizing the hydrologic response of many basins at a variety of spatial scales. For most of these networks, a delay of several hours between data acquisition and transmission is acceptable.

25.6.1 Real-Time Networks

A major application of real-time networks is flood forecasting and warning. The scale of most of these networks and their requirement for reliable real-time data access on a continuing basis commonly make the use of some combination of land-line and radio communications desirable. An additional requirement of this type of network is microprocessor-controlled sensors at the remote sites to enable the analysis of sensor response to determine if preselected thresholds for variables such as precipitation or stream stage have been exceeded. When thresholds are exceeded, a message is sent from the remote site to the central control station to initiate flood warnings or other actions.

Many cities and water management agencies, particularly in the western United States, have established real-time networks for flood detection under a general National Weather Service program titled Automated Local Evaluation in Real Time (ALERT). A similar program titled the Integrated Flood Observing and Warning System (IFLOWS) has been established in the Appalachian states of Kentucky, Pennsylvania, Virginia, and West Virginia.

The ALERT network in the Denver–Boulder, Colo., area is typical of most ALERT Systems.[9,28] Several major streams in the Denver–Boulder area are instrumented with automated stream gauges, and self-reporting tipping-bucket rain gauges are installed within each basin. Figure 25.6.1 shows instrumentation in Lena Gulch,

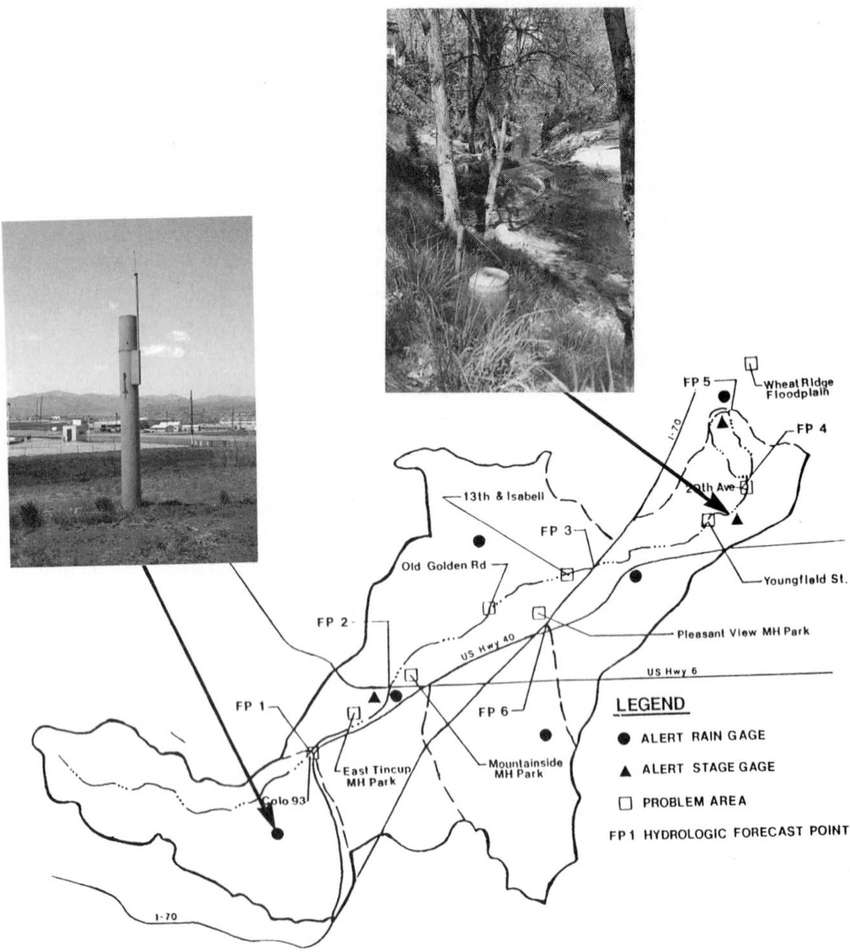

FIGURE 25.6.1 The ALERT system network, Lena Gulch, Colorado. *(Modified from Stewart.[28] Used with permission.)*

a 13.8-mi^2 basin that contains three stream gauges and six rain gauges. Each rain gauge bucket tip triggers a transmission as does any change in stream stage during each 4-min monitoring period.

All gauges transmit data to a central base station using battery-powered VHF radio transmitters for use in the development of flash-flood forecasts. These forecasts and the real-time rainfall and stream-flow data are made available to affected local government agencies via an electronic mail distribution system.

Larger-area real-time networks may include use of the full range of data-transmission technologies. Orwig[22] reported on the joint application of radio, satellite, and meteor-burst telemetry for flood forecasting in the area of Mount St. Helens following its 1980 eruption. In this application, an emergency channel feature of the GOES satellite was used to provide 5-min-interval data to forecasters when stream stage exceeded a preset threshold.

25.6.2 Regional Networks

Regional networks normally are designed for data collection from areas ranging from 1000 to 100,000 km^2. Applications range from real-time forecasting to monitoring the hydrologic response of a number of basins to precipitation. Real-time forecasting normally is done for much larger basins than those in ALERT networks and thus a delay of 3 to 6 h between data collection and data application is not critical for most applications. Thus satellite and meteor-burst systems generally are used for data acquisition and transmission.

Some networks interface VHF/UHF radio systems with satellite and meteor-burst telemetry systems to optimize network cost and efficiency. A number of remote sites have been established that use VHF/UHF radio to transmit data to a satellite DCP or meteor-burst site. Data are collected at these sites for periodic transmission to a central network data-processing station. Kleppe and Yori[19] describe the application of this technology which they term *random adaptive sub-telemetry systems.* Curtis[6] discusses the benefits of this approach for coupling real-time event reporting technologies with satellite and meteor-burst telemetry.

25.7 REFERENCES

1. Billings, R. H., U.S. Geological Survey, written communication, 1991.
2. Buchanan, T. J., and W. P. Somers, *USGS Stage Measurement at Gaging Stations,* Book 3, Chap. A7, *Techniques of Water-Resources Investigations,* U.S. Geological Survey, 1968.
3. Chow, V. T., D. R. Maidment, and L. W. Mays, *Applied Hydrology,* Chap. 6, McGraw-Hill, New York, pp. 175–200, 1988.
4. Craig, J. D., *USGS Installation and Service Manual for USGS Manometers,* Book 8, Chap. A2, "Techniques of Water-Resources Investigations," U.S. Geological Survey, 1983.
5. Crook, A. G., "Operational Experiences in Meteor Burst Telemetry—Eight Years of SNOTEL Project Observations," in *Hydrologic Applications of Space Technology,* A. I. Johnson, ed., IAHS publication no. 160, pp. 411–418, 1986.
6. Curtis, D. C., "On Merging Satellite and Meteor Burst Communications with Real-Time Event Reporting Technologies," in *Hydrologic Applications of Space Technology,* A. I. Johnson, ed., IAHS publication no. 160, pp. 419–426, 1986.
7. Flanders, A. F., "General Report," pp. 39–56 in *Proc. Int. Conf. on Telemetry and Data Transmission for Hydrology,* Toulouse, France, March 23–27, 1987.

8. Halliday, R. A., "The Use of Satellites in Hydrometry," in *Hydrometry—Principles and Practices,* R. W. Herschy, ed., Wiley, London, p. 427–451, 1978.

9. Henz, J. F., and R. A. Kelly, "Innovative Uses of Rainfall Prediction in Urban Flash Flood Forecasting," in *Computerized Decision Support Systems for Water Managers,* J. W. Labadie, L. E. Brazil, I. Corbu, and L. W. Johnson, ed., *Proc. 3rd Resources Operations Management Workshop,* ASCE, Fort Collins, Colo., pp. 714–722, 1988.

10. Herschy, R. W., "Towards a Satellite-Based Hydrometric Data Collection System," in *Advances in Hydrometry,* J. A. Cole, ed., IAHS publication no. 134, pp. 285–296, 1982.

11. Herschy, R. W., "Hydrometric Data Transmission by the European Space Agency Satellite METEOSAT," *New Technology in Hydrometry,* R. W. Herschy, ed., Adam Hilger, Bristol, U.K., pp. 123–137, 1986.

12. Herschy, R. W., "Satellite Data Transmission as an Aid to Hydrological Telemetry," in *Hydrologic Applications of Space Technology,* A. I. Johnson, ed., IAHS publication no. 160, pp. 369–376, 1986.

13. Herschy, R. W., *Streamflow Measurement,* Chap. 3, Elsevier Applied Science, London, pp. 143–153, 1985.

14. Hess, A. E., "Thermal Pulse Flowmeter for Measuring Slow Water Velocities in Boreholes," U.S. Geological Survey open-file report 87-121, 1990.

15. Hydrologic Instrumentation Facility, "U.S. Geological Survey Instrumentation Documentation and Drawing Files," Stennis Space Center, Miss., 1991.

16. India Meteorological Department, "Details of the INSAT Meteorological Applications Programme of India," in *Hydrological Applications of Remote Sensing and Remote Data Transmission,* B. E. Goodison, ed., IAHS publication no. 145, pp. 61–64, 1985.

17. Johnson, R. A., and J. I. Rorabaugh, *USGS Operating Manual for the R200 Downhole Recorder with Tandy 102 Retriever,* U.S. Geological Survey open-file report 88-488, 1988.

18. Kalske, R., and I. Lahteenoja, "Use of Meteor Scatter Telemetry for Hydrometeorological Data Collection Networks," in *Hydrologic Applications of Space Technology,* A. I. Johnson, ed., IAHS publication no. 160, pp. 403–409, 1986.

19. Kleppe, J. A., and L. G. Yori, "The Use of Random Adaptive Subtelemetry for Satellite and Meteor-Burst Applications," in *Hydrological Applications of Remote Sensing and Remote Data Transmission,* B. E. Goodison, ed., IAHS publication no. 145, pp. 189–195, 1985.

20. MacCallum, D. H., "Operational Satellite Data Collection Systems Operated by the National Oceanic and Atmospheric Administration," in *Hydrological Applications of Remote Sensing and Remote Data Transmission,* B. E. Goodison, ed., IAHS publication no. 145, pp. 89–97, 1985.

21. National Space Agency of Japan, "Development of Satellite Remote Sensing Systems in Japan," in *Hydrological Applications of Remote Sensing and Remote Data Transmission,* B. E. Goodison, ed., IAHS publication no. 145, pp. 45–60, 1985.

22. Orwig, C. E., "Real Time Remote Sensing in the Mount St. Helens Drainage Basin," in *Hydrological Applications of Remote Sensing and Remote Data Transmission,* B. E. Goodison, ed., IAHS publication no. 145, pp. 197–205, 1985.

23. Rantz, S. E., et al., *USGS Measurement and Computation of Streamflow,* vol. 1: *Measurement of Stage and Discharge,* U.S. Geological Survey water-supply paper 2175, 1982.

24. Rodda, J. C., *Supporting World Hydrology—Activities of International Hydrological Programs,* p. 996, EOS, July 24, 1990.

25. Sharp, K. V., and W. L. Rapp, *USGS Operating Manual for the Current Meter Digitizer,* U.S. Geological Survey report 6-84-03, 1984; revised 1989.

26. Shope, W. G., Jr., and R. W. Paulson, "Development of a Distributed System for Handling Real Time Hydrological Data Collected by the U.S. Geological Survey," in *Hydrological Applications of Remote Sensing and Remote Data Transmission,* B. E. Goodison, ed., IAHS publication no. 145, pp. 99–108, 1985.

27. Shope, W. G., Jr., and R. W. Paulson, "Development of a National Real Time Hydrologic Information System Using GOES Satellite Technology," in *Hydrologic Applications of Space Technology,* A. I. Johnson, ed., IAHS publication no. 160, pp. 13–21, 1986.
28. Stewart, K. G., "Effective Timely Responses to Urban Flash Floods," pp. 1–15 in *Proc. 12th Annual Meeting of the National Weather Association,* Denver, Colo., 1988.
29. Taillade, M., "Data Collection and Location by Satellite—the Argos System," in *Hydrological Applications of Remote Sensing and Remote Data Transmission,* B. E. Goodison, ed., IAHS publication no. 145, pp. 33–44, 1985.

CHAPTER 26
HYDROLOGIC FORECASTING

Dennis P. Lettenmaier
Department of Civil Engineering
University of Washington
Seattle, Washington

Eric F. Wood
Department of Civil Engineering
and Operations Research
Princeton University
Princeton, New Jersey

26.1 WORTH AND USES OF FORECASTING

26.1.1 Definitions

Forecasting means the estimation of conditions at a specific future time, or during a specific time interval. It is distinguished from *prediction,* which is the estimation of future conditions, without reference to a specific time. The stage of a river at 1200 hours tomorrow is forecasted, as is the next 5 months' reservoir inflow. Conversely, the 100-year flood is predicted, as is the 10-year, 7-day low flow. As the forecasting *lead time* (the time or time interval for which the forecast is made) increases, forecast accuracy usually decreases. For very long lead times, the distinction between forecasts and predictions is blurred, and most forecasts are no more accurate than those made by using the long-term statistical mean.

26.1.2 Forecast Uses and Classification

Forecasts are used for warning of extreme events (e.g., floods and droughts); for operation of water resource systems such as reservoirs, run-of-the-river diversions, and hydropower generation projects; and for contract negotiation, especially for hydropower sales. Contingency planning functions, especially related to droughts, depend on stream-flow forecasts. These functions include thermal electric power generation, which may depend on the availability of cooling water and thermal discharge limits related to stream-flow level; crop selection, which depends on the forecasted availability of irrigation water; implementation of conservation programs for municipal and industrial water users during droughts; and industrial production, where raw materials or products are transported via waterways.

Although any classification of forecasts inevitably is subject to some overlap, there are substantial differences in forecasting methods, depending on forecast lead time, so some distinction is helpful. For the purposes of this chapter, *short-term forecasts* are those with lead times less than 7 days, and *long-term forecasts* are those with longer lead times, usually up to several months. At present, little forecast skill is possible for hydrologic variables when forecast lead times extend beyond several months to a year; for longer lead times, prediction techniques are more applicable. Seven days is a convenient division between short- and long-term forecasts because it is slightly longer than the current maximum lead time of 96 h for the U.S. National Weather Service's Quantitative Precipitation Forecasts, but within the realm of likely future improvements in weather forecasting that might affect stream-flow forecasting methods.

Short-term forecasts are most often used for flood warning purposes and for real-time operation of water resources systems (for example, hydropower scheduling). Long-term forecasts are most often used for water management, such as allocation of irrigation water, negotiation of hydropower sales contracts, and, during droughts, for evaluation and implementation of mitigation measures such as water conservation.

26.1.3 Forecast Accuracy

Forecast accuracy is a measure of forecast error, that is, the difference between the amount forecasted, and the value that actually occurs. Forecast errors can be either *systematic* (recurring), or *random* (due to case-specific conditions, such as errors in the meteorological forecast on which the hydrologic forecast is based). Forecast accuracy is best assessed by retrospective comparison of forecasts actually made or that might have been made, and the values observed during the forecast period.[32] Let $Q_f(i)$ be the forecasted stream flow for forecast period T in year i, $i = 1, 2, \ldots, n$, and $Q_o(i)$ be the observed stream flow during the same period, and define M_f and M_o, the means of the forecasts and observations for the same period, as follows:

$$M_f = \frac{1}{n} \sum_{i=1}^{n} Q_f(i)$$

$$M_o = \frac{1}{n} \sum_{i=1}^{n} Q_o(i)$$

The following are widely used measures of forecast error:

$$\text{Bias: } B = M_f - M_o \tag{26.1.1}$$

$$\text{Mean squared error: MSE} = \frac{1}{n} \sum_{i=1}^{n} [Q_f(i) - Q_o(i)]^2 \tag{26.1.2}$$

$$\text{Root mean square error: RMSE} = (\text{MSE})^{0.5} \tag{26.1.3}$$

$$\text{Variance: } V = \text{MSE} - B^2 \tag{26.1.4}$$

$$\text{Relative bias: RB} = \frac{B}{M_o} \tag{26.1.5}$$

$$\text{Mean absolute error: MAE} = \frac{1}{n} \sum_{i=1}^{n} |Q_f(i) - Q_o(i)| \tag{26.1.6}$$

Relative mean absolute error: $\text{RMAE} = \dfrac{\text{MAE}}{M_o}$ (26.1.7)

Forecast efficiency: $E = 1 - \dfrac{\text{MSE}}{\text{V}}$ (26.1.8)

R squared: $$R^2 = \left[\frac{\frac{1}{n}\sum_{i=1}^{n} Q_o(i)Q_f(i) - M_o M_f}{\left(\frac{1}{n}\sum_{i=1}^{n} Q_o^2 - M_o^2\right)\left(\frac{1}{n}\sum_{i=1}^{n} Q_f^2 - M_f^2\right)} \right]^2 \qquad (26.1.9)$$

Bias and relative bias are measures of *systematic error* in the forecast; that is, over a period of many years, they measure the degree to which the forecast is consistently above or below the actual value. Variance is a measure of the variability, or scatter, of a number of forecasts about the true value, and is therefore a measure of the *random error.* Mean square error, root mean square error, mean absolute error, relative mean absolute error, and forecast efficiency are all measures that incorporate both systematic and random errors. A perfect forecast exists only if both the bias and the variance are zero, which occurs only when all forecasted values are identical to the observations. R^2 is the square of the correlation coefficient between the observed and forecasted values. Although R^2 is a widely used measure of forecast accuracy, care must be taken if appreciable bias is present, since R^2 evaluates the accuracy of a forecast with respect to random error only. The highest value of R^2, 1.0, can be achieved for cases where there is a constant bias in the forecasts; that is, the forecasted value is equal to the observation plus or minus a constant. For this reason, instead of using R^2, forecast accuracy is better assessed by using the bias and the variance, or the bias and the mean absolute error. MAE or RMAE is preferred to MSE because, when compared to squared error measures, absolute error measures are less dominated by a small number of large errors, and are thus a more reliable indicator of typical error magnitudes.

26.1.4 Forecast Worth

Forecast worth is the economic or other value that results from a forecast, that is, the net benefits that accrue if a forecast is available, less those that occur in the absence of a forecast. In some cases, the baseline, or no-forecast, case can be evaluated by using information derivable from past conditions, for instance, long-term stream-flow records. A forecast has positive worth if, over the long-term—that is, in statistical expectation—it results in better water utilization, or alternatively, less severe damages, than would otherwise occur.

26.2 SHORT-TERM FORECASTING METHODS

26.2.1 Choice of Models

There are two basic types of short-term forecasting models: those based on *channel routing* and those based on *rainfall-runoff modeling.* For any particular application, one may base a forecast on channel routing models, rainfall-runoff models, or a combination depending on the problem classification. The classification of short-term forecasting problems depends on two criteria. The first relates the required forecast lead time T_f to the time of concentration of the catchment at the forecast point. This includes both the *hydrologic response time* of the catchment, T_c (often measured by the *time of concentration,* which is the time of travel from the farthest point in the catchment to the forecast point), and the travel time through the channel/river system, T_r. The second criterion is the ratio of the spatial scale of the meteorological event to the spatial scale of the catchment above the forecast point. We will refer to this ratio as R_s, the *spatial scale* of the forecast problem.

Case 1. $T_f > T_c + T_r$. If the required forecast lead time is larger than the time of concentration, then meteorological forecasting of the precipitation is required. This is because some of the water which is included in the flow forecast has yet to fall as precipitation on the watershed at the time the forecast is made. In such cases, a forecast procedure that includes both precipitation forecasting (see Chap. 3) and hydrologic (precipitation-runoff) forecasting is needed.

Case 2. $T_f < T_c + T_r$ and $T_c \ll T_r$. If the forecast lead time is shorter than the time of concentration, and the total time of concentration is dominated by the routing time of the flood wave through the channel system, then stream-flow forecasts can be based on observed flows at upstream gauge locations. For such applications, the upstream observed flows must be transmitted to a forecast control center as the flows occur. This is the situation for large river systems such as the lower Colorado River in Texas and the Mississippi River and its major tributaries. For such systems, stream-flow forecasts can be based on channel routing models alone.

Case 3. $T_f < T_c + T_r$ and $T_r \ll T_c$. If the forecast lead time is shorter than the time of concentration, and the time of concentration of the catchment is dominated by the hydrologic response time of the catchment, then stream-flow forecasts should be based on observed rainfall from a network of rain gauges whose data are transmitted to the forecast center. This is the situation for small catchments and urban areas. Such forecasts must incorporate a rainfall-runoff model.

Case 4. $R_s \lesssim 0.7$. When the ratio of the spatial scale of the meteorological event to the spatial scale of the catchment is less than about 0.7, then there is partial coverage of the catchment by the rain event. In this case, the accuracy of forecasts based on rainfall-runoff models that assume spatially uniform rainfall will suffer. This is often a problem for large catchments [e.g., drainage area larger than about 3000 mi^2 (7500 km^2)], and for smaller catchments during intense convective storms. For these situations, the basin should be partitioned with stream gauging stations on the major tributaries and an extensive rain gauge telemetry system. The upstream channel inflows can be forecasted by using a rainfall-runoff model based on observed rainfall. The downstream channel flows can then be forecasted by a channel routing model based on either observed or forecast upstream channel flows as follows: if T_r is greater than T_f, then use observed upstream channel flows; if T_f is less than T_r and Case 1 doesn't hold, then use forecasted upstream channel flows.

26.2.2 Forecasts Based on Channel Routing

Dynamic Wave Method. The routing of water down a river channel is described by one-dimensional hydrodynamic equations of unsteady flow known as the St. Venant equations, which can be expressed in *nonconservation form* (independent variables water depth y and velocity U) as

$$U\frac{\partial y}{\partial x} + y\frac{\partial U}{\partial x} + \frac{\partial y}{\partial t} = 0 \tag{26.2.1}$$

$$\frac{\partial U}{\partial t} + U\frac{\partial U}{\partial x} + g\frac{\partial y}{\partial x} - g(S_0 - S_f) = 0 \tag{26.2.2}$$

where x is the distance along the channel, g is the gravitational constant, S_0 is the channel slope, and S_f is the friction slope (see Chap. 10). Equations (26.2.1) and (26.2.2) can also be written in *conservation form,* where the independent variables are water surface elevation and discharge. Forecasting models based on the full St. Venant equations are referred to as dynamic wave models and can be solved using finite difference numerical schemes. They are most applicable for the following cases of unsteady flow:[47]

1. Upstream movement of waves such as tidal action or storm surges
2. Backwater effects caused by downstream reservoirs or tributary flows
3. Rivers with extremely flat bottom slopes, e.g., $S_0 \lesssim 0.0005$
4. Abrupt waves caused by rapid reservoir releases or dam failures

An example of a dynamic wave model is the U.S. National Weather Service (NWS) model described in Chap. 10 (see also Refs. 12 and 13).

Typical time steps for dynamic wave routing of spring snowmelt floods on large river systems like the Mississippi-Ohio-Cumberland-Tennessee system may be on the order of 24 h. For forecasting floods due to hurricanes on smaller systems like the Susquehanna or the Connecticut rivers in the Eastern United States, time steps on the order of several hours are used, and for forecasting dam failure impacts, considerably shorter time steps (on the order of minutes) are employed. Typical root mean square (rms) errors for forecasting channel water surface elevations where the channel cross sections and roughness characteristics are well-known are on the order of 0.60 ft (Ref. 12).

Diffusion Equation Method. If the inertial terms in the momentum equation [the first two terms on the left-hand side of Eq. (26.2.2)] are ignored, then the continuity and momentum equations describing gradually varied unsteady flow can be written in conservation form as

$$\frac{\partial Q}{\partial x} + \frac{\partial A}{\partial t} = q \tag{26.2.3}$$

$$\frac{\partial y}{\partial x} = S_0 - S_f \tag{26.2.4}$$

where Q is discharge at location x along the channel, A is channel cross-sectional area, q is lateral inflow, and the friction slope, S_f, can be parameterized using either Manning's equation:

$$S_f = \frac{n^2|Q|Q}{A^2R^{1.33}} \tag{26.2.5}$$

where n is Manning's roughness coefficient and R is the hydraulic radius, or using

$$S_f = \frac{|Q|Q}{A^2C^2R} \tag{26.2.6}$$

where C is the Chézy coefficient. The continuity and momentum equations can be rewritten for a rectangular channel of width B, with a conveyance K_c which is a single-valued function of depth as[29]

$$\frac{\partial Q}{\partial t} = D\frac{\partial^2 Q}{\partial x^2} - c\frac{\partial Q}{\partial x} + cq \tag{26.2.7}$$

where D is a diffusion coefficient estimated as

$$D = \frac{K_c^2}{2QB} \tag{26.2.8}$$

and c is an advective velocity estimated as

$$c = \frac{Q}{BK_c}\frac{dK_c}{dy} \tag{26.2.9}$$

The conveyance $K_c = Q\sqrt{S_f}$. The diffusion equation is a second-order parabolic equation, similar to a heat-conduction equation for which solutions can be found in reference books, for example, Carslaw and Jaeger.[7] Diffusion wave routing can be used for flood routing in rivers without significant backwater effects so long as the channel slope S_0 is greater than about 0.05, although a sample computation of the first two terms in Eq. (26.2.2) to assure that they are negligible relative to the channel and friction slopes is advisable.

Kinematic Wave Method. A further simplification of the dynamic wave equations occurs when the first three terms in the momentum equation (Eq. 26.2.2) are ignored, so that the equation is reduced to a balance between gravitational forces S_0 and frictional forces S_f. This formulation is known as the kinematic wave model, and is applicable to forecasting in cases without significant backwater effects and where the channel slope is greater than about 0.001 (Ref. 47). A single-valued function relating flow area and discharge of the form

$$Q = \alpha A^m \tag{26.2.10}$$

is usually specified.[11] The forms for α and m suggested by Ref. 6 for wide rectangular channels are

$$\alpha = \frac{1.49}{nB^{2/3}}S^{0.5} \qquad \text{and} \qquad m = 1.67$$

and for triangular (isosceles) channels

$$\alpha = \frac{0.94}{n}S^{0.5}\frac{y}{1+y^2} \qquad \text{and} \qquad m = 1.33$$

where the variables are expressed in English units.

Muskingum Method. Simpler models than those based on the St. Venant equations have been successfully used for river forecasting. An example is the Muskingum method in which a lumped form of the channel continuity equation is combined with a linear storage discharge relationship. The continuity equation is

$$\frac{dS}{dt} = I - Q \tag{26.2.11}$$

where dS/dt represents change in storage with time, I represents inflow into the river section, and Q represents outflow from the river section. The linear storage-discharge relationship relates the storage in the river section to I, Q, and a parameter k, the travel time of the flood through the reach, in the following manner:

$$S = k[fI + (1 - f)Q] \tag{26.2.12}$$

where f is a weighting factor less than 0.5; typical values of f lie between 0 and 0.3 with a mean of about 0.2 (Ref. 8). Smaller values of f are applicable to rivers with reservoir-type storage and large values for rivers with wedge storage. Section 10.3.2 of Chap. 10 describes techniques to estimate k and f from channel characteristics and flow rates. The value for k depends on the time step for the flood routing and f, and is estimated by using measured inflows and outflows for the river section. The parameter k has units of time, as does the time step; f is dimensionless.

The Muskingum method can be recast in a form ideally suited for flood forecasting. If the storage relationship is considered at two subsequent time steps [$t_j = j\,\Delta t$ and $t_{j+1} = (j+1)\Delta t$] and is combined with the continuity equation, then the outflow from the reach at time step $j+1$ can be related to the previous outflows and current and previous inflows as follows:

$$Q_{j+1} = C_1 I_{j+1} + C_2 I_j + C_3 Q_j \tag{26.2.13}$$

where C_1, C_2, and C_3 are parameters which depend on k, f, and the chosen time step Δt, and are given by Eqs. (10.2.8) to (10.2.10) in Chap. 10.

Impulse Response Function Method. A third approach to river routing is to derive the impulse response function for the river reach by analysis of observed hydrographs at upstream and downstream points. The impulse response function relates the outflow from the reach to a unit input at the upper boundary of the reach in exactly the same way that the instantaneous unit hydrograph (IUH) relates a catchment hydrograph to a unit input of excess rainfall (see Sec. 9.5). The outflow hydrograph for a river reach is the convolution of the upstream flood hydrograph entering the river section with the impulse response function for the river and is computed in the same way that rainfall is convoluted through an IUH. Goring[16] describes an application of this approach for flood forecasting for New Zealand rivers in the tidal zone, where the impulse response function is estimated via spectral analysis,[34] and separating the harmonic tidal component from the river response.

Application of the impulse response function requires a continuous input, from which the continuous output can be derived for any time t. In almost all forecasting applications, data are available at discrete times, and forecasts are computed at specified time intervals, $j\,\Delta t$, rather than for continuous time. The convolution for discrete times used in forecasting is

$$Q_j = \sum_{l=0}^{L} I_{j-l} H_l \tag{26.2.14}$$

where Q_j is the forecasted flow at time $j\,\Delta t$, H_1 is the *discrete pulse response function*, and I_{j-1} is the input at time $(j-1)\,\Delta t$. The discrete pulse response function H_1 can be computed from the continuous impulse response function $h(t)$ for an arbitrary time step Δt as described by Chow et al. (Ref. 8, pp. 204–213). The length of the discrete pulse response, L, is equivalent to the time base T_b of $h(t)$ [the time at which $h(t)$ can be taken] and can be computed as $L = T_b/\Delta t$.

One complication in application of impulse response methods for forecasting is that the forecast may make use of unobserved inputs. If j^* is the forecast time, then for $j > j^*$, the forecast Q_j depends on values of unobserved inputs. However, the weights H_1 corresponding to unobserved inputs will generally be small so long as the channel travel time is much less than the forecast lead time, and may be set to zero. As the forecast lead time $(j - j^*)\,\Delta t$ approaches the channel travel time T_r, the values of the unobserved inputs will exert more influence on the forecast, and greater care will need to be taken in their selection. Options are to extrapolate the upstream hydrograph, or to develop a forecasting procedure (e.g., using rainfall-runoff models as described in Sec. 26.2.3) for the future inputs. In any event, it is important to evaluate the magnitude of the weights to determine whether forecasts of the upstream inputs are needed. It is also important that, as observed inputs become available, they are substituted for extrapolated or forecasted values.

Diffusion Equation Model. For the case when the parameters of the diffusion equation model are constant, the impulse response function for the system can be written as[17]

$$h(x, t) = \frac{1}{2(\pi D)^{0.5}} \frac{x}{t^{1.5}} \exp\left[-\frac{(ct - x)^2}{4Dt}\right] \tag{26.2.15}$$

where c and D are defined in Eqs. (26.2.8) and (26.2.9), respectively, and x is the location of the forecast point downstream from the upstream boundary.

Muskingum Model. Dooge[10] presents the impulse response function for a river section assuming that the storage is related to inflows and outflows as described in the Muskingum model. The resulting impulse response function is

$$h(t) = \frac{1}{k(1-f)^2} \exp\left[-\frac{t}{k(1-f)}\right] - \frac{f}{1-f}\delta(0) \tag{26.2.16}$$

where $\delta(0)$ is the Kronecker delta function, equal to 1 when the argument is zero (at $t = 0$), and zero otherwise. Note that, unlike the diffusion model, the Muskingum model (like variations of the linear reservoir method described below) is applied in an input-output sense to observations at an upstream and a downstream gauge; hence the forecast point is the downstream gauge location, and there is no explicit channel location argument in the impulse response function. Because the second term on the right-hand side of Eq. (26.2.15) is negative, it is possible to forecast negative downstream flows using this method. This can occur when the inflow hydrograph is such that its contribution to the first term in the convolution for the outflow calculations is insufficient to counteract the effect from the second term. In this case, a common procedure is to set the second term to zero.

Harley[17] and Dooge[10] describe a method to obtain parameter estimates for the impulse response function for a uniform channel. Their results for the Muskingum method are

$$k = \frac{X}{1.5U_0} \tag{26.2.17a}$$

and

$$f = 0.5 - \frac{1}{3}\left(1 - \frac{F^2}{4}\right)\left(\frac{Y_0}{S_0 X}\right) \tag{26.2.17b}$$

where X is the river section length, U_0 and Y_0 are the reference channel velocity and depth respectively, S_0 is the channel slope, and F is the Froude number. The reference values U_0 and Y_0 can be computed in the following manner. First choose the reference velocity U_0. Using a single-valued stage-discharge relationship (e.g., Manning's equation), the corresponding reference depth Y_0 is calculated. Dooge[10] analyzed the sensitivity of the routed flood wave computation to different reference discharges. Using smaller reference discharges resulted in flood waves that had slightly lower peaks and slightly higher wave velocities. A reasonable value for U_0 is the velocity corresponding to the average discharge of the input hydrograph. For short river sections, the Muskingum method described above can give negative values of f, which implies (physically) a system with no wedge storage ($f = 0$). In such cases, a linear reservoir model may be a better approach.

Linear Reservoir Model. A simpler model for flow routing occurs when f in Eq. (26.2.12) is assumed to be zero; this case is also known as the linear reservoir model. For this case the outflow and storage at time t are related linearly as $S(t) = kQ(t)$, where k may be interpreted as a storage constant with units of time. The impulse response is just

$$h(t) = \frac{1}{k} \exp\left(-\frac{t}{k}\right) \tag{26.2.18}$$

Cascade of Linear Reservoirs. The concept of a linear reservoir can be extended to a cascade of linear reservoirs. This method has also been referred to as successive routing through a characteristic reach whose development has been attributed to Kalinin and Milyukov.[23] If the number of reservoirs is taken as n and the storage constant k is equal for all reservoirs, then the impulse response function is mathematically the same as a gamma probability distribution function:

$$h(t) = \frac{1}{k\Gamma(n)}\left(\frac{t}{k}\right)^{n-1} \exp\left(-\frac{t}{k}\right) \tag{26.2.19}$$

where $\Gamma()$ is the gamma function, tabulated values of which are readily available in mathematical handbooks (for example, Ref. 1) and as computer library functions. For the linear reservoir model of size n, Dooge[10] found

$$k = \frac{4}{9}\left(1 - \frac{F^2}{4}\right)\frac{Y_0}{S_0 U_0} \tag{26.2.20a}$$

and

$$n = \frac{6}{4 - F^2}\,\frac{S_0 X}{Y_0} \tag{26.2.20b}$$

where k is the time constant for the single reservoir of the cascade, and U_0 and Y_0 are the reference velocity and depth.

Lag-and-Route Method. The linear reservoir method can be extended by considering the storage at any time to be proportional to the outflow which occurs after the lapse of time t^*; this is often referred to as a *lag-and-route procedure*. The relationship is

$$S(t) = kQ(t + t^*) \tag{26.2.21}$$

which leads to the impulse response function

$$h(t) = 0 \qquad t < t^*$$
$$h(t) = \frac{1}{k} \exp\left(-\frac{t - t^*}{k}\right) \qquad t \geq t^* \tag{26.2.22}$$

For the lag-and-route method, estimates of the parameters can be obtained by using

$$k = \frac{X}{(1.5U_0)}\left[\frac{2}{3}\left(1 - \frac{F^2}{4}\right)\left(\frac{Y_0}{S_0 X}\right)\right]^{0.5} \tag{26.2.23a}$$

and

$$t^* = \frac{X}{1.5U_0} - k \tag{26.2.23b}$$

where U_0 and Y_0 are reference velocity and depth, as discussed earlier. Unrealistic values for t^* may arise from short channels, implying that there is no lag in the system and t^* should be set to zero.

Lagged Cascade of Linear Reservoirs Method. The lag-and-route method can be combined with a cascade of n linear reservoirs to yield a model that works well for long river reaches. The model first translates the input hydrograph by the lag time t^*. The resulting impulse response function is

$$h(t) = 0 \qquad t < t^*$$
$$h(t) = \frac{1}{k}\,\Gamma(n)\left(\frac{t - t^*}{k}\right)^{n-1} \exp\left(-\frac{t - t^*}{k}\right) \qquad t \geq t^* \tag{26.2.24}$$

which is a function of three parameters: the lag time t^*; the time constant for a single reservoir, k; and the number of reservoirs in the cascade, n. This leads to the following parameter estimates:

$$k = \frac{1}{3}(2 + F^2)\frac{Y_0}{S_0 U_0} \tag{26.2.25a}$$

$$n = \frac{2}{3}\frac{(4 - F^2)}{(2 + F^2)^2}\frac{S_0 X}{Y_0} \tag{26.2.25b}$$

$$t^* = \frac{X}{1.5U_0} - nk \tag{26.2.25c}$$

Comparisons of Methods. Harley[17] and Dooge[10] compared the Muskingum method and variations of the linear reservoir method and found that for relatively short river lengths of 10 to 25 mi (16 to 40 km) the accuracy of the different methods was comparable. For longer channel lengths, on the order of 200 mi (300 km), the Muskingum method gave poor results, but the cascade of linear reservoirs and the diffusion model performed much better. The lag-and-route method predicted the travel time quite well but underestimated the attenuation of the flood hydrograph. The lagged cascade model produced the best forecasts, especially for the longest channels. These comparisons were made for channels with relatively small Froude numbers, that is, small channel velocities relative to the square root of channel depth.

26.2.3 Forecasts Based on Rainfall-Runoff Modeling

There is a range of forecasting models for stream discharge based on rainfall inputs. The most complex of the rainfall-runoff models currently being used are conceptual storage models, for which the U.S. National Weather Service River Forecast System is a representative example. Complex rainfall-runoff models represent the various water storage terms (interception, soil moisture, and surface storage) and flux terms (infiltration, evapotranspiration, snowmelt, interflow, groundwater baseflow, and surface runoff from rainfall and snowmelt) in varying levels of complexity. Singh (Ref. 36, Chap. 23) provides a list of conceptual models. Chapter 21 discusses the structure of some conceptual models in greater detail.

Relatively simple models for the representation of the rainfall-runoff processes have been successfully used for forecasting purposes. For example, if the response of the catchment is assumed linear with respect to effective rainfall, then methods based on convoluting the impulse response for the catchment with the effective rainfall can be used to forecast stream discharge. The instantaneous unit hydrograph model is an example of this approach. Effective rainfall is the total rainfall less any losses due to infiltration, interception, and surface ponding; the impulse response function of the basin is the hydrograph response of the catchment to a unit amount (usually taken as 1 in or 1 cm) of effective rainfall over the catchment. This approach requires the user to overcome two problems. The first is the estimation of the form of the unit hydrograph. The second is the estimation of the effective rainfall from measurements of gauge measurements for rainfall.

Unit Hydrograph Models. A variety of analytical forms and estimation procedures for the impulse response function have been proposed. Given data for effective rainfall and stream discharge, a least squares estimate for the impulse response function can be found as described in Sec. 9.5. An alternative method for the estimation of the impulse response function is through optimization procedures. The constrained linear system (CLS) model[44] uses a quadratic programming approach where the squared error between the estimated discharge and the observed discharge is minimized under the constraint that the ordinates of the unit hydrograph must be nonnegative.

A second approach is to relate the time to peak for the unit hydrograph and the peak of the unit hydrograph to basin characteristics. Snyder[37] carried out some of the first systematic studies in this area and his relationships are still used today. More recently, these unit hydrograph characteristics have been estimated from the geomorphological characteristics of the catchments. The geomorphological approach is summarized in Sec. 9.5 and in Ref. 6.

A common functional form for the impulse response function is the same as that used earlier for flood routing, namely the gamma function:

$$h(t) = \frac{1}{k\Gamma(n)}\left(\frac{t}{k}\right)^{n-1} \exp\left(\frac{t}{k}\right) \tag{26.2.26}$$

where $\Gamma(n)$ is the gamma function, which is equal to $(n-1)!$ for n integer and is tabulated, or can be evaluated by numerical approximations, given in Abramowitz and Stegun.[1] The parameters n and k can be fitted either by using the observed rainfall and streamflow data or by using the basin characteristics to estimate time to peak and the peak value of the impulse response function. Application of this method for forecasting purposes is exactly as for use of the impulse response method for channel routing; the discrete time unit hydrograph at the time interval of interest (generally some fraction of the required forecast lead time) must first be derived from the continuous time impulse response function [e.g., Eq. (26.2.26)]. The method described in Ref. 8, Chap. 7, is suggested. Then, the unit hydrograph is applied to the precipitation sequence. Where precipitation values are needed during the forecast period, simple approximations will suffice so long as the forecast lead time T_f is less than T_c, since they will receive little weight in the convolution. If $T_f > T_c$, forecast period precipitation will strongly affect the stream-flow forecast, and quantitative precipitation forecasts or related methods (see Chap. 3) are required.

Effective Rainfall Estimation. Estimating effective rainfall from gauged rainfall can be accomplished by preprocessing the rainfall data so as to estimate the infiltration, interception, and surface ponding losses. A variety of models can be used for this purpose, including the Soil Conservation Service (SCS) method for rural catchments (Sec. 9.4). The SCS method uses both catchment characteristics (soil and land cover) and the wetness of the catchment to estimate the effective rainfall.

The use of *antecedent precipitation* is quite effective as a surrogate for soil wetness and the subsequent effect on effective rainfall. The standard *antecedent precipitation index A* (for time period t) is

$$A_t = A_{t-1}K + p_t \tag{26.2.27}$$

where K is a decay rate ($K < 1$) that represents the drying of the soil and p_t is the rainfall during the time period t. The parameter K can be related to the average values of A_t and p_t by

$$K = 1.0 - \frac{E(p_t)}{E(A_t)} \tag{26.2.28}$$

where $E(p_t)$ is the average rainfall for the season of interest during a time period (for instance, one day) and $E(A_t)$ is the average value of the index A. Typical values of K are in the range 0.85 to 0.95 for a daily time step.

Bras (Ref. 6, Chap. 8) combines a 30-day antecedent precipitation index with Horton's infiltration equation to get an infiltration equation that depends on soil wetness. The equation is

$$f = f_c + (f_0 - f_c) \exp\left[-B(A_{30} + P)\right] \tag{26.2.29}$$

where P is the accumulated rainfall during the rainfall event for which the infiltration is being computed, f_0 is the initial infiltration rate at the beginning of the storm, f_c is the limiting (wet condition) infiltration rate, and B is a parameter to be estimated. After preprocessing, the resulting rainfall can be considered as the direct or effective rainfall that produces the flood event, and is used as the input into an impulse response model of the type discussed earlier.

Storage Accounting Models. There are a wide variety of hydrologic models that fall in between the complex conceptual models, like that of the U.S. National Weather Service system, and the linear impulse response models with a rainfall preprocessor. The International Association of Hydrological Sciences (IAHS) Oxford Symposium

Proceedings[21] provides a good overview of these models which generally use a storage function based formulation with either an infiltration loss function or a rainfall scaling factor to account for prestorm catchment wetness. When the model parameters have been estimated over a range of conditions (usually 10 to 20 storms), these models have demonstrated good forecasting performance.

26.2.4 Short-Term Forecast Updating

Forecasting models can be used in one of two modes: they can take in measured inputs like rainfall and make stream-flow forecasts, or they can use both measured inputs and measured stream flows when making the stream-flow forecasts. The second case is usually referred to as *updating* because the stream-flow forecast is adjusted as measurements of the stream flow become available. These measurements allow estimation of the forecast error which is the difference between the true and the forecasted stream flow, and can be approximated by the difference between the observed stream flow Q_o and the forecast Q_f.

$$e = Q_f - Q_o \tag{26.2.30}$$

Kalman Filtering. Updating increases the complexity of forecasting, but can significantly improve forecast accuracy. Updating is usually referred to as *filtering* in the statistics and estimation literature because the errors in measurements and models are filtered out through special algorithms. The most common algorithm is the Kalman filter which can be applied to linear models (like the impulse response models discussed earlier) with errors in both the model and measurements. One requirement of these updating algorithms is that the stream flow for period t be written explicitly as a function of both the stream flows from earlier time periods and the rainfall inputs. This requirement is difficult to satisfy for complex conceptual models but easier to satisfy for linear impulse response models. Many widely available mathematical computer packages (like the International Mathematical and Statistical Library package) contain Kalman filtering and related algorithms. The theory of forecast updating using Kalman filtering is described by Gelb,[14] Anderson and Moore,[2] and Jazwinski.[22]

A number of investigators have evaluated the benefits of forecast updating by Kalman filtering. Georgakakos[15] studied the performance of a hydrometeorological model for stream-flow forecasting using 6-hourly data for Bird Creek, a 2344 km^2 catchment in Oklahoma. A precipitation forecasting model was developed and coupled to a modified version of the U.S. National Weather Service rainfall-runoff model to produce the stream-flow forecast. Variables such as soil moisture storages and stream flow were updated through a Kalman filter. Eight 2-month forecast periods were examined. The results show that the nonupdating forecasting model produced forecasts whose time to peak was on average 10 h later than that observed and whose peak discharges were on average 37 percent lower than those observed. The incorporation of the real-time data and updating improved these statistics so that the time to peak was forecast within 0.5 h on average, and the average error in the peak discharge was reduced to 22 percent (see Figs. 26.2.1 and 26.2.2). Figure 26.2.3 shows the typical performance of the model for the period March 1 to April 30, 1958.

Takasao and Shiiba[39] and Takasao et al.[40] developed a simple nonlinear storage model for forecasting stream discharge on the Haze River, a 370 km^2 basin in Japan. Figure 26.2.4 shows their forecast performance for a flood in September 1965 with and without updating. The forecast errors without updating are consistently larger

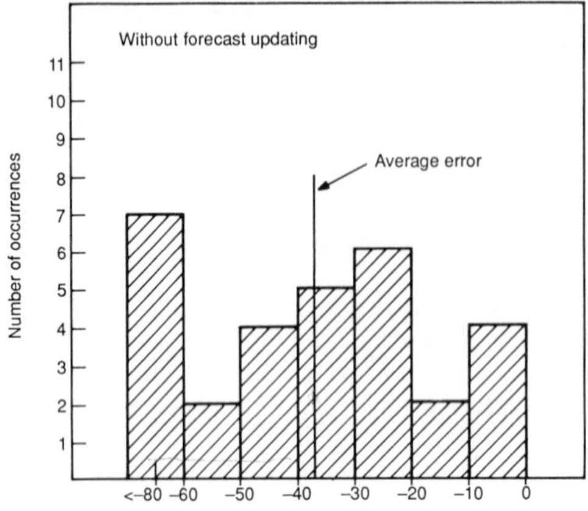

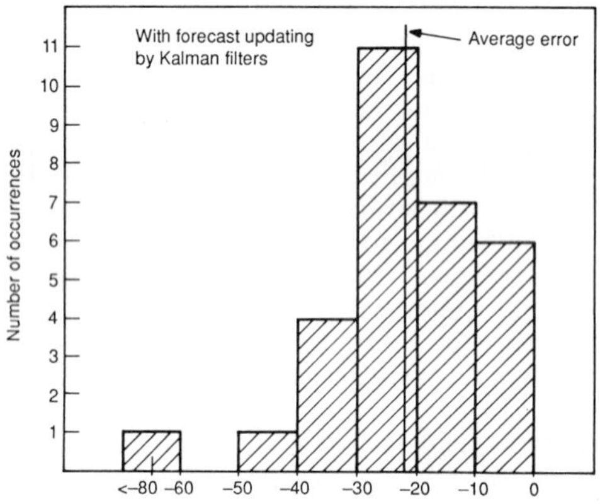

FIGURE 26.2.1 The effect of forecast updating on the error in forecasting peak discharge for Bird Creek, Oklahoma. The variable shown is the percent difference between the observed peak discharge and the discharge forecast at that time. *(From Georgakakos.[15])*

than when updating takes place. This can be seen more clearly in Fig. 26.2.5 which shows the autocorrelations of the forecast errors for the September 1965 flood. The autocorrelations show that the forecast errors persist longer when no updating is done, i.e., the errors are highly autocorrelated.

Forecast Error Correction. Updating of short-term forecasts is most effective when the forecast errors are highly autocorrelated. This persistence in the errors can be used to develop a forecasting procedure that consists of an initial forecast based on the hydrologic model plus a correction based on the errors between the forecast and measurement:

$$Q_f(t) = M(P, C, Q) + e_f(t) \tag{26.2.31}$$

where $Q_f(t)$ is the forecast for time t, $M(P, C, Q)$ is the hydrologic model which depends on precipitation P up to the forecast time, catchment characteristics, and variables represented collectively by C, and possibly previously observed stream discharge represented by Q; and $e_f(t)$ is a predicted forecast error for time t based on the previous forecast errors.

A time-series model such as an autoregressive model can be constructed for the prediction of the forecast error using the techniques described in Chap. 19 and Sec. 26.3.5. Using a number of calibration storms, a time series of forecast errors is

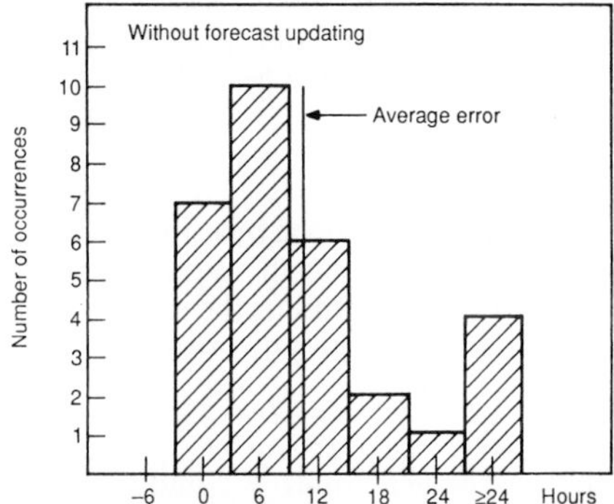

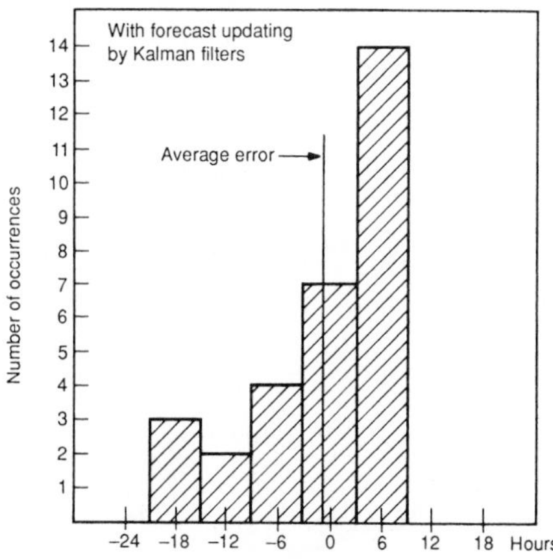

FIGURE 26.2.2 Error of forecasts of time of peak flow for Bird Creek, Oklahoma, with and without forecast updating. *(From Georgakakos.[15])*

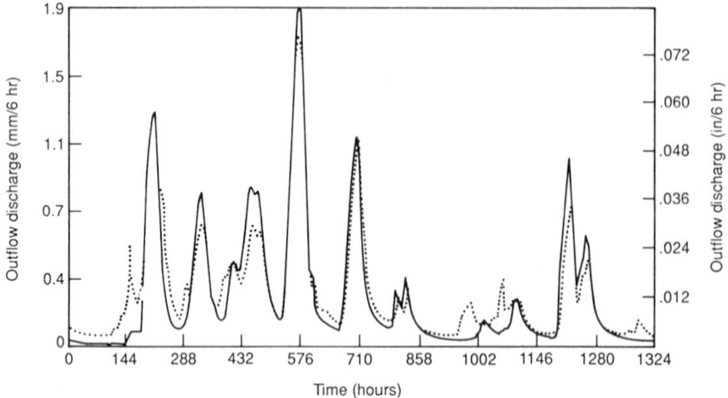

FIGURE 26.2.3 Six-hour stream-flow forecasts, in mm/6 h, for Bird Creek, Oklahoma; observations are solid curves, forecasts are dashed. *(From Georgakakos.[15])*

constructed. A statistical model is developed by using the identification and parameter estimation techniques described in Chap. 19, which is then used to correct the forecast with Eq. (26.2.31). A typical form for the error model is lag-one autoregressive, which results in an estimated forecast error $e_f(t)$ at time t:

$$\hat{e}_f(t) = \rho e_f(t-1) \qquad (26.2.32)$$

where $e_f(t-1)$ is the observed error at time $t-1$, and ρ is the lag-one autocorrelation coefficient of the forecast errors without updating.

When an operational forecast for time t is being made, the stream flow for the previous period has been measured and the forecast error calculated. Using the measured inputs (precipitation and any other observed values required by the model), an initial hydrologic-model-based forecast is made. Using the calculated forecast errors from the previous time periods, a forecast error for time t is predicted.

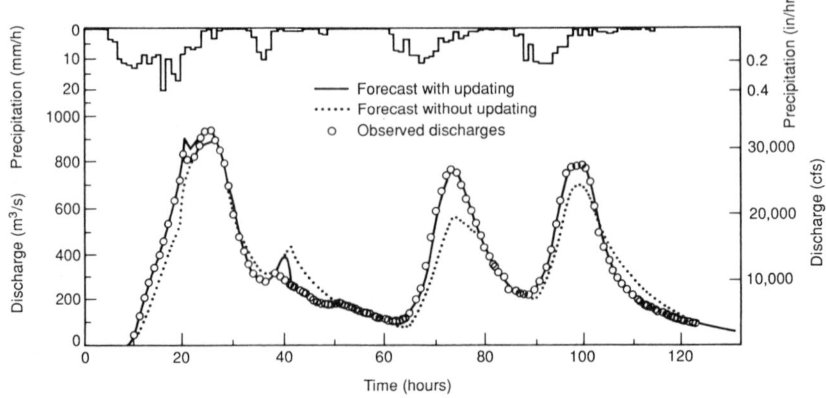

FIGURE 26.2.4 One-hour forecasts with and without forecast updating by Kalman filtering for flood of September 1965 on the Haze River, Japan. *(From Takasao and Shiiba.[39])*

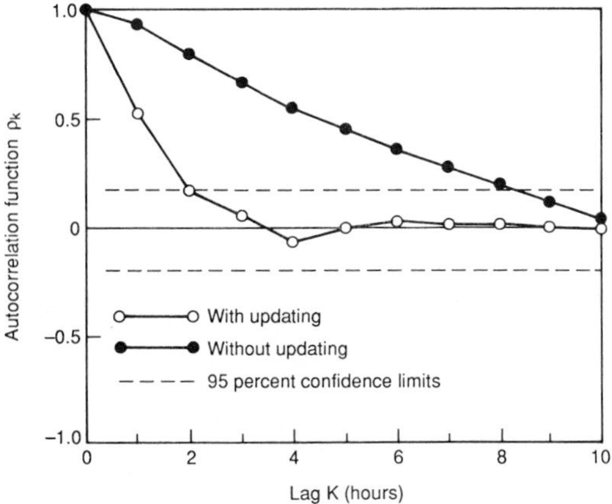

FIGURE 26.2.5 Autocorrelations of 1-h forecasts with and without updating for the Haze River flood of September 1965. *(From Takasao and Shiiba.[39])*

The final forecast for time *t* is the sum of the hydrologic-model-based forecast and the forecast error prediction. This procedure is a simple, but effective, way of updating operational forecast procedures.

The technique can be evaluated for any particular application by computing the forecast error autocorrelation function. If the forecast errors are uncorrelated, this is an indication that the forecasting procedure is extracting as much information as possible (for that particular model) from the measurements, and no further accuracy improvements can be achieved by updating. It should be noted that different forecasting models and procedures can be compared by comparing the variances and autocorrelation functions of the forecast error time series. The best models have the lowest forecast error variances; the best forecasting procedures have the smallest autocorrelations. Choosing between various combinations of models and forecasting procedures often requires that a tradeoff be made between low forecast error variances with some autocorrelation in the forecast errors, and higher variances with lower forecast error autocorrelations. The specific application must guide the hydrologist in making a tradeoff. Nonetheless, it must be emphasized that selection of the correct form for the forecasting model dynamics is critical; updating cannot be used as a substitute for a sound hydrologic model.

26.3 LONG-TERM FORECASTING METHODS

Long-term forecasting models fall into three general classes: *index-variable, storage accounting,* and *conceptual simulation.* Index-variable methods are the simplest, and oldest, methods of forecasting long-term runoff. They relate forecast period runoff to one or more index variables that are likely to affect stream flow during the forecast period, such as accumulated precipitation prior to the time of forecast and soil moisture at the time of forecast. Storage accounting models estimate the amount of water stored in the catchment (either as surface, subsurface, or snowpack storage) that is available to contribute to future runoff; the forecast is then some function (usually linear) of the storage estimate. The conceptual simulation approach uses precipitation-runoff simulation whose input data up to the time of forecast are the observed meteorology, and during the forecast period are estimates (e.g., forecasts) of the relevant meteorological variables.

Errors in long-term forecasts arise from three sources: *model error,* an incorrect conceptualization of the relationship between future runoff and the controlling variables such as precipitation; *data error,* or errors in the forecast model inputs such as precipitation due, for instance, to sparseness of precipitation gauge coverage; and *meteorological forecast error,* such as the error in forecast precipitation or error temperature. Meteorological forecasts are only marginally more accurate than climatological averages for forecast lead times of more than a few days, so for practical purposes, accurate long-term runoff forecasts are possible only in situations where the future runoff is more strongly affected by catchment water storage at the time of the forecast than by meteorological conditions (especially precipitation) during the forecast period. This is the case for forecasting of snowmelt runoff in the western United States and similar areas, where most of the precipitation falling in the winter that will contribute to spring and summer runoff is stored (and measurable) as snow water equivalent at the beginning of the spring snowmelt period. It may also be true in drought conditions in basins where forecast period runoff will result primarily from base flow, and in cases (e.g., forecasting of the runoff from the Laurentian Great Lakes) where forecast period runoff is largely controlled by surface water storage. While the methods described in this section are not particular to snowmelt forecasting, it should be recognized that in situations other than those mentioned, stream-flow forecast accuracy is usually limited by the accuracy of long-term weather forecasts, the state of the art of which, at present, represents only a slight improvement over historical climatology. Therefore, except in cases where forecast period runoff is controlled by initial catchment water storage, long-term stream-flow forecasts cannot be expected to be much better than a climatological average, and there is little to distinguish a long-term forecast from a prediction as defined in Sec. 26.1.

26.3.1 Snowmelt Forecasting

The oldest and most established long-term forecasting application is for snowmelt runoff, especially in areas such as the western United States, where a large fraction of the annual runoff is derived from winter precipitation that occurs as snow. The accuracy of snowmelt forecasts is determined by

1. The accuracy with which snow water storage can be measured at the beginning of the forecast period
2. Knowledge of soil moisture and its spatial distribution at the beginning of the snowmelt period
3. The availability and timing of energy available for melting the snowpack during the forecast period
4. Precipitation and its form (rain or snow) during the forecast period
5. The accuracy of the relationship (model) used to trace snowmelt through the soil and channel network to the stream gauge

Generally, the most accurate snowmelt forecasts are obtainable in situations where the snow accumulation and melt periods are well-defined, and there is relatively little precipitation during the melt (forecast) period. This is the case in much of the Sierra Nevada and the western slopes of the Rocky Mountains, where the most accurate forecasts are made at the end of the snow accumulation period (usually around April 1) for a forecast period through the end of the melt season (usually from May to July, depending on catchment elevation and latitude). Under such conditions, long-term snowmelt forecast accuracy is determined primarily by the accuracy with which the amount of snow water storage, and its spatial distribution, can be determined at the beginning of the forecast season. If the catchment is fully snow-covered at the beginning of the forecast period, and remains nearly so, fairly accurate forecasts can be made solely on the basis of an index of the energy available for snowmelt during the forecast period. The most widely used index for this purpose is the degree-day factor (see Chap. 7). When the catchment becomes partially bare at some time during the forecast period, forecast methods can make use of information about the areal extent of snowcover, which can be measured either via satellite remote sensing or aircraft.[31] Except for very large catchments, the channel routing time is much less than the forecast period length, and simple routing schemes can be used, such as an average time of travel from the point of melting to the catchment outlet.

Accurate information about catchment snow water storage is critical to the success of long-term snowmelt forecasts. Snow deposition and ablation processes are highly variable spatially. Methods presently used to estimate snow water storage are

1. *Manual surveys.* Manual snow surveys have been used in the mountainous areas of the western United States for over 50 years. The measurement device most often used is the standard federal snow sampler. The usual approach is to identify a clear area, and to take multiple cores across transects. The final reported value is the average snow water equivalent over the transect. The U.S. Soil Conservation Service[46] and Atmospheric Environment Service[4] define the procedure used in manual snow surveys. Because manual snow surveys are often used in connection with index-variable forecasting methods, it is important that the location of the snow survey site, and the procedure used for sample collection (location of the transects, number of replicates) not be changed without careful consideration of how prior reported observations can be adjusted so that comparability of data from one year to the next is assured.

2. *Automated recording devices.* The U.S. Soil Conservation Service now uses an automated system known as SNOTEL, which incorporates snow water equivalent measurement via snow pillows, as well as precipitation and air temperature measurements. Data transmission is via meteor-burst telemetry (see Chap. 25). The SNOTEL snow pillows have a surface which is a 1.22×1.52-m stainless-steel plate. The pillow is filled with an antifreeze solution. The pressure in the pillow is related to the water-equivalent depth of the snow on the platform. The great advantage of snow pillows used in connection with automated recording devices, as compared with snow surveys, is the greater frequency of data acquisition, typically twice per day for SNOTEL as compared with twice per month at most for manual surveys. Three common problems with snow pillows are (1) bridging, that is, partial suspension of the pillow via connection of the snow on the weighing platform to the surrounding snow pack, usually caused by the formation of ice layers; (2) edge effects, that is, erosion of the snow pack near the edge of the measuring platform; and (3) differential thermal effects at the base of the measurement platform as compared with the surrounding ground (the snow pillow conducts heat differently than the surrounding soil). Bridging can usually be observed visually, but requires periodic visits to the sites, or overflights. Edge and thermal base effects represent long-term biases for which corrections are required if comparability with snow survey data is important, as in the case of index-variable methods.

3. *Aircraft and satellite remote sensing.* Although remote sensing has been successfully used for estimation of snow areal extent,[35] estimation of catchment snow water storage is much more difficult. The greatest success has been achieved with aircraft gamma-radiation measurements, which perform best for relatively thin snowpacks, less than 30-cm snow water equivalent, in nonforested areas. The U.S. National Weather Service routinely uses airborne gamma-radiation observations for its forecasts of spring runoff in the Upper Mississippi River basin, as does the Geological Survey of Canada for forecasting of spring runoff of several rivers in the northern Great Plains. Remote sensing of mountain snowpacks is a more difficult problem because of the greater depth of snowpacks, topographic variability, and difficulties in measurement of snow under forest canopies. Experimental methods using microwave radars are being developed for remote sensing of snow water storage in mountainous areas.

26.3.2 Forecasts Based on Index Variables

An index variable is a surrogate, readily measured prior to the time of forecast, that can be related to forecast period runoff. Commonly used index variables are snow water storage, accumulated precipitation up to the forecast date, accumulated runoff prior to the forecast date, and current soil moisture. Index-variable methods are of the general form

$$Q_f = f(X_1, X_2, \ldots, X_n) \tag{26.3.1}$$

where X_i is the ith of n index variables. X_i might indicate, for instance, values of the same variable at different measurement locations, such as at several precipitation gauges or snow courses. Generally, a separate relationship f is developed for each forecast beginning and ending date. Many early index methods used graphical techniques.[28] Later, statistical techniques such as multiple regression, principal components, and pattern search were introduced.[30,48] A common practice in snowmelt forecasting is to include among the index variables X_i the total precipitation from a specified initial date (e.g., the beginning of the water year) through the end of the forecast period, rather than just the precipitation occurring from the initial date to the beginning of the forecast period. Stedinger et al.[38] have shown that this approach is undesirable because it artificially inflates the apparent forecast accuracy, but more importantly, when several index variables are considered, inclusion of estimated future precipitation may result in selection of an inferior set of index variables, and hence reduce forecast accuracy. Stedinger et al.[38] also found that the use of nonlinear transformations of observed variables (e.g., logarithms) resulted in little or no improvement in forecast accuracy.

Index methods have the major advantage that they are simple and easily implemented. Virtually all statistical packages available for small computers have multiple regression and other applicable routines (e.g., principal components) necessary to implement index-variable forecasts. Further, if implemented via statistical methods such as regression, confidence bounds on the forecasted volumes can be derived. Where water-use decisions must be based on forecasted runoff volumes, knowledge of the confidence bounds about a forecast, or, equivalently, the range within which the actual runoff is expected to lie, is usually more important than the "best" forecast.[20,27] Knowledge of the statistical properties of the forecast also permits determination of the risk that the actual value will exceed a trigger for a special response even though the best forecast (expected value of the forecast probability distribution) does not lie in the special response region.

26.3.3 Storage Accounting Schemes

The concept of storage accounting, as outlined above, is that runoff in a future forecast period is determined by the amount of water presently in storage in a catchment (either as soil moisture, groundwater, or snow), as well as forecast period precipitation. The idea of storage accounting was introduced by Tangborn and Rasmussen[43] for forecasting snowmelt runoff. They suggested that basin storage could be estimated as a linear function of basin precipitation from the beginning of an accounting period (generally about the beginning of the water year) up to the forecast date, less runoff:

$$S_t = a + bP_t - R_t \tag{26.3.2}$$

where P_t is the total precipitation (up to the forecast date t) at one or more precipitation gauges and R_t is the total runoff in the same period. The forecast runoff for the period T is then a linear function of S_t,

$$Q_f(t, T) = \alpha(t, T)S_t + \beta(t, T) = A(t, T) + B(t, T) - C(t, T)R_t \tag{26.3.3}$$

Tangborn and Rasmussen[43] and later Tangborn[41] estimated the coefficients A and B by regression, and constrained $C(t, T)$ to 1.0. The above model can be viewed as a special case of the index-variable method with regression estimates of parameters.

Figure 26.3.1 compares R^2 values for index variable forecasts for the Stanislaus and Sacramento Rivers, California, and storage accounting forecasts with and without snow course data for the Clark Fork Yellowstone River, Montana, and the Cedar River, Washington. The California forecast results are taken from Stedinger et al.,[38] while the storage accounting results are from Lettenmaier and Garen.[26] Figure 26.3.1 shows several important characteristics of snowmelt runoff forecasts in general, and index-variable and storage accounting forecasts in particular. First, forecast accuracy increases throughout the snow accumulation season, as more of the water that will ultimately contribute to runoff is stored in the snowpack. Once the date of maximum snow accumulation has passed, forecast accuracy decreases, because forecasts become increasingly sensitive to precipitation during the forecast period. This decline is evident from the Clark Fork Yellowstone River results but not in the other two graphs because the date of maximum forecast accuracy is not until after May 1 for the Stanislaus, Sacramento, and Cedar Rivers. Both index and storage accounting forecasts are quite sensitive to the variables used in the forecast. For instance, for the Cedar River, the best forecasts occurred when only low-elevation precipitation information was used to estimate basin water storage; inclusion of snow course information degraded the forecast accuracy. For the Clark Fork Yellowstone River, the reverse occurred, and forecast accuracy increases when precipitation and snow course data are used together. Similar results for other streams are reported by Lettenmaier and Garen[26] and Lettenmaier.[24] These differences are site-specific. Therefore, a careful screening analysis using statistical methods such as stepwise regression must be performed to assure that the proper variables are included in the forecast equation.

Although the storage accounting method can be implemented via regression, other estimates of basin storage can be used as well. Tangborn[42] has proposed the use of a spatially lumped conceptual simulation model for the Columbia River basin to simulate soil moisture, groundwater, and snow water storage. Storage accounting forecasts of future runoff are then based on the total basin storage (soil moisture, groundwater, and snow water) at time t. A forecast for lead time T beginning at t is

$$Q_f(t, T) = A(t, T)S_t + B(t, T) \tag{26.3.4}$$

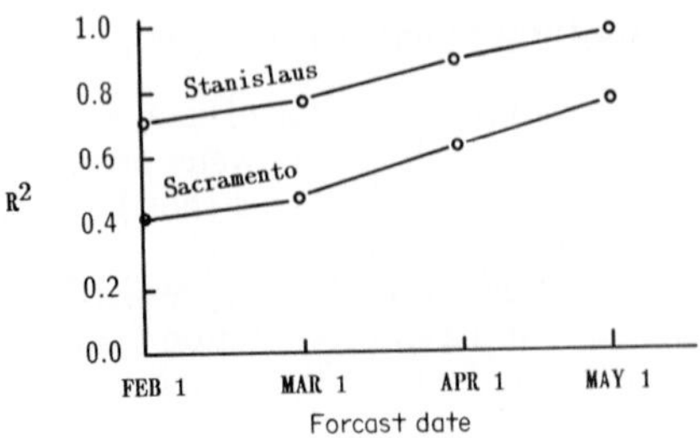

(a) Stanislaus & Sacramento R., California

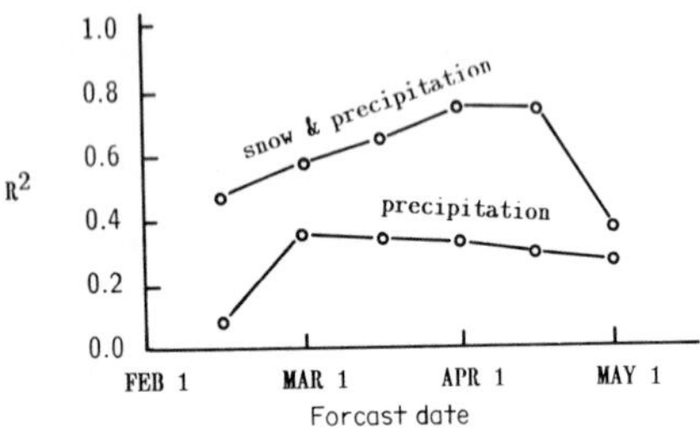

(b) Clark Fork Yellowstone R., Montana

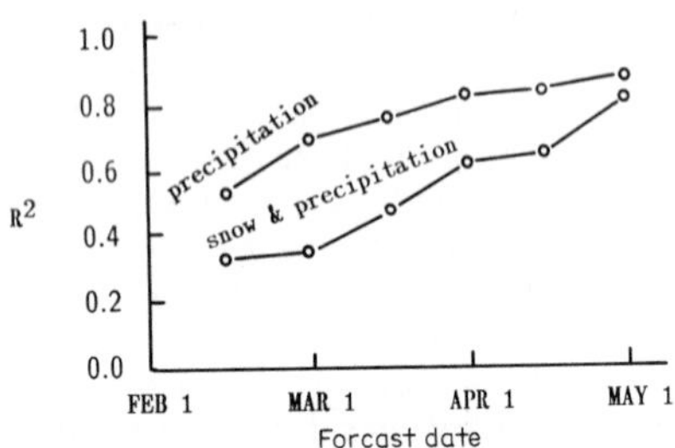

(c) Cedar R., Washington

FIGURE 26.3.1 Comparison of R^2 for snowmelt forecasts using (*a*) index-variable method and (*b*) storage accounting models for selected western U.S. catchments. In (*a*), the forecast period is from the date indicated through June 30; in (*b*) and (*c*), through July 31. The two lines in (*b*) and (*c*) indicate runoff forecasts based on low-level precipitation measurements alone, or precipitation measurements combined with snow survey data. *[Data for* (a) *are from Stedinger et al.*;[38] (b) *and* (c) *are from Lettenmaier and Garen.*[26] *Used with permission.]*

with coefficients $A(t, T)$ and $B(t, T)$ estimated via regression. Considerable improvements in forecast accuracy may be achievable using this method, as compared with index method forecasts. Tangborn's application of the storage accounting method forecasts incorporates a test season correction as follows. First, Eq. (26.3.4) is applied to a test period of length T_t, usually of length one week to one month, beginning at $t - T_t$, resulting in a forecast $Q_f(t - T_t, T_t)$. Because the flow in the test period has been observed at time t, the test period error $e_t(t - T_t, T_t) = Q_f(t - T_t, T_t) - Q_o(t - T_t, T_t)$ is available. The forecast period error $e_f(t, T) = Q_f(t, T) - Q_o(t, T)$ can then be regressed on the test period error, and a correction developed:

$$Q_f^*(t, T) = Q_f(t, T) + \gamma(t, T)e_t(t - T_t, T_t) \tag{26.3.5}$$

where $\gamma(t, T)$ is the coefficient from the regression of $e_f(t, T)$ on $e_t(t - T_t, T_t)$, and $Q_f^*(t, T)$ is the corrected forecast. The test period correction can be viewed as a form of updating to correct for misestimation of basin water storage by the conceptual simulation model.

26.3.4 Extended Stream-Flow Prediction

Extended stream-flow prediction (ESP) is a method for deriving runoff forecasts based on a spatially lumped precipitation-runoff model.[9,45] The model is run up to the time of forecast with observed meteorological inputs. During the forecast period, the meteorological inputs can be of several forms: (1) a forecast of future conditions; (2) a historical data set which is similar to conditions up to the time of forecast; (3) all observed historical sequences of meteorological inputs, possibly weighted to reflect prior probabilities; and (4) synthetically generated realizations of possible future meteorology, possibly also weighted according to their prior probabilities. Figure 26.3.2 shows an example application of the ESP procedure to the Rex River, Washington. The forecasting model consisted of the National Weather Service River Forecast System snow accumulation and ablation model[3] coupled to a simplified soil moisture accounting model known as the Nanjing model.[33] The model was run in simulation mode with observed data up to the forecast date, March 1, 1977. From March 1 through the end of the forecast period (July 31), precipitation and temperature from each of the 12 years in the period 1968–1980, excluding 1977, were used as input to the model. Figure 26.3.2 shows the results of forecasts of cumulative flow corresponding to each of the 12 years. From the ensemble of forecasts a best estimate can be determined (usually the mean of the alternative forecasts).

In the case where several series of meteorological inputs are used, the resulting vector of forecasts can be analyzed to produce an estimate of the probability distribution of the forecast. The mean, or other statistic of the empirical probability distribution, such as the median, can be used as the "best" forecast. In addition, worst-case forecasts can be made by entering extreme precipitation and/or temperature during the forecast period (e.g., zero precipitation for drought contingency analysis).

One major advantage of extended stream-flow prediction is the ability to conduct alternative scenario analyses. It can be argued that the ESP procedure may be better than index-variable methods in extreme years, since the nonlinearities in the precipitation-runoff process are incorporated in the conceptual simulation model on which the procedure is based. In addition, ESP produces a time sequence of runoff during the forecast period, rather than just the total forecast period volume, which is especially useful if reservoir management decisions are to be made. The disadvantage of ESP is that it is much more cumbersome to implement than either the index

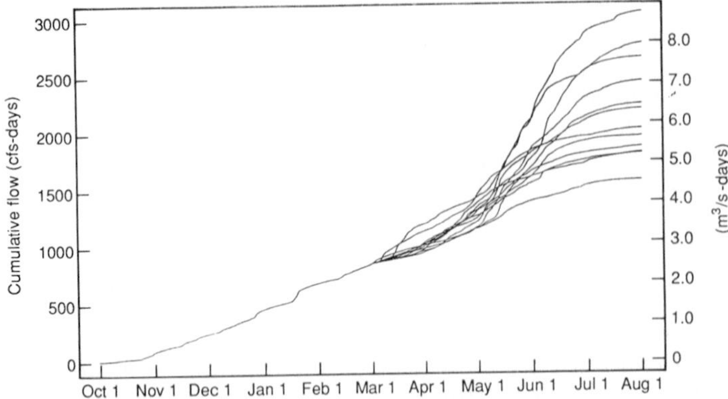

FIGURE 26.3.2 An example of extended stream-flow prediction (ESP) for the Rex River, Washington. The simulation was run to March 1 with observed precipitation and temperature for water year 1977; twelve alternative scenarios were prepared for the forecast period March 1 – July 31, using precipitation and temperature records for this period for water years 1968 – 1980, excluding 1977.

variable or storage accounting methods. In addition, the conceptual simulation models on which ESP is based usually require estimation of a large number of parameters, some of which may not be accurately estimated because of errors in precipitation observations, model errors, and data record length. Lettenmaier[25] has shown that parameter error provides a lower bound below which forecast error cannot be reduced in the absence of forecast updating procedures. This lower bound for ESP was found to be similar to the observed forecast accuracy of storage accounting models, which suggests that, if ESP methods are to be implemented, the use of forecast updating methods will be essential to assure that the time of forecast conditions on which the ESP forecasts are based, such as basin soil and snow water storage, are properly represented.

26.3.5 Time-Series Forecasting Methods

Time-series models can be used to describe the stochastic structure of the time sequence of stream flows and precipitation. Time series models are usually of the form

$$X_t = \alpha_0 + \alpha_1 X_{t-1} + \cdots \alpha_p X_{t-p} + \epsilon_t + \beta_1 \epsilon_{t-1} + \cdots \beta_q \epsilon_{t-q} \qquad (26.3.6)$$

where $X_t = x_t - \mu_t$, x_t is the observed flow, μ_t is the mean flow at time t, ϵ_t is a normally distributed error term that is independent of the previous observations and error terms, and α_i and β_i are parameters to be estimated.

For practical purposes, the autoregressive and moving average order, p and q, of such models is usually quite small, typically $p = 1$ or 2, and often $q = 0$. More complex models that account for seasonal nonstationarity in a more sophisticated manner than can be modeled by a time-dependent mean may be appropriate under some circumstances.[5,18,19]

For forecasting purposes, the terms in $\epsilon_{t-1}, \epsilon_{t-2}, \ldots$ can be eliminated by back substitution of expressions for $X_{t-1}, X_{t-2}, \ldots$, and Eq. (26.3.6) can be rewritten as[5]

$$\hat{X}_t = \pi_0 + \sum_{i=1}^{\infty} \pi_i \tilde{X}_{t-i} + \epsilon_t \tag{26.3.7}$$

where the coefficients $\pi_0, \pi_1, \pi_2, \ldots$ can be calculated from the α's and β's in Eq. (26.3.6) and $\tilde{X}_{t-i}$ is the conditional expectation of X_{t-i}, defined as

$$\tilde{X}_{t-i} = E(X_{t-i} | X_{t-i-1}, X_{t-i-2}, \ldots) \tag{26.3.8}$$

In practice, the infinite sum of terms in $\tilde{X}_{t-i}$ can be approximated by a small finite number of terms. The procedure for generating forecasts is as follows. The conditional expectations $\tilde{X}_{t-i}$ in Eq. (26.3.7) are computed iteratively from Eq. (26.3.6) by substituting the forecasted flows for those $\tilde{X}_{t-i}$ which have not occurred, substituting the observed flows for those which have occurred, and inserting the conditional expectations of the error terms $\tilde{\epsilon}_{t-i}$ (usually $\tilde{\epsilon}_t = 0$). The resulting methodology is rigorous and has well-defined forecast error properties which are useful for risk analysis.

Time-series forecasts have the property that they approach the long-term mean, μ_t, as the forecast lead time increases. The rate of decay of the forecasts to the mean depends on how far the initial value (most recent observed flow) is from the long-term mean, and the time series persistence or autocorrelation, which determines the parameters α_i and β_i. In general, higher levels of persistence, such as are found during those parts of the year dominated by base flow recession, lead to more accurate forecasts, and lower persistence, such as are present in those times of year dominated by future precipitation, have less accurate forecasts. Univariate time series methods are limited in that the only information they incorporate is that which is present in the past flows. In mountainous regions dominated by snowmelt runoff, the index variable, storage accounting, and ESP methods, which make direct use of snow water storage, are usually preferable. However, in situations where snow storage is not a factor, time series methods have the advantage that they are easily implemented, and the forecast error variance can be readily computed.[5]

Forecasting Reservoir Storage. Figure 26.3.3 illustrates a hypothetical use of a time-series forecast procedure described by Hirsch[20] for the Potomac River using the drought period of the early 1930s with an approximation of the current total system reservoir storage and demands. September 1, 1930 was assumed to be the forecast origin; the simulated reservoir storage on this date was about 50 percent of capacity using observed data from water year 1897 through the forecast date. A variation of time-series model was used to simulate 100 stochastic realizations of reservoir inflows for the period October 1930 to September 1932, conditioned on the observed initial flows through September 1930. The simulation procedure was similar to extended stream-flow prediction, except that a stochastic model, which made use only of information in the historic stream-flow record, rather than a deterministic precipitation-runoff model, was used to simulate alternative stream-flow sequences. As in ESP, the reservoir storage sequences from the 100 alternative stream-flow sequences formed an ensemble from which the risk analysis could be performed. Figure 26.3.3 shows that the risk of reservoir failure (zero storage at some time in the future) was a little more than 10 percent, since this curve falls to zero storage for about 3 months in 1932.

Forecasting Stream-Flow Volume. Figure 26.3.4 illustrates a slightly different use of the time-series method for one component of the City of Seattle water supply system

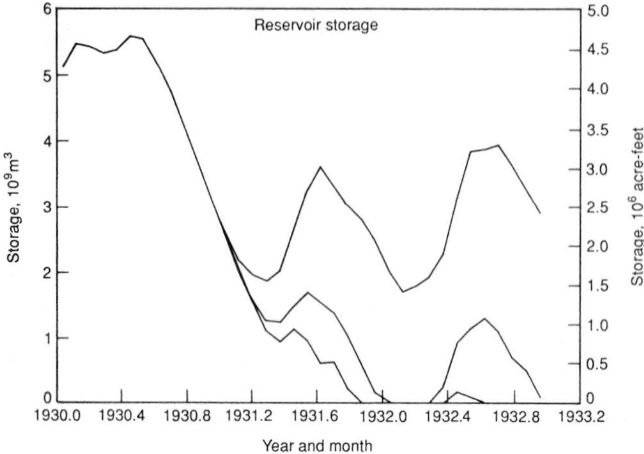

FIGURE 26.3.3 Example of position analysis for hydrologic data from September 1930 showing storage up to September 1930 with 2 percent (lower curve), 10 percent (middle curve), and 50 percent (upper curve) risk storage levels through September 1932. *(From Hirsch.[20] Used with permission.)*

(South Fork Tolt River) during the 1987 drought.[27] The forecasting model was of a form similar to that used by Hirsch, although it was somewhat simpler in that no moving average term was included [$q = 0$ in Eq. (26.3.6)]. Rather than simulate alternative sequences, a procedure was developed to estimate the empirical cumulative probability distribution of stream flows at each forecast time simultaneously for each of four system inflow locations. The forecasts were made at weekly intervals

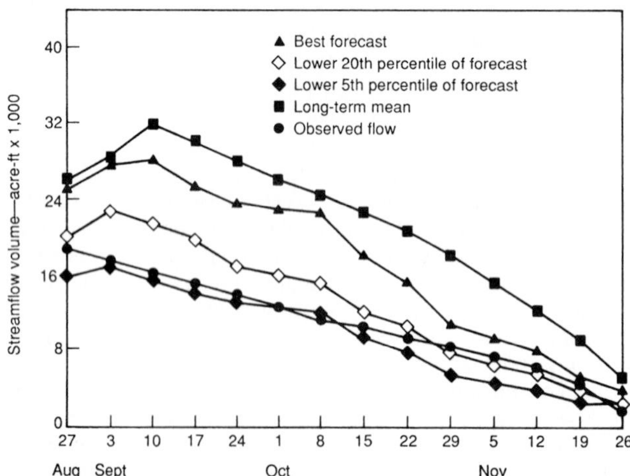

FIGURE 26.3.4 Seasonal stream-flow volume forecasts and corresponding observed and long-term mean flow for forecasts beginning on the given dates of 1987, South Fork Tolt River, Washington. *(From Lettenmaier et al.[27] Used with permission.)*

throughout the period August–November, 1987; on each forecast date stream-flow scenarios corresponding to given risk levels were produced for each of four system inflow points. Figure 26.3.4 shows forecasts of stream-flow volume through the end of the study period. As time passes, the forecast lead time grows shorter so the forecasted and observed flow volumes decrease with time. The study period was during an extreme drought; for this reason, most of the "best" forecasts were much larger than the observed flows, and for most of the sites for much of the period, the observed flows were near the lower 5th percentile of the forecast distribution. This occurred because under normal conditions there is little persistence in flows during the fall season; however, during a period of severe drought, both meteorological and hydrologic conditions can display much more persistence than usual. This persistence is difficult to capture in time-series forecasts, however, and this argues in favor of extended stream-flow prediction methods for drought forecasting.

26.3.6 Updating Long-Term Forecasts

Long-term forecast accuracy can often be improved through implementation of simple forecast updating procedures. If the objective of the long-term forecast is to provide a time series of future flows, the updating methods described in Sec. 26.2.4, which are based on the application of time-series models to extract information from the persistence of the forecast error time series, can be applied directly. Often, though, long-term forecasts are desired to provide an estimate of the total volume of runoff over the forecast time horizon, and the specific sequence of flows within that horizon is of less interest. In such cases, regression-based updating schemes like Eq. (26.3.5) are more appropriate. The reason that updating may be successful, particularly where forecasts make use of conceptual rainfall-runoff or snowmelt models, is that errors in estimating the conceptual moisture storages in the model prior to the time of forecast tend to persist through the forecast period. These errors most commonly are the result of misestimation of precipitation or snowmelt prior to the forecast date. The easiest approach to forecast updating is simply to develop an error correction equation similar to Eq. (26.2.31) to adjust the forecasted stream flows:

$$Q_f(t, T) = M(P, C, Q) + e_f(t, T) \tag{26.3.9}$$

where $Q_f(t, T)$ represents the updated forecast made at time t for lead time T, $M(P, C, Q)$ is the model estimate of the stream flow in the forecast period, and $e_f(t, T)$ is the error adjustment. By rearranging Eq. (26.3.9), find that the error adjustment is just the difference between the unadjusted model forecast and the observed stream flow; by retrospectively applying the forecasting model to previous years' data, a series of forecast errors can be obtained. This error series can then be related to quantities that are observable at the time of the forecast. For instance, one can construct an estimate of the change in catchment water storage over the period preceding the forecast date, say, from the beginning of the water year, as

$$\Delta S(t_0, t) = P(t_0, t) - Q(t_0, t) - E(t_0, t) \tag{26.3.10}$$

where $P(t_0, t) - Q(t_0, t) - E(t_0, t)$ are the precipitation, runoff, and evaporation up to the forecast date. An equivalent quantity, $\Delta S^*(t_0, t)$ can be estimated from the model storage change over the period (t_0, t), and a sequence of the apparent storage change errors $\Delta S(t_0, t) - \Delta S^*(t_0, t)$ (one value for each year) can be constructed. Linear regression can then be used to estimate the forecast error as a function of the storage error, and this relationship can be used operationally to provide the error adjustment $e_f(t, T)$ in Eq. (26.3.9).

REFERENCES

1. Abramowitz, M., and I. A. Stegun, *Handbook of Mathematical Functions,* Dover, New York, 1972.
2. Anderson, B. D. O., and J. B. Moore, *Optimal Filtering,* Prentice-Hall, Englewood Cliffs, N. J., 1979.
3. Anderson, E. A., "National Weather Service River Forecast System—Snow Accumulation and Ablation Model," NOAA technical memorandum NWS HYDRO-17, November 1973.
4. Atmospheric Environment Service, *Snow Surveying* (2d ed.), Environment Canada, Downsview, Ontario, 1973.
5. Box, G. E. P., and G. M. Jenkins, *Time Series Analysis Forecasting and Control.* Holden-Day, Oakland, Calif., 1976.
6. Bras, R. L., *Hydrology: An Introduction to Hydrologic Science,* Addison-Wesley, Reading, Mass., 1990.
7. Carslaw, H. S., and C. Jaeger, *Conduction of Heat in Solids* (2d ed.), Oxford University Press, Fair Lawn, N. J., 1959.
8. Chow, V. T., D. R. Maidment, and L. W. Mays, *Applied Hydrology,* McGraw-Hill, New York, 1988.
9. Day, G. N., "Extended Streamflow Forecasting Using NWSRFS," *J. Water Resour. Plann. Manage.,* vol. 111, no. WR2, pp. 157–170, April 1985.
10. Dooge, J. C. I., "Linear Theory of Hydrologic Systems," technical bulletin no. 1468, U.S. Department of Agriculture, Washington, D.C., 1973.
11. Eagleson, P. S., *Dynamic Hydrology,* McGraw-Hill, New York, 1970.
12. Fread, D. L., "NWS Operational Dynamic Wave Model," *Proc. 25th Annual Hydraulics Division Specialty Conference,* American Society of Civil Engineers, pp. 455–464, August 1978.
13. Fread, D. L., "Channel Routing," Chap. 14 in *Hydrological Forecasting,* M. G. Anderson and T. P. Burt, eds. Wiley, Chichester, U.K., 1985.
14. Gelb, A., ed., *Applied Optimal Estimation,* MIT Press, Cambridge, Mass., 1974.
15. Georgakakos, K. P., "A Generalized Stochastic Hydrometeorological Model for Flood and Flash-Flood Forecasting: 2. Case Studies," *Water Resour. Res.,* vol. 22, no. 13, pp. 2096–2106, December 1986.
16. Goring, D. G., "Analysis of Tidal River Records by a Harmonic Regressive Technique," *J. Hydrol.,* vol. 73, pp. 21–37, 1984.
17. Harley, B. M., "Linear Routing in Uniform Open Channels," thesis submitted for the M.Eng.Sc. degree, Department of Civil Engineering, National University of Ireland, University College, Cork, September 1967.
18. Hipel, K. W., A. I. McLeod, and W. C. Lennox, "Advances in Box-Jenkins Modeling: 1. Model Construction," *Water Resour. Res.,* vol. 13, no. 3, pp. 567–576, June 1977*a*.
19. Hipel, K. W., A. I. McLeod, and W. C. Lennox, "Advances in Box-Jenkins Modeling: 2. Applications," *Water Resour. Res.,* vol. 13, no. 3, pp. 577–586, June 1977*b*.
20. Hirsch, R. M., "Stochastic Hydrologic Model for Drought Management," *J. Water Resour. Plann. Manage. Div., Am. Soc. Civ. Eng.* (now *J. Water Resour. Plann. Manage.*), vol. 107, no. WR2, pp. 303–313, October, 1981.
21. International Association of Hydrological Sciences, "Hydrological Forecasting Symposium," IAHS-AISH Publication No. 129, *Proceedings of the Oxford Symposium,* International Association of Hydrological Sciences, Wallingford, U.K., 1980.
22. Jazwinski, A. H., *Stochastic Processes and Filtering Theory,* Academic Press, New York, 1970.

23. Kalinin, G. P., and P. I. Milyukov, "O Raschete Neustanovivshegosya Dvizheniya Vody V Otkrytykh Ruslakh" ["On the Computation of Unsteady Flow in Open Channels"], *Meteorologiya i Gidrologiya Zhuzurnal,* vol. 10, pp. 10–18, Leningrad, 1957.

24. Lettenmaier, D. P., "Value of Snow Course Data in Forecasting Snowmelt Runoff," *Proc. Workshop on Modeling of Snowcover Runoff,* S. C. Colbeck and M. Ray, eds., U.S. Army Cold Regions Research and Engineering Laboratory, pp. 44–55, September 1978.

25. Lettenmaier, D. P., "Limitations on Seasonal Snowmelt Forecast Accuracy," *J. Water Resour. Plann. Manage.,* vol. 108, no. WR3, pp. 255–269, 1984.

26. Lettenmaier, D. P., and D. C. Garen, "Evaluation of Streamflow Forecasting Methods," *Proc. 47th Western Snow Conference,* Sparks, Nev., pp. 48–55, 1979.

27. Lettenmaier, D. P., E. F. Wood, and D. B. Parkinson, "Operating the Seattle Water System during the 1987 Drought," *J. Am. Water Works Assoc.,* vol. 82, no. 5, pp. 55–60, May 1990.

28. Linsley, R. K., M. A. Kohler, and J. L. H. Paulhus, *Hydrology for Engineers* (2d ed.), McGraw-Hill, New York, 1958.

29. Mahmood, K., and V. Yevjevich, *Unsteady Flow in Open Channels,* vol. 1, Water Resources Publications, Fort Collins, Colo., 1975.

30. Marsden, M. A., and R. T. Davis, "Regression on Principle Components as a Tool in Water Supply Forecasting," *Proc. 36th Western Snow Conference,* pp. 33–40, April 1968.

31. Martinec, J., and A. Rango, "Application of a Snowmelt-Runoff Model Using LANDSAT Data," *Nordic Hydrology,* vol. 10, pp. 225–238, 1979.

32. Nash, J. E., and J. V. Sutcliffe, "River Flow Forecasting Through Conceptual Models: 1. A Discussion of Principles," *J. Hydrol.,* vol. 10, pp. 282–290, 1970.

33. Pandolfi, C., A. Gabos, and E. Todini, "Real-Time Flood Forecasting and Management for the Han River in China," paper presented at IUGG Symposium, Hamburg, Germany, 1983.

34. Priestley, M. B., *Spectral Analysis and Time Series,* Academic Press, London, 1981.

35. Rango, A., "Operational Applications of Satellite Snow Cover Observations," *Water Resour. Bull.,* vol. 16, no. 6, pp. 1066–1073, 1980.

36. Singh, V. P., *Elementary Hydrology,* Prentice-Hall, Englewood Cliffs, N. J., 1992.

37. Snyder, F. F., "Synthetic Unit-graphs," *Trans. AGU,* vol. 19, pp. 447–454, 1938.

38. Stedinger, J. R., J. Grygier, and H. Yin, "Seasonal Streamflow Forecasts Based Upon Regression," in *Computerized Decision Support Systems for Water Managers,* J. W. Labadie, L. E. Brazil, I. Corbu, and L. E. Johnson, eds., *Proc. 3d Water Resources Operations Management Workshop,* American Society of Civil Engineers, 1989.

39. Takasao, T., and M. Shiiba "Development of Techniques for On-Line Forecasting of Rainfall and Flood Runoff," *Natural Disaster Science,* vol. 6, no. 2, pp. 83–112, 1984.

40. Takasao, T., M. Shiiba, and K. Takara, "Stochastic State-Space Techniques for Flood Runoff Forecasting," *Proc. Pacific Int. Seminar on Water Resources Systems,* Tomamu, Japan, pp. 117–132, 1989.

41. Tangborn, W. V., "Application of a New Hydrometeorological Streamflow Prediction Model," *Proc. 45th Western Snow Conference,* Albuquerque, N. Mex., April 1977.

42. Tangborn, W. V., "A Basin Water Storage Model to Forecast Columbia River Discharge," unpublished document, HyMet Co., Seattle, Wash., 1990.

43. Tangborn, W. V., and L. A. Rasmussen, "Hydrology of the North Cascades Region, Washington: 2. A Proposed Hydrometeorological Streamflow Prediction Method," *Water Resour. Res.,* vol. 12, no. 2, pp. 203–216, 1976.

44. Todini, E., and J. R. Wallis "Using CLS for Daily or Longer Period Rainfall-Runoff Modeling," pp. 149–168 in *Mathematical Models for Surface Water Hydrology,* T. A. Ciriani, U. Maione, and J. R. Wallis, eds., Wiley, London, 1977.

45. Twedt, T. M., J. C. Schaake, Jr., and E. L. Peck, "National Weather Service Extended Streamflow Prediction," *Proc. 45th Western Snow Conference,* Albuquerque, N. Mex., pp. 52–57, April 1977.

46. U.S. Soil Conservation Service, "Snow Survey and Water Supply Forecasting," Sec. 22 in *SCS National Engineering Handbook,* U.S. Department of Agriculture, Washington, D.C., 1972.

47. World Meteorological Organization, "Hydrological Forecasting," Chap. 6 in *Guide to Hydrological Practices,* vol. 1: *Data Acquisition and Processing,* report no. 168 (4th ed.), Geneva, Switzerland, 1981.

48. Zuzel, J. F., D. L. Robertson, and W. J. Rawls, "Optimizing Long-Term Streamflow Forecasts," *J. Soil Water Conserv.,* vol. 30, no. 2, pp. 76–78, 1975.

CHAPTER 27

HYDROLOGIC DESIGN FOR WATER USE

T. A. McMahon
Department of Civil and Agricultural Engineering
University of Melbourne, Australia

During the past 25 years, since the publication of Chow's *Handbook of Applied Hydrology,* there have been important developments in analytical and computational procedures for estimating the hydrologic design parameters of water-supply systems. This chapter reflects these developments and recognizes that, in addition to the needs of people, industry, hydropower, and irrigation, new demands associated with aquatic habitats must be addressed.

The first section in this chapter sets the design procedure in context. Such issues as the design of new systems versus more efficient management of present systems, competing uses, and the design procedure are briefly discussed. This leads into the substantive part of the chapter, which deals first with the design of water supply. Yield of surface water reservoirs and forecasting demands are considered in detail. (Yield of groundwater basis is treated in Chap. 6.) This section concludes with brief comments on design life of system components. Design for hydropower follows next. The final section addresses the question of in-stream use. Compared to the techniques outlined in the earlier sections of the chapter, the procedures for assessing water needs of aquatic habitats have not been widely applied and should be used with caution.

27.1 THE DESIGN CONTEXT

27.1.1 New Systems versus Better Management of Present Systems

In developed countries, the traditional approach to meeting increased water demands has been to seek new sources of supply or to complete the next stage of a staged project. With increased pressure on limited financial and natural resources, there has been a marked shift away from capital works toward examining the current supply system with the objective of increasing its efficiency through better management. Most suitable reservoir sites have been utilized, and the next phase of water-supply expansion will be relatively more expensive. In developing countries, the pressures are different; often water-supply systems are nonexistent and cheaper dam sites are available; nevertheless, better management is also a key to the successful operation of a new system. Consequently, the procedures outlined in this chapter take account of this new emphasis on more effective management.

27.1.2 Competing Issues

Competing Uses. In considering the design or management of a water-supply system, many competing issues must be addressed.

The years after World War II saw the conflict between allocating reservoir space for flood mitigation and utilizing more of the capacity for conservation purposes. While the flood control controversy[33] was debated in the United States, it was also an issue in many other countries. With the continued expansion of urbanization, it is inevitable that there will be urban/rural conflicts. For example, in Australia 85 percent of the population lives in major cities, yet 80 percent of total water use is for agriculture. This is also true for the United States. Consequently, in a country with scarce water resources and urban dwellers who wish to maintain large home gardens and green open spaces, competing uses take on a new dimension. Recreational use of reservoirs also competes with the operating needs of conservation and/or flood management because these needs require the surface water elevation of the reservoir to be varied, while recreational pursuits require a stable water surface elevation. During the past decade, aquatic habitat and in-stream flow requirements have become important factors. The requirements for maintenance of specific habitats conflict with water level variations that occur with hydropower operation. Of equal importance are the effects on habitat of river regulation, in which natural flood peaks, which are the basis of the maintenance of a healthy ecosystem, are reduced in magnitude or duration and periods of naturally occurring low flow are considerably augmented. Navigation of inland waterways requires a minimum depth of water. All these management issues must be considered during the design process and the competing demands must be balanced within the legal and administrative framework.

Conjunctive Use/Cyclic Storage. In order to maximize the efficient use of both surface and groundwater systems, a coordinated and planned management scheme is required. Todd[57] defines this process as *conjunctive use.* Lettenmaier and Burges[34] distinguish *consumptive use* which deals with short-term use from the more important long-term pumping and recharging known as *cyclic storage.* An important element of such a system is the recharge of the groundwater storage using excess runoff during periods of above-normal precipitation. During dry periods, the groundwater storage is drawn down again. Correctly managed, the integrated systems will yield more water at more economic rates than separately managed surface and groundwater systems.

Water Rights. Engineering aspects of water law are described by Rice and White.[48] A *water right* is the legal right to use water. In the eastern United States, this right has existed only for land that is adjacent or *riparian* to a stream. Known as the *riparian doctrine,* it has worked satisfactorily in regions where water was plentiful and use was relatively small and nonconsumptive. In areas where water is scarce, riparian rights were unsatisfactory and the doctrine of *prior appropriation* developed, in which a water right was created simply by taking (or appropriating) the water and using it, regardless of location or ownership of adjacent land. Rights under the appropriation doctrine are ordered by the time sequence in which the right was applied for: the "first in time is first in right" method. In modern times, appropriation rights may be obtained by filing an application with a state regulatory office. In Australia, another system exists where ownership of all water is vested in the state and, in government-supplied irrigation areas, a water right is associated with each unit area of irrigable land. A new element in this picture is the enactment of legislation allowing irrigators to sell their water right separately from the land.[32] Various United States state laws have provided for fair market transfers, subject to the constraint of state regulatory approval. Although these issues are not described further here, they do affect the efficient management of water resources and need to be taken into account as part of the design process. Readers are referred to a general discussion of water rights in Linsley and Franzini[36] and a U.S. case study of water rights modeling and analysis by Wurbs and Walls.[64]

Water Quality and Habitat Needs. Return flow from water application (e.g., irrigation) finds its way back to the source, but is normally of lower quality than the source, so slowly the source degrades. *Leaching,* which is required when poor-quality water is used for irrigation, often aggravates the overall problem. Some areas irrigated for many years have become waterlogged through inadequate drainage, and *soil salinization* has increased because irrigation water brought in salts that were deposited in soils as irrigation water evaporated.

Downstream habitat needs, both in-stream and those associated with riparian wetlands, must also be taken into account. In addition, both existing and future downstream rights need to be included in estimating water demand from a reservoir.

27.1.3 Design Procedure

Design or Management. An important part of the design process relates to better management of present water-supply systems rather than building new ones. Thus some of the procedures that follow need to be considered with these two strategies in mind. In most cases, techniques for analysis will be the same, but the objective will be to estimate firm yield for a given reservoir or system rather than estimating the size of a new reservoir.

Preliminary or Final Design of Surface Reservoirs. For surface water reservoirs, a two-step design process is adopted. In the first step, a number of alternative dam sites are investigated, not only for construction requirements, but also in terms of the hydrology. Generally, simple and quick techniques are used to establish reservoir capacity-yield relationships. Such methods are called *preliminary design* techniques.

In preliminary techniques, simplifying assumptions are normally made—reservoir releases are assumed constant, evaporation is ignored in temperate and humid regions, seasonal flows may not be taken into account—but the *final design* procedure should take into account all factors affecting the design outcome, including system inflows, variations of releases with season, the possibility of release restriction during periods of low storage, the net evaporation loss from the reservoir, minimum pool requirements, and the probability of not being able to meet the demand.

27.1.4 Other Considerations

Climate Change. Historically, water-supply systems have been designed to take account of climate variability. Extremes are expected and the analysis of representative historical stream flows makes provision for these. However, until recently, the long-term climate was regarded as unchanging during the normal life of water systems. But observed rates of change of greenhouse-effect gases and their likely effects on temperature and other climate variables would suggest that climate change needs to be considered in planning. Titus et al.[56] argue that, rather than ignore the effect until its consequences are firmly established, water engineers and planners should evaluate whether new and old systems are vulnerable to the risks of climate change.

It is important, therefore, that in applying the techniques that follow, consideration be given to examining the impact on the design of climate change. The major difficulty is to adopt a scenario of climate change appropriate to the area under investigation. The only feasible approach at present is to use a sensitivity analysis and examine, for example, the effect of an increase or a decrease in system inflow on firm yield. This could be done by changing mean annual flows by, say, ± 20 and ± 40 percent. This is approximately equivalent to changing precipitation or potential evaporation by ± 10 and ± 20 percent, respectively.

Characterizing Droughts. Drought is a popular term in arid regions, but strictly it is a relative term applicable to any climate. There are many definitions of drought, but the one adopted herein is that of Palmer:[46] a drought is ". . . a prolonged and abnormal moisture deficiency." Palmer used a water balance approach to characterize drought, and his index is widely known as the *Palmer Drought index.*

The Palmer index was developed for semiarid (western Kansas, United States) to subhumid (Iowa) regions for assessing drought severity and is now applied over the whole of the United States. The analysis is based on a weekly or monthly water balance, and output from the analysis is an index value ranging from -4 (extreme drought) through 0 (normal condition) to $+4$ (extreme wet period). The index is computed as a function of the difference, accumulated through time, of the actual rainfall and the CAFEC rainfall (climatically appropriate for existing conditions of temperature, evaporation, and other components of the water balance). Evaporation is computed by the Thornthwaite method, and runoff, percolation, and soil-water levels, by a simple soil-water balance model. An example map of the Palmer index for the United States is given as Fig. 27.1.1.

In applying the Palmer index to New South Wales, Australia, McDonald[38] offered several comments:

1. The minimum data set is preferably at least 30 years of daily precipitation and temperature data and an estimate of soil field capacity.
2. The method computes the incidence and severity of both wet and dry conditions.
3. Calculations are normally made for regions, but smaller areas can be analyzed.
4. The computational procedure developed in the midwest United States incorporates a number of empirical constants, which are also appropriate for New South Wales.

Uncertainty. *Uncertainty* is a major element in the design process. This is inherent not only in the flow records where temporal and spatial variability is significant, but also in the forecasts of demand. In water-supply systems, storages and water-transfer channels are introduced to reduce the effects of flow and demand variability. Thus variability is a key parameter in the design process for water supplies.

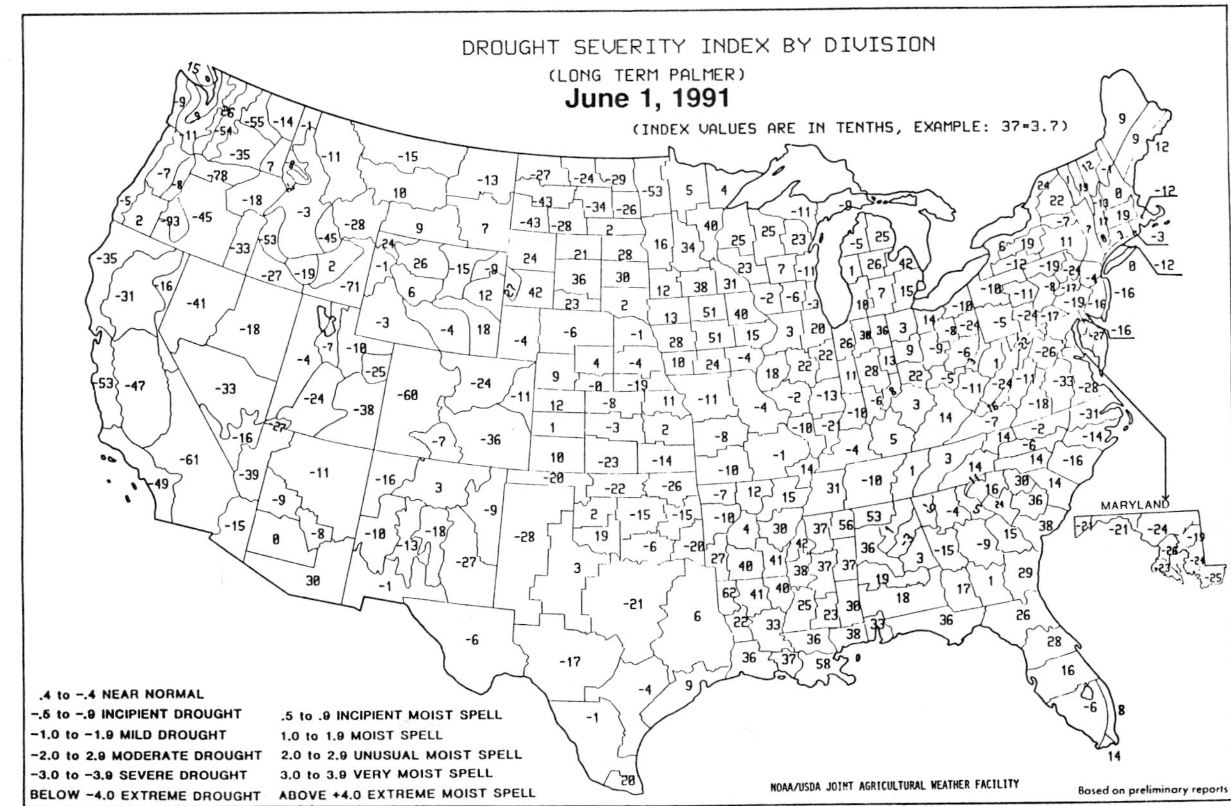

FIGURE 27.1.1 Palmer drought severity index for the United States for June 1, 1991. *(By permission of NOAA/USDA Joint Agricultural Weather Facility.)*

27.2 DESIGN FOR WATER SUPPLY

Water-supply needs are usually met from surface water and groundwater. Surface supplies are of two types (Fig. 27.2.1): (1) *on-stream storages,* which are necessary for streams that exhibit significant flow variability or where the percentage utilization of the stream flow will be large and (2) *unregulated streams,* for which two alternatives may be considered—*direct supply source,* i.e., a supply directly from the unregulated stream (also called run-of-the-river supply), or pumping to an *off-stream storage* and then on to the demand center (Fig. 27.2.2).

27.2.1 Yield of Surface Water Resources

Background. In the hydrologic design process, a distinction is made between low- and high-cost dams. In the case of low-cost dams, which are those associated with small water uses like on-farm dams for stock and irrigation use, a site is chosen and then a simple analysis is carried out to determine its capacity. It would be unusual for the time spent to be more than a few worker hours. For larger reservoirs for irrigation or municipal water supply, the two processes of preliminary screening followed by detailed final design are carried out.

A useful summary result of either a preliminary or final design analysis is illustrated in Fig. 27.2.3, where draft or withdrawal (or regulated outflow) from the reservoir, expressed as a percentage of mean annual flow, is plotted against the required storage for several levels of probability of failure. From such a diagram, the storage at a specific site required to meet a given draft and probability of failure can be obtained. Alternatively, Fig. 27.2.3 shows, for a fixed reservoir size, the relationship of withdrawal or firm yield to probability of failure (or reliability). In the United States, the probability-of-failure concept is replaced by *no-failure yield;* that is, one is interested in the yield which can be consistently supplied over an N-year planning horizon.

The material in this section is treated in considerable detail in the text by McMahon and Mein,[39] which contains many worked examples.

Definition of Terms. The *active storage* of a reservoir is the volume of water stored above the level of the lowest valve or offtake, or above some legally imposed minimum level. The storage below the lowest offtake is known as *dead* storage and should be sufficient to store anticipated sediment accumulation for the design life of the reservoir.

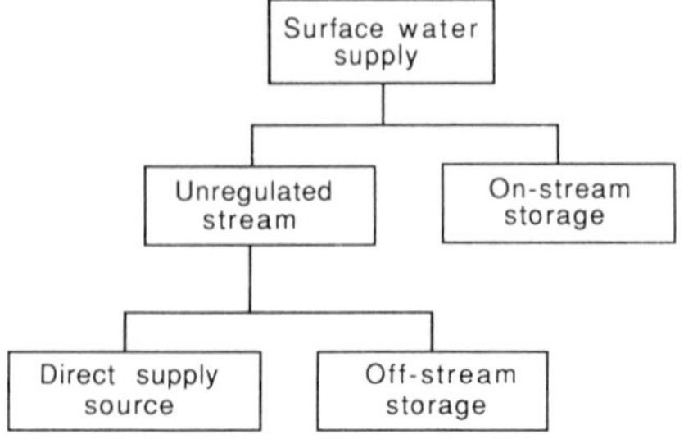

FIGURE 27.2.1 Sources of surface water supply.

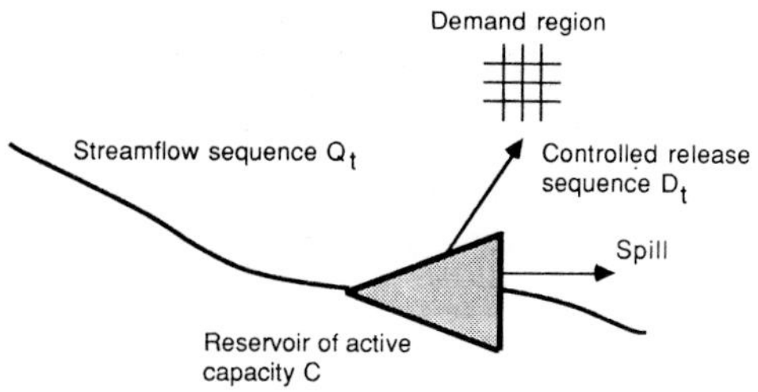

(a) On-stream storage (Eqn. 27.2.14)

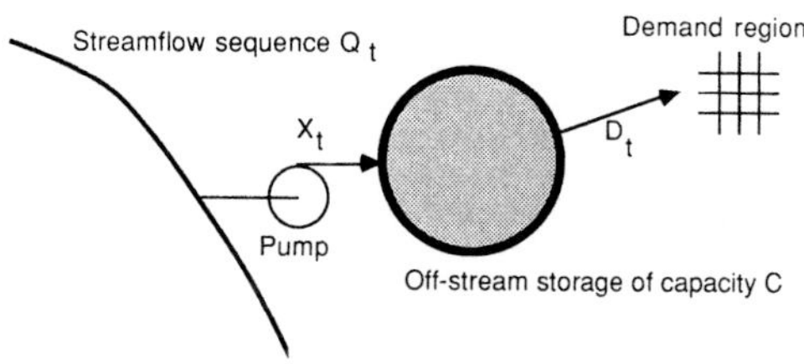

(b) Off-stream storage (Eqn. 27.2.17)

FIGURE 27.2.2 Types of reservoir storage.

Within-year storage is the storage capacity required to smooth out seasonal fluctuations in stream flow within a year. In humid regions, like the eastern United States, most reservoirs refill every winter and are drawn down for water supply in the summer. Thus they require only sufficient capacity to meet within-year fluctuations in stream flow and demand.

Carry-over storage refers to the additional capacity of a reservoir to compensate for year-to-year variations in stream flow and demand for reservoirs that do not refill every year. Reservoirs located in more arid climates are normally of this type.

The terms *yield, release, draft, withdrawal, outflow,* and *regulation* are used synonymously and refer to the regulated outflow from a reservoir. *Spill* is uncontrolled flow from a reservoir and will take place when the stored water is above full storage capacity level. Yield is often expressed as a percentage of mean flow during the project life, with typical values being in the range 50 to 70 percent. Data from Australian water authorities indicate that the median value of regulation in that country is about 65 percent. Hardison[23] has shown that for mainland United States regulation varies from 97 percent in Tennessee to 57 percent in the lower Colorado basin.

Standardized inflow m is a term introduced by Hurst[26] and defined as

$$m = \frac{(1 - D)\bar{x}}{s} \tag{27.2.1}$$

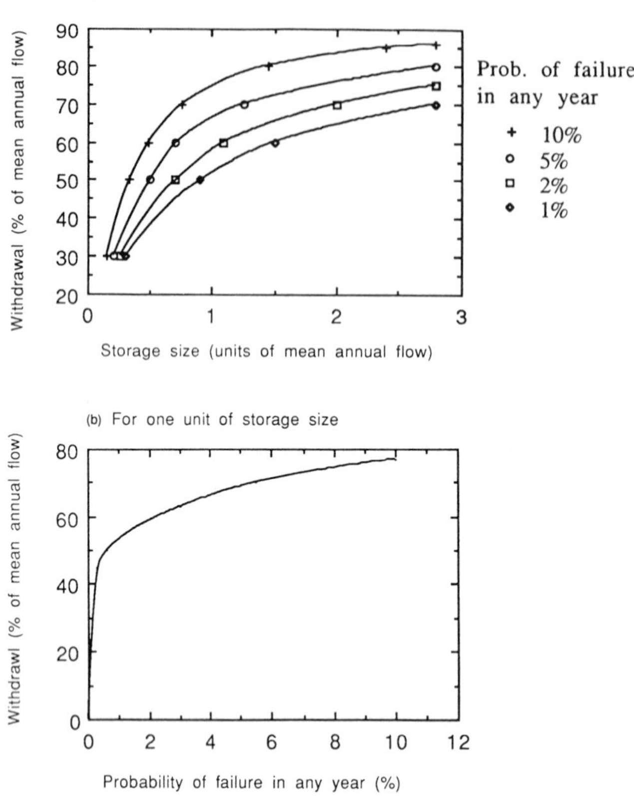

FIGURE 27.2.3 Examples of the relationship between storage size, withdrawal, and probability of failure, where failure is defined as the inability, at any time during the year, of the reservoir to supply the required draft.

where D = draft expressed as a ratio of mean annual flow
$\bar{x}$ = mean annual flow
s = standard deviation of annual flows

According to Vogel and Stedinger,[61] over-year storage would be required when m is approximately less than unity.

Firm yield is a term used in the United States and is the draft or withdrawal that lowers the water content in a reservoir from a full condition to a minimum level just once during the critical historical drought. It is essentially the no-failure yield.

Release or *operating rules.* Usually the water supplied to the consumer is equal to that demanded. However, during periods when the reservoir contents are so low that the water required cannot be supplied, or when prudence suggests that only part of the demand should be met from storage, withdrawals may be a function of water stored where the function relating release and storage is the release rule. These concepts are illustrated in Fig. 27.2.4.

A number of definitions are used of *probability of failure* of a reservoir from a

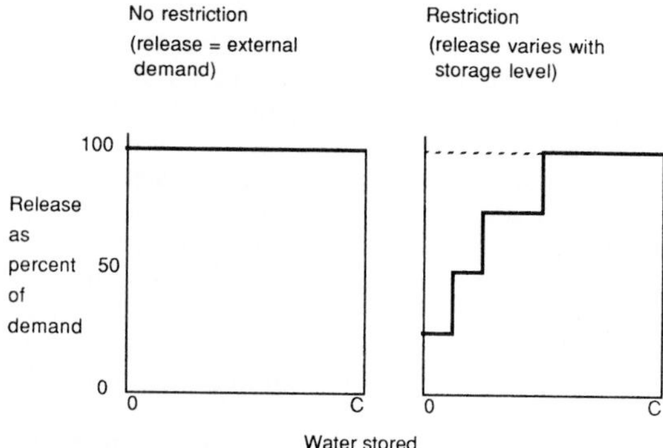

FIGURE 27.2.4 Release rules (C = reservoir capacity).

hydrologic point of view. A common definition of failure in theoretical analysis is the ratio of the time the reservoir is empty, n_f, to the total time period used in the analysis, n:

$$P_f = \frac{n_f}{n} \tag{27.2.2}$$

This definition is unrealistic. For example, a city water-supply reservoir is never allowed to empty; restrictions on releases are applied before this situation is reached. Hence a more appropriate definition of failure is

$$P_r = \frac{n_r}{n} \tag{27.2.3}$$

where n_r = time during which the reservoir is unable to meet the full demand. In other words, P_r is the proportion of time restrictions are imposed.

Both P_f and P_r are really a percent of time that the specified service is not fulfilled. They are not probabilities in the strict sense, which usually are expressed as percent chance of failure in a year.

Reliability is the term used to represent the proportion of time the reservoir is able to meet the consumer demand. Hence, it is the complement of probability of failure:

$$R_f = 1 - P_f \tag{27.2.4}$$

$$R_r = 1 - P_r \tag{27.2.5}$$

Volumetric reliability is equivalent to Fiering's[17] performance index and is defined as the ratio of the volume of water supplied, V_s, to the volume of water demanded, V_d:

$$R_v = \frac{V_s}{V_d} \tag{27.2.6}$$

Two other criteria are useful in assessing reservoir performance, namely *resilience* (the rate of recovery from failure) and *vulnerability* (the severity of failure). The probability of recovery from failure to some acceptable state within a specified time interval is defined as resilience. Hashimoto et al.[24] argued that resilience can be a measure of the probability of being in a period of no failure, given that there was a failure in the last period. Thus a resilient system is one that is capable of recovering from a deficit state to a normal state in a short time. For mathematical modeling, Moy et al.[42] defined resilience as the maximum number of consecutive periods of shortage that occur prior to recovery.

Vulnerability is a measure of the likely magnitude or significance of a failure. Moy et al.[42] defined it as the largest deficit during the period of operation of a reservoir.

A *critical period* is defined as the period from a full condition through emptiness (or a condition defined as failure) to the next full condition (Fig. 27.2.5). On the other hand, a *critical drawdown period* is the period during which the reservoir contents fall from an initially full condition to an empty condition, without spilling during the intervening period. In the United States, these terms refer only to the worst historical drought.

In a critical drawdown period, only one failure occurs and that signals the end of the period. In Fig. 27.2.5, two critical drawdown periods are illustrated. In the second one, the period beyond the initial empty point in 1987 is not part of the critical drawdown period. Critical drawdown periods can be as short as a few months or as long as 10 years or more.

Example of Reservoir Water-Supply Behavior (Routing). Although a behavior (routing), procedure is discussed in detail later, a tabulated water-supply routing example using monthly data is presented as Table 27.2.1. This illustrates the various elements that need to be considered in developing reservoir size–withdrawal–probability of failure relationships. All demands should be taken into account; they include water-supply needs of urban areas, irrigation, in-stream uses, and of other water rights. Evaporation losses from the reservoir are often very important, especially in semiarid and arid regions.

The procedures that follow have been developed to overcome various limitations in this sort of analysis. The key issues include computing probability of failure, usefulness of rapid techniques for preliminary analysis and adequacy of historical flows as representative of future flows.

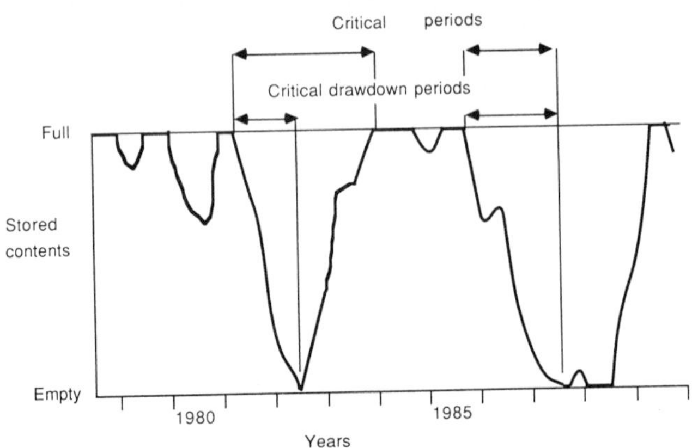

FIGURE 27.2.5 Behavior diagram illustrating critical periods. (In the United States, the critical period refers only to the period that yielded the lowest sustainable draft.)

TABLE 27.2.1 Reservoir Water-Supply Routing Example Using Monthly Data in Units of Volume

Data in acre-feet; 1 acre · ft = 1233.5 m³

Year/ month	Storage content at start of month	Inflow	Net evaporation loss	Other losses	Withdrawal*	Storage content at end of month	Spill
1974							
February	1300	370	70	—	100	1500	
March	1500	700	50	—	100	2000†	50
April	2000	0	60	—	100	1840	
⋮							
1980							
June	150	25	5	—	100	70	
July	70	20	1	—	89‡	0	
August	0	250	5	—	100	145	

* Withdrawals normally vary seasonally, depending on the enterprise.
† Full supply level is 2000 acre · ft.
‡ July 1980 demand was 100, but only 89 available.

Stream-Flow Record. One of the critical elements in evaluating the adequacy of surface water supplies, whether reservoir or run-of-river, is developing a stream-flow record that is suitable for analysis. Often little or no data are available at the site in question. One simple method for transferring records from gauged to ungauged watersheds is to use the ratio of their respective drainage areas. In estimating available flows, it is important that proper account be taken not only of historical upstream depletions, but also estimates must be made of future depletions affecting the availability of supply.

Usable Reservoir Storage Volume. The critical elements in determining usable reservoir storage volume are minimum pool elevations and loss of storage because of sedimentation. Minimum pool elevations are not defined solely by hydraulic limitations of the outlet or diversion works. More severe constraints may be imposed by recreational interests, habitat values in the reservoir, or adverse water-quality effects if the pool is drawn too low. The loss of storage because of sedimentation may be significant and must be accounted for in the yield evaluation. Most planning is based on projections of 50 or 100 years of sediment accumulation, although one can show changes in yield over time, so that current water-supply needs are not constrained by yield estimates that reflect conditions in the distant future. As estimating reservoir sediment accumulations is difficult at best, regular sediment surveys are important (at least once every 10 years).

Mass Curve Procedure (Rippl Diagram). One of the earliest methods for estimating the size of storage to meet a given draft is Rippl's method.[49] The method (as it is now used) is illustrated in Fig. 27.2.6. The steps are

1. For the proposed location, construct a cumulative curve of stream flow; monthly data are preferred. Determine the slope of the cumulative draft appropriate for the graphical scales adopted.

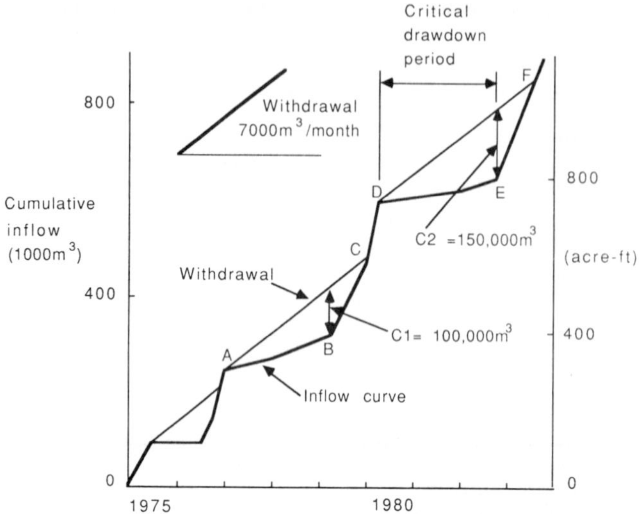

FIGURE 27.2.6 Reservoir capacity-yield estimation by mass curve procedure.

2. On the mass curve diagram, superimpose the cumulative draft line such that it is tangential to the mass inflow curve, as shown in Fig. 27.2.6.
3. Measure the largest intercept between the inflow curve and draft line.

For the example in Fig. 27.2.6, the intercept C_2 is greater than C_1, and therefore the design capacity would be taken as C_2 (=150,000 m³). From the figure, it can be seen that the reservoir is full at A, begins to empty from A to B, and refills from B to C. From C to D, the reservoir spills, and, again, the content falls until it just empties at E. The critical drawdown period, D to E, for this example is approximately 28 months.

In the mass curve procedure, two important assumptions are made:

1. The reservoir is full at the beginning of the critical drawdown period.
2. As the analysis utilizes historical stream-flow data, it is implicit that future sequences of inflow will not contain a more severe drought than the historical sequence.

The procedure exhibits two important attributes:

1. It is simple and widely understood.
2. Because the analysis uses historical data, seasonality, autocorrelation, and other flow parameters are taken into account.

Rippl's mass curve method is equivalent to the computer-efficient sequent peak algorithm (see Loucks et al.[37]). However, the procedure does have several deficiencies, especially where variable drafts, evaporation losses, and multireservoir systems are involved. Because of these limitations, the mass curve procedure is not recommended as a design technique. Nevertheless, it is included herein because the capacity of many reservoirs has been estimated using this technique, and analysts reviewing the yield of older storages should be aware of its limitations.

Preliminary Design Procedure. Following an extensive review of preliminary reservoir capacity-yield techniques and their application to Australian and Malaysian streams (that cover the range of hydrologic characteristics throughout the world) Teoh and McMahon[55] recommend the Gould gamma method for establishing reservoir storage–draft–probability of failure relationships. It should be noted, however, that the Gould gamma method applies to reservoirs whose storage volume fluctuates from year to year and that no reliable techniques except local site-specific empirical ones are available to deal with the within-year storage problem.

Gould Gamma Method. Derivation of the Gould gamma method is set out in McMahon and Mein.[39] It is based on the following assumptions:

1. The critical drawdown period is sufficiently long for the sum of n-year inflows to be normally distributed.
2. For nonnormally distributed flows, a small correction allows gamma distributed flows to be modeled.
3. Annual flows are assumed independent.
4. The draft rate or withdrawal is uniform from year to year.

Computed storage need at the site in question is related to the mean and variability of annual stream flow, draft rate, and probability of nonexceedance of the inflow during the critical period, as follows:

$$\tau = \left[\frac{z_p^2}{4(1-D)} - d\right] C_v^2 \qquad (27.2.7)$$

where τ = required storage expressed as a ratio of mean annual flow
C_v = coefficient of variation of the annual stream flow
D = constant draft rate or withdrawal expressed as a ratio of mean annual flow ($D < 1$)
z_p = standardized normal variate corresponding to a p percent risk of failure in any year
p = probability (in percent) of nonexceedance of inflow during the critical drawdown period
d = factor to adjust annual stream flows to a gamma distribution; this factor is always used unless annual flows are normally distributed

Values of z_p and d for given values of p are given in Table 27.2.2. The critical drawdown period (years) may be estimated from

$$C_p = \frac{z_p^2}{4(1-D)^2} C_v^2 \qquad (27.2.8)$$

In addition to the above assumptions, this technique assumes that only one failure occurs during the critical drawdown period; that is, in Fig. 27.2.5, for example, the 1988 failure before refilling is not taken into account. Although the method is based on annual flows, it provides reliable estimates of storage for reservoirs as small as 0.1 times mean annual flow.

Notwithstanding the above limitations, as a preliminary procedure the Gould gamma method provides very good estimates of carryover storage over the whole range of practical interest.

The complementary use of Eq. (27.2.7) is to make a preliminary estimate of reservoir yield. Equation (27.2.7) can be recast so that it is expressed in terms of

TABLE 27.2.2 Standardized Normal Variate z_p and Gould Gamma Correction Factor d

p percentile risk of failure in any year	z_p	d*
0.5	3.30	†
1.0	2.33	1.5
2.0	2.05	1.1
5.0	1.64	0.6
10.0	1.28	0.3‡

* Adapted from McMahon and Mein.[39] Used with permission.

† d is not available. Assumptions in method break down.

‡ Method is not recommended for use in this range.

withdrawal or reservoir yield:

$$D = 1 - \frac{z_p^2 C_v^2}{4(\tau + dC_v^2)} \tag{27.2.9}$$

To be of practical use, net effective lake evaporation needs to be subtracted from the gross yield to provide an effective yield estimate. An appropriate adjustment procedure is discussed in the next subsection.

Also, given τ, C_v, and D, one can obtain a quick estimate of reliability:

$$Z_p = 2\sqrt{(1 - D)\left[\left(\frac{\tau}{C_v^2}\right) + d\right]} \tag{27.2.10}$$

Adjustment for Evaporation. If the draft rate has not been modified to reflect the effects of net lake evaporation, then an adjustment to storage size is required:

$$\Delta S_E = cA_F \, \Delta E \, C_p \tag{27.2.11}$$

where S_E = volume that the computed reservoir storage should be increased to account for net evaporation loss ΔE, m^3 or acre · ft

c = constant

A_F = area of reservoir at full supply or conservation pool, m^2 or acre · ft

ΔE = net evaporation loss [see Eq. (27.2.15)]

C_p = critical drawdown period [Eq. (27.2.8)]

Based on an analysis[39] of only six reservoirs in Australia, $c = 0.7$.

An alternative approach is to first determine the value of the effective area ($A_F + A_E$), where A_E is the area of the minimum pool to be maintained. By multiplying this effective area by the net effective evaporation (lake evaporation minus rainfall), with appropriate unit conversion, a quick estimate of evaporation volume loss can be found for preliminary design purposes. This adjustment also allows the gross yield curves in Fig. 27.2.3 to be converted to a family of net yield curves obtained for alternative minimum pool levels.

Sampling Error. Many water-supply managers have unrealistic expectations of the accuracy of hydrologic estimates. In Australia, for example, many streams have coefficients of variation of annual flows greater than 0.5. To achieve a standard error of estimation of the mean annual flow of ±5 percent, such streams would require at least 100 years of data. Thus, with shorter records, say 40 years, errors in estimates of mean annual flows are quite large.

With the Gould gamma method [Eq. (27.2.7)] as a basis for analysis, the sensitivity of storage estimates to errors in the mean and coefficient of variation of annual flows is

$$\frac{\Delta C}{C} = \left\{ \frac{-z_p^2 + 4d(1-D)^2}{z_p^2(1-D) - 4d(1-D)^2} \right\} \frac{\Delta \bar{x}}{\bar{x}} \qquad (27.2.12)$$

or

$$\frac{\Delta C}{C} = 2\frac{\Delta C_v}{C_v} \qquad (27.2.13)$$

where ΔC = error in storage capacity C
$\Delta \bar{x}$ = error in mean annual flow $\bar{x}$
ΔC_v = error in coefficient of variation of annual C_v

Thus, from Eqs. (27.2.12) and (27.2.13), a standard error of 10 percent in the mean and C_v results separately in a standard error of approximately 35 and 20 percent in the respective storage estimates (for D approximately 0.7).

Final Design Procedure. By considering factors in addition to those used in preliminary techniques, final design procedures are intended to provide more accurate estimates of the storage-yield relationship. Such factors should include seasonal inflows, drafts as a function of season and contents of the reservoir, and net evaporation and other losses. Procedures that take into account these aspects are discussed next—behavior (or simulation) analysis, simulation with stochastically generated data, and multireservoir systems.

Behavior (or Simulation) Analysis. In a *behavior* (or *simulation*) *analysis,* as illustrated in Fig. 27.2.7, the changes in storage content of a finite reservoir are calculated by the continuity equation:

$$S_{t+1} = S_t + Q_t - D_t - \Delta \mathrm{SE}_t - L_t \qquad (27.2.14)$$

subject to the constraint $0 \geq S_{t+1} \geq C$,

where S_{t+1}, S_t = reservoir contents at the beginning of the $(t+1)$th and tth period, respectively
Q_t = inflow during the tth period
D_t = draft (required by the demand center, plus an allocation for in-stream use) during the tth period
$\Delta \mathrm{SE}_t$ = net evaporation loss from the reservoir during the tth period
L_t = other losses during the tth period
C = active reservoir capacity

For most studies, a time period of one month is adopted, but shorter time intervals, e.g., daily or weekly, may be used, especially for within-year storage estimates. If the reservoir capacity C is set so that the reservoir empties just once during simulation using all the historical data, the storage size will be the same as that found for the mass curve procedure (assuming similar draft and related conditions).

To estimate probability of failure from Fig. 27.2.7, the number of time periods for which the reservoir is empty is divided by the total number of periods used in the analysis. Alternatively, the number of years in which failure occurs divided by the number of years used in the analysis expressed as a percentage is the percentage chance of failure in any year.

In applying a behavior analysis to estimating storage size or firm yield, two assumptions are made: (1) the reservoir is initially full and (2) the historical data are representative of future inflows.

Behavior analysis has a number of limitations:

1. Assuming the reservoir is initially full can significantly affect the probability of failure if low flows occur at the beginning of the historical record. Of course, this effect can be checked by carrying out simulations for different starting conditions. Klemes (personal communication) has suggested an alternative approach: Use the final reservoir level as the starting level for a second (or concatenated) sequence where the hydrologic data are used in the same sequence twice, end to end. This ensures a water balance over the period of record.

2. Because the analysis is based on historical data, the sequencing of inflows may not be representative of the population of flows.

3. Broken records cannot be satisfactorily handled because of the difficulty of assigning the correct reservoir condition after a break in the flow record. However, under normal circumstances gaps in the record would be filled by using information from a nearby gauging station.

Notwithstanding the above deficiencies in behavior analysis, the method has several important attributes:

1. It is a simple procedure and displays clearly the behavior of the reservoir contents.
2. Because historical data are used, all flow parameters are modeled insofar as they are incorporated in the historical record, and the critical period is identified with known years.
3. A behavior analysis can be based on any time interval (daily, monthly, annual).
4. Seasonal withdrawals and complicated release rules can be taken into account.

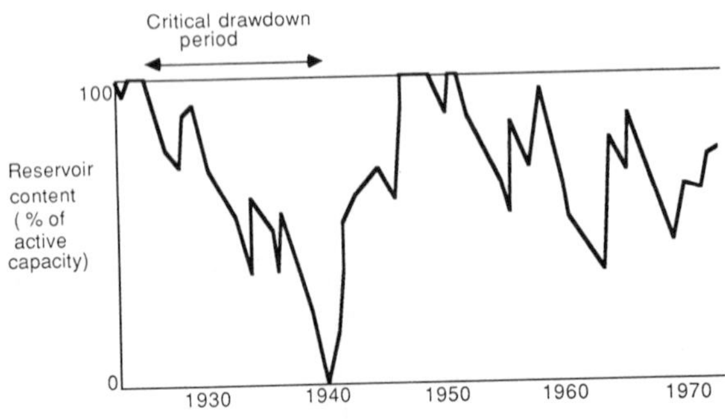

FIGURE 27.2.7 Example of behavior or simulation diagram.

In using the method to estimate yield from a known reservoir and river system, a trial-and-error procedure must also be adopted. Reservoir size is known, but demands on the system are assumed. These are varied for successive simulations using the whole historical record and appropriate starting conditions in Eq. (27.2.14). Firm yield is found by determining, for the given storage, the withdrawal rate that lowers the water content of the reservoir to a specified minimum level just once during the critical historical drought. For monthly simulation, in Eq. (27.2.14), $D_t = Yd_t$, where Y is the annual firm yield and d_t is the fraction of the annual demand normally required in month t.

Adjustment for Reservoir Evaporation. Evaporation from a reservoir needs to be taken into account. It will depend on the surface area of the reservoir, the potential evaporation at the site, and the net increase in evaporation over the evapotranspiration that occurred prior to the reservoir area being inundated. The derivation of the adjustment factor is given in McMahon and Mein,[39] and the factor is represented by the following equation:

$$\Delta E = E_o - \mathrm{ET} \tag{27.2.15}$$

$$\Delta \mathrm{SE}_t = \Delta E_t A_t \tag{27.2.16}$$

where ΔE = net evaporation loss per unit area per unit time from the reservoir water surface

ΔE_t = net evaporation loss during the tth period

E_o = open surface water evaporation from the reservoir after dam construction

ET = evapotranspiration that would have occurred from the reservoir area if it had not been inundated

$\Delta \mathrm{SE}_t$ = storage correction of evaporation loss

A_t = reservoir surface area during the tth period

E_o is usually estimated from pan evaporation data. Details, including appropriate pan coefficients, are given in Chap. 4. ET is difficult to estimate, and so a long-term value is computed as the average annual rainfall less the average annual runoff for the inundated area. Net evaporation should not be computed as the difference between open surface water evaporation less precipitation unless the reservoir inflow is derived from an area equal to the catchment area less the lake surface area. In the United States, a common practice is to adjust the inflow data for the impact of evapotranspiration from the natural reservoir terrain. Thus net evaporation loss is then the difference between lake evaporation and lake precipitation.

Simulation with Stochastically Generated Data. Stochastic data generation is used to understand the sensitivity of a storage-yield estimate to other data sequences that could have been observed besides the historical one.

The approach recommended is shown diagrammatically in Fig. 27.2.8. The first step is to use the historical flow record to estimate the magnitude of the variable under consideration, for example, the reservoir yield denoted by Y_H from a given storage for a specific condition of probability of failure. For each stochastically generated replicate (there are n shown in the figure), an estimate of the variable is made. Next, values of the n estimates are plotted as a frequency diagram on which the historical estimate is shown. Fig. 27.2.8 shows the percentage chance (hatched area) that a larger yield would be obtained in any year.

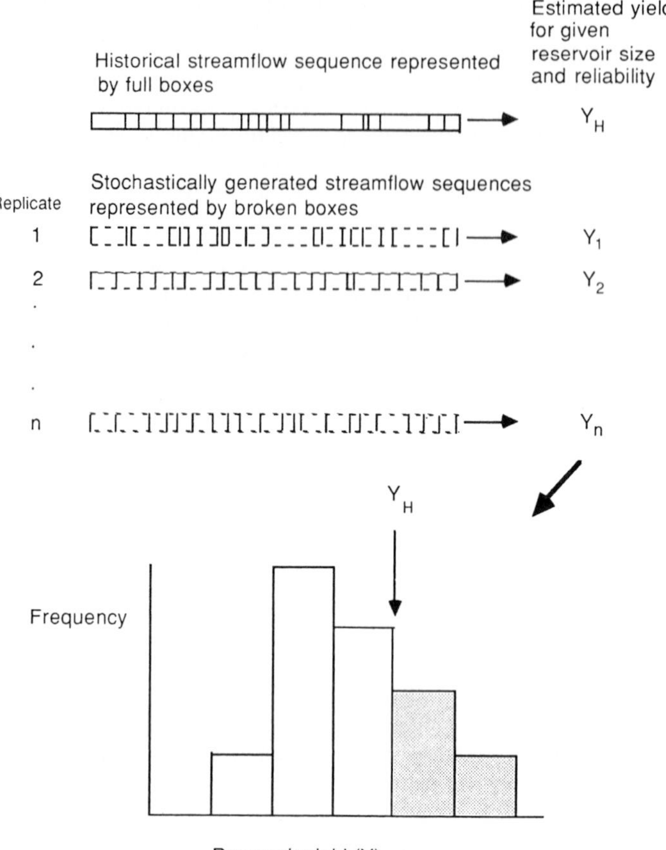

FIGURE 27.2.8 Schematic illustration of the use of stochastically generated data in storage-yield analysis. The shaded region gives the proportion of the stochastic sequences whose yields are greater than that derived from the historical data.

Multireservoir Systems. The previous material deals with the storage capacity–draft–probability relationship for a single reservoir. However, many water-supply systems have more than one reservoir, and the techniques discussed so far provide unsatisfactory estimates of the storage-draft relationship for a multireservoir system.

The traditional technique for solving multireservoir problems, whether for design or for management, is to develop a computer simulation of the system, to which is applied the historical record, normally on a monthly basis.

A variation of this approach that has been used in practice is to carry out n simulations using the N-year historical record concatenated with itself and beginning each separate simulation in a new year of the historical sequence and continuing for N successive years. In this way, n analyses are completed that are based on N years of record and provide n estimates of the variable under consideration. A serious difficulty with this approach is that, because of the overlapping nature of the sequence between each simulation, their outcomes are not independent.

As outlined in Chap. 19, the most appropriate technique to generate multisite data, whether it be stream-flow, rainfall, and/or climatic variables, is to use Matalas' residual approach[40] at an annual level and then disaggregate the annual figures into monthly values. This procedure preserves all the important statistical properties of the historical data—means, standard deviations, lag-1 autocorrelations, and cross-correlations. Relevant practical issues in this approach are discussed in Srikanthan et al.[52]

Computer Models for Reservoir Simulation. A comprehensive summary and review of available computer models for optimization by mathematical programming—linear programming, dynamic programming, nonlinear programming, and simulation—is given by Yeh.[65] Simulation is a modeling approach that represents the behavior of a system by mathematic or algebraic descriptors. Unlike mathematical programming, simulation does not find an optimum decision for system operation, but allows the analyst to explore the consequences of various scenarios of an existing water resources system or a new system.

A number of computer packages are available to perform multireservoir simulation. HEC-3, developed by the United States Army Corps of Engineers Hydrologic Engineering Center, is one such model that operates on a monthly time step.[27] This model can accommodate reservoirs, diversions, and power plants. The use of daily simulation programs, e.g., HEC-5,[28] is becoming more widespread with the need to operate and manage water resource systems more efficiently. The HEC models can be obtained from the Hydrologic Engineering Center, 609 Second St., Davis, CA 95616, U.S.A.

Within-Year Storage. For reservoirs that are likely to spill at least once every year, two approaches are available: *frequency mass curves* and a behavior analysis, as described previously.

The steps in the frequency mass curve procedure are as follows:

1. Low-flow frequency curves are constructed for durations of 1, 7, 15, 30, 60, 120, and 183 consecutive days. An example is given in Fig. 27.2.9.
2. Flows for a given annual probability of nonexceedance or average recurrence interval are read from the low-flow frequency curves and replotted as inflows against duration (Fig. 27.2.10*a*).
3. A constant draft or yield line is superimposed on the diagram.
4. As shown in Fig. 27.2.10, the largest intercept between the draft line and the inflow curve is the reservoir capacity to meet the draft at the design level of reliability (taken as the complement of the annual probability of nonexceedance).

In situations where the storage capacity is known and the reservoir yield is the unknown variable, the procedure is modified as follows:

1. As above.
2. As above.
3. As shown in Fig. 27.2.10*b*, locate a point *A* on the negative side of the vertical axis to represent an origin equal to the storage size.
4. Draw a line from this point tangential to the inflow curve. The slope of this line is the reservoir yield.

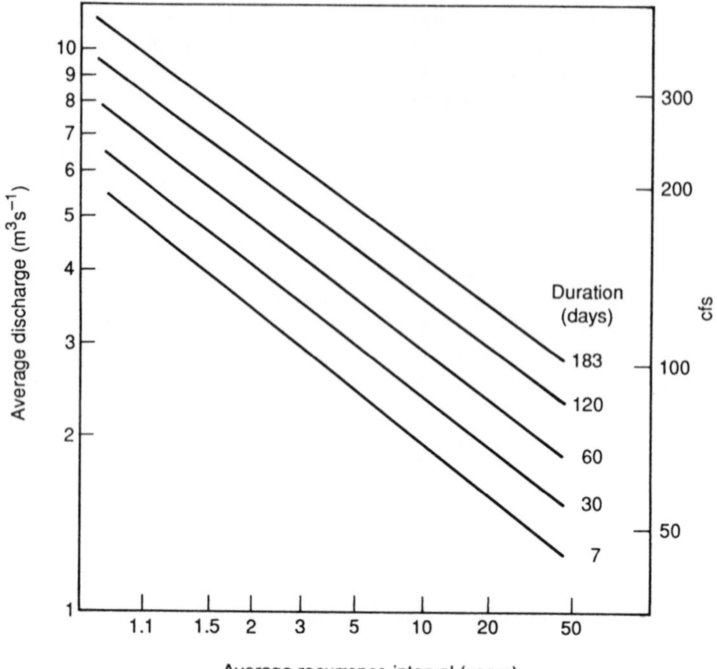

FIGURE 27.2.9 Schematic diagram of low-flow frequency curves.

The method is based on two assumptions: (1) the reservoir is assumed to be full at the beginning of the critical drawdown period and (2) failures that occur after the end of the critical drawdown period, but before the next spill, are neglected.

The method has several limitations:

1. Variable draft rates cannot be easily taken into account.
2. The frequency–mass curve procedure introduces a bias into the storage estimate[39] which requires storage estimates to be increased by 10 percent.[22]
3. Low-flow frequency curves are required. Where these are not available and cannot be produced by a computer, then considerable manual computation is required.
4. Evaporation losses are not taken into account directly, although the withdrawal rate can be modified to reflect the effects of net lake evaporation or the storage size can be increased appropriately as described previously for the Gould gamma method.

Despite these inadequacies, the method is a quick, simple, and reliable procedure to estimate within-year storage need if low-flow frequency curves are available.

Adjustment for Autocorrelation. For some storage-yield procedures, it is assumed that annual flow volumes are independent; i.e., annual autocorrelation is zero. For streams in which there is a positive autocorrelation, a storage adjustment needs to be made. This can be achieved either through increasing the required storage or decreasing the reservoir yield. A selection of results for storage adjustment presented by Srikanthan[53] is included as Fig. 27.2.11, and an adjustment for draft developed by Hardison[22] is presented in Fig. 27.2.12. Figure 27.2.11 is applied by multiplying the storage size for autocorrelation $r = 0$ by the adjustment factor. It has been shown by Phatarfod[47] that the autocorrelation effect on required storage size asymptotically reaches the value $(1 + r)/(1 - r)$ as storage size increases. In Fig. 27.2.12, the adjustment should be subtracted from the allowable draft computed with no autocorrelation between years.

Off-Stream Storages. Figure 27.2.2*b* illustrates the off-stream storage problem. A flow duration curve (developed normally by using daily data) provides a first estimate of the required storage size. Typically, a low-head, high-capacity pump is used to lift water to the off-stream storage. Assuming that a regular wet season occurs each year, an approximate estimate of the available water (volume pumped) on an annual basis can be made (Fig. 27.2.13). Allowance for evaporation loss from the storage and for downstream water requirements may also be included in this analysis.

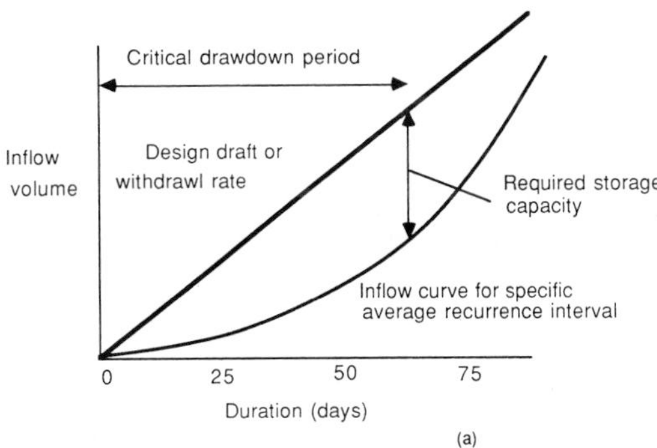

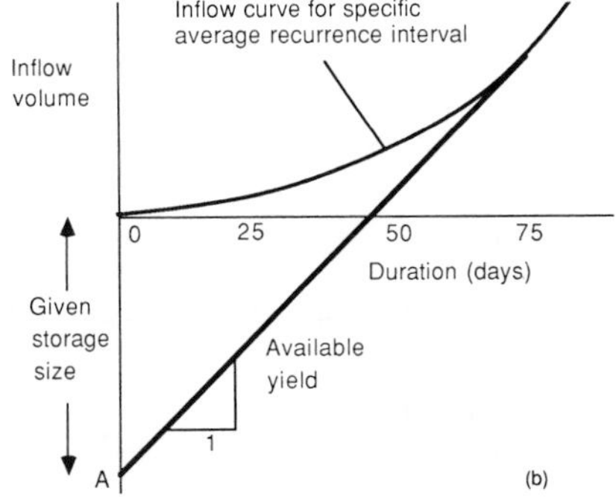

FIGURE 27.2.10 Schematic diagram of mass inflow-duration curve for estimating within-year required storage or available yield.

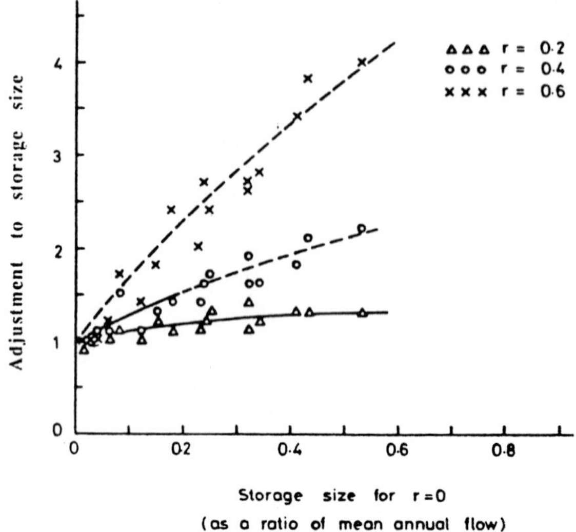

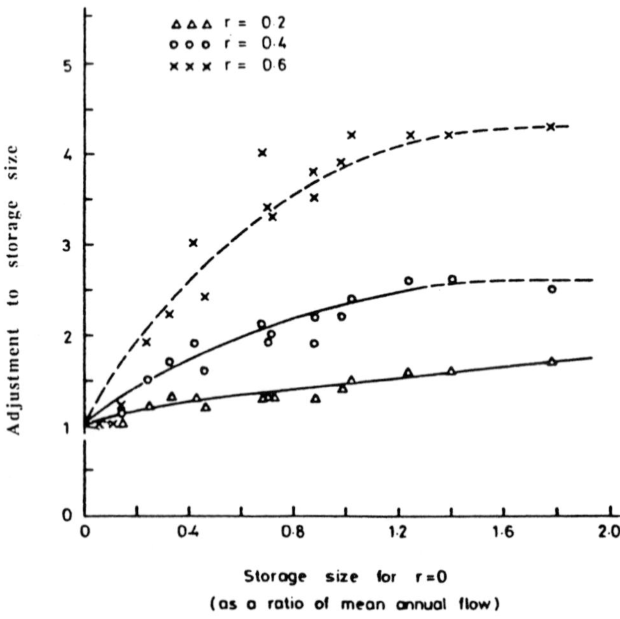

FIGURE 27.2.11 Adjustment of storage size for autocorrelation effects. *(Adapted from Srikanthan, 1985.[53] Used with permission.)*

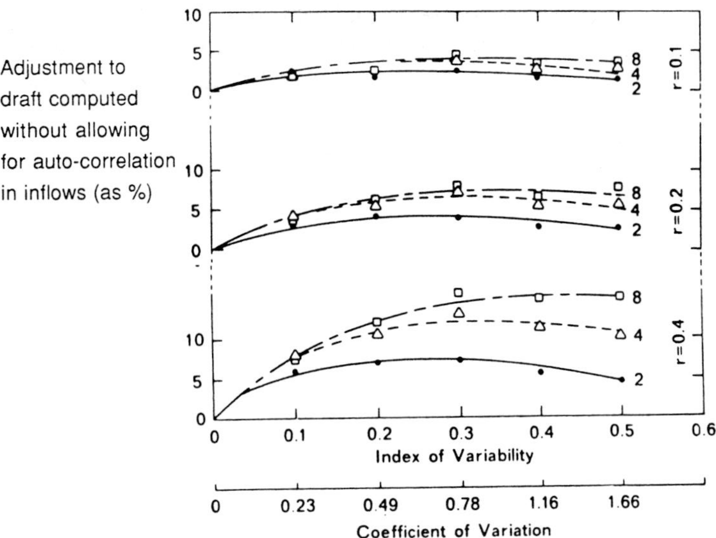

FIGURE 27.2.12 Adjustment of draft for autocorrelation effects. *(Adapted from Hardison.[22] Used with permission.)*

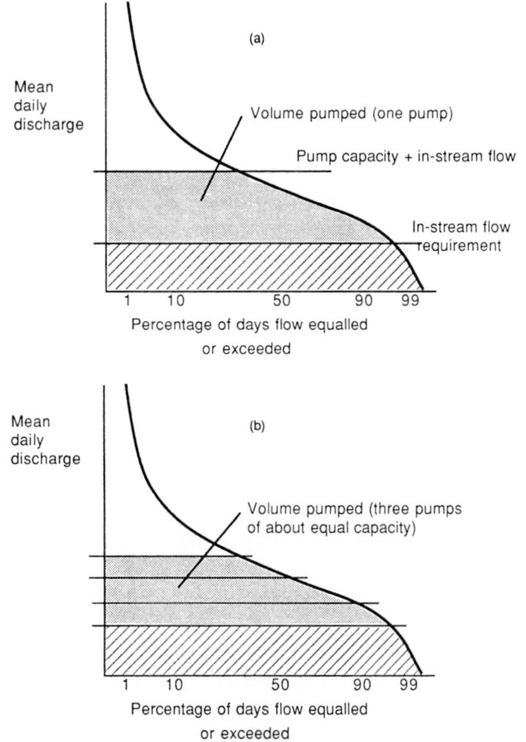

FIGURE 27.2.13 Use of flow duration curve for estimating off-stream storage capacity: (*a*) one pump; (*b*) three pumps.

Where more detailed analysis is warranted, simulation of the system should be carried out. Depending on the scale of operation, a daily, weekly, or monthly time step should be adopted. The approach is similar to the behavior analysis, except that the inflow is limited:

$$S'_{t+1} = S'_t + X_t - D_t - \Delta E'_t - L'_t \tag{27.2.17}$$

subject to the constraints

$$S'_{t+1} \le C$$

$$X_t \le P$$

$$> Q_t$$

$$\not> C - (S'_t - D_t - \Delta E'_t - L'_t)$$

where S'_{t+1}, S'_t = off-stream reservoir contents at the beginning of the $(t + 1)$th and tth period, respectively,
X_t = pumped inflow to the storage during the tth period
D_t = draft during the tth period
Q_t = stream flow during the tth period
ΔE_t = evaporation less rainfall on the reservoir during the tth period
L'_t = seepage losses during the tth period
C = active capacity of the off-stream reservoir
P = maximum discharge capacity of the pumps

The last constraint on X_t related to capacity C ensures that the storage is not overfilled.

27.2.2 Yield of Groundwater Basins

This chapter would be incomplete without some reference to groundwater as a source of supply. Groundwater can be considered as a mineable resource, with management goals ranging from complete depletion, through a continuum to a situation where complete conservation is justified. In addition to the relation of discharge to recharge, consolidation of the aquifer, and penetration of saline water or other forms of pollution must be taken into account. Suffice it to say, however, that in optimum development of a water resources system both surface and subsurface sources need to be studied both in isolation and as a complementary surface/subsurface storage system.

In arid regions, evaporation losses from surface reservoirs make long-term carryover storage less effective than in humid regions. It is therefore best to use groundwater for long-term storage, drawing on it mostly after surface supplies reach a critical stage and allowing replenishment in the meantime.

27.2.3 Forecasting Demand

Urban Demand. Forecasting urban water demand has traditionally consisted of estimating population, multiplying by an average daily per capita use to obtain mean annual demand and then applying peak-to-average ratios to estimate peak hourly and daily demands. Variations in this approach are either to extrapolate the trends in past demand or to use the coefficient method in which the demand per unit is estimated (per household demand, water use per ton of industrial production, etc.) and multiplied by the number of units projected into the future. Such approaches do not contain any allowance for seasonality in climate, variations in economic activity or price elasticity of demand, all of which have been found to be important parameters in forecasting demand. Some guidance on population forecasts and per capita water use estimates is given in Gupta.[21]

In this section, techniques are discussed that overcome some of the limitations of the traditional method and that can be used as tools for management and operation.

Cross-sectional data are defined as data collected from a number of geographic areas or individual households at a single point in time. A second set of data known as *time-series data* involves the collection of information at a point in time for a single region or city. By noting changes in consumption in response to price changes, values of *price elasticity of demand* can be computed. The price elasticity is the percent change in demand for a 1 percent increase in price and is usually about -0.2 for urban water demand. A third category of data, known as *pooled cross-sectional time-series data,* comprises information for a number of geographic areas over time.

Water demand forecasts are used for several purposes. In design, they are necessary in planning new developments or system expansion, for example, the scale, sequencing, and timing of headwater storage capacity extension. At a daily or an hourly time scale, peak values are needed for estimating the size and operation of local balancing reservoirs, pumping stations and pipe capacities. Short-term forecasting data are also required in demand management, for example, in determining an appropriate pricing policy, for providing short-term forecasts of demand for headwater/service basin operation, and for estimating appropriate levels of water-use restrictions.

Regression Models. From data collected during the seminal study on U.S. residential water use,[35] Howe and Linaweaver[25] in 1967 developed a range of regression models covering domestic demands, summer sprinkler demands, and maximum day sprinkler demands. To illustrate the range of models, three equations developed by Howe and Linaweaver follow:

1. Domestic demand in a metered area with public sewer:

$$q_{a,d} = a + bv - cp_w \tag{27.2.18}$$

2. Summer sprinkling demand in a metered area with public sewer in eastern United States:

$$q_{s,s} = dB^e(w_s - fr_s)^g p_s^h v^j \tag{27.2.19}$$

3. Maximum day sprinkling demand in a metered area with public sewer in western United States:

$$q_{\max,s} = kW_{\max}^m v^n \tag{27.2.20}$$

where $q_{a,d}$ = average annual quantity of water demanded for domestic purposes
$q_{s,s}$ = average summer sprinkling demand
$q_{\max,s}$ = maximum day sprinkling demand
v = market value of dwelling unit
p_w = sum of water and sewer charges
B = irrigable area per dwelling unit

w_s = summer potential evapotranspiration
r_s = summer precipitation
p_s = marginal commodity charge applicable to average summer total rates of use
W_{max} = maximum day potential evapotranspiration
$a, \ldots, n$ = coefficients

In estimating the above models, Howe and Linaweaver found R^2 (coefficient of determination) varying between 72 and 93 percent. Such models allow water planners and managers to determine water demand, taking into account economic and pricing factors as well as climatic ones. They also permit estimates of elasticity of demand to be computed, such as parameter h in Eq. (27.2.19).

Agthe and Billings[1] observed that most water demand analysts use static models, like those above, of the form:

$$Q = a_0 + a_1 P + a_2 D + a_3 Y + a_4 W + u \tag{27.2.21}$$

where
Q = monthly water consumption of the average household
P = marginal price facing the average household
D = difference between what the typical consumer actually pays for water and what would be paid if all the water were purchased at the marginal rate
Y = personal income per household
W = evapotranspiration for Bermuda grass minus rainfall
u = random error term
$a_0, \ldots, a_4$ = coefficients

They argued, however, that if current water use is strongly influenced by past water use, a dynamic model should be used. Their dynamic model was of the form:

$$Q_t = C_0 + C_1 P + C_2 D + C_3 Y + C_4 W + C_5 Q_{t-1} + u \tag{27.2.22}$$

where Q_t, Q_{t-1} = water consumption of the average household in months t and $t-1$ and $C_0, \ldots, C_5$ = coefficients.

The above static and dynamic models of Agthe and Billings,[1] which were used to examine short- and long-term price elasticities of demand, produced satisfactory theoretical and statistical results. Coefficients of determination ranged between 80 and 86 percent.

In an extensive study of water use in Adelaide, Australia, Dandy[11] recommended the adoption of two models—a dynamic model for short-term forecasts and a static model for long-term forecasts—as follows:

$$Q'_t = a + bQ_{t-1} + cA + dN + eW \tag{27.2.23}$$

and

$$Q'_t = f + gA + hN + jW \tag{27.2.24}$$

where Q'_t, Q'_{t-1} = annual total water consumption in years t, $t-1$ respectively,
A = annual water allowance below which there is no financial penalty for use
N = number of residents
W = watered area of plot
$a, \ldots, j$ = coefficients

Approximately 68 and 32 percent of the variance was accounted for in the models, respectively.

Disaggregated Econometric Model. This model[16] considers urban water use as an aggregation of a large number of water use categories, taking into account factors that determine the need and intensity of use. The approach is packaged into a computerized forecasting model known as IWR-MAIN. The model is available as a public domain software program supported by a user's manual.[13]

Estimation of water use is given by the following summation:

$$Q_{t,d} = \sum_{s=1}^{k} \sum_{i=1}^{n} Q_{t,d,s,i} \tag{27.2.25}$$

$$Q_{t,d,s,i} = f(P,V,H,W,C,N,E) \tag{27.2.26}$$

where $Q_{t,d}$ = total urban water use for year t, and for use dimension d (e.g., summer seasonal use)

$Q_{t,d,s,i}$ = average daily water use in year t, and for use dimension d in user sector s (e.g., single-family residential) by user category i (e.g., homes with a specific market value range),

P = price of water and sewer service

V = market value of housing units (by residential categories)

H = number of persons per housing unit (by residential categories)

W = weather conditions or climate (by residential categories)

C = conservation programs

N = number of users

E = number of employees (by nonresidential categories)

Figure 27.2.14 outlines the input data requirements, water use models, and disaggregated water-use data. Residential water use is estimated using the Howe-Linaweaver equations that utilize price, housing unit value, homestead size, and weather conditions.

Nonresidential uses in IWR-MAIN are disaggregated into 280 categories and average water requirements in each are determined on the basis of water use per employee per day. Price elasticity of nonresidential water demand is given by

$$q_2 = q_1 \left(\frac{P_2}{P_1}\right)^e \tag{27.2.27}$$

where q_2 = modified employee use coefficient representing use at price level P_2

q_1 = employee use coefficient at price level P_1

e = price elasticity of nonresidential water demand

The impacts of conservation measures are taken into account by a conservation adjustment defined as

$$P_{s,d} = \sum_{m=1}^{L} (R_{m,s,d} C_{m,s,t} I_{m,m+j,d}) \tag{27.2.28}$$

where $P_{s,d}$ = adjustment factor for effect of all conservation measures affecting sector s and use dimension d

$R_{m,s,d}$ = fractional reduction in use of water by sector s in use dimension d expected as a result of implementing measure m

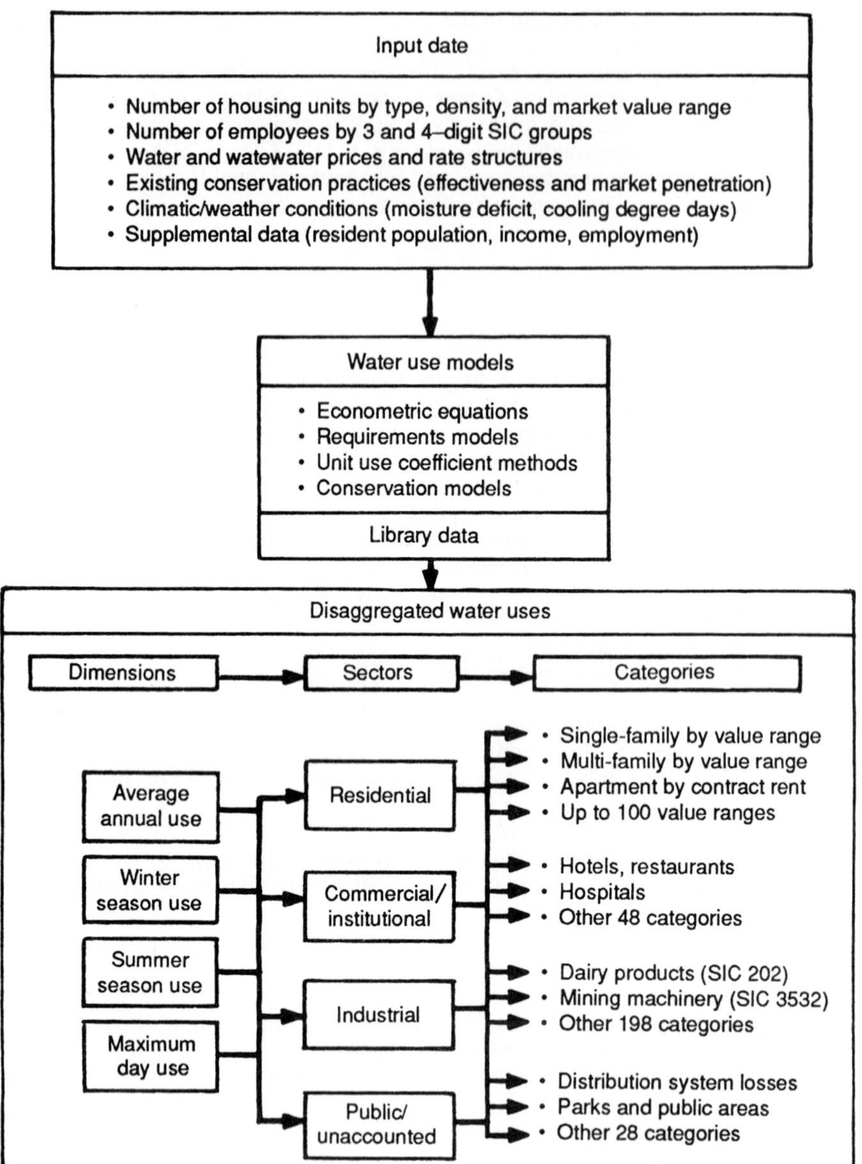

FIGURE 27.2.14 Structure of IWR-MAIN model. *(Reproduced with permission from Dziegielewski and Boland.[16])*

$C_{m,s,t}$ = coverage of measure m in use sector s at time t, expressed as a fraction of sectional water use
$I_{m,m+j,d}$ = interaction factor for combinations of individual pairs of measures m and $m+j$ for dimension d
L = number of measures implemented

The forecast of the restricted water use for each use sector and dimension is

$$Q_{c,s,d} = Q_{u,s,d}(1 - P_{s,d}) \tag{27.2.29}$$

where $Q_{c,s,d}$ = restricted water use (with conservation) for sector s and dimension d and $Q_{u,s,d}$ = unrestricted water use.

Irrigation Demand. For many purposes, it is sufficient to estimate total irrigation demand by adopting a *water duty* in m³/ha (or acre · ft/acre) and using that value as the design or demand forecasting criterion. In other words,

$$\text{Demand} = \text{irrigated area} \times \text{water applied per unit area} \tag{27.2.30}$$

The irrigated area is less than or equal to the irrigable area and is estimated from surveys of existing irrigated areas or by design estimates of new ones. The water applied per unit area is estimated from experience in the area or by the detailed method that follows.

Procedures for estimating irrigation demand are outlined in a number of publications.[2,7]

The steps in estimating irrigation demand are as follows:

1. Calculation of reference crop evapotranspiration
2. Selection of crop coefficient
3. Adjustment for climate, soil water, method of irrigation, and cultural practices
4. Computation of net irrigation requirements, taking into account rainfall, groundwater, and stored soil water
5. Determination of irrigation demand after allowing for leaching requirements and irrigation efficiency

These steps follow those given in Doorenbos and Pruitt,[15] and the material presented below is from that reference unless otherwise indicated.

Reference Crop Evaporation. In Sec. 4.4.5 of Chap. 4, the procedures for calculating reference crop evapotranspiration (ET_o) are set out. Reference crop evapotranspiration is defined as "the rate of evapotranspiration from an extensive surface of 8 to 15 cm (3.1 to 5.9 in) tall, green grass cover of uniform height, actively growing, completely shading the ground and not short of water."[15] According to Doorenbos and Pruitt,[15] the relative errors in estimating ET_o could be as low as ±10 percent and as high as ±25 percent.

Selection of Crop Coefficient. In order to take account of the particular crop in estimating its water requirement, a crop coefficient k_c is introduced:

$$ET_{crop} = k_c ET_o \tag{27.2.31}$$

where ET_{crop} = crop evapotranspiration
ET_o = reference crop evapotranspiration
k_c = crop coefficient

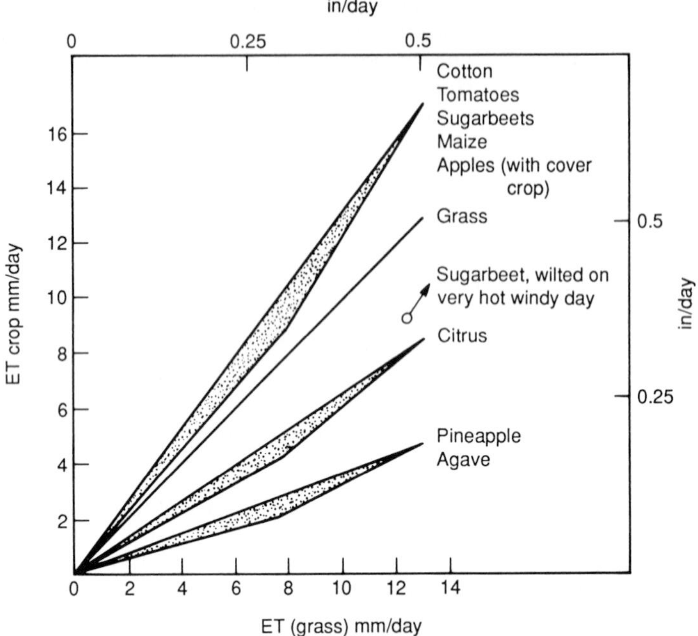

FIGURE 27.2.15 Relationship between ET_{crop} and ET_o. *(Reproduced with permission from Doorenbos and Pruitt.[15])*

The effect of crop characteristics on the relationship between ET_{crop} and ET_o is illustrated in Fig. 27.2.15. In addition to crop characteristics, other factors affecting the value of k_c are crop sowing and planting date, rate of crop development, and climatic conditions. Also during the early stages of growth, the frequency of rain or irrigation is important. The effect of length of growing season is also important. This is illustrated conceptually in Fig. 27.2.16. Consequently, in selecting the appropriate k_c for each period or month in the growing season for a given crop, the rate of crop development must be considered.

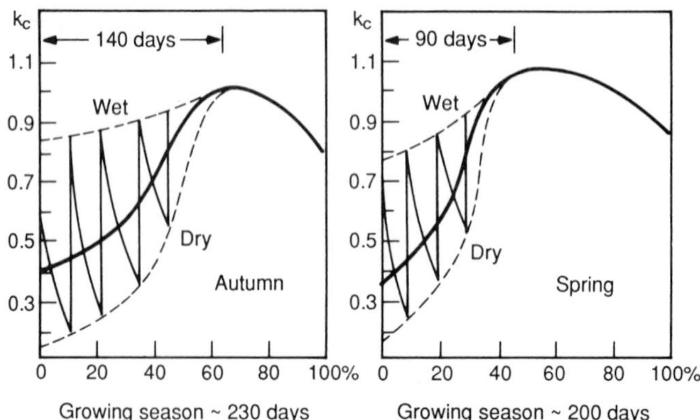

FIGURE 27.2.16 Example of k_c values for sugarcane for different growing seasons. *(Reproduced with permission from Doorenbos and Pruitt.[15])*

ET_{crop} is the sum of crop transpiration and soil evaporation. During full crop cover, soil evaporation is negligible. However, during initial crop development, soil evaporation may be significant and is taken into account through an appropriate k_c.

Values of k_c can be found in a number of references; a comprehensive list is given in Doorenbos and Pruitt.[15]

Factors Affecting ET_{crop}. ET_{crop} calculated from Eq. (27.2.31) refers to evapotranspiration of a disease-free crop, grown in very large fields and not short of water and fertilizers. In reality, a number of factors need to be taken into account.

Climate. Crop water needs vary from year to year, with values differing by up to 25 percent for midcontinental climates. Monthly variations show even greater variations. Such variations are illustrated in Fig. 27.2.17, where mean daily values of ET for rye grass are plotted as a frequency distribution by month.

Advection needs to be taken into account. Often the meteorological data used in the analysis were collected prior to irrigation development and consequently a different microclimate to that prior to irrigation develops. As an air mass moves into an irrigated field, particularly in an arid or semiarid climate, the air mass gives up heat as it passes over the irrigated area, resulting in increased ET at the upward edge and lower evaporation as it moves further into the irrigation area. As seen in Fig. 27.2.18, these effects can be significant and the correction factor given is applied (by multiplication) to the estimated ET value for the condition prior to irrigation development.

Soil Water. Immediately after irrigation, the crop transpires at the predicted rate, but as the soil dries out, this rate decreases. However, in planning and design, the predicted ET_{crop} should be used unless specific objectives are pursued that require a lower value.

Both high water tables and soil salinities can have adverse effects on plant growth

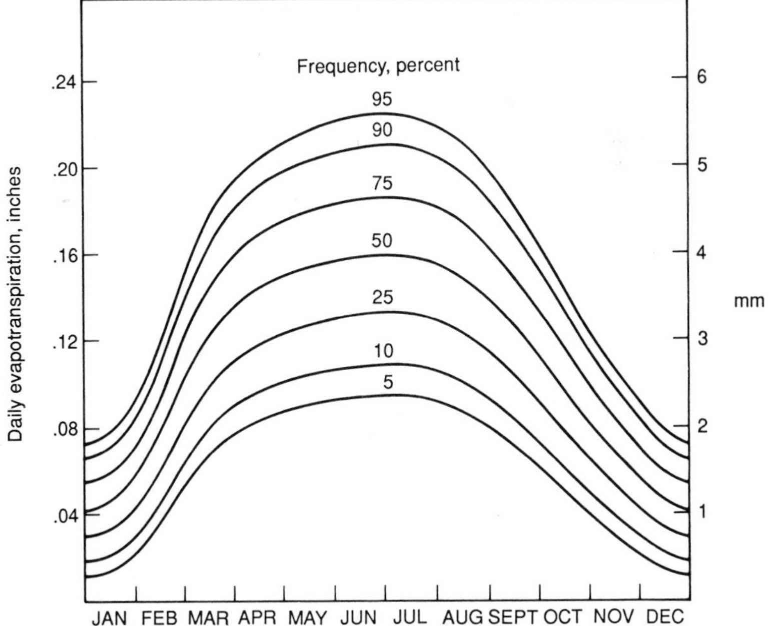

FIGURE 27.2.17 Monthly frequency distribution of mean daily ET for ryegrass in coastal California valley. *(Reproduced with permission from Nixon et al.[45])*

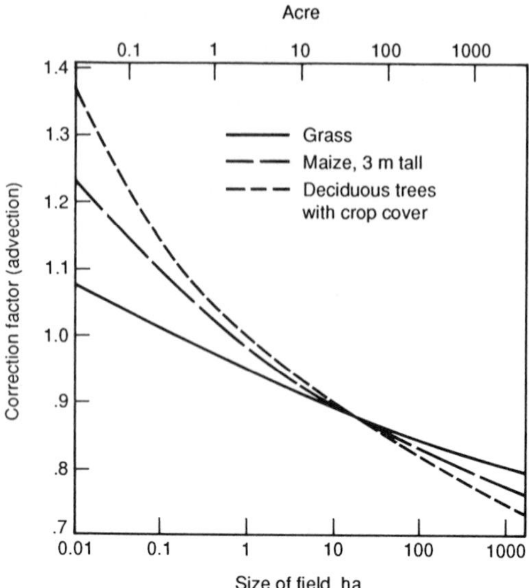

FIGURE 27.2.18 Correction factor for advection effects on evapotranspiration from irrigated fields. *(Reproduced with permission from Doorenbos and Pruitt.[15])*

and hence ET. Specialist advice should be sought on these matters. Many factors affect the relationship between crop yields and ET. The effect of time and duration of water shortages can be very pronounced. Some issues are treated in Doorenbos and Pruitt,[15] while others are dealt with in Slatyer.[50]

Irrigation Methods. The evapotranspiration of a crop is little affected by the method of irrigation, if the system is properly designed and installed. It is affected, however, by the system management. A particular system will be chosen such that the crop water requirements can be best met, at the same time minimizing soil structure deterioration. Readers are referred to Jensen[29] for guidance in the design and operation of farm irrigation systems.

Cultural Practices. Other factors that also need to be considered include the use of fertilizers, plant population, tillage, mulching, and windbreaks. These issues are treated in Doorenbos and Pruitt.[15]

Net Irrigation Requirements. The net irrigation requirements of a crop are based on a field water balance, which includes evapotranspiration ET_{crop}, rainfall P, groundwater contribution G, and stored water at the beginning of each period W:

$$I_n = \underset{\text{losses}}{ET_{crop}} - \underset{\text{gains}}{(P + G + W)} \tag{27.2.32}$$

where I_n = net irrigation requirement of the crop during a given period. For preliminary studies, monthly data are frequently used. The sum of I_n for the different crops over the entire irrigation area forms the basis for determining the required supply.

Instead of using a mean rainfall value in Eq. (27.2.32), a dependable level of rainfall would be used, e.g., the depth of rainfall that can be expected 4 out of 5 years or 9 out of 10 years. However, it should be noted that not all the rainfall is effective, as some may run off or evaporate. Guidelines for adjusting these and related effects are given in Doorenbos and Pruitt[15] and Dastane.[12]

Groundwater contribution G may be important where water tables are high. However, detailed experiments will be required to determine the groundwater contribution under field conditions.

The water stored in the root zone, W, prior to the first irrigation needs to be taken into account. This may be equivalent to one full irrigation. But for design this factor should not be considered.

Irrigation Requirement. In addition to meeting the net irrigation requirement, water is required for leaching accumulated salts from the root zone and to compensate for water losses during conveyance and application.

The *leaching requirement* is the maximum amount of irrigation water supplied that must be drained through the root zone to control soil salinity at a defined specific level. A detailed account of leaching requirements is given by Ayers and Westcot;[3] the following should be used only as a guide. For sandy loam to clay loam soils with good drainage and where rainfall is low, the leaching requirement can be obtained from:[15]

- For surface irrigation methods (including sprinklers):

$$\mathrm{LR} = \frac{\mathrm{EC}_w}{5\ \mathrm{EC}_e - \mathrm{EC}_w} \tag{27.2.33}$$

- For drip and high-frequency sprinklers (near daily):

$$\mathrm{LR} = \frac{\mathrm{EC}_w}{2 \max \mathrm{EC}_e} \tag{27.2.34}$$

where LR = leaching requirement (expressed as a fraction of the irrigation water application)
EC_w = electrical conductivity of the irrigation water,
EC_e = electrical conductivity of the soil saturation extract for a given crop appropriate to the tolerable degree of yield reduction
$\max \mathrm{EC}_e$ = maximum tolerable electrical conductivity of the soil saturation extract for a given crop

Some typical values of the above variables are given in Table 27.2.3. A complete set can be found in Doorenbos and Pruitt[15] or Ayers and Westcot.[3]

To account for losses during conveyance and application, an efficiency factor should be included when estimating project irrigation requirements. These vary widely:

Conveyance efficiency E_c	0.65–0.9
Field channel efficiency E_b	0.7–0.9
Field application efficiency E_a	0.5–0.8

Together these efficiencies define project efficiency E_p as

$$E_p = E_a E_b E_c \tag{27.2.35}$$

Further details are given in Doorenbos and Pruitt[15] and Bos and Nugteren.[4] A typical value of E_p is 0.4.

TABLE 27.2.3 Crop Tolerance Levels for Selected Crops, mmho/cm*

	Yield potential				
	100%		75%		Max.
Crop	EC_e	EC_w	EC_e	EC_w	EC_e
Alfalfa	2.0	1.3	5.4	3.6	16
Apple, pear	1.7	1.0	3.3	2.2	8
Corn	1.7	1.1	3.8	2.5	10
Cotton	7.7	5.1	13.0	8.4	27
Orange	1.7	1.1	3.2	2.2	8
Perennial rye	5.6	3.7	8.9	5.9	19
Tomato	2.5	1.7	5.0	3.4	13
Wheat†	6.0	4.0	9.5	6.4	20

* 1 mmho/cm = 2.54 mmho/in.

† During germination and seedling stage, EC_e should not exceed 4 or 5 mmho/cm. Data may not apply to new semidwarf varieties of wheat.

Source: Reproduced with permission from Doorenbos and Pruitt[15] as modified from Ayers and Westcot.[3]

From all the above material, project irrigation supply requirements can be estimated by:

$$V = C\frac{f}{E_p}\sum_{i=1}^{cr}\left(\frac{AI_n}{1-\text{LR}}\right)_i \tag{27.2.36}$$

where V = project irrigation supply requirements, m³/month or acre · ft/month
E_p = project irrigation efficiency, fraction
A = area under a given crop, ha or acre
I_n = net irrigation requirement of a given crop, mm/month or in/month
LR = leaching requirements (fraction)
i = index for crop
cr = number of crops
f = flexibility factor
C = 10 for SI units or 0.083 for U.S. units

For preliminary planning, monthly data are used.

To determine the capacity of the engineering works, the magnitude of the peak-water-use month would be used. Doorenbos and Pruitt[15] recommend including flexibility factor f to allow for future intensification and diversification of crop production. A typical value is $f = 1.2$.

27.2.4 Design Life

The design life of a component of a water-supply system needs to be understood in the context of the economic life and the expected useful life of the system.

The *useful life* of an engineering component is the period during which the component can successfully perform the task for which it was designed. On the other

TABLE 27.2.4 Estimated Useful Life of Elements of Water Resources Projects, years

Canals and ditches	75	Penstock	50
Dams:		Pipes:	
Earthen, concrete	150	Concrete	20
Loose rock	60	Steel <100 mm	30
Steel	40	>100 mm	40
Flumes:		Pumps	18–25
Concrete	75	Water meters	30
Steel	50	Wells	40–50
Hydraulic turbines	35		

Source: Reproduced with permission from Linsley and Franzini.[36]

hand, the *economic life* is of duration less than or equal to the useful life and is defined as the period during which the equivalent cost of retaining the component is less than or equal to installing a new one.[14] In other words, it is that part of the useful life that results in a minimum equivalent uniform annual cost, where the cost includes depreciation of the usefulness of the component and the increase in operation and maintenance costs.[44] Finally, *design life* is the period adopted in the design for replacement of the item.

Table 27.2.4 lists the estimated average useful lives of a number of elements of a water resources project.

The choice of an appropriate design life or design period is difficult, as there are many factors to be considered and of these some are subjective. However, the most satisfactory method to select the optimum design life is to analyze and compare the costs and benefits for different design periods, especially when there are a number of subjective factors involved.

Groundwater systems pose a special problem. As a general rule, if recharge ≥ discharge, the useful life is unlimited. However, under some design conditions, discharge > recharge, that is, the resource is overdrafted or mined, and if the water table or piezometric surface is then lowered so that polluted water enters the aquifer, the life of the system is severely reduced.

27.3 DESIGN FOR HYDROPOWER

The purpose of this section is to estimate the energy potential of a hydropower site, whether it be a run-of-river type, a single-purpose storage, or part of a system. The discussion is limited to preliminary studies associated with the site hydrology. The material that follows is extracted in the main from a U.S. Army Corps of Engineers manual.[60]

27.3.1 Definitions

Definition of Terms. *Hydroelectric energy* is produced by a hydraulic turbine using the energy of water flowing from a higher to a lower elevation. The energy is usually measured in multiples of watt-hours (for example, kilowatt-hours, kWh). Three classes of energy need to be defined:

- *Average annual energy* is the estimate of the energy that could be generated in a year by the project.
- *Firm energy* (or *primary energy*) is the electrical energy that could be generated during the critical period (Sec. 27.2.1).
- Energy that is in excess of firm energy is defined as *secondary energy*.

Other terms that need to be defined follow. *Electrical power* is defined as

$$P = CQHe \tag{27.3.1}$$

where P = electrical power, kW
Q = discharge, m^3/s or cfs
H = net available head, m or ft
e = overall efficiency (usually 80 to 85 percent)
C = 9.81 for SI units or 0.0847 for U.S. units

To convert power output to *energy E*, Eq. (27.3.1) is integrated over time:

$$E = C \int_{t=0}^{n} Q_t H_t e \, dt \tag{27.3.2}$$

The integration is accomplished using either a flow-duration curve analysis or stream-flow routing.

Static head is the difference between the water surface elevation in the forebay and the water level in the tailwater of the powerhouse (Fig. 27.3.1).

Net head represents the actual head available for power generation, calculated as follows:

$$\mathrm{NH} = \mathrm{FB} - \mathrm{TW} - \text{losses} \tag{27.3.3}$$

where NH = net head
FB = forebay elevation
TW = tailwater elevation
losses = trashrack and penstock head losses as defined in the next subsection

The term *efficiency* usually refers to the combined (or overall) efficiencies of the turbine and generator efficiencies; i.e.,

$$e = e_t e_g \tag{27.3.4}$$

where e = overall efficiency
e_t = turbine efficiency
e_g = generator efficiency

Run-of-river plant has very limited storage capacity and uses water only as it becomes available.

Turbine Characteristics. Although the process of selection of turbine characteristics is not included in this handbook, several characteristics about turbines need to be noted. For preliminary power studies, it is usual to ignore the minimum discharge constraint and the head range limitation. However, for feasibility studies, these

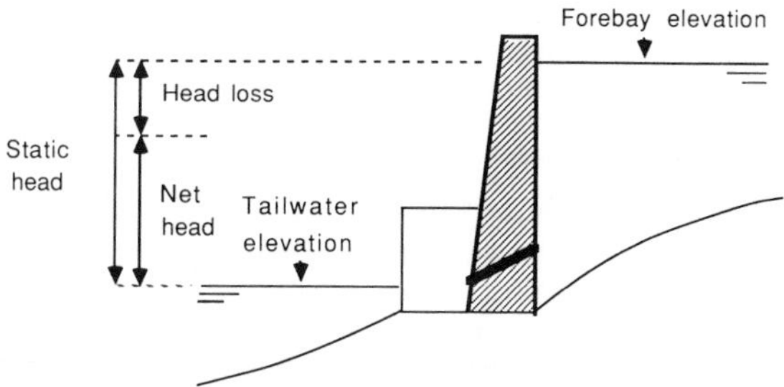

FIGURE 27.3.1 Definition of head and elevation terms.

characteristics should be accounted for. Turbine efficiencies vary widely. However, for reconnaissance-level studies, a fixed value of 80 to 85 percent may be used.[60]

27.3.2 Data Requirements

Stream-Flow Data

1. For sequential stream-flow routing studies (to be discussed in Sec. 27.3.4), only complete years of stream-flow data should be used.
2. The data should reflect upstream regulation and diversion, and take into account other operational procedures like flood control.
3. For low-variability streams (coefficient of variation of annual stream flows < 0.25), 30 years of historical stream flows is generally considered sufficient to assure statistical reliability. Where a shorter record exists, the stream-flow sequence and/or the flow duration curve should be extended by standard procedures.

Stream-Flow Losses. Not all the stream flow passing a dam site will be available for power generation. Consumption losses include net evaporation and diversions. There are potentially many nonconsumptive losses, e.g., navigation, lock requirements, fish passage requirements.[60]

Routing Interval. A daily time interval should be used for the duration curve method (to be discussed in Sec. 27.3.3). Day-to-day variations tend to be masked by weekly or monthly average flows.

For sequential stream-flow routing, the appropriate time interval depends on the type of project. For those projects with seasonal power storage, a weekly or monthly interval is adequate. For run-of-river systems with no upstream storage regulation, daily flow must be used. However, for peaking projects, pumpback projects, and off-stream pumped storages, hourly data may be required.

Reservoir and Tailwater Characteristics. For storage projects, storage volume and reservoir surface area versus reservoir elevation data are needed.

In run-of-river systems where constant reservoir levels are not specified, the forebay elevation versus discharge relationship will have to be developed for hourly modeling. For daily or longer time intervals, an average pool elevation is satisfactory for energy potential assessment.

In the flow-duration procedure, either a constant average reservoir elevation or an elevation versus discharge relationship must be assumed for all projects.

Three types of tailwater data need to be considered:

1. A tailwater rating curve (elevation versus discharge)
2. A weighted average tailwater elevation
3. Elevation of a downstream reservoir if it encroaches on the project being studied

Head Loss. In order to determine the net head available for power generation, head losses must be estimated. These include friction losses in the trashrack, intake structure, and penstock. (The turbine efficiency factor accounts for entrance losses to the turbine and draft tube exit losses.) A typical trashrack head loss is 0.3 m (1 ft).

For preliminary studies and for analysis of projects with short penstocks, it is usual to adopt a fixed penstock head loss based on the average discharge:

$$h_f = Ck_s \frac{V^{1.9}}{D^{1.1}} \tag{27.3.5}$$

where h_f = friction loss in penstock, m per 1000 m, or ft per 1000 ft
D = penstock diameter, m or ft
V = average velocity of flow in penstock, m/s or ft/s
k_s = a friction loss coefficient (for steel penstocks, assume $k_s = 0.34$)
C = 2.58 for SI units or 1 for U.S. units

During preliminary studies and for situations in which penstock costs are not a major cost component, a preliminary penstock diameter can be estimated by using a velocity of 17 percent of the exit velocity:

$$V_R = 0.17(2gH)^{0.5} \tag{27.3.6}$$

where V_R = velocity of flow in the penstock at the rated discharge, m/s or ft/s
g = gravitation constant, 9.81 m/s^2 or 32 ft/s^2
H = gross head, m or ft

Normally, $V_R \leq 7.6$ m/s (or 25 ft/s) and $D \leq 12.2$ m (or 40 ft/s).

Other Aspects

1. The power plant installed capacity establishes an upper limit on the amount of energy that can be generated. For preliminary studies, energy estimates are sometimes made without applying an installed capacity constraint.
2. Operating constraints other than those directly related to power may need to be defined, e.g., storage releases scheduled for downstream uses, optimum pool elevation for reservoir recreation.
3. Channel routing characteristics may also need to be defined, especially for hourly studies.

27.3.3 Flow Duration Method

Two analytical techniques are available to determine the energy potential of a hydropower site, namely the flow duration method and sequential stream-flow routing. Which type is adopted will depend on the kind of hydropower plant being studied.

A flow-duration curve (Sec. 8.6), which is the basis of one method, is converted to a power-duration curve through the application of the water power equation. From this, an estimate can be made of the site's energy potential.

Advantages and Disadvantages. A major advantage of the flow-duration curve method is that it is relatively simple and fast, once the flow duration curve has been computed. However, it is restricted to run-of-river situations where the head varies with the flow.

Steps in Procedure

1. Develop the daily flow-duration curve (Fig. 27.3.2*a*).
2. Estimate net head from Eq. (27.3.3) and head-versus-discharge curve (Fig. 27.3.2*b*).
3. Select 20 to 30 points on the flow-duration curve and compute the power at each flow using Eq. (27.3.1). Figure 27.3.2 graphically illustrates the method. It should be noted that this procedure provides an estimate of the theoretical energy potential of the site. It overestimates the actual power that can be developed at the site because:
   a. Installed capacity constraint is not taken into account.
   b. Turbine efficiencies are assumed constant, although these may vary substantially with both head and discharge.
   c. At stream flows greater than the rated discharge, the full gate discharge decreases with the reduced head.
   d. Operational constraints on pool levels or release rates can significantly reduce power generation.
4. Compute average annual energy from the power-duration curve (Fig. 27.3.2*c*) as follows:

$$\text{Average annual energy (kWh)} = 87.6 \int_0^{100} P\,dp \tag{27.3.7}$$

   where P = power, kW, and p = percentage of time.
5. The final step is to compute dependable capacity, which is defined as the load-carrying ability of a power plant under adverse load and flow conditions. For a run-of-river project, this involves developing a generation-duration curve based on stream flows occurring in the peak demand months. Thus

$$\begin{aligned}\text{Dependable capacity (kWh)} &= \text{average generation}\\ &= \frac{1}{100}\int_0^{100} P\,dp\end{aligned} \tag{27.3.8}$$

27.3.4 Sequential Stream-Flow Routing Method

The sequential stream-flow routing method uses the continuity equation to evaluate storage projects and systems:

$$\Delta S = I - O - L \tag{27.3.9}$$

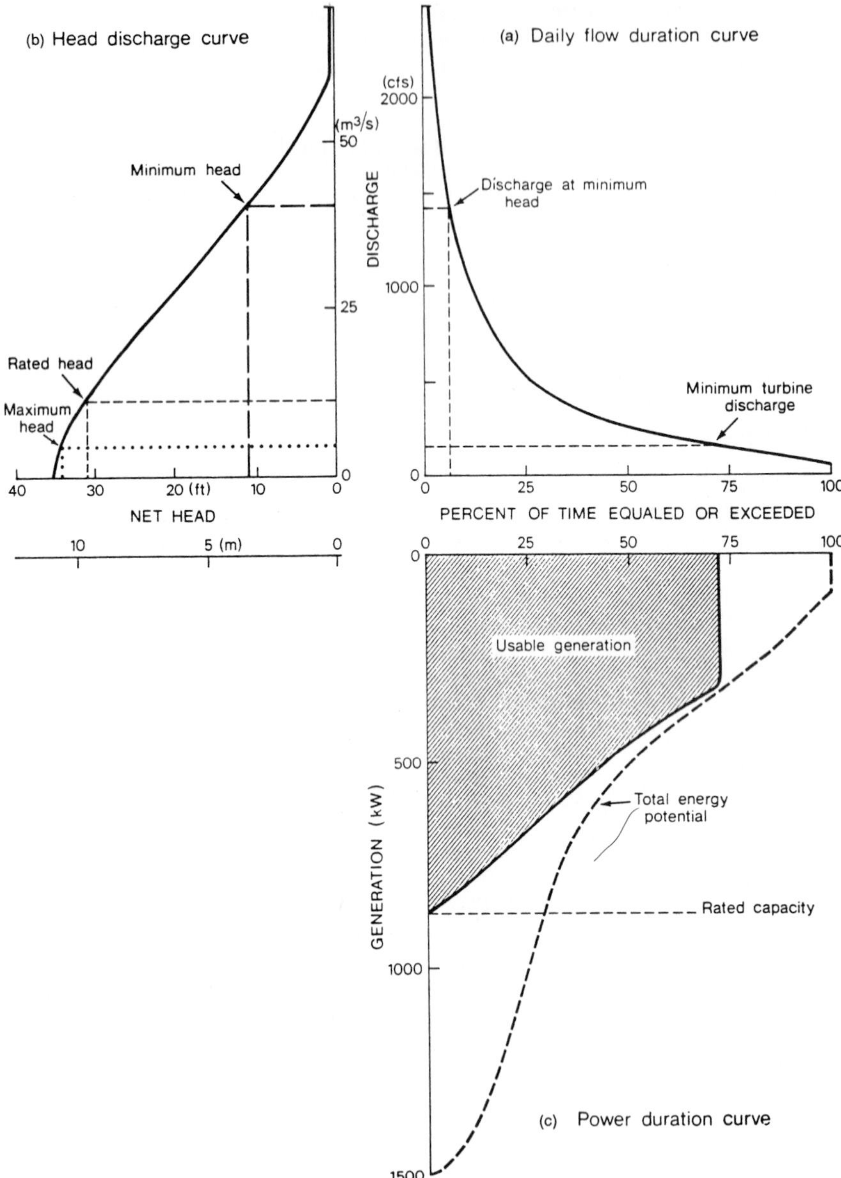

FIGURE 27.3.2 Flow-duration method for estimating potential energy at hydropower site: (*a*) daily flow-duration curve, (*b*) net head discharge curve, (*c*) power-duration curve. *(Adapted from U.S. Army Corps of Engineers.[60])*

where ΔS = change in reservoir storage
I = reservoir inflow
O = reservoir outflow
L = losses (diversions, evaporation)

The routing interval (discussed earlier) depends on the type of system under review.

Energy can be estimated at the hydroplant by applying the reservoir outflow values to the water power equation. Head, discharge, and efficiency may be affected by the operation of the continuity equation through the change in reservoir storage.

Steps in the Procedure

1. Select plant capacity. For preliminary studies, it is usual to base rated capacity for pondage and for storage projects on a head about equal to the average head.
2. Compute stream flow available for power generation. For other than preliminary studies, losses need to be taken into account.
3. Determine average pool elevation.
4. Compute net head.
5. Estimate efficiency. For preliminary analysis, a fixed value is normally adopted.
6. Compute generation. Use the water power equation to compute the average power output (in kW) for each time interval. (If the installed capacity is known or assumed, the power output is reduced to this value.) Multiply the average power output by the number of hours in the time interval to obtain energy.
7. Compute average annual energy. Carry through the above process for each time interval until all the flow data have been analyzed. From this analysis, the annual energy production for each year, the average annual energy, and the average monthly energy output can be estimated.

Application of Sequential Stream-Flow Routing to Storage Projects. Where the primary objective of the analysis is to maximize firm energy output, the following steps would be undertaken:

1. Identify the critical period.
2. Make a preliminary estimate of firm energy potential.
3. Make one or more critical period sequential stream-flow routings to determine the actual firm energy capability and to define operating criteria for the remainder of the period of record.
4. Carry through routing for the period of record to determine average annual energy.
5. Use alternative operating strategies to maximize power benefits.

27.4 DESIGN FOR IN-STREAM HABITATS

27.4.1 General

Section 27.4 is about estimating in-stream flows to enhance, or at least maintain, an aquatic river ecosystem. *In-stream flows* are those which remain in the stream channel after diversions for off-stream uses such as urban and irrigation needs and, in some configurations, hydropower, are appropriately catered for. Compared with the techniques for estimating yield of surface water resources, the methods that follow to assess the hydrologic need of aquatic habitats have not been widely applied and should be used with caution. Up-to-date techniques on the hydrology-habitat link can be found in Gordon et al.[19]

As a general observation, it is only in the last decade or so that streams and rivers have been considered to be more than natural conveyance systems for water supply, or as sinks for disposing, sometimes through dilution, of high-salinity returns from irrigation districts, or of effluents containing high concentrations of nitrogen and phosphorus released from sewage treatment plants. In addition, for centuries, dams, reservoirs, and diversions have had a major influence on the natural patterns of stream flow and stream morphology.[10,62] Flood hydrographs have been attenuated and, in some cases, completely dissipated, low flows have been augmented and, as a result of hydropower, surges of flow pass rapidly down the stream system. Dams act as physical barriers to the movement of fish; desnagging and channelization have reduced native fish cover and spawning areas. Increased sediment has altered stream bed characteristics and caused filling in of deep holes. How then do we measure the effects of river regulation on aquatic ecosystems? How much will the hydrologic change affect stream habitat, and what flows are required and how should they be distributed to maintain or enhance a stream ecosystem?

Recommendations for in-stream flows may include minimum discharges necessary to allow the passage of fish, to ensure acceptable levels of temperature and dissolved oxygen, or to maintain an appropriate habitat, including living space, for the flora and fauna in the stream under consideration. In addition, periodic high flows may be required to remove fine sediment from the stream bed, or flush anoxic or highly saline water from stratified pools. For the Gellibrand River in Victoria, Australia, Tunbridge and Glenane[59] recommended three levels of environmental flow and a flushing flow after assessing the habitat requirements for rearing, passage, resting, and spawning of the particular fish species in the river:

1. An *optimum environmental flow,* to allow the full production of fish, especially for recovery after a period of stress
2. A *minimum environmental flow,* which would result in little or no reduction in numbers of fish, for average rainfall years
3. A *survival environmental flow,* which may cause a reduction in fish numbers, but no loss of species, for low rainfall years
4. A short-term *flushing flow,* to displace the salt wedge in the estuary

Techniques for determining in-stream flow requirements can be assigned to one of three classes: historical discharges or rule-of-thumb methods, threshold methods, and in-stream habitat simulation methods.

27.4.2 Historical Discharges or Rule-of-Thumb Methods

Techniques included under this heading utilize only stream-flow data to define in-stream flow requirements. For example, in regard to water-quality standards, the 7-day–10-year average recurrence interval flow has been used, but is usually considered to be too low for habitat maintenance. However, more sophisticated procedures include the variation of required flow with season. A well-known procedure is the Tennant method.

Tennant Method. The Tennant method[54] (also referred to as the *Montana method*) is based on field observations over a 10-year period in Montana, Nebraska, and Wyoming (U.S.A.) and uses a correlation between stream flow and fish habitat quality in cold and warm waters. Table 27.4.1 summarizes the in-stream recommendations for a range of stream conditions. Tennant[54] found that at flows equivalent to 10 percent of the mean annual discharge, fish were crowded into the deeper pools and riffles were too shallow for larger fish to pass. At 30 percent discharge, satisfactory widths, depths, and velocities were maintained.

A fundamental assumption in applying the technique to other streams requires that they be morphologically similar to those for which the technique was developed. However, such information was not provided by Tennant.[54] The method should be restricted to streams with low variability. Wesche and Rechard[63] suggested that the recommended flows be compared with the average 10- and 30-day natural low flows to check whether the required flows are available naturally during low flow periods.

27.4.3 Threshold Methods

Threshold methods take into account the habitat requirements of fish species and the availability of habitat at various discharge levels. These methods specify a flow below which the habitat is not considered to be adequate for the in-stream flow needs. However, the need for field data makes the techniques more time-consuming and costly than the rule-of-thumb methods.

Idaho Method. This technique, presented by Cochnauer,[8] was developed for large unwadable rivers and is based on the prediction of habitat loss for key fish species at reduced discharges. Critical areas for spawning, rearing, and passage of fish are identified in the field and measurements, at the lowest practical flow, are made of cross-sectional profiles, water surface elevation, velocity and substrate type. Up to 100 transects in each study reach are taken, including one at the control section, which is the downstream section controlling the water surface profile in the reach. A

TABLE 27.4.1 Tennant Method: In-Stream Flow Recommendations for Fish, Wildlife, Recreation, and Related Environmental Resources

	Recommended base flow	
Stream condition	October–March (dry)	April–September (wet)
Flushing or maximum	200% mean annual discharge	
Optimum range	60–100% mean annual discharge	
Outstanding	40%	60%
Excellent	30%	50%
Good	20%	40%
Fair or degrading	10%	30%
Poor or minimum	10%	10%
Severe degradation	10% to zero discharge	

Source: Reproduced with permission from Tennant.[54] Numbers given are applicable in northwestern United States.

TABLE 27.4.2 Habitat Requirements for Rearing Fish of Southwestern Victoria, Australia

Species	Depth,* m	Velocity, m/s	Substrate type
Blackfish	>0.20	0–0.30	All types
Brown trout (adult) (juveniles are common in shallow, fast water with boulders and rubble cover)	>0.20	0–0.50	All types
Redfin and common carp (juveniles also found in faster water in gravel/rubble sites)	>1.0	0–0.20	Mud/sand
Short-finned eel	>0.20	0–0.30	All types

* 1 m = 3.28 ft.
Source: Reproduced with permission from Tunbridge.[58]

water surface profile computer program (Sec. 21.4.5) is used to generate hydraulic characteristics for the reach over a range of flows, although data collection is made at a single discharge. Curves are then constructed (for example, water depth versus discharge) which are compared with known biological criteria like the information given in Table 27.4.2.

Washington Department of Fisheries Method. In contrast to the Idaho method, the Washington method outlined by Collings et al.[9] requires measurements of stream depth and velocity at a number of transects, taken at several different discharges, to determine the amount of spawning and rearing habitats for salmon. If species criteria are not available, then a wetted perimeter method is used, which is based on the premise that the inflection point on a graph of wetted perimeter versus discharge represents the quantity of water preferred by salmon.

An advantage of the Washington method over the Idaho approach is that it produces a graph showing the change in habitat with measured discharge. But the method is considerably more labor-intensive than the Idaho method.

Summary. Compared with rule-of-thumb methods, threshold methods are able to take into account habitat requirements of fish species at various levels of discharge. But the techniques are more time-consuming and costly and, although developed for salmon, could be modified to cater for other fish species. However, it has been argued[31] that for almost the same amount of effort as for the threshold models, it may be preferable to use the more sophisticated stream habitat modeling methods.

27.4.4 In-Stream Habitat Simulation Methods

These methods combine the hydraulic characteristics of a reach defined by the changes in such variables as velocity, depth, wetted perimeter, and substrate, with data on habitat preferences of a given species to estimate the amount of habitat available over a range of discharges. From this analysis, the optimum discharges for a number of individual species can be determined and the results used as a guide for recommending in-stream flows.

The In-Stream Flow Incremental Method (IFIM). IFIM is a conceptual framework for the assessment of riverine habitat and may be thought of as a collection of computer models and analytical procedures designed to predict changes in fish or invertebrate habitats due to flow changes. The methodology was developed by the Cooperative Instream Flow Group of the United States Fish and Wildlife Service[5] and is discussed in detail by Nestler et al.[43]

The decision variable generated by IFIM is total habitat area for fish or invertebrates before and after the completion of a project affecting stream flows. The steps in the method are as follows:

1. *Outline the problem* to be addressed and define rigorously the objectives of the study.
2. Determine the *geographical data* of the study area and the actual length of the river reach to be considered.
3. Determine the *environmental variables* that must be analyzed and those that can be safely ignored. This will require an understanding of the potential disturbance caused by the project, for example, water yield, sediment and chemical loadings to the river, flow regime, and channel morphology.
4. Select an appropriate *evaluation species.* All interpretations regarding the significance of an environmental change are based on the consequences of the change to the evaluation species.
5. Compute the *total habitat* area as a function of discharge. The habitat area is the area of stream having suitable conditions for the species at a particular life stage (fry, juvenile, or adult) or for a particular activity (spawning, resting, feeding). This habitat-discharge function may be utilized in a number of ways, including optimization, habitat time series and duration curves:
   a. *Optimization* involves finding a flow for each month that minimizes habitat reductions for all life stages occupying the stream during the month.
   b. A *habitat time series* is obtained by combining the habitat discharge function with a time series of discharge, which may be either historical (reflecting conditions before a project) or simulated (to reflect flow conditions after the completion of a project).
   c. A *habitat duration curve* summarizes the habitat time series in terms of the percent of time a certain amount of habitat is equaled or exceeded with or without the project.

PHABSIM. A major component of IFIM is the Physical Habitat Simulation system (PHABSIM).[6] This is a large suite of computer programs used to relate changes in in-stream variables (e.g., depth, velocity, substrate, and cover) to changes in physical habitat availability (for a specific species stratified by life stages).

PHABSIM is a computer package by which total habitat area is obtained as a function of discharge. It uses predefined ranges of preferred and tolerable physical habitat conditions for a specific species at a specific life stage, called *habitat preference criteria.* The method of PHABSIM is summarized by Mosley and Jowett[41] and Gan and McMahon,[18] as follows. Observations of fish distribution are employed to determine the probability of use of a location, having a given depth, velocity, cover, and substrate (or other variables for a particular fish life stage). For each controlling variable, a *habitat preference curve,* as in Fig. 27.4.1, is defined, which provides an index of the suitability of use of the habitat over a possible range of values of the variable. The habitat preference index varies between 0 and 1. The peaks of the curve represent the optimal range of values and are given a weighting factor of 1. The tails of the curve represent zero suitability.

The net suitability of use for a specific life stage of a species at a given location is determined by the values of the habitat preference indices for the pertinent variables that control habitat quality. Thus, for example, a 10 m^2 (108 ft^2) area of stream bed

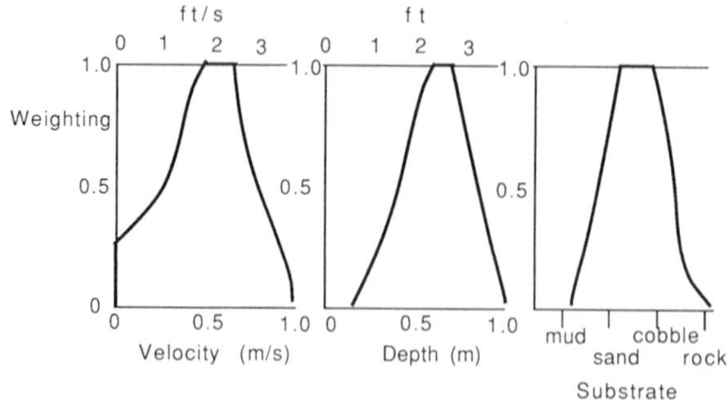

FIGURE 27.4.1 Hypothetical velocity, depth, and substrate habitat preference curves for an aquatic species. *(Adapted from Gore and Nestler.[20] Reprinted by permission of John Wiley & Sons, Ltd.)*

with preference index values of 0.90, 0.85, and 1.0 for depth, velocity, and substrate, respectively, would have a net suitability of use of $0.90 \times 0.85 \times 1.0 = 0.76$. Thus, 7.6 m^2 (82 ft^2) of the 10 m^2 (108 ft^2) stream bed may be regarded as being suitable for use and is called the *weighted usable area* (WUA) for the particular section of the stream. By measuring depth, velocity, and substrate throughout a stream reach and obtaining habitat preference index values at each measurement point from the appropriate curve, the WUA for the entire stream reach may be determined by examining the WUA for each stream cell, where each cell has a characteristic depth, velocity, and substrate. The computation is performed for a particular stream discharge. When the flow is changed, the physical habitat conditions for the stream reach must be redefined and the computation repeated for the new conditions. This leads to the WUA for a stream reach as a function of discharge.

Apart from the preference curves shown in Fig. 27.4.1, other types of criteria may be used. The binary criterion is simplest, taking a value of either 0 or 1. Preference criteria may be derived in three ways—by using professional judgment, from the literature, or by field studies. The last method is the most reliable; however, it is extremely labor-intensive. For field determination of preference curves, a stream is selected which encompasses the entire range of conditions which the species might occupy at various life stages. This normally requires some *a priori* knowledge about the habitat preferences of the species being investigated.

Time-Series Simulation (TSLIB). The habitat-discharge functions of PHABSIM may be combined with stream-flow data to produce monthly or daily habitat time series and habitat duration curves using the program TSLIB, a module of PHABSIM. The time series is useful for comparing pre- and postproject habitat availability.

RHYHABSIM. The River Hydraulics and Habitat Simulation Program (RHYHABSIM), 1988 version, was developed by Jowett[30] in New Zealand. It is a computer model used for modeling the changes in in-stream habitat with changes in flow and is based on IFIM methodology. It may be regarded as an alternative to PHABSIM for computer implementation of this methodology.

Comparison between PHABSIM and RHYHABSIM. A comparison between the two models has been published.[18] The key points are

1. RHYHABSIM consists of a single program, including both habitat and hydraulic simulation, which in executable form is only 254 KB in size. On the other hand, PHABSIM is a collection of 240 separate programs, consisting of executable and batch files about 9000 KB in size. In PHABSIM, there is a choice of four programs for hydraulic simulation of depths and velocities and it can be cojoined to one external program, HEC2. In addition, there are five habitat simulation programs, which have a range of options.

2. Both models may be run on IBM-compatible microcomputers, but hardware requirements for RHYHABSIM are simpler than those for PHABSIM.

3. User manuals are available for both models, but because of the many possible alternatives within PHABSIM, between and within programs, considerably theoretical background knowledge is required for choices to be made, as the manual offers very little guidance.

4. The basic data requirements are similar for both models: cross-sectional bank and bed profiles, water surface levels and point velocities for all cross-sections at one discharge, substrate data at each cross section, and species preference curves. PHABSIM can also use stage-discharge data.

5. RHYHABSIM is user-friendly and easy to run with an internal HELP facility. A major difficulty experienced by the users of PHABSIM is the choice of program, as each must be selected and run separately.

6. Despite the size of PHABSIM compared with RHYHABSIM, the range of application of both models is about the same, although PHABSIM does provide a choice of five habitat and five hydraulic simulation programs, where RHYHABSIM has only one habitat and one hydraulic simulation program.

Limitations. There have been a number of criticisms of both the hydraulics and habitat simulation program of PHABSIM (and hence RHYHABSIM). Some are listed below.

1. The standard step backwater methods used in these programs require that the flow is steady, gradually varied, and subcritical. Under some conditions, e.g., in steep-gradient streams, during hydro peaking operations, or at low flows when bedrocks are exposed, the above criteria are not met.
2. An underlying assumption is that, if the habitat is maintained, the fish population will also be maintained—in other words, the WUA index is a proxy for species biomass or abundance. Some field studies have established such a relationship, while others have found WUA and fish biomass to be uncorrelated. Better relationships between biomass and habitat measures have been found in cold water streams, which have simpler ecosystems than in streams with warmer water.
3. Factors other than habitat, e.g., food supply, competition and predation, nutrients, dissolved oxygen, temperature, and flow regimes, may be of greater importance than the physical habitat in limiting species biomass. Thus it has been suggested that the prediction of available habitat, rather than biomass, is the appropriate level of use of PHABSIM as a management tool.
4. A critical limitation in the use of either model is the lack of well-defined or standardized habitat suitability curves.

Further information on the models described above can be obtained from:

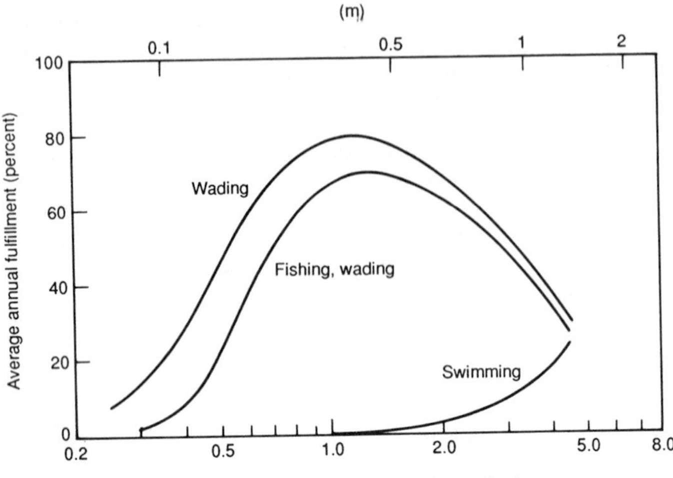

FIGURE 27.4.2 Average annual fulfillment for water contact uses as function of depth at median recreational season discharge. *(Reproduced with permission from Smith and Carswell.[51])*

- PHABSIM: Aquatic Systems Research
  National Ecology Research Center
  U.S. Fish and Wildlife Service
  2627 Redwing Road
  Fort Collins, CO 80526-2899
  U.S.A.
- RHYHABSIM: Freshwater Fisheries Centre
  P.O. Box 8324
  Riccarton
  New Zealand

Annual Fulfillment of In-Stream Uses. Smith and Carswell[51] have developed a methodology based on regionalized parameters of the hydraulic geometry of streams to quantify the relative fulfillment of selected in-stream uses. Figure 27.4.2 shows the average annual fulfillment for water contact uses as a function of stream depth at the median recreational season discharge. The method provides an objective evaluation of relative in-stream flows to be achieved on different gauged streams.

REFERENCES

1. Agthe, D. E., and R. B. Billings, "Dynamic Models of Residential Water Demand," *Water Resour. Res.,* vol. 16, no. 3, pp. 476–480, 1980.
2. American Society of Civil Engineers, *Evapotranspiration and Irrigation Water Requirements,* M. E. Jensen, ed., manual no. 70, American Society of Civil Engineers, New York, 1990.
3. Ayers, R. S., and D. W. Westcot, *Water Quality for Agriculture,* FAO irrigation and drainage paper no. 29, Food and Agriculture Organization of the United Nations, 1976.

4. Bos, M. G. and J. Nugteren, *On Irrigation Efficiencies,* International Institute for Land Reclamation and Improvement Publication no. 19, 1974.
5. Bovee, K. D., *A Guide to Stream Habitat Analysis Using the Instream Flow Incremental Methodology,* United States Fish and Wildlife Service, Cooperative Instream Flow Group, instream flow information paper no. 12, 1982.
6. Bovee, K. D., *Development and Evaluation of Habitat Suitability Criteria for Use in Instream Flow Incremental Methodology,* United States Fish and Wildlife Service Cooperative Instream Flow Group, instream flow information paper no. 21, 1986.
7. Burman, R. D., R. H. Cuenca, and A. Weiss, "Techniques for Estimating Irrigation Water Requirements," *Advances in Irrigation,* D. Hillel, ed., vol. 2, Academic Press, New York, pp. 335–394, 1983.
8. Cochnauer, T., "Instream Flow Techniques for Large Rivers," in *Instream Flow Needs,* J. F. Osborne and C. H. Allman, eds., vol. 2, pp. 387–400, Amer. Fish. Soc., Bethesda, Md., 1976.
9. Collings, M. R., R. W. Smith, and G. T. Higgins, *The Hydrology of Four Streams in Western Washington as Related to Several Salmon Species,* United States Geological Survey open-file, Tacoma, Wash., 1970.
10. Craig, J. F., and J. B. Kemper, eds., *Regulated Streams: Advances in Ecology,* Plenum Press, New York, 1987.
11. Dandy, G. C., *A Study of the Factors Which Affect Residential Water Consumption in Adelaide.* The University of Adelaide, Department of Civil Engineering, 1987.
12. Dastane, N. G., *Effective Rainfall,* FAO irrigation and drainage paper no. 25, Food and Agriculture Organization of the United Nations, Rome, 1975.
13. Davis, W. U., et al., *IWR-MAIN Water Use Forecasting System, Version 5.1: User's Manual and System Description,* IWR-Report 88-R-6, U.S. Army Corps of Engineers Institute of Water Resources, Fort Belvoir, Va., 1981.
14. De Garmo, E. P., W. G. Sullivan, and J. R. Canada, *Engineering Economy,* Macmillan, New York, 1984.
15. Doorenbos, J., and W. O. Pruitt, *Guidelines for Predicting Crop Water Requirements,* FAO irrigation and drainage paper no. 24, revised 1977, Food and Agriculture Organization of the United Nations, Rome, 1984.
16. Dziegielewski, B., and J. J. Boland, "Forecasting Urban Water Use: The IWR-MAIN Model," *Water Resour. Bul.,* vol. 25, no. 1, pp. 101–109, 1989.
17. Fiering, M. B., *Streamflow Synthesis,* Macmillan, London, 1967.
18. Gan, K. C., and T. A. McMahon, *Comparison of Two Computer Models for Assessing Environmental Flow Requirements,* report for Office of Water Resources, Department of Conservation and Environment, Victoria, Australia.
19. Gordon, N. D., T. A. McMahon, and B. L. Finlayson, *Stream Hydrology: An Introduction for Ecologists,* Wiley, Chichester, U.K., 1992.
20. Gore, J. A., and J. M. Nestler, "Instream Flow Studies in Perspective," *Regulated Rivers,* vol. 2, pp. 93–101, 1988.
21. Gupta, R. S., *Hydrology and Hydraulic Systems,* Prentice-Hall, Englewood Cliffs, N.J., 1989.
22. Hardison, C. H., "Storage to Augment Low Flows," *Proc. Reservoir Yield Symp.,* Water Research Assoc. (UK), 1965.
23. Hardison, C. H., "Potential United States Water-Supply Development," *J. Irrig. Drain. Div., Am. Soc. Civ. Eng.* (now *J. Irrig. Drain Eng.*), vol. 98, no. IR3, pp. 479–492, 1972.
24. Hashimoto, T., J. R. Stedinger, and D. P. Loucks, "Reliability, Resiliency, and Vulnerability Criteria for Water Resources System Performance Evaluation," *Water Resour. Res.,* vol. 18, no. 1, pp. 14–20, 1982.
25. Howe, C. W., and F. P. Linaweaver, "The Impact of Price on Residential Water Demand

and Its Relation to System Design and Price Structure," *Water Resour. Res.*, vol. 3, no. 1, pp. 13–32, 1967.

26. Hurst, H. E., "Long-Term Storage Capacities for Reservoirs," *Trans. Amer. Soc. Civil Eng.*, vol. 116, p. 776, 1951.
27. Hydrologic Engineering Center, *HEC-3 Reservoir System Analysis for Conservation,* U.S. Army Corps of Engineers, 1976.
28. Hydrologic Engineering Center, *HEC-5 Simulation of Flood Control and Conservation Systems,* U.S. Army Corps of Engineers, 1982.
29. Jensen, M. E., "Design and Operations of Farm Irrigation Systems," monograph no. 3, American Society of Agricultural Engineers, St. Joseph, Mich., 1980.
30. Jowett, I. G., *River Hydraulics and Habitat Simulation, RHYHABSIM,* Computer Manual, Freshwater Fisheries Centre, Riccarton, New Zealand, 1988.
31. Kinhill Engineers Pty. Ltd., *Techniques for Determining Environmental Water Requirements—A Review,* Department of Water Resources, Victoria, Australia, technical report series no. 40, 1988.
32. Langford, K. J., and B. E. Foley, "Transferable Water Entitlements—Victorian Perspective," Int. Seminar and Workshop on Water Allocation and Transfer Systems, University of New England, Australia, 1990.
33. Leopold, L. B., and T. Maddock, *The Flood Control Controversy,* Ronald Press, New York, 1954.
34. Lettenmaier, D. P., and S. J. Burges, "Cyclic Storage: A Preliminary Assessment," *Ground Water,* vol. 20, no. 3, pp. 278–288, 1982.
35. Linaweaver, R. P., J. C. Beebe, and F. A. Skrivan, *Data Report of the Residential Water Use Research Project,* Johns Hopkins University, Department of Environmental Engineering Science, Baltimore, 1966.
36. Linsley, R. K., and J. B. Franzini, *Water Resources Engineering* (2d ed.), McGraw-Hill Kogakusha, Tokyo, 1972.
37. Loucks, D. P., J. R. Stedinger, and D. A. Haith, *Water Resource Systems Planning and Analysis,* Prentice-Hall, Englewood Cliffs, N.J., 1981.
38. McDonald, N. S., "Decision Making Using a Drought Severity Index," *Proc. United Nations University Workshop, Need for Climate and Hydrologic Data in Agriculture in Southeast Asia,* CSIRO Division of Water Research, technical memo 89/5, 1989.
39. McMahon, T. A., and R. G. Mein, *River and Reservoir Yield,* Water Resources Publications, Littleton, Colo., 1986.
40. Matalas, N. C., "Mathematical Assessment of Synthetic Hydrology," *Water Resour. Res.*, vol. 3, no. 4, pp. 937–945, 1967.
41. Mosley, M. P., and I. G. Jowett, "Fish Habitat Analysis Using River Flow Simulation," *New Zealand J. Mar. and Freshwater Res.*, vol. 19, pp. 293–309, 1985.
42. Moy, W.-S., J. L. Cohon, and C. S. ReVelle, "A Programming Model for Analysis of the Reliability, Resilience, and Vulnerability of a Water Supply Reservoir," *Water Resour. Res.*, vol. 22, no. 4, pp. 489–498, 1986.
43. Nestler, J. M., R. T. Milhous, and J. B. Layzer, "Instream Habitat Modeling Techniques," in *Alternatives in Regulative River Management,* J. A. Gore and G. E. Petts, eds., CRC Press, Baca Raton, Fla., pp. 295–315, 1989.
44. Newnan, D. G., *Engineering Economic Analysis,* Engineering Press, San Jose, Calif., 1980.
45. Nixon, P. R., G. P. Lawless, and G. V. Richardson, "Coastal California Evapotranspiration Frequencies," *J. Irrig. Drainage Div., Am. Soc. Civ. Eng.* (now *J. Irrig. Drain. Eng.*), vol. 98, pp. 185–191, 1972.
46. Palmer, W. C., "Meteorological Drought," U.S. Department of Commerce, Weather Bureau research paper no. 45, 1965.

47. Phatarfod, R. M., *The Effect of Annual Serial Correlation on Reservoir Size,* statistical research report no. 95, Department of Mathematics, Monash University, Australia, 1984.
48. Rice, L., and M. D. White, *Engineering Aspects of Water Law,* Wiley, New York, 1987.
49. Rippl, W., "Capacity of Storage Reservoirs for Water Supply," *Minutes Proc. Inst. Civ. Eng.,* vol. 71, pp. 270–278, 1883.
50. Slatyer, R. O., *Plant-Water Relationships,* Academic Press, New York, 1967.
51. Smith, R. L., and W. J. Carswell, "Average Annual Fulfillment of Instream Uses," *J. Water Resour. Plann. Manage.,* vol. 110, no. 4, pp. 497–510, 1984.
52. Srikanthan, R., T. A. McMahon, G. P. Codner, and R. G. Mein, "Theory and Application of Multisite Streamflow Generation Models," *Civil Eng. Trans. Inst. Eng. Aust.,* vol. CE26, no. 4, pp. 272–279, 1984.
53. Srikanthan, R., "The Effect of Annual Auto-Correlation on Storage Size," *Civil Eng. Trans. Inst. Eng. Aust.,* vol. CE27, no. 2, pp. 225–229, 1985.
54. Tennant, D. L., "Instream Flow Regimens for Fish, Wildlife, Recreation and Related Environmental Resources," *Fisheries,* vol. 1, pp. 6–10, 1976.
55. Teoh, C. H., and T. A. McMahon, "Evaluation of Rapid Reservoir Storage-Yield Procedures," *Adv. in Water Resources,* vol. 5, pp. 208–215, 1982.
56. Titus, J. G., et al., "Greenhouse Effect, Sea Level Rise, and Coastal Drainage Systems," *J. Water Resour. Plann. Manage.,* vol. 113, no. 2, pp. 216–227, 1985.
57. Todd, D., *Groundwater Hydrology,* Wiley, New York, 1959.
58. Tunbridge, B. R., *Environmental Flows and Fish Populations of Waters in the South-Western Region of Victoria,* Department of Conservation, Forests and Lands, Victoria, Australia, Arthur Rylah Institute for Environmental Research technical report no. 65, 1988.
59. Tunbridge, B. R., and T. J. Glenane, *A Study of Environmental Flows Necessary to Maintain Fish Populations in the Gellibrand River and Estuary,* Department of Conservation, Forests and Lands, Victoria, Australia, Arthur Rylah Institute for Environmental Research technical report no. 25, 1988.
60. U.S. Army Corps of Engineers, *Hydropower: Engineering and Design,* engineer manual EM 1110-2-1701, U.S. Army Corps of Engineers, Washington D.C., 31 December 1985.
61. Vogel, R. M., and J. R. Stedinger, "Generalized Storage-Reliability-Yield Relationships," *J. Hydrol.,* vol. 89, pp. 303–327, 1987.
62. Ward, J. V., and J. A. Stanford, eds., *The Ecology of Regulated Streams,* Plenum Press, New York, 1979.
63. Wesche, T. A., and P. A. Rechard, *A Summary of Instream Flow Methods for Fisheries and Related Needs,* Eisenhower Consortium bulletin no. 9, 1980.
64. Wurbs, R. A., and W. B. Walls, "Water Rights Modelling and Analysis," *J. Water Resour. Plann. Manage.,* vol. 115, no. 4, pp. 416–430, 1989.
65. Yeh, W. W.-G., "Reservoir Management and Operations Model: A State-of-the-Art Review," *Water Resour. Res.,* vol. 21, no. 12, pp. 1797–1818, 1985.

CHAPTER 28

HYDROLOGIC DESIGN FOR URBAN DRAINAGE AND FLOOD CONTROL

Ben R. Urbonas
Urban Drainage and Flood Control District
Denver, Colorado

Larry A. Roesner
Camp Dresser and McKee
Maitland, Florida

Urban hydrologic design offers unique challenges for the hydrologist. Often the principles of watershed hydrology cannot be directly applied to urban hydrology. One's thinking has to shift to very small watersheds having highly variable watershedding surfaces. In addition, where water would normally run over land as sheet flow, in an urban setting it is concentrated into swales, open channels, and storm sewers, all of which accelerate the flow. As a result, the hydrologist needs to consider that local flooding can occur in a matter of minutes instead of hours or days.

The goal of this chapter is to describe some of the unique problems encountered in urban drainage and flood damage management. How urbanization affects runoff is briefly described. This is followed by a discussion of the hydrologic aspects of policies, criteria, and systems planning for drainage and flood control problems. The very important subject of floodplain management, which differs somewhat from stormwater drainage, is discussed. Last, but not least, guidance is provided for the design of storm sewers, detention and retention facilities, and water-quality enhancement facilities.

28.1 IMPACTS OF URBANIZATION ON STORMWATER RUNOFF

Runoff Volume. Urbanization increases surface stormwater runoff and modifies its quality. As land urbanizes, it is covered by impervious surfaces such as paved roads, parking lots, and roofs which prevent rainfall or snowmelt from infiltrating into the ground. The U.S. Nationwide Urban Runoff Program (NURP) reported its findings

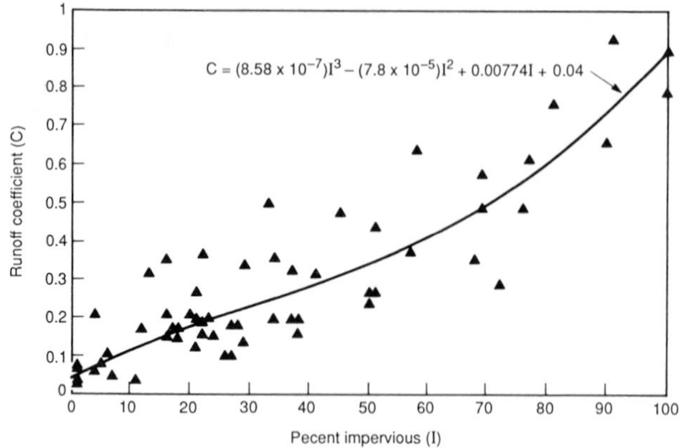

FIGURE 28.1.1 Runoff coefficient versus percent impervious. *(Used with permission from Ref. 71.)*

in 1983[77] that the observed runoff coefficient C increases as the percentage of imperviousness in a watershed increases, as illustrated in Fig. 28.1.1. Since these data were collected over a 2-year period, they are indicative of urban runoff from frequently occurring storm events, namely, a 2-year storm or less. The Urban Drainage and Flood Control District in Denver, after a 12-year study,[66] found that the runoff coefficient increases as the volume of rainfall increases, so C is not a constant value for a given land-use surface.

Peak Rates of Flow. Surface runoff in urban areas has a higher velocity than in nonurban areas because impervious surfaces are smoother than meadow, range land,

TABLE 28.1.1 Ratio of Peak Runoff Rates Before and After Development at Three Single-Family Residential Sites

	Post/Preurbanization ratios of runoff peaks		
Return period, years	New Jersey* site	Denver, Colorado site†	Canberra, Australia site‡
2		57.0	9.0
10		3.10	4.7
15	3.0		
100		1.85	1.9

* 33-in (840-mm) annual precipitation; based on modeling pre- and postdevelopment conditions using SCS TR-55 model and Type II storm distribution.

† 15-in (380-mm) annual precipitation; based on 8-year rainfall-runoff data record and 73-year simulation of pre- and postdevelopment conditions.

‡ 22-in (550-mm) annual precipitation; based on statistical analysis of similar-size adjacent developed and undeveloped tracts of land.

forest, or farm fields. This increase in velocity, along with the increase in runoff volume and the concentration of the runoff into pipes and channels, results in quicker concentration of flows from various parts of the watershed. The end result is an increase in the observed peak rate of flow as the area urbanizes.

Considerable differences in the effects of urbanization on peak runoff rates were reported in three studies,[9,68,91] summarized in Table 28.1.1. The largest changes in runoff due to urbanization are seen in the frequently occurring 2-year storm. The change in runoff and the differences between the three studies are smallest for the 100-year storms. Possible explanations why the increase in runoff with urbanization differed between the three studies include the fact that different hydrologic regions were studied and that different study techniques were employed, which points out the need for caution in using such studies to justify a point of view. Results from one watershed, or when using a specific hydrologic model may not be the same for another watershed or when a different methodology is used. Nevertheless, the underlying conclusion is that urbanization increases stormwater runoff volumes and peak rates, which is supported by the findings of Driver and Tasker,[16] based on the study of rainfall and runoff data from 30 different cities throughout the United States, and studies by others.[57,86]

28.2 POLICY, CRITERIA, AND DRAINAGE SYSTEM PLANNING

28.2.1 Need for Statement of Policy

Storm drainage and management systems are constructed to cope with an increase in runoff as land is developed. Communities with successfully functioning drainage systems generally have a set of written policy statements concerning the management of this increase in stormwater runoff. Such statements can be found in local ordinances, rules, and regulations which specify basic goals and objectives that need to be followed by professional staff, land developers, and the community's planning and advisory boards.

Since each city and county is unique in its geographic, hydrologic, social, economic, and political makeup, it is not possible to state a single policy that should be adopted by all communities. Thus the following discussion provides some ideas to consider in formulating local drainage policy.

Drainage System's Role in the Community. The stormwater drainage system is a part of the total urban infrastructure, and space or right-of-way has to be reserved for this system. Storm sewers require little space and can often be located within the street right-of-way. Major open channels, natural or manmade, require their own rights-of-way, preferably under public ownership or control.

The drainage system serves several vital community functions:

1. It removes stormwater from the streets and permits the transportation arteries to function during bad weather—when this is done efficiently, the life expectancy of street pavement is extended
2. The drainage system controls the rate and velocity of runoff along gutters and other surfaces in a manner that reduces the hazard to local residents and potential for damage to pavement

3. The drainage system conveys runoff to natural or manmade major drainageways
4. The system can be designed to control the mass of pollutants arriving at receiving waterways
5. Major open drainageways and detention facilities offer opportunities for multiple use such as recreation, parks, and wildlife preserves.

Minor versus Major Runoff Events. When establishing local policy and criteria, each community defines for itself what constitutes the *initial drainage system,* sometimes called the *minor drainage system,* which serves as the surface drainage system, and the *major drainage system,* which is the flood control system.

The initial system is intended to handle the frequent runoff events and nuisance flows. It is generally designed to serve runoff peak flows and volumes not exceeding the 2- or 5-year return period. The definition of the initial system includes depth and velocity limitations for roadside swales, gutters, storm sewers, and open channels during the selected design event.

The major drainage system captures, conveys, and stores runoff from large storms. Current practice by the Federal Emergency Management Agency (FEMA) of the United States requires all cities and counties to locate new homes and businesses above the 100-year flood level. As a result, the major system is often designed for the 100-year flood, although there are instances where the major system is being designed for only a 10-, 25-, or 50-year flood.

The major system is made up of the initial system plus the flow-carrying capacity of streets, gutters, borrow ditches, parks, open spaces, and portions of individual lawns, as well as the major drainageways and natural waterways. Each community needs to define what constitutes its formal, dedicated, major drainage system.

28.2.2 Elements of Local Urban Storm Drainage Criteria

Local *storm drainage criteria* are the backbone of a developing stormwater drainage system. They set the limits on development that a city or a county is willing to accept; provide guidance and methods of design and the details for key components of an acceptable drainage and flood control system; and ensure consistency in function, longevity, safety, aesthetics, and maintainability of the system. References 7, 8, 51, and 66 provide examples of such documents. Table 28.2.1 contains a checklist of items to consider when developing stormwater drainage criteria.

28.2.3 Master Planning Studies—Watershed Approach

If local urban storm drainage criteria are the backbone of a developing stormwater drainage system, *master plans* add the rest of the skeleton to support the body. A drainage system master plan provides a municipality with a road map of how to develop this vital part of its infrastructure. Such a plan addresses system needs and design in a coherent manner. Without a master plan, each new land development project adds its drainage subsystem piecemeal to the total system. This is done with no assurance that each subsystem is consistent with the neighboring elements of the completed drainage network.

TABLE 28.2.1 Checklist for Developing Local Storm Drainage Criteria

Governing legislation and statements of policy and procedure
- Legal basis for criteria
- Define what constitutes the drainage system
- Benefits of the drainage system
- Policy for dedication of right-of-way
- Compatible multipurpose uses
- Review and approval procedures
- Procedures for obtaining variances or waivers of criteria.

Initial and major drainage system provisions
- Definitions of initial system and major system
- Where should a separate formal major drainageway begin
- Allowable flow capacities in streets for initial and major storms
- Maximum and minimum velocities in pipes and channels
- Maximum flow depths in channels and freeboard requirements

Data required for design, such as:
- Watershed boundaries
- Local rainfall and runoff data
- History of flooding in the area
- Defined regulatory flood plains and floodways
- Existing and projected land use for project site
- Existing and planned future land uses upstream
- Existing and planned drainage systems off-site
- Tabulation of previous studies affecting site
- Conflicts with existing utilities
- Design storms, intensity-duration-frequency data
- Hydrologic methods and/or models
- Storm sewer design criteria, including materials
- Manhole details and spacing, inlet details and spacing, types of inlets, trenching, bedding, backfill, etc.
- Street flow calculations and limitations
- Details of major system components such as channels, drop structures, erosion checks, transitions, major culverts and pipes, bridges, bends, energy dissipators, riprap, sediment transport

Detention requirements
- When and where to use detention
- Design storms
- Hydrologic sizing criteria and/or procedures
- Safety, aesthetics, and maintainability criteria
- Multipurpose uses and design details for each

Water-quality criteria
- Goals and objectives
- Minimum capture volumes
- Required or acceptable best management practices
- Technical design criteria for each best management practice (BMP)

Special considerations
- Right-of-way dedication requirements
- Use of irrigation ditches
- Floodproofing and when it is accepted
- Any other items reflecting local needs

List of technical references

A master planning study for a watershed looks at the entire system. As a result, projected future urban land uses, rather than the existing development, should be used as the basis for runoff computations. This avoids undersizing drainage facilities. Master plans address not only physical and hydrologic concerns, but also consider environmental, safety, aesthetic, recreational, economic, fiscal, and maintenance issues.[23] These studies may also have to resolve interjurisdictional drainage issues and potential regulatory roadblocks.

The major advantage a master-planned watershed has over a watershed that develops piecemeal is that each element of the major system is based on systemwide modeling. An urban watershed is a composite of multiple land uses which produce differing runoff volumes, times to peak, and flow rates. Watershed modeling permits the integration of all these varying features. Thus, alternative drainage and flood management plans can be tested and judged on how they affect the hydrologic response of the entire system.

Various types of *stormwater detention* are often investigated during planning studies to mitigate the effects of increased runoff resulting from urbanization, to reduce the size of facilities needed for the conveyance of the runoff, to mitigate water-quality concerns, or for a combination of any of these reasons. Barring other overriding considerations, economics can help decide if detention is cost-effective and if it should be used. However, civil law principles and statutory mandates, as well as a number of federal, state, and local rules and ordinances, often require that on-site detention be used to ensure that runoff peaks from a development site do not exceed those existing before development and that urban stormwater quality is improved before it leaves the site.

On-site and regional detention can be considered during planning studies. *On-site detention* provides stormwater detention at each new land development site, sometimes at each legal lot or section. It is virtually impossible to quantify the hydrologic impacts of a very large number of existing and future on-site facilities located randomly within a watershed.

Several studies[32,43,69] have shown that on-site detention has the potential for maintaining peak flow rates at or below their predevelopment levels along major drainageways only for larger storm events. It appears that restricting peak flows to their predevelopment levels along downstream major drainageways is not possible for storms with a return period of less than 5 years using on-site detention only, unless the soils, geology, groundwater, and meteorologic conditions at the development site permit successful infiltration of stormwater into the ground.

On-site detention is most effective in controlling peak flow rates along major drainageways during large storms when all on-site detention is sized uniformly, with the goal of limiting the peak flows along major drainageways to predevelopment levels.[68,69] Restricting rates of runoff to their predevelopment levels at each site often fails to accomplish this wider goal.[12] Watershed planning permits the development of on-site detention schemes that address this wider purpose.

A preplanned regional detention system is much easier to design than a random collection of on-site detention facilities. As a result, master planning studies are good vehicles for identifying the locations, sizes, and required release rates of regional detention facilities. On the other hand, regional detention is often difficult to finance and, as a result, difficult to implement as planned. What may appear to be an ideal master plan may not be implemented without a fiscal commitment by a city or a county, a commitment which has to include a financing plan and appropriations of funds over a number of years.

28.3 FLOODPLAIN MANAGEMENT

28.3.1 Introduction

Flooding is a natural and certain phenomena. It can occur on any land surface, ranging in size from a street intersection or a home lot to the extensive floodplain areas inundated by large rivers such as the Nile River. Flooding generally results in damage to property and negative impacts on human welfare. *Floodplain management* is the process of minimizing property damage and reducing the threat to human life and welfare when major storms occur.

Floodplain management employs many of the same methods of analysis used for stormwater management. While stormwater management usually deals with one to a few jurisdictions and with more frequently occurring storms, i.e. 2-, 5-, and 10-year storms, floodplain management deals with larger watersheds and large storms. Floodplain delineation studies are typically done for watersheds of about 1 mi^2 (2.6 km^2) or larger and such studies focus on floods having return periods of 25 to 100 years, and sometimes 500 years.

This section presents an overview of floodplain management. It addresses both the regulatory aspects of floodplain management in the United States and the procedures for determining floodplains along rivers, lakes, and coastal areas. Floodplain management, while mandated by federal or state government, is very much a local issue. Anyone engaged in floodplain management should first check with the local agency responsible for drainage or flood control to determine the ground rules to follow.

28.3.2 Regulatory Framework

In the United States, the Federal Emergency Management Agency (FEMA) has promulgated minimum floodplain management standards for local governments to follow. This agency has also developed flood insurance rate maps (FIRM) for all urban areas of United States showing the floodway for the 100-year flood and the areas of inundation for the 100- and 500-year floods. Also included are flood stage profiles for the 10-, 50-, 100-, and 500-year storms. Figure 28.3.1 shows a typical flood profile along a stream in the FIRM format. FIRM maps and profiles for the watershed are determined from land use and conditions at the time of the study. They do not consider land-use changes that will result from future development within the watershed. As a result, these maps become outdated as urbanization occurs. Some local governments have developed maps based on projected future land uses, and these maps should remain valid for the foreseeable future.

Figure 28.3.2 shows a *riverine floodplain* which is composed of a floodway and the flood fringe. The *floodway* is that part of the floodplain considered to be the zone of highest hazard and the zone to be reserved for the passage of larger floods. The *flood fringe* is a zone of floodwater storage where water moves slowly, or is ponded, thus attenuating the flood peak as the flood wave moves downstream.

Federal regulations do not allow any structure to be built in the floodway. Any structure built in the flood fringe must have its finished floor elevations at or above the 100-year flood elevation. Local regulations, on the other hand, often require finished floor elevations be 1 to 2 ft (0.305 to 0.61 m) above the 100-year flood

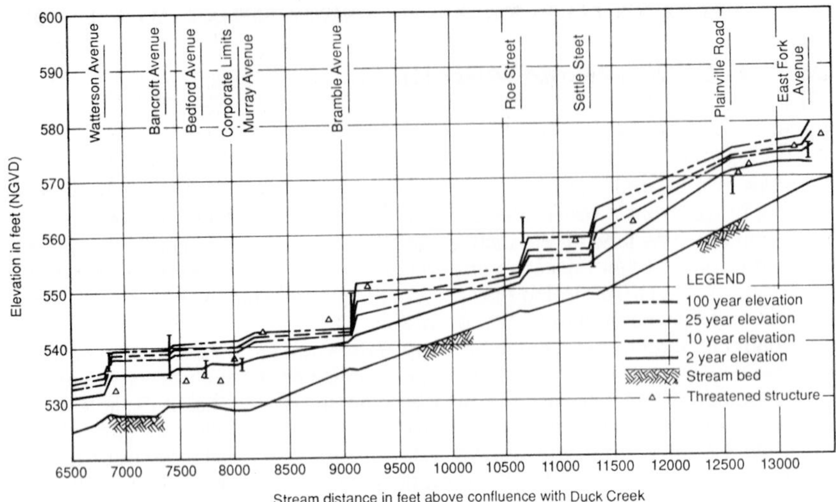

FIGURE 28.3.1 Example of flood profile drawing in FEMA format. *(Courtesy of Camp Dresser and McKee, Cambridge, Mass.)*

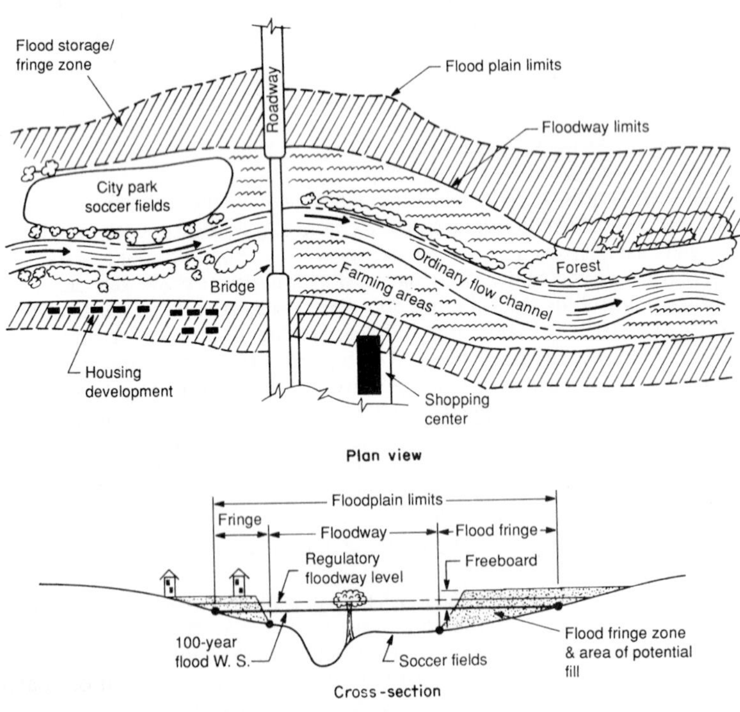

FIGURE 28.3.2 Floodplain, floodway, and flood fringe.

elevation. Properties constructed in the floodplain prior to the publication of the final FIRM map, and properties constructed outside the FEMA-defined 100-year floodplain, are eligible for FEMA-subsidized flood insurance. Federally insured lenders cannot issue mortgage loans for properties within the floodplain without proof of flood insurance. Properties constructed within the delineated floodplain after the FIRM map is published are not eligible for FEMA subsidized insurance.

Many FEMA floodplain studies have used approximate methods to estimate peak flows. As a result, some FIRM floodplain maps may not be very accurate. Regardless, the FIRM flood profile elevation governs the determination of whether the properties are in the floodplain or not. Thus, care should be used when dealing with properties just outside the delineated floodplain to be sure that they are, in fact, above the 100-year flood profile.

FEMA will amend the FIRM maps if it can be shown that the new data are more accurate than the original data, or that the analytical approach used for the new definition is more accurate than the original analysis. Maps are revised when a physical facility such as a levee, floodwall, or channel improvement changes the floodplain boundaries, provided minimum FEMA standards of design, construction, and maintenance of the facility are met.

Floodplain and floodway boundaries, once published by FEMA, are very difficult to change. An appeal must be made both to the affected community and to FEMA, and must be based on sound technical information. Details of the appeals process are described by FEMA.[19] Local ordinances also need to be followed during this appeal process.

28.3.3 Delineation of River Floodways and Floodplains

The first step in floodplain delineation is to compute the flood runoff rates and volumes, route them through the drainageway network, and to compute the corresponding water surface profiles, such as those shown in Fig. 28.3.1, using the techniques described in Chaps. 9, 10, and 21. The floodplain boundaries are then determined by transcribing the water surface profile elevations onto a topographic map.

Floodway limits are determined by blocking off flow conveyance areas in equal increments on both sides of the floodplain. This is done by starting at the edges of the floodplain at each cross section and extending the blockage or encroachment incrementally inward until a specified rise in the computed flood profile occurs or the top of the bank for a defined channel, river, or stream is reached. The floodway limit is then defined by connecting along the channel the maximum encroachment points found at each cross section. However, encroachment into an existing formal channel is not permitted, since it is already considered to be an existing floodway. In steep terrain, where the velocity head is significant [i.e., velocity head greater than ½ ft (0.15 m)], the encroachment producing a specified rise in energy grade line is often used to define the floodway limits instead of a rise in the water surface profile.

Floodplain Maps. FEMA defines floodways on its FIRM maps using the encroachment necessary to produce a 1-ft (0.305-m) maximum rise in the water surface or energy grade line elevation. Some cities have used a 0.5-ft (0.15-m) rise in the energy grade line to define floodway limits, while others use criteria based on maximum velocity and/or depth. All of these approaches are valid.

In urban areas, the floodway and the floodplain limits are plotted on maps having a horizontal scale no less than 1 in to 200 ft (1:2400), and preferably 1 in to 100 ft (1:1200). The contour interval used will depend on the amount of relief within the floodplain. Preferably, the horizontal distance between adjacent contours should be less than 200 ft (61 m). In very flat areas such as south Florida, this can mean contour intervals of 1 ft (0.3 m) or less. In most cases, however, a 2-ft (0.6-m) contour map is adequate.

Floodplain Modification. Floodplain limits can be altered by filling in the flood fringe or by constructing levees, dikes, or floodwalls. The floodplain can also be reduced through channel modifications which increase the conveyance of the floodway so that water moves faster through the floodplain, resulting in less backwater and floodplain storage. Since modifications of conveyance capacity may also increase the peak flow downstream, such modifications should be approached with caution. Similarly, filling the flood fringe also decreases floodplain storage, which results in increased peak flow and flood depth downstream.

Altering the floodway from its existing or natural placement should also be approached with caution. Newspaper files are replete with stories of structures that failed during large storms. Often when floodways are modified, new structures are constructed within the historic floodway, and some form of channelization, levee, or floodwall is designed to modify the floodway limits to protect them. Reducing floodway limits results in a reduced ability of the floodplain to absorb and transport floods that exceed the design discharge.

Whenever a floodway is modified, the hydrologist should practice the highest level of accepted hydrologic and hydraulic design. Sufficient safety factors need to be used to ensure that the new, manmade conveyance facilities actually provide the protection they are designed to provide. A simple sensitivity analysis can reveal how the new design will function under a larger flood.[6] It will also reveal the effectiveness of the design safety factors. Be sure to use actual field cross sections and longitudinal slopes when testing sensitivity, and not idealized rectangular or trapezoidal sections.

28.3.4 Floodplain Delineation around Lakes

For lakes whose surface area is less than 1000 acres (400 ha), delineation of the floodplain around the lake is a simple exercise because the water surface elevation is assumed to be horizontal throughout the lake at any point in time. To determine the peak water surface elevation of the lake, the runoff hydrographs from all watersheds tributary to the lake are determined and hydrographs are routed through the lake using a storage-routing method, with the stage-discharge curve for the lake or reservoir used to determine the outflow. The contour on the topographic map of the maximum water surface elevation is then traced around the lake and it becomes the floodplain boundary.

There are two cases when the assumption of a level water surface may not be valid. The first is if the lake is very long, narrow, and shallow. In this case the lake may act more like a stream segment than a lake. If such a possibility exists, use the riverine approach or dynamic routing to determine the maximum flood elevations along the long axis of the reservoir (see Chap. 10). If the lake is relatively wide, this method will not result in significantly different flood elevations along its axis. As a rule of thumb, consider the possibility of a varying water surface profile along the lake's axis if its length-to-width ratio is greater than 4. See Chap. 10 for more details on when dynamic routing should be used instead of storage routing for lakes.

A second case is where a chain of lakes is connected by a short channel or an outlet that controls the flow between the two water bodies. Under some hydrologic conditions, the lower lake may rise sufficiently to diminish the outflow from the upper

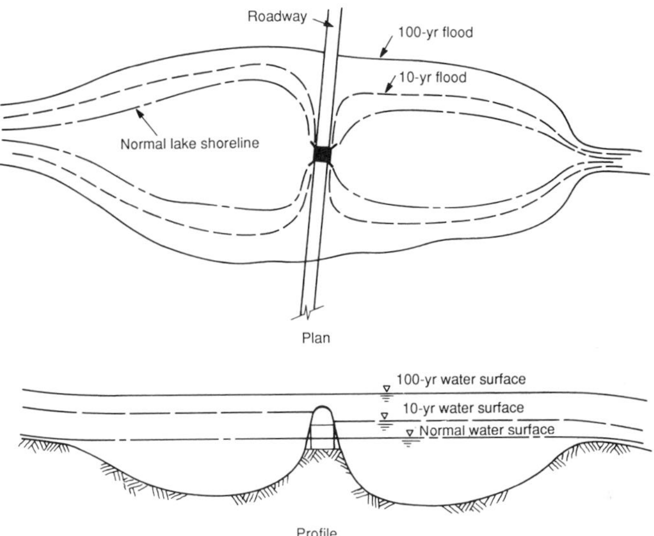

FIGURE 28.3.3 Sequential lakes behave as a single lake during a flood.

lake, thus requiring either a dynamic or diffusion routing model for analysis of backwater effects. On the other hand, it is possible that the two lakes may function as a single lake, as shown in Fig. 28.3.3. Each case has to be analyzed to determine how the water surface reacts dynamically to the inflow hydrograph.

28.3.5 Floodplain Delineation in Tidal Areas

In coastal areas and along the shoreline of very large lakes, such as the Great Lakes, flooding is caused by two phenomena: tidal surges and rainfall runoff. Except for tsunamis caused by undersea earthquakes, tidal flooding is caused by large-scale atmospheric disturbances over the open ocean (or a very large lake) that produce a surge wave which eventually reaches the shore. Tidal flooding is primarily a concern along shallow coastlines similar to those found along the Atlantic and Gulf coasts of the United States. In the Northern Hemisphere, winds blow counterclockwise about the eye of a hurricane, thus producing strong wave surge to the right of the place where the eye of the hurricane crosses the coast. In the Southern Hemisphere the reverse is true and the surge wave occurs to the left of the crossing point.

Tidal elevations for the 100-year recurrence interval can be determined by extrapolating tidal stage records, but more commonly, mathematical models are used that translate atmospheric variables into a tidal wave and then propagate the wave overland in the near shore area using a two-dimensional flow routing model. This analysis requires special routing models that can move water from wet cells or elements to dry ones as the flood wave moves inland.[34]

To delineate the floodplain, the tidal surge of the desired recurrence interval, usually the 100-year event, is mathematically propagated over the land surface. The maximum landward extent of the surge is mapped and becomes the tidal surge floodplain limit.

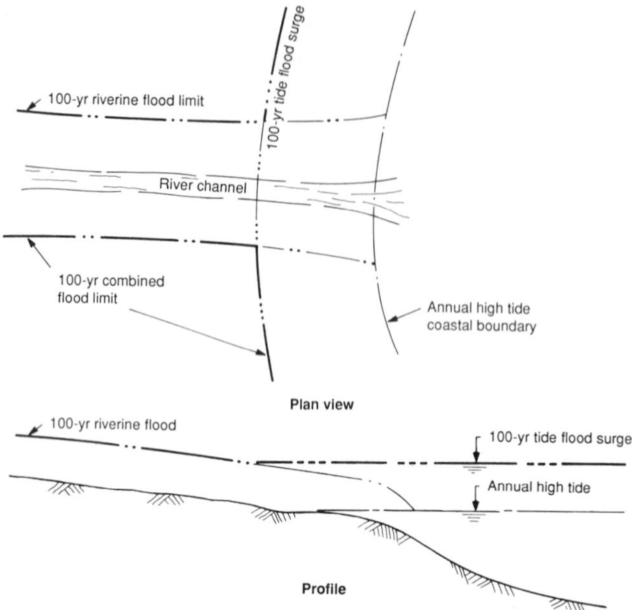

FIGURE 28.3.4 Floodplain limits for a combined riverine and tidal flood plain.

In areas where streams or rivers drain to a tidal water body, riverine flooding is computed in the normal manner by using the mean high tide level at the coast as the downstream starting water surface elevation for backwater calculations. It is assumed that the occurrences of the 100-year tidal surge and the 100-year rainstorm-generated floods are independent phenomena. Whether or not they are independent is debatable.[44,96] The riverine floodplain so determined is overlaid on the tidal surge map, and the outside boundary, shown in Fig. 28.3.4, is defined as the floodplain limit.

28.4 STORM SEWER DESIGN

Storm sewer design is described in many publications.[1,6,18,30,50,55,84,85] Often local storm drainage criteria manuals[7,8,10,31,51,66] contain specific design procedures and/or guidance. The most widely used reference in the United States on this topic is a joint publication by the American Society of Civil Engineers (ASCE) and the Water Pollution Control Federation (WPCF),[3] which is being updated as a stormwater management manual of practice.[4] As a result, the emphasis in this chapter is on practical guidance concerning the sizing and the design of storm sewers.

28.4.1 Information Needs

A common problem in designing storm sewers for areas of new development is the lack of an adequate information base. Site data collection should not end at the

TABLE 28.4.1 Condensed Checklist of Information Needs for Storm Sewer Design

- Local storm drainage criteria and design standards
- Maps, preferably topographic, of the subbasin in which the new system is to be located
- Detailed topographic map of the design area
- Locations, sizes, and types of existing storm sewers and channels located upstream and downstream of design area
- Locations, depths, and types of all existing and proposed utilities
- Layout of design area including existing and planned street patterns and profiles, types of street cross sections, street intersection elevations, grades of any irrigation and drainage ditches, and elevations of all other items that may pose physical constraints to the new system
- Soil borings, soil mechanical properties, and soil chemistry to help select appropriate pipe materials and strength classes
- Seasonal water table levels
- Intensity-duration-frequency and design storm data for the locally required design return periods
- Pipe vendor information for the types of storm sewer pipe materials accepted by local jurisdiction

project's limits. The designer needs also to obtain information from the local government about the changes planned for the site and all adjacent lands. This information helps to integrate the new development's drainage system into the total drainage system. Table 28.4.1 provides a condensed checklist of information that a hydrologist needs to collect when designing storm sewers.

28.4.2 Storm Sewer Design Criteria

Storm sewer design criteria depend on the pipe materials being used, local terrain, and local climate and hydrology. Table 28.4.2 contains many of the technical items and limitations to consider when designing a new storm sewer system. The hydrologist begins the design process by first finding the theoretical curb-full gutter flow capacity during the minor storm. Using the variables shown in Fig. 28.4.1, Manning's equation can be rewritten in the form of Eq. (28.4.1). This equation contains an adjustment factor of 1.2 to account for the effects of the very small hydraulic radius of flow in the street. Without this factor the theoretical gutter capacity is underestimated.[25,47]

$$Q_t = 1.2\, K_g \left(\frac{z}{n}\right) S^{1/2}\, y^{8/3} \qquad (28.4.1)$$

where Q_t = flow rate, ft³/s (m³/s)
K_g = 0.47 for U.S. standard units (0.32 for SI units)
$z = T/y$; T = top width of flow, ft (m)
n = Manning's coefficient (usually 0.016 for paved streets)
S = longitudinal slope of the gutter, ft/ft (m/m)
y = depth of flow at the curb, ft (m)

The first stormwater inlet in the system, and each subsequent inlet along a gutter, are located where the allowable gutter capacity is reached during the minor storm.

TABLE 28.4.2 Technical Items and Limitations to Consider in Storm Sewer Design

Velocity:	
Minimum design velocity	2–3 ft/s (0.6–0.9 m/s)
Maximum design velocity	
Rigid pipe	15–21 ft/s (4.6–6.4 m/s)
Flexible pipe	10–15 ft/s (3.0–4.6 m/s)
Maximum manhole spacing: (function of pipe size)	400–600 ft (122–183 m)
Minimum size of pipe	12–24 in (0.3–0.6 m)
Vertical alignment at manholes:	
Different size pipe	Match crown of pipe or 80 to 85% depth lines
Same size pipe	Minimum of 0.1–0.2 ft (0.03–0.06 m) in invert drop
Minimum depth of soil cover	12–24 in (0.3–0.6 m)
Final hydraulic design	Check design for surcharge and junction losses by using backwater analysis
Location of inlets	In street where the allowable gutter flow capacity is exceeded

The allowable gutter capacity Q_a is computed from the theoretical curb-full flow capacity, using a reduction factor F:

$$Q_a = F\,Q_t \qquad (28.4.2)$$

Figure 28.4.2 suggests values of F for various gutter slopes. This figure represents the policies of the Denver metropolitan area.[66] Other communities may have different policies with respect to F. A similar approach is used for the major storm, except the theoretical gutter flow is based on a depth greater than the curb-full flow.

28.4.3 Rational Method

The *rational method* is a storm sewer design method based on the *rational formula* for computation of the design discharge in each pipe. The validity of the rational formula is based on the assumption that the rainfall intensity for any given duration is uniform over the entire tributary watershed. The rational formula is

$$Q = K_u CiA \qquad (28.4.3)$$

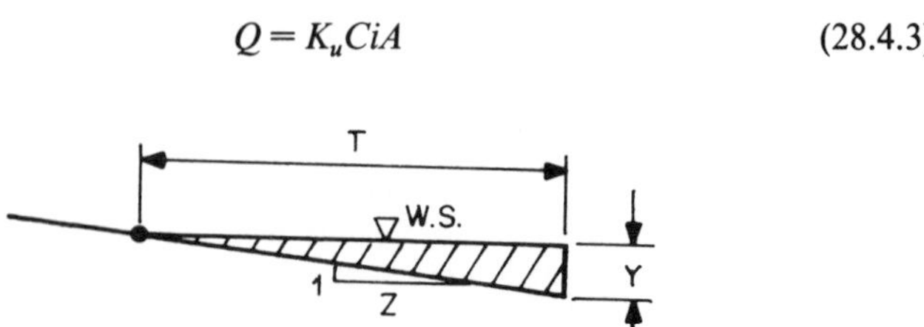

FIGURE 28.4.1 Elements of shallow street flow.

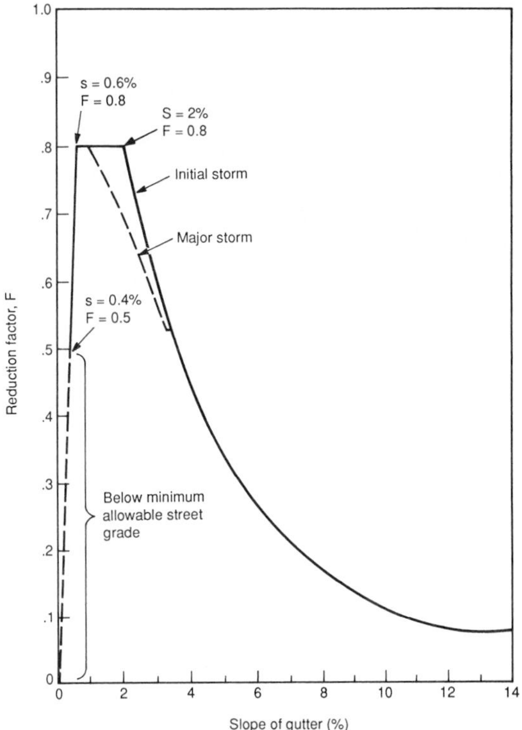

FIGURE 28.4.2 Reduction factor for allowable gutter capacity in an arterial street. *(Used with permission from Ref. 66.)*

where Q = peak flow rate, ft³/s (m³/s)
K_u = 1.0 for U.S. standard units (0.28 for SI units)
C = runoff coefficient for the watershed
i = intensity of rainfall taken from the intensity-duration-frequency curves for the specified design return period at the time of concentration T_c, in/h (mm/h)
A = area of tributary watershed, acres (km²)

Despite many critics, the rational method continues to dominate hydrologic practice for the design of storm sewers, especially in the field of land development. Other, more complicated design tools may, in fact, not offer more accurate results unless rainfall and runoff data are available for the design subbasin. At the same time, many of the more complicated procedures and models incorporate some aspects of the rational method. As an example, another procedure that is ingrained in hydrologic practice, the SCS method, utilizes the time of concentration and an empirically redistributed intensity-duration-frequency curve as input hyetograph. The U.S. Army Corps of Engineers' STORM model uses runoff coefficients to estimate runoff rates and volumes. Thus, what may appear to be a more sophisticated model is often built by using some rational formula components to estimate runoff rates.

Selecting Rational Method Design Parameters

Time of Concentration. The use of appropriate values for time of concentration T_c is very important, although it is hard sometimes to judge what is the correct value. Local data collection efforts can be used to calibrate the T_c so that it works properly with the calibrated runoff coefficients. When this was done in Denver, the resulting peak flow estimates were found to be consistently within 20 percent of those obtained with calibrated distributed routing models.[66]

Equation (28.4.4) states that T_c is the sum of two flow times. The first is the initial time required for the surface runoff to reach the first swale, gutter, sewer, or channel. The second is the travel time in the conveyance elements.

$$T_c = t_i + t_t \tag{28.4.4}$$

where T_c = time of concentration, min
t_i = initial inlet or sheet flow time, min
t_t = travel time in a conveyance element, min

Finding the initial time seems to create the greatest amount of confusion and conflict. It can be estimated by Eq. (28.4.5). This equation was developed for estimating the initial surface sheet flow time over short distances. This equation should not be used for distances larger than 200 to 300 ft under urban conditions. Surface flows beyond the initial sheet flow distance are likely to concentrate into either a swale, a gutter, or another type of conveyance element.

$$t_i = \frac{K_u\,[1.8\,(1.1 - C_5)\,L^{0.5}]}{S^{1/3}} \tag{28.4.5}$$

where t_i = initial overland flow time, min
K_u = 1.0 for U.S. standard units (0.552 for SI units)
C_5 = runoff coefficient used for the 5-year storm
L = length of overland flow, ft (m)
S = average basin slope, %

Travel time is the time it takes the flow to travel through the various conveyance elements to the next inlet or design point. It is calculated by adding the time it takes the water to travel in each segment of the system from the point where the sheet flow (i.e., initial time t_i) enters a recognizable conveyance element.

Before the calculated value of T_c is used, it must be determined that the calculated value makes sense for the site. Figure 28.4.3 was developed by the U.S. Soil Conservation Service[82] and can be used as one of the checks to gauge if the final T_c makes sense. If the value of T_c calculated by any of the available equations differs significantly from the one calculated using the average flow velocity obtained from this figure, the calculations must be double-checked.

Often, limitations on the time of concentration are imposed by local governing agencies. For example, after a 10-year data-gathering effort, the maximum value of T_c for urban basins in Denver is limited by Eq. (28.4.6).[66] It is speculated that this is a reasonable limitation for urban basins having at least a 0.4 percent slope. Thus, if the calculated time of concentration exceeds this limiting value, the Eq. (28.4.6) value of T_c is used.

$$T_c = K_u \left(\frac{L}{180}\right) + 10 \tag{28.4.6}$$

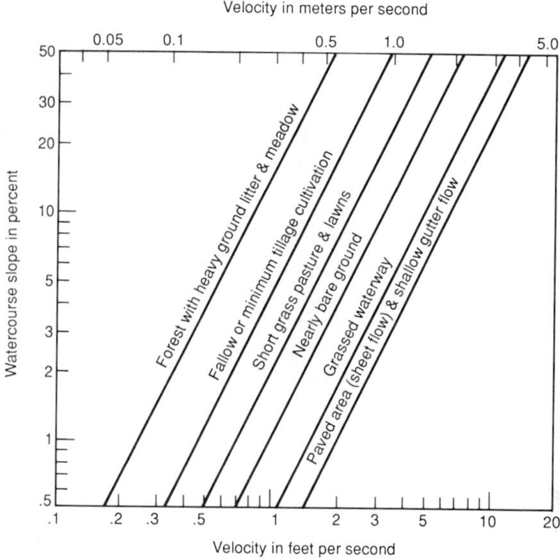

FIGURE 28.4.3 Estimate of average flow velocity for use with the rational formula. *(From Ref. 82.)*

where T_c = maximum time of concentration, min, for basins having a slope greater or equal to 0.4 percent

L = maximum distance from inlet (design point) to watershed boundary, ft (m)

K_u = 1.0 for U.S. standard units (0.305 for SI units).

Runoff Coefficient. The runoff coefficient C varies with the type of surface area and the total depth and intensity of rainfall. Studies by Schaake and others[58,66] showed that the runoff coefficient increases as the return period of the design storm increases. Table 28.4.3 lists some of the values suggested in the Urban Storm Drainage Criteria Manual for Denver.[66] These may or may not be appropriate for other localities. However, C for the smaller rainfall values in this table correspond well to the values

TABLE 28.4.3 Runoff Coefficients Suggested for Denver, Colo., Area

	2-h rainfall depth, in (mm)			
Land use or surface type	1.2 (27)	1.7 (38)	2.0 (43)	3.1 (83)
Lawns, sandy soil	0.00	0.05	0.10	0.20
Lawns, clayey soil	0.05	0.15	0.25	0.50
Paved areas	0.87	0.88	0.90	0.93
Gravel streets	0.15	0.25	0.35	0.65
Roofs	0.80	0.85	0.90	0.93

Values in this table may be combined by area weighting. The coefficients may not be valid for watershed areas exceeding 160 to 200 acres (65 to 80 ha).

Source: After Ref. 66.

presented in Fig. 28.1.1, which was derived using National Urban Runoff Program data.[77]

What Part of the Watershed Controls Design? A criticism of the rational method is that it does not account for which portion of the watershed controls the storm sewer size. An example would be an office complex where T_c is calculated by starting with overland flow T_i over a grass surface near the top of the basin, which is then added to the travel time T_t in gutters. The value T_i in such a case can be large. As a result, the rainfall intensity taken from the intensity-duration-frequency (I-D-F) curves will be low.

On the other hand, if T_c is calculated by starting just downstream of the grass areas, namely, at the upstream edge of the parking lot surrounding the buildings, the I-D-F rainfall intensity will be higher. The more intensely developed lower area, even if it is only a part of the total tributary area, can produce a higher calculated peak flow than the one calculated using the entire tributary area. Figure 28.4.4 illustrates typical I-D-F curves for the 2-, 5-, and 100-year return periods.[66] Note how rapidly the rainfall intensity diminishes as the time of concentration or rainfall duration increases from 5 to 60 min. As a result, one should check which portion of the tributary watershed (i.e., T_c) governs the size of each storm sewer.

Computer Programs. There are many desktop computer programs to help the designer size storm sewers by the rational method. Most of them are distributed by private vendors. It is not possible to list all of them, nor is it possible to pass judgment on how well each program performs this task. As a result, this discussion is limited to three programs, two supported by public agencies and one distributed by a professional nonprofit association:

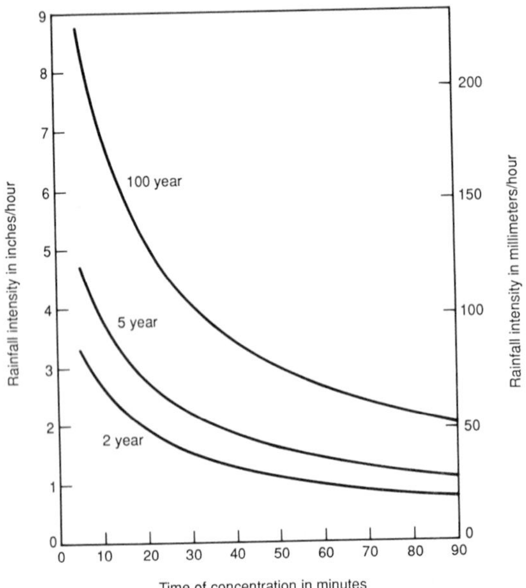

FIGURE 28.4.4 An example of idealized intensity-duration-frequency curves. *(From Ref. 66.)*

- *UDSEWER.* Distributed by Civil Engineering Department, University of Colorado at Denver, P.O. Box 113, Denver, CO 80204-5300; Attn.: Prof. C. Y. Guo.
- *CESTORM.* Distributed by the American Public Works Association, 1313 E. 60th St., Chicago, IL 60637.
- *HYDRA.* Distributed by McTrans, Department of Civil Engineering, University of Florida, Gainesville, FL 32611, under a contract to U.S. Federal Highway Administration, as part of HYDRAIN—Integrated Drainage Design Computer System.[20]

All three programs provide for storm sewer design using rational method. UDSEWER and HYDRA can be used to check the design by using backwater analysis, which accounts for manhole and juncture losses. UDSEWER, when using the rational method, checks which portion of the watershed controls the runoff peak. HYDRA can also generate hydrographs, route overflows down a gutter, store water in ponds, and size stormwater inlets. All three programs also permit user input of design flows and have the capability either to size new sewers or to check whether existing sewers are adequate.

28.4.4 Hydrograph Routing Methods for Design

There are also many good desktop computer programs not based on the rational formula that can assist with the design of storm sewers. Again, most are supported by private vendors and it is not possible to list all of them. Two programs, ILLUDAS and UDSWM2-PC, both supported by public agencies, are briefly described here.

ILLUDAS. The title stands for Illinois Urban Drainage Area Simulator, and this program has its hydrologic roots in the British Road Research Laboratory model.[61] It calculates runoff hydrographs using the time-area curve method and routes them kinematically through any user-specified conveyance system. It has an option to size storm sewers, but it does not account for backwater effects or junction losses. It is available from the Illinois State Water Survey, 2204 Griffith Dr., Champaign, IL 61820.

UDSWM2-PC. The name stands for Urban Drainage Stormwater Model,[65] and this program is a modified version of the runoff block of EPA's Storm Water Management Model (SWMM). It can use, as input, stormwater hydrographs generated by the Colorado Unit Hydrograph Procedure software, or it can generate its own subbasin hydrographs using the EPA SWMM runoff block algorithms.[75] The latest version has an option to size circular storm sewers, but, like ILLUDAS, it does not account for backwater effects or junction losses. It is available from the University of Colorado at Denver (see UDSEWER in Sec. 28.4.3 for address).

Other Models for Analysis and Checking of Design. A very comprehensive distributed routing model developed for the analysis of urban stormwater systems in the Environmental Protection Agency's Storm Water Management Model.[78] It is described in detail in Chap. 21 of this handbook. Its runoff block and transport block may be used to check storm sewer system adequacy based on kinematic routing algorithms.

Its EXTRAN block, however, offers full dynamic routing capabilities and, as a result, can be used to check sewer system adequacy while accounting for backwater and flow momentum effects. This is an important feature in areas with very flat terrain, interconnected sewer loops, and storm sewers affected by downstream tidal fluctuations or the filling of detention basins. This is because nondynamic routing models do not account for the effects of a time-variable energy grade line on flow. Instead, they assume all flow is occurring under normal depth conditions, which is not always valid.

EXTRAN, SWMM Transport, ILLUDAS, CESTORM, UDSWM2-PC, and many other storm sewer design and analysis software systems on the market often do not account for junction losses or losses that occur at hydraulic jumps, where the sewer flow shifts between supercritical flow and subcritical flow. In systems that do not have supercritical flow, the effects of junction losses can be somewhat compensated by increasing the pipe friction factors by approximately 10 percent. In sewer systems that have a mix of subcritical and supercritical flow, hydraulic losses at all junctions and potential zones of hydraulic jumps within the pipes require special attention by the designer.

28.4.5 Data Limitations

Except for some of the larger urban centers, reliable and accurate rainfall/runoff data are not available for the calibration of models. Even in areas where such data exist, they are rarely available for the specific watershed being studied. As a result, the hydrologist is faced with a plethora of hydrologic methods without the benefit of data for calibration. Without calibration, it is difficult to know the accuracy of any hydrologic method in calculating runoff for a given site.

Although some hydrologists estimate runoff using several methods and then compare results, this does not ensure a correct answer. This approach guarantees only consistency with previous experience of the hydrologist. Also, similar answers among several methods may imply only that the methods are based on the same technology, even when different models are used. Collection and analysis of local rainfall, runoff, and water-quality data is the only reliable way to improve the reliability of hydrologic calculations.

28.4.6 Precision versus Accuracy

The ASCE Task Committee on the Design of Stormwater Detention Outlet Structures[2] and McCuen[42] addressed the topic of accuracy in design of detention basin outlets. The committee pointed out that

> Design accuracy is a function of the stormwater management policy as well as the hydrologic and hydraulic models used in design. These factors may either introduce a bias or cause a lack of precision, with bias and precision being components of net accuracy. *Bias* is defined as a systematic variation from the true value, and lack of *precision* results from non-systematic, random, or chance error.

The committee concluded that hydraulic calculations introduce the least inaccuracy and hydrologic calculations, the most. For example, the U.S. Geological Survey[80] reported errors in estimating runoff peaks, using regression equations in South Dakota, to be +152 to −60 percent. While this is an example of the randomness of runoff data, precipitation measurements can routinely be 15 percent low and, for extreme events, 40 percent low, which indicates a bias in rainfall data.[35]

Inaccuracies exist when surface runoff is calculated, and this fact needs to be recognized in design of surface drainage systems. Often, resources are expended in refining hydraulic calculations to a degree rarely justified by the accuracy of hydrologic information. As a result, refinement in calculations beyond two significant figures is rarely realistic. Instead, include a reasonable safety factor when sizing drainage and flood control systems.

28.4.7 Design Procedure and Example

Actual design of a storm sewer system is an involved process. The following steps identify some of the steps that need to be followed when computer software is used to perform pipe sizing calculations:

1. Locate and develop information about the design site. Use Table 28.4.1 as a checklist.
2. Evaluate the site and lay out a preliminary storm sewer system that appears to be cost-effective—that is, uses the least amount of pipe. Do not size the pipe yet.
3. Sketch the preliminary profiles for the top of each pipe segment. Maintain the minimum depth of cover while avoiding interferences with utilities such as sanitary sewers, water lines, and gas lines. Do not forget to account for sanitary sewer service lines that flow under gravity into the collector sewers. There can be large numbers of these lines that control both the vertical and horizontal alignments of storm sewer pipes.
4. Identify the tributary subbasins to the system at major points along the storm sewer such as the location of major land-use changes, juncture manholes, changes in direction or pipe, street intersections, and major inlets.
5. Measure the subbasin areas and determine runoff coefficients, slopes, flow paths, gutter capacities, and potential inlet locations.
6. Locate the most upstream points of the storm sewer network on city streets at points where the allowable gutter capacity may be exceeded. Adjust the preliminary sewer layout to extend to these points, or to other upstream points that need to have inlets to intercept flow.
7. Code the information as input to the design software.
8. Size the storm sewers using software for the design storm.
9. Evaluate the system's hydraulics using backwater calculations, accounting for junction losses.
10. If any pipe segment has excessive surcharge, namely, the hydraulic grade line is at or near the street level, increase size of downstream pipes and repeat steps 8 and 9.
11. Examine the system and try another storm sewer layout that you think may be cost-effective. Then repeat steps 3 through 9.
12. Estimate the cost of each system by comparing cost of pipe, trenching, bedding, and backfill.
13. Determine which layout is most cost-effective and examine this layout to see if there are any unusual items not accounted for, such as the need to relocate a major water line, sewer line, or telephone vault; to excavate in rock; to tunnel under a major highway; or any other factor that can make an otherwise best layout impractical to build.

These steps were followed to design a storm sewer for the basin shown in Fig. 28.4.5 to intercept and convey runoff from a 2-year storm. Selected output from the

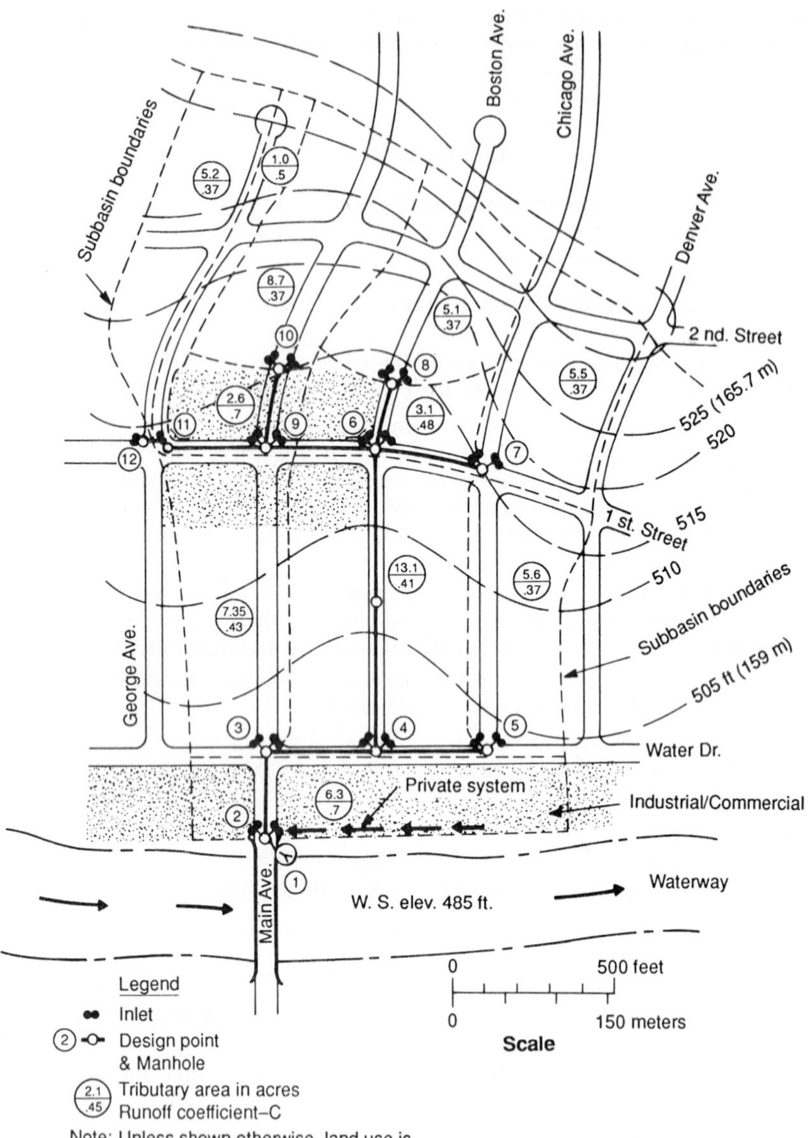

FIGURE 28.4.5 Storm sewer design example watershed.

UDSEWER[63] software are presented in Table 28.4.4. In addition to sizing all pipes, the software also evaluates one existing storm sewer, number 102. Note that the smallest pipe in the network is 18 in (0.46 m) in diameter, which was specified by the user as the minimum size for this design. The software also calculates its own I-D-F curve for the design storm using the 1-h depth for a 2-year storm as input.

28.5 DESIGN OF DETENTION AND RETENTION FACILITIES

Stormwater detention, as an urban stormwater runoff management component, began to gain broad recognition in the early 1970s. Before long it was being used throughout the United States, Canada, Sweden, Australia, and many other countries around the world. Unfortunately, in many instances it was used indiscriminately, or as a political expediency. This was particularly true of on-site detention, which was often required by local governments at each new land-development site.

Typically, the developer is required to install on-site detention which limits peak runoff rates after development to those present before new development took place. This criterion was applied in the past to a single-recurrence-frequency flood such as the 5-, 10-, or 100-year flood. More recently, multilevel control of a range of storm frequencies is often required. Although the concept of multilevel control appears straightforward, it may not always adequately control flooding along natural waterways or control stream erosion.[32,45,59,69,95]

Limiting hydrograph peaks from a single design storm, or even two or three design storms at an individual development site does not address the consequences of the increases in the runoff volume. What works for a single development site may not work well for a large urban watershed with many randomly designed and located on-site detention facilities. Nevertheless, on-site detention will continue to be used.

One should not conclude that use of on-site detention is a poor policy. It is not. On-site detention can be beneficial when used properly, and it does help mitigate some of the impacts of urbanization. It is the easiest way for a local authority to implement runoff controls for new development sites, since on-site detention is financed by the land developer. From a political perspective, a city or a county can point to its on-site detention policy as evidence that it requires new development to control increases in runoff from the site. This can provide a defense, albeit not a perfect one, against claims resulting from increased flooding or environmental damages that may be occurring downstream.

28.5.1 Locating Detention Facilities

Stahre and Urbonas[59] summarized the findings of three independent studies between 1974 and 1983 and concluded the following about randomly located on-site detention facilities:

> . . . a system of detention basins sized using uniform volume and release requirements can, in fact, be more effective in controlling peaks along the drainageways than a system of random designs.

Lakatos,[36] after conducting model studies in New Jersey using the Soil Conservation Service (SCS) TR-20 method concluded that locating on-site detention basins in

TABLE 28.4.4 Example of Storm Sewer Design Using Rational Method

Design conditions at manholes

Manhole ID number	Subbasin area C	Time of concentration: Overland T_o (min)	Time of concentration: Gutter T_f (min)	Time of concentration: Subbasin T_c (min)	Cumulative area C	Cumulative T_c (min)	Rainfall intensity, in/h	Design peak flow, ft^3/s	Ground elevation, ft	Water elevation, ft	Comments
1.00											
2.00	4.41	7.92	4.61	12.53	27.89	21.58	1.89	52.69	502.00	489.61	OK
3.00	3.16	12.97	5.28	18.25	23.48	21.18	1.91	44.80	503.00	492.46	OK
4.00	5.37	14.24	4.72	18.96	20.32	20.57	1.94	39.38	503.00	496.40	OK
5.00	2.07	13.35	3.95	16.28	2.07	16.28	2.18	4.52	504.00	500.32	OK
6.00	1.49	10.59	1.94	12.54	12.87	18.68	2.04	26.23	512.00	503.07	OK
7.00	2.04	6.10	4.17	10.26	2.04	10.26	2.68	5.45	515.00	510.40	OK
8.00	1.89	10.56	2.70	13.25	1.89	13.25	2.40	4.53	514.00	508.52	OK
9.00	1.82	7.12	1.77	8.88	7.46	17.02	2.14	15.69	513.00	505.62	OK
10.00	3.22	16.08	3.92	16.11	3.22	16.11	2.19	7.06	515.00	510.53	OK
11.00	0.50	5.25	5.64	10.89	2.42	16.80	2.15	5.21	514.00	508.78	OK
12.00	1.92	14.93	4.41	16.67	1.92	16.67	2.16	4.15	514.00	509.29	OK

Sewer design and hydraulics

Sewer ID number	Manhole number		Sewer shape	Required dia. high, in	Suggested dia. high, in	Pipes existing	Slope, %	Invert elevation		Sewer length, ft	Pipe surchrge length, ft
	Upstream ID no.	Downstream ID no.				Dia. high, in		Upstream, ft	Downstream, ft		
102	2	1	ROUND	28.94	30.00	33.00	2.00	487.25	484.65	130.00	0.00
203	3	2	ROUND	30.47	33.00		1.10	490.25	487.50	250.00	0.00
304	4	3	ROUND	29.55	30.00		1.00	494.30	490.80	350.00	0.00
405	5	4	ROUND	13.13	18.00		1.00	499.50	496.00	350.00	0.00
406	6	4	ROUND	26.46	27.00		0.80	500.75	493.55	900.00	75.37
607	7	6	ROUND	13.21	18.00		1.40	509.50	504.46	360.00	0.00
608	8	6	ROUND	13.70	18.00		0.80	507.70	504.90	350.00	0.00
609	9	6	ROUND	21.82	24.00		0.80	504.20	501.40	350.00	0.00
910	10	9	ROUND	15.51	18.00		1.00	509.50	506.00	350.00	0.00
911	11	9	ROUND	14.43	18.00		0.80	507.90	505.10	350.00	0.00
1112	12	11	ROUND	13.25	18.00		0.80	508.50	508.18	40.00	0.00

the central zone of the watershed was most effective, while locating detention in the lower zone may actually result in peak flow increases downstream because the downstream detention holds back water until the upstream peaks arrive.

In many cases, runoff from new city streets is not considered a part of the developer's land, and is not captured by on-site detention. This simple oversight is commonplace. Yet, runoff from streets is a significant portion of the total runoff that occurs from urbanized areas, and needs to be accounted for in modeling on-site detention.

28.5.2 Sizing Detention Volume

The volume of stormwater that is detained during a storm is a function of the volume of runoff, the detention facility's outlet characteristics and the available storage volume in the facility. The designer's goal is to find a storage volume that will match the pond's discharge limits for the specified set of design storms. In essence, Eq. (28.5.1) has to be solved, which states that the storage volume is the time integral of the difference between the inflow and outflow hydrographs:

$$V_{max} = \int_0^{t_*} (Q_{in} - Q_{out})\, dt \tag{28.5.1}$$

where V_{max} = storage volume
t_* = time from beginning of runoff to a point of maximum storage, where $Q_{in} = Q_{out}$ on the hydrograph recession limb
Q_{in} = inflow rate
Q_{out} = outflow rate

Figure 28.5.1 illustrates what this calculation represents. To determine the required storage volume, the designer routes one or more runoff hydrographs through a preliminarily sized detention facility using one of the computer programs available for this purpose. Computer design programs, such as HYDROPOND,[64] solve Eq. (28.5.1) using finite-difference numerical calculating procedures. Many of these are described in detail in the SCS *National Hydrology Handbook*.[81] Reservoir storage routing is also described in Chap. 10 of this handbook.

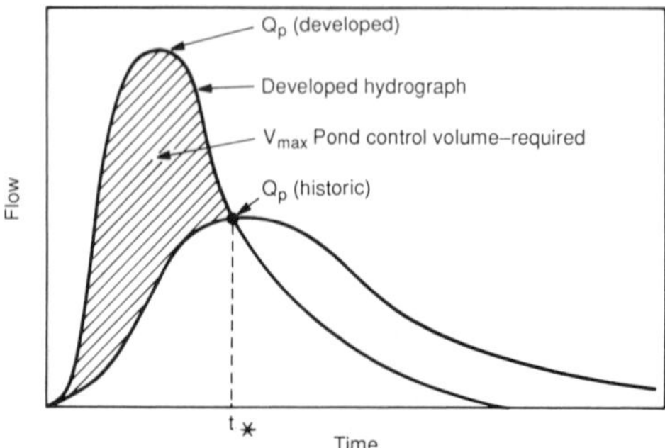

FIGURE 28.5.1 Illustration of detention storage volume computation.

Design Process Using Hydrographs. Briefly stated, the classic detention sizing procedure consists of the following steps:

1. Estimate the storage volume V_{est} needed, using a simplified assumption such as that suggested by Eq. (28.5.2).
2. Using site topography, prepare a preliminary layout of a detention basin that has the desired volume and outlet configuration.
3. Code the stage-storage-outflow characteristics of the trial pond as input into the detention design software you are using.
4. Route the input hydrographs through the pond model.
5. If the trial size and outlet assumptions do not satisfy the desired goals, resize the basin and/or reconfigure the outlets and repeat steps 3 through 5 until the design goals are achieved.

$$V_{est} = 1.1\, V_{in} \left(\frac{Q_{pin}}{Q_{pout}} \right) \tag{28.5.2}$$

where V_{est} = first estimate of required storage volume
V_{in} = volume of inflow hydrograph
Q_{pin} = peak inflow rate
Q_{pout} = desired peak outflow rate

Inflow hydrographs can be generated by using any surface runoff model, including the ones mentioned in Sec. 28.4.4 and in Chap. 9. Detention design using reservoir storage routing of hydrographs is the preferred technique for large detention basins. This approach is mandatory for designing a system of many detention basins. Storage routing is time-consuming and not as accurate for watersheds less than 160 acres (65 ha), as may be implied in the calculation rigor. As a result, less rigorous I-D-F-based procedures can be used to quickly size storage volumes of on-site detention facilities.

I-D-F-Based Detention Design Procedure. I-D-F-based (i.e., rational method) procedures can be used to find storage volumes needed to control a single runoff event such as 2-, 10-, or 100-year storm. This procedure was first introduced by Kropf in 1957 and was adopted by the Federal Aviation Agency (FAA).[18] This simple mass balance technique, when properly used, gives reliable estimates of detention volumes needed for small urban watersheds. Other rational method-based procedures are described in the literature.[1,4,6,59]

The FAA procedure assumes that rainfall volume accumulates with time and is a time integral of the desired I-D-F curve. Equation (28.5.3) transforms the rainfall volume-duration curve into a runoff volume-duration curve using the watershed's runoff coefficient and its tributary area. It is used to estimate the cumulative runoff volume entering a detention basin. The cumulative volume leaving the basin is estimated by Eq. (28.5.4).

$$V_{in} = K_u CiAT \tag{28.5.3}$$

where V_{in} = cumulative runoff volume, ft³ (m³)
C = runoff coefficient
K_u = 1.0 for U.S. standard units (0.28 for SI units)
i = storm's intensity taken from the I-D-F curve at time T, in/h (mm/h)
A = tributary area, acres (km²)
T = storm duration, s

$$V_{out} = k\, Q_{out}\, T \tag{28.5.4}$$

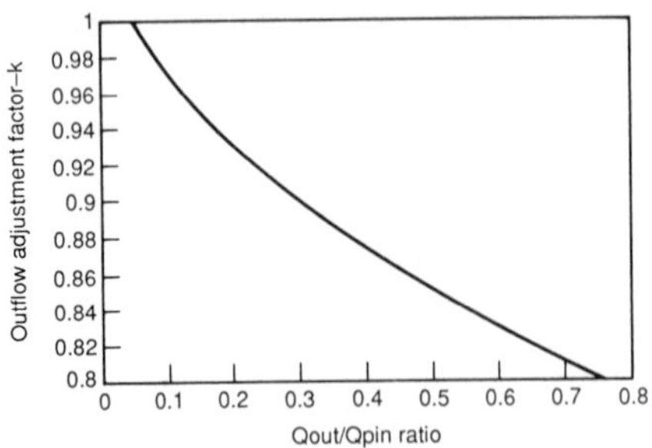

FIGURE 28.5.2 Outflow adjustment factor versus outflow rate/inflow peak ratio. *(Used with permission from Ref. 64.)*

in which T is defined above and

V_{out} = cumulative volume of outflow, ft³ (m³)
Q_{out} = maximum outflow rate, ft³/s (m³/s)
k = outflow adjustment coefficient from Fig. 28.5.2; find Q_{pin} (peak inflow rate) from the rational formula, Eq. (28.4.3)

The required detention volume is then found by using Eq. (28.5.5), which states that the required storage volume is the maximum difference between the cumulative inflow and the cumulative outflow volumes.

$$V = \max(V_{in} - V_{out}) \tag{28.5.5}$$

This procedure assumes a constant outflow rate Q_{out}, which is the rate of discharge when the detention basin is full. Discharge varies, however, with the depth of water. This fact is compensated for by the outflow adjustment factor k.

Example. Determine the size of an on-site detention basin to limit the peak runoff rate for a 10-year storm from a 100-acre (40-ha) single-family residential site to no more then 40 ft³/s (1.1 m³/s). The solution is as follows. From a runoff analysis and the desired release rate, it is determined that $Q_{out}/Q_{pin} = 0.3$. From Fig. 28.5.2, the value of $k = 0.90$. The entire calculating procedure is set up on a spreadsheet, and its printout is reproduced as Table 28.5.1. We see that the needed detention basin volume is estimated as 2.85 acre · ft (35,100 m³). The procedure is illustrated graphically in Fig. 28.5.3.

28.5.3 Design Considerations

Inflow Structures. Erosion and deposition occur at inflow points to detention ponds. Inflow structures to convey the flow into the basin and to control erosion are needed whenever flow has to drop into a pond or into an empty detention basin. These structures can be drop manholes or surface rundowns. For large flows that have to drop several feet in elevation, more elaborate energy-dissipating structures have to be used such as a standard U.S. Bureau of Reclamation (USBR) baffle chute.[73,75] Figure 28.5.4 is a photograph of an inflow structure constructed for a detention basin in Denver which also serves as a park. This structure is a variation of the USBR baffle chute, is attractive, and fits the urban park setting very nicely while dissipating inflow energy.

TABLE 28.5.1 Example of Rational Formula Method for Detention Basin Sizing

Project title: Example Problem

Basin size A:	100 Acres
% impervious:	40 percent
Runoff coefficient C:	0.4
Design frequency:	10 year
1-h rainfall:	1.65 inches
Peak discharge Q:	40 cfs
Outflow rate factor k:	0.90

Storm duration, min	Rainfall intensity, in/h	Runoff volume, ft³	Outflow volume, ft³	Storage volume, ft³	Storage volume, acre · ft	
T (1)	*I* (2)	*CIAT* (3)	*kQT* (4)	(3) − (4) (5)	(6)	
0.0	7.70	0	0	0	0.00	
5.0	5.60	67,718	10,800	56,918	1.31	
10.0	4.46	108,028	21,600	86,428	1.98	
15.0	3.75	135,974	32,400	103,574	2.38	
20.0	3.25	157,093	43,200	113,893	2.61	
25.0	2.88	173,959	54,000	119,959	2.75	
30.0	2.59	187,952	64,800	123,152	2.83	
35.0	2.36	199,888	75,600	124,288	2.85	*
40.0	2.17	210,288	86,400	123,888	2.84	
45.0	2.02	219,499	97,200	122,299	2.81	
50.0	1.88	227,765	108,000	119,765	2.75	
60.0	1.67	242,130	129,600	112,530	2.58	

* Required detention volume.

Note: 1 acre · ft = 43,560 ft³ = 1233.5 m³; 1 ft³ = 0.0283 m³; 1 in. = 25.4 mm.

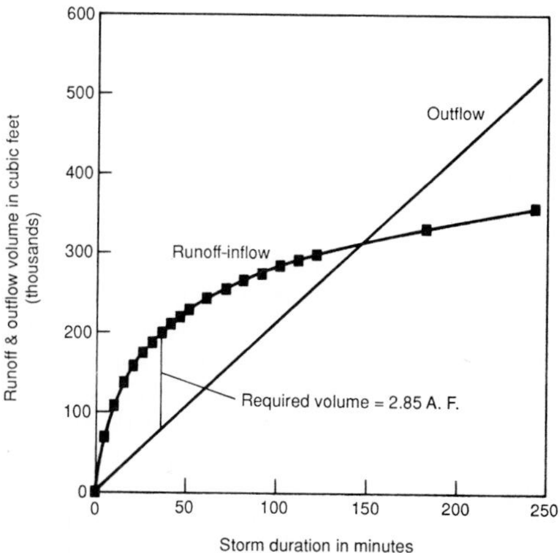

FIGURE 28.5.3 I-D-F-based method for sizing an on-site detention basin. *(After Ref. 18.)*

FIGURE 28.5.4 An inflow control structure in Denver. *(Photo by Eric Stiles, U.S. Bureau of Reclamation, Denver, Colo. Used with permission.)*

While erosion can be reduced through the use of inflow structures, nothing can be done to prevent sediment deposition. In fact, deposition of sediments should be encouraged to take place in areas that are accessible to maintenance equipment. Sediment will eventually need to be removed from all detention facilities, and encouraging deposition at selected locations reduces removal cost.

Outlet Structures. Outlet structures have greater incidence of maintenance than other structures in a detention basin, especially the very small outlets susceptible to clogging. They can be an operational headache. A clogged outlet, or one that has been otherwise modified (i.e., through vandalism, improper maintenance, etc.), will invalidate even the most elaborate and sophisticated hydrologic design. Trash, debris, vandalism, children plugging outlets to see what happens, and other factors are all at work in urban areas to modify the characteristics of detention outlets.[2,4,13,56,59]

Each outlet serves a unique flow control function and has to be designed with clogging, vandalism, maintenance, aesthetics, and safety in mind. Triangular weirs, while offering a wide range of hydraulic control, are notorious for catching trash and debris in the bottom of the V. When conditions permit, rectangular weirs, hooded overflow risers, or vertical slits can reduce operational problems. For small on-site facilities (ones with very small outflow rates) normal outlet designs, such as suggested by the U.S. Bureau of Reclamation,[74] will not provide sufficient flow throttling. Instead, the designer has to use small orifices, perforated risers, and oversized trash racks to mitigate outlet clogging.

Figure 28.5.5 is a photograph of a vertical slit weir type of an outlet control structure located in front of a large outlet pipe. This structure works well when the outflow rate is more than 1 ft^3/s (0.093 m^3/s) because it is less susceptible to clogging

FIGURE 28.5.5 A vertical slit outlet from a detention basin. *(Used with permission from Urban Drainage and Flood Control District, Denver, Colo.)*

than a simple orifice. When the inflow exceeds the primary capacity of this structure, water spills over the top of the weir and continues to discharge through the outlet pipe. Very large floods overtop a saddle spillway and leave the basin along an adjacent street.

Figure 28.5.6 is a chart that can be used as a guide to size trash racks that can be placed in front of orifice-type outlets. When the outflow rate is very small, such as for extended detention basins used for water quality, trash racks can become impractical. In such cases the outlet, often a perforated riser, is packed in coarse gravel. Figure 28.5.7 depicts a standard design for such an outlet used in Littleton, Colo.[8]

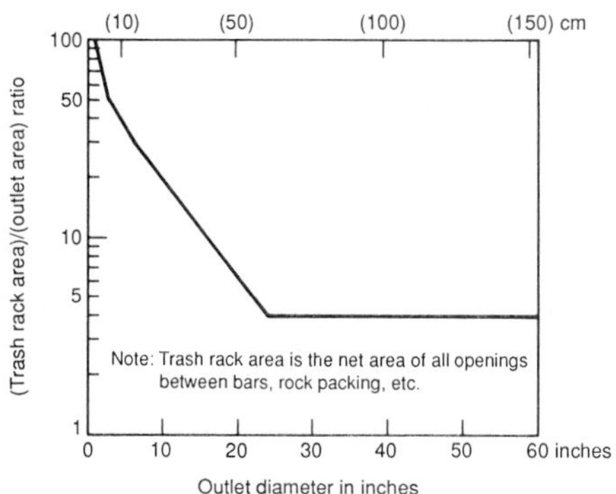

FIGURE 28.5.6 Minimum trash rack size versus outlet diameter. *(Used with permission from Ref. 66.)*

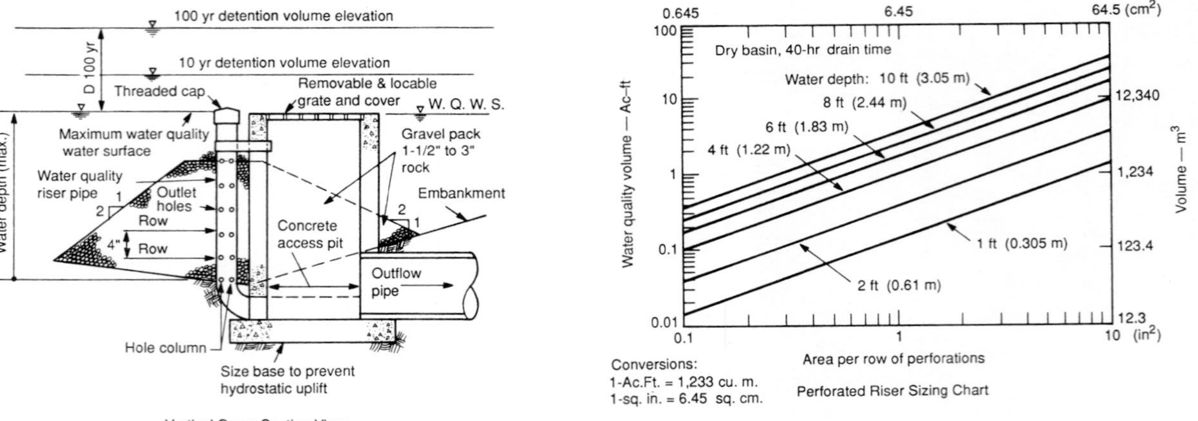

FIGURE 28.5.7 Example of a flood control and water-quality outlet structure. *(Used with permission from Ref. 8.)*

Aesthetics. Although there are examples of unattractive "holes in the ground," most new detention facilities are tastefully incorporated into the urban setting in which they reside. This is not a hydrologic consideration, but it is a consideration which will be used by the public to judge these facilities. Aesthetics of the finished facility are important, and the services of a landscape architect can help with this aspect of the design.

Safety. Safety is only partially a hydrologic design issue, and it also includes the structural integrity of the water-impounding embankment and its ability to withstand floods greater than the nominal design. Safety to the public when the facility is in operation and when the facility is dormant, namely, between runoff events, is very important. The designer needs to consider flow velocities, water depth, and how to prevent and to discourage the public from being exposed to high-hazard areas during periods of storm runoff. In addition, the designer needs to size an emergency spillway and/or design the embankment so it will not fail catastrophically during a very large event.

When the facility is not operational, which is most of the time, its layout should minimize the use of high vertical drops, deep water near the shore, and steep side slopes above and below the permanent water level. Also, outlet and inflow structures require special attention. Use of flat side slopes, flat benches above and below permanent pool water level, planting thorny shrubs around inflow and outflow structures, and the use of trash/safety racks at all outlet orifices and pipes all help to enhance the safety of detention facilities.[2,4,26,49,59]

28.5.4 Additional Considerations for Water Quality

When designing detention to enhance urban stormwater quality, the designer has to modify ordinary thinking and focus on the capture of small, frequently occurring runoff events. Stormwater quality detention facilities need to facilitate the settling of small sediment particles found in urban runoff.[33,37,38,39,52,59,60,92] These small particles, less than 60 μm in size, account for approximately 80 percent of the particulates in the water column.

Average residence times of 12 to 60 h in a pond are needed to achieve significant removal rates of these small suspended sediments. As a result, outflow rates for water-quality control are much smaller than for control of peak runoff rates for drainage and flood control purposes. The designer is faced with the difficult problem of providing for aesthetics, maintenance, operation, and compatible recreational uses for a facility experiencing frequent ponding and sediment deposition.

28.6 DESIGN FOR WATER-QUALITY ENHANCEMENT

28.6.1 Introduction

In 1987 the United States Congress passed the Clean Water Act. Section 402(p) of this act requires that stormwater discharges from industries and municipal separate storm sewer systems be regulated through permits issued by the U.S. Environmental Protection Agency (EPA). These permits are to include requirements for

> . . . controls to reduce the discharge of pollutants to the maximum extent practicable, including management practices, control technologies and system, design and engineering methods, and such other provisions as . . . appropriate for the control of such pollutants.

The good intent of the Congress at the time notwithstanding, the technological state of the art with respect to the cost-effective removal of pollutants from urban runoff is in its infancy. While studies have been conducted to quantify pollutant concentrations in urban runoff,[53,60,62,77] the data collected so far are insufficient to develop definitive cause-effect relationships between pollutant sources and their impacts on streams and rivers. This probably is a result of the complexing nature of stormwater affecting the toxicology of many pollutants.[11] The only identified, yet inconsistent, effects of separate stormwater urban runoff on receiving waters were reported for some tidal estuaries and lakes.[5,77] Furthermore, the pollutant removal efficiencies of control measures and devices are described by approximate and uncertain empirical relationships, so urban runoff quality control is still very much an educated art. Effective stormwater quality management requires a working knowledge of hydrology, water chemistry, and aquatic ecology.

28.6.2 The Hydrology of Urban Runoff Quality Control

In contrast to flood control and drainage projects, which are based on large infrequent storms, design storms for runoff quality control are small, frequently occurring events—generally less than the 1-year storm. The reason for this is seen in Fig. 28.6.1, which shows the percent of runoff capture that can be achieved for a watershed with a runoff coefficient of 0.5 using various sized detention basins. This curve was derived by Roesner for the City of Cincinnati by continuous modeling for 40 years of hourly rainfall data at the Cincinnati airport and a drawdown time of 24 h for a brimful detention basin. It shows that almost 95 percent of the runoff events can be captured with a storage volume of 0.5 watershed inches, 0.5 acre · in/acre (12.7

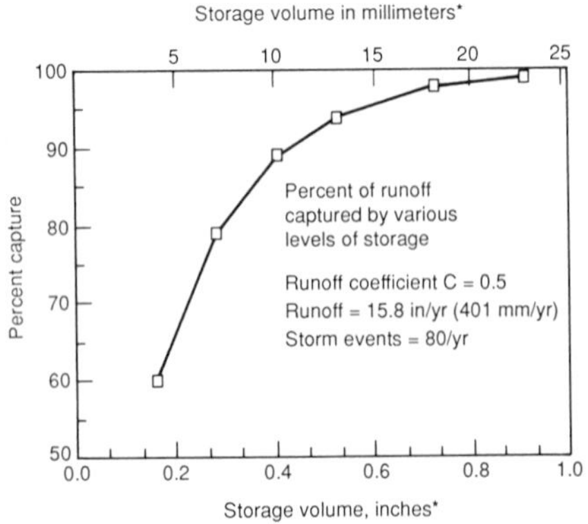

FIGURE 28.6.1 Percent runoff capture versus storage volume in Cincinnati.

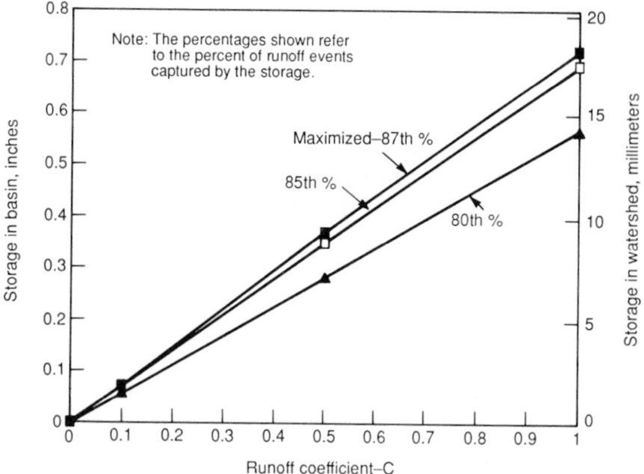

FIGURE 28.6.2 Capture volume in Denver with 40-h drain time. *(Used with permission from Ref. 71.)*

watershed millimeters). Stating it differently, such a basin overflows during only 5 percent of runoff events, or four times a year on the average.

Similar results may be found at other locations. For example, Fig. 28.6.2 shows similar findings for Denver using a slightly different set of curves.[70,71] It shows that for a catchment with a 0.5 runoff coefficient, 87 percent of runoff events can be totally captured with a basin having 0.37 in of storage. Despite the differences in climate between Cincinnati and Denver, the storage volume required to achieve this capture rate is practically identical at both locations.

For the Cincinnati example described above, capturing the 1-month storm (12 overflows per year) will totally treat almost 80 percent of the runoff events. At the same time, capturing the 2-month storm (6 overflows per year) will totally treat almost 90 percent of the runoff events. Regardless of the number of overflows that will occur, the combined sewage entering these basins during the storms that cause them to overflow will still receive primary treatment as the flow passes through the basins.

The slope of the curve in Fig. 28.6.1 indicates that the increase in storage needed to capture one additional percent of runoff events beyond the 90 percent level is in an area of rapidly diminishing returns. The point of diminishing returns appears to be similar at other locations and appears to occur when 80 to 90 percent of all runoff events are captured. However, analysis of local rainfall and runoff should be performed to develop similar detention sizing criteria for water-quality control.

28.6.3 Basic Principles for Planning Urban Runoff Quality Controls

There are five basic factors involved in the control of urban runoff quality. They are

- Prevention or reduction of pollutant deposition in urban areas
- Source control of pollutants
- Minimizing directly connected impervious area to the stormwater conveyance system

- Designing controls for the small storm, generally less than 1 in rainfall
- Using the treatment train concept

Preventing Deposition of Pollutants. Preventing the deposition of pollutants in urban areas is a best-management practice that, through good housekeeping measures, eliminates or reduces pollutants entering the urban stormwater systems. This simple concept requires the cooperation of the general public.

People need to become aware of the problem they may be causing. For example, draining crankcase oil onto an empty lot or down a gutter results in this oil polluting streams and lakes. Public education and the installation of public facilities are needed for the disposal of household products such as crankcase oil, antifreeze, gasoline, pesticides, herbicides, paints, and solvents. Public awareness and prompt reporting to local public health authorities can help to quickly locate and deal with illegal dumping of pollutants and hazardous wastes.

Source Controls. Source controls are measures that prevent pollutants from ever coming into contact with the rainfall and/or runoff. They are the best and cheapest controls and include: covering of chemical storage areas, diking around chemical unloading and potential spill areas, minimizing the use of deicing chemicals, careful use and handling of household herbicides and pesticides, and disconnecting illicit wastewater connections to storm sewers. Examples of illicit connections to separate storm sewers are: wastewater sewers from buildings, floor drains from car wash areas, drains from chemical storage areas, all of which should be connected to a sanitary sewer system. Hydraulic relief connections from the sanitary sewer to a storm sewer should also be disconnected.

A study in Michigan[46] revealed that dry weather discharge of pollutants from storm drains serving commercial and industrial areas is a major contributor of toxic chemicals to a river. Hubbard and Sample[29] found that in an industrial area of Seattle, Wash., a major source of pollutants in the stormwater came from old, pollutant-laden sediments deposited in the storm sewer. These sediments were contaminated by illegal dumping of industrial wastes over a number of years. In both instances, illegal dumping and illicit connections to storm sewers were the primary causes of toxic effects observed in the receiving waters. Simply being alert to source control opportunities and practicing good housekeeping on the urban landscape can appreciably reduce the amount of pollutants entering receiving waters through separate storm sewers (those not carrying municipal sewage).

Minimizing Directly Connected Impervious Areas. After prevention and source control, the next most cost-effective runoff quality control is the minimizing of directly connected impervious area (DCIA). DCIA is the impermeable area that drains directly to an improved drainage component, such as a street gutter, ditch, or pipe. Minimizing DCIA is effective because it results in slowing of stormwater runoff as it travels to the drainage system. This maximizes the opportunity for rainfall to infiltrate where it falls. Moreover, contact with grassed areas filters the stormwater and removes some of the pollutants it contains.

Minimizing DCIA is effective only during small storms, but these are the storms of greatest concern for water quality. In some situations it may be possible to divert runoff from smaller storms and the initial runoff from larger storms into a treatment device where the water is removed from the main drainage system. In other cases the stormwater management quantity and quality controls may be combined in the same facility using dual controls, namely, one set of controls for regulating small storms and a second set designed to function during the larger, flood-producing storms.[4,17,41,56,59]

Treatment Train. The treatment train concept of runoff quality management was first introduced by Livingston in Florida.[40] This concept views runoff quality management as a set of treatment practices in series, as illustrated in Fig. 28.6.3. The first process in the train is, of course, source control followed by individual lot or parcel controls, such as minimizing of DCIA, that reduce and filter runoff. Further down in the drainage system, larger controls that serve areas of 5 to 50 acres (2 to 20 ha) may be added; finally, regional controls that can treat runoff from areas ranging from 100 to 600 acres (40 to 240 ha) can be installed.

The number of processes required in the treatment train depends on the availability of land, economics, the degree of pollutant removal desired/required for the system, and the type of practices planned for the system. With the exception of infiltration practices, a single-treatment practice can eliminate, at best, 80 percent of the suspended solids and much lower percentages of nutrients from runoff. Also, removal during any one storm event may be highly variable, depending on the size of the storm, the time since the last storm, the working condition of the treatment practices, and other factors not yet fully understood. Table 28.6.1 summarizes the findings of an extensive literature study by the Stormwater Management Task Force established in 1989 by the Colorado Department of Health.[67] As can be seen, the range in performance varies substantially and depends much on the site-specific details of each facility.

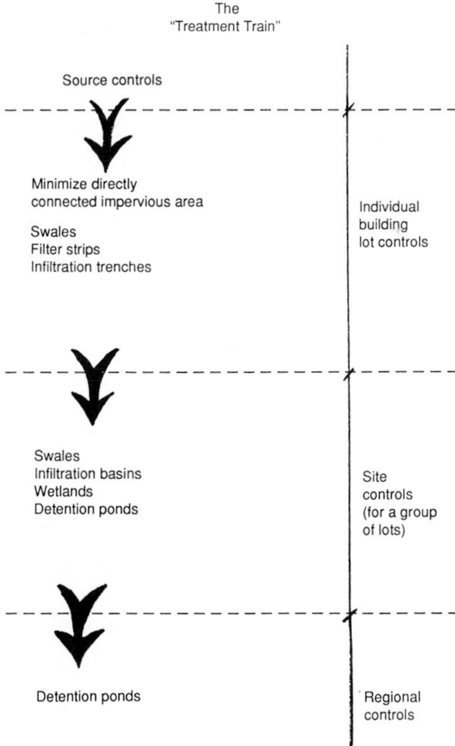

FIGURE 28.6.3 The treatment train for control of urban stormwater runoff quality.

The treatment practices identified in Table 28.6.1 can be grouped into two basic categories: infiltration practices and detention. Wetlands[89] can also be used as another form of stormwater detention, but this practice currently lacks sound, field-verified design criteria.

28.6.4 Infiltration Practices

Infiltration practices generally fall into four classes: (1) *swales* and *filter strips,* (2) *porous pavement* and *modular pavement,* (3) *percolation trenches,* and (4) *infiltration basins.* Since infiltrated water ultimately percolates into the groundwater, care should be taken with these practices when applied near well fields, or when gasoline stations, chemical storage areas, or other industrial activities drain into them.

Swales and Filter Strips. These are among the oldest stormwater control measures, having been used along streets and highways and on farms for many years. A *swale* is a shallow vegetated trench with a flat longitudinal slope and mild side slopes. A *filter strip* is simply a strip of land across which stormwater from a street, parking lot, rooftop, etc., sheet-flows before entering an adjacent conveyance system. Swales and filter strips serve a similar function to the minimizing of directly connected impervious area. They reduce the runoff velocity and provide an opportunity for runoff to infiltrate into the soil. They also filter the flow from small storms, removing some of the suspended solids and other pollutants attached to the solids. Where the soil underlying the swales and filter strips has a high infiltration rate, the removal efficiency of these controls can be very good—in excess of 80 percent removal,[90] however, in most cases, the removal efficiency of a swale is very low.

Sizing of grass-lined swales and channels for flow conveyance can be done using standard open-channel equations.[14] The length of a swale or filter strip needed to infiltrate the water-quality design storm can be found from Eq. (28.6.1), derived by

TABLE 28.6.1 Potential Pollutant Removal Rates, in Percent, of Various Treatment Practices[67]

Type of practice	Total Suspended solids	Total P	Total N	Zinc	Lead	Biological oxygen demand (BOD)	Bacteria
Porous pavement[1]	85–95	65	75–85	98	80	80	n/a
Infiltration[1]	0–99	0–75	0–70	0–99	0–99	0–90	75–98
Percolation trench[1]	99	65–75	60–70	95–99	N/A	90	98
Retention ponds[2]	91	0–79	0–80	0–71	9–95	0–69	n/a
Extended detention[3]	50–70	10–20	10–20	30–60	75–90		50–90
Wetland[4]	41	9–58	21	56	73	18	n/a
Sand filters[5]	60–80	60–80	−110–0[6]	10–80	60–80	60–80	n/a

[1] Schueler.[56] Estimates based on assumed removal efficiencies and modelling

[2] EPA.[76] Based on NURP data without regard to design.

[3] EPA,[77] Grizzard et al.,[24] Whipple and Hunter.[92] Based on field and laboratory data.

[4] USGS[79] for all constituents except total P. Based on average performance from 13 sampled runoff events in Orlando, Fla. Lakatos and McNemer[37] for total P only. Summary of field data from eight study sites in eastern United States.

[5] Veenhuis et al.,[83] based on field data in Austin, Texas. Wanielista et al.[88]

[6] Sand filters contribute rather than remove Total N.

Wanielista.[87] Most of the time it is not possible to design a swale or filter strip of sufficient length to infiltrate 100 percent of the design storm runoff.

$$L = \frac{K_u(KQ^{5/8}\, S^{3/16})}{(n^{3/8}\, i)} \tag{28.6.1}$$

where L = length of swale to infiltrate the flow, ft (m)
K_u = 1.0 for U.S. standard units, 77.3 for SI units
K = 10,000(V/H), where V and H are the vertical and horizontal distances of the side slope of the swale
Q = average flow rate in the swale, ft³/s (m³/s)
S = longitudinal slope in ft/ft (m/m) of the swale or filter strip
n = Manning's coefficient
i = saturated infiltration rate, in/h (cm/h)

Equation 28.6.1 appears to have physical inconsistencies, since, as the slope S approaches zero, the length of the swale approaches zero. However, since the flow rate also approaches zero with a diminishing longitudinal slope, the integrity of this equation is maintained. When the slope becomes zero, the designer needs to treat the swale as a long and narrow infiltration basin by balancing the inflow with infiltration into the ground.

Other design criteria for swales and filter strips include: the longitudinal slope should be less than 0.02 ft/ft (m/m) to minimize erosion (use grade check structures to achieve this); side slopes for swales must be no steeper than 4H:1V for easy mowing, but preferably 8:1 or 10:1 to maximize the amount of water coming into contact with the soil and vegetation; the seasonal high groundwater level, or bedrock, must be at least 4 ft below an infiltration swale's bottom; and the filter strip must be designed so that runoff flows across it in a sheet and does not channelize. The filter strip should be at least 20 feet (6 m) long.

Porous Pavement. Porous pavement, especially porous concrete, and modular pavement have a potential for use in parking, or overflow parking areas. Porous pavement can help reduce the amount of land required for runoff quality control and preserve the natural water balance at a given site. Porous concrete can also provide a safer driving surface in the rain. It is necessary that the underlying soil be permeable, the ground be fairly flat, and the seasonal high groundwater be at least 3 ft (0.9 m) and bedrock at least 4 ft (1.2 m) below the paved surface.

Porous concrete pavement is used extensively in the state of Florida,[21] but caution should be exercised when applying this practice in regions with colder climate. Questions regarding the integrity of the pavement when subjected to winter freeze-thaw conditions still need to be resolved. On the other hand, modular porous block pavement, as shown in Fig. 28.6.4, has the potential of working in all climates.

Percolation Trenches. These devices can be located on the surface of the ground or buried beneath the surface. They are usually designed to serve tributary areas less than 5 acres (2 ha), but can serve areas as large as 10 acres (4 ha). They must be located in good, high-permeability soils and in areas with a low groundwater table. An infiltration trench generally consists of a long trench, ranging from 3 to 12 ft (0.9 to 3.7 m) in depth, which is backfilled with stone aggregate, allowing for the temporary storage of stormwater in the pores of the aggregate material (see Fig. 28.6.5). Stored runoff then infiltrates into the surrounding soil. Since the bottom of the trench will seal first with the fine particulates brought in by the stormwater, it is best to assume that exfiltration occurs only through the trench sidewalls.

FIGURE 28.6.4 Typical modular pavement parking area. *(Photo by Eric Livingston, Department of Environmental Regulation, State of Florida. Used with permission.)*

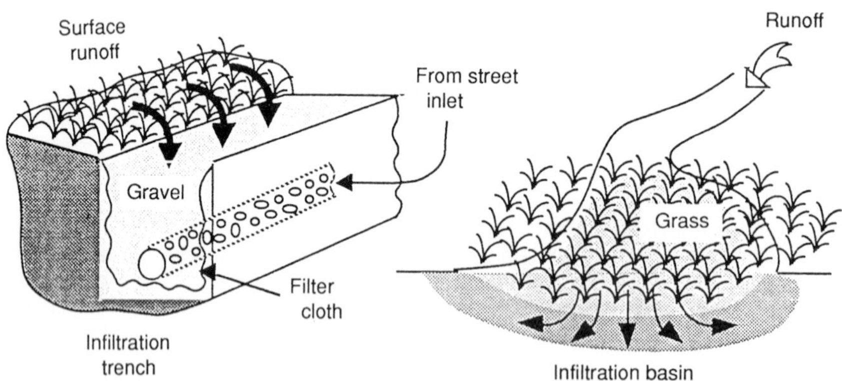

- Efficiency: Excellent (Small depressional infiltration basins are good onsite controls)
- Function: Infiltrates runoff to groundwater; soil filter pollutants
- Maintenance-intensive (mowing, upstream erosion control)
- Nonfuntional if plugged
- Soil must be highly permeable, and wet season water table 4 feet below bottom
- Experience with underdrained infiltration basins is poor

FIGURE 28.6.5 Infiltration devices.

There are two basic types of *percolation trenches,* the *open surface* type and the *underground* or *covered* type.[56,59] There is also a *dry well,* a device used primarily in Maryland to store and infiltrate rooftop runoff. Infiltration trenches are appropriate in highway medians, at parking lot edges, and in narrow landscaped areas.

Clogging is a major problem in percolation trenches. Once clogged, the device will no longer dispose of the accumulated stormwater. Some form of runoff pretreatment, such as swale drainage, use of filter strips, and/or the use of filter devices must precede the introduction of stormwater into the trench. It is also recommended that the infiltration trench not be installed in a new construction area until construction is complete and the soils in the tributary drainage area are stabilized by sod or seeding and mulching to prevent erosion. Clogging during construction is a common cause of failure. Rehabilitation of a clogged trench is a major and very expensive effort.

To compute the required trench size, the hydraulic conductivity of the adjacent soils and the porosity of the gravel media must be known.[59] Low unit loading rates must be used so that the facility can "rest" between runoff events to permit soil pores to drain fully. This practice in septic leaching fields appears to reduce failure rates.

The maximum time to drain the water from a trench should be 48 to 72 h. In addition, the trench bottom must be at least 4 ft (1.2 m) above the seasonal high groundwater and bedrock levels. Filter cloth or granular filter media must be placed around the trench walls, top, and bottom to prevent the adjacent soils from entering the aggregate. Aggregate for filling the trench must be free of fines, and be 1 to 3 in (2 to 7 cm) in diameter. Lightweight excavation equipment should be used so that the soil around the trench is not compacted, and observation wells should be installed to permit periodic inspection for standing water, which may indicate failure.

Infiltration Basins. An *infiltration basin* is a retention facility in which captured runoff is infiltrated into the ground. An infiltration basin may be made by constructing an embankment and impounding water behind it or by excavating a depression to capture the runoff from the design-water-quality storm. If properly designed, the basin can be integrated into the landscaping as a depressed open area. Trees, plants, and grass can be planted within the basin if water is not stored longer than 24 to 36 h. An infiltration basin can serve an area as small as a front yard and up to tributary areas having 50 acres (20 ha); however, its reliability diminishes as the tributary area increases.

Infiltration basins are most efficient when they are constructed so that during larger storms the basin captures the first portion of the runoff and the remaining flow is diverted around the basin. A schematic of a dual-pond off-line system with a smart weir designed to divert the initial runoff volume to the infiltration basin is shown in Fig. 28.6.6.

Infiltration basins must be located on very porous soils and the basin bottom must be at least 4 ft (1.2 m) above the seasonal high groundwater or bedrock. Exfiltration drain time should never exceed 72 h except for landscaped basins, which should drain in no more than 24 to 36 h. These basins need to be vegetated for aesthetic reasons. In addition, grass and other vegetative roots break up the surface soils and keep the infiltrating surfaces from getting plugged. It is suggested that basin side slopes be less than 4H:1V to facilitate maintenance and reduce erosion of the banks. Other design criteria include the use of light equipment for excavation and maintenance and a need for an emergency spillway to pass larger floods when the basin is on-line. For additional guidance and insight into the design and use of infiltration basins, see Refs. 28, 54, 56, 59, and 72. Field observations of older installations indicate a very high rate of failure for infiltration basins and percolation trenches.

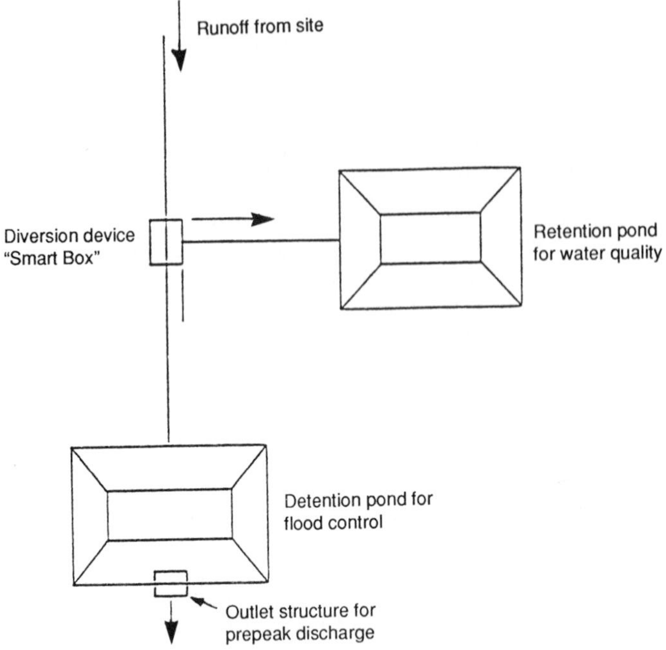

"Smart Box" schematic

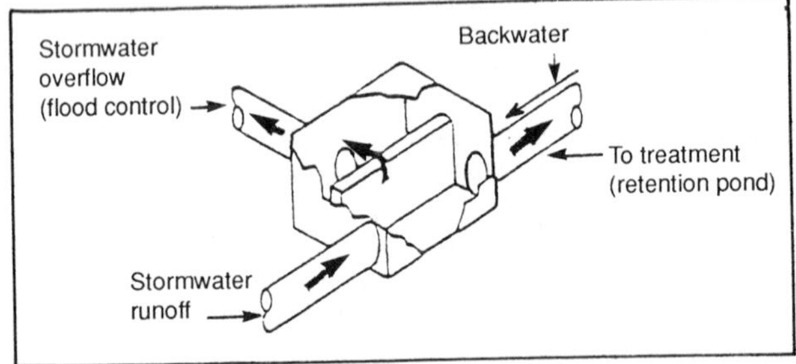

FIGURE 28.6.6 Dual-pond off-line, in-line system with a smart weir. *(Used with permission from Ref. 40.)*

28.6.5 Detention Practices Overview

The stormwater treatment practices described in the previous section are effective and efficient for reducing pollutants in stormwater runoff. However, unless the soils are very permeable and the tributary drainage area is small, these controls are not practical, since they become very large, requiring large areas of land, and will tend to develop groundwater mounds that will not drain off laterally rapidly enough to keep the groundwater from surfacing. As a result, in most stormwater management systems, some form of detention is required.

Detention basins are more widely used for urban stormwater management than any other form of stormwater control. However, until the late 1980s their primary application was for drainage control, namely, peak flow attenuation of the 2-, 10-, 25, or 100-year storms, rather than for water-quality control. The state of practice for pollutant removal has been more qualitative than quantitative because the relationships between the design parameters and the pollutant removal efficiencies of the basins are not yet reliably established. However, enough empirical evidence has been gained from operating prototypes to develop sound guidelines for the design of these facilities to control runoff quality.

There are two basic types of detention for water quality: *extended detention basins* (sometimes called *dry detention basins*), and *ponds* (sometimes called *retention ponds* or *wet detention ponds*). An *extended detention basin* is a facility that empties completely between storms. This emptying time, even for very small runoff events, needs to be 24 to 60 h. A *retention pond,* on the other hand, has a permanent pool of water as part of the facility and detains new stormwater above its permanent water surface. Field data summarized in Table 28.6.1 and also reported by Hartigan,[27] EPA,[76] and the Northern Virginia Planning District Commission[48] indicate that retention ponds are more effective than detention basins in removing pollutants from stormwater. This is especially true for nutrients. Properly designed retention ponds should remove 2 to 3 times more phosphorus (i.e., 50 to 60 percent versus 20 to 30 percent) and 1.3 to 2 times more total nitrogen (i.e., 30 to 40 percent versus 20 to 30 percent) than extended detention basins.[27] See Table 28.6.1 for a comparison of the removal efficiencies of other constituents by the two types of detention.

Retention ponds may need 2 to 7 times more permanent pool volume than the temporary storage volume needed for an extended detention basin. This permanent pool for wet detention can be excavated below the outlet level, while the bottom of an extended detention basin is above the outlet level. The pond volume is designed by using a eutrophication model for the removal of dissolved phosphorus and other nutrients.[27] On the other hand, if the goal is to remove suspended solids, total lead, or other particulates, sizing of ponds or basins can be based on sedimentation theory.

Driscoll developed a procedure for prediction of annual sediment removal rates by detention ponds.[15,76] The curves in Fig. 28.6.7 illustrate an example of Driscoll's work for a single-family residential areas (i.e., runoff coefficient of 0.2). The curves relate the long-term suspended solids removal rate to the pond's surface area and its tributary watershed area for several regions in the United States. For example, annual sediment removal rate of 85 percent is predicted for the southeast region if a total of 0.25 in (6.4 mm) of runoff is captured by such a pond.

28.6.6 Design of (Dry) Extended Detention Basins

Dry detention basins are the most common type of detention used in the United States, Canada, Australia, and possibly other countries. They are primarily designed for peak-flow attenuation. Emptying time is usually less than 6 h and pollutant removal is virtually nonexistent. Laboratory tests show that 80 percent of the pollutants associated with suspended solids are attached to particles less than 60 μm in size, namely fine silts and clays.[59,60] As a result, extended detention periods of 24 to 60 h are needed in dry basins to remove pollutants by sedimentation. Even this extended detention will not remove dissolved pollutants.

A schematic representation of an extended detention basin is in Fig. 28.6.8. In this case, water entering the basin is impounded behind the embankment and is released slowly through the perforated riser outlet. Coarse aggregate around the perforated

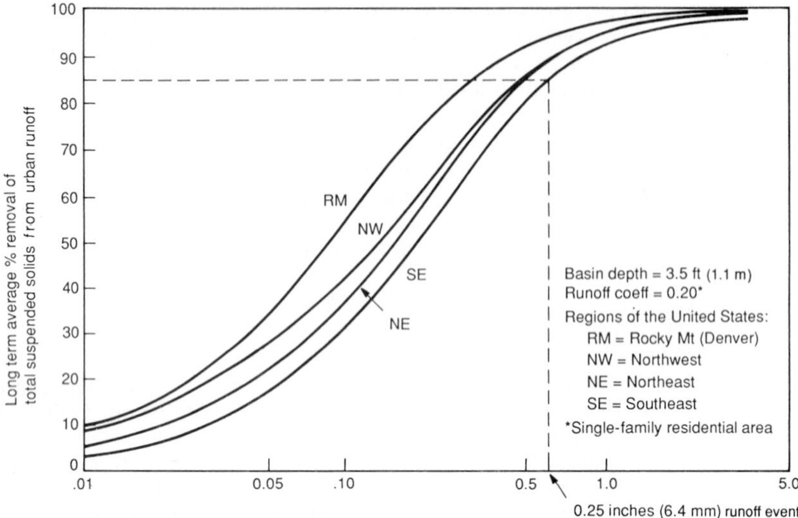

FIGURE 28.6.7 Geographically based design curves for solids settling model. *(From Ref. 76.)*

riser reduces plugging by leaves, paper, plastic bags, and other debris. When the water-quality volume is filled, additional inflow is either diverted around the basin upstream of the inlet, or the pond overflows through the primary spillway. Most of the sediment in the incoming stormwater settles to the bottom of the basin.

The capture volume required in the basin is a function of the tributary land use and local meteorology. The determination of a cost-effective capture volume was

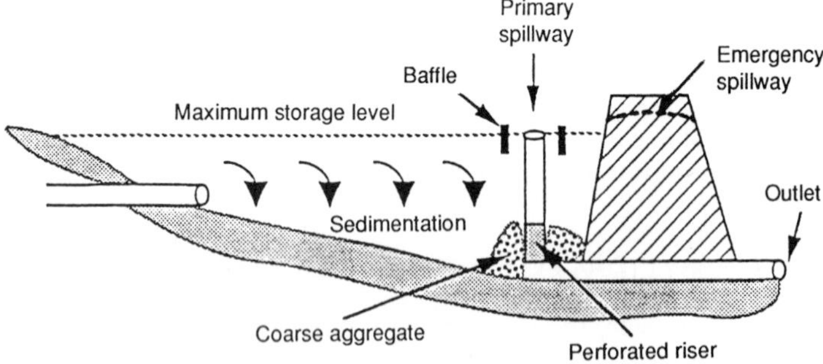

- Efficiency: Poor for detention times under 12 hours
  Good for detention times greater than 24 hours
- Function: Settle pollutants out; soluble pollutants pass through
- Maintenance is moderate if properly designed
- Improper design can make facilities an eyesore and a mosquito-breeding mudhole
- Newer designs are incorporating a shallow marsh around outlet
  Result: Better removal efficiency and no mosquito nuisance
- Regional detention facilities serving 100-200 acres can be aesthetically developed
  Result: Lower maintenance costs

FIGURE 28.6.8 Design of an extended detention basin.

discussed in Sec. 28.6.2. Outlet controls should be designed to drain the basin in 20 to 60 h (the longer the better for water quality, but poorer for the maintenance of healthy vegetation), with no more than 50 percent of the brim-full volume released during the first quarter of the design draw down period. The basin bottom should be designed to drain fully. In areas of permeable soils, some of the drainage will be into the groundwater. When this is the case, keep the bottom at least 3 ft (0.9 m) above the annual high water table to help the basin drain dry between storms. To control erosion, the basin bottom and side slopes should be vegetated with turf-forming grasses that tolerate prolonged inundation. Basin side slopes should be less than 4H:1V so that a mechanical mowers can operate safely on them.

28.6.7 Design of Retention Ponds

A retention pond is a small lake designed to remove pollutants from urban runoff. The basic treatment processes at work in a retention pond are illustrated schematically in Fig. 28.6.9. In the main body of the lake, pollutants are removed by settling and nutrients are removed by phytoplankton growth in the water column. Shallow marsh plants around the perimeter of the basin also remove nutrients. A small forebay between the pond's inflow and the main body of the pond will help to remove coarse sediments at a location where they can be readily extracted for maintenance purposes.[22]

The outlet is designed to slowly release flows from small storms, resulting in temporary surcharge storage above the permanent pool which captures the runoff from all smaller storms and the first part of runoff from larger storms. In some instances, retention ponds are designed with an outlet that discharges directly through the primary drainage and a flood control spillway designed without a specific water-quality capture volume. Some basins are designed to go off-line after they fill, with additional inflow diverted around the basin as illustrated in Fig. 28.6.6.

The permanent pool size can be computed as the average runoff volume over the wettest 14-day period of the rainy season, a criterion suggested by Hartigan[27] to

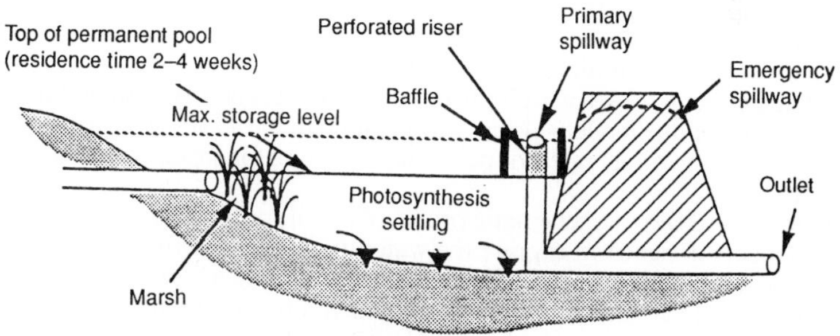

- Efficiency: Excellent if properly designed.
  Can be poor if bottom goes anoxic.
- Function: Removes pollutants by settling, dissolved pollutants biochemically.
- Maintenance: Relatively free after first year except for major cleanout at about ten years.
- Aesthetic design can make pond an asset to community.
- Excellent as a regional facility.

FIGURE 28.6.9 Design of a retention pond.

control dissolved phosphates. If phosphate removal is not an issue, the permanent pool can be much smaller. Below the water line the soil should have a low permeability, or the pond should be lined. When groundwater is at the same elevation as the permanent pool, exfiltration is not a concern, but the potential for groundwater contamination may exist.

Water in the permanent pool should be deep enough to minimize daylight penetration to the bottom, thereby reducing weed growth. It should be sufficiently shallow to allow wind mixing and aeration of the water. When the bottom is allowed to become depleted of oxygen, sediments become anoxic and release nutrients and trace metals back into the water column. This defeats the purpose of the pond. A depth of 3.5 to 12 ft (1 to 4 m) appears to work best.

The pond should have a length-to-width ratio greater than 2 to 1, with the inlet and outlet located at opposite ends of the long axis. This enhances the possibility of plug flow through the pond and provides an opportunity for a wind fetch to set up. Shorelines should slope less than 4H:1V and be planted in native or domestic grasses. A 10-ft-wide littoral bench along the shore, up to 2 ft in depth, will aid with nutrient uptake, filter direct runoff, and enhance public safety. For a balanced pond ecology, the littoral zone should be 25 to 50 percent of the pond's surface area, with the remaining 75 to 50 percent being open water. Outlets should be protected from clogging by trash racks, surface skimmers, or other devices.

In arid and semiarid areas, wet ponds may not be feasible in some cases because of the inability to maintain a permanent pool. As part of the preliminary design of a pond in these areas, a water budget analysis should be performed to ensure that the pond does not lose too much water through evapotranspiration and exfiltration during dry periods.

28.6.8 Design of Multipurpose Detention Basins for Quantity and Quality Control

As discussed in Sec. 28.6.2, runoff quality controls are designed to capture and treat small storms which have a return interval on the order of 1 to 6 months. Stream bank erosion control, on the other hand, also requires control of up to the 2-year storm,[91,95] while drainage and flood control practices generally attempt to regulate runoff from 5-year and larger storms. To conserve land, all of these functions can be combined into a single, on-line, multipurpose facility.[93]

First, the outlet works of a multipurpose facility must be staged so that the water-quality design volume is released very slowly. Subsequent stages provide the required storage volume and outlet rates for peak runoff capture for erosion and flood control functions.

Figure 28.6.10 shows a schematic profile of a multipurpose detention pond. The stage 1 outlet is designed to capture the water-quality design volume and release it slowly. The stage 2 outlet controls the runoff volume from a 5-year storm so that the postdevelopment release rate does not exceed the predevelopment flow during a 5-year event. The stage 3 outlet limits the sum of the flows over the spillway and through the stage 2 outlet so as not to exceed the predevelopment flow for runoff events such as the 10- to 100-year floods. The spillway is designed so that it will pass the postdevelopment 100-year and/or larger floods safely without overtopping the embankment.

The embankment and its spillways should be designed to prevent catastrophic failure of the embankment. Dam failure is often judged by the courts as an absolute liability of the owner when it can be shown that the failure caused more damage than

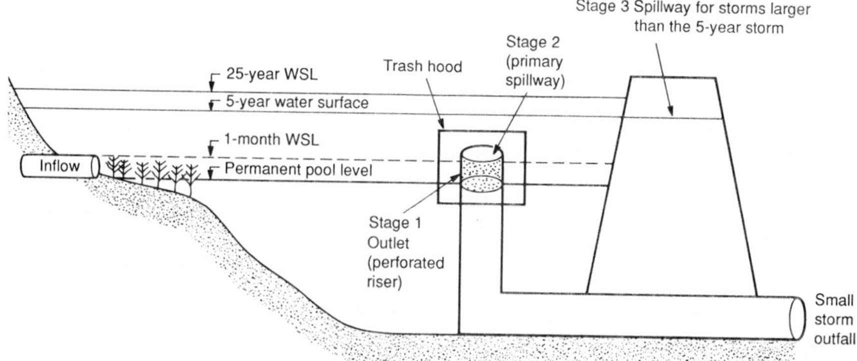

FIGURE 28.6.10 Conceptual design of a multipurpose pond.

would have otherwise occurred if the dam was never built. This simple principle of common law always has to be considered in design of any type of detention facility in an urbanized setting.

28.6.9 Comparison of Regional and on-Site Detention Facilities

The question of who is responsible for maintenance of detention facilities is often left unanswered, but this is extremely important for water-quality facilities. For example, if the average size of a development is 40 acres (16 ha), and a detention facility or other form of water-quality control is constructed for each development, the result will be 16 facilities per square mile (7 basins per km^2). Because water-quality outlets for such small facilities can be only 1- to 2-in orifices, they will plug easily. Often, on-site basins serve less than 10 acres (4 ha) and the outlets are even smaller. Also, areas dedicated for such small facilities are too small for other public uses.

A more logical approach is to use regional detention facilities serving 100 to 600 acres (40 to 240 ha). This minimizes the hydrologic uncertainties of cumulative effects of large numbers of on-site detention facilities. Wiegand et al.[94] showed that fewer regional facilities are also most cost-effective to build and maintain than a large number of on-site facilities. An added bonus is that the water-quality control outlets are larger, which makes them easier to design, build, operate, and maintain. Since regional facilities are generally maintained by some form of public body, the chances that they will receive continuing operation and maintenance are much better than for the multitudes of individually owned on-site facilities.

When properly sited within a watershed, regional basins provide treatment for existing development as well as for new developments and capture all of the street runoff, which is often missed by on-site ponds. The land area for regional facilities is sufficiently large to permit other compatible uses such as recreation, wildlife habitat, aesthetic open space, and other public uses. The major disadvantage of regional facilities is that they require advanced planning by local jurisdictions and need up-front financing to be built. Financing early in the watershed's development and before sufficient developer contributions are available can preclude their timely implementation.

28.7 REFERENCES

1. APWA, *Urban Stormwater Management,* special report no. 49. American Public Works Association, Chicago, 1981.
2. ASCE, *Stormwater Detention Outlet Control Structures,* prepared by the Task Committee of the American Society of Civil Engineers, New York, 1985.
3. ASCE and WPCF, *Design and Construction of Sanitary and Storm Sewers.* ASCE Manual of Hydrologisting Practice no. 37, WPCF Manual of Practice no. 9, prepared by a Joint Committee of the American Society of Civil Engineers and Water Pollution Control Federation, New York, 1960.
4. ASCE and WPCF, *Stormwater Management Manual of Practice,* Water Pollution Control Federation and American Society of Civil Engineers, 1992.
5. Athayde, D., "Nationwide Urban Runoff Program," *APWA Reporter,* Chicago, 1984.
6. Chow, V. T., D. R. Maidment, and L. W. Mays, *Applied Hydrology,* McGraw-Hill, New York, 1988.
7. City of Fort Worth, *Storm Drainage Criteria and Design Manual,* Public Works Department, Fort Worth, Texas, 1967.
8. City of Littleton, *Storm Drainage Design and Technical Criteria,* Engineering Department, Littleton, Colo., October 1986.
9. Corder, G. P., E. M. Laurenson, and R. G. Mein, "Hydrologic Effects of Urbanization: Case Study," *Proc. Hydraulic and Water Resources Symposium,* Australian National University, Canberra, February 1988.
10. County of Los Angeles, *Hydrology Manual,* Los Angeles County Department of Public Works, Alhambra, Calif., 1989.
11. Davies, P. E., "Toxicology and Chemistry in Urban Runoff," *Urban Runoff Quality—Impacts and Quality Enhancement Technology,* American Society of Civil Engineers, New York, 1986.
12. Debo, T. N., "Detention Ordinances—Solving or Causing Problems," *Stormwater Detention Facilities,* American Society of Civil Engineers, New York, 1982.
13. DeGroot, W. G., *Stormwater Detention Facilities,* American Society of Civil Engineers, New York, 1982.
14. *Design of Roadside Drainage Channels,* U.S. Department of Commerce, Bureau of Public Roads, Washington, D.C., 1967.
15. Driscoll, E. D., "Performance of Detention Basins for Control of Urban Runoff," *1983 Int. Symp. Urban Hydrology, Hydraulics and Sediment Control,* University of Kentucky, Lexington, 1983.
16. Driver, N., and G. D. Tasker, "Techniques for Estimation of Storm-Runoff Loads, Volumes, and Selected Constituent Concentrations," USGS open-file report 88-191, U.S. Geological Survey, Denver, Colo., 1988.
17. Fairfax County, *Preliminary Design Manual for BMP Facilities,* Department of Environmental Management, Fairfax County, Va., 1980.
18. Federal Aviation Agency, *Airport Drainage,* Washington, D.C., 1966.
19. FEMA, *Conditions and Criteria for Appeals of Proposed Base Flood Evaluation Determination,* Federal Emergency Management Agency, Washington, D.C., June 1984.
20. FHWA, *HYDRAIN—Integrated Drainage Design Computer System,* under development for the Federal Highway Administration and 29-state coalition by GKY and Associates Inc., 5411-E Backlick Rd., Springfield, VA 22151, February 1990.
21. Florida Concrete Products Association, *Pervious Pavement Manual,* Orlando, Fla., 1988.
22. Goyen, A. G., B. C. Phillips, and J. F. Neal, "Urban Stormwater Quality—Australian Control Structures," *Proc. 4th Int. Conf. on Urban Storm Drainage,* Lausanne, Switzerland, 1987.

23. Grigg, N. S., et al., *"Urban Drainage and Flood Control Projects: Economic, Legal and Financial Aspects,"* Colorado State University, Environmental Research Center report no. 65, Fort Collins, Colo., July 1975.

24. Grizzard, T. J., C. W. Randall, B. L. Weand, and K. L. Ellis, "Effectiveness of Extended Detention Ponds," *Urban Runoff Quality—Impact and Quality Enhancement Technology,* American Society of Civil Engineers, New York, 1986.

25. Hamlin, H., and J. Bautista, "On-the-Spot Tests Check Gutter Capacity," *The American City,* April 1965.

26. *Handbook for Stormwater Detention Basins,* Somerset County, N.J., 1982.

27. Hartigan, J. P., "Basis for Design of Wet Detention Basin BMP's," *Design of Urban Runoff Quality Controls,* American Society of Civil Engineers, New York, 1989.

28. Heise, P., "Infiltration Systems" (in Danish), *Proc. Seminar in Surface Water Technology,* Fagernes, Denmark, 1977.

29. Hubbard, T. P., and T. E. Sample, "Source Tracing of Toxicants in Storm Drains," *Design of Urban Runoff Quality Controls,* American Society of Civil Engineers, New York, 1989.

30. Izzard, C. F., "Hydraulic Capacity of Curb Opening Inlets," *Flood Hazard News,* Urban Drainage and Flood Control District, Denver, Colo., 1977.

31. Jefferson County, *Storm Drainage Design and Technical Criteria,* Engineering Department, Jefferson County, Golden, Colo., 1987.

32. Kamedulski, G. E., and R. H. McCuen, "Evaluation of Alternative Stormwater Detention Policies," *J. Water Resour. Plann. Manage. Div., Am. Soc. Civ. Eng.* (now *J. Water Resour. Plann. Manage.*), vol. 105, no. WR2, pp. 171–186, September 1979.

33. Kropp, R. H., "Water Quality Enhancement Techniques," *Stormwater Detention Facilities,* American Society of Civil Engineers, New York, 1982.

34. Kuo, C. Y., editor, *Urban Stormwater Management in Urban Coastal Areas,* American Society of Civil Engineers, New York, June 1980.

35. Kurtyka, J. C., "Precipitation Measurement Study," State Water Survey Division report, Illinois State Water Survey, Champaign, 1953.

36. Lakatos, D. F., and R. H. Kropp, "Stormwater Detention—Downstream Effects on Peak Flow Rates," *Stormwater Detention Facilities,* American Society of Civil Engineers, New York, 1982.

37. Lakatos, D. F., and L. J. McNemer, "Wetlands and Stormwater Pollution Management," *Wetland Hydrology—Proc. National Wetland Symposium,* Chicago, 1987.

38. Lawrence, A. I., and A. G. Goyen, "Improving Urban Stormwater Quality—An Australian Strategy," *Proc. 4th Int. Conf. on Urban Storm Drainage,* Lausanne, Switzerland, 1987.

39. Livingston, E. H., "Use of Wetlands for Urban Stormwater Management," *Design of Urban Runoff Quality Controls,* American Society of Civil Engineers, New York, 1989.

40. Livingston, E. H., et al., *The Florida Development Manual: A Guide to Sound Land and Water Management,* Department of Environmental Regulation, Tallahassee, Fla., June 1988.

41. *Local Disposal of Storm Water—Design Manual,* Swedish Association of Water and Sewage Works, publication VAV (in Swedish), 1983.

42. McCuen, R. H., "Design Accuracy of Stormwater Detention Basins," technical report, Department of Civil Engineering, University of Maryland, 1983.

43. McCuen, R. H., S. G. Walesh, and W. J. Rawls, "Control of Urban Stormwater Runoff by Detention and Retention," *U.S. Department of Agriculture Research Service miscellaneous publication no. 1428,* March 1983.

44. Meyers, V. A., and F. P. Ho, "Coastal Storm Tide Frequency, Virginia to Delaware," *Urban Stormwater Measurement in Urban Coastal Areas,* American Society of Civil Engineers, New York, June 1980.

45. Miller, A. C., "Hydraulic Effects on Streams," *Stormwater Detention Facilities,* American Society of Civil Engineers, New York, 1982.

46. Murray, J., S. D. Schmidt, and D. R. Spencer, "Nonpoint Pollution First Step in Control," *Design of Urban Runoff Quality Controls,* American Society of Civil Engineers, New York, 1989.

47. National Research Council, *A Study of Inverted Crown Residential Streets and Alleys,* Conducted for Federal Housing Administration by Building Advisory Research Board, Washington, D.C., 1957.

48. Northern Virginia District Planning Commission, *Washington Metropolitan Area Urban Runoff Demonstration Project,* Annapolis, Md., April 1983.

49. O'Loughlin, G. G., and H. T. B. Corderoy, "Safety and Retarding Basins," *Proc. Seminar on Retarding Basins,* The Institute of Engineers Australia, Sidney Division, Australia, 1983.

50. O'Loughlin, G. G., and D. K. Robinson, "Urban Storm Drainage," Chap. 14 in *Australian Rainfall and Runoff, A Guide to Flood Estimation,* The Institute of Engineers Australia, Barton, Australia, 1987.

51. Orange County, *Orange County Hydrology Manual,* Orange County EMA, Santa Ana, Calif., October 1986.

52. Randall, C. W., Ellis, K., Grizzard, T. J., and W. R. Knocke, "Urban Runoff Pollutant Removal by Sedimentation," *Stormwater Detention Facilities,* American Society of Civil Engineers, New York, 1982.

53. Rinella, J. F. and McKenzie, S. W., *Methods for Relating Suspended-Chemical Concentrations to Suspended-Sediment Particle Size Classes in Storm-Water Runoff,* U.S. Geological Survey Water Resources Investigation 82-39, Portland, Ore., 1983.

54. Roesner, L. A., B. R. Urbonas, and M. B. Sonnen, eds., *Design of Urban Runoff Quality Controls, Proc. Engineering Foundation Conference on Current Practice and Design Criteria for Urban Quality Control,* American Society of Civil Engineers, New York, 1989.

55. Schaefer, J. R., K. R. Wright, W. E. Taggart, and R. M. Wright, *Urban Storm Drainage Management,* Marcel Dekker, New York, 1982.

56. Schueler, T. R., *Controlling Urban Runoff: A Practical Manual for Planning and Designing Urban Best Management Practices,* Metropolitan Washington Water Resources Planning Board, Washington, D.C., 1987.

57. Seaburn, G. E., "Effects of Urban Development on Direct Runoff to East Meadow Brook," professional paper 627-B, U.S. Geological Survey, Washington, D.C., 1969.

58. Schaake, J. C., J. C. Geyer, and J. W. Knapp, "Experimental Examination of the Rational Method," *J. Hydraul. Div., Am. Soc. Civ. Eng.* (now *J. Hydraul. Eng.*), vol. 93, no. 6, pp. 353–370, November 1967.

59. Stahre, P., and B. Urbonas, *Stormwater Detention for Drainage, Water Quality and CSO Management,* Prentice-Hall, Englewood Cliffs, N. J., 1990.

60. Stockholm Water and Sewer Works, "Stormwater Studies at Jarvafaltet" (in Swedish), 1978.

61. Terstriep, M. L., and J. B. Stall, *The Illinois Drainage Area Simulator, ILLUDAS,* Illinois State Water Survey, Urbana, 1974.

62. Torno, H. C., J. Marsalek, and M. Desbores, editors, *Urban Runoff Pollution, Proc. NATO Advanced Research Workshop,* Springer-Verlag, Berlin, 1986.

63. *UDSEWER, A Personal Computer Digital Model, Ver. 3.01,* Civil Engineering Department, University of Colorado, Denver, 1990.

64. University of Colorado, *HYDRO POND User's Manual,* Department of Civil Engineering, University of Colorado at Denver, 1991.

65. Urban Drainage and Flood Control District, *Urban Drainage Stormwater Management Model—PC Version (UDSWM2-PC),* distributed by Department of Civil Engineering, University of Colorado at Denver, 1985.

66. Urban Drainage and Flood Control District, *Urban Storm Drainage Criteria Manual,* Denver, Colo., revised 1991.

67. Urbonas, B. R., ed., "BMP Practices Assessment for the Development of Colorado's Stormwater Management Program," final report of the Assessment Subcommittee of Colorado's Stormwater Task Force to Colorado Water Quality Control Division, Denver, 1990.

68. Urbonas, B. R., B. Benik, and M. Hunter, "Stream Stability Under a Changing Environment," *Proc. Stream Bank Erosion Symposium,* Snowmass, Colo., August 1989.

69. Urbonas, B. R., and M. W. Glidden, "Potential Effects of Detention Policies," *Proc. 2d Southwest Regional Symposium on Stormwater Management,* Texas A&M University, November 1983.

70. Urbonas, B. R., C. Y. Guo, and L. S. Tucker, "Optimization of Stormwater Quality Capture Volume," *Urban Stormwater Quality Enhancement—Source Control, Retrofitting and Combined Sewer Technology,* American Society of Civil Engineers, New York, 1990.

71. Urbonas, B. R., C. Y. Guo, and L. S. Tucker, "Sizing Capture Volume for Stormwater Quality Enhancement," *Flood Hazard News,* Urban Drainage and Flood Control District, Denver, Colo., 1990.

72. Urbonas, B. R., and L. A. Roesner, editors, *Urban Runoff Quality—Impact and Quality Enhancement Technology, Proc. Engineering Foundation Conference,* American Society of Civil Engineers, New York, 1986.

73. U.S. Bureau of Reclamation, *Design of Small Canal Structures,* U.S. Department of Interior, Denver, Colo., 1974.

74. U.S. Bureau of Reclamation, *Design of Small Dams,* Denver, Colo., 1973.

75. U.S. Bureau of Reclamation, *Hydraulic Design of Stilling Basins and Energy Dissipators,* U.S. Department of the Interior, Denver, Colo., 1964.

76. U.S. Environmental Protection Agency, *Methodology for Analysis of Detention Basins for Control of Urban Runoff Quality,* EPA440/5-87-001, Washington, D.C., September 1986.

77. U.S. Environmental Protection Agency, *Results of the Nationwide Urban Runoff Program,* final report, NTIS access no. PB84-18552, Washington, D.C., 1983.

78. U.S. Environmental Protection Agency, *Stormwater Management Model User Manual,* User's Manual, ver. IV, Athens, Ga., 1987.

79. U.S. Geological Survey, "Constituent-Load Changes in Urban Stormwater Runoff Routed through a Detention Pool–Wetland System in Central Florida," Water Resources Investigations 85-4310, U.S. Geological Survey, Tallahassee, Fla., 1986.

80. U.S. Geological Survey, "Techniques for Estimating Flood Peaks, Volumes and Hydrographs for Small Watersheds in South Dakota," water resources investigations 80-80, September 1980.

81. U.S. Soil Conservation Service, *National Engineering Handbook,* Sec. 4, "Hydrology," U.S. Department of Agriculture, Soil Conservation Service, Washington, D.C., 1972.

82. U.S. Soil Conservation Service, *"Urban Hydrology for Small Watersheds,"* technical report no. 55, Washington, D.C., 1975.

83. Veenhuis, J. E., J. H. Parish, and M. E. Jennings, "Monitoring and Design of Stormwater Control Basins," *Design of Urban Runoff Quality Controls,* American Society of Civil Engineers, New York, 1989.

84. Viessman, W., Jr., G. L. Lewis, and J. W. Knapp, *Introduction to Hydrology* (3d ed.), Harper and Row, New York, 1989.

85. Walesh, S. G., *Urban Surface Water Management,* Wiley, New York, 1989.

86. Wallace, J. R., "The Effects of Land Use Changes on the Hydrology of an Urban Watershed," OWRR report, project C-1786, School of Civil Engineering, Georgia Institute of Technology, Atlanta, 1971.

87. Wanielista, M. P., "Best Management Practices Overview," *Urban Runoff Quality—Impact and Quality Enhancement Technology,* American Society of Civil Engineers, New York, 1986.

88. Wanielista, M. P., Y. A. Yousef, H. H. Harper, and C. L. Cassagnol, "Detention with Effluent Filtration for Stormwater Management," *Proc. 2d Int. Conf. on Urban Storm Drainage,* Urbana, Ill., 1981.

89. Water Resources Administration, *Guidelines for Constructing Wetland Stormwater Basins,* Maryland Department of Natural Resources, Annapolis, March 1987.

90. Whalen, P. J., and M. G. Callum, "An Assessment of Urban Land Use/Stormwater Runoff Quality Relationships and Treatment Efficiencies of Selected Stormwater Management Systems," South Florida Water Management District, Technical Publication 88-9, 1988.

91. Whipple, W., "Coping with Increased Stream Erosion in Urban Streams," *Water Resour. Res.,* vol. 17, no. 5, pp. 1561–1564, October 1981.

92. Whipple, W., and J. V. Hunter, "Settleability of Urban Runoff Pollution," *J. Water Pollut. Control Fed.,* vol. 53, no. 12, pp. 1726–1731, December 1981.

93. Whipple, W., R. Kropp, and S. Burke, "Implementing Dual-Purpose Stormwater Detention Programs," *J. Water Res. Plann. Manage.,* vol. 113, no. 6, pp. 779–792, November 1987.

94. Wiegand, C., T. Schueler, W. Chittenden, and D. Jellick, "Cost of Urban Quality Controls," *Design of Urban Runoff Quality Controls,* American Society of Civil Engineers, New York, 1989.

95. Williams, L. H., "Effectiveness of Stormwater Detention," *Stormwater Detention Facilities,* American Society of Civil Engineers, New York, 1982.

96. Yeh, F.-F., "A Note on Joint Probability of Surge and Rainfall," *Urban Stormwater Management in Urban Coastal Areas,* American Society of Civil Engineers, New York, June 1980.

CHAPTER 29
HYDROLOGIC DESIGN FOR GROUNDWATER POLLUTION CONTROL

Philip B. Bedient
Department of Environmental Science and Engineering
Rice University
Houston, Texas

Frank W. Schwartz
Department of Geological Sciences
Ohio State University
Columbus, Ohio

Hanadi S. Rifai
Department of Environmental Science and Engineering
Rice University
Houston, Texas

29.1 OVERVIEW OF REMEDIAL TECHNOLOGIES

Chapter 29 is devoted to various aspects of hydrologic design for groundwater pollution control. Prior to the passage of comprehensive environmental legislation in 1980 and 1984, detailed monitoring of groundwater at industrial and waste disposal sites was rare in the United States. There has been an explosion of information and new technology in aquifer investigation and remediation since 1984, mainly due to requirements under environmental legislation related to the Safe Drinking Water Act, the Resource Conservation and Recovery Act (RCRA), and Superfund legislation for cleaning up hazardous waste sites. While many methods for remediation exist in the literature, the emphasis of this chapter is on proven technology for which there exist a number of case studies where groundwater cleanup has been successfully demonstrated.

The first step in selecting any remediation scheme involves a detailed evaluation and characterization of the subsurface hydrogeology of a site combined with an assessment of contaminant sources and the extent and rate of the contaminant plume. Then, analytical or numerical methods can be used to evaluate the effectiveness of a set of pumping and/or injecting wells designed to slowly remove the soluble contaminants from the saturated zone beneath the water table or within a confined aquifer. Section 29.2 in this chapter reviews the primary methods of containment or hydrodynamic control including excavation, pump and treat, and source control. Section 29.6 reviews and evaluates a number of these systems at actual sites on the basis of the results of an extensive U.S. Environmental Protection Agency (EPA) investigation.[74]

Knowledge of the nature and extent of organic contamination of groundwater has advanced considerably during the past decade because of the expenditure of more than a billion dollars on site investigations and cleanup activities.[52] Hundreds of plumes of organic contaminants in various hydrogeologic settings have now been delineated by networks of monitoring wells. Classic studies at Conroe, Texas, the Borden landfill in Canada, and Moffitt Field in California have contributed significantly to the understanding of subsurface transport processes.

It has become clear that simple pump-and-treat systems provide some measure of cleanup, but at many sites, once the wells are turned off, contaminants reappear in monitoring wells at elevated concentrations. Pump-and-treat systems can best be thought of as a management tool to prevent, by hydraulic controls, the continued migration of contaminants in an aquifer.[52] Conferences such as the annual National Water Well Association Meeting on Petroleum Hydrocarbons and Organic Chemicals in Ground Water have stressed the need for better and more comprehensive remediation methods. Recent research has resulted in several new methods to help address the difficult cleanup of both soluble organic compounds and nonaqueous-phase liquids (NAPLs) which can persist in a separate phase in the subsurface.

Vacuum extraction or *soil venting* is an important new technology that has emerged to treat spills of volatile organics in the unsaturated zone, and organics that may be trapped at residual saturation. A practical approach to the design, operation, and monitoring of an in situ soil-venting system is described by Johnson et al.[43] This method offers a useful alternative for soils contaminated with gasoline, solvents, and other volatile compounds as described in Sec. 29.3.

Bioremediation or *biorestoration* can clean up aquifers contaminated with biodegradable organic compounds. By injecting nutrients, oxygen, or microbes into the subsurface of a spill site, microbial degradation of selected compounds can be enhanced. This methodology provides a cost-effective alternative to standard pump-and-treat scenarios for biodegradable contaminants. Section 29.4 reviews the theory behind the bioremediation process and provides the necessary design details and case studies for implementing such a system in the field.

Many sites have NAPLs floating on the water table or in a denser form (DNAPL) which resides near the bottom of an aquifer layer. Some sites have residual oils trapped in the unsaturated zone near the water table. Section 29.5 treats the flow of NAPLs and describes several methods for pumping NAPLs from the subsurface. Movement of a contaminant as a separate, immiscible phase is currently not well understood in either the saturated or unsaturated zones. A nonaqueous liquid phase moves in response to pressure gradients and gravity. Its movement and, hence, recovery, is also influenced by interfacial tension and by the process of volatilization and dissolution. The presence of NAPLs at a site greatly complicates the transport process and the ultimate choice of a remediation method (Sec. 29.5).

The following sections of Chap. 29 present details on the theory and design of systems for groundwater pollution control. A large number of references have been included in this chapter, but several major review articles provide the most in-depth coverage of the specific topics.[15,21,32,43,51,52,54,61,68,69,73,74]

29.2 PUMP-AND-TREAT SYSTEMS FOR DISSOLVED GROUNDWATER CONTAMINANTS

29.2.1 Introduction

The nature and scope of a groundwater contamination problem must be understood before an appropriate remedial action can be determined. Usually, a group of boreholes and wells are drilled and sampled at a site to define the hydrogeology and the source and extent of contaminant migration. A history of contamination events is prepared to define the types of waste and to quantify their loadings to the system. Data collection from monitoring wells is an iterative process performed in phases so that information is constantly updated as more wells are added and results are analyzed. The source definition is one of the most important factors affecting groundwater contamination, and site characterization with accurate hydrogeologic data and contaminant data is important for successful remedial design.

29.2.2 Data Requirements

Hydrogeological Data. The drilling of soil borings and monitoring wells yields information to describe the strata underlying the site to at least the maximum depth of known or potential contamination and generates a reliable description of the subsurface geology. Continuous core samples are collected by auger or rotary drilling methods. The information obtained during the geologic investigation can be presented in geologic cross sections and fence diagrams (sections plotted at oblique angles to one another). Laboratory analysis of sediment or rock samples includes grain size analysis, hydraulic conductivity, plasticity, moisture content, dry density, clay mineralogy identification, and partitioning coefficients for pertinent chemicals. In addition to laboratory analysis, in situ analysis can be made of the geology through borehole and other geophysical methods which provide information on the extent of contaminant plumes, areas of buried trenching operations, and abandoned well locations.

Another key element affecting any remedial design is an accurate characterization of the groundwater flow system. This includes defining the physical parameters of the contaminated region, such as the hydraulic conductivity, storage coefficient, and aquifer thickness. Recharge rates and pumping rates are required along with any physical and hydraulic boundaries which affect the rate and direction of flow. Groundwater movement can be analyzed through the measurement of water levels in wells and piezometers. It is helpful to categorize wells according to the elevation and geologic formation of the screened interval so that the horizontal and vertical gradients of hydraulic potential can be analyzed separately. If there are enough measuring points, a contour map of the piezometric surface of each separate aquifer can be prepared. The contour map can be evaluated to determine possible areas of groundwater recharge and discharge and to identify the direction of groundwater movement. Water level data collected from all the wells on the same day provide the most representative information for producing a potentiometric surface map (Fig. 29.2.1).

The rate of groundwater flow is controlled by the porosity and hydraulic conductivity of the media through which it travels and by hydraulic gradients, which are influenced by recharge and discharge.[24,26] Because contaminants are advected with moving groundwater, it is important to characterize both the rate and direction of groundwater flow from areas of recharge (commonly via rainfall, surface pits or

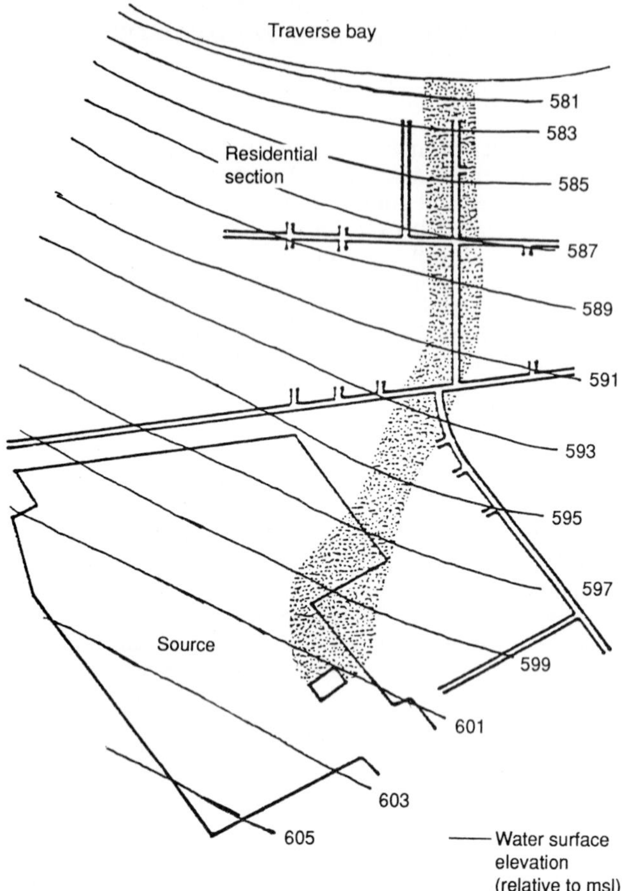

FIGURE 29.2.1 Water table and direction of groundwater flow at Traverse City, Mich.

ponds, or irrigation) to areas of discharge (surface water or wells). Groundwater flows are greatly affected by subsurface heterogeneities, both vertical and horizontal, subsurface fractures, and other features which alter the average hydraulic conductivity of the aquifer. A typical plume of contamination is shown in Fig. 29.2.2.

It is important to conduct a site characterization in a short time period, since groundwater flow systems can vary with time. Seasonal variations in water levels as high as several feet can adversely impact a remediation scheme. For example, at a fuel spill site in Traverse City, Mich., the reduction of the water table by 3 ft during the summer drought of 1988 caused a contaminated zone of residual fuel oil near the original water table to not be properly flushed by an injection well system.[61]

Source Characterization. Source characterization consists of defining the following: (1) the total mass or volume of chemicals released, (2) the total surface area affected by infiltration of chemicals, and (3) the time sequence of the release. Often, the release occurred so long ago that information is difficult to obtain, since many sites have been abandoned and records often have not been kept. In these cases, one is forced to use hydrogeologic data and plume contaminant data to compute what the release must have been by using a mass balance approach. In cases involving leaking tanks or pits, one can often estimate the volume of waste from known pit or tank volumes and estimated infiltration rates. However, without detailed records, contaminant source definition is still the greatest source of error in designing a groundwater remedial scheme.

Contaminant Plume Data. Information about the contaminant mix and spatial distribution of the plume is generally needed to select and analyze remedial alternatives during screening and detailed analysis phases. Physical and chemical properties of contaminants, such as density, solubility, and partitioning should be assessed because they influence plume movement. It should be recognized that some contaminants may not be detected by using routine analytical services, though they are present at levels that would be above cleanup levels, and special analytical services may have to used.

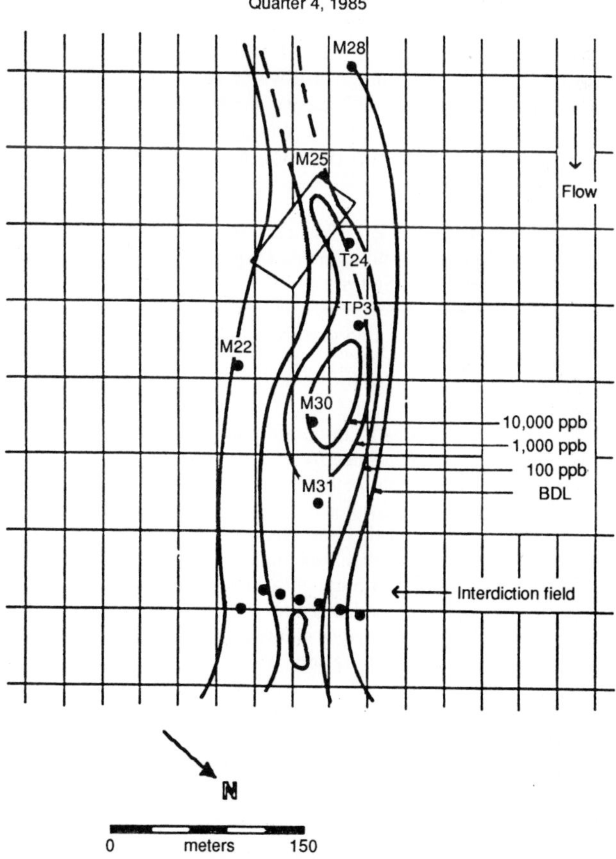

FIGURE 29.2.2 Traverse City field site—BTX plume (quarter 4, 1985). *(Used with permission of ASCE from Ref. 61.)*

Indicator chemicals are those site contaminants that are generally the most mobile and toxic in relation to their concentration; consequently, they reflect most of the risk posed by the site. Generally indicator chemicals are selected on the basis of toxicity, mobility, persistence, treatability, and volume of contaminants at the site. By initially identifying these constituents during the investigation, analytical costs can be reduced. During initial testing of the remedial action, however, samples should be analyzed for all contaminants present to ensure that indicator chemicals have been appropriately selected. Samples are generally analyzed once for total metals, cyanide, semivolatiles, volatiles, and major anions and cations; periodically for those contaminants found at the site; and more frequently (e.g., during aquifer tests) for indicator chemicals. Before completing the remedial action, samples should be analyzed for all contaminants originally detected. Section 15.2 of Chap. 15 tabulates a number of properties of organic chemicals commonly found in groundwater.

Quantitative characterization of the subsurface chemistry includes sampling both the vadose and saturated zones to determine the concentration distributions in groundwater, soil, and vadose water. A network of properly constructed monitoring wells needs to be installed to collect depth-discrete groundwater samples.[71] Wells should be located in areas that will supply information on ambient (background) groundwater chemistry and on contamination in the plume. In the case of layered systems, it may be necessary to sample from wells properly completed in more than one vertical unit. Generally, soil and groundwater samples should be analyzed for the inorganic and organic parameters of concern from the waste stream. Of the various contaminants found in groundwater, the widely used industrial solvents and aromatic hydrocarbons from petroleum products are most common (Table 29.2.1).

TABLE 29.2.1 Typical Organic Compounds Found in Waste Disposal Sites in the United States

Groundwater contaminant
Acetone
Benzene
bis-(2-ethylhexyl)phthalate
Chlorobenzene
Chloroethane
Chloroform
1,1-dichloroethane
1,2-dichloroethane
Di-n-butyl phthalate
Ethyl benzene
Methylene chloride
Naphthalene
Phenol
Tetrachloroethene
Toluene
1,2-trans-dichloroethane
1,1,1-trichloroethane
Trichloroethylene
Vinyl chloride
Xylene

Source: Mercer et al.[54]

After analysis of the samples, the resulting concentration data should be mapped in two or three dimensions to determine the spatial distribution of contamination. These plume delineation maps and the results from aquifer slug and pump tests will yield estimates on plume migration and identify possible locations for injection or extraction wells to be used for remediation (Fig. 29.2.2).

29.2.3 Solute Transport Processes

Transport processes of major concern in ground waste include advection, dispersion, adsorption, biodegradation, and chemical reaction. These processes are described in some detail in Chaps. 15, 16, and 22 of this handbook. The incorporation of transport mechanisms into groundwater models is described by Refs. 2, 3, 8, 9, 27, 36, and 77.

Advection represents the movement of a contaminant with the bulk fluid according to the seepage velocity in the pore space. Capture zone approaches rely on concepts from well mechanics and assume that advection is dominant during pumping or injection.[40] The dispersion process is a mixing mechanism due mainly to heterogeneities in the medium that cause variations in flow velocities and flow paths.

Chemical properties of the plume are necessary to characterize the transport mechanisms of the chemicals and to evaluate the feasibility of a remedial system. A number of properties influence the mobility of dissolved chemicals in groundwater and should be considered in detail for plume migration and cleanup. They include aqueous solubility, density, absorption, Henry's law constant, and biodegradability. These are described in more detail in Chap. 16 of this handbook.

29.2.4 Site Characterization and Monitoring

Subsurface conditions can be studied only by indirect techniques or by using point well or borehole data. Table 29.2.2, from Mercer et al.,[54] lists common data collection methods in the subsurface. Monitoring wells are described by Refs. 20, 30, and 62; geophysical techniques are described in Refs. 28, 45, and 48 and in Chap. 6 (Sec. 6.9) in this handbook. The choice of appropriate sampling methods depends on the overall objectives of the project. Throughout the field study, it is essential to document all well construction details, sampling episodes, and field tests in order to arrive at an accurate evaluation of the entire site.

Methods for determining hydraulic properties of subsurface units primarily consist of aquifer tests such as pump tests or slug tests. In a pump test, a well is pumped at a constant rate and water-level responses are measured in surrounding wells. The slug test method involves inducing a rapid water-level change within a well and measuring the head changes as the water in the well returns to its initial level. The initial water-level change can be induced by either introducing or withdrawing a volume of water or a displacement device into or out of the well. The rate of recovery is related to the hydraulic conductivity of the surrounding aquifer material.[17,26,39,58]

To determine flow directions and vertical and horizontal gradients, water levels must be measured and converted to elevations relative to a datum, usually mean sea level. Water-level measurements may be taken by several different means including chalk and tape, electrical water-level probe, and pressure transducer.[34,66] Horizontal gradients are determined by using water-level data from wells that are screened into the same aquifer layer and/or at the same elevation but separated vertically. Water

TABLE 29.2.2 Data Collection Methods

Category	Commonly used methods	Advantages/disadvantages
Geophysics	Electromagnetics	Good for delineation of high-conductivity plumes
	Resistivity	Useful in locating fractures
	Seismic	Limited use in shallow soil studies
	Ground-penetrating radar	Useful in very shallow soil studies
Drilling	Augering	Poor stratigraphic data
	Augering with split-spoon sampling	Good soil samples
	Air/water rotary	Rock sample formation
	Mud rotary	Fills fractures—needs intensive development
	Coring	Complete details on bedrock
	Jetting/driving	No subsurface data
Groundwater sampling	Bailer	Allows escape of volatiles (operator-dependent)
	Centrifugal pump	Can produce turbid samples, increasing chance of misrepresentation contamination
	Peristaltic/bladder pump	Gives more representative samples
Soil sampling	Soil boring	Restricted to shallow depths
Aquifer tests	Pump test	Samples a large aquifer section
	Slug test	Does not require liquid disposal

elevations are often plotted on a map as contours so that flow direction can be established, as shown in Fig. 29.2.1.

Vertical hydraulic gradients are determined by using water-level data from wells in the same location but screened to different elevations. The gradient is the difference in water levels divided by the distance between the measurement locations. Vertical gradients often occur across confining beds of clays or silts. Because the water levels constitute a complex three-dimensional surface, care must be taken in computing the hydraulic gradient, which determines the direction of flow.

To ensure proper quality assurance (QA) and quality control (QC) of groundwater samples, strict protocols must be followed in the field. The pH, temperature, and specific conductance of a sample should be measured. Ideally, before a sample is gathered, water should be extracted from the well until these parameters have stabilized. This will help ensure that the sample is from the formation rather than from residual water in the well. Proper sample storage and shipment to a qualified laboratory is also important. A sampling plan should address issues such as sampling frequency, locations, and statistical relevance of samples.[72] For more details on sampling guidance, see Refs. 5, 6, and 22.

29.2.5 Analysis of Plume Movement and Remediation

Once a site has been accurately characterized, the first step in data interpretation is making preliminary calculations such as using the hydraulic gradient, hydraulic conductivity, and porosity in Darcy's equation to estimate advective transport. Next, one may compare these velocity calculations with estimates of the mean rate of plume movement. If the two are not at all comparable, this indicates uncertainty in the source release rate or location, or that processes such as sorption or biodegradation are important. There are numerous tools that can be used to interpret data, including geostatistical analyses and computer modeling.

Geostatistics. Methods such as kriging can be used to quantify the spatial variability inherent in the hydraulic conductivity field or the concentration plume of an aquifer.[44] De Marsily[27] presents a concise review of geostatistical approaches in hydrogeology. Kriging is an optimal estimation method which can be used to estimate maps of transmissivity, heads, and concentrations from existing spatial measurements at scattered locations. See Chap. 20 for more details.

Groundwater Models. Groundwater flow and transport modeling performed during the remedial investigation can be used as a tool to estimate plume movement and response to various remedies. The model should be calibrated to a measured plume of contamination to the extent possible. However, caution should be used when applying models at waste sites because there is uncertainty whenever subsurface movement is modeled, particularly when the results of the model are based on estimated parameters. The purposes of modeling groundwater flow and transport include the following:

1. To guide the placement of monitoring wells and hydrogeologic characterization when the remedial study is conducted in phases.
2. To predict concentrations of contaminants at receptor locations.
3. To estimate the effect of source-control actions (e.g., remove the leaking tank) on groundwater remediation.
4. To evaluate expected remedy performance during the feasibility study so that the rate of restoration can be predicted and the cost of alternative methods can be compared.

A variety of subsurface models are reviewed in Chap. 22 of this handbook.

Selection of Remedial Alternatives. Once a site has been well-characterized for hydrogeology and contaminant concentrations, alternatives for control and remediation can be selected and combined to provide an overall strategy for cleanup. Choosing a remedial technology is a function of the contaminant, site characteristics, and the location of the contaminant with respect to the water table. Hydraulic conductivity or transmissivity of a formation is the most important parameter, since pumping rates and velocities are directly related. The reactivity or biodegradability of the contaminant is vital for determining whether an in-situ treatment process will work. If pure product exists at or near the water table in the form of separate-phase fluid, the problem of removal may be greatly complicated, as described in Sec. 29.5. Thus, depending on the situation, it may be necessary to combine a pumping system with other techniques (bioderemediation, soil venting, skimming of hydrocarbon product) in order to complete remediation in the saturated and vadose zones. Alternatives for handling groundwater contamination include: (1) containment, (2) excavation, (3) pumped removal of product or contaminated water, (4) in situ treatment (chemical or biological), and (5) vacuum extraction.

29.2.6 Containment Methods

Containment methods control the spread of contaminants in the subsurface by the use of physical barriers or by hydrodynamic methods which require pumping of groundwater via a series of wells surrounding or in the plume. Isolation techniques for the surface and subsurface include excavation and removal of the contaminated soil and groundwater, barriers to groundwater flow, and surface water controls. Historical barrier approaches include slurry walls, grout curtains, sheet piling, and liners or geomembranes.[21]

Excavation Methods. A pit is dug to remove shallow contaminated soils, pumping wells are installed to control the plume, and the excavated soil is transported to a landfill or surface impoundment for disposal. The groundwater is pumped out and can be treated by a variety of techniques. The inherent problem in excavation and removal to another location is that a new hazardous waste site is often created.

Barriers to Groundwater Flow. Physical barriers used to prevent the flow of groundwater include slurry walls, grout curtains, sheet piling, and various manmade liners.[21,32] These barriers may be used in an effort to contain contaminated groundwater or leachate or slow the flow of clean groundwater into a zone of contamination. Some general characteristics and advantages and disadvantages of slurry walls, grout curtains, and sheet pilings are compared in Table 29.2.3 as adapted from Thomas et al.[68]

Aquifers with sandy surficial soil less than 60 ft (18 m) in depth and underlaid by an impermeable layer of fine-grain deposits or bedrock are most amenable to slurry wall construction.[55] Construction of a *slurry wall* entails excavating a narrow trench (2 to 5 ft; 0.6 to 1.5 m) surrounding the contaminated zone. The slurry acts to maintain the trench during excavation and is usually a mixture of soil or cement, bentonite, and water.[32] The trench is generally excavated through the aquifer and into the bedrock. Installation of a slurry wall at depths greater than 60 ft (18 m) is difficult. However, slurry walls have not been as successful in recent years because of leakage and failure problems associated with contaminated groundwater.

Liners represent another type of physical barrier and are often used in conjunction with surface water controls and caps.[21] Liners may be used to protect groundwater from leachate from landfills containing hazardous materials. The type of liner depends on the type of soil and contaminants which are present. Liners include polyethylene, polyvinyl chloride (PVC), many asphalt-based materials, and soil-bentonite or -cement mixtures. Polyvinyl chloride liners have permeabilities of less than 3.0×10^{-11} ft/s (9.15×10^{-10} cm/s); however, little is known about the service life of the PVC membranes.[70] The membrane should be installed over fine-grained soil to prevent punctures.

Grout curtains are another type of physical barrier which are constructed by injecting grout (liquid, slurry, or emulsion) under pressure into the ground through well points.[21] Groundwater flow is impeded by the grout that solidifies in the interstitial pore space. The curtain is made continuous by injecting the grout into staggered well points that form a two- or three-row grid pattern.[32] Spacing of the well points for grout injection and the rate of injection are critical to construction. Premature solidi-

TABLE 29.2.3 Advantages and Disadvantages of Some Physical Barriers to Groundwater Flow

Methods	Advantages	Disadvantages
Slurry walls	Very low upkeep Fairly effective Simple construction Long service life Minimal environmental impact	Expensive Bentonite deteriorates in concentrated ionic solutions Must be anchored to impermeable strata Some construction procedures patented Bentonite availability limited
Grout curtains	Very low upkeep Fairly effective Increases soil's bearing capacity Long service life Minimal environmental impact Used in consolidated and unconsolidated material Versatile Specific targets can be reached A variety of fluids can be injected	Very expensive Hard to place Difficult to determine completeness of wall Some techniques are proprietary Limited to soils with permeabilities of 10^{-5} cm/s or greater Cannot be used at shallow depths (1.5 m) Some applications can create additional pollution
Sheet pilings	Inexpensive for small projects Fairly effective Minimal environmental impact Long service life Sections of steel piling are reusable	Difficult to form effective barrier in coarse, dense material Steel may corrode from contamination Not initially watertight

Source: Adapted from Thomas et al.[68]

fication occurs when the injection rate is too slow, whereas the soil formation is fractured when the rate is too fast. Soil permeability is decreased and soil-bearing capacity is increased after the grout properly solidifies. Chemical or particulate grouting is most effective in soils which are of sand-sized grains or larger. The expense of grouting and its potential for contamination-related problems in the grout limit its usefulness.

Construction of *sheet piling* involves driving interlocking sections of steel sheet piling into the ground.[21] The sheets are assembled before use by slotted or ball-and-socket connections and are driven into the soil in sections. The connections between the steel sheets are not initially watertight; however, fine-grained soil particles eventually fill the gaps and the barrier generally becomes impermeable to groundwater flow. Sheet piling may be ineffective in coarse, dense material because the interlocking web may be disrupted during construction.[56] There are not many examples of their use in remedial work.

Surface Water Controls. Caps, dikes, terraces, channels, chutes, downpipes, grading, vegetation, seepage basins, and ditches are used to divert uncontaminated surface water away from waste sites and reduce the amount of infiltration and resulting hazardous leachate, or direct contaminated water away from clean areas.[32,68] Many of these techniques may be used in combination with each other. Surface capping involves covering the contaminated area with an impermeable material, regrading the land surface to minimize infiltration of surface water, and revegetating the site.[21]

Surface caps are constructed using natural soil, commercially designed materials, or waste materials. Examples include clay, concrete, asphalt, lime, fly ash, and mixed layers or synthetic liners. Fine-textured soils, often from the site, are most commonly used. The soil is compacted into a cap that covers the waste site and minimizes infiltration of surface water. Blending the soil with additives and cements may increase the effectiveness of the resulting cap.[21]

Dikes are used to control runoff or flood water whereas terraces are designed to divert surface water away from an area or to control erosion.[32,68] Downdrain structures, such as chutes and pipes, are used to channel surface water around the site. Channels, or diversion ditches, are used to intercept runoff upstream of a location and to direct the flow along a different course. Surface water intercepted by channels may be diverted into ditches or seepage basins from which the water infiltrates into the ground. Surface water controls only isolate the contamination instead of providing active treatment or management of plumes.[32]

29.2.7 Hydrodynamic Controls and Contaminant Withdrawal Systems

Hydrodynamic controls of groundwater contamination lower the water table to prevent groundwater discharge to a river or lake, to reduce the rate of migration by dewatering the waste, or to confine the plume to a region of low piezometric head created by a combination of pumping and injection wells. Pumping wells are used to extract water from the saturated zone by creating a capture zone for migrating contaminants. A major problem is the proper treatment and disposal of the contaminated water. On-site treatment facilities are usually required before water can be reinjected to the aquifer or released to the surface water system. The number of wells, their locations, and the required pumping rates are the key design parameters of interest, and methods of analysis are described in more detail in Sec. 29.2.8. Chapter 6 in the handbook discusses well mechanics in detail.

Interceptor systems use drains, a line of buried perforated pipe, and/or trenches (open excavations usually backfilled with gravel, to collect contaminated groundwater close to the water table). These systems operate as an infinite line of wells near the shallow water table and are sometimes efficient at removing contamination near the surface. Trenches are often used to collect nonaqueous-phase liquids like crude oil and gasoline, which are light and tend to move near the capillary fringe just above the water table.

Soil venting removes volatile organic contaminants from the unsaturated zone by vacuum pumping. When operated properly, soil venting, or vapor extraction, can be one of the most cost-effective remediation methods for soils contaminated with volatile compounds such as gasoline or solvents. The design of a soil venting system is complex, and design criteria include the number of wells, well spacing, well location, construction, and vapor treatment systems.[43] Soil venting is presented in more detail in Sec. 29.3.

29.2.8 Examples of Capture Zone and Pump-and-Treat Systems

Pumping water that contains dissolved contaminants can be addressed by using standard well mechanics and capture zone theory, which is well understood. If the hydraulic conductivity is too low or the geology is very complex and heterogeneous, then pumping may not be a feasible alternative for hazardous waste cleanup. If the hydrogeology is conducive to an injection pumping system, then several design approaches can be used to develop an efficient and reliable system for contaminant removal.

Pilot-scale systems or small field demonstration projects have been used at a number of sites to evaluate required pumping rates and the placement of wells in a small area of the site, before expanding to the entire site. In this way, operational policies, mechanical problems, and costs can be evaluated before the larger cleanup is attempted. Careful monitoring of the system is the key to understanding how the injection pumping pattern will respond over time.

Javandel and Tsang[40] developed a useful analytical method for the design of recovery well systems, based on the concept of a capture zone (Fig. 29.2.3). The capture zone for a well depends on the pumping rate and the aquifer conditions. Ideally, the capture zone should be somewhat larger than the plume to be cleaned up, and so wells can be added until sufficient pumping capacity is provided to create a useful capture zone. However, even with more wells, some contaminants may pass between the wells, and well spacing becomes an important parameter, in addition to pumping rate. The greater the pumping rate, the larger the capture zone, and the closer the wells are placed, the better the chance of complete plume capture. Overall, the method minimizes the pumping injection rates through a proper choice of well location and distance between wells.

Javandel and Tsang[40] use complex potential theory as the basis for a simple graphical procedure to determine the pumping rate, the number of wells, and the distance between wells. The method requires the type curves for one to four wells presented in Fig. 29.2.4, and values for two parameters: B the aquifer thickness (assumed to be constant) and U the specific discharge or background Darcy velocity (also assumed constant) for the regional flow system. The method involves the following five steps:

1. Construct a map of the contaminant plume at the same scale as the type curves.

2. Superimpose the type curve for one well on the plume, keeping the x axis parallel to the direction of regional groundwater flow and along the midline of the plume so that approximately equal proportions of the plume lie on each side of the x

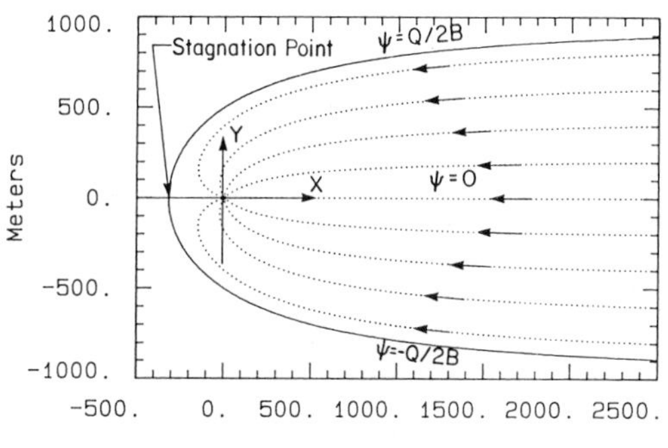

FIGURE 29.2.3 The paths of some water particles within the capture zone with $Q/BU = 2000$, leading to the pumping well located at the origin. *(Reproduced from Ref. 40 with permission.)*

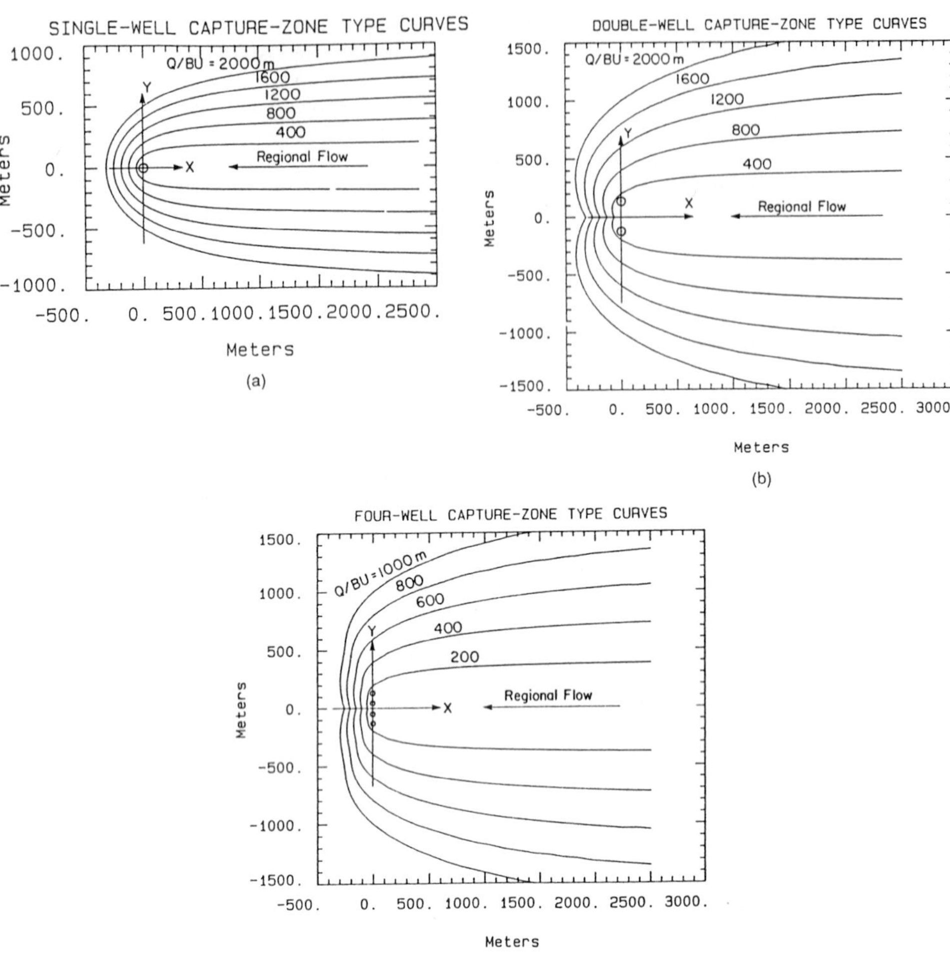

FIGURE 29.2.4 *(a)* A set of type curves showing the capture zones of a single pumping well located at the origin for various values of Q/BU. *(b)* A set of type curves showing the capture zones of two pumping wells located on the y axis for various values of Q/BU. *(c)* A set of type curves showing capture zones of four pumping wells, all located on the y axis for several values of Q/BU. *(Reproduced from Ref. 40 with permission.)*

axis. The pumped well on the type curve will be at the downstream end of the plume. The type curve is adjusted so that the plume is enclosed by a single Q/BU curve.

3. The single-well pumping rate Q is calculated from the known values of aquifer thickness B and the Darcy velocity for regional flow U, along with the value of Q/BU indicated on the type curve (TCV) with the equation

$$Q = B \cdot U \cdot \text{TCV} \tag{29.2.1}$$

4. If the pumping rate is feasible, one well with pumping rate Q is required for cleanup. If the required production is not feasible because of a lack of available drawdown, it will be necessary to continue adding wells (see step 5).

5. Repeat steps 2, 3, and 4 using the two-, three-, and four-well type curves in that order, until a single-well pumping rate is calculated that the aquifer can support. The only extra difficulty comes from having to calculate the optimum spacing between wells using the following simple rules:

Number of wells	2	3	4
Spacing	$\dfrac{Q}{\pi BU}$	$\dfrac{1.26Q}{\pi BU}$	$\dfrac{1.2Q}{\pi BU}$

and to account for the interfacing among the pumped wells in checking on the feasibility of the pumping rates. The wells are always located symmetrically around the x axis, as the type curves show.

Reinjecting the treated water produced by the wells will accelerate the rate of aquifer cleanup. The procedure is essentially the same as that just discussed except the type curves are reversed and the wells are injecting instead of pumping. The authors suggest that the injection wells should be moved slightly upstream of the calculated location to avoid causing parts of plume to follow a long flow path. Their rule of thumb is to place wells half the distance between the theoretical location and the tail of the plume. The following example taken from Javandel and Tsang[40] illustrates how the technique is used.

Example. Shown on Fig. 29.2.5 is a plume of trichloroethylene (TCE) present in a shallow confined aquifer having a thickness of 10 m, a hydraulic conductivity of 10^{-4} m/s, an effective porosity of 0.2, and a storativity of 3×10^{-5}. The hydraulic gradient for the regional flow system is 0.002, and the available drawdown for wells in the aquifer is 7 m. Given this information, design an optimum collection system.

Values of B and U are required for the calculation. B is given as 10 m, but U needs to be calculated from the Darcy equation

$$U = K \text{ grad } (h) \qquad \text{or} \qquad U = 10^{-4} \times 0.002 = 2 \times 10^{-7} \text{ m/s} \qquad (29.2.2)$$

Superposition of the type curve for one well on the plume provides a Q/BU curve of about 2500. Using this number and the values of B and U, the single-well pumping rate is

$$Q = B \cdot U \cdot \text{TCV} \qquad \text{or} \qquad 10 \times (2 \times 10^{-7}) \times 2500 = 5 \times 10^{-3} \text{ m}^3/\text{s} \qquad (29.2.3)$$

A check is required to determine whether this pumping rate can be supported for the aquifer. The Cooper-Jacob[40] equation provides the drawdown at the well, assuming $r = 0.2$ m and the pumping period is 1 year:

$$s = \frac{2.3Q}{4\pi T} \log \frac{2.25Tt}{r^2 S} \qquad (29.2.4)$$

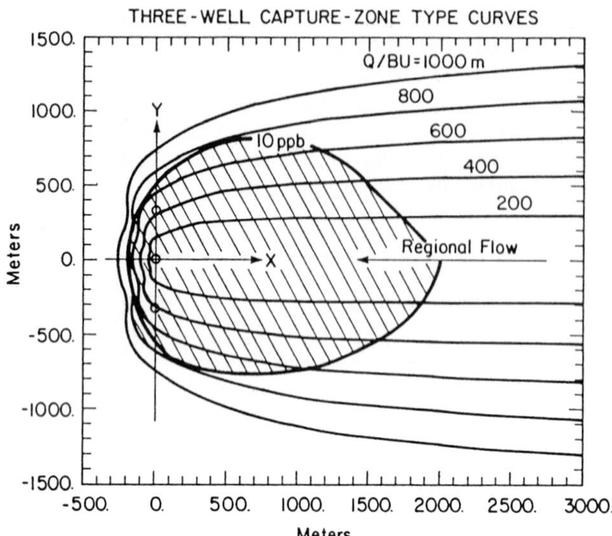

FIGURE 29.2.5 The 10-ppb contour line of TCE at the matching position with the capture-zone type curve of $Q/BU = 800$. *(Reproduced from Ref. 40 with permission.)*

where $Q = 5 \times 10^{-3}$ m³/s
$T = KB = 10^{-3}$ m²/s
$t = 1$ year or 3.15×10^{7} s
$r = 0.2$ m
$S = 3 \times 10^{-5}$

The pumping period represents some preselected planning horizon for cleanup. Substitution of the known values into the Cooper-Jacob equation gives a drawdown of $S = 9.85$ m. Even without accounting for well loss, the calculated drawdown exceeds the 7 m available. Thus, a multiwell system is necessary. Superimposing the plume on the two-well type curve gives $Q/BU = 1200$ or $Q = 2.4 \times 10^{-3}$ m³/s in each well. This provides a predicted drawdown of 6.57 m at each well after 1 year compared to the available 7 m. Because of symmetry, the drawdown in each well is the same. The total drawdown at one of the wells includes the contribution of that well pumping plus the second one 382 m away, based on the principal of superposition. The well separation of 382 m is computed from $Q/\pi BU$ for two wells where $Q = 2.4 \times 10^{-3}$ m³/s in each well. Considering well losses for the two-well scheme makes it unacceptable.

Moving to a three-well scheme, Q/BU is 800 (Fig. 29.2.5), which translates to a pumping rate of 1.6×10^{-3} m³/s for each well. Carrying out the drawdown calculation for three wells located $1.26Q/\pi BU$ or 320 m apart provides an estimate of 5.7 m for the center well, which is comfortably less than the available drawdown of 7 m. Thus, the design requires three wells, located 320 m apart, each pumped at 1.6×10^{-3} m³/s.

This design method is extremely useful but cannot be applied to every design problem encountered in practice. There are assumptions built into the formulation, such as constant aquifer transmissivity, fully penetrating wells, no recharge, and isotropic hydraulic conductivity, which have to be satisfied by the field problem or the method will not necessarily yield a correct result. Actual field sites where boundary conditions and site variabilities are important issues may require analysis by numerical models (Chap. 22).

29.3 VACUUM EXTRACTION SYSTEMS

Vacuum extraction is an important new technology that has emerged to treat spills of volatile organic compounds in the unsaturated zone. It is suited to removing volatile nonaqueous-phase liquids in the unsaturated zone that are trapped at residual saturation, and free product layers. Experience has shown that in comparison to remedial strategies like excavation or pump and treat, soil venting is more efficient and cost-effective.[42] Vacuum extraction involves passing large volumes of air through or close to a contaminated spill by using an air circulation system. The organic compounds or various fractions of a mixture of organic compounds volatilize or evaporate into the air, and are transported to the surface. Thus, just like the other contaminant transport problems considered so far in this chapter, the vapor extraction method involves fluid flow (air in this case) and the transport of mass dissolved in the fluid.

29.3.1 Air Circulation

An example of a vapor extraction system, with a simple representation of the gas flow process, is depicted in Fig. 29.3.1. Vacuum pumping of soil gas creates a gradient in pressure with an accompanying gas flow to the vapor extraction well. In this case, with no air injection wells, the air circulates from the ground surface to the extraction well and back to the surface, where the air is treated to remove the contaminants.

Patterns of air circulation to extraction wells have been studied in the field by direct measurements,[7] and more recently by mathematical and experimental modeling.[42,43,46] Most of the theoretical work to date has assumed that any density differences in the vapor at various locations in the soil can be neglected under the forced convective conditions created by the vacuum extraction.

Johnson et al.[42] have exploited the analogy to groundwater flow in developing simple screening models to describe the distribution of pressure around venting

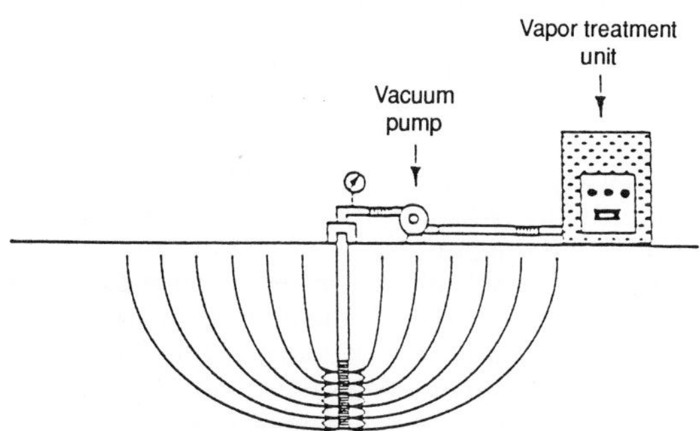

FIGURE 29.3.1 A simple vapor extraction system and the resulting pattern of air flow. *(Modified from Ref. 43. Used with permission.)*

wells. For conditions of radial flow, the governing equation can be written

$$\frac{1}{r}\frac{\partial}{\partial r}\left(r\frac{\partial P'}{\partial r}\right)=\left(\frac{\epsilon\mu}{kP_{\text{atm}}}\right)\frac{\partial P'}{\partial t} \tag{29.3.1}$$

where P' = deviation of pressure from the reference pressure P_{atm}
k = soil permeability
μ = vapor viscosity
ϵ = porosity
t = time

When Eq. (29.3.1) is solved with appropriate boundary conditions, with m as the thickness of the unconfined zone and r as the radial distance from the well to the point of interest, the

$$P'=\frac{Q}{4\pi m(k/\mu)}W(u) \tag{29.3.2}$$

where $u = r^2\,\epsilon\mu/4kP_{\text{atm}}t$ and $W(u)$ is the well function of u, which is a commonly tabulated function.

Calculations with Eq. (29.3.2) show that for sandy soils ($10 < k < 100$ darcys), the pressure distribution approximates a steady state in 1 day to 1 week. Thus, it is appropriate to model pressure distributions by using a steady-state solution to the governing flow equation for the following set of boundary conditions: $P = P_w$ at $r = R_w$ and $P = P_{\text{atm}}$ at $r = R_I$, where P_w is the pressure at the well with radius R_w and P_{atm} is the ambient pressure at the radius of influence R_I. Johnson et al.[42] provide the following solution to the steady-state equation for radial flow

$$P(r)=P_w\left\{1+\left[1-\left(\frac{P_{\text{atm}}}{P_w}\right)^2\right]\frac{\ln(r/R_w)}{\ln(R_w/R_I)}\right\}^{1/2} \tag{29.3.3}$$

As Johnson et al.[42] point out, while not explicitly represented in Eq. (29.3.3), the properties of the soil do influence the steady-state pressure distribution because the radius of influence R_I does vary as a function of permeability and layering. Johnson et al.[42] develop corresponding solutions for radial Darcy velocity and volumetric flow rate for this steady-state case. The second of these solutions provides a useful way to determine what the theoretical maximum air flow is to a vapor extraction well. This solution is written

$$Q=H\pi\frac{k}{\mu}P_w\frac{[1-(P_{\text{atm}}/P_w)^2]}{\ln(R_w/R_I)} \tag{29.3.4}$$

where H is the total length of the screen. Just as with well hydraulics, these analytical equations form not only the basis for predictive analysis, but also for well-testing methods for permeability estimation.

Analytical solutions are most useful for screening purposes and for exploring the relationships among variables; however, their practical applicability is limited to simple problems. An alternative way of solving differential equations like Eq. (29.3.1) is with powerful numerical models such as the one developed by Krishnayya et al.[46] Numerical models are effective in modeling complicated problems of the type commonly encountered in practice.

By analogy with problems involving groundwater flow to wells, the general pattern of air circulation will be influenced by features of the fluid being circulated, the geologic system, and the well system used to withdraw and inject air. The permeability is the most important of all the parameters which influence flow. Ultimately, it is the permeability which determines the efficacy of vapor extraction because flow rates at steady state for a well under a specified vacuum are a direct function of permeability. To be practically useful, vapor extraction requires some minimal rate of air circulation, which may not be feasible in some low-permeability units.

From an analytical model, Johnson et al.[43] developed a series of relationships between permeability and flow rate (Fig. 29.3.2). For a given vacuum in the extraction well, the steady-state rate of air flow is a linear function of permeability (when both variables are plotted on log scales). An increase in the vacuum (smaller P_w) at a given permeability will increase the air flow. However, the maximum change that might be expected by varying P_w is about an order of magnitude in flow rate (Fig. 29.3.2).

It is well-known from groundwater flow theory that variability in permeability plays an important role in controlling the pattern of flow. This is also the case for patterns of air circulation. Consider the situation in Fig. 29.3.3, where a high-permeability zone, 1.5 m thick, overlies a lower-permeability zone, 4.5 m thick, and

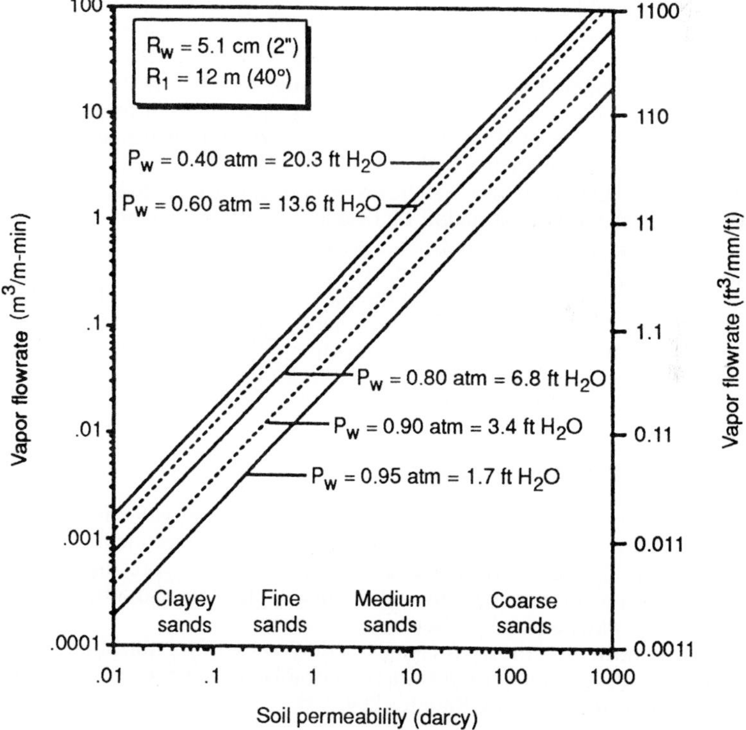

FIGURE 29.3.2 Predicted steady-state rates of air flow per unit length of well screen from a vapor extraction well for a range of permeabilities and applied vacuums (P_w). *(Reproduced from Ref. 43 with permission.)*

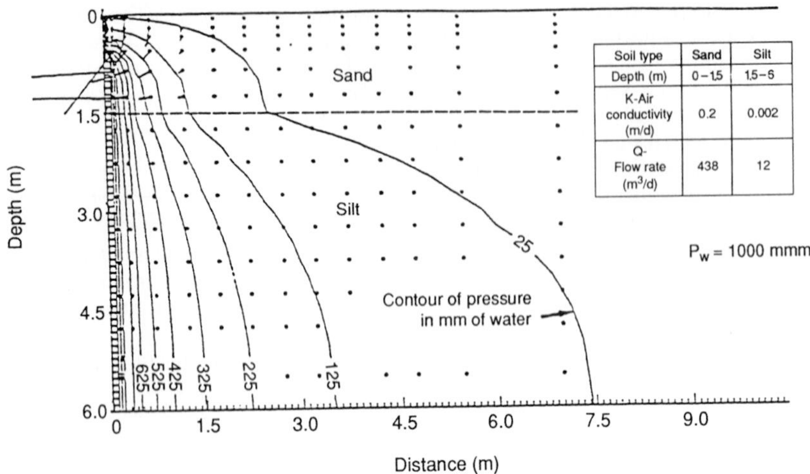

FIGURE 29.3.3 Pressure distribution around a vapor extraction well sealed to a depth of 0.7 m in a layered soil. *(Modified from Ref. 46. Used with permission.)*

the vapor extraction well is screened in both units. The distribution in pressures around the well shows a relatively active pattern of circulation from the surface through the shallow sand and into the well. In fact, as the table on the figure indicates, 438 m^3/day or approximately 97 percent of the air circulated by the well flows through the permeable sand unit.[46] If the lower-permeability silt layer was the target for the cleanup, this vapor extraction well would be very inefficient.

Features of the design also control air circulation. The most important factors include: (1) the flow rates in the injection/extraction wells, (2) the types and locations of wells in a multiwell system, and (3) the presence of a surface seal. For a single vacuum well, Fig. 29.3.2 illustrates how reducing P_w increases the air withdrawal rate. Adding more wells to the system will do the same. Furthermore, the pattern of air circulation can be controlled by the types and locations of the wells.

29.3.2 Organic Vapor Transport

The vapor-phase transport of mass has much in common with the transport of dissolved contaminants in groundwater. The processes of advection and dispersion operate to physically transport the mass, while the chemical processes are involved with the generation of the contaminants through volatilization and subsequent interactions with the water and solid phases in the system.

Removal of mass from the spill depends on the process of volatilization, which is a phase partitioning between a liquid and a gas. For solvents, the process is described in terms of equilibrium theory by Raoult's law

$$P_i = x_i P_i^0 \tag{29.3.5}$$

where P_i is the vapor pressure of component i (atm) in the soil gas, x_i is the mole fraction of the component in the solvent, and P_i^0 is the vapor pressure of the pure solvent at the temperature of interest. Values of P_i^0 are listed in Table 15.2.5 for some organic compounds; see also Ref. 75. For dealing with complex solvent mixtures like gasoline, Eq. (29.3.5) can be written

$$P_i = x_i \tau_i P_i^0 \tag{29.3.6}$$

where an activity coefficient τ_i for the ith component in the mixture is added to account for nonidealities.[37] For gasoline, τ values are generally close to 1 except for the aromatic components, which can approach 1.3 (Ref. 37). By applying the ideal gas equation, the partial pressures calculated in Eq. (29.3.5) can be expressed in terms of concentration:

$$C_s = \frac{PV}{RT} \sum \text{MW} \tag{29.3.7}$$

where C_s = saturation concentration, mg/L
V = volume, assumed equal to 1000 cm³ to calculate contraction
T = temperature, kelvins
R = universal gas constant (82.04 atm · cm³ · mole⁻¹)
MW = number of milligrams per mole
P = partial pressure of the compound.

The equilibrium model provides the basis for most analyses carried out in practice. The equilibrium approach applies in cases where the rate of volatilization is large relative to the rate of physical transport through the medium. Where the flow of air contacts residual product in every pore, phase equilibrium is achieved quite rapidly, at least for the more volatile compounds at the start of venting. Conditions may exist when this assumption will not hold, meaning that the vapor concentration in the air flow will be less than its theoretical equilibrium concentration. At the pore scale, this effect can develop, for example, because of difficulties experienced by compounds in moving to the liquid interfaces. A more important reason for less-than-equilibrium concentrations of vapor is geologic heterogeneities that cause most of the air flow to avoid most of the contaminated zones.

With spills of complex solvent mixtures such as gasoline, soil venting removes the components with higher vapor pressure first. The residual contamination thus becomes progressively enriched with the less volatile compounds. Because of this compositional change, the overall rate of mass removal decreases with time. This process has been described mathematically by Marley and Hoag[53] and Johnson et al.[42] by using forms of Eq. (29.3.7) and information on the most volatile components of gasoline. Given some starting composition of the gasoline, the saturation concentration for each component in the vapor is determined. The mass loss of the components from the spill over some small time increment is

$$\text{ML}_i(\Delta t) = QC_i \, \Delta t \tag{29.3.8}$$

where ML_i is the mass loss rate of component i, mg; Q is the air flow rate, L/min; and C_i is the concentration of component i in the vapor from Eq. (29.3.7), mg/L. This mass loss causes the total mass of each component in the liquid phase to change along with the mole fraction. By taking small steps in time, one can simulate the composition changes and hence the progress of soil venting. Marley and Hoag[53] provide a computer code to facilitate these calculations.

Vapor pressures for organic compounds increase significantly as a function of increasing temperature. For example, the vapor pressure for a volatile compound like benzene increases from 0.037 atm at 0°C to 0.137 atm at 26.7°C (Ref. 42). Over the same temperature range, the vapor pressure for n-dodecane increases from 2.8×10^{-5} atm to 2.3×10^{-4} atm. The main implication of this result is that the overall time required for cleanup will change, depending on temperature. Johnson et al.[42] show, for example, that a cleanup operation by soil venting in Michigan in winter might take 5 times longer than the identical problem in California. This temperature dependence does not apply so strongly to other remediation or cleanup methods.

The physical transport processes, advection and dispersion, play an important role with respect to vapor transport. The simplest and most extensively analyzed model of physical transport assumes that advection transports contaminants from the point of generation to the vapor extraction well (Fig. 29.3.4*a*). The distribution of contaminants is reasonably homogeneous and much of the air flow has the opportunity to move through the bulk of the spill. Ideally, the vapor concentration in that fraction of the air moving through the spill is the equilibrium concentration deter-

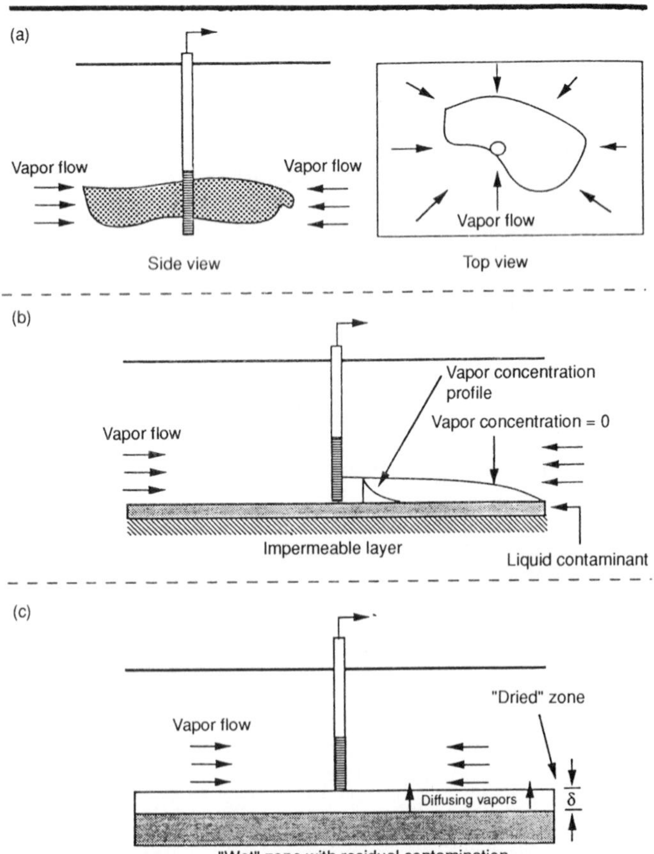

FIGURE 29.3.4 Different ways in which the circulating air interacts with a volatile contaminant source. *(Reproduced from Ref. 43 with permission.)*

mined by Raoult's law. In Fig. 29.3.4*a*, about 25 percent of the air passes through the spill, giving a vapor concentration in the well that is 25 percent of equilibrium value and a removal rate of $0.25QC_{tot}$.

However, in heterogeneous media or when pure product is present, the air flow may not pass directly through the spill. Specific examples cited by Johnson et al.[42,43] include: (1) air flow across the surface of a free-liquid floating on the water table or low-permeability layer (Fig. 29.3.4*b*) and (2) product trapped in a lower permeability lens (Fig. 29.3.4*c*). In both of these cases, the mass loss rate of contaminants is controlled by the rate at which mass can diffuse into the moving vapor stream. Thus, when flowing air bypasses the spill, the rate of mass removal may be much lower than for the homogeneous case. The result may be a vapor extraction system that circulates considerable quantities of air without removing much of the contaminant. Mathematical approaches for estimating the contaminant removal rates under these more complex conditions are presented by Johnson et al.[43]

When vapor diffusion is an important process, the diffusion coefficient becomes an important parameter controlling the process. Diffusion coefficients do vary slightly as a function of temperature, but much more significantly as a function of moisture content. Thus, soil venting operations with very wet soils are less efficient than with the same soils in a dry state.

29.3.3 Implementing Vapor Extraction Systems

Johnson et al.[43] present a detailed and comprehensive approach that can be followed in practice for implementing a vapor extraction scheme. Having arrived at vapor extraction as a candidate strategy for remediating a contamination problem, the following four major activities constitute a vapor extraction project:

1. Feasibility analysis based on available data to establish whether a vapor system is an appropriate remedial strategy
2. Testing with a pilot vapor extraction system and groundwater wells to determine physical parameters and confirm the accuracy of the above analysis
3. Design of the complete system
4. Monitoring to confirm that the designed level of cleanup is being achieved

The feasibility analysis (step 1) requires preliminary information about the geology and hydrogeology of the site, the chemical and physical characteristics of the liquid contaminant, and any regulations applicable to the spill. As Table 29.3.1 indicates, there are five questions that need to be answered by this analysis. Most of the approaches to answering the questions have been discussed in this section to some extent.

In terms of ongoing monitoring activities, Johnson et al.[43] recommend that the following data be collected on an ongoing basis: (1) air flow rates, (2) pressure at each extraction and injection well, (3) temperature (ambient and soil), (4) water table elevation, (5) vapor concentrations and composition at each extraction well, and (6) soil-gas vapor concentrations at various distances from the extraction wells. The vapor extraction system is shut down when target levels of a cleanup are reached. These targets are often site-specific, depending on particular water-quality and health standards, and safety considerations. Johnson et al.[42] discuss in detail how the monitoring data are evaluated before making this decision.

TABLE 29.3.1 A Summary of the Basic Issues to be Addressed in Planning for the Development of a Vapor Extraction System

Major activity/issue	Approach to resolving issue
Feasibility analysis:	
1. What is the expected range in rates of air flow for a single well?	Estimate Q with the flow equation
2. What is the estimated removal rate with a single well?	Estimate total connection of vapor in the air flow C_{tot}
	Estimate single-well removal rate (percent air flow contacting spill $\times Q \times C$)
3. Is the single-well removal rate acceptable?	Compare removal rate for a single well with the rate required (i.e., total mass of spill × target time for recovering spill)
4. What residual contamination will remain at the end, and will the level of cleanup meet regulatory requirements?	Given the air flow rates and a model describing the removal rates as a function of time, determine soil concentrations of key constituents at the end of the remedial period
	Compare these levels with requirements of the regulations
5. What negative effect may develop as a consequence of vapor extraction?	Evaluate the possibility of offsite contributions from other sources
	Evaluate the possibility of a water table rise accompanying vapor extraction
Testing with vapor extraction and groundwater wells:	Evaluation of pumping tests gives site-specific data necessary for finalizing the design of the vapor extraction and water control systems
	Monitoring of vapor concentrations during the well test helps to validate earlier calculations and establish whether efficiencies are affected by heterogeneities
Final system design:	Site-specific data is used in conjunction with flow equations to determine the single-well removal rate
	The required removal rate is divided by the single-well rate to establish the number of wells
Monitoring:	Monitoring documents the progress of the cleanup and will assist in determining whether changes to the system are required
	Monitoring helps to establish when the system can be shut off

Source: Johnson et al.[43] Used with permission.

29.4 BIOREMEDIATION

29.4.1 Introduction

Bioremediation (also called *biorestoration* or *biological treatment*) involves the treatment of subsurface pollutants by stimulating the native microbial population to biodegrade organic pollutants into simple carbon dioxide and water. Biodegradation of organic contaminants has been utilized in municipal wastewater treatment processes for years, including activated sludge reactors, lagoons, waste-stabilization ponds, and fluidized-bed reactors. Fixed-film processes include trickling filters, rotating biological disks, and sequencing batch reactors.[50]

The natural biodegradation process is a biochemical reaction which is mediated by microorganisms, which have been found to exist naturally in the subsurface. An organic compound is oxidized (loses hydrogen electrons) by an electron acceptor which in itself is reduced (gains hydrogen electrons). Electron acceptors include oxygen (O_2), nitrate (NO_3^-), sulfate (SO_4^{2-}), and carbon dioxide (CO_2). The utilization of oxygen as an electron acceptor is termed *aerobic biodegradation* and that of nitrate is called *anaerobic biodegradation.* An example of the aerobic biodegradation reaction for benzene is

$$C_6H_6 + 7.5\,O_2 \longrightarrow 6\,CO_2 + 3\,H_2O \tag{29.4.1}$$

The ultimate goal of biodegradation is to convert organic wastes into biomass, CO_2, CH_4, and inorganic salts. Two criteria must be satisfied before biodegradation or bioremediation can occur: (1) the subsurface geology must have a relatively large hydraulic conductivity to allow the transport of oxygen and nutrients through the aquifer and (2) microorganisms must be present in sufficient numbers and types to degrade the contaminants of interest. Because of the heterogeneous nature of the subsurface environment, hydraulic conductivity should be evaluated at multiple locations on a site. Formations with hydraulic conductivity K values greater than 10^{-4} cm/s are most amenable to in situ bioremediation.

Enhanced aerobic bioremediation for a petroleum spill is an engineered delivery of nutrients and oxygen to the contaminated zone in an aquifer. Oxygen sources can be created in wells or infiltration galleries and include air, pure oxygen (gaseous and liquid forms), and hydrogen peroxide. *Sparging* is a procedure in which air or a gas is injected as bubbles into water. Sparging the groundwater with air and pure oxygen can supply only 8 to 40 mg/L of oxygen, depending on the temperature of the injection field.[50]

29.4.2 Bioremediation of Organics in the Subsurface

The interest in bioremediation as a cleanup technology was generated by early studies such as that presented by Dunlap and McNabb,[31] which indicated the presence of active microbial populations in the subsurface. The microbial populations were metabolically active, and nutritionally diverse. The main advantage of biological treatment is that it offers partial or complete breakdown of the contaminant instead of simply transferring the contaminant from one phase in the environment to another. Lee et al.[50] and Thomas et al.[68] have discussed in detail other advantages and disadvantages of bioremediation (listed in Table 29.4.1).

Unlike many aquifer remediation techniques, bioremediation can be used to treat contaminants that are sorbed to soil or trapped in pore spaces. In addition to treat-

TABLE 29.4.1 Advantages and Disadvantages of Bioremediation

Advantages
1. Can be used to treat hydrocarbons and certain organic compounds, especially water-soluble pollutants and low levels of other compounds that would be difficult to remove by other methods
2. Environmentally sound because it does not usually generate waste products and typically results in complete degradation of the contaminants
3. Utilizes the indigenous microbial flora and does not introduce potentially harmful organisms
4. Fast, safe, and generally economical
5. Treatment moves with the groundwater
6. Good for short-term treatment of organic contaminated ground water
Disadvantages
1. Can be inhibited by heavy metals and toxic organics
2. Bacteria can plug the soil and reduce circulation
3. Introduction of nutrients could adversely affect nearby surface water
4. Residues may cause taste and odor problems
5. Labor and maintenance requirements may be high, especially for long-term treatment
6. May not work for aquifers with low permeabilities that do not permit adequate circulation of nutrients

Source: Thomas et al.[68]

ment of the saturated zone, organics held in the unsaturated and capillary zone can be treated when an infiltration gallery or soil flushing is used. The time required to treat subsurface pollution using bioremediation can be faster than some withdrawal and treatment procedures.

Bioremediation can cost less than other remedial options, although data are generally lacking at this time to compare pump-and-treat alone versus bioremediation. Flathman and Githens[35] estimated that the cost of bioremediation would be one-fifth of that for excavation and disposal of soil contaminated with isopropanol and tetrahydrofuran and, in addition, would provide an ultimate disposal solution.

There are several factors that may limit or inhibit the biodegradation of subsurface organic pollutants, even in the presence of microorganisms. Many organic compounds in the subsurface are resistant to degradation, and heavy metals and toxic concentrations of pollutants may inhibit microbial activity and preclude the use of the indigenous microorganisms for bioremediation at some sites. The pumping and injection wells may clog from excessive microbial growth which results from the addition of oxygen and nutrients. In addition, the hydrodynamics of the restoration program must be properly managed. The nutrients added must be contained within the treatment zone to prevent eutrophication of untargeted areas. High concentrations of nitrate can render groundwater unpotable. Metabolites of partial degradation of organic compounds may impart objectionable tastes and odors.

Bioremediation projects require extensive monitoring and maintenance for successful treatment; whether these requirements are greater than those for other remedial actions is debatable. Bioremediation is difficult to implement in low-permeability aquifers; however, many other physical and chemical remediation processes are subject to the same restrictions. Overall, the long-term effects of bioremediation are unknown.

To date, the most aerobically biodegradable compounds in the subsurface have been petroleum hydrocarbons such as gasoline, crude oil, heating oil, fuel oil, lube oil waste, and mineral oil.[50] Other compounds such as alcohols (isopropanol, methanol, ethanol), ketones (acetone, methyl ethyl ketone), and glycols (ethylene glycol) are also aerobically biodegradable. Recent studies have expanded the list of aerobically degraded compounds to include methylate benzenes, chlorinated benzenes,[47] chlorinated phenols,[67] methylene chloride,[41] naphthalene, methylnapthalenes, dibenzofuran, fluorene, and phenanthrene.[47,79]

By increasing the rates at which required oxygen and nutrients are delivered to microorganisms, the main concept behind aerobic bioremediation, one can enhance the natural biodegradation process and achieve faster removal rates.

29.4.3 Design of a Bioremediation Process

The steps involved in an in situ bioremediation program[50] are (1) site investigation, (2) free product recovery, (3) microbial degradation enhancement study, (4) system design, (5) operation, and (6) maintenance. It is important to define the hydrogeology and the extent of contamination at the site prior to the initiation of any in situ effort (see Sec. 29.2).

Monitoring wells should be sampled for presence of hydrocarbon contamination, and a plume should be mapped for the site. The pumping rate that can be sustained in the aquifer is very important because it limits the amount of water that can be delivered to the system during the bioremediation process. Usually, hydraulic conductivity should exceed about 10^{-4} cm/s for a site to be amenable to bioremediation.

After defining the hydrogeology, recovery of the free product (NAPL), if any, at the site should be completed (see Sec. 29.5). The pure product can be removed by physical recovery techniques such as a single-pump system that produces water and hydrocarbon or a two-pump, two-well system that steepens the hydraulic gradient and recovers the accumulating hydrocarbon.

Prior to the initiation of bioremediation activity, it is important to conduct a feasibility study for the biodegradation of the contaminants present at the site. First, contaminant-degrading microorganisms must be present, and, second, the response of these native microorganisms to the proposed treatment method must be evaluated. In addition, the feasibility study is conducted to determine the nutrient requirements of the microorganisms. These laboratory studies provide a reliable basis for performance at the field level only if they are performed under conditions that simulate the field.

The chemistry of a field site will affect the types and amounts of nutrients that are required. Limestone and high-mineral-content soils, for example, will affect nutrient availability by reacting with phosphorus. Silts and clays at sites may induce nutrient sorption on the soil matrix, and hence decrease the amount of nutrients available for growth. In general, a chemical analysis of the groundwater provides little information about the nutrient requirements at a field site; it is mostly the soil composition that is significant.

In situ bioremediation requires the addition of an electron receptor such as oxygen, and the rate of aerobic biodegradation is limited by the amount of oxygen that can be transported to the organisms in the zone of contamination. A number of methods are available to supply oxygen to groundwater, including the addition of air, pure oxygen, or hydrogen peroxide, with increasing concentrations, respectively. Addition of hydrogen peroxide must be carefully monitored to avoid toxicity to microorganisms at elevated concentrations.[69] Aquifer plugging due to precipitation of inorganic nutrients is a possible side effect.

A system for injection of nutrients into the formation and circulating them through the contaminated portion of the aquifer must be designed and constructed.[49] The system usually includes injection and production wells and equipment for the addition and mixing of the nutrient solution.[60] A typical system is shown in Fig. 29.4.1. Well installation should be performed under the direction of a hydrogeologist to ensure adequate circulation of the groundwater.[49] Produced water can be recycled to recirculate unused nutrients, avoid disposal of potentially contaminated groundwater, and avoid the need for makeup water. Inorganic nutrients can be added to the subsurface once the system is constructed. Nutrients also can be circulated by an infiltration gallery (Fig. 29.4.2); this method provides the additional advantage of treating the residual gasoline that may be trapped in the pore spaces of the unsaturated zone.[18]

Bioremediation is not without its problems, however, the most important being the lack of well-documented field demonstrations that show the effectiveness of the technology and what, if any, are the long-term effects of this treatment on groundwater systems. Other problems include the possibility of generating undesirable intermediate compounds during the biodegradation process which are more persistent in the environment than is the parent compound. Hydrogen peroxide, which dissociates to form water and one-half molecule of oxygen, is infinitely soluble in water;[69] however, hydrogen peroxide can be toxic to microorganisms at concentrations as low as 100 ppm. A stepping-up procedure is usually utilized to allow the microorganisms to adapt to the higher concentrations of the oxidant. If hydrogen peroxide is destabilized, oxygen will come out of solution as a gas, and the process becomes less efficient.[38] Proprietary techniques have been developed to stabilize

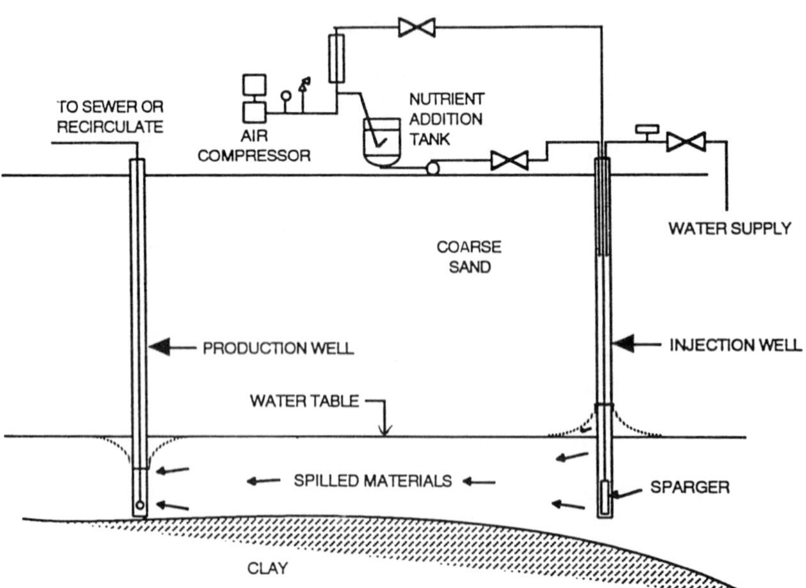

FIGURE 29.4.1 Injection system for oxygen.

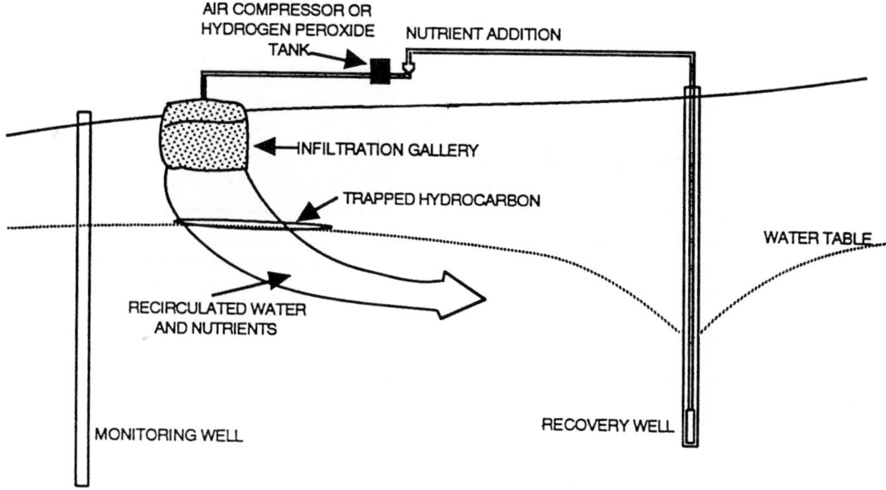

FIGURE 29.4.2 Infiltration gallery.

hydrogen peroxide. In severe cases, gas production (both O_2 and CO_2) can lead to a reduction in hydraulic conductivity.[19] One undesirable effect of using hydrogen peroxide is that other redox reactions may also be enhanced, such as iron precipitation.

29.4.4 Application of Models to Biodegradation Processes

The problem of quantifying biodegradation in the subsurface can be addressed by using models which combine physical, chemical, and biological processes (see Chap. 22). Developing such models is not simple, however, because of the complex nature of microbial kinetics, the limitations of computer resources, the lack of field data on biodegradation, and the lack of robust numerical schemes that can simulate the physical, chemical, and biological processes accurately. Several researchers have developed groundwater biodegradation models. The main approaches utilized for modeling biodegradation kinetics are

1. First-order decay
2. Biofilm models (including kinetic expressions)
3. Instantaneous reaction model
4. Dual-substrate monod model

Borden and Bedient[15] developed the first version of the BIOPLUME model to simulate the simultaneous growth, decay, and transport of microorganisms combined with the transport and removal of hydrocarbons and oxygen. Simulation results indicated that any available oxygen in the region near the hydrocarbon source will be rapidly consumed. In the body of the hydrocarbon plume, the oxygen transport will be rate limiting and the consumption of oxygen and hydrocarbon can be approximated as an instantaneous reaction. The major sources of oxygen, these

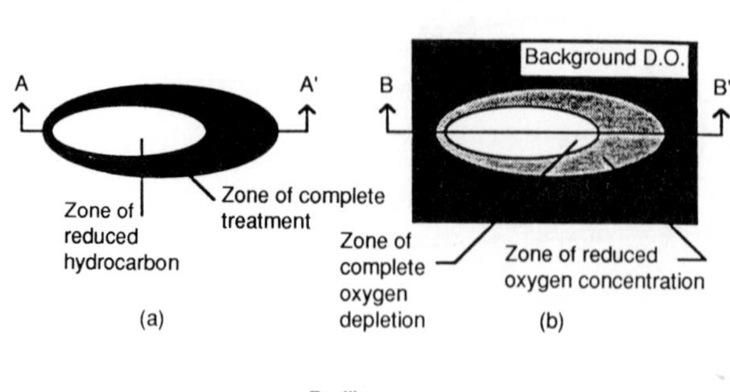

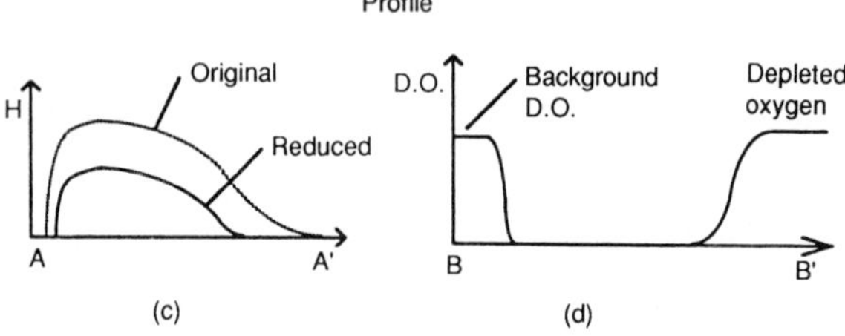

FIGURE 29.4.3 Use of principle of superposition for hydrocarbon and oxygen in BIOPLUME II: *(a)* hydrocarbon plume; *(b)* oxygen plume; *(c)* hydrocarbon; *(d)* oxygen. *(Reproduced from Ref. 61 with permission of ASCE.)*

researchers concluded, are transverse mixing, advective fluxes, and vertical exchange with the unsaturated zone (Fig. 29.4.3).

The only model input parameters to BIOPLUME II which are required to simulate biodegradation include the amount of dissolved oxygen in the aquifer prior to contamination and the oxygen demand of the contaminant determined from a stoichiometric relationship.

29.4.5 Examples and Detailed Tabulation of Case Studies

In a controlled field experiment at the Canada Forces Base Borden site, two plumes of gasoline-contaminated groundwater were introduced into the aquifer. Immediately upgradient of one plume, groundwater spiked with nitrate was added so that a nitrate plume would overtake the organic plume.[10] The success of the field experiment was limited.[11] The dissolved organic contaminant mass (BTEX) decreased rapidly because of residual oxygen concentrations in the aquifer prior to the nitrate overlap. The organic mass left in the aquifer was not adequate to evaluate anaerobic biotransformation.

Semprini et al.[65] presented the results from a field evaluation of in situ biodegradation of trichloroethylene and related compounds. The method that was used in the field demonstration relied on the ability of methane-oxidizing bacteria to degrade these contaminants to stable, nontoxic end products. The field site is located at the Moffitt Naval Air Station, Mountain View, Calif., and the test zone is a shallow confined aquifer composed of coarse-grained alluvial sediments. Results from the biotransformation experiments at the site indicate that biodegradation of TCE was on the order of 30 percent of the mass injected.

Borden and Bedient[15,16] conducted a three-well injection-production test at the United Creosoting Co. (UCC) site in Conroe, Texas, to evaluate the significance of biotransformation in limiting the transport of polycyclic aromatics present in the shallow aquifer. During the test, chloride, a nonreactive tracer, and two organic compounds, naphthalene and paradichlorobenzene (pDCB), were injected into a center well for 24 h; flushing by clean groundwater followed for 6 days. Groundwater was continuously produced from two adjoining wells and monitored to observe the breakthrough of these compounds. A significant loss of naphthalene and pDCB attributed to biotransformation processes was observed during the test.

Rifai et al.[61] studied the naturally occurring biodegradation at an aviation fuel spill in Traverse City, Mich. Contamination data from approximately 25 wells at the site were utilized to define the dissolved benzene, toluene, and xylene plume over a 2-year period. Rifai et al.[61] calculated a biodegradation rate of 1.0 percent a day at the Traverse City site (Fig. 29.4.3) and indicated that the pump-and-treat system at the site was removing a very small percentage of the total dissolved mass present at the site. A modeling effort was completed by using the BIOPLUME II model, and results matched the field observations reasonably well.

Chiang et al.[24] characterized soluble hydrocarbon and dissolved oxygen in a shallow aquifer beneath a field site by sampling groundwater at 42 monitoring wells. Results from 10 sampling periods over 3 years showed a significant reduction in total benzene mass with time in groundwater. The natural attenuation rate was calculated to be 0.95 percent per day. Spatial relationships between DO and total benzene, toluene, and xylene (BTX) were shown to be strongly correlated by statistical analyses and solute transport modeling.

29.5 PUMPING LNAPLs AND DNAPLs

Pumping is a favored method for removing both light and dense nonaqueous-phase liquids (LNAPLs and DNAPLs) present as free liquids in the subsurface. Experience with a large number of contamination problems resulting from stored petroleum products indicates that LNAPL removal by pumping is invariably one of the primary elements of a comprehensive remedial strategy. In the case of problems with DNAPLs, there is much less information and experience available on the efficacy of pumping technologies. However, in the absence of other alternatives, pumping is often selected for evaluation.

29.5.1 The Occurrence of NAPLs and DNAPLs

Figure 29.5.1 reviews some of the important features of the occurrence of NAPLs and DNAPLs in a simple, layered aquifer. In both cases illustrated in the figure, some finite volume of free product, or separate-phase liquid, spilled at the ground surface, moved downward through the unsaturated zone. Because LNAPLs are less dense than water, they accumulate at the top of the capillary fringe and spread laterally (Fig. 29.5.1*a*). The direction of spreading coincides with the gradient of the water table. DNAPLs, which are more dense than water, spread somewhat as they begin to encounter the capillary fringe, but move downward and accumulate at the low-

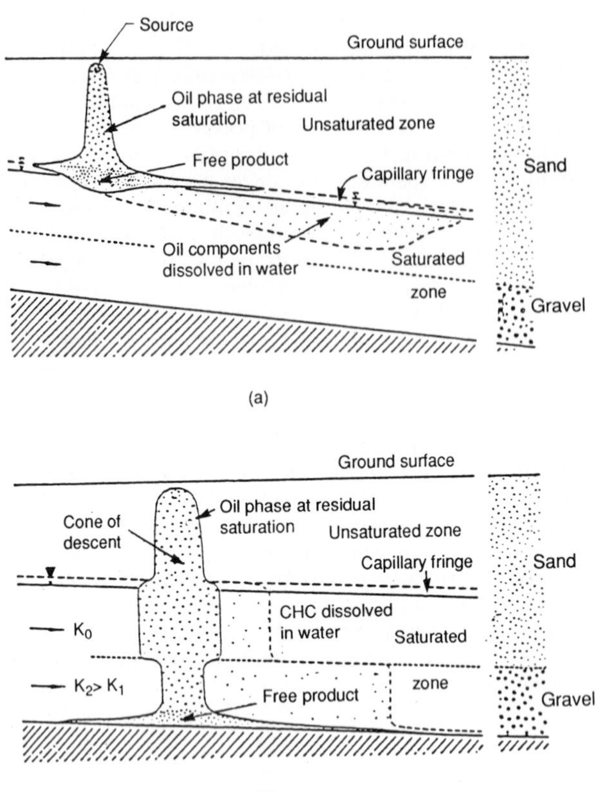

FIGURE 29.5.1 Idealized patterns of *(a)* LNAPL and *(b)* DNAPL flow in the subsurface. Partitioning of the liquid and gas phases provides a source of secondary contamination. *(Reproduced from Ref. 64 with permission.)*

permeability interfaces they encounter. The direction of lateral flow of a DNAPL follows the gradient of the surface of the low-permeability barrier.

As both Fig. 29.5.1*a* and *b* illustrate, a small fraction of the NAPL and DNAPL is left behind in every pore as the free product flows. In this so-called *cone of descent,* the NAPL is at *residual saturation,* a condition in which the NAPL is trapped in the pore space and cannot flow. This residually saturated fluid will not directly respond to pumping except for slow dissolution over a very long period of time. As the size of a NAPL plume increases, the volume of *free product* decreases and the volume of fluid held at residual saturation increases (free product is the hydrocarbon that can move through the porous medium). Thus, in the case of NAPLs, pumping has to be used with other alternatives like vacuum extraction or bioremediation to remediate both the free and residual components of the spill.

The figures also illustrate the tendency for soluble components to dissolve in flowing groundwater. Similarly, volatile components will partition or evaporate into the soil gas. Thus, the problem of cleanup may be compounded by the need to control and remediate vapor contamination in the soil zone and plumes of dissolved contaminants in the groundwater.

Experience from many different studies has shown that real problems are more complex than the idealized situations represented in Fig. 29.5.1. The main source of complexity is heterogeneity in the geologic and hydrogeologic setting that ultimately produces much more irregular fluid distributions. This heterogeneity may be related, for example, to spatial variability caused by layering, lenses, fractures, or combinations of all these. Figure 29.5.2 presents a series of conceptual models for NAPL distributions. The first (Fig. 29.5.2*a*) is a case of fuel leakage through a fractured clay overlying a porous sand aquifer. It is designed to display the unique role played by fractures in channeling the free product downward through a limited number of fracture pathways. What makes this case complicated is the uncertainty concerning where the fractures go and how they are interconnected. Schwartz et al.,[63] in examining a problem of NAPL contamination in this type of geologic setting, found that contaminant distributions within a clay unit were extremely heterogeneous and difficult to quantify. This heterogeneity produced a complicated NAPL distribution in the underlying sand aquifer.

The second example (Fig. 29.5.2*b*) highlights the tendency of DNAPLs to occur in a number of small unconnected depressions. This pattern is a consequence of the heterogeneity in the hydraulic conductivity distribution and the small-scale topography that exists along the tops of low-permeability units. Figure 29.5.2*c* illustrates another type of heterogeneity that also contributes to isolating small volumes of free product. In this case, the complexity is caused by an irregular and ill-defined fracture network. Invariably, the more widely distributed the free product becomes, the more difficult it becomes even to find the NAPL,[52] let alone clean it up. If NAPLs in the subsurface are the source of a dissolved contaminant plume, they will have to be removed before a site can be permanently cleaned up. Mackay and Cherry[52] raise the possibility of pump-and-treat systems for dissolved contaminants that may be required to be operated in perpetuity if NAPLs remain undiscovered and unrecovered.

29.5.2 Technical Considerations in Pumping

The most successful use of wells for NAPL removal is with LNAPLs such as oil, gasoline, or other hydrocarbons that occur at the top of the capillary fringe. Experience with DNAPL removal, which is reviewed later in this section, paints a much less optimistic picture of the suitability of pumping methodologies.

LNAPLs. To remove LNAPLs, wells need to accomplish two things: create a cone of depression in the water table centered on the pumping well and remove all the free product that enters the well. The gradient of the water table, produced by the well, causes free product within the cone of depression to flow toward the well.

Several different pumping arrangements are possible depending on the geologic setting, the equipment available, the extent of the spill, and the availability of existing wells.[12] Shown in Fig. 29.5.3 are examples of three types of installations for LNAPL recovery: (a) a one-pump, one-well system, (b) a two-pump, two-well system, and (c) a two-pump, one-well system.

The first system is similar in design to a traditional water well. During its operation, the pump cycles on and off, pumping both water and free product. Normally, a simple float system is used to maintain a pumping level close to the pump intake (Fig. 29.5.3*a*), and to maximize the pumping of product at the expense of water. The advantages of this system include: (1) a lower cost per system as compared to the others because of fewer wells and/or pumps and the feasibility of using smaller-diameter wells and (2) a simple control system for operating the pump. The disad-

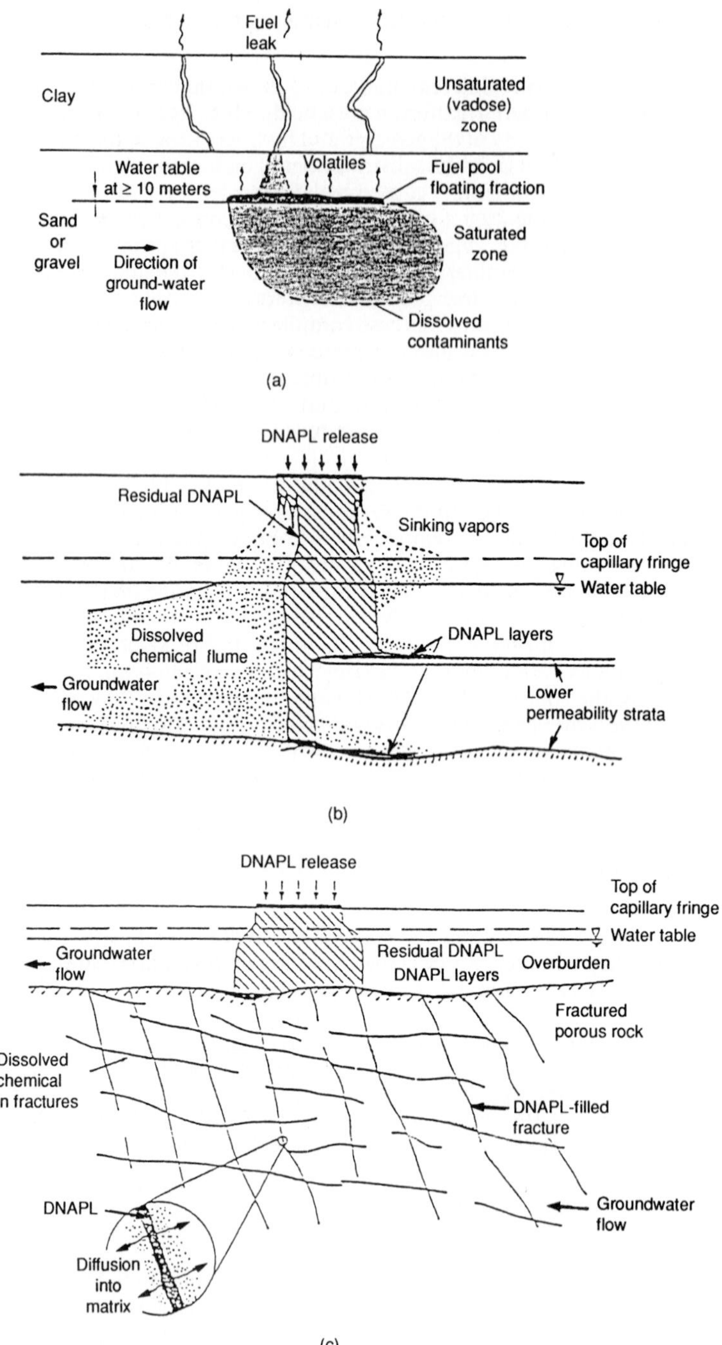

FIGURE 29.5.2 Examples of NAPL and DNAPL occurrence in complex geologic and hydrogeologic settings. *(a)* NAPL—fuel lack over fissured clay. *(Reproduced from Ref. 78 with permission.)* *(b)* DNAPL—leak into a heterogeneous aquifer. *(c)* DNAPL—migration into a fractured rock. *(b and c are from Ref. 33.)*

vantages are (1) reduced pumping efficiencies as compared to the other systems because the pump cycles on and off, (2) the tendency for some pumps to emulsify the product, which makes it necessary for a separator on the surface, and which increases the quantity of dissolved hydrocarbons in the water. This load of dissolved contaminants is usually great enough to require surface treatment of the water before release.

As Blake and Fryberger[12] point out, the two-pump, two-well systems (Fig. 29.5.3*b*) are often used when existing wells at the site are not large enough to contain two pumps. Specific two-well designs that involve both drilled and bored wells have also evolved to take advantage of readily obtainable pumping equipment and the larger capacity of bored wells.[57] The main advantages of the two-pump, two-well system are that (1) it pumps continuously and is more efficient than the one-pump, one-well system and (2) much less surface treatment is required because the oil and water are pumped separately and do not require separation at the site. Further, the large columns of water being pumped to control the water table position should contain no dissolved hydrocarbons. The disadvantages of the two-pump, two-well system are increased costs, if both sets of wells and pumps need to be purchased, and a more complex control system that, for example, will shut down the system if there is a threat that the water well could start to pump free product.

The two-pump, one-well system (Fig. 29.5.3*c*) has the same advantages and disadvantages as the two-pump, two-well system, but may be somewhat more cost-efficient. Normally, a larger-diameter well (minimum 10 in) is required to contain the two pumps and associated hardware. With this system the pump intake for the water well is set near the bottom of the well, while the product pump is set at or slightly below the oil/water interface (Fig. 29.5.3*c*). A control device is set between the two pumps to trigger a shutdown if the product for some reason accumulates in the well and threatens to enter the water pump.

Blake and Fryberger[12] provide the following advice on designing and operating these wells:

1. Make the wells as efficient as possible through the proper selection of materials—the rate of product recovery is directly related to the efficiency of the pumping well.
2. Keep the length of the screen longer than normal to be sure that the product has access to the casing in the event that drawdown is not as great as expected.
3. Use continuous-slot wire-wrapped screens to maximize the open area—bacterial growth and clogging is accelerated in recovery wells. The greater the open area, the longer period of time that the well can operate without servicing.
4. Keep the cone of depression of the water table at the minimum size necessary to control the spill—overpumping could reduce the overall product yield by forcing large volumes of product into residual saturation as the water table declines.

Certain modifications to these basic pumping systems have been developed to handle special situations. For example, hydrocarbon cleanups from low-permeability units often pose a problem because the small radius of influence produced by one well would ordinarily require that a large number of wells be installed to clean up a site. Agar et al.[1] describe a slim hole system involving a piston pump that is much less expensive than conventional systems. Thus, more wells can be installed for a given expenditure. Blake and Gates[13] attack this same problem by applying a vacuum to a conventional well system to increase the local hydraulic gradient. This not only increases the size of the capture zone for the well, but is thought to remobilize some of the product trapped at residual saturation because of increasing flow velocities.

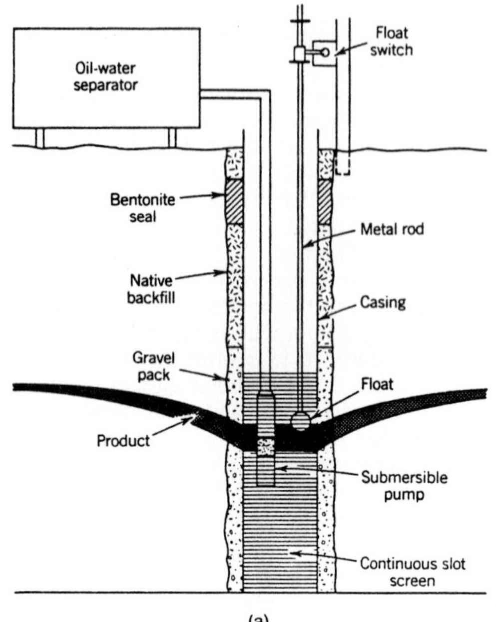
Oil-water separator
Float switch
Bentonite seal
Metal rod
Native backfill
Casing
Gravel pack
Float
Product
Submersible pump
Continuous slot screen

(a)

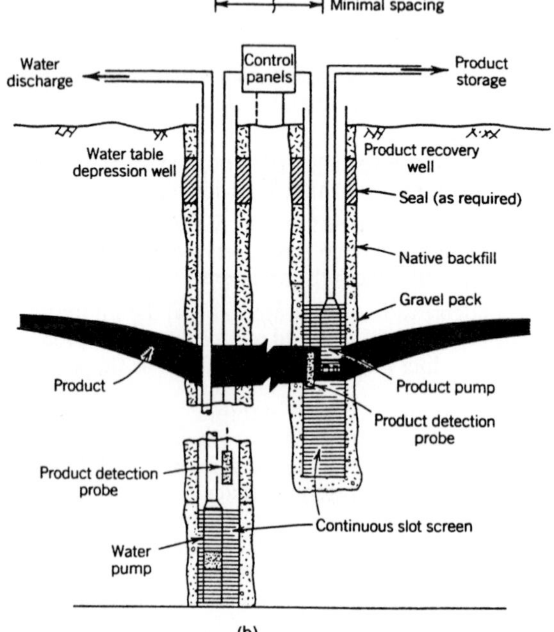
Minimal spacing
Water discharge
Control panels
Product storage
Water table depression well
Product recovery well
Seal (as required)
Native backfill
Gravel pack
Product
Product pump
Product detection probe
Product detection probe
Continuous slot screen
Water pump

(b)

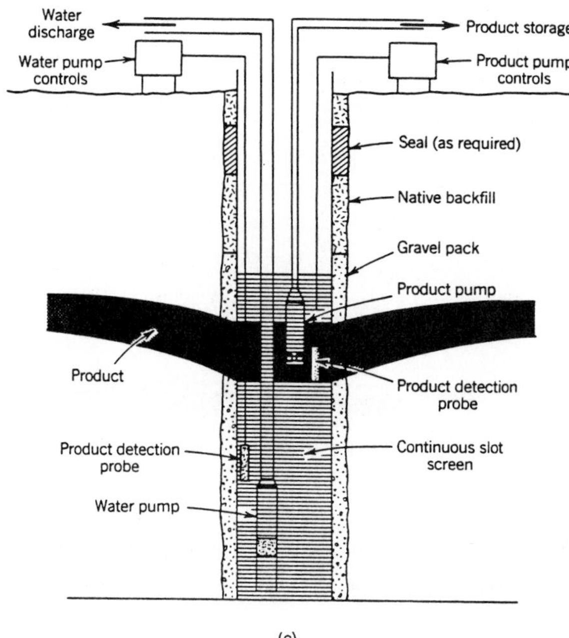

FIGURE 29.5.3 Examples: *(a)* one-well, one-pump system for recovering NAPLs, *(b)* two-well, two-pump system for the recovery of NAPLs, *(c)* one-well, two-pump system for the recovery of NAPLS. *(Reproduced from Ref. 14 with permission.)*

DNAPLs. Methods for removing DNAPLs with wells have not progressed as far technically, nor have recovery schemes met with as much success as has been the case with LNAPLs. Although the processes of flow for both LNAPLs and DNAPLs are similar, the major difference between the two types of fluids that affects their recovery is the driving forces that can be marshaled to move the NAPL to the collection well. With LNAPLs, the ability to provide a cone of depression in the water table is the key to promoting large volume recovery. With DNAPLs, there is no comparable way to marshal gravity-flow forces because the DNAPL accumulates along geologic boundaries and not the water table.

Experience with DNAPL collection is that a single well producing from a small zone saturated and free product is not very efficient. For example, several studies[33,76] have tested one-pump, one-well systems in the field. The work by Villaume et al.[76] looked at a problem of coal-tar contamination of a gravel aquifer in Pennsylvania. The experience with the sand and gravel aquifer was that initial DNAPL yields from four wells, pumped at a very slow rate, started out at about 100 gal/day (0.37 m^3/day) and decreased dramatically with time as the volume of coal tar in the immediate vicinity of the well was depleted. Given an estimated volume of product in the ground of possibly several hundred thousand gallons, a remedial scheme based on pumping of only free product would not be successful.

Figure 29.5.4 illustrates the two-pump, one-well system used by Villaume et al.[76] This modified pumping scheme doubled the yield of product and avoided the abrupt

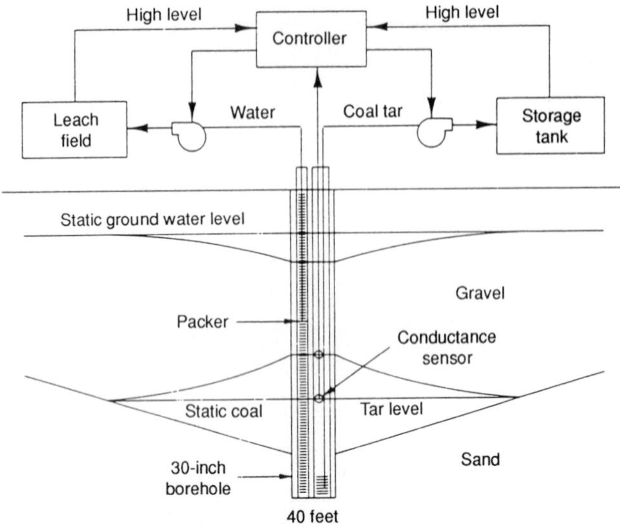

FIGURE 29.5.4 Two-pump, one-well system for DNAPL recovery where deliberate upwelling of the coal-tar interface is used to increase the flow of product into the recovery well. *(Reproduced from Ref. 76 with permission.)*

decline in rates with time. Over a 9-month period, approximately 7000 gal of coal tar was recovered. Villaume et al.[76] explain the increase in yields as an effect related to upcoming, which is shown on the figure. An observation first made in removing freshwater lying above seawater is that the position of the interface between the two fluids of differing density will rise due to pumping, according to the Ghyben-Herzberg principle. A sharp interface approximation for the water/DNAPL interface should theoretically result in the same behavior. This process provides an additional driving force to move the DNAPL to the well.

More recently, Connor et al.[25] examined three pumping strategies for removing DNAPLs from a silty sand unit at the Motco Superfund site near Houston, Texas. During a test of a recovery pilot system, they examined withdrawal rates for 4 days by pumping alone, and for 4 days with vacuum-enhanced pumping coupled with freshwater injection. Throughout the test, the DNAPL concentration averaged 1 to 2 percent of the total volume of produced water. However, there was a significant difference in the quantities of water produced with the various pumping schemes. Pumping alone yielded a maximum continuous flow rate of 1.7 gal/min. Vacuum-enhanced pumping produced a continuous rate of 2.5 gal/min, and, finally, vacuum-enhanced pumping augmented by a 1.5 gal/min injection of freshwater yielded groundwater at a continuous rate of 3.4 gal/min. Thus, the series of technical refinements in the pumping scheme double the maximum production of water and, in this case, the DNAPL.

29.5.3 Considerations in Pumping System Design

The state-of-the-art for designing pumping systems at the present time appears to involve what Domenico and Schwartz[29] refer to as an expanding pilot-scale system.

This approach involves selection of one of the standard pumping strategies and installing a system whose size reflects the certainty of knowledge about its expected performance. For example, in the case of LNAPLs, for which the experience base is well-developed, the initial system could involve a relatively large number of withdrawal wells. In the case of DNAPLs where the experience base is not as well-developed, this pilot system usually involves a small number of wells.

The basis for design with this type of approach is experience and careful observations of how the initial system deployed in the field actually performs. In the case of the large pumping system for LNAPL removal, a series of careful observations can pinpoint any gaps in the recovery system that could require additional wells. In the case of the DNAPL systems, one can decide on the basis of the small-scale test how successful a larger-scale operation would be.

Mathematical techniques are not used generally to optimize a NAPL recovery system to nearly the same extent as with pump-and-treat systems for recovering a plume of contaminants dissolved in water. Part of the reason is that software tools with the capability of simulating multiphase flow to a large number of pumping wells were not available for many years. Now, however, it appears that the development of simpler approaches and the availability of more powerful personal computers are making these analyses more feasible. For example, Charbeneau et al.[23] demonstrate the use of a NAPL/water flow model to evaluate pumping strategies for a site in California. Their model analysis suggested that for the site under consideration a strategy of boundary containment would be feasible, as opposed to a large system of interior recovery wells. Parker and Kaluarachchi[59] describe the theory of a water/LNAPL code called ARMOS and consider its application to a hypothetical LNAPL spill. They evaluated the efficiency of four possible schemes for single-well pumping in which the location and/or pattern of well operation change.

29.5.4 Case Studies of LNAPL and DNAPL Recovery by Pumping

The literature contains many case studies on the successful application of pumping for LNAPL recovery, especially those relating to the cleanup of spills of petroleum products. A very good source of information is the *Proceedings of the Conference on Petroleum Hydrocarbons and Organic Chemicals—Prevention, Detection and Restoration,* published by the National Water Well Association. This conference has been held annually since 1984 and provides a wealth of information on problems of contamination involving both LNAPLs and DNAPLs. In many cases, collection of the free product is simply the first step in an ongoing process of site remediation that involves other alternatives like vapor extraction and enhanced bioremediation.

In terms of DNAPLs, the discussion here has looked at the few cases of remediation by pumping that appear in the open literature. Clearly, both theoretical and practical work of all kinds is required to improve the understanding of how to remediate sites contaminated by DNAPLs.

29.6 EVALUATION OF GROUNDWATER EXTRACTION REMEDIES

Groundwater extraction is the most commonly used remedial technology for contaminated aquifers. In a detailed investigation performed by U.S. Environmental Protection Agency (EPA),[74] information was assembled from hazardous waste sites throughout the United States to show how groundwater extraction systems are being used, how their performance compares with expectations, and what factors are affecting their success. General data describing the site locations, the types of contaminants involved, the geologic nature of the sites, and the status of the remediations were collected for a total of 112 sites.[74]

29.6.1 Remediation Objectives

Two different remedial objectives were identified for the groundwater extraction systems described in the case studies: aquifer restoration and migration control. *Aquifer restoration* means that the contaminant concentrations in the aquifer are to be reduced below specified levels that have been determined to be protective for the site. In the case of Superfund sites, the cleanup levels are either the regulatory maximum contaminant levels (MCLs) or 10^{-4} to 10^{-6} of the excess cancer risk concentrations. Of the 19 sites studied in detail by EPA,[74] 13 had aquifer restoration as their primary goal, and only one restoration has been successful so far. Several of the other systems show promise of eventual aquifer restoration, but typically progress toward this goal is behind schedule. Concentrations often decline rapidly when the extraction system is first turned on, but, after the initial decrease, continued reductions are usually slower than expected.

Sites that are favorable for aquifer restoration have relatively simple stratigraphy with fairly homogeneous unconsolidated aquifer materials and constituents in the groundwater. Departures from these ideal conditions impede the progress of aquifer restoration. However, even if the concentrations are not rapidly reduced to cleanup goals, the extraction systems may still significantly reduce contaminant mass in the aquifer.

Migration control, or *plume containment,* was the primary goal at seven of the 19 EPA case study sites. In some of these cases the systems were initially intended for aquifer restoration, but operating experience indicated that this goal was not feasible. Migration control is a less ambitious objective, but one that can be more easily attained. In all but three of the 19 case study sites, successful plume containment has been demonstrated.

Failure to contain the contaminant plume has been noted at two of the sites described in the case studies. These are sites where water-supply wells have become contaminated, but the wells have not been taken out of service. Instead, treatment systems have been installed at the wells to make the groundwater produced suitable for its intended use. Both of the systems studied are successfully delivering water for domestic supply, and it appears that wellhead treatment can be operated successfully. In addition to serving the primary need for water supply, these systems can help to prevent the spread of contamination in the aquifer beyond the region already affected.

29.6.2 Extraction System Design

Table 29.6.1 presents a summary of information concerning the design of the extraction systems at the 19 case study sites. The first several columns in the table indicate the kinds of site data and the analytical methods that were used in designing the extraction systems. The remaining columns indicate the extraction capacity of the systems, whether the system configuration has been adjusted in response to operating experience, and the expected time required to complete the remediation. This infor-

TABLE 29.6.1 Summary of Design Information for Case Study Sites

Site No.	Site name	Aquifer tests	Flow model	Travel time analysis	Multilevel sampling	Soil sampling	Sorption considered	Pilot testing	System modifications	Maximum pumping rate, gal/min	Projected cleanup period
1	Amphenol Corp.	Yes	Yes	Yes	Yes	Yes	Yes	No	No	200	5–10 years
2	Black and Decker Inc.	Yes	No	No	Yes	No	No	No	No	15–20	Not projected
3	Des Moines TCE	Yes	Yes	Yes	Yes	Yes	Yes	No	No	1300	Not projected
4	Du Pont Mobile plant	Yes	No	No	No	Yes	Yes	No	Yes	150–180	N/A[1]
5	Emerson Electric Co.	No	Yes	Yes	Yes	Yes	Yes	No	No	30	9 months
6	Fairchild Semiconductor	Yes	No	No	Yes	Yes	No	No	Yes	9200	N/A[1]
7	General Mills Inc.	Yes	Yes	No	Yes	Yes	No	No	No	370	Not projected
8	GenRad Corp.	Yes	Yes	No	Yes	Yes	Yes	Yes	No	30	>5 years
9	Harris Corp.	Yes	Yes	No	Yes	Yes	No	No	Yes	1000[2] 300[3]	Not projected
10	IBM Dayton	Yes	Yes	Yes	Yes	Yes	Yes	Yes	Yes	1000	6 to 11 years
11	IBM San Jose	Yes	Yes	No	Yes	Yes	Yes	No	Yes	6000	10 years[4]
12	Nichols Engineering	Yes	No	No	Yes	Yes	Yes	Yes	Yes	65	2.25 years
13	Olin Corp.	Yes	Yes	No	No	Yes	No	No	No	2600–3600	N/A[1]
14	Ponders Corner	Yes	No	No	Yes	Yes	Yes	No	No	1200	10 years
15	Savannah River Plant	Yes	Yes	Yes	Yes	Yes	Yes	No	No[5]	440	30 years
16	Site A	No	Yes	Yes	Yes	Yes	No	No	Yes	50	25–60 days
17	Utah Power and Light	Yes	Yes	No	Yes	Yes	Yes	Yes	Yes	200	N/A[1]
18	Verona Well Field	Yes	Yes	No	Yes	Yes	Yes	No	Yes	2000[2] 400[3]	Not projected
19	Ville Mercier[6]	—	—	—	Yes	—		—	No	750	N/A[1]

1. Projection of cleanup period is not applicable to containment systems.
2. Containment or wellhead treatment portion.
3. Restoration portion.
4. Not explicitly projected; figure based on interpretation of design information.
5. Increase in extraction rates is planned.
6. Little design information is given in the case study for this site.

mation was taken from the design and operating reports collected for the case study sites. For some of the sites, the available reports may not have provided complete descriptions of the site data collected or the design methods used.

At all but two of the sites, aquifer testing was done to help determine the hydrogeologic characteristics of the aquifer to be remediated. *Aquifer testing,* as referred to here, means that a well on the site was pumped at a controlled rate for a certain period of the time, and the resulting water-level drawdown was measured in separate observation wells.

Numerical modeling was used to analyze groundwater flow at most of the case study sites. The flow models were generally used to determine capture zones of the planned extraction wells. At six of the sites, the flow patterns predicted by the models were used to estimate the travel times required for contaminants at the edge of the contaminant plume to reach the extraction wells. However, numerical contaminant transport modeling was not used at any of the sites.

At all but two of the sites, the subsurface exploration program included multilevel groundwater sampling to determine the vertical distribution of contaminants. This was done at all of the sites involving more than one aquifer and also at many of the single-aquifer sites.

Subsurface soil samples were taken and analyzed at all but one of the case study sites. In most cases, soil sampling was conducted for the purpose of identifying the contaminants present and for characterizing the contaminant source. Only rarely were soils sampled over a wider area to study the contaminant sorption characteristics of the aquifer or to search for contaminants in the nonaqueous phase. However, sorption was considered, in one way or another, at all but six of the sites.

Pilot testing, as referred to in Table 29.6.1, means that a small-scale extraction system was initially installed to provide data for the design of the final system. This was done at four of the case study sites. *System modification* means that the configuration of the full-scale system was adjusted on the basis of information gathered during performance monitoring. This happened at about half the sites. In some cases, wells were added or pumping rates were increased because monitoring showed that the initial system was inadequate. In other cases, extraction at some wells was terminated as a result of the progress of the remediation.

The last column of Table 29.6.1 lists the projected cleanup period as estimated by the designers of the extraction system. At sites where the remedial objective is not aquifer restoration, such a projection is not applicable, and the extraction system is expected to continue operating indefinitely. Even for several of the aquifer restoration sites, no projection of the cleanup time has been made.

29.6.3 General Findings from the EPA Study

The following general conclusions can be drawn from the case study data collected in the EPA project:

1. The groundwater extraction systems were generally effective in maintaining hydraulic containment of contaminant plumes, thus preventing further migration of contaminants. Even so, the design of successful containment systems requires careful study of site hydrogeology.
2. Significant removal of contaminant mass from the subsurface can often be achieved by the operation of groundwater extraction systems. Where site conditions are favorable, and the extraction system is properly designed and operated, it may be possible to remediate the aquifer to health-based levels. However, the time required for complete remediation is usually longer than initially estimated.

3. The contaminant concentrations observed at sites where aquifer restoration systems are in operation typically show a rapid initial decrease and then level off or decrease at a greatly reduced rate. This effect is to be expected because the mass reductions are greatest when contaminants are being removed at high concentration.
4. The success of groundwater extraction systems is highly dependent on site hydrogeology and contaminant characteristics. Site conditions must be taken into account in the selection of realistic system objectives.
5. Wellhead treatment systems can provide plume containment while maintaining water supply from contaminated aquifers. Such systems are usually installed as a cost-effective remedy for water-supply needs.
6. It is important to consider the contaminated subsurface, both saturated and unsaturated zones, as an integrated whole, and to design remediation programs to address source areas and unsaturated zone contamination in addition to the groundwater contamination in the saturated part of the aquifer. Enhanced methods such as pulsed pumping and vapor extraction can potentially improve the effectiveness of groundwater extraction systems by extracting residual contamination in the vadose zone.

REFERENCES

1. Agar, J. G., S. K. Ray, and W. R. Allan, "SHP pumps—A New Approach to the Recovery of Subsurface Petroleum Hydrocarbons," *Proc. Petroleum Hydrocarbons and Organic Chemicals in Ground Water,* NWWA, Dublin, Ohio, pp. 770–784, 1986.
2. Anderson, M. P., "Using Models to Simulate the Movement of Contaminants through Ground Water Flow Systems," *CRC Critical Rev. Environ. Control,* Chemical Rubber Co., vol. 9, p. 96, 1979.
3. Anderson, M. P., "Movement of Contaminants in Groundwater: Groundwater Transport —Advection and Dispersion," *Studies in Geophysics, Groundwater Contamination,* National Academy Press, Washington, D.C., pp. 37–45, 1984.
4. Anderson, M. P., D. S. Ward, E. G. Lappala, and T. A. Prickett, "Computer Models for Subsurface Water," Chap. 22, *Handbook of Hydrology,* McGraw-Hill, New York, 1993.
5. Barcelona, M. J., J. P. Gibb, and R. A. Miller, "A Guide to the Selection of Materials for Monitoring Well Construction and Ground-Water Sampling," Illinois State Water Survey contract report no. 327, USEPA-RSKERL, EPA-600/52-84/024, U.S. Environmental Protection Agency, 1983.
6. Barcelona, M. J., J. P. Gibb, J. A. Helfrich, and E. E. Garske, "Practical Guide for Ground-Water Sampling," Illinois State Water Survey contrast report no. 374, USEPA-RSKERL under cooperative agreement CR-809966-01, U.S. Environmental Protection Agency, Ada, Okla., 1985.
7. Batchelder, G. V., W. A. Panzeri, and H. T. Phillips, "Soil Ventilation for the Removal of Adsorbed Liquid Hydrocarbons in the Subsurface," *Proc. Petroleum Hydrocarbons and Organic Chemicals in Ground Water,* NWWA, Dublin, Ohio, pp. 672–688, 1986.
8. Bear, J., *Hydraulics of Ground Water,* McGraw-Hill, New York, 1979.
9. Bedient, P. B., R. C. Borden, and D. I. Leib, "Basic Concepts for Ground Water Modeling," Chap. 28 in *Ground Water Quality,* C. H. Ward, W. Giger, and P. L. McCarty, eds., Wiley, pp. 512–531, 1985.
10. Berry-Spark, K., J. F. Barker, D. Major, and C. I. Mayfield, "Remediation of Gasoline-Contaminated Ground-Waters: A Controlled Field Experiment," *Proc. Petroleum Hydrocarbons and Organic Chemicals in Ground Water,* NWWA, pp. 613–623, 1986.

11. Berry-Spark, K., and J. F. Barker, "Nitrate Remediation of Gasoline Contaminated Ground Waters: Results of a Controlled Field Experiment," *Proc. Petroleum Hydrocarbons and Organic Chemicals in Ground Water,* NWWA/API, pp. 3–10, 1987.

12. Blake, S. B., and J. S. Fryberger, "Containment and Recovery of Refined Hydrocarbons from Groundwater," Seminar on Groundwater and Petroleum Hydrocarbons Protection, Detection, Restoration, Petroleum Association for Conservation of the Canadian Environment, Ottawa, Canada, pp. V1–V47, 1983.

13. Blake, S. B., and M. M. Gates, "Vacuum Enhanced Hydrocarbon Recovery: A Case Study," *Proc. Petroleum Hydrocarbons and Organic Chemicals in Ground Water,* NWWA, Dublin, Ohio, pp. 709–721, 1986.

14. Blake, S. B., and R. W. Lewis, "Underground Oil Recovery," *Proc. 2d National Symposium on Aquifer Restoration and Ground Water Monitoring,* D. M. Nielsen, ed., NWWA, Dublin, Ohio, pp. 69–75, 1982.

15. Borden, R. C., and P. B. Bedient, "Transport of Dissolved Hydrocarbons Influenced by Reaeration and Oxygen Limited Biodegradation: 1. Theoretical Development," *Water Resour. Res.,* vol. 22, pp. 1973–1982, 1986.

16. Borden, R. C., and P. B. Bedient, "*In Situ* Measurement of Adsorption and Biotransformation at a Hazardous Waste Site," *Water Resour. Bull.,* vol. 23, pp. 629–636, 1987.

17. Bouwer, H., and R. C. Rice, "A Slug Test for Determining Hydraulic Conductivity of Unconfined Aquifers and Completely or Partially Penetrating Wells," *Water Resour. Res.,* vol. 12, pp. 423–428, 1976.

18. Brenoel, M., and R. A. Brown, "Remediation of a Leaking Underground Storage Tank with Enhanced Bioreclamation, *Proc. Nat. Symp. Exp. on Aquifer Restoration and Ground Water Monitoring,* NWWA, Worthington, Ohio, p. 527, 1985.

19. Brown, R. A., R. D. Norris, and R. L. Raymond, "Oxygen Transport in Contaminated Aquifers," *Proc. Conference on Petroleum Hydrocarbons and Organic Chemicals in Ground Water,* NWWA, Worthington, Ohio, p. 421, 1984.

20. Campbell, M. D., and J. H. Lehr, *Water Well Technology,* McGraw-Hill, New York, 1973.

21. Canter, L. W., and R. C. Knox, *Ground Water Pollution Control,* Lewis Publishers, Chelsea, Mich., 1986.

22. Cartwright, K., and J. M. Shafer, "Selected Technical Considerations for Data Collection and Interpretation—Groundwater," in *National Water Quality Monitoring and Assessment,* Washington, D.C., 1987.

23. Charbeneau, R. J., et al., "A Two-Layer Model to Simulate Floating Free Product Recovery: Formulation and Applications," *Proc. Petroleum Hydrocarbons and Organic Chemicals in Ground Water,* NWWA, Dublin, Ohio, pp. 333–345, 1989.

24. Chiang, C. Y., et al., "Aerobic Biodegradation of Benzene, Xylene in a Sandy Aquifer—Data Analysis and Computer Modeling," *Ground Water,* vol. 27, no. 6, pp. 823–834, 1989.

25. Connor, J. A., C. J. Newell, and D. K. Wilson, "Assessment, Field Testing, and Conceptual Design for Managing Dense Non-Aqueous Phase Liquids (DNAPL) at a Superfund Site," *Proc. Petroleum Hydrocarbons and Organic Chemicals in Ground Water,* NWWA, Dublin, Ohio, pp. 519–533, 1989.

26. Cooper, H. H., Jr., J. D. Bredehoeft, and S. S. Papadopulos, "Response of a Finite Diameter Well to an Instantaneous Charge of Water," *Water Resour. Res.,* vol. 3, no. 1, pp. 263–269, 1967.

27. De Marsily, G., *Quantitative Hydrogeology,* Academic Press, Harcourt Brace Jovanovich, Orlando, Fla., 1986.

28. Dobrin, M. B., *Introduction to Geophysical Prospecting* (3d ed.), McGraw-Hill, New York, 1976.

29. Domenico, P. A., and F. W. Schwartz, *Physical and Chemical Hydrogeology,* Wiley, New York, 1990.

30. Driscoll, F. G., *Ground Water and Wells* (2d ed.), Johnson Division, UOP, St. Paul, Minn., 1986.

31. Dunlap, W. J., and J. F. McNabb, "Subsurface Biological Activity in Relation to Ground Water Pollution," EPA-660/2-73-014, U.S. Environmental Protection Agency, Ada, Okla., p. 60, 1973.

32. Ehrenfeld, J., and J. Bass, *Evaluation of Remedial Action Unit Operations at Hazardous Waste Disposal Sites,* Noyes Publications, Park Ridge, N.J., 1984.

33. Feenstra, S., and J. A. Cherry, "Subsurface Contamination by Dense Non-Aqueous Phase (DNAPL) Chemicals," Figs. 3 and 4, International Groundwater Symposium, International Association of Hydrogeologists, Halifax, Nova Scotia, pp. 61–69, 1988.

34. Fetter, C. W., *Applied Hydrogeology,* Merrill, Columbus, Ohio, 1988.

35. Flathman, P. E., and G. D. Githens, "In Situ Biological Treatment of Isopropanol, Acetone, and Tetrahydrofuran in the Soil/Groundwater Environment," *Groundwater Treatment Technology,* E. K. Nyer, ed., Van Nostrand Reinhold, New York, p. 173, 1985.

36. Freeze, R. A., and J. A. Cherry, *Groundwater,* Prentice-Hall, Englewood Cliffs, N.J., 1979.

37. Hinchee, R. E., et al., "Underground Fuel Contamination, Investigation, and Remediation a Risk Assessment Approach to How Clean Is Clean," *Proc. Petroleum Hydrocarbons and Organic Chemicals in Ground Water,* NWWA, Dublin, Ohio, pp. 539–563, 1986.

38. Hinchee, R. E., D. C. Downey, and E. J. Coleman, "Enhanced Bioremediation, Soil Venting, and Ground-Water Extraction; a Cost-Effectiveness and Feasibility Comparison," *Proc. Petroleum Hydrocarbons and Organic Chemicals in Ground Water,* NWWA, Dublin, Ohio, pp. 147–164, 1987.

39. Hvorslev, M. J., "Time Lag and Soil Permeability in Groundwater Observations," U. S. Army Corps of Engineers, *Waterways Exp. Sta. Bull.,* vol. 36, Vicksburg, Miss., 1951.

40. Javandel, I., and C. F. Tsang, "Capture-Zone Type Curves: A Tool for Aquifer Cleanup," *Ground Water,* vol. 24, no. 5, pp. 616–625, 1986.

41. Jhaveri, V., and A. J. Mazzacca, "Bio-Reclamation of Ground and Groundwater: A Case History," *Proc. 4th Nat. Conf. Management of Uncontrolled Hazardous Waste Sites,* Washington, D.C., p. 242, October 1983.

42. Johnson, P. C., M. W. Kemblowski, and J. D. Colthart, "Practical Screening Models for Soil Venting Applications," *Proc. Petroleum Hydrocarbons and Organic Chemicals in Ground Water,* NWWA, Dublin, Ohio, pp. 521–546, 1988.

43. Johnson, P. C., et al., "A Practical Approach to the Design, Operation, and Monitoring of In-Situ Soil-Venting Systems," *Ground Water Monitoring Review,* vol. 10, no. 2, pp. 159–178, 1990.

44. Journel, A. G., and C. J. Huijbregts, *Mining Geostatistics,* Academic Press, London, 1978.

45. Keys, W. S., and L. M. MacCary, "Application of Borehole Geophysics to Water-Resources Investigations," in *Techniques of Water-Resources Investigations,*" U.S. Geological Survey, Book 2, Chap. E1, 1971.

46. Krishnayya, A. V., M. J. O'Connor, J. G. Agar, and R. D. King, "Vapor Extraction Systems: Factors Affecting Their Design and Performance," *Petroleum Hydrocarbons and Organic Chemicals in Ground Water,* NWWA, Dublin, Ohio, pp. 547–569, 1988.

47. Kuhn, E. P., et al., "Microbial Transformation of Substituted Benzenes during Infiltration of River Water to Ground Water: Laboratory Column Studies," *Environ. Sci. Technol.,* vol. 19, p. 961, 1985.

48. Kwader, T., "The Use of Geophysical Logs for Determining Formation Water Quality," *Ground Water,* vol. 24, pp. 11–15, 1986.

49. Lee, M. D., and C. H. Ward, "Biological Methods for the Restoration of Contaminated Aquifers," *Environ. Toxicol. Chem.,* vol. 4, p. 743, 1985.

50. Lee, M. D., et al., "Biorestoration of Aquifers Contaminated with Organic Compounds," NCGWR, R. S. Kerr Environmental Research Laboratory, U.S. Environmental Protection Agency, Ada, Okla., vol. 18, no. 1, pp. 29–89, 1988.

51. Mackay, D. M., D. L. Freyberg, P. V. Roberts, and J. A. Cherry, "A Natural Gradient

Experiment on Solute Transport in a Sand Aquifer: 1. Approach and Overview of Plume Movement," *Water Resour. Res.*, vol. 22, no. 13, pp. 2017–2029, December 1986.

52. Mackay, D. M., and J. A. Cherry, "Groundwater Contamination: Pump-and-Treat Remediation," *Environ. Sci. Technol.*, vol. 23, no. 6, pp. 630–636, 1989.

53. Marley, M. C., and G. E. Hoag, "Induced Soil Venting for the Recovery/Restoration of Gasoline Hydrocarbons in the Vadose Zone," *Proc. Petroleum Hydrocarbons and Organic Chemicals in Ground Water,* NWWA, Dublin, Ohio, pp. 473–503, 1984.

54. Mercer, J. W., D. C. Skipp, and D. Giffin, *Basics of Pump-and-Treat Ground-Water Remediation Technology,* Robert S. Kerr Environmental Research Laboratory, U.S. Environmental Protection Agency, Ada, Okla., EPA-600/8-90/003, March 1990.

55. Need, E. A., and M. J. Costello, "Hydrogeologic Aspects of Slurry Wall Isolation Systems in Areas of High Downward Gradients," *Proc. 4th Nat. Symp. on Aquifer Restoration and Ground Water Monitoring,* D. M. Nielsen and M. Curl, eds., Columbus, Ohio, May 1984, National Water Well Association, Worthington, Ohio, p. 18, 1984.

56. Nielsen, D. M., "Remedial Methods Available in Areas of Groundwater Contamination," *Proc. 6th Nat. Ground Water Quality Symp.,* Atlanta, September 1982, National Water Well Association, Worthington, Ohio, p. 219, 1983.

57. O'Connor, M. J., A. M. Wofford, and S. K. Ray, "Recovery of Subsurface Hydrocarbons at an Asphalt Plant—Results of a Five-Year Monitoring Program," *Proc. Petroleum Hydrocarbons and Organic Chemicals in Ground Water,* NWWA, Dublin, Ohio, pp. 359–376, 1984.

58. Papadopulos, I. S., J. D. Bredehoeft, and H. H. Cooper, Jr., "On the Analysis of 'Slug Test' Data," *Water Resour. Res.,* vol. 9, no. 4, pp. 1087–1089, 1973.

59. Parker, J. C., and J. J. Kaluarachchi, "A Numerical Model for Design of Free Product Recovery Systems at Hydrocarbon Spill Sites," *Proc. Solving Ground Water Problems with Models,* National Water Well Association, Dublin, Ohio, pp. 271–281, 1989.

60. Raymond, R. L., "Environmental Bioreclamation," 1978 Mid-Continent Conference and Exhibition on Control of Chemicals and Oil Spills, Detroit, September 1978.

61. Rifai, H. S., et al., "Biodegradation Modeling at a Jet Fuel Spill Site," *J. Environ. Eng.,* vol. 114, no. 5, pp. 1007–1019, 1988.

62. Scalf, M. R., et al., "Manual of Groundwater Quality Sampling Procedures," Robert S. Kerr Environmental Research Laboratory, U.S. EPA, Ada, Okla., 1981.

63. Schwartz, F. W., J. A. Cherry, and J. R. Roberts, "A Case Study of a Chemical Spill: Polychlorinated Biphenyls (PCBs): 2. Hydrogeologic Conditions and Contaminant Migration," *Water Resour. Res.,* vol. 18, no. 3, pp. 535–545, 1982.

64. Schwille, F., "Migration of Organic Fluids Immiscible with Water in the Unsaturated and Saturated Zones," Figs. 2 and 3, *Second Canadian/American Conference on Hydrogeology,* B. Hitchon and M. Trudell, eds., National Water Well Association, Dublin, Ohio, p. 34, 1985.

65. Semprini, L., et al., "A Field Evaluation of In Situ Biodegradation for Aquifer Restoration," U.S. Environmental Protection Agency, EPA/600/S2/87/096, U.S. Government Printing Office, Ada, Okla., 1988.

66. Streltsova, T. D., *Well Testing in Heterogeneous Formations,* Wiley, New York, 1988.

67. Suflita, J. M., and G. D. Miller, "Microbial Metabolism of Chlorophenolic Compounds in Ground Water Aquifers, *Environ. Toxicol. Chem.,* vol. 4, p. 751, 1985.

68. Thomas, J. M., et al., "Leaking Underground Storage Tanks: Remediation with Emphasis on In Situ Biorestoration," NCGWR, Robert S. Kerr Environmental Research Laboratory, U.S. Environmental Protection Agency, Ada, OK, EPA/600/2-87/008, January 1987.

69. Thomas, J. M., and C. H. Ward, "In Situ Biorestoration of Organic Contaminants in the Subsurface," *Environ. Sci. Technol.,* vol. 23, no. 7, pp.760–766, November 7, 1989.

70. Threlfall, D., and M. J. Dowiak, "Remedial Options for Ground Water Protection at

Abandoned Solid Waste Disposal Facilities," U.S. Environmental Protection Agency National Conference on Management of Uncontrolled Hazardous Waste Sites, Washington, D.C., October 1980.

71. U.S. Environmental Protection Agency, "RCRA Ground-Water Monitoring Technical Enforcement Guidance Document," OSWER-9950.1, Washington, D.C., 1986.
72. U.S. Environmental Protection Agency, "Handbook of Ground Water," EPA/625/6-87/016, Cincinnati, Ohio, 1987.
73. U.S. Environmental Protection Agency, *Guidance on Remedial Actions for Contaminated Ground Water at Superfund Sites,* EPA/540/G-88/003, December 1988.
74. U.S. Environmental Protection Agency, "Evaluation of Ground-Water Extraction Remedies," vol. 1, summary report, EPA/540/2-89/054, Washington, D.C., September 1989.
75. Verschueren, K., *Handbook of Environmental Data on Organic Chemicals,* Van Nostrand Reinhold, New York, 1983.
76. Villaume, J. F., P. C. Lowe, and D. F. Unites, "Recovery of Coal Gasification Wastes: An Innovative Approach," *Proc. 3d Nat. Symposium on Aquifer Restoration and Ground Water Monitoring,* National Water Well Association, Dublin, Ohio, pp. 434–444, 1983.
77. Wang, H. F., and M. P. Anderson, *Introduction to Groundwater Modeling, Finite Difference and Finite Element Methods,* Freeman, San Francisco, 1982.
78. Walther, E. G., A. M. Pitchford, and G. R. Olhoeft, "A Strategy for Detecting Subsurface Organic Contaminants," *Proc. Petroleum Hydrocarbons and Organic Chemicals in Ground Water—Prevention, Detection and Restoration,* Fig. 9, National Water Well Association, Dublin, Ohio, p. 371, 1986.
79. Wilson, J. T., et al., "Influence of Microbial Adaptation on the Fate of Organic Pollutants in Ground Water," *Environ. Toxicol. Chem.,* vol. 4, p. 721, 1985.

APPENDIX A
Hydrologic Unit Conversions

The following tables contain conversion factors for units of hydrologic measurement. These factors are accurate to five significant figures.*

Table	Quantity	Dimensions
A.1	Length	L
A.2	Area	L^2
A.3	Volume	L^3
A.4	Velocity or rate	LT^{-1}
A.5	Discharge	L^3T^{-1}
A.6	Mass	M
A.7	Force	$F = MLT^{-2}$
A.8	Pressure or stress	FL^{-2}
A.9	Energy	$E = FL$
A.10	Energy flux	$EL^{-2}T^{-1}$
A.11	Power	ET^{-1}
A.12	Time	T
A.13	Temperature	θ

SI units are preceded by a prefix indicating the order of magnitude of the unit in powers of 10. These prefixes are

Multiplying factor	Prefix	Symbol
10^{12}	tera	T
10^{9}	giga	G
10^{6}	mega	M
10^{3}	kilo	k
10^{2}	hecto	h
10^{1}	deka (deca)	da
10^{-1}	deci	d
10^{-2}	centi	c
10^{-3}	milli	m
10^{-6}	micro	μ
10^{-9}	nano	n
10^{-12}	pico	p

* The editor appreciated the assistance of Pawel Mizgalewicz in preparing these tables.

TABLE A.1 Length Conversion Factors

To convert *from* these units		*To* these units, multiply by the tabulated factor						
		mm	cm	m	km	in	ft	mi
millimeter	mm	1	0.1	0.001	10^{-6}	0.039370	0.0032808	6.2137×10^{-7}
centimeter	cm	10	1	0.01	10^{-5}	0.39370	0.032808	6.2137×10^{-6}
meter	m	1000	100	1	0.001	39.370	3.2808	6.2137×10^{-4}
kilometer	km	10^{6}	10^{5}	1000	1	39,370	3280.8	0.62137
inch	in	25.4	2.54	0.0254	2.54×10^{-5}	1	0.083333	1.5783×10^{-5}
foot	ft	304.8	30.48	0.3048	3.048×10^{-4}	12	1	1.8939×10^{-4}
mile	mi	1.6093×10^{6}	1.6093×10^{5}	1609.3	1.6093	63,360	5280	1

Example: in $\times$ 2.54 = cm

Other units: 1 yard = 3 ft; 1 micrometer (μm) = 10^{-6} m; 1 angstrom (Å) = 10^{-10} m.

TABLE A.2 Area Conversion Factors

To convert *from* these units		***To*** these units, multiply by the tabulated factor: cm^2	m^2	ha	km^2	ft^2	ac	mi^2
square centimeter	cm^2	1	10^{-4}	10^{-8}	10^{-10}	0.0010764	2.4711×10^{-8}	3.8610×10^{-11}
square meter	m^2	10^4	1	10^{-4}	10^{-6}	10.764	2.4711×10^{-4}	3.8610×10^{-7}
hectare	ha	10^8	10^4	1	0.01	107,639	2.4711	0.0038610
square kilometer	km^2	10^{10}	10^6	100	1	1.0764×10^7	247.11	0.38610
square foot	ft^2	929.03	0.092903	9.2903×10^{-6}	9.2903×10^{-8}	1	2.2957×10^{-5}	3.5870×10^{-8}
acre	acre	4.0469×10^7	4046.9	0.40469	0.0040469	43,560	1	0.0015625
square mile	mi^2	2.5900×10^{10}	2.5900×10^6	259.00	2.5900	2.7878×10^7	640	1

Example: acre $\times$ 0.40469 = hectare

TABLE A.3 Volume Conversion Factors

To convert *from* these units		*To* these units, multiply by the tabulated factor						
		cm^3	L	m^3	ha·m	ft^3	gal	acre·ft
cubic centimeter	cm^3	1	0.001	10^{-6}	10^{-10}	3.5315×10^{-5}	2.6417×10^{-4}	8.1071×10^{-10}
liter	L	1000	1	0.001	10^{-7}	0.035315	0.26417	8.1071×10^{-7}
cubic meter	m^3	10^6	1000	1	10^{-4}	35.315	264.17	8.1071×10^{-4}
hectare-meter	ha·m	10^{10}	10^7	10^4	1	353,147	2.6417×10^6	8.1071
cubic foot	ft^3	28,317	28.317	0.028317	2.8317×10^{-6}	1	7.4805	2.2957×10^{-5}
U.S. gallon	gal	3785.4	3.7854	0.0037854	3.7854×10^{-7}	0.13368	1	3.0689×10^{-6}
acre-foot	acre·ft	1.2335×10^9	1.2335×10^6	1233.5	0.12335	43,560	325,851	1

Example: $m^3 \times 35.315 = ft^3$

Other conversions: 1 milliliter (mL) = 1 cm^3; 1 liter = 1 dm^3; U.S. gallon × 1.2 = 1 imperial gallon; 1 ft^3/s·day = 1.9835 acre-ft = 86,400 ft^3; 1 m^3/s·day = 8.6400 ha·m = 86,400 m^3.

TABLE A.4 Velocity or Rate Conversion Factors

To convert *from* these units		*To* these units, multiply by the tabulated factor							
		cm/s	m/s	mm/h	cm/h	m/d	in/h	ft/s	gal/day·ft^2
centimeter/second	cm/s	1	0.01	36,000	3600	864	1417.3	0.032808	21,205
meter/second	m/s	100	1	3.6×10^6	3.6×10^5	86,400	1.4173×10^5	3.2808	2.1205×10^6
millimeter/hour	mm/h	2.7778×10^{-5}	2.7778×10^{-7}	1	0.1	0.024	0.039370	9.1134×10^{-7}	0.58902
centimeter/hour	cm/h	2.7778×10^{-4}	2.7778×10^{-6}	10	1	0.24	0.39370	9.1134×10^{-6}	5.8902
meter/day	m/d	0.0011574	1.1574×10^{-5}	41.667	4.1667	1	1.6404	3.7973×10^{-5}	24.542
inch/hour	in/h	7.0556×10^{-4}	7.0556×10^{-6}	25.4	2.54	0.6096	1	2.3148×10^{-5}	14.961
feet/second	ft/s	30.48	0.3048	1.0973×10^6	1.0973×10^5	26,335	43,200	1	6.4632×10^5
gal/day·square foot	gal/day·ft^2	4.7160×10^{-5}	4.7160×10^{-7}	1.6997	0.16977	0.040746	0.066840	1.5472×10^{-6}	1

Example: ft/s × 0.3048 = m/s

TABLE A.5 Discharge Conversion Factors

To convert *from* these units		*To* these units, multiply by the tabulated factor						
		L/s	m^3/s	m^3/day	ft^3/s	gal/min	MGD	acre-ft/year
liter/second	L/s	1	0.001	86.4	0.035315	15.850	0.022824	25.567
cubic meter/second	m^3/s	1000	1	86,400	35.315	15,850	22.824	25,567
cubic meter/day	m^3/day	0.011574	1.1574×10^{-5}	1	4.0873×10^{-4}	0.18345	2.6417×10^{-4}	0.29591
cubic feet/second	ft^3/s	28.317	0.028317	2446.6	1	448.83	0.64632	723.97
gallon/minute	gal/min	0.063090	6.3090×10^{-5}	5.4510	0.0022280	1	0.00144	1.6130
million gallons/day	MGD	43.813	0.043813	3785.4	1.5472	694.44	1	1120.1
acre-feet/year	acre·ft/year	0.039113	3.9113×10^{-5}	3.3794	0.0013813	0.61996	8.9274×10^{-4}	1

Example: gal/min $\times$ 5.4510 = m^3/day

TABLE A.6 Mass Conversion Factors

To convert *from* these units		*To* these units, multiply by the tabulated factor				
		mg	g	kg	slug	lb_m
milligram	mg	1	0.001	10^{-6}	6.8522×10^{-8}	2.2046×10^{-6}
gram	g	1000	1	0.001	6.8522×10^{-5}	0.0022046
kilogram	kg	10^6	1000	1	0.068522	2.2046
slug	slug	1.4594×10^7	14,594	14.594	1	32.174
pound mass	lb_m	4.5359×10^5	453.59	0.45359	0.031081	1

Example: $lb_m \times 0.45359 = kg$
Other units: 1 ton (metric) = 1000 kg; 1 U.S. ton = 2000 lb_m = 907.18 kg.

TABLE A.7 Force Conversion Factors

To convert *from* these units		*To* these units, multiply by the tabulated factor			
		N	kg_f	dyne	lb_f
newton	N	1	0.10197	10^5	0.22481
kilogram-force	kg_f	9.8067	1	9.8067×10^5	2.2046
dyne	dyne	10^{-5}	1.0197×10^{-6}	1	2.2481×10^{-6}
pound-force	lb_f	4.4482	0.45359	4.4482×10^5	1

Example: $lb_f \times 4.4482 = N$
Note: Standard conditions of gravity $g = 9.086650$ m/s^2.

TABLE A.8 Pressure or Stress Conversion Factors

To convert *from* these units		***To*** these units, multiply by the tabulated factor						
		atm	mb	Pa	mmHg	inHg	inH_2O	lb/in^2
atmosphere standard	atm	1	1013.3	1.0133×10^5	760	30.006	407.19	14.696
millibar	mb	9.8692×10^{-4}	1	100	0.75006	0.029613	0.40186	0.014504
pascal	Pa	9.8692×10^{-6}	0.01	1	0.0075006	2.9613×10^{-4}	0.0040186	1.4504×10^{-4}
torr = millimeter of mercury 0°C	mmHg	0.0013158	1.3332	133.32	1	0.039481	0.53577	0.019337
inch of mercury, 60°F	inHg	0.033327	33.769	3376.9	25.329	1	13.570	0.48977
inch of water, 60°F	inH_2O	0.0024559	2.4884	248.84	1.8665	0.073690	1	0.036091
pound force per square inch	lb/in^2	0.068046	68.948	6894.8	51.715	2.0418	27.708	1

Example: inHg $\times$ 3376.9 = Pa
Note: $lb_f/ft^2 \times 47.880$ = Pa

TABLE A.9 Energy Conversion Factors

To convert ***from*** these units		***To*** these units, multiply by the tabulated factor					
		J	kWh	cal	Btu	erg	$ft \cdot lb_f$
joule	J	1	2.7778×10^{-7}	0.23885	9.4782×10^{-4}	10^{7}	0.73756
kilowatt-hour	kWh	3.6×10^{6}	1	8.5985×10^{5}	3412.1	3.6×10^{13}	2.6552×10^{6}
calorie	cal	4.1868	1.1630×10^{-6}	1	0.0039683	4.1868×10^{7}	3.0880
British thermal unit	Btu	1055.1	2.9307×10^{-4}	252.00	1	1.0551×10^{10}	778.17
erg	erg	10^{-7}	2.7778×10^{-14}	2.3885×10^{-8}	9.4782×10^{-11}	1	7.3756×10^{-8}
foot-pound force	$ft \cdot lb_f$	1.3558	3.7662×10^{-7}	0.32383	0.0012851	1.3558×10^{7}	1

British thermal unit and calorie are from the International Table.
Example: cal × 4.1868 = J

TABLE A.10 Energy Flux Conversion Factors

To convert ***from*** these units		***To*** these units, multiply by the tabulated factor					
		W/m^2	W/cm^2	$MJ/day \cdot m^2$	ly/s	ly/min	ly/day
watt/square meter	W/m^2	1	10^{-4}	0.086400	2.3885×10^{-5}	0.0014331	2.0636
watt/square cm	W/cm^2	10,000	1	864	0.23885	14.331	20,636
Megajoule/day/square meter	$MJ/day \cdot m^2$	11.574	0.0011574	1	2.7644×10^{-4}	0.016587	23.885
langley/second	ly/s	41,868	4.1868	3617.4	1	60	86,400
langley/minute	ly/min	697.80	0.069780	60.290	0.016667	1	1440
langley/day	ly/day	0.48458	4.8458×10^{-5}	0.041868	1.1574×10^{-5}	6.9444×10^{-4}	1

Example: ly/day $\times$ 0.041868 = $MJ/day \cdot m^2$
Note: 1 MJ = 10^6 J; 1 ly = 1 cal/cm^2

TABLE A.11 Power Conversion Factors

To convert *from* these units		*To* these units, multiply by the tabulated factor: W	kW	Btu/h	ft·lb_f/s	hp
watt	W	1	0.001	3.4121	0.73756	0.0013410
kilowatt	kW	1000	1	3412.1	737.56	1.3410
Btu* per hour	Btu/h	0.29307	2.9307×10^{-4}	1	0.21616	3.9301×10^{-4}
foot-pound force per second	ft·lb_f/s	1.3558	0.0013558	4.6262	1	0.0018182
horsepower	hp	745.70	0.74570	2544.4	550	1

*British thermal unit (International Table).
Example: Btu/h $\times$ 2.9307×10^{-4} = kW

TABLE A.12 Time Conversion Factors

To convert *from* these units		*To* these units, multiply by the tabulated factor: s	min	h	day	year
second	s	1	0.016667	2.7778×10^{-4}	1.1574×10^{-5}	3.1710×10^{-8}
minute	min	60	1	0.016667	6.9444×10^{-4}	1.9026×10^{-6}
hour	h	3600	60	1	0.041667	1.1416×10^{-4}
day	day	86,400	1440	24	1	0.0027397
year	year	31,536,000	525,600	8760	365	1

Example: day $\times$ 86,400 = seconds

TABLE A.13 Temperature Conversion Relationships

To convert *from* these units		*To* these units, use the tabulated formula: K	°F	°C	°R
kelvin	K	1	$\frac{9}{5}$K − 459.67	K − 273.15	$\frac{9}{5}$K
degrees Fahrenheit	°F	$\frac{5}{9}$(°F + 459.67)	1	$\frac{5}{9}$(°F − 32)	°F + 459.67
degrees Celsius	°C	°C + 273.15	$\frac{9}{5}$°C + 32	1	$\frac{9}{5}$(°C + 273.15)
degrees Rankine	°R	$\frac{5}{9}$°R	°R − 459.67	$\frac{5}{9}$°R − 273.15	1

Example: K = °C + 273.15

INDEX